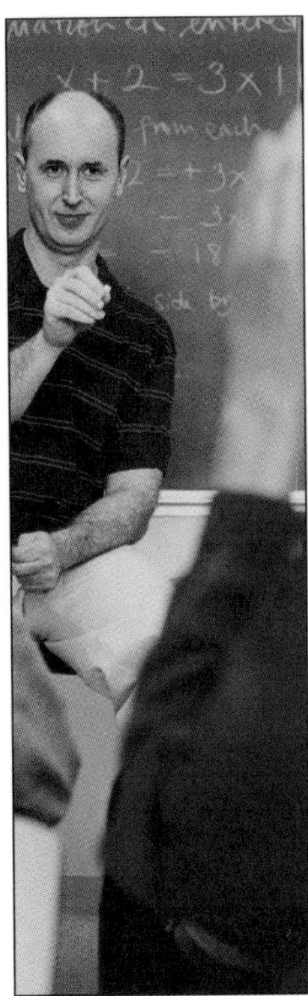

The first person to invent a car that runs on water...

... may be sitting right in your classroom! Every one of your students has the potential to make a difference. And realizing that potential starts right here, in your course.

When students succeed in your course—when they stay on-task and make the breakthrough that turns confusion into confidence—they are empowered to realize the possibilities for greatness that lie within each of them. We know your goal is to create an environment where students reach their full potential and experience the exhilaration of academic success that will last them a lifetime. *WileyPLUS* can help you reach that goal.

Wiley**PLUS** is an online suite of resources—including the complete text—that will help your students:

- come to class better prepared for your lectures
- get immediate feedback and context-sensitive help on assignments and quizzes
- track their progress throughout the course

"I just wanted to say how much this program helped me in studying... I was able to actually see my mistakes and correct them. ... I really think that other students should have the chance to use *WileyPLUS*."

Ashlee Krisko, *Oakland University*

---

www.wiley.com/college/wileyplus

80% of students surveyed said it improved their understanding of the material. *

TO THE INSTRUCTOR

WileyPLUS is built around the activities you perform

Prepare & Present

Create outstanding class presentations using a wealth of resources, such as PowerPoint™ slides, image galleries, interactive simulations, demonstration videos, and more. Plus you can easily upload any materials you have created into your course, and combine them with the resources Wiley provides you with.

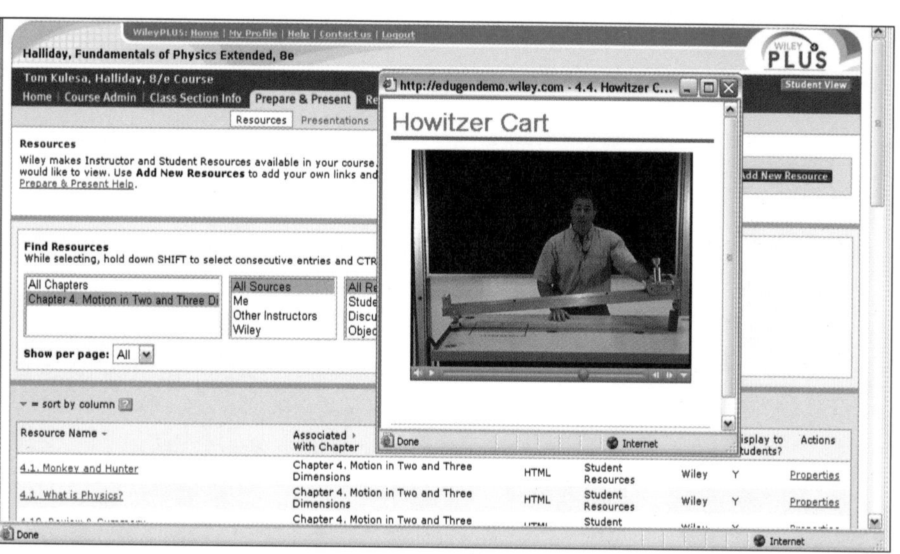

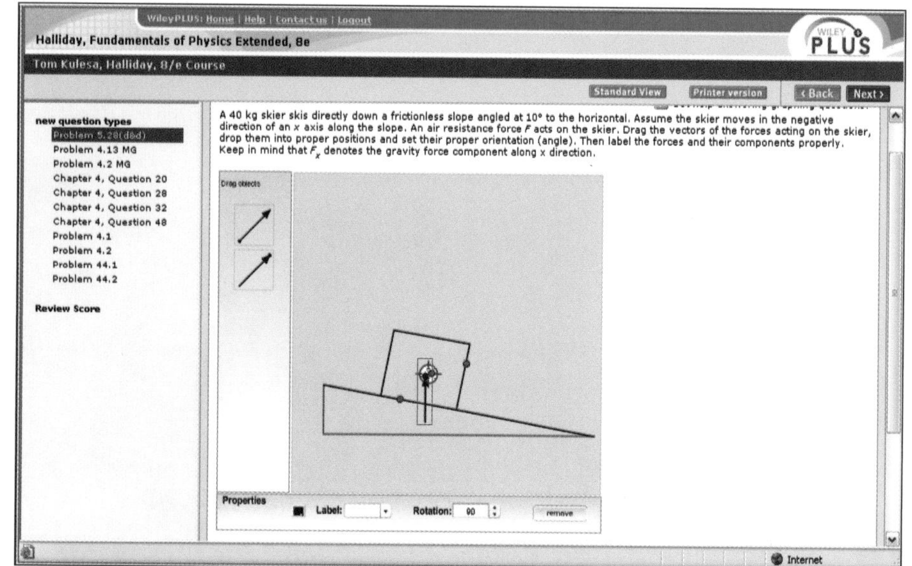

Create Assignments

All of the end-of-chapter problems from the 8th edition of *Fundamentals of Physics* are available for assignment in an algorithmic format. Each problem has a link to the relevant portion of the text and a Hint or other problem solving help that can be made available to the student at the instructor's discretion. New to this edition are GO (Guided Online) problems that provide problem solving guidance in an interactive tutorial format. Also new are vector drawing and vector diagram problems and problems built around interactive simulations.

*Based on a spring 2005 survey of 972 student users of *WileyPLUS*

TO THE STUDENT

You have the potential to make a difference!

Will you be the first person to land on Mars? Will you invent a car that runs on water? But, first and foremost, will you get through this course?

WileyPLUS is a powerful online system packed with features to help you make the most of your potential, and get the best grade you can!

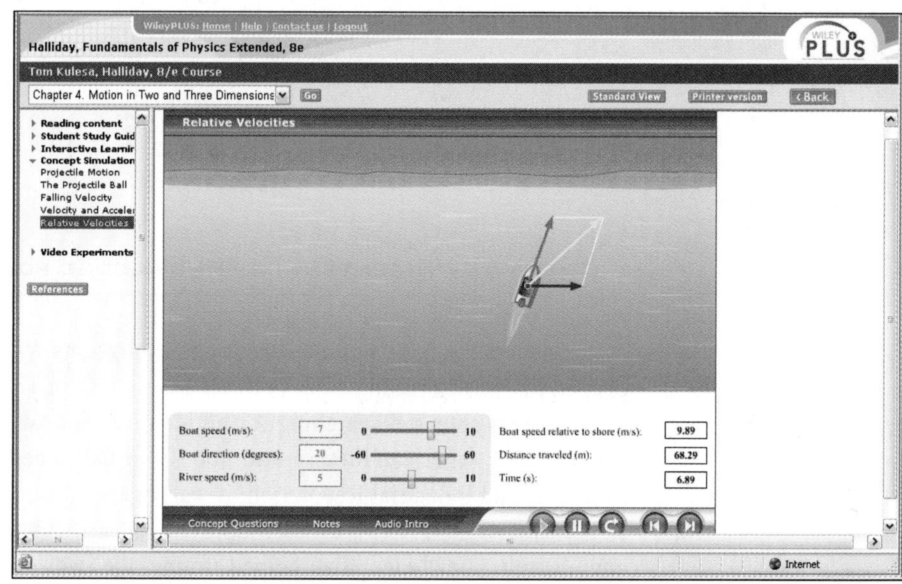

With WileyPLUS you get:

A complete online version of your text and other study resources

Study more effectively and get instant feedback when you practice on your own. Resources like self-assessment quizzes, concept simulations, and Interactive LearningWare bring the subject matter to life, and help you master the material.

Problem-solving help, instant grading, and feedback on your homework and quizzes

You can keep all of your assigned work in one location, making it easy for you to stay on task. Plus, many homework problems contain direct links to the relevant portion of your text to help you deal with problem-solving obstacles at the moment they come up.

The ability to track your progress and grades throughout the term.

A personal gradebook allows you to monitor your results from past assignments at any time. You'll always know exactly where you stand.

If your instructor uses *WileyPLUS*, you will receive a URL for your class. If not, your instructor can get more information about *WileyPLUS* by visiting www.wiley.com/college/wileyplus

"It has been a great help, and I believe it has helped me to achieve a better grade."
Michael Morris, *Columbia Basin College*

69% of students surveyed said it helped them get a better grade. *

The Wiley Bicentennial–Knowledge for Generations

Each generation has its unique needs and aspirations. When Charles Wiley first opened his small printing shop in lower Manhattan in 1807, it was a generation of boundless potential searching for an identity. And we were there, helping to define a new American literary tradition. Over half a century later, in the midst of the Second Industrial Revolution, it was a generation focused on building the future. Once again, we were there, supplying the critical scientific, technical, and engineering knowledge that helped frame the world. Throughout the 20th Century, and into the new millennium, nations began to reach out beyond their own borders and a new international community was born. Wiley was there, expanding its operations around the world to enable a global exchange of ideas, opinions, and know-how.

For 200 years, Wiley has been an integral part of each generation's journey, enabling the flow of information and understanding necessary to meet their needs and fulfill their aspirations. Today, bold new technologies are changing the way we live and learn. Wiley will be there, providing you the must-have knowledge you need to imagine new worlds, new possibilities, and new opportunities.

Generations come and go, but you can always count on Wiley to provide you the knowledge you need, when and where you need it!

William J. Pesce
President and Chief Executive Officer

Peter Booth Wiley
Chairman of the Board

HALLIDAY / RESNICK

Fundamentals of Physics

Jearl Walker

Cleveland State University

John Wiley & Sons, Inc.

ACQUISITIONS EDITOR Stuart Johnson
PROJECT EDITOR Geraldine Osnato
EDITORIAL ASSISTANT Aly Rentrop
SENIOR MARKETING MANAGER Amanda Wygal
SENIOR PRODUCTION EDITOR Elizabeth Swain
TEXT DESIGNER Madelyn Lesure
COVER DESIGNER Norm Christiansen
DUMMY DESIGNER Lee Goldstein
PHOTO EDITOR Hilary Newman
ILLUSTRATION EDITOR Anna Melhorn
ILLUSTRATION STUDIO Radiant Illustrations Inc.
SENIOR MEDIA EDITOR Thomas Kulesa
MEDIA PROJECT MANAGER Bridget O'Lavin
COVER IMAGE ©Eric Heller/Photo Researchers

This book was set in 10/12 Times Ten by Progressive Information Technologies and printed and bound by Von Hoffmann Press. The cover was printed by Von Hoffmann Press.

This book is printed on acid free paper.

To order books or for customer service please call 1-800-CALL WILEY (225-5945).

Library of Congress Cataloging-in-Publication Data
Halliday, David
Fundamentals of physics.—8th ed., Volume 1/David Halliday, Robert Resnick, Jearl Walker.
p. cm.
Includes index.
Volume 1: ISBN 13: 978-0-470-04473-5 (acid-free paper)
ISBN 10: 0-470-04473-X (acid-free paper)
Also catalogued as
Extended version: ISBN-13: 978-0-471-75801-3 (acid-free paper)
ISBN-10: 0-471-75801-9 (acid-free paper)
1. Physics—Textbooks. I. Resnick, Robert II. Walker, Jearl III. Title.
QC21.3.H35 2008
530—dc22

2006041375

Printed in the United States of America

10 9 8 7 6 5 4 3

Brief Contents

Volume 1

PART 1

PART 2

Volume 2

PART 3

PART 4

PART 5

Contents

7 Kinetic Energy and Work 140

What property of a funny car determines the winning time in a drag race?

8 Potential Energy and Conservation of Energy 166

Why can a large landslide move almost 30 times as far across a valley as a small landslide?

9 Center of Mass and Linear Momentum 201

How can a male bighorn sheep survive the severe head collision with which it gains dominance?

10 Rotation 241

How can a small snapping shrimp snap its claw so firmly that the sound stuns its prey?

11 Rolling, Torque, and Angular Momentum 275

What is the magic behind the onset of a ballet dancer's midair rotation during a tour jeté?

12 Equilibrium and Elasticity 305

Why is even the slight tilt of the leaning tower in Pisa dangerous?

13 Gravitation 330

What "monster" lurks at the center of our Milky Way Galaxy?

14 Fluids 359

How does a surfer surf?

15 Oscillations 386

How can you stop the natural but annoying oscillations of a tall building in a wind?

16 Waves—I 413

What causes the sometimes dangerous oscillations of footbridges and mosh pits?

17 Waves—II 445

What causes the musical echo from the stairs at an ancient Mayan pyramid?

18 Temperature, Heat, and the First Law of Thermodynamics 476

How can a beetle detect a forest fire at a large distance without seeing or smelling it?

Preface

Fun with a big challenge. That is how I have regarded physics since the day when Sharon, one of the students in a class I taught as a graduate student, suddenly demanded of me, "What has any of this got to do with my life?" Of course I immediately responded, "Sharon, this has everything to do with your life—this is physics."

She asked me for an example. I thought and thought but could not come up with a single one. That night I created *The Flying Circus of Physics* for Sharon but also for me because I realized her complaint was mine. I had spent six years slugging my way through many dozens of physics textbooks that were carefully written with the best of pedagogical plans, but there was something missing. Physics is the most interesting subject in the world because it is about how the world works, and yet the textbooks had been thoroughly wrung of any connection with the real world. The fun was missing.

I have packed a lot of real-world physics into this HRW book, connecting it with the new edition of *The Flying Circus of Physics*. Much of the material comes from the HRW classes I teach, where I can judge from the faces and blunt comments what material and presentations work and what do not. The notes I make on my successes and failures there help form the basis of this book. My message here is the same as I had with every student I've met since Sharon so long ago: "Yes, you *can* reason from basic physics concepts all the way to valid conclusions about the real world, and that understanding of the real world is where the fun is."

I have many goals in writing this book but the overriding one is to provide instructors with a tool by which they can teach students how to effectively read scientific material, identify fundamental concepts, reason through scientific questions, and solve quantitative problems. This process is not easy for either students or instructors. Indeed, the course associated with this book may be one of the most challenging of all the courses taken by a student. However, it can also be one of the most rewarding because it reveals the world's fundamental clockwork from which all scientific and engineering applications spring.

Many users of the seventh edition (both instructors and students) sent in comments and suggestions to improve the book. These improvements are now incorporated into the narrative and problems throughout the book. The publisher John Wiley & Sons and I regard the book as an ongoing project and encourage more input from users. You can send suggestions, corrections, and positive or negative comments to John Wiley & Sons (http:www.wiley.com/college/halliday) or Jearl Walker (mail address: Physics Department, Cleveland State University, Cleveland, OH 44115 USA; fax number: (USA) 216 687 2424; or email address: physics@wiley.com; or the blog site at www.flyingcircusofphysics.com). We may not be able to respond to all suggestions, but we keep and study each of them.

Major Content Changes

- Flying Circus material has been incorporated into the text in several ways: chapter opening Puzzlers, Sample Problems, text examples, and end-of-chapter Problems. The purpose of this is two-fold: (1) make the subject more interesting and engaging, (2) show the student that the world around them can be examined and understood using the fundamental principles of physics.

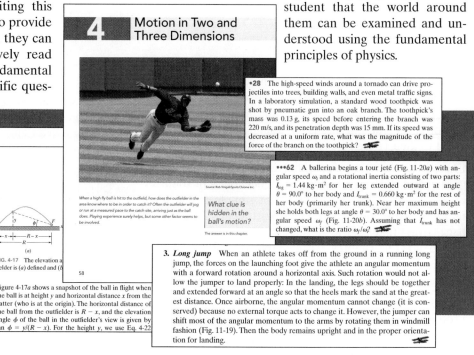

4 Motion in Two and Three Dimensions

Source: Rob Tringali/Sports Chrome Inc.

When a high fly ball is hit to the outfield, how does the outfielder in the area know where to be in order to catch it? Often the outfielder will jog or run at a measured pace to the catch site, arriving just as the ball does. Playing experience surely helps, but some other factor seems to be involved.

What clue is hidden in the ball's motion?

The answer is in this chapter.

58

Sample Problem 4-8

Suppose a baseball batter B hits a high fly ball to the outfield, directly toward an outfielder F and with a launch speed of $v_0 = 40$ m/s and a launch angle of $\theta_0 = 35°$. During the flight, a line from the outfielder to the ball makes an angle ϕ with the ground. Plot elevation angle ϕ versus time t, assuming that the outfielder is already positioned to catch the ball, is 6.0 m too close to the batter, and is 6.0 m too far away.

KEY IDEAS (1) If we neglect air drag, the ball is a projectile for which the vertical motion and the horizontal motion can be analyzed separately. (2) Assuming the ball is caught at approximately the height it is hit, the horizontal distance traveled by the ball is the range R, given by Eq. 4-26 ($R = (v_0^2/g) \sin 2\theta_0$).

Calculations: The ball can be caught if the outfielder's distance from the batter is equal to the range R of the ball. Using Eq. 4-26, we find

$$R = \frac{v_0^2}{g} \sin 2\theta_0 = \frac{(40\text{ m/s})^2}{9.8\text{ m/s}^2} \sin (70°) = 153.42\text{ m}.$$

y B F ϕ x R − x R (a)

FIG. 4-17 The elevation a
fielder is (a) defined and (b

Figure 4-17*a* shows a snapshot of the ball in flight when the ball is at height y and horizontal distance x from the batter (who is at the origin). The horizontal distance of the ball from the outfielder is $R - x$, and the elevation angle ϕ of the ball in the outfielder's view is given by $\tan \phi = y/(R - x)$. For the height y, we use Eq. 4-22

•28 The high-speed winds around a tornado can drive projectiles into trees, building walls, and even metal traffic signs. In a laboratory simulation, a standard wood toothpick was shot by pneumatic gun into an oak branch. The toothpick's mass was 0.13 g, its speed before entering the branch was 220 m/s, and its penetration depth was 15 mm. If its speed was decreased at a uniform rate, what was the magnitude of the force of the branch on the toothpick?

•••62 A ballerina begins a tour jeté (Fig. 11-20*a*) with angular speed ω_i and a rotational inertia consisting of two parts: $I_{leg} = 1.44\text{ kg}\cdot\text{m}^2$ for her leg extended outward at angle $\theta = 90.0°$ to her body and $I_{trunk} = 0.660\text{ kg}\cdot\text{m}^2$ for the rest of her body (primarily her trunk). Near her maximum height she holds both legs at angle $\theta = 30.0°$ to her body and has angular speed ω_f (Fig. 11-20*b*). Assuming that I_{trunk} has not changed, what is the ratio ω_f/ω_i?

3. ***Long jump*** When an athlete takes off from the ground in a running long jump, the forces on the launching foot give the athlete an angular momentum with a forward rotation around a horizontal axis. Such rotation would not allow the jumper to land properly: In the landing, the legs should be together and extended forward at an angle so that the heels mark the sand at the greatest distance. Once airborne, the angular momentum cannot change (it is conserved) because no external torque acts to change it. However, the jumper can shift most of the angular momentum to the arms by rotating them in windmill fashion (Fig. 11-19). Then the body remains upright and in the proper orientation for landing.

- Links to *The Flying Circus of Physics* are shown throughout the text material and end-of-chapter problems with a biplane icon. In the electronic version of this book, clicking on the icon takes you to the corresponding item in *Flying Circus*. The bibliography of *Flying Circus* (over 10 000 references to scientific and engineering journals) is located at www.flyingcircusofphysics.com.

- The Newtonian gravitational law, the Coulomb law, and the Biot-Savart law are now introduced in unit-vector notation.

- Most of the chapter-opening puzzlers (the examples of applied physics designed to entice a reader into each chapter) are new and come straight from research journals in many different fields.

- Several thousand of the end-of-chapter problems have been rewritten to streamline both the presentation and the answer. Many new problems of the moderate and difficult categories have been included.

Chapter Features

Opening puzzlers. A curious puzzling situation opens each chapter and is explained somewhere within the chapter, to entice a student to read the chapter. These features, which are a hallmark of *Fundamentals of Physics*, are based on current research as reported in scientific, engineering, medical, and legal journals.

What is physics? The narrative of every chapter now begins with this question, and with an answer that pertains to the subject of the chapter. (A plumber once asked me, "What do you do for a living?" I replied, "I teach physics." He thought for several minutes and then asked, "What is physics?" The plumber's career was entirely based on physics, yet he did not even know what physics is. Many students in introductory physics do not know what physics is but assume that it is irrelevant to their chosen career.)

Checkpoints are stopping points that effectively ask the student, "Can you answer this question with some reasoning based on the narrative or sample problem that you just read?" If not, then the student should go back over that previous material before traveling deeper into the chapter. For example, see Checkpoint 1 on page 62 and Checkpoint 2 on page 280. *Answers to all checkpoints are in the back of the book.*

Sample problems are chosen to demonstrate how problems can be solved with reasoned solutions rather than quick and simplistic plugging of numbers into an equation with no regard for what the equation means. The sample problems with the label "Build your skill" are typically longer, with more guidance.

Key Ideas in the sample problems focus a student on the basic concepts at the root of the solution to a problem. In effect, these key ideas say, "We start our solution by using this basic concept, a procedure that prepares us for solving many other problems. We don't start by grabbing an equation for a quick plug-and-chug, a procedure that prepares us for nothing."

Problem-solving tactics contain helpful instructions to guide the beginning physics student as to how to solve problems and avoid common errors.

Review & Summary is a brief outline of the chapter contents that contains the essential concepts but which is not a substitute for reading the chapter.

Questions are like the checkpoints and require reasoning and understanding rather than calculations. *Answers to the odd-numbered questions are in the back of the book.*

Problems are grouped under section titles and are labeled according to difficulty. *Answers to the odd-numbered problems are in the back of the book.*

Icons for additional help. When worked-out solutions are provided either in print or electronically for certain of the odd-numbered problems, the statements for those problems include a trailing icon to alert both student and instructor as to where the solutions are located. An icon guide is provided here and at the beginning of each set of problems:

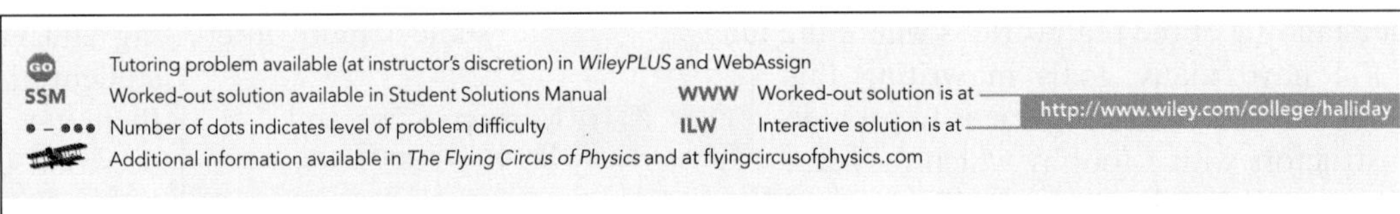

GO Tutoring problem available (at instructor's discretion) in WileyPLUS and WebAssign
SSM Worked-out solution available in Student Solutions Manual
• – ••• Number of dots indicates level of problem difficulty
Additional information available in The Flying Circus of Physics and at flyingcircusofphysics.com
WWW Worked-out solution is at
ILW Interactive solution is at
http://www.wiley.com/college/halliday

SSM Solution is in the Student Solutions Manual.
WWW Solution is at
http://www.wiley.com/college/halliday
ILW Interactive LearningWare solution is at
http://www.wiley.com/college/halliday

Additional problems. These problems are not ordered or sorted in any way so that a student must determine which parts of the chapter apply to any given problem.

Additional Features

Reasoning versus plug-and-chug. A primary goal of this book is to teach students to reason through challenging situations, from basic principles to a solution. Although some plug-and-chug homework problems remain in the book (on purpose), most homework problems emphasize reasoning.

Chapters of reasonable length. To avoid producing a book thick enough to stop a bullet (and thus also a student), I have made the chapters of reasonable length. I explain enough to get a student going but not so much that a student no longer must analyze and fuse ideas. After all, a student will need the skill of analyzing and fusing ideas long after this book is read and the course is completed.

Use of vector-capable calculators. When vector calculations in a sample problem can be performed directly on-screen with a vector-capable calculator, the solution of the sample problem indicates that fact but still carries through the traditional component analysis. When vector calculations cannot be performed directly on-screen, the solution explains why.

Graphs as puzzles. These are problems that give a graph and ask for a result that requires much more than just reading off a data point from the graph. Rather, the solution requires an understanding of the physical arrangement in a problem and the principles behind the associated equations. These problems are more like Sherlock Holmes puzzles because a student must decide what data are important. For examples, see problem 50 on page 80, problem 12 on page 108, and problem 22 on page 231.

Problems with applied physics, based on published research, appear in many places, either as the opening puzzler of a chapter, a sample problem, or a homework problem. For example, see the opening puzzler for Chapter 4 on page 58, Sample Problem 4-8 on pages 69-70, and homework problem 62 on page 302. For an example of homework problems that build on a continuing story, see problems 2, 39, and 61 on pages 131, 134, and 136.

Problems with novel situations. Here is one of several hundred such problems: Problem 69 on page 113 relates a true story of how Air Canada flight 143 ran out of fuel at an altitude of 7.9 km because the crew and airport personnel did not consider the units for the fuel (an important lesson for students who tend to "blow off" units).

Versions of the Text

To accommodate the individual needs of instructors and students, the eighth edition of *Fundamentals of Physics* is available in a number of different versions.

The **Regular Edition** consists of Chapters 1 through 37 (ISBN 978-0-470-04472-8).

The **Extended Edition** contains six additional chapters on quantum physics and cosmology, Chapters 1–44 (ISBN 978-0-471-75801-3).

Both editions are available as single, hard-cover books, or in the following alternative versions:

- **Volume 1** - Chapters 1–20 (Mechanics and Thermodynamics), hardcover, ISBN 978-0-47004473-5
- **Volume 2** - Chapters 21–44 (E&M, Optics, and Quantum Physics), hardcover, ISBN 978-0-470-04474-2
- **Part 1** - Chapters 1–11, paperback, ISBN 978-0-470-04475-9
- **Part 2** - Chapters 12–20, paperback, ISBN 978-0-470-04476-6
- **Part 3** - Chapters 21–32, paperback, ISBN 978-0-470-04477-3
- **Part 4** - Chapters 33–37, paperback, ISBN 978-0-470-04478-0
- **Part 5** - Chapters 38–44, paperback, ISBN 978-0-470-04479-7

Wiley*PLUS*

There have been several significant additions to the Wiley*PLUS* course that accompanies *Fundamentals of Physics:*

- All of the end-of-chapter problems have been coded and are now available for assignment.
- Every problem has an associated Hint that can made available to the students at the instructor's discretion.
- There are approximately 400 additional Sample Problems available to the student **at the instructor's discretion.** The Sample Problems are written in the same style and format as those in the text, i.e., they are intended to give the student transferable problem-solving skills rather than specific recipes.
- For every chapter, approximately 6 problems are available in a tutorial format that provides step-by-step, interactive problem-solving guidance. Most are marked here in this book with the icon GO ("Guided Online") but more are being added.
- For every chapter, approximately 6 problems are available in a version that requires the student to enter an algebraic answer.
- There are vector drawing and vector diagram problems that use "drag 'n drop" functionality to assess the students' ability to draw vectors and vector diagrams.
- There are simulation problems that require the student to work with a java applet.

The overall purpose of this new material is to move the on-line homework experience beyond simple Right/Wrong grading and provide meaningful problem solving guidance and support.

Instructor's Supplements

Instructor's Solutions Manual by Sen-Ben Liao, Lawrence Livermore National Laboratory. This manual provides worked-out solutions for all problems found at the end of each chapter.

Instructor Companion Site

http://www.wiley.com/college/halliday

- **Instructor's Manual** by J. Richard Christman, U.S. Coast Guard Academy. This resource contains lecture notes outlining the most important topics of each chapter; demonstration experiments; laboratory and computer projects; film and video sources; answers to all Questions, Exercises, Problems, and Checkpoints; and a correlation guide to the Questions, Exercises, and Problems in the previous edition. It also contains a complete list of all problems for which solutions are available to students (SSM, WWW, and ILW).
- **Lecture PowerPoint Slides** by Athos Petrou and John Cerne of the University of Buffalo. These PowerPoints cover the entire book and are heavily illustrated with figures from the text.
- **Classroom Response Systems ("Clicker") Questions** by David Marx, Illinois State University. There are two sets of questions available: Reading Quiz and Interactive Lecture. The Reading Quiz questions are intended to be relatively straightforward for any student who read the assigned material. The Interactive Lecture questions are intended to for use in an interactive lecture setting.
- **Wiley Physics Simulations** by Andrew Duffy, Boston University. 50 interactive simulations (Java applets) that can be used for classroom demonstrations.
- **Wiley Physics Demonstrations** by David Maiullo, Rutgers University. This is a collection of digital videos of 80 standard physics demonstrations. They can be shown in class or accessed from the student companion site. There is an accompanying Instructor's Guide that includes "clicker" questions.
- **Test Bank** by J. Richard Christman, U.S. Coast Guard Academy. The Test Bank includes more than 2200 multiple-choice questions. These items are also available in the Computerized Test Bank which provides full editing features to help you customize tests (available in both IBM and Macintosh versions).
- All of the *Instructor's Solutions Manual* in MSWord, and pdf files
- All text illustrations, suitable for both classroom projection and printing.

On-line homework and quizzing. In addition to Wiley*PLUS, Fundamentals of Physics*, eighth edition also supports WebAssignPLUS and CAPA, which are other programs that give instructors the ability to deliver and grade homework and quizzes on-line.

WebCT and Blackboard. A variety of materials have been prepared for easy incorporation in either WebCT or Blackboard. WebCT and Blackboard are powerful and easy-to-use web-based course-management systems that allow instructors to set up complete on-line courses with chat rooms, bulletin boards, quizzing, student tracking, etc.

Student's Supplements

Student Companion site. This web site

http://www.wiley.com/college/halliday

was developed specifically for *Fundamentals of Physics,* eighth edition, and is designed to further assist students in the study of physics. The site includes solutions to selected end-of-chapter problems (which are identified with a www icon in the text); self-quizzes; simulation exercises; tips on how to make best use of a programmable calculator; and the Interactive LearningWare tutorials that are described below.

Student Study Guide. The student study guide consists of an overview of the chapter's important concepts, hints for solving end-of-chapter questions/problems, and practice quizzes.

Student's Solutions Manual by J. Richard Christman, U.S. Coast Guard Academy and Edward Derringh, Wentworth Institute. This manual provides student with complete worked-out solutions to 15 percent of the problems found at the end of each chapter within the text. These problems are indicated with an ssm icon.

Interactive LearningWare. This software guides students through solutions to 200 of the end-of-chapter problems. These problems are indicated with an ilw icon. The solutions process is developed interactively, with appropriate feedback and access to error-specific help for the most common mistakes.

Wiley Desktop Edition. An electronic version of *Fundamentals of Physics,* eighth edition containing the complete, extended version of the text is available for download at:

www.wiley.com/college/desktop

Wiley Desktop Editions are a cost effective alternative to the printed text.

Physics as a Second Language: *Mastering Problem Solving* by Thomas Barrett, of Ohio State University. This brief paperback teaches the student how to approach problems more efficiently and effectively. The student will learn how to recognize common patterns in physics problems, break problems down into manageable steps, and apply appropriate techniques. The book takes the student step-by-step through the solutions to numerous examples.

Acknowledgments

A great many people have contributed to this book. J. Richard Christman, of the U.S. Coast Guard Academy, has once again created many fine supplements; his recommendations to this book have been invaluable. Sen-Ben Liao of Lawrence Livermore National Laboratory, James Whitenton of Southern Polytechnic State University, and Jerry Shi, of Pasadena City College, performed the Herculean task of working out solutions for every one of the homework problems in the book. At John Wiley publishers, the book received support from Stuart Johnson, the editor who oversaw the entire project, Tom Kulesa, who coordinated the state-of-the-art media package, and Geraldine Osnato, who managed a super team to create an impressive supplements package. We thank Elizabeth Swain, the production editor, for pulling all the pieces together during the complex production process. We also thank Maddy Lesure, for her design of both the text and the book cover; Lee Goldstein for her page make-up; Helen Walden for her copyediting; Anna Melhorn for managing the illustration program; and Lilian Brady for her proofreading. Hilary Newman was inspired in the search for unusual and interesting photographs. Both the publisher John Wiley & Sons, Inc. and Jearl Walker would like to thank the following for comments and ideas about the 7th edition: Richard Woodard, University of Florida; David Wick, Clarkson University; Patrick Rapp, University of Puerto Rico at Mayagüez; Nora Thornber, Raritan Valley Community College; Laurence I. Gould, University of Hartford; Greg Childers, California State University at Fullerton; Asha Khakpour of California State University at Fullerton; Joe F. McCullough, Cabrillo College. Finally, our external reviewers have been outstanding and we acknowledge here our debt to each member of that team.

Maris A. Abolins
Michigan State University

Edward Adelson
Ohio State University

Nural Akchurin
Texas Tech

Barbara Andereck
Ohio Wesleyan University

Mark Arnett
Kirkwood Community College

Arun Bansil
Northeastern University

Richard Barber
Santa Clara University

Neil Basecu
Westchester Community College

Anand Batra
Howard University

Richard Bone
Florida International University

Michael E. Browne
University of Idaho

Timothy J. Burns
Leeward Community College

Joseph Buschi
Manhattan College

Philip A. Casabella
Rensselaer Polytechnic Institute

Randall Caton
Christopher Newport College

Roger Clapp
University of South Florida

W. R. Conkie
Queen's University

Renate Crawford
University of Massachusetts-Dartmouth

Mike Crivello
San Diego State University

Robert N. Davie, Jr.
St. Petersburg Junior College

Cheryl K. Dellai
Glendale Community College

Eric R. Dietz
California State University at Chico

N. John DiNardo
Drexel University

Eugene Dunnam
University of Florida

Robert Endorf
University of Cincinnati

F. Paul Esposito
University of Cincinnati

Jerry Finkelstein
San Jose State University

Robert H. Good
California State University-Hayward

John B. Gruber
San Jose State University

Ann Hanks
American River College

Randy Harris
University of California-Davis

Samuel Harris
Purdue University

Harold B. Hart
Western Illinois University

Rebecca Hartzler
Seattle Central Community College

John Hubisz
North Carolina State University

Joey Huston
Michigan State University

David Ingram
Ohio University

Shawn Jackson
University of Tulsa

Hector Jimenez
University of Puerto Rico

Sudhakar B. Joshi
York University

Leonard M. Kahn
University of Rhode Island

Leonard Kleinman
University of Texas at Austin

Craig Kletzing
University of Iowa

Arthur Z. Kovacs
Rochester Institute of Technology

Kenneth Krane
Oregon State University

Priscilla Laws
Dickinson College

Edbertho Leal
Polytechnic University of Puerto Rico

Vern Lindberg
Rochester Institute of Technology

Peter Loly
University of Manitoba

Andreas Mandelis
University of Toronto

Robert R. Marchini
Memphis State University

Paul Marquard
Caspar College

David Marx
Illinois State University

James H. McGuire
Tulane University

David M. McKinstry
Eastern Washington University

Eugene Mosca
United States Naval Academy

James Napolitano
Rensselaer Polytechnic Institute

Michael O'Shea
Kansas State University

Patrick Papin
San Diego State University

Kiumars Parvin
San Jose State University

Robert Pelcovits
Brown University

Oren P. Quist
South Dakota State University

Joe Redish
University of Maryland

Timothy M. Ritter
University of North Carolina at Pembroke

Gerardo A. Rodriguez
Skidmore College

John Rosendahl
University of California at Irvine

Todd Ruskell
Colorado School of Mines

Michael Schatz
Georgia Institute of Technology

Darrell Seeley
Milwaukee School of Engineering

Bruce Arne Sherwood
North Carolina State University

Ross L. Spencer
Brigham Young University

Paul Stanley
Beloit College

Harold Stokes
Brigham Young University

Michael G. Strauss
University of Oklahoma

Jay D. Strieb
Villanova University

Dan Styer
Oberlin College

Michael Tammaro
University of Rhode Island

Marshall Thomsen
Eastern Michigan University

David Toot
Alfred University

Tung Tsang
Howard University

J. S. Turner
University of Texas at Austin

T. S. Venkataraman
Drexel University

Gianfranco Vidali
Syracuse University

Fred Wang
Prairie View A & M

Robert C. Webb
Texas A & M University

William M. Whelan
Ryerson Polytechnic University

George Williams
University of Utah

David Wolfe
University of New Mexico

Measurement

When an earthquake strikes a populated region, it can shake apart buildings and other structures or cause them to topple over. However, in some regions it can cause structures to sink into the ground until they are significantly submerged, as if the structures were on a dense fluid instead of solid ground.

How can a building sink into the ground?

The answer is in this chapter.

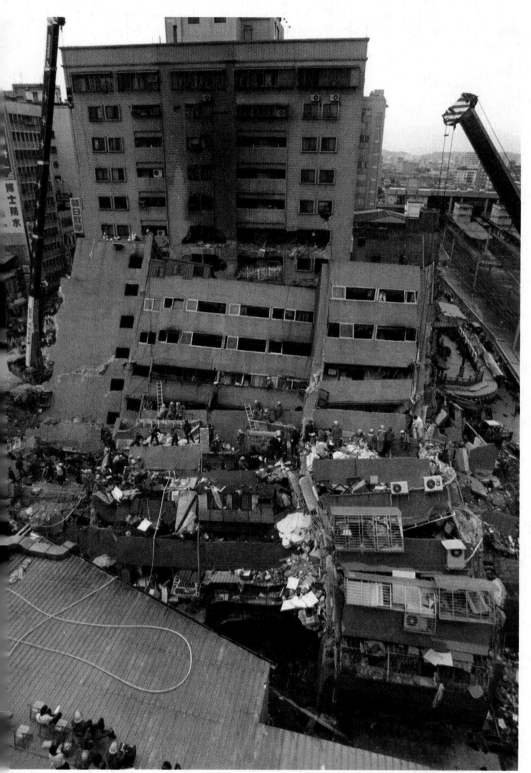

©AP/Wide World Photos

1-1 WHAT IS PHYSICS?

Science and engineering are based on measurements and comparisons. Thus, we need rules about how things are measured and compared, and we need experiments to establish the units for those measurements and comparisons. One purpose of physics (and engineering) is to design and conduct those experiments.

For example, physicists strive to develop clocks of extreme accuracy so that any time or time interval can be precisely determined and compared. You may wonder whether such accuracy is actually needed or worth the effort. Here is one example of the worth: Without clocks of extreme accuracy, the Global Positioning System (GPS) that is now vital to worldwide navigation would be useless.

1-2 | Measuring Things

We discover physics by learning how to measure the quantities involved in physics. Among these quantities are length, time, mass, temperature, pressure, and electric current.

We measure each physical quantity in its own units, by comparison with a **standard.** The **unit** is a unique name we assign to measures of that quantity—for example, meter (m) for the quantity length. The standard corresponds to exactly 1.0 unit of the quantity. As you will see, the standard for length, which corresponds to exactly 1.0 m, is the distance traveled by light in a vacuum during a certain fraction of a second. We can define a unit and its standard in any way we care to. However, the important thing is to do so in such a way that scientists around the world will agree that our definitions are both sensible and practical.

Once we have set up a standard—say, for length—we must work out procedures by which any length whatever, be it the radius of a hydrogen atom, the wheelbase of a skateboard, or the distance to a star, can be expressed in terms of the standard. Rulers, which approximate our length standard, give us one such procedure for measuring length. However, many of our comparisons must be indirect. You cannot use a ruler, for example, to measure the radius of an atom or the distance to a star.

There are so many physical quantities that it is a problem to organize them. Fortunately, they are not all independent; for example, speed is the ratio of a length to a time. Thus, what we do is pick out—by international agreement—a small number of physical quantities, such as length and time, and assign standards to them alone. We then define all other physical quantities in terms of these *base quantities* and their standards (called *base standards*). Speed, for example, is defined in terms of the base quantities length and time and their base standards.

Base standards must be both accessible and invariable. If we define the length standard as the distance between one's nose and the index finger on an outstretched arm, we certainly have an accessible standard—but it will, of course, vary from person to person. The demand for precision in science and engineering pushes us to aim first for invariability. We then exert great effort to make duplicates of the base standards that are accessible to those who need them.

1-3 | The International System of Units

In 1971, the 14th General Conference on Weights and Measures picked seven quantities as base quantities, thereby forming the basis of the International System of Units, abbreviated SI from its French name and popularly known as the *metric system.* Table 1-1 shows the units for the three base quantities—length, mass, and time—that we use in the early chapters of this book. These units were defined to be on a "human scale."

TABLE 1-1

Units for Three SI Base Quantities

Quantity	Unit Name	Unit Symbol
Length	meter	m
Time	second	s
Mass	kilogram	kg

Many SI *derived units* are defined in terms of these base units. For example, the SI unit for power, called the **watt** (W), is defined in terms of the base units for mass, length, and time. Thus, as you will see in Chapter 7,

$$1 \text{ watt} = 1 \text{ W} = 1 \text{ kg} \cdot \text{m}^2/\text{s}^3, \quad (1\text{-}1)$$

where the last collection of unit symbols is read as kilogram-meter squared per second cubed.

To express the very large and very small quantities we often run into in physics, we use *scientific notation,* which employs powers of 10. In this notation,

$$3\,560\,000\,000 \text{ m} = 3.56 \times 10^9 \text{ m} \quad (1\text{-}2)$$

and

$$0.000\,000\,492 \text{ s} = 4.92 \times 10^{-7} \text{ s}. \quad (1\text{-}3)$$

Scientific notation on computers sometimes takes on an even briefer look, as in 3.56 E9 and 4.92 E–7, where E stands for "exponent of ten." It is briefer still on some calculators, where E is replaced with an empty space.

As a further convenience when dealing with very large or very small measurements, we use the prefixes listed in Table 1-2. As you can see, each prefix represents a certain power of 10, to be used as a multiplication factor. Attaching a prefix to an SI unit has the effect of multiplying by the associated factor. Thus, we can express a particular electric power as

$$1.27 \times 10^9 \text{ watts} = 1.27 \text{ gigawatts} = 1.27 \text{ GW} \quad (1\text{-}4)$$

or a particular time interval as

$$2.35 \times 10^{-9} \text{ s} = 2.35 \text{ nanoseconds} = 2.35 \text{ ns}. \quad (1\text{-}5)$$

Some prefixes, as used in milliliter, centimeter, kilogram, and megabyte, are probably familiar to you.

TABLE 1-2

Prefixes for SI Units

Factor	Prefix[a]	Symbol
10^{24}	yotta-	Y
10^{21}	zetta-	Z
10^{18}	exa-	E
10^{15}	peta-	P
10^{12}	tera-	T
$\mathbf{10^{9}}$	**giga-**	**G**
$\mathbf{10^{6}}$	**mega-**	**M**
$\mathbf{10^{3}}$	**kilo-**	**k**
10^{2}	hecto-	h
10^{1}	deka-	da
10^{-1}	deci-	d
$\mathbf{10^{-2}}$	**centi-**	**c**
$\mathbf{10^{-3}}$	**milli-**	**m**
$\mathbf{10^{-6}}$	**micro-**	$\boldsymbol{\mu}$
$\mathbf{10^{-9}}$	**nano-**	**n**
$\mathbf{10^{-12}}$	**pico-**	**p**
10^{-15}	femto-	f
10^{-18}	atto-	a
10^{-21}	zepto-	z
10^{-24}	yocto-	y

[a]The most frequently used prefixes are shown in bold type.

1-4 | Changing Units

We often need to change the units in which a physical quantity is expressed. We do so by a method called *chain-link conversion.* In this method, we multiply the original measurement by a **conversion factor** (a ratio of units that is equal to unity). For example, because 1 min and 60 s are identical time intervals, we have

$$\frac{1 \text{ min}}{60 \text{ s}} = 1 \quad \text{and} \quad \frac{60 \text{ s}}{1 \text{ min}} = 1.$$

Thus, the ratios (1 min)/(60 s) and (60 s)/(1 min) can be used as conversion factors. This is *not* the same as writing $\frac{1}{60} = 1$ or $60 = 1$; each *number* and its *unit* must be treated together.

Because multiplying any quantity by unity leaves the quantity unchanged, we can introduce conversion factors wherever we find them useful. In chain-link conversion, we use the factors to cancel unwanted units. For example, to convert 2 min to seconds, we have

$$2 \text{ min} = (2 \text{ min})(1) = (2 \cancel{\text{min}})\left(\frac{60 \text{ s}}{1 \cancel{\text{min}}}\right) = 120 \text{ s}. \quad (1\text{-}6)$$

If you introduce a conversion factor in such a way that unwanted units do *not* cancel, invert the factor and try again. In conversions, the units obey the same algebraic rules as variables and numbers.

Appendix D gives conversion factors between SI and other systems of units, including non-SI units still used in the United States. However, the conversion factors are written in the style of "1 min = 60 s" rather than as a ratio. The following sample problem gives an example of how to set up such ratios.

Sample Problem 1-1

When, according to legend, Pheidippides ran from Marathon to Athens in 490 B.C. to bring word of the Greek victory over the Persians, he probably ran at a speed of about 23 rides per hour (rides/h). The ride is an ancient Greek unit for length, as are the stadium and the plethron: 1 ride was defined to be 4 stadia, 1 stadium was defined to be 6 plethra, and, in terms of a modern unit, 1 plethron is 30.8 m. How fast did Pheidippides run in kilometers per second (km/s)?

KEY IDEA In chain-link conversions, we write the conversion factors as ratios that will eliminate unwanted units.

Calculation: Here we write

$$23 \text{ rides/h} = \left(23 \frac{\cancel{\text{rides}}}{\cancel{\text{h}}}\right)\left(\frac{4 \cancel{\text{ stadia}}}{1 \cancel{\text{ ride}}}\right)\left(\frac{6 \cancel{\text{ plethra}}}{1 \cancel{\text{ stadium}}}\right)$$
$$\times \left(\frac{30.8 \cancel{\text{ m}}}{1 \cancel{\text{ plethron}}}\right)\left(\frac{1 \text{ km}}{1000 \cancel{\text{ m}}}\right)\left(\frac{1 \cancel{\text{ h}}}{3600 \text{ s}}\right)$$
$$= 4.7227 \times 10^{-3} \text{ km/s} \approx 4.7 \times 10^{-3} \text{ km/s.} \quad \text{(Answer)}$$

Sample Problem 1-2

The cran is a British volume unit for freshly caught herrings: 1 cran = 170.474 liters (L) of fish, about 750 herrings. Suppose that, to be cleared through customs in Saudi Arabia, a shipment of 1255 crans must be declared in terms of cubic covidos, where the covido is an Arabic unit of length: 1 covido = 48.26 cm. What is the required declaration?

KEY IDEA From Appendix D we see that 1 L is equivalent to 1000 cm^3. To convert from *cubic* centimeters to *cubic* covidos, we must *cube* the conversion ratio between centimeters and covidos.

Calculation: We write the following chain-link conversion:

$$1255 \text{ crans}$$
$$= (1255 \text{ crans})\left(\frac{170.474 \text{ L}}{1 \text{ cran}}\right)\left(\frac{1000 \text{ cm}^3}{1 \text{ L}}\right)\left(\frac{1 \text{ covido}}{48.26 \text{ cm}}\right)^3$$
$$= 1.903 \times 10^3 \text{ covidos}^3. \quad \text{(Answer)}$$

PROBLEM-SOLVING TACTICS

Tactic 1: Significant Figures and Decimal Places If you calculated the answer to Sample Problem 1-1 without your calculator automatically rounding it off, the number $4.722\,666\,666\,67 \times 10^{-3}$ might have appeared in the display. The precision implied by this number is meaningless. We rounded the answer to 4.7×10^{-3} km/s so as not to imply that it is more precise than the given data. The given speed of 23 rides/h consists of two digits, called **significant figures.** Thus, we rounded the answer to two significant figures. In this book, final results of calculations are often rounded to match the least number of significant figures in the given data. (However, sometimes an extra significant figure is kept.) When the leftmost of the digits to be discarded is 5 or more, the last remaining digit is rounded up; otherwise it is retained as is. For example, 11.3516 is rounded to three significant figures as 11.4 and 11.3279 is rounded to three significant figures as 11.3. (The answers to sample problems in this book are usually presented with the symbol = instead of ≈ even if rounding is involved.)

When a number such as 3.15 or 3.15×10^3 is provided in a problem, the number of significant figures is apparent, but how about the number 3000? Is it known to only one significant figure (3×10^3)? Or is it known to as many as four significant figures (3.000×10^3)? In this book, we assume that all the zeros in such given numbers as 3000 are significant, but you had better not make that assumption elsewhere.

Don't confuse *significant figures* with *decimal places.* Consider the lengths 35.6 mm, 3.56 m, and 0.00356 m. They all have three significant figures but they have one, two, and five decimal places, respectively.

1-5 | Length

In 1792, the newborn Republic of France established a new system of weights and measures. Its cornerstone was the meter, defined to be one ten-millionth of the distance from the north pole to the equator. Later, for practical reasons, this Earth standard was abandoned and the meter came to be defined as the distance between two fine lines engraved near the ends of a platinum–iridium bar, the **standard meter bar,** which was kept at the International Bureau of Weights and Measures near Paris. Accurate copies of the bar were sent to standardizing laboratories throughout the world. These **secondary standards** were used to produce other, still more accessible standards, so that ultimately every measuring device derived its authority from the standard meter bar through a complicated chain of comparisons.

Eventually, a standard more precise than the distance between two fine scratches on a metal bar was required. In 1960, a new standard for the meter, based on the wavelength of light, was adopted. Specifically, the standard for the meter was redefined to be 1 650 763.73 wavelengths of a particular orange-red light emitted by atoms of krypton-86 (a particular isotope, or type, of krypton) in a gas discharge tube. This awkward number of wavelengths was chosen so that the new standard would be close to the old meter-bar standard.

By 1983, however, the demand for higher precision had reached such a point that even the krypton-86 standard could not meet it, and in that year a bold step was taken. The meter was redefined as the distance traveled by light in a specified time interval. In the words of the 17th General Conference on Weights and Measures:

> The meter is the length of the path traveled by light in a vacuum during a time interval of 1/299 792 458 of a second.

This time interval was chosen so that the speed of light c is exactly

$$c = 299\,792\,458 \text{ m/s}.$$

Measurements of the speed of light had become extremely precise, so it made sense to adopt the speed of light as a defined quantity and to use it to redefine the meter.

Table 1-3 shows a wide range of lengths, from that of the universe (top line) to those of some very small objects.

TABLE 1-3

Some Approximate Lengths

Measurement	Length in Meters
Distance to the first galaxies formed	2×10^{26}
Distance to the Andromeda galaxy	2×10^{22}
Distance to the nearby star Proxima Centauri	4×10^{16}
Distance to Pluto	6×10^{12}
Radius of Earth	6×10^{6}
Height of Mt. Everest	9×10^{3}
Thickness of this page	1×10^{-4}
Length of a typical virus	1×10^{-8}
Radius of a hydrogen atom	5×10^{-11}
Radius of a proton	1×10^{-15}

PROBLEM-SOLVING TACTICS

***Tactic 2:* Order of Magnitude** The *order of magnitude* of a number is the power of ten when the number is expressed in scientific notation. For example, if $A = 2.3 \times 10^4$ and $B = 7.8 \times 10^4$, then the orders of magnitude of both A and B are 4.

Often, engineering and science professionals will estimate the result of a calculation to the *nearest* order of magnitude. For our example, the nearest order of magnitude is 4 for A and 5 for B. Such estimation is common when detailed or precise data required in the calculation are not known or easily found. Sample Problem 1-3 gives an example.

Sample Problem 1-3 Build your skill

The world's largest ball of string is about 2 m in radius. To the nearest order of magnitude, what is the total length L of the string in the ball?

KEY IDEA We could, of course, take the ball apart and measure the total length L, but that would take great effort and make the ball's builder most unhappy. Instead, because we want only the nearest order of magnitude, we can estimate any quantities required in the calculation.

Calculations: Let us assume the ball is spherical with radius $R = 2$ m. The string in the ball is not closely packed (there are uncountable gaps between adjacent sections of string). To allow for these gaps, let us somewhat overestimate the cross-sectional area of the string by assuming the cross section is square, with an edge length $d = 4$ mm. Then, with a cross-sectional area of d^2 and a length L, the string occupies a total volume of

$$V = (\text{cross-sectional area})(\text{length}) = d^2L.$$

This is approximately equal to the volume of the ball, given by $\frac{4}{3}\pi R^3$, which is about $4R^3$ because π is about 3. Thus, we have

$$d^2L = 4R^3,$$

or

$$L = \frac{4R^3}{d^2} = \frac{4(2\text{ m})^3}{(4 \times 10^{-3}\text{ m})^2}$$
$$= 2 \times 10^6\text{ m} \approx 10^6\text{ m} = 10^3\text{ km}. \quad \text{(Answer)}$$

(Note that you do not need a calculator for such a simplified calculation.) To the nearest order of magnitude, the ball contains about 1000 km of string!

1-6 | Time

Time has two aspects. For civil and some scientific purposes, we want to know the time of day so that we can order events in sequence. In much scientific work, we want to know how long an event lasts. Thus, any time standard must be able to answer two questions: "*When* did it happen?" and "What is its *duration*?" Table 1-4 shows some time intervals.

TABLE 1-4

Some Approximate Time Intervals

Measurement	Time Interval in Seconds
Lifetime of the proton (predicted)	3×10^{40}
Age of the universe	5×10^{17}
Age of the pyramid of Cheops	1×10^{11}
Human life expectancy	2×10^{9}
Length of a day	9×10^{4}
Time between human heartbeats	8×10^{-1}
Lifetime of the muon	2×10^{-6}
Shortest lab light pulse	1×10^{-16}
Lifetime of the most unstable particle	1×10^{-23}
The Planck time[a]	1×10^{-43}

[a]This is the earliest time after the big bang at which the laws of physics as we know them can be applied.

FIG. 1-1 When the metric system was proposed in 1792, the hour was redefined to provide a 10-hour day. The idea did not catch on. The maker of this 10-hour watch wisely provided a small dial that kept conventional 12-hour time. Do the two dials indicate the same time? *(Steven Pitkin)*

Any phenomenon that repeats itself is a possible time standard. Earth's rotation, which determines the length of the day, has been used in this way for centuries; Fig. 1-1 shows one novel example of a watch based on that rotation. A quartz clock, in which a quartz ring is made to vibrate continuously, can be calibrated against Earth's rotation via astronomical observations and used to measure time intervals in the laboratory. However, the calibration cannot be carried out with the accuracy called for by modern scientific and engineering technology.

To meet the need for a better time standard, atomic clocks have been developed. An atomic clock at the National Institute of Standards and Technology (NIST) in Boulder, Colorado, is the standard for Coordinated Universal Time (UTC) in the United States. Its time signals are available by shortwave radio (stations WWV and WWVH) and by telephone (303-499-7111). Time signals (and related information) are also available from the United States Naval Observatory at website http://tycho.usno.navy.mil/time.html. (To set a clock extremely accurately at your particular location, you would have to account for the travel time required for these signals to reach you.)

Figure 1-2 shows variations in the length of one day on Earth over a 4-year period, as determined by comparison with a cesium (atomic) clock. Because the variation displayed by Fig. 1-2 is seasonal and repetitious, we suspect the rotating Earth when there is a difference between Earth and atom as timekeepers. The variation is due to tidal effects caused by the Moon and to large-scale winds.

The 13th General Conference on Weights and Measures in 1967 adopted a standard second based on the cesium clock:

> One second is the time taken by 9 192 631 770 oscillations of the light (of a specified wavelength) emitted by a cesium-133 atom.

Atomic clocks are so consistent that, in principle, two cesium clocks would have to run for 6000 years before their readings would differ by more than 1 s. Even such accuracy pales in comparison with that of clocks currently being developed; their precision may be 1 part in 10^{18}—that is, 1 s in 1×10^{18} s (which is about 3×10^{10} y).

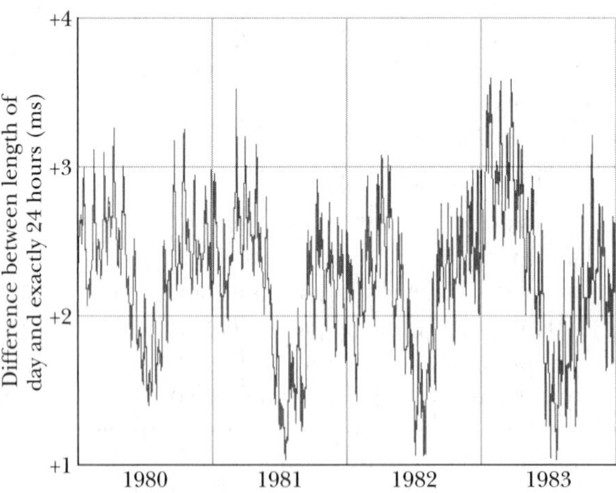

FIG. 1-2 Variations in the length of the day over a 4-year period. Note that the entire vertical scale amounts to only 3 ms (= 0.003 s).

1-7 | Mass

The Standard Kilogram

The SI standard of mass is a platinum–iridium cylinder (Fig. 1-3) kept at the International Bureau of Weights and Measures near Paris and assigned, by international agreement, a mass of 1 kilogram. Accurate copies have been sent to standardizing laboratories in other countries, and the masses of other bodies can be determined by balancing them against a copy. Table 1-5 shows some masses expressed in kilograms, ranging over about 83 orders of magnitude.

The U.S. copy of the standard kilogram is housed in a vault at NIST. It is removed, no more than once a year, for the purpose of checking duplicate copies that are used elsewhere. Since 1889, it has been taken to France twice for recomparison with the primary standard.

FIG. 1-3 The international 1 kg standard of mass, a platinum–iridium cylinder 3.9 cm in height and in diameter. *(Courtesy Bureau International des Poids et Mesures, France)*

A Second Mass Standard

The masses of atoms can be compared with one another more precisely than they can be compared with the standard kilogram. For this reason, we have a second mass standard. It is the carbon-12 atom, which, by international agreement, has been assigned a mass of 12 **atomic mass units** (u). The relation between the two units is

$$1\text{ u} = 1.660\,538\,86 \times 10^{-27}\text{ kg}, \tag{1-7}$$

with an uncertainty of ± 10 in the last two decimal places. Scientists can, with reasonable precision, experimentally determine the masses of other atoms relative to the mass of carbon-12. What we presently lack is a reliable means of extending that precision to more common units of mass, such as a kilogram.

TABLE 1-5

Some Approximate Masses

Object	Mass in Kilograms
Known universe	1×10^{53}
Our galaxy	2×10^{41}
Sun	2×10^{30}
Moon	7×10^{22}
Asteroid Eros	5×10^{15}
Small mountain	1×10^{12}
Ocean liner	7×10^{7}
Elephant	5×10^{3}
Grape	3×10^{-3}
Speck of dust	7×10^{-10}
Penicillin molecule	5×10^{-17}
Uranium atom	4×10^{-25}
Proton	2×10^{-27}
Electron	9×10^{-31}

Density

As we shall discuss further in Chapter 14, the **density** ρ of a material is the mass per unit volume:

$$\rho = \frac{m}{V}. \tag{1-8}$$

Densities are typically listed in kilograms per cubic meter or grams per cubic centimeter. The density of water (1.00 gram per cubic centimeter) is often used as a comparison. Fresh snow has about 10% of that density; platinum has a density that is about 21 times that of water.

Sample Problem 1-4

A heavy object can sink into the ground during an earthquake if the shaking causes the ground to undergo *liquefaction,* in which the soil grains experience little friction as they slide over one another. The ground is then effectively quicksand. The possibility of liquefaction in sandy ground can be predicted in terms of the *void ratio e* for a sample of the ground:

$$e = \frac{V_{\text{voids}}}{V_{\text{grains}}}. \tag{1-9}$$

Here, V_{grains} is the total volume of the sand grains in the sample and V_{voids} is the total volume between the grains (in the *voids*). If e exceeds a critical value of 0.80, liquefaction can occur during an earthquake. What is the corresponding sand density ρ_{sand}? Solid silicon dioxide (the primary component of sand) has a density of $\rho_{SiO_2} = 2.600 \times 10^3\text{ kg/m}^3$.

KEY IDEA The density of the sand ρ_{sand} in a sample is the mass per unit volume—that is, the ratio of the total mass m_{sand} of the sand grains to the total volume V_{total} of the sample:

$$\rho_{\text{sand}} = \frac{m_{\text{sand}}}{V_{\text{total}}}. \tag{1-10}$$

Calculations: The total volume V_{total} of a sample is

$$V_{total} = V_{grains} + V_{voids}.$$

Substituting for V_{voids} from Eq. 1-9 and solving for V_{grains} lead to

$$V_{grains} = \frac{V_{total}}{1 + e}. \qquad (1\text{-}11)$$

From Eq. 1-8, the total mass m_{sand} of the sand grains is the product of the density of silicon dioxide and the total volume of the sand grains:

$$m_{sand} = \rho_{SiO_2} V_{grains}. \qquad (1\text{-}12)$$

Substituting this expression into Eq. 1-10 and then substituting for V_{grains} from Eq. 1-11 lead to

$$\rho_{sand} = \frac{\rho_{SiO_2}}{V_{total}} \frac{V_{total}}{1 + e} = \frac{\rho_{SiO_2}}{1 + e}. \qquad (1\text{-}13)$$

Substituting $\rho_{SiO_2} = 2.600 \times 10^3$ kg/m^3 and the critical value of $e = 0.80$, we find that liquefaction occurs when the sand density exceeds

$$\rho_{sand} = \frac{2.600 \times 10^3 \text{ kg/m}^3}{1.80} = 1.4 \times 10^3 \text{ kg/m}^3.$$

(Answer)

REVIEW & SUMMARY

Measurement in Physics Physics is based on measurement of physical quantities. Certain physical quantities have been chosen as **base quantities** (such as length, time, and mass); each has been defined in terms of a **standard** and given a **unit** of measure (such as meter, second, and kilogram). Other physical quantities are defined in terms of the base quantities and their standards and units.

SI Units The unit system emphasized in this book is the International System of Units (SI). The three physical quantities displayed in Table 1-1 are used in the early chapters. Standards, which must be both accessible and invariable, have been established for these base quantities by international agreement. These standards are used in all physical measurement, for both the base quantities and the quantities derived from them. Scientific notation and the prefixes of Table 1-2 are used to simplify measurement notation.

Changing Units Conversion of units may be performed by using *chain-link conversions* in which the original data are multiplied successively by conversion factors written as unity and the units are manipulated like algebraic quantities until only the desired units remain.

Length The meter is defined as the distance traveled by light during a precisely specified time interval.

Time The second is defined in terms of the oscillations of light emitted by an atomic (cesium-133) source. Accurate time signals are sent worldwide by radio signals keyed to atomic clocks in standardizing laboratories.

Mass The kilogram is defined in terms of a platinum–iridium standard mass kept near Paris. For measurements on an atomic scale, the atomic mass unit, defined in terms of the atom carbon-12, is usually used.

Density The density ρ of a material is the mass per unit volume:

$$\rho = \frac{m}{V}. \qquad (1\text{-}8)$$

PROBLEMS

Tutoring problem available (at instructor's discretion) in *WileyPLUS* and WebAssign

SSM Worked-out solution available in Student Solutions Manual

• – ••• Number of dots indicates level of problem difficulty

Additional information available in *The Flying Circus of Physics* and at flyingcircusofphysics.com

WWW Worked-out solution is at http://www.wiley.com/college/halliday

ILW Interactive solution is at http://www.wiley.com/college/halliday

sec. 1-5 Length

•1 The micrometer (1 μm) is often called the *micron.* (a) How many microns make up 1.0 km? (b) What fraction of a centimeter equals 1.0 μm? (c) How many microns are in 1.0 yd?

•2 Spacing in this book was generally done in units of points and picas: 12 points = 1 pica, and 6 picas = 1 inch. If a figure was misplaced in the page proofs by 0.80 cm, what was the misplacement in (a) picas and (b) points?

•3 Horses are to race over a certain English meadow for a distance of 4.0 furlongs. What is the race distance in (a) rods and (b) chains? (1 furlong = 201.168 m, 1 rod = 5.0292 m, and 1 chain = 20.117 m.) SSM WWW

•4 A *gry* is an old English measure for length, defined as 1/10 of a line, where *line* is another old English measure for length, defined as 1/12 inch. A common measure for length in the publishing business is a *point,* defined as 1/72 inch. What is an area of 0.50 gry^2 in points squared (points^2)?

•5 Earth is approximately a sphere of radius 6.37×10^6 m. What are (a) its circumference in kilometers, (b) its surface area in square kilometers, and (c) its volume in cubic kilometers? SSM

••6 Harvard Bridge, which connects MIT with its fraternities across the Charles River, has a length of 364.4 Smoots plus one ear. The unit of one Smoot is based on the length of Oliver Reed Smoot, Jr., class of 1962, who was carried or dragged length by length across the bridge so that other pledge members of the Lambda Chi Alpha fraternity could mark off (with paint) 1-Smoot lengths along the bridge. The marks have been repainted biannually by fraternity pledges since the initial measurement, usually during times of traffic congestion so that the police cannot easily interfere. (Presumably, the police were originally upset because the Smoot is not an SI base unit, but these days they seem to have accepted the unit.) Figure 1-4 shows three parallel paths, measured in Smoots (S), Willies (W), and Zeldas (Z). What is the length of 50.0 Smoots in (a) Willies and (b) Zeldas?

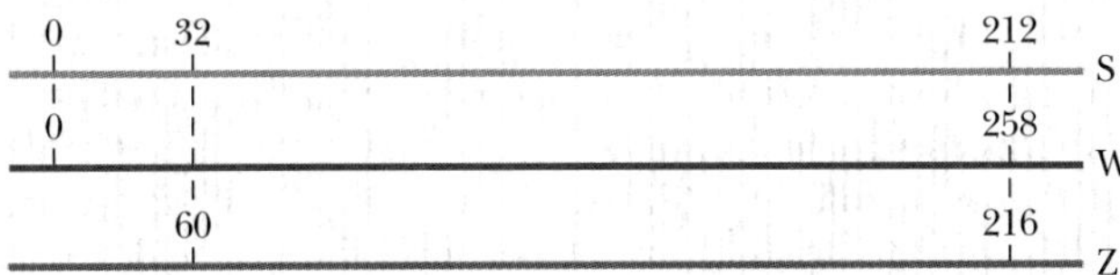

FIG. 1-4 Problem 6.

••7 Antarctica is roughly semicircular, with a radius of 2000 km (Fig. 1-5). The average thickness of its ice cover is 3000 m. How many cubic centimeters of ice does Antarctica contain? (Ignore the curvature of Earth.)

2000 km

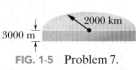

3000 m

FIG. 1-5 Problem 7.

••8 You can easily convert common units and measures electronically, but you still should be able to use a conversion table, such as those in Appendix D. Table 1-6 is part of a conversion table for a system of volume measures once common in Spain; a volume of 1 fanega is equivalent to 55.501 dm^3 (cubic decimeters). To complete the table, what numbers (to three significant figures) should be entered in (a) the cahiz column, (b) the fanega column, (c) the cuartilla column, and (d) the almude column, starting with the top blank? Express 7.00 almudes in (e) medios, (f) cahizes, and (g) cubic centimeters (cm^3).

TABLE 1-6

Problem 8

	cahiz	fanega	cuartilla	almude	medio
1 cahiz =	1	12	48	144	288
1 fanega =		1	4	12	24
1 cuartilla =			1	3	6
1 almude =				1	2
1 medio =					1

••9 Hydraulic engineers in the United States often use, as a unit of volume of water, the *acre-foot,* defined as the volume of water that will cover 1 acre of land to a depth of 1 ft. A severe thunderstorm dumped 2.0 in. of rain in 30 min on a town of area 26 km^2. What volume of water, in acre-feet, fell on the town? ILW

sec. 1-6 Time

•10 The fastest growing plant on record is a *Hesperoyucca whipplei* that grew 3.7 m in 14 days. What was its growth rate in micrometers per second?

•11 A fortnight is a charming English measure of time equal to 2.0 weeks (the word is a contraction of "fourteen nights"). That is a nice amount of time in pleasant company but perhaps a painful string of microseconds in unpleasant company. How many microseconds are in a fortnight?

•12 A lecture period (50 min) is close to 1 microcentury. (a) How long is a microcentury in minutes? (b) Using

$$\text{percentage difference} = \left(\frac{\text{actual} - \text{approximation}}{\text{actual}}\right)100,$$

find the percentage difference from the approximation.

•13 For about 10 years after the French Revolution, the French government attempted to base measures of time on multiples of ten: One week consisted of 10 days, one day consisted of 10 hours, one hour consisted of 100 minutes, and one minute consisted of 100 seconds. What are the ratios of (a) the French decimal week to the standard week and (b) the French decimal second to the standard second?

•14 Time standards are now based on atomic clocks. A promising second standard is based on *pulsars,* which are rotating neutron stars (highly compact stars consisting only of neutrons). Some rotate at a rate that is highly stable, sending out a radio beacon that sweeps briefly across Earth once with each rotation, like a lighthouse beacon. Pulsar PSR 1937+21 is an example; it rotates once every 1.557 806 448 872 75 ± 3 ms, where the trailing ±3 indicates the uncertainty in the last decimal place (it does *not* mean ±3 ms). (a) How many rotations does PSR 1937+21 make in 7.00 days? (b) How much time does the pulsar take to rotate exactly one million times and (c) what is the associated uncertainty?

•15 Three digital clocks *A, B,* and *C* run at different rates and do not have simultaneous readings of zero. Figure 1-6 shows simultaneous readings on pairs of the clocks for four occasions. (At the earliest occasion, for example, *B* reads 25.0 s and *C* reads 92.0 s.) If two events are 600 s apart on clock *A*, how far apart are they on (a) clock *B* and (b) clock *C*? (c) When clock *A* reads 400 s, what does clock *B* read? (d) When clock *C* reads 15.0 s, what does clock *B* read? (Assume negative readings for prezero times.)

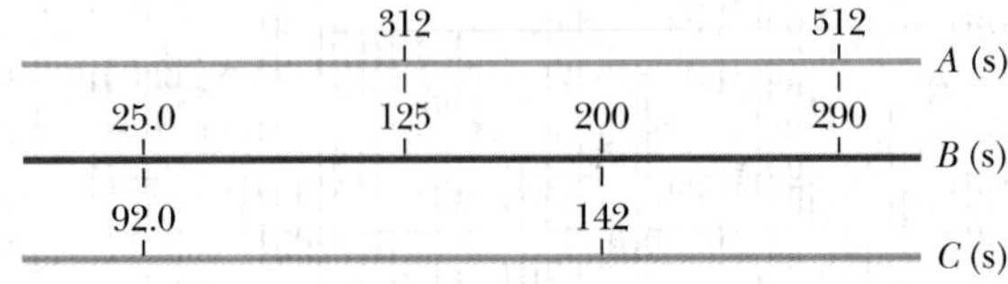

FIG. 1-6 Problem 15.

•16 Until 1883, every city and town in the United States kept its own local time. Today, travelers reset their watches only when the time change equals 1.0 h. How far, on the average, must you travel in degrees of longitude between the time-zone boundaries at which your watch must be reset by 1.0 h? (*Hint:* Earth rotates 360° in about 24 h.)

•17 Five clocks are being tested in a laboratory. Exactly at noon, as determined by the WWV time signal, on successive days of a week the clocks read as in the following table. Rank the five clocks according to their relative value as good timekeepers, best to worst. Justify your choice. SSM

Clock	Sun.	Mon.	Tues.	Wed.	Thurs.	Fri.	Sat.
A	12:36:40	12:36:56	12:37:12	12:37:27	12:37:44	12:37:59	12:38:14
B	11:59:59	12:00:02	11:59:57	12:00:07	12:00:02	11:59:56	12:00:03
C	15:50:45	15:51:43	15:52:41	15:53:39	15:54:37	15:55:35	15:56:33
D	12:03:59	12:02:52	12:01:45	12:00:38	11:59:31	11:58:24	11:57:17
E	12:03:59	12:02:49	12:01:54	12:01:52	12:01:32	12:01:22	12:01:12

••18 Because Earth's rotation is gradually slowing, the length of each day increases: The day at the end of 1.0 century is 1.0 ms longer than the day at the start of the century. In 20 centuries, what is the total of the daily increases in time?

•••19 Suppose that, while lying on a beach near the equator watching the Sun set over a calm ocean, you start a stopwatch just as the top of the Sun disappears. You then stand, elevating your eyes by a height $H = 1.70$ m, and stop the watch when the top of the Sun again disappears. If the elapsed time is $t = 11.1$ s, what is the radius r of Earth?

sec. 1-7 Mass

•20 Gold, which has a density of 19.32 g/cm^3, is the most ductile metal and can be pressed into a thin leaf or drawn out into a long fiber. (a) If a sample of gold, with a mass of 27.63 g, is pressed into a leaf of 1.000 μm thickness, what is the area of the leaf? (b) If, instead, the gold is drawn out into a cylindrical fiber of radius 2.500 μm, what is the length of the fiber?

•21 (a) Assuming that water has a density of exactly 1 g/cm^3, find the mass of one cubic meter of water in kilograms. (b) Suppose that it takes 10.0 h to drain a container of 5700 m^3 of water. What is the "mass flow rate," in kilograms per second, of water from the container? SSM

•22 The record for the largest glass bottle was set in 1992 by a team in Millville, New Jersey—they blew a bottle with a volume of 193 U.S. fluid gallons. (a) How much short of 1.0 million cubic centimeters is that? (b) If the bottle were filled with water at the leisurely rate of 1.8 g/min, how long would the filling take? Water has a density of 1000 kg/m^3. GO

•23 Earth has a mass of 5.98×10^{24} kg. The average mass of the atoms that make up Earth is 40 u. How many atoms are there in Earth?

••24 One cubic centimeter of a typical cumulus cloud contains 50 to 500 water drops, which have a typical radius of 10 μm. For that range, give the lower value and the higher value, respectively, for the following. (a) How many cubic meters of water are in a cylindrical cumulus cloud of height 3.0 km and radius 1.0 km? (b) How many 1-liter pop bottles would that water fill? (c) Water has a density of 1000 kg/m^3. How much mass does the water in the cloud have?

••25 Iron has a density of 7.87 g/cm^3, and the mass of an iron atom is 9.27×10^{-26} kg. If the atoms are spherical and tightly packed, (a) what is the volume of an iron atom and (b) what is the distance between the centers of adjacent atoms?

••26 A mole of atoms is 6.02×10^{23} atoms. To the nearest order of magnitude, how many moles of atoms are in a large domestic cat? The masses of a hydrogen atom, an oxygen atom, and a carbon atom are 1.0 u, 16 u, and 12 u, respectively. (*Hint:* Cats are sometimes known to kill a mole.)

••27 On a spending spree in Malaysia, you buy an ox with a weight of 28.9 piculs in the local unit of weights: 1 picul = 100 gins, 1 gin = 16 tahils, 1 tahil = 10 chees, and 1 chee = 10 hoons. The weight of 1 hoon corresponds to a mass of 0.3779 g. When you arrange to ship the ox home to your astonished family, how much mass in kilograms must you declare on the shipping manifest? (*Hint:* Set up multiple chain-link conversions.)

••28 Grains of fine California beach sand are approximately spheres with an average radius of 50 μm and are made of silicon dioxide, which has a density of 2600 kg/m^3. What mass of sand grains would have a total surface area (the total area of all the individual spheres) equal to the surface area of a cube 1.00 m on an edge? GO

••29 During heavy rain, a section of a mountainside measuring 2.5 km horizontally, 0.80 km up along the slope, and 2.0 m deep slips into a valley in a mud slide. Assume that the mud ends up uniformly distributed over a surface area of the valley measuring 0.40 km × 0.40 km and that mud has a density of 1900 kg/m^3. What is the mass of the mud sitting above a 4.0 m^2 area of the valley floor?

••30 Water is poured into a container that has a leak. The mass m of the water is given as a function of time t by $m = 5.00t^{0.8} - 3.00t + 20.00$, with $t \geq 0$, m in grams, and t in seconds. (a) At what time is the water mass greatest, and (b) what is that greatest mass? In kilograms per minute, what is the rate of mass change at (c) $t = 2.00$ s and (d) $t = 5.00$ s?

•••31 A vertical container with base area measuring 14.0 cm by 17.0 cm is being filled with identical pieces of candy, each with a volume of 50.0 mm^3 and a mass of 0.0200 g. Assume that the volume of the empty spaces between the candies is negligible. If the height of the candies in the container increases at the rate of 0.250 cm/s, at what rate (kilograms per minute) does the mass of the candies in the container increase?

Additional Problems

32 Table 1-7 shows some old measures of liquid volume. To complete the table, what numbers (to three significant figures) should be entered in (a) the wey column, (b) the chaldron column, (c) the bag column, (d) the pottle column, and (e) the gill column, starting with the top blank? (f) The volume of 1 bag is equal to 0.1091 m^3. If an old story has a witch cooking up some vile liquid in a cauldron of volume 1.5 chaldrons, what is the volume in cubic meters?

TABLE 1-7

Problem 32

	wey	chaldron	bag	pottle	gill
1 wey =	1	10/9	40/3	640	120 240
1 chaldron =					
1 bag =					
1 pottle =					
1 gill =					

33 An old English children's rhyme states, "Little Miss Muffet sat on a tuffet, eating her curds and whey, when along came a spider who sat down beside her. . . ." The spider sat down not because of the curds and whey but because Miss Muffet had a stash of 11 tuffets of dried flies. The volume measure of a tuffet is given by 1 tuffet = 2 pecks = 0.50 Imperial bushel, where 1 Imperial bushel = 36.3687 liters (L). What was Miss Muffet's stash in (a) pecks, (b) Imperial bushels, and (c) liters?

34 An old manuscript reveals that a landowner in the time of King Arthur held 3.00 acres of plowed land plus a livestock area of 25.0 perches by 4.00 perches. What was the total area in (a) the old unit of roods and (b) the more modern unit of square meters? Here, 1 acre is an area of 40 perches by 4 perches, 1 rood is an area of 40 perches by 1 perch, and 1 perch is the length 16.5 ft.

35 A tourist purchases a car in England and ships it home to the United States. The car sticker advertised that the car's fuel consumption was at the rate of 40 miles per gallon on the open road. The tourist does not realize that the U.K. gallon differs from the U.S. gallon:

$$1 \text{ U.K. gallon} = 4.545\,963\,1 \text{ liters}$$
$$1 \text{ U.S. gallon} = 3.785\,306\,0 \text{ liters.}$$

For a trip of 750 miles (in the United States), how many gallons of fuel does (a) the mistaken tourist believe she needs and (b) the car actually require? SSM

36 Two types of *barrel* units were in use in the 1920s in the United States. The apple barrel had a legally set volume of 7056 cubic inches; the cranberry barrel, 5826 cubic inches. If a merchant sells 20 cranberry barrels of goods to a customer who thinks he is receiving apple barrels, what is the discrepancy in the shipment volume in liters?

37 The description for a certain brand of house paint claims a coverage of 460 ft^2/gal. (a) Express this quantity in square meters per liter. (b) Express this quantity in an SI unit (see Appendices A and D). (c) What is the inverse of the original quantity, and (d) what is its physical significance?

38 In the United States, a doll house has the scale of 1:12 of a real house (that is, each length of the doll house is $\frac{1}{12}$ that of the real house) and a miniature house (a doll house to fit within a doll house) has the scale of 1:144 of a real house. Suppose a real house (Fig. 1-7) has a front length of 20 m, a depth of 12 m, a height of 6.0 m, and a standard sloped roof (vertical triangular faces on the ends) of height 3.0 m. In cubic meters, what are the volumes of the corresponding (a) doll house and (b) miniature house?

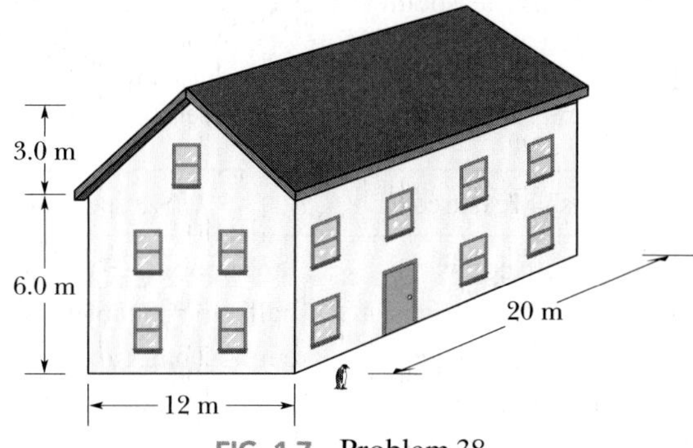

FIG. 1-7 Problem 38.

39 A *cord* is a volume of cut wood equal to a stack 8 ft long, 4 ft wide, and 4 ft high. How many cords are in 1.0 m^3? SSM

40 One molecule of water (H_2O) contains two atoms of hydrogen and one atom of oxygen. A hydrogen atom has a mass of 1.0 u and an atom of oxygen has a mass of 16 u, approximately. (a) What is the mass in kilograms of one molecule of water? (b) How many molecules of water are in the world's oceans, which have an estimated total mass of 1.4×10^{21} kg?

41 A ton is a measure of volume frequently used in shipping, but that use requires some care because there are at least three types of tons: A *displacement ton* is equal to 7 barrels bulk, a *freight ton* is equal to 8 barrels bulk, and a *register ton* is equal to 20 barrels bulk. A *barrel bulk* is another measure of volume: 1 barrel bulk = 0.1415 m^3. Suppose you spot a shipping order for "73 tons" of M&M candies, and you are certain that the client who sent the order intended "ton" to refer to volume (instead of weight or mass, as discussed in Chapter 5). If the client actually meant displacement tons, how many extra U.S. bushels of the candies will you erroneously ship if you interpret the order as (a) 73 freight tons and (b) 73 register tons? (1 m^3 = 28.378 U.S. bushels.) SSM

42 Strangely, the wine for a large wedding reception is to be served in a stunning cut-glass receptacle with the interior dimensions of 40 cm $\times$ 40 cm $\times$ 30 cm (height). The receptacle is to be initially filled to the top. The wine can be purchased in bottles of the sizes given in the following table. Purchasing a larger bottle instead of multiple smaller bottles decreases the overall cost of the wine. To minimize the cost, (a) which bottle sizes should be purchased and how many of each should be purchased and, once the receptacle is filled, how much wine is left over in terms of (b) standard bottles and (c) liters?

1 standard bottle
1 magnum = 2 standard bottles
1 jeroboam = 4 standard bottles
1 rehoboam = 6 standard bottles
1 methuselah = 8 standard bottles
1 salmanazar = 12 standard bottles
1 balthazar = 16 standard bottles = 11.356 L
1 nebuchadnezzar = 20 standard bottles

43 A typical sugar cube has an edge length of 1 cm. If you had a cubical box that contained a mole of sugar cubes, what would its edge length be? (One mole = 6.02×10^{23} units.)

44 Using conversions and data in the chapter, determine the number of hydrogen atoms required to obtain 1.0 kg of hydrogen. A hydrogen atom has a mass of 1.0 u.

45 An astronomical unit (AU) is the average distance between Earth and the Sun, approximately 1.50×10^8 km. The speed of light is about 3.0×10^8 m/s. Express the speed of light in astronomical units per minute. SSM

46 What mass of water fell on the town in Problem 9? Water has a density of 1.0×10^3 kg/m^3.

47 A person on a diet might lose 2.3 kg per week. Express the mass loss rate in milligrams per second, as if the dieter could sense the second-by-second loss.

48 The *corn–hog ratio* is a financial term used in the pig market and presumably is related to the cost of feeding a pig until it is large enough for market. It is defined as the ratio of the market price of a pig with a mass of 3.108 slugs to the market price of a U.S. bushel of corn. (The word "slug" is derived from an old German word that means "to hit"; we have the same meaning for "slug" as a verb in modern English.) A U.S. bushel is equal to 35.238 L. If the corn–hog ratio is listed as 5.7 on the market exchange, what is it in the metric units of

$$\frac{\text{price of 1 kilogram of pig}}{\text{price of 1 liter of corn}}?$$

(*Hint:* See the Mass table in Appendix D.)

49 You are to fix dinners for 400 people at a convention of Mexican food fans. Your recipe calls for 2 jalapeño peppers per serving (one serving per person). However, you have only habanero peppers on hand. The spiciness of peppers is measured in terms of the *scoville heat unit* (SHU). On average, one jalapeño pepper has a spiciness of 4000 SHU and one habanero pepper has a spiciness of 300 000 SHU. To get the desired spiciness, how many habanero peppers should you substitute for the jalapeño peppers in the recipe for the 400 dinners?

50 A unit of area often used in measuring land areas is the *hectare,* defined as 10^4 m^2. An open-pit coal mine consumes 75 hectares of land, down to a depth of 26 m, each year. What volume of earth, in cubic kilometers, is removed in this time?

51 (a) A unit of time sometimes used in microscopic physics is the *shake.* One shake equals 10^{-8} s. Are there more shakes in a second than there are seconds in a year? (b) Humans have existed for about 10^6 years, whereas the universe is about 10^{10} years old. If the age of the universe is defined as 1 "universe day," where a universe day consists of "universe seconds" as a normal day consists of normal seconds, how many universe seconds have humans existed?

52 As a contrast between the old and the modern and between the large and the small, consider the following: In old rural England 1 hide (between 100 and 120 acres) was the area of land needed to sustain one family with a single plough for one year. (An area of 1 acre is equal to 4047 m^2.) Also, 1 wapentake was the area of land needed by 100 such families. In quantum physics, the cross-sectional area of a nucleus (defined in terms of the chance of a particle hitting and being absorbed by it) is measured in units of barns, where 1 barn is 1×10^{-28} m^2. (In nuclear physics jargon, if a nucleus is "large," then shooting a particle at it is like shooting a bullet at a barn door, which can hardly be missed.) What is the ratio of 25 wapentakes to 11 barns?

53 A traditional unit of length in Japan is the ken (1 ken = 1.97 m). What are the ratios of (a) square kens to square meters and (b) cubic kens to cubic meters? What is the volume of a cylindrical water tank of height 5.50 kens and radius 3.00 kens in (c) cubic kens and (d) cubic meters?

54 You receive orders to sail due east for 24.5 mi to put your salvage ship directly over a sunken pirate ship. However, when your divers probe the ocean floor at that location and find no evidence of a ship, you radio back to your source of information, only to discover that the sailing distance was supposed to be 24.5 *nautical miles,* not regular miles. Use the Length table in Appendix D.

55 A standard interior staircase has steps each with a rise (height) of 19 cm and a run (horizontal depth) of 23 cm. Research suggests that the stairs would be safer for descent if the run were, instead, 28 cm. For a particular staircase of total height 4.57 m, how much farther into the room would the staircase extend if this change in run were made?

56 The common Eastern mole, a mammal, typically has a mass of 75 g, which corresponds to about 7.5 moles of atoms. (A mole of atoms is 6.02×10^{23} atoms.) In atomic mass units (u), what is the average mass of the atoms in the common Eastern mole?

57 An *astronomical unit* (AU) is equal to the average distance from Earth to the Sun, about 92.9×10^6 mi. A *parsec* (pc) is the distance at which a length of 1 AU would subtend an angle of exactly 1 second of arc (Fig. 1-8). A *light-year* (ly) is the distance that light, traveling through a vacuum with a speed of 186 000 mi/s, would cover in 1.0 year. Express the Earth–Sun distance in (a) parsecs and (b) light-years. SSM

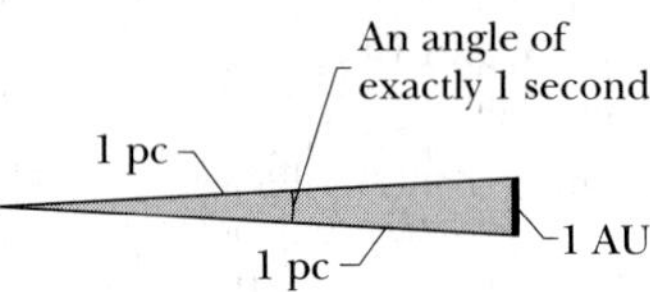

FIG. 1-8 Problem 57.

58 In purchasing food for a political rally, you erroneously order shucked medium-size Pacific oysters (which come 8 to 12 per U.S. pint) instead of shucked medium-size Atlantic oysters (which come 26 to 38 per U.S. pint). The filled oyster container shipped to you has the interior measure of 1.0 m × 12 cm × 20 cm, and a U.S. pint is equivalent to 0.4732 liter. By how many oysters is the order short of your anticipated count?

59 The cubit is an ancient unit of length based on the distance between the elbow and the tip of the middle finger of the measurer. Assume that the distance ranged from 43 to 53 cm, and suppose that ancient drawings indicate that a cylindrical pillar was to have a length of 9 cubits and a diameter of 2 cubits. For the stated range, what are the lower value and the upper value, respectively, for (a) the cylinder's length in meters, (b) the cylinder's length in millimeters, and (c) the cylinder's volume in cubic meters?

60 An old English cookbook carries this recipe for cream of nettle soup: "Boil stock of the following amount: 1 breakfastcup plus 1 teacup plus 6 tablespoons plus 1 dessertspoon. Using gloves, separate nettle tops until you have 0.5 quart; add the tops to the boiling stock. Add 1 tablespoon of cooked rice and 1 saltspoon of salt. Simmer for 15 min." The following table gives some of the conversions among old (premetric) British measures and among common (still premetric) U.S. measures. (These measures scream for metrication.) For liquid measures, 1 British teaspoon = 1 U.S. teaspoon. For dry measures, 1 British teaspoon = 2 U.S. teaspoons and 1 British quart = 1 U.S. quart. In U.S. measures, how much (a) stock, (b) nettle tops, (c) rice, and (d) salt are required in the recipe?

Old British Measures	U.S. Measures
teaspoon = 2 saltspoons	tablespoon = 3 teaspoons
dessertspoon = 2 teaspoons	half cup = 8 tablespoons
tablespoon = 2 dessertspoons	cup = 2 half cups
teacup = 8 tablespoons	
breakfastcup = 2 teacups	

Motion Along a Straight Line 2

Jeremy Woodhouse/Masterfile

A woodpecker hammers its beak into the limb of a tree to search for insects to eat, to create storage space, or to audibly advertise for a mate. The motion toward the limb may be very rapid, but the stopping once the limb is reached is extremely rapid and would be fatal to a human. Thus, a woodpecker should seemingly fall from the tree either dead or unconscious every time it slams its beak into the tree. Not only does it survive, but it rapidly repeats the motion, sending out a rat-tat-tat signal through the air.

Why can a woodpecker survive the severe impacts with a tree limb?

The answer is in this chapter.

2-1 WHAT IS PHYSICS?

One purpose of physics is to study the motion of objects—how fast they move, for example, and how far they move in a given amount of time. NASCAR engineers are fanatical about this aspect of physics as they determine the performance of their cars before and during a race. Geologists use this physics to measure tectonic-plate motion as they attempt to predict earthquakes. Medical researchers need this physics to map the blood flow through a patient when diagnosing a partially closed artery, and motorists use it to determine how they might slow sufficiently when their radar detector sounds a warning. There are countless other examples. In this chapter, we study the basic physics of motion where the object (race car, tectonic plate, blood cell, or any other object) moves along a single axis. Such motion is called *one-dimensional motion.*

2-2 | Motion

The world, and everything in it, moves. Even seemingly stationary things, such as a roadway, move with Earth's rotation, Earth's orbit around the Sun, the Sun's orbit around the center of the Milky Way galaxy, and that galaxy's migration relative to other galaxies. The classification and comparison of motions (called **kinematics**) is often challenging. What exactly do you measure, and how do you compare?

Before we attempt an answer, we shall examine some general properties of motion that is restricted in three ways.

1. The motion is along a straight line only. The line may be vertical, horizontal, or slanted, but it must be straight.
2. Forces (pushes and pulls) cause motion but will not be discussed until Chapter 5. In this chapter we discuss only the motion itself and changes in the motion. Does the moving object speed up, slow down, stop, or reverse direction? If the motion does change, how is time involved in the change?
3. The moving object is either a **particle** (by which we mean a point-like object such as an electron) or an object that moves like a particle (such that every portion moves in the same direction and at the same rate). A stiff pig slipping down a straight playground slide might be considered to be moving like a particle; however, a tumbling tumbleweed would not.

2-3 | Position and Displacement

To locate an object means to find its position relative to some reference point, often the **origin** (or zero point) of an axis such as the x axis in Fig. 2-1. The **positive direction** of the axis is in the direction of increasing numbers (coordinates), which is to the right in Fig. 2-1. The opposite is the **negative direction.**

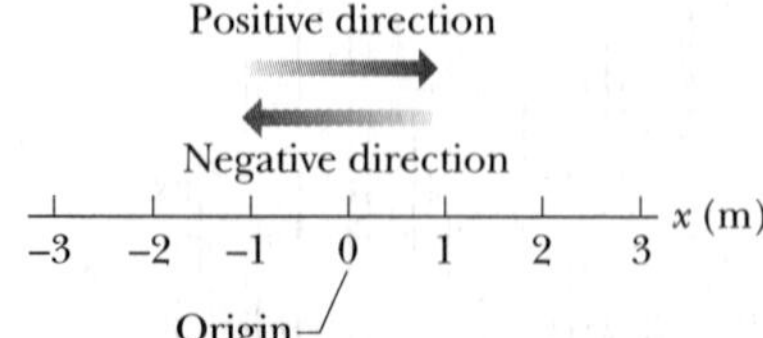

FIG. 2-1 Position is determined on an axis that is marked in units of length (here meters) and that extends indefinitely in opposite directions. The axis name, here x, is always on the positive side of the origin.

For example, a particle might be located at $x = 5$ m, which means it is 5 m in the positive direction from the origin. If it were at $x = -5$ m, it would be just as far from the origin but in the opposite direction. On the axis, a coordinate of -5 m is less than a coordinate of -1 m, and both coordinates are less than a coordinate of $+5$ m. A plus sign for a coordinate need not be shown, but a minus sign must always be shown.

A change from position x_1 to position x_2 is called a **displacement** Δx, where

$$\Delta x = x_2 - x_1. \tag{2-1}$$

(The symbol Δ, the Greek uppercase delta, represents a change in a quantity, and it means the final value of that quantity minus the initial value.) When numbers are inserted for the position values x_1 and x_2 in Eq. 2-1, a displacement in the positive direction (to the right in Fig. 2-1) always comes out positive, and a displacement in the opposite direction (left in the figure) always comes out negative. For example, if the particle moves from $x_1 = 5$ m to $x_2 = 12$ m, then $\Delta x = (12 \text{ m}) - (5 \text{ m}) = +7$ m. The positive result indicates that the motion is in the positive direction. If, instead, the particle moves from $x_1 = 5$ m to $x_2 = 1$ m, then $\Delta x = (1 \text{ m}) - (5 \text{ m}) = -4$ m. The negative result indicates that the motion is in the negative direction.

The actual number of meters covered for a trip is irrelevant; displacement involves only the original and final positions. For example, if the particle moves from $x = 5$ m out to $x = 200$ m and then back to $x = 5$ m, the displacement from start to finish is $\Delta x = (5 \text{ m}) - (5 \text{ m}) = 0$.

A plus sign for a displacement need not be shown, but a minus sign must always be shown. If we ignore the sign (and thus the direction) of a displacement, we are left with the **magnitude** (or absolute value) of the displacement. For example, a displacement of $\Delta x = -4$ m has a magnitude of 4 m.

Displacement is an example of a **vector quantity,** which is a quantity that has both a direction and a magnitude. We explore vectors more fully in Chapter 3 (in fact, some of you may have already read that chapter), but here all we need is the idea that displacement has two features: (1) Its *magnitude* is the distance (such as the number of meters) between the original and final positions. (2) Its *direction,* from an original position to a final position, can be represented by a plus sign or a minus sign if the motion is along a single axis.

What follows is the first of many checkpoints you will see in this book. Each consists of one or more questions whose answers require some reasoning or a mental calculation, and each gives you a quick check of your understanding of a point just discussed. The answers are listed in the back of the book.

CHECKPOINT 1 Here are three pairs of initial and final positions, respectively, along an x axis. Which pairs give a negative displacement: (a) -3 m, $+5$ m; (b) -3 m, -7 m; (c) 7 m, -3 m?

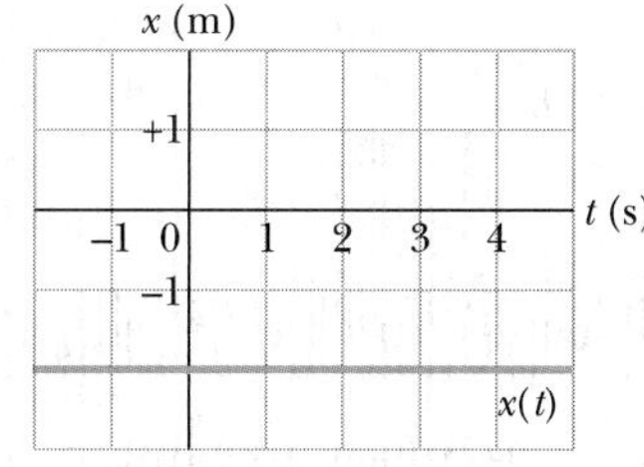

FIG. 2-2 The graph of $x(t)$ for an armadillo that is stationary at $x = -2$ m. The value of x is -2 m for all times t.

2-4 | Average Velocity and Average Speed

A compact way to describe position is with a graph of position x plotted as a function of time t—a graph of $x(t)$. (The notation $x(t)$ represents a function x of t, not the product x times t.) As a simple example, Fig. 2-2 shows the position function $x(t)$ for a stationary armadillo (which we treat as a particle) over a 7 s time interval. The animal's position stays at $x = -2$ m.

Figure 2-3*a* is more interesting, because it involves motion. The armadillo is apparently first noticed at $t = 0$ when it is at the position $x = -5$ m. It moves toward $x = 0$, passes through that point at $t = 3$ s, and then moves on to increasingly larger positive values of x. Figure 2-3*b* depicts the straight line motion of the armadillo and is something like what you would see. The graph in Fig. 2-3*a* is more abstract and quite unlike what you would see, but it is richer in information. It also reveals how fast the armadillo moves.

Actually, several quantities are associated with the phrase "how fast." One of them is the **average velocity** v_{avg}, which is the ratio of the displacement Δx that occurs during a particular time interval Δt to that interval:

$$v_{avg} = \frac{\Delta x}{\Delta t} = \frac{x_2 - x_1}{t_2 - t_1}. \tag{2-2}$$

The notation means that the position is x_1 at time t_1 and then x_2 at time t_2. A common unit for v_{avg} is the meter per second (m/s). You may see other units in the problems, but they are always in the form of length/time.

On a graph of x versus t, v_{avg} is the **slope** of the straight line that connects two particular points on the $x(t)$ curve: one is the point that corresponds to x_2 and t_2,

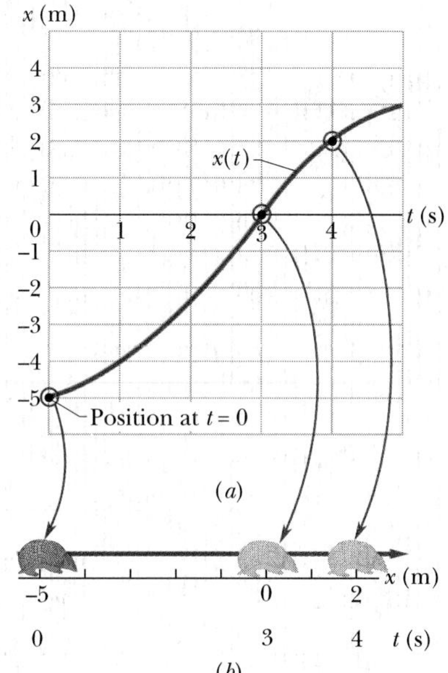

FIG. 2-3 (*a*) The graph of $x(t)$ for a moving armadillo. (*b*) The path associated with the graph. The scale below the x axis shows the times at which the armadillo reaches various x values.

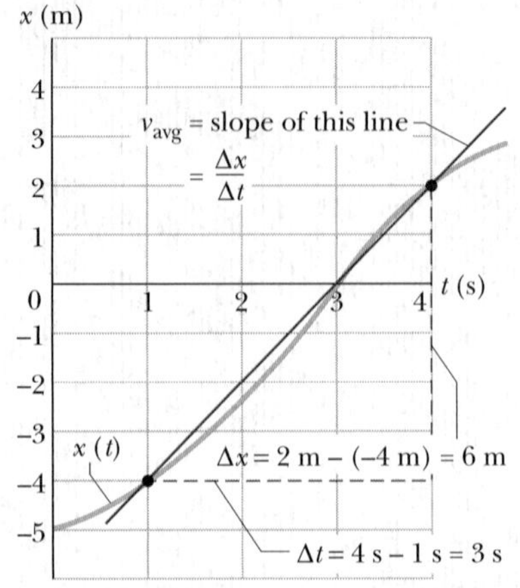

FIG. 2-4 Calculation of the average velocity between $t = 1$ s and $t = 4$ s as the slope of the line that connects the points on the $x(t)$ curve representing those times.

and the other is the point that corresponds to x_1 and t_1. Like displacement, v_{avg} has both magnitude and direction (it is another vector quantity). Its magnitude is the magnitude of the line's slope. A positive v_{avg} (and slope) tells us that the line slants upward to the right; a negative v_{avg} (and slope) tells us that the line slants downward to the right. The average velocity v_{avg} always has the same sign as the displacement Δx because Δt in Eq. 2-2 is always positive.

Figure 2-4 shows how to find v_{avg} in Fig. 2-3 for the time interval $t = 1$ s to $t = 4$ s. We draw the straight line that connects the point on the position curve at the beginning of the interval and the point on the curve at the end of the interval. Then we find the slope $\Delta x/\Delta t$ of the straight line. For the given time interval, the average velocity is

$$v_{avg} = \frac{6 \text{ m}}{3 \text{ s}} = 2 \text{ m/s}.$$

Average speed s_{avg} is a different way of describing "how fast" a particle moves. Whereas the average velocity involves the particle's displacement Δx, the average speed involves the total distance covered (for example, the number of meters moved), independent of direction; that is,

$$s_{avg} = \frac{\text{total distance}}{\Delta t}. \tag{2-3}$$

Because average speed does *not* include direction, it lacks any algebraic sign. Sometimes s_{avg} is the same (except for the absence of a sign) as v_{avg}. However, as is demonstrated in Sample Problem 2-1, the two can be quite different.

Sample Problem 2-1

You drive a beat-up pickup truck along a straight road for 8.4 km at 70 km/h, at which point the truck runs out of gasoline and stops. Over the next 30 min, you walk another 2.0 km farther along the road to a gasoline station.

(a) What is your overall displacement from the beginning of your drive to your arrival at the station?

KEY IDEA Assume, for convenience, that you move in the positive direction of an x axis, from a first position of $x_1 = 0$ to a second position of x_2 at the station. That second position must be at $x_2 = 8.4$ km $+ 2.0$ km $=$ 10.4 km. Then your displacement Δx along the x axis is the second position minus the first position.

Calculation: From Eq. 2-1, we have

$$\Delta x = x_2 - x_1 = 10.4 \text{ km} - 0 = 10.4 \text{ km}. \quad \text{(Answer)}$$

Thus, your overall displacement is 10.4 km in the positive direction of the x axis.

(b) What is the time interval Δt from the beginning of your drive to your arrival at the station?

KEY IDEA We already know the walking time interval Δt_{wlk} (= 0.50 h), but we lack the driving time interval Δt_{dr}. However, we know that for the drive the displacement Δx_{dr} is 8.4 km and the average velocity $v_{avg,dr}$ is 70 km/h. Thus, this average velocity is the ratio of the displacement for the drive to the time interval for the drive.

Calculations: We first write

$$v_{avg,dr} = \frac{\Delta x_{dr}}{\Delta t_{dr}}.$$

Rearranging and substituting data then give us

$$\Delta t_{dr} = \frac{\Delta x_{dr}}{v_{avg,dr}} = \frac{8.4 \text{ km}}{70 \text{ km/h}} = 0.12 \text{ h}.$$

So,

$$\Delta t = \Delta t_{dr} + \Delta t_{wlk}$$
$$= 0.12 \text{ h} + 0.50 \text{ h} = 0.62 \text{ h}. \quad \text{(Answer)}$$

(c) What is your average velocity v_{avg} from the beginning of your drive to your arrival at the station? Find it both numerically and graphically.

KEY IDEA From Eq. 2-2 we know that v_{avg} *for the entire trip* is the ratio of the displacement of 10.4 km *for the entire trip* to the time interval of 0.62 h *for the entire trip.*

Calculation: Here we find

$$v_{avg} = \frac{\Delta x}{\Delta t} = \frac{10.4 \text{ km}}{0.62 \text{ h}}$$
$$= 16.8 \text{ km/h} \approx 17 \text{ km/h}. \quad \text{(Answer)}$$

To find v_{avg} graphically, first we graph the function $x(t)$ as shown in Fig. 2-5, where the beginning and arrival points on the graph are the origin and the point labeled as "Station." Your average velocity is the slope of the straight line connecting those points; that is, v_{avg} is the

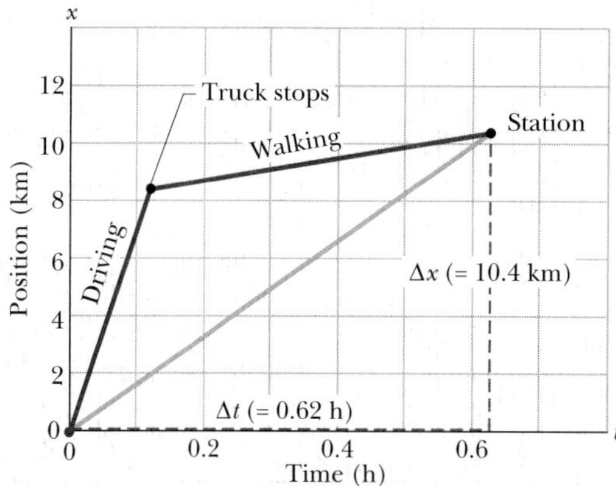

FIG. 2-5 The lines marked "Driving" and "Walking" are the position–time plots for the driving and walking stages. (The plot for the walking stage assumes a constant rate of walking.) The slope of the straight line joining the origin and the point labeled "Station" is the average velocity for the trip, from the beginning to the station.

ratio of the *rise* ($\Delta x = 10.4$ km) to the *run* ($\Delta t = 0.62$ h), which gives us $v_{avg} = 16.8$ km/h.

(d) Suppose that to pump the gasoline, pay for it, and walk back to the truck takes you another 45 min. What is your average speed from the beginning of your drive to your return to the truck with the gasoline?

KEY IDEA Your average speed is the ratio of the total distance you move to the total time interval you take to make that move.

Calculation: The total distance is 8.4 km + 2.0 km + 2.0 km = 12.4 km. The total time interval is 0.12 h + 0.50 h + 0.75 h = 1.37 h. Thus, Eq. 2-3 gives us

$$s_{avg} = \frac{12.4 \text{ km}}{1.37 \text{ h}} = 9.1 \text{ km/h}. \qquad \text{(Answer)}$$

PROBLEM-SOLVING TACTICS

Tactic 1: ***Do You Understand the Problem?*** The common difficulty is simply not understanding the problem. The best test of understanding is this: Can *you* explain the problem?

Write down the given data, with units, using the symbols of the chapter. (In Sample Problem 2-1, the given data allow you to find your net displacement Δx in part (a) and the corresponding time interval Δt in part (b).) Identify the unknown and its symbol. (In the sample problem, the unknown in part (c) is your average velocity v_{avg}.) Then find the connection between the unknown and the data. (The connection is provided by Eq. 2-2, the definition of average velocity.)

Tactic 2: ***Are the Units OK?*** Be sure to use a consistent set of units when putting numbers into the equations. In Sample Problem 2-1, the logical units in terms of the given data are kilometers for distances, hours for time intervals, and kilometers per hour for velocities. You may sometimes need to convert units.

Tactic 3: ***Is Your Answer Reasonable?*** Does your answer make sense, or is it far too large or far too small? Is the sign correct? Are the units appropriate? In part (c) of Sample Problem 2-1, for example, the correct answer is 17 km/h. If you find 0.00017 km/h, −17 km/h, 17 km/s, or 17 000 km/h, you should realize at once that you have done something wrong. The error may lie in your method, in your algebra, or in your keystroking of numbers on a calculator.

Tactic 4: ***Reading a Graph*** Figures 2-2, 2-3*a*, 2-4, and 2-5 are graphs you should be able to read easily. In each graph, the variable on the horizontal axis is the time *t*, with the direction of increasing time to the right. In each, the variable on the vertical axis is the position *x* of the moving particle with respect to the origin, with the positive direction of *x* upward. Always note the units (seconds or minutes; meters or kilometers) in which the variables are expressed.

2-5 | Instantaneous Velocity and Speed

You have now seen two ways to describe how fast something moves: average velocity and average speed, both of which are measured over a time interval Δt. However, the phrase "how fast" more commonly refers to how fast a particle is moving at a given instant—its **instantaneous velocity** (or simply **velocity**) v.

The velocity at any instant is obtained from the average velocity by shrinking the time interval Δt closer and closer to 0. As Δt dwindles, the average velocity approaches a limiting value, which is the velocity at that instant:

$$v = \lim_{\Delta t \to 0} \frac{\Delta x}{\Delta t} = \frac{dx}{dt}. \qquad (2\text{-}4)$$

Note that v is the rate at which position x is changing with time at a given instant; that is, v is the derivative of x with respect to t. Also note that v at any instant is the slope of the position–time curve at the point representing that instant. Velocity is another vector quantity and thus has an associated direction.

Speed is the magnitude of velocity; that is, speed is velocity that has been stripped of any indication of direction, either in words or via an algebraic sign. (*Caution:* Speed and average speed can be quite different.) A velocity of +5 m/s and one of −5 m/s both have an associated speed of 5 m/s. The speedometer in a car measures speed, not velocity (it cannot determine the direction).

CHECKPOINT 2 The following equations give the position $x(t)$ of a particle in four situations (in each equation, x is in meters, t is in seconds, and $t > 0$): (1) $x = 3t - 2$; (2) $x = -4t^2 - 2$; (3) $x = 2/t^2$; and (4) $x = -2$. (a) In which situation is the velocity v of the particle constant? (b) In which is v in the negative x direction?

Sample Problem 2-2

Figure 2-6*a* is an $x(t)$ plot for an elevator cab that is initially stationary, then moves upward (which we take to be the positive direction of x), and then stops. Plot $v(t)$.

KEY IDEA We can find the velocity at any time from the slope of the $x(t)$ curve at that time.

Calculations: The slope of $x(t)$, and so also the velocity, is zero in the intervals from 0 to 1 s and from 9 s on, so then the cab is stationary. During the interval *bc*, the slope is constant and nonzero, so then the cab moves with constant velocity. We calculate the slope of $x(t)$ then as

$$\frac{\Delta x}{\Delta t} = v = \frac{24\text{ m} - 4.0\text{ m}}{8.0\text{ s} - 3.0\text{ s}} = +4.0\text{ m/s}.$$

The plus sign indicates that the cab is moving in the positive x direction. These intervals (where $v = 0$ and $v = 4$ m/s) are plotted in Fig. 2-6*b*. In addition, as the cab initially begins to move and then later slows to a stop, v varies as indicated in the intervals 1 s to 3 s and 8 s to 9 s. Thus, Fig. 2-6*b* is the required plot. (Figure 2-6*c* is considered in Section 2-6.)

Given a $v(t)$ graph such as Fig. 2-6*b*, we could "work backward" to produce the shape of the associated $x(t)$ graph (Fig. 2-6*a*). However, we would not know the actual values for x at various times, because the $v(t)$ graph indicates only *changes* in x. To find such a change in x during any interval, we must, in the language of calculus, calculate the area "under the curve" on the $v(t)$ graph for that interval. For example, during the interval 3 s to 8 s in which the cab has a velocity of 4.0 m/s, the change in x is

$$\Delta x = (4.0\text{ m/s})(8.0\text{ s} - 3.0\text{ s}) = +20\text{ m}.$$

(This area is positive because the $v(t)$ curve is above the t axis.) Figure 2-6*a* shows that x does indeed increase by 20 m in that interval. However, Fig. 2-6*b* does not tell us the *values* of x at the beginning and end of the interval. For that, we need additional information, such as the value of x at some instant.

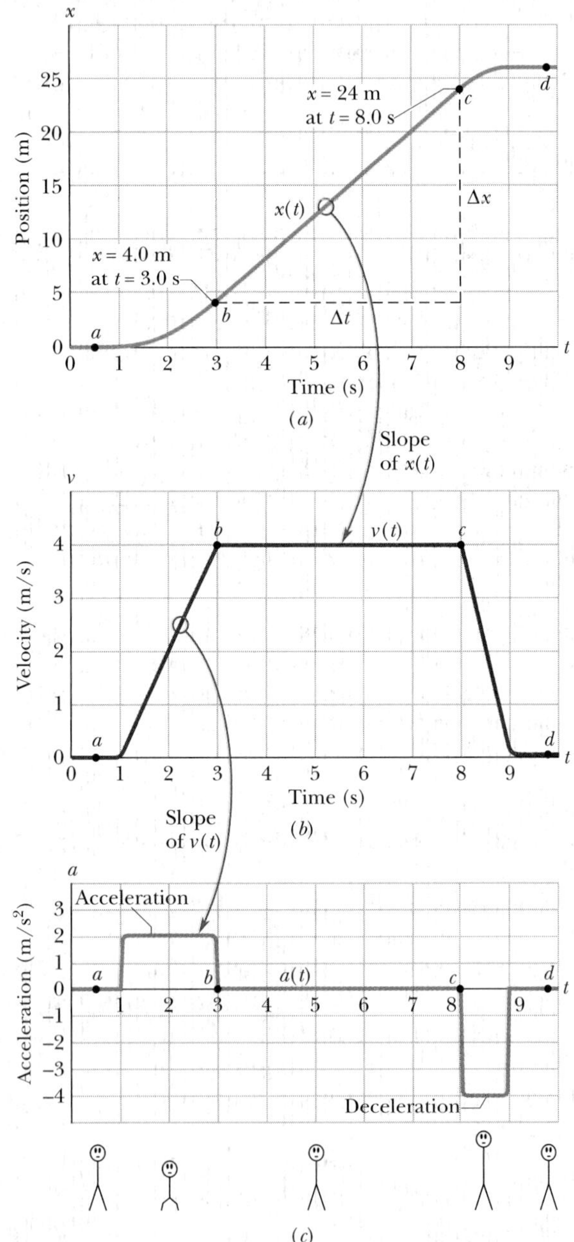

FIG. 2-6 (*a*) The $x(t)$ curve for an elevator cab that moves upward along an x axis. (*b*) The $v(t)$ curve for the cab. Note that it is the derivative of the $x(t)$ curve ($v = dx/dt$). (*c*) The $a(t)$ curve for the cab. It is the derivative of the $v(t)$ curve ($a = dv/dt$). The stick figures along the bottom suggest how a passenger's body might feel during the accelerations.

Sample Problem 2-3

The position of a particle moving on an x axis is given by

$$x = 7.8 + 9.2t - 2.1t^3, \qquad (2\text{-}5)$$

with x in meters and t in seconds. What is its velocity at $t = 3.5$ s? Is the velocity constant, or is it continuously changing?

KEY IDEA Velocity is the first derivative (with respect to time) of the position function $x(t)$.

Calculations: For simplicity, the units have been omitted from Eq. 2-5, but you can insert them if you like by changing the coefficients to 7.8 m, 9.2 m/s, and -2.1 m/s^3. Taking the derivative of Eq. 2-5, we write

$$v = \frac{dx}{dt} = \frac{d}{dt}(7.8 + 9.2t - 2.1t^3),$$

which becomes

$$v = 0 + 9.2 - (3)(2.1)t^2 = 9.2 - 6.3t^2. \qquad (2\text{-}6)$$

At $t = 3.5$ s,

$$v = 9.2 - (6.3)(3.5)^2 = -68 \text{ m/s}. \qquad \text{(Answer)}$$

At $t = 3.5$ s, the particle is moving in the negative direction of x (note the minus sign) with a speed of 68 m/s. Since the quantity t appears in Eq. 2-6, the velocity v depends on t and so is continuously changing.

2-6 | Acceleration

When a particle's velocity changes, the particle is said to undergo **acceleration** (or to accelerate). For motion along an axis, the **average acceleration** a_{avg} over a time interval Δt is

$$a_{avg} = \frac{v_2 - v_1}{t_2 - t_1} = \frac{\Delta v}{\Delta t}, \qquad (2\text{-}7)$$

where the particle has velocity v_1 at time t_1 and then velocity v_2 at time t_2. The **instantaneous acceleration** (or simply **acceleration**) is

$$a = \frac{dv}{dt}. \qquad (2\text{-}8)$$

In words, the acceleration of a particle at any instant is the rate at which its velocity is changing at that instant. Graphically, the acceleration at any point is the slope of the curve of $v(t)$ at that point. We can combine Eq. 2-8 with Eq. 2-4 to write

$$a = \frac{dv}{dt} = \frac{d}{dt}\left(\frac{dx}{dt}\right) = \frac{d^2x}{dt^2}. \qquad (2\text{-}9)$$

In words, the acceleration of a particle at any instant is the second derivative of its position $x(t)$ with respect to time.

A common unit of acceleration is the meter per second per second: m/(s · s) or m/s^2. Other units are in the form of length/(time · time) or length/time^2. Acceleration has both magnitude and direction (it is yet another vector quantity). Its algebraic sign represents its direction on an axis just as for displacement and velocity; that is, acceleration with a positive value is in the positive direction of an axis, and acceleration with a negative value is in the negative direction.

Figure 2-6*c* is a plot of the acceleration of the elevator cab discussed in Sample Problem 2-2. Compare this $a(t)$ curve with the $v(t)$ curve—each point on the $a(t)$ curve shows the derivative (slope) of the $v(t)$ curve at the corresponding time. When v is constant (at either 0 or 4 m/s), the derivative is zero and so also is the acceleration. When the cab first begins to move, the $v(t)$ curve has a positive derivative (the slope is positive), which means that $a(t)$ is positive. When the cab slows to a stop, the derivative and slope of the $v(t)$ curve are negative; that is, $a(t)$ is negative.

Next compare the slopes of the $v(t)$ curve during the two acceleration periods. The slope associated with the cab's slowing down (commonly called "deceleration") is steeper because the cab stops in half the time it took to get up to speed. The steeper slope means that the magnitude of the deceleration is larger than that of the acceleration, as indicated in Fig. 2-6*c*.

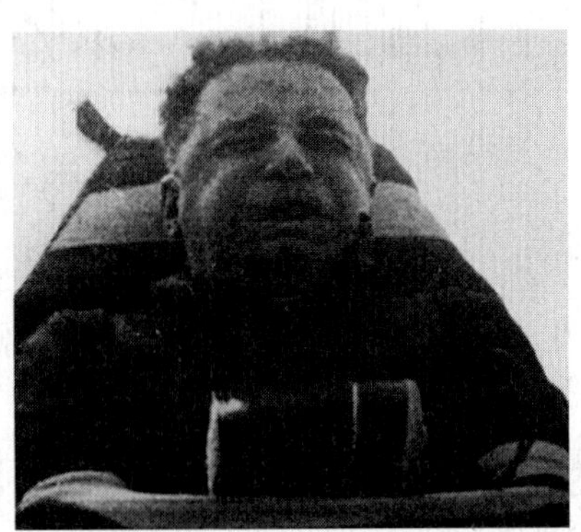

FIG. 2-7 Colonel J. P. Stapp in a rocket sled as it is brought up to high speed (acceleration out of the page) and then very rapidly braked (acceleration into the page). *(Courtesy U.S. Air Force)*

The sensations you would feel while riding in the cab of Fig. 2-6 are indicated by the sketched figures at the bottom. When the cab first accelerates, you feel as though you are pressed downward; when later the cab is braked to a stop, you seem to be stretched upward. In between, you feel nothing special. In other words, your body reacts to accelerations (it is an accelerometer) but not to velocities (it is not a speedometer). When you are in a car traveling at 90 km/h or an airplane traveling at 900 km/h, you have no bodily awareness of the motion. However, if the car or plane quickly changes velocity, you may become keenly aware of the change, perhaps even frightened by it. Part of the thrill of an amusement park ride is due to the quick changes of velocity that you undergo (you pay for the accelerations, not for the speed). A more extreme example is shown in the photographs of Fig. 2-7, which were taken while a rocket sled was rapidly accelerated along a track and then rapidly braked to a stop.

Large accelerations are sometimes expressed in terms of g units, with

$$1g = 9.8 \text{ m/s}^2 \qquad (g \text{ unit}). \tag{2-10}$$

(As we shall discuss in Section 2-9, g is the magnitude of the acceleration of a falling object near Earth's surface.) On a roller coaster, you may experience brief accelerations up to $3g$, which is $(3)(9.8 \text{ m/s}^2)$, or about 29 m/s^2, more than enough to justify the cost of the ride.

PROBLEM-SOLVING TACTICS

***Tactic 5:* An Acceleration's Sign** In common language, the sign of an acceleration has a nonscientific meaning: positive acceleration means that the speed of an object is increasing, and negative acceleration means that the speed is decreasing (the object is decelerating). In this book, however, the sign of an acceleration indicates a direction, not whether an object's speed is increasing or decreasing.

For example, if a car with an initial velocity $v = -25$ m/s is braked to a stop in 5.0 s, then $a_{avg} = +5.0 \text{ m/s}^2$. The acceleration is *positive*, but the car's speed has decreased. The reason is the difference in signs: the direction of the acceleration is opposite that of the velocity.

Here then is the proper way to interpret the signs:

If the signs of the velocity and acceleration of a particle are the same, the speed of the particle increases. If the signs are opposite, the speed decreases.

CHECKPOINT 3 A wombat moves along an x axis. What is the sign of its acceleration if it is moving (a) in the positive direction with increasing speed, (b) in the positive direction with decreasing speed, (c) in the negative direction with increasing speed, and (d) in the negative direction with decreasing speed?

Sample Problem 2-4 Build your skill

A particle's position on the x axis of Fig. 2-1 is given by

$$x = 4 - 27t + t^3,$$

with x in meters and t in seconds.

(a) Because position x depends on time t, the particle must be moving. Find the particle's velocity function $v(t)$ and acceleration function $a(t)$.

KEY IDEAS (1) To get the velocity function $v(t)$, we differentiate the position function $x(t)$ with respect to time. (2) To get the acceleration function $a(t)$, we differentiate the velocity function $v(t)$ with respect to time.

Calculations: Differentiating the position function, we find

$$v = -27 + 3t^2, \qquad \text{(Answer)}$$

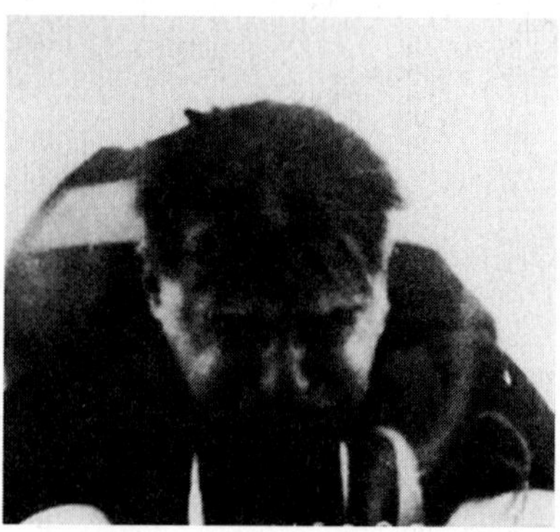

FIG. 2-7 *Continued*

with v in meters per second. Differentiating the velocity function then gives us

$$a = +6t, \qquad \text{(Answer)}$$

with a in meters per second squared.

(b) Is there ever a time when $v = 0$?

Calculation: Setting $v(t) = 0$ yields

$$0 = -27 + 3t^2,$$

which has the solution

$$t = \pm 3 \text{ s.} \qquad \text{(Answer)}$$

Thus, the velocity is zero both 3 s before and 3 s after the clock reads 0.

(c) Describe the particle's motion for $t \geq 0$.

Reasoning: We need to examine the expressions for $x(t)$, $v(t)$, and $a(t)$.

At $t = 0$, the particle is at $x(0) = +4$ m and is moving with a velocity of $v(0) = -27$ m/s—that is, in the negative direction of the x axis. Its acceleration is $a(0) = 0$ because just then the particle's velocity is not changing.

For $0 < t < 3$ s, the particle still has a negative velocity, so it continues to move in the negative direction. However, its acceleration is no longer 0 but is increasing and positive. Because the signs of the velocity and the acceleration are opposite, the particle must be slowing.

Indeed, we already know that it stops momentarily at $t = 3$ s. Just then the particle is as far to the left of the origin in Fig. 2-1 as it will ever get. Substituting $t = 3$ s into the expression for $x(t)$, we find that the particle's position just then is $x = -50$ m. Its acceleration is still positive.

For $t > 3$ s, the particle moves to the right on the axis. Its acceleration remains positive and grows progressively larger in magnitude. The velocity is now positive, and it too grows progressively larger in magnitude.

2-7 | Constant Acceleration: A Special Case

In many types of motion, the acceleration is either constant or approximately so. For example, you might accelerate a car at an approximately constant rate when a traffic light turns from red to green. Then graphs of your position, velocity, and acceleration would resemble those in Fig. 2-8. (Note that $a(t)$ in Fig. 2-8*c* is constant, which requires that $v(t)$ in Fig. 2-8*b* have a constant slope.) Later when you brake the car to a stop, the acceleration (or deceleration in common language) might also be approximately constant.

Such cases are so common that a special set of equations has been derived for dealing with them. One approach to the derivation of these equations is given in this section. A second approach is given in the next section. Throughout both sections and later when you work on the homework problems, keep in mind that *these equations are valid only for constant acceleration (or situations in which you can approximate the acceleration as being constant).*

When the acceleration is constant, the average acceleration and instantaneous acceleration are equal and we can write Eq. 2-7, with some changes in notation, as

$$a = a_{\text{avg}} = \frac{v - v_0}{t - 0}.$$

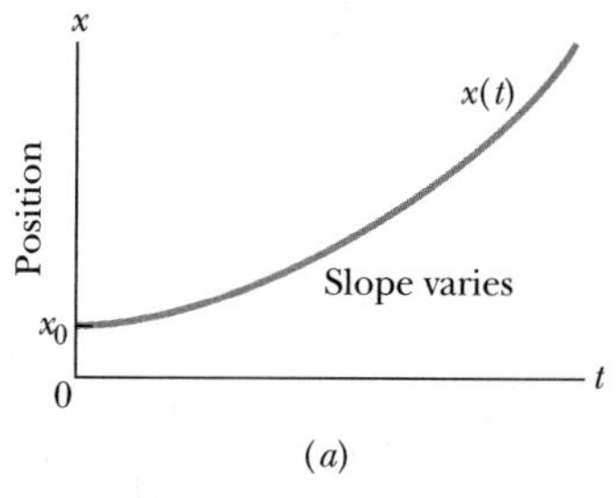

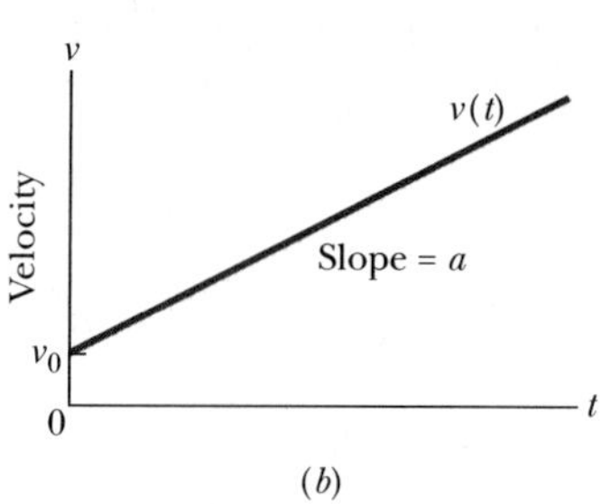

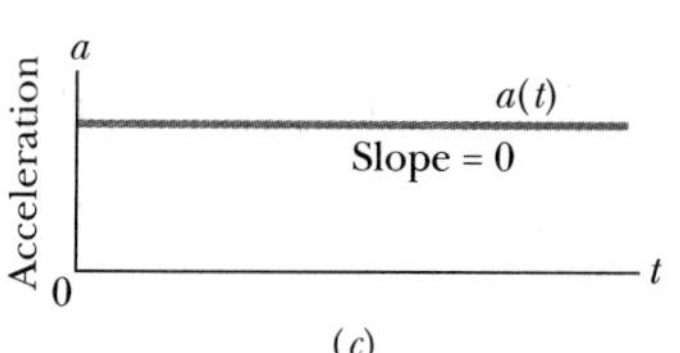

FIG. 2-8 (*a*) The position $x(t)$ of a particle moving with constant acceleration. (*b*) Its velocity $v(t)$, given at each point by the slope of the curve of $x(t)$. (*c*) Its (constant) acceleration, equal to the (constant) slope of the curve of $v(t)$.

Here v_0 is the velocity at time $t = 0$ and v is the velocity at any later time t. We can recast this equation as

$$v = v_0 + at. \tag{2-11}$$

As a check, note that this equation reduces to $v = v_0$ for $t = 0$, as it must. As a further check, take the derivative of Eq. 2-11. Doing so yields $dv/dt = a$, which is the definition of a. Figure 2-8*b* shows a plot of Eq. 2-11, the $v(t)$ function; the function is linear and thus the plot is a straight line.

In a similar manner, we can rewrite Eq. 2-2 (with a few changes in notation) as

$$v_{avg} = \frac{x - x_0}{t - 0}$$

and then as

$$x = x_0 + v_{avg}t, \tag{2-12}$$

in which x_0 is the position of the particle at $t = 0$ and v_{avg} is the average velocity between $t = 0$ and a later time t.

For the linear velocity function in Eq. 2-11, the *average* velocity over any time interval (say, from $t = 0$ to a later time t) is the average of the velocity at the beginning of the interval ($= v_0$) and the velocity at the end of the interval ($= v$). For the interval from $t = 0$ to the later time t then, the average velocity is

$$v_{avg} = \tfrac{1}{2}(v_0 + v). \tag{2-13}$$

Substituting the right side of Eq. 2-11 for v yields, after a little rearrangement,

$$v_{avg} = v_0 + \tfrac{1}{2}at. \tag{2-14}$$

Finally, substituting Eq. 2-14 into Eq. 2-12 yields

$$x - x_0 = v_0 t + \tfrac{1}{2}at^2. \tag{2-15}$$

As a check, note that putting $t = 0$ yields $x = x_0$, as it must. As a further check, taking the derivative of Eq. 2-15 yields Eq. 2-11, again as it must. Figure 2-8*a* shows a plot of Eq. 2-15; the function is quadratic and thus the plot is curved.

Equations 2-11 and 2-15 are the *basic equations for constant acceleration;* they can be used to solve any constant acceleration problem in this book. However, we can derive other equations that might prove useful in certain specific situations. First, note that as many as five quantities can possibly be involved in any problem about constant acceleration—namely, $x - x_0$, v, t, a, and v_0. Usually, one of these quantities is *not* involved in the problem, *either as a given or as an unknown.* We are then presented with three of the remaining quantities and asked to find the fourth.

Equations 2-11 and 2-15 each contain four of these quantities, but not the same four. In Eq. 2-11, the "missing ingredient" is the displacement $x - x_0$. In Eq. 2-15, it is the velocity v. These two equations can also be combined in three ways to yield three additional equations, each of which involves a different "missing variable." First, we can eliminate t to obtain

$$v^2 = v_0^2 + 2a(x - x_0). \tag{2-16}$$

This equation is useful if we do not know t and are not required to find it. Second, we can eliminate the acceleration a between Eqs. 2-11 and 2-15 to produce an equation in which a does not appear:

$$x - x_0 = \tfrac{1}{2}(v_0 + v)t. \tag{2-17}$$

Finally, we can eliminate v_0, obtaining

$$x - x_0 = vt - \tfrac{1}{2}at^2. \tag{2-18}$$

Note the subtle difference between this equation and Eq. 2-15. One involves the initial velocity v_0; the other involves the velocity v at time t.

Table 2-1 lists the basic constant acceleration equations (Eqs. 2-11 and 2-15) as well as the specialized equations that we have derived. To solve a simple constant acceleration problem, you can usually use an equation from this list (*if* you have the list with you). Choose an equation for which the only unknown variable is the variable requested in the problem. A simpler plan is to remember only Eqs. 2-11 and 2-15, and then solve them as simultaneous equations whenever needed.

TABLE 2-1

Equations for Motion with Constant Acceleration[a]

Equation Number	Equation	Missing Quantity
2-11	$v = v_0 + at$	$x - x_0$
2-15	$x - x_0 = v_0 t + \frac{1}{2}at^2$	v
2-16	$v^2 = v_0^2 + 2a(x - x_0)$	t
2-17	$x - x_0 = \frac{1}{2}(v_0 + v)t$	a
2-18	$x - x_0 = vt - \frac{1}{2}at^2$	v_0

[a]Make sure that the acceleration is indeed constant before using the equations in this table.

CHECKPOINT 4 The following equations give the position $x(t)$ of a particle in four situations: (1) $x = 3t - 4$; (2) $x = -5t^3 + 4t^2 + 6$; (3) $x = 2/t^2 - 4/t$; (4) $x = 5t^2 - 3$. To which of these situations do the equations of Table 2-1 apply?

Sample Problem 2-5

The head of a woodpecker is moving forward at a speed of 7.49 m/s when the beak makes first contact with a tree limb. The beak stops after penetrating the limb by 1.87 mm. Assuming the acceleration to be constant, find the acceleration magnitude in terms of g.

KEY IDEA We can use the constant-acceleration equations; in particular, we can use Eq. 2-16 ($v^2 = v_0^2 + 2a(x - x_0)$), which relates velocity and displacement.

Calculations: Because the woodpecker's head stops, the final velocity is $v = 0$. The initial velocity is $v_0 = 7.49$ m/s, and the displacement during the constant acceleration is $x - x_0 = 1.87 \times 10^{-3}$ m. Substituting these values into Eq. 2-16, we have

$$0^2 = (7.49 \text{ m/s})^2 + 2a(1.87 \times 10^{-3} \text{ m}),$$

or

$$a = -1.500 \times 10^4 \text{ m/s}^2.$$

Dividing by $g = 9.8$ m/s² and taking the absolute value, we find that the magnitude of the head's acceleration is

$$a = (1.53 \times 10^3)g. \quad \text{(Answer)}$$

Comment: This typical acceleration magnitude for a woodpecker is about 70 times the acceleration magnitude of Colonel Stapp in Fig. 2-7 and certainly would have been lethal to him. The ability of a woodpecker to withstand such huge acceleration magnitudes is not well understood, but there are two main arguments. (1) The woodpecker's motion is almost along a straight line. Some researchers believe that concussion can occur in humans and animals when the head is rapidly rotated around the neck (and brain stem), but that it is less likely in straight-line motion. (2) The woodpecker's brain is attached so well to the skull that there is little residual movement or oscillation of the brain just after the impact and no chance for the tissue connecting the skull and brain to tear.

Sample Problem 2-6 Build your skill

Figure 2-9 gives a particle's velocity v versus its position as it moves along an x axis with constant acceleration. What is its velocity at position $x = 0$?

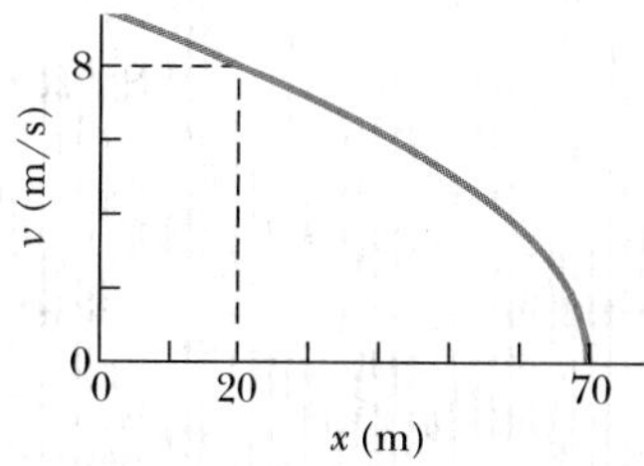

FIG. 2-9 Velocity versus position.

KEY IDEA We can use the constant-acceleration equations; in particular, we can use Eq. 2-16 ($v^2 = v_0^2 + 2a(x - x_0)$), which relates velocity and position.

First try: Normally we want to use an equation that includes the requested variable. In Eq. 2-16, we can identify x_0 as 0 and v_0 as being the requested variable. Then we can identify a second pair of values as being v and x. From the graph, we have two such pairs: (1) $v = 8$ m/s and $x = 20$ m, and (2) $v = 0$ and $x = 70$ m. For example, we can write Eq. 2-16 as

$$(8 \text{ m/s})^2 = v_0^2 + 2a(20 \text{ m} - 0). \quad (2\text{-}19)$$

However, we know neither v_0 nor a.

Second try: Instead of directly involving the requested variable, let's use Eq. 2-16 with the two pairs of known data, identifying $v_0 = 8$ m/s and $x_0 = 20$ m as the first pair and $v = 0$ m/s and $x = 70$ m as the second pair. Then we can write

$$(0 \text{ m/s})^2 = (8 \text{ m/s})^2 + 2a(70 \text{ m} - 20 \text{ m}),$$

which gives us $a = -0.64$ m/s². Substituting this value into Eq. 2-19 and solving for v_0 (the velocity associated with the position of $x = 0$), we find

$$v_0 = 9.5 \text{ m/s}. \quad \text{(Answer)}$$

Comment: Some problems involve an equation that includes the requested variable. A more challenging problem requires you to first use an equation that does *not* include the requested variable but that gives you a value needed to find it. Sometimes that procedure takes *physics courage* because it is so indirect. However, if you build your solving skills by solving lots of problems, the procedure gradually requires less courage and may even become obvious. Solving problems of any kind, whether physics or social, requires practice.

2-8 | Another Look at Constant Acceleration*

The first two equations in Table 2-1 are the basic equations from which the others are derived. Those two can be obtained by integration of the acceleration with the condition that a is constant. To find Eq. 2-11, we rewrite the definition of acceleration (Eq. 2-8) as

$$dv = a\,dt.$$

We next write the *indefinite integral* (or *antiderivative*) of both sides:

$$\int dv = \int a\,dt.$$

Since acceleration a is a constant, it can be taken outside the integration. We obtain

$$\int dv = a \int dt$$

or

$$v = at + C. \tag{2-20}$$

To evaluate the constant of integration C, we let $t = 0$, at which time $v = v_0$. Substituting these values into Eq. 2-20 (which must hold for all values of t, including $t = 0$) yields

$$v_0 = (a)(0) + C = C.$$

Substituting this into Eq. 2-20 gives us Eq. 2-11.

To derive Eq. 2-15, we rewrite the definition of velocity (Eq. 2-4) as

$$dx = v\,dt$$

and then take the indefinite integral of both sides to obtain

$$\int dx = \int v\,dt.$$

Next, we substitute for v with Eq. 2-11:

$$\int dx = \int (v_0 + at)\,dt.$$

Since v_0 is a constant, as is the acceleration a, this can be rewritten as

$$\int dx = v_0 \int dt + a \int t\,dt.$$

Integration now yields

$$x = v_0 t + \tfrac{1}{2}at^2 + C', \tag{2-21}$$

where C' is another constant of integration. At time $t = 0$, we have $x = x_0$. Substituting these values in Eq. 2-21 yields $x_0 = C'$. Replacing C' with x_0 in Eq. 2-21 gives us Eq. 2-15.

2-9 | Free-Fall Acceleration

If you tossed an object either up or down and could somehow eliminate the effects of air on its flight, you would find that the object accelerates downward at a certain constant rate. That rate is called the **free-fall acceleration,** and its magnitude is represented by g. The acceleration is independent of the object's characteristics, such as mass, density, or shape; it is the same for all objects.

*This section is intended for students who have had integral calculus.

Two examples of free-fall acceleration are shown in Fig. 2-10, which is a series of stroboscopic photos of a feather and an apple. As these objects fall, they accelerate downward—both at the same rate g. Thus, their speeds increase at the same rate, and they fall together.

The value of g varies slightly with latitude and with elevation. At sea level in Earth's midlatitudes the value is 9.8 m/s^2 (or 32 ft/s^2), which is what you should use as an exact number for the problems in this book unless otherwise noted.

The equations of motion in Table 2-1 for constant acceleration also apply to free fall near Earth's surface; that is, they apply to an object in vertical flight, either up or down, when the effects of the air can be neglected. However, note that for free fall: (1) The directions of motion are now along a vertical y axis instead of the x axis, with the positive direction of y upward. (This is important for later chapters when combined horizontal and vertical motions are examined.) (2) The free-fall acceleration is negative—that is, downward on the y axis, toward Earth's center—and so it has the value $-g$ in the equations.

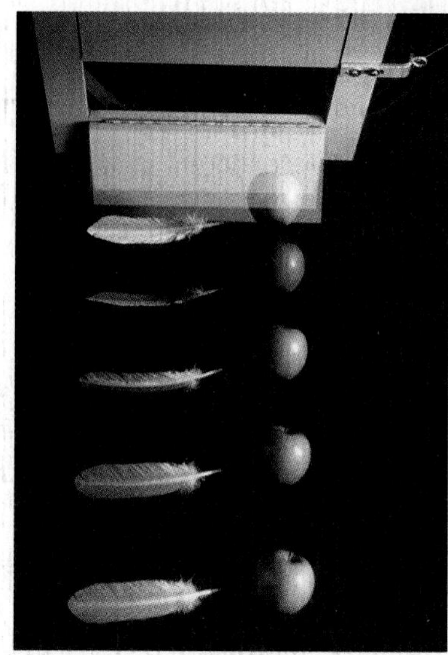

FIG. 2-10 A feather and an apple free fall in vacuum at the same magnitude of acceleration g. The acceleration increases the distance between successive images. In the absence of air, the feather and apple fall together. *(Jim Sugar/Corbis Images)*

The free-fall acceleration near Earth's surface is $a = -g = -9.8$ m/s^2, and the *magnitude* of the acceleration is $g = 9.8$ m/s^2. Do not substitute -9.8 m/s^2 for g.

Suppose you toss a tomato directly upward with an initial (positive) velocity v_0 and then catch it when it returns to the release level. During its *free-fall flight* (from just after its release to just before it is caught), the equations of Table 2-1 apply to its motion. The acceleration is always $a = -g = -9.8$ m/s^2, negative and thus downward. The velocity, however, changes, as indicated by Eqs. 2-11 and 2-16: during the ascent, the magnitude of the positive velocity decreases, until it momentarily becomes zero. Because the tomato has then stopped, it is at its maximum height. During the descent, the magnitude of the (now negative) velocity increases.

CHECKPOINT 5 (a) If you toss a ball straight up, what is the sign of the ball's displacement for the ascent, from the release point to the highest point? (b) What is it for the descent, from the highest point back to the release point? (c) What is the ball's acceleration at its highest point?

Sample Problem 2-7

On September 26, 1993, Dave Munday went over the Canadian edge of Niagara Falls in a steel ball equipped with an air hole and then fell 48 m to the water (and rocks). Assume his initial velocity was zero, and neglect the effect of the air on the ball during the fall.

(a) How long did Munday fall to reach the water surface?

KEY IDEA Because Munday's fall was a free fall, the constant-acceleration equations of Table 2-1 apply.

Calculations: Let us place a y axis along the path of his fall, with $y = 0$ at his starting point and the positive direction up the axis (Fig. 2-11). Then the acceleration is $a = -g$ along that axis, and the water level is at $y = -48$ m (negative because it is below $y = 0$). Let the fall begin at time $t = 0$, with initial velocity $v_0 = 0$.

From Table 2-1 we choose Eq. 2-15 (but in y notation) because it contains the requested time t. We find

$$y - y_0 = v_0t - \tfrac{1}{2}gt^2,$$

$$-48 \text{ m} - 0 = 0t - \tfrac{1}{2}(9.8 \text{ m/s}^2)t^2,$$

$$t^2 = 48/4.9,$$

and $$t = 3.1 \text{ s.} \qquad \text{(Answer)}$$

Note that Munday's displacement $y - y_0$ is a negative quantity—Munday fell down, in the *negative direction* of the y axis (he did not fall up!). Also note that 48/4.9 has two square roots: 3.1 and -3.1. Here we choose the positive root because Munday obviously reaches the water surface *after* he begins to fall at $t = 0$.

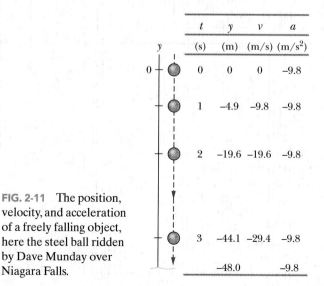

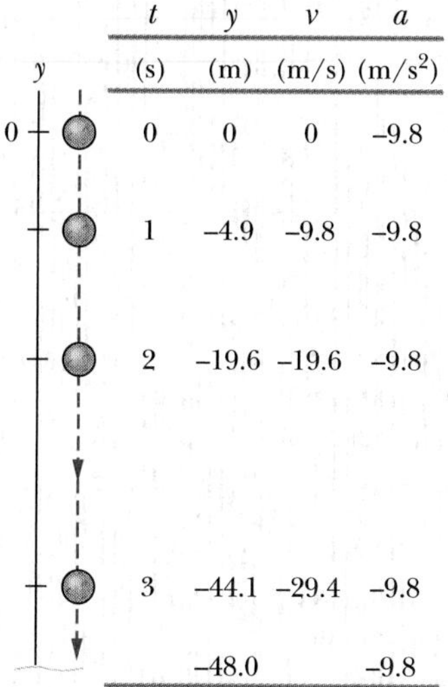

y
0
t y v a
(s) (m) (m/s) (m/s²)
0 0 0 -9.8
1 -4.9 -9.8 -9.8
2 -19.6 -19.6 -9.8
3 -44.1 -29.4 -9.8
-48.0 -9.8

FIG. 2-11 The position, velocity, and acceleration of a freely falling object, here the steel ball ridden by Dave Munday over Niagara Falls.

(b) Munday could count off the three seconds of free fall but could not see how far he had fallen with each count. Determine his position at each full second.

Calculations: We again use Eq. 2-15 but now we substitute, in turn, the values $t = 1.0$ s, 2.0 s, and 3.0 s, and solve for Munday's position y. The results are shown in Fig. 2-11.

(c) What was Munday's velocity as he reached the water surface?

Calculation: To find the velocity from the original data without using the time of fall from (a), we rewrite Eq. 2-16 in y notation and then substitute known data:

$$v^2 = v_0^2 - 2g(y - y_0) = 0 - (2)(9.8 \text{ m/s}^2)(-48 \text{ m}),$$

so $$v = -30.67 \text{ m/s} \approx -31 \text{ m/s} = -110 \text{ km/h}. \quad \text{(Answer)}$$

We chose the negative root here because the velocity was in the negative direction.

(d) What was Munday's velocity at each count of one full second? Was he aware of his increasing speed?

Calculations: To find the velocities from the original data without using the positions from (b), we let $a = -g$ in Eq. 2-11 and then substitute, in turn, the values $t = 1.0$ s, 2.0 s, and 3.0 s. Here is an example:

$$v = v_0 - gt$$
$$= 0 - (9.8 \text{ m/s}^2)(1.0 \text{ s}) = -9.8 \text{ m/s}. \quad \text{(Answer)}$$

The other results are shown in Fig. 2-11.

Once he was in free fall, Munday was unaware of the increasing speed because the acceleration during the fall was always -9.8 m/s^2, as noted in the last column of Fig. 2-11. He was, of course, sharply aware of hitting the water because then the acceleration abruptly changed. (Munday survived the fall but then faced stiff legal fines for his daredevil action.)

Sample Problem 2-8

In Fig. 2-12, a pitcher tosses a baseball up along a y axis, with an initial speed of 12 m/s.

(a) How long does the ball take to reach its maximum height?

KEY IDEAS (1) Once the ball leaves the pitcher and before it returns to his hand, its acceleration is the free-fall acceleration $a = -g$. Because this is constant, Table 2-1 applies to the motion. (2) The velocity v at the maximum height must be 0.

Calculation: Knowing v, a, and the initial velocity $v_0 = 12$ m/s, and seeking t, we solve Eq. 2-11, which contains those four variables. This yields

$$t = \frac{v - v_0}{a} = \frac{0 - 12 \text{ m/s}}{-9.8 \text{ m/s}^2} = 1.2 \text{ s}. \quad \text{(Answer)}$$

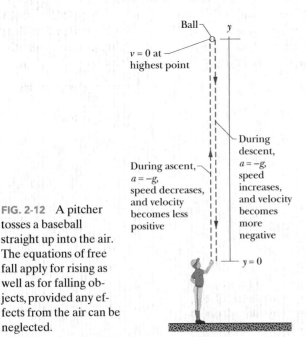

FIG. 2-12 A pitcher tosses a baseball straight up into the air. The equations of free fall apply for rising as well as for falling objects, provided any effects from the air can be neglected.

(b) What is the ball's maximum height above its release point?

Calculation: We can take the ball's release point to be $y_0 = 0$. We can then write Eq. 2-16 in y notation, set $y - y_0 = y$ and $v = 0$ (at the maximum height), and solve for y. We get

$$y = \frac{v^2 - v_0^2}{2a} = \frac{0 - (12 \text{ m/s})^2}{2(-9.8 \text{ m/s}^2)} = 7.3 \text{ m}. \quad \text{(Answer)}$$

(c) How long does the ball take to reach a point 5.0 m above its release point?

Calculations: We know v_0, $a = -g$, and displacement $y - y_0 = 5.0$ m, and we want t, so we choose Eq. 2-15. Rewriting it for y and setting $y_0 = 0$ give us

$$y = v_0 t - \tfrac{1}{2}gt^2,$$

or $$5.0 \text{ m} = (12 \text{ m/s})t - (\tfrac{1}{2})(9.8 \text{ m/s}^2)t^2.$$

If we temporarily omit the units (having noted that they are consistent), we can rewrite this as

$$4.9t^2 - 12t + 5.0 = 0.$$

Solving this quadratic equation for t yields

$$t = 0.53 \text{ s} \quad \text{and} \quad t = 1.9 \text{ s}. \quad \text{(Answer)}$$

During ascent, a = −g, speed decreases, and velocity becomes less positive
During descent, a = −g, speed increases, and velocity becomes more negative

There are two such times! This is not really surprising because the ball passes twice through $y = 5.0$ m, once on the way up and once on the way down.

PROBLEM-SOLVING TACTICS

Tactic 6: Meanings of Minus Signs In Sample Problems 2-7 and 2-8, we established a vertical axis (the y axis) and we chose—quite arbitrarily—its upward direction to be positive. We then chose the origin of the y axis (that is, the $y = 0$ position) to suit the problem. In Sample Problem 2-7, the origin was at the top of the falls, and in Sample Problem 2-8 it was at the pitcher's hand. A negative value of y then means that the body is below the chosen origin. A negative velocity means that the body is moving in the negative direction of the y axis—that is, downward. This is true no matter where the body is located.

We take the acceleration to be negative ($-9.8\ \text{m/s}^2$) in all problems dealing with falling bodies. A negative acceleration means that, as time goes on, the velocity of the body becomes either less positive or more negative. This is true no matter where the body is located and no matter how fast or in what direction it is moving. In Sample Problem 2-8, the acceleration of the ball is negative (downward) throughout its flight, whether the ball is rising or falling.

Tactic 7: Unexpected Answers Mathematics often generates answers that you might not have thought of as possibilities, as in Sample Problem 2-8c. If you get more answers than you expect, do not automatically discard the ones that do not seem to fit. Examine them carefully for physical meaning. If time is your variable, even a negative value can mean something; negative time simply refers to time before $t = 0$, the (arbitrary) time at which you decided to start your stopwatch.

2-10 | Graphical Integration in Motion Analysis

When we have a graph of an object's acceleration versus time, we can integrate on the graph to find the object's velocity at any given time. Because acceleration a is defined in terms of velocity as $a = dv/dt$, the Fundamental Theorem of Calculus tells us that

$$v_1 - v_0 = \int_{t_0}^{t_1} a\,dt. \tag{2-22}$$

The right side of the equation is a definite integral (it gives a numerical result rather than a function), v_0 is the velocity at time t_0, and v_1 is the velocity at later time t_1. The definite integral can be evaluated from an $a(t)$ graph, such as in Fig. 2-13a. In particular,

$$\int_{t_0}^{t_1} a\,dt = \begin{pmatrix}\text{area between acceleration curve}\\ \text{and time axis, from } t_0 \text{ to } t_1\end{pmatrix}. \tag{2-23}$$

If a unit of acceleration is $1\ \text{m/s}^2$ and a unit of time is 1 s, then the corresponding unit of area on the graph is

$$(1\ \text{m/s}^2)(1\ \text{s}) = 1\ \text{m/s},$$

which is (properly) a unit of velocity. When the acceleration curve is above the time axis, the area is positive; when the curve is below the time axis, the area is negative.

Similarly, because velocity v is defined in terms of the position x as $v = dx/dt$, then

$$x_1 - x_0 = \int_{t_0}^{t_1} v\,dt, \tag{2-24}$$

where x_0 is the position at time t_0 and x_1 is the position at time t_1. The definite integral on the right side of Eq. 2-24 can be evaluated from a $v(t)$ graph, like that shown in Fig. 2-13b. In particular,

$$\int_{t_0}^{t_1} v\,dt = \begin{pmatrix}\text{area between velocity curve}\\ \text{and time axis, from } t_0 \text{ to } t_1\end{pmatrix}. \tag{2-25}$$

If the unit of velocity is 1 m/s and the unit of time is 1 s, then the corresponding unit of area on the graph is

$$(1\ \text{m/s})(1\ \text{s}) = 1\ \text{m},$$

which is (properly) a unit of position and displacement. Whether this area is positive or negative is determined as described for the $a(t)$ curve of Fig. 2-13a.

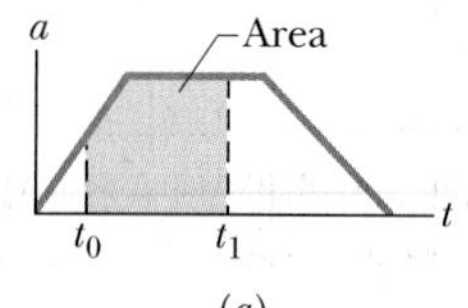

(*a*)

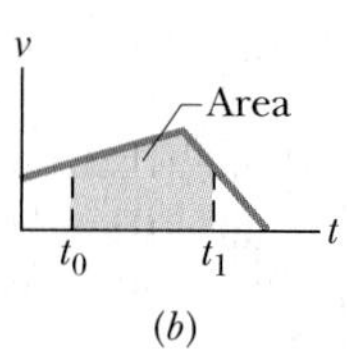

(*b*)

FIG. 2-13 The area between a plotted curve and the horizontal time axis, from time t_0 to time t_1, is indicated for (a) a graph of acceleration a versus t and (b) a graph of velocity v versus t.

Sample Problem 2-9

"Whiplash injury" commonly occurs in a rear-end collision where a front car is hit from behind by a second car. In the 1970s, researchers concluded that the injury was due to the occupant's head being whipped back over the top of the seat as the car was slammed forward. As a result of this finding, head restraints were built into cars, yet neck injuries in rear-end collisions continued to occur.

In a recent test to study neck injury in rear-end collisions, a volunteer was strapped to a seat that was then moved abruptly to simulate a collision by a rear car moving at 10.5 km/h. Figure 2-14*a* gives the accelerations of the volunteer's torso and head during the collision, which began at time $t = 0$. The torso acceleration was delayed by 40 ms because during that time interval the seat back had to compress against the volunteer. The head acceleration was delayed by an additional 70 ms. What was the torso speed when the head began to accelerate?

KEY IDEA We can calculate the torso speed at any time by finding an area on the torso $a(t)$ graph.

Calculations: We know that the initial torso speed is $v_0 = 0$ at time $t_0 = 0$, at the start of the "collision." We want the torso speed v_1 at time $t_1 = 110$ ms, which is when the head begins to accelerate.

Combining Eqs. 2-22 and 2-23, we can write

$$v_1 - v_0 = \begin{pmatrix}\text{area between acceleration curve}\\ \text{and time axis, from } t_0 \text{ to } t_1\end{pmatrix}. \quad (2\text{-}26)$$

For convenience, let us separate the area into three regions (Fig. 2-14*b*). From 0 to 40 ms, region A has no area:

$$\text{area}_A = 0.$$

From 40 ms to 100 ms, region B has the shape of a triangle, with area

$$\text{area}_B = \tfrac{1}{2}(0.060 \text{ s})(50 \text{ m/s}^2) = 1.5 \text{ m/s}.$$

From 100 ms to 110 ms, region C has the shape of a rectangle, with area

$$\text{area}_C = (0.010 \text{ s})(50 \text{ m/s}^2) = 0.50 \text{ m/s}.$$

Substituting these values and $v_0 = 0$ into Eq. 2-26 gives us

$$v_1 - 0 = 0 + 1.5 \text{ m/s} + 0.50 \text{ m/s},$$

or

$$v_1 = 2.0 \text{ m/s} = 7.2 \text{ km/h}. \qquad \text{(Answer)}$$

Comments: When the head is just starting to move forward, the torso already has a speed of 7.2 km/h. Researchers argue that it is this difference in speeds during the early stage of a rear-end collision that injures the neck. The backward whipping of the head happens later and could, especially if there is no head restraint, increase the injury.

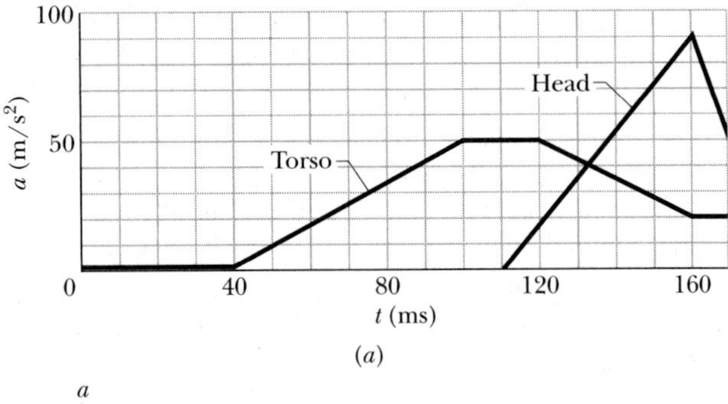

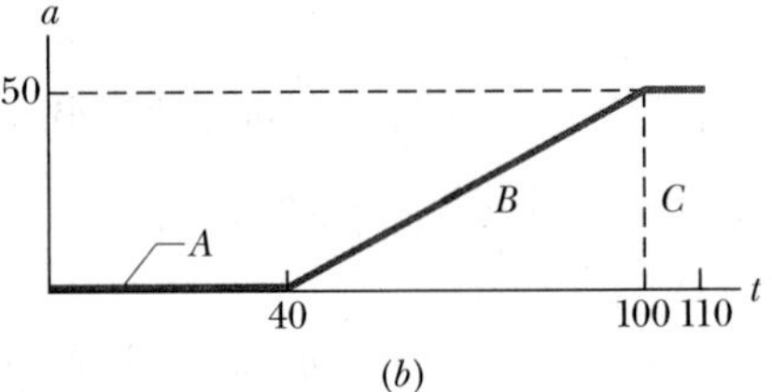

FIG. 2-14 (*a*) The $a(t)$ curve of the torso and head of a volunteer in a simulation of a rear-end collision. (*b*) Breaking up the region between the plotted curve and the time axis to calculate the area.

REVIEW & SUMMARY

Position The *position* x of a particle on an x axis locates the particle with respect to the **origin,** or zero point, of the axis. The position is either positive or negative, according to which side of the origin the particle is on, or zero if the particle is at the origin. The **positive direction** on an axis is the direction of increasing positive numbers; the opposite direction is the **negative direction.**

Displacement The *displacement* Δx of a particle is the change in its position:

$$\Delta x = x_2 - x_1. \quad (2\text{-}1)$$

Displacement is a vector quantity. It is positive if the particle has moved in the positive direction of the x axis and negative if the particle has moved in the negative direction.

Average Velocity When a particle has moved from position x_1 to position x_2 during a time interval $\Delta t = t_2 - t_1$, its *average velocity* during that interval is

$$v_{avg} = \frac{\Delta x}{\Delta t} = \frac{x_2 - x_1}{t_2 - t_1}. \quad (2\text{-}2)$$

The algebraic sign of v_{avg} indicates the direction of motion (v_{avg} is a vector quantity). Average velocity does not depend on the actual distance a particle moves, but instead depends on its original and final positions.

On a graph of x versus t, the average velocity for a time interval Δt is the slope of the straight line connecting the points on the curve that represent the two ends of the interval.

Average Speed The *average speed* s_{avg} of a particle during a time interval Δt depends on the total distance the particle moves in that time interval:

$$s_{avg} = \frac{\text{total distance}}{\Delta t}. \tag{2-3}$$

Instantaneous Velocity The *instantaneous velocity* (or simply **velocity**) v of a moving particle is

$$v = \lim_{\Delta t \to 0} \frac{\Delta x}{\Delta t} = \frac{dx}{dt}, \tag{2-4}$$

where Δx and Δt are defined by Eq. 2-2. The instantaneous velocity (at a particular time) may be found as the slope (at that particular time) of the graph of x versus t. **Speed** is the magnitude of instantaneous velocity.

Average Acceleration *Average acceleration* is the ratio of a change in velocity Δv to the time interval Δt in which the change occurs:

$$a_{avg} = \frac{\Delta v}{\Delta t}. \tag{2-7}$$

The algebraic sign indicates the direction of a_{avg}.

Instantaneous Acceleration *Instantaneous acceleration* (or simply **acceleration**) a is the first time derivative of velocity $v(t)$ and the second time derivative of position $x(t)$:

$$a = \frac{dv}{dt} = \frac{d^2x}{dt^2}. \tag{2-8, 2-9}$$

On a graph of v versus t, the acceleration a at any time t is the slope of the curve at the point that represents t.

Constant Acceleration The five equations in Table 2-1 describe the motion of a particle with constant acceleration:

$$v = v_0 + at, \tag{2-11}$$

$$x - x_0 = v_0 t + \tfrac{1}{2}at^2, \tag{2-15}$$

$$v^2 = v_0^2 + 2a(x - x_0), \tag{2-16}$$

$$x - x_0 = \tfrac{1}{2}(v_0 + v)t, \tag{2-17}$$

$$x - x_0 = vt - \tfrac{1}{2}at^2. \tag{2-18}$$

These are *not* valid when the acceleration is not constant.

Free-Fall Acceleration An important example of straight-line motion with constant acceleration is that of an object rising or falling freely near Earth's surface. The constant acceleration equations describe this motion, but we make two changes in notation: (1) we refer the motion to the vertical y axis with $+y$ vertically *up*; (2) we replace a with $-g$, where g is the magnitude of the free-fall acceleration. Near Earth's surface, $g = 9.8\ \text{m/s}^2$ ($= 32\ \text{ft/s}^2$).

QUESTIONS

1 Figure 2-15 shows four paths along which objects move from a starting point to a final point, all in the same time interval. The paths pass over a grid of equally spaced straight lines. Rank the paths according to (a) the average velocity of the objects and (b) the average speed of the objects, greatest first.

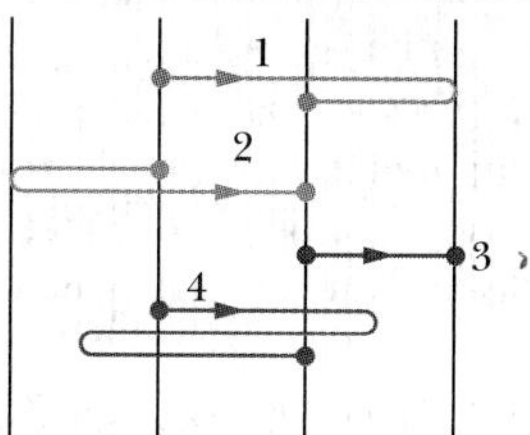

FIG. 2-15 Question 1.

2 Figure 2-16 is a graph of a particle's position along an x axis versus time. (a) At time $t = 0$, what is the sign of the particle's position? Is the particle's velocity positive, negative, or 0 at (b) $t = 1$ s, (c) $t = 2$ s, and (d) $t = 3$ s? (e) How many times does the particle go through the point $x = 0$?

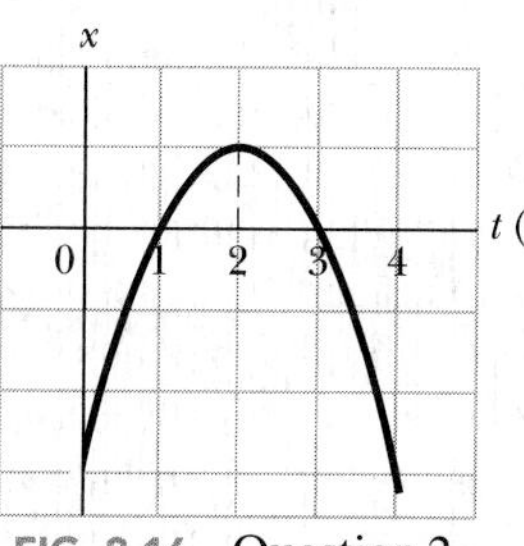

FIG. 2-16 Question 2.

3 Figure 2-17 gives the velocity of a particle moving on an x axis. What are (a) the initial and (b) the final directions of travel? (c) Does the particle stop momentarily? (d) Is the acceleration positive or negative? (e) Is it constant or varying?

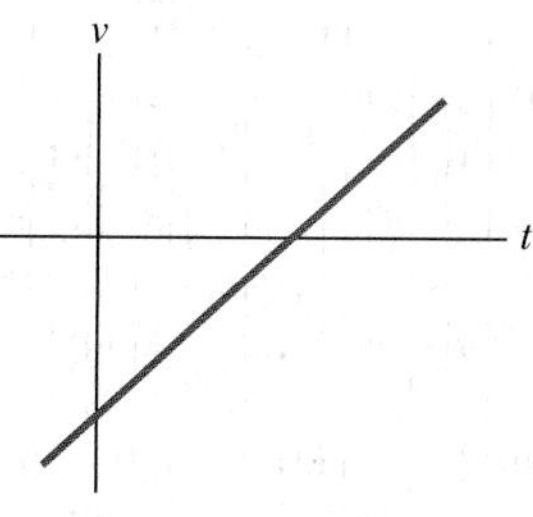

FIG. 2-17 Question 3.

4 Figure 2-18 gives the acceleration $a(t)$ of a Chihuahua as it chases a German shepherd along an axis. In which of the time periods indicated does the Chihuahua move at constant speed?

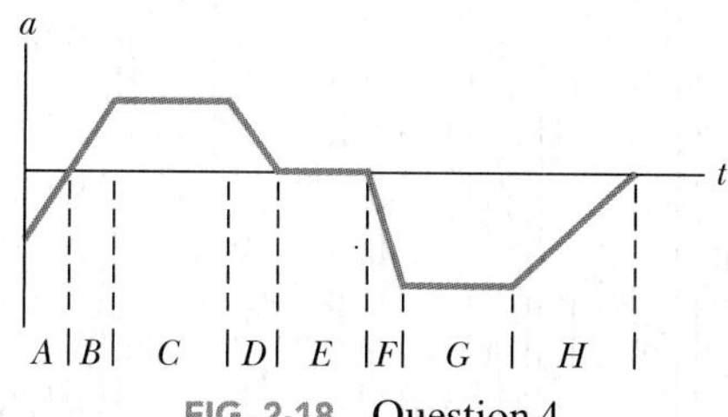

FIG. 2-18 Question 4.

5 Figure 2-19 gives the velocity of a particle moving along an axis. Point 1 is at the highest point on the curve; point 4 is at the lowest point; and points 2 and 6 are at the same height. What is the direction of travel at (a) time $t = 0$ and (b) point 4? (c) At which of the six numbered points does the particle reverse its direction of travel? (d) Rank the six points according to the magnitude of the acceleration, greatest first.

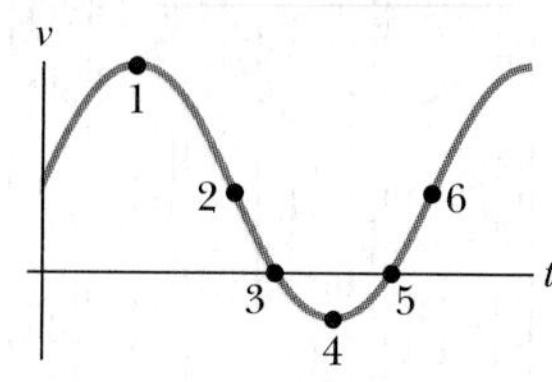

FIG. 2-19 Question 5.

6 The following equations give the velocity $v(t)$ of a particle in four situations: (a) $v = 3$; (b) $v = 4t^2 + 2t - 6$; (c) $v = 3t - 4$; (d) $v = 5t^2 - 3$. To which of these situations do the equations of Table 2-1 apply?

7 In Fig. 2-20, a cream tangerine is thrown directly upward past three evenly spaced windows of equal heights. Rank the windows according to (a) the average speed of the cream tangerine while passing them, (b) the time the cream tangerine takes to pass them, (c) the magnitude of the acceleration of the cream tangerine while passing them, and (d) the change Δv in the speed of the cream tangerine during the passage, greatest first.

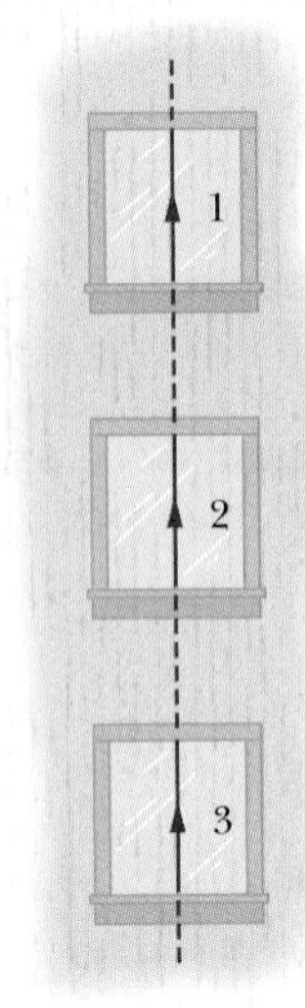

FIG. 2-20 Question 7.

8 At $t = 0$, a particle moving along an x axis is at position $x_0 = -20$ m. The signs of the particle's initial velocity v_0 (at time t_0) and constant acceleration a are, respectively, for four situations: (1) +, +; (2) +, −; (3) −, +; (4) −, −. In which situations will the particle (a) stop momentarily, (b) pass through the origin, and (c) never pass through the origin?

9 Hanging over the railing of a bridge, you drop an egg (no initial velocity) as you throw a second egg downward. Which curves in Fig. 2-21 give the velocity $v(t)$ for (a) the dropped egg and (b) the thrown egg? (Curves A and B are parallel; so are C, D, and E; so are F and G.)

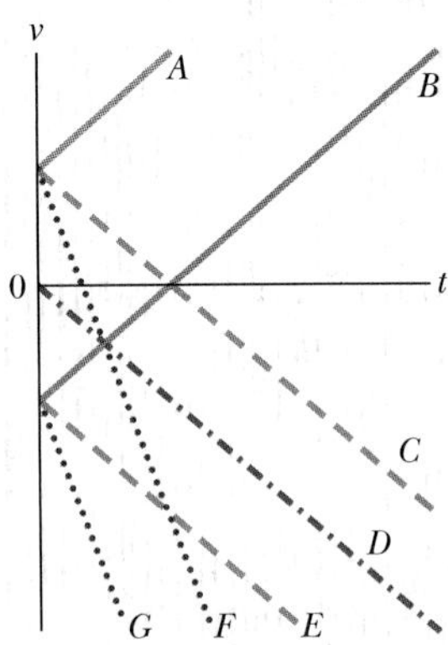

FIG. 2-21 Question 9.

PROBLEMS

GO Tutoring problem available (at instructor's discretion) in *WileyPLUS* and WebAssign
SSM Worked-out solution available in Student Solutions Manual
WWW Worked-out solution is at
• – ••• Number of dots indicates level of problem difficulty
ILW Interactive solution is at
http://www.wiley.com/college/halliday
Additional information available in *The Flying Circus of Physics* and at flyingcircusofphysics.com

sec. 2-4 Average Velocity and Average Speed

•1 An automobile travels on a straight road for 40 km at 30 km/h. It then continues in the same direction for another 40 km at 60 km/h. (a) What is the average velocity of the car during this 80 km trip? (Assume that it moves in the positive x direction.) (b) What is the average speed? (c) Graph x versus t and indicate how the average velocity is found on the graph. SSM WWW

•2 A car travels up a hill at a constant speed of 40 km/h and returns down the hill at a constant speed of 60 km/h. Calculate the average speed for the round trip.

•3 During a hard sneeze, your eyes might shut for 0.50 s. If you are driving a car at 90 km/h during such a sneeze, how far does the car move during that time?

•4 The 1992 world speed record for a bicycle (human powered vehicle) was set by Chris Huber. His time through the measured 200 m stretch was a sizzling 6.509 s, at which he commented, "Cogito ergo zoom!" (I think, therefore I go fast!). In 2001, Sam Whittingham beat Huber's record by 19.0 km/h. What was Whittingham's time through the 200 m?

•5 The position of an object moving along an x axis is given by $x = 3t - 4t^2 + t^3$, where x is in meters and t in seconds. Find the position of the object at the following values of t: (a) 1 s, (b) 2 s, (c) 3 s, and (d) 4 s. (e) What is the object's displacement between $t = 0$ and $t = 4$ s? (f) What is its average velocity for the time interval from $t = 2$ s to $t = 4$ s? (g) Graph x versus t for $0 \le t \le 4$ s and indicate how the answer for (f) can be found on the graph. SSM

•6 Compute your average velocity in the following two cases: (a) You walk 73.2 m at a speed of 1.22 m/s and then run 73.2 m at a speed of 3.05 m/s along a straight track. (b) You walk for 1.00 min at a speed of 1.22 m/s and then run for 1.00 min at 3.05 m/s along a straight track. (c) Graph x versus t for both cases and indicate how the average velocity is found on the graph.

••7 In 1 km races, runner 1 on track 1 (with time 2 min, 27.95 s) appears to be faster than runner 2 on track 2 (2 min, 28.15 s). However, length L_2 of track 2 might be slightly greater than length L_1 of track 1. How large can $L_2 - L_1$ be for us still to conclude that runner 1 is faster? ILW

••8 To set a speed record in a measured (straight-line) distance d, a race car must be driven first in one direction (in time t_1) and then in the opposite direction (in time t_2). (a) To eliminate the effects of the wind and obtain the car's speed v_c in a windless situation, should we find the average of d/t_1 and d/t_2 (method 1) or should we divide d by the average of t_1 and t_2? (b) What is the fractional difference in the two methods when a steady wind blows along the car's route and the ratio of the wind speed v_w to the car's speed v_c is 0.0240?

••9 You are to drive to an interview in another town, at a distance of 300 km on an expressway. The interview is at 11:15 A.M. You plan to drive at 100 km/h, so you leave at 8:00 A.M. to allow some extra time. You drive at that speed for the first 100 km, but then construction work forces you to slow to 40 km/h for 40 km. What would be the least speed needed for the rest of the trip to arrive in time for the interview?

••10 *Panic escape*. Figure 2-22 shows a general situation in which a stream of people attempt to escape through an exit door that turns out to be locked. The people move toward the door at speed $v_s = 3.50$ m/s, are

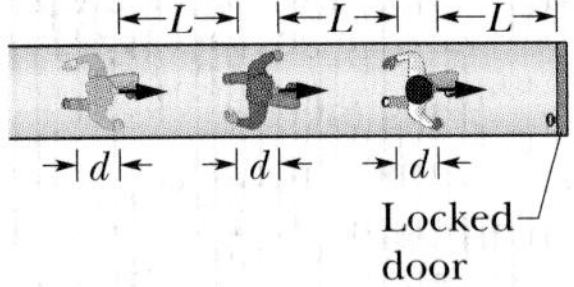

FIG. 2-22 Problem 10.

each $d = 0.25$ m in depth, and are separated by $L = 1.75$ m. The arrangement in Fig. 2-22 occurs at time $t = 0$. (a) At what average rate does the layer of people at the door increase? (b) At what time does the layer's depth reach 5.0 m? (The answers reveal how quickly such a situation becomes dangerous.)

••11 Two trains, each having a speed of 30 km/h, are headed at each other on the same straight track. A bird that can fly 60 km/h flies off the front of one train when they are 60 km apart and heads directly for the other train. On reaching the other train, the bird flies directly back to the first train, and so forth. (We have no idea *why* a bird would behave in this way.) What is the total distance the bird travels before the trains collide?

•••12 *Traffic shock wave.* An abrupt slowdown in concentrated traffic can travel as a pulse, termed a *shock wave,* along the line of cars, either downstream (in the traffic direction) or upstream, or it can be stationary. Figure 2-23 shows a uniformly spaced line of cars moving at speed $v = 25.0$ m/s toward a uniformly spaced line of slow cars moving at speed $v_s = 5.00$ m/s. Assume that each faster car adds length $L = 12.0$ m (car length plus buffer zone) to the line of slow cars when it joins the line, and assume it slows abruptly at the last instant. (a) For what separation distance d between the faster cars does the shock wave remain stationary? If the separation is twice that amount, what are the (b) speed and (c) direction (upstream or downstream) of the shock wave?

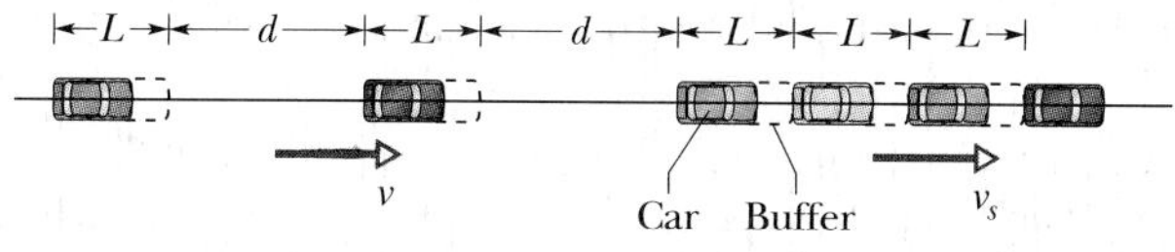

FIG. 2-23 Problem 12.

•••13 You drive on Interstate 10 from San Antonio to Houston, half the *time* at 55 km/h and the other half at 90 km/h. On the way back you travel half the *distance* at 55 km/h and the other half at 90 km/h. What is your average speed (a) from San Antonio to Houston, (b) from Houston back to San Antonio, and (c) for the entire trip? (d) What is your average velocity for the entire trip? (e) Sketch x versus t for (a), assuming the motion is all in the positive x direction. Indicate how the average velocity can be found on the sketch. ILW

sec. 2-5 Instantaneous Velocity and Speed

•14 The position function $x(t)$ of a particle moving along an x axis is $x = 4.0 - 6.0t^2$, with x in meters and t in seconds. (a) At what time and (b) where does the particle (momentarily) stop? At what (c) negative time and (d) positive time does the particle pass through the origin? (e) Graph x versus t for the range -5 s to $+5$ s. (f) To shift the curve rightward on the graph, should we include the term $+20t$ or the term $-20t$ in $x(t)$? (g) Does that inclusion increase or decrease the value of x at which the particle momentarily stops?

•15 (a) If a particle's position is given by $x = 4 - 12t + 3t^2$ (where t is in seconds and x is in meters), what is its velocity at $t = 1$ s? (b) Is it moving in the positive or negative direction of x just then? (c) What is its speed just then? (d) Is the speed increasing or decreasing just then? (Try answering the next two questions without further calculation.) (e) Is there ever an instant when the velocity is zero? If so, give the time t; if not, answer no. (f) Is there a time after $t = 3$ s when the particle is moving in the negative direction of x? If so, give the time t; if not, answer no. GO

•16 An electron moving along the x axis has a position given by $x = 16te^{-t}$ m, where t is in seconds. How far is the electron from the origin when it momentarily stops?

••17 The position of a particle moving along the x axis is given in centimeters by $x = 9.75 + 1.50t^3$, where t is in seconds. Calculate (a) the average velocity during the time interval $t = 2.00$ s to $t = 3.00$ s; (b) the instantaneous velocity at $t = 2.00$ s; (c) the instantaneous velocity at $t = 3.00$ s; (d) the instantaneous velocity at $t = 2.50$ s; and (e) the instantaneous velocity when the particle is midway between its positions at $t = 2.00$ s and $t = 3.00$ s. (f) Graph x versus t and indicate your answers graphically.

sec. 2-6 Acceleration

•18 (a) If the position of a particle is given by $x = 20t - 5t^3$, where x is in meters and t is in seconds, when, if ever, is the particle's velocity zero? (b) When is its acceleration a zero? (c) For what time range (positive or negative) is a negative? (d) Positive? (e) Graph $x(t)$, $v(t)$, and $a(t)$.

•19 At a certain time a particle had a speed of 18 m/s in the positive x direction, and 2.4 s later its speed was 30 m/s in the opposite direction. What is the average acceleration of the particle during this 2.4 s interval? SSM

•20 The position of a particle moving along an x axis is given by $x = 12t^2 - 2t^3$, where x is in meters and t is in seconds. Determine (a) the position, (b) the velocity, and (c) the acceleration of the particle at $t = 3.0$ s. (d) What is the maximum positive coordinate reached by the particle and (e) at what time is it reached? (f) What is the maximum positive velocity reached by the particle and (g) at what time is it reached? (h) What is the acceleration of the particle at the instant the particle is not moving (other than at $t = 0$)? (i) Determine the average velocity of the particle between $t = 0$ and $t = 3$ s.

••21 The position of a particle moving along the x axis depends on the time according to the equation $x = ct^2 - bt^3$, where x is in meters and t in seconds. What are the units of (a) constant c and (b) constant b? Let their numerical values be 3.0 and 2.0, respectively. (c) At what time does the particle reach its maximum positive x position? From $t = 0.0$ s to $t = 4.0$ s, (d) what distance does the particle move and (e) what is its displacement? Find its velocity at times (f) 1.0 s, (g) 2.0 s, (h) 3.0 s, and (i) 4.0 s. Find its acceleration at times (j) 1.0 s, (k) 2.0 s, (l) 3.0 s, and (m) 4.0 s.

••22 From $t = 0$ to $t = 5.00$ min, a man stands still, and from $t = 5.00$ min to $t = 10.0$ min, he walks briskly in a straight line at a constant speed of 2.20 m/s. What are (a) his average velocity v_{avg} and (b) his average acceleration a_{avg} in the time interval 2.00 min to 8.00 min? What are (c) v_{avg} and (d) a_{avg} in the time interval 3.00 min to 9.00 min? (e) Sketch x versus t and v versus t, and indicate how the answers to (a) through (d) can be obtained from the graphs.

sec. 2-7 Constant Acceleration: A Special Case

•23 An electron has a constant acceleration of $+3.2$ m/s^2. At a certain instant its velocity is $+9.6$ m/s. What is its velocity (a) 2.5 s earlier and (b) 2.5 s later?

•**24** A muon (an elementary particle) enters a region with a speed of 5.00×10^6 m/s and then is slowed at the rate of 1.25×10^{14} m/s^2. (a) How far does the muon take to stop? (b) Graph x versus t and v versus t for the muon.

•**25** Suppose a rocket ship in deep space moves with constant acceleration equal to 9.8 m/s^2, which gives the illusion of normal gravity during the flight. (a) If it starts from rest, how long will it take to acquire a speed one-tenth that of light, which travels at 3.0×10^8 m/s? (b) How far will it travel in so doing? SSM

•**26** On a dry road, a car with good tires may be able to brake with a constant deceleration of 4.92 m/s^2. (a) How long does such a car, initially traveling at 24.6 m/s, take to stop? (b) How far does it travel in this time? (c) Graph x versus t and v versus t for the deceleration.

•**27** An electron with an initial velocity $v_0 = 1.50 \times 10^5$ m/s enters a region of length L = 1.00 cm where it is electrically accelerated (Fig. 2-24). It emerges with $v = 5.70 \times 10^6$ m/s. What is its acceleration, assumed constant? SSM

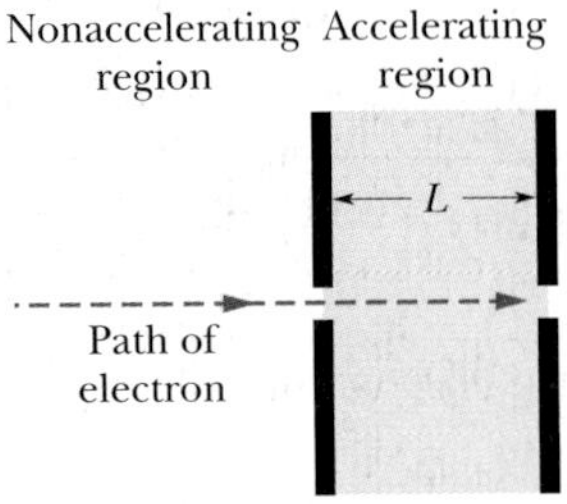

Nonaccelerating region
Accelerating region
L
Path of electron

FIG. 2-24 Problem 27.

•**28** *Catapulting mushrooms.* Certain mushrooms launch their spores by a catapult mechanism. As water condenses from the air onto a spore that is attached to the mushroom, a drop grows on one side of the spore and a film grows on the other side. The spore is bent over by the drop's weight, but when the film reaches the drop, the drop's water suddenly spreads into the film and the spore springs upward so rapidly that it is slung off into the air. Typically, the spore reaches a speed of 1.6 m/s in a 5.0 μm launch; its speed is then reduced to zero in 1.0 mm by the air. Using that data and assuming constant accelerations, find the acceleration in terms of g during (a) the launch and (b) the speed reduction.

•**29** An electric vehicle starts from rest and accelerates at a rate of 2.0 m/s^2 in a straight line until it reaches a speed of 20 m/s. The vehicle then slows at a constant rate of 1.0 m/s^2 until it stops. (a) How much time elapses from start to stop? (b) How far does the vehicle travel from start to stop?

•**30** A world's land speed record was set by Colonel John P. Stapp when in March 1954 he rode a rocket-propelled sled that moved along a track at 1020 km/h. He and the sled were brought to a stop in 1.4 s. (See Fig. 2-7.) In terms of g, what acceleration did he experience while stopping?

•**31** A certain elevator cab has a total run of 190 m and a maximum speed of 305 m/min, and it accelerates from rest and then back to rest at 1.22 m/s^2. (a) How far does the cab move while accelerating to full speed from rest? (b) How long does it take to make the nonstop 190 m run, starting and ending at rest? ILW

•**32** The brakes on your car can slow you at a rate of 5.2 m/s^2. (a) If you are going 137 km/h and suddenly see a state trooper, what is the minimum time in which you can get your car under the 90 km/h speed limit? (The answer reveals the futility of braking to keep your high speed from being detected with a radar or laser gun.) (b) Graph x versus t and v versus t for such a slowing.

•**33** A car traveling 56.0 km/h is 24.0 m from a barrier when the driver slams on the brakes. The car hits the barrier 2.00 s later. (a) What is the magnitude of the car's constant acceleration before impact? (b) How fast is the car traveling at impact? SSM ILW

••**34** A car moves along an x axis through a distance of 900 m, starting at rest (at $x = 0$) and ending at rest (at $x = 900$ m). Through the first $\frac{1}{4}$ of that distance, its acceleration is $+2.25$ m/s^2. Through the next $\frac{3}{4}$ of that distance, its acceleration is -0.750 m/s^2. What are (a) its travel time through the 900 m and (b) its maximum speed? (c) Graph position x, velocity v, and acceleration a versus time t for the trip.

••**35** Figure 2-25 depicts the motion of a particle moving along an x axis with a constant acceleration. The figure's vertical scaling is set by x_s = 6.0 m. What are the (a) magnitude and (b) direction of the particle's acceleration?

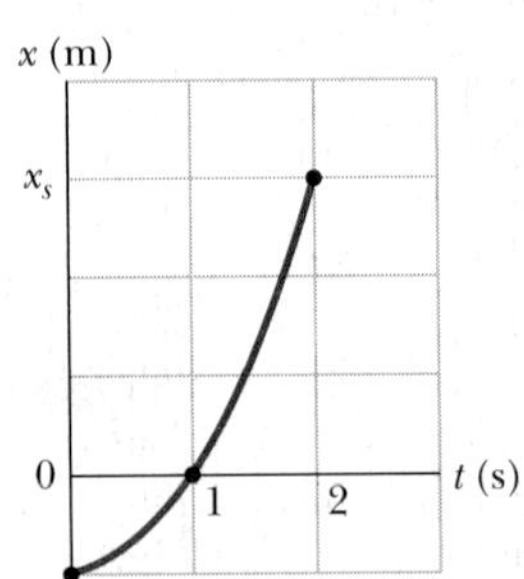

FIG. 2-25 Problem 35.

••**36** (a) If the maximum acceleration that is tolerable for passengers in a subway train is 1.34 m/s^2 and subway stations are located 806 m apart, what is the maximum speed a subway train can attain between stations? (b) What is the travel time between stations? (c) If a subway train stops for 20 s at each station, what is the maximum average speed of the train, from one start-up to the next? (d) Graph x, v, and a versus t for the interval from one start-up to the next.

••**37** Cars A and B move in the same direction in adjacent lanes. The position x of car A is given in Fig. 2-26, from time $t = 0$ to $t = 7.0$ s. The figure's vertical scaling is set by x_s = 32.0 m. At t = 0, car B is at x = 0, with a velocity of 12 m/s and a negative constant acceleration a_B. (a) What must a_B be such that the cars are (momentarily) side by side (momentarily at the same value of x) at t = 4.0 s? (b) For that value of a_B, how many times are the cars side by side? (c) Sketch the position x of car B versus time t on Fig. 2-26. How many times will the cars be side by side if the magnitude of acceleration a_B is (d) more than and (e) less than the answer to part (a)?

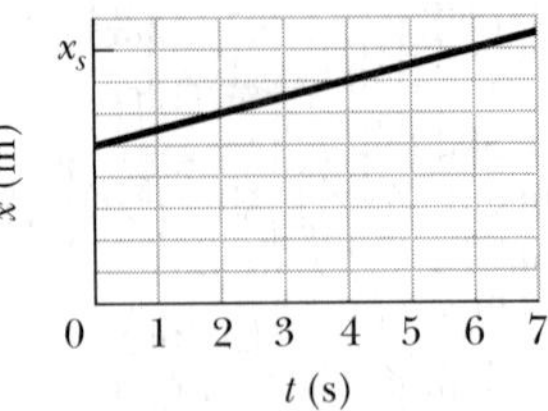

FIG. 2-26 Problem 37.

••**38** You are driving toward a traffic signal when it turns yellow. Your speed is the legal speed limit of v_0 = 55 km/h; your best deceleration rate has the magnitude a = 5.18 m/s^2. Your best reaction time to begin braking is T = 0.75 s. To avoid having the front of your car enter the intersection after the light turns red, should you brake to a stop or continue to move at 55 km/h if the distance to the intersection and the duration of the yellow light are (a) 40 m and 2.8 s, and (b) 32 m and 1.8 s? Give an answer of brake, continue, either (if either strategy works), or neither (if neither strategy works and the yellow duration is inappropriate).

••**39** As two trains move along a track, their conductors suddenly notice that they are headed toward each other. Figure 2-27

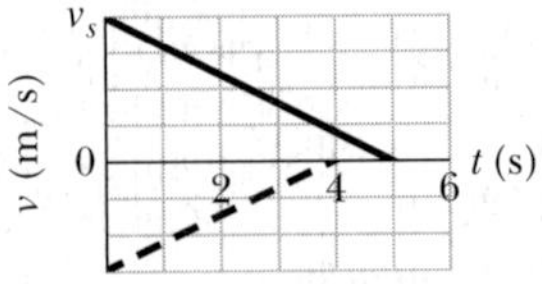

FIG. 2-27 Problem 39.

gives their velocities v as functions of time t as the conductors slow the trains. The figure's vertical scaling is set by $v_s = 40.0$ m/s. The slowing processes begin when the trains are 200 m apart. What is their separation when both trains have stopped?

••40 In Fig. 2-28, a red car and a green car, identical except for the color, move toward each other in adjacent lanes and parallel to an x axis. At time $t = 0$, the red car is at $x_r = 0$ and the green car is at $x_g = 220$ m. If the red car has a constant velocity of 20 km/h, the cars pass each other at $x = 44.5$ m, and if it has a constant velocity of 40 km/h, they pass each other at $x = 76.6$ m. What are (a) the initial velocity and (b) the acceleration of the green car?

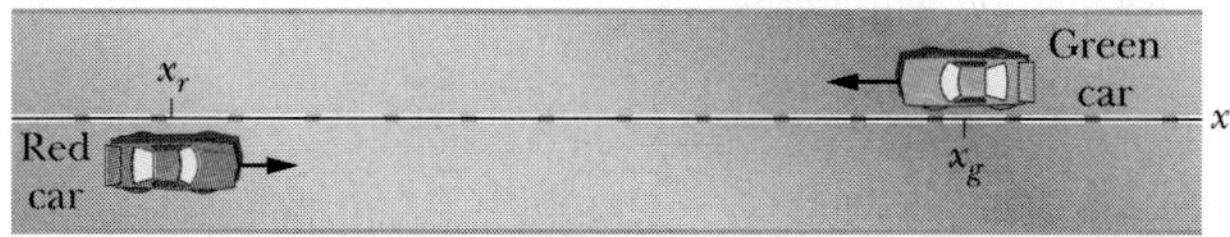

FIG. 2-28 Problems 40 and 41.

••41 Figure 2-28 shows a red car and a green car that move toward each other. Figure 2-29 is a graph of their motion, showing the positions $x_{g0} = 270$ m and $x_{r0} = -35.0$ m at time $t = 0$. The green car has a constant speed of 20.0 m/s and the red car begins from rest. What is the acceleration magnitude of the red car?

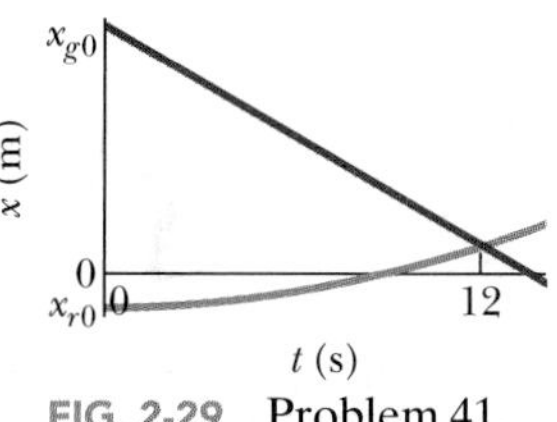

FIG. 2-29 Problem 41.

•••42 When a high-speed passenger train traveling at 161 km/h rounds a bend, the engineer is shocked to see that a locomotive has improperly entered onto the track from a siding and is a distance $D = 676$ m ahead (Fig. 2-30). The locomotive is moving at 29.0 km/h. The engineer of the high-speed train immediately applies the brakes. (a) What must be the magnitude of the resulting constant deceleration if a collision is to be just avoided? (b) Assume that the engineer is at $x = 0$ when, at $t = 0$, he first spots the locomotive. Sketch $x(t)$ curves for the locomotive and high-speed train for the cases in which a collision is just avoided and is not quite avoided. GO

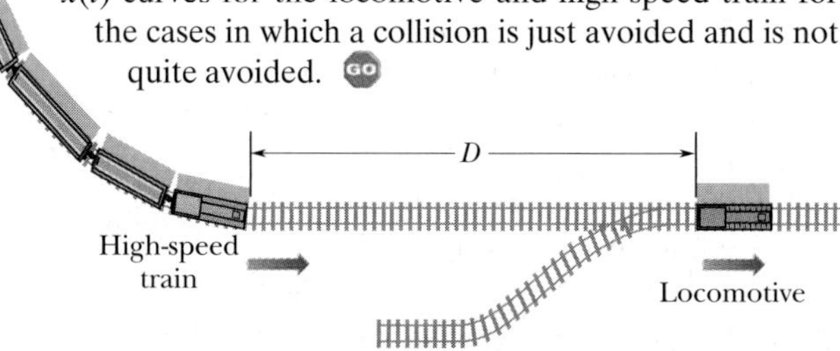

FIG. 2-30 Problem 42.

•••43 You are arguing over a cell phone while trailing an unmarked police car by 25 m; both your car and the police car are traveling at 110 km/h. Your argument diverts your attention from the police car for 2.0 s (long enough for you to look at the phone and yell, "I won't do that!"). At the beginning of that 2.0 s, the police officer begins braking suddenly at 5.0 m/s². (a) What is the separation between the two cars when your attention finally returns? Suppose that you take another 0.40 s to realize your danger and begin braking. (b) If you too brake at 5.0 m/s², what is your speed when you hit the police car?

sec. 2-9 Free-Fall Acceleration

•44 Raindrops fall 1700 m from a cloud to the ground. (a) If they were not slowed by air resistance, how fast would the drops be moving when they struck the ground? (b) Would it be safe to walk outside during a rainstorm?

•45 At a construction site a pipe wrench struck the ground with a speed of 24 m/s. (a) From what height was it inadvertently dropped? (b) How long was it falling? (c) Sketch graphs of y, v, and a versus t for the wrench. SSM

•46 A hoodlum throws a stone vertically downward with an initial speed of 12.0 m/s from the roof of a building, 30.0 m above the ground. (a) How long does it take the stone to reach the ground? (b) What is the speed of the stone at impact?

•47 (a) With what speed must a ball be thrown vertically from ground level to rise to a maximum height of 50 m? (b) How long will it be in the air? (c) Sketch graphs of y, v, and a versus t for the ball. On the first two graphs, indicate the time at which 50 m is reached. SSM WWW

•48 When startled, an armadillo will leap upward. Suppose it rises 0.544 m in the first 0.200 s. (a) What is its initial speed as it leaves the ground? (b) What is its speed at the height of 0.544 m? (c) How much higher does it go?

•49 A hot-air balloon is ascending at the rate of 12 m/s and is 80 m above the ground when a package is dropped over the side. (a) How long does the package take to reach the ground? (b) With what speed does it hit the ground? SSM

••50 A bolt is dropped from a bridge under construction, falling 90 m to the valley below the bridge. (a) In how much time does it pass through the last 20% of its fall? What is its speed (b) when it begins that last 20% of its fall and (c) when it reaches the valley beneath the bridge?

••51 A key falls from a bridge that is 45 m above the water. It falls directly into a model boat, moving with constant velocity, that is 12 m from the point of impact when the key is released. What is the speed of the boat? SSM ILW

••52 At time $t = 0$, apple 1 is dropped from a bridge onto a roadway beneath the bridge; somewhat later, apple 2 is thrown down from the same height. Figure 2-31 gives the vertical positions y of the apples versus t during the falling, until both apples have hit the roadway. With approximately what speed is apple 2 thrown down?

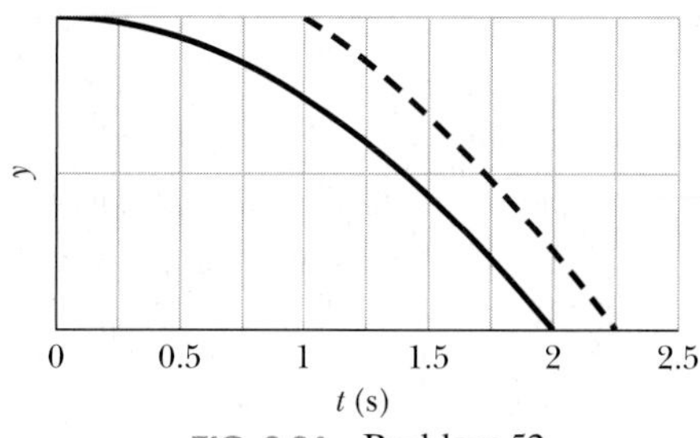

FIG. 2-31 Problem 52.

••53 As a runaway scientific balloon ascends at 19.6 m/s, one of its instrument packages breaks free of a harness and free-falls. Figure 2-32 gives the vertical velocity of the package versus time, from before it breaks free to when it reaches the ground. (a) What maximum

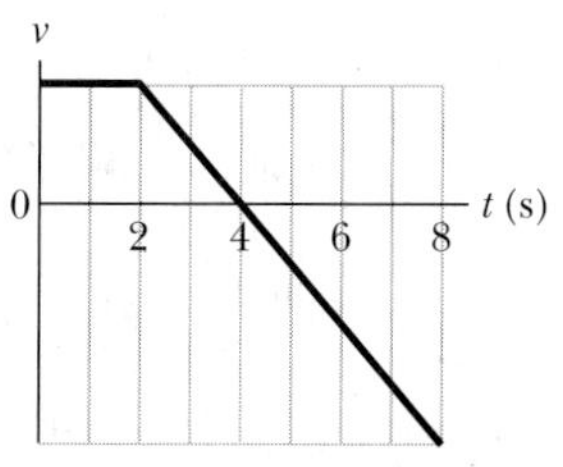

FIG. 2-32 Problem 53.

height above the break-free point does it rise? (b) How high is the break-free point above the ground?

••54 Figure 2-33 shows the speed v versus height y of a ball tossed directly upward, along a y axis. Distance d is 0.40 m. The speed at height y_A is v_A. The speed at height y_B is $\frac{1}{3}v_A$. What is speed v_A?

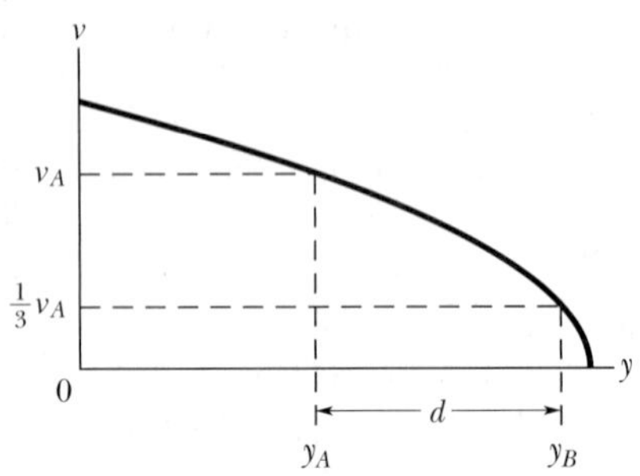

FIG. 2-33 Problem 54.

••55 A ball of moist clay falls 15.0 m to the ground. It is in contact with the ground for 20.0 ms before stopping. (a) What is the magnitude of the average acceleration of the ball during the time it is in contact with the ground? (Treat the ball as a particle.) (b) Is the average acceleration up or down? SSM

••56 A stone is dropped into a river from a bridge 43.9 m above the water. Another stone is thrown vertically down 1.00 s after the first is dropped. The stones strike the water at the same time. (a) What is the initial speed of the second stone? (b) Plot velocity versus time on a graph for each stone, taking zero time as the instant the first stone is released.

••57 To test the quality of a tennis ball, you drop it onto the floor from a height of 4.00 m. It rebounds to a height of 2.00 m. If the ball is in contact with the floor for 12.0 ms, (a) what is the magnitude of its average acceleration during that contact and (b) is the average acceleration up or down?

••58 A rock is thrown vertically upward from ground level at time $t = 0$. At $t = 1.5$ s it passes the top of a tall tower, and 1.0 s later it reaches its maximum height. What is the height of the tower?

••59 Water drips from the nozzle of a shower onto the floor 200 cm below. The drops fall at regular (equal) intervals of time, the first drop striking the floor at the instant the fourth drop begins to fall. When the first drop strikes the floor, how far below the nozzle are the (a) second and (b) third drops?

••60 An object falls a distance h from rest. If it travels $0.50h$ in the last 1.00 s, find (a) the time and (b) the height of its fall. (c) Explain the physically unacceptable solution of the quadratic equation in t that you obtain.

••61 A drowsy cat spots a flowerpot that sails first up and then down past an open window. The pot is in view for a total of 0.50 s, and the top-to-bottom height of the window is 2.00 m. How high above the window top does the flowerpot go?

•••62 A ball is shot vertically upward from the surface of another planet. A plot of y versus t for the ball is shown in Fig. 2-34,

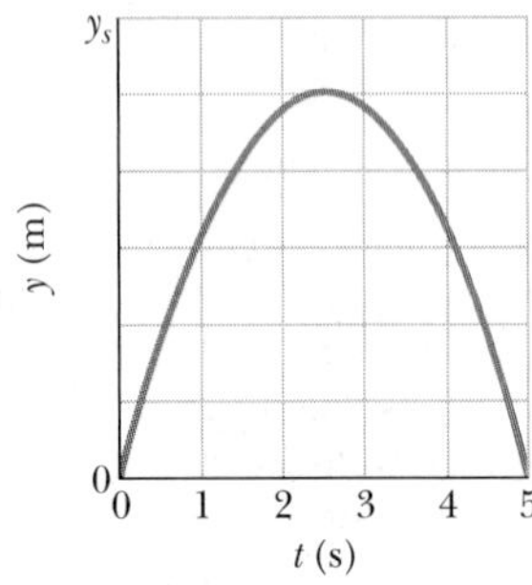

FIG. 2-34 Problem 62.

where y is the height of the ball above its starting point and $t = 0$ at the instant the ball is shot. The figure's vertical scaling is set by $y_s = 30.0$ m. What are the magnitudes of (a) the free-fall acceleration on the planet and (b) the initial velocity of the ball?

•••63 A steel ball is dropped from a building's roof and passes a window, taking 0.125 s to fall from the top to the bottom of the window, a distance of 1.20 m. It then falls to a sidewalk and bounces back past the window, moving from bottom to top in 0.125 s. Assume that the upward flight is an exact reverse of the fall. The time the ball spends below the bottom of the window is 2.00 s. How tall is the building? GO

•••64 A basketball player grabbing a rebound jumps 76.0 cm vertically. How much total time (ascent and descent) does the player spend (a) in the top 15.0 cm of this jump and (b) in the bottom 15.0 cm? Do your results explain why such players seem to hang in the air at the top of a jump?

sec. 2-10 Graphical Integration in Motion Analysis

•65 In Sample Problem 2-9, at maximum head acceleration, what is the speed of (a) the head and (b) the torso?

•66 A salamander of the genus *Hydromantes* captures prey by launching its tongue as a projectile: The skeletal part of the tongue is shot forward, unfolding the rest of the tongue, until the outer portion lands on the prey, sticking to it. Figure 2-35 shows the acceleration magnitude a versus time t for the acceleration phase of the launch in a typical situation. The indicated accelerations are $a_2 = 400$ m/s^2 and $a_1 = 100$ m/s^2. What is the outward speed of the tongue at the end of the acceleration phase?

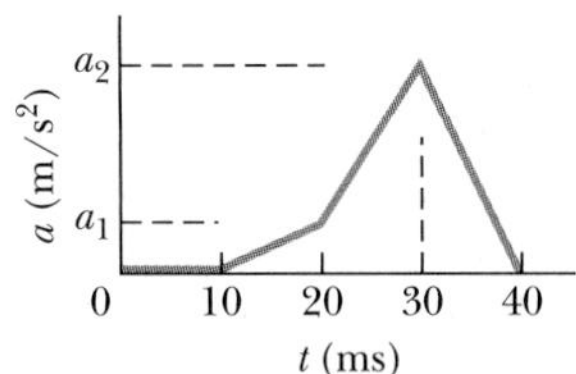

FIG. 2-35 Problem 66.

••67 How far does the runner whose velocity–time graph is shown in Fig. 2-36 travel in 16 s? The figure's vertical scaling is set by $v_s = 8.0$ m/s. ILW

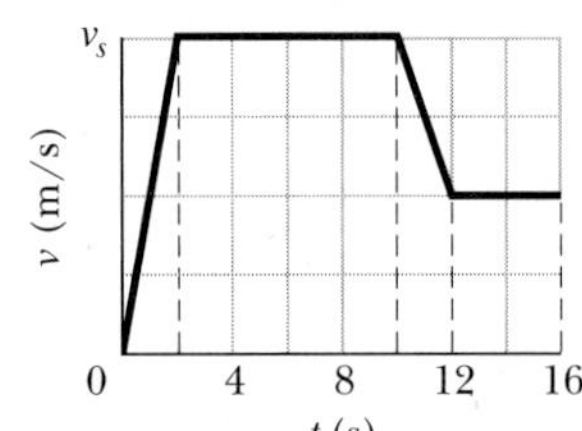

FIG. 2-36 Problem 67.

••68 In a forward punch in karate, the fist begins at rest at the waist and is brought rapidly forward until the arm is fully extended. The speed $v(t)$ of the fist is given in Fig. 2-37 for someone skilled in karate. How far has the fist moved at (a) time $t = 50$ ms and (b) when the speed of the fist is maximum?

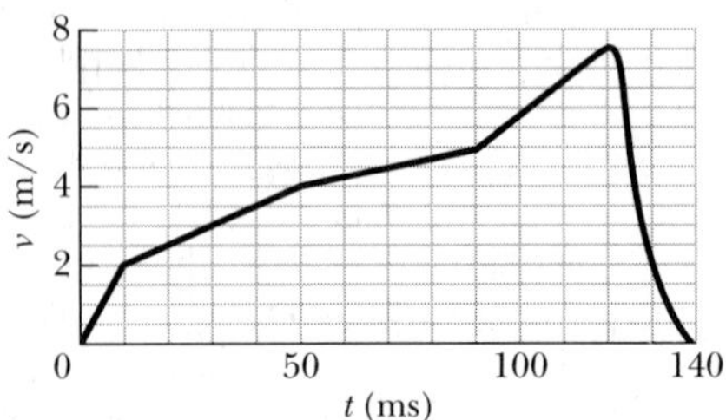

FIG. 2-37 Problem 68.

••69 When a soccer ball is kicked toward a player and the player deflects the ball by "heading" it, the acceleration of the head during the collision can be significant. Figure 2-38 gives the measured acceleration $a(t)$ of a soccer player's head for a bare head and a helmeted head, starting from rest. At time $t = 7.0$ ms, what is the difference in the speed acquired by the bare head and the speed acquired by the helmeted head?

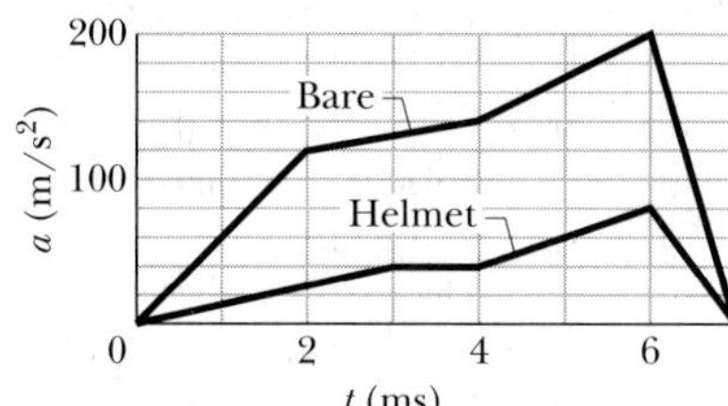

FIG. 2-38 Problem 69.

•••**70** Two particles move along an x axis. The position of particle 1 is given by $x = 6.00t^2 + 3.00t + 2.00$ (in meters and seconds); the acceleration of particle 2 is given by $a = -8.00t$ (in meters per seconds squared and seconds) and, at $t = 0$, its velocity is 20 m/s. When the velocities of the particles match, what is their velocity?

Additional Problems

71 At the instant the traffic light turns green, an automobile starts with a constant acceleration a of 2.2 m/s^2. At the same instant a truck, traveling with a constant speed of 9.5 m/s, overtakes and passes the automobile. (a) How far beyond the traffic signal will the automobile overtake the truck? (b) How fast will the automobile be traveling at that instant? GO

72 Figure 2-39 shows part of a street where traffic flow is to be controlled to allow a *platoon* of cars to move smoothly along the street. Suppose that the platoon leaders have just reached intersection 2, where the green appeared when they were distance d from the intersection. They continue to travel at a certain speed v_p (the speed limit) to reach intersection 3, where the green appears when they are distance d from it. The intersections are separated by distances D_{23} and D_{12}. (a) What should be the time delay of the onset of green at intersection 3 relative to that at intersection 2 to keep the platoon moving smoothly?

Suppose, instead, that the platoon had been stopped by a red light at intersection 1. When the green comes on there, the leaders require a certain time t_r to respond to the change and an additional time to accelerate at some rate a to the cruising speed v_p. (b) If the green at intersection 2 is to appear when the leaders are distance d from that intersection, how long after the light at intersection 1 turns green should the light at intersection 2 turn green?

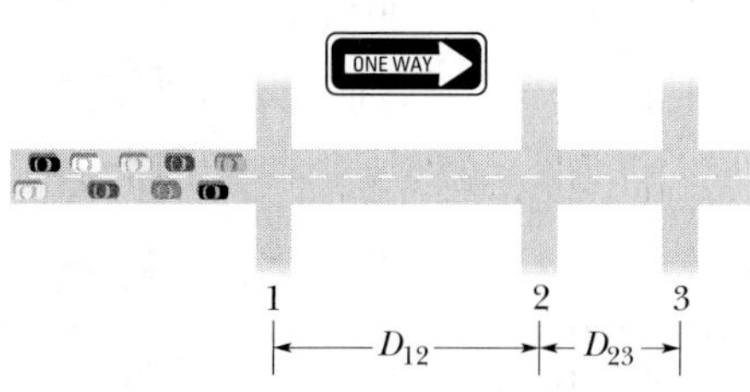

FIG. 2-39 Problem 72.

73 In an arcade video game, a spot is programmed to move across the screen according to $x = 9.00t - 0.750t^3$, where x is distance in centimeters measured from the left edge of the screen and t is time in seconds. When the spot reaches a screen edge, at either $x = 0$ or $x = 15.0$ cm, t is reset to 0 and the spot starts moving again according to $x(t)$. (a) At what time after starting is the spot instantaneously at rest? (b) At what value of x does this occur? (c) What is the spot's acceleration (including sign) when this occurs? (d) Is it moving right or left just prior to coming to rest? (e) Just after? (f) At what time $t > 0$ does it first reach an edge of the screen?

74 A lead ball is dropped in a lake from a diving board 5.20 m above the water. It hits the water with a certain velocity and then sinks to the bottom with this same constant velocity. It reaches the bottom 4.80 s after it is dropped. (a) How deep is the lake? What are the (b) magnitude and (c) direction (up or down) of the average velocity of the ball for the entire fall? Suppose that all the water is drained from the lake. The ball is now thrown from the diving board so that it again reaches the bottom in 4.80 s. What are the (d) magnitude and (e) direction of the initial velocity of the ball?

75 The single cable supporting an unoccupied construction elevator breaks when the elevator is at rest at the top of a 120-m-high building. (a) With what speed does the elevator strike the ground? (b) How long is it falling? (c) What is its speed when it passes the halfway point on the way down? (d) How long has it been falling when it passes the halfway point?

76 Two diamonds begin a free fall from rest from the same height, 1.0 s apart. How long after the first diamond begins to fall will the two diamonds be 10 m apart?

77 If a baseball pitcher throws a fastball at a horizontal speed of 160 km/h, how long does the ball take to reach home plate 18.4 m away?

78 A proton moves along the x axis according to the equation $x = 50t + 10t^2$, where x is in meters and t is in seconds. Calculate (a) the average velocity of the proton during the first 3.0 s of its motion, (b) the instantaneous velocity of the proton at $t = 3.0$ s, and (c) the instantaneous acceleration of the proton at $t = 3.0$ s. (d) Graph x versus t and indicate how the answer to (a) can be obtained from the plot. (e) Indicate the answer to (b) on the graph. (f) Plot v versus t and indicate on it the answer to (c).

79 A motorcycle is moving at 30 m/s when the rider applies the brakes, giving the motorcycle a constant deceleration. During the 3.0 s interval immediately after braking begins, the speed decreases to 15 m/s. What distance does the motorcycle travel from the instant braking begins until the motorcycle stops?

80 A pilot flies horizontally at 1300 km/h, at height $h = 35$ m above initially level ground. However, at time $t = 0$, the pilot begins to fly over ground sloping upward at angle $\theta = 4.3°$ (Fig. 2-40). If the pilot does not change the airplane's heading, at what time t does the plane strike the ground?

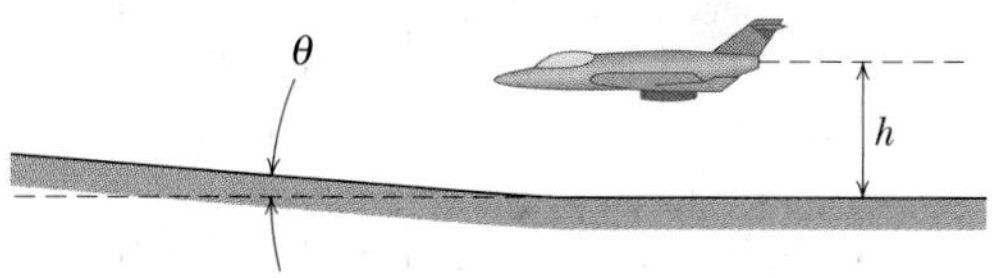

FIG. 2-40 Problem 80.

81 A shuffleboard disk is accelerated at a constant rate from rest to a speed of 6.0 m/s over a 1.8 m distance by a player using a cue. At this point the disk loses contact with the cue and slows at a constant rate of 2.5 m/s^2 until it stops. (a) How much time elapses from when the disk begins to accelerate until it stops? (b) What total distance does the disk travel?

82 The head of a rattlesnake can accelerate at 50 m/s^2 in striking a victim. If a car could do as well, how long would it take to reach a speed of 100 km/h from rest?

83 A jumbo jet must reach a speed of 360 km/h on the runway for takeoff. What is the lowest constant acceleration needed for takeoff from a 1.80 km runway?

84 An automobile driver increases the speed at a constant rate from 25 km/h to 55 km/h in 0.50 min. A bicycle rider speeds up at a constant rate from rest to 30 km/h in 0.50 min. What are the magnitudes of (a) the driver's acceleration and (b) the rider's acceleration?

85 To stop a car, first you require a certain reaction time to begin braking; then the car slows at a constant rate. Suppose that the total distance moved by your car during these two phases is 56.7 m when its initial speed is 80.5 km/h, and 24.4 m when its initial speed is 48.3 km/h. What are (a) your reaction time and (b) the magnitude of the acceleration?

86 A red train traveling at 72 km/h and a green train traveling at 144 km/h are headed toward each other along a straight, level track. When they are 950 m apart, each engineer sees the other's train and applies the brakes. The brakes slow each train at the rate of 1.0 m/s^2. Is there a collision? If so, answer yes and give the speed of the red train and the speed of the green train at impact, respectively. If not, answer no and give the separation between the trains when they stop.

87 At time $t = 0$, a rock climber accidentally allows a piton to fall freely from a high point on the rock wall to the valley below him. Then, after a short delay, his climbing partner, who is 10 m higher on the wall, throws a piton downward. The positions y of the pitons versus t during the falling are given in Fig. 2-41. With what speed is the second piton thrown?

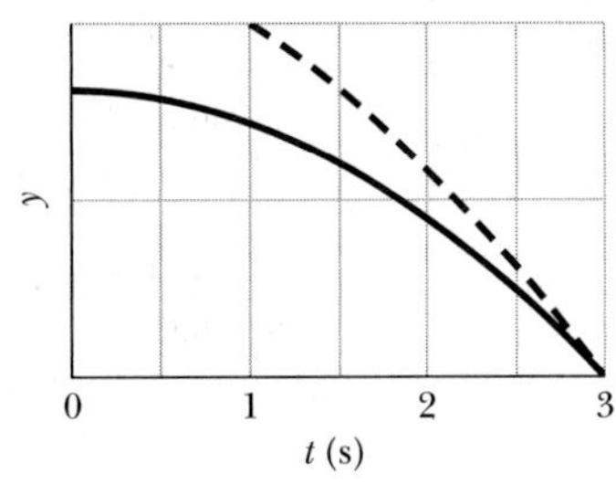

FIG. 2-41 Problem 87.

88 A rock is shot vertically upward from the edge of the top of a tall building. The rock reaches its maximum height above the top of the building 1.60 s after being shot. Then, after barely missing the edge of the building as it falls downward, the rock strikes the ground 6.00 s after it is launched. In SI units: (a) with what upward velocity is the rock shot, (b) what maximum height above the top of the building is reached by the rock, and (c) how tall is the building?

89 A particle's acceleration along an x axis is $a = 5.0t$, with t in seconds and a in meters per second squared. At $t = 2.0$ s, its velocity is +17 m/s. What is its velocity at $t = 4.0$ s? SSM

90 A train started from rest and moved with constant acceleration. At one time it was traveling 30 m/s, and 160 m farther on it was traveling 50 m/s. Calculate (a) the acceleration, (b) the time required to travel the 160 m mentioned, (c) the time required to attain the speed of 30 m/s, and (d) the distance moved from rest to the time the train had a speed of 30 m/s. (e) Graph x versus t and v versus t for the train, from rest.

91 A hot rod can accelerate from 0 to 60 km/h in 5.4 s. (a) What is its average acceleration, in m/s^2, during this time? (b) How far will it travel during the 5.4 s, assuming its acceleration is constant? (c) From rest, how much time would it require to go a distance of 0.25 km if its acceleration could be maintained at the value in (a)? SSM

92 A rocket-driven sled running on a straight, level track is used to investigate the effects of large accelerations on humans. One such sled can attain a speed of 1600 km/h in 1.8 s, starting from rest. Find (a) the acceleration (assumed constant) in terms of g and (b) the distance traveled.

93 Figure 2-42 shows a simple device for measuring your reaction time. It consists of a cardboard strip marked with a scale and two large dots. A friend holds the strip *vertically,* with thumb and forefinger at the dot on the right in Fig. 2-42. You then position your thumb and forefinger at the other dot (on the left in Fig. 2-42), being careful not to touch the strip. Your friend releases the strip, and you try to pinch it as soon as possible after you see it begin to fall. The mark at the place where you pinch the strip gives your reaction time. (a) How far from the lower dot should you place the 50.0 ms mark? How much higher should you place the marks for (b) 100, (c) 150, (d) 200, and (e) 250 ms? (For example, should the 100 ms marker be 2 times as far from the dot as the 50 ms marker? If so, give an answer of 2 times. Can you find any pattern in the answers?)

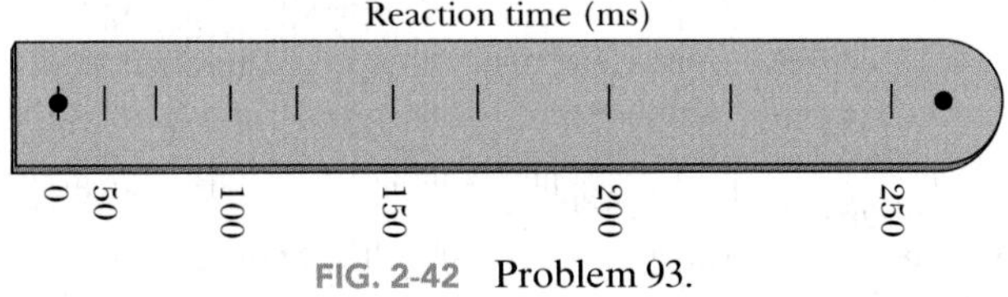

FIG. 2-42 Problem 93.

94 Figure 2-43 gives the acceleration a versus time t for a particle moving along an x axis. The a-axis scale is set by $a_s = 12.0$ m/s^2. At $t = -2.0$ s, the particle's velocity is 7.0 m/s. What is its velocity at $t = 6.0$ s?

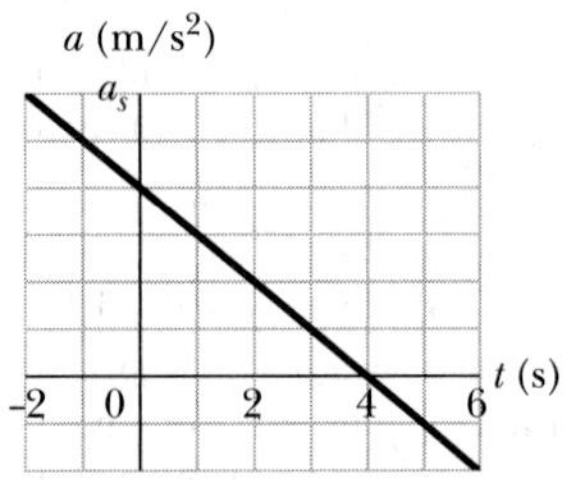

FIG. 2-43 Problem 94.

95 A mining cart is pulled up a hill at 20 km/h and then pulled back down the hill at 35 km/h through its original level. (The time required for the cart's reversal at the top of its climb is negligible.) What is the average speed of the cart for its round trip, from its original level back to its original level?

96 On average, an eye blink lasts about 100 ms. How far does a MiG-25 "Foxbat" fighter travel during a pilot's blink if the plane's average velocity is 3400 km/h?

97 When the legal speed limit for the New York Thruway was increased from 55 mi/h to 65 mi/h, how much time was saved by a motorist who drove the 700 km between the Buffalo entrance and the New York City exit at the legal speed limit? SSM

98 A motorcyclist who is moving along an x axis directed toward the east has an acceleration given by $a = (6.1 - 1.2t)$ m/s^2 for $0 \leq t \leq 6.0$ s. At $t = 0$, the velocity and position of the cyclist are 2.7 m/s and 7.3 m. (a) What is the maximum speed achieved by the cyclist? (b) What total distance does the cyclist travel between $t = 0$ and 6.0 s?

99 A certain juggler usually tosses balls vertically to a height H. To what height must they be tossed if they are to spend twice as much time in the air? SSM

100 A car moving with constant acceleration covered the distance between two points 60.0 m apart in 6.00 s. Its speed as it passed the second point was 15.0 m/s. (a) What was the speed at the first point? (b) What was the magnitude of the acceleration? (c) At what prior distance from the first point was the car at rest? (d) Graph x versus t and v versus t for the car, from rest ($t = 0$).

101 A rock is dropped from a 100-m-high cliff. How long does it take to fall (a) the first 50 m and (b) the second 50 m?

102 Two subway stops are separated by 1100 m. If a subway train accelerates at $+1.2\ m/s^2$ from rest through the first half of the distance and decelerates at $-1.2\ m/s^2$ through the second half, what are (a) its travel time and (b) its maximum speed? (c) Graph x, v, and a versus t for the trip.

103 A certain sprinter has a top speed of 11.0 m/s. If the sprinter starts from rest and accelerates at a constant rate, he is able to reach his top speed in a distance of 12.0 m. He is then able to maintain this top speed for the remainder of a 100 m race. (a) What is his time for the 100 m race? (b) In order to improve his time, the sprinter tries to decrease the distance required for him to reach his top speed. What must this distance be if he is to achieve a time of 10.0 s for the race?

104 A particle starts from the origin at $t = 0$ and moves along the positive x axis. A graph of the velocity of the particle as a function of the time is shown in Fig. 2-44; the v-axis scale is set by $v_s = 4.0$ m/s. (a) What is the coordinate of the particle at $t = 5.0$ s? (b) What is the velocity of the particle at $t = 5.0$ s? (c) What is the acceleration of the particle at $t = 5.0$ s? (d) What is the average velocity of the particle between $t = 1.0$ s and $t = 5.0$ s? (e) What is the average acceleration of the particle between $t = 1.0$ s and $t = 5.0$ s?

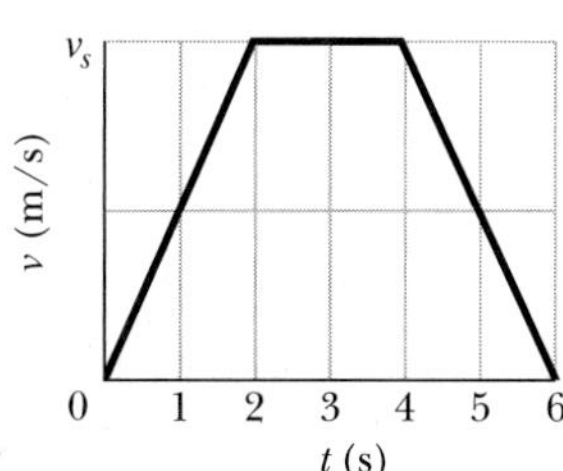

FIG. 2-44 Problem 104.

105 A stone is thrown vertically upward. On its way up it passes point A with speed v, and point B, 3.00 m higher than A, with speed $\frac{1}{2}v$. Calculate (a) the speed v and (b) the maximum height reached by the stone above point B.

106 A rock is dropped (from rest) from the top of a 60-m-tall building. How far above the ground is the rock 1.2 s before it reaches the ground?

107 An iceboat has a constant velocity toward the east when a sudden gust of wind causes the iceboat to have a constant acceleration toward the east for a period of 3.0 s. A plot of x versus t is shown in Fig. 2-45, where $t = 0$ is taken to be the instant the wind starts to blow and the positive x axis is toward the east. (a) What is the acceleration of the iceboat during the 3.0 s interval? (b) What is the velocity of the iceboat at the end of the 3.0 s interval? (c) If the acceleration remains constant for an additional 3.0 s, how far does the iceboat travel during this second 3.0 s interval? SSM

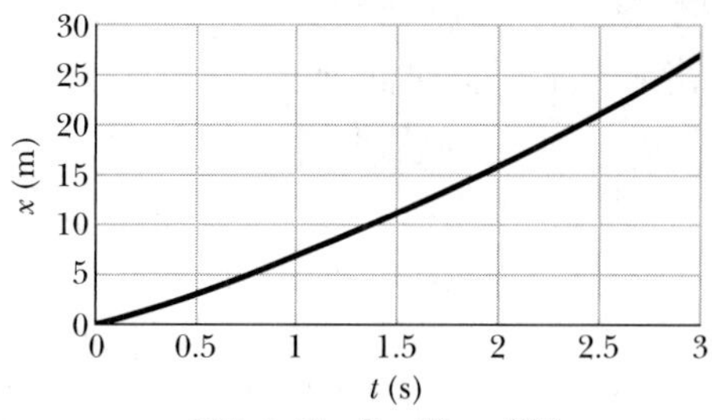

FIG. 2-45 Problem 107.

108 A ball is thrown vertically downward from the top of a 36.6-m-tall building. The ball passes the top of a window that is 12.2 m above the ground 2.00 s after being thrown. What is the speed of the ball as it passes the top of the window?

109 The speed of a bullet is measured to be 640 m/s as the bullet emerges from a barrel of length 1.20 m. Assuming constant acceleration, find the time that the bullet spends in the barrel after it is fired.

110 A parachutist bails out and freely falls 50 m. Then the parachute opens, and thereafter she decelerates at $2.0\ m/s^2$. She reaches the ground with a speed of 3.0 m/s. (a) How long is the parachutist in the air? (b) At what height does the fall begin?

111 The Zero Gravity Research Facility at the NASA Glenn Research Center includes a 145 m drop tower. This is an evacuated vertical tower through which, among other possibilities, a 1 m diameter sphere containing an experimental package can be dropped. (a) How long is the sphere in free fall? (b) What is its speed just as it reaches a catching device at the bottom of the tower? (c) When caught, the sphere experiences an average deceleration of $25g$ as its speed is reduced to zero. Through what distance does it travel during the deceleration?

112 A ball is thrown *down* vertically with an initial *speed* of v_0 from a height of h. (a) What is its speed just before it strikes the ground? (b) How long does the ball take to reach the ground? What would be the answers to (c) part a and (d) part b if the ball were thrown *upward* from the same height and with the same initial speed? Before solving any equations, decide whether the answers to (c) and (d) should be greater than, less than, or the same as in (a) and (b).

113 A car can be braked to a stop from the autobahn-like speed of 200 km/h in 170 m. Assuming the acceleration is constant, find its magnitude in (a) SI units and (b) in terms of g. (c) How much time T_b is required for the braking? Your *reaction time* T_r is the time you require to perceive an emergency, move your foot to the brake, and begin the braking. If $T_r = 400$ ms, then (d) what is T_b in terms of T_r, and (e) is most of the full time required to stop spent in reacting or braking? Dark sunglasses delay the visual signals sent from the eyes to the visual cortex in the brain, increasing T_r. (f) In the extreme case in which T_r is increased by 100 ms, how much farther does the car travel during your reaction time?

114 The sport with the fastest moving ball is jai alai, where measured speeds have reached 303 km/h. If a professional jai alai player faces a ball at that speed and involuntarily blinks, he blacks out the scene for 100 ms. How far does the ball move during the blackout?

3 Vectors

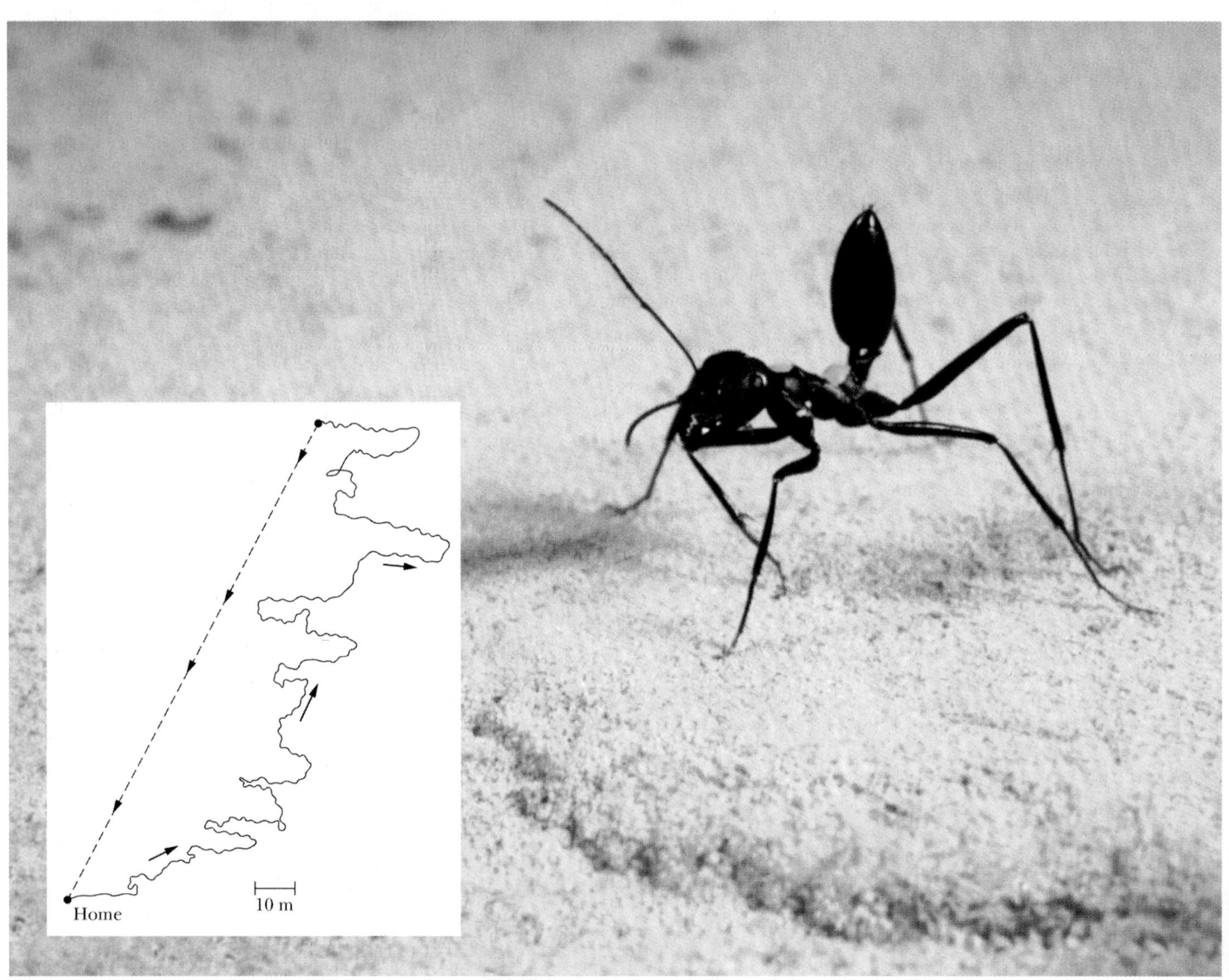

Courtesy Rudiger Wehner, Zoologisches Institut der Universitat Zurich

The desert ant Cataglyphis fortis *lives in the plains of the Sahara desert. When one of the ants forages for food, it travels from its home nest along a haphazard search path like the one shown here. The ant may travel more than 500 m along such a complicated path over flat, featureless sand that contains no landmarks. Yet, when the ant decides to return home, it turns and then runs directly home.*

How does the ant know the way home with no guiding clues on the desert plain?

The answer is in this chapter.

3-1 WHAT IS PHYSICS?

Physics deals with a great many quantities that have both size and direction, and it needs a special mathematical language—the language of vectors—to describe those quantities. This language is also used in engineering, the other sciences, and even in common speech. If you have ever given directions such as "Go five blocks down this street and then hang a left," you have used the language of vectors. In fact, navigation of any sort is based on vectors, but physics and engineering also need vectors in special ways to explain phenomena involving rotation and magnetic forces, which we get to in later chapters. In this chapter, we focus on the basic language of vectors.

3-2 | Vectors and Scalars

A particle moving along a straight line can move in only two directions. We can take its motion to be positive in one of these directions and negative in the other. For a particle moving in three dimensions, however, a plus sign or minus sign is no longer enough to indicate a direction. Instead, we must use a *vector.*

A **vector** has magnitude as well as direction, and vectors follow certain (vector) rules of combination, which we examine in this chapter. A **vector quantity** is a quantity that has both a magnitude and a direction and thus can be represented with a vector. Some physical quantities that are vector quantities are displacement, velocity, and acceleration. You will see many more throughout this book, so learning the rules of vector combination now will help you greatly in later chapters.

Not all physical quantities involve a direction. Temperature, pressure, energy, mass, and time, for example, do not "point" in the spatial sense. We call such quantities **scalars,** and we deal with them by the rules of ordinary algebra. A single value, with a sign (as in a temperature of $-40°F$), specifies a scalar.

The simplest vector quantity is displacement, or change of position. A vector that represents a displacement is called, reasonably, a **displacement vector.** (Similarly, we have velocity vectors and acceleration vectors.) If a particle changes its position by moving from A to B in Fig. 3-1*a*, we say that it undergoes a displacement from A to B, which we represent with an arrow pointing from A to B. The arrow specifies the vector graphically. To distinguish vector symbols from other kinds of arrows in this book, we use the outline of a triangle as the arrowhead.

In Fig. 3-1*a*, the arrows from A to B, from A' to B', and from A'' to B'' have the same magnitude and direction. Thus, they specify identical displacement vectors and represent the same *change of position* for the particle. A vector can be shifted without changing its value *if* its length and direction are not changed.

The displacement vector tells us nothing about the actual path that the particle takes. In Fig. 3-1*b*, for example, all three paths connecting points A and B correspond to the same displacement vector, that of Fig. 3-1*a*. Displacement vectors represent only the overall effect of the motion, not the motion itself.

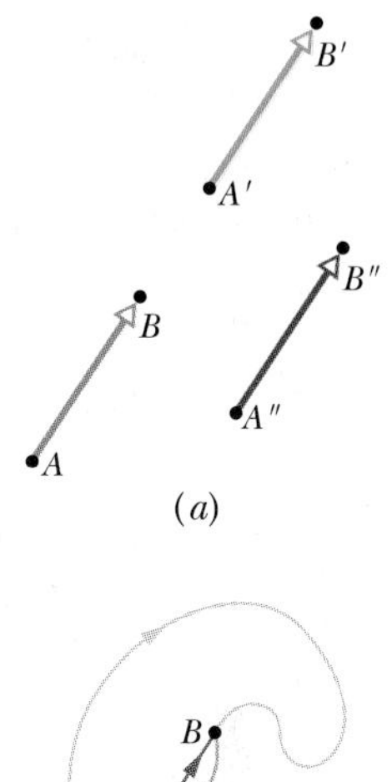

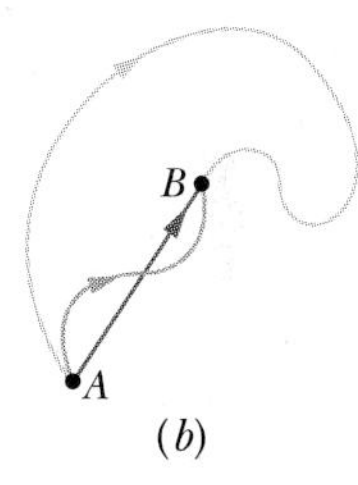

FIG. 3-1 (*a*) All three arrows have the same magnitude and direction and thus represent the same displacement. (*b*) All three paths connecting the two points correspond to the same displacement vector.

3-3 | Adding Vectors Geometrically

Suppose that, as in the vector diagram of Fig. 3-2*a*, a particle moves from A to B and then later from B to C. We can represent its overall displacement (no matter what its actual path) with two successive displacement vectors, AB and BC. The *net* displacement of these two displacements is a single displacement from A to C. We call AC the **vector sum** (or **resultant**) of the vectors AB and BC. This sum is not the usual algebraic sum.

In Fig. 3-2*b*, we redraw the vectors of Fig. 3-2*a* and relabel them in the way that we shall use from now on, namely, with an arrow over an italic symbol, as

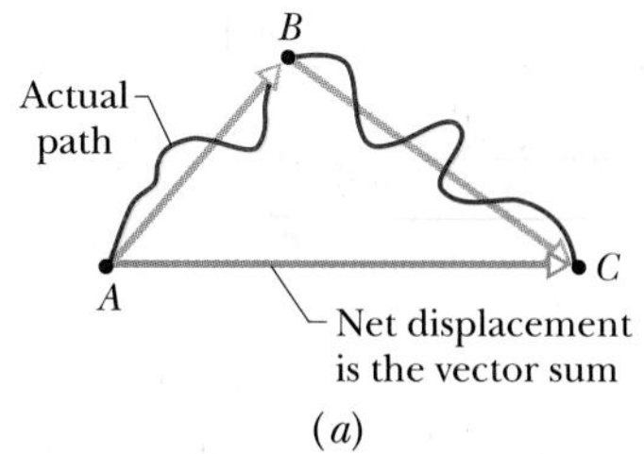

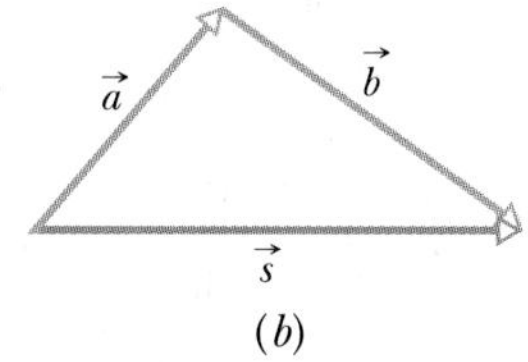

FIG. 3-2 (*a*) AC is the vector sum of the vectors AB and BC. (*b*) The same vectors relabeled.

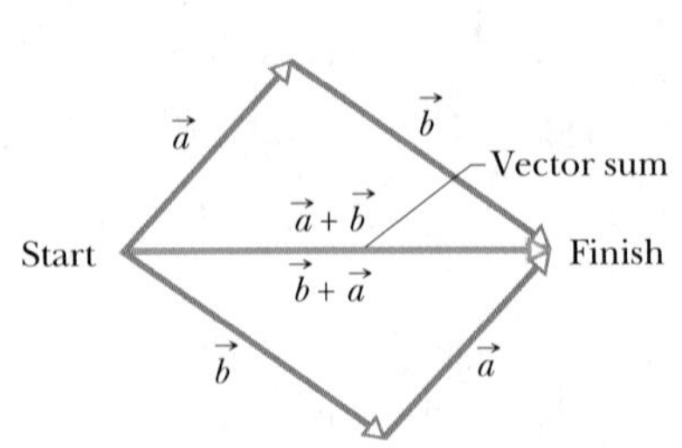

FIG. 3-3 The two vectors $\vec{a}$ and $\vec{b}$ can be added in either order; see Eq. 3-2.

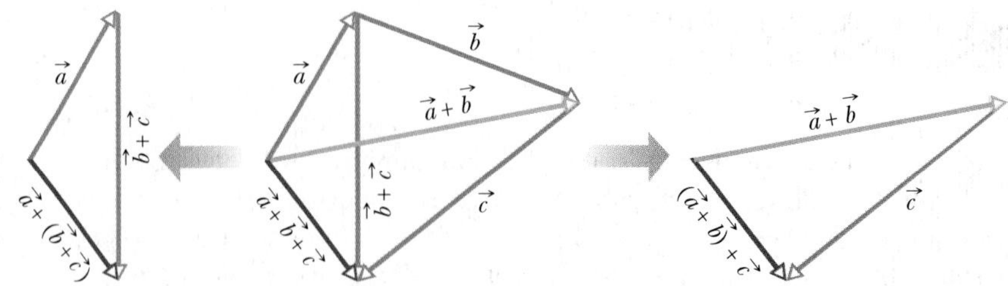

FIG. 3-4 The three vectors $\vec{a}$, $\vec{b}$, and $\vec{c}$ can be grouped in any way as they are added; see Eq. 3-3.

in $\vec{a}$. If we want to indicate only the magnitude of the vector (a quantity that lacks a sign or direction), we shall use the italic symbol, as in a, b, and s. (You can use just a handwritten symbol.) A symbol with an overhead arrow always implies both properties of a vector, magnitude and direction.

We can represent the relation among the three vectors in Fig. 3-2*b* with the *vector equation*

$$\vec{s} = \vec{a} + \vec{b}, \tag{3-1}$$

which says that the vector $\vec{s}$ is the vector sum of vectors $\vec{a}$ and $\vec{b}$. The symbol + in Eq. 3-1 and the words "sum" and "add" have different meanings for vectors than they do in the usual algebra because they involve both magnitude *and* direction.

Figure 3-2 suggests a procedure for adding two-dimensional vectors $\vec{a}$ and $\vec{b}$ geometrically. (1) On paper, sketch vector $\vec{a}$ to some convenient scale and at the proper angle. (2) Sketch vector $\vec{b}$ to the same scale, with its tail at the head of vector $\vec{a}$, again at the proper angle. (3) The vector sum $\vec{s}$ is the vector that extends from the tail of $\vec{a}$ to the head of $\vec{b}$.

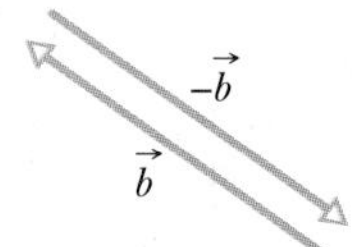

FIG. 3-5 The vectors $\vec{b}$ and $-\vec{b}$ have the same magnitude and opposite directions.

Vector addition, defined in this way, has two important properties. First, the order of addition does not matter. Adding $\vec{a}$ to $\vec{b}$ gives the same result as adding $\vec{b}$ to $\vec{a}$ (Fig. 3-3); that is,

$$\vec{a} + \vec{b} = \vec{b} + \vec{a} \quad \text{(commutative law)}. \tag{3-2}$$

Second, when there are more than two vectors, we can group them in any order as we add them. Thus, if we want to add vectors $\vec{a}$, $\vec{b}$, and $\vec{c}$, we can add $\vec{a}$ and $\vec{b}$ first and then add their vector sum to $\vec{c}$. We can also add $\vec{b}$ and $\vec{c}$ first and then add *that* sum to $\vec{a}$. We get the same result either way, as shown in Fig. 3-4. That is,

$$(\vec{a} + \vec{b}) + \vec{c} = \vec{a} + (\vec{b} + \vec{c}) \quad \text{(associative law)}. \tag{3-3}$$

The vector $-\vec{b}$ is a vector with the same magnitude as $\vec{b}$ but the opposite direction (see Fig. 3-5). Adding the two vectors in Fig. 3-5 would yield

$$\vec{b} + (-\vec{b}) = 0.$$

Thus, adding $-\vec{b}$ has the effect of subtracting $\vec{b}$. We use this property to define the difference between two vectors: let $\vec{d} = \vec{a} - \vec{b}$. Then

$$\vec{d} = \vec{a} - \vec{b} = \vec{a} + (-\vec{b}) \quad \text{(vector subtraction)}; \tag{3-4}$$

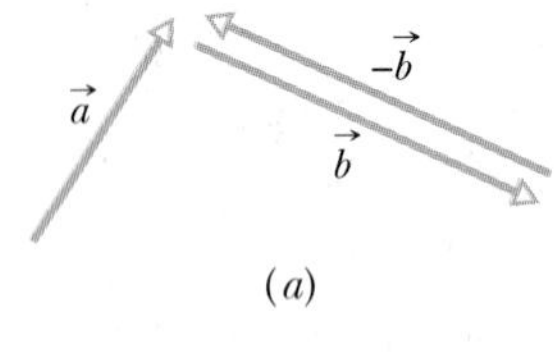

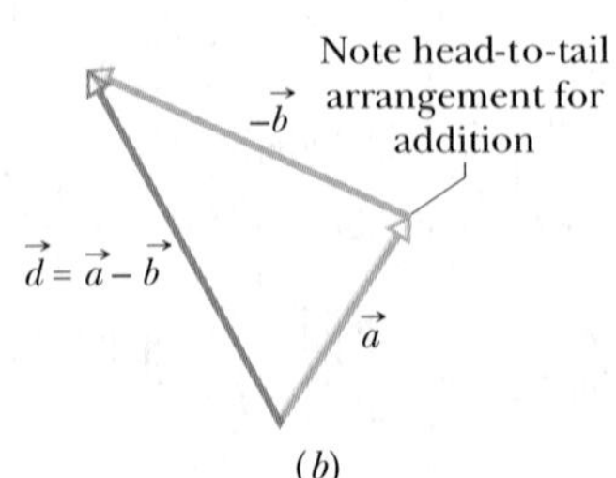

FIG. 3-6 (*a*) Vectors $\vec{a}$, $\vec{b}$, and $-\vec{b}$. (*b*) To subtract vector $\vec{b}$ from vector $\vec{a}$, add vector $-\vec{b}$ to vector $\vec{a}$.

that is, we find the difference vector $\vec{d}$ by adding the vector $-\vec{b}$ to the vector $\vec{a}$. Figure 3-6 shows how this is done geometrically.

As in the usual algebra, we can move a term that includes a vector symbol from one side of a vector equation to the other, but we must change its sign. For example, if we are given Eq. 3-4 and need to solve for $\vec{a}$, we can rearrange the equation as

$$\vec{d} + \vec{b} = \vec{a} \quad \text{or} \quad \vec{a} = \vec{d} + \vec{b}.$$

Remember that, although we have used displacement vectors here, the rules for addition and subtraction hold for vectors of all kinds, whether they represent velocities, accelerations, or any other vector quantity. However, we can add only vectors of the same kind. For example, we can add two displacements, or two velocities, but adding a displacement and a velocity makes no sense. In the arithmetic of scalars, that would be like trying to add 21 s and 12 m.

CHECKPOINT 1 The magnitudes of displacements $\vec{a}$ and $\vec{b}$ are 3 m and 4 m, respectively, and $\vec{c} = \vec{a} + \vec{b}$. Considering various orientations of $\vec{a}$ and $\vec{b}$, what is (a) the maximum possible magnitude for $\vec{c}$ and (b) the minimum possible magnitude?

Sample Problem 3-1

In an orienteering class, you have the goal of moving as far (straight-line distance) from base camp as possible by making three straight-line moves. You may use the following displacements in any order: (a) $\vec{a}$, 2.0 km due east (directly toward the east); (b) $\vec{b}$, 2.0 km 30° north of east (at an angle of 30° toward the north from due east); (c) $\vec{c}$, 1.0 km due west. Alternatively, you may substitute either $-\vec{b}$ for $\vec{b}$ or $-\vec{c}$ for $\vec{c}$. What is the greatest distance you can be from base camp at the end of the third displacement?

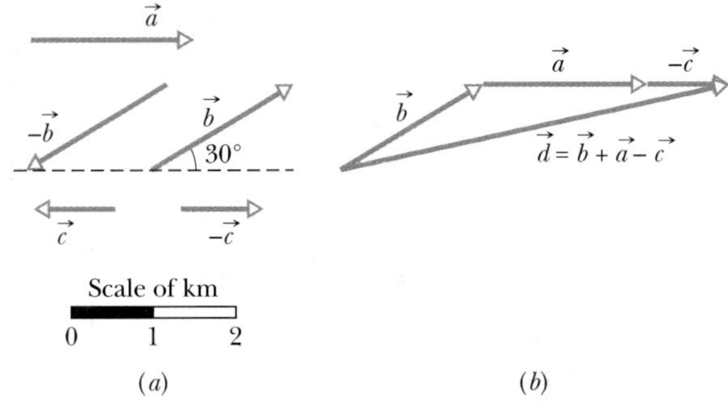

FIG. 3-7 (*a*) Displacement vectors; three are to be used. (*b*) Your distance from base camp is greatest if you undergo displacements $\vec{a}$, $\vec{b}$, and $-\vec{c}$, in any order.

Reasoning: Using a convenient scale, we draw vectors $\vec{a}$, $\vec{b}$, $\vec{c}$, $-\vec{b}$, and $-\vec{c}$ as in Fig. 3-7*a*. We then mentally slide the vectors over the page, connecting three of them at a time in head-to-tail arrangements to find their vector sum $\vec{d}$. The tail of the first vector represents base camp. The head of the third vector represents the point at which you stop. The vector sum $\vec{d}$ extends from the tail of the first vector to the head of the third vector. Its magnitude d is your distance from base camp.

We find that distance d is greatest for a head-to-tail arrangement of vectors $\vec{a}$, $\vec{b}$, and $-\vec{c}$. They can be in any order, because their vector sum is the same for any order. The order shown in Fig. 3-7*b* is for the vector sum

$$\vec{d} = \vec{b} + \vec{a} + (-\vec{c}).$$

Using the scale given in Fig. 3-7*a*, we measure the length d of this vector sum, finding

$$d = 4.8 \text{ m}. \qquad \text{(Answer)}$$

3-4 | Components of Vectors

Adding vectors geometrically can be tedious. A neater and easier technique involves algebra but requires that the vectors be placed on a rectangular coordinate system. The x and y axes are usually drawn in the plane of the page, as shown in Fig. 3-8*a*. The z axis comes directly out of the page at the origin; we ignore it for now and deal only with two-dimensional vectors.

A **component** of a vector is the projection of the vector on an axis. In Fig. 3-8*a*, for example, a_x is the component of vector $\vec{a}$ on (or along) the x axis and a_y is the component along the y axis. To find the projection of a vector along an axis, we draw perpendicular lines from the two ends of the vector to the axis, as shown. The projection of a vector on an x axis is its *x component*, and similarly the projection on the y axis is the *y component*. The process of finding the components of a vector is called **resolving the vector.**

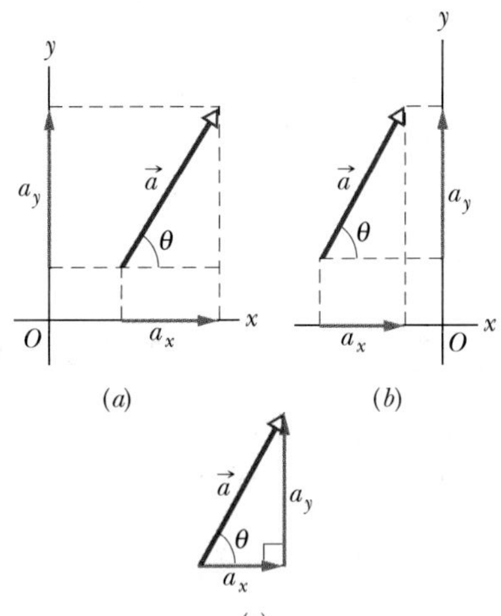

FIG. 3-8 (*a*) The components a_x and a_y of vector $\vec{a}$. (*b*) The components are unchanged if the vector is shifted, as long as the magnitude and orientation are maintained. (*c*) The components form the legs of a right triangle whose hypotenuse is the magnitude of the vector.

A component of a vector has the same direction (along an axis) as the vector. In Fig. 3-8, a_x and a_y are both positive because $\vec{a}$ extends in the positive direction of both axes. (Note the small arrowheads on the components, to indicate their direction.) If we were to reverse vector $\vec{a}$, then both components would be negative and their arrowheads would point toward negative x and y. Resolving vector $\vec{b}$ in Fig. 3-9 yields a positive component b_x and a negative component b_y.

In general, a vector has three components, although for the case of Fig. 3-8*a*

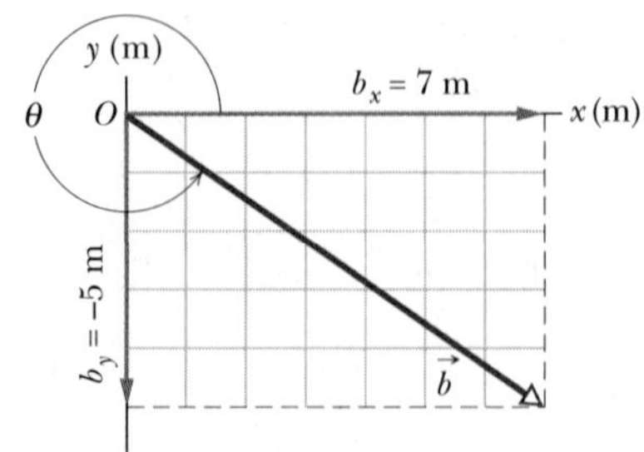

FIG. 3-9 The component of $\vec{b}$ on the x axis is positive, and that on the y axis is negative.

the component along the z axis is zero. As Figs. 3-8*a* and *b* show, if you shift a vector without changing its direction, its components do not change.

We can find the components of $\vec{a}$ in Fig. 3-8*a* geometrically from the right triangle there:

$$a_x = a\cos\theta \quad \text{and} \quad a_y = a\sin\theta, \tag{3-5}$$

where θ is the angle that the vector $\vec{a}$ makes with the positive direction of the x axis, and a is the magnitude of $\vec{a}$. Figure 3-8*c* shows that $\vec{a}$ and its x and y components form a right triangle. It also shows how we can reconstruct a vector from its components: we arrange those components *head to tail.* Then we complete a right triangle with the vector forming the hypotenuse, from the tail of one component to the head of the other component.

Once a vector has been resolved into its components along a set of axes, the components themselves can be used in place of the vector. For example, $\vec{a}$ in Fig. 3-8*a* is given (completely determined) by a and θ. It can also be given by its components a_x and a_y. Both pairs of values contain the same information. If we know a vector in *component notation* (a_x and a_y) and want it in *magnitude-angle notation* (a and θ), we can use the equations

$$a = \sqrt{a_x^2 + a_y^2} \quad \text{and} \quad \tan\theta = \frac{a_y}{a_x} \tag{3-6}$$

to transform it.

In the more general three-dimensional case, we need a magnitude and two angles (say, a, θ, and ϕ) or three components (a_x, a_y, and a_z) to specify a vector.

CHECKPOINT 2

In the figure, which of the indicated methods for combining the x and y components of vector $\vec{a}$ are proper to determine that vector?

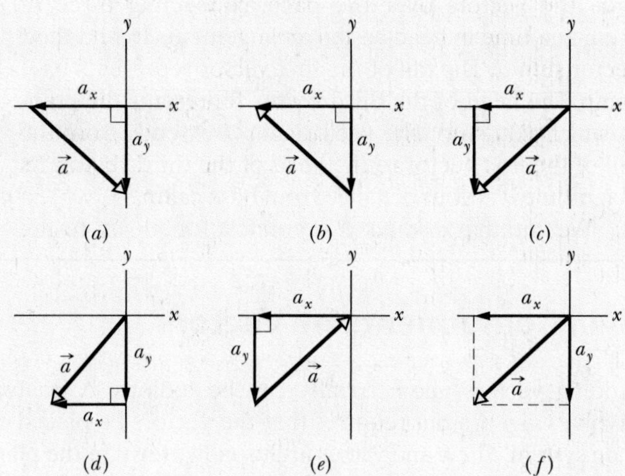

Sample Problem 3-2

A small airplane leaves an airport on an overcast day and is later sighted 215 km away, in a direction making an angle of 22° east of due north. How far east and north is the airplane from the airport when sighted?

KEY IDEA We are given the magnitude (215 km) and the angle (22° east of due north) of a vector and need to find the components of the vector.

Calculations: We draw an xy coordinate system with the positive direction of x due east and that of y due north (Fig. 3-10). For convenience, the origin is placed at

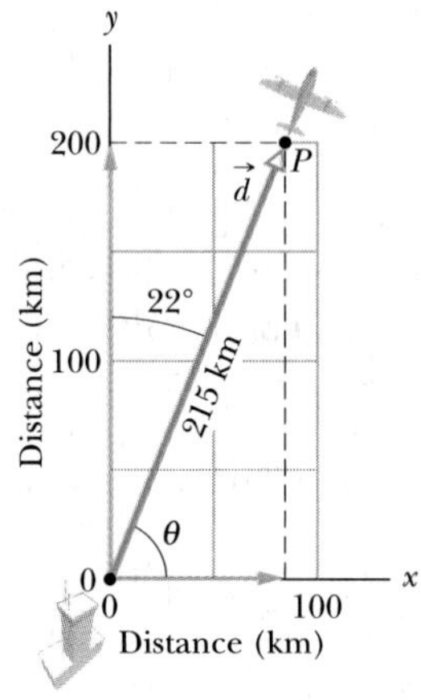

FIG. 3-10 A plane takes off from an airport at the origin and is later sighted at P.

the airport. The airplane's displacement $\vec{d}$ points from the origin to where the airplane is sighted.

To find the components of $\vec{d}$, we use Eq. 3-5 with $\theta = 68°\ (= 90° - 22°)$:

$$d_x = d\cos\theta = (215\text{ km})(\cos 68°) = 81\text{ km} \qquad \text{(Answer)}$$

$$d_y = d\sin\theta = (215\text{ km})(\sin 68°) = 199\text{ km} \approx 2.0 \times 10^2\text{ km}. \qquad \text{(Answer)}$$

Thus, the airplane is 81 km east and 2.0×10^2 km north of the airport.

Sample Problem 3-3

For two decades, spelunking teams sought a connection between the Flint Ridge cave system and Mammoth Cave, which are in Kentucky. When the connection was finally discovered, the combined system was declared the world's longest cave (more than 200 km long). The team that found the connection had to crawl, climb, and squirm through countless passages, traveling a net 2.6 km westward, 3.9 km southward, and 25 m upward. What was their displacement from start to finish?

KEY IDEA We have the components of a three-dimensional vector, and we need to find the vector's magnitude and two angles to specify the vector's direction.

Horizontal Components: We first draw the components as in Fig. 3-11*a*. The horizontal components (2.6 km west and 3.9 km south) form the legs of a horizontal right triangle. The team's horizontal displacement forms the hypotenuse of the triangle, and its magnitude d_h is given by the Pythagorean theorem:

$$d_h = \sqrt{(2.6\text{ km})^2 + (3.9\text{ km})^2} = 4.69\text{ km}.$$

Also from the horizontal triangle in Fig. 3-11*a*, we see that this horizontal displacement is directed south of due west by an angle θ_h given by

$$\tan\theta_h = \frac{3.9\text{ km}}{2.6\text{ km}},$$

so

$$\theta_h = \tan^{-1}\frac{3.9\text{ km}}{2.6\text{ km}} = 56°, \qquad \text{(Answer)}$$

which is one of the two angles we need to specify the direction of the overall displacement.

Overall Displacement: To include the vertical component (25 m = 0.025 km), we now take a side view of Fig. 3-11*a*, looking northwest. We get Fig. 3-11*b*, where the vertical component and the horizontal displacement d_h form the legs of another right triangle. Now the team's overall displacement forms the hypotenuse of that triangle, with a magnitude d given by

$$d = \sqrt{(4.69\text{ km})^2 + (0.025\text{ km})^2} = 4.69\text{ km} \approx 4.7\text{ km}. \qquad \text{(Answer)}$$

This displacement is directed upward from the horizontal displacement by the angle

$$\theta_v = \tan^{-1}\frac{0.025\text{ km}}{4.69\text{ km}} = 0.3°. \qquad \text{(Answer)}$$

Thus, the team's displacement vector had a magnitude of 4.7 km and was at an angle of 56° south of west and at an angle of 0.3° upward. The net vertical motion was, of course, insignificant compared with the horizontal motion. However, that fact would have been of no comfort to the team, which had to climb up and down countless times to get through the cave. The route they actually covered was quite different from the displacement vector.

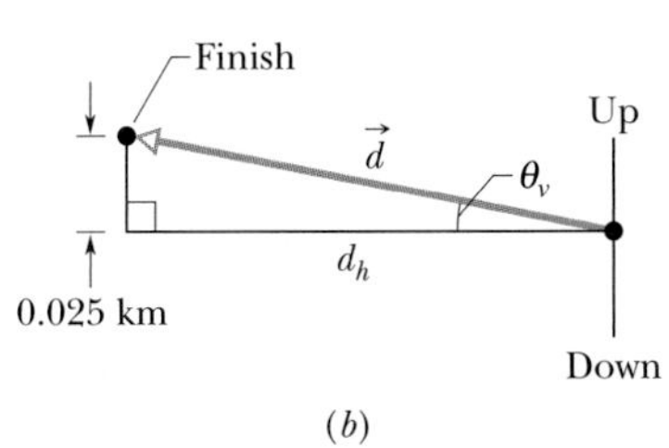

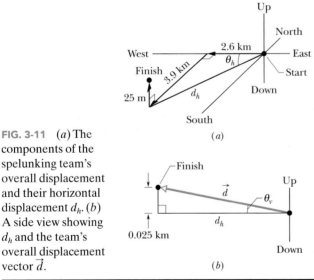

FIG. 3-11 (*a*) The components of the spelunking team's overall displacement and their horizontal displacement d_h. (*b*) A side view showing d_h and the team's overall displacement vector $\vec{d}$.

PROBLEM-SOLVING TACTICS

Tactic 1: Angles—Degrees and Radians Angles that are measured relative to the positive direction of the x axis are positive if they are measured in the counterclockwise direction and negative if measured clockwise. For example, 210° and −150° are the same angle.

Angles may be measured in degrees or radians (rad). To relate the two measures, recall that a full circle is 360° and 2π rad. To convert, say, 40° to radians, write

$$40° \frac{2\pi\text{ rad}}{360°} = 0.70\text{ rad}.$$

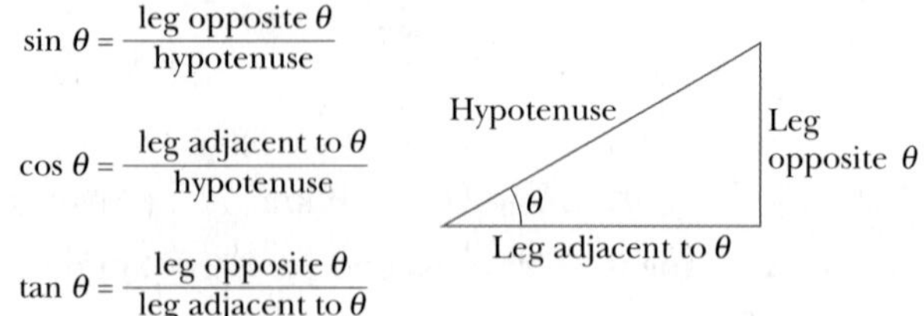

FIG. 3-12 A triangle used to define the trigonometric functions. See also Appendix E.

Tactic 2: Trig Functions You need to know the definitions of the common trigonometric functions—sine, cosine, and tangent—because they are part of the language of science and engineering. They are given in Fig. 3-12 in a form that does not depend on how the triangle is labeled.

You should also be able to sketch how the trig functions vary with angle, as in Fig. 3-13, in order to be able to judge whether a calculator result is reasonable. Even knowing the signs of the functions in the various quadrants can be of help.

Tactic 3: Inverse Trig Functions When the inverse trig functions $\sin^{-1}$, $\cos^{-1}$, and $\tan^{-1}$ are taken on a calculator, you must consider the reasonableness of the answer you get, because there is usually another possible answer that the calculator does not give. The range of operation for a calculator in taking each inverse trig function is indicated in Fig. 3-13. As an example, $\sin^{-1} 0.5$ has associated angles of 30° (which is displayed by the calculator, since 30° falls within its range of operation) and 150°. To see both values, draw a horizontal line through 0.5 in Fig. 3-13*a* and note where it cuts the sine curve.

How do you distinguish a correct answer? It is the one that seems more reasonable for the given situation. As an example, reconsider the calculation of θ_h in Sample Problem 3-3, where $\tan \theta_h = 3.9/2.6 = 1.5$. Taking $\tan^{-1} 1.5$ on your calculator tells you that $\theta_h = 56°$, but $\theta_h = 236°$ (= 180° + 56°) also has a tangent of 1.5. Which is correct? From the physical situation (Fig. 3-11*a*), 56° is reasonable and 236° is clearly not.

Tactic 4: Measuring Vector Angles The equations for $\cos \theta$ and $\sin \theta$ in Eq. 3-5 and for $\tan \theta$ in Eq. 3-6 are valid only if the angle is measured from the positive direction of the x axis. If it is measured relative to some other direction, then the trig functions in Eq. 3-5 may have to be interchanged and the ratio in Eq. 3-6 may have to be inverted. A safer method is to convert the angle to one measured from the positive direction of the x axis.

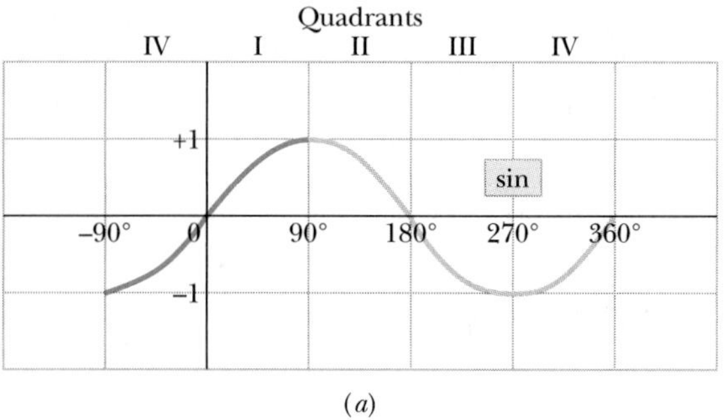

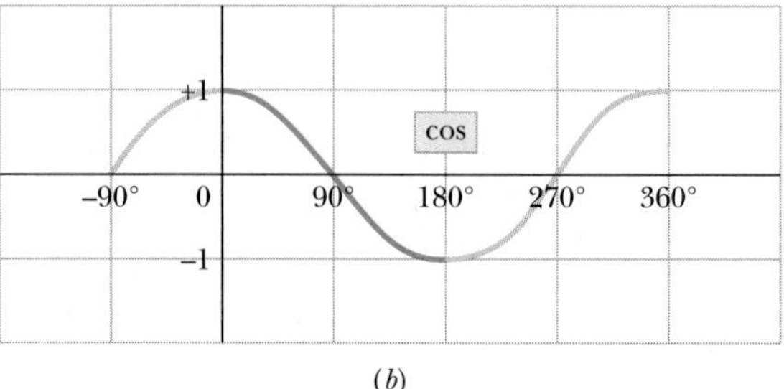

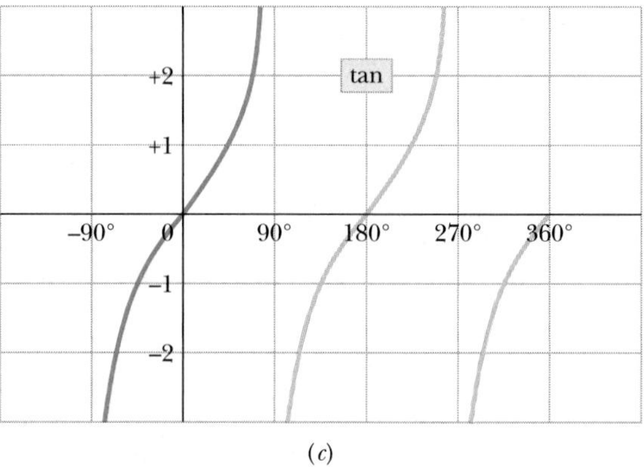

FIG. 3-13 Three useful curves to remember. A calculator's range of operation for taking *inverse* trig functions is indicated by the darker portions of the colored curves.

3-5 | Unit Vectors

A **unit vector** is a vector that has a magnitude of exactly 1 and points in a particular direction. It lacks both dimension and unit. Its sole purpose is to point—that is, to specify a direction. The unit vectors in the positive directions of the x, y, and z axes are labeled $\hat{\text{i}}$, $\hat{\text{j}}$, and $\hat{\text{k}}$, where the hat ˆ is used instead of an overhead arrow as for other vectors (Fig. 3-14). The arrangement of axes in Fig. 3-14 is said to be a **right-handed coordinate system.** The system remains right-handed if it is rotated rigidly. We use such coordinate systems exclusively in this book.

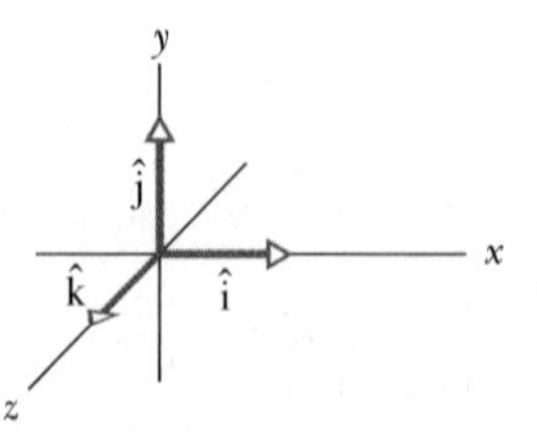

FIG. 3-14 Unit vectors $\hat{\text{i}}$, $\hat{\text{j}}$, and $\hat{\text{k}}$ define the directions of a right-handed coordinate system.

Unit vectors are very useful for expressing other vectors; for example, we can express $\vec{a}$ and $\vec{b}$ of Figs. 3-8 and 3-9 as

$$\vec{a} = a_x\hat{\text{i}} + a_y\hat{\text{j}} \tag{3-7}$$

and

$$\vec{b} = b_x\hat{\text{i}} + b_y\hat{\text{j}}. \tag{3-8}$$

These two equations are illustrated in Fig. 3-15. The quantities $a_x\hat{\mathrm{i}}$ and $a_y\hat{\mathrm{j}}$ are vectors, called the **vector components** of $\vec{a}$. The quantities a_x and a_y are scalars, called the **scalar components** of $\vec{a}$ (or, as before, simply its **components**).

As an example, let us write the displacement $\vec{d}$ of the spelunking team of Sample Problem 3-3 in terms of unit vectors. First, superimpose the coordinate system of Fig. 3-14 on the one shown in Fig. 3-11*a*. Then the directions of $\hat{\mathrm{i}}$, $\hat{\mathrm{j}}$, and $\hat{\mathrm{k}}$ are toward the east, up, and toward the south, respectively. Thus, displacement $\vec{d}$ from start to finish is neatly expressed in unit-vector notation as

$$\vec{d} = -(2.6\text{ km})\hat{\mathrm{i}} + (0.025\text{ km})\hat{\mathrm{j}} + (3.9\text{ km})\hat{\mathrm{k}}. \tag{3-9}$$

Here $-(2.6\text{ km})\hat{\mathrm{i}}$ is the vector component $d_x\hat{\mathrm{i}}$ along the x axis, and $-(2.6\text{ km})$ is the x component d_x.

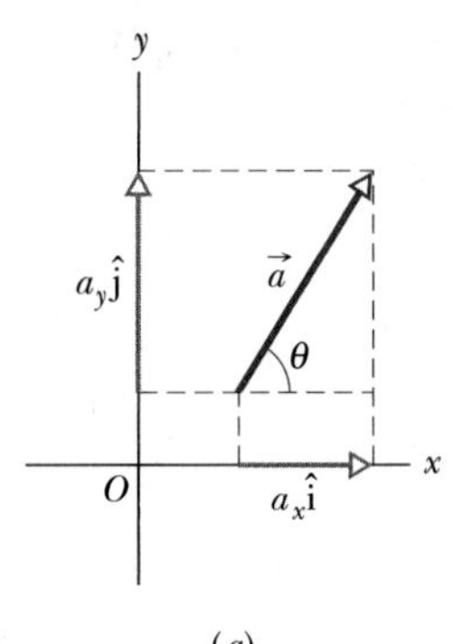

(*a*)

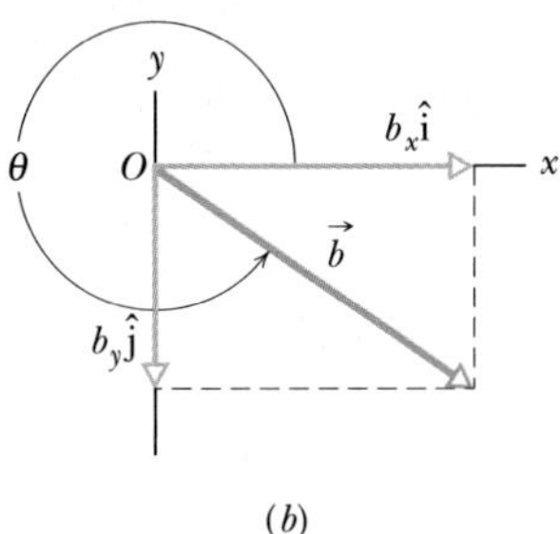

(*b*)

FIG. 3-15 (*a*) The vector components of vector $\vec{a}$. (*b*) The vector components of vector $\vec{b}$.

3-6 | Adding Vectors by Components

Using a sketch, we can add vectors geometrically. On a vector-capable calculator, we can add them directly on the screen. A third way to add vectors is to combine their components axis by axis, which is the way we examine here.

To start, consider the statement

$$\vec{r} = \vec{a} + \vec{b}, \tag{3-10}$$

which says that the vector $\vec{r}$ is the same as the vector $(\vec{a} + \vec{b})$. Thus, each component of $\vec{r}$ must be the same as the corresponding component of $(\vec{a} + \vec{b})$:

$$r_x = a_x + b_x \tag{3-11}$$

$$r_y = a_y + b_y \tag{3-12}$$

$$r_z = a_z + b_z. \tag{3-13}$$

In other words, two vectors must be equal if their corresponding components are equal. Equations 3-10 to 3-13 tell us that to add vectors $\vec{a}$ and $\vec{b}$, we must (1) resolve the vectors into their scalar components; (2) combine these scalar components, axis by axis, to get the components of the sum $\vec{r}$; and (3) combine the components of $\vec{r}$ to get $\vec{r}$ itself. We have a choice in step 3. We can express $\vec{r}$ in unit-vector notation (as in Eq. 3-9) or in magnitude-angle notation (as in the answer to Sample Problem 3-3).

This procedure for adding vectors by components also applies to vector subtractions. Recall that a subtraction such as $\vec{d} = \vec{a} - \vec{b}$ can be rewritten as an addition $\vec{d} = \vec{a} + (-\vec{b})$. To subtract, we add $\vec{a}$ and $-\vec{b}$ by components, to get

$$d_x = a_x - b_x, \quad d_y = a_y - b_y, \quad \text{and} \quad d_z = a_z - b_z,$$

where

$$\vec{d} = d_x\hat{\mathrm{i}} + d_y\hat{\mathrm{j}} + d_z\hat{\mathrm{k}}.$$

CHECKPOINT 3 (a) In the figure here, what are the signs of the x components of $\vec{d}_1$ and $\vec{d}_2$? (b) What are the signs of the y components of $\vec{d}_1$ and $\vec{d}_2$? (c) What are the signs of the x and y components of $\vec{d}_1 + \vec{d}_2$?

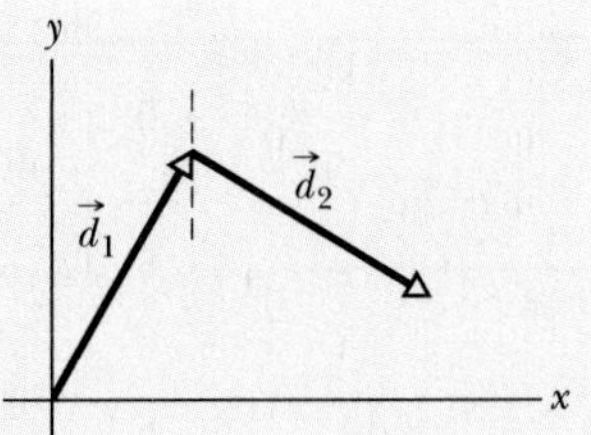

Sample Problem 3-4

Figure 3-16*a* shows the following three vectors:

$$\vec{a} = (4.2\text{ m})\hat{\mathrm{i}} - (1.5\text{ m})\hat{\mathrm{j}},$$

$$\vec{b} = (-1.6\text{ m})\hat{\mathrm{i}} + (2.9\text{ m})\hat{\mathrm{j}},$$

and

$$\vec{c} = (-3.7\text{ m})\hat{\mathrm{j}}.$$

What is their vector sum $\vec{r}$ which is also shown?

KEY IDEA We can add the three vectors by components, axis by axis, and then combine the components to write the vector sum $\vec{r}$.

Calculations: For the x axis, we add the x components of $\vec{a}$, $\vec{b}$, and $\vec{c}$, to get the x component of the vector sum $\vec{r}$:

$$\begin{aligned} r_x &= a_x + b_x + c_x \\ &= 4.2\text{ m} - 1.6\text{ m} + 0 = 2.6\text{ m}. \end{aligned}$$

Similarly, for the y axis,

$$\begin{aligned} r_y &= a_y + b_y + c_y \\ &= -1.5\text{ m} + 2.9\text{ m} - 3.7\text{ m} = -2.3\text{ m}. \end{aligned}$$

We then combine these components of $\vec{r}$ to write the vector in unit-vector notation:

$$\vec{r} = (2.6\text{ m})\hat{\text{i}} - (2.3\text{ m})\hat{\text{j}}, \qquad \text{(Answer)}$$

where $(2.6\text{ m})\hat{\text{i}}$ is the vector component of $\vec{r}$ along the x axis and $-(2.3\text{ m})\hat{\text{j}}$ is that along the y axis. Figure 3-16b shows one way to arrange these vector components to form $\vec{r}$. (Can you sketch the other way?)

We can also answer the question by giving the magnitude and an angle for $\vec{r}$. From Eq. 3-6, the magnitude is

$$r = \sqrt{(2.6\text{ m})^2 + (-2.3\text{ m})^2} \approx 3.5\text{ m} \qquad \text{(Answer)}$$

and the angle (measured from the $+x$ direction) is

$$\theta = \tan^{-1}\left(\frac{-2.3\text{ m}}{2.6\text{ m}}\right) = -41°, \qquad \text{(Answer)}$$

where the minus sign means clockwise.

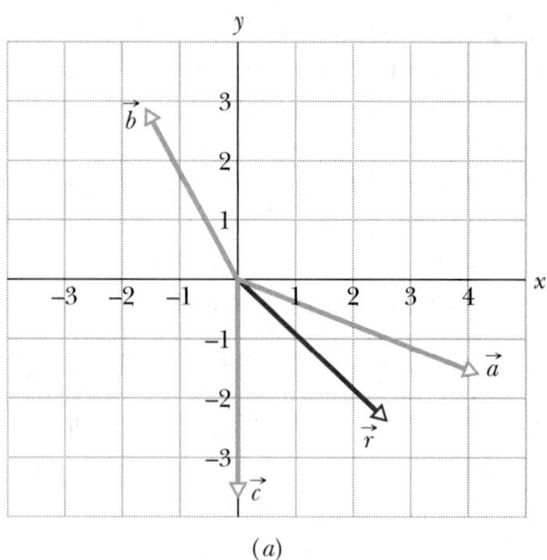

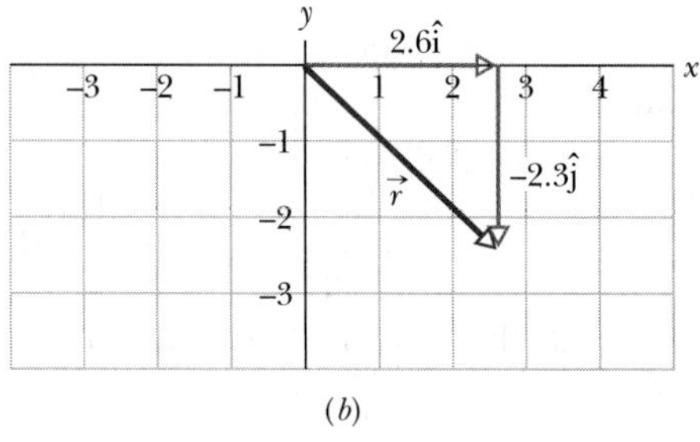

FIG. 3-16 Vector $\vec{r}$ is the vector sum of the other three vectors.

Sample Problem 3-5

According to experiments, the desert ant shown in the chapter opening photograph keeps track of its movements along a mental coordinate system. When it wants to return to its home nest, it effectively sums its displacements along the axes of the system to calculate a vector that points directly home. As an example of the calculation, let's consider an ant making five runs of 6.0 cm each on an xy coordinate system, in the directions shown in Fig. 3-17a, starting from home. At the end of the fifth run, what are the magnitude and angle of the ant's net displacement vector $\vec{d}_{\text{net}}$, and what are those of the homeward vector $\vec{d}_{\text{home}}$ that extends from the ant's final position back to home?

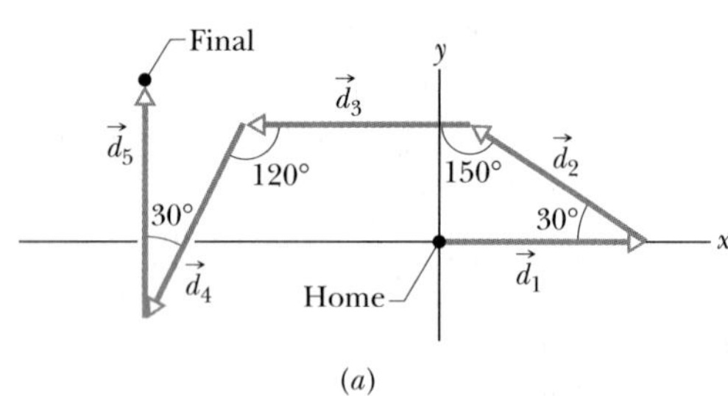

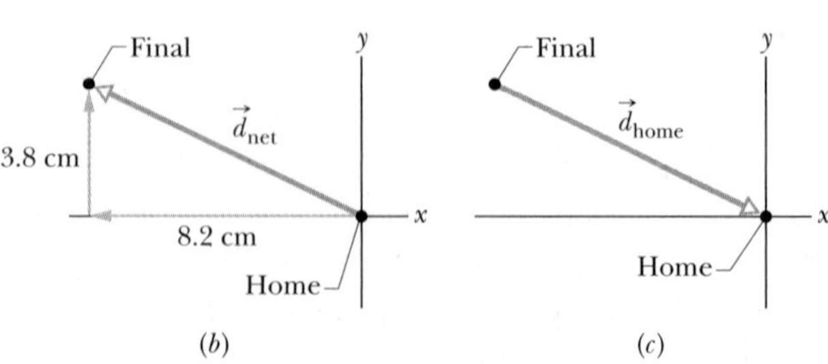

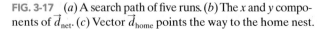

FIG. 3-17 (a) A search path of five runs. (b) The x and y components of $\vec{d}_{\text{net}}$. (c) Vector $\vec{d}_{\text{home}}$ points the way to the home nest.

KEY IDEAS (1) To find the net displacement $\vec{d}_{\text{net}}$, we need to sum the five individual displacement vectors:

$$\vec{d}_{\text{net}} = \vec{d}_1 + \vec{d}_2 + \vec{d}_3 + \vec{d}_4 + \vec{d}_5.$$

(2) We evaluate this sum for the x components alone,

$$d_{\text{net},x} = d_{1x} + d_{2x} + d_{3x} + d_{4x} + d_{5x}, \qquad (3\text{-}14)$$

and for the y components alone,

$$d_{\text{net},y} = d_{1y} + d_{2y} + d_{3y} + d_{4y} + d_{5y}. \qquad (3\text{-}15)$$

(3) We construct $\vec{d}_{\text{net}}$ from its x and y components.

Calculations: To evaluate Eq. 3-14, we apply the x part of Eq. 3-5 to each run:

$$\begin{aligned} d_{1x} &= (6.0\text{ cm})\cos 0° = +6.0\text{ cm} \\ d_{2x} &= (6.0\text{ cm})\cos 150° = -5.2\text{ cm} \\ d_{3x} &= (6.0\text{ cm})\cos 180° = -6.0\text{ cm} \\ d_{4x} &= (6.0\text{ cm})\cos(-120°) = -3.0\text{ cm} \\ d_{5x} &= (6.0\text{ cm})\cos 90° = 0. \end{aligned}$$

Equation 3-14 then gives us

$$d_{\text{net},x} = +6.0 \text{ cm} + (-5.2 \text{ cm}) + (-6.0 \text{ cm}) + (-3.0 \text{ cm}) + 0 = -8.2 \text{ cm}.$$

Similarly, we evaluate the individual y components of the five runs using the y part of Eq. 3-5. The results are shown in Table 3-1. Substituting the results into Eq. 3-15 then gives us

$$d_{\text{net},y} = +3.8 \text{ cm}.$$

Vector $\vec{d}_{\text{net}}$ and its x and y components are shown in Fig. 3-17*b*. To find the magnitude and angle of $\vec{d}_{\text{net}}$ from its components, we use Eq. 3-6. The magnitude is

$$d_{\text{net}} = \sqrt{d^2_{\text{net},x} + d^2_{\text{net},y}} = \sqrt{(-8.2 \text{ cm})^2 + (3.8 \text{ cm})^2} = 9.0 \text{ cm}.$$

To find the angle (measured from the positive direction of x), we take an inverse tangent:

$$\theta = \tan^{-1}\left(\frac{d_{\text{net},y}}{d_{\text{net},x}}\right) = \tan^{-1}\left(\frac{3.8 \text{ cm}}{-8.2 \text{ cm}}\right) = -24.86°.$$

Caution: Recall from Problem-Solving Tactic 3 that taking an inverse tangent on a calculator may not give the correct answer. The answer $-24.86°$ indicates that the direction of $\vec{d}_{\text{net}}$ is in the fourth quadrant of our xy coordinate system. However, when we construct the vector from its components (Fig. 3-17*b*), we see that the direction of $\vec{d}_{\text{net}}$ is in the second quadrant. Thus, we must "fix" the calculator's answer by adding 180°:

TABLE 3-1

Run	d_x (cm)	d_y (cm)
1	+6.0	0
2	−5.2	+3.0
3	−6.0	0
4	−3.0	−5.2
5	0	+6.0
net	−8.2	+3.8

$$\theta = -24.86° + 180° = 155.14° \approx 155°.$$

Thus, the ant's displacement $\vec{d}_{\text{net}}$ has magnitude and angle

$$d_{\text{net}} = 9.0 \text{ cm at } 155°. \quad \text{(Answer)}$$

Vector $\vec{d}_{\text{home}}$ directed from the ant to its home has the same magnitude as $\vec{d}_{\text{net}}$ but the opposite direction (Fig. 3-17*c*). We already have the angle ($-24.86° \approx -25°$) for the direction opposite $\vec{d}_{\text{net}}$. Thus, $\vec{d}_{\text{home}}$ has magnitude and angle

$$d_{\text{home}} = 9.0 \text{ cm at } -25°. \quad \text{(Answer)}$$

A desert ant traveling more than 500 m from its home will actually make thousands of individual runs. Yet, it somehow knows how to calculate $\vec{d}_{\text{home}}$ (without studying this chapter).

Sample Problem 3-6 Build your skill

Here is a problem involving vector addition that *cannot* be solved directly on a vector-capable calculator, using the vector notation of the calculator. A fellow camper is to walk away from you in a straight line (vector $\vec{A}$), turn, walk in a second straight line (vector $\vec{B}$) and then stop. How far must you walk in a straight line (vector $\vec{C}$) to reach her?

The three vectors (shown in Fig. 3-18) are related by

$$\vec{C} = \vec{A} + \vec{B}. \quad (3\text{-}16)$$

$\vec{A}$ has a magnitude of 22.0 m and is directed at an angle of $-47.0°$ (clockwise) from the positive direction of an x axis. $\vec{B}$ has a magnitude of 17.0 m and is directed counterclockwise from the positive direction of the x axis by angle ϕ. $\vec{C}$ is in the positive direction of the x axis. What is the magnitude of $\vec{C}$?

KEY IDEA We cannot answer the question by adding $\vec{A}$ and $\vec{B}$ directly on a vector-capable calculator, say, in the generic form of

[magnitude A ∠ angle A] + [magnitude B ∠ angle B]

because we do not know the value for the angle ϕ of $\vec{B}$. However, we *can* express Eq. 3-16 in terms of components for either the x axis or the y axis.

Calculations: Since $\vec{C}$ is directed along the x axis, we choose that axis and write

$$C_x = A_x + B_x.$$

We next express each x component in the form of the x part of Eq. 3-5 and substitute known data. We then have

$$C \cos 0° = 22.0 \cos(-47.0°) + 17.0 \cos \phi. \quad (3\text{-}17)$$

However, this hardly seems to help, because we still cannot solve for C without knowing ϕ.

Let us now express Eq. 3-16 in terms of components along the y axis:

$$C_y = A_y + B_y.$$

We then cast these y components in the form of the y part of Eq. 3-5 and substitute known data, to write

$$C \sin 0° = 22.0 \sin(-47.0°) + 17.0 \sin \phi,$$

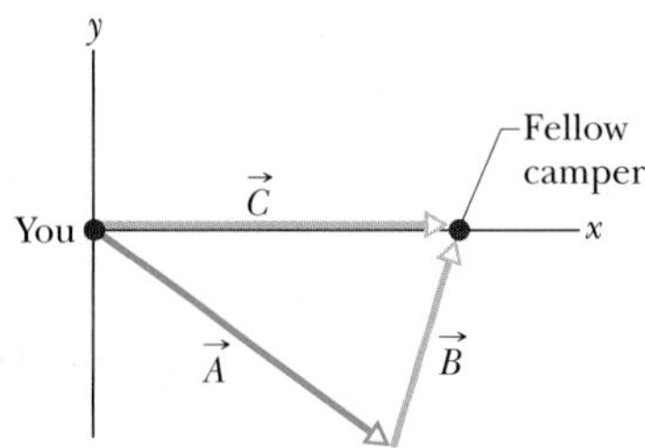

FIG. 3-18 $\vec{C}$ equals the sum $\vec{A} + \vec{B}$.

which yields

$$0 = 22.0 \sin(-47.0°) + 17.0 \sin \phi.$$

Solving for ϕ then gives us

$$\phi = \sin^{-1} - \frac{22.0 \sin(-47.0°)}{17.0} = 71.17°.$$

Substituting this result into Eq. 3-17 leads us to

$$C = 20.5 \text{ m}. \qquad \text{(Answer)}$$

Note the technique of solution: When we got stuck with components on the x axis, we worked with components on the y axis, to evaluate ϕ. We next moved back to the x axis, to evaluate C.

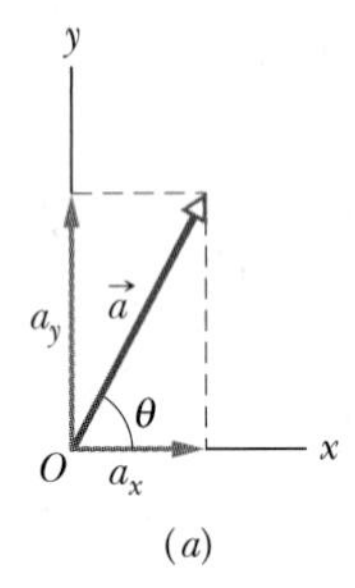

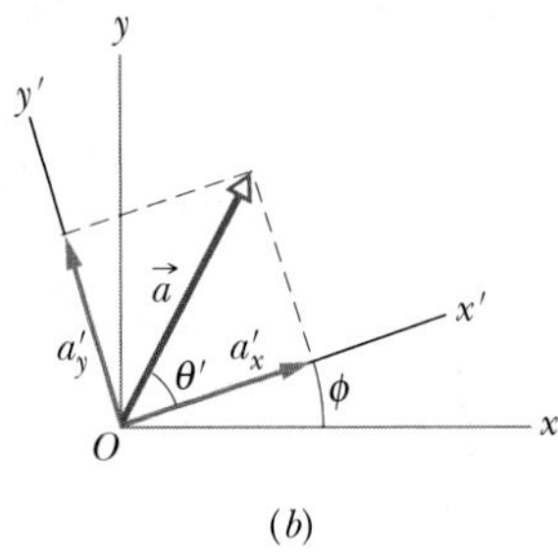

FIG. 3-19 (*a*) The vector $\vec{a}$ and its components. (*b*) The same vector, with the axes of the coordinate system rotated through an angle ϕ.

3-7 | Vectors and the Laws of Physics

So far, in every figure that includes a coordinate system, the x and y axes are parallel to the edges of the book page. Thus, when a vector $\vec{a}$ is included, its components a_x and a_y are also parallel to the edges (as in Fig. 3-19*a*). The only reason for that orientation of the axes is that it looks "proper"; there is no deeper reason. We could, instead, rotate the axes (but not the vector $\vec{a}$) through an angle ϕ as in Fig. 3-19*b*, in which case the components would have new values, call them a'_x and a'_y. Since there are an infinite number of choices of ϕ, there are an infinite number of different pairs of components for $\vec{a}$.

Which then is the "right" pair of components? The answer is that they are all equally valid because each pair (with its axes) just gives us a different way of describing the same vector $\vec{a}$; all produce the same magnitude and direction for the vector. In Fig. 3-19 we have

$$a = \sqrt{a_x^2 + a_y^2} = \sqrt{a_x'^2 + a_y'^2} \qquad (3\text{-}18)$$

and

$$\theta = \theta' + \phi. \qquad (3\text{-}19)$$

The point is that we have great freedom in choosing a coordinate system, because the relations among vectors do not depend on the location of the origin or on the orientation of the axes. This is also true of the relations of physics; they are all independent of the choice of coordinate system. Add to that the simplicity and richness of the language of vectors and you can see why the laws of physics are almost always presented in that language: one equation, like Eq. 3-10, can represent three (or even more) relations, like Eqs. 3-11, 3-12, and 3-13.

3-8 | Multiplying Vectors*

There are three ways in which vectors can be multiplied, but none is exactly like the usual algebraic multiplication. As you read this section, keep in mind that a vector-capable calculator will help you multiply vectors only if you understand the basic rules of that multiplication.

Multiplying a Vector by a Scalar

If we multiply a vector $\vec{a}$ by a scalar s, we get a new vector. Its magnitude is the product of the magnitude of $\vec{a}$ and the absolute value of s. Its direction is the direction of $\vec{a}$ if s is positive but the opposite direction if s is negative. To divide $\vec{a}$ by s, we multiply $\vec{a}$ by $1/s$.

Multiplying a Vector by a Vector

There are two ways to multiply a vector by a vector: one way produces a scalar (called the *scalar product*), and the other produces a new vector (called the *vector product*). (Students commonly confuse the two ways.)

*This material will not be employed until later (Chapter 7 for scalar products and Chapter 11 for vector products), and so your instructor may wish to postpone assignment of this section.

The Scalar Product

The **scalar product** of the vectors $\vec{a}$ and $\vec{b}$ in Fig. 3-20a is written as $\vec{a}\cdot\vec{b}$ and defined to be

$$\vec{a}\cdot\vec{b} = ab\cos\phi, \qquad (3\text{-}20)$$

where a is the magnitude of $\vec{a}$, b is the magnitude of $\vec{b}$, and ϕ is the angle between $\vec{a}$ and $\vec{b}$ (or, more properly, between the directions of $\vec{a}$ and $\vec{b}$). There are actually two such angles: ϕ and $360° - \phi$. Either can be used in Eq. 3-20, because their cosines are the same.

Note that there are only scalars on the right side of Eq. 3-20 (including the value of $\cos\phi$). Thus $\vec{a}\cdot\vec{b}$ on the left side represents a *scalar* quantity. Because of the notation, $\vec{a}\cdot\vec{b}$ is also known as the **dot product** and is spoken as "a dot b."

A dot product can be regarded as the product of two quantities: (1) the magnitude of one of the vectors and (2) the scalar component of the second vector along the direction of the first vector. For example, in Fig. 3-20b, $\vec{a}$ has a scalar component $a\cos\phi$ along the direction of $\vec{b}$; note that a perpendicular dropped from the head of $\vec{a}$ onto $\vec{b}$ determines that component. Similarly, $\vec{b}$ has a scalar component $b\cos\phi$ along the direction of $\vec{a}$.

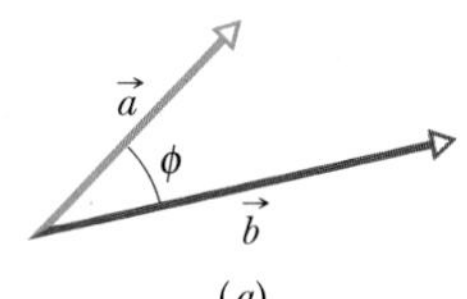

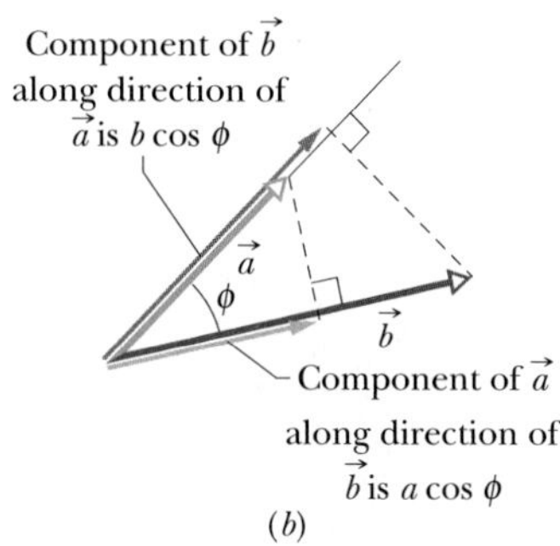

FIG. 3-20 (a) Two vectors $\vec{a}$ and $\vec{b}$, with an angle ϕ between them. (b) Each vector has a component along the direction of the other vector.

If the angle ϕ between two vectors is 0°, the component of one vector along the other is maximum, and so also is the dot product of the vectors. If, instead, ϕ is 90°, the component of one vector along the other is zero, and so is the dot product.

Equation 3-20 can be rewritten as follows to emphasize the components:

$$\vec{a}\cdot\vec{b} = (a\cos\phi)(b) = (a)(b\cos\phi). \qquad (3\text{-}21)$$

The commutative law applies to a scalar product, so we can write

$$\vec{a}\cdot\vec{b} = \vec{b}\cdot\vec{a}.$$

When two vectors are in unit-vector notation, we write their dot product as

$$\vec{a}\cdot\vec{b} = (a_x\hat{i} + a_y\hat{j} + a_z\hat{k})\cdot(b_x\hat{i} + b_y\hat{j} + b_z\hat{k}), \qquad (3\text{-}22)$$

which we can expand according to the distributive law: Each vector component of the first vector is to be dotted with each vector component of the second vector. By doing so, we can show that

$$\vec{a}\cdot\vec{b} = a_xb_x + a_yb_y + a_zb_z. \qquad (3\text{-}23)$$

CHECKPOINT 4 Vectors $\vec{C}$ and $\vec{D}$ have magnitudes of 3 units and 4 units, respectively. What is the angle between the directions of $\vec{C}$ and $\vec{D}$ if $\vec{C}\cdot\vec{D}$ equals (a) zero, (b) 12 units, and (c) −12 units?

Sample Problem 3-7

What is the angle ϕ between $\vec{a} = 3.0\hat{i} - 4.0\hat{j}$ and $\vec{b} = -2.0\hat{i} + 3.0\hat{k}$? (*Caution:* Although many of the following steps can be bypassed with a vector-capable calculator, you will learn more about scalar products if, at least here, you use these steps.)

KEY IDEA The angle between the directions of two vectors is included in the definition of their scalar product (Eq. 3-20):

$$\vec{a}\cdot\vec{b} = ab\cos\phi. \qquad (3\text{-}24)$$

Calculations: In Eq. 3-24, a is the magnitude of $\vec{a}$, or

$$a = \sqrt{3.0^2 + (-4.0)^2} = 5.00, \qquad (3\text{-}25)$$

and b is the magnitude of $\vec{b}$, or

$$b = \sqrt{(-2.0)^2 + 3.0^2} = 3.61. \qquad (3\text{-}26)$$

We can separately evaluate the left side of Eq. 3-24 by writing the vectors in unit-vector notation and using the distributive law:

$$\begin{aligned}\vec{a}\cdot\vec{b} &= (3.0\hat{i} - 4.0\hat{j})\cdot(-2.0\hat{i} + 3.0\hat{k})\\ &= (3.0\hat{i})\cdot(-2.0\hat{i}) + (3.0\hat{i})\cdot(3.0\hat{k})\\ &\quad + (-4.0\hat{j})\cdot(-2.0\hat{i}) + (-4.0\hat{j})\cdot(3.0\hat{k}).\end{aligned}$$

We next apply Eq. 3-20 to each term in this last expression. The angle between the unit vectors in the first term ($\hat{i}$ and $\hat{i}$) is 0°, and in the other terms it is 90°. We then have

$$\vec{a}\cdot\vec{b} = -(6.0)(1) + (9.0)(0) + (8.0)(0) - (12)(0)$$
$$= -6.0.$$

Substituting this result and the results of Eqs. 3-25 and 3-26 into Eq. 3-24 yields

$$-6.0 = (5.00)(3.61)\cos\phi,$$

so $$\phi = \cos^{-1}\frac{-6.0}{(5.00)(3.61)} = 109^\circ \approx 110^\circ. \quad \text{(Answer)}$$

The Vector Product

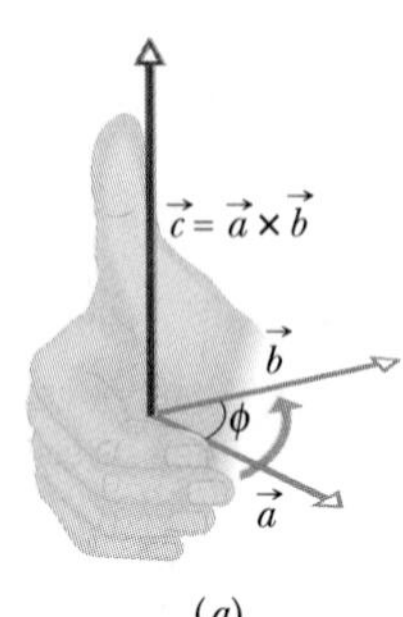

(a)

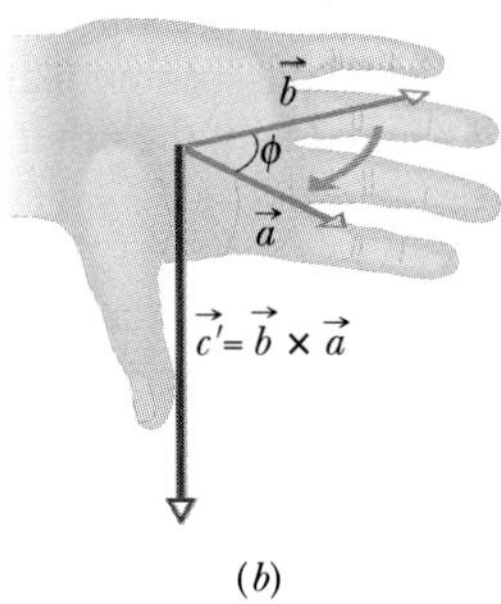

(b)

FIG. 3-21 Illustration of the right-hand rule for vector products. (a) Sweep vector $\vec{a}$ into vector $\vec{b}$ with the fingers of your right hand. Your outstretched thumb shows the direction of vector $\vec{c} = \vec{a} \times \vec{b}$. (b) Showing that $\vec{b} \times \vec{a}$ is the reverse of $\vec{a} \times \vec{b}$.

The **vector product** of $\vec{a}$ and $\vec{b}$, written $\vec{a} \times \vec{b}$, produces a third vector $\vec{c}$ whose magnitude is

$$c = ab\sin\phi, \tag{3-27}$$

where ϕ is the *smaller* of the two angles between $\vec{a}$ and $\vec{b}$. (You must use the smaller of the two angles between the vectors because $\sin\phi$ and $\sin(360^\circ - \phi)$ differ in algebraic sign.) Because of the notation, $\vec{a} \times \vec{b}$ is also known as the **cross product,** and in speech it is "a cross b."

If $\vec{a}$ and $\vec{b}$ are parallel or antiparallel, $\vec{a} \times \vec{b} = 0$. The magnitude of $\vec{a} \times \vec{b}$, which can be written as $|\vec{a} \times \vec{b}|$, is maximum when $\vec{a}$ and $\vec{b}$ are perpendicular to each other.

The direction of $\vec{c}$ is perpendicular to the plane that contains $\vec{a}$ and $\vec{b}$. Figure 3-21*a* shows how to determine the direction of $\vec{c} = \vec{a} \times \vec{b}$ with what is known as a **right-hand rule.** Place the vectors $\vec{a}$ and $\vec{b}$ tail to tail without altering their orientations, and imagine a line that is perpendicular to their plane where they meet. Pretend to place your *right* hand around that line in such a way that your fingers would sweep $\vec{a}$ into $\vec{b}$ through the smaller angle between them. Your outstretched thumb points in the direction of $\vec{c}$.

The order of the vector multiplication is important. In Fig. 3-21*b*, we are determining the direction of $\vec{c}' = \vec{b} \times \vec{a}$, so the fingers are placed to sweep $\vec{b}$ into $\vec{a}$ through the smaller angle. The thumb ends up in the opposite direction from previously, and so it must be that $\vec{c}' = -\vec{c}$; that is,

$$\vec{b} \times \vec{a} = -(\vec{a} \times \vec{b}). \tag{3-28}$$

In other words, the commutative law does not apply to a vector product.

In unit-vector notation, we write

$$\vec{a} \times \vec{b} = (a_x\hat{i} + a_y\hat{j} + a_z\hat{k}) \times (b_x\hat{i} + b_y\hat{j} + b_z\hat{k}), \tag{3-29}$$

which can be expanded according to the distributive law; that is, each component of the first vector is to be crossed with each component of the second vector. The cross products of unit vectors are given in Appendix E (see "Products of Vectors"). For example, in the expansion of Eq. 3-29, we have

$$a_x\hat{i} \times b_x\hat{i} = a_xb_x(\hat{i} \times \hat{i}) = 0,$$

because the two unit vectors $\hat{i}$ and $\hat{i}$ are parallel and thus have a zero cross product. Similarly, we have

$$a_x\hat{i} \times b_y\hat{j} = a_xb_y(\hat{i} \times \hat{j}) = a_xb_y\hat{k}.$$

In the last step we used Eq. 3-27 to evaluate the magnitude of $\hat{i} \times \hat{j}$ as unity. (These vectors $\hat{i}$ and $\hat{j}$ each have a magnitude of unity, and the angle between them is 90°.) Also, we used the right-hand rule to get the direction of $\hat{i} \times \hat{j}$ as being in the positive direction of the z axis (thus in the direction of $\hat{k}$).

Continuing to expand Eq. 3-29, you can show that

$$\vec{a} \times \vec{b} = (a_yb_z - b_ya_z)\hat{i} + (a_zb_x - b_za_x)\hat{j} + (a_xb_y - b_xa_y)\hat{k}. \tag{3-30}$$

A determinant (Appendix E) or a vector-capable calculator can also be used.

To check whether any xyz coordinate system is a right-handed coordinate system, use the right-hand rule for the cross product $\hat{\mathrm{i}} \times \hat{\mathrm{j}} = \hat{\mathrm{k}}$ with that system. If your fingers sweep $\hat{\mathrm{i}}$ (positive direction of x) into $\hat{\mathrm{j}}$ (positive direction of y) with the outstretched thumb pointing in the positive direction of z, then the system is right-handed.

CHECKPOINT 5 Vectors $\vec{C}$ and $\vec{D}$ have magnitudes of 3 units and 4 units, respectively. What is the angle between the directions of $\vec{C}$ and $\vec{D}$ if the magnitude of the vector product $\vec{C} \times \vec{D}$ is (a) zero and (b) 12 units?

Sample Problem 3-8

In Fig. 3-22, vector $\vec{a}$ lies in the xy plane, has a magnitude of 18 units and points in a direction 250° from the $+x$ direction. Also, vector $\vec{b}$ has a magnitude of 12 units and points in the $+z$ direction. What is the vector product $\vec{c} = \vec{a} \times \vec{b}$?

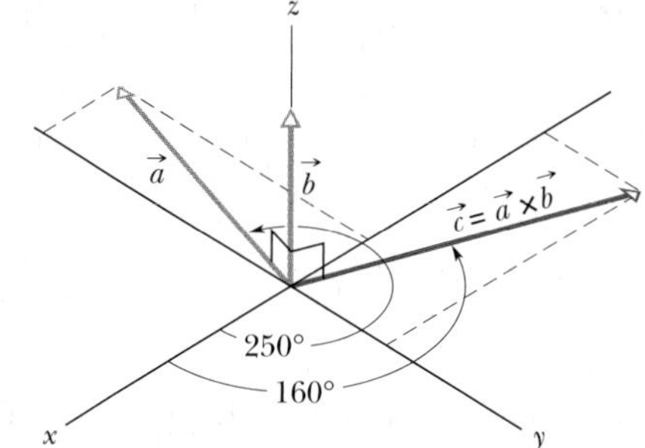

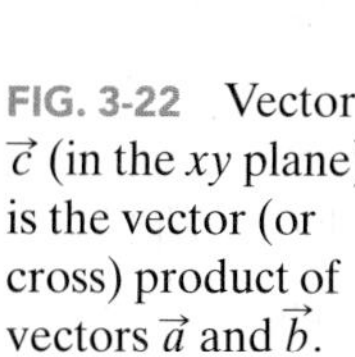

FIG. 3-22 Vector $\vec{c}$ (in the xy plane) is the vector (or cross) product of vectors $\vec{a}$ and $\vec{b}$.

KEY IDEA When we have two vectors in magnitude-angle notation, we find the magnitude of their cross product with Eq. 3-27 and the direction of their cross product with the right-hand rule of Fig. 3-21.

Calculations: For the magnitude we write

$$c = ab \sin \phi = (18)(12)(\sin 90°) = 216. \quad \text{(Answer)}$$

To determine the direction in Fig. 3-22, imagine placing the fingers of your right hand around a line perpendicular to the plane of $\vec{a}$ and $\vec{b}$ (the line on which $\vec{c}$ is shown) such that your fingers sweep $\vec{a}$ into $\vec{b}$. Your outstretched thumb then gives the direction of $\vec{c}$. Thus, as shown in the figure, $\vec{c}$ lies in the xy plane. Because its direction is perpendicular to the direction of $\vec{a}$, it is at an angle of

$$250° - 90° = 160° \quad \text{(Answer)}$$

from the positive direction of the x axis.

Sample Problem 3-9

If $\vec{a} = 3\hat{\mathrm{i}} - 4\hat{\mathrm{j}}$ and $\vec{b} = -2\hat{\mathrm{i}} + 3\hat{\mathrm{k}}$, what is $\vec{c} = \vec{a} \times \vec{b}$?

KEY IDEA When two vectors are in unit-vector notation, we can find their cross product by using the distributive law.

Calculations: Here we write

$$\begin{aligned}\vec{c} &= (3\hat{\mathrm{i}} - 4\hat{\mathrm{j}}) \times (-2\hat{\mathrm{i}} + 3\hat{\mathrm{k}}) \\ &= 3\hat{\mathrm{i}} \times (-2\hat{\mathrm{i}}) + 3\hat{\mathrm{i}} \times 3\hat{\mathrm{k}} + (-4\hat{\mathrm{j}}) \times (-2\hat{\mathrm{i}}) \\ &\quad + (-4\hat{\mathrm{j}}) \times 3\hat{\mathrm{k}}.\end{aligned}$$

We next evaluate each term with Eq. 3-27, finding the direction with the right-hand rule. For the first term here, the angle ϕ between the two vectors being crossed is 0. For the other terms, ϕ is 90°. We find

$$\begin{aligned}\vec{c} &= -6(0) + 9(-\hat{\mathrm{j}}) + 8(-\hat{\mathrm{k}}) - 12\hat{\mathrm{i}} \\ &= -12\hat{\mathrm{i}} - 9\hat{\mathrm{j}} - 8\hat{\mathrm{k}}. \quad \text{(Answer)}\end{aligned}$$

This vector $\vec{c}$ is perpendicular to both $\vec{a}$ and $\vec{b}$, a fact you can check by showing that $\vec{c} \cdot \vec{a} = 0$ and $\vec{c} \cdot \vec{b} = 0$; that is, there is no component of $\vec{c}$ along the direction of either $\vec{a}$ or $\vec{b}$.

PROBLEM-SOLVING TACTICS

Tactic 5: Common Errors with Cross Products Several errors are common in finding a cross product. (1) Failure to arrange vectors tail to tail is tempting when an illustration presents them head to tail; you must mentally shift (or better, redraw) one vector to the proper arrangement without changing its orientation. (2) Failing to use the right hand in applying the right-hand rule is easy when the right hand is occupied with a calculator or pencil. (3) Failure to sweep the first vector of the product into the second vector can occur when the orientations of the vectors require an awkward twisting of your hand to apply the right-hand rule. Sometimes that happens when you try to make the sweep mentally rather than actually using your hand. (4) Failure to work with a right-handed coordinate system results when you forget how to draw such a system. See Fig. 3-14 for one perspective. Practice drawing other perspectives, such as the (correct ones) shown in Fig. 3-25 on page 53.

REVIEW & SUMMARY

Scalars and Vectors *Scalars,* such as temperature, have magnitude only. They are specified by a number with a unit (10°C) and obey the rules of arithmetic and ordinary algebra. *Vectors,* such as displacement, have both magnitude and direction (5 m, north) and obey the rules of vector algebra.

Adding Vectors Geometrically Two vectors $\vec{a}$ and $\vec{b}$ may be added geometrically by drawing them to a common scale and placing them head to tail. The vector connecting the tail of the first to the head of the second is the vector sum $\vec{s}$. To subtract $\vec{b}$ from $\vec{a}$, reverse the direction of $\vec{b}$ to get $-\vec{b}$; then add $-\vec{b}$ to $\vec{a}$. Vector addition is commutative and obeys the associative law.

Components of a Vector The (scalar) *components* a_x and a_y of any two-dimensional vector $\vec{a}$ along the coordinate axes are found by dropping perpendicular lines from the ends of $\vec{a}$ onto the coordinate axes. The components are given by

$$a_x = a\cos\theta \quad \text{and} \quad a_y = a\sin\theta, \tag{3-5}$$

where θ is the angle between the positive direction of the x axis and the direction of $\vec{a}$. The algebraic sign of a component indicates its direction along the associated axis. Given its components, we can find the magnitude and orientation of the vector $\vec{a}$ with

$$a = \sqrt{a_x^2 + a_y^2} \quad \text{and} \quad \tan\theta = \frac{a_y}{a_x}. \tag{3-6}$$

Unit-Vector Notation *Unit vectors* $\hat{i}, \hat{j}$, and $\hat{k}$ have magnitudes of unity and are directed in the positive directions of the x, y, and z axes, respectively, in a right-handed coordinate system. We can write a vector $\vec{a}$ in terms of unit vectors as

$$\vec{a} = a_x\hat{i} + a_y\hat{j} + a_z\hat{k}, \tag{3-7}$$

in which $a_x\hat{i}$, $a_y\hat{j}$, and $a_z\hat{k}$ are the **vector components** of $\vec{a}$ and a_x, a_y, and a_z are its **scalar components.**

Adding Vectors in Component Form To add vectors in component form, we use the rules

$$r_x = a_x + b_x \quad r_y = a_y + b_y \quad r_z = a_z + b_z. \tag{3-11 to 3-13}$$

Here $\vec{a}$ and $\vec{b}$ are the vectors to be added, and $\vec{r}$ is the vector sum.

Product of a Scalar and a Vector The product of a scalar s and a vector $\vec{v}$ is a new vector whose magnitude is sv and whose direction is the same as that of $\vec{v}$ if s is positive, and opposite that of $\vec{v}$ if s is negative. To divide $\vec{v}$ by s, multiply $\vec{v}$ by $1/s$.

The Scalar Product The **scalar** (or **dot**) **product** of two vectors $\vec{a}$ and $\vec{b}$ is written $\vec{a}\cdot\vec{b}$ and is the *scalar* quantity given by

$$\vec{a}\cdot\vec{b} = ab\cos\phi, \tag{3-20}$$

in which ϕ is the angle between the directions of $\vec{a}$ and $\vec{b}$. A scalar product is the product of the magnitude of one vector and the scalar component of the second vector along the direction of the first vector. In unit-vector notation,

$$\vec{a}\cdot\vec{b} = (a_x\hat{i} + a_y\hat{j} + a_z\hat{k})\cdot(b_x\hat{i} + b_y\hat{j} + b_z\hat{k}), \tag{3-22}$$

which may be expanded according to the distributive law. Note that $\vec{a}\cdot\vec{b} = \vec{b}\cdot\vec{a}$.

The Vector Product The **vector** (or **cross**) **product** of two vectors $\vec{a}$ and $\vec{b}$ is written $\vec{a}\times\vec{b}$ and is a *vector* $\vec{c}$ whose magnitude c is given by

$$c = ab\sin\phi, \tag{3-27}$$

in which ϕ is the smaller of the angles between the directions of $\vec{a}$ and $\vec{b}$. The direction of $\vec{c}$ is perpendicular to the plane defined by $\vec{a}$ and $\vec{b}$ and is given by a right-hand rule, as shown in Fig. 3-21. Note that $\vec{a}\times\vec{b} = -(\vec{b}\times\vec{a})$. In unit-vector notation,

$$\vec{a}\times\vec{b} = (a_x\hat{i} + a_y\hat{j} + a_z\hat{k})\times(b_x\hat{i} + b_y\hat{j} + b_z\hat{k}), \tag{3-29}$$

which we may expand with the distributive law.

QUESTIONS

1 Being part of the "Gators," the University of Florida golfing team must play on a putting green with an alligator pit. Figure 3-23 shows an overhead view of one putting challenge of the team; an xy coordinate system is superimposed. Team members must putt from the origin to the hole, which is at xy coordinates (8 m, 12 m), but they can putt the golf ball using only one or more of the following displacements, one or more times:

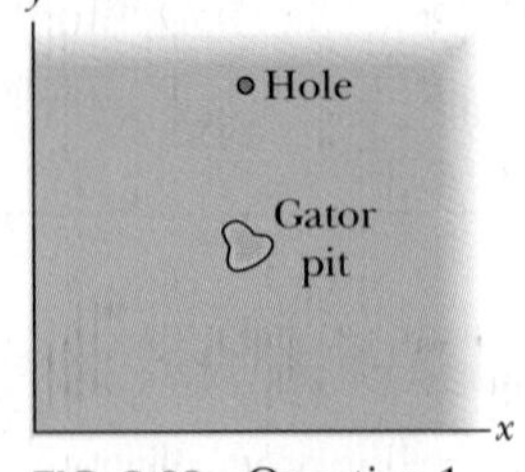

FIG. 3-23 Question 1.

$$\vec{d}_1 = (8\text{ m})\hat{i} + (6\text{ m})\hat{j}, \quad \vec{d}_2 = (6\text{ m})\hat{j}, \quad \vec{d}_3 = (8\text{ m})\hat{i}.$$

The pit is at coordinates (8 m, 6 m). If a team member putts the ball into or through the pit, the member is automatically transferred to Florida State University, the arch rival. What sequence of displacements should a team member use to avoid the pit?

2 Equation 3-2 shows that the addition of two vectors $\vec{a}$ and $\vec{b}$ is commutative. Does that mean subtraction is commutative, so that $\vec{a} - \vec{b} = \vec{b} - \vec{a}$?

3 Can the sum of the magnitudes of two vectors ever be equal to the magnitude of the sum of the same two vectors? If no, why not? If yes, when?

4 The two vectors shown in Fig. 3-24 lie in an xy plane. What are the signs of the x and y components, respectively, of (a) $\vec{d}_1 + \vec{d}_2$, (b) $\vec{d}_1 - \vec{d}_2$, and (c) $\vec{d}_2 - \vec{d}_1$?

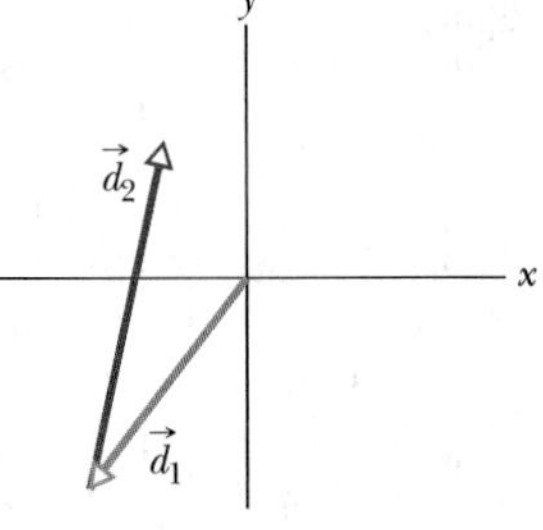

FIG. 3-24 Question 4.

5 If $\vec{d} = \vec{a} + \vec{b} + (-\vec{c})$, does (a) $\vec{a} + (-\vec{d}) = \vec{c} + (-\vec{b})$, (b) $\vec{a} = (-\vec{b}) + \vec{d} + \vec{c}$, and (c) $\vec{c} + (-\vec{d}) = \vec{a} + \vec{b}$?

6 Describe two vectors $\vec{a}$ and $\vec{b}$ such that

(a) $\vec{a} + \vec{b} = \vec{c}$ and $a + b = c$;

(b) $\vec{a} + \vec{b} = \vec{a} - \vec{b}$;

(c) $\vec{a} + \vec{b} = \vec{c}$ and $a^2 + b^2 = c^2$.

7 Which of the arrangements of axes in Fig. 3-25 can be labeled "right-handed coordinate system"? As usual, each axis label indicates the positive side of the axis.

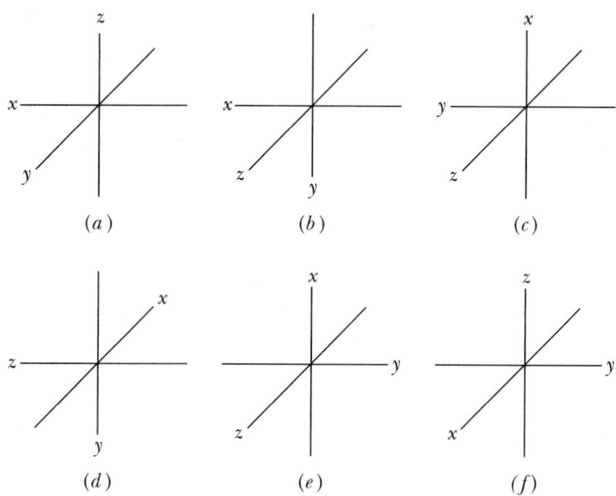

FIG. 3-25 Question 7.

8 Figure 3-26 shows vector $\vec{A}$ and four other vectors that have the same magnitude but differ in orientation. (a) Which of those other four vectors have the same dot product with $\vec{A}$? (b) Which have a negative dot product with $\vec{A}$?

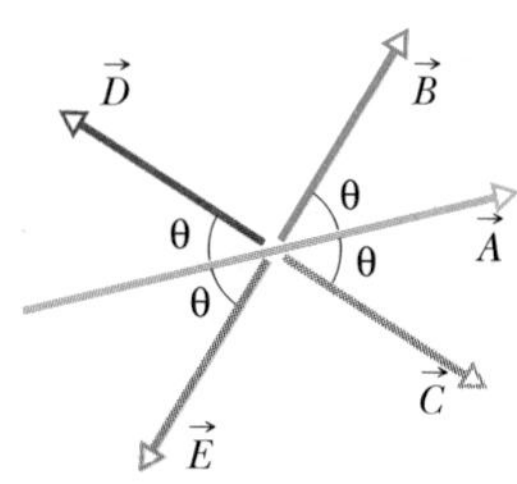

FIG. 3-26 Question 8.

9 If $\vec{F} = q(\vec{v} \times \vec{B})$ and $\vec{v}$ is perpendicular to $\vec{B}$, then what is the direction of $\vec{B}$ in the three situations shown in Fig. 3-27 when constant q is (a) positive and (b) negative?

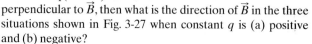

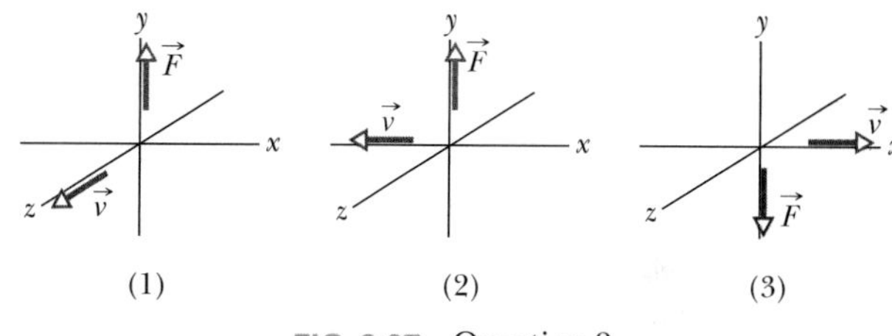

FIG. 3-27 Question 9.

10 If $\vec{a} \cdot \vec{b} = \vec{a} \cdot \vec{c}$, must $\vec{b}$ equal $\vec{c}$?

PROBLEMS

GO Tutoring problem available (at instructor's discretion) in *WileyPLUS* and WebAssign
SSM Worked-out solution available in Student Solutions Manual
WWW Worked-out solution is at http://www.wiley.com/college/halliday
• – ••• Number of dots indicates level of problem difficulty
ILW Interactive solution is at http://www.wiley.com/college/halliday
Additional information available in *The Flying Circus of Physics* and at flyingcircusofphysics.com

sec. 3-4 Components of Vectors

•**1** The x component of vector $\vec{A}$ is -25.0 m and the y component is $+40.0$ m. (a) What is the magnitude of $\vec{A}$? (b) What is the angle between the direction of $\vec{A}$ and the positive direction of x? SSM

•**2** Express the following angles in radians: (a) 20.0°, (b) 50.0°, (c) 100°. Convert the following angles to degrees: (d) 0.330 rad, (e) 2.10 rad, (f) 7.70 rad.

•**3** What are (a) the x component and (b) the y component of a vector $\vec{a}$ in the xy plane if its direction is 250° counterclockwise from the positive direction of the x axis and its magnitude is 7.3 m? SSM

•**4** In Fig. 3-28, a heavy piece of machinery is raised by sliding it a distance $d = 12.5$ m along a plank oriented at angle $\theta = 20.0°$ to the horizontal. How far is it moved (a) vertically and (b) horizontally?

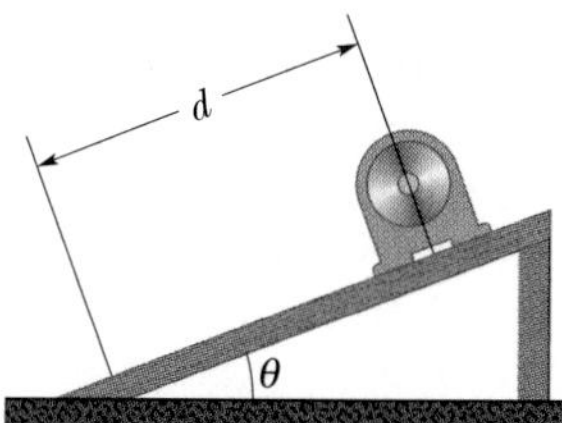

FIG. 3-28 Problem 4.

•**5** A ship sets out to sail to a point 120 km due north. An unexpected storm blows the ship to a point 100 km due east of its starting point. (a) How far and (b) in what direction must it now sail to reach its original destination?

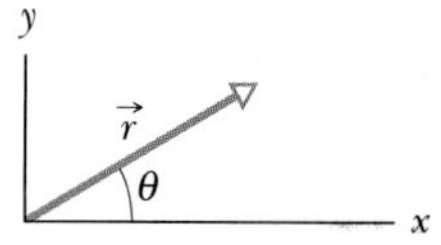

•**6** A displacement vector $\vec{r}$ in the xy plane is 15 m long and directed at angle $\theta = 30°$ in Fig. 3-29. Determine (a) the x component and (b) the y component of the vector.

FIG. 3-29 Problem 6.

••**7** A room has dimensions 3.00 m (height) × 3.70 m × 4.30 m. A fly starting at one corner flies around, ending up at the diagonally opposite corner. (a) What is the magnitude of its displacement? (b) Could the length of its path be less than this magnitude? (c) Greater? (d) Equal? (e) Choose a suitable coordinate system and express the components of the displacement vector in that system in unit-vector notation. (f) If the fly walks, what is the length of the shortest path? (*Hint:* This can be answered without calculus. The room is like a box. Unfold its walls to flatten them into a plane.) SSM WWW

sec. 3-6 Adding Vectors by Components

•**8** A car is driven east for a distance of 50 km, then north for 30 km, and then in a direction 30° east of north for 25 km. Sketch the vector diagram and determine (a) the magnitude and (b) the angle of the car's total displacement from its starting point.

•9 (a) In unit-vector notation, what is the sum $\vec{a} + \vec{b}$ if $\vec{a} = (4.0\text{ m})\hat{i} + (3.0\text{ m})\hat{j}$ and $\vec{b} = (-13.0\text{ m})\hat{i} + (7.0\text{ m})\hat{j}$? What are the (b) magnitude and (c) direction of $\vec{a} + \vec{b}$? SSM

•10 A person walks in the following pattern: 3.1 km north, then 2.4 km west, and finally 5.2 km south. (a) Sketch the vector diagram that represents this motion. (b) How far and (c) in what direction would a bird fly in a straight line from the same starting point to the same final point?

•11 A person desires to reach a point that is 3.40 km from her present location and in a direction that is 35.0° north of east. However, she must travel along streets that are oriented either north–south or east–west. What is the minimum distance she could travel to reach her destination?

•12 For the vectors $\vec{a} = (3.0\text{ m})\hat{i} + (4.0\text{ m})\hat{j}$ and $\vec{b} = (5.0\text{ m})\hat{i} + (-2.0\text{ m})\hat{j}$, give $\vec{a} + \vec{b}$ in (a) unit-vector notation, and as (b) a magnitude and (c) an angle (relative to $\hat{i}$). Now give $\vec{b} - \vec{a}$ in (d) unit-vector notation, and as (e) a magnitude and (f) an angle.

•13 Two vectors are given by

$$\vec{a} = (4.0\text{ m})\hat{i} - (3.0\text{ m})\hat{j} + (1.0\text{ m})\hat{k}$$

and

$$\vec{b} = (-1.0\text{ m})\hat{i} + (1.0\text{ m})\hat{j} + (4.0\text{ m})\hat{k}.$$

In unit-vector notation, find (a) $\vec{a} + \vec{b}$, (b) $\vec{a} - \vec{b}$, and (c) a third vector $\vec{c}$ such that $\vec{a} - \vec{b} + \vec{c} = 0$.

•14 Find the (a) x, (b) y, and (c) z components of the sum $\vec{r}$ of the displacements $\vec{c}$ and $\vec{d}$ whose components in meters along the three axes are $c_x = 7.4$, $c_y = -3.8$, $c_z = -6.1$; $d_x = 4.4$, $d_y = -2.0$, $d_z = 3.3$.

•15 An ant, crazed by the Sun on a hot Texas afternoon, darts over an xy plane scratched in the dirt. The x and y components of four consecutive darts are the following, all in centimeters: (30.0, 40.0), (b_x, −70.0), (−20.0, c_y), (−80.0, −70.0). The overall displacement of the four darts has the xy components (−140, −20.0). What are (a) b_x and (b) c_y? What are the (c) magnitude and (d) angle (relative to the positive direction of the x axis) of the overall displacement? GO

•16 In the sum $\vec{A} + \vec{B} = \vec{C}$, vector $\vec{A}$ has a magnitude of 12.0 m and is angled 40.0° counterclockwise from the $+x$ direction, and vector $\vec{C}$ has a magnitude of 15.0 m and is angled 20.0° counterclockwise from the $-x$ direction. What are (a) the magnitude and (b) the angle (relative to $+x$) of $\vec{B}$?

•17 The two vectors $\vec{a}$ and $\vec{b}$ in Fig. 3-30 have equal magnitudes of 10.0 m and the angles are $\theta_1 = 30°$ and $\theta_2 = 105°$. Find the (a) x and (b) y components of their vector sum $\vec{r}$, (c) the magnitude of $\vec{r}$, and (d) the angle $\vec{r}$ makes with the positive direction of the x axis. SSM ILW WWW

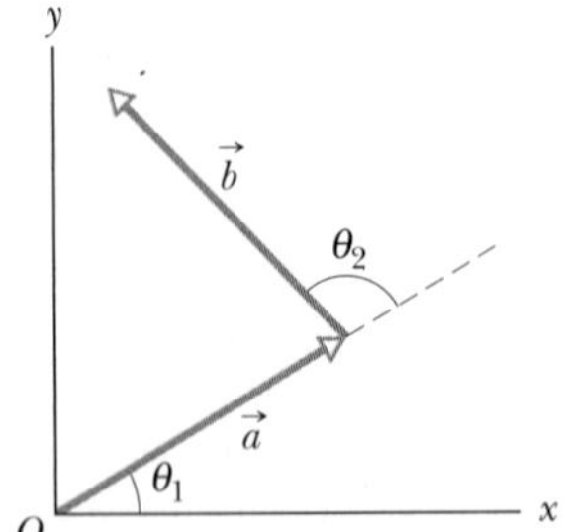

FIG. 3-30 Problem 17.

•18 You are to make four straight-line moves over a flat desert floor, starting at the origin of an xy coordinate system and ending at the xy coordinates (−140 m, 30 m). The x component and y component of your moves are the following, respectively, in meters: (20 and 60), then (b_x and −70), then (−20 and c_y), then (−60 and −70). What are (a) component b_x and (b) component c_y? What are (c) the magnitude and (d) the angle (relative to the positive direction of the x axis) of the overall displacement?

•19 Three vectors $\vec{a}$, $\vec{b}$, and $\vec{c}$ each have a magnitude of 50 m and lie in an xy plane. Their directions relative to the positive direction of the x axis are 30°, 195°, and 315°, respectively. What are (a) the magnitude and (b) the angle of the vector $\vec{a} + \vec{b} + \vec{c}$, and (c) the magnitude and (d) the angle of $\vec{a} - \vec{b} + \vec{c}$? What are the (e) magnitude and (f) angle of a fourth vector $\vec{d}$ such that $(\vec{a} + \vec{b}) - (\vec{c} + \vec{d}) = 0$? ILW

•20 (a) What is the sum of the following four vectors in unit-vector notation? For that sum, what are (b) the magnitude, (c) the angle in degrees, and (d) the angle in radians?

$\vec{E}$: 6.00 m at +0.900 rad $\vec{F}$: 5.00 m at −75.0°

$\vec{G}$: 4.00 m at +1.20 rad $\vec{H}$: 6.00 m at −210°

•21 In a game of lawn chess, where pieces are moved between the centers of squares that are each 1.00 m on edge, a knight is moved in the following way: (1) two squares forward, one square rightward; (2) two squares leftward, one square forward; (3) two squares forward, one square leftward. What are (a) the magnitude and (b) the angle (relative to "forward") of the knight's overall displacement for the series of three moves?

••22 An explorer is caught in a whiteout (in which the snowfall is so thick that the ground cannot be distinguished from the sky) while returning to base camp. He was supposed to travel due north for 5.6 km, but when the snow clears, he discovers that he actually traveled 7.8 km at 50° north of due east. (a) How far and (b) in what direction must he now travel to reach base camp?

••23 Oasis B is 25 km due east of oasis A. Starting from oasis A, a camel walks 24 km in a direction 15° south of east and then walks 8.0 km due north. How far is the camel then from oasis B? GO

••24 Two beetles run across flat sand, starting at the same point. Beetle 1 runs 0.50 m due east, then 0.80 m at 30° north of due east. Beetle 2 also makes two runs; the first is 1.6 m at 40° east of due north. What must be (a) the magnitude and (b) the direction of its second run if it is to end up at the new location of beetle 1?

••25 If $\vec{B}$ is added to $\vec{C} = 3.0\hat{i} + 4.0\hat{j}$, the result is a vector in the positive direction of the y axis, with a magnitude equal to that of $\vec{C}$. What is the magnitude of $\vec{B}$?

••26 Vector $\vec{A}$, which is directed along an x axis, is to be added to vector $\vec{B}$, which has a magnitude of 7.0 m. The sum is a third vector that is directed along the y axis, with a magnitude that is 3.0 times that of $\vec{A}$. What is that magnitude of $\vec{A}$?

••27 Typical backyard ants often create a network of chemical trails for guidance. Extending outward from the nest, a trail branches (*bifurcates*) repeatedly, with 60° between the branches. If a roaming ant chances upon a trail, it can tell the way to the nest at any branch point: If it is moving away from the nest, it has two choices of path requiring a small turn in its travel direction, either 30° leftward or 30° rightward. If it is moving toward the nest, it has only one such choice. Figure 3-31 shows a typical ant trail, with lettered straight sec-

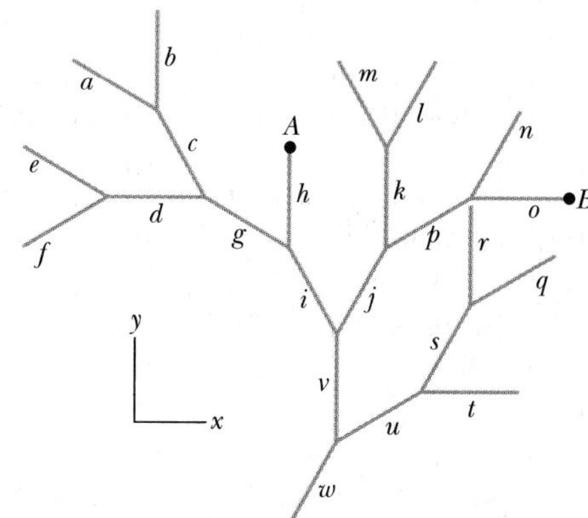

FIG. 3-31 Problem 27.

tions of 2.0 cm length and symmetric bifurcation of 60°. What are the (a) magnitude and (b) angle (relative to the positive direction of the superimposed x axis) of an ant's displacement from the nest (find it in the figure) if the ant enters the trail at point A? What are the (c) magnitude and (d) angle if it enters at point B?

••28 Here are two vectors:

$$\vec{a} = (4.0\text{ m})\hat{i} - (3.0\text{ m})\hat{j} \quad\text{and}\quad \vec{b} = (6.0\text{ m})\hat{i} + (8.0\text{ m})\hat{j}.$$

What are (a) the magnitude and (b) the angle (relative to $\hat{i}$) of $\vec{a}$? What are (c) the magnitude and (d) the angle of $\vec{b}$? What are (e) the magnitude and (f) the angle of $\vec{a} + \vec{b}$; (g) the magnitude and (h) the angle of $\vec{b} - \vec{a}$; and (i) the magnitude and (j) the angle of $\vec{a} - \vec{b}$? (k) What is the angle between the directions of $\vec{b} - \vec{a}$ and $\vec{a} - \vec{b}$?

••29 If $\vec{d}_1 + \vec{d}_2 = 5\vec{d}_3$, $\vec{d}_1 - \vec{d}_2 = 3\vec{d}_3$, and $\vec{d}_3 = 2\hat{i} + 4\hat{j}$, then what are, in unit-vector notation, (a) $\vec{d}_1$ and (b) $\vec{d}_2$?

••30 What is the sum of the following four vectors in (a) unit-vector notation, and as (b) a magnitude and (c) an angle?

$\vec{A} = (2.00\text{ m})\hat{i} + (3.00\text{ m})\hat{j}$ $\quad$ $\vec{B}$: 4.00 m, at +65.0°

$\vec{C} = (-4.00\text{ m})\hat{i} + (-6.00\text{ m})\hat{j}$ $\quad$ $\vec{D}$: 5.00 m, at −235°

•••31 In Fig. 3-32, a cube of edge length a sits with one corner at the origin of an xyz coordinate system. A *body diagonal* is a line that extends from one corner to another through the center. In unit-vector notation, what is the body diagonal that extends from the corner at (a) coordinates (0, 0, 0), (b) coordinates (a, 0, 0), (c) coordinates (0, a, 0), and (d) coordinates (a, a, 0)? (e) Determine the angles that the body diagonals make with the adjacent edges. (f) Determine the length of the body diagonals in terms of a.

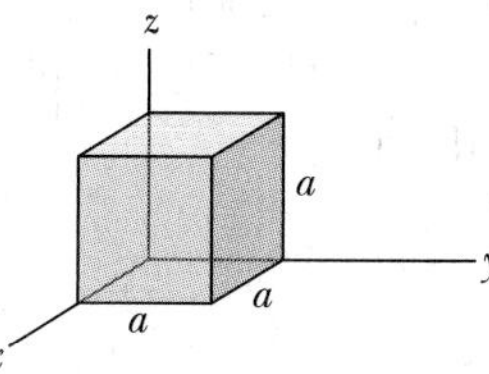

FIG. 3-32 Problem 31.

sec. 3-7 Vectors and the Laws of Physics

•32 In Fig. 3-33, a vector $\vec{a}$ with a magnitude of 17.0 m is directed at angle $\theta = 56.0°$ counterclockwise from the $+x$ axis. What are the components (a) a_x and (b) a_y of the vector? A second coordinate system is inclined by angle $\theta' = 18.0°$ with respect to the first. What are the components (c) a'_x and (d) a'_y in this primed coordinate system?

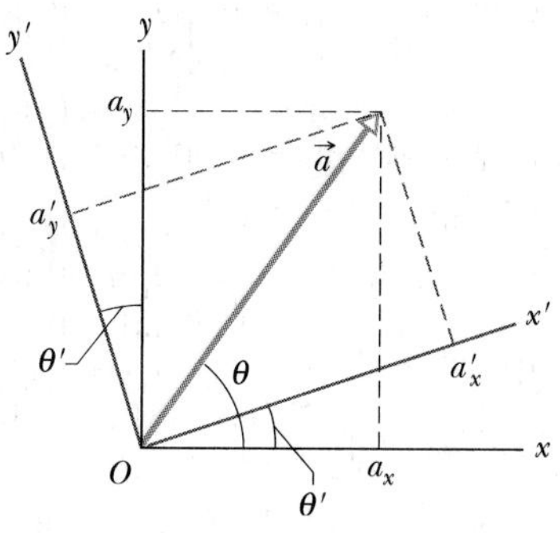

FIG. 3-33 Problem 32.

sec. 3-8 Multiplying Vectors

•33 Two vectors, $\vec{r}$ and $\vec{s}$, lie in the xy plane. Their magnitudes are 4.50 and 7.30 units, respectively, and their directions are 320° and 85.0°, respectively, as measured counterclockwise from the positive x axis. What are the values of (a) $\vec{r}\cdot\vec{s}$ and (b) $\vec{r}\times\vec{s}$?

•34 If $\vec{d}_1 = 3\hat{i} - 2\hat{j} + 4\hat{k}$ and $\vec{d}_2 = -5\hat{i} + 2\hat{j} - \hat{k}$, then what is $(\vec{d}_1 + \vec{d}_2)\cdot(\vec{d}_1 \times 4\vec{d}_2)$?

•35 Three vectors are given by $\vec{a} = 3.0\hat{i} + 3.0\hat{j} - 2.0\hat{k}$, $\vec{b} = -1.0\hat{i} - 4.0\hat{j} + 2.0\hat{k}$, and $\vec{c} = 2.0\hat{i} + 2.0\hat{j} + 1.0\hat{k}$. Find (a) $\vec{a}\cdot(\vec{b}\times\vec{c})$, (b) $\vec{a}\cdot(\vec{b}+\vec{c})$, and (c) $\vec{a}\times(\vec{b}+\vec{c})$.

•36 Two vectors are given by $\vec{a} = 3.0\hat{i} + 5.0\hat{j}$ and $\vec{b} = 2.0\hat{i} + 4.0\hat{j}$. Find (a) $\vec{a}\times\vec{b}$, (b) $\vec{a}\cdot\vec{b}$, (c) $(\vec{a}+\vec{b})\cdot\vec{b}$, and (d) the component of $\vec{a}$ along the direction of $\vec{b}$. (*Hint:* For (d), consider Eq. 3-20 and Fig. 3-20.)

•37 For the vectors in Fig. 3-34, with $a = 4$, $b = 3$, and $c = 5$, what are (a) the magnitude and (b) the direction of $\vec{a}\times\vec{b}$, (c) the magnitude and (d) the direction of $\vec{a}\times\vec{c}$, and (e) the magnitude and (f) the direction of $\vec{b}\times\vec{c}$? (The z axis is not shown.)

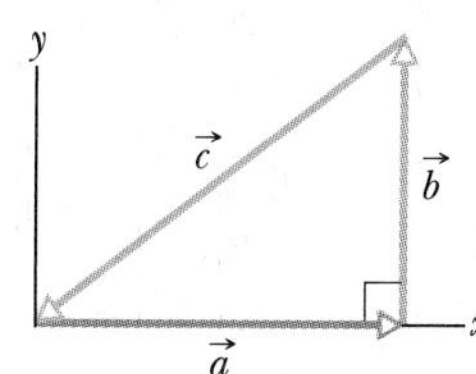

FIG. 3-34 Problems 37 and 50.

••38 Displacement $\vec{d}_1$ is in the yz plane 63.0° from the positive direction of the y axis, has a positive z component, and has a magnitude of 4.50 m. Displacement $\vec{d}_2$ is in the xz plane 30.0° from the positive direction of the x axis, has a positive z component, and has magnitude 1.40 m. What are (a) $\vec{d}_1\cdot\vec{d}_2$, (b) $\vec{d}_1\times\vec{d}_2$, and (c) the angle between $\vec{d}_1$ and $\vec{d}_2$?

••39 Use the definition of scalar product, $\vec{a}\cdot\vec{b} = ab\cos\theta$, and the fact that $\vec{a}\cdot\vec{b} = a_xb_x + a_yb_y + a_zb_z$ to calculate the angle between the two vectors given by $\vec{a} = 3.0\hat{i} + 3.0\hat{j} + 3.0\hat{k}$ and $\vec{b} = 2.0\hat{i} + 1.0\hat{j} + 3.0\hat{k}$. SSM ILW WWW

••40 For the following three vectors, what is $3\vec{C}\cdot(2\vec{A}\times\vec{B})$?

$$\vec{A} = 2.00\hat{i} + 3.00\hat{j} - 4.00\hat{k}$$
$$\vec{B} = -3.00\hat{i} + 4.00\hat{j} + 2.00\hat{k} \qquad \vec{C} = 7.00\hat{i} - 8.00\hat{j}$$

••41 Vector $\vec{A}$ has a magnitude of 6.00 units, vector $\vec{B}$ has a magnitude of 7.00 units, and $\vec{A}\cdot\vec{B}$ has a value of 14.0. What is the angle between the directions of $\vec{A}$ and $\vec{B}$?

••42 In the product $\vec{F} = q\vec{v}\times\vec{B}$, take $q = 2$,

$$\vec{v} = 2.0\hat{i} + 4.0\hat{j} + 6.0\hat{k} \quad\text{and}\quad \vec{F} = 4.0\hat{i} - 20\hat{j} + 12\hat{k}.$$

What then is $\vec{B}$ in unit-vector notation if $B_x = B_y$? GO

••43 The three vectors in Fig. 3-35 have magnitudes $a = 3.00$ m, $b = 4.00$ m, and $c = 10.0$ m and angle $\theta = 30.0°$. What are (a) the x component and (b) the y component of $\vec{a}$; (c) the x component and (d) the y com-

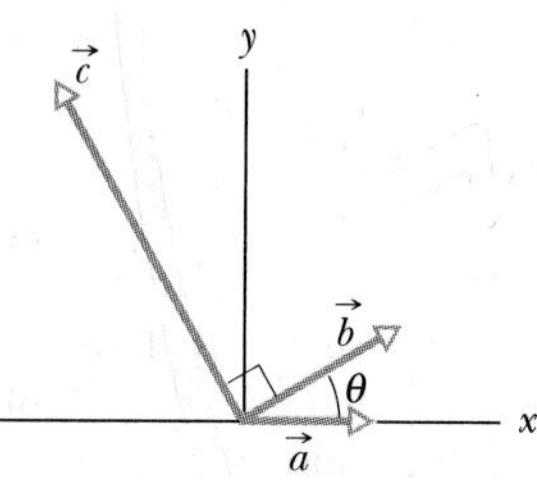

FIG. 3-35 Problem 43.

ponent of $\vec{b}$; and (e) the x component and (f) the y component of $\vec{c}$? If $\vec{c} = p\vec{a} + q\vec{b}$, what are the values of (g) p and (h) q? SSM ILW

••44 In a meeting of mimes, mime 1 goes through a displacement $\vec{d}_1 = (4.0 \text{ m})\hat{i} + (5.0 \text{ m})\hat{j}$ and mime 2 goes through a displacement $\vec{d}_2 = (-3.0 \text{ m})\hat{i} + (4.0 \text{ m})\hat{j}$. What are (a) $\vec{d}_1 \times \vec{d}_2$, (b) $\vec{d}_1 \cdot \vec{d}_2$, (c) $(\vec{d}_1 + \vec{d}_2) \cdot \vec{d}_2$, and (d) the component of $\vec{d}_1$ along the direction of $\vec{d}_2$? (*Hint:* For (d), see Eq. 3-20 and Fig. 3-20.)

Additional Problems

45 Rock *faults* are ruptures along which opposite faces of rock have slid past each other. In Fig. 3-36, points A and B coincided before the rock in the foreground slid down to the right. The net displacement $\overrightarrow{AB}$ is along the plane of the fault. The horizontal component of $\overrightarrow{AB}$ is the *strike-slip AC*. The component of $\overrightarrow{AB}$ that is directed down the plane of the fault is the *dip-slip AD*. (a) What is the magnitude of the net displacement $\overrightarrow{AB}$ if the strike-slip is 22.0 m and the dip-slip is 17.0 m? (b) If the plane of the fault is inclined at angle $\phi = 52.0°$ to the horizontal, what is the vertical component of $\overrightarrow{AB}$?

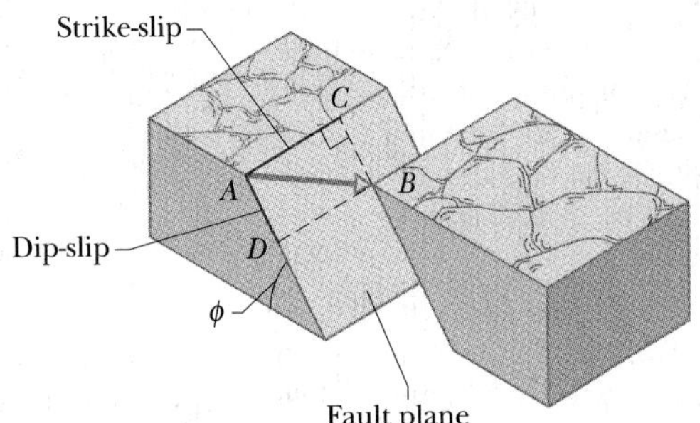

FIG. 3-36 Problem 45.

46 Two vectors $\vec{a}$ and $\vec{b}$ have the components, in meters, $a_x = 3.2$, $a_y = 1.6$, $b_x = 0.50$, $b_y = 4.5$. (a) Find the angle between the directions of $\vec{a}$ and $\vec{b}$. There are two vectors in the xy plane that are perpendicular to $\vec{a}$ and have a magnitude of 5.0 m. One, vector $\vec{c}$, has a positive x component and the other, vector $\vec{d}$, a negative x component. What are (b) the x component and (c) the y component of $\vec{c}$, and (d) the x component and (e) the y component of vector $\vec{d}$?

47 A vector $\vec{a}$ of magnitude 10 units and another vector $\vec{b}$ of magnitude 6.0 units differ in directions by 60°. Find (a) the scalar product of the two vectors and (b) the magnitude of the vector product $\vec{a} \times \vec{b}$. SSM

48 Vector $\vec{a}$ has a magnitude of 5.0 m and is directed east. Vector $\vec{b}$ has a magnitude of 4.0 m and is directed 35° west of due north. What are (a) the magnitude and (b) the direction of $\vec{a} + \vec{b}$? What are (c) the magnitude and (d) the direction of $\vec{b} - \vec{a}$? (e) Draw a vector diagram for each combination.

49 A particle undergoes three successive displacements in a plane, as follows: $\vec{d}_1$, 4.00 m southwest; then $\vec{d}_2$, 5.00 m east; and finally $\vec{d}_3$, 6.00 m in a direction 60.0° north of east. Choose a coordinate system with the y axis pointing north and the x axis pointing east. What are (a) the x component and (b) the y component of $\vec{d}_1$? What are (c) the x component and (d) the y component of $\vec{d}_2$? What are (e) the x component and (f) the y component of $\vec{d}_3$? Next, consider the *net* displacement of the particle for the three successive displacements. What are (g) the x component, (h) the y component, (i) the magnitude, and (j) the direction of the net displacement? If the particle is to return directly to the starting point, (k) how far and (l) in what direction should it move?

50 For the vectors in Fig. 3-34, with $a = 4$, $b = 3$, and $c = 5$, calculate (a) $\vec{a} \cdot \vec{b}$, (b) $\vec{a} \cdot \vec{c}$, and (c) $\vec{b} \cdot \vec{c}$.

51 A sailboat sets out from the U.S. side of Lake Erie for a point on the Canadian side, 90.0 km due north. The sailor, however, ends up 50.0 km due east of the starting point. (a) How far and (b) in what direction must the sailor now sail to reach the original destination? SSM

52 Find the sum of the following four vectors in (a) unit-vector notation, and as (b) a magnitude and (c) an angle relative to $+x$.

$\vec{P}$: 10.0 m, at 25.0° counterclockwise from $+x$

$\vec{Q}$: 12.0 m, at 10.0° counterclockwise from $+y$

$\vec{R}$: 8.00 m, at 20.0° clockwise from $-y$

$\vec{S}$: 9.00 m, at 40.0° counterclockwise from $-y$

53 Vectors $\vec{A}$ and $\vec{B}$ lie in an xy plane. $\vec{A}$ has magnitude 8.00 and angle 130°; $\vec{B}$ has components $B_x = -7.72$ and $B_y = -9.20$. What are the angles between the negative direction of the y axis and (a) the direction of $\vec{A}$, (b) the direction of the product $\vec{A} \times \vec{B}$, and (c) the direction of $\vec{A} \times (\vec{B} + 3.00\hat{k})$?

54 Here are three displacements, each in meters: $\vec{d}_1 = 4.0\hat{i} + 5.0\hat{j} - 6.0\hat{k}$, $\vec{d}_2 = -1.0\hat{i} + 2.0\hat{j} + 3.0\hat{k}$, and $\vec{d}_3 = 4.0\hat{i} + 3.0\hat{j} + 2.0\hat{k}$. (a) What is $\vec{r} = \vec{d}_1 - \vec{d}_2 + \vec{d}_3$? (b) What is the angle between $\vec{r}$ and the positive z axis? (c) What is the component of $\vec{d}_1$ along the direction of $\vec{d}_2$? (d) What is the component of $\vec{d}_1$ that is perpendicular to the direction of $\vec{d}_2$ and in the plane of $\vec{d}_1$ and $\vec{d}_2$? (*Hint:* For (c), consider Eq. 3-20 and Fig. 3-20; for (d), consider Eq. 3-27.)

55 Vectors $\vec{A}$ and $\vec{B}$ lie in an xy plane. $\vec{A}$ has magnitude 8.00 and angle 130°; $\vec{B}$ has components $B_x = -7.72$ and $B_y = -9.20$. (a) What is $5\vec{A} \cdot \vec{B}$? What is $4\vec{A} \times 3\vec{B}$ in (b) unit-vector notation and (c) magnitude-angle notation with spherical coordinates (see Fig. 3-37)? (d) What is the angle between the directions of $\vec{A}$ and $4\vec{A} \times 3\vec{B}$? (*Hint:* Think a bit before you resort to a calculation.) What is $\vec{A} + 3.00\hat{k}$ in (e) unit-vector notation and (f) magnitude-angle notation with spherical coordinates?

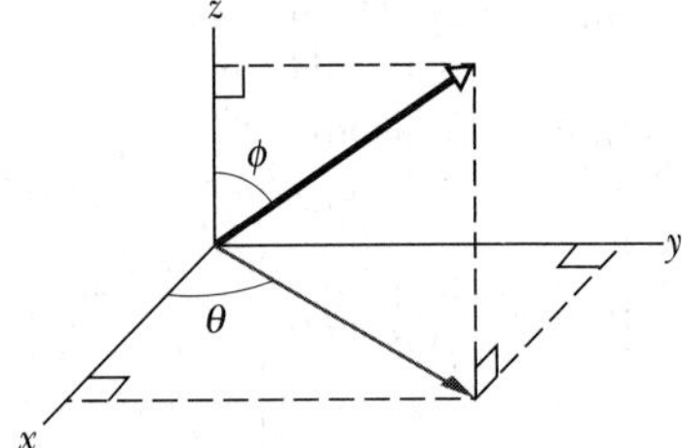

FIG. 3-37 Problem 55.

56 Vector $\vec{d}_1$ is in the negative direction of a y axis, and vector $\vec{d}_2$ is in the positive direction of an x axis. What are the directions of (a) $\vec{d}_2/4$ and (b) $\vec{d}_1/(-4)$? What are the magnitudes of products (c) $\vec{d}_1 \cdot \vec{d}_2$ and (d) $\vec{d}_1 \cdot (\vec{d}_2/4)$? What is the direction of the vector resulting from (e) $\vec{d}_1 \times \vec{d}_2$ and (f) $\vec{d}_2 \times \vec{d}_1$? What is the magnitude of the vector product in (g) part (e) and (h) part (f)? What are the (i) magnitude and (j) direction of $\vec{d}_1 \times (\vec{d}_2/4)$?

57 Here are three vectors in meters:

$$\vec{d}_1 = -3.0\hat{\mathrm{i}} + 3.0\hat{\mathrm{j}} + 2.0\hat{\mathrm{k}}$$
$$\vec{d}_2 = -2.0\hat{\mathrm{i}} - 4.0\hat{\mathrm{j}} + 2.0\hat{\mathrm{k}}$$
$$\vec{d}_3 = 2.0\hat{\mathrm{i}} + 3.0\hat{\mathrm{j}} + 1.0\hat{\mathrm{k}}.$$

What results from (a) $\vec{d}_1 \cdot (\vec{d}_2 + \vec{d}_3)$, (b) $\vec{d}_1 \cdot (\vec{d}_2 \times \vec{d}_3)$, and (c) $\vec{d}_1 \times (\vec{d}_2 + \vec{d}_3)$?

58 A golfer takes three putts to get the ball into the hole. The first putt displaces the ball 3.66 m north, the second 1.83 m southeast, and the third 0.91 m southwest. What are (a) the magnitude and (b) the direction of the displacement needed to get the ball into the hole on the first putt?

59 Consider $\vec{a}$ in the positive direction of x, $\vec{b}$ in the positive direction of y, and a scalar d. What is the direction of $\vec{b}/d$ if d is (a) positive and (b) negative? What is the magnitude of (c) $\vec{a} \cdot \vec{b}$ and (d) $\vec{a} \cdot \vec{b}/d$? What is the direction of the vector resulting from (e) $\vec{a} \times \vec{b}$ and (f) $\vec{b} \times \vec{a}$? (g) What is the magnitude of the vector product in (e)? (h) What is the magnitude of the vector product in (f)? What are (i) the magnitude and (j) the direction of $\vec{a} \times \vec{b}/d$ if d is positive?

60 A vector $\vec{d}$ has a magnitude of 2.5 m and points north. What are (a) the magnitude and (b) the direction of $4.0\vec{d}$? What are (c) the magnitude and (d) the direction of $-3.0\vec{d}$?

61 Let $\hat{\mathrm{i}}$ be directed to the east, $\hat{\mathrm{j}}$ be directed to the north, and $\hat{\mathrm{k}}$ be directed upward. What are the values of products (a) $\hat{\mathrm{i}} \cdot \hat{\mathrm{k}}$, (b) $(-\hat{\mathrm{k}}) \cdot (-\hat{\mathrm{j}})$, and (c) $\hat{\mathrm{j}} \cdot (-\hat{\mathrm{j}})$? What are the directions (such as east or down) of products (d) $\hat{\mathrm{k}} \times \hat{\mathrm{j}}$, (e) $(-\hat{\mathrm{i}}) \times (-\hat{\mathrm{j}})$, and (f) $(-\hat{\mathrm{k}}) \times (-\hat{\mathrm{j}})$?

62 Consider two displacements, one of magnitude 3 m and another of magnitude 4 m. Show how the displacement vectors may be combined to get a resultant displacement of magnitude (a) 7 m, (b) 1 m, and (c) 5 m.

63 A bank in downtown Boston is robbed (see the map in Fig. 3-38). To elude police, the robbers escape by helicopter, making three successive flights described by the following displacements: 32 km, 45° south of east; 53 km, 26° north of west; 26 km, 18° east of south. At the end of the third flight they are captured. In what town are they apprehended?

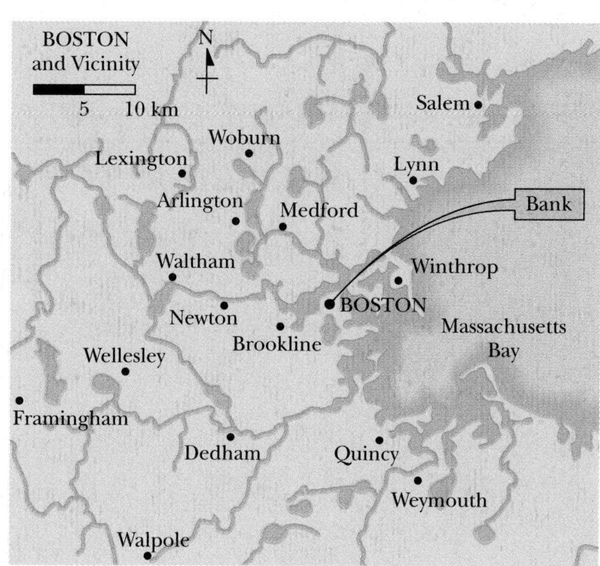

FIG. 3-38 Problem 63.

64 A wheel with a radius of 45.0 cm rolls without slipping along a horizontal floor (Fig. 3-39). At time t_1, the dot P painted on the rim of the wheel is at the point of contact between the wheel and the floor. At a later time t_2, the wheel has rolled through one-half of a revolution. What are (a) the magnitude and (b) the angle (relative to the floor) of the displacement of P?

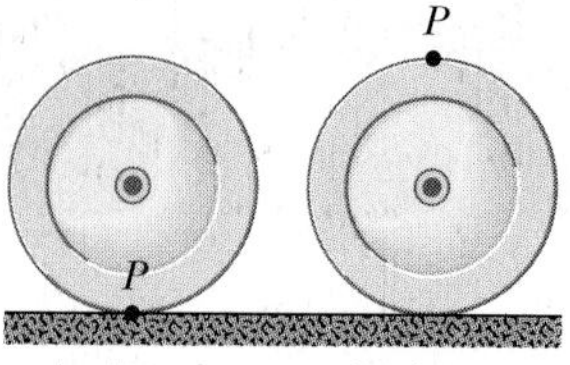

FIG. 3-39 Problem 64.

65 $\vec{A}$ has the magnitude 12.0 m and is angled 60.0° counterclockwise from the positive direction of the x axis of an xy coordinate system. Also, $\vec{B} = (12.0\text{ m})\hat{\mathrm{i}} + (8.00\text{ m})\hat{\mathrm{j}}$ on that same coordinate system. We now rotate the system counterclockwise about the origin by 20.0° to form an $x'y'$ system. On this new system, what are (a) $\vec{A}$ and (b) $\vec{B}$, both in unit-vector notation?

66 A woman walks 250 m in the direction 30° east of north, then 175 m directly east. Find (a) the magnitude and (b) the angle of her final displacement from the starting point. (c) Find the distance she walks. (d) Which is greater, that distance or the magnitude of her displacement?

67 (a) In unit-vector notation, what is $\vec{r} = \vec{a} - \vec{b} + \vec{c}$ if $\vec{a} = 5.0\hat{\mathrm{i}} + 4.0\hat{\mathrm{j}} - 6.0\hat{\mathrm{k}}$, $\vec{b} = -2.0\hat{\mathrm{i}} + 2.0\hat{\mathrm{j}} + 3.0\hat{\mathrm{k}}$, and $\vec{c} = 4.0\hat{\mathrm{i}} + 3.0\hat{\mathrm{j}} + 2.0\hat{\mathrm{k}}$? (b) Calculate the angle between $\vec{r}$ and the positive z axis. (c) What is the component of $\vec{a}$ along the direction of $\vec{b}$? (d) What is the component of $\vec{a}$ perpendicular to the direction of $\vec{b}$ but in the plane of $\vec{a}$ and $\vec{b}$? (*Hint:* For (c), see Eq. 3-20 and Fig. 3-20; for (d), see Eq. 3-27.)

68 If $\vec{a} - \vec{b} = 2\vec{c}$, $\vec{a} + \vec{b} = 4\vec{c}$, and $\vec{c} = 3\hat{\mathrm{i}} + 4\hat{\mathrm{j}}$, then what are (a) $\vec{a}$ and (b) $\vec{b}$?

69 A protester carries his sign of protest, starting from the origin of an xyz coordinate system, with the xy plane horizontal. He moves 40 m in the negative direction of the x axis, then 20 m along a perpendicular path to his left, and then 25 m up a water tower. (a) In unit-vector notation, what is the displacement of the sign from start to end? (b) The sign then falls to the foot of the tower. What is the magnitude of the displacement of the sign from start to this new end?

70 A vector $\vec{d}$ has a magnitude 3.0 m and is directed south. What are (a) the magnitude and (b) the direction of the vector $5.0\vec{d}$? What are (c) the magnitude and (d) the direction of the vector $-2.0\vec{d}$?

71 If $\vec{B}$ is added to $\vec{A}$, the result is $6.0\hat{\mathrm{i}} + 1.0\hat{\mathrm{j}}$. If $\vec{B}$ is subtracted from $\vec{A}$, the result is $-4.0\hat{\mathrm{i}} + 7.0\hat{\mathrm{j}}$. What is the magnitude of $\vec{A}$? SSM

72 A fire ant, searching for hot sauce in a picnic area, goes through three displacements along level ground: $\vec{d}_1$ for 0.40 m southwest (that is, at 45° from directly south and from directly west), $\vec{d}_2$ for 0.50 m due east, $\vec{d}_3$ for 0.60 m at 60° north of east. Let the positive x direction be east and the positive y direction be north. What are (a) the x component and (b) the y component of $\vec{d}_1$? What are (c) the x component and (d) the y component of $\vec{d}_2$? What are (e) the x component and (f) the y component of $\vec{d}_3$?

What are (g) the x component, (h) the y component, (i) the magnitude, and (j) the direction of the ant's net displacement? If the ant is to return directly to the starting point, (k) how far and (l) in what direction should it move?

4 Motion in Two and Three Dimensions

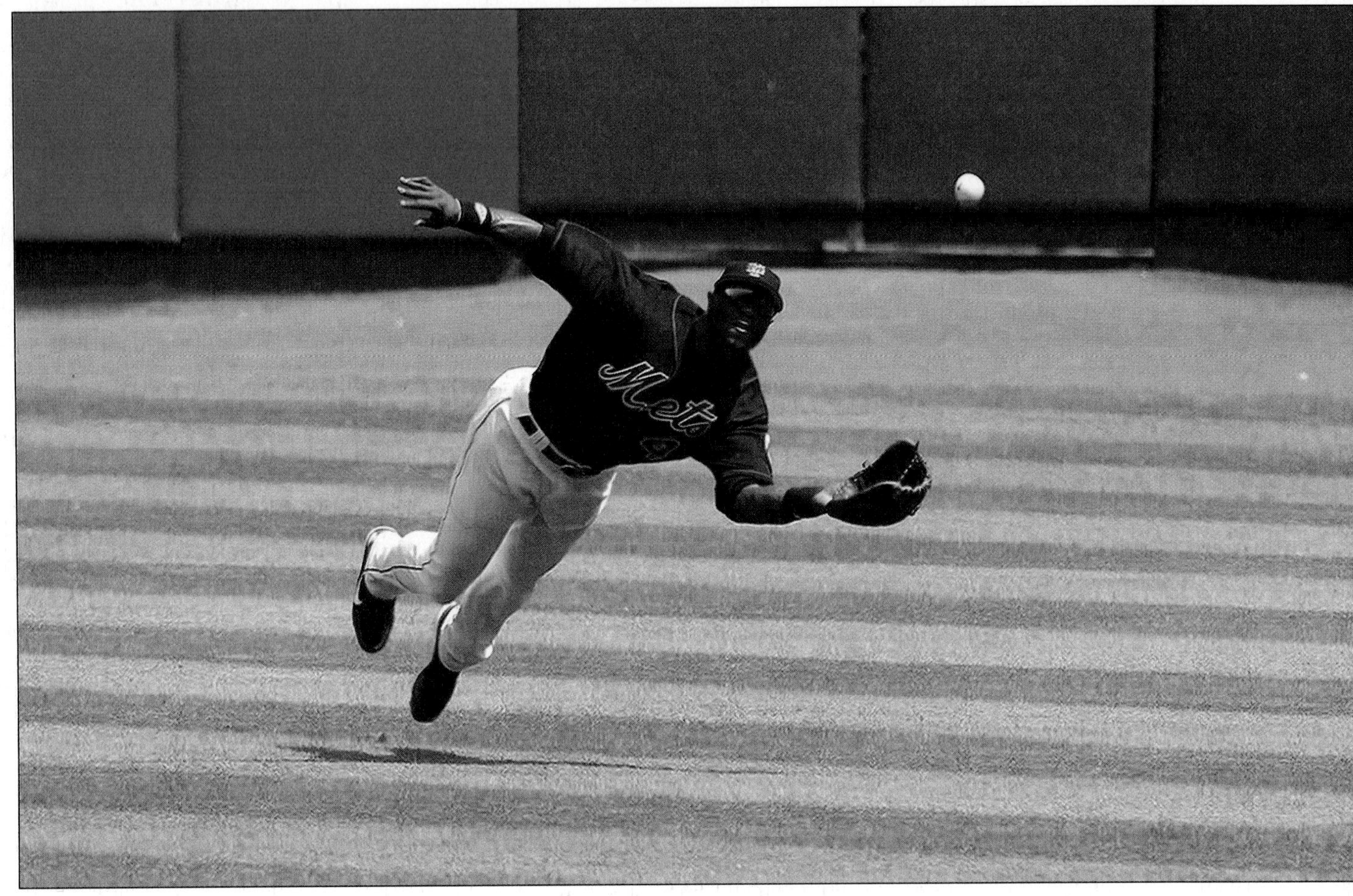

Source: Rob Tringali/Sports Chrome Inc.

When a high fly ball is hit to the outfield, how does the outfielder in the area know where to be in order to catch it? Often the outfielder will jog or run at a measured pace to the catch site, arriving just as the ball does. Playing experience surely helps, but some other factor seems to be involved.

What clue is hidden in the ball's motion?

The answer is in this chapter.

4-1 WHAT IS PHYSICS?

In this chapter we continue looking at the aspect of physics that analyzes motion, but now the motion can be in two or three dimensions. For example, medical researchers and aeronautical engineers might concentrate on the physics of the two- and three-dimensional turns taken by fighter pilots in dogfights because a modern high-performance jet can take a tight turn so quickly that the pilot immediately loses consciousness. A sports engineer might focus on the physics of basketball. For example, in a *free throw* (where a player gets an uncontested shot at the basket from about 4.3 m), a player might employ the *overhand push shot,* in which the ball is pushed away from about shoulder height and then released. Or the player might use an *underhand loop shot,* in which the ball is brought upward from about the belt-line level and released. The first technique is the overwhelming choice among professional players, but the legendary Rick Barry set the record for free-throw shooting with the underhand technique.

Motion in three dimensions is not easy to understand. For example, you are probably good at driving a car along a freeway (one-dimensional motion) but would probably have a difficult time in landing an airplane on a runway (three-dimensional motion) without a lot of training.

In our study of two- and three-dimensional motion, we start with position and displacement.

4-2 | Position and Displacement

One general way of locating a particle (or particle-like object) is with a **position vector** $\vec{r}$, which is a vector that extends from a reference point (usually the origin) to the particle. In the unit-vector notation of Section 3-5, $\vec{r}$ can be written

$$\vec{r} = x\hat{\text{i}} + y\hat{\text{j}} + z\hat{\text{k}}, \tag{4-1}$$

where $x\hat{\text{i}}$, $y\hat{\text{j}}$, and $z\hat{\text{k}}$ are the vector components of $\vec{r}$ and the coefficients x, y, and z are its scalar components.

The coefficients x, y, and z give the particle's location along the coordinate axes and relative to the origin; that is, the particle has the rectangular coordinates (x, y, z). For instance, Fig. 4-1 shows a particle with position vector

$$\vec{r} = (-3\text{ m})\hat{\text{i}} + (2\text{ m})\hat{\text{j}} + (5\text{ m})\hat{\text{k}}$$

and rectangular coordinates $(-3\text{ m}, 2\text{ m}, 5\text{ m})$. Along the x axis the particle is 3 m from the origin, in the $-\hat{\text{i}}$ direction. Along the y axis it is 2 m from the origin, in the $+\hat{\text{j}}$ direction. Along the z axis it is 5 m from the origin, in the $+\hat{\text{k}}$ direction.

As a particle moves, its position vector changes in such a way that the vector always extends to the particle from the reference point (the origin). If the position vector changes—say, from $\vec{r}_1$ to $\vec{r}_2$ during a certain time interval—then the particle's **displacement** $\Delta\vec{r}$ during that time interval is

$$\Delta\vec{r} = \vec{r}_2 - \vec{r}_1. \tag{4-2}$$

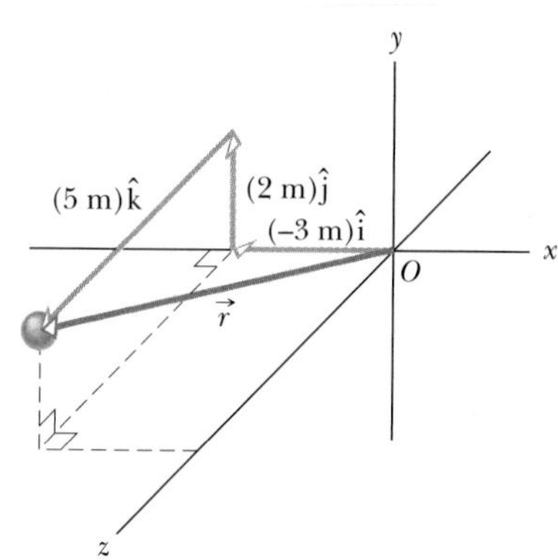

FIG. 4-1 The position vector $\vec{r}$ for a particle is the vector sum of its vector components.

Using the unit-vector notation of Eq. 4-1, we can rewrite this displacement as

$$\Delta\vec{r} = (x_2\hat{\text{i}} + y_2\hat{\text{j}} + z_2\hat{\text{k}}) - (x_1\hat{\text{i}} + y_1\hat{\text{j}} + z_1\hat{\text{k}})$$

or as

$$\Delta\vec{r} = (x_2 - x_1)\hat{\text{i}} + (y_2 - y_1)\hat{\text{j}} + (z_2 - z_1)\hat{\text{k}}, \tag{4-3}$$

where coordinates (x_1, y_1, z_1) correspond to position vector $\vec{r}_1$ and coordinates (x_2, y_2, z_2) correspond to position vector $\vec{r}_2$. We can also rewrite the displacement by substituting Δx for $(x_2 - x_1)$, Δy for $(y_2 - y_1)$, and Δz for $(z_2 - z_1)$:

$$\Delta\vec{r} = \Delta x\hat{\text{i}} + \Delta y\hat{\text{j}} + \Delta z\hat{\text{k}}. \tag{4-4}$$

Sample Problem 4-1

In Fig. 4-2, the position vector for a particle initially is

$$\vec{r}_1 = (-3.0\text{ m})\hat{\text{i}} + (2.0\text{ m})\hat{\text{j}} + (5.0\text{ m})\hat{\text{k}}$$

and then later is

$$\vec{r}_2 = (9.0\text{ m})\hat{\text{i}} + (2.0\text{ m})\hat{\text{j}} + (8.0\text{ m})\hat{\text{k}}.$$

What is the particle's displacement $\Delta\vec{r}$ from $\vec{r}_1$ to $\vec{r}_2$?

KEY IDEA The displacement $\Delta\vec{r}$ is obtained by subtracting the initial $\vec{r}_1$ from the later $\vec{r}_2$.

Calculation: The subtraction gives us

$$\begin{aligned}\Delta\vec{r} &= \vec{r}_2 - \vec{r}_1\\ &= [9.0 - (-3.0)]\hat{\text{i}} + [2.0 - 2.0]\hat{\text{j}} + [8.0 - 5.0]\hat{\text{k}}\\ &= (12\text{ m})\hat{\text{i}} + (3.0\text{ m})\hat{\text{k}}. \qquad \text{(Answer)}\end{aligned}$$

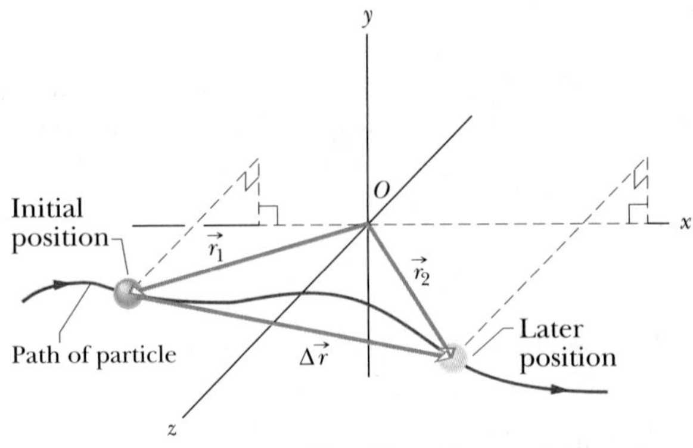

FIG. 4-2 The displacement $\Delta\vec{r} = \vec{r}_2 - \vec{r}_1$ extends from the head of the initial position vector $\vec{r}_1$ to the head of the later position vector $\vec{r}_2$.

This displacement vector is parallel to the xz plane because it lacks a y component.

Sample Problem 4-2

A rabbit runs across a parking lot on which a set of coordinate axes has, strangely enough, been drawn. The coordinates (meters) of the rabbit's position as functions of time t (seconds) are given by

$$x = -0.31t^2 + 7.2t + 28 \qquad (4\text{-}5)$$

and

$$y = 0.22t^2 - 9.1t + 30. \qquad (4\text{-}6)$$

(a) At $t = 15$ s, what is the rabbit's position vector $\vec{r}$ in unit-vector notation and in magnitude-angle notation?

KEY IDEA The x and y coordinates of the rabbit's position, as given by Eqs. 4-5 and 4-6, are the scalar components of the rabbit's position vector $\vec{r}$.

Calculations: We can write

$$\vec{r}(t) = x(t)\hat{\text{i}} + y(t)\hat{\text{j}}. \qquad (4\text{-}7)$$

(We write $\vec{r}(t)$ rather than $\vec{r}$ because the components are functions of t, and thus $\vec{r}$ is also.)

At $t = 15$ s, the scalar components are

$$x = (-0.31)(15)^2 + (7.2)(15) + 28 = 66\text{ m}$$

and

$$y = (0.22)(15)^2 - (9.1)(15) + 30 = -57\text{ m},$$

so

$$\vec{r} = (66\text{ m})\hat{\text{i}} - (57\text{ m})\hat{\text{j}}, \qquad \text{(Answer)}$$

which is drawn in Fig. 4-3a. To get the magnitude and angle of $\vec{r}$, we use Eq. 3-6:

$$\begin{aligned}r &= \sqrt{x^2 + y^2} = \sqrt{(66\text{ m})^2 + (-57\text{ m})^2}\\ &= 87\text{ m}, \qquad \text{(Answer)}\end{aligned}$$

and

$$\theta = \tan^{-1}\frac{y}{x} = \tan^{-1}\left(\frac{-57\text{ m}}{66\text{ m}}\right) = -41°. \qquad \text{(Answer)}$$

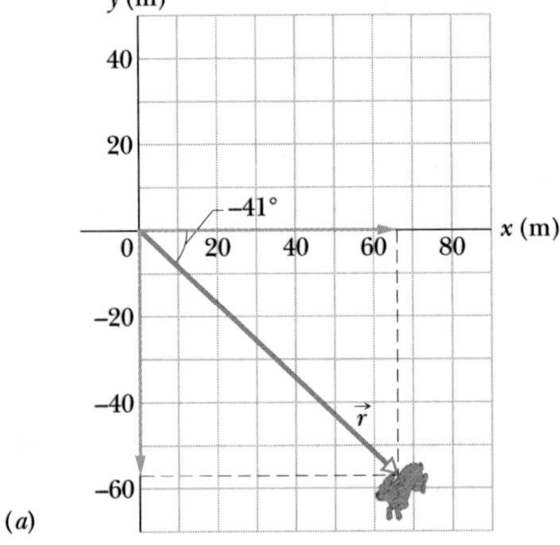

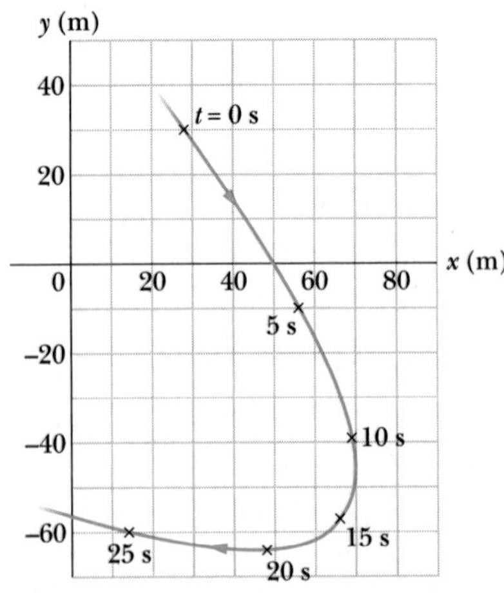

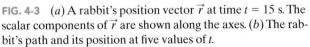

FIG. 4-3 (a) A rabbit's position vector $\vec{r}$ at time $t = 15$ s. The scalar components of $\vec{r}$ are shown along the axes. (b) The rabbit's path and its position at five values of t.

Check: Although $\theta = 139°$ has the same tangent as $-41°$, the components of $\vec{r}$ indicate that the desired angle is $139° - 180° = -41°$.

(b) Graph the rabbit's path for $t = 0$ to $t = 25$ s.

Graphing: We can repeat part (a) for several values of t and then plot the results. Figure 4-3*b* shows the plots for five values of t and the path connecting them. We can also plot Eqs. 4-5 and 4-6 on a calculator.

4-3 | Average Velocity and Instantaneous Velocity

If a particle moves from one point to another, we might need to know how fast it moves. Just as in Chapter 2, we can define two quantities that deal with "how fast": *average velocity* and *instantaneous velocity.* However, here we must consider these quantities as vectors and use vector notation.

If a particle moves through a displacement $\Delta\vec{r}$ in a time interval Δt, then its **average velocity** $\vec{v}_{avg}$ is

$$\text{average velocity} = \frac{\text{displacement}}{\text{time interval}},$$

or

$$\vec{v}_{avg} = \frac{\Delta\vec{r}}{\Delta t}. \tag{4-8}$$

This tells us that the direction of $\vec{v}_{avg}$ (the vector on the left side of Eq. 4-8) must be the same as that of the displacement $\Delta\vec{r}$ (the vector on the right side). Using Eq. 4-4, we can write Eq. 4-8 in vector components as

$$\vec{v}_{avg} = \frac{\Delta x\hat{\mathrm{i}} + \Delta y\hat{\mathrm{j}} + \Delta z\hat{\mathrm{k}}}{\Delta t} = \frac{\Delta x}{\Delta t}\hat{\mathrm{i}} + \frac{\Delta y}{\Delta t}\hat{\mathrm{j}} + \frac{\Delta z}{\Delta t}\hat{\mathrm{k}}. \tag{4-9}$$

For example, if the particle in Sample Problem 4-1 moves from its initial position to its later position in 2.0 s, then its average velocity during that move is

$$\vec{v}_{avg} = \frac{\Delta\vec{r}}{\Delta t} = \frac{(12\text{ m})\hat{\mathrm{i}} + (3.0\text{ m})\hat{\mathrm{k}}}{2.0\text{ s}} = (6.0\text{ m/s})\hat{\mathrm{i}} + (1.5\text{ m/s})\hat{\mathrm{k}}.$$

That is, the average velocity (a vector quantity) has a component of 6.0 m/s along the x axis and a component of 1.5 m/s along the z axis.

When we speak of the **velocity** of a particle, we usually mean the particle's **instantaneous velocity** $\vec{v}$ at some instant. This $\vec{v}$ is the value that $\vec{v}_{avg}$ approaches in the limit as we shrink the time interval Δt to 0 about that instant. Using the language of calculus, we may write $\vec{v}$ as the derivative

$$\vec{v} = \frac{d\vec{r}}{dt}. \tag{4-10}$$

Figure 4-4 shows the path of a particle that is restricted to the xy plane. As the particle travels to the right along the curve, its position vector sweeps to the right. During time interval Δt, the position vector changes from $\vec{r}_1$ to $\vec{r}_2$ and the particle's displacement is $\Delta\vec{r}$.

To find the instantaneous velocity of the particle at, say, instant t_1 (when the particle is at position 1), we shrink interval Δt to 0 about t_1. Three things happen as we do so. (1) Position vector $\vec{r}_2$ in Fig. 4-4 moves toward $\vec{r}_1$ so that $\Delta\vec{r}$ shrinks toward zero. (2) The direction of $\Delta\vec{r}/\Delta t$ (and thus of $\vec{v}_{avg}$) approaches the direction of the line tangent to the particle's path at position 1. (3) The average velocity $\vec{v}_{avg}$ approaches the instantaneous velocity $\vec{v}$ at t_1.

In the limit as $\Delta t \to 0$, we have $\vec{v}_{avg} \to \vec{v}$ and, most important here, $\vec{v}_{avg}$ takes on the direction of the tangent line. Thus, $\vec{v}$ has that direction as well:

> The direction of the instantaneous velocity $\vec{v}$ of a particle is always tangent to the particle's path at the particle's position.

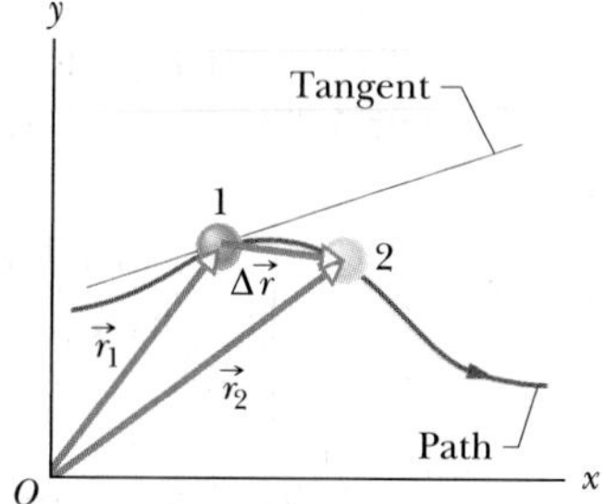

FIG. 4-4 The displacement $\Delta\vec{r}$ of a particle during a time interval Δt, from position 1 with position vector $\vec{r}_1$ at time t_1 to position 2 with position vector $\vec{r}_2$ at time t_2. The tangent to the particle's path at position 1 is shown.

The result is the same in three dimensions: $\vec{v}$ is always tangent to the particle's path.

To write Eq. 4-10 in unit-vector form, we substitute for $\vec{r}$ from Eq. 4-1:

$$\vec{v} = \frac{d}{dt}(x\hat{\text{i}} + y\hat{\text{j}} + z\hat{\text{k}}) = \frac{dx}{dt}\hat{\text{i}} + \frac{dy}{dt}\hat{\text{j}} + \frac{dz}{dt}\hat{\text{k}}.$$

This equation can be simplified somewhat by writing it as

$$\vec{v} = v_x\hat{\text{i}} + v_y\hat{\text{j}} + v_z\hat{\text{k}}, \tag{4-11}$$

where the scalar components of $\vec{v}$ are

$$v_x = \frac{dx}{dt}, \quad v_y = \frac{dy}{dt}, \quad \text{and} \quad v_z = \frac{dz}{dt}. \tag{4-12}$$

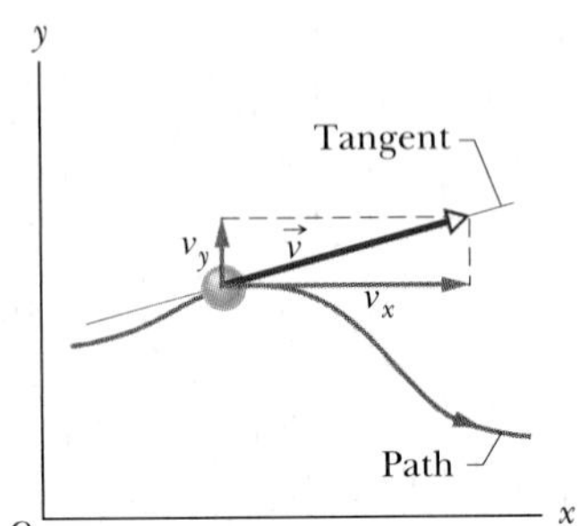

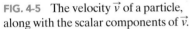

FIG. 4-5 The velocity $\vec{v}$ of a particle, along with the scalar components of $\vec{v}$.

For example, dx/dt is the scalar component of $\vec{v}$ along the x axis. Thus, we can find the scalar components of $\vec{v}$ by differentiating the scalar components of $\vec{r}$.

Figure 4-5 shows a velocity vector $\vec{v}$ and its scalar x and y components. Note that $\vec{v}$ is tangent to the particle's path at the particle's position. *Caution:* When a position vector is drawn, as in Figs. 4-1 through 4-4, it is an arrow that extends from one point (a "here") to another point (a "there"). However, when a velocity vector is drawn, as in Fig. 4-5, it does *not* extend from one point to another. Rather, it shows the instantaneous direction of travel of a particle at the tail, and its length (representing the velocity magnitude) can be drawn to any scale.

CHECKPOINT 1 The figure shows a circular path taken by a particle. If the instantaneous velocity of the particle is $\vec{v} = (2\text{ m/s})\hat{\text{i}} - (2\text{ m/s})\hat{\text{j}}$, through which quadrant is the particle moving at that instant if it is traveling (a) clockwise and (b) counterclockwise around the circle? For both cases, draw $\vec{v}$ on the figure.

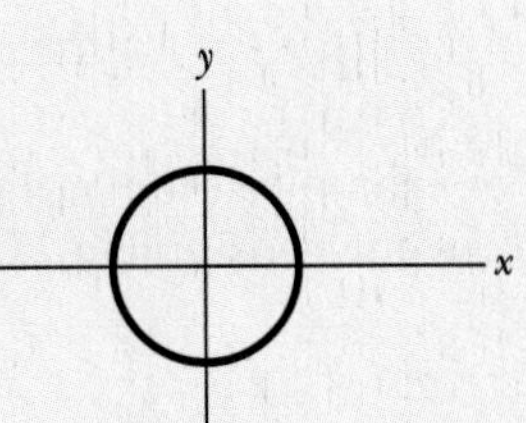

Sample Problem 4-3

For the rabbit in Sample Problem 4-2 find the velocity $\vec{v}$ at time $t = 15$ s.

KEY IDEA We can find $\vec{v}$ by taking derivatives of the components of the rabbit's position vector.

Calculations: Applying the v_x part of Eq. 4-12 to Eq. 4-5, we find the x component of $\vec{v}$ to be

$$v_x = \frac{dx}{dt} = \frac{d}{dt}(-0.31t^2 + 7.2t + 28)$$
$$= -0.62t + 7.2. \tag{4-13}$$

At $t = 15$ s, this gives $v_x = -2.1$ m/s. Similarly, applying the v_y part of Eq. 4-12 to Eq. 4-6, we find

$$v_y = \frac{dy}{dt} = \frac{d}{dt}(0.22t^2 - 9.1t + 30)$$
$$= 0.44t - 9.1. \tag{4-14}$$

At $t = 15$ s, this gives $v_y = -2.5$ m/s. Equation 4-11 then yields

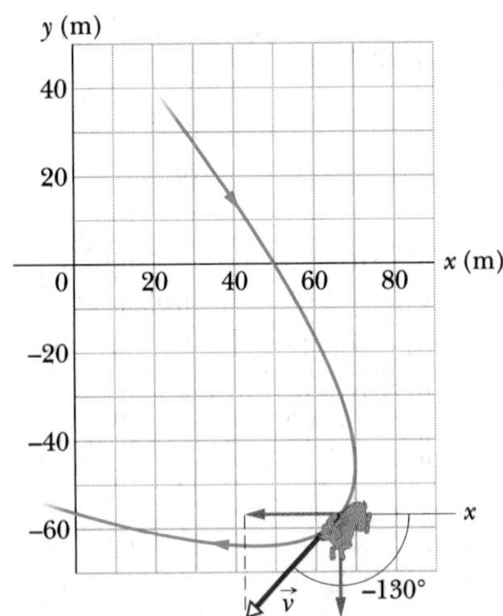

FIG. 4-6 The rabbit's velocity $\vec{v}$ at $t = 15$ s.

$$\vec{v} = (-2.1\text{ m/s})\hat{\text{i}} + (-2.5\text{ m/s})\hat{\text{j}}, \quad \text{(Answer)}$$

which is shown in Fig. 4-6, tangent to the rabbit's path and in the direction the rabbit is running at $t = 15$ s.

To get the magnitude and angle of $\vec{v}$, either we use a vector-capable calculator or we follow Eq. 3-6 to write

$$v = \sqrt{v_x^2 + v_y^2} = \sqrt{(-2.1\text{ m/s})^2 + (-2.5\text{ m/s})^2}$$
$$= 3.3\text{ m/s} \qquad \text{(Answer)}$$

and

$$\theta = \tan^{-1}\frac{v_y}{v_x} = \tan^{-1}\left(\frac{-2.5\text{ m/s}}{-2.1\text{ m/s}}\right)$$
$$= \tan^{-1}1.19 = -130°. \qquad \text{(Answer)}$$

Check: Is the angle $-130°$ or $-130° + 180° = 50°$?

4-4 | Average Acceleration and Instantaneous Acceleration

When a particle's velocity changes from $\vec{v}_1$ to $\vec{v}_2$ in a time interval Δt, its **average acceleration** $\vec{a}_{\text{avg}}$ during Δt is

$$\text{average acceleration} = \frac{\text{change in velocity}}{\text{time interval}},$$

or

$$\vec{a}_{\text{avg}} = \frac{\vec{v}_2 - \vec{v}_1}{\Delta t} = \frac{\Delta\vec{v}}{\Delta t}. \qquad (4\text{-}15)$$

If we shrink Δt to zero about some instant, then in the limit $\vec{a}_{\text{avg}}$ approaches the **instantaneous acceleration** (or **acceleration**) $\vec{a}$ at that instant; that is,

$$\vec{a} = \frac{d\vec{v}}{dt}. \qquad (4\text{-}16)$$

If the velocity changes in *either* magnitude *or* direction (or both), the particle must have an acceleration.

We can write Eq. 4-16 in unit-vector form by substituting Eq. 4-11 for $\vec{v}$ to obtain

$$\vec{a} = \frac{d}{dt}(v_x\hat{\imath} + v_y\hat{\jmath} + v_z\hat{k})$$
$$= \frac{dv_x}{dt}\hat{\imath} + \frac{dv_y}{dt}\hat{\jmath} + \frac{dv_z}{dt}\hat{k}.$$

We can rewrite this as

$$\vec{a} = a_x\hat{\imath} + a_y\hat{\jmath} + a_z\hat{k}, \qquad (4\text{-}17)$$

where the scalar components of $\vec{a}$ are

$$a_x = \frac{dv_x}{dt}, \quad a_y = \frac{dv_y}{dt}, \quad \text{and} \quad a_z = \frac{dv_z}{dt}. \qquad (4\text{-}18)$$

To find the scalar components of $\vec{a}$, we differentiate the scalar components of $\vec{v}$.

Figure 4-7 shows an acceleration vector $\vec{a}$ and its scalar components for a particle moving in two dimensions. *Caution:* When an acceleration vector is drawn, as in Fig. 4-7, it does *not* extend from one position to another. Rather, it shows the direction of acceleration for a particle located at its tail, and its length (representing the acceleration magnitude) can be drawn to any scale.

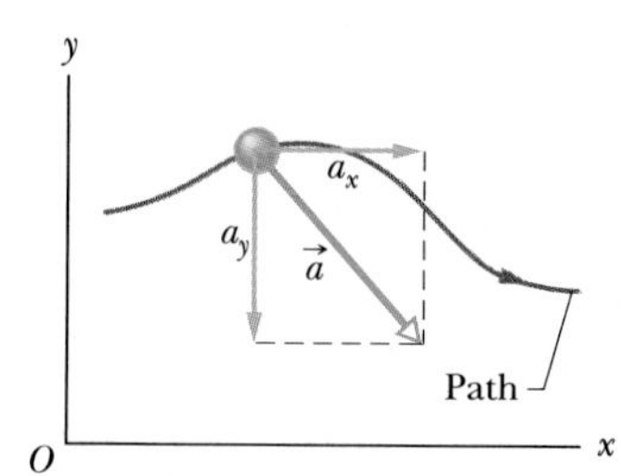

FIG. 4-7 The acceleration $\vec{a}$ of a particle and the scalar components of $\vec{a}$.

CHECKPOINT 2 Here are four descriptions of the position (in meters) of a puck as it moves in an xy plane:

(1) $x = -3t^2 + 4t - 2$ and $y = 6t^2 - 4t$

(2) $x = -3t^3 - 4t$ and $y = -5t^2 + 6$

(3) $\vec{r} = 2t^2\hat{\imath} - (4t + 3)\hat{\jmath}$

(4) $\vec{r} = (4t^3 - 2t)\hat{\imath} + 3\hat{\jmath}$

Are the x and y acceleration components constant? Is acceleration $\vec{a}$ constant?

Sample Problem 4-4

For the rabbit in Sample Problems 4-2 and 4-3, find the acceleration $\vec{a}$ at time $t = 15$ s.

KEY IDEA We can find $\vec{a}$ by taking derivatives of the rabbit's velocity components.

Calculations: Applying the a_x part of Eq. 4-18 to Eq. 4-13, we find the x component of $\vec{a}$ to be

$$a_x = \frac{dv_x}{dt} = \frac{d}{dt}(-0.62t + 7.2) = -0.62 \text{ m/s}^2.$$

Similarly, applying the a_y part of Eq. 4-18 to Eq. 4-14 yields the y component as

$$a_y = \frac{dv_y}{dt} = \frac{d}{dt}(0.44t - 9.1) = 0.44 \text{ m/s}^2.$$

We see that the acceleration does not vary with time (it is a constant) because the time variable t does not appear in the expression for either acceleration component. Equation 4-17 then yields

$$\vec{a} = (-0.62 \text{ m/s}^2)\hat{\text{i}} + (0.44 \text{ m/s}^2)\hat{\text{j}}, \quad \text{(Answer)}$$

which is superimposed on the rabbit's path in Fig. 4-8.

To get the magnitude and angle of $\vec{a}$, either we use a vector-capable calculator or we follow Eq. 3-6. For the magnitude we have

$$a = \sqrt{a_x^2 + a_y^2} = \sqrt{(-0.62 \text{ m/s}^2)^2 + (0.44 \text{ m/s}^2)^2}$$
$$= 0.76 \text{ m/s}^2. \quad \text{(Answer)}$$

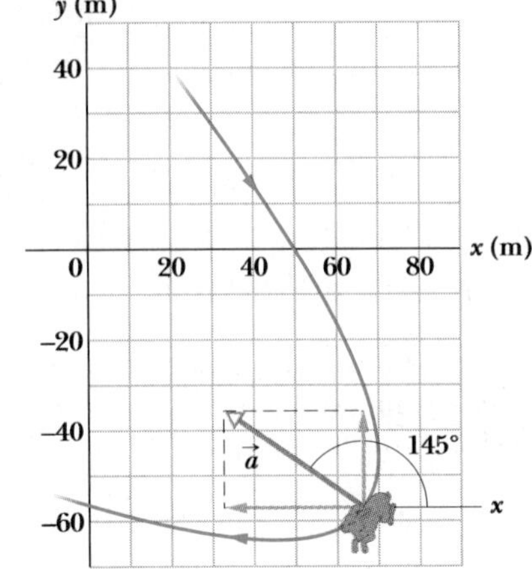

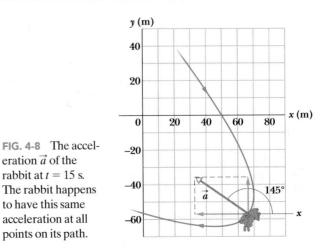

FIG. 4-8 The acceleration $\vec{a}$ of the rabbit at $t = 15$ s. The rabbit happens to have this same acceleration at all points on its path.

For the angle we have

$$\theta = \tan^{-1}\frac{a_y}{a_x} = \tan^{-1}\left(\frac{0.44 \text{ m/s}^2}{-0.62 \text{ m/s}^2}\right) = -35°.$$

However, this angle, which is the one displayed on a calculator, indicates that $\vec{a}$ is directed to the right and downward in Fig. 4-8. Yet, we know from the components that $\vec{a}$ must be directed to the left and upward. To find the other angle that has the same tangent as $-35°$ but is not displayed on a calculator, we add 180°:

$$-35° + 180° = 145°. \quad \text{(Answer)}$$

This *is* consistent with the components of $\vec{a}$. Note that $\vec{a}$ has the same magnitude and direction throughout the rabbit's run because the acceleration is constant.

Sample Problem 4-5

A particle with velocity $\vec{v}_0 = -2.0\hat{\text{i}} + 4.0\hat{\text{j}}$ (in meters per second) at $t = 0$ undergoes a constant acceleration $\vec{a}$ of magnitude $a = 3.0$ m/s² at an angle $\theta = 130°$ from the positive direction of the x axis. What is the particle's velocity $\vec{v}$ at $t = 5.0$ s?

KEY IDEA Because the acceleration is constant, Eq. 2-11 ($v = v_0 + at$) applies, but we must use it separately for motion parallel to the x axis and motion parallel to the y axis.

Calculations: We find the velocity components v_x and v_y from the equations

$$v_x = v_{0x} + a_x t \quad \text{and} \quad v_y = v_{0y} + a_y t.$$

In these equations, v_{0x} ($= -2.0$ m/s) and v_{0y} ($= 4.0$ m/s) are the x and y components of $\vec{v}_0$, and a_x and a_y are the x and y components of $\vec{a}$. To find a_x and a_y, we resolve $\vec{a}$ either with a vector-capable calculator or with Eq. 3-5:

$$a_x = a\cos\theta = (3.0 \text{ m/s}^2)(\cos 130°) = -1.93 \text{ m/s}^2,$$
$$a_y = a\sin\theta = (3.0 \text{ m/s}^2)(\sin 130°) = +2.30 \text{ m/s}^2.$$

When these values are inserted into the equations for v_x and v_y, we find that, at time $t = 5.0$ s,

$$v_x = -2.0 \text{ m/s} + (-1.93 \text{ m/s}^2)(5.0 \text{ s}) = -11.65 \text{ m/s},$$
$$v_y = 4.0 \text{ m/s} + (2.30 \text{ m/s}^2)(5.0 \text{ s}) = 15.50 \text{ m/s}.$$

Thus, at $t = 5.0$ s, we have, after rounding,

$$\vec{v} = (-12 \text{ m/s})\hat{\text{i}} + (16 \text{ m/s})\hat{\text{j}}. \quad \text{(Answer)}$$

Either using a vector-capable calculator or following Eq. 3-6, we find that the magnitude and angle of $\vec{v}$ are

$$v = \sqrt{v_x^2 + v_y^2} = 19.4 \approx 19 \text{ m/s} \quad \text{(Answer)}$$

and

$$\theta = \tan^{-1}\frac{v_y}{v_x} = 127° \approx 130°. \quad \text{(Answer)}$$

Check: Does 127° appear on your calculator's display, or does $-53°$ appear? Now sketch the vector $\vec{v}$ with its components to see which angle is reasonable.

4-5 | Projectile Motion

We next consider a special case of two-dimensional motion: A particle moves in a vertical plane with some initial velocity $\vec{v}_0$ but its acceleration is always the free-fall acceleration $\vec{g}$, which is downward. Such a particle is called a **projectile** (meaning that it is projected or launched), and its motion is called **projectile motion.** A projectile might be a tennis ball (Fig. 4-9) or baseball in flight, but it is not an airplane or a duck in flight. Many sports (from golf and football to lacrosse and racquetball) involve the projectile motion of a ball, and much effort is spent in trying to control that motion for an advantage. For example, the racquetball player who discovered the Z-shot in the 1970s easily won his games because the ball's peculiar flight to the rear of the court always perplexed his opponents.

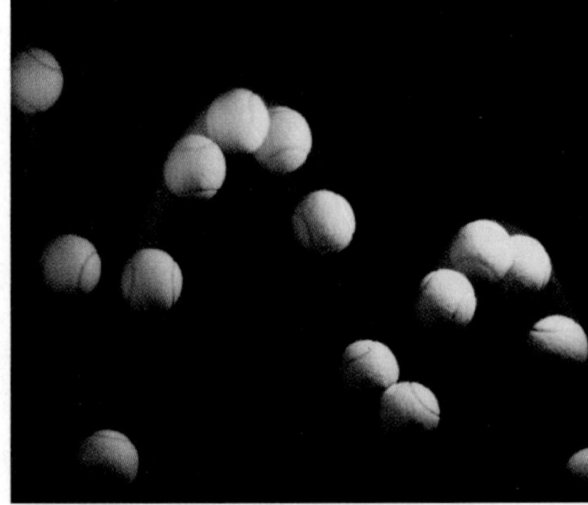

FIG. 4-9 A stroboscopic photograph of a yellow tennis ball bouncing off a hard surface. Between impacts, the ball has projectile motion. *Source:* Richard Megna/Fundamental Photographs.

Our goal here is to analyze projectile motion using the tools for two-dimensional motion described in Sections 4-2 through 4-4 and making the assumption that air has no effect on the projectile. Figure 4-10, which is analyzed in the next section, shows the path followed by a projectile when the air has no effect. The projectile is launched with an initial velocity $\vec{v}_0$ that can be written as

$$\vec{v}_0 = v_{0x}\hat{\mathrm{i}} + v_{0y}\hat{\mathrm{j}}. \tag{4-19}$$

The components v_{0x} and v_{0y} can then be found if we know the angle θ_0 between $\vec{v}_0$ and the positive x direction:

$$v_{0x} = v_0 \cos\theta_0 \quad \text{and} \quad v_{0y} = v_0 \sin\theta_0. \tag{4-20}$$

During its two-dimensional motion, the projectile's position vector $\vec{r}$ and velocity vector $\vec{v}$ change continuously, but its acceleration vector $\vec{a}$ is constant and *always* directed vertically downward. The projectile has *no* horizontal acceleration.

Projectile motion, like that in Figs. 4-9 and 4-10, looks complicated, but we have the following simplifying feature (known from experiment):

> In projectile motion, the horizontal motion and the vertical motion are independent of each other; that is, neither motion affects the other.

This feature allows us to break up a problem involving two-dimensional motion into two separate and easier one-dimensional problems, one for the horizontal motion (with *zero acceleration*) and one for the vertical motion (with *constant downward acceleration*). Here are two experiments that show that the horizontal motion and the vertical motion are independent.

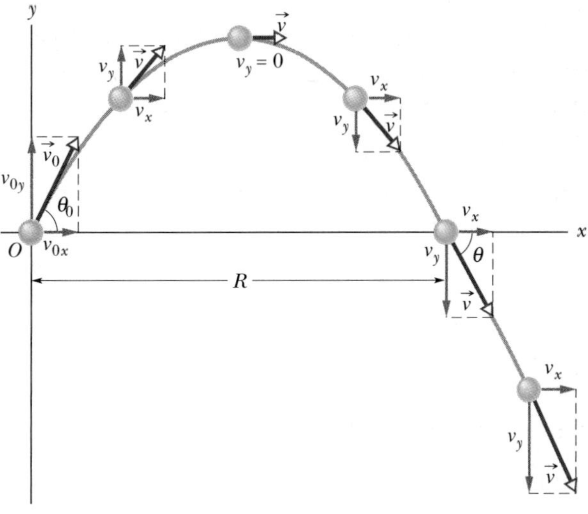

FIG. 4-10 The path of a projectile that is launched at $x_0 = 0$ and $y_0 = 0$, with an initial velocity $\vec{v}_0$. The initial velocity and the velocities at various points along its path are shown, along with their components. Note that the horizontal velocity component remains constant but the vertical velocity component changes continuously. The *range R* is the horizontal distance the projectile has traveled *when it returns to its launch height.*

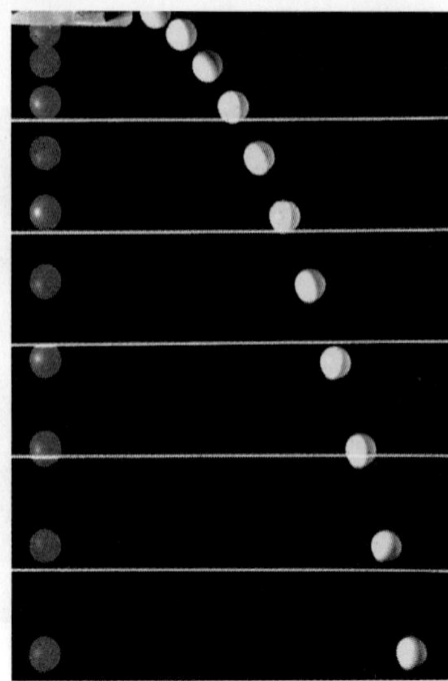

FIG. 4-11 One ball is released from rest at the same instant that another ball is shot horizontally to the right. Their vertical motions are identical. *Source:* Richard Megna/ Fundamental Photographs.

Two Golf Balls

Figure 4-11 is a stroboscopic photograph of two golf balls, one simply released and the other shot horizontally by a spring. The golf balls have the same vertical motion, both falling through the same vertical distance in the same interval of time. *The fact that one ball is moving horizontally while it is falling has no effect on its vertical motion;* that is, the horizontal and vertical motions are independent of each other.

A Great Student Rouser

Figure 4-12 shows a demonstration that has enlivened many a physics lecture. It involves a blowgun G, using a ball as a projectile. The target is a can suspended from a magnet M, and the tube of the blowgun is aimed directly at the can. The experiment is arranged so that the magnet releases the can just as the ball leaves the blowgun.

If g (the magnitude of the free-fall acceleration) were zero, the ball would follow the straight-line path shown in Fig. 4-12 and the can would float in place after the magnet released it. The ball would certainly hit the can.

However, g is *not* zero, but the ball *still* hits the can! As Fig. 4-12 shows, during the time of flight of the ball, both ball and can fall the same distance h from their zero-g locations. The harder the demonstrator blows, the greater is the ball's initial speed, the shorter the flight time, and the smaller the value of h.

CHECKPOINT 3 At a certain instant, a fly ball has velocity $\vec{v} = 25\hat{\imath} - 4.9\hat{\jmath}$ (the x axis is horizontal, the y axis is upward, and $\vec{v}$ is in meters per second). Has the ball passed its highest point?

4-6 | Projectile Motion Analyzed

Now we are ready to analyze projectile motion, horizontally and vertically.

The Horizontal Motion

Because there is *no acceleration* in the horizontal direction, the horizontal component v_x of the projectile's velocity remains unchanged from its initial value v_{0x} throughout the motion, as demonstrated in Fig. 4-13. At any time t, the projectile's horizontal displacement $x - x_0$ from an initial position x_0 is given by Eq. 2-15 with $a = 0$, which we write as

$$x - x_0 = v_{0x}t.$$

Because $v_{0x} = v_0 \cos \theta_0$, this becomes

$$x - x_0 = (v_0 \cos \theta_0)t. \tag{4-21}$$

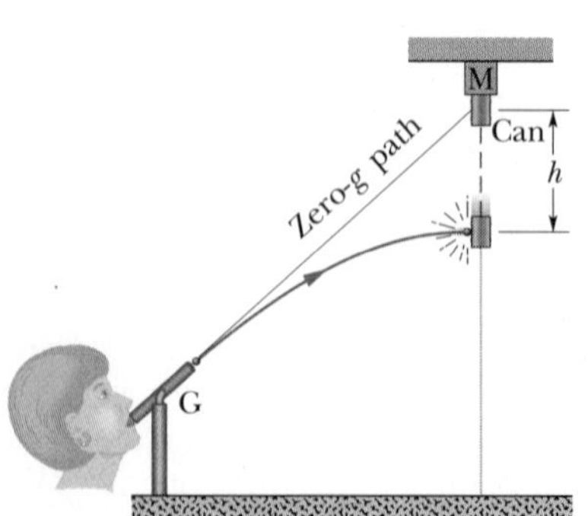

FIG. 4-12 The projectile ball always hits the falling can. Each falls a distance h from where it would be were there no free-fall acceleration.

The Vertical Motion

The vertical motion is the motion we discussed in Section 2-9 for a particle in free fall. Most important is that the acceleration is constant. Thus, the equations of Table 2-1 apply, provided we substitute $-g$ for a and switch to y notation. Then, for example, Eq. 2-15 becomes

$$\begin{aligned} y - y_0 &= v_{0y}t - \tfrac{1}{2}gt^2 \\ &= (v_0 \sin \theta_0)t - \tfrac{1}{2}gt^2, \end{aligned} \tag{4-22}$$

where the initial vertical velocity component v_{0y} is replaced with the equivalent $v_0 \sin \theta_0$. Similarly, Eqs. 2-11 and 2-16 become

$$v_y = v_0 \sin \theta_0 - gt \tag{4-23}$$

and

$$v_y^2 = (v_0 \sin \theta_0)^2 - 2g(y - y_0). \tag{4-24}$$

As is illustrated in Fig. 4-10 and Eq. 4-23, the vertical velocity component behaves just as for a ball thrown vertically upward. It is directed upward initially, and its magnitude steadily decreases to zero, *which marks the maximum height of the path.* The vertical velocity component then reverses direction, and its magnitude becomes larger with time.

FIG. 4-13 The vertical component of this skateboarder's velocity is changing but not the horizontal component, which matches the skateboard's velocity. As a result, the skateboard stays underneath him, allowing him to land on it. *Source:* Jamie Budge/Liaison/ Getty Images, Inc.

The Equation of the Path

We can find the equation of the projectile's path (its **trajectory**) by eliminating time t between Eqs. 4-21 and 4-22. Solving Eq. 4-21 for t and substituting into Eq. 4-22, we obtain, after a little rearrangement,

$$y = (\tan\theta_0)x - \frac{gx^2}{2(v_0\cos\theta_0)^2} \quad \text{(trajectory)}. \tag{4-25}$$

This is the equation of the path shown in Fig. 4-10. In deriving it, for simplicity we let $x_0 = 0$ and $y_0 = 0$ in Eqs. 4-21 and 4-22, respectively. Because g, θ_0, and v_0 are constants, Eq. 4-25 is of the form $y = ax + bx^2$, in which a and b are constants. This is the equation of a parabola, so the path is *parabolic*.

The Horizontal Range

The *horizontal range* R of the projectile, as Fig. 4-10 shows, is the *horizontal* distance the projectile has traveled when it returns to its initial (launch) height. To find range R, let us put $x - x_0 = R$ in Eq. 4-21 and $y - y_0 = 0$ in Eq. 4-22, obtaining

$$R = (v_0\cos\theta_0)t$$

and

$$0 = (v_0\sin\theta_0)t - \tfrac{1}{2}gt^2.$$

Eliminating t between these two equations yields

$$R = \frac{2v_0^2}{g}\sin\theta_0\cos\theta_0.$$

Using the identity $\sin 2\theta_0 = 2\sin\theta_0\cos\theta_0$ (see Appendix E), we obtain

$$R = \frac{v_0^2}{g}\sin 2\theta_0. \tag{4-26}$$

Caution: This equation does *not* give the horizontal distance traveled by a projectile when the final height is not the launch height.

Note that R in Eq. 4-26 has its maximum value when $\sin 2\theta_0 = 1$, which corresponds to $2\theta_0 = 90°$ or $\theta_0 = 45°$.

The horizontal range R is maximum for a launch angle of 45°.

However, when the launch and landing heights differ, as in shot put, hammer throw, and basketball, a launch angle of 45° does not yield the maximum horizontal distance.

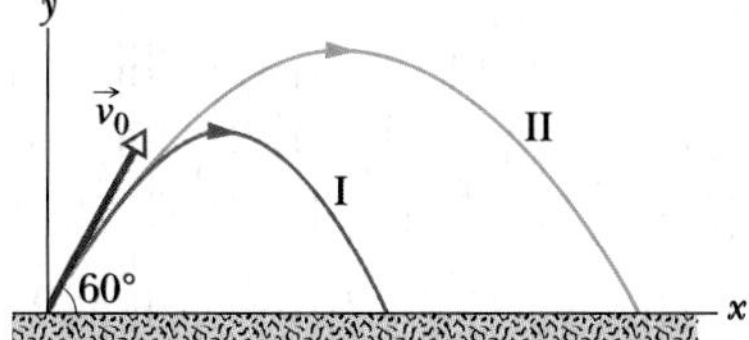

FIG. 4-14 (I) The path of a fly ball calculated by taking air resistance into account. (II) The path the ball would follow in a vacuum, calculated by the methods of this chapter. See Table 4-1 for corresponding data. (Adapted from "The Trajectory of a Fly Ball," by Peter J. Brancazio, *The Physics Teacher,* January 1985.)

The Effects of the Air

We have assumed that the air through which the projectile moves has no effect on its motion. However, in many situations, the disagreement between our calculations and the actual motion of the projectile can be large because the air resists (opposes) the motion. Figure 4-14, for example, shows two paths for a fly ball that leaves the bat at an angle of 60° with the horizontal and an initial speed of 44.7 m/s. Path I (the baseball player's fly ball) is a calculated path that approximates normal conditions of play, in air. Path II (the physics professor's fly ball) is the path the ball would follow in a vacuum.

TABLE 4-1

Two Fly Balls[a]

	Path I (Air)	Path II (Vacuum)
Range	98.5 m	177 m
Maximum height	53.0 m	76.8 m
Time of flight	6.6 s	7.9 s

[a]See Fig. 4-14. The launch angle is 60° and the launch speed is 44.7 m/s.

CHECKPOINT 4 A fly ball is hit to the outfield. During its flight (ignore the effects of the air), what happens to its (a) horizontal and (b) vertical components of velocity? What are the (c) horizontal and (d) vertical components of its acceleration during ascent, during descent, and at the topmost point of its flight?

Sample Problem 4-6 Build your skill

In Fig. 4-15, a rescue plane flies at 198 km/h (= 55.0 m/s) and constant height $h = 500$ m toward a point directly over a victim, where a rescue capsule is to land.

(a) What should be the angle ϕ of the pilot's line of sight to the victim when the capsule release is made?

KEY IDEAS Once released, the capsule is a projectile, so its horizontal and vertical motions can be considered separately (we need not consider the actual curved path of the capsule).

Calculations: In Fig. 4-15, we see that ϕ is given by

$$\phi = \tan^{-1}\frac{x}{h}, \tag{4-27}$$

where x is the horizontal coordinate of the victim (and of the capsule when it hits the water) and $h = 500$ m. We should be able to find x with Eq. 4-21:

$$x - x_0 = (v_0 \cos \theta_0)t. \tag{4-28}$$

Here we know that $x_0 = 0$ because the origin is placed at the point of release. Because the capsule is *released* and not shot from the plane, its initial velocity $\vec{v}_0$ is equal to the plane's velocity. Thus, we know also that the initial velocity has magnitude $v_0 = 55.0$ m/s and angle $\theta_0 = 0°$ (measured relative to the positive direction of the x axis). However, we do not know the time t the capsule takes to move from the plane to the victim.

To find t, we next consider the *vertical* motion and specifically Eq. 4-22:

$$y - y_0 = (v_0 \sin \theta_0)t - \tfrac{1}{2}gt^2. \tag{4-29}$$

Here the vertical displacement $y - y_0$ of the capsule is -500 m (the negative value indicates that the capsule moves *downward*). So,

$$-500 \text{ m} = (55.0 \text{ m/s})(\sin 0°)t - \tfrac{1}{2}(9.8 \text{ m/s}^2)t^2.$$

Solving for t, we find $t = 10.1$ s. Using that value in Eq. 4-28 yields

$$x - 0 = (55.0 \text{ m/s})(\cos 0°)(10.1 \text{ s}),$$

or

$$x = 555.5 \text{ m}.$$

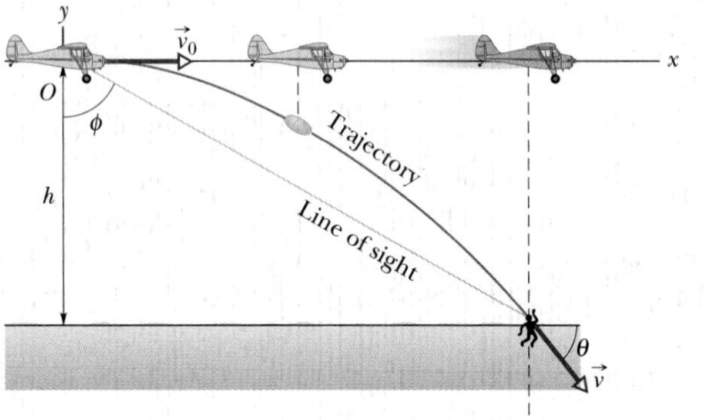

FIG. 4-15 A plane drops a rescue capsule while moving at constant velocity in level flight. While falling, the capsule remains under the plane.

Then Eq. 4-27 gives us

$$\phi = \tan^{-1}\frac{555.5 \text{ m}}{500 \text{ m}} = 48.0°. \quad \text{(Answer)}$$

(b) As the capsule reaches the water, what is its velocity $\vec{v}$ in unit-vector notation and in magnitude-angle notation?

KEY IDEAS (1) The horizontal and vertical components of the capsule's velocity are independent. (2) Component v_x does not change from its initial value $v_{0x} = v_0 \cos \theta_0$ because there is no horizontal acceleration. (3) Component v_y changes from its initial value $v_{0y} = v_0 \sin \theta_0$ because there is a vertical acceleration.

Calculations: When the capsule reaches the water,

$$v_x = v_0 \cos \theta_0 = (55.0 \text{ m/s})(\cos 0°) = 55.0 \text{ m/s}.$$

Using Eq. 4-23 and the capsule's time of fall $t = 10.1$ s, we also find that when the capsule reaches the water,

$$\begin{aligned} v_y &= v_0 \sin \theta_0 - gt \qquad (4\text{-}30)\\ &= (55.0 \text{ m/s})(\sin 0°) - (9.8 \text{ m/s}^2)(10.1 \text{ s})\\ &= -99.0 \text{ m/s}. \end{aligned}$$

Thus, at the water

$$\vec{v} = (55.0 \text{ m/s})\hat{\text{i}} - (99.0 \text{ m/s})\hat{\text{j}}. \quad \text{(Answer)}$$

Using Eq. 3-6 as a guide, we find that the magnitude and the angle of $\vec{v}$ are

$$v = 113 \text{ m/s} \quad \text{and} \quad \theta = -60.9°. \quad \text{(Answer)}$$

Sample Problem 4-7

Figure 4-16 shows a pirate ship 560 m from a fort defending a harbor entrance. A defense cannon, located at sea level, fires balls at initial speed $v_0 = 82$ m/s.

(a) At what angle θ_0 from the horizontal must a ball be fired to hit the ship?

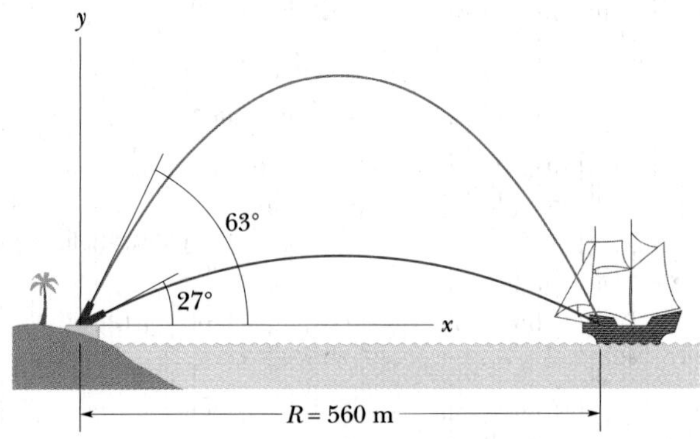

FIG. 4-16 A pirate ship under fire.

KEY IDEAS (1) A fired cannonball is a projectile. We want an equation that relates the launch angle θ_0 to the ball's horizontal displacement as it moves from cannon to ship. (2) Because the cannon and the ship are at the same height, the horizontal displacement is the range.

Calculations: We can relate the launch angle θ_0 to the range R with Eq. 4-26 ($R = (v_0^2/g)\sin 2\theta_0$), which, after rearrangement, gives

$$\theta_0 = \frac{1}{2}\sin^{-1}\frac{gR}{v_0^2} = \frac{1}{2}\sin^{-1}\frac{(9.8\text{ m/s}^2)(560\text{ m})}{(82\text{ m/s})^2}$$

$$= \frac{1}{2}\sin^{-1} 0.816. \qquad (4\text{-}31)$$

One solution of $\sin^{-1}$ (54.7°) is displayed by a calculator; we subtract it from 180° to get the other solution (125.3°). Thus, Eq. 4-31 gives us

$$\theta_0 = 27° \quad \text{and} \quad \theta_0 = 63°. \qquad \text{(Answer)}$$

(b) What is the maximum range of the cannonballs?

Calculations: We have seen that maximum range corresponds to an elevation angle θ_0 of 45°. Thus,

$$R = \frac{v_0^2}{g}\sin 2\theta_0 = \frac{(82\text{ m/s})^2}{9.8\text{ m/s}^2}\sin(2\times 45°)$$

$$= 686\text{ m} \approx 690\text{ m}. \qquad \text{(Answer)}$$

As the pirate ship sails away, the two elevation angles at which the ship can be hit draw together, eventually merging at $\theta_0 = 45°$ when the ship is 690 m away. Beyond that distance the ship is safe.

Sample Problem 4-8

Suppose a baseball batter B hits a high fly ball to the outfield, dirctly toward an outfielder F and with a launch speed of $v_0 = 40$ m/s and a launch angle of $\theta_0 = 35°$. During the flight, a line from the outfielder to the ball makes an angle ϕ with the ground. Plot elevation angle ϕ versus time t, assuming that the outfielder is already positioned to catch the ball, is 6.0 m too close to the batter, and is 6.0 m too far away.

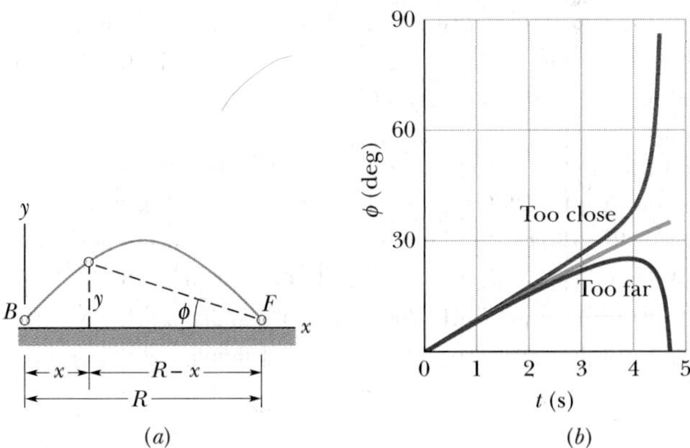

FIG. 4-17 The elevation angle ϕ for a ball hit toward an outfielder is (*a*) defined and (*b*) plotted versus time t.

KEY IDEAS (1) If we neglect air drag, the ball is a projectile for which the vertical motion and the horizontal motion can be analyzed separately. (2) Assuming the ball is caught at approximately the height it is hit, the horizontal distance traveled by the ball is the range R, given by Eq. 4-26 ($R = (v_0^2/g)\sin 2\theta_0$).

Calculations: The ball can be caught if the outfielder's distance from the batter is equal to the range R of the ball. Using Eq. 4-26, we find

$$R = \frac{v_0^2}{g}\sin 2\theta_0 = \frac{(40\text{ m/s})^2}{9.8\text{ m/s}^2}\sin(70°) = 153.42\text{ m}. \qquad (4\text{-}32)$$

Figure 4-17*a* shows a snapshot of the ball in flight when the ball is at height y and horizontal distance x from the batter (who is at the origin). The horizontal distance of the ball from the outfielder is $R - x$, and the elevation angle ϕ of the ball in the outfielder's view is given by $\tan\phi = y/(R - x)$. For the height y, we use Eq. 4-22 ($y - y_0 = (v_0 \sin\theta_0)t - \frac{1}{2}gt^2$), setting $y_0 = 0$. For the horizontal distance x, we substitute with Eq. 4-21 $(x - x_0 = (v_0 \cos \theta_0)t)$, setting $x_0 = 0$. Thus, using $v_0 = 40$ m/s and $\theta_0 = 35°$, we have

$$\phi = \tan^{-1} \frac{(40 \sin 35°)t - 4.9t^2}{153.42 - (40 \cos 35°)t}. \quad (4\text{-}33)$$

Graphing this function versus t gives the middle plot in Fig. 4-17b. We see that the ball's angle in the outfielder's view increases at an almost steady rate throughout the flight.

If the outfielder is 6.0 m too close to the batter, we replace the distance of 153.42 m in Eq. 4-33 with 153.42 m − 6.0 m = 147.42 m. Regraphing the function gives the "Too close" plot in Fig. 4-17b. Now the elevation angle of the ball rapidly increases toward the end of the flight as the ball soars over the outfielder's head. If the outfielder is 6.0 m too far away from the batter, we replace the distance of 153.42 m in Eq. 4-33 with 159.42 m. The resulting plot is labeled "Too far" in the figure: The angle first increases and then rapidly decreases. Thus, if a ball is hit directly toward an outfielder, the player can tell from the change in the ball's elevation angle ϕ whether to stay put, run toward the batter, or back away from the batter.

Sample Problem 4-9 Build your skill

At time $t = 0$, a golf ball is shot from ground level into the air, as indicated in Fig. 4-18a. The angle θ between the ball's direction of travel and the positive direction of the x axis is given in Fig. 4-18b as a function of time t. The ball lands at $t = 6.00$ s. What is the magnitude v_0 of the ball's launch velocity, at what height $(y - y_0)$ above the launch level does the ball land, and what is the ball's direction of travel just as it lands?

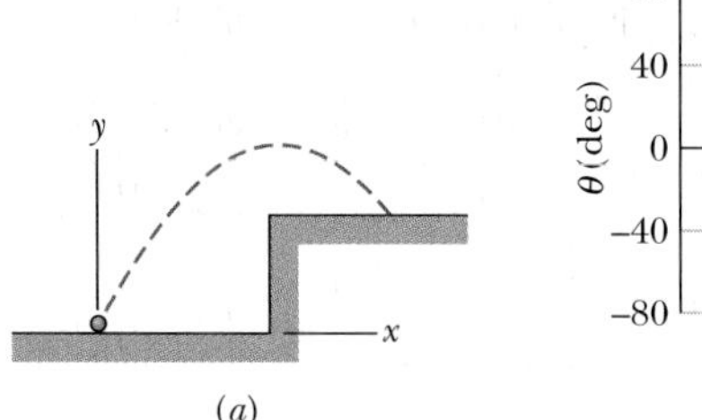

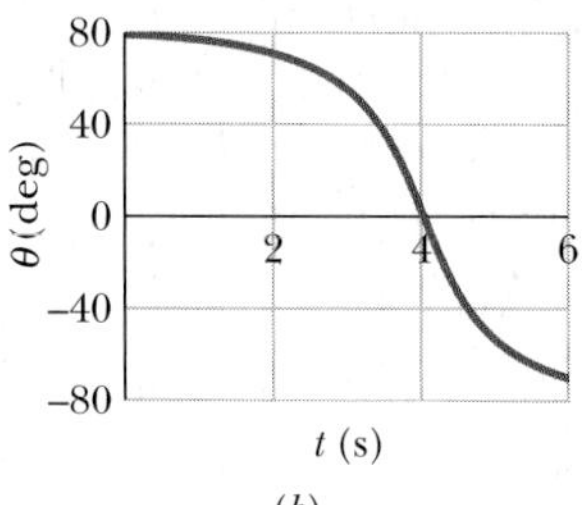

FIG. 4-18 (a) Path of a golf ball shot onto a plateau. (b) The angle θ that gives the ball's direction of motion during the flight is plotted versus time t.

KEY IDEAS (1) The ball is a projectile, and so its horizontal and vertical motions can be considered separately. (2) The horizontal component v_x $(= v_0 \cos \theta_0)$ of the ball's velocity does not change during the flight. (3) The vertical component v_y of its velocity *does* change and is zero when the ball reaches maximum height. (4) The ball's direction of travel at any time during the flight is at the angle of its velocity vector $\vec{v}$ just then. That angle is given by $\tan \theta = v_y/v_x$, with the velocity components evaluated at that time.

Calculations: When the ball reaches its maximum height, $v_y = 0$. So, the direction of the velocity $\vec{v}$ is horizontal, at angle $\theta = 0°$. From the graph, we see that this condition occurs at $t = 4.0$ s. We also see that the launch angle θ_0 (at $t = 0$) is 80°. Using Eq. 4-23 $(v_y = v_0 \sin \theta_0 - gt)$, with $t = 4.0$ s, $g = 9.8$ m/s², $\theta_0 = 80°$, and $v_y = 0$, we find

$$v_0 = 39.80 \approx 40 \text{ m/s}. \quad \text{(Answer)}$$

The ball lands at $t = 6.00$ s. Using Eq. 4-22 $(y - y_0 = (v_0 \sin \theta_0)t - \frac{1}{2}gt^2)$ with $t = 6.00$ s, we obtain

$$y - y_0 = 58.77 \text{ m} \approx 59 \text{ m}. \quad \text{(Answer)}$$

Just as the ball lands, its horizontal velocity v_x is still $v_0 \cos \theta_0$; substituting for v_0 and θ_0 gives us $v_x = 6.911$ m/s. We find its vertical velocity just then by using Eq. 4-23 $(v_y = v_0 \sin \theta_0 - gt)$ with $t = 6.00$ s, which yields $v_y = -19.60$ m/s. Thus, the angle of the ball's direction of travel at landing is

$$\theta = \tan^{-1} \frac{v_y}{v_x} = \tan^{-1} \frac{-19.60 \text{ m/s}}{6.911 \text{ m/s}} \approx -71°. \quad \text{(Answer)}$$

4-7 | Uniform Circular Motion

A particle is in **uniform circular motion** if it travels around a circle or a circular arc at constant (*uniform*) speed. Although the speed does not vary, *the particle is accelerating* because the velocity changes in direction.

Figure 4-19 shows the relationship between the velocity and acceleration vectors at various stages during uniform circular motion. Both vectors have constant magnitude, but their directions change continuously. The velocity is always directed tangent to the circle in the direction of motion. The acceleration is always directed *radially inward.* Because of this, the acceleration associated with uniform circular motion is called a **centripetal** (meaning "center seeking")

acceleration. As we prove next, the magnitude of this acceleration $\vec{a}$ is

$$a = \frac{v^2}{r} \quad \text{(centripetal acceleration),} \tag{4-34}$$

where r is the radius of the circle and v is the speed of the particle.

In addition, during this acceleration at constant speed, the particle travels the circumference of the circle (a distance of $2\pi r$) in time

$$T = \frac{2\pi r}{v} \quad \text{(period).} \tag{4-35}$$

T is called the *period of revolution,* or simply the *period,* of the motion. It is, in general, the time for a particle to go around a closed path exactly once.

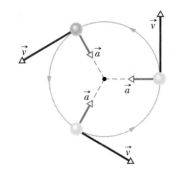

FIG. 4-19 Velocity and acceleration vectors for uniform circular motion.

Proof of Eq. 4-34

To find the magnitude and direction of the acceleration for uniform circular motion, we consider Fig. 4-20. In Fig. 4-20*a*, particle p moves at constant speed v around a circle of radius r. At the instant shown, p has coordinates x_p and y_p.

Recall from Section 4-3 that the velocity $\vec{v}$ of a moving particle is always tangent to the particle's path at the particle's position. In Fig. 4-20*a*, that means $\vec{v}$ is perpendicular to a radius r drawn to the particle's position. Then the angle θ that $\vec{v}$ makes with a vertical at p equals the angle θ that radius r makes with the x axis.

The scalar components of $\vec{v}$ are shown in Fig. 4-20*b*. With them, we can write the velocity $\vec{v}$ as

$$\vec{v} = v_x\hat{\mathrm{i}} + v_y\hat{\mathrm{j}} = (-v \sin\theta)\hat{\mathrm{i}} + (v\cos\theta)\hat{\mathrm{j}}. \tag{4-36}$$

Now, using the right triangle in Fig. 4-20*a*, we can replace $\sin\theta$ with y_p/r and $\cos\theta$ with x_p/r to write

$$\vec{v} = \left(-\frac{vy_p}{r}\right)\hat{\mathrm{i}} + \left(\frac{vx_p}{r}\right)\hat{\mathrm{j}}. \tag{4-37}$$

To find the acceleration $\vec{a}$ of particle p, we must take the time derivative of this equation. Noting that speed v and radius r do not change with time, we obtain

$$\vec{a} = \frac{d\vec{v}}{dt} = \left(-\frac{v}{r}\frac{dy_p}{dt}\right)\hat{\mathrm{i}} + \left(\frac{v}{r}\frac{dx_p}{dt}\right)\hat{\mathrm{j}}. \tag{4-38}$$

Now note that the rate dy_p/dt at which y_p changes is equal to the velocity component v_y. Similarly, $dx_p/dt = v_x$, and, again from Fig. 4-20*b*, we see that $v_x = -v\sin\theta$ and $v_y = v\cos\theta$. Making these substitutions in Eq. 4-38, we find

$$\vec{a} = \left(-\frac{v^2}{r}\cos\theta\right)\hat{\mathrm{i}} + \left(-\frac{v^2}{r}\sin\theta\right)\hat{\mathrm{j}}. \tag{4-39}$$

This vector and its components are shown in Fig. 4-20*c*. Following Eq. 3-6, we find

$$a = \sqrt{a_x^2 + a_y^2} = \frac{v^2}{r}\sqrt{(\cos\theta)^2 + (\sin\theta)^2} = \frac{v^2}{r}\sqrt{1} = \frac{v^2}{r},$$

as we wanted to prove. To orient $\vec{a}$, we find the angle ϕ shown in Fig. 4-20*c*:

$$\tan\phi = \frac{a_y}{a_x} = \frac{-(v^2/r)\sin\theta}{-(v^2/r)\cos\theta} = \tan\theta.$$

Thus, $\phi = \theta$, which means that $\vec{a}$ is directed along the radius r of Fig. 4-20*a*, toward the circle's center, as we wanted to prove.

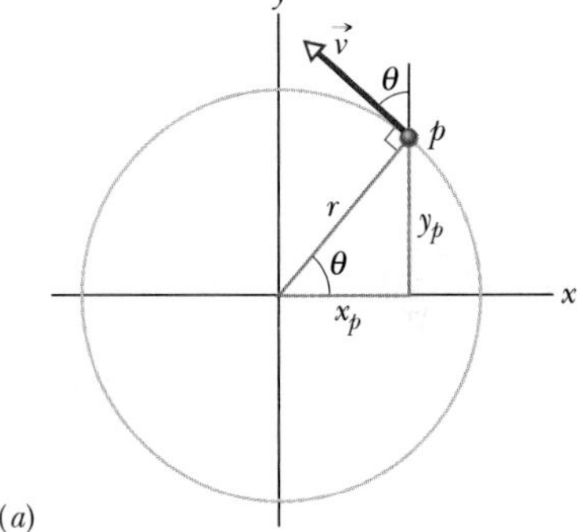

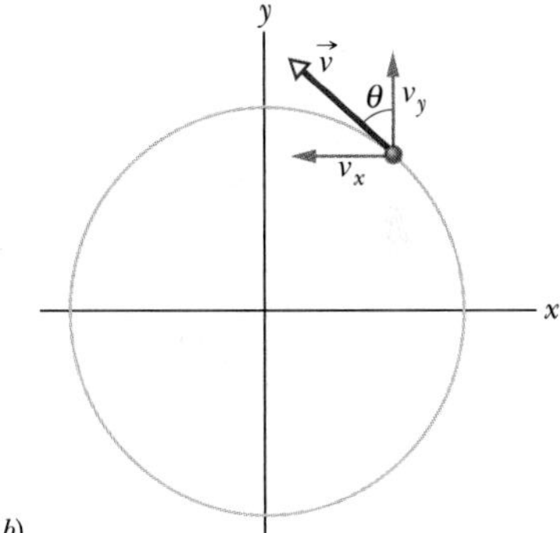

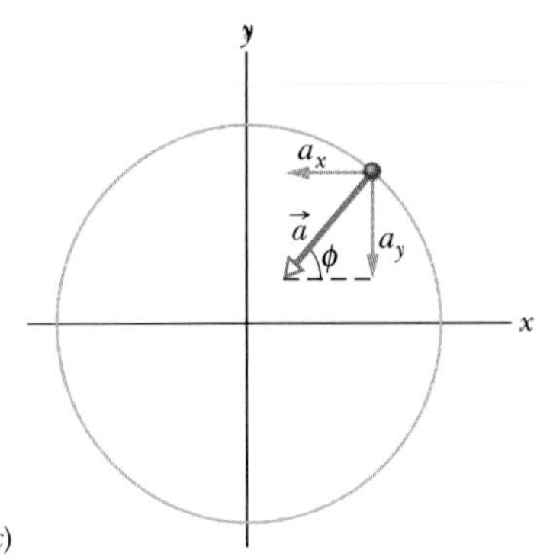

FIG. 4-20 Particle p moves in counterclockwise uniform circular motion. (*a*) Its position and velocity $\vec{v}$ at a certain instant. (*b*) Velocity $\vec{v}$. (*c*) Acceleration $\vec{a}$.

CHECKPOINT 5 An object moves at constant speed along a circular path in a horizontal xy plane, with the center at the origin. When the object is at $x = -2$ m, its velocity is $-(4 \text{ m/s})\hat{\mathrm{j}}$. Give the object's (a) velocity and (b) acceleration at $y = 2$ m.

Sample Problem 4-10

"Top gun" pilots have long worried about taking a turn too tightly. As a pilot's body undergoes centripetal acceleration, with the head toward the center of curvature, the blood pressure in the brain decreases, leading to loss of brain function.

There are several warning signs. When the centripetal acceleration is $2g$ or $3g$, the pilot feels heavy. At about $4g$, the pilot's vision switches to black and white and narrows to "tunnel vision." If that acceleration is sustained or increased, vision ceases and, soon after, the pilot is unconscious—a condition known as g-LOC for "g-induced loss of consciousness."

What is the magnitude of the acceleration, in g units, of a pilot whose aircraft enters a horizontal circular turn with a velocity of $\vec{v}_i = (400\hat{\text{i}} + 500\hat{\text{j}})$ m/s and 24.0 s later leaves the turn with a velocity of $\vec{v}_f = (-400\hat{\text{i}} - 500\hat{\text{j}})$ m/s?

KEY IDEAS We assume the turn is made with uniform circular motion. Then the pilot's acceleration is centripetal and has magnitude a given by Eq. 4-34 ($a = v^2/R$), where R is the circle's radius. Also, the time required to complete a full circle is the period given by Eq. 4-35 ($T = 2\pi R/v$).

Calculations: Because we do not know radius R, let's solve Eq. 4-35 for R and substitute into Eq. 4-34. We find

$$a = \frac{2\pi v}{T}.$$

Speed v here is the (constant) magnitude of the velocity during the turning. Let's substitute the components of the initial velocity into Eq. 3-6:

$$v = \sqrt{(400 \text{ m/s})^2 + (500 \text{ m/s})^2} = 640.31 \text{ m/s}.$$

To find the period T of the motion, first note that the final velocity is the reverse of the initial velocity. This means the aircraft leaves on the opposite side of the circle from the initial point and must have completed half a circle in the given 24.0 s. Thus a full circle would have taken $T = 48.0$ s. Substituting these values into our equation for a, we find

$$a = \frac{2\pi(640.31 \text{ m/s})}{48.0 \text{ s}} = 83.81 \text{ m/s}^2 \approx 8.6g. \quad \text{(Answer)}$$

4-8 | Relative Motion in One Dimension

Suppose you see a duck flying north at 30 km/h. To another duck flying alongside, the first duck seems to be stationary. In other words, the velocity of a particle depends on the **reference frame** of whoever is observing or measuring the velocity. For our purposes, a reference frame is the physical object to which we attach our coordinate system. In everyday life, that object is the ground. For example, the speed listed on a speeding ticket is always measured relative to the ground. The speed relative to the police officer would be different if the officer were moving while making the speed measurement.

Suppose that Alex (at the origin of frame A in Fig. 4-21) is parked by the side of a highway, watching car P (the "particle") speed past. Barbara (at the origin of frame B) is driving along the highway at constant speed and is also watching car P. Suppose that they both measure the position of the car at a given moment. From Fig. 4-21 we see that

$$x_{PA} = x_{PB} + x_{BA}. \tag{4-40}$$

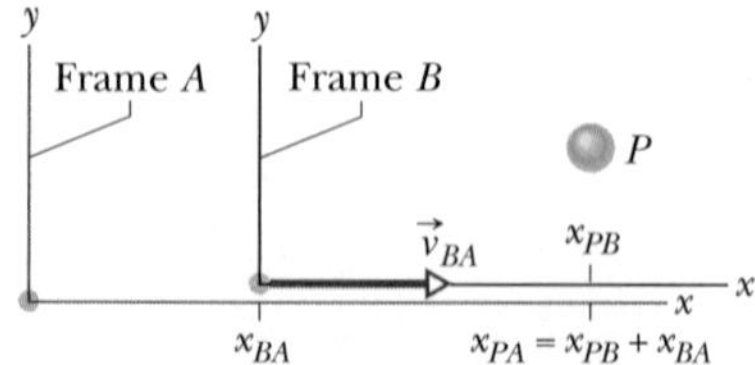

FIG. 4-21 Alex (frame A) and Barbara (frame B) watch car P, as both B and P move at different velocities along the common x axis of the two frames. At the instant shown, x_{BA} is the coordinate of B in the A frame. Also, P is at coordinate x_{PB} in the B frame and coordinate $x_{PA} = x_{PB} + x_{BA}$ in the A frame.

The equation is read: "The coordinate x_{PA} of P as measured by A *is equal to* the coordinate x_{PB} of P as measured by B *plus* the coordinate x_{BA} of B as measured by A." Note how this reading is supported by the sequence of the subscripts.

Taking the time derivative of Eq. 4-40, we obtain

$$\frac{d}{dt}(x_{PA}) = \frac{d}{dt}(x_{PB}) + \frac{d}{dt}(x_{BA}).$$

Thus, the velocity components are related by

$$v_{PA} = v_{PB} + v_{BA}. \tag{4-41}$$

This equation is read: "The velocity v_{PA} of P as measured by A *is equal to* the velocity v_{PB} of P as measured by B *plus* the velocity v_{BA} of B as measured by A." The term v_{BA} is the velocity of frame B relative to frame A.

Here we consider only frames that move at constant velocity relative to each other. In our example, this means that Barbara (frame B) drives always at constant velocity v_{BA} relative to Alex (frame A). Car P (the moving particle), however, can change speed and direction (that is, it can accelerate).

To relate an acceleration of P as measured by Barbara and by Alex, we take the time derivative of Eq. 4-41:

$$\frac{d}{dt}(v_{PA}) = \frac{d}{dt}(v_{PB}) + \frac{d}{dt}(v_{BA}).$$

Because v_{BA} is constant, the last term is zero and we have

$$a_{PA} = a_{PB}. \qquad (4\text{-}42)$$

In other words,

> Observers on different frames of reference that move at constant velocity relative to each other will measure the same acceleration for a moving particle.

Sample Problem 4-11

In Fig. 4-21, suppose that Barbara's velocity relative to Alex is a constant $v_{BA} = 52$ km/h and car P is moving in the negative direction of the x axis.

(a) If Alex measures a constant $v_{PA} = -78$ km/h for car P, what velocity v_{PB} will Barbara measure?

KEY IDEAS We can attach a frame of reference A to Alex and a frame of reference B to Barbara. Because the frames move at constant velocity relative to each other along one axis, we can use Eq. 4-41 ($v_{PA} = v_{PB} + v_{BA}$) to relate v_{PB} to v_{PA} and v_{BA}.

Calculation: We find

$$-78 \text{ km/h} = v_{PB} + 52 \text{ km/h}.$$

Thus,

$$v_{PB} = -130 \text{ km/h}. \qquad \text{(Answer)}$$

Comment: If car P were connected to Barbara's car by a cord wound on a spool, the cord would be unwinding at a speed of 130 km/h as the two cars separated.

(b) If car P brakes to a stop relative to Alex (and thus relative to the ground) in time $t = 10$ s at constant acceleration, what is its acceleration a_{PA} relative to Alex?

KEY IDEAS To calculate the acceleration of car P *relative to Alex,* we must use the car's velocities *relative to Alex.* Because the acceleration is constant, we can use Eq. 2-11 ($v = v_0 + at$) to relate the acceleration to the initial and final velocities of P.

Calculation: The initial velocity of P relative to Alex is $v_{PA} = -78$ km/h and the final velocity is 0. Thus,

$$a_{PA} = \frac{v - v_0}{t} = \frac{0 - (-78 \text{ km/h})}{10 \text{ s}} \frac{1 \text{ m/s}}{3.6 \text{ km/h}}$$
$$= 2.2 \text{ m/s}^2. \qquad \text{(Answer)}$$

(c) What is the acceleration a_{PB} of car P relative to Barbara during the braking?

KEY IDEA To calculate the acceleration of car P *relative to Barbara,* we must use the car's velocities *relative to Barbara.*

Calculation: We know the initial velocity of P relative to Barbara from part (a) ($v_{PB} = -130$ km/h). The final velocity of P relative to Barbara is -52 km/h (this is the velocity of the stopped car relative to the moving Barbara). Thus,

$$a_{PB} = \frac{v - v_0}{t} = \frac{-52 \text{ km/h} - (-130 \text{ km/h})}{10 \text{ s}} \frac{1 \text{ m/s}}{3.6 \text{ km/h}}$$
$$= 2.2 \text{ m/s}^2. \qquad \text{(Answer)}$$

Comment: We should have foreseen this result: Because Alex and Barbara have a constant relative velocity, they must measure the same acceleration for the car.

4-9 | Relative Motion in Two Dimensions

Our two observers are again watching a moving particle P from the origins of reference frames A and B, while B moves at a constant velocity $\vec{v}_{BA}$ relative to A. (The corresponding axes of these two frames remain parallel.) Figure 4-22 shows a certain instant during the motion. At that instant, the position vector of the origin of B

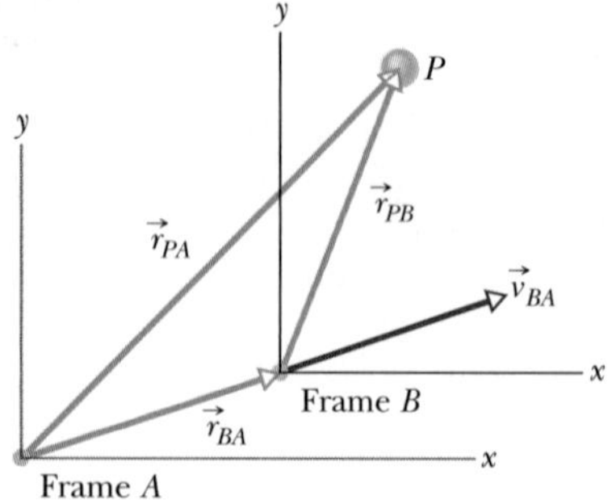

FIG. 4-22 Frame B has the constant two-dimensional velocity $\vec{v}_{BA}$ relative to frame A. The position vector of B relative to A is $\vec{r}_{BA}$. The position vectors of particle P are $\vec{r}_{PA}$ relative to A and $\vec{r}_{PB}$ relative to B.

relative to the origin of A is $\vec{r}_{BA}$. Also, the position vectors of particle P are $\vec{r}_{PA}$ relative to the origin of A and $\vec{r}_{PB}$ relative to the origin of B. From the arrangement of heads and tails of those three position vectors, we can relate the vectors with

$$\vec{r}_{PA} = \vec{r}_{PB} + \vec{r}_{BA}. \tag{4-43}$$

By taking the time derivative of this equation, we can relate the velocities $\vec{v}_{PA}$ and $\vec{v}_{PB}$ of particle P relative to our observers:

$$\vec{v}_{PA} = \vec{v}_{PB} + \vec{v}_{BA}. \tag{4-44}$$

By taking the time derivative of this relation, we can relate the accelerations $\vec{a}_{PA}$ and $\vec{a}_{PB}$ of the particle P relative to our observers. However, note that because $\vec{v}_{BA}$ is constant, its time derivative is zero. Thus, we get

$$\vec{a}_{PA} = \vec{a}_{PB}. \tag{4-45}$$

As for one-dimensional motion, we have the following rule: Observers on different frames of reference that move at constant velocity relative to each other will measure the *same* acceleration for a moving particle.

Sample Problem 4-12

In Fig. 4-23*a*, a plane moves due east while the pilot points the plane somewhat south of east, toward a steady wind that blows to the northeast. The plane has velocity $\vec{v}_{PW}$ relative to the wind, with an airspeed (speed relative to the wind) of 215 km/h, directed at angle θ south of east. The wind has velocity $\vec{v}_{WG}$ relative to the ground with speed 65.0 km/h, directed 20.0° east of north. What is the magnitude of the velocity $\vec{v}_{PG}$ of the plane relative to the ground, and what is θ?

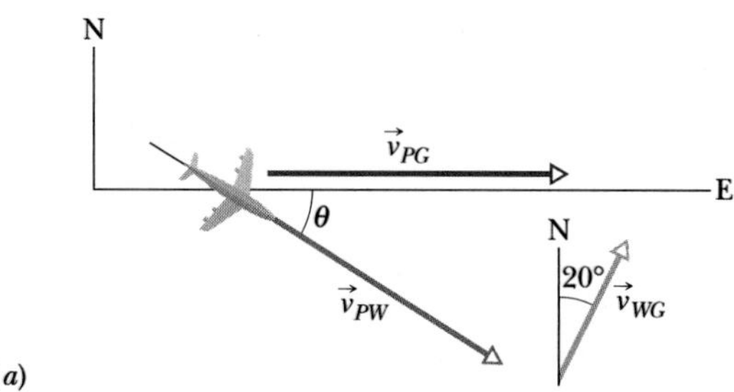

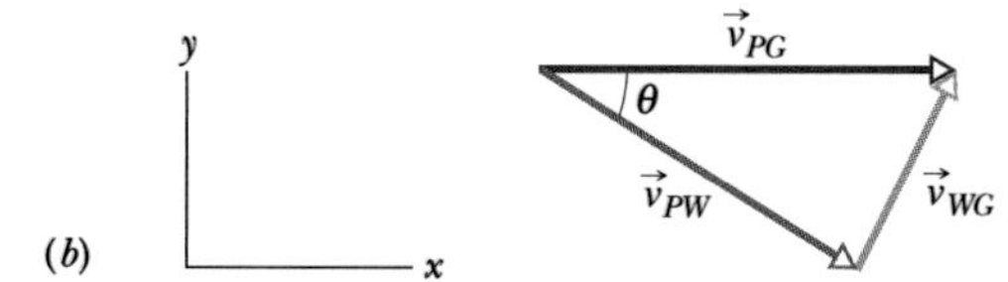

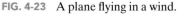

FIG. 4-23 A plane flying in a wind.

KEY IDEAS The situation is like the one in Fig. 4-22. Here the moving particle P is the plane, frame A is attached to the ground (call it G), and frame B is "attached" to the wind (call it W). We need a vector diagram like Fig. 4-22 but with three velocity vectors.

Calculations: First we construct a sentence that relates the three vectors shown in Fig. 4-23*b*:

velocity of plane relative to ground (*PG*) = velocity of plane relative to wind (*PW*) + velocity of wind relative to ground. (*WG*)

This relation is written in vector notation as

$$\vec{v}_{PG} = \vec{v}_{PW} + \vec{v}_{WG}. \tag{4-46}$$

We need to resolve the vectors into components on the coordinate system of Fig. 4-23*b* and then solve Eq. 4-46 axis by axis. For the y components, we find

$$v_{PG,y} = v_{PW,y} + v_{WG,y}$$

or $$0 = -(215 \text{ km/h}) \sin\theta + (65.0 \text{ km/h})(\cos 20.0°).$$

Solving for θ gives us

$$\theta = \sin^{-1}\frac{(65.0 \text{ km/h})(\cos 20.0°)}{215 \text{ km/h}} = 16.5°. \quad \text{(Answer)}$$

Similarly, for the x components we find

$$v_{PG,x} = v_{PW,x} + v_{WG,x}.$$

Here, because $\vec{v}_{PG}$ is parallel to the x axis, the component $v_{PG,x}$ is equal to the magnitude v_{PG}. Substituting this notation and the value $\theta = 16.5°$, we find

$$v_{PG} = (215 \text{ km/h})(\cos 16.5°) + (65.0 \text{ km/h})(\sin 20.0°) = 228 \text{ km/h}. \quad \text{(Answer)}$$

REVIEW & SUMMARY

Position Vector The location of a particle relative to the origin of a coordinate system is given by a *position vector* $\vec{r}$, which in unit-vector notation is

$$\vec{r} = x\hat{i} + y\hat{j} + z\hat{k}. \tag{4-1}$$

Here $x\hat{i}$, $y\hat{j}$, and $z\hat{k}$ are the vector components of position vector $\vec{r}$, and x, y, and z are its scalar components (as well as the coordinates of the particle). A position vector is described either by a magnitude and one or two angles for orientation, or by its vector or scalar components.

Displacement If a particle moves so that its position vector changes from $\vec{r}_1$ to $\vec{r}_2$, the particle's *displacement* $\Delta\vec{r}$ is

$$\Delta\vec{r} = \vec{r}_2 - \vec{r}_1. \tag{4-2}$$

The displacement can also be written as

$$\Delta\vec{r} = (x_2 - x_1)\hat{i} + (y_2 - y_1)\hat{j} + (z_2 - z_1)\hat{k} \tag{4-3}$$

$$= \Delta x\hat{i} + \Delta y\hat{j} + \Delta z\hat{k}. \tag{4-4}$$

Average Velocity and Instantaneous Velocity If a particle undergoes a displacement $\Delta\vec{r}$ in time interval Δt, its *average velocity* $\vec{v}_{avg}$ for that time interval is

$$\vec{v}_{avg} = \frac{\Delta\vec{r}}{\Delta t}. \tag{4-8}$$

As Δt in Eq. 4-8 is shrunk to 0, $\vec{v}_{avg}$ reaches a limit called either the *velocity* or the *instantaneous velocity* $\vec{v}$:

$$\vec{v} = \frac{d\vec{r}}{dt}, \tag{4-10}$$

which can be rewritten in unit-vector notation as

$$\vec{v} = v_x\hat{i} + v_y\hat{j} + v_z\hat{k}, \tag{4-11}$$

where $v_x = dx/dt$, $v_y = dy/dt$, and $v_z = dz/dt$. The instantaneous velocity $\vec{v}$ of a particle is always directed along the tangent to the particle's path at the particle's position.

Average Acceleration and Instantaneous Acceleration If a particle's velocity changes from $\vec{v}_1$ to $\vec{v}_2$ in time interval Δt, its *average acceleration* during Δt is

$$\vec{a}_{avg} = \frac{\vec{v}_2 - \vec{v}_1}{\Delta t} = \frac{\Delta\vec{v}}{\Delta t}. \tag{4-15}$$

As Δt in Eq. 4-15 is shrunk to 0, $\vec{a}_{avg}$ reaches a limiting value called either the *acceleration* or the *instantaneous acceleration* $\vec{a}$:

$$\vec{a} = \frac{d\vec{v}}{dt}. \tag{4-16}$$

In unit-vector notation,

$$\vec{a} = a_x\hat{i} + a_y\hat{j} + a_z\hat{k}, \tag{4-17}$$

where $a_x = dv_x/dt$, $a_y = dv_y/dt$, and $a_z = dv_z/dt$.

Projectile Motion *Projectile motion* is the motion of a particle that is launched with an initial velocity $\vec{v}_0$. During its flight, the particle's horizontal acceleration is zero and its vertical acceleration is the free-fall acceleration $-g$. (Upward is taken to be a positive direction.) If $\vec{v}_0$ is expressed as a magnitude (the speed v_0) and an angle θ_0 (measured from the horizontal), the particle's equations of motion along the horizontal x axis and vertical y axis are

$$x - x_0 = (v_0 \cos\theta_0)t, \tag{4-21}$$

$$y - y_0 = (v_0 \sin\theta_0)t - \tfrac{1}{2}gt^2, \tag{4-22}$$

$$v_y = v_0 \sin\theta_0 - gt, \tag{4-23}$$

$$v_y^2 = (v_0 \sin\theta_0)^2 - 2g(y - y_0). \tag{4-24}$$

The **trajectory** (path) of a particle in projectile motion is parabolic and is given by

$$y = (\tan\theta_0)x - \frac{gx^2}{2(v_0 \cos\theta_0)^2}, \tag{4-25}$$

if x_0 and y_0 of Eqs. 4-21 to 4-24 are zero. The particle's **horizontal range** R, which is the horizontal distance from the launch point to the point at which the particle returns to the launch height, is

$$R = \frac{v_0^2}{g}\sin 2\theta_0. \tag{4-26}$$

Uniform Circular Motion If a particle travels along a circle or circular arc of radius r at constant speed v, it is said to be in *uniform circular motion* and has an acceleration $\vec{a}$ of constant magnitude

$$a = \frac{v^2}{r}. \tag{4-34}$$

The direction of $\vec{a}$ is toward the center of the circle or circular arc, and $\vec{a}$ is said to be *centripetal*. The time for the particle to complete a circle is

$$T = \frac{2\pi r}{v}. \tag{4-35}$$

T is called the *period of revolution*, or simply the *period*, of the motion.

Relative Motion When two frames of reference A and B are moving relative to each other at constant velocity, the velocity of a particle P as measured by an observer in frame A usually differs from that measured from frame B. The two measured velocities are related by

$$\vec{v}_{PA} = \vec{v}_{PB} + \vec{v}_{BA}, \tag{4-44}$$

where $\vec{v}_{BA}$ is the velocity of B with respect to A. Both observers measure the same acceleration for the particle:

$$\vec{a}_{PA} = \vec{a}_{PB}. \tag{4-45}$$

QUESTIONS

1 Figure 4-24 shows the initial position i and the final position f of a particle. What are the (a) initial position vector $\vec{r}_i$ and (b) final position vector $\vec{r}_f$, both in unit-vector notation? (c) What is the x component of displacement $\Delta\vec{r}$?

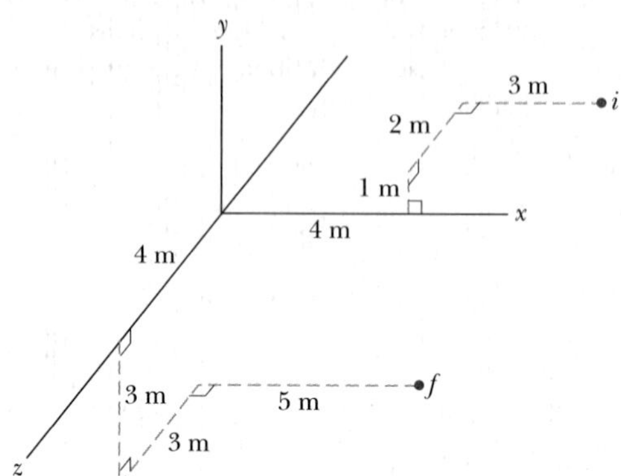

FIG. 4-24 Question 1.

2 Figure 4-25 shows the path taken by a skunk foraging for trash food, from initial point i. The skunk took the same time T to go from each labeled point to the next along its path. Rank points a, b, and c according to the magnitude of the average velocity of the skunk to reach them from initial point i, greatest first.

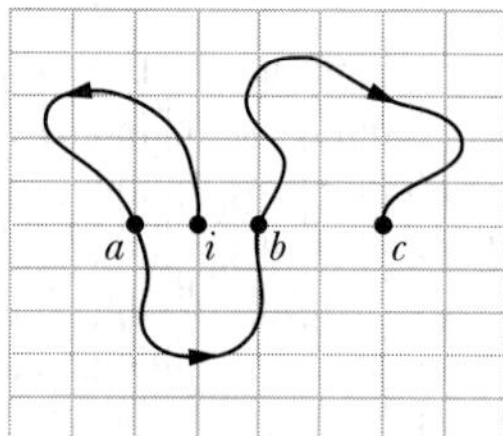

FIG. 4-25 Question 2.

3 You are to launch a rocket, from just above the ground, with one of the following initial velocity vectors: (1) $\vec{v}_0 = 20\hat{i} + 70\hat{j}$, (2) $\vec{v}_0 = -20\hat{i} + 70\hat{j}$, (3) $\vec{v}_0 = 20\hat{i} - 70\hat{j}$, (4) $\vec{v}_0 = -20\hat{i} - 70\hat{j}$. In your coordinate system, x runs along level ground and y increases upward. (a) Rank the vectors according to the launch speed of the projectile, greatest first. (b) Rank the vectors according to the time of flight of the projectile, greatest first.

4 Figure 4-26 shows three situations in which identical projectiles are launched (at the same level) at identical initial speeds and angles. The projectiles do not land on the same terrain, however. Rank the situations according to the final speeds of the projectiles just before they land, greatest first.

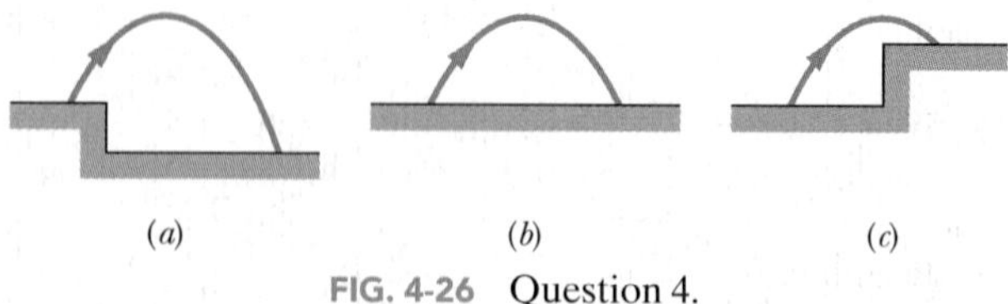

FIG. 4-26 Question 4.

5 When Paris was shelled from 100 km away with the WWI long-range artillery piece "Big Bertha," the shells were fired at an angle greater than 45° to give them a greater range, possibly even twice as long as at 45°. Does that result mean that the air density at high altitudes increases with altitude or decreases?

6 In Fig. 4-27, a cream tangerine is thrown up past windows 1, 2, and 3, which are identical in size and regularly spaced vertically. Rank those three windows according to (a) the time the cream tangerine takes to pass them and (b) the average speed of the cream tangerine during the passage, greatest first.

The cream tangerine then moves down past windows 4, 5, and 6, which are identical in size and irregularly spaced horizontally. Rank those three windows according to (c) the time the cream tangerine takes to pass them and (d) the average speed of the cream tangerine during the passage, greatest first.

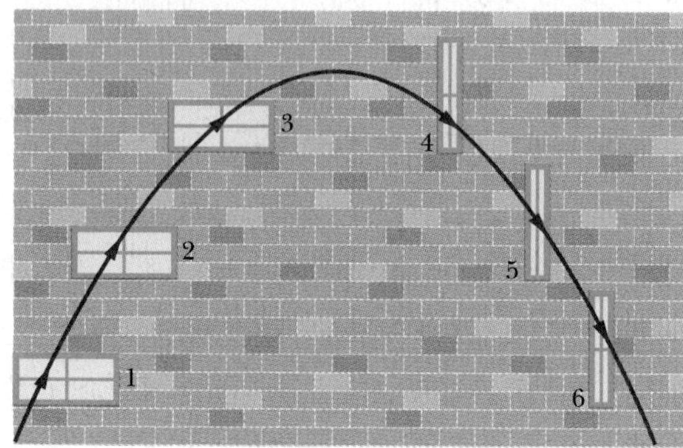

FIG. 4-27 Question 6.

7 Figure 4-28 shows three paths for a football kicked from ground level. Ignoring the effects of air, rank the paths according to (a) time of flight, (b) initial vertical velocity component, (c) initial horizontal velocity component, and (d) initial speed, greatest first.

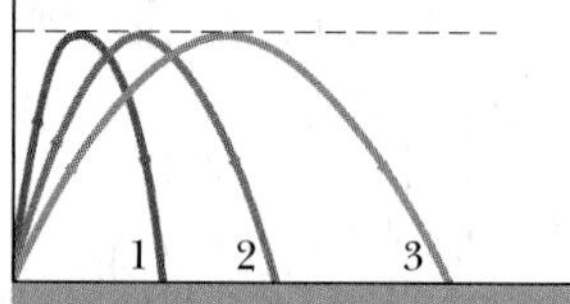

FIG. 4-28 Question 7.

8 The only good use of a fruitcake is in catapult practice. Curve 1 in Fig. 4-29 gives the height y of a catapulted fruitcake versus the angle θ between its velocity vector and its acceleration vector during flight. (a) Which of the lettered points on that curve corresponds to the landing of the fruitcake on the ground? (b) Curve 2 is a similar plot for the same launch speed but for a different launch angle. Does the fruitcake now land farther away or closer to the launch point?

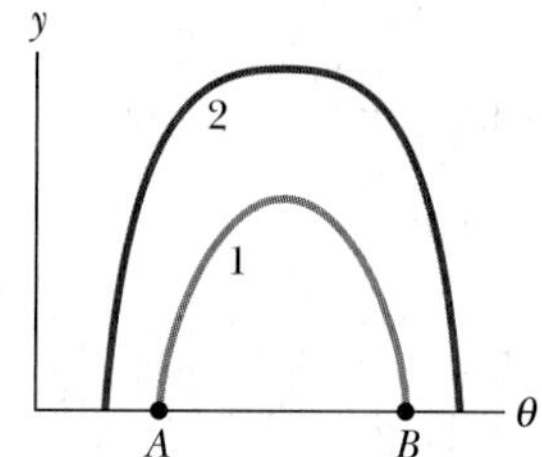

FIG. 4-29 Question 8.

9 An airplane flying horizontally at a constant speed of 350 km/h over level ground releases a bundle of food supplies. Ignore the effect of the air on the bundle. What are the bundle's initial (a) vertical and (b) horizontal components of velocity? (c) What is its horizontal component of velocity just before hitting the ground? (d) If the airplane's speed were, instead, 450 km/h, would the time of fall be longer, shorter, or the same?

10 A ball is shot from ground level over level ground at a certain initial speed. Figure 4-30 gives the range R

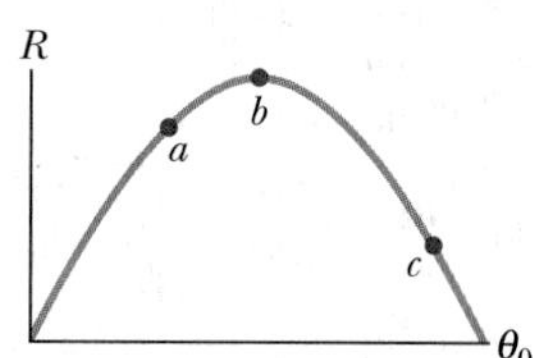

FIG. 4-30 Question 10.

of the ball versus its launch angle θ_0. Rank the three lettered points on the plot according to (a) the total flight time of the ball and (b) the ball's speed at maximum height, greatest first.

11 In Fig. 4-31, particle P is in uniform circular motion, centered on the origin of an xy coordinate system. (a) At what values of θ is the vertical component r_y of the position vector greatest in magnitude? (b) At what values of θ is the vertical component v_y of the particle's velocity greatest in magnitude? (c) At what values of θ is the vertical component a_y of the particle's acceleration greatest in magnitude?

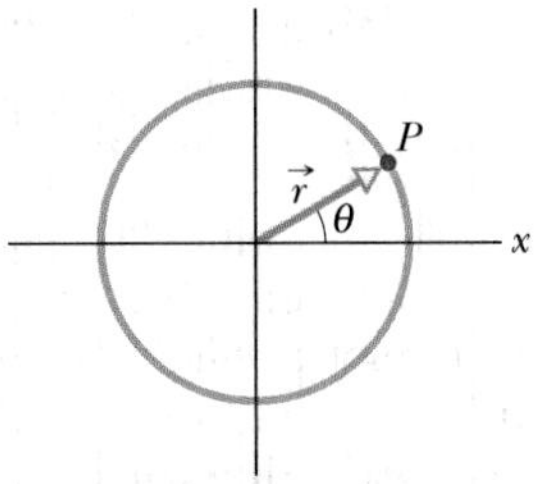

FIG. 4-31 Question 11.

12 (a) Is it possible to be accelerating while traveling at constant speed? Is it possible to round a curve with (b) zero acceleration and (c) a constant magnitude of acceleration?

13 Figure 4-32 shows four tracks (either half- or quarter-circles) that can be taken by a train, which moves at a constant speed. Rank the tracks according to the magnitude of a train's acceleration on the curved portion, greatest first.

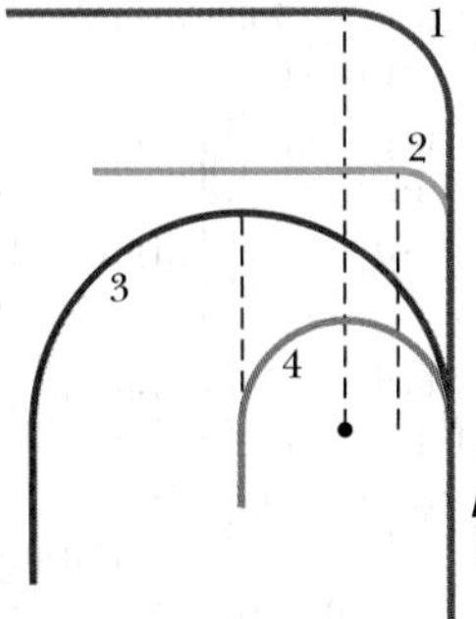

FIG. 4-32 Question 13.

PROBLEMS

 Tutoring problem available (at instructor's discretion) in *WileyPLUS* and WebAssign
SSM Worked-out solution available in Student Solutions Manual
WWW Worked-out solution is at
ILW Interactive solution is at
http://www.wiley.com/college/halliday
• – ••• Number of dots indicates level of problem difficulty
 Additional information available in *The Flying Circus of Physics* and at flyingcircusofphysics.com

sec. 4-2 Position and Displacement

•1 A positron undergoes a displacement $\Delta\vec{r} = 2.0\hat{i} - 3.0\hat{j} + 6.0\hat{k}$, ending with the position vector $\vec{r} = 3.0\hat{j} - 4.0\hat{k}$, in meters. What was the positron's initial position vector?

•2 A watermelon seed has the following coordinates: $x = -5.0$ m, $y = 8.0$ m, and $z = 0$ m. Find its position vector (a) in unit-vector notation and as (b) a magnitude and (c) an angle relative to the positive direction of the x axis. (d) Sketch the vector on a right-handed coordinate system. If the seed is moved to the xyz coordinates (3.00 m, 0 m, 0 m), what is its displacement (e) in unit-vector notation and as (f) a magnitude and (g) an angle relative to the positive x direction?

•3 The position vector for an electron is $\vec{r} = (5.0 \text{ m})\hat{i} - (3.0 \text{ m})\hat{j} + (2.0 \text{ m})\hat{k}$. (a) Find the magnitude of $\vec{r}$. (b) Sketch the vector on a right-handed coordinate system.

••4 The minute hand of a wall clock measures 10 cm from its tip to the axis about which it rotates. The magnitude and angle of the displacement vector of the tip are to be determined for three time intervals. What are the (a) magnitude and (b) angle from a quarter after the hour to half past, the (c) magnitude and (d) angle for the next half hour, and the (e) magnitude and (f) angle for the hour after that?

sec. 4-3 Average Velocity and Instantaneous Velocity

•5 An ion's position vector is initially $\vec{r} = 5.0\hat{i} - 6.0\hat{j} + 2.0\hat{k}$, and 10 s later it is $\vec{r} = -2.0\hat{i} + 8.0\hat{j} - 2.0\hat{k}$, all in meters. In unit-vector notation, what is its $\vec{v}_{avg}$ during the 10 s?

•6 An electron's position is given by $\vec{r} = 3.00t\hat{i} - 4.00t^2\hat{j} + 2.00\hat{k}$, with t in seconds and $\vec{r}$ in meters. (a) In unit-vector notation, what is the electron's velocity $\vec{v}(t)$? At $t = 2.00$ s, what is $\vec{v}$ (b) in unit-vector notation and as (c) a magnitude and (d) an angle relative to the positive direction of the x axis?

•7 A train at a constant 60.0 km/h moves east for 40.0 min, then in a direction 50.0° east of due north for 20.0 min, and then west for 50.0 min. What are the (a) magnitude and (b) angle of its average velocity during this trip? SSM

••8 A plane flies 483 km east from city A to city B in 45.0 min and then 966 km south from city B to city C in 1.50 h. For the total trip, what are the (a) magnitude and (b) direction of the plane's displacement, the (c) magnitude and (d) direction of its average velocity, and (e) its average speed?

••9 Figure 4-33 gives the path of a squirrel moving about on level ground, from point A (at time $t = 0$), to points B (at $t = 5.00$ min), C (at $t = 10.0$ min), and finally D (at $t = 15.0$ min). Consider the average velocities of the squirrel from point A to each of the other three points. Of them, what are the (a) magnitude and (b) angle of the one with the least magnitude and the (c) magnitude and (d) angle of the one with the greatest magnitude?

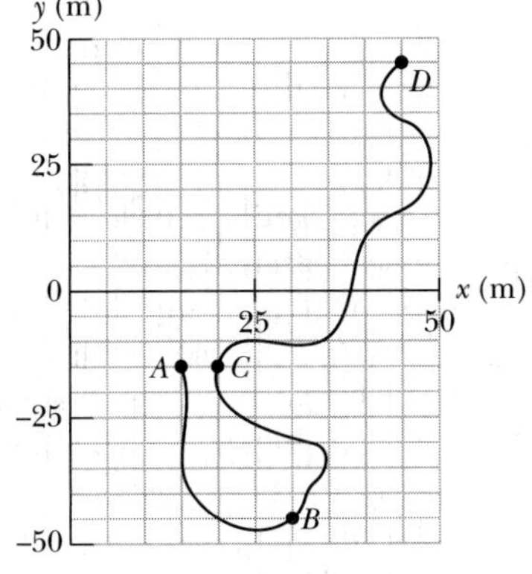

FIG. 4-33 Problem 9.

•••10 The position vector $\vec{r} = 5.00t\hat{i} + (et + ft^2)\hat{j}$ locates a particle as a function of time t. Vector $\vec{r}$ is in meters, t is in seconds, and factors e and f are constants. Figure 4-34 gives the angle θ of the particle's direction of travel as a function of t (θ is measured from the positive x direction). What are (a) e and (b) f, including units?

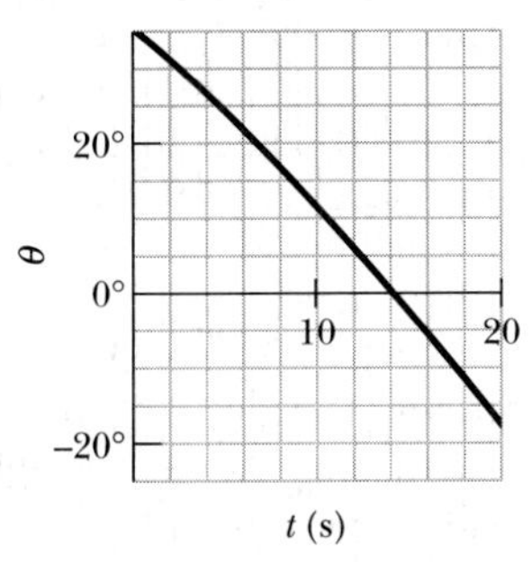

FIG. 4-34 Problem 10.

sec. 4-4 Average Acceleration and Instantaneous Acceleration

•**11** A particle moves so that its position (in meters) as a function of time (in seconds) is $\vec{r} = \hat{i} + 4t^2\hat{j} + t\hat{k}$. Write expressions for (a) its velocity and (b) its acceleration as functions of time. SSM

•**12** A proton initially has $\vec{v} = 4.0\hat{i} - 2.0\hat{j} + 3.0\hat{k}$ and then 4.0 s later has $\vec{v} = -2.0\hat{i} - 2.0\hat{j} + 5.0\hat{k}$ (in meters per second). For that 4.0 s, what are (a) the proton's average acceleration $\vec{a}_{avg}$ in unit-vector notation, (b) the magnitude of $\vec{a}_{avg}$, and (c) the angle between $\vec{a}_{avg}$ and the positive direction of the x axis?

•**13** The position $\vec{r}$ of a particle moving in an xy plane is given by $\vec{r} = (2.00t^3 - 5.00t)\hat{i} + (6.00 - 7.00t^4)\hat{j}$, with $\vec{r}$ in meters and t in seconds. In unit-vector notation, calculate (a) $\vec{r}$, (b) $\vec{v}$, and (c) $\vec{a}$ for $t = 2.00$ s. (d) What is the angle between the positive direction of the x axis and a line tangent to the particle's path at $t = 2.00$ s? GO

•**14** At one instant a bicyclist is 40.0 m due east of a park's flagpole, going due south with a speed of 10.0 m/s. Then 30.0 s later, the cyclist is 40.0 m due north of the flagpole, going due east with a speed of 10.0 m/s. For the cyclist in this 30.0 s interval, what are the (a) magnitude and (b) direction of the displacement, the (c) magnitude and (d) direction of the average velocity, and the (e) magnitude and (f) direction of the average acceleration?

••**15** A cart is propelled over an xy plane with acceleration components $a_x = 4.0$ m/s^2 and $a_y = -2.0$ m/s^2. Its initial velocity has components $v_{0x} = 8.0$ m/s and $v_{0y} = 12$ m/s. In unit-vector notation, what is the velocity of the cart when it reaches its greatest y coordinate?

••**16** A moderate wind accelerates a pebble over a horizontal xy plane with a constant acceleration $\vec{a} = (5.00 \text{ m/s}^2)\hat{i} + (7.00 \text{ m/s}^2)\hat{j}$. At time $t = 0$, the velocity is $(4.00 \text{ m/s})\hat{i}$. What are the (a) magnitude and (b) angle of its velocity when it has been displaced by 12.0 m parallel to the x axis?

••**17** A particle leaves the origin with an initial velocity $\vec{v} = (3.00\hat{i})$ m/s and a constant acceleration $\vec{a} = (-1.00\hat{i} - 0.500\hat{j})$ m/s^2. When it reaches its maximum x coordinate, what are its (a) velocity and (b) position vector? SSM ILW

••**18** The velocity $\vec{v}$ of a particle moving in the xy plane is given by $\vec{v} = (6.0t - 4.0t^2)\hat{i} + 8.0\hat{j}$, with $\vec{v}$ in meters per second and t (> 0) in seconds. (a) What is the acceleration when $t = 3.0$ s? (b) When (if ever) is the acceleration zero? (c) When (if ever) is the velocity zero? (d) When (if ever) does the speed equal 10 m/s?

•••**19** The acceleration of a particle moving only on a horizontal xy plane is given by $\vec{a} = 3t\hat{i} + 4t\hat{j}$, where $\vec{a}$ is in meters per second-squared and t is in seconds. At $t = 0$, the position vector $\vec{r} = (20.0 \text{ m})\hat{i} + (40.0 \text{ m})\hat{j}$ locates the particle, which then has the velocity vector $\vec{v} = (5.00 \text{ m/s})\hat{i} + (2.00 \text{ m/s})\hat{j}$. At $t = 4.00$ s, what are (a) its position vector in unit-vector notation and (b) the angle between its direction of travel and the positive direction of the x axis?

•••**20** In Fig. 4-35, particle A moves along the line $y = 30$ m with a constant velocity $\vec{v}$ of magnitude 3.0 m/s and parallel to the x axis. At the instant particle A passes the y axis, particle B leaves the origin with zero initial speed and constant acceleration $\vec{a}$ of magnitude 0.40 m/s^2. What angle θ between $\vec{a}$ and the positive direction of the y axis would result in a collision?

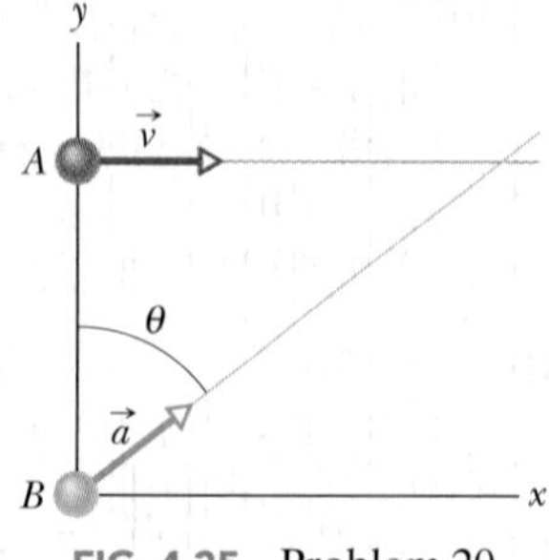

FIG. 4-35 Problem 20.

sec. 4-6 Projectile Motion Analyzed

•**21** A projectile is fired horizontally from a gun that is 45.0 m above flat ground, emerging from the gun with a speed of 250 m/s. (a) How long does the projectile remain in the air? (b) At what horizontal distance from the firing point does it strike the ground? (c) What is the magnitude of the vertical component of its velocity as it strikes the ground?

•**22** In the 1991 World Track and Field Championships in Tokyo, Mike Powell jumped 8.95 m, breaking by a full 5 cm the 23-year long-jump record set by Bob Beamon. Assume that Powell's speed on takeoff was 9.5 m/s (about equal to that of a sprinter) and that $g = 9.80$ m/s^2 in Tokyo. How much less was Powell's range than the maximum possible range for a particle launched at the same speed?

•**23** The current world-record motorcycle jump is 77.0 m, set by Jason Renie. Assume that he left the take-off ramp at 12.0° to the horizontal and that the take-off and landing heights are the same. Neglecting air drag, determine his take-off speed.

•**24** A small ball rolls horizontally off the edge of a tabletop that is 1.20 m high. It strikes the floor at a point 1.52 m horizontally from the table edge. (a) How long is the ball in the air? (b) What is its speed at the instant it leaves the table?

•**25** A dart is thrown horizontally with an initial speed of 10 m/s toward point P, the bull's-eye on a dart board. It hits at point Q on the rim, vertically below P, 0.19 s later. (a) What is the distance PQ? (b) How far away from the dart board is the dart released?

•**26** In Fig. 4-36, a stone is projected at a cliff of height h with an initial speed of 42.0 m/s directed at angle $\theta_0 = 60.0°$ above the horizontal. The stone strikes at A, 5.50 s after launching. Find (a) the height h of the cliff, (b) the speed of the stone just before impact at A, and (c) the maximum height H reached above the ground.

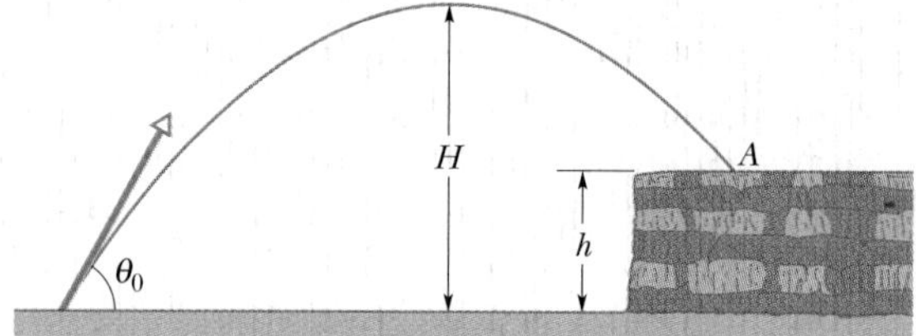

FIG. 4-36 Problem 26.

•**27** A certain airplane has a speed of 290.0 km/h and is diving at an angle of $\theta = 30.0°$ below the horizontal when the pilot releases a radar decoy (Fig. 4-37). The horizontal distance between the release point and the point where the decoy strikes the ground is $d = 700$ m. (a) How long is the decoy in the air? (b) How high was the release point? ILW

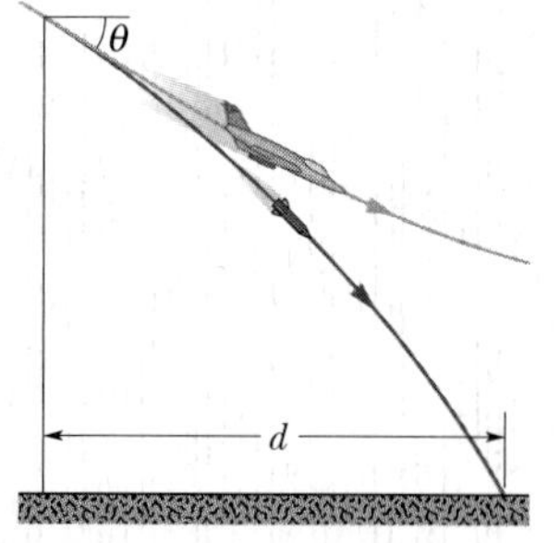

FIG. 4-37 Problem 27.

•28 A stone is catapulted at time $t = 0$, with an initial velocity of magnitude 20.0 m/s and at an angle of 40.0° above the horizontal. What are the magnitudes of the (a) horizontal and (b) vertical components of its displacement from the catapult site at $t = 1.10$ s? Repeat for the (c) horizontal and (d) vertical components at $t = 1.80$ s, and for the (e) horizontal and (f) vertical components at $t = 5.00$ s.

••29 A lowly high diver pushes off horizontally with a speed of 2.00 m/s from the platform edge 10.0 m above the surface of the water. (a) At what horizontal distance from the edge is the diver 0.800 s after pushing off? (b) At what vertical distance above the surface of the water is the diver just then? (c) At what horizontal distance from the edge does the diver strike the water? SSM WWW

••30 A trebuchet was a hurling machine built to attack the walls of a castle under siege. A large stone could be hurled against a wall to break apart the wall. The machine was not placed near the wall because then arrows could reach it from the castle wall. Instead, it was positioned so that the stone hit the wall during the second half of its flight. Suppose a stone is launched with a speed of $v_0 = 28.0$ m/s and at an angle of $\theta_0 = 40.0°$. What is the speed of the stone if it hits the wall (a) just as it reaches the top of its parabolic path and (b) when it has descended to half that height? (c) As a percentage, how much faster is it moving in part (b) than in part (a)?

••31 A plane, diving with constant speed at an angle of 53.0° with the vertical, releases a projectile at an altitude of 730 m. The projectile hits the ground 5.00 s after release. (a) What is the speed of the plane? (b) How far does the projectile travel horizontally during its flight? What are the (c) horizontal and (d) vertical components of its velocity just before striking the ground? SSM

••32 During a tennis match, a player serves the ball at 23.6 m/s, with the center of the ball leaving the racquet horizontally 2.37 m above the court surface. The net is 12 m away and 0.90 m high. When the ball reaches the net, (a) does the ball clear it and (b) what is the distance between the center of the ball and the top of the net? Suppose that, instead, the ball is served as before but now it leaves the racquet at 5.00° below the horizontal. When the ball reaches the net, (c) does the ball clear it and (d) what now is the distance between the center of the ball and the top of the net?

••33 In a jump spike, a volleyball player slams the ball from overhead and toward the opposite floor. Controlling the angle of the spike is difficult. Suppose a ball is spiked from a height of 2.30 m with an initial speed of 20.0 m/s at a downward angle of 18.00°. How much farther on the opposite floor would it have landed if the downward angle were, instead, 8.00°?

••34 A soccer ball is kicked from the ground with an initial speed of 19.5 m/s at an upward angle of 45°. A player 55 m away in the direction of the kick starts running to meet the ball at that instant. What must be his average speed if he is to meet the ball just before it hits the ground?

••35 A projectile's launch speed is five times its speed at maximum height. Find launch angle θ_0.

••36 Suppose that a shot putter can put a shot at the world-class speed $v_0 = 15.00$ m/s and at a height of 2.160 m. What horizontal distance would the shot travel if the launch angle θ_0 is (a) 45.00° and (b) 42.00°? The answers indicate that the angle of 45°, which maximizes the range of projectile motion, does not maximize the horizontal distance when the launch and landing are at different heights.

••37 A ball is shot from the ground into the air. At a height of 9.1 m, its velocity is $\vec{v} = (7.6\hat{i} + 6.1\hat{j})$ m/s, with $\hat{i}$ horizontal and $\hat{j}$ upward. (a) To what maximum height does the ball rise? (b) What total horizontal distance does the ball travel? What are the (c) magnitude and (d) angle (below the horizontal) of the ball's velocity just before it hits the ground? ILW

••38 You throw a ball toward a wall at speed 25.0 m/s and at angle $\theta_0 = 40.0°$ above the horizontal (Fig. 4-38). The wall is distance $d = 22.0$ m from the release point of the ball. (a) How far above the release point does the ball hit the wall? What are the (b) horizontal and (c) vertical components of its velocity as it hits the wall? (d) When it hits, has it passed the highest point on its trajectory?

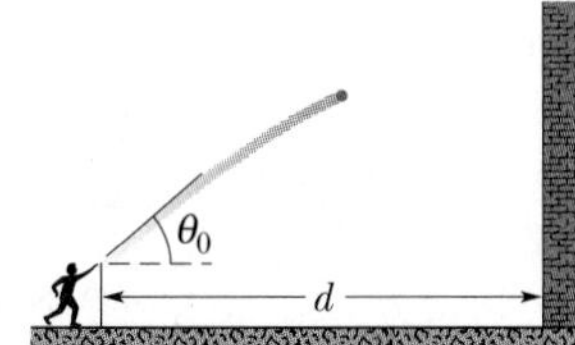

FIG. 4-38 Problem 38.

••39 A rifle that shoots bullets at 460 m/s is to be aimed at a target 45.7 m away. If the center of the target is level with the rifle, how high above the target must the rifle barrel be pointed so that the bullet hits dead center? SSM

••40 A baseball leaves a pitcher's hand horizontally at a speed of 161 km/h. The distance to the batter is 18.3 m. (a) How long does the ball take to travel the first half of that distance? (b) The second half? (c) How far does the ball fall freely during the first half? (d) During the second half? (e) Why aren't the quantities in (c) and (d) equal?

••41 In Fig. 4-39, a ball is thrown leftward from the left edge of the roof, at height h above the ground. The ball hits the ground 1.50 s later, at distance $d = 25.0$ m from the building and at angle $\theta = 60.0°$ with the horizontal. (a) Find h. (*Hint:* One way is to reverse the motion, as if on videotape.) What are the (b) magnitude and (c) angle relative to the horizontal of the velocity at which the ball is thrown? (d) Is the angle above or below the horizontal?

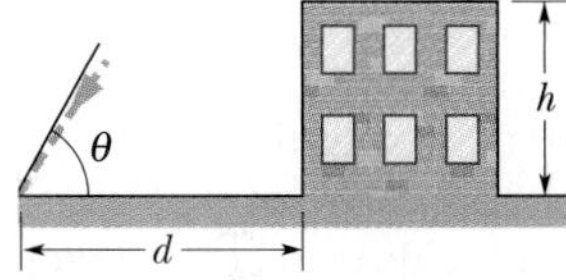

FIG. 4-39 Problem 41.

••42 A golf ball is struck at ground level. The speed of the golf ball as a function of the time is shown in Fig. 4-40, where $t = 0$ at the instant the ball is struck. (a) How far does the golf ball travel horizontally before returning to ground level? (b) What is the maximum height above ground level attained by the ball?

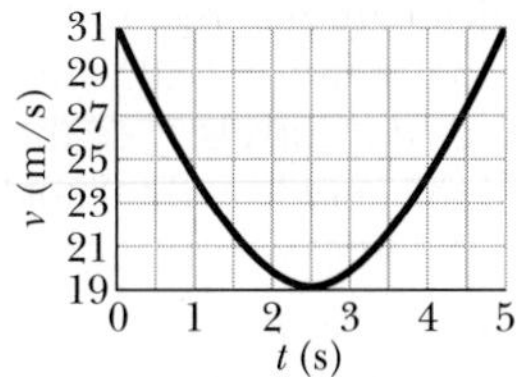

FIG. 4-40 Problem 42.

••43 In Fig. 4-41, a ball is launched with a velocity of magnitude 10.0 m/s, at an angle of 50.0° to the horizontal. The launch point is at the base of a ramp of horizontal length $d_1 = 6.00$ m and height $d_2 = 3.60$ m. A plateau is located at the top of the ramp. (a) Does the ball land on the ramp or

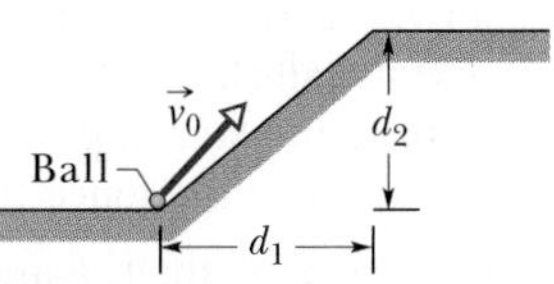

FIG. 4-41 Problem 43.

the plateau? When it lands, what are the (b) magnitude and (c) angle of its displacement from the launch point?

••44 In 1939 or 1940, Emanuel Zacchini took his human-cannonball act to an extreme: After being shot from a cannon, he soared over three Ferris wheels and into a net (Fig. 4-42). (a) Treating him as a particle, calculate his clearance over the first wheel. (b) If he reached maximum height over the middle wheel, by how much did he clear it? (c) How far from the cannon should the net's center have been positioned (neglect air drag)?

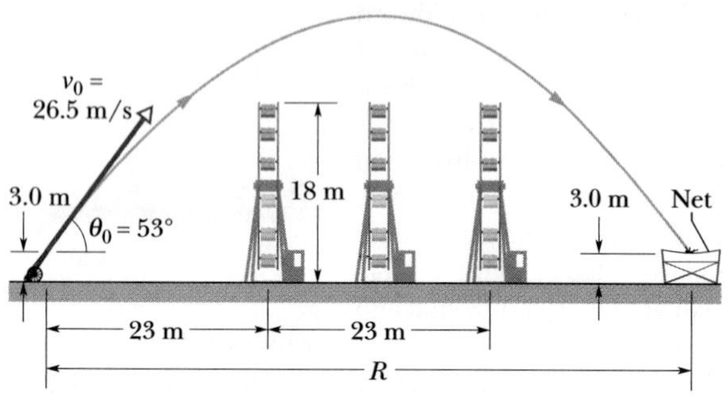

FIG. 4-42 Problem 44.

••45 Upon spotting an insect on a twig overhanging water, an archer fish squirts water drops at the insect to knock it into the water (Fig. 4-43). Although the fish sees the insect along a straight-line path at angle ϕ and distance d, a drop must be launched at a different angle θ_0 if its parabolic path is to intersect the insect. If $\phi = 36.0°$, $d =$ 0.900 m, and the launch speed is 3.56 m/s, what θ_0 is required for the drop to be at the top of the parabolic path when it reaches the insect?

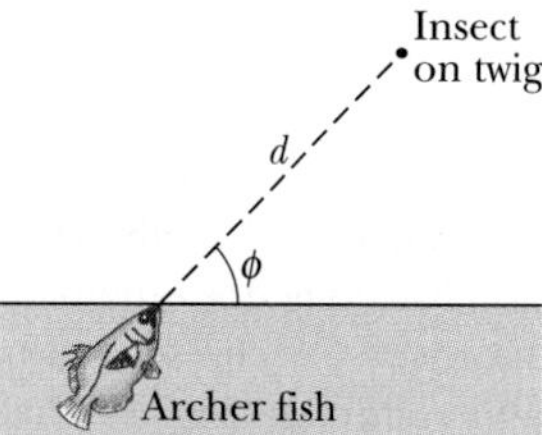

FIG. 4-43 Problem 45.

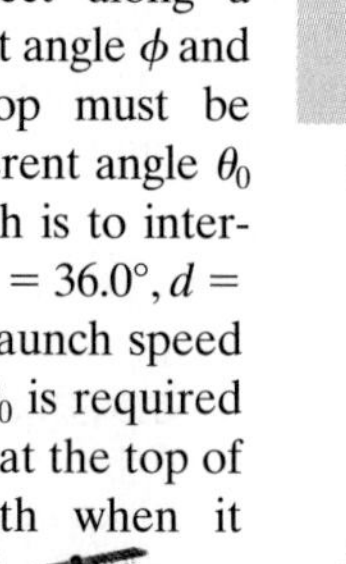

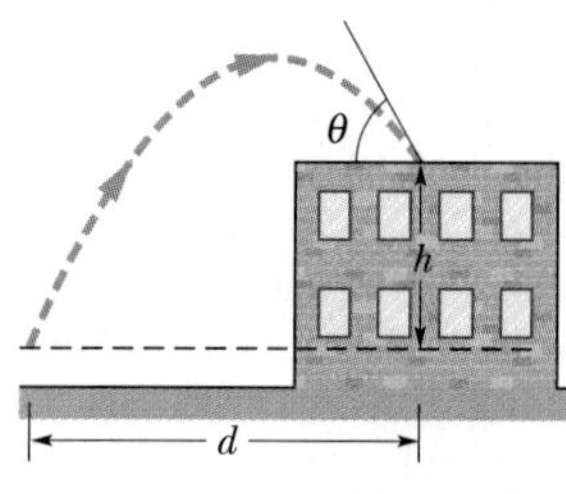

FIG. 4-44 Problem 46.

••46 In Fig. 4-44, a ball is thrown up onto a roof, landing 4.00 s later at height $h = 20.0$ m above the release level. The ball's path just before landing is angled at $\theta = 60.0°$ with the roof. (a) Find the horizontal distance d it travels. (See the hint to Problem 41.) What are the (b) magnitude and (c) angle (relative to the horizontal) of the ball's initial velocity?

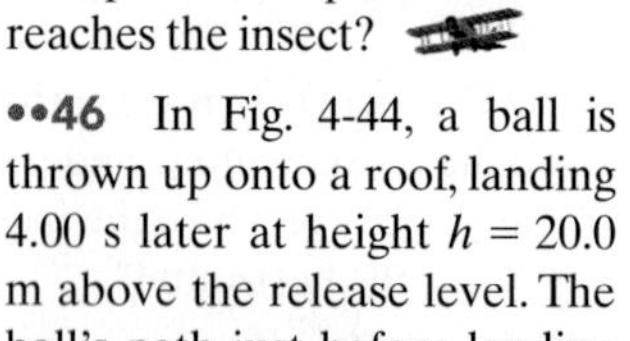

••47 A batter hits a pitched ball when the center of the ball is 1.22 m above the ground. The ball leaves the bat at an angle of 45° with the ground. With that launch, the ball should have a horizontal range (returning to the *launch* level) of 107 m. (a) Does the ball clear a 7.32-m-high fence that is 97.5 m horizontally from the launch point? (b) At the fence, what is the distance between the fence top and the ball center? SSM WWW

••48 In basketball, *hang* is an illusion in which a player seems to weaken the gravitational acceleration while in midair. The illusion depends much on a skilled player's ability to rapidly shift the ball between hands during the flight, but it might also be supported by the longer horizontal distance the player travels in the upper part of the jump than in the lower part. If a player jumps with an initial speed of $v_0 = 7.00$ m/s at an angle of $\theta_0 = 35.0°$, what percent of the jump's range does the player spend in the upper half of the jump (between maximum height and half maximum height)?

•••49 A skilled skier knows to jump upward before reaching a downward slope. Consider a jump in which the launch speed is $v_0 = 10$ m/s, the launch angle is $\theta_0 = 9.0°$, the initial course is approximately flat, and the steeper track has a slope of 11.3°. Figure 4-45*a* shows a *prejump* that allows the skier to land on the top portion of the steeper track. Figure 4-45*b* shows a jump at the edge of the steeper track. In Fig. 4-45*a*, the skier lands at approximately the launch level. (a) In the landing, what is the angle ϕ between the skier's path and the slope? In Fig. 4-45*b*, (b) how far below the launch level does the skier land and (c) what is ϕ? (The greater fall and greater ϕ can result in loss of control in the landing.)

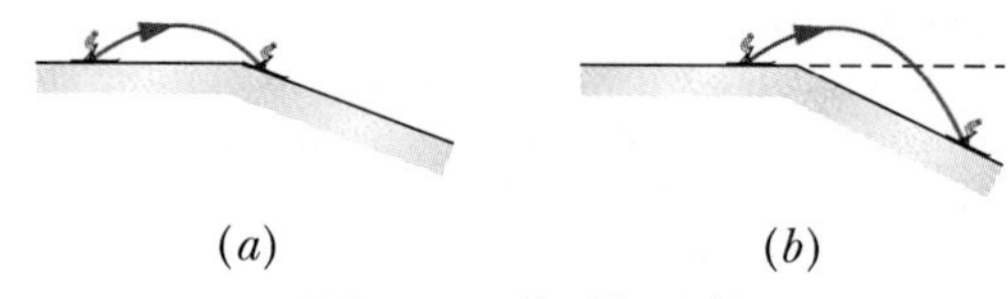

FIG. 4-45 Problem 49.

•••50 A ball is to be shot from level ground toward a wall at distance x (Fig. 4-46*a*). Figure 4-46*b* shows the y component v_y of the ball's velocity just as it would reach the wall, as a function of that distance x. What is the launch angle?

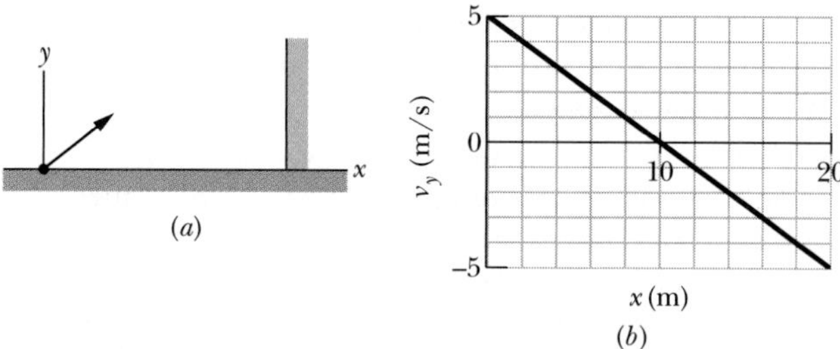

FIG. 4-46 Problem 50.

•••51 A football kicker can give the ball an initial speed of 25 m/s. What are the (a) least and (b) greatest elevation angles at which he can kick the ball to score a field goal from a point 50 m in front of goalposts whose horizontal bar is 3.44 m above the ground? SSM

•••52 A ball is to be shot from level ground with a certain speed. Figure 4-47 shows the range R it will have versus the launch angle θ_0. The value of θ_0 determines the flight time; let t_{max} represent the maximum flight time. What is the least speed the ball will have during its flight if θ_0 is chosen such that the flight time is $0.500t_{max}$?

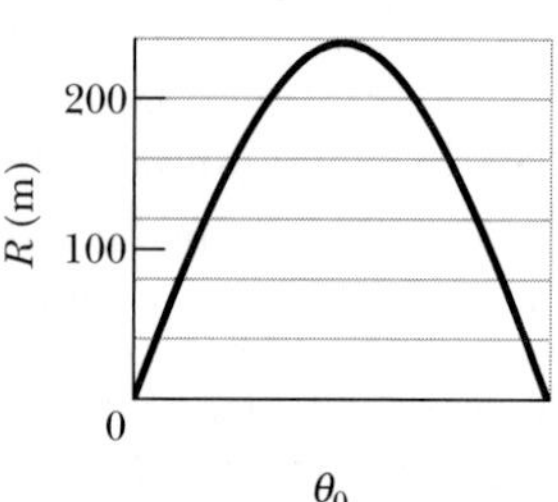

FIG. 4-47 Problem 52.

•••53 A ball rolls horizontally off the top of a stairway with a speed of 1.52 m/s. The steps are 20.3 cm high and 20.3 cm wide. Which step does the ball hit first? SSM

•••54 Two seconds after being projected from ground level, a projectile is displaced 40 m horizontally and 53 m vertically above its launch point. What are the (a) horizontal and (b) vertical components of the initial velocity of the projectile? (c) At the instant the projectile achieves its maximum height above ground level, how far is it displaced horizontally from the launch point? GO

•••55 In Fig. 4-48, a baseball is hit at a height $h = 1.00$ m and then caught at the same height. It travels alongside a wall, moving up past the top of the wall 1.00 s after it is hit and then down past the top of the wall 4.00 s later, at distance $D = 50.0$ m farther along the wall. (a) What horizontal distance is traveled by the ball from hit to catch? What are the (b) magnitude and (c) angle (relative to the horizontal) of the ball's velocity just after being hit? (d) How high is the wall?

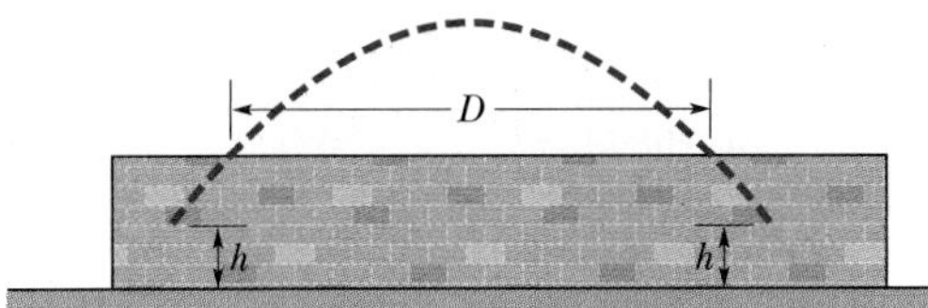

FIG. 4-48 Problem 55.

sec. 4-7 Uniform Circular Motion

•56 A centripetal-acceleration addict rides in uniform circular motion with period $T = 2.0$ s and radius $r = 3.00$ m. At t_1 his acceleration is $\vec{a} = (6.00 \text{ m/s}^2)\hat{i} + (-4.00 \text{ m/s}^2)\hat{j}$. At that instant, what are the values of (a) $\vec{v} \cdot \vec{a}$ and (b) $\vec{r} \times \vec{a}$?

•57 A woman rides a carnival Ferris wheel at radius 15 m, completing five turns about its horizontal axis every minute. What are (a) the period of the motion, the (b) magnitude and (c) direction of her centripetal acceleration at the highest point, and the (d) magnitude and (e) direction of her centripetal acceleration at the lowest point? ILW

•58 What is the magnitude of the acceleration of a sprinter running at 10 m/s when rounding a turn of a radius 25 m?

•59 When a large star becomes a *supernova,* its core may be compressed so tightly that it becomes a *neutron star,* with a radius of about 20 km (about the size of the San Francisco area). If a neutron star rotates once every second, (a) what is the speed of a particle on the star's equator and (b) what is the magnitude of the particle's centripetal acceleration? (c) If the neutron star rotates faster, do the answers to (a) and (b) increase, decrease, or remain the same?

•60 An Earth satellite moves in a circular orbit 640 km above Earth's surface with a period of 98.0 min. What are the (a) speed and (b) magnitude of the centripetal acceleration of the satellite?

•61 A carnival merry-go-round rotates about a vertical axis at a constant rate. A man standing on the edge has a constant speed of 3.66 m/s and a centripetal acceleration $\vec{a}$ of magnitude 1.83 m/s². Position vector $\vec{r}$ locates him relative to the rotation axis. (a) What is the magnitude of $\vec{r}$? What is the direction of $\vec{r}$ when $\vec{a}$ is directed (b) due east and (c) due south?

•62 A rotating fan completes 1200 revolutions every minute. Consider the tip of a blade, at a radius of 0.15 m. (a) Through what distance does the tip move in one revolution? What are (b) the tip's speed and (c) the magnitude of its acceleration? (d) What is the period of the motion?

••63 A purse at radius 2.00 m and a wallet at radius 3.00 m travel in uniform circular motion on the floor of a merry-go-round as the ride turns. They are on the same radial line. At one instant, the acceleration of the purse is $(2.00 \text{ m/s}^2)\hat{i} + (4.00 \text{ m/s}^2)\hat{j}$. At that instant and in unit-vector notation, what is the acceleration of the wallet?

••64 A particle moves along a circular path over a horizontal xy coordinate system, at constant speed. At time $t_1 = 4.00$ s, it is at point (5.00 m, 6.00 m) with velocity $(3.00 \text{ m/s})\hat{j}$ and acceleration in the positive x direction. At time $t_2 = 10.0$ s, it has velocity $(-3.00 \text{ m/s})\hat{i}$ and acceleration in the positive y direction. What are the (a) x and (b) y coordinates of the center of the circular path if $t_2 - t_1$ is less than one period?

••65 At $t_1 = 2.00$ s, the acceleration of a particle in counterclockwise circular motion is $(6.00 \text{ m/s}^2)\hat{i} + (4.00 \text{ m/s}^2)\hat{j}$. It moves at constant speed. At time $t_2 = 5.00$ s, its acceleration is $(4.00 \text{ m/s}^2)\hat{i} + (-6.00 \text{ m/s}^2)\hat{j}$. What is the radius of the path taken by the particle if $t_2 - t_1$ is less than one period? GO

••66 A particle moves horizontally in uniform circular motion, over a horizontal xy plane. At one instant, it moves through the point at coordinates (4.00 m, 4.00 m) with a velocity of $-5.00\hat{i}$ m/s and an acceleration of $+12.5\hat{j}$ m/s². What are the (a) x and (b) y coordinates of the center of the circular path?

•••67 A boy whirls a stone in a horizontal circle of radius 1.5 m and at height 2.0 m above level ground. The string breaks, and the stone flies off horizontally and strikes the ground after traveling a horizontal distance of 10 m. What is the magnitude of the centripetal acceleration of the stone during the circular motion? SSM WWW

•••68 A cat rides a merry-go-round turning with uniform circular motion. At time $t_1 = 2.00$ s, the cat's velocity is $\vec{v}_1 = (3.00 \text{ m/s})\hat{i} + (4.00 \text{ m/s})\hat{j}$, measured on a horizontal xy coordinate system. At $t_2 = 5.00$ s, its velocity is $\vec{v}_2 = (-3.00 \text{ m/s})\hat{i} + (-4.00 \text{ m/s})\hat{j}$. What are (a) the magnitude of the cat's centripetal acceleration and (b) the cat's average acceleration during the time interval $t_2 - t_1$, which is less than one period?

sec. 4-8 Relative Motion in One Dimension

•69 A cameraman on a pickup truck is traveling westward at 20 km/h while he videotapes a cheetah that is moving westward 30 km/h faster than the truck. Suddenly, the cheetah stops, turns, and then runs at 45 km/h eastward, as measured by a suddenly nervous crew member who stands alongside the cheetah's path. The change in the animal's velocity takes 2.0 s. What are the (a) magnitude and (b) direction of the animal's acceleration according to the cameraman and the (c) magnitude and (d) direction according to the nervous crew member?

•70 A boat is traveling upstream in the positive direction of an x axis at 14 km/h with respect to the water of a river. The water is flowing at 9.0 km/h with respect to the ground. What are the (a) magnitude and (b) direction of the boat's velocity with respect to the ground? A child on the boat walks from front to rear at 6.0 km/h with respect to the boat. What are the (c) magnitude and (d) direction of the child's velocity with respect to the ground?

••71 A suspicious-looking man runs as fast as he can along a moving sidewalk from one end to the other, taking 2.50 s. Then security agents appear, and the man runs as fast as he can back along the sidewalk to his starting point, taking 10.0 s. What is the ratio of the man's running speed to the sidewalk's speed?

sec. 4-9 Relative Motion in Two Dimensions

•72 A rugby player runs with the ball directly toward his opponent's goal, along the positive direction of an x axis. He can legally pass the ball to a teammate as long as the ball's velocity relative to the field does not have a positive x component. Suppose the player runs at speed 4.0 m/s relative to the field while he passes the ball with velocity $\vec{v}_{BP}$ relative to himself. If $\vec{v}_{BP}$ has magnitude 6.0 m/s, what is the smallest angle it can have for the pass to be legal?

••73 Two ships, A and B, leave port at the same time. Ship A travels northwest at 24 knots, and ship B travels at 28 knots in a direction 40° west of south. (1 knot = 1 nautical mile per hour; see Appendix D.) What are the (a) magnitude and (b) direction of the velocity of ship A relative to B? (c) After what time will the ships be 160 nautical miles apart? (d) What will be the bearing of B (the direction of B's position) relative to A at that time? SSM ILW

••74 A light plane attains an airspeed of 500 km/h. The pilot sets out for a destination 800 km due north but discovers that the plane must be headed 20.0° east of due north to fly there directly. The plane arrives in 2.00 h. What were the (a) magnitude and (b) direction of the wind velocity?

••75 Snow is falling vertically at a constant speed of 8.0 m/s. At what angle from the vertical do the snowflakes appear to be falling as viewed by the driver of a car traveling on a straight, level road with a speed of 50 km/h? SSM

••76 After flying for 15 min in a wind blowing 42 km/h at an angle of 20° south of east, an airplane pilot is over a town that is 55 km due north of the starting point. What is the speed of the airplane relative to the air?

••77 A train travels due south at 30 m/s (relative to the ground) in a rain that is blown toward the south by the wind. The path of each raindrop makes an angle of 70° with the vertical, as measured by an observer stationary on the ground. An observer on the train, however, sees the drops fall perfectly vertically. Determine the speed of the raindrops relative to the ground. SSM

••78 A 200-m-wide river flows due east at a uniform speed of 2.0 m/s. A boat with a speed of 8.0 m/s relative to the water leaves the south bank pointed in a direction 30° west of north. What are the (a) magnitude and (b) direction of the boat's velocity relative to the ground? (c) How long does the boat take to cross the river? GO

••79 Two highways intersect as shown in Fig. 4-49. At the instant shown, a police car P is distance $d_P = 800$ m from the intersection and moving at speed $v_P = 80$ km/h. Motorist M is distance $d_M = 600$ m from the intersection and moving at speed $v_M = 60$ km/h. (a) In unit-vector notation, what is the velocity of the motorist with respect to the police car? (b) For the instant shown in Fig. 4-49, what is the angle between the velocity found in (a) and the line of sight between the two cars? (c) If the cars maintain their velocities, do the answers to (a) and (b) change as the cars move nearer the intersection?

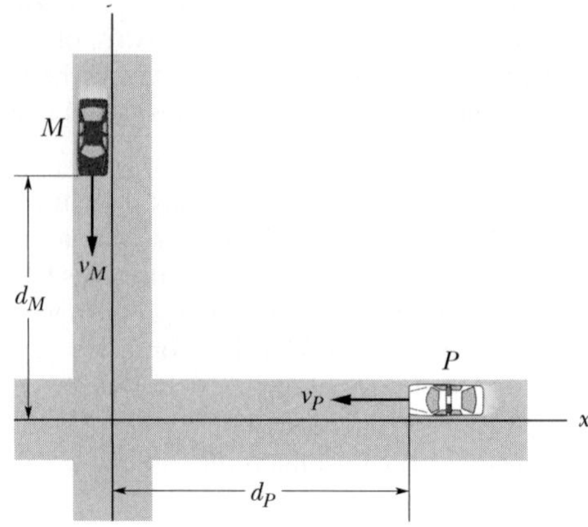

FIG. 4-49 Problem 79.

••80 In the overhead view of Fig. 4-50, Jeeps P and B race along straight lines, across flat terrain, and past stationary border guard A. Relative to the guard, B travels at a constant speed of 20.0 m/s, at the angle $\theta_2 = 30.0°$. Relative to the guard, P has accelerated from rest at a constant rate of 0.400 m/s² at the angle $\theta_1 = 60.0°$. At a certain time during the acceleration, P has a speed of 40.0 m/s. At that time, what are the (a) magnitude and (b) direction of the velocity of P relative to B and the (c) magnitude and (d) direction of the acceleration of P relative to B?

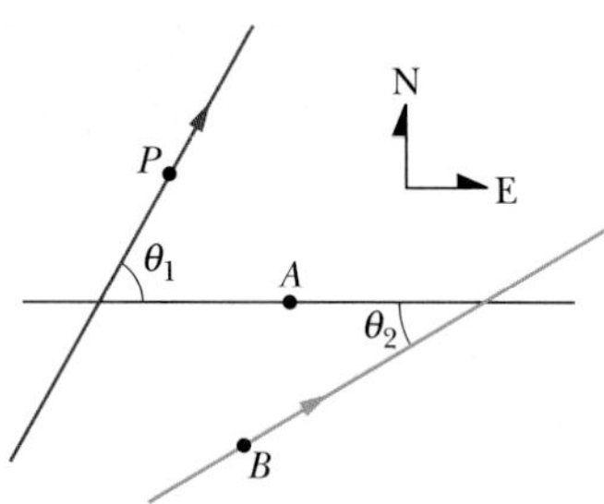

FIG. 4-50 Problem 80.

•••81 Ship A is located 4.0 km north and 2.5 km east of ship B. Ship A has a velocity of 22 km/h toward the south, and ship B has a velocity of 40 km/h in a direction 37° north of east. (a) What is the velocity of A relative to B in unit-vector notation with $\hat{i}$ toward the east? (b) Write an expression (in terms of $\hat{i}$ and $\hat{j}$) for the position of A relative to B as a function of t, where $t = 0$ when the ships are in the positions described above. (c) At what time is the separation between the ships least? (d) What is that least separation?

•••82 A 200-m-wide river has a uniform flow speed of 1.1 m/s through a jungle and toward the east. An explorer wishes to leave a small clearing on the south bank and cross the river in a powerboat that moves at a constant speed of 4.0 m/s with respect to the water. There is a clearing on the north bank 82 m upstream from a point directly opposite the clearing on the south bank. (a) In what direction must the boat be pointed in order to travel in a straight line and land in the clearing on the north bank? (b) How long will the boat take to cross the river and land in the clearing?

Additional Problems

83 You are kidnapped by political-science majors (who are upset because you told them political science is not a real science). Although blindfolded, you can tell the speed of their car (by the whine of the engine), the time of travel (by men-

tally counting off seconds), and the direction of travel (by turns along the rectangular street system). From these clues, you know that you are taken along the following course: 50 km/h for 2.0 min, turn 90° to the right, 20 km/h for 4.0 min, turn 90° to the right, 20 km/h for 60 s, turn 90° to the left, 50 km/h for 60 s, turn 90° to the right, 20 km/h for 2.0 min, turn 90° to the left, 50 km/h for 30 s. At that point, (a) how far are you from your starting point, and (b) in what direction relative to your initial direction of travel are you?

84 *Curtain of death.* A large metallic asteroid strikes Earth and quickly digs a crater into the rocky material below ground level by launching rocks upward and outward. The following table gives five pairs of launch speeds and angles (from the horizontal) for such rocks, based on a model of crater formation. (Other rocks, with intermediate speeds and angles, are also launched.) Suppose that you are at $x = 20$ km when the asteroid strikes the ground at time $t = 0$ and position $x = 0$ (Fig. 4-51). (a) At $t = 20$ s, what are the x and y coordinates of the rocks headed in your direction from launches A through E? (b) Plot these coordinates and then sketch a curve through the points to include rocks with intermediate launch speeds and angles. The curve should indicate what you would see as you look up into the approaching rocks and what dinosaurs must have seen during asteroid strikes long ago.

Launch	Speed (m/s)	Angle (degrees)
A	520	14.0
B	630	16.0
C	750	18.0
D	870	20.0
E	1000	22.0

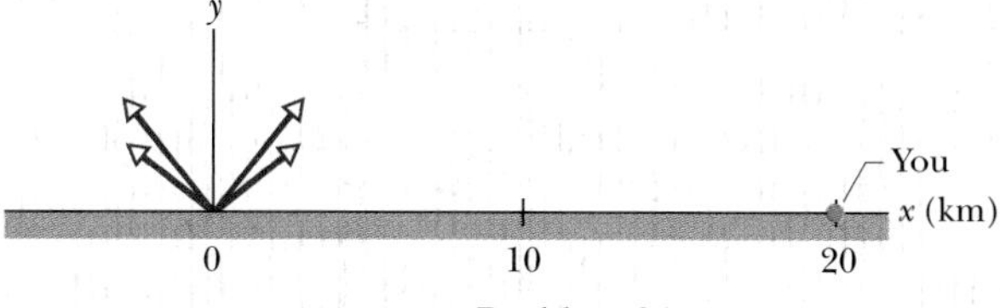

FIG. 4-51 Problem 84.

85 In Fig. 4-52, a lump of wet putty moves in uniform circular motion as it rides at a radius of 20.0 cm on the rim of a wheel rotating counterclockwise with a period of 5.00 ms. The lump then happens to fly off the rim at the 5 o'clock position (as if on a clock face). It leaves the rim at a height of $h = 1.20$ m from the floor and at a distance $d = 2.50$ m from a wall. At what height on the wall does the lump hit?

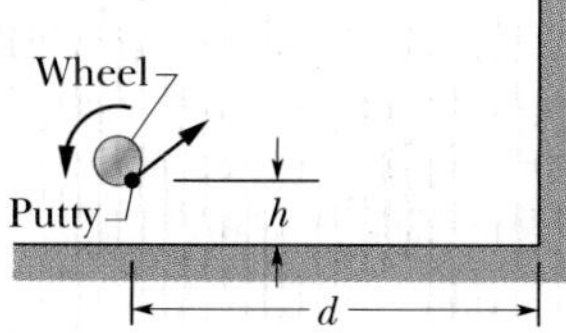

FIG. 4-52 Problem 85.

86 A particle is in uniform circular motion about the origin of an xy coordinate system, moving clockwise with a period of 7.00 s. At one instant, its position vector (from the origin) is $\vec{r} = (2.00\text{ m})\hat{i} - (3.00\text{ m})\hat{j}$. At that instant, what is its velocity in unit-vector notation?

87 In Fig. 4-53, a ball is shot directly upward from the ground with an initial speed of $v_0 = 7.00$ m/s. Simultaneously, a construction elevator cab begins to move upward from the ground with a constant speed of $v_c = 3.00$ m/s. What maximum height does the ball reach relative to (a) the ground and (b) the cab floor? At what rate does the speed of the ball change relative to (c) the ground and (d) the cab floor?

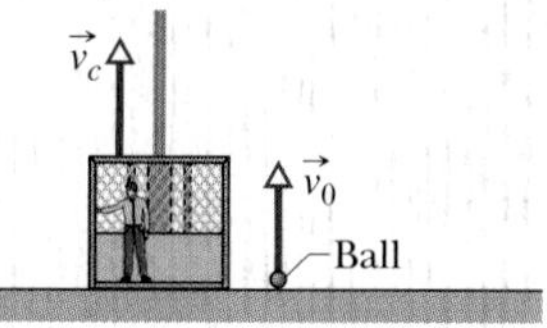

FIG. 4-53 Problem 87.

88 In Fig. 4-54a, a sled moves in the negative x direction at constant speed v_s while a ball of ice is shot from the sled with a velocity $\vec{v}_0 = v_{0x}\hat{i} + v_{0y}\hat{j}$ relative to the sled. When the ball lands, its horizontal displacement Δx_{bg} relative to the ground (from its launch position to its landing position) is measured. Figure 4-54b gives Δx_{bg} as a function of v_s. Assume the ball lands at approximately its launch height. What are the values of (a) v_{0x} and (b) v_{0y}? The ball's displacement Δx_{bs} relative to the sled can also be measured. Assume that the sled's velocity is not changed when the ball is shot. What is Δx_{bs} when v_s is (c) 5.0 m/s and (d) 15 m/s?

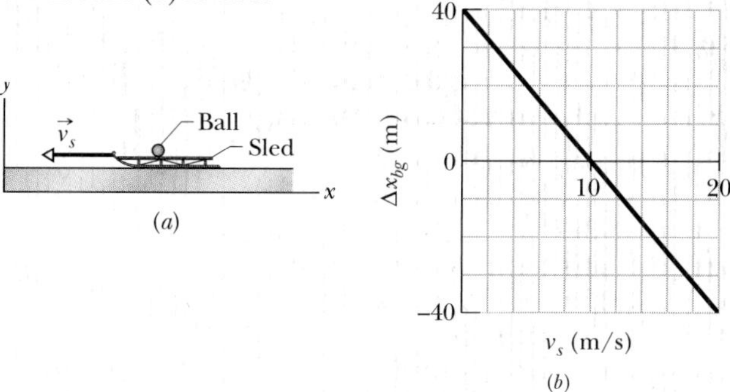

FIG. 4-54 Problem 88.

89 A woman who can row a boat at 6.4 km/h in still water faces a long, straight river with a width of 6.4 km and a current of 3.2 km/h. Let $\hat{i}$ point directly across the river and $\hat{j}$ point directly downstream. If she rows in a straight line to a point directly opposite her starting position, (a) at what angle to $\hat{i}$ must she point the boat and (b) how long will she take? (c) How long will she take if, instead, she rows 3.2 km *down* the river and then back to her starting point? (d) How long if she rows 3.2 km *up* the river and then back to her starting point? (e) At what angle to $\hat{i}$ should she point the boat if she wants to cross the river in the shortest possible time? (f) How long is that shortest time?

90 In Fig. 4-55, a radar station detects an airplane approaching directly from the east. At first observation, the airplane is at distance $d_1 = 360$ m from the station and at angle $\theta_1 = 40°$ above the horizon. The airplane is tracked through an angular change $\Delta\theta = 123°$ in the vertical east–west plane; its distance is then $d_2 = 790$ m. Find the (a) magnitude and (b) direction of the airplane's displacement during this period.

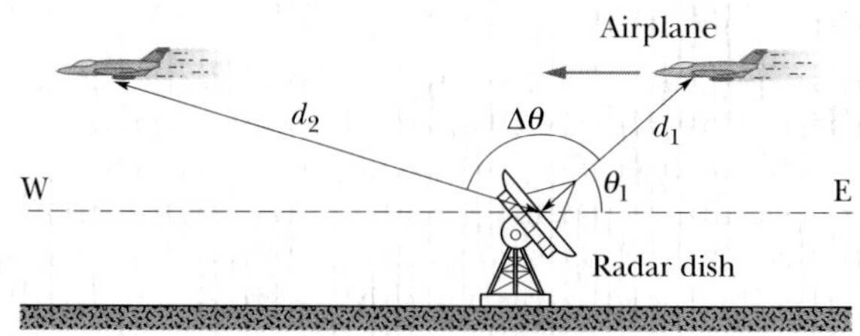

FIG. 4-55 Problem 90.

91 A rifle is aimed horizontally at a target 30 m away. The bullet hits the target 1.9 cm below the aiming point. What are (a) the bullet's time of flight and (b) its speed as it emerges from the rifle? SSM

92 The fast French train known as the TGV (Train à Grande Vitesse) has a scheduled average speed of 216 km/h. (a) If the train goes around a curve at that speed and the magnitude of the acceleration experienced by the passengers is to be limited to 0.050g, what is the smallest radius of curvature for the track that can be tolerated? (b) At what speed must the train go around a curve with a 1.00 km radius to be at the acceleration limit?

93 A magnetic field can force a charged particle to move in a circular path. Suppose that an electron moving in a circle experiences a radial acceleration of magnitude 3.0×10^{14} m/s^2 in a particular magnetic field. (a) What is the speed of the electron if the radius of its circular path is 15 cm? (b) What is the period of the motion?

94 The position vector for a proton is initially $\vec{r} = 5.0\hat{i} - 6.0\hat{j} + 2.0\hat{k}$ and then later is $\vec{r} = -2.0\hat{i} + 6.0\hat{j} + 2.0\hat{k}$, all in meters. (a) What is the proton's displacement vector, and (b) to what plane is that vector parallel?

95 A particle P travels with constant speed on a circle of radius $r = 3.00$ m (Fig. 4-56) and completes one revolution in 20.0 s. The particle passes through O at time $t = 0$. State the following vectors in magnitude-angle notation (angle relative to the positive direction of x). With respect to O, find the particle's position vector at the times t of (a) 5.00 s, (b) 7.50 s, and (c) 10.0 s.

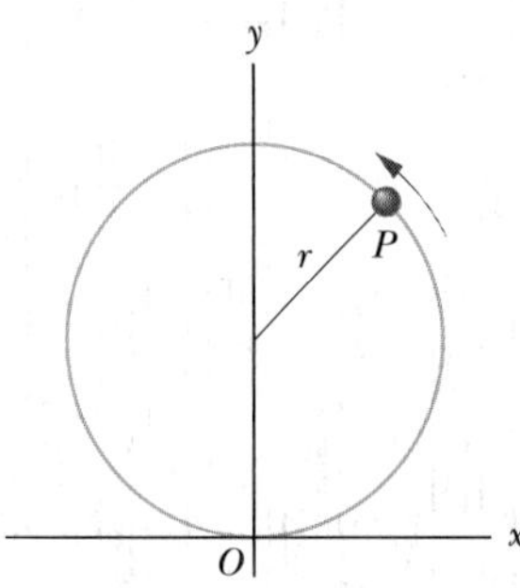

FIG. 4-56 Problem 95.

(d) For the 5.00 s interval from the end of the fifth second to the end of the tenth second, find the particle's displacement. For that interval, find (e) its average velocity and its velocity at the (f) beginning and (g) end. Next, find the acceleration at the (h) beginning and (i) end of that interval.

96 An iceboat sails across the surface of a frozen lake with constant acceleration produced by the wind. At a certain instant the boat's velocity is $(6.30\hat{i} - 8.42\hat{j})$ m/s. Three seconds later, because of a wind shift, the boat is instantaneously at rest. What is its average acceleration for this 3 s interval?

97 In 3.50 h, a balloon drifts 21.5 km north, 9.70 km east, and 2.88 km upward from its release point on the ground. Find (a) the magnitude of its average velocity and (b) the angle its average velocity makes with the horizontal.

98 A ball is thrown horizontally from a height of 20 m and hits the ground with a speed that is three times its initial speed. What is the initial speed?

99 A projectile is launched with an initial speed of 30 m/s at an angle of 60° above the horizontal. What are the (a) magnitude and (b) angle of its velocity 2.0 s after launch, and (c) is the angle above or below the horizontal? What are the (d) magnitude and (e) angle of its velocity 5.0 s after launch, and (f) is the angle above or below the horizontal?

100 An airport terminal has a moving sidewalk to speed passengers through a long corridor. Larry does not use the moving sidewalk; he takes 150 s to walk through the corridor. Curly, who simply stands on the moving sidewalk, covers the same distance in 70 s. Moe boards the sidewalk and walks along it. How long does Moe take to move through the corridor? Assume that Larry and Moe walk at the same speed.

101 A football player punts the football so that it will have a "hang time" (time of flight) of 4.5 s and land 46 m away. If the ball leaves the player's foot 150 cm above the ground, what must be the (a) magnitude and (b) angle (relative to the horizontal) of the ball's initial velocity?

102 For women's volleyball the top of the net is 2.24 m above the floor and the court measures 9.0 m by 9.0 m on each side of the net. Using a jump serve, a player strikes the ball at a point that is 3.0 m above the floor and a horizontal distance of 8.0 m from the net. If the initial velocity of the ball is horizontal, (a) what minimum magnitude must it have if the ball is to clear the net and (b) what maximum magnitude can it have if the ball is to strike the floor inside the back line on the other side of the net?

103 Figure 4-57 shows the straight path of a particle across an xy coordinate system as the particle is accelerated from rest during time interval Δt_1. The acceleration is constant. The xy coordinates for point A are (4.00 m, 6.00 m); those for point B are (12.0 m, 18.0 m). (a) What is the ratio a_y/a_x of the acceleration components? (b) What are the coordinates of the particle if the motion is continued for another interval equal to Δt_1?

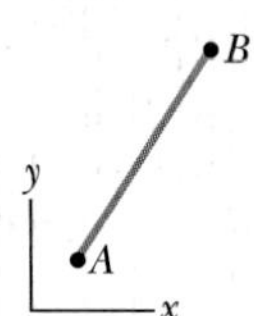

FIG. 4-57 Problem 103.

104 An astronaut is rotated in a horizontal centrifuge at a radius of 5.0 m. (a) What is the astronaut's speed if the centripetal acceleration has a magnitude of 7.0g? (b) How many revolutions per minute are required to produce this acceleration? (c) What is the period of the motion?

105 (a) What is the magnitude of the centripetal acceleration of an object on Earth's equator due to the rotation of Earth? (b) What would Earth's rotation period have to be for objects on the equator to have a centripetal acceleration of magnitude 9.8 m/s^2?

106 A person walks up a stalled 15-m-long escalator in 90 s. When standing on the same escalator, now moving, the person is carried up in 60 s. How much time would it take that person to walk up the moving escalator? Does the answer depend on the length of the escalator?

107 A baseball is hit at ground level. The ball reaches its maximum height above ground level 3.0 s after being hit. Then 2.5 s after reaching its maximum height, the ball barely clears a fence that is 97.5 m from where it was hit. Assume the ground is level. (a) What maximum height above ground level is reached by the ball? (b) How high is the fence? (c) How far beyond the fence does the ball strike the ground? SSM

108 The range of a projectile depends not only on v_0 and θ_0 but also on the value g of the free-fall acceleration, which varies from place to place. In 1936, Jesse Owens established a world's running broad jump record of 8.09 m at the Olympic Games at Berlin (where $g = 9.8128$ m/s^2). Assuming the same values of v_0 and θ_0, by how much would his record have differed if he had competed instead in 1956 at Melbourne (where $g = 9.7999$ m/s^2)?

109 During volcanic eruptions, chunks of solid rock can be blasted out of the volcano; these projectiles are called *volcanic bombs.* Figure 4-58 shows a cross section of Mt. Fuji, in Japan. (a) At what initial speed would a bomb have to be ejected, at angle $\theta_0 = 35°$ to the horizontal, from the vent at A in order to fall at the foot of the volcano at B, at vertical distance $h = 3.30$ km and horizontal distance $d = 9.40$ km? Ignore, for the moment, the effects of air on the bomb's travel. (b) What would be the time of flight? (c) Would the effect of the air increase or decrease your answer in (a)?

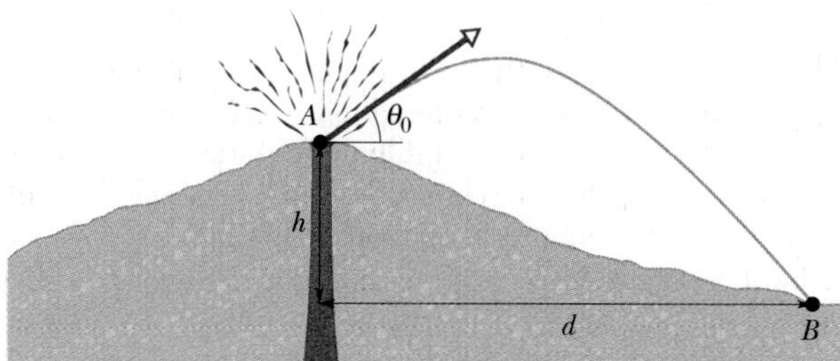

FIG. 4-58 Problem 109.

110 Long flights at midlatitudes in the Northern Hemisphere encounter the jet stream, an eastward airflow that can affect a plane's speed relative to Earth's surface. If a pilot maintains a certain speed relative to the air (the plane's *airspeed*), the speed relative to the surface (the plane's *ground speed*) is more when the flight is in the direction of the jet stream and less when the flight is opposite the jet stream. Suppose a round-trip flight is scheduled between two cities separated by 4000 km, with the outgoing flight in the direction of the jet stream and the return flight opposite it. The airline computer advises an airspeed of 1000 km/h, for which the difference in flight times for the outgoing and return flights is 70.0 min. What jet-stream speed is the computer using?

111 A particle starts from the origin at $t = 0$ with a velocity of $8.0\hat{j}$ m/s and moves in the xy plane with constant acceleration $(4.0\hat{i} + 2.0\hat{j})$ m/s^2. When the particle's x coordinate is 29 m, what are its (a) y coordinate and (b) speed? SSM

112 A sprinter running on a circular track has a velocity of constant magnitude 9.2 m/s and a centripetal acceleration of magnitude 3.8 m/s^2. What are (a) the track radius and (b) the period of the circular motion?

113 An electron having an initial horizontal velocity of magnitude 1.00×10^9 cm/s travels into the region between two horizontal metal plates that are electrically charged. In that region, the electron travels a horizontal distance of 2.00 cm and has a constant downward acceleration of magnitude 1.00×10^{17} cm/s^2 due to the charged plates. Find (a) the time the electron takes to travel the 2.00 cm, (b) the vertical distance it travels during that time, and the magnitudes of its (c) horizontal and (d) vertical velocity components as it emerges from the region.

114 An elevator without a ceiling is ascending with a constant speed of 10 m/s. A boy on the elevator shoots a ball directly upward, from a height of 2.0 m above the elevator floor, just as the elevator floor is 28 m above the ground. The initial speed of the ball with respect to the elevator is 20 m/s. (a) What maximum height above the ground does the ball reach? (b) How long does the ball take to return to the elevator floor?

115 Suppose that a space probe can withstand the stresses of a $20g$ acceleration. (a) What is the minimum turning radius of such a craft moving at a speed of one-tenth the speed of light? (b) How long would it take to complete a 90° turn at this speed?

116 At what initial speed must the basketball player in Fig. 4-59 throw the ball, at angle $\theta_0 = 55°$ above the horizontal, to make the foul shot? The horizontal distances are $d_1 = 1.0$ ft and $d_2 = 14$ ft, and the heights are $h_1 = 7.0$ ft and $h_2 = 10$ ft.

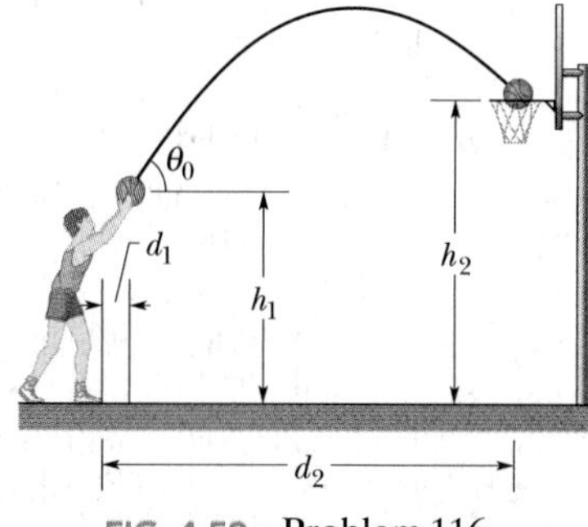

FIG. 4-59 Problem 116.

117 A wooden boxcar is moving along a straight railroad track at speed v_1. A sniper fires a bullet (initial speed v_2) at it from a high-powered rifle. The bullet passes through both lengthwise walls of the car, its entrance and exit holes being exactly opposite each other as viewed from within the car. From what direction, relative to the track, is the bullet fired? Assume that the bullet is not deflected upon entering the car, but that its speed decreases by 20%. Take $v_1 = 85$ km/h and $v_2 = 650$ m/s. (Why don't you need to know the width of the boxcar?)

118 You are to throw a ball with a speed of 12.0 m/s at a target that is height $h = 5.00$ m above the level at which you release the ball (Fig. 4-60). You want the ball's velocity to be horizontal at the instant it reaches the target. (a) At what angle θ above the horizontal must you throw the ball? (b) What is the horizontal distance from the release point to the target? (c) What is the speed of the ball just as it reaches the target?

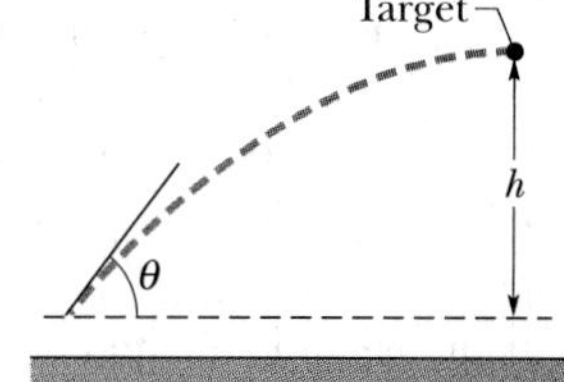

FIG. 4-60 Problem 118.

119 Figure 4-61 shows the path taken by a drunk skunk over level ground, from initial point i to final point f. The angles are $\theta_1 = 30.0°$, $\theta_2 = 50.0°$, and $\theta_3 = 80.0°$, and the distances are $d_1 = 5.00$ m, $d_2 = 8.00$ m, and $d_3 = 12.0$ m. What are the (a) magnitude and (b) angle of the skunk's displacement from i to f?

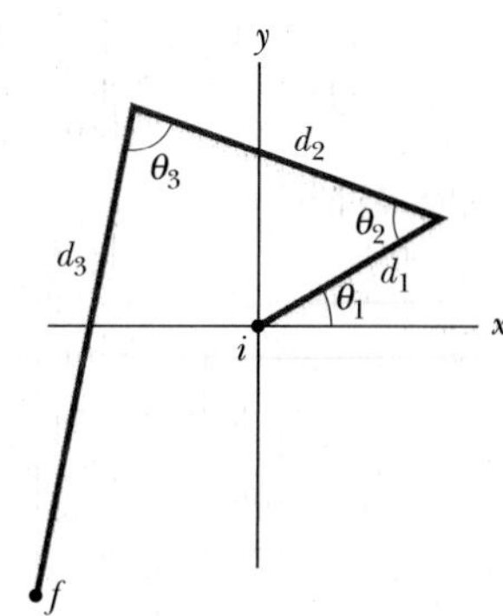

FIG. 4-61 Problem 119.

120 A projectile is fired with an initial speed $v_0 = 30.0$ m/s from level ground at a target that is on the ground, at distance $R = 20.0$ m, as shown in Fig. 4-62. What are the (a) least and (b) greatest launch angles that will allow the projectile to hit the target?

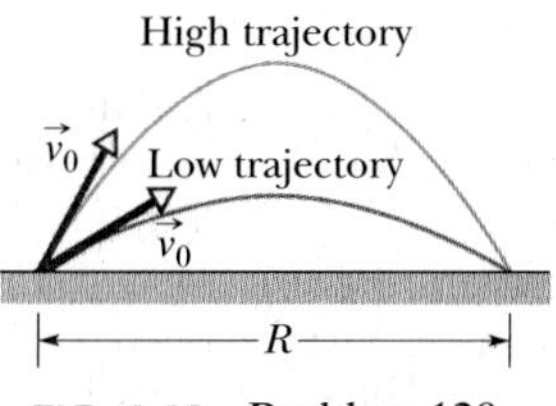

FIG. 4-62 Problem 120.

121 Oasis A is 90 km due west of oasis B. A desert camel leaves A and takes 50 h to walk 75 km at 37° north of due east. Next it takes 35 h to walk 65 km due south. Then it rests for 5.0 h. What are the (a) magnitude and (b) direction of the camel's displacement relative to A at the resting point? From the time the camel leaves A until the end of the rest period, what are the (c) magnitude and (d) direction of its average velocity and (e) its average speed? The camel's last drink was at A; it must be at B no more than 120 h later for its next drink. If it is to reach B just in time, what must be the (f) magnitude and (g) direction of its average velocity after the rest period? SSM

122 *A graphing surprise.* At time $t = 0$, a burrito is launched from level ground, with an initial speed of 16.0 m/s and launch angle θ_0. Imagine a position vector $\vec{r}$ continuously directed from the launching point to the burrito during the flight. Graph the magnitude r of the position vector for (a) $\theta_0 = 40.0°$ and (b) $\theta_0 = 80.0°$. For $\theta_0 = 40.0°$, (c) when does r reach its maximum value, (d) what is that value, and how far (e) horizontally and (f) vertically is the burrito from the launch point? For $\theta_0 = 80.0°$, (g) when does r reach its maximum value, (h) what is that value, and how far (i) horizontally and (j) vertically is the burrito from the launch point?

123 In Sample Problem 4-7b, a ball is shot through a horizontal distance of 686 m by a cannon located at sea level and angled at 45° from the horizontal. How much greater would the horizontal distance have been had the cannon been 30 m higher?

124 (a) If an electron is projected horizontally with a speed of 3.0×10^6 m/s, how far will it fall in traversing 1.0 m of horizontal distance? (b) Does the answer increase or decrease if the initial speed is increased?

125 The magnitude of the velocity of a projectile when it is at its maximum height above ground level is 10 m/s. (a) What is the magnitude of the velocity of the projectile 1.0 s before it achieves its maximum height? (b) What is the magnitude of the velocity of the projectile 1.0 s after it achieves its maximum height? If we take $x = 0$ and $y = 0$ to be at the point of maximum height and positive x to be in the direction of the velocity there, what are the (c) x coordinate and (d) y coordinate of the projectile 1.0 s before it reaches its maximum height and the (e) x coordinate and (f) y coordinate 1.0 s after it reaches its maximum height?

126 A frightened rabbit moving at 6.0 m/s due east runs onto a large area of level ice of negligible friction. As the rabbit slides across the ice, the force of the wind causes it to have a constant acceleration of 1.4 m/s^2, due north. Choose a coordinate system with the origin at the rabbit's initial position on the ice and the positive x axis directed toward the east. In unit-vector notation, what are the rabbit's (a) velocity and (b) position when it has slid for 3.0 s?

127 The pilot of an aircraft flies due east relative to the ground in a wind blowing 20 km/h toward the south. If the speed of the aircraft in the absence of wind is 70 km/h, what is the speed of the aircraft relative to the ground?

128 The pitcher in a slow-pitch softball game releases the ball at a point 3.0 ft above ground level. A stroboscopic plot of the position of the ball is shown in Fig. 4-63, where the readings are 0.25 s apart and the ball is released at $t = 0$. (a) What is the initial speed of the ball? (b) What is the speed of the ball at the instant it reaches its maximum height above ground level? (c) What is that maximum height?

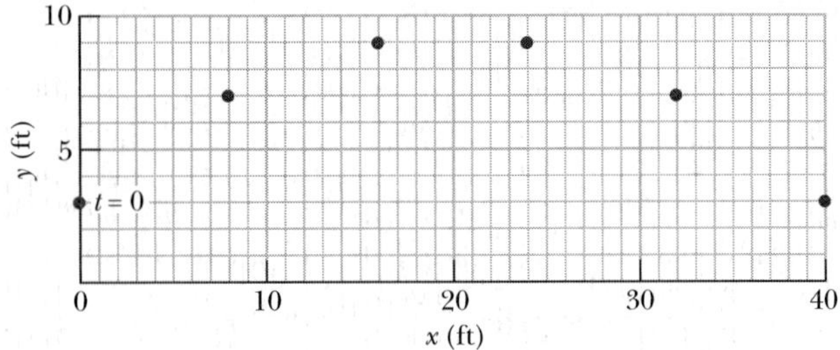

FIG. 4-63 Problem 128.

129 The New Hampshire State Police use aircraft to enforce highway speed limits. Suppose that one of the airplanes has a speed of 135 mi/h in still air. It is flying straight north so that it is at all times directly above a north–south highway. A ground observer tells the pilot by radio that a 70.0 mi/h wind is blowing but neglects to give the wind direction. The pilot observes that in spite of the wind the plane can travel 135 mi along the highway in 1.00 h. In other words, the ground speed is the same as if there were no wind. (a) From what direction is the wind blowing? (b) What is the heading of the plane; that is, in what direction does it point?

130 The position $\vec{r}$ of a particle moving in the xy plane is given by $\vec{r} = 2t\hat{i} + 2\sin[(\pi/4 \text{ rad/s})t]\hat{j}$, where $\vec{r}$ is in meters and t is in seconds. (a) Calculate the x and y components of the particle's position at $t = 0, 1.0, 2.0, 3.0$, and 4.0 s and sketch the particle's path in the xy plane for the interval $0 \leq t \leq 4.0$ s. (b) Calculate the components of the particle's velocity at $t = 1.0$, 2.0, and 3.0 s. Show that the velocity is tangent to the path of the particle and in the direction the particle is moving at each time by drawing the velocity vectors on the plot of the particle's path in part (a). (c) Calculate the components of the particle's acceleration at $t = 1.0, 2.0$, and 3.0 s.

131 A golfer tees off from the top of a rise, giving the golf ball an initial velocity of 43 m/s at an angle of 30° above the horizontal. The ball strikes the fairway a horizontal distance of 180 m from the tee. Assume the fairway is level. (a) How high is the rise above the fairway? (b) What is the speed of the ball as it strikes the fairway?

132 A track meet is held on a planet in a distant solar system. A shot-putter releases a shot at a point 2.0 m above ground level. A stroboscopic plot of the position of the shot is shown in Fig. 4-64, where the readings are 0.50 s apart and the shot is released at time $t = 0$. (a) What is the initial velocity of the shot in unit-vector notation? (b) What is the magnitude of the free-fall acceleration on the planet? (c) How long after it is released does the shot reach the ground? (d) If an identical throw of the shot is made on the surface of Earth, how long after it is released does it reach the ground?

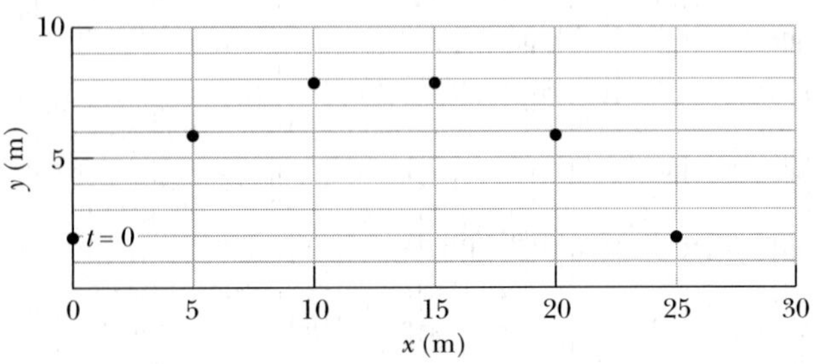

FIG. 4-64 Problem 132.

Force and Motion—I

5

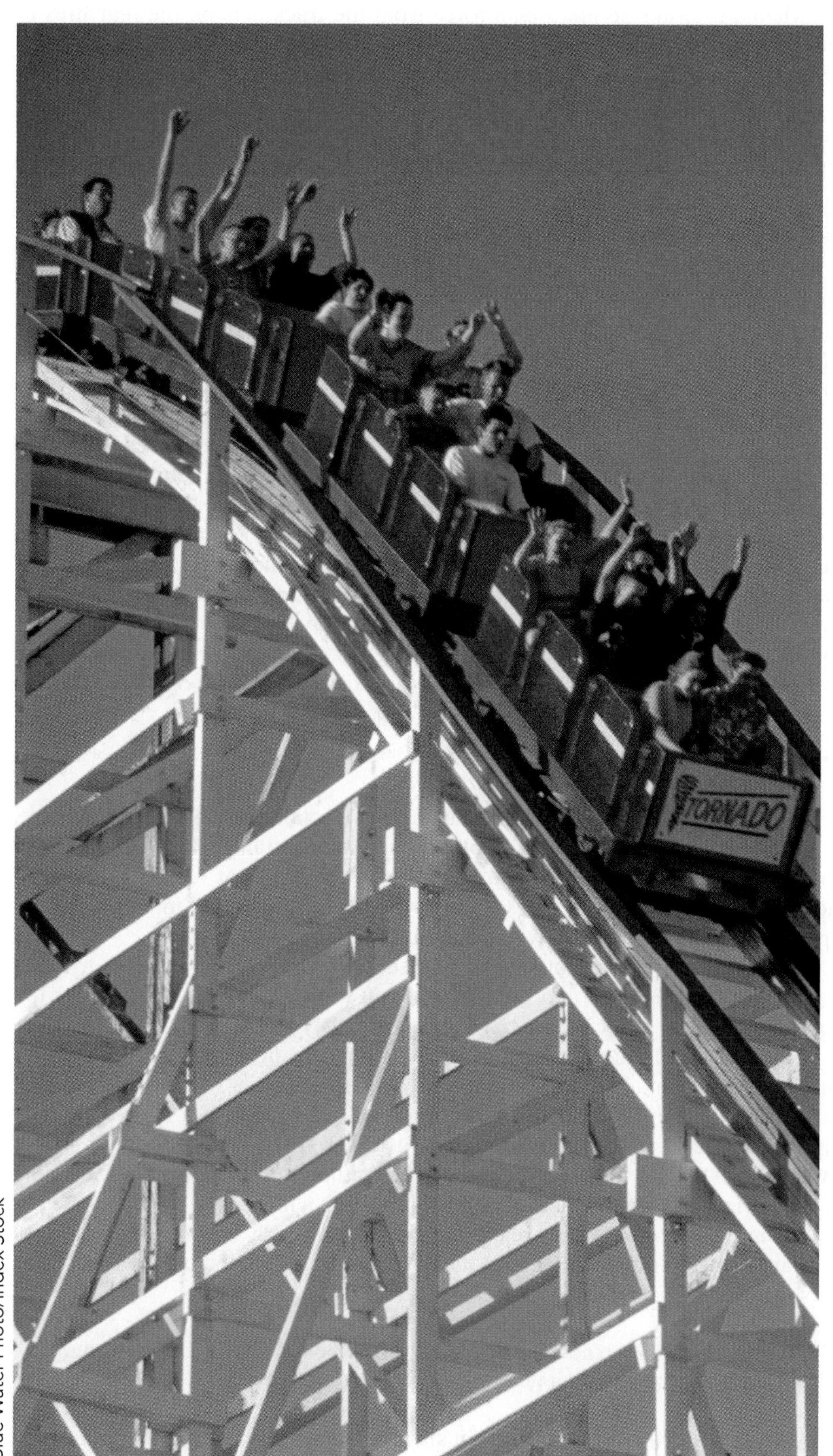

Blue Water Photo/Index Stock

Many roller-coaster enthusiasts prefer riding in the first car because they enjoy being the first to go over an "edge" and onto a downward slope. However, many other enthusiasts prefer the rear car—they claim that going over the edge is far more frightening there. The roller coaster is certainly moving faster when the last car is dragged over the edge by the rest of the roller coaster. But there seems to be some other, more subtle element that brings out the fear as that last car comes to the edge.

What is the subtle fear factor in riding the last car in a roller coaster?

The answer is in this chapter.

5-1 WHAT IS PHYSICS?

We have seen that part of physics is a study of motion, including accelerations, which are changes in velocities. Physics is also a study of what can *cause* an object to accelerate. That cause is a **force,** which is, loosely speaking, a push or pull on the object. The force is said to *act* on the object to change its velocity. For example, when a dragster accelerates, a force from the track acts on the rear tires to cause the dragster's acceleration. When a defensive guard knocks down a quarterback, a force from the guard acts on the quarterback to cause the quarterback's backward acceleration. When a car slams into a telephone pole, a force on the car from the pole causes the car to stop. Science, engineering, legal, and medical journals are filled with articles about forces on objects, including people.

5-2 | Newtonian Mechanics

The relation between a force and the acceleration it causes was first understood by Isaac Newton (1642–1727) and is the subject of this chapter. The study of that relation, as Newton presented it, is called *Newtonian mechanics.* We shall focus on its three primary laws of motion.

Newtonian mechanics does not apply to all situations. If the speeds of the interacting bodies are very large—an appreciable fraction of the speed of light—we must replace Newtonian mechanics with Einstein's special theory of relativity, which holds at any speed, including those near the speed of light. If the interacting bodies are on the scale of atomic structure (for example, they might be electrons in an atom), we must replace Newtonian mechanics with quantum mechanics. Physicists now view Newtonian mechanics as a special case of these two more comprehensive theories. Still, it is a very important special case because it applies to the motion of objects ranging in size from the very small (almost on the scale of atomic structure) to astronomical (galaxies and clusters of galaxies).

5-3 | Newton's First Law

Before Newton formulated his mechanics, it was thought that some influence, a "force," was needed to keep a body moving at constant velocity. Similarly, a body was thought to be in its "natural state" when it was at rest. For a body to move with constant velocity, it seemingly had to be propelled in some way, by a push or a pull. Otherwise, it would "naturally" stop moving.

These ideas were reasonable. If you send a puck sliding across a wooden floor, it does indeed slow and then stop. If you want to make it move across the floor with constant velocity, you have to continuously pull or push it.

Send a puck sliding over the ice of a skating rink, however, and it goes a lot farther. You can imagine longer and more slippery surfaces, over which the puck would slide farther and farther. In the limit you can think of a long, extremely slippery surface (said to be a **frictionless surface**), over which the puck would hardly slow. (We can in fact come close to this situation by sending a puck sliding over a horizontal air table, across which it moves on a film of air.)

From these observations, we can conclude that a body will keep moving with constant velocity if no force acts on it. That leads us to the first of Newton's three laws of motion:

> **Newton's First Law:** If no force acts on a body, the body's velocity cannot change; that is, the body cannot accelerate.

In other words, if the body is at rest, it stays at rest. If it is moving, it continues to move with the same velocity (same magnitude *and* same direction).

5-4 | Force

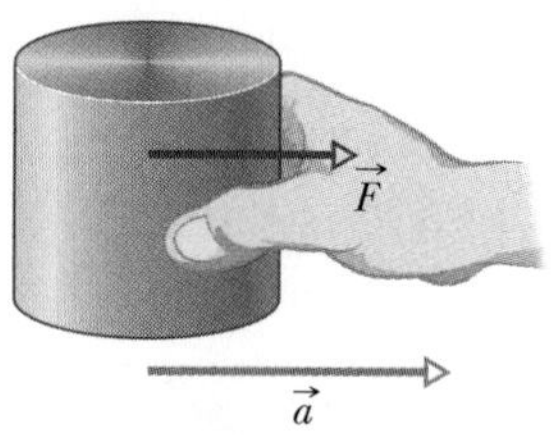

FIG. 5-1 A force $\vec{F}$ on the standard kilogram gives that body an acceleration $\vec{a}$.

We now wish to define the unit of force. We know that a force can cause the acceleration of a body. Thus, we shall define the unit of force in terms of the acceleration that a force gives to a standard reference body, which we take to be the standard kilogram of Fig. 1-3. This body has been assigned, exactly and by definition, a mass of 1 kg.

We put the standard body on a horizontal frictionless table and pull the body to the right (Fig. 5-1) so that, by trial and error, it eventually experiences a measured acceleration of 1 m/s^2. We then declare, as a matter of definition, that the force we are exerting on the standard body has a magnitude of 1 newton (abbreviated N).

We can exert a 2 N force on our standard body by pulling it so that its measured acceleration is 2 m/s^2, and so on. Thus in general, if our standard body of 1 kg mass has an acceleration of magnitude a, we know that a force F must be acting on it and that the magnitude of the force (in newtons) is equal to the magnitude of the acceleration (in meters per second per second).

Thus, a force is measured by the acceleration it produces. However, acceleration is a vector quantity, with both magnitude and direction. Is force also a vector quantity? We can easily assign a direction to a force (just assign the direction of the acceleration), but that is not sufficient. We must prove by experiment that forces are vector quantities. Actually, that has been done: forces are indeed vector quantities; they have magnitudes and directions, and they combine according to the vector rules of Chapter 3.

This means that when two or more forces act on a body, we can find their **net force,** or **resultant force,** by adding the individual forces vectorially. A single force that has the magnitude and direction of the net force has the same effect on the body as all the individual forces together. This fact is called the **principle of superposition for forces.** The world would be quite strange if, for example, you and a friend were to pull on the standard body in the same direction, each with a force of 1 N, and yet somehow the net pull was 14 N.

In this book, forces are most often represented with a vector symbol such as $\vec{F}$, and a net force is represented with the vector symbol $\vec{F}_{net}$. As with other vectors, a force or a net force can have components along coordinate axes. When forces act only along a single axis, they are single-component forces. Then we can drop the overhead arrows on the force symbols and just use signs to indicate the directions of the forces along that axis.

Instead of the wording used in Section 5-3, the more proper statement of Newton's First Law is in terms of a *net* force:

Newton's First Law: If no *net* force acts on a body ($\vec{F}_{net} = 0$), the body's velocity cannot change; that is, the body cannot accelerate.

There may be multiple forces acting on a body, but if their net force is zero, the body cannot accelerate.

Inertial Reference Frames

Newton's first law is not true in all reference frames, but we can always find reference frames in which it (as well as the rest of Newtonian mechanics) is true. Such frames are called **inertial reference frames,** or simply **inertial frames.**

An inertial reference frame is one in which Newton's laws hold.

For example, we can assume that the ground is an inertial frame provided we can neglect Earth's astronomical motions (such as its rotation).

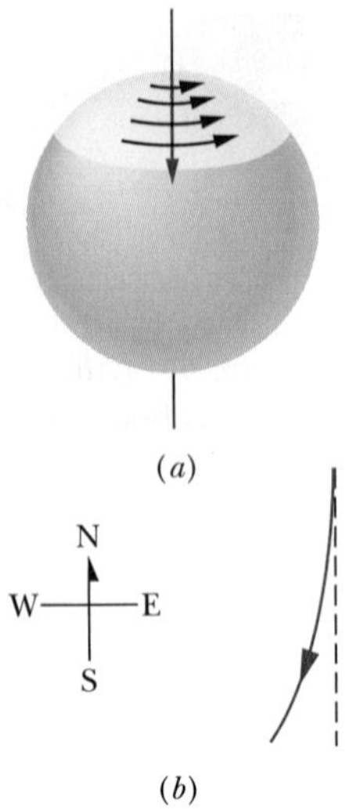

FIG. 5-2 (*a*) The path of a puck sliding from the north pole as seen from a stationary point in space. Earth rotates to the east. (*b*) The path of the puck as seen from the ground.

That assumption works well if, say, a puck is sent sliding along a *short* strip of frictionless ice—we would find that the puck's motion obeys Newton's laws. However, suppose the puck is sent sliding along a *long* ice strip extending from the north pole (Fig. 5-2*a*). If we view the puck from a stationary frame in space, the puck moves south along a simple straight line because Earth's rotation around the north pole merely slides the ice beneath the puck. However, if we view the puck from a point on the ground so that we rotate with Earth, the puck's path is not a simple straight line. Because the eastward speed of the ground beneath the puck is greater the farther south the puck slides, from our ground-based view the puck appears to be deflected westward (Fig. 5-2*b*). However, this apparent deflection is caused not by a force as required by Newton's laws but by the fact that we see the puck from a rotating frame. In this situation, the ground is a **noninertial frame.**

In this book we usually assume that the ground is an inertial frame and that measured forces and accelerations are from this frame. If measurements are made in, say, an elevator that is accelerating relative to the ground, then the measurements are being made in a noninertial frame and the results can be surprising. We see an example of this in Sample Problem 5-8.

CHECKPOINT 1 Which of the figure's six arrangements correctly show the vector addition of forces $\vec{F}_1$ and $\vec{F}_2$ to yield the third vector, which is meant to represent their net force $\vec{F}_{net}$?

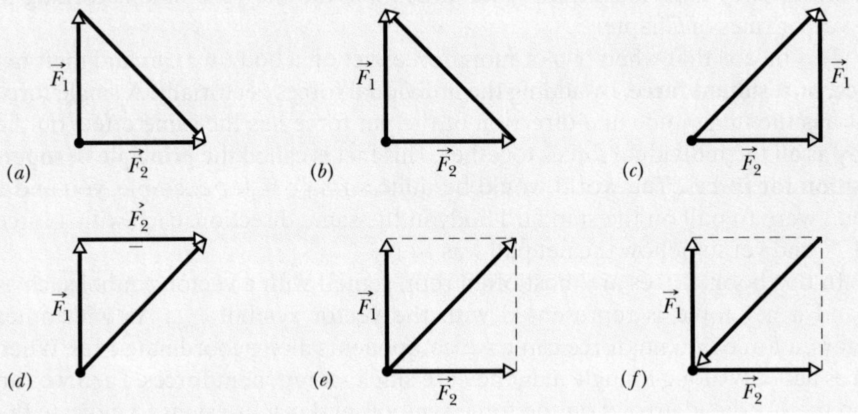

5-5 | Mass

Everyday experience tells us that a given force produces different magnitudes of acceleration for different bodies. Put a baseball and a bowling ball on the floor and give both the same sharp kick. Even if you don't actually do this, you know the result: The baseball receives a noticeably larger acceleration than the bowling ball. The two accelerations differ because the mass of the baseball differs from the mass of the bowling ball—but what, exactly, is mass?

We can explain how to measure mass by imagining a series of experiments in an inertial frame. In the first experiment we exert a force on a standard body, whose mass m_0 is defined to be 1.0 kg. Suppose that the standard body accelerates at 1.0 m/s^2. We can then say the force on that body is 1.0 N.

We next apply that same force (we would need some way of being certain it is the same force) to a second body, body X, whose mass is not known. Suppose we find that this body X accelerates at 0.25 m/s^2. We know that a *less massive* baseball receives a *greater acceleration* than a more massive bowling ball when the same force (kick) is applied to both. Let us then make the following conjecture: The ratio of the masses of two bodies is equal to the inverse of the ratio of their accelerations when the same force is applied to both. For body X and the standard body, this tells us that

$$\frac{m_X}{m_0} = \frac{a_0}{a_X}.$$

Solving for m_X yields

$$m_X = m_0 \frac{a_0}{a_X} = (1.0\ \text{kg}) \frac{1.0\ \text{m/s}^2}{0.25\ \text{m/s}^2} = 4.0\ \text{kg}.$$

Our conjecture will be useful, of course, only if it continues to hold when we change the applied force to other values. For example, if we apply an 8.0 N force to the standard body, we obtain an acceleration of 8.0 m/s^2. When the 8.0 N force is applied to body X, we obtain an acceleration of 2.0 m/s^2. Our conjecture then gives us

$$m_X = m_0 \frac{a_0}{a_X} = (1.0\ \text{kg}) \frac{8.0\ \text{m/s}^2}{2.0\ \text{m/s}^2} = 4.0\ \text{kg},$$

consistent with our first experiment. Many experiments yielding similar results indicate that our conjecture provides a consistent and reliable means of assigning a mass to any given body.

Our measurement experiments indicate that mass is an *intrinsic* characteristic of a body—that is, a characteristic that automatically comes with the existence of the body. They also indicate that mass is a scalar quantity. However, the nagging question remains: What, exactly, is mass?

Since the word *mass* is used in everyday English, we should have some intuitive understanding of it, maybe something that we can physically sense. Is it a body's size, weight, or density? The answer is no, although those characteristics are sometimes confused with mass. We can say only that *the mass of a body is the characteristic that relates a force on the body to the resulting acceleration.* Mass has no more familiar definition; you can have a physical sensation of mass only when you try to accelerate a body, as in the kicking of a baseball or a bowling ball.

5-6 | Newton's Second Law

All the definitions, experiments, and observations we have discussed so far can be summarized in one neat statement:

Newton's Second Law: The net force on a body is equal to the product of the body's mass and its acceleration.

In equation form,

$$\vec{F}_{\text{net}} = m\vec{a} \quad \text{(Newton's second law).} \tag{5-1}$$

This equation is simple, but we must use it cautiously. First, we must be certain about which body we are applying it to. Then $\vec{F}_{\text{net}}$ must be the vector sum of *all* the forces that act on *that* body. Only forces that act on *that* body are to be included in the vector sum, not forces acting on other bodies that might be involved in the given situation. For example, if you are in a rugby scrum, the net force on *you* is the vector sum of all the pushes and pulls on *your* body. It does not include any push or pull on another player from you.

Like other vector equations, Eq. 5-1 is equivalent to three component equations, one for each axis of an xyz coordinate system:

$$F_{\text{net},x} = ma_x, \quad F_{\text{net},y} = ma_y, \quad \text{and} \quad F_{\text{net},z} = ma_z. \tag{5-2}$$

Each of these equations relates the net force component along an axis to the acceleration along that same axis. For example, the first equation tells us that the

TABLE 5-1

Units in Newton's Second Law (Eqs. 5-1 and 5-2)

System	Force	Mass	Acceleration
SI	newton (N)	kilogram (kg)	m/s^2
CGS[a]	dyne	gram (g)	cm/s^2
British[b]	pound (lb)	slug	ft/s^2

[a] 1 dyne = $1\ g \cdot cm/s^2$.
[b] 1 lb = $1\ slug \cdot ft/s^2$.

sum of all the force components along the x axis causes the x component a_x of the body's acceleration, but causes no acceleration in the y and z directions. Turned around, the acceleration component a_x is caused only by the sum of the force components along the x axis. In general,

The acceleration component along a given axis is caused *only by* the sum of the force components along that *same* axis, and not by force components along any other axis.

Equation 5-1 tells us that if the net force on a body is zero, the body's acceleration $\vec{a} = 0$. If the body is at rest, it stays at rest; if it is moving, it continues to move at constant velocity. In such cases, any forces on the body *balance* one another, and both the forces and the body are said to be in *equilibrium*. Commonly, the forces are also said to *cancel* one another, but the term "cancel" is tricky. It does *not* mean that the forces cease to exist (canceling forces is not like canceling dinner reservations). The forces still act on the body.

For SI units, Eq. 5-1 tells us that

$$1\ \text{N} = (1\ \text{kg})(1\ \text{m/s}^2) = 1\ \text{kg} \cdot \text{m/s}^2. \tag{5-3}$$

Some force units in other systems of units are given in Table 5-1 and Appendix D.

To solve problems with Newton's second law, we often draw a **free-body diagram** in which the only body shown is the one for which we are summing forces. A sketch of the body itself is preferred by some teachers but, to save space in these chapters, we shall usually represent the body with a dot. Also, each force on the body is drawn as a vector arrow with its tail on the body. A coordinate system is usually included, and the acceleration of the body is sometimes shown with a vector arrow (labeled as an acceleration).

A **system** consists of one or more bodies, and any force on the bodies inside the system from bodies outside the system is called an **external force.** If the bodies making up a system are rigidly connected to one another, we can treat the system as one composite body, and the net force $\vec{F}_{net}$ on it is the vector sum of all external forces. (We do not include **internal forces**—that is, forces between two bodies inside the system.) For example, a connected railroad engine and car form a system. If, say, a tow line pulls on the front of the engine, the force due to the tow line acts on the whole engine–car system. Just as for a single body, we can relate the net external force on a system to its acceleration with Newton's second law, $\vec{F}_{net} = m\vec{a}$, where m is the total mass of the system.

CHECKPOINT 2 The figure here shows two horizontal forces acting on a block on a frictionless floor. If a third horizontal force $\vec{F}_3$ also acts on the block, what are the magnitude and direction of $\vec{F}_3$ when the block is (a) stationary and (b) moving to the left with a constant speed of 5 m/s?

3 N
5 N

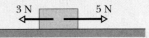

Sample Problem 5-1

Figures 5-3*a* to *c* show three situations in which one or two forces act on a puck that moves over frictionless ice along an x axis, in one-dimensional motion. The puck's mass is $m = 0.20$ kg. Forces $\vec{F}_1$ and $\vec{F}_2$ are directed along the axis and have magnitudes $F_1 = 4.0$ N and $F_2 = 2.0$ N. Force $\vec{F}_3$ is directed at angle $\theta = 30°$ and has magnitude $F_3 = 1.0$ N. In each situation, what is the acceleration of the puck?

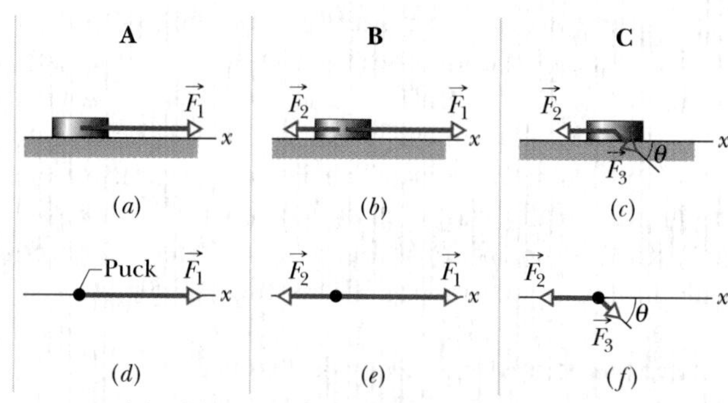

FIG. 5-3 (*a*)–(*c*) In three situations, forces act on a puck that moves along an x axis. (*d*)–(*f*) Free-body diagrams.

KEY IDEA In each situation we can relate the acceleration $\vec{a}$ to the net force $\vec{F}_{net}$ acting on the puck with Newton's second law, $\vec{F}_{net} = m\vec{a}$. However, because the motion is along only the x axis, we can simplify each situation by writing the second law for x components only:

$$F_{net,x} = ma_x. \tag{5-4}$$

The free-body diagrams for the three situations are given in Figs. 5-3*d* to *f* with the puck represented by a dot.

Situation A: For Fig. 5-3*d*, where only one horizontal force acts, Eq. 5-4 gives us

$$F_1 = ma_x,$$

which, with given data, yields

$$a_x = \frac{F_1}{m} = \frac{4.0 \text{ N}}{0.20 \text{ kg}} = 20 \text{ m/s}^2. \quad \text{(Answer)}$$

The positive answer indicates that the acceleration is in the positive direction of the x axis.

Situation B: In Fig. 5-3*e*, two horizontal forces act on the puck, $\vec{F}_1$ in the positive direction of x and $\vec{F}_2$ in the negative direction. Now Eq. 5-4 gives us

$$F_1 - F_2 = ma_x,$$

which, with given data, yields

$$a_x = \frac{F_1 - F_2}{m} = \frac{4.0 \text{ N} - 2.0 \text{ N}}{0.20 \text{ kg}} = 10 \text{ m/s}^2.$$

(Answer)

Thus, the net force accelerates the puck in the positive direction of the x axis.

Situation C: In Fig. 5-3*f*, force $\vec{F}_3$ is not directed along the direction of the puck's acceleration; only x component $F_{3,x}$ is. (Force $\vec{F}_3$ is two-dimensional but the motion is only one-dimensional.) Thus, we write Eq. 5-4 as

$$F_{3,x} - F_2 = ma_x. \tag{5-5}$$

From the figure, we see that $F_{3,x} = F_3 \cos \theta$. Solving for the acceleration and substituting for $F_{3,x}$ yield

$$a_x = \frac{F_{3,x} - F_2}{m} = \frac{F_3 \cos \theta - F_2}{m}$$

$$= \frac{(1.0 \text{ N})(\cos 30°) - 2.0 \text{ N}}{0.20 \text{ kg}} = -5.7 \text{ m/s}^2.$$

(Answer)

Thus, the net force accelerates the puck in the negative direction of the x axis.

Sample Problem 5-2

In the overhead view of Fig. 5-4*a*, a 2.0 kg cookie tin is accelerated at 3.0 m/s² in the direction shown by $\vec{a}$, over a frictionless horizontal surface. The acceleration is caused by three horizontal forces, only two of which are shown: $\vec{F}_1$ of magnitude 10 N and $\vec{F}_2$ of magnitude 20 N. What is the third force $\vec{F}_3$ in unit-vector notation and in magnitude-angle notation?

KEY IDEA The net force $\vec{F}_{net}$ on the tin is the sum of the three forces and is related to the acceleration $\vec{a}$ via Newton's second law ($\vec{F}_{net} = m\vec{a}$). Thus,

$$\vec{F}_1 + \vec{F}_2 + \vec{F}_3 = m\vec{a}, \tag{5-6}$$

which gives us

$$\vec{F}_3 = m\vec{a} - \vec{F}_1 - \vec{F}_2. \tag{5-7}$$

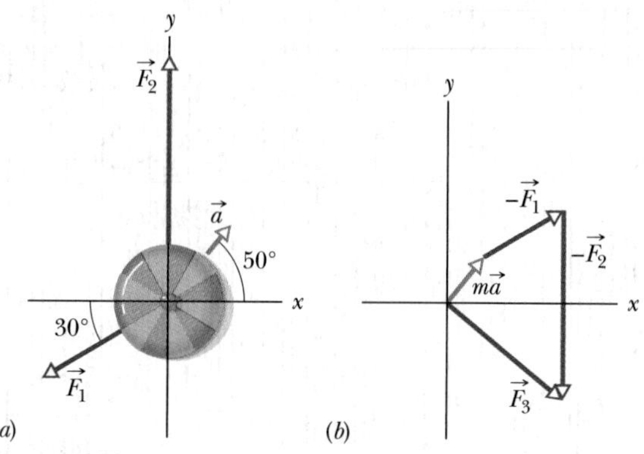

FIG. 5-4 (*a*) An overhead view of two of three horizontal forces that act on a cookie tin, resulting in acceleration $\vec{a}$. $\vec{F}_3$ is not shown. (*b*) An arrangement of vectors $m\vec{a}$, $-\vec{F}_1$, and $-\vec{F}_2$ to find force $\vec{F}_3$.

Calculations: Because this is a two-dimensional problem, we *cannot* find $\vec{F}_3$ merely by substituting the magnitudes for the vector quantities on the right side of Eq. 5-7. Instead, we must vectorially add $m\vec{a}$, $-\vec{F}_1$ (the reverse of $\vec{F}_1$), and $-\vec{F}_2$ (the reverse of $\vec{F}_2$), as shown in Fig. 5-4*b*. This addition can be done directly on a vector-capable calculator because we know both magnitude and angle for all three vectors. However, here we shall evaluate the right side of Eq. 5-7 in terms of components, first along the *x* axis and then along the *y* axis.

x components: Along the *x* axis we have

$$F_{3,x} = ma_x - F_{1,x} - F_{2,x}$$
$$= m(a\cos 50°) - F_1\cos(-150°) - F_2\cos 90°.$$

Then, substituting known data, we find

$$F_{3,x} = (2.0\text{ kg})(3.0\text{ m/s}^2)\cos 50° - (10\text{ N})\cos(-150°)$$
$$- (20\text{ N})\cos 90°$$
$$= 12.5\text{ N}.$$

y components: Similarly, along the *y* axis we find

$$F_{3,y} = ma_y - F_{1,y} - F_{2,y}$$
$$= m(a\sin 50°) - F_1\sin(-150°) - F_2\sin 90°$$
$$= (2.0\text{ kg})(3.0\text{ m/s}^2)\sin 50° - (10\text{ N})\sin(-150°)$$
$$- (20\text{ N})\sin 90°$$
$$= -10.4\text{ N}.$$

Vector: In unit-vector notation, we can write

$$\vec{F}_3 = F_{3,x}\hat{\text{i}} + F_{3,y}\hat{\text{j}} = (12.5\text{ N})\hat{\text{i}} - (10.4\text{ N})\hat{\text{j}}$$
$$\approx (13\text{ N})\hat{\text{i}} - (10\text{ N})\hat{\text{j}}. \qquad \text{(Answer)}$$

We can now use a vector-capable calculator to get the magnitude and the angle of $\vec{F}_3$. We can also use Eq. 3-6 to obtain the magnitude and the angle (from the positive direction of the *x* axis) as

$$F_3 = \sqrt{F_{3,x}^2 + F_{3,y}^2} = 16\text{ N}$$

and

$$\theta = \tan^{-1}\frac{F_{3,y}}{F_{3,x}} = -40°. \qquad \text{(Answer)}$$

PROBLEM-SOLVING TACTICS

Tactic 1: Dimensions and Vectors When you are dealing with forces, you cannot just add or subtract their magnitudes to find their net force unless they happen to be directed *along the same axis.* If they are not, you must use vector addition, either by means of a vector-capable calculator or by finding components along axes, one axis at a time, as is done in Sample Problem 5-2.

Tactic 2: Reading Force Problems Read the problem statement several times until you have a good mental picture of what the situation is, what data are given, and what is requested. If you know what the problem is about but don't know what to do next, put the problem aside and reread the text. If you are hazy about Newton's second law, reread that section. Study the sample problems. And remember that solving physics problems (like repairing cars and designing computer chips) takes training.

Tactic 3: Draw Two Types of Figures You may need two figures. One is a rough sketch of the actual situation. When you draw the forces, place the tail of each force vector either on the boundary of or within the body on which that force acts. The other figure is a free-body diagram: the forces on a *single* body are drawn, with the body represented by a dot or a sketch. Place the tail of each force vector on the dot or sketch.

Tactic 4: What Is Your System? If you are using Newton's second law, you must know what body or system you are applying it to. In Sample Problem 5-1 it is the puck (not the ice). In Sample Problem 5-2, it is the cookie tin.

Tactic 5: Choose Your Axes Wisely Often, we can save a lot of work by choosing one of our coordinate axes to coincide with one of the forces.

5-7 | Some Particular Forces

The Gravitational Force

A **gravitational force** $\vec{F}_g$ on a body is a certain type of pull that is directed toward a second body. In these early chapters, we do not discuss the nature of this force and usually consider situations in which the second body is Earth. Thus, when we speak of *the* gravitational force $\vec{F}_g$ on a body, we usually mean a force that pulls on it directly toward the center of Earth—that is, directly down toward the ground. We shall assume that the ground is an inertial frame.

Suppose a body of mass m is in free fall with the free-fall acceleration of magnitude g. Then, if we neglect the effects of the air, the only force acting on the body is the gravitational force $\vec{F}_g$. We can relate this downward force and downward acceleration with Newton's second law ($\vec{F} = m\vec{a}$). We place a vertical y axis along the body's path, with the positive direction upward. For this axis, Newton's second law can be written in the form $F_{\text{net},y} = ma_y$, which, in our situation, becomes

$$-F_g = m(-g)$$

or $$F_g = mg. \qquad (5\text{-}8)$$

In words, the magnitude of the gravitational force is equal to the product mg.

This same gravitational force, with the same magnitude, still acts on the body even when the body is not in free fall but is, say, at rest on a pool table or moving across the table. (For the gravitational force to disappear, Earth would have to disappear.)

We can write Newton's second law for the gravitational force in these vector forms:

$$\vec{F}_g = -F_g\hat{\text{j}} = -mg\hat{\text{j}} = m\vec{g}, \qquad (5\text{-}9)$$

where $\hat{\text{j}}$ is the unit vector that points upward along a y axis, directly away from the ground, and $\vec{g}$ is the free-fall acceleration (written as a vector), directed downward.

Weight

The **weight** W of a body is the magnitude of the net force required to prevent the body from falling freely, as measured by someone on the ground. For example, to keep a ball at rest in your hand while you stand on the ground, you must provide an upward force to balance the gravitational force on the ball from Earth. Suppose the magnitude of the gravitational force is 2.0 N. Then the magnitude of your upward force must be 2.0 N, and thus the weight W of the ball is 2.0 N. We also say that the ball *weighs* 2.0 N and speak about the ball *weighing* 2.0 N.

A ball with a weight of 3.0 N would require a greater force from you—namely, a 3.0 N force—to keep it at rest. The reason is that the gravitational force you must balance has a greater magnitude—namely, 3.0 N. We say that this second ball is *heavier* than the first ball.

Now let us generalize the situation. Consider a body that has an acceleration $\vec{a}$ of zero relative to the ground, which we again assume to be an inertial frame. Two forces act on the body: a downward gravitational force $\vec{F}_g$ and a balancing upward force of magnitude W. We can write Newton's second law for a vertical y axis, with the positive direction upward, as

$$F_{\text{net},y} = ma_y.$$

In our situation, this becomes

$$W - F_g = m(0) \qquad (5\text{-}10)$$

or $$W = F_g \quad \text{(weight, with ground as inertial frame).} \qquad (5\text{-}11)$$

This equation tells us (assuming the ground is an inertial frame) that

> The weight W of a body is equal to the magnitude F_g of the gravitational force on the body.

Substituting mg for F_g from Eq. 5-8, we find

$$W = mg \quad \text{(weight)}, \qquad (5\text{-}12)$$

which relates a body's weight to its mass.

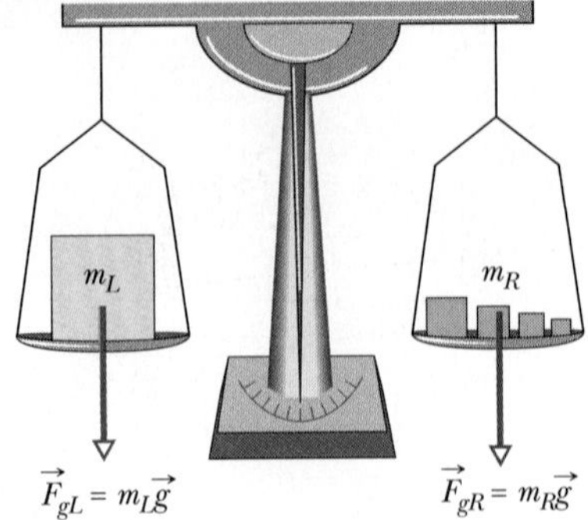

FIG. 5-5 An equal-arm balance. When the device is in balance, the gravitational force $\vec{F}_{gL}$ on the body being weighed (on the left pan) and the total gravitational force $\vec{F}_{gR}$ on the reference bodies (on the right pan) are equal. Thus, the mass m_L of the body being weighed is equal to the total mass m_R of the reference bodies.

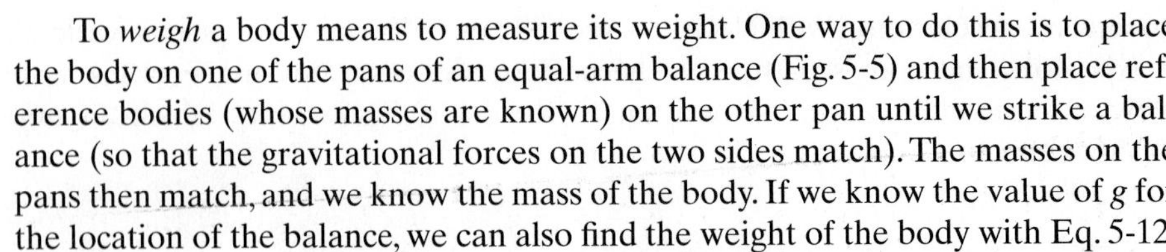

To *weigh* a body means to measure its weight. One way to do this is to place the body on one of the pans of an equal-arm balance (Fig. 5-5) and then place reference bodies (whose masses are known) on the other pan until we strike a balance (so that the gravitational forces on the two sides match). The masses on the pans then match, and we know the mass of the body. If we know the value of g for the location of the balance, we can also find the weight of the body with Eq. 5-12.

We can also weigh a body with a spring scale (Fig. 5-6). The body stretches a spring, moving a pointer along a scale that has been calibrated and marked in either mass or weight units. (Most bathroom scales in the United States work this way and are marked in the force unit pounds.) If the scale is marked in mass units, it is accurate only where the value of g is the same as where the scale was calibrated.

The weight of a body must be measured when the body is not accelerating vertically relative to the ground. For example, you can measure your weight on a scale in your bathroom or on a fast train. However, if you repeat the measurement with the scale in an accelerating elevator, the reading differs from your weight because of the acceleration. Such a measurement is called an *apparent weight.*

Caution: A body's weight is not its mass. Weight is the magnitude of a force and is related to mass by Eq. 5-12. If you move a body to a point where the value of g is different, the body's mass (an intrinsic property) is not different but the weight is. For example, the weight of a bowling ball having a mass of 7.2 kg is 71 N on Earth but only 12 N on the Moon. The mass is the same on Earth and Moon, but the free-fall acceleration on the Moon is only 1.6 m/s^2.

The Normal Force

If you stand on a mattress, Earth pulls you downward, but you remain stationary. The reason is that the mattress, because it deforms downward due to you, pushes up on you. Similarly, if you stand on a floor, it deforms (it is compressed, bent, or buckled ever so slightly) and pushes up on you. Even a seemingly rigid concrete floor does this (if it is not sitting directly on the ground, enough people on the floor could break it).

The push on you from the mattress or floor is a **normal force** $\vec{F}_N$. The name comes from the mathematical term *normal,* meaning perpendicular: The force on you from, say, the floor is perpendicular to the floor.

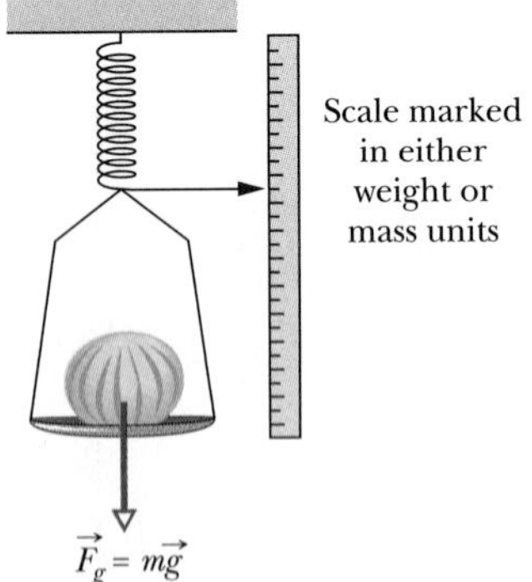

FIG. 5-6 A spring scale. The reading is proportional to the *weight* of the object on the pan, and the scale gives that weight if marked in weight units. If, instead, it is marked in mass units, the reading is the object's weight only if the value of g at the location where the scale is being used is the same as the value of g at the location where the scale was calibrated.

> When a body presses against a surface, the surface (even a seemingly rigid one) deforms and pushes on the body with a normal force $\vec{F}_N$ that is perpendicular to the surface.

Figure 5-7*a* shows an example. A block of mass m presses down on a table, deforming it somewhat because of the gravitational force $\vec{F}_g$ on the block. The table pushes up on the block with normal force $\vec{F}_N$. The free-body diagram for the block is given in Fig. 5-7*b*. Forces $\vec{F}_g$ and $\vec{F}_N$ are the only two forces on the block and they are both vertical. Thus, for the block we can write Newton's second law for a positive-upward y axis ($F_{\text{net},y} = ma_y$) as

$$F_N - F_g = ma_y.$$

From Eq. 5-8, we substitute mg for F_g, finding

$$F_N - mg = ma_y.$$

Then the magnitude of the normal force is

$$F_N = mg + ma_y = m(g + a_y) \qquad (5\text{-}13)$$

for any vertical acceleration a_y of the table and block (they might be in an accelerating elevator). If the table and block are not accelerating relative to the ground, then $a_y = 0$ and Eq. 5-13 yields

$$F_N = mg. \tag{5-14}$$

CHECKPOINT 3 In Fig. 5-7, is the magnitude of the normal force $\vec{F}_N$ greater than, less than, or equal to mg if the block and table are in an elevator moving upward (a) at constant speed and (b) at increasing speed?

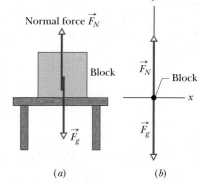

FIG. 5-7 (*a*) A block resting on a table experiences a normal force $\vec{F}_N$ perpendicular to the tabletop. (*b*) The free-body diagram for the block.

Friction

If we either slide or attempt to slide a body over a surface, the motion is resisted by a bonding between the body and the surface. (We discuss this bonding more in the next chapter.) The resistance is considered to be a single force $\vec{f}$, called either the **frictional force** or simply **friction.** This force is directed along the surface, opposite the direction of the intended motion (Fig. 5-8). Sometimes, to simplify a situation, friction is assumed to be negligible (the surface is *frictionless*).

Tension

When a cord (or a rope, cable, or other such object) is attached to a body and pulled taut, the cord pulls on the body with a force $\vec{T}$ directed away from the body and along the cord (Fig. 5-9*a*). The force is often called a *tension force* because the cord is said to be in a state of *tension* (or to be *under tension*), which means that it is being pulled taut. The *tension in the cord* is the magnitude T of the force on the body. For example, if the force on the body from the cord has magnitude $T = 50$ N, the tension in the cord is 50 N.

A cord is often said to be *massless* (meaning its mass is negligible compared to the body's mass) and *unstretchable.* The cord then exists only as a connection between two bodies. It pulls on both bodies with the same force magnitude T, even if the bodies and the cord are accelerating and even if the cord runs around a *massless, frictionless pulley* (Figs. 5-9*b* and *c*). Such a pulley has negligible mass compared to the bodies and negligible friction on its axle opposing its rotation. If the cord wraps halfway around a pulley, as in Fig. 5-9*c*, the net force on the pulley from the cord has the magnitude $2T$.

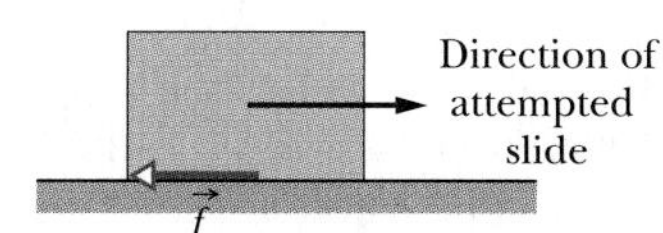

FIG. 5-8 A frictional force $\vec{f}$ opposes the attempted slide of a body over a surface.

CHECKPOINT 4 The suspended body in Fig. 5-9*c* weighs 75 N. Is T equal to, greater than, or less than 75 N when the body is moving upward (a) at constant speed, (b) at increasing speed, and (c) at decreasing speed?

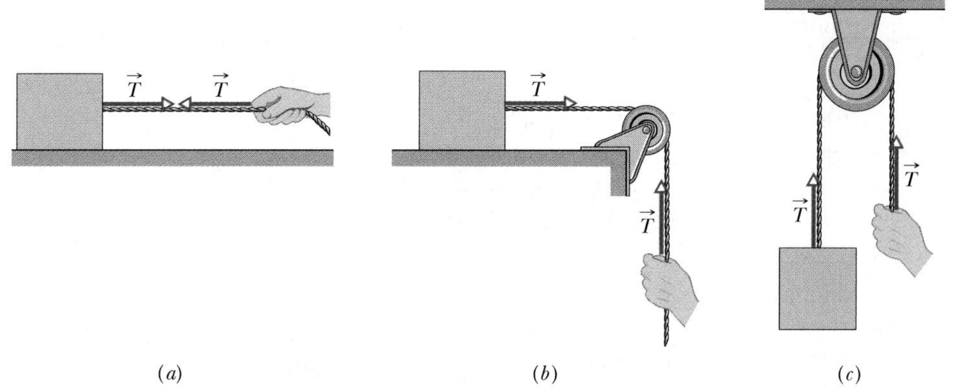

FIG. 5-9 (*a*) The cord, pulled taut, is under tension. If its mass is negligible, the cord pulls on the body and the hand with force $\vec{T}$, even if the cord runs around a massless, frictionless pulley as in (*b*) and (*c*).

PROBLEM-SOLVING TACTICS

Tactic 6: Normal Force Equation 5-14 for the normal force on a body holds only when $\vec{F}_N$ is directed upward and the body's vertical acceleration is zero; so we do *not* apply it for other orientations of $\vec{F}_N$ or when the vertical acceleration is not zero. Instead, we must derive a new expression for $\vec{F}_N$ from Newton's second law.

We are free to move $\vec{F}_N$ around in a figure as long as we maintain its orientation. For example, in Fig. 5-7*a* we can slide it downward so that its head is at the boundary between block and tabletop. However, $\vec{F}_N$ is least likely to be misinterpreted when its tail is either at that boundary or somewhere within the block (as shown). An even better technique is to draw a free-body diagram as in Fig. 5-7*b*, with the tail of $\vec{F}_N$ directly on the dot or sketch representing the block.

Sample Problem 5-3

Takeoff illusion. A jet plane taking off from an aircraft carrier is propelled by its powerful engines while being thrown forward by a catapult mechanism installed in the carrier deck. The resulting high acceleration allows the plane to reach takeoff speed in a short distance on the deck. However, that high acceleration also compels the pilot to angle the plane sharply nose-down as it leaves the deck. Pilots are trained to ignore this compulsion, but occasionally a plane is flown straight into the ocean. Let's explore the physics behind the compulsion.

Your sense of vertical depends on visual clues and on the vestibular system located in your inner ear. That system contains tiny hair cells in a fluid. When you hold your head upright, the hairs are vertically in line with the gravitational force $\vec{F}_g$ on you and the system signals your brain that your head is upright. When you tilt your head backward by some angle ϕ, the hairs are bent and the system signals your brain about the tilt. The hairs are also bent when you are accelerated forward by an applied horizontal force $\vec{F}_{app}$. The signal sent to your brain then indicates, erroneously, that your head is tilted back, to be in line with an extension through the vector sum $\vec{F}_{sum} = \vec{F}_g + \vec{F}_{app}$ (Fig. 5-10*a*). However, the erroneous signal is ignored when visual clues clearly indicate no tilt, such as when you are accelerated in a car.

A pilot being hurled along the deck of an aircraft carrier at night has almost no visual clues. The illusion of tilt is strong and very convincing, with the result that the pilot feels as though the plane leaves the deck headed sharply upward. Without proper training, a pilot will attempt to level the plane by bringing its nose sharply down, sending the plane into the ocean.

Suppose that, starting from rest, a pilot undergoes constant horizontal acceleration to reach a takeoff speed of 85 m/s in 90 m. What is the angle ϕ of the illusionary tilt experienced by the pilot?

KEY IDEAS (1) We can use Newton's second law to relate the magnitude F_{app} of the force on the pilot (from the seatback) to the resulting acceleration a_x: $F_{app} = ma_x$, where m is the mass of the pilot. (2) Because the acceleration is constant, we can use the equations of Table 2-1 to find a_x.

Calculations: We need to find the tilt angle ϕ of the line that extends through $\vec{F}_{sum}$, the vector sum of the vertical gravitational force $\vec{F}_g$ acting on the pilot and the horizontal applied force $\vec{F}_{app}$. We can find ϕ by rearranging the force vectors as in Fig. 5-10*b* and then writing

$$\tan\phi = \frac{F_{app}}{F_g},$$

or

$$\phi = \tan^{-1}\left(\frac{F_{app}}{F_g}\right). \quad (5\text{-}15)$$

Since we know the initial speed ($v_0 = 0$), the final speed ($v_x = 85$ m/s), and the displacement ($x - x_0 = 90$ m), we use Eq. 2-16 ($v^2 = v_0^2 + 2a(x - x_0)$) to write

$$(85 \text{ m/s})^2 = 0^2 + 2a_x(90 \text{ m}),$$

or

$$a_x = 40.1 \text{ m/s}^2.$$

Then, by Newton's second law, $F_{app} = m(40.1 \text{ m/s}^2)$. Substituting this result and the result $F_g = m(9.8 \text{ m/s}^2)$ in Eq. 5-15 gives us

$$\phi = \tan^{-1}\left(\frac{m(40.1 \text{ m/s}^2)}{m(9.8 \text{ m/s}^2)}\right) = 76°. \quad \text{(Answer)}$$

Thus, as the plane is accelerated along the carrier deck, the pilot feels an illusion of a backward tilt of 76°, as though the plane is angled nose-up by 76°. The illusion may compel the pilot to put the plane nose-down by 76° just after takeoff.

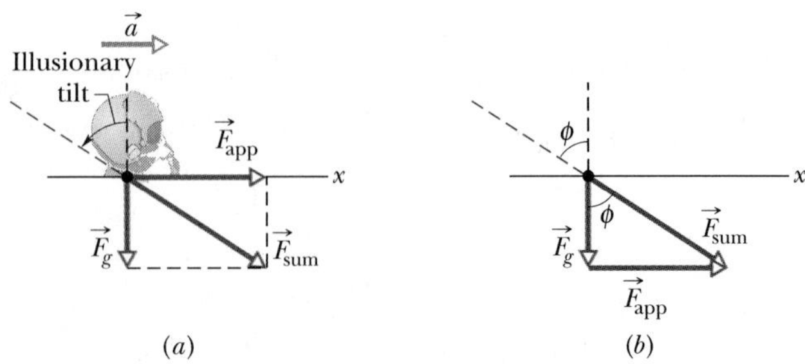

FIG. 5-10 (*a*) Force $\vec{F}_{app}$, directed to the right, is applied to the pilot during takeoff. The pilot's head feels as though it is tilted back along the red dashed line. (*b*) The vector sum $\vec{F}_{sum}$ $(= \vec{F}_g + \vec{F}_{app})$ is at angle ϕ from the vertical.

5-8 | Newton's Third Law

Two bodies are said to *interact* when they push or pull on each other—that is, when a force acts on each body due to the other body. For example, suppose you position a book B so it leans against a crate C (Fig. 5-11*a*). Then the book and crate interact: There is a horizontal force $\vec{F}_{BC}$ on the book from the crate (or due to the crate) and a horizontal force $\vec{F}_{CB}$ on the crate from the book (or due to the book). This pair of forces is shown in Fig. 5-11*b*. Newton's third law states that

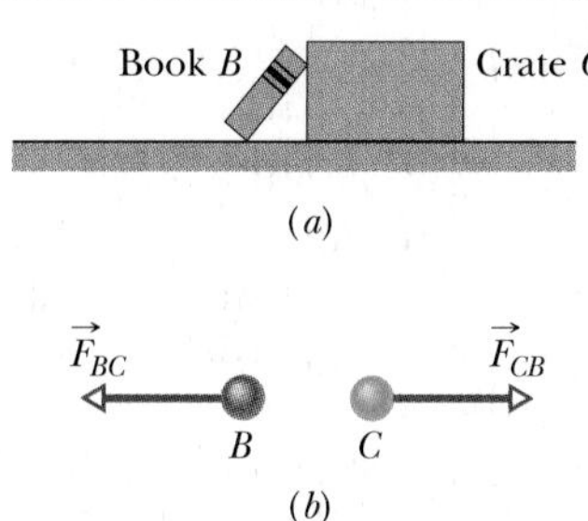

FIG. 5-11 (*a*) Book B leans against crate C. (*b*) Forces $\vec{F}_{BC}$ (the force on the book from the crate) and $\vec{F}_{CB}$ (the force on the crate from the book) have the same magnitude and are opposite in direction.

Newton's Third Law: When two bodies interact, the forces on the bodies from each other are always equal in magnitude and opposite in direction.

For the book and crate, we can write this law as the scalar relation

$$F_{BC} = F_{CB} \quad \text{(equal magnitudes)}$$

or as the vector relation

$$\vec{F}_{BC} = -\vec{F}_{CB} \quad \text{(equal magnitudes and opposite directions)},$$

where the minus sign means that these two forces are in opposite directions. We can call the forces between two interacting bodies a **third-law force pair.** When any two bodies interact in any situation, a third-law force pair is present. The book and crate in Fig. 5-11*a* are stationary, but the third law would still hold if they were moving and even if they were accelerating.

As another example, let us find the third-law force pairs involving the cantaloupe in Fig. 5-12*a*, which lies on a table that stands on Earth. The cantaloupe interacts with the table and with Earth (this time, there are three bodies whose interactions we must sort out).

Let's first focus on the forces acting on the cantaloupe (Fig. 5-12*b*). Force $\vec{F}_{CT}$ is the normal force on the cantaloupe from the table, and force $\vec{F}_{CE}$ is the gravitational force on the cantaloupe due to Earth. Are they a third-law force pair? No, because they are forces on a single body, the cantaloupe, and not on two interacting bodies.

To find a third-law pair, we must focus not on the cantaloupe but on the interaction between the cantaloupe and one other body. In the cantaloupe–Earth interaction (Fig. 5-12*c*), Earth pulls on the cantaloupe with a gravitational force $\vec{F}_{CE}$ and the cantaloupe pulls on Earth with a gravitational force $\vec{F}_{EC}$. Are these forces a third-law force pair? Yes, because they are forces on two interacting bodies, the force on each due to the other. Thus, by Newton's third law,

$$\vec{F}_{CE} = -\vec{F}_{EC} \quad \text{(cantaloupe – Earth interaction)}.$$

Next, in the cantaloupe–table interaction, the force on the cantaloupe from the table is $\vec{F}_{CT}$ and, conversely, the force on the table from the cantaloupe is $\vec{F}_{TC}$ (Fig. 5-12*d*). These forces are also a third-law force pair, and so

$$\vec{F}_{CT} = -\vec{F}_{TC} \quad \text{(cantaloupe–table interaction)}.$$

CHECKPOINT 5 Suppose that the cantaloupe and table of Fig. 5-12 are in an elevator cab that begins to accelerate upward. (a) Do the magnitudes of $\vec{F}_{TC}$ and $\vec{F}_{CT}$ increase, decrease, or stay the same? (b) Are those two forces still equal in magnitude and opposite in direction? (c) Do the magnitudes of $\vec{F}_{CE}$ and $\vec{F}_{EC}$ increase, decrease, or stay the same? (d) Are those two forces still equal in magnitude and opposite in direction?

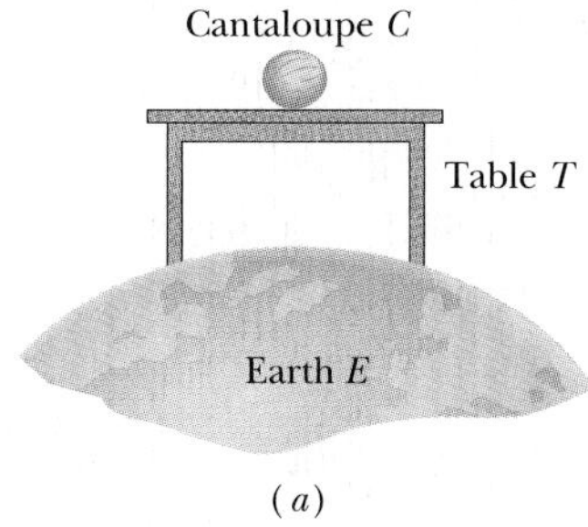

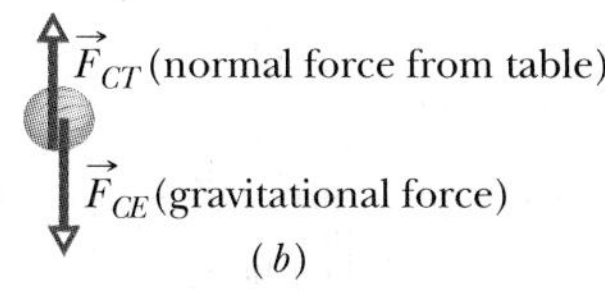

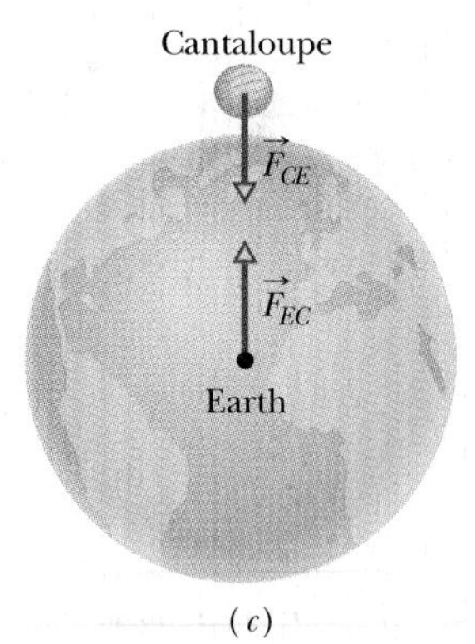

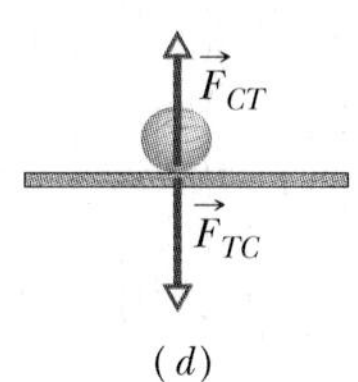

FIG. 5-12 (*a*) A cantaloupe lies on a table that stands on Earth. (*b*) The forces *on the cantaloupe* are $\vec{F}_{CT}$ and $\vec{F}_{CE}$. (*c*) The third-law force pair for the cantaloupe–Earth interaction. (*d*) The third-law force pair for the cantaloupe–table interaction.

5-9 | Applying Newton's Laws

The rest of this chapter consists of sample problems. You should pore over them, learning their procedures for attacking a problem. Especially important is knowing how to translate a sketch of a situation into a free-body diagram with appropriate axes, so that Newton's laws can be applied.

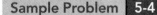

Sample Problem 5-4 Build your skill

Figure 5-13 shows a block S (the *sliding block*) with mass $M = 3.3$ kg. The block is free to move along a horizontal frictionless surface and connected, by a cord that wraps over a frictionless pulley, to a second block H (the *hanging block*), with mass $m = 2.1$ kg. The cord and pulley have negligible masses compared to the blocks (they are "massless"). The hanging block H falls as the sliding block S accelerates to the right. Find (a) the acceleration of block S, (b) the acceleration of block H, and (c) the tension in the cord.

Q *What is this problem all about?*

You are given two bodies—sliding block and hanging block—but must also consider *Earth,* which pulls on both bodies. (Without Earth, nothing would happen here.) A total of five forces act on the blocks, as shown in Fig. 5-14:

1. The cord pulls to the right on sliding block S with a force of magnitude T.
2. The cord pulls upward on hanging block H with a force of the same magnitude T. This upward force keeps block H from falling freely.
3. Earth pulls down on block S with the gravitational force $\vec{F}_{gS}$, which has a magnitude equal to Mg.
4. Earth pulls down on block H with the gravitational force $\vec{F}_{gH}$, which has a magnitude equal to mg.
5. The table pushes up on block S with a normal force $\vec{F}_N$.

There is another thing you should note. We assume that the cord does not stretch, so that if block H falls 1 mm in a certain time, block S moves 1 mm to the right in that same time. This means that the blocks move together and their accelerations have the same magnitude a.

Q *How do I classify this problem? Should it suggest a particular law of physics to me?*

Yes. Forces, masses, and accelerations are involved, and they should suggest Newton's second law of motion, $\vec{F}_{\text{net}} = m\vec{a}$. That is our starting **Key Idea**.

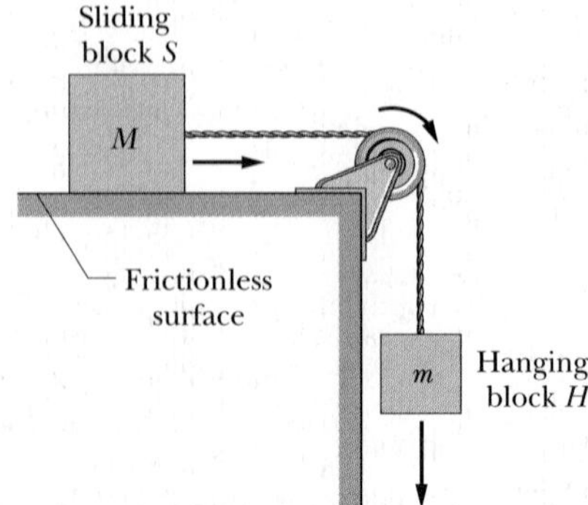

FIG. 5-13 A block S of mass M is connected to a block H of mass m by a cord that wraps over a pulley.

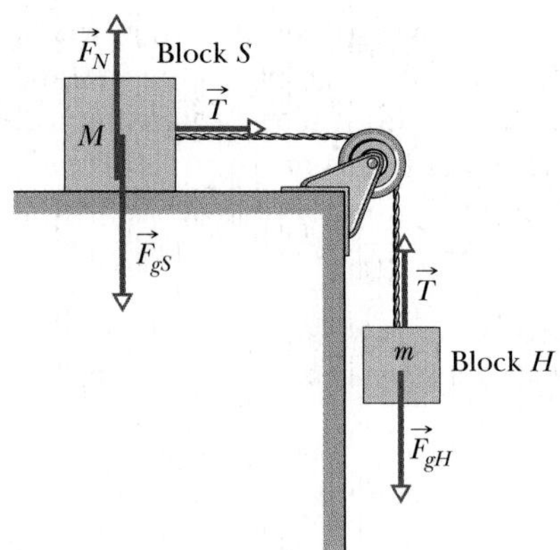

FIG. 5-14 The forces acting on the two blocks of Fig. 5-13.

Q *If I apply Newton's second law to this problem, to which body should I apply it?*

We focus on two bodies, the sliding block and the hanging block. Although they are *extended objects* (they are not points), we can still treat each block as a particle because every part of it moves in exactly the same way. A second **Key Idea** is to apply Newton's second law separately to each block.

Q *What about the pulley?*

We cannot represent the pulley as a particle because different parts of it move in different ways. When we discuss rotation, we shall deal with pulleys in detail. Meanwhile, we eliminate the pulley from consideration by assuming its mass to be negligible compared with the masses of the two blocks. Its only function is to change the cord's orientation.

Q *OK. Now how do I apply $\vec{F}_{\text{net}} = m\vec{a}$ to the sliding block?*

Represent block S as a particle of mass M and draw *all* the forces that act *on* it, as in Fig. 5-15*a*. This is the block's free-body diagram. Next, draw a set of axes. It makes sense to draw the x axis parallel to the table, in the direction in which the block moves.

Q *Thanks, but you still haven't told me how to apply $\vec{F}_{\text{net}} = m\vec{a}$ to the sliding block. All you've done is explain how to draw a free-body diagram.*

You are right, and here's the third **Key Idea**: The expression $\vec{F}_{\text{net}} = M\vec{a}$ is a vector equation, so we can write it as three component equations:

$$F_{\text{net},x} = Ma_x \quad F_{\text{net},y} = Ma_y \quad F_{\text{net},z} = Ma_z \qquad (5\text{-}16)$$

in which $F_{\text{net},x}$, $F_{\text{net},y}$, and $F_{\text{net},z}$ are the components of the net force along the three axes. Now we apply each component equation to its corresponding direction. Because block S does not accelerate vertically, $F_{\text{net},y} = Ma_y$ becomes

$$F_N - F_{gS} = 0 \quad \text{or} \quad F_N = F_{gS}.$$

Thus in the y direction, the magnitude of the normal force is equal to the magnitude of the gravitational force.

No force acts in the z direction, which is perpendicular to the page.

In the x direction, there is only one force component, which is T. Thus, $F_{\text{net},x} = Ma_x$ becomes

$$T = Ma. \tag{5-17}$$

This equation contains two unknowns, T and a; so we cannot yet solve it. Recall, however, that we have not said anything about the hanging block.

Q *I agree. How do I apply $\vec{F}_{\text{net}} = m\vec{a}$ to the hanging block?*

We apply it just as we did for block S: Draw a free-body diagram for block H, as in Fig. 5-15*b*. Then apply $\vec{F}_{\text{net}} = m\vec{a}$ in component form. This time, because the acceleration is along the y axis, we use the y part of Eq. 5-16 ($F_{\text{net},y} = ma_y$) to write

$$T - F_{gH} = ma_y.$$

We can now substitute mg for F_{gH} and $-a$ for a_y (negative because block H accelerates in the negative direction of the y axis). We find

$$T - mg = -ma. \tag{5-18}$$

Now note that Eqs. 5-17 and 5-18 are simultaneous equations with the same two unknowns, T and a. Subtracting these equations eliminates T. Then solving for a yields

$$a = \frac{m}{M + m}g. \tag{5-19}$$

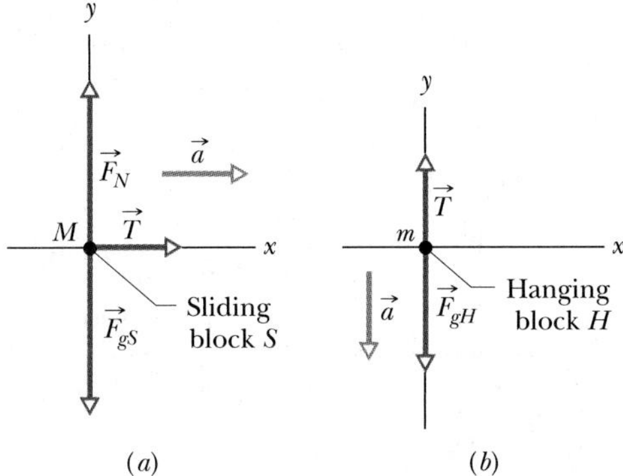

FIG. 5-15 (*a*) A free-body diagram for block S of Fig. 5-13. (*b*) A free-body diagram for block H of Fig. 5-13.

Substituting this result into Eq. 5-17 yields

$$T = \frac{Mm}{M + m}g. \tag{5-20}$$

Putting in the numbers gives, for these two quantities,

$$a = \frac{m}{M + m}g = \frac{2.1\text{ kg}}{3.3\text{ kg} + 2.1\text{ kg}}(9.8\text{ m/s}^2)$$
$$= 3.8\text{ m/s}^2 \qquad \text{(Answer)}$$

and
$$T = \frac{Mm}{M + m}g = \frac{(3.3\text{ kg})(2.1\text{ kg})}{3.3\text{ kg} + 2.1\text{ kg}}(9.8\text{ m/s}^2)$$
$$= 13\text{ N}. \qquad \text{(Answer)}$$

Q *The problem is now solved, right?*

That's a fair question, but the problem is not really finished until we have examined the results to see whether they make sense. (If you made these calculations on the job, wouldn't you want to see whether they made sense before you turned them in?)

Look first at Eq. 5-19. Note that it is dimensionally correct and that the acceleration a will always be less than g. This is as it must be, because the hanging block is not in free fall. The cord pulls upward on it.

Look now at Eq. 5-20, which we can rewrite in the form

$$T = \frac{M}{M + m}mg. \tag{5-21}$$

In this form, it is easier to see that this equation is also dimensionally correct, because both T and mg have dimensions of forces. Equation 5-21 also lets us see that the tension in the cord is always less than mg, and thus is always less than the gravitational force on the hanging block. That is a comforting thought because, if T were *greater* than mg, the hanging block would accelerate upward.

We can also check the results by studying special cases, in which we can guess what the answers must be. A simple example is to put $g = 0$, as if the experiment were carried out in interstellar space. We know that in that case, the blocks would not move from rest, there would be no forces on the ends of the cord, and so there would be no tension in the cord. Do the formulas predict this? Yes, they do. If you put $g = 0$ in Eqs. 5-19 and 5-20, you find $a = 0$ and $T = 0$. Two more special cases you might try are $M = 0$ and $m \to \infty$.

Sample Problem 5-5

In Fig. 5-16*a*, a cord pulls on a box of sea biscuits up along a frictionless plane inclined at $\theta = 30°$. The box has mass $m = 5.00$ kg, and the force from the cord has magnitude $T = 25.0$ N. What is the box's acceleration component a along the inclined plane?

KEY IDEA The acceleration along the plane is set by the force components along the plane (not by force components perpendicular to the plane), as expressed by Newton's second law (Eq. 5-1).

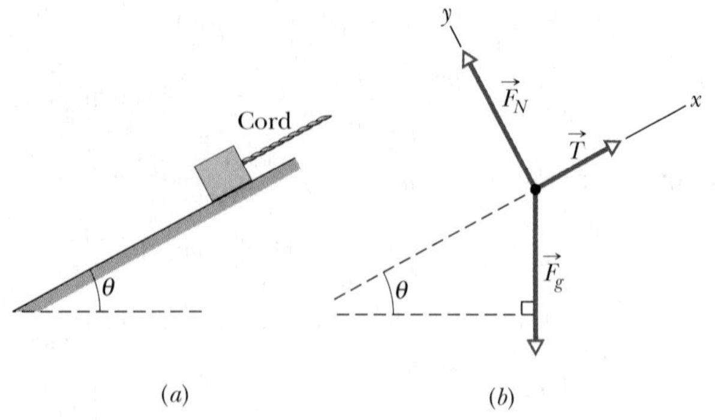

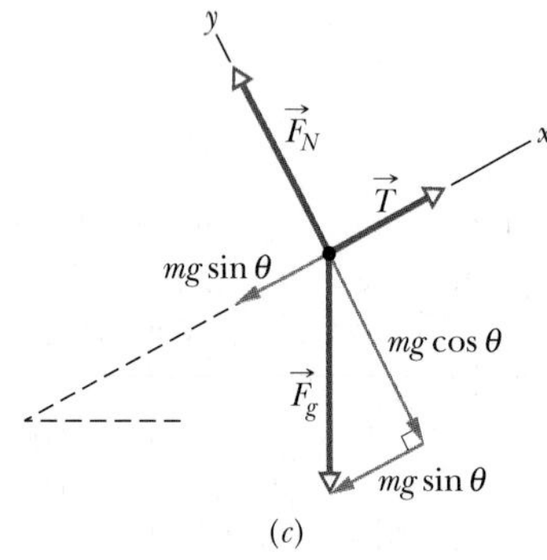

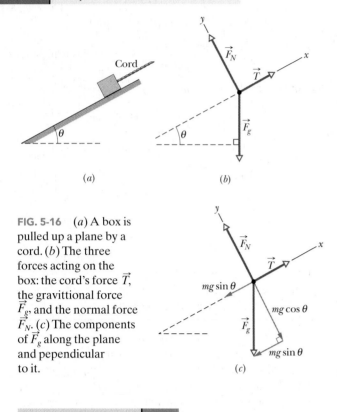

FIG. 5-16 (*a*) A box is pulled up a plane by a cord. (*b*) The three forces acting on the box: the cord's force $\vec{T}$, the gravittional force $\vec{F}_g$, and the normal force $\vec{F}_N$. (*c*) The components of $\vec{F}_g$ along the plane and perpendicular to it.

Calculation: For convenience, we draw a coordinate system and a free-body diagram as shown in Fig. 5-16*b*. The positive direction of the *x* axis is up the plane. Force $\vec{T}$ from the cord is up the plane and has magnitude $T = 25.0$ N. The gravitational force $\vec{F}_g$ is downward and has magnitude $mg = (5.00 \text{ kg})(9.8 \text{ m/s}^2) = 49.0$ N. More important, its component along the plane is down the plane and has magnitude $mg \sin\theta$ as indicated in Fig. 5-16*c*. (To see why that trig function is involved, compare the right triangles in Figs. 5-16*b* and *c*.) To indicate the direction, we can write the component as $-mg \sin\theta$. The normal force $\vec{F}_N$ is perpendicular to the plane and thus does not determine acceleration along the plane.

We write Newton's second law ($\vec{F}_{net} = m\vec{a}$) for motion along the *x* axis as

$$T - mg \sin\theta = ma. \tag{5-22}$$

Substituting data and solving for *a*, we find

$$a = 0.100 \text{ m/s}^2, \qquad \text{(Answer)}$$

where the positive result indicates that the box accelerates up the plane.

Sample Problem 5-6

Let's return to the chapter opening question: What produces the fear factor in the last car of a traditional gravity-driven roller coaster? Let's consider a coaster having 10 identical cars with total mass *M* and massless interconnections. Figure 5-17*a* shows the coaster just after the first car has begun its descent along a frictionless slope with an angle θ. Figure 5-17*b* shows the coaster just before the last car begins its descent. What is the acceleration of the coaster in these two situations?

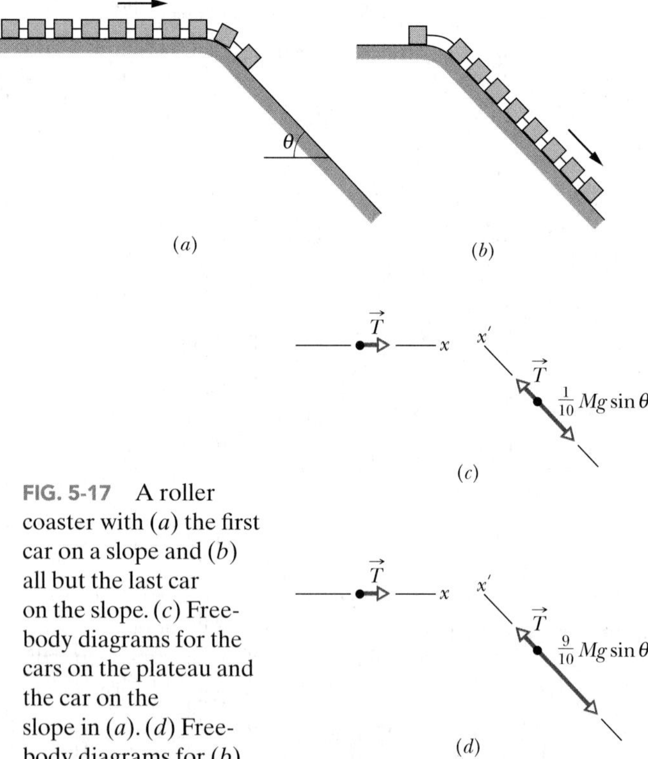

FIG. 5-17 A roller coaster with (*a*) the first car on a slope and (*b*) all but the last car on the slope. (*c*) Free-body diagrams for the cars on the plateau and the car on the slope in (*a*). (*d*) Free-body diagrams for (*b*).

KEY IDEAS (1) The net force on an object causes the object's acceleration, as related by Newton's second law (Eq. 5-1, $\vec{F}_{net} = m\vec{a}$). (2) When the motion is along a single axis, we write that law in component form (such as $F_{net,x} = ma_x$) and we use only force components along that axis. (3) When several objects move together at the same velocity and with the same acceleration, they can be regarded as a single composite object. *Internal forces* act between the individual objects, but only *external forces* can cause the composite object to accelerate.

Calculations for Fig. 5-17a: Figure 5-17*c* shows free-body diagrams associated with Fig. 5-17*a*, with convenient axes superimposed. The tilted *x*′ axis has its positive direction up the slope. *T* is the magnitude of the interconnection force between the car on the slope and the cars still on the plateau. Because the coaster consists of 10 identical cars with total mass *M*, the mass of the car on the slope is $\frac{1}{10}M$ and the mass of the cars on the plateau is $\frac{9}{10}M$. Only a single *external* force acts along the *x* axis on the nine-car composite—namely, the interconnection force with magnitude *T*. (The forces between the nine cars are internal forces.) Thus, Newton's second law for motion along the x axis ($F_{\text{net},x} = ma_x$) becomes

$$T = \tfrac{9}{10}Ma, \qquad (5\text{-}23)$$

where a is the magnitude of the acceleration a_x along the x axis.

Along the tilted x' axis, two forces act on the car on the slope: the interconnection force with magnitude T (in the positive direction of the axis) and the x' component of the gravitational force (in the negative direction of the axis). From Sample Problem 5-5, we know to write that gravitational component as $-mg \sin \theta$, where m is the mass. Because we know that the car accelerates *down* the slope in the negative x' direction with magnitude a, we can write the acceleration as $-a$. Thus, for this car, with mass $\tfrac{1}{10}M$ we write Newton's second law for motion along the x' axis as

$$T - \tfrac{1}{10}Mg \sin \theta = \tfrac{1}{10}M(-a). \qquad (5\text{-}24)$$

Substituting for T from Eq. 5-23 and solving for a, we have

$$a = \tfrac{1}{10}g \sin \theta. \qquad \text{(Answer)}$$

Calculations for Fig. 5-17b: Figure 5-17d shows free-body diagrams associated with Fig. 5-17b. For the car still on the plateau, we rewrite Eq. 5-23 as

$$T = \tfrac{1}{10}Ma.$$

For the nine cars on the slope, we rewrite Eq. 5-24 as

$$T - \tfrac{9}{10}Mg \sin \theta = \tfrac{9}{10}M(-a).$$

Again solving for a, we now find

$$a = \tfrac{9}{10}g \sin \theta. \qquad \text{(Answer)}$$

The fear factor: This last answer is 9 times the first answer. Thus, in general, the acceleration of the cars greatly increases as more of them go over the edge and onto the slope. That increase in acceleration occurs regardless of your car choice, but your interpretation of the acceleration depends on the choice. In the first car, most of the acceleration occurs on the slope and is due to the component of the gravitational force along the slope, which is reasonable. In the last car, most of the acceleration occurs on the plateau and is due to the push on you from the back of your seat. That push rapidly increases as you approach the edge, giving you the frightening sensation that you are about to be hurled off the plateau and into the air.

Sample Problem 5-7 Build your skill

Figure 5-18a shows the general arrangement in which two forces are applied to a 4.00 kg block on a frictionless floor, but only force $\vec{F}_1$ is indicated. That force has a fixed magnitude but can be applied at angle θ to the positive direction of the x axis. Force $\vec{F}_2$ is horizontal and fixed in both magnitude and angle. Figure 5-18b gives the horizontal acceleration a_x of the block for any given value of θ from 0° to 90°. What is the value of a_x for $\theta = 180°$?

KEY IDEAS (1) The horizontal acceleration a_x depends on the net horizontal force $F_{\text{net},x}$, as given by Newton's second law. (2) The net horizontal force is the sum of the horizontal components of forces $\vec{F}_1$ and $\vec{F}_2$.

Calculations: The x component of $\vec{F}_2$ is F_2 because the vector is horizontal. The x component of $\vec{F}_1$ is $F_1 \cos \theta$. Using these expressions and a mass m of 4.00 kg, we can write Newton's second law ($\vec{F}_{\text{net}} = m\vec{a}$) for motion along the x axis as

$$F_1 \cos \theta + F_2 = 4.00a_x. \qquad (5\text{-}25)$$

From this equation we see that when $\theta = 90°$, $F_1 \cos \theta$ is zero and $F_2 = 4.00a_x$. From the graph we see that the corresponding acceleration is 0.50 m/s². Thus, $F_2 = 2.00$ N and $\vec{F}_2$ must be in the positive direction of the x axis.

From Eq. 5-25, we find that when $\theta = 0°$,

$$F_1 \cos 0° + 2.00 = 4.00a_x. \qquad (5\text{-}26)$$

From the graph we see that the corresponding acceleration is 3.0 m/s². From Eq. 5-26, we then find that $F_1 = 10$ N.

Substituting $F_1 = 10$ N, $F_2 = 2.00$ N, and $\theta = 180°$ into Eq. 5-25 leads to

$$a_x = -2.00 \text{ m/s}^2. \qquad \text{(Answer)}$$

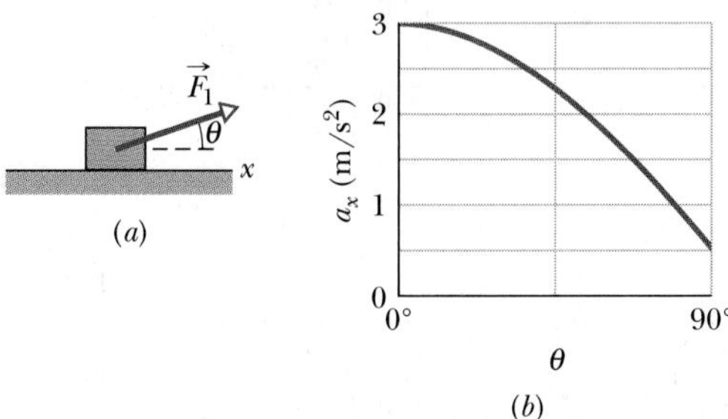

FIG. 5-18 (a) One of the two forces applied to a block is shown. Its angle θ can be varied. (b) The block's acceleration component a_x versus θ.

Sample Problem 5-8 Build your skill

In Fig. 5-19*a*, a passenger of mass $m = 72.2$ kg stands on a platform scale in an elevator cab. We are concerned with the scale readings when the cab is stationary and when it is moving up or down.

(a) Find a general solution for the scale reading, whatever the vertical motion of the cab.

KEY IDEAS (1) The reading is equal to the magnitude of the normal force $\vec{F}_N$ on the passenger from the scale. The only other force acting on the passenger is the gravitational force $\vec{F}_g$, as shown in the free-body diagram of Fig. 5-19*b*. (2) We can relate the forces on the passenger to his acceleration $\vec{a}$ by using Newton's second law ($\vec{F}_{net} = m\vec{a}$). However, recall that we can use this law only in an inertial frame. If the cab accelerates, then it is *not* an inertial frame. So we choose the ground to be our inertial frame and make any measure of the passenger's acceleration relative to it.

Calculations: Because the two forces on the passenger and his acceleration are all directed vertically, along the y axis in Fig. 5-19*b*, we can use Newton's second law written for y components ($F_{net,y} = ma_y$) to get

$$F_N - F_g = ma$$

or

$$F_N = F_g + ma. \tag{5-27}$$

This tells us that the scale reading, which is equal to F_N, depends on the vertical acceleration. Substituting mg for F_g gives us

$$F_N = m(g + a) \quad \text{(Answer)} \tag{5-28}$$

for any choice of acceleration a.

(b) What does the scale read if the cab is stationary or moving upward at a constant 0.50 m/s?

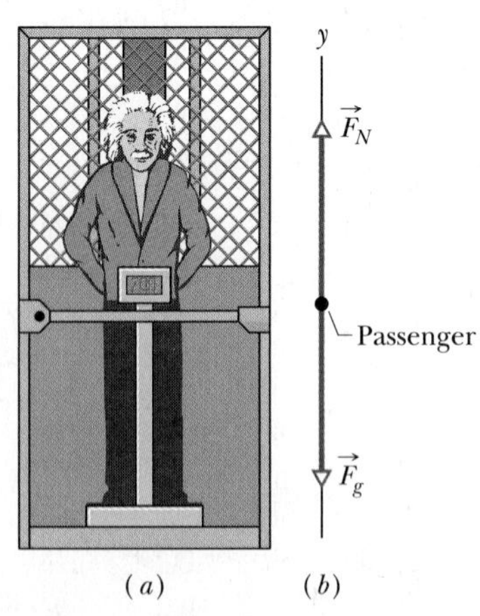

FIG. 5-19 (*a*) A passenger stands on a platform scale that indicates either his weight or his apparent weight. (*b*) The free-body diagram for the passenger, showing the normal force $\vec{F}_N$ on him from the scale and the gravitational force $\vec{F}_g$.

KEY IDEA For any constant velocity (zero or otherwise), the acceleration a of the passenger is zero.

Calculation: Substituting this and other known values into Eq. 5-28, we find

$$F_N = (72.2 \text{ kg})(9.8 \text{ m/s}^2 + 0) = 708 \text{ N}. \quad \text{(Answer)}$$

This is the weight of the passenger and is equal to the magnitude F_g of the gravitational force on him.

(c) What does the scale read if the cab accelerates upward at 3.20 m/s² and downward at 3.20 m/s²?

Calculations: For $a = 3.20$ m/s², Eq. 5-28 gives

$$F_N = (72.2 \text{ kg})(9.8 \text{ m/s}^2 + 3.20 \text{ m/s}^2) = 939 \text{ N}, \quad \text{(Answer)}$$

and for $a = -3.20$ m/s², it gives

$$F_N = (72.2 \text{ kg})(9.8 \text{ m/s}^2 - 3.20 \text{ m/s}^2) = 477 \text{ N}. \quad \text{(Answer)}$$

For an upward acceleration (either the cab's upward speed is increasing or its downward speed is decreasing), the scale reading is greater than the passenger's weight. That reading is a measurement of an apparent weight, because it is made in a noninertial frame. For a downward acceleration (either decreasing upward speed or increasing downward speed), the scale reading is less than the passenger's weight.

(d) During the upward acceleration in part (c), what is the magnitude F_{net} of the net force on the passenger, and what is the magnitude $a_{p,cab}$ of his acceleration as measured in the frame of the cab? Does $\vec{F}_{net} = m\vec{a}_{p,cab}$?

Calculation: The magnitude F_g of the gravitational force on the passenger does not depend on the motion of the passenger or the cab; so, from part (b), F_g is 708 N. From part (c), the magnitude F_N of the normal force on the passenger during the upward acceleration is the 939 N reading on the scale. Thus, the net force on the passenger is

$$F_{net} = F_N - F_g = 939 \text{ N} - 708 \text{ N} = 231 \text{ N}, \quad \text{(Answer)}$$

during the upward acceleration. However, his acceleration $a_{p,cab}$ relative to the frame of the cab is zero. Thus, in the noninertial frame of the accelerating cab, F_{net} is not equal to $ma_{p,cab}$, and Newton's second law does not hold.

Sample Problem 5-9 Build your skill

In Fig. 5-20*a*, a constant horizontal force $\vec{F}_{app}$ of magnitude 20 N is applied to block A of mass $m_A = 4.0$ kg, which pushes against block B of mass $m_B = 6.0$ kg. The blocks slide over a frictionless surface, along an x axis.

(a) What is the acceleration of the blocks?

Serious Error: Because force $\vec{F}_{app}$ is applied directly to block A, we use Newton's second law to relate that force to the acceleration $\vec{a}$ of block A. Because the motion is along the x axis, we use that law for x components ($F_{net,x} = ma_x$), writing it as

$$F_{app} = m_A a.$$

However, this is seriously wrong because $\vec{F}_{app}$ is not the only horizontal force acting on block A. There is also the force $\vec{F}_{AB}$ from block B (Fig. 5-20*b*).

Dead-End Solution: Let us now include force $\vec{F}_{AB}$ by writing, again for the x axis,

$$F_{app} - F_{AB} = m_A a.$$

(We use the minus sign to include the direction of $\vec{F}_{AB}$.) Because F_{AB} is a second unknown, we cannot solve this equation for a.

Successful Solution: Because of the direction in which force $\vec{F}_{app}$ is applied, the two blocks form a rigidly connected system. We can relate the net force *on the system* to the acceleration *of the system* with Newton's second law. Here, once again for the x axis, we can write that law as

$$F_{app} = (m_A + m_B)a,$$

where now we properly apply $\vec{F}_{app}$ to the system with total mass $m_A + m_B$. Solving for a and substituting known values, we find

$$a = \frac{F_{app}}{m_A + m_B} = \frac{20 \text{ N}}{4.0 \text{ kg} + 6.0 \text{ kg}} = 2.0 \text{ m/s}^2. \quad \text{(Answer)}$$

Thus, the acceleration of the system and of each block is in the positive direction of the x axis and has the magnitude 2.0 m/s².

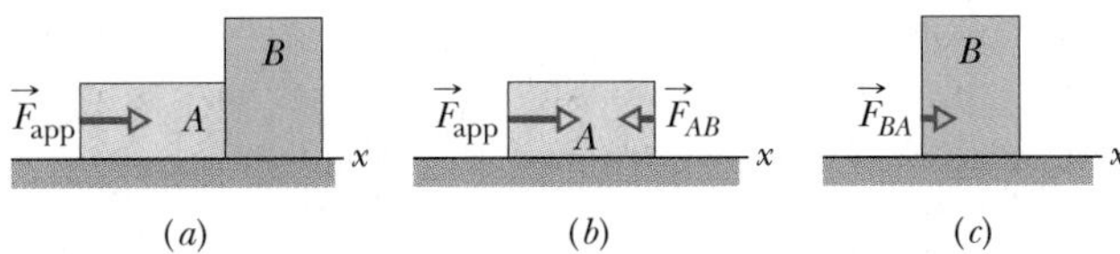

FIG. 5-20 (*a*) A constant horizontal force $\vec{F}_{app}$ is applied to block A, which pushes against block B. (*b*) Two horizontal forces act on block A. (*c*) Only one horizontal force acts on block B.

(b) What is the (horizontal) force $\vec{F}_{BA}$ on block B from block A (Fig. 5-20*c*)?

KEY IDEA We can relate the net force on block B to the block's acceleration with Newton's second law.

Calculation: Here we can write that law, still for components along the x axis, as

$$F_{BA} = m_B a,$$

which, with known values, gives

$$F_{BA} = (6.0 \text{ kg})(2.0 \text{ m/s}^2) = 12 \text{ N}. \quad \text{(Answer)}$$

Thus, force $\vec{F}_{BA}$ is in the positive direction of the x axis and has a magnitude of 12 N.

REVIEW & SUMMARY

Newtonian Mechanics The velocity of an object can change (the object can accelerate) when the object is acted on by one or more **forces** (pushes or pulls) from other objects. *Newtonian mechanics* relates accelerations and forces.

Force Forces are vector quantities. Their magnitudes are defined in terms of the acceleration they would give the standard kilogram. A force that accelerates that standard body by exactly 1 m/s² is defined to have a magnitude of 1 N. The direction of a force is the direction of the acceleration it causes. Forces are combined according to the rules of vector algebra. The **net force** on a body is the vector sum of all the forces acting on the body.

Newton's First Law If there is no net force on a body, the body remains at rest if it is initially at rest or moves in a straight line at constant speed if it is in motion.

Inertial Reference Frames Reference frames in which Newtonian mechanics holds are called *inertial reference frames* or *inertial frames.* Reference frames in which Newtonian mechanics does not hold are called *noninertial reference frames* or *noninertial frames.*

Mass The **mass** of a body is the characteristic of that body that relates the body's acceleration to the net force causing the acceleration. Masses are scalar quantities.

Newton's Second Law The net force $\vec{F}_{net}$ on a body with mass m is related to the body's acceleration $\vec{a}$ by

$$\vec{F}_{net} = m\vec{a}, \quad (5\text{-}1)$$

which may be written in the component versions

$$F_{net,x} = ma_x \quad F_{net,y} = ma_y \quad \text{and} \quad F_{net,z} = ma_z. \quad (5\text{-}2)$$

The second law indicates that in SI units

$$1 \text{ N} = 1 \text{ kg} \cdot \text{m/s}^2. \quad (5\text{-}3)$$

A **free-body diagram** is a stripped-down diagram in which only *one* body is considered. That body is represented by either a sketch or a dot. The external forces on the body are drawn, and a coordinate system is superimposed, oriented so as to simplify the solution.

Some Particular Forces A **gravitational force** $\vec{F}_g$ on a body is a pull by another body. In most situations in this book, the other body is Earth or some other astronomical body. For Earth, the force is directed down toward the ground, which is assumed to be an inertial frame. With that assumption, the magnitude of $\vec{F}_g$ is

$$F_g = mg, \tag{5-8}$$

where m is the body's mass and g is the magnitude of the free-fall acceleration.

The **weight** W of a body is the magnitude of the upward force needed to balance the gravitational force on the body. A body's weight is related to the body's mass by

$$W = mg. \tag{5-12}$$

A **normal force** $\vec{F}_N$ is the force on a body from a surface against which the body presses. The normal force is always perpendicular to the surface.

A **frictional force** $\vec{f}$ is the force on a body when the body slides or attempts to slide along a surface. The force is always parallel to the surface and directed so as to oppose the sliding. On a *frictionless surface,* the frictional force is negligible.

When a cord is under **tension,** each end of the cord pulls on a body. The pull is directed along the cord, away from the point of attachment to the body. For a *massless* cord (a cord with negligible mass), the pulls at both ends of the cord have the same magnitude T, even if the cord runs around a *massless, frictionless pulley* (a pulley with negligible mass and negligible friction on its axle to oppose its rotation).

Newton's Third Law If a force $\vec{F}_{BC}$ acts on body B due to body C, then there is a force $\vec{F}_{CB}$ on body C due to body B:

$$\vec{F}_{BC} = -\vec{F}_{CB}.$$

QUESTIONS

1 In Fig. 5-21, forces $\vec{F}_1$ and $\vec{F}_2$ are applied to a lunchbox as it slides at constant velocity over a frictionless floor. We are to decrease angle θ without changing the magnitude of $\vec{F}_1$. For constant velocity, should we increase, decrease, or maintain the magnitude of $\vec{F}_2$?

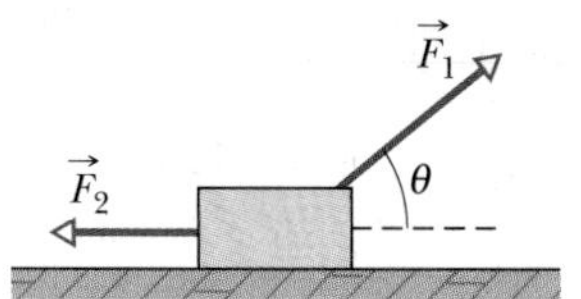

FIG. 5-21 Question 1.

2 At time $t = 0$, constant $\vec{F}$ begins to act on a rock moving through deep space in the $+x$ direction. (a) For time $t > 0$, which are possible functions $x(t)$ for the rock's position: (1) $x = 4t - 3$, (2) $x = -4t^2 + 6t - 3$, (3) $x = 4t^2 + 6t - 3$? (b) For which function is $\vec{F}$ directed opposite the rock's initial direction of motion?

3 Figure 5-22 shows overhead views of four situations in which forces act on a block that lies on a frictionless floor. If the force magnitudes are chosen properly, in which situations is it possible that the block is (a) stationary and (b) moving with a constant velocity?

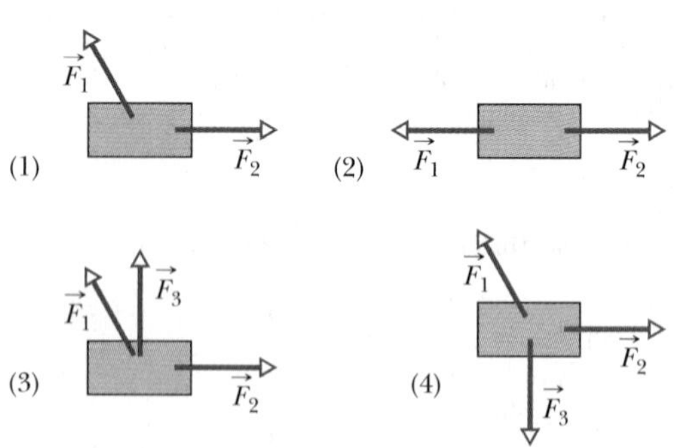

FIG. 5-22 Question 3.

4 Two horizontal forces,

$$\vec{F}_1 = (3\text{ N})\hat{\imath} - (4\text{ N})\hat{\jmath} \quad \text{and} \quad \vec{F}_2 = -(1\text{ N})\hat{\imath} - (2\text{ N})\hat{\jmath}$$

pull a banana split across a frictionless lunch counter. Without using a calculator, determine which of the vectors in the free-body diagram of Fig. 5-23 best represent (a) $\vec{F}_1$ and (b) $\vec{F}_2$. What is the net-force component along (c) the x axis and (d) the y axis? Into which quadrants do (e) the net-force vector and (f) the split's acceleration vector point?

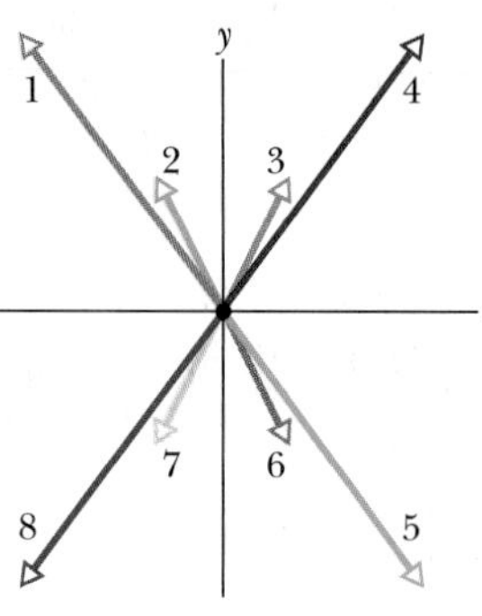

FIG. 5-23 Question 4.

5 Figure 5-24 gives the free-body diagram for four situations in which an object is pulled by several forces across a frictionless floor, as seen from overhead. In which situations does the object's acceleration $\vec{a}$ have (a) an x component and (b) a y component? (c) In each situation, give the direction of $\vec{a}$ by naming either a quadrant or a direction along an axis. (This can be done with a few mental calculations.)

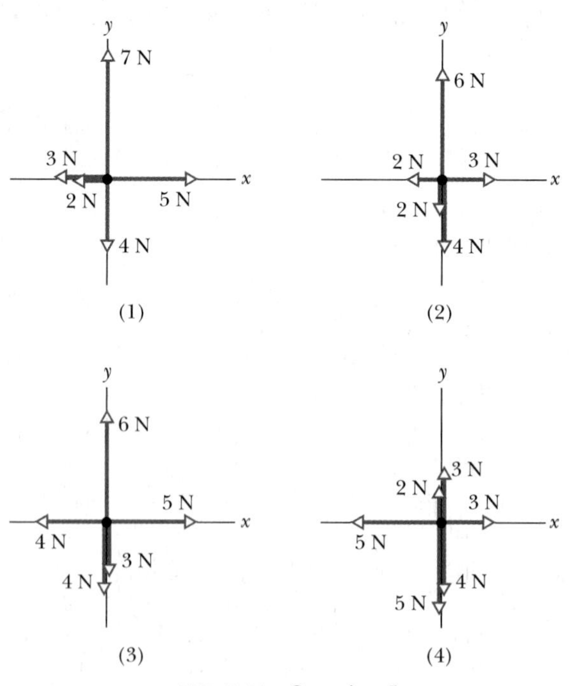

FIG. 5-24 Question 5.

6 Figure 5-25 gives three graphs of velocity component $v_x(t)$ and three graphs of velocity component $v_y(t)$. The graphs are not to scale. Which $v_x(t)$ graph and which $v_y(t)$ graph best correspond to each of the four situations in Question 5 and Fig. 5-24?

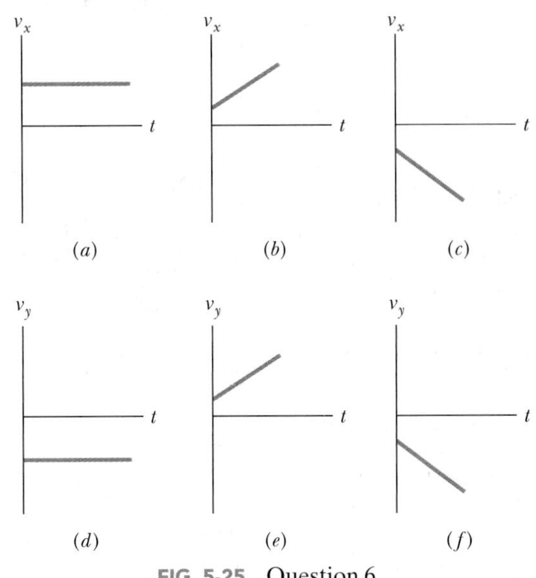

FIG. 5-25 Question 6.

7 Figure 5-26 shows a train of four blocks being pulled across a frictionless floor by force $\vec{F}$. What total mass is accelerated to the right by (a) force $\vec{F}$, (b) cord 3, and (c) cord 1? (d) Rank the blocks according to their accelerations, greatest first. (e) Rank the cords according to their tension, greatest first. (Warm-up for Problems 50 and 51)

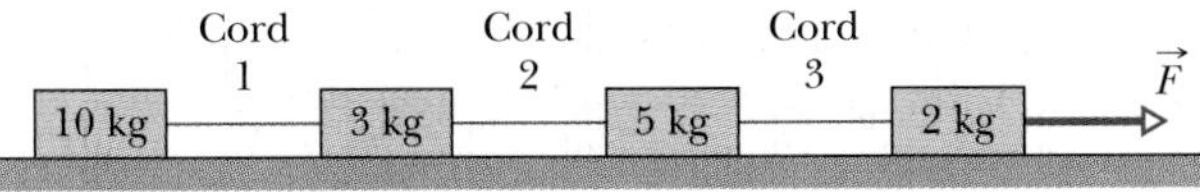

FIG. 5-26 Question 7.

8 Figure 5-27 shows the same breadbox in four situations where horizontal forces are applied. Rank the situations according to the magnitude of the box's acceleration, greatest first.

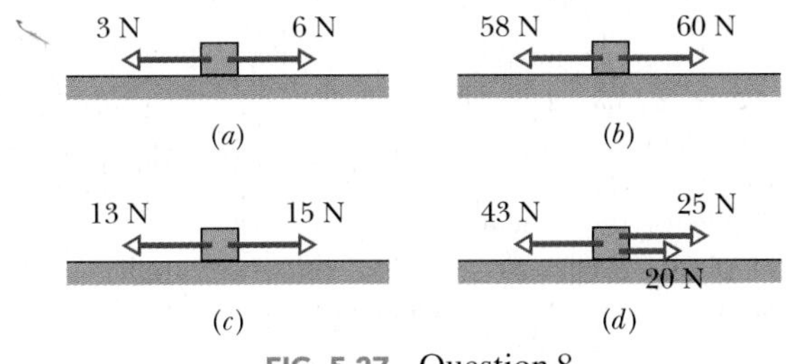

FIG. 5-27 Question 8.

9 A vertical force $\vec{F}$ is applied to a block of mass m that lies on a floor. What happens to the magnitude of the normal force $\vec{F}_N$ on the block from the floor as magnitude F is increased from zero if force $\vec{F}$ is (a) downward and (b) upward?

10 Figure 5-28 shows four choices for the direction of a force of magnitude F to be applied to a block on an inclined plane. The directions are either horizontal or vertical. (For choices a and b, the force is not enough to lift the block off the plane.) Rank the choices according to the magnitude of the normal force on the block from the plane, greatest first.

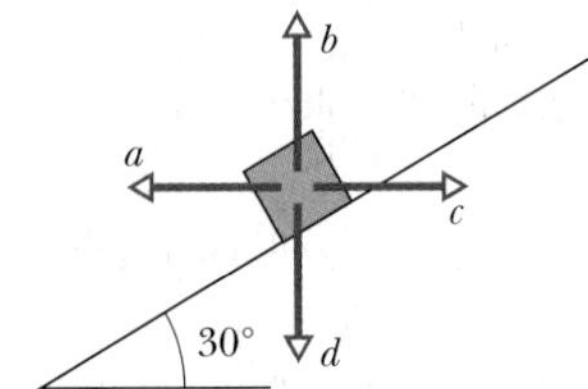

FIG. 5-28 Question 10.

11 July 17, 1981, Kansas City: The newly opened Hyatt Regency is packed with people listening and dancing to a band playing favorites from the 1940s. Many of the people are crowded onto the walkways that hang like bridges across the wide atrium. Suddenly two of the walkways collapse, falling onto the merrymakers on the main floor.

The walkways were suspended one above another on vertical rods and held in place by nuts threaded onto the rods. In the original design, only two long rods were to be used, each extending through all three walkways (Fig. 5-29a). If each walkway and the merrymakers on it have a combined mass of M, what is the total mass supported by the threads and two nuts on (a) the lowest walkway and (b) the highest walkway?

Threading nuts on a rod is impossible except at the ends, so the design was changed: Instead, six rods were used, each connecting two walkways (Fig. 5-29b). What now is the total mass supported by the threads and two nuts on (c) the lowest walkway, (d) the upper side of the highest walkway, and (e) the lower side of the highest walkway? It was this design that failed.

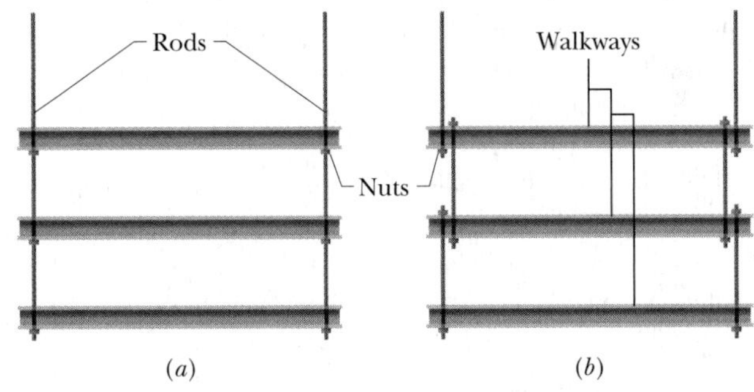

FIG. 5-29 Question 11.

12 Figure 5-30 shows three blocks being pushed across a frictionless floor by horizontal force $\vec{F}$. What total mass is accelerated to the right by (a) force $\vec{F}$, (b) force $\vec{F}_{21}$ on block 2 from block 1, and (c) force $\vec{F}_{32}$ on block 3 from block 2? (d) Rank the blocks according to their acceleration magnitudes, greatest first. (e) Rank forces $\vec{F}$, $\vec{F}_{21}$, and $\vec{F}_{32}$ according to magnitude, greatest first. (Warm-up for Problem 53)

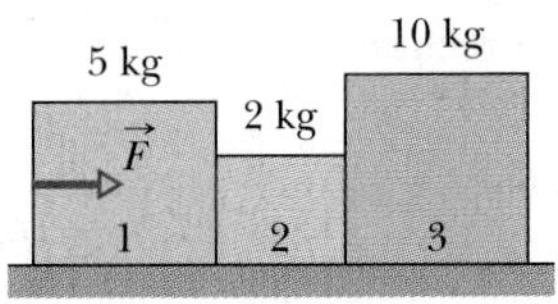

FIG. 5-30 Question 12.

PROBLEMS

GO Tutoring problem available (at instructor's discretion) in *WileyPLUS* and WebAssign

SSM Worked-out solution available in Student Solutions Manual

• – ••• Number of dots indicates level of problem difficulty

Additional information available in *The Flying Circus of Physics* and at flyingcircusofphysics.com

WWW Worked-out solution is at

ILW Interactive solution is at

http://www.wiley.com/college/halliday

sec. 5-6 Newton's Second Law

•1 If the 1 kg standard body has an acceleration of 2.00 m/s^2 at 20.0° to the positive direction of an x axis, what are (a) the x component and (b) the y component of the net force acting on the body, and (c) what is the net force in unit-vector notation?

•2 Two horizontal forces act on a 2.0 kg chopping block that can slide over a frictionless kitchen counter, which lies in an xy plane. One force is $\vec{F}_1 = (3.0\text{ N})\hat{i} + (4.0\text{ N})\hat{j}$. Find the acceleration of the chopping block in unit-vector notation when the other force is (a) $\vec{F}_2 = (-3.0\text{ N})\hat{i} + (-4.0\text{ N})\hat{j}$, (b) $\vec{F}_2 = (-3.0\text{ N})\hat{i} + (4.0\text{ N})\hat{j}$, and (c) $\vec{F}_2 = (3.0\text{ N})\hat{i} + (-4.0\text{ N})\hat{j}$.

•3 Only two horizontal forces act on a 3.0 kg body that can move over a frictionless floor. One force is 9.0 N, acting due east, and the other is 8.0 N, acting 62° north of west. What is the magnitude of the body's acceleration?

••4 A 2.00 kg object is subjected to three forces that give it an acceleration $\vec{a} = -(8.00\text{ m/s}^2)\hat{i} + (6.00\text{ m/s}^2)\hat{j}$. If two of the three forces are $\vec{F}_1 = (30.0\text{ N})\hat{i} + (16.0\text{ N})\hat{j}$ and $\vec{F}_2 = -(12.0\text{ N})\hat{i} + (8.00\text{ N})\hat{j}$, find the third force.

••5 There are two forces on the 2.00 kg box in the overhead view of Fig. 5-31, but only one is shown. For $F_1 = 20.0$ N, $a = 12.0$ m/s^2, and $\theta = 30.0°$, find the second force (a) in unit-vector notation and as (b) a magnitude and (c) an angle relative to the positive direction of the x axis. SSM

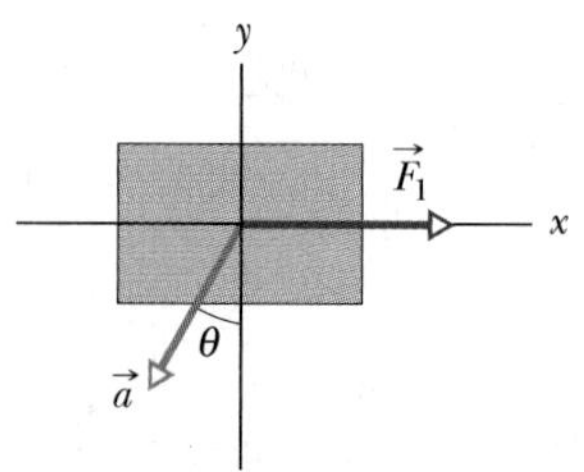

FIG. 5-31 Problem 5.

••6 While two forces act on it, a particle is to move at the constant velocity $\vec{v} = (3\text{ m/s})\hat{i} - (4\text{ m/s})\hat{j}$. One of the forces is $\vec{F}_1 = (2\text{ N})\hat{i} + (-6\text{ N})\hat{j}$. What is the other force?

••7 Three astronauts, propelled by jet backpacks, push and guide a 120 kg asteroid toward a processing dock, exerting the forces shown in Fig. 5-32, with $F_1 = 32$ N, $F_2 = 55$ N, $F_3 = 41$ N, $\theta_1 = 30°$, and $\theta_3 = 60°$. What is the asteroid's acceleration (a) in unit-vector notation and as (b) a magnitude and (c) a direction relative to the positive direction of the x axis? GO

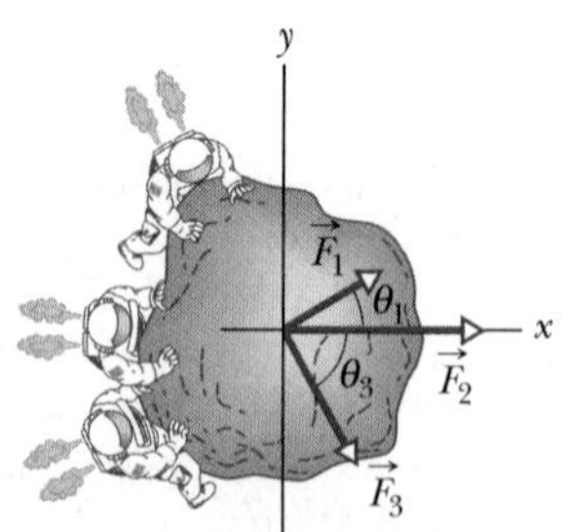

FIG. 5-32 Problem 7.

••8 In a two-dimensional tug-of-war, Alex, Betty, and Charles pull horizontally on an automobile tire at the angles shown in the overhead view of Fig. 5-33. The tire remains stationary in spite of the three pulls. Alex pulls with force $\vec{F}_A$ of magnitude 220 N, and Charles pulls with force $\vec{F}_C$ of magnitude 170 N. Note that the direction of $\vec{F}_C$ is not given. What is the magnitude of Betty's force $\vec{F}_B$?

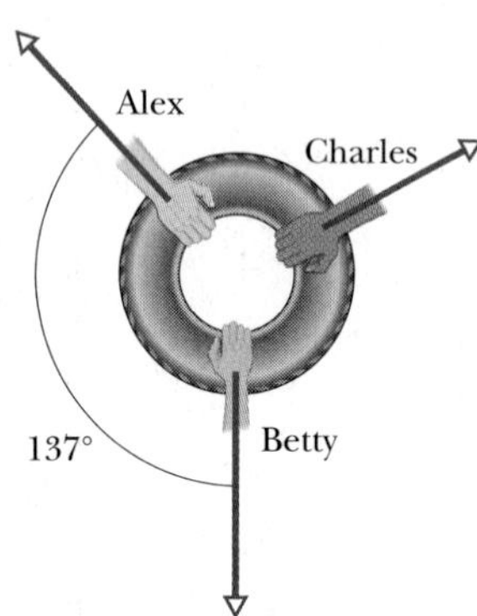

FIG. 5-33 Problem 8.

••9 A 2.0 kg particle moves along an x axis, being propelled by a variable force directed along that axis. Its position is given by $x = 3.0\text{ m} + (4.0\text{ m/s})t + ct^2 - (2.0\text{ m/s}^3)t^3$, with x in meters and t in seconds. The factor c is a constant. At $t = 3.0$ s, the force on the particle has a magnitude of 36 N and is in the negative direction of the axis. What is c?

••10 A 0.150 kg particle moves along an x axis according to $x(t) = -13.00 + 2.00t + 4.00t^2 - 3.00t^3$, with x in meters and t in seconds. In unit-vector notation, what is the net force acting on the particle at $t = 3.40$ s?

••11 A 0.340 kg particle moves in an xy plane according to $x(t) = -15.00 + 2.00t - 4.00t^3$ and $y(t) = 25.00 + 7.00t - 9.00t^2$, with x and y in meters and t in seconds. At $t = 0.700$ s, what are (a) the magnitude and (b) the angle (relative to the positive direction of the x axis) of the net force on the particle, and (c) what is the angle of the particle's direction of travel?

•••12 Two horizontal forces $\vec{F}_1$ and $\vec{F}_2$ act on a 4.0 kg disk that slides over frictionless ice, on which an xy coordinate system is laid out. Force $\vec{F}_1$ is in the positive direction of the x axis and has a magnitude of 7.0 N. Force $\vec{F}_2$ has a magnitude of 9.0 N. Figure 5-34 gives the x component v_x of the velocity of the disk as a function of time t during the sliding. What is the angle between the constant directions of forces $\vec{F}_1$ and $\vec{F}_2$?

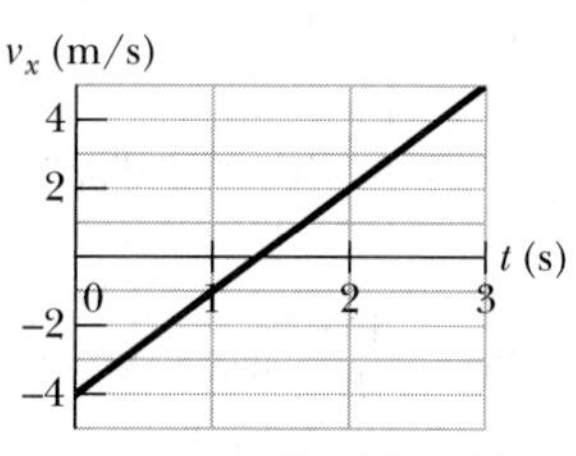

FIG. 5-34 Problem 12.

sec. 5-7 Some Particular Forces

•13 (a) An 11.0 kg salami is supported by a cord that runs to a spring scale, which is supported by a cord hung from the ceiling (Fig. 5-35a). What is the reading on the scale, which is marked in weight units? (b) In Fig. 5-35b the salami is supported by a cord that runs around a pulley and to a scale. The opposite end of the scale is attached by a cord to a wall. What is the reading on the scale? (c) In Fig. 5-35c the wall has been replaced with a second 11.0 kg salami, and the assembly is stationary. What is the reading on the scale? SSM

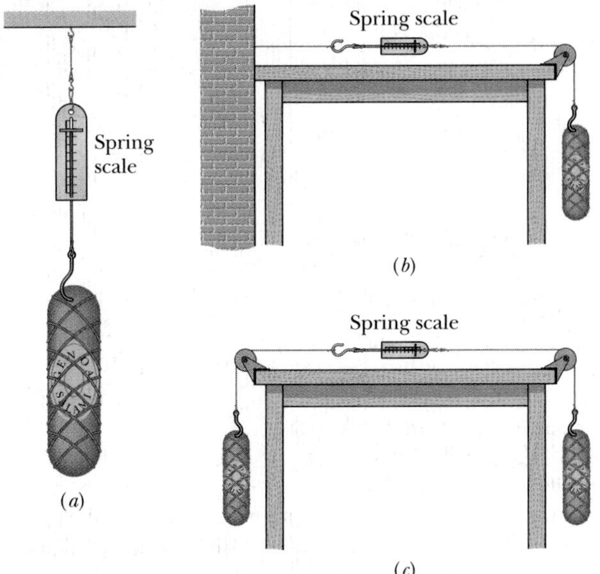

FIG. 5-35 Problem 13.

•**14** A block with a weight of 3.0 N is at rest on a horizontal surface. A 1.0 N upward force is applied to the block by means of an attached vertical string. What are the (a) magnitude and (b) direction of the force of the block on the horizontal surface?

•**15** Figure 5-36 shows an arrangement in which four disks are suspended by cords. The longer, top cord loops over a frictionless pulley and pulls with a force of magnitude 98 N on the wall to which it is attached. The tensions in the shorter cords are $T_1 = 58.8$ N, $T_2 = 49.0$ N, and $T_3 = 9.8$ N. What are the masses of (a) disk A, (b) disk B, (c) disk C, and (d) disk D?

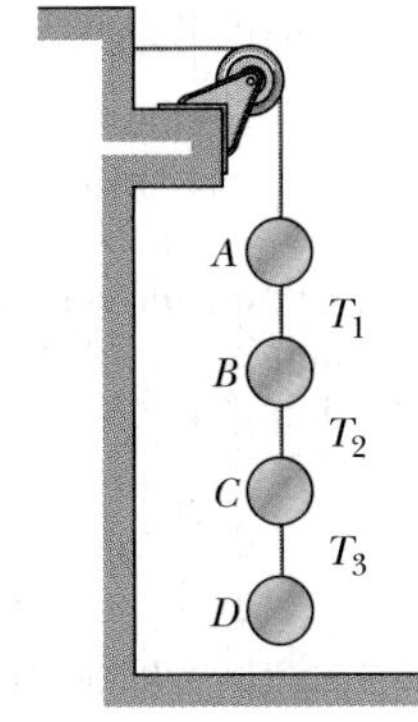

FIG. 5-36 Problem 15.

••**16** Some insects can walk below a thin rod (such as a twig) by hanging from it. Suppose that such an insect has mass m and hangs from a horizontal rod as shown in Fig. 5-37, with angle $\theta = 40°$. Its six legs are all under the same tension, and the leg sections nearest the body are horizontal. (a) What is the ratio of the tension in each tibia (forepart of a leg) to the insect's weight? (b) If the insect straightens out its legs somewhat, does the tension in each tibia increase, decrease, or stay the same?

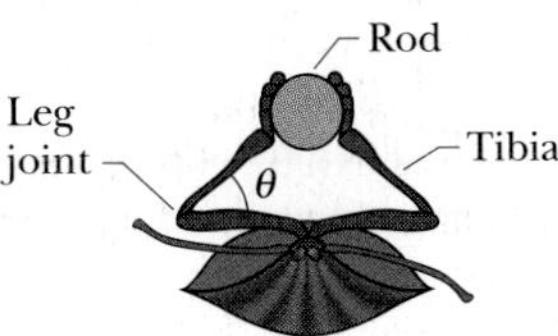

FIG. 5-37 Problem 16.

sec. 5-9 Applying Newton's Laws

•**17** A customer sits in an amusement park ride in which the compartment is to be pulled downward in the negative direction of a y axis with an acceleration magnitude of $1.24g$, with $g = 9.80$ m/s^2. A 0.567 g coin rests on the customer's knee. Once the motion begins and in unit-vector notation, what is the coin's acceleration relative to (a) the ground and (b) the customer? (c) How long does the coin take to reach the compartment ceiling, 2.20 m above the knee? In unit-vector notation, what are (d) the actual force on the coin and (e) the apparent force according to the customer's measure of the coin's acceleration?

•**18** Tarzan, who weighs 820 N, swings from a cliff at the end of a 20.0 m vine that hangs from a high tree limb and initially makes an angle of 22.0° with the vertical. Assume that an x axis extends horizontally away from the cliff edge and a y axis extends upward. Immediately after Tarzan steps off the cliff, the tension in the vine is 760 N. Just then, what are (a) the force on him from the vine in unit-vector notation and the net force on him (b) in unit-vector notation and as (c) a magnitude and (d) an angle relative to the positive direction of the x axis? What are the (e) magnitude and (f) angle of Tarzan's acceleration just then?

•**19** In Fig. 5-38, let the mass of the block be 8.5 kg and the angle θ be 30°. Find (a) the tension in the cord and (b) the normal force acting on the block. (c) If the cord is cut, find the magnitude of the resulting acceleration of the block. SSM WWW

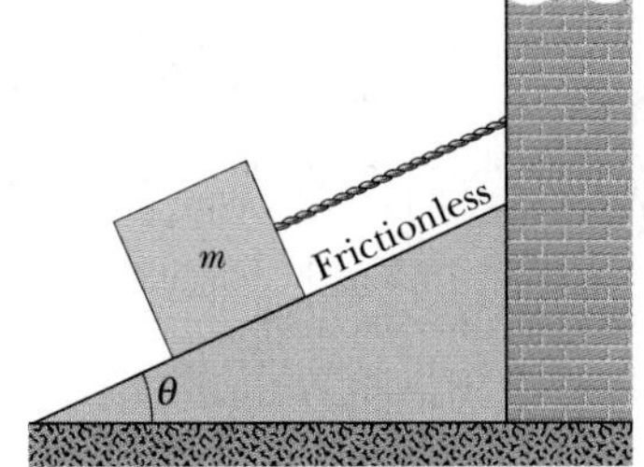

FIG. 5-38 Problem 19.

•**20** There are two horizontal forces on the 2.0 kg box in the overhead view of Fig. 5-39 but only one (of magnitude $F_1 = 20$ N) is shown. The box moves along the x axis. For each of the following values for the acceleration a_x of the box, find the second force in unit-vector notation: (a) 10 m/s^2, (b) 20 m/s^2, (c) 0, (d) -10 m/s^2, and (e) -20 m/s^2.

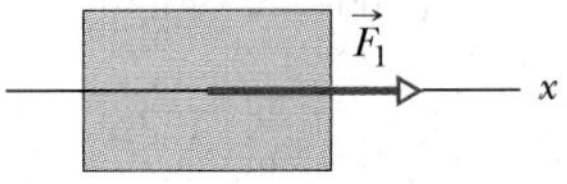

FIG. 5-39 Problem 20.

•**21** A constant horizontal force $\vec{F}_a$ pushes a 2.00 kg FedEx package across a frictionless floor on which an xy coordinate system has been drawn. Figure 5-40 gives the package's x and y velocity components versus time t. What are the (a) magnitude and (b) direction of $\vec{F}_a$?

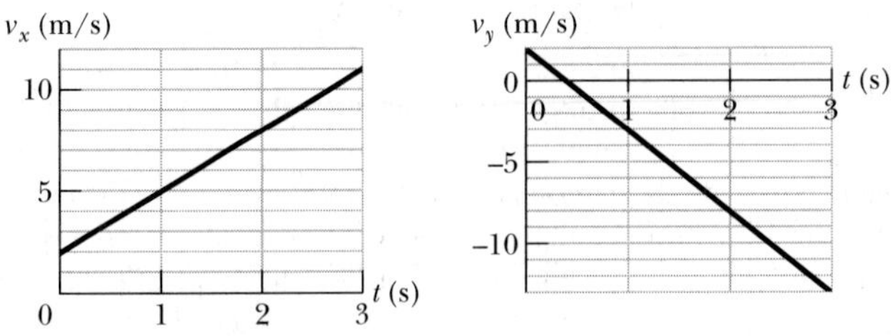

FIG. 5-40 Problem 21.

•**22** In April 1974, John Massis of Belgium managed to move two passenger railroad cars. He did so by clamping his teeth down on a bit that was attached to the cars with a rope and then leaning backward while pressing his feet against the railway ties. The cars together weighed 700 kN (about 80 tons). Assume that he pulled with a constant force that was 2.5 times his body weight, at an upward angle θ of 30° from the horizontal. His mass was 80 kg, and he moved the cars by 1.0 m. Neglecting any retarding force from the wheel rotation, find the speed of the cars at the end of the pull.

•**23** *Sunjamming.* A "sun yacht" is a spacecraft with a large sail that is pushed by sunlight. Although such a push is tiny in everyday circumstances, it can be large enough to send the spacecraft outward from the Sun on a cost-free but slow trip. Suppose that the spacecraft has a mass of 900 kg and receives a push of 20 N. (a) What is the magnitude of the resulting acceleration? If the craft starts from rest, (b) how far will it travel in 1 day and (c) how fast will it then be moving?

•**24** The tension at which a fishing line snaps is commonly called the line's "strength." What minimum strength is needed for a line that is to stop a salmon of weight 85 N in 11 cm if the fish is initially drifting at 2.8 m/s? Assume a constant deceleration.

•**25** A 500 kg rocket sled can be accelerated at a constant rate from rest to 1600 km/h in 1.8 s. What is the magnitude of the required net force? SSM

•**26** A car traveling at 53 km/h hits a bridge abutment. A passenger in the car moves forward a distance of 65 cm (with respect to the road) while being brought to rest by an inflated air bag. What magnitude of force (assumed constant) acts on the passenger's upper torso, which has a mass of 41 kg?

•**27** A firefighter who weighs 712 N slides down a vertical pole with an acceleration of 3.00 m/s^2, directed downward. What are the (a) magnitude and (b) direction (up or down) of the vertical force on the firefighter from the pole and the (c) magnitude and (d) direction of the vertical force of the pole on the firefighter?

•**28** The high-speed winds around a tornado can drive projectiles into trees, building walls, and even metal traffic signs. In a laboratory simulation, a standard wood toothpick was shot by pneumatic gun into an oak branch. The toothpick's mass was 0.13 g, its speed before entering the branch was 220 m/s, and its penetration depth was 15 mm. If its speed was decreased at a uniform rate, what was the magnitude of the force of the branch on the toothpick?

•**29** An electron with a speed of 1.2×10^7 m/s moves horizontally into a region where a constant vertical force of 4.5×10^{-16} N acts on it. The mass of the electron is 9.11×10^{-31} kg. Determine the vertical distance the electron is deflected during the time it has moved 30 mm horizontally. SSM

•**30** A car that weighs 1.30×10^4 N is initially moving at 40 km/h when the brakes are applied and the car is brought to a stop in 15 m. Assuming the force that stops the car is constant, find (a) the magnitude of that force and (b) the time required for the change in speed. If the initial speed is doubled, and the car experiences the same force during the braking, by what factors are (c) the stopping distance and (d) the stopping time multiplied? (There could be a lesson here about the danger of driving at high speeds.)

••**31** The velocity of a 3.00 kg particle is given by $\vec{v} = (8.00t\hat{i} + 3.00t^2\hat{j})$ m/s, with time t in seconds. At the instant the net force on the particle has a magnitude of 35.0 N, what are the direction (relative to the positive direction of the x axis) of (a) the net force and (b) the particle's direction of travel?

••**32** In Fig. 5-41, a crate of mass $m = 100$ kg is pushed at constant speed up a frictionless ramp ($\theta = 30.0°$) by a horizontal force $\vec{F}$. What are the magnitudes of (a) $\vec{F}$ and (b) the force on the crate from the ramp? GO

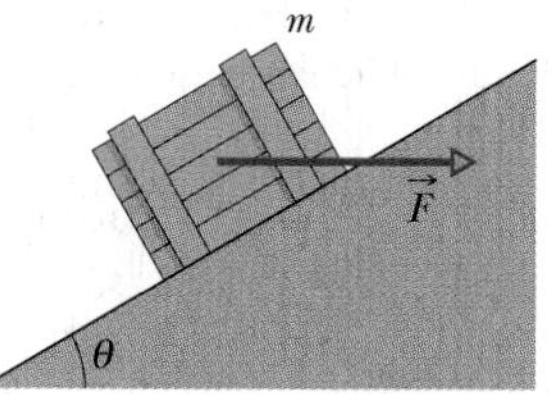

FIG. 5-41 Problem 32.

••**33** A 40 kg girl and an 8.4 kg sled are on the frictionless ice of a frozen lake, 15 m apart but connected by a rope of negligible mass. The girl exerts a horizontal 5.2 N force on the rope. What are the acceleration magnitudes of (a) the sled and (b) the girl? (c) How far from the girl's initial position do they meet?

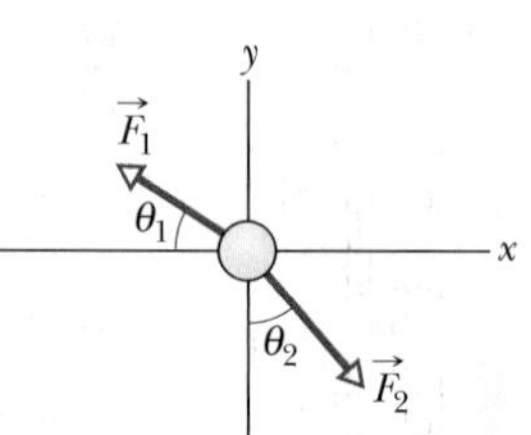

FIG. 5-42 Problem 34.

••**34** Figure 5-42 shows an overhead view of a 0.0250 kg lemon half and two of the three horizontal forces that act on it as it is on a frictionless table. Force $\vec{F}_1$ has a magnitude of 6.00 N and is at $\theta_1 = 30.0°$. Force $\vec{F}_2$ has a magnitude of 7.00 N and is at $\theta_2 = 30.0°$. In unit-vector notation, what is the third force if the lemon half (a) is stationary, (b) has constant velocity $\vec{v} = (13.0\hat{i} - 14.0\hat{j})$ m/s, and (c) has varying velocity $\vec{v} = (13.0t\hat{i} - 14.0t\hat{j})$ m/s^2, where t is time?

••**35** A block is projected up a frictionless inclined plane with initial speed $v_0 = 3.50$ m/s. The angle of incline is $\theta = 32.0°$. (a) How far up the plane does the block go? (b) How long does it take to get there? (c) What is its speed when it gets back to the bottom? SSM WWW

••**36** A 40 kg skier skis directly down a frictionless slope angled at 10° to the horizontal. Assume the skier moves in the negative direction of an x axis along the slope. A wind force with component F_x acts on the skier. What is F_x if the magnitude of the skier's velocity is (a) constant, (b) increasing at a rate of 1.0 m/s^2, and (c) increasing at a rate of 2.0 m/s^2?

••**37** A sphere of mass 3.0×10^{-4} kg is suspended from a cord. A steady horizontal breeze pushes the sphere so that the cord makes a constant angle of 37° with the vertical. Find (a) the push magnitude and (b) the tension in the cord. ILW

••**38** A dated box of dates, of mass 5.00 kg, is sent sliding up a frictionless ramp at an angle of θ to the horizontal. Figure 5-43 gives, as a function of time t, the component v_x of the box's velocity along an x axis that extends directly up the ramp. What is the magnitude of the normal force on the box from the ramp?

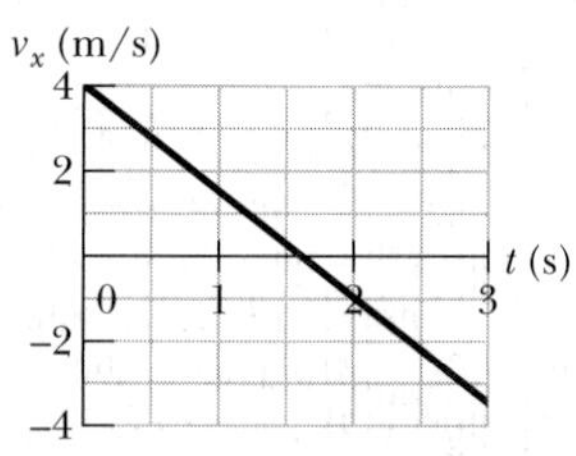

FIG. 5-43 Problem 38.

••**39** An elevator cab and its load have a combined mass of 1600 kg. Find the tension in the supporting cable when the cab, originally moving downward at 12 m/s, is brought to rest with constant acceleration in a distance of 42 m.

••**40** Holding on to a towrope moving parallel to a frictionless ski slope, a 50 kg skier is pulled up the slope, which is at an angle of 8.0° with the horizontal. What is the magnitude F_{rope} of the force on the skier from the rope when (a) the magnitude v of the skier's velocity is constant at 2.0 m/s and (b) $v =$ 2.0 m/s as v increases at a rate of 0.10 m/s^2?

••41 An elevator cab that weighs 27.8 kN moves upward. What is the tension in the cable if the cab's speed is (a) increasing at a rate of $1.22\ m/s^2$ and (b) decreasing at a rate of $1.22\ m/s^2$?

••42 A lamp hangs vertically from a cord in a descending elevator that decelerates at $2.4\ m/s^2$. (a) If the tension in the cord is 89 N, what is the lamp's mass? (b) What is the cord's tension when the elevator ascends with an upward acceleration of $2.4\ m/s^2$?

••43 Using a rope that will snap if the tension in it exceeds 387 N, you need to lower a bundle of old roofing material weighing 449 N from a point 6.1 m above the ground. (a) What magnitude of the bundle's acceleration will put the rope on the verge of snapping? (b) At that acceleration, with what speed would the bundle hit the ground?

••44 An elevator cab is pulled upward by a cable. The cab and its single occupant have a combined mass of 2000 kg. When that occupant drops a coin, its acceleration relative to the cab is $8.00\ m/s^2$ downward. What is the tension in the cable?

••45 In Fig. 5-44, a chain consisting of five links, each of mass 0.100 kg, is lifted vertically with constant acceleration of magnitude $a = 2.50\ m/s^2$. Find the magnitudes of (a) the force on link 1 from link 2, (b) the force on link 2 from link 3, (c) the force on link 3 from link 4, and (d) the force on link 4 from link 5. Then find the magnitudes of (e) the force $\vec{F}$ on the top link from the person lifting the chain and (f) the *net* force accelerating each link. SSM

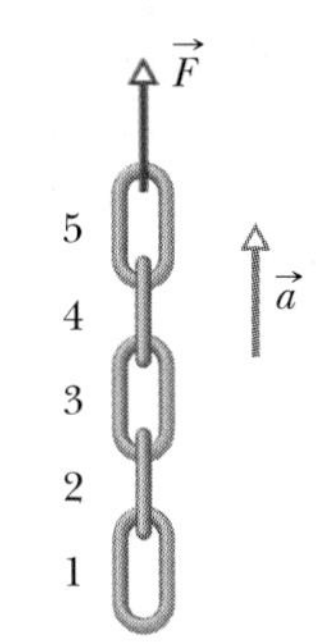

FIG. 5-44 Problem 45.

••46 In Fig. 5-45, elevator cabs *A* and *B* are connected by a short cable and can be pulled upward or lowered by the cable above cab *A*. Cab *A* has mass 1700 kg; cab *B* has mass 1300 kg. A 12.0 kg box of catnip lies on the floor of cab *A*. The tension in the cable connecting the cabs is 1.91×10^4 N. What is the magnitude of the normal force on the box from the floor?

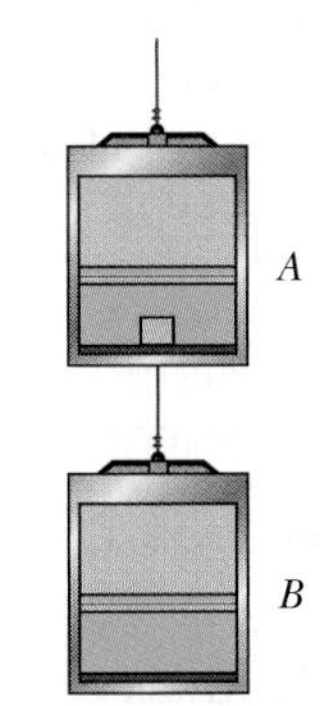

FIG. 5-45 Problem 46.

••47 In Fig. 5-46, a block of mass $m = 5.00$ kg is pulled along a horizontal frictionless floor by a cord that exerts a force of magnitude $F = 12.0$ N at an angle $\theta = 25.0°$. (a) What is the magnitude of the block's acceleration? (b) The force magnitude F is slowly increased. What is its value just before the block is lifted (completely) off the floor? (c) What is the magnitude of the block's acceleration just before it is lifted (completely) off the floor?

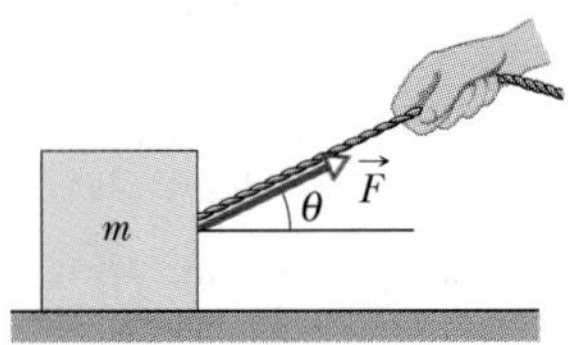

FIG. 5-46 Problems 47 and 62.

••48 In earlier days, horses pulled barges down canals in the manner shown in Fig. 5-47. Suppose the horse pulls on the rope with a force of 7900 N at an angle of $\theta = 18°$ to the direction of motion of the barge, which is headed straight along the positive direction of an x axis. The mass of the barge is 9500 kg, and the magnitude of its acceleration is $0.12\ m/s^2$. What are the (a) magnitude and (b) direction (relative to positive x) of the force on the barge from the water? GO

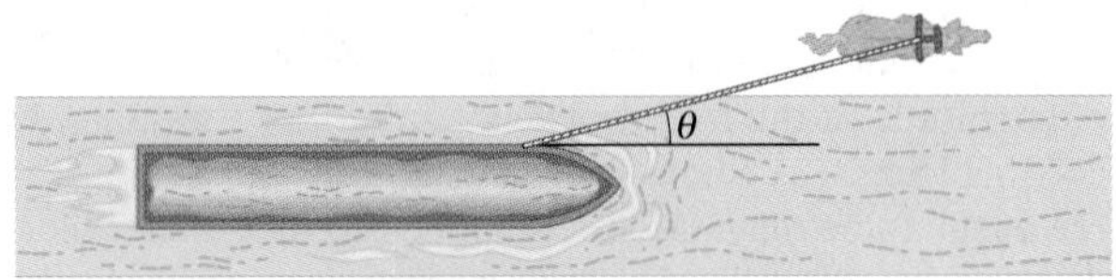

FIG. 5-47 Problem 48.

••49 The Zacchini family was renowned for their human-cannonball act in which a family member was shot from a cannon using either elastic bands or compressed air. In one version of the act, Emanuel Zacchini was shot over three Ferris wheels to land in a net at the same height as the open end of the cannon and at a range of 69 m. He was propelled inside the barrel for 5.2 m and launched at an angle of 53°. If his mass was 85 kg and he underwent constant acceleration inside the barrel, what was the magnitude of the force propelling him? (*Hint:* Treat the launch as though it were along a ramp at 53°. Neglect air drag.)

••50 Figure 5-48 shows four penguins that are being playfully pulled along very slippery (frictionless) ice by a curator. The masses of three penguins and the tension in two of the cords are $m_1 = 12$ kg, $m_3 = 15$ kg, $m_4 = 20$ kg, $T_2 = 111$ N, and $T_4 = 222$ N. Find the penguin mass m_2 that is not given.

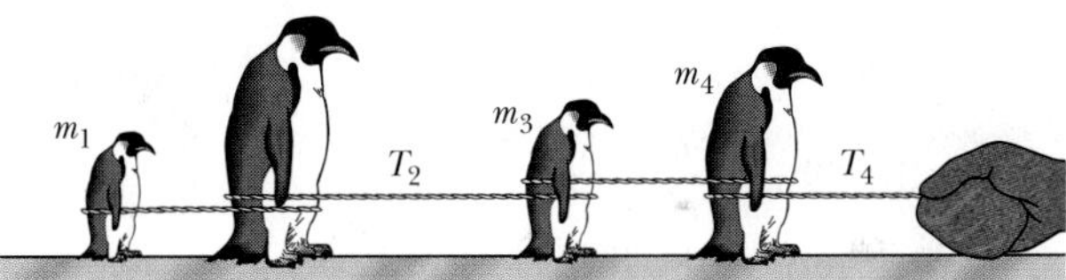

FIG. 5-48 Problem 50.

••51 In Fig. 5-49, three connected blocks are pulled to the right on a horizontal frictionless table by a force of magnitude $T_3 = 65.0$ N. If $m_1 = 12.0$ kg, $m_2 = 24.0$ kg, and $m_3 = 31.0$ kg, calculate (a) the magnitude of the system's acceleration, (b) the tension T_1, and (c) the tension T_2.

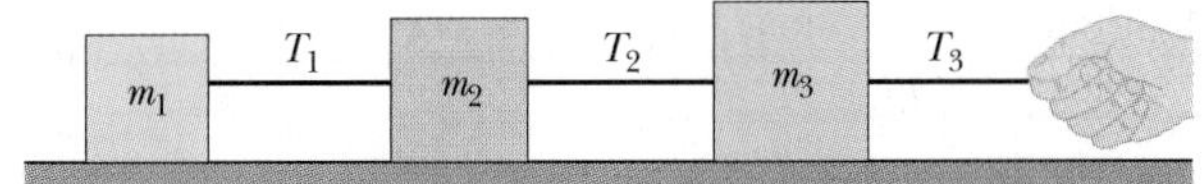

FIG. 5-49 Problem 51.

••52 In Fig. 5-50*a*, a constant horizontal force $\vec{F}_a$ is applied to block *A*, which pushes against block *B* with a 20.0 N force directed horizontally to the right. In Fig. 5-50*b*, the same force $\vec{F}_a$ is applied to block *B*; now block *A* pushes on block *B* with a

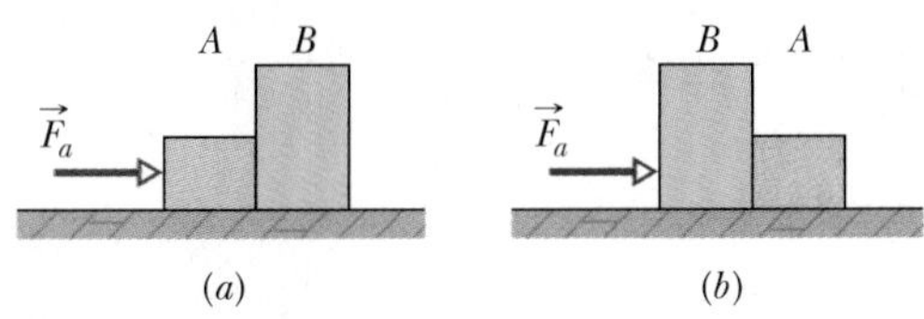

FIG. 5-50 Problem 52.

10.0 N force directed horizontally to the left. The blocks have a combined mass of 12.0 kg. What are the magnitudes of (a) their acceleration in Fig. 5-50*a* and (b) force $\vec{F}_a$?

••**53** Two blocks are in contact on a frictionless table. A horizontal force is applied to the larger block, as shown in Fig. 5-51. (a) If $m_1 = 2.3$ kg, $m_2 = 1.2$ kg, and $F = 3.2$ N, find the magnitude of the force between the two blocks. (b) Show that if a force of the same magnitude F is applied to the smaller block but in the opposite direction, the magnitude of the force between the blocks is 2.1 N, which is not the same value calculated in (a). (c) Explain the difference. SSM ILW WWW

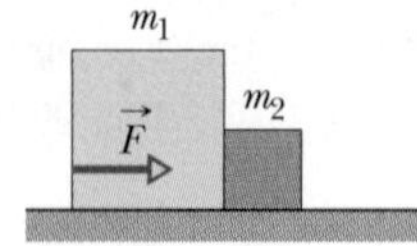

FIG. 5-51 Problem 53.

••**54** In Fig. 5-52, three ballot boxes are connected by cords, one of which wraps over a pulley having negligible friction on its axle and negligible mass. The three masses are $m_A = 30.0$ kg, $m_B = 40.0$ kg, and $m_C = 10.0$ kg. When the assembly is released from rest, (a) what is the tension in the cord connecting B and C, and (b) how far does A move in the first 0.250 s (assuming it does not reach the pulley)?

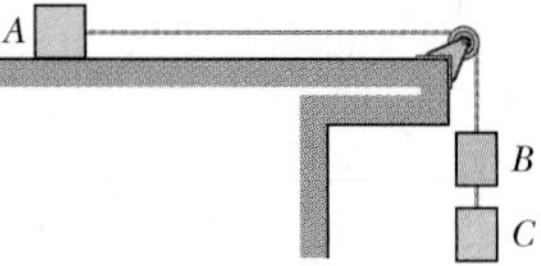

FIG. 5-52 Problem 54.

••**55** Figure 5-53 shows two blocks connected by a cord (of negligible mass) that passes over a frictionless pulley (also of negligible mass). The arrangement is known as *Atwood's machine*. One block has mass $m_1 = 1.30$ kg; the other has mass $m_2 = 2.80$ kg. What are (a) the magnitude of the blocks' acceleration and (b) the tension in the cord? GO

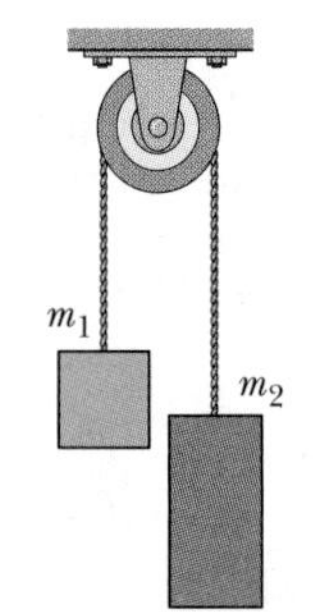

FIG. 5-53 Problems 55 and 63.

••**56** In shot putting, many athletes elect to launch the shot at an angle that is smaller than the theoretical one (about 42°) at which the distance of a projected ball at the same speed and height is greatest. One reason has to do with the speed the athlete can give the shot during the acceleration phase of the throw. Assume that a 7.260 kg shot is accelerated along a straight path of length 1.650 m by a constant applied force of magnitude 380.0 N, starting with an initial speed of 2.500 m/s (due to the athlete's preliminary motion). What is the shot's speed at the end of the acceleration phase if the angle between the path and the horizontal is (a) 30.00° and (b) 42.00°? (*Hint:* Treat the motion as though it were along a ramp at the given angle.) (c) By what percent is the launch speed decreased if the athlete increases the angle from 30.00° to 42.00°?

••**57** A 10 kg monkey climbs up a massless rope that runs over a frictionless tree limb and back down to a 15 kg package on the ground (Fig. 5-54). (a) What is the magnitude of the least acceleration the monkey must have if it is to lift the package off the ground? If, after the package has been lifted, the monkey stops its climb and holds onto the rope, what are the (b) magnitude and (c) direction of the monkey's acceleration and (d) the tension in the rope? SSM

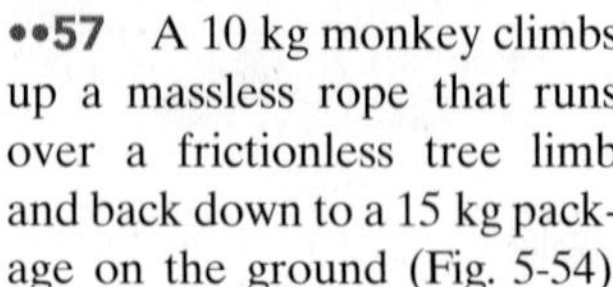

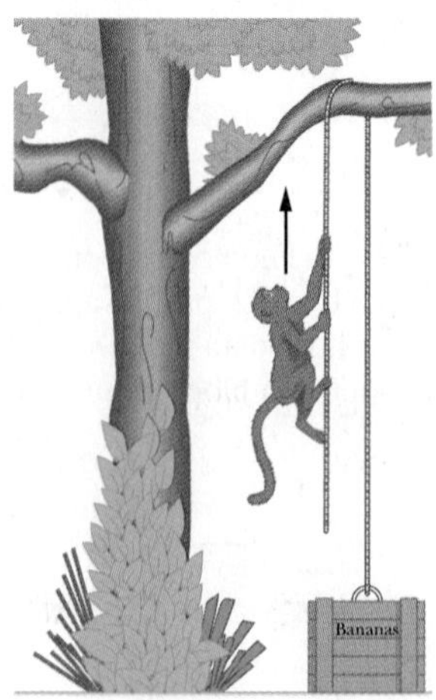

FIG. 5-54 Problem 57.

••**58** An 85 kg man lowers himself to the ground from a height of 10.0 m by holding onto a rope that runs over a frictionless pulley to a 65 kg sandbag. With what speed does the man hit the ground if he started from rest?

••**59** A block of mass $m_1 = 3.70$ kg on a frictionless plane inclined at angle $\theta = 30.0°$ is connected by a cord over a massless, frictionless pulley to a second block of mass $m_2 = 2.30$ kg (Fig. 5-55). What are (a) the magnitude of the acceleration of each block, (b) the direction of the acceleration of the hanging block, and (c) the tension in the cord? ILW

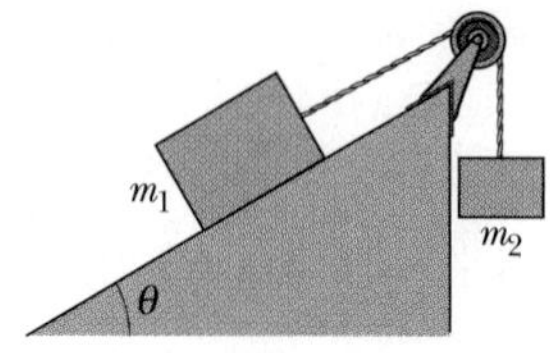

FIG. 5-55 Problem 59.

••**60** Figure 5-56 shows a man sitting in a bosun's chair that dangles from a massless rope, which runs over a massless, frictionless pulley and back down to the man's hand. The combined mass of man and chair is 95.0 kg. With what force magnitude must the man pull on the rope if he is to rise (a) with a constant velocity and (b) with an upward acceleration of 1.30 m/s²? (*Hint:* A free-body diagram can really help.) If the rope on the right extends to the ground and is pulled by a co-worker, with what force magnitude must the co-worker pull for the man to rise (c) with a constant velocity and (d) with an upward acceleration of 1.30 m/s²? What is the magnitude of the force on the ceiling from the pulley system in (e) part a, (f) part b, (g) part c, and (h) part d?

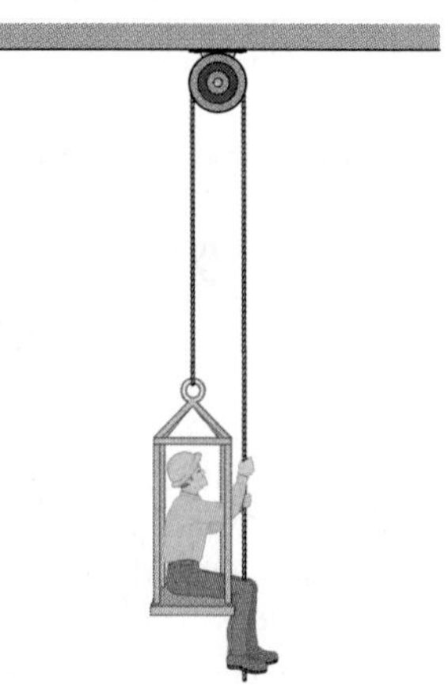
FIG. 5-56 Problem 60.

••**61** A hot-air balloon of mass M is descending vertically with downward acceleration of magnitude a. How much mass (ballast) must be thrown out to give the balloon an upward acceleration of magnitude a? Assume that the upward force from the air (the lift) does not change because of the decrease in mass. SSM ILW

••**62** Figure 5-46 shows a 5.00 kg block being pulled along a frictionless floor by a cord that applies a force of constant magnitude 20.0 N but with an angle $\theta(t)$ that varies with time. When angle $\theta = 25.0°$, at what rate is the block's acceleration changing if (a) $\theta(t) = (2.00 \times 10^{-2}$ deg/s$)t$ and (b) $\theta(t) = -(2.00 \times 10^{-2}$ deg/s$)t$? (*Hint:* Switch to radians.)

•••**63** Figure 5-53 shows *Atwood's machine*, in which two containers are connected by a cord (of negligible mass) passing over a frictionless pulley (also of negligible mass). At time $t = 0$, container 1 has mass 1.30 kg and container 2 has mass 2.80 kg, but container 1 is losing mass (through a leak) at the constant rate of 0.200 kg/s. At what rate is the acceleration magnitude of the containers changing at (a) $t = 0$ and (b) $t = 3.00$ s? (c) When does the acceleration reach its maximum value?

•••64 A shot putter launches a 7.260 kg shot by pushing it along a straight line of length 1.650 m and at an angle of 34.10° from the horizontal, accelerating the shot to the launch speed from its initial speed of 2.500 m/s (which is due to the athlete's preliminary motion). The shot leaves the hand at a height of 2.110 m and at an angle of 34.10°, and it lands at a horizontal distance of 15.90 m. What is the magnitude of the athlete's average force on the shot during the acceleration phase? (*Hint:* Treat the motion during the acceleration phase as though it were along a ramp at the given angle.)

•••65 Figure 5-57 shows three blocks attached by cords that loop over frictionless pulleys. Block B lies on a frictionless table; the masses are m_A = 6.00 kg, m_B = 8.00 kg, and m_C = 10.0 kg. When the blocks are released, what is the tension in the cord at the right?

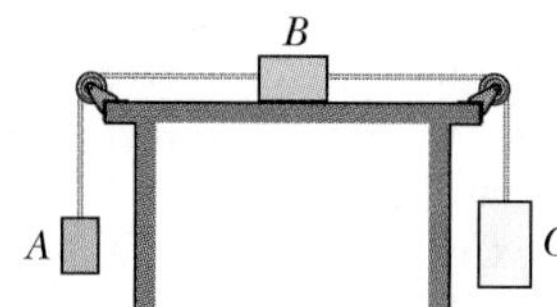

FIG. 5-57 Problem 65.

•••66 Figure 5-58 shows a box of mass m_2 = 1.0 kg on a frictionless plane inclined at angle θ = 30°. It is connected by a cord of negligible mass to a box of mass m_1 = 3.0 kg on a horizontal frictionless surface. The pulley is frictionless and massless. (a) If the magnitude of horizontal force $\vec{F}$ is 2.3 N, what is the tension in the connecting cord? (b) What is the largest value the magnitude of $\vec{F}$ may have without the cord becoming slack?

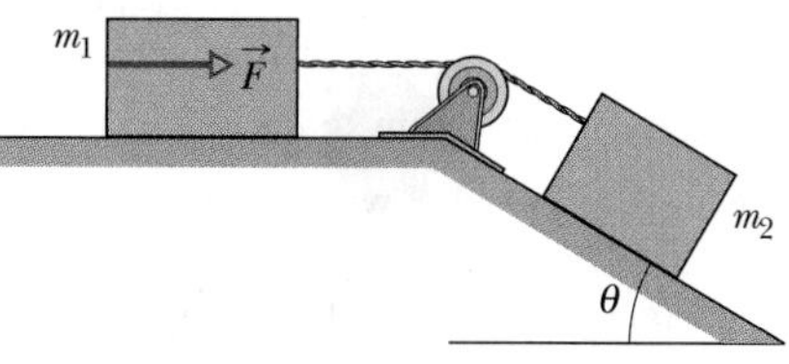

FIG. 5-58 Problem 66.

•••67 Figure 5-59 gives, as a function of time t, the force component F_x that acts on a 3.00 kg ice block that can move only along the x axis. At t = 0, the block is moving in the positive direction of the axis, with a speed of 3.0 m/s. What are its (a) speed and (b) direction of travel at t = 11 s?

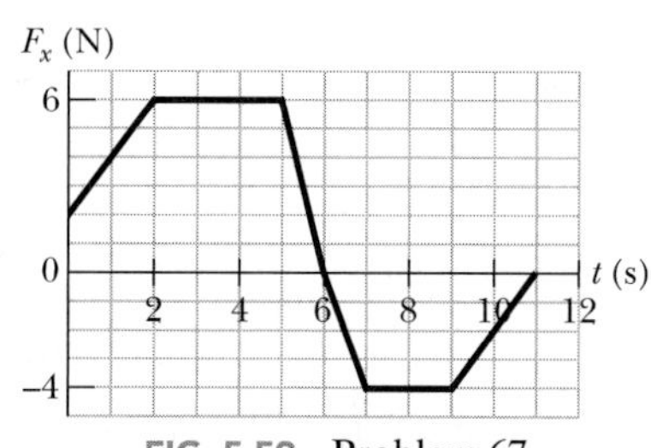

FIG. 5-59 Problem 67.

•••68 Figure 5-60 shows a section of a cable-car system. The maximum permissible mass of each car with occupants is 2800 kg. The cars, riding on a support cable, are pulled by a second cable attached to the support tower on each car. Assume that the cables are taut and inclined at angle θ = 35°. What is the difference in tension between adjacent sections of pull cable if the cars are at the maximum permissible mass and are being accelerated up the incline at 0.81 m/s²?

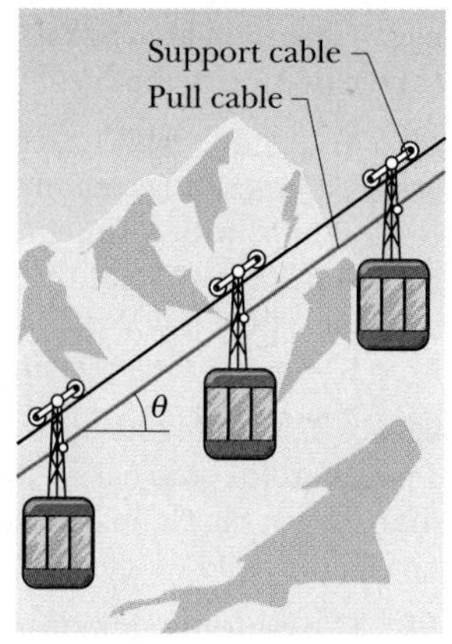

FIG. 5-60 Problem 68.

Additional Problems

69 *Blowing off units.* Throughout your physics course, your instructor will expect you to be careful with units in your calculations. Yet some students tend to neglect them and just trust that they always work out properly. Maybe this real-world example will keep you from such a sloppy habit.

On July 23, 1983, Air Canada Flight 143 was being readied for its long trip from Montreal to Edmonton when the flight crew asked the ground crew to determine how much fuel was already on board. The flight crew knew they needed to begin the trip with 22 300 kg of fuel. They knew that amount in kilograms because Canada had recently switched to the metric system; previously fuel had been measured in pounds. The ground crew could measure the onboard fuel only in liters, which they reported as 7682 L. Thus, to determine how much fuel was on board and how much additional fuel was needed, the flight crew asked the ground crew for the conversion factor from liters to kilograms of fuel. The response was 1.77, which the flight crew used (1.77 kg corresponds to 1 L). (a) How many kilograms of fuel did the flight crew think they had? (In this problem, take all given data as being exact.) (b) How many liters did they ask to be added?

Unfortunately, the response from the ground crew was based on pre-metric habits—1.77 was the conversion factor not from liters to kilograms but rather from liters to *pounds* of fuel (1.77 lb corresponds to 1 L). (c) How many kilograms of fuel were actually on board? (Except for the given 1.77, use four significant figures for other conversion factors.) (d) How many liters of additional fuel were actually needed? (e) When the airplane left Montreal, what percentage of the required fuel did it have?

En route to Edmonton, at an altitude of 7.9 km, the airplane ran out of fuel and began to fall. Although the airplane had no power, the pilot managed to put it into a downward glide. Because the nearest working airport was too far to reach by gliding only, the pilot angled the glide toward an old, nonworking airport.

Unfortunately, the runway at that airport had been converted to a track for race cars, and a steel barrier had been constructed across it. Fortunately, as the airplane hit the runway, the front landing gear collapsed, dropping the nose of the airplane onto the runway. The skidding slowed the airplane so that it stopped just short of the steel barrier, with stunned race drivers and fans looking on. All on board the airplane emerged safely. The point here is this: Take care of the units.

70 The only two forces acting on a body have magnitudes of 20 N and 35 N and directions that differ by 80°. The resulting acceleration has a magnitude of 20 m/s². What is the mass of the body?

71 Figure 5-61 is an overhead view of a 12 kg tire that is to be pulled by three horizontal ropes. One rope's force (F_1 = 50 N) is indicated. The forces from the other ropes are to be oriented such that the tire's acceleration magnitude a is least. What is that least a if (a) F_2 = 30 N, F_3 = 20 N; (b) F_2 = 30 N, F_3 = 10 N; and (c) $F_2 = F_3$ = 30 N?

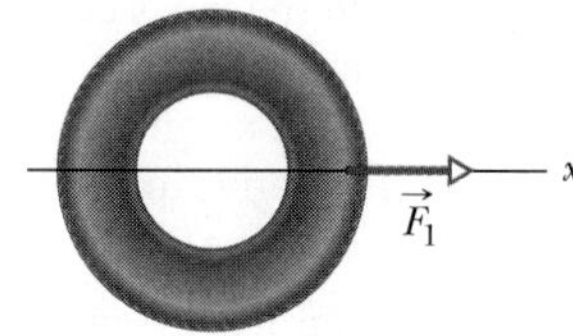

FIG. 5-61 Problem 71.

72 A block of mass M is pulled along a horizontal frictionless surface by a rope of mass m, as shown in Fig. 5-62. A horizontal force $\vec{F}$ acts on one end of the rope. (a) Show that the rope *must* sag, even if only by an imperceptible amount. Then, assuming that the sag is negligible, find (b) the acceleration of rope and block, (c) the force on the block from the rope, and (d) the tension in the rope at its midpoint.

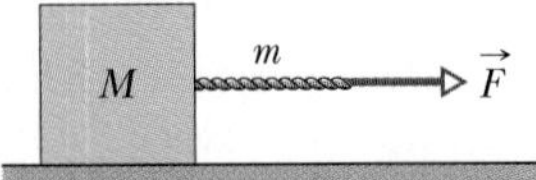

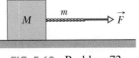

FIG. 5-62 Problem 72.

73 A worker drags a crate across a factory floor by pulling on a rope tied to the crate. The worker exerts a force of magnitude $F = 450$ N on the rope, which is inclined at an upward angle $\theta = 38°$ to the horizontal, and the floor exerts a horizontal force of magnitude $f = 125$ N that opposes the motion. Calculate the magnitude of the acceleration of the crate if (a) its mass is 310 kg and (b) its weight is 310 N. SSM

74 Three forces act on a particle that moves with unchanging velocity $\vec{v} = (2 \text{ m/s})\hat{i} - (7 \text{ m/s})\hat{j}$. Two of the forces are $\vec{F}_1 = (2 \text{ N})\hat{i} + (3 \text{ N})\hat{j} + (-2 \text{ N})\hat{k}$ and $\vec{F}_2 = (-5 \text{ N})\hat{i} + (8 \text{ N})\hat{j} + (-2 \text{ N})\hat{k}$. What is the third force?

75 A 52 kg circus performer is to slide down a rope that will break if the tension exceeds 425 N. (a) What happens if the performer hangs stationary on the rope? (b) At what magnitude of acceleration does the performer just avoid breaking the rope?

76 An 80 kg man drops to a concrete patio from a window 0.50 m above the patio. He neglects to bend his knees on landing, taking 2.0 cm to stop. (a) What is his average acceleration from when his feet first touch the patio to when he stops? (b) What is the magnitude of the average stopping force exerted on him by the patio?

77 In Fig. 5-63, 4.0 kg block A and 6.0 kg block B are connected by a string of negligible mass. Force $\vec{F}_A = (12 \text{ N})\hat{i}$ acts on block A; force $\vec{F}_B = (24 \text{ N})\hat{i}$ acts on block B. What is the tension in the string?

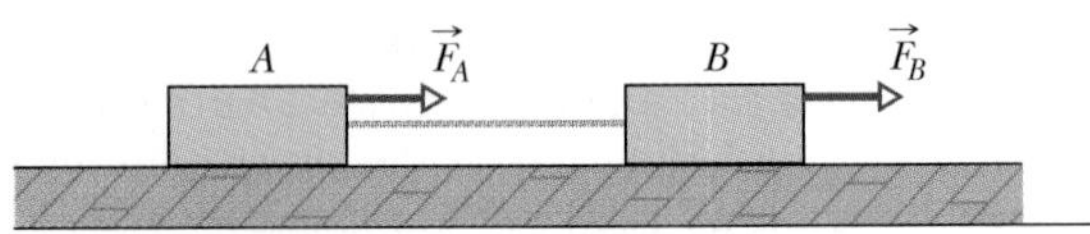

FIG. 5-63 Problem 77.

78 In the overhead view of Fig. 5-64, five forces pull on a box of mass $m = 4.0$ kg. The force magnitudes are $F_1 = 11$ N, $F_2 = 17$ N, $F_3 = 3.0$ N, $F_4 = 14$ N, and $F_5 = 5.0$ N, and angle θ_4 is 30°. Find the box's acceleration (a) in unit-vector notation and as (b) a magnitude and (c) an angle relative to the positive direction of the x axis.

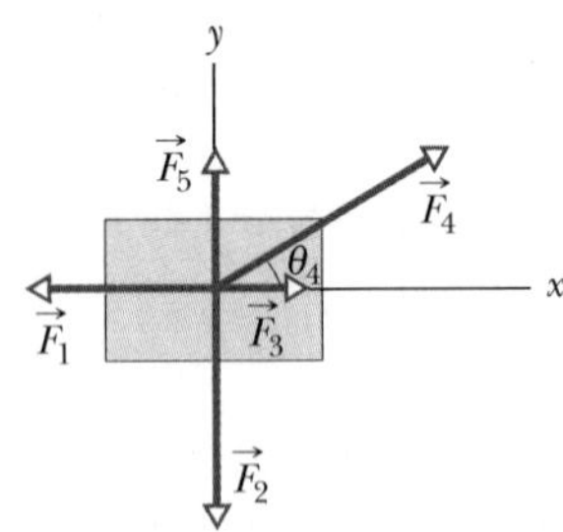

FIG. 5-64 Problem 78.

79 A certain force gives an object of mass m_1 an acceleration of 12.0 m/s² and an object of mass m_2 an acceleration of 3.30 m/s². What acceleration would the force give to an object of mass (a) $m_2 - m_1$ and (b) $m_2 + m_1$? SSM

80 Imagine a landing craft approaching the surface of Callisto, one of Jupiter's moons. If the engine provides an upward force (thrust) of 3260 N, the craft descends at constant speed; if the engine provides only 2200 N, the craft accelerates downward at 0.39 m/s². (a) What is the weight of the landing craft in the vicinity of Callisto's surface? (b) What is the mass of the craft? (c) What is the magnitude of the free-fall acceleration near the surface of Callisto?

81 An object is hung from a spring balance attached to the ceiling of an elevator cab. The balance reads 65 N when the cab is standing still. What is the reading when the cab is moving upward (a) with a constant speed of 7.6 m/s and (b) with a speed of 7.6 m/s while decelerating at a rate of 2.4 m/s²?

82 In Fig. 5-65, a force $\vec{F}$ of magnitude 12 N is applied to a FedEx box of mass $m_2 = 1.0$ kg. The force is directed up a plane tilted by $\theta = 37°$. The box is connected by a cord to a UPS box of mass $m_1 = 3.0$ kg on the floor. The floor, plane, and pulley are frictionless, and the masses of the pulley and cord are negligible. What is the tension in the cord?

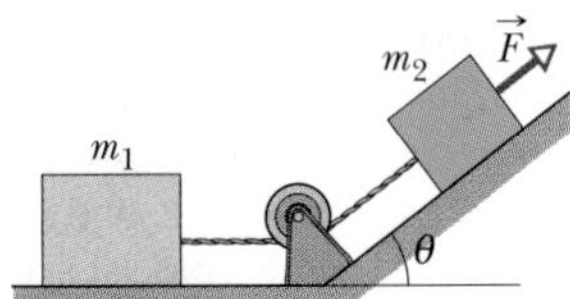

FIG. 5-65 Problem 82.

83 A certain particle has a weight of 22 N at a point where $g = 9.8$ m/s². What are its (a) weight and (b) mass at a point where $g = 4.9$ m/s²? What are its (c) weight and (d) mass if it is moved to a point in space where $g = 0$?

84 Compute the weight of a 75 kg space ranger (a) on Earth, (b) on Mars, where $g = 3.7$ m/s², and (c) in interplanetary space, where $g = 0$. (d) What is the ranger's mass at each location?

85 A 1400 kg jet engine is fastened to the fuselage of a passenger jet by just three bolts (this is the usual practice). Assume that each bolt supports one-third of the load. (a) Calculate the force on each bolt as the plane waits in line for clearance to take off. (b) During flight, the plane encounters turbulence, which suddenly imparts an upward vertical acceleration of 2.6 m/s² to the plane. Calculate the force on each bolt now.

86 An 80 kg person is parachuting and experiencing a downward acceleration of 2.5 m/s². The mass of the parachute is 5.0 kg. (a) What is the upward force on the open parachute from the air? (b) What is the downward force on the parachute from the person?

87 Suppose that in Fig. 5-13, the masses of the blocks are 2.0 kg and 4.0 kg. (a) Which mass should the hanging block have if the magnitude of the acceleration is to be as large as possible? What then are (b) the magnitude of the acceleration and (c) the tension in the cord?

88 You pull a short refrigerator with a constant force $\vec{F}$ across a greased (frictionless) floor, either with $\vec{F}$ horizontal (case 1) or with $\vec{F}$ tilted upward at an angle θ (case 2). (a) What is the ratio of the refrigerator's speed in case 2 to its speed in case 1 if you pull for a certain time t? (b) What is this ratio if you pull for a certain distance d?

89 A spaceship lifts off vertically from the Moon, where $g = 1.6$ m/s². If the ship has an upward acceleration of 1.0 m/s² as it lifts off, what is the magnitude of the force exerted by the ship on its pilot, who weighs 735 N on Earth?

90 Compute the initial upward acceleration of a rocket of mass 1.3×10^4 kg if the initial upward force produced by its engine (the thrust) is 2.6×10^5 N. Do not neglect the gravitational force on the rocket.

91 Figure 5-66*a* shows a mobile hanging from a ceiling; it consists of two metal pieces ($m_1 = 3.5$ kg and $m_2 = 4.5$ kg) that are strung together by cords of negligible mass. What is the tension in (a) the bottom cord and (b) the top cord? Figure 5-66*b* shows a mobile consisting of three metal pieces. Two of the masses are $m_3 = 4.8$ kg and $m_5 = 5.5$ kg. The tension in the top cord is 199 N. What is the tension in (c) the lowest cord and (d) the middle cord? SSM

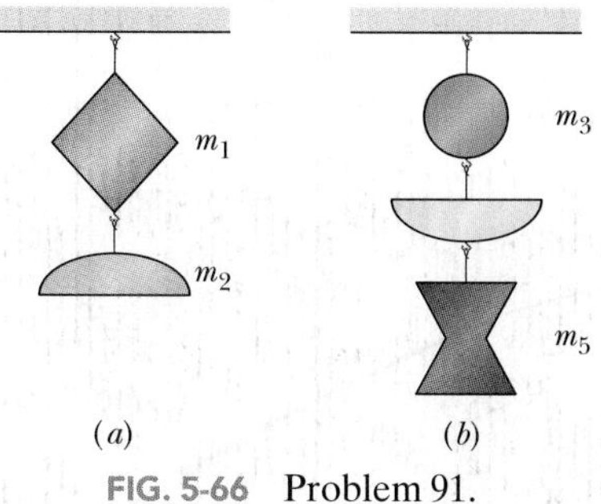

FIG. 5-66 Problem 91.

92 If the 1 kg standard body is accelerated by only $\vec{F}_1 = (3.0\text{ N})\hat{i} + (4.0\text{ N})\hat{j}$ and $\vec{F}_2 = (-2.0\text{ N})\hat{i} + (-6.0\text{ N})\hat{j}$, then what is $\vec{F}_{net}$ (a) in unit-vector notation and as (b) a magnitude and (c) an angle relative to the positive x direction? What are the (d) magnitude and (e) angle of $\vec{a}$?

93 A nucleus that captures a stray neutron must bring the neutron to a stop within the diameter of the nucleus by means of the *strong force.* That force, which "glues" the nucleus together, is approximately zero outside the nucleus. Suppose that a stray neutron with an initial speed of 1.4×10^7 m/s is just barely captured by a nucleus with diameter $d = 1.0 \times 10^{-14}$ m. Assuming the strong force on the neutron is constant, find the magnitude of that force. The neutron's mass is 1.67×10^{-27} kg.

94 A 15 000 kg helicopter lifts a 4500 kg truck with an upward acceleration of 1.4 m/s². Calculate (a) the net upward force on the helicopter blades from the air and (b) the tension in the cable between helicopter and truck.

95 A motorcycle and 60.0 kg rider accelerate at 3.0 m/s² up a ramp inclined 10° above the horizontal. What are the magnitudes of (a) the net force on the rider and (b) the force on the rider from the motorcycle? SSM

96 An interstellar ship has a mass of 1.20×10^6 kg and is initially at rest relative to a star system. (a) What constant acceleration is needed to bring the ship up to a speed of $0.10c$ (where c is the speed of light, 3.0×10^8 m/s) relative to the star system in 3.0 days? (b) What is that acceleration in g units? (c) What force is required for the acceleration? (d) If the engines are shut down when $0.10c$ is reached (the speed then remains constant), how long does the ship take (start to finish) to journey 5.0 light-months, the distance that light travels in 5.0 months?

97 For sport, a 12 kg armadillo runs onto a large pond of level, frictionless ice. The armadillo's initial velocity is 5.0 m/s along the positive direction of an x axis. Take its initial position on the ice as being the origin. It slips over the ice while being pushed by a wind with a force of 17 N in the positive direction of the y axis. In unit-vector notation, what are the animal's (a) velocity and (b) position vector when it has slid for 3.0 s?

98 A 50 kg passenger rides in an elevator cab that starts from rest on the ground floor of a building at $t = 0$ and rises to the top floor during a 10 s interval. The cab's acceleration as a function of the time is shown in Fig. 5-67, where positive values of the acceleration mean that it is directed upward. What are the (a) magnitude and (b) direction (up or down) of the maximum force on the passenger from the floor, the (c) magnitude and (d) direction of the minimum force on the passenger from the floor, and the (e) magnitude and (f) direction of the maximum force on the floor from the passenger?

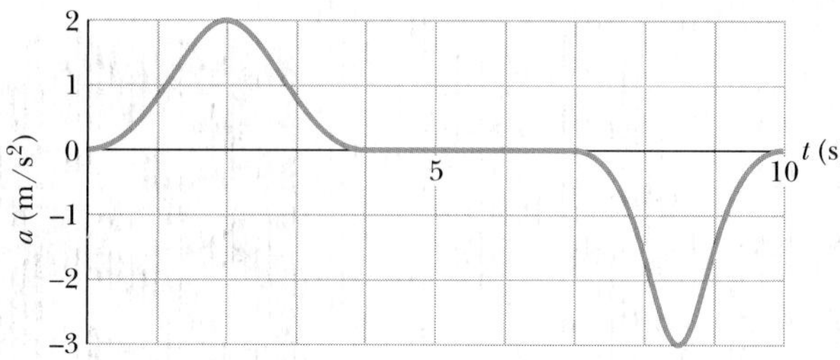

FIG. 5-67 Problem 98.

99 Figure 5-68 shows a box of dirty money (mass $m_1 = 3.0$ kg) on a frictionless plane inclined at angle $\theta_1 = 30°$. The box is connected via a cord of negligible mass to a box of laundered money (mass $m_2 = 2.0$ kg) on a frictionless plane inclined at angle $\theta_2 = 60°$. The pulley is frictionless and has negligible mass. What is the tension in the cord? SSM

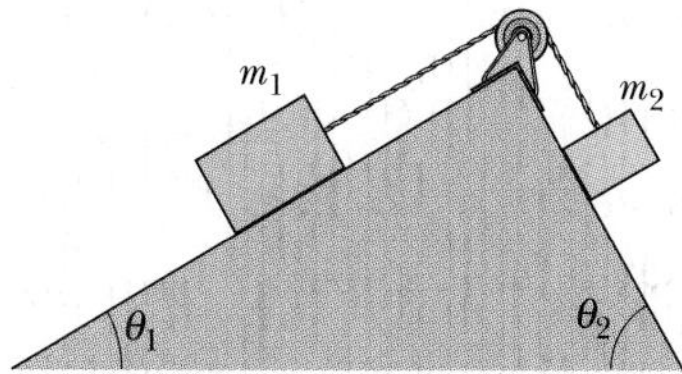

FIG. 5-68 Problem 99.

100 Suppose the 1 kg standard body accelerates at 4.00 m/s² at 160° from the positive direction of an x axis due to two forces; one is $\vec{F}_1 = (2.50\text{ N})\hat{i} + (4.60\text{ N})\hat{j}$. What is the other force (a) in unit-vector notation and as (b) a magnitude and (c) an angle?

101 In Fig. 5-69, a tin of antioxidants ($m_1 = 1.0$ kg) on a frictionless inclined surface is connected to a tin of corned beef ($m_2 = 2.0$ kg). The pulley is massless and frictionless. An upward force of magnitude $F = 6.0$ N acts on the corned beef tin, which has a downward acceleration of 5.5 m/s². What are (a) the tension in the connecting cord and (b) angle β? SSM

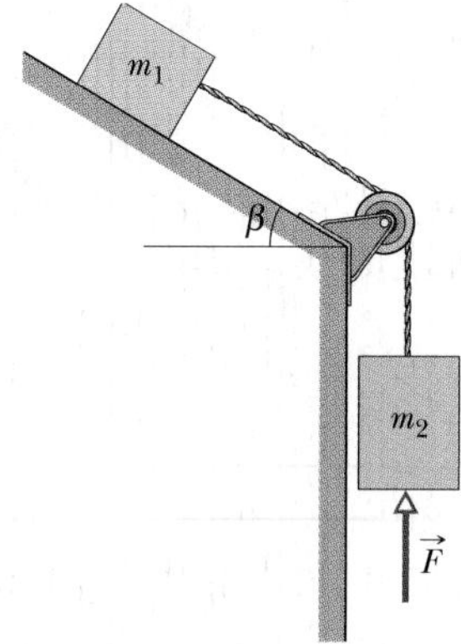

FIG. 5-69 Problem 101.

102 A rocket and its payload have a total mass of 5.0×10^4 kg. How large is the force produced by the engine (the thrust) when the rocket is (a) "hovering" over the launchpad just after ignition and (b) accelerating upward at 20 m/s²?

103 A motorcycle of weight 2.0 kN accelerates from 0 to 88.5 km/h in 6.0 s. What are the magnitudes of (a) the constant acceleration and (b) the net force causing the acceleration?

104 An initially stationary electron (mass $= 9.11 \times 10^{-31}$ kg) undergoes a constant acceleration through 1.5 cm, reaching 6.0×10^6 m/s. What are (a) the magnitude of the force accelerating the electron and (b) the electron's weight?

6 Force and Motion—II

The Great Pyramid, built about 4500 years ago, consists of about 2 300 000 stone blocks, most with a mass of 2000 to 3000 kg. How did the engineers and workers manage to lift the stones into place to construct this pyramid, which is over 140 m high? Some researchers argue that during the construction a large team of men would pull a block up a giant earthen ramp that ran at a modest angle up one side of the pyramid. However, no evidence (such as rubble or painted pictures) exists to support this theory. Other researchers argue that a spiral ramp ran around the pyramid. However, such a ramp would have been highly unstable and, besides, maneuvering a 2000 kg stone around the 90° corners along the ramp would have been daunting, if not impossible.

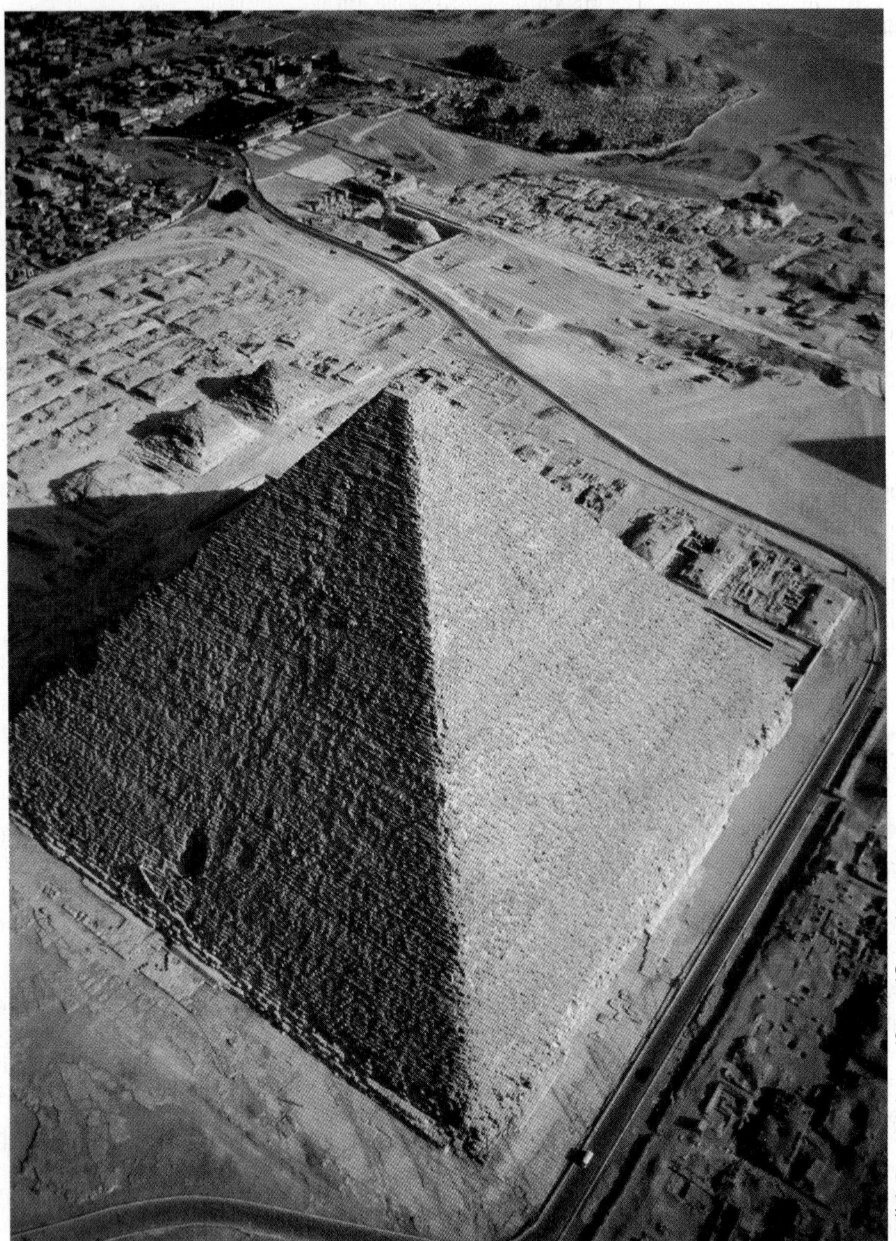

Will & Deni McIntyre/Stone/Getty Images

How did the ancient people move the blocks up and into position?

The answer is in this chapter.

6-1 WHAT IS PHYSICS?

In this chapter we focus on the physics of three common types of force: frictional force, drag force, and centripetal force. An engineer preparing a car for the Indianapolis 500 must consider all three types. Frictional forces acting on the tires are crucial to the car's acceleration out of the pit and out of a curve (if the car hits an oil slick, the friction is lost and so is the car). Drag forces acting on the car from the passing air must be minimized or else the car will consume too much fuel and have to pit too early (even one 14 s pit stop can cost a driver the race). Centripetal forces are crucial in the turns (if there is insufficient centripetal force, the car slides into the wall). We start our discussion with frictional forces.

6-2 | Friction

Frictional forces are unavoidable in our daily lives. If we were not able to counteract them, they would stop every moving object and bring to a halt every rotating shaft. About 20% of the gasoline used in an automobile is needed to counteract friction in the engine and in the drive train. On the other hand, if friction were totally absent, we could not get an automobile to go anywhere, and we could not walk or ride a bicycle. We could not hold a pencil, and, if we could, it would not write. Nails and screws would be useless, woven cloth would fall apart, and knots would untie.

Here we deal with the frictional forces that exist between dry solid surfaces, either stationary relative to each other or moving across each other at slow speeds. Consider three simple thought experiments:

1. Send a book sliding across a long horizontal counter. As expected, the book slows and then stops. This means the book must have an acceleration parallel to the counter surface, in the direction opposite the book's velocity. From Newton's second law, then, a force must act on the book parallel to the counter surface, in the direction opposite its velocity. That force is a frictional force.
2. Push horizontally on the book to make it travel at constant velocity along the counter. Can the force from you be the only horizontal force on the book? No, because then the book would accelerate. From Newton's second law, there must be a second force, directed opposite your force but with the same magnitude, so that the two forces balance. That second force is a frictional force, directed parallel to the counter.
3. Push horizontally on a heavy crate. The crate does not move. From Newton's second law, a second force must also be acting on the crate to counteract your force. Moreover, this second force must be directed opposite your force and have the same magnitude as your force, so that the two forces balance. That second force is a frictional force. Push even harder. The crate still does not move. Apparently the frictional force can change in magnitude so that the two forces still balance. Now push with all your strength. The crate begins to slide. Evidently, there is a maximum magnitude of the frictional force. When you exceed that maximum magnitude, the crate slides.

Figure 6-1 shows a similar situation. In Fig. 6-1a, a block rests on a tabletop, with the gravitational force $\vec{F}_g$ balanced by a normal force $\vec{F}_N$. In Fig. 6-1b, you exert a force $\vec{F}$ on the block, attempting to pull it to the left. In response, a frictional force $\vec{f}_s$ is directed to the right, exactly balancing your force. The force $\vec{f}_s$ is called the **static frictional force.** The block does not move.

Figures 6-1c and 6-1d show that as you increase the magnitude of your applied force, the magnitude of the static frictional force $\vec{f}_s$ also increases and the block remains at rest. When the applied force reaches a certain magnitude, however, the block "breaks away" from its intimate contact with the tabletop and

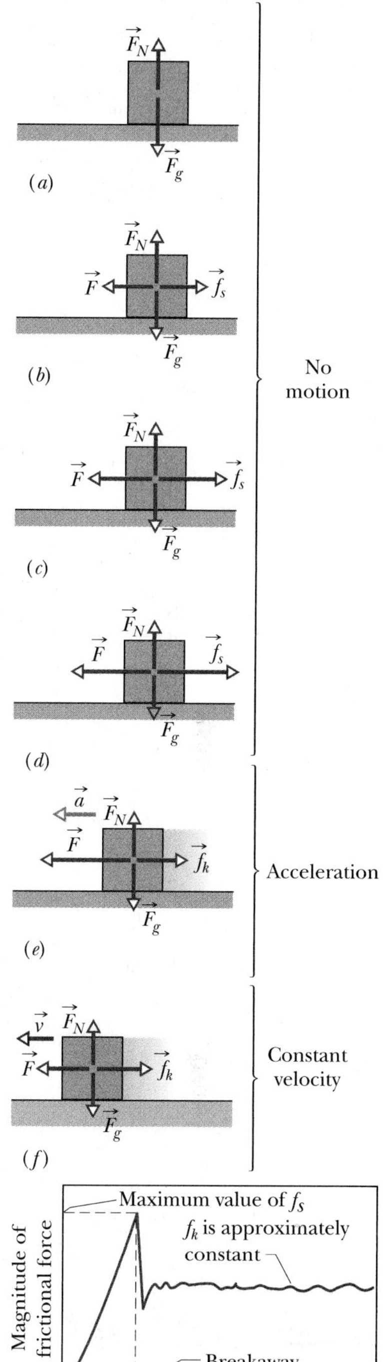

FIG. 6-1 (a) The forces on a stationary block. (b–d) An external force $\vec{F}$, applied to the block, is balanced by a static frictional force $\vec{f}_s$. As F is increased, f_s also increases, until f_s reaches a certain maximum value. (e) The block then "breaks away," accelerating suddenly in the direction of $\vec{F}$. (f) If the block is now to move with constant velocity, F must be reduced from the maximum value it had just before the block broke away. (g) Some experimental results for the sequence (a) through (f).

accelerates leftward (Fig. 6-1e). The frictional force that then opposes the motion is called the **kinetic frictional force** $\vec{f}_k$.

Usually, the magnitude of the kinetic frictional force, which acts when there is motion, is less than the maximum magnitude of the static frictional force, which acts when there is no motion. Thus, if you wish the block to move across the surface with a constant speed, you must usually decrease the magnitude of the applied force once the block begins to move, as in Fig. 6-1f. As an example, Fig. 6-1g shows the results of an experiment in which the force on a block was slowly increased until breakaway occurred. Note the reduced force needed to keep the block moving at constant speed after breakaway.

A frictional force is, in essence, the vector sum of many forces acting between the surface atoms of one body and those of another body. If two highly polished and carefully cleaned metal surfaces are brought together in a very good vacuum (to keep them clean), they cannot be made to slide over each other. Because the surfaces are so smooth, many atoms of one surface contact many atoms of the other surface, and the surfaces *cold-weld* together instantly, forming a single piece of metal. If a machinist's specially polished gage blocks are brought together in air, there is less atom-to-atom contact, but the blocks stick firmly to each other and can be separated only by means of a wrenching motion. Usually, however, this much atom-to-atom contact is not possible. Even a highly polished metal surface is far from being flat on the atomic scale. Moreover, the surfaces of everyday objects have layers of oxides and other contaminants that reduce cold-welding.

When two ordinary surfaces are placed together, only the high points touch each other. (It is like having the Alps of Switzerland turned over and placed down on the Alps of Austria.) The actual *micro*scopic area of contact is much less than the apparent *macro*scopic contact area, perhaps by a factor of 10^4. Nonetheless, many contact points do cold-weld together. These welds produce static friction when an applied force attempts to slide the surfaces relative to each other.

If the applied force is great enough to pull one surface across the other, there is first a tearing of welds (at breakaway) and then a continuous re-forming and tearing of welds as movement occurs and chance contacts are made (Fig. 6-2). The kinetic frictional force $\vec{f}_k$ that opposes the motion is the vector sum of the forces at those many chance contacts.

If the two surfaces are pressed together harder, many more points cold-weld. Now getting the surfaces to slide relative to each other requires a greater applied force: The static frictional force $\vec{f}_s$ has a greater maximum value. Once the surfaces are sliding, there are many more points of momentary cold-welding, so the kinetic frictional force $\vec{f}_k$ also has a greater magnitude.

Often, the sliding motion of one surface over another is "jerky" because the two surfaces alternately stick together and then slip. Such repetitive *stick-and-slip* can produce squeaking or squealing, as when tires skid on dry pavement, fingernails scratch along a chalkboard, or a rusty hinge is opened. It can also produce beautiful sounds, as when a bow is drawn properly across a violin string.

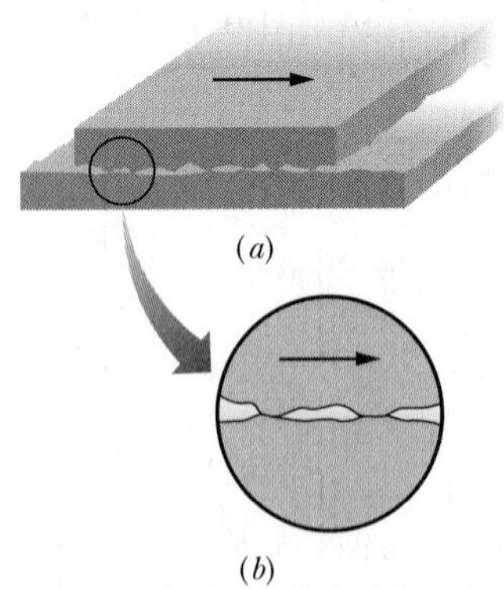

FIG. 6-2 The mechanism of sliding friction. (a) The upper surface is sliding to the right over the lower surface in this enlarged view. (b) A detail, showing two spots where cold-welding has occurred. Force is required to break the welds and maintain the motion.

6-3 | Properties of Friction

Experiment shows that when a dry and unlubricated body presses against a surface in the same condition and a force $\vec{F}$ attempts to slide the body along the surface, the resulting frictional force has three properties:

Property 1. If the body does not move, then the static frictional force $\vec{f}_s$ and the component of $\vec{F}$ that is parallel to the surface balance each other. They are equal in magnitude, and $\vec{f}_s$ is directed opposite that component of $\vec{F}$.

Property 2. The magnitude of $\vec{f}_s$ has a maximum value $f_{s,\max}$ that is given by

$$f_{s,\max} = \mu_s F_N, \tag{6-1}$$

where μ_s is the **coefficient of static friction** and F_N is the magnitude of the normal force on the body from the surface. If the magnitude of the component of $\vec{F}$ that is parallel to the surface exceeds $f_{s,\max}$, then the body begins to slide along the surface.

Property 3. If the body begins to slide along the surface, the magnitude of the frictional force rapidly decreases to a value f_k given by

$$f_k = \mu_k F_N, \tag{6-2}$$

where μ_k is the **coefficient of kinetic friction.** Thereafter, during the sliding, a kinetic frictional force $\vec{f}_k$ with magnitude given by Eq. 6-2 opposes the motion.

The magnitude F_N of the normal force appears in properties 2 and 3 as a measure of how firmly the body presses against the surface. If the body presses harder, then, by Newton's third law, F_N is greater. Properties 1 and 2 are worded in terms of a single applied force $\vec{F}$, but they also hold for the net force of several applied forces acting on the body. Equations 6-1 and 6-2 are *not* vector equations; the direction of $\vec{f}_s$ or $\vec{f}_k$ is always parallel to the surface and opposed to the attempted sliding, and the normal force $\vec{F}_N$ is perpendicular to the surface.

The coefficients μ_s and μ_k are dimensionless and must be determined experimentally. Their values depend on certain properties of both the body and the surface; hence, they are usually referred to with the preposition "between," as in "the value of μ_s *between* an egg and a Teflon-coated skillet is 0.04, but that *between* rock-climbing shoes and rock is as much as 1.2." We assume that the value of μ_k does not depend on the speed at which the body slides along the surface.

CHECKPOINT 1 A block lies on a floor. (a) What is the magnitude of the frictional force on it from the floor? (b) If a horizontal force of 5 N is now applied to the block, but the block does not move, what is the magnitude of the frictional force on it? (c) If the maximum value $f_{s,\max}$ of the static frictional force on the block is 10 N, will the block move if the magnitude of the horizontally applied force is 8 N? (d) If it is 12 N? (e) What is the magnitude of the frictional force in part (c)?

Sample Problem 6-1

If a car's wheels are "locked" (kept from rolling) during emergency braking, the car slides along the road. Ripped-off bits of tire and small melted sections of road form the "skid marks" that reveal that cold-welding occurred during the slide. The record for the longest skid marks on a public road was reportedly set in 1960 by a Jaguar on the M1 highway in England (Fig. 6-3*a*)—the marks were 290 m long! Assuming that $\mu_k = 0.60$ and the car's acceleration was constant during the braking, how fast was the car going when the wheels became locked?

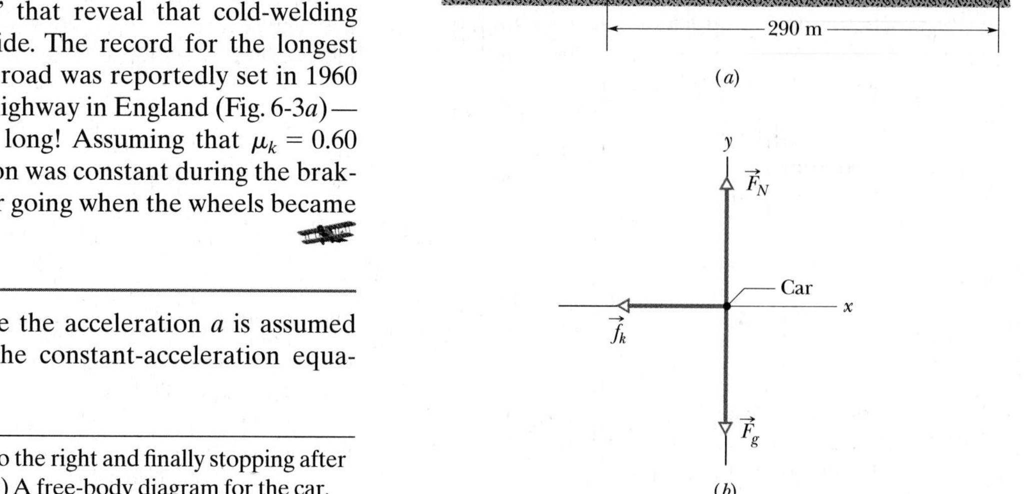

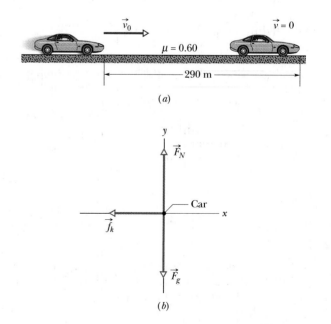

FIG. 6-3 (*a*) A car sliding to the right and finally stopping after a displacement of 290 m. (*b*) A free-body diagram for the car.

KEY IDEAS (1) Because the acceleration a is assumed constant, we can use the constant-acceleration equations of Table 2-1 to find the car's initial speed v_0. (2) If we neglect the effects of the air on the car, acceleration a was due only to a kinetic frictional force $\vec{f}_k$ on the car from the road, directed opposite the direction of the car's motion, assumed to be in the positive direction of an x axis (Fig. 6-3*b*). We can relate this force to the acceleration by writing Newton's second law for x components ($F_{\text{net},x} = ma_x$) as

$$-f_k = ma, \tag{6-3}$$

where m is the car's mass. The minus sign indicates the direction of the kinetic frictional force.

Calculations: From Eq. 6-2, the frictional force has the magnitude $f_k = \mu_k F_N$, where F_N is the magnitude of the normal force on the car from the road. Because the car is not accelerating vertically, we know from Fig. 6-3*b* and Newton's second law that the magnitude of $\vec{F}_N$ is equal to the magnitude of the gravitational force $\vec{F}_g$ on the car, which is mg. Thus, $F_N = mg$.

Now solving Eq. 6-3 for a and substituting $f_k = \mu_k F_N = \mu_k mg$ for f_k yield

$$a = -\frac{f_k}{m} = -\frac{\mu_k mg}{m} = -\mu_k g, \tag{6-4}$$

where the minus sign indicates that the acceleration is in the negative direction of the x axis, opposite the direction of the velocity. Next, let's use Eq. 2-16,

$$v^2 = v_0^2 + 2a(x - x_0),$$

from the constant-acceleration equations of Chapter 2. We know that the displacement $x - x_0$ was 290 m and assume that the final speed v was 0. Substituting for a from Eq. 6-4 and solving for v_0 give

$$v_0 = \sqrt{2\mu_k g(x - x_0)} = \sqrt{(2)(0.60)(9.8 \text{ m/s}^2)(290 \text{ m})}$$
$$= 58 \text{ m/s} = 210 \text{ km/h}. \quad \text{(Answer)}$$

We assumed that $v = 0$ at the far end of the skid marks. Actually, the marks ended only because the Jaguar left the road after 290 m. So v_0 was at *least* 210 km/h.

Sample Problem 6-2

In Fig. 6-4*a*, a block of mass $m = 3.0$ kg slides along a floor while a force $\vec{F}$ of magnitude 12.0 N is applied to it at an upward angle θ. The coefficient of kinetic friction between the block and the floor is $\mu_k = 0.40$. We can vary θ from 0 to 90° (the block remains on the floor). What θ gives the maximum value of the block's acceleration magnitude a?

KEY IDEAS Because the block is moving, a *kinetic* frictional force acts on it. The magnitude is given by Eq. 6-2 ($f_k = \mu_k F_N$, where F_N is the normal force). The direction is opposite the motion (the friction opposes the sliding).

Calculating F_N: Because we need the magnitude f_k of the frictional force, we first must calculate the magnitude F_N of the normal force. Figure 6-4*b* is a free-body diagram showing the forces along the vertical y axis. The normal force is upward, the gravitational force $\vec{F}_g$ with magnitude mg is downward, and (note) the vertical component F_y of the applied force is upward. That component is shown in Fig. 6-4*c*, where we can see that $F_y = F \sin\theta$. We can write Newton's second law ($\vec{F}_{\text{net}} = m\vec{a}$) for those forces along the y axis as

$$F_N + F\sin\theta - mg = m(0), \tag{6-5}$$

where we substituted zero for the acceleration along the y axis (the block does not even move along that axis). Thus,

$$F_N = mg - F\sin\theta. \tag{6-6}$$

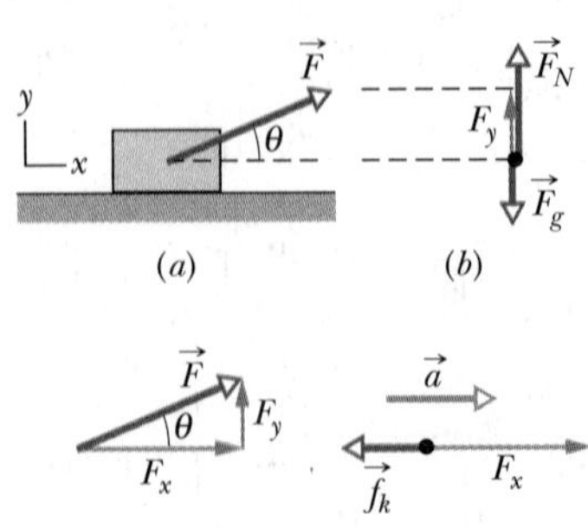

FIG. 6-4 (*a*) A force is applied to a moving block. (*b*) The vertical forces. (*c*) The components of the applied force. (*d*) The horizontal forces and acceleration.

Calculating acceleration a: Figure 6-4*d* is a free-body diagram for motion along the x axis. The horizontal component F_x of the applied force is rightward; from Fig. 6-4*c*, we see that $F_x = F\cos\theta$. The frictional force has magnitude f_k ($= \mu_k F_N$) and is leftward. Writing Newton's second law for motion along the x axis gives us

$$F\cos\theta - \mu_k F_N = ma. \tag{6-7}$$

Substituting for F_N from Eq. 6-6 and solving for a lead to

$$a = \frac{F}{m}\cos\theta - \mu_k\left(g - \frac{F}{m}\sin\theta\right). \tag{6-8}$$

Finding a maximum: To find the value of θ that maximizes a, we take the derivative of a with respect to θ and set the result equal to zero:

$$\frac{da}{d\theta} = -\frac{F}{m}\sin\theta + \mu_k\frac{F}{m}\cos\theta = 0.$$

Rearranging and using the identity $(\sin\theta)/(\cos\theta) = \tan\theta$ give us

$$\tan\theta = \mu_k.$$

Solving for θ and substituting the given $\mu_k = 0.40$, we find that the acceleration will be maximum if

$$\theta = \tan^{-1}\mu_k = 21.8^\circ \approx 22^\circ. \quad \text{(Answer)}$$

Comment: As we increase θ from 0, more of the applied force $\vec{F}$ is upward, relieving the normal force. The decrease in the normal force causes a decrease in the frictional force, which opposes the block's motion. Thus, the block's acceleration tends to increase. However, the increase in θ also decreases the horizontal component of $\vec{F}$, and so the block's acceleration tends to decrease. These opposing tendencies produce a maximum acceleration at $\theta = 22^\circ$.

Sample Problem 6-3

Although many ingenious schemes have been attributed to the building of the Great Pyramid, the stone blocks were probably hauled up the side of the pyramid by men pulling on ropes. Figure 6-5*a* represents a 2000 kg stone block in the process of being pulled up the finished (smooth) side of the Great Pyramid, which forms a plane inclined at angle $\theta = 52^\circ$. The block is secured to a wood sled and is pulled by multiple ropes (only one is shown). The sled's track is lubricated with water to decrease the coefficient of static friction to 0.40. Assume negligible friction at the (lubricated) point where the ropes pass over the edge at the top of the side. If each man on top of the pyramid pulls with a (reasonable) force of 686 N, how many men are needed to put the block on the verge of moving?

KEY IDEAS (1) Because the block is on the *verge* of moving, the static frictional force must be at its maximum possible value; that is, $f_s = f_{s,\max}$. (2) Because the block is on the verge of moving *up* the plane, the frictional force must be *down* the plane (to oppose the pending motion). (3) From Sample Problem 5-5, we know that the component of the gravitational force down the plane is $mg\sin\theta$ and the component perpendicular to (and inward from) the plane is $mg\cos\theta$ (Fig. 6-5*b*).

Calculations: Figure 6-5*c* is a free-body diagram for the block, showing the force $\vec{F}$ applied by the ropes, the static frictional force $\vec{f}_s$, and the two components of the gravitational force. We can write Newton's second law ($\vec{F}_{\text{net}} = m\vec{a}$) for forces along the x axis as

$$F - mg\sin\theta - f_s = m(0). \quad (6\text{-}9)$$

Because the block is on the verge of sliding and the frictional force is at the maximum possible value $f_{s,\max}$, we use Eq. 6-1 to replace f_s with $\mu_s F_N$:

$$f_s = f_{s,\max} = \mu_s F_N. \quad (6\text{-}10)$$

From Figure 6-5*c*, we see that along the y axis Newton's second law becomes

$$F_N - mg\cos\theta = m(0). \quad (6\text{-}11)$$

Solving Eq. 6-11 for F_N and substituting the result into Eq. 6-10, we have

$$f_s = \mu_s mg\cos\theta. \quad (6\text{-}12)$$

Substituting this expression into Eq. 6-9 and solving for F lead to

$$F = \mu_s mg\cos\theta + mg\sin\theta. \quad (6\text{-}13)$$

Substituting $m = 2000$ kg, $\theta = 52^\circ$, and $\mu_s = 0.40$, we find that the force required to put the stone block on the verge of moving is 2.027×10^4 N. Dividing this by the assumed pulling force of 686 N from each man, we find that the required number of men is

$$N = \frac{2.027 \times 10^4\ \text{N}}{686\ \text{N}} = 29.5 \approx 30\ \text{men}. \quad \text{(Answer)}$$

Comment: Once the stone block began to move, the friction was kinetic friction and the coefficient was about 0.20. You can show that the required number of men was then 26 or 27. Thus, the huge stone blocks of the Great Pyramid could be pulled up into position by reasonably small teams of men.

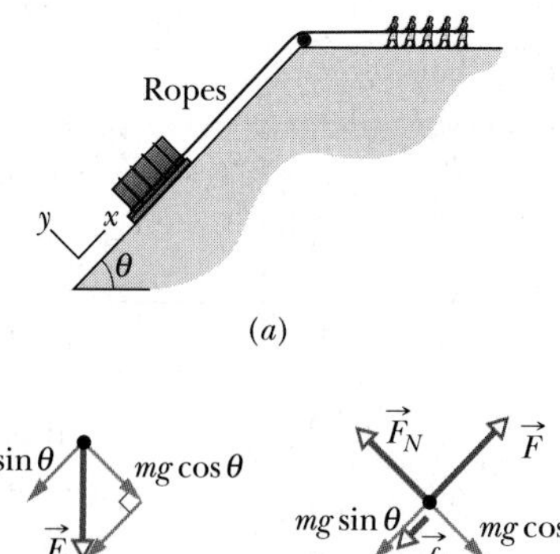

FIG. 6-5 (*a*) A stone block on the verge of being pulled up the side of the Great Pyramid. (*b*) The components of the gravitational force. (*c*) A free-body diagram for the block.

FIG. 6-6 This skier crouches in an "egg position" so as to minimize her effective cross-sectional area and thus minimize the air drag acting on her. *(Karl-Josef Hildenbrand/dpa/Landov LLC)*

6-4 | The Drag Force and Terminal Speed

A **fluid** is anything that can flow—generally either a gas or a liquid. When there is a relative velocity between a fluid and a body (either because the body moves through the fluid or because the fluid moves past the body), the body experiences a **drag force** $\vec{D}$ that opposes the relative motion and points in the direction in which the fluid flows relative to the body.

Here we examine only cases in which air is the fluid, the body is blunt (like a baseball) rather than slender (like a javelin), and the relative motion is fast enough so that the air becomes turbulent (breaks up into swirls) behind the body. In such cases, the magnitude of the drag force $\vec{D}$ is related to the relative speed v by an experimentally determined **drag coefficient** C according to

$$D = \tfrac{1}{2}C\rho Av^2, \tag{6-14}$$

where ρ is the air density (mass per volume) and A is the **effective cross-sectional area** of the body (the area of a cross section taken perpendicular to the velocity $\vec{v}$). The drag coefficient C (typical values range from 0.4 to 1.0) is not truly a constant for a given body because if v varies significantly, the value of C can vary as well. Here, we ignore such complications.

Downhill speed skiers know well that drag depends on A and v^2. To reach high speeds a skier must reduce D as much as possible by, for example, riding the skis in the "egg position" (Fig. 6-6) to minimize A.

When a blunt body falls from rest through air, the drag force $\vec{D}$ is directed upward; its magnitude gradually increases from zero as the speed of the body increases. This upward force $\vec{D}$ opposes the downward gravitational force $\vec{F}_g$ on the body. We can relate these forces to the body's acceleration by writing Newton's second law for a vertical y axis ($F_{\text{net},y} = ma_y$) as

$$D - F_g = ma, \tag{6-15}$$

where m is the mass of the body. As suggested in Fig. 6-7, if the body falls long enough, D eventually equals F_g. From Eq. 6-15, this means that $a = 0$, and so the body's speed no longer increases. The body then falls at a constant speed, called the **terminal speed** v_t.

To find v_t, we set $a = 0$ in Eq. 6-15 and substitute for D from Eq. 6-14, obtaining

$$\tfrac{1}{2}C\rho Av_t^2 - F_g = 0,$$

which gives

$$v_t = \sqrt{\frac{2F_g}{C\rho A}}. \tag{6-16}$$

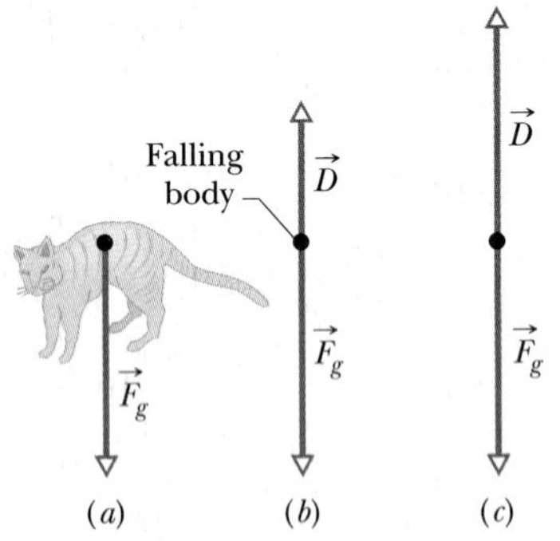

FIG. 6-7 The forces that act on a body falling through air: (*a*) the body when it has just begun to fall and (*b*) the free-body diagram a little later, after a drag force has developed. (*c*) The drag force has increased until it balances the gravitational force on the body. The body now falls at its constant terminal speed.

Table 6-1 gives values of v_t for some common objects.

According to calculations* based on Eq. 6-14, a cat must fall about six floors to reach terminal speed. Until it does so, $F_g > D$ and the cat accelerates downward because of the net downward force. Recall from Chapter 2 that your body is an accelerometer, not a speedometer. Because the cat also senses the acceleration, it is frightened and keeps its feet underneath its body, its head tucked in, and its spine bent upward, making A small, v_t large, and injury likely.

However, if the cat does reach v_t during a longer fall, the acceleration vanishes and the cat relaxes somewhat, stretching its legs and neck horizontally outward and straightening its spine (it then resembles a flying squirrel). These actions increase area A and thus also, by Eq. 6-14, the drag D. The cat begins to slow because now $D > F_g$ (the net force is upward), until a new, smaller v_t is reached. The decrease

*W. O. Whitney and C. J. Mehlhaff, "High-Rise Syndrome in Cats." *The Journal of the American Veterinary Medical Association*, 1987.

TABLE 6-1

Some Terminal Speeds in Air

Object	Terminal Speed (m/s)	95% Distance[a] (m)
Shot (from shot put)	145	2500
Sky diver (typical)	60	430
Baseball	42	210
Tennis ball	31	115
Basketball	20	47
Ping-Pong ball	9	10
Raindrop (radius = 1.5 mm)	7	6
Parachutist (typical)	5	3

[a]This is the distance through which the body must fall from rest to reach 95% of its terminal speed.
Source: Adapted from Peter J. Brancazio, *Sport Science*, 1984, Simon & Schuster, New York.

in v_t reduces the possibility of serious injury on landing. Just before the end of the fall, when it sees it is nearing the ground, the cat pulls its legs back beneath its body to prepare for the landing.

Humans often fall from great heights for the fun of skydiving. However, in April 1987, during a jump, sky diver Gregory Robertson noticed that fellow sky diver Debbie Williams had been knocked unconscious in a collision with a third sky diver and was unable to open her parachute. Robertson, who was well above Williams at the time and who had not yet opened his parachute for the 4 km plunge, reoriented his body head-down so as to minimize A and maximize his downward speed. Reaching an estimated v_t of 320 km/h, he caught up with Williams and then went into a horizontal "spread eagle" (as in Fig. 6-8) to increase D so that he could grab her. He opened her parachute and then, after releasing her, his own, a scant 10 s before impact. Williams received extensive internal injuries due to her lack of control on landing but survived.

FIG. 6-8 Sky divers in a horizontal "spread eagle" maximize air drag. *(Steve Fitchett/Taxi/Getty Images)*

Sample Problem 6-4

If a falling cat reaches a first terminal speed of 97 km/h while it is tucked in and then stretches out, doubling A, how fast is it falling when it reaches a new terminal speed?

KEY IDEA The terminal speeds of the cat depend on (among other things) the effective cross-sectional areas A of the cat, according to Eq. 6-16. Thus, we can use that equation to set up a ratio of speeds. We let v_{to} and v_{tn} represent the original and new terminal speeds, and A_o and A_n the original and new areas. Then by Eq. 6-16,

$$\frac{v_{tn}}{v_{to}} = \frac{\sqrt{2F_g/C\rho A_n}}{\sqrt{2F_g/C\rho A_o}} = \sqrt{\frac{A_o}{A_n}} = \sqrt{\frac{A_o}{2A_o}} = \sqrt{0.5} \approx 0.7,$$

which means that $v_{tn} \approx 0.7v_{to}$, or about 68 km/h.

Sample Problem 6-5

A raindrop with radius $R = 1.5$ mm falls from a cloud that is at height $h = 1200$ m above the ground. The drag coefficient C for the drop is 0.60. Assume that the drop is spherical throughout its fall. The density of water ρ_w is 1000 kg/m^3, and the density of air ρ_a is 1.2 kg/m^3.

(a) What is the terminal speed of the drop?

KEY IDEA The drop reaches a terminal speed v_t when the gravitational force on it is balanced by the air drag force on it, so its acceleration is zero. We could then apply Newton's second law and the drag force equation to find v_t, but Eq. 6-16 does all that for us.

Calculations: To use Eq. 6-16, we need the drop's effective cross-sectional area A and the magnitude F_g of the gravitational force. Because the drop is spherical, A is the area of a circle (πR^2) that has the same radius as the sphere. To find F_g, we use three facts: (1) $F_g = mg$, where m is the drop's mass; (2) the (spherical) drop's volume is $V = \frac{4}{3}\pi R^3$; and (3) the density of the water in the drop is the mass per volume, or $\rho_w = m/V$. Thus, we find

$$F_g = V\rho_w g = \tfrac{4}{3}\pi R^3 \rho_w g.$$

We next substitute this, the expression for A, and the given data into Eq. 6-16. Being careful to distinguish between the air density ρ_a and the water density ρ_w, we obtain

$$v_t = \sqrt{\frac{2F_g}{C\rho_a A}} = \sqrt{\frac{8\pi R^3 \rho_w g}{3C\rho_a \pi R^2}} = \sqrt{\frac{8R\rho_w g}{3C\rho_a}}$$

$$= \sqrt{\frac{(8)(1.5 \times 10^{-3}\text{ m})(1000\text{ kg/m}^3)(9.8\text{ m/s}^2)}{(3)(0.60)(1.2\text{ kg/m}^3)}}$$

$$= 7.4\text{ m/s} \approx 27\text{ km/h.} \qquad \text{(Answer)}$$

Note that the height of the cloud does not enter into the calculation. As Table 6-1 indicates, the raindrop reaches terminal speed after falling just a few meters.

(b) What would be the drop's speed just before impact if there were no drag force?

KEY IDEA With no drag force to reduce the drop's speed during the fall, the drop would fall with the constant free-fall acceleration g, so the constant-acceleration equations of Table 2-1 apply.

Calculation: Because we know the acceleration is g, the initial velocity v_0 is 0, and the displacement $x - x_0$ is $-h$, we use Eq. 2-16 to find v:

$$v = \sqrt{2gh} = \sqrt{(2)(9.8\text{ m/s}^2)(1200\text{ m})}$$

$$= 153\text{ m/s} \approx 550\text{ km/h.} \qquad \text{(Answer)}$$

Had he known this, Shakespeare would scarcely have written, "it droppeth as the gentle rain from heaven, upon the place beneath." In fact, the speed is close to that of a bullet from a large-caliber handgun!

6-5 | Uniform Circular Motion

From Section 4-7, recall that when a body moves in a circle (or a circular arc) at constant speed v, it is said to be in uniform circular motion. Also recall that the body has a centripetal acceleration (directed toward the center of the circle) of constant magnitude given by

$$a = \frac{v^2}{R} \quad \text{(centripetal acceleration),} \qquad (6\text{-}17)$$

where R is the radius of the circle.

Let us examine two examples of uniform circular motion:

1. *Rounding a curve in a car.* You are sitting in the center of the rear seat of a car moving at a constant high speed along a flat road. When the driver suddenly turns left, rounding a corner in a circular arc, you slide across the seat toward the right and then jam against the car wall for the rest of the turn. What is going on?

   While the car moves in the circular arc, it is in uniform circular motion; that is, it has an acceleration that is directed toward the center of the circle. By Newton's second law, a force must cause this acceleration. Moreover, the force must also be directed toward the center of the circle. Thus, it is a **centripetal force,** where the adjective indicates the direction. In this example, the centripetal force is a frictional force on the tires from the road; it makes the turn possible.

   If you are to move in uniform circular motion along with the car, there must also be a centripetal force on you. However, apparently the frictional force on you from the seat was not great enough to make you go in a circle with the car. Thus, the seat slid beneath you, until the right wall of the car jammed into you. Then its push on you provided the needed centripetal force on you, and you joined the car's uniform circular motion.

2. *Orbiting Earth.* This time you are a passenger in the space shuttle *Atlantis.* As it and you orbit Earth, you float through your cabin. What is going on?

   Both you and the shuttle are in uniform circular motion and have accelerations directed toward the center of the circle. Again by Newton's second law, centripetal forces must cause these accelerations. This time the centripetal forces are gravitational pulls (the pull on you and the pull on the shuttle) exerted by Earth and directed radially inward, toward the center of Earth.

In both car and shuttle you are in uniform circular motion, acted on by a centripetal force—yet your sensations in the two situations are quite different. In the car, jammed up against the wall, you are aware of being compressed by the wall. In the orbiting shuttle, however, you are floating around with no sensation of any force acting on you. Why this difference?

The difference is due to the nature of the two centripetal forces. In the car, the centripetal force is the push on the part of your body touching the car wall. You can sense the compression on that part of your body. In the shuttle, the centripetal force is Earth's gravitational pull on every atom of your body. Thus, there is no compression (or pull) on any one part of your body and no sensation of a force acting on you. (The sensation is said to be one of "weightlessness," but that description is tricky. The pull on you by Earth has certainly not disappeared and, in fact, is only a little less than it would be with you on the ground.)

Another example of a centripetal force is shown in Fig. 6-9. There a hockey puck moves around in a circle at constant speed v while tied to a string looped around a central peg. This time the centripetal force is the radially inward pull on the puck from the string. Without that force, the puck would slide off in a straight line instead of moving in a circle.

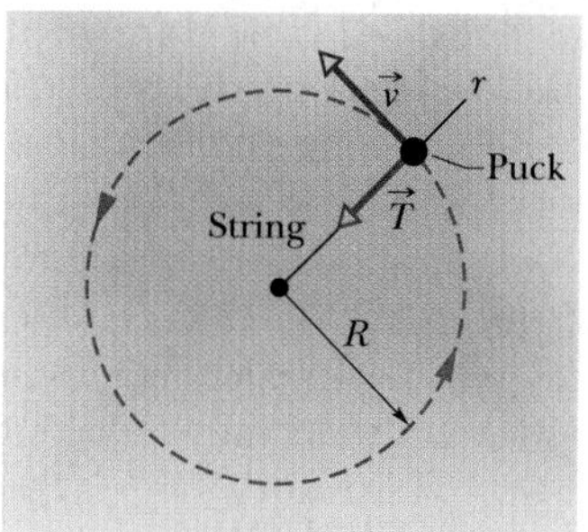

FIG. 6-9 An overhead view of a hockey puck moving with constant speed v in a circular path of radius R on a horizontal frictionless surface. The centripetal force on the puck is $\vec{T}$, the pull from the string, directed inward along the radial axis r extending through the puck.

Note again that a centripetal force is not a new kind of force. The name merely indicates the direction of the force. It can, in fact, be a frictional force, a gravitational force, the force from a car wall or a string, or any other force. For any situation:

> A centripetal force accelerates a body by changing the direction of the body's velocity without changing the body's speed.

From Newton's second law and Eq. 6-17 ($a = v^2/R$), we can write the magnitude F of a centripetal force (or a net centripetal force) as

$$F = m\frac{v^2}{R} \quad \text{(magnitude of centripetal force).} \tag{6-18}$$

Because the speed v here is constant, the magnitudes of the acceleration and the force are also constant.

However, the directions of the centripetal acceleration and force are not constant; they vary continuously so as to always point toward the center of the circle. For this reason, the force and acceleration vectors are sometimes drawn along a radial axis r that moves with the body and always extends from the center of the circle to the body, as in Fig. 6-9. The positive direction of the axis is radially outward, but the acceleration and force vectors point radially inward.

CHECKPOINT 2 When you ride in a Ferris wheel at constant speed, what are the directions of your acceleration $\vec{a}$ and the normal force $\vec{F}_N$ on you (from the always upright seat) as you pass through (a) the highest point and (b) the lowest point of the ride?

Sample Problem 6-6

Igor is a cosmonaut on the International Space Station, in a circular orbit around Earth, at an altitude h of 520 km and with a constant speed v of 7.6 km/s. Igor's mass m is 79 kg.

(a) What is his acceleration?

KEY IDEA Igor is in uniform circular motion and thus has a centripetal acceleration of magnitude given by Eq. 6-17 ($a = v^2/R$).

Calculation: The radius R of Igor's motion is $R_E + h$, where R_E is Earth's radius (6.37×10^6 m, from Appendix C). Thus,

$$a = \frac{v^2}{R} = \frac{v^2}{R_E + h}$$

$$= \frac{(7.6 \times 10^3\ \text{m/s})^2}{6.37 \times 10^6\ \text{m} + 0.52 \times 10^6\ \text{m}}$$

$$= 8.38\ \text{m/s}^2 \approx 8.4\ \text{m/s}^2. \qquad \text{(Answer)}$$

This is the value of the free-fall acceleration at Igor's altitude. If he were lifted to that altitude and released, instead of being put into orbit there, he would fall toward Earth's center, starting out with that value for his acceleration. The difference in the two situations is that when he orbits Earth, he always has a "sideways" motion as well: As he falls, he also moves to the side, so that he ends up moving along a curved path around Earth.

(b) What force does Earth exert on Igor?

KEY IDEAS (1) There must be a centripetal force on Igor if he is to be in uniform circular motion. (2) That force is the gravitational force $\vec{F}_g$ on him from Earth, directed toward his center of rotation (at the center of Earth).

Calculation: From Newton's second law, written along the radial axis r, this force has the magnitude

$$F_g = ma = (79 \text{ kg})(8.38 \text{ m/s}^2)$$
$$= 662 \text{ N} \approx 660 \text{ N}. \qquad \text{(Answer)}$$

If Igor were to stand on a scale placed on the top of a tower of height $h = 520$ km, the scale would read 660 N. In orbit, the scale (if Igor could "stand" on it) would read zero because he and the scale are in free fall together, and therefore his feet do not actually press against it.

Sample Problem 6-7

In a 1901 circus performance, Allo "Dare Devil" Diavolo introduced the stunt of riding a bicycle in a loop-the-loop (Fig. 6-10a). Assuming that the loop is a circle with radius $R = 2.7$ m, what is the least speed v Diavolo could have at the top of the loop to remain in contact with it there?

KEY IDEA We can assume that Diavolo and his bicycle travel through the top of the loop as a single particle in uniform circular motion. Thus, at the top, the acceleration $\vec{a}$ of this particle must have the magnitude $a = v^2/R$ given by Eq. 6-17 and be directed downward, toward the center of the circular loop.

(a)

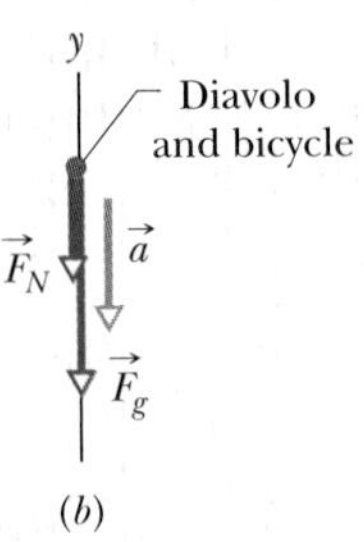

(b)

FIG. 6-10 (a) Contemporary advertisement for Diavolo and (b) free-body diagram for the performer at the top of the loop. *(Photograph in part a reproduced with permission of Circus World Museum)*

Calculations: The forces on the particle when it is at the top of the loop are shown in the free-body diagram of Fig 6-10b. The gravitational force $\vec{F}_g$ is directed downward along a y axis; so is the normal force $\vec{F}_N$ on the particle from the loop; so also is the centripetal acceleration of the particle. Thus, Newton's second law for y components ($F_{\text{net},y} = ma_y$) gives us

$$-F_N - F_g = m(-a)$$

and

$$-F_N - mg = m\left(-\frac{v^2}{R}\right). \qquad (6\text{-}19)$$

If the particle has the *least speed* v needed to remain in contact, then it is on the *verge of losing contact* with the loop (falling away from the loop), which means that $F_N = 0$ at the top of the loop (the particle and loop touch but without any normal force). Substituting 0 for F_N in Eq. 6-19, solving for v, and then substituting known values give us

$$v = \sqrt{gR} = \sqrt{(9.8 \text{ m/s}^2)(2.7 \text{ m})}$$
$$= 5.1 \text{ m/s}. \qquad \text{(Answer)}$$

Comments: Diavolo made certain that his speed at the top of the loop was greater than 5.1 m/s so that he did not lose contact with the loop and fall away from it. Note that this speed requirement is independent of the mass of Diavolo and his bicycle. Had he feasted on, say, pierogies before his performance, he still would have had to exceed only 5.1 m/s to maintain contact as he passed through the top of the loop.

Sample Problem 6-8 Build your skill

Even some seasoned roller-coaster riders blanch at the thought of riding the Rotor, which is essentially a large, hollow cylinder that is rotated rapidly around its central axis (Fig. 6-11). Before the ride begins, a rider enters the cylinder through a door on the side and stands on a floor, up against a canvas-covered wall. The door is closed, and as the cylinder begins to turn, the rider, wall, and floor move in unison. When the rider's speed reaches some predetermined value, the floor abruptly and alarmingly falls away. The rider does not fall with it but instead is pinned to the wall while the cylinder rotates, as if an unseen (and somewhat unfriendly) agent is pressing the body to the wall. Later, the floor is eased back to the rider's feet, the cylinder slows, and the rider sinks a few centimeters to regain footing on the floor. (Some riders consider all this to be fun.)

Suppose that the coefficient of static friction μ_s between the rider's clothing and the canvas is 0.40 and that the cylinder's radius R is 2.1 m.

(a) What minimum speed v must the cylinder and rider have if the rider is not to fall when the floor drops?

KEY IDEAS

1. The gravitational force $\vec{F}_g$ on the rider tends to slide her down the wall, but she does not move because a frictional force from the wall acts upward on her (Fig. 6-11).
2. If she is to be on the verge of sliding down, that upward force must be a *static* frictional force $\vec{f}_s$ at its maximum value $\mu_s F_N$, where F_N is the magnitude of the normal force $\vec{F}_N$ on her from the cylinder (Fig. 6-11).
3. This normal force is directed horizontally toward the central axis of the cylinder and is the centripetal force that causes the rider to move in a circular path, with centripetal acceleration of magnitude $a = v^2/R$ and directed toward the center of the circle.

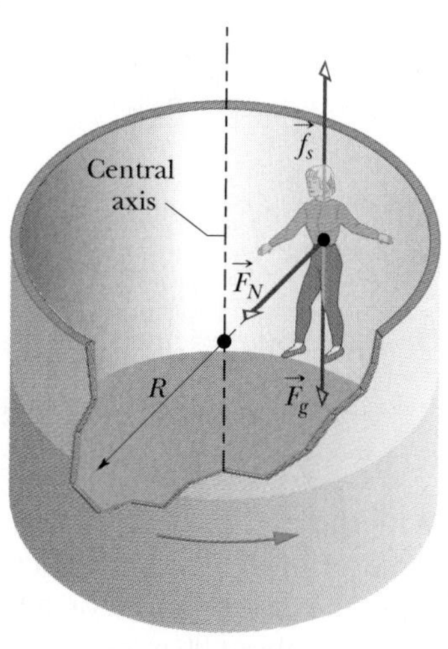

FIG. 6-11 A Rotor in an amusement park, showing the forces on a rider. The centripetal force is the normal force $\vec{F}_N$ with which the wall pushes inward on the rider.

We want speed v in that last expression, for the condition that the rider is on the verge of sliding.

Vertical calculations: We first place a vertical y axis through the rider, with the positive direction upward. We can then apply Newton's second law to the rider, writing it for y components ($F_{\text{net},y} = ma_y$) as

$$f_s - mg = m(0),$$

where m is the rider's mass and mg is the magnitude of $\vec{F}_g$. Because the rider is on the verge of sliding, we substitute the maximum value $\mu_s F_N$ for f_s in this equation, getting

$$\mu_s F_N - mg = 0,$$

or

$$F_N = \frac{mg}{\mu_s}. \tag{6-20}$$

Radial calculations: Next we place a radial r axis through the rider, with the positive direction outward. We can then write Newton's second law for components along that axis as

$$-F_N = m\left(-\frac{v^2}{R}\right). \tag{6-21}$$

Substituting Eq. 6-20 for F_N and then solving for v, we find

$$v = \sqrt{\frac{gR}{\mu_s}} = \sqrt{\frac{(9.8 \text{ m/s}^2)(2.1 \text{ m})}{0.40}}$$
$$= 7.17 \text{ m/s} \approx 7.2 \text{ m/s}. \quad \text{(Answer)}$$

Note that the result is independent of the rider's mass; it holds for anyone riding the Rotor, from a child to a sumo wrestler, which is why no one has to "weigh in" to ride the Rotor.

(b) If the rider's mass is 49 kg, what is the magnitude of the centripetal force on her?

Calculation: According to Eq. 6-21,

$$F_N = m\frac{v^2}{R} = (49 \text{ kg})\frac{(7.17 \text{ m/s})^2}{2.1 \text{ m}}$$
$$\approx 1200 \text{ N}. \quad \text{(Answer)}$$

Although this force is directed toward the central axis, the rider has an overwhelming sensation that the force pinning her against the wall is directed radially outward. Her sensation stems from the fact that she is in a noninertial frame (she and it are accelerating). As measured from such frames, forces can be illusionary. The illusion is part of the Rotor's attraction.

Sample Problem 6-9 Build your skill

Upside-down racing: A modern race car is designed so that the passing air pushes down on it, allowing the car to travel much faster through a flat turn in a Grand Prix without friction failing. This downward push is called *negative lift.* Can a race car have so much negative lift that it could be driven upside down on a long ceiling, as done fictionally by a sedan in the first *Men in Black* movie?

Figure 6-12*a* represents a Grand Prix race car of mass $m = 600$ kg as it travels on a flat track in a circular arc of radius $R = 100$ m. Because of the shape of the car and the wings on it, the passing air exerts a negative lift $\vec{F}_L$ downward on the car. The coefficient of static friction between the tires and the track is 0.75. (Assume that the forces on the four tires are identical.)

(a) If the car is on the verge of sliding out of the turn when its speed is 28.6 m/s, what is the magnitude of $\vec{F}_L$?

KEY IDEAS

1. A centripetal force must act on the car because the car is moving around a circular arc; that force must be directed toward the center of curvature of the arc (here, that is horizontally).
2. The only horizontal force acting on the car is a frictional force on the tires from the road. So the required centripetal force is a frictional force.
3. Because the car is not sliding, the frictional force must be a *static* frictional force $\vec{f}_s$ (Fig. 6-12*a*).
4. Because the car is on the verge of sliding, the magnitude f_s is equal to the maximum value $f_{s,\max} = \mu_s F_N$, where F_N is the magnitude of the normal force $\vec{F}_N$ acting on the car from the track.

Radial calculations: The frictional force $\vec{f}_s$ is shown in the free-body diagram of Fig. 6-12*b*. It is in the negative direction of a radial axis *r* that always extends from the center of curvature through the car as the car moves. The force produces a centripetal acceleration of magnitude v^2/R. We can relate the force and acceleration by writing Newton's second law for components along the *r* axis ($F_{\text{net},r} = ma_r$) as

$$-f_s = m\left(-\frac{v^2}{R}\right). \tag{6-22}$$

Substituting $f_{s,\max} = \mu_s F_N$ for f_s leads us to

$$\mu_s F_N = m\left(\frac{v^2}{R}\right). \tag{6-23}$$

Vertical calculations: Next, let's consider the vertical forces on the car. The normal force $\vec{F}_N$ is directed up, in the positive direction of the *y* axis in Fig. 6-12*b*. The gravitational force $\vec{F}_g = m\vec{g}$ and the negative lift $\vec{F}_L$ are directed down. The acceleration of the car along the *y* axis is zero. Thus we can write Newton's second law for components along the *y* axis ($F_{\text{net},y} = ma_y$) as

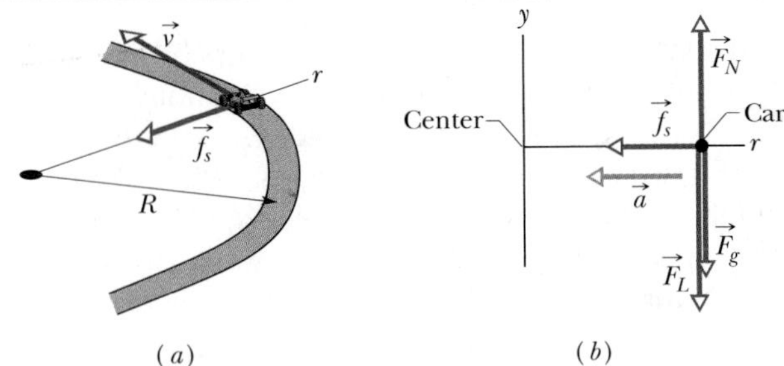

FIG. 6-12 (*a*) A race car moves around a flat curved track at constant speed *v*. The frictional force $\vec{f}_s$ provides the necessary centripetal force along a radial axis *r*. (*b*) A free-body diagram (not to scale) for the car, in the vertical plane containing *r*.

$$F_N - mg - F_L = 0,$$

or

$$F_N = mg + F_L. \tag{6-24}$$

Combining results: Now we can combine our results along the two axes by substituting Eq. 6-24 for F_N in Eq. 6-23. Doing so and then solving for F_L lead to

$$\begin{aligned} F_L &= m\left(\frac{v^2}{\mu_s R} - g\right) \\ &= (600 \text{ kg})\left(\frac{(28.6 \text{ m/s})^2}{(0.75)(100 \text{ m})} - 9.8 \text{ m/s}^2\right) \\ &= 663.7 \text{ N} \approx 660 \text{ N}. \end{aligned} \qquad \text{(Answer)}$$

(b) The magnitude F_L of the negative lift on a car depends on the square of the car's speed v^2, just as the drag force does (Eq. 6-14). Thus, the negative lift on the car here is greater when the car travels faster, as it does on a straight section of track. What is the magnitude of the negative lift for a speed of 90 m/s?

KEY IDEA F_L is proportional to v^2.

Calculations: Thus we can write a ratio of the negative lift $F_{L,90}$ at $v = 90$ m/s to our result for the negative lift F_L at $v = 28.6$ m/s as

$$\frac{F_{L,90}}{F_L} = \frac{(90 \text{ m/s})^2}{(28.6 \text{ m/s})^2}.$$

Substituting $F_L = 663.7$ N and solving for $F_{L,90}$ give us

$$F_{L,90} = 6572 \text{ N} \approx 6600 \text{ N}. \qquad \text{(Answer)}$$

Upside-down racing: The gravitational force is

$$\begin{aligned} F_g &= mg = (600 \text{ kg})(9.8 \text{ m/s}^2) \\ &= 5880 \text{ N}. \end{aligned}$$

With the car upside down, the negative lift is an *upward* force of 6600 N, which exceeds the downward 5880 N. Thus, the car could run on a long ceiling *provided* that it moves at about 90 m/s (= 324 km/h = 201 mi/h).

Sample Problem 6-10

Curved portions of highways are always banked (tilted) to prevent cars from sliding off the highway. When a highway is dry, the frictional force between the tires and the road surface may be enough to prevent sliding. When the highway is wet, however, the frictional force may be negligible, and banking is then essential. Figure 6-13a represents a car of mass m as it moves at a constant speed v of 20 m/s around a banked circular track of radius $R = 190$ m. (It is a normal car, rather than a race car, which means any vertical force from the passing air is negligible.) If the frictional force from the track is negligible, what bank angle θ prevents sliding?

KEY IDEA Unlike Sample Problem 6-9, the track is banked so as to tilt the normal force $\vec{F}_N$ on the car toward the center of the circle (Fig. 6-13b). Thus, $\vec{F}_N$ now has a centripetal component of magnitude F_{Nr}, directed inward along a radial axis r. We want to find the value of the bank angle θ such that this centripetal component keeps the car on the circular track without need of friction.

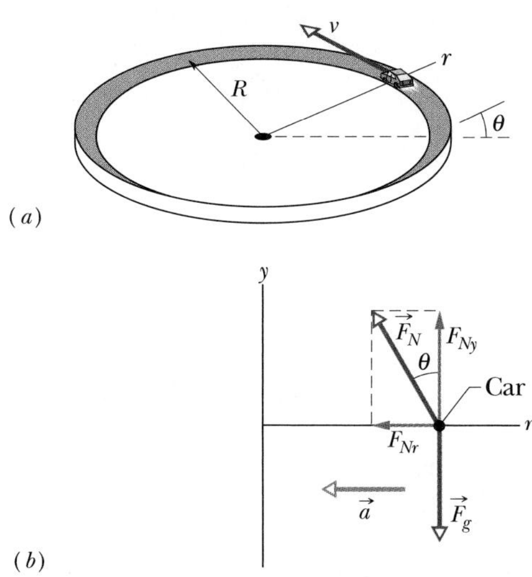

FIG. 6-13 (a) A car moves around a curved banked road at constant speed v. The bank angle is exaggerated for clarity. (b) A free-body diagram for the car, assuming that friction between tires and road is zero and that the car lacks negative lift. The radially inward component F_{Nr} of the normal force (along radial axis r) provides the necessary centripetal force and radial acceleration.

Radial calculation: As Fig. 6-13b shows (and as you should verify), the angle that force $\vec{F}_N$ makes with the vertical is equal to the bank angle θ of the track. Thus, the radial component F_{Nr} is equal to $F_N \sin\theta$. We can now write Newton's second law for components along the r axis ($F_{\text{net},r} = ma_r$) as

$$-F_N \sin\theta = m\left(-\frac{v^2}{R}\right). \tag{6-25}$$

We cannot solve this equation for the value of θ because it also contains the unknowns F_N and m.

Vertical calculations: We next consider the forces and acceleration along the y axis in Fig. 6-13b. The vertical component of the normal force is $F_{Ny} = F_N \cos\theta$, the gravitational force $\vec{F}_g$ on the car has the magnitude mg, and the acceleration of the car along the y axis is zero. Thus we can write Newton's second law for components along the y axis ($F_{\text{net},y} = ma_y$) as

$$F_N \cos\theta - mg = m(0),$$

from which

$$F_N \cos\theta = mg. \tag{6-26}$$

Combining results: Equation 6-26 also contains the unknowns F_N and m, but note that dividing Eq. 6-25 by Eq. 6-26 neatly eliminates both those unknowns. Doing so, replacing $(\sin\theta)/(\cos\theta)$ with $\tan\theta$, and solving for θ then yield

$$\theta = \tan^{-1}\frac{v^2}{gR}$$

$$= \tan^{-1}\frac{(20\ \text{m/s})^2}{(9.8\ \text{m/s}^2)(190\ \text{m})} = 12°. \quad \text{(Answer)}$$

REVIEW & SUMMARY

Friction When a force $\vec{F}$ tends to slide a body along a surface, a **frictional force** from the surface acts on the body. The frictional force is parallel to the surface and directed so as to oppose the sliding. It is due to bonding between the body and the surface.

If the body does not slide, the frictional force is a **static frictional force** $\vec{f}_s$. If there is sliding, the frictional force is a **kinetic frictional force** $\vec{f}_k$.

1. If a body does not move, the static frictional force $\vec{f}_s$ and the component of $\vec{F}$ parallel to the surface are equal in magnitude, and $\vec{f}_s$ is directed opposite that component. If the component increases, f_s also increases.
2. The magnitude of $\vec{f}_s$ has a maximum value $f_{s,\text{max}}$ given by

$$f_{s,\text{max}} = \mu_s F_N, \tag{6-1}$$

where μ_s is the **coefficient of static friction** and F_N is the magnitude of the normal force. If the component of $\vec{F}$ parallel to the surface exceeds $f_{s,\text{max}}$, the body slides on the surface.

3. If the body begins to slide on the surface, the magnitude of the frictional force rapidly decreases to a constant value f_k given by

$$f_k = \mu_k F_N, \tag{6-2}$$

where μ_k is the **coefficient of kinetic friction.**

Drag Force When there is relative motion between air (or some other fluid) and a body, the body experiences a **drag force** $\vec{D}$ that opposes the relative motion and points in the direction in which the fluid flows relative to the body. The magnitude of $\vec{D}$ is related to the relative speed v by an experimentally determined **drag coefficient** C according to

$$D = \tfrac{1}{2}C\rho A v^2, \qquad (6\text{-}14)$$

where ρ is the fluid density (mass per unit volume) and A is the **effective cross-sectional area** of the body (the area of a cross section taken perpendicular to the relative velocity $\vec{v}$).

Terminal Speed When a blunt object has fallen far enough through air, the magnitudes of the drag force $\vec{D}$ and the gravitational force $\vec{F}_g$ on the body become equal. The body then falls at a constant **terminal speed** v_t given by

$$v_t = \sqrt{\frac{2F_g}{C\rho A}}. \qquad (6\text{-}16)$$

Uniform Circular Motion If a particle moves in a circle or a circular arc of radius R at constant speed v, the particle is said to be in **uniform circular motion.** It then has a **centripetal acceleration** $\vec{a}$ with magnitude given by

$$a = \frac{v^2}{R}. \qquad (6\text{-}17)$$

This acceleration is due to a net **centripetal force** on the particle, with magnitude given by

$$F = \frac{mv^2}{R}, \qquad (6\text{-}18)$$

where m is the particle's mass. The vector quantities $\vec{a}$ and $\vec{F}$ are directed toward the center of curvature of the particle's path.

QUESTIONS

1 In Fig. 6-14, horizontal force $\vec{F}_1$ of magnitude 10 N is applied to a box on a floor, but the box does not slide. Then, as the magnitude of vertical force $\vec{F}_2$ is increased from zero, do the following quantities increase, decrease, or stay the same: (a) the magnitude of the frictional force $\vec{f}_s$ on the box; (b) the magnitude of the normal force $\vec{F}_N$ on the box from the floor; (c) the maximum value $f_{s,\max}$ of the magnitude of the static frictional force on the box? (d) Does the box eventually slide?

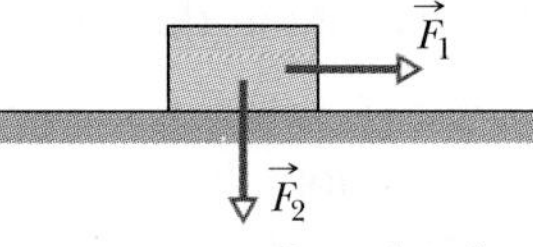

FIG. 6-14 Question 1.

2 In three experiments, three different horizontal forces are applied to the same block lying on the same countertop. The force magnitudes are $F_1 = 12$ N, $F_2 = 8$ N, and $F_3 = 4$ N. In each experiment, the block remains stationary in spite of the applied force. Rank the forces according to (a) the magnitude f_s of the static frictional force on the block from the countertop and (b) the maximum value $f_{s,\max}$ of that force, greatest first.

3 In Fig. 6-15, if the box is stationary and the angle θ between the horizontal and force $\vec{F}$ is increased somewhat, do the following quantities increase, decrease, or remain the same: (a) F_x; (b) f_s; (c) F_N; (d) $f_{s,\max}$? (e) If, instead, the box is sliding and θ is increased, does the magnitude of the frictional force on the box increase, decrease, or remain the same?

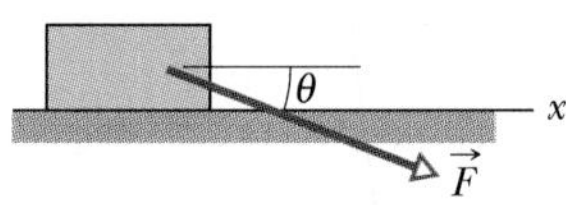

FIG. 6-15 Question 3.

4 Repeat Question 3 for force $\vec{F}$ angled upward instead of downward as drawn.

5 If you press an apple crate against a wall so hard that the crate cannot slide down the wall, what is the direction of (a) the static frictional force $\vec{f}_s$ on the crate from the wall and (b) the normal force $\vec{F}_N$ on the crate from the wall? If you increase your push, what happens to (c) f_s, (d) F_N, and (e) $f_{s,\max}$?

6 In Fig. 6-16, a block of mass m is held stationary on a ramp by the frictional force on it from the ramp. A force $\vec{F}$, directed up the ramp, is then applied to the block and gradually increased in magnitude from zero. During the increase, what happens to the direction and magnitude of the frictional force on the block?

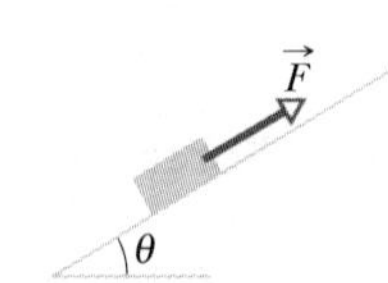

FIG. 6-16 Question 6.

7 Reconsider Question 6 but with the force $\vec{F}$ now directed down the ramp. As the magnitude of $\vec{F}$ is increased from zero, what happens to the direction and magnitude of the frictional force on the block?

8 In Fig. 6-17, a horizontal force of 100 N is to be applied to a 10 kg slab that is initially stationary on a frictionless floor, to accelerate the slab. A 10 kg block lies on top of the slab; the coefficient of friction μ between the block and the slab is not known, and the block might slip. (a) Considering that possibility, what is the possible range of values for the magnitude of the slab's acceleration a_{slab}? (*Hint:* You don't need written calculations; just consider extreme values for μ.) (b) What is the possible range for the magnitude a_{block} of the block's acceleration?

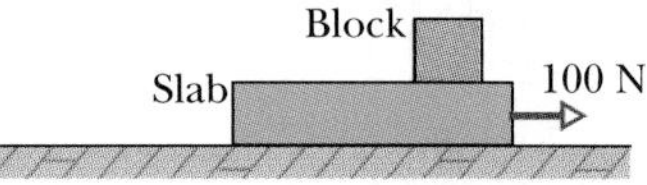

FIG. 6-17 Question 8.

9 A person riding a Ferris wheel moves through positions at (1) the top, (2) the bottom, and (3) midheight. If the wheel rotates at a constant rate, rank these three positions according to (a) the magnitude of the person's centripetal acceleration, (b) the magnitude of the net centripetal force on the person, and (c) the magnitude of the normal force on the person, greatest first.

10 In 1987, as a Halloween stunt, two sky divers passed a pumpkin back and forth between them while they were in free fall just west of Chicago. The stunt was great fun until the last sky diver with the pumpkin opened his parachute. The pumpkin broke free from his grip, plummeted about 0.5 km, ripped through the roof of a house, slammed into the kitchen floor, and splattered all over the newly remodeled kitchen. From the sky diver's viewpoint and from the pumpkin's viewpoint, why did the sky diver lose control of the pumpkin?

11 Figure 6-18 shows the path of a park ride that travels at constant speed through five circular arcs of radii R_0, $2R_0$, and $3R_0$. Rank the arcs according to the magnitude of the centripetal force on a rider traveling in the arcs, greatest first.

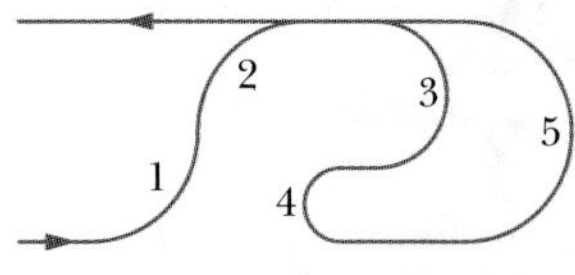

FIG. 6-18 Question 11.

PROBLEMS

GO Tutoring problem available (at instructor's discretion) in *WileyPLUS* and WebAssign

SSM Worked-out solution available in Student Solutions Manual

WWW Worked-out solution is at http://www.wiley.com/college/halliday

• – ••• Number of dots indicates level of problem difficulty

ILW Interactive solution is at http://www.wiley.com/college/halliday

Additional information available in *The Flying Circus of Physics* and at flyingcircusofphysics.com

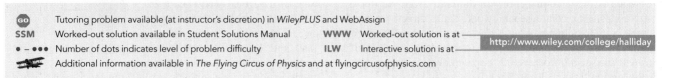

sec. 6-3 Properties of Friction

•1 A bedroom bureau with a mass of 45 kg, including drawers and clothing, rests on the floor. (a) If the coefficient of static friction between the bureau and the floor is 0.45, what is the magnitude of the minimum horizontal force that a person must apply to start the bureau moving? (b) If the drawers and clothing, with 17 kg mass, are removed before the bureau is pushed, what is the new minimum magnitude? SSM WWW

•2 *The mysterious sliding stones.* Along the remote Racetrack Playa in Death Valley, California, stones sometimes gouge out prominent trails in the desert floor, as if the stones had been migrating (Fig. 6-19). For years curiosity mounted about why the stones moved. One explanation was that strong winds during occasional rainstorms would drag the rough stones over ground softened by rain. When the desert dried out, the trails behind the stones were hard-baked in place. According to measurements, the coefficient of kinetic friction between the stones and the wet playa ground is about 0.80. What horizontal force must act on a 20 kg stone (a typical mass) to maintain the stone's motion once a gust has started it moving? (Story continues with Problem 39.)

FIG. 6-19 Problem 2. What moved the stone? *(Jerry Schad/Photo Researchers)*

•3 A person pushes horizontally with a force of 220 N on a 55 kg crate to move it across a level floor. The coefficient of kinetic friction is 0.35. What is the magnitude of (a) the frictional force and (b) the crate's acceleration? SSM ILW

•4 A baseball player with mass $m = 79$ kg, sliding into second base, is retarded by a frictional force of magnitude 470 N. What is the coefficient of kinetic friction μ_k between the player and the ground?

•5 The floor of a railroad flatcar is loaded with loose crates having a coefficient of static friction of 0.25 with the floor. If the train is initially moving at a speed of 48 km/h, in how short a distance can the train be stopped at constant acceleration without causing the crates to slide over the floor?

•6 A slide-loving pig slides down a certain 35° slide in twice the time it would take to slide down a frictionless 35° slide. What is the coefficient of kinetic friction between the pig and the slide?

•7 A 3.5 kg block is pushed along a horizontal floor by a force $\vec{F}$ of magnitude 15 N at an angle $\theta = 40°$ with the horizontal (Fig. 6-20). The coefficient of kinetic friction between the block and the floor is 0.25. Calculate the magnitudes of (a) the frictional force on the block from the floor and (b) the block's acceleration. GO

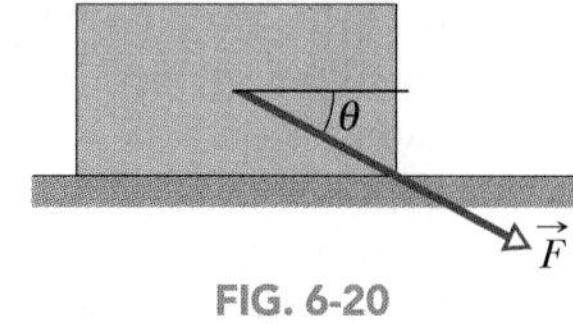

FIG. 6-20 Problems 7 and 24.

•8 In a pickup game of dorm shuffleboard, students crazed by final exams use a broom to propel a calculus book along the dorm hallway. If the 3.5 kg book is pushed from rest through a distance of 0.90 m by the horizontal 25 N force from the broom and then has a speed of 1.60 m/s, what is the coefficient of kinetic friction between the book and floor?

•9 A 2.5 kg block is initially at rest on a horizontal surface. A horizontal force $\vec{F}$ of magnitude 6.0 N and a vertical force $\vec{P}$ are then applied to the block (Fig. 6-21). The coefficients of friction for the block and surface are $\mu_s = 0.40$ and $\mu_k = 0.25$. Determine the magnitude of the frictional force acting on the block if the magnitude of $\vec{P}$ is (a) 8.0 N, (b) 10 N, and (c) 12 N. GO

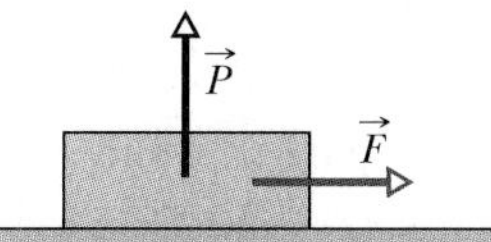

FIG. 6-21 Problem 9.

•10 In about 1915, Henry Sincosky of Philadelphia suspended himself from a rafter by gripping the rafter with the thumb of each hand on one side and the fingers on the opposite side (Fig. 6-22). Sincosky's mass was 79 kg. If the coefficient of static friction between hand and rafter was 0.70, what was the least magnitude of the normal force on the rafter from each thumb or opposite fingers? (After suspending himself, Sincosky chinned himself on the rafter and then moved hand-over-hand along the rafter. If you do not think Sincosky's grip was remarkable, try to repeat his stunt.)

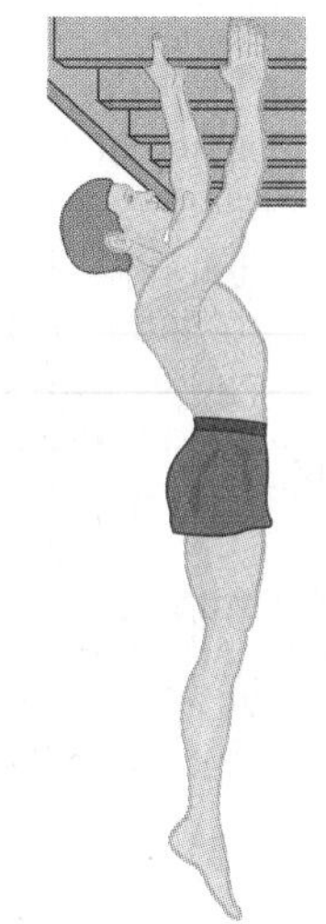

FIG. 6-22 Problem 10.

•11 A worker pushes horizontally on a 35 kg crate with a force of magnitude 110 N. The coefficient of static friction between the crate and the floor is 0.37. (a) What is the value of $f_{s,max}$ under the circumstances? (b) Does the crate move? (c) What is the frictional force on the crate from the floor? (d) Suppose, next, that a second worker pulls directly upward on the crate to help out. What is the least vertical pull that will allow the first worker's 110 N push to move the crate? (e) If, instead, the second worker pulls horizontally to help out, what is the least pull that will get the crate moving?

•12 Figure 6-23 shows the cross section of a road cut into the side of a mountain. The solid line AA' represents a weak bedding plane along which sliding is possible. Block B directly above the highway is separated from uphill rock by a large crack (called a *joint*), so that only friction between the block and the bedding plane prevents sliding. The mass of the block is 1.8×10^7 kg, the *dip angle* θ of the bedding plane is 24°, and the coefficient of static friction between block and plane is 0.63. (a) Show that the block will not slide. (b) Water seeps into the joint and expands upon freezing, exerting on the block a force $\vec{F}$ parallel to AA'. What minimum value of force magnitude F will trigger a slide down the plane?

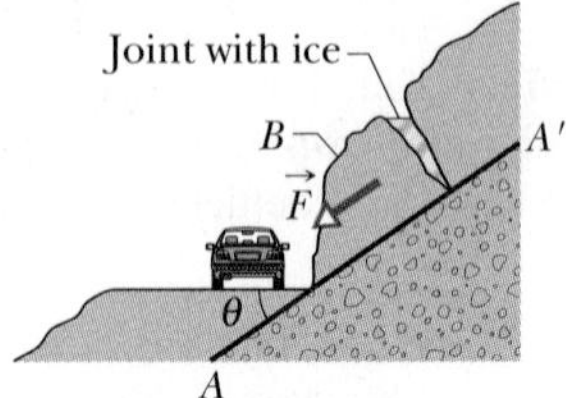

FIG. 6-23 Problem 12.

•13 A 68 kg crate is dragged across a floor by pulling on a rope attached to the crate and inclined 15° above the horizontal. (a) If the coefficient of static friction is 0.50, what minimum force magnitude is required from the rope to start the crate moving? (b) If $\mu_k = 0.35$, what is the magnitude of the initial acceleration of the crate? SSM

•14 Figure 6-24 shows an initially stationary block of mass m on a floor. A force of magnitude $0.500mg$ is then applied at upward angle $\theta = 20°$. What is the magnitude of the acceleration of the block across the floor if (a) $\mu_s = 0.600$ and $\mu_k = 0.500$ and (b) $\mu_s = 0.400$ and $\mu_k = 0.300$?

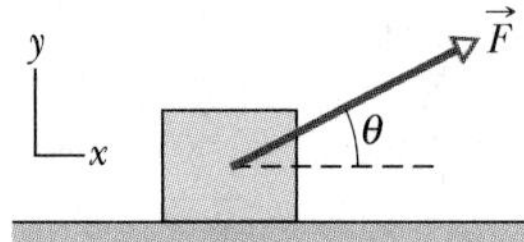

FIG. 6-24 Problem 14.

•15 The coefficient of static friction between Teflon and scrambled eggs is about 0.04. What is the smallest angle from the horizontal that will cause the eggs to slide across the bottom of a Teflon-coated skillet?

••16 You testify as an *expert witness* in a case involving an accident in which car A slid into the rear of car B, which was stopped at a red light along a road headed down a hill (Fig. 6-25). You find that the slope of the hill is $\theta = 12.0°$, that the cars were separated by distance $d = 24.0$ m when the driver of car A put the car into a slide (it lacked any automatic anti-brake-lock system), and that the speed of car A at the onset of braking was $v_0 = 18.0$ m/s. With what speed did car A hit car B if the coefficient of kinetic friction was (a) 0.60 (dry road surface) and (b) 0.10 (road surface covered with wet leaves)?

FIG. 6-25 Problem 16.

••17 A 12 N horizontal force $\vec{F}$ pushes a block weighing 5.0 N against a vertical wall (Fig. 6-26). The coefficient of static friction between the wall and the block is 0.60, and the coefficient of kinetic friction is 0.40. Assume that the block is not moving initially. (a) Will the block move? (b) In unit-vector notation, what is the force on the block from the wall?

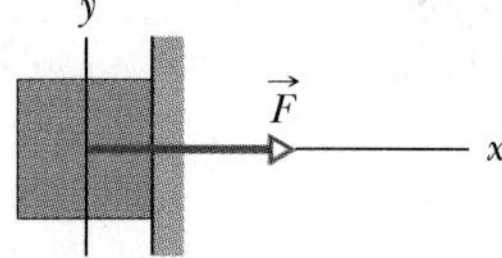

FIG. 6-26 Problem 17.

••18 A 4.10 kg block is pushed along a floor by a constant applied force that is horizontal and has a magnitude of 40.0 N. Figure 6-27 gives the block's speed v versus time t as the block moves along an x axis on the floor. The scale of the figure's vertical axis is set by $v_s = 5.0$ m/s. What is the coefficient of kinetic friction between the block and the floor?

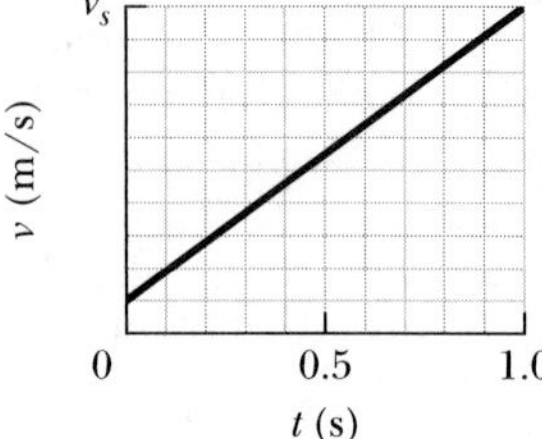

FIG. 6-27 Problem 18.

••19 An initially stationary box of sand is to be pulled across a floor by means of a cable in which the tension should not exceed 1100 N. The coefficient of static friction between the box and the floor is 0.35. (a) What should be the angle between the cable and the horizontal in order to pull the greatest possible amount of sand, and (b) what is the weight of the sand and box in that situation?

••20 A loaded penguin sled weighing 80 N rests on a plane inclined at angle $\theta = 20°$ to the horizontal (Fig. 6-28). Between the sled and the plane, the coefficient of static friction is 0.25, and the coefficient of kinetic friction is 0.15. (a) What is the least magnitude of the force $\vec{F}$, parallel to the plane, that will prevent the sled from slipping down the plane? (b) What is the minimum magnitude F that will start the sled moving up the plane? (c) What value of F is required to move the sled up the plane at constant velocity?

FIG. 6-28 Problems 20 and 26.

••21 In Fig. 6-29, a force $\vec{P}$ acts on a block weighing 45 N. The block is initially at rest on a plane inclined at angle $\theta = 15°$ to the horizontal. The positive direction of the x axis is up the plane. The coefficients of friction between block and plane are $\mu_s = 0.50$ and $\mu_k = 0.34$. In unit-vector notation, what is the frictional force on the block from the plane when $\vec{P}$ is (a) $(-5.0\text{ N})\hat{i}$, (b) $(-8.0\text{ N})\hat{i}$, and (c) $(-15\text{ N})\hat{i}$?

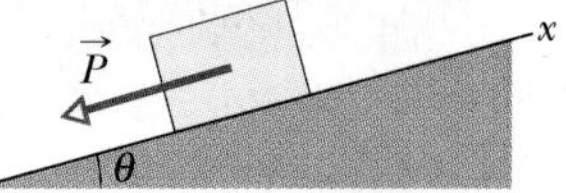

FIG. 6-29 Problem 21.

••22 In Fig. 6-30, a box of Cheerios (mass $m_C = 1.0$ kg) and a box of Wheaties (mass $m_W = 3.0$ kg) are accelerated across a horizontal surface by a horizontal force $\vec{F}$ applied to the Cheerios box. The magnitude of the frictional force on the Cheerios box is 2.0 N, and the magnitude of the frictional force on the Wheaties box is 4.0 N. If the magnitude of $\vec{F}$ is

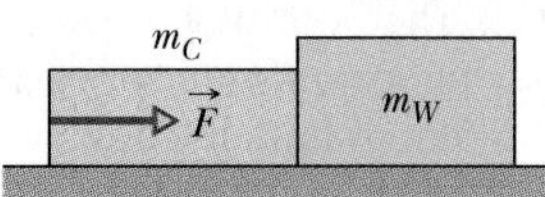

FIG. 6-30 Problem 22.

12 N, what is the magnitude of the force on the Wheaties box from the Cheerios box?

••23 Block B in Fig. 6-31 weighs 711 N. The coefficient of static friction between block and table is 0.25; angle θ is 30°; assume that the cord between B and the knot is horizontal. Find the maximum weight of block A for which the system will be stationary. SSM WWW

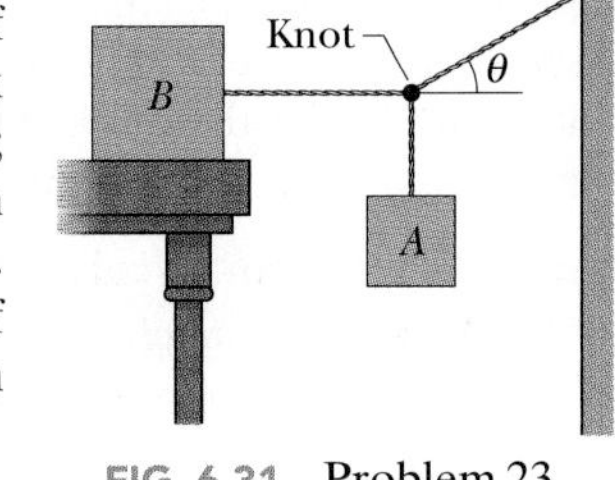

FIG. 6-31 Problem 23.

••24 A block is pushed across a floor by a constant force that is applied at downward angle θ (Fig. 6-20). Figure 6-32 gives the acceleration magnitude a versus a range of values for the coefficient of kinetic friction μ_k between block and floor: $a_1 = 3.0\text{ m/s}^2$, $\mu_{k2} = 0.20$, and $\mu_{k3} = 0.40$. What is the value of θ?

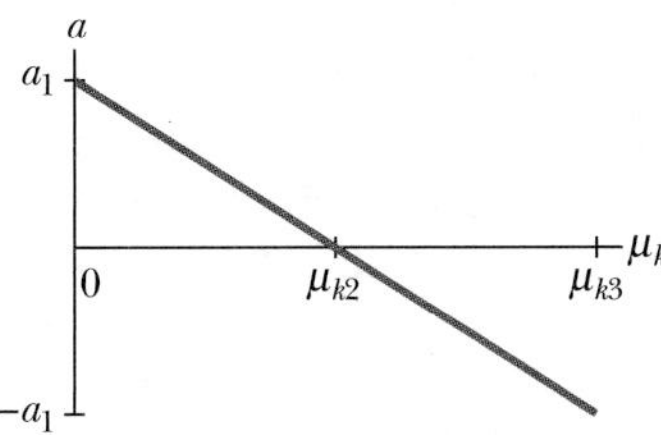

FIG. 6-32 Problem 24.

••25 When the three blocks in Fig. 6-33 are released from rest, they accelerate with a magnitude of 0.500 m/s². Block 1 has mass M, block 2 has $2M$, and block 3 has $2M$. What is the coefficient of kinetic friction between block 2 and the table?

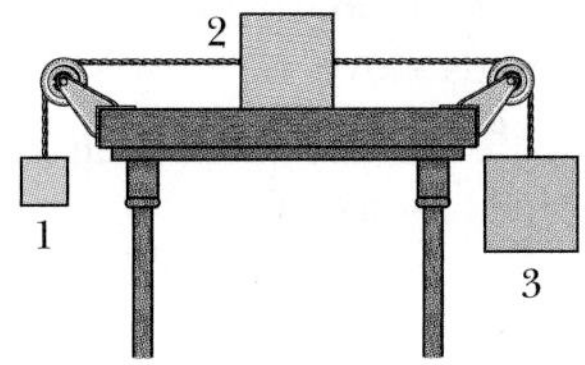

FIG. 6-33 Problem 25

••26 In Fig. 6-28, a sled is held on an inclined plane by a cord pulling directly up the plane. The sled is to be on the verge of moving up the plane. In Fig. 6-34, the magnitude F required of the cord's force on the sled is plotted versus a range of values for the coefficient of static friction μ_s between sled and plane: $F_1 = 2.0$ N, $F_2 = 5.0$ N, and $\mu_2 = 0.50$. At what angle θ is the plane inclined?

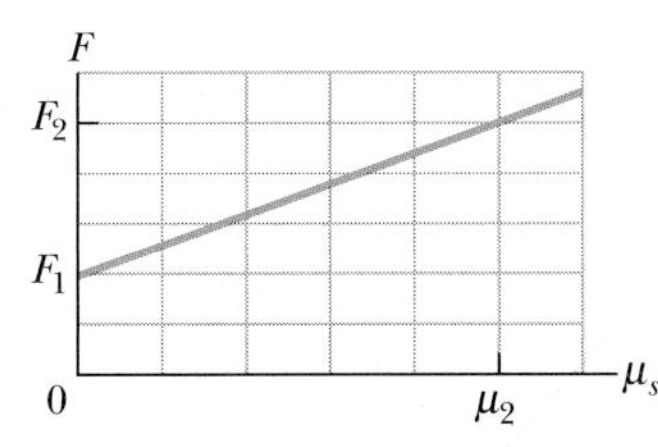

FIG. 6-34 Problem 26.

••27 Two blocks, of weights 3.6 N and 7.2 N, are connected by a massless string and slide down a 30° inclined plane. The coefficient of kinetic friction between the lighter block and the plane is 0.10; that between the heavier block and the plane is 0.20. Assuming that the lighter block leads, find (a) the magnitude of the acceleration of the blocks and (b) the tension in the string. SSM

••28 Figure 6-35 shows three crates being pushed over a concrete floor by a horizontal force $\vec{F}$ of magnitude 440 N. The masses of the crates are $m_1 = 30.0$ kg, $m_2 = 10.0$ kg, and $m_3 = 20.0$ kg. The coefficient of kinetic friction between the floor and each of the crates is 0.700. (a) What is the magnitude F_{32} of the force on crate 3 from crate 2? (b) If the crates then slide onto a polished floor, where the coefficient of kinetic friction is less than 0.700, is magnitude F_{32} more than, less than, or the same as it was when the coefficient was 0.700? GO

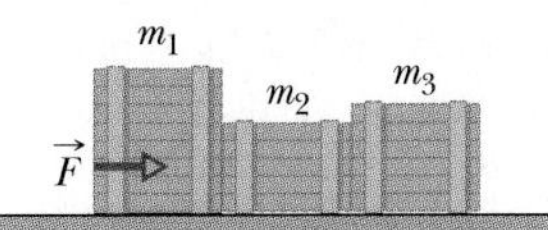

FIG. 6-35 Problem 28.

••29 Body A in Fig. 6-36 weighs 102 N, and body B weighs 32 N. The coefficients of friction between A and the incline are $\mu_s = 0.56$ and $\mu_k = 0.25$. Angle θ is 40°. Let the positive direction of an x axis be up the incline. In unit-vector notation, what is the acceleration of A if A is initially (a) at rest, (b) moving up the incline, and (c) moving down the incline?

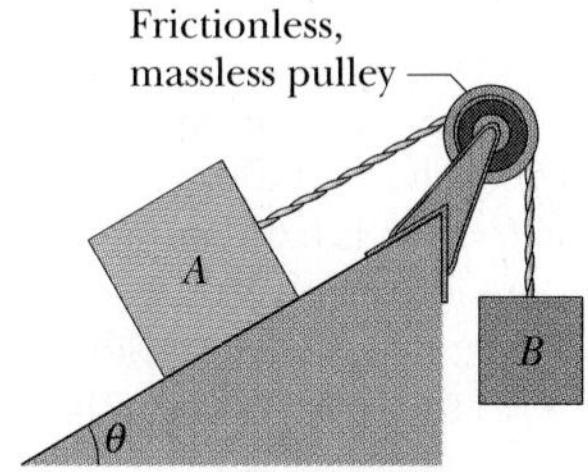

FIG. 6-36 Problems 29 and 30.

••30 In Fig. 6-36, two blocks are connected over a pulley. The mass of block A is 10 kg, and the coefficient of kinetic friction between A and the incline is 0.20. Angle θ of the incline is 30°. Block A slides down the incline at constant speed. What is the mass of block B?

••31 In Fig. 6-37, blocks A and B have weights of 44 N and 22 N, respectively. (a) Determine the minimum weight of block C to keep A from sliding if μ_s between A and the table is 0.20. (b) Block C suddenly is lifted off A. What is the acceleration of block A if μ_k between A and the table is 0.15?

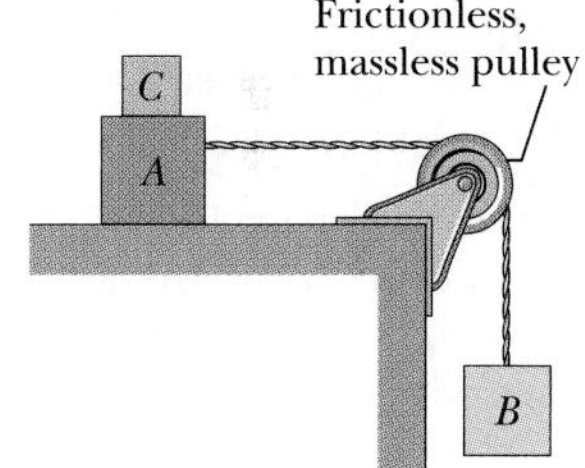

FIG. 6-37 Problem 31.

••32 A toy chest and its contents have a combined weight of 180 N. The coefficient of static friction between toy chest and floor is 0.42. The child in Fig. 6-38 attempts to move the chest across the floor by pulling on an attached rope. (a) If θ is 42°, what is the magnitude of the force $\vec{F}$ that the child must exert on the rope to put the chest on the verge of moving? (b) Write an expression for the magnitude F required to put the chest on the verge of moving as a function of the angle θ. Determine (c) the value of θ for which F is a minimum and (d) that minimum magnitude.

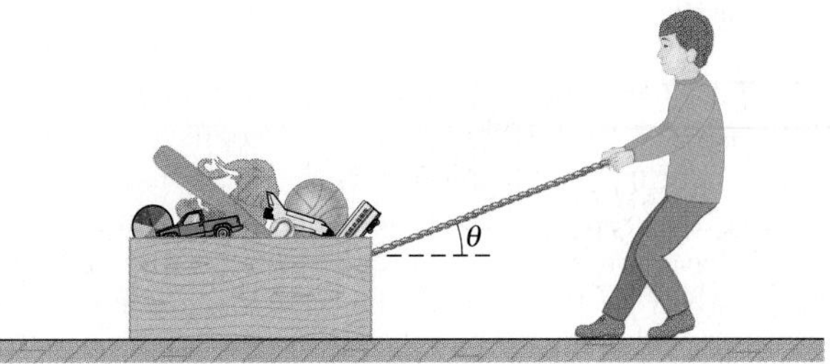

FIG. 6-38 Problem 32.

••33 The two blocks ($m = 16$ kg and $M = 88$ kg) in Fig. 6-39 are not attached to each other. The coefficient of static friction between the blocks is $\mu_s = 0.38$, but the surface beneath the larger block is frictionless.

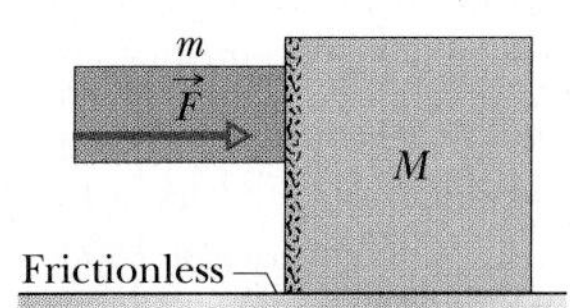

FIG. 6-39 Problem 33.

What is the minimum magnitude of the horizontal force $\vec{F}$ required to keep the smaller block from slipping down the larger block? ILW

•••34 In Fig. 6-40, a slab of mass $m_1 = 40$ kg rests on a frictionless floor, and a block of mass $m_2 = 10$ kg rests on top of the slab. Between block and slab, the coefficient of static friction is 0.60, and the coefficient of kinetic friction is 0.40. The block is pulled by a horizontal force $\vec{F}$ of magnitude 100 N. In unit-vector notation, what are the resulting accelerations of (a) the block and (b) the slab? GO

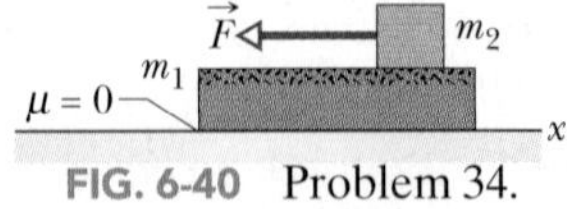

FIG. 6-40 Problem 34.

•••35 A 1000 kg boat is traveling at 90 km/h when its engine is shut off. The magnitude of the frictional force $\vec{f}_k$ between boat and water is proportional to the speed v of the boat: $f_k = 70v$, where v is in meters per second and f_k is in newtons. Find the time required for the boat to slow to 45 km/h. SSM

sec. 6-4 The Drag Force and Terminal Speed

•36 The terminal speed of a sky diver is 160 km/h in the spread-eagle position and 310 km/h in the nosedive position. Assuming that the diver's drag coefficient C does not change from one position to the other, find the ratio of the effective cross-sectional area A in the slower position to that in the faster position.

••37 Calculate the ratio of the drag force on a jet flying at 1000 km/h at an altitude of 10 km to the drag force on a prop-driven transport flying at half that speed and altitude. The density of air is 0.38 kg/m^3 at 10 km and 0.67 kg/m^3 at 5.0 km. Assume that the airplanes have the same effective cross-sectional area and drag coefficient C.

••38 In downhill speed skiing a skier is retarded by both the air drag force on the body and the kinetic frictional force on the skis. (a) Suppose the slope angle is $\theta = 40.0°$, the snow is dry snow with a coefficient of kinetic friction $\mu_k = 0.0400$, the mass of the skier and equipment is $m = 85.0$ kg, the cross-sectional area of the (tucked) skier is $A = 1.30$ m^2, the drag coefficient is $C = 0.150$, and the air density is 1.20 kg/m^3. (a) What is the terminal speed? (b) If a skier can vary C by a slight amount dC by adjusting, say, the hand positions, what is the corresponding variation in the terminal speed?

••39 *Continuation of Problem 2.* Now assume that Eq. 6-14 gives the magnitude of the air drag force on the typical 20 kg stone, which presents to the wind a vertical cross-sectional area of 0.040 m^2 and has a drag coefficient C of 0.80. Take the air density to be 1.21 kg/m^3, and the coefficient of kinetic friction to be 0.80. (a) In kilometers per hour, what wind speed V along the ground is needed to maintain the stone's motion once it has started moving? Because winds along the ground are retarded by the ground, the wind speeds reported for storms are often measured at a height of 10 m. Assume wind speeds are 2.00 times those along the ground. (b) For your answer to (a), what wind speed would be reported for the storm? (c) Is that value reasonable for a high-speed wind in a storm? (Story continues with Problem 61.)

••40 Assume Eq. 6-14 gives the drag force on a pilot plus ejection seat just after they are ejected from a plane traveling horizontally at 1300 km/h. Assume also that the mass of the seat is equal to the mass of the pilot and that the drag coefficient is that of a sky diver. Making a reasonable guess of the pilot's mass and using the appropriate v_t value from Table 6-1, estimate the magnitudes of (a) the drag force on the *pilot + seat* and (b) their horizontal deceleration (in terms of g), both just after ejection. (The result of (a) should indicate an engineering requirement: The seat must include a protective barrier to deflect the initial wind blast away from the pilot's head.)

sec. 6-5 Uniform Circular Motion

•41 What is the smallest radius of an unbanked (flat) track around which a bicyclist can travel if her speed is 29 km/h and the μ_s between tires and track is 0.32? ILW

•42 During an Olympic bobsled run, the Jamaican team makes a turn of radius 7.6 m at a speed of 96.6 km/h. What is their acceleration in terms of g?

•43 A cat dozes on a stationary merry-go-round, at a radius of 5.4 m from the center of the ride. Then the operator turns on the ride and brings it up to its proper turning rate of one complete rotation every 6.0 s. What is the least coefficient of static friction between the cat and the merry-go-round that will allow the cat to stay in place, without sliding?

•44 Suppose the coefficient of static friction between the road and the tires on a car is 0.60 and the car has no negative lift. What speed will put the car on the verge of sliding as it rounds a level curve of 30.5 m radius?

••45 A circular-motion addict of mass 80 kg rides a Ferris wheel around in a vertical circle of radius 10 m at a constant speed of 6.1 m/s. (a) What is the period of the motion? What is the magnitude of the normal force on the addict from the seat when both go through (b) the highest point of the circular path and (c) the lowest point?

••46 A roller-coaster car has a mass of 1200 kg when fully loaded with passengers. As the car passes over the top of a circular hill of radius 18 m, its speed is not changing. At the top of the hill, what are the (a) magnitude F_N and (b) direction (up or down) of the normal force on the car from the track if the car's speed is $v = 11$ m/s? What are (c) F_N and (d) the direction if $v = 14$ m/s?

••47 In Fig. 6-41, a car is driven at constant speed over a circular hill and then into a circular valley with the same radius. At the top of the hill, the normal force on the driver from the car seat is 0. The driver's mass is 70.0 kg. What is the magnitude of the normal force on the driver from the seat when the car passes through the bottom of the valley?

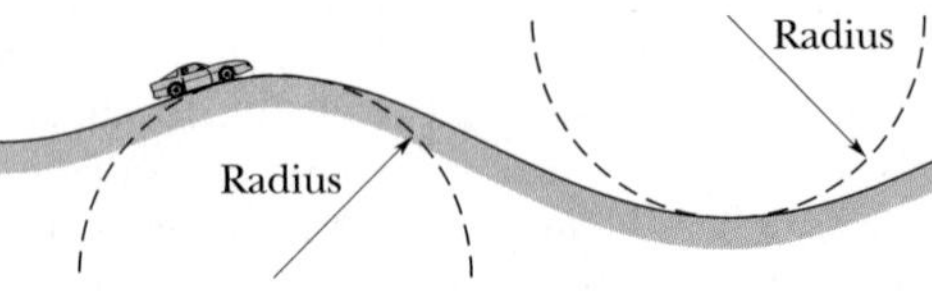

FIG. 6-41 Problem 47.

••48 A police officer in hot pursuit drives her car through a circular turn of radius 300 m with a constant speed of 80.0 km/h. Her mass is 55.0 kg. What are (a) the magnitude and (b) the angle (relative to vertical) of the *net* force of the officer on the car seat? (*Hint:* Consider both horizontal and vertical forces.)

••49 A student of weight 667 N rides a steadily rotating Ferris wheel (the student sits upright). At the highest point, the magnitude of the normal force $\vec{F}_N$ on the student from the seat is 556 N. (a) Does the student feel "light" or "heavy" there? (b) What is the magnitude of $\vec{F}_N$ at the lowest point? If the wheel's speed is doubled, what is the magnitude F_N at the (c) highest and (d) lowest point? SSM ILW

••50 An amusement park ride consists of a car moving in a vertical circle on the end of a rigid boom of negligible mass. The combined weight of the car and riders is 5.0 kN, and the circle's radius is 10 m. At the top of the circle, what are the (a) magnitude F_B and (b) direction (up or down) of the force on the car from the boom if the car's speed is $v = 5.0$ m/s? What are (c) F_B and (d) the direction if $v = 12$ m/s?

••51 An old streetcar rounds a flat corner of radius 9.1 m, at 16 km/h. What angle with the vertical will be made by the loosely hanging hand straps?

••52 In designing circular rides for amusement parks, mechanical engineers must consider how small variations in certain parameters can alter the net force on a passenger. Consider a passenger of mass m riding around a horizontal circle of radius r at speed v. What is the variation dF in the net force magnitude for (a) a variation dr in the radius with v held constant, (b) a variation dv in the speed with r held constant, and (c) a variation dT in the period with r held constant?

••53 An airplane is flying in a horizontal circle at a speed of 480 km/h (Fig. 6-42). If its wings are tilted at angle $\theta = 40°$ to the horizontal, what is the radius of the circle in which the plane is flying? Assume that the required force is provided entirely by an "aerodynamic lift" that is perpendicular to the wing surface. SSM WWW

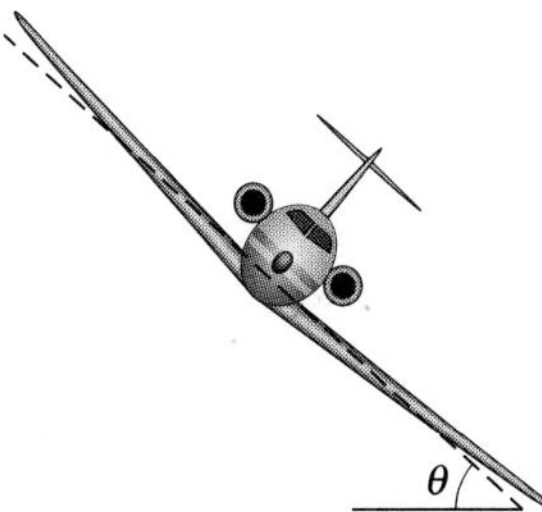

FIG. 6-42 Problem 53.

••54 An 85.0 kg passenger is made to move along a circular path of radius $r = 3.50$ m in uniform circular motion. (a) Figure 6-43a is a plot of the required magnitude F of the net centripetal force for a range of possible values of the passenger's speed v. What is the plot's slope at $v = 8.30$ m/s? (b) Figure 6-43b is a plot of F for a range of possible values of T, the period of the motion. What is the plot's slope at $T = 2.50$ s?

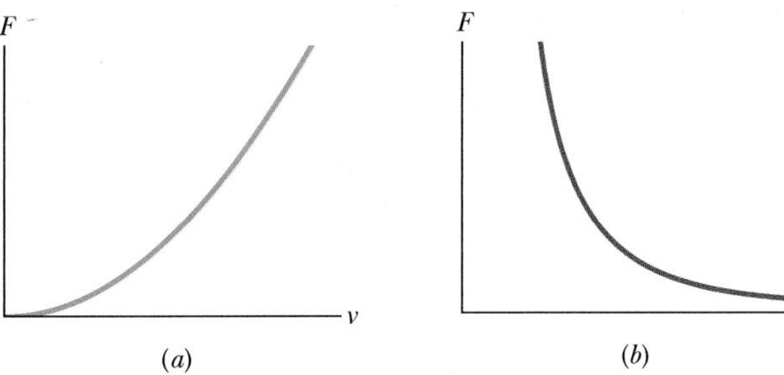

FIG. 6-43 Problem 54.

••55 A puck of mass $m = 1.50$ kg slides in a circle of radius $r = 20.0$ cm on a frictionless table while attached to a hanging cylinder of mass $M = 2.50$ kg by a cord through a hole in the table (Fig. 6-44). What speed keeps the cylinder at rest? GO

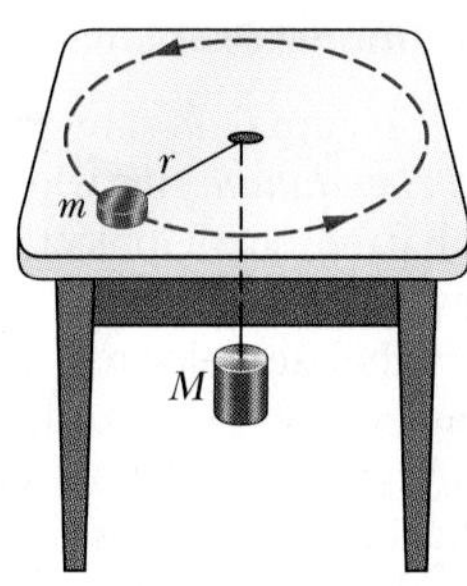

FIG. 6-44 Problem 55.

••56 *Brake or turn?* Figure 6-45 depicts an overhead view of a car's path as the car travels toward a wall. Assume that the driver begins to brake the car when the distance to the wall is $d = 107$ m, and take the car's mass as $m = 1400$ kg, its initial speed as $v_0 = 35$ m/s, and the coefficient of static friction as $\mu_s = 0.50$. Assume that the car's weight is distributed evenly on the four wheels, even during braking. (a) What magnitude of static friction is needed (between tires and road) to stop the car just as it reaches the wall? (b) What is the maximum possible static friction $f_{s,\max}$? (c) If the coefficient of kinetic friction between the (sliding) tires and the road is $\mu_k = 0.40$, at what speed will the car hit the wall? To avoid the crash, a driver could elect to turn the car so that it just barely misses the wall, as shown in the figure. (d) What magnitude of frictional force would be required to keep the car in a circular path of radius d and at the given speed v_0? (e) Is the required force less than $f_{s,\max}$ so that a circular path is possible?

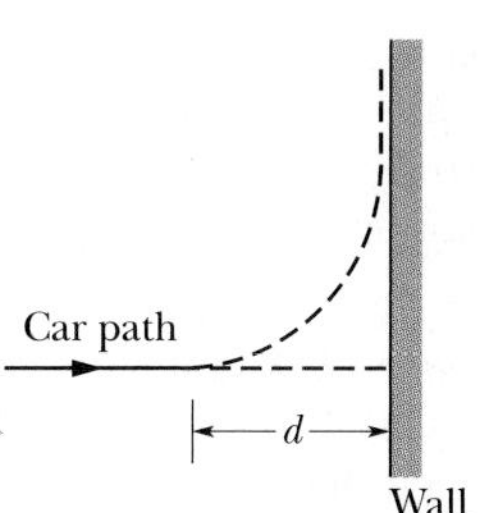

FIG. 6-45 Problem 56.

••57 A bolt is threaded onto one end of a thin horizontal rod, and the rod is then rotated horizontally about its other end. An engineer monitors the motion by flashing a strobe lamp onto the rod and bolt, adjusting the strobe rate until the bolt appears to be in the same eight places during each full rotation of the rod (Fig. 6-46). The strobe rate is 2000 flashes per second; the bolt has mass 30 g and is at radius 3.5 cm. What is the magnitude of the force on the bolt from the rod?

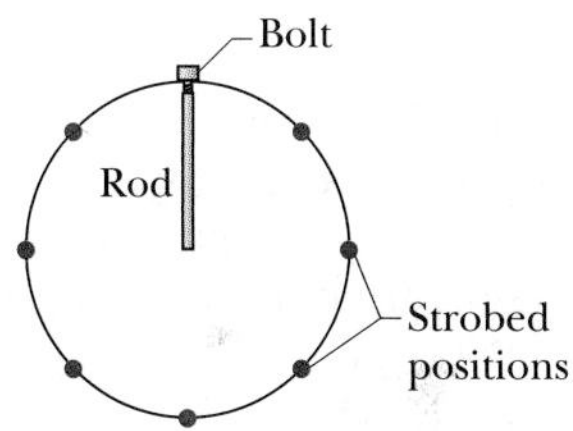

FIG. 6-46 Problem 57.

••58 A banked circular highway curve is designed for traffic moving at 60 km/h. The radius of the curve is 200 m. Traffic is moving along the highway at 40 km/h on a rainy day. What is the minimum coefficient of friction between tires and road that will allow cars to take the turn without sliding off the road? (Assume the cars do not have negative lift.)

•••59 In Fig. 6-47, a 1.34 kg ball is connected by means of two massless strings, each of length $L = 1.70$ m, to a vertical, rotating rod. The strings are tied to the rod with separation $d = 1.70$ m and are taut. The tension in the upper string is 35 N. What are the (a) tension in the lower string, (b) magnitude of the net force $\vec{F}_{net}$ on the ball, and (c) speed of the ball? (d) What is the direction of $\vec{F}_{net}$? SSM ILW

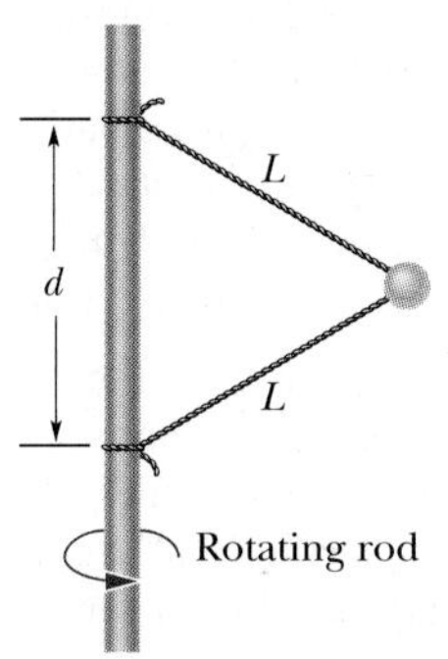

FIG. 6-47 Problem 59.

Additional Problems

60 Figure 6-48 shows a *conical pendulum,* in which the bob (the small object at the lower end of the cord) moves in a horizontal circle at constant speed. (The cord sweeps out a cone as the bob rotates.) The bob has a mass of 0.040 kg, the string has length L = 0.90 m and negligible mass, and the bob follows a circular path of circumference 0.94 m. What are (a) the tension in the string and (b) the period of the motion?

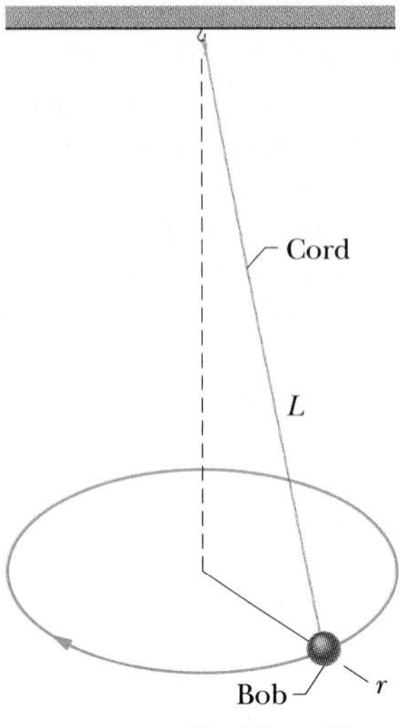

FIG. 6-48 Problem 60.

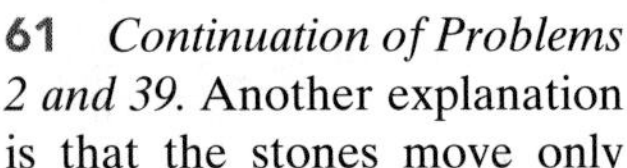

61 *Continuation of Problems 2 and 39.* Another explanation is that the stones move only when the water dumped on the playa during a storm freezes into a large, thin sheet of ice. The stones are trapped in place in the ice. Then, as air flows across the ice during a wind, the air-drag forces on the ice and stones move them both, with the stones gouging out the trails. The magnitude of the air-drag force on this horizontal "ice sail" is given by $D_{ice} = 4C_{ice}\rho A_{ice}v^2$, where C_{ice} is the drag coefficient (2.0×10^{-3}), ρ is the air density (1.21 kg/m^3), A_{ice} is the horizontal area of the ice, and v is the wind speed along the ice.

Assume the following: The ice sheet measures 400 m by 500 m by 4.0 mm and has a coefficient of kinetic friction of 0.10 with the ground and a density of 917 kg/m^3. Also assume that 100 stones identical to the one in Problem 2 are trapped in the ice. To maintain the motion of the sheet, what are the required wind speeds (a) near the sheet and (b) at a height of 10 m? (c) Are these reasonable values for high-speed winds in a storm?

62 *Engineering a highway curve.* If a car goes through a curve too fast, the car tends to slide out of the curve. For a banked curve with friction, a frictional force acts on a fast car to oppose the tendency to slide out of the curve; the force is directed down the bank (in the direction water would drain). Consider a circular curve of radius R = 200 m and bank angle θ, where the coefficient of static friction between tires and pavement is μ_s. A car (without negative lift) is driven around the curve as shown in Fig. 6-13. (a) Find an expression for the car speed v_{max} that puts the car on the verge of sliding out. (b) On the same graph, plot v_{max} versus angle θ for the range 0° to 50°, first for $\mu_s = 0.60$ (dry pavement) and then for $\mu_s = 0.050$ (wet or icy pavement). In kilometers per hour, evaluate v_{max} for a bank angle of $\theta = 10°$ and for (c) $\mu_s = 0.60$ and (d) $\mu_s = 0.050$. (Now you can see why accidents occur in highway curves when icy conditions are not obvious to drivers, who tend to drive at normal speeds.)

63 In Fig. 6-49, the coefficient of kinetic friction between the block and inclined plane is 0.20, and angle θ is 60°. What are the (a) magnitude a and (b) direction (up or down the plane) of the block's acceleration if the block is sliding down the plane? What are (c) a and (d) the direction if the block is sent sliding up the plane?

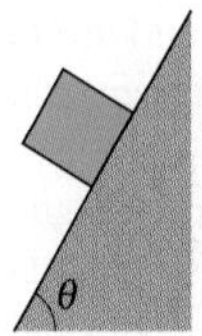

FIG. 6-49 Problem 63.

64 In Fig. 6-50, block 1 of mass m_1 = 2.0 kg and block 2 of mass m_2 = 3.0 kg are connected by a string of negligible mass and are initially held in place. Block 2 is on a frictionless surface tilted at $\theta = 30°$. The coefficient of kinetic friction between block 1 and the horizontal surface is 0.25. The pulley has negligible mass and friction. Once they are released, the blocks move. What then is the tension in the string?

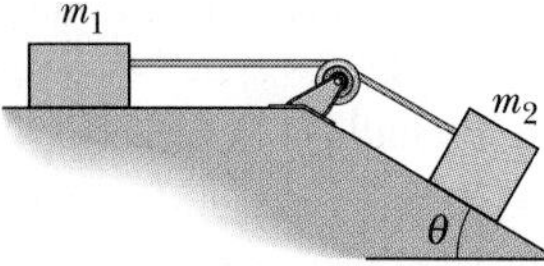

FIG. 6-50 Problem 64.

65 A block of mass m_t = 4.0 kg is put on top of a block of mass m_b = 5.0 kg. To cause the top block to slip on the bottom one while the bottom one is held fixed, a horizontal force of at least 12 N must be applied to the top block. The assembly of blocks is now placed on a horizontal, frictionless table (Fig. 6-51). Find the magnitudes of (a) the maximum horizontal force $\vec{F}$ that can be applied to the lower block so that the blocks will move together and (b) the resulting acceleration of the blocks. SSM

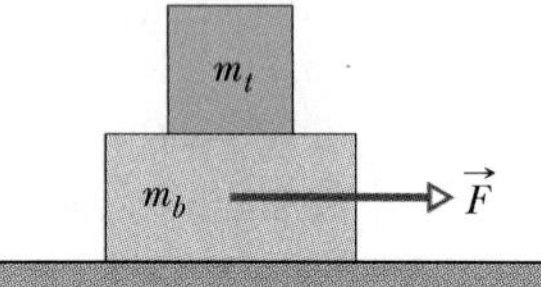

FIG. 6-51 Problem 65.

66 A box of canned goods slides down a ramp from street level into the basement of a grocery store with acceleration 0.75 m/s^2 directed down the ramp. The ramp makes an angle of 40° with the horizontal. What is the coefficient of kinetic friction between the box and the ramp?

67 An 8.00 kg block of steel is at rest on a horizontal table. The coefficient of static friction between the block and the table is 0.450. A force is to be applied to the block. To three significant figures, what is the magnitude of that applied force if it puts the block on the verge of sliding when the force is directed (a) horizontally, (b) upward at 60.0° from the horizontal, and (c) downward at 60.0° from the horizontal?

68 In Fig. 6-52, a box of ant aunts (total mass m_1 = 1.65 kg) and a box of ant uncles (total mass m_2 = 3.30 kg) slide down an inclined plane while attached by a massless rod parallel to the plane. The angle of incline is $\theta = 30.0°$. The coefficient of kinetic friction between the aunt box and the incline is $\mu_1 = 0.226$; that between the uncle box and the incline is $\mu_2 = 0.113$. Compute (a) the tension in the rod and (b) the magnitude of the common acceleration of the two boxes. (c) How would the answers to (a) and (b) change if the uncles trailed the aunts?

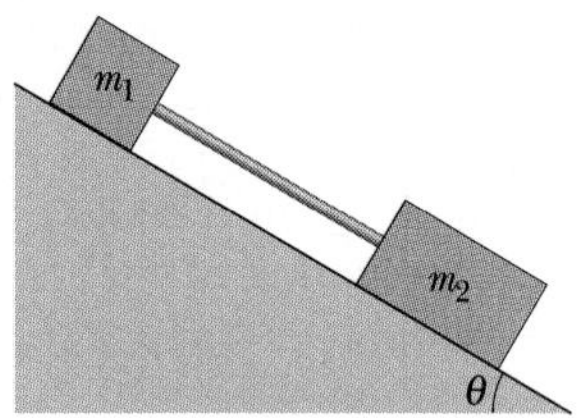

FIG. 6-52 Problem 68.

69 In Fig. 6-53, a crate slides down an inclined right-angled

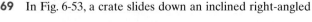

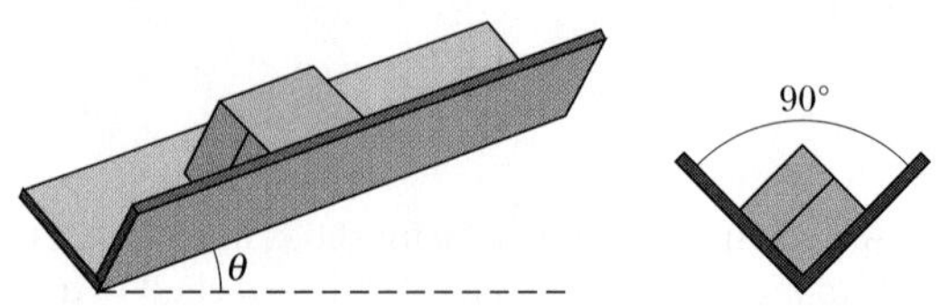

FIG. 6-53 Problem 69.

trough. The coefficient of kinetic friction between the crate and the trough is μ_k. What is the acceleration of the crate in terms of μ_k, θ, and g?

70 A student wants to determine the coefficients of static friction and kinetic friction between a box and a plank. She places the box on the plank and gradually raises one end of the plank. When the angle of inclination with the horizontal reaches 30°, the box starts to slip, and it then slides 2.5 m down the plank in 4.0 s at constant acceleration. What are (a) the coefficient of static friction and (b) the coefficient of kinetic friction between the box and the plank?

71 A locomotive accelerates a 25-car train along a level track. Every car has a mass of 5.0×10^4 kg and is subject to a frictional force $f = 250v$, where the speed v is in meters per second and the force f is in newtons. At the instant when the speed of the train is 30 km/h, the magnitude of its acceleration is 0.20 m/s^2. (a) What is the tension in the coupling between the first car and the locomotive? (b) If this tension is equal to the maximum force the locomotive can exert on the train, what is the steepest grade up which the locomotive can pull the train at 30 km/h?

72 A house is built on the top of a hill with a nearby slope at angle $\theta = 45°$ (Fig. 6-54). An engineering study indicates that the slope angle should be reduced because the top layers of soil along the slope might slip past the lower layers. If the coefficient of static friction between two such layers is 0.5, what is the least angle ϕ through which the present slope should be reduced to prevent slippage?

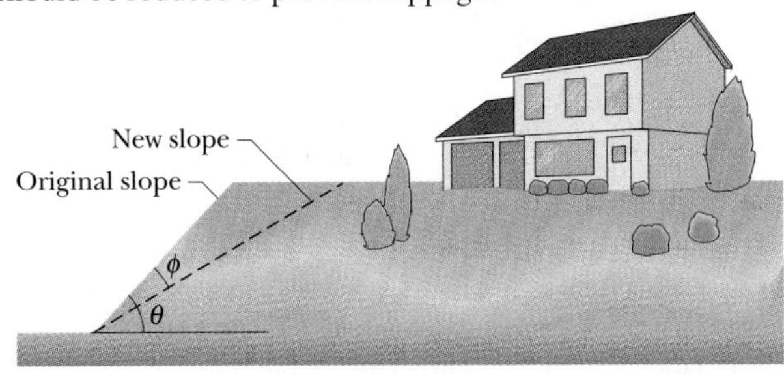

FIG. 6-54 Problem 72.

73 What is the terminal speed of a 6.00 kg spherical ball that has a radius of 3.00 cm and a drag coefficient of 1.60? The density of the air through which the ball falls is 1.20 kg/m^3.

74 A high-speed railway car goes around a flat, horizontal circle of radius 470 m at a constant speed. The magnitudes of the horizontal and vertical components of the force of the car on a 51.0 kg passenger are 210 N and 500 N, respectively. (a) What is the magnitude of the net force (of *all* the forces) on the passenger? (b) What is the speed of the car?

75 An 11 kg block of steel is at rest on a horizontal table. The coefficient of static friction between block and table is 0.52. (a) What is the magnitude of the horizontal force that will put the block on the verge of moving? (b) What is the magnitude of a force acting upward 60° from the horizontal that will put the block on the verge of moving? (c) If the force acts downward at 60° from the horizontal, how large can its magnitude be without causing the block to move?

76 Calculate the magnitude of the drag force on a missile 53 cm in diameter cruising at 250 m/s at low altitude, where the density of air is 1.2 kg/m^3. Assume $C = 0.75$.

77 A bicyclist travels in a circle of radius 25.0 m at a constant speed of 9.00 m/s. The bicycle–rider mass is 85.0 kg. Calculate the magnitudes of (a) the force of friction on the bicycle from the road and (b) the *net* force on the bicycle from the road. SSM

78 A 110 g hockey puck sent sliding over ice is stopped in 15 m by the frictional force on it from the ice. (a) If its initial speed is 6.0 m/s, what is the magnitude of the frictional force? (b) What is the coefficient of friction between the puck and the ice?

79 In Fig. 6-55, a 49 kg rock climber is climbing a "chimney." The coefficient of static friction between her shoes and the rock is 1.2; between her back and the rock is 0.80. She has reduced her push against the rock until her back and her shoes are on the verge of slipping. (a) Draw a free-body diagram of her. (b) What is the magnitude of her push against the rock? (c) What fraction of her weight is supported by the frictional force on her shoes? 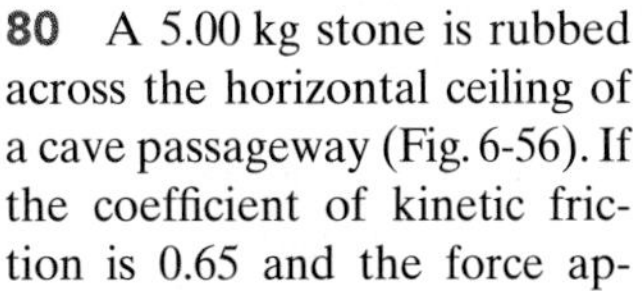

FIG. 6-55 Problem 79.

80 A 5.00 kg stone is rubbed across the horizontal ceiling of a cave passageway (Fig. 6-56). If the coefficient of kinetic friction is 0.65 and the force applied to the stone is angled at $\theta = 70.0°$, what must the magnitude of the force be for the stone to move at constant velocity?

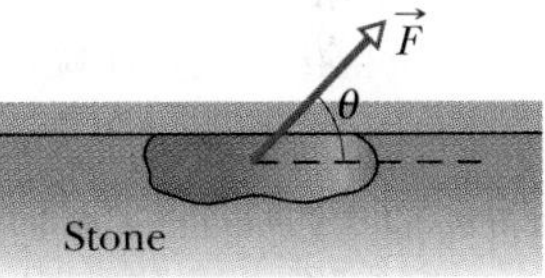

FIG. 6-56 Problem 80.

81 Block A in Fig. 6-57 has mass $m_A = 4.0$ kg, and block B has mass $m_B = 2.0$ kg. The coefficient of kinetic friction between block B and the horizontal plane is $\mu_k = 0.50$. The inclined plane is frictionless and at angle $\theta = 30°$. The pulley serves only to change the direction of the cord connecting the blocks. The cord has negligible mass. Find (a) the tension in the cord and (b) the magnitude of the acceleration of the blocks. SSM

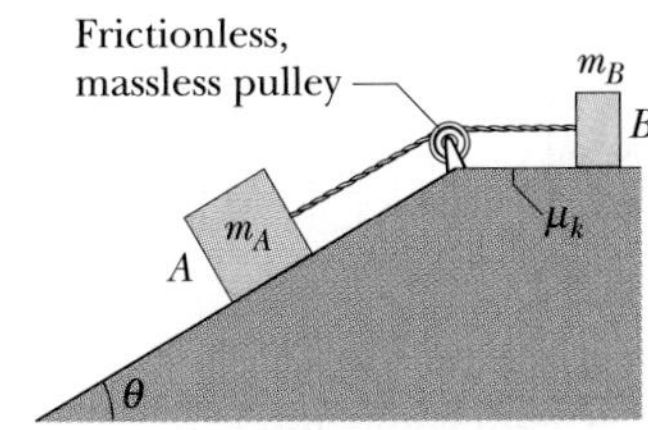

FIG. 6-57 Problem 81.

82 A ski that is placed on snow will stick to the snow. However, when the ski is moved along the snow, the rubbing warms and partially melts the snow, reducing the coefficient of kinetic friction and promoting sliding. Waxing the ski makes it water repellent and reduces friction with the resulting layer of water. A magazine reports that a new type of plastic ski is especially water repellent and that, on a gentle 200 m slope in the Alps, a skier reduced his top-to-bottom time from 61 s with standard skis to 42 s with the new skis. Determine the magnitude of his average acceleration with (a) the standard skis and (b) the new skis. Assuming a 3.0° slope, compute the coefficient of kinetic friction for (c) the standard skis and (d) the new skis.

83 Playing near a road construction site, a child falls over a barrier and down onto a dirt slope that is angled downward at 35° to the horizontal. As the child slides *down* the slope, he has an acceleration that has a magnitude of 0.50 m/s² and that is directed *up* the slope. What is the coefficient of kinetic friction between the child and the slope?

84 In Fig. 6-58, a stuntman drives a car (without negative lift) over the top of a hill, the cross section of which can be approximated by a circle of radius $R = 250$ m. What is the greatest speed at which he can drive without the car leaving the road at the top of the hill?

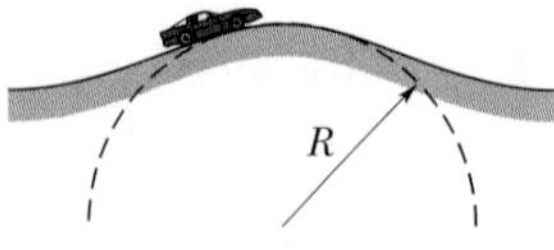

FIG. 6-58 Problem 84.

85 A car weighing 10.7 kN and traveling at 13.4 m/s without negative lift attempts to round an unbanked curve with a radius of 61.0 m. (a) What magnitude of the frictional force on the tires is required to keep the car on its circular path? (b) If the coefficient of static friction between the tires and the road is 0.350, is the attempt at taking the curve successful? SSM

86 A 100 N force, directed at an angle θ above a horizontal floor, is applied to a 25.0 kg chair sitting on the floor. If $\theta = 0°$, what are (a) the horizontal component F_h of the applied force and (b) the magnitude F_N of the normal force of the floor on the chair? If $\theta = 30.0°$, what are (c) F_h and (d) F_N? If $\theta = 60.0°$, what are (e) F_h and (f) F_N? Now assume that the coefficient of static friction between chair and floor is 0.420. Does the chair slide or remain at rest if θ is (g) 0°, (h) 30.0°, and (i) 60.0°?

87 A student, crazed by final exams, uses a force $\vec{P}$ of magnitude 80 N and angle $\theta = 70°$ to push a 5.0 kg block across the ceiling of his room (Fig. 6-59). If the coefficient of kinetic friction between the block and the ceiling is 0.40, what is the magnitude of the block's acceleration?

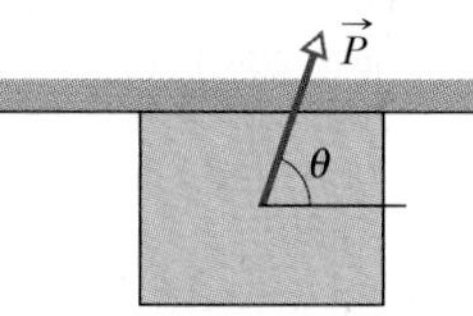

FIG. 6-59 Problem 87.

88 A certain string can withstand a maximum tension of 40 N without breaking. A child ties a 0.37 kg stone to one end and, holding the other end, whirls the stone in a vertical circle of radius 0.91 m, slowly increasing the speed until the string breaks. (a) Where is the stone on its path when the string breaks? (b) What is the speed of the stone as the string breaks?

89 You must push a crate across a floor to a docking bay. The crate weighs 165 N. The coefficient of static friction between crate and floor is 0.510, and the coefficient of kinetic friction is 0.32. Your force on the crate is directed horizontally. (a) What magnitude of your push puts the crate on the verge of sliding? (b) With what magnitude must you then push to keep the crate moving at a constant velocity? (c) If, instead, you then push with the same magnitude as the answer to (a), what is the magnitude of the crate's acceleration?

90 A child weighing 140 N sits at rest at the top of a playground slide that makes an angle of 25° with the horizontal. The child keeps from sliding by holding onto the sides of the slide. After letting go of the sides, the child has a constant acceleration of 0.86 m/s² (down the slide, of course). (a) What is the coefficient of kinetic friction between the child and the slide? (b) What maximum and minimum values for the coefficient of static friction between the child and the slide are consistent with the information given here?

91 A filing cabinet weighing 556 N rests on the floor. The coefficient of static friction between it and the floor is 0.68, and the coefficient of kinetic friction is 0.56. In four different attempts to move it, it is pushed with horizontal forces of magnitudes (a) 222 N, (b) 334 N, (c) 445 N, and (d) 556 N. For each attempt, calculate the magnitude of the frictional force on it from the floor. (The cabinet is initially at rest.) (e) In which of the attempts does the cabinet move? SSM

92 A sling-thrower puts a stone (0.250 kg) in the sling's pouch (0.010 kg) and then begins to make the stone and pouch move in a vertical circle of radius 0.650 m. The cord between the pouch and the person's hand has negligible mass and will break when the tension in the cord is 33.0 N or more. Suppose the sling-thrower could gradually increase the speed of the stone. (a) Will the breaking occur at the lowest point of the circle or at the highest point? (b) At what speed of the stone will that breaking occur?

93 A four-person bobsled (total mass = 630 kg) comes down a straightaway at the start of a bobsled run. The straightaway is 80.0 m long and is inclined at a constant angle of 10.2° with the horizontal. Assume that the combined effects of friction and air drag produce on the bobsled a constant force of 62.0 N that acts parallel to the incline and up the incline. Answer the following questions to three significant digits. (a) If the speed of the bobsled at the start of the run is 6.20 m/s, how long does the bobsled take to come down the straightaway? (b) Suppose the crew is able to reduce the effects of friction and air drag to 42.0 N. For the same initial velocity, how long does the bobsled now take to come down the straightaway?

94 In Fig. 6-60, force $\vec{F}$ is applied to a crate of mass m on a floor where the coefficient of static friction between crate and floor is μ_s. Angle θ is initially 0° but is gradually increased so that the force vector rotates clockwise in the figure. During the rotation, the magnitude F of the force is continuously adjusted so that the crate is always on the verge of sliding. For $\mu_s = 0.70$, (a) plot the ratio F/mg versus θ and (b) determine the angle θ_{inf} at which the ratio approaches an infinite value. (c) Does lubricating the floor increase or decrease θ_{inf}, or is the value unchanged? (d) What is θ_{inf} for $\mu_s = 0.60$?

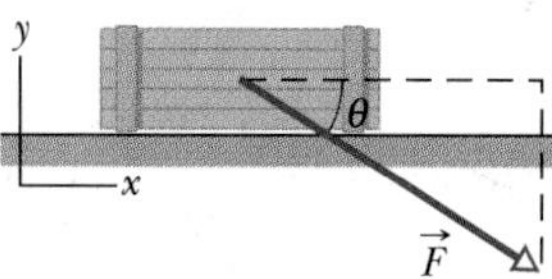

FIG. 6-60 Problem 94.

95 In the early afternoon, a car is parked on a street that runs down a steep hill, at an angle of 35.0° relative to the horizontal. Just then the coefficient of static friction between the tires and the street surface is 0.725. Later, after nightfall, a sleet storm hits the area, and the coefficient decreases due to both the ice and a chemical change in the road surface because of the temperature decrease. By what percentage must the coefficient decrease if the car is to be in danger of sliding down the street?

96 In Fig. 6-61, block 1 of mass $m_1 = 2.0$ kg and block 2 of mass $m_2 = 1.0$ kg are connected by a string of negligible mass. Block 2 is pushed by force $\vec{F}$ of magnitude 20 N and angle $\theta = 35°$. The coefficient of kinetic friction between

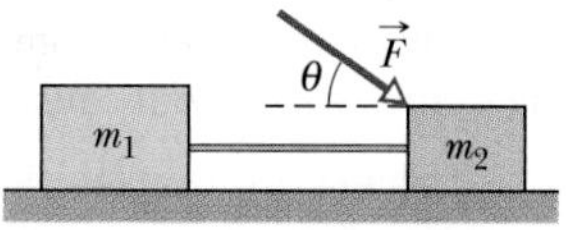

FIG. 6-61 Problem 96.

each block and the horizontal surface is 0.20. What is the tension in the string?

97 In Fig. 6-62 a fastidious worker pushes directly along the handle of a mop with a force $\vec{F}$. The handle is at an angle θ with the vertical, and μ_s and μ_k are the coefficients of static and kinetic friction between the head of the mop and the floor. Ignore the mass of the handle and assume that all the mop's mass m is in its head. (a) If the mop head moves along the floor with a constant velocity, then what is F? (b) Show that if θ is less than a certain value θ_0, then $\vec{F}$ (still directed along the handle) is unable to move the mop head. Find θ_0.

FIG. 6-62 Problem 97.

98 A circular curve of highway is designed for traffic moving at 60 km/h. Assume the traffic consists of cars without negative lift. (a) If the radius of the curve is 150 m, what is the correct angle of banking of the road? (b) If the curve were not banked, what would be the minimum coefficient of friction between tires and road that would keep traffic from skidding out of the turn when traveling at 60 km/h?

99 A block slides with constant velocity down an inclined plane that has slope angle θ. The block is then projected up the same plane with an initial speed v_0. (a) How far up the plane will it move before coming to rest? (b) After the block comes to rest, will it slide down the plane again? Give an argument to back your answer. SSM

100 In Fig. 6-63, a block weighing 22 N is held at rest against a vertical wall by a horizontal force $\vec{F}$ of magnitude 60 N. The coefficient of static friction between the wall and the block is 0.55, and the coefficient of kinetic friction between them is 0.38. In six experiments, a second force $\vec{P}$ is applied to the block and directed parallel to the wall with these magnitudes and directions: (a) 34 N, up, (b) 12 N, up, (c) 48 N, up, (d) 62 N, up, (e) 10 N, down, and (f) 18 N, down. In each experiment, what is the magnitude of the frictional force on the block? In which does the block move (g) up the wall and (h) down the wall? (i) In which is the frictional force directed down the wall?

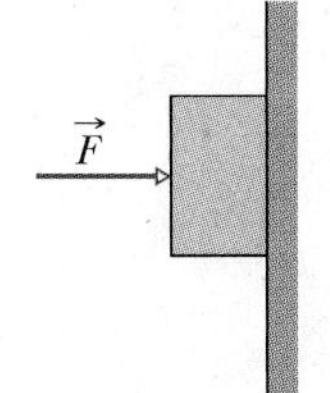

FIG. 6-63 Problem 100.

101 When a small 2.0 g coin is placed at a radius of 5.0 cm on a horizontal turntable that makes three full revolutions in 3.14 s, the coin does not slip. What are (a) the coin's speed, the (b) magnitude and (c) direction (radially inward or outward) of the coin's acceleration, and the (d) magnitude and (e) direction (inward or outward) of the frictional force on the coin? The coin is on the verge of slipping if it is placed at a radius of 10 cm. (f) What is the coefficient of static friction between coin and turntable?

102 A child places a picnic basket on the outer rim of a merry-go-round that has a radius of 4.6 m and revolves once every 30 s. (a) What is the speed of a point on that rim? (b) What is the lowest value of the coefficient of static friction between basket and merry-go-round that allows the basket to stay on the ride?

103 A 1.5 kg box is initially at rest on a horizontal surface when at $t = 0$ a horizontal force $\vec{F} = (1.8t)\hat{\mathrm{i}}$ N (with t in seconds) is applied to the box. The acceleration of the box as a function of time t is given by $\vec{a} = 0$ for $0 \le t \le 2.8$ s and $\vec{a} = (1.2t - 2.4)\hat{\mathrm{i}}$ m/s² for $t > 2.8$ s. (a) What is the coefficient of static friction between the box and the surface? (b) What is the coefficient of kinetic friction between the box and the surface?

104 A trunk with a weight of 220 N rests on the floor. The coefficient of static friction between the trunk and the floor is 0.41, and the coefficient of kinetic friction is 0.32. (a) What is the magnitude of the minimum horizontal force with which a person must push on the trunk to start it moving? (b) Once the trunk is moving, what magnitude of horizontal force must the person apply to keep it moving with constant velocity? (c) If the person continued to push with the force used to start the motion, what would be the magnitude of the trunk's acceleration?

105 A warehouse worker exerts a constant horizontal force of magnitude 85 N on a 40 kg box that is initially at rest on the horizontal floor of the warehouse. When the box has moved a distance of 1.4 m, its speed is 1.0 m/s. What is the coefficient of kinetic friction between the box and the floor? SSM

106 Imagine that the standard kilogram is placed on Earth's equator, where it moves in a circle of radius 6.40×10^6 m (Earth's radius) at a constant speed of 465 m/s due to Earth's rotation. (a) What is the magnitude of the centripetal force on the standard kilogram during the rotation? Imagine that the standard kilogram hangs from a spring balance at that location and assume that it would weigh exactly 9.80 N if Earth did not rotate. (b) What is the reading on the spring balance; that is, what is the magnitude of the force on the spring balance from the standard kilogram?

107 As a 40 N block slides down a plane that is inclined at 25° to the horizontal, its acceleration is 0.80 m/s², directed up the plane. What is the coefficient of kinetic friction between the block and the plane?

108 Luggage is transported from one location to another in an airport by a conveyor belt. At a certain location, the belt moves down an incline that makes an angle of 2.5° with the horizontal. Assume that with such a slight angle there is no slipping of the luggage. Determine the magnitude of the frictional force by the belt on a box weighing 69 N when the box is on the inclined portion of the belt and the belt speed is (a) 0 and constant, (b) 0.65 m/s and constant, (c) 0.65 m/s and increasing at a rate of 0.20 m/s², (d) 0.65 m/s and decreasing at a rate of 0.20 m/s², and (e) 0.65 m/s and increasing at a rate of 0.57 m/s². (f) For which of these five situations is the frictional force directed down the incline?

109 In Fig. 6-64, a 5.0 kg block is sent sliding up a plane inclined at $\theta = 37°$ while a horizontal force $\vec{F}$ of magnitude 50 N acts on it. The coefficient of kinetic friction between block and plane is 0.30. What are the (a) magnitude and (b) direction (up or down the plane) of the block's acceleration? The block's initial speed is 4.0 m/s. (c) How far up the plane does the block go? (d) When it reaches its highest point, does it remain at rest or slide back down the plane?

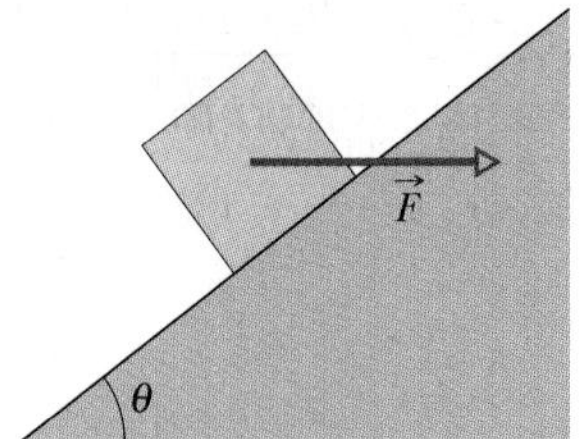

FIG. 6-64 Problem 109.

7 Kinetic Energy and Work

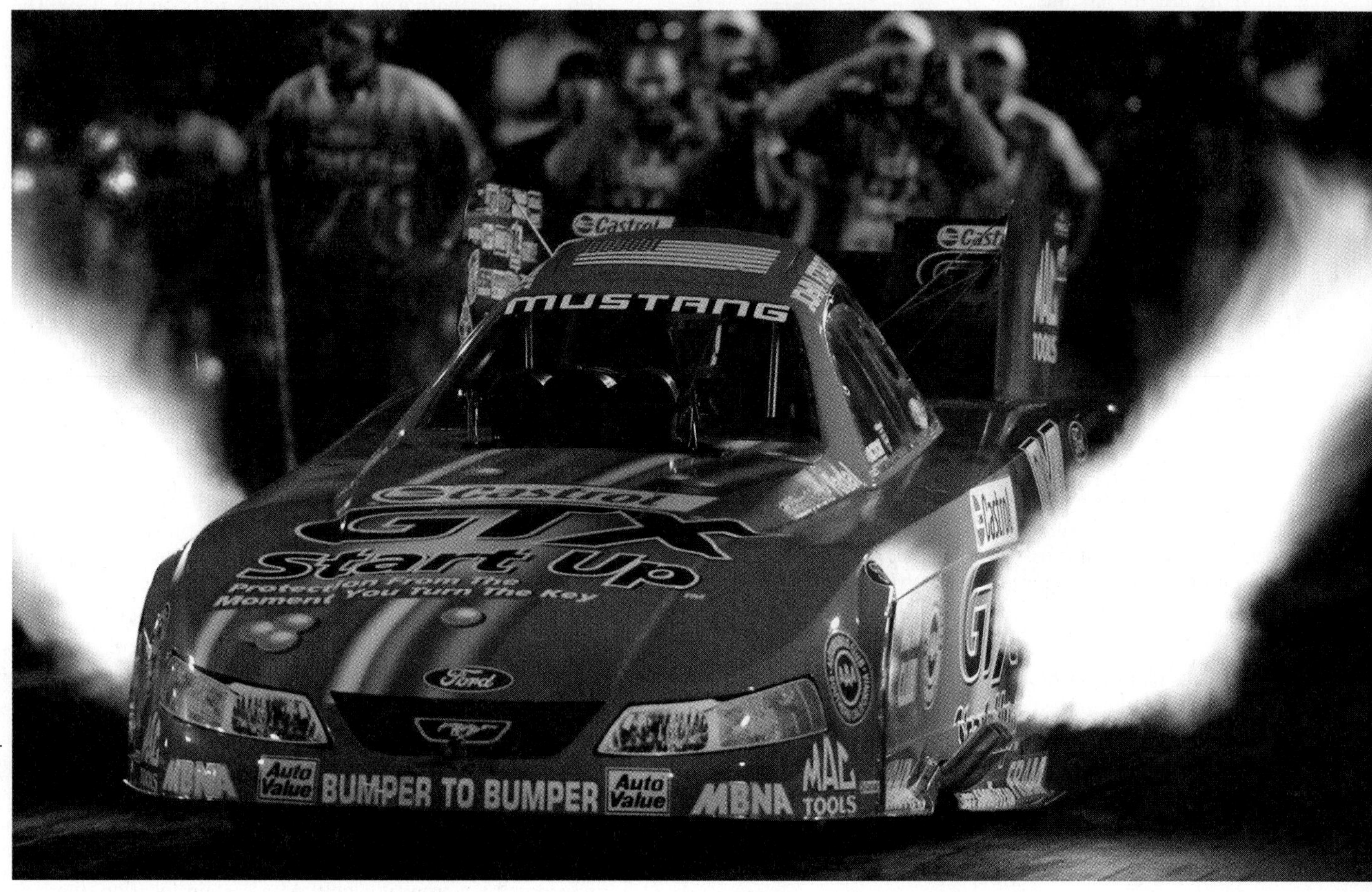

UPI Photo/Dilip Vishwanat/Landov LLC

The driver of a funny car prepares for a timed run along a quarter-mile track by spinning the wheels, to make the tires and track sticky so that traction is high. Then the driver waits at the starting line until the countdown on the Christmas tree lights *reaches green. The car's forward surge is so powerful that the car is effectively launched like a horizontal rocket. The science and engineering of funny cars is now so advanced that winning and losing is often determined by elapsed times differing by only 1 ms.*

What property of a car determines the winning time?

The answer is in this chapter.

7-1 WHAT IS PHYSICS?

One of the fundamental goals of physics is to investigate something that everyone talks about: energy. The topic is obviously important. Indeed, our civilization is based on acquiring and effectively using energy.

For example, everyone knows that any type of motion requires energy: Flying across the Pacific Ocean requires it. Lifting material to the top floor of an office building or to an orbiting space station requires it. Throwing a fastball requires it. We spend a tremendous amount of money to acquire and use energy. Wars have been started because of energy resources. Wars have been ended because of a sudden, overpowering use of energy by one side. Everyone knows many examples of energy and its use, but what does the term *energy* really mean?

7-2 | What Is Energy?

The term *energy* is so broad that a clear definition is difficult to write. Technically, energy is a scalar quantity associated with the state (or condition) of one or more objects. However, this definition is too vague to be of help to us now.

A looser definition might at least get us started. Energy is a number that we associate with a system of one or more objects. If a force changes one of the objects by, say, making it move, then the energy number changes. After countless experiments, scientists and engineers realized that if the scheme by which we assign energy numbers is planned carefully, the numbers can be used to predict the outcomes of experiments and, even more important, to build machines, such as flying machines. This success is based on a wonderful property of our universe: Energy can be transformed from one type to another and transferred from one object to another, but the total amount is always the same (energy is *conserved*). No exception to this *principle of energy conservation* has ever been found.

Think of the many types of energy as being numbers representing money in many types of bank accounts. Rules have been made about what such money numbers mean and how they can be changed. You can transfer money numbers from one account to another or from one system to another, perhaps electronically with nothing material actually moving. However, the total amount (the total of all the money numbers) can always be accounted for: It is always conserved.

In this chapter we focus on only one type of energy (*kinetic energy*) and on only one way in which energy can be transferred (*work*). In the next chapter we examine a few other types of energy and how the principle of energy conservation can be written as equations to be solved.

7-3 | Kinetic Energy

Kinetic energy K is energy associated with the *state of motion* of an object. The faster the object moves, the greater is its kinetic energy. When the object is stationary, its kinetic energy is zero.

For an object of mass m whose speed v is well below the speed of light,

$$K = \tfrac{1}{2}mv^2 \qquad \text{(kinetic energy).} \tag{7-1}$$

For example, a 3.0 kg duck flying past us at 2.0 m/s has a kinetic energy of $6.0\ \text{kg}\cdot\text{m}^2/\text{s}^2$; that is, we associate that number with the duck's motion.

The SI unit of kinetic energy (and every other type of energy) is the **joule** (J), named for James Prescott Joule, an English scientist of the 1800s. It is defined directly from Eq. 7-1 in terms of the units for mass and velocity:

$$1\ \text{joule} = 1\ \text{J} = 1\ \text{kg}\cdot\text{m}^2/\text{s}^2. \tag{7-2}$$

Thus, the flying duck has a kinetic energy of 6.0 J.

Sample Problem 7-1

In 1896 in Waco, Texas, William Crush parked two locomotives at opposite ends of a 6.4-km-long track, fired them up, tied their throttles open, and then allowed them to crash head-on at full speed (Fig. 7-1) in front of 30,000 spectators. Hundreds of people were hurt by flying debris; several were killed. Assuming each locomotive weighed 1.2×10^6 N and its acceleration was a constant 0.26 m/s^2, what was the total kinetic energy of the two locomotives just before the collision?

KEY IDEAS (1) We need to find the kinetic energy of each locomotive with Eq. 7-1, but that means we need each locomotive's speed just before the collision and its mass. (2) Because we can assume each locomotive had constant acceleration, we can use the equations in Table 2-1 to find its speed v just before the collision.

Calculations: We choose Eq. 2-16 because we know values for all the variables except v:

$$v^2 = v_0^2 + 2a(x - x_0).$$

With $v_0 = 0$ and $x - x_0 = 3.2 \times 10^3$ m (half the initial separation), this yields

$$v^2 = 0 + 2(0.26 \text{ m/s}^2)(3.2 \times 10^3 \text{ m}),$$

or

$$v = 40.8 \text{ m/s}$$

(about 150 km/h).

We can find the mass of each locomotive by dividing its given weight by g:

$$m = \frac{1.2 \times 10^6 \text{ N}}{9.8 \text{ m/s}^2} = 1.22 \times 10^5 \text{ kg}.$$

Now, using Eq. 7-1, we find the total kinetic energy of the two locomotives just before the collision as

$$K = 2(\tfrac{1}{2}mv^2) = (1.22 \times 10^5 \text{ kg})(40.8 \text{ m/s})^2$$
$$= 2.0 \times 10^8 \text{ J.} \quad \text{(Answer)}$$

This collision was like an exploding bomb.

FIG. 7-1 The aftermath of an 1896 crash of two locomotives. *(Courtesy Library of Congress)*

7-4 | Work

If you accelerate an object to a greater speed by applying a force to the object, you increase the kinetic energy K $(= \frac{1}{2}mv^2)$ of the object. Similarly, if you decelerate the object to a lesser speed by applying a force, you decrease the kinetic energy of the object. We account for these changes in kinetic energy by saying that your force has transferred energy *to* the object from yourself or *from* the object to yourself. In such a transfer of energy via a force, **work** W is said to be *done on the object by the force.* More formally, we define work as follows:

> Work W is energy transferred to or from an object by means of a force acting on the object. Energy transferred to the object is positive work, and energy transferred from the object is negative work.

"Work," then, is transferred energy; "doing work" is the act of transferring the energy. Work has the same units as energy and is a scalar quantity.

The term *transfer* can be misleading. It does not mean that anything material flows into or out of the object; that is, the transfer is not like a flow of water. Rather, it is like the electronic transfer of money between two bank accounts: The number in one account goes up while the number in the other account goes down, with nothing material passing between the two accounts.

Note that we are not concerned here with the common meaning of the word "work," which implies that *any* physical or mental labor is work. For example, if you push hard against a wall, you tire because of the continuously repeated muscle contractions that are required, and you are, in the common sense, working.

However, such effort does not cause an energy transfer to or from the wall and thus is not work done on the wall as defined here.

To avoid confusion in this chapter, we shall use the symbol W only for work and shall represent a weight with its equivalent mg.

7-5 | Work and Kinetic Energy

Finding an Expression for Work

Let us find an expression for work by considering a bead that can slide along a frictionless wire that is stretched along a horizontal x axis (Fig. 7-2). A constant force $\vec{F}$, directed at an angle ϕ to the wire, accelerates the bead along the wire. We can relate the force and the acceleration with Newton's second law, written for components along the x axis:

$$F_x = ma_x, \tag{7-3}$$

where m is the bead's mass. As the bead moves through a displacement $\vec{d}$, the force changes the bead's velocity from an initial value $\vec{v}_0$ to some other value $\vec{v}$. Because the force is constant, we know that the acceleration is also constant. Thus, we can use Eq. 2-16 to write, for components along the x axis,

$$v^2 = v_0^2 + 2a_x d. \tag{7-4}$$

Solving this equation for a_x, substituting into Eq. 7-3, and rearranging then give us

$$\tfrac{1}{2}mv^2 - \tfrac{1}{2}mv_0^2 = F_x d. \tag{7-5}$$

The first term on the left side of the equation is the kinetic energy K_f of the bead at the end of the displacement d, and the second term is the kinetic energy K_i of the bead at the start of the displacement. Thus, the left side of Eq. 7-5 tells us the kinetic energy has been changed by the force, and the right side tells us the change is equal to $F_x d$. Therefore, the work W done on the bead by the force (the energy transfer due to the force) is

$$W = F_x d. \tag{7-6}$$

If we know values for F_x and d, we can use this equation to calculate the work W done on the bead by the force.

> To calculate the work a force does on an object as the object moves through some displacement, we use only the force component along the object's displacement. The force component perpendicular to the displacement does zero work.

From Fig. 7-2, we see that we can write F_x as $F \cos \phi$, where ϕ is the angle between the directions of the displacement $\vec{d}$ and the force $\vec{F}$. Thus,

$$W = Fd \cos \phi \quad \text{(work done by a constant force).} \tag{7-7}$$

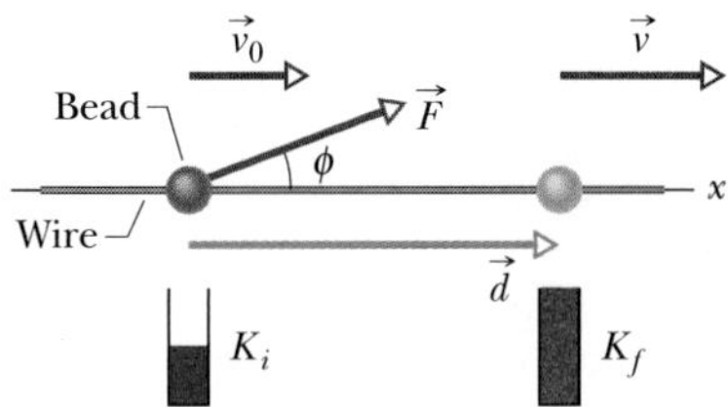

FIG. 7-2 A constant force $\vec{F}$ directed at angle ϕ to the displacement $\vec{d}$ of a bead on a wire accelerates the bead along the wire, changing the velocity of the bead from $\vec{v}_0$ to $\vec{v}$. A "kinetic energy gauge" indicates the resulting change in the kinetic energy of the bead, from the value K_i to the value K_f.

Because the right side of this equation is equivalent to the scalar (dot) product $\vec{F} \cdot \vec{d}$, we can also write

$$W = \vec{F} \cdot \vec{d} \quad \text{(work done by a constant force),} \tag{7-8}$$

where F is the magnitude of $\vec{F}$. (You may wish to review the discussion of scalar products in Section 3-8.) Equation 7-8 is especially useful for calculating the work when $\vec{F}$ and $\vec{d}$ are given in unit-vector notation.

Cautions: There are two restrictions to using Eqs. 7-6 through 7-8 to calculate work done on an object by a force. First, the force must be a *constant force;* that is, it must not change in magnitude or direction as the object moves. (Later, we shall discuss what to do with a *variable force* that changes in magnitude.) Second, the object must be *particle-like.* This means that the object must be *rigid;* all parts of it must move together, in the same direction. In this chapter we consider only particle-like objects, such as the bed and its occupant being pushed in Fig. 7-3.

Signs for work. The work done on an object by a force can be either positive work or negative work. For example, if the angle ϕ in Eq. 7-7 is less than 90°, then cos ϕ is positive and thus so is the work. If ϕ is greater than 90° (up to 180°), then cos ϕ is negative and thus so is the work. (Can you see that the work is zero when $\phi = 90°$?) These results lead to a simple rule. To find the sign of the work done by a force, consider the force vector component that is parallel to the displacement:

> A force does positive work when it has a vector component in the same direction as the displacement, and it does negative work when it has a vector component in the opposite direction. It does zero work when it has no such vector component.

Units for work. Work has the SI unit of the joule, the same as kinetic energy. However, from Eqs. 7-6 and 7-7 we can see that an equivalent unit is the newton-meter (N·m). The corresponding unit in the British system is the foot-pound (ft·lb). Extending Eq. 7-2, we have

$$1\text{ J} = 1\text{ kg}\cdot\text{m}^2/\text{s}^2 = 1\text{ N}\cdot\text{m} = 0.738\text{ ft}\cdot\text{lb}. \tag{7-9}$$

Net work done by several forces. When two or more forces act on an object, the **net work** done on the object is the sum of the works done by the individual forces. We can calculate the net work in two ways. (1) We can find the work done by each force and then sum those works. (2) Alternatively, we can first find the net force $\vec{F}_{net}$ of those forces. Then we can use Eq. 7-7, substituting the magnitude F_{net} for F and also the angle between the directions of $\vec{F}_{net}$ and $\vec{d}$ for ϕ. Similarly, we can use Eq. 7-8 with $\vec{F}_{net}$ substituted for $\vec{F}$.

Work–Kinetic Energy Theorem

Equation 7-5 relates the change in kinetic energy of the bead (from an initial $K_i = \frac{1}{2}mv_0^2$ to a later $K_f = \frac{1}{2}mv^2$) to the work W $(= F_x d)$ done on the bead. For such particle-like objects, we can generalize that equation. Let ΔK be the change in the kinetic energy of the object, and let W be the net work done on it. Then

$$\Delta K = K_f - K_i = W, \tag{7-10}$$

which says that

$$\begin{pmatrix}\text{change in the kinetic}\\ \text{energy of a particle}\end{pmatrix} = \begin{pmatrix}\text{net work done on}\\ \text{the particle}\end{pmatrix}.$$

We can also write

$$K_f = K_i + W, \tag{7-11}$$

which says that

$$\begin{pmatrix}\text{kinetic energy after}\\ \text{the net work is done}\end{pmatrix} = \begin{pmatrix}\text{kinetic energy}\\ \text{before the net work}\end{pmatrix} + \begin{pmatrix}\text{the net}\\ \text{work done}\end{pmatrix}.$$

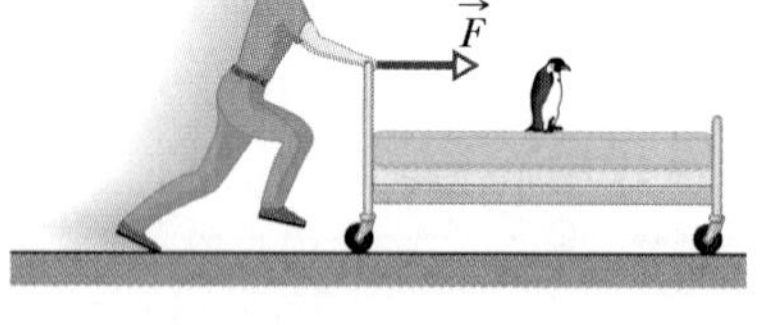

FIG. 7-3 A contestant in a bed race. We can approximate the bed and its occupant as being a particle for the purpose of calculating the work done on them by the force applied by the student.

These statements are known traditionally as the **work–kinetic energy theorem** for particles. They hold for both positive and negative work: If the net work done on a particle is positive, then the particle's kinetic energy increases by the amount of the work. If the net work done is negative, then the particle's kinetic energy decreases by the amount of the work.

For example, if the kinetic energy of a particle is initially 5 J and there is a net transfer of 2 J to the particle (positive net work), the final kinetic energy is 7 J. If, instead, there is a net transfer of 2 J from the particle (negative net work), the final kinetic energy is 3 J.

CHECKPOINT 1 A particle moves along an x axis. Does the kinetic energy of the particle increase, decrease, or remain the same if the particle's velocity changes (a) from −3 m/s to −2 m/s and (b) from −2 m/s to 2 m/s? (c) In each situation, is the work done on the particle positive, negative, or zero?

Sample Problem 7-2

Figure 7-4*a* shows two industrial spies sliding an initially stationary 225 kg floor safe a displacement $\vec{d}$ of magnitude 8.50 m, straight toward their truck. The push $\vec{F}_1$ of spy 001 is 12.0 N, directed at an angle of 30.0° downward from the horizontal; the pull $\vec{F}_2$ of spy 002 is 10.0 N, directed at 40.0° above the horizontal. The magnitudes and directions of these forces do not change as the safe moves, and the floor and safe make frictionless contact.

(a) What is the net work done on the safe by forces $\vec{F}_1$ and $\vec{F}_2$ during the displacement $\vec{d}$?

KEY IDEAS (1) The net work W done on the safe by the two forces is the sum of the works they do individually. (2) Because we can treat the safe as a particle and the forces are constant in both magnitude and direction, we can use either Eq. 7-7 ($W = Fd\cos\phi$) or Eq. 7-8 ($W = \vec{F}\cdot\vec{d}$) to calculate those works. Since we know the magnitudes and directions of the forces, we choose Eq. 7-7.

Calculations: From Eq. 7-7 and the free-body diagram for the safe in Fig. 7-4*b*, the work done by $\vec{F}_1$ is

$$W_1 = F_1 d\cos\phi_1 = (12.0\text{ N})(8.50\text{ m})(\cos 30.0°)$$
$$= 88.33\text{ J},$$

and the work done by $\vec{F}_2$ is

$$W_2 = F_2 d\cos\phi_2 = (10.0\text{ N})(8.50\text{ m})(\cos 40.0°)$$
$$= 65.11\text{ J}.$$

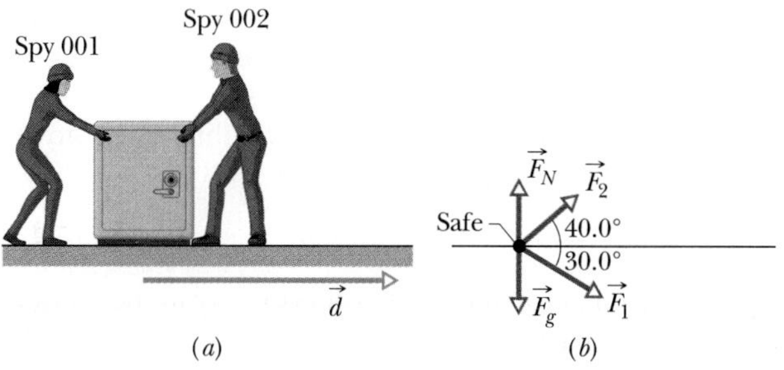

FIG. 7-4 (*a*) Two spies move a floor safe through a displacement $\vec{d}$. (*b*) A free-body diagram for the safe.

Thus, the net work W is

$$W = W_1 + W_2 = 88.33\text{ J} + 65.11\text{ J}$$
$$= 153.4\text{ J} \approx 153\text{ J}. \quad \text{(Answer)}$$

During the 8.50 m displacement, therefore, the spies transfer 153 J of energy to the kinetic energy of the safe.

(b) During the displacement, what is the work W_g done on the safe by the gravitational force $\vec{F}_g$ and what is the work W_N done on the safe by the normal force $\vec{F}_N$ from the floor?

KEY IDEA Because these forces are constant in both magnitude and direction, we can find the work they do with Eq. 7-7.

Calculations: Thus, with mg as the magnitude of the gravitational force, we write

$$W_g = mgd\cos 90° = mgd(0) = 0 \quad \text{(Answer)}$$

and

$$W_N = F_N d\cos 90° = F_N d(0) = 0. \quad \text{(Answer)}$$

We should have known this result. Because these forces are perpendicular to the displacement of the safe, they do zero work on the safe and do not transfer any energy to or from it.

(c) The safe is initially stationary. What is its speed v_f at the end of the 8.50 m displacement?

KEY IDEA The speed of the safe changes because its kinetic energy is changed when energy is transferred to it by $\vec{F}_1$ and $\vec{F}_2$.

Calculations: We relate the speed to the work done by combining Eqs. 7-10 and 7-1:

$$W = K_f - K_i = \tfrac{1}{2}mv_f^2 - \tfrac{1}{2}mv_i^2.$$

The initial speed v_i is zero, and we now know that the work done is 153.4 J. Solving for v_f and then substituting known data, we find that

$$v_f = \sqrt{\frac{2W}{m}} = \sqrt{\frac{2(153.4\text{ J})}{225\text{ kg}}}$$
$$= 1.17\text{ m/s}. \quad \text{(Answer)}$$

Sample Problem 7-3

During a storm, a crate of crepe is sliding across a slick, oily parking lot through a displacement $\vec{d} = (-3.0\text{ m})\hat{i}$ while a steady wind pushes against the crate with a force $\vec{F} = (2.0\text{ N})\hat{i} + (-6.0\text{ N})\hat{j}$. The situation and coordinate axes are shown in Fig. 7-5.

(a) How much work does this force do on the crate during the displacement?

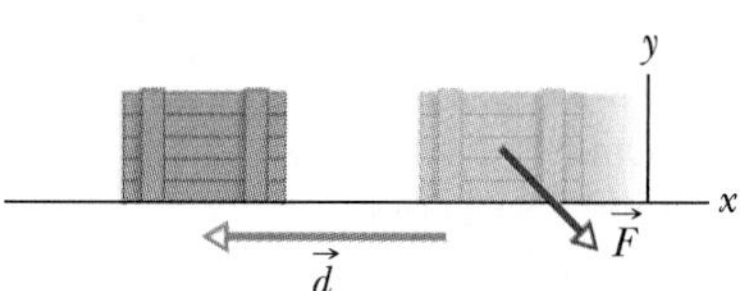

FIG. 7-5 Force $\vec{F}$ slows a crate during displacement $\vec{d}$.

KEY IDEA Because we can treat the crate as a particle and because the wind force is constant ("steady") in both magnitude and direction during the displacement, we can use either Eq. 7-7 ($W = Fd\cos\phi$) or Eq. 7-8 ($W = \vec{F}\cdot\vec{d}$) to calculate the work. Since we know $\vec{F}$ and $\vec{d}$ in unit-vector notation, we choose Eq. 7-8.

Calculations: We write

$$W = \vec{F}\cdot\vec{d} = [(2.0\text{ N})\hat{\text{i}} + (-6.0\text{ N})\hat{\text{j}}]\cdot[(-3.0\text{ m})\hat{\text{i}}].$$

Of the possible unit-vector dot products, only $\hat{\text{i}}\cdot\hat{\text{i}}$, $\hat{\text{j}}\cdot\hat{\text{j}}$, and $\hat{\text{k}}\cdot\hat{\text{k}}$ are nonzero (see Appendix E). Here we obtain

$$W = (2.0\text{ N})(-3.0\text{ m})\hat{\text{i}}\cdot\hat{\text{i}} + (-6.0\text{ N})(-3.0\text{ m})\hat{\text{j}}\cdot\hat{\text{i}}$$
$$= (-6.0\text{ J})(1) + 0 = -6.0\text{ J}. \qquad \text{(Answer)}$$

Thus, the force does a negative 6.0 J of work on the crate, transferring 6.0 J of energy from the kinetic energy of the crate.

(b) If the crate has a kinetic energy of 10 J at the beginning of displacement $\vec{d}$, what is its kinetic energy at the end of $\vec{d}$?

KEY IDEA Because the force does negative work on the crate, it reduces the crate's kinetic energy.

Calculation: Using the work–kinetic energy theorem in the form of Eq. 7-11, we have

$$K_f = K_i + W = 10\text{ J} + (-6.0\text{ J}) = 4.0\text{ J}. \qquad \text{(Answer)}$$

Less kinetic energy means that the crate has been slowed.

7-6 | Work Done by the Gravitational Force

We next examine the work done on an object by the gravitational force acting on it. Figure 7-6 shows a particle-like tomato of mass m that is thrown upward with initial speed v_0 and thus with initial kinetic energy $K_i = \frac{1}{2}mv_0^2$. As the tomato rises, it is slowed by a gravitational force $\vec{F}_g$; that is, the tomato's kinetic energy decreases because $\vec{F}_g$ does work on the tomato as it rises. Because we can treat the tomato as a particle, we can use Eq. 7-7 ($W = Fd\cos\phi$) to express the work done during a displacement $\vec{d}$. For the force magnitude F, we use mg as the magnitude of $\vec{F}_g$. Thus, the work W_g done by the gravitational force $\vec{F}_g$ is

$$W_g = mgd\cos\phi \quad \text{(work done by gravitational force).} \qquad (7\text{-}12)$$

For a rising object, force $\vec{F}_g$ is directed opposite the displacement $\vec{d}$, as indicated in Fig. 7-6. Thus, $\phi = 180°$ and

$$W_g = mgd\cos 180° = mgd(-1) = -mgd. \qquad (7\text{-}13)$$

The minus sign tells us that during the object's rise, the gravitational force acting on the object transfers energy in the amount mgd from the kinetic energy of the object. This is consistent with the slowing of the object as it rises.

After the object has reached its maximum height and is falling back down, the angle ϕ between force $\vec{F}_g$ and displacement $\vec{d}$ is zero. Thus,

$$W_g = mgd\cos 0° = mgd(+1) = +mgd. \qquad (7\text{-}14)$$

The plus sign tells us that the gravitational force now transfers energy in the amount mgd to the kinetic energy of the object. This is consistent with the speeding up of the object as it falls. (Actually, as we shall see in Chapter 8, energy transfers associated with lifting and lowering an object involve the full object–Earth system.)

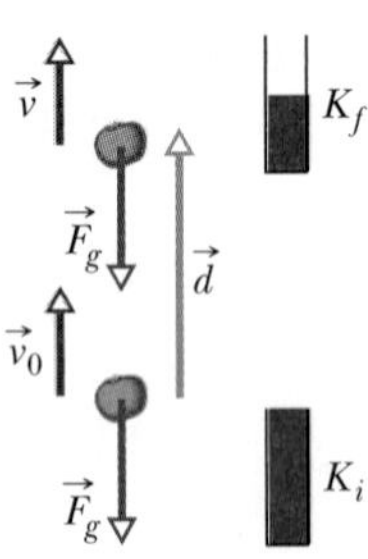

FIG. 7-6 Because the gravitational force $\vec{F}_g$ acts on it, a particle-like tomato of mass m thrown upward slows from velocity $\vec{v}_0$ to velocity $\vec{v}$ during displacement $\vec{d}$. A kinetic energy gauge indicates the resulting change in the kinetic energy of the tomato, from $K_i (= \frac{1}{2}mv_0^2)$ to $K_f (= \frac{1}{2}mv^2)$.

Work Done in Lifting and Lowering an Object

Now suppose we lift a particle-like object by applying a vertical force $\vec{F}$ to it. During the upward displacement, our applied force does positive work W_a on the object while the gravitational force does negative work W_g on it. Our applied force tends to transfer energy to the object while the gravitational force tends to transfer energy from it. By Eq. 7-10, the change ΔK in the kinetic energy of the object due to these two energy transfers is

$$\Delta K = K_f - K_i = W_a + W_g, \qquad (7\text{-}15)$$

in which K_f is the kinetic energy at the end of the displacement and K_i is that at the start of the displacement. This equation also applies if we lower the object, but then the gravitational force tends to transfer energy *to* the object while our force tends to transfer energy *from* it.

In one common situation, the object is stationary before and after the lift—for example, when you lift a book from the floor to a shelf. Then K_f and K_i are both zero, and Eq. 7-15 reduces to

$$W_a + W_g = 0$$

or

$$W_a = -W_g. \tag{7-16}$$

Note that we get the same result if K_f and K_i are not zero but are still equal. Either way, the result means that the work done by the applied force is the negative of the work done by the gravitational force; that is, the applied force transfers the same amount of energy to the object as the gravitational force transfers from the object. Using Eq. 7-12, we can rewrite Eq. 7-16 as

$$W_a = -mgd\cos\phi \quad \text{(work done in lifting and lowering; } K_f = K_i\text{)}, \tag{7-17}$$

with ϕ being the angle between $\vec{F}_g$ and $\vec{d}$. If the displacement is vertically upward (Fig. 7-7*a*), then $\phi = 180°$ and the work done by the applied force equals mgd. If the displacement is vertically downward (Fig. 7-7*b*), then $\phi = 0°$ and the work done by the applied force equals $-mgd$.

Equations 7-16 and 7-17 apply to any situation in which an object is lifted or lowered, with the object stationary before and after the lift. They are independent of the magnitude of the force used. For example, if you lift a mug from the floor to over your head, your force on the mug varies considerably during the lift. Still, because the mug is stationary before and after the lift, the work your force does on the mug is given by Eqs. 7-16 and 7-17, where, in Eq. 7-17, mg is the weight of the mug and d is the distance you lift it.

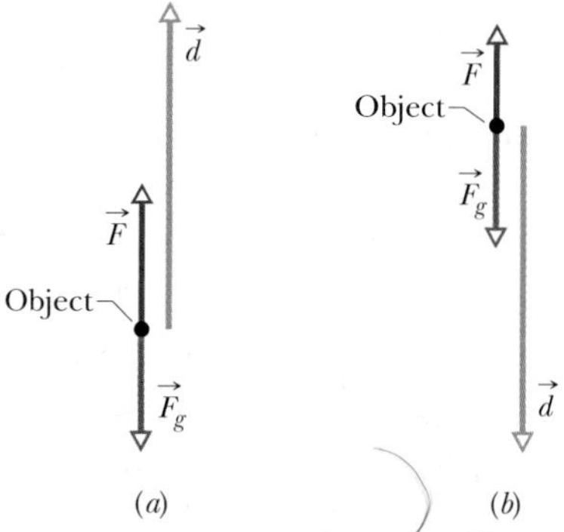

FIG. 7-7 (*a*) An applied force $\vec{F}$ lifts an object. The object's displacement $\vec{d}$ makes an angle $\phi = 180°$ with the gravitational force $\vec{F}_g$ on the object. The applied force does positive work on the object. (*b*) An applied force $\vec{F}$ lowers an object. The displacement $\vec{d}$ of the object makes an angle $\phi = 0°$ with the gravitational force $\vec{F}_g$. The applied force does negative work on the object.

Sample Problem 7-4

One of the lifts of Paul Anderson (Fig. 7-8) in the 1950s remains a record: Anderson stooped beneath a reinforced wood platform, placed his hands on a short stool to brace himself, and then pushed upward on the platform with his back, lifting the platform straight up by 1.0 cm. The platform held automobile parts and a safe filled with lead, with a total weight of 27 900 N (6270 lb).

(a) As Anderson lifted the load, how much work was done on it by the gravitational force $\vec{F}_g$?

KEY IDEA We can treat the load as a single particle because the components moved rigidly together. Thus we can use Eq. 7-12 ($W_g = mgd\cos\phi$) to find the work W_g done on the load by $\vec{F}_g$.

Calculation: The angle ϕ between the directions of the downward gravitational force and the upward displacement was 180°. Substituting this and the given data into Eq. 7-12, we find

$$W_g = mgd\cos\phi = (27\,900\text{ N})(0.010\text{ m})(\cos 180°)$$
$$= -280\text{ J}. \quad \text{(Answer)}$$

(b) How much work was done by the force Anderson applied to make the lift?

FIG. 7-8 Using a harness across his back, Paul Anderson lifted a platform and a scout troop off the ground. (*©AP/WideWorld Photos*)

KEY IDEAS Anderson's force was certainly not constant. Thus, we *cannot* just substitute a force magnitude into Eq. 7-7 to find the work done. However, we know that the load was stationary both at the start and at the end of the lift. Therefore, we know that the work W_A done by Anderson's applied force was the negative of the work W_g done by the gravitational force $\vec{F}_g$.

Calculation: Equation 7-16 gives us

$$W_A = -W_g = +280 \text{ J}. \qquad \text{(Answer)}$$

Comments: This is hardly more than the work needed to lift a stuffed school backpack from the floor to shoulder level. So, why was Anderson's lift so amazing? Work (energy transfer) and force are different quantities; although Anderson's lift required an unremarkable energy transfer, it required a truly remarkable force.

Sample Problem 7-5

An initially stationary 15.0 kg crate of cheese wheels is pulled, via a cable, a distance $d = 5.70$ m up a frictionless ramp to a height h of 2.50 m, where it stops (Fig. 7-9*a*).

(a) How much work W_g is done on the crate by the gravitational force $\vec{F}_g$ during the lift?

KEY IDEA We treat the crate as a particle and use Eq. 7-12 ($W_g = mgd \cos \phi$) to find the work W_g done by $\vec{F}_g$.

Calculations: We do not know the angle ϕ between the directions of $\vec{F}_g$ and displacement $\vec{d}$. However, from the crate's free-body diagram in Fig. 7-9*b*, we find that ϕ is $\theta + 90°$, where θ is the (unknown) angle of the ramp. Equation 7-12 then gives us

$$W_g = mgd \cos(\theta + 90°) = -mgd \sin \theta, \qquad (7\text{-}18)$$

where we have used a trigonometic identity to simplify the expression. The result seems to be useless because θ is unknown. But (continuing with physics courage) we see from Fig. 7-9*a* that $d \sin \theta = h$, where h is a known quantity. With this substitution, Eq. 7-18 gives us

$$W_g = -mgh \qquad (7\text{-}19)$$
$$= -(15.0 \text{ kg})(9.8 \text{ m/s}^2)(2.50 \text{ m})$$
$$= -368 \text{ J}. \qquad \text{(Answer)}$$

Note that Eq. 7-19 tells us that the work W_g done by the gravitational force depends on the vertical displacement but (surprisingly) not on the horizontal displacement. (We return to this point in Chapter 8.)

(b) How much work W_T is done on the crate by the force $\vec{T}$ from the cable during the lift?

KEY IDEA We cannot just substitute the force magnitude T for F in Eq. 7-7 ($W = Fd \cos \phi$) because we do not know the value of T. However, to get us going we can treat the crate as a particle and then apply the work–kinetic energy theorem ($\Delta K = W$) to it.

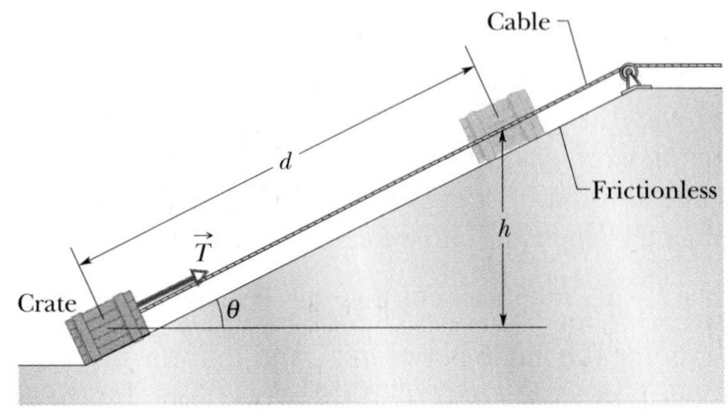

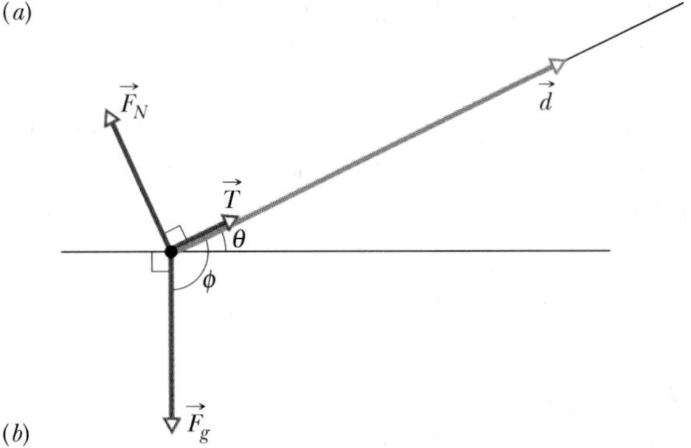

FIG. 7-9 (*a*) A crate is pulled up a frictionless ramp by a force $\vec{T}$ parallel to the ramp. (*b*) A free-body diagram for the crate, showing also the displacement $\vec{d}$.

Calculations: Because the crate is stationary before and after the lift, the change ΔK in its kinetic energy is zero. For the net work W done on the crate, we must sum the works done by all three forces acting on the crate. From (a), the work W_g done by the gravitational force $\vec{F}_g$ is -368 J. The work W_N done by the normal force $\vec{F}_N$ on the crate from the ramp is zero because $\vec{F}_N$ is perpendicular to the displacement. We want the work W_T done by $\vec{T}$. Thus, the work–kinetic energy theorem gives us

$$\Delta K = W_T + W_g + W_N$$

or

$$0 = W_T - 368 \text{ J} + 0,$$

and so

$$W_T = 368 \text{ J}. \qquad \text{(Answer)}$$

Sample Problem 7-6 Build your skill

An elevator cab of mass $m = 500$ kg is descending with speed $v_i = 4.0$ m/s when its supporting cable begins to slip, allowing it to fall with constant acceleration $\vec{a} = \vec{g}/5$ (Fig. 7-10*a*).

(a) During the fall through a distance $d = 12$ m, what is the work W_g done on the cab by the gravitational force $\vec{F}_g$?

KEY IDEA We can treat the cab as a particle and thus use Eq. 7-12 ($W_g = mgd \cos \phi$) to find the work W_g.

Calculation: From Fig. 7-10*b*, we see that the angle between the directions of $\vec{F}_g$ and the cab's displacement $\vec{d}$ is 0°. Then, from Eq. 7-12, we find

$$W_g = mgd\cos 0° = (500\text{ kg})(9.8\text{ m/s}^2)(12\text{ m})(1)$$
$$= 5.88 \times 10^4\text{ J} \approx 59\text{ kJ}. \quad \text{(Answer)}$$

(b) During the 12 m fall, what is the work W_T done on the cab by the upward pull $\vec{T}$ of the elevator cable?

KEY IDEAS (1) We can calculate the work W_T with Eq. 7-7 ($W = Fd\cos\phi$) if we first find an expression for the magnitude T of the cable's pull. (2) We can find that expression by writing Newton's second law for components along the y axis in Fig. 7-10*b* ($F_{\text{net},y} = ma_y$).

Calculations: We get

$$T - F_g = ma.$$

Solving for T, substituting mg for F_g, and then substituting the result in Eq. 7-7, we obtain

$$W_T = Td\cos\phi = m(a + g)d\cos\phi.$$

Next, substituting $-g/5$ for the (downward) acceleration a and then 180° for the angle ϕ between the directions of forces $\vec{T}$ and $m\vec{g}$, we find

$$W_T = m\left(-\frac{g}{5} + g\right)d\cos\phi = \frac{4}{5}mgd\cos\phi$$
$$= \frac{4}{5}(500\text{ kg})(9.8\text{ m/s}^2)(12\text{ m})\cos 180°$$
$$= -4.70 \times 10^4\text{ J} \approx -47\text{ kJ}. \quad \text{(Answer)}$$

Caution: Note that W_T is not simply the negative of W_g. The reason is that, because the cab accelerates during the fall, its speed changes during the fall, and thus its kinetic energy also changes. Therefore, Eq. 7-16 (which assumes that the initial and final kinetic energies are equal) does *not* apply here.

(c) What is the net work W done on the cab during the fall?

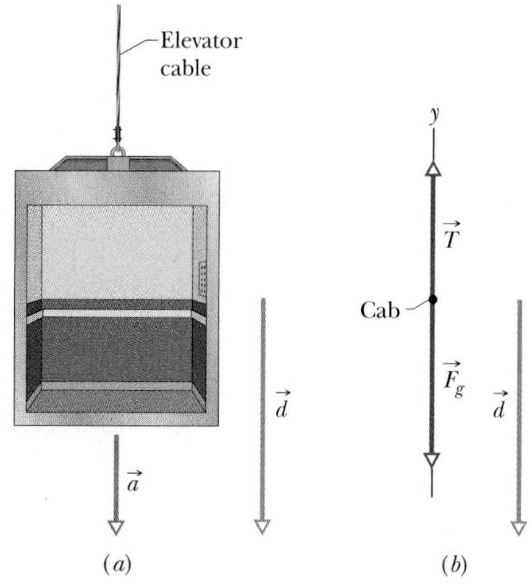

FIG. 7-10 An elevator cab, descending with speed v_i, suddenly begins to accelerate downward. (*a*) It moves through a displacement $\vec{d}$ with constant acceleration $\vec{a} = \vec{g}/5$. (*b*) A free-body diagram for the cab, displacement included.

Calculation: The net work is the sum of the works done by the forces acting on the cab:

$$W = W_g + W_T = 5.88 \times 10^4\text{ J} - 4.70 \times 10^4\text{ J}$$
$$= 1.18 \times 10^4\text{ J} \approx 12\text{ kJ}. \quad \text{(Answer)}$$

(d) What is the cab's kinetic energy at the end of the 12 m fall?

KEY IDEA The kinetic energy changes *because* of the net work done on the cab, according to Eq. 7-11 ($K_f = K_i + W$).

Calculation: From Eq. 7-1, we can write the kinetic energy at the start of the fall as $K_i = \frac{1}{2}mv_i^2$. We can then write Eq. 7-11 as

$$K_f = K_i + W = \tfrac{1}{2}mv_i^2 + W$$
$$= \tfrac{1}{2}(500\text{ kg})(4.0\text{ m/s})^2 + 1.18 \times 10^4\text{ J}$$
$$= 1.58 \times 10^4\text{ J} \approx 16\text{ kJ}. \quad \text{(Answer)}$$

7-7 | Work Done by a Spring Force

We next want to examine the work done on a particle-like object by a particular type of *variable force*—namely, a **spring force**, the force from a spring. Many forces in nature have the same mathematical form as the spring force. Thus, by examining this one force, you can gain an understanding of many others.

The Spring Force

Figure 7-11*a* shows a spring in its **relaxed state**—that is, neither compressed nor extended. One end is fixed, and a particle-like object—a block, say—is attached to the other, free end. If we stretch the spring by pulling the block to the right as in Fig. 7-11*b*, the spring pulls on the block toward the left. (Because a spring

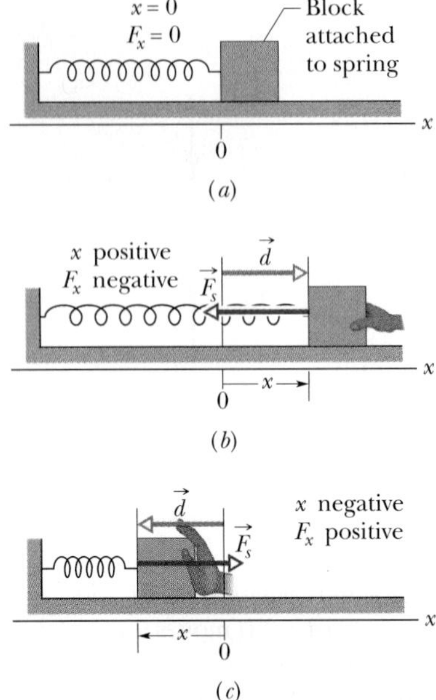

FIG. 7-11 (*a*) A spring in its relaxed state. The origin of an x axis has been placed at the end of the spring that is attached to a block. (*b*) The block is displaced by $\vec{d}$, and the spring is stretched by a positive amount x. Note the restoring force $\vec{F}_s$ exerted by the spring. (*c*) The spring is compressed by a negative amount x. Again, note the restoring force.

force acts to restore the relaxed state, it is sometimes said to be a *restoring force.*) If we compress the spring by pushing the block to the left as in Fig. 7-11*c*, the spring now pushes on the block toward the right.

To a good approximation for many springs, the force $\vec{F}_s$ from a spring is proportional to the displacement $\vec{d}$ of the free end from its position when the spring is in the relaxed state. The *spring force* is given by

$$\vec{F}_s = -k\vec{d} \quad \text{(Hooke's law)}, \tag{7-20}$$

which is known as **Hooke's law** after Robert Hooke, an English scientist of the late 1600s. The minus sign in Eq. 7-20 indicates that the direction of the spring force is always opposite the direction of the displacement of the spring's free end. The constant k is called the **spring constant** (or **force constant**) and is a measure of the stiffness of the spring. The larger k is, the stiffer the spring; that is, the larger k is, the stronger the spring's pull or push for a given displacement. The SI unit for k is the newton per meter.

In Fig. 7-11 an x axis has been placed parallel to the length of the spring, with the origin ($x = 0$) at the position of the free end when the spring is in its relaxed state. For this common arrangement, we can write Eq. 7-20 as

$$F_x = -kx \quad \text{(Hooke's law)}, \tag{7-21}$$

where we have changed the subscript. If x is positive (the spring is stretched toward the right on the x axis), then F_x is negative (it is a pull toward the left). If x is negative (the spring is compressed toward the left), then F_x is positive (it is a push toward the right). Note that a spring force is a *variable force* because it is a function of x, the position of the free end. Thus F_x can be symbolized as $F(x)$. Also note that Hooke's law is a *linear* relationship between F_x and x.

The Work Done by a Spring Force

To find the work done by the spring force as the block in Fig. 7-11*a* moves, let us make two simplifying assumptions about the spring. (1) It is *massless;* that is, its mass is negligible relative to the block's mass. (2) It is an *ideal spring;* that is, it obeys Hooke's law exactly. Let us also assume that the contact between the block and the floor is frictionless and that the block is particle-like.

We give the block a rightward jerk to get it moving and then leave it alone. As the block moves rightward, the spring force F_x does work on the block, decreasing the kinetic energy and slowing the block. However, we *cannot* find this work by using Eq. 7-7 ($W = Fd \cos \phi$) because that equation assumes a constant force. The spring force is a variable force.

To find the work done by the spring, we use calculus. Let the block's initial position be x_i and its later position x_f. Then divide the distance between those two positions into many segments, each of tiny length Δx. Label these segments, starting from x_i, as segments 1, 2, and so on. As the block moves through a segment, the spring force hardly varies because the segment is so short that x hardly varies. Thus, we can approximate the force magnitude as being constant within the segment. Label these magnitudes as F_{x1} in segment 1, F_{x2} in segment 2, and so on.

With the force now constant in each segment, we *can* find the work done within each segment by using Eq. 7-7. Here $\phi = 180°$, and so $\cos \phi = -1$. Then the work done is $-F_{x1}\,\Delta x$ in segment 1, $-F_{x2}\,\Delta x$ in segment 2, and so on. The net work W_s done by the spring, from x_i to x_f, is the sum of all these works:

$$W_s = \sum -F_{xj}\,\Delta x, \tag{7-22}$$

where j labels the segments. In the limit as Δx goes to zero, Eq. 7-22 becomes

$$W_s = \int_{x_i}^{x_f} -F_x\,dx. \tag{7-23}$$

From Eq. 7-21, the force magnitude F_x is kx. Thus, substitution leads to

$$W_s = \int_{x_i}^{x_f} -kx\,dx = -k\int_{x_i}^{x_f} x\,dx$$

$$= (-\tfrac{1}{2}k)[x^2]_{x_i}^{x_f} = (-\tfrac{1}{2}k)(x_f^2 - x_i^2). \tag{7-24}$$

Multiplied out, this yields

$$W_s = \tfrac{1}{2}kx_i^2 - \tfrac{1}{2}kx_f^2 \quad \text{(work by a spring force).} \tag{7-25}$$

This work W_s done by the spring force can have a positive or negative value, depending on whether the *net* transfer of energy is to or from the block as the block moves from x_i to x_f. *Caution:* The final position x_f appears in the *second* term on the right side of Eq. 7-25. Therefore, Eq. 7-25 tells us:

Work W_s is positive if the block ends up closer to the relaxed position ($x = 0$) than it was initially. It is negative if the block ends up farther away from $x = 0$. It is zero if the block ends up at the same distance from $x = 0$.

If $x_i = 0$ and if we call the final position x, then Eq. 7-25 becomes

$$W_s = -\tfrac{1}{2}kx^2 \quad \text{(work by a spring force).} \tag{7-26}$$

The Work Done by an Applied Force

Now suppose that we displace the block along the x axis while continuing to apply a force $\vec{F}_a$ to it. During the displacement, our applied force does work W_a on the block while the spring force does work W_s. By Eq. 7-10, the change ΔK in the kinetic energy of the block due to these two energy transfers is

$$\Delta K = K_f - K_i = W_a + W_s, \tag{7-27}$$

in which K_f is the kinetic energy at the end of the displacement and K_i is that at the start of the displacement. If the block is stationary before and after the displacement, then K_f and K_i are both zero and Eq. 7-27 reduces to

$$W_a = -W_s. \tag{7-28}$$

If a block that is attached to a spring is stationary before and after a displacement, then the work done on it by the applied force displacing it is the negative of the work done on it by the spring force.

Caution: If the block is not stationary before and after the displacement, then this statement is *not* true.

CHECKPOINT 2 For three situations, the initial and final positions, respectively, along the x axis for the block in Fig. 7-11 are (a) −3 cm, 2 cm; (b) 2 cm, 3 cm; and (c) −2 cm, 2 cm. In each situation, is the work done by the spring force on the block positive, negative, or zero?

Sample Problem 7-7

A package of spicy Cajun pralines lies on a frictionless floor, attached to the free end of a spring in the arrangement of Fig. 7-11*a*. A rightward applied force of magnitude $F_a = 4.9$ N would be needed to hold the package at $x_1 = 12$ mm.

(a) How much work does the spring force do on the package if the package is pulled rightward from $x_0 = 0$ to $x_2 = 17$ mm?

KEY IDEA As the package moves from one position to another, the spring force does work on it as given by Eq. 7-25 or Eq. 7-26.

Calculations: We know that the initial position x_i is 0 and the final position x_f is 17 mm, but we do not know the spring constant k. We can probably find k with Eq. 7-21 (Hooke's law), but we need this fact to use it: Were the package held stationary at $x_1 = 12$ mm, the spring force would have to balance the applied force (according to Newton's second law). Thus, the spring force F_x would have to be -4.9 N (toward the left in Fig. 7-11b); so Eq. 7-21 ($F_x = -kx$) gives us

$$k = -\frac{F_x}{x_1} = -\frac{-4.9\text{ N}}{12 \times 10^{-3}\text{ m}} = 408\text{ N/m}.$$

Now, with the package at $x_2 = 17$ mm, Eq. 7-26 yields

$$W_s = -\tfrac{1}{2}kx_2^2 = -\tfrac{1}{2}(408\text{ N/m})(17 \times 10^{-3}\text{ m})^2$$
$$= -0.059\text{ J}. \qquad \text{(Answer)}$$

(b) Next, the package is moved leftward to $x_3 = -12$ mm. How much work does the spring force do on the package during this displacement? Explain the sign of this work.

Calculation: Now $x_i = +17$ mm and $x_f = -12$ mm, and Eq. 7-25 yields

$$W_s = \tfrac{1}{2}kx_i^2 - \tfrac{1}{2}kx_f^2 = \tfrac{1}{2}k(x_i^2 - x_f^2)$$
$$= \tfrac{1}{2}(408\text{ N/m})[(17 \times 10^{-3}\text{ m})^2 - (-12 \times 10^{-3}\text{ m})^2]$$
$$= 0.030\text{ J} = 30\text{ mJ}. \qquad \text{(Answer)}$$

This work done on the block by the spring force is positive because the spring force does more positive work as the block moves from $x_i = +17$ mm to the spring's relaxed position than it does negative work as the block moves from the spring's relaxed position to $x_f = -12$ mm.

Sample Problem 7-8

In Fig. 7-12, a cumin canister of mass $m = 0.40$ kg slides across a horizontal frictionless counter with speed $v = 0.50$ m/s. It then runs into and compresses a spring of spring constant $k = 750$ N/m. When the canister is momentarily stopped by the spring, by what distance d is the spring compressed?

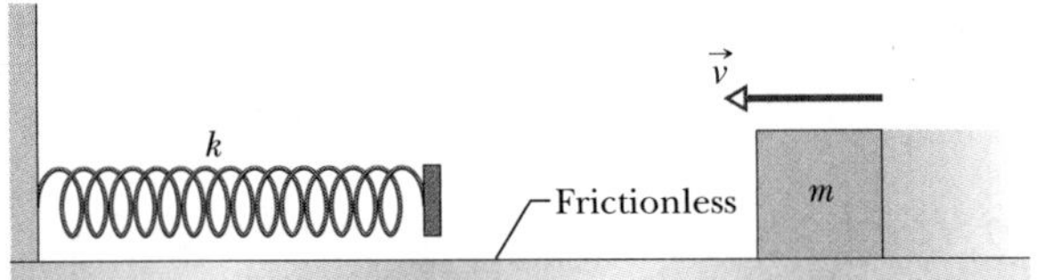

FIG. 7-12 A canister of mass m moves at velocity $\vec{v}$ toward a spring that has spring constant k.

KEY IDEAS

1. The work W_s done on the canister by the spring force is related to the requested distance d by Eq. 7-26 ($W_s = -\tfrac{1}{2}kx^2$), with d replacing x.
2. The work W_s is also related to the kinetic energy of the canister by Eq. 7-10 ($K_f - K_i = W$).
3. The canister's kinetic energy has an initial value of $K = \tfrac{1}{2}mv^2$ and a value of zero when the canister is momentarily at rest.

Calculations: Putting the first two of these ideas together, we write the work–kinetic energy theorem for the canister as

$$K_f - K_i = -\tfrac{1}{2}kd^2.$$

Substituting according to the third idea makes this expression

$$0 - \tfrac{1}{2}mv^2 = -\tfrac{1}{2}kd^2.$$

Simplifying, solving for d, and substituting known data then give us

$$d = v\sqrt{\frac{m}{k}} = (0.50\text{ m/s})\sqrt{\frac{0.40\text{ kg}}{750\text{ N/m}}}$$
$$= 1.2 \times 10^{-2}\text{ m} = 1.2\text{ cm}. \qquad \text{(Answer)}$$

7-8 | Work Done by a General Variable Force

One-Dimensional Analysis

Let us return to the situation of Fig. 7-2 but now consider the force to be in the positive direction of the x axis and the force magnitude to vary with position x. Thus, as the bead (particle) moves, the magnitude $F(x)$ of the force doing work on it changes. Only the magnitude of this variable force changes, not its direction, and the magnitude at any position does not change with time.

Figure 7-13a shows a plot of such a *one-dimensional variable force.* We want an expression for the work done on the particle by this force as the particle moves from an initial point x_i to a final point x_f. However, we *cannot* use Eq. 7-7 ($W = Fd\cos\phi$) because it applies only for a constant force $\vec{F}$. Here, again, we shall use calculus. We divide the area under the curve of Fig. 7-13a into a number of narrow strips of width Δx (Fig. 7-13b). We choose Δx small enough to permit us to take the force $F(x)$ as being reasonably constant over that interval. We let $F_{j,\text{avg}}$ be the average value of $F(x)$ within the jth interval. Then in Fig. 7-13b, $F_{j,\text{avg}}$ is the height of the jth strip.

With $F_{j,\text{avg}}$ considered constant, the increment (small amount) of work ΔW_j done by the force in the jth interval is now approximately given by Eq. 7-7 and is

$$\Delta W_j = F_{j,\text{avg}}\,\Delta x. \tag{7-29}$$

In Fig. 7-13b, ΔW_j is then equal to the area of the jth rectangular, shaded strip.

To approximate the total work W done by the force as the particle moves from x_i to x_f, we add the areas of all the strips between x_i and x_f in Fig. 7-13b:

$$W = \sum \Delta W_j = \sum F_{j,\text{avg}}\,\Delta x. \tag{7-30}$$

Equation 7-30 is an approximation because the broken "skyline" formed by the tops of the rectangular strips in Fig. 7-13b only approximates the actual curve of $F(x)$.

We can make the approximation better by reducing the strip width Δx and using more strips (Fig. 7-13c). In the limit, we let the strip width approach zero; the number of strips then becomes infinitely large and we have, as an exact result,

$$W = \lim_{\Delta x \to 0} \sum F_{j,\text{avg}}\,\Delta x. \tag{7-31}$$

This limit is exactly what we mean by the integral of the function $F(x)$ between the limits x_i and x_f. Thus, Eq. 7-31 becomes

$$W = \int_{x_i}^{x_f} F(x)\,dx \quad \text{(work: variable force).} \tag{7-32}$$

If we know the function $F(x)$, we can substitute it into Eq. 7-32, introduce the proper limits of integration, carry out the integration, and thus find the work. (Appendix E contains a list of common integrals.) Geometrically, the work is equal to the area between the $F(x)$ curve and the x axis, between the limits x_i and x_f (shaded in Fig. 7-13d).

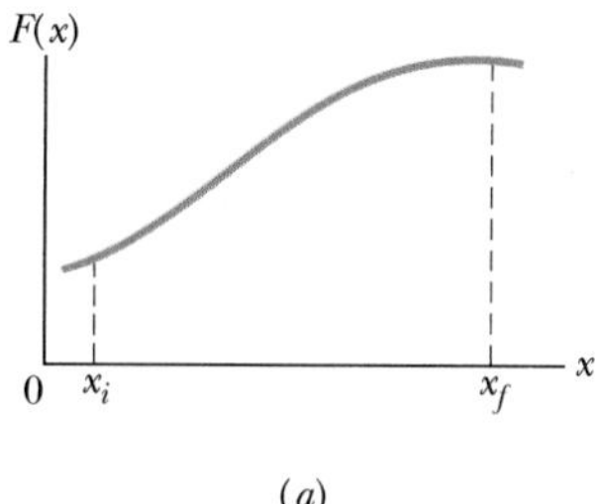

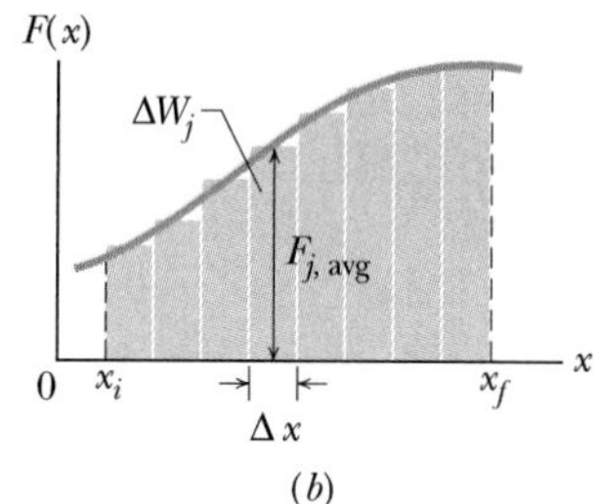

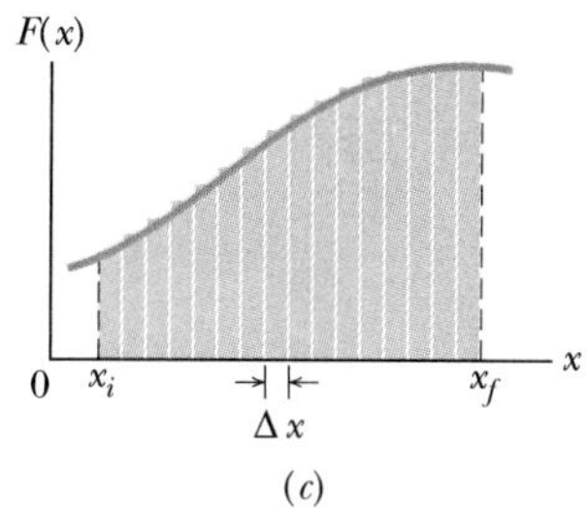

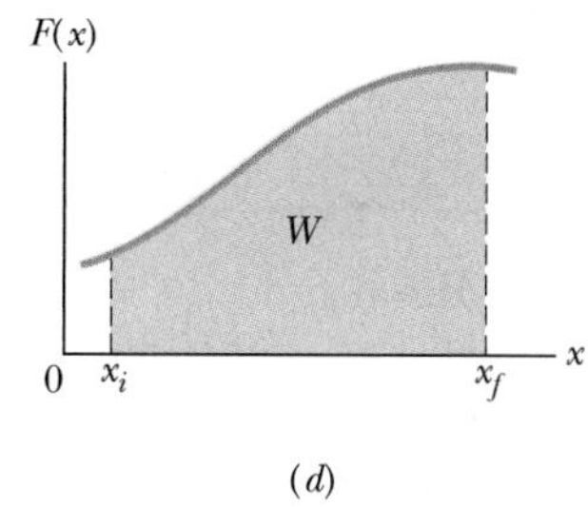

FIG. 7-13 (a) A one-dimensional force $\vec{F}(x)$ plotted against the displacement x of a particle on which it acts. The particle moves from x_i to x_f. (b) Same as (a) but with the area under the curve divided into narrow strips. (c) Same as (b) but with the area divided into narrower strips. (d) The limiting case. The work done by the force is given by Eq. 7-32 and is represented by the shaded area between the curve and the x axis and between x_i and x_f.

Three-Dimensional Analysis

Consider now a particle that is acted on by a three-dimensional force

$$\vec{F} = F_x\hat{\text{i}} + F_y\hat{\text{j}} + F_z\hat{\text{k}}, \tag{7-33}$$

in which the components F_x, F_y, and F_z can depend on the position of the particle; that is, they can be functions of that position. However, we make three simplifications: F_x may depend on x but not on y or z, F_y may depend on y but not on x or z, and F_z may depend on z but not on x or y. Now let the particle move through an incremental displacement

$$d\vec{r} = dx\hat{\text{i}} + dy\hat{\text{j}} + dz\hat{\text{k}}. \tag{7-34}$$

The increment of work dW done on the particle by $\vec{F}$ during the displacement $d\vec{r}$ is, by Eq. 7-8,

$$dW = \vec{F}\cdot d\vec{r} = F_x\,dx + F_y\,dy + F_z\,dz. \tag{7-35}$$

The work W done by $\vec{F}$ while the particle moves from an initial position r_i having coordinates (x_i, y_i, z_i) to a final position r_f having coordinates (x_f, y_f, z_f) is then

$$W = \int_{r_i}^{r_f} dW = \int_{x_i}^{x_f} F_x\,dx + \int_{y_i}^{y_f} F_y\,dy + \int_{z_i}^{z_f} F_z\,dz. \tag{7-36}$$

If $\vec{F}$ has only an x component, then the y and z terms in Eq. 7-36 are zero and the equation reduces to Eq. 7-32.

Work–Kinetic Energy Theorem with a Variable Force

Equation 7-32 gives the work done by a variable force on a particle in a one-dimensional situation. Let us now make certain that the calculated work is indeed equal to the change in kinetic energy of the particle, as the work–kinetic energy theorem states.

Consider a particle of mass m, moving along an x axis and acted on by a net force $F(x)$ that is directed along that axis. The work done on the particle by this force as the particle moves from position x_i to position x_f is given by Eq. 7-32 as

$$W = \int_{x_i}^{x_f} F(x)\,dx = \int_{x_i}^{x_f} ma\,dx, \tag{7-37}$$

in which we use Newton's second law to replace $F(x)$ with ma. We can write the quantity $ma\,dx$ in Eq. 7-37 as

$$ma\,dx = m\frac{dv}{dt}\,dx. \tag{7-38}$$

From the chain rule of calculus, we have

$$\frac{dv}{dt} = \frac{dv}{dx}\frac{dx}{dt} = \frac{dv}{dx}v, \tag{7-39}$$

and Eq. 7-38 becomes

$$ma\,dx = m\frac{dv}{dx}v\,dx = mv\,dv. \tag{7-40}$$

Substituting Eq. 7-40 into Eq. 7-37 yields

$$W = \int_{v_i}^{v_f} mv\,dv = m\int_{v_i}^{v_f} v\,dv$$

$$= \tfrac{1}{2}mv_f^2 - \tfrac{1}{2}mv_i^2. \tag{7-41}$$

Note that when we change the variable from x to v we are required to express the limits on the integral in terms of the new variable. Note also that because the mass m is a constant, we are able to move it outside the integral.

Recognizing the terms on the right side of Eq. 7-41 as kinetic energies allows us to write this equation as

$$W = K_f - K_i = \Delta K,$$

which is the work–kinetic energy theorem.

Sample Problem 7-9

In an epidural procedure, as used in childbirth, a surgeon or an anesthetist must run a needle through the skin on the patient's back, through various tissue layers and into a narrow region called the epidural space that lies within the spinal canal surrounding the spinal cord. The needle is intended to deliver an anesthetic fluid. This tricky procedure requires much practice so that the doctor knows when the needle has reached the epidural space and not overshot it, a mistake that could result in serious complications.

The feel a doctor has for the needle's penetration is the variable force that must be applied to advance the needle through the tissues. Figure 7-14a is a graph of the force magnitude F versus displacement x of the needle tip in a typical epidural procedure. (The line segments have been straightened somewhat from the original

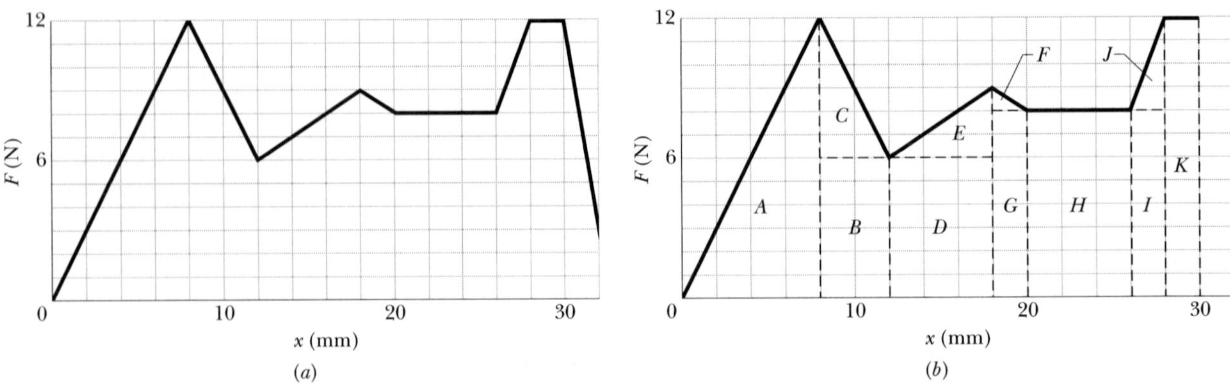

FIG. 7-14 (a) The force magnitude F versus the displacement x of the needle in an epidural procedure. (b) Breaking up the region between the plotted curve and the displacement axis to calculate the area.

data.) As x increases from 0, the skin resists the needle, but at $x = 8.0$ mm the force is finally great enough to pierce the skin, and then the required force decreases. Similarly, the needle finally pierces the interspinous ligament at $x = 18$ mm and the relatively tough ligamentum flavum at $x = 30$ mm. The needle then enters the epidural space (where it is to deliver the anesthetic fluid), and the force drops sharply. A new doctor must learn this pattern of force versus displacement to recognize when to stop pushing on the needle. (This is the pattern to be programmed into a virtual-reality simulation of an epidural procedure.) How much work W is done by the force exerted on the needle to get the needle to the epidural space at $x = 30$ mm?

KEY IDEAS (1) We can calculate the work W done by a variable force $F(x)$ by integrating the force versus position x. Equation 7-32 tells us that

$$W = \int_{x_i}^{x_f} F(x)\,dx.$$

We want the work done by the force during the displacement from $x_i = 0$ to $x_f = 0.030$ m. (2) We can evaluate the integral by finding the area under the curve on the graph of Fig. 7-14*a*.

$$W = \begin{pmatrix}\text{area between force curve}\\ \text{and } x \text{ axis, from } x_i \text{ to } x_f\end{pmatrix}.$$

Calculations: Because our graph consists of straight-line segments, we can find the area by splitting the region below the curve into rectangular and triangular regions, as shown in Fig. 7-14*b*. For example, the area in triangular region A is

$$\text{area}_A = \tfrac{1}{2}(0.0080\text{ m})(12\text{ N}) = 0.048\text{ N}\cdot\text{m} = 0.048\text{ J}.$$

Once we've calculated the areas for all the labeled regions in Fig. 7-14*b*, we find that the total work is

$$\begin{aligned} W &= (\text{sum of the areas of regions } A \text{ through } K)\\ &= 0.048 + 0.024 + 0.012 + 0.036 + 0.009 + 0.001\\ &\quad + 0.016 + 0.048 + 0.016 + 0.004 + 0.024\\ &= 0.238\text{ J}. \qquad \text{(Answer)}\end{aligned}$$

Sample Problem 7-10

Force $\vec{F} = (3x^2\text{ N})\hat{\imath} + (4\text{ N})\hat{\jmath}$, with x in meters, acts on a particle, changing only the kinetic energy of the particle. How much work is done on the particle as it moves from coordinates (2 m, 3 m) to (3 m, 0 m)? Does the speed of the particle increase, decrease, or remain the same?

KEY IDEA The force is a variable force because its x component depends on the value of x. Thus, we cannot use Eqs. 7-7 and 7-8 to find the work done. Instead, we must use Eq. 7-36 to integrate the force.

Calculation: We set up two integrals, one along each axis:

$$\begin{aligned} W &= \int_2^3 3x^2\,dx + \int_3^0 4\,dy = 3\int_2^3 x^2\,dx + 4\int_3^0 dy\\ &= 3[\tfrac{1}{3}x^3]_2^3 + 4[y]_3^0 = [3^3 - 2^3] + 4[0-3]\\ &= 7.0\text{ J}. \qquad \text{(Answer)}\end{aligned}$$

The positive result means that energy is transferred to the particle by force $\vec{F}$. Thus, the kinetic energy of the particle increases and, because $K = \frac{1}{2}mv^2$, its speed must also increase.

7-9 | Power

The time rate at which work is done by a force is said to be the **power** due to the force. If a force does an amount of work W in an amount of time Δt, the **average power** due to the force during that time interval is

$$P_{avg} = \frac{W}{\Delta t} \quad \text{(average power).} \tag{7-42}$$

The **instantaneous power** P is the instantaneous time rate of doing work, which we can write as

$$P = \frac{dW}{dt} \quad \text{(instantaneous power).} \tag{7-43}$$

Suppose we know the work $W(t)$ done by a force as a function of time. Then to get the instantaneous power P at, say, time $t = 3.0$ s during the work, we would first take the time derivative of $W(t)$ and then evaluate the result for $t = 3.0$ s.

The SI unit of power is the joule per second. This unit is used so often that it has a special name, the **watt** (W), after James Watt, who greatly improved the rate at which

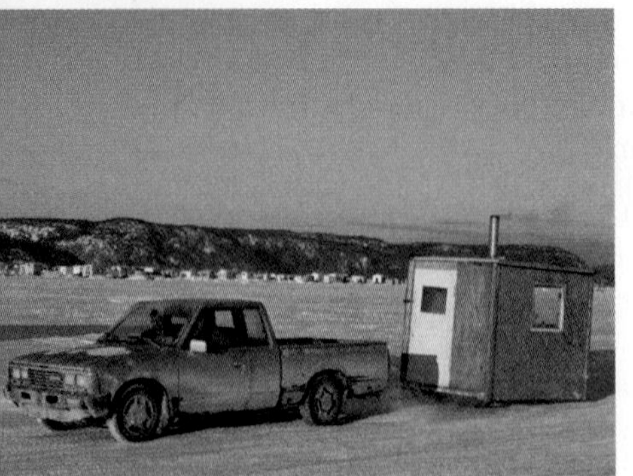

FIG. 7-15 The power due to the truck's applied force on the trailing load is the rate at which that force does work on the load. *(REGLAIN FREDERIC/Gamma-Presse, Inc.)*

steam engines could do work. In the British system, the unit of power is the foot-pound per second. Often the horsepower is used. These are related by

$$1 \text{ watt} = 1 \text{ W} = 1 \text{ J/s} = 0.738 \text{ ft} \cdot \text{lb/s} \tag{7-44}$$

and

$$1 \text{ horsepower} = 1 \text{ hp} = 550 \text{ ft} \cdot \text{lb/s} = 746 \text{ W}. \tag{7-45}$$

Inspection of Eq. 7-42 shows that work can be expressed as power multiplied by time, as in the common unit kilowatt-hour. Thus,

$$1 \text{ kilowatt-hour} = 1 \text{ kW} \cdot \text{h} = (10^3 \text{ W})(3600 \text{ s}) = 3.60 \times 10^6 \text{ J} = 3.60 \text{ MJ}. \tag{7-46}$$

Perhaps because they appear on our utility bills, the watt and the kilowatt-hour have become identified as electrical units. They can be used equally well as units for other examples of power and energy. Thus, if you pick up a book from the floor and put it on a tabletop, you are free to report the work that you have done as, say, 4×10^{-6} kW · h (or more conveniently as 4 mW · h).

We can also express the rate at which a force does work on a particle (or particle-like object) in terms of that force and the particle's velocity. For a particle that is moving along a straight line (say, an x axis) and is acted on by a constant force $\vec{F}$ directed at some angle ϕ to that line, Eq. 7-43 becomes

$$P = \frac{dW}{dt} = \frac{F \cos \phi \, dx}{dt} = F \cos \phi \left(\frac{dx}{dt}\right),$$

or

$$P = Fv \cos \phi. \tag{7-47}$$

Reorganizing the right side of Eq. 7-47 as the dot product $\vec{F} \cdot \vec{v}$, we may also write the equation as

$$P = \vec{F} \cdot \vec{v} \quad \text{(instantaneous power)}. \tag{7-48}$$

For example, the truck in Fig. 7-15 exerts a force $\vec{F}$ on the trailing load, which has velocity $\vec{v}$ at some instant. The instantaneous power due to $\vec{F}$ is the rate at which $\vec{F}$ does work on the load at that instant and is given by Eqs. 7-47 and 7-48. Saying that this power is "the power of the truck" is often acceptable, but keep in mind what is meant: Power is the rate at which the applied *force* does work.

CHECKPOINT 3 A block moves with uniform circular motion because a cord tied to the block is anchored at the center of a circle. Is the power due to the force on the block from the cord positive, negative, or zero?

Sample Problem 7-11

Figure 7-16 shows constant forces $\vec{F}_1$ and $\vec{F}_2$ acting on a box as the box slides rightward across a frictionless floor. Force $\vec{F}_1$ is horizontal, with magnitude 2.0 N; force $\vec{F}_2$ is angled upward by 60° to the floor and has magnitude 4.0 N. The speed v of the box at a certain instant is 3.0 m/s. What is the power due to each force acting on the box at that instant, and what is the net power? Is the net power changing at that instant?

KEY IDEA We want an instantaneous power, not an average power over a time period. Also, we know the box's velocity (rather than the work done on it).

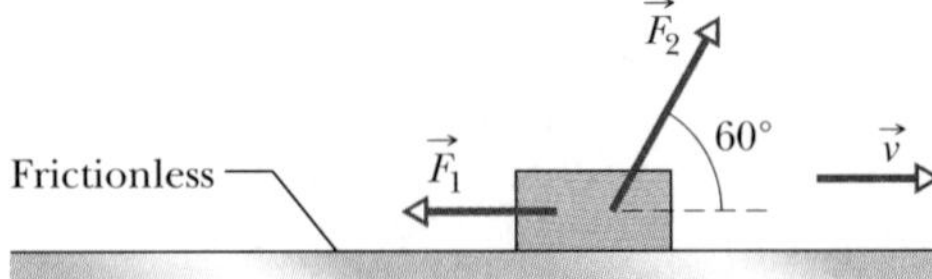

FIG. 7-16 Two forces $\vec{F}_1$ and $\vec{F}_2$ act on a box that slides rightward across a frictionless floor. The velocity of the box is $\vec{v}$.

Calculation: We use Eq. 7-47 for each force. For force $\vec{F}_1$, at angle $\phi_1 = 180°$ to velocity $\vec{v}$, we have

$$P_1 = F_1 v \cos \phi_1 = (2.0 \text{ N})(3.0 \text{ m/s}) \cos 180° = -6.0 \text{ W}. \quad \text{(Answer)}$$

This negative result tells us that force $\vec{F}_1$ is transferring energy *from* the box at the rate of 6.0 J/s.

For force $\vec{F}_2$, at angle $\phi_2 = 60°$ to velocity $\vec{v}$, we have

$$P_2 = F_2 v \cos \phi_2 = (4.0 \text{ N})(3.0 \text{ m/s}) \cos 60° = 6.0 \text{ W}. \quad \text{(Answer)}$$

This positive result tells us that force $\vec{F}_2$ is transferring energy *to* the box at the rate of 6.0 J/s.

The net power is the sum of the individual powers:

$$P_{net} = P_1 + P_2 = -6.0 \text{ W} + 6.0 \text{ W} = 0, \quad \text{(Answer)}$$

which tells us that the net rate of transfer of energy to or from the box is zero. Thus, the kinetic energy ($K = \frac{1}{2}mv^2$) of the box is not changing, and so the speed of the box will remain at 3.0 m/s. With neither the forces $\vec{F}_1$ and $\vec{F}_2$ nor the velocity $\vec{v}$ changing, we see from Eq. 7-48 that P_1 and P_2 are constant and thus so is P_{net}.

Sample Problem 7-12

Provided a funny car does not lose traction, the time it takes to race from rest through a distance D depends primarily on the engine's power P. Assuming the power is constant, derive the time in terms of D and P.

KEY IDEAS (1) The power of an engine is the rate at which it can do work, as expressed by Eq. 7-43 ($P = dW/dt$). (2) We can relate the work done during the race to the kinetic energy with Eq. 7-10, the work–kinetic energy theorem ($W = K_f - K_i$).

Power and kinetic energy: From the work–kinetic energy theorem, a small amount of work dW results in a small change dK of kinetic energy: $dW = dK$. Substituting this into Eq. 7-43 and rearranging give us

$$dK = P\, dt.$$

Integrating both sides and substituting that the kinetic energy is $K = 0$ when the race starts at $t = 0$, we find

$$\int_0^K dK = \int_0^t P\, dt$$

and

$$K = Pt.$$

After substituting $\frac{1}{2}mv^2$ for K, we solve for v, the speed at the end of the race:

$$v = \left(\frac{2Pt}{m}\right)^{1/2}. \quad (7\text{-}49)$$

Distance and speed: From the definition of velocity in Chapter 2, we know that $v = dx/dt$. Rearranging the definition and setting up integration on both sides, we find

$$\int_0^D dx = \int_0^t v\, dt.$$

Substituting from Eq. 7-49, we have

$$\int_0^D dx = \int_0^t \left(\frac{2Pt}{m}\right)^{1/2} dt = \left(\frac{2P}{m}\right)^{1/2} \int_0^t t^{1/2}\, dt.$$

Integrating then yields

$$D = \left(\frac{2P}{m}\right)^{1/2} \frac{2}{3} t^{3/2}.$$

Solving for t tells us that a funny car's elapsed time t depends on D and P as given by

$$t = \left(\frac{3}{2}D\right)^{2/3} \left(\frac{m}{2P}\right)^{1/3}. \quad \text{(Answer)}$$

Comments: In words, the elapsed time depends on the inverse cube root of the power. If the racing crew can coax more power out of the engine, the elapsed time decreases because of the *inverse* dependence, but only modestly because of the *cube root* dependence.

REVIEW & SUMMARY

Kinetic Energy The **kinetic energy** K associated with the motion of a particle of mass m and speed v, where v is well below the speed of light, is

$$K = \tfrac{1}{2}mv^2 \quad \text{(kinetic energy)}. \quad (7\text{-}1)$$

Work **Work** W is energy transferred to or from an object via a force acting on the object. Energy transferred to the object is positive work, and from the object, negative work.

Work Done by a Constant Force The work done on a particle by a constant force $\vec{F}$ during displacement $\vec{d}$ is

$$W = Fd \cos \phi = \vec{F} \cdot \vec{d} \quad \text{(work, constant force)}, \quad (7\text{-}7, 7\text{-}8)$$

in which ϕ is the constant angle between the directions of $\vec{F}$ and $\vec{d}$. Only the component of $\vec{F}$ that is along the displacement $\vec{d}$ can do work on the object. When two or more forces act on an object, their **net work** is the sum of the individual works done by the forces, which is also equal to the work that would be done on the object by the net force $\vec{F}_{net}$ of those forces.

Work and Kinetic Energy For a particle, a change ΔK in the kinetic energy equals the net work W done on the particle:

$$\Delta K = K_f - K_i = W \quad \text{(work–kinetic energy theorem)}, \quad (7\text{-}10)$$

in which K_i is the initial kinetic energy of the particle and K_f is the kinetic energy after the work is done. Equation 7-10 rearranged gives us

$$K_f = K_i + W. \quad (7\text{-}11)$$

Work Done by the Gravitational Force The work W_g done by the gravitational force $\vec{F}_g$ on a particle-like object of mass m as the object moves through a displacement $\vec{d}$ is given by

$$W_g = mgd\cos\phi, \quad (7\text{-}12)$$

in which ϕ is the angle between $\vec{F}_g$ and $\vec{d}$.

Work Done in Lifting and Lowering an Object The work W_a done by an applied force as a particle-like object is either lifted or lowered is related to the work W_g done by the gravitational force and the change ΔK in the object's kinetic energy by

$$\Delta K = K_f - K_i = W_a + W_g. \quad (7\text{-}15)$$

If $K_f = K_i$, then Eq. 7-15 reduces to

$$W_a = -W_g, \quad (7\text{-}16)$$

which tells us that the applied force transfers as much energy to the object as the gravitational force transfers from it.

Spring Force The force $\vec{F}_s$ from a spring is

$$\vec{F}_s = -k\vec{d} \quad \text{(Hooke's law)}, \quad (7\text{-}20)$$

where $\vec{d}$ is the displacement of the spring's free end from its position when the spring is in its **relaxed state** (neither compressed nor extended), and k is the **spring constant** (a measure of the spring's stiffness). If an x axis lies along the spring, with the origin at the location of the spring's free end when the spring is in its relaxed state, Eq. 7-20 can be written as

$$F_x = -kx \quad \text{(Hooke's law)}. \quad (7\text{-}21)$$

A spring force is thus a variable force: It varies with the displacement of the spring's free end.

Work Done by a Spring Force If an object is attached to the spring's free end, the work W_s done on the object by the spring force when the object is moved from an initial position x_i to a final position x_f is

$$W_s = \tfrac{1}{2}kx_i^2 - \tfrac{1}{2}kx_f^2. \quad (7\text{-}25)$$

If $x_i = 0$ and $x_f = x$, then Eq. 7-25 becomes

$$W_s = -\tfrac{1}{2}kx^2. \quad (7\text{-}26)$$

Work Done by a Variable Force When the force $\vec{F}$ on a particle-like object depends on the position of the object, the work done by $\vec{F}$ on the object while the object moves from an initial position r_i with coordinates (x_i, y_i, z_i) to a final position r_f with coordinates (x_f, y_f, z_f) must be found by integrating the force. If we assume that component F_x may depend on x but not on y or z, component F_y may depend on y but not on x or z, and component F_z may depend on z but not on x or y, then the work is

$$W = \int_{x_i}^{x_f} F_x\,dx + \int_{y_i}^{y_f} F_y\,dy + \int_{z_i}^{z_f} F_z\,dz. \quad (7\text{-}36)$$

If $\vec{F}$ has only an x component, then Eq. 7-36 reduces to

$$W = \int_{x_i}^{x_f} F(x)\,dx. \quad (7\text{-}32)$$

Power The **power** due to a force is the *rate* at which that force does work on an object. If the force does work W during a time interval Δt, the *average power* due to the force over that time interval is

$$P_{\text{avg}} = \frac{W}{\Delta t}. \quad (7\text{-}42)$$

Instantaneous power is the instantaneous rate of doing work:

$$P = \frac{dW}{dt}. \quad (7\text{-}43)$$

For a force $\vec{F}$ at an angle ϕ to the direction of travel of the instantaneous velocity $\vec{v}$, the instantaneous power is

$$P = Fv\cos\phi = \vec{F}\cdot\vec{v}. \quad (7\text{-}47, 7\text{-}48)$$

QUESTIONS

1 Is positive or negative work done by a constant force $\vec{F}$ on a particle during a straight-line displacement $\vec{d}$ if (a) the angle between $\vec{F}$ and $\vec{d}$ is 30°; (b) the angle is 100°; (c) $\vec{F} = 2\hat{i} - 3\hat{j}$ and $\vec{d} = -4\hat{i}$?

2 In three situations, a briefly applied horizontal force changes the velocity of a hockey puck that slides over frictionless ice. The overhead views of Fig. 7-17 indicate, for each situation, the puck's initial speed v_i, its final speed v_f, and the directions of the corresponding velocity vectors. Rank the situations according to the work done on the puck by the applied force, most positive first and most negative last.

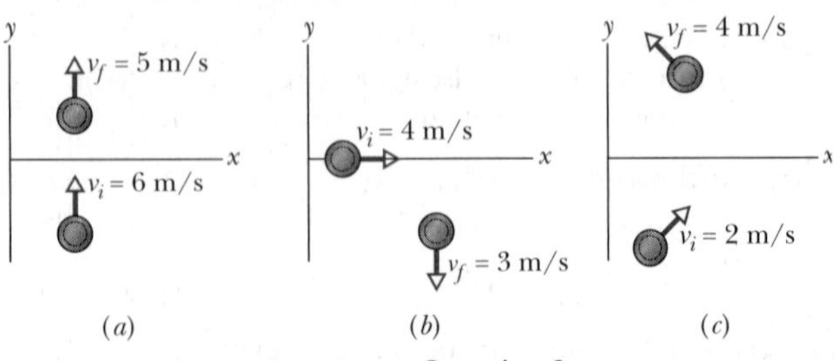

FIG. 7-17 Question 2.

3 Rank the following velocities according to the kinetic energy a particle will have with each velocity, greatest first: (a) $\vec{v} = 4\hat{i} + 3\hat{j}$, (b) $\vec{v} = -4\hat{i} + 3\hat{j}$, (c) $\vec{v} = -3\hat{i} + 4\hat{j}$, (d) $\vec{v} = 3\hat{i} - 4\hat{j}$, (e) $\vec{v} = 5\hat{i}$, and (f) v = 5 m/s at 30° to the horizontal.

4 Figure 7-18*a* shows two horizontal forces that act on a block that is sliding to the right across a frictionless floor. Figure 7-18*b* shows three plots of the block's kinetic energy K

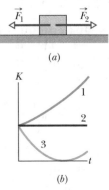

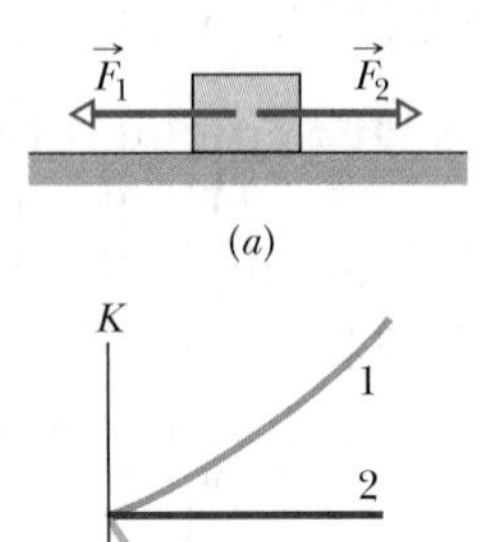

FIG. 7-18 Question 4.

versus time t. Which of the plots best corresponds to the following three situations: (a) $F_1 = F_2$, (b) $F_1 > F_2$, (c) $F_1 < F_2$?

5 In Fig. 7-19, a greased pig has a choice of three frictionless slides along which to slide to the ground. Rank the slides according to how much work the gravitational force does on the pig during the descent, greatest first.

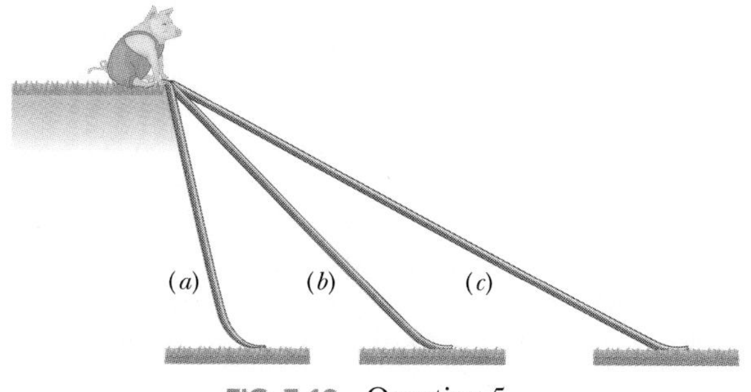

FIG. 7-19 Question 5.

6 Figure 7-20a shows four situations in which a horizontal force acts on the same block, which is initially at rest. The force magnitudes are $F_2 = F_4 = 2F_1 = 2F_3$. The horizontal component v_x of the block's velocity is shown in Fig. 7-20b for the four situations. (a) Which plot in Fig. 7-20b best corresponds to which force in Fig. 7-20a? (b) Which plot in Fig. 7-20c (for kinetic energy K versus time t) best corresponds to which plot in Fig. 7-20b?

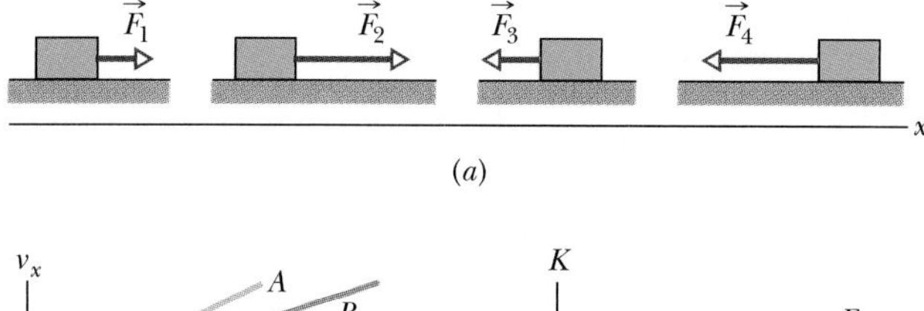

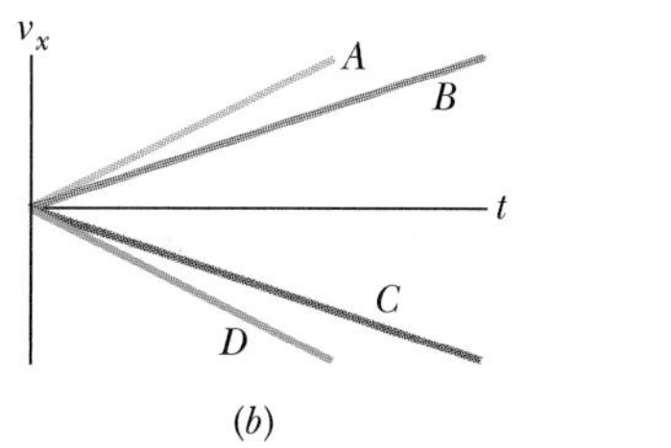

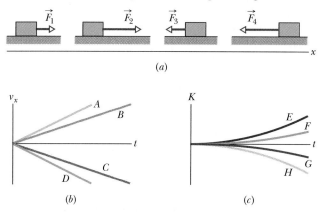

FIG. 7-20 Question 6.

7 Figure 7-21 shows four graphs (drawn to the same scale) of the x component F_x of a variable force (directed along an x axis) versus the position x of a particle on which the force acts. Rank the graphs according to the work done by the force on the particle from $x = 0$ to $x = x_1$, from most positive work first to most negative work last.

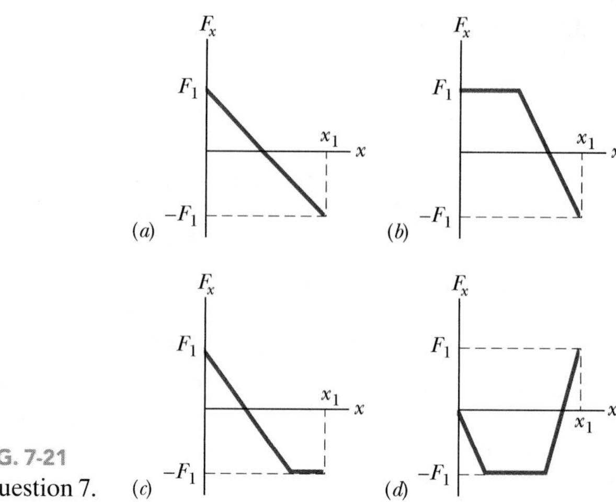

FIG. 7-21 Question 7.

8 Figure 7-22 gives the x component F_x of a force that can act on a particle. If the particle begins at rest at $x = 0$, what is its coordinate when it has (a) its greatest kinetic energy, (b) its greatest speed, and (c) zero speed? (d) What is the particle's direction of travel after it reaches $x = 6$ m?

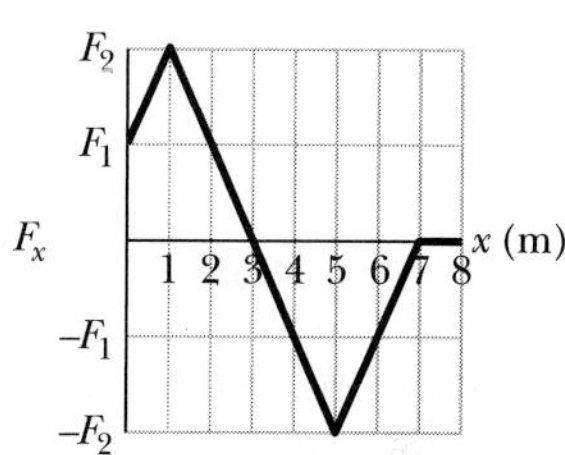

FIG. 7-22 Question 8.

9 Spring A is stiffer than spring B ($k_A > k_B$). The spring force of which spring does more work if the springs are compressed (a) the same distance and (b) by the same applied force?

10 A glob of slime is launched or dropped from the edge of a cliff. Which of the graphs in Fig. 7-23 could possibly show how the kinetic energy of the glob changes during its flight?

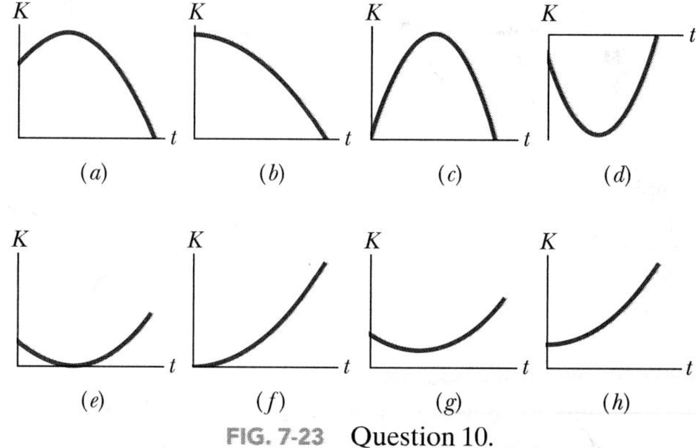

FIG. 7-23 Question 10.

PROBLEMS

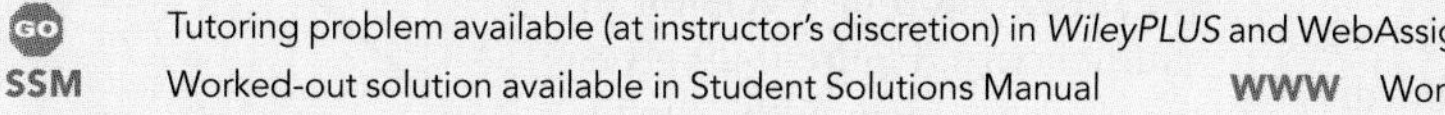

GO Tutoring problem available (at instructor's discretion) in *WileyPLUS* and WebAssign

SSM Worked-out solution available in Student Solutions Manual

WWW Worked-out solution is at http://www.wiley.com/college/halliday

• – ••• Number of dots indicates level of problem difficulty

ILW Interactive solution is at http://www.wiley.com/college/halliday

Additional information available in *The Flying Circus of Physics* and at flyingcircusofphysics.com

sec. 7-3 Kinetic Energy

•**1** On August 10, 1972, a large meteorite skipped across the atmosphere above the western United States and western Canada, much like a stone skipped across water. The accompanying fireball was so bright that it could be seen in the daytime sky and was brighter than the usual meteorite trail. The meteorite's mass was about 4×10^6 kg; its speed was about 15 km/s. Had it entered the atmosphere vertically, it would have hit Earth's surface with about the same speed. (a) Calculate the meteorite's loss of kinetic energy (in joules) that would have been associated with the vertical impact. (b) Express the energy as a multiple of the explosive energy of 1 megaton of TNT, which is 4.2×10^{15} J. (c) The energy associated with the atomic bomb explosion over Hiroshima was equivalent to 13 kilotons of TNT. To how many Hiroshima bombs would the meteorite impact have been equivalent?

•2 If a Saturn V rocket with an Apollo spacecraft attached had a combined mass of 2.9×10^5 kg and reached a speed of 11.2 km/s, how much kinetic energy would it then have?

•3 A proton (mass $m = 1.67 \times 10^{-27}$ kg) is being accelerated along a straight line at 3.6×10^{15} m/s² in a machine. If the proton has an initial speed of 2.4×10^7 m/s and travels 3.5 cm, what then is (a) its speed and (b) the increase in its kinetic energy? SSM

•4 A force $\vec{F}_a$ is applied to a bead as the bead is moved along a straight wire through displacement +5.0 cm. The magnitude of $\vec{F}_a$ is set at a certain value, but the angle ϕ between $\vec{F}_a$ and the bead's displacement can be chosen. Figure 7-24 gives the work W done by $\vec{F}_a$ on the bead for a range of ϕ values; $W_0 =$ 25 J. How much work is done by $\vec{F}_a$ if ϕ is (a) 64° and (b) 147°?

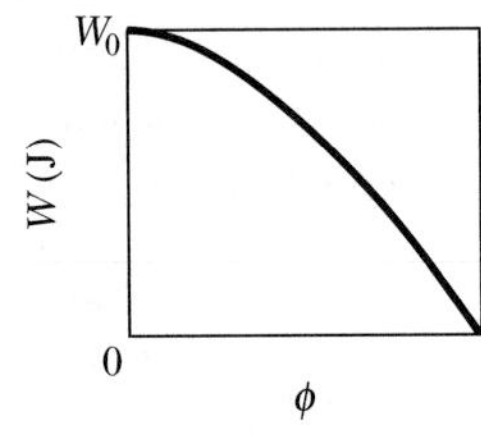

FIG. 7-24 Problem 4.

••5 A father racing his son has half the kinetic energy of the son, who has half the mass of the father. The father speeds up by 1.0 m/s and then has the same kinetic energy as the son. What are the original speeds of (a) the father and (b) the son?

••6 A bead with mass 1.8×10^{-2} kg is moving along a wire in the positive direction of an x axis. Beginning at time $t = 0$, when the bead passes through $x = 0$ with speed 12 m/s, a constant force acts on the bead. Figure 7-25 indicates the bead's position at times $t_0 = 0$, $t_1 = 1.0$ s, $t_2 = 2.0$ s, and $t_3 = 3.0$ s. The bead momentarily stops at $t = 3.0$ s. What is the kinetic energy of the bead at $t = 10$ s?

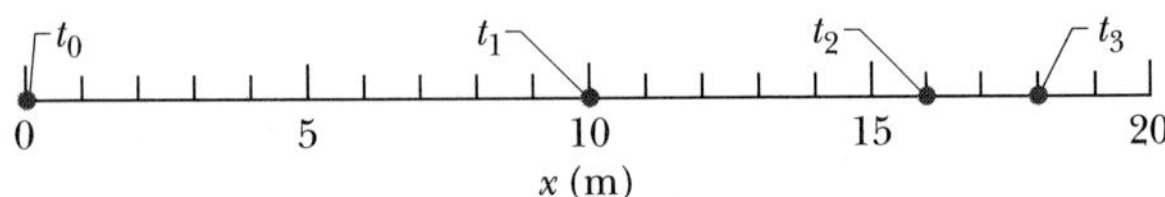

FIG. 7-25 Problem 6.

sec. 7-5 Work and Kinetic Energy

•7 The only force acting on a 2.0 kg canister that is moving in an xy plane has a magnitude of 5.0 N. The canister initially has a velocity of 4.0 m/s in the positive x direction and some time later has a velocity of 6.0 m/s in the positive y direction. How much work is done on the canister by the 5.0 N force during this time?

•8 A coin slides over a frictionless plane and across an xy coordinate system from the origin to a point with xy coordinates (3.0 m, 4.0 m) while a constant force acts on it. The force has magnitude 2.0 N and is directed at a counterclockwise angle of 100° from the positive direction of the x axis. How much work is done by the force on the coin during the displacement?

•9 A 3.0 kg body is at rest on a frictionless horizontal air track when a constant horizontal force $\vec{F}$ acting in the positive direction of an x axis along the track is applied to the body. A stroboscopic graph of the position of the body as it slides to the right is shown in Fig. 7-26. The force $\vec{F}$ is applied to the body at $t = 0$, and the graph records the position of the body at 0.50 s intervals. How much work is done on the body by the applied force $\vec{F}$ between $t = 0$ and $t = 2.0$ s?

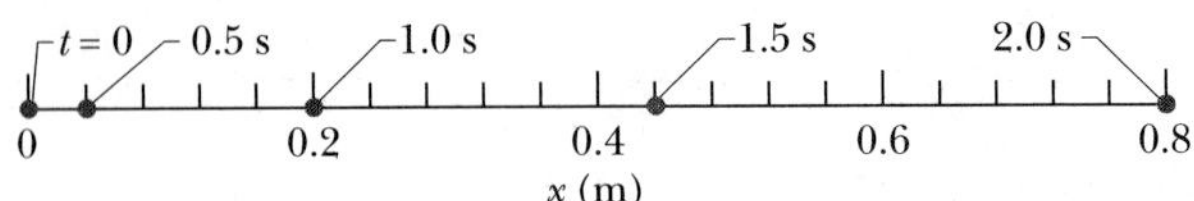

FIG. 7-26 Problem 9.

•10 A floating ice block is pushed through a displacement $\vec{d} = (15\text{ m})\hat{i} - (12\text{ m})\hat{j}$ along a straight embankment by rushing water, which exerts a force $\vec{F} = (210\text{ N})\hat{i} - (150\text{ N})\hat{j}$ on the block. How much work does the force do on the block during the displacement?

••11 A luge and its rider, with a total mass of 85 kg, emerge from a downhill track onto a horizontal straight track with an initial speed of 37 m/s. If a force slows them to a stop at a constant rate of 2.0 m/s², (a) what magnitude F is required for the force, (b) what distance d do they travel while slowing, and (c) what work W is done on them by the force? What are (d) F, (e) d, and (f) W if they, instead, slow at 4.0 m/s²?

••12 An 8.0 kg object is moving in the positive direction of an x axis. When it passes through $x = 0$, a constant force directed along the axis begins to act on it. Figure 7-27 gives its kinetic energy K versus position x as it moves from $x = 0$ to $x = 5.0$ m; $K_0 = 30.0$ J. The force continues to act. What is v when the object moves back through $x = -3.0$ m?

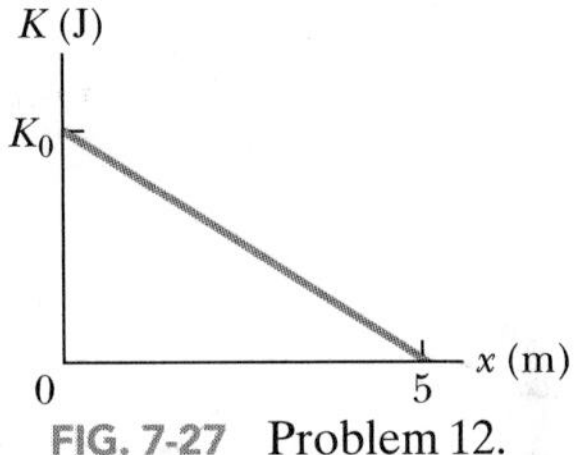

FIG. 7-27 Problem 12.

••13 Figure 7-28 shows three forces applied to a trunk that moves leftward by 3.00 m over a frictionless floor. The force magnitudes are $F_1 = 5.00$ N, $F_2 = 9.00$ N, and $F_3 = 3.00$ N, and the indicated angle is $\theta = 60.0°$. During the displacement, (a) what is the net work done on the trunk by the three forces and (b) does the kinetic energy of the trunk increase or decrease? GO

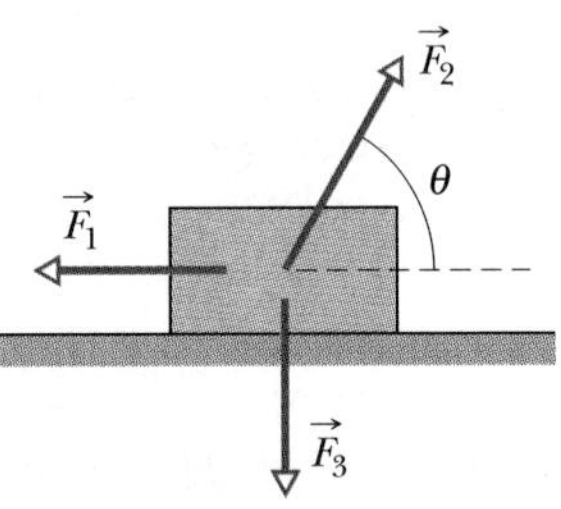

FIG. 7-28 Problem 13.

••14 A can of bolts and nuts is pushed 2.00 m along an x axis by a broom along the greasy (frictionless) floor of a car repair shop in a version of shuffleboard. Figure 7-29 gives the work W done on the can by the constant horizontal force from the broom, versus the can's position x. The scale of the figure's vertical axis is set by $W_s = 6.0$ J. (a) What is the magnitude of that force? (b) If the can had an initial kinetic energy of 3.00 J, moving in the positive direction of the x axis, what is its kinetic energy at the end of the 2.00 m?

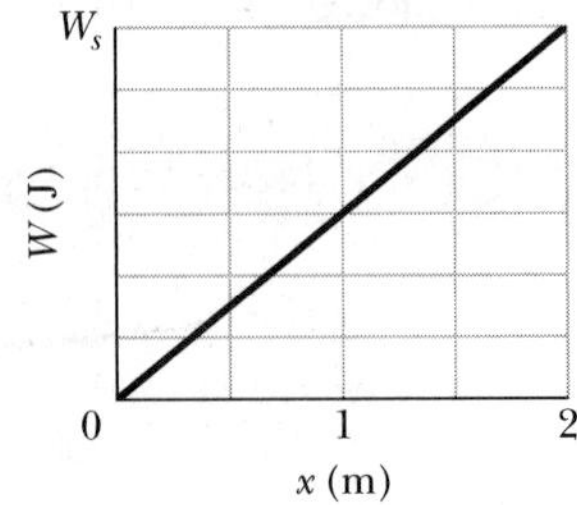

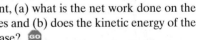

FIG. 7-29 Problem 14.

••15 A 12.0 N force with a fixed orientation does work on a particle as the particle moves through displacement $\vec{d} = (2.00\hat{i} - 4.00\hat{j} + 3.00\hat{k})$ m. What is the angle between the force and the displacement if the change in the particle's kinetic energy is (a) +30.0 J and (b) −30.0 J?

••16 Figure 7-30 shows an overhead view of three horizontal forces acting on a cargo canister that was initially stationary but now moves across a frictionless floor. The force magnitudes are $F_1 = 3.00$ N, $F_2 = 4.00$ N, and $F_3 = 10.0$ N, and the indicated angles are $\theta_2 = 50.0°$ and $\theta_3 = 35.0°$. What is the net work done on the canister by the three forces during the first 4.00 m of displacement?

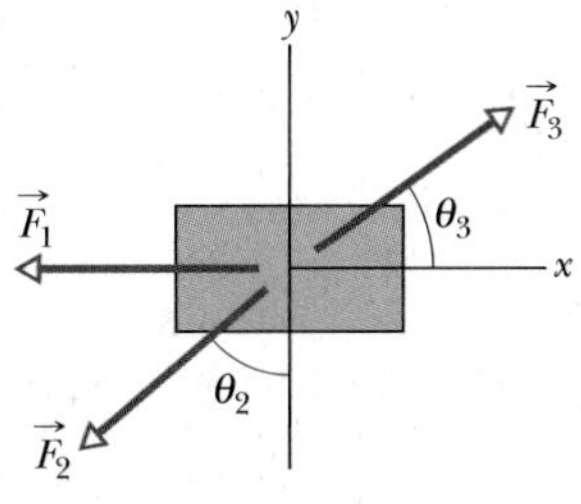

FIG. 7-30 Problem 16.

sec. 7-6 Work Done by the Gravitational Force

•17 A helicopter lifts a 72 kg astronaut 15 m vertically from the ocean by means of a cable. The acceleration of the astronaut is $g/10$. How much work is done on the astronaut by (a) the force from the helicopter and (b) the gravitational force on her? Just before she reaches the helicopter, what are her (c) kinetic energy and (d) speed? SSM WWW

•18 (a) In 1975 the roof of Montreal's Velodrome, with a weight of 360 kN, was lifted by 10 cm so that it could be centered. How much work was done on the roof by the forces making the lift? (b) In 1960 a Tampa, Florida, mother reportedly raised one end of a car that had fallen onto her son when a jack failed. If her panic lift effectively raised 4000 N (about $\frac{1}{4}$ of the car's weight) by 5.0 cm, how much work did her force do on the car?

••19 A cord is used to vertically lower an initially stationary block of mass M at a constant downward acceleration of $g/4$. When the block has fallen a distance d, find (a) the work done by the cord's force on the block, (b) the work done by the gravitational force on the block, (c) the kinetic energy of the block, and (d) the speed of the block. SSM

••20 In Fig. 7-31, a horizontal force $\vec{F}_a$ of magnitude 20.0 N is applied to a 3.00 kg psychology book as the book slides a distance $d = 0.500$ m up a frictionless ramp at angle $\theta = 30.0°$. (a) During the displacement, what is the net work done on the book by $\vec{F}_a$, the gravitational force on the book, and the normal force on the book? (b) If the book has zero kinetic energy at the start of the displacement, what is its speed at the end of the displacement? GO

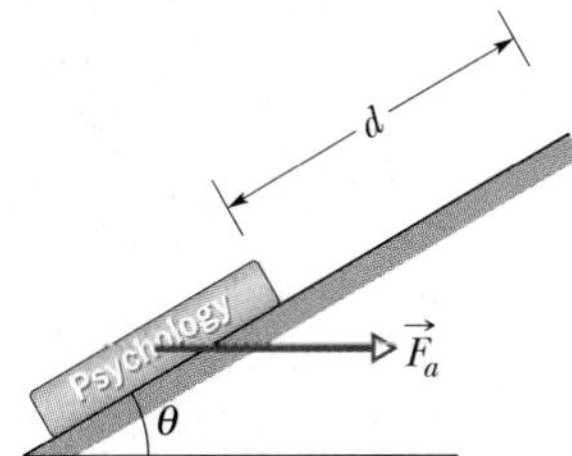

FIG. 7-31 Problem 20.

••21 In Fig. 7-32, a constant force $\vec{F}_a$ of magnitude 82.0 N is applied to a 3.00 kg shoe box at angle $\phi = 53.0°$, causing the box to move up a frictionless ramp at constant speed. How much work is done on the box by $\vec{F}_a$ when the box has moved through vertical distance $h = 0.150$ m?

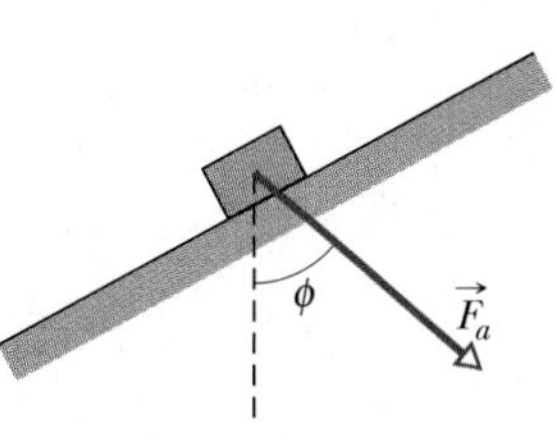

FIG. 7-32 Problem 21.

••22 A block is sent up a frictionless ramp along which an x axis extends upward. Figure 7-33 gives the kinetic energy of the block as a function of position x; the scale of the figure's vertical axis is set by $K_s = 40.0$ J. If the block's initial speed is 4.00 m/s, what is the normal force on the block?

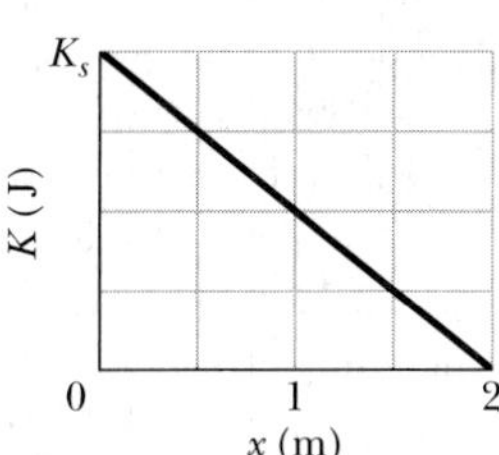

FIG. 7-33 Problem 22.

••23 In Fig. 7-34, a block of ice slides down a frictionless ramp at angle $\theta = 50°$ while an ice worker pulls on the block (via a rope) with a force $\vec{F}_r$ that has a magnitude of 50 N and is directed up the ramp. As the block slides through distance $d = 0.50$ m along the ramp, its kinetic energy increases by 80 J. How much greater would its kinetic energy have been if the rope had not been attached to the block?

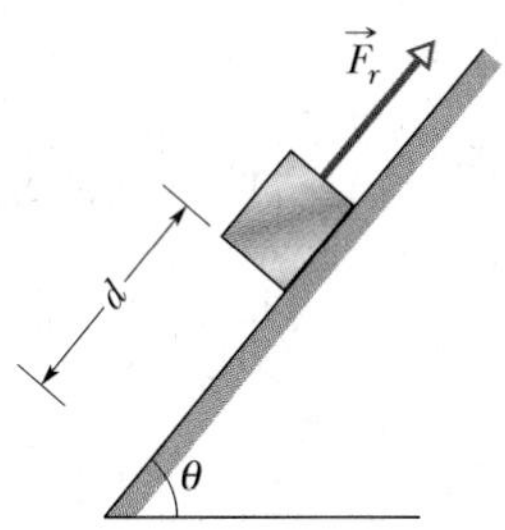

FIG. 7-34 Problem 23.

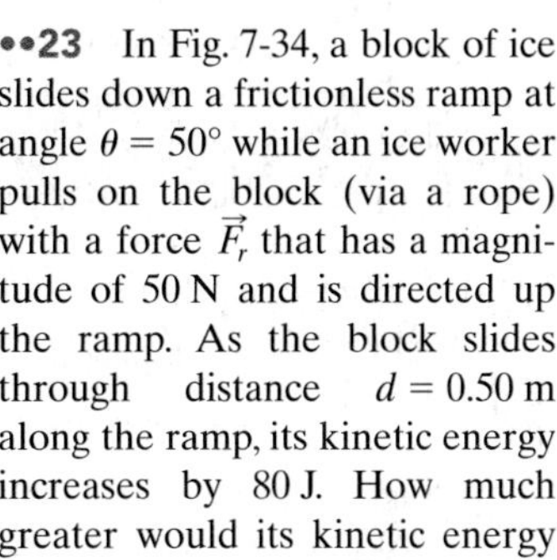

••24 A cave rescue team lifts an injured spelunker directly upward and out of a sinkhole by means of a motor-driven cable. The lift is performed in three stages, each requiring a vertical distance of 10.0 m: (a) the initially stationary spelunker is accelerated to a speed of 5.00 m/s; (b) he is then lifted at the constant speed of 5.00 m/s; (c) finally he is decelerated to zero speed. How much work is done on the 80.0 kg rescuee by the force lifting him during each stage?

•••25 In Fig. 7-35, a 0.250 kg block of cheese lies on the floor of a 900 kg elevator cab that is being pulled upward by a cable through distance $d_1 = 2.40$ m and then through distance $d_2 = 10.5$ m. (a) Through d_1, if the normal force on the block from the floor has constant magnitude $F_N = 3.00$ N, how much work is done on the cab by the force from the cable? (b) Through d_2, if the work done on the cab by the (constant) force from the cable is 92.61 kJ, what is the magnitude of F_N? GO

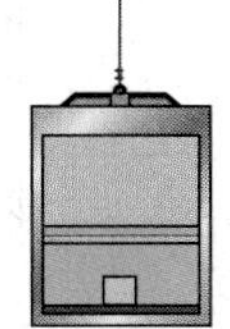
FIG. 7-35 Problem 25.

sec. 7-7 Work Done by a Spring Force

•26 During spring semester at MIT, residents of the parallel buildings of the East Campus dorms battle one another with large catapults that are made with surgical hose mounted on a window frame. A balloon filled with dyed water is placed in a pouch attached to the hose, which is then stretched through the width of the room. Assume that the stretching of the hose obeys Hooke's law with a spring constant of 100 N/m. If the hose is stretched by 5.00 m and then released, how much work does the force from the hose do on the balloon in the pouch by the time the hose reaches its relaxed length?

•27 A spring and block are in the arrangement of Fig. 7-11. When the block is pulled out to $x = +4.0$ cm, we must apply a force of magnitude 360 N to hold it there. We pull the block to $x = 11$ cm and then release it. How much work does the spring do on the block as the block moves from $x_i = +5.0$ cm to (a) $x = +3.0$ cm, (b) $x = -3.0$ cm, (c) $x = -5.0$ cm, and (d) $x = -9.0$ cm?

•28 In Fig. 7-11, we must apply a force of magnitude 80 N to hold the block stationary at $x = -2.0$ cm. From that position, we then slowly move the block so that our force does +4.0 J of work on the spring–block system; the block is then again stationary. What is the block's position? (*Hint:* There are two answers.)

••29 The only force acting on a 2.0 kg body as it moves along a positive x axis has an x component $F_x = -6x$ N, with x in meters. The velocity at $x = 3.0$ m is 8.0 m/s. (a) What is the velocity of the body at $x = 4.0$ m? (b) At what positive value of x will the body have a velocity of 5.0 m/s? SSM WWW

••30 Figure 7-36 gives spring force F_x versus position x for the spring–block arrangement of Fig. 7-11. The scale is set by $F_s = 160.0$ N. We release the block at $x = 12$ cm. How much work does the spring do on the block when the block moves from $x_i = +8.0$ cm to (a) $x = +5.0$ cm, (b) $x = -5.0$ cm, (c) $x = -8.0$ cm, and (d) $x = -10.0$ cm?

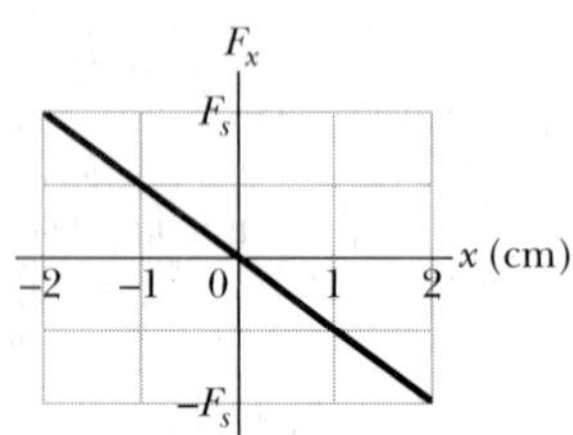

FIG. 7-36 Problem 30.

••31 In the arrangement of Fig. 7-11, we gradually pull the block from $x = 0$ to $x = +3.0$ cm, where it is stationary. Figure 7-37 gives the work that our force does on the block. The scale of the figure's vertical axis is set by $W_s = 1.0$ J. We then pull the block out to $x = +5.0$ cm and release it from rest. How much work does the spring do on the block when the block moves from $x_i = +5.0$ cm to (a) $x = +4.0$ cm, (b) $x = -2.0$ cm, and (c) $x = -5.0$ cm?

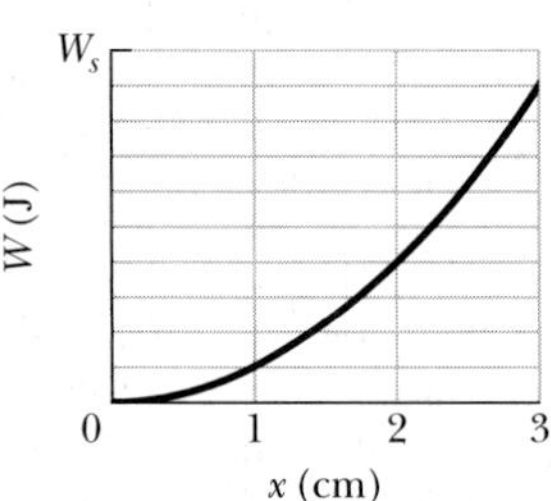

FIG. 7-37 Problem 31.

••32 In Fig. 7-11a, a block of mass m lies on a horizontal frictionless surface and is attached to one end of a horizontal spring (spring constant k) whose other end is fixed. The block is initially at rest at the position where the spring is unstretched ($x = 0$) when a constant horizontal force $\vec{F}$ in the positive direction of the x axis is applied to it. A plot of the resulting kinetic energy of the block versus its position x is shown in Fig. 7-38. The scale of the figure's vertical axis is set by $K_s = 4.0$ J. (a) What is the magnitude of $\vec{F}$? (b) What is the value of k?

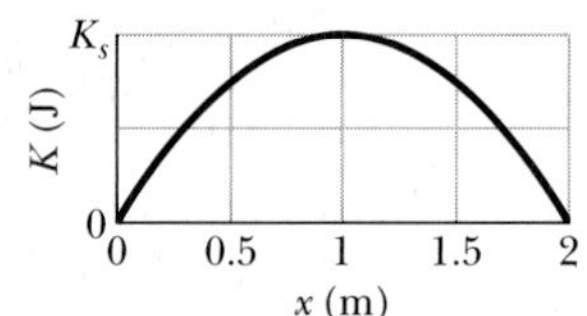

FIG. 7-38 Problem 32.

•••33 The block in Fig. 7-11a lies on a horizontal frictionless surface, and the spring constant is 50 N/m. Initially, the spring is at its relaxed length and the block is stationary at position $x = 0$. Then an applied force with a constant magnitude of 3.0 N pulls the block in the positive direction of the x axis, stretching the spring until the block stops. When that stopping point is reached, what are (a) the position of the block, (b) the work that has been done on the block by the applied force, and (c) the work that has been done on the block by the spring force? During the block's displacement, what are (d) the block's position when its kinetic energy is maximum and (e) the value of that maximum kinetic energy?

sec. 7-8 Work Done by a General Variable Force

•34 A 5.0 kg block moves in a straight line on a horizontal frictionless surface under the influence of a force that varies with position as shown in Fig. 7-39. The scale of the figure's vertical axis is set by $F_s = 10.0$ N. How much work is done by the force as the block moves from the origin to $x = 8.0$ m?

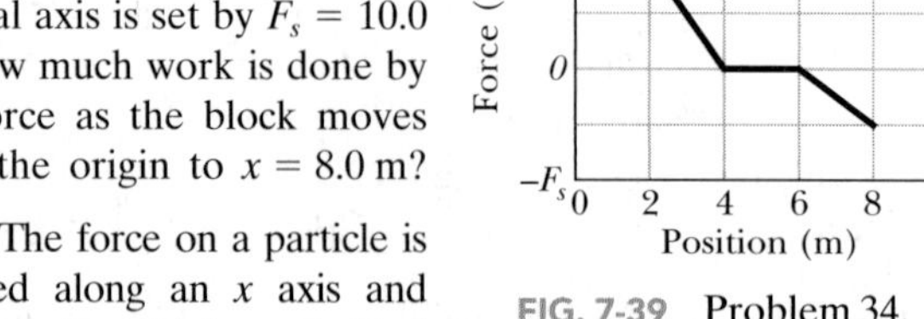

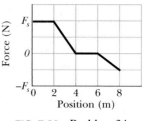

FIG. 7-39 Problem 34.

•35 The force on a particle is directed along an x axis and given by $F = F_0(x/x_0 - 1)$. Find the work done by the force in moving the particle from $x = 0$ to $x = 2x_0$ by (a) plotting $F(x)$ and measuring the work from the graph and (b) integrating $F(x)$. SSM WWW

•36 A 10 kg brick moves along an x axis. Its acceleration as a function of its position is shown in Fig. 7-40. The scale of the figure's vertical axis is set by $a_s = 20.0$ m/s². What is the net work performed on the brick by the force causing the acceleration as the brick moves from $x = 0$ to $x = 8.0$ m? ILW

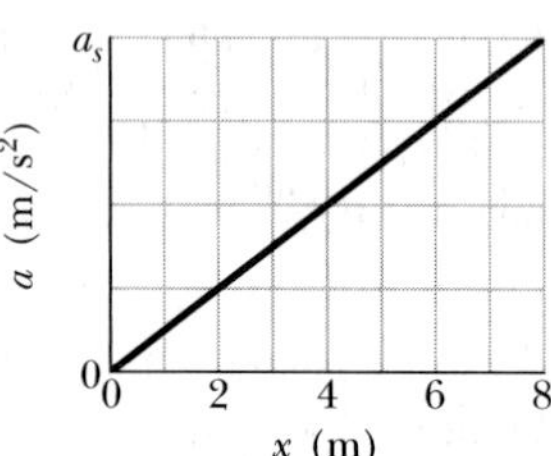

FIG. 7-40 Problem 36.

••37 A single force acts on a 3.0 kg particle-like object whose position is given by $x = 3.0t - 4.0t^2 + 1.0t^3$, with x in meters and t in seconds. Find the work done on the object by the force from $t = 0$ to $t = 4.0$ s.

••38 A can of sardines is made to move along an x axis from $x = 0.25$ m to $x = 1.25$ m by a force with a magnitude given by $F = \exp(-4x^2)$, with x in meters and F in newtons. (Here exp is the exponential function.) How much work is done on the can by the force?

••39 Figure 7-41 gives the acceleration of a 2.00 kg particle as an applied force $\vec{F}_a$ moves it from rest along an x axis from $x = 0$ to $x = 9.0$ m. The scale of the figure's vertical axis is set by $a_s = 6.0$ m/s². How much work has the force done on the particle when the particle reaches (a) $x = 4.0$ m, (b) $x = 7.0$ m, and (c) $x = 9.0$ m? What is the particle's speed and direction of travel when it reaches (d) $x = 4.0$ m, (e) $x = 7.0$ m, and (f) $x = 9.0$ m?

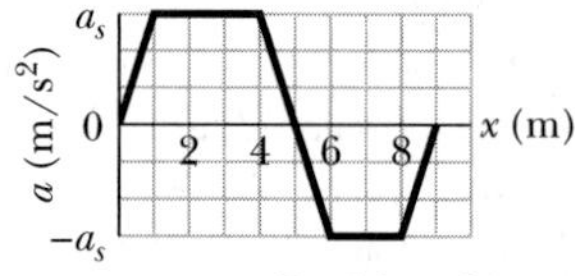

FIG. 7-41 Problem 39.

••40 A 1.5 kg block is initially at rest on a horizontal frictionless surface when a horizontal force along an x axis is applied to the block. The force is given by $\vec{F}(x) = (2.5 - x^2)\hat{i}$ N, where x is in meters and the initial position of the block is $x = 0$. (a) What is the kinetic energy of the block as it passes through $x = 2.0$ m? (b) What is the maximum kinetic energy of the block between $x = 0$ and $x = 2.0$ m?

••41 A force $\vec{F} = (cx - 3.00x^2)\hat{i}$ acts on a particle as the particle moves along an x axis, with $\vec{F}$ in newtons, x in meters, and c a constant. At $x = 0$, the particle's kinetic energy is 20.0 J; at $x = 3.00$ m, it is 11.0 J. Find c. GO

•••42 Figure 7-42 shows a cord attached to a cart that can slide along a frictionless horizontal rail aligned along an x axis. The left end of the cord is pulled over a pulley, of negligible mass and friction and at cord height $h = 1.20$ m, so the cart slides from $x_1 = 3.00$ m to $x_2 = 1.00$ m. During the move, the tension in the cord is a constant 25.0 N. What is the change in the kinetic energy of the cart during the move?

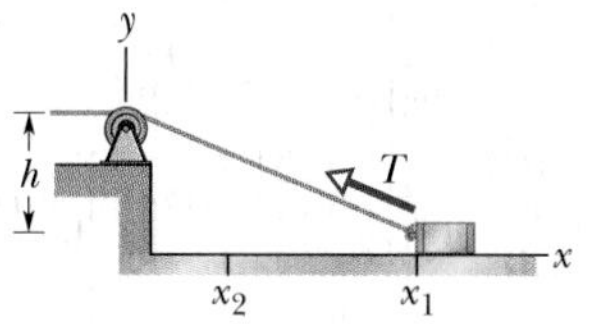

FIG. 7-42 Problem 42.

sec. 7-9 Power

•**43** A 100 kg block is pulled at a constant speed of 5.0 m/s across a horizontal floor by an applied force of 122 N directed 37° above the horizontal. What is the rate at which the force does work on the block? SSM ILW

•**44** The loaded cab of an elevator has a mass of 3.0×10^3 kg and moves 210 m up the shaft in 23 s at constant speed. At what average rate does the force from the cable do work on the cab?

•**45** A force of 5.0 N acts on a 15 kg body initially at rest. Compute the work done by the force in (a) the first, (b) the second, and (c) the third seconds and (d) the instantaneous power due to the force at the end of the third second. SSM

•**46** A skier is pulled by a towrope up a frictionless ski slope that makes an angle of 12° with the horizontal. The rope moves parallel to the slope with a constant speed of 1.0 m/s. The force of the rope does 900 J of work on the skier as the skier moves a distance of 8.0 m up the incline. (a) If the rope moved with a constant speed of 2.0 m/s, how much work would the force of the rope do on the skier as the skier moved a distance of 8.0 m up the incline? At what rate is the force of the rope doing work on the skier when the rope moves with a speed of (b) 1.0 m/s and (c) 2.0 m/s?

••**47** A fully loaded, slow-moving freight elevator has a cab with a total mass of 1200 kg, which is required to travel upward 54 m in 3.0 min, starting and ending at rest. The elevator's counterweight has a mass of only 950 kg, and so the elevator motor must help. What average power is required of the force the motor exerts on the cab via the cable? SSM

••**48** (a) At a certain instant, a particle-like object is acted on by a force $\vec{F} = (4.0\text{ N})\hat{i} - (2.0\text{ N})\hat{j} + (9.0\text{ N})\hat{k}$ while the object's velocity is $\vec{v} = -(2.0\text{ m/s})\hat{i} + (4.0\text{ m/s})\hat{k}$. What is the instantaneous rate at which the force does work on the object? (b) At some other time, the velocity consists of only a y component. If the force is unchanged and the instantaneous power is -12 W, what is the velocity of the object?

••**49** A machine carries a 4.0 kg package from an initial position of $\vec{d}_i = (0.50\text{ m})\hat{i} + (0.75\text{ m})\hat{j} + (0.20\text{ m})\hat{k}$ at $t = 0$ to a final position of $\vec{d}_f = (7.50\text{ m})\hat{i} + (12.0\text{ m})\hat{j} + (7.20\text{ m})\hat{k}$ at $t = 12$ s. The constant force applied by the machine on the package is $\vec{F} = (2.00\text{ N})\hat{i} + (4.00\text{ N})\hat{j} + (6.00\text{ N})\hat{k}$. For that displacement, find (a) the work done on the package by the machine's force and (b) the average power of the machine's force on the package.

••**50** A 0.30 kg ladle sliding on a horizontal frictionless surface is attached to one end of a horizontal spring (k = 500 N/m) whose other end is fixed. The ladle has a kinetic energy of 10 J as it passes through its equilibrium position (the point at which the spring force is zero). (a) At what rate is the spring doing work on the ladle as the ladle passes through its equilibrium position? (b) At what rate is the spring doing work on the ladle when the spring is compressed 0.10 m and the ladle is moving away from the equilibrium position?

••**51** A force $\vec{F} = (3.00\text{ N})\hat{i} + (7.00\text{ N})\hat{j} + (7.00\text{ N})\hat{k}$ acts on a 2.00 kg mobile object that moves from an initial position of $\vec{d}_i = (3.00\text{ m})\hat{i} - (2.00\text{ m})\hat{j} + (5.00\text{ m})\hat{k}$ to a final position of $\vec{d}_f = -(5.00\text{ m})\hat{i} + (4.00\text{ m})\hat{j} + (7.00\text{ m})\hat{k}$ in 4.00 s. Find (a) the work done on the object by the force in the 4.00 s interval, (b) the average power due to the force during that interval, and (c) the angle between vectors $\vec{d}_i$ and $\vec{d}_f$.

•••**52** A funny car accelerates from rest through a measured track distance in time T with the engine operating at a constant power P. If the track crew can increase the engine power by a differential amount dP, what is the change in the time required for the run?

Additional Problems

53 An explosion at ground level leaves a crater with a diameter that is proportional to the energy of the explosion raised to the $\frac{1}{3}$ power; an explosion of 1 megaton of TNT leaves a crater with a 1 km diameter. Below Lake Huron in Michigan there appears to be an ancient impact crater with a 50 km diameter. What was the kinetic energy associated with that impact, in terms of (a) megatons of TNT (1 megaton yields 4.2×10^{15} J) and (b) Hiroshima bomb equivalents (13 kilotons of TNT each)? (Ancient meteorite or comet impacts may have significantly altered Earth's climate and contributed to the extinction of the dinosaurs and other life-forms.)

54 A 250 g block is dropped onto a relaxed vertical spring that has a spring constant of k = 2.5 N/cm (Fig. 7-43). The block becomes attached to the spring and compresses the spring 12 cm before momentarily stopping. While the spring is being compressed, what work is done on the block by (a) the gravitational force on it and (b) the spring force? (c) What is the speed of the block just before it hits the spring? (Assume that friction is negligible.) (d) If the speed at impact is doubled, what is the maximum compression of the spring?

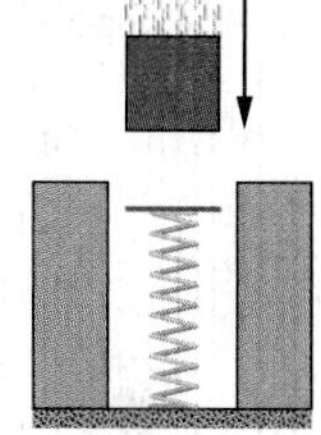

FIG. 7-43 Problem 54.

55 How much work is done by a force $\vec{F} = (2x\text{ N})\hat{i} + (3\text{ N})\hat{j}$, with x in meters, that moves a particle from a position $\vec{r}_i = (2\text{ m})\hat{i} + (3\text{ m})\hat{j}$ to a position $\vec{r}_f = -(4\text{ m})\hat{i} - (3\text{ m})\hat{j}$?

56 To pull a 50 kg crate across a horizontal frictionless floor, a worker applies a force of 210 N, directed 20° above the horizontal. As the crate moves 3.0 m, what work is done on the crate by (a) the worker's force, (b) the gravitational force on the crate, and (c) the normal force on the crate from the floor? (d) What is the total work done on the crate?

57 In Fig. 7-44, a cord runs around two massless, frictionless pulleys. A canister with mass m = 20 kg hangs from one pulley, and you exert a force $\vec{F}$ on the free end of the cord. (a) What must be the magnitude of $\vec{F}$ if you are to lift the canister at a constant speed? (b) To lift the canister by 2.0 cm, how far must you

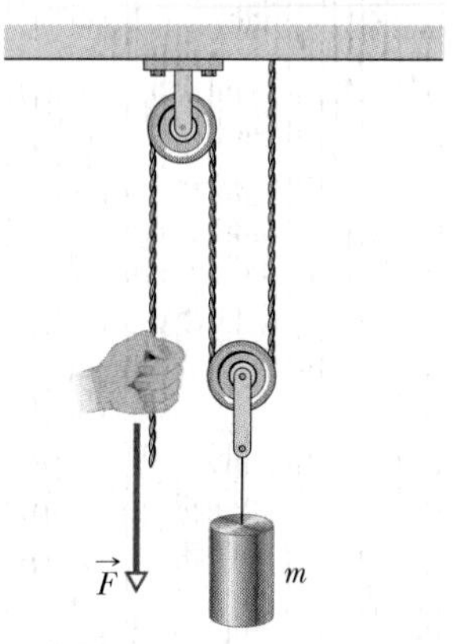

FIG. 7-44 Problem 57.

pull the free end of the cord? During that lift, what is the work done on the canister by (c) your force (via the cord) and (d) the gravitational force? (*Hint:* When a cord loops around a pulley as shown, it pulls on the pulley with a net force that is twice the tension in the cord.)

58 A force $\vec{F} = (4.0\text{ N})\hat{\text{i}} + c\hat{\text{j}}$ acts on a particle as the particle goes through displacement $\vec{d} = (3.0\text{ m})\hat{\text{i}} - (2.0\text{ m})\hat{\text{j}}$. (Other forces also act on the particle.) What is c if the work done on the particle by force $\vec{F}$ is (a) 0, (b) 17 J, and (c) −18 J?

59 A constant force of magnitude 10 N makes an angle of 150° (measured counterclockwise) with the positive x direction as it acts on a 2.0 kg object moving in an xy plane. How much work is done on the object by the force as the object moves from the origin to the point having position vector $(2.0\text{ m})\hat{\text{i}} - (4.0\text{ m})\hat{\text{j}}$?

60 An initially stationary 2.0 kg object accelerates horizontally and uniformly to a speed of 10 m/s in 3.0 s. (a) In that 3.0 s interval, how much work is done on the object by the force accelerating it? What is the instantaneous power due to that force (b) at the end of the interval and (c) at the end of the first half of the interval?

61 If a ski lift raises 100 passengers averaging 660 N in weight to a height of 150 m in 60.0 s, at constant speed, what average power is required of the force making the lift?

62 Boxes are transported from one location to another in a warehouse by means of a conveyor belt that moves with a constant speed of 0.50 m/s. At a certain location the conveyor belt moves for 2.0 m up an incline that makes an angle of 10° with the horizontal, then for 2.0 m horizontally, and finally for 2.0 m down an incline that makes an angle of 10° with the horizontal. Assume that a 2.0 kg box rides on the belt without slipping. At what rate is the force of the conveyor belt doing work on the box as the box moves (a) up the 10° incline, (b) horizontally, and (c) down the 10° incline?

63 A horse pulls a cart with a force of 40 lb at an angle of 30° above the horizontal and moves along at a speed of 6.0 mi/h. (a) How much work does the force do in 10 min? (b) What is the average power (in horsepower) of the force? SSM

64 An iceboat is at rest on a frictionless frozen lake when a sudden wind exerts a constant force of 200 N, toward the east, on the boat. Due to the angle of the sail, the wind causes the boat to slide in a straight line for a distance of 8.0 m in a direction 20° north of east. What is the kinetic energy of the iceboat at the end of that 8.0 m?

65 A 230 kg crate hangs from the end of a rope of length $L = 12.0$ m. You push horizontally on the crate with a varying force $\vec{F}$ to move it distance $d = 4.00$ m to the side (Fig. 7-45). (a) What is the magnitude of $\vec{F}$ when the crate is in this final position? During the crate's displacement, what are (b) the total work done on it, (c) the work done by the gravitational force on the crate, and (d) the work done by the pull on the crate from the rope? (e) Knowing that the crate is motionless before and after its displacement, use the answers to (b), (c), and (d) to find the work your force $\vec{F}$ does on the crate. (f) Why is the work of your force not equal to the product of the horizontal displacement and the answer to (a)?

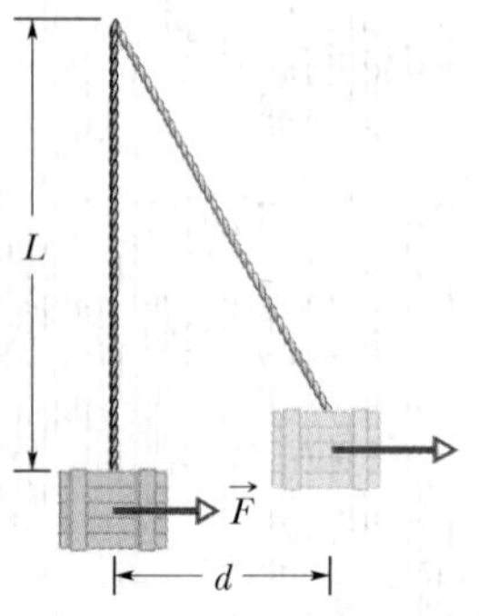

FIG. 7-45 Problem 65.

66 The only force acting on a 2.0 kg body as the body moves along an x axis varies as shown in Fig. 7-46. The scale of the figure's vertical axis is set by F_s = 4.0 N. The velocity of the body at $x = 0$ is 4.0 m/s. (a) What is the kinetic energy of the body at $x = 3.0$ m? (b) At what value of x will the body have a kinetic energy of 8.0 J? (c) What is the maximum kinetic energy of the body between $x = 0$ and $x = 5.0$ m?

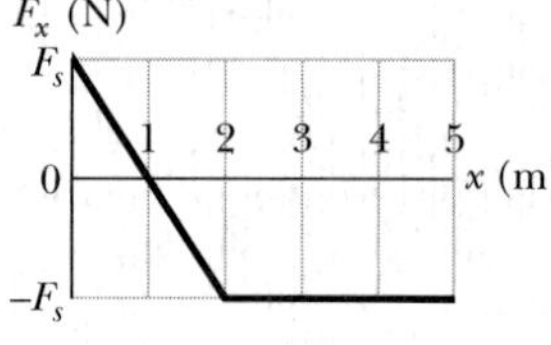

FIG. 7-46 Problem 66.

67 Figure 7-47 shows a cold package of hot dogs sliding rightward across a frictionless floor through a distance d = 20.0 cm while three forces act on the package. Two of them are horizontal and have the magnitudes F_1 = 5.00 N and F_2 = 1.00 N; the third is angled down by θ = 60.0° and has the magnitude F_3 = 4.00 N. (a) For the 20.0 cm displacement, what is the *net* work done on the package by the three applied forces, the gravitational force on the package, and the normal force on the package? (b) If the package has a mass of 2.0 kg and an initial kinetic energy of 0, what is its speed at the end of the displacement?

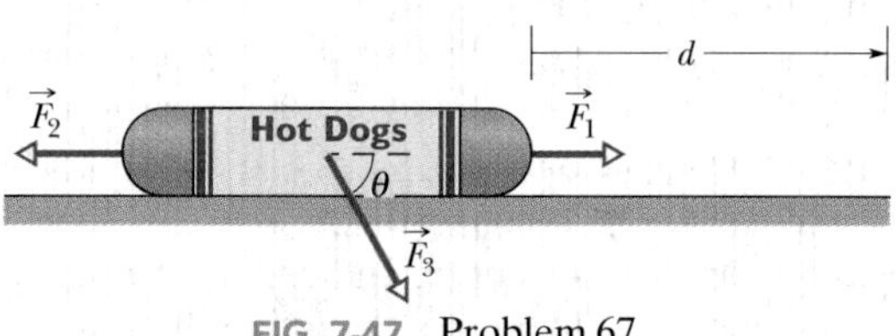

FIG. 7-47 Problem 67.

68 A frightened child is restrained by her mother as the child slides down a frictionless playground slide. If the force on the child from the mother is 100 N up the slide, the child's kinetic energy increases by 30 J as she moves down the slide a distance of 1.8 m. (a) How much work is done on the child by the gravitational force during the 1.8 m descent? (b) If the child is not restrained by her mother, how much will the child's kinetic energy increase as she comes down the slide that same distance of 1.8 m?

69 To push a 25.0 kg crate up a frictionless incline, angled at 25.0° to the horizontal, a worker exerts a force of 209 N parallel to the incline. As the crate slides 1.50 m, how much work is done on the crate by (a) the worker's applied force, (b) the gravitational force on the crate, and (c) the normal force exerted by the incline on the crate? (d) What is the total work done on the crate? SSM

70 If a car of mass 1200 kg is moving along a highway at 120 km/h, what is the car's kinetic energy as determined by someone standing alongside the highway?

71 A spring with a pointer attached is hanging next to a scale marked in millimeters. Three different packages are hung from the spring, in turn, as shown in Fig. 7-48. (a) Which mark on the scale will the pointer indicate when no package is hung from the spring? (b) What is the weight W of the third package? SSM

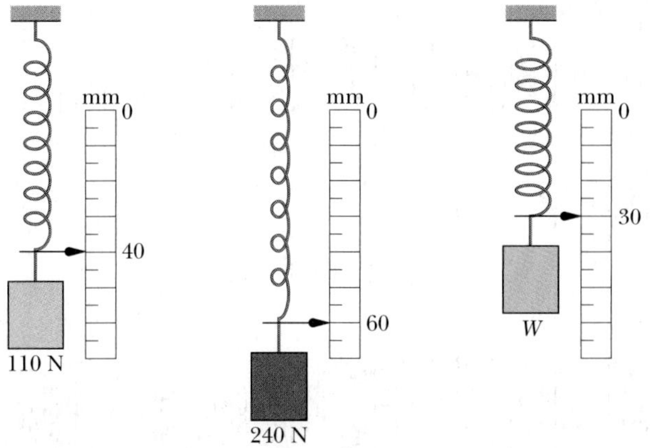

mm 0
40
110 N
mm 0
60
240 N
mm 0
30
W

FIG. 7-48 Problem 71.

72 A particle moves along a straight path through displacement $\vec{d} = (8\text{ m})\hat{\text{i}} + c\hat{\text{j}}$ while force $\vec{F} = (2\text{ N})\hat{\text{i}} - (4\text{ N})\hat{\text{j}}$ acts on it. (Other forces also act on the particle.) What is the value of c if the work done by $\vec{F}$ on the particle is (a) zero, (b) positive, and (c) negative?

73 An elevator cab has a mass of 4500 kg and can carry a maximum load of 1800 kg. If the cab is moving upward at full load at 3.80 m/s, what power is required of the force moving the cab to maintain that speed? SSM

74 A 45 kg block of ice slides down a frictionless incline 1.5 m long and 0.91 m high. A worker pushes up against the ice, parallel to the incline, so that the block slides down at constant speed. (a) Find the magnitude of the worker's force. How much work is done on the block by (b) the worker's force, (c) the gravitational force on the block, (d) the normal force on the block from the surface of the incline, and (e) the net force on the block?

75 A force $\vec{F}$ in the positive direction of an x axis acts on an object moving along the axis. If the magnitude of the force is $F = 10e^{-x/2.0}$ N, with x in meters, find the work done by $\vec{F}$ as the object moves from $x = 0$ to $x = 2.0$ m by (a) plotting $F(x)$ and estimating the area under the curve and (b) integrating to find the work analytically.

76 In Fig. 7-49*a*, a 2.0 N force is applied to a 4.0 kg block at a downward angle θ as the block moves rightward through 1.0 m across a frictionless floor. Find an expression for the speed v_f of the block at the end of that distance if the block's initial velocity is (a) 0 and (b) 1.0 m/s to the right. (c) The situation in Fig. 7-49*b* is similar in that the block is initially moving at 1.0 m/s to the right, but now the 2.0 N force is directed downward to the left. Find an expression for the speed v_f of the block at the end of the 1.0 m distance. (d) Graph all three expressions for v_f versus downward angle θ for $\theta = 0°$ to $\theta = 90°$. Interpret the graphs.

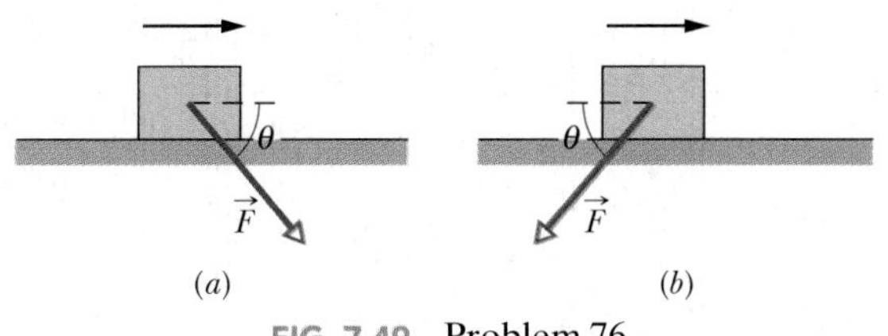

FIG. 7-49 Problem 76.

77 A 2.0 kg lunchbox is sent sliding over a frictionless surface, in the positive direction of an x axis along the surface. Beginning at time $t = 0$, a steady wind pushes on the lunchbox in the negative direction of the x axis. Figure 7-50 shows the position x of the lunchbox as a function of time t as the wind pushes on the lunchbox. From the graph, estimate the kinetic energy of the lunchbox at (a) $t = 1.0$ s and (b) $t = 5.0$ s. (c) How much work does the force from the wind do on the lunchbox from $t = 1.0$ s to $t = 5.0$ s? SSM

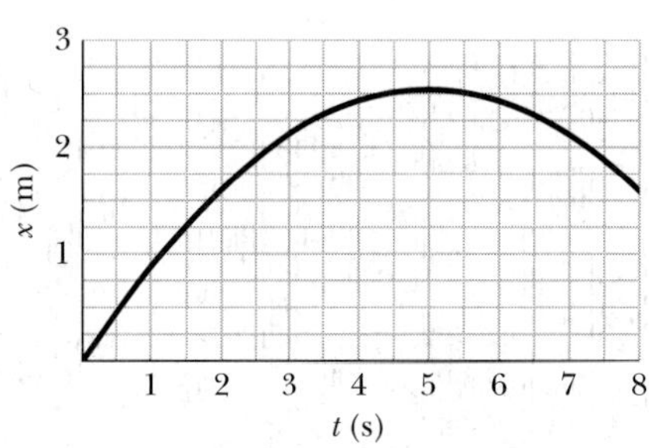

FIG. 7-50 Problem 77.

78 *Numerical integration.* A breadbox is made to move along an x axis from $x = 0.15$ m to $x = 1.20$ m by a force with a magnitude given by $F = \exp(-2x^2)$, with x in meters and F in newtons. (Here exp is the exponential function.) How much work is done on the breadbox by the force?

79 As a particle moves along an x axis, a force in the positive direction of the axis acts on it. Figure 7-51 shows the magnitude F of the force versus position x of the particle. The curve is given by $F = a/x^2$, with $a = 9.0\text{ N}\cdot\text{m}^2$. Find the work done on the particle by the force as the particle moves from $x = 1.0$ m to $x = 3.0$ m by (a) estimating the work from the graph and (b) integrating the force function.

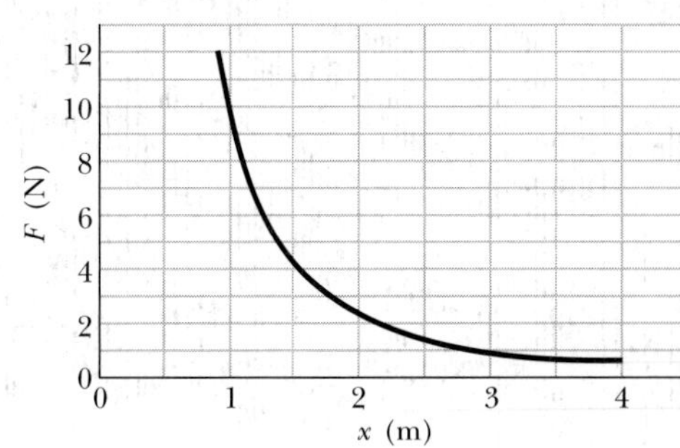

FIG. 7-51 Problem 79.

80 A CD case slides along a floor in the positive direction of an x axis while an applied force $\vec{F}_a$ acts on the case. The force is directed along the x axis and has the x component $F_{ax} = 9x - 3x^2$, with x in meters and F_{ax} in newtons. The case starts at rest at the position $x = 0$, and it moves until it is again at rest. (a) Plot the work $\vec{F}_a$ does on the case as a function of x. (b) At what position is the work maximum, and (c) what is that maximum value? (d) At what position has the work decreased to zero? (e) At what position is the case again at rest?

8 Potential Energy and Conservation of Energy

Courtesy Mark Reid, USGS

When an avalanche of rocks moves down a mountainside and into a valley, friction between the moving rocks and the land slow and eventually stop the flow. The runout (the distance the rocks can move across a valley) is typically 2/3 of the height from which the rocks descended. However, in large avalanches, where a large amount of material comes down a mountainside, the runout can be 30 times as much, which can take a seemingly safe community by fatal surprise.

Why can large avalanches have such long runouts?

The answer is in this chapter.

8-1 WHAT IS PHYSICS?

One job of physics is to identify the different types of energy in the world, especially those that are of common importance. One general type of energy is **potential energy** U. Technically, potential energy is energy that can be associated with the configuration (arrangement) of a system of objects that exert forces on one another.

This is a pretty formal definition of something that is actually familiar to you. An example might help better than the definition: A bungee-cord jumper plunges from a staging platform (Fig. 8-1). The system of objects consists of Earth and the jumper. The force between the objects is the gravitational force. The configuration of the system changes (the separation between the jumper and Earth decreases—that is, of course, the thrill of the jump). We can account for the jumper's motion and increase in kinetic energy by defining a **gravitational potential energy** U. This is the energy associated with the state of separation between two objects that attract each other by the gravitational force, here the jumper and Earth.

When the jumper begins to stretch the bungee cord near the end of the plunge, the system of objects consists of the cord and the jumper. The force between the objects is an elastic (spring-like) force. The configuration of the system changes (the cord stretches). We can account for the jumper's decrease in kinetic energy and the cord's increase in length by defining an **elastic potential energy** U. This is the energy associated with the state of compression or extension of an elastic object, here the bungee cord.

Physics determines how the potential energy of a system can be calculated so that energy might be stored or put to use. For example, before any particular bungee-cord jumper takes the plunge, someone (probably a mechanical engineer) must determine the correct cord to be used by calculating the gravitational and elastic potential energies that can be expected. Then the jump is only thrilling and not fatal.

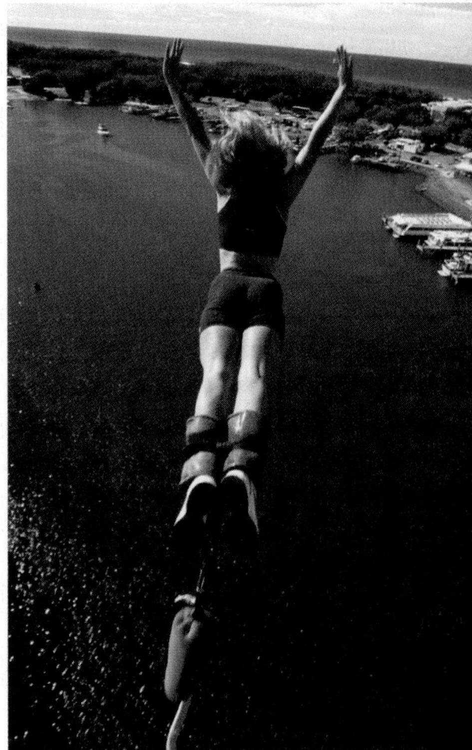

FIG. 8-1 The kinetic energy of a bungee-cord jumper increases during the free fall, and then the cord begins to stretch, slowing the jumper. *(KOFUJIWARA/amana images/ Getty Images News and Sport Services)*

8-2 | Work and Potential Energy

In Chapter 7 we discussed the relation between work and a change in kinetic energy. Here we discuss the relation between work and a change in potential energy.

Let us throw a tomato upward (Fig. 8-2). We already know that as the tomato rises, the work W_g done on the tomato by the gravitational force is negative because the force transfers energy *from* the kinetic energy of the tomato. We can now finish the story by saying that this energy is transferred by the gravitational force *to* the gravitational potential energy of the tomato–Earth system.

The tomato slows, stops, and then begins to fall back down because of the gravitational force. During the fall, the transfer is reversed: The work W_g done on the tomato by the gravitational force is now positive—that force transfers energy *from* the gravitational potential energy of the tomato–Earth system *to* the kinetic energy of the tomato.

For either rise or fall, the change ΔU in gravitational potential energy is defined as being equal to the negative of the work done on the tomato by the gravitational force. Using the general symbol W for work, we write this as

$$\Delta U = -W. \tag{8-1}$$

This equation also applies to a block–spring system, as in Fig. 8-3. If we abruptly shove the block to send it moving rightward, the spring force acts leftward and thus does negative work on the block, transferring energy from the kinetic energy of the block to the elastic potential energy of the spring–block

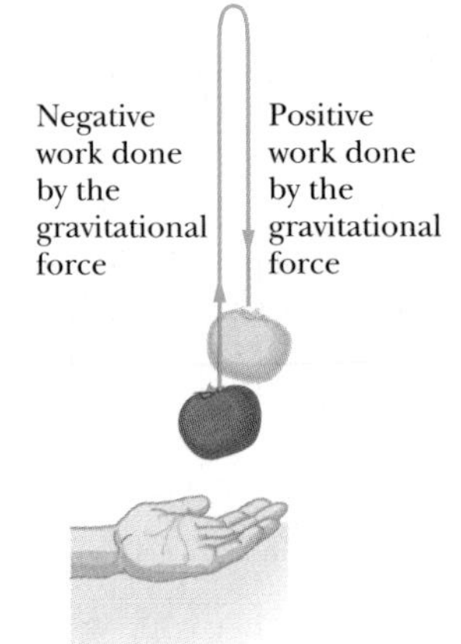

FIG. 8-2 A tomato is thrown upward. As it rises, the gravitational force does negative work on it, decreasing its kinetic energy. As the tomato descends, the gravitational force does positive work on it, increasing its kinetic energy.

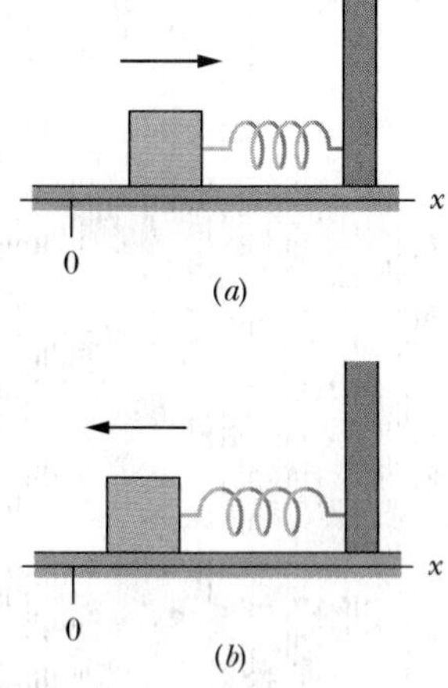

FIG. 8-3 A block, attached to a spring and initially at rest at $x = 0$, is set in motion toward the right. (*a*) As the block moves rightward (as indicated by the arrow), the spring force does negative work on it. (*b*) Then, as the block moves back toward $x = 0$, the spring force does positive work on it.

system. The block slows and eventually stops, and then begins to move leftward because the spring force is still leftward. The transfer of energy is then reversed—it is from potential energy of the spring–block system to kinetic energy of the block.

Conservative and Nonconservative Forces

Let us list the key elements of the two situations we just discussed:

1. The *system* consists of two or more objects.
2. A *force* acts between a particle-like object (tomato or block) in the system and the rest of the system.
3. When the system configuration changes, the force does *work* (call it W_1) on the particle-like object, transferring energy between the kinetic energy K of the object and some other type of energy of the system.
4. When the configuration change is reversed, the force reverses the energy transfer, doing work W_2 in the process.

In a situation in which $W_1 = -W_2$ is always true, the other type of energy is a potential energy and the force is said to be a **conservative force.** As you might suspect, the gravitational force and the spring force are both conservative (since otherwise we could not have spoken of gravitational potential energy and elastic potential energy, as we did previously).

A force that is not conservative is called a **nonconservative force.** The kinetic frictional force and drag force are nonconservative. For an example, let us send a block sliding across a floor that is not frictionless. During the sliding, a kinetic frictional force from the floor slows the block by transferring energy from its kinetic energy to a type of energy called *thermal energy* (which has to do with the random motions of atoms and molecules). We know from experiment that this energy transfer cannot be reversed (thermal energy cannot be transferred back to kinetic energy of the block by the kinetic frictional force). Thus, although we have a system (made up of the block and the floor), a force that acts between parts of the system, and a transfer of energy by the force, the force is not conservative. Therefore, thermal energy is not a potential energy.

When only conservative forces act on a particle-like object, we can greatly simplify otherwise difficult problems involving motion of the object. The next section, in which we develop a test for identifying conservative forces, provides one means for simplifying such problems.

8-3 | Path Independence of Conservative Forces

The primary test for determining whether a force is conservative or nonconservative is this: Let the force act on a particle that moves along any *closed path,* beginning at some initial position and eventually returning to that position (so that the particle makes a *round trip* beginning and ending at the initial position). The force is conservative only if the total energy it transfers to and from the particle during the round trip along this and any other closed path is zero. In other words:

> The net work done by a conservative force on a particle moving around any closed path is zero.

We know from experiment that the gravitational force passes this *closed-path test.* An example is the tossed tomato of Fig. 8-2. The tomato leaves the launch point with speed v_0 and kinetic energy $\frac{1}{2}mv_0^2$. The gravitational force acting on the tomato slows it, stops it, and then causes it to fall back down. When the tomato returns to the launch point, it again has speed v_0 and kinetic energy $\frac{1}{2}mv_0^2$. Thus, the gravitational force transfers as much energy *from* the tomato during the ascent as it transfers *to* the tomato during the descent back to the launch point. The net work done on the tomato by the gravitational force during the round trip is zero.

An important result of the closed-path test is that:

> The work done by a conservative force on a particle moving between two points does not depend on the path taken by the particle.

For example, suppose that a particle moves from point a to point b in Fig. 8-4a along either path 1 or path 2. If only a conservative force acts on the particle, then the work done on the particle is the same along the two paths. In symbols, we can write this result as

$$W_{ab,1} = W_{ab,2}, \tag{8-2}$$

where the subscript ab indicates the initial and final points, respectively, and the subscripts 1 and 2 indicate the path.

This result is powerful because it allows us to simplify difficult problems when only a conservative force is involved. Suppose you need to calculate the work done by a conservative force along a given path between two points, and the calculation is difficult or even impossible without additional information. You can find the work by substituting some other path between those two points for which the calculation is easier and possible. Sample Problem 8-1 gives an example, but first we need to prove Eq. 8-2.

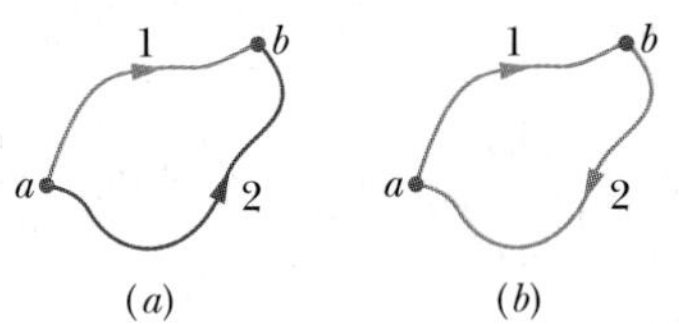

FIG. 8-4 (a) As a conservative force acts on it, a particle can move from point a to point b along either path 1 or path 2. (b) The particle moves in a round trip, from point a to point b along path 1 and then back to point a along path 2.

Proof of Equation 8-2

Figure 8-4b shows an arbitrary round trip for a particle that is acted upon by a single force. The particle moves from an initial point a to point b along path 1 and then back to point a along path 2. The force does work on the particle as the particle moves along each path. Without worrying about where positive work is done and where negative work is done, let us just represent the work done from a to b along path 1 as $W_{ab,1}$ and the work done from b back to a along path 2 as $W_{ba,2}$. If the force is conservative, then the net work done during the round trip must be zero:

$$W_{ab,1} + W_{ba,2} = 0,$$

and thus

$$W_{ab,1} = -W_{ba,2}. \tag{8-3}$$

In words, the work done along the outward path must be the negative of the work done along the path back.

Let us now consider the work $W_{ab,2}$ done on the particle by the force when the particle moves from a to b along path 2, as indicated in Fig. 8-4a. If the force is conservative, that work is the negative of $W_{ba,2}$:

$$W_{ab,2} = -W_{ba,2}. \tag{8-4}$$

Substituting $W_{ab,2}$ for $-W_{ba,2}$ in Eq. 8-3, we obtain

$$W_{ab,1} = W_{ab,2},$$

which is what we set out to prove.

CHECKPOINT 1 The figure shows three paths connecting points a and b. A single force $\vec{F}$ does the indicated work on a particle moving along each path in the indicated direction. On the basis of this information, is force $\vec{F}$ conservative?

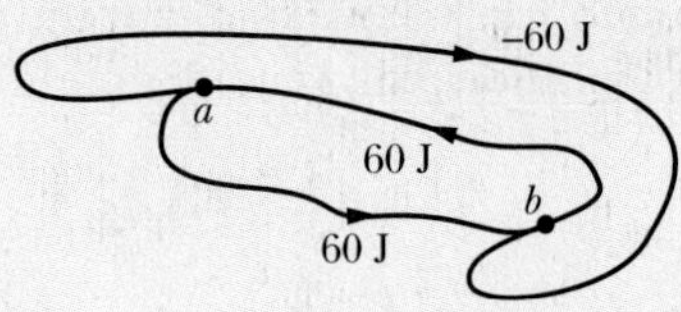

Sample Problem 8-1

Figure 8-5*a* shows a 2.0 kg block of slippery cheese that slides along a frictionless track from point *a* to point *b*. The cheese travels through a total distance of 2.0 m along the track, and a net vertical distance of 0.80 m. How much work is done on the cheese by the gravitational force during the slide?

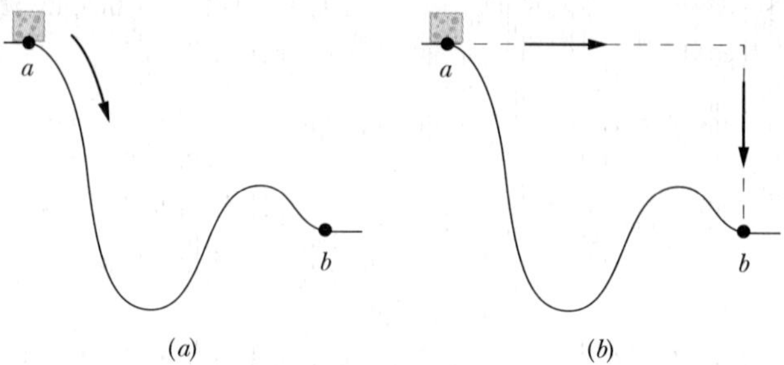

FIG. 8-5 (*a*) A block of cheese slides along a frictionless track from point *a* to point *b*. (*b*) Finding the work done on the cheese by the gravitational force is easier along the dashed path than along the actual path taken by the cheese; the result is the same for both paths.

KEY IDEAS (1) We *cannot* calculate the work by using Eq. 7-12 ($W_g = mgd\cos\phi$). The reason is that the angle ϕ between the directions of the gravitational force $\vec{F}_g$ and the displacement $\vec{d}$ varies along the track in an unknown way. (Even if we did know the shape of the track and could calculate ϕ along it, the calculation could be very difficult.) (2) Because $\vec{F}_g$ is a conservative force, we can find the work by choosing some other path between *a* and *b*—one that makes the calculation easy.

Calculations: Let us choose the dashed path in Fig. 8-5*b*; it consists of two straight segments. Along the horizontal segment, the angle ϕ is a constant 90°. Even though we do not know the displacement along that horizontal segment, Eq. 7-12 tells us that the work W_h done there is

$$W_h = mgd\cos 90^\circ = 0.$$

Along the vertical segment, the displacement d is 0.80 m and, with $\vec{F}_g$ and $\vec{d}$ both downward, the angle ϕ is a constant 0°. Thus, Eq. 7-12 gives us, for the work W_v done along the vertical part of the dashed path,

$$\begin{aligned} W_v &= mgd\cos 0^\circ \\ &= (2.0\text{ kg})(9.8\text{ m/s}^2)(0.80\text{ m})(1) = 15.7\text{ J}. \end{aligned}$$

The total work done on the cheese by $\vec{F}_g$ as the cheese moves from point *a* to point *b* along the dashed path is then

$$W = W_h + W_v = 0 + 15.7\text{ J} \approx 16\text{ J}. \quad \text{(Answer)}$$

This is also the work done as the cheese slides along the track from *a* to *b*.

8-4 | Determining Potential Energy Values

Here we find equations that give the value of the two types of potential energy discussed in this chapter: gravitational potential energy and elastic potential energy. However, first we must find a general relation between a conservative force and the associated potential energy.

Consider a particle-like object that is part of a system in which a conservative force $\vec{F}$ acts. When that force does work W on the object, the change ΔU in the potential energy associated with the system is the negative of the work done. We wrote this fact as Eq. 8-1 ($\Delta U = -W$). For the most general case, in which the force may vary with position, we may write the work W as in Eq. 7-32:

$$W = \int_{x_i}^{x_f} F(x)\,dx. \tag{8-5}$$

This equation gives the work done by the force when the object moves from point x_i to point x_f, changing the configuration of the system. (Because the force is conservative, the work is the same for all paths between those two points.)

Substituting Eq. 8-5 into Eq. 8-1, we find that the change in potential energy due to the change in configuration is, in general notation,

$$\Delta U = -\int_{x_i}^{x_f} F(x)\,dx. \tag{8-6}$$

Gravitational Potential Energy

We first consider a particle with mass m moving vertically along a y axis (the positive direction is upward). As the particle moves from point y_i to point y_f, the gravitational force $\vec{F}_g$ does work on it. To find the corresponding change in the gravitational potential energy of the particle–Earth system, we use Eq. 8-6 with two changes: (1) We integrate along the y axis instead of the x axis, because the gravitational force acts vertically. (2) We substitute $-mg$ for the force symbol F, because $\vec{F}_g$ has the magnitude mg and is directed down the y axis. We then have

$$\Delta U = -\int_{y_i}^{y_f} (-mg)\,dy = mg\int_{y_i}^{y_f} dy = mg\Big[y\Big]_{y_i}^{y_f},$$

which yields

$$\Delta U = mg(y_f - y_i) = mg\,\Delta y. \tag{8-7}$$

Only *changes* ΔU in gravitational potential energy (or any other type of potential energy) are physically meaningful. However, to simplify a calculation or a discussion, we sometimes would like to say that a certain gravitational potential value U is associated with a certain particle–Earth system when the particle is at a certain height y. To do so, we rewrite Eq. 8-7 as

$$U - U_i = mg(y - y_i). \tag{8-8}$$

Then we take U_i to be the gravitational potential energy of the system when it is in a **reference configuration** in which the particle is at a **reference point** y_i. Usually we take $U_i = 0$ and $y_i = 0$. Doing this changes Eq. 8-8 to

$$U(y) = mgy \quad \text{(gravitational potential energy).} \tag{8-9}$$

This equation tells us:

The gravitational potential energy associated with a particle–Earth system depends only on the vertical position y (or height) of the particle relative to the reference position $y = 0$, not on the horizontal position.

Elastic Potential Energy

We next consider the block–spring system shown in Fig. 8-3, with the block moving on the end of a spring of spring constant k. As the block moves from point x_i to point x_f, the spring force $F_x = -kx$ does work on the block. To find the corresponding change in the elastic potential energy of the block–spring system, we substitute $-kx$ for $F(x)$ in Eq. 8-6. We then have

$$\Delta U = -\int_{x_i}^{x_f} (-kx)\,dx = k\int_{x_i}^{x_f} x\,dx = \tfrac{1}{2}k\Big[x^2\Big]_{x_i}^{x_f},$$

or

$$\Delta U = \tfrac{1}{2}kx_f^2 - \tfrac{1}{2}kx_i^2. \tag{8-10}$$

To associate a potential energy value U with the block at position x, we choose the reference configuration to be when the spring is at its relaxed length and the block is at $x_i = 0$. Then the elastic potential energy U_i is 0, and Eq. 8-10 becomes

$$U - 0 = \tfrac{1}{2}kx^2 - 0,$$

which gives us

$$U(x) = \tfrac{1}{2}kx^2 \quad \text{(elastic potential energy).} \tag{8-11}$$

CHECKPOINT 2 A particle is to move along an x axis from $x = 0$ to x_1 while a conservative force, directed along the x axis, acts on the particle. The figure shows three situations in which the x component of that force varies with x. The force has the same maximum magnitude F_1 in all three situations. Rank the situations according to the change in the associated potential energy during the particle's motion, most positive first.

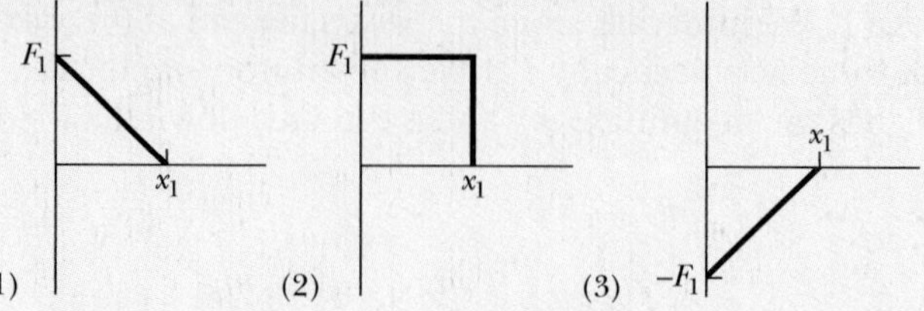

PROBLEM-SOLVING TACTICS

Tactic 1: ***Using the Term "Potential Energy"*** A potential energy is associated with a system as a whole. However, you might see statements that associate it with only part of the system. For example, you might read, "An apple hanging in a tree has a gravitational potential energy of 30 J." Such statements are often acceptable, but you should always keep in mind that the potential energy is actually associated with a system—here the apple–Earth system. Also keep in mind that assigning a particular potential energy value, such as 30 J here, to an object or even a system makes sense *only* if the reference potential energy value is known, as explored in Sample Problem 8-2.

Sample Problem 8-2

A 2.0 kg sloth hangs 5.0 m above the ground (Fig. 8-6).

(a) What is the gravitational potential energy U of the sloth–Earth system if we take the reference point $y = 0$ to be (1) at the ground, (2) at a balcony floor that is 3.0 m above the ground, (3) at the limb, and (4) 1.0 m above the limb? Take the gravitational potential energy to be zero at $y = 0$.

KEY IDEA Once we have chosen the reference point for $y = 0$, we can calculate the gravitational potential energy U of the system *relative to that reference point* with Eq. 8-9.

Calculations: For choice (1) the sloth is at $y = 5.0$ m, and

$$U = mgy = (2.0 \text{ kg})(9.8 \text{ m/s}^2)(5.0 \text{ m}) = 98 \text{ J}. \quad \text{(Answer)}$$

For the other choices, the values of U are

(2) $U = mgy = mg(2.0 \text{ m}) = 39$ J,

(3) $U = mgy = mg(0) = 0$ J,

(4) $U = mgy = mg(-1.0 \text{ m}) = -19.6 \text{ J} \approx -20$ J. (Answer)

(b) The sloth drops to the ground. For each choice of reference point, what is the change ΔU in the potential energy of the sloth–Earth system due to the fall?

KEY IDEA The *change* in potential energy does not depend on the choice of the reference point for $y = 0$; instead, it depends on the change in height Δy.

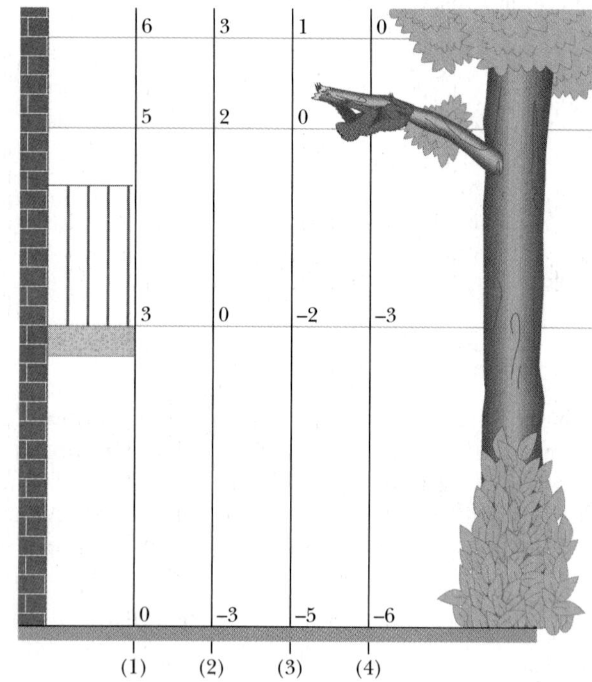

FIG. 8-6 Four choices of reference point $y = 0$. Each y axis is marked in units of meters. The choice affects the value of the potential energy U of the sloth–Earth system. However, it does not affect the change ΔU in potential energy of the system if the sloth moves by, say, falling.

Calculation: For all four situations, we have the same $\Delta y = -5.0$ m. Thus, for (1) to (4), Eq. 8-7 tells us that

$$\Delta U = mg\,\Delta y = (2.0 \text{ kg})(9.8 \text{ m/s}^2)(-5.0 \text{ m}) = -98 \text{ J}. \quad \text{(Answer)}$$

8-5 | Conservation of Mechanical Energy

The **mechanical energy** E_{mec} of a system is the sum of its potential energy U and the kinetic energy K of the objects within it:

$$E_{mec} = K + U \quad \text{(mechanical energy).} \tag{8-12}$$

In this section, we examine what happens to this mechanical energy when only conservative forces cause energy transfers within the system—that is, when frictional and drag forces do not act on the objects in the system. Also, we shall assume that the system is *isolated* from its environment; that is, no *external force* from an object outside the system causes energy changes inside the system.

When a conservative force does work W on an object within the system, that force transfers energy between kinetic energy K of the object and potential energy U of the system. From Eq. 7-10, the change ΔK in kinetic energy is

$$\Delta K = W \tag{8-13}$$

and from Eq. 8-1, the change ΔU in potential energy is

$$\Delta U = -W. \tag{8-14}$$

Combining Eqs. 8-13 and 8-14, we find that

$$\Delta K = -\Delta U. \tag{8-15}$$

In words, one of these energies increases exactly as much as the other decreases.

We can rewrite Eq. 8-15 as

$$K_2 - K_1 = -(U_2 - U_1), \tag{8-16}$$

where the subscripts refer to two different instants and thus to two different arrangements of the objects in the system. Rearranging Eq. 8-16 yields

$$K_2 + U_2 = K_1 + U_1 \quad \text{(conservation of mechanical energy).} \tag{8-17}$$

In words, this equation says:

$$\begin{pmatrix}\text{the sum of } K \text{ and } U \text{ for} \\ \text{any state of a system}\end{pmatrix} = \begin{pmatrix}\text{the sum of } K \text{ and } U \text{ for} \\ \text{any other state of the system}\end{pmatrix},$$

when the system is isolated and only conservative forces act on the objects in the system. In other words:

In an isolated system where only conservative forces cause energy changes, the kinetic energy and potential energy can change, but their sum, the mechanical energy E_{mec} of the system, cannot change.

This result is called the **principle of conservation of mechanical energy.** (Now you can see where *conservative* forces got their name.) With the aid of Eq. 8-15, we can write this principle in one more form, as

$$\Delta E_{mec} = \Delta K + \Delta U = 0. \tag{8-18}$$

The principle of conservation of mechanical energy allows us to solve problems that would be quite difficult to solve using only Newton's laws:

When the mechanical energy of a system is conserved, we can relate the sum of kinetic energy and potential energy at one instant to that at another instant *without considering the intermediate motion* and *without finding the work done by the forces involved.*

Figure 8-7 shows an example in which the principle of conservation of mechanical energy can be applied: As a pendulum swings, the energy of the

In olden days, a person would be tossed via a blanket to be able to see farther over the flat terrain. Nowadays, it is done just for fun. During the ascent of the person in the photograph, energy is transferred from kinetic energy to gravitational potential energy. The maximum height is reached when that transfer is complete. Then the transfer is reversed during the fall. *(©AP/Wide World Photos)*

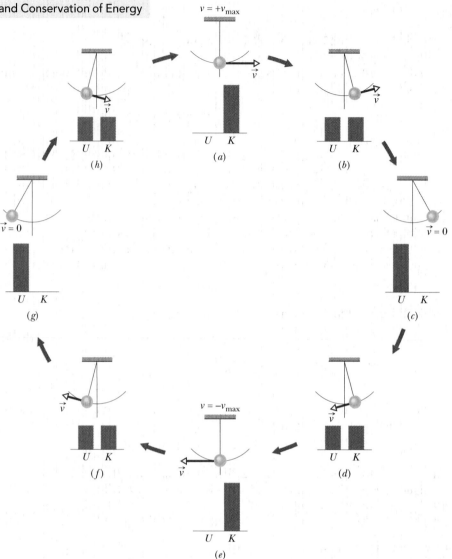

FIG. 8-7 A pendulum, with its mass concentrated in a bob at the lower end, swings back and forth. One full cycle of the motion is shown. During the cycle the values of the potential and kinetic energies of the pendulum–Earth system vary as the bob rises and falls, but the mechanical energy E_{mec} of the system remains constant. The energy E_{mec} can be described as continuously shifting between the kinetic and potential forms. In stages (*a*) and (*e*), all the energy is kinetic energy. The bob then has its greatest speed and is at its lowest point. In stages (*c*) and (*g*), all the energy is potential energy. The bob then has zero speed and is at its highest point. In stages (*b*), (*d*), (*f*), and (*h*), half the energy is kinetic energy and half is potential energy. If the swinging involved a frictional force at the point where the pendulum is attached to the ceiling, or a drag force due to the air, then E_{mec} would not be conserved, and eventually the pendulum would stop.

pendulum–Earth system is transferred back and forth between kinetic energy K and gravitational potential energy U, with the sum $K + U$ being constant. If we know the gravitational potential energy when the pendulum bob is at its highest point (Fig. 8-7*c*), Eq. 8-17 gives us the kinetic energy of the bob at the lowest point (Fig. 8-7*e*).

For example, let us choose the lowest point as the reference point, with the gravitational potential energy $U_2 = 0$. Suppose then that the potential energy at the highest point is $U_1 = 20$ J relative to the reference point. Because the bob momentarily stops at its highest point, the kinetic energy there is $K_1 = 0$. Putting these values into Eq. 8-17 gives us the kinetic energy K_2 at the lowest point:

$$K_2 + 0 = 0 + 20 \text{ J} \quad \text{or} \quad K_2 = 20 \text{ J}.$$

Note that we get this result without considering the motion between the highest and lowest points (such as in Fig. 8-7*d*) and without finding the work done by any forces involved in the motion.

CHECKPOINT 3 The figure shows four situations—one in which an initially stationary block is dropped and three in which the block is allowed to slide down frictionless ramps. (a) Rank the situations according to the kinetic energy of the block at point *B*, greatest first. (b) Rank them according to the speed of the block at point *B*, greatest first.

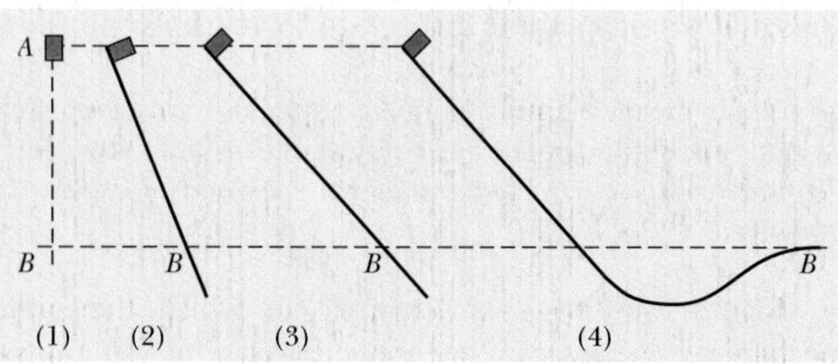

Sample Problem 8-3 Build your skill

In Fig. 8-8, a child of mass m is released from rest at the top of a water slide, at height $h = 8.5$ m above the bottom of the slide. Assuming that the slide is frictionless because of the water on it, find the child's speed at the bottom of the slide.

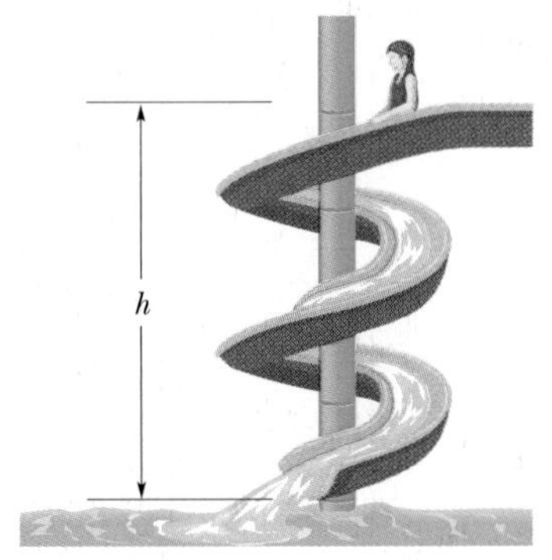

FIG. 8-8 A child slides down a water slide as she descends a height h.

KEY IDEAS (1) We cannot find her speed at the bottom by using her acceleration along the slide as we might have in earlier chapters because we do not know the slope (angle) of the slide. However, because that speed is related to her kinetic energy, perhaps we can use the principle of conservation of mechanical energy to get the speed. Then we would not need to know the slope. (2) Mechanical energy is conserved in a system *if* the system is isolated and *if* only conservative forces cause energy transfers within it. Let's check.

Forces: Two forces act on the child. The *gravitational force,* a conservative force, does work on her. The *normal force* on her from the slide does no work because its direction at any point during the descent is always perpendicular to the direction in which the child moves.

System: Because the only force doing work on the child is the gravitational force, we choose the child–Earth system as our system, which we can take to be isolated.

Thus, we have only a conservative force doing work in an isolated system, so we *can* use the principle of conservation of mechanical energy.

Calculations: Let the mechanical energy be $E_{mec,t}$ when the child is at the top of the slide and $E_{mec,b}$ when she is at the bottom. Then the conservation principle tells us

$$E_{mec,b} = E_{mec,t}. \tag{8-19}$$

To show both kinds of mechanical energy, we have

$$K_b + U_b = K_t + U_t, \tag{8-20}$$

or

$$\tfrac{1}{2}mv_b^2 + mgy_b = \tfrac{1}{2}mv_t^2 + mgy_t.$$

Dividing by m and rearranging yield

$$v_b^2 = v_t^2 + 2g(y_t - y_b).$$

Putting $v_t = 0$ and $y_t - y_b = h$ leads to

$$v_b = \sqrt{2gh} = \sqrt{(2)(9.8 \text{ m/s}^2)(8.5 \text{ m})}$$
$$= 13 \text{ m/s}. \qquad \text{(Answer)}$$

This is the same speed that the child would reach if she fell 8.5 m vertically. On an actual slide, some frictional forces would act and the child would not be moving quite so fast.

Comments: Although this problem is hard to solve directly with Newton's laws, using conservation of mechanical energy makes the solution much easier. However, if we were asked to find the time taken for the child to reach the bottom of the slide, energy methods would be of no use; we would need to know the shape of the slide, and we would have a difficult problem.

PROBLEM-SOLVING TACTICS

Tactic 2: ***Conservation of Mechanical Energy*** Asking the following questions will help you to solve problems involving the principle of conservation of mechanical energy.

For what system is mechanical energy conserved? You should be able to separate your system from its environment. Imagine drawing a closed surface such that whatever is inside is your system and whatever is outside is the environment of that system.

Is friction or drag present? If friction or drag is present, mechanical energy is not conserved.

Is your system isolated? The principle of conservation of mechanical energy applies only to isolated systems. That means that no *external forces* (forces exerted by objects outside the system) should do work on the objects in the system.

What are the initial and final states of your system? The system changes from some initial configuration to some final configuration. You apply the principle of conservation of mechanical energy by saying that E_{mec} has the same value in both these configurations. Be very clear about what these two configurations are.

8-6 | Reading a Potential Energy Curve

Once again we consider a particle that is part of a system in which a conservative force acts. This time suppose that the particle is constrained to move along an

x axis while the conservative force does work on it. We can learn a lot about the motion of the particle from a plot of the system's potential energy $U(x)$. However, before we discuss such plots, we need one more relationship.

Finding the Force Analytically

Equation 8-6 tells us how to find the change ΔU in potential energy between two points in a one-dimensional situation if we know the force $F(x)$. Now we want to go the other way; that is, we know the potential energy function $U(x)$ and want to find the force.

For one-dimensional motion, the work W done by a force that acts on a particle as the particle moves through a distance Δx is $F(x)\,\Delta x$. We can then write Eq. 8-1 as

$$\Delta U(x) = -W = -F(x)\,\Delta x. \tag{8-21}$$

Solving for $F(x)$ and passing to the differential limit yield

$$F(x) = -\frac{dU(x)}{dx} \quad \text{(one-dimensional motion)}, \tag{8-22}$$

which is the relation we sought.

We can check this result by putting $U(x) = \frac{1}{2}kx^2$, which is the elastic potential energy function for a spring force. Equation 8-22 then yields, as expected, $F(x) = -kx$, which is Hooke's law. Similarly, we can substitute $U(x) = mgx$, which is the gravitational potential energy function for a particle–Earth system, with a particle of mass m at height x above Earth's surface. Equation 8-22 then yields $F = -mg$, which is the gravitational force on the particle.

The Potential Energy Curve

Figure 8-9a is a plot of a potential energy function $U(x)$ for a system in which a particle is in one-dimensional motion while a conservative force $F(x)$ does work

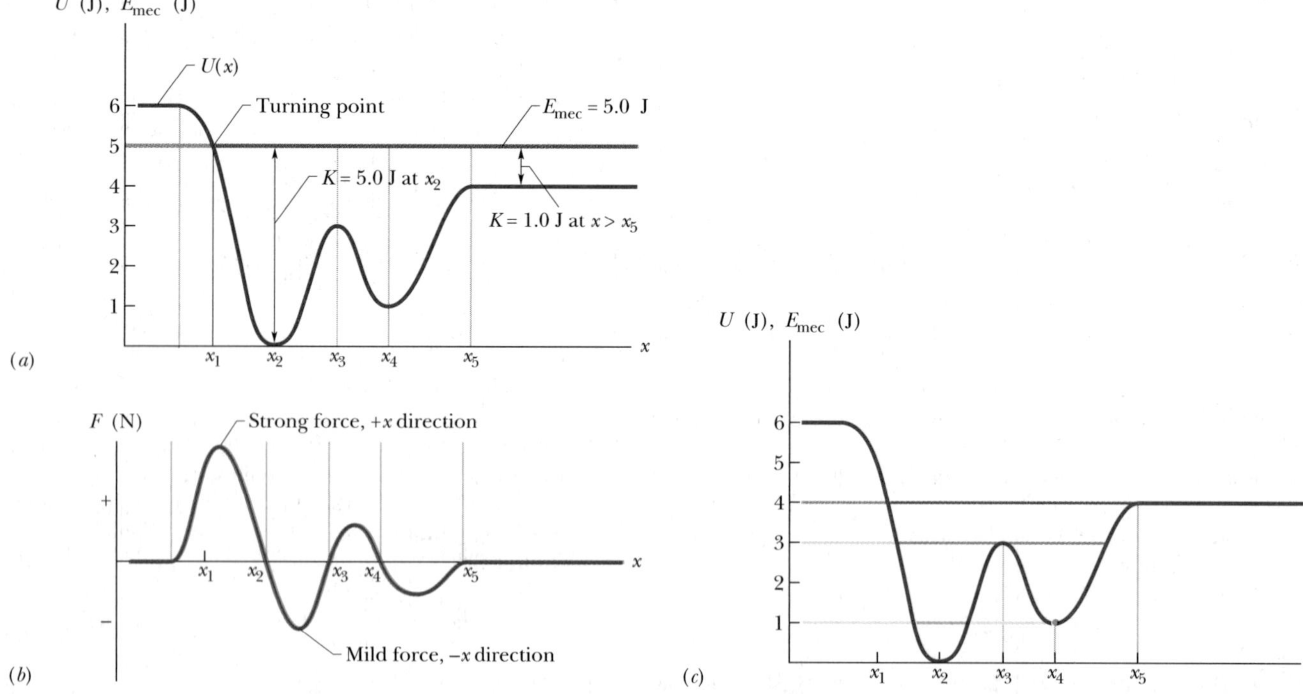

FIG. 8-9 (a) A plot of $U(x)$, the potential energy function of a system containing a particle confined to move along an x axis. There is no friction, so mechanical energy is conserved. (b) A plot of the force $F(x)$ acting on the particle, derived from the potential energy plot by taking its slope at various points. (c) The $U(x)$ plot of (a) with three possible values of E_{mec} shown.

on it. We can easily find $F(x)$ by (graphically) taking the slope of the $U(x)$ curve at various points. (Equation 8-22 tells us that $F(x)$ is the negative of the slope of the $U(x)$ curve.) Figure 8-9*b* is a plot of $F(x)$ found in this way.

Turning Points

In the absence of a nonconservative force, the mechanical energy E of a system has a constant value given by

$$U(x) + K(x) = E_{mec}. \tag{8-23}$$

Here $K(x)$ is the *kinetic energy function* of a particle in the system (this $K(x)$ gives the kinetic energy as a function of the particle's location x). We may rewrite Eq. 8-23 as

$$K(x) = E_{mec} - U(x). \tag{8-24}$$

Suppose that E_{mec} (which has a constant value, remember) happens to be 5.0 J. It would be represented in Fig. 8-9*a* by a horizontal line that runs through the value 5.0 J on the energy axis. (It is, in fact, shown there.)

Equation 8-24 tells us how to determine the kinetic energy K for any location x of the particle: On the $U(x)$ curve, find U for that location x and then subtract U from E_{mec}. For example, if the particle is at any point to the right of x_5, then $K =$ 1.0 J. The value of K is greatest (5.0 J) when the particle is at x_2 and least (0 J) when the particle is at x_1.

Since K can never be negative (because v^2 is always positive), the particle can never move to the left of x_1, where $E_{mec} - U$ is negative. Instead, as the particle moves toward x_1 from x_2, K decreases (the particle slows) until $K = 0$ at x_1 (the particle stops there).

Note that when the particle reaches x_1, the force on the particle, given by Eq. 8-22, is positive (because the slope dU/dx is negative). This means that the particle does not remain at x_1 but instead begins to move to the right, opposite its earlier motion. Hence x_1 is a **turning point,** a place where $K = 0$ (because $U = E$) and the particle changes direction. There is no turning point (where $K = 0$) on the right side of the graph. When the particle heads to the right, it will continue indefinitely.

Equilibrium Points

Figure 8-9*c* shows three different values for E_{mec} superposed on the plot of the potential energy function $U(x)$ of Fig. 8-9*a*. Let us see how they change the situation. If $E_{mec} = 4.0$ J (purple line), the turning point shifts from x_1 to a point between x_1 and x_2. Also, at any point to the right of x_5, the system's mechanical energy is equal to its potential energy; thus, the particle has no kinetic energy and (by Eq. 8-22) no force acts on it, and so it must be stationary. A particle at such a position is said to be in **neutral equilibrium.** (A marble placed on a horizontal tabletop is in that state.)

If $E_{mec} = 3.0$ J (pink line), there are two turning points: One is between x_1 and x_2, and the other is between x_4 and x_5. In addition, x_3 is a point at which $K = 0$. If the particle is located exactly there, the force on it is also zero, and the particle remains stationary. However, if it is displaced even slightly in either direction, a nonzero force pushes it farther in the same direction, and the particle continues to move. A particle at such a position is said to be in **unstable equilibrium.** (A marble balanced on top of a bowling ball is an example.)

Next consider the particle's behavior if $E_{mec} = 1.0$ J (green line). If we place it at x_4, it is stuck there. It cannot move left or right on its own because to do so would require a negative kinetic energy. If we push it slightly left or right, a restoring force appears that moves it back to x_4. A particle at such a position is said to be in **stable equilibrium.** (A marble placed at the bottom of a hemispherical bowl is an example.) If we place the particle in the cup-like *potential well* centered at x_2, it is between two turning points. It can still move somewhat, but only partway to x_1 or x_3.

CHECKPOINT 4 The figure gives the potential energy function $U(x)$ for a system in which a particle is in one-dimensional motion. (a) Rank regions AB, BC, and CD according to the magnitude of the force on the particle, greatest first. (b) What is the direction of the force when the particle is in region AB?

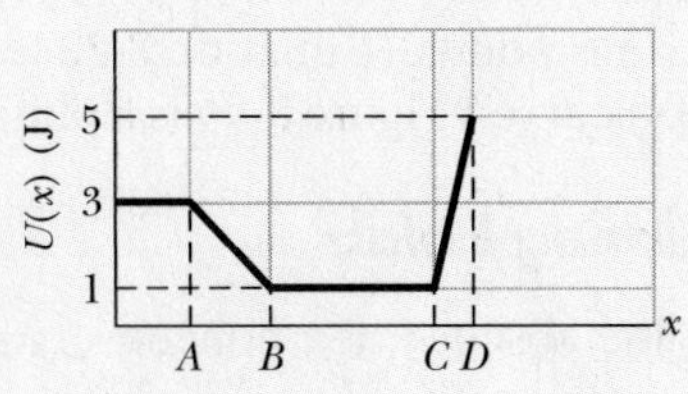

Sample Problem 8-4

A 2.00 kg particle moves along an x axis in one-dimensional motion while a conservative force along that axis acts on it. The potential energy $U(x)$ associated with the force is plotted in Fig. 8-10a. That is, if the particle were placed at any position between $x = 0$ and $x = 7.00$ m, it would have the plotted value of U. At $x = 6.5$ m, the particle has velocity $\vec{v}_0 = (-4.00 \text{ m/s})\hat{\text{i}}$.

(a) From Fig. 8-10a, determine the particle's speed at $x_1 = 4.5$ m.

KEY IDEAS (1) The particle's kinetic energy is given by Eq. 7-1 ($K = \frac{1}{2}mv^2$). (2) Because only a conservative force acts on the particle, the mechanical energy E_{mec} $(= K + U)$ is conserved as the particle moves. (3) Therefore, on a plot of $U(x)$ such as Fig. 8-10a, the kinetic energy is equal to the difference between E_{mec} and U.

Calculations: At $x = 6.5$ m, the particle has kinetic energy

$$K_0 = \tfrac{1}{2}mv_0^2 = \tfrac{1}{2}(2.00 \text{ kg})(4.00 \text{ m/s})^2 = 16.0 \text{ J}.$$

Because the potential energy there is $U = 0$, the mechanical energy is

$$E_{mec} = K_0 + U_0 = 16.0 \text{ J} + 0 = 16.0 \text{ J}.$$

This value for E_{mec} is plotted as a horizontal line in Fig. 8-10a. From that figure we see that at $x = 4.5$ m, the potential energy is $U_1 = 7.0$ J. The kinetic energy K_1 is the difference between E_{mec} and U_1:

$$K_1 = E_{mec} - U_1 = 16.0 \text{ J} - 7.0 \text{ J} = 9.0 \text{ J}.$$

Because $K_1 = \frac{1}{2}mv_1^2$, we find

$$v_1 = 3.0 \text{ m/s}. \qquad \text{(Answer)}$$

(b) Where is the particle's turning point located?

KEY IDEA The turning point is where the force momentarily stops and then reverses the particle's motion. That is, it is where the particle momentarily has $v = 0$ and thus $K = 0$.

Calculations: Because K is the difference between E_{mec} and U, we want the point in Fig. 8-10a where the plot of U rises to meet the horizontal line of E_{mec}, as shown in Fig. 8-10b. Because the plot of U is a straight line in Fig. 8-10b, we can draw nested right triangles as shown and then write the proportionality of distances

$$\frac{16 - 7.0}{d} = \frac{20 - 7.0}{4.0 - 1.0},$$

which gives us $d = 2.08$ m. Thus, the turning point is at

$$x = 4.0 \text{ m} - d = 1.9 \text{ m}. \qquad \text{(Answer)}$$

(c) Evaluate the force acting on the particle when it is in the region $1.9 \text{ m} < x < 4.0 \text{ m}$.

KEY IDEA The force is given by Eq. 8-22 ($F(x) = -dU(x)/dx$). The equation states that the force is equal to the negative of the slope on a graph of $U(x)$.

Calculations: For the graph of Fig. 8-10b, we see that for the range $1.0 \text{ m} < x < 4.0 \text{ m}$ the force is

$$F = -\frac{20 \text{ J} - 7.0 \text{ J}}{1.0 \text{ m} - 4.0 \text{ m}} = 4.3 \text{ N}. \qquad \text{(Answer)}$$

Thus, the force has magnitude 4.3 N and is in the positive direction of the x axis. This result is consistent with the fact that the initially leftward-moving particle is stopped by the force and then sent rightward.

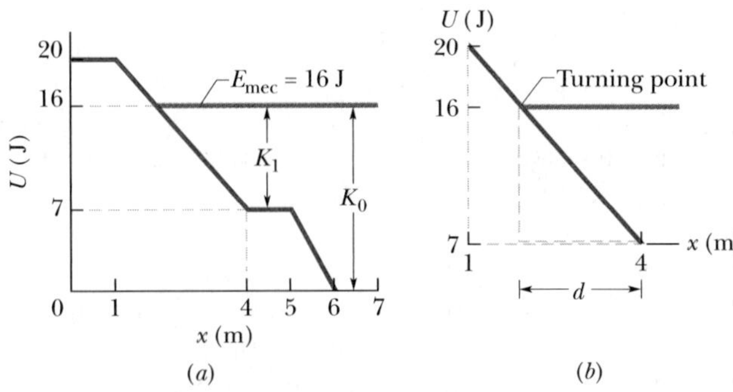

FIG. 8-10 (a) A plot of potential energy U versus position x. (b) A section of the plot used to find where the particle turns around.

8-7 | Work Done on a System by an External Force

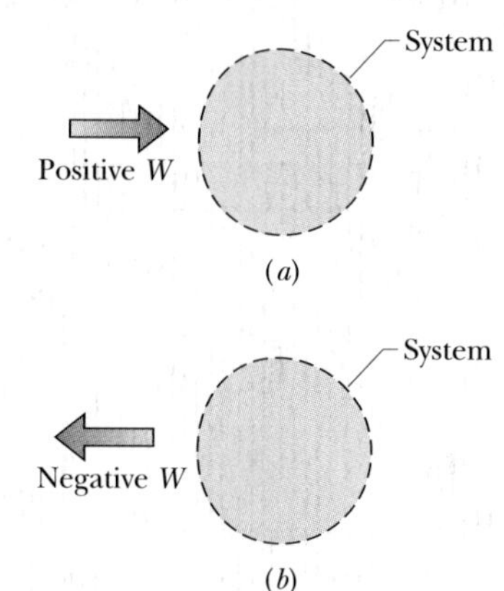

FIG. 8-11 (a) Positive work W done on an arbitrary system means a transfer of energy to the system. (b) Negative work W means a transfer of energy from the system.

In Chapter 7, we defined work as being energy transferred to or from an object by means of a force acting on the object. We can now extend that definition to an external force acting on a system of objects.

Work is energy transferred to or from a system by means of an external force acting on that system.

Figure 8-11*a* represents positive work (a transfer of energy *to* a system), and Fig. 8-11*b* represents negative work (a transfer of energy *from* a system). When more than one force acts on a system, their *net work* is the energy transferred to or from the system.

These transfers are like transfers of money to and from a bank account. If a system consists of a single particle or particle-like object, as in Chapter 7, the work done on the system by a force can change only the kinetic energy of the system. The energy statement for such transfers is the work–kinetic energy theorem of Eq. 7-10 ($\Delta K = W$); that is, a single particle has only one energy account, called kinetic energy. External forces can transfer energy into or out of that account. If a system is more complicated, however, an external force can change other forms of energy (such as potential energy); that is, a more complicated system can have multiple energy accounts.

Let us find energy statements for such systems by examining two basic situations, one that does not involve friction and one that does.

No Friction Involved

To compete in a bowling-ball-hurling contest, you first squat and cup your hands under the ball on the floor. Then you rapidly straighten up while also pulling your hands up sharply, launching the ball upward at about face level. During your upward motion, your applied force on the ball obviously does work; that is, it is an external force that transfers energy, but to what system?

To answer, we check to see which energies change. There is a change ΔK in the ball's kinetic energy and, because the ball and Earth become more separated, there is a change ΔU in the gravitational potential energy of the ball–Earth system. To include both changes, we need to consider the ball–Earth system. Then your force is an external force doing work on that system, and the work is

$$W = \Delta K + \Delta U, \tag{8-25}$$

or

$$W = \Delta E_{\text{mec}} \quad \text{(work done on system, no friction involved)}, \tag{8-26}$$

where ΔE_{mec} is the change in the mechanical energy of the system. These two equations, which are represented in Fig. 8-12, are equivalent energy statements for work done on a system by an external force when friction is not involved.

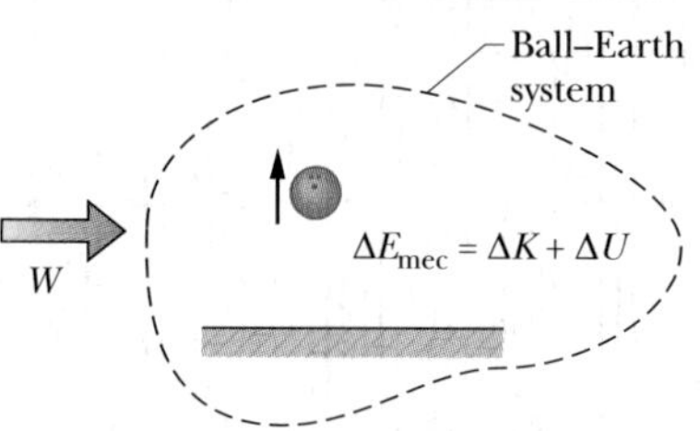

FIG. 8-12 Positive work W is done on a system of a bowling ball and Earth, causing a change ΔE_{mec} in the mechanical energy of the system, a change ΔK in the ball's kinetic energy, and a change ΔU in the system's gravitational potential energy.

Friction Involved

We next consider the example in Fig. 8-13*a*. A constant horizontal force $\vec{F}$ pulls a block along an x axis and through a displacement of magnitude d, increasing the block's velocity from $\vec{v}_0$ to $\vec{v}$. During the motion, a constant kinetic frictional force $\vec{f}_k$ from the floor acts on the block. Let us first choose the block as our system and apply Newton's second law to it. We can write that law for components along the x axis ($F_{\text{net},x} = ma_x$) as

$$F - f_k = ma. \tag{8-27}$$

Because the forces are constant, the acceleration $\vec{a}$ is also constant. Thus, we can use Eq. 2-16 to write

$$v^2 = v_0^2 + 2ad.$$

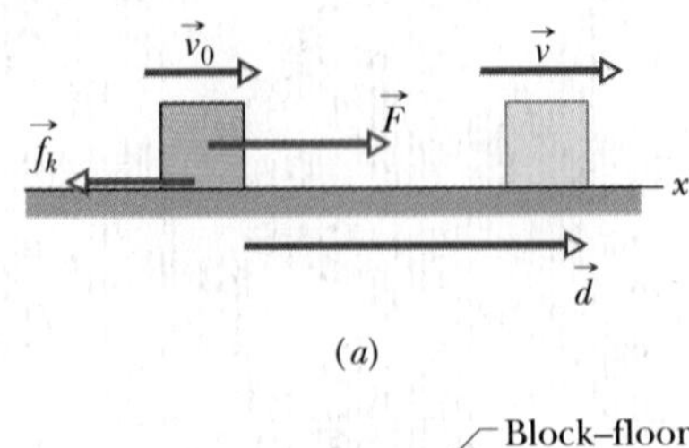

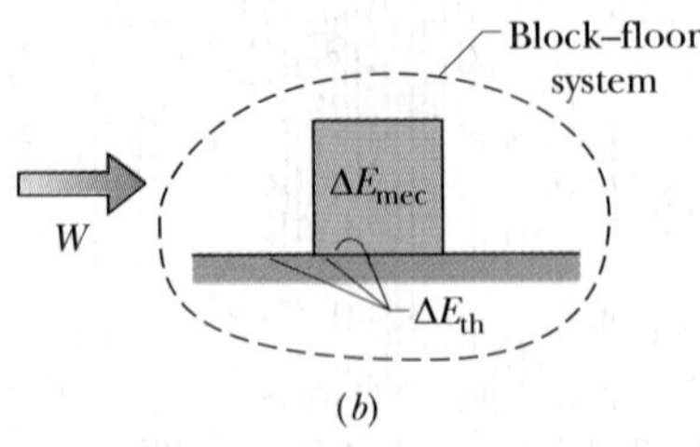

FIG. 8-13 (*a*) A block is pulled across a floor by force $\vec{F}$ while a kinetic frictional force $\vec{f}_k$ opposes the motion. The block has velocity $\vec{v}_0$ at the start of a displacement $\vec{d}$ and velocity $\vec{v}$ at the end of the displacement. (*b*) Positive work W is done on the block–floor system by force $\vec{F}$, resulting in a change ΔE_{mec} in the block's mechanical energy and a change ΔE_{th} in the thermal energy of the block and floor.

Solving this equation for a, substituting the result into Eq. 8-27, and rearranging then give us

$$Fd = \tfrac{1}{2}mv^2 - \tfrac{1}{2}mv_0^2 + f_k d \tag{8-28}$$

or, because $\frac{1}{2}mv^2 - \frac{1}{2}mv_0^2 = \Delta K$ for the block,

$$Fd = \Delta K + f_k d. \tag{8-29}$$

In a more general situation (say, one in which the block is moving up a ramp), there can be a change in potential energy. To include such a possible change, we generalize Eq. 8-29 by writing

$$Fd = \Delta E_{mec} + f_k d. \tag{8-30}$$

By experiment we find that the block and the portion of the floor along which it slides become warmer as the block slides. As we shall discuss in Chapter 18, the temperature of an object is related to the object's thermal energy E_{th} (the energy associated with the random motion of the atoms and molecules in the object). Here, the thermal energy of the block and floor increases because (1) there is friction between them and (2) there is sliding. Recall that friction is due to the cold-welding between two surfaces. As the block slides over the floor, the sliding causes repeated tearing and re-forming of the welds between the block and the floor, which makes the block and floor warmer. Thus, the sliding increases their thermal energy E_{th}.

Through experiment, we find that the increase ΔE_{th} in thermal energy is equal to the product of the magnitudes f_k and d:

$$\Delta E_{th} = f_k d \quad \text{(increase in thermal energy by sliding).} \tag{8-31}$$

Thus, we can rewrite Eq. 8-30 as

$$Fd = \Delta E_{mec} + \Delta E_{th}. \tag{8-32}$$

Fd is the work W done by the external force $\vec{F}$ (the energy transferred by the force), but on which system is the work done (where are the energy transfers made)? To answer, we check to see which energies change. The block's mechanical energy changes, and the thermal energies of the block and floor also change. Therefore, the work done by force $\vec{F}$ is done on the block–floor system. That work is

$$W = \Delta E_{mec} + \Delta E_{th} \quad \text{(work done on system, friction involved).} \tag{8-33}$$

This equation, which is represented in Fig. 8-13*b*, is the energy statement for the work done on a system by an external force when friction is involved.

CHECKPOINT 5 In three trials, a block is pushed by a horizontal applied force across a floor that is not frictionless, as in Fig. 8-13*a*. The magnitudes F of the applied force and the results of the pushing on the block's speed are given in the table. In all three trials, the block is pushed through the same distance d. Rank the three trials according to the change in the thermal energy of the block and floor that occurs in that distance d, greatest first.

Trial	F	Result on Block's Speed
a	5.0 N	decreases
b	7.0 N	remains constant
c	8.0 N	increases

Sample Problem 8-5

The prehistoric people of Easter Island carved hundreds of gigantic stone statues in a quarry and then moved them to sites all over the island (Fig. 8-14). How they managed to move the statues by as much as 10 km without the use

FIG. 8-14 Easter Island stone statues. *(©LMR Group/Alamy Images)*

of sophisticated machines has been hotly debated. They most likely cradled each statue in a wooden sled and then pulled the sled over a "runway" consisting of almost identical logs acting as rollers. In a modern reenactment of this technique, 25 men were able to move a 9000 kg Easter Island–type statue 45 m over level ground in 2 min.

(a) Estimate the work the net force $\vec{F}$ from the men did during the 45 m displacement of the statue, and determine the system on which that force did the work.

KEY IDEAS (1) We can calculate the work done with Eq. 7-7 ($W = Fd \cos \phi$). (2) To determine the system on which the work is done we see which energies change.

Calculations: In Eq. 7-7, d is the distance 45 m, F is the magnitude of the net force on the statue from the 25 men, and $\phi = 0°$. Let us estimate that each man pulled with a force magnitude equal to twice his weight, which we take to be the same value mg for all the men. Thus, the magnitude of the net force from the men was $F = (25)(2mg) = 50mg$. Estimating a man's mass as 80 kg, we can then write Eq. 7-7 as

$$W = Fd \cos \phi = 50mgd \cos \phi$$
$$= (50)(80 \text{ kg})(9.8 \text{ m/s}^2)(45 \text{ m}) \cos 0°$$
$$= 1.8 \times 10^6 \text{ J} \approx 2 \text{ MJ}. \qquad \text{(Answer)}$$

Because the statue moved, there was certainly a change ΔK in its kinetic energy during the motion. We can easily guess that there must have been considerable kinetic friction between the sled, logs, and ground, resulting in a change ΔE_{th} in their thermal energies. Thus, the system on which the work was done consisted of the statue, sled, logs, and ground.

(b) What was the increase ΔE_{th} in the thermal energy of the system during the 45 m displacement?

KEY IDEA We can relate ΔE_{th} to the work W done by $\vec{F}$ with the energy statement of Eq. 8-33 for a system that involves friction:

$$W = \Delta E_{mec} + \Delta E_{th}.$$

Calculations: We know the value of W from (a). The change ΔE_{mec} in the statue's mechanical energy was zero because the statue was stationary at the beginning and the end of the move and its elevation did not change. Thus, we find

$$\Delta E_{th} = W = 1.8 \times 10^6 \text{ J} \approx 2 \text{ MJ}. \quad \text{(Answer)}$$

(c) Estimate the work that would have been done by the 25 men if they had moved the statue 10 km across level ground on Easter Island. Also estimate the total change ΔE_{th} that would have occurred in the statue–sled–logs–ground system.

Calculation: We calculate W as in (a), but with 1×10^4 m substituted for d. Also, we again equate ΔE_{th} to W. We get

$$W = \Delta E_{th} = 3.9 \times 10^8 \text{ J} \approx 400 \text{ MJ}. \quad \text{(Answer)}$$

This would have been a staggering amount of energy for the men to have transferred during the movement of a statue. Still, the 25 men *could* have moved the statue 10 km without some mysterious energy source.

Sample Problem 8-6

A food shipper pushes a wood crate of cabbage heads (total mass $m = 14$ kg) across a concrete floor with a constant horizontal force $\vec{F}$ of magnitude 40 N. In a straight-line displacement of magnitude $d = 0.50$ m, the speed of the crate decreases from $v_0 = 0.60$ m/s to $v = 0.20$ m/s.

(a) How much work is done by force $\vec{F}$, and on what system does it do the work?

KEY IDEA Because the applied force $\vec{F}$ is constant, we can calculate the work it does by using Eq. 7-7 ($W = Fd \cos \phi$).

Calculation: Substituting given data, including the fact that force $\vec{F}$ and displacement $\vec{d}$ are in the same direction, we find

$$W = Fd \cos \phi = (40 \text{ N})(0.50 \text{ m}) \cos 0°$$
$$= 20 \text{ J}. \qquad \text{(Answer)}$$

Reasoning: We can determine the system on which the work is done to see which energies change. Because the crate's speed changes, there is certainly a change ΔK in the crate's kinetic energy. Is there friction between the floor and the crate, and thus a change in thermal energy? Note that $\vec{F}$ and the crate's velocity have the same direction. Thus, if there is no friction, then $\vec{F}$ should be accelerating the crate to a *greater* speed. However, the crate is *slowing,* so there must be friction and a change ΔE_{th} in thermal energy of the crate and the floor. Therefore, the system on which the work is done is the crate–floor system, because both energy changes occur in that system.

(b) What is the increase ΔE_{th} in the thermal energy of the crate and floor?

KEY IDEA We can relate ΔE_{th} to the work W done by $\vec{F}$ with the energy statement of Eq. 8-33 for a system that involves friction:

$$W = \Delta E_{mec} + \Delta E_{th}. \tag{8-34}$$

Calculations: We know the value of W from (a). The change ΔE_{mec} in the crate's mechanical energy is just the change in its kinetic energy because no potential energy changes occur, so we have

$$\Delta E_{mec} = \Delta K = \tfrac{1}{2}mv^2 - \tfrac{1}{2}mv_0^2.$$

Substituting this into Eq. 8-34 and solving for ΔE_{th}, we find

$$\begin{aligned}\Delta E_{th} &= W - (\tfrac{1}{2}mv^2 - \tfrac{1}{2}mv_0^2) = W - \tfrac{1}{2}m(v^2 - v_0^2)\\ &= 20\text{ J} - \tfrac{1}{2}(14\text{ kg})[(0.20\text{ m/s})^2 - (0.60\text{ m/s})^2]\\ &= 22.2\text{ J} \approx 22\text{ J}.\end{aligned} \quad \text{(Answer)}$$

8-8 | Conservation of Energy

We now have discussed several situations in which energy is transferred to or from objects and systems, much like money is transferred between accounts. In each situation we assume that the energy that was involved could always be accounted for; that is, energy could not magically appear or disappear. In more formal language, we assumed (correctly) that energy obeys a law called the **law of conservation of energy,** which is concerned with the **total energy** E of a system. That total is the sum of the system's mechanical energy, thermal energy, and any type of *internal energy* in addition to thermal energy. (We have not yet discussed other types of internal energy.) The law states that

> The total energy E of a system can change only by amounts of energy that are transferred to or from the system.

The only type of energy transfer that we have considered is work W done on a system. Thus, for us at this point, this law states that

$$W = \Delta E = \Delta E_{mec} + \Delta E_{th} + \Delta E_{int}, \tag{8-35}$$

where ΔE_{mec} is any change in the mechanical energy of the system, ΔE_{th} is any change in the thermal energy of the system, and ΔE_{int} is any change in any other type of internal energy of the system. Included in ΔE_{mec} are changes ΔK in kinetic energy and changes ΔU in potential energy (elastic, gravitational, or any other type we might find).

This law of conservation of energy is *not* something we have derived from basic physics principles. Rather, it is a law based on countless experiments. Scientists and engineers have never found an exception to it.

FIG. 8-15 To descend, the rock climber must transfer energy from the gravitational potential energy of a system consisting of him, his gear, and Earth. He has wrapped the rope around metal rings so that the rope rubs against the rings. This allows most of the transferred energy to go to the thermal energy of the rope and rings rather than to his kinetic energy. *(Tyler Stableford/The Image Bank/ Getty Images)*

Isolated System

If a system is isolated from its environment, there can be no energy transfers to or from it. For that case, the law of conservation of energy states:

> The total energy E of an isolated system cannot change.

Many energy transfers may be going on *within* an isolated system—between, say, kinetic energy and a potential energy or between kinetic energy and thermal energy. However, the total of all the types of energy in the system cannot change.

We can use the rock climber in Fig. 8-15 as an example, approximating him, his gear, and Earth as an isolated system. As he rappels down the rock face, changing the configuration of the system, he needs to control the transfer of energy from the gravitational potential energy of the system. (That energy cannot just disappear.) Some of it is transferred to his kinetic energy. However, he obviously does not want very much transferred to that type or he will be moving too quickly, so he has wrapped the rope around metal rings to produce friction between the rope and the rings as he moves down. The sliding of the rings on the rope then transfers the gravitational potential energy of the system to thermal energy of the rings and rope in a way that he can control. The total energy of the climber–gear–Earth system (the total of its gravitational potential energy, kinetic energy, and thermal energy) does not change during his descent.

For an isolated system, the law of conservation of energy can be written in two ways. First, by setting $W = 0$ in Eq. 8-35, we get

$$\Delta E_{mec} + \Delta E_{th} + \Delta E_{int} = 0 \quad \text{(isolated system).} \tag{8-36}$$

We can also let $\Delta E_{mec} = E_{mec,2} - E_{mec,1}$, where the subscripts 1 and 2 refer to two different instants—say, before and after a certain process has occurred. Then Eq. 8-36 becomes

$$E_{mec,2} = E_{mec,1} - \Delta E_{th} - \Delta E_{int}. \tag{8-37}$$

Equation 8-37 tells us:

In an isolated system, we can relate the total energy at one instant to the total energy at another instant *without considering the energies at intermediate times.*

This fact can be a very powerful tool in solving problems about isolated systems when you need to relate energies of a system before and after a certain process occurs in the system.

In Section 8-5, we discussed a special situation for isolated systems—namely, the situation in which nonconservative forces (such as a kinetic frictional force) do not act within them. In that special situation, ΔE_{th} and ΔE_{int} are both zero, and so Eq. 8-37 reduces to Eq. 8-18. In other words, the mechanical energy of an isolated system is conserved when nonconservative forces do not act in it.

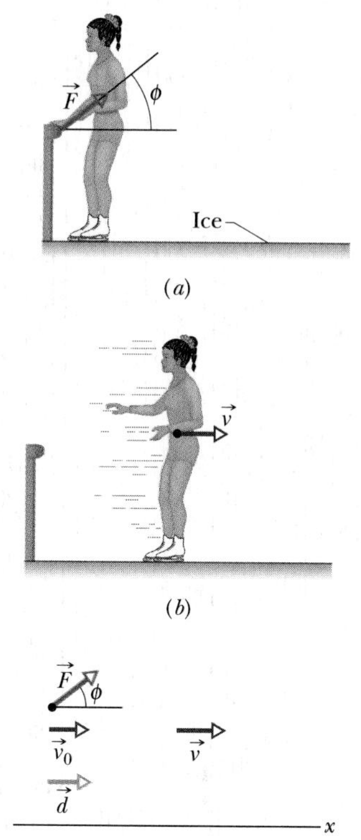

FIG. 8-16 (*a*) As a skater pushes herself away from a railing, the force on her from the railing is $\vec{F}$. (*b*) After the skater leaves the railing, she has velocity $\vec{v}$. (*c*) External force $\vec{F}$ acts on the skater, at angle ϕ with a horizontal x axis. When the skater goes through displacement $\vec{d}$, her velocity is changed from $\vec{v}_0$ (= 0) to $\vec{v}$ by the horizontal component of $\vec{F}$.

External Forces and Internal Energy Transfers

An external force can change the kinetic energy or potential energy of an object without doing work on the object—that is, without transferring energy to the object. Instead, the force is responsible for transfers of energy from one type to another inside the object.

Figure 8-16 shows an example. An initially stationary ice-skater pushes away from a railing and then slides over the ice (Figs. 8-16*a* and *b*). Her kinetic energy increases because of an external force $\vec{F}$ on her from the rail. However, that force does not transfer energy from the rail to her. Thus, the force does no work on her. Rather, her kinetic energy increases as a result of internal transfers from the biochemical energy in her muscles.

Figure 8-17 shows another example. An engine increases the speed of a car with four-wheel drive (all four wheels are made to turn by the engine). During the acceleration, the engine causes the tires to push backward on the road surface. This push produces frictional forces $\vec{f}$ that act on each tire in the forward direction. The net external force $\vec{F}$ from the road, which is the sum of these frictional forces, accelerates the car, increasing its kinetic energy. However, $\vec{F}$ does not transfer energy from the road to the car and so does no work on the car.

FIG. 8-17 A vehicle accelerates to the right using four-wheel drive. The road exerts four frictional forces (two of them shown) on the bottom surfaces of the tires. Taken together, these four forces make up the net external force $\vec{F}$ acting on the car.

Rather, the car's kinetic energy increases as a result of internal transfers from the energy stored in the fuel.

In situations like these two, we can sometimes relate the external force $\vec{F}$ on an object to the change in the object's mechanical energy if we can simplify the situation. Consider the ice-skater example. During her push through distance d in Fig. 8-16c, we can simplify by assuming that the acceleration is constant, her speed changing from $v_0 = 0$ to v. (That is, we assume $\vec{F}$ has constant magnitude F and angle ϕ.) After the push, we can simplify the skater as being a particle and neglect the fact that the exertions of her muscles have increased the thermal energy in her muscles and changed other physiological features. Then we can apply Eq. 7-5 ($\frac{1}{2}mv^2 - \frac{1}{2}mv_0^2 = F_xd$) to write

$$K - K_0 = (F\cos\phi)d,$$

or

$$\Delta K = Fd\cos\phi. \tag{8-38}$$

If the situation also involves a change in the elevation of an object, we can include the change ΔU in gravitational potential energy by writing

$$\Delta U + \Delta K = Fd\cos\phi. \tag{8-39}$$

The force on the right side of this equation does no work on the object but is still responsible for the changes in energy shown on the left side.

Power

Now that you have seen how energy can be transferred from one type to another, we can expand the definition of power given in Section 7-9. There power is defined as the rate at which work is done by a force. In a more general sense, power P is the rate at which energy is transferred by a force from one type to another. If an amount of energy ΔE is transferred in an amount of time Δt, the **average power** due to the force is

$$P_{\text{avg}} = \frac{\Delta E}{\Delta t}. \tag{8-40}$$

Similarly, the **instantaneous power** due to the force is

$$P = \frac{dE}{dt}. \tag{8-41}$$

Sample Problem 8-7

In Fig. 8-18, a 2.0 kg package of tamales slides along a floor with speed $v_1 = 4.0$ m/s. It then runs into and compresses a spring, until the package momentarily stops. Its path to the initially relaxed spring is frictionless, but as it compresses the spring, a kinetic frictional force from the floor, of magnitude 15 N, acts on the package. If $k = 10\,000$ N/m, by what distance d is the spring compressed when the package stops?

KEY IDEAS We need to examine all the forces and then to determine whether we have an isolated system or a system on which an external force is doing work.

Forces: The normal force on the package from the floor does no work on the package because the direction of this force is always perpendicular to the direction of the package's displacement. For the same reason, the gravitational force on the package does no work. As the spring is compressed, however, a spring force does work on the package, transferring energy to elastic potential energy of the spring. The spring force also pushes against a rigid wall. Because there is friction between the package and the floor, the sliding of

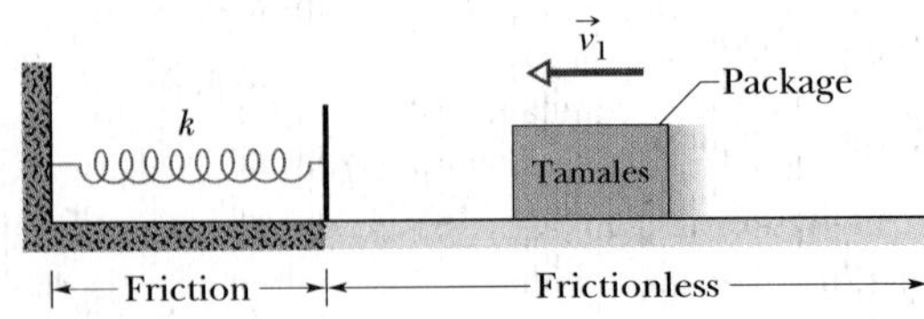

FIG. 8-18 A package slides across a frictionless floor with velocity $\vec{v}_1$ toward a spring of spring constant k. When the package reaches the spring, a frictional force from the floor acts on the package.

the package across the floor increases their thermal energies.

System: The package–spring–floor–wall system includes all these forces and energy transfers in one isolated system. Therefore, because the system is isolated, its total energy cannot change. We can then apply the law of conservation of energy in the form of Eq. 8-37 to the system:

$$E_{mec,2} = E_{mec,1} - \Delta E_{th}. \tag{8-42}$$

Calculations: In Eq. 8-42, let subscript 1 correspond to the initial state of the sliding package and subscript 2 correspond to the state in which the package is momentarily stopped and the spring is compressed by distance d. For both states the mechanical energy of the system is the sum of the package's kinetic energy ($K = \frac{1}{2}mv^2$) and the spring's potential energy ($U = \frac{1}{2}kx^2$). For state 1, $U = 0$ (because the spring is not compressed), and the package's speed is v_1. Thus, we have

$$E_{mec,1} = K_1 + U_1 = \tfrac{1}{2}mv_1^2 + 0.$$

For state 2, $K = 0$ (because the package is stopped), and the compression distance is d. Therefore, we have

$$E_{mec,2} = K_2 + U_2 = 0 + \tfrac{1}{2}kd^2.$$

Finally, by Eq. 8-31, we can substitute $f_k d$ for the change ΔE_{th} in the thermal energy of the package and the floor. We can now rewrite Eq. 8-42 as

$$\tfrac{1}{2}kd^2 = \tfrac{1}{2}mv_1^2 - f_k d.$$

Rearranging and substituting known data give us

$$5000d^2 + 15d - 16 = 0.$$

Solving this quadratic equation yields

$$d = 0.055 \text{ m} = 5.5 \text{ cm}. \quad \text{(Answer)}$$

Sample Problem 8-8

Figure 8-19a shows the mountain slope and the valley along which a rock avalanche moves. The rocks have a total mass m, fall from a height $y = H$, move a distance d_1 along a slope of angle $\theta = 45°$, and then move a distance d_2 along a flat valley. What is the ratio d_2/H of the runout to the fall height if the coefficient of kinetic friction has the reasonable value of 0.60?

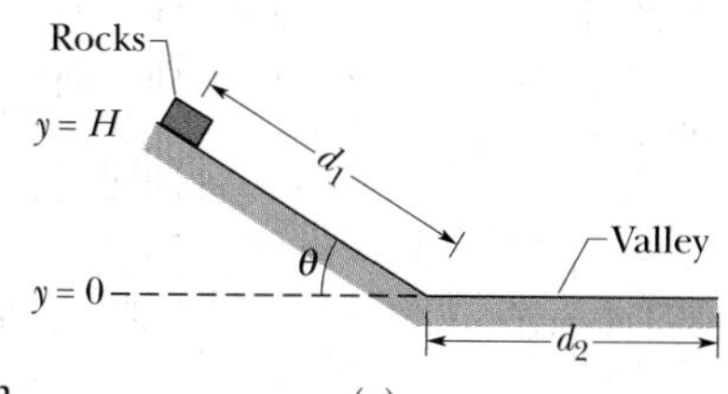

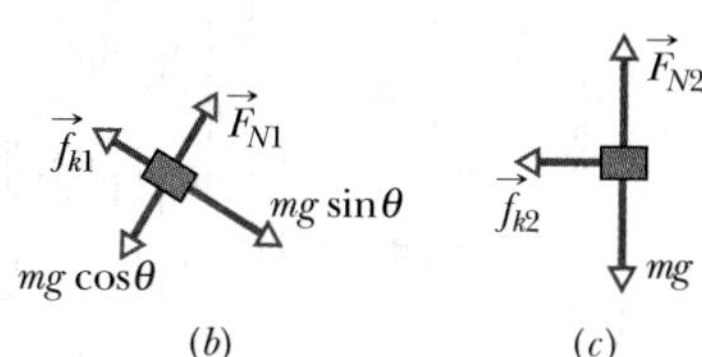

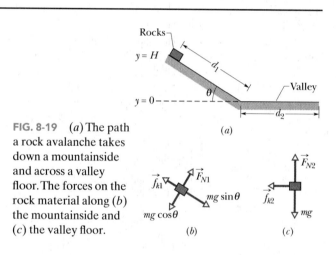

FIG. 8-19 (a) The path a rock avalanche takes down a mountainside and across a valley floor. The forces on the rock material along (b) the mountainside and (c) the valley floor.

KEY IDEAS (1) The mechanical energy E_{mec} of the rocks–Earth system is the sum of the kinetic energy ($K = \frac{1}{2}mv^2$) and the gravitational potential energy ($U = mgy$). (2) The mechanical energy is not conserved during the slide because a (nonconservative) frictional force acts on the rocks, transferring an amount of energy ΔE_{th} to the thermal energy of the rocks and ground. (3) The transferred energy ΔE_{th} is related to the magnitude of the kinetic frictional force and the distance of sliding by Eq. 8-31 ($\Delta E_{th} = f_k d$). (4) The mechanical energy $E_{mec,2}$ at any point during the slide is related to the initial mechanical energy $E_{mec,1}$ and the transferred energy ΔE_{th} by Eq. 8-37, which can be rewritten as $E_{mec,2} = E_{mec,1} - \Delta E_{th}$.

Calculations: The final mechanical energy $E_{mec,2}$ is equal to the initial mechanical energy $E_{mec,1}$ minus the amount ΔE_{th} lost to thermal energy:

$$E_{mec,2} = E_{mec,1} - \Delta E_{th}. \tag{8-43}$$

Initially the rocks have potential energy $U = mgH$ and kinetic energy $K = 0$, and so the initial mechanical energy is $E_{mec,1} = mgH$. Finally (when the rocks stop) the rocks have potential energy $U = 0$ and kinetic energy $K = 0$, and so $E_{mec,2} = 0$. The amount of energy transferred to thermal energy is $\Delta E_{th,1} = f_{k1}d_1$ during the slide down the slope and $\Delta E_{th,2} = f_{k2}d_2$ during the runout across the valley. Substituting these expressions into Eq. 8-43, we have

$$0 = mgH - f_{k1}d_1 - F_{k2}d_2. \tag{8-44}$$

From Fig. 8-19a, we see that $d_1 = H/(\sin\theta)$. To obtain expressions for the kinetic frictional forces, we use Eq. 6-2 ($f_k = \mu_k F_N$). Recall from Chapter 6 that on an inclined plane the normal force offsets the component $mg\cos\theta$ of the gravitational force (Fig. 8-19b). Similarly, recall from Chapter 5 that on a horizontal surface the normal force offsets the full magnitude mg of the gravitational force (Fig. 8-19c). Substituting these expressions into Eq. 8-44 and solving for the ratio d_2/H, we find

$$0 = mgH - \mu_k(mg\cos\theta)\frac{H}{\sin\theta} - \mu_k mgd_2$$

and

$$\frac{d_2}{H} = \left(\frac{1}{\mu_k} - \frac{1}{\tan\theta}\right). \tag{8-45}$$

Substituting $\mu_k = 0.60$ and $\theta = 45°$, we find

$$\frac{d_2}{H} = 0.67. \qquad \text{(Answer)}$$

Comments: Our result is typical for a small avalanche. However, for a large avalanche, the ratio d_2/H may be as large as 20. If you substitute this ratio into Eq. 8-45 and solve for the coefficient of kinetic friction, you find $\mu_k = 0.05$. Researchers do not understand how a large avalanche of jagged, tumbling rocks can have a value of μ_k small enough to rival that of very slippery ice. One of the most promising ideas is that the material is continuously levitated by a thin layer of small oscillating debris and thus almost never touches the mountain slope or valley floor until the avalanche stops.

Sample Problem 8-9

In Fig. 8-20*a*, a 20 kg block is about to collide with a spring at its relaxed length. As the block compresses the spring, a kinetic frictional force between the block and floor acts on the block. Figure 8-20*b* gives the block's kinetic energy $K(x)$ and the spring's potential energy $U(x)$ as functions of the block's position x as the spring is compressed. What is the coefficient of kinetic friction μ_k between the block and floor?

KEY IDEAS (1) The mechanical energy $E_{\text{mec}}\,(= K + U)$ is not conserved during the compression because the nonconservative frictional force acts on the block, transferring an amount of energy ΔE_{th} to the thermal energy of the block and floor. (2) The transferred energy ΔE_{th} is related to the magnitude of the kinetic frictional force and the distance of sliding by Eq. 8-31 ($\Delta E_{\text{th}} = f_k d$). (3) The mechanical energy $E_{\text{mec},2}$ at any point during the compression is related to the initial mechanical energy $E_{\text{mec},1}$ and ΔE_{th} by Eq. 8-37, which can be rewritten as $E_{\text{mec},2} = E_{\text{mec},1} - \Delta E_{\text{th}}$.

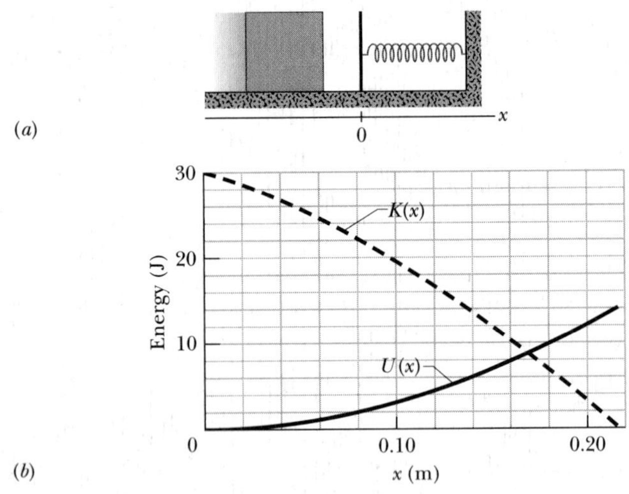

FIG. 8-20 (*a*) Block on the verge of colliding with a spring. (*b*) The change of kinetic energy K and potential energy U as the spring is compressed and the block slows to a stop.

Finding ΔE_{th}: From Fig. 8-20*b*, we see that when the block is at $x = 0$ and is on the verge of compressing the spring, its kinetic energy is $K = 30$ J and the spring's potential energy is $U = 0$. Thus, the sum of K and U is

$$E_{\text{mec},1} = 30 \text{ J}.$$

The spring is at its maximum compression when the block stops—that is, when the kinetic energy drops to 0. From the figure we see that this occurs at $x \approx 0.215$ m, at which $K = 0$ and $U = 14$ J. Thus, at the stopping point, the sum of K and U is

$$E_{\text{mec},2} = 14 \text{ J}.$$

To find the amount of energy transferred to thermal energy, we write $E_{\text{mec},2} = E_{\text{mec},1} - \Delta E_{\text{th}}$ as

$$14 \text{ J} = 30 \text{ J} - \Delta E_{\text{th}}$$

or

$$\Delta E_{\text{th}} = 16 \text{ J}.$$

Finding μ_k: From Eq. 6-2, we know that a kinetic frictional force is given by $f_k = \mu_k F_N$, where the normal force is given by Eq. 5-14 ($F_N = mg$). Here, the frictional force f_k transfers 16 J to thermal energy over distance $d = 0.215$ m, according to $\Delta E_{\text{th}} = f_k d$. Pulling these several expressions together, we write

$$\Delta E_{\text{th}} = f_k d = \mu_k F_N d = \mu_k mgd$$

and then substitute the data $\Delta E_{\text{th}} = 16$ J, $m = 20$ kg, $g = 9.8$ m/s^2, and $d = 0.215$ m, finding

$$\mu_k = 0.38. \qquad \text{(Answer)}$$

REVIEW & SUMMARY

Conservative Forces A force is a **conservative force** if the net work it does on a particle moving around any closed path, from an initial point and then back to that point, is zero. Equivalently, a force is conservative if the net work it does on a particle moving between two points does not depend on the path taken by the particle. The gravitational force and the spring force are conservative forces; the kinetic frictional force is a **nonconservative force.**

Potential Energy A **potential energy** is energy that is associated with the configuration of a system in which a conservative force acts. When the conservative force does work W

on a particle within the system, the change ΔU in the potential energy of the system is

$$\Delta U = -W. \tag{8-1}$$

If the particle moves from point x_i to point x_f, the change in the potential energy of the system is

$$\Delta U = -\int_{x_i}^{x_f} F(x)\,dx. \tag{8-6}$$

Gravitational Potential Energy The potential energy associated with a system consisting of Earth and a nearby particle is **gravitational potential energy.** If the particle moves from height y_i to height y_f, the change in the gravitational potential energy of the particle–Earth system is

$$\Delta U = mg(y_f - y_i) = mg\,\Delta y. \tag{8-7}$$

If the **reference point** of the particle is set as $y_i = 0$ and the corresponding gravitational potential energy of the system is set as $U_i = 0$, then the gravitational potential energy U when the particle is at any height y is

$$U(y) = mgy. \tag{8-9}$$

Elastic Potential Energy **Elastic potential energy** is the energy associated with the state of compression or extension of an elastic object. For a spring that exerts a spring force $F = -kx$ when its free end has displacement x, the elastic potential energy is

$$U(x) = \tfrac{1}{2}kx^2. \tag{8-11}$$

The **reference configuration** has the spring at its relaxed length, at which $x = 0$ and $U = 0$.

Mechanical Energy The **mechanical energy** E_{mec} of a system is the sum of its kinetic energy K and potential energy U:

$$E_{mec} = K + U. \tag{8-12}$$

An *isolated system* is one in which no *external force* causes energy changes. If only conservative forces do work within an isolated system, then the mechanical energy E_{mec} of the system cannot change. This **principle of conservation of mechanical energy** is written as

$$K_2 + U_2 = K_1 + U_1, \tag{8-17}$$

in which the subscripts refer to different instants during an energy transfer process. This conservation principle can also be written as

$$\Delta E_{mec} = \Delta K + \Delta U = 0. \tag{8-18}$$

Potential Energy Curves If we know the potential energy function $U(x)$ for a system in which a one-dimensional force $F(x)$ acts on a particle, we can find the force as

$$F(x) = -\frac{dU(x)}{dx}. \tag{8-22}$$

If $U(x)$ is given on a graph, then at any value of x, the force $F(x)$ is the negative of the slope of the curve there and the kinetic energy of the particle is given by

$$K(x) = E_{mec} - U(x), \tag{8-24}$$

where E_{mec} is the mechanical energy of the system. A **turning point** is a point x at which the particle reverses its motion (there, $K = 0$). The particle is in **equilibrium** at points where the slope of the $U(x)$ curve is zero (there, $F(x) = 0$).

Work Done on a System by an External Force Work W is energy transferred to or from a system by means of an external force acting on the system. When more than one force acts on a system, their *net work* is the transferred energy. When friction is not involved, the work done on the system and the change ΔE_{mec} in the mechanical energy of the system are equal:

$$W = \Delta E_{mec} = \Delta K + \Delta U. \tag{8-26, 8-25}$$

When a kinetic frictional force acts within the system, then the thermal energy E_{th} of the system changes. (This energy is associated with the random motion of atoms and molecules in the system.) The work done on the system is then

$$W = \Delta E_{mec} + \Delta E_{th}. \tag{8-33}$$

The change ΔE_{th} is related to the magnitude f_k of the frictional force and the magnitude d of the displacement caused by the external force by

$$\Delta E_{th} = f_k d. \tag{8-31}$$

Conservation of Energy The **total energy** E of a system (the sum of its mechanical energy and its internal energies, including thermal energy) can change only by amounts of energy that are transferred to or from the system. This experimental fact is known as the **law of conservation of energy.** If work W is done on the system, then

$$W = \Delta E = \Delta E_{mec} + \Delta E_{th} + \Delta E_{int}. \tag{8-35}$$

If the system is isolated ($W = 0$), this gives

$$\Delta E_{mec} + \Delta E_{th} + \Delta E_{int} = 0 \tag{8-36}$$

and

$$E_{mec,2} = E_{mec,1} - \Delta E_{th} - \Delta E_{int}, \tag{8-37}$$

where the subscripts 1 and 2 refer to two different instants.

Power The **power** due to a force is the *rate* at which that force transfers energy. If an amount of energy ΔE is transferred by a force in an amount of time Δt, the **average power** of the force is

$$P_{avg} = \frac{\Delta E}{\Delta t}. \tag{8-40}$$

The **instantaneous power** due to a force is

$$P = \frac{dE}{dt}. \tag{8-41}$$

QUESTIONS

1 Figure 8-21 shows one direct path and four indirect paths from point i to point f. Along the direct path and three of the indirect paths, only a conservative force F_c acts on a certain object. Along the fourth indirect path, both F_c and a nonconservative force F_{nc} act on the object. The change ΔE_{mec} in the object's mechanical energy (in joules) in going from i to f is indicated along each straight-line segment of the indirect paths. What is ΔE_{mec} (a) from i to f along the direct path and

(b) due to F_{nc} along the one path where it acts?

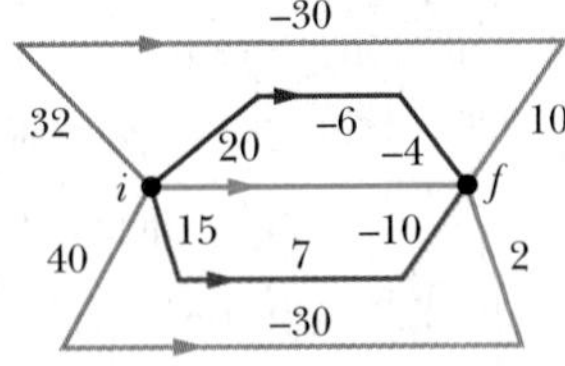

FIG. 8-21 Question 1.

2 In Fig. 8-22, a small, initially stationary block is released on a frictionless ramp at a height of 3.0 m. Hill heights along the ramp are as shown. The hills have identical circular tops, and the block does not fly off any hill. (a) Which hill is the first the block cannot cross? (b) What does the block do after failing to cross that hill? On which hilltop is (c) the centripetal acceleration of the block greatest and (d) the normal force on the block least?

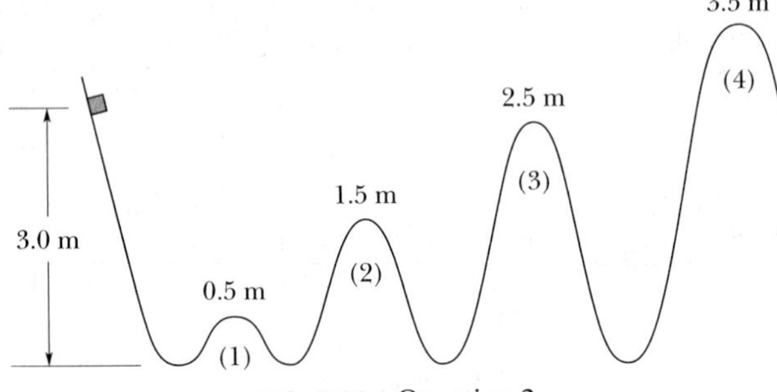

FIG. 8-22 Question 2.

3 In Fig. 8-23, a horizontally moving block can take three frictionless routes, differing only in elevation, to reach the dashed finish line. Rank the routes according to (a) the speed of the block at the finish line and (b) the travel time of the block to the finish line, greatest first.

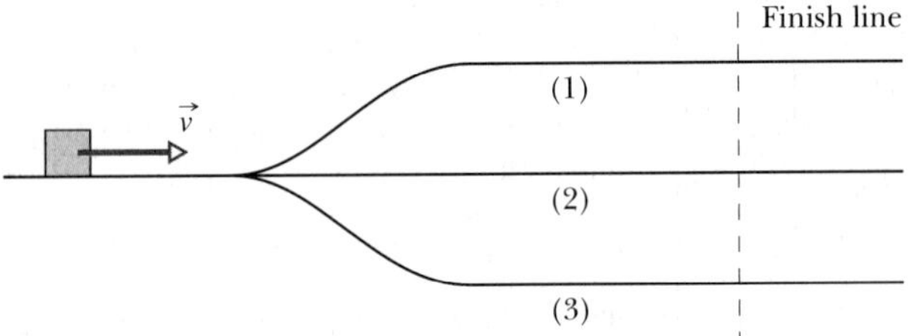

FIG. 8-23 Question 3.

4 Figure 8-24 gives the potential energy function of a particle. (a) Rank regions *AB*, *BC*, *CD*, and *DE* according to the magnitude of the force on the particle, greatest first. What value must the mechanical energy E_{mec} of the particle not exceed if the particle is to be (b) trapped in the potential well at the left, (c) trapped in the potential well at the right, and (d) able to move between the two potential wells but not to the right of point *H*? For the situation of (d), in which of regions *BC*, *DE*, and *FG* will the particle have (e) the greatest kinetic energy and (f) the least speed?

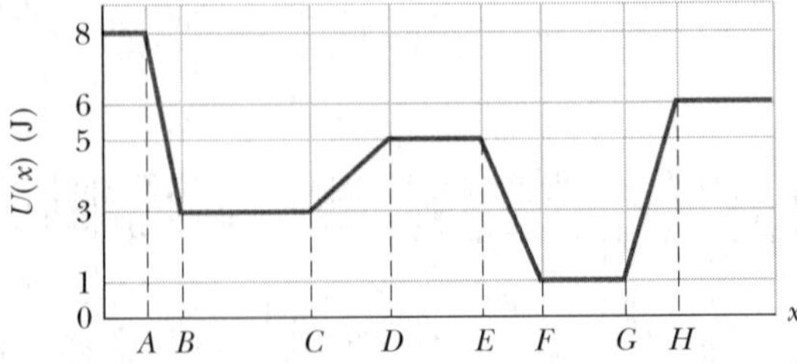

FIG. 8-24 Question 4.

5 Figure 8-25 shows three situations involving a plane that is not frictionless and a block sliding along the plane. The block begins with the same speed in all three situations and slides until the kinetic frictional force has stopped it. Rank the situations according to the increase in thermal energy due to the sliding, greatest first.

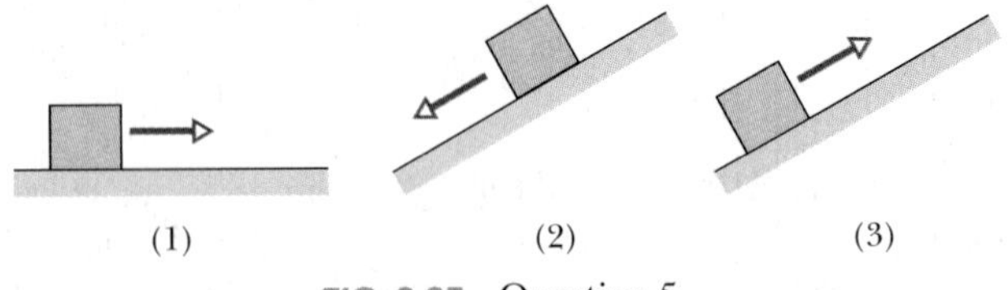

FIG. 8-25 Question 5.

6 In Fig. 8-26*a*, you pull upward on a rope that is attached to a cylinder on a vertical rod. Because the cylinder fits tightly on the rod, the cylinder slides along the rod with considerable friction. Your force does work $W = +100$ J on the cylinder–rod–Earth system (Fig. 8-26*b*). An "energy statement" for the system is shown in Fig. 8-26*c*: the kinetic energy *K* increases by 50 J, and the gravitational potential energy U_g increases by 20 J. The only other change in energy within the system is for the thermal energy E_{th}. What is the change ΔE_{th}?

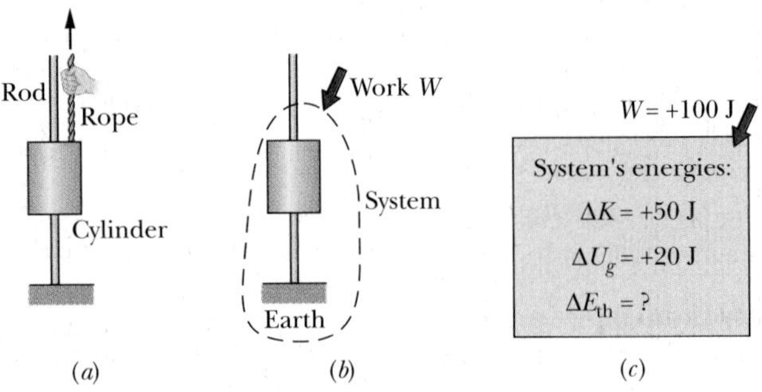

FIG. 8-26 Question 6.

7 The arrangement shown in Fig. 8-27 is similar to that in Question 6. Here you pull downward on the rope that is attached to the cylinder, which fits tightly on the rod. Also, as the cylinder descends, it pulls on a block via a second rope, and the block slides over a lab table. Again consider the cylinder–rod–Earth system, similar to that shown in Fig. 8-26*b*. Your work on the system is 200 J. The system does work of 60 J on the block. Within the system, the kinetic energy increases by 130 J and the gravitational potential energy decreases by 20 J. (a) Draw an "energy statement" for the system, as in Fig. 8-26*c*. (b) What is the change in the thermal energy within the system?

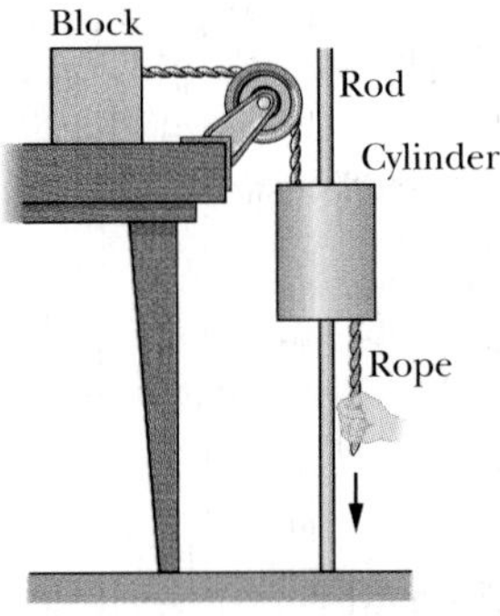

FIG. 8-27 Question 7.

8 In Fig. 8-28, a block slides along a track that descends through distance *h*. The track is frictionless except for the lower section. There the block slides to a stop in a certain distance *D* because of friction. (a) If we decrease *h*, will the block now slide to a stop in a distance that is greater than, less than, or equal to D? (b) If, instead, we increase the mass of the block, will the stopping distance now be greater than, less than, or equal to D?

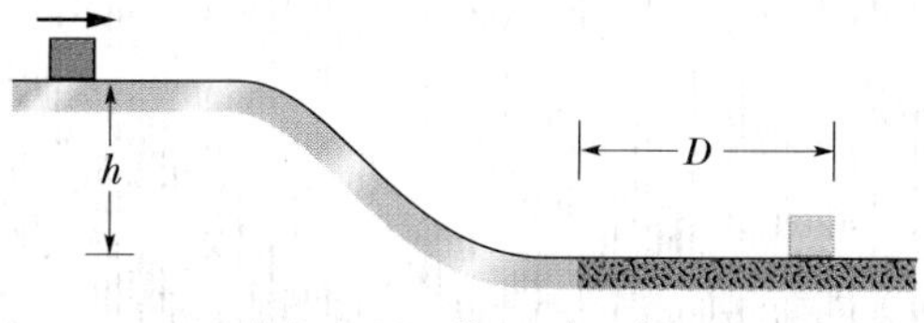

FIG. 8-28 Question 8.

9 In Fig. 8-29, a block slides from A to C along a frictionless ramp, and then it passes through horizontal region CD, where a frictional force acts on it. Is the block's kinetic energy increasing, decreasing, or constant in (a) region AB, (b) region BC, and (c) region CD? (d) Is the block's mechanical energy increasing, decreasing, or constant in those regions?

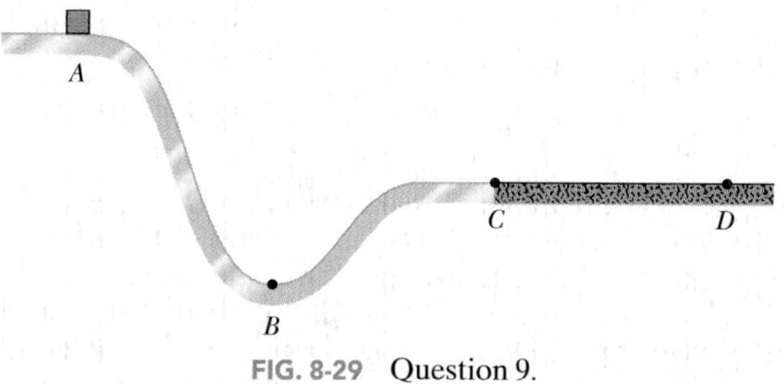

FIG. 8-29 Question 9.

PROBLEMS

GO Tutoring problem available (at instructor's discretion) in *WileyPLUS* and WebAssign
SSM Worked-out solution available in Student Solutions Manual
• – ••• Number of dots indicates level of problem difficulty
Additional information available in *The Flying Circus of Physics* and at flyingcircusofphysics.com
WWW Worked-out solution is at
ILW Interactive solution is at
http://www.wiley.com/college/halliday

sec. 8-4 Determining Potential Energy Values

•1 You drop a 2.00 kg book to a friend who stands on the ground at distance $D = 10.0$ m below. If your friend's outstretched hands are at distance $d = 1.50$ m above the ground (Fig. 8-30), (a) how much work W_g does the gravitational force do on the book as it drops to her hands? (b) What is the change ΔU in the gravitational potential energy of the book–Earth system during the drop? If the gravitational potential energy U of that system is taken to be zero at ground level, what is U (c) when the book is released and (d) when it reaches her hands? Now take U to be 100 J at ground level and again find (e) W_g, (f) ΔU, (g) U at the release point, and (h) U at her hands.

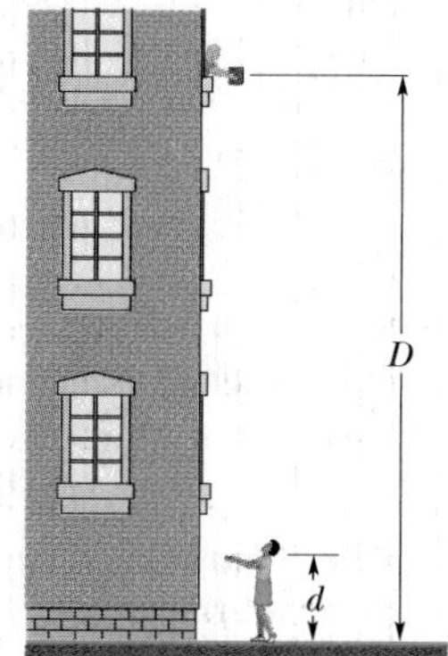

FIG. 8-30 Problems 1 and 10.

•2 Figure 8-31 shows a ball with mass $m = 0.341$ kg attached to the end of a thin rod with length $L = 0.452$ m and negligible mass. The other end of the rod is pivoted so that the ball can move in a vertical circle. The rod is held horizontally as shown and then given enough of a downward push to cause the ball to swing down and around and just reach the vertically up position, with zero speed there. How much work is done on the ball by the gravitational force from the initial point to (a) the lowest point, (b) the highest point, and (c) the point on the right level with the initial point? If the gravitational potential energy of the ball–Earth system is taken to be zero at the initial point, what is it when the ball reaches (d) the lowest point, (e) the highest point, and (f) the point on the right level with the initial point? (g) Suppose the rod were pushed harder so that the ball passed through the highest point with a nonzero speed. Would ΔU_g from the lowest point to the highest point then be greater than, less than, or the same as it was when the ball stopped at the highest point?

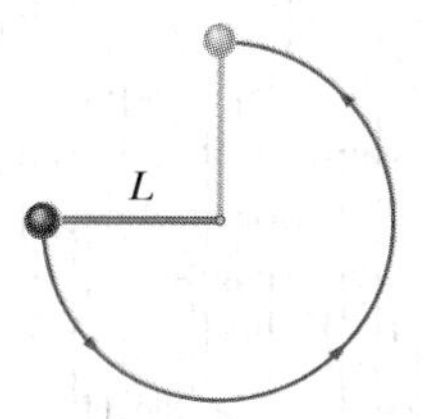

FIG. 8-31 Problems 2 and 12.

•3 In Fig. 8-32, a 2.00 g ice flake is released from the edge of a hemispherical bowl whose radius r is 22.0 cm. The flake–bowl contact is frictionless. (a) How much work is done on the flake by the gravitational force during the flake's descent to the bottom of the bowl? (b) What is the change in the potential energy of the flake–Earth system during that descent? (c) If that potential energy is taken to be zero at the bottom of the bowl, what is its value when the flake is released? (d) If, instead, the potential energy is taken to be zero at the release point, what is its value when the flake reaches the bottom of the bowl? (e) If the mass of the flake were doubled, would the magnitudes of the answers to (a) through (d) increase, decrease, or remain the same? SSM

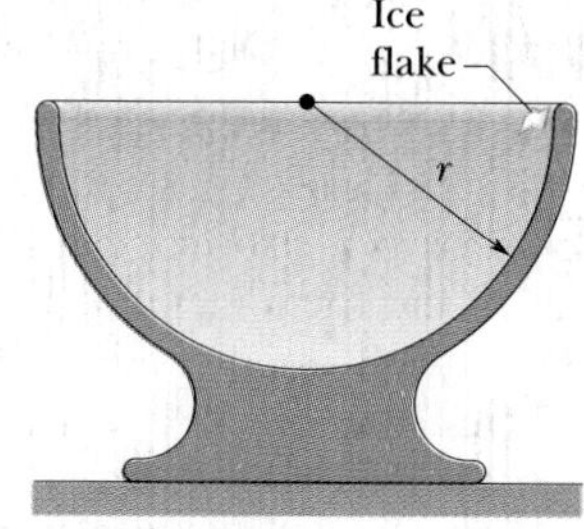

FIG. 8-32 Problems 3 and 11.

•4 In Fig. 8-33, a frictionless roller-coaster car of mass $m = 825$ kg tops the first hill with speed $v_0 = 17.0$ m/s at height $h = 42.0$ m. How much work does the gravitational

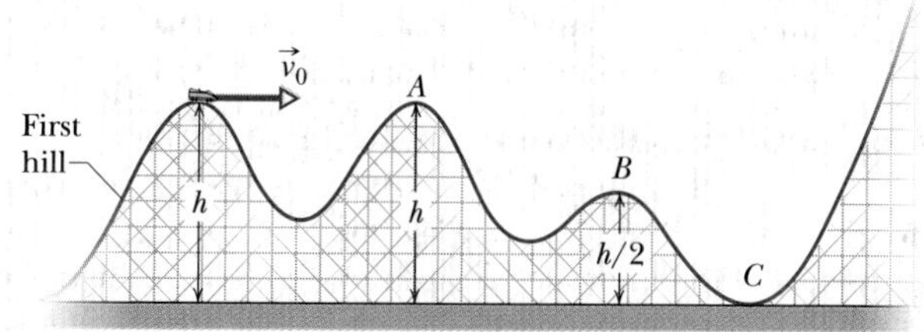

FIG. 8-33 Problems 4 and 13.

force do on the car from that point to (a) point A, (b) point B, and (c) point C? If the gravitational potential energy of the car–Earth system is taken to be zero at C, what is its value when the car is at (d) B and (e) A? (f) If mass m were doubled, would the change in the gravitational potential energy of the system between points A and B increase, decrease, or remain the same?

•5 What is the spring constant of a spring that stores 25 J of elastic potential energy when compressed by 7.5 cm? SSM

•6 A 1.50 kg snowball is fired from a cliff 12.5 m high. The snowball's initial velocity is 14.0 m/s, directed 41.0° above the horizontal. (a) How much work is done on the snowball by the gravitational force during its flight to the flat ground below the cliff? (b) What is the change in the gravitational potential energy of the snowball–Earth system during the flight? (c) If that gravitational potential energy is taken to be zero at the height of the cliff, what is its value when the snowball reaches the ground?

••7 Figure 8-34 shows a thin rod, of length $L = 2.00$ m and negligible mass, that can pivot about one end to rotate in a vertical circle. A ball of mass $m = 5.00$ kg is attached to the other end. The rod is pulled aside to angle $\theta_0 = 30.0°$ and released with initial velocity $\vec{v}_0 = 0$. As the ball descends to its lowest point, (a) how much work does the gravitational force do on it and (b) what is the change in the gravitational potential energy of the ball–Earth system? (c) If the gravitational potential energy is taken to be zero at the lowest point, what is its value just as the ball is released? (d) Do the magnitudes of the answers to (a) through (c) increase, decrease, or remain the same if angle θ is increased?

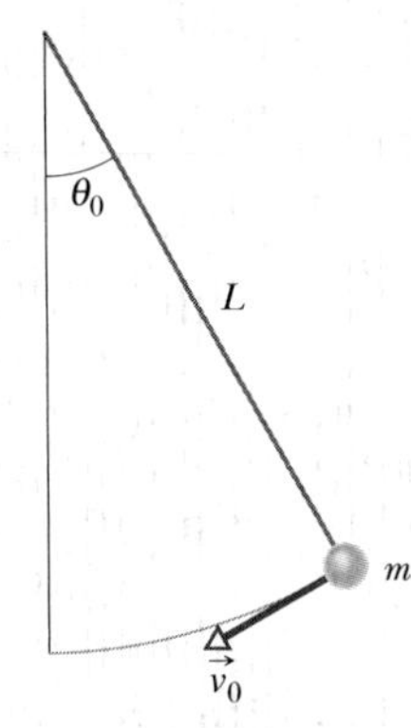

FIG. 8-34 Problems 7, 16, and 17.

••8 In Fig. 8-35, a small block of mass $m = 0.032$ kg can slide along the frictionless loop-the-loop, with loop radius $R = 12$ cm. The block is released from rest at point P, at height $h = 5.0R$ above the bottom of the loop. How much work does the gravitational force do on the block as the block travels from point P to (a) point Q and (b) the top of the loop? If the gravitational potential energy of the block–Earth system is taken to be zero at the bottom of the loop, what is that potential energy when the block is (c) at point P, (d) at point Q, and (e) at the top of the loop? (f) If, instead of merely being released, the block is given some initial speed downward along the track, do the answers to (a) through (e) increase, decrease, or remain the same?

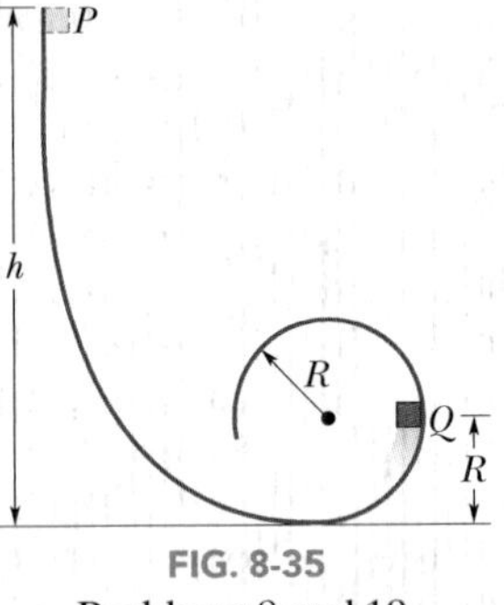

FIG. 8-35 Problems 8 and 19.

sec. 8-5 Conservation of Mechanical Energy

•9 In Fig. 8-36, a runaway truck with failed brakes is moving downgrade at 130 km/h just before the driver steers the truck up a frictionless emergency escape ramp with an inclination of $\theta = 15°$. The truck's mass is 1.2×10^4 kg. (a) What minimum length L must the ramp have if the truck is to stop (momentarily) along it? (Assume the truck is a particle, and justify that assumption.) Does the minimum length L increase, decrease, or remain the same if (b) the truck's mass is decreased and (c) its speed is decreased? SSM

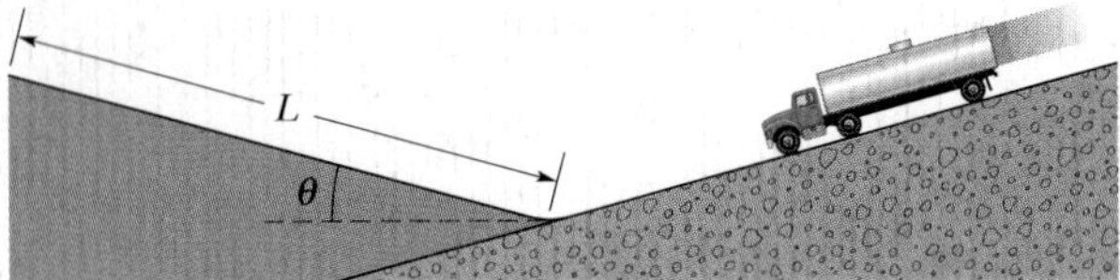

FIG. 8-36 Problem 9.

•10 (a) In Problem 1, what is the speed of the book when it reaches the hands? (b) If we substituted a second book with twice the mass, what would its speed be? (c) If, instead, the book were thrown down, would the answer to (a) increase, decrease, or remain the same?

•11 (a) In Problem 3, what is the speed of the flake when it reaches the bottom of the bowl? (b) If we substituted a second flake with twice the mass, what would its speed be? (c) If, instead, we gave the flake an initial downward speed along the bowl, would the answer to (a) increase, decrease, or remain the same? SSM WWW

•12 (a) In Problem 2, what initial speed must be given the ball so that it reaches the vertically upward position with zero speed? What then is its speed at (b) the lowest point and (c) the point on the right at which the ball is level with the initial point? (d) If the ball's mass were doubled, would the answers to (a) through (c) increase, decrease, or remain the same?

•13 In Problem 4, what is the speed of the car at (a) point A, (b) point B, and (c) point C? (d) How high will the car go on the last hill, which is too high for it to cross? (e) If we substitute a second car with twice the mass, what then are the answers to (a) through (d)? GO

•14 (a) In Problem 6, using energy techniques rather than the techniques of Chapter 4, find the speed of the snowball as it reaches the ground below the cliff. What is that speed (b) if the launch angle is changed to 41.0° *below* the horizontal and (c) if the mass is changed to 2.50 kg?

•15 A 5.0 g marble is fired vertically upward using a spring gun. The spring must be compressed 8.0 cm if the marble is to just reach a target 20 m above the marble's position on the compressed spring. (a) What is the change ΔU_g in the gravitational potential energy of the marble–Earth system during the 20 m ascent? (b) What is the change ΔU_s in the elastic potential energy of the spring during its launch of the marble? (c) What is the spring constant of the spring? SSM

••16 (a) In Problem 7, what is the speed of the ball at the lowest point? (b) Does the speed increase, decrease, or remain the same if the mass is increased?

••17 Figure 8-34 shows a pendulum of length $L = 1.25$ m. Its bob (which effectively has all the mass) has speed v_0 when the cord makes an angle $\theta_0 = 40.0°$ with the vertical. (a) What is the speed of the bob when it is in its lowest position if $v_0 = 8.00$ m/s? What is the least value that v_0 can have if the pendulum is to swing down and then up (b) to a horizontal position, and (c) to a vertical position with the cord remaining straight? (d) Do the answers to (b) and (c) increase, decrease, or remain the same if θ_0 is increased by a few degrees?

••18 A 700 g block is released from rest at height h_0 above a vertical spring with spring constant $k = 400$ N/m and negligible mass. The block sticks to the spring and momentarily stops after compressing the spring 19.0 cm. How much work is done (a) by the block on the spring and (b) by the spring on the block? (c) What is the value of h_0? (d) If the block were released from height $2.00h_0$ above the spring, what would be the maximum compression of the spring?

••19 In Problem 8, what are the magnitudes of (a) the horizontal component and (b) the vertical component of the *net* force acting on the block at point Q? (c) At what height h should the block be released from rest so that it is on the verge of losing contact with the track at the top of the loop? (*On the verge of losing contact* means that the normal force on the block from the track has just then become zero.) (d) Graph the magnitude of the normal force on the block at the top of the loop versus initial height h, for the range $h = 0$ to $h = 6R$.

••20 A single conservative force $\vec{F} = (6.0x - 12)\hat{\imath}$ N, where x is in meters, acts on a particle moving along an x axis. The potential energy U associated with this force is assigned a value of 27 J at $x = 0$. (a) Write an expression for U as a function of x, with U in joules and x in meters. (b) What is the maximum positive potential energy? At what (c) negative value and (d) positive value of x is the potential energy equal to zero?

••21 The string in Fig. 8-37 is $L = 120$ cm long, has a ball attached to one end, and is fixed at its other end. The distance d from the fixed end to a fixed peg at point P is 75.0 cm. When the initially stationary ball is released with the string horizontal as shown, it will swing along the dashed arc. What is its speed when it reaches (a) its lowest point and (b) its highest point after the string catches on the peg? ILW

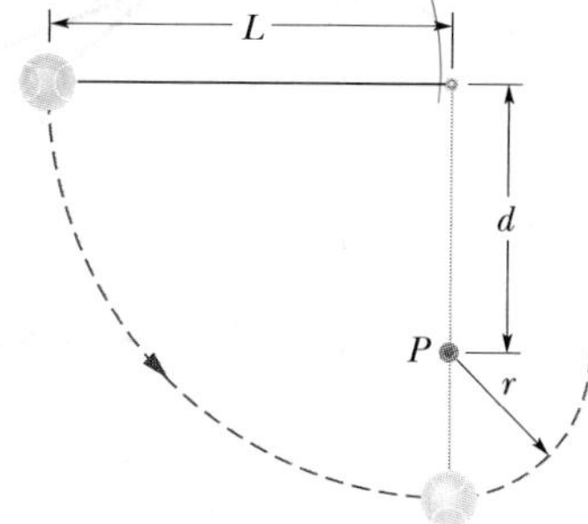

FIG. 8-37 Problems 21 and 68.

••22 A block of mass $m = 2.0$ kg is dropped from height $h = 40$ cm onto a spring of spring constant $k = 1960$ N/m (Fig. 8-38). Find the maximum distance the spring is compressed.

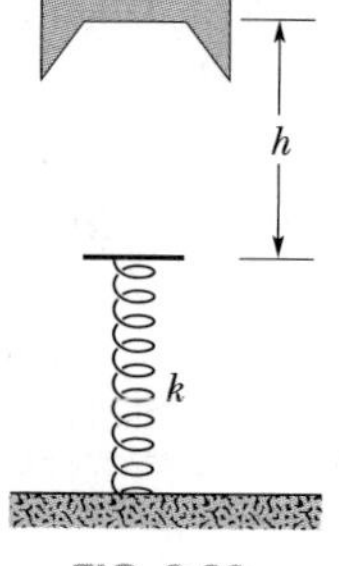

FIG. 8-38 Problem 22.

••23 At $t = 0$ a 1.0 kg ball is thrown from a tall tower with $\vec{v} = (18\text{ m/s})\hat{\imath} + (24\text{ m/s})\hat{\jmath}$. What is ΔU of the ball–Earth system between $t = 0$ and $t = 6.0$ s (still free fall)?

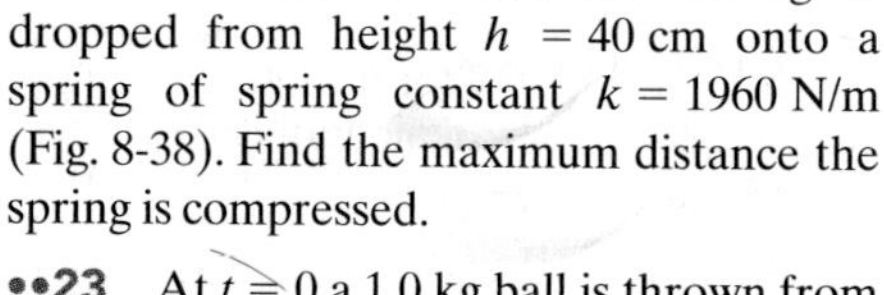

••24 A 60 kg skier starts from rest at height $H = 20$ m above the end of a ski-jump ramp (Fig. 8-39) and leaves the ramp at angle $\theta = 28°$. Neglect the effects of air resistance and assume the ramp is frictionless. (a) What is the maximum height h of his jump above the end of the ramp? (b) If he increased his weight by putting on a backpack, would h then be greater, less, or the same?

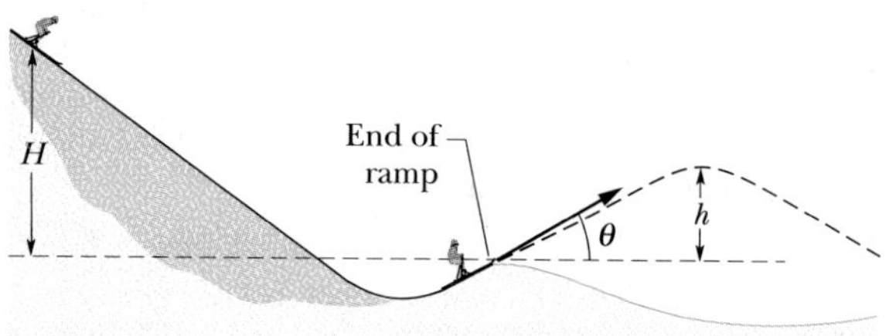

FIG. 8-39 Problem 24.

••25 Tarzan, who weighs 688 N, swings from a cliff at the end of a vine 18 m long (Fig. 8-40). From the top of the cliff to the bottom of the swing, he descends by 3.2 m. The vine will break if the force on it exceeds 950 N. (a) Does the vine break? (b) If no, what is the greatest force on it during the swing? If yes, at what angle with the vertical does it break?

FIG. 8-40 Problem 25.

••26 A pendulum consists of a 2.0 kg stone swinging on a 4.0 m string of negligible mass. The stone has a speed of 8.0 m/s when it passes its lowest point. (a) What is the speed when the string is at 60° to the vertical? (b) What is the greatest angle with the vertical that the string will reach during the stone's motion? (c) If the potential energy of the pendulum–Earth system is taken to be zero at the stone's lowest point, what is the total mechanical energy of the system?

••27 Figure 8-41 shows an 8.00 kg stone at rest on a spring. The spring is compressed 10.0 cm by the stone. (a) What is the spring constant? (b) The stone is pushed down an additional 30.0 cm and released. What is the elastic potential energy of the compressed spring just before that release? (c) What is the change in the gravitational potential energy of the stone–Earth system when the stone moves from the release point to its maximum height? (d) What is that maximum height, measured from the release point? GO

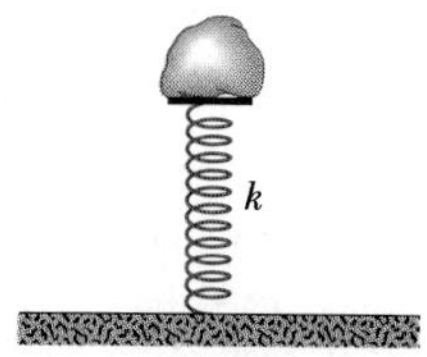

FIG. 8-41 Problem 27.

••28 A 2.0 kg breadbox on a frictionless incline of angle $\theta = 40°$ is connected, by a cord that runs over a pulley, to a light spring of spring constant $k = 120$ N/m, as shown in Fig. 8-42. The box is released from rest when the spring is unstretched. Assume that the pulley is massless and frictionless. (a) What is the speed of the box when it has moved 10 cm down the incline? (b) How far down the incline from its point of release does the box slide before momentarily stopping, and what are the (c) magnitude and (d) direction (up or down the incline) of the box's acceleration at the instant the box momentarily stops?

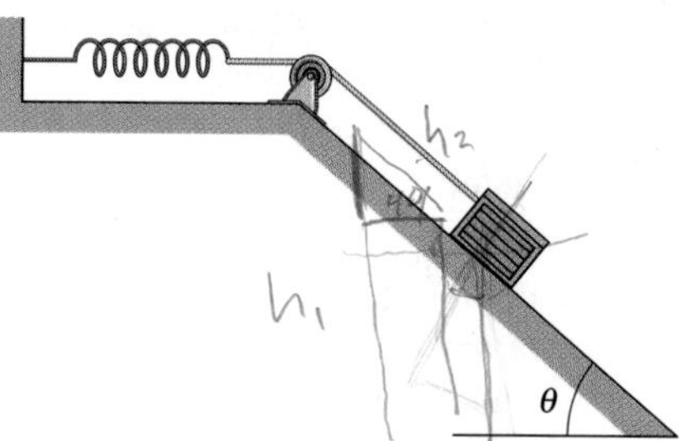

FIG. 8-42 Problem 28.

••29 A block with mass $m = 2.00$ kg is placed against a spring on a frictionless incline with angle $\theta = 30.0°$ (Fig. 8-43). (The block is not attached to the spring.) The spring, with spring constant $k = 19.6$ N/cm, is compressed 20.0 cm and then released. (a) What is the elastic potential energy of the compressed spring? (b) What is the change in the gravitational potential energy of the block–Earth system as the block moves from the release point to its highest point on the incline? (c) How far along the incline is the highest point from the release point? ILW

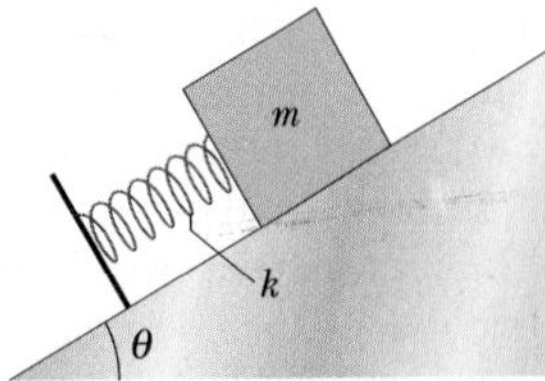

FIG. 8-43 Problem 29.

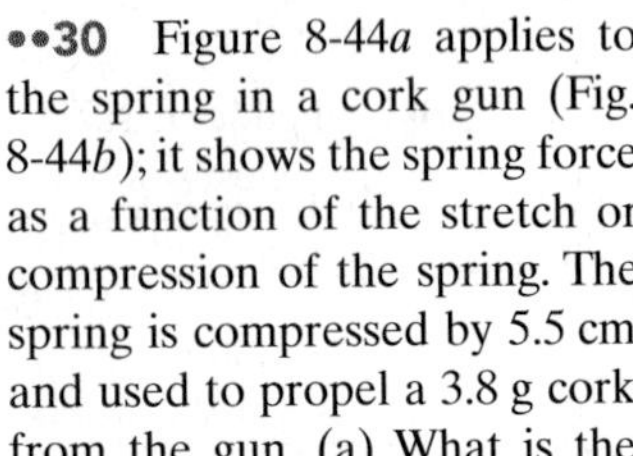

••30 Figure 8-44*a* applies to the spring in a cork gun (Fig. 8-44*b*); it shows the spring force as a function of the stretch or compression of the spring. The spring is compressed by 5.5 cm and used to propel a 3.8 g cork from the gun. (a) What is the speed of the cork if it is released as the spring passes through its relaxed position? (b) Suppose, instead, that the cork sticks to the spring and stretches it 1.5 cm before separation occurs. What now is the speed of the cork at the time of release?

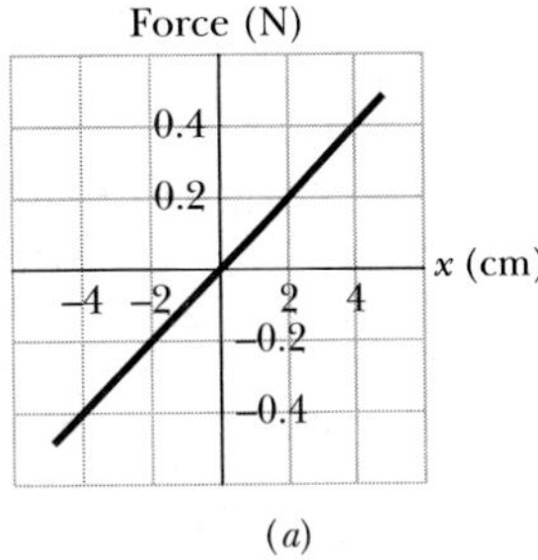

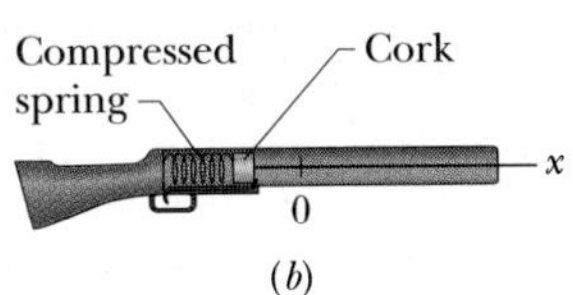

FIG. 8-44 Problem 30.

••31 In Fig. 8-45, a block of mass $m = 12$ kg is released from rest on a frictionless incline of angle $\theta = 30°$. Below the block is a spring that can be compressed 2.0 cm by a force of 270 N. The block momentarily stops when it compresses the spring by 5.5 cm. (a) How far does the block move down the incline from its rest position to this stopping point? (b) What is the speed of the block just as it touches the spring? SSM WWW

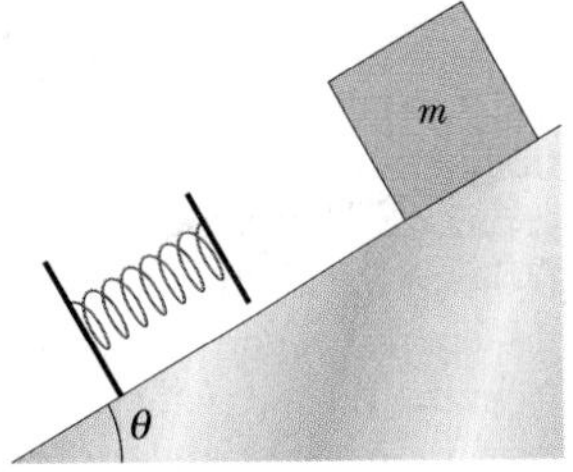

FIG. 8-45 Problems 31 and 37.

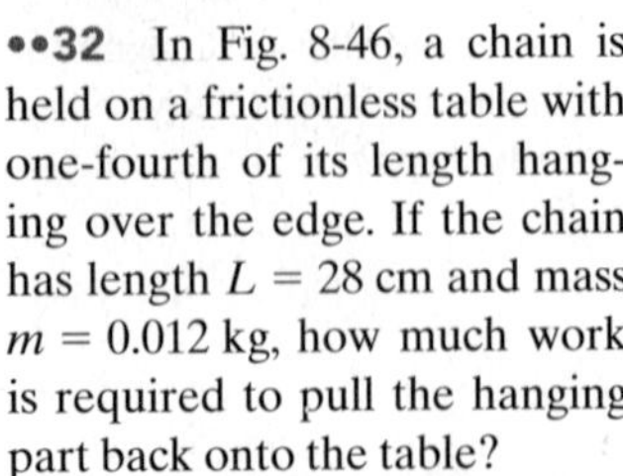

••32 In Fig. 8-46, a chain is held on a frictionless table with one-fourth of its length hanging over the edge. If the chain has length $L = 28$ cm and mass $m = 0.012$ kg, how much work is required to pull the hanging part back onto the table?

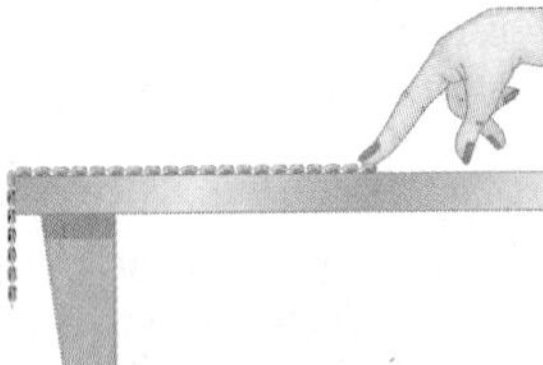
FIG. 8-46 Problem 32.

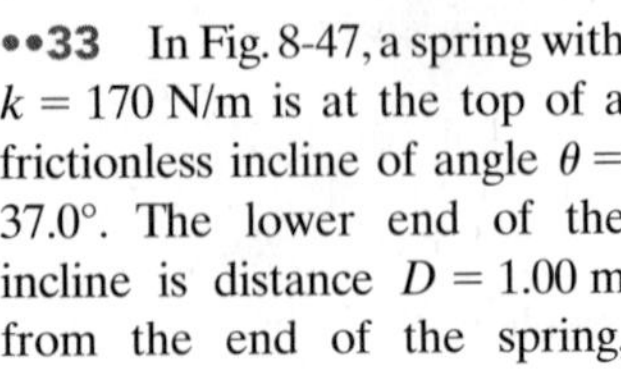

••33 In Fig. 8-47, a spring with $k = 170$ N/m is at the top of a frictionless incline of angle $\theta = 37.0°$. The lower end of the incline is distance $D = 1.00$ m from the end of the spring, which is at its relaxed length. A 2.00 kg canister is pushed against the spring until the spring is compressed 0.200 m and released from rest. (a) What is the speed of the canister at the instant the spring returns to its relaxed length (which is when the canister loses contact with the spring)? (b) What is the speed of the canister when it reaches the lower end of the incline? GO

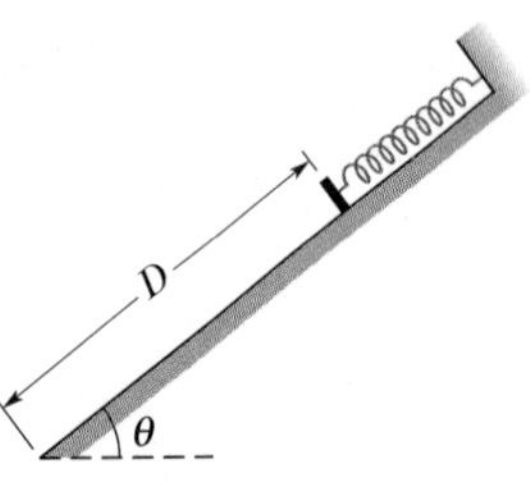

FIG. 8-47 Problem 33.

•••34 Two children are playing a game in which they try to hit a small box on the floor with a marble fired from a spring-loaded gun that is mounted on a table. The target box is horizontal distance $D = 2.20$ m from the edge of the table; see Fig. 8-48. Bobby compresses the spring 1.10 cm, but the center of the marble falls 27.0 cm short of the center of the box. How far should Rhoda compress the spring to score a direct hit? Assume that neither the spring nor the ball encounters friction in the gun.

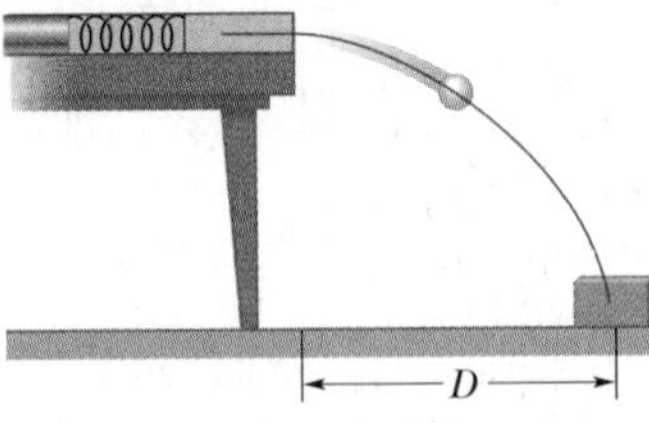

FIG. 8-48 Problem 34.

•••35 A uniform cord of length 25 cm and mass 15 g is initially stuck to a ceiling. Later, it hangs vertically from the ceiling with only one end still stuck. What is the change in the gravitational potential energy of the cord with this change in orientation? (*Hint:* Consider a differential slice of the cord and then use integral calculus.)

•••36 A boy is initially seated on the top of a hemispherical ice mound of radius $R = 13.8$ m. He begins to slide down the ice, with a negligible initial speed (Fig. 8-49). Approximate the ice as being frictionless. At what height does the boy lose contact with the ice?

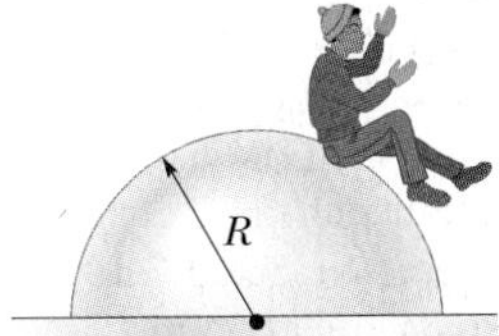

FIG. 8-49 Problem 36.

•••37 In Fig. 8-45, a block of mass $m = 3.20$ kg slides from rest a distance d down a frictionless incline at angle $\theta = 30.0°$ where it runs into a spring of spring constant 431 N/m. When the block momentarily stops, it has compressed the spring by 21.0 cm. What are (a) distance d and (b) the distance between the point of the first block–spring contact and the point where the block's speed is greatest?

sec. 8-6 Reading a Potential Energy Curve

••38 The potential energy of a diatomic molecule (a two-atom system like H_2 or O_2) is given by

$$U = \frac{A}{r^{12}} - \frac{B}{r^6},$$

where r is the separation of the two atoms of the molecule and A and B are positive constants. This potential energy is associated with the force that binds the two atoms together. (a) Find the *equilibrium separation*—that is, the distance between the atoms at which the force on each atom is zero. Is the force repulsive (the atoms are pushed apart) or attractive (they are pulled together) if their separation is (b) smaller and (c) larger than the equilibrium separation?

••39 Figure 8-50 shows a plot of potential energy U versus position x of a 0.90 kg particle that can travel only along an x axis. (Nonconservative forces are not involved.) Three values are $U_A = 15.0$ J, $U_B = 35.0$ J, and $U_C = 45.0$ J. The particle is released at $x = 4.5$ m with an initial speed of 7.0 m/s, headed in the negative x direction. (a) If the particle can reach $x = 1.0$ m, what is its speed there, and if it cannot, what is its turning point? What are the (b) magnitude and (c) direction of the force on the particle as it begins to move to the left of $x = 4.0$ m? Suppose, instead, the particle is headed in the positive x direction when it is released at $x = 4.5$ m at speed 7.0 m/s. (d) If the particle can reach $x = 7.0$ m, what is its speed there, and if it cannot, what is its turning point? What are the (e) magnitude and (f) direction of the force on the particle as it begins to move to the right of $x = 5.0$ m?

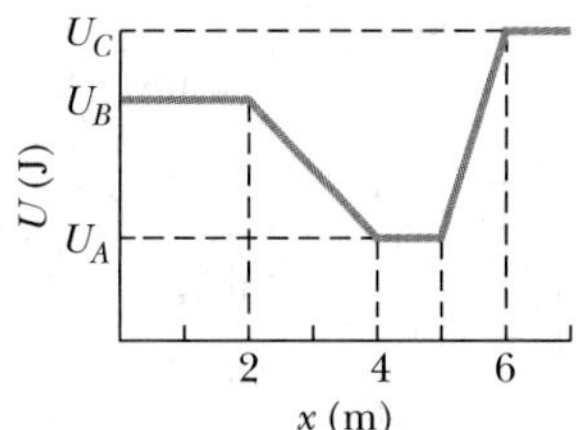

FIG. 8-50 Problem 39.

••40 Figure 8-51 shows a plot of potential energy U versus position x of a 0.200 kg particle that can travel only along an x axis under the influence of a conservative force. The graph has these values: $U_A = 9.00$ J, $U_C = 20.00$ J, and $U_D = 24.00$ J. The particle is released at the point where U forms a "potential hill" of "height" $U_B = 12.00$ J, with kinetic energy 4.00 J. What is the speed of the particle at (a) $x = 3.5$ m and (b) $x = 6.5$ m? What is the position of the turning point on (c) the right side and (d) the left side?

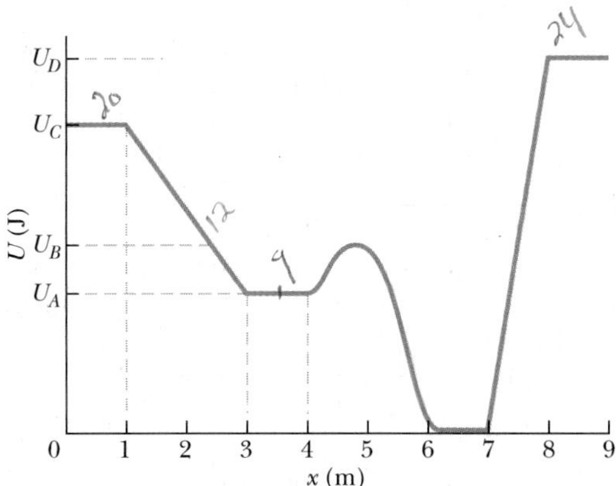

FIG. 8-51 Problem 40.

•••41 A single conservative force $F(x)$ acts on a 1.0 kg particle that moves along an x axis. The potential energy $U(x)$ associated with $F(x)$ is given by

$$U(x) = -4x\,e^{-x/4}\ \text{J},$$

where x is in meters. At $x = 5.0$ m the particle has a kinetic energy of 2.0 J. (a) What is the mechanical energy of the system? (b) Make a plot of $U(x)$ as a function of x for $0 \le x \le 10$ m, and on the same graph draw the line that represents the mechanical energy of the system. Use part (b) to determine (c) the least value of x the particle can reach and (d) the greatest value of x the particle can reach. Use part (b) to determine (e) the maximum kinetic energy of the particle and (f) the value of x at which it occurs. (g) Determine an expression in newtons and meters for $F(x)$ as a function of x. (h) For what (finite) value of x does $F(x) = 0$?

sec. 8-7 Work Done on a System by an External Force

•42 A worker pushed a 27 kg block 9.2 m along a level floor at constant speed with a force directed 32° below the horizontal. If the coefficient of kinetic friction between block and floor was 0.20, what were (a) the work done by the worker's force and (b) the increase in thermal energy of the block–floor system?

•43 A collie drags its bed box across a floor by applying a horizontal force of 8.0 N. The kinetic frictional force acting on the box has magnitude 5.0 N. As the box is dragged through 0.70 m along the way, what are (a) the work done by the collie's applied force and (b) the increase in thermal energy of the bed and floor?

••44 A horizontal force of magnitude 35.0 N pushes a block of mass 4.00 kg across a floor where the coefficient of kinetic friction is 0.600. (a) How much work is done by that applied force on the block–floor system when the block slides through a displacement of 3.00 m across the floor? (b) During that displacement, the thermal energy of the block increases by 40.0 J. What is the increase in thermal energy of the floor? (c) What is the increase in the kinetic energy of the block?

••45 A rope is used to pull a 3.57 kg block at constant speed 4.06 m along a horizontal floor. The force on the block from the rope is 7.68 N and directed 15.0° above the horizontal. What are (a) the work done by the rope's force, (b) the increase in thermal energy of the block–floor system, and (c) the coefficient of kinetic friction between the block and floor? SSM

sec. 8-8 Conservation of Energy

•46 A 60 kg skier leaves the end of a ski-jump ramp with a velocity of 24 m/s directed 25° above the horizontal. Suppose that as a result of air drag the skier returns to the ground with a speed of 22 m/s, landing 14 m vertically below the end of the ramp. From the launch to the return to the ground, by how much is the mechanical energy of the skier–Earth system reduced because of air drag?

•47 A 25 kg bear slides, from rest, 12 m down a lodgepole pine tree, moving with a speed of 5.6 m/s just before hitting the ground. (a) What change occurs in the gravitational potential energy of the bear–Earth system during the slide? (b) What is the kinetic energy of the bear just before hitting the ground? (c) What is the average frictional force that acts on the sliding bear? SSM ILW

•48 An outfielder throws a baseball with an initial speed of 81.8 mi/h. Just before an infielder catches the ball at the same level, the ball's speed is 110 ft/s. In foot-pounds, by how much is the mechanical energy of the ball–Earth system reduced because of air drag? (The weight of a baseball is 9.0 oz.)

•49 A 75 g Frisbee is thrown from a point 1.1 m above the ground with a speed of 12 m/s. When it has reached a height of 2.1 m, its speed is 10.5 m/s. What was the reduction in E_{mec} of the Frisbee–Earth system because of air drag?

•50 In Fig. 8-52, a block slides down an incline. As it moves from point A to point B, which are 5.0 m apart, force $\vec{F}$ acts on the block, with magnitude 2.0 N and directed down the incline. The magnitude of the frictional force acting on the block is 10 N. If the kinetic energy of the block increases by 35 J between A and B, how much work is done on the block by the gravitational force as the block moves from A to B?

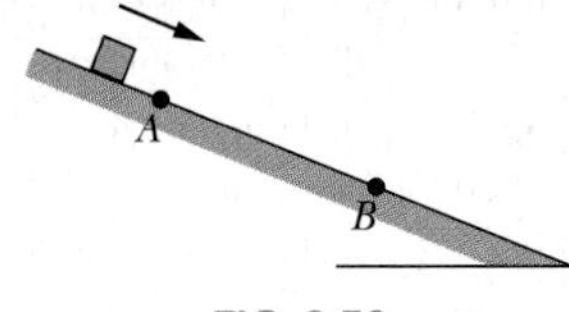

FIG. 8-52 Problems 50 and 69.

•51 During a rockslide, a 520 kg rock slides from rest down

a hillside that is 500 m long and 300 m high. The coefficient of kinetic friction between the rock and the hill surface is 0.25. (a) If the gravitational potential energy U of the rock–Earth system is zero at the bottom of the hill, what is the value of U just before the slide? (b) How much energy is transferred to thermal energy during the slide? (c) What is the kinetic energy of the rock as it reaches the bottom of the hill? (d) What is its speed then?

••52 You push a 2.0 kg block against a horizontal spring, compressing the spring by 15 cm. Then you release the block, and the spring sends it sliding across a tabletop. It stops 75 cm from where you released it. The spring constant is 200 N/m. What is the block–table coefficient of kinetic friction?

••53 In Fig. 8-53, a block slides along a track from one level to a higher level after passing through an intermediate valley. The track is frictionless until the block reaches the higher level. There a frictional force stops the block in a distance d. The block's initial speed v_0 is 6.0 m/s, the height difference h is 1.1 m, and μ_k is 0.60. Find d. GO

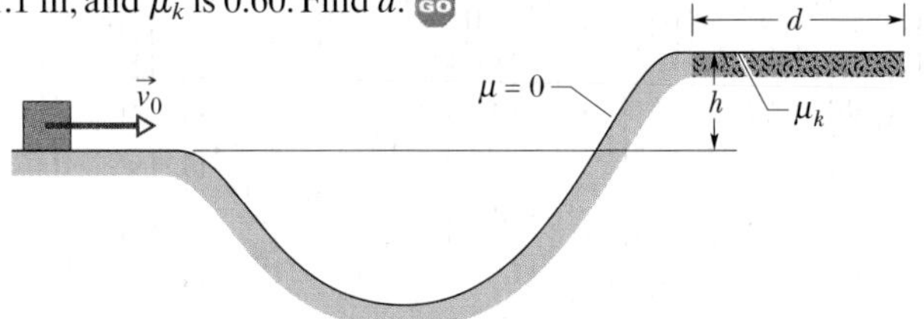

FIG. 8-53 Problem 53.

••54 A large fake cookie sliding on a horizontal surface is attached to one end of a horizontal spring with spring constant $k = 400$ N/m; the other end of the spring is fixed in place. The cookie has a kinetic energy of 20.0 J as it passes through the spring's equilibrium position. As the cookie slides, a frictional force of magnitude 10.0 N acts on it. (a) How far will the cookie slide from the equilibrium position before coming momentarily to rest? (b) What will be the kinetic energy of the cookie as it slides back through the equilibrium position?

••55 In Fig. 8-54, a 3.5 kg block is accelerated from rest by a compressed spring of spring constant 640 N/m. The block leaves the spring at the spring's relaxed length and then travels over a horizontal floor with a coefficient of kinetic friction $\mu_k = 0.25$. The frictional force stops the block in distance $D = 7.8$ m. What are (a) the increase in the thermal energy of the block–floor system, (b) the maximum kinetic energy of the block, and (c) the original compression distance of the spring? GO

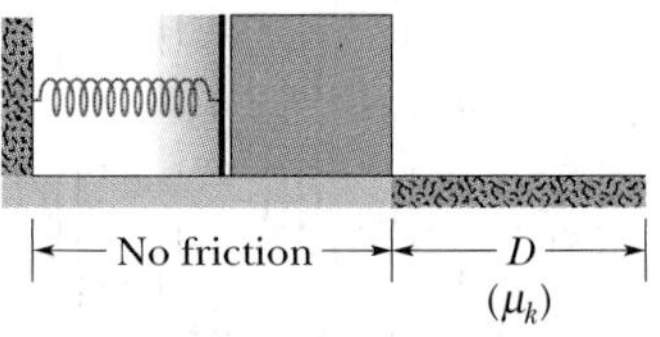

FIG. 8-54 Problem 55.

••56 A 4.0 kg bundle starts up a 30° incline with 128 J of kinetic energy. How far will it slide up the incline if the coefficient of kinetic friction between bundle and incline is 0.30?

••57 When a click beetle is upside down on its back, it jumps upward by suddenly arching its back, transferring energy stored in a muscle to mechanical energy. This launching mechanism produces an audible click, giving the beetle its name. Videotape of a certain click-beetle jump shows that a beetle of mass $m = 4.0 \times 10^{-6}$ kg moved directly upward by 0.77 mm during the launch and then to a maximum height of $h = 0.30$ m. During the launch, what are the average magnitudes of (a) the external force on the beetle's back from the floor and (b) the acceleration of the beetle in terms of g?

••58 A child whose weight is 267 N slides down a 6.1 m playground slide that makes an angle of 20° with the horizontal. The coefficient of kinetic friction between slide and child is 0.10. (a) How much energy is transferred to thermal energy? (b) If she starts at the top with a speed of 0.457 m/s, what is her speed at the bottom?

••59 In Fig. 8-55, a block of mass $m = 2.5$ kg slides head on into a spring of spring constant $k = 320$ N/m. When the block stops, it has compressed the spring by 7.5 cm. The coefficient of kinetic friction between block and floor is 0.25. While the block is in contact with the spring and being brought to rest, what are (a) the work done by the spring force and (b) the increase in thermal energy of the block–floor system? (c) What is the block's speed just as it reaches the spring? ILW

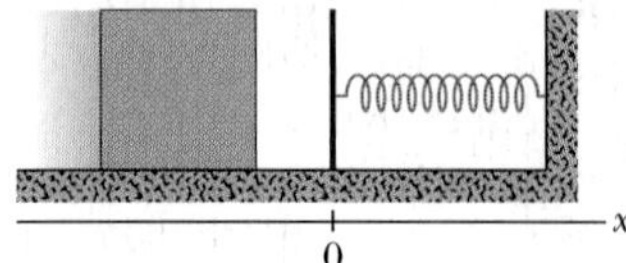

FIG. 8-55 Problem 59.

••60 A cookie jar is moving up a 40° incline. At a point 55 cm from the bottom of the incline (measured along the incline), the jar has a speed of 1.4 m/s. The coefficient of kinetic friction between jar and incline is 0.15. (a) How much farther up the incline will the jar move? (b) How fast will it be going when it has slid back to the bottom of the incline? (c) Do the answers to (a) and (b) increase, decrease, or remain the same if we decrease the coefficient of kinetic friction (but do not change the given speed or location)?

••61 A stone with a weight of 5.29 N is launched vertically from ground level with an initial speed of 20.0 m/s, and the air drag on it is 0.265 N throughout the flight. What are (a) the maximum height reached by the stone and (b) its speed just before it hits the ground?

•••62 In Fig. 8-56, a block is released from rest at height $d =$ 40 cm and slides down a frictionless ramp and onto a first plateau, which has length d and where the coefficient of kinetic friction is 0.50. If the block is still moving, it then slides down a second frictionless ramp through height $d/2$ and onto a lower plateau, which has length $d/2$ and where the coefficient of kinetic friction is again 0.50. If the block is still moving, it then slides up a frictionless ramp until it (momentarily) stops. Where does the block stop? If its final stop is on a plateau, state which one and give the distance L from the left edge of that plateau. If the block reaches the ramp, give the height H above the lower plateau where it momentarily stops.

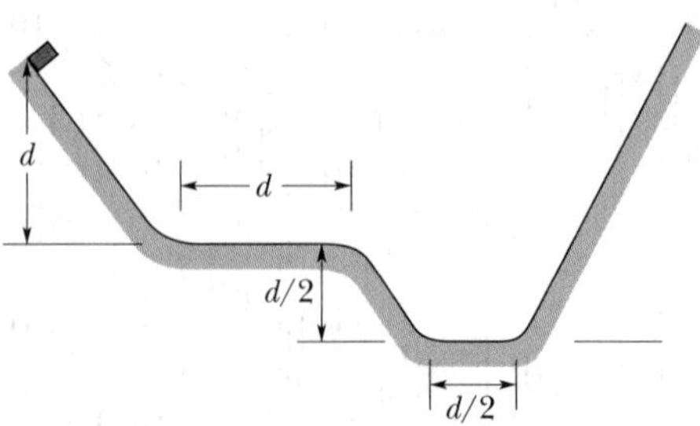

FIG. 8-56 Problem 62.

•••63 A particle can slide along a track with elevated ends and a flat central part, as shown in Fig. 8-57. The flat part has length $L = 40$ cm. The curved portions of the track are frictionless, but for the flat part the coefficient of kinetic friction is $\mu_k = 0.20$. The particle is released from rest at point A, which is at height $h = L/2$. How far from the left edge of the flat part does the particle finally stop?

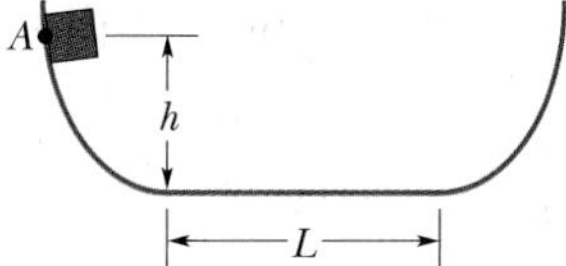

FIG. 8-57 Problem 63.

•••64 In Fig. 8-58, a block slides along a path that is without friction until the block reaches the section of length $L = 0.75$ m, which begins at height $h = 2.0$ m on a ramp of angle $\theta = 30°$. In that section, the coefficient of kinetic friction is 0.40. The block passes through point A with a speed of 8.0 m/s. If the block can reach point B (where the friction ends), what is its speed there, and if it cannot, what is its greatest height above A?

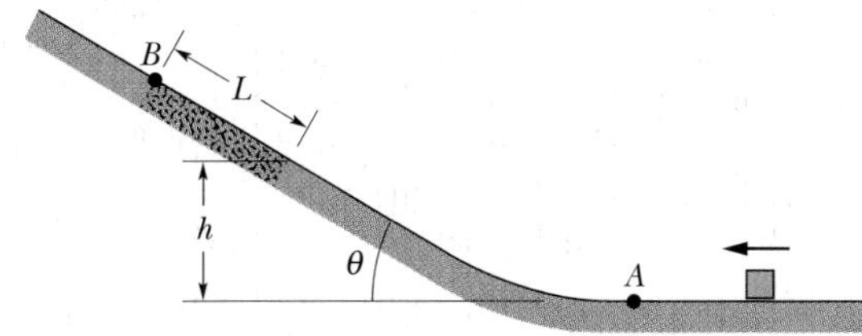

FIG. 8-58 Problem 64.

•••65 The cable of the 1800 kg elevator cab in Fig. 8-59 snaps when the cab is at rest at the first floor, where the cab bottom is a distance $d = 3.7$ m above a spring of spring constant $k = 0.15$ MN/m. A safety device clamps the cab against guide rails so that a constant frictional force of 4.4 kN opposes the cab's motion. (a) Find the speed of the cab just before it hits the spring. (b) Find the maximum distance x that the spring is compressed (the frictional force still acts during this compression). (c) Find the distance that the cab will bounce back up the shaft. (d) Using conservation of energy, find the approximate total distance that the cab will move before coming to rest. (Assume that the frictional force on the cab is negligible when the cab is stationary.)

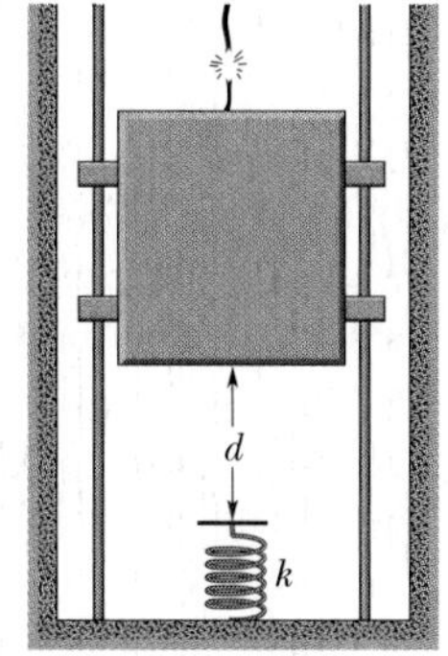

FIG. 8-59 Problem 65.

Additional Problems

66 At a certain factory, 300 kg crates are dropped vertically from a packing machine onto a conveyor belt moving at 1.20 m/s (Fig. 8-60). (A motor maintains the belt's constant speed.) The coefficient of kinetic friction between the belt and each crate is 0.400. After a short time, slipping between the belt and the crate ceases, and the crate then moves along with the belt. For the period of time during which the crate is being brought to rest relative to the belt, calculate, for a coordinate system at rest in the factory, (a) the kinetic energy supplied to the crate, (b) the magnitude of the kinetic frictional force acting on the crate, and (c) the energy supplied by the motor. (d) Explain why answers (a) and (c) differ.

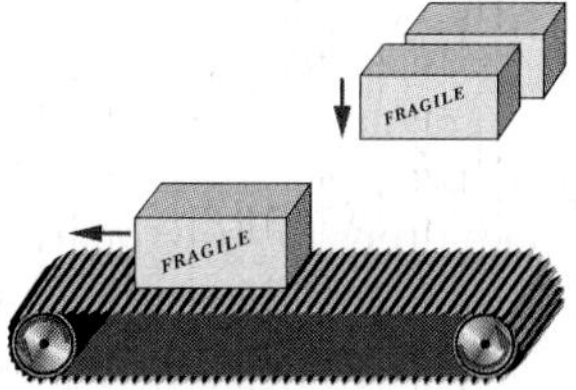

FIG. 8-60 Problem 66.

67 A playground slide is in the form of an arc of a circle that has a radius of 12 m. The maximum height of the slide is $h = 4.0$ m, and the ground is tangent to the circle (Fig. 8-61). A 25 kg child starts from rest at the top of the slide and has a speed of 6.2 m/s at the bottom. (a) What is the length of the slide? (b) What average frictional force acts on the child over this distance? If, instead of the ground, a vertical line through the *top of the slide* is tangent to the circle, what are (c) the length of the slide and (d) the average frictional force on the child?

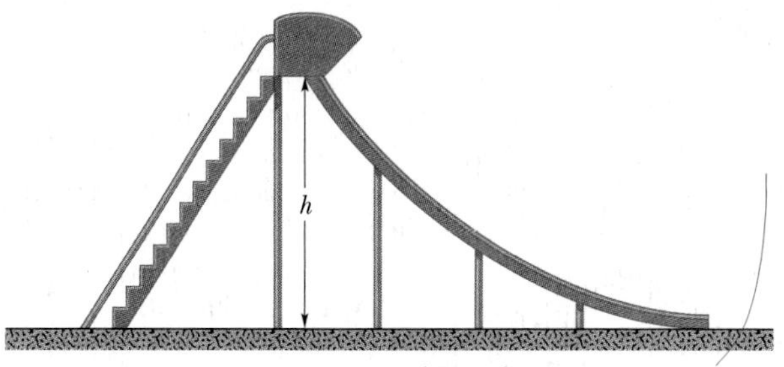

FIG. 8-61 Problem 67.

68 In Fig. 8-37, the string is $L = 120$ cm long, has a ball attached to one end, and is fixed at its other end. A fixed peg is at point P. Released from rest, the ball swings down until the string catches on the peg; then the ball swings up, around the peg. If the ball is to swing completely around the peg, what value must distance d exceed? (*Hint:* The ball must still be moving at the top of its swing. Do you see why?)

69 In Fig. 8-52, a block is sent sliding down a frictionless ramp. Its speeds at points A and B are 2.00 m/s and 2.60 m/s, respectively. Next, it is again sent sliding down the ramp, but this time its speed at point A is 4.00 m/s. What then is its speed at point B? SSM

70 A certain spring is found *not* to conform to Hooke's law. The force (in newtons) it exerts when stretched a distance x (in meters) is found to have magnitude $52.8x + 38.4x^2$ in the direction opposing the stretch. (a) Compute the work required to stretch the spring from $x = 0.500$ m to $x = 1.00$ m. (b) With one end of the spring fixed, a particle of mass 2.17 kg is attached to the other end of the spring when it is stretched by an amount $x = 1.00$ m. If the particle is then released from rest, what is its speed at the instant the stretch in the spring is $x = 0.500$ m? (c) Is the force exerted by the spring conservative or nonconservative? Explain.

71 A factory worker accidentally releases a 180 kg crate that was being held at rest at the top of a ramp that is 3.7 m long and inclined at 39° to the horizontal. The coefficient of kinetic friction between the crate and the ramp, and between the crate and the horizontal factory floor, is 0.28. (a) How fast is the crate moving as it reaches the bottom of the ramp? (b) How far will it subsequently slide across the floor? (Assume that the crate's kinetic energy does not change as it moves from the ramp onto the floor.) (c) Do the answers to (a) and (b) increase, decrease, or remain the same if we halve the mass of the crate?

72 In Fig. 8-62, a small block is sent through point A with a speed of 7.0 m/s. Its path is without friction until it reaches the section of length $L = 12$ m, where the coefficient of kinetic fric-

tion is 0.70. The indicated heights are $h_1 = 6.0$ m and $h_2 = 2.0$ m. What are the speeds of the block at (a) point B and (b) point C? (c) Does the block reach point D? If so, what is its speed there; if not, how far through the section of friction does it travel?

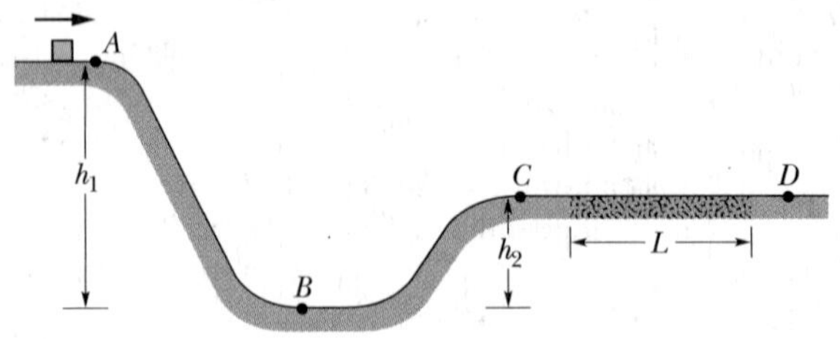

FIG. 8-62 Problem 72.

73 A 2.50 kg beverage can is thrown directly downward from a height of 4.00 m, with an initial speed of 3.00 m/s. The air drag on the can is negligible. What is the kinetic energy of the can (a) as it reaches the ground at the end of its fall and (b) when it is halfway to the ground? What are (c) the kinetic energy of the can and (d) the gravitational potential energy of the can–Earth system 0.200 s before the can reaches the ground? For the latter, take the reference point $y = 0$ to be at the ground.

74 A 1.50 kg water balloon is shot straight up with an initial speed of 3.00 m/s. (a) What is the kinetic energy of the balloon just as it is launched? (b) How much work does the gravitational force do on the balloon during the balloon's full ascent? (c) What is the change in the gravitational potential energy of the balloon–Earth system during the full ascent? (d) If the gravitational potential energy is taken to be zero at the launch point, what is its value when the balloon reaches its maximum height? (e) If, instead, the gravitational potential energy is taken to be zero at the maximum height, what is its value at the launch point? (f) What is the maximum height?

75 In Fig. 8-63, the pulley has negligible mass, and both it and the inclined plane are frictionless. Block A has a mass of 1.0 kg, block B has a mass of 2.0 kg, and angle θ is 30°. If the blocks are released from rest with the connecting cord taut, what is their total kinetic energy when block B has fallen 25 cm? SSM

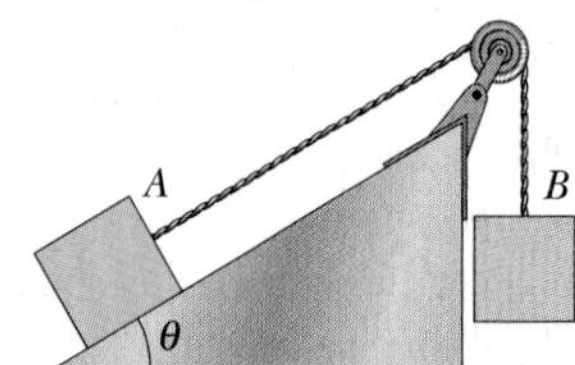

FIG. 8-63 Problem 75.

76 From the edge of a cliff, a 0.55 kg projectile is launched with an initial kinetic energy of 1550 J. The projectile's maximum upward displacement from the launch point is +140 m. What are the (a) horizontal and (b) vertical components of its launch velocity? (c) At the instant the vertical component of its velocity is 65 m/s, what is its vertical displacement from the launch point?

77 The only force acting on a particle is conservative force $\vec{F}$. If the particle is at point A, the potential energy of the system associated with $\vec{F}$ and the particle is 40 J. If the particle moves from point A to point B, the work done on the particle by $\vec{F}$ is +25 J. What is the potential energy of the system with the particle at B?

78 A constant horizontal force moves a 50 kg trunk 6.0 m up a 30° incline at constant speed. The coefficient of kinetic friction between the trunk and the incline is 0.20. What are (a) the work done by the applied force and (b) the increase in the thermal energy of the trunk and incline?

79 Two blocks, of masses $M =$ 2.0 kg and $2M$, are connected to a spring of spring constant $k = 200$ N/m that has one end fixed, as shown in Fig. 8-64. The horizontal surface and the pulley are frictionless, and the pulley has negligible mass. The blocks are released from rest with the spring relaxed. (a) What is the combined kinetic energy of the two blocks when the hanging block has fallen 0.090 m? (b) What is the kinetic energy of the hanging block when it has fallen that 0.090 m? (c) What maximum distance does the hanging block fall before momentarily stopping?

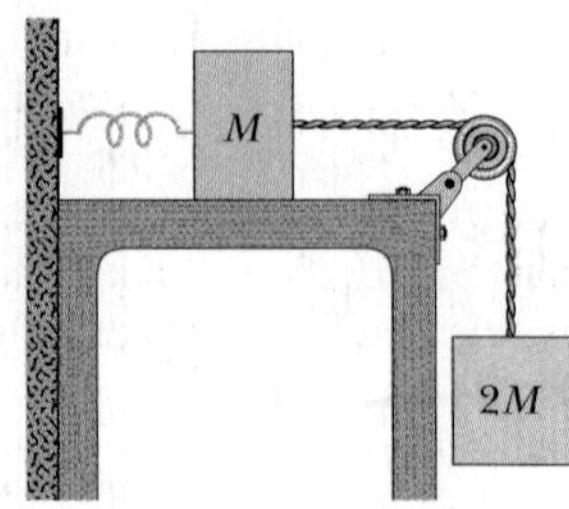

FIG. 8-64 Problem 79.

80 A volcanic ash flow is moving across horizontal ground when it encounters a 10° upslope. The front of the flow then travels 920 m up the slope before stopping. Assume that the gases entrapped in the flow lift the flow and thus make the frictional force from the ground negligible; assume also that the mechanical energy of the front of the flow is conserved. What was the initial speed of the front of the flow?

81 A 0.50 kg banana is thrown directly upward with an initial speed of 4.00 m/s and reaches a maximum height of 0.80 m. What change does air drag cause in the mechanical energy of the banana–Earth system during the ascent?

82 If a 70 kg baseball player steals home by sliding into the plate with an initial speed of 10 m/s just as he hits the ground, (a) what is the decrease in the player's kinetic energy and (b) what is the increase in the thermal energy of his body and the ground along which he slides?

83 A spring ($k = 200$ N/m) is fixed at the top of a frictionless plane inclined at angle $\theta = 40°$ (Fig. 8-65). A 1.0 kg block is projected up the plane, from an initial position that is distance $d = 0.60$ m from the end of the relaxed spring, with an initial kinetic energy of 16 J. (a) What is the kinetic energy of the block at the instant it has compressed the spring 0.20 m? (b) With what kinetic energy must the block be projected up the plane if it is to stop momentarily when it has compressed the spring by 0.40 m? SSM

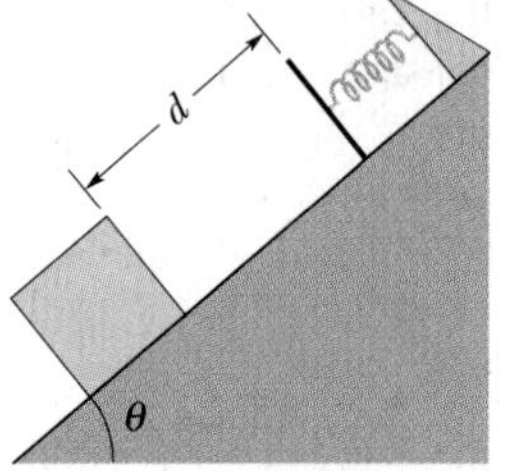

FIG. 8-65 Problem 83.

84 A 3.2 kg sloth hangs 3.0 m above the ground. (a) What is the gravitational potential energy of the sloth–Earth system if we take the reference point $y = 0$ to be at the ground? If the sloth drops to the ground and air drag on it is assumed to be negligible, what are the (b) kinetic energy and (c) speed of the sloth just before it reaches the ground?

85 A machine pulls a 40 kg trunk 2.0 m up a 40° ramp at constant velocity, with the machine's force on the trunk directed parallel to the ramp. The coefficient of kinetic friction between the trunk and the ramp is 0.40. What are (a) the work done on the trunk by the machine's force and (b) the increase in thermal energy of the trunk and the ramp?

86 The luxury liner *Queen Elizabeth 2* has a diesel-electric power plant with a maximum power of 92 MW at a cruising speed of 32.5 knots. What forward force is exerted on the ship at this speed? (1 knot = 1.852 km/h.)

87 The temperature of a plastic cube is monitored while the cube is pushed 3.0 m across a floor at constant speed by a horizontal force of 15 N. The thermal energy of the cube increases by 20 J. What is the increase in the thermal energy of the floor along which the cube slides? SSM

88 Two snowy peaks are at heights $H = 850$ m and $h = 750$ m above the valley between them. A ski run extends between the peaks, with a total length of 3.2 km and an average slope of $\theta = 30°$ (Fig. 8-66). (a) A skier starts from rest at the top of the higher peak. At what speed will he arrive at the top of the lower peak if he coasts without using ski poles? Ignore friction. (b) Approximately what coefficient of kinetic friction between snow and skis would make him stop just at the top of the lower peak?

FIG. 8-66 Problem 88.

89 A swimmer moves through the water at an average speed of 0.22 m/s. The average drag force is 110 N. What average power is required of the swimmer?

90 An automobile with passengers has weight 16 400 N and is moving at 113 km/h when the driver brakes, sliding to a stop. The frictional force on the wheels from the road has a magnitude of 8230 N. Find the stopping distance.

91 A 0.63 kg ball thrown directly upward with an initial speed of 14 m/s reaches a maximum height of 8.1 m. What is the change in the mechanical energy of the ball–Earth system during the ascent of the ball to that maximum height?

92 The summit of Mount Everest is 8850 m above sea level. (a) How much energy would a 90 kg climber expend against the gravitational force on him in climbing to the summit from sea level? (b) How many candy bars, at 1.25 MJ per bar, would supply an energy equivalent to this? Your answer should suggest that work done against the gravitational force is a very small part of the energy expended in climbing a mountain.

93 A sprinter who weighs 670 N runs the first 7.0 m of a race in 1.6 s, starting from rest and accelerating uniformly. What are the sprinter's (a) speed and (b) kinetic energy at the end of the 1.6 s? (c) What average power does the sprinter generate during the 1.6 s interval?

94 In Fig. 8-67, a 1400 kg block of granite is pulled up an incline at a constant speed of 1.34 m/s by a cable and winch. The indicated distances are $d_1 = 40$ m and $d_2 = 30$ m. The coefficient of kinetic friction between the block and the incline is 0.40. What is the power due to the force applied to the block by the cable?

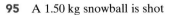

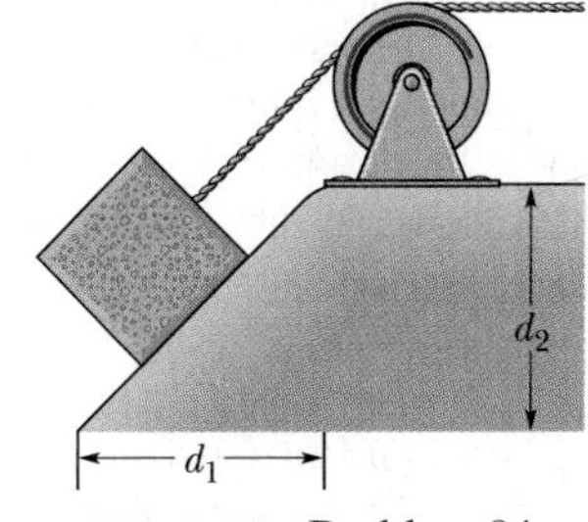

FIG. 8-67 Problem 94.

95 A 1.50 kg snowball is shot upward at an angle of 34.0° to the horizontal with an initial speed of 20.0 m/s. (a) What is its initial kinetic energy? (b) By how much does the gravitational potential energy of the snowball–Earth system change as the snowball moves from the launch point to the point of maximum height? (c) What is that maximum height?

96 A 20 kg object is acted on by a conservative force given by $F = -3.0x - 5.0x^2$, with F in newtons and x in meters. Take the potential energy associated with the force to be zero when the object is at $x = 0$. (a) What is the potential energy of the system associated with the force when the object is at $x = 2.0$ m? (b) If the object has a velocity of 4.0 m/s in the negative direction of the x axis when it is at $x = 5.0$ m, what is its speed when it passes through the origin? (c) What are the answers to (a) and (b) if the potential energy of the system is taken to be -8.0 J when the object is at $x = 0$?

97 A 9.40 kg projectile is fired vertically upward. Air drag decreases the mechanical energy of the projectile–Earth system by 68.0 kJ during the projectile's ascent. How much higher would the projectile have gone were air drag negligible?

98 A metal tool is sharpened by being held against the rim of a wheel on a grinding machine by a force of 180 N. The frictional forces between the rim and the tool grind off small pieces of the tool. The wheel has a radius of 20.0 cm and rotates at 2.50 rev/s. The coefficient of kinetic friction between the wheel and the tool is 0.320. At what rate is energy being transferred from the motor driving the wheel to the thermal energy of the wheel and tool and to the kinetic energy of the material thrown from the tool?

99 A spring with a spring constant of 3200 N/m is initially stretched until the elastic potential energy of the spring is 1.44 J. ($U = 0$ for the relaxed spring.) What is ΔU if the initial stretch is changed to (a) a stretch of 2.0 cm, (b) a compression of 2.0 cm, and (c) a compression of 4.0 cm?

100 The spring in the muzzle of a child's spring gun has a spring constant of 700 N/m. To shoot a ball from the gun, first the spring is compressed and then the ball is placed on it. The gun's trigger then releases the spring, which pushes the ball through the muzzle. The ball leaves the spring just as it leaves the outer end of the muzzle. When the gun is inclined upward by 30° to the horizontal, a 57 g ball is shot to a maximum height of 1.83 m above the gun's muzzle. Assume air drag on the ball is negligible. (a) At what speed does the spring launch the ball? (b) Assuming that friction on the ball within the gun can be neglected, find the spring's initial compression distance.

101 A 60.0 kg circus performer slides 4.00 m down a pole to the circus floor, starting from rest. What is the kinetic energy of the performer as she reaches the floor if the frictional force on her from the pole (a) is negligible (she will be hurt) and (b) has a magnitude of 500 N?

102 In 1981, Daniel Goodwin climbed 443 m up the *exterior* of the Sears Building in Chicago using suction cups and metal clips. (a) Approximate his mass and then compute how much energy he had to transfer from biomechanical (internal) energy to the gravitational potential energy of the Earth–Goodwin system to lift himself to that height. (b) How much energy would he have had to transfer if he had, instead, taken the stairs inside the building (to the same height)?

103 A 30 g bullet moving a horizontal velocity of 500 m/s comes to a stop 12 cm within a solid wall. (a) What is the change in the bullet's mechanical energy? (b) What is the magnitude of the average force from the wall stopping it?

104 Resistance to the motion of an automobile consists of road friction, which is almost independent of speed, and air drag, which is proportional to speed-squared. For a certain car with a weight of 12 000 N, the total resistant force F is given by $F = 300 + 1.8v^2$, with F in newtons and v in meters per second. Calculate the power (in horsepower) required to accelerate the car at 0.92 m/s^2 when the speed is 80 km/h.

105 A locomotive with a power capability of 1.5 MW can accelerate a train from a speed of 10 m/s to 25 m/s in 6.0 min. (a) Calculate the mass of the train. Find (b) the speed of the train and (c) the force accelerating the train as functions of time (in seconds) during the 6.0 min interval. (d) Find the distance moved by the train during the interval.

106 A 5.0 kg block is projected at 5.0 m/s up a plane that is inclined at 30° with the horizontal. How far up along the plane does the block go (a) if the plane is frictionless and (b) if the coefficient of kinetic friction between the block and the plane is 0.40? (c) In the latter case, what is the increase in thermal energy of block and plane during the block's ascent? (d) If the block then slides back down against the frictional force, what is the block's speed when it reaches the original projection point?

107 A 20 kg block on a horizontal surface is attached to a horizontal spring of spring constant $k = 4.0$ kN/m. The block is pulled to the right so that the spring is stretched 10 cm beyond its relaxed length, and the block is then released from rest. The frictional force between the sliding block and the surface has a magnitude of 80 N. (a) What is the kinetic energy of the block when it has moved 2.0 cm from its point of release? (b) What is the kinetic energy of the block when it first slides back through the point at which the spring is relaxed? (c) What is the maximum kinetic energy attained by the block as it slides from its point of release to the point at which the spring is relaxed?

108 A 70.0 kg man jumping from a window lands in an elevated fire rescue net 11.0 m below the window. He momentarily stops when he has stretched the net by 1.50 m. Assuming that mechanical energy is conserved during this process and that the net functions like an ideal spring, find the elastic potential energy of the net when it is stretched by 1.50 m.

109 To form a pendulum, a 0.092 kg ball is attached to one end of a rod of length 0.62 m and negligible mass, and the other end of the rod is mounted on a pivot. The rod is rotated until it is straight up, and then it is released from rest so that it swings down around the pivot. When the ball reaches its lowest point, what are (a) its speed and (b) the tension in the rod? Next, the rod is rotated until it is horizontal, and then it is again released from rest. (c) At what angle from the vertical does the tension in the rod equal the weight of the ball? (d) If the mass of the ball is increased, does the answer to (c) increase, decrease, or remain the same? SSM

110 A skier weighing 600 N goes over a frictionless circular hill of radius $R = 20$ m (Fig. 8-68). Assume that the effects of air resistance on the skier are negligible. As she comes up the hill, her speed is 8.0 m/s at point B, at angle $\theta = 20°$. (a) What is her speed at the hilltop (point A) if she coasts without using her poles? (b) What minimum speed can she have at B and still coast to the hilltop? (c) Do the answers to these two questions increase, decrease, or remain the same if the skier weighs 700 N?

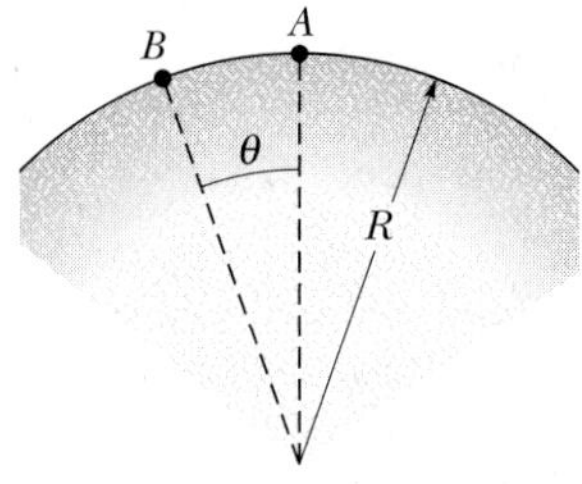

FIG. 8-68 Problem 110.

111 A 50 g ball is thrown from a window with an initial velocity of 8.0 m/s at an angle of 30° above the horizontal. Using energy methods, determine (a) the kinetic energy of the ball at the top of its flight and (b) its speed when it is 3.0 m below the window. Does the answer to (b) depend on either (c) the mass of the ball or (d) the initial angle? SSM

112 A 68 kg sky diver falls at a constant terminal speed of 59 m/s. (a) At what rate is the gravitational potential energy of the Earth–sky diver system being reduced? (b) At what rate is the system's mechanical energy being reduced?

113 A river descends 15 m through rapids. The speed of the water is 3.2 m/s upon entering the rapids and 13 m/s upon leaving. What percentage of the gravitational potential energy of the water–Earth system is transferred to kinetic energy during the descent? (*Hint:* Consider the descent of, say, 10 kg of water.)

114 The magnitude of the gravitational force between a particle of mass m_1 and one of mass m_2 is given by

$$F(x) = G\frac{m_1 m_2}{x^2},$$

where G is a constant and x is the distance between the particles. (a) What is the corresponding potential energy function $U(x)$? Assume that $U(x) \to 0$ as $x \to \infty$ and that x is positive. (b) How much work is required to increase the separation of the particles from $x = x_1$ to $x = x_1 + d$?

115 Approximately 5.5×10^6 kg of water falls 50 m over Niagara Falls each second. (a) What is the decrease in the gravitational potential energy of the water–Earth system each second? (b) If all this energy could be converted to electrical energy (it cannot be), at what rate would electrical energy be supplied? (The mass of 1 m^3 of water is 1000 kg.) (c) If the electrical energy were sold at 1 cent/kW·h, what would be the yearly income?

116 A 1500 kg car starts from rest on a horizontal road and gains a speed of 72 km/h in 30 s. (a) What is its kinetic energy at the end of the 30 s? (b) What is the average power required of the car during the 30 s interval? (c) What is the instantaneous power at the end of the 30 s interval, assuming that the acceleration is constant?

117 A particle can move along only an x axis, where conservative forces act on it (Fig. 8-69 and the following table). The particle is released at $x = 5.00$ m with a kinetic energy of $K = 14.0$ J and a potential energy of $U = 0$. If its motion is in the negative direction of the x axis, what are its (a) K and (b) U at $x = 2.00$ m and its (c) K and (d) U at $x = 0$? If its motion is in the positive direction of the x axis, what are its (e) K and (f) U at $x = 11.0$ m, its (g) K and (h) U at $x = 12.0$ m, and its (i) K and (j) U at $x = 13.0$ m? (k) Plot $U(x)$ versus x for the range $x = 0$ to $x = 13.0$ m.

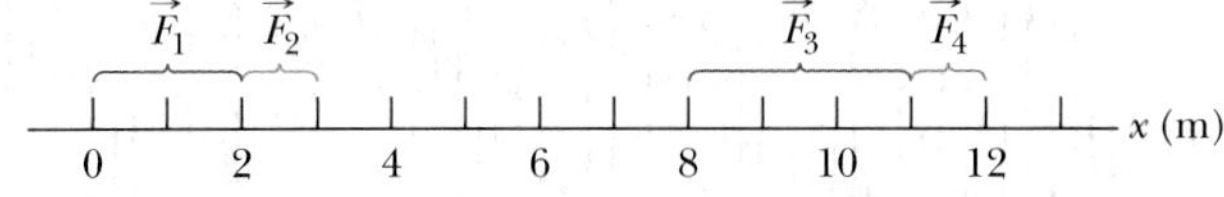

FIG. 8-69 Problems 117 and 118.

Next, the particle is released from rest at $x = 0$. What are (l) its kinetic energy at $x = 5.0$ m and (m) the maximum positive position x_{max} it reaches? (n) What does the particle do after it reaches x_{max}?

Range	Force
0 to 2.00 m	$\vec{F}_1 = +(3.00\text{ N})\hat{\text{i}}$
2.00 m to 3.00 m	$\vec{F}_2 = +(5.00\text{ N})\hat{\text{i}}$
3.00 m to 8.00 m	$F = 0$
8.00 m to 11.0 m	$\vec{F}_3 = -(4.00\text{ N})\hat{\text{i}}$
11.0 m to 12.0 m	$\vec{F}_4 = -(1.00\text{ N})\hat{\text{i}}$
12.0 m to 15.0 m	$F = 0$

118 For the arrangement of forces in Problem 117, a 2.00 kg particle is released at $x = 5.00$ m with an initial velocity of 3.45 m/s in the negative direction of the x axis. (a) If the particle can reach $x = 0$ m, what is its speed there, and if it cannot, what is its turning point? Suppose, instead, the particle is headed in the positive x direction when it is released at $x = 5.00$ m at speed 3.45 m/s. (b) If the particle can reach $x = 13.0$ m, what is its speed there, and if it cannot, what is its turning point?

119 A 0.42 kg shuffleboard disk is initially at rest when a player uses a cue to increase its speed to 4.2 m/s at constant acceleration. The acceleration takes place over a 2.0 m distance, at the end of which the cue loses contact with the disk. Then the disk slides an additional 12 m before stopping. Assume that the shuffleboard court is level and that the force of friction on the disk is constant. What is the increase in the thermal energy of the disk–court system (a) for that additional 12 m and (b) for the entire 14 m distance? (c) How much work is done on the disk by the cue? SSM

120 We move a particle along an x axis, first outward from $x = 1.0$ m to $x = 4.0$ m and then back to $x = 1.0$ m, while an external force acts on it. That force is directed along the x axis, and its x component can have different values for the outward trip and for the return trip. Here are the values (in newtons) for four situations, where x is in meters:

Outward	Inward
(a) +3.0	−3.0
(b) +5.0	+5.0
(c) $+2.0x$	$-2.0x$
(d) $+3.0x^2$	$+3.0x^2$

Find the net work done on the particle by the external force *for the round trip* for each of the four situations. (e) For which, if any, is the external force conservative?

121 A conservative force $F(x)$ acts on a 2.0 kg particle that moves along an x axis. The potential energy $U(x)$ associated with $F(x)$ is graphed in Fig. 8-70. When the particle is at $x = 2.0$ m, its velocity is −1.5 m/s. What are the (a) magnitude and (b) direction of $F(x)$ at this position? Between what positions on the (c) left and (d) right does the particle move? (e) What is the particle's speed at $x = 7.0$ m? SSM

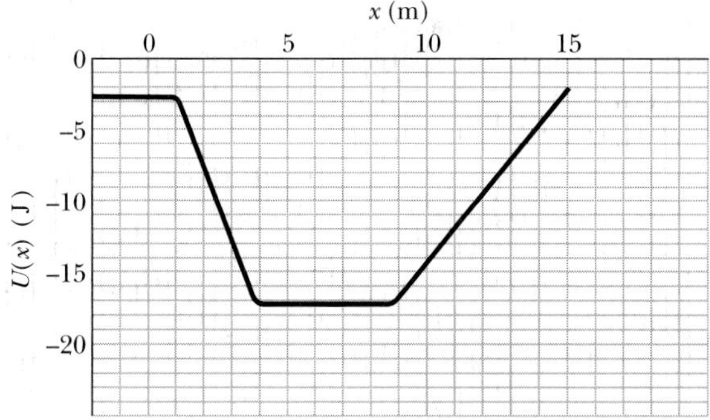

FIG. 8-70 Problem 121.

122 To make a pendulum, a 300 g ball is attached to one end of a string that has a length of 1.4 m and negligible mass. (The other end of the string is fixed.) The ball is pulled to one side until the string makes an angle of 30.0° with the vertical; then (with the string taut) the ball is released from rest. Find (a) the speed of the ball when the string makes an angle of 20.0° with the vertical and (b) the maximum speed of the ball. (c) What is the angle between the string and the vertical when the speed of the ball is one-third its maximum value?

123 A 1500 kg car begins sliding down a 5.0° inclined road with a speed of 30 km/h. The engine is turned off, and the only forces acting on the car are a net frictional force from the road and the gravitational force. After the car has traveled 50 m along the road, its speed is 40 km/h. (a) How much is the mechanical energy of the car reduced because of the net frictional force? (b) What is the magnitude of that net frictional force? SSM

124 In a circus act, a 60 kg clown is shot from a cannon with an initial velocity of 16 m/s at some unknown angle above the horizontal. A short time later the clown lands in a net that is 3.9 m vertically above the clown's initial position. Disregard air drag. What is the kinetic energy of the clown as he lands in the net?

125 The maximum force you can exert on an object with one of your back teeth is about 750 N. Suppose that as you gradually bite on a clump of licorice, the licorice resists compression by one of your teeth by acting like a spring for which $k = 2.5 \times 10^5$ N/m. Find (a) the distance the licorice is com-

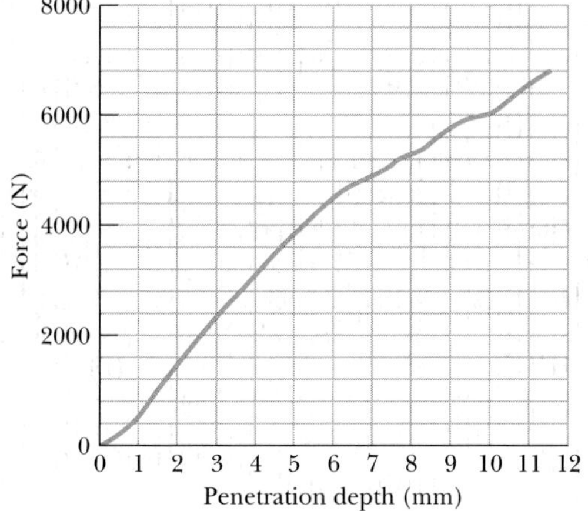

FIG. 8-71 Problem 125.

pressed by your tooth and (b) the work the tooth does on the licorice during the compression. (c) Plot the magnitude of your force versus the compression distance. (d) If there is a potential energy associated with this compression, plot it versus compression distance.

In the 1990s the pelvis of a particular *Triceratops* dinosaur was found to have deep bite marks. The shape of the marks suggested that they were made by a *Tyrannosaurus rex* dinosaur. To test the idea, researchers made a replica of a *T. rex* tooth from bronze and aluminum and then used a hydraulic press to gradually drive the replica into cow bone to the depth seen in the *Triceratops* bone. A graph of the force required versus depth of penetration is given in Fig. 8-71 for one trial; the required force increased with depth because, as the nearly conical tooth penetrated the bone, more of the tooth came in contact with the bone. (e) How much work was done by the hydraulic press—and thus presumably by the *T. rex*—in such a penetration? (f) Is there a potential energy associated with this penetration? (The large biting force and energy expenditure attributed to the *T. rex* by this research suggest that the animal was a predator and not a scavenger.)

126 A 70 kg firefighter slides, from rest, 4.3 m down a vertical pole. (a) If the firefighter holds onto the pole lightly, so that the frictional force of the pole on her is negligible, what is her speed just before reaching the ground floor? (b) If the firefighter grasps the pole more firmly as she slides, so that the average frictional force of the pole on her is 500 N upward, what is her speed just before reaching the ground floor?

127 A 15 kg block is accelerated at 2.0 m/s^2 along a horizontal frictionless surface, with the speed increasing from 10 m/s to 30 m/s. What are (a) the change in the block's mechanical energy and (b) the average rate at which energy is transferred to the block? What is the instantaneous rate of that transfer when the block's speed is (c) 10 m/s and (d) 30 m/s? SSM

128 Repeat Problem 127, but now with the block accelerated up a frictionless plane inclined at 5.0° to the horizontal.

129 The surface of the continental United States has an area of about 8×10^6 km^2 and an average elevation of about 500 m (above sea level). The average yearly rainfall is 75 cm. The fraction of this rainwater that returns to the atmosphere by evaporation is $\frac{2}{3}$; the rest eventually flows into the ocean. If the decrease in gravitational potential energy of the water–Earth system associated with that flow could be fully converted to electrical energy, what would be the average power? (The mass of 1 m^3 of water is 1000 kg.)

130 A spring with spring constant $k = 200$ N/m is suspended vertically with its upper end fixed to the ceiling and its lower end at position $y = 0$. A block of weight 20 N is attached to the lower end, held still for a moment, and then released. What are (a) the kinetic energy K, (b) the change (from the initial value) in the gravitational potential energy ΔU_g, and (c) the change in the elastic potential energy ΔU_e of the spring–block system when the block is at $y = -5.0$ cm? What are (d) K, (e) ΔU_g, and (f) ΔU_e when $y = -10$ cm, (g) K, (h) ΔU_g, and (i) ΔU_e when $y = -15$ cm, and (j) K, (k) ΔU_g, and (l) ΔU_e when $y = -20$ cm?

131 Each second, 1200 m^3 of water passes over a waterfall 100 m high. Three-fourths of the kinetic energy gained by the water in falling is transferred to electrical energy by a hydroelectric generator. At what rate does the generator produce electrical energy? (The mass of 1 m^3 of water is 1000 kg.) SSM

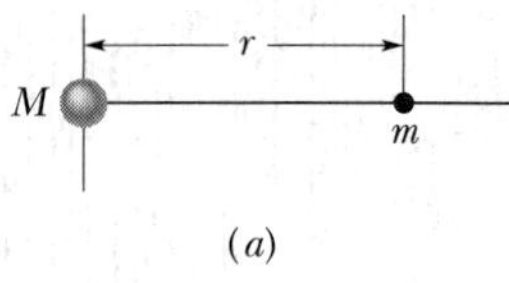

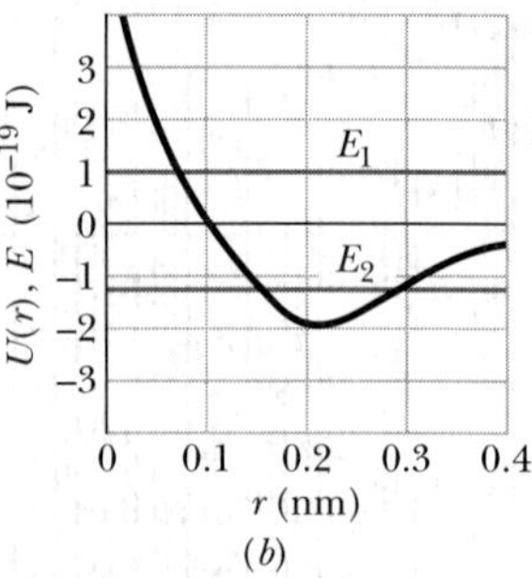

FIG. 8-72 Problem 132.

132 Figure 8-72*a* shows a molecule consisting of two atoms of masses m and M (with $m \ll M$) and separation r. Figure 8-72*b* shows the potential energy $U(r)$ of the molecule as a function of r. Describe the motion of the atoms (a) if the total mechanical energy E of the two-atom system is greater than zero (as is E_1), and (b) if E is less than zero (as is E_2). For $E_1 = 1 \times 10^{-19}$ J and $r = 0.3$ nm, find (c) the potential energy of the system, (d) the total kinetic energy of the atoms, and (e) the force (magnitude and direction) acting on each atom. For what values of r is the force (f) repulsive, (g) attractive, and (h) zero?

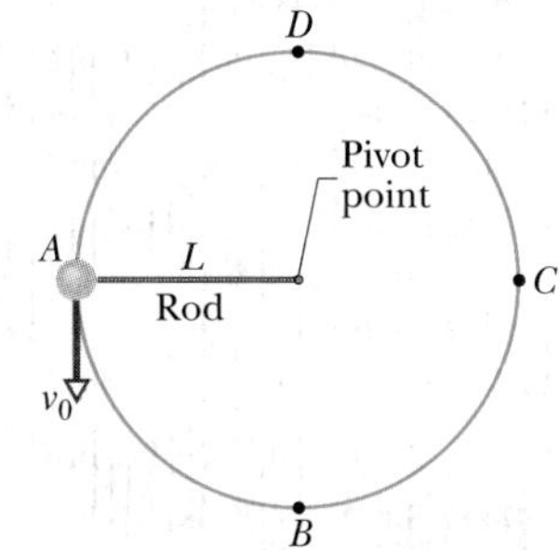

FIG. 8-73 Problem 133.

133 A massless rigid rod of length L has a ball of mass m attached to one end (Fig. 8-73). The other end is pivoted in such a way that the ball will move in a vertical circle. First, assume that there is no friction at the pivot. The system is launched downward from the horizontal position A with initial speed v_0. The ball just barely reaches point D and then stops. (a) Derive an expression for v_0 in terms of L, m, and g. (b) What is the tension in the rod when the ball passes through B? (c) A little grit is placed on the pivot to increase the friction there. Then the ball just barely reaches C when launched from A with the same speed as before. What is the decrease in the mechanical energy during this motion? (d) What is the decrease in the mechanical energy by the time the ball finally comes to rest at B after several oscillations? SSM

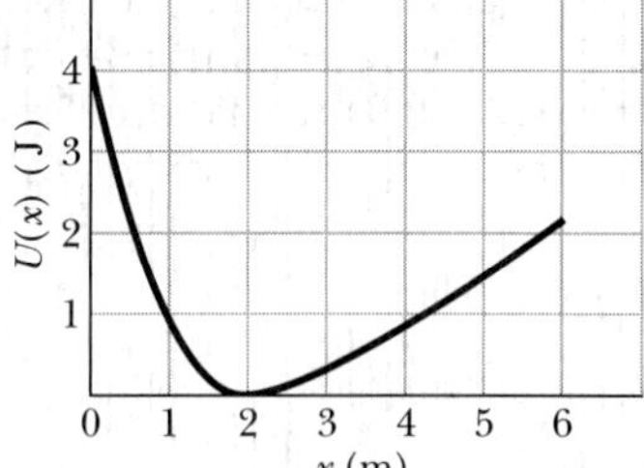

FIG. 8-74 Problem 134.

134 Conservative force $F(x)$ acts on a particle that moves along an x axis. Figure 8-74 shows how the potential energy $U(x)$ associated with force $F(x)$ varies with the position of the particle. (a) Plot $F(x)$ for the range $0 < x < 6$ m. (b) The mechanical energy E of the system is 4.0 J. Plot the kinetic energy $K(x)$ of the particle directly on Fig. 8-74.

135 Fasten one end of a vertical spring to a ceiling, attach a cabbage to the other end, and then slowly lower the cabbage until the upward force on it from the spring balances the gravitational force on it. Show that the loss of gravitational potential energy of the cabbage–Earth system equals twice the gain in the spring's potential energy.

Center of Mass and Linear Momentum 9

©franzfoto.com/Alamy Images

The males of the bighorn sheep (Ovis canadensis) fight one another to gain the attention of the females. Two males repeatedly run at each other at full speed with their heads lowered so that the horns collide, until one of them finally gives up. Such clashes can be costly because if a horn breaks, the male will likely be seriously hurt or killed in the next collision. But even without a broken horn, both sheep should seemingly fall to the ground with concussions.

How can the head-banging sheep withstand such violent clashes?

The answer is in this chapter.

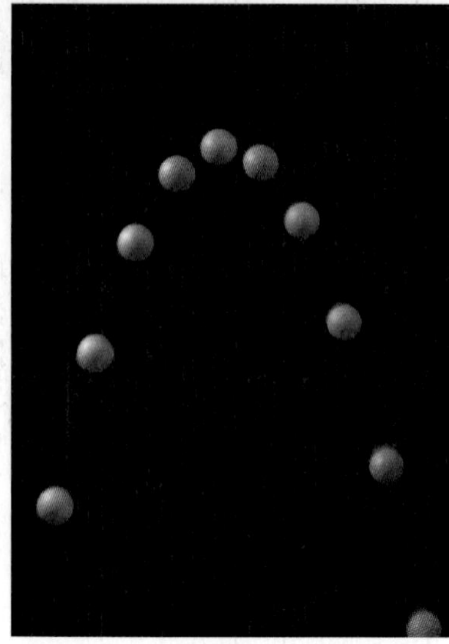

(a)

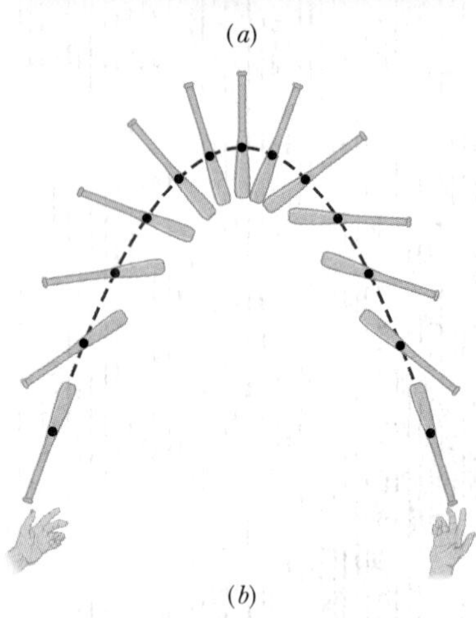

(b)

FIG. 9-1 (a) A ball tossed into the air follows a parabolic path. (b) The center of mass (black dot) of a baseball bat flipped into the air follows a parabolic path, but all other points of the bat follow more complicated curved paths. *(a: Richard Megna/Fundamental Photographs)*

9-1 WHAT IS PHYSICS?

Every mechanical engineer hired as an expert witness to reconstruct a traffic accident uses physics. Every trainer who coaches a ballerina on how to leap uses physics. Indeed, analyzing complicated motion of any sort requires simplification via an understanding of physics. In this chapter we discuss how the complicated motion of a system of objects, such as a car or a ballerina, can be simplified if we determine a special point of the system—the *center of mass* of that system.

Here is a quick example. If you toss a ball into the air without much spin on the ball (Fig. 9-1*a*), its motion is simple—it follows a parabolic path, as we discussed in Chapter 4, and the ball can be treated as a particle. If, instead, you flip a baseball bat into the air (Fig. 9-1*b*), its motion is more complicated. Because every part of the bat moves differently, along paths of many different shapes, you cannot represent the bat as a particle. Instead, it is a system of particles each of which follows its own path through the air. However, the bat has one special point—the center of mass—that *does* move in a simple parabolic path. The other parts of the bat move around the center of mass. (To locate the center of mass, balance the bat on an outstretched finger; the point is above your finger, on the bat's central axis.)

You cannot make a career of flipping baseball bats into the air, but you can make a career of advising long-jumpers or dancers on how to leap properly into the air while either moving their arms and legs or rotating their torso. Your starting point would be the person's center of mass because of its simple motion.

9-2 | The Center of Mass

We define the **center of mass** (com) of a system of particles (such as a person) in order to predict the possible motion of the system.

> The center of mass of a system of particles is the point that moves as though (1) all of the system's mass were concentrated there and (2) all external forces were applied there.

In this section we discuss how to determine where the center of mass of a system of particles is located. We start with a system of only a few particles, and then we consider a system of a great many particles (a solid body, such as a baseball bat). Later in the chapter, we discuss how the center of mass of a system moves when external forces act on the system.

Systems of Particles

Figure 9-2*a* shows two particles of masses m_1 and m_2 separated by distance d. We have arbitrarily chosen the origin of an x axis to coincide with the particle of mass m_1. We *define* the position of the center of mass (com) of this two-particle system to be

$$x_{\text{com}} = \frac{m_2}{m_1 + m_2}d. \tag{9-1}$$

Suppose, as an example, that $m_2 = 0$. Then there is only one particle, of mass m_1, and the center of mass must lie at the position of that particle; Eq. 9-1 dutifully reduces to $x_{\text{com}} = 0$. If $m_1 = 0$, there is again only one particle (of mass m_2), and we have, as we expect, $x_{\text{com}} = d$. If $m_1 = m_2$, the center of mass should be halfway between the two particles; Eq. 9-1 reduces to $x_{\text{com}} = \frac{1}{2}d$, again as we expect. Finally, Eq. 9-1 tells us that if neither m_1 nor m_2 is zero, x_{com} can have only values that lie between zero and d; that is, the center of mass must lie somewhere between the two particles.

Figure 9-2*b* shows a more generalized situation, in which the coordinate system has been shifted leftward. The position of the center of mass is now defined as

$$x_{\text{com}} = \frac{m_1x_1 + m_2x_2}{m_1 + m_2}. \tag{9-2}$$

Note that if we put $x_1 = 0$, then x_2 becomes d and Eq. 9-2 reduces to Eq. 9-1, as it must. Note also that in spite of the shift of the coordinate system, the center of mass is still the same distance from each particle.

We can rewrite Eq. 9-2 as

$$x_{com} = \frac{m_1x_1 + m_2x_2}{M}, \tag{9-3}$$

in which M is the total mass of the system. (Here, $M = m_1 + m_2$.) We can extend this equation to a more general situation in which n particles are strung out along the x axis. Then the total mass is $M = m_1 + m_2 + \cdots + m_n$, and the location of the center of mass is

$$x_{com} = \frac{m_1x_1 + m_2x_2 + m_3x_3 + \cdots + m_nx_n}{M}$$

$$= \frac{1}{M}\sum_{i=1}^{n} m_ix_i. \tag{9-4}$$

The subscript i is an index that takes on all integer values from 1 to n.

If the particles are distributed in three dimensions, the center of mass must be identified by three coordinates. By extension of Eq. 9-4, they are

$$x_{com} = \frac{1}{M}\sum_{i=1}^{n} m_ix_i, \qquad y_{com} = \frac{1}{M}\sum_{i=1}^{n} m_iy_i, \qquad z_{com} = \frac{1}{M}\sum_{i=1}^{n} m_iz_i. \tag{9-5}$$

We can also define the center of mass with the language of vectors. First recall that the position of a particle at coordinates x_i, y_i, and z_i is given by a position vector:

$$\vec{r}_i = x_i\hat{\imath} + y_i\hat{\jmath} + z_i\hat{k}. \tag{9-6}$$

Here the index identifies the particle, and $\hat{\imath}$, $\hat{\jmath}$, and $\hat{k}$ are unit vectors pointing, respectively, in the positive direction of the x, y, and z axes. Similarly, the position of the center of mass of a system of particles is given by a position vector:

$$\vec{r}_{com} = x_{com}\hat{\imath} + y_{com}\hat{\jmath} + z_{com}\hat{k}. \tag{9-7}$$

The three scalar equations of Eq. 9-5 can now be replaced by a single vector equation,

$$\vec{r}_{com} = \frac{1}{M}\sum_{i=1}^{n} m_i\vec{r}_i, \tag{9-8}$$

where again M is the total mass of the system. You can check that this equation is correct by substituting Eqs. 9-6 and 9-7 into it, and then separating out the x, y, and z components. The scalar relations of Eq. 9-5 result.

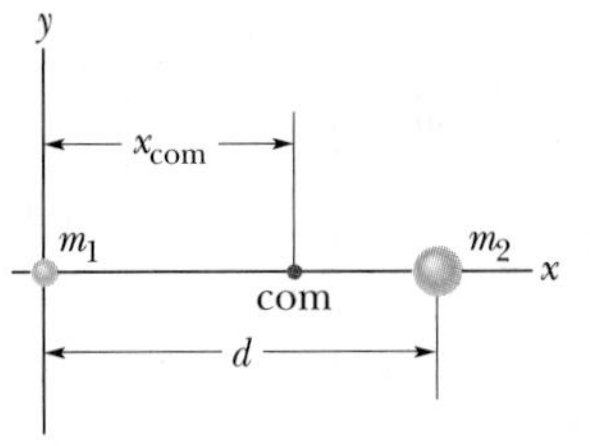

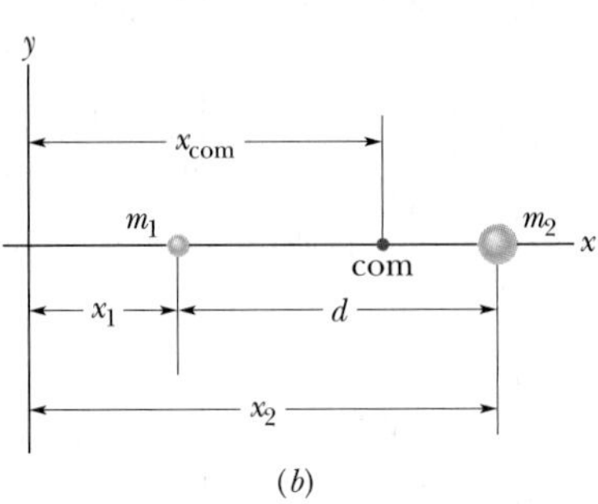

FIG. 9-2 (*a*) Two particles of masses m_1 and m_2 are separated by distance d. The dot labeled com shows the position of the center of mass, calculated from Eq. 9-1. (*b*) The same as (*a*) except that the origin is located farther from the particles. The position of the center of mass is calculated from Eq. 9-2. The location of the center of mass with respect to the particles is the same in both cases.

Solid Bodies

An ordinary object, such as a baseball bat, contains so many particles (atoms) that we can best treat it as a continuous distribution of matter. The "particles" then become differential mass elements dm, the sums of Eq. 9-5 become integrals, and the coordinates of the center of mass are defined as

$$x_{com} = \frac{1}{M}\int x\,dm, \qquad y_{com} = \frac{1}{M}\int y\,dm, \qquad z_{com} = \frac{1}{M}\int z\,dm, \tag{9-9}$$

where M is now the mass of the object.

Evaluating these integrals for most common objects (such as a television set or a moose) would be difficult, so here we consider only *uniform* objects. Such objects have uniform *density,* or mass per unit volume; that is, the density ρ

(Greek letter rho) is the same for any given element of an object as for the whole object. From Eq. 1-8, we can write

$$\rho = \frac{dm}{dV} = \frac{M}{V}, \qquad (9\text{-}10)$$

where dV is the volume occupied by a mass element dm, and V is the total volume of the object. Substituting $dm = (M/V)\,dV$ from Eq. 9-10 into Eq. 9-9 gives

$$x_{com} = \frac{1}{V}\int x\,dV, \qquad y_{com} = \frac{1}{V}\int y\,dV, \qquad z_{com} = \frac{1}{V}\int z\,dV. \qquad (9\text{-}11)$$

You can bypass one or more of these integrals if an object has a point, a line, or a plane of symmetry. The center of mass of such an object then lies at that point, on that line, or in that plane. For example, the center of mass of a uniform sphere (which has a point of symmetry) is at the center of the sphere (which is the point of symmetry). The center of mass of a uniform cone (whose axis is a line of symmetry) lies on the axis of the cone. The center of mass of a banana (which has a plane of symmetry that splits it into two equal parts) lies somewhere in that plane.

The center of mass of an object need not lie within the object. There is no dough at the com of a doughnut, and no iron at the com of a horseshoe.

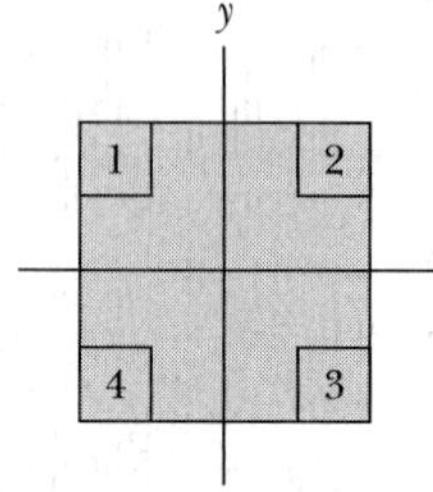

CHECKPOINT 1 The figure shows a uniform square plate from which four identical squares at the corners will be removed. (a) Where is the center of mass of the plate originally? Where is it after the removal of (b) square 1; (c) squares 1 and 2; (d) squares 1 and 3; (e) squares 1, 2, and 3; (f) all four squares? Answer in terms of quadrants, axes, or points (without calculation, of course).

Sample Problem 9-1

Three particles of masses $m_1 = 1.2$ kg, $m_2 = 2.5$ kg, and $m_3 = 3.4$ kg form an equilateral triangle of edge length $a = 140$ cm. Where is the center of mass of this system?

KEY IDEA We are dealing with particles instead of an extended solid body, so we can use Eq. 9-5 to locate their center of mass. The particles are in the plane of the equilateral triangle, so we need only the first two equations.

Calculations: We can simplify the calculations by choosing the x and y axes so that one of the particles is located at the origin and the x axis coincides with one of the triangle's sides (Fig. 9-3). The three particles then have the following coordinates:

Particle	Mass (kg)	x (cm)	y (cm)
1	1.2	0	0
2	2.5	140	0
3	3.4	70	120

The total mass M of the system is 7.1 kg.

From Eq. 9-5, the coordinates of the center of mass are

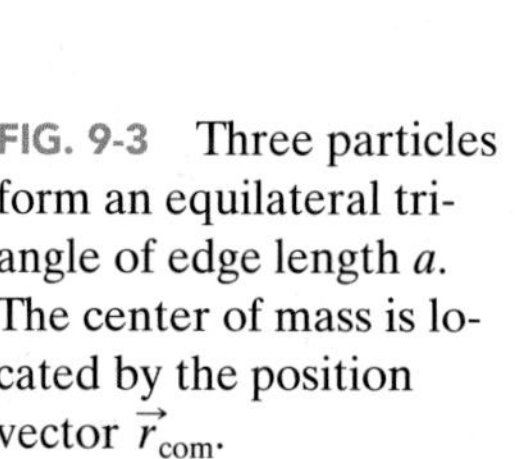
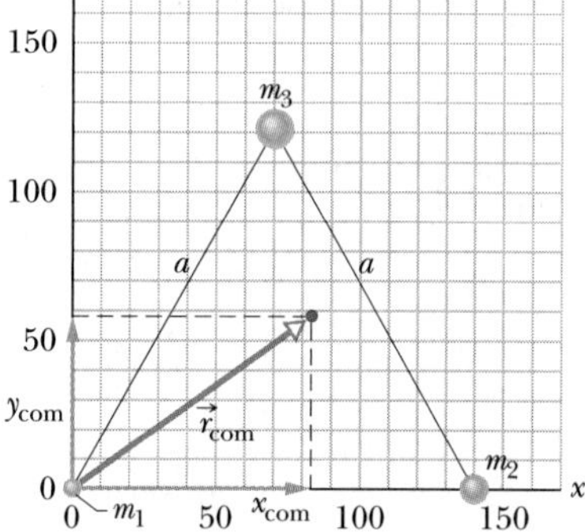

FIG. 9-3 Three particles form an equilateral triangle of edge length a. The center of mass is located by the position vector $\vec{r}_{com}$.

$$x_{com} = \frac{1}{M}\sum_{i=1}^{3} m_i x_i = \frac{m_1x_1 + m_2x_2 + m_3x_3}{M}$$

$$= \frac{(1.2\text{ kg})(0) + (2.5\text{ kg})(140\text{ cm}) + (3.4\text{ kg})(70\text{ cm})}{7.1\text{ kg}}$$

$$= 83\text{ cm} \qquad \text{(Answer)}$$

and

$$y_{com} = \frac{1}{M}\sum_{i=1}^{3} m_i y_i = \frac{m_1y_1 + m_2y_2 + m_3y_3}{M}$$

$$= \frac{(1.2\text{ kg})(0) + (2.5\text{ kg})(0) + (3.4\text{ kg})(120\text{ cm})}{7.1\text{ kg}}$$

$$= 58\text{ cm}. \qquad \text{(Answer)}$$

In Fig. 9-3, the center of mass is located by the position vector $\vec{r}_{com}$, which has components x_{com} and y_{com}.

Sample Problem 9-2 Build your skill

Figure 9-4*a* shows a uniform metal plate P of radius $2R$ from which a disk of radius R has been stamped out (removed) in an assembly line. Using the xy coordinate system shown, locate the center of mass com_P of the plate.

KEY IDEAS (1) Let us roughly locate the center of plate P by using symmetry. We note that the plate is symmetric about the x axis (we get the portion below that axis by rotating the upper portion about the axis). Thus, com_P must be on the x axis. The plate (with the disk removed) is not symmetric about the y axis. However, because there is somewhat more mass on the right of the y axis, com_P must be somewhat to the right of that axis. Thus, the location of com_P should be roughly as indicated in Fig. 9-4*a*.

(2) Plate P is an extended solid body, so we can use Eqs. 9-11 to find the actual coordinates of com_P. However, that procedure is difficult. Here is a much easier way: In working with centers of mass, we can assume that the mass of a *uniform* object is concentrated in a particle at the object's center of mass.

Calculations: First, put the stamped-out disk (call it disk S) back into place (Fig. 9-4*b*) to form the original composite plate (call it plate C). Because of its circular symmetry, the center of mass com_S for disk S is at the center of S, at $x = -R$ (as shown). Similarly, the center of mass com_C for composite plate C is at the center of C, at the origin (as shown). We then have the following:

Plate	Center of Mass	Location of com	Mass
P	com_P	$x_P = ?$	m_P
S	com_S	$x_S = -R$	m_S
C	com_C	$x_C = 0$	$m_C = m_S + m_P$

Assume that mass m_S of disk S is concentrated in a particle at $x_S = -R$, and mass m_P is concentrated in a particle at x_P (Fig. 9-4*c*). Next treat these two particles as a two-particle system, using Eq. 9-2 to find their center of mass x_{S+P}. We get

$$x_{S+P} = \frac{m_S x_S + m_P x_P}{m_S + m_P}. \tag{9-12}$$

Next note that the combination of disk S and plate P is composite plate C. Thus, the position x_{S+P} of com_{S+P} must coincide with the position x_C of com_C, which is at the origin; so $x_{S+P} = x_C = 0$. Substituting this into Eq. 9-12 and solving for x_P, we get

$$x_P = -x_S \frac{m_S}{m_P}. \tag{9-13}$$

We can relate these masses to the face areas of S and P by noting that

$$\begin{aligned}\text{mass} &= \text{density} \times \text{volume}\\ &= \text{density} \times \text{thickness} \times \text{area}.\end{aligned}$$

Then
$$\frac{m_S}{m_P} = \frac{\text{density}_S}{\text{density}_P} \times \frac{\text{thickness}_S}{\text{thickness}_P} \times \frac{\text{area}_S}{\text{area}_P}.$$

Because the plate is uniform, the densities and thicknesses are equal; we are left with

$$\begin{aligned}\frac{m_S}{m_P} &= \frac{\text{area}_S}{\text{area}_P} = \frac{\text{area}_S}{\text{area}_C - \text{area}_S}\\ &= \frac{\pi R^2}{\pi(2R)^2 - \pi R^2} = \frac{1}{3}.\end{aligned}$$

Substituting this and $x_S = -R$ into Eq. 9-13, we have

$$x_P = \tfrac{1}{3}R. \qquad \text{(Answer)}$$

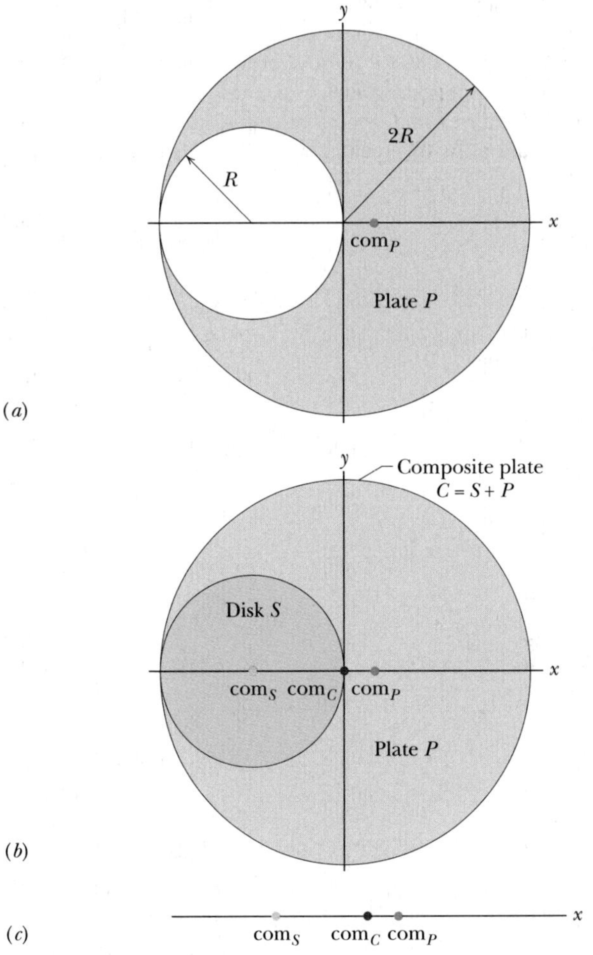

FIG. 9-4 (*a*) Plate P is a metal plate of radius $2R$, with a circular hole of radius R. The center of mass of P is at point com_P. (*b*) Disk S has been put back into place to form a composite plate C. The center of mass com_S of disk S and the center of mass com_C of plate C are shown. (*c*) The center of mass com_{S+P} of the combination of S and P coincides with com_C, which is at $x = 0$.

PROBLEM-SOLVING TACTICS

Tactic 1: Center-of-Mass Problems Sample Problems 9-1 and 9-2 provide three strategies for simplifying center-of-mass problems. (1) Make full use of the symmetry of the object, be it about a point, a line, or a plane. (2) If the object can be divided into several parts, treat each of these parts as a particle, located at its own center of mass. (3) Choose your axes wisely: If your system is a group of particles, choose one of the particles as your origin. If your system is a body with a line of symmetry, let that be your x or y axis. The choice of origin can be completely arbitrary because the location of the center of mass is the same regardless of the origin from which it is measured.

9-3 | Newton's Second Law for a System of Particles

Now that we know how to locate the center of mass of a system of particles, we discuss how external forces can move a center of mass. Let us start with a simple system of two billiard balls.

If you roll a cue ball at a second billiard ball that is at rest, you expect that the two-ball system will continue to have some forward motion after impact. You would be surprised, for example, if both balls came back toward you or if both moved to the right or to the left.

What continues to move forward, its steady motion completely unaffected by the collision, is the center of mass of the two-ball system. If you focus on this point—which is always halfway between these bodies because they have identical masses—you can easily convince yourself by trial at a billiard table that this is so. No matter whether the collision is glancing, head-on, or somewhere in between, the center of mass continues to move forward, as if the collision had never occurred. Let us look into this center-of-mass motion in more detail.

To do so, we replace the pair of billiard balls with an assemblage of n particles of (possibly) different masses. We are interested not in the individual motions of these particles but *only* in the motion of the center of mass of the assemblage. Although the center of mass is just a point, it moves like a particle whose mass is equal to the total mass of the system; we can assign a position, a velocity, and an acceleration to it. We state (and shall prove next) that the vector equation that governs the motion of the center of mass of such a system of particles is

$$\vec{F}_{net} = M\vec{a}_{com} \quad \text{(system of particles).} \tag{9-14}$$

This equation is Newton's second law for the motion of the center of mass of a system of particles. Note that its form is the same as the form of the equation ($\vec{F}_{net} = m\vec{a}$) for the motion of a single particle. However, the three quantities that appear in Eq. 9-14 must be evaluated with some care:

1. $\vec{F}_{net}$ is the net force of *all external forces* that act on the system. Forces on one part of the system from another part of the system (*internal forces*) are not included in Eq. 9-14.
2. M is the *total mass* of the system. We assume that no mass enters or leaves the system as it moves, so that M remains constant. The system is said to be **closed.**
3. $\vec{a}_{com}$ is the acceleration of the *center of mass* of the system. Equation 9-14 gives no information about the acceleration of any other point of the system.

Equation 9-14 is equivalent to three equations involving the components of $\vec{F}_{net}$ and $\vec{a}_{com}$ along the three coordinate axes. These equations are

$$F_{net,x} = Ma_{com,x} \qquad F_{net,y} = Ma_{com,y} \qquad F_{net,z} = Ma_{com,z}. \tag{9-15}$$

Now we can go back and examine the behavior of the billiard balls. Once the cue ball has begun to roll, no net external force acts on the (two-ball) system. Thus, because $\vec{F}_{net} = 0$, Eq. 9-14 tells us that $\vec{a}_{com} = 0$ also. Because acceleration is the rate of change of velocity, we conclude that the velocity of the center of mass of the system of two balls does not change. When the two balls collide, the forces that come into play are *internal* forces, on one ball from the other. Such forces do not contribute to the net force $\vec{F}_{net}$, which remains zero. Thus, the center of mass of the system, which was moving forward before the collision, must continue to move forward after the collision, with the same speed and in the same direction.

Equation 9-14 applies not only to a system of particles but also to a solid body, such as the bat of Fig. 9-1*b*. In that case, M in Eq. 9-14 is the mass of the bat and $\vec{F}_{net}$ is the gravitational force on the bat. Equation 9-14 then tells us that $\vec{a}_{com} = \vec{g}$. In other words, the center of mass of the bat moves as if the bat were a single particle of mass M, with force $\vec{F}_g$ acting on it.

FIG. 9-5 A fireworks rocket explodes in flight. In the absence of air drag, the center of mass of the fragments would continue to follow the original parabolic path, until fragments began to hit the ground.

Figure 9-5 shows another interesting case. Suppose that at a fireworks display, a rocket is launched on a parabolic path. At a certain point, it explodes into fragments. If the explosion had not occurred, the rocket would have continued along the trajectory shown in the figure. The forces of the explosion are *internal* to the system (at first the system is just the rocket, and later it is its fragments); that is, they are forces on parts of the system from other parts. If we ignore air drag, the net *external* force $\vec{F}_{net}$ acting on the system is the gravitational force on the system, regardless of whether the rocket explodes. Thus, from Eq. 9-14, the acceleration $\vec{a}_{com}$ of the center of mass of the fragments (while they are in flight) remains equal to $\vec{g}$. This means that the center of mass of the fragments follows the same parabolic trajectory that the rocket would have followed had it not exploded.

When a ballet dancer leaps across the stage in a grand jeté, she raises her arms and stretches her legs out horizontally as soon as her feet leave the stage (Fig. 9-6). These actions shift her center of mass upward through her body. Although the shifting center of mass faithfully follows a parabolic path across the stage, its movement relative to the body decreases the height that is attained by her head and torso, relative to that of a normal jump. The result is that the head and torso follow a nearly horizontal path, giving an illusion that the dancer is floating.

Proof of Equation 9-14

Now let us prove this important equation. From Eq. 9-8 we have, for a system of n particles,

$$M\vec{r}_{com} = m_1\vec{r}_1 + m_2\vec{r}_2 + m_3\vec{r}_3 + \cdots + m_n\vec{r}_n, \tag{9-16}$$

FIG. 9-6 A grand jeté. (Adapted from *The Physics of Dance,* by Kenneth Laws, Schirmer Books, 1984.)

in which M is the system's total mass and $\vec{r}_{com}$ is the vector locating the position of the system's center of mass.

Differentiating Eq. 9-16 with respect to time gives

$$M\vec{v}_{com} = m_1\vec{v}_1 + m_2\vec{v}_2 + m_3\vec{v}_3 + \cdots + m_n\vec{v}_n. \tag{9-17}$$

Here $\vec{v}_i$ $(= d\vec{r}_i/dt)$ is the velocity of the ith particle, and $\vec{v}_{com}$ $(= d\vec{r}_{com}/dt)$ is the velocity of the center of mass.

Differentiating Eq. 9-17 with respect to time leads to

$$M\vec{a}_{com} = m_1\vec{a}_1 + m_2\vec{a}_2 + m_3\vec{a}_3 + \cdots + m_n\vec{a}_n. \tag{9-18}$$

Here $\vec{a}_i$ $(= d\vec{v}_i/dt)$ is the acceleration of the ith particle, and $\vec{a}_{com}$ $(= d\vec{v}_{com}/dt)$ is the acceleration of the center of mass. Although the center of mass is just a geometrical point, it has a position, a velocity, and an acceleration, as if it were a particle.

From Newton's second law, $m_i\vec{a}_i$ is equal to the resultant force $\vec{F}_i$ that acts on the ith particle. Thus, we can rewrite Eq. 9-18 as

$$M\vec{a}_{com} = \vec{F}_1 + \vec{F}_2 + \vec{F}_3 + \cdots + \vec{F}_n. \tag{9-19}$$

Among the forces that contribute to the right side of Eq. 9-19 will be forces that the particles of the system exert on each other (internal forces) and forces exerted on the particles from outside the system (external forces). By Newton's third law, the internal forces form third-law force pairs and cancel out in the sum that appears on the right side of Eq. 9-19. What remains is the vector sum of all the *external* forces that act on the system. Equation 9-19 then reduces to Eq. 9-14, the relation that we set out to prove.

CHECKPOINT 2 Two skaters on frictionless ice hold opposite ends of a pole of negligible mass. An axis runs along the pole, and the origin of the axis is at the center of mass of the two-skater system. One skater, Fred, weighs twice as much as the other skater, Ethel. Where do the skaters meet if (a) Fred pulls hand over hand along the pole so as to draw himself to Ethel, (b) Ethel pulls hand over hand to draw herself to Fred, and (c) both skaters pull hand over hand?

Sample Problem 9-3

The three particles in Fig. 9-7*a* are initially at rest. Each experiences an *external* force due to bodies outside the three-particle system. The directions are indicated, and the magnitudes are $F_1 = 6.0$ N, $F_2 = 12$ N, and $F_3 = 14$ N. What is the acceleration of the center of mass of the system, and in what direction does it move?

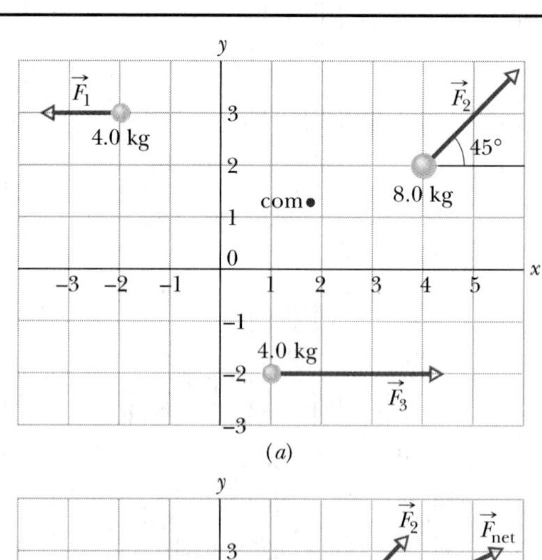

KEY IDEAS The position of the center of mass, calculated by the method of Sample Problem 9-1, is marked by a dot in the figure. We can treat the center of mass as if it were a real particle, with a mass equal to the system's total mass $M = 16$ kg. We can also treat the

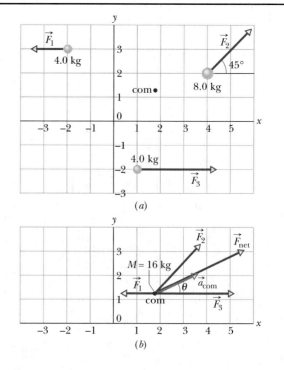

FIG. 9-7 (*a*) Three particles, initially at rest in the positions shown, are acted on by the external forces shown. The center of mass (com) of the system is marked. (*b*) The forces are now transferred to the center of mass of the system, which behaves like a particle with a mass M equal to the total mass of the system. The net external force $\vec{F}_{net}$ and the acceleration $\vec{a}_{com}$ of the center of mass are shown.

three external forces as if they act at the center of mass (Fig. 9-7*b*).

Calculations: We can now apply Newton's second law ($\vec{F}_{net} = m\vec{a}$) to the center of mass, writing

$$\vec{F}_{net} = M\vec{a}_{com} \tag{9-20}$$

or

$$\vec{F}_1 + \vec{F}_2 + \vec{F}_3 = M\vec{a}_{com}$$

so

$$\vec{a}_{com} = \frac{\vec{F}_1 + \vec{F}_2 + \vec{F}_3}{M}. \tag{9-21}$$

Equation 9-20 tells us that the acceleration $\vec{a}_{com}$ of the center of mass is in the same direction as the net external force $\vec{F}_{net}$ on the system (Fig. 9-7*b*). Because the particles are initially at rest, the center of mass must also be at rest. As the center of mass then begins to accelerate, it must move off in the common direction of $\vec{a}_{com}$ and $\vec{F}_{net}$.

We can evaluate the right side of Eq. 9-21 directly on a vector-capable calculator, or we can rewrite Eq. 9-21 in component form, find the components of $\vec{a}_{com}$, and then find $\vec{a}_{com}$. Along the x axis, we have

$$a_{com,x} = \frac{F_{1x} + F_{2x} + F_{3x}}{M}$$
$$= \frac{-6.0\text{ N} + (12\text{ N})\cos 45° + 14\text{ N}}{16\text{ kg}} = 1.03\text{ m/s}^2.$$

Along the y axis, we have

$$a_{com,y} = \frac{F_{1y} + F_{2y} + F_{3y}}{M}$$
$$= \frac{0 + (12\text{ N})\sin 45° + 0}{16\text{ kg}} = 0.530\text{ m/s}^2.$$

From these components, we find that $\vec{a}_{com}$ has the magnitude

$$a_{com} = \sqrt{(a_{com,x})^2 + (a_{com,y})^2}$$
$$= 1.16\text{ m/s}^2 \approx 1.2\text{ m/s}^2 \qquad \text{(Answer)}$$

and the angle (from the positive direction of the x axis)

$$\theta = \tan^{-1}\frac{a_{com,y}}{a_{com,x}} = 27°. \qquad \text{(Answer)}$$

9-4 | Linear Momentum

In this section, we discuss only a single particle instead of a system of particles, in order to define two important quantities. Then in Section 9-5, we extend those definitions to systems of many particles.

The first definition concerns a familiar word—*momentum*—that has several meanings in everyday language but only a single precise meaning in physics and engineering. The **linear momentum** of a particle is a vector quantity $\vec{p}$ that is defined as

$$\vec{p} = m\vec{v} \qquad \text{(linear momentum of a particle)}, \tag{9-22}$$

in which m is the mass of the particle and $\vec{v}$ is its velocity. (The adjective *linear* is often dropped, but it serves to distinguish $\vec{p}$ from *angular* momentum, which is introduced in Chapter 11 and which is associated with rotation.) Since m is always a positive scalar quantity, Eq. 9-22 tells us that $\vec{p}$ and $\vec{v}$ have the same direction. From Eq. 9-22, the SI unit for momentum is the kilogram-meter per second (kg · m/s).

Newton expressed his second law of motion in terms of momentum:

> The time rate of change of the momentum of a particle is equal to the net force acting on the particle and is in the direction of that force.

In equation form this becomes

$$\vec{F}_{net} = \frac{d\vec{p}}{dt}. \tag{9-23}$$

In words, Eq. 9-23 says that the net external force $\vec{F}_{net}$ on a particle changes the particle's linear momentum $\vec{p}$. Conversely, the linear momentum can be changed only by a net external force. If there is no net external force, $\vec{p}$ *cannot* change. As we shall see in Section 9-7, this last fact can be an extremely powerful tool in solving problems.

Manipulating Eq. 9-23 by substituting for $\vec{p}$ from Eq. 9-22 gives, for constant mass m,

$$\vec{F}_{net} = \frac{d\vec{p}}{dt} = \frac{d}{dt}(m\vec{v}) = m\frac{d\vec{v}}{dt} = m\vec{a}.$$

Thus, the relations $\vec{F}_{net} = d\vec{p}/dt$ and $\vec{F}_{net} = m\vec{a}$ are equivalent expressions of Newton's second law of motion for a particle.

CHECKPOINT 3 The figure gives the magnitude p of the linear momentum versus time t for a particle moving along an axis. A force directed along the axis acts on the particle. (a) Rank the four regions indicated according to the magnitude of the force, greatest first. (b) In which region is the particle slowing?

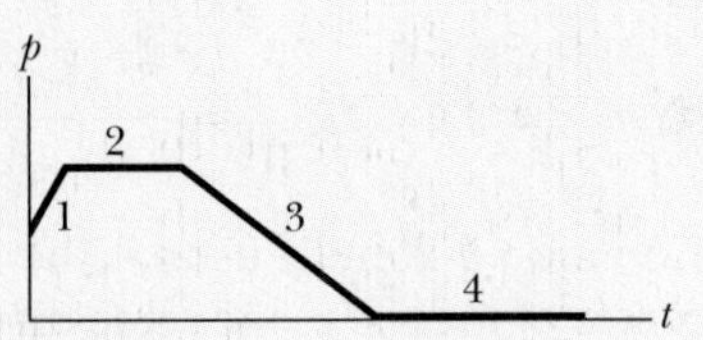

9-5 | The Linear Momentum of a System of Particles

Let's extend the definition of linear momentum to a system of particles. Consider a system of n particles, each with its own mass, velocity, and linear momentum. The particles may interact with each other, and external forces may act on them. The system as a whole has a total linear momentum $\vec{P}$, which is defined to be the vector sum of the individual particles' linear momenta. Thus,

$$\vec{P} = \vec{p}_1 + \vec{p}_2 + \vec{p}_3 + \cdots + \vec{p}_n$$
$$= m_1\vec{v}_1 + m_2\vec{v}_2 + m_3\vec{v}_3 + \cdots + m_n\vec{v}_n. \quad (9\text{-}24)$$

If we compare this equation with Eq. 9-17, we see that

$$\vec{P} = M\vec{v}_{com} \quad \text{(linear momentum, system of particles)}, \quad (9\text{-}25)$$

which is another way to define the linear momentum of a system of particles:

The linear momentum of a system of particles is equal to the product of the total mass M of the system and the velocity of the center of mass.

If we take the time derivative of Eq. 9-25, we find

$$\frac{d\vec{P}}{dt} = M\frac{d\vec{v}_{com}}{dt} = M\vec{a}_{com}. \quad (9\text{-}26)$$

Comparing Eqs. 9-14 and 9-26 allows us to write Newton's second law for a system of particles in the equivalent form

$$\vec{F}_{net} = \frac{d\vec{P}}{dt} \quad \text{(system of particles)}, \quad (9\text{-}27)$$

where $\vec{F}_{net}$ is the net external force acting on the system. This equation is the generalization of the single-particle equation $\vec{F}_{net} = d\vec{p}/dt$ to a system of many particles. In words, the equation says that the net external force $\vec{F}_{net}$ on a system of particles changes the linear momentum $\vec{P}$ of the system. Conversely, the linear momentum can be changed only by a net external force. If there is no net external force, $\vec{P}$ *cannot* change.

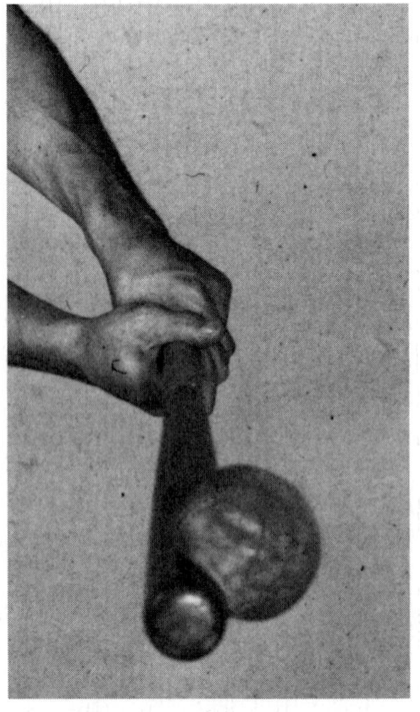

The collision of a ball with a bat collapses part of the ball. *(Photo by Harold E. Edgerton. ©The Harold and Esther Edgerton Family Trust, courtesy of Palm Press, Inc.)*

9-6 | Collision and Impulse

The momentum $\vec{p}$ of any particle-like body cannot change unless a net external force changes it. For example, we could push on the body to change its momentum. More dramatically, we could arrange for the body to collide with a baseball bat. In such a *collision* (or *crash*), the external force on the body is brief, has large magnitude, and suddenly changes the body's momentum. Collisions occur commonly in our world, but before we get to them, we need to consider a simple collision in which a moving particle-like body (a *projectile*) collides with some other body (a *target*).

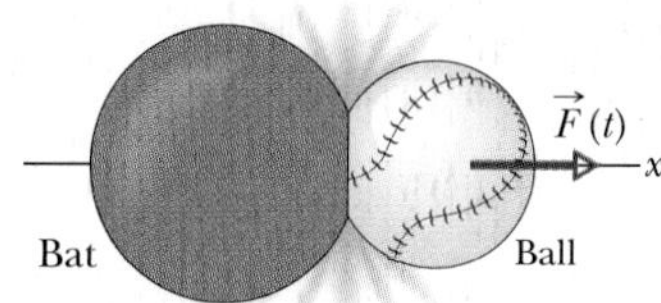

FIG. 9-8 Force $\vec{F}(t)$ acts on a ball as the ball and a bat collide.

Single Collision

Let the projectile be a ball and the target be a bat. The collision is brief, and the ball experiences a force that is great enough to slow, stop, or even reverse its motion. Figure 9-8 depicts the collision at one instant. The ball experiences a force $\vec{F}(t)$ that varies during the collision and changes the linear momentum $\vec{p}$ of the ball. That change is related to the force by Newton's second law written in the form $\vec{F} = d\vec{p}/dt$. Thus, in time interval dt, the change in the ball's momentum is

$$d\vec{p} = \vec{F}(t)\,dt. \tag{9-28}$$

We can find the net change in the ball's momentum due to the collision if we integrate both sides of Eq. 9-28 from a time t_i just before the collision to a time t_f just after the collision:

$$\int_{t_i}^{t_f} d\vec{p} = \int_{t_i}^{t_f} \vec{F}(t)\,dt. \tag{9-29}$$

The left side of this equation gives us the change in momentum: $\vec{p}_f - \vec{p}_i = \Delta\vec{p}$. The right side, which is a measure of both the magnitude and the duration of the collision force, is called the **impulse** $\vec{J}$ of the collision:

$$\vec{J} = \int_{t_i}^{t_f} \vec{F}(t)\,dt \quad \text{(impulse defined).} \tag{9-30}$$

Thus, the change in an object's momentum is equal to the impulse on the object:

$$\Delta\vec{p} = \vec{J} \quad \text{(linear momentum–impulse theorem).} \tag{9-31}$$

This expression can also be written in the vector form

$$\vec{p}_f - \vec{p}_i = \vec{J} \tag{9-32}$$

and in such component forms as

$$\Delta p_x = J_x \tag{9-33}$$

and

$$p_{fx} - p_{ix} = \int_{t_i}^{t_f} F_x\,dt. \tag{9-34}$$

If we have a function for $\vec{F}(t)$, we can evaluate $\vec{J}$ (and thus the change in momentum) by integrating the function. If we have a plot of $\vec{F}$ versus time t, we can evaluate $\vec{J}$ by finding the area between the curve and the t axis, such as in Fig. 9-9*a*. In many situations we do not know how the force varies with time but we do know the average magnitude F_{avg} of the force and the duration Δt $(= t_f - t_i)$ of the collision. Then we can write the magnitude of the impulse as

$$J = F_{avg}\,\Delta t. \tag{9-35}$$

The average force is plotted versus time as in Fig. 9-9*b*. The area under that curve is equal to the area under the curve for the actual force $F(t)$ in Fig. 9-9*a* because both areas are equal to impulse magnitude J.

Instead of the ball, we could have focused on the bat in Fig. 9-8. At any instant, Newton's third law tells us that the force on the bat has the same magnitude but the opposite direction as the force on the ball. From Eq. 9-30, this means that the impulse on the bat has the same magnitude but the opposite direction as the impulse on the ball.

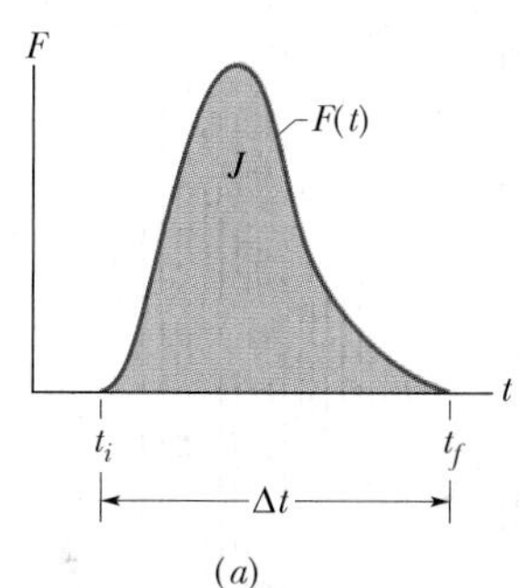

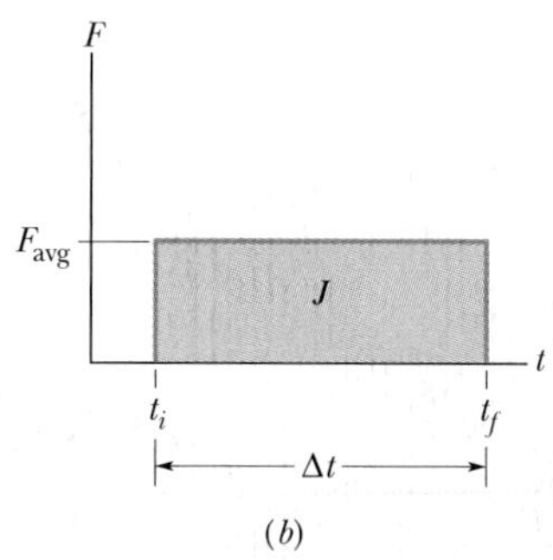

FIG. 9-9 (*a*) The curve shows the magnitude of the time-varying force $F(t)$ that acts on the ball in the collision of Fig. 9-8. The area under the curve is equal to the magnitude of the impulse $\vec{J}$ on the ball in the collision. (*b*) The height of the rectangle represents the average force F_{avg} acting on the ball over the time interval Δt. The area within the rectangle is equal to the area under the curve in (*a*) and thus is also equal to the magnitude of the impulse $\vec{J}$ in the collision.

CHECKPOINT 4 A paratrooper whose chute fails to open lands in snow; he is hurt slightly. Had he landed on bare ground, the stopping time would have been 10 times shorter and the collision lethal. Does the presence of the snow increase, decrease, or leave unchanged the values of (a) the paratrooper's change in momentum, (b) the impulse stopping the paratrooper, and (c) the force stopping the paratrooper?

Series of Collisions

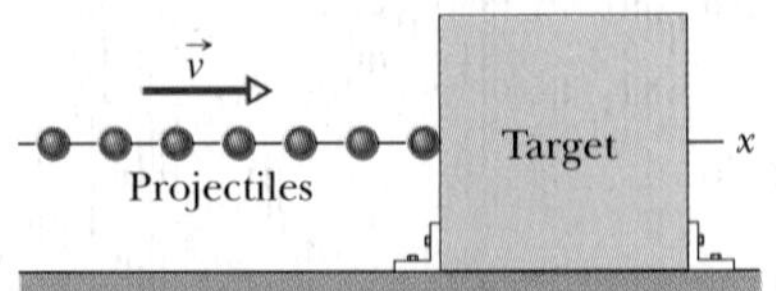

FIG. 9-10 A steady stream of projectiles, with identical linear momenta, collides with a target, which is fixed in place. The average force F_{avg} on the target is to the right and has a magnitude that depends on the rate at which the projectiles collide with the target or, equivalently, the rate at which mass collides with the target.

Now let's consider the force on a body when it undergoes a series of identical, repeated collisions. For example, as a prank, we might adjust one of those machines that fire tennis balls to fire them at a rapid rate directly at a wall. Each collision would produce a force on the wall, but that is not the force we are seeking. We want the average force F_{avg} on the wall during the bombardment—that is, the average force during a large number of collisions.

In Fig. 9-10, a steady stream of projectile bodies, with identical mass m and linear momenta $m\vec{v}$, moves along an x axis and collides with a target body that is fixed in place. Let n be the number of projectiles that collide in a time interval Δt. Because the motion is along only the x axis, we can use the components of the momenta along that axis. Thus, each projectile has initial momentum mv and undergoes a change Δp in linear momentum because of the collision. The total change in linear momentum for n projectiles during interval Δt is $n\,\Delta p$. The resulting impulse $\vec{J}$ on the target during Δt is along the x axis and has the same magnitude of $n\,\Delta p$ but is in the opposite direction. We can write this relation in component form as

$$J = -n\,\Delta p, \tag{9-36}$$

where the minus sign indicates that J and Δp have opposite directions.

By rearranging Eq. 9-35 and substituting Eq. 9-36, we find the average force F_{avg} acting on the target during the collisions:

$$F_{avg} = \frac{J}{\Delta t} = -\frac{n}{\Delta t}\Delta p = -\frac{n}{\Delta t} m\,\Delta v. \tag{9-37}$$

This equation gives us F_{avg} in terms of $n/\Delta t$, the rate at which the projectiles collide with the target, and Δv, the change in the velocity of those projectiles.

If the projectiles stop upon impact, then in Eq. 9-37 we can substitute, for Δv,

$$\Delta v = v_f - v_i = 0 - v = -v, \tag{9-38}$$

where v_i $(= v)$ and v_f $(= 0)$ are the velocities before and after the collision, respectively. If, instead, the projectiles bounce (rebound) directly backward from the target with no change in speed, then $v_f = -v$ and we can substitute

$$\Delta v = v_f - v_i = -v - v = -2v. \tag{9-39}$$

In time interval Δt, an amount of mass $\Delta m = nm$ collides with the target. With this result, we can rewrite Eq. 9-37 as

$$F_{avg} = -\frac{\Delta m}{\Delta t}\Delta v. \tag{9-40}$$

This equation gives the average force F_{avg} in terms of $\Delta m/\Delta t$, the rate at which mass collides with the target. Here again we can substitute for Δv from Eq. 9-38 or 9-39 depending on what the projectiles do.

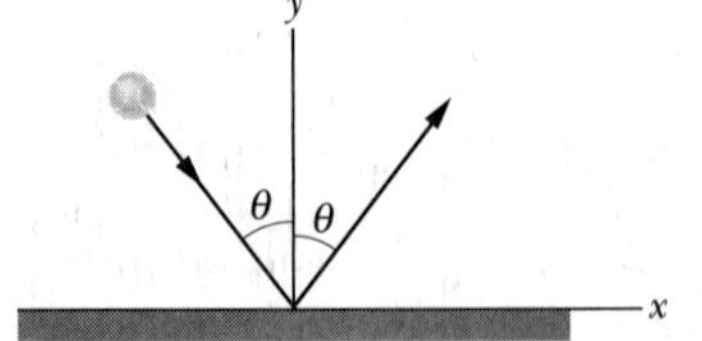

CHECKPOINT 5 The figure shows an overhead view of a ball bouncing from a vertical wall without any change in its speed. Consider the change $\Delta\vec{p}$ in the ball's linear momentum. (a) Is Δp_x positive, negative, or zero? (b) Is Δp_y positive, negative, or zero? (c) What is the direction of $\Delta\vec{p}$?

Sample Problem 9-4

When a male bighorn sheep runs head-first into another male, the rate at which its speed drops to zero is dramatic. Figure 9-11 gives a typical graph of the acceleration a versus time t for such a collision, with the acceleration taken as negative to correspond to an initially positive velocity. The peak acceleration has magnitude 34 m/s^2 and the duration of the collision is 0.27 s. Assume that the sheep's mass is 90.0 kg. What are the magnitudes of the impulse and average force due to the collision?

KEY IDEAS (1) Impulse is defined as being the integration of force with respect to time, according to Eq. 9-30 ($\vec{J} = \int \vec{F}(t)\,dt$). (2) Average force is related to the impulse and the elapsed time by Eq. 9-35 ($J = F_{avg}\,\Delta t$).

Calculations: We do not have a function for the force that we can integrate. However, we do have a graph of a versus t that we can transform to be F versus t by multiplying the scale on the acceleration axis by the mass of 90.0 kg. Then we can graphically integrate by finding the area between the plot and the time axis. Since the plot is in the shape of a triangle, we have for the impulse magnitude

$$J = \text{area} = \tfrac{1}{2}(0.27\text{ s})(90.0\text{ kg})(34.0\text{ m/s}^2)$$
$$= 4.13 \times 10^2\,\text{N}\cdot\text{s} \approx 4.1 \times 10^2\,\text{N}\cdot\text{s}. \quad \text{(Answer)}$$

For the magnitude of the average force, we can write

$$F_{avg} = \frac{J}{\Delta t} = \frac{4.13 \times 10^2\,\text{N}\cdot\text{s}}{0.27\text{ s}}$$
$$= 1.5 \times 10^3\,\text{N}. \quad \text{(Answer)}$$

Comment: The impulse is equal to the change in the sheep's momentum during the collision. So, the size of the impulse depends on the sheep's mass and its speed right before the collision. To win the fight, a male wants a large momentum magnitude. However, if the sheep were to hit skull-to-skull or skull-to-horn, the collision duration would be 1/10 of what we just used and thus the average force would be 10 times what we just calculated. Such a large force would result in concussion or even death; neither result would win the favors of onlooking female sheep. A male avoids such results by having flexible horns that yield somewhat during the collision. Such yielding prolongs the collision and decreases the force to about 1500 N, which the skull, brain, and muscles can withstand. Thus, if a horn breaks during a collision, the next collision could be fatal.

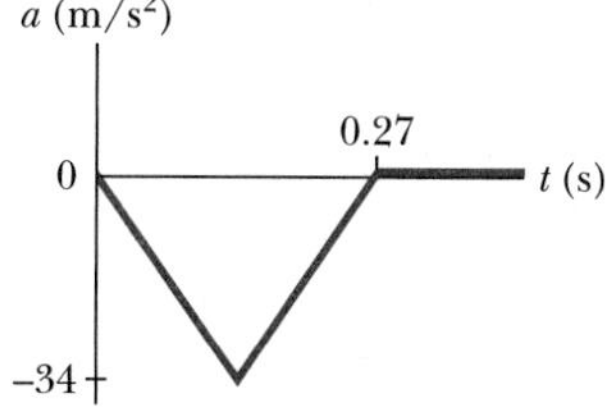

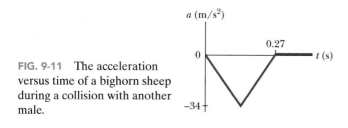

FIG. 9-11 The acceleration versus time of a bighorn sheep during a collision with another male.

Sample Problem 9-5

Race-car wall collision. Figure 9-12a is an overhead view of the path taken by a race car driver as his car collides with the racetrack wall. Just before the collision, he is traveling at speed $v_i = 70$ m/s along a straight line at 30° from the wall. Just after the collision, he is traveling at speed $v_f = 50$ m/s along a straight line at 10° from the wall. His mass m is 80 kg.

(a) What is the impulse $\vec{J}$ on the driver due to the collision?

KEY IDEAS We can treat the driver as a particle-like body and thus apply the physics of this section. However, we cannot calculate $\vec{J}$ directly from Eq. 9-30 because we do not know anything about the force $\vec{F}(t)$ on the driver during the collision. That is, we do not have a function of $\vec{F}(t)$ or a plot for it and thus cannot integrate to find $\vec{J}$. However, we *can* find $\vec{J}$ from the change in the driver's linear momentum $\vec{p}$ via Eq. 9-32 ($\vec{J} = \vec{p}_f - \vec{p}_i$).

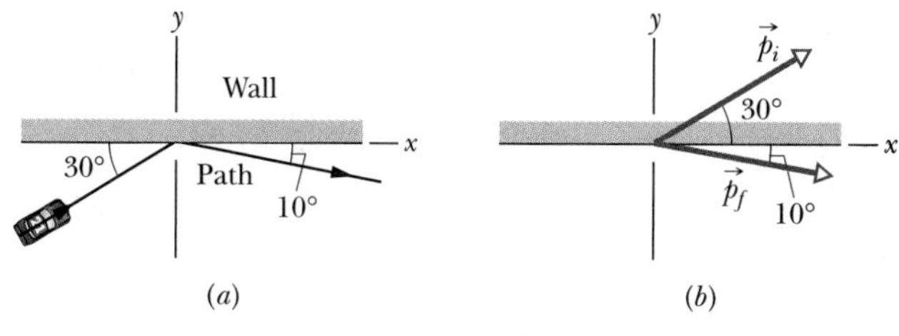

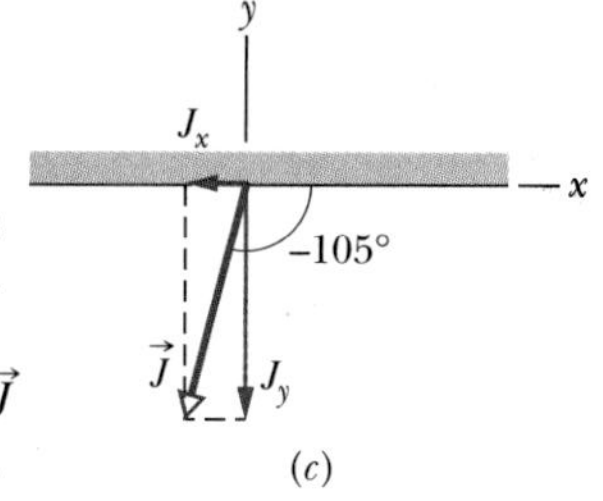

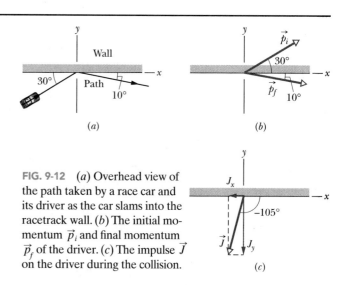

FIG. 9-12 (a) Overhead view of the path taken by a race car and its driver as the car slams into the racetrack wall. (b) The initial momentum $\vec{p}_i$ and final momentum $\vec{p}_f$ of the driver. (c) The impulse $\vec{J}$ on the driver during the collision.

Calculations: Figure 9-12b shows the driver's momentum $\vec{p}_i$ before the collision (at angle 30° from the positive x direction) and his momentum $\vec{p}_f$ after the collision (at angle −10°). From Eqs. 9-32 and 9-22 ($\vec{p} =$

$m\vec{v}$), we can write

$$\vec{J} = \vec{p}_f - \vec{p}_i = m\vec{v}_f - m\vec{v}_i = m(\vec{v}_f - \vec{v}_i). \quad (9\text{-}41)$$

We could evaluate the right side of this equation directly on a vector-capable calculator because we know m is 80 kg, $\vec{v}_f$ is 50 m/s at $-10°$, and $\vec{v}_i$ is 70 m/s at 30°. Instead, here we evaluate Eq. 9-41 in component form.

***x* component:** Along the x axis we have

$$\begin{aligned} J_x &= m(v_{fx} - v_{ix}) \\ &= (80\text{ kg})[(50\text{ m/s})\cos(-10°) - (70\text{ m/s})\cos 30°] \\ &= -910\text{ kg}\cdot\text{m/s}. \end{aligned}$$

***y* component:** Along the y axis,

$$\begin{aligned} J_y &= m(v_{fy} - v_{iy}) \\ &= (80\text{ kg})[(50\text{ m/s})\sin(-10°) - (70\text{ m/s})\sin 30°] \\ &= -3495\text{ kg}\cdot\text{m/s} \approx -3500\text{ kg}\cdot\text{m/s}. \end{aligned}$$

Impulse: The impulse is then

$$\vec{J} = (-910\hat{i} - 3500\hat{j})\text{ kg}\cdot\text{m/s}, \quad \text{(Answer)}$$

which means the impulse magnitude is

$$J = \sqrt{J_x^2 + J_y^2} = 3616\text{ kg}\cdot\text{m/s} \approx 3600\text{ kg}\cdot\text{m/s}.$$

The angle of $\vec{J}$ is given by

$$\theta = \tan^{-1}\frac{J_y}{J_x}, \quad \text{(Answer)}$$

which a calculator evaluates as 75.4°. Recall that the physically correct result of an inverse tangent might be the displayed answer plus 180°. We can tell which is correct here by drawing the components of $\vec{J}$ (Fig. 9-12c). We find that θ is actually $75.4° + 180° = 255.4°$, which we can write as

$$\theta = -105°. \quad \text{(Answer)}$$

(b) The collision lasts for 14 ms. What is the magnitude of the average force on the driver during the collision?

KEY IDEA From Eq. 9-35 ($J = F_{avg}\,\Delta t$), the magnitude F_{avg} of the average force is the ratio of the impulse magnitude J to the duration Δt of the collision.

Calculations: We have

$$\begin{aligned} F_{avg} &= \frac{J}{\Delta t} = \frac{3616\text{ kg}\cdot\text{m/s}}{0.014\text{ s}} \\ &= 2.583 \times 10^5\text{ N} \approx 2.6 \times 10^5\text{ N}. \quad \text{(Answer)} \end{aligned}$$

Using $F = ma$ with $m = 80$ kg, you can show that the magnitude of the driver's average acceleration during the collision is about $3.22 \times 10^3\text{ m/s}^2 = 329g$. We can guess that the collision will probably be fatal.

Surviving: Mechanical engineers attempt to reduce the chances of a fatality by designing and building racetrack walls with more "give," so that a collision lasts longer. For example, if the collision here lasted 10 times longer and the other data remained the same, the magnitudes of the average force and average acceleration would be 10 times less and probably survivable.

9-7 | Conservation of Linear Momentum

Suppose that the net external force $\vec{F}_{net}$ (and thus the net impulse $\vec{J}$) acting on a system of particles is zero (the system is isolated) and that no particles leave or enter the system (the system is closed). Putting $\vec{F}_{net} = 0$ in Eq. 9-27 then yields $d\vec{P}/dt = 0$, or

$$\vec{P} = \text{constant} \quad \text{(closed, isolated system)}. \quad (9\text{-}42)$$

In words,

> If no net external force acts on a system of particles, the total linear momentum $\vec{P}$ of the system cannot change.

This result is called the **law of conservation of linear momentum.** It can also be written as

$$\vec{P}_i = \vec{P}_f \quad \text{(closed, isolated system)}. \quad (9\text{-}43)$$

In words, this equation says that, for a closed, isolated system,

$$\begin{pmatrix}\text{total linear momentum}\\ \text{at some initial time } t_i\end{pmatrix} = \begin{pmatrix}\text{total linear momentum}\\ \text{at some later time } t_f\end{pmatrix}.$$

Caution: Momentum should not be confused with energy. In the sample problems of this section, momentum is conserved but energy is definitely not.

Equations 9-42 and 9-43 are vector equations and, as such, each is equivalent to three equations corresponding to the conservation of linear momentum in three mutually perpendicular directions as in, say, an xyz coordinate system. Depending on the forces acting on a system, linear momentum might be conserved in one or two directions but not in all directions. However,

> If the component of the net *external* force on a closed system is zero along an axis, then the component of the linear momentum of the system along that axis cannot change.

As an example, suppose that you toss a grapefruit across a room. During its flight, the only external force acting on the grapefruit (which we take as the system) is the gravitational force $\vec{F}_g$, which is directed vertically downward. Thus, the vertical component of the linear momentum of the grapefruit changes, but since no horizontal external force acts on the grapefruit, the horizontal component of the linear momentum cannot change.

Note that we focus on the external forces acting on a closed system. Although internal forces can change the linear momentum of portions of the system, they cannot change the total linear momentum of the entire system.

The sample problems in this section involve explosions that are either one-dimensional (meaning that the motions before and after the explosion are along a single axis) or two-dimensional (meaning that they are in a plane containing two axes). In the following sections we consider one-dimensional and two-dimensional collisions.

CHECKPOINT 6 An initially stationary device lying on a frictionless floor explodes into two pieces, which then slide across the floor. One piece slides in the positive direction of an x axis. (a) What is the sum of the momenta of the two pieces after the explosion? (b) Can the second piece move at an angle to the x axis? (c) What is the direction of the momentum of the second piece?

Sample Problem 9-6

One-dimensional explosion: A ballot box with mass $m = 6.0$ kg slides with speed $v = 4.0$ m/s across a frictionless floor in the positive direction of an x axis. The box explodes into two pieces. One piece, with mass $m_1 = 2.0$ kg, moves in the positive direction of the x axis at $v_1 = 8.0$ m/s. What is the velocity of the second piece, with mass m_2?

KEY IDEAS (1) We could get the velocity of the second piece if we knew its momentum, because we already know its mass is $m_2 = m - m_1 = 4.0$ kg. (2) We can relate the momenta of the two pieces to the original momentum of the box if momentum is conserved.

Calculations: Our reference frame will be that of the floor. Our system, which consists initially of the box and then of the two pieces, is closed but is not isolated, because the box and pieces each experience a normal force from the floor and a gravitational force. However, those forces are both vertical and thus cannot change the horizontal component of the momentum of the system. Neither can the forces produced by the explosion, because those forces are internal to the system. Thus, the horizontal component of the momentum of the system is conserved, and we can apply Eq. 9-43 along the x axis.

The initial momentum of the system is that of the box:

$$\vec{P}_i = m\vec{v}.$$

Similarly, we can write the final momenta of the two pieces as

$$\vec{P}_{f1} = m_1\vec{v}_1 \quad \text{and} \quad \vec{P}_{f2} = m_2\vec{v}_2.$$

The final total momentum $\vec{P}_f$ of the system is the vector sum of the momenta of the two pieces:

$$\vec{P}_f = \vec{P}_{f1} + \vec{P}_{f2} = m_1\vec{v}_1 + m_2\vec{v}_2.$$

Since all the velocities and momenta in this problem are vectors along the x axis, we can write them in terms of their x components. Doing so while applying Eq. 9-43, we now obtain

$$P_i = P_f$$

or

$$mv = m_1v_1 + m_2v_2.$$

Inserting known data, we find

$$(6.0\text{ kg})(4.0\text{ m/s}) = (2.0\text{ kg})(8.0\text{ m/s}) + (4.0\text{ kg})v_2$$

and thus

$$v_2 = 2.0\text{ m/s}. \qquad \text{(Answer)}$$

Since the result is positive, the second piece moves in the positive direction of the x axis.

Sample Problem 9-7

One-dimensional explosion: Figure 9-13*a* shows a space hauler and cargo module, of total mass M, traveling along an x axis in deep space. They have an initial velocity $\vec{v}_i$ of magnitude 2100 km/h relative to the Sun. With a small explosion, the hauler ejects the cargo module, of mass $0.20M$ (Fig. 9-13*b*). The hauler then travels 500 km/h faster than the module along the x axis; that is, the relative speed v_{rel} between the hauler and the module is 500 km/h. What then is the velocity $\vec{v}_{HS}$ of the hauler relative to the Sun?

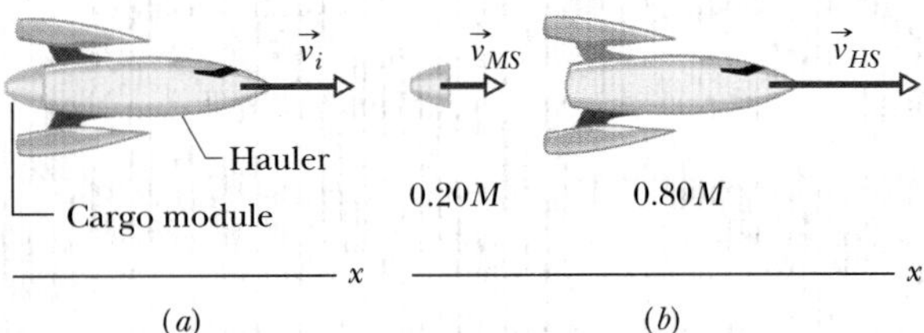

FIG. 9-13 (*a*) A space hauler, with a cargo module, moving at initial velocity $\vec{v}_i$. (*b*) The hauler has ejected the cargo module. Now the velocities relative to the Sun are $\vec{v}_{MS}$ for the module and $\vec{v}_{HS}$ for the hauler.

KEY IDEA Because the hauler–module system is closed and isolated, its total linear momentum is conserved; that is,

$$\vec{P}_i = \vec{P}_f, \tag{9-44}$$

where the subscripts i and f refer to values before and after the ejection, respectively.

Calculations: Because the motion is along a single axis, we can write momenta and velocities in terms of their x components, using a sign to indicate direction. Before the ejection, we have

$$P_i = Mv_i. \tag{9-45}$$

Let v_{MS} be the velocity of the ejected module relative to the Sun. The total linear momentum of the system after the ejection is then

$$P_f = (0.20M)v_{MS} + (0.80M)v_{HS}, \tag{9-46}$$

where the first term on the right is the linear momentum of the module and the second term is that of the hauler.

We do not know the velocity v_{MS} of the module relative to the Sun, but we can relate it to the known velocities with

$$\begin{pmatrix}\text{velocity of}\\ \text{hauler relative}\\ \text{to Sun}\end{pmatrix} = \begin{pmatrix}\text{velocity of}\\ \text{hauler relative}\\ \text{to module}\end{pmatrix} + \begin{pmatrix}\text{velocity of}\\ \text{module relative}\\ \text{to Sun}\end{pmatrix}.$$

In symbols, this gives us

$$v_{HS} = v_{rel} + v_{MS} \tag{9-47}$$

or

$$v_{MS} = v_{HS} - v_{rel}.$$

Substituting this expression for v_{MS} into Eq. 9-46, and then substituting Eqs. 9-45 and 9-46 into Eq. 9-44, we find

$$Mv_i = 0.20M(v_{HS} - v_{rel}) + 0.80Mv_{HS},$$

which gives us

$$v_{HS} = v_i + 0.20v_{rel},$$

or

$$v_{HS} = 2100 \text{ km/h} + (0.20)(500 \text{ km/h})$$
$$= 2200 \text{ km/h}. \quad \text{(Answer)}$$

Sample Problem 9-8

Two-dimensional explosion: A firecracker placed inside a coconut of mass M, initially at rest on a frictionless floor, blows the coconut into three pieces that slide across the floor. An overhead view is shown in Fig. 9-14*a*. Piece C, with mass $0.30M$, has final speed $v_{fC} = 5.0$ m/s.

(a) What is the speed of piece B, with mass $0.20M$?

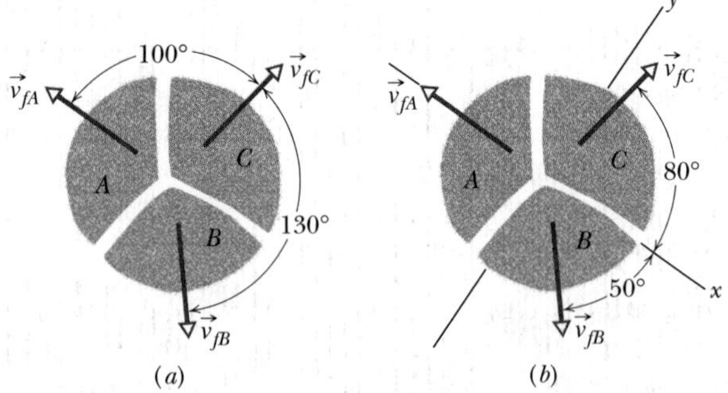

FIG. 9-14 Three pieces of an exploded coconut move off in three directions along a frictionless floor. (*a*) An overhead view of the event. (*b*) The same with a two-dimensional axis system imposed.

KEY IDEA First we need to see whether linear momentum is conserved. We note that (1) the coconut and its pieces form a closed system, (2) the explosion forces are internal to that system, and (3) no net external force acts on the system. Therefore, the linear momentum of the system is conserved.

Calculations: To get started, we superimpose an xy coordinate system as shown in Fig. 9-14*b*, with the negative direction of the x axis coinciding with the direction of $\vec{v}_{fA}$. The x axis is at 80° with the direction of $\vec{v}_{fC}$ and 50° with the direction of $\vec{v}_{fB}$.

Linear momentum is conserved separately along

each axis. Let's use the y axis and write

$$P_{iy} = P_{fy}, \tag{9-48}$$

where subscript i refers to the initial value (before the explosion), and subscript y refers to the y component of $\vec{P}_i$ or $\vec{P}_f$.

The component P_{iy} of the initial linear momentum is zero, because the coconut is initially at rest. To get an expression for P_{fy}, we find the y component of the final linear momentum of each piece, using the y-component version of Eq. 9-22 ($p_y = mv_y$):

$$p_{fA,y} = 0,$$
$$p_{fB,y} = -0.20Mv_{fB,y} = -0.20Mv_{fB}\sin 50°,$$
$$p_{fC,y} = 0.30Mv_{fC,y} = 0.30Mv_{fC}\sin 80°.$$

(Note that $p_{fA,y} = 0$ because of our choice of axes.) Equation 9-48 can now be written as

$$P_{iy} = P_{fy} = p_{fA,y} + p_{fB,y} + p_{fC,y}.$$

Then, with $v_{fC} = 5.0$ m/s, we have

$$0 = 0 - 0.20Mv_{fB}\sin 50° + (0.30M)(5.0 \text{ m/s})\sin 80°,$$

from which we find

$$v_{fB} = 9.64 \text{ m/s} \approx 9.6 \text{ m/s}. \quad \text{(Answer)}$$

(b) What is the speed of piece A?

Calculations: Because linear momentum is also conserved along the x axis, we have

$$P_{ix} = P_{fx}, \tag{9-49}$$

where $P_{ix} = 0$ because the coconut is initially at rest. To get P_{fx}, we find the x components of the final momenta, using the fact that piece A must have a mass of $0.50M$ $(= M - 0.20M - 0.30M)$:

$$p_{fA,x} = -0.50Mv_{fA},$$
$$p_{fB,x} = 0.20Mv_{fB,x} = 0.20Mv_{fB}\cos 50°,$$
$$p_{fC,x} = 0.30Mv_{fC,x} = 0.30Mv_{fC}\cos 80°.$$

Equation 9-49 can now be written as

$$P_{ix} = P_{fx} = p_{fA,x} + p_{fB,x} + p_{fC,x}.$$

Then, with $v_{fC} = 5.0$ m/s and $v_{fB} = 9.64$ m/s, we have

$$0 = -0.50Mv_{fA} + 0.20M(9.64 \text{ m/s})\cos 50° + 0.30M(5.0 \text{ m/s})\cos 80°,$$

from which we find

$$v_{fA} = 3.0 \text{ m/s}. \quad \text{(Answer)}$$

PROBLEM-SOLVING TACTICS

Tactic 2: *Conservation of Linear Momentum* For problems involving the conservation of linear momentum, first make sure that you have chosen a closed, isolated system. *Closed* means that no matter (no particles) passes through the system boundary in any direction. *Isolated* means that the net external force acting on the system is zero. If it is not isolated, then remember that each component of linear momentum is conserved separately if the corresponding component of the net external force is zero. So, you might conserve one component and not another.

Next, select two appropriate states of the system (which you may choose to call the initial state and the final state) and write expressions for the linear momentum of the system in each of these two states. In writing these expressions, make sure that you know what inertial reference frame you are using, and make sure also that you include the entire system, not missing any part of it and not including objects that do not belong to your system.

Finally, set your expressions for $\vec{P}_i$ and $\vec{P}_f$ equal to each other and solve for what is requested.

9-8 | Momentum and Kinetic Energy in Collisions

In Section 9-6, we considered the collision of two particle-like bodies but focused on only one of the bodies at a time. For the next several sections we switch our focus to the system itself, with the assumption that the system is closed and isolated. In Section 9-7, we discussed a rule about such a system: The total linear momentum $\vec{P}$ of the system cannot change because there is no net external force to change it. This is a very powerful rule because it can allow us to determine the results of a collision *without* knowing the details of the collision (such as how much damage is done).

We shall also be interested in the total kinetic energy of a system of two colliding bodies. If that total happens to be unchanged by the collision, then the kinetic energy of the system is *conserved* (it is the same before and after the collision). Such a collision is called an **elastic collision.** In everyday collisions of common bodies, such as two cars or a ball and a bat, some energy is always transferred from kinetic energy to other forms of energy, such as thermal energy or energy of sound. Thus, the kinetic energy of the system is *not* conserved. Such a collision is called an **inelastic collision.**

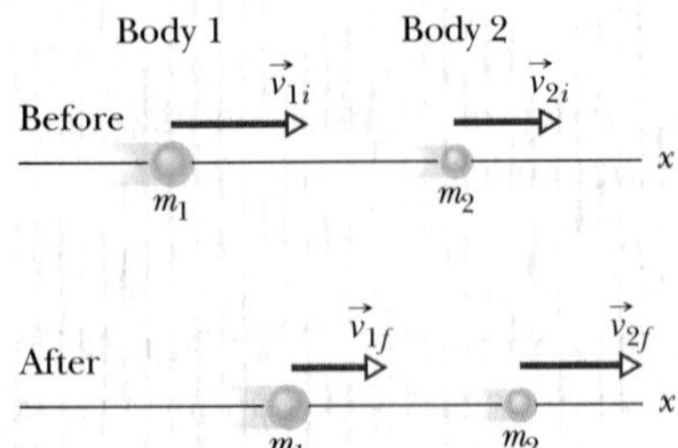

FIG. 9-15 Bodies 1 and 2 move along an x axis, before and after they have an inelastic collision.

However, in some situations, we can *approximate* a collision of common bodies as elastic. Suppose that you drop a Superball onto a hard floor. If the collision between the ball and floor (or Earth) were elastic, the ball would lose no kinetic energy because of the collision and would rebound to its original height. However, the actual rebound height is somewhat short, showing that at least some kinetic energy is lost in the collision and thus that the collision is somewhat inelastic. Still, we might choose to neglect that small loss of kinetic energy to approximate the collision as elastic.

The inelastic collision of two bodies always involves a loss in the kinetic energy of the system. The greatest loss occurs if the bodies stick together, in which case the collision is called a **completely inelastic collision.** The collision of a baseball and a bat is inelastic. However, the collision of a wet putty ball and a bat is completely inelastic because the putty sticks to the bat.

9-9 | Inelastic Collisions in One Dimension

One-Dimensional Inelastic Collision

Figure 9-15 shows two bodies just before and just after they have a one-dimensional collision. The velocities before the collision (subscript i) and after the collision (subscript f) are indicated. The two bodies form our system, which is closed and isolated. We can write the law of conservation of linear momentum for this two-body system as

$$\begin{pmatrix}\text{total momentum } \vec{P}_i \\ \text{before the collision}\end{pmatrix} = \begin{pmatrix}\text{total momentum } \vec{P}_f \\ \text{after the collision}\end{pmatrix},$$

which we can symbolize as

$$\vec{p}_{1i} + \vec{p}_{2i} = \vec{p}_{1f} + \vec{p}_{2f} \quad \text{(conservation of linear momentum).} \tag{9-50}$$

Because the motion is one-dimensional, we can drop the overhead arrows for vectors and use only components along the axis, indicating direction with a sign. Thus, from $p = mv$, we can rewrite Eq. 9-50 as

$$m_1v_{1i} + m_2v_{2i} = m_1v_{1f} + m_2v_{2f}. \tag{9-51}$$

If we know values for, say, the masses, the initial velocities, and one of the final velocities, we can find the other final velocity with Eq. 9-51.

One-Dimensional Completely Inelastic Collision

Figure 9-16 shows two bodies before and after they have a completely inelastic collision (meaning they stick together). The body with mass m_2 happens to be initially at rest ($v_{2i} = 0$). We can refer to that body as the *target* and to the incoming body as the *projectile.* After the collision, the stuck-together bodies move with velocity V. For this situation, we can rewrite Eq. 9-51 as

$$m_1v_{1i} = (m_1 + m_2)V \tag{9-52}$$

or

$$V = \frac{m_1}{m_1 + m_2}v_{1i}. \tag{9-53}$$

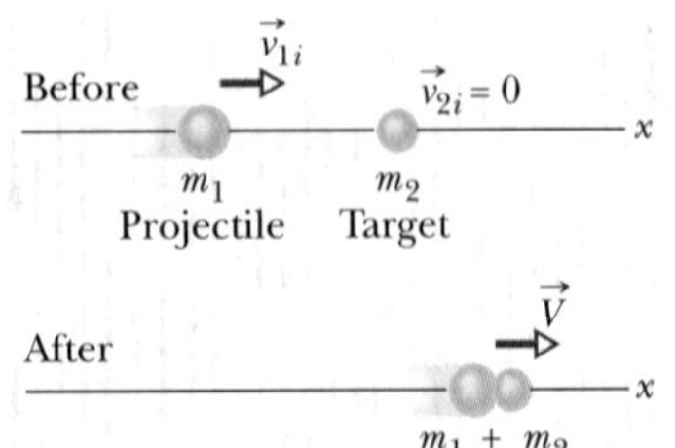

FIG. 9-16 A completely inelastic collision between two bodies. Before the collision, the body with mass m_2 is at rest and the body with mass m_1 moves directly toward it. After the collision, the stuck-together bodies move with the same velocity $\vec{V}$.

If we know values for, say, the masses and the initial velocity v_{1i} of the projectile, we can find the final velocity V with Eq. 9-53. Note that V must be less than v_{1i} because the mass ratio $m_1/(m_1 + m_2)$ must be less than unity.

Velocity of the Center of Mass

In a closed, isolated system, the velocity $\vec{v}_{com}$ of the center of mass of the system cannot be changed by a collision because, with the system isolated, there is no net external force to change it. To get an expression for $\vec{v}_{com}$, let us return to the two-body system and one-dimensional collision of Fig. 9-15. From Eq. 9-25 ($\vec{P} = M\vec{v}_{com}$), we can relate $\vec{v}_{com}$ to the total linear momentum $\vec{P}$ of that two-body system by writing

$$\vec{P} = M\vec{v}_{com} = (m_1 + m_2)\vec{v}_{com}. \tag{9-54}$$

The total linear momentum $\vec{P}$ is conserved during the collision; so it is given by either side of Eq. 9-50. Let us use the left side to write

$$\vec{P} = \vec{p}_{1i} + \vec{p}_{2i}. \tag{9-55}$$

Substituting this expression for $\vec{P}$ in Eq. 9-54 and solving for $\vec{v}_{com}$ give us

$$\vec{v}_{com} = \frac{\vec{P}}{m_1 + m_2} = \frac{\vec{p}_{1i} + \vec{p}_{2i}}{m_1 + m_2}. \tag{9-56}$$

The right side of this equation is a constant, and $\vec{v}_{com}$ has that same constant value before and after the collision.

For example, Fig. 9-17 shows, in a series of freeze-frames, the motion of the center of mass for the completely inelastic collision of Fig. 9-16. Body 2 is the target, and its initial linear momentum in Eq. 9-56 is $\vec{p}_{2i} = m_2\vec{v}_{2i} = 0$. Body 1 is the projectile, and its initial linear momentum in Eq. 9-56 is $\vec{p}_{1i} = m_1\vec{v}_{1i}$. Note that as the series of freeze-frames progresses to and then beyond the collision, the center of mass moves at a constant velocity to the right. After the collision, the common final speed V of the bodies is equal to $\vec{v}_{com}$ because then the center of mass travels with the stuck-together bodies.

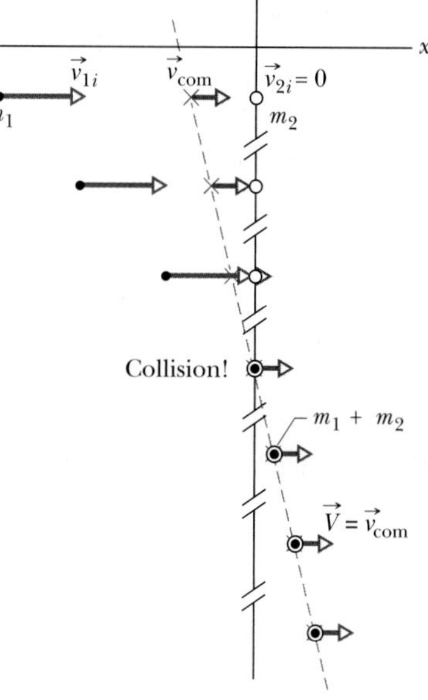

FIG. 9-17 Some freeze-frames of the two-body system in Fig. 9-16, which undergoes a completely inelastic collision. The system's center of mass is shown in each freeze-frame. The velocity $\vec{v}_{com}$ of the center of mass is unaffected by the collision. Because the bodies stick together after the collision, their common velocity $\vec{V}$ must be equal to $\vec{v}_{com}$.

CHECKPOINT 7 Body 1 and body 2 are in a completely inelastic one-dimensional collision. What is their final momentum if their initial momenta are, respectively, (a) 10 kg · m/s and 0; (b) 10 kg · m/s and 4 kg · m/s; (c) 10 kg · m/s and −4 kg · m/s?

Sample Problem 9-9 Build your skill

The *ballistic pendulum* was used to measure the speeds of bullets before electronic timing devices were developed. The version shown in Fig. 9-18 consists of a large block of wood of mass $M = 5.4$ kg, hanging from two long cords. A bullet of mass $m = 9.5$ g is fired into the block, coming quickly to rest. The *block + bullet* then swing upward, their center of mass rising a vertical distance $h = 6.3$ cm before the pendulum comes momentarily to rest at the end of its arc. What is the speed of the bullet just prior to the collision?

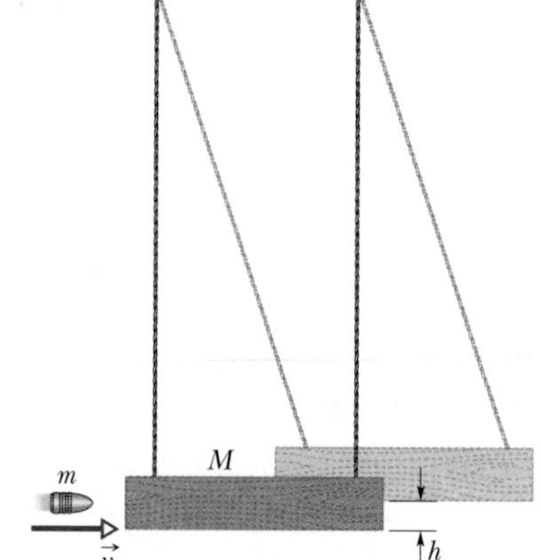

FIG. 9-18 A ballistic pendulum, used to measure the speeds of bullets.

KEY IDEAS We can see that the bullet's speed v must determine the rise height h. However, we cannot use the conservation of mechanical energy to relate these two quantities because surely energy is transferred from mechanical energy to other forms (such as thermal energy and energy to break apart the wood) as the bullet penetrates the block. Nevertheless, we can split this complicated motion into two steps that we can separately analyze: (1) the bullet–block collision and (2) the bullet–block rise, during which mechanical energy *is* conserved.

Reasoning step 1: Because the collision within the bullet–block system is so brief, we can make two important assumptions: (1) During the collision, the gravitational force on the block and the force on the block from the cords are still balanced. Thus, during the collision, the net external impulse on the bullet–block system is zero. Therefore, the system is isolated and its total linear momentum is conserved. (2) The collision is one-dimensional in the sense that the direction of the bullet and block *just after the collision* is in the bullet's original direction of motion.

Because the collision is one-dimensional, the block is initially at rest, and the bullet sticks in the block, we use Eq. 9-53 to express the conservation of linear momentum. Replacing the symbols there with the corresponding symbols here, we have

$$V = \frac{m}{m + M} v. \tag{9-57}$$

Reasoning step 2: As the bullet and block now swing up together, the mechanical energy of the bullet–block–Earth system is conserved. (This mechanical energy is not changed by the force of the cords on the block, because that force is always directed perpendicular to the block's direction of travel.) Let's take the block's initial level as our reference level of zero gravitational potential energy. Then conservation of mechanical energy means that the system's kinetic energy at the start of the swing must equal its gravitational potential energy at the highest point of the swing. Because the speed of the bullet and block at the start of the swing is the speed V immediately after the collision, we may write this conservation as

$$\tfrac{1}{2}(m + M)V^2 = (m + M)gh.$$

Combining steps: Substituting for V from Eq. 9-57 leads to

$$v = \frac{m + M}{m}\sqrt{2gh}$$
$$= \left(\frac{0.0095 \text{ kg} + 5.4 \text{ kg}}{0.0095 \text{ kg}}\right)\sqrt{(2)(9.8 \text{ m/s}^2)(0.063 \text{ m})}$$
$$= 630 \text{ m/s}. \qquad \text{(Answer)}$$

The ballistic pendulum is a kind of "transformer," exchanging the high speed of a light object (the bullet) for the low—and thus more easily measurable—speed of a massive object (the block).

Sample Problem 9-10 Build your skill

The most dangerous type of collision between two cars is a head-on collision. Surprisingly, data suggest that the risk of fatality to a driver is less if that driver has a passenger in the car. Let's see why.

Figure 9-19 represents two identical cars about to collide head-on in a completely inelastic, one-dimensional collision along an x axis. During the collision, the two cars form a closed system. Let's make the reasonable assumption that during the collision the impulse between the cars is so great that we can neglect the relatively minor impulses due to the frictional forces on the tires from the road. Then we can assume that there is no net external force on the two-car system.

The x component of the initial velocity of car 1 along the x axis is $v_{1i} = +25$ m/s, and that of car 2 is $v_{2i} = -25$ m/s. During the collision, the force (and thus the impulse) on each car causes a change Δv in the car's velocity. The probability of a driver being killed depends on the magnitude of Δv for that driver's car. We want to calculate the changes Δv_1 and Δv_2 in the velocities of the two cars.

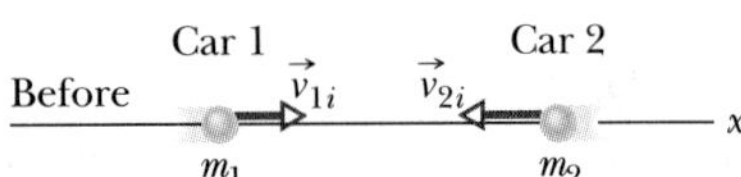

FIG. 9-19 Two cars about to collide head-on.

(a) First, let each car carry only a driver. The total mass of car 1 (including driver 1) is $m_1 = 1400$ kg, and the total mass of car 2 (including driver 2) is $m_2 = 1400$ kg. What are the changes Δv_1 and Δv_2 in the velocities of the cars?

KEY IDEA Because the system is closed and isolated, its total linear momentum is conserved.

Calculations: From Eq. 9-51, we can write this as

$$m_1 v_{1i} + m_2 v_{2i} = m_1 v_{1f} + m_2 v_{2f}. \tag{9-58}$$

Since the collision is completely inelastic, the two cars stick together and thus have the same velocity V after the collision. Substituting V for v_{1f} and v_{2f} into Eq. 9-58 and solving for V, we have

$$V = \frac{m_1 v_{1i} + m_2 v_{2i}}{m_1 + m_2}. \tag{9-59}$$

Substitution of the given data then results in

$$V = \frac{(1400 \text{ kg})(+25 \text{ m/s}) + (1400 \text{ kg})(-25 \text{ m/s})}{1400 \text{ kg} + 1400 \text{ kg}} = 0.$$

Thus, the change in the velocity of car 1 is

$$\Delta v_1 = v_{1f} - v_{1i} = V - v_{1i}$$
$$= 0 - (+25 \text{ m/s}) = -25 \text{ m/s}, \qquad \text{(Answer)}$$

and the change in the velocity of car 2 is

$$\Delta v_2 = v_{2f} - v_{2i} = V - v_{2i}$$
$$= 0 - (-25 \text{ m/s}) = +25 \text{ m/s}. \qquad \text{(Answer)}$$

(b) Next, we reconsider the collision, but this time with an 80 kg passenger in car 1. What are Δv_1 and Δv_2 now?

Calculations: Repeating our steps but now substituting $m_1 = 1480$ kg, we find that

$$V = 0.694 \text{ m/s},$$

which gives

$$\Delta v_1 = -24.3 \text{ m/s}$$

and

$$\Delta v_2 = +25.7 \text{ m/s}. \qquad \text{(Answer)}$$

(c) The magnitude of Δv_1 is less with the passenger in the car. Because the probability of a driver being killed depends on Δv_1, we can reason that the probability is less for driver 1.

The data on head-on car collisions do not include values of Δv, but they do include the car masses and whether or not a collision was fatal. Fitting a function to the collected data, researchers have found that the fatality risk r_1 of driver 1 is given by

$$r_1 = c\left(\frac{m_2}{m_1}\right)^{1.79}, \qquad (9\text{-}60)$$

where c is a constant. Justify why the ratio m_2/m_1 appears in this equation, and then use the equation to compare the fatality risks for driver 1 with and without the passenger.

Calculations: We first rewrite Eq. 9-58 as

$$m_1(v_{1f} - v_{1i}) = -m_2(v_{2f} - v_{2i}).$$

Substituting $\Delta v_1 = v_{1f} - v_{1i}$ and $\Delta v_2 = v_{2f} - v_{2i}$ and rearranging give us

$$\frac{m_2}{m_1} = -\frac{\Delta v_1}{\Delta v_2}. \qquad (9\text{-}61)$$

A driver's fatality risk depends on the change Δv for that driver. In Eq. 9-61, we see that the ratio of Δv values in a collision is the inverse of the ratio of the masses, and this is the reason researchers can link fatality risk to the ratio of masses in Eq. 9-60.

From part (a) and Eq. 9-60, *without* the passenger, driver 1 has a fatality risk of

$$r_1 = c\left(\frac{1400\text{ kg}}{1400\text{ kg}}\right)^{1.79} = c. \qquad (9\text{-}62)$$

From part (b) and Eq. 9-60, *with* the passenger, driver 1 has a fatality risk of

$$r_1' = c\left(\frac{1400\text{ kg}}{1400\text{ kg} + 80\text{ kg}}\right)^{1.79} = 0.9053c.$$

Substituting for c from Eq. 9-62, we find

$$r_1' = 0.9053r_1 \approx 0.91r_1. \qquad \text{(Answer)}$$

In words, the fatality risk for driver 1 is about 9% less when a passenger is in the car.

9-10 | Elastic Collisions in One Dimension

As we discussed in Section 9-8, everyday collisions are inelastic but we can approximate some of them as being elastic; that is, we can approximate that the total kinetic energy of the colliding bodies is conserved and is not transferred to other forms of energy:

$$\begin{pmatrix}\text{total kinetic energy}\\ \text{before the collision}\end{pmatrix} = \begin{pmatrix}\text{total kinetic energy}\\ \text{after the collision}\end{pmatrix}.$$

This does not mean that the kinetic energy of each colliding body cannot change. Rather, it means this:

> In an elastic collision, the kinetic energy of each colliding body may change, but the total kinetic energy of the system does not change.

For example, the collision of a cue ball with an object ball in a game of pool can be approximated as being an elastic collision. If the collision is head-on (the cue ball heads directly toward the object ball), the kinetic energy of the cue ball can be transferred almost entirely to the object ball. (Still, the fact that the collision makes a sound means that at least a little of the kinetic energy is transferred to the energy of the sound.)

Stationary Target

Figure 9-20 shows two bodies before and after they have a one-dimensional collision, like a head-on collision between pool balls. A projectile body of mass m_1 and initial velocity v_{1i} moves toward a target body of mass m_2 that is initially at rest ($v_{2i} = 0$). Let's assume that this two-body system is closed and isolated. Then the net linear momentum of the system is conserved, and from Eq. 9-51 we can write that conservation as

$$m_1v_{1i} = m_1v_{1f} + m_2v_{2f} \qquad \text{(linear momentum)}. \qquad (9\text{-}63)$$

If the collision is also elastic, then the total kinetic energy is conserved and we can write that conservation as

$$\tfrac{1}{2}m_1v_{1i}^2 = \tfrac{1}{2}m_1v_{1f}^2 + \tfrac{1}{2}m_2v_{2f}^2 \qquad \text{(kinetic energy)}. \qquad (9\text{-}64)$$

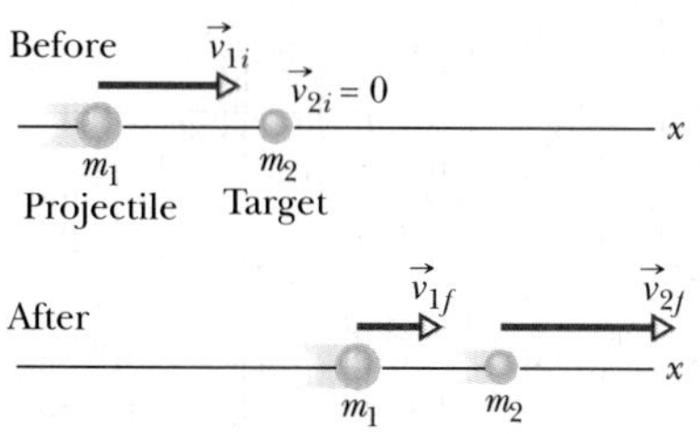

FIG. 9-20 Body 1 moves along an x axis before having an elastic collision with body 2, which is initially at rest. Both bodies move along that axis after the collision.

In each of these equations, the subscript i identifies the initial velocities and the subscript f the final velocities of the bodies. If we know the masses of the bodies and if we also know v_{1i}, the initial velocity of body 1, the only unknown quantities are v_{1f} and v_{2f}, the final velocities of the two bodies. With two equations at our disposal, we should be able to find these two unknowns.

To do so, we rewrite Eq. 9-63 as

$$m_1(v_{1i} - v_{1f}) = m_2 v_{2f} \tag{9-65}$$

and Eq. 9-64 as*

$$m_1(v_{1i} - v_{1f})(v_{1i} + v_{1f}) = m_2 v_{2f}^2. \tag{9-66}$$

After dividing Eq. 9-66 by Eq. 9-65 and doing some more algebra, we obtain

$$v_{1f} = \frac{m_1 - m_2}{m_1 + m_2} v_{1i} \tag{9-67}$$

and

$$v_{2f} = \frac{2m_1}{m_1 + m_2} v_{1i}. \tag{9-68}$$

We note from Eq. 9-68 that v_{2f} is always positive (the initially stationary target body with mass m_2 always moves forward). From Eq. 9-67 we see that v_{1f} may be of either sign (the projectile body with mass m_1 moves forward if $m_1 > m_2$ but rebounds if $m_1 < m_2$).

Let us look at a few special situations.

1. ***Equal masses*** If $m_1 = m_2$, Eqs. 9-67 and 9-68 reduce to

$$v_{1f} = 0 \quad \text{and} \quad v_{2f} = v_{1i},$$

which we might call a pool player's result. It predicts that after a head-on collision of bodies with equal masses, body 1 (initially moving) stops dead in its tracks and body 2 (initially at rest) takes off with the initial speed of body 1. In head-on collisions, bodies of equal mass simply exchange velocities. This is true even if body 2 is not initially at rest.

2. ***A massive target*** In Fig. 9-20, a massive target means that $m_2 \gg m_1$. For example, we might fire a golf ball at a stationary cannonball. Equations 9-67 and 9-68 then reduce to

$$v_{1f} \approx -v_{1i} \quad \text{and} \quad v_{2f} \approx \left(\frac{2m_1}{m_2}\right) v_{1i}. \tag{9-69}$$

This tells us that body 1 (the golf ball) simply bounces back along its incoming path, its speed essentially unchanged. Initially stationary body 2 (the cannonball) moves forward at a low speed, because the quantity in parentheses in Eq. 9-69 is much less than unity. All this is what we should expect.

3. ***A massive projectile*** This is the opposite case; that is, $m_1 \gg m_2$. This time, we fire a cannonball at a stationary golf ball. Equations 9-67 and 9-68 reduce to

$$v_{1f} \approx v_{1i} \quad \text{and} \quad v_{2f} \approx 2v_{1i}. \tag{9-70}$$

Equation 9-70 tells us that body 1 (the cannonball) simply keeps on going, scarcely slowed by the collision. Body 2 (the golf ball) charges ahead at twice the speed of the cannonball.

You may wonder: Why twice the speed? As a starting point in thinking about the matter, recall the collision described by Eq. 9-69, in which the velocity of the incident light body (the golf ball) changed from $+v$ to $-v$, a velocity *change* of $2v$. The same *change* in velocity (but now from zero to $2v$) occurs in this example also.

*In this step, we use the identity $a^2 - b^2 = (a - b)(a + b)$. It reduces the amount of algebra needed to solve the simultaneous equations Eqs. 9-65 and 9-66.

Moving Target

Now that we have examined the elastic collision of a projectile and a stationary target, let us examine the situation in which both bodies are moving before they undergo an elastic collision.

For the situation of Fig. 9-21, the conservation of linear momentum is written as

$$m_1 v_{1i} + m_2 v_{2i} = m_1 v_{1f} + m_2 v_{2f}, \tag{9-71}$$

and the conservation of kinetic energy is written as

$$\tfrac{1}{2} m_1 v_{1i}^2 + \tfrac{1}{2} m_2 v_{2i}^2 = \tfrac{1}{2} m_1 v_{1f}^2 + \tfrac{1}{2} m_2 v_{2f}^2. \tag{9-72}$$

To solve these simultaneous equations for v_{1f} and v_{2f}, we first rewrite Eq. 9-71 as

$$m_1(v_{1i} - v_{1f}) = -m_2(v_{2i} - v_{2f}), \tag{9-73}$$

and Eq. 9-72 as

$$m_1(v_{1i} - v_{1f})(v_{1i} + v_{1f}) = -m_2(v_{2i} - v_{2f})(v_{2i} + v_{2f}). \tag{9-74}$$

After dividing Eq. 9-74 by Eq. 9-73 and doing some more algebra, we obtain

$$v_{1f} = \frac{m_1 - m_2}{m_1 + m_2} v_{1i} + \frac{2m_2}{m_1 + m_2} v_{2i} \tag{9-75}$$

and

$$v_{2f} = \frac{2m_1}{m_1 + m_2} v_{1i} + \frac{m_2 - m_1}{m_1 + m_2} v_{2i}. \tag{9-76}$$

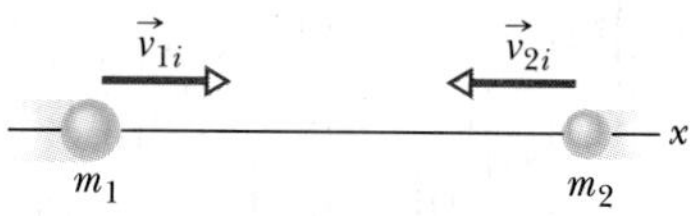

FIG. 9-21 Two bodies headed for a one-dimensional elastic collision.

Note that the assignment of subscripts 1 and 2 to the bodies is arbitrary. If we exchange those subscripts in Fig. 9-21 and in Eqs. 9-75 and 9-76, we end up with the same set of equations. Note also that if we set $v_{2i} = 0$, body 2 becomes a stationary target as in Fig. 9-20, and Eqs. 9-75 and 9-76 reduce to Eqs. 9-67 and 9-68, respectively.

CHECKPOINT 8 What is the final linear momentum of the target in Fig. 9-20 if the initial linear momentum of the projectile is 6 kg·m/s and the final linear momentum of the projectile is (a) 2 kg·m/s and (b) −2 kg·m/s? (c) What is the final kinetic energy of the target if the initial and final kinetic energies of the projectile are, respectively, 5 J and 2 J?

Sample Problem 9-11

Two metal spheres, suspended by vertical cords, initially just touch, as shown in Fig. 9-22. Sphere 1, with mass $m_1 = 30$ g, is pulled to the left to height $h_1 = 8.0$ cm, and then released from rest. After swinging down, it undergoes an elastic collision with sphere 2, whose mass $m_2 = 75$ g. What is the velocity v_{1f} of sphere 1 just after the collision?

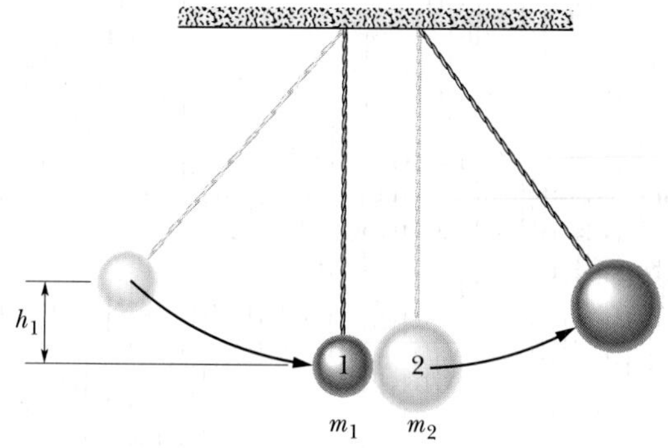

FIG. 9-22 Two metal spheres suspended by cords just touch when they are at rest. Sphere 1, with mass m_1, is pulled to the left to height h_1 and then released.

KEY IDEA We can split this complicated motion into two steps that we can analyze separately: (1) the descent of sphere 1 (in which mechanical energy is conserved) and (2) the two-sphere collision (in which momentum is conserved).

Step 1: As sphere 1 swings down, the mechanical energy of the sphere–Earth system is conserved. (The mechanical energy is not changed by the force of the cord on sphere 1 because that force is always directed perpendicular to the sphere's direction of travel.)

Calculation: Let's take the lowest level as our reference level of zero gravitational potential energy. Then the kinetic energy of sphere 1 at the lowest level must equal the gravitational potential energy of the system when sphere 1 is at height h_1. Thus,

$$\tfrac{1}{2}m_1 v_{1i}^2 = m_1 g h_1,$$

which we solve for the speed v_{1i} of sphere 1 just before the collision:

$$v_{1i} = \sqrt{2gh_1} = \sqrt{(2)(9.8\text{ m/s}^2)(0.080\text{ m})} = 1.252\text{ m/s}.$$

Step 2: Here we can make two assumptions in addition to the assumption that the collision is elastic. First, we can assume that the collision is one-dimensional because the motions of the spheres are approximately horizontal from just before the collision to just after it. Second, because the collision is so brief, we can assume that the two-sphere system is closed and isolated. This means that the total linear momentum of the system is conserved.

Calculation: Thus, we can use Eq. 9-67 to find the velocity of sphere 1 just after the collision:

$$v_{1f} = \frac{m_1 - m_2}{m_1 + m_2} v_{1i} = \frac{0.030\text{ kg} - 0.075\text{ kg}}{0.030\text{ kg} + 0.075\text{ kg}}(1.252\text{ m/s}) = -0.537\text{ m/s} \approx -0.54\text{ m/s}. \quad \text{(Answer)}$$

The minus sign tells us that sphere 1 moves to the left just after the collision.

9-11 | Collisions in Two Dimensions

When two bodies collide, the impulse between them determines the directions in which they then travel. In particular, when the collision is not head-on, the bodies do not end up traveling along their initial axis. For such two-dimensional collisions in a closed, isolated system, the total linear momentum must still be conserved:

$$\vec{P}_{1i} + \vec{P}_{2i} = \vec{P}_{1f} + \vec{P}_{2f}. \quad (9\text{-}77)$$

If the collision is also elastic (a special case), then the total kinetic energy is also conserved:

$$K_{1i} + K_{2i} = K_{1f} + K_{2f}. \quad (9\text{-}78)$$

Equation 9-77 is often more useful for analyzing a two-dimensional collision if we write it in terms of components on an xy coordinate system. For example, Fig. 9-23 shows a *glancing collision* (it is not head-on) between a projectile body and a target body initially at rest. The impulses between the bodies have sent the bodies off at angles θ_1 and θ_2 to the x axis, along which the projectile initially traveled. In this situation we would rewrite Eq. 9-77 for components along the x axis as

$$m_1 v_{1i} = m_1 v_{1f} \cos\theta_1 + m_2 v_{2f} \cos\theta_2, \quad (9\text{-}79)$$

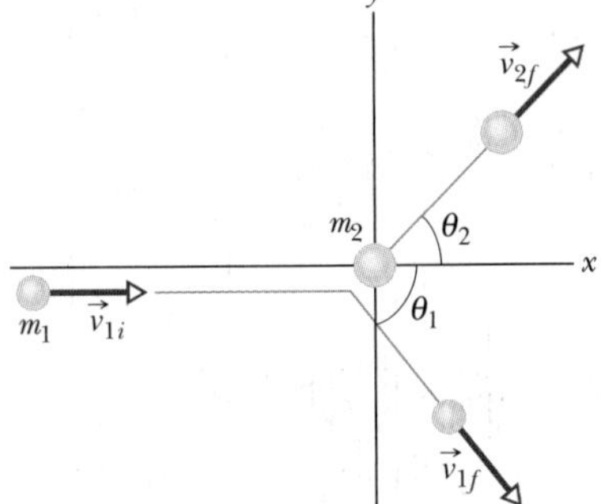

FIG. 9-23 An elastic collision between two bodies in which the collision is not head-on. The body with mass m_2 (the target) is initially at rest.

and along the y axis as

$$0 = -m_1 v_{1f} \sin\theta_1 + m_2 v_{2f} \sin\theta_2. \quad (9\text{-}80)$$

We can also write Eq. 9-78 (for the special case of an elastic collision) in terms of speeds:

$$\tfrac{1}{2}m_1 v_{1i}^2 = \tfrac{1}{2}m_1 v_{1f}^2 + \tfrac{1}{2}m_2 v_{2f}^2 \quad \text{(kinetic energy)}. \quad (9\text{-}81)$$

Equations 9-79 to 9-81 contain seven variables: two masses, m_1 and m_2; three speeds, v_{1i}, v_{1f}, and v_{2f}; and two angles, θ_1 and θ_2. If we know any four of these quantities, we can solve the three equations for the remaining three quantities.

CHECKPOINT 9 In Fig. 9-23, suppose that the projectile has an initial momentum of 6 kg · m/s, a final x component of momentum of 4 kg · m/s, and a final y component of momentum of −3 kg · m/s. For the target, what then are (a) the final x component of momentum and (b) the final y component of momentum?

9-12 | Systems with Varying Mass: A Rocket

In the systems we have dealt with so far, we have assumed that the total mass of the system remains constant. Sometimes, as in a rocket, it does not. Most of the mass of a rocket on its launching pad is fuel, all of which will eventually be burned and ejected from the nozzle of the rocket engine.

We handle the variation of the mass of the rocket as the rocket accelerates by applying Newton's second law, not to the rocket alone but to the rocket and its ejected combustion products taken together. The mass of *this* system does *not* change as the rocket accelerates.

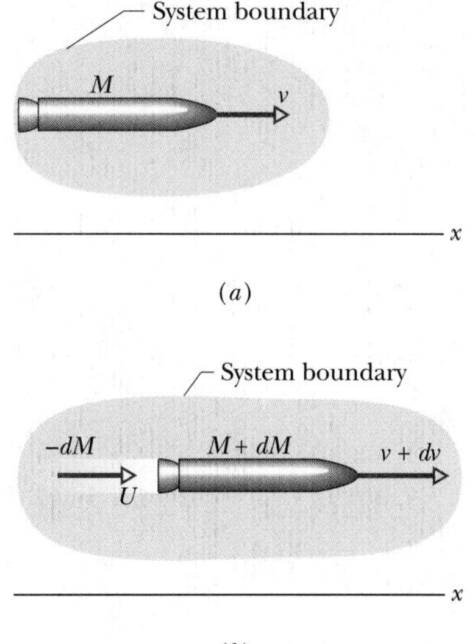

FIG. 9-24 (*a*) An accelerating rocket of mass M at time t, as seen from an inertial reference frame. (*b*) The same but at time $t + dt$. The exhaust products released during interval dt are shown.

Finding the Acceleration

Assume that we are at rest relative to an inertial reference frame, watching a rocket accelerate through deep space with no gravitational or atmospheric drag forces acting on it. For this one-dimensional motion, let M be the mass of the rocket and v its velocity at an arbitrary time t (see Fig. 9-24*a*).

Figure 9-24*b* shows how things stand a time interval dt later. The rocket now has velocity $v + dv$ and mass $M + dM$, where the change in mass dM is a *negative quantity.* The exhaust products released by the rocket during interval dt have mass $-dM$ and velocity U relative to our inertial reference frame.

Our system consists of the rocket and the exhaust products released during interval dt. The system is closed and isolated, so the linear momentum of the system must be conserved during dt; that is,

$$P_i = P_f, \qquad (9\text{-}82)$$

where the subscripts i and f indicate the values at the beginning and end of time interval dt. We can rewrite Eq. 9-82 as

$$Mv = -dM\,U + (M + dM)(v + dv), \qquad (9\text{-}83)$$

where the first term on the right is the linear momentum of the exhaust products released during interval dt and the second term is the linear momentum of the rocket at the end of interval dt.

We can simplify Eq. 9-83 by using the relative speed v_{rel} between the rocket and the exhaust products, which is related to the velocities relative to the frame with

$$\begin{pmatrix}\text{velocity of rocket}\\ \text{relative to frame}\end{pmatrix} = \begin{pmatrix}\text{velocity of rocket}\\ \text{relative to products}\end{pmatrix} + \begin{pmatrix}\text{velocity of products}\\ \text{relative to frame}\end{pmatrix}.$$

In symbols, this means

$$(v + dv) = v_{rel} + U,$$

or

$$U = v + dv - v_{rel}. \qquad (9\text{-}84)$$

Substituting this result for U into Eq. 9-83 yields, with a little algebra,

$$-dM\,v_{rel} = M\,dv. \qquad (9\text{-}85)$$

Dividing each side by dt gives us

$$-\frac{dM}{dt}v_{rel} = M\frac{dv}{dt}. \qquad (9\text{-}86)$$

We replace dM/dt (the rate at which the rocket loses mass) by $-R$, where R is the (positive) mass rate of fuel consumption, and we recognize that dv/dt is the acceleration of the rocket. With these changes, Eq. 9-86 becomes

$$Rv_{rel} = Ma \quad \text{(first rocket equation).} \qquad (9\text{-}87)$$

Equation 9-87 holds at any instant, with the mass M, the fuel consumption rate R, and the acceleration a evaluated at that instant.

Note the left side of Eq. 9-87 has the dimensions of force ($\text{kg/s} \cdot \text{m/s} = \text{kg} \cdot \text{m/s}^2 = \text{N}$) and depends only on design characteristics of the rocket engine—namely, the rate R at which it consumes fuel mass and the speed v_{rel} with which that mass is ejected relative to the rocket. We call this term Rv_{rel} the **thrust** of the rocket engine and represent it with T. Newton's second law emerges clearly if we write Eq. 9-87 as $T = Ma$, in which a is the acceleration of the rocket at the time that its mass is M.

Finding the Velocity

How will the velocity of a rocket change as it consumes its fuel? From Eq. 9-85 we have

$$dv = -v_{rel}\frac{dM}{M}.$$

Integrating leads to

$$\int_{v_i}^{v_f} dv = -v_{rel}\int_{M_i}^{M_f}\frac{dM}{M},$$

in which M_i is the initial mass of the rocket and M_f its final mass. Evaluating the integrals then gives

$$v_f - v_i = v_{rel}\ln\frac{M_i}{M_f} \quad \text{(second rocket equation)} \qquad (9\text{-}88)$$

for the increase in the speed of the rocket during the change in mass from M_i to M_f. (The symbol "ln" in Eq. 9-88 means the *natural logarithm.*) We see here the advantage of multistage rockets, in which M_f is reduced by discarding successive stages when their fuel is depleted. An ideal rocket would reach its destination with only its payload remaining.

Sample Problem 9-12

A rocket whose initial mass M_i is 850 kg consumes fuel at the rate $R = 2.3$ kg/s. The speed v_{rel} of the exhaust gases relative to the rocket engine is 2800 m/s. What thrust does the rocket engine provide?

KEY IDEA Thrust T is equal to the product of the fuel consumption rate R and the relative speed v_{rel} at which exhaust gases are expelled, as given by Eq. 9-87.

Calculation: Here we find

$$T = Rv_{rel} = (2.3\ \text{kg/s})(2800\ \text{m/s}) = 6440\ \text{N} \approx 6400\ \text{N}. \quad \text{(Answer)}$$

(b) What is the initial acceleration of the rocket?

KEY IDEA We can relate the thrust T of a rocket to the magnitude a of the resulting acceleration with $T = Ma$, where M is the rocket's mass. However, M decreases and a increases as fuel is consumed. Because we want the initial value of a here, we must use the intial value M_i of the mass.

Calculation: We find

$$a = \frac{T}{M_i} = \frac{6440\ \text{N}}{850\ \text{kg}} = 7.6\ \text{m/s}^2. \quad \text{(Answer)}$$

To be launched from Earth's surface, a rocket must have an initial acceleration greater than $g = 9.8$ m/s^2. Put another way, the thrust T of the rocket engine must exceed the initial gravitational force on the rocket, which here has the magnitude $M_i g$, which gives us (850 kg)(9.8 m/s^2), or 8330 N. Because the acceleration or thrust requirement is not met (here $T = 6400$ N), our rocket could not be launched from Earth's surface by itself; it would require another, more powerful, rocket.

REVIEW & SUMMARY

Center of Mass The **center of mass** of a system of n particles is defined to be the point whose coordinates are given by

$$x_{com} = \frac{1}{M}\sum_{i=1}^{n} m_i x_i, \quad y_{com} = \frac{1}{M}\sum_{i=1}^{n} m_i y_i, \quad z_{com} = \frac{1}{M}\sum_{i=1}^{n} m_i z_i, \qquad (9\text{-}5)$$

or

$$\vec{r}_{com} = \frac{1}{M}\sum_{i=1}^{n} m_i \vec{r}_i, \qquad (9\text{-}8)$$

where M is the total mass of the system.

Newton's Second Law for a System of Particles The motion of the center of mass of any system of particles is governed by **Newton's second law for a system of particles,** which is

$$\vec{F}_{net} = M\vec{a}_{com}. \qquad (9\text{-}14)$$

Here $\vec{F}_{net}$ is the net force of all the *external* forces acting on the system, M is the total mass of the system, and $\vec{a}_{com}$ is the acceleration of the system's center of mass.

Linear Momentum and Newton's Second Law For a single particle, we define a quantity $\vec{p}$ called its **linear momentum** as

$$\vec{p} = m\vec{v}, \tag{9-22}$$

and can write Newton's second law in terms of this momentum:

$$\vec{F}_{net} = \frac{d\vec{p}}{dt}. \tag{9-23}$$

For a system of particles these relations become

$$\vec{P} = M\vec{v}_{com} \quad \text{and} \quad \vec{F}_{net} = \frac{d\vec{P}}{dt}. \tag{9-25, 9-27}$$

Collision and Impulse Applying Newton's second law in momentum form to a particle-like body involved in a collision leads to the **impulse–linear momentum theorem:**

$$\vec{p}_f - \vec{p}_i = \Delta\vec{p} = \vec{J}, \tag{9-31, 9-32}$$

where $\vec{p}_f - \vec{p}_i = \Delta\vec{p}$ is the change in the body's linear momentum, and $\vec{J}$ is the **impulse** due to the force $\vec{F}(t)$ exerted on the body by the other body in the collision:

$$\vec{J} = \int_{t_i}^{t_f} \vec{F}(t)\,dt. \tag{9-30}$$

If F_{avg} is the average magnitude of $\vec{F}(t)$ during the collision and Δt is the duration of the collision, then for one-dimensional motion

$$J = F_{avg}\,\Delta t. \tag{9-35}$$

When a steady stream of bodies, each with mass m and speed v, collides with a body whose position is fixed, the average force on the fixed body is

$$F_{avg} = -\frac{n}{\Delta t}\Delta p = -\frac{n}{\Delta t} m\,\Delta v, \tag{9-37}$$

where $n/\Delta t$ is the rate at which the bodies collide with the fixed body, and Δv is the change in velocity of each colliding body. This average force can also be written as

$$F_{avg} = -\frac{\Delta m}{\Delta t}\Delta v, \tag{9-40}$$

where $\Delta m/\Delta t$ is the rate at which mass collides with the fixed body. In Eqs. 9-37 and 9-40, $\Delta v = -v$ if the bodies stop upon impact and $\Delta v = -2v$ if they bounce directly backward with no change in their speed.

Conservation of Linear Momentum If a system is isolated so that no net *external* force acts on it, the linear momentum $\vec{P}$ of the system remains constant:

$$\vec{P} = \text{constant} \quad \text{(closed, isolated system)}. \tag{9-42}$$

This can also be written as

$$\vec{P}_i = \vec{P}_f \quad \text{(closed, isolated system)}, \tag{9-43}$$

where the subscripts refer to the values of $\vec{P}$ at some initial time and at a later time. Equations 9-42 and 9-43 are equivalent statements of the **law of conservation of linear momentum.**

Inelastic Collision in One Dimension In an *inelastic collision* of two bodies, the kinetic energy of the two-body system is not conserved. If the system is closed and isolated, the total linear momentum of the system *must* be conserved, which we can write in vector form as

$$\vec{p}_{1i} + \vec{p}_{2i} = \vec{p}_{1f} + \vec{p}_{2f}, \tag{9-50}$$

where subscripts i and f refer to values just before and just after the collision, respectively.

If the motion of the bodies is along a single axis, the collision is one-dimensional and we can write Eq. 9-50 in terms of velocity components along that axis:

$$m_1v_{1i} + m_2v_{2i} = m_1v_{1f} + m_2v_{2f}. \tag{9-51}$$

If the bodies stick together, the collision is a *completely inelastic collision* and the bodies have the same final velocity V (because they *are* stuck together).

Motion of the Center of Mass The center of mass of a closed, isolated system of two colliding bodies is not affected by a collision. In particular, the velocity $\vec{v}_{com}$ of the center of mass cannot be changed by the collision.

Elastic Collisions in One Dimension An *elastic collision* is a special type of collision in which the kinetic energy of a system of colliding bodies is conserved. If the system is closed and isolated, its linear momentum is also conserved. For a one-dimensional collision in which body 2 is a target and body 1 is an incoming projectile, conservation of kinetic energy and linear momentum yield the following expressions for the velocities immediately after the collision:

$$v_{1f} = \frac{m_1 - m_2}{m_1 + m_2}v_{1i} \tag{9-67}$$

and

$$v_{2f} = \frac{2m_1}{m_1 + m_2}v_{1i}. \tag{9-68}$$

Collisions in Two Dimensions If two bodies collide and their motion is not along a single axis (the collision is not head-on), the collision is two-dimensional. If the two-body system is closed and isolated, the law of conservation of momentum applies to the collision and can be written as

$$\vec{P}_{1i} + \vec{P}_{2i} = \vec{P}_{1f} + \vec{P}_{2f}. \tag{9-77}$$

In component form, the law gives two equations that describe the collision (one equation for each of the two dimensions). If the collision is also elastic (a special case), the conservation of kinetic energy during the collision gives a third equation:

$$K_{1i} + K_{2i} = K_{1f} + K_{2f}. \tag{9-78}$$

Variable-Mass Systems In the absence of external forces a rocket accelerates at an instantaneous rate given by

$$Rv_{rel} = Ma \quad \text{(first rocket equation)}, \tag{9-87}$$

in which M is the rocket's instantaneous mass (including unexpended fuel), R is the fuel consumption rate, and v_{rel} is the fuel's exhaust speed relative to the rocket. The term Rv_{rel} is the **thrust** of the rocket engine. For a rocket with constant R and v_{rel}, whose speed changes from v_i to v_f when its mass changes from M_i to M_f,

$$v_f - v_i = v_{rel}\ln\frac{M_i}{M_f} \quad \text{(second rocket equation)}. \tag{9-88}$$

QUESTIONS

1 Figure 9-25 shows an overhead view of three particles on which external forces act. The magnitudes and directions of the forces on two of the particles are indicated. What are the magnitude and direction of the force acting on the third particle if the center of mass of the three-particle system is (a) stationary, (b) moving at a constant velocity rightward, and (c) accelerating rightward?

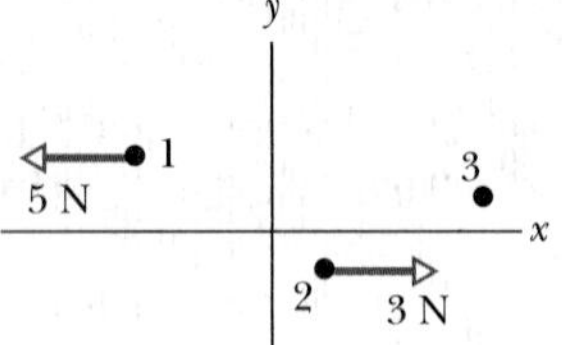

FIG. 9-25 Question 1.

2 Figure 9-26 shows an overhead view of four particles of equal mass sliding over a frictionless surface at constant velocity. The directions of the velocities are indicated; their magnitudes are equal. Consider pairing the particles. Which pairs form a system with a center of mass that (a) is stationary, (b) is stationary and at the origin, and (c) passes through the origin?

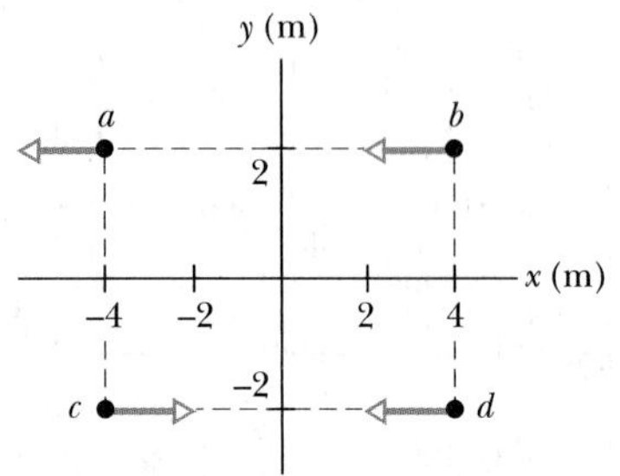

FIG. 9-26 Question 2.

3 The free-body diagrams in Fig. 9-27 give, from overhead views, the horizontal forces acting on three boxes of chocolates as the boxes move over a frictionless confectioner's counter. For each box, is its linear momentum conserved along the x axis and the y axis?

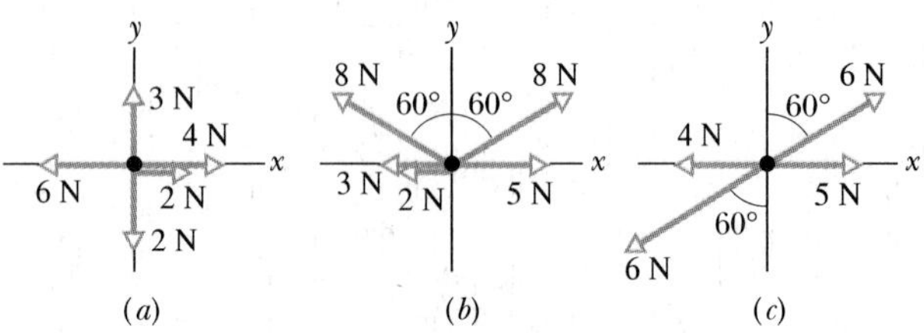

FIG. 9-27 Question 3.

4 Figure 9-28 shows four groups of three or four identical particles that move parallel to either the x axis or the y axis, at identical speeds. Rank the groups according to center-of-mass speed, greatest first.

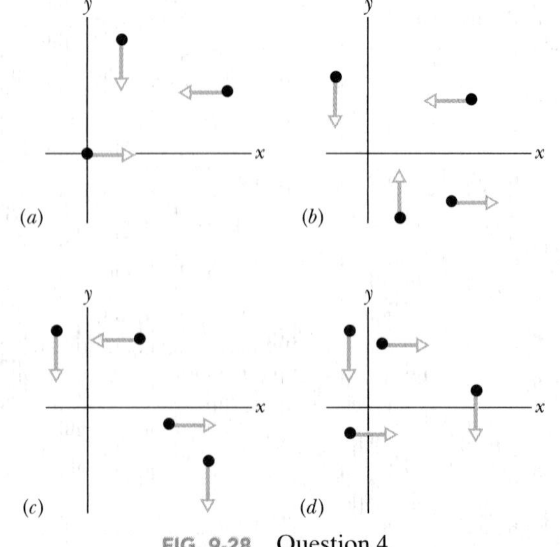

FIG. 9-28 Question 4.

5 Consider a box, like that in Sample Problem 9-6, which explodes into two pieces while moving with a constant positive velocity along an x axis. If one piece, with mass m_1, ends up with positive velocity $\vec{v}_1$, then the second piece, with mass m_2, could end up with (a) a positive velocity $\vec{v}_2$ (Fig. 9-29a), (b) a negative velocity $\vec{v}_2$ (Fig. 9-29b), or (c) zero velocity (Fig. 9-29c). Rank those three possible results for the second piece according to the corresponding magnitude of $\vec{v}_1$, greatest first.

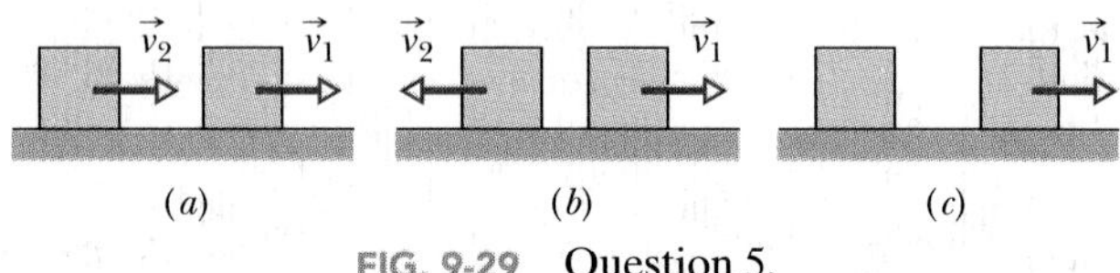

FIG. 9-29 Question 5.

6 Figure 9-30 shows graphs of force magnitude versus time for a body involved in a collision. Rank the graphs according to the magnitude of the impulse on the body, greatest first.

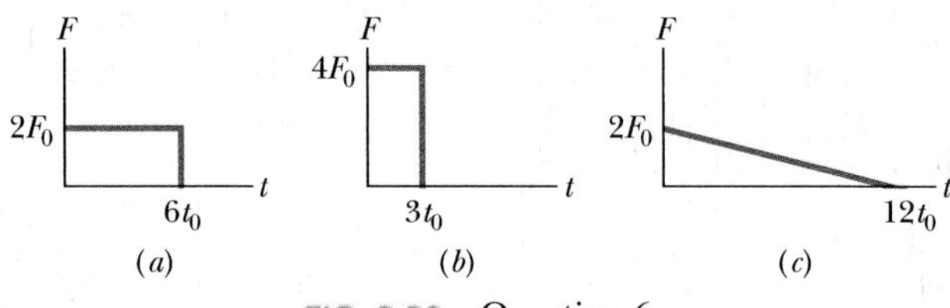

FIG. 9-30 Question 6.

7 Two bodies have undergone an elastic one-dimensional collision along an x axis. Figure 9-31 is a graph of position versus time for those bodies and for their center of mass. (a) Were both bodies initially moving, or was one initially stationary? Which line segment corresponds to the motion of the center of mass (b) before the collision and (c) after the collision? (d) Is the mass of the body that was moving faster before the collision greater than, less than, or equal to that of the other body?

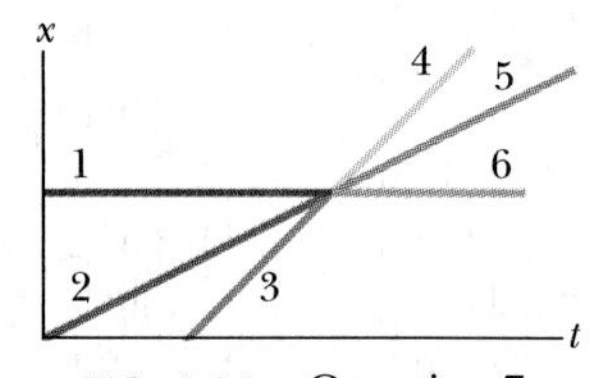

FIG. 9-31 Question 7.

8 Figure 9-32: A block on a horizontal floor is initially either stationary, sliding in the positive direction of an x axis, or sliding in the negative direction of that axis. Then the block ex-

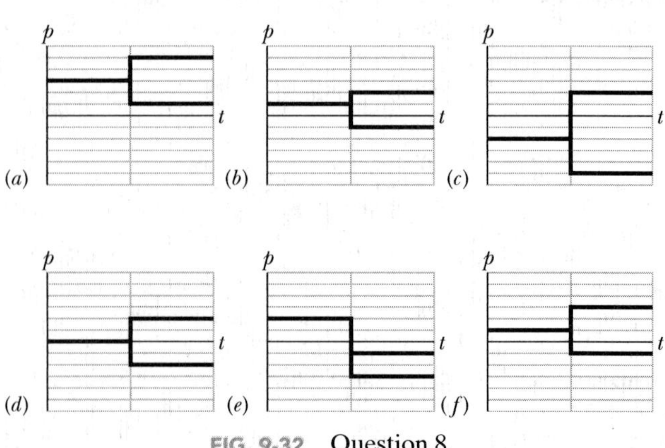

FIG. 9-32 Question 8.

plodes into two pieces that slide along the x axis. Assume the block and the two pieces form a closed, isolated system. Six choices for a graph of the momenta of the block and the pieces are given, all versus time t. Determine which choices represent physically impossible situations and explain why.

9 Block 1 with mass m_1 slides along an x axis across a frictionless floor and then undergoes an elastic collision with a stationary block 2 with mass m_2. Figure 9-33 shows a plot of position x versus time t of block 1 until the collision occurs at position x_c and time t_c. In which of the lettered regions on the graph will the plot be continued (after the collision) if (a) $m_1 < m_2$ and (b) $m_1 > m_2$? (c) Along which of the numbered dashed lines will the plot be continued if $m_1 = m_2$?

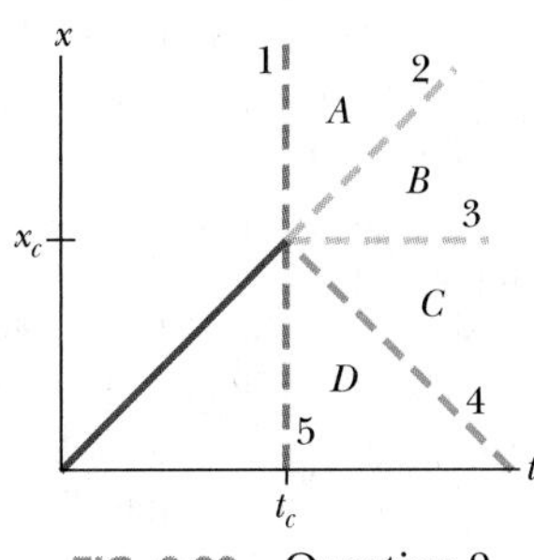

FIG. 9-33 Question 9.

10 Figure 9-34 shows four graphs of position versus time for two bodies and their center of mass. The two bodies form a closed, isolated system and undergo a completely inelastic, one-dimensional collision on an x axis. In graph 1, are (a) the two bodies and (b) the center of mass moving in the positive or negative direction of the x axis? (c) Which graphs correspond to a physically impossible situation? Explain.

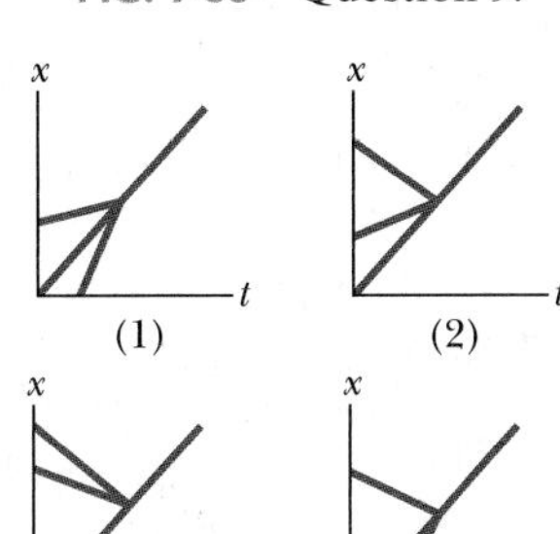

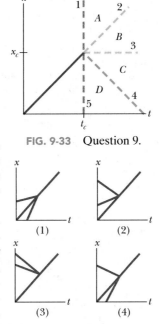

FIG. 9-34 Question 10.

11 A block slides along a frictionless floor and into a stationary second block with the same mass. Figure 9-35 shows four choices for a graph of the kinetic energies K of the blocks. (a) Determine which represent physically impossible situations. Of the others, which best represents (b) an elastic collision and (c) an inelastic collision?

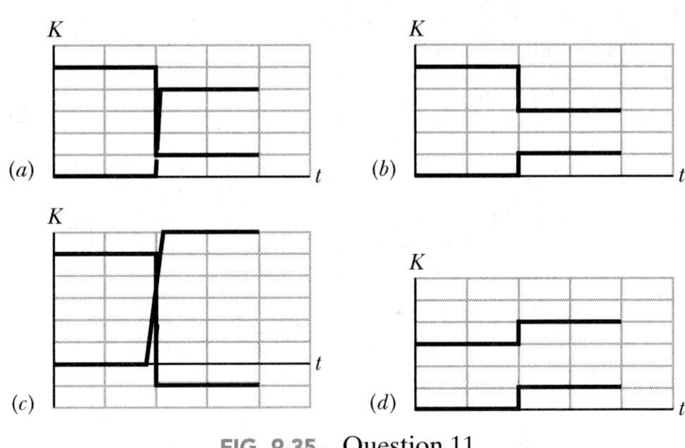

FIG. 9-35 Question 11.

12 Figure 9-36 shows a snapshot of block 1 as it slides along an x axis on a frictionless floor, before it undergoes an elastic collision with stationary block 2. The figure also shows three possible positions of the center of mass (com) of the two-block system at the time of the snapshot. (Point B is halfway between the centers of the two blocks.) Is block 1 stationary, moving forward, or moving backward after the collision if the com is located in the snapshot at (a) A, (b) B, and (c) C?

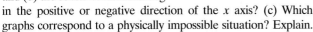

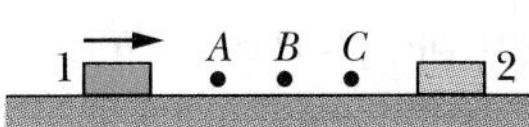

FIG. 9-36 Question 12.

PROBLEMS

GO Tutoring problem available (at instructor's discretion) in *WileyPLUS* and WebAssign

SSM Worked-out solution available in Student Solutions Manual

WWW Worked-out solution is at http://www.wiley.com/college/halliday

• – ••• Number of dots indicates level of problem difficulty

ILW Interactive solution is at http://www.wiley.com/college/halliday

Additional information available in *The Flying Circus of Physics* and at flyingcircusofphysics.com

sec. 9-2 The Center of Mass

•**1** A 2.00 kg particle has the xy coordinates (-1.20 m, 0.500 m), and a 4.00 kg particle has the xy coordinates (0.600 m, -0.750 m). Both lie on a horizontal plane. At what (a) x and (b) y coordinates must you place a 3.00 kg particle such that the center of mass of the three-particle system has the coordinates (-0.500 m, -0.700 m)?

•**2** Figure 9-37 shows a three-particle system, with masses $m_1 = 3.0$ kg, $m_2 = 4.0$ kg, and $m_3 = 8.0$ kg. The scales on the axes are set by $x_s = 2.0$ m and $y_s = 2.0$ m. What are (a) the x coordinate and (b) the y coordinate of the system's center of mass? (c) If m_3 is gradually increased, does the center of mass of the system shift toward or away from that particle, or does it remain stationary?

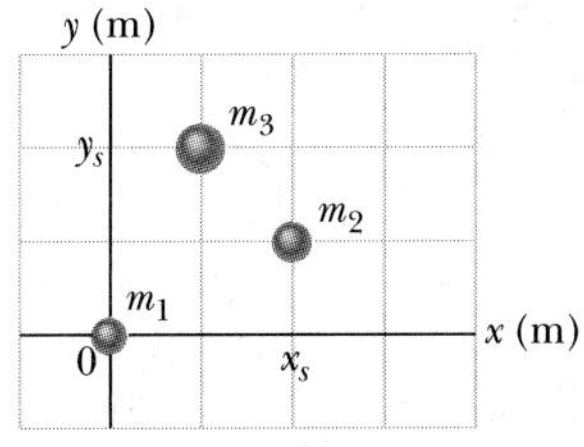

FIG. 9-37 Problem 2.

••**3** What are (a) the x coordinate and (b) the y coordinate of the center of mass for the uniform plate shown in Fig. 9-38 if $L = 5.0$ cm? GO

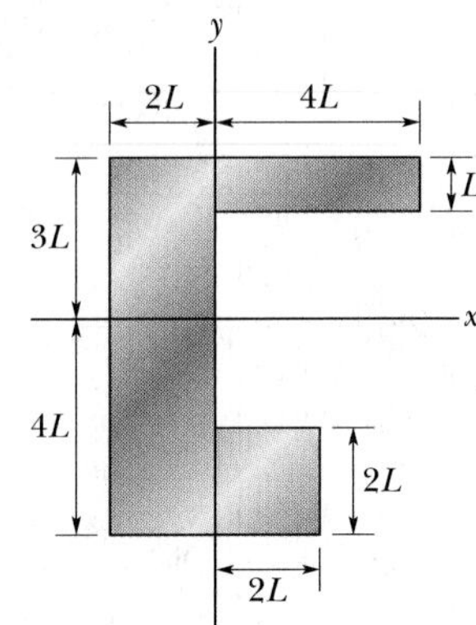

FIG. 9-38 Problem 3.

••**4** In Fig. 9-39, three uniform thin rods, each of length $L = 22$ cm, form an inverted U. The vertical rods each have a mass of 14 g; the horizontal rod has a mass of 42 g. What are (a) the x coordinate and (b) the

y coordinate of the system's center of mass?

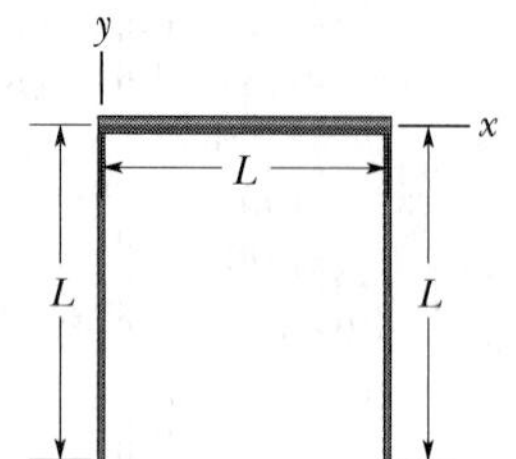

FIG. 9-39 Problem 4.

••5 In the ammonia (NH_3) molecule of Fig. 9-40, three hydrogen (H) atoms form an equilateral triangle, with the center of the triangle at distance $d = 9.40 \times 10^{-11}$ m from each hydrogen atom. The nitrogen (N) atom is at the apex of a pyramid, with the three hydrogen atoms forming the base. The nitrogen-to-hydrogen atomic mass ratio is 13.9, and the nitrogen-to-hydrogen distance is $L = 10.14 \times 10^{-11}$ m. What are the (a) x and (b) y coordinates of the molecule's center of mass? ILW

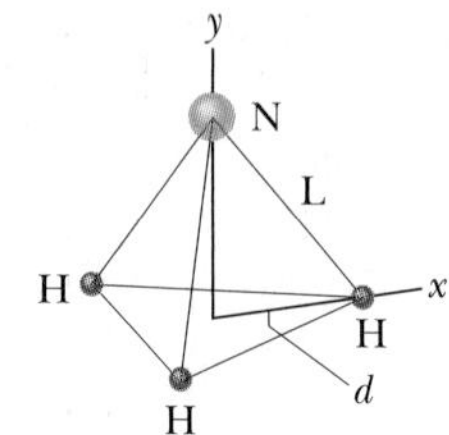

FIG. 9-40 Problem 5.

••6 Figure 9-41 shows a cubical box that has been constructed from uniform metal plate of negligible thickness. The box is open at the top and has edge length L = 40 cm. Find (a) the x coordinate, (b) the y coordinate, and (c) the z coordinate of the center of mass of the box.

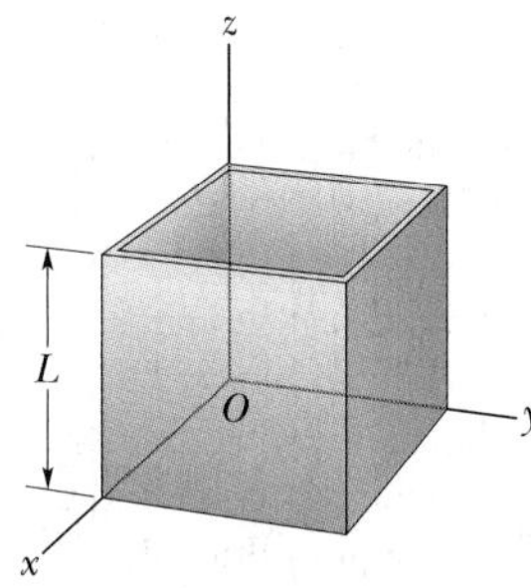

FIG. 9-41 Problem 6.

••7 Figure 9-42 shows a slab with dimensions $d_1 = 11.0$ cm, $d_2 = 2.80$ cm, and $d_3 = 13.0$ cm. Half the slab consists of aluminum (density = 2.70 g/cm³) and half consists of iron (density = 7.85 g/cm³). What are (a) the x coordinate, (b) the y coordinate, and (c) the z coordinate of the slab's center of mass?

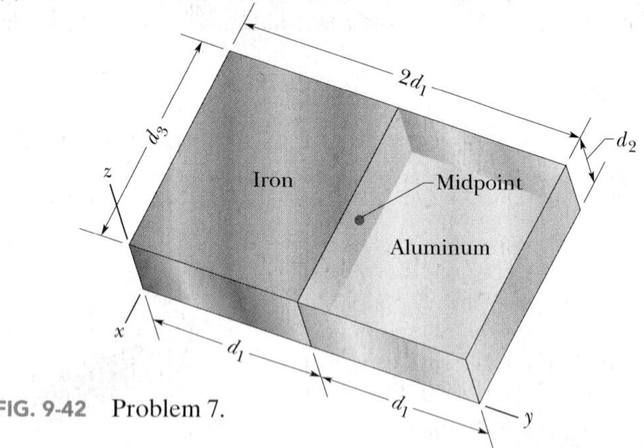

FIG. 9-42 Problem 7.

•••8 A uniform soda can of mass 0.140 kg is 12.0 cm tall and filled with 1.31 kg of soda (Fig. 9-43). Then small holes are drilled in the top and bottom (with negligible loss of metal) to drain the soda. What is the height h of the com of the can and contents (a) initially and (b) after the can loses all the soda? (c) What happens to h as the soda drains out? (d) If x is the height of the remaining soda at any given instant, find x when the com reaches its lowest point.

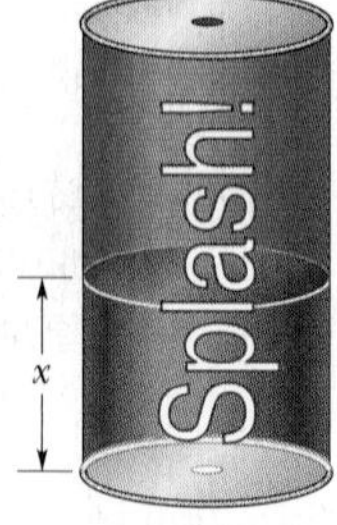

FIG. 9-43 Problem 8.

sec. 9-3 Newton's Second Law for a System of Particles

•9 A big olive ($m = 0.50$ kg) lies at the origin of an xy coordinate system, and a big Brazil nut ($M = 1.5$ kg) lies at the point (1.0, 2.0) m. At $t = 0$, a force $\vec{F}_o = (2.0\hat{i} + 3.0\hat{j})$ N begins to act on the olive, and a force $\vec{F}_n = (-3.0\hat{i} - 2.0\hat{j})$ N begins to act on the nut. In unit-vector notation, what is the displacement of the center of mass of the olive–nut system at $t = 4.0$ s, with respect to its position at $t = 0$?

•10 Two skaters, one with mass 65 kg and the other with mass 40 kg, stand on an ice rink holding a pole of length 10 m and negligible mass. Starting from the ends of the pole, the skaters pull themselves along the pole until they meet. How far does the 40 kg skater move?

•11 A stone is dropped at $t = 0$. A second stone, with twice the mass of the first, is dropped from the same point at $t = 100$ ms. (a) How far below the release point is the center of mass of the two stones at $t = 300$ ms? (Neither stone has yet reached the ground.) (b) How fast is the center of mass of the two-stone system moving at that time? ILW

•12 A 1000 kg automobile is at rest at a traffic signal. At the instant the light turns green, the automobile starts to move with a constant acceleration of 4.0 m/s². At the same instant a 2000 kg truck, traveling at a constant speed of 8.0 m/s, overtakes and passes the automobile. (a) How far is the com of the automobile–truck system from the traffic light at $t = 3.0$ s? (b) What is the speed of the com then?

••13 Figure 9-44 shows an arrangement with an air track, in which a cart is connected by a cord to a hanging block. The cart has mass $m_1 = 0.600$ kg, and its center is initially at xy coordinates (−0.500 m, 0 m); the block has mass $m_2 = 0.400$ kg, and its center is initially at xy coordinates (0, −0.100 m). The mass of the cord and pulley are negligible. The cart is released from rest, and both cart and block move until the cart hits the pulley. The friction between the cart and the air track and between the pulley and its axle is negligible. (a) In unit-vector notation, what is the acceleration of the center of mass of the cart–block system? (b) What is the velocity of the com as a function of time t? (c) Sketch the path taken by the com. (d) If the path is curved, determine whether it bulges upward to the right or downward to the left, and if it is straight, find the angle between it and the x axis.

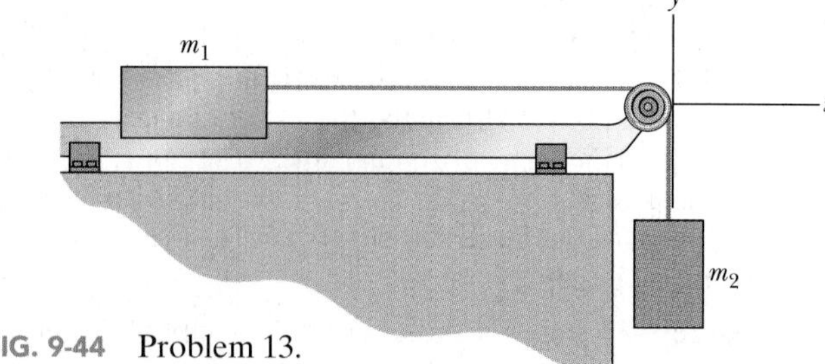

FIG. 9-44 Problem 13.

••14 In Figure 9-45, two particles are launched from the origin of the coordinate system at time $t = 0$. Particle 1 of mass $m_1 = 5.00$ g is shot directly along the x axis on a frictionless floor, with constant speed 10.0 m/s. Particle 2 of mass $m_2 = 3.00$ g is shot with a velocity of magnitude 20.0 m/s, at an upward angle such that it

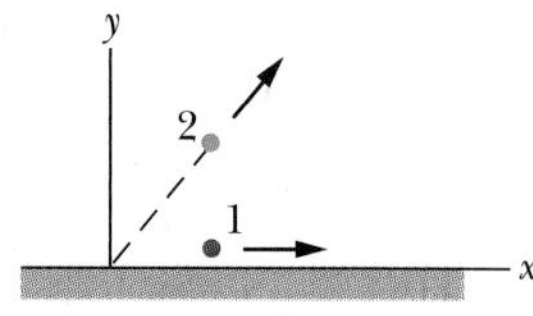

FIG. 9-45 Problem 14.

always stays directly above particle 1. (a) What is the maximum height H_{max} reached by the com of the two-particle system? In unit-vector notation, what are the (b) velocity and (c) acceleration of the com when the com reaches H_{max}?

••15 A shell is shot with an initial velocity $\vec{v}_0$ of 20 m/s, at an angle of $\theta_0 = 60°$ with the horizontal. At the top of the trajectory, the shell explodes into two fragments of equal mass (Fig. 9-46). One fragment, whose speed immediately after the explosion is zero, falls vertically. How far from the gun does the other fragment land, assuming that the terrain is level and that air drag is negligible? SSM

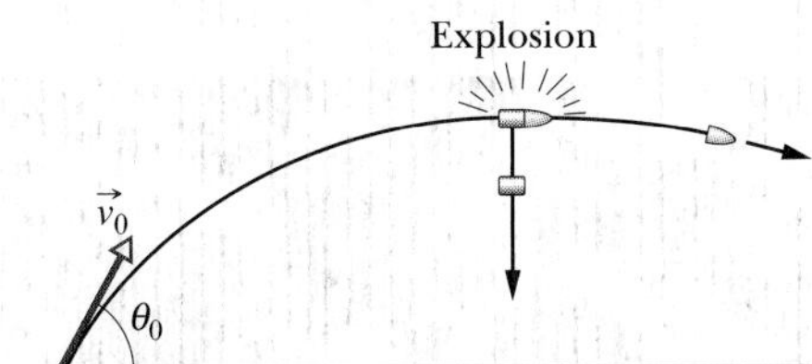

FIG. 9-46 Problem 15.

•••16 Ricardo, of mass 80 kg, and Carmelita, who is lighter, are enjoying Lake Merced at dusk in a 30 kg canoe. When the canoe is at rest in the placid water, they exchange seats, which are 3.0 m apart and symmetrically located with respect to the canoe's center. If the canoe moves 40 cm horizontally relative to a pier post, what is Carmelita's mass?

•••17 In Fig. 9-47*a*, a 4.5 kg dog stands on an 18 kg flatboat at distance $D = 6.1$ m from the shore. It walks 2.4 m along the boat toward shore and then stops. Assuming no friction between the boat and the water, find how far the dog is then from the shore. (*Hint:* See Fig. 9-47*b*.)

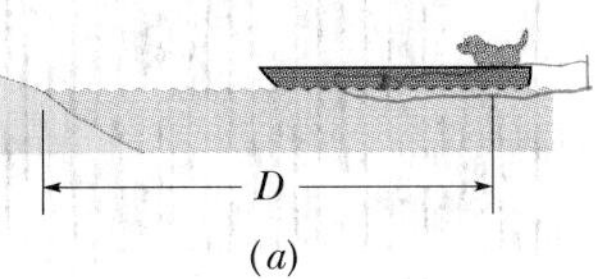

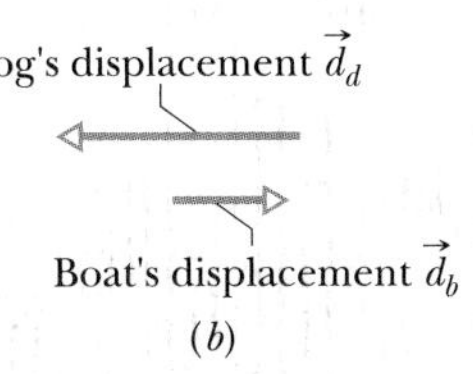

FIG. 9-47 Problem 17.

sec. 9-5 The Linear Momentum of a System of Particles

•18 A 0.70 kg ball moving horizontally at 5.0 m/s strikes a vertical wall and rebounds with speed 2.0 m/s. What is the magnitude of the change in its linear momentum?

•19 A 2100 kg truck traveling north at 41 km/h turns east and accelerates to 51 km/h. (a) What is the change in the truck's kinetic energy? What are the (b) magnitude and (c) direction of the change in its momentum? ILW

••20 Figure 9-48 gives an overhead view of the path taken by a 0.165 kg cue ball as it bounces from a rail of a pool table. The ball's initial speed is 2.00 m/s, and the angle θ_1 is 30.0°. The bounce reverses the y component of the ball's velocity but does not alter the x component. What are (a) angle θ_2 and (b) the change in the ball's linear momentum in unit-vector notation? (The fact that the ball rolls is irrelevant to the problem.)

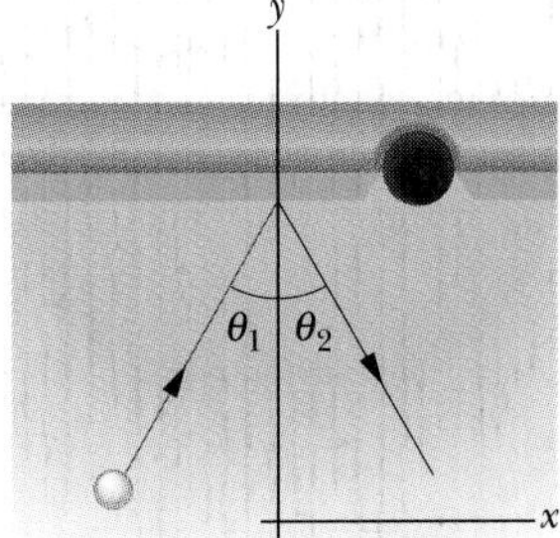

FIG. 9-48 Problem 20.

••21 A 0.30 kg softball has a velocity of 15 m/s at an angle of 35° below the horizontal just before making contact with the bat. What is the magnitude of the change in momentum of the ball while in contact with the bat if the ball leaves with a velocity of (a) 20 m/s, vertically downward, and (b) 20 m/s, horizontally back toward the pitcher?

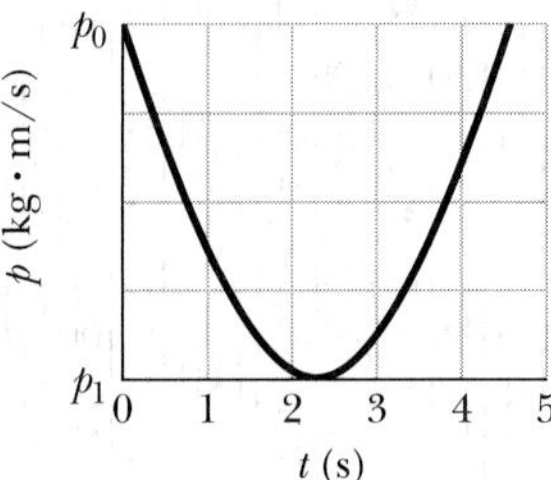

FIG. 9-49 Problem 22.

••22 At time $t = 0$, a ball is struck at ground level and sent over level ground. Figure 9-49 gives momentum p versus t during the flight ($p_0 = 6.0$ kg·m/s and $p_1 = 4.0$ kg·m/s). At what initial angle is the ball launched?

sec. 9-6 Collision and Impulse

•23 A force in the negative direction of an x axis is applied for 27 ms to a 0.40 kg ball initially moving at 14 m/s in the positive direction of the axis. The force varies in magnitude, and the impulse has magnitude 32.4 N·s. What are the ball's (a) speed and (b) direction of travel just after the force is applied? What are (c) the average magnitude of the force and (d) the direction of the impulse on the ball? SSM

•24 In a common but dangerous prank, a chair is pulled away as a person is moving downward to sit on it, causing the victim to land hard on the floor. Suppose the victim falls by 0.50 m, the mass that moves downward is 70 kg, and the collision on the floor lasts 0.082 s. What are the magnitudes of the (a) impulse and (b) average force acting on the victim from the floor during the collision?

•25 Until his seventies, Henri LaMothe (Fig. 9-50) excited

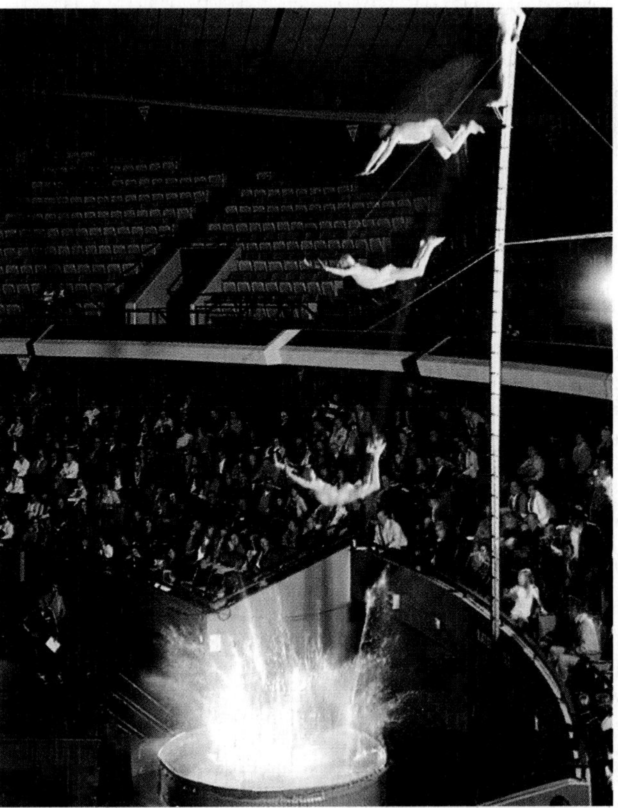

FIG. 9-50 Problem 25. Belly-flopping into 30 cm of water. (*George Long/ Sports Illustrated/©Time, Inc.*)

audiences by belly-flopping from a height of 12 m into 30 cm of water. Assuming that he stops just as he reaches the bottom of the water and estimating his mass, find the magnitude of the impulse on him from the water.

•**26** In February 1955, a paratrooper fell 370 m from an airplane without being able to open his chute but happened to land in snow, suffering only minor injuries. Assume that his speed at impact was 56 m/s (terminal speed), that his mass (including gear) was 85 kg, and that the magnitude of the force on him from the snow was at the survivable limit of 1.2×10^5 N. What are (a) the minimum depth of snow that would have stopped him safely and (b) the magnitude of the impulse on him from the snow?

•**27** A 1.2 kg ball drops vertically onto a floor, hitting with a speed of 25 m/s. It rebounds with an initial speed of 10 m/s. (a) What impulse acts on the ball during the contact? (b) If the ball is in contact with the floor for 0.020 s, what is the magnitude of the average force on the floor from the ball?

•**28** In tae-kwon-do, a hand is slammed down onto a target at a speed of 13 m/s and comes to a stop during the 5.0 ms collision. Assume that during the impact the hand is independent of the arm and has a mass of 0.70 kg. What are the magnitudes of the (a) impulse and (b) average force on the hand from the target?

•**29** Suppose a gangster sprays Superman's chest with 3 g bullets at the rate of 100 bullets/min, and the speed of each bullet is 500 m/s. Suppose too that the bullets rebound straight back with no change in speed. What is the magnitude of the average force on Superman's chest?

••**30** A 5.0 kg toy car can move along an x axis; Fig. 9-51 gives F_x of the force acting on the car, which begins at rest at time $t = 0$. The scale on the F_x axis is set by $F_{xs} = 5.0$ N. In unit-vector notation, what is $\vec{p}$ at (a) $t = 4.0$ s and (b) $t = 7.0$ s, and (c) what is $\vec{v}$ at $t = 9.0$ s?

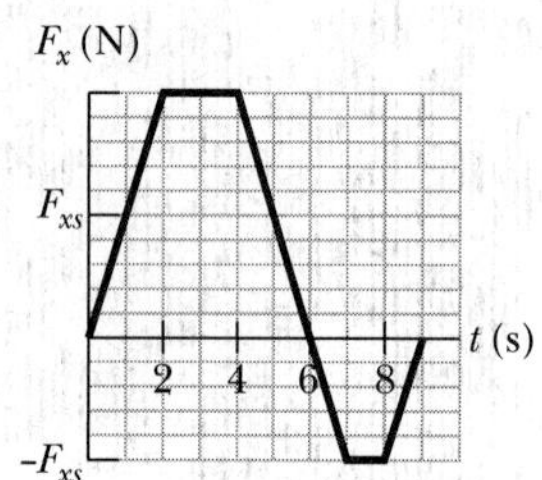

FIG. 9-51 Problem 30.

••**31** Figure 9-52 shows a 0.300 kg baseball just before and just after it collides with a bat. Just before, the ball has velocity $\vec{v}_1$ of magnitude 12.0 m/s and angle $\theta_1 = 35.0°$. Just after, it is traveling directly upward with velocity $\vec{v}_2$ of magnitude 10.0 m/s. The duration of the collision is 2.00 ms. What are the (a) magnitude and (b) direction (relative to the positive direction of the x axis) of the impulse on the ball from the bat? What are the (c) magnitude and (d) direction of the average force on the ball from the bat?

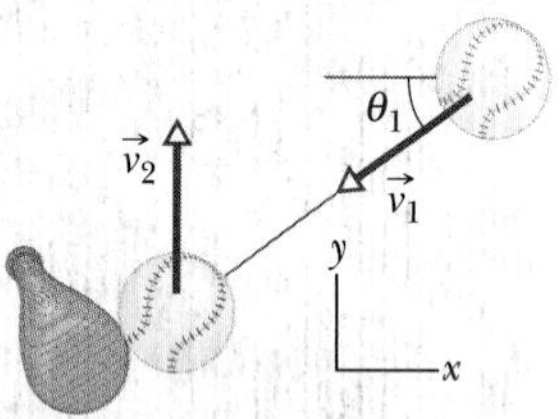

FIG. 9-52 Problem 31.

••**32** Basilisk lizards can run across the top of a water surface (Fig. 9-53). With each step, a lizard first slaps its foot against the water and then pushes it down into the water rapidly enough to form an air cavity around the top of the foot. To avoid having to pull the foot back up against water drag in order to complete the step, the lizard withdraws the foot before water can flow into the air cavity. If the lizard is not to sink, the average upward impulse on the lizard during this full action of slap, downward push, and withdrawal must match the downward impulse due to the gravitational force. Suppose the mass of a basilisk lizard is 90.0 g, the mass of each foot is 3.00 g, the speed of a foot as it slaps the water is 1.50 m/s, and the time for a single step is 0.600 s. (a) What is the magnitude of the impulse on the lizard during the slap? (Assume this impulse is directly upward.) (b) During the 0.600 s duration of a step, what is the downward impulse on the lizard due to the gravitational force? (c) Which action, the slap or the push, provides the primary support for the lizard, or are they approximately equal in their support?

FIG. 9-53 Problem 32. Lizard running across water. *(Stephen Dalton/Photo Researchers)*

••**33** *Jumping up before the elevator hits.* After the cable snaps and the safety system fails, an elevator cab free-falls from a height of 36 m. During the collision at the bottom of the elevator shaft, a 90 kg passenger is stopped in 5.0 ms. (Assume that neither the passenger nor the cab rebounds.) What are the magnitudes of the (a) impulse and (b) average force on the passenger during the collision? If the passenger were to jump upward with a speed of 7.0 m/s relative to the cab floor just before the cab hits the bottom of the shaft, what are the magnitudes of the (c) impulse and (d) average force (assuming the same stopping time)?

••**34** *Two average forces.* A steady stream of 0.250 kg snowballs is shot perpendicularly into a wall at a speed of 4.00 m/s. Each ball sticks to the wall. Figure 9-54 gives the magnitude F of the force on the wall as a function of time t for two of the snowball impacts. Impacts occur with a repetition time interval $\Delta t_r = 50.0$ ms, last a duration time interval $\Delta t_d = 10$ ms, and produce isosceles triangles on the graph, with each impact reaching a force maximum $F_{max} = 200$ N. During each impact, what are the magnitudes of (a) the impulse and (b) the average force on the wall? (c) During a time interval

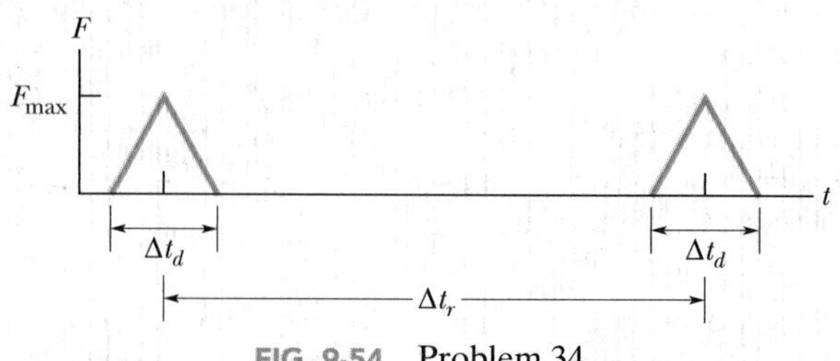

FIG. 9-54 Problem 34.

of many impacts, what is the magnitude of the average force on the wall?

••**35** A soccer player kicks a soccer ball of mass 0.45 kg that is initially at rest. The player's foot is in contact with the ball for 3.0×10^{-3} s, and the force of the kick is given by

$$F(t) = [(6.0 \times 10^6)t - (2.0 \times 10^9)t^2]\ \text{N}$$

for $0 \le t \le 3.0 \times 10^{-3}$ s, where t is in seconds. Find the magnitudes of (a) the impulse on the ball due to the kick, (b) the average force on the ball from the player's foot during the period of contact, (c) the maximum force on the ball from the player's foot during the period of contact, and (d) the ball's velocity immediately after it loses contact with the player's foot. SSM

••**36** In the overhead view of Fig. 9-55, a 300 g ball with a speed v of 6.0 m/s strikes a wall at an angle θ of 30° and then rebounds with the same speed and angle. It is in contact with the wall for 10 ms. In unit-vector notation, what are (a) the impulse on the ball from the wall and (b) the average force on the wall from the ball?

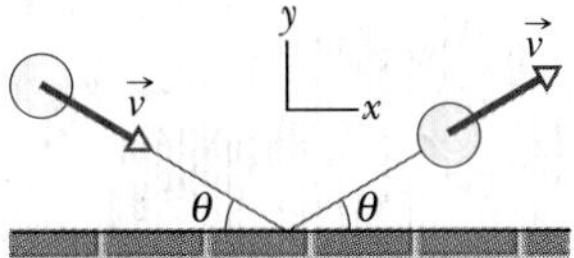

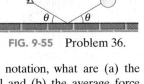

FIG. 9-55 Problem 36.

••**37** Figure 9-56 shows an approximate plot of force magnitude F versus time t during the collision of a 58 g Superball with a wall. The initial velocity of the ball is 34 m/s perpendicular to the wall; the ball rebounds directly back with approximately the same speed, also perpendicular to the wall. What is F_{max}, the maximum magnitude of the force on the ball from the wall during the collision? GO

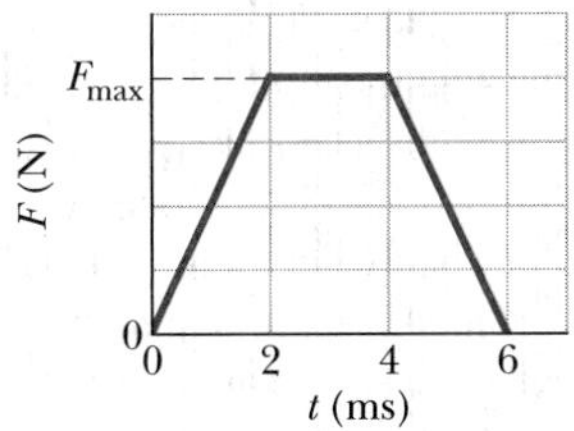

FIG. 9-56 Problem 37.

••**38** A 0.25 kg puck is initially stationary on an ice surface with negligible friction. At time $t = 0$, a horizontal force begins to move the puck. The force is given by $\vec{F} = (12.0 - 3.00t^2)\hat{\text{i}}$, with $\vec{F}$ in newtons and t in seconds, and it acts until its magnitude is zero. (a) What is the magnitude of the impulse on the puck from the force between $t = 0.500$ s and $t = 1.25$ s? (b) What is the change in momentum of the puck between $t = 0$ and the instant at which $F = 0$?

sec. 9-7 Conservation of Linear Momentum

•**39** A 91 kg man lying on a surface of negligible friction shoves a 68 g stone away from himself, giving it a speed of 4.0 m/s. What speed does the man acquire as a result? SSM

•**40** A space vehicle is traveling at 4300 km/h relative to Earth when the exhausted rocket motor (mass $4m$) is disengaged and sent backward with a speed of 82 km/h relative to the command module (mass m). What is the speed of the command module relative to Earth just after the separation?

••**41** In the Olympiad of 708 B.C., some athletes competing in the standing long jump used handheld weights called *halteres* to lengthen their jumps (Fig. 9-57). The weights were swung up in front just before liftoff and then swung down and thrown backward during the flight. Suppose a modern 78 kg long jumper similarly uses two 5.50 kg halteres, throwing them horizontally to the rear at his maximum height such that their horizontal velocity is zero relative to the ground. Let his liftoff velocity be $\vec{v} = (9.5\hat{\text{i}} + 4.0\hat{\text{j}})$ m/s with or without the halteres, and assume that he lands at the liftoff level. What distance would the use of the halteres add to his range?

FIG. 9-57 Problem 41. *(Réunion des Musées Nationaux/Art Resource)*

••**42** A 4.0 kg mess kit sliding on a frictionless surface explodes into two 2.0 kg parts: 3.0 m/s, due north, and 5.0 m/s, 30° north of east. What is the original speed of the mess kit?

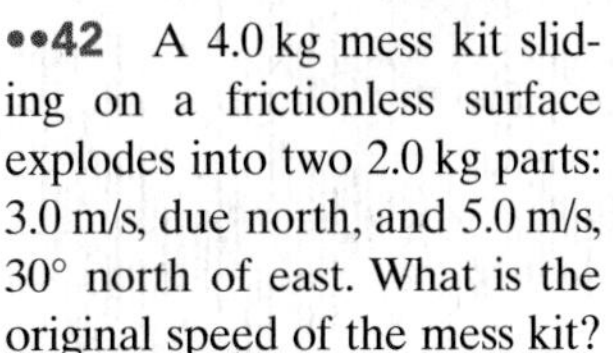

••**43** Figure 9-58 shows a two-ended "rocket" that is initially stationary on a frictionless floor, with its center at the origin of an x axis. The rocket consists of a central block C (of mass $M = 6.00$ kg) and blocks L and R (each of mass $m = 2.00$ kg) on the left and right sides. Small explosions can shoot either of the side blocks away from block C and along the x axis. Here is the sequence: (1) At time $t = 0$, block L is shot to the left with a speed of 3.00 m/s *relative* to the velocity that the explosion gives the rest of the rocket. (2) Next, at time $t = 0.80$ s, block R is shot to the right with a speed of 3.00 m/s *relative* to the velocity that block C then has. At $t = 2.80$ s, what are (a) the velocity of block C and (b) the position of its center?

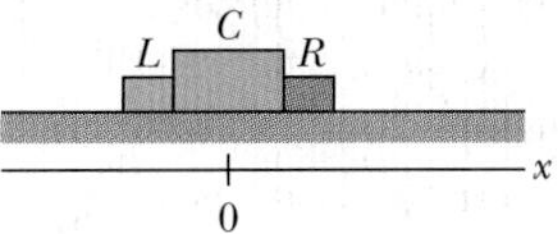

FIG. 9-58 Problem 43.

••**44** An object, with mass m and speed v relative to an observer, explodes into two pieces, one three times as massive as the other; the explosion takes place in deep space. The less massive piece stops relative to the observer. How much kinetic energy is added to the system during the explosion, as measured in the observer's reference frame?

••**45** A vessel at rest at the origin of an xy coordinate system explodes into three pieces. Just after the explosion, one piece, of mass m, moves with velocity $(-30\text{ m/s})\hat{\text{i}}$ and a second piece, also of mass m, moves with velocity $(-30\text{ m/s})\hat{\text{j}}$. The third piece has mass $3m$. Just after the explosion, what are the (a) magnitude and (b) direction of the velocity of the third piece?

••**46** In Fig. 9-59, a stationary block explodes into two pieces L and R that slide across a frictionless floor and then into regions with friction, where they stop. Piece L, with a mass of 2.0 kg, encounters a coefficient of kinetic friction $\mu_L = 0.40$ and slides to a stop in distance $d_L = 0.15$ m. Piece R encounters a coefficient of kinetic friction $\mu_R = 0.50$ and slides to a stop in distance $d_R = 0.25$ m. What was the mass of the block?

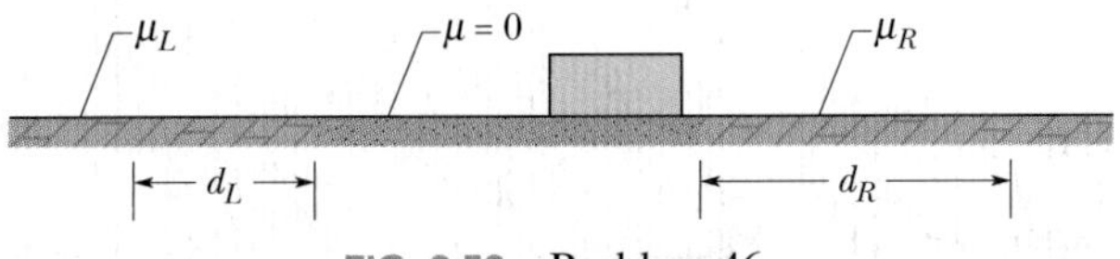

FIG. 9-59 Problem 46.

••**47** A 20.0 kg body is moving through space in the positive direction of an x axis with a speed of 200 m/s when, due to an internal explosion, it breaks into three parts. One part, with a mass of 10.0 kg, moves away from the point of explosion with a speed of 100 m/s in the positive y direction. A second part, with a mass of 4.00 kg, moves in the negative x direction with a speed of 500 m/s. (a) In unit-vector notation, what is the velocity of the third part? (b) How much energy is released in the explosion? Ignore effects due to the gravitational force. SSM WWW

•••48 Particle A and particle B are held together with a compressed spring between them. When they are released, the spring pushes them apart, and they then fly off in opposite directions, free of the spring. The mass of A is 2.00 times the mass of B, and the energy stored in the spring was 60 J. Assume that the spring has negligible mass and that all its stored energy is transferred to the particles. Once that transfer is complete, what are the kinetic energies of (a) particle A and (b) particle B?

sec. 9-9 Inelastic Collisions in One Dimension

•49 A bullet of mass 10 g strikes a ballistic pendulum of mass 2.0 kg. The center of mass of the pendulum rises a vertical distance of 12 cm. Assuming that the bullet remains embedded in the pendulum, calculate the bullet's initial speed.

•50 A 5.20 g bullet moving at 672 m/s strikes a 700 g wooden block at rest on a frictionless surface. The bullet emerges, traveling in the same direction with its speed reduced to 428 m/s. (a) What is the resulting speed of the block? (b) What is the speed of the bullet–block center of mass?

••51 In Anchorage, collisions of a vehicle with a moose are so common that they are referred to with the abbreviation MVC. Suppose a 1000 kg car slides into a stationary 500 kg moose on a very slippery road, with the moose being thrown through the windshield (a common MVC result). (a) What percent of the original kinetic energy is lost in the collision to other forms of energy? A similar danger occurs in Saudi Arabia because of camel–vehicle collisions (CVC). (b) What percent of the original kinetic energy is lost if the car hits a 300 kg camel? (c) Generally, does the percent loss increase or decrease if the animal mass decreases?

••52 In the "before" part of Fig. 9-60, car A (mass 1100 kg) is stopped at a traffic light when it is rear-ended by car B (mass 1400 kg). Both cars then slide with locked wheels until the frictional force from the slick road (with a low μ_k of 0.13) stops them, at distances $d_A = 8.2$ m and $d_B = 6.1$ m. What are the speeds of (a) car A and (b) car B at the start of the sliding, just after the collision? (c) Assuming that linear momentum is conserved during the collision, find the speed of car B just before the collision. (d) Explain why this assumption may be invalid.

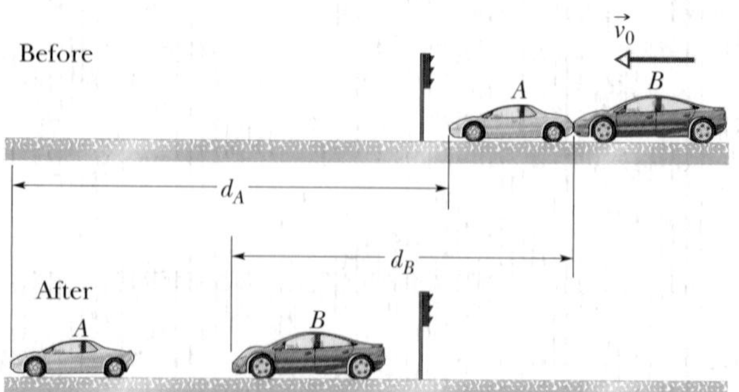

FIG. 9-60 Problem 52.

••53 In Fig. 9-61a, a 3.50 g bullet is fired horizontally at two blocks at rest on a frictionless table. The bullet passes through block 1 (mass 1.20 kg) and embeds itself in block 2 (mass 1.80 kg). The blocks end up with speeds $v_1 = 0.630$ m/s and $v_2 = 1.40$ m/s (Fig. 9-61b). Neglecting the material removed from block 1 by the bullet, find the speed of the bullet as it (a) leaves and (b) enters block 1. GO

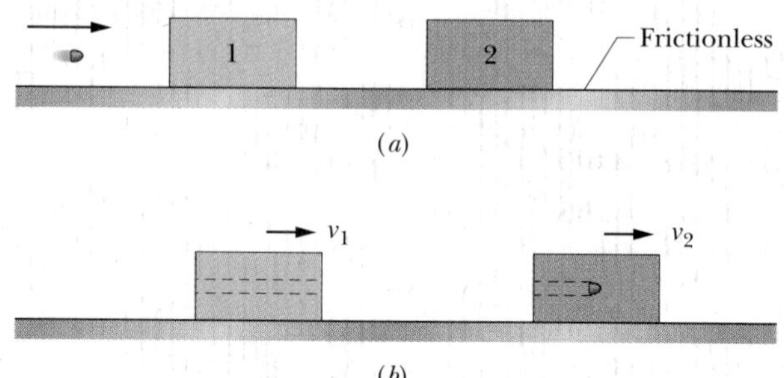

FIG. 9-61 Problem 53.

••54 In Fig. 9-62, a 10 g bullet moving directly upward at 1000 m/s strikes and passes through the center of mass of a 5.0 kg block initially at rest. The bullet emerges from the block moving directly upward at 400 m/s. To what maximum height does the block then rise above its initial position?

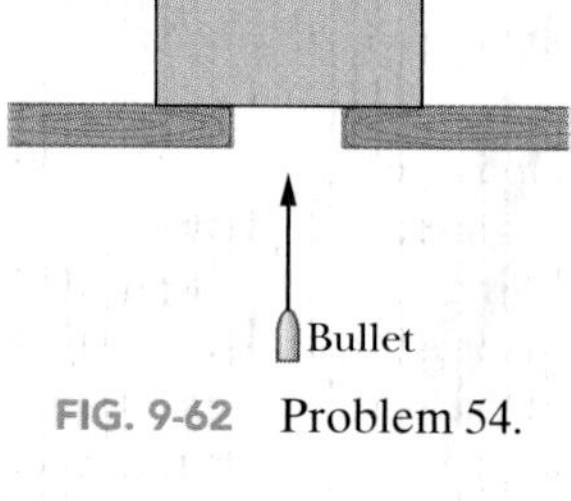

FIG. 9-62 Problem 54.

••55 In Fig. 9-63, a ball of mass $m = 60$ g is shot with speed $v_i = 22$ m/s into the barrel of a spring gun of mass $M = 240$ g initially at rest on a frictionless surface. The ball sticks in the barrel at the point of maximum compression of the spring. Assume that the increase in thermal energy due to friction between the ball and the barrel is negligible. (a) What is the speed of the spring gun after the ball stops in the barrel? (b) What fraction of the initial kinetic energy of the ball is stored in the spring? GO

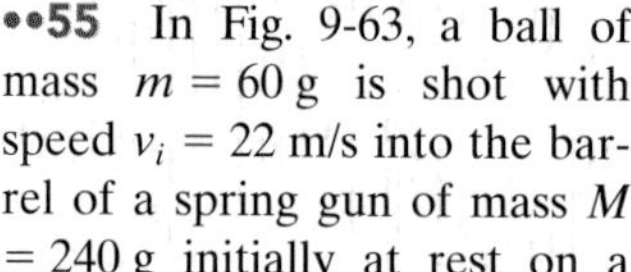

FIG. 9-63 Problem 55.

••56 A completely inelastic collision occurs between two balls of wet putty that move directly toward each other along a vertical axis. Just before the collision, one ball, of mass 3.0 kg, is moving upward at 20 m/s and the other ball, of mass 2.0 kg, is moving downward at 12 m/s. How high do the combined two balls of putty rise above the collision point? (Neglect air drag.)

••57 A 5.0 kg block with a speed of 3.0 m/s collides with a 10 kg block that has a speed of 2.0 m/s in the same direction. After the collision, the 10 kg block travels in the original direction with a speed of 2.5 m/s. (a) What is the velocity of the 5.0 kg block immediately after the collision? (b) By how much does the total kinetic energy of the system of two blocks change because of the collision? (c) Suppose, instead, that the 10 kg block ends up with a speed of 4.0 m/s. What then is the change in the total kinetic energy? (d) Account for the result you obtained in (c). ILW

•••58 In Fig. 9-64, block 2 (mass 1.0 kg) is at rest on a frictionless surface and touching the end of an unstretched spring of spring constant 200 N/m. The other end of the spring is fixed to a wall. Block 1 (mass 2.0 kg), traveling at speed $v_1 = 4.0$ m/s, collides with block 2, and the two blocks stick together. When the blocks momentarily stop, by what distance is the spring compressed?

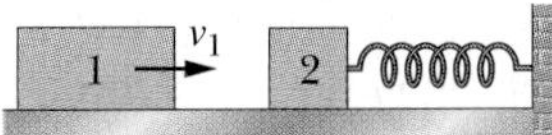

FIG. 9-64 Problem 58.

•••59 In Fig. 9-65, block 1 (mass 2.0 kg) is moving rightward at 10 m/s and block 2 (mass 5.0 kg) is moving rightward at 3.0 m/s. The surface is frictionless, and a spring with a spring constant of 1120 N/m is fixed to block 2. When the blocks collide, the compression of the spring is maximum at the instant the blocks have the same velocity. Find the maximum compression. ILW

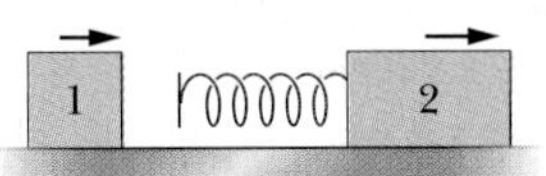

FIG. 9-65 Problems 59 and 126.

sec. 9-10 Elastic Collisions in One Dimension

•60 Two titanium spheres approach each other head-on with the same speed and collide elastically. After the collision, one of the spheres, whose mass is 300 g, remains at rest. (a) What is the mass of the other sphere? (b) What is the speed of the two-sphere center of mass if the initial speed of each sphere is 2.00 m/s?

•61 A cart with mass 340 g moving on a frictionless linear air track at an initial speed of 1.2 m/s undergoes an elastic collision with an initially stationary cart of unknown mass. After the collision, the first cart continues in its original direction at 0.66 m/s. (a) What is the mass of the second cart? (b) What is its speed after impact? (c) What is the speed of the two-cart center of mass? SSM

•62 In Fig. 9-66, block A (mass 1.6 kg) slides into block B (mass 2.4 kg), along a frictionless surface. The directions of three velocities before (i) and after (f) the collision are indicated; the corresponding speeds are $v_{Ai} = 5.5$ m/s, $v_{Bi} = 2.5$ m/s, and $v_{Bf} = 4.9$ m/s. What are the (a) speed and(b) direction (left or right) of velocity $\vec{v}_{Af}$? (c) Is the collision elastic?

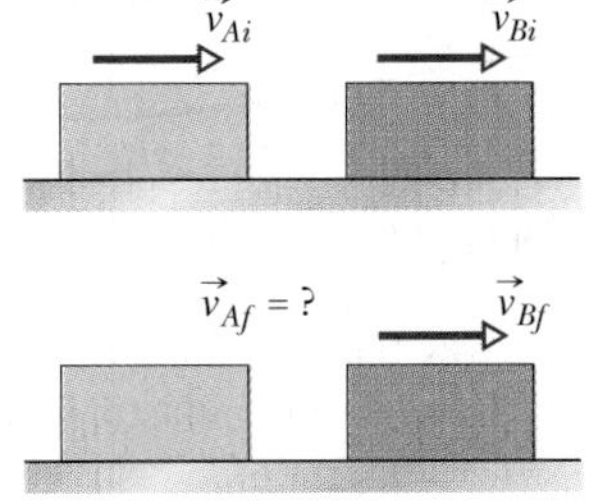

FIG. 9-66 Problem 62.

••63 A body of mass 2.0 kg makes an elastic collision with another body at rest and continues to move in the original direction but with one-fourth of its original speed. (a) What is the mass of the other body? (b) What is the speed of the two-body center of mass if the initial speed of the 2.0 kg body was 4.0 m/s? SSM

••64 Block 1, with mass m_1 and speed 4.0 m/s, slides along an x axis on a frictionless floor and then undergoes a one-dimensional elastic collision with stationary block 2, with mass $m_2 = 0.40m_1$. The two blocks then slide into a region where the coefficient of kinetic friction is 0.50; there they stop. How far into that region do (a) block 1 and (b) block 2 slide?

••65 In Fig. 9-67, particle 1 of mass $m_1 = 0.30$ kg slides rightward along an x axis on a frictionless floor with a speed of 2.0 m/s. When it reaches $x = 0$, it undergoes a one-dimensional elastic collision with stationary particle 2 of mass $m_2 = 0.40$ kg. When particle 2 then reaches a wall at $x_w = 70$ cm, it bounces from the wall with no loss of speed. At what position on the x axis does particle 2 then collide with particle 1?

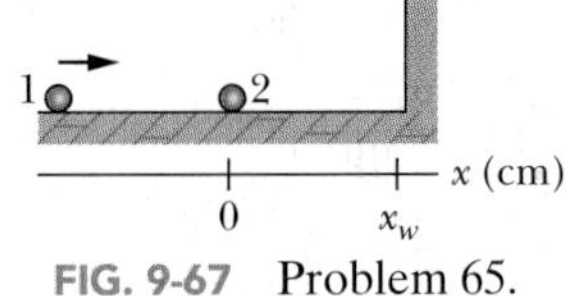

FIG. 9-67 Problem 65.

••66 A steel ball of mass 0.500 kg is fastened to a cord that is 70.0 cm long and fixed at the far end. The ball is then released when the cord is horizontal (Fig. 9-68). At the bottom of its path, the ball strikes a 2.50 kg steel block initially at rest on a frictionless surface. The collision is elastic. Find (a) the speed of the ball and (b) the speed of the block, both just after the collision.

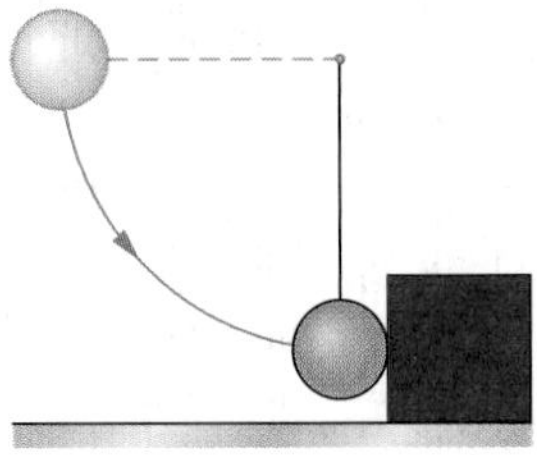

FIG. 9-68 Problem 66.

••67 Block 1 of mass m_1 slides along a frictionless floor and into a one-dimensional elastic collision with stationary block 2 of mass $m_2 = 3m_1$. Prior to the collision, the center of mass of the two-block system had a speed of 3.00 m/s. Afterward, what are the speeds of (a) the center of mass and (b) block 2?

••68 In Fig. 9-69, block 1 of mass m_1 slides from rest along a frictionless ramp from height $h = 2.50$ m and then collides with stationary block 2, which has mass $m_2 = 2.00m_1$. After the collision, block 2 slides into a region where the coefficient of kinetic friction μ_k is 0.500 and comes to a stop in distance d within that region. What is the value of distance d if the collision is (a) elastic and (b) completely inelastic?

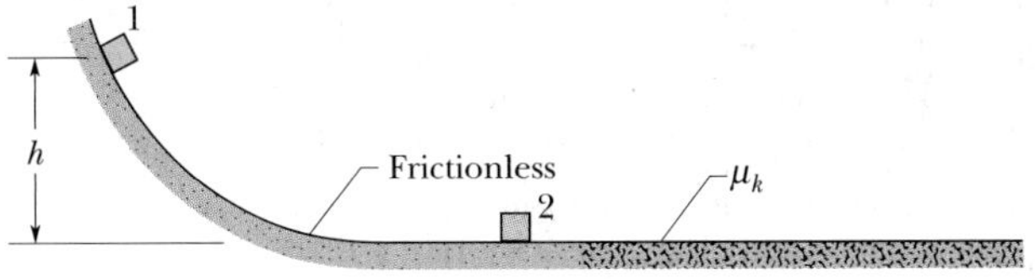

FIG. 9-69 Problem 68.

•••69 A small ball of mass m is aligned above a larger ball of mass $M = 0.63$ kg (with a slight separation, as with the baseball and basketball of Fig. 9-70a), and the two are dropped simultaneously from a height of $h = 1.8$ m. (Assume the radius of each ball is negligible relative to h.) (a) If the larger ball rebounds elastically from the floor and then the small ball rebounds elastically from the larger ball, what value of m results in the larger ball stopping when it collides with the small ball? (b) What height does the small ball then reach (Fig. 9-70b)?

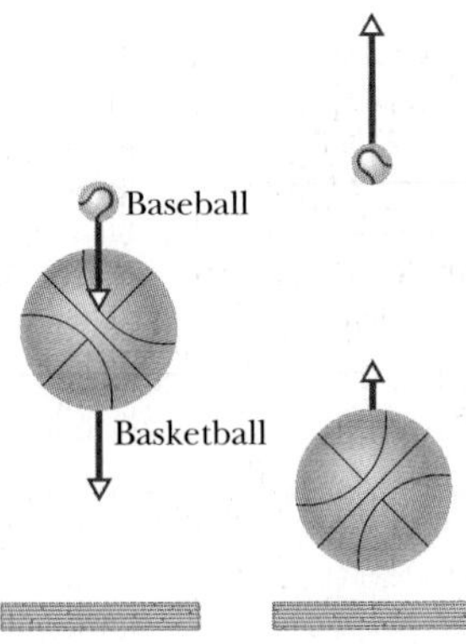

FIG. 9-70 Problem 69.

•••70 In Fig. 9-71, puck 1 of mass $m_1 = 0.20$ kg is sent sliding across a frictionless lab bench, to undergo a one-dimensional elastic collision with stationary puck 2. Puck 2 then slides off the bench and lands a distance d from the base of the bench. Puck 1 rebounds from the collision and slides off the opposite edge of the bench, landing a distance $2d$ from the base of the bench. What is the mass of puck 2? (*Hint:* Be careful with signs.)

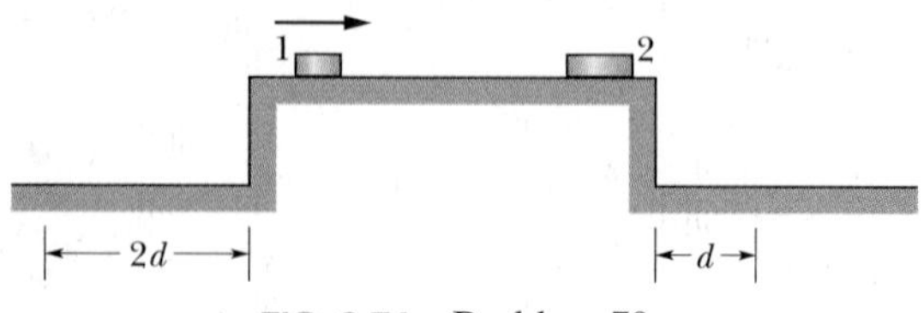

FIG. 9-71 Problem 70.

sec. 9-11 Collisions in Two Dimensions

••71 A projectile proton with a speed of 500 m/s collides elastically with a target proton initially at rest. The two protons then move along perpendicular paths, with the projectile path at 60° from the original direction. After the collision, what are the speeds of (a) the target proton and (b) the projectile proton?

••72 Two 2.0 kg bodies, A and B, collide. The velocities before the collision are $\vec{v}_A = (15\hat{i} + 30\hat{j})$ m/s and $\vec{v}_B = (-10\hat{i} + 5.0\hat{j})$ m/s. After the collision, $\vec{v}'_A = (-5.0\hat{i} + 20\hat{j})$ m/s. What are (a) the final velocity of B and (b) the change in the total kinetic energy (including sign)?

••73 In Fig. 9-23, projectile particle 1 is an alpha particle and target particle 2 is an oxygen nucleus. The alpha particle is scattered at angle $\theta_1 = 64.0°$ and the oxygen nucleus recoils with speed 1.20×10^5 m/s and at angle $\theta_2 = 51.0°$. In atomic mass units, the mass of the alpha particle is 4.00 u and the mass of the oxygen nucleus is 16.0 u. What are the (a) final and (b) initial speeds of the alpha particle? ILW

••74 Ball B, moving in the positive direction of an x axis at speed v, collides with stationary ball A at the origin. A and B have different masses. After the collision, B moves in the negative direction of the y axis at speed $v/2$. (a) In what direction does A move? (b) Show that the speed of A cannot be determined from the given information.

•••75 After a completely inelastic collision, two objects of the same mass and same initial speed move away together at half their initial speed. Find the angle between the initial velocities of the objects.

sec. 9-12 Systems with Varying Mass: A Rocket

•76 Consider a rocket that is in deep space and at rest relative to an inertial reference frame. The rocket's engine is to be fired for a certain interval. What must be the rocket's *mass ratio* (ratio of initial to final mass) over that interval if the rocket's original speed relative to the inertial frame is to be equal to (a) the exhaust speed (speed of the exhaust products relative to the rocket) and (b) 2.0 times the exhaust speed?

•77 A rocket that is in deep space and initially at rest relative to an inertial reference frame has a mass of 2.55×10^5 kg, of which 1.81×10^5 kg is fuel. The rocket engine is then fired for 250 s while fuel is consumed at the rate of 480 kg/s. The speed of the exhaust products relative to the rocket is 3.27 km/s. (a) What is the rocket's thrust? After the 250 s firing, what are (b) the mass and (c) the speed of the rocket? SSM ILW

•78 A 6090 kg space probe moving nose-first toward Jupiter at 105 m/s relative to the Sun fires its rocket engine, ejecting 80.0 kg of exhaust at a speed of 253 m/s relative to the space probe. What is the final velocity of the probe?

•79 In Fig. 9-72, two long barges are moving in the same direction in still water, one with a speed of 10 km/h and the other with a speed of 20 km/h. While they are passing each other, coal is shoveled from the slower to the faster one at a rate of 1000 kg/min. How much additional force must be provided by the driving engines of (a) the faster barge and (b) the slower barge if neither is to change speed? Assume that the shoveling is always perfectly sideways and that the frictional forces between the barges and the water do not depend on the mass of the barges. SSM

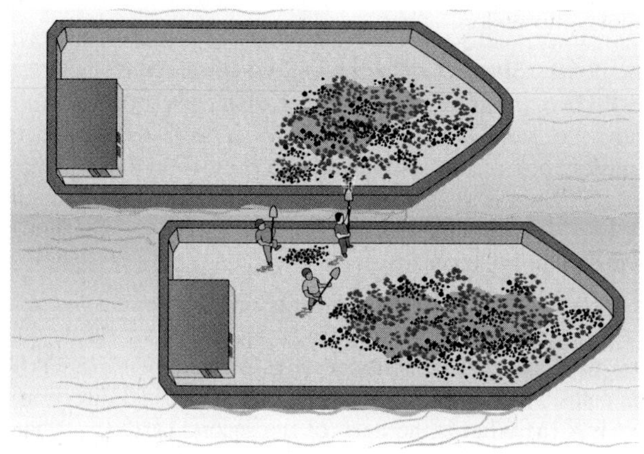

FIG. 9-72 Problem 79.

Additional Problems

80 *Speed amplifier.* In Fig. 9-73, block 1 of mass m_1 slides along an x axis on a frictionless floor with a speed of $v_{1i} = 4.00$ m/s. Then it undergoes a one-dimensional elastic collision with stationary block 2 of mass $m_2 = 0.500m_1$. Next, block 2 undergoes a one-dimensional elastic collision with stationary block 3 of mass $m_3 = 0.500m_2$. (a) What then is the speed of block 3? Are (b) the speed, (c) the kinetic energy, and (d) the momentum of block 3 greater than, less than, or the same as the initial values for block 1?

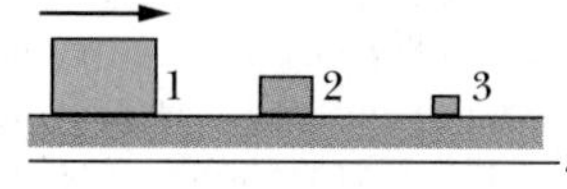

FIG. 9-73 Problem 80.

81 *Speed deamplifier.* In Fig. 9-74, block 1 of mass m_1 slides along an x axis on a frictionless floor at speed 4.00 m/s. Then it undergoes a one-dimensional elastic collision with stationary block 2 of mass $m_2 = 2.00m_1$. Next, block 2 undergoes a one-dimensional elastic collision with stationary block 3 of mass $m_3 = 2.00m_2$. (a) What then is the speed of block 3? Are (b) the speed, (c) the kinetic energy, and (d) the momentum of block 3 greater than, less than, or the same as the initial values for block 1?

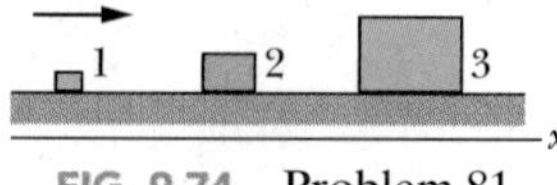

FIG. 9-74 Problem 81.

82 Figure 9-75 shows an overhead view of two particles sliding at constant velocity over a frictionless surface. The particles have the same mass and the same initial speed $v = 4.00$ m/s, and they collide where their paths intersect. An x axis is arranged to bisect the angle between their incoming paths, so that $\theta = 40.0°$. The region to the right of the collision is divided into four lettered sections by the x axis and four numbered dashed lines. In what region or along what line do the particles travel if the collision is (a) completely inelastic, (b) elastic, and (c) inelastic? What are their final speeds if the collision is (d) completely inelastic and (e) elastic?

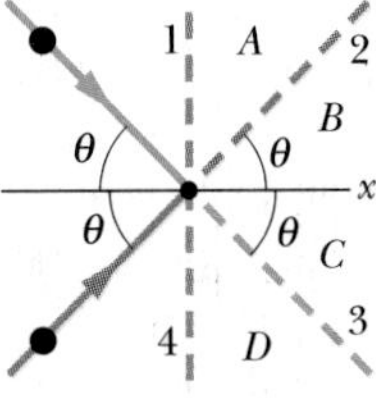

FIG. 9-75 Problem 82.

83 *"Relative" is an important word.* In Fig. 9-76, block L of mass $m_L = 1.00$ kg and block R of mass $m_R = 0.500$ kg are held in place with a compressed spring between them. When the blocks are released, the spring sends them sliding across a frictionless floor. (The spring has negligible mass and falls to the floor after the blocks leave it.) (a) If the spring gives block L a release speed of 1.20 m/s *relative* to the floor, how far does block R travel in the next 0.800 s? (b) If, instead, the spring gives block L a release speed of 1.20 m/s *relative* to the velocity that the spring gives block R, how far does block R travel in the next 0.800 s?

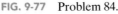

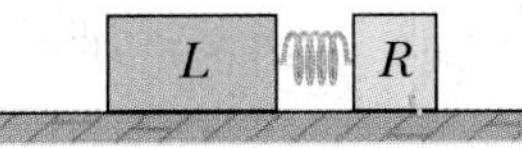

FIG. 9-76 Problem 83.

84 *Pancake collapse of a tall building*. In the section of a tall building shown in Fig. 9-77*a*, the infrastructure of any given floor K must support the weight W of all higher floors. Normally the infrastructure is constructed with a safety factor s so that it can withstand an even greater downward force of sW. If, however, the support columns between K and L suddenly collapse and allow the higher floors to free-fall together onto floor K (Fig. 9-77*b*), the force in the collision can exceed sW and, after a brief pause, cause K to collapse onto floor J, which collapses on floor I, and so on until the ground is reached. Assume that the floors are separated by $d = 4.0$ m and have the same mass. Also assume that when the floors above K free-fall onto K, the collision lasts 1.5 ms. Under these simplified conditions, what value must the safety factor s exceed to prevent pancake collapse of the building?

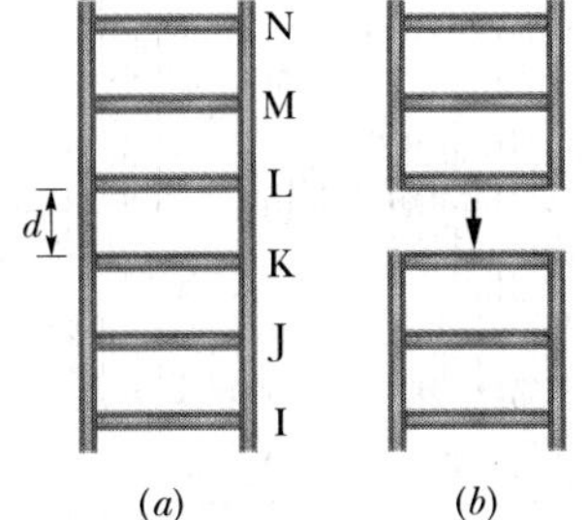

FIG. 9-77 Problem 84.

85 A railroad car moves under a grain elevator at a constant speed of 3.20 m/s. Grain drops into the car at the rate of 540 kg/min. What is the magnitude of the force needed to keep the car moving at constant speed if friction is negligible?

86 *Tyrannosaurus rex* may have known from experience not to run particularly fast because of the danger of tripping, in which case its short forearms would have been no help in cushioning the fall. Suppose a *T. rex* of mass m trips while walking, toppling over, with its center of mass falling freely a distance of 1.5 m. Then its center of mass descends an additional 0.30 m due to compression of its body and the ground. (a) In multiples of the dinosaur's weight, what is the approximate magnitude of the average vertical force on the dinosaur during its collision with the ground (during the descent of 0.30 m)? Now assume that the dinosaur is running at a speed of 19 m/s (fast) when it trips, falls to the ground, and then slides to a stop with a coefficient of kinetic friction of 0.6. Assume also that the average vertical force during the collision and sliding is that in (a). What, approximately, are (b) the magnitude of the average total force on the dinosaur from the ground (again in multiples of its weight) and (c) the sliding distance? The force magnitudes of (a) and (b) strongly suggest that the collision would injure the torso of the dinosaur. The head, which would fall farther, would suffer even greater injury.

87 A man (weighing 915 N) stands on a long railroad flatcar (weighing 2415 N) as it rolls at 18.2 m/s in the positive direction of an x axis, with negligible friction. Then the man runs along the flatcar in the negative x direction at 4.00 m/s relative to the flatcar. What is the resulting increase in the speed of the flatcar?

88 Figure 9-78 shows a uniform square plate of edge length $6d = 6.0$ m from which a square piece of edge length $2d$ has been removed. What are (a) the x coordinate and (b) the y coordinate of the center of mass of the remaining piece?

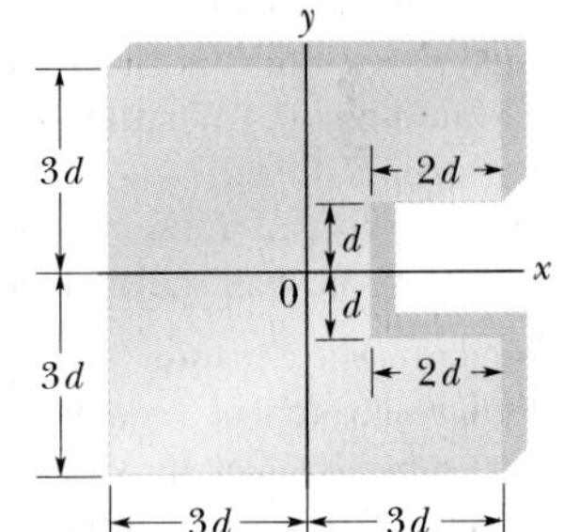

FIG. 9-78 Problem 88.

89 The last stage of a rocket, which is traveling at a speed of 7600 m/s, consists of two parts that are clamped together: a rocket case with a mass of 290.0 kg and a payload capsule with a mass of 150.0 kg. When the clamp is released, a compressed spring causes the two parts to separate with a relative speed of 910.0 m/s. What are the speeds of (a) the rocket case and (b) the payload after they have separated? Assume that all velocities are along the same line. Find the total kinetic energy of the two parts (c) before and (d) after they separate. (e) Account for the difference.

90 An object is tracked by a radar station and found to have a position vector given by $\vec{r} = (3500 - 160t)\hat{i} + 2700\hat{j} + 300\hat{k}$, with $\vec{r}$ in meters and t in seconds. The radar station's x axis points east, its y axis north, and its z axis vertically up. If the object is a 250 kg meteorological missile, what are (a) its linear momentum, (b) its direction of motion, and (c) the net force on it?

91 A pellet gun fires ten 2.0 g pellets per second with a speed of 500 m/s. The pellets are stopped by a rigid wall. What are (a) the magnitude of the momentum of each pellet, (b) the kinetic energy of each pellet, and (c) the magnitude of the average force on the wall from the stream of pellets? (d) If each pellet is in contact with the wall for 0.60 ms, what is the magnitude of the average force on the wall from each pellet during contact? (e) Why is this average force so different from the average force calculated in (c)? SSM

92 A body is traveling at 2.0 m/s along the positive direction of an x axis; no net force acts on the body. An internal explosion separates the body into two parts, each of 4.0 kg, and increases the total kinetic energy by 16 J. The forward part continues to move in the original direction of motion. What are the speeds of (a) the rear part and (b) the forward part?

93 A 1400 kg car moving at 5.3 m/s is initially traveling north along the positive direction of a y axis. After completing a 90° right-hand turn in 4.6 s, the inattentive operator drives into a tree, which stops the car in 350 ms. In unit-vector notation, what is the impulse on the car (a) due to the turn and (b) due to the collision? What is the magnitude of the average force that acts on the car (c) during the turn and (d) during the collision? (e) What is the direction of the average force during the turn? SSM

94 A spacecraft is separated into two parts by detonating the explosive bolts that hold them together. The masses of the parts are 1200 kg and 1800 kg; the magnitude of the impulse on each part from the bolts is 300 N·s. With what relative speed do the two parts separate because of the detonation?

95 A ball having a mass of 150 g strikes a wall with a speed of 5.2 m/s and rebounds with only 50% of its initial kinetic energy. (a) What is the speed of the ball immediately after rebounding? (b) What is the magnitude of the impulse on the wall from the ball? (c) If the ball is in contact with the wall for 7.6 ms, what is the magnitude of the average force on the ball from the wall during this time interval?

96 An old Chrysler with mass 2400 kg is moving along a straight stretch of road at 80 km/h. It is followed by a Ford with mass 1600 kg moving at 60 km/h. How fast is the center of mass of the two cars moving?

97 A railroad freight car of mass 3.18×10^4 kg collides with a stationary caboose car. They couple together, and 27.0% of the initial kinetic energy is transferred to thermal energy, sound, vibrations, and so on. Find the mass of the caboose. SSM

98 Two blocks of masses 1.0 kg and 3.0 kg are connected by a spring and rest on a frictionless surface. They are given velocities toward each other such that the 1.0 kg block travels initially at 1.7 m/s toward the center of mass, which remains at rest. What is the initial speed of the other block?

99 A 75 kg man is riding on a 39 kg cart traveling at a velocity of 2.3 m/s. He jumps off with zero horizontal velocity relative to the ground. What is the resulting change in the cart's velocity, including sign?

100 A certain radioactive (parent) nucleus transforms to a different (daughter) nucleus by emitting an electron and a neutrino. The parent nucleus was at rest at the origin of an xy coordinate system. The electron moves away from the origin with linear momentum $(-1.2 \times 10^{-22}\ \text{kg} \cdot \text{m/s})\hat{i}$; the neutrino moves away from the origin with linear momentum $(-6.4 \times 10^{-23}\ \text{kg} \cdot \text{m/s})\hat{j}$. What are the (a) magnitude and (b) direction of the linear momentum of the daughter nucleus? (c) If the daughter nucleus has a mass of 5.8×10^{-26} kg, what is its kinetic energy? ILW

101 In the arrangement of Fig. 9-23, billiard ball 1 moving at a speed of 2.2 m/s undergoes a glancing collision with identical billiard ball 2 that is at rest. After the collision, ball 2 moves at speed 1.1 m/s, at an angle of $\theta_2 = 60°$. What are (a) the magnitude and (b) the direction of the velocity of ball 1 after the collision? (c) Do the given data suggest the collision is elastic or inelastic? SSM

102 A rocket is moving away from the solar system at a speed of 6.0×10^3 m/s. It fires its engine, which ejects exhaust with a speed of 3.0×10^3 m/s relative to the rocket. The mass of the rocket at this time is 4.0×10^4 kg, and its acceleration is 2.0 m/s². (a) What is the thrust of the engine? (b) At what rate, in kilograms per second, is exhaust ejected during the firing?

103 The three balls in the overhead view of Fig. 9-79 are identical. Balls 2 and 3 touch each other and are aligned perpendicular to the path of ball 1. The velocity of ball 1 has magnitude $v_0 = 10$ m/s and is directed at the contact point of balls 1 and 2. After the collision, what are the (a) speed and (b) direction of the velocity of ball 2, the (c) speed and (d) direction of the velocity of ball 3, and the (e) speed and (f) direction of the velocity of ball 1? (*Hint:* With friction absent, each impulse is directed along the line connecting the centers of the colliding balls, normal to the colliding surfaces.)

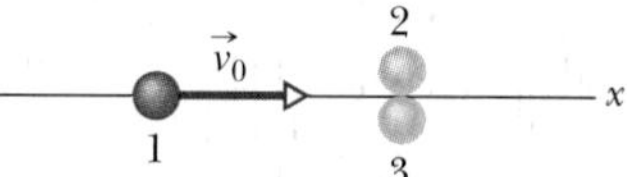

FIG. 9-79 Problem 103.

104 In a game of pool, the cue ball strikes another ball of the same mass and initially at rest. After the collision, the cue ball moves at 3.50 m/s along a line making an angle of 22.0° with the cue ball's original direction of motion, and the second ball has a speed of 2.00 m/s. Find (a) the angle between the direction of motion of the second ball and the original direction of motion of the cue ball and (b) the original speed of the cue ball. (c) Is kinetic energy (of the centers of mass, don't consider the rotation) conserved?

105 In Fig. 9-80, two identical containers of sugar are connected by a cord that passes over a frictionless pulley. The cord and pulley have negligible mass, each container and its sugar together have a mass of 500 g, the centers of the containers are separated by 50 mm, and the containers are held fixed at the same height. What is the horizontal distance between the center of container 1 and the center of mass of the two-container system (a) initially and (b) after 20 g of sugar is transferred from container 1 to container 2? After the transfer and after the containers are released, (c) in what direction and (d) at what acceleration magnitude does the center of mass move?

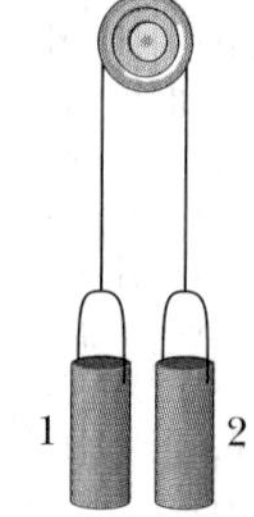

FIG. 9-80 Problem 105.

106 A 0.15 kg ball hits a wall with a velocity of $(5.00\ \text{m/s})\hat{i} + (6.50\ \text{m/s})\hat{j} + (4.00\ \text{m/s})\hat{k}$. It rebounds from the wall with a velocity of $(2.00\ \text{m/s})\hat{i} + (3.50\ \text{m/s})\hat{j} + (-3.20\ \text{m/s})\hat{k}$. What are (a) the change in the ball's momentum, (b) the impulse on the ball, and (c) the impulse on the wall?

107 At time $t = 0$, force $\vec{F}_1 = (-4.00\hat{i} + 5.00\hat{j})$ N acts on an initially stationary particle of mass 2.00×10^{-3} kg and force $\vec{F}_2 = (2.00\hat{i} - 4.00\hat{j})$ N acts on an initially stationary particle of mass 4.00×10^{-3} kg. From time $t = 0$ to $t = 2.00$ ms, what are the (a) magnitude and (b) angle (relative to the positive direction of the x axis) of the displacement of the center of mass of the two-particle system? (c) What is the kinetic energy of the center of mass at $t = 2.00$ ms? SSM

108 A 0.550 kg ball falls directly down onto concrete, hitting it with a speed of 12.0 m/s and rebounding directly upward with a speed of 3.00 m/s. Extend a y axis upward. In unit-vector notation, what are (a) the change in the ball's momentum, (b) the impulse on the ball, and (c) the impulse on the concrete?

109 A collision occurs between a 2.00 kg particle traveling with velocity $\vec{v}_1 = (-4.00 \text{ m/s})\hat{i} + (-5.00 \text{ m/s})\hat{j}$ and a 4.00 kg particle traveling with velocity $\vec{v}_2 = (6.00 \text{ m/s})\hat{i} + (-2.00 \text{ m/s})\hat{j}$. The collision connects the two particles. What then is their velocity in (a) unit-vector notation and as a (b) magnitude and (c) angle?

110 An atomic nucleus at rest at the origin of an xy coordinate system transforms into three particles. Particle 1, mass 16.7×10^{-27} kg, moves away from the origin at velocity $(6.00 \times 10^6 \text{ m/s})\hat{i}$; particle 2, mass 8.35×10^{-27} kg, moves away at velocity $(-8.00 \times 10^6 \text{ m/s})\hat{j}$. (a) In unit-vector notation, what is the linear momentum of the third particle, mass 11.7×10^{-27} kg? (b) How much kinetic energy appears in this transformation?

111 An electron undergoes a one-dimensional elastic collision with an initially stationary hydrogen atom. What percentage of the electron's initial kinetic energy is transferred to kinetic energy of the hydrogen atom? (The mass of the hydrogen atom is 1840 times the mass of the electron.)

112 The script for an action movie calls for a small race car (of mass 1500 kg and length 3.0 m) to accelerate along a flattop boat (of mass 4000 kg and length 14 m), from one end of the boat to the other, where the car will then jump the gap between the boat and a somewhat lower dock. You are the technical advisor for the movie. The boat will initially touch the dock, as in Fig. 9-81; the boat can slide through the water without significant resistance; both the car and the boat can be approximated as uniform in their mass distribution. Determine what the width of the gap will be just as the car is about to make the jump.

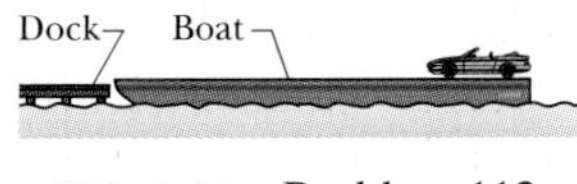

FIG. 9-81 Problem 112.

113 A rocket sled with a mass of 2900 kg moves at 250 m/s on a set of rails. At a certain point, a scoop on the sled dips into a trough of water located between the tracks and scoops water into an empty tank on the sled. By applying the principle of conservation of linear momentum, determine the speed of the sled after 920 kg of water has been scooped up. Ignore any retarding force on the scoop. SSM

114 A 140 g ball with speed 7.8 m/s strikes a wall perpendicularly and rebounds in the opposite direction with the same speed. The collision lasts 3.80 ms. What are the magnitudes of the (a) impulse and (b) average force on the wall from the ball?

115 (a) How far is the center of mass of the Earth–Moon system from the center of Earth? (Appendix C gives the masses of Earth and the Moon and the distance between the two.) (b) What percentage of Earth's radius is that distance? SSM

116 A 500.0 kg module is attached to a 400.0 kg shuttle craft, which moves at 1000 m/s relative to the stationary main spaceship. Then a small explosion sends the module backward with speed 100.0 m/s relative to the new speed of the shuttle craft. As measured by someone on the main spaceship, by what fraction did the kinetic energy of the module and shuttle craft increase because of the explosion?

117 A 6100 kg rocket is set for vertical firing from the ground. If the exhaust speed is 1200 m/s, how much gas must be ejected each second if the thrust (a) is to equal the magnitude of the gravitational force on the rocket and (b) is to give the rocket an initial upward acceleration of 21 m/s^2? SSM

118 A 2140 kg railroad flatcar, which can move with negligible friction, is motionless next to a platform. A 242 kg sumo wrestler runs at 5.3 m/s along the platform (parallel to the track) and then jumps onto the flatcar. What is the speed of the flatcar if he then (a) stands on it, (b) runs at 5.3 m/s relative to it in his original direction, and (c) turns and runs at 5.3 m/s relative to the flatcar opposite his original direction?

119 In Fig. 9-82, block 1 slides along an x axis on a frictionless floor with a speed of 0.75 m/s. When it reaches stationary block 2, the two blocks undergo an elastic collision. The following table gives the mass and length of the (uniform) blocks and also the locations of their centers at time $t = 0$. Where is the center of mass of the two-block system located (a) at $t = 0$, (b) when the two blocks first touch, and (c) at $t = 4.0$ s?

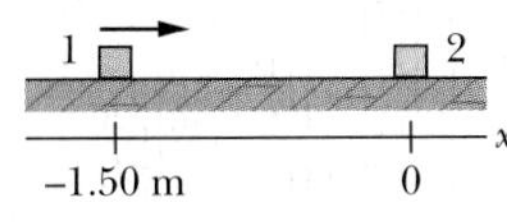

FIG. 9-82 Problem 119.

Block	Mass (kg)	Length (cm)	Center at $t = 0$
1	0.25	5.0	$x = -1.50$ m
2	0.50	6.0	$x = 0$

120 In Fig. 9-83, an 80 kg man is on a ladder hanging from a balloon that has a total mass of 320 kg (including the basket passenger). The balloon is initially stationary relative to the ground. If the man on the ladder begins to climb at 2.5 m/s relative to the ladder, (a) in what direction and (b) at what speed does the balloon move? (c) If the man then stops climbing, what is the speed of the balloon?

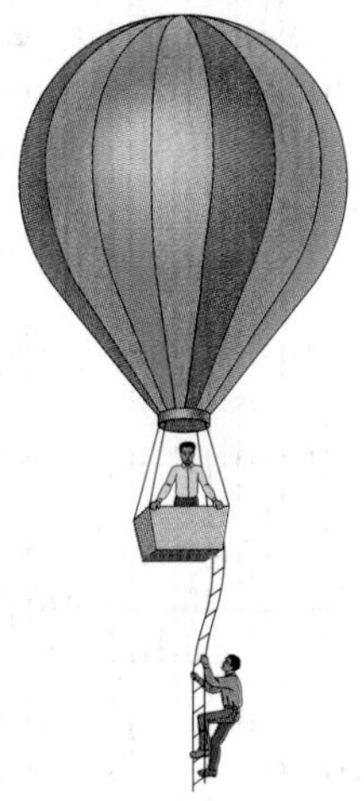
FIG. 9-83 Problem 120.

121 Particle 1 of mass 200 g and speed 3.00 m/s undergoes a one-dimensional collision with stationary particle 2 of mass 400 g. What is the magnitude of the impulse on particle 1 if the collision is (a) elastic and (b) completely inelastic?

122 During a lunar mission, it is necessary to increase the speed of a spacecraft by 2.2 m/s when it is moving at 400 m/s relative to the Moon. The speed of the exhaust products from the rocket engine is 1000 m/s relative to the spacecraft. What fraction of the initial mass of the spacecraft must be burned and ejected to accomplish the speed increase?

123 In Fig. 9-84, a 3.2 kg box of running shoes slides on a horizontal frictionless table and collides with a 2.0 kg box of ballet slippers initially at rest on the edge of the table, at height $h = 0.40$ m. The speed of the 3.2 kg box is 3.0 m/s just before the collision. If the two boxes stick together because of packing tape on their sides, what is their kinetic energy just before they strike the floor?

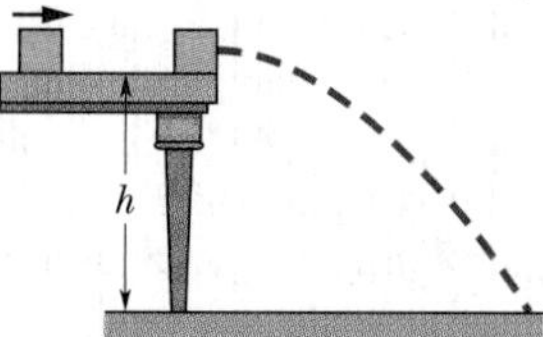

FIG. 9-84 Problem 123.

124 In the two-sphere arrangement of Sample Problem 9-11, assume that sphere 1 has a mass of 50 g and an initial height of $h_1 = 9.0$ cm, and that sphere 2 has a mass of 85 g. After sphere 1 is released and collides elastically with sphere 2, what height is reached by (a) sphere 1 and (b) sphere 2? After the next (elastic) collision, what height is reached by (c) sphere 1 and (d) sphere 2? (*Hint:* Do not use rounded-off values.)

125 A 3000 kg block falls vertically through 6.0 m and then collides with a 500 kg pile, driving it 3.0 cm into bedrock. Assuming that the block–pile collision is completely inelastic, find the magnitude of the average force on the pile from the bedrock during the 3.0 cm descent.

126 In Fig. 9-65, block 1 (mass 6.0 kg) is moving rightward at 8.0 m/s and block 2 (mass 4.0 kg) is moving rightward at 2.0 m/s. The surface is frictionless, and a spring with a spring constant of 8000 N/m is fixed to block 2. Eventually block 1 overtakes block 2. At the instant block 1 is moving rightward at 6.4 m/s, what are (a) the speed of block 2 and (b) the elastic potential energy of the spring?

127 An electron (mass $m_1 = 9.11 \times 10^{-31}$ kg) and a proton (mass $m_2 = 1.67 \times 10^{-27}$ kg) attract each other via an electrical force. Suppose that an electron and a proton are released from rest with an initial separation $d = 3.0 \times 10^{-6}$ m. When their separation has decreased to 1.0×10^{-6} m, what is the ratio of (a) the electron's linear momentum magnitude to the proton's linear momentum magnitude, (b) the electron's speed to the proton's speed, and (c) the electron's kinetic energy to the proton's kinetic energy? (d) As the separation continues to decrease, do the answers to (a) through (c) increase, decrease, or remain the same?

128 A railroad freight car weighing 280 kN and traveling at 1.52 m/s overtakes one weighing 210 kN and traveling at 0.914 m/s in the same direction. If the cars couple together, find (a) the speed of the cars after the collision and (b) the loss of kinetic energy during the collision. If instead, as is very unlikely, the collision is elastic, find the after-collision speed of (c) the lighter car and (d) the heavier car.

129 A 3.0 kg object moving at 8.0 m/s in the positive direction of an x axis has a one-dimensional elastic collision with an object of mass M, initially at rest. After the collision the object of mass M has a velocity of 6.0 m/s in the positive direction of the axis. What is mass M? SSM

130 Two particles P and Q are released from rest 1.0 m apart. P has a mass of 0.10 kg, and Q a mass of 0.30 kg. P and Q attract each other with a constant force of 1.0×10^{-2} N. No external forces act on the system. (a) What is the speed of the center of mass of P and Q when the separation is 0.50 m? (b) At what distance from P's original position do the particles collide?

131 In Fig. 9-85, block 1 of mass $m_1 = 6.6$ kg is at rest on a long frictionless table that is up against a wall. Block 2 of mass m_2 is placed between block 1 and the wall and sent sliding to the left, toward block 1, with constant speed v_{2i}. Find the value of m_2 for which both blocks move with the same velocity after block 2 has collided once with block 1 and once with the wall. Assume all collisions are elastic (the collision with the wall does not change the speed of block 2).

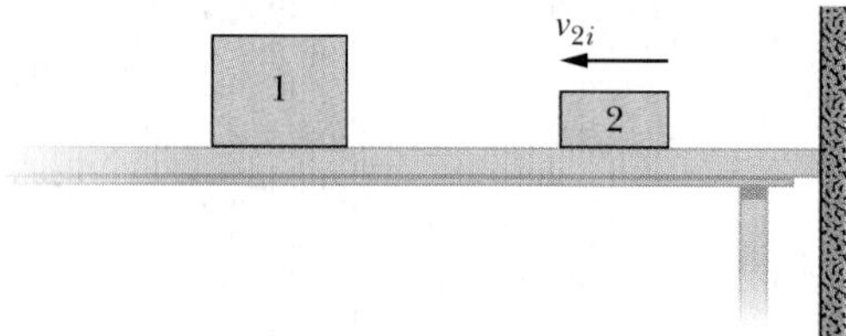

FIG. 9-85 Problem 131.

132 A rocket of mass M moves along an x axis at the constant speed $v_i = 40$ m/s. A small explosion separates the rocket into a rear section (of mass m_1) and a front section; both sections move along the x axis. The relative speed between the rear and front sections is 20 m/s. What are (a) the minimum possible value of final speed v_f of the front section and (b) for what limiting value of m_1 does it occur? (c) What is the maximum possible value of v_f and (d) for what limiting value of m_1 does it occur?

133 A 2.65 kg stationary package explodes into three parts that then slide across a frictionless floor. The package had been at the origin of a coordinate system. Part 1 has mass $m_1 = 0.500$ kg and velocity $(10.0\hat{i} + 12.0\hat{j})$ m/s. Part 2 has mass $m_2 = 0.750$ kg and a speed of 14.0 m/s, and travels at an angle 110° (counterclockwise from the positive direction of the x axis). (a) What is the speed of part 3? (b) In what direction does it travel?

134 Particle 1 with mass 3.0 kg and velocity $(5.0 \text{ m/s})\hat{i}$ undergoes a one-dimensional elastic collision with particle 2 with mass 2.0 kg and velocity $(-6.0 \text{ m/s})\hat{i}$. After the collision, what are the velocities of (a) particle 1 and (b) particle 2?

135 At a certain instant, four particles have the xy coordinates and velocities given in the following table. At that instant, what are the (a) x and (b) y coordinates of their center of mass and (c) the velocity of their center of mass?

Particle	Mass (kg)	Position (m)	Velocity (m/s)
1	2.0	0, 3.0	$-9.0\hat{j}$
2	4.0	3.0, 0	$6.0\hat{i}$
3	3.0	0, −2.0	$6.0\hat{j}$
4	12	−1.0, 0	$-2.0\hat{i}$

136 Figure 9-86 shows two 22.7 kg ice sleds that are placed a short distance apart, one directly behind the other. A 3.63 kg cat initially standing on one sled jumps to the other one and then back to the first. Both jumps are made at a speed of 3.05 m/s relative to the ice. What are the final speeds of (a) the first sled and (b) the other sled?

FIG. 9-86 Problem 136.

Rotation 10

David Wrobel/Visuals Unlimited

The snapping shrimp stuns its prey (tiny crabs) by clamping its oversized claw shut but not on the prey itself. Instead, the prey is stunned by a powerful sound wave generated by the moving part of the claw as it nears the stationary part of the claw. The sound (a snap resembling the pop of popcorn) can be heard by a diver and, with enough shrimp, can be loud enough to hide a submarine from sonar detection. The sound wave can also produce very faint flashes of light, what is generally called sound-produced light or sonoluminescence. *However, some researchers have dubbed the shrimp-produced light* shrimpoluminescence.

How can a shrimp claw generate such a powerful sound wave?

The answer is in this chapter.

(a)

(b)

FIG. 10-1 Figure skater Sasha Cohen in motion of (*a*) pure translation in a fixed direction and (*b*) pure rotation about a vertical axis. *(a: Mike Segar/Reuters/Landov LLC; b: Elsa/Getty Images, Inc.)*

10-1 WHAT IS PHYSICS?

As we have discussed, one focus of physics is motion. However, so far we have examined only the motion of **translation,** in which an object moves along a straight or curved line, as in Fig. 10-1*a*. We now turn to the motion of **rotation,** in which an object turns about an axis, as in Fig. 10-1*b*.

You see rotation in nearly every machine, you use it every time you open a beverage can with a pull tab, and you pay to experience it every time you go to an amusement park. Rotation is the key to many fun activities, such as hitting a long drive in golf (the ball needs to rotate in order for the air to keep it aloft longer) and throwing a curveball in baseball (the ball needs to rotate in order for the air to push it left or right). Rotation is also the key to more serious matters, such as metal failure in aging airplanes.

We begin our discussion of rotation by defining the variables for the motion, just as we did for translation in Chapter 2.

10-2 | The Rotational Variables

We wish to examine the rotation of a rigid body about a fixed axis. A **rigid body** is a body that can rotate with all its parts locked together and without any change in its shape. A **fixed axis** means that the rotation occurs about an axis that does not move. Thus, we shall not examine an object like the Sun, because the parts of the Sun (a ball of gas) are not locked together. We also shall not examine an object like a bowling ball rolling along a lane, because the ball rotates about a moving axis (the ball's motion is a mixture of rotation and translation).

Figure 10-2 shows a rigid body of arbitrary shape in rotation about a fixed axis, called the **axis of rotation** or the **rotation axis.** In pure rotation (*angular motion*), every point of the body moves in a circle whose center lies on the axis of rotation, and every point moves through the same angle during a particular time interval. In pure translation (*linear motion*), every point of the body moves in a straight line, and every point moves through the same *linear distance* during a particular time interval.

We deal now—one at a time—with the angular equivalents of the linear quantities position, displacement, velocity, and acceleration.

Angular Position

Figure 10-2 shows a *reference line*, fixed in the body, perpendicular to the rotation axis and rotating with the body. The **angular position** of this line is the angle of the line relative to a fixed direction, which we take as the **zero angular position.** In Fig. 10-3, the angular position θ is measured relative to the positive direction of the x axis. From geometry, we know that θ is given by

$$\theta = \frac{s}{r} \quad \text{(radian measure).} \tag{10-1}$$

Here s is the length of a circular arc that extends from the x axis (the zero angular position) to the reference line, and r is the radius of the circle.

An angle defined in this way is measured in **radians** (rad) rather than in revolutions (rev) or degrees. The radian, being the ratio of two lengths, is a pure number and thus has no dimension. Because the circumference of a circle of radius r is $2\pi r$, there are 2π radians in a complete circle:

$$1 \text{ rev} = 360° = \frac{2\pi r}{r} = 2\pi \text{ rad}, \tag{10-2}$$

and thus

$$1 \text{ rad} = 57.3° = 0.159 \text{ rev}. \tag{10-3}$$

We do *not* reset θ to zero with each complete rotation of the reference line about the rotation axis. If the reference line completes two revolutions from the zero angular position, then the angular position θ of the line is $\theta = 4\pi$ rad.

For pure translation along an x axis, we can know all there is to know about a moving body if we know $x(t)$, its position as a function of time. Similarly, for pure rotation, we can know all there is to know about a rotating body if we know $\theta(t)$, the angular position of the body's reference line as a function of time.

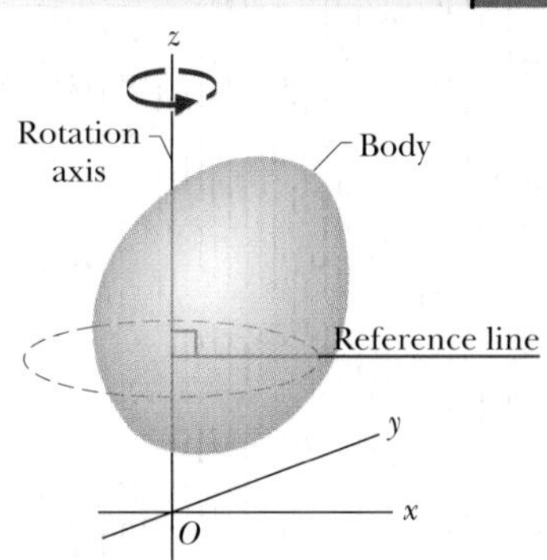

FIG. 10-2 A rigid body of arbitrary shape in pure rotation about the z axis of a coordinate system. The position of the *reference line* with respect to the rigid body is arbitrary, but it is perpendicular to the rotation axis. It is fixed in the body and rotates with the body.

Angular Displacement

If the body of Fig. 10-3 rotates about the rotation axis as in Fig. 10-4, changing the angular position of the reference line from θ_1 to θ_2, the body undergoes an **angular displacement** $\Delta\theta$ given by

$$\Delta\theta = \theta_2 - \theta_1. \tag{10-4}$$

This definition of angular displacement holds not only for the rigid body as a whole but also for *every particle within that body.*

If a body is in translational motion along an x axis, its displacement Δx is either positive or negative, depending on whether the body is moving in the positive or negative direction of the axis. Similarly, the angular displacement $\Delta\theta$ of a rotating body is either positive or negative, according to the following rule:

An angular displacement in the counterclockwise direction is positive, and one in the clockwise direction is negative.

The phrase "*clocks are negative*" can help you remember this rule (they certainly are negative when their alarms sound off early in the morning).

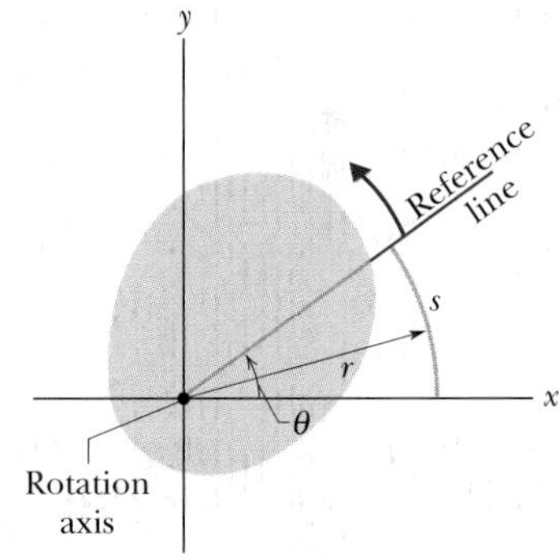

FIG. 10-3 The rotating rigid body of Fig. 10-2 in cross section, viewed from above. The plane of the cross section is perpendicular to the rotation axis, which now extends out of the page, toward you. In this position of the body, the reference line makes an angle θ with the x axis.

CHECKPOINT 1 A disk can rotate about its central axis like a merry-go-round. Which of the following pairs of values for its initial and final angular positions, respectively, give a negative angular displacement: (a) -3 rad, $+5$ rad, (b) -3 rad, -7 rad, (c) 7 rad, -3 rad?

Angular Velocity

Suppose that our rotating body is at angular position θ_1 at time t_1 and at angular position θ_2 at time t_2 as in Fig. 10-4. We define the **average angular velocity** of the body in the time interval Δt from t_1 to t_2 to be

$$\omega_{avg} = \frac{\theta_2 - \theta_1}{t_2 - t_1} = \frac{\Delta\theta}{\Delta t}, \tag{10-5}$$

in which $\Delta\theta$ is the angular displacement that occurs during Δt (ω is the lowercase Greek letter omega).

The **(instantaneous) angular velocity** ω, with which we shall be most concerned, is the limit of the ratio in Eq. 10-5 as Δt approaches zero. Thus,

$$\omega = \lim_{\Delta t \to 0} \frac{\Delta\theta}{\Delta t} = \frac{d\theta}{dt}. \tag{10-6}$$

If we know $\theta(t)$, we can find the angular velocity ω by differentiation.

Equations 10-5 and 10-6 hold not only for the rotating rigid body as a whole but also for *every particle of that body* because the particles are all locked together. The unit of angular velocity is commonly the radian per second (rad/s) or the revolution per second (rev/s). Another measure of angular velocity was used during at least the first three decades of rock: Music was produced by vinyl (phonograph) records that were played on turntables at "$33\frac{1}{3}$ rpm" or "45 rpm," meaning at $33\frac{1}{3}$ rev/min or 45 rev/min.

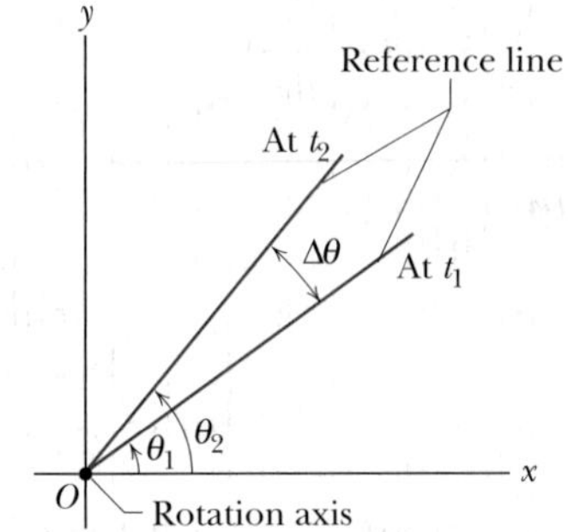

FIG. 10-4 The reference line of the rigid body of Figs. 10-2 and 10-3 is at angular position θ_1 at time t_1 and at angular position θ_2 at a later time t_2. The quantity $\Delta\theta$ ($= \theta_2 - \theta_1$) is the angular displacement that occurs during the interval Δt ($= t_2 - t_1$). The body itself is not shown.

If a particle moves in translation along an x axis, its linear velocity v is either positive or negative, depending on whether the particle is moving in the positive or negative direction of the axis. Similarly, the angular velocity ω of a rotating rigid body is either positive or negative, depending on whether the body is rotating counterclockwise (positive) or clockwise (negative). ("Clocks are negative" still works.) The magnitude of an angular velocity is called the **angular speed,** which is also represented with ω.

Angular Acceleration

If the angular velocity of a rotating body is not constant, then the body has an angular acceleration. Let ω_2 and ω_1 be its angular velocities at times t_2 and t_1, respectively. The **average angular acceleration** of the rotating body in the interval from t_1 to t_2 is defined as

$$\alpha_{avg} = \frac{\omega_2 - \omega_1}{t_2 - t_1} = \frac{\Delta\omega}{\Delta t}, \qquad (10\text{-}7)$$

in which $\Delta\omega$ is the change in the angular velocity that occurs during the time interval Δt. The **(instantaneous) angular acceleration** α, with which we shall be most concerned, is the limit of this quantity as Δt approaches zero. Thus,

$$\alpha = \lim_{\Delta t \to 0} \frac{\Delta\omega}{\Delta t} = \frac{d\omega}{dt}. \qquad (10\text{-}8)$$

Equations 10-7 and 10-8 also hold for *every particle of that body.* The unit of angular acceleration is commonly the radian per second-squared (rad/s^2) or the revolution per second-squared (rev/s^2).

Sample Problem 10-1 Build your skill

The disk in Fig. 10-5*a* is rotating about its central axis like a merry-go-round. The angular position $\theta(t)$ of a reference line on the disk is given by

$$\theta = -1.00 - 0.600t + 0.250t^2, \qquad (10\text{-}9)$$

with t in seconds, θ in radians, and the zero angular position as indicated in the figure.

(a) Graph the angular position of the disk versus time from $t = -3.0$ s to $t = 5.4$ s. Sketch the disk and its angular position reference line at $t = -2.0$ s, 0 s, and 4.0 s, and when the curve crosses the t axis.

KEY IDEA The angular position of the disk is the angular position $\theta(t)$ of its reference line, which is given by Eq. 10-9 as a function of time t. So we graph Eq. 10-9; the result is shown in Fig. 10-5*b*.

Calculations: To sketch the disk and its reference line at a particular time, we need to determine θ for that time. To do so, we substitute the time into Eq. 10-9. For $t = -2.0$ s, we get

$$\theta = -1.00 - (0.600)(-2.0) + (0.250)(-2.0)^2$$
$$= 1.2 \text{ rad} = 1.2 \text{ rad}\,\frac{360°}{2\pi \text{ rad}} = 69°.$$

This means that at $t = -2.0$ s the reference line on the disk is rotated counterclockwise from the zero position by 1.2 rad = 69° (counterclockwise because θ is positive). Sketch 1 in Fig. 10-5*b* shows this position of the reference line.

Similarly, for $t = 0$, we find $\theta = -1.00$ rad $= -57°$, which means that the reference line is rotated clockwise from the zero angular position by 1.0 rad, or 57°, as shown in sketch 3. For $t = 4.0$ s, we find $\theta = 0.60$ rad $=$ 34° (sketch 5). Drawing sketches for when the curve crosses the t axis is easy, because then $\theta = 0$ and the reference line is momentarily aligned with the zero angular position (sketches 2 and 4).

(b) At what time t_{min} does $\theta(t)$ reach the minimum value shown in Fig. 10-5*b*? What is that minimum value?

KEY IDEA To find the extreme value (here the minimum) of a function, we take the first derivative of the function and set the result to zero.

Calculations: The first derivative of $\theta(t)$ is

$$\frac{d\theta}{dt} = -0.600 + 0.500t. \qquad (10\text{-}10)$$

Setting this to zero and solving for t give us the time at

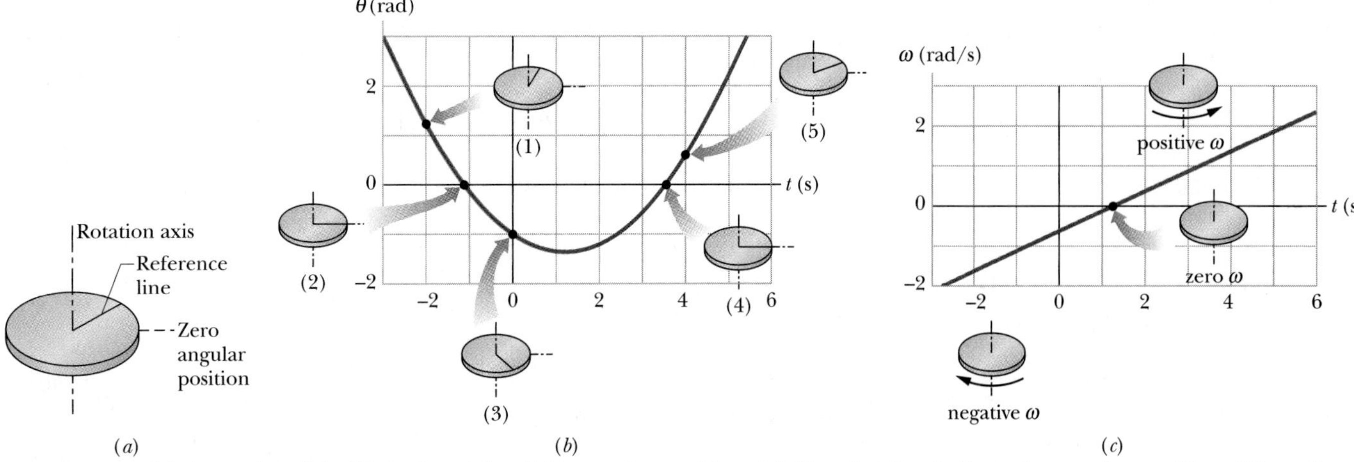

(*a*) (*b*) (*c*)

FIG. 10-5 (*a*) A rotating disk. (*b*) A plot of the disk's angular position $\theta(t)$. Five sketches indicate the angular position of the reference line on the disk for five points on the curve. (*c*) A plot of the disk's angular velocity $\omega(t)$. Positive values of ω correspond to counterclockwise rotation, and negative values to clockwise rotation.

which $\theta(t)$ is minimum:

$$t_{\min} = 1.20 \text{ s.} \qquad \text{(Answer)}$$

To get the minimum value of θ, we next substitute $t_{\min}$ into Eq. 10-9, finding

$$\theta = -1.36 \text{ rad} \approx -77.9°. \qquad \text{(Answer)}$$

This *minimum* of $\theta(t)$ (the bottom of the curve in Fig. 10-5*b*) corresponds to the *maximum clockwise* rotation of the disk from the zero angular position, somewhat more than is shown in sketch 3.

(c) Graph the angular velocity ω of the disk versus time from $t = -3.0$ s to $t = 6.0$ s. Sketch the disk and indicate the direction of turning and the sign of ω at $t = -2.0$ s, 4.0 s, and $t_{\min}$.

KEY IDEA From Eq. 10-6, the angular velocity ω is equal to $d\theta/dt$ as given in Eq. 10-10. So, we have

$$\omega = -0.600 + 0.500t. \qquad (10\text{-}11)$$

The graph of this function $\omega(t)$ is shown in Fig. 10-5*c*.

Calculations: To sketch the disk at $t = -2.0$ s, we substitute that value into Eq. 10-11, obtaining

$$\omega = -1.6 \text{ rad/s.} \qquad \text{(Answer)}$$

The minus sign tells us that at $t = -2.0$ s, the disk is turning clockwise (the lowest sketch in Fig. 10-5*c*).

Substituting $t = 4.0$ s into Eq. 10-11 gives us

$$\omega = 1.4 \text{ rad/s.} \qquad \text{(Answer)}$$

The implied plus sign tells us that at $t = 4.0$ s, the disk is turning counterclockwise (the highest sketch in Fig. 10-5*c*).

For $t_{\min}$, we already know that $d\theta/dt = 0$. So, we must also have $\omega = 0$. That is, the disk momentarily stops when the reference line reaches the minimum value of θ in Fig. 10-5*b*, as suggested by the center sketch in Fig. 10-5*c*.

(d) Use the results in parts (a) through (c) to describe the motion of the disk from $t = -3.0$ s to $t = 6.0$ s.

Description: When we first observe the disk at $t = -3.0$ s, it has a positive angular position and is turning clockwise but slowing. It stops at angular position $\theta = -1.36$ rad and then begins to turn counterclockwise, with its angular position eventually becoming positive again.

Sample Problem 10-2

A child's top is spun with angular acceleration

$$\alpha = 5t^3 - 4t,$$

with t in seconds and α in radians per second-squared. At $t = 0$, the top has angular velocity 5 rad/s, and a reference line on it is at angular position $\theta = 2$ rad.

(a) Obtain an expression for the angular velocity $\omega(t)$ of the top.

KEY IDEA By definition, $\alpha(t)$ is the derivative of $\omega(t)$ with respect to time. Thus, we can find $\omega(t)$ by integrating $\alpha(t)$ with respect to time.

Calculations: Equation 10-8 tells us

$$d\omega = \alpha\, dt,$$

so

$$\int d\omega = \int \alpha\, dt.$$

From this we find

$$\omega = \int (5t^3 - 4t)\, dt = \tfrac{5}{4}t^4 - \tfrac{4}{2}t^2 + C.$$

To evaluate the constant of integration C, we note that $\omega = 5$ rad/s at $t = 0$. Substituting these values in our expression for ω yields

$$5 \text{ rad/s} = 0 - 0 + C,$$

so $C = 5$ rad/s. Then

$$\omega = \tfrac{5}{4}t^4 - 2t^2 + 5. \qquad \text{(Answer)}$$

(b) Obtain an expression for the angular position $\theta(t)$ of the top.

KEY IDEA By definition, $\omega(t)$ is the derivative of $\theta(t)$ with respect to time. Therefore, we can find $\theta(t)$ by integrating $\omega(t)$ with respect to time.

Calculations: Since Eq. 10-6 tells us that

$$d\theta = \omega\, dt,$$

we can write

$$\begin{aligned}\theta &= \int \omega\, dt = \int (\tfrac{5}{4}t^4 - 2t^2 + 5)\, dt \\ &= \tfrac{1}{4}t^5 - \tfrac{2}{3}t^3 + 5t + C' \\ &= \tfrac{1}{4}t^5 - \tfrac{2}{3}t^3 + 5t + 2, \qquad \text{(Answer)}\end{aligned}$$

where C' has been evaluated by noting that $\theta = 2$ rad at $t = 0$.

10-3 | Are Angular Quantities Vectors?

We can describe the position, velocity, and acceleration of a single particle by means of vectors. If the particle is confined to a straight line, however, we do not really need vector notation. Such a particle has only two directions available to it, and we can indicate these directions with plus and minus signs.

In the same way, a rigid body rotating about a fixed axis can rotate only clockwise or counterclockwise as seen along the axis, and again we can select between the two directions by means of plus and minus signs. The question arises: "Can we treat the angular displacement, velocity, and acceleration of a rotating body as vectors?" The answer is a qualified "yes" (see the caution below, in connection with angular displacements).

Consider the angular velocity. Figure 10-6*a* shows a vinyl record rotating on a turntable. The record has a constant angular speed ω (= $33\tfrac{1}{3}$ rev/min) in the clockwise direction. We can represent its angular velocity as a vector $\vec{\omega}$ pointing along the axis of rotation, as in Fig. 10-6*b*. Here's how: We choose the length of this vector according to some convenient scale, for example, with 1 cm corresponding to 10 rev/min. Then we establish a direction for the vector $\vec{\omega}$ by using a **right-hand rule,** as Fig. 10-6*c* shows: Curl your right hand about the rotating record, your fingers pointing *in the direction of rotation.* Your extended thumb will then point in the direction of the angular velocity vector. If the record were to rotate in the opposite sense, the right-hand rule would tell you that the angular velocity vector then points in the opposite direction.

It is not easy to get used to representing angular quantities as vectors. We instinctively expect that something should be moving *along* the direction of a vector. That is not the case here. Instead, something (the rigid body) is rotating *around* the direction of the vector. In the world of pure rotation, a vector defines

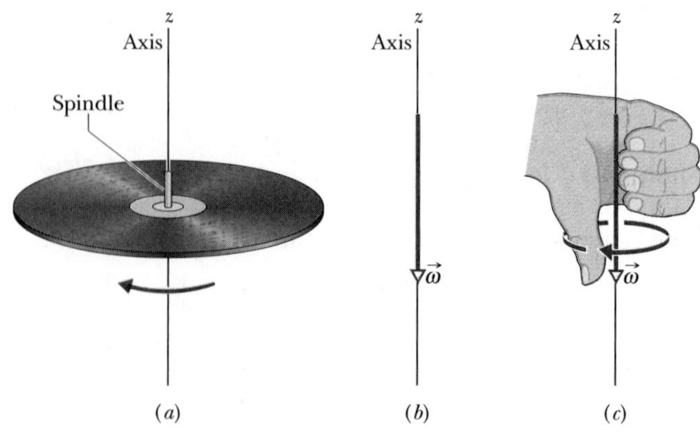

FIG. 10-6 (*a*) A record rotating about a vertical axis that coincides with the axis of the spindle. (*b*) The angular velocity of the rotating record can be represented by the vector $\vec{\omega}$, lying along the axis and pointing down, as shown. (*c*) We establish the direction of the angular velocity vector as downward by using a right-hand rule. When the fingers of the right hand curl around the record and point the way it is moving, the extended thumb points in the direction of $\vec{\omega}$.

an axis of rotation, not a direction in which something moves. Nonetheless, the vector also defines the motion. Furthermore, it obeys all the rules for vector manipulation discussed in Chapter 3. The angular acceleration $\vec{\alpha}$ is another vector, and it too obeys those rules.

In this chapter we consider only rotations that are about a fixed axis. For such situations, we need not consider vectors—we can represent angular velocity with ω and angular acceleration with α, and we can indicate direction with an implied plus sign for counterclockwise or an explicit minus sign for clockwise.

Now for the caution: Angular *displacements* (unless they are very small) *cannot* be treated as vectors. Why not? We can certainly give them both magnitude and direction, as we did for the angular velocity vector in Fig. 10-6. However, to be represented as a vector, a quantity must *also* obey the rules of vector addition, one of which says that if you add two vectors, the order in which you add them does not matter. Angular displacements fail this test.

Figure 10-7 gives an example. An initially horizontal book is given two 90° angular displacements, first in the order of Fig. 10-7*a* and then in the order of Fig. 10-7*b*. Although the two angular displacements are identical, their order is not, and the book ends up with different orientations. Here's another example. Hold your right arm downward, palm toward your thigh. Keeping your wrist rigid, (1) lift the arm forward until it is horizontal, (2) move it horizontally until it points toward the right, and (3) then bring it down to your side. Your palm faces forward. If you start over, but reverse the steps, which way does your palm end up facing? From either example, we must conclude that the addition of two angular displacements depends on their order and they cannot be vectors.

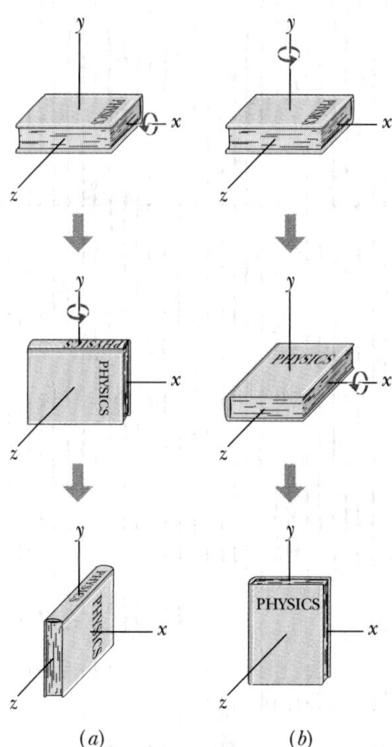

FIG. 10-7 (*a*) From its initial position, at the top, the book is given two successive 90° rotations, first about the (horizontal) x axis and then about the (vertical) y axis. (*b*) The book is given the same rotations, but in the reverse order.

10-4 | Rotation with Constant Angular Acceleration

In pure translation, motion with a *constant linear acceleration* (for example, that of a falling body) is an important special case. In Table 2-1, we displayed a series of equations that hold for such motion.

In pure rotation, the case of *constant angular acceleration* is also important, and a parallel set of equations holds for this case also. We shall not derive them here, but simply write them from the corresponding linear equations, substituting equivalent angular quantities for the linear ones. This is done in Table 10-1, which lists both sets of equations (Eqs. 2-11 and 2-15 to 2-18; 10-12 to 10-16).

Recall that Eqs. 2-11 and 2-15 are basic equations for constant linear acceleration—the other equations in the Linear list can be derived from them. Similarly, Eqs. 10-12 and 10-13 are the basic equations for constant angular acceleration, and the other equations in the Angular list can be derived from them. To solve a simple problem involving constant angular acceleration, you can usually use an equation from the Angular list (*if* you have the list). Choose an equation for which the only unknown variable will be the variable requested in the problem. A better plan is to remember only Eqs. 10-12 and 10-13, and then solve them as simultaneous equations whenever needed. An example is given in Sample Problem 10-4.

TABLE 10-1

Equations of Motion for Constant Linear Acceleration and for Constant Angular Acceleration

Equation Number	Linear Equation	Missing Variable		Angular Equation	Equation Number
(2-11)	$v = v_0 + at$	$x - x_0$	$\theta - \theta_0$	$\omega = \omega_0 + \alpha t$	(10-12)
(2-15)	$x - x_0 = v_0 t + \frac{1}{2}at^2$	v	ω	$\theta - \theta_0 = \omega_0 t + \frac{1}{2}\alpha t^2$	(10-13)
(2-16)	$v^2 = v_0^2 + 2a(x - x_0)$	t	t	$\omega^2 = \omega_0^2 + 2\alpha(\theta - \theta_0)$	(10-14)
(2-17)	$x - x_0 = \frac{1}{2}(v_0 + v)t$	a	α	$\theta - \theta_0 = \frac{1}{2}(\omega_0 + \omega)t$	(10-15)
(2-18)	$x - x_0 = vt - \frac{1}{2}at^2$	v_0	ω_0	$\theta - \theta_0 = \omega t - \frac{1}{2}\alpha t^2$	(10-16)

CHECKPOINT 2 In four situations, a rotating body has angular position $\theta(t)$ given by (a) $\theta = 3t - 4$, (b) $\theta = -5t^3 + 4t^2 + 6$, (c) $\theta = 2/t^2 - 4/t$, and (d) $\theta = 5t^2 - 3$. To which situations do the angular equations of Table 10-1 apply?

Sample Problem 10-3

A grindstone (Fig. 10-8) rotates at constant angular acceleration $\alpha = 0.35\ \text{rad/s}^2$. At time $t = 0$, it has an angular velocity of $\omega_0 = -4.6$ rad/s and a reference line on it is horizontal, at the angular position $\theta_0 = 0$.

(a) At what time after $t = 0$ is the reference line at the angular position $\theta = 5.0$ rev?

KEY IDEA The angular acceleration is constant, so we can use the rotation equations of Table 10-1. We choose Eq. 10-13,

$$\theta - \theta_0 = \omega_0 t + \tfrac{1}{2}\alpha t^2,$$

because the only unknown variable it contains is the desired time t.

Calculations: Substituting known values and setting $\theta_0 = 0$ and $\theta = 5.0\ \text{rev} = 10\pi$ rad give us

$$10\pi\ \text{rad} = (-4.6\ \text{rad/s})t + \tfrac{1}{2}(0.35\ \text{rad/s}^2)t^2.$$

(We converted 5.0 rev to 10π rad to keep the units consistent.) Solving this quadratic equation for t, we find

$$t = 32\ \text{s}. \qquad \text{(Answer)}$$

(b) Describe the grindstone's rotation between $t = 0$ and $t = 32$ s.

Description: The wheel is initially rotating in the negative (clockwise) direction with angular velocity $\omega_0 = -4.6$ rad/s, but its angular acceleration α is positive. This

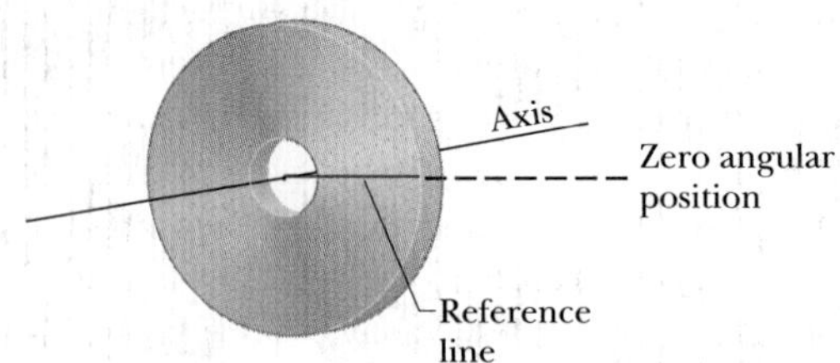

FIG. 10-8 A grindstone. At $t = 0$ the reference line (which we imagine to be marked on the stone) is horizontal.

initial opposition of the signs of angular velocity and angular acceleration means that the wheel slows in its rotation in the negative direction, stops, and then reverses to rotate in the positive direction. After the reference line comes back through its initial orientation of $\theta = 0$, the wheel turns an additional 5.0 rev by time $t = 32$ s.

(c) At what time t does the grindstone momentarily stop?

Calculation: We again go to the table of equations for constant angular acceleration, and again we need an equation that contains only the desired unknown variable t. However, now the equation must also contain the variable ω, so that we can set it to 0 and then solve for the corresponding time t. We choose Eq. 10-12, which yields

$$t = \frac{\omega - \omega_0}{\alpha} = \frac{0 - (-4.6\ \text{rad/s})}{0.35\ \text{rad/s}^2} = 13\ \text{s}. \qquad \text{(Answer}$$

Sample Problem 10-4

While you are operating a Rotor (the rotating cylindrical ride discussed in Sample Problem 6-8), you spot a passenger in acute distress and decrease the angular velocity of the cylinder from 3.40 rad/s to 2.00 rad/s in 20.0 rev, at constant angular acceleration. (The passenger is obviously more of a "translation person" than a "rotation person.")

(a) What is the constant angular acceleration during this decrease in angular speed?

KEY IDEA Because the cylinder's angular acceleration is constant, we can relate it to the angular velocity and angular displacement via the basic equations for constant angular acceleration (Eqs. 10-12 and 10-13).

Calculations: The initial angular velocity is $\omega_0 = 3.40$ rad/s, the angular displacement is $\theta - \theta_0 = 20.0$ rev, and the angular velocity at the end of that displacement is $\omega = 2.00$ rad/s. But we do not know the angular acceleration α and time t, which are in both basic equations.

To eliminate the unknown t, we use Eq. 10-12 to write

$$t = \frac{\omega - \omega_0}{\alpha},$$

which we then substitute into Eq. 10-13 to write

$$\theta - \theta_0 = \omega_0\left(\frac{\omega - \omega_0}{\alpha}\right) + \tfrac{1}{2}\alpha\left(\frac{\omega - \omega_0}{\alpha}\right)^2.$$

Solving for α, substituting known data, and converting 20 rev to 125.7 rad, we find

$$\alpha = \frac{\omega^2 - \omega_0^2}{2(\theta - \theta_0)} = \frac{(2.00 \text{ rad/s})^2 - (3.40 \text{ rad/s})^2}{2(125.7 \text{ rad})}$$

$$= -0.0301 \text{ rad/s}^2. \qquad \text{(Answer)}$$

(b) How much time did the speed decrease take?

Calculation: Now that we know α, we can use Eq. 10-12 to solve for t:

$$t = \frac{\omega - \omega_0}{\alpha} = \frac{2.00 \text{ rad/s} - 3.40 \text{ rad/s}}{-0.0301 \text{ rad/s}^2}$$

$$= 46.5 \text{ s}. \qquad \text{(Answer)}$$

10-5 | Relating the Linear and Angular Variables

In Section 4-7, we discussed uniform circular motion, in which a particle travels at constant linear speed v along a circle and around an axis of rotation. When a rigid body, such as a merry-go-round, rotates around an axis, each particle in the body moves in its own circle around that axis. Since the body is rigid, all the particles make one revolution in the same amount of time; that is, they all have the same angular speed ω.

However, the farther a particle is from the axis, the greater the circumference of its circle is, and so the faster its linear speed v must be. You can notice this on a merry-go-round. You turn with the same angular speed ω regardless of your distance from the center, but your linear speed v increases noticeably if you move to the outside edge of the merry-go-round.

We often need to relate the linear variables s, v, and a for a particular point in a rotating body to the angular variables θ, ω, and α for that body. The two sets of variables are related by r, the *perpendicular distance* of the point from the rotation axis. This perpendicular distance is the distance between the point and the rotation axis, measured along a perpendicular to the axis. It is also the radius r of the circle traveled by the point around the axis of rotation.

The Position

If a reference line on a rigid body rotates through an angle θ, a point within the body at a position r from the rotation axis moves a distance s along a circular arc, where s is given by Eq. 10-1:

$$s = \theta r \qquad \text{(radian measure).} \qquad (10\text{-}17)$$

This is the first of our linear–angular relations. *Caution:* The angle θ here must be measured in radians because Eq. 10-17 is itself the definition of angular measure in radians.

The Speed

Differentiating Eq. 10-17 with respect to time—with r held constant—leads to

$$\frac{ds}{dt} = \frac{d\theta}{dt} r.$$

However, ds/dt is the linear speed (the magnitude of the linear velocity) of the point in question, and $d\theta/dt$ is the angular speed ω of the rotating body. So

$$v = \omega r \qquad \text{(radian measure).} \qquad (10\text{-}18)$$

Caution: The angular speed ω must be expressed in radian measure.

Equation 10-18 tells us that since all points within the rigid body have the same angular speed ω, points with greater radius r have greater linear speed v. Figure 10-9*a* reminds us that the linear velocity is always tangent to the circular path of the point in question.

If the angular speed ω of the rigid body is constant, then Eq. 10-18 tells us that the linear speed v of any point within it is also constant. Thus, each point within the body undergoes uniform circular motion. The period of revolution T

for the motion of each point and for the rigid body itself is given by Eq. 4-35:

$$T = \frac{2\pi r}{v}. \tag{10-19}$$

This equation tells us that the time for one revolution is the distance $2\pi r$ traveled in one revolution divided by the speed at which that distance is traveled. Substituting for v from Eq. 10-18 and canceling r, we find also that

$$T = \frac{2\pi}{\omega} \quad \text{(radian measure).} \tag{10-20}$$

This equivalent equation says that the time for one revolution is the angular distance 2π rad traveled in one revolution divided by the angular speed (or rate) at which that angle is traveled.

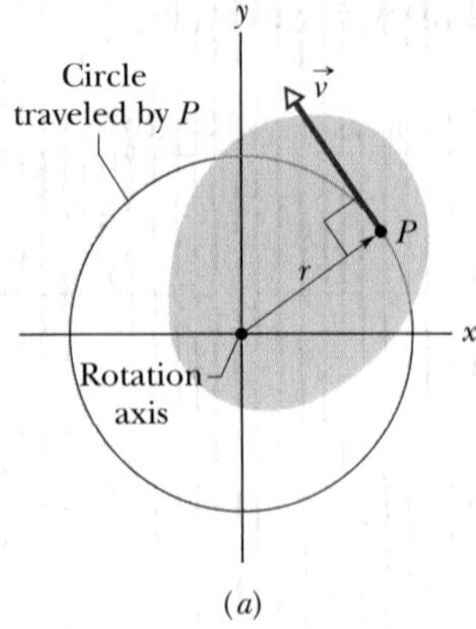

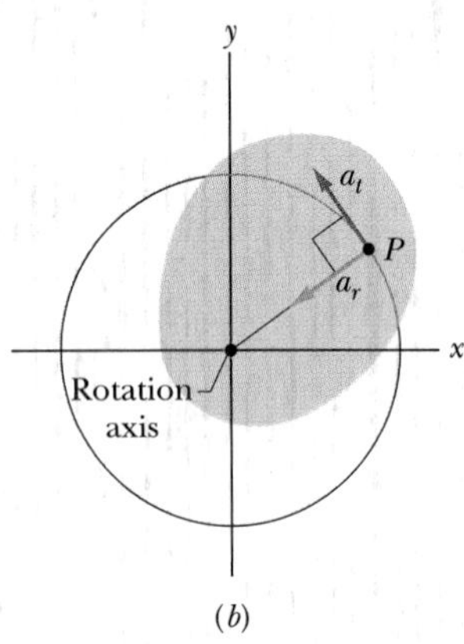

FIG. 10-9 The rotating rigid body of Fig. 10-2, shown in cross section viewed from above. Every point of the body (such as P) moves in a circle around the rotation axis. (a) The linear velocity $\vec{v}$ of every point is tangent to the circle in which the point moves. (b) The linear acceleration $\vec{a}$ of the point has (in general) two components: tangential a_t and radial a_r.

The Acceleration

Differentiating Eq. 10-18 with respect to time—again with r held constant—leads to

$$\frac{dv}{dt} = \frac{d\omega}{dt} r. \tag{10-21}$$

Here we run up against a complication. In Eq. 10-21, dv/dt represents only the part of the linear acceleration that is responsible for changes in the *magnitude* v of the linear velocity $\vec{v}$. Like $\vec{v}$, that part of the linear acceleration is tangent to the path of the point in question. We call it the *tangential component* a_t of the linear acceleration of the point, and we write

$$a_t = \alpha r \quad \text{(radian measure),} \tag{10-22}$$

where $\alpha = d\omega/dt$. *Caution:* The angular acceleration α in Eq. 10-22 must be expressed in radian measure.

In addition, as Eq. 4-34 tells us, a particle (or point) moving in a circular path has a *radial component* of linear acceleration, $a_r = v^2/r$ (directed radially inward), that is responsible for changes in the *direction* of the linear velocity $\vec{v}$. By substituting for v from Eq. 10-18, we can write this component as

$$a_r = \frac{v^2}{r} = \omega^2 r \quad \text{(radian measure).} \tag{10-23}$$

Thus, as Fig. 10-9b shows, the linear acceleration of a point on a rotating rigid body has, in general, two components. The radially inward component a_r (given by Eq. 10-23) is present whenever the angular velocity of the body is not zero. The tangential component a_t (given by Eq. 10-22) is present whenever the angular acceleration is not zero.

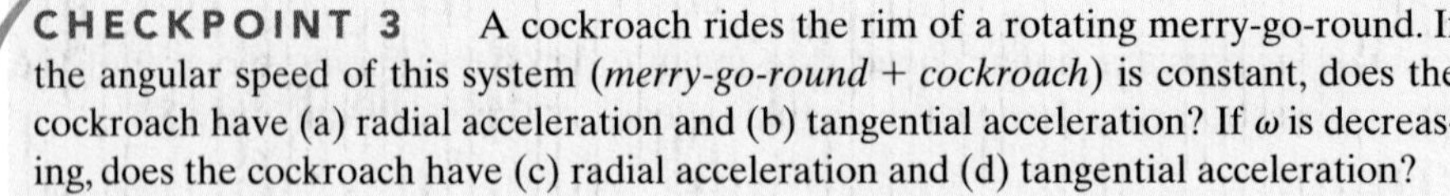
CHECKPOINT 3 A cockroach rides the rim of a rotating merry-go-round. If the angular speed of this system (*merry-go-round + cockroach*) is constant, does the cockroach have (a) radial acceleration and (b) tangential acceleration? If ω is decreasing, does the cockroach have (c) radial acceleration and (d) tangential acceleration?

Sample Problem 10-5

In spite of the extreme care taken in engineering a roller coaster, an unlucky few of the millions of people who ride roller coasters each year end up with a medical condition called *roller-coaster headache.* Symptoms, which might not appear for several days, include vertigo and headache, both severe enough to require medical treatment.

Let's investigate the probable cause by designing the track for our own *induction roller coaster* (which can be accelerated by magnetic forces even on a hori-

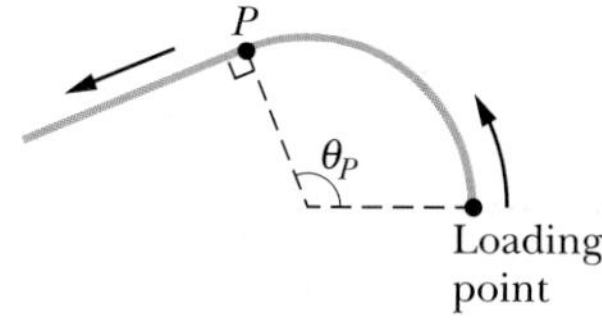

FIG. 10-10 An overhead view of a horizontal track for a roller coaster. The track begins as a circular arc at the loading point and then, at point P, continues along a tangent to the arc.

zontal track). To create an initial thrill, we want each passenger to leave the loading point with acceleration g along the horizontal track. To increase the thrill, we also want that first section of track to form a circular arc (Fig. 10-10), so that the passenger also experiences a centripetal acceleration. As the passenger accelerates along the arc, the magnitude of this centripetal acceleration increases alarmingly. When the magnitude a of the net acceleration reaches $4g$ at some point P and angle θ_P along the arc, we want the passenger then to move in a straight line, along a tangent to the arc.

(a) What angle θ_P should the arc subtend so that a is $4g$ at point P?

KEY IDEAS (1) At any given time, the passenger's net acceleration $\vec{a}$ is the vector sum of the tangential acceleration $\vec{a}_t$ along the track and the radial acceleration $\vec{a}_r$ toward the arc's center of curvature (as in Fig. 10-9*b*). (2) The value of a_r at any given time depends on the angular speed ω according to Eq. 10-23 ($a_r = \omega^2 r$, where r is the radius of the circular arc). (3) An angular acceleration α around the arc is associated with the tangential acceleration a_t along the track according to Eq. 10-22 ($a_t = \alpha r$). (4) Because a_t and r are constant, so is α and thus we can use the constant angular-acceleration equations.

Calculations: Because we are trying to determine a value for angular position θ, let's choose Eq. 10-14 from among the constant angular-acceleration equations:

$$\omega^2 = \omega_0^2 + 2\alpha(\theta - \theta_0). \tag{10-24}$$

For the angular acceleration α, we substitute from Eq. 10-22:

$$\alpha = \frac{a_t}{r}. \tag{10-25}$$

We also substitute $\omega_0 = 0$ and $\theta_0 = 0$, and we find

$$\omega^2 = \frac{2a_t\theta}{r}. \tag{10-26}$$

Substituting this result for ω^2 into

$$a_r = \omega^2 r \tag{10-27}$$

gives a relation between the radial acceleration, the tangential acceleration, and the angular position θ:

$$a_r = 2a_t\theta. \tag{10-28}$$

Because $\vec{a}_t$ and $\vec{a}_r$ are perpendicular vectors, their sum has the magnitude

$$a = \sqrt{a_t^2 + a_r^2}. \tag{10-29}$$

Substituting for a_r from Eq. 10-28 and solving for θ lead to

$$\theta = \tfrac{1}{2}\sqrt{\frac{a^2}{a_t^2} - 1}. \tag{10-30}$$

When a reaches the design value of $4g$, angle θ is the angle θ_P we want. Substituting $a = 4g$, $\theta = \theta_P$, and $a_t = g$ into Eq. 10-30, we find

$$\theta_P = \tfrac{1}{2}\sqrt{\frac{(4g)^2}{g^2} - 1} = 1.94 \text{ rad} = 111°. \quad \text{(Answer)}$$

(b) What is the magnitude a of the passenger's net acceleration at point P and after point P?

Reasoning: At P, a has the design value of $4g$. Just after P is reached, the passenger moves in a straight line and no longer has centripetal acceleration. Thus, the passenger has only the acceleration magnitude g along the track. Hence,

$$a = 4g \text{ at } P \quad \text{and} \quad a = g \text{ after } P. \quad \text{(Answer)}$$

Roller-coaster headache can occur when a passenger's head undergoes an abrupt change in acceleration, with the acceleration magnitude large before or after the change. The reason is that the change can cause the brain to move relative to the skull, tearing the veins that bridge the brain and skull. Our design to increase the acceleration from g to $4g$ along the path to P might harm the passenger, but the abrupt change in acceleration as the passenger passes through point P is more likely to cause roller-coaster headache.

PROBLEM-SOLVING TACTICS

Tactic 1: *Units for Angular Variables* In Eq. 10-1 ($\theta = s/r$), we began the use of radian measure for all angular variables whenever we are using equations that contain both angular and linear variables. Thus, we must express angular displacements in radians, angular velocities in rad/s and rad/min, and angular accelerations in rad/s^2 and rad/min^2. Equations 10-17, 10-18, 10-20, 10-22, and 10-23 are marked to emphasize this. The only exceptions to this rule are equations that involve *only* angular variables, such as the angular equations listed in Table 10-1. Here you are free to use any unit you wish for the angular variables; that is, you may use radians, degrees, or revolutions, as long as you use them consistently.

In equations where radian measure must be used, you need not keep track of the unit "radian" (rad) algebraically, as you must do for other units. You can add or delete it at will, to suit the context.

10-6 | Kinetic Energy of Rotation

The rapidly rotating blade of a table saw certainly has kinetic energy due to that rotation. How can we express the energy? We cannot apply the familiar formula $K = \frac{1}{2}mv^2$ to the saw as a whole because that would give us the kinetic energy only of the saw's center of mass, which is zero.

Instead, we shall treat the table saw (and any other rotating rigid body) as a collection of particles with different speeds. We can then add up the kinetic energies of all the particles to find the kinetic energy of the body as a whole. In this way we obtain, for the kinetic energy of a rotating body,

$$K = \tfrac{1}{2}m_1v_1^2 + \tfrac{1}{2}m_2v_2^2 + \tfrac{1}{2}m_3v_3^2 + \cdots$$
$$= \sum \tfrac{1}{2}m_iv_i^2, \qquad (10\text{-}31)$$

in which m_i is the mass of the ith particle and v_i is its speed. The sum is taken over all the particles in the body.

The problem with Eq. 10-31 is that v_i is not the same for all particles. We solve this problem by substituting for v from Eq. 10-18 ($v = \omega r$), so that we have

$$K = \sum \tfrac{1}{2}m_i(\omega r_i)^2 = \tfrac{1}{2}\left(\sum m_ir_i^2\right)\omega^2, \qquad (10\text{-}32)$$

in which ω *is* the same for all particles.

The quantity in parentheses on the right side of Eq. 10-32 tells us how the mass of the rotating body is distributed about its axis of rotation. We call that quantity the **rotational inertia** (or **moment of inertia**) I of the body with respect to the axis of rotation. It is a constant for a particular rigid body and a particular rotation axis. (That axis must always be specified if the value of I is to be meaningful.)

We may now write

$$I = \sum m_ir_i^2 \quad \text{(rotational inertia)} \qquad (10\text{-}33)$$

and substitute into Eq. 10-32, obtaining

$$K = \tfrac{1}{2}I\omega^2 \quad \text{(radian measure)} \qquad (10\text{-}34)$$

as the expression we seek. Because we have used the relation $v = \omega r$ in deriving Eq. 10-34, ω must be expressed in radian measure. The SI unit for I is the kilogram–square meter ($\text{kg}\cdot\text{m}^2$).

Equation 10-34, which gives the kinetic energy of a rigid body in pure rotation, is the angular equivalent of the formula $K = \frac{1}{2}Mv_{\text{com}}^2$, which gives the kinetic energy of a rigid body in pure translation. In both formulas there is a factor of $\frac{1}{2}$. Where mass M appears in one equation, I (which involves both mass and its distribution) appears in the other. Finally, each equation contains as a factor the square of a speed—translational or rotational as appropriate. The kinetic energies of translation and of rotation are not different kinds of energy. They are both kinetic energy, expressed in ways that are appropriate to the motion at hand.

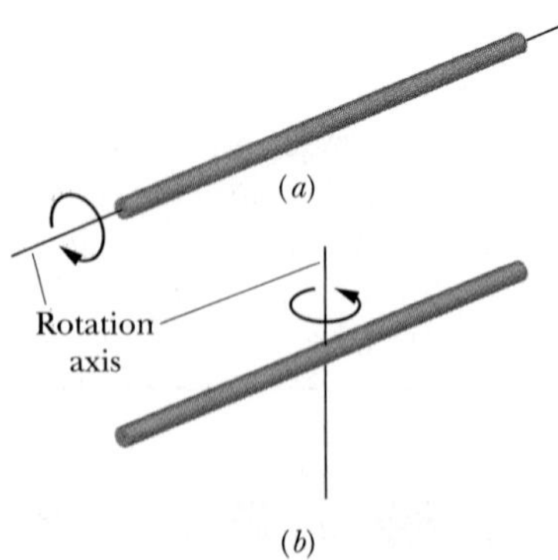

FIG. 10-11 A long rod is much easier to rotate about (*a*) its central (longitudinal) axis than about (*b*) an axis through its center and perpendicular to its length. The reason for the difference is that the mass is distributed closer to the rotation axis in (*a*) than in (*b*).

We noted previously that the rotational inertia of a rotating body involves not only its mass but also how that mass is distributed. Here is an example that you can literally feel. Rotate a long, fairly heavy rod (a pole, a length of lumber, or something similar), first around its central (longitudinal) axis (Fig. 10-11*a*) and then around an axis perpendicular to the rod and through the center (Fig. 10-11*b*). Both rotations involve the very same mass, but the first rotation is much easier than the second. The reason is that the mass is distributed much closer to the rotation axis in the first rotation. As a result, the rotational inertia of the rod is much smaller in Fig. 10-11*a* than in Fig. 10-11*b*. In general, smaller rotational inertia means easier rotation.

CHECKPOINT 4 The figure shows three small spheres that rotate about a vertical axis. The perpendicular distance between the axis and the center of each sphere is given. Rank the three spheres according to their rotational inertia about that axis, greatest first.

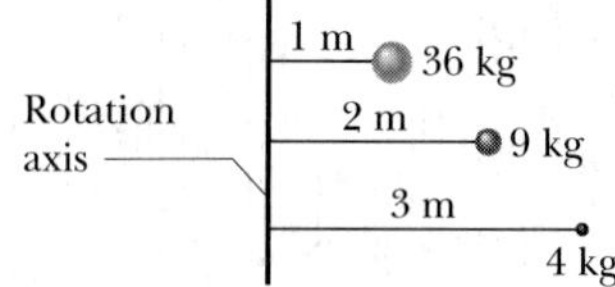

10-7 | Calculating the Rotational Inertia

If a rigid body consists of a few particles, we can calculate its rotational inertia about a given rotation axis with Eq. 10-33 ($I = \Sigma\, m_i r_i^2$); that is, we can find the product mr^2 for each particle and then sum the products. (Recall that r is the perpendicular distance a particle is from the given rotation axis.)

If a rigid body consists of a great many adjacent particles (it is *continuous,* like a Frisbee), using Eq. 10-33 would require a computer. Thus, instead, we replace the sum in Eq. 10-33 with an integral and define the rotational inertia of the body as

$$I = \int r^2\, dm \qquad \text{(rotational inertia, continuous body).} \qquad (10\text{-}35)$$

Table 10-2 gives the results of such integration for nine common body shapes and the indicated axes of rotation.

Parallel-Axis Theorem

Suppose we want to find the rotational inertia I of a body of mass M about a given axis. In principle, we can always find I with the integration of Eq. 10-35. However, there is a shortcut if we happen to already know the rotational inertia I_{com} of the body about a *parallel* axis that extends through the body's center of mass. Let h be the perpendicular distance between the given axis and the axis

TABLE 10-2

Some Rotational Inertias

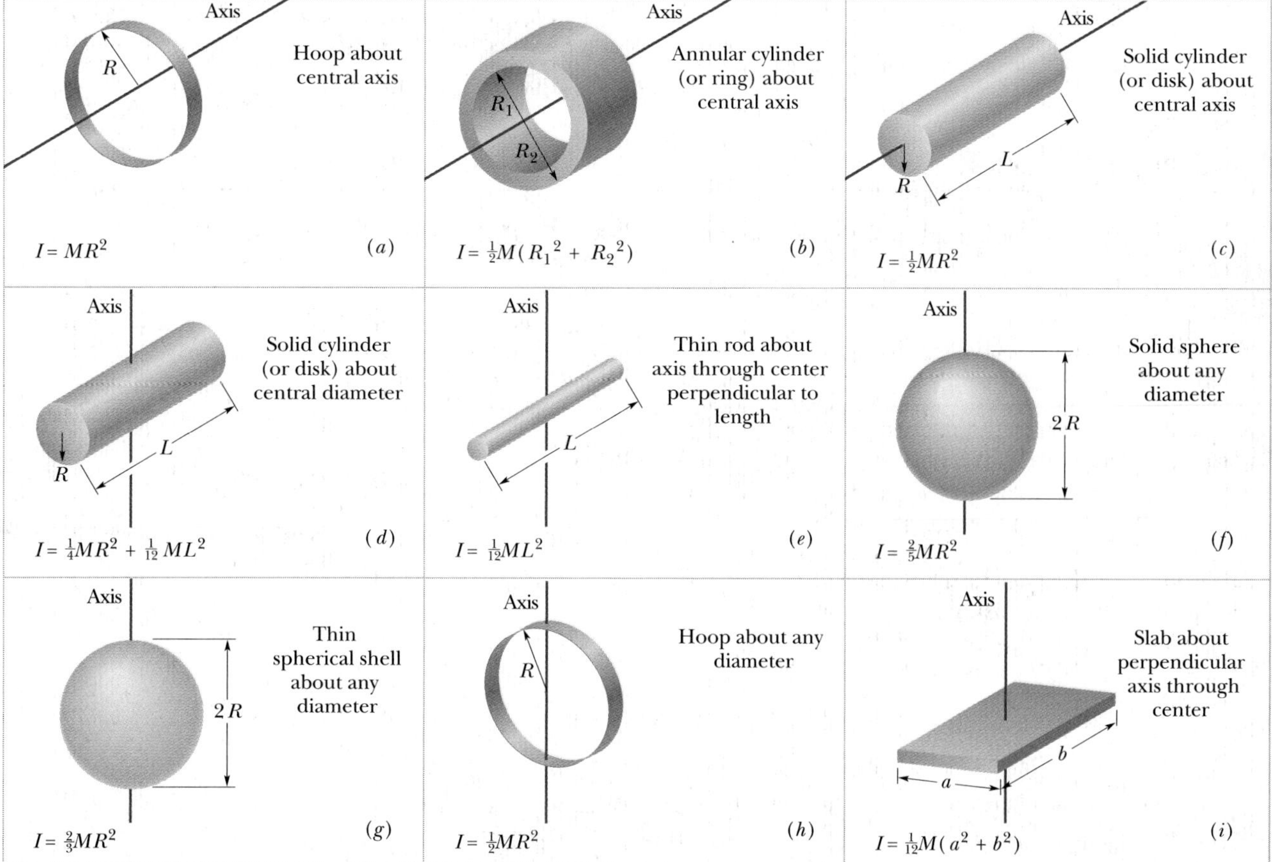

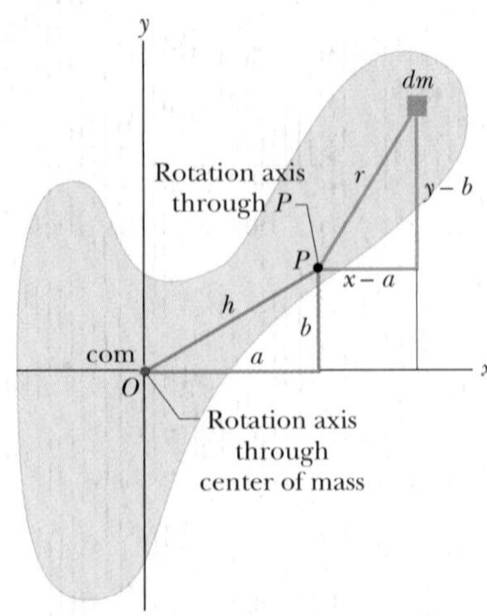

FIG. 10-12 A rigid body in cross section, with its center of mass at O. The parallel-axis theorem (Eq. 10-36) relates the rotational inertia of the body about an axis through O to that about a parallel axis through a point such as P, a distance h from the body's center of mass. Both axes are perpendicular to the plane of the figure.

through the center of mass (remember these two axes must be parallel). Then the rotational inertia I about the given axis is

$$I = I_{com} + Mh^2 \qquad \text{(parallel-axis theorem).} \tag{10-36}$$

This equation is known as the **parallel-axis theorem.** We shall now prove it and then put it to use in Checkpoint 5 and Sample Problem 10-6.

Proof of the Parallel-Axis Theorem

Let O be the center of mass of the arbitrarily shaped body shown in cross section in Fig. 10-12. Place the origin of the coordinates at O. Consider an axis through O perpendicular to the plane of the figure, and another axis through point P parallel to the first axis. Let the x and y coordinates of P be a and b.

Let dm be a mass element with the general coordinates x and y. The rotational inertia of the body about the axis through P is then, from Eq. 10-35,

$$I = \int r^2\,dm = \int [(x-a)^2 + (y-b)^2]\,dm,$$

which we can rearrange as

$$I = \int (x^2 + y^2)\,dm - 2a\int x\,dm - 2b\int y\,dm + \int (a^2 + b^2)\,dm. \tag{10-37}$$

From the definition of the center of mass (Eq. 9-9), the middle two integrals of Eq. 10-37 give the coordinates of the center of mass (multiplied by a constant) and thus must each be zero. Because $x^2 + y^2$ is equal to R^2, where R is the distance from O to dm, the first integral is simply I_{com}, the rotational inertia of the body about an axis through its center of mass. Inspection of Fig. 10-12 shows that the last term in Eq. 10-37 is Mh^2, where M is the body's total mass. Thus, Eq. 10-37 reduces to Eq. 10-36, which is the relation that we set out to prove.

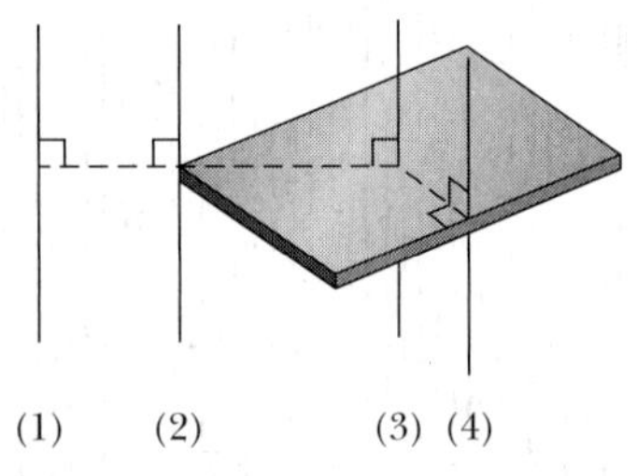

CHECKPOINT 5 The figure shows a book-like object (one side is longer than the other) and four choices of rotation axes, all perpendicular to the face of the object. Rank the choices according to the rotational inertia of the object about the axis, greatest first.

Sample Problem 10-6

Figure 10-13a shows a rigid body consisting of two particles of mass m connected by a rod of length L and negligible mass.

(a) What is the rotational inertia I_{com} about an axis through the center of mass, perpendicular to the rod as shown?

KEY IDEA Because we have only two particles with mass, we can find the body's rotational inertia I_{com} by using Eq. 10-33 rather than by integration.

Calculations: For the two particles, each at perpendicular distance $\frac{1}{2}L$ from the rotation axis, we have

$$I = \sum m_i r_i^2 = (m)(\tfrac{1}{2}L)^2 + (m)(\tfrac{1}{2}L)^2$$
$$= \tfrac{1}{2}mL^2. \qquad \text{(Answer)}$$

(b) What is the rotational inertia I of the body about an axis through the left end of the rod and parallel to the first axis (Fig. 10-13b)?

KEY IDEAS This situation is simple enough that we can find I using either of two techniques. The first is similar to the one used in part (a). The other, more powerful one is to apply the parallel-axis theorem.

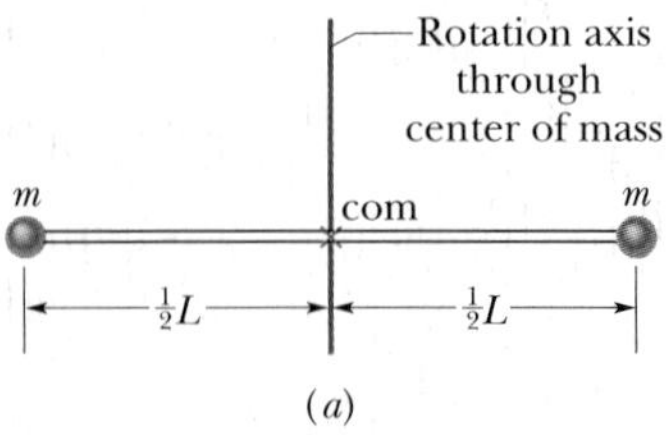

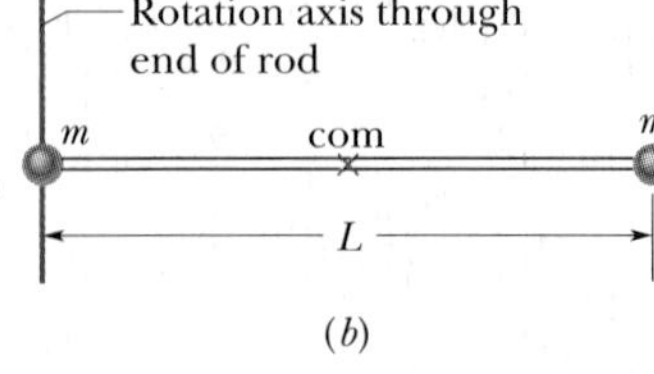

FIG. 10-13 A rigid body consisting of two particles of mass m joined by a rod of negligible mass.

First technique: We calculate I as in part (a), except here the perpendicular distance r_i is zero for the particle on the left and L for the particle on the right. Now Eq. 10-33 gives us

$$I = m(0)^2 + mL^2 = mL^2. \qquad \text{(Answer)}$$

Second technique: Because we already know I_{com} about an axis through the center of mass and because the axis here is parallel to that "com axis," we can apply the parallel-axis theorem (Eq. 10-36). We find

$$I = I_{com} + Mh^2 = \tfrac{1}{2}mL^2 + (2m)(\tfrac{1}{2}L)^2$$
$$= mL^2. \qquad \text{(Answer)}$$

Sample Problem 10-7

Figure 10-14 shows a thin, uniform rod of mass M and length L, on an x axis with the origin at the rod's center.

(a) What is the rotational inertia of the rod about the perpendicular rotation axis through the center?

KEY IDEAS (1) Because the rod is uniform, its center of mass is at its center. Therefore, we are looking for I_{com}. (2) Because the rod is a continuous object, we must use the integral of Eq. 10-35,

$$I = \int r^2\,dm, \qquad (10\text{-}38)$$

to find the rotational inertia.

Calculations: We want to integrate with respect to coordinate x (not mass m as indicated in the integral), so we must relate the mass dm of an element of the rod to its length dx along the rod. (Such an element is shown in Fig. 10-14.) Because the rod is uniform, the ratio of mass to length is the same for all the elements and for the rod as a whole. Thus, we can write

$$\frac{\text{element's mass } dm}{\text{element's length } dx} = \frac{\text{rod's mass } M}{\text{rod's length } L}$$

or

$$dm = \frac{M}{L}\,dx.$$

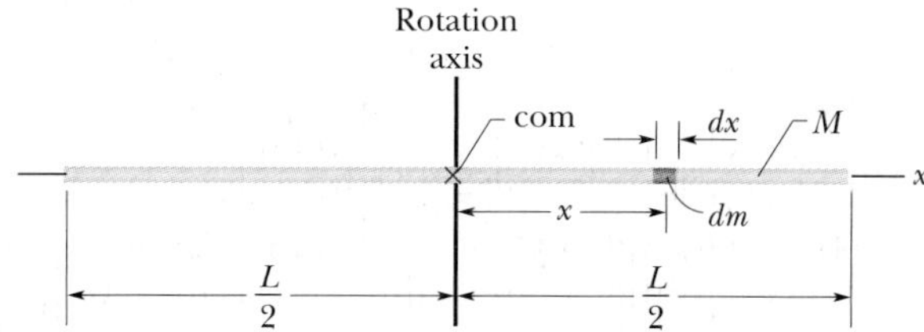

FIG. 10-14 A uniform rod of length L and mass M. An element of mass dm and length dx is represented.

We can now substitute this result for dm and x for r in Eq. 10-38. Then we integrate from end to end of the rod (from $x = -L/2$ to $x = L/2$) to include all the elements. We find

$$I = \int_{x=-L/2}^{x=+L/2} x^2\left(\frac{M}{L}\right)dx$$
$$= \frac{M}{3L}\left[x^3\right]_{-L/2}^{+L/2} = \frac{M}{3L}\left[\left(\frac{L}{2}\right)^3 - \left(-\frac{L}{2}\right)^3\right]$$
$$= \tfrac{1}{12}ML^2. \qquad \text{(Answer)}$$

This agrees with the result given in Table 10-2*e*.

(b) What is the rod's rotational inertia I about a new rotation axis that is perpendicular to the rod and through the left end?

KEY IDEAS We can find I by shifting the origin of the x axis to the left end of the rod and then integrating from $x = 0$ to $x = L$. However, here we shall use a more powerful (and easier) technique by applying the parallel-axis theorem (Eq. 10-36).

Calculations: If we place the axis at the rod's end so that it is parallel to the axis through the center of mass, then we can use the parallel-axis theorem (Eq. 10-36). We know from part (a) that I_{com} is $\frac{1}{12}ML^2$. From Fig. 10-14, the perpendicular distance h between the new rotation axis and the center of mass is $\frac{1}{2}L$. Equation 10-36 then gives us

$$I = I_{com} + Mh^2 = \tfrac{1}{12}ML^2 + (M)(\tfrac{1}{2}L)^2$$
$$= \tfrac{1}{3}ML^2. \qquad \text{(Answer)}$$

Actually, this result holds for any axis through the left or right end that is perpendicular to the rod, whether it is parallel to the axis shown in Fig. 10-14 or not.

Sample Problem 10-8

Large machine components that undergo prolonged, high-speed rotation are first examined for the possibility of failure in a *spin test system.* In this system, a component is *spun up* (brought up to high speed) while inside a cylindrical arrangement of lead bricks and containment liner, all within a steel shell that is closed by a lid clamped into place. If the rotation causes the component to shatter, the soft lead bricks are supposed to catch the pieces for later analysis.

In 1985, Test Devices, Inc. (www.testdevices.com) was spin testing a sample of a solid steel rotor (a disk) of mass $M = 272$ kg and radius $R = 38.0$ cm. When the

FIG. 10-15 Some of the destruction caused by the explosion of a rapidly rotating steel disk. *(Courtesy Test Devices, Inc)*

sample reached an angular speed ω of 14 000 rev/min, the test engineers heard a dull thump from the test system, which was located one floor down and one room over from them. Investigating, they found that lead bricks had been thrown out in the hallway leading to the test room, a door to the room had been hurled into the adjacent parking lot, one lead brick had shot from the test site through the wall of a neighbor's kitchen, the structural beams of the test building had been damaged, the concrete floor beneath the spin chamber had been shoved downward by about 0.5 cm, and the 900 kg lid had been blown upward through the ceiling and had then crashed back onto the test equipment (Fig. 10-15). The exploding pieces had not penetrated the room of the test engineers only by luck.

How much energy was released in the explosion of the rotor?

KEY IDEA The released energy was equal to the rotational kinetic energy K of the rotor just as it reached the angular speed of 14 000 rev/min.

Calculations: We can find K with Eq. 10-34 ($K = \frac{1}{2}I\omega^2$), but first we need an expression for the rotational inertia I. Because the rotor was a disk that rotated like a merry-go-round, I is given by the expression in Table 10-2c ($I = \frac{1}{2}MR^2$). Thus, we have

$$I = \tfrac{1}{2}MR^2 = \tfrac{1}{2}(272 \text{ kg})(0.38 \text{ m})^2 = 19.64 \text{ kg}\cdot\text{m}^2.$$

The angular speed of the rotor was

$$\omega = (14\,000 \text{ rev/min})(2\pi \text{ rad/rev})\left(\frac{1 \text{ min}}{60 \text{ s}}\right)$$
$$= 1.466 \times 10^3 \text{ rad/s}.$$

Now we can use Eq. 10-34 to write

$$K = \tfrac{1}{2}I\omega^2 = \tfrac{1}{2}(19.64 \text{ kg}\cdot\text{m}^2)(1.466 \times 10^3 \text{ rad/s})^2$$
$$= 2.1 \times 10^7 \text{ J}. \qquad \text{(Answer)}$$

Being near this explosion was quite dangerous.

10-8 | Torque

A doorknob is located as far as possible from the door's hinge line for a good reason. If you want to open a heavy door, you must certainly apply a force; that alone, however, is not enough. Where you apply that force and in what direction you push are also important. If you apply your force nearer to the hinge line than the knob, or at any angle other than 90° to the plane of the door, you must use a greater force to move the door than if you apply the force at the knob and perpendicular to the door's plane.

Figure 10-16a shows a cross section of a body that is free to rotate about an axis passing through O and perpendicular to the cross section. A force $\vec{F}$ is applied at point P, whose position relative to O is defined by a position vector $\vec{r}$. The directions of vectors $\vec{F}$ and $\vec{r}$ make an angle ϕ with each other. (For simplicity, we consider only forces that have no component parallel to the rotation axis; thus, $\vec{F}$ is in the plane of the page.)

To determine how $\vec{F}$ results in a rotation of the body around the rotation axis, we resolve $\vec{F}$ into two components (Fig. 10-16b). One component, called the *radial component* F_r, points along $\vec{r}$. This component does not cause rotation, because it acts along a line that extends through O. (If you pull on a door parallel to the plane of the door, you do not rotate the door.) The other component of $\vec{F}$, called the *tangential component* F_t, is perpendicular to $\vec{r}$ and has magnitude $F_t = F \sin \phi$. This component *does* cause rotation. (If you pull on a door perpendicular to its plane, you can rotate the door.)

The ability of $\vec{F}$ to rotate the body depends not only on the magnitude of its tangential component F_t, but also on just how far from O the force is applied. To include both these factors, we define a quantity called **torque** τ as the product of the two factors and write it as

$$\tau = (r)(F \sin \phi). \tag{10-39}$$

Two equivalent ways of computing the torque are

$$\tau = (r)(F \sin \phi) = rF_t \tag{10-40}$$

and

$$\tau = (r \sin \phi)(F) = r_\perp F, \tag{10-41}$$

where $r_\perp$ is the perpendicular distance between the rotation axis at O and an extended line running through the vector $\vec{F}$ (Fig. 10-16c). This extended line is called the **line of action** of $\vec{F}$, and $r_\perp$ is called the **moment arm** of $\vec{F}$. Figure 10-16b shows that we can describe r, the magnitude of $\vec{r}$, as being the moment arm of the force component F_t.

Torque, which comes from the Latin word meaning "to twist," may be loosely identified as the turning or twisting action of the force $\vec{F}$. When you apply a force to an object—such as a screwdriver or torque wrench—with the purpose of turning that object, you are applying a torque. The SI unit of torque is the newton-meter (N·m). *Caution:* The newton-meter is also the unit of work. Torque and work, however, are quite different quantities and must not be confused. Work is often expressed in joules (1 J = 1 N·m), but torque never is.

In the next chapter we shall discuss torque in a general way as being a vector quantity. Here, however, because we consider only rotation around a single axis, we do not need vector notation. Instead, a torque has either a positive or negative value depending on the direction of rotation it would give a body initially at rest: If the body would rotate counterclockwise, the torque is positive. If the object would rotate clockwise, the torque is negative. (The phrase "clocks are negative" from Section 10-2 still works.)

Torques obey the superposition principle that we discussed in Chapter 5 for forces: When several torques act on a body, the **net torque** (or **resultant torque**) is the sum of the individual torques. The symbol for net torque is τ_{net}.

CHECKPOINT 6 The figure shows an overhead view of a meter stick that can pivot about the dot at the position marked 20 (for 20 cm). All five forces on the stick are horizontal and have the same magnitude. Rank the forces according to the magnitude of the torque they produce, greatest first.

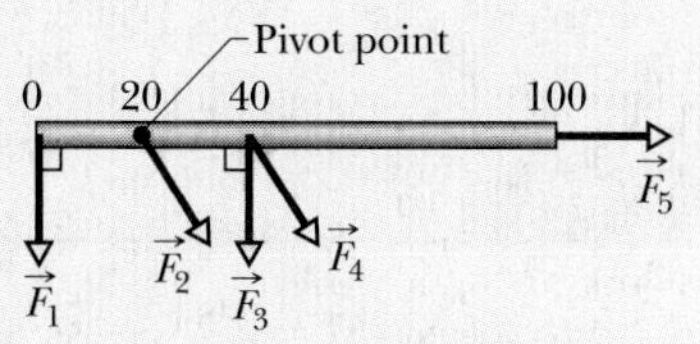

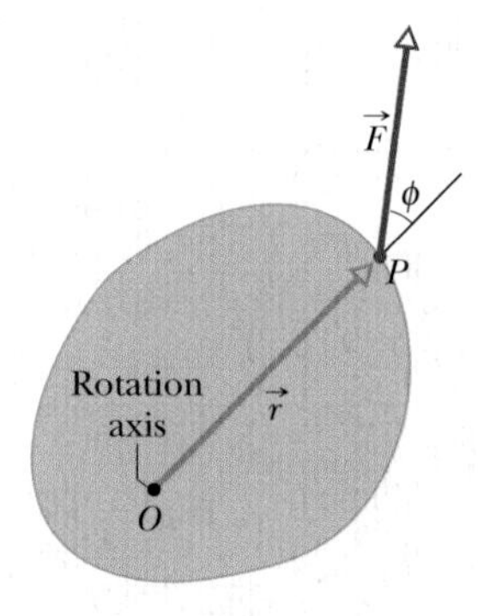

(a)

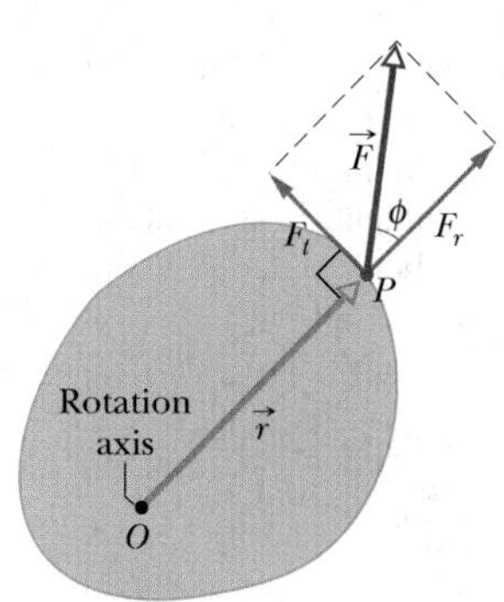

(b)

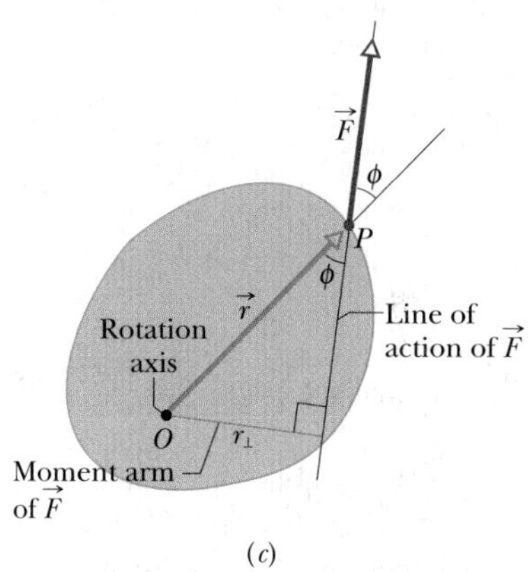

(c)

FIG. 10-16 (a) A force $\vec{F}$ acts at point P on a rigid body that is free to rotate about an axis through O; the axis is perpendicular to the plane of the cross section shown here. (b) The torque due to this force is $(r)(F \sin \phi)$. We can also write it as rF_t, where F_t is the tangential component of $\vec{F}$. (c) The torque can also be written as $r_\perp F$, where $r_\perp$ is the moment arm of $\vec{F}$.

10-9 | Newton's Second Law for Rotation

A torque can cause rotation of a rigid body, as when you use a torque to rotate a door. Here we want to relate the net torque τ_{net} on a rigid body to the angular acceleration α that torque causes about a rotation axis. We do so by analogy with Newton's second law ($F_{net} = ma$) for the acceleration a of a body of mass m due to a net force F_{net} along a coordinate axis. We replace F_{net} with τ_{net}, m with I, and a with α in radian measure, writing

$$\tau_{net} = I\alpha \quad \text{(Newton's second law for rotation).} \tag{10-42}$$

Proof of Equation 10-42

We prove Eq. 10-42 by first considering the simple situation shown in Fig. 10-17. The rigid body there consists of a particle of mass m on one end of a massless rod of length r. The rod can move only by rotating about its other end, around a rotation axis (an axle) that is perpendicular to the plane of the page. Thus, the particle can move only in a circular path that has the rotation axis at its center.

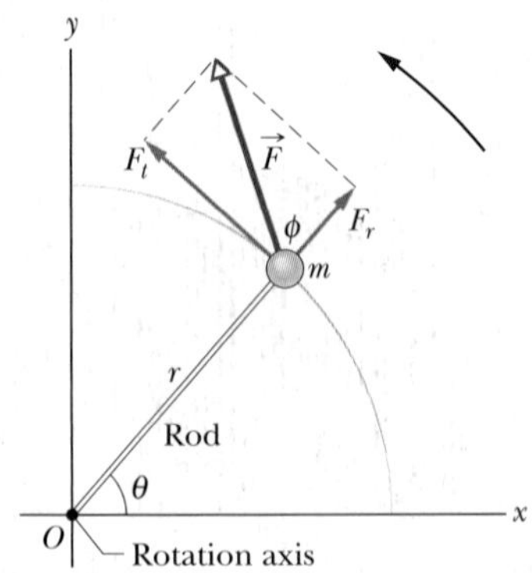

FIG. 10-17 A simple rigid body, free to rotate about an axis through O, consists of a particle of mass m fastened to the end of a rod of length r and negligible mass. An applied force $\vec{F}$ causes the body to rotate.

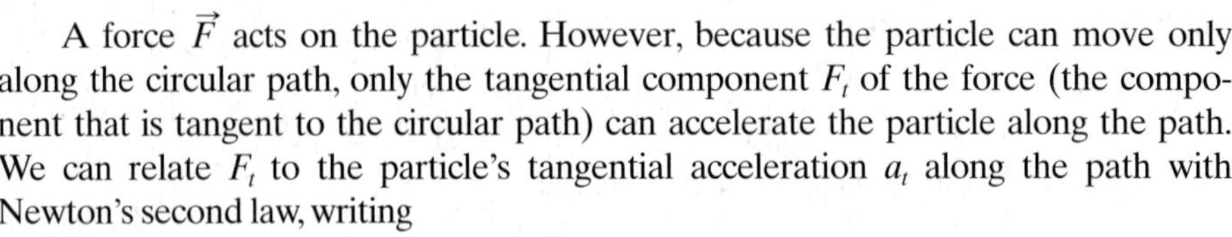

A force $\vec{F}$ acts on the particle. However, because the particle can move only along the circular path, only the tangential component F_t of the force (the component that is tangent to the circular path) can accelerate the particle along the path. We can relate F_t to the particle's tangential acceleration a_t along the path with Newton's second law, writing

$$F_t = ma_t.$$

The torque acting on the particle is, from Eq. 10-40,

$$\tau = F_t r = ma_t r.$$

From Eq. 10-22 ($a_t = \alpha r$) we can write this as

$$\tau = m(\alpha r)r = (mr^2)\alpha. \quad (10\text{-}43)$$

The quantity in parentheses on the right is the rotational inertia of the particle about the rotation axis (see Eq. 10-33). Thus, Eq. 10-43 reduces to

$$\tau = I\alpha \quad \text{(radian measure).} \quad (10\text{-}44)$$

For the situation in which more than one force is applied to the particle, we can generalize Eq. 10-44 as

$$\tau_{\text{net}} = I\alpha \quad \text{(radian measure),} \quad (10\text{-}45)$$

which we set out to prove. We can extend this equation to any rigid body rotating about a fixed axis, because any such body can always be analyzed as an assembly of single particles.

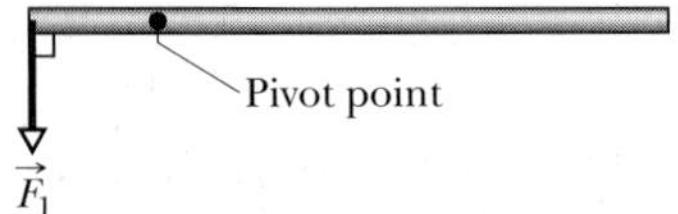

CHECKPOINT 7 The figure shows an overhead view of a meter stick that can pivot about the point indicated, which is to the left of the stick's midpoint. Two horizontal forces, $\vec{F}_1$ and $\vec{F}_2$, are applied to the stick. Only $\vec{F}_1$ is shown. Force $\vec{F}_2$ is perpendicular to the stick and is applied at the right end. If the stick is not to turn, (a) what should be the direction of $\vec{F}_2$, and (b) should F_2 be greater than, less than, or equal to F_1?

Sample Problem 10-9 Build your skill

Figure 10-18*a* shows a uniform disk, with mass $M = 2.5$ kg and radius $R = 20$ cm, mounted on a fixed horizontal axle. A block with mass $m = 1.2$ kg hangs from a massless cord that is wrapped around the rim of the disk. Find the acceleration of the falling block, the angular acceleration of the disk, and the tension in the cord. The cord does not slip, and there is no friction at the axle.

KEY IDEAS (1) Taking the block as a system, we can relate its acceleration a to the forces acting on it with Newton's second law ($\vec{F}_{\text{net}} = m\vec{a}$). (2) Taking the disk as a system, we can relate its angular acceleration α to the torque acting on it with Newton's second law for rotation ($\tau_{\text{net}} = I\alpha$). (3) To combine the motions of block and disk, we use the fact that the linear acceleration a of the block and the (tangential) linear acceleration a_t of the disk rim are equal.

Forces on block: The forces are shown in the block's free-body diagram in Fig. 10-18*b*: The force from the cord is $\vec{T}$, and the gravitational force is $\vec{F}_g$, of magnitude mg. We can now write Newton's second law for components along a vertical y axis ($F_{\text{net},y} = ma_y$) as

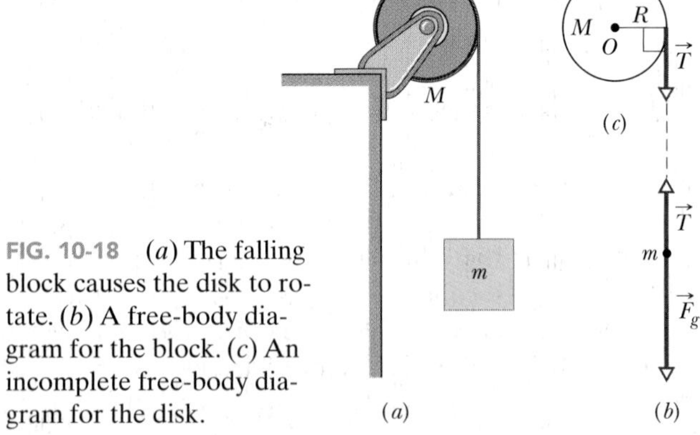

FIG. 10-18 (*a*) The falling block causes the disk to rotate. (*b*) A free-body diagram for the block. (*c*) An incomplete free-body diagram for the disk.

$$T - mg = ma. \quad (10\text{-}46)$$

However, we cannot solve this equation for a because it also contains the unknown T.

Torque on disk: Previously, when we got stuck on the y axis, we switched to the x axis. Here, we switch to the rotation of the disk. To calculate the torques and the rotational inertia I, we take the rotation axis to be perpendicular to the disk and through its center, at point O in Fig. 10-18c.

The torques are then given by Eq. 10-40 ($\tau = rF_t$). The gravitational force on the disk and the force on the disk from the axle both act at the center of the disk and thus at distance $r = 0$, so their torques are zero. The force $\vec{T}$ on the disk due to the cord acts at distance $r = R$ and is tangent to the rim of the disk. Therefore, its torque is $-RT$, negative because the torque rotates the disk clockwise from rest. From Table 10-2c, the rotational inertia I of the disk is $\frac{1}{2}MR^2$. Thus we can write $\tau_{\text{net}} = I\alpha$ as

$$-RT = \tfrac{1}{2}MR^2\alpha. \qquad (10\text{-}47)$$

This equation seems useless because it has two unknowns, α and T, neither of which is the desired a. However, mustering physics courage, we can make it useful with this fact: Because the cord does not slip, the linear acceleration a of the block and the (tangential) linear acceleration a_t of the rim of the disk are equal. Then, by Eq. 10-22 ($a_t = \alpha r$) we see that here $\alpha = a/R$. Substituting this in Eq. 10-47 yields

$$T = -\tfrac{1}{2}Ma. \qquad (10\text{-}48)$$

Combining results: Combining Eqs. 10-46 and 10-48 leads to

$$a = -g\frac{2m}{M + 2m} = -(9.8 \text{ m/s}^2)\frac{(2)(1.2 \text{ kg})}{2.5 \text{ kg} + (2)(1.2 \text{ kg})}$$

$$= -4.8 \text{ m/s}^2. \qquad \text{(Answer)}$$

We then use Eq. 10-48 to find T:

$$T = -\tfrac{1}{2}Ma = -\tfrac{1}{2}(2.5 \text{ kg})(-4.8 \text{ m/s}^2)$$

$$= 6.0 \text{ N}. \qquad \text{(Answer)}$$

As we should expect, acceleration a of the falling block is less than g, and tension T in the cord ($= 6.0$ N) is less than the gravitational force on the hanging block ($= mg = 11.8$ N). We see also that a and T depend on the mass of the disk but not on its radius. As a check, we note that the formulas derived above predict $a = -g$ and $T = 0$ for the case of a massless disk ($M = 0$). This is what we would expect; the block simply falls as a free body. From Eq. 10-22, the angular acceleration of the disk is

$$\alpha = \frac{a}{R} = \frac{-4.8 \text{ m/s}^2}{0.20 \text{ m}} = -24 \text{ rad/s}^2. \qquad \text{(Answer)}$$

Sample Problem 10-10

To throw an 80 kg opponent with a basic judo hip throw, you intend to pull his uniform with a force $\vec{F}$ and a moment arm $d_1 = 0.30$ m from a pivot point (rotation axis) on your right hip (Fig. 10-19). You wish to rotate him about the pivot point with an angular acceleration α of -6.0 rad/s^2—that is, with an angular acceleration that is *clockwise* in the figure. Assume that his rotational inertia I relative to the pivot point is 15 kg · m^2.

(a) What must the magnitude of $\vec{F}$ be if, before you throw him, you bend your opponent forward to bring his center of mass to your hip (Fig. 10-19a)?

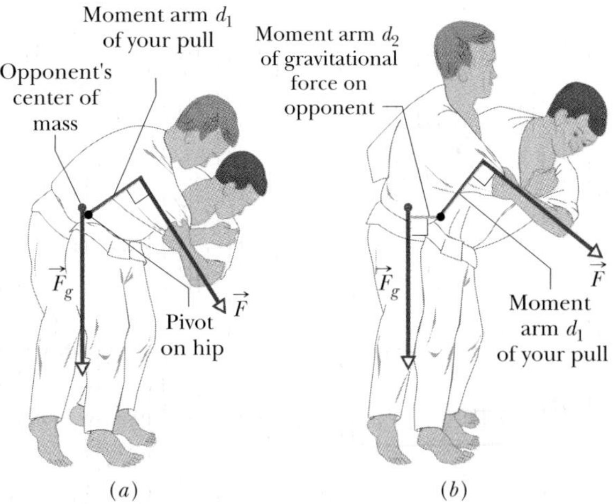

FIG. 10-19 A judo hip throw (a) correctly executed and (b) incorrectly executed.

KEY IDEA We can relate your pull $\vec{F}$ on your opponent to the given angular acceleration α via Newton's second law for rotation ($\tau_{\text{net}} = I\alpha$).

Calculations: As his feet leave the floor, we can assume that only three forces act on him: your pull $\vec{F}$, a force $\vec{N}$ on him from you at the pivot point (this force is not indicated in Fig. 10-19), and the gravitational force $\vec{F}_g$. To use $\tau_{\text{net}} = I\alpha$, we need the corresponding three torques, each about the pivot point.

From Eq. 10-41 ($\tau = r_\perp F$), the torque due to your pull $\vec{F}$ is equal to $-d_1F$, where d_1 is the moment arm $r_\perp$ and the sign indicates the clockwise rotation this torque tends to cause. The torque due to $\vec{N}$ is zero, because $\vec{N}$ acts at the pivot point and thus has moment arm $r_\perp = 0$.

To evaluate the torque due to $\vec{F}_g$, we can assume that $\vec{F}_g$ acts at your opponent's center of mass. With the center of mass at the pivot point, $\vec{F}_g$ has moment arm $r_\perp = 0$ and thus the torque due to $\vec{F}_g$ is zero. Thus, the only torque on your opponent is due to your pull $\vec{F}$, and we can write $\tau_{\text{net}} = I\alpha$ as

$$-d_1F = I\alpha.$$

We then find

$$F = \frac{-I\alpha}{d_1} = \frac{-(15\ \text{kg}\cdot\text{m}^2)(-6.0\ \text{rad/s}^2)}{0.30\ \text{m}}$$

$$= 300\ \text{N}. \qquad \text{(Answer)}$$

(b) What must the magnitude of $\vec{F}$ be if your opponent remains upright before you throw him, so that $\vec{F}_g$ has a moment arm $d_2 = 0.12$ m (Fig. 10-19*b*)?

KEY IDEA Because the moment arm for $\vec{F}_g$ is no longer zero, the torque due to $\vec{F}_g$ is now equal to d_2mg and is positive because the torque attempts counterclockwise rotation.

Calculations: Now we write $\tau_{net} = I\alpha$ as

$$-d_1F + d_2mg = I\alpha,$$

which gives

$$F = -\frac{I\alpha}{d_1} + \frac{d_2mg}{d_1}.$$

From (a), we know that the first term on the right is equal to 300 N. Substituting this and the given data, we have

$$F = 300\ \text{N} + \frac{(0.12\ \text{m})(80\ \text{kg})(9.8\ \text{m/s}^2)}{0.30\ \text{m}}$$

$$= 613.6\ \text{N} \approx 610\ \text{N}. \qquad \text{(Answer)}$$

The results indicate that you will have to pull much harder if you do not initially bend your opponent to bring his center of mass to your hip. A good judo fighter knows this lesson from physics.

10-10 | Work and Rotational Kinetic Energy

As we discussed in Chapter 7, when a force F causes a rigid body of mass m to accelerate along a coordinate axis, the force does work W on the body. Thus, the body's kinetic energy ($K = \frac{1}{2}mv^2$) can change. Suppose it is the only energy of the body that changes. Then we relate the change ΔK in kinetic energy to the work W with the work–kinetic energy theorem (Eq. 7-10), writing

$$\Delta K = K_f - K_i = \tfrac{1}{2}mv_f^2 - \tfrac{1}{2}mv_i^2 = W \qquad \text{(work–kinetic energy theorem).} \qquad (10\text{-}49)$$

For motion confined to an x axis, we can calculate the work with Eq. 7-32,

$$W = \int_{x_i}^{x_f} F\,dx \qquad \text{(work, one-dimensional motion).} \qquad (10\text{-}50)$$

This reduces to $W = Fd$ when F is constant and the body's displacement is d. The rate at which the work is done is the power, which we can find with Eqs. 7-43 and 7-48,

$$P = \frac{dW}{dt} = Fv \qquad \text{(power, one-dimensional motion).} \qquad (10\text{-}51)$$

Now let us consider a rotational situation that is similar. When a torque accelerates a rigid body in rotation about a fixed axis, the torque does work W on the body. Therefore, the body's rotational kinetic energy ($K = \frac{1}{2}I\omega^2$) can change. Suppose that it is the only energy of the body that changes. Then we can still relate the change ΔK in kinetic energy to the work W with the work–kinetic energy theorem, except now the kinetic energy is a rotational kinetic energy:

$$\Delta K = K_f - K_i = \tfrac{1}{2}I\omega_f^2 - \tfrac{1}{2}I\omega_i^2 = W \qquad \text{(work–kinetic energy theorem).} \qquad (10\text{-}52)$$

Here, I is the rotational inertia of the body about the fixed axis and ω_i and ω_f are the angular speeds of the body before and after the work is done, respectively.

Also, we can calculate the work with a rotational equivalent of Eq. 10-50,

$$W = \int_{\theta_i}^{\theta_f} \tau\,d\theta \qquad \text{(work, rotation about fixed axis),} \qquad (10\text{-}53)$$

where τ is the torque doing the work W, and θ_i and θ_f are the body's angular positions before and after the work is done, respectively. When τ is constant, Eq. 10-53 reduces to

$$W = \tau(\theta_f - \theta_i) \qquad \text{(work, constant torque).} \qquad (10\text{-}54)$$

TABLE 10-3

Some Corresponding Relations for Translational and Rotational Motion

Pure Translation (Fixed Direction)		Pure Rotation (Fixed Axis)	
Position	x	Angular position	θ
Velocity	$v = dx/dy$	Angular velocity	$\omega = d\theta/dt$
Acceleration	$a = dv/dt$	Angular acceleration	$\alpha = d\omega/dt$
Mass	m	Rotational inertia	I
Newton's second law	$F_{\text{net}} = ma$	Newton's second law	$\tau_{\text{net}} = I\alpha$
Work	$W = \int F\,dx$	Work	$W = \int \tau\,d\theta$
Kinetic energy	$K = \frac{1}{2}mv^2$	Kinetic energy	$K = \frac{1}{2}I\omega^2$
Power (constant force)	$P = Fv$	Power (constant torque)	$P = \tau\omega$
Work–kinetic energy theorem	$W = \Delta K$	Work–kinetic energy theorem	$W = \Delta K$

The rate at which the work is done is the power, which we can find with the rotational equivalent of Eq. 10-51,

$$P = \frac{dW}{dt} = \tau\omega \quad \text{(power, rotation about fixed axis).} \tag{10-55}$$

Table 10-3 summarizes the equations that apply to the rotation of a rigid body about a fixed axis and the corresponding equations for translational motion.

Proof of Eqs. 10-52 through 10-55

Let us again consider the situation of Fig. 10-17, in which force $\vec{F}$ rotates a rigid body consisting of a single particle of mass m fastened to the end of a massless rod. During the rotation, force $\vec{F}$ does work on the body. Let us assume that the only energy of the body that is changed by $\vec{F}$ is the kinetic energy. Then we can apply the work–kinetic energy theorem of Eq. 10-49:

$$\Delta K = K_f - K_i = W. \tag{10-56}$$

Using $K = \frac{1}{2}mv^2$ and Eq. 10-18 ($v = \omega r$), we can rewrite Eq. 10-56 as

$$\Delta K = \tfrac{1}{2}mr^2\omega_f^2 - \tfrac{1}{2}mr^2\omega_i^2 = W. \tag{10-57}$$

From Eq. 10-33, the rotational inertia for this one-particle body is $I = mr^2$. Substituting this into Eq. 10-57 yields

$$\Delta K = \tfrac{1}{2}I\omega_f^2 - \tfrac{1}{2}I\omega_i^2 = W,$$

which is Eq. 10-52. We derived it for a rigid body with one particle, but it holds for any rigid body rotated about a fixed axis.

We next relate the work W done on the body in Fig. 10-17 to the torque τ on the body due to force $\vec{F}$. When the particle moves a distance ds along its circular path, only the tangential component F_t of the force accelerates the particle along the path. Therefore, only F_t does work on the particle. We write that work dW as $F_t\,ds$. However, we can replace ds with $r\,d\theta$, where $d\theta$ is the angle through which the particle moves. Thus we have

$$dW = F_t r\,d\theta. \tag{10-58}$$

From Eq. 10-40, we see that the product $F_t r$ is equal to the torque τ, so we can rewrite Eq. 10-58 as

$$dW = \tau\,d\theta. \tag{10-59}$$

The work done during a finite angular displacement from θ_i to θ_f is then

$$W = \int_{\theta_i}^{\theta_f} \tau\,d\theta,$$

which is Eq. 10-53. It holds for any rigid body rotating about a fixed axis. Equation 10-54 comes directly from Eq. 10-53.

We can find the power P for rotational motion from Eq. 10-59:

$$P = \frac{dW}{dt} = \tau \frac{d\theta}{dt} = \tau\omega,$$

which is Eq. 10-55

Sample Problem 10-11

Let the disk in Sample Problem 10-9 and Fig. 10-18 start from rest at time $t = 0$. What is its rotational kinetic energy K at $t = 2.5$ s?

KEY IDEA We can find K with Eq. 10-34 ($K = \frac{1}{2}I\omega^2$). We already know that $I = \frac{1}{2}MR^2$, but we do not yet know ω at $t = 2.5$ s. However, because the angular acceleration α has the constant value of -24 rad/s^2, we can apply the equations for constant angular acceleration in Table 10-1.

Calculations: Because we want ω and know α and ω_0 ($= 0$), we use Eq. 10-12:

$$\omega = \omega_0 + \alpha t = 0 + \alpha t = \alpha t.$$

Substituting $\omega = \alpha t$ and $I = \frac{1}{2}MR^2$ into Eq. 10-34, we find

$$K = \tfrac{1}{2}I\omega^2 = \tfrac{1}{2}(\tfrac{1}{2}MR^2)(\alpha t)^2 = \tfrac{1}{4}M(R\alpha t)^2$$
$$= \tfrac{1}{4}(2.5 \text{ kg})[(0.20 \text{ m})(-24 \text{ rad/s}^2)(2.5 \text{ s})]^2$$
$$= 90 \text{ J.} \qquad \text{(Answer)}$$

KEY IDEA We can also get this answer by finding the disk's kinetic energy from the work done on the disk.

Calculations: First, we relate the *change* in the kinetic energy of the disk to the net work W done on the disk, using the work–kinetic energy theorem of Eq. 10-52 ($K_f - K_i = W$). With K substituted for K_f and 0 for K_i, we get

$$K = K_i + W = 0 + W = W. \qquad (10\text{-}60)$$

Next we want to find the work W. We can relate W to the torques acting on the disk with Eq. 10-53 or 10-54. The only torque causing angular acceleration and doing work is the torque due to force $\vec{T}$ on the disk from the cord. From Sample Problem 10-9, this torque is equal to $-TR$. Because α is constant, this torque also must be constant. Thus, we can use Eq. 10-54 to write

$$W = \tau(\theta_f - \theta_i) = -TR(\theta_f - \theta_i). \qquad (10\text{-}61)$$

Because α is constant, we can use Eq. 10-13 to find $\theta_f - \theta_i$. With $\omega_i = 0$, we have

$$\theta_f - \theta_i = \omega_i t + \tfrac{1}{2}\alpha t^2 = 0 + \tfrac{1}{2}\alpha t^2 = \tfrac{1}{2}\alpha t^2.$$

Now we substitute this into Eq. 10-61 and then substitute the result into Eq. 10-60. With $T = 6.0$ N and $\alpha = -24$ rad/s^2 (from Sample Problem 10-9), we have

$$K = W = -TR(\theta_f - \theta_i) = -TR(\tfrac{1}{2}\alpha t^2) = -\tfrac{1}{2}TR\alpha t^2$$
$$= -\tfrac{1}{2}(6.0 \text{ N})(0.20 \text{ m})(-24 \text{ rad/s}^2)(2.5 \text{ s})^2$$
$$= 90 \text{ J.} \qquad \text{(Answer)}$$

Sample Problem 10-12

A tall, cylindrical chimney will fall over when its base is ruptured. Treat the chimney as a thin rod of length $L = 55.0$ m (Fig. 10-20*a*). At the instant it makes an angle of $\theta = 35.0°$ with the vertical, what is its angular speed ω_f?

KEY IDEAS (1) During the rotation, the mechanical energy (the sum of the rotational kinetic energy K and the gravitational potential energy U) does not change. (2) The rotational kinetic energy is given by Eq. 10-34 ($K = \frac{1}{2}I\omega^2$).

Conservation of mechanical energy: As the center of mass of the chimney falls, energy is transferred from gravitational potential energy U to rotational kinetic energy K but the total amount does not change. We can write this fact as

$$K_f + U_f = K_i + U_i. \qquad (10\text{-}62)$$

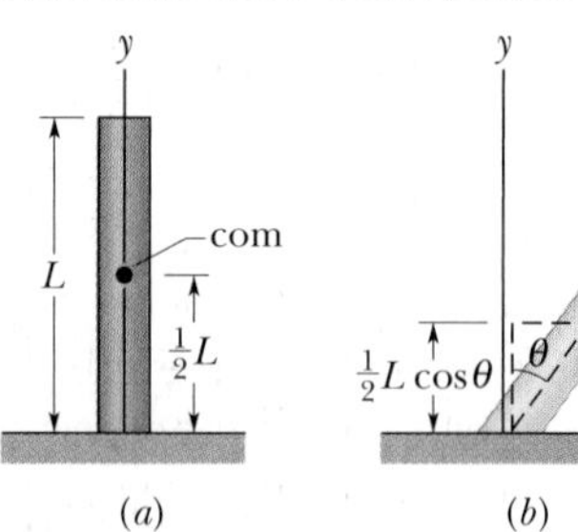

FIG. 10-20 (*a*) A cylindrical chimney. (*b*) The height of its center of mass is determined with the right triangle.

Rotational kinetic energy: The kinetic energy K is initially zero but its value thereafter ($= \frac{1}{2}I\omega^2$) depends on the rotational inertia I. If we had a thin rod rotating around its center of mass (at its center), we know from Table 10-2 that $I_{com} = \frac{1}{12}mL^2$, where m is the rod's mass and L is the rod's length. However, our rod-like chimney rotates around one end, at a distance of $L/2$ from the center, and so we use the parallel-axis theorem to find

$$I = \tfrac{1}{12}mL + m\left(\frac{L}{2}\right)^2 = \tfrac{1}{3}mL^2. \qquad (10\text{-}63)$$

Substituting this into $K = \frac{1}{2}I\omega^2$ tells us

$$K_f = \tfrac{1}{2}(\tfrac{1}{3}mL^2)\omega^2. \qquad (10\text{-}64)$$

Potential energy: The potential energy $U\,(= mgy)$ depends on the height of each segment of the chimney. However, we can calculate U by assuming that all of the mass is concentrated at the chimney's com, which is initially at height $\frac{1}{2}L$. So the initial potential energy is

$$U_i = \tfrac{1}{2}mgL. \qquad (10\text{-}65)$$

When the chimney has rotated through angle θ, Fig. 10-20*b* tells us that the center is at height $\frac{1}{2}L\cos\theta$. So, the potential energy is now

$$U_f = \tfrac{1}{2}mgL\cos\theta. \qquad (10\text{-}66)$$

Angular speed: After substituting Eqs. 10-66, 10-65, and 10-64 into Eq. 10-62 and setting $K_i = 0$, we solve for ω_f and find

$$\omega = \sqrt{\frac{3g}{L}(1-\cos\theta)} = \sqrt{\frac{3(9.8\text{ m/s}^2)}{55.0\text{ m}}(1-\cos 35.0°)}$$
$$= 0.311\text{ rad/s}. \qquad \text{(Answer)}$$

Comments: The bottom portion of a chimney tends to rotate around the base faster than the top portion, and the chimney is likely to break apart during the rotation, with the top half lagging behind the bottom half.

Sample Problem 10-13

In the oversized claw of a snapping shrimp, the dactylus (the large, mobile section of the claw) is drawn away from the propodius (the opposing, stationary part of the claw) by a muscle that is gradually put under tension (Fig. 10-21). Energy stored in the muscle increases as the tension increases. The sudden release of the dactylus allows it to rotate about a pivot point, to slam shut on the propodius in a time interval Δt of only 290 μs. In particular, the *plunger* on the dactylus runs into a cavity on the propodius, causing water to squirt out of the cavity so quickly that the water undergoes *cavitation*. That is, the water vaporizes to form bubbles of water vapor. These bubbles rapidly grow as they enter the surrounding water and then they suddenly collapse, emitting an intense sound wave. The combination of these sound waves from many bubbles can stun the shrimp's prey.

The peak angular speed ω of the dactylus is about 2×10^3 rad/s and its rotational inertia I is about 3×10^{-11} kg·m². At what akverage rate is energy transferred from the muscle to the rotation?

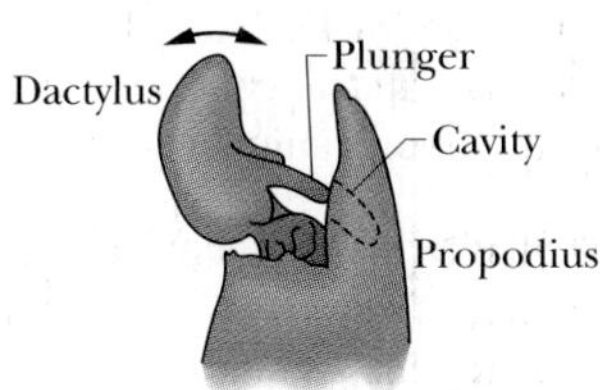

FIG. 10-21 The oversized claw of a snapping shrimp. The dactylus is first pulled away from the opposing section of the propodius and then allowed to snap back to it, thrusting the plunger into the cavity.

KEY IDEA (1) Rotational kinetic energy is given by Eq. 10-34 ($K = \frac{1}{2}I\omega^2$). (2) Average power is given by Eq. 8-40 ($P_{avg} = \Delta E/\Delta t$).

Calculations: When the angular speed reaches its peak value, the rotational kinetic energy is

$$K = \tfrac{1}{2}I\omega^2 = \tfrac{1}{2}(3 \times 10^{-11}\text{ kg}\cdot\text{m}^2)(2 \times 10^3\text{ rad/s})^2$$
$$= 6 \times 10^{-5}\text{ J}.$$

The average power is then

$$P_{avg} = \frac{\Delta E}{\Delta t} = \frac{6 \times 10^{-5}\text{ J}}{290 \times 10^{-6}\text{ s}}$$
$$= 0.2\text{ W}. \qquad \text{(Answer)}$$

This power greatly exceeds what any fast-acting muscle in the shrimp can produce. However, in the claw the shrimp effectively locks the dactylus against a spring so that it can gradually increase thc tension and stored energy (the power of this stage is low). Then, once the stored energy is high, the dactylus is released and the spring-like muscle slams it shut (the power is now very high). Many other animals make use of such low-power storing of energy and then a high-power release of the energy that allows them to capture lunch or to avoid becoming lunch.

REVIEW & SUMMARY

Angular Position To describe the rotation of a rigid body about a fixed axis, called the **rotation axis,** we assume a **reference line** is fixed in the body, perpendicular to that axis and rotating with the body. We measure the **angular position** θ of this line relative to a fixed direction. When θ is measured in **radians,**

$$\theta = \frac{s}{r} \qquad \text{(radian measure)}, \qquad (10\text{-}1)$$

where s is the arc length of a circular path of radius r and angle θ. Radian measure is related to angle measure in revolutions and degrees by

$$1 \text{ rev} = 360° = 2\pi \text{ rad}. \tag{10-2}$$

Angular Displacement A body that rotates about a rotation axis, changing its angular position from θ_1 to θ_2, undergoes an **angular displacement**

$$\Delta\theta = \theta_2 - \theta_1, \tag{10-4}$$

where $\Delta\theta$ is positive for counterclockwise rotation and negative for clockwise rotation.

Angular Velocity and Speed If a body rotates through an angular displacement $\Delta\theta$ in a time interval Δt, its **average angular velocity** ω_{avg} is

$$\omega_{avg} = \frac{\Delta\theta}{\Delta t}. \tag{10-5}$$

The **(instantaneous) angular velocity** ω of the body is

$$\omega = \frac{d\theta}{dt}. \tag{10-6}$$

Both ω_{avg} and ω are vectors, with directions given by the **right-hand rule** of Fig. 10-6. They are positive for counterclockwise rotation and negative for clockwise rotation. The magnitude of the body's angular velocity is the **angular speed.**

Angular Acceleration If the angular velocity of a body changes from ω_1 to ω_2 in a time interval $\Delta t = t_2 - t_1$, the **average angular acceleration** α_{avg} of the body is

$$\alpha_{avg} = \frac{\omega_2 - \omega_1}{t_2 - t_1} = \frac{\Delta\omega}{\Delta t}. \tag{10-7}$$

The **(instantaneous) angular acceleration** α of the body is

$$\alpha = \frac{d\omega}{dt}. \tag{10-8}$$

Both α_{avg} and α are vectors.

The Kinematic Equations for Constant Angular Acceleration *Constant angular acceleration* (α = constant) is an important special case of rotational motion. The appropriate kinematic equations, given in Table 10-1, are

$$\omega = \omega_0 + \alpha t, \tag{10-12}$$

$$\theta - \theta_0 = \omega_0 t + \tfrac{1}{2}\alpha t^2, \tag{10-13}$$

$$\omega^2 = \omega_0^2 + 2\alpha(\theta - \theta_0), \tag{10-14}$$

$$\theta - \theta_0 = \tfrac{1}{2}(\omega_0 + \omega)t, \tag{10-15}$$

$$\theta - \theta_0 = \omega t - \tfrac{1}{2}\alpha t^2. \tag{10-16}$$

Linear and Angular Variables Related A point in a rigid rotating body, at a *perpendicular distance* r from the rotation axis, moves in a circle with radius r. If the body rotates through an angle θ, the point moves along an arc with length s given by

$$s = \theta r \quad \text{(radian measure)}, \tag{10-17}$$

where θ is in radians.

The linear velocity $\vec{v}$ of the point is tangent to the circle; the point's linear speed v is given by

$$v = \omega r \quad \text{(radian measure)}, \tag{10-18}$$

where ω is the angular speed (in radians per second) of the body.

The linear acceleration $\vec{a}$ of the point has both *tangential* and *radial* components. The tangential component is

$$a_t = \alpha r \quad \text{(radian measure)}, \tag{10-22}$$

where α is the magnitude of the angular acceleration (in radians per second-squared) of the body. The radial component of $\vec{a}$ is

$$a_r = \frac{v^2}{r} = \omega^2 r \quad \text{(radian measure)}. \tag{10-23}$$

If the point moves in uniform circular motion, the period T of the motion for the point and the body is

$$T = \frac{2\pi r}{v} = \frac{2\pi}{\omega} \quad \text{(radian measure)}. \tag{10-19, 10-20}$$

Rotational Kinetic Energy and Rotational Inertia The kinetic energy K of a rigid body rotating about a fixed axis is given by

$$K = \tfrac{1}{2}I\omega^2 \quad \text{(radian measure)}, \tag{10-34}$$

in which I is the **rotational inertia** of the body, defined as

$$I = \sum m_i r_i^2 \tag{10-33}$$

for a system of discrete particles and defined as

$$I = \int r^2\, dm \tag{10-35}$$

for a body with continuously distributed mass. The r and r_i in these expressions represent the perpendicular distance from the axis of rotation to each mass element in the body.

The Parallel-Axis Theorem The *parallel-axis theorem* relates the rotational inertia I of a body about any axis to that of the same body about a parallel axis through the center of mass:

$$I = I_{com} + Mh^2. \tag{10-36}$$

Here h is the perpendicular distance between the two axes.

Torque *Torque* is a turning or twisting action on a body about a rotation axis due to a force $\vec{F}$. If $\vec{F}$ is exerted at a point given by the position vector $\vec{r}$ relative to the axis, then the magnitude of the torque is

$$\tau = rF_t = r_\perp F = rF \sin\phi, \tag{10-40, 10-41, 10-39}$$

where F_t is the component of $\vec{F}$ perpendicular to $\vec{r}$ and ϕ is the angle between $\vec{r}$ and $\vec{F}$. The quantity $r_\perp$ is the perpendicular distance between the rotation axis and an extended line running through the $\vec{F}$ vector. This line is called the **line of action** of $\vec{F}$, and $r_\perp$ is called the **moment arm** of $\vec{F}$. Similarly, r is the moment arm of F_t.

The SI unit of torque is the newton-meter (N·m). A torque τ is positive if it tends to rotate a body at rest counterclockwise and negative if it tends to rotate the body in the clockwise direction.

Newton's Second Law in Angular Form The rotational analog of Newton's second law is

$$\tau_{net} = I\alpha, \tag{10-45}$$

where τ_{net} is the net torque acting on a particle or rigid body, I is the rotational inertia of the particle or body about the rotation axis, and α is the resulting angular acceleration about that axis.

Work and Rotational Kinetic Energy The equations used for calculating work and power in rotational motion correspond to equations used for translational motion and are

$$W = \int_{\theta_i}^{\theta_f} \tau \, d\theta \qquad (10\text{-}53)$$

and

$$P = \frac{dW}{dt} = \tau\omega. \qquad (10\text{-}55)$$

When τ is constant, Eq. 10-53 reduces to

$$W = \tau(\theta_f - \theta_i). \qquad (10\text{-}54)$$

The form of the work–kinetic energy theorem used for rotating bodies is

$$\Delta K = K_f - K_i = \tfrac{1}{2}I\omega_f^2 - \tfrac{1}{2}I\omega_i^2 = W. \qquad (10\text{-}52)$$

QUESTIONS

1 Figure 10-22 is a graph of the angular velocity versus time for a disk rotating like a merry-go-round. For a point on the disk rim, rank the instants a, b, c, and d according to the magnitude of the (a) tangential and (b) radial acceleration, greatest first.

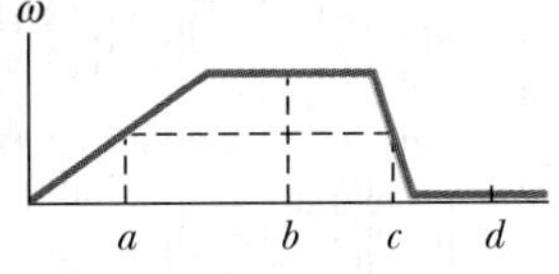

FIG. 10-22 Question 1.

2 Figure 10-23b is a graph of the angular position of the rotating disk of Fig. 10-23a. Is the angular velocity of the disk positive, negative, or zero at (a) $t = 1$ s, (b) $t = 2$ s, and (c) $t = 3$ s? (d) Is the angular acceleration positive or negative?

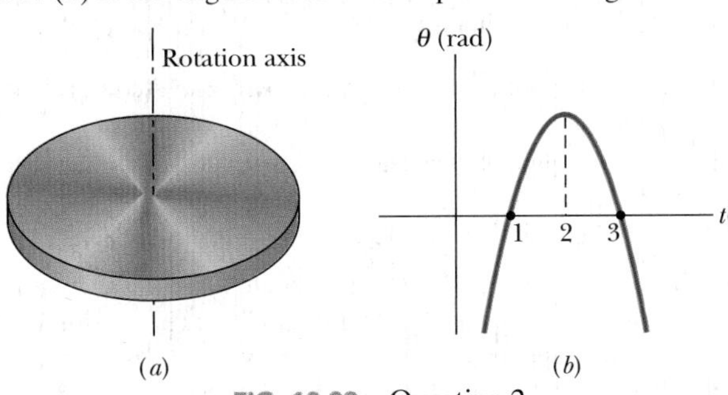

FIG. 10-23 Question 2.

3 Figure 10-24 shows a uniform metal plate that had been square before 25% of it was snipped off. Three lettered points are indicated. Rank them according to the rotational inertia of the plate around a perpendicular axis through them, greatest first.

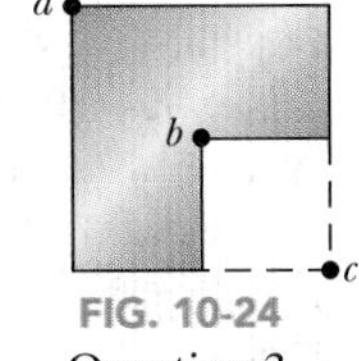

FIG. 10-24 Question 3.

4 Figure 10-25 shows plots of angular position θ versus time t for three cases in which a disk is rotated like a merry-go-round. In each case, the rotation direction changes at a certain angular position θ_{change}. (a) For each case, determine whether θ_{change} is clockwise or counterclockwise from $\theta = 0$, or whether it is at $\theta = 0$. For each case, determine (b) whether ω is zero before, after, or at $t = 0$ and (c) whether α is positive, negative, or zero.

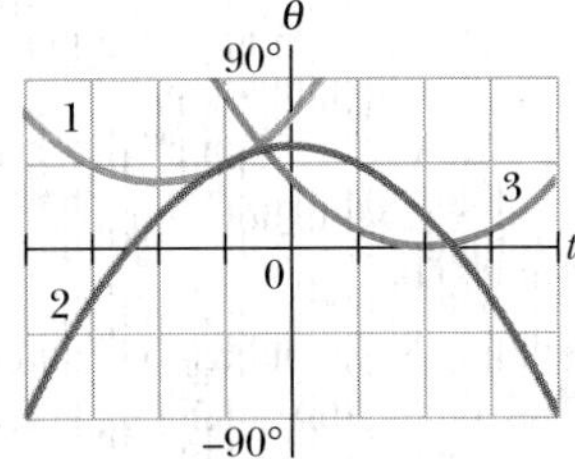

FIG. 10-25 Question 4.

5 Figure 10-26a is an overhead view of a horizontal bar that can pivot; two horizontal forces act on the bar, but it is stationary. If the angle between the bar and $\vec{F}_2$ is now decreased from 90° and the bar is still not to turn, should F_2 be made larger, made smaller, or left the same?

6 Figure 10-26b shows an overhead view of a horizontal bar that is rotated about the pivot point by two horizontal forces, $\vec{F}_1$ and $\vec{F}_2$, with $\vec{F}_2$ at angle ϕ to the bar. Rank the following values of ϕ according to the magnitude of the angular acceleration of the bar, greatest first: 90°, 70°, and 110°.

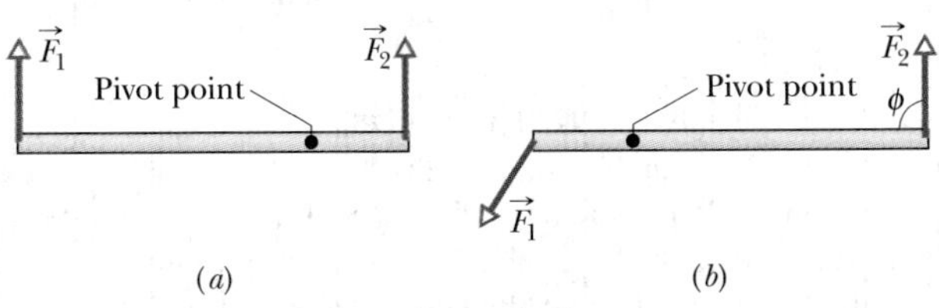

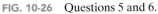

FIG. 10-26 Questions 5 and 6.

7 In Fig. 10-27, two forces $\vec{F}_1$ and $\vec{F}_2$ act on a disk that turns about its center like a merry-go-round. The forces maintain the indicated angles during the rotation, which is counterclockwise and at a constant rate. However, we are to decrease the angle θ of $\vec{F}_1$ without changing the magnitude of $\vec{F}_1$. (a) To keep the angular speed constant, should we increase, decrease, or maintain the magnitude of $\vec{F}_2$? Do forces (b) $\vec{F}_1$ and (c) $\vec{F}_2$ tend to rotate the disk clockwise or counterclockwise?

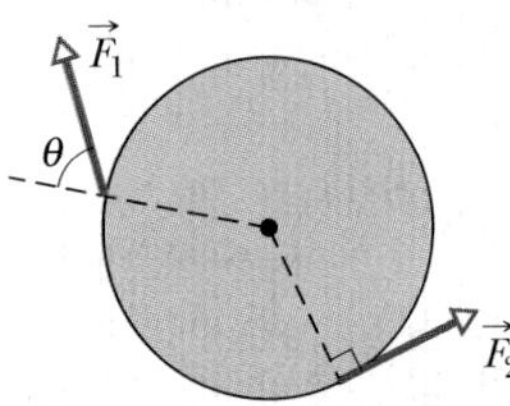

FIG. 10-27 Question 7.

8 In the overhead view of Fig. 10-28, five forces of the same magnitude act on a strange merry-go-round; it is a square that can rotate about point P, at midlength along one of the edges. Rank the forces according to the magnitude of the torque they create about point P, greatest first.

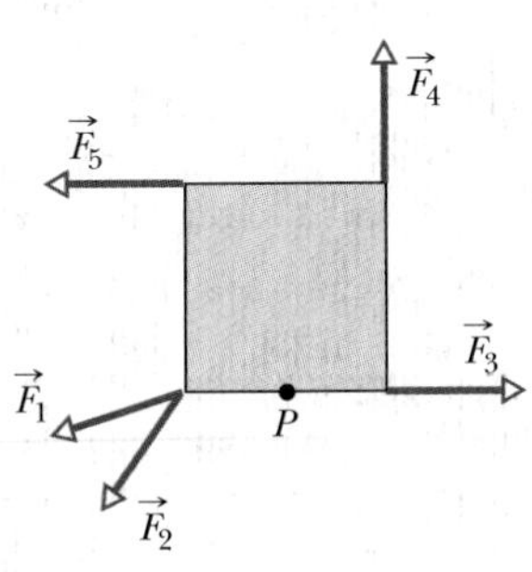

FIG. 10-28 Question 8.

9 A force is applied to the rim of a disk that can rotate like a merry-go-round, so as to change its angular velocity. Its initial and final angular velocities, respectively, for four situations are: (a) -2 rad/s, 5 rad/s; (b) 2 rad/s, 5 rad/s; (c) -2 rad/s, -5 rad/s; and (d) 2 rad/s, -5 rad/s. Rank the situations according to the work done by the torque due to the force, greatest first.

10 Figure 10-29 shows three flat disks (of the same radius) that can rotate about their centers like merry-go-rounds. Each disk consists of the same two materials, one denser than the other (density is mass per unit volume). In disks 1 and 3, the denser material forms the outer half of the disk area. In disk 2, it forms the inner half of the disk area. Forces with identical magnitudes are applied tangentially to the disk, either at the outer edge or at the interface of the two materials, as shown. Rank the disks according to (a) the torque about the disk center, (b) the rotational inertia about the disk center, and (c) the angular acceleration of the disk, greatest first.

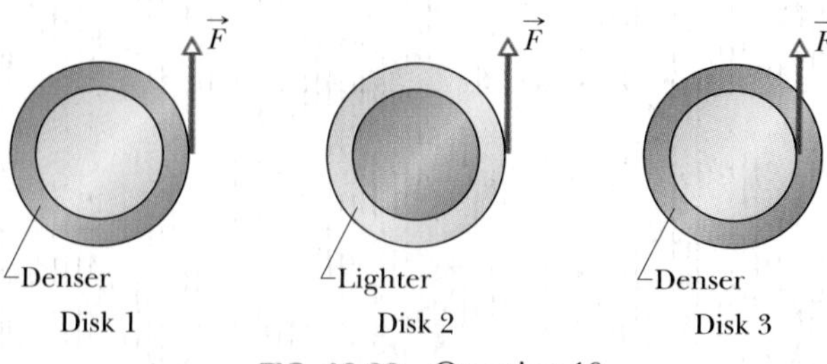

FIG. 10-29 Question 10.

PROBLEMS

GO Tutoring problem available (at instructor's discretion) in *WileyPLUS* and WebAssign
SSM Worked-out solution available in Student Solutions Manual
• – ••• Number of dots indicates level of problem difficulty
Additional information available in *The Flying Circus of Physics* and at flyingcircusofphysics.com
WWW Worked-out solution is at
ILW Interactive solution is at
http://www.wiley.com/college/halliday

sec. 10-2 The Rotational Variables

•1 A good baseball pitcher can throw a baseball toward home plate at 85 mi/h with a spin of 1800 rev/min. How many revolutions does the baseball make on its way to home plate? For simplicity, assume that the 60 ft path is a straight line.

•2 What is the angular speed of (a) the second hand, (b) the minute hand, and (c) the hour hand of a smoothly running analog watch? Answer in radians per second.

••3 A diver makes 2.5 revolutions on the way from a 10-m-high platform to the water. Assuming zero initial vertical velocity, find the average angular velocity during the dive. ILW

••4 The angular position of a point on the rim of a rotating wheel is given by $\theta = 4.0t - 3.0t^2 + t^3$, where θ is in radians and t is in seconds. What are the angular velocities at (a) $t = 2.0$ s and (b) $t = 4.0$ s? (c) What is the average angular acceleration for the time interval that begins at $t = 2.0$ s and ends at $t = 4.0$ s? What are the instantaneous angular accelerations at (d) the beginning and (e) the end of this time interval?

••5 When a slice of buttered toast is accidentally pushed over the edge of a counter, it rotates as it falls. If the distance to the floor is 76 cm and for rotation less than 1 rev, what are the (a) smallest and (b) largest angular speeds that cause the toast to hit and then topple to be butter-side down?

••6 The angular position of a point on a rotating wheel is given by $\theta = 2.0 + 4.0t^2 + 2.0t^3$, where θ is in radians and t is in seconds. At $t = 0$, what are (a) the point's angular position and (b) its angular velocity? (c) What is its angular velocity at $t = 4.0$ s? (d) Calculate its angular acceleration at $t = 2.0$ s. (e) Is its angular acceleration constant?

•••7 The wheel in Fig. 10-30 has eight equally spaced spokes and a radius of 30 cm. It is mounted on a fixed axle and is spinning at 2.5 rev/s. You want to shoot a 20-cm-long arrow parallel to this axle and through the wheel without hitting any of the spokes. Assume that the arrow and the spokes are very thin. (a) What minimum speed must the arrow have? (b) Does it matter where between the axle and rim of the wheel you aim? If so, what is the best location?

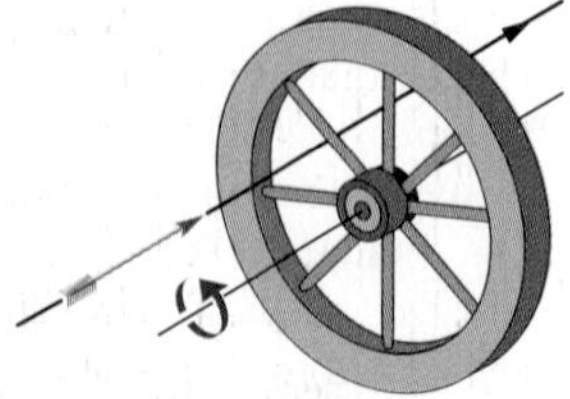
FIG. 10-30 Problem 7.

•••8 The angular acceleration of a wheel is $\alpha = 6.0t^4 - 4.0t^2$, with α in radians per second-squared and t in seconds. At time $t = 0$, the wheel has an angular velocity of $+2.0$ rad/s and an angular position of $+1.0$ rad. Write expressions for (a) the angular velocity (rad/s) and (b) the angular position (rad) as functions of time (s).

sec. 10-4 Rotation with Constant Angular Acceleration

•9 A disk, initially rotating at 120 rad/s, is slowed down with a constant angular acceleration of magnitude 4.0 rad/s^2. (a) How much time does the disk take to stop? (b) Through what angle does the disk rotate during that time?

•10 The angular speed of an automobile engine is increased at a constant rate from 1200 rev/min to 3000 rev/min in 12 s. (a) What is its angular acceleration in revolutions per minute-squared? (b) How many revolutions does the engine make during this 12 s interval?

•11 A drum rotates around its central axis at an angular velocity of 12.60 rad/s. If the drum then slows at a constant rate of 4.20 rad/s^2, (a) how much time does it take and (b) through what angle does it rotate in coming to rest?

•12 Starting from rest, a disk rotates about its central axis with constant angular acceleration. In 5.0 s, it rotates 25 rad. During that time, what are the magnitudes of (a) the angular acceleration and (b) the average angular velocity? (c) What is the instantaneous angular velocity of the disk at the end of the 5.0 s? (d) With the angular acceleration unchanged, through what additional angle will the disk turn during the next 5.0 s?

••13 A wheel has a constant angular acceleration of 3.0 rad/s^2. During a certain 4.0 s interval, it turns through an angle of 120 rad. Assuming that the wheel started from rest, how long has it been in motion at the start of this 4.0 s interval? SSM

••14 A merry-go-round rotates from rest with an angular acceleration of 1.50 rad/s^2. How long does it take to rotate through (a) the first 2.00 rev and (b) the next 2.00 rev?

••15 At $t = 0$, a flywheel has an angular velocity of 4.7 rad/s, a constant angular acceleration of -0.25 rad/s^2, and a reference line at $\theta_0 = 0$. (a) Through what maximum angle θ_{max} will the reference line turn in the positive direction? What are the (b) first and (c) second times the reference line will be at $\theta = \frac{1}{2}\theta_{max}$? At what (d) negative time and (e) positive time will the reference line be at $\theta = -10.5$ rad? (f) Graph θ versus t, and indicate the answers to (a) through (e) on the graph.

••16 A disk rotates about its central axis starting from rest and accelerates with constant angular acceleration. At one time it is rotating at 10 rev/s; 60 revolutions later, its angular speed is 15 rev/s. Calculate (a) the angular acceleration, (b) the time required to complete the 60 revolutions, (c) the time required to reach the 10 rev/s angular speed, and (d) the number of revolutions from rest until the time the disk reaches the 10 rev/s angular speed.

••17 A flywheel turns through 40 rev as it slows from an angular speed of 1.5 rad/s to a stop. (a) Assuming a constant angular acceleration, find the time for it to come to rest. (b) What is its angular acceleration? (c) How much time is required for it to complete the first 20 of the 40 revolutions? ILW

sec. 10-5 Relating the Linear and Angular Variables

•18 A vinyl record is played by rotating the record so that an approximately circular groove in the vinyl slides under a stylus. Bumps in the groove run into the stylus, causing it to oscillate. The equipment converts those oscillations to electrical signals and then to sound. Suppose that a record turns at the rate of $33\frac{1}{3}$ rev/min, the groove being played is at a radius of 10.0 cm, and the bumps in the groove are uniformly separated by 1.75 mm. At what rate (hits per second) do the bumps hit the stylus?

•19 Between 1911 and 1990, the top of the leaning bell tower at Pisa, Italy, moved toward the south at an average rate of 1.2 mm/y. The tower is 55 m tall. In radians per second, what is the average angular speed of the tower's top about its base?

•20 An astronaut is being tested in a centrifuge. The centrifuge has a radius of 10 m and, in starting, rotates according to $\theta = 0.30t^2$, where t is in seconds and θ is in radians. When $t = 5.0$ s, what are the magnitudes of the astronaut's (a) angular velocity, (b) linear velocity, (c) tangential acceleration, and (d) radial acceleration?

•21 A flywheel with a diameter of 1.20 m is rotating at an angular speed of 200 rev/min. (a) What is the angular speed of the flywheel in radians per second? (b) What is the linear speed of a point on the rim of the flywheel? (c) What constant angular acceleration (in revolutions per minute-squared) will increase the wheel's angular speed to 1000 rev/min in 60.0 s? (d) How many revolutions does the wheel make during that 60.0 s? SSM WWW

•22 If an airplane propeller rotates at 2000 rev/min while the airplane flies at a speed of 480 km/h relative to the ground, what is the linear speed of a point on the tip of the propeller, at radius 1.5 m, as seen by (a) the pilot and (b) an observer on the ground? The plane's velocity is parallel to the propeller's axis of rotation.

•23 What are the magnitudes of (a) the angular velocity, (b) the radial acceleration, and (c) the tangential acceleration of a spaceship taking a circular turn of radius 3220 km at a speed of 29 000 km/h?

•24 An object rotates about a fixed axis, and the angular position of a reference line on the object is given by $\theta = 0.40e^{2t}$, where θ is in radians and t is in seconds. Consider a point on the object that is 4.0 cm from the axis of rotation. At $t =$ 0, what are the magnitudes of the point's (a) tangential component of acceleration and (b) radial component of acceleration?

••25 A disk, with a radius of 0.25 m, is to be rotated like a merry-go-round through 800 rad, starting from rest, gaining angular speed at the constant rate α_1 through the first 400 rad and then losing angular speed at the constant rate $-\alpha_1$ until it is again at rest. The magnitude of the centripetal acceleration of any portion of the disk is not to exceed 400 m/s^2. (a) What is the least time required for the rotation? (b) What is the corresponding value of α_1?

••26 A gyroscope flywheel of radius 2.83 cm is accelerated from rest at 14.2 rad/s^2 until its angular speed is 2760 rev/min. (a) What is the tangential acceleration of a point on the rim of the flywheel during this spin-up process? (b) What is the radial acceleration of this point when the flywheel is spinning at full speed? (c) Through what distance does a point on the rim move during the spin-up?

••27 An early method of measuring the speed of light makes use of a rotating slotted wheel. A beam of light passes through one of the slots at the outside edge of the wheel, as in Fig. 10-31, travels to a distant mirror, and returns to the wheel just in time to pass through the next slot in the wheel. One such slotted wheel has a radius of 5.0 cm and 500 slots around its edge. Measurements taken when the mirror is $L = 500$ m from the wheel indicate a speed of light of 3.0×10^5 km/s. (a) What is the (constant) angular speed of the wheel? (b) What is the linear speed of a point on the edge of the wheel?

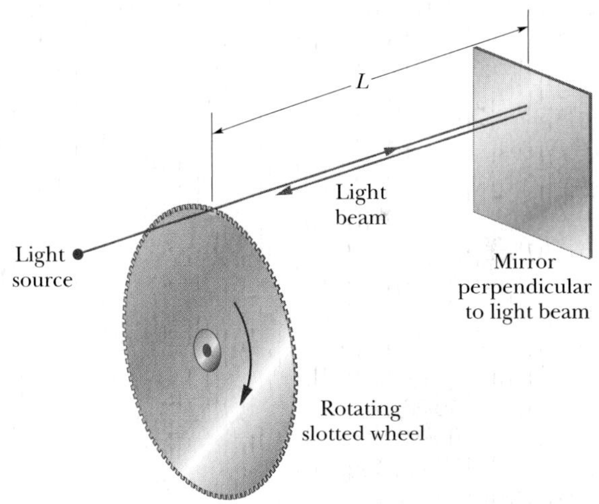

FIG. 10-31 Problem 27.

••28 The flywheel of a steam engine runs with a constant angular velocity of 150 rev/min. When steam is shut off, the friction of the bearings and of the air stops the wheel in 2.2 h. (a) What is the constant angular acceleration, in revolutions per minute-squared, of the wheel during the slowdown? (b) How many revolutions does the wheel make before stopping? (c) At the instant the flywheel is turning at 75 rev/min, what is the tangential component of the linear acceleration of a flywheel particle that is 50 cm from the axis of rotation? (d) What is the magnitude of the net linear acceleration of the particle in (c)?

••29 (a) What is the angular speed ω about the polar axis of a point on Earth's surface at latitude 40° N? (Earth rotates about that axis.) (b) What is the linear speed v of the point? What are (c) ω and (d) v for a point at the equator? SSM

••30 In Fig. 10-32, wheel A of radius $r_A = 10$ cm is coupled by belt B to wheel C of radius $r_C = 25$ cm. The angular speed of wheel A is increased from rest at a constant rate of 1.6 rad/s^2. Find the time needed for wheel C to reach an angular speed of 100 rev/min, assuming the belt does not slip. (*Hint:* If the belt does not slip, the linear speeds at the two rims must be equal.)

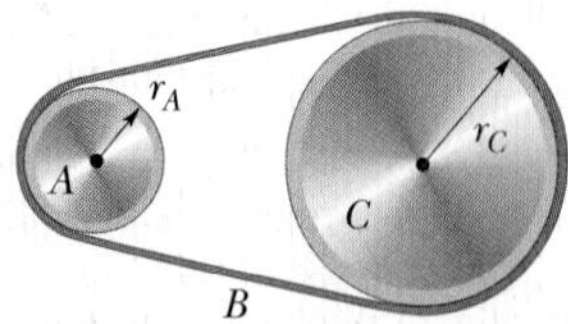

FIG. 10-32 Problem 30.

••31 A record turntable is rotating at $33\frac{1}{3}$ rev/min. A watermelon seed is on the turntable 6.0 cm from the axis of rotation. (a) Calculate the acceleration of the seed, assuming that it does not slip. (b) What is the minimum value of the coefficient of static friction between the seed and the turntable if the seed is not to slip? (c) Suppose that the turntable achieves its angular speed by starting from rest and undergoing a constant angular acceleration for 0.25 s. Calculate the minimum coefficient of static friction required for the seed not to slip during the acceleration period.

•••32 A pulsar is a rapidly rotating neutron star that emits a radio beam the way a lighthouse emits a light beam. We receive a radio pulse for each rotation of the star. The period T of rotation is found by measuring the time between pulses. The pulsar in the Crab nebula has a period of rotation of $T = 0.033$ s that is increasing at the rate of 1.26×10^{-5} s/y. (a) What is the pulsar's angular acceleration α? (b) If α is constant, how many years from now will the pulsar stop rotating? (c) The pulsar originated in a supernova explosion seen in the year 1054. Assuming constant α, find the initial T.

sec. 10-6 Kinetic Energy of Rotation

•33 Calculate the rotational inertia of a wheel that has a kinetic energy of 24 400 J when rotating at 602 rev/min. SSM

•34 Figure 10-33 gives angular speed versus time for a thin rod that rotates around one end. The scale on the ω axis is set by $\omega_s = 6.0$ rad/s. (a) What is the magnitude of the rod's angular acceleration? (b) At $t = 4.0$ s, the rod has a rotational kinetic energy of 1.60 J. What is its kinetic energy at $t = 0$?

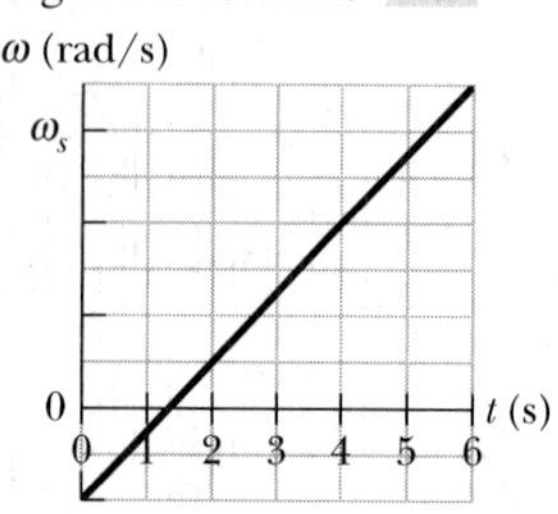

FIG. 10-33 Problem 34.

sec. 10-7 Calculating the Rotational Inertia

•35 Calculate the rotational inertia of a meter stick, with mass 0.56 kg, about an axis perpendicular to the stick and located at the 20 cm mark. (Treat the stick as a thin rod.) SSM

•36 Figure 10-34 shows three 0.0100 kg particles that have been glued to a rod of length $L = 6.00$ cm and negligible mass. The assembly can rotate around a perpendicular axis through point O at the left end. If we remove one particle (that is, 33% of the mass), by what percentage does the rotational inertia of the assembly around the rotation axis decrease when that removed particle is (a) the innermost one and (b) the outermost one?

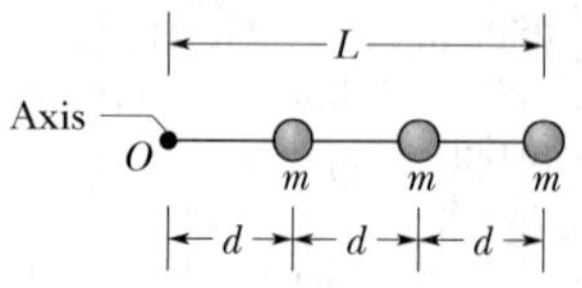

FIG. 10-34 Problems 36 and 64.

•37 Two uniform solid cylinders, each rotating about its central (longitudinal) axis at 235 rad/s, have the same mass of 1.25 kg but differ in radius. What is the rotational kinetic energy of (a) the smaller cylinder, of radius 0.25 m, and (b) the larger cylinder, of radius 0.75 m? SSM

•38 Figure 10-35*a* shows a disk that can rotate about an axis at a radial distance h from the center of the disk. Figure 10-35*b* gives the rotational inertia I of the disk about the axis as a function of that distance h, from the center out to the edge of the disk. The scale on the I axis is set by $I_A = 0.050$ kg·m^2 and $I_B = 0.150$ kg·m^2. What is the mass of the disk?

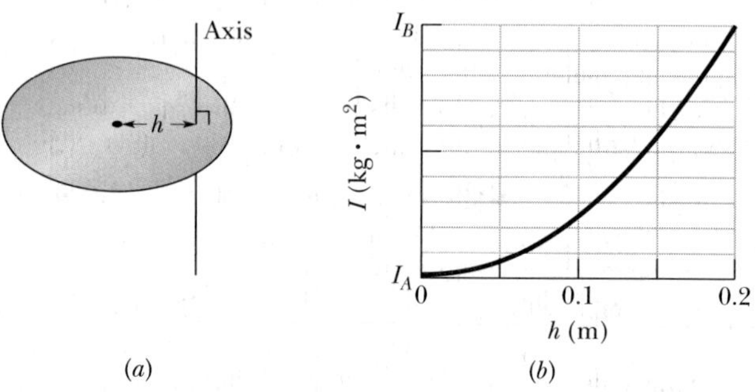

FIG. 10-35 Problem 38.

••39 In Fig. 10-36, two particles, each with mass $m = 0.85$ kg, are fastened to each other, and to a rotation axis at O, by two thin rods, each with length $d = 5.6$ cm and mass $M = 1.2$ kg. The combination rotates around the rotation axis with angular speed $\omega = 0.30$ rad/s. Measured about O, what are the combination's (a) rotational inertia and (b) kinetic energy? GO

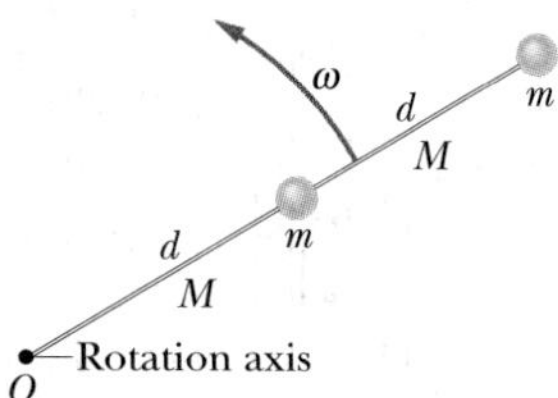

FIG. 10-36 Problem 39.

••40 Four identical particles of mass 0.50 kg each are placed at the vertices of a 2.0 m × 2.0 m square and held there by four massless rods, which form the sides of the square. What is the rotational inertia of this rigid body about an axis that (a) passes through the midpoints of opposite sides and lies in the plane of the square, (b) passes through the midpoint of one of the sides and is perpendicular to the plane of the square, and (c) lies in the plane of the square and passes through two diagonally opposite particles?

••41 The uniform solid block in Fig. 10-37 has mass 0.172 kg and edge lengths $a = 3.5$ cm, $b = 8.4$ cm, and $c = 1.4$ cm. Calculate its rotational inertia about an axis through one corner and perpendicular to the large faces. SSM WWW

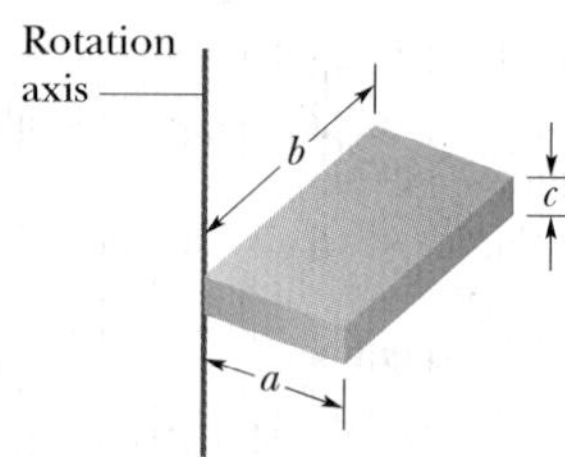

FIG. 10-37 Problem 41.

••42 Figure 10-38 shows an arrangement of 15 identical disks that have been glued together in a rod-like shape of length $L = 1.0000$ m and (total) mass $M = 100.0$ mg. The disk arrangement can rotate about a perpendicular axis through its central disk at point O. (a) What is the rotational inertia of the arrangement about that axis? (b) If we approximated the arrangement as being a uniform rod of mass M and length L, what percentage error would we make in using the formula in Table 10-2e to calculate the rotational inertia?

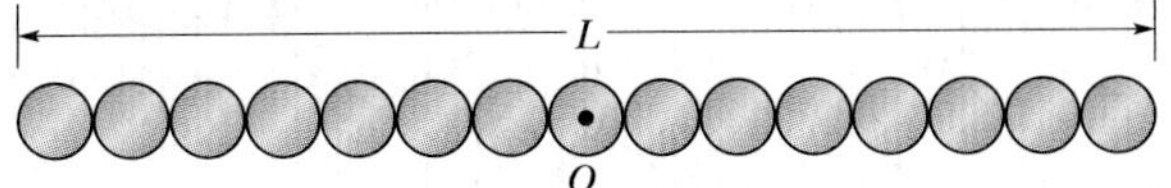

FIG. 10-38 Problem 42.

••43 Trucks can be run on energy stored in a rotating flywheel, with an electric motor getting the flywheel up to its top speed of 200π rad/s. One such flywheel is a solid, uniform cylinder with a mass of 500 kg and a radius of 1.0 m. (a) What is the kinetic energy of the flywheel after charging? (b) If the truck uses an average power of 8.0 kW, for how many minutes can it operate between chargings?

••44 The masses and coordinates of four particles are as follows: 50 g, $x = 2.0$ cm, $y = 2.0$ cm; 25 g, $x = 0$, $y = 4.0$ cm; 25 g, $x = -3.0$ cm, $y = -3.0$ cm; 30 g, $x = -2.0$ cm, $y = 4.0$ cm. What are the rotational inertias of this collection about the (a) x, (b) y, and (c) z axes? (d) Suppose the answers to (a) and (b) are A and B, respectively. Then what is the answer to (c) in terms of A and B?

sec. 10-8 Torque

•45 A small ball of mass 0.75 kg is attached to one end of a 1.25-m-long massless rod, and the other end of the rod is hung from a pivot. When the resulting pendulum is 30° from the vertical, what is the magnitude of the gravitational torque calculated about the pivot? SSM

•46 The length of a bicycle pedal arm is 0.152 m, and a downward force of 111 N is applied to the pedal by the rider. What is the magnitude of the torque about the pedal arm's pivot when the arm is at angle (a) 30°, (b) 90°, and (c) 180° with the vertical?

•47 The body in Fig. 10-39 is pivoted at O, and two forces act on it as shown. If $r_1 = 1.30$ m, $r_2 = 2.15$ m, $F_1 = 4.20$ N, $F_2 = 4.90$ N, $\theta_1 = 75.0°$, and $\theta_2 = 60.0°$, what is the net torque about the pivot? SSM ILW

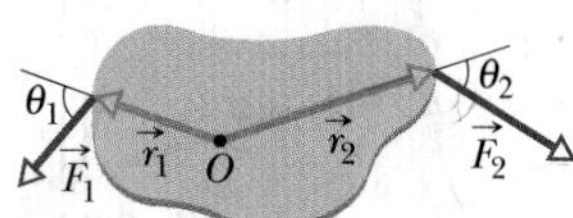

FIG. 10-39 Problem 47.

•48 The body in Fig. 10-40 is pivoted at O. Three forces act on it: $F_A = 10$ N at point A, 8.0 m from O; $F_B = 16$ N at B, 4.0 m from O; and $F_C = 19$ N at C, 3.0 m from O. What is the net torque about O?

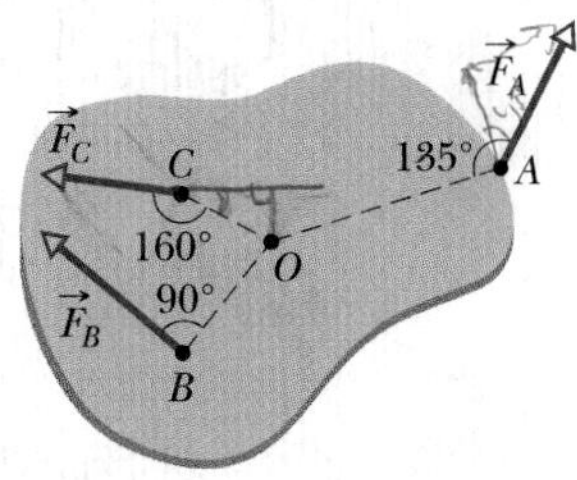

FIG. 10-40 Problem 48.

sec. 10-9 Newton's Second Law for Rotation

•49 During the launch from a board, a diver's angular speed about her center of mass changes from zero to 6.20 rad/s in 220 ms. Her rotational inertia about her center of mass is 12.0 kg·m^2. During the launch, what are the magnitudes of (a) her average angular acceleration and (b) the average external torque on her from the board? SSM ILW

•50 If a 32.0 N·m torque on a wheel causes angular acceleration 25.0 rad/s^2, what is the wheel's rotational inertia?

••51 Figure 10-41 shows a uniform disk that can rotate around its center like a merry-go-round. The disk has a radius of 2.00 cm and a mass of 20.0 grams and is initially at rest. Starting at time $t = 0$, two forces are to be applied tangentially to the rim as indicated, so that at time $t = 1.25$ s the disk has an angular velocity of 250 rad/s counterclockwise. Force $\vec{F}_1$ has a magnitude of 0.100 N. What is magnitude F_2? GO

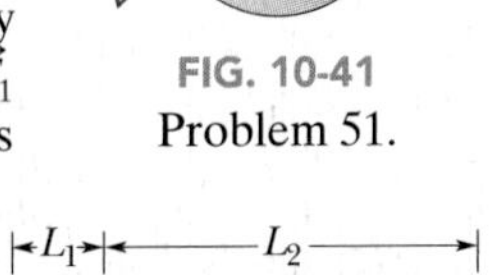

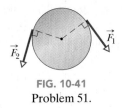

FIG. 10-41 Problem 51.

••52 Figure 10-42 shows particles 1 and 2, each of mass m, attached to the ends of a rigid massless rod of length $L_1 + L_2$, with $L_1 = 20$ cm and $L_2 = 80$ cm. The rod is held horizontally on the fulcrum and then released. What are the magnitudes of the initial accelerations of (a) particle 1 and (b) particle 2?

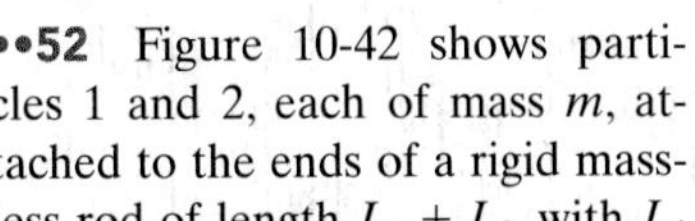

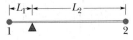

FIG. 10-42 Problem 52.

••53 In Fig. 10-43a, an irregularly shaped plastic plate with uniform thickness and density (mass per unit volume) is to be rotated around an axle that is perpendicular to the plate face and through point O. The rotational inertia of the plate about that axle is measured with the following method. A circular disk of mass 0.500 kg and radius 2.00 cm is glued to the plate, with its center aligned with point O (Fig. 10-43b). A string is wrapped around the edge of the disk the way a string is wrapped around a top. Then the string is pulled for 5.00 s. As a result, the disk and plate are rotated by a constant force of 0.400 N that is applied by the string tangentially to the edge of the disk. The resulting angular speed is 114 rad/s. What is the rotational inertia of the plate about the axle? GO

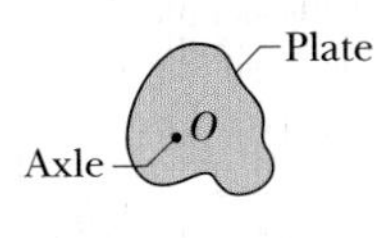

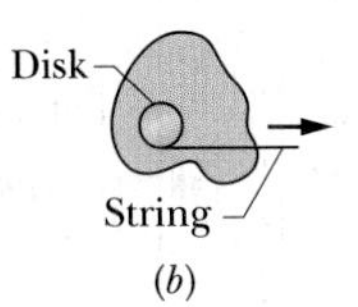

FIG. 10-43 Problem 53.

••54 In Fig. 10-44, a cylinder having a mass of 2.0 kg can rotate about its central axis through point O. Forces are applied as shown: $F_1 = 6.0$ N, $F_2 = 4.0$ N, $F_3 = 2.0$ N, and $F_4 = 5.0$ N. Also, $r = 5.0$ cm and $R = 12$ cm. Find the (a) magnitude and (b) direction of the angular acceleration of the cylinder. (During the rotation, the forces maintain their same angles relative to the cylinder.)

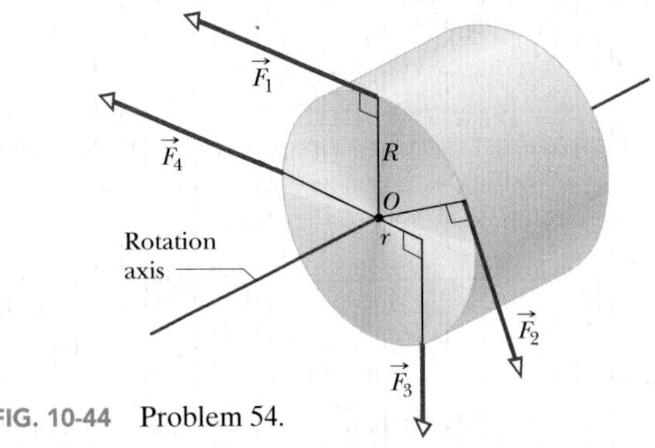

FIG. 10-44 Problem 54.

••55 In Fig. 10-45, block 1 has mass $m_1 = 460$ g, block 2 has mass $m_2 = 500$ g, and the pulley, which is mounted on a horizontal axle with negligible friction, has radius $R = 5.00$ cm. When released from rest, block 2 falls 75.0 cm in 5.00 s without the cord slipping on the pulley. (a) What is the magnitude of the acceleration of the blocks? What are (b) tension T_2 and (c) tension T_1? (d) What is the magnitude of the pulley's angular acceleration? (e) What is its rotational inertia? GO

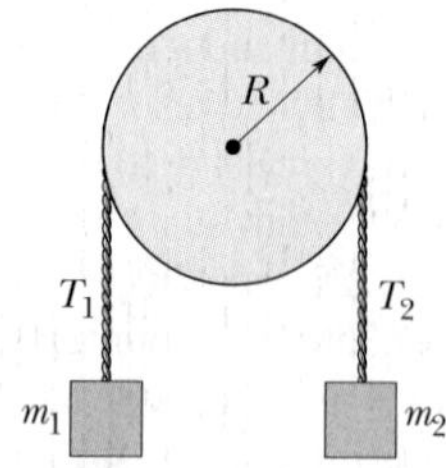

FIG. 10-45 Problems 55 and 73.

••56 In a judo foot-sweep move, you sweep your opponent's left foot out from under him while pulling on his gi (uniform) toward that side. As a result, your opponent rotates around his right foot and onto the mat. Figure 10-46 shows a simplified diagram of your opponent as you face him, with his left foot swept out. The rotational axis is through point O. The gravitational force $\vec{F}_g$ on him effectively acts at his center of mass, which is a horizontal distance $d = 28$ cm from point O. His mass is 70 kg, and his rotational inertia about point O is 65 kg · m². What is the magnitude of his initial angular acceleration about point O if your pull $\vec{F}_a$ on his gi is (a) negligible and (b) horizontal with a magnitude of 300 N and applied at height $h = 1.4$ m?

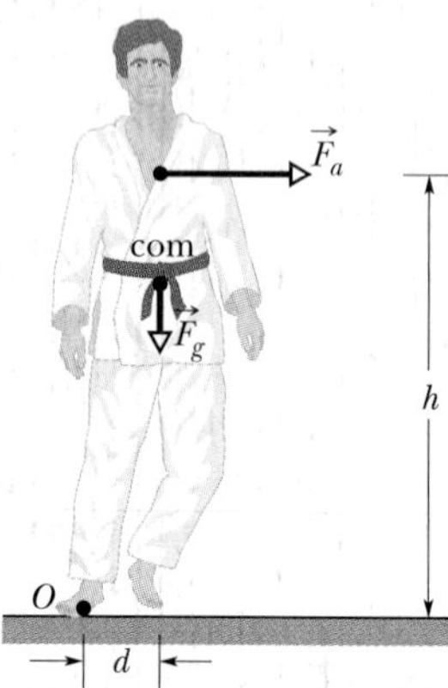

FIG. 10-46 Problem 56.

•••57 A pulley, with a rotational inertia of 1.0×10^{-3} kg · m² about its axle and a radius of 10 cm, is acted on by a force applied tangentially at its rim. The force magnitude varies in time as $F = 0.50t + 0.30t^2$, with F in newtons and t in seconds. The pulley is initially at rest. At $t = 3.0$ s what are (a) its angular acceleration and (b) its angular speed?

sec. 10-10 Work and Rotational Kinetic Energy

•58 A thin rod of length 0.75 m and mass 0.42 kg is suspended freely from one end. It is pulled to one side and then allowed to swing like a pendulum, passing through its lowest position with angular speed 4.0 rad/s. Neglecting friction and air resistance, find (a) the rod's kinetic energy at its lowest position and (b) how far above that position the center of mass rises.

•59 A 32.0 kg wheel, essentially a thin hoop with radius 1.20 m, is rotating at 280 rev/min. It must be brought to a stop in 15.0 s. (a) How much work must be done to stop it? (b) What is the required average power?

•60 (a) If $R = 12$ cm, $M = 400$ g, and $m = 50$ g in Fig. 10-18, find the speed of the block after it has descended 50 cm starting from rest. Solve the problem using energy conservation principles. (b) Repeat (a) with $R = 5.0$ cm.

•61 An automobile crankshaft transfers energy from the engine to the axle at the rate of 100 hp (= 74.6 kW) when rotating at a speed of 1800 rev/min. What torque (in newton-meters) does the crankshaft deliver?

••62 A uniform cylinder of radius 10 cm and mass 20 kg is mounted so as to rotate freely about a horizontal axis that is parallel to and 5.0 cm from the central longitudinal axis of the cylinder. (a) What is the rotational inertia of the cylinder about the axis of rotation? (b) If the cylinder is released from rest with its central longitudinal axis at the same height as the axis about which the cylinder rotates, what is the angular speed of the cylinder as it passes through its lowest position?

••63 A meter stick is held vertically with one end on the floor and is then allowed to fall. Find the speed of the other end just before it hits the floor, assuming that the end on the floor does not slip. (*Hint:* Consider the stick to be a thin rod and use the conservation of energy principle.) SSM ILW

••64 In Fig. 10-34, three 0.0100 kg particles have been glued to a rod of length $L = 6.00$ cm and negligible mass and can rotate around a perpendicular axis through point O at one end. How much work is required to change the rotational rate (a) from 0 to 20.0 rad/s, (b) from 20.0 rad/s to 40.0 rad/s, and (c) from 40.0 rad/s to 60.0 rad/s? (d) What is the slope of a plot of the assembly's kinetic energy (in joules) versus the square of its rotation rate (in radians-squared per second-squared)?

•••65 Figure 10-47 shows a rigid assembly of a thin hoop (of mass m and radius $R = 0.150$ m) and a thin radial rod (of mass m and length $L = 2.00R$). The assembly is upright, but if we give it a slight nudge, it will rotate around a horizontal axis in the plane of the rod and hoop, through the lower end of the rod. Assuming that the energy given to the assembly in such a nudge is negligible, what would be the assembly's angular speed about the rotation axis when it passes through the upside-down (inverted) orientation? GO

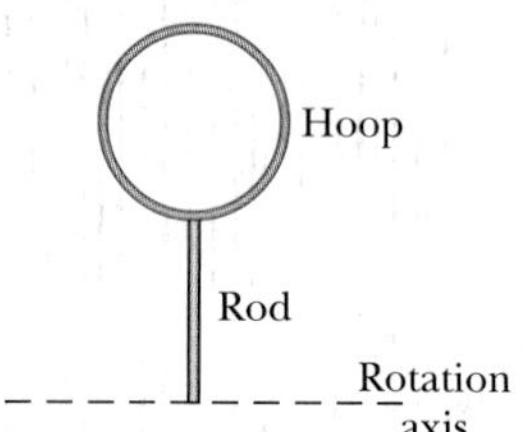

FIG. 10-47 Problem 65.

•••66 A uniform spherical shell of mass $M = 4.5$ kg and radius $R = 8.5$ cm can rotate about a vertical axis on frictionless bearings (Fig. 10-48). A massless cord passes around the equator of the shell, over a pulley of rotational inertia $I = 3.0 \times 10^{-3}$ kg · m² and radius $r = 5.0$ cm, and is attached to a small object of mass $m = 0.60$ kg. There is no friction on the pulley's axle; the cord does not slip on the pulley. What is the speed of the object when it has fallen 82 cm after being released from rest? Use energy considerations.

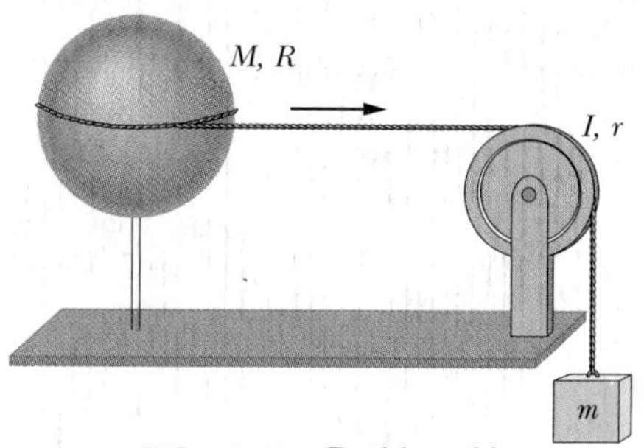

FIG. 10-48 Problem 66.

•••67 A tall, cylindrical chimney falls over when its base is ruptured. Treat the chimney as a thin rod of length 55.0 m. At the instant it makes an angle of 35.0° with the vertical as it falls, what are (a) the radial acceleration of the top, and (b) the tangential acceleration of the top. (*Hint:* Use energy considerations, not a torque.) (c) At what angle θ is the tangential acceleration equal to g?

Additional Problems

68 George Washington Gale Ferris, Jr., a civil engineering graduate from Rensselaer Polytechnic Institute, built the original Ferris wheel for the 1893 World's Columbian Exposition in Chicago. The wheel, an astounding engineering construc-

tion at the time, carried 36 wooden cars, each holding up to 60 passengers, around a circle 76 m in diameter. The cars were loaded 6 at a time, and once all 36 cars were full, the wheel made a complete rotation at constant angular speed in about 2 min. Estimate the amount of work that was required of the machinery to rotate the passengers alone.

69 In Fig. 10-49, two 6.20 kg blocks are connected by a massless string over a pulley of radius 2.40 cm and rotational inertia 7.40×10^{-4} kg·m². The string does not slip on the pulley; it is not known whether there is friction between the table and the sliding block; the pulley's axis is frictionless. When this system is released from rest, the pulley turns through 1.30 rad in 91.0 ms and the acceleration of the blocks is constant. What are (a) the magnitude of the pulley's angular acceleration, (b) the magnitude of either block's acceleration, (c) string tension T_1, and (d) string tension T_2? SSM

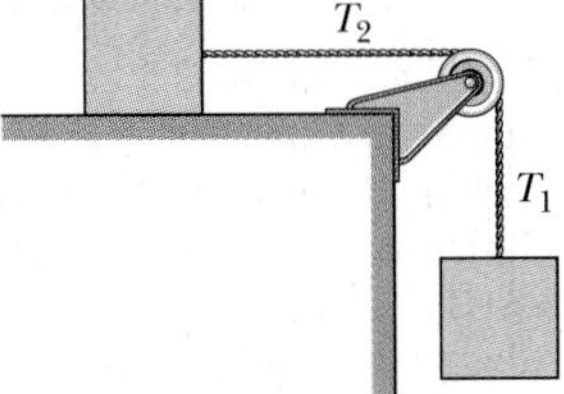

FIG. 10-49 Problem 69.

70 Figure 10-50 shows a flat construction of two circular rings that have a common center and are held together by three rods of negligible mass. The construction, which is initially at rest, can rotate around the common center (like a merry-go-round), where another rod of negligible mass lies. The mass, inner radius, and outer radius of the rings are given in the following table. A tangential force of magnitude 12.0 N is applied to the outer edge of the outer ring for 0.300 s. What is the change in the angular speed of the construction during that time interval?

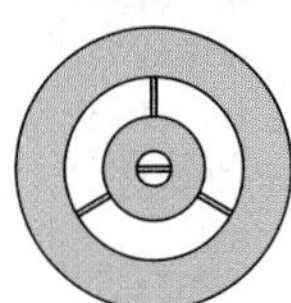
FIG. 10-50 Problem 70.

Ring	Mass (kg)	Inner Radius (m)	Outer Radius (m)
1	0.120	0.0160	0.0450
2	0.240	0.0900	0.1400

71 In Fig. 10-51, a small disk of radius $r = 2.00$ cm has been glued to the edge of a larger disk of radius $R = 4.00$ cm so that the disks lie in the same plane. The disks can be rotated around a perpendicular axis through point O at the center of the larger disk. The disks both have a uniform density (mass per unit volume) of 1.40×10^3 kg/m³ and a uniform thickness of 5.00 mm. What is the rotational inertia of the two-disk assembly about the rotation axis through O?

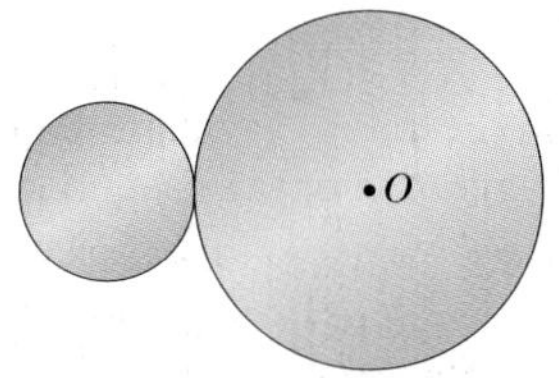

FIG. 10-51 Problem 71.

72 At 7:14 A.M. on June 30, 1908, a huge explosion occurred above remote central Siberia, at latitude 61° N and longitude 102° E; the fireball thus created was the brightest flash seen by anyone before nuclear weapons. The *Tunguska Event,* which according to one chance witness "covered an enormous part of the sky," was probably the explosion of a *stony asteroid* about 140 m wide. (a) Considering only Earth's rotation, determine how much later the asteroid would have had to arrive to put the explosion above Helsinki at longitude 25° E. This would have obliterated the city. (b) If the asteroid had, instead, been a *metallic asteroid,* it could have reached Earth's surface. How much later would such an asteroid have had to arrive to put the impact in the Atlantic Ocean at longitude 20° W? (The resulting tsunamis would have wiped out coastal civilization on both sides of the Atlantic.)

73 In Fig. 10-45, two blocks, of mass $m_1 = 400$ g and $m_2 = 600$ g, are connected by a massless cord that is wrapped around a uniform disk of mass $M = 500$ g and radius $R = 12.0$ cm. The disk can rotate without friction about a fixed horizontal axis through its center; the cord cannot slip on the disk. The system is released from rest. Find (a) the magnitude of the acceleration of the blocks, (b) the tension T_1 in the cord at the left, and (c) the tension T_2 in the cord at the right.

74 Attached to each end of a thin steel rod of length 1.20 m and mass 6.40 kg is a small ball of mass 1.06 kg. The rod is constrained to rotate in a horizontal plane about a vertical axis through its midpoint. At a certain instant, it is rotating at 39.0 rev/s. Because of friction, it slows to a stop in 32.0 s. Assuming a constant retarding torque due to friction, compute (a) the angular acceleration, (b) the retarding torque, (c) the total energy transferred from mechanical energy to thermal energy by friction, and (d) the number of revolutions rotated during the 32.0 s. (e) Now suppose that the retarding torque is known not to be constant. If any of the quantities (a), (b), (c), and (d) can still be computed without additional information, give its value.

75 A uniform helicopter rotor blade is 7.80 m long, has a mass of 110 kg, and is attached to the rotor axle by a single bolt. (a) What is the magnitude of the force on the bolt from the axle when the rotor is turning at 320 rev/min? (*Hint:* For this calculation the blade can be considered to be a point mass at its center of mass. Why?) (b) Calculate the torque that must be applied to the rotor to bring it to full speed from rest in 6.70 s. Ignore air resistance. (The blade cannot be considered to be a point mass for this calculation. Why not? Assume the mass distribution of a uniform thin rod.) (c) How much work does the torque do on the blade in order for the blade to reach a speed of 320 rev/min?

76 A wheel, starting from rest, rotates with a constant angular acceleration of 2.00 rad/s². During a certain 3.00 s interval, it turns through 90.0 rad. (a) What is the angular velocity of the wheel at the start of the 3.00 s interval? (b) How long has the wheel been turning before the start of the 3.00 s interval?

77 A golf ball is launched at an angle of 20° to the horizontal, with a speed of 60 m/s and a rotation rate of 90 rad/s. Neglecting air drag, determine the number of revolutions the ball makes by the time it reaches maximum height.

78 Two uniform solid spheres have the same mass of 1.65 kg, but one has a radius of 0.226 m and the other has a radius of 0.854 m. Each can rotate about an axis through its center. (a) What is the magnitude τ of the torque required to bring the smaller sphere from rest to an angular speed of 317 rad/s in 15.5 s? (b) What is the magnitude F of the force that must be applied tangentially at the sphere's equator to give that torque? What are the corresponding values of (c) τ and (d) F for the larger sphere?

79 The thin uniform rod in Fig. 10-52 has length 2.0 m and can pivot about a horizontal, frictionless pin through one end. It is released from rest at angle $\theta = 40°$ above the horizontal. Use the principle of conservation of energy to determine the angular speed of the rod as it passes through the horizontal position.

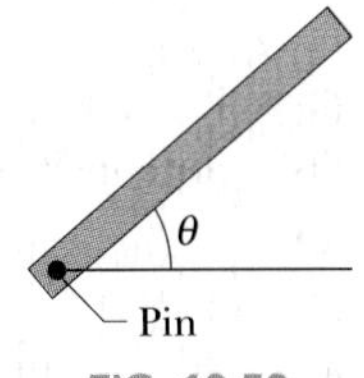

FIG. 10-52 Problem 79.

80 The flywheel of an engine is rotating at 25.0 rad/s. When the engine is turned off, the flywheel slows at a constant rate and stops in 20.0 s. Calculate (a) the angular acceleration of the flywheel, (b) the angle through which the flywheel rotates in stopping, and (c) the number of revolutions made by the flywheel in stopping.

81 A small ball with mass 1.30 kg is mounted on one end of a rod 0.780 m long and of negligible mass. The system rotates in a horizontal circle about the other end of the rod at 5010 rev/min. (a) Calculate the rotational inertia of the system about the axis of rotation. (b) There is an air drag of 2.30×10^{-2} N on the ball, directed opposite its motion. What torque must be applied to the system to keep it rotating at constant speed?

82 Starting from rest at $t = 0$, a wheel undergoes a constant angular acceleration. When $t = 2.0$ s, the angular velocity of the wheel is 5.0 rad/s. The acceleration continues until $t = 20$ s, when it abruptly ceases. Through what angle does the wheel rotate in the interval $t = 0$ to $t = 40$ s?

83 A high-wire walker always attempts to keep his center of mass over the wire (or rope). He normally carries a long, heavy pole to help: If he leans, say, to his right (his com moves to the right) and is in danger of rotating around the wire, he moves the pole to his left (its com moves to the left) to slow the rotation and allow himself time to adjust his balance. Assume that the walker has a mass of 70.0 kg and a rotational inertia of 15.0 kg $\cdot$ m^2 about the wire. What is the magnitude of his angular acceleration about the wire if his com is 5.0 cm to the right of the wire and (a) he carries no pole and (b) the 14.0 kg pole he carries has its com 10 cm to the left of the wire?

84 *Racing disks.* Figure 10-53 shows two disks that can rotate about their centers like a merry-go-round. At time $t = 0$, the reference lines of the two disks have the same orientation. Disk A is already rotating, with a constant angular velocity of 9.5 rad/s. Disk B has been stationary but now begins to rotate at a constant angular acceleration of 2.2 rad/s^2. (a) At what time t will the reference lines of the two disks momentarily have the same angular displacement θ? (b) Will that time t be the first time since $t = 0$ that the reference lines are momentarily aligned?

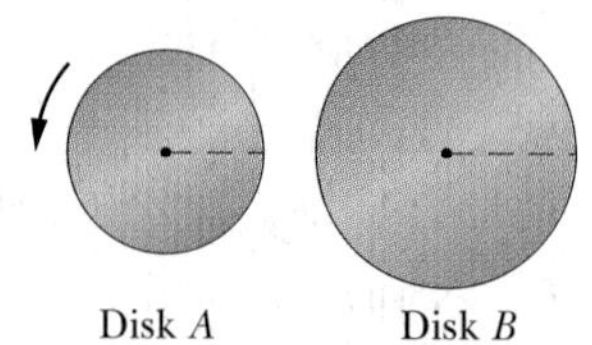

FIG. 10-53 Problem 84.

85 A bicyclist of mass 70 kg puts all his mass on each downward-moving pedal as he pedals up a steep road. Take the diameter of the circle in which the pedals rotate to be 0.40 m, and determine the magnitude of the maximum torque he exerts about the rotation axis of the pedals.

86 A disk rotates at constant angular acceleration, from angular position $\theta_1 = 10.0$ rad to angular position $\theta_2 = 70.0$ rad in 6.00 s. Its angular velocity at θ_2 is 15.0 rad/s. (a) What was its angular velocity at θ_1? (b) What is the angular acceleration? (c) At what angular position was the disk initially at rest? (d) Graph θ versus time t and angular speed ω versus t for the disk, from the beginning of the motion (let $t = 0$ then).

87 A wheel of radius 0.20 m is mounted on a frictionless horizontal axis. The rotational inertia of the wheel about the axis is 0.050 kg $\cdot$ m^2. A massless cord wrapped around the wheel is attached to a 2.0 kg block that slides on a horizontal frictionless surface. If a horizontal force of magnitude $P = 3.0$ N is applied to the block as shown in Fig. 10-54, what is the magnitude of the angular acceleration of the wheel? Assume the cord does not slip on the wheel. SSM

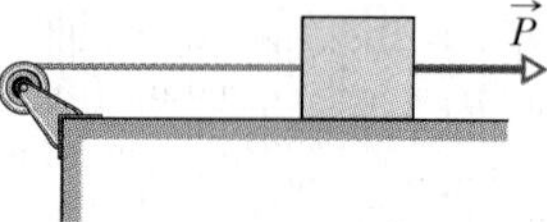

FIG. 10-54 Problem 87.

88 Our Sun is 2.3×10^4 ly (light-years) from the center of our Milky Way galaxy and is moving in a circle around that center at a speed of 250 km/s. (a) How long does it take the Sun to make one revolution about the galactic center? (b) How many revolutions has the Sun completed since it was formed about 4.5×10^9 years ago?

89 A record turntable rotating at $33\frac{1}{3}$ rev/min slows down and stops in 30 s after the motor is turned off. (a) Find its (constant) angular acceleration in revolutions per minute-squared. (b) How many revolutions does it make in this time? SSM

90 A rigid body is made of three identical thin rods, each with length $L = 0.600$ m, fastened together in the form of a letter **H** (Fig. 10-55). The body is free to rotate about a horizontal axis that runs along the length of one of the legs of the **H**. The body is allowed to fall from rest from a position in which the plane of the **H** is horizontal. What is the angular speed of the body when the plane of the **H** is vertical?

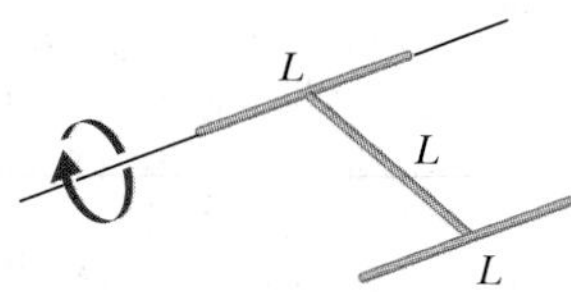

FIG. 10-55 Problem 90.

91 (a) Show that the rotational inertia of a solid cylinder of mass M and radius R about its central axis is equal to the rotational inertia of a thin hoop of mass M and radius $R/\sqrt{2}$ about its central axis. (b) Show that the rotational inertia I of any given body of mass M about any given axis is equal to the rotational inertia of an *equivalent hoop* about that axis, if the hoop has the same mass M and a radius k given by

$$k = \sqrt{\frac{I}{M}}.$$

The radius k of the equivalent hoop is called the *radius of gyration* of the given body. SSM

92 A thin spherical shell has a radius of 1.90 m. An applied torque of 960 N $\cdot$ m gives the shell an angular acceleration of 6.20 rad/s^2 about an axis through the center of the shell. What are (a) the rotational inertia of the shell about that axis and (b) the mass of the shell?

93 In Fig. 10-56, a wheel of radius 0.20 m is mounted on a frictionless horizontal axle. A massless cord is wrapped around the wheel and attached to a 2.0 kg box that slides on a frictionless surface inclined at angle $\theta = 20°$ with the horizontal. The box accelerates down the surface at 2.0 m/s^2. What is the rotational inertia of the wheel about the axle?

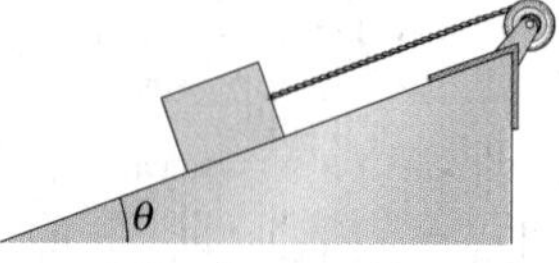

FIG. 10-56 Problem 93.

94 The method by which the massive lintels (top stones) were lifted to the top of the upright stones at Stonehenge has long been debated. One possible method was tested in a small Czech town. A concrete block of mass 5124 kg was pulled up along two oak beams whose top surfaces had been debarked and then lubricated with fat (Fig. 10-57). The beams were 10 m long, and each extended from the ground to the top of one of the two upright pillars onto which the block was to be raised. The pillars were 3.9 m high; the coefficient of static friction between block and beams was 0.22. The pull on the block was via ropes wrapped around it and around the top ends of two spruce logs of length 4.5 m. A platform was strung at the opposite end of each log. When enough workers sat or stood on a platform, the attached spruce log would pivot about the top of its upright pillar and pull one end of the block a short distance up a beam. For each log, the rope holding the block was approximately perpendicular to the log; the distance between the pivot point and the point where the rope wrapped around the log was 0.70 m. Assuming that each worker had a mass of 85 kg, find the smallest number of workers needed on the two platforms so that the block begins to move up along the beams. (About half this number could actually move the block by moving first one end of it and then the other.)

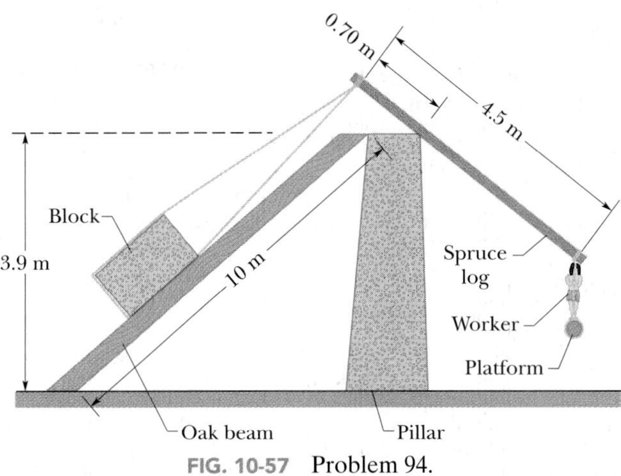

FIG. 10-57 Problem 94.

95 Figure 10-58 shows a propeller blade that rotates at 2000 rev/min about a perpendicular axis at point *B*. Point *A* is at the outer tip of the blade, at radial distance 1.50 m. (a) What is the difference in the magnitudes *a* of the centripetal acceleration of point *A* and of a point at radial distance 0.150 m? (b) Find the slope of a plot of *a* versus radial distance along the blade.

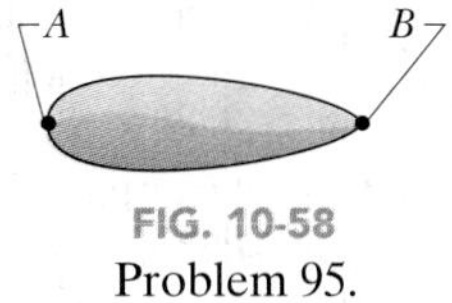

FIG. 10-58 Problem 95.

96 A yo-yo-shaped device mounted on a horizontal frictionless axis is used to lift a 30 kg box as shown in Fig. 10-59. The outer radius *R* of the device is 0.50 m, and the radius *r* of the hub is 0.20 m. When a constant horizontal force $\vec{F}_{app}$ of magnitude 140 N is applied to a rope wrapped around the outside of the device, the box, which is suspended from a rope wrapped around the hub, has an upward acceleration of magnitude 0.80 m/s². What is the rotational inertia of the device about its axis of rotation?

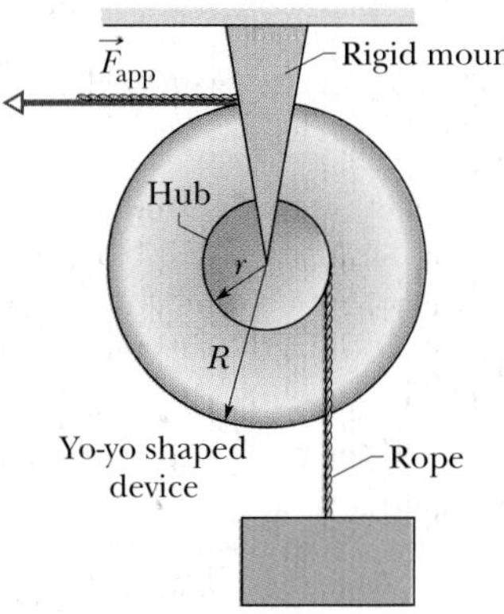

FIG. 10-59 Problem 96.

97 The rigid body shown in Fig. 10-60 consists of three particles connected by massless rods. It is to be rotated about an axis perpendicular to its plane through point *P*. If $M = 0.40$ kg, $a = 30$ cm, and $b = 50$ cm, how much work is required to take the body from rest to an angular speed of 5.0 rad/s?

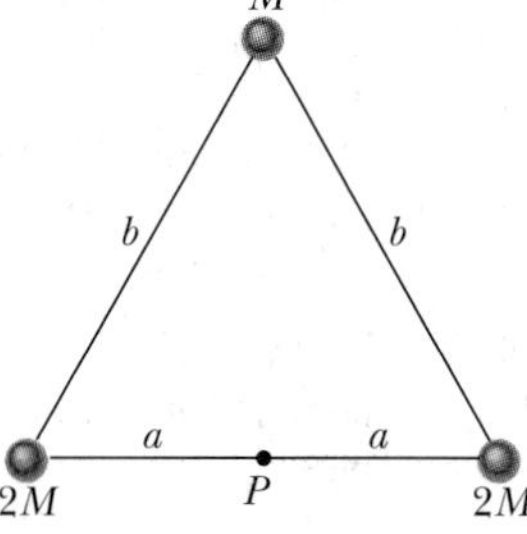

FIG. 10-60 Problem 97.

98 *Beverage engineering.* The pull tab was a major advance in the engineering design of beverage containers. The tab pivots on a central bolt in the can's top. When you pull upward on one end of the tab, the other end presses downward on a portion of the can's top that has been scored. If you pull upward with a 10 N force, approximately what is the magnitude of the force applied to the scored section? (You will need to examine a can with a pull tab.)

99 Cheetahs running at top speed have been reported at an astounding 114 km/h (about 71 mi/h) by observers driving alongside the animals. Imagine trying to measure a cheetah's speed by keeping your vehicle abreast of the animal while also glancing at your speedometer, which is registering 114 km/h. You keep the vehicle a constant 8.0 m from the cheetah, but the noise of the vehicle causes the cheetah to continuously veer away from you along a circular path of radius 92 m. Thus, you travel along a circular path of radius 100 m. (a) What is the angular speed of you and the cheetah around the circular paths? (b) What is the linear speed of the cheetah along its path? (If you did not account for the circular motion, you would conclude erroneously that the cheetah's speed is 114 km/h, and that type of error was apparently made in the published reports.)

100 A point on the rim of a 0.75-m-diameter grinding wheel changes speed at a constant rate from 12 m/s to 25 m/s in 6.2 s. What is the average angular acceleration of the wheel?

101 In Fig. 10-61, a thin uniform rod (mass 3.0 kg, length 4.0 m) rotates freely about a horizontal axis *A* that is perpendicular to the rod and passes through a point at distance $d = 1.0$ m from the end of the rod. The kinetic energy of the rod as it passes through the vertical position is 20 J. (a) What is the rotational inertia of the rod about axis *A*? (b) What is the (linear) speed of the end *B* of the rod as the rod passes through the vertical position? (c) At what angle θ will the rod momentarily stop in its upward swing?

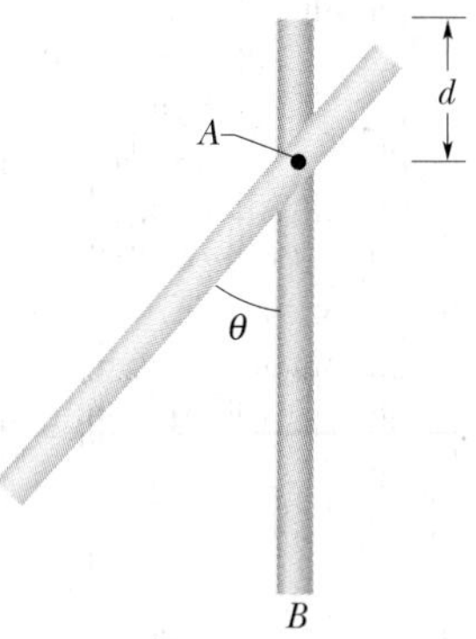

FIG. 10-61 Problem 101.

102 A car starts from rest and moves around a circular track of radius 30.0 m. Its speed increases at the constant rate of 0.500 m/s². (a) What is the magnitude of its *net* linear acceleration 15.0 s later? (b) What angle does this net acceleration vector make with the car's velocity at this time?

103 A pulley wheel that is 8.0 cm in diameter has a 5.6-m-long cord wrapped around its periphery. Starting from rest, the wheel is given a constant angular acceleration of 1.5 rad/s^2. (a) Through what angle must the wheel turn for the cord to unwind completely? (b) How long will this take?

104 A heavy flywheel rotating on its central axis is slowing down because of friction in its bearings. At the end of the first minute of slowing, its angular speed is 0.900 of its initial angular speed of 250 rev/min. Assuming a constant angular acceleration, find its angular speed at the end of the second minute.

105 Figure 10-62 shows a communications satellite that is a solid cylinder with mass 1210 kg, diameter 1.21 m, and length 1.75 m. Prior to launch from the shuttle cargo bay, the satellite is set spinning at 1.52 rev/s about its long axis. What are (a) its rotational inertia about the rotation axis and (b) its rotational kinetic energy?

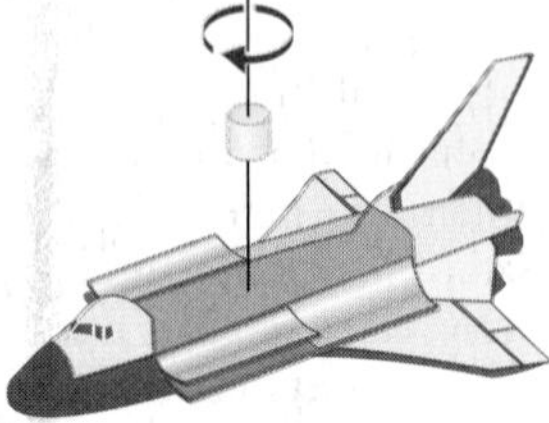

FIG. 10-62 Problem 105.

106 A vinyl record on a turntable rotates at $33\frac{1}{3}$ rev/min. (a) What is its angular speed in radians per second? What is the linear speed of a point on the record (b) 15 cm and (c) 7.4 cm from the turntable axis?

107 What is the angular speed of a car traveling at 50 km/h and rounding a circular turn of radius 110 m?

108 Calculate (a) the torque, (b) the energy, and (c) the average power required to accelerate Earth in 1 day from rest to its present angular speed about its axis.

109 The oxygen molecule O_2 has a mass of 5.30×10^{-26} kg and a rotational inertia of 1.94×10^{-46} kg·m^2 about an axis through the center of the line joining the atoms and perpendicular to that line. Suppose the center of mass of an O_2 molecule in a gas has a translational speed of 500 m/s and the molecule has a rotational kinetic energy that is $\frac{2}{3}$ of the translational kinetic energy of its center of mass. What then is the molecule's angular speed about the center of mass?

110 The rigid object shown in Fig. 10-63 consists of three balls and three connecting rods, with $M = 1.6$ kg, $L = 0.60$ m, and $\theta = 30°$. The balls may be treated as particles, and the connecting rods have negligible mass. Determine the rotational kinetic energy of the object if it has an angular speed of 1.2 rad/s about (a) an axis that passes through point P and is perpendicular to the plane of the figure and (b) an axis that passes through point P, is perpendicular to the rod of length $2L$, and lies in the plane of the figure.

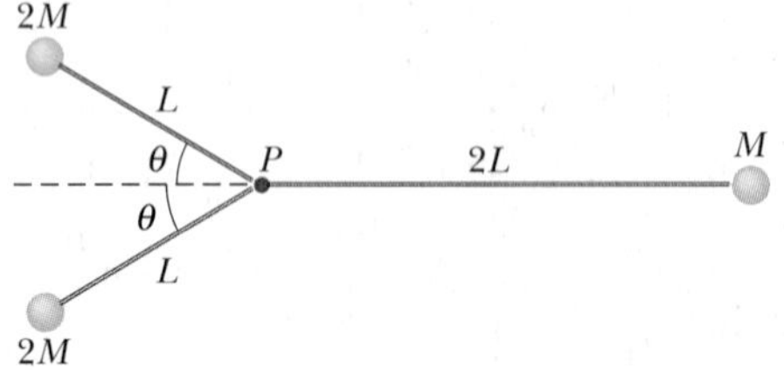

FIG. 10-63 Problem 110.

111 In Fig. 10-64, four pulleys are connected by two belts. Pulley A (radius 15 cm) is the drive pulley, and it rotates at 10 rad/s. Pulley B (radius 10 cm) is connected by belt 1 to pulley A. Pulley B' (radius 5 cm) is concentric with pulley B and is rigidly attached to it. Pulley C (radius 25 cm) is connected by belt 2 to pulley B'. Calculate (a) the linear speed of a point on belt 1, (b) the angular speed of pulley B, (c) the angular speed of pulley B', (d) the linear speed of a point on belt 2, and (e) the angular speed of pulley C. (*Hint:* If the belt between two pulleys does not slip, the linear speeds at the rims of the two pulleys must be equal.)

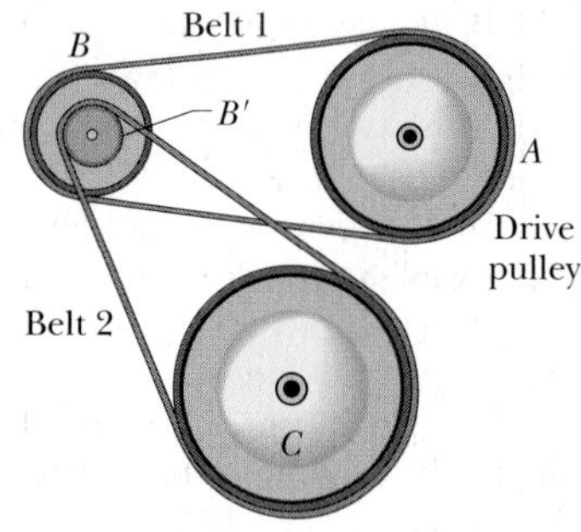

FIG. 10-64 Problem 111.

112 Four particles, each of mass 0.20 kg, are placed at the vertices of a square with sides of length 0.50 m. The particles are connected by rods of negligible mass. This rigid body can rotate in a vertical plane about a horizontal axis A that passes through one of the particles. The body is released from rest with rod AB horizontal (Fig. 10-65). (a) What is the rotational inertia of the body about axis A? (b) What is the angular speed of the body about axis A when rod AB swings through the vertical position?

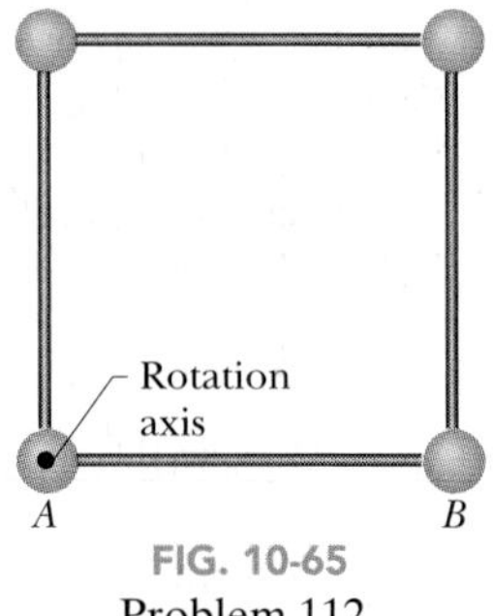

FIG. 10-65 Problem 112.

113 The turntable of a record player has an angular speed of 8.0 rad/s at the instant it is switched off. Three seconds later, the turntable has an angular speed of 2.6 rad/s. Through how many radians does the turntable rotate from the time it is turned off until it stops? (Assume constant α.)

114 Two thin rods (each of mass 0.20 kg) are joined together to form a rigid body as shown in Fig. 10-66. One of the rods has length $L_1 = 0.40$ m, and the other has length $L_2 = 0.50$ m. What is the rotational inertia of this rigid body about (a) an axis that is perpendicular to the plane of the paper and passes through the center of the shorter rod and (b) an axis that is perpendicular to the plane of the paper and passes through the center of the longer rod?

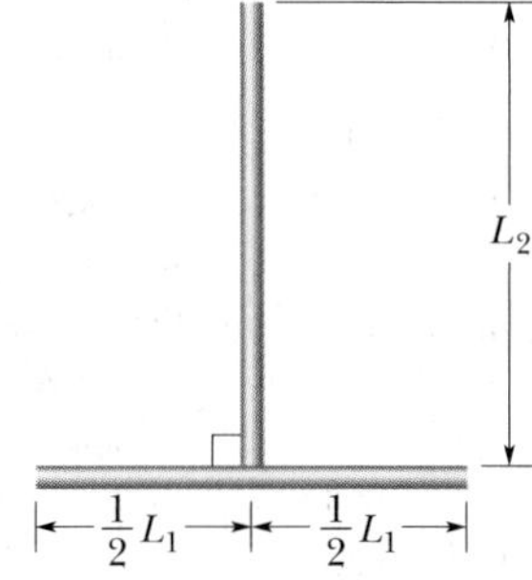

FIG. 10-66 Problem 114.

115 In Fig. 10-18a, a wheel of radius 0.20 m is mounted on a frictionless horizontal axis. The rotational inertia of the wheel about the axis is 0.40 kg·m^2. A massless cord wrapped around the wheel's circumference is attached to a 6.0 kg box. The system is released from rest. When the box has a kinetic energy of 6.0 J, what are (a) the wheel's rotational kinetic energy and (b) the distance the box has fallen? SSM

116 Three 0.50 kg particles form an equilateral triangle with 0.60 m sides. The particles are connected by rods of negligible mass. What is the rotational inertia of this rigid body about (a) an axis that passes through one of the particles and is parallel to the rod connecting the other two, (b) an axis that passes through the midpoint of one of the sides and is perpendicular to the plane of the triangle, and (c) an axis that is parallel to one side of the triangle and passes through the midpoints of the other two sides?

Rolling, Torque, and Angular Momentum

Ballet has several types of jumps but a tour jeté is the most enchanting. After leaping straight up in that jump, a ballet performer suddenly begins to rotate as if spun by an invisible hand. After half a turn, the rotation vanishes and then the performer lands. Even if an audience knows nothing of Newton's laws, they know that rotation cannot suddenly turn on and off while the performer is in midair. Hence, what they see is magical.

What accounts for the magic of a tour jeté?

The answer is in this chapter.

FIG. 11-1 The self-righting Segway Human Transporter. *(Justin Sullivan/ Getty Images News and Sport Services)*

11-1 WHAT IS PHYSICS?

As we discussed in Chapter 10, physics includes the study of rotation. Arguably, the most important application of that physics is in the rolling motion of wheels and wheel-like objects. This applied physics has long been used. For example, when the prehistoric people of Easter Island moved their gigantic stone statues from the quarry and across the island, they dragged them over logs acting as rollers. Much later, when settlers moved westward across America in the 1800s, they rolled their possessions first by wagon and then later by train. Today, like it or not, the world is filled with cars, trucks, motorcycles, bicycles, and other rolling vehicles.

The physics and engineering of rolling have been around for so long that you might think no fresh ideas remain to be developed. However, skateboards and in-line skates were invented and engineered fairly recently, to become huge financial successes. Street luge is now catching on, and the self-righting Segway (Fig. 11-1) may change the way people move around in large cities. Applying the physics of rolling can still lead to surprises and rewards. Our starting point in exploring that physics is to simplify rolling motion.

11-2 | Rolling as Translation and Rotation Combined

FIG. 11-2 A time-exposure photograph of a rolling disk. Small lights have been attached to the disk, one at its center and one at its edge. The latter traces out a curve called a cycloid. *(Richard Megna/ Fundamental Photographs)*

Here we consider only objects that *roll smoothly* along a surface; that is, the objects roll without slipping or bouncing on the surface. Figure 11-2 shows how complicated smooth rolling motion can be: Although the center of the object moves in a straight line parallel to the surface, a point on the rim certainly does not. However, we can study this motion by treating it as a combination of translation of the center of mass and rotation of the rest of the object around that center.

To see how we do this, pretend you are standing on a sidewalk watching the bicycle wheel of Fig. 11-3 as it rolls along a street. As shown, you see the center of mass O of the wheel move forward at constant speed v_{com}. The point P on the street where the wheel makes contact with the street surface also moves forward at speed v_{com}, so that P always remains directly below O.

During a time interval t, you see both O and P move forward by a distance s. The bicycle rider sees the wheel rotate through an angle θ about the center of the wheel, with the point of the wheel that was touching the street at the beginning of t moving through arc length s. Equation 10-17 relates the arc length s to the rotation angle θ:

$$s = \theta R, \quad (11\text{-}1)$$

where R is the radius of the wheel. The linear speed v_{com} of the center of the wheel (the center of mass of this uniform wheel) is ds/dt. The angular speed ω of the wheel about its center is $d\theta/dt$. Thus, differentiating Eq. 11-1 with respect to time (with R held constant) gives us

$$v_{com} = \omega R \quad \text{(smooth rolling motion).} \quad (11\text{-}2)$$

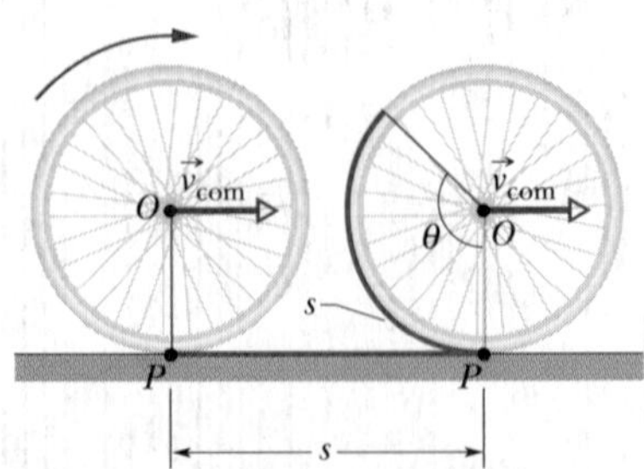

FIG. 11-3 The center of mass O of a rolling wheel moves a distance s at velocity $\vec{v}_{com}$ while the wheel rotates through angle θ. The point P at which the wheel makes contact with the surface over which the wheel rolls also moves a distance s.

Figure 11-4 shows that the rolling motion of a wheel is a combination of purely translational and purely rotational motions. Figure 11-4*a* shows the purely rotational motion (as if the rotation axis through the center were stationary): Every point on the wheel rotates about the center with angular speed ω. (This is the type of motion we considered in Chapter 10.) Every point on the outside edge of the wheel has linear speed v_{com} given by Eq. 11-2. Figure 11-4*b* shows the purely translational motion (as if the wheel did not rotate at all): Every point on the wheel moves to the right with speed v_{com}.

The combination of Figs. 11-4*a* and 11-4*b* yields the actual rolling motion of the wheel, Fig. 11-4*c*. Note that in this combination of motions, the portion of the

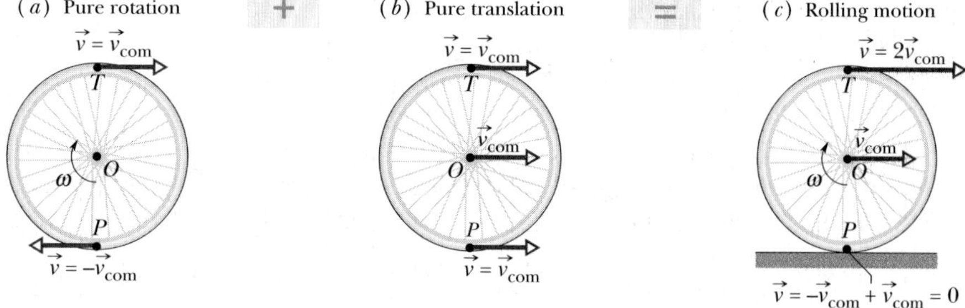

FIG. 11-4 Rolling motion of a wheel as a combination of purely rotational motion and purely translational motion. (*a*) The purely rotational motion: All points on the wheel move with the same angular speed ω. Points on the outside edge of the wheel all move with the same linear speed $v = v_{com}$. The linear velocities $\vec{v}$ of two such points, at top (T) and bottom (P) of the wheel, are shown. (*b*) The purely translational motion: All points on the wheel move to the right with the same linear velocity $\vec{v}_{com}$. (*c*) The rolling motion of the wheel is the combination of (*a*) and (*b*).

wheel at the bottom (at point P) is stationary and the portion at the top (at point T) is moving at speed $2v_{com}$, faster than any other portion of the wheel. These results are demonstrated in Fig. 11-5, which is a time exposure of a rolling bicycle wheel. You can tell that the wheel is moving faster near its top than near its bottom because the spokes are more blurred at the top than at the bottom.

The motion of any round body rolling smoothly over a surface can be separated into purely rotational and purely translational motions, as in Figs. 11-4*a* and 11-4*b*.

FIG. 11-5 A photograph of a rolling bicycle wheel. The spokes near the wheel's top are more blurred than those near the bottom because the top ones are moving faster, as Fig. 11-4*c* shows. *(Courtesy Alice Halliday)*

Rolling as Pure Rotation

Figure 11-6 suggests another way to look at the rolling motion of a wheel—namely, as pure rotation about an axis that always extends through the point where the wheel contacts the street as the wheel moves. We consider the rolling motion to be pure rotation about an axis passing through point P in Fig. 11-4*c* and perpendicular to the plane of the figure. The vectors in Fig. 11-6 then represent the instantaneous velocities of points on the rolling wheel.

Question: What angular speed about this new axis will a stationary observer assign to a rolling bicycle wheel?

Answer: The same ω that the rider assigns to the wheel as she or he observes it in pure rotation about an axis through its center of mass.

To verify this answer, let us use it to calculate the linear speed of the top of the rolling wheel from the point of view of a stationary observer. If we call the wheel's radius R, the top is a distance $2R$ from the axis through P in Fig. 11-6, so the linear speed at the top should be (using Eq. 11-2)

$$v_{top} = (\omega)(2R) = 2(\omega R) = 2v_{com},$$

in exact agreement with Fig. 11-4*c*. You can similarly verify the linear speeds shown for the portions of the wheel at points O and P in Fig. 11-4*c*.

CHECKPOINT 1 The rear wheel on a clown's bicycle has twice the radius of the front wheel. (a) When the bicycle is moving, is the linear speed at the very top of the rear wheel greater than, less than, or the same as that of the very top of the front wheel? (b) Is the angular speed of the rear wheel greater than, less than, or the same as that of the front wheel?

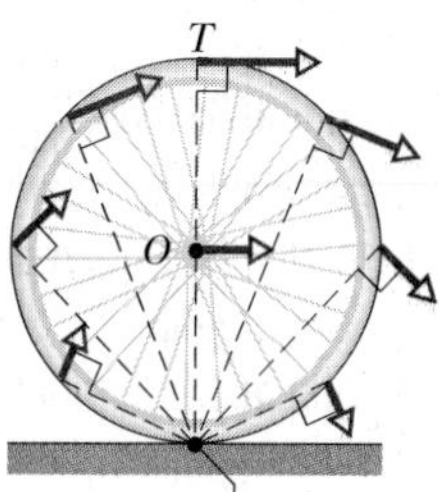

FIG. 11-6 Rolling can be viewed as pure rotation, with angular speed ω, about an axis that always extends through P. The vectors show the instantaneous linear velocities of selected points on the rolling wheel. You can obtain the vectors by combining the translational and rotational motions as in Fig. 11-4.

11-3 | The Kinetic Energy of Rolling

Let us now calculate the kinetic energy of the rolling wheel as measured by the stationary observer. If we view the rolling as pure rotation about an axis through P in Fig. 11-6, then from Eq. 10-34 we have

$$K = \tfrac{1}{2}I_P\omega^2, \tag{11-3}$$

in which ω is the angular speed of the wheel and I_P is the rotational inertia of the wheel about the axis through P. From the parallel-axis theorem of Eq. 10-36 ($I = I_{com} + Mh^2$), we have

$$I_P = I_{com} + MR^2, \tag{11-4}$$

in which M is the mass of the wheel, I_{com} is its rotational inertia about an axis through its center of mass, and R (the wheel's radius) is the perpendicular distance h. Substituting Eq. 11-4 into Eq. 11-3, we obtain

$$K = \tfrac{1}{2}I_{com}\omega^2 + \tfrac{1}{2}MR^2\omega^2,$$

and using the relation $v_{com} = \omega R$ (Eq. 11-2) yields

$$K = \tfrac{1}{2}I_{com}\omega^2 + \tfrac{1}{2}Mv_{com}^2. \tag{11-5}$$

We can interpret the term $\frac{1}{2}I_{com}\omega^2$ as the kinetic energy associated with the rotation of the wheel about an axis through its center of mass (Fig. 11-4*a*), and the term $\frac{1}{2}Mv_{com}^2$ as the kinetic energy associated with the translational motion of the wheel's center of mass (Fig. 11-4*b*). Thus, we have the following rule:

A rolling object has two types of kinetic energy: a rotational kinetic energy ($\frac{1}{2}I_{com}\omega^2$) due to its rotation about its center of mass and a translational kinetic energy ($\frac{1}{2}Mv_{com}^2$) due to translation of its center of mass.

Sample Problem 11-1

The current land-speed record was set in the Black Rock Desert of Nevada in 1997 by the jet-powered car *Thrust SSC*. The car's speed was 1222 km/h in one direction and 1233 km/h in the opposite direction. Both speeds exceeded the speed of sound at that location (1207 km/h).

Setting the land-speed record was obviously very dangerous for many reasons. One of them had to do with the car's wheels. Approximate each wheel on the car *Thrust SSC* as a disk of uniform thickness and mass $M = 170$ kg, and assume smooth rolling. When the car's speed was 1233 km/h, what was the kinetic energy of each wheel?

KEY IDEAS Equation 11-5 gives the kinetic energy of a rolling object, but we need three ideas to use it:

1. When we speak of the speed of a rolling object, we always mean the speed of the center of mass, so here $v_{com} = 1233$ km/h $= 342.5$ m/s.
2. Equation 11-5 requires the angular speed ω of the rolling object, which we can relate to v_{com} with Eq. 11-2, writing $\omega = v_{com}/R$, where R is the wheel's radius.
3. Equation 11-5 also requires the rotational inertia I_{com} of the object about its center of mass. From Table 10-2*c*, we find that, for a uniform disk, $I_{com} = \frac{1}{2}MR^2$.

Calculations: Now Eq. 11-5 gives us

$$\begin{aligned} K &= \tfrac{1}{2}I_{com}\omega^2 + \tfrac{1}{2}Mv_{com}^2 \\ &= (\tfrac{1}{2})(\tfrac{1}{2}MR^2)(v_{com}/R)^2 + \tfrac{1}{2}Mv_{com}^2 = \tfrac{3}{4}Mv_{com}^2 \\ &= \tfrac{3}{4}(170\text{ kg})(342.5\text{ m/s})^2 \\ &= 1.50 \times 10^7\text{ J}. \qquad \text{(Answer)} \end{aligned}$$

(Note that the wheel's radius R cancels out of the calculation.)

This answer gives one measure of the danger when the land-speed record was set by *Thrust SSC:* The kinetic energy of each (cast aluminum) wheel on the car was huge, almost as much as the kinetic energy (2.1×10^7 J) of the spinning steel disk that exploded in Sample Problem 10-8. Had a wheel hit any hard obstacle along the car's path, the wheel would have exploded the way the steel disk did, with the car and driver moving faster than sound!

11-4 | The Forces of Rolling

Friction and Rolling

If a wheel rolls at constant speed, as in Fig. 11-3, it has no tendency to slide at the point of contact P, and thus no frictional force acts there. However, if a net force acts on the rolling wheel to speed it up or to slow it, then that net force causes acceleration $\vec{a}_{com}$ of the center of mass along the direction of travel. It also causes the wheel to rotate faster or slower, which means it causes an angular acceleration α. These accelerations tend to make the wheel slide at P. Thus, a frictional force must act on the wheel at P to oppose that tendency.

If the wheel *does not* slide, the force is a *static* frictional force $\vec{f}_s$ and the motion is smooth rolling. We can then relate the magnitudes of the linear acceleration $\vec{a}_{com}$ and the angular acceleration α by differentiating Eq. 11-2 with respect to time (with R held constant). On the left side, dv_{com}/dt is a_{com}, and on the right side $d\omega/dt$ is α. So, for smooth rolling we have

$$a_{com} = \alpha R \quad \text{(smooth rolling motion).} \tag{11-6}$$

If the wheel *does* slide when the net force acts on it, the frictional force that acts at P in Fig. 11-3 is a *kinetic* frictional force $\vec{f}_k$. The motion then is not smooth rolling, and Eq. 11-6 does not apply to the motion. In this chapter we discuss only smooth rolling motion.

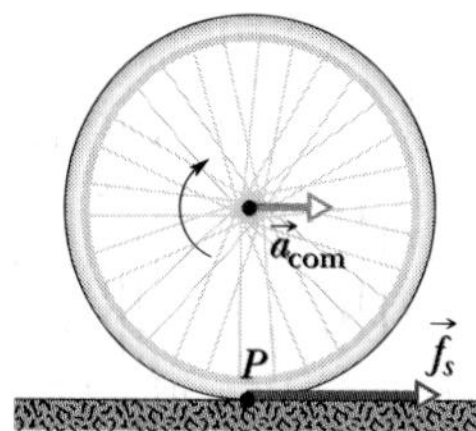

FIG. 11-7 A wheel rolls horizontally without sliding while accelerating with linear acceleration $\vec{a}_{com}$. A static frictional force $\vec{f}_s$ acts on the wheel at P, opposing its tendency to slide.

Figure 11-7 shows an example in which a wheel is being made to rotate faster while rolling to the right along a flat surface, as on a bicycle at the start of a race. The faster rotation tends to make the bottom of the wheel slide to the left at point P. A frictional force at P, directed to the right, opposes this tendency to slide. If the wheel does not slide, that frictional force is a static frictional force $\vec{f}_s$ (as shown), the motion is smooth rolling, and Eq. 11-6 applies to the motion. (Without friction, bicycle races would be stationary and very boring.)

If the wheel in Fig. 11-7 were made to rotate slower, as on a slowing bicycle, we would change the figure in two ways: The directions of the center-of-mass acceleration $\vec{a}_{com}$ and the frictional force $\vec{f}_s$ at point P would now be to the left.

Rolling Down a Ramp

Figure 11-8 shows a round uniform body of mass M and radius R rolling smoothly down a ramp at angle θ, along an x axis. We want to find an expression for the body's acceleration $a_{com,x}$ down the ramp. We do this by using Newton's second law in both its linear version ($F_{net} = Ma$) and its angular version ($\tau_{net} = I\alpha$).

We start by drawing the forces on the body as shown in Fig. 11-8:

1. The gravitational force $\vec{F}_g$ on the body is directed downward. The tail of the vector is placed at the center of mass of the body. The component along the ramp is $F_g \sin\theta$, which is equal to $Mg \sin\theta$.
2. A normal force $\vec{F}_N$ is perpendicular to the ramp. It acts at the point of contact P, but in Fig. 11-8 the vector has been shifted along its direction until its tail is at the body's center of mass.
3. A static frictional force $\vec{f}_s$ acts at the point of contact P and is directed up the ramp. (Do you see why? If the body were to slide at P, it would slide *down* the ramp. Thus, the frictional force opposing the sliding must be *up* the ramp.)

We can write Newton's second law for components along the x axis in Fig. 11-8 ($F_{net,x} = ma_x$) as

$$f_s - Mg\sin\theta = Ma_{com,x}. \tag{11-7}$$

This equation contains two unknowns, f_s and $a_{com,x}$. (We should *not* assume that f_s is at its maximum value $f_{s,max}$. All we know is that the value of f_s is just right for the body to roll smoothly down the ramp, without sliding.)

We now wish to apply Newton's second law in angular form to the body's rotation about its center of mass. First, we shall use Eq. 10-41 ($\tau = r_\perp F$) to write the torques on the body about that point. The frictional force $\vec{f}_s$ has moment arm R and thus produces a torque Rf_s, which is positive because it tends to rotate the body counterclockwise in Fig. 11-8. Forces $\vec{F}_g$ and $\vec{F}_N$ have zero moment arms about the center of mass and thus produce zero torques. So we can write the angular form of Newton's second law ($\tau_{net} = I\alpha$) about an axis through the body's center of mass as

$$Rf_s = I_{com}\alpha. \tag{11-8}$$

This equation contains two unknowns, f_s and α.

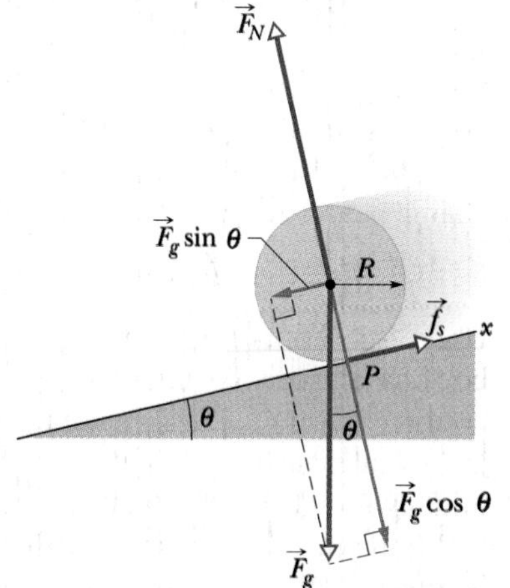

FIG. 11-8 A round uniform body of radius R rolls down a ramp. The forces that act on it are the gravitational force $\vec{F}_g$, a normal force $\vec{F}_N$, and a frictional force $\vec{f}_s$ pointing up the ramp. (For clarity, vector $\vec{F}_N$ has been shifted in the direction it points until its tail is at the center of the body.)

Because the body is rolling smoothly, we can use Eq. 11-6 ($a_{com} = \alpha R$) to relate the unknowns $a_{com,x}$ and α. But we must be cautious because here $a_{com,x}$ is negative (in the negative direction of the x axis) and α is positive (counter-clockwise). Thus we substitute $-a_{com,x}/R$ for α in Eq. 11-8. Then, solving for f_s, we obtain

$$f_s = -I_{com}\frac{a_{com,x}}{R^2}. \qquad (11\text{-}9)$$

Substituting the right side of Eq. 11-9 for f_s in Eq. 11-7, we then find

$$a_{com,x} = -\frac{g \sin \theta}{1 + I_{com}/MR^2}. \qquad (11\text{-}10)$$

We can use this equation to find the linear acceleration $a_{com,x}$ of any body rolling along an incline of angle θ with the horizontal.

CHECKPOINT 2 Disks A and B are identical and roll across a floor with equal speeds. Then disk A rolls up an incline, reaching a maximum height h, and disk B moves up an incline that is identical except that it is frictionless. Is the maximum height reached by disk B greater than, less than, or equal to h?

Sample Problem 11-2 Build your skill

A uniform ball, of mass $M = 6.00$ kg and radius R, rolls smoothly from rest down a ramp at angle $\theta = 30.0°$ (Fig. 11-8).

(a) The ball descends a vertical height $h = 1.20$ m to reach the bottom of the ramp. What is its speed at the bottom?

KEY IDEAS The mechanical energy E of the ball–Earth system is conserved as the ball rolls down the ramp. The reason is that the only force doing work on the ball is the gravitational force, a conservative force. The normal force on the ball from the ramp does zero work because it is perpendicular to the ball's path. The frictional force on the ball from the ramp does not transfer any energy to thermal energy because the ball does not slide (it *rolls smoothly*).

Therefore, we can write the conservation of mechanical energy ($E_f = E_i$) as

$$K_f + U_f = K_i + U_i, \qquad (11\text{-}11)$$

where subscripts f and i refer to the final values (at the bottom) and initial values (at rest), respectively. The gravitational potential energy is initially $U_i = Mgh$ (where M is the ball's mass) and finally $U_f = 0$. The kinetic energy is initially $K_i = 0$. For the final kinetic energy K_f, we need an additional idea: Because the ball rolls, the kinetic energy involves both translation *and* rotation, so we include them both by using the right side of Eq. 11-5.

Calculations: Substituting into Eq. 11-11 gives us

$$(\tfrac{1}{2}I_{com}\omega^2 + \tfrac{1}{2}Mv_{com}^2) + 0 = 0 + Mgh, \qquad (11\text{-}12)$$

where I_{com} is the ball's rotational inertia about an axis through its center of mass, v_{com} is the requested speed at the bottom, and ω is the angular speed there.

Because the ball rolls smoothly, we can use Eq. 11-2 to substitute v_{com}/R for ω to reduce the unknowns in Eq. 11-12. Doing so, substituting $\frac{2}{5}MR^2$ for I_{com} (from Table 10-2f), and then solving for v_{com} give us

$$v_{com} = \sqrt{(\tfrac{10}{7})gh} = \sqrt{(\tfrac{10}{7})(9.8 \text{ m/s}^2)(1.20 \text{ m})}$$
$$= 4.10 \text{ m/s}. \qquad \text{(Answer)}$$

Note that the answer does not depend on M or R.

(b) What are the magnitude and direction of the frictional force on the ball as it rolls down the ramp?

KEY IDEA Because the ball rolls smoothly, Eq. 11-9 gives the frictional force on the ball.

Calculations: Before we can use Eq. 11-9, we need the ball's acceleration $a_{com,x}$ from Eq. 11-10:

$$a_{com,x} = -\frac{g \sin \theta}{1 + I_{com}/MR^2} = -\frac{g \sin \theta}{1 + \frac{2}{5}MR^2/MR^2}$$
$$= -\frac{(9.8 \text{ m/s}^2) \sin 30.0°}{1 + \frac{2}{5}} = -3.50 \text{ m/s}^2.$$

Note that we needed neither mass M nor radius R to find $a_{com,x}$. Thus, any size ball with any uniform mass would have this acceleration down a 30.0° ramp, provided the ball rolls smoothly.

We can now solve Eq. 11-9 as

$$f_s = -I_{com}\frac{a_{com,x}}{R^2} = -\tfrac{2}{5}MR^2\frac{a_{com,x}}{R^2} = -\tfrac{2}{5}Ma_{com,x}$$
$$= -\tfrac{2}{5}(6.00 \text{ kg})(-3.50 \text{ m/s}^2) = 8.40 \text{ N}. \qquad \text{(Answer)}$$

Note that we needed mass M but not radius R. Thus, the frictional force on any 6.00 kg ball rolling smoothly down a 30.0° ramp would be 8.40 N regardless of the ball's radius.

11-5 | The Yo-Yo

A yo-yo is a physics lab that you can fit in your pocket. If a yo-yo rolls down its string for a distance h, it loses potential energy in amount mgh but gains kinetic energy in both translational ($\frac{1}{2}Mv_{com}^2$) and rotational ($\frac{1}{2}I_{com}\omega^2$) forms. As it climbs back up, it loses kinetic energy and regains potential energy.

In a modern yo-yo, the string is not tied to the axle but is looped around it. When the yo-yo "hits" the bottom of its string, an upward force on the axle from the string stops the descent. The yo-yo then spins, axle inside loop, with only rotational kinetic energy. The yo-yo keeps spinning ("sleeping") until you "wake it" by jerking on the string, causing the string to catch on the axle and the yo-yo to climb back up. The rotational kinetic energy of the yo-yo at the bottom of its string (and thus the sleeping time) can be considerably increased by throwing the yo-yo downward so that it starts down the string with initial speeds v_{com} and ω instead of rolling down from rest.

To find an expression for the linear acceleration a_{com} of a yo-yo rolling down a string, we could use Newton's second law just as we did for the body rolling down a ramp in Fig. 11-8. The analysis is the same except for the following:

1. Instead of rolling down a ramp at angle θ with the horizontal, the yo-yo rolls down a string at angle $\theta = 90°$ with the horizontal.
2. Instead of rolling on its outer surface at radius R, the yo-yo rolls on an axle of radius R_0 (Fig. 11-9*a*).
3. Instead of being slowed by frictional force $\vec{f}_s$, the yo-yo is slowed by the force $\vec{T}$ on it from the string (Fig. 11-9*b*).

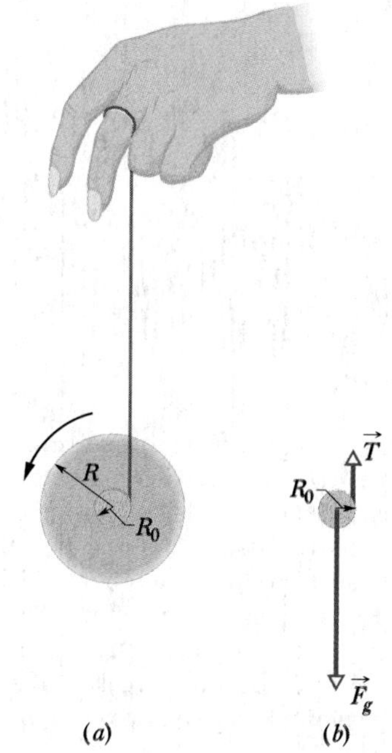

FIG. 11-9 (*a*) A yo-yo, shown in cross section. The string, of assumed negligible thickness, is wound around an axle of radius R_0. (*b*) A free-body diagram for the falling yo-yo. Only the axle is shown.

The analysis would again lead us to Eq. 11-10. Therefore, let us just change the notation in Eq. 11-10 and set $\theta = 90°$ to write the linear acceleration as

$$a_{com} = -\frac{g}{1 + I_{com}/MR_0^2}, \tag{11-13}$$

where I_{com} is the yo-yo's rotational inertia about its center and M is its mass. A yo-yo has the same downward acceleration when it is climbing back up.

11-6 | Torque Revisited

In Chapter 10 we defined torque τ for a rigid body that can rotate around a fixed axis, with each particle in the body forced to move in a path that is a circle centered on that axis. We now expand the definition of torque to apply it to an individual particle that moves along any path relative to a fixed *point* (rather than a fixed axis). The path need no longer be a circle, and we must write the torque as a vector $\vec{\tau}$ that may have any direction.

Figure 11-10*a* shows such a particle at point A in an xy plane. A single force $\vec{F}$ in that plane acts on the particle, and the particle's position relative to the origin O is given by position vector $\vec{r}$. The torque $\vec{\tau}$ acting on the particle relative to the fixed point O is a vector quantity defined as

$$\vec{\tau} = \vec{r} \times \vec{F} \quad \text{(torque defined)}. \tag{11-14}$$

We can evaluate the vector (or cross) product in this definition of $\vec{\tau}$ by using the rules for such products given in Section 3-8. To find the direction of $\vec{\tau}$, we slide the vector $\vec{F}$ (without changing its direction) until its tail is at the origin O, so that the two vectors in the vector product are tail to tail as in Fig. 11-10*b*. We then use the right-hand rule for vector products in Fig. 3-21*a*, sweeping the fingers of the right hand from $\vec{r}$ (the first vector in the product) into $\vec{F}$ (the second vector). The outstretched right thumb then gives the direction of $\vec{\tau}$. In Fig. 11-10*b*, the direction of $\vec{\tau}$ is in the positive direction of the z axis.

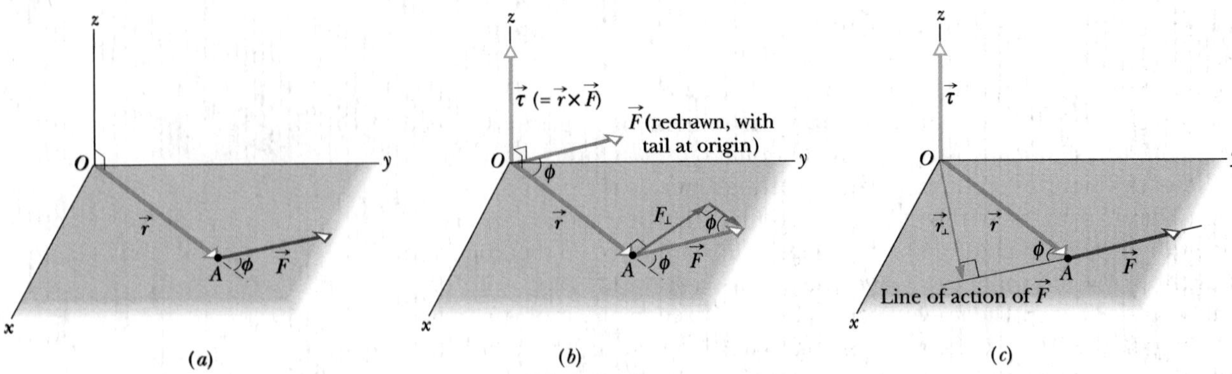

FIG. 11-10 Defining torque. (*a*) A force $\vec{F}$, lying in an *xy* plane, acts on a particle at point *A*. (*b*) This force produces a torque $\vec{\tau}$ (= $\vec{r} \times \vec{F}$) on the particle with respect to the origin *O*. By the right-hand rule for vector (cross) products, the torque vector points in the positive direction of *z*. Its magnitude is given by $rF_\perp$ in (*b*) and by $r_\perp F$ in (*c*).

To determine the magnitude of $\vec{\tau}$, we apply the general result of Eq. 3-27 ($c = ab \sin \phi$), finding

$$\tau = rF \sin \phi, \tag{11-15}$$

where ϕ is the smaller angle between the directions of $\vec{r}$ and $\vec{F}$ when the vectors are tail to tail. From Fig. 11-10*b*, we see that Eq. 11-15 can be rewritten as

$$\tau = rF_\perp, \tag{11-16}$$

where $F_\perp$ (= $F \sin \phi$) is the component of $\vec{F}$ perpendicular to $\vec{r}$. From Fig. 11-10*c*, we see that Eq. 11-15 can also be rewritten as

$$\tau = r_\perp F, \tag{11-17}$$

where $r_\perp$ (= $r \sin \phi$) is the moment arm of $\vec{F}$ (the perpendicular distance between *O* and the line of action of $\vec{F}$).

CHECKPOINT 3 The position vector $\vec{r}$ of a particle points along the positive direction of a *z* axis. If the torque on the particle is (a) zero, (b) in the negative direction of *x*, and (c) in the negative direction of *y*, in what direction is the force causing the torque?

Sample Problem 11-3

In Fig. 11-11*a*, three forces, each of magnitude 2.0 N, act on a particle. The particle is in the *xz* plane at point *A* given by position vector $\vec{r}$, where $r = 3.0$ m and $\theta = 30°$. Force $\vec{F}_1$ is parallel to the *x* axis, force $\vec{F}_2$ is parallel to the *z* axis, and force $\vec{F}_3$ is parallel to the *y* axis. What is the torque, about the origin *O*, due to each force?

KEY IDEA Because the three force vectors do not lie in a plane, we cannot evaluate their torques as in Chapter 10. Instead, we must use vector (or cross) products, with magnitudes given by Eq. 11-15 ($\tau = rF \sin \phi$) and directions given by the right-hand rule for vector products.

Calculations: Because we want the torques with respect to the origin *O*, the vector $\vec{r}$ required for each cross product is the given position vector. To determine the angle ϕ between the direction of $\vec{r}$ and the direction of each force, we shift the force vectors of Fig. 11-11*a*, each in

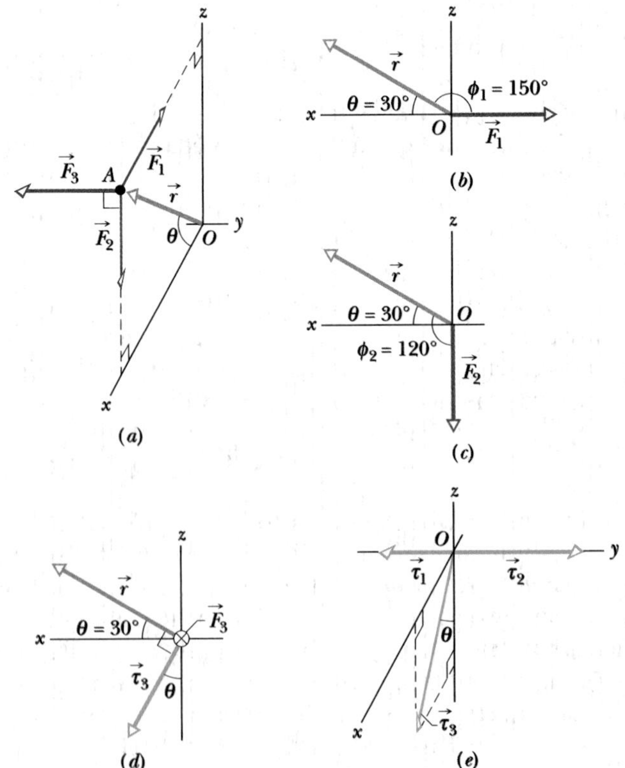

FIG. 11-11 (*a*) A particle at point *A* is acted on by three forces, each parallel to a coordinate axis. The angle ϕ (used in finding torque) is shown (*b*) for $\vec{F}_1$ and (*c*) for $\vec{F}_2$. (*d*) Torque $\vec{\tau}_3$ is perpendicular to both $\vec{r}$ and $\vec{F}_3$ (force $\vec{F}_3$ is directed into the plane of the figure). (*e*) The torques (relative to the origin *O*) acting on the particle.

turn, so that their tails are at the origin. Figures 11-11*b*, *c*, and *d*, which are direct views of the *xz* plane, show the shifted force vectors $\vec{F}_1$, $\vec{F}_2$, and $\vec{F}_3$, respectively. (Note how much easier the angles are to see.) In Fig. 11-11*d*, the angle between the directions of $\vec{r}$ and $\vec{F}_3$ is 90° and the symbol ⊗ means $\vec{F}_3$ is directed into the page. If it were directed out of the page, it would be represented with the symbol ⊙.

Now, applying Eq. 11-15 for each force, we find the magnitudes of the torques to be

$$\tau_1 = rF_1 \sin \phi_1 = (3.0 \text{ m})(2.0 \text{ N})(\sin 150°) = 3.0 \text{ N} \cdot \text{m},$$

$$\tau_2 = rF_2 \sin \phi_2 = (3.0 \text{ m})(2.0 \text{ N})(\sin 120°) = 5.2 \text{ N} \cdot \text{m},$$

and $$\tau_3 = rF_3 \sin \phi_3 = (3.0 \text{ m})(2.0 \text{ N})(\sin 90°) = 6.0 \text{ N} \cdot \text{m}. \quad \text{(Answer)}$$

To find the directions of these torques, we use the right-hand rule, placing the fingers of the right hand so as to rotate $\vec{r}$ into $\vec{F}$ through the *smaller* of the two angles between their directions. The thumb points in the direction of the torque. Thus $\vec{\tau}_1$ is directed into the page in Fig. 11-11*b*; $\vec{\tau}_2$ is directed out of the page in Fig. 11-11*c*; and $\vec{\tau}_3$ is directed as shown in Fig. 11-11*d*. All three torque vectors are shown in Fig. 11-11*e*.

PROBLEM-SOLVING TACTICS

Tactic 1: ***Vector Products and Torques*** Equation 11-15 for torques is our first application of the vector (or cross) product. Section 3-8, where the rules for the vector product are given, lists many common errors in finding the direction of a vector product.

Keep in mind that a torque is calculated *with respect to* (or *about*) a point, which must be known if the value of the torque is to be meaningful. Changing the point can change the torque in both magnitude and direction. For example, in Sample Problem 11-3, the torques due to the three forces are calculated about the origin *O*. You can show that the torques due to the same three forces are all zero if they are calculated about point *A* (at the position of the particle), because then $r = 0$ for each force.

11-7 | Angular Momentum

Recall that the concept of linear momentum $\vec{p}$ and the principle of conservation of linear momentum are extremely powerful tools. They allow us to predict the outcome of, say, a collision of two cars without knowing the details of the collision. Here we begin a discussion of the angular counterpart of $\vec{p}$, winding up in Section 11-11 with the angular counterpart of the conservation principle.

Figure 11-12 shows a particle of mass *m* with linear momentum $\vec{p}\,(= m\vec{v})$ as it passes through point *A* in an *xy* plane. The **angular momentum** $\vec{\ell}$ of this particle with respect to the origin *O* is a vector quantity defined as

$$\vec{\ell} = \vec{r} \times \vec{p} = m(\vec{r} \times \vec{v}) \quad \text{(angular momentum defined)}, \tag{11-18}$$

where $\vec{r}$ is the position vector of the particle with respect to *O*. As the particle moves relative to *O* in the direction of its momentum $\vec{p}\,(= m\vec{v})$, position vector $\vec{r}$ rotates around *O*. Note carefully that to have angular momentum about *O*, the particle does *not* itself have to rotate around *O*. Comparison of Eqs. 11-14 and 11-18 shows that angular momentum bears the same relation to linear momentum that torque does to force. The SI unit of angular momentum is the kilogram-meter-squared per second ($\text{kg} \cdot \text{m}^2/\text{s}$), equivalent to the joule-second ($\text{J} \cdot \text{s}$).

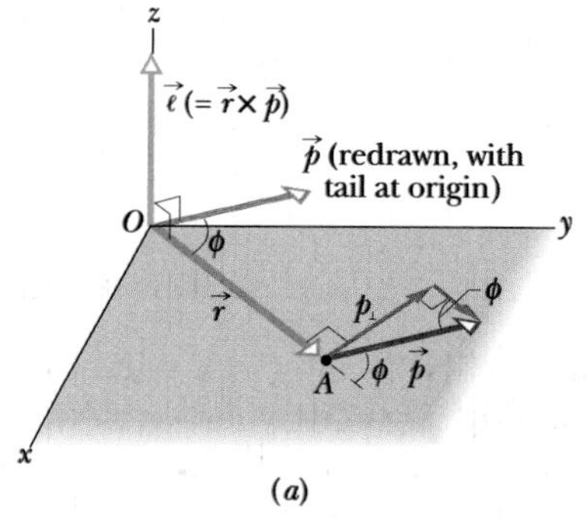

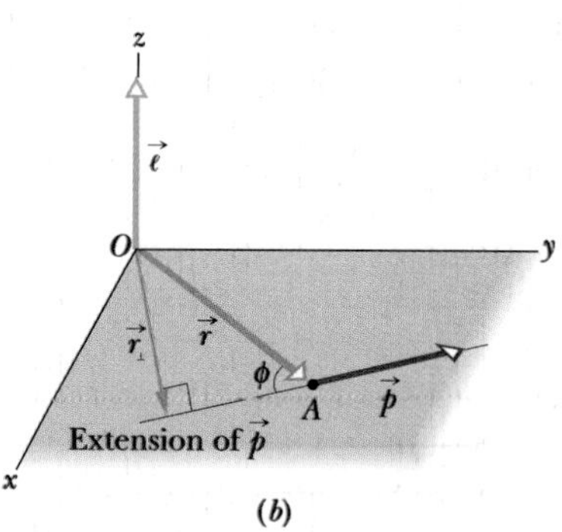

FIG. 11-12 Defining angular momentum. A particle passing through point *A* has linear momentum $\vec{p}\,(= m\vec{v})$, with the vector $\vec{p}$ lying in an *xy* plane. The particle has angular momentum $\vec{\ell}\,(= \vec{r} \times \vec{p})$ with respect to the origin *O*. By the right-hand rule, the angular momentum vector points in the positive direction of *z*. (*a*) The magnitude of $\vec{\ell}$ is given by $\ell = rp_\perp = rmv_\perp$. (*b*) The magnitude of $\vec{\ell}$ is also given by $\ell = r_\perp p = r_\perp mv$.

To find the direction of the angular momentum vector $\vec{\ell}$ in Fig. 11-12, we slide the vector $\vec{p}$ until its tail is at the origin *O*. Then we use the right-hand rule for vector products, sweeping the fingers from $\vec{r}$ into $\vec{p}$. The outstretched thumb then shows that the direction of $\vec{\ell}$ is in the positive direction of the *z* axis in Fig. 11-12. This positive direction is consistent with the counterclockwise rotation of position vector $\vec{r}$ about the *z* axis, as the particle moves. (A negative direction of $\vec{\ell}$ would be consistent with a clockwise rotation of $\vec{r}$ about the *z* axis.)

To find the magnitude of $\vec{\ell}$, we use the general result of Eq. 3-27 to write

$$\ell = rmv \sin \phi, \tag{11-19}$$

where ϕ is the smaller angle between $\vec{r}$ and $\vec{p}$ when these two vectors are tail to tail. From Fig. 11-12*a*, we see that Eq. 11-19 can be rewritten as

$$\ell = rp_\perp = rmv_\perp, \tag{11-20}$$

where $p_\perp$ is the component of $\vec{p}$ perpendicular to $\vec{r}$ and $v_\perp$ is the component of $\vec{v}$ perpendicular to $\vec{r}$. From Fig. 11-12*b*, we see that Eq. 11-19 can also be rewritten as

$$\ell = r_\perp p = r_\perp mv, \tag{11-21}$$

where $r_\perp$ is the perpendicular distance between O and the extension of $\vec{p}$.

Just as is true for torque, angular momentum has meaning only with respect to a specified origin. Moreover, if the particle in Fig. 11-12 did not lie in the xy plane, or if the linear momentum $\vec{p}$ of the particle did not also lie in that plane, the angular momentum $\vec{\ell}$ would not be parallel to the z axis. The direction of the angular momentum vector is always perpendicular to the plane formed by the position and linear momentum vectors $\vec{r}$ and $\vec{p}$.

CHECKPOINT 4 In part *a* of the figure, particles 1 and 2 move around point O in opposite directions, in circles with radii 2 m and 4 m. In part *b*, particles 3 and 4 travel in the same direction, along straight lines at perpendicular distances of 4 m and 2 m from point O. Particle 5 moves directly away from O. All five particles have the same mass and the same constant speed. (a) Rank the particles according to the magnitudes of their angular momentum about point O, greatest first. (b) Which particles have negative angular momentum about point O?

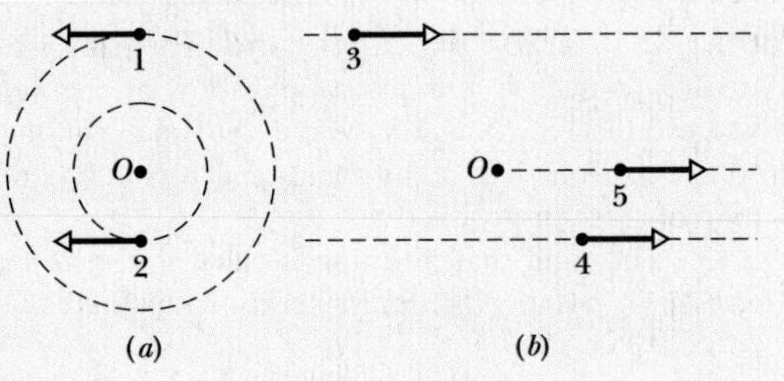

Sample Problem 11-4

Figure 11-13 shows an overhead view of two particles moving at constant momentum along horizontal paths. Particle 1, with momentum magnitude $p_1 = 5.0\ \text{kg}\cdot\text{m/s}$, has position vector $\vec{r}_1$ and will pass 2.0 m from point O. Particle 2, with momentum magnitude $p_2 = 2.0\ \text{kg}\cdot\text{m/s}$, has position vector $\vec{r}_2$ and will pass 4.0 m from point O. What are the magnitude and direction of the net angular momentum $\vec{L}$ about point O of the two-particle system?

KEY IDEA To find $\vec{L}$, we can first find the individual angular momenta $\vec{\ell}_1$ and $\vec{\ell}_2$ and then add them. To evaluate their magnitudes, we can use any one of Eqs. 11-18 through 11-21. However, Eq. 11-21 is easiest, because we are given the perpendicular distances $r_{1\perp}$ (= 2.0 m) and $r_{2\perp}$ (= 4.0 m) and the momentum magnitudes p_1 and p_2.

Calculations: For particle 1, Eq. 11-21 yields

$$\ell_1 = r_{\perp 1}p_1 = (2.0\ \text{m})(5.0\ \text{kg}\cdot\text{m/s})$$
$$= 10\ \text{kg}\cdot\text{m}^2/\text{s}.$$

To find the direction of vector $\vec{\ell}_1$, we use Eq. 11-18 and the right-hand rule for vector products. For $\vec{r}_1 \times \vec{p}_1$, the vector product is out of the page, perpendicular to the plane of Fig. 11-13. This is the positive direction, consistent with the counterclockwise rotation of the particle's

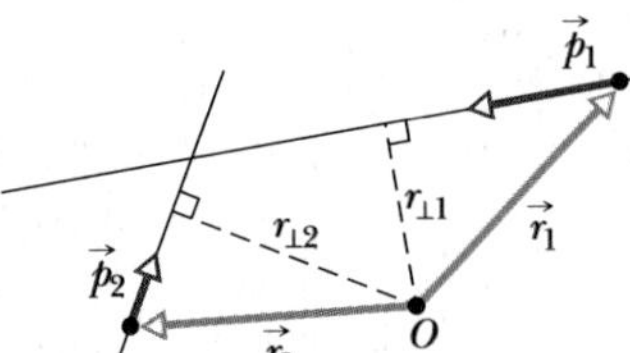

FIG. 11-13 Two particles pass near point O.

position vector $\vec{r}_1$ around O as particle 1 moves. Thus, the angular momentum vector for particle 1 is

$$\ell_1 = +10\ \text{kg}\cdot\text{m}^2/\text{s}.$$

Similarly, the magnitude of $\vec{\ell}_2$ is

$$\ell_2 = r_{\perp 2}p_2 = (4.0\ \text{m})(2.0\ \text{kg}\cdot\text{m/s})$$
$$= 8.0\ \text{kg}\cdot\text{m}^2/\text{s},$$

and the vector product $\vec{r}_2 \times \vec{p}_2$ is into the page, which is the negative direction, consistent with the clockwise rotation of $\vec{r}_2$ around O as particle 2 moves. Thus, the angular momentum vector for particle 2 is

$$\ell_2 = -8.0\ \text{kg}\cdot\text{m}^2/\text{s}.$$

The net angular momentum for the two-particle system is

$$L = \ell_1 + \ell_2 = +10\ \text{kg}\cdot\text{m}^2/\text{s} + (-8.0\ \text{kg}\cdot\text{m}^2/\text{s})$$
$$= +2.0\ \text{kg}\cdot\text{m}^2/\text{s}. \qquad \text{(Answer)}$$

The plus sign means that the system's net angular momentum about point O is out of the page.

11-8 | Newton's Second Law in Angular Form

Newton's second law written in the form

$$\vec{F}_{\text{net}} = \frac{d\vec{p}}{dt} \quad \text{(single particle)} \tag{11-22}$$

expresses the close relation between force and linear momentum for a single particle. We have seen enough of the parallelism between linear and angular quantities to be pretty sure that there is also a close relation between torque and angular momentum. Guided by Eq. 11-22, we can even guess that it must be

$$\vec{\tau}_{\text{net}} = \frac{d\vec{\ell}}{dt} \quad \text{(single particle).} \tag{11-23}$$

Equation 11-23 is indeed an angular form of Newton's second law for a single particle:

> The (vector) sum of all the torques acting on a particle is equal to the time rate of change of the angular momentum of that particle.

Equation 11-23 has no meaning unless the torques $\vec{\tau}$ and the angular momentum $\vec{\ell}$ are defined with respect to the same origin.

Proof of Equation 11-23

We start with Eq. 11-18, the definition of the angular momentum of a particle:

$$\vec{\ell} = m(\vec{r} \times \vec{v}),$$

where $\vec{r}$ is the position vector of the particle and $\vec{v}$ is the velocity of the particle. Differentiating* each side with respect to time t yields

$$\frac{d\vec{\ell}}{dt} = m\left(\vec{r} \times \frac{d\vec{v}}{dt} + \frac{d\vec{r}}{dt} \times \vec{v}\right). \tag{11-24}$$

However, $d\vec{v}/dt$ is the acceleration $\vec{a}$ of the particle, and $d\vec{r}/dt$ is its velocity $\vec{v}$. Thus, we can rewrite Eq. 11-24 as

$$\frac{d\vec{\ell}}{dt} = m(\vec{r} \times \vec{a} + \vec{v} \times \vec{v}).$$

Now $\vec{v} \times \vec{v} = 0$ (the vector product of any vector with itself is zero because the angle between the two vectors is necessarily zero). This leads to

$$\frac{d\vec{\ell}}{dt} = m(\vec{r} \times \vec{a}) = \vec{r} \times m\vec{a}.$$

We now use Newton's second law ($\vec{F}_{\text{net}} = m\vec{a}$) to replace $m\vec{a}$ with its equal, the vector sum of the forces that act on the particle, obtaining

$$\frac{d\vec{\ell}}{dt} = \vec{r} \times \vec{F}_{\text{net}} = \sum(\vec{r} \times \vec{F}). \tag{11-25}$$

Here the symbol Σ indicates that we must sum the vector products $\vec{r} \times \vec{F}$ for all the forces. However, from Eq. 11-14, we know that each one of those vector products is the torque associated with one of the forces. Therefore, Eq. 11-25 tells us that

$$\vec{\tau}_{\text{net}} = \frac{d\vec{\ell}}{dt}.$$

This is Eq. 11-23, the relation that we set out to prove.

*In differentiating a vector product, be sure not to change the order of the two quantities (here $\vec{r}$ and $\vec{v}$) that form that product. (See Eq. 3-28.)

CHECKPOINT 5 The figure shows the position vector $\vec{r}$ of a particle at a certain instant, and four choices for the direction of a force that is to accelerate the particle. All four choices lie in the xy plane. (a) Rank the choices according to the magnitude of the time rate of change ($d\vec{\ell}/dt$) they produce in the angular momentum of the particle about point O, greatest first. (b) Which choice results in a negative rate of change about O?

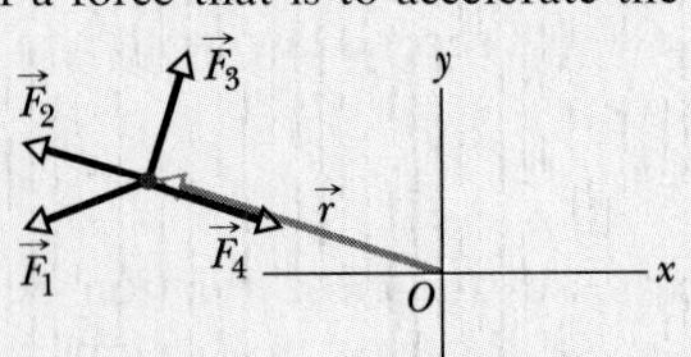

Sample Problem 11-5 Build your skill

In Fig. 11-14, a penguin of mass m falls from rest at point A, a horizontal distance D from the origin O of an xyz coordinate system. (The positive direction of the z axis is directly outward from the plane of the figure.)

(a) What is the angular momentum $\vec{\ell}$ of the falling penguin about O?

KEY IDEA We can treat the penguin as a particle, and thus its angular momentum $\vec{\ell}$ is given by Eq. 11-18 ($\vec{\ell} = \vec{r} \times \vec{p}$), where $\vec{r}$ is the penguin's position vector (extending from O to the penguin) and $\vec{p}$ is the penguin's linear momentum. (The penguin has *angular* momentum about O even though it moves in a straight line, because vector $\vec{r}$ rotates about O as the penguin falls.)

Calculations: To find the magnitude of $\vec{\ell}$, we can use any one of the scalar equations derived from Eq. 11-18—namely, Eqs. 11-19 through 11-21. However, Eq. 11-21 ($\ell = r_\perp mv$) is easiest because the perpendicular distance $r_\perp$ between O and an extension of vector $\vec{p}$ is the given distance D. The speed of an object that has fallen from rest for a time t is $v = gt$. We can now write Eq. 11-21 in terms of given quantities as

$$\ell = r_\perp mv = Dmgt. \qquad \text{(Answer)}$$

To find the direction of $\vec{\ell}$, we use the right-hand rule for the vector product $\vec{r} \times \vec{p}$ in Eq. 11-18. Mentally shift $\vec{p}$ until its tail is at the origin, and then use the fingers of your right hand to rotate $\vec{r}$ into $\vec{p}$ through the smaller angle between the two vectors. Your outstretched thumb then points into the plane of the figure, indicating that the product $\vec{r} \times \vec{p}$ and thus also $\vec{\ell}$ are directed into that plane, in the negative direction of the z axis. We represent $\vec{\ell}$ with an encircled cross ⊗ at O. The vector $\vec{\ell}$ changes with time in magnitude only; its direction remains unchanged.

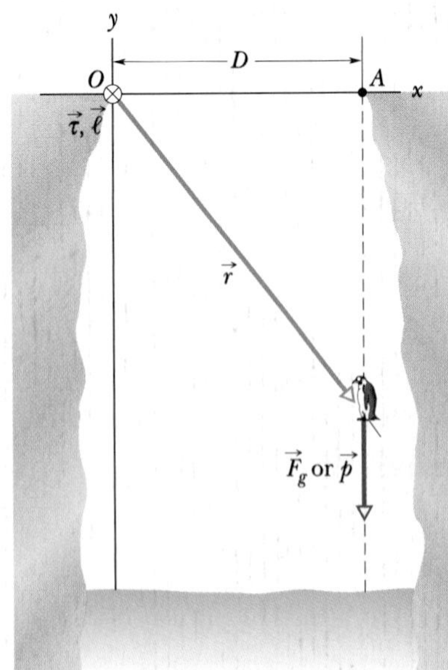

FIG. 11-14 A penguin falls vertically from point A. The torque $\vec{\tau}$ and the angular momentum $\vec{\ell}$ of the falling penguin with respect to the origin O are directed into the plane of the figure at O.

(b) About the origin O, what is the torque $\vec{\tau}$ on the penguin due to the gravitational force $\vec{F}_g$?

KEY IDEAS (1) The torque is given by Eq. 11-14 ($\vec{\tau} = \vec{r} \times \vec{F}$), where now the force is $\vec{F}_g$. (2) Force $\vec{F}_g$ causes a torque on the penguin, even though the penguin moves in a straight line, because $\vec{r}$ rotates about O as the penguin moves.

Calculations: To find the magnitude of $\vec{\tau}$, we can use any one of the scalar equations derived from Eq. 11-14—namely, Eqs. 11-15 through 11-17. However, Eq. 11-17 ($\tau = r_\perp F$) is easiest because the perpendicular distance $r_\perp$ between O and the line of action of $\vec{F}_g$ is the given distance D. So, substituting D and using mg for the magnitude of $\vec{F}_g$, we can write Eq. 11-17 as

$$\tau = DF_g = Dmg. \qquad \text{(Answer)}$$

Using the right-hand rule for the vector product $\vec{r} \times \vec{F}$ in Eq. 11-14, we find that the direction of $\vec{\tau}$ is the negative direction of the z axis, the same as $\vec{\ell}$.

The results we obtained in parts (a) and (b) must be consistent with Newton's second law in the angular form of Eq. 11-23 ($\vec{\tau}_{net} = d\vec{\ell}/dt$). To check the magnitudes we got, we write Eq. 11-23 in component form for the z axis and then substitute our result $\ell = Dmgt$. We find

$$\tau = \frac{d\ell}{dt} = \frac{d(Dmgt)}{dt} = Dmg,$$

which is the magnitude we found for $\vec{\tau}$. To check the directions, we note that Eq. 11-23 tells us that $\vec{\tau}$ and $d\vec{\ell}/dt$ must have the same direction. So $\vec{\tau}$ and $\vec{\ell}$ must also have the same direction, which is what we found.

11-9 | The Angular Momentum of a System of Particles

Now we turn our attention to the angular momentum of a system of particles with respect to an origin. The total angular momentum $\vec{L}$ of the system is the (vector) sum of the angular momenta $\vec{\ell}$ of the individual particles (here with label i):

$$\vec{L} = \vec{\ell}_1 + \vec{\ell}_2 + \vec{\ell}_3 + \cdots + \vec{\ell}_n = \sum_{i=1}^{n} \vec{\ell}_i. \tag{11-26}$$

With time, the angular momenta of individual particles may change because of interactions between the particles or with the outside. We can find the resulting change in $\vec{L}$ by taking the time derivative of Eq. 11-26. Thus,

$$\frac{d\vec{L}}{dt} = \sum_{i=1}^{n} \frac{d\vec{\ell}_i}{dt}. \tag{11-27}$$

From Eq. 11-23, we see that $d\vec{\ell}_i/dt$ is equal to the net torque $\vec{\tau}_{\text{net},i}$ on the ith particle. We can rewrite Eq. 11-27 as

$$\frac{d\vec{L}}{dt} = \sum_{i=1}^{n} \vec{\tau}_{\text{net},i}. \tag{11-28}$$

That is, the rate of change of the system's angular momentum $\vec{L}$ is equal to the vector sum of the torques on its individual particles. Those torques include *internal torques* (due to forces between the particles) and *external torques* (due to forces on the particles from bodies external to the system). However, the forces between the particles always come in third-law force pairs so their torques sum to zero. Thus, the only torques that can change the total angular momentum $\vec{L}$ of the system are the external torques acting on the system.

Let $\vec{\tau}_{\text{net}}$ represent the net external torque, the vector sum of all external torques on all particles in the system. Then we can write Eq. 11-28 as

$$\vec{\tau}_{\text{net}} = \frac{d\vec{L}}{dt} \quad \text{(system of particles)}, \tag{11-29}$$

which is Newton's second law in angular form. It says:

The net external torque $\vec{\tau}_{\text{net}}$ acting on a system of particles is equal to the time rate of change of the system's total angular momentum $\vec{L}$.

Equation 11-29 is analogous to $\vec{F}_{\text{net}} = d\vec{P}/dt$ (Eq. 9-27) but requires extra caution: Torques and the system's angular momentum must be measured relative to the same origin. If the center of mass of the system is not accelerating relative to an inertial frame, that origin can be any point. However, if the center of mass of the system *is* accelerating, the origin can be only at that center of mass. As an example, consider a wheel as the system of particles. If the wheel is rotating about an axis that is fixed relative to the ground, then the origin for applying Eq. 11-29 can be any point that is stationary relative to the ground. However, if the wheel is rotating about an axis that is accelerating (such as when the wheel rolls down a ramp), then the origin can be only at the wheel's center of mass.

11-10 | The Angular Momentum of a Rigid Body Rotating About a Fixed Axis

We next evaluate the angular momentum of a system of particles that form a rigid body that rotates about a fixed axis. Figure 11-15a shows such a body. The fixed axis of rotation is a z axis, and the body rotates about it with constant angular speed ω. We wish to find the angular momentum of the body about that axis.

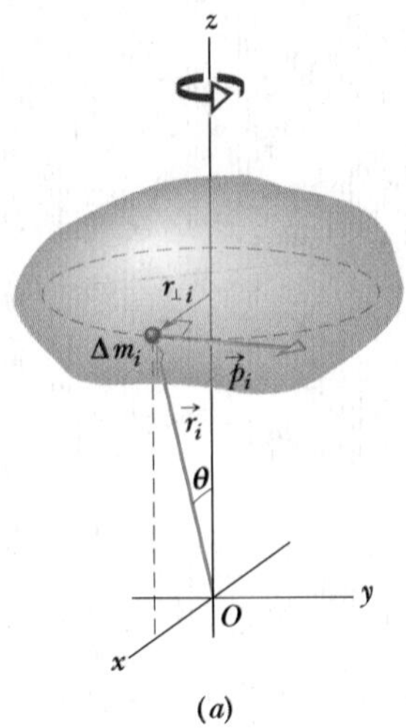

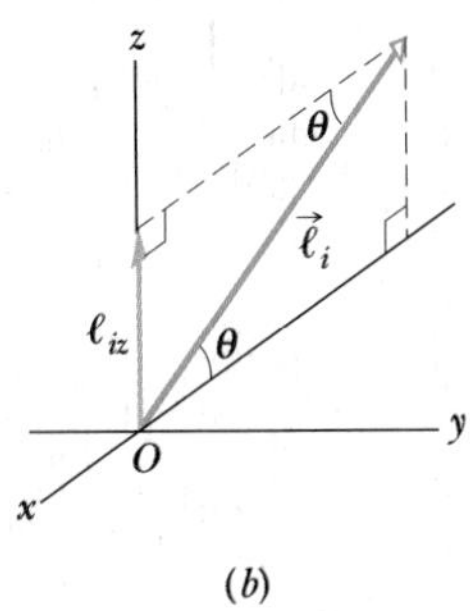

FIG. 11-15 (*a*) A rigid body rotates about a z axis with angular speed ω. A mass element of mass Δm_i within the body moves about the z axis in a circle with radius $r_{\perp i}$. The mass element has linear momentum $\vec{p}_i$, and it is located relative to the origin O by position vector $\vec{r}_i$. Here the mass element is shown when $r_{\perp i}$ is parallel to the x axis. (*b*) The angular momentum $\vec{\ell}_i$, with respect to O, of the mass element in (*a*). The z component ℓ_{iz} is also shown.

We can find the angular momentum by summing the z components of the angular momenta of the mass elements in the body. In Fig. 11-15*a*, a typical mass element, of mass Δm_i, moves around the z axis in a circular path. The position of the mass element is located relative to the origin O by position vector $\vec{r}_i$. The radius of the mass element's circular path is $r_{\perp i}$, the perpendicular distance between the element and the z axis.

The magnitude of the angular momentum $\vec{\ell}_i$ of this mass element, with respect to O, is given by Eq. 11-19:

$$\ell_i = (r_i)(p_i)(\sin 90^\circ) = (r_i)(\Delta m_i\, v_i),$$

where p_i and v_i are the linear momentum and linear speed of the mass element, and 90° is the angle between $\vec{r}_i$ and $\vec{p}_i$. The angular momentum vector $\vec{\ell}_i$ for the mass element in Fig. 11-15*a* is shown in Fig. 11-15*b*; its direction must be perpendicular to those of $\vec{r}_i$ and $\vec{p}_i$.

We are interested in the component of $\vec{\ell}_i$ that is parallel to the rotation axis, here the z axis. That z component is

$$\ell_{iz} = \ell_i \sin\theta = (r_i \sin\theta)(\Delta m_i\, v_i) = r_{\perp i}\,\Delta m_i\, v_i.$$

The z component of the angular momentum for the rotating rigid body as a whole is found by adding up the contributions of all the mass elements that make up the body. Thus, because $v = \omega r_\perp$, we may write

$$L_z = \sum_{i=1}^{n} \ell_{iz} = \sum_{i=1}^{n} \Delta m_i\, v_i r_{\perp i} = \sum_{i=1}^{n} \Delta m_i(\omega r_{\perp i}) r_{\perp i}$$
$$= \omega\left(\sum_{i=1}^{n} \Delta m_i\, r_{\perp i}^2\right). \tag{11-30}$$

We can remove ω from the summation here because it has the same value for all points of the rotating rigid body.

The quantity $\Sigma\,\Delta m_i\, r_{\perp i}^2$ in Eq. 11-30 is the rotational inertia I of the body about the fixed axis (see Eq. 10-33). Thus Eq. 11-30 reduces to

$$L = I\omega \quad \text{(rigid body, fixed axis).} \tag{11-31}$$

We have dropped the subscript z, but you must remember that the angular momentum defined by Eq. 11-31 is the angular momentum about the rotation axis. Also, I in that equation is the rotational inertia about that same axis.

Table 11-1, which supplements Table 10-3, extends our list of corresponding linear and angular relations.

TABLE 11-1

More Corresponding Variables and Relations for Translational and Rotational Motion[a]

Translational		Rotational	
Force	$\vec{F}$	Torque	$\vec{\tau}\,(=\vec{r}\times\vec{F})$
Linear momentum	$\vec{p}$	Angular momentum	$\vec{\ell}\,(=\vec{r}\times\vec{p})$
Linear momentum[b]	$\vec{P}\,(=\Sigma\vec{p}_i)$	Angular momentum[b]	$\vec{L}\,(=\Sigma\vec{\ell}_i)$
Linear momentum[b]	$\vec{P} = M\vec{v}_{com}$	Angular momentum[c]	$L = I\omega$
Newton's second law[b]	$\vec{F}_{net} = \dfrac{d\vec{P}}{dt}$	Newton's second law[b]	$\vec{\tau}_{net} = \dfrac{d\vec{L}}{dt}$
Conservation law[d]	$\vec{P}$ = a constant	Conservation law[d]	$\vec{L}$ = a constant

[a]See also Table 10-3.

[b]For systems of particles, including rigid bodies.

[c]For a rigid body about a fixed axis, with L being the component along that axis.

[d]For a closed, isolated system.

CHECKPOINT 6 In the figure, a disk, a hoop, and a solid sphere are made to spin about fixed central axes (like a top) by means of strings wrapped around them, with the strings producing the same constant tangential force $\vec{F}$ on all three objects. The three objects have the same mass and radius, and they are initially stationary. Rank the objects according to (a) their angular momentum about their central axes and (b) their angular speed, greatest first, when the strings have been pulled for a certain time t.

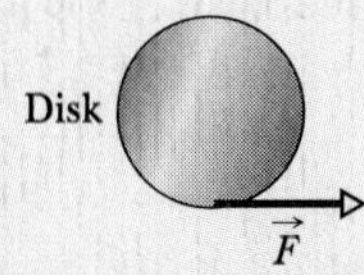

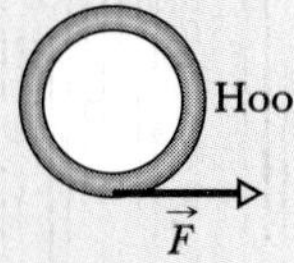

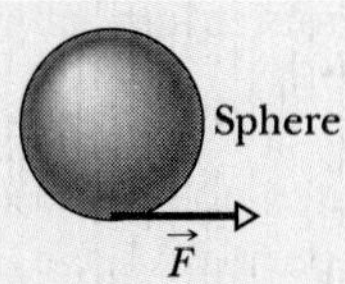

Sample Problem 11-6

George Washington Gale Ferris, Jr., a civil engineering graduate from Rensselaer Polytechnic Institute, built the original Ferris wheel (Fig. 11-16) for the 1893 World's Columbian Exposition in Chicago. The wheel, an astounding engineering construction at the time, carried 36 wooden cars, each holding as many as 60 passengers, around a circle of radius $R = 38$ m. The mass of each car was about 1.1×10^4 kg. The mass of the wheel's structure was about 6.0×10^5 kg, which was mostly in the circular grid from which the cars were suspended. The wheel made a complete rotation at an angular speed ω_F in about 2 min.

(a) Estimate the magnitude L of the angular momentum of the wheel and its passengers while the wheel rotated at ω_F.

KEY IDEA We can treat the wheel, cars, and passengers as a rigid object rotating about a fixed axis, at the wheel's axle. Then Eq. 11-31 ($L = I\omega$) gives the magnitude of the angular momentum of that object. We need to find ω_F and the rotational inertia I of this object.

Rotational inertia: To find I, let us start with the loaded cars. Because we can treat them as particles, at distance R from the axis of rotation, we know from Eq. 10-33 that their rotational inertia is $I_{pc} = M_{pc}R^2$, where M_{pc} is their total mass. Let us assume that the 36 cars are each filled with 60 passengers, each of mass 70 kg. Then their total mass is

$$M_{pc} = 36[1.1 \times 10^4 \text{ kg} + 60(70 \text{ kg})] = 5.47 \times 10^5 \text{ kg}$$

and their rotational inertia is

$$I_{pc} = M_{pc}R^2 = (5.47 \times 10^5 \text{ kg})(38 \text{ m})^2 = 7.90 \times 10^8 \text{ kg}\cdot\text{m}^2.$$

Next we consider the structure of the wheel. Let us assume that the rotational inertia of the structure is due mainly to the circular grid suspending the cars. Further, let us assume that the grid forms a hoop of radius R, with a mass M_{hoop} of 3.0×10^5 kg (half the wheel's mass). From Table 10-2*a*, the rotational inertia of the hoop is

$$I_{hoop} = M_{hoop}R^2 = (3.0 \times 10^5 \text{ kg})(38 \text{ m})^2 = 4.33 \times 10^8 \text{ kg}\cdot\text{m}^2.$$

The combined rotational inertia I of the cars, passengers, and hoop is then

FIG. 11-16 The original Ferris wheel. *(From "Shepp's World's Fair Photographed" by James W. Shepp and Daniel P. Shepp, Globe Publishing Co., Chicago and Philadelphia, 1893)*

$$I = I_{pc} + I_{hoop} = 7.90 \times 10^8 \text{ kg}\cdot\text{m}^2 + 4.33 \times 10^8 \text{ kg}\cdot\text{m}^2 = 1.22 \times 10^9 \text{ kg}\cdot\text{m}^2.$$

Angular speed: To find the rotational speed ω_F, we use Eq. 10-5 ($\omega_{avg} = \Delta\theta/\Delta t$). Here the wheel goes through an angular displacement of $\Delta\theta = 2\pi$ rad in a time period $\Delta t = 2$ min. Thus, we have

$$\omega_F = \frac{2\pi \text{ rad}}{(2 \text{ min})(60 \text{ s/min})} = 0.0524 \text{ rad/s}.$$

Angular momentum: Now we can find the magnitude L of the angular momentum with Eq. 11-31:

$$L = I\omega_F = (1.22 \times 10^9 \text{ kg}\cdot\text{m}^2)(0.0524 \text{ rad/s}) = 6.39 \times 10^7 \text{ kg}\cdot\text{m}^2/\text{s} \approx 6.4 \times 10^7 \text{ kg}\cdot\text{m}^2/\text{s}. \quad \text{(Answer)}$$

(b) If the fully loaded wheel is rotated from rest to ω_F in a time period $\Delta t_1 = 5.0$ s, what is the magnitude τ_{avg} of the average net external torque acting on it?

KEY IDEA The average net external torque is related

to the change ΔL in the angular momentum of the loaded wheel by Eq. 11-29 ($\vec{\tau}_{net} = d\vec{L}/dt$).

Calculation: Because the wheel rotates about a fixed axis to reach angular speed ω_F in time period Δt_1, we can rewrite Eq. 11-29 as $\tau_{avg} = \Delta L/\Delta t_1$. The change ΔL is from zero to the answer for part (a). Thus, we have

$$\tau_{avg} = \frac{\Delta L}{\Delta t_1} = \frac{6.39 \times 10^7 \text{ kg}\cdot\text{m}^2/\text{s} - 0}{5.0 \text{ s}}$$

$$\approx 1.3 \times 10^7 \text{ N}\cdot\text{m}. \qquad \text{(Answer)}$$

11-11 | Conservation of Angular Momentum

So far we have discussed two powerful conservation laws, the conservation of energy and the conservation of linear momentum. Now we meet a third law of this type, involving the conservation of angular momentum. We start from Eq. 11-29 ($\vec{\tau}_{net} = d\vec{L}/dt$), which is Newton's second law in angular form. If no net external torque acts on the system, this equation becomes $d\vec{L}/dt = 0$, or

$$\vec{L} = \text{a constant} \quad \text{(isolated system)}. \tag{11-32}$$

This result, called the **law of conservation of angular momentum,** can also be written as

$$\begin{pmatrix}\text{net angular momentum}\\ \text{at some initial time } t_i\end{pmatrix} = \begin{pmatrix}\text{net angular momentum}\\ \text{at some later time } t_f\end{pmatrix},$$

or

$$\vec{L}_i = \vec{L}_f \quad \text{(isolated system)}. \tag{11-33}$$

Equations 11-32 and 11-33 tell us:

> If the net external torque acting on a system is zero, the angular momentum $\vec{L}$ of the system remains constant, no matter what changes take place within the system.

Equations 11-32 and 11-33 are vector equations; as such, they are equivalent to three component equations corresponding to the conservation of angular momentum in three mutually perpendicular directions. Depending on the torques acting on a system, the angular momentum of the system might be conserved in only one or two directions but not in all directions:

> If the component of the net *external* torque on a system along a certain axis is zero, then the component of the angular momentum of the system along that axis cannot change, no matter what changes take place within the system.

We can apply this law to the isolated body in Fig. 11-15, which rotates around the z axis. Suppose that the initially rigid body somehow redistributes its mass relative to that rotation axis, changing its rotational inertia about that axis. Equations 11-32 and 11-33 state that the angular momentum of the body cannot change. Substituting Eq. 11-31 (for the angular momentum along the rotational axis) into Eq. 11-33, we write this conservation law as

$$I_i\omega_i = I_f\omega_f. \tag{11-34}$$

Here the subscripts refer to the values of the rotational inertia I and angular speed ω before and after the redistribution of mass.

Like the other two conservation laws that we have discussed, Eqs. 11-32 and 11-33 hold beyond the limitations of Newtonian mechanics. They hold for particles whose speeds approach that of light (where the theory of special relativity reigns), and they remain true in the world of subatomic particles (where quantum physics reigns). No exceptions to the law of conservation of angular momentum have ever been found.

We now discuss four examples involving this law.

1. *The spinning volunteer* Figure 11-17 shows a student seated on a stool that

(a)

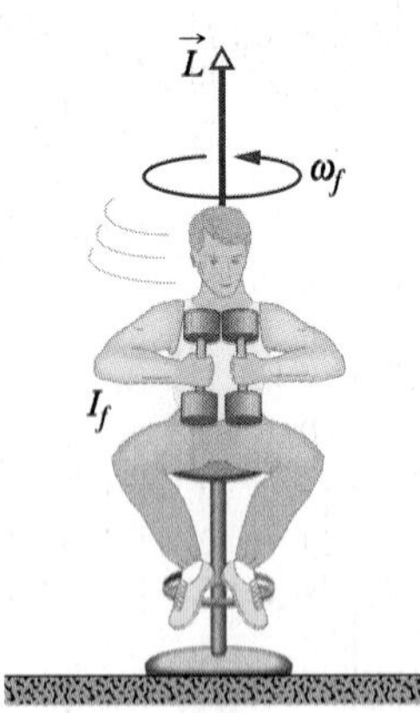

(b)

FIG. 11-17 (*a*) The student has a relatively large rotational inertia about the rotation axis and a relatively small angular speed. (*b*) By decreasing his rotational inertia, the student automatically increases his angular speed. The angular momentum $\vec{L}$ of the rotating system remains unchanged.

can rotate freely about a vertical axis. The student, who has been set into rotation at a modest initial angular speed ω_i, holds two dumbbells in his outstretched hands. His angular momentum vector $\vec{L}$ lies along the vertical rotation axis, pointing upward.

The instructor now asks the student to pull in his arms; this action reduces his rotational inertia from its initial value I_i to a smaller value I_f because he moves mass closer to the rotation axis. His rate of rotation increases markedly, from ω_i to ω_f. The student can then slow down by extending his arms once more, moving the dumbbells outward.

No net external torque acts on the system consisting of the student, stool, and dumbbells. Thus, the angular momentum of that system about the rotation axis must remain constant, no matter how the student maneuvers the dumbbells. In Fig. 11-17*a*, the student's angular speed ω_i is relatively low and his rotational inertia I_i is relatively high. According to Eq. 11-34, his angular speed in Fig. 11-17*b* must be greater to compensate for the decreased I_f.

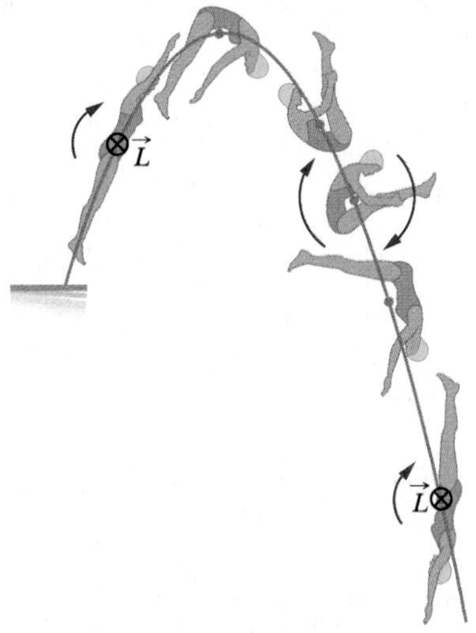

FIG. 11-18 The diver's angular momentum $\vec{L}$ is constant throughout the dive, being represented by the tail ⊗ of an arrow that is perpendicular to the plane of the figure. Note also that her center of mass (see the dots) follows a parabolic path.

2. ***The springboard diver*** Figure 11-18 shows a diver doing a forward one-and-a-half-somersault dive. As you should expect, her center of mass follows a parabolic path. She leaves the springboard with a definite angular momentum $\vec{L}$ about an axis through her center of mass, represented by a vector pointing into the plane of Fig. 11-18, perpendicular to the page. When she is in the air, no net external torque acts on her about her center of mass, so her angular momentum about her center of mass cannot change. By pulling her arms and legs into the closed *tuck position,* she can considerably reduce her rotational inertia about the same axis and thus, according to Eq. 11-34, considerably increase her angular speed. Pulling out of the tuck position (into the *open layout position*) at the end of the dive increases her rotational inertia and thus slows her rotation rate so she can enter the water with little splash. Even in a more complicated dive involving both twisting and somersaulting, the angular momentum of the diver must be conserved, in both magnitude *and* direction, throughout the dive.

FIG. 11-19 Windmill motion of the arms during a long jump helps maintain body orientation for a proper landing.

3. ***Long jump*** When an athlete takes off from the ground in a running long jump, the forces on the launching foot give the athlete an angular momentum with a forward rotation around a horizontal axis. Such rotation would not allow the jumper to land properly: In the landing, the legs should be together and extended forward at an angle so that the heels mark the sand at the greatest distance. Once airborne, the angular momentum cannot change (it is conserved) because no external torque acts to change it. However, the jumper can shift most of the angular momentum to the arms by rotating them in windmill fashion (Fig. 11-19). Then the body remains upright and in the proper orientation for landing.

4. ***Tour jeté*** In a tour jeté, a ballet performer leaps with a small twisting motion on the floor with one foot while holding the other leg perpendicular to the body (Fig. 11-20*a*). The angular speed is so small that it may not be perceptible to the audience. As the performer ascends, the outstretched leg is brought down and the other leg is brought up, with both ending up at angle θ to the

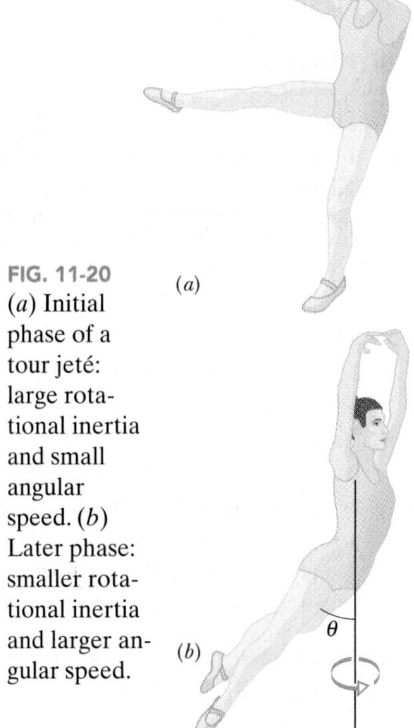

FIG. 11-20 (*a*) Initial phase of a tour jeté: large rotational inertia and small angular speed. (*b*) Later phase: smaller rotational inertia and larger angular speed.

body (Fig. 11-20*b*). The motion is graceful, but it also serves to increase the rotation because bringing in the initially outstretched leg decreases the performer's rotational inertia. Since no external torque acts on the airborne performer, the angular momentum cannot change. Thus, with a decrease in rotational inertia, the angular speed must increase. When the jump is well executed, the performer seems to suddenly begin to spin and rotates 180° before the initial leg orientations are reversed in preparation for the landing. Once a leg is again outstretched, the rotation seems to vanish.

CHECKPOINT 7 A rhinoceros beetle rides the rim of a small disk that rotates like a merry-go-round. If the beetle crawls toward the center of the disk, do the following (each relative to the central axis) increase, decrease, or remain the same for the beetle–disk system: (a) rotational inertia, (b) angular momentum, and (c) angular speed?

Sample Problem 11-7

Figure 11-21*a* shows a student, again sitting on a stool that can rotate freely about a vertical axis. The student, initially at rest, is holding a bicycle wheel whose rim is loaded with lead and whose rotational inertia I_{wh} about its central axis is 1.2 kg · m^2. The wheel is rotating at an angular speed ω_{wh} of 3.9 rev/s; as seen from overhead, the rotation is counterclockwise. The axis of the wheel is vertical, and the angular momentum $\vec{L}_{wh}$ of the wheel points vertically upward. The student now inverts the wheel (Fig. 11-21*b*) so that, as seen from overhead, it is rotating clockwise. Its angular momentum is now $-\vec{L}_{wh}$. The inversion results in the student, the stool, and the wheel's center rotating together as a composite rigid body about the stool's rotation axis, with rotational inertia $I_b = 6.8$ kg · m^2. (The fact that the wheel is also rotating about its center does not affect the mass distribution of this composite body; thus, I_b has the same value whether or not the wheel rotates.) With what angular speed ω_b and in what direction does the composite body rotate after the inversion of the wheel?

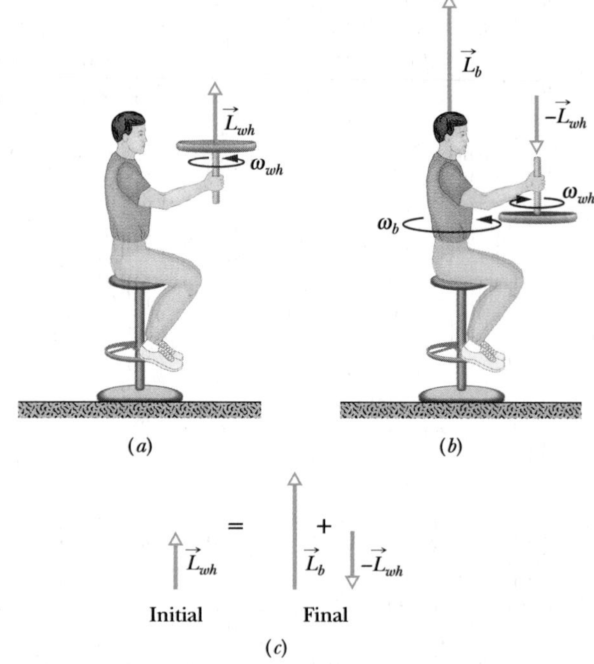

FIG. 11-21 (*a*) A student holds a bicycle wheel rotating around a vertical axis. (*b*) The student inverts the wheel, setting himself into rotation. (*c*) The net angular momentum of the system must remain the same in spite of the inversion.

KEY IDEAS

1. The angular speed ω_b we seek is related to the final angular momentum $\vec{L}_b$ of the composite body about the stool's rotation axis by Eq. 11-31 ($L = I\omega$).
2. The initial angular speed ω_{wh} of the wheel is related to the angular momentum $\vec{L}_{wh}$ of the wheel's rotation about its center by the same equation.
3. The vector addition of $\vec{L}_b$ and $\vec{L}_{wh}$ gives the total angular momentum $\vec{L}_{tot}$ of the system of student, stool, and wheel.
4. As the wheel is inverted, no net *external* torque acts on that system to change $\vec{L}_{tot}$ about any vertical axis. (Torques due to forces between the student and the wheel as the student inverts the wheel are *internal* to the system.) So, the system's total angular momentum is conserved about any vertical axis.

Calculations: The conservation of $\vec{L}_{tot}$ is represented with vectors in Fig. 11-21*c*. We can also write this conservation in terms of components along a vertical axis as

$$L_{b,f} + L_{wh,f} = L_{b,i} + L_{wh,i}, \qquad (11\text{-}35)$$

where i and f refer to the initial state (before inversion of the wheel) and the final state (after inversion). Because inversion of the wheel inverted the angular momentum vector of the wheel's rotation, we substitute $-L_{wh,i}$ for $L_{wh,f}$. Then, if we set $L_{b,i} = 0$ (because the student, the stool, and the wheel's center were initially at rest), Eq. 11-35 yields

$$L_{b,f} = 2L_{wh,i}.$$

Using Eq. 11-31, we next substitute $I_b\omega_b$ for $L_{b,f}$ and

$I_{wh}\omega_{wh}$ for $L_{wh,i}$ and solve for ω_b, finding

$$\omega_b = \frac{2I_{wh}}{I_b}\,\omega_{wh}$$

$$= \frac{(2)(1.2\text{ kg}\cdot\text{m}^2)(3.9\text{ rev/s})}{6.8\text{ kg}\cdot\text{m}^2} = 1.4\text{ rev/s.} \quad \text{(Answer)}$$

This positive result tells us that the student rotates counterclockwise about the stool axis as seen from overhead. If the student wishes to stop rotating, he has only to invert the wheel once more.

Sample Problem 11-8

In Fig. 11-22, a cockroach with mass m rides on a disk of mass $6.00m$ and radius R. The disk rotates like a merry-go-round around its central axis at angular speed $\omega_i = 1.50$ rad/s. The cockroach is initially at radius $r = 0.800R$, but then it crawls out to the rim of the disk. Treat the cockroach as a particle. What then is the angular speed?

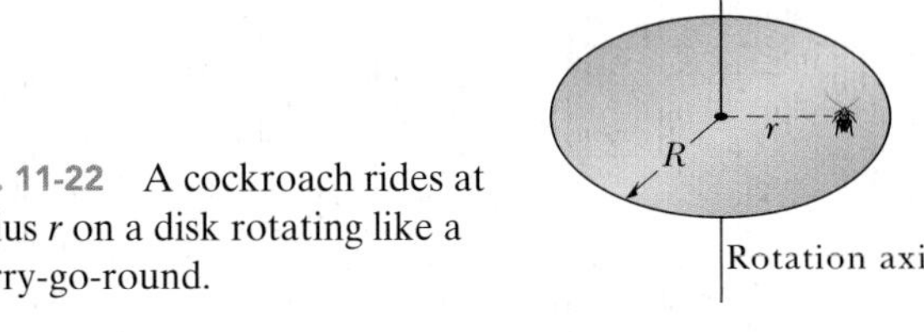

FIG. 11-22 A cockroach rides at radius r on a disk rotating like a merry-go-round.

KEY IDEAS (1) The cockroach's crawl changes the mass distribution (and thus the rotational inertia) of the cockroach–disk system. (2) The angular momentum of the system does not change because there is no external torque to change it. (The forces and torques due to the cockroach's crawl are internal to the system.) (3) The magnitude of the angular momentum of a rigid body or a particle is given by Eq. 11-31 ($L = I\omega$).

Calculations: We want to find the final angular speed. Our key is to equate the final angular momentum L_f to the initial angular momentum L_i, because both involve angular speed. They also involve rotational inertia I. So, let's start by finding the rotational inertia of the system of cockroach and disk before and after the crawl.

The rotational inertia of a disk rotating about its central axis is given by Table 10-2*c* as $\frac{1}{2}MR^2$. Substituting $6.00m$ for the mass M, our disk here has rotational inertia

$$I_d = 3.00mR^2. \qquad (11\text{-}36)$$

(We don't have values for m and R, but we shall continue with physics courage.)

From Eq. 10-33, we know that the rotational inertia of the cockroach (a particle) is equal to mr^2. Substituting the cockroach's initial radius ($r = 0.800R$) and final radius ($r = R$), we find that its initial rotational inertia about the rotation axis is

$$I_{ci} = 0.64mR^2 \qquad (11\text{-}37)$$

and its final rotational inertia about the rotation axis is

$$I_{cf} = mR^2. \qquad (11\text{-}38)$$

So, the cockroach–disk system initially has the rotational inertia

$$I_i = I_d + I_{ci} = 3.64mR^2, \qquad (11\text{-}39)$$

and finally has the rotational inertia

$$I_f = I_d + I_{cf} = 4.00mR^2. \qquad (11\text{-}40)$$

Next, we use Eq. 11-31 ($L = I\omega$) to write the fact that the system's final angular momentum L_f is equal to the system's initial angular momentum L_i:

$$I_f\omega_f = I_i\omega_i$$

or

$$4.00mR^2\omega_f = 3.64mR^2(1.50\text{ rad/s}).$$

After canceling the unknowns m and R, we come to

$$\omega_f = 1.37\text{ rad/s.} \qquad \text{(Answer)}$$

Note that the angular speed decreased because part of the mass moved outward from the rotation axis.

11-12 | Precession of a Gyroscope

A simple gyroscope consists of a wheel fixed to a shaft and free to spin about the axis of the shaft. If one end of the shaft of a *nonspinning* gyroscope is placed on a support as in Fig. 11-23*a* and the gyroscope is released, the gyroscope falls by rotating downward about the tip of the support. Since the fall involves rotation, it is governed by Newton's second law in angular form, which is given by Eq. 11-29:

$$\vec{\tau} = \frac{d\vec{L}}{dt}. \qquad (11\text{-}41)$$

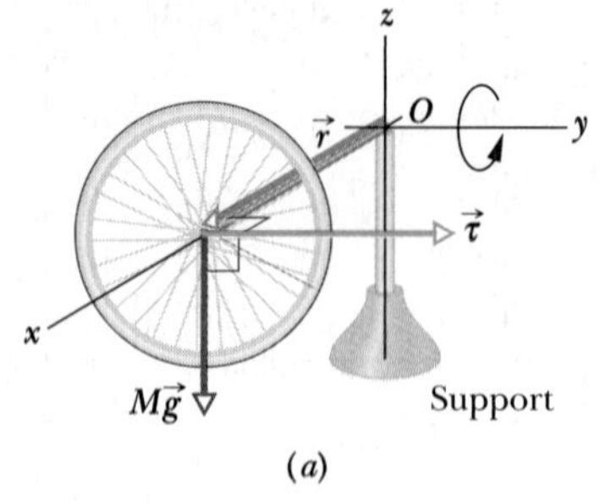

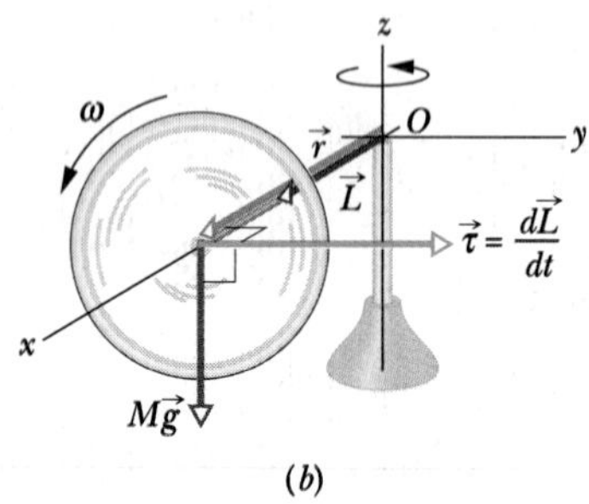

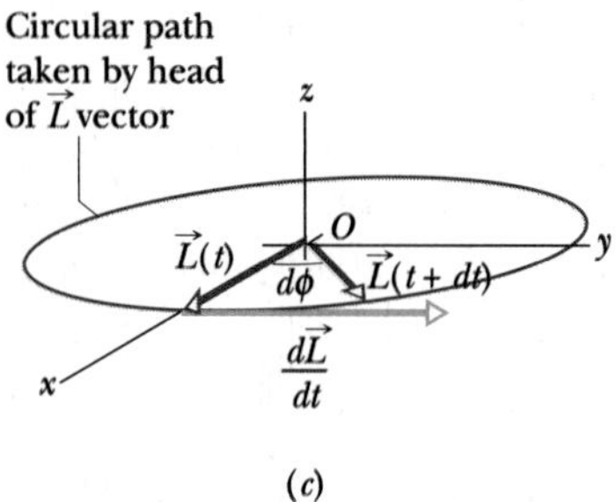

FIG. 11-23 (*a*) A nonspinning gyroscope falls by rotating in an *xz* plane because of torque $\vec{\tau}$. (*b*) A rapidly spinning gyroscope, with angular momentum $\vec{L}$, precesses around the *z* axis. Its precessional motion is in the *xy* plane. (*c*) The change $d\vec{L}/dt$ in angular momentum leads to a rotation of $\vec{L}$ about *O*.

This equation tells us that the torque causing the downward rotation (the fall) changes the angular momentum $\vec{L}$ of the gyroscope from its initial value of zero. The torque $\vec{\tau}$ is due to the gravitational force $M\vec{g}$ acting at the gyroscope's center of mass, which we take to be at the center of the wheel. The moment arm relative to the support tip, located at O in Fig. 11-23*a*, is $\vec{r}$. The magnitude of $\vec{\tau}$ is

$$\tau = Mgr \sin 90° = Mgr \tag{11-42}$$

(because the angle between $M\vec{g}$ and $\vec{r}$ is 90°), and its direction is as shown in Fig. 11-23*a*.

A rapidly spinning gyroscope behaves differently. Assume it is released with the shaft angled slightly upward. It first rotates slightly downward but then, while it is still spinning about its shaft, it begins to rotate horizontally about a vertical axis through support point O in a motion called **precession.**

Why does the spinning gyroscope stay aloft instead of falling over like the nonspinning gyroscope? The clue is that when the spinning gyroscope is released, the torque due to $M\vec{g}$ must change not an initial angular momentum of zero but rather some already existing nonzero angular momentum due to the spin.

To see how this nonzero initial angular momentum leads to precession, we first consider the angular momentum $\vec{L}$ of the gyroscope due to its spin. To simplify the situation, we assume the spin rate is so rapid that the angular momentum due to precession is negligible relative to $\vec{L}$. We also assume the shaft is horizontal when precession begins, as in Fig. 11-23*b*. The magnitude of $\vec{L}$ is given by Eq. 11-31:

$$L = I\omega, \tag{11-43}$$

where I is the rotational moment of the gyroscope about its shaft and ω is the angular speed at which the wheel spins about the shaft. The vector $\vec{L}$ points along the shaft, as in Fig. 11-23*b*. Since $\vec{L}$ is parallel to $\vec{r}$, torque $\vec{\tau}$ must be perpendicular to $\vec{L}$.

According to Eq. 11-41, torque $\vec{\tau}$ causes an incremental change $d\vec{L}$ in the angular momentum of the gyroscope in an incremental time interval dt; that is,

$$d\vec{L} = \vec{\tau}\,dt. \tag{11-44}$$

However, for a *rapidly spinning* gyroscope, the magnitude of $\vec{L}$ is fixed by Eq. 11-43. Thus the torque can change only the direction of $\vec{L}$, not its magnitude.

From Eq. 11-44 we see that the direction of $d\vec{L}$ is in the direction of $\vec{\tau}$, perpendicular to $\vec{L}$. The only way that $\vec{L}$ can be changed in the direction of $\vec{\tau}$ without the magnitude L being changed is for $\vec{L}$ to rotate around the z axis as shown in Fig. 11-23*c*. $\vec{L}$ maintains its magnitude, the head of the $\vec{L}$ vector follows a circular path, and $\vec{\tau}$ is always tangent to that path. Since $\vec{L}$ must always point along the shaft, the shaft must rotate about the z axis in the direction of $\vec{\tau}$. Thus we have precession. Because the spinning gyroscope must obey Newton's law in angular form in response to any change in its initial angular momentum, it must precess instead of merely toppling over.

We can find the **precession rate** Ω by first using Eqs. 11-44 and 11-42 to get the magnitude of $d\vec{L}$:

$$dL = \tau\,dt = Mgr\,dt. \tag{11-45}$$

As $\vec{L}$ changes by an incremental amount in an incremental time interval dt, the shaft and $\vec{L}$ precess around the z axis through incremental angle $d\phi$. (In Fig. 11-23*c*, angle $d\phi$ is exaggerated for clarity.) With the aid of Eqs. 11-43 and 11-45, we find that $d\phi$ is given by

$$d\phi = \frac{dL}{L} = \frac{Mgr\,dt}{I\omega}.$$

Dividing this expression by dt and setting the rate $\Omega = d\phi/dt$, we obtain

$$\Omega = \frac{Mgr}{I\omega} \quad \text{(precession rate).} \tag{11-46}$$

This result is valid under the assumption that the spin rate ω is rapid. Note that Ω decreases as ω is increased. Note also that there would be no precession if the gravitational force $M\vec{g}$ did not act on the gyroscope, but because I is a function of M, mass cancels from Eq. 11-46; thus Ω is independent of the mass.

Equation 11-46 also applies if the shaft of a spinning gyroscope is at an angle to the horizontal. It holds as well for a spinning top, which is essentially a spinning gyroscope at an angle to the horizontal.

REVIEW & SUMMARY

Rolling Bodies For a wheel of radius R rolling smoothly,

$$v_{com} = \omega R, \tag{11-2}$$

where v_{com} is the linear speed of the wheel's center of mass and ω is the angular speed of the wheel about its center. The wheel may also be viewed as rotating instantaneously about the point P of the "road" that is in contact with the wheel. The angular speed of the wheel about this point is the same as the angular speed of the wheel about its center. The rolling wheel has kinetic energy

$$K = \tfrac{1}{2}I_{com}\omega^2 + \tfrac{1}{2}Mv_{com}^2, \tag{11-5}$$

where I_{com} is the rotational moment of the wheel about its center of mass and M is the mass of the wheel. If the wheel is being accelerated but is still rolling smoothly, the acceleration of the center of mass $\vec{a}_{com}$ is related to the angular acceleration α about the center with

$$a_{com} = \alpha R. \tag{11-6}$$

If the wheel rolls smoothly down a ramp of angle θ, its acceleration along an x axis extending up the ramp is

$$a_{com,x} = -\frac{g\sin\theta}{1 + I_{com}/MR^2}. \tag{11-10}$$

Torque as a Vector In three dimensions, *torque* $\vec{\tau}$ is a vector quantity defined relative to a fixed point (usually an origin); it is

$$\vec{\tau} = \vec{r} \times \vec{F}, \tag{11-14}$$

where $\vec{F}$ is a force applied to a particle and $\vec{r}$ is a position vector locating the particle relative to the fixed point. The magnitude of $\vec{\tau}$ is given by

$$\tau = rF\sin\phi = rF_\perp = r_\perp F, \qquad (11\text{-}15, 11\text{-}16, 11\text{-}17)$$

where ϕ is the angle between $\vec{F}$ and $\vec{r}$, $F_\perp$ is the component of $\vec{F}$ perpendicular to $\vec{r}$, and $r_\perp$ is the moment arm of $\vec{F}$. The direction of $\vec{\tau}$ is given by the right-hand rule.

Angular Momentum of a Particle The *angular momentum* $\vec{\ell}$ of a particle with linear momentum $\vec{p}$, mass m, and linear velocity $\vec{v}$ is a vector quantity defined relative to a fixed point (usually an origin) as

$$\vec{\ell} = \vec{r} \times \vec{p} = m(\vec{r} \times \vec{v}). \tag{11-18}$$

The magnitude of $\vec{\ell}$ is given by

$$\ell = rmv\sin\phi \tag{11-19}$$

$$= rp_\perp = rmv_\perp \tag{11-20}$$

$$= r_\perp p = r_\perp mv, \tag{11-21}$$

where ϕ is the angle between $\vec{r}$ and $\vec{p}$, $p_\perp$ and $v_\perp$ are the components of $\vec{p}$ and $\vec{v}$ perpendicular to $\vec{r}$, and $r_\perp$ is the perpendicular distance between the fixed point and the extension of $\vec{p}$. The direction of $\vec{\ell}$ is given by the right-hand rule for cross products.

Newton's Second Law in Angular Form Newton's second law for a particle can be written in angular form as

$$\vec{\tau}_{net} = \frac{d\vec{\ell}}{dt}, \tag{11-23}$$

where $\vec{\tau}_{net}$ is the net torque acting on the particle and $\vec{\ell}$ is the angular momentum of the particle.

Angular Momentum of a System of Particles The angular momentum $\vec{L}$ of a system of particles is the vector sum of the angular momenta of the individual particles:

$$\vec{L} = \vec{\ell}_1 + \vec{\ell}_2 + \cdots + \vec{\ell}_n = \sum_{i=1}^{n} \vec{\ell}_i. \tag{11-26}$$

The time rate of change of this angular momentum is equal to the net external torque on the system (the vector sum of the torques due to interactions of the particles of the system with particles external to the system):

$$\vec{\tau}_{net} = \frac{d\vec{L}}{dt} \quad \text{(system of particles)}. \tag{11-29}$$

Angular Momentum of a Rigid Body For a rigid body rotating about a fixed axis, the component of its angular momentum parallel to the rotation axis is

$$L = I\omega \quad \text{(rigid body, fixed axis)}. \tag{11-31}$$

Conservation of Angular Momentum The angular momentum $\vec{L}$ of a system remains constant if the net external torque acting on the system is zero:

$$\vec{L} = \text{a constant} \quad \text{(isolated system)} \tag{11-32}$$

or

$$\vec{L}_i = \vec{L}_f \quad \text{(isolated system)}. \tag{11-33}$$

This is the **law of conservation of angular momentum.**

Precession of a Gyroscope A spinning gyroscope can precess about a vertical axis through its support at the rate

$$\Omega = \frac{Mgr}{I\omega}, \tag{11-46}$$

where M is the gyroscope's mass, r is the moment arm, I is the rotational inertia, and ω is the spin rate.

QUESTIONS

1 In Fig. 11-24, three forces of the same magnitude are applied to a particle at the origin ($\vec{F}_1$ acts directly into the plane of the figure). Rank the forces according to the magnitudes of the torques they create about (a) point P_1, (b) point P_2, and (c) point P_3, greatest first.

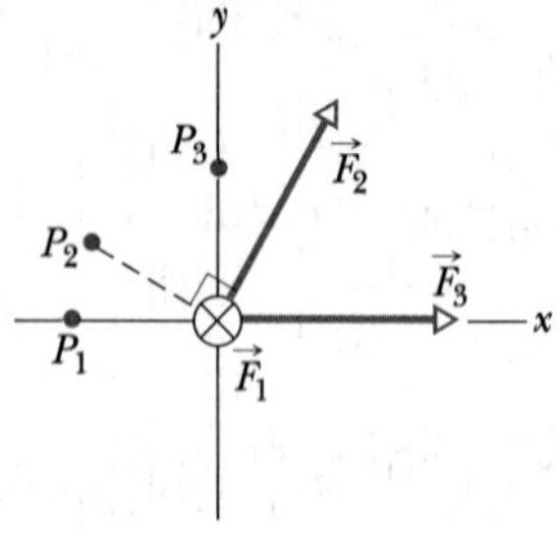

FIG. 11-24 Question 1.

2 The position vector $\vec{r}$ of a particle relative to a certain point has a magnitude of 3 m, and the force $\vec{F}$ on the particle has a magnitude of 4 N. What is the angle between the directions of $\vec{r}$ and $\vec{F}$ if the magnitude of the associated torque equals (a) zero and (b) 12 N·m?

3 What happens to the initially stationary yo-yo in Fig. 11-25 if you pull it via its string with (a) force $\vec{F}_2$ (the line of action passes through the point of contact on the table, as indicated), (b) force $\vec{F}_1$ (the line of action passes above the point of contact), and (c) force $\vec{F}_3$ (the line of action passes to the right of the point of contact)?

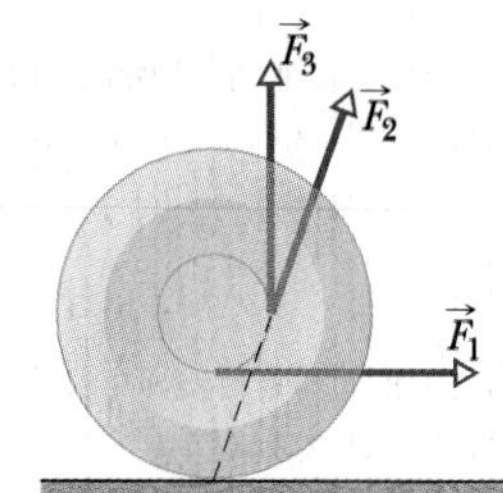

FIG. 11-25 Question 3.

4 Figure 11-26 shows two particles A and B at xyz coordinates (1 m, 1 m, 0) and (1 m, 0, 1 m). Acting on each particle are three numbered forces, all of the same magnitude and each directed parallel to an axis. (a) Which of the forces produce a torque about the origin that is directed parallel to y? (b) Rank the forces according to the magnitudes of the torques they produce on the particles about the origin, greatest first.

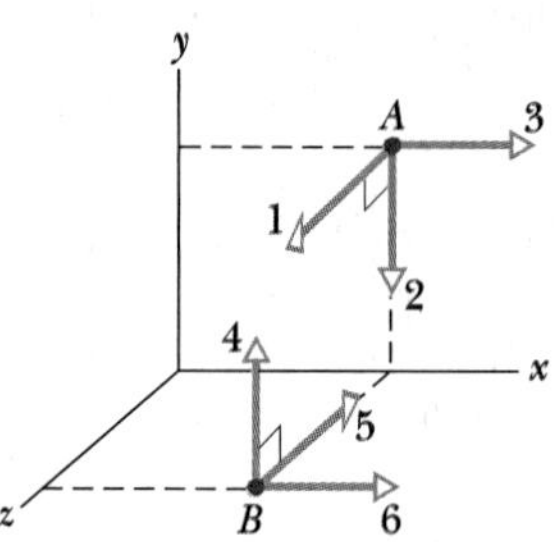

FIG. 11-26 Question 4.

5 Figure 11-27 shows three particles of the same mass and the same constant speed moving as indicated by the velocity vectors. Points a, b, c, and d form a square, with point e at the center. Rank the points according to the magnitude of the net angular momentum of the three-particle system when measured about the points, greatest first.

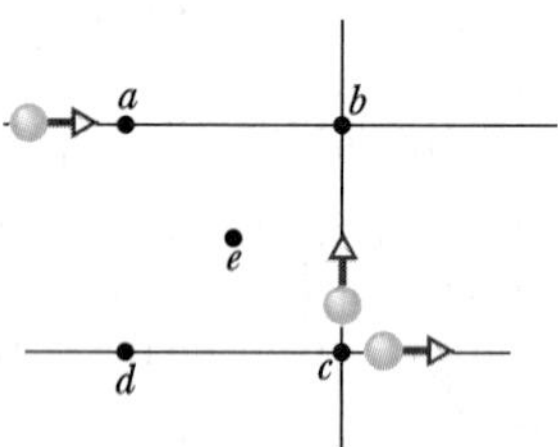

FIG. 11-27 Question 5.

6 Figure 11-28 shows a particle moving at constant velocity $\vec{v}$ and five points with their xy coordinates. Rank the points according to the magnitude of the angular momentum of the particle measured about them, greatest first.

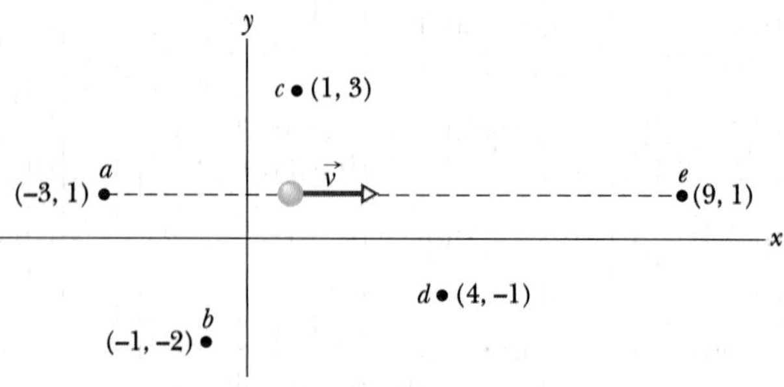

FIG. 11-28 Question 6.

7 Figure 11-29 gives the angular momentum magnitude L of a wheel versus time t. Rank the four lettered time intervals according to the magnitude of the torque acting on the wheel, greatest first.

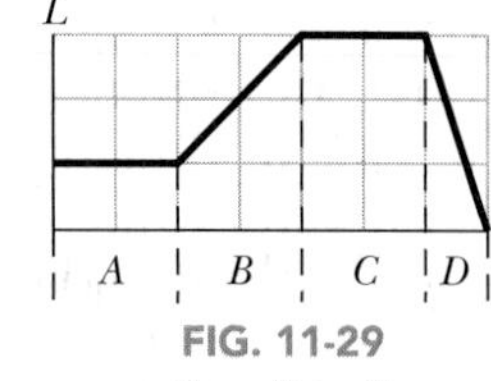

FIG. 11-29 Question 7.

8 The angular momenta $\ell(t)$ of a particle in four situations are (1) $\ell = 3t + 4$; (2) $\ell = -6t^2$; (3) $\ell = 2$; (4) $\ell = 4/t$. In which situation is the net torque on the particle (a) zero, (b) positive and constant, (c) negative and increasing in magnitude ($t > 0$), and (d) negative and decreasing in magnitude ($t > 0$)?

9 A rhinoceros beetle rides the rim of a horizontal disk rotating counterclockwise like a merry-go-round. If the beetle then walks along the rim in the direction of the rotation, will the magnitudes of the following quantities (each measured about the rotation axis) increase, decrease, or remain the same (the disk is still rotating in the counterclockwise direction): (a) the angular momentum of the beetle–disk system, (b) the angular momentum and angular velocity of the beetle, and (c) the angular momentum and angular velocity of the disk? (d) What are your answers if the beetle walks in the direction opposite the rotation?

10 Figure 11-30 shows an overhead view of a rectangular slab that can spin like a merry-go-round about its center at O. Also shown are seven paths along which wads of bubble gum can be thrown (all with the same speed and mass) to stick onto the stationary slab. (a) Rank the paths according to the angular speed that the slab (and gum) will have after the gum sticks, greatest first. (b) For which paths will the angular momentum of the slab (and gum) about O be negative from the view of Fig. 11-30?

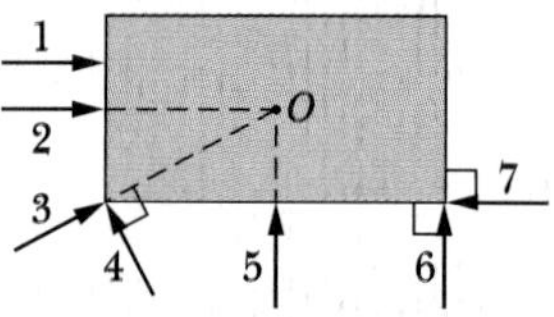

FIG. 11-30 Question 10.

PROBLEMS

GO Tutoring problem available (at instructor's discretion) in *WileyPLUS* and WebAssign
SSM Worked-out solution available in Student Solutions Manual
WWW Worked-out solution is at
• – ••• Number of dots indicates level of problem difficulty
ILW Interactive solution is at
http://www.wiley.com/college/halliday
Additional information available in *The Flying Circus of Physics* and at flyingcircusofphysics.com

sec. 11-2 Rolling as Translation and Rotation Combined

•1 A car travels at 80 km/h on a level road in the positive direction of an x axis. Each tire has a diameter of 66 cm. Relative to a woman riding in the car and in unit-vector notation, what are the velocity $\vec{v}$ at the (a) center, (b) top, and (c) bottom of the tire and the magnitude a of the acceleration at the (d) center, (e) top, and (f) bottom of each tire? Relative to a hitchhiker sitting next to the road and in unit-vector notation, what are the velocity $\vec{v}$ at the (g) center, (h) top, and (i) bottom of the tire and the magnitude a of the acceleration at the (j) center, (k) top, and (l) bottom of each tire?

•2 An automobile traveling at 80.0 km/h has tires of 75.0 cm diameter. (a) What is the angular speed of the tires about their axles? (b) If the car is brought to a stop uniformly in 30.0 complete turns of the tires (without skidding), what is the magnitude of the angular acceleration of the wheels? (c) How far does the car move during the braking?

sec. 11-4 The Forces of Rolling

•3 A 1000 kg car has four 10 kg wheels. When the car is moving, what fraction of its total kinetic energy is due to rotation of the wheels about their axles? Assume that the wheels have the same rotational inertia as uniform disks of the same mass and size. Why do you not need to know the radius of the wheels? ILW

•4 A uniform solid sphere rolls down an incline. (a) What must be the incline angle if the linear acceleration of the center of the sphere is to have a magnitude of $0.10g$? (b) If a frictionless block were to slide down the incline at that angle, would its acceleration magnitude be more than, less than, or equal to $0.10g$? Why?

•5 A 140 kg hoop rolls along a horizontal floor so that the hoop's center of mass has a speed of 0.150 m/s. How much work must be done on the hoop to stop it? SSM

••6 A hollow sphere of radius 0.15 m, with rotational inertia $I = 0.040\ \text{kg}\cdot\text{m}^2$ about a line through its center of mass, rolls without slipping up a surface inclined at 30° to the horizontal. At a certain initial position, the sphere's total kinetic energy is 20 J. (a) How much of this initial kinetic energy is rotational? (b) What is the speed of the center of mass of the sphere at the initial position? When the sphere has moved 1.0 m up the incline from its initial position, what are (c) its total kinetic energy and (d) the speed of its center of mass?

••7 In Fig. 11-31, a constant horizontal force $\vec{F}_{app}$ of magnitude 10 N is applied to a wheel of mass 10 kg and radius 0.30 m. The wheel rolls smoothly on the horizontal surface, and the acceleration of its center of mass has magnitude 0.60 m/s². (a) In unit-vector notation, what is the frictional force on the wheel? (b) What is the rotational inertia of the wheel about the rotation axis through its center of mass?

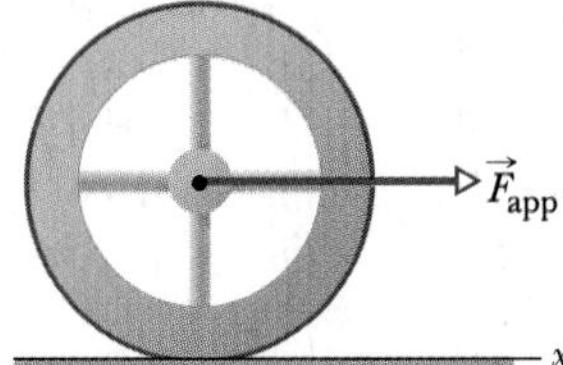

FIG. 11-31 Problem 7.

••8 In Fig. 11-32, a solid brass ball of mass 0.280 g will roll smoothly along a loop-the-loop track when released from rest along the straight section. The circular loop has radius R = 14.0 cm, and the ball has radius $r \ll R$. (a) What is h if the ball is on the verge of leaving the track when it reaches the top of the loop? If the ball is released at height $h = 6.00R$, what are the (b) magnitude and (c) direction of the horizontal force component acting on the ball at point Q?

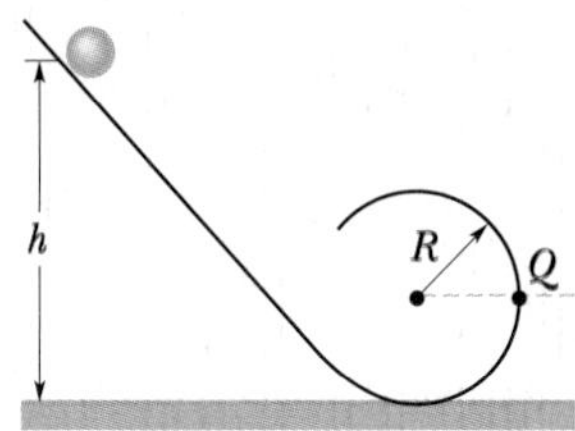

FIG. 11-32 Problem 8.

••9 In Fig. 11-33, a solid cylinder of radius 10 cm and mass 12 kg starts from rest and rolls without slipping a distance L = 6.0 m down a roof that is inclined at the angle $\theta = 30°$. (a) What is the angular speed of the cylinder about its center as it leaves the roof? (b) The roof's edge is at height H = 5.0 m. How far horizontally from the roof's edge does the cylinder hit the level ground? ILW

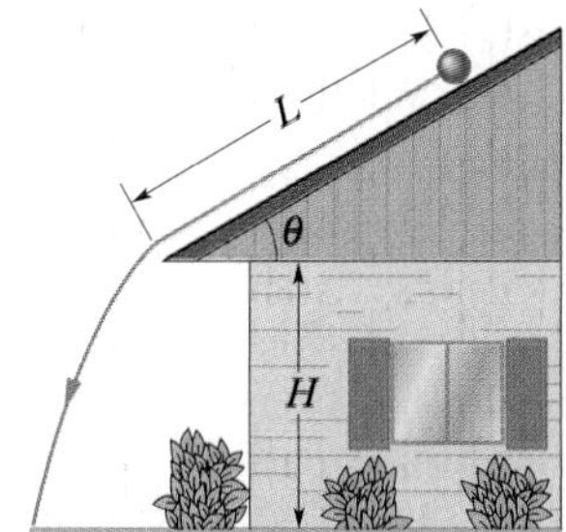

FIG. 11-33 Problem 9.

••10 Figure 11-34 gives the speed v versus time t for a 0.500 kg object of radius 6.00 cm that rolls smoothly down a 30° ramp. The scale on the velocity axis is set by v_s = 4.0 m/s. What is the rotational inertia of the object?

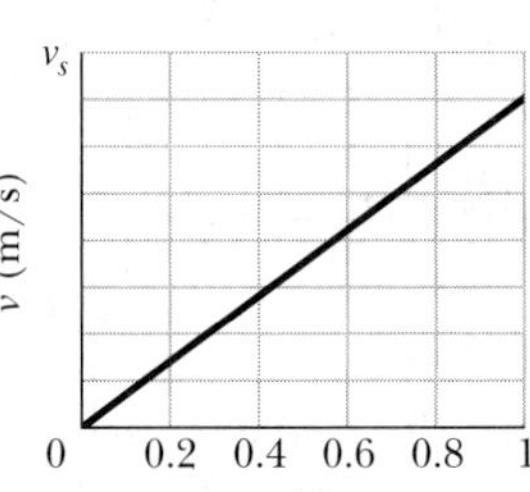

FIG. 11-34 Problem 10.

••11 In Fig. 11-35, a solid ball rolls smoothly from rest (starting at height H = 6.0 m) until it leaves the horizontal section at the end of the track, at height h = 2.0 m. How far horizontally from point A does the ball hit the floor? GO

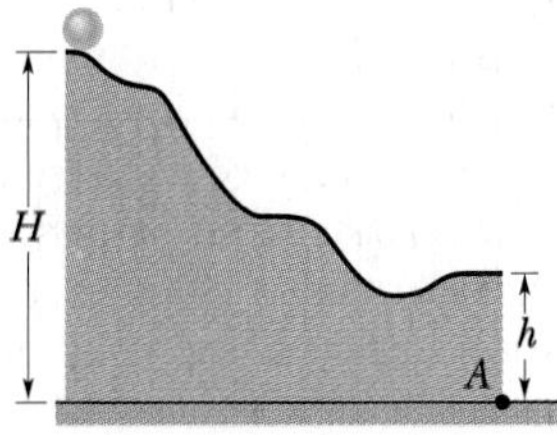

FIG. 11-35 Problem 11.

••12 Figure 11-36 shows the potential energy $U(x)$ of a solid ball that can roll along an x axis. The scale on the U axis is set by U_s = 100 J. The ball is uniform, rolls smoothly, and has a mass of 0.400 kg. It is released at x = 7.0 m headed in the negative direction of the x axis with a mechanical energy of 75 J.

(a) If the ball can reach $x = 0$ m, what is its speed there, and if it cannot, what is its turning point? Suppose, instead, it is headed in the positive direction of the x axis when it is released at $x = 7.0$ m with 75 J. (b) If the ball can reach $x = 13$ m, what is its speed there, and if it cannot, what is its turning point?

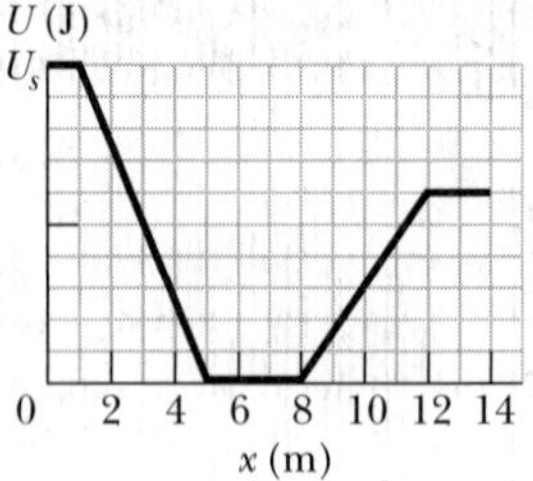

FIG. 11-36 Problem 12.

••13 A bowler throws a bowling ball of radius $R = 11$ cm along a lane. The ball (Fig. 11-37) slides on the lane with initial speed $v_{com,0} = 8.5$ m/s and initial angular speed $\omega_0 = 0$. The coefficient of kinetic friction between the ball and the lane is 0.21. The kinetic frictional force $\vec{f}_k$ acting on the ball causes a linear acceleration of the ball while producing a torque that causes an angular acceleration of the ball. When speed v_{com} has decreased enough and angular speed ω has increased enough, the ball stops sliding and then rolls smoothly. (a) What then is v_{com} in terms of ω? During the sliding, what are the ball's (b) linear acceleration and (c) angular acceleration? (d) How long does the ball slide? (e) How far does the ball slide? (f) What is the linear speed of the ball when smooth rolling begins?

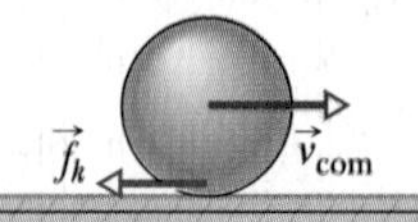

FIG. 11-37 Problem 13.

••14 In Fig. 11-38, a small, solid, uniform ball is to be shot from point P so that it rolls smoothly along a horizontal path, up along a ramp, and onto a plateau. Then it leaves the plateau horizontally to land on a game board, at a horizontal distance d from the right edge of the plateau. The vertical heights are $h_1 = 5.00$ cm and $h_2 = 1.60$ cm. With what speed must the ball be shot at point P for it to land at $d = 6.00$ cm?

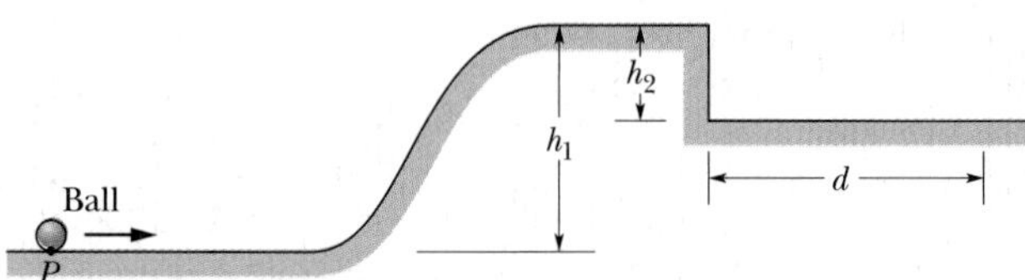

FIG. 11-38 Problem 14.

•••15 *Nonuniform ball.* In Fig. 11-39, a ball of mass M and radius R rolls smoothly from rest down a ramp and onto a circular loop of radius 0.48 m. The initial height of the ball is $h = 0.36$ m. At the loop bottom, the magnitude of the normal force on the ball is $2.00Mg$. The ball consists of an outer spherical shell (of a certain uniform density) that is glued to a central sphere (of a different uniform density). The rotational inertia of the ball can be expressed in the general form $I = \beta MR^2$, but β is not 0.4 as it is for a ball of uniform density. Determine β.

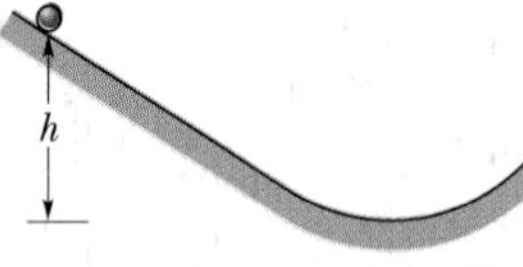

FIG. 11-39 Problem 15.

•••16 *Nonuniform cylindrical object.* In Fig. 11-40, a cylindrical object of mass M and radius R rolls smoothly from rest down a ramp and onto a horizontal section. From there it rolls off the ramp and onto the floor, landing a horizontal distance $d = 0.506$ m from the end of the ramp. The initial height of the object is $H = 0.90$ m; the end of the ramp is at height $h = 0.10$ m. The object consists of an outer cylindrical shell (of a certain uniform density) that is glued to a central cylinder (of a different uniform density). The rotational inertia of the object can be expressed in the general form $I = \beta MR^2$, but β is not 0.5 as it is for a cylinder of uniform density. Determine β.

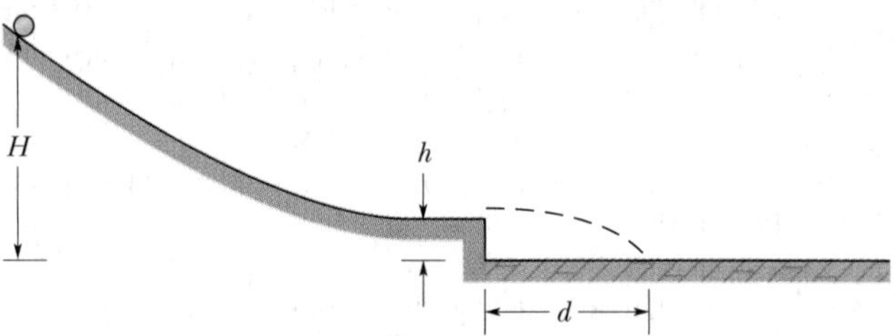

FIG. 11-40 Problem 16.

sec. 11-5 The Yo-Yo

•17 A yo-yo has a rotational inertia of 950 g·cm^2 and a mass of 120 g. Its axle radius is 3.2 mm, and its string is 120 cm long. The yo-yo rolls from rest down to the end of the string. (a) What is the magnitude of its linear acceleration? (b) How long does it take to reach the end of the string? As it reaches the end of the string, what are its (c) linear speed, (d) translational kinetic energy, (e) rotational kinetic energy, and (f) angular speed? SSM

•18 In 1980, over San Francisco Bay, a large yo-yo was released from a crane. The 116 kg yo-yo consisted of two uniform disks of radius 32 cm connected by an axle of radius 3.2 cm. What was the magnitude of the acceleration of the yo-yo during (a) its fall and (b) its rise? (c) What was the tension in the cord on which it rolled? (d) Was that tension near the cord's limit of 52 kN? Suppose you build a scaled-up version of the yo-yo (same shape and materials but larger). (e) Will the magnitude of your yo-yo's acceleration as it falls be greater than, less than, or the same as that of the San Francisco yo-yo? (f) How about the tension in the cord?

sec. 11-6 Torque Revisited

•19 In unit-vector notation, what is the torque about the origin on a particle located at coordinates (0, −4.0 m, 3.0 m) if that torque is due to (a) force $\vec{F}_1$ with components $F_{1x} = 2.0$ N, $F_{1y} = F_{1z} = 0$, and (b) force $\vec{F}_2$ with components $F_{2x} = 0$, $F_{2y} = 2.0$ N, $F_{2z} = 4.0$ N?

•20 A plum is located at coordinates (−2.0 m, 0, 4.0 m). In unit-vector notation, what is the torque about the origin on the plum if that torque is due to a force $\vec{F}$ whose only component is (a) $F_x = 6.0$ N, (b) $F_x = -6.0$ N, (c) $F_z = 6.0$ N, and (d) $F_z = -6.0$ N?

•21 In unit-vector notation, what is the net torque about the origin on a flea located at coordinates (0, −4.0 m, 5.0 m) when forces $\vec{F}_1 = (3.0\text{ N})\hat{k}$ and $\vec{F}_2 = (-2.0\text{ N})\hat{j}$ act on the flea?

••22 In unit-vector notation, what is the torque about the origin on a jar of jalapeño peppers located at coordinates (3.0 m, −2.0 m, 4.0 m) due to (a) force $\vec{F}_1 = (3.0\text{ N})\hat{i} - (4.0\text{ N})\hat{j} + (5.0\text{ N})\hat{k}$, (b) force $\vec{F}_2 = (-3.0\text{ N})\hat{i} - (4.0\text{ N})\hat{j} - (5.0\text{ N})\hat{k}$, and (c) the vector sum of $\vec{F}_1$ and $\vec{F}_2$? (d) Repeat part

(c) for the torque about the point with coordinates (3.0 m, 2.0 m, 4.0 m).

••23 Force $\vec{F} = (-8.0\text{ N})\hat{i} + (6.0\text{ N})\hat{j}$ acts on a particle with position vector $\vec{r} = (3.0\text{ m})\hat{i} + (4.0\text{ m})\hat{j}$. What are (a) the torque on the particle about the origin, in unit-vector notation, and (b) the angle between the directions of $\vec{r}$ and $\vec{F}$? SSM

••24 A particle moves through an xyz coordinate system while a force acts on the particle. When the particle has the position vector $\vec{r} = (2.00\text{ m})\hat{i} - (3.00\text{ m})\hat{j} + (2.00\text{ m})\hat{k}$, the force is $\vec{F} = F_x\hat{i} + (7.00\text{ N})\hat{j} - (6.00\text{ N})\hat{k}$ and the corresponding torque about the origin is $\vec{\tau} = (4.00\text{ N}\cdot\text{m})\hat{i} + (2.00\text{ N}\cdot\text{m})\hat{j} - (1.00\text{ N}\cdot\text{m})\hat{k}$. Determine F_x.

••25 Force $\vec{F} = (2.0\text{ N})\hat{i} - (3.0\text{ N})\hat{k}$ acts on a pebble with position vector $\vec{r} = (0.50\text{ m})\hat{j} - (2.0\text{ m})\hat{k}$ relative to the origin. In unit-vector notation, what is the resulting torque on the pebble about (a) the origin and (b) the point (2.0 m, 0, −3.0 m)?

sec. 11-7 Angular Momentum

•26 A 2.0 kg particle-like object moves in a plane with velocity components $v_x = 30$ m/s and $v_y = 60$ m/s as it passes through the point with (x, y) coordinates of (3.0, −4.0) m. Just then, in unit-vector notation, what is its angular momentum relative to (a) the origin and (b) the point (−2.0, −2.0) m?

•27 In the instant of Fig. 11-41, two particles move in an xy plane. Particle P_1 has mass 6.5 kg and speed $v_1 = 2.2$ m/s, and it is at distance $d_1 = 1.5$ m from point O. Particle P_2 has mass 3.1 kg and speed $v_2 = 3.6$ m/s, and it is at distance $d_2 = 2.8$ m from point O. What are the (a) magnitude and (b) direction of the net angular momentum of the two particles about O? ILW

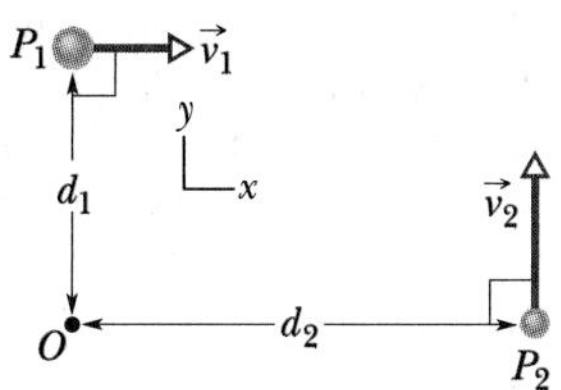

FIG. 11-41 Problem 27.

•28 At the instant of Fig. 11-42, a 2.0 kg particle P has a position vector $\vec{r}$ of magnitude 3.0 m and angle $\theta_1 = 45°$ and a velocity vector $\vec{v}$ of magnitude 4.0 m/s and angle $\theta_2 = 30°$. Force $\vec{F}$, of magnitude 2.0 N and angle $\theta_3 = 30°$, acts on P. All three vectors lie in the xy plane. About the origin, what are the (a) magnitude and (b) direction of the angular momentum of P and the (c) magnitude and (d) direction of the torque acting on P?

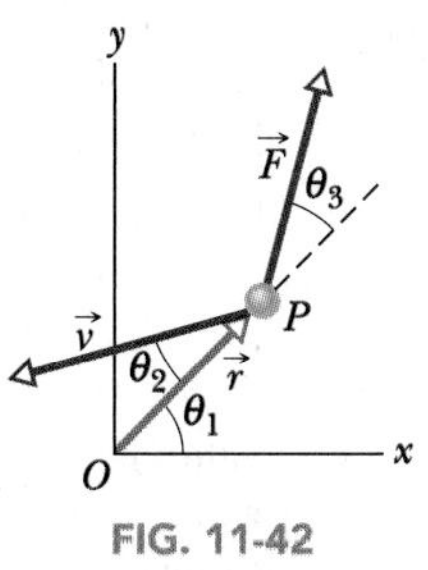

FIG. 11-42 Problem 28.

•29 At one instant, force $\vec{F} = 4.0\hat{j}$ N acts on a 0.25 kg object that has position vector $\vec{r} = (2.0\hat{i} - 2.0\hat{k})$ m and velocity vector $\vec{v} = (-5.0\hat{i} + 5.0\hat{k})$ m/s. About the origin and in unit-vector notation, what are (a) the object's angular momentum and (b) the torque acting on the object? SSM

••30 At the instant the displacement of a 2.00 kg object relative to the origin is $\vec{d} = (2.00\text{ m})\hat{i} + (4.00\text{ m})\hat{j} - (3.00\text{ m})\hat{k}$, its velocity is $\vec{v} = -(6.00\text{ m/s})\hat{i} + (3.00\text{ m/s})\hat{j} + (3.00\text{ m/s})\hat{k}$ and it is subject to a force $\vec{F} = (6.00\text{ N})\hat{i} - (8.00\text{ N})\hat{j} + (4.00\text{ N})\hat{k}$. Find (a) the acceleration of the object, (b) the angular momentum of the object about the origin, (c) the torque about the origin acting on the object, and (d) the angle between the velocity of the object and the force acting on the object.

••31 In Fig. 11-43, a 0.400 kg ball is shot directly upward at initial speed 40.0 m/s. What is its angular momentum about P, 2.00 m horizontally from the launch point, when the ball is (a) at maximum height and (b) halfway back to the ground? What is the torque on the ball about P due to the gravitational force when the ball is (c) at maximum height and (d) halfway back to the ground?

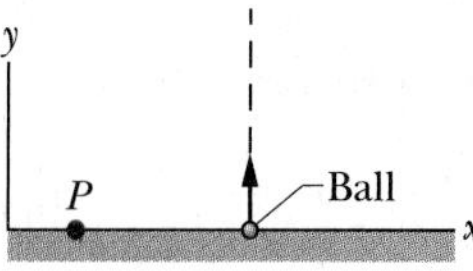

FIG. 11-43 Problem 31.

sec. 11-8 Newton's Second Law in Angular Form

•32 A particle is to move in an xy plane, clockwise around the origin as seen from the positive side of the z axis. In unit-vector notation, what torque acts on the particle if the magnitude of its angular momentum about the origin is (a) 4.0 kg·m²/s, (b) $4.0t^2$ kg·m²/s, (c) $4.0\sqrt{t}$ kg·m²/s, and (d) $4.0/t^2$ kg·m²/s?

•33 A 3.0 kg particle with velocity $\vec{v} = (5.0\text{ m/s})\hat{i} - (6.0\text{ m/s})\hat{j}$ is at $x = 3.0$ m, $y = 8.0$ m. It is pulled by a 7.0 N force in the negative x direction. About the origin, what are (a) the particle's angular momentum, (b) the torque acting on the particle, and (c) the rate at which the angular momentum is changing? SSM ILW WWW

•34 A particle is acted on by two torques about the origin: $\vec{\tau}_1$ has a magnitude of 2.0 N·m and is directed in the positive direction of the x axis, and $\vec{\tau}_2$ has a magnitude of 4.0 N·m and is directed in the negative direction of the y axis. In unit-vector notation, find $d\vec{\ell}/dt$, where $\vec{\ell}$ is the angular momentum of the particle about the origin.

••35 At time t, $\vec{r} = 4.0t^2\hat{i} - (2.0t + 6.0t^2)\hat{j}$ gives the position of a 3.0 kg particle relative to the origin of an xy coordinate system ($\vec{r}$ is in meters and t is in seconds). (a) Find an expression for the torque acting on the particle relative to the origin. (b) Is the magnitude of the particle's angular momentum relative to the origin increasing, decreasing, or unchanging?

sec. 11-10 The Angular Momentum of a Rigid Body Rotating About a Fixed Axis

•36 A sanding disk with rotational inertia 1.2×10^{-3} kg·m² is attached to an electric drill whose motor delivers a torque of magnitude 16 N·m about the central axis of the disk. About that axis and with the torque applied for 33 ms, what is the magnitude of the (a) angular momentum and (b) angular velocity of the disk?

•37 The angular momentum of a flywheel having a rotational inertia of 0.140 kg·m² about its central axis decreases from 3.00 to 0.800 kg·m²/s in 1.50 s. (a) What is the magnitude of the average torque acting on the flywheel about its central axis during this period? (b) Assuming a constant angular acceleration, through what angle does the flywheel turn? (c) How much work is done on the wheel? (d) What is the average power of the flywheel? SSM

•38 Figure 11-44 shows three rotating, uniform disks that are coupled by belts. One belt runs around the rims of disks A and C. Another belt runs around a central hub on disk A and the rim of disk B. The belts move smoothly without slippage on the rims and hub. Disk A has radius R; its hub has radius $0.5000R$; disk B has radius $0.2500R$; and disk C has radius $2.000R$. Disks B and C have the same density (mass per unit volume) and thickness. What is the ratio of the magnitude of the angular momentum of disk C to that of disk B?

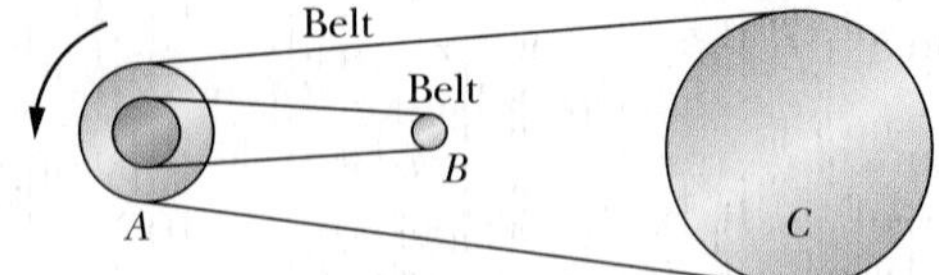

FIG. 11-44 Problem 38.

•**39** In Fig. 11-45, three particles of mass $m = 23$ g are fastened to three rods of length $d = 12$ cm and negligible mass. The rigid assembly rotates around point O at angular speed $\omega = 0.85$ rad/s. About O, what are (a) the rotational inertia of the assembly, (b) the magnitude of the angular momentum of the middle particle, and (c) the magnitude of the angular momentum of the asssembly?

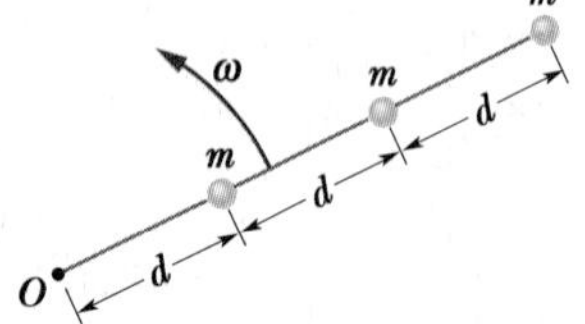

FIG. 11-45 Problem 39.

••**40** Figure 11-46 gives the torque τ that acts on an initially stationary disk that can rotate about its center like a merry-go-round. The scale on the τ axis is set by $\tau_s = 4.0$ N·m. What is the angular momentum of the disk about the rotation axis at times (a) $t = 7.0$ s and (b) $t = 20$ s?

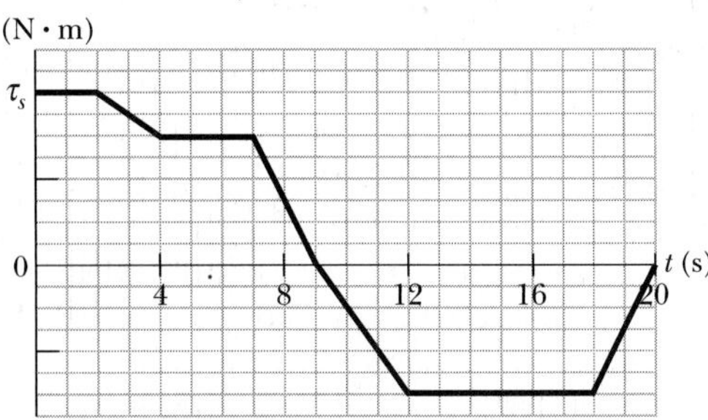

FIG. 11-46 Problem 40.

••**41** Figure 11-47 shows a rigid structure consisting of a circular hoop of radius R and mass m, and a square made of four thin bars, each of length R and mass m. The rigid structure rotates at a constant speed about a vertical axis, with a period of rotation of 2.5 s. Assuming $R = 0.50$ m and $m = 2.0$ kg, calculate (a) the structure's rotational inertia about the axis of rotation and (b) its angular momentum about that axis. GO

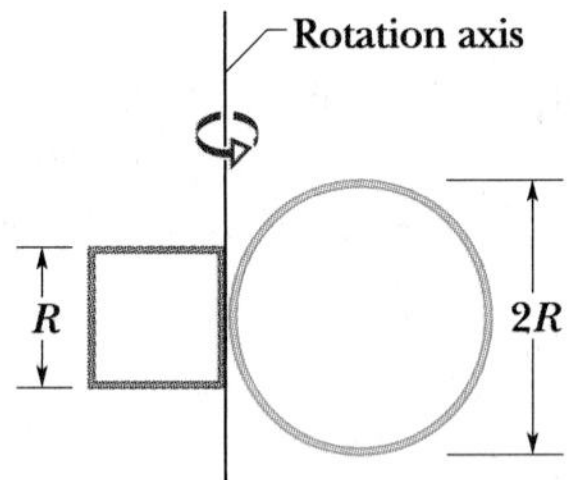

FIG. 11-47 Problem 41.

••**42** A disk with a rotational inertia of 7.00 kg·m^2 rotates like a merry-go-round while undergoing a torque given by $\tau = (5.00 + 2.00t)$ N·m. At time $t = 1.00$ s, its angular momentum is 5.00 kg·m^2/s. What is its angular momentum at $t = 3.00$ s?

sec. 11-11 Conservation of Angular Momentum

•**43** A man stands on a platform that is rotating (without friction) with an angular speed of 1.2 rev/s; his arms are outstretched and he holds a brick in each hand. The rotational inertia of the system consisting of the man, bricks, and platform about the central vertical axis of the platform is 6.0 kg·m^2. If by moving the bricks the man decreases the rotational inertia of the system to 2.0 kg·m^2, what are (a) the resulting angular speed of the platform and (b) the ratio of the new kinetic energy of the system to the original kinetic energy? (c) What source provided the added kinetic energy? SSM WWW

•**44** The rotor of an electric motor has rotational inertia $I_m = 2.0 \times 10^{-3}$ kg·m^2 about its central axis. The motor is used to change the orientation of the space probe in which it is mounted. The motor axis is mounted along the central axis of the probe; the probe has rotational inertia $I_p = 12$ kg·m^2 about this axis. Calculate the number of revolutions of the rotor required to turn the probe through 30° about its central axis.

•**45** A wheel is rotating freely at angular speed 800 rev/min on a shaft whose rotational inertia is negligible. A second wheel, initially at rest and with twice the rotational inertia of the first, is suddenly coupled to the same shaft. (a) What is the angular speed of the resultant combination of the shaft and two wheels? (b) What fraction of the original rotational kinetic energy is lost? SSM ILW

•**46** A Texas cockroach first rides at the center of a circular disk that rotates freely like a merry-go-round without external torques. The cockroach then walks out to the edge of the disk, at radius R. Figure 11-48 gives the angular speed ω of the cockroach–disk system during the walk. The scale on the ω axis is set by $\omega_a = 5.0$ rad/s and $\omega_b = 6.0$ rad/s. When the cockroach is on the edge at radius R, what is the ratio of the bug's rotational inertia to that of the disk, both calculated about the rotation axis?

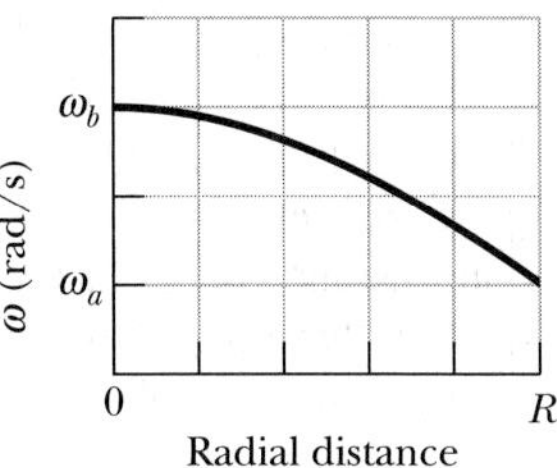

FIG. 11-48 Problem 46.

•**47** Two disks are mounted (like a merry-go-round) on low-friction bearings on the same axle and can be brought together so that they couple and rotate as one unit. The first disk, with rotational inertia 3.30 kg·m^2 about its central axis, is set spinning counterclockwise at 450 rev/min. The second disk, with rotational inertia 6.60 kg·m^2 about its central axis, is set spinning counterclockwise at 900 rev/min. They then couple together. (a) What is their angular speed after coupling? If instead the second disk is set spinning clockwise at 900 rev/min, what are their (b) angular speed and (c) direction of rotation after they couple together?

•**48** The rotational inertia of a collapsing spinning star drops to $\frac{1}{3}$ its initial value. What is the ratio of the new rotational kinetic energy to the initial rotational kinetic energy?

•**49** A track is mounted on a large wheel that is free to turn with negligible friction about a vertical axis (Fig. 11-49). A toy train of mass m is placed on the track and, with the system initially at rest, the train's electrical power is turned on. The train reaches speed 0.15 m/s with respect to the track. What is the angular speed of the wheel if its mass is 1.1m and its radius is 0.43 m? (Treat the wheel as a hoop, and neglect

the mass of the spokes and hub.) SSM

FIG. 11-49 Problem 49.

•50 A Texas cockroach of mass 0.17 kg runs counterclockwise around the rim of a lazy Susan (a circular disk mounted on a vertical axle) that has radius 15 cm, rotational inertia 5.0×10^{-3} kg·m², and frictionless bearings. The cockroach's speed (relative to the ground) is 2.0 m/s, and the lazy Susan turns clockwise with angular velocity $\omega_0 = 2.8$ rad/s. The cockroach finds a bread crumb on the rim and, of course, stops. (a) What is the angular speed of the lazy Susan after the cockroach stops? (b) Is mechanical energy conserved as it stops?

•51 In Fig. 11-50, two skaters, each of mass 50 kg, approach each other along parallel paths separated by 3.0 m. They have opposite velocities of 1.4 m/s each. One skater carries one end of a long pole of negligible mass, and the other skater grabs the other end as she passes. The skaters then rotate around the center of the pole. Assume that the friction between skates and ice is negligible. What are (a) the radius of the circle, (b) the angular speed of the skaters, and (c) the kinetic energy of the two-skater system? Next, the skaters pull along the pole until they are separated by 1.0 m. What then are (d) their angular speed and (e) the kinetic energy of the system? (f) What provided the energy for the increased kinetic energy?

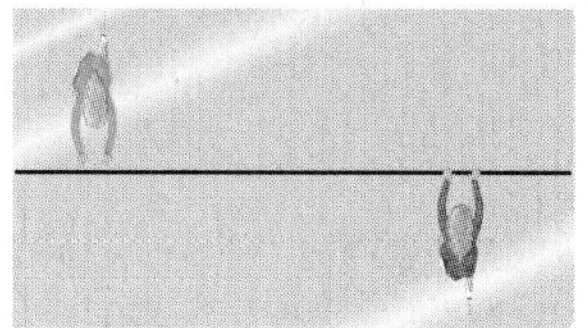
FIG. 11-50 Problem 51.

•52 A bola consists of three massive, identical spheres connected to a common point by identical lengths of sturdy string (Fig. 11-51*a*). To launch this native South American weapon, you hold one of the spheres overhead and then rotate that hand about its wrist so as to rotate the other two spheres in a horizontal path about the hand. Once you manage sufficient rotation, you cast the weapon at a target. Initially the bola rotates around the previously held sphere at angular speed ω_i but then quickly changes so that the spheres rotate around the common connection point at angular speed ω_f (Fig. 11-51*b*). (a) What is the ratio ω_f/ω_i? (b) In the center-of-mass frame, what is the ratio K_f/K_i of the corresponding rotational kinetic energies?

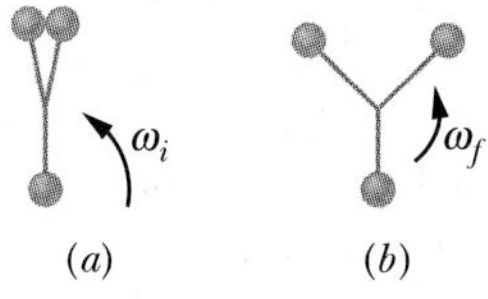

FIG. 11-51 Problem 52.

•53 A horizontal vinyl record of mass 0.10 kg and radius 0.10 m rotates freely about a vertical axis through its center with an angular speed of 4.7 rad/s. The rotational inertia of the record about its axis of rotation is 5.0×10^{-4} kg·m². A wad of wet putty of mass 0.020 kg drops vertically onto the record from above and sticks to the edge of the record. What is the angular speed of the record immediately after the putty sticks to it?

••54 In a long jump, an athlete leaves the ground with an initial angular momentum that tends to rotate her body forward, threatening to ruin her landing. To counter this tendency, she rotates her outstretched arms to "take up" the angular momentum (Fig. 11-19). In 0.700 s, one arm sweeps through 0.500 rev and the other arm sweeps through 1.000 rev. Treat each arm as a thin rod of mass 4.0 kg and length 0.60 m, rotating around one end. In the athlete's reference frame, what is the magnitude of the total angular momentum of the arms around the common rotation axis through the shoulders?

••55 A uniform thin rod of length 0.500 m and mass 4.00 kg can rotate in a horizontal plane about a vertical axis through its center. The rod is at rest when a 3.00 g bullet traveling in the rotation plane is fired into one end of the rod. As viewed from above, the bullet's path makes angle $\theta = 60.0°$ with the rod (Fig. 11-52). If the bullet lodges in the rod and the angular velocity of the rod is 10 rad/s immediately after the collision, what is the bullet's speed just before impact? GO

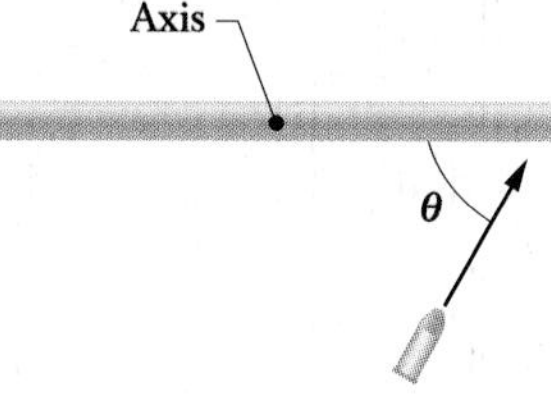

FIG. 11-52 Problem 55.

••56 A cockroach of mass m lies on the rim of a uniform disk of mass $4.00m$ that can rotate freely about its center like a merry-go-round. Initially the cockroach and disk rotate together with an angular velocity of 0.260 rad/s. Then the cockroach walks halfway to the center of the disk. (a) What then is the angular velocity of the cockroach–disk system? (b) What is the ratio K/K_0 of the new kinetic energy of the system to its initial kinetic energy? (c) What accounts for the change in the kinetic energy?

••57 Figure 11-53 is an overhead view of a thin uniform rod of length 0.800 m and mass M rotating horizontally at angular speed 20.0 rad/s about an axis through its center. A particle of mass $M/3.00$ initially attached to one end is ejected from the rod and travels along a path that is perpendicular to the rod at the instant of ejection. If the particle's speed v_p is 6.00 m/s greater than the speed of the rod end just after ejection, what is the value of v_p?

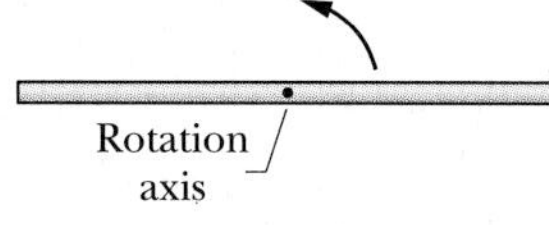

FIG. 11-53 Problem 57.

••58 In Fig. 11-54, a 1.0 g bullet is fired into a 0.50 kg block attached to the end of a 0.60 m nonuniform rod of mass 0.50 kg. The block–rod–bullet system then rotates in the plane of the figure, about a fixed axis at A. The rotational inertia of the rod alone about that axis at A is 0.060 kg·m². Treat the block as a particle. (a) What then is the rotational inertia of the block–rod–bullet system about point A? (b) If the angular speed of the system about A just after impact is 4.5 rad/s, what is the bullet's speed just before impact?

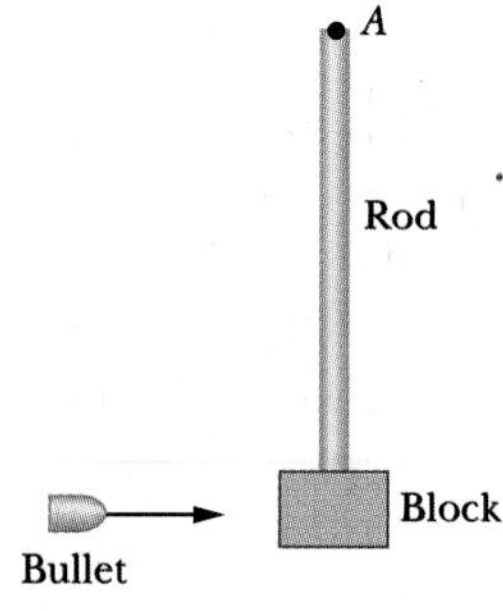

FIG. 11-54 Problem 58.

••59 A uniform disk of mass $10m$ and radius $3.0r$ can rotate freely about its fixed center like a merry-go-round. A smaller uniform disk of mass m and radius r lies on top of the larger disk, concentric with it. Initially the two disks rotate together with an angular velocity of 20 rad/s. Then a slight disturbance causes the smaller disk to slide outward across the larger disk, until the outer edge of the smaller disk catches on the outer edge of the larger disk. Afterward, the two disks again rotate together (without further sliding). (a) What then is their angular velocity about the center of the larger disk? (b) What is the ratio K/K_0 of the new kinetic energy of the two-disk system to the system's initial kinetic energy?

••**60** A horizontal platform in the shape of a circular disk rotates on a frictionless bearing about a vertical axle through the center of the disk. The platform has a mass of 150 kg, a radius of 2.0 m, and a rotational inertia of 300 kg · m² about the axis of rotation. A 60 kg student walks slowly from the rim of the platform toward the center. If the angular speed of the system is 1.5 rad/s when the student starts at the rim, what is the angular speed when she is 0.50 m from the center?

••**61** The uniform rod (length 0.60 m, mass 1.0 kg) in Fig. 11-55 rotates in the plane of the figure about an axis through one end, with a rotational inertia of 0.12 kg · m². As the rod swings through its lowest position, it collides with a 0.20 kg putty wad that sticks to the end of the rod. If the rod's angular speed just before collision is 2.4 rad/s, what is the angular speed of the rod–putty system immediately after collision?

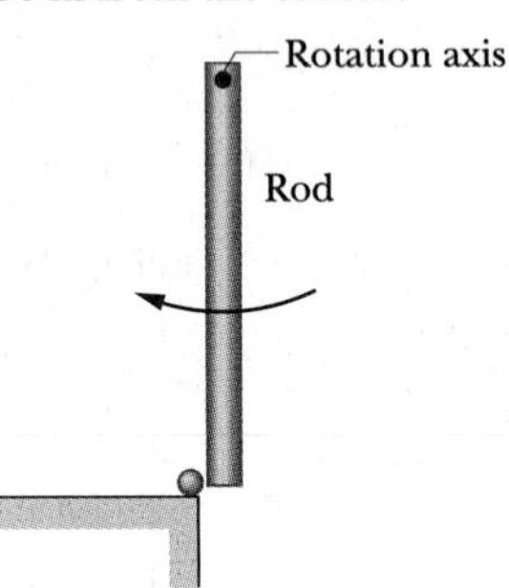

FIG. 11-55 Problem 61.

•••**62** A ballerina begins a tour jeté (Fig. 11-20a) with angular speed ω_i and a rotational inertia consisting of two parts: I_{leg} = 1.44 kg · m² for her leg extended outward at angle θ = 90.0° to her body and I_{trunk} = 0.660 kg · m² for the rest of her body (primarily her trunk). Near her maximum height she holds both legs at angle θ = 30.0° to her body and has angular speed ω_f (Fig. 11-20b). Assuming that I_{trunk} has not changed, what is the ratio ω_f/ω_i?

•••**63** Figure 11-56 is an overhead view of a thin uniform rod of length 0.600 m and mass M rotating horizontally at 80.0 rad/s counterclockwise about an axis through its center. A particle of mass M/3.00 and traveling horizontally at speed 40.0 m/s hits the rod and sticks. The particle's path is perpendicular to the rod at the instant of the hit, at a distance d from the rod's center. (a) At what value of d are rod and particle stationary after the hit? (b) In which direction do rod and particle rotate if d is greater than this value?

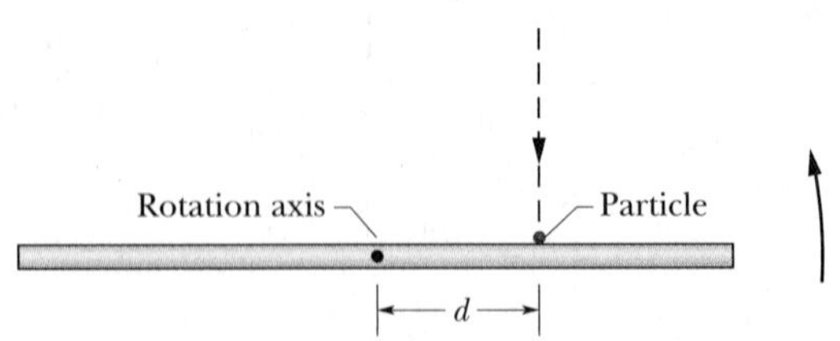

FIG. 11-56 Problem 63.

•••**64** During a jump to his partner, an aerialist is to make a quadruple somersault lasting a time t = 1.87 s. For the first and last quarter-revolution, he is in the extended orientation shown in Fig. 11-57, with rotational inertia I_1 = 19.9 kg · m² around his center of mass (the dot). During the rest of the flight he is in a tight tuck, with rotational inertia I_2 = 3.93 kg · m². What must be his angular speed ω_2 around his center of mass during the tuck?

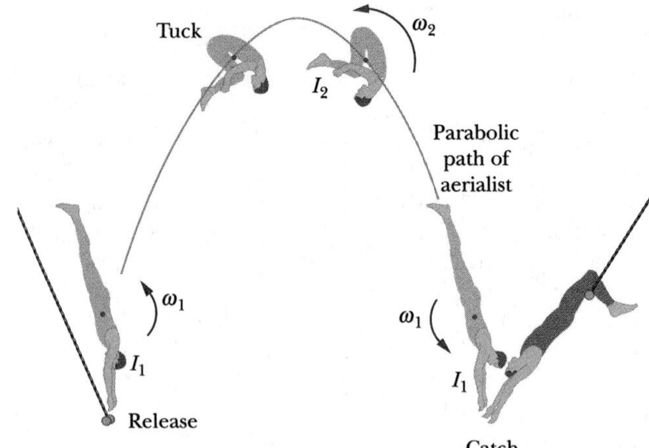

FIG. 11-57 Problem 64.

•••**65** In Fig. 11-58, a 30 kg child stands on the edge of a stationary merry-go-round of mass 100 kg and radius 2.0 m. The rotational inertia of the merry-go-round about its rotation axis is 150 kg · m². The child catches a ball of mass 1.0 kg thrown by a friend. Just before the ball is caught, it has a horizontal velocity $\vec{v}$ of magnitude 12 m/s, at angle ϕ = 37° with a line tangent to the outer edge of the merry-go-round, as shown. What is the angular speed of the merry-go-round just after the ball is caught?

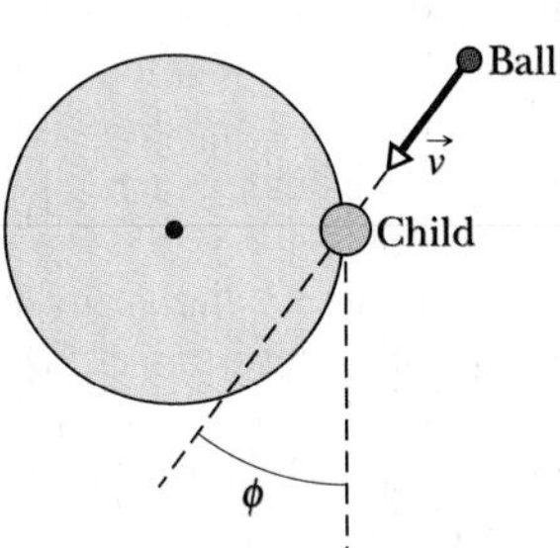

FIG. 11-58 Problem 65.

•••**66** In Fig. 11-59, a small 50 g block slides down a frictionless surface through height h = 20 cm and then sticks to a uniform rod of mass 100 g and length 40 cm. The rod pivots about point O through angle θ before momentarily stopping. Find θ.

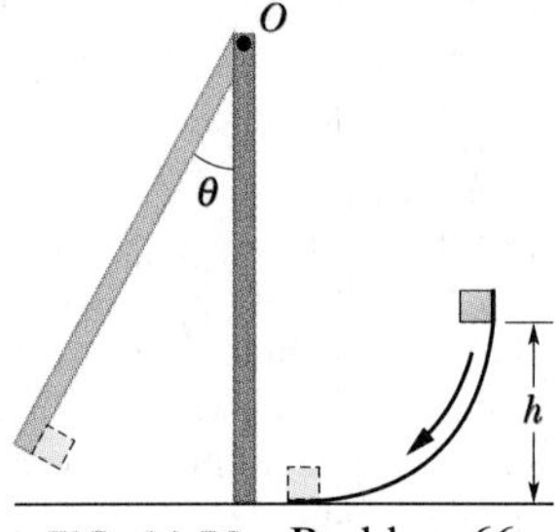

FIG. 11-59 Problem 66.

•••**67** Two 2.00 kg balls are attached to the ends of a thin rod of length 50.0 cm and negligible mass. The rod is free to rotate in a vertical plane without friction about a horizontal axis through its center. With the rod initially horizontal (Fig. 11-60), a 50.0 g wad of wet putty drops onto one of the balls, hitting it with a speed of 3.00 m/s and then sticking to it. (a) What is the angular speed of the system just after the putty wad hits? (b) What is the ratio of the kinetic energy of the system after the collision to that of the putty wad just before? (c) Through what angle will the system rotate before it momentarily stops? SSM WWW

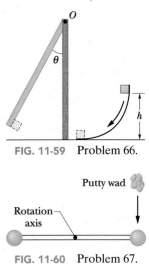

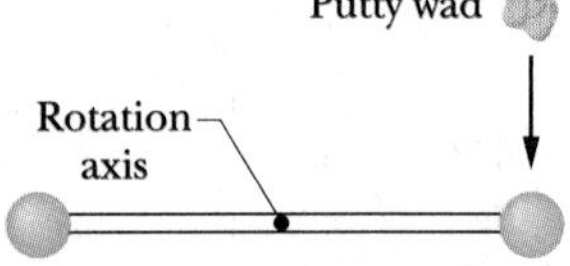

Putty wad

Rotation axis

FIG. 11-60 Problem 67.

sec. 11-12 Precession of a Gyroscope

••**68** A top spins at 30 rev/s about an axis that makes an angle of 30° with the vertical. The mass of the top is 0.50 kg, its rotational inertia about its central axis is 5.0 × 10⁻⁴ kg · m², and its center of mass is 4.0 cm from the pivot point.

If the spin is clockwise from an overhead view, what are the (a) precession rate and (b) direction of the precession as viewed from overhead?

••69 A certain gyroscope consists of a uniform disk with a 50 cm radius mounted at the center of an axle that is 11 cm long and of negligible mass. The axle is horizontal and supported at one end. If the disk is spinning around the axle at 1000 rev/min, what is the precession rate?

Additional Problems

70 A uniform block of granite in the shape of a book has face dimensions of 20 cm and 15 cm and a thickness of 1.2 cm. The density (mass per unit volume) of granite is 2.64 g/cm^3. The block rotates around an axis that is perpendicular to its face and halfway between its center and a corner. Its angular momentum about that axis is 0.104 kg·m^2/s. What is its rotational kinetic energy about that axis?

71 Figure 11-61 shows an overhead view of a ring that can rotate about its center like a merry-go-round. Its outer radius R_2 is 0.800 m, its inner radius R_1 is $R_2/2.00$, its mass M is 8.00 kg, and the mass of the crossbars at its center is negligible. It initially rotates at an angular speed of 8.00 rad/s with a cat of mass $m = M/4.00$ on its outer edge, at radius R_2. By how much does the cat increase the kinetic energy of the cat–ring system if the cat crawls to the inner edge, at radius R_1? GO

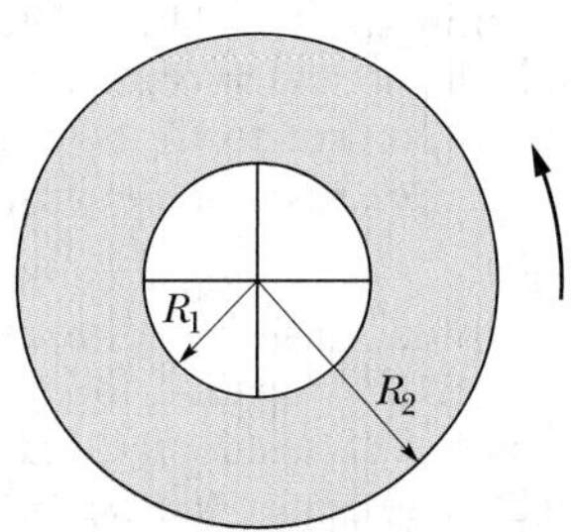

FIG. 11-61 Problem 71.

72 A 2.50 kg particle that is moving horizontally over a floor with velocity $(-3.00 \text{ m/s})\hat{j}$ undergoes a completely inelastic collision with a 4.00 kg particle that is moving horizontally over the floor with velocity $(4.50 \text{ m/s})\hat{i}$. The collision occurs at xy coordinates $(-0.500 \text{ m}, -0.100 \text{ m})$. After the collision and in unit-vector notation, what is the angular momentum of the stuck-together particles with respect to the origin?

73 Two particles, each of mass 2.90×10^{-4} kg and speed 5.46 m/s, travel in opposite directions along parallel lines separated by 4.20 cm. (a) What is the magnitude L of the angular momentum of the two-particle system around a point midway between the two lines? (b) Does the value of L change if the point about which it is calculated is not midway between the lines? If the direction of travel for one of the particles is reversed, what would be (c) the answer to part (a) and (d) the answer to part (b)? SSM

74 A uniform rod rotates in a horizontal plane about a vertical axis through one end. The rod is 6.00 m long, weighs 10.0 N, and rotates at 240 rev/min. Calculate (a) its rotational inertia about the axis of rotation and (b) the magnitude of its angular momentum about that axis.

75 Wheels A and B in Fig. 11-62 are connected by a belt that does not slip. The radius of B is 3.00 times the radius of A. What would be the ratio of the rotational inertias I_A/I_B if the two wheels had (a) the same angular momentum about their central axes and (b) the same rotational kinetic energy?

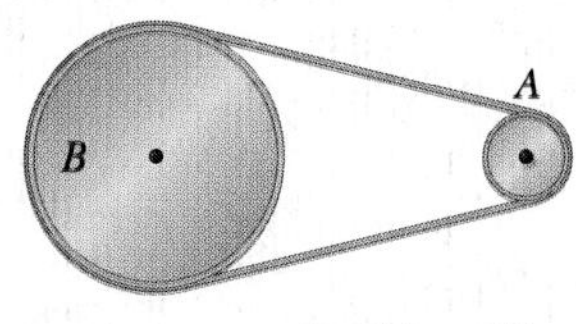

FIG. 11-62 Problem 75.

76 At time $t = 0$, a 2.0 kg particle has position vector $\vec{r} = (4.0 \text{ m})\hat{i} - (2.0 \text{ m})\hat{j}$ relative to the origin. Its velocity is given by $\vec{v} = (-6.0t^2 \text{ m/s})\hat{i}$ for $t \geq 0$ in seconds. About the origin, what are (a) the particle's angular momentum $\vec{L}$ and (b) the torque $\vec{\tau}$ acting on the particle, both in unit-vector notation and for $t > 0$? About the point $(-2.0 \text{ m}, -3.0 \text{ m}, 0)$, what are (c) $\vec{L}$ and (d) $\vec{\tau}$ for $t > 0$?

77 A uniform wheel of mass 10.0 kg and radius 0.400 m is mounted rigidly on an axle through its center (Fig. 11-63). The radius of the axle is 0.200 m, and the rotational inertia of the wheel–axle combination about its central axis is 0.600 kg·m^2. The wheel is initially at rest at the top of a surface that is inclined at angle $\theta = 30.0°$ with the horizontal; the axle rests on the surface while the wheel extends into a groove in the surface without touching the surface. Once released, the axle rolls down along the surface smoothly and without slipping. When the wheel–axle combination has moved down the surface by 2.00 m, what are (a) its rotational kinetic energy and (b) its translational kinetic energy? SSM

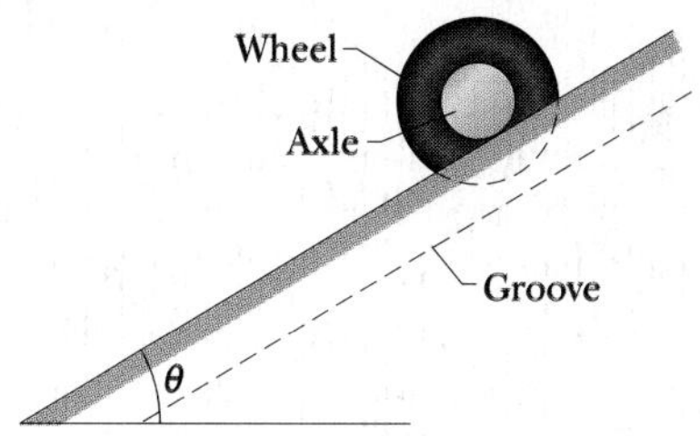

FIG. 11-63 Problem 77.

78 Suppose that the yo-yo in Problem 17, instead of rolling from rest, is thrown so that its initial speed down the string is 1.3 m/s. (a) How long does the yo-yo take to reach the end of the string? As it reaches the end of the string, what are its (b) total kinetic energy, (c) linear speed, (d) translational kinetic energy, (e) angular speed, and (f) rotational kinetic energy?

79 A small solid sphere with radius 0.25 cm and mass 0.56 g rolls without slipping on the inside of a large fixed hemisphere with radius 15 cm and a vertical axis of symmetry. The sphere starts at the top from rest. (a) What is its kinetic energy at the bottom? (b) What fraction of its kinetic energy at the bottom is associated with rotation about an axis through its com? (c) What is the magnitude of the normal force on the hemisphere from the sphere when the sphere reaches the bottom?

80 A uniform solid ball rolls smoothly along a floor, then up a ramp inclined at 15.0°. It momentarily stops when it has rolled 1.50 m along the ramp. What was its initial speed?

81 A body of radius R and mass m is rolling smoothly with speed v on a horizontal surface. It then rolls up a hill to a maximum height h. (a) If $h = 3v^2/4g$, what is the body's rotational inertia about the rotational axis through its center of mass? (b) What might the body be?

82 A wheel of radius 0.250 m, which is moving initially at 43.0 m/s, rolls to a stop in 225 m. Calculate the magnitudes of (a) its linear acceleration and (b) its angular acceleration. (c) The wheel's rotational inertia is 0.155 kg·m^2 about its central axis. Calculate the magnitude of the torque about the central axis due to friction on the wheel.

83 If Earth's polar ice caps fully melted and the water returned to the oceans, the oceans would be deeper by about 30 m. What effect would this have on Earth's rotation? Make an estimate of the resulting change in the length of the day.

84 A 1200 kg airplane is flying in a straight line at 80 m/s, 1.3 km above the ground. What is the magnitude of its angular momentum with respect to a point on the ground directly under the path of the plane?

85 In a playground, there is a small merry-go-round of radius 1.20 m and mass 180 kg. Its radius of gyration (see Problem 91 of Chapter 10) is 91.0 cm. A child of mass 44.0 kg runs at a speed of 3.00 m/s along a path that is tangent to the rim of the initially stationary merry-go-round and then jumps on. Neglect friction between the bearings and the shaft of the merry-go-round. Calculate (a) the rotational inertia of the merry-go-round about its axis of rotation, (b) the magnitude of the angular momentum of the running child about the axis of rotation of the merry-go-round, and (c) the angular speed of the merry-go-round and child after the child has jumped onto the merry-go-round. SSM

86 A wheel rotates clockwise about its central axis with an angular momentum of 600 kg · m²/s. At time $t = 0$, a torque of magnitude 50 N · m is applied to the wheel to reverse the rotation. At what time t is the angular speed zero?

87 A 3.0 kg toy car moves along an x axis with a velocity given by $\vec{v} = -2.0t^3\hat{i}$ m/s, with t in seconds. For $t > 0$, what are (a) the angular momentum $\vec{L}$ of the car and (b) the torque $\vec{\tau}$ on the car, both calculated about the origin? What are (c) $\vec{L}$ and (d) $\vec{\tau}$ about the point (2.0 m, 5.0 m, 0)? What are (e) $\vec{L}$ and (f) $\vec{\tau}$ about the point (2.0 m, −5.0 m, 0)? SSM

88 A thin-walled pipe rolls along the floor. What is the ratio of its translational kinetic energy to its rotational kinetic energy about the central axis parallel to its length?

89 A solid sphere of weight 36.0 N rolls up an incline at an angle of 30.0°. At the bottom of the incline the center of mass of the sphere has a translational speed of 4.90 m/s. (a) What is the kinetic energy of the sphere at the bottom of the incline? (b) How far does the sphere travel up along the incline? (c) Does the answer to (b) depend on the sphere's mass?

90 An automobile has a total mass of 1700 kg. It accelerates from rest to 40 km/h in 10 s. Assume each wheel is a uniform 32 kg disk. Find, for the end of the 10 s interval, (a) the rotational kinetic energy of each wheel about its axle, (b) the total kinetic energy of each wheel, and (c) the total kinetic energy of the automobile.

91 With axle and spokes of negligible mass and a thin rim, a certain bicycle wheel has a radius of 0.350 m and weighs 37.0 N; it can turn on its axle with negligible friction. A man holds the wheel above his head with the axle vertical while he stands on a turntable that is free to rotate without friction; the wheel rotates clockwise, as seen from above, with an angular speed of 57.7 rad/s, and the turntable is initially at rest. The rotational inertia of *wheel + man + turntable* about the common axis of rotation is 2.10 kg · m². The man's free hand suddenly stops the rotation of the wheel (relative to the turntable). Determine the resulting (a) angular speed and (b) direction of rotation of the system.

92 For an 84 kg person standing at the equator, what is the magnitude of the angular momentum about Earth's center due to Earth's rotation?

93 A girl of mass M stands on the rim of a frictionless merry-go-round of radius R and rotational inertia I that is not moving. She throws a rock of mass m horizontally in a direction that is tangent to the outer edge of the merry-go-round. The speed of the rock, relative to the ground, is v. Afterward, what are (a) the angular speed of the merry-go-round and (b) the linear speed of the girl?

94 A 4.0 kg particle moves in an xy plane. At the instant when the particle's position and velocity are $\vec{r} = (2.0\hat{i} + 4.0\hat{j})$ m and $\vec{v} = -4.0\hat{j}$ m/s, the force on the particle is $\vec{F} = -3.0\hat{i}$ N. At this instant, determine (a) the particle's angular momentum about the origin, (b) the particle's angular momentum about the point $x = 0, y = 4.0$ m, (c) the torque acting on the particle about the origin, and (d) the torque acting on the particle about the point $x = 0, y = 4.0$ m.

95 In Fig. 11-64, a constant horizontal force $\vec{F}_{app}$ of magnitude 12 N is applied to a uniform solid cylinder by fishing line wrapped around the cylinder. The mass of the cylinder is 10 kg, its radius is 0.10 m, and the cylinder rolls smoothly on the horizontal surface. (a) What is the magnitude of the acceleration of the center of mass of the cylinder? (b) What is the magnitude of the angular acceleration of the cylinder about the center of mass? (c) In unit-vector notation, what is the frictional force acting on the cylinder? SSM

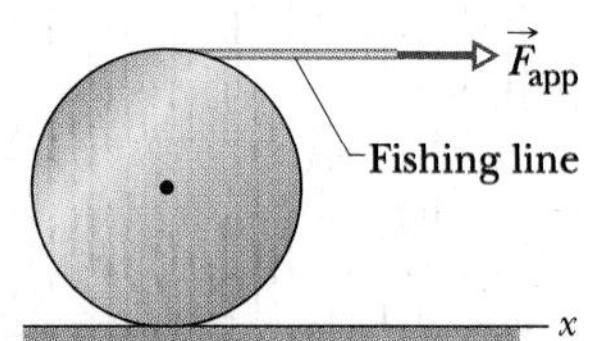

FIG. 11-64 Problem 95.

96 (a) In Sample Problem 10-8, when the rotor exploded, how much angular momentum, calculated about the rotation axis, was released to the surroundings? (b) If we assume that most of the pieces of the rotor were stopped within 0.025 s after the explosion, what was the magnitude of the average torque acting on those pieces, calculated about the rotation axis?

97 A particle of mass M = 0.25 kg is dropped from a point that is at height h = 1.80 m above the ground and horizontal distance s = 0.45 m from an observation point O, as shown in Fig. 11-65. What is the magnitude of the angular momentum of the particle with respect to point O when the particle has fallen half the distance to the ground?

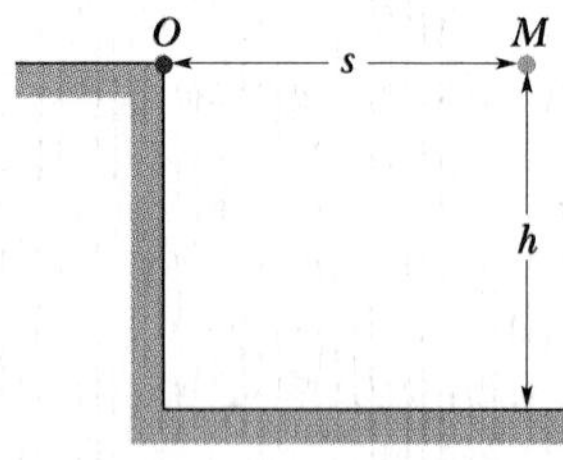

FIG. 11-65 Problem 97.

98 At one instant, a 0.80 kg particle is located at the position $\vec{r} = (2.0\text{ m})\hat{i} + (3.0\text{ m})\hat{j}$. The linear momentum of the particle lies in the xy plane and has a magnitude of 2.4 kg · m/s and a direction of 115° measured counterclockwise from the positive direction of x. What is the angular momentum of the particle about the origin, in unit-vector notation?

Equilibrium and Elasticity

The famous tower in Pisa, Italy, began to lean toward the south even during its construction, which spanned two centuries. The leaning increased with time but only at the snail's pace of 0.001° per year. In recent years, when the tilt reached 5.5°, the tower was closed to tourists because authorities feared that it would soon collapse. But doesn't collapse require that the tower's center of mass move out beyond the base of the tower? That would not have happened for many more years.

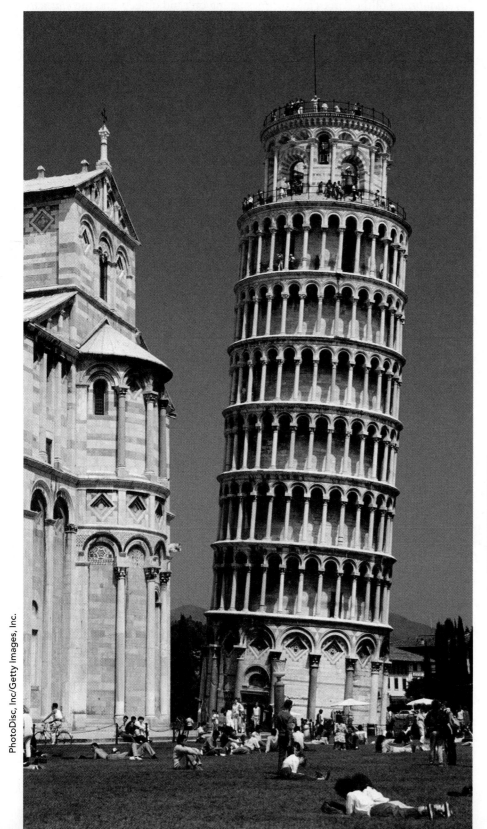

PhotoDisc, Inc/Getty Images, Inc.

So, what was the danger to the tower?

The answer is in this chapter.

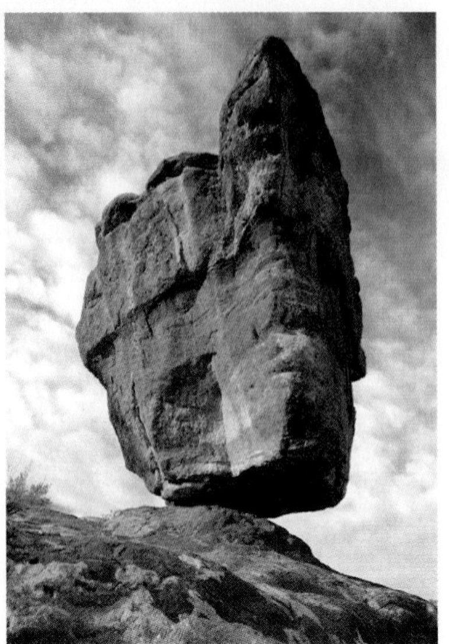

FIG. 12-1 A balancing rock. Although its perch seems precarious, the rock is in static equilibrium. *(Symon Lobsang/Photis/Jupiter Images Corp.)*

12-1 WHAT IS PHYSICS?

Human constructions are supposed to be stable in spite of the forces that act on them. A building, for example, should be stable in spite of the gravitational force and wind forces on it, and a bridge should be stable in spite of the gravitational force pulling it downward and the repeated jolting it receives from cars and trucks.

One focus of physics is on what allows an object to be stable in spite of any forces acting on it. In this chapter we examine the two main aspects of stability: the *equilibrium* of the forces and torques acting on rigid objects and the *elasticity* of nonrigid objects, a property that governs how such objects can deform. When this physics is done correctly, it is the subject of countless articles in physics and engineering journals; when it is done incorrectly, it is the subject of countless articles in newspapers and legal journals.

12-2 | Equilibrium

Consider these objects: (1) a book resting on a table, (2) a hockey puck sliding with constant velocity across a frictionless surface, (3) the rotating blades of a ceiling fan, and (4) the wheel of a bicycle that is traveling along a straight path at constant speed. For each of these four objects,

1. The linear momentum $\vec{P}$ of its center of mass is constant.
2. Its angular momentum $\vec{L}$ about its center of mass, or about any other point, is also constant.

We say that such objects are in **equilibrium.** The two requirements for equilibrium are then

$$\vec{P} = \text{a constant} \quad \text{and} \quad \vec{L} = \text{a constant}. \tag{12-1}$$

Our concern in this chapter is with situations in which the constants in Eq. 12-1 are zero; that is, we are concerned largely with objects that are not moving in any way—either in translation or in rotation—in the reference frame from which we observe them. Such objects are in **static equilibrium.** Of the four objects mentioned at the beginning of this section, only one—the book resting on the table—is in static equilibrium.

The balancing rock of Fig. 12-1 is another example of an object that, for the present at least, is in static equilibrium. It shares this property with countless other structures, such as cathedrals, houses, filing cabinets, and taco stands, that remain stationary over time.

As we discussed in Section 8-6, if a body returns to a state of static equilibrium after having been displaced from that state by a force, the body is said to be in *stable* static equilibrium. A marble placed at the bottom of a hemispherical bowl is an example. However, if a small force can displace the body and end the equilibrium, the body is in *unstable* static equilibrium.

For example, suppose we balance a domino with the domino's center of mass vertically above the supporting edge, as in Fig. 12-2*a*. The torque about the supporting edge due to the gravitational force $\vec{F}_g$ on the domino is zero because the line of action of $\vec{F}_g$ is through that edge. Thus, the domino is in equilibrium. Of course, even a slight force on it due to some chance disturbance ends the equilibrium. As the line of action of $\vec{F}_g$ moves to one side of the supporting edge (as in Fig. 12-2*b*), the torque due to $\vec{F}_g$ increases the rotation of the domino. Therefore, the domino in Fig. 12-2*a* is in unstable static equilibrium.

The domino in Fig. 12-2*c* is not quite as unstable. To topple this domino, a force would have to rotate it through and then beyond the balance position of Fig. 12-2*a*, in which the center of mass is above a supporting edge. A slight force will not topple this domino, but a vigorous flick of the finger against the domino

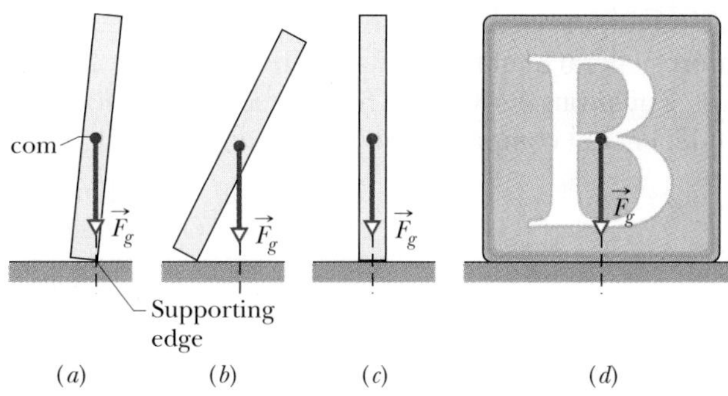

FIG. 12-2 (*a*) A domino balanced on one edge, with its center of mass vertically above that edge. The gravitational force $\vec{F}_g$ on the domino is directed through the supporting edge. (*b*) If the domino is rotated even slightly from the balanced orientation, then $\vec{F}_g$ causes a torque that increases the rotation. (*c*) A domino upright on a narrow side is somewhat more stable than the domino in (*a*). (*d*) A square block is even more stable.

certainly will. (If we arrange a chain of such upright dominos, a finger flick against the first can cause the whole chain to fall.)

The child's square block in Fig. 12-2*d* is even more stable because its center of mass would have to be moved even farther to get it to pass above a supporting edge. A flick of the finger may not topple the block. (This is why you never see a chain of toppling square blocks.) The worker in Fig. 12-3 is like both the domino and the square block: Parallel to the beam, his stance is wide and he is stable; perpendicular to the beam, his stance is narrow and he is unstable (and at the mercy of a chance gust of wind).

The analysis of static equilibrium is very important in engineering practice. The design engineer must isolate and identify all the external forces and torques that may act on a structure and, by good design and wise choice of materials, ensure that the structure will remain stable under these loads. Such analysis is necessary to ensure, for example, that bridges do not collapse under their traffic and wind loads and that the landing gear of aircraft will function after the shock of rough landings.

FIG. 12-3 A construction worker balanced on a steel beam is in static equilibrium but is more stable parallel to the beam than perpendicular to it. *(Robert Brenner/PhotoEdit)*

12-3 | The Requirements of Equilibrium

The translational motion of a body is governed by Newton's second law in its linear momentum form, given by Eq. 9-27 as

$$\vec{F}_{\text{net}} = \frac{d\vec{P}}{dt}. \tag{12-2}$$

If the body is in translational equilibrium—that is, if $\vec{P}$ is a constant—then $d\vec{P}/dt = 0$ and we must have

$$\vec{F}_{\text{net}} = 0 \quad \text{(balance of forces).} \tag{12-3}$$

The rotational motion of a body is governed by Newton's second law in its angular momentum form, given by Eq. 11-29 as

$$\vec{\tau}_{\text{net}} = \frac{d\vec{L}}{dt}. \tag{12-4}$$

If the body is in rotational equilibrium—that is, if $\vec{L}$ is a constant—then $d\vec{L}/dt = 0$ and we must have

$$\vec{\tau}_{\text{net}} = 0 \quad \text{(balance of torques).} \tag{12-5}$$

Thus, the two requirements for a body to be in equilibrium are as follows:

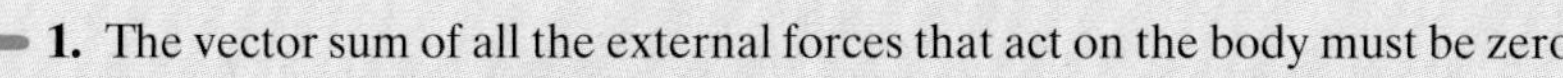

1. The vector sum of all the external forces that act on the body must be zero.
2. The vector sum of all external torques that act on the body, measured about *any* possible point, must also be zero.

These requirements obviously hold for *static* equilibrium. They also hold for the more general equilibrium in which $\vec{P}$ and $\vec{L}$ are constant but not zero.

Equations 12-3 and 12-5, as vector equations, are each equivalent to three independent component equations, one for each direction of the coordinate axes:

Balance of forces	Balance of torques
$F_{\text{net},x} = 0$	$\tau_{\text{net},x} = 0$
$F_{\text{net},y} = 0$	$\tau_{\text{net},y} = 0$
$F_{\text{net},z} = 0$	$\tau_{\text{net},z} = 0$

(12-6)

We shall simplify matters by considering only situations in which the forces that act on the body lie in the xy plane. This means that the only torques that can act on the body must tend to cause rotation around an axis parallel to the z axis. With this assumption, we eliminate one force equation and two torque equations from Eqs. 12-6, leaving

$$F_{\text{net},x} = 0 \quad \text{(balance of forces)}, \tag{12-7}$$

$$F_{\text{net},y} = 0 \quad \text{(balance of forces)}, \tag{12-8}$$

$$\tau_{\text{net},z} = 0 \quad \text{(balance of torques)}. \tag{12-9}$$

Here, $\tau_{\text{net},z}$ is the net torque that the external forces produce either about the z axis or about *any* axis parallel to it.

A hockey puck sliding at constant velocity over ice satisfies Eqs. 12-7, 12-8, and 12-9 and is thus in equilibrium *but not in static equilibrium.* For static equilibrium, the linear momentum $\vec{P}$ of the puck must be not only constant but also zero; the puck must be at rest on the ice. Thus, there is another requirement for static equilibrium:

3. The linear momentum $\vec{P}$ of the body must be zero.

CHECKPOINT 1 The figure gives six overhead views of a uniform rod on which two or more forces act perpendicularly to the rod. If the magnitudes of the forces are adjusted properly (but kept nonzero), in which situations can the rod be in static equilibrium?

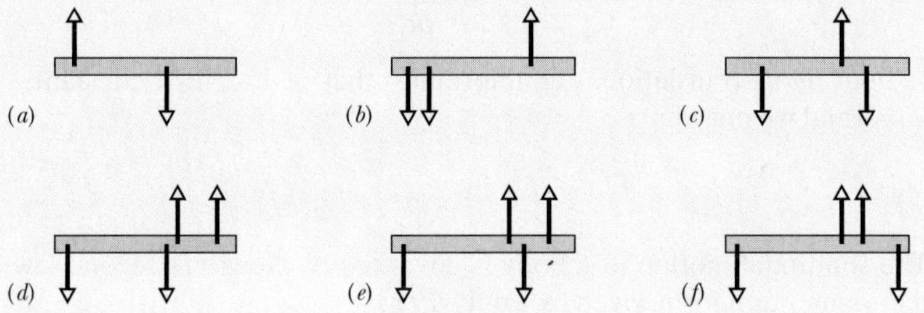

12-4 | The Center of Gravity

The gravitational force on an extended body is the vector sum of the gravitational forces acting on the individual elements (the atoms) of the body. Instead of considering all those individual elements, we can say that

The gravitational force $\vec{F}_g$ on a body effectively acts at a single point, called the **center of gravity** (cog) of the body.

Here the word "effectively" means that if the forces on the individual elements were somehow turned off and force $\vec{F}_g$ at the center of gravity were turned on, the net force and the net torque (about any point) acting on the body would not change.

Until now, we have assumed that the gravitational force $\vec{F}_g$ acts at the center of mass (com) of the body. This is equivalent to assuming that the center of gravity is at the center of mass. Recall that, for a body of mass M, the force $\vec{F}_g$ is equal to $M\vec{g}$, where $\vec{g}$ is the acceleration that the force would produce if the body were to fall freely. In the proof that follows, we show that

> If $\vec{g}$ is the same for all elements of a body, then the body's center of gravity (cog) is coincident with the body's center of mass (com).

This is approximately true for everyday objects because $\vec{g}$ varies only a little along Earth's surface and decreases in magnitude only slightly with altitude. Thus, for objects like a mouse or a moose, we have been justified in assuming that the gravitational force acts at the center of mass. After the following proof, we shall resume that assumption.

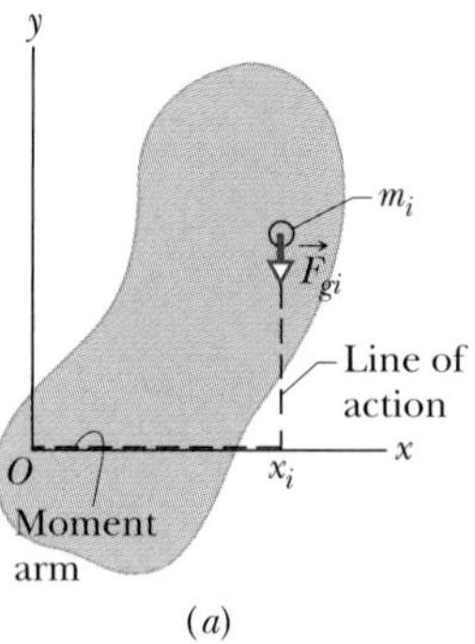

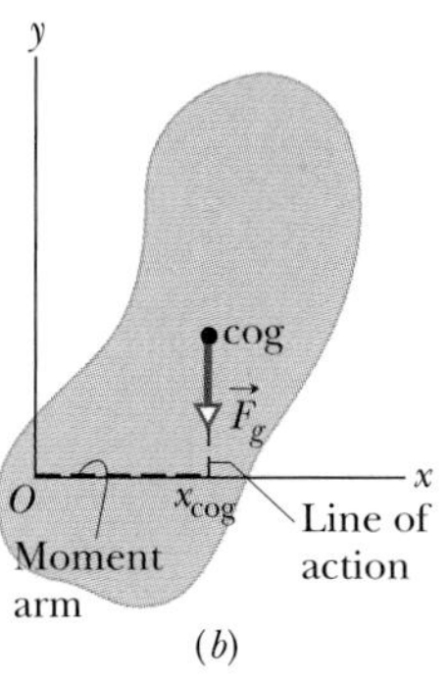

FIG. 12-4 (*a*) An element of mass m_i in an extended body. The gravitational force $\vec{F}_{gi}$ on the element has moment arm x_i about the origin O of the coordinate system. (*b*) The gravitational force $\vec{F}_g$ on a body is said to act at the center of gravity (cog) of the body. Here $\vec{F}_g$ has moment arm x_{cog} about origin O.

Proof

First, we consider the individual elements of the body. Figure 12-4*a* shows an extended body, of mass M, and one of its elements, of mass m_i. A gravitational force $\vec{F}_{gi}$ acts on each such element and is equal to $m_i\vec{g}_i$. The subscript on $\vec{g}_i$ means $\vec{g}_i$ is the gravitational acceleration *at the location of the element i* (it can be different for other elements).

In Fig. 12-4*a*, each force $\vec{F}_{gi}$ produces a torque τ_i on the element about the origin O, with moment arm x_i. Using Eq. 10-41 ($\tau = r_\perp F$), we can write torque τ_i as

$$\tau_i = x_i F_{gi}. \tag{12-10}$$

The net torque on all the elements of the body is then

$$\tau_{net} = \sum \tau_i = \sum x_i F_{gi}. \tag{12-11}$$

Next, we consider the body as a whole. Figure 12-4*b* shows the gravitational force $\vec{F}_g$ acting at the body's center of gravity. This force produces a torque τ on the body about O, with moment arm x_{cog}. Again using Eq. 10-41, we can write this torque as

$$\tau = x_{cog} F_g. \tag{12-12}$$

The gravitational force $\vec{F}_g$ on the body is equal to the sum of the gravitational forces $\vec{F}_{gi}$ on all its elements, so we can substitute ΣF_{gi} for F_g in Eq. 12-12 to write

$$\tau = x_{cog} \sum F_{gi}. \tag{12-13}$$

Now recall that the torque due to force $\vec{F}_g$ acting at the center of gravity is equal to the net torque due to all the forces $\vec{F}_{gi}$ acting on all the elements of the body. (That is how we defined the center of gravity.) Thus, τ in Eq. 12-13 is equal to τ_{net} in Eq. 12-11. Putting those two equations together, we can write

$$x_{cog} \sum F_{gi} = \sum x_i F_{gi}.$$

Substituting $m_i g_i$ for F_{gi} gives us

$$x_{cog} \sum m_i g_i = \sum x_i m_i g_i. \tag{12-14}$$

Now here is a key idea: If the accelerations g_i at all the locations of the elements are the same, we can cancel g_i from this equation to write

$$x_{cog} \sum m_i = \sum x_i m_i. \tag{12-15}$$

The sum Σm_i of the masses of all the elements is the mass M of the body. Therefore, we can rewrite Eq. 12-15 as

$$x_{cog} = \frac{1}{M} \sum x_i m_i. \tag{12-16}$$

The right side of this equation gives the coordinate x_{com} of the body's center of mass (Eq. 9-4). We now have what we sought to prove:

$$x_{cog} = x_{com}. \tag{12-17}$$

12-5 | Some Examples of Static Equilibrium

In this section we examine four sample problems involving static equilibrium. In each, we select a system of one or more objects to which we apply the equations of equilibrium (Eqs. 12-7, 12-8, and 12-9). The forces involved in the equilibrium are all in the xy plane, which means that the torques involved are parallel to the z axis. Thus, in applying Eq. 12-9, the balance of torques, we select an axis parallel to the z axis about which to calculate the torques. Although Eq. 12-9 is satisfied for *any* such choice of axis, you will see that certain choices simplify the application of Eq. 12-9 by eliminating one or more unknown force terms.

CHECKPOINT 2 The figure gives an overhead view of a uniform rod in static equilibrium. (a) Can you find the magnitudes of unknown forces $\vec{F}_1$ and $\vec{F}_2$ by balancing the forces? (b) If you wish to find the magnitude of force $\vec{F}_2$ by using a balance of torques equation, where should you place a rotational axis to eliminate $\vec{F}_1$ from the equation? (c) The magnitude of $\vec{F}_2$ turns out to be 65 N. What then is the magnitude of $\vec{F}_1$?

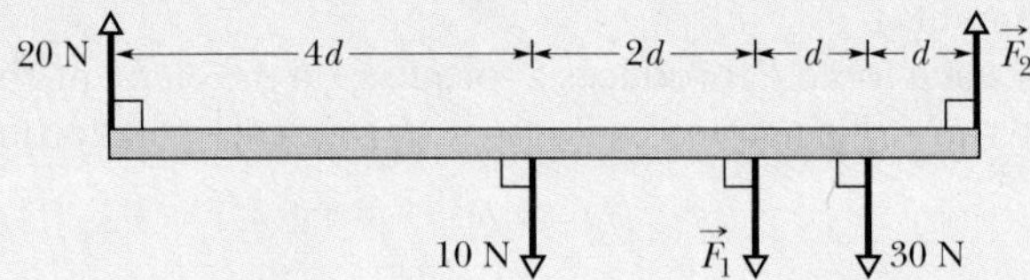

Sample Problem 12-1 Build your skill

In Fig. 12-5a, a uniform beam, of length L and mass m = 1.8 kg, is at rest on two scales. A uniform block, with mass M = 2.7 kg, is at rest on the beam, with its center a distance $L/4$ from the beam's left end. What do the scales read?

KEY IDEAS The first steps in the solution of *any* problem about static equilibrium are these: Clearly define the system to be analyzed and then draw a free-body diagram of it, indicating all the forces on the system. Here, let us choose the system as the beam and block taken together. Then the forces on the system are shown in the free-body diagram of Fig. 12-5b. (Choosing the system takes experience, and often there can be more than one good choice; see item 2 of Problem-Solving Tactic 1 below.) Because the system is in static equilibrium, we can apply the balance of forces equations (Eqs. 12-7 and 12-8) and the balance of torques equation (Eq. 12-9) to it.

Calculations: The normal forces on the beam from the scales are $\vec{F}_l$ on the left and $\vec{F}_r$ on the right. The scale readings that we want are equal to the magnitudes of those forces. The gravitational force $\vec{F}_{g,beam}$ on the beam acts at the beam's center of mass and is equal to $m\vec{g}$. Similarly, the gravitational force $\vec{F}_{g,block}$ on the block acts at the block's center of mass and is equal to $M\vec{g}$. However, to simplify Fig. 12-5b, the block is represented by a dot within the boundary of the beam and $\vec{F}_{g,block}$ is

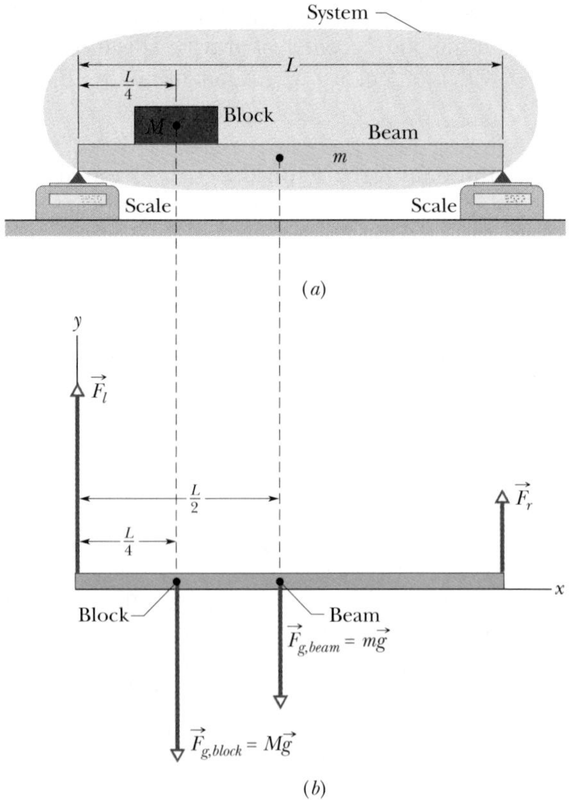

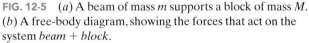

FIG. 12-5 (a) A beam of mass m supports a block of mass M. (b) A free-body diagram, showing the forces that act on the system *beam + block*.

drawn with its tail on that dot. (This shift of vector $\vec{F}_{g,block}$ along its line of action does not alter the torque due to $\vec{F}_{g,block}$ about any axis perpendicular to the figure.)

The forces have no x components, so Eq. 12-7 ($F_{\text{net},x} = 0$) provides no information. For the y components, Eq. 12-8 ($F_{\text{net},y} = 0$) gives us

$$F_l + F_r - Mg - mg = 0. \qquad (12\text{-}18)$$

This equation contains two unknowns, the forces F_l and F_r, so we also need to use Eq. 12-9, the balance of torques equation. We can apply it to *any* rotation axis perpendicular to the plane of Fig. 12-5. Let us choose a rotation axis through the left end of the beam. We shall also use our general rule for assigning signs to torques: If a torque would cause an initially stationary body to rotate clockwise about the rotation axis, the torque is negative. If the rotation would be counterclockwise, the torque is positive. Finally, we shall write the torques in the form $r_\perp F$, where the moment arm $r_\perp$ is 0 for $\vec{F}_l$, $L/4$ for $M\vec{g}$, $L/2$ for $m\vec{g}$, and L for $\vec{F}_r$.

We now can write the balancing equation ($\tau_{\text{net},z} = 0$) as

$$(0)(F_l) - (L/4)(Mg) - (L/2)(mg) + (L)(F_r) = 0,$$

which gives us

$$\begin{aligned} F_r &= \tfrac{1}{4}Mg + \tfrac{1}{2}mg \\ &= \tfrac{1}{4}(2.7\text{ kg})(9.8\text{ m/s}^2) + \tfrac{1}{2}(1.8\text{ kg})(9.8\text{ m/s}^2) \\ &= 15.44\text{ N} \approx 15\text{ N}. \qquad \text{(Answer)} \end{aligned}$$

Now, solving Eq. 12-18 for F_l and substituting this result, we find

$$\begin{aligned} F_l &= (M + m)g - F_r \\ &= (2.7\text{ kg} + 1.8\text{ kg})(9.8\text{ m/s}^2) - 15.44\text{ N} \\ &= 28.66\text{ N} \approx 29\text{ N}. \qquad \text{(Answer)} \end{aligned}$$

Notice the strategy in the solution: When we wrote an equation for the balance of force components, we got stuck with two unknowns. If we had written an equation for the balance of torques around some *arbitrary* axis, we would have again gotten stuck with those two unknowns. However, because we chose the axis to pass through the point of application of one of the unknown forces, here $\vec{F}_l$, we did not get stuck. Our choice neatly eliminated that force from the torque equation, allowing us to solve for the other unknown force magnitude F_r. Then we returned to the equation for the balance of force components to find the remaining unknown force magnitude.

Sample Problem 12-2

In Fig. 12-6*a*, a ladder of length $L = 12$ m and mass $m = 45$ kg leans against a slick (frictionless) wall. The ladder's upper end is at height $h = 9.3$ m above the pavement on which the lower end rests (the pavement is not frictionless). The ladder's center of mass is $L/3$ from the lower end. A firefighter of mass $M = 72$ kg climbs the ladder until her center of mass is $L/2$ from the lower end. What then are the magnitudes of the forces on the ladder from the wall and the pavement?

KEY IDEAS First, we choose our system as being the firefighter and ladder, together, and then we draw the free-body diagram of Fig. 12-6*b*. Because the system is in static equilibrium, the balancing equations (Eqs. 12-7 through 12-9) apply to it.

Calculations: In Fig. 12-6*b*, the firefighter is represented with a dot within the boundary of the ladder. The gravitational force on her is represented with its equivalent $M\vec{g}$, and that vector has been shifted along its line of action, so that its tail is on the dot. (The shift does not alter a torque due to $M\vec{g}$ about any axis perpendicular to the figure.)

The only force on the ladder from the wall is the horizontal force $\vec{F}_w$ (there cannot be a frictional force along a frictionless wall). The force $\vec{F}_p$ on the ladder from the pavement has a horizontal component $\vec{F}_{px}$ that is a static frictional force and a vertical component $\vec{F}_{py}$ that is a normal force.

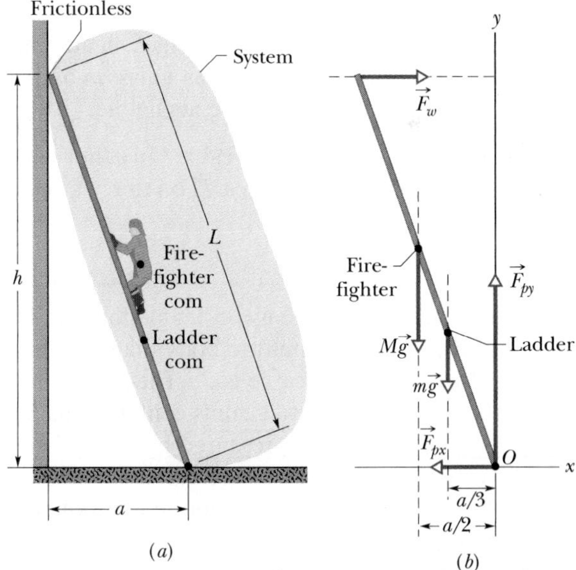

FIG. 12-6 (*a*) A firefighter climbs halfway up a ladder that is leaning against a frictionless wall. The pavement beneath the ladder is not frictionless. (*b*) A free-body diagram, showing the forces that act on the *firefighter + ladder* system. The origin O of a coordinate system is placed at the point of application of the unknown force $\vec{F}_p$ (whose vector components $\vec{F}_{px}$ and $\vec{F}_{py}$ are shown).

To apply the balancing equations, let's start with Eq. 12-9 ($\tau_{\text{net},z} = 0$). To choose an axis about which to calculate the torques, note that we have unknown forces

($\vec{F}_w$ and $\vec{F}_p$) at the two ends of the ladder. To eliminate, say, $\vec{F}_p$ from the calculation, we place the axis at point O, perpendicular to the figure. We also place the origin of an xy coordinate system at O. We can find torques about O with any of Eqs. 10-39 through 10-41, but Eq. 10-41 ($\tau = r_\perp F$) is easiest to use here.

To find the moment arm $r_\perp$ of $\vec{F}_w$, we draw a line of action through that vector (horizontal dashed line in Fig. 12-6*b*). Then $r_\perp$ is the perpendicular distance between O and the line of action. In Fig. 12-6*b*, $r_\perp$ extends along the y axis and is equal to the height h. We similarly draw lines of action for $M\vec{g}$ and $m\vec{g}$ and see that their moment arms extend along the x axis. For the distance a shown in Fig. 12-6*a*, the moment arms are $a/2$ (the firefighter is halfway up the ladder) and $a/3$ (the ladder's center of mass is one-third of the way up the ladder), respectively. The moment arms for $\vec{F}_{px}$ and $\vec{F}_{py}$ are zero.

Now, with torques written in the form $r_\perp F$, the balancing equation $\tau_{\text{net},z} = 0$ becomes

$$-(h)(F_w) + (a/2)(Mg) + (a/3)(mg) + (0)(F_{px}) + (0)(F_{py}) = 0. \quad (12\text{-}19)$$

(Recall our rule: A positive torque corresponds to counterclockwise rotation and a negative torque corresponds to clockwise rotation.)

Using the Pythagorean theorem, we find that

$$a = \sqrt{L^2 - h^2} = 7.58 \text{ m}.$$

Then Eq. 12-19 gives us

$$F_w = \frac{ga(M/2 + m/3)}{h} = \frac{(9.8 \text{ m/s}^2)(7.58 \text{ m})(72/2 \text{ kg} + 45/3 \text{ kg})}{9.3 \text{ m}} = 407 \text{ N} \approx 410 \text{ N}. \quad \text{(Answer)}$$

Now we need to use the force balancing equations. The equation $F_{\text{net},x} = 0$ gives us

$$F_w - F_{px} = 0,$$

so

$$F_{px} = F_w = 410 \text{ N}. \quad \text{(Answer)}$$

The equation $F_{\text{net},y} = 0$ gives us

$$F_{py} - Mg - mg = 0,$$

so

$$F_{py} = (M + m)g = (72 \text{ kg} + 45 \text{ kg})(9.8 \text{ m/s}^2) = 1146.6 \text{ N} \approx 1100 \text{ N}. \quad \text{(Answer)}$$

Sample Problem 12-3

Figure 12-7*a* shows a safe, of mass $M = 430$ kg, hanging by a rope from a boom with dimensions $a = 1.9$ m and $b = 2.5$ m. The boom consists of a hinged beam and a horizontal cable. The uniform beam has a mass m of 85 kg; the masses of the cable and rope are negligible.

(a) What is the tension T_c in the cable? In other words, what is the magnitude of the force $\vec{T}_c$ on the beam from the cable?

KEY IDEAS The system here is the beam alone, and the forces on it are shown in the free-body diagram of Fig. 12-7*b*. The force from the cable is $\vec{T}_c$. The gravitational force on the beam acts at the beam's center of mass (at the beam's center) and is represented by its equivalent $m\vec{g}$. The vertical component of the force on the beam from the hinge is $\vec{F}_v$, and the horizontal component of the force from the hinge is $\vec{F}_h$. The force from the rope supporting the safe is $\vec{T}_r$. Because beam, rope, and safe are stationary, the magnitude of $\vec{T}_r$ is equal to the weight of the safe: $T_r = Mg$. We place the origin O of an xy coordinate system at the hinge. Because the system is in static equilibrium, the balancing equations apply to it.

Calculations: Let us start with Eq. 12-9 ($\tau_{\text{net},z} = 0$). Note that we are asked for the magnitude of force $\vec{T}_c$ and not of forces $\vec{F}_h$ and $\vec{F}_v$ acting at the hinge, at point O. To eliminate $\vec{F}_h$ and $\vec{F}_v$ from the torque calculation,

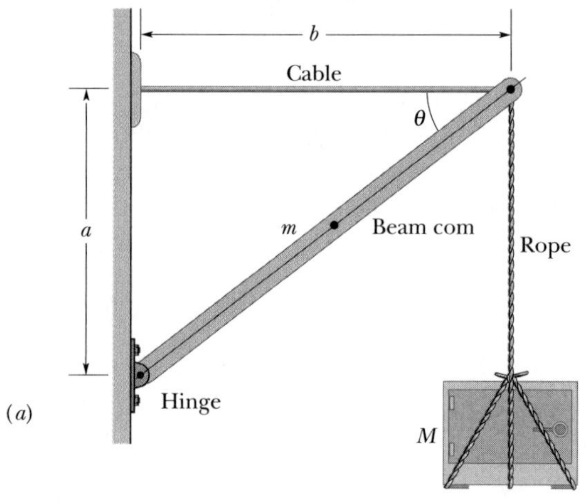

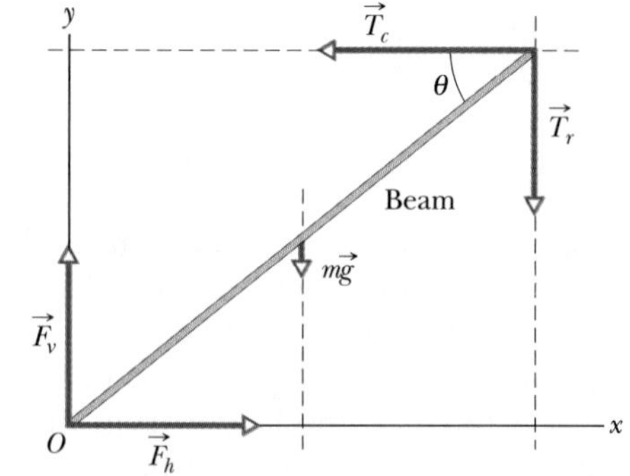

FIG. 12-7 (*a*) A heavy safe is hung from a boom consisting of a horizontal steel cable and a uniform beam. (*b*) A free-body diagram for the beam.

we should calculate torques about an axis that is perpendicular to the figure at point O. Then $\vec{F}_h$ and $\vec{F}_v$ will have moment arms of zero. The lines of action for $\vec{T}_c$, $\vec{T}_r$, and $m\vec{g}$ are dashed in Fig. 12-7*b*. The corresponding moment arms are a, b, and $b/2$.

Writing torques in the form of $r_\perp F$ and using our rule about signs for torques, the balancing equation $\tau_{\text{net},z} = 0$ becomes

$$(a)(T_c) - (b)(T_r) - (\tfrac{1}{2}b)(mg) = 0.$$

Substituting Mg for T_r and solving for T_c, we find that

$$T_c = \frac{gb(M + \frac{1}{2}m)}{a}$$

$$= \frac{(9.8 \text{ m/s}^2)(2.5 \text{ m})(430 \text{ kg} + 85/2 \text{ kg})}{1.9 \text{ m}}$$

$$= 6093 \text{ N} \approx 6100 \text{ N}. \qquad \text{(Answer)}$$

(b) Find the magnitude F of the net force on the beam from the hinge.

KEY IDEA Now we want F_h and F_v so we can combine them to get F. Because we know T_c, we apply the force balancing equations to the beam.

Calculations: For the horizontal balance, we write $F_{\text{net},x} = 0$ as

$$F_h - T_c = 0,$$

and so $\qquad F_h = T_c = 6093 \text{ N}.$

For the vertical balance, we write $F_{\text{net},y} = 0$ as

$$F_v - mg - T_r = 0.$$

Substituting Mg for T_r and solving for F_v, we find that

$$F_v = (m + M)g = (85 \text{ kg} + 430 \text{ kg})(9.8 \text{ m/s}^2)$$
$$= 5047 \text{ N}.$$

From the Pythagorean theorem, we now have

$$F = \sqrt{F_h^2 + F_v^2}$$
$$= \sqrt{(6093 \text{ N})^2 + (5047 \text{ N})^2} \approx 7900 \text{ N}. \qquad \text{(Answer)}$$

Note that F is substantially greater than either the combined weights of the safe and the beam, 5000 N, or the tension in the horizontal wire, 6100 N.

Sample Problem 12-4

Let's assume that the Tower of Pisa is a uniform hollow cylinder of radius $R = 9.8$ m and height $h = 60$ m. The center of mass is located at height $h/2$, along the cylinder's central axis. In Fig. 12-8*a*, the cylinder is upright. In Fig. 12-8*b*, it leans rightward (toward the tower's southern wall) by $\theta = 5.5°$, which shifts the com by a distance d. Let's assume that the ground exerts only two forces on the tower. A normal force $\vec{F}_{NL}$ acts on the left (northern) wall, and a normal force $\vec{F}_{NR}$ acts on the right (southern) wall. By what percent does the magnitude F_{NR} increase because of the leaning?

KEY IDEA Because the tower is still standing, it is in equilibrium and thus the sum of torques calculated around any point must be zero.

Calculations: Because we want to calculate F_{NR} on the right side and do not know or want F_{NL} on the left side, we use a pivot point on the left side to calculate torques. The forces on the upright tower are represented in Fig. 12-8*c*. The gravitational force $m\vec{g}$, taken to act at the com, has a vertical line of action and a moment arm of R (the perpendicular distance from the pivot to the line of action). About the pivot, the torque associated with this force would tend to create clockwise rotation and thus is negative. The normal force $\vec{F}_{NR}$ on the southern wall also has a vertical line of action, and its moment arm is $2R$. About the pivot, the torque associated with this force would tend to create counterclockwise rotation and thus is positive. We now can write the torque-balancing equation ($\tau_{\text{net},z} = 0$) as

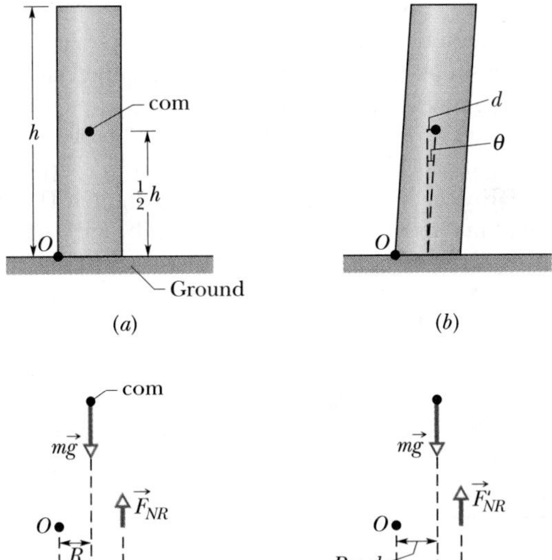

FIG. 12-8 A cylinder modeling the Tower of Pisa: (*a*) upright and (*b*) leaning, with the center of mass shifted rightward. The forces and moment arms to find torques about a pivot at point O for the cylinder (*c*) upright and (*d*) leaning.

$$-(R)(mg) + (2R)(F_{NR}) = 0,$$

which yields

$$F_{NR} = \tfrac{1}{2}mg. \qquad (12\text{-}20)$$

We should have been able to guess this result: With the center of mass located on the central axis (the cylinder's line of symmetry), the right side supports half the cylinder's weight.

In Fig. 12-8*b*, the com is shifted rightward by distance

$$d = \tfrac{1}{2}h \tan \theta .$$

The only change in the balance of torques equation is that the moment arm for the gravitational force is now $R + d$ and the normal force at the right has a new magnitude F'_{NR} (Fig. 12-8*d*). Thus, we write

$$-(R + d)(mg) + (2R)(F'_{NR}) = 0,$$

which gives us

$$F'_{NR} = \frac{(R + d)}{2R}\, mg. \qquad (12\text{-}21)$$

Dividing Eq. 12-21 by Eq. 12-20 and then substituting for d, we obtain

$$\frac{F'_{NR}}{F_{NR}} = \frac{R + d}{R} = 1 + \frac{d}{R} = 1 + \frac{0.5h \tan \theta}{R}.$$

Substituting the values of $h = 60$ m, $R = 9.8$ m, and $\theta = 5.5°$ leads to

$$\frac{F'_{NR}}{F_{NR}} = 1.29.$$

Thus, our simple model predicts that, although the tilt is modest, the normal force on the tower's southern wall has increased by about 30%. One danger to the tower is that the force may cause the southern wall to buckle and explode outward.

PROBLEM-SOLVING TACTICS

Tactic 1: ***Static Equilibrium Problems*** Here is a list of steps for solving static equilibrium problems:

1. Draw a *sketch* of the problem.
2. Select the *system* to which you will apply the laws of equilibrium, drawing a closed curve around the system on your sketch to fix it clearly in your mind. In some situations you can select a single object as the system; it is the object you wish to be in equilibrium. In other situations, you might include additional objects in the system *if* their inclusion simplifies the calculations for equilibrium. For example, suppose in Sample Problem 12-2 you select only the ladder as the system. Then in Fig. 12-6*b* you will have to account for additional unknown forces exerted on the ladder by the hands and feet of the firefighter. These additional unknowns complicate the equilibrium calculations. The system of Fig. 12-6 was chosen to include the firefighter so that those unknown forces are *internal* to the system and thus need not be found in order to solve Sample Problem 12-2.
3. Draw a *free-body diagram* of the system. Show all the forces that act on the system, labeling them and making sure that their points of application and lines of action are correctly shown.
4. Draw in the *x and y axes* of a coordinate system with at least one axis parallel to one or more unknown force. Resolve into components the forces that do not lie along one of the axes. In all our sample problems it made sense to choose the x axis horizontal and the y axis vertical.
5. Write the two *balance of forces equations,* using symbols throughout.
6. Choose one or more rotation axes perpendicular to the plane of the figure and write the *balance of torques equation* for each axis. If you choose an axis that passes through the line of action of an unknown force, the equation will be simplified because that force will not appear in it.
7. *Solve* your equations *algebraically* for the unknowns. Some students feel more confident in substituting numbers with units in the independent equations at this stage, especially if the algebra is particularly involved. However, experienced problem solvers prefer the algebra, which reveals the dependence of solutions on the various variables.
8. Finally, *substitute numbers* with units in your algebraic solutions, obtaining numerical values for the unknowns.
9. Look at your answer—does it make sense? Is it obviously too large or too small? Is the sign correct? Are the units appropriate?

12-6 | Indeterminate Structures

For the problems of this chapter, we have only three independent equations at our disposal, usually two balance of forces equations and one balance of torques equation about a given rotation axis. Thus, if a problem has more than three unknowns, we cannot solve it.

It is easy to find such problems. In Sample Problem 12-2, for example, we could have assumed that there is friction between the wall and the top of the ladder. Then there would have been a vertical frictional force acting where the ladder touches the wall, making a total of four unknown forces. With only three equations, we could not have solved this problem.

Consider also an unsymmetrically loaded car. What are the forces—all different—on the four tires? Again, we cannot find them because we have only three independent equations. Similarly, we can solve an equilibrium problem for a table with three legs but not for one with four legs. Problems like these, in which there are more unknowns than equations, are called **indeterminate.**

Yet solutions to indeterminate problems exist in the real world. If you rest the tires of the car on four platform scales, each scale will register a definite reading, the sum of the readings being the weight of the car. What is eluding us in our efforts to find the individual forces by solving equations?

The problem is that we have assumed—without making a great point of it—that the bodies to which we apply the equations of static equilibrium are perfectly rigid. By this we mean that they do not deform when forces are applied to them. Strictly, there are no such bodies. The tires of the car, for example, deform easily under load until the car settles into a position of static equilibrium.

We have all had experience with a wobbly restaurant table, which we usually level by putting folded paper under one of the legs. If a big enough elephant sat on such a table, however, you may be sure that if the table did not collapse, it would deform just like the tires of a car. Its legs would all touch the floor, the forces acting upward on the table legs would all assume definite (and different) values as in Fig. 12-9, and the table would no longer wobble. How do we find the values of those forces acting on the legs?

To solve such indeterminate equilibrium problems, we must supplement equilibrium equations with some knowledge of *elasticity,* the branch of physics and engineering that describes how real bodies deform when forces are applied to them. The next section provides an introduction to this subject.

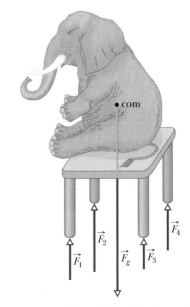

FIG. 12-9 The table is an indeterminate structure. The four forces on the table legs differ from one another in magnitude and cannot be found from the laws of static equilibrium alone.

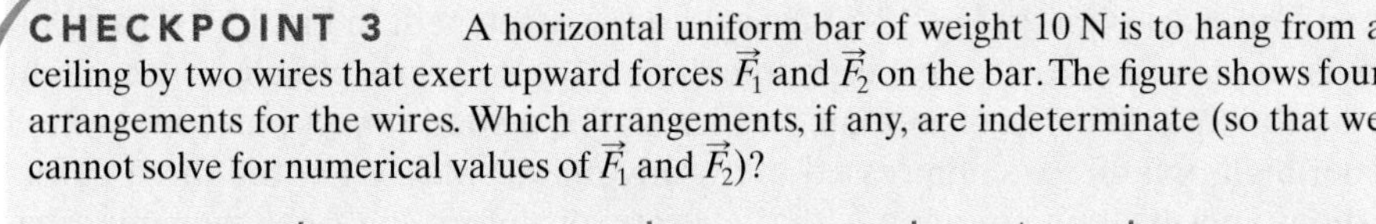

CHECKPOINT 3 A horizontal uniform bar of weight 10 N is to hang from a ceiling by two wires that exert upward forces $\vec{F}_1$ and $\vec{F}_2$ on the bar. The figure shows four arrangements for the wires. Which arrangements, if any, are indeterminate (so that we cannot solve for numerical values of $\vec{F}_1$ and $\vec{F}_2$)?

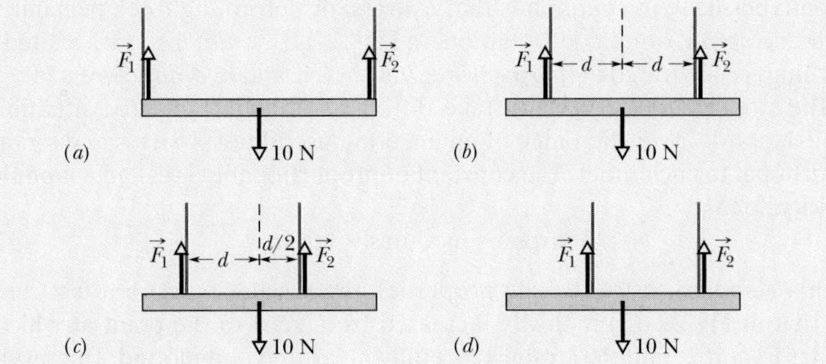

12-7 | Elasticity

When a large number of atoms come together to form a metallic solid, such as an iron nail, they settle into equilibrium positions in a three-dimensional *lattice,* a repetitive arrangement in which each atom is a well-defined equilibrium distance from its nearest neighbors. The atoms are held together by interatomic forces that are modeled as tiny springs in Fig. 12-10. The lattice is remarkably rigid, which is another way of saying that the "interatomic springs" are extremely stiff. It is for this reason that we perceive many ordinary objects, such as metal ladders, tables, and spoons, as perfectly rigid. Of course, some ordinary objects, such as garden hoses or rubber gloves, do not strike us as rigid at all. The atoms that make up these objects *do not* form a rigid lattice like that of Fig. 12-10 but are aligned in long, flexible molecular chains, each chain being only loosely bound to its neighbors.

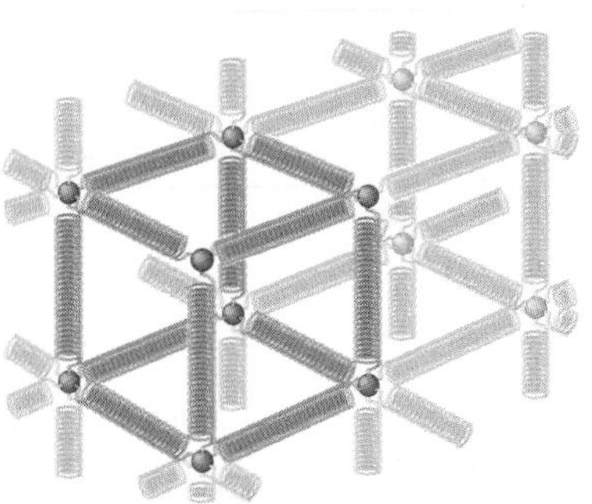

FIG. 12-10 The atoms of a metallic solid are distributed on a repetitive three-dimensional lattice. The springs represent interatomic forces.

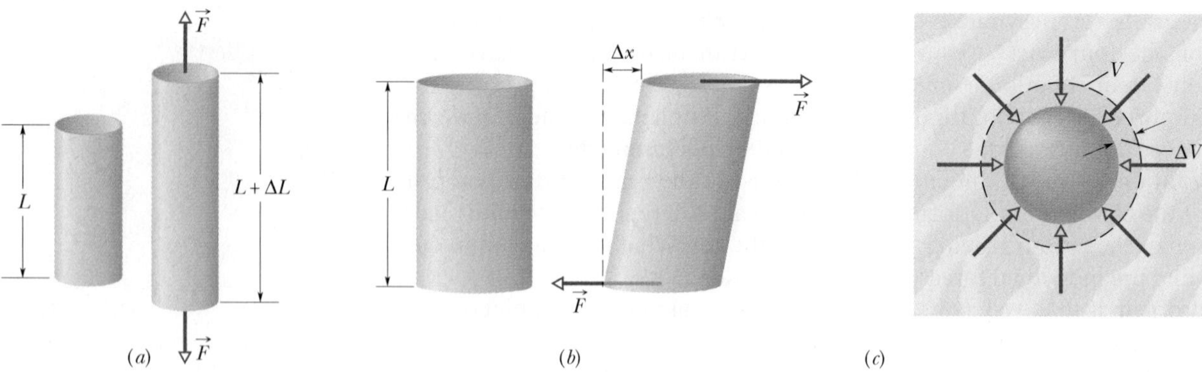

FIG. 12-11 (*a*) A cylinder subject to *tensile stress* stretches by an amount ΔL. (*b*) A cylinder subject to *shearing stress* deforms by an amount Δx, somewhat like a pack of playing cards would. (*c*) A solid sphere subject to uniform *hydraulic stress* from a fluid shrinks in volume by an amount ΔV. All the deformations shown are greatly exaggerated.

All real "rigid" bodies are to some extent **elastic,** which means that we can change their dimensions slightly by pulling, pushing, twisting, or compressing them. To get a feeling for the orders of magnitude involved, consider a vertical steel rod 1 m long and 1 cm in diameter attached to a factory ceiling. If you hang a subcompact car from the free end of such a rod, the rod will stretch but only by about 0.5 mm, or 0.05%. Furthermore, the rod will return to its original length when the car is removed.

If you hang two cars from the rod, the rod will be permanently stretched and will not recover its original length when you remove the load. If you hang three cars from the rod, the rod will break. Just before rupture, the elongation of the rod will be less than 0.2%. Although deformations of this size seem small, they are important in engineering practice. (Whether a wing under load will stay on an airplane is obviously important.)

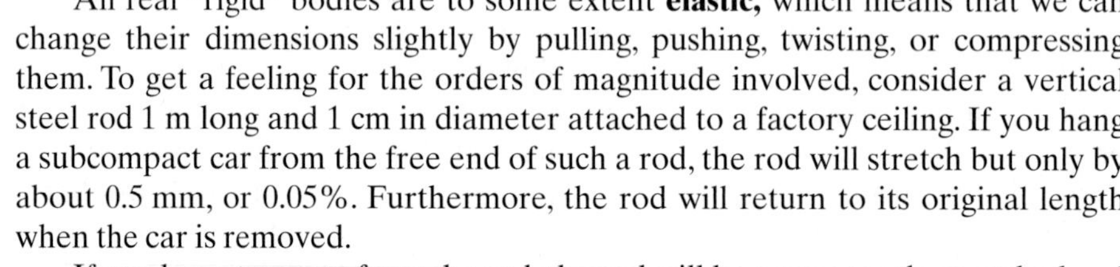

FIG. 12-12 A test specimen used to determine a stress–strain curve such as that of Fig. 12-13. The change ΔL that occurs in a certain length L is measured in a tensile stress–strain test.

Figure 12-11 shows three ways in which a solid might change its dimensions when forces act on it. In Fig. 12-11*a*, a cylinder is stretched. In Fig. 12-11*b*, a cylinder is deformed by a force perpendicular to its long axis, much as we might deform a pack of cards or a book. In Fig. 12-11*c*, a solid object placed in a fluid under high pressure is compressed uniformly on all sides. What the three deformation types have in common is that a **stress,** or deforming force per unit area, produces a **strain,** or unit deformation. In Fig. 12-11, *tensile stress* (associated with stretching) is illustrated in (*a*), *shearing stress* in (*b*), and *hydraulic stress* in (*c*).

The stresses and the strains take different forms in the three situations of Fig. 12-11, but—over the range of engineering usefulness—stress and strain are proportional to each other. The constant of proportionality is called a **modulus of elasticity,** so that

$$\text{stress} = \text{modulus} \times \text{strain}. \tag{12-22}$$

In a standard test of tensile properties, the tensile stress on a test cylinder (like that in Fig. 12-12) is slowly increased from zero to the point at which the cylinder fractures, and the strain is carefully measured and plotted. The result is a graph of stress versus strain like that in Fig. 12-13. For a substantial range of applied stresses, the stress–strain relation is linear, and the specimen recovers its original dimensions when the stress is removed; it is here that Eq. 12-22 applies. If the stress is increased beyond the **yield strength** S_y of the specimen, the specimen becomes permanently deformed. If the stress continues to increase, the specimen eventually ruptures, at a stress called the **ultimate strength** S_u.

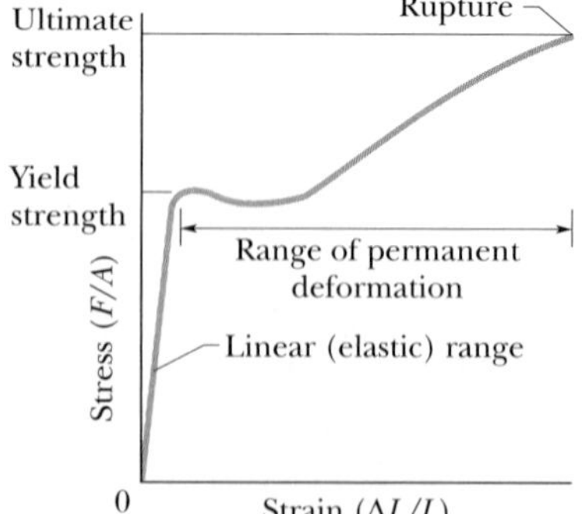

FIG. 12-13 A stress–strain curve for a steel test specimen such as that of Fig. 12-12. The specimen deforms permanently when the stress is equal to the *yield strength* of the specimen's material. It ruptures when the stress is equal to the *ultimate strength* of the material.

Tension and Compression

For simple tension or compression, the stress on the object is defined as F/A, where F is the magnitude of the force applied perpendicularly to an area A on the object. The strain, or unit deformation, is then the dimensionless quantity $\Delta L/L$, the fractional (or sometimes percentage) change in a length of the specimen. If the specimen is a long rod and the stress does not exceed the yield strength, then not only the entire rod but also every section of it experiences the same strain when a given stress is applied. Because the strain is dimensionless, the modulus in Eq. 12-22 has the same dimensions as the stress—namely, force per unit area.

The modulus for tensile and compressive stresses is called the **Young's modulus** and is represented in engineering practice by the symbol E. Equation 12-22 becomes

$$\frac{F}{A} = E\frac{\Delta L}{L}. \tag{12-23}$$

The strain $\Delta L/L$ in a specimen can often be measured conveniently with a *strain gage* (Fig. 12-14). This simple and useful device, which can be attached directly to operating machinery with an adhesive, is based on the principle that its electrical properties are dependent on the strain it undergoes.

Although the Young's modulus for an object may be almost the same for tension and compression, the object's ultimate strength may well be different for the two types of stress. Concrete, for example, is very strong in compression but is so weak in tension that it is almost never used in that manner. Table 12-1 shows the Young's modulus and other elastic properties for some materials of engineering interest.

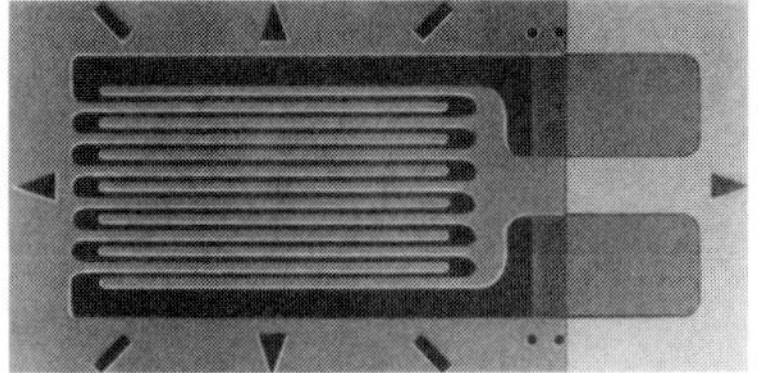

FIG. 12-14 A strain gage of overall dimensions 9.8 mm by 4.6 mm. The gage is fastened with adhesive to the object whose strain is to be measured; it experiences the same strain as the object. The electrical resistance of the gage varies with the strain, permitting strains up to 3% to be measured. *(Courtesy Vishay Micro-Measurements Group, Raleigh, NC)*

Shearing

In the case of shearing, the stress is also a force per unit area, but the force vector lies in the plane of the area rather than perpendicular to it. The strain is the dimensionless ratio $\Delta x/L$, with the quantities defined as shown in Fig. 12-11*b*. The corresponding modulus, which is given the symbol G in engineering practice, is called the **shear modulus.** For shearing, Eq. 12-22 is written as

$$\frac{F}{A} = G\frac{\Delta x}{L}. \tag{12-24}$$

Shearing stresses play a critical role in the buckling of shafts that rotate under load and in bone fractures caused by bending.

Hydraulic Stress

In Fig. 12-11*c*, the stress is the fluid pressure p on the object, and, as you will see in Chapter 14, pressure is a force per unit area. The strain is $\Delta V/V$, where V is the original volume of the specimen and ΔV is the absolute value of the change in volume. The corresponding modulus, with symbol B, is called the **bulk modulus** of the material. The object is said to be under *hydraulic compression,* and the pressure can be called the *hydraulic stress.* For this situation, we write Eq. 12-22 as

$$p = B\frac{\Delta V}{V}. \tag{12-25}$$

TABLE 12-1

Some Elastic Properties of Selected Materials of Engineering Interest

Material	Density ρ (kg/m^3)	Young's Modulus E (10^9 N/m^2)	Ultimate Strength S_u (10^6 N/m^2)	Yield Strength S_y (10^6 N/m^2)
Steel[a]	7860	200	400	250
Aluminum	2710	70	110	95
Glass	2190	65	50[b]	—
Concrete[c]	2320	30	40[b]	—
Wood[d]	525	13	50[b]	—
Bone	1900	9[b]	170[b]	—
Polystyrene	1050	3	48	—

[a]Structural steel (ASTM-A36).
[b]In compression.
[c]High strength.
[d]Douglas fir.

The bulk modulus is 2.2×10^9 N/m^2 for water and 1.6×10^{11} N/m^2 for steel. The pressure at the bottom of the Pacific Ocean, at its average depth of about 4000 m, is 4.0×10^7 N/m^2. The fractional compression $\Delta V/V$ of a volume of water due to this pressure is 1.8%; that for a steel object is only about 0.025%. In general, solids—with their rigid atomic lattices—are less compressible than liquids, in which the atoms or molecules are less tightly coupled to their neighbors.

Sample Problem 12-5

One end of a steel rod of radius $R = 9.5$ mm and length $L = 81$ cm is held in a vise. A force of magnitude $F = 62$ kN is then applied perpendicularly to the end face (uniformly across the area) at the other end. What are the stress on the rod and the elongation ΔL and strain of the rod?

KEY IDEAS (1) The stress is the ratio of the magnitude F of the perpendicular force to the area A. The ratio is the left side of Eq. 12-23. (2) The elongation ΔL is related to the stress and Young's modulus E by Eq. 12-23 ($F/A = E\,\Delta L/L$). (3) Strain is the ratio of the elongation to the initial length L.

Calculations: To find the stress, we write

$$\text{stress} = \frac{F}{A} = \frac{F}{\pi R^2} = \frac{6.2 \times 10^4 \text{ N}}{(\pi)(9.5 \times 10^{-3} \text{ m})^2}$$
$$= 2.2 \times 10^8 \text{ N/m}^2. \qquad \text{(Answer)}$$

The yield strength for structural steel is 2.5×10^8 N/m^2, so this rod is dangerously close to its yield strength.

We find the value of Young's modulus for steel in Table 12-1. Then from Eq. 12-23 we find the elongation:

$$\Delta L = \frac{(F/A)L}{E} = \frac{(2.2 \times 10^8 \text{ N/m}^2)(0.81 \text{ m})}{2.0 \times 10^{11} \text{ N/m}^2}$$
$$= 8.9 \times 10^{-4} \text{ m} = 0.89 \text{ mm}. \qquad \text{(Answer)}$$

For the strain, we have

$$\frac{\Delta L}{L} = \frac{8.9 \times 10^{-4} \text{ m}}{0.81 \text{ m}}$$
$$= 1.1 \times 10^{-3} = 0.11\%. \qquad \text{(Answer)}$$

Sample Problem 12-6

A table has three legs that are 1.00 m in length and a fourth leg that is longer by $d = 0.50$ mm, so that the table wobbles slightly. A steel cylinder with mass $M = 290$ kg is placed on the table (which has a mass much less than M) so that all four legs are compressed but unbuckled and the table is level but no longer wobbles. The legs are wooden cylinders with cross-sectional area $A = 1.0$ cm^2; Young's modulus is $E = 1.3 \times 10^{10}$ N/m^2. What are the magnitudes of the forces on the legs from the floor?

KEY IDEAS We take the table plus steel cylinder as our system. The situation is like that in Fig. 12-9, except now we have a steel cylinder on the table. If the tabletop remains level, the legs must be compressed in the following ways: Each of the short legs must be compressed by the same amount (call it ΔL_3) and thus by the same force of magnitude F_3. The single long leg must be compressed by a larger amount ΔL_4 and thus by a force with a larger magnitude F_4. In other words, for a level tabletop, we must have

$$\Delta L_4 = \Delta L_3 + d. \qquad (12\text{-}26)$$

From Eq. 12-23, we can relate a change in length to the force causing the change with $\Delta L = FL/AE$, where L is the original length of a leg. We can use this relation to replace ΔL_4 and ΔL_3 in Eq. 12-26. However, note that we can approximate the original length L as being the same for all four legs.

Calculations: Making those replacements and that approximation gives us

$$\frac{F_4 L}{AE} = \frac{F_3 L}{AE} + d. \qquad (12\text{-}27)$$

We cannot solve this equation because it has two unknowns, F_4 and F_3.

To get a second equation containing F_4 and F_3, we can use a vertical y axis and then write the balance of vertical forces ($F_{\text{net},y} = 0$) as

$$3F_3 + F_4 - Mg = 0, \qquad (12\text{-}28)$$

where Mg is equal to the magnitude of the gravitational force on the system. (*Three* legs have force $\vec{F}_3$ on them.) To solve the simultaneous equations 12-27 and 12-28 for, say, F_3, we first use Eq. 12-28 to find that $F_4 = Mg - 3F_3$. Substituting that into Eq. 12-27 then yields, after some algebra,

$$F_3 = \frac{Mg}{4} - \frac{dAE}{4L}$$
$$= \frac{(290 \text{ kg})(9.8 \text{ m/s}^2)}{4} - \frac{(5.0 \times 10^{-4} \text{ m})(10^{-4} \text{ m}^2)(1.3 \times 10^{10} \text{ N/m}^2)}{(4)(1.00 \text{ m})}$$
$$= 548 \text{ N} \approx 5.5 \times 10^2 \text{ N}. \qquad \text{(Answer)}$$

From Eq. 12-28 we then find

$$F_4 = Mg - 3F_3 = (290 \text{ kg})(9.8 \text{ m/s}^2) - 3(548 \text{ N})$$
$$\approx 1.2 \text{ kN}. \qquad \text{(Answer)}$$

You can show that to reach their equilibrium configuration, the three short legs are each compressed by 0.42 mm and the single long leg by 0.92 mm.

REVIEW & SUMMARY

Static Equilibrium A rigid body at rest is said to be in **static equilibrium.** For such a body, the vector sum of the external forces acting on it is zero:

$$\vec{F}_{\text{net}} = 0 \quad \text{(balance of forces)}. \qquad (12\text{-}3)$$

If all the forces lie in the xy plane, this vector equation is equivalent to two component equations:

$$F_{\text{net},x} = 0 \quad \text{and} \quad F_{\text{net},y} = 0 \quad \text{(balance of forces)}. \quad (12\text{-}7, 12\text{-}8)$$

Static equilibrium also implies that the vector sum of the external torques acting on the body about *any* point is zero, or

$$\vec{\tau}_{\text{net}} = 0 \quad \text{(balance of torques)}. \qquad (12\text{-}5)$$

If the forces lie in the xy plane, all torque vectors are parallel to the z axis, and Eq. 12-5 is equivalent to the single component equation

$$\tau_{\text{net},z} = 0 \quad \text{(balance of torques)}. \qquad (12\text{-}9)$$

Center of Gravity The gravitational force acts individually on each element of a body. The net effect of all individual actions may be found by imagining an equivalent total gravitational force $\vec{F}_g$ acting at the **center of gravity.** If the gravitational acceleration $\vec{g}$ is the same for all the elements of the body, the center of gravity is at the center of mass.

Elastic Moduli Three **elastic moduli** are used to describe the elastic behavior (deformations) of objects as they respond to forces that act on them. The **strain** (fractional change in length) is linearly related to the applied **stress** (force per unit area) by the proper modulus, according to the general relation

$$\text{stress} = \text{modulus} \times \text{strain}. \qquad (12\text{-}22)$$

Tension and Compression When an object is under tension or compression, Eq. 12-22 is written as

$$\frac{F}{A} = E\frac{\Delta L}{L}, \qquad (12\text{-}23)$$

where $\Delta L/L$ is the tensile or compressive strain of the object, F is the magnitude of the applied force $\vec{F}$ causing the strain, A is the cross-sectional area over which $\vec{F}$ is applied (perpendicular to A, as in Fig. 12-11*a*), and E is the **Young's modulus** for the object. The stress is F/A.

Shearing When an object is under a shearing stress, Eq. 12-22 is written as

$$\frac{F}{A} = G\frac{\Delta x}{L}, \qquad (12\text{-}24)$$

where $\Delta x/L$ is the shearing strain of the object, Δx is the displacement of one end of the object in the direction of the applied force $\vec{F}$ (as in Fig. 12-11*b*), and G is the **shear modulus** of the object. The stress is F/A.

Hydraulic Stress When an object undergoes *hydraulic compression* due to a stress exerted by a surrounding fluid, Eq. 12-22 is written as

$$p = B\frac{\Delta V}{V}, \qquad (12\text{-}25)$$

where p is the pressure (*hydraulic stress*) on the object due to the fluid, $\Delta V/V$ (the strain) is the absolute value of the fractional change in the object's volume due to that pressure, and B is the **bulk modulus** of the object.

QUESTIONS

1 Figure 12-15 shows four overhead views of rotating uniform disks that are sliding across a frictionless floor. Three forces, of magnitude F, $2F$, or $3F$, act on each disk, either at the rim, at the center, or halfway between rim and center. The force vectors rotate along with the disks, and, in the "snapshots" of Fig. 12-15, point left or right. Which disks are in equilibrium?

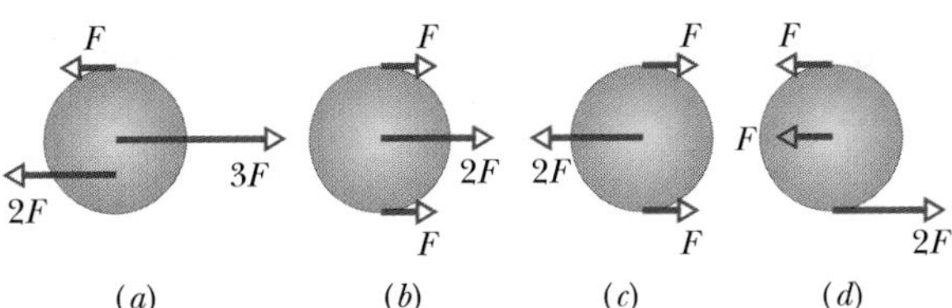

FIG. 12-15 Question 1.

2 Figure 12-16 shows an overhead view of a uniform stick on which four forces act. Suppose we choose a rotational axis through point O, calculate the torques about that axis due to the forces, and find that these torques balance. Will the torques balance if, instead, the rotational axis is chosen to be at (a) point A (on the stick), (b) point B (on line with the stick), or (c) point C (off to one side of the stick)? (d) Suppose, instead, that we find that the torques about point O do not balance. Is there another point about which the torques will balance?

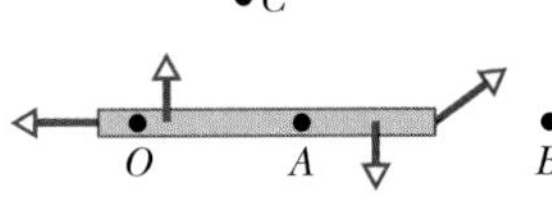

FIG. 12-16 Question 2.

3 Figure 12-17 shows a mobile of toy penguins hanging from a ceiling. Each crossbar is horizontal, has negligible mass, and extends three times as far to the right of the wire supporting it as to the left. Penguin 1 has mass $m_1 = 48$ kg. What are the masses of (a) penguin 2, (b) penguin 3, and (c) penguin 4?

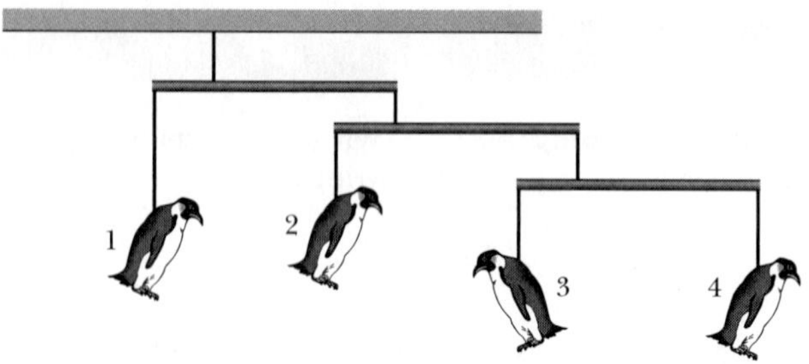

FIG. 12-17 Question 3.

4 In Fig. 12-18, a rigid beam is attached to two posts that are fastened to a floor. A small but heavy safe is placed at the six positions indicated, in turn. Assume that the mass of the beam is negligible compared to that of the safe. (a) Rank the positions according to the force on post A due to the safe, greatest compression first, greatest tension last, and indicate where, if anywhere, the force is zero. (b) Rank them according to the force on post B.

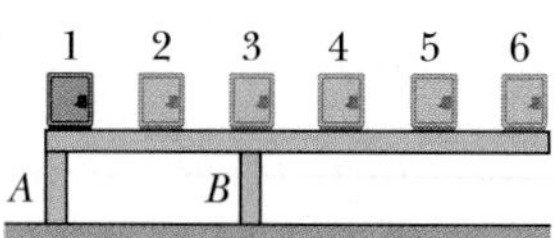

FIG. 12-18 Question 4.

5 Figure 12-19 shows three situations in which the same horizontal rod is supported by a hinge on a wall at one end and a cord at its other end. Without written calculation, rank the situations according to the magnitudes of (a) the force on the rod from the cord, (b) the vertical force on the rod from the hinge, and (c) the horizontal force on the rod from the hinge, greatest first.

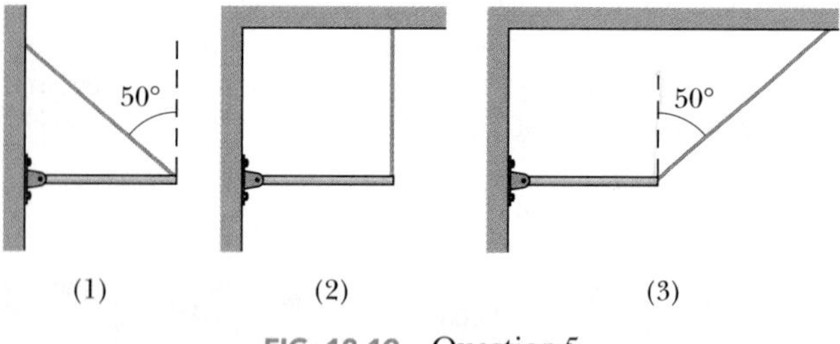

FIG. 12-19 Question 5.

6 A ladder leans against a frictionless wall but is prevented from falling because of friction between it and the ground. Suppose you shift the base of the ladder toward the wall. Determine whether the following become larger, smaller, or stay the same (in magnitude): (a) the normal force on the ladder from the ground, (b) the force on the ladder from the wall, (c) the static frictional force on the ladder from the ground, and (d) the maximum value $f_{s,\max}$ of the static frictional force.

7 In Fig. 12-20, a vertical rod is hinged at its lower end and attached to a cable at its upper end. A horizontal force $\vec{F}_a$ is to be applied to the rod as shown. If the point at which the force is applied is moved up the rod, does the tension in the cable increase, decrease, or remain the same?

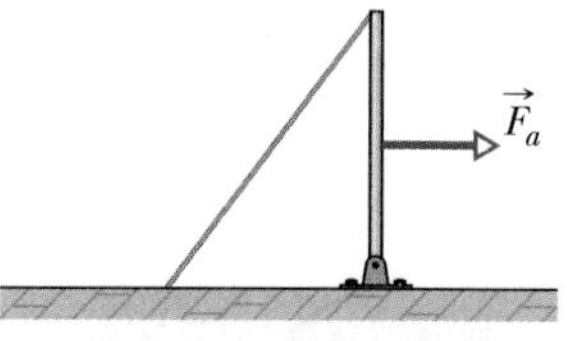

FIG. 12-20 Question 7.

8 Three piñatas hang from the (stationary) assembly of massless pulleys and cords seen in Fig. 12-21. One long cord runs from the ceiling at the right to the lower pulley at the left, looping halfway around all the pulleys. Several shorter cords suspend pulleys from the ceiling or piñatas from the pulleys. The weights (in newtons) of two piñatas are given. (a) What is the weight of the third piñata? (*Hint:* A cord that loops halfway around a pulley pulls on the pulley with a net force that is twice the tension in the cord.) (b) What is the tension in the short cord labeled with T?

FIG. 12-21 Question 8.

9 In Fig. 12-22, a stationary 5 kg rod AC is held against a wall by a rope and friction between rod and wall. The uniform rod is 1 m long, and angle $\theta = 30^\circ$. (a) If you are to find the magnitude of the force $\vec{T}$ on the rod from the rope with a single equation, at what labeled point should a rotational axis be placed? With that choice of axis and counterclockwise torques positive, what is the sign of (b) the torque τ_w due to the rod's weight and (c) the torque τ_r due to the pull on the rod by the rope? (d) Is the magnitude of τ_r greater than, less than, or equal to the magnitude of τ_w?

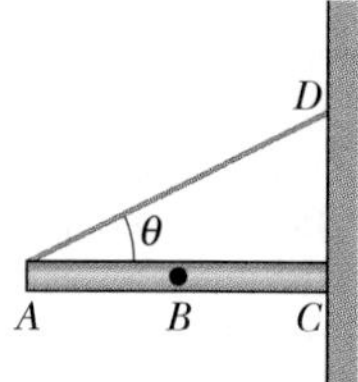

FIG. 12-22 Question 9.

10 Figure 12-23 shows a horizontal block that is suspended by two wires, A and B, which are identical except for their original lengths. The center of mass of the block is closer to wire B than to wire A. (a) Measuring torques about the block's center of mass, state whether the magnitude of the torque due to wire A is greater than, less than, or equal to the magnitude of the torque due to wire B. (b) Which wire exerts more force on the block? (c) If the wires are now equal in length, which one was originally shorter (before the block was suspended)?

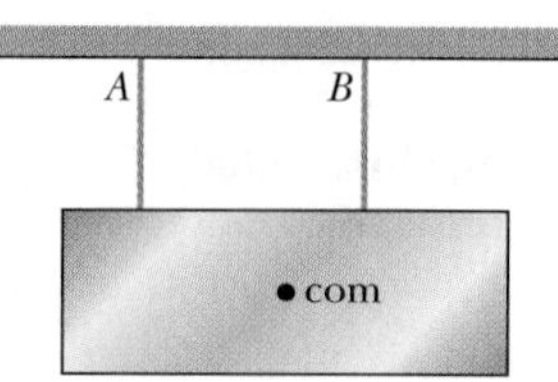

FIG. 12-23 Question 10.

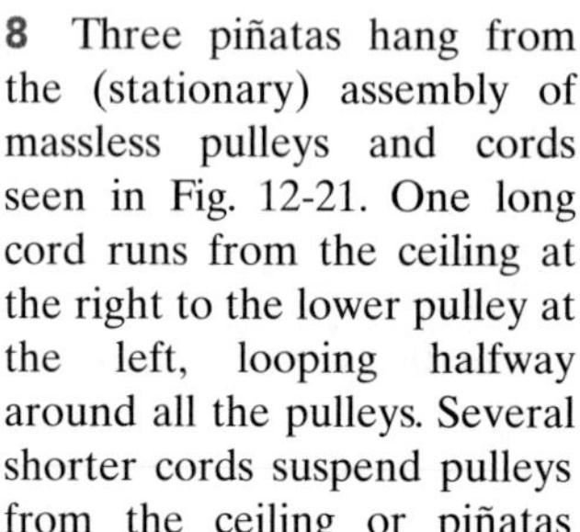

PROBLEMS

GO Tutoring problem available (at instructor's discretion) in *WileyPLUS* and WebAssign
SSM Worked-out solution available in Student Solutions Manual
WWW Worked-out solution is at http://www.wiley.com/college/halliday
• – ••• Number of dots indicates level of problem difficulty
ILW Interactive solution is at http://www.wiley.com/college/halliday
Additional information available in *The Flying Circus of Physics* and at flyingcircusofphysics.com

sec. 12-4 The Center of Gravity

•**1** Because g varies so little over the extent of most structures, any structure's center of gravity effectively coincides with its center of mass. Here is a fictitious example where g varies more significantly. Figure 12-24 shows an array of six particles, each with mass m, fixed to the edge of a rigid structure of negligible mass. The distance between adjacent particles along the edge is 2.00 m. The following table gives the value of g (m/s^2) at each particle's location. Using the coordinate system shown, find (a) the x coordinate x_{com} and (b) the y coordinate y_{com} of the center of mass of the six-particle system. Then find (c) the x coordinate x_{cog} and (d) the y coordinate y_{cog} of the center of gravity of the six-particle system.

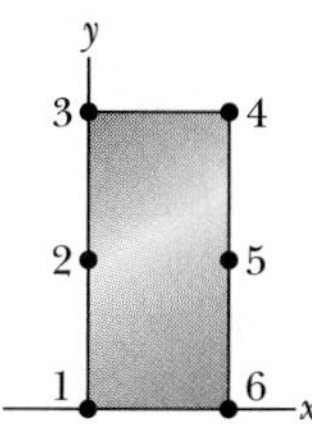

FIG. 12-24 Problem 1.

Particle	g	Particle	g
1	8.00	4	7.40
2	7.80	5	7.60
3	7.60	6	7.80

sec. 12-5 Some Examples of Static Equilibrium

•**2** An archer's bow is drawn at its midpoint until the tension in the string is equal to the force exerted by the archer. What is the angle between the two halves of the string?

•**3** A rope of negligible mass is stretched horizontally between two supports that are 3.44 m apart. When an object of weight 3160 N is hung at the center of the rope, the rope is observed to sag by 35.0 cm. What is the tension in the rope? ILW

•**4** A physics Brady Bunch, whose weights in newtons are indicated in Fig. 12-25, is balanced on a seesaw. What is the number of the person who causes the largest torque about the rotation axis at *fulcrum f* directed (a) out of the page and (b) into the page?

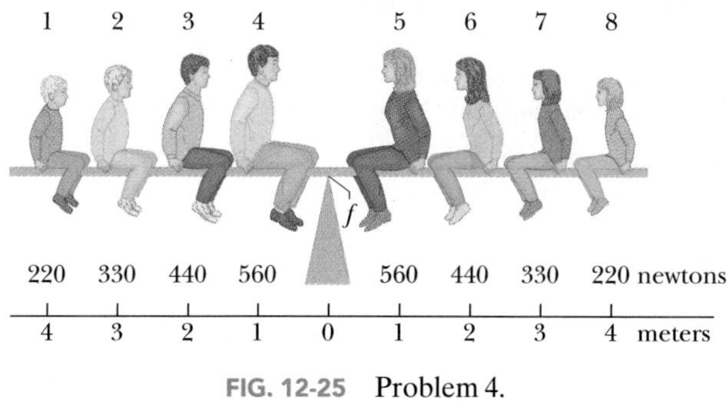

FIG. 12-25 Problem 4.

•**5** In Fig. 12-26, a uniform sphere of mass $m = 0.85$ kg and radius $r = 4.2$ cm is held in place by a massless rope attached to a frictionless wall a distance $L = 8.0$ cm above the center of the sphere. Find (a) the tension in the rope and (b) the force on the sphere from the wall. SSM WWW

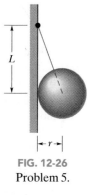

FIG. 12-26 Problem 5.

•**6** An automobile with a mass of 1360 kg has 3.05 m between the front and rear axles. Its center of gravity is located 1.78 m behind the front axle. With the automobile on level ground, determine the magnitude of the force from the ground on (a) each front wheel (assuming equal forces on the front wheels) and (b) each rear wheel (assuming equal forces on the rear wheels).

•**7** A diver of weight 580 N stands at the end of a diving board of length $L = 4.5$ m and negligible mass (Fig. 12-27). The board is fixed to two pedestals separated by distance $d = 1.5$ m. Of the forces acting on the board, what are the (a) magnitude and (b) direction (up or down) of the force from the left pedestal and the (c) magnitude and (d) direction (up or down) of the force from the right pedestal? (e) Which pedestal (left or right) is being stretched, and (f) which is being compressed? SSM

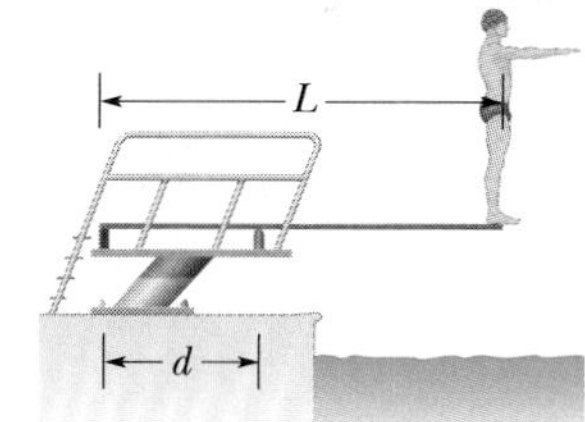

FIG. 12-27 Problem 7.

•**8** A scaffold of mass 60 kg and length 5.0 m is supported in a horizontal position by a vertical cable at each end. A window washer of mass 80 kg stands at a point 1.5 m from one end. What is the tension in (a) the nearer cable and (b) the farther cable?

•**9** A 75 kg window cleaner uses a 10 kg ladder that is 5.0 m long. He places one end on the ground 2.5 m from a wall, rests the upper end against a cracked window, and climbs the ladder. He is 3.0 m up along the ladder when the window breaks. Neglect friction between the ladder and window and assume that the base of the ladder does not slip. When the window is on the verge of breaking, what are (a) the magnitude of the force on the window from the ladder, (b) the magnitude of the force on the ladder from the ground, and (c) the angle (relative to the horizontal) of that force on the ladder?

•**10** In Fig. 12-28, a man is trying to get his car out of mud on the shoulder of a road. He ties one end of a rope tightly around the front bumper and the other end tightly around a utility pole 18 m away. He then pushes sideways on the rope at its midpoint with a force of 550 N, displacing the center of the rope 0.30 m from its previous position, and the car barely moves. What is the magnitude of the force on the car from the rope? (The rope stretches somewhat.)

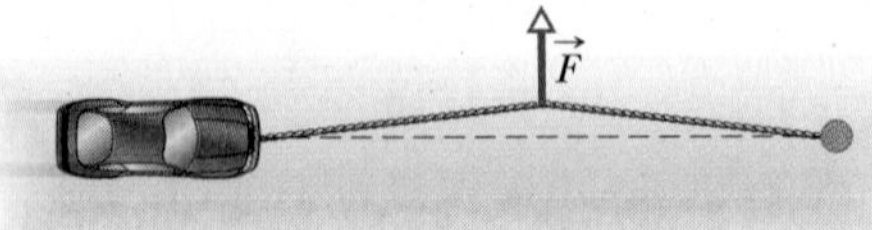

FIG. 12-28 Problem 10.

•**11** A meter stick balances horizontally on a knife-edge at the 50.0 cm mark. With two 5.00 g coins stacked over the 12.0 cm mark, the stick is found to balance at the 45.5 cm mark. What is the mass of the meter stick? SSM

•**12** The system in Fig. 12-29 is in equilibrium, with the string in the center exactly horizontal. Block A weighs 40 N, block B weighs 50 N, and angle ϕ is 35°. Find (a) tension T_1, (b) tension T_2, (c) tension T_3, and (d) angle θ. GO

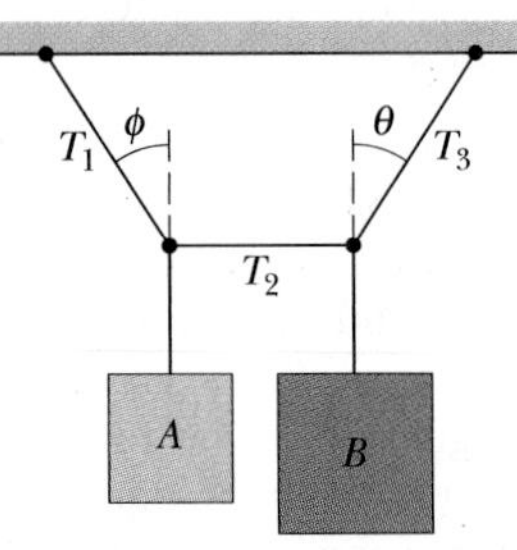

FIG. 12-29 Problem 12.

•**13** Forces $\vec{F}_1$, $\vec{F}_2$, and $\vec{F}_3$ act on the structure of Fig. 12-30, shown in an overhead view. We wish to put the structure in equilibrium by applying a fourth force, at a point such as P. The fourth force has vector components $\vec{F}_h$ and $\vec{F}_v$. We are given that $a = 2.0$ m, $b = 3.0$ m, $c = 1.0$ m, $F_1 = 20$ N, $F_2 = 10$ N, and $F_3 = 5.0$ N. Find (a) F_h, (b) F_v, and (c) d. ILW

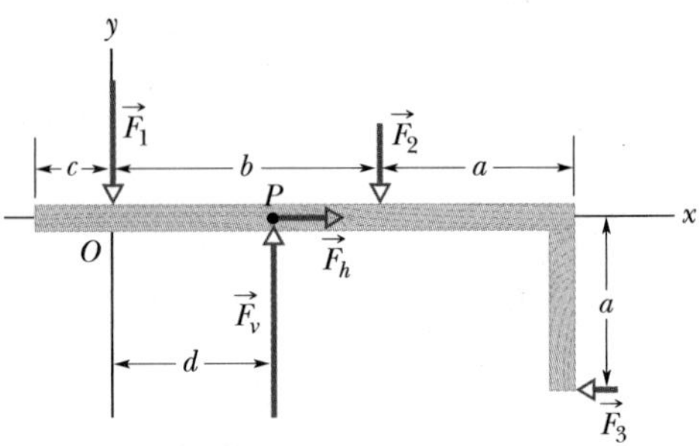

FIG. 12-30 Problem 13.

•**14** A uniform cubical crate is 0.750 m on each side and weighs 500 N. It rests on a floor with one edge against a very small, fixed obstruction. At what least height above the floor must a horizontal force of magnitude 350 N be applied to the crate to tip it?

•**15** To crack a certain nut in a nutcracker, forces with magnitudes of at least 40 N must act on its shell from both sides. For the nutcracker of Fig. 12-31, with distances $L = 12$ cm and $d = 2.6$ cm, what are the force components $F_\perp$ (perpendicular to the handles) corresponding to that 40 N?

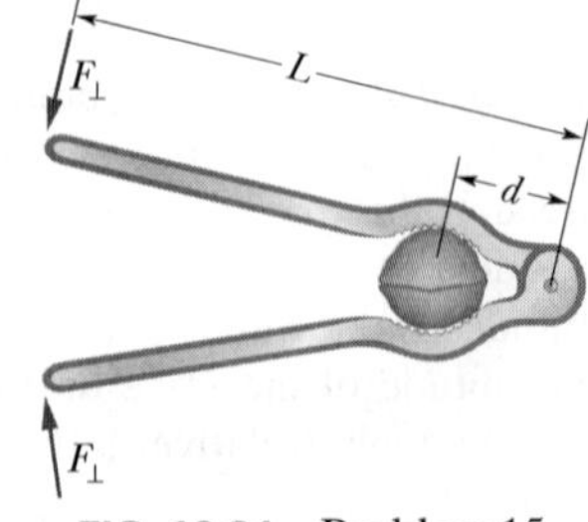

FIG. 12-31 Problem 15.

•**16** In Fig. 12-32, a horizontal scaffold, of length 2.00 m and uniform mass 50.0 kg, is suspended from a building by two cables. The scaffold has dozens of paint cans stacked on it at various points. The total mass of the paint cans is 75.0 kg. The tension in the cable at the right is 722 N. How far horizontally from *that* cable is the center of mass of the system of paint cans?

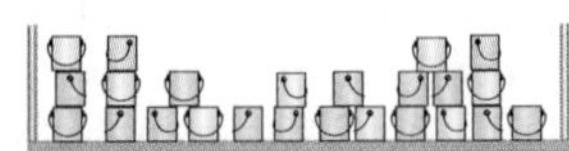

FIG. 12-32 Problem 16.

•**17** Figure 12-33 shows the anatomical structures in the lower leg and foot that are involved in standing on tiptoe, with the heel raised slightly off the floor so that the foot effectively contacts the floor only at point P. Assume distance $a = 5.0$ cm, distance $b = 15$ cm, and the person's weight $W = 900$ N. Of the forces acting on the foot, what are the (a) magnitude and (b) direction (up or down) of the force at point A from the calf muscle and the (c) magnitude and (d) direction (up or down) of the force at point B from the lower leg bones?

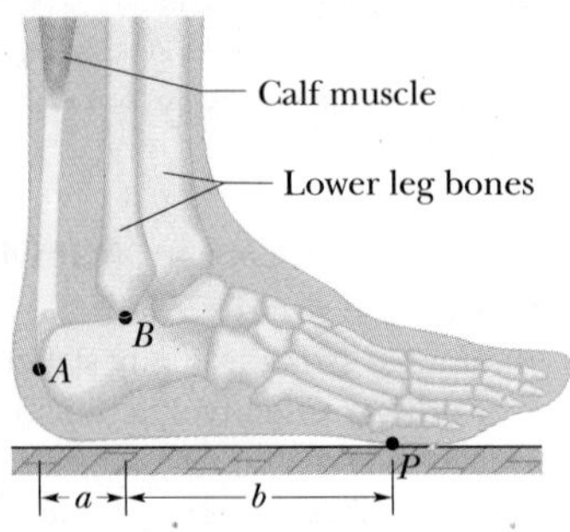

FIG. 12-33 Problem 17.

•**18** A bowler holds a bowling ball ($M = 7.2$ kg) in the palm of his hand (Fig. 12-34). His upper arm is vertical, his lower arm (1.8 kg) is horizontal. What is the magnitude of (a) the force of the biceps muscle on the lower arm and (b) the force between the bony structures at the elbow contact point?

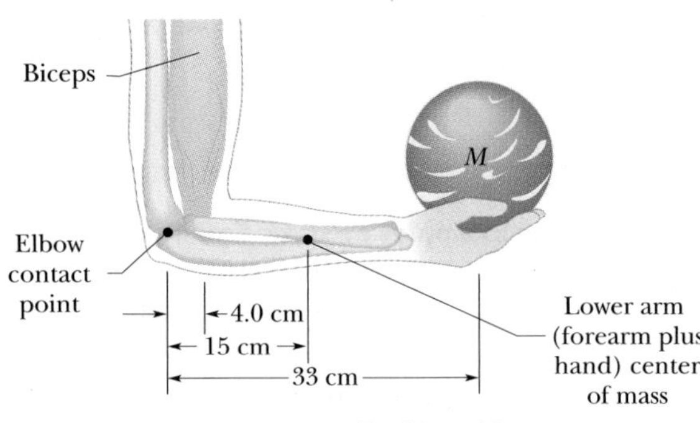

FIG. 12-34 Problem 18.

•**19** In Fig. 12-35, a uniform beam of weight 500 N and length 3.0 m is suspended horizontally. On the left it is hinged to a wall; on the right it is supported by a cable bolted to the wall at distance D above the beam. The least tension that will snap the cable is 1200 N. (a) What value of D corresponds to that tension? (b) To prevent the cable from snapping, should D be increased or decreased from that value?

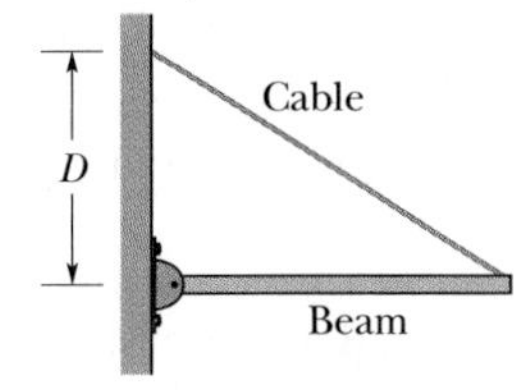

FIG. 12-35 Problem 19.

•**20** In Fig. 12-36, horizontal scaffold 2, with uniform mass $m_2 = 30.0$ kg and length $L_2 = 2.00$ m, hangs from horizontal scaffold 1, with uniform mass $m_1 = 50.0$ kg. A 20.0 kg box of nails lies on scaffold 2, centered at distance $d = 0.500$ m from the left end. What is the tension T in the cable indicated? GO

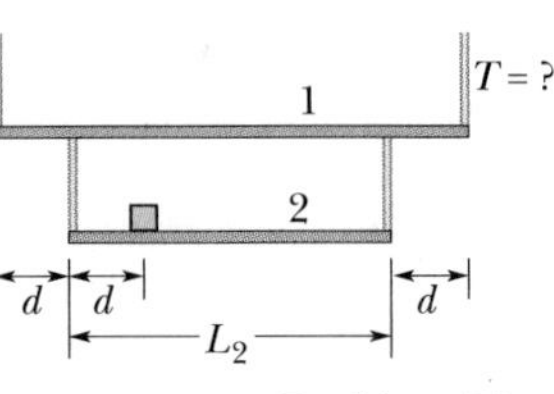

FIG. 12-36 Problem 20.

••21 In Fig. 12-37, what magnitude of (constant) force $\vec{F}$ applied horizontally at the axle of the wheel is necessary to raise the wheel over an obstacle of height $h = 3.00$ cm? The wheel's radius is $r = 6.00$ cm, and its mass is $m = 0.800$ kg. SSM WWW

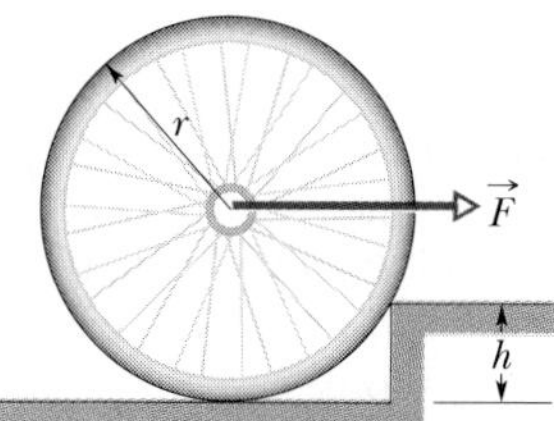

FIG. 12-37 Problem 21.

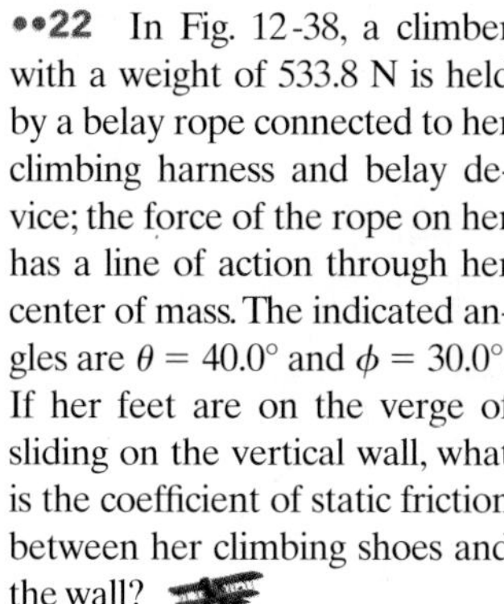

••22 In Fig. 12-38, a climber with a weight of 533.8 N is held by a belay rope connected to her climbing harness and belay device; the force of the rope on her has a line of action through her center of mass. The indicated angles are $\theta = 40.0°$ and $\phi = 30.0°$. If her feet are on the verge of sliding on the vertical wall, what is the coefficient of static friction between her climbing shoes and the wall?

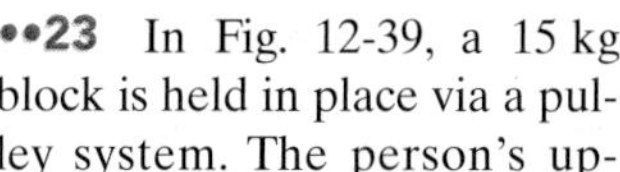

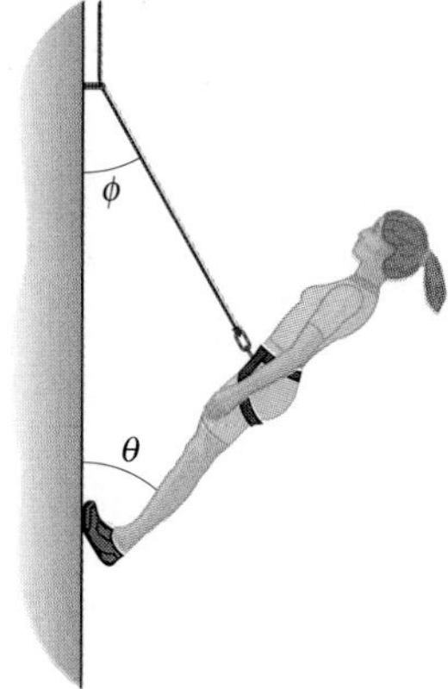

FIG. 12-38 Problem 22.

••23 In Fig. 12-39, a 15 kg block is held in place via a pulley system. The person's upper arm is vertical; the forearm is at angle $\theta = 30°$ with the horizontal. Forearm and hand together have a mass of 2.0 kg, with a center of mass at distance $d_1 = 15$ cm from the contact point of the forearm bone and the upper-arm bone (humerus). The triceps muscle pulls vertically upward on the forearm at distance $d_2 = 2.5$ cm behind that contact point. Distance d_3 is 35 cm. What are the (a) magnitude and (b) direction (up or down) of the force on the forearm from the triceps muscle and the (c) magnitude and (d) direction (up or down) of the force on the forearm from the humerus?

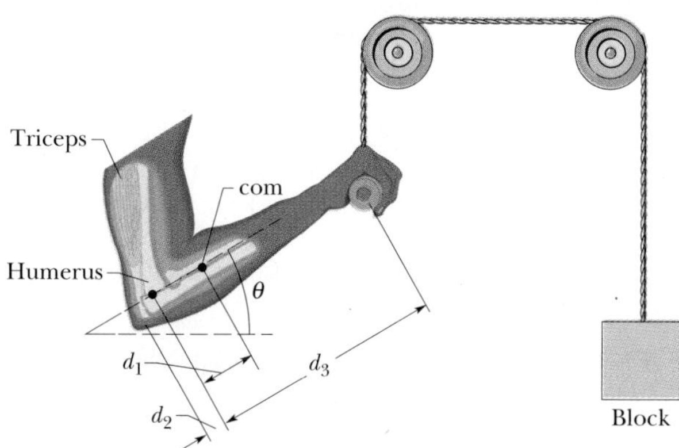

FIG. 12-39 Problem 23.

••24 In Fig. 12-40, a climber leans out against a vertical ice wall that has negligible friction. Distance a is 0.914 m and distance L is 2.10 m. His center of mass is distance $d = 0.940$ m from the feet–ground contact point. If he is on the verge of sliding, what is the coefficient of static friction between feet and ground?

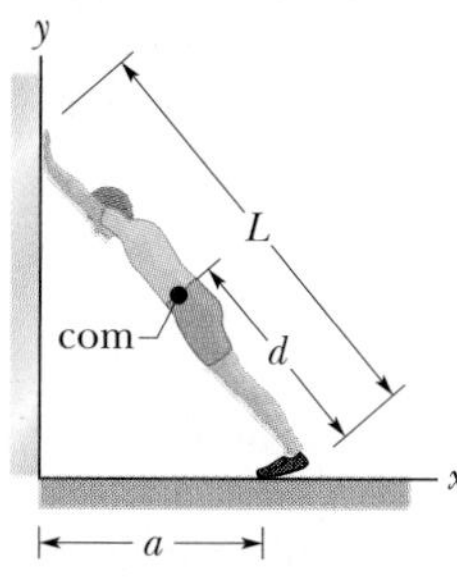

FIG. 12-40 Problem 24.

••25 In Fig. 12-41, one end of a uniform beam of weight 222 N is hinged to a wall; the other end is supported by a wire that makes angles $\theta = 30.0°$ with both wall and beam. Find (a) the tension in the wire and the (b) horizontal and (c) vertical components of the force of the hinge on the beam.

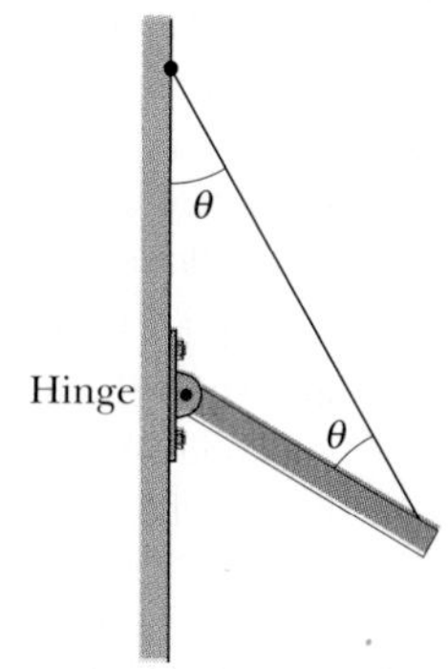

FIG. 12-41 Problem 25.

••26 In Fig. 12-42, a 55 kg rock climber is in a lie-back climb along a fissure, with hands pulling on one side of the fissure and feet pressed against the opposite side. The fissure has width $w = 0.20$ m, and the center of mass of the climber is a horizontal distance $d = 0.40$ m from the fissure. The coefficient of static friction between hands and rock is $\mu_1 = 0.40$, and between boots and rock it is $\mu_2 = 1.2$. (a) What is the least horizontal pull by the hands and push by the feet that will keep the climber stable? (b) For the horizontal pull of (a), what must be the vertical distance h between hands and feet? If the climber encounters wet rock, so that μ_1 and μ_2 are reduced, what happens to (c) the answer to (a) and (d) the answer to (b)? GO

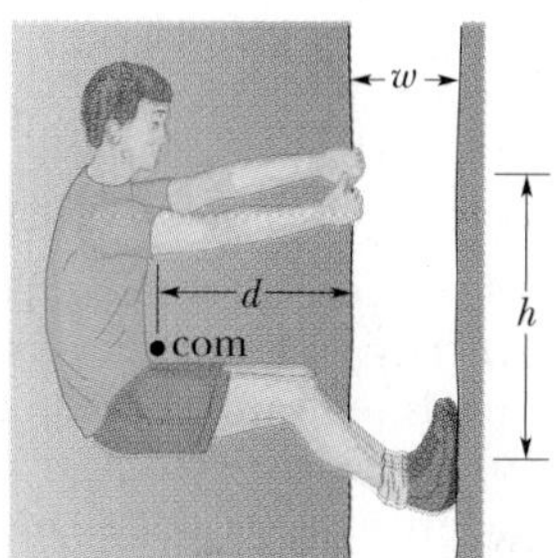

FIG. 12-42 Problem 26.

••27 The system in Fig. 12-43 is in equilibrium. A concrete block of mass 225 kg hangs from the end of the uniform strut of mass 45.0 kg. For angles $\phi = 30.0°$ and $\theta = 45.0°$, find (a) the tension T in the cable and the (b) horizontal and (c) vertical components of the force on the strut from the hinge. ILW

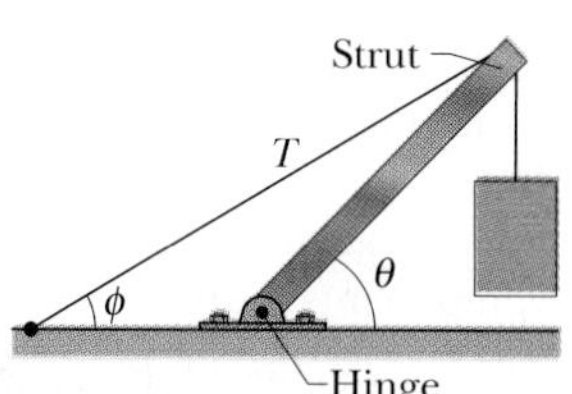

FIG. 12-43 Problem 27.

••28 In Fig. 12-44, a 50.0 kg uniform square sign, of edge length $L = 2.00$ m, is hung from a horizontal rod of length $d_h = 3.00$ m and negligible mass. A cable is attached to the end of the rod and to a point on the wall at distance $d_v = 4.00$ m above the point where the rod is hinged to the wall. (a) What is the tension in the cable? What are the (b) magnitude and (c) direction (left or right) of the horizontal component of the force on the rod from the wall, and the (d) magnitude and (e) direction (up or down) of the vertical component of this force?

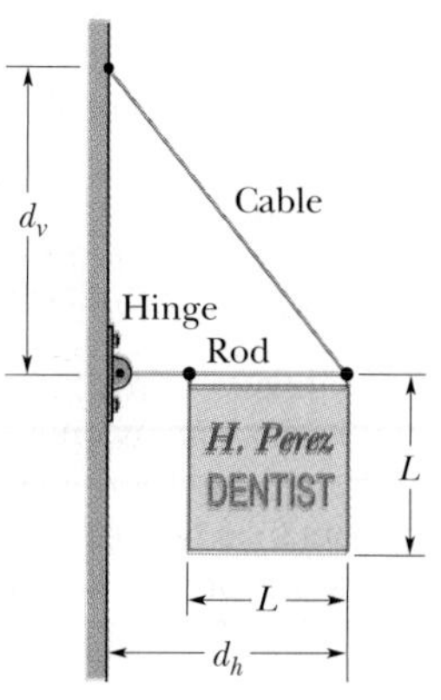

FIG. 12-44 Problem 28.

••29 In Fig. 12-45, a nonuniform bar is suspended at rest in a horizontal position by two

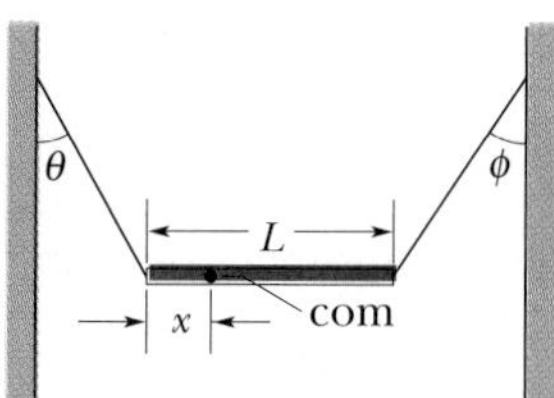

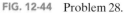

FIG. 12-45 Problem 29.

massless cords. One cord makes the angle $\theta = 36.9°$ with the vertical; the other makes the angle $\phi = 53.1°$ with the vertical. If the length L of the bar is 6.10 m, compute the distance x from the left end of the bar to its center of mass.

••30 In Fig. 12-46, suppose the length L of the uniform bar is 3.00 m and its weight is 200 N. Also, let the block's weight W = 300 N and the angle $\theta = 30.0°$. The wire can withstand a maximum tension of 500 N. (a) What is the maximum possible distance x before the wire breaks? With the block placed at this maximum x, what are the (b) horizontal and (c) vertical components of the force on the bar from the hinge at A? GO

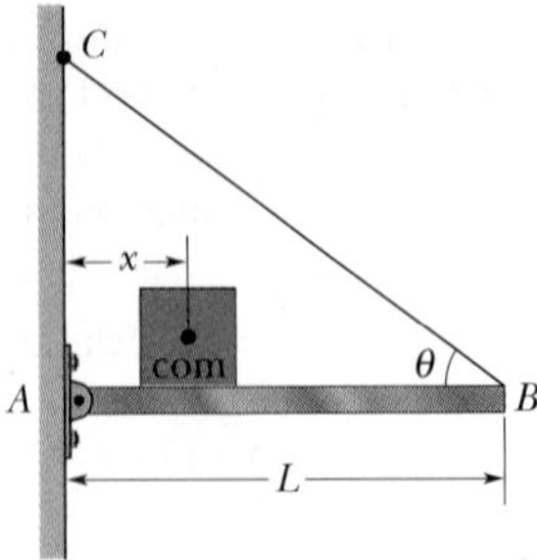

FIG. 12-46 Problems 30 and 32.

••31 A door has a height of 2.1 m along a y axis that extends vertically upward and a width of 0.91 m along an x axis that extends outward from the hinged edge of the door. A hinge 0.30 m from the top and a hinge 0.30 m from the bottom each support half the door's mass, which is 27 kg. In unit-vector notation, what are the forces on the door at (a) the top hinge and (b) the bottom hinge?

••32 In Fig. 12-46, a thin horizontal bar AB of negligible weight and length L is hinged to a vertical wall at A and supported at B by a thin wire BC that makes an angle θ with the horizontal. A block of weight W can be moved anywhere along the bar; its position is defined by the distance x from the wall to its center of mass. As a function of x, find (a) the tension in the wire, and the (b) horizontal and (c) vertical components of the force on the bar from the hinge at A.

••33 A cubical box is filled with sand and weighs 890 N. We wish to "roll" the box by pushing horizontally on one of the upper edges. (a) What minimum force is required? (b) What minimum coefficient of static friction between box and floor is required? (c) If there is a more efficient way to roll the box, find the smallest possible force that would have to be applied directly to the box to roll it. (*Hint:* At the onset of tipping, where is the normal force located?) SSM WWW

••34 Figure 12-47 shows a 70 kg climber hanging by only the *crimp hold* of one hand on the edge of a shallow horizontal ledge in a rock wall. (The fingers are pressed down to gain purchase.) Her feet touch the rock wall at distance H = 2.0 m directly below her crimped fingers but do not provide any support. Her center of mass is distance a = 0.20 m from the wall. Assume that the force from the ledge supporting her fingers is equally shared by the four fingers. What are the values of the (a) horizontal component F_h and (b) vertical component F_v of the force on *each* fingertip?

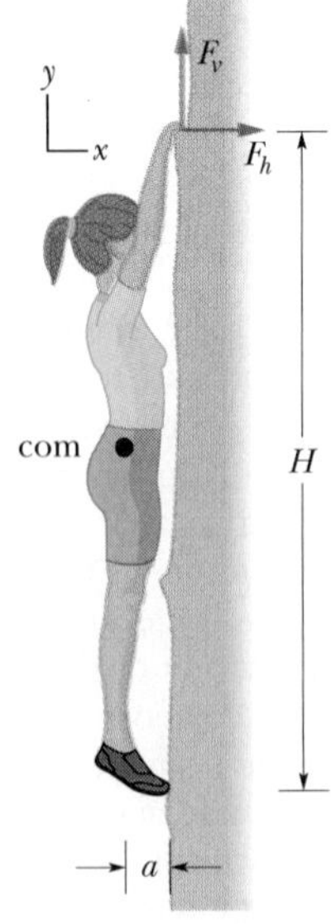

FIG. 12-47 Problem 34.

••35 Figure 12-48*a* shows a vertical uniform beam of length L that is hinged at its lower end. A horizontal force $\vec{F}_a$ is applied to the beam at distance y from the lower end. The beam remains vertical because of a cable attached at the upper end, at angle θ with the horizontal. Figure 12-48*b* gives the tension T in the cable as a function of the position of the applied force given as a fraction y/L of the beam length. The scale of the T axis is set by T_s = 600 N. Figure 12-48*c* gives the magnitude F_h of the horizontal force on the beam from the hinge, also as a function of y/L. Evaluate (a) angle θ and (b) the magnitude of $\vec{F}_a$.

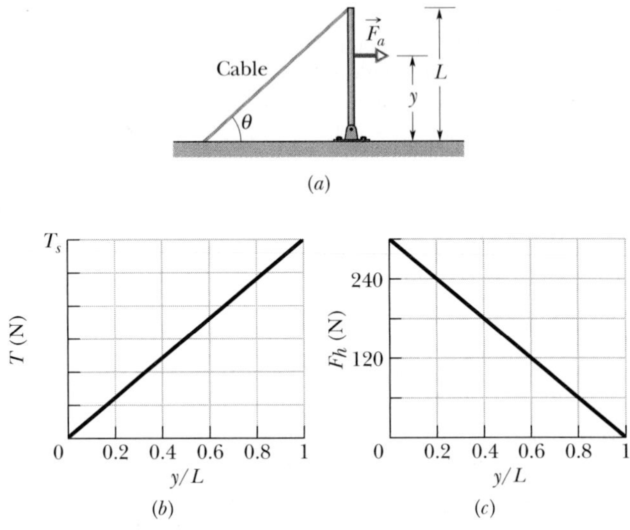

FIG. 12-48 Problem 35.

••36 In Fig. 12-49, the driver of a car on a horizontal road makes an emergency stop by applying the brakes so that all four wheels lock and skid along the road. The coefficient of kinetic friction between tires and road is 0.40. The separation between the front and rear axles is L = 4.2 m, and the center of mass of the car is located at distance d = 1.8 m behind the front axle and distance h = 0.75 m above the road. The car weighs 11 kN. Find the magnitude of (a) the braking acceleration of the car, (b) the normal force on each rear wheel, (c) the normal force on each front wheel, (d) the braking force on each rear wheel, and (e) the braking force on each front wheel. (*Hint:* Although the car is not in translational equilibrium, it *is* in rotational equilibrium.)

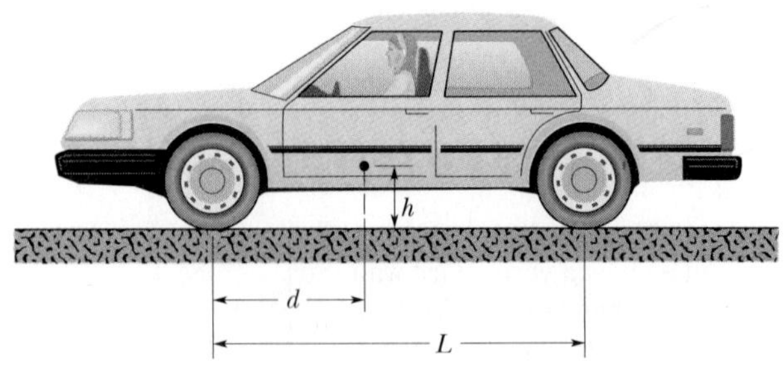

FIG. 12-49 Problem 36.

••37 In Fig. 12-50, a uniform plank, with a length L of 6.10 m and a weight of 445 N, rests on the ground and against a frictionless roller at the top of a wall of height h = 3.05 m. The plank remains in equilibrium for any value of $\theta \geq 70°$ but slips if $\theta < 70°$. Find the coefficient of static friction between the plank and the ground.

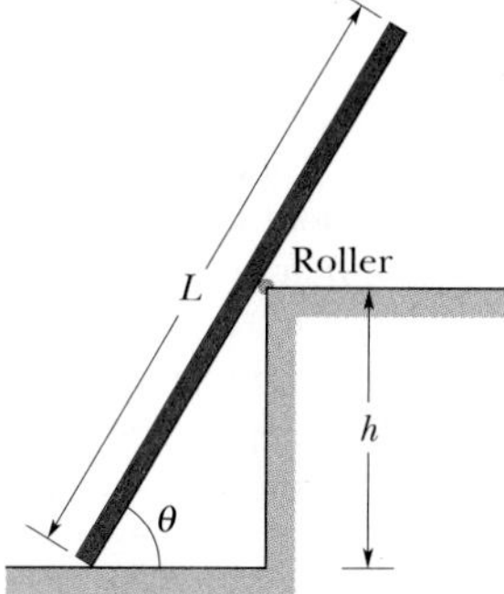

FIG. 12-50 Problem 37.

••**38** In Fig. 12-51, uniform beams A and B are attached to a wall with hinges and loosely bolted together (there is no torque of one on the other). Beam A has length $L_A = 2.40$ m and mass 54.0 kg; beam B has mass 68.0 kg. The two hinge points are separated by distance $d = 1.80$ m. In unit-vector notation, what is the force on (a) beam A due to its hinge, (b) beam A due to the bolt, (c) beam B due to its hinge, and (d) beam B due to the bolt?

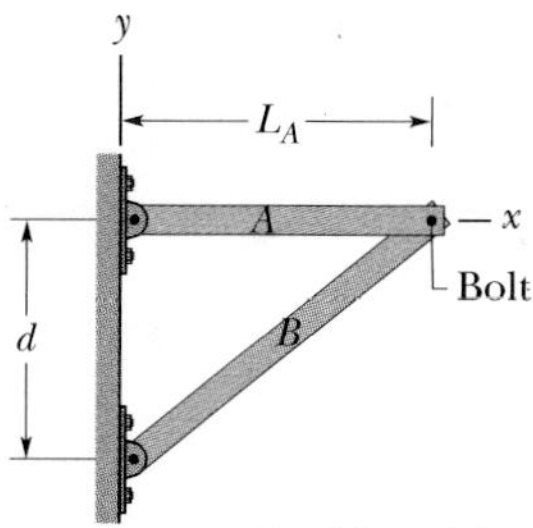

FIG. 12-51 Problem 38.

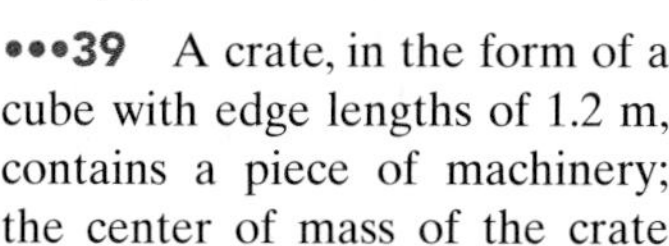

•••**39** A crate, in the form of a cube with edge lengths of 1.2 m, contains a piece of machinery; the center of mass of the crate and its contents is located 0.30 m above the crate's geometrical center. The crate rests on a ramp that makes an angle θ with the horizontal. As θ is increased from zero, an angle will be reached at which the crate will either tip over or start to slide down the ramp. If the coefficient of static friction μ_s between ramp and crate is 0.60, (a) does the crate tip or slide and (b) at what angle θ does this occur? If $\mu_s = 0.70$, (c) does the crate tip or slide and (d) at what angle θ does this occur? (*Hint:* At the onset of tipping, where is the normal force located?)

•••**40** In Sample Problem 12-2, let the coefficient of static friction μ_s between the ladder and the pavement be 0.53. How far (in percent) up the ladder must the firefighter go to put the ladder on the verge of sliding?

•••**41** For the stepladder shown in Fig. 12-52, sides AC and CE are each 2.44 m long and hinged at C. Bar BD is a tie-rod 0.762 m long, halfway up. A man weighing 854 N climbs 1.80 m along the ladder. Assuming that the floor is frictionless and neglecting the mass of the ladder, find (a) the tension in the tie-rod and the magnitudes of the forces on the ladder from the floor at (b) A and (c) E. (*Hint:* Isolate parts of the ladder in applying the equilibrium conditions.)

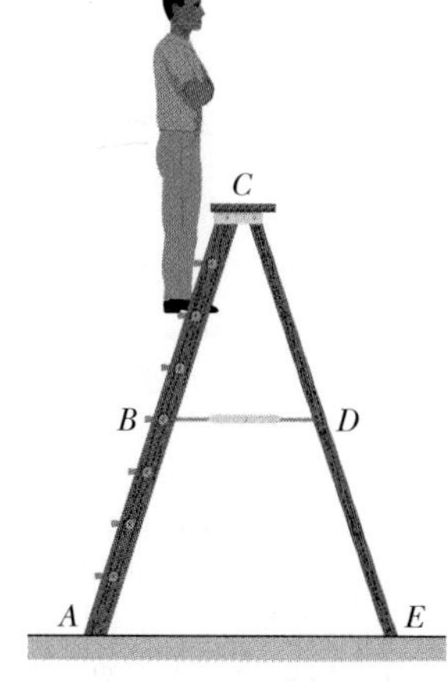

FIG. 12-52 Problem 41.

•••**42** Figure 12-53a shows a horizontal uniform beam of mass m_b and length L that is supported on the left by a hinge attached to a wall and on the right by a cable at angle θ with the horizontal. A package of mass m_p is positioned on the beam at a distance x from the left end. The total mass is $m_b + m_p = 61.22$ kg. Figure 12-53b gives the tension T in the cable as a function of the package's position given as a fraction x/L of the beam length. The scale of the T axis is set by $T_a = 500$ N and $T_b = 700$ N. Evaluate (a) angle θ, (b) mass m_b, and (c) mass m_p.

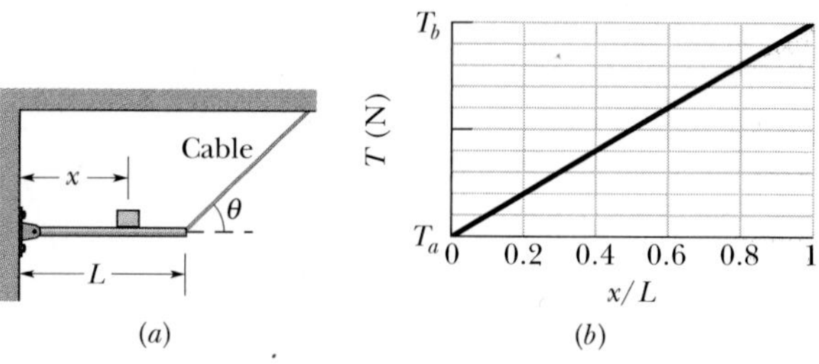

FIG. 12-53 Problem 42.

sec. 12-7 Elasticity

•**43** A horizontal aluminum rod 4.8 cm in diameter projects 5.3 cm from a wall. A 1200 kg object is suspended from the end of the rod. The shear modulus of aluminum is 3.0×10^{10} N/m^2. Neglecting the rod's mass, find (a) the shear stress on the rod and (b) the vertical deflection of the end of the rod. SSM ILW

•**44** Figure 12-54 shows the stress–strain curve for a material. The scale of the stress axis is set by $s = 300$, in units of 10^6 N/m^2. What are (a) the Young's modulus and (b) the approximate yield strength for this material?

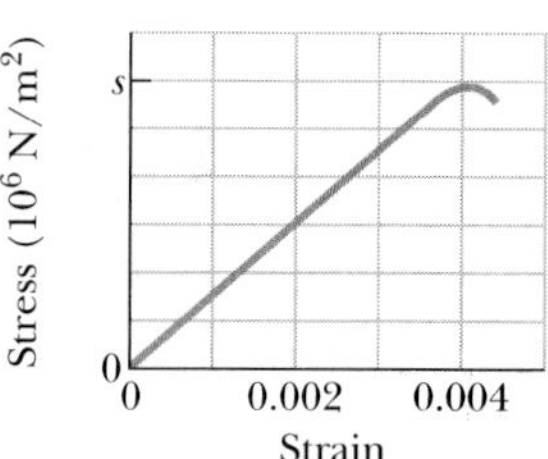

FIG. 12-54 Problem 44.

••**45** In Fig. 12-55, a 103 kg uniform log hangs by two steel wires, A and B, both of radius 1.20 mm. Initially, wire A was 2.50 m long and 2.00 mm shorter than wire B. The log is now horizontal. What are the magnitudes of the forces on it from (a) wire A and (b) wire B? (c) What is the ratio d_A/d_B? GO

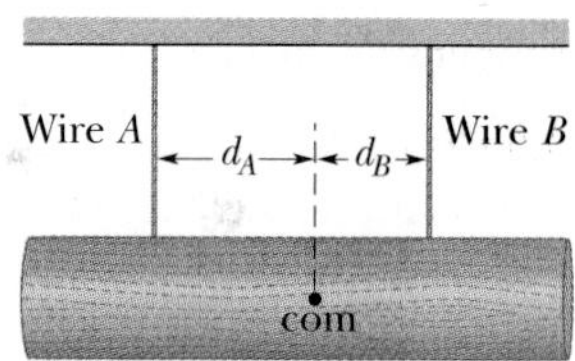

FIG. 12-55 Problem 45.

••**46** Figure 12-56 shows the stress versus strain plot for an aluminum wire that is stretched by a machine pulling in opposite directions at the two ends of the wire. The scale of the stress axis is set by $s = 7.0$, in units of 10^7 N/m^2. The wire has an initial length of 0.800 m and an initial cross-sectional area of 2.00×10^{-6} m^2. How much work does the force from the machine do on the wire to produce a strain of 1.00×10^{-3}?

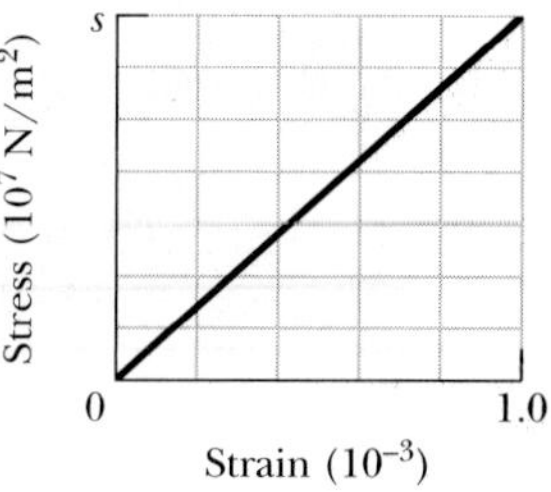

FIG. 12-56 Problem 46.

••**47** In Fig. 12-57, a lead brick rests horizontally on cylinders A and B. The areas of the top faces of the cylinders are related by $A_A = 2A_B$; the Young's moduli of the cylinders are related by $E_A = 2E_B$. The

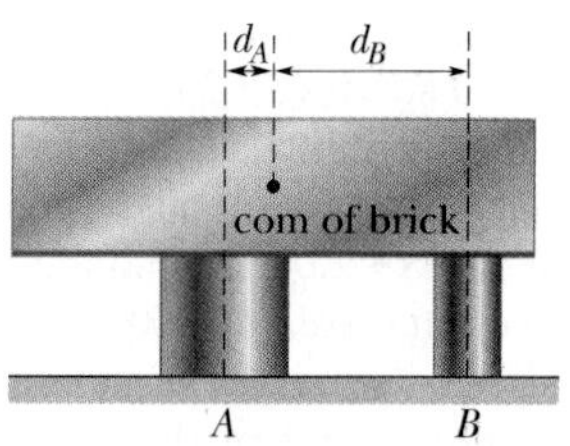

FIG. 12-57 Problem 47.

cylinders had identical lengths before the brick was placed on them. What fraction of the brick's mass is supported (a) by cylinder A and (b) by cylinder B? The horizontal distances between the center of mass of the brick and the centerlines of the cylinders are d_A for cylinder A and d_B for cylinder B. (c) What is the ratio d_A/d_B?

••48 Figure 12-58 shows an approximate plot of stress versus strain for a spider-web thread, out to the point of breaking at a strain of 2.00. The vertical axis scale is set by $a = 0.12\ \text{GN/m}^2$, $b = 0.30\ \text{GN/m}^2$, and $c = 0.80\ \text{GN/m}^2$. Assume that the thread has an initial length of 0.80 cm, an initial cross-sectional area of $8.0 \times 10^{-12}\ \text{m}^2$, and (during stretching) a constant volume. Assume also that when the single thread snares a flying insect, the insect's kinetic energy is transferred to the stretching of the thread. (a) How much kinetic energy would put the thread on the verge of breaking? What is the kinetic energy of (b) a fruit fly of mass 6.00 mg and speed 1.70 m/s and (c) a bumble bee of mass 0.388 g and speed 0.420 m/s? Would (d) the fruit fly and (e) the bumble bee break the thread?

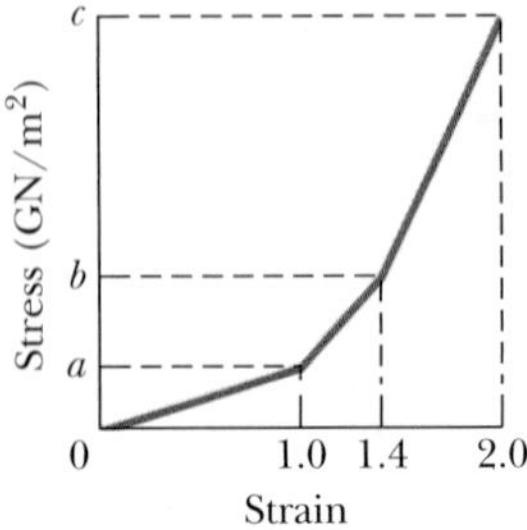

FIG. 12-58 Problem 48.

••49 A tunnel of length $L = 150$ m, height $H = 7.2$ m, and width 5.8 m (with a flat roof) is to be constructed at distance $d = 60$ m beneath the ground. (See Fig. 12-59.) The tunnel roof is to be supported entirely by square steel columns, each with a cross-sectional area of 960 cm². The mass of 1.0 cm³ of the ground material is 2.8 g. (a) What is the total weight of the ground material the columns must support? (b) How many columns are needed to keep the compressive stress on each column at one-half its ultimate strength?

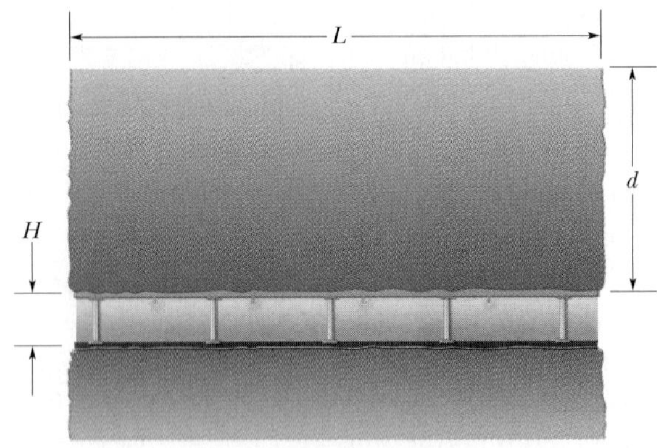

FIG. 12-59 Problem 49.

•••50 Figure 12-60 represents an insect caught at the midpoint of a spider-web thread. The thread breaks under a stress of $8.20 \times 10^8\ \text{N/m}^2$ and a strain of 2.00. Initially, it was horizontal and had a length of 2.00 cm and a cross-sectional area of $8.00 \times 10^{-12}\ \text{m}^2$. As the thread was stretched under the weight of the insect, its volume remained constant. If the weight of the insect puts the thread on the verge of breaking, what is the insect's mass? (A spider's web is built to break if a potentially harmful insect, such as a bumble bee, becomes snared in the web.)

FIG. 12-60 Problem 50.

•••51 Figure 12-61 is an overhead view of a rigid rod that turns about a vertical axle until the identical rubber stoppers A and B are forced against rigid walls at distances $r_A = 7.0$ cm and $r_B = 4.0$ cm from the axle. Initially the stoppers touch the walls without being compressed. Then force $\vec{F}$ of magnitude 220 N is applied perpendicular to the rod at a distance $R = 5.0$ cm from the axle. Find the magnitude of the force compressing (a) stopper A and (b) stopper B.

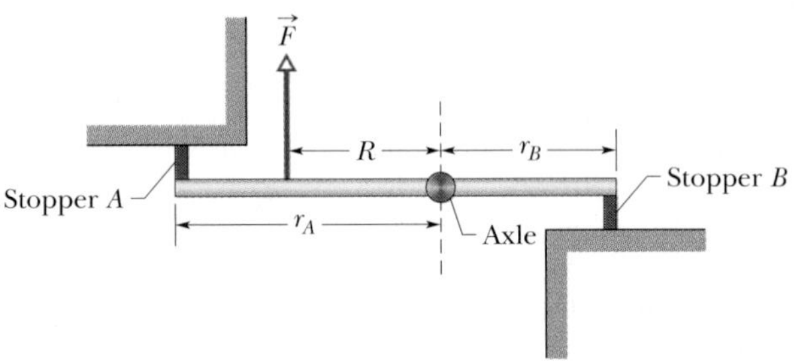

FIG. 12-61 Problem 51.

Additional Problems

52 Figure 12-62a shows a uniform ramp between two buildings that allows for motion between the buildings due to strong winds. At its left end, it is hinged to the building wall; at its right end, it has a roller that can roll along the building wall. There is no vertical force on the roller from the building, only a horizontal force with magnitude F_h. The horizontal distance between the buildings is $D = 4.00$ m. The rise of the ramp is $h = 0.490$ m. A man walks across the ramp from the left. Figure 12-62b gives F_h as a function of the horizontal distance x of the man from the building at the left. The scale of the F_h axis is set by $a = 20$ kN and $b = 25$ kN. What are the masses of (a) the ramp and (b) the man?

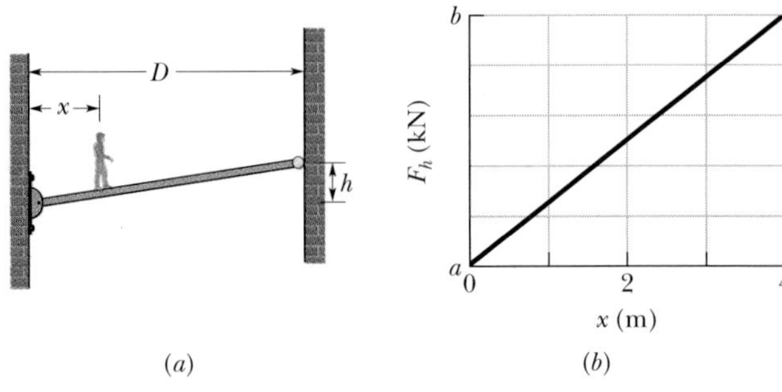

FIG. 12-62 Problem 52.

53 In Fig. 12-63, a 10 kg sphere is supported on a frictionless plane inclined at angle $\theta = 45°$ from the horizontal. Angle ϕ is 25°. Calculate the tension in the cable.

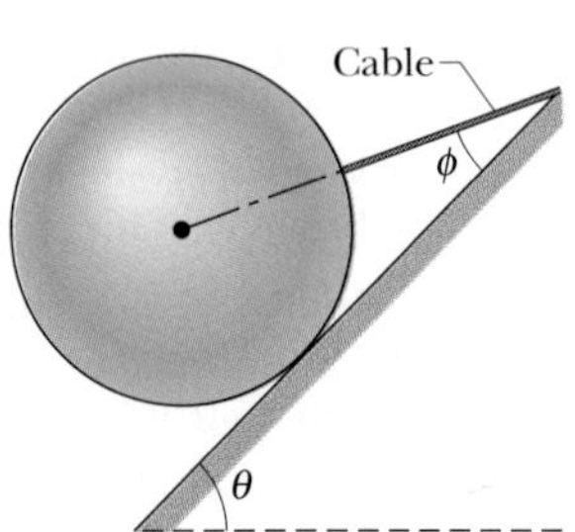

FIG. 12-63 Problem 53.

54 In Fig. 12-64a, a uniform 40.0 kg beam is centered over two rollers. Vertical lines across the beam mark off equal lengths. Two of the lines are centered over the rollers; a 10.0 kg package of tamales is centered over roller B. What

are the magnitudes of the forces on the beam from (a) roller A and (b) roller B? The beam is then rolled to the left until the right-hand end is centered over roller B (Fig. 12-64b). What now are the magnitudes of the forces on the beam from (c) roller A and (d) roller B? Next, the beam is rolled to the right. Assume that it has a length of 0.800 m. (e) What horizontal distance between the package and roller B puts the beam on the verge of losing contact with roller A?

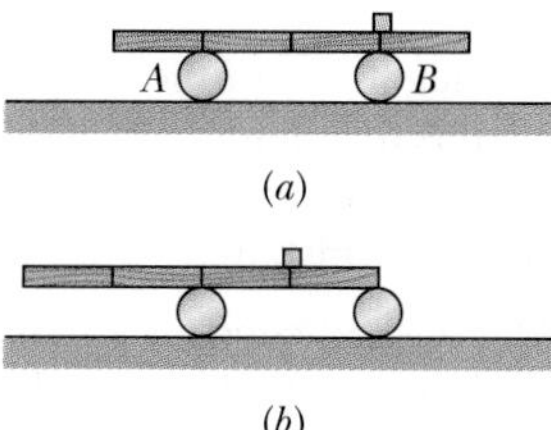

FIG. 12-64 Problem 54.

55 In Fig. 12-65, an 817 kg construction bucket is suspended by a cable A that is attached at O to two other cables B and C, making angles $\theta_1 = 51.0°$ and $\theta_2 = 66.0°$ with the horizontal. Find the tensions in (a) cable A, (b) cable B, and (c) cable C. (*Hint:* To avoid solving two equations in two unknowns, position the axes as shown in the figure.) SSM

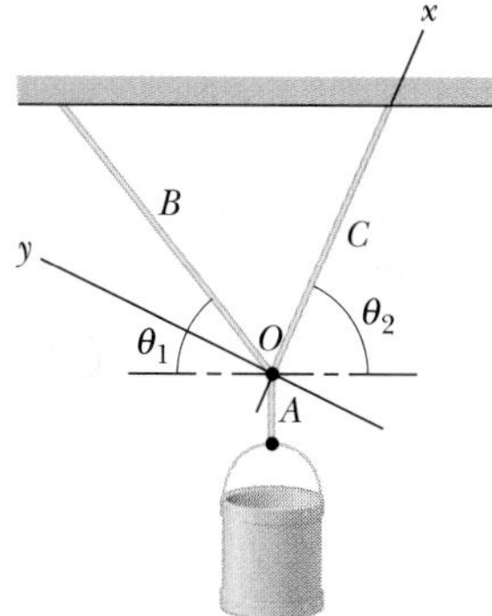

FIG. 12-65 Problem 55.

56 In Fig. 12-66, a package of mass m hangs from a short cord that is tied to the wall via cord 1 and to the ceiling via cord 2. Cord 1 is at angle $\phi = 40°$ with the horizontal; cord 2 is at angle θ. (a) For what value of θ is the tension in cord 2 minimized? (b) In terms of mg, what is the minimum tension in cord 2?

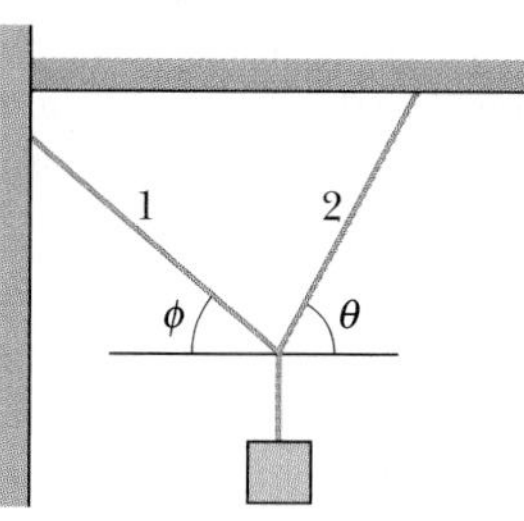

FIG. 12-66 Problem 56.

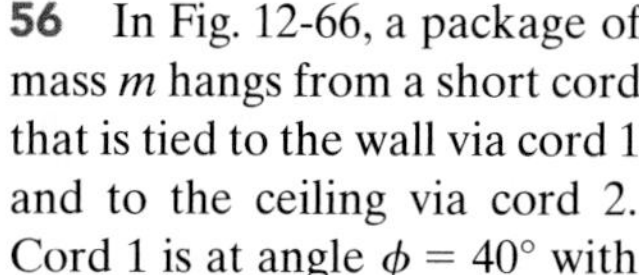

57 The force $\vec{F}$ in Fig. 12-67 keeps the 6.40 kg block and the pulleys in equilibrium. The pulleys have negligible mass and friction. Calculate the tension T in the upper cable. (*Hint:* When a cable wraps halfway around a pulley as here, the magnitude of its net force on the pulley is twice the tension in the cable.) ILW

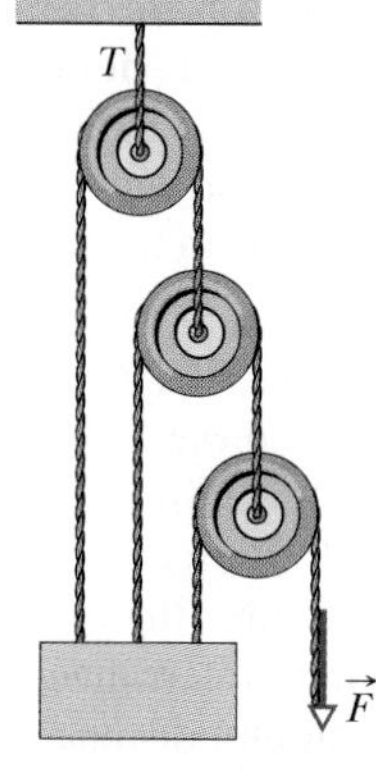

FIG. 12-67 Problem 57.

58 In Fig. 12-68, two identical, uniform, and frictionless spheres, each of mass m, rest in a rigid rectangular container. A line connecting their centers is at 45° to the horizontal. Find the magnitudes of the forces on the spheres from (a) the bottom of the container, (b) the left side of the container, (c) the right side of the container, and (d) each other. (*Hint:* The force of one sphere on the other is directed along the center–center line.)

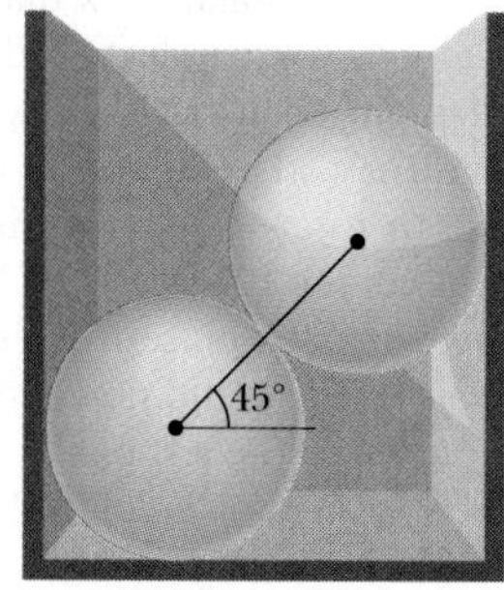

FIG. 12-68 Problem 58.

59 Four bricks of length L, identical and uniform, are stacked on top of one another (Fig. 12-69) in such a way that part of each extends beyond the one beneath. Find, in terms of L, the maximum values of (a) a_1, (b) a_2, (c) a_3, (d) a_4, and (e) h, such that the stack is in equilibrium.

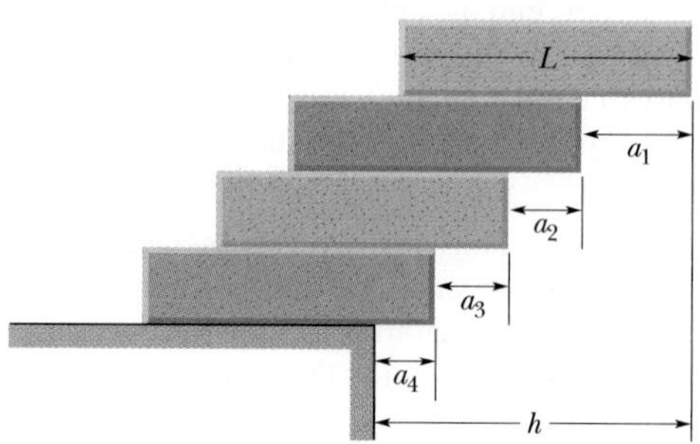

FIG. 12-69 Problem 59.

60 After a fall, a 95 kg rock climber finds himself dangling from the end of a rope that had been 15 m long and 9.6 mm in diameter but has stretched by 2.8 cm. For the rope, calculate (a) the strain, (b) the stress, and (c) the Young's modulus.

61 In Fig. 12-70, a rectangular slab of slate rests on a bedrock surface inclined at angle $\theta = 26°$. The slab has length $L = 43$ m, thickness $T = 2.5$ m, and width $W = 12$ m, and 1.0 cm^3 of it has a mass of 3.2 g. The coefficient of static friction between slab and bedrock is 0.39. (a) Calculate the component of the gravitational force on the slab parallel to the bedrock surface. (b) Calculate the magnitude of the static frictional force on the slab. By comparing (a) and (b), you can see that the slab is in danger of sliding. This is prevented only by chance protrusions of bedrock. (c) To stabilize the slab, bolts are to be driven perpendicular to the bedrock surface (two bolts are shown). If each bolt has a cross-sectional area of 6.4 cm^2 and will snap under a shearing stress of 3.6×10^8 N/m^2, what is the minimum number of bolts needed? Assume that the bolts do not affect the normal force. SSM

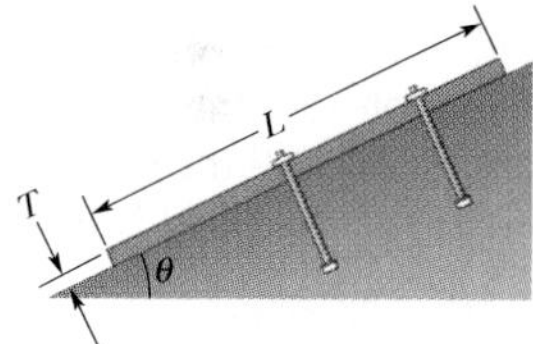

FIG. 12-70 Problem 61.

62 A uniform ladder whose length is 5.0 m and whose weight is 400 N leans against a frictionless vertical wall. The coefficient of static friction between the level ground and the foot of the ladder is 0.46. What is the greatest distance the foot of the ladder can be placed from the base of the wall without the ladder immediately slipping?

63 In Fig. 12-71, block A (mass 10 kg) is in equilibrium, but it would slip if block B (mass 5.0 kg) were any heavier. For angle $\theta = 30°$, what is the coefficient of static friction between block A and the surface below it? SSM

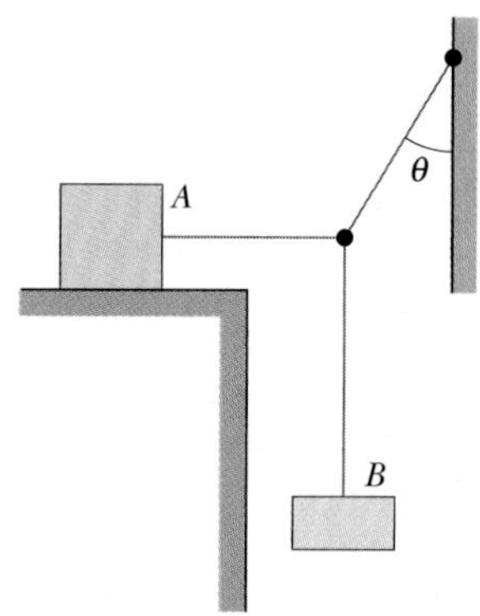

FIG. 12-71 Problem 63.

64 A mine elevator is supported by a single steel cable 2.5 cm in diameter. The total mass of the elevator cage and occupants is 670 kg. By how much does the cable stretch when the elevator hangs by (a) 12 m of cable and (b) 362 m of cable? (Neglect the mass of the cable.)

65 In Fig. 12-72, a uniform rod of mass m is hinged to a building at its lower end, while its upper end is held in place by a rope attached to the wall. If angle $\theta_1 = 60°$, what value must angle θ_2 have so that the tension in the rope is equal to $mg/2$? SSM

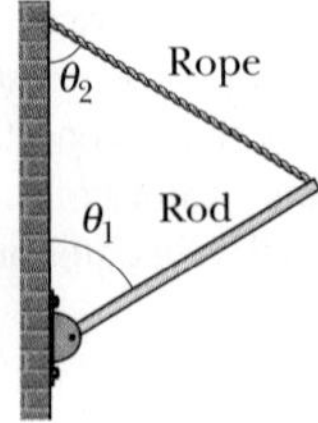

FIG. 12-72 Problem 65.

66 A 73 kg man stands on a level bridge of length L. He is at distance $L/4$ from one end. The bridge is uniform and weighs 2.7 kN. What are the magnitudes of the vertical forces on the bridge from its supports at (a) the end farther from him and (b) the nearer end?

67 A makeshift swing is constructed by making a loop in one end of a rope and tying the other end to a tree limb. A child is sitting in the loop with the rope hanging vertically when the child's father pulls on the child with a horizontal force and displaces the child to one side. Just before the child is released from rest, the rope makes an angle of 15° with the vertical and the tension in the rope is 280 N. (a) How much does the child weigh? (b) What is the magnitude of the (horizontal) force of the father on the child just before the child is released? (c) If the maximum horizontal force the father can exert on the child is 93 N, what is the maximum angle with the vertical the rope can make while the father is pulling horizontally?

68 The system in Fig. 12-73 is in equilibrium. The angles are $\theta_1 = 60°$ and $\theta_2 = 20°$, and the ball has mass $M = 2.0$ kg. What is the tension in (a) string ab and (b) string bc?

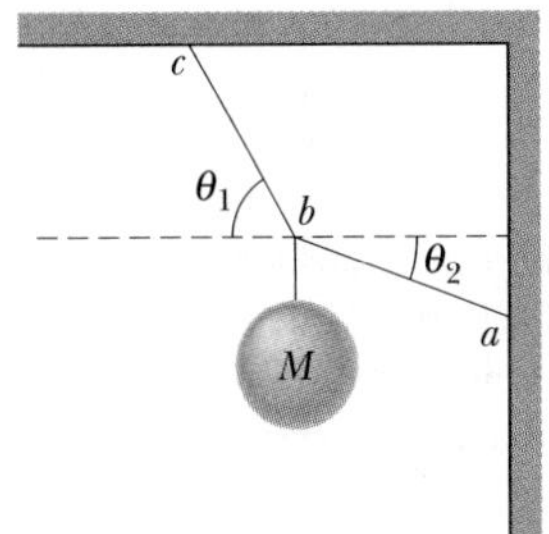

FIG. 12-73 Problem 68.

69 Figure 12-74 shows a stationary arrangement of two crayon boxes and three cords. Box A has a mass of 11.0 kg and is on a ramp at angle $\theta = 30.0°$; box B has a mass of 7.00 kg and hangs on a cord. The cord connected to box A is parallel to the ramp, which is frictionless. (a) What is the tension in the upper cord, and (b) what angle does that cord make with the horizontal?

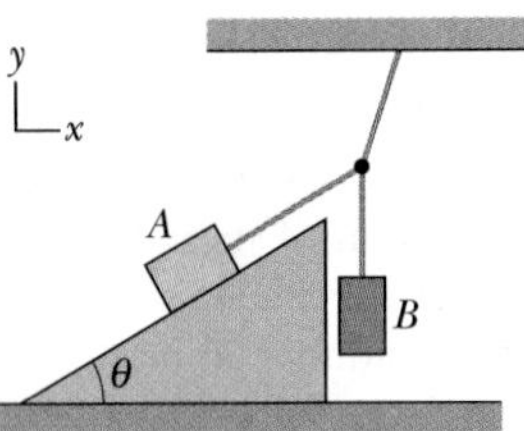

FIG. 12-74 Problem 69.

70 A construction worker attempts to lift a uniform beam off the floor and raise it to a vertical position. The beam is 2.50 m long and weighs 500 N. At a certain instant the worker holds the beam momentarily at rest with one end at distance $d = 1.50$ m above the floor, as shown in Fig. 12-75, by exerting a force $\vec{P}$ on the beam, perpendicular to the beam. (a) What is the magnitude P? (b) What is the magnitude of the (net) force of the floor on the beam? (c) What is the minimum value the coefficient of static friction between beam and floor can have in order for the beam not to slip at this instant?

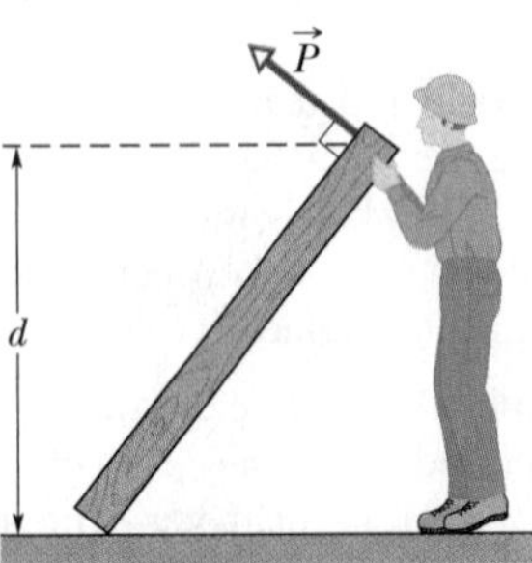

FIG. 12-75 Problem 70.

71 A solid copper cube has an edge length of 85.5 cm. How much stress must be applied to the cube to reduce the edge length to 85.0 cm? The bulk modulus of copper is 1.4×10^{11} N/m^2.

72 A uniform beam is 5.0 m long and has a mass of 53 kg. In Fig. 12-76, the beam is supported in a horizontal position by a hinge and a cable, with angle $\theta = 60°$. In unit-vector notation, what is the force on the beam from the hinge?

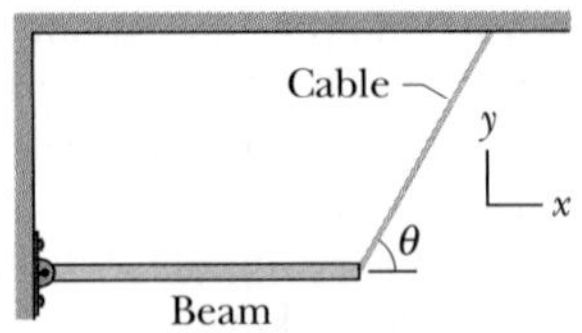

FIG. 12-76 Problem 72.

73 In Fig. 12-77, a uniform beam with a weight of 60 N and a length of 3.2 m is hinged at its lower end, and a horizontal force $\vec{F}$ of magnitude 50 N acts at its upper end. The beam is held vertical by a cable that makes angle $\theta = 25°$ with the ground and is attached to the beam at height $h = 2.0$ m. What are (a) the tension in the cable and (b) the force on the beam from the hinge in unit-vector notation?

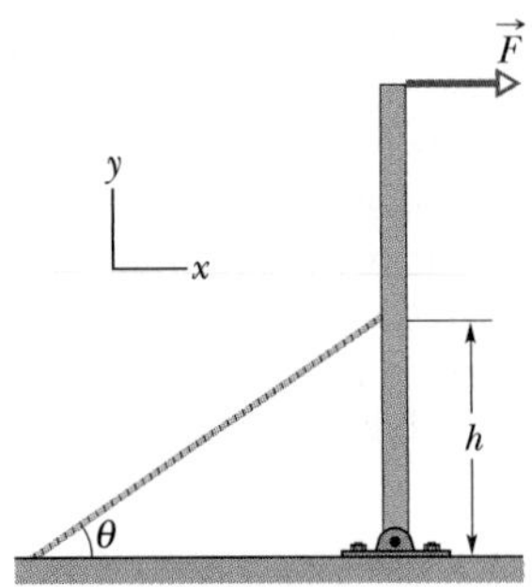

FIG. 12-77 Problem 73.

74 In Fig. 12-78, a uniform beam of length 12.0 m is supported by a horizontal cable and a hinge at angle $\theta = 50.0°$. The tension in the cable is 400 N. In unit-vector notation, what are (a) the gravitational force on the beam and (b) the force on the beam from the hinge?

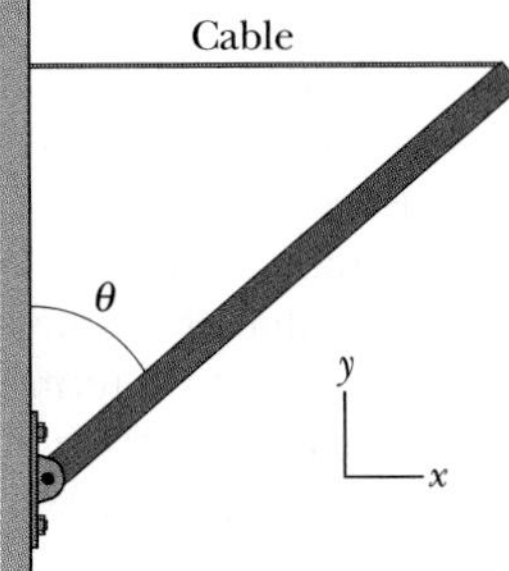

FIG. 12-78 Problem 74.

75 Four bricks of length L, identical and uniform, are stacked on a table in two ways, as shown in Fig. 12-79 (compare with Problem 59). We seek to maximize the overhang distance h in both arrangements. Find the optimum distances a_1, a_2, b_1, and b_2, and calculate h for the two arrangements.

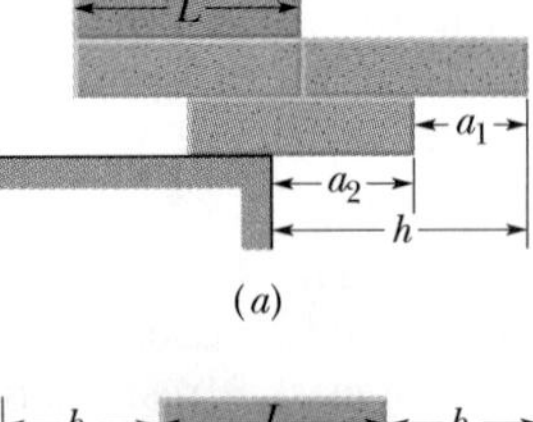

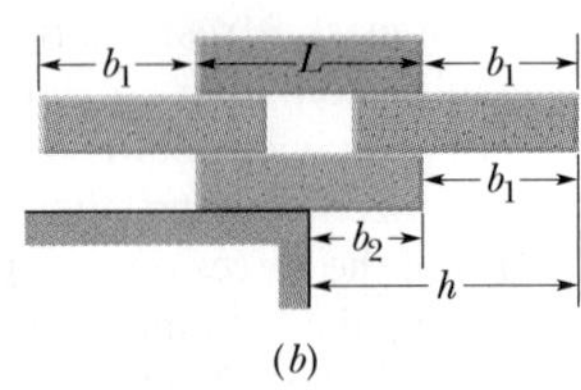

FIG. 12-79 Problem 75.

76 A pan balance is made up of a rigid, massless rod with a hanging pan attached at each end. The rod is supported at and free to rotate about a point not at its center. It is balanced by unequal masses placed in the two pans. When an unknown mass m is placed in the left pan, it is balanced by a mass m_1 placed in the right pan; when the mass m is placed in the right pan, it is balanced by a mass m_2 in the left pan. Show that $m = \sqrt{m_1 m_2}$.

77 The rigid square frame in Fig. 12-80 consists of the four side bars AB, BC, CD, and DA plus two diagonal bars AC and BD, which pass each other freely at E. By means of the turnbuckle G, bar AB is put under tension, as if its ends were subject to horizontal, outward forces $\vec{T}$ of magnitude 535 N. (a) Which of the other bars are in tension? What are the magnitudes of (b) the forces causing the tension in those bars and (c) the forces causing compression in the other bars? (*Hint:* Symmetry considerations can lead to considerable simplification in this problem.)

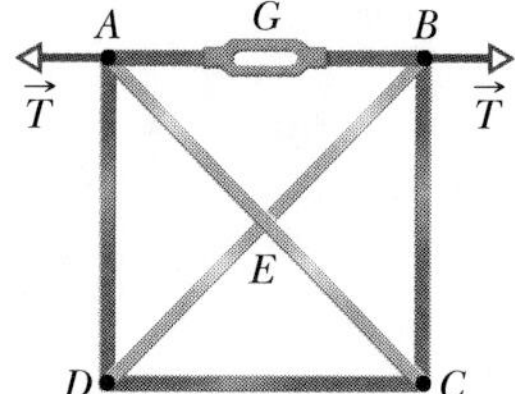

FIG. 12-80 Problem 77.

78 A gymnast with mass 46.0 kg stands on the end of a uniform balance beam as shown in Fig. 12-81. The beam is 5.00 m long and has a mass of 250 kg (excluding the mass of the two supports). Each support is 0.540 m from its end of the beam. In unit-vector notation, what are the forces on the beam due to (a) support 1 and (b) support 2?

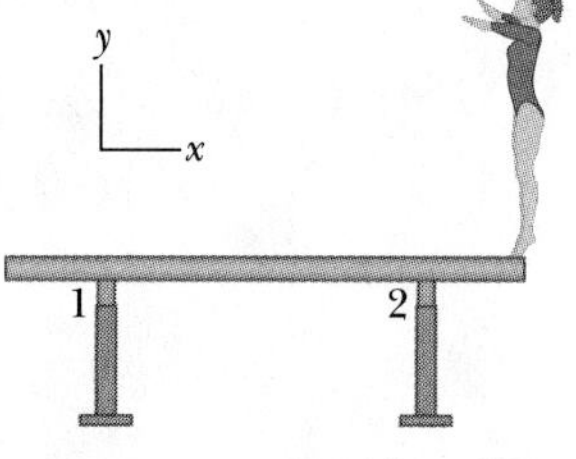

FIG. 12-81 Problem 78.

79 Figure 12-82 shows a 300 kg cylinder that is horizontal. Three steel wires support the cylinder from a ceiling. Wires 1 and 3 are attached at the ends of the cylinder, and wire 2 is attached at the center. The wires each have a cross-sectional area of 2.00×10^{-6} m^2. Initially (before the cylinder was put in place) wires 1 and 3 were 2.0000 m long and wire 2 was 6.00 mm longer than that. Now (with the cylinder in place) all three wires have been stretched. What is the tension in (a) wire 1 and (b) wire 2?

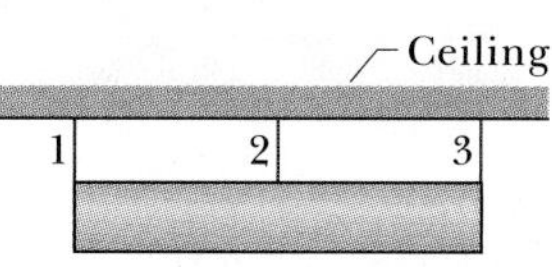

FIG. 12-82 Problem 79.

80 Figure 12-83*a* shows details of a finger in the crimp hold of the climber in Fig. 12-47. A tendon that runs from muscles in the forearm is attached to the far bone in the finger. Along the way, the tendon runs through several guiding sheaths called *pulleys.* The A2 pulley is attached to the first finger bone; the A4 pulley is attached to the second finger bone. To pull the finger toward the palm, the forearm muscles pull the tendon through the pulleys, much like strings on a marionette can be pulled to move parts of the marionette. Figure 12-83*b* is a simplified diagram of the second finger bone, which has length d. The tendon's pull $\vec{F}_t$ on the bone acts at the point where the tendon enters the A4 pulley, at distance $d/3$ along the bone. If the force components on each of the four crimped fingers in Fig. 12-47 are $F_h = 13.4$ N and $F_v = 162.4$ N, what is the magnitude of $\vec{F}_t$? The result is probably tolerable, but if the climber hangs by only one or two fingers, the A2 and A4 pulleys can be ruptured, a common ailment among rock climbers.

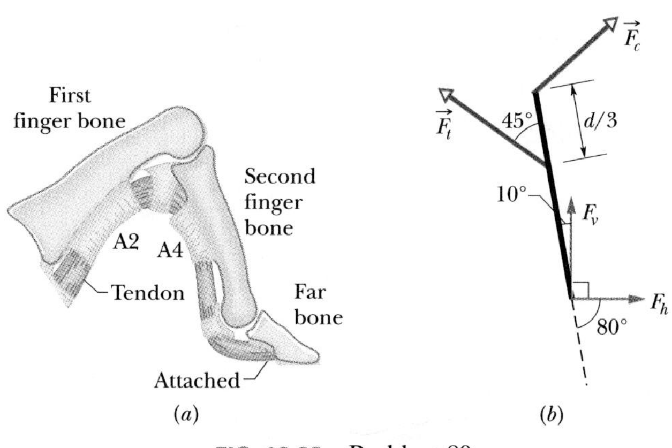

FIG. 12-83 Problem 80.

81 A uniform cube of side length 8.0 cm rests on a horizontal floor. The coefficient of static friction between cube and floor is μ. A horizontal pull $\vec{P}$ is applied perpendicular to one of the vertical faces of the cube, at a distance 7.0 cm above the floor on the vertical midline of the cube face. The magnitude of $\vec{P}$ is gradually increased. During that increase, for what values of μ will the cube eventually (a) begin to slide and (b) begin to tip? (*Hint:* At the onset of tipping, where is the normal force located?) SSM

82 A cylindrical aluminum rod, with an initial length of 0.8000 m and radius 1000.0 μm, is clamped in place at one end and then stretched by a machine pulling parallel to its length at its other end. Assuming that the rod's density (mass per unit volume) does not change, find the force magnitude that is required of the machine to decrease the radius to 999.9 μm. (The yield strength is not exceeded.)

83 A beam of length L is carried by three men, one man at one end and the other two supporting the beam between them on a crosspiece placed so that the load of the beam is equally divided among the three men. How far from the beam's free end is the crosspiece placed? (Neglect the mass of the crosspiece.)

84 A trap door in a ceiling is 0.91 m square, has a mass of 11 kg, and is hinged along one side, with a catch at the opposite side. If the center of gravity of the door is 10 cm toward the hinged side from the door's center, what are the magnitudes of the forces exerted by the door on (a) the catch and (b) the hinge?

85 A uniform ladder is 10 m long and weighs 200 N. In Fig. 12-84, the ladder leans against a vertical, frictionless wall at height $h = 8.0$ m above the ground. A horizontal force $\vec{F}$ is applied to the ladder at distance $d = 2.0$ m from its base (measured along the ladder). (a) If force magnitude $F = 50$ N, what is the force of the ground on the ladder, in unit-vector notation? (b) If $F = 150$ N, what is the force of the ground on the ladder, also in unit-vector notation? (c) Suppose the coefficient of static friction between the ladder and the ground is 0.38; for what minimum value of the force magnitude F will the base of the ladder just barely start to move toward the wall? SSM

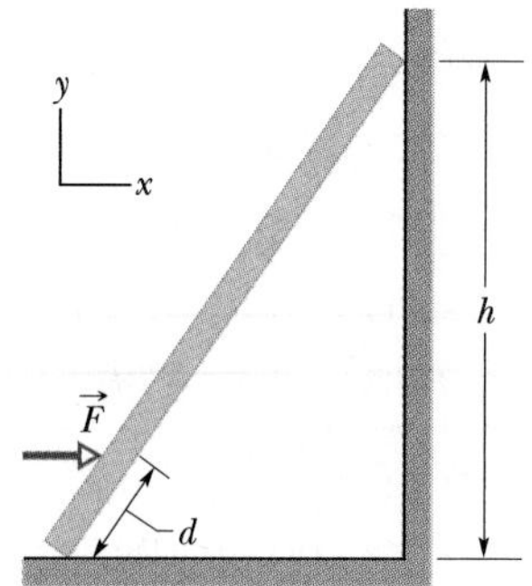

FIG. 12-84 Problem 85.

86 If the (square) beam in Fig. 12-7*a* is of Douglas fir, what must be its thickness to keep the compressive stress on it to $\frac{1}{6}$ of its ultimate strength? (See Sample Problem 12-3.)

13 Gravitation

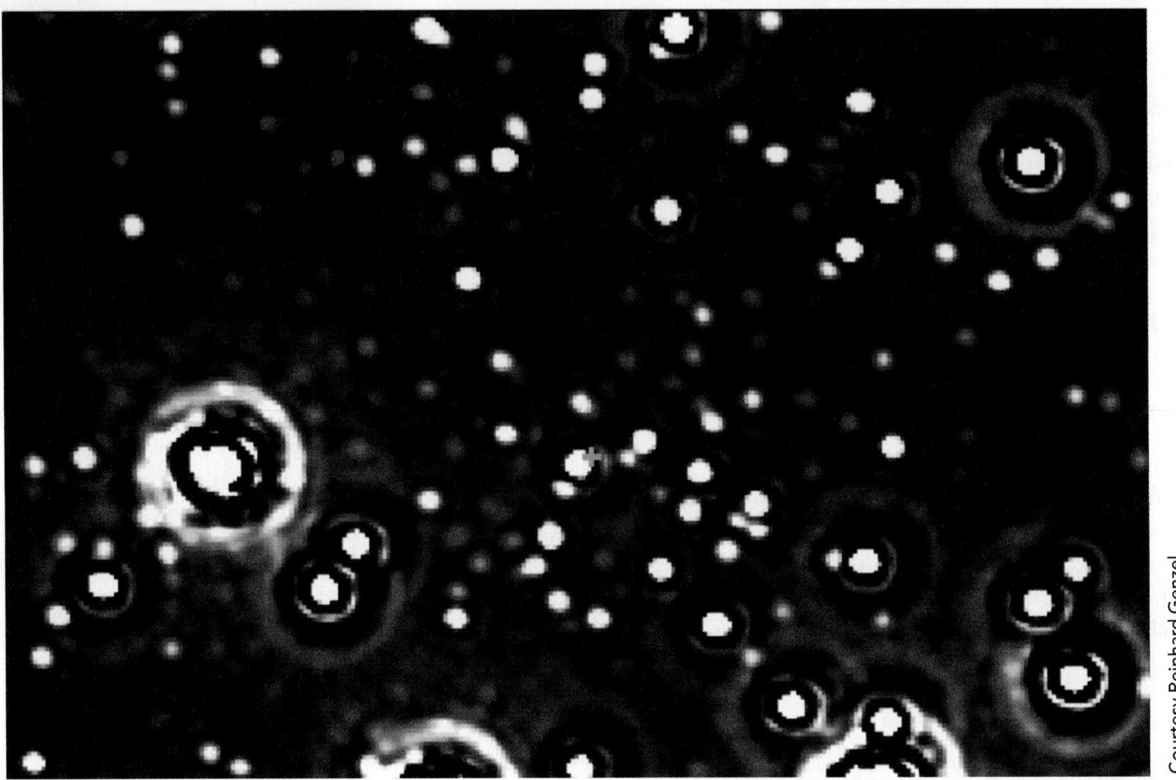

Courtesy Reinhard Genzel

This is an image of the stars near our Milky Way galaxy's center, which is marked with a small cross. Note that nothing shows up exactly at the center, but slightly off center (at the 8:00 position) there is a small circle. That circle is the image of a star known as S2. The other circles are also images of stars (the halos around them are artificially produced by the method of processing the images). Most stars in our galaxy move so slowly that we cannot actually see them move relative to one another, not even over a lifetime of observations. However, S2 is very different—we can see it move. In fact, it is moving so rapidly that it makes a complete trip around the Galaxy's center in only 15.2 years. There must be something huge at the center, yet we see nothing there.

What monster lies at our galaxy's center?

The answer is in this chapter.

13-1 WHAT IS PHYSICS?

One of the long-standing goals of physics is to understand the gravitational force—the force that holds you to Earth, holds the Moon in orbit around Earth, and holds Earth in orbit around the Sun. It also reaches out through the whole of our Milky Way galaxy, holding together the billions and billions of stars in the Galaxy and the countless molecules and dust particles between stars. We are located somewhat near the edge of this disk-shaped collection of stars and other matter, 2.6×10^4 light-years (2.5×10^{20} m) from the galactic center, around which we slowly revolve.

The gravitational force also reaches across intergalactic space, holding together the Local Group of galaxies, which includes, in addition to the Milky Way, the Andromeda Galaxy (Fig. 13-1) at a distance of 2.3×10^6 light-years away from Earth, plus several closer dwarf galaxies, such as the Large Magellanic Cloud. The Local Group is part of the Local Supercluster of galaxies that is being drawn by the gravitational force toward an exceptionally massive region of space called the Great Attractor. This region appears to be about 3.0×10^8 light-years from Earth, on the opposite side of the Milky Way. And the gravitational force is even more far-reaching because it attempts to hold together the entire universe, which is expanding.

FIG. 13-1 The Andromeda Galaxy. Located 2.3×10^6 light-years from us, and faintly visible to the naked eye, it is very similar to our home galaxy, the Milky Way. *(Courtesy NASA)*

This force is also responsible for some of the most mysterious structures in the universe: *black holes.* When a star considerably larger than our Sun burns out, the gravitational force between all its particles can cause the star to collapse in on itself and thereby to form a black hole. The gravitational force at the surface of such a collapsed star is so strong that neither particles nor light can escape from the surface (thus the term "black hole"). Any star coming too near a black hole can be ripped apart by the strong gravitational force and pulled into the hole. Enough captures like this yields a *supermassive black hole.* Such mysterious monsters appear to be common in the universe.

Although the gravitational force is still not fully understood, the starting point in our understanding of it lies in the *law of gravitation* of Isaac Newton.

13-2 | Newton's Law of Gravitation

Physicists like to study seemingly unrelated phenomena to show that a relationship can be found if the phenomena are examined closely enough. This search for unification has been going on for centuries. In 1665, the 23-year-old Isaac Newton made a basic contribution to physics when he showed that the force that holds the Moon in its orbit is the same force that makes an apple fall. We take this knowledge so much for granted now that it is not easy for us to comprehend the ancient belief that the motions of earthbound bodies and heavenly bodies were different in kind and were governed by different laws.

Newton concluded not only that Earth attracts both apples and the Moon but also that every body in the universe attracts every other body; this tendency of bodies to move toward each other is called **gravitation.** Newton's conclusion takes a little getting used to, because the familiar attraction of Earth for earthbound bodies is so great that it overwhelms the attraction that earthbound bodies have for each other. For example, Earth attracts an apple with a force magnitude of about 0.8 N. You also attract a nearby apple (and it attracts you), but the force of attraction has less magnitude than the weight of a speck of dust.

Newton proposed a *force law* that we call **Newton's law of gravitation:** Every particle attracts any other particle with a **gravitational force** of magnitude

$$F = G\frac{m_1 m_2}{r^2} \quad \text{(Newton's law of gravitation).} \qquad (13\text{-}1)$$

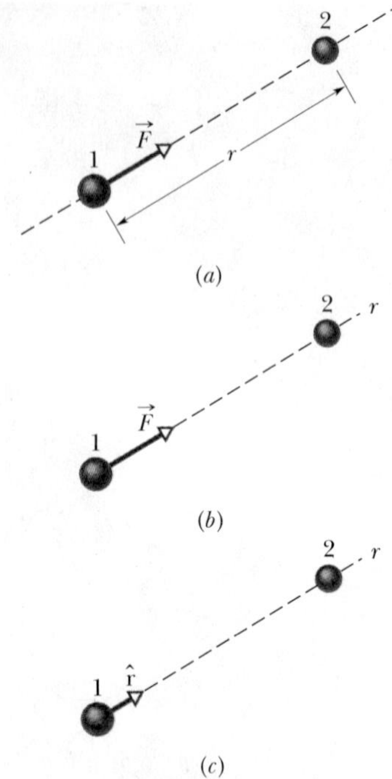

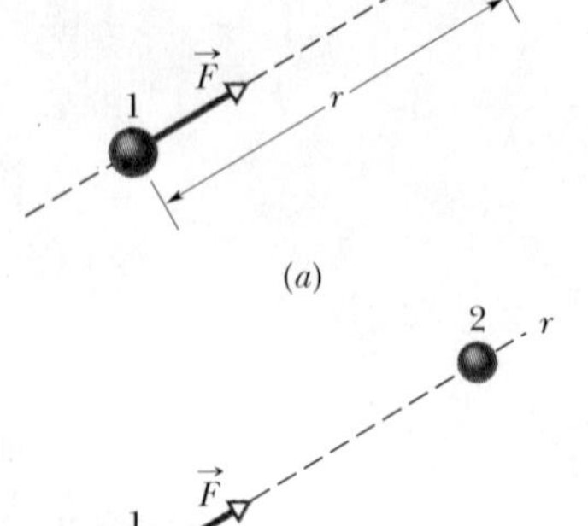

FIG. 13-2 (*a*) The gravitational force $\vec{F}$ on particle 1 due to particle 2 is an attractive force because particle 1 is attracted to particle 2. (*b*) Force $\vec{F}$ is directed along a radial coordinate axis r extending from particle 1 through particle 2. (*c*) $\vec{F}$ is in the direction of a unit vector $\hat{r}$ along the r axis.

Here m_1 and m_2 are the masses of the particles, r is the distance between them, and G is the **gravitational constant,** with a value that is now known to be

$$G = 6.67 \times 10^{-11}\ \text{N}\cdot\text{m}^2/\text{kg}^2$$
$$= 6.67 \times 10^{-11}\ \text{m}^3/\text{kg}\cdot\text{s}^2. \quad (13\text{-}2)$$

In Fig. 13-2*a*, $\vec{F}$ is the gravitational force acting on particle 1 (mass m_1) due to particle 2 (mass m_2). The force is directed toward particle 2 and is said to be an *attractive force* because particle 1 is attracted toward particle 2. The magnitude of the force is given by Eq. 13-1.

We can describe $\vec{F}$ as being in the positive direction of an r axis extending radially from particle 1 through particle 2 (Fig. 13-2*b*). We can also describe $\vec{F}$ by using a radial unit vector $\hat{r}$ (a dimensionless vector of magnitude 1) that is directed away from particle 1 along the r axis (Fig. 13-2*c*). From Eq. 13-1, the force on particle 1 is then

$$\vec{F} = G\frac{m_1 m_2}{r^2}\hat{r}. \quad (13\text{-}3)$$

The gravitational force on particle 2 due to particle 1 has the same magnitude as the force on particle 1 but the opposite direction. These two forces form a third-law force pair, and we can speak of the gravitational force *between* the two particles as having a magnitude given by Eq. 13-1. This force between two particles is not altered by other objects, even if they are located between the particles. Put another way, no object can shield either particle from the gravitational force due to the other particle.

The strength of the gravitational force—that is, how strongly two particles with given masses at a given separation attract each other—depends on the value of the gravitational constant G. If G—by some miracle—were suddenly multiplied by a factor of 10, you would be crushed to the floor by Earth's attraction. If G were divided by this factor, Earth's attraction would be so weak that you could jump over a building.

Although Newton's law of gravitation applies strictly to particles, we can also apply it to real objects as long as the sizes of the objects are small relative to the distance between them. The Moon and Earth are far enough apart so that, to a good approximation, we can treat them both as particles—but what about an apple and Earth? From the point of view of the apple, the broad and level Earth, stretching out to the horizon beneath the apple, certainly does not look like a particle.

Newton solved the apple–Earth problem by proving an important theorem called the *shell theorem:*

> A uniform spherical shell of matter attracts a particle that is outside the shell as if all the shell's mass were concentrated at its center.

Earth can be thought of as a nest of such shells, one within another and each shell attracting a particle outside Earth's surface as if the mass of that shell were at the center of the shell. Thus, from the apple's point of view, Earth *does* behave like a particle, one that is located at the center of Earth and has a mass equal to that of Earth.

Suppose that, as in Fig. 13-3, Earth pulls down on an apple with a force of magnitude 0.80 N. The apple must then pull up on Earth with a force of magnitude 0.80 N, which we take to act at the center of Earth. Although the forces are matched in magnitude, they produce different accelerations when the apple is released. The accelerations of the apple is about 9.8 m/s^2, the familiar acceleration of a falling body near Earth's surface. The acceleration of Earth, however, measured in a reference frame attached to the center of mass of the apple–Earth system, is only about 1×10^{-25} m/s^2.

FIG. 13-3 The apple pulls up on Earth just as hard as Earth pulls down on the apple.

CHECKPOINT 1 A particle is to be placed, in turn, outside four objects, each of mass m: (1) a large uniform solid sphere, (2) a large uniform spherical shell, (3) a small uniform solid sphere, and (4) a small uniform shell. In each situation, the distance between the particle and the center of the object is d. Rank the objects according to the magnitude of the gravitational force they exert on the particle, greatest first.

13-3 | Gravitation and the Principle of Superposition

Given a group of particles, we find the net (or resultant) gravitational force on any one of them from the others by using the **principle of superposition.** This is a general principle that says a net effect is the sum of the individual effects. Here, the principle means that we first compute the individual gravitational forces that act on our selected particle due to each of the other particles. We then find the net force by adding these forces vectorially, as usual.

For n interacting particles, we can write the principle of superposition for the gravitational forces on particle 1 as

$$\vec{F}_{1,\mathrm{net}} = \vec{F}_{12} + \vec{F}_{13} + \vec{F}_{14} + \vec{F}_{15} + \cdots + \vec{F}_{1n}. \tag{13-4}$$

Here $\vec{F}_{1,\mathrm{net}}$ is the net force on particle 1 and, for example, $\vec{F}_{13}$ is the force on particle 1 from particle 3. We can express this equation more compactly as a vector sum:

$$\vec{F}_{1,\mathrm{net}} = \sum_{i=2}^{n} \vec{F}_{1i}. \tag{13-5}$$

What about the gravitational force on a particle from a real (extended) object? This force is found by dividing the object into parts small enough to treat as particles and then using Eq. 13-5 to find the vector sum of the forces on the particle from all the parts. In the limiting case, we can divide the extended object into differential parts each of mass dm and each producing a differential force $d\vec{F}$ on the particle. In this limit, the sum of Eq. 13-5 becomes an integral and we have

$$\vec{F}_1 = \int d\vec{F}, \tag{13-6}$$

in which the integral is taken over the entire extended object and we drop the subscript "net." If the extended object is a uniform sphere or a spherical shell, we can avoid the integration of Eq. 13-6 by assuming that the object's mass is concentrated at the object's center and using Eq. 13-1.

CHECKPOINT 2 The figure shows four arrangements of three particles of equal masses. (a) Rank the arrangements according to the magnitude of the net gravitational force on the particle labeled m, greatest first. (b) In arrangement 2, is the direction of the net force closer to the line of length d or to the line of length D?

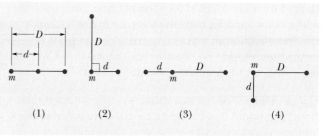

D d m (1)
D d m (2)
d D m (3)
m D d (4)

Sample Problem 13-1

Figure 13-4a shows an arrangement of three particles, particle 1 of mass $m_1 = 6.0$ kg and particles 2 and 3 of mass $m_2 = m_3 = 4.0$ kg, and distance $a = 2.0$ cm. What is the net gravitational force $\vec{F}_{1,\mathrm{net}}$ on particle 1 due to the other particles?

KEY IDEAS (1) Because we have particles, the magnitude of the gravitational force on particle 1 due to either of the other particles is given by Eq. 13-1 ($F = Gm_1m_2/r^2$). (2) The direction of either gravitational force on particle 1 is toward the particle responsible for it. (3) Because the forces are not along a single axis, we *cannot* simply add or subtract their magnitudes or their components to get the net force. Instead, we must add them as vectors.

Calculations: From Eq. 13-1, the magnitude of the force $\vec{F}_{12}$ on particle 1 from particle 2 is

$$F_{12} = \frac{Gm_1m_2}{a^2} = \frac{(6.67 \times 10^{-11}\ \text{m}^3/\text{kg}\cdot\text{s}^2)(6.0\ \text{kg})(4.0\ \text{kg})}{(0.020\ \text{m})^2} = 4.00 \times 10^{-6}\ \text{N}.$$

Similarly, the magnitude of force $\vec{F}_{13}$ on particle 1 from particle 3 is

$$F_{13} = \frac{Gm_1m_3}{(2a)^2} = \frac{(6.67 \times 10^{-11}\ \text{m}^3/\text{kg}\cdot\text{s}^2)(6.0\ \text{kg})(4.0\ \text{kg})}{(0.040\ \text{m})^2} = 1.00 \times 10^{-6}\ \text{N}.$$

Force $\vec{F}_{12}$ is directed in the positive direction of the y axis (Fig. 13-4*b*) and has only the y component F_{12}. Similarly, $\vec{F}_{13}$ is directed in the negative direction of the x axis and has only the x component $-F_{13}$.

To find the net force $\vec{F}_{1,\text{net}}$ on particle 1, we must add the two forces as vectors. We can do so on a vector-capable calculator. However, here we note that $-F_{13}$ and F_{12} are actually the x and y components of $\vec{F}_{1,\text{net}}$.

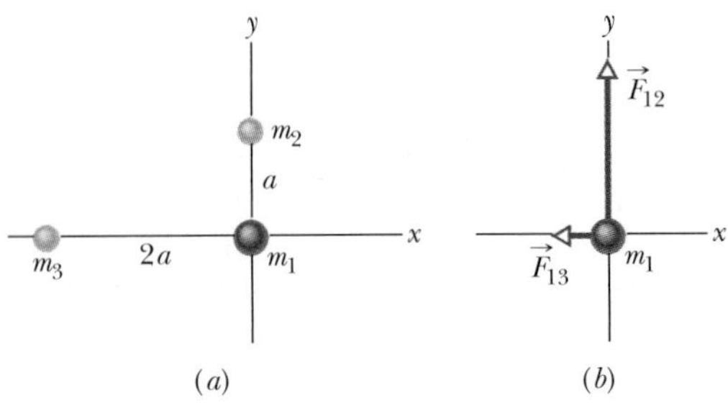

FIG. 13-4 (*a*) An arrangement of three particles. (*b*) The forces acting on the particle of mass m_1 due to the other particles.

Therefore, we can use Eq. 3-6 to find first the magnitude and then the direction of $\vec{F}_{1,\text{net}}$. The magnitude is

$$F_{1,\text{net}} = \sqrt{(F_{12})^2 + (-F_{13})^2} = \sqrt{(4.00 \times 10^{-6}\ \text{N})^2 + (-1.00 \times 10^{-6}\ \text{N})^2} = 4.1 \times 10^{-6}\ \text{N}. \quad \text{(Answer)}$$

Relative to the positive direction of the x axis, Eq. 3-6 gives the direction of $\vec{F}_{1,\text{net}}$ as

$$\theta = \tan^{-1}\frac{F_{12}}{-F_{13}} = \tan^{-1}\frac{4.00 \times 10^{-6}\ \text{N}}{-1.00 \times 10^{-6}\ \text{N}} = -76°.$$

Is this a reasonable direction? No, because the direction of $\vec{F}_{1,\text{net}}$ must be between the directions of $\vec{F}_{12}$ and $\vec{F}_{13}$. Recall from Chapter 3 (Problem-Solving Tactic 3) that a calculator displays only one of the two possible answers to a $\tan^{-1}$ function. We find the other answer by adding 180°:

$$-76° + 180° = 104°, \quad \text{(Answer)}$$

which *is* a reasonable direction for $\vec{F}_{1,\text{net}}$.

Sample Problem 13-2 Build your skill

Figure 13-5*a* shows an arrangement of five particles, with masses $m_1 = 8.0$ kg, $m_2 = m_3 = m_4 = m_5 = 2.0$ kg, and with $a = 2.0$ cm and $\theta = 30°$. What is the net gravitational force $\vec{F}_{1,\text{net}}$ on particle 1 due to the other particles?

KEY IDEAS (1) Because we have particles, the magnitude of the gravitational force on particle 1 due to either of the other particles is given by Eq. 13-1 ($F = Gm_1m_2/r^2$). (2) The direction of a gravitational force on particle 1 is toward the particle responsible for the force. (3) We can use symmetry to eliminate unneeded calculations.

Calculations: For the magnitudes of the forces on particle 1, first note that particles 2 and 4 have equal masses and equal distances of $r = 2a$ from particle 1. Thus, from Eq. 13-1, we find

$$F_{12} = F_{14} = \frac{Gm_1m_2}{(2a)^2}. \tag{13-7}$$

Similarly, since particles 3 and 5 have equal masses and are both distance $r = a$ from particle 1, we find

$$F_{13} = F_{15} = \frac{Gm_1m_3}{a^2}. \tag{13-8}$$

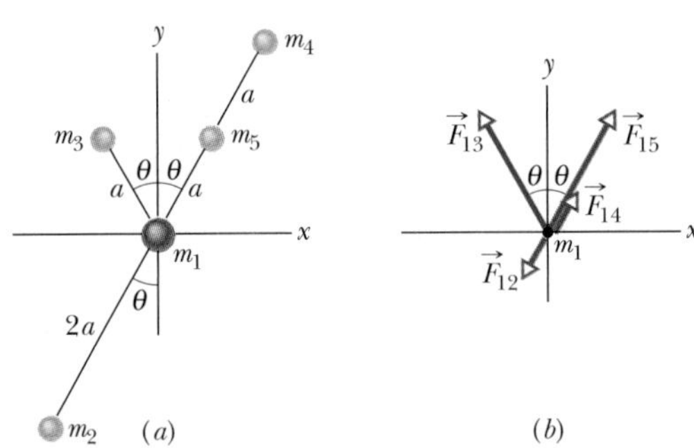

FIG. 13-5 (*a*) An arrangement of five particles. (*b*) The forces acting on the particle of mass m_1 due to the other four particles.

We could now substitute known data into these two equations to evaluate the magnitudes of the forces, indicate the directions of the forces on the free-body diagram of Fig. 13-5*b*, and then find the net force either (1) by resolving the vectors into x and y components, finding the net x and net y components, and then vectorially combining them or (2) by adding the vectors directly on a vector-capable calculator.

Instead, however, we shall make further use of the symmetry of the problem. First, we note that $\vec{F}_{12}$ and $\vec{F}_{14}$ are equal in magnitude but opposite in direction; thus, those forces *cancel*. Inspection of Fig. 13-5*b* and Eq. 13-8 reveals that the *x* components of $\vec{F}_{13}$ and $\vec{F}_{15}$ also *cancel*, and that their *y* components are identical in magnitude and both act in the positive direction of the *y* axis. Thus, $\vec{F}_{1,\text{net}}$ acts in that same direction, and its magnitude is twice the *y* component of $\vec{F}_{13}$:

$$F_{1,\text{net}} = 2F_{13}\cos\theta = 2\frac{Gm_1m_3}{a^2}\cos\theta$$

$$= \frac{2(6.67\times10^{-11}\ \text{m}^3/\text{kg}\cdot\text{s}^2)(8.0\ \text{kg})(2.0\ \text{kg})}{(0.020\ \text{m})^2}\cos 30^\circ$$

$$= 4.6\times10^{-6}\ \text{N}. \qquad \text{(Answer)}$$

Note that the presence of particle 5 along the line between particles 1 and 4 does not alter the gravitational force on particle 1 from particle 4.

PROBLEM-SOLVING TACTICS

Tactic 1: ***Drawing Gravitational Force Vectors*** When you are given a diagram of particles, such as Fig. 13-4*a*, and asked to find the net gravitational force on one of them, you should usually draw a free-body diagram showing only the particle of concern and the forces on *it alone*, as in Fig. 13-4*b*. If, instead, you choose to superimpose the force vectors on the given diagram, be sure to draw the vectors with either all tails (preferably) or all heads on the particle experiencing those forces. If you draw the vectors elsewhere, you invite confusion—and confusion is guaranteed if you draw the vectors on the particles *causing* the forces.

Tactic 2: ***Simplifying a Sum of Forces with Symmetry*** In Sample Problem 13-2 we used the symmetry of the situation: By realizing that particles 2 and 4 are positioned symmetrically about particle 1, and thus that $\vec{F}_{12}$ and $\vec{F}_{14}$ cancel, we avoided calculating either force. By realizing that the *x* components of $\vec{F}_{13}$ and $\vec{F}_{15}$ cancel and that their *y* components are identical and add, we saved even more effort.

In problems with symmetry, you can save much effort and reduce the chance of error by identifying which calculations are not needed because of the symmetry. Such identification is a skill acquired only by doing many homework problems.

13-4 | Gravitation Near Earth's Surface

Let us assume that Earth is a uniform sphere of mass *M*. The magnitude of the gravitational force from Earth on a particle of mass *m*, located outside Earth a distance *r* from Earth's center, is then given by Eq. 13-1 as

$$F = G\frac{Mm}{r^2}. \qquad (13\text{-}9)$$

If the particle is released, it will fall toward the center of Earth, as a result of the gravitational force $\vec{F}$, with an acceleration we shall call the **gravitational acceleration** $\vec{a}_g$. Newton's second law tells us that magnitudes *F* and a_g are related by

$$F = ma_g. \qquad (13\text{-}10)$$

Now, substituting *F* from Eq. 13-9 into Eq. 13-10 and solving for a_g, we find

$$a_g = \frac{GM}{r^2}. \qquad (13\text{-}11)$$

Table 13-1 shows values of a_g computed for various altitudes above Earth's surface. Notice a_g is significant even at 400 km.

TABLE 13-1

Variation of a_g with Altitude

Altitude (km)	a_g (m/s^2)	Altitude Example
0	9.83	Mean Earth surface
8.8	9.80	Mt. Everest
36.6	9.71	Highest crewed balloon
400	8.70	Space shuttle orbit
35 700	0.225	Communications satellite

Since Section 5-4, we have assumed that Earth is an inertial frame by neglecting its rotation. This simplification has allowed us to assume that the free-fall acceleration *g* of a particle is the same as the particle's gravitational acceleration (which we now call a_g). Furthermore, we assumed that *g* has the constant value 9.8 m/s^2 any place on Earth's surface. However, any *g* value measured at a given location will differ from the a_g value calculated with Eq. 13-11 for that location for three reasons: (1) Earth's mass is not distributed uniformly, (2) Earth is not a perfect sphere, and (3) Earth rotates. Moreover, because *g* differs from a_g, the same three reasons mean that the measured weight *mg* of a particle differs from the magnitude of the gravitational force on the particle as given by Eq. 13-9. Let us now examine those reasons.

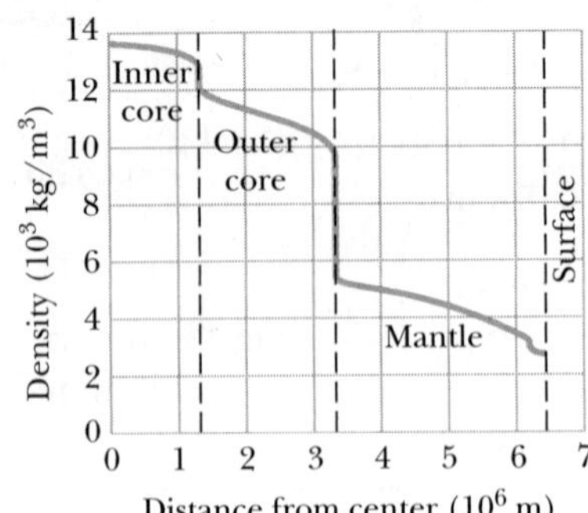

FIG. 13-6 The density of Earth as a function of distance from the center. The limits of the solid inner core, the largely liquid outer core, and the solid mantle are shown, but the crust of Earth is too thin to show clearly on this plot.

1. ***Earth's mass is not uniformly distributed.*** The density (mass per unit volume) of Earth varies radially as shown in Fig. 13-6, and the density of the crust (outer section) varies from region to region over Earth's surface. Thus, g varies from region to region over the surface.
2. ***Earth is not a sphere.*** Earth is approximately an ellipsoid, flattened at the poles and bulging at the equator. Its equatorial radius is greater than its polar radius by 21 km. Thus, a point at the poles is closer to the dense core of Earth than is a point on the equator. This is one reason the free-fall acceleration g increases as one proceeds, at sea level, from the equator toward either pole.
3. ***Earth is rotating.*** The rotation axis runs through the north and south poles of Earth. An object located on Earth's surface anywhere except at those poles must rotate in a circle about the rotation axis and thus must have a centripetal acceleration directed toward the center of the circle. This centripetal acceleration requires a centripetal net force that is also directed toward that center.

To see how Earth's rotation causes g to differ from a_g, let us analyze a simple situation in which a crate of mass m is on a scale at the equator. Figure 13-7*a* shows this situation as viewed from a point in space above the north pole.

Figure 13-7*b*, a free-body diagram for the crate, shows the two forces on the crate, both acting along a radial r axis that extends from Earth's center. The normal force $\vec{F}_N$ on the crate from the scale is directed outward, in the positive direction of the r axis. The gravitational force, represented with its equivalent $m\vec{a}_g$, is directed inward. Because it travels in a circle about the center of Earth as Earth turns, the crate has a centripetal acceleration $\vec{a}$ directed toward Earth's center. From Eq. 10-23 ($a_r = \omega^2 r$), we know this acceleration is equal to $\omega^2 R$, where ω is Earth's angular speed and R is the circle's radius (approximately Earth's radius). Thus, we can write Newton's second law for forces along the r axis ($F_{\text{net},r} = ma_r$) as

$$F_N - ma_g = m(-\omega^2 R). \tag{13-12}$$

The magnitude F_N of the normal force is equal to the weight mg read on the scale. With mg substituted for F_N, Eq. 13-12 gives us

$$mg = ma_g - m(\omega^2 R), \tag{13-13}$$

which says

$$\begin{pmatrix}\text{measured}\\ \text{weight}\end{pmatrix} = \begin{pmatrix}\text{magnitude of}\\ \text{gravitational force}\end{pmatrix} - \begin{pmatrix}\text{mass times}\\ \text{centripetal acceleration}\end{pmatrix}.$$

Thus, the measured weight is less than the magnitude of the gravitational force on the crate, because of Earth's rotation.

To find a corresponding expression for g and a_g, we cancel m from Eq. 13-13 to write

$$g = a_g - \omega^2 R, \tag{13-14}$$

which says

$$\begin{pmatrix}\text{free-fall}\\ \text{acceleration}\end{pmatrix} = \begin{pmatrix}\text{gravitational}\\ \text{acceleration}\end{pmatrix} - \begin{pmatrix}\text{centripetal}\\ \text{acceleration}\end{pmatrix}.$$

Thus, the measured free-fall acceleration is less than the gravitational acceleration because of Earth's rotation.

The difference between accelerations g and a_g is equal to $\omega^2 R$ and is greatest on the equator (for one reason, the radius of the circle traveled by the crate is

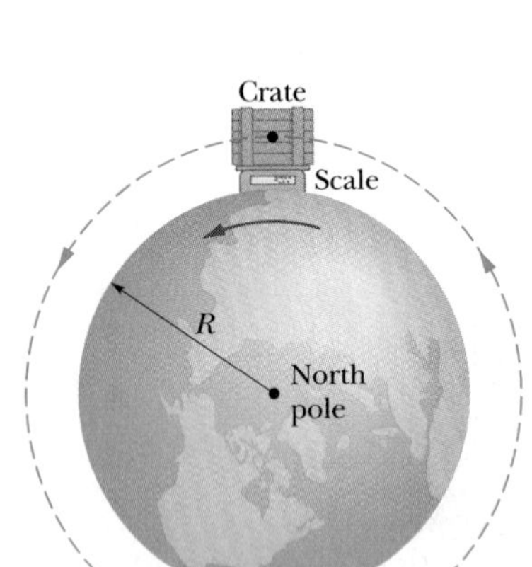

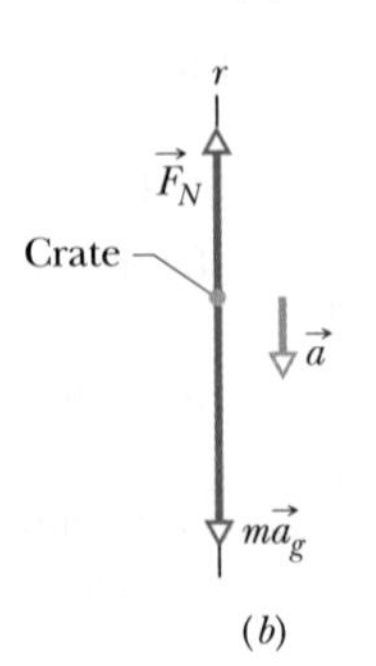

FIG. 13-7 (*a*) A crate sitting on a scale at Earth's equator, as seen by an observer positioned on Earth's rotation axis at some point above the north pole. (*b*) A free-body diagram for the crate, with a radial r axis extending from Earth's center. The gravitational force on the crate is represented with its equivalent $m\vec{a}_g$. The normal force on the crate from the scale is $\vec{F}_N$. Because of Earth's rotation, the crate has a centripetal acceleration $\vec{a}$ that is directed toward Earth's center.

greatest there). To find the difference, we can use Eq. 10-5 ($\omega = \Delta\theta/\Delta t$) and Earth's radius $R = 6.37 \times 10^6$ m. For one rotation of Earth, θ is 2π rad and the time period Δt is about 24 h. Using these values (and converting hours to seconds), we find that g is less than a_g by only about 0.034 m/s^2 (small compared to 9.8 m/s^2). Therefore, neglecting the difference in accelerations g and a_g is often justified. Similarly, neglecting the difference between weight and the magnitude of the gravitational force is also often justified.

Sample Problem 13-3

(a) An astronaut whose height h is 1.70 m floats "feet down" in an orbiting space shuttle at distance $r = 6.77 \times 10^6$ m away from the center of Earth. What is the difference between the gravitational acceleration at her feet and at her head?

KEY IDEAS We can approximate Earth as a uniform sphere of mass M_E. Then, from Eq. 13-11, the gravitational acceleration at any distance r from the center of Earth is

$$a_g = \frac{GM_E}{r^2}. \tag{13-15}$$

We might simply apply this equation twice, first with $r = 6.77 \times 10^6$ m for the feet and then with $r = 6.77 \times 10^6$ m $+ 1.70$ m for the head. However, a calculator may give us the same value for a_g twice, and thus a difference of zero, because h is so much smaller than r. Here's a more promising approach: Because we have a differential change dr in r between the astronaut's feet and head, we should differentiate Eq. 13-15 with respect to r.

Calculations: The differentiation gives us

$$da_g = -2\frac{GM_E}{r^3}\,dr, \tag{13-16}$$

where da_g is the differential change in the gravitational acceleration due to the differential change dr in r. For the astronaut, $dr = h$ and $r = 6.77 \times 10^6$ m. Substituting data into Eq. 13-16, we find

$$da_g = -2\frac{(6.67 \times 10^{-11}\ \text{m}^3/\text{kg}\cdot\text{s}^2)(5.98 \times 10^{24}\ \text{kg})}{(6.77 \times 10^6\ \text{m})^3}(1.70\ \text{m})$$

$$= -4.37 \times 10^{-6}\ \text{m/s}^2, \qquad \text{(Answer)}$$

where the M_E value is taken from Appendix C. This result means that the gravitational acceleration of the astronaut's feet toward Earth is slightly greater than the gravitational acceleration of her head toward Earth. This difference in acceleration tends to stretch her body, but the difference is so small that the stretching is unnoticeable.

(b) If the astronaut is now "feet down" at the same orbital radius $r = 6.77 \times 10^6$ m about a black hole of mass $M_h = 1.99 \times 10^{31}$ kg (10 times our Sun's mass), what is the difference between the gravitational acceleration at her feet and at her head? The black hole has a mathematical surface (*event horizon*) of radius $R_h = 2.95 \times 10^4$ m. Nothing, not even light, can escape from that surface or anywhere inside it. Note that the astronaut is well outside the surface (at $r = 229R_h$).

Calculations: We again have a differential change dr in r between the astronaut's feet and head, so we can again use Eq. 13-16. However, now we substitute $M_h = 1.99 \times 10^{31}$ kg for M_E. We find

$$da_g = -2\frac{(6.67 \times 10^{-11}\ \text{m}^3/\text{kg}\cdot\text{s}^2)(1.99 \times 10^{31}\ \text{kg})}{(6.77 \times 10^6\ \text{m})^3}(1.70\ \text{m})$$

$$= -14.5\ \text{m/s}^2. \qquad \text{(Answer)}$$

This means that the gravitational acceleration of the astronaut's feet toward the black hole is noticeably larger than that of her head. The resulting tendency to stretch her body would be bearable but quite painful. If she drifted closer to the black hole, the stretching tendency would increase drastically.

13-5 | Gravitation Inside Earth

Newton's shell theorem can also be applied to a situation in which a particle is located *inside* a uniform shell, to show the following:

> A uniform shell of matter exerts no *net* gravitational force on a particle located inside it.

Caution: This statement does *not* mean that the gravitational forces on the particle from the various elements of the shell magically disappear. Rather, it means that the *sum* of the force vectors on the particle from all the elements is zero.

If Earth's mass were uniformly distributed, the gravitational force acting on a particle would be a maximum at Earth's surface and would decrease as the particle moved outward, away from the planet. If the particle were to move inward, perhaps down a deep mine shaft, the gravitational force would change for two reasons. (1) It would tend to increase because the particle would be moving closer to the center of Earth. (2) It would tend to decrease because the thickening shell of material lying outside the particle's radial position would not exert any net force on the particle.

For a uniform Earth, the second influence would prevail and the force on the particle would steadily decrease to zero as the particle approached the center of Earth. However, for the real (nonuniform) Earth, the force on the particle actually increases as the particle begins to descend. The force reaches a maximum at a certain depth and then decreases as the particle descends farther.

Sample Problem 13-4

In *Pole to Pole,* an early science fiction story by George Griffith, three explorers attempt to travel by capsule through a naturally formed (and, of course, fictional) tunnel directly from the south pole to the north pole (Fig. 13-8). According to the story, as the capsule approaches Earth's center, the gravitational force on the explorers becomes alarmingly large and then, exactly at the center, it suddenly but only momentarily disappears. Then the capsule travels through the second half of the tunnel, to the north pole.

Check Griffith's description by finding the gravitational force on the capsule of mass m when it reaches a distance r from Earth's center. Assume that Earth is a sphere of uniform density ρ (mass per unit volume).

KEY IDEAS Newton's shell theorem gives us three ideas:

1. When the capsule is at radius r from Earth's center, the portion of Earth that lies outside a sphere of radius r does *not* produce a net gravitational force on the capsule.
2. The portion of Earth that lies inside that sphere *does* produce a net gravitational force on the capsule.
3. We can treat the mass M_{ins} of that inside portion of Earth as being the mass of a particle located at Earth's center.

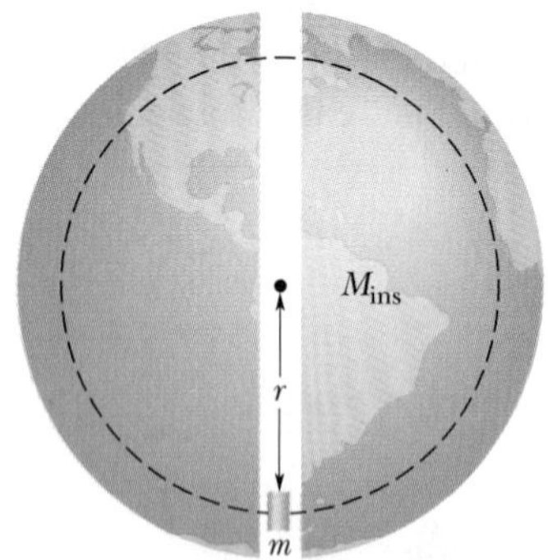

FIG. 13-8 A capsule of mass m falls from rest through a tunnel that connects Earth's south and north poles. When the capsule is at distance r from Earth's center, the portion of Earth's mass that is contained in a sphere of that radius is M_{ins}.

Calculations: All three ideas tell us that we can write Eq. 13-1, for the magnitude of the gravitational force on the capsule, as

$$F = \frac{GmM_{ins}}{r^2}. \quad (13\text{-}17)$$

To write the mass M_{ins} in terms of the radius r, we note that the volume V_{ins} containing this mass is $\frac{4}{3}\pi r^3$. Also, because we're assuming an Earth of uniform density, the density $\rho_{ins} = M_{ins}/V_{ins}$ is Earth's density ρ. Thus, we have

$$M_{ins} = \rho V_{ins} = \rho\frac{4\pi r^3}{3}. \quad (13\text{-}18)$$

Then, after substituting this expression into Eq. 13-17 and canceling, we have

$$F = \frac{4\pi Gm\rho}{3}r. \quad \text{(Answer)} \quad (13\text{-}19)$$

This equation tells us that the force magnitude F depends linearly on the capsule's distance r from Earth's center. Thus, as r decreases, F also decreases (opposite of Griffith's description), until it is zero at Earth's center. At least Griffith got the zero-at-the-center detail correct.

Equation 13-19 can also be written in terms of the force vector $\vec{F}$ and the capsule's position vector $\vec{r}$ along a radial axis extending from Earth's center. Let K represent the collection of constants $4\pi Gm\rho/3$. Then, Eq. 13-19 becomes

$$\vec{F} = -K\vec{r}, \quad (13\text{-}20)$$

in which we have inserted a minus sign to indicate that $\vec{F}$ and $\vec{r}$ have opposite directions. Equation 13-20 has the form of Hooke's law (Eq. 7-20, $\vec{F} = -k\vec{d}$). Thus, under the idealized conditions of the story, the capsule would oscillate like a block on a spring, with the center of the oscillation at Earth's center. After the capsule had fallen from the south pole to Earth's center, it would travel from the center to the north pole (as Griffith said) and then back again, repeating the cycle forever.

13-6 | Gravitational Potential Energy

In Section 8-4, we discussed the gravitational potential energy of a particle–Earth system. We were careful to keep the particle near Earth's surface, so that we could regard the gravitational force as constant. We then chose some reference configuration of the system as having a gravitational potential energy of zero. Often, in this configuration the particle was on Earth's surface. For particles not on Earth's surface, the gravitational potential energy decreased when the separation between the particle and Earth decreased.

Here, we broaden our view and consider the gravitational potential energy U of two particles, of masses m and M, separated by a distance r. We again choose a reference configuration with U equal to zero. However, to simplify the equations, the separation distance r in the reference configuration is now large enough to be approximated as *infinite*. As before, the gravitational potential energy decreases when the separation decreases. Since $U = 0$ for $r = \infty$, the potential energy is negative for any finite separation and becomes progressively more negative as the particles move closer together.

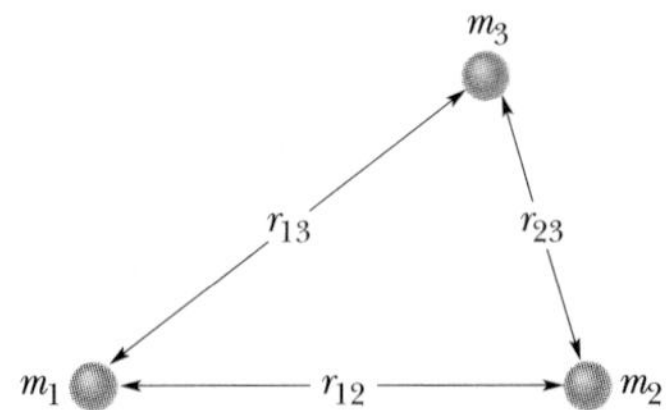

FIG. 13-9 A system consisting of three particles. The gravitational potential energy *of the system* is the sum of the gravitational potential energies of all three pairs of particles.

With these facts in mind and as we shall justify next, we take the gravitational potential energy of the two-particle system to be

$$U = -\frac{GMm}{r} \quad \text{(gravitational potential energy).} \tag{13-21}$$

Note that $U(r)$ approaches zero as r approaches infinity and that for any finite value of r, the value of $U(r)$ is negative.

The potential energy given by Eq. 13-21 is a property of the system of two particles rather than of either particle alone. There is no way to divide this energy and say that so much belongs to one particle and so much to the other. However, if $M \gg m$, as is true for Earth (mass M) and a baseball (mass m), we often speak of "the potential energy of the baseball." We can get away with this because, when a baseball moves in the vicinity of Earth, changes in the potential energy of the baseball–Earth system appear almost entirely as changes in the kinetic energy of the baseball, since changes in the kinetic energy of Earth are too small to be measured. Similarly, in Section 13-8 we shall speak of "the potential energy of an artificial satellite" orbiting Earth, because the satellite's mass is so much smaller than Earth's mass. When we speak of the potential energy of bodies of comparable mass, however, we have to be careful to treat them as a system.

If our system contains more than two particles, we consider each pair of particles in turn, calculate the gravitational potential energy of that pair with Eq. 13-21 as if the other particles were not there, and then algebraically sum the results. Applying Eq. 13-21 to each of the three pairs of Fig. 13-9, for example, gives the potential energy of the system as

$$U = -\left(\frac{Gm_1m_2}{r_{12}} + \frac{Gm_1m_3}{r_{13}} + \frac{Gm_2m_3}{r_{23}}\right). \tag{13-22}$$

Proof of Equation 13-21

Let us shoot a baseball directly away from Earth along the path in Fig. 13-10. We want to find an expression for the gravitational potential energy U of the ball at point P along its path, at radial distance R from Earth's center. To do so, we first find the work W done on the ball by the gravitational force as the ball travels from point P to a great (infinite) distance from Earth. Because the gravitational force $\vec{F}(r)$ is a variable force (its magnitude depends on r), we must use the techniques of Section 7-8 to find the work. In vector notation, we can write

$$W = \int_R^\infty \vec{F}(r) \cdot d\vec{r}. \tag{13-23}$$

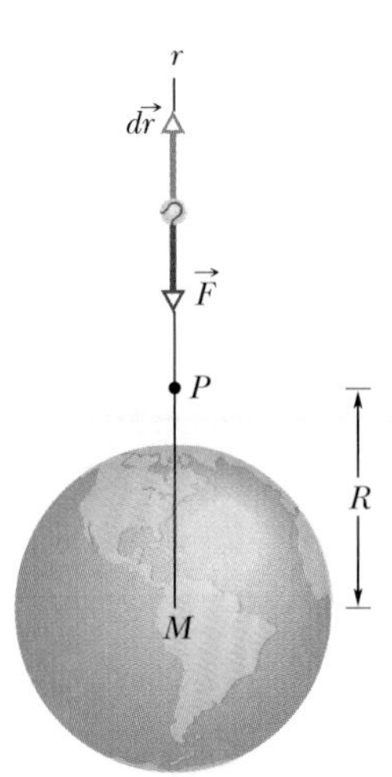

FIG. 13-10 A baseball is shot directly away from Earth, through point P at radial distance R from Earth's center. The gravitational force $\vec{F}$ on the ball and a differential displacement vector $d\vec{r}$ are shown, both directed along a radial r axis.

The integral contains the scalar (or dot) product of the force $\vec{F}(r)$ and the differential displacement vector $d\vec{r}$ along the ball's path. We can expand that product as

$$\vec{F}(r) \cdot d\vec{r} = F(r)\, dr \cos \phi, \tag{13-24}$$

where ϕ is the angle between the directions of $\vec{F}(r)$ and $d\vec{r}$. When we substitute

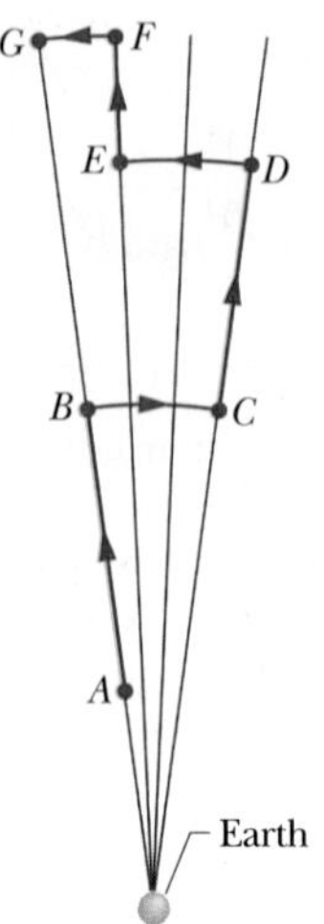

FIG. 13-11 Near Earth, a baseball is moved from point A to point G along a path consisting of radial lengths and circular arcs.

180° for ϕ and Eq. 13-1 for $F(r)$, Eq. 13-24 becomes

$$\vec{F}(r) \cdot d\vec{r} = -\frac{GMm}{r^2}\, dr,$$

where M is Earth's mass and m is the mass of the ball.

Substituting this into Eq. 13-23 and integrating give us

$$W = -GMm \int_R^\infty \frac{1}{r^2}\, dr = \left[\frac{GMm}{r}\right]_R^\infty$$

$$= 0 - \frac{GMm}{R} = -\frac{GMm}{R}, \qquad (13\text{-}25)$$

where W is the work required to move the ball from point P (at distance R) to infinity. Equation 8-1 ($\Delta U = -W$) tells us that we can also write that work in terms of potential energies as

$$U_\infty - U = -W.$$

Because the potential energy U_∞ at infinity is zero, U is the potential energy at P, and W is given by Eq. 13-25, this equation becomes

$$U = W = -\frac{GMm}{R}.$$

Switching R to r gives us Eq. 13-21, which we set out to prove.

Path Independence

In Fig. 13-11, we move a baseball from point A to point G along a path consisting of three radial lengths and three circular arcs (centered on Earth). We are interested in the total work W done by Earth's gravitational force $\vec{F}$ on the ball as it moves from A to G. The work done along each circular arc is zero, because the direction of $\vec{F}$ is perpendicular to the arc at every point. Thus, W is the sum of only the works done by $\vec{F}$ along the three radial lengths.

Now, suppose we mentally shrink the arcs to zero. We would then be moving the ball directly from A to G along a single radial length. Does that change W? No. Because no work was done along the arcs, eliminating them does not change the work. The path taken from A to G now is clearly different, but the work done by $\vec{F}$ is the same.

We discussed such a result in a general way in Section 8-3. Here is the point: The gravitational force is a conservative force. Thus, the work done by the gravitational force on a particle moving from an initial point i to a final point f is independent of the path taken between the points. From Eq. 8-1, the change ΔU in the gravitational potential energy from point i to point f is given by

$$\Delta U = U_f - U_i = -W. \qquad (13\text{-}26)$$

Since the work W done by a conservative force is independent of the actual path taken, the change ΔU in gravitational potential energy is *also independent* of the path taken.

Potential Energy and Force

In the proof of Eq. 13-21, we derived the potential energy function $U(r)$ from the force function $\vec{F}(r)$. We should be able to go the other way—that is, to start from the potential energy function and derive the force function. Guided by Eq. 8-22 ($F(x) = -dU(x)/dx$), we can write

$$F = -\frac{dU}{dr} = -\frac{d}{dr}\left(-\frac{GMm}{r}\right)$$

$$= -\frac{GMm}{r^2}. \qquad (13\text{-}27)$$

This is Newton's law of gravitation (Eq. 13-1). The minus sign indicates that the force on mass m points radially inward, toward mass M.

Escape Speed

If you fire a projectile upward, usually it will slow, stop momentarily, and return to Earth. There is, however, a certain minimum initial speed that will cause it to move upward forever, theoretically coming to rest only at infinity. This minimum initial speed is called the (Earth) **escape speed.**

Consider a projectile of mass m, leaving the surface of a planet (or some other astronomical body or system) with escape speed v. The projectile has a kinetic energy K given by $\frac{1}{2}mv^2$ and a potential energy U given by Eq. 13-21:

$$U = -\frac{GMm}{R},$$

in which M is the mass of the planet and R is its radius.

When the projectile reaches infinity, it stops and thus has no kinetic energy. It also has no potential energy because an infinite separation between two bodies is our zero-potential-energy configuration. Its total energy at infinity is therefore zero. From the principle of conservation of energy, its total energy at the planet's surface must also have been zero, and so

$$K + U = \tfrac{1}{2}mv^2 + \left(-\frac{GMm}{R}\right) = 0.$$

This yields

$$v = \sqrt{\frac{2GM}{R}}. \tag{13-28}$$

Note that v does not depend on the direction in which a projectile is fired from a planet. However, attaining that speed is easier if the projectile is fired in the direction the launch site is moving as the planet rotates about its axis. For example, rockets are launched eastward at Cape Canaveral to take advantage of the Cape's eastward speed of 1500 km/h due to Earth's rotation.

Equation 13-28 can be applied to find the escape speed of a projectile from any astronomical body, provided we substitute the mass of the body for M and the radius of the body for R. Table 13-2 shows some escape speeds.

CHECKPOINT 3 You move a ball of mass m away from a sphere of mass M. (a) Does the gravitational potential energy of the ball–sphere system increase or decrease? (b) Is positive or negative work done by the gravitational force between the ball and the sphere?

TABLE 13-2

Some Escape Speeds

Body	Mass (kg)	Radius (m)	Escape Speed (km/s)
Ceres[a]	1.17×10^{21}	3.8×10^5	0.64
Earth's moon[a]	7.36×10^{22}	1.74×10^6	2.38
Earth	5.98×10^{24}	6.37×10^6	11.2
Jupiter	1.90×10^{27}	7.15×10^7	59.5
Sun	1.99×10^{30}	6.96×10^8	618
Sirius B[b]	2×10^{30}	1×10^7	5200
Neutron star[c]	2×10^{30}	1×10^4	2×10^5

[a]The most massive of the asteroids.

[b]A *white dwarf* (a star in a final stage of evolution) that is a companion of the bright star Sirius.

[c]The collapsed core of a star that remains after that star has exploded in a *supernova* event.

Sample Problem 13-5

An asteroid, headed directly toward Earth, has a speed of 12 km/s relative to the planet when the asteroid is 10 Earth radii from Earth's center. Neglecting the effects of Earth's atmosphere on the asteroid, find the asteroid's speed v_f when it reaches Earth's surface.

KEY IDEAS Because we are to neglect the effects of the atmosphere on the asteroid, the mechanical energy of the asteroid–Earth system is conserved during the fall. Thus, the final mechanical energy (when the asteroid reaches Earth's surface) is equal to the initial mechanical energy. With kinetic energy K and gravitational potential energy U, we can write this as

$$K_f + U_f = K_i + U_i. \tag{13-29}$$

Also, if we assume the system is isolated, the system's linear momentum must be conserved during the fall. Therefore, the momentum change of the asteroid and that of Earth must be equal in magnitude and opposite in sign. However, because Earth's mass is so much greater than the asteroid's mass, the change in Earth's speed is negligible relative to the change in the asteroid's speed. So, the change in Earth's kinetic energy is also negligible. Thus, we can assume that the kinetic energies in Eq. 13-29 are those of the asteroid alone.

Calculations: Let m represent the asteroid's mass and M represent Earth's mass (5.98×10^{24} kg). The asteroid is initially at distance $10R_E$ and finally at distance R_E, where R_E is Earth's radius (6.37×10^6 m). Substituting Eq. 13-21 for U and $\frac{1}{2}mv^2$ for K, we rewrite Eq. 13-29 as

$$\tfrac{1}{2}mv_f^2 - \frac{GMm}{R_E} = \tfrac{1}{2}mv_i^2 - \frac{GMm}{10R_E}.$$

Rearranging and substituting known values, we find

$$\begin{aligned} v_f^2 &= v_i^2 + \frac{2GM}{R_E}\left(1 - \frac{1}{10}\right) \\ &= (12 \times 10^3 \text{ m/s})^2 \\ &\quad + \frac{2(6.67 \times 10^{-11} \text{ m}^3/\text{kg}\cdot\text{s}^2)(5.98 \times 10^{24} \text{ kg})}{6.37 \times 10^6 \text{ m}} 0.9 \\ &= 2.567 \times 10^8 \text{ m}^2/\text{s}^2, \end{aligned}$$

and

$$v_f = 1.60 \times 10^4 \text{ m/s} = 16 \text{ km/s}. \quad \text{(Answer)}$$

At this speed, the asteroid would not have to be particularly large to do considerable damage at impact. If it were only 5 m across, the impact could release about as much energy as the nuclear explosion at Hiroshima. Alarmingly, about 500 million asteroids of this size are near Earth's orbit, and in 1994 one of them apparently penetrated Earth's atmosphere and exploded 20 km above the South Pacific (setting off nuclear-explosion warnings on six military satellites). The impact of an asteroid 500 m across (there may be a million of them near Earth's orbit) could end modern civilization and almost eliminate humans worldwide.

13-7 | Planets and Satellites: Kepler's Laws

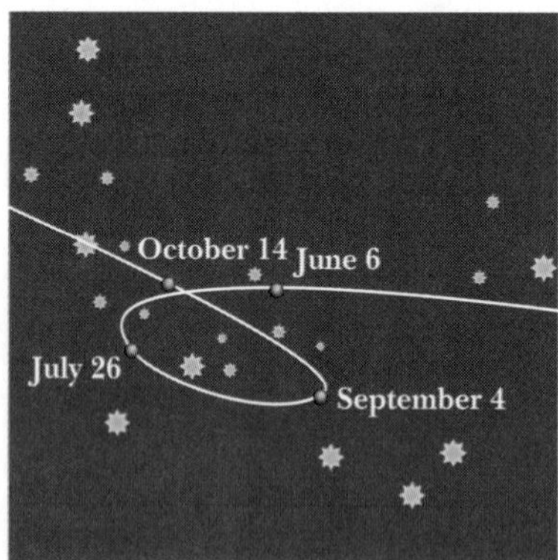

FIG. 13-12 The path seen from Earth for the planet Mars as it moved against a background of the constellation Capricorn during 1971. The planet's position on four days is marked. Both Mars and Earth are moving in orbits around the Sun so that we see the position of Mars relative to us; this relative motion sometimes results in an apparent loop in the path of Mars.

The motions of the planets, as they seemingly wander against the background of the stars, have been a puzzle since the dawn of history. The "loop-the-loop" motion of Mars, shown in Fig. 13-12, was particularly baffling. Johannes Kepler (1571–1630), after a lifetime of study, worked out the empirical laws that govern these motions. Tycho Brahe (1546–1601), the last of the great astronomers to make observations without the help of a telescope, compiled the extensive data from which Kepler was able to derive the three laws of planetary motion that now bear Kepler's name. Later, Newton (1642–1727) showed that his law of gravitation leads to Kepler's laws.

In this section we discuss each of Kepler's three laws. Although here we apply the laws to planets orbiting the Sun, they hold equally well for satellites, either natural or artificial, orbiting Earth or any other massive central body.

> **1. THE LAW OF ORBITS:** All planets move in elliptical orbits, with the Sun at one focus.

Figure 13-13 shows a planet of mass m moving in such an orbit around the Sun, whose mass is M. We assume that $M \gg m$, so that the center of mass of the planet–Sun system is approximately at the center of the Sun.

The orbit in Fig. 13-13 is described by giving its **semimajor axis** a and its **eccentricity** e, the latter defined so that ea is the distance from the center of the ellipse to either focus F or F'. *An eccentricity of zero corresponds to a circle,* in which the two foci merge to a single central point. The eccentricities of the planetary orbits are not large; so if the orbits are drawn to scale, they look circular. The eccentricity of the ellipse of Fig. 13-13, which has been exaggerated for clarity, is 0.74. The eccentricity of Earth's orbit is only 0.0167.

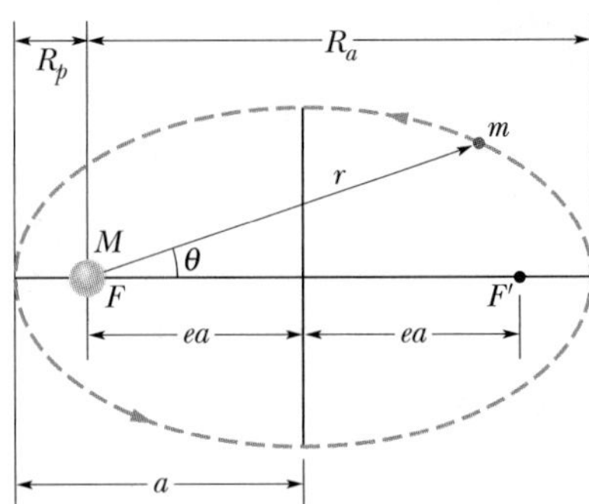

FIG. 13-13 A planet of mass m moving in an elliptical orbit around the Sun. The Sun, of mass M, is at one focus F of the ellipse. The other focus is F', which is located in empty space. Each focus is a distance ea from the ellipse's center, with e being the eccentricity of the ellipse. The semimajor axis a of the ellipse, the perihelion (nearest the Sun) distance R_p, and the aphelion (farthest from the Sun) distance R_a are also shown.

2. THE LAW OF AREAS: A line that connects a planet to the Sun sweeps out equal areas in the plane of the planet's orbit in equal time intervals; that is, the rate dA/dt at which it sweeps out area A is constant.

Qualitatively, this second law tells us that the planet will move most slowly when it is farthest from the Sun and most rapidly when it is nearest to the Sun. As it turns out, Kepler's second law is totally equivalent to the law of conservation of angular momentum. Let us prove it.

The area of the shaded wedge in Fig. 13-14*a* closely approximates the area swept out in time Δt by a line connecting the Sun and the planet, which are separated by distance r. The area ΔA of the wedge is approximately the area of a triangle with base $r\,\Delta\theta$ and height r. Since the area of a triangle is one-half of the base times the height, $\Delta A \approx \frac{1}{2}r^2\,\Delta\theta$. This expression for ΔA becomes more exact as Δt (hence $\Delta\theta$) approaches zero. The instantaneous rate at which area is being swept out is then

$$\frac{dA}{dt} = \tfrac{1}{2}r^2\frac{d\theta}{dt} = \tfrac{1}{2}r^2\omega, \tag{13-30}$$

in which ω is the angular speed of the rotating line connecting Sun and planet.

Figure 13-14*b* shows the linear momentum $\vec{p}$ of the planet, along with the radial and perpendicular components of $\vec{p}$. From Eq. 11-20 ($L = rp_\perp$), the magnitude of the angular momentum $\vec{L}$ of the planet about the Sun is given by the product of r and $p_\perp$, the component of $\vec{p}$ perpendicular to r. Here, for a planet of mass m,

$$\begin{aligned} L &= rp_\perp = (r)(mv_\perp) = (r)(m\omega r) \\ &= mr^2\omega, \end{aligned} \tag{13-31}$$

where we have replaced $v_\perp$ with its equivalent ωr (Eq. 10-18). Eliminating $r^2\omega$ between Eqs. 13-30 and 13-31 leads to

$$\frac{dA}{dt} = \frac{L}{2m}. \tag{13-32}$$

If dA/dt is constant, as Kepler said it is, then Eq. 13-32 means that L must also be constant—angular momentum is conserved. Kepler's second law is indeed equivalent to the law of conservation of angular momentum.

3. THE LAW OF PERIODS: The square of the period of any planet is proportional to the cube of the semimajor axis of its orbit.

To see this, consider the circular orbit of Fig. 13-15, with radius r (the radius of a circle is equivalent to the semimajor axis of an ellipse). Applying Newton's

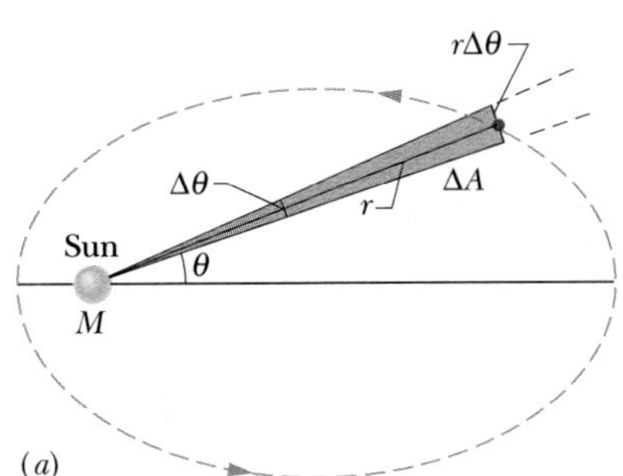

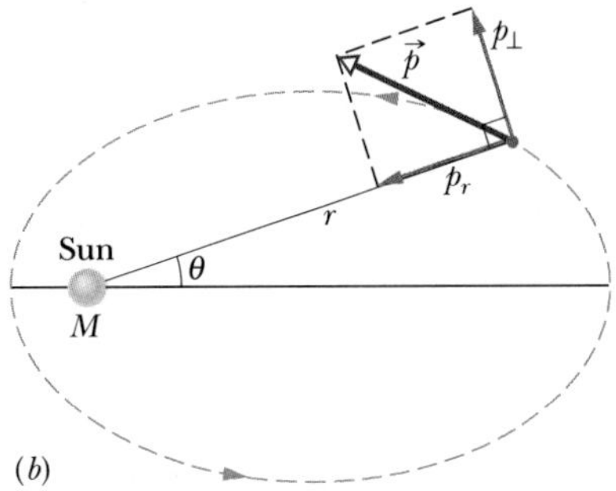

FIG. 13-14 (*a*) In time Δt, the line r connecting the planet to the Sun moves through an angle $\Delta\theta$, sweeping out an area ΔA (shaded). (*b*) The linear momentum $\vec{p}$ of the planet and the components of $\vec{p}$.

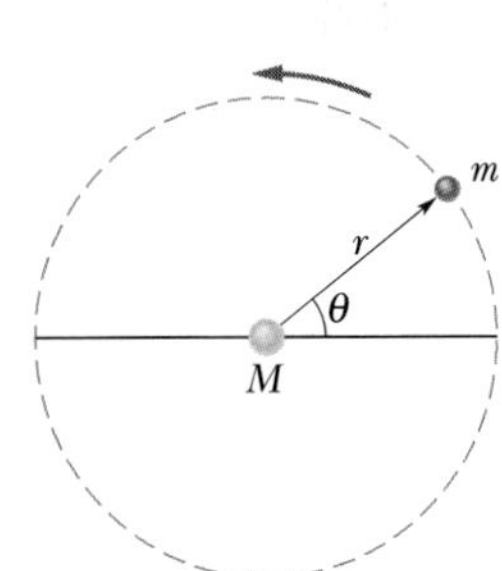

FIG. 13-15 A planet of mass m moving around the Sun in a circular orbit of radius r.

TABLE 13-3

Kepler's Law of Periods for the Solar System

Planet	Semimajor Axis a (10^{10} m)	Period T (y)	T^2/a^3 (10^{-34} y^2/m^3)
Mercury	5.79	0.241	2.99
Venus	10.8	0.615	3.00
Earth	15.0	1.00	2.96
Mars	22.8	1.88	2.98
Jupiter	77.8	11.9	3.01
Saturn	143	29.5	2.98
Uranus	287	84.0	2.98
Neptune	450	165	2.99
Pluto	590	248	2.99

second law ($F = ma$) to the orbiting planet in Fig. 13-15 yields

$$\frac{GMm}{r^2} = (m)(\omega^2 r). \qquad (13\text{-}33)$$

Here we have substituted from Eq. 13-1 for the force magnitude F and used Eq. 10-23 to substitute $\omega^2 r$ for the centripetal acceleration. If we now use Eq. 10-20 to replace ω with $2\pi/T$, where T is the period of the motion, we obtain Kepler's third law:

$$T^2 = \left(\frac{4\pi^2}{GM}\right)r^3 \quad \text{(law of periods).} \qquad (13\text{-}34)$$

The quantity in parentheses is a constant that depends only on the mass M of the central body about which the planet orbits.

Equation 13-34 holds also for elliptical orbits, provided we replace r with a, the semimajor axis of the ellipse. This law predicts that the ratio T^2/a^3 has essentially the same value for every planetary orbit around a given massive body. Table 13-3 shows how well it holds for the orbits of the planets of the solar system.

CHECKPOINT 4 Satellite 1 is in a certain circular orbit around a planet, while satellite 2 is in a larger circular orbit. Which satellite has (a) the longer period and (b) the greater speed?

Sample Problem 13-6

Comet Halley orbits the Sun with a period of 76 years and, in 1986, had a distance of closest approach to the Sun, its *perihelion distance* R_p, of 8.9×10^{10} m. Table 13-3 shows that this is between the orbits of Mercury and Venus.

(a) What is the comet's farthest distance from the Sun, which is called its *aphelion distance* R_a?

KEY IDEAS From Fig. 13-13, we see that $R_a + R_p = 2a$, where a is the semimajor axis of the orbit. Thus, we can find R_a if we first find a. We can relate a to the given period via the law of periods (Eq. 13-34) if we simply substitute the semimajor axis a for r.

Calculations: Making that substitution and then solving for a, we have

$$a = \left(\frac{GMT^2}{4\pi^2}\right)^{1/3}. \qquad (13\text{-}35)$$

If we substitute the mass M of the Sun, 1.99×10^{30} kg, and the period T of the comet, 76 years or 2.4×10^9 s, into Eq. 13-35, we find that $a = 2.7 \times 10^{12}$ m. Now we have

$$\begin{aligned} R_a &= 2a - R_p \\ &= (2)(2.7 \times 10^{12}\text{ m}) - 8.9 \times 10^{10}\text{ m} \\ &= 5.3 \times 10^{12}\text{ m}. \end{aligned} \qquad \text{(Answer)}$$

Table 13-3 shows that this is a little less than the semimajor axis of the orbit of Pluto. Thus, the comet does not get farther from the Sun than Pluto.

(b) What is the eccentricity e of the orbit of comet Halley?

KEY IDEA We can relate e, a, and R_p via Fig. 13-13, in which we see that $ea = a - R_p$.

Calculation: We have

$$\begin{aligned} e &= \frac{a - R_p}{a} = 1 - \frac{R_p}{a} \\ &= 1 - \frac{8.9 \times 10^{10}\text{ m}}{2.7 \times 10^{12}\text{ m}} = 0.97. \end{aligned} \qquad \text{(Answer)}$$

This tells us that, with an eccentricity approaching unity, this orbit must be a long thin ellipse.

Sample Problem 13-7

Let's return to the story that opens this chapter. Figure 13-16 shows the observed orbit of the star S2 as the star moves around a mysterious and unobserved object called Sagittarius A* (pronounced "A star"), which is at the center of the Milky Way galaxy. S2 orbits Sagittarius A* with a period of $T = 15.2$ y and with a semimajor axis of $a = 5.50$ light-days ($= 1.42 \times 10^{14}$ m). What is the mass M of Sagittarius A*? What is Sagittarius A*?

KEY IDEA The period T and the semimajor axis a of the orbit are related to the mass M of Sagittarius A* according to Kepler's law of periods. From Eq. 13-34, with a replacing the radius r of a circular orbit, we have

$$T^2 = \left(\frac{4\pi^2}{GM}\right)a^3. \tag{13-36}$$

Calculations: Solving Eq. 13-36 for M and substituting the given data lead us to

$$M = \frac{4\pi^2 a^3}{GT^2}$$

$$= \frac{4\pi^2(1.42 \times 10^{14}\text{ m})^3}{(6.67 \times 10^{-11}\text{ N}\cdot\text{m}^2/\text{kg}^2)[(15.2\text{ y})(3.16 \times 10^7\text{ s/y})]^2}$$

$$= 7.35 \times 10^{36}\text{ kg}. \qquad \text{(Answer)}$$

To figure out what Sagittarius A* might be, let's divide this mass by the mass of our Sun ($M_{\text{Sun}} = 1.99 \times 10^{30}$ kg) to find that

$$M = (3.7 \times 10^6)M_{\text{Sun}}.$$

Sagittarius A* has a mass of 3.7 million Suns! However, it cannot be seen. Thus, it is an extremely compact object. Such a huge mass in such a small object leads to the reasonable conclusion that this object is a *supermassive* black hole. In fact, evidence is mounting that a supermassive black hole lurks at the center of most galaxies. (Movies of the stars orbiting Sagittarius A* are available on the Web; search under "black hole galactic center.")

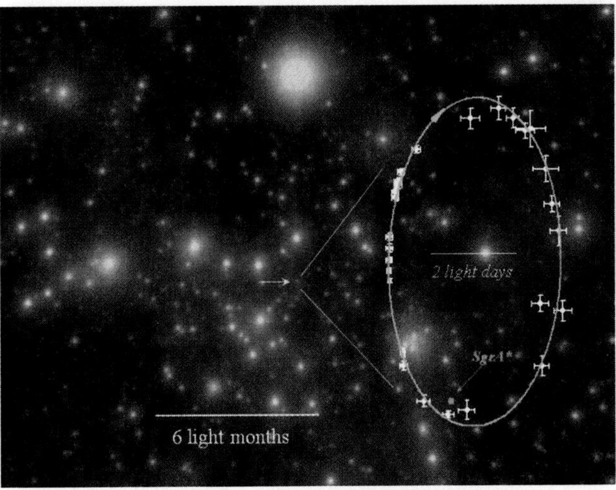

FIG. 13-16 The orbit of star S2 about Sagittarius A* (Sgr A*). The elliptical orbit appears skewed because we do not see it from directly above the orbital plane. Uncertainties in the location of S2 are indicated by the crossbars. *(Courtesy Reinhard Genzel)*

13-8 | Satellites: Orbits and Energy

As a satellite orbits Earth in an elliptical path, both its speed, which fixes its kinetic energy K, and its distance from the center of Earth, which fixes its gravitational potential energy U, fluctuate with fixed periods. However, the mechanical energy E of the satellite remains constant. (Since the satellite's mass is so much smaller than Earth's mass, we assign U and E for the Earth–satellite system to the satellite alone.)

The potential energy of the system is given by Eq. 13-21:

$$U = -\frac{GMm}{r}$$

(with $U = 0$ for infinite separation). Here r is the radius of the satellite's orbit, assumed for the time being to be circular, and M and m are the masses of Earth and the satellite, respectively.

To find the kinetic energy of a satellite in a circular orbit, we write Newton's second law ($F = ma$) as

$$\frac{GMm}{r^2} = m\frac{v^2}{r}, \tag{13-37}$$

where v^2/r is the centripetal acceleration of the satellite. Then, from Eq. 13-37, the kinetic energy is

$$K = \tfrac{1}{2}mv^2 = \frac{GMm}{2r}, \tag{13-38}$$

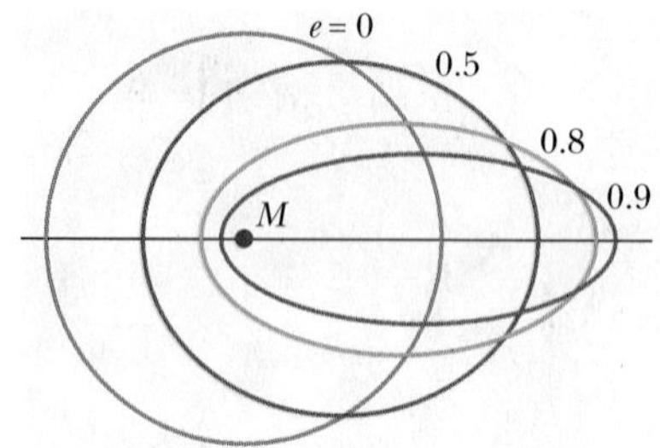

FIG. 13-17 Four orbits with different eccentricities e about an object of mass M. All four orbits have the same semimajor axis a and thus correspond to the same total mechanical energy E.

which shows us that for a satellite in a circular orbit,

$$K = -\frac{U}{2} \quad \text{(circular orbit).} \tag{13-39}$$

The total mechanical energy of the orbiting satellite is

$$E = K + U = \frac{GMm}{2r} - \frac{GMm}{r}$$

or

$$E = -\frac{GMm}{2r} \quad \text{(circular orbit).} \tag{13-40}$$

This tells us that for a satellite in a circular orbit, the total energy E is the negative of the kinetic energy K:

$$E = -K \quad \text{(circular orbit).} \tag{13-41}$$

For a satellite in an elliptical orbit of semimajor axis a, we can substitute a for r in Eq. 13-40 to find the mechanical energy:

$$E = -\frac{GMm}{2a} \quad \text{(elliptical orbit).} \tag{13-42}$$

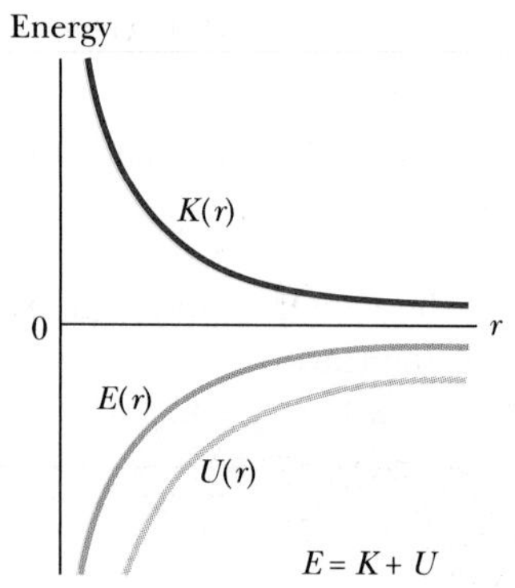

FIG. 13-18 The variation of kinetic energy K, potential energy U, and total energy E with radius r for a satellite in a circular orbit. For any value of r, the values of U and E are negative, the value of K is positive, and $E = -K$. As $r \rightarrow \infty$, all three energy curves approach a value of zero.

Equation 13-42 tells us that the total energy of an orbiting satellite depends only on the semimajor axis of its orbit and not on its eccentricity e. For example, four orbits with the same semimajor axis are shown in Fig. 13-17; the same satellite would have the same total mechanical energy E in all four orbits. Figure 13-18 shows the variation of K, U, and E with r for a satellite moving in a circular orbit about a massive central body.

CHECKPOINT 5 In the figure here, a space shuttle is initially in a circular orbit of radius r about Earth. At point P, the pilot briefly fires a forward-pointing thruster to decrease the shuttle's kinetic energy K and mechanical energy E. (a) Which of the dashed elliptical orbits shown in the figure will the shuttle then take? (b) Is the orbital period T of the shuttle (the time to return to P) then greater than, less than, or the same as in the circular orbit?

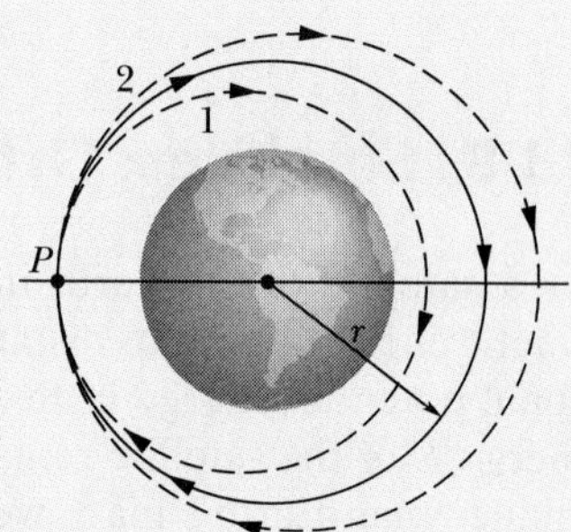

Sample Problem 13-8

A playful astronaut releases a bowling ball, of mass $m = 7.20$ kg, into circular orbit about Earth at an altitude h of 350 km.

(a) What is the mechanical energy E of the ball in its orbit?

KEY IDEA We can get E from the orbital energy, given by Eq. 13-40 ($E = -GMm/2r$), if we first find the orbital radius r.

Calculations: The orbital radius must be

$$r = R + h = 6370 \text{ km} + 350 \text{ km} = 6.72 \times 10^6 \text{ m},$$

in which R is the radius of Earth. Then, from Eq. 13-40, the mechanical energy is

$$E = -\frac{GMm}{2r}$$
$$= -\frac{(6.67 \times 10^{-11} \text{ N}\cdot\text{m}^2/\text{kg}^2)(5.98 \times 10^{24} \text{ kg})(7.20 \text{ kg})}{(2)(6.72 \times 10^6 \text{ m})}$$
$$= -2.14 \times 10^8 \text{ J} = -214 \text{ MJ}. \quad \text{(Answer)}$$

(b) What is the mechanical energy E_0 of the ball on the launchpad at Cape Canaveral? From there to the orbit, what is the change ΔE in the ball's mechanical energy?

KEY IDEA On the launchpad, the ball is *not* in orbit and thus Eq. 13-40 does *not* apply. Instead, we must find

$E_0 = K_0 + U_0$, where K_0 is the ball's kinetic energy and U_0 is the gravitational potential energy of the ball–Earth system.

Calculations: To find U_0, we use Eq. 13-21 to write

$$U_0 = -\frac{GMm}{R}$$

$$= -\frac{(6.67 \times 10^{-11}\ \text{N}\cdot\text{m}^2/\text{kg}^2)(5.98 \times 10^{24}\ \text{kg})(7.20\ \text{kg})}{6.37 \times 10^6\ \text{m}}$$

$$= -4.51 \times 10^8\ \text{J} = -451\ \text{MJ}.$$

The kinetic energy K_0 of the ball is due to the ball's motion with Earth's rotation. You can show that K_0 is less than 1 MJ, which is negligible relative to U_0. Thus, the mechanical energy of the ball on the launchpad is

$$E_0 = K_0 + U_0 \approx 0 - 451\ \text{MJ} = -451\ \text{MJ}. \quad \text{(Answer)}$$

The *increase* in the mechanical energy of the ball from launchpad to orbit is

$$\Delta E = E - E_0 = (-214\ \text{MJ}) - (-451\ \text{MJ})$$

$$= 237\ \text{MJ}. \quad \text{(Answer)}$$

This is worth a few dollars at your utility company. Obviously the high cost of placing objects into orbit is not due to their required mechanical energy.

13-9 | Einstein and Gravitation

Principle of Equivalence

Albert Einstein once said: "I was . . . in the patent office at Bern when all of a sudden a thought occurred to me: 'If a person falls freely, he will not feel his own weight.' I was startled. This simple thought made a deep impression on me. It impelled me toward a theory of gravitation."

Thus Einstein tells us how he began to form his **general theory of relativity.** The fundamental postulate of this theory about gravitation (the gravitating of objects toward each other) is called the **principle of equivalence,** which says that gravitation and acceleration are equivalent. If a physicist were locked up in a small box as in Fig. 13-19, he would not be able to tell whether the box was at rest on Earth (and subject only to Earth's gravitational force), as in Fig. 13-19*a*, or accelerating through interstellar space at 9.8 m/s² (and subject only to the force producing that acceleration), as in Fig. 13-19*b*. In both situations he would feel the same and would read the same value for his weight on a scale. Moreover, if he watched an object fall past him, the object would have the same acceleration relative to him in both situations.

FIG. 13-19 (*a*) A physicist in a box resting on Earth sees a cantaloupe falling with acceleration $a = 9.8\ \text{m/s}^2$. (*b*) If he and the box accelerate in deep space at 9.8 m/s², the cantaloupe has the same acceleration relative to him. It is not possible, by doing experiments within the box, for the physicist to tell which situation he is in. For example, the platform scale on which he stands reads the same weight in both situations.

Curvature of Space

We have thus far explained gravitation as due to a force between masses. Einstein showed that, instead, gravitation is due to a curvature of space that is caused by the masses. (As is discussed later in this book, space and time are entangled, so the curvature of which Einstein spoke is really a curvature of *spacetime,* the combined four dimensions of our universe.)

Picturing how space (such as vacuum) can have curvature is difficult. An analogy might help: Suppose that from orbit we watch a race in which two boats begin on Earth's equator with a separation of 20 km and head due south (Fig. 13-20*a*). To the sailors, the boats travel along flat, parallel paths. However, with time the boats draw together until, nearer the south pole, they touch. The sailors in the boats can interpret this drawing together in terms of a force acting on the boats. Looking on from space, however, we can see that the boats draw together simply because of the curvature of Earth's surface. We can see this because we are viewing the race from "outside" that surface.

Figure 13-20*b* shows a similar race: Two horizontally separated apples are dropped from the same height above Earth. Although the apples may appear to travel along parallel paths, they actually move toward each other because they both fall toward Earth's center. We can interpret the motion of the apples in terms of the gravitational force on the apples from Earth. We can also interpret the motion in terms of a curvature of the space near Earth, a curvature due to the

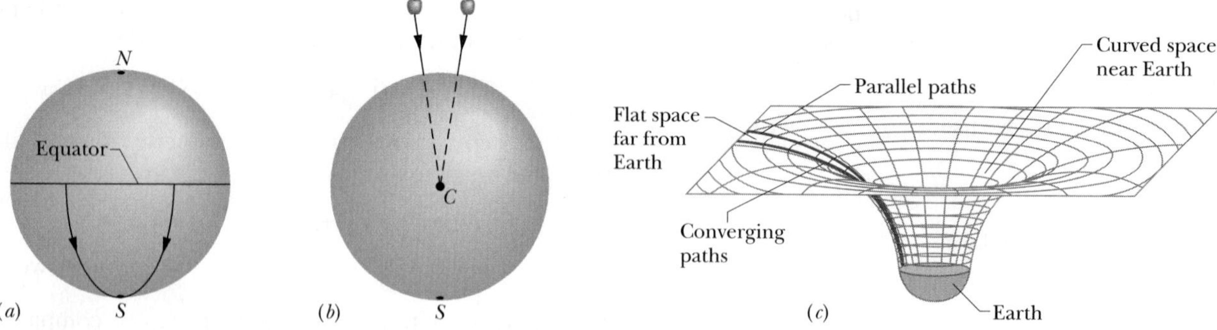

FIG. 13-20 (*a*) Two objects moving along lines of longitude toward the south pole converge because of the curvature of Earth's surface. (*b*) Two objects falling freely near Earth move along lines that converge toward the center of Earth because of the curvature of space near Earth. (*c*) Far from Earth (and other masses), space is flat and parallel paths remain parallel. Close to Earth, the parallel paths begin to converge because space is curved by Earth's mass.

presence of Earth's mass. This time we cannot see the curvature because we cannot get "outside" the curved space, as we got "outside" the curved Earth in the boat example. However, we can depict the curvature with a drawing like Fig. 13-20*c*; there the apples would move along a surface that curves toward Earth because of Earth's mass.

When light passes near Earth, the path of the light bends slightly because of the curvature of space there, an effect called *gravitational lensing.* When light passes a more massive structure, like a galaxy or a black hole having large mass, its path can be bent more. If such a massive structure is between us and a quasar (an extremely bright, extremely distant source of light), the light from the quasar can bend around the massive structure and toward us (Fig. 13-21*a*). Then, because the light seems to be coming to us from a number of slightly different directions in the sky, we see the same quasar in all those different directions. In some situations, the quasars we see blend together to form a giant luminous arc, which is called an *Einstein ring* (Fig. 13-21*b*).

Should we attribute gravitation to the curvature of spacetime due to the presence of masses or to a force between masses? Or should we attribute it to the actions of a type of fundamental particle called a *graviton,* as conjectured in some modern physics theories? We just don't know.

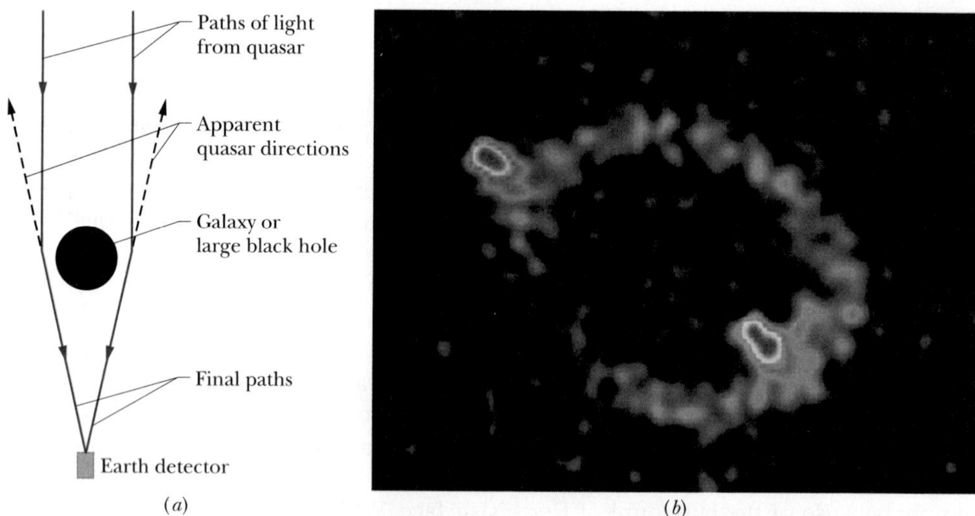

FIG. 13-21 (*a*) Light from a distant quasar follows curved paths around a galaxy or a large black hole because the mass of the galaxy or black hole has curved the adjacent space. If the light is detected, it appears to have originated along the backward extensions of the final paths (dashed lines). (*b*) The Einstein ring known as MG1131+0456 on the computer screen of a telescope. The source of the light (actually, radio waves, which are a form of invisible light) is far behind the large, unseen galaxy that produces the ring; a portion of the source appears as the two bright spots seen along the ring. (*Courtesy National Radio Astronomy Observatory*)

REVIEW & SUMMARY

The Law of Gravitation Any particle in the universe attracts any other particle with a **gravitational force** whose magnitude is

$$F = G\frac{m_1 m_2}{r^2} \quad \text{(Newton's law of gravitation)}, \tag{13-1}$$

where m_1 and m_2 are the masses of the particles, r is their separation, and G $(= 6.67 \times 10^{-11}\ \text{N} \cdot \text{m}^2/\text{kg}^2)$ is the *gravitational constant.*

Gravitational Behavior of Uniform Spherical Shells Equation 13-1 holds only for particles. The gravitational force between extended bodies must generally be found by adding (integrating) the individual forces on individual particles within the bodies. However, if either of the bodies is a uniform spherical shell or a spherically symmetric solid, the net gravitational force it exerts on an *external* object may be computed as if all the mass of the shell or body were located at its center.

Superposition Gravitational forces obey the **principle of superposition;** that is, if n particles interact, the net force $\vec{F}_{1,\text{net}}$ on a particle labeled particle 1 is the sum of the forces on it from all the other particles taken one at a time:

$$\vec{F}_{1,\text{net}} = \sum_{i=2}^{n} \vec{F}_{1i}, \tag{13-5}$$

in which the sum is a vector sum of the forces $\vec{F}_{1i}$ on particle 1 from particles 2, 3, . . . , n. The gravitational force $\vec{F}_1$ on a particle from an extended body is found by dividing the body into units of differential mass dm, each of which produces a differential force $d\vec{F}$ on the particle, and then integrating to find the sum of those forces:

$$\vec{F}_1 = \int d\vec{F}. \tag{13-6}$$

Gravitational Acceleration The *gravitational acceleration* a_g of a particle (of mass m) is due solely to the gravitational force acting on it. When the particle is at distance r from the center of a uniform, spherical body of mass M, the magnitude F of the gravitational force on the particle is given by Eq. 13-1. Thus, by Newton's second law,

$$F = ma_g, \tag{13-10}$$

which gives

$$a_g = \frac{GM}{r^2}. \tag{13-11}$$

Free-Fall Acceleration and Weight Because Earth's mass is not distributed uniformly, because the planet is not perfectly spherical, and because it rotates, the actual free-fall acceleration $\vec{g}$ of a particle near Earth differs slightly from the gravitational acceleration $\vec{a}_g$, and the particle's weight (equal to mg) differs from the magnitude of the gravitational force acting on the particle as computed with Eq. 13-1.

Gravitation Within a Spherical Shell A uniform shell of matter exerts no net gravitational force on a particle located inside it. This means that if a particle is located inside a uniform solid sphere at distance r from its center, the gravitational force exerted on the particle is due only to the mass M_{ins} that lies inside a sphere of radius r. This mass is given by

$$M_{\text{ins}} = \rho\frac{4\pi r^3}{3}, \tag{13-18}$$

where ρ is the density of the sphere.

Gravitational Potential Energy The gravitational potential energy $U(r)$ of a system of two particles, with masses M and m and separated by a distance r, is the negative of the work that would be done by the gravitational force of either particle acting on the other if the separation between the particles were changed from infinite (very large) to r. This energy is

$$U = -\frac{GMm}{r} \quad \text{(gravitational potential energy)}. \tag{13-21}$$

Potential Energy of a System If a system contains more than two particles, its total gravitational potential energy U is the sum of terms representing the potential energies of all the pairs. As an example, for three particles, of masses m_1, m_2, and m_3,

$$U = -\left(\frac{Gm_1m_2}{r_{12}} + \frac{Gm_1m_3}{r_{13}} + \frac{Gm_2m_3}{r_{23}}\right). \tag{13-22}$$

Escape Speed An object will escape the gravitational pull of an astronomical body of mass M and radius R (that is, it will reach an infinite distance) if the object's speed near the body's surface is at least equal to the **escape speed,** given by

$$v = \sqrt{\frac{2GM}{R}}. \tag{13-28}$$

Kepler's Laws Gravitational attraction holds the solar system together and makes possible orbiting Earth satellites, both natural and artificial. Such motions are governed by Kepler's three laws of planetary motion, all of which are direct consequences of Newton's laws of motion and gravitation:

1. ***The law of orbits.*** All planets move in elliptical orbits with the Sun at one focus.
2. ***The law of areas.*** A line joining any planet to the Sun sweeps out equal areas in equal time intervals. (This statement is equivalent to conservation of angular momentum.)
3. ***The law of periods.*** The square of the period T of any planet is proportional to the cube of the semimajor axis a of its orbit. For circular orbits with radius r,

$$T^2 = \left(\frac{4\pi^2}{GM}\right)r^3 \quad \text{(law of periods)}, \tag{13-34}$$

where M is the mass of the attracting body—the Sun in the case of the solar system. For elliptical planetary orbits, the semimajor axis a is substituted for r.

Energy in Planetary Motion When a planet or satellite with mass m moves in a circular orbit with radius r, its potential energy U and kinetic energy K are given by

$$U = -\frac{GMm}{r} \quad \text{and} \quad K = \frac{GMm}{2r}. \qquad (13\text{-}21, 13\text{-}38)$$

The mechanical energy $E = K + U$ is then

$$E = -\frac{GMm}{2r}. \qquad (13\text{-}40)$$

For an elliptical orbit of semimajor axis a,

$$E = -\frac{GMm}{2a}. \qquad (13\text{-}42)$$

Einstein's View of Gravitation Einstein pointed out that gravitation and acceleration are equivalent. This **principle of equivalence** led him to a theory of gravitation (the **general theory of relativity**) that explains gravitational effects in terms of a curvature of space.

QUESTIONS

1 In Fig. 13-22, a central particle is surrounded by two circular rings of particles, at radii r and R, with $R > r$. All the particles have mass m. What are the magnitude and direction of the net gravitational force on the central particle due to the particles in the rings?

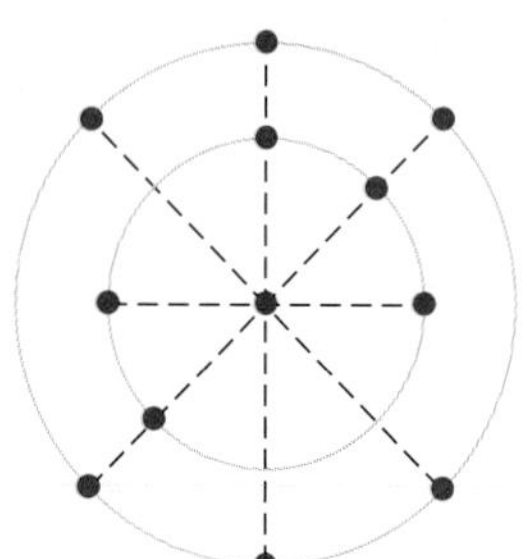

FIG. 13-22 Question 1.

2 In Fig. 13-23, two particles, of masses m and $2m$, are fixed in place on an axis. (a) Where on the axis can a third particle of mass $3m$ be placed (other than at infinity) so that the net gravitational force on it from the first two particles is zero: to the left of the first two particles, to their right, between them but closer to the more massive particle, or between them but closer to the less massive particle? (b) Does the answer change if the third particle has, instead, a mass of $16m$? (c) Is there a point off the axis (other than infinity) at which the net force on the third particle would be zero?

FIG. 13-23 Question 2.

3 Figure 13-24 shows three situations involving a point particle P with mass m and a spherical shell with a uniformly distributed mass M. The radii of the shells are given. Rank the situations according to the magnitude of the gravitational force on particle P due to the shell, greatest first.

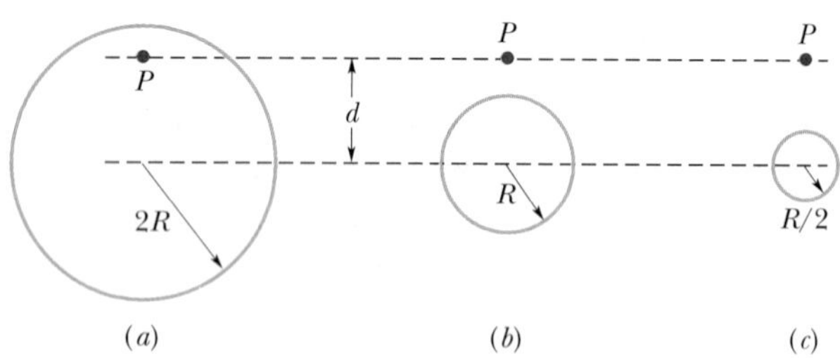

FIG. 13-24 Question 3.

4 Figure 13-25 shows three arrangements of the same identical particles, with three of them placed on a circle of radius 0.20 m and the fourth one placed at the center of the circle. (a) Rank the arrangements according to the magnitude of the net gravitational force on the central particle due to the other three particles, greatest first. (b) Rank them according to the gravitational potential energy of the four-particle system, least negative first.

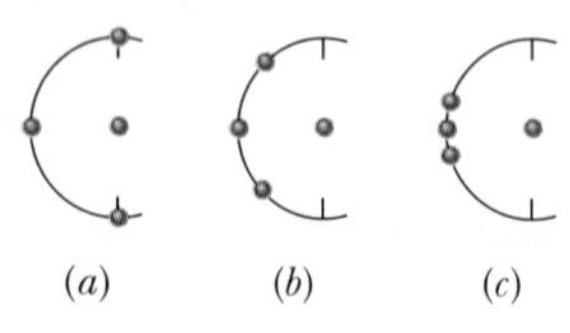

FIG. 13-25 Question 4.

5 In Fig. 13-26, a central particle of mass M is surrounded by a square array of other particles, separated by either distance d or distance $d/2$ along the perimeter of the square. What are the magnitude and direction of the net gravitational force on the central particle due to the other particles?

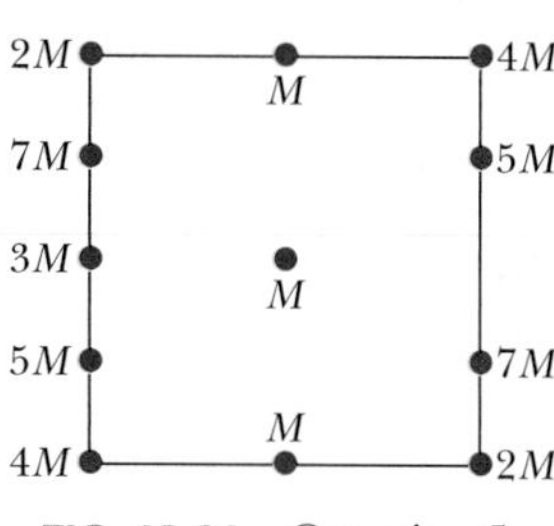

FIG. 13-26 Question 5.

6 Figure 13-27 gives the gravitational acceleration a_g for four planets as a function of the radial distance r from the center of the planet, starting at the surface of the planet (at radius R_1, R_2, R_3, or R_4). Plots 1 and 2 coincide for $r \geq R_2$; plots 3 and 4 coincide for $r \geq R_4$. Rank the four planets according to (a) mass and (b) mass per unit volume, greatest first.

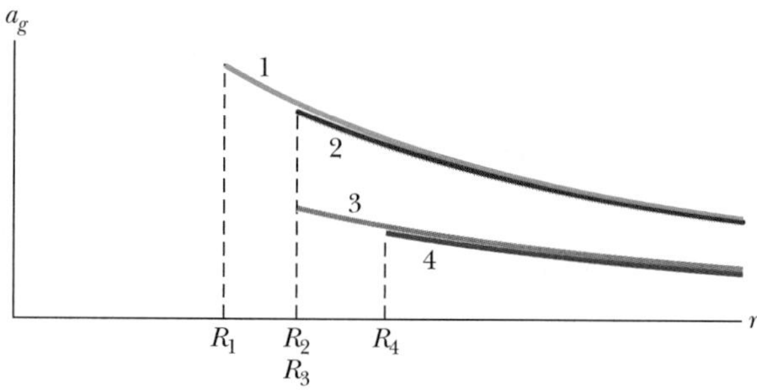

FIG. 13-27 Question 6.

7 Figure 13-28 shows three particles initially fixed in place, with B and C identical and positioned symmetrically about the y axis, at distance d from A. (a) In what direction is the net gravitational force $\vec{F}_{net}$ on A? (b) If we move C directly away from the origin, does $\vec{F}_{net}$ change in direction? If so, how and what is the limit of the change?

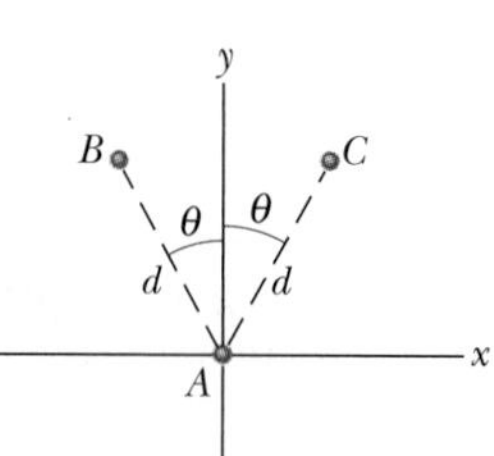

FIG. 13-28 Question 7.

8 In Fig. 13-29, three particles are fixed in place. The mass of B is greater than the mass of C. Can a fourth particle (particle D) be placed somewhere so that the net gravitational force on particle A from particles B, C, and D is zero? If so, in which quadrant should it be placed and which axis should it be near?

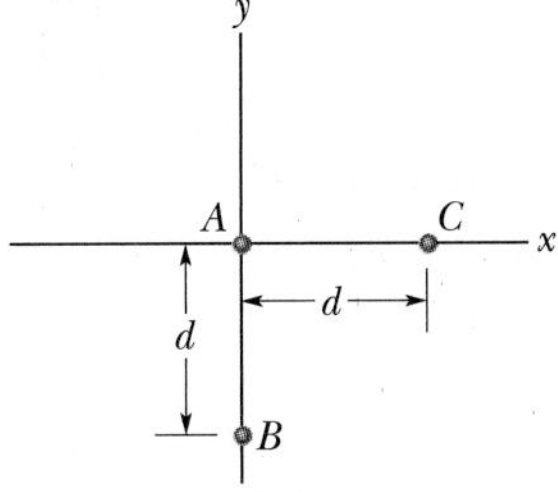

FIG. 13-29 Question 8.

9 Rank the four systems of equal-mass particles shown in Checkpoint 2 according to the absolute value of the gravitational potential energy of the system, greatest first.

10 In Fig. 13-30, a particle of mass m (not shown) is to be moved from an infinite distance to one of the three possible locations a, b, and c. Two other particles, of masses m and $2m$, are fixed in place. Rank the three possible locations according to the work done by the net gravitational force on the moving particle due to the fixed particles, greatest first.

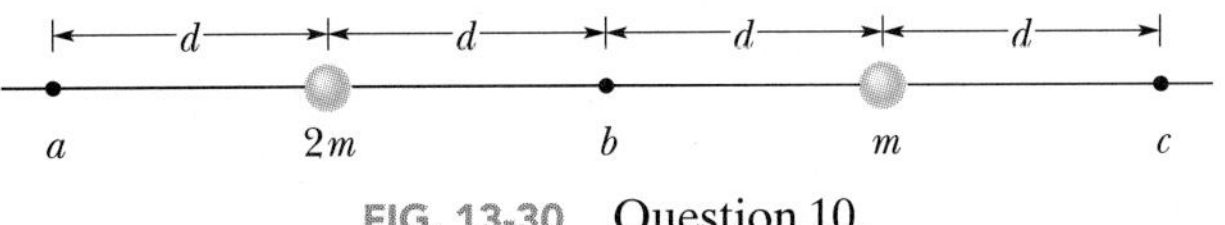

FIG. 13-30 Question 10.

11 Figure 13-31 shows three uniform spherical planets that are identical in size and mass. The periods of rotation T for the planets are given, and six lettered points are indicated—three points are on the equators of the planets and three points are on the north poles. Rank the points according to the value of the free-fall acceleration g at them, greatest first.

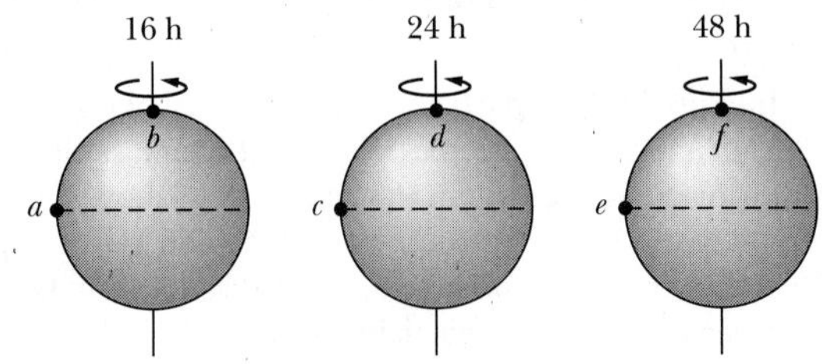

FIG. 13-31 Question 11.

12 Figure 13-32 shows six paths by which a rocket orbiting a moon might move from point a to point b. Rank the paths according to (a) the corresponding change in the gravitational potential energy of the rocket–moon system and (b) the net work done on the rocket by the gravitational force from the moon, greatest first.

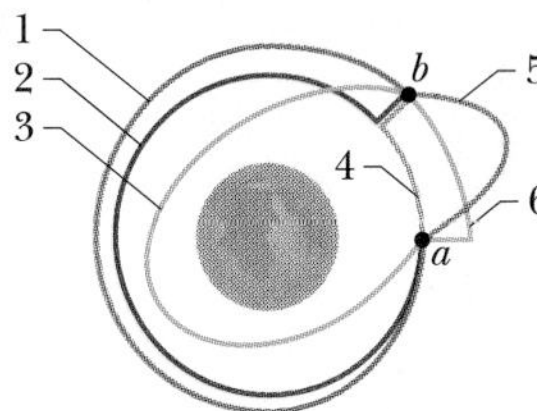

FIG. 13-32 Question 12.

PROBLEMS

GO Tutoring problem available (at instructor's discretion) in *WileyPLUS* and WebAssign

SSM Worked-out solution available in Student Solutions Manual

WWW Worked-out solution is at http://www.wiley.com/college/halliday

• – ••• Number of dots indicates level of problem difficulty

ILW Interactive solution is at http://www.wiley.com/college/halliday

Additional information available in *The Flying Circus of Physics* and at flyingcircusofphysics.com

sec. 13-2 Newton's Law of Gravitation

•**1** What must the separation be between a 5.2 kg particle and a 2.4 kg particle for their gravitational attraction to have a magnitude of 2.3×10^{-12} N? SSM

•**2** The Sun and Earth each exert a gravitational force on the Moon. What is the ratio F_{Sun}/F_{Earth} of these two forces? (The average Sun–Moon distance is equal to the Sun–Earth distance.)

•**3** A mass M is split into two parts, m and $M - m$, which are then separated by a certain distance. What ratio m/M maximizes the magnitude of the gravitational force between the parts? ILW

•**4** *Moon effect.* Some people believe that the Moon controls their activities. If the Moon moves from being directly on the opposite side of Earth from you to being directly overhead, by what percent does (a) the Moon's gravitational pull on you increase and (b) your weight (as measured on a scale) decrease? Assume that the Earth–Moon (center-to-center) distance is 3.82×10^8 m and Earth's radius is 6.37×10^6 m.

sec. 13-3 Gravitation and the Principle of Superposition

•**5** *One dimension.* In Fig. 13-33, two point particles are fixed on an x axis separated by distance d. Particle A has mass m_A and particle B has mass $3.00m_A$. A third particle C, of mass $75.0m_A$, is to be placed on the x axis and near particles A and B. In terms of distance d, at what x coordinate should C be placed so that the net gravitational force on particle A from particles B and C is zero?

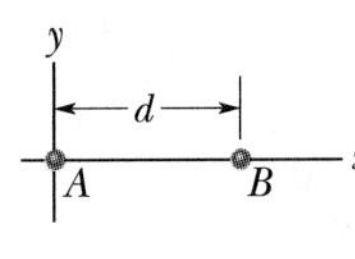

FIG. 13-33 Problem 5.

•**6** In Fig. 13-34, three 5.00 kg spheres are located at distances $d_1 = 0.300$ m and $d_2 = 0.400$ m. What are the (a) magnitude and (b) direction (relative to the positive direction of the x axis) of the net gravitational force on sphere B due to spheres A and C?

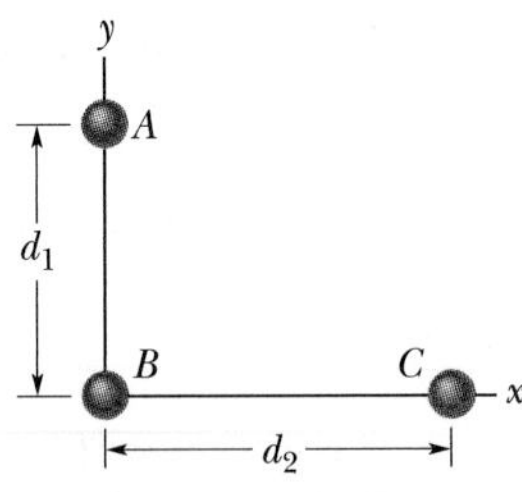

FIG. 13-34 Problem 6.

•**7** How far from Earth must a space probe be along a line toward the Sun so that the Sun's gravitational pull on the probe balances Earth's pull? SSM WWW

•**8** In Fig. 13-35, a square of edge length 20.0 cm is formed by four spheres of masses $m_1 = 5.00$ g, $m_2 = 3.00$ g, $m_3 = 1.00$ g, and $m_4 = 5.00$ g. In unit-vector notation, what is the net gravitational force from them on a central sphere with mass $m_5 = 2.50$ g? GO

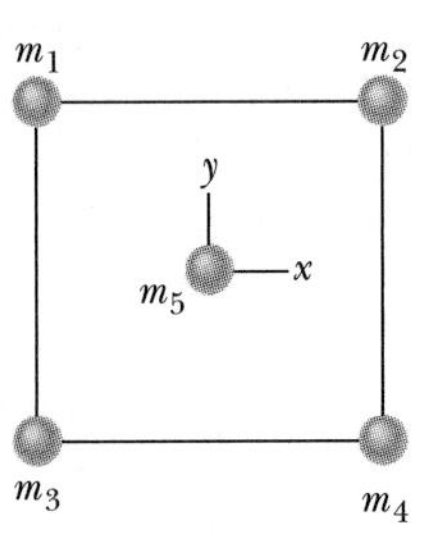

FIG. 13-35 Problem 8.

•9 *Miniature black holes.* Left over from the big-bang beginning of the universe, tiny black holes might still wander through the universe. If one with a mass of 1×10^{11} kg (and a radius of only 1×10^{-16} m) reached Earth, at what distance from your head would its gravitational pull on you match that of Earth's?

••10 In Fig. 13-36*a*, particle *A* is fixed in place at $x = -0.20$ m on the *x* axis and particle *B*, with a mass of 1.0 kg, is fixed in place at the origin. Particle *C* (not shown) can be moved along the *x* axis, between particle *B* and $x = \infty$. Figure 13-36*b* shows the *x* component $F_{net,x}$ of the net gravitational force on particle *B* due to particles *A* and *C*, as a function of position *x* of particle *C*. The plot actually extends to the right, approaching an asymptote of -4.17×10^{-10} N as $x \to \infty$. What are the masses of (a) particle *A* and (b) particle *C*?

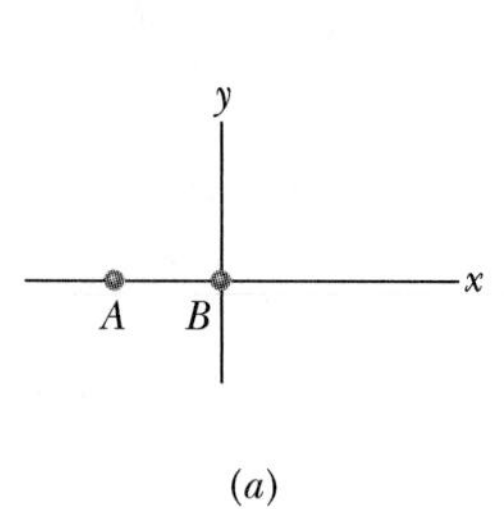

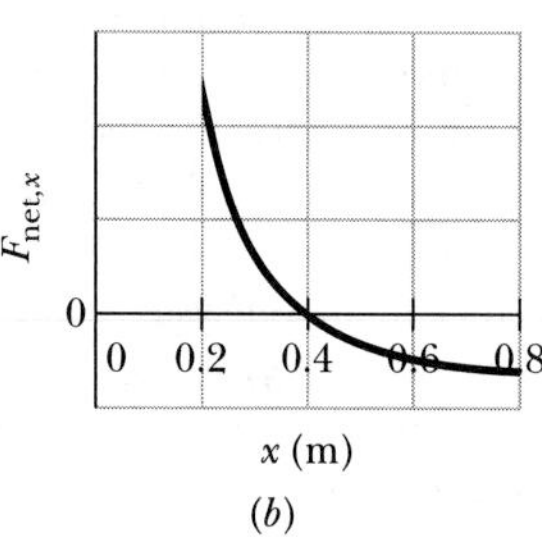

FIG. 13-36 Problem 10.

••11 As seen in Fig. 13-37, two spheres of mass *m* and a third sphere of mass *M* form an equilateral triangle, and a fourth sphere of mass m_4 is at the center of the triangle. The net gravitational force on that central sphere from the three other spheres is zero. (a) What is *M* in terms of *m*? (b) If we double the value of m_4, what then is the magnitude of the net gravitational force on the central sphere?

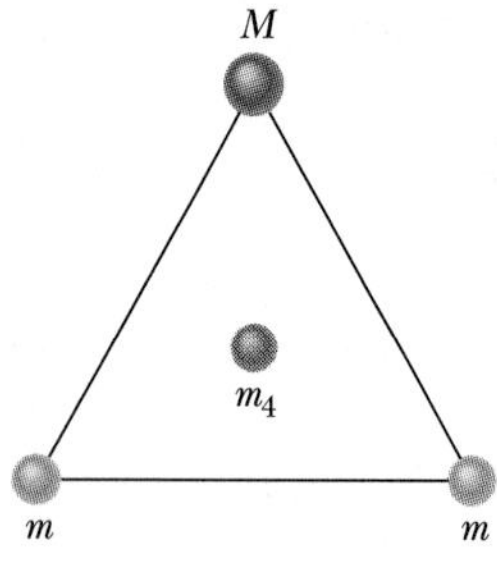

FIG. 13-37 Problem 11.

••12 Three point particles are fixed in position in an *xy* plane. Two of them, particle *A* of mass 6.00 g and particle *B* of mass 12.0 g, are shown in Fig. 13-38, with a separation of $d_{AB} = 0.500$ m at angle $\theta = 30°$. Particle *C*, with mass 8.00 g, is not shown. The net gravitational force acting on particle *A* due to particles *B* and *C* is 2.77×10^{-14} N at an angle of $-163.8°$ from the positive direction of the *x* axis. What are (a) the *x* coordinate and (b) the *y* coordinate of particle *C*?

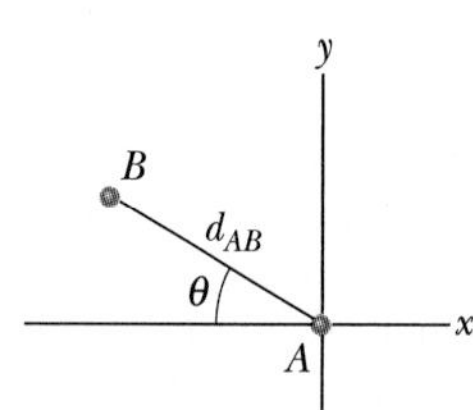

FIG. 13-38 Problem 12.

••13 Figure 13-39 shows a spherical hollow inside a lead sphere of radius $R = 4.00$ cm; the surface of the hollow passes through the center of the sphere and "touches" the right side of the sphere. The mass of the sphere before hollowing was $M = 2.95$ kg. With what gravitational force does the hollowed-out lead sphere attract a small sphere of mass $m = 0.431$ kg that lies at a distance $d = 9.00$ cm from the center of the lead sphere, on the straight line connecting the centers of the spheres and of the hollow?

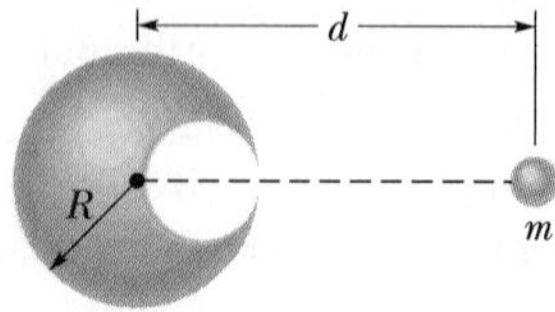

FIG. 13-39 Problem 13.

••14 *Two dimensions.* In Fig. 13-40, three point particles are fixed in place in an *xy* plane. Particle *A* has mass m_A, particle *B* has mass $2.00m_A$, and particle *C* has mass $3.00m_A$. A fourth particle *D*, with mass $4.00m_A$, is to be placed near the other three particles. In terms of distance *d*, at what (a) *x* coordinate and (b) *y* coordinate should particle *D* be placed so that the net gravitational force on particle *A* from particles *B*, *C*, and *D* is zero? GO

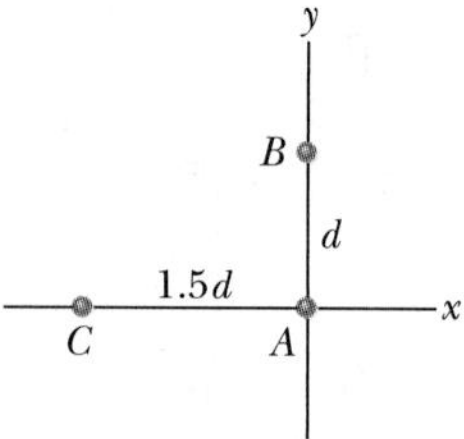

FIG. 13-40 Problem 14.

•••15 *Three dimensions.* Three point particles are fixed in place in an *xyz* coordinate system. Particle *A*, at the origin, has mass m_A. Particle *B*, at *xyz* coordinates $(2.00d, 1.00d, 2.00d)$, has mass $2.00m_A$, and particle *C*, at coordinates $(-1.00d, 2.00d, -3.00d)$, has mass $3.00m_A$. A fourth particle *D*, with mass $4.00m_A$, is to be placed near the other particles. In terms of distance *d*, at what (a) *x*, (b) *y*, and (c) *z* coordinate should *D* be placed so that the net gravitational force on *A* from *B*, *C*, and *D* is zero?

•••16 In Fig. 13-41, a particle of mass $m_1 = 0.67$ kg is a distance $d = 23$ cm from one end of a uniform rod with length $L = 3.0$ m and mass $M = 5.0$ kg. What is the magnitude of the gravitational force $\vec{F}$ on the particle from the rod?

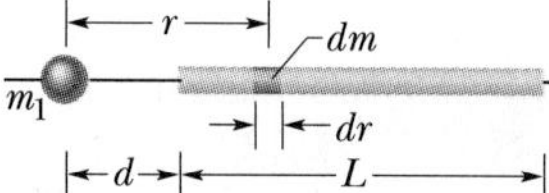

FIG. 13-41 Problem 16.

sec. 13-4 Gravitation Near Earth's Surface

•17 At what altitude above Earth's surface would the gravitational acceleration be 4.9 m/s^2? SSM

•18 *Mile-high building.* In 1956, Frank Lloyd Wright proposed the construction of a mile-high building in Chicago. Suppose the building had been constructed. Ignoring Earth's rotation, find the change in your weight if you were to ride an elevator from the street level, where you weigh 600 N, to the top of the building.

•19 (a) What will an object weigh on the Moon's surface if it weighs 100 N on Earth's surface? (b) How many Earth radii must this same object be from the center of Earth if it is to weigh the same as it does on the Moon?

•20 *Mountain pull.* A large mountain can slightly affect the direction of "down" as determined by a plumb line. Assume that we can model a mountain as a sphere of radius $R = 2.00$ km and density (mass per unit volume) 2.6×10^3 kg/m^3. Assume also that we hang a 0.50 m plumb line at a distance of 3*R* from the sphere's center and such that the sphere pulls horizontally on the lower end. How far would the lower end move toward the sphere?

••21 One model for a certain planet has a core of radius *R* and mass *M* surrounded by an outer shell of inner radius *R*, outer radius 2*R*, and mass 4*M*. If $M = 4.1 \times 10^{24}$ kg and $R = 6.0 \times 10^6$ m, what is the gravitational acceleration of a particle at points (a) *R* and (b) 3*R* from the center of the planet?

••22 The radius R_h and mass M_h of a black hole are related by $R_h = 2GM_h/c^2$, where c is the speed of light. Assume that the gravitational acceleration a_g of an object at a distance $r_o = 1.001R_h$ from the center of a black hole is given by Eq. 13-11 (it is, for large black holes). (a) In terms of M_h, find a_g at r_o. (b) Does a_g at r_o increase or decrease as M_h increases? (c) What is a_g at r_o for a very large black hole whose mass is 1.55×10^{12} times the solar mass of 1.99×10^{30} kg? (d) If the astronaut of Sample Problem 13-3 is at r_o with her feet toward this black hole, what is the difference in gravitational acceleration between her head and her feet? (e) Is the tendency to stretch the astronaut severe?

••23 Certain neutron stars (extremely dense stars) are believed to be rotating at about 1 rev/s. If such a star has a radius of 20 km, what must be its minimum mass so that material on its surface remains in place during the rapid rotation? ILW

sec. 13-5 Gravitation Inside Earth

•24 Two concentric spherical shells with uniformly distributed masses M_1 and M_2 are situated as shown in Fig. 13-42. Find the magnitude of the net gravitational force on a particle of mass m, due to the shells, when the particle is located at radial distance (a) a, (b) b, and (c) c.

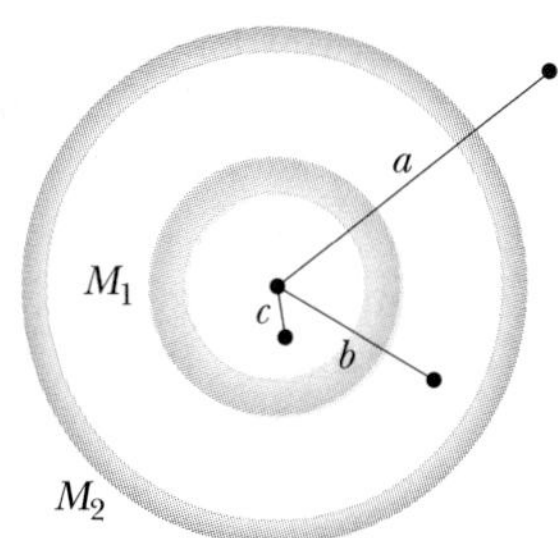

FIG. 13-42 Problem 24.

••25 Figure 13-43 shows, not to scale, a cross section through the interior of Earth. Rather than being uniform throughout, Earth is divided into three zones: an outer *crust,* a *mantle,* and an inner *core.* The dimensions of these zones and the masses contained within them are shown on the figure. Earth has a total mass of 5.98×10^{24} kg and a radius of 6370 km. Ignore rotation and assume that Earth is spherical. (a) Calculate a_g at the surface. (b) Suppose that a bore hole (the *Mohole*) is driven to the crust–mantle interface at a depth of 25.0 km; what would be the value of a_g at the bottom of the hole? (c) Suppose that Earth were a uniform sphere with the same total mass and size. What would be the value of a_g at a depth of 25.0 km? (Precise measurements of a_g are sensitive probes of the interior structure of Earth, although results can be clouded by local variations in mass distribution.)

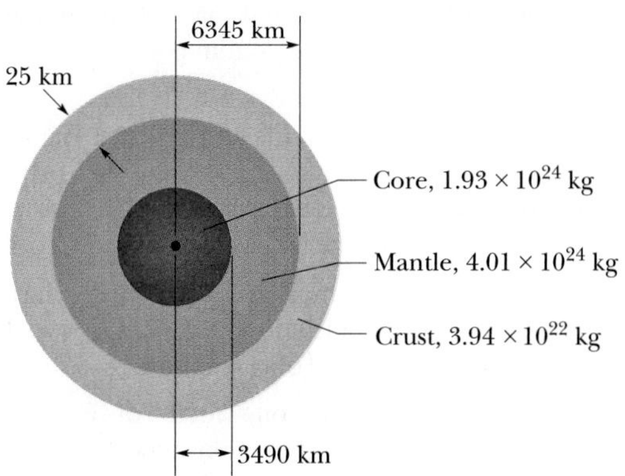

FIG. 13-43 Problem 25.

••26 Assume a planet is a uniform sphere of radius R that (somehow) has a narrow radial tunnel through its center (Fig. 13-8). Also assume we can position an apple anywhere along the tunnel or outside the sphere. Let F_R be the magnitude of the gravitational force on the apple when it is located at the planet's surface. How far from the surface is there a point where the magnitude of the gravitational force on the apple is $\frac{1}{2}F_R$ if we move the apple (a) away from the planet and (b) into the tunnel?

••27 A solid uniform sphere has a mass of 1.0×10^4 kg and a radius of 1.0 m. What is the magnitude of the gravitational force due to the sphere on a particle of mass m located at a distance of (a) 1.5 m and (b) 0.50 m from the center of the sphere? (c) Write a general expression for the magnitude of the gravitational force on the particle at a distance $r \leq 1.0$ m from the center of the sphere.

••28 Consider a pulsar, a collapsed star of extremely high density, with a mass M equal to that of the Sun (1.98×10^{30} kg), a radius R of only 12 km, and a rotational period T of 0.041 s. By what percentage does the free-fall acceleration g differ from the gravitational acceleration a_g at the equator of this spherical star?

sec. 13-6 Gravitational Potential Energy

•29 The mean diameters of Mars and Earth are 6.9×10^3 km and 1.3×10^4 km, respectively. The mass of Mars is 0.11 times Earth's mass. (a) What is the ratio of the mean density (mass per unit volume) of Mars to that of Earth? (b) What is the value of the gravitational acceleration on Mars? (c) What is the escape speed on Mars? SSM

•30 (a) What is the gravitational potential energy of the two-particle system in Problem 1? If you triple the separation between the particles, how much work is done (b) by the gravitational force between the particles and (c) by you?

•31 What multiple of the energy needed to escape from Earth gives the energy needed to escape from (a) the Moon and (b) Jupiter?

•32 Figure 13-44 gives the potential energy function $U(r)$ of a projectile, plotted outward from the surface of a planet of radius R_s. If the projectile is launched radially outward from the surface with a mechanical energy of -2.0×10^9 J, what are (a) its kinetic energy at radius $r = 1.25R_s$ and (b) its *turning point* (see Section 8-6) in terms of R_s?

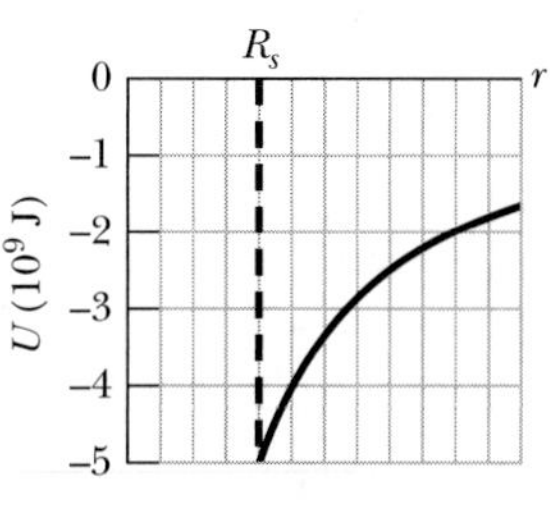

FIG. 13-44 Problems 32 and 33.

•33 Figure 13-44 gives the potential energy function $U(r)$ of a projectile, plotted outward from the surface of a planet of radius R_s. What least kinetic energy is required of a projectile launched at the surface if the projectile is to "escape" the planet?

•34 In Problem 3, what ratio m/M gives the least gravitational potential energy for the system?

••35 The three spheres in Fig. 13-45, with masses $m_A = 80$ g, $m_B = 10$ g, and $m_C = 20$ g, have their centers on a common line, with $L = 12$ cm and $d = 4.0$ cm. You move sphere

B along the line until its center-to-center separation from C is $d = 4.0$ cm. How much work is done on sphere B (a) by you and (b) by the net gravitational force on B due to spheres A and C? GO

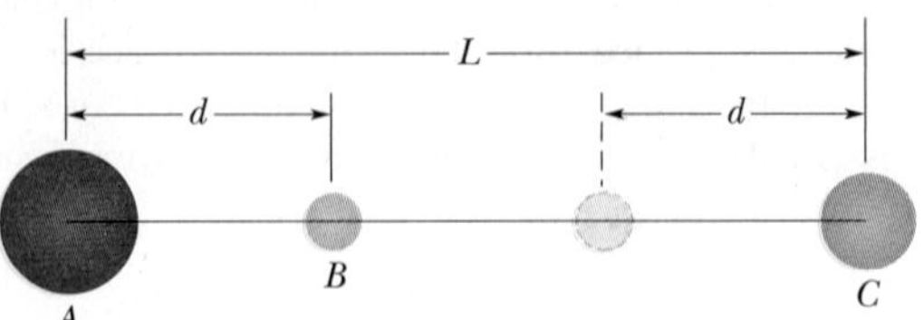

FIG. 13-45 Problem 35.

••**36** A projectile is shot directly away from Earth's surface. Neglect the rotation of Earth. What multiple of Earth's radius R_E gives the radial distance a projectile reaches if (a) its initial speed is 0.500 of the escape speed from Earth and (b) its initial kinetic energy is 0.500 of the kinetic energy required to escape Earth? (c) What is the least initial mechanical energy required at launch if the projectile is to escape Earth?

••**37** (a) What is the escape speed on a spherical asteroid whose radius is 500 km and whose gravitational acceleration at the surface is 3.0 m/s²? (b) How far from the surface will a particle go if it leaves the asteroid's surface with a radial speed of 1000 m/s? (c) With what speed will an object hit the asteroid if it is dropped from 1000 km above the surface? SSM

••**38** Zero, a hypothetical planet, has a mass of 5.0×10^{23} kg, a radius of 3.0×10^6 m, and no atmosphere. A 10 kg space probe is to be launched vertically from its surface. (a) If the probe is launched with an initial energy of 5.0×10^7 J, what will be its kinetic energy when it is 4.0×10^6 m from the center of Zero? (b) If the probe is to achieve a maximum distance of 8.0×10^6 m from the center of Zero, with what initial kinetic energy must it be launched from the surface of Zero?

••**39** Two neutron stars are separated by a distance of 1.0×10^{10} m. They each have a mass of 1.0×10^{30} kg and a radius of 1.0×10^5 m. They are initially at rest with respect to each other. As measured from that rest frame, how fast are they moving when (a) their separation has decreased to one-half its initial value and (b) they are about to collide? SSM

••**40** In deep space, sphere A of mass 20 kg is located at the origin of an x axis and sphere B of mass 10 kg is located on the axis at $x = 0.80$ m. Sphere B is released from rest while sphere A is held at the origin. (a) What is the gravitational potential energy of the two-sphere system just as B is released? (b) What is the kinetic energy of B when it has moved 0.20 m toward A?

••**41** Figure 13-46 shows four particles, each of mass 20.0 g, that form a square with an edge length of $d = 0.600$ m. If d is reduced to 0.200 m, what is the change in the gravitational potential energy of the four-particle system? GO

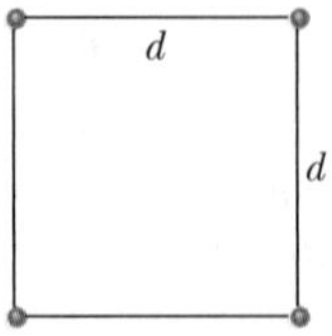

FIG. 13-46 Problem 41.

••**42** Figure 13-47a shows a particle A that can be moved along a y axis from an infinite distance to the origin. That origin lies at the midpoint between particles B and C, which have identical masses, and the y axis is a perpendicular bisector between them. Distance D is 0.3057 m. Figure 13-47b shows the potential energy U of the three-particle system as a function of the position of particle A along the y axis. The curve actually extends rightward and approaches an asymptote of -2.7×10^{-11} J as $y \to \infty$. What are the masses of (a) particles B and C and (b) particle A?

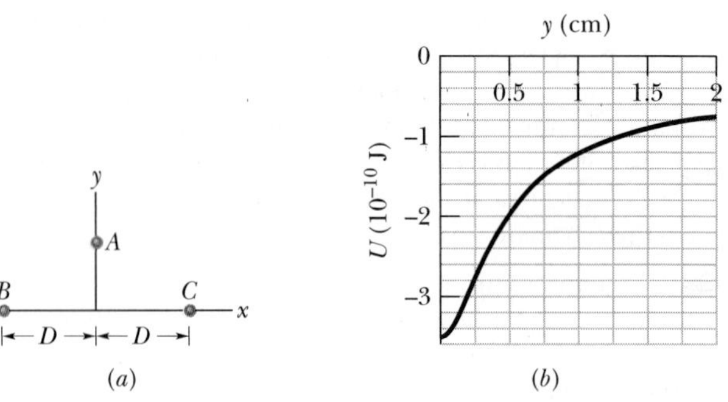

FIG. 13-47 Problem 42.

sec. 13-7 Planets and Satellites: Kepler's Laws

•**43** The Martian satellite Phobos travels in an approximately circular orbit of radius 9.4×10^6 m with a period of 7 h 39 min. Calculate the mass of Mars from this information.

•**44** The first known collision between space debris and a functioning satellite occurred in 1996: At an altitude of 700 km, a year-old French spy satellite was hit by a piece of an Ariane rocket that had been in orbit for 10 years. A stabilizing boom on the satellite was demolished, and the satellite was sent spinning out of control. Just before the collision and in kilometers per hour, what was the speed of the rocket piece relative to the satellite if both were in circular orbits and the collision was (a) head-on and (b) along perpendicular paths?

•**45** The Sun, which is 2.2×10^{20} m from the center of the Milky Way galaxy, revolves around that center once every 2.5×10^8 years. Assuming each star in the Galaxy has a mass equal to the Sun's mass of 2.0×10^{30} kg, the stars are distributed uniformly in a sphere about the galactic center, and the Sun is at the edge of that sphere, estimate the number of stars in the Galaxy. SSM WWW

•**46** The mean distance of Mars from the Sun is 1.52 times that of Earth from the Sun. From Kepler's law of periods, calculate the number of years required for Mars to make one revolution around the Sun; compare your answer with the value given in Appendix C.

•**47** A satellite, moving in an elliptical orbit, is 360 km above Earth's surface at its farthest point and 180 km above at its closest point. Calculate (a) the semimajor axis and (b) the eccentricity of the orbit. SSM

•**48** A satellite is put in a circular orbit about Earth with a radius equal to one-half the radius of the Moon's orbit. What is its period of revolution in lunar months? (A lunar month is the period of revolution of the Moon.)

•**49** (a) What linear speed must an Earth satellite have to be in a circular orbit at an altitude of 160 km above Earth's surface? (b) What is the period of revolution?

•**50** The Sun's center is at one focus of Earth's orbit. How far from this focus is the other focus, (a) in meters and (b) in terms of the solar radius, 6.96×10^8 m? The eccentricity is 0.0167, and the semimajor axis is 1.50×10^{11} m.

•**51** A comet that was seen in April 574 by Chinese astronomers on a day known by them as the Woo Woo day was spotted again in May 1994. Assume the time between observations is the period of the Woo Woo day comet and take its eccentricity as 0.11. What are (a) the semimajor axis of the comet's orbit and (b) its greatest distance from the Sun in terms of the mean orbital radius R_P of Pluto?

•**52** An orbiting satellite stays over a certain spot on the equator of (rotating) Earth. What is the altitude of the orbit (called a *geosynchronous orbit*)?

••**53** In 1610, Galileo used his telescope to discover four prominent moons around Jupiter. Their mean orbital radii a and periods T are as follows:
(a) Plot log a (y axis) against log T (x axis) and show that you get a straight line. (b) Measure the slope of the line and compare it with the value that you expect from Kepler's third law. (c) Find the mass of Jupiter from the intercept of this line with the y axis.

Name	a (10^8 m)	T (days)
Io	4.22	1.77
Europa	6.71	3.55
Ganymede	10.7	7.16
Callisto	18.8	16.7

••**54** In 1993 the spacecraft *Galileo* sent home an image (Fig. 13-48) of asteroid 243 Ida and a tiny orbiting moon (now known as Dactyl), the first confirmed example of an asteroid–moon system. In the image, the moon, which is 1.5 km wide, is 100 km from the center of the asteroid, which is 55 km long. The shape of the moon's orbit is not well known; assume it is circular with a period of 27 h. (a) What is the mass of the asteroid? (b) The volume of the asteroid, measured from the *Galileo* images, is 14 100 km^3. What is the density (mass per unit volume) of the asteroid?

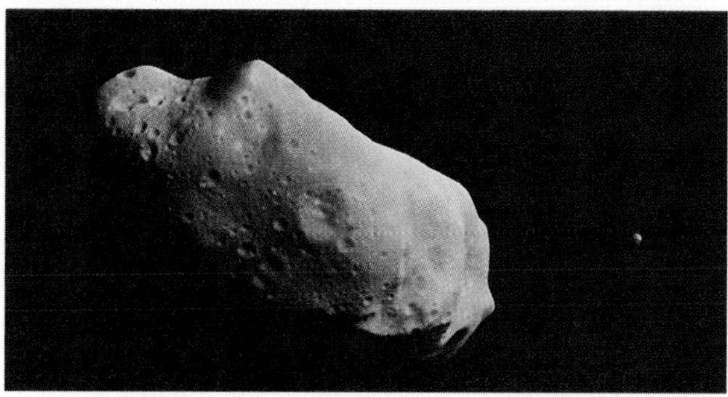

FIG. 13-48 Problem 54. A tiny moon (at right) orbits asteroid 243 Ida. *(Courtesy NASA)*

••**55** In a certain binary-star system, each star has the same mass as our Sun, and they revolve about their center of mass. The distance between them is the same as the distance between Earth and the Sun. What is their period of revolution in years? ILW

••**56** *Hunting a black hole.* Observations of the light from a certain star indicate that it is part of a binary (two-star) system. This visible star has orbital speed $v = 270$ km/s, orbital period $T = 1.70$ days, and approximate mass $m_1 = 6M_s$, where M_s is the Sun's mass, 1.99×10^{30} kg. Assume that the visible star and its companion star, which is dark and unseen, are both in circular orbits (Fig. 13-49). What multiple of M_s gives the approximate mass m_2 of the dark star?

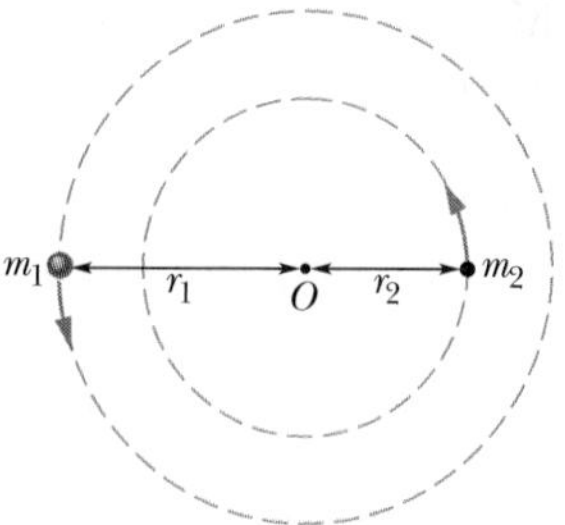

FIG. 13-49 Problem 56.

••**57** A 20 kg satellite has a circular orbit with a period of 2.4 h and a radius of 8.0×10^6 m around a planet of unknown mass. If the magnitude of the gravitational acceleration on the surface of the planet is 8.0 m/s^2, what is the radius of the planet?

•••**58** The presence of an unseen planet orbiting a distant star can sometimes be inferred from the motion of the star as we see it. As the star and planet orbit the center of mass of the star–planet system, the star moves toward and away from us with what is called the *line of sight velocity,* a motion that can be detected. Figure 13-50 shows a graph of the line of sight velocity versus time for the star 14 Herculis. The star's mass is believed to be 0.90 of the mass of our Sun. Assume that only one planet orbits the star and that our view is along the plane of the orbit. Then approximate (a) the planet's mass in terms of Jupiter's mass m_J and (b) the planet's orbital radius in terms of Earth's orbital radius r_E.

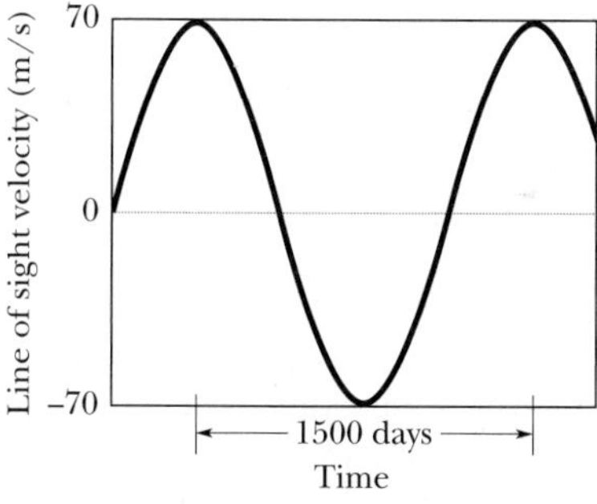

FIG. 13-50 Problem 58.

•••**59** Three identical stars of mass M form an equilateral triangle that rotates around the triangle's center as the stars move in a common circle about that center. The triangle has edge length L. What is the speed of the stars?

sec. 13-8 Satellites: Orbits and Energy

•**60** Two Earth satellites, A and B, each of mass m, are to be launched into circular orbits about Earth's center. Satellite A is to orbit at an altitude of 6370 km. Satellite B is to orbit at an altitude of 19 110 km. The radius of Earth R_E is 6370 km. (a) What is the ratio of the potential energy of satellite B to that of satellite A, in orbit? (b) What is the ratio of the kinetic energy of satellite B to that of satellite A, in orbit? (c) Which satellite has the greater total energy if each has a mass of 14.6 kg? (d) By how much?

•**61** An asteroid, whose mass is 2.0×10^{-4} times the mass of Earth, revolves in a circular orbit around the Sun at a distance that is twice Earth's distance from the Sun. (a) Calculate the period of revolution of the asteroid in years. (b) What is the ratio of the kinetic energy of the asteroid to the kinetic energy of Earth? SSM WWW

•62 A satellite orbits a planet of unknown mass in a circle of radius 2.0×10^7 m. The magnitude of the gravitational force on the satellite from the planet is $F = 80$ N. (a) What is the kinetic energy of the satellite in this orbit? (b) What would F be if the orbit radius were increased to 3.0×10^7 m?

•63 (a) At what height above Earth's surface is the energy required to lift a satellite to that height equal to the kinetic energy required for the satellite to be in orbit at that height? (b) For greater heights, which is greater, the energy for lifting or the kinetic energy for orbiting?

•64 In Fig. 13-51, two satellites, A and B, both of mass $m = 125$ kg, move in the same circular orbit of radius $r = 7.87 \times 10^6$ m around Earth but in opposite senses of rotation and therefore on a collision course. (a) Find the total mechanical energy $E_A + E_B$ of the *two satellites + Earth* system before the collision. (b) If the collision is completely inelastic so that the wreckage remains as one piece of tangled material (mass $= 2m$), find the total mechanical energy immediately after the collision. (c) Just after the collision, is the wreckage falling directly toward Earth's center or orbiting around Earth?

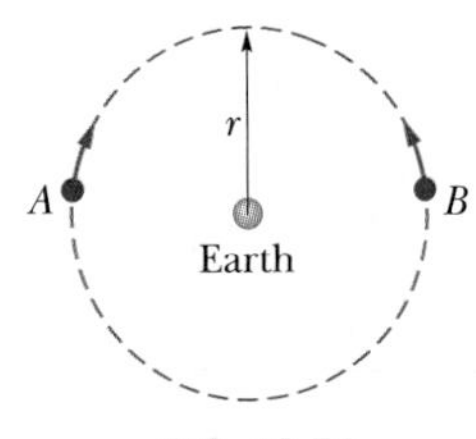

FIG. 13-51 Problem 64.

••65 A satellite is in a circular Earth orbit of radius r. The area A enclosed by the orbit depends on r^2 because $A = \pi r^2$. Determine how the following properties of the satellite depend on r: (a) period, (b) kinetic energy, (c) angular momentum, and (d) speed.

••66 One way to attack a satellite in Earth orbit is to launch a swarm of pellets in the same orbit as the satellite but in the opposite direction. Suppose a satellite in a circular orbit 500 km above Earth's surface collides with a pellet having mass 4.0 g. (a) What is the kinetic energy of the pellet in the reference frame of the satellite just before the collision? (b) What is the ratio of this kinetic energy to the kinetic energy of a 4.0 g bullet from a modern army rifle with a muzzle speed of 950 m/s?

•••67 What are (a) the speed and (b) the period of a 220 kg satellite in an approximately circular orbit 640 km above the surface of Earth? Suppose the satellite loses mechanical energy at the average rate of 1.4×10^5 J per orbital revolution. Adopting the reasonable approximation that the satellite's orbit becomes a "circle of slowly diminishing radius," determine the satellite's (c) altitude, (d) speed, and (e) period at the end of its 1500th revolution. (f) What is the magnitude of the average retarding force on the satellite? Is angular momentum around Earth's center conserved for (g) the satellite and (h) the satellite–Earth system (assuming that system is isolated)?

•••68 Two small spaceships, each with mass $m = 2000$ kg, are in the circular Earth orbit of Fig. 13-52, at an altitude h of 400 km. Igor, the commander of one of the ships, arrives at any fixed point in the orbit 90 s ahead of Picard, the commander of the other ship. What are the (a) period T_0 and (b) speed v_0 of the ships? At point P in Fig. 13-52, Picard fires an instantaneous burst in the forward direction, *reducing* his ship's speed by 1.00%. After this burst, he follows the elliptical orbit shown dashed in the figure. What are the (c) kinetic energy and (d) potential energy of his ship immediately after the burst? In Picard's new elliptical orbit, what are (e) the total energy E, (f) the semimajor axis a, and (g) the orbital period T? (h) How much earlier than Igor will Picard return to P? GO

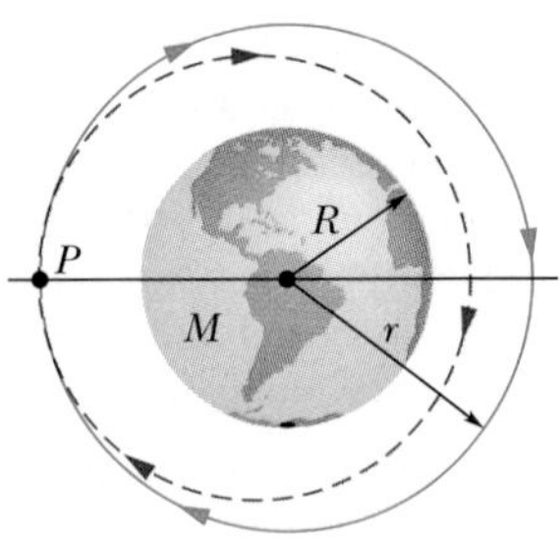

FIG. 13-52 Problem 68.

sec. 13-9 Einstein and Gravitation

•69 In Fig. 13-19b, the scale on which the 60 kg physicist stands reads 220 N. How long will the cantaloupe take to reach the floor if the physicist drops it (from rest relative to himself) at a height of 2.1 m above the floor?

Additional Problems

70 The mysterious visitor that appears in the enchanting story *The Little Prince* was said to come from a planet that "was scarcely any larger than a house!" Assume that the mass per unit volume of the planet is about that of Earth and that the planet does not appreciably spin. Approximate (a) the free-fall acceleration on the planet's surface and (b) the escape speed from the planet.

71 Figure 13-53 is a graph of the kinetic energy K of an asteroid versus its distance r from Earth's center, as the asteroid falls directly in toward that center. (a) What is the (approximate) mass of the asteroid? (b) What is its speed at $r = 1.945 \times 10^7$ m?

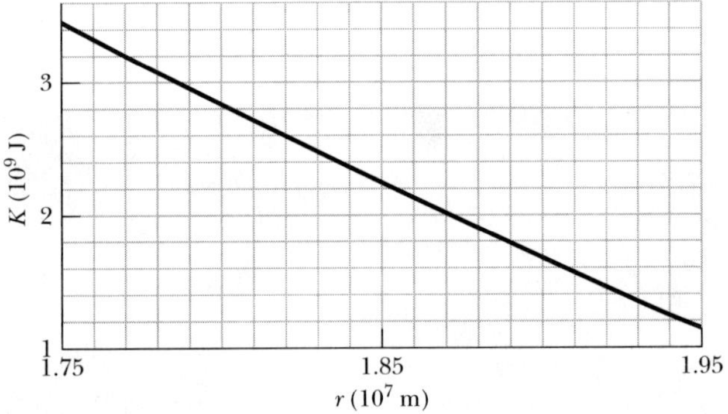

FIG. 13-53 Problem 71.

72 The radius R_h of a black hole is the radius of a mathematical sphere, called the event horizon, that is centered on the black hole. Information from events inside the event horizon cannot reach the outside world. According to Einstein's general theory of relativity, $R_h = 2GM/c^2$, where M is the mass of the black hole and c is the speed of light.

Suppose that you wish to study a black hole near it, at a radial distance of $50R_h$. However, you do not want the difference in gravitational acceleration between your feet and your head to exceed 10 m/s² when you are feet down (or head down) toward the black hole. (a) As a multiple of our Sun's mass M_S, approximately what is the limit to the mass of the black hole you can tolerate at the given radial distance? (You need to estimate your height.) (b) Is the limit an upper limit (you can tolerate smaller masses) or a lower limit (you can tolerate larger masses)?

73 Sphere A with mass 80 kg is located at the origin of an xy coordinate system; sphere B with mass 60 kg is located at coordinates (0.25 m, 0); sphere C with mass 0.20 kg is located in the first quadrant 0.20 m from A and 0.15 m from B. In unit-vector notation, what is the gravitational force on C due to A and B?

74 A satellite is in elliptical orbit with a period of 8.00×10^4 s about a planet of mass 7.00×10^{24} kg. At aphelion, at radius 4.5×10^7 m, the satellite's angular speed is 7.158×10^{-5} rad/s. What is its angular speed at perihelion?

75 In a shuttle craft of mass $m = 3000$ kg, Captain Janeway orbits a planet of mass $M = 9.50 \times 10^{25}$ kg, in a circular orbit of radius $r = 4.20 \times 10^7$ m. What are (a) the period of the orbit and (b) the speed of the shuttle craft? Janeway briefly fires a forward-pointing thruster, reducing her speed by 2.00%. Just then, what are (c) the speed, (d) the kinetic energy, (e) the gravitational potential energy, and (f) the mechanical energy of the shuttle craft? (g) What is the semi-major axis of the elliptical orbit now taken by the craft? (h) What is the difference between the period of the original circular orbit and that of the new elliptical orbit? (i) Which orbit has the smaller period? SSM

76 A typical neutron star may have a mass equal to that of the Sun but a radius of only 10 km. (a) What is the gravitational acceleration at the surface of such a star? (b) How fast would an object be moving if it fell from rest through a distance of 1.0 m on such a star? (Assume the star does not rotate.)

77 Four uniform spheres, with masses $m_A = 40$ kg, $m_B = 35$ kg, $m_C = 200$ kg, and $m_D = 50$ kg, have (x, y) coordinates of (0, 50 cm), (0, 0), (−80 cm, 0), and (40 cm, 0), respectively. In unit-vector notation, what is the net gravitational force on sphere B due to the other spheres? GO

78 (a) In Problem 77, remove sphere A and calculate the gravitational potential energy of the remaining three-particle system. (b) If A is then put back in place, is the potential energy of the four-particle system more or less than that of the system in (a)? (c) In (a), is the work done by you to remove A positive or negative? (d) In (b), is the work done by you to replace A positive or negative?

79 A very early, simple satellite consisted of an inflated spherical aluminum balloon 30 m in diameter and of mass 20 kg. Suppose a meteor having a mass of 7.0 kg passes within 3.0 m of the surface of the satellite. What is the magnitude of the gravitational force on the meteor from the satellite at the closest approach? SSM

80 A uniform solid sphere of radius R produces a gravitational acceleration of a_g on its surface. At what distance from the sphere's center are there points (a) inside and (b) outside the sphere where the gravitational acceleration is $a_g/3$?

81 A projectile is fired vertically from Earth's surface with an initial speed of 10 km/s. Neglecting air drag, how far above the surface of Earth will it go? ILW

82 A 50 kg satellite circles planet Cruton every 6.0 h. The magnitude of the gravitational force exerted on the satellite by Cruton is 80 N. (a) What is the radius of the orbit? (b) What is the kinetic energy of the satellite? (c) What is the mass of planet Cruton?

83 In a double-star system, two stars of mass 3.0×10^{30} kg each rotate about the system's center of mass at radius 1.0×10^{11} m. (a) What is their common angular speed? (b) If a meteoroid passes through the system's center of mass perpendicular to their orbital plane, what minimum speed must it have at the center of mass if it is to escape to "infinity" from the two-star system? SSM

84 An object lying on Earth's equator is accelerated (a) toward the center of Earth because Earth rotates, (b) toward the Sun because Earth revolves around the Sun in an almost circular orbit, and (c) toward the center of our galaxy because the Sun moves around the galactic center. For the latter, the period is 2.5×10^8 y and the radius is 2.2×10^{20} m. Calculate these three accelerations as multiples of $g = 9.8$ m/s^2.

85 The masses and coordinates of three spheres are as follows: 20 kg, $x = 0.50$ m, $y = 1.0$ m; 40 kg, $x = -1.0$ m, $y = -1.0$ m; 60 kg, $x = 0$ m, $y = -0.50$ m. What is the magnitude of the gravitational force on a 20 kg sphere located at the origin due to these three spheres? ILW

86 With what speed would mail pass through the center of Earth if falling in the tunnel of Sample Problem 13-4?

87 The orbit of Earth around the Sun is *almost* circular: The closest and farthest distances are 1.47×10^8 km and 1.52×10^8 km respectively. Determine the corresponding variations in (a) total energy, (b) gravitational potential energy, (c) kinetic energy, and (d) orbital speed. (*Hint:* Use conservation of energy and conservation of angular momentum.) SSM

88 A spaceship is on a straight-line path between Earth and the Moon. At what distance from Earth is the net gravitational force on the spaceship zero?

89 An object of mass m is initially held in place at radial distance $r = 3R_E$ from the center of Earth, where R_E is the radius of Earth. Let M_E be the mass of Earth. A force is applied to the object to move it to a radial distance $r = 4R_E$, where it again is held in place. Calculate the work done by the applied force during the move by integrating the force magnitude.

90 The fastest possible rate of rotation of a planet is that for which the gravitational force on material at the equator just barely provides the centripetal force needed for the rotation. (Why?) (a) Show that the corresponding shortest period of rotation is

$$T = \sqrt{\frac{3\pi}{G\rho}},$$

where ρ is the uniform density (mass per unit volume) of the spherical planet. (b) Calculate the rotation period assuming a density of 3.0 g/cm^3, typical of many planets, satellites, and asteroids. No astronomical object has ever been found to be spinning with a period shorter than that determined by this analysis.

91 (a) If the legendary apple of Newton could be released from rest at a height of 2 m from the surface of a neutron star with a mass 1.5 times that of our Sun and a radius of 20 km, what would be the apple's speed when it reached the surface of the star? (b) If the apple could rest on the surface of the star, what would be the approximate difference between the gravitational acceleration at the top and at the bottom of the apple? (Choose a reasonable size for an apple; the answer indicates that an apple would never survive near a neutron star.)

92 Some people believe that the positions of the planets at the time of birth influence the newborn. Others deride this belief and claim that the gravitational force exerted on a baby by the obstetrician is greater than the force exerted by the planets. To check this claim, calculate the magnitude of the gravitational force exerted on a 3 kg baby (a) by a 70 kg obstetrician who is 1 m away and roughly approximated as a point mass, (b) by the massive planet Jupiter ($m = 2 \times 10^{27}$ kg) at its closest approach to Earth ($= 6 \times 10^{11}$ m), and (c) by Jupiter at its greatest distance from Earth ($= 9 \times 10^{11}$ m). (d) Is the claim correct?

93 A certain triple-star system consists of two stars, each of mass m, revolving in the same circular orbit of radius r around a central star of mass M (Fig. 13-54). The two orbiting stars are always at opposite ends of a diameter of the orbit. Derive an expression for the period of revolution of the stars. SSM

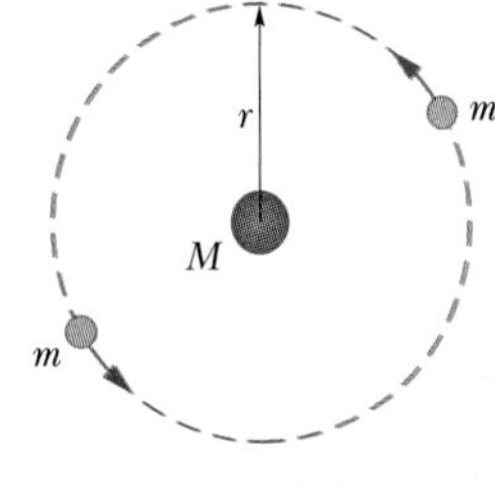

FIG. 13-54 Problem 93.

94 A 150.0 kg rocket moving radially outward from Earth has a speed of 3.70 km/s when its engine shuts off 200 km above Earth's surface. (a) Assuming negligible air drag, find the rocket's kinetic energy when the rocket is 1000 km above Earth's surface. (b) What maximum height above the surface is reached by the rocket?

95 Planet Roton, with a mass of 7.0×10^{24} kg and a radius of 1600 km, gravitationally attracts a meteorite that is initially at rest relative to the planet, at a distance great enough to take as infinite. The meteorite falls toward the planet. Assuming the planet is airless, find the speed of the meteorite when it reaches the planet's surface.

96 Two 20 kg spheres are fixed in place on a y axis, one at $y = 0.40$ m and the other at $y = -0.40$ m. A 10 kg ball is then released from rest at a point on the x axis that is at a great distance (effectively infinite) from the spheres. If the only forces acting on the ball are the gravitational forces from the spheres, then when the ball reaches the (x, y) point (0.30 m, 0), what are (a) its kinetic energy and (b) the net force on it from the spheres, in unit-vector notation?

97 A satellite of mass 125 kg is in a circular orbit of radius 7.00×10^6 m around a planet. If the period is 8050 s, what is the mechanical energy of the satellite?

98 In his 1865 science fiction novel *From the Earth to the Moon,* Jules Verne described how three astronauts are shot to the Moon by means of a huge gun. According to Verne, the aluminum capsule containing the astronauts is accelerated by ignition of nitrocellulose to a speed of 11 km/s along the gun barrel's length of 220 m. (a) In g units, what is the average acceleration of the capsule and astronauts in the gun barrel? (b) Is that acceleration tolerable or deadly to the astronauts?

A modern version of such gun-launched spacecraft (although without passengers) has been proposed. In this modern version, called the SHARP (Super High Altitude Research Project) gun, ignition of methane and air shoves a piston along the gun's tube, compressing hydrogen gas that then launches a rocket. During this launch, the rocket moves 3.5 km and reaches a speed of 7.0 km/s. Once launched, the rocket can be fired to gain additional speed. (c) In g units, what would be the average acceleration of the rocket within the launcher? (d) How much additional speed is needed (via the rocket engine) if the rocket is to orbit Earth at an altitude of 700 km?

99 Several planets (Jupiter, Saturn, Uranus) are encircled by rings, perhaps composed of material that failed to form a satellite. In addition, many galaxies contain ring-like structures. Consider a homogeneous thin ring of mass M and outer radius R (Fig. 13-55). (a) What gravitational attraction does it exert on a particle of mass m located on the ring's central axis a distance x from the ring center? (b) Suppose the particle falls from rest as a result of the attraction of the ring of matter. What is the speed with which it passes through the center of the ring?

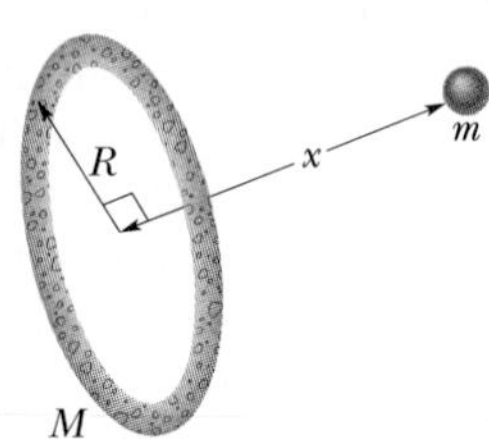

FIG. 13-55 Problem 99.

100 Four identical 1.5 kg particles are placed at the corners of a square with sides equal to 20 cm. What is the magnitude of the net gravitational force on any one of the particles due to the others?

101 We watch two identical astronomical bodies A and B, each of mass m, fall toward each other from rest because of the gravitational force on each from the other. Their initial center-to-center separation is R_i. Assume that we are in an inertial reference frame that is stationary with respect to the center of mass of this two-body system. Use the principle of conservation of mechanical energy ($K_f + U_f = K_i + U_i$) to find the following when the center-to-center separation is $0.5R_i$: (a) the total kinetic energy of the system, (b) the kinetic energy of each body, (c) the speed of each body relative to us, and (d) the speed of body B relative to body A.

Next assume that we are in a reference frame attached to body A (we ride on the body). Now we see body B fall from rest toward us. From this reference frame, again use $K_f + U_f = K_i + U_i$ to find the following when the center-to-center separation is $0.5R_i$: (e) the kinetic energy of body B and (f) the speed of body B relative to body A. (g) Why are the answers to (d) and (f) different? Which answer is correct?

102 What is the percentage change in the acceleration of Earth toward the Sun when the alignment of Earth, Sun, and Moon changes from an eclipse of the Sun (with the Moon between Earth and Sun) to an eclipse of the Moon (Earth between Moon and Sun)?

103 A planet requires 300 (Earth) days to complete its circular orbit around its sun, which has a mass of 6.0×10^{30} kg. What are the planet's (a) orbital radius and (b) orbital speed?

104 A particle of comet dust with mass m is a distance R from Earth's center and a distance r from the Moon's center. If Earth's mass is M_E and the Moon's mass is M_m, what is the sum of the gravitational potential energy of the particle–Earth system and the gravitational potential energy of the particle–Moon system?

Fluids 

Warren Bolster/Stone/Getty Images

A surfer patiently kneels on his surfboard to catch the next big wave. When he spots a wave building in height as it heads his way, he rapidly paddles toward shore so that he is moving almost as fast as the wave as the wave begins to sweep under him. Then he stands up on the board, continuously adjusting his balance. How does he manage to continue riding the wave? How, instead, can he move up or down the front of the wave?

In short, how does a surfer surf?

The answer is in this chapter.

14-1 WHAT IS PHYSICS?

The physics of fluids is the basis of hydraulic engineering, a branch of engineering that is applied in a great many fields. A nuclear engineer might study the fluid flow in the hydraulic system of an aging nuclear reactor, while a medical engineer might study the blood flow in the arteries of an aging patient. An environmental engineer might be concerned about the drainage from waste sites or the efficient irrigation of farmlands. A naval engineer might be concerned with the dangers faced by a deep-sea diver or with the possibility of a crew escaping from a downed submarine. An aeronautical engineer might design the hydraulic systems controlling the wing flaps that allow a jet airplane to land. Hydraulic engineering is also applied in many Broadway and Las Vegas shows, where huge sets are quickly put up and brought down by hydraulic systems.

Before we can study any such application of the physics of fluids, we must first answer the question "What is a fluid?"

14-2 | What Is a Fluid?

A **fluid,** in contrast to a solid, is a substance that can flow. Fluids conform to the boundaries of any container in which we put them. They do so because a fluid cannot sustain a force that is tangential to its surface. (In the more formal language of Section 12-7, a fluid is a substance that flows because it cannot withstand a shearing stress. It can, however, exert a force in the direction perpendicular to its surface.) Some materials, such as pitch, take a long time to conform to the boundaries of a container, but they do so eventually; thus, we classify even those materials as fluids.

You may wonder why we lump liquids and gases together and call them fluids. After all (you may say), liquid water is as different from steam as it is from ice. Actually, it is not. Ice, like other crystalline solids, has its constituent atoms organized in a fairly rigid three-dimensional array called a crystalline lattice. In neither steam nor liquid water, however, is there any such orderly long-range arrangement.

14-3 | Density and Pressure

When we discuss rigid bodies, we are concerned with particular lumps of matter, such as wooden blocks, baseballs, or metal rods. Physical quantities that we find useful, and in whose terms we express Newton's laws, are *mass* and *force.* We might speak, for example, of a 3.6 kg block acted on by a 25 N force.

With fluids, we are more interested in the extended substance and in properties that can vary from point to point in that substance. It is more useful to speak of **density** and **pressure** than of mass and force.

Density

To find the density ρ of a fluid at any point, we isolate a small volume element ΔV around that point and measure the mass Δm of the fluid contained within that element. The **density** is then

$$\rho = \frac{\Delta m}{\Delta V}. \tag{14-1}$$

In theory, the density at any point in a fluid is the limit of this ratio as the volume element ΔV at that point is made smaller and smaller. In practice, we assume that a fluid sample is large relative to atomic dimensions and thus is "smooth" (with uni-

TABLE 14-1

Some Densities

Material or Object	Density (kg/m³)	Material or Object	Density (kg/m³)
Interstellar space	10^{-20}	Iron	7.9×10^3
Best laboratory vacuum	10^{-17}	Mercury (the metal, not the planet)	13.6×10^3
Air: 20°C and 1 atm pressure	1.21	Earth: average	5.5×10^3
20°C and 50 atm	60.5	core	9.5×10^3
Styrofoam	1×10^2	crust	2.8×10^3
Ice	0.917×10^3	Sun: average	1.4×10^3
Water: 20°C and 1 atm	0.998×10^3	core	1.6×10^5
20°C and 50 atm	1.000×10^3	White dwarf star (core)	10^{10}
Seawater: 20°C and 1 atm	1.024×10^3	Uranium nucleus	3×10^{17}
Whole blood	1.060×10^3	Neutron star (core)	10^{18}

form density), rather than "lumpy" with atoms. This assumption allows us to write Eq. 14-1 as

$$\rho = \frac{m}{V} \quad \text{(uniform density)}, \tag{14-2}$$

where m and V are the mass and volume of the sample.

Density is a scalar property; its SI unit is the kilogram per cubic meter. Table 14-1 shows the densities of some substances and the average densities of some objects. Note that the density of a gas (see Air in the table) varies considerably with pressure, but the density of a liquid (see Water) does not; that is, gases are readily *compressible* but liquids are not.

Pressure

Let a small pressure-sensing device be suspended inside a fluid-filled vessel, as in Fig. 14-1*a*. The sensor (Fig. 14-1*b*) consists of a piston of surface area ΔA riding in a close-fitting cylinder and resting against a spring. A readout arrangement allows us to record the amount by which the (calibrated) spring is compressed by the surrounding fluid, thus indicating the magnitude ΔF of the force that acts normal to the piston. We define the **pressure** on the piston from the fluid as

$$p = \frac{\Delta F}{\Delta A}. \tag{14-3}$$

In theory, the pressure at any point in the fluid is the limit of this ratio as the surface area ΔA of the piston, centered on that point, is made smaller and smaller. However, if the force is uniform over a flat area A, we can write Eq. 14-3 as

$$p = \frac{F}{A} \quad \text{(pressure of uniform force on flat area)}, \tag{14-4}$$

where F is the magnitude of the normal force on area A. (When we say a force is uniform over an area, we mean that the force is evenly distributed over every point of the area.)

We find by experiment that at a given point in a fluid at rest, the pressure p defined by Eq. 14-4 has the same value no matter how the pressure sensor is oriented. Pressure is a scalar, having no directional properties. It is true that the force acting on the piston of our pressure sensor is a vector quantity, but Eq. 14-4 involves only the *magnitude* of that force, a scalar quantity.

The SI unit of pressure is the newton per square meter, which is given a special name, the **pascal** (Pa). In metric countries, tire pressure gauges are calibrated in kilopascals. The pascal is related to some other common (non-SI)

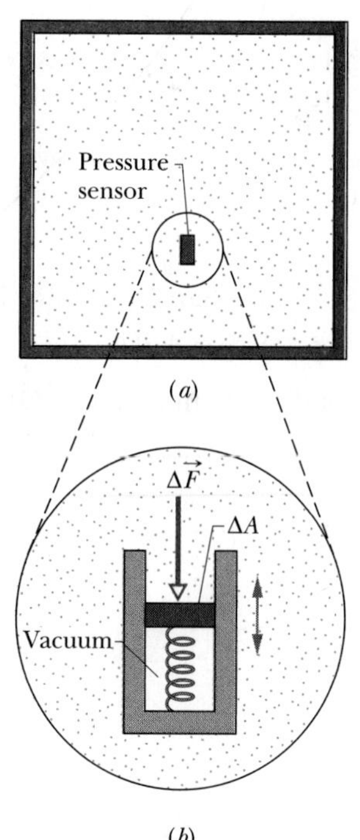

FIG. 14-1 (*a*) A fluid-filled vessel containing a small pressure sensor, shown in (*b*). The pressure is measured by the relative position of the movable piston in the sensor.

TABLE 14-2

Some Pressures

	Pressure (Pa)		Pressure (Pa)
Center of the Sun	2×10^{16}	Automobile tire[a]	2×10^{5}
Center of Earth	4×10^{11}	Atmosphere at sea level	1.0×10^{5}
Highest sustained laboratory pressure	1.5×10^{10}	Normal blood systolic pressure[a,b]	1.6×10^{4}
Deepest ocean trench (bottom)	1.1×10^{8}	Best laboratory vacuum	10^{-12}
Spike heels on a dance floor	10^{6}		

[a]Pressure in excess of atmospheric pressure.

[b]Equivalent to 120 torr on the physician's pressure gauge.

pressure units as follows:

$$1 \text{ atm} = 1.01 \times 10^5 \text{ Pa} = 760 \text{ torr} = 14.7 \text{ lb/in.}^2.$$

The *atmosphere* (atm) is, as the name suggests, the approximate average pressure of the atmosphere at sea level. The *torr* (named for Evangelista Torricelli, who invented the mercury barometer in 1674) was formerly called the *millimeter of mercury* (mm Hg). The pound per square inch is often abbreviated psi. Table 14-2 shows some pressures.

Sample Problem 14-1

A living room has floor dimensions of 3.5 m and 4.2 m and a height of 2.4 m.

(a) What does the air in the room weigh when the air pressure is 1.0 atm?

KEY IDEAS (1) The air's weight is equal to mg, where m is its mass. (2) Mass m is related to the air density ρ and the air volume V by Eq. 14-2 ($\rho = m/V$).

Calculation: Putting the two ideas together and taking the density of air at 1.0 atm from Table 14-1, we find

$$\begin{aligned} mg &= (\rho V)g \\ &= (1.21 \text{ kg/m}^3)(3.5 \text{ m} \times 4.2 \text{ m} \times 2.4 \text{ m})(9.8 \text{ m/s}^2) \\ &= 418 \text{ N} \approx 420 \text{ N}. \qquad \text{(Answer)} \end{aligned}$$

This is the weight of about 110 cans of Pepsi.

(b) What is the magnitude of the atmosphere's downward force on the top of your head, which we take to have an area of 0.040 m²?

KEY IDEA When the fluid pressure p on a surface of area A is uniform, the fluid force on the surface can be obtained from Eq. 14-4 ($p = F/A$).

Calculation: Although air pressure varies daily, we can approximate that $p = 1.0$ atm. Then Eq. 14-4 gives

$$\begin{aligned} F = pA &= (1.0 \text{ atm})\left(\frac{1.01 \times 10^5 \text{ N/m}^2}{1.0 \text{ atm}}\right)(0.040 \text{ m}^2) \\ &= 4.0 \times 10^3 \text{ N}. \qquad \text{(Answer)} \end{aligned}$$

This large force is equal to the weight of the column of air that covers the top of your head and extends all the way to the top of the atmosphere.

14-4 | Fluids at Rest

Figure 14-2*a* shows a tank of water—or other liquid—open to the atmosphere. As every diver knows, the pressure *increases* with depth below the air–water interface. The diver's depth gauge, in fact, is a pressure sensor much like that of Fig. 14-1*b*. As every mountaineer knows, the pressure *decreases* with altitude as one ascends into the atmosphere. The pressures encountered by the diver and the mountaineer are usually called *hydrostatic pressures,* because they are due to fluids that are static (at rest). Here we want to find an expression for hydrostatic pressure as a function of depth or altitude.

Let us look first at the increase in pressure with depth below the water's surface. We set up a vertical y axis in the tank, with its origin at the air–water interface and the positive direction upward. We next consider a water sample contained in an imaginary right circular cylinder of horizontal base (or face) area A, such that y_1 and y_2 (both of which are *negative* numbers) are the depths below the surface of the upper and lower cylinder faces, respectively.

Figure 14-2*b* shows a free-body diagram for the water in the cylinder. The water is in *static equilibrium;* that is, it is stationary and the forces on it balance. Three forces act on it vertically: Force $\vec{F}_1$ acts at the top surface of the cylinder and is due to the water above the cylinder. Similarly, force $\vec{F}_2$ acts at the bottom surface of the cylinder and is due to the water below the cylinder. The gravitational force on the water in the cylinder is represented by $m\vec{g}$, where m is the mass of the water in the cylinder. The balance of these forces is written as

$$F_2 = F_1 + mg. \qquad (14\text{-}5)$$

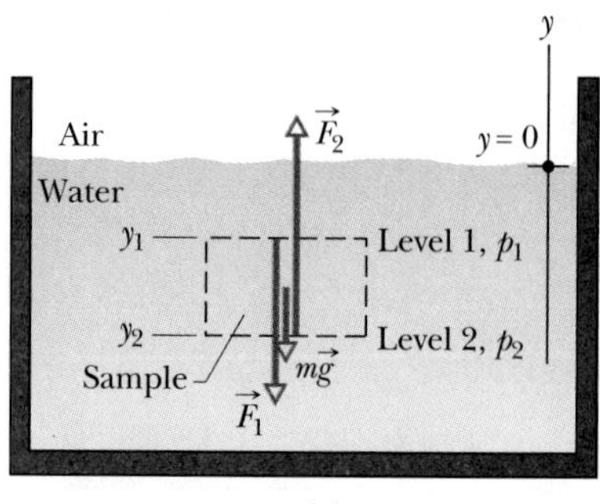

(*a*)

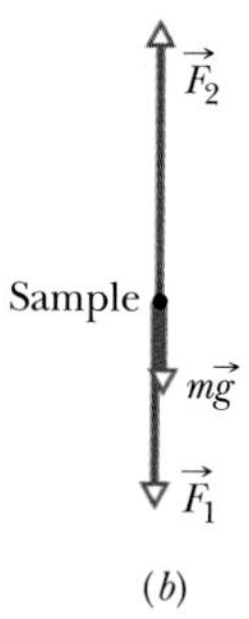

(*b*)

FIG. 14-2 (*a*) A tank of water in which a sample of water is contained in an imaginary cylinder of horizontal base area A. Force $\vec{F}_1$ acts at the top surface of the cylinder; force $\vec{F}_2$ acts at the bottom surface of the cylinder; the gravitational force on the water in the cylinder is represented by $m\vec{g}$. (*b*) A free-body diagram of the water sample.

We want to transform Eq. 14-5 into an equation involving pressures. From Eq. 14-4, we know that

$$F_1 = p_1 A \quad \text{and} \quad F_2 = p_2 A. \qquad (14\text{-}6)$$

The mass m of the water in the cylinder is, from Eq. 14-2, $m = \rho V$, where the cylinder's volume V is the product of its face area A and its height $y_1 - y_2$. Thus, m is equal to $\rho A(y_1 - y_2)$. Substituting this and Eq. 14-6 into Eq. 14-5, we find

$$p_2 A = p_1 A + \rho A g(y_1 - y_2)$$

or

$$p_2 = p_1 + \rho g(y_1 - y_2). \qquad (14\text{-}7)$$

This equation can be used to find pressure both in a liquid (as a function of depth) and in the atmosphere (as a function of altitude or height). For the former, suppose we seek the pressure p at a depth h below the liquid surface. Then we choose level 1 to be the surface, level 2 to be a distance h below it (as in Fig. 14-3), and p_0 to represent the atmospheric pressure on the surface. We then substitute

$$y_1 = 0, \quad p_1 = p_0 \quad \text{and} \quad y_2 = -h, \quad p_2 = p$$

into Eq. 14-7, which becomes

$$p = p_0 + \rho g h \quad \text{(pressure at depth } h). \qquad (14\text{-}8)$$

Note that the pressure at a given depth in the liquid depends on that depth but not on any horizontal dimension.

The pressure at a point in a fluid in static equilibrium depends on the depth of that point but not on any horizontal dimension of the fluid or its container.

Thus, Eq. 14-8 holds no matter what the shape of the container. If the bottom surface of the container is at depth h, then Eq. 14-8 gives the pressure p there.

In Eq. 14-8, p is said to be the total pressure, or **absolute pressure,** at level 2. To see why, note in Fig. 14-3 that the pressure p at level 2 consists of two contributions: (1) p_0, the pressure due to the atmosphere, which bears down on the liquid, and (2) ρgh, the pressure due to the liquid above level 2, which bears down on level 2. In general, the difference between an absolute pressure and an atmospheric pressure is called the **gauge pressure.** (The name comes from the use of a gauge to measure this difference in pressures.) For the situation of Fig. 14-3, the gauge pressure is ρgh.

Equation 14-7 also holds above the liquid surface: It gives the atmospheric pressure at a given distance above level 1 in terms of the atmospheric pressure p_1 at level 1 (*assuming* that the atmospheric density is uniform over that distance). For example, to find the atmospheric pressure at a distance d above level 1 in Fig. 14-3, we substitute

$$y_1 = 0, \quad p_1 = p_0 \quad \text{and} \quad y_2 = d, \quad p_2 = p.$$

Then with $\rho = \rho_{air}$, we obtain

$$p = p_0 - \rho_{air} g d.$$

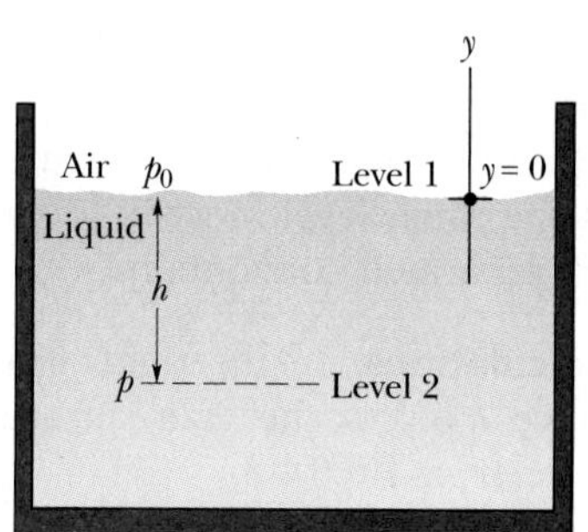

FIG. 14-3 The pressure p increases with depth h below the liquid surface according to Eq. 14-8.

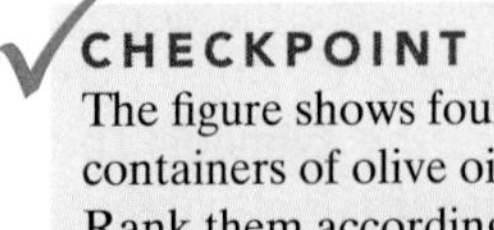

CHECKPOINT 1
The figure shows four containers of olive oil. Rank them according to the pressure at depth h, greatest first.

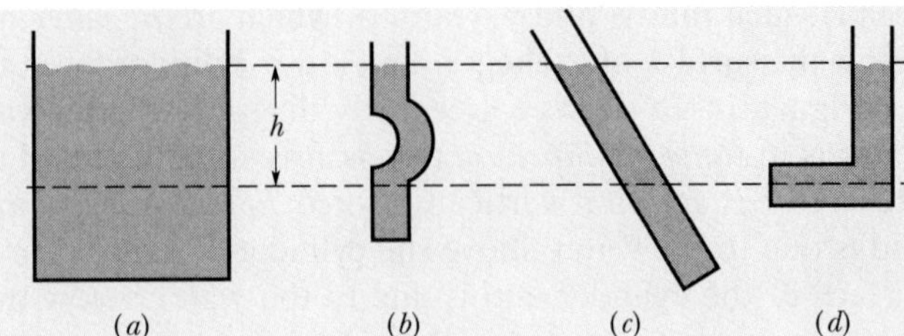

Sample Problem 14-2

A novice scuba diver practicing in a swimming pool takes enough air from his tank to fully expand his lungs before abandoning the tank at depth L and swimming to the surface. He ignores instructions and fails to exhale during his ascent. When he reaches the surface, the difference between the external pressure on him and the air pressure in his lungs is 9.3 kPa. From what depth does he start? What potentially lethal danger does he face?

KEY IDEA The pressure at depth h in a liquid of density ρ is given by Eq. 14-8 ($p = p_0 + \rho g h$), where the gauge pressure $\rho g h$ is added to the atmospheric pressure p_0.

Calculations: Here, when the diver fills his lungs at depth L, the external pressure on him (and thus the air pressure within his lungs) is greater than normal and given by Eq. 14-8 as

$$p = p_0 + \rho g L,$$

where p_0 is atmospheric pressure and ρ is the water's density (998 kg/m^3, from Table 14-1). As he ascends, the external pressure on him decreases, until it is atmospheric pressure p_0 at the surface. His blood pressure also decreases, until it is normal. However, because he does not exhale, the air pressure in his lungs remains at the value it had at depth L. At the surface, the pressure difference between the higher pressure in his lungs and the lower pressure on his chest is

$$\Delta p = p - p_0 = \rho g L,$$

from which we find

$$L = \frac{\Delta p}{\rho g} = \frac{9300 \text{ Pa}}{(998 \text{ kg/m}^3)(9.8 \text{ m/s}^2)} = 0.95 \text{ m.} \quad \text{(Answer)}$$

This is not deep! Yet, the pressure difference of 9.3 kPa (about 9% of atmospheric pressure) is sufficient to rupture the diver's lungs and force air from them into the depressurized blood, which then carries the air to the heart, killing the diver. If the diver follows instructions and gradually exhales as he ascends, he allows the pressure in his lungs to equalize with the external pressure, and then there is no danger.

Sample Problem 14-3

The U-tube in Fig. 14-4 contains two liquids in static equilibrium: Water of density ρ_w (= 998 kg/m^3) is in the right arm, and oil of unknown density ρ_x is in the left. Measurement gives $l = 135$ mm and $d = 12.3$ mm. What is the density of the oil?

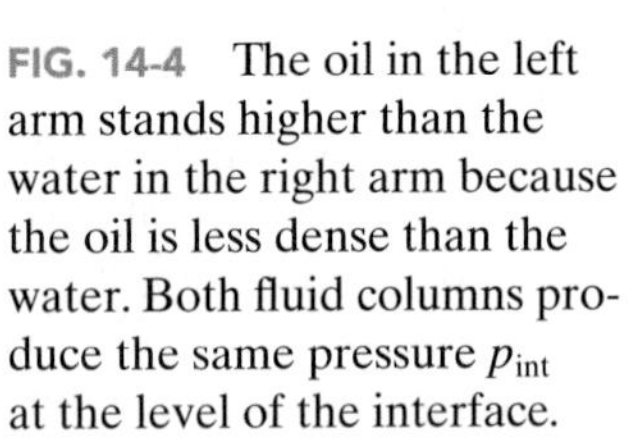

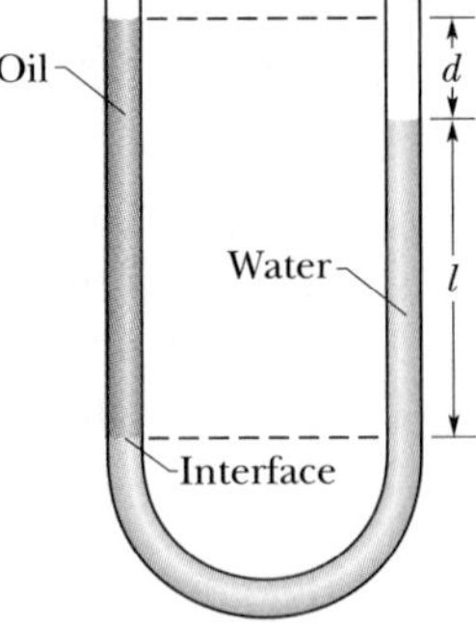

FIG. 14-4 The oil in the left arm stands higher than the water in the right arm because the oil is less dense than the water. Both fluid columns produce the same pressure p_{int} at the level of the interface.

KEY IDEAS (1) The pressure p_{int} at the level of the oil–water interface in the left arm depends on the density ρ_x and height of the oil above the interface. (2) The water in the right arm *at the same level* must be at the same pressure p_{int}. The reason is that, because the water is in static equilibrium, pressures at points in the water at the same level must be the same even if the points are separated horizontally.

Calculations: In the right arm, the interface is a distance l below the free surface of the *water*, and we have, from Eq. 14-8,

$$p_{int} = p_0 + \rho_w g l \quad \text{(right arm).}$$

In the left arm, the interface is a distance $l + d$ below the free surface of the *oil*, and we have, again from Eq. 14-8,

$$p_{int} = p_0 + \rho_x g(l + d) \quad \text{(left arm).}$$

Equating these two expressions and solving for the unknown density yield

$$\rho_x = \rho_w \frac{l}{l + d} = (998 \text{ kg/m}^3)\frac{135 \text{ mm}}{135 \text{ mm} + 12.3 \text{ mm}} = 915 \text{ kg/m}^3. \quad \text{(Answer)}$$

Note that the answer does not depend on the atmospheric pressure p_0 or the free-fall acceleration g.

14-5 I Measuring Pressure

The Mercury Barometer

Figure 14-5*a* shows a very basic *mercury barometer,* a device used to measure the pressure of the atmosphere. The long glass tube is filled with mercury and inverted with its open end in a dish of mercury, as the figure shows. The space above the mercury column contains only mercury vapor, whose pressure is so small at ordinary temperatures that it can be neglected.

We can use Eq. 14-7 to find the atmospheric pressure p_0 in terms of the height h of the mercury column. We choose level 1 of Fig. 14-2 to be that of the air–mercury interface and level 2 to be that of the top of the mercury column, as labeled in Fig. 14-5*a*. We then substitute

$$y_1 = 0, \quad p_1 = p_0 \quad \text{and} \quad y_2 = h, \quad p_2 = 0$$

into Eq. 14-7, finding that

$$p_0 = \rho g h, \tag{14-9}$$

where ρ is the density of the mercury.

For a given pressure, the height h of the mercury column does not depend on the cross-sectional area of the vertical tube. The fanciful mercury barometer of Fig. 14-5*b* gives the same reading as that of Fig. 14-5*a*; all that counts is the vertical distance h between the mercury levels.

Equation 14-9 shows that, for a given pressure, the height of the column of mercury depends on the value of g at the location of the barometer and on the density of mercury, which varies with temperature. The height of the column (in millimeters) is numerically equal to the pressure (in torr) *only* if the barometer is at a place where g has its accepted standard value of 9.80665 m/s^2 *and* the temperature of the mercury is 0°C. If these conditions do not prevail (and they rarely do), small corrections must be made before the height of the mercury column can be transformed into a pressure.

The Open-Tube Manometer

An *open-tube manometer* (Fig. 14-6) measures the gauge pressure p_g of a gas. It consists of a **U**-tube containing a liquid, with one end of the tube connected to the vessel whose gauge pressure we wish to measure and the other end open to the atmosphere. We can use Eq. 14-7 to find the gauge pressure in terms of the height h shown in Fig. 14-6. Let us choose levels 1 and 2 as shown in Fig. 14-6. We then substitute

$$y_1 = 0, \quad p_1 = p_0 \quad \text{and} \quad y_2 = -h, \quad p_2 = p$$

into Eq. 14-7, finding that

$$p_g = p - p_0 = \rho g h, \tag{14-10}$$

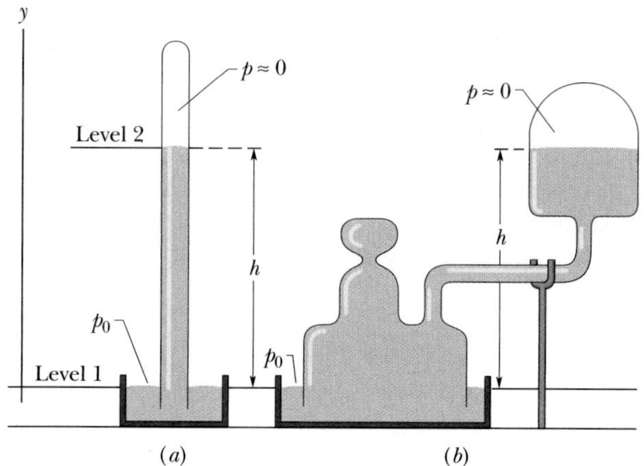

FIG. 14-5 (*a*) A mercury barometer. (*b*) Another mercury barometer. The distance h is the same in both cases.

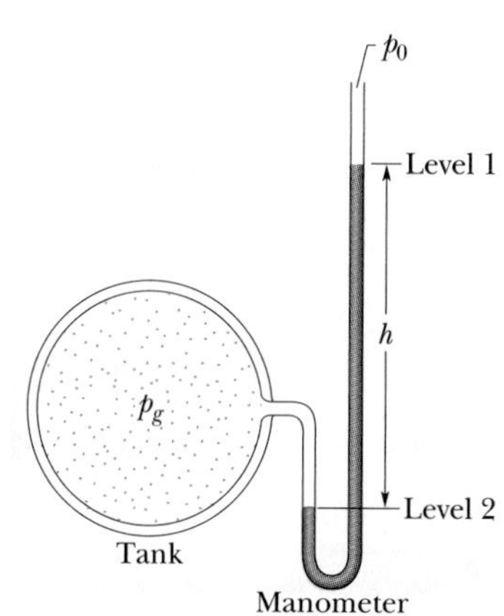

FIG. 14-6 An open-tube manometer, connected to measure the gauge pressure of the gas in the tank on the left. The right arm of the **U**-tube is open to the atmosphere.

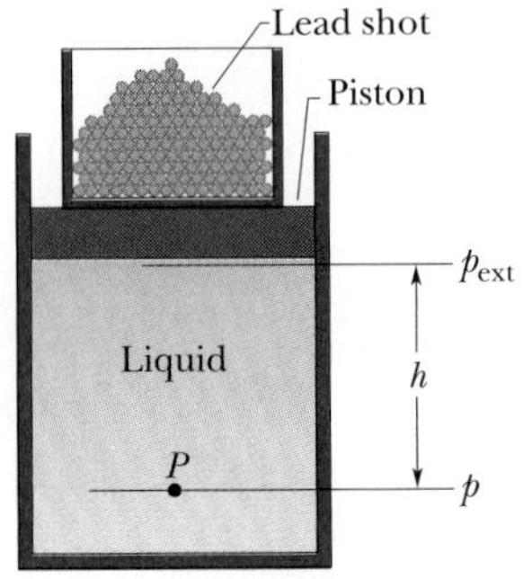

FIG. 14-7 Lead shot (small balls of lead) loaded onto the piston create a pressure p_{ext} at the top of the enclosed (incompressible) liquid. If p_{ext} is increased, by adding more lead shot, the pressure increases by the same amount at all points within the liquid.

where ρ is the density of the liquid in the tube. The gauge pressure p_g is directly proportional to h.

The gauge pressure can be positive or negative, depending on whether $p > p_0$ or $p < p_0$. In inflated tires or the human circulatory system, the (absolute) pressure is greater than atmospheric pressure, so the gauge pressure is a positive quantity, sometimes called the *overpressure.* If you suck on a straw to pull fluid up the straw, the (absolute) pressure in your lungs is actually less than atmospheric pressure. The gauge pressure in your lungs is then a negative quantity.

14-6 | Pascal's Principle

When you squeeze one end of a tube to get toothpaste out the other end, you are watching **Pascal's principle** in action. This principle is also the basis for the Heimlich maneuver, in which a sharp pressure increase properly applied to the abdomen is transmitted to the throat, forcefully ejecting food lodged there. The principle was first stated clearly in 1652 by Blaise Pascal (for whom the unit of pressure is named):

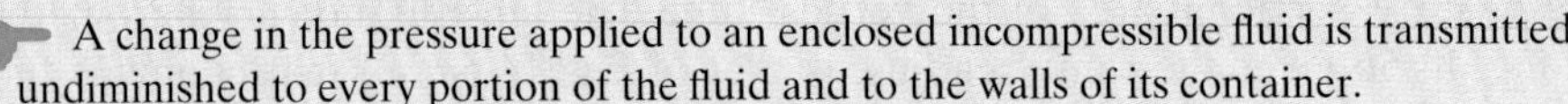

A change in the pressure applied to an enclosed incompressible fluid is transmitted undiminished to every portion of the fluid and to the walls of its container.

Demonstrating Pascal's Principle

Consider the case in which the incompressible fluid is a liquid contained in a tall cylinder, as in Fig. 14-7. The cylinder is fitted with a piston on which a container of lead shot rests. The atmosphere, container, and shot exert pressure p_{ext} on the piston and thus on the liquid. The pressure p at any point P in the liquid is then

$$p = p_{ext} + \rho g h. \qquad (14\text{-}11)$$

Let us add a little more lead shot to the container to increase p_{ext} by an amount Δp_{ext}. The quantities ρ, g, and h in Eq. 14-11 are unchanged, so the pressure change at P is

$$\Delta p = \Delta p_{ext}. \qquad (14\text{-}12)$$

This pressure change is independent of h, so it must hold for all points within the liquid, as Pascal's principle states.

Pascal's Principle and the Hydraulic Lever

Figure 14-8 shows how Pascal's principle can be made the basis of a hydraulic lever. In operation, let an external force of magnitude F_i be directed downward on the left-hand (or input) piston, whose surface area is A_i. An incompressible liquid in the device then produces an upward force of magnitude F_o on the right-hand (or output) piston, whose surface area is A_o. To keep the system in equilibrium, there must be a downward force of magnitude F_o on the output piston from an external load (not shown). The force $\vec{F}_i$ applied on the left and the downward force $\vec{F}_o$ from the load on the right produce a change Δp in the pressure of the liquid that is given by

$$\Delta p = \frac{F_i}{A_i} = \frac{F_o}{A_o},$$

so

$$F_o = F_i \frac{A_o}{A_i}. \qquad (14\text{-}13)$$

Equation 14-13 shows that the output force F_o on the load must be greater than the input force F_i if $A_o > A_i$, as is the case in Fig. 14-8.

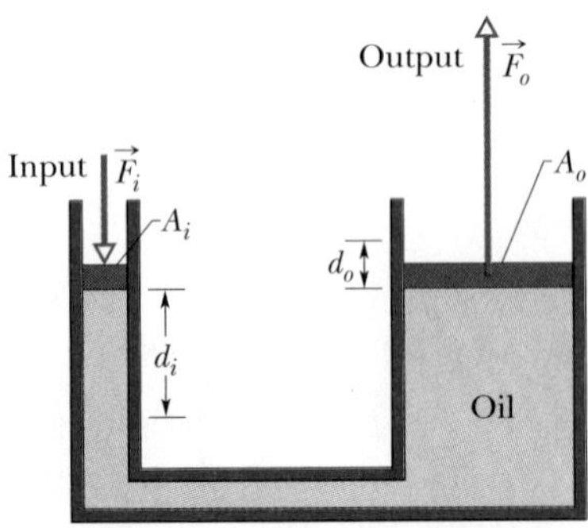

FIG. 14-8 A hydraulic arrangement that can be used to magnify a force $\vec{F}_i$. The work done is, however, not magnified and is the same for both the input and output forces.

If we move the input piston downward a distance d_i, the output piston moves upward a distance d_o, such that the same volume V of the incompressible liquid is displaced at both pistons. Then

$$V = A_i d_i = A_o d_o,$$

which we can write as

$$d_o = d_i \frac{A_i}{A_o}. \qquad (14\text{-}14)$$

This shows that, if $A_o > A_i$ (as in Fig. 14-8), the output piston moves a smaller distance than the input piston moves.

From Eqs. 14-13 and 14-14 we can write the output work as

$$W = F_o d_o = \left(F_i \frac{A_o}{A_i}\right)\left(d_i \frac{A_i}{A_o}\right) = F_i d_i, \qquad (14\text{-}15)$$

which shows that the work W done *on* the input piston by the applied force is equal to the work W done *by* the output piston in lifting the load placed on it.

The advantage of a hydraulic lever is this:

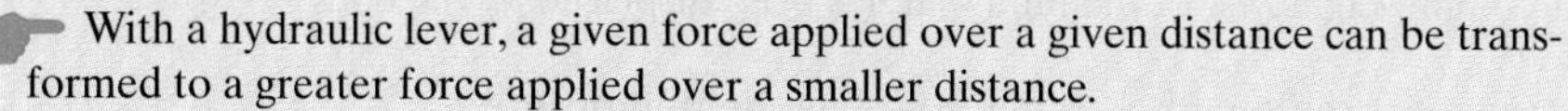

With a hydraulic lever, a given force applied over a given distance can be transformed to a greater force applied over a smaller distance.

The product of force and distance remains unchanged so that the same work is done. However, there is often tremendous advantage in being able to exert the larger force. Most of us, for example, cannot lift an automobile directly but can with a hydraulic jack, even though we have to pump the handle farther than the automobile rises and in a series of small strokes.

FIG. 14-9 A thin-walled plastic sack of water is in static equilibrium in the pool. The gravitational force on the sack must be balanced by a net upward force on it from the surrounding water.

14-7 | Archimedes' Principle

Figure 14-9 shows a student in a swimming pool, manipulating a very thin plastic sack (of negligible mass) that is filled with water. She finds that the sack and its contained water are in static equilibrium, tending neither to rise nor to sink. The downward gravitational force $\vec{F}_g$ on the contained water must be balanced by a net upward force from the water surrounding the sack.

This net upward force is a **buoyant force** $\vec{F}_b$. It exists because the pressure in the surrounding water increases with depth below the surface. Thus, the pressure near the bottom of the sack is greater than the pressure near the top, which means the forces on the sack due to this pressure are greater in magnitude near the bottom of the sack than near the top. Some of the forces are represented in Fig. 14-10*a*, where the space occupied by the sack has been left empty. Note that the force vectors drawn near the bottom of that space (with upward components) have longer lengths than those drawn near the top of the sack (with downward components). If we vectorially add all the forces on the sack from the water, the horizontal components cancel and the vertical components add to yield the upward buoyant force $\vec{F}_b$ on the sack. (Force $\vec{F}_b$ is shown to the right of the pool in Fig. 14-10*a*.)

Because the sack of water is in static equilibrium, the magnitude of $\vec{F}_b$ is equal to the magnitude $m_f g$ of the gravitational force $\vec{F}_g$ on the sack of water: $F_b = m_f g$. (Subscript f refers to *fluid,* here the water.) In words, the magnitude of the buoyant force is equal to the weight of the water in the sack.

In Fig. 14-10*b*, we have replaced the sack of water with a stone that exactly fills the hole in Fig. 14-10*a*. The stone is said to *displace* the water, meaning that the stone occupies space that would otherwise be occupied by water. We have changed nothing about the shape of the hole, so the forces at the hole's surface must be the same as when the water-filled sack was in place. Thus, the same upward buoyant force that acted on the water-filled sack now acts on the stone;

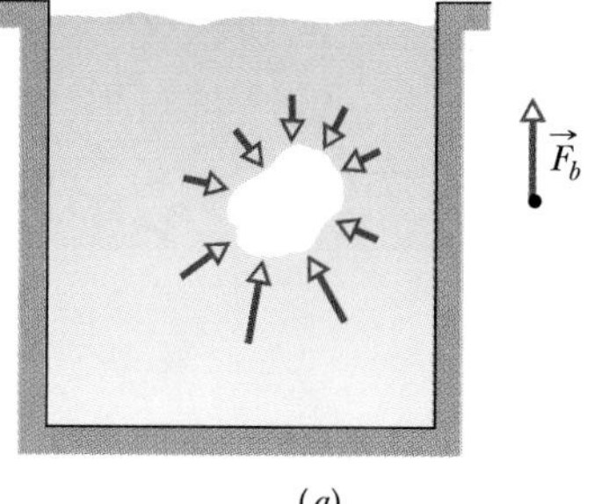

(*a*)

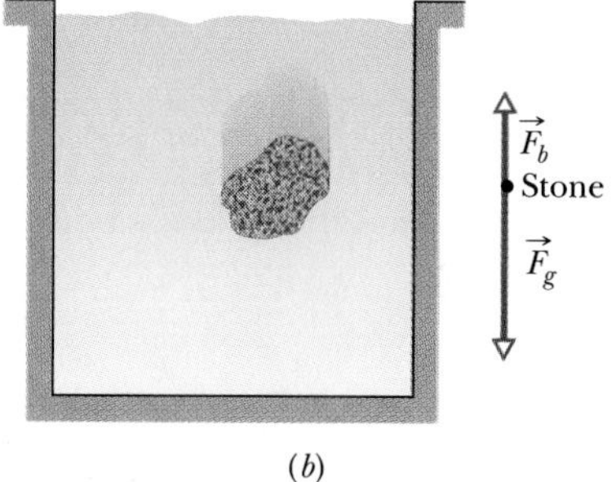

(*b*)

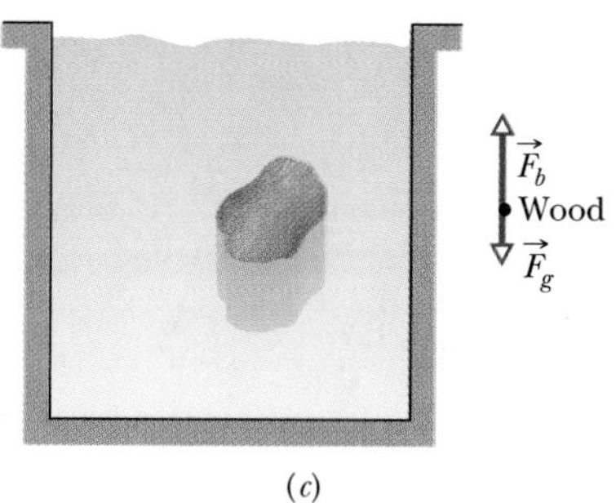

(*c*)

FIG. 14-10 (*a*) The water surrounding the hole in the water produces a net upward buoyant force on whatever fills the hole. (*b*) For a stone of the same volume as the hole, the gravitational force exceeds the buoyant force in magnitude. (*c*) For a lump of wood of the same volume, the gravitational force is less than the buoyant force in magnitude.

that is, the magnitude F_b of the buoyant force is equal to $m_f g$, the weight of the water displaced by the stone.

Unlike the water-filled sack, the stone is not in static equilibrium. The downward gravitational force $\vec{F}_g$ on the stone is greater in magnitude than the upward buoyant force, as is shown in the free-body diagram in Fig. 14-10*b*. The stone thus accelerates downward, sinking to the bottom of the pool.

Let us next exactly fill the hole in Fig. 14-10*a* with a block of lightweight wood, as in Fig. 14-10*c*. Again, nothing has changed about the forces at the hole's surface, so the magnitude F_b of the buoyant force is still equal to $m_f g$, the weight of the displaced water. Like the stone, the block is not in static equilibrium. However, this time the gravitational force $\vec{F}_g$ is lesser in magnitude than the buoyant force (as shown to the right of the pool), and so the block accelerates upward, rising to thc top surface of the water.

Our results with the sack, stone, and block apply to all fluids and are summarized in **Archimedes' principle:**

> When a body is fully or partially submerged in a fluid, a buoyant force $\vec{F}_b$ from the surrounding fluid acts on the body. The force is directed upward and has a magnitude equal to the weight $m_f g$ of the fluid that has been displaced by the body.

The buoyant force on a body in a fluid has the magnitude

$$F_b = m_f g \quad \text{(buoyant force)}, \tag{14-16}$$

where m_f is the mass of the fluid that is displaced by the body.

Floating

When we release a block of lightweight wood just above the water in a pool, the block moves into the water because the gravitational force on it pulls it downward. As the block displaces more and more water, the magnitude F_b of the upward buoyant force acting on it increases. Eventually, F_b is large enough to equal the magnitude F_g of the downward gravitational force on the block, and the block comes to rest. The block is then in static equilibrium and is said to be *floating* in the water. In general,

> When a body floats in a fluid, the magnitude F_b of the buoyant force on the body is equal to the magnitude F_g of the gravitational force on the body.

We can write this statement as

$$F_b = F_g \quad \text{(floating)}. \tag{14-17}$$

From Eq. 14-16, we know that $F_b = m_f g$. Thus,

> When a body floats in a fluid, the magnitude F_g of the gravitational force on the body is equal to the weight $m_f g$ of the fluid that has been displaced by the body.

We can write this statement as

$$F_g = m_f g \quad \text{(floating)}. \tag{14-18}$$

In other words, a floating body displaces its own weight of fluid.

Apparent Weight in a Fluid

If we place a stone on a scale that is calibrated to measure weight, then the reading on the scale is the stone's weight. However, if we do this underwater, the upward buoyant force on the stone from the water decreases the reading. That reading is then an apparent weight. In general, an **apparent weight** is related to the actual weight of a body and the buoyant force on the body by

$$\begin{pmatrix}\text{apparent}\\ \text{weight}\end{pmatrix} = \begin{pmatrix}\text{actual}\\ \text{weight}\end{pmatrix} - \begin{pmatrix}\text{magnitude of}\\ \text{buoyant force}\end{pmatrix},$$

which we can write as

$$\text{weight}_{\text{app}} = \text{weight} - F_b \quad \text{(apparent weight).} \tag{14-19}$$

If, in some test of strength, you had to lift a heavy stone, you could do it more easily with the stone underwater. Then your applied force would need to exceed only the stone's apparent weight, not its larger actual weight, because the upward buoyant force would help you lift the stone.

The magnitude of the buoyant force on a floating body is equal to the body's weight. Equation 14-19 thus tells us that a floating body has an apparent weight of zero—the body would produce a reading of zero on a scale. (When astronauts prepare to perform a complex task in space, they practice the task floating underwater, where their apparent weight is zero as it is in space.)

CHECKPOINT 2 A penguin floats first in a fluid of density ρ_0, then in a fluid of density $0.95\rho_0$, and then in a fluid of density $1.1\rho_0$. (a) Rank the densities according to the magnitude of the buoyant force on the penguin, greatest first. (b) Rank the densities according to the amount of fluid displaced by the penguin, greatest first.

Sample Problem 14-4

In Fig. 14-11*a*, a surfer rides on the front side of a wave, at a point where a tangent to the wave has a slope of $\theta = 30.0°$. The combined mass of surfer and surfboard is $m = 83.0$ kg, and the board has a submerged volume of $V = 2.50 \times 10^{-2}$ m^3. The surfer maintains his position on the wave as the wave moves at constant speed toward shore. What are the magnitude and direction (relative to the positive direction of the x axis in Fig. 14-11*b*) of the drag force on the surfboard from the water?

KEY IDEAS (1) The buoyancy force on the surfer has a magnitude F_b equal to the weight of the seawater displaced by the submerged volume of the surfboard. The direction of the force is perpendicular to the surface at the surfer's location. (2) By Newton's second law, because the surfer moves at constant speed toward the shore, the (vector) sum of the buoyancy force $\vec{F}_b$, the gravitational force $\vec{F}_g$, and the drag force $\vec{F}_d$ must be 0.

Calculations: The forces and their components are shown in the free-body diagram of Fig. 14-11*b*. The gravitational force $\vec{F}_g$ is downward and (as we saw in Chapter 5) has a component of $mg \sin\theta$ down the slope and a component of $mg \cos\theta$ perpendicular to the slope. A drag force $\vec{F}_d$ from the water acts on the surfboard because water is continuously forced up into the wave as the wave continues to move toward the shore. This push on the surfboard is upward and to the rear, at angle ϕ to the x axis. The buoyancy force $\vec{F}_b$ is perpendicular to the water surface; its magnitude depends on the mass m_f of the water displaced by the surfboard, as given by Eq. 14-16 ($F_b = m_f g$). From Eq. 14-2 ($\rho = m/V$), we can write the mass in terms of the seawater density ρ_w and the submerged volume V of the surfboard: $m_f = \rho_w V$. From Table 14-1, seawater density ρ_w is 1.024×10^3 kg/m^3. Thus, the magnitude of the buoyant force is

$$\begin{aligned} F_b &= m_f g = \rho_w V g \\ &= (1.024 \times 10^3 \text{ kg/m}^3)(2.50 \times 10^{-2} \text{ m}^3)(9.8 \text{ m/s}^2) \\ &= 2.509 \times 10^2 \text{ N.} \end{aligned}$$

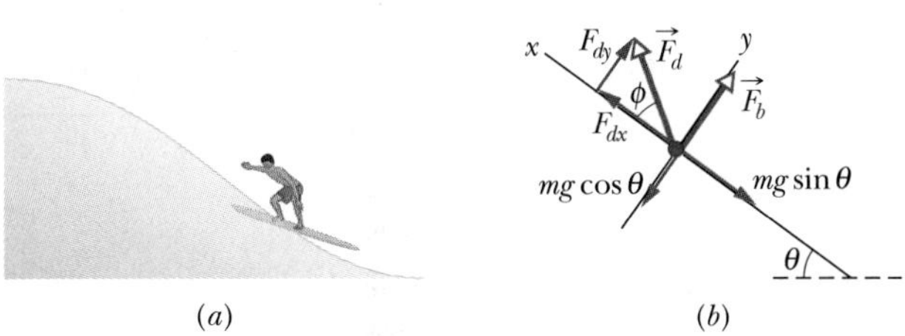

FIG. 14-11 (*a*) Surfer. (*b*) Free-body diagram showing the forces on the surfer–surfboard system.

So, Newton's second law for the y axis,

$$F_{dy} + F_b - mg\cos\theta = m(0),$$

becomes

$$F_{dy} + 2.509 \times 10^2 \text{ N} - (83 \text{ kg})(9.8 \text{ m/s}^2)\cos 30.0° = 0,$$

yielding

$$F_{dy} = 453.5 \text{ N}.$$

Similarly, Newton's second law $\vec{F}_{\text{net}} = m\vec{a}$ for the x axis,

$$F_{dx} - mg\sin\theta = m(0),$$

yields

$$F_{dx} = 406.7 \text{ N}.$$

Combining the two components of the drag force tells us that the force has magnitude

$$F_d = \sqrt{(406.7\text{ N})^2 + (453.5\text{ N})^2} = 609\text{ N} \qquad \text{(Answer)}$$

and angle

$$\phi = \tan^{-1}\left(\frac{453.5\text{ N}}{406.7\text{ N}}\right) = 48.1°. \qquad \text{(Answer)}$$

Wipeout avoided: If the surfer tilts the board slightly forward, the magnitude of the drag force decreases and angle ϕ changes. The result is that the net force is no longer zero and the surfer moves down the face of the wave. The descent is somewhat self-adjusting because, as the surfer descends, the tilt angle θ of the wave surface decreases and thus so does the component of the gravitational force $mg \sin \theta$ pulling the surfer down the slope. So, the surfer can adjust the board to re-establish equilibrium, now lower on the wave. Similarly, by tilting the board slightly backward, the surfer increases the drag and moves up the face of the wave. If the surfer is still on the lower part of the wave, then both θ and $mg \sin \theta$ increase and again the surfer can control the forces and re-establish equilibrium.

Sample Problem 14-5

In Fig. 14-12, a block of density $\rho = 800\text{ kg/m}^3$ floats face down in a fluid of density $\rho_f = 1200\text{ kg/m}^3$. The block has height $H = 6.0$ cm.

(a) By what depth h is the block submerged?

KEY IDEA (1) Floating requires that the upward buoyant force on the block match the downward gravitational force on the block. (2) The buoyant force is equal to the weight $m_f g$ of the fluid displaced by the submerged portion of the block.

Calculations: From Eq. 14-16, we know that the buoyant force has the magnitude $F_b = m_f g$, where m_f is the mass of the fluid displaced by the block's submerged volume V_f. From Eq. 14-2 ($\rho = m/V$), we know that the mass of the displaced fluid is $m_f = \rho_f V_f$. We don't know V_f but if we symbolize the block's face length as L and its width as W, then from Fig. 14-12 we see that the submerged volume must be $V_f = LWh$. If we now combine our three expressions, we find that the upward buoyant force has magnitude

$$F_b = m_f g = \rho_f V_f g = \rho_f LWhg. \qquad (14\text{-}20)$$

Similarly, we can write the magnitude F_g of the gravitational force on the block, first in terms of the block's mass m, then in terms of the block's density ρ and (full) volume V, and then in terms of the block's dimensions L, W, and H (the full height):

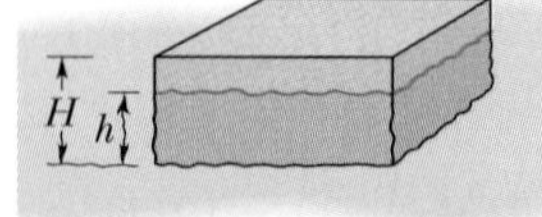

FIG. 14-12 Block of height H floats in a fluid, to a depth of h.

$$F_g = mg = \rho Vg = \rho_f LWHg. \qquad (14\text{-}21)$$

The floating block is stationary. Thus, writing Newton's second law for components along a vertical y axis ($F_{\text{net},y} = ma_y$), we have

$$F_b - F_g = m(0),$$

or from Eqs. 14-20 and 14-21,

$$\rho_f LWhg - \rho LWHg = 0,$$

which gives us

$$h = \frac{\rho}{\rho_f} H = \frac{800\text{ kg/m}^3}{1200\text{ kg/m}^3}(6.0\text{ cm}) = 4.0\text{ cm}. \qquad \text{(Answer)}$$

(b) If the block is held fully submerged and then released, what is the magnitude of its acceleration?

Calculations: The gravitational force on the block is the same but now, with the block fully submerged, the volume of the displaced water is $V = LWH$. (The full height of the block is used.) This means that the value of F_b is now larger, and the block will no longer be stationary but will accelerate upward. Now Newton's second law yields

$$F_b - F_g = ma,$$

or

$$\rho_f LWHg - \rho LWHg = \rho LWHa,$$

where we inserted ρLWH for the mass m of the block. Solving for a leads to

$$a = \left(\frac{\rho_f}{\rho} - 1\right)g = \left(\frac{1200\text{ kg/m}^3}{800\text{ kg/m}^3} - 1\right)(9.8\text{ m/s}^2) = 4.9\text{ m/s}^2. \qquad \text{(Answer)}$$

14-8 | Ideal Fluids in Motion

The motion of *real fluids* is very complicated and not yet fully understood. Instead, we shall discuss the motion of an **ideal fluid,** which is simpler to handle mathematically and yet provides useful results. Here are four assumptions that we make about our ideal fluid; they all are concerned with *flow:*

1. ***Steady flow*** In *steady* (or *laminar*) *flow,* the velocity of the moving fluid at any fixed point does not change with time, either in magnitude or in direction. The gentle flow of water near the center of a quiet stream is steady; the flow in a chain of rapids is not. Figure 14-13 shows a transition from steady flow to *nonsteady* (or *nonlaminar* or *turbulent*) *flow* for a rising stream of smoke. The speed of the smoke particles increases as they rise and, at a certain critical speed, the flow changes from steady to nonsteady.
2. ***Incompressible flow*** We assume, as for fluids at rest, that our ideal fluid is incompressible; that is, its density has a constant, uniform value.
3. ***Nonviscous flow*** Roughly speaking, the viscosity of a fluid is a measure of how resistive the fluid is to flow. For example, thick honey is more resistive to flow than water, and so honey is said to be more viscous than water. Viscosity is the fluid analog of friction between solids; both are mechanisms by which the kinetic energy of moving objects can be transferred to thermal energy. In the absence of friction, a block could glide at constant speed along a horizontal surface. In the same way, an object moving through a nonviscous fluid would experience no *viscous drag force*—that is, no resistive force due to viscosity; it could move at constant speed through the fluid. The British scientist Lord Rayleigh noted that in an ideal fluid a ship's propeller would not work, but, on the other hand, in an ideal fluid a ship (once set into motion) would not need a propeller!
4. ***Irrotational flow*** Although it need not concern us further, we also assume that the flow is *irrotational.* To test for this property, let a tiny grain of dust move with the fluid. Although this test body may (or may not) move in a circular path, in irrotational flow the test body will not rotate about an axis through its own center of mass. For a loose analogy, the motion of a Ferris wheel is rotational; that of its passengers is irrotational.

FIG. 14-13 At a certain point, the rising flow of smoke and heated gas changes from steady to turbulent. *(Will McIntyre/Photo Researchers)*

We can make the flow of a fluid visible by adding a *tracer.* This might be a dye injected into many points across a liquid stream (Fig. 14-14) or smoke particles added to a gas flow (Fig. 14-13). Each bit of a tracer follows a *streamline,* which is the path that a tiny element of the fluid would take as the fluid flows. Recall from Chapter 4 that the velocity of a particle is always tangent to the path taken by the particle. Here the particle is the fluid element, and its velocity $\vec{v}$ is always tangent to a streamline (Fig. 14-15). For this reason, two streamlines can never intersect; if they did, then an element arriving at their intersection would have two different velocities simultaneously—an impossibility.

14-9 | The Equation of Continuity

You may have noticed that you can increase the speed of the water emerging from a garden hose by partially closing the hose opening with your thumb. Apparently the speed v of the water depends on the cross-sectional area A through which the water flows.

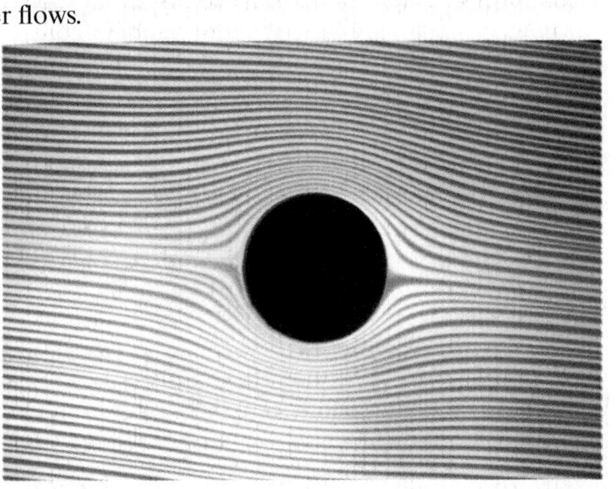

FIG. 14-14 The steady flow of a fluid around a cylinder, as revealed by a dye tracer that was injected into the fluid upstream of the cylinder. *(Courtesy D.H. Peregrine, University of Bristol)*

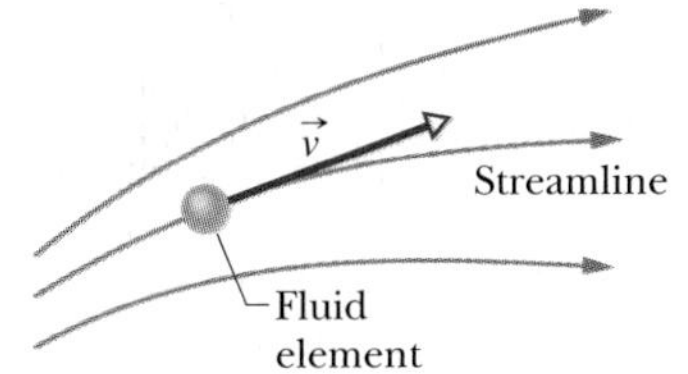

FIG. 14-15 A fluid element traces out a streamline as it moves. The velocity vector of the element is tangent to the streamline at every point.

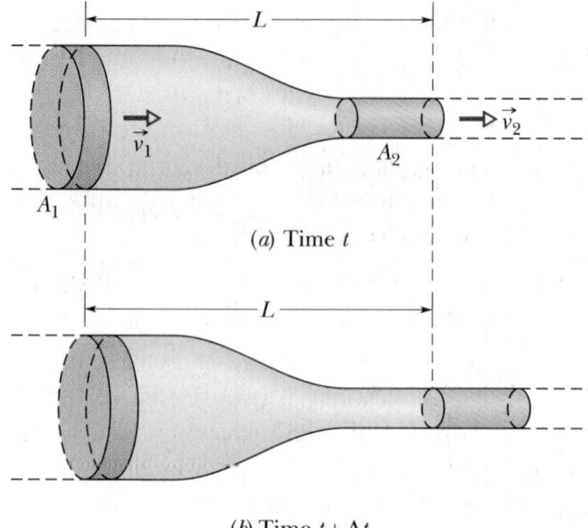

FIG. 14-16 Fluid flows from left to right at a steady rate through a tube segment of length L. The fluid's speed is v_1 at the left side and v_2 at the right side. The tube's cross-sectional area is A_1 at the left side and A_2 at the right side. From time t in (a) to time $t + \Delta t$ in (b), the amount of fluid shown in purple enters at the left side and the equal amount of fluid shown in green emerges at the right side.

Here we wish to derive an expression that relates v and A for the steady flow of an ideal fluid through a tube with varying cross section, like that in Fig. 14-16. The flow there is toward the right, and the tube segment shown (part of a longer tube) has length L. The fluid has speeds v_1 at the left end of the segment and v_2 at the right end. The tube has cross-sectional areas A_1 at the left end and A_2 at the right end. Suppose that in a time interval Δt a volume ΔV of fluid enters the tube segment at its left end (that volume is colored purple in Fig. 14-16). Then, because the fluid is incompressible, an identical volume ΔV must emerge from the right end of the segment (it is colored green in Fig. 14-16).

We can use this common volume ΔV to relate the speeds and areas. To do so, we first consider Fig. 14-17, which shows a side view of a tube of *uniform* cross-sectional area A. In Fig. 14-17a, a fluid element e is about to pass through the dashed line drawn across the tube width. The element's speed is v, so during a time interval Δt, the element moves along the tube a distance $\Delta x = v\,\Delta t$. The volume ΔV of fluid that has passed through the dashed line in that time interval Δt is

$$\Delta V = A\,\Delta x = Av\,\Delta t. \tag{14-22}$$

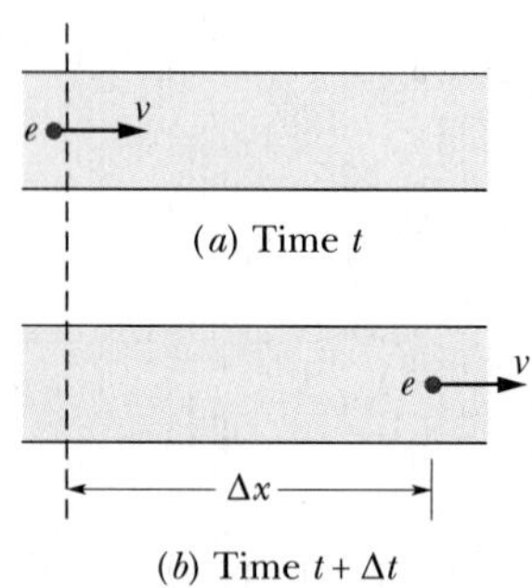

FIG. 14-17 Fluid flows at a constant speed v through a tube. (a) At time t, fluid element e is about to pass the dashed line. (b) At time $t + \Delta t$, element e is a distance $\Delta x = v\,\Delta t$ from the dashed line.

Applying Eq. 14-22 to both the left and right ends of the tube segment in Fig. 14-16, we have

$$\Delta V = A_1v_1\,\Delta t = A_2v_2\,\Delta t$$

or

$$A_1v_1 = A_2v_2 \quad \text{(equation of continuity).} \tag{14-23}$$

This relation between speed and cross-sectional area is called the **equation of continuity** for the flow of an ideal fluid. It tells us that the flow speed increases when we decrease the cross-sectional area through which the fluid flows (as when we partially close off a garden hose with a thumb).

Equation 14-23 applies not only to an actual tube but also to any so-called *tube of flow,* or imaginary tube whose boundary consists of streamlines. Such a tube acts like a real tube because no fluid element can cross a streamline; thus, all the fluid within a tube of flow must remain within its boundary. Figure 14-18 shows a tube of flow in which the cross-sectional area increases from area A_1 to area A_2 along the flow direction. From Eq. 14-23 we know that, with the increase in area, the speed must decrease, as is indicated by the greater spacing between streamlines at the right in Fig. 14-18. Similarly, you can see that in Fig. 14-14 the speed of the flow is greatest just above and just below the cylinder.

We can rewrite Eq. 14-23 as

$$R_V = Av = \text{a constant} \quad \text{(volume flow rate, equation of continuity),} \tag{14-24}$$

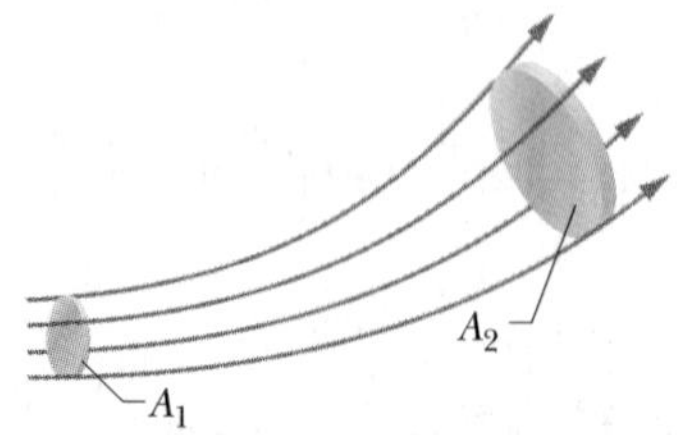

FIG. 14-18 A tube of flow is defined by the streamlines that form the boundary of the tube. The volume flow rate must be the same for all cross sections of the tube of flow.

in which R_V is the **volume flow rate** of the fluid (volume past a given point per unit time). Its SI unit is the cubic meter per second (m^3/s). If the density ρ of the fluid is uniform, we can multiply Eq. 14-24 by that density to get the **mass flow rate** R_m (mass per unit time):

$$R_m = \rho R_V = \rho A v = \text{a constant} \quad \text{(mass flow rate)}. \qquad (14\text{-}25)$$

The SI unit of mass flow rate is the kilogram per second (kg/s). Equation 14-25 says that the mass that flows into the tube segment of Fig. 14-16 each second must be equal to the mass that flows out of that segment each second.

CHECKPOINT 3

The figure shows a pipe and gives the volume flow rate (in cm³/s) and the direction of flow for all but one section. What are the volume flow rate and the direction of flow for that section?

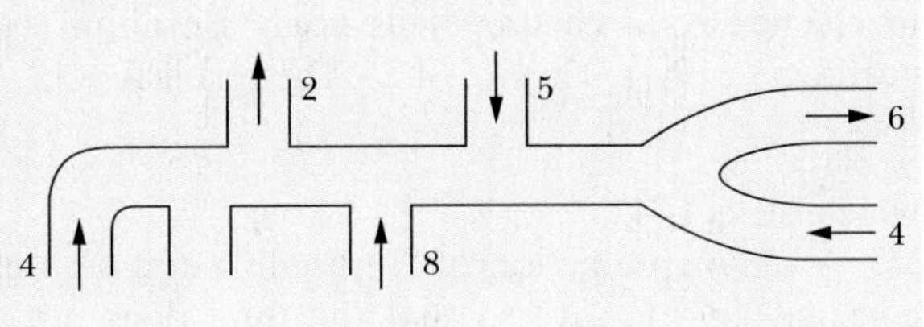

Sample Problem 14-6

Figure 14-19 shows how the stream of water emerging from a faucet "necks down" as it falls. The indicated cross-sectional areas are $A_0 = 1.2\ \text{cm}^2$ and $A = 0.35\ \text{cm}^2$. The two levels are separated by a vertical distance $h = 45$ mm. What is the volume flow rate from the tap?

KEY IDEA The volume flow rate through the higher cross section must be the same as that through the lower cross section.

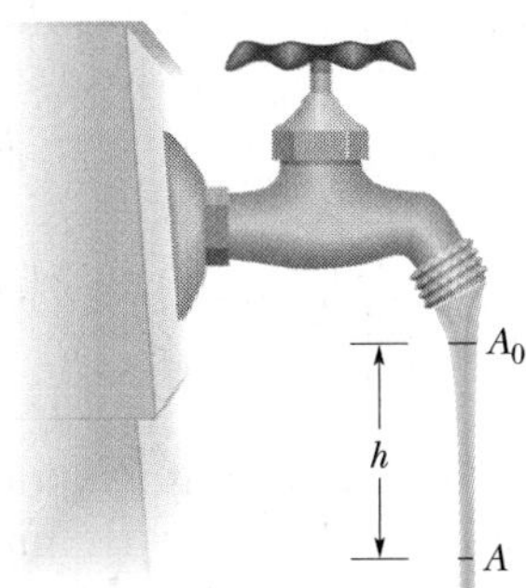

FIG. 14-19 As water falls from a tap, its speed increases. Because the volume flow rate must be the same at all horizontal cross sections of the stream, the stream must "neck down" (narrow).

Calculations: From Eq. 14-24, we have

$$A_0 v_0 = A v, \qquad (14\text{-}26)$$

where v_0 and v are the water speeds at the levels corresponding to A_0 and A. From Eq. 2-16 we can also write, because the water is falling freely with acceleration g,

$$v^2 = v_0^2 + 2gh. \qquad (14\text{-}27)$$

Eliminating v between Eqs. 14-26 and 14-27 and solving for v_0, we obtain

$$v_0 = \sqrt{\frac{2ghA^2}{A_0^2 - A^2}}$$

$$= \sqrt{\frac{(2)(9.8\ \text{m/s}^2)(0.045\ \text{m})(0.35\ \text{cm}^2)^2}{(1.2\ \text{cm}^2)^2 - (0.35\ \text{cm}^2)^2}}$$

$$= 0.286\ \text{m/s} = 28.6\ \text{cm/s}.$$

From Eq. 14-24, the volume flow rate R_V is then

$$R_V = A_0 v_0 = (1.2\ \text{cm}^2)(28.6\ \text{cm/s})$$

$$= 34\ \text{cm}^3/\text{s}. \qquad \text{(Answer)}$$

14-10 | Bernoulli's Equation

Figure 14-20 represents a tube through which an ideal fluid is flowing at a steady rate. In a time interval Δt, suppose that a volume of fluid ΔV, colored purple in Fig. 14-20, enters the tube at the left (or input) end and an identical volume, colored green in Fig. 14-20, emerges at the right (or output) end. The emerging volume must be the same as the entering volume because the fluid is incompressible, with an assumed constant density ρ.

Let y_1, v_1, and p_1 be the elevation, speed, and pressure of the fluid entering at the left, and y_2, v_2, and p_2 be the corresponding quantities for the fluid emerging at the right. By applying the principle of conservation of energy to the fluid, we shall show that these quantities are related by

$$p_1 + \tfrac{1}{2}\rho v_1^2 + \rho g y_1 = p_2 + \tfrac{1}{2}\rho v_2^2 + \rho g y_2. \qquad (14\text{-}28)$$

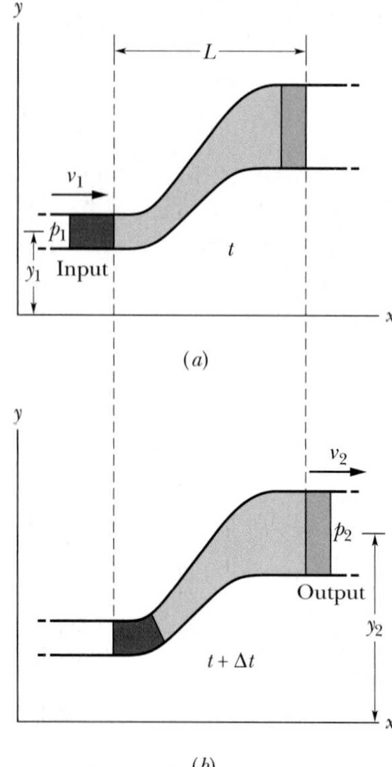

FIG. 14-20 Fluid flows at a steady rate through a length L of a tube, from the input end at the left to the output end at the right. From time t in (a) to time $t + \Delta t$ in (b), the amount of fluid shown in purple enters the input end and the equal amount shown in green emerges from the output end.

In general, the term $\frac{1}{2}\rho v^2$ is called the fluid's **kinetic energy density** (kinetic energy per unit volume). We can also write Eq. 14-28 as

$$p + \tfrac{1}{2}\rho v^2 + \rho g y = \text{a constant} \quad \text{(Bernoulli's equation).} \tag{14-29}$$

Equations 14-28 and 14-29 are equivalent forms of **Bernoulli's equation,** after Daniel Bernoulli, who studied fluid flow in the 1700s.* Like the equation of continuity (Eq. 14-24), Bernoulli's equation is not a new principle but simply the reformulation of a familiar principle in a form more suitable to fluid mechanics. As a check, let us apply Bernoulli's equation to fluids at rest, by putting $v_1 = v_2 = 0$ in Eq. 14-28. The result is

$$p_2 = p_1 + \rho g(y_1 - y_2),$$

which is Eq. 14-7.

A major prediction of Bernoulli's equation emerges if we take y to be a constant ($y = 0$, say) so that the fluid does not change elevation as it flows. Equation 14-28 then becomes

$$p_1 + \tfrac{1}{2}\rho v_1^2 = p_2 + \tfrac{1}{2}\rho v_2^2, \tag{14-30}$$

which tells us that:

If the speed of a fluid element increases as the element travels along a horizontal streamline, the pressure of the fluid must decrease, and conversely.

Put another way, where the streamlines are relatively close together (where the velocity is relatively great), the pressure is relatively low, and conversely.

The link between a change in speed and a change in pressure makes sense if you consider a fluid element. When the element nears a narrow region, the higher pressure behind it accelerates it so that it then has a greater speed in the narrow region. When it nears a wide region, the higher pressure ahead of it decelerates it so that it then has a lesser speed in the wide region.

Bernoulli's equation is strictly valid only to the extent that the fluid is ideal. If viscous forces are present, thermal energy will be involved. We take no account of this in the derivation that follows.

Proof of Bernoulli's Equation

Let us take as our system the entire volume of the (ideal) fluid shown in Fig. 14-20. We shall apply the principle of conservation of energy to this system as it moves from its initial state (Fig. 14-20*a*) to its final state (Fig. 14-20*b*). The fluid lying between the two vertical planes separated by a distance L in Fig. 14-20 does not change its properties during this process; we need be concerned only with changes that take place at the input and output ends.

First, we apply energy conservation in the form of the work–kinetic energy theorem,

$$W = \Delta K, \tag{14-31}$$

which tells us that the change in the kinetic energy of our system must equal the net work done on the system. The change in kinetic energy results from the change in speed between the ends of the tube and is

$$\begin{aligned} \Delta K &= \tfrac{1}{2}\Delta m\, v_2^2 - \tfrac{1}{2}\Delta m\, v_1^2 \\ &= \tfrac{1}{2}\rho\, \Delta V(v_2^2 - v_1^2), \end{aligned} \tag{14-32}$$

in which Δm ($= \rho\, \Delta V$) is the mass of the fluid that enters at the input end and leaves at the output end during a small time interval Δt.

*For irrotational flow (which we assume), the constant in Eq. 14-29 has the same value for all points within the tube of flow; the points do not have to lie along the same streamline. Similarly, the points 1 and 2 in Eq. 14-28 can lie anywhere within the tube of flow.

The work done on the system arises from two sources. The work W_g done by the gravitational force ($\Delta m\,\vec{g}$) on the fluid of mass Δm during the vertical lift of the mass from the input level to the output level is

$$W_g = -\Delta m\, g(y_2 - y_1) = -\rho g\, \Delta V(y_2 - y_1). \quad (14\text{-}33)$$

This work is negative because the upward displacement and the downward gravitational force have opposite directions.

Work must also be done *on* the system (at the input end) to push the entering fluid into the tube and *by* the system (at the output end) to push forward the fluid that is located ahead of the emerging fluid. In general, the work done by a force of magnitude F, acting on a fluid sample contained in a tube of area A to move the fluid through a distance Δx, is

$$F\,\Delta x = (pA)(\Delta x) = p(A\,\Delta x) = p\,\Delta V.$$

The work done on the system is then $p_1\,\Delta V$, and the work done by the system is $-p_2\,\Delta V$. Their sum W_p is

$$W_p = -p_2\,\Delta V + p_1\,\Delta V = -(p_2 - p_1)\,\Delta V. \quad (14\text{-}34)$$

The work–kinetic energy theorem of Eq. 14-31 now becomes

$$W = W_g + W_p = \Delta K.$$

Substituting from Eqs. 14-32, 14-33, and 14-34 yields

$$-\rho g\,\Delta V(y_2 - y_1) - \Delta V(p_2 - p_1) = \tfrac{1}{2}\rho\,\Delta V(v_2^2 - v_1^2).$$

This, after a slight rearrangement, matches Eq. 14-28, which we set out to prove.

CHECKPOINT 4 Water flows smoothly through the pipe shown in the figure, descending in the process. Rank the four numbered sections of pipe according to (a) the volume flow rate R_V through them, (b) the flow speed v through them, and (c) the water pressure p within them, greatest first.

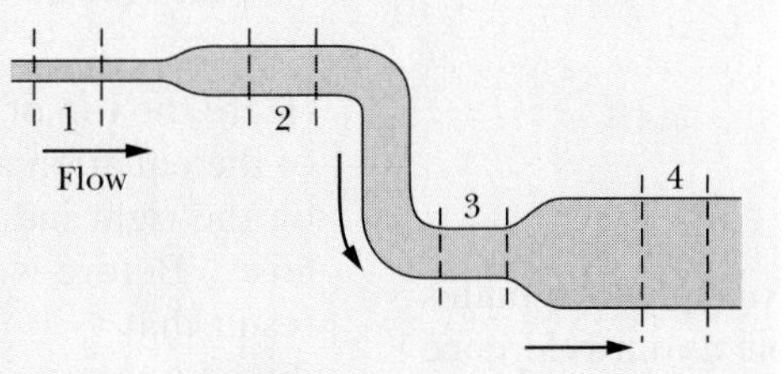

Sample Problem 14-7

Ethanol of density $\rho = 791$ kg/m^3 flows smoothly through a horizontal pipe that tapers (as in Fig. 14-16) in cross-sectional area from $A_1 = 1.20 \times 10^{-3}$ m^2 to $A_2 = A_1/2$. The pressure difference between the wide and narrow sections of pipe is 4120 Pa. What is the volume flow rate R_V of the ethanol?

KEY IDEAS (1) Because the fluid flowing through the wide section of pipe must entirely pass through the narrow section, the volume flow rate R_V must be the same in the two sections. Thus, from Eq. 14-24,

$$R_V = v_1 A_1 = v_2 A_2. \quad (14\text{-}35)$$

However, with two unknown speeds, we cannot evaluate this equation for R_V. (2) Because the flow is smooth, we can apply Bernoulli's equation. From Eq. 14-28, we can write

$$p_1 + \tfrac{1}{2}\rho v_1^2 + \rho g y = p_2 + \tfrac{1}{2}\rho v_2^2 + \rho g y, \quad (14\text{-}36)$$

where subscripts 1 and 2 refer to the wide and narrow sections of pipe, respectively, and y is their common elevation. This equation hardly seems to help because it does not contain the desired R_V and it contains the unknown speeds v_1 and v_2.

Calculations: There is a neat way to make Eq. 14-36 work for us: First, we can use Eq. 14-35 and the fact that $A_2 = A_1/2$ to write

$$v_1 = \frac{R_V}{A_1} \quad \text{and} \quad v_2 = \frac{R_V}{A_2} = \frac{2R_V}{A_1}. \quad (14\text{-}37)$$

Then we can substitute these expressions into Eq. 14-36 to eliminate the unknown speeds and introduce the desired volume flow rate. Doing this and solving for R_V yield

$$R_V = A_1\sqrt{\frac{2(p_1 - p_2)}{3\rho}}. \quad (14\text{-}38)$$

We still have a decision to make: We know that the pressure difference between the two sections is 4120 Pa, but does that mean that $p_1 - p_2$ is 4120 Pa or -4120 Pa? We could guess the former is true, or otherwise the square root in Eq. 14-38 would give us an imaginary number. Instead of guessing, however, let's try some reasoning. From Eq. 14-35 we see that speed v_2 in the narrow section (small A_2) must be greater than speed v_1 in the wider section (larger A_1). Recall that if the speed of a fluid increases as the fluid travels along a horizontal path (as here), the pressure of the fluid must decrease. Thus, p_1 is greater than p_2, and $p_1 - p_2 = 4120$ Pa. Inserting this and known data into Eq. 14-38 gives

$$R_V = 1.20 \times 10^{-3}\,\text{m}^2 \sqrt{\frac{(2)(4120\ \text{Pa})}{(3)(791\ \text{kg/m}^3)}}$$

$$= 2.24 \times 10^{-3}\ \text{m}^3/\text{s}. \qquad \text{(Answer)}$$

Sample Problem 14-8

In the old West, a desperado fires a bullet into an open water tank (Fig. 14-21), creating a hole a distance h below the water surface. What is the speed v of the water exiting the tank?

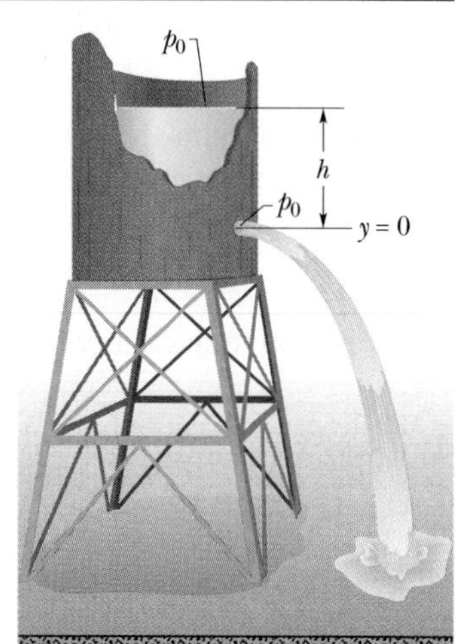

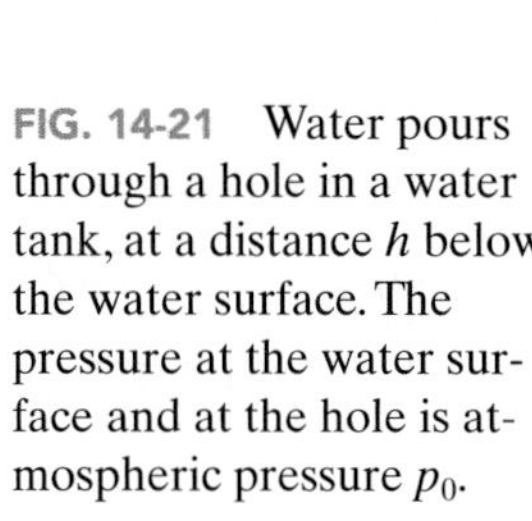

FIG. 14-21 Water pours through a hole in a water tank, at a distance h below the water surface. The pressure at the water surface and at the hole is atmospheric pressure p_0.

KEY IDEAS (1) This situation is essentially that of water moving (downward) with speed v_0 through a wide pipe (the tank) of cross-sectional area A and then moving (horizontally) with speed v through a narrow pipe (the hole) of cross-sectional area a. (2) Because the water flowing through the wide pipe must entirely pass through the narrow pipe, the volume flow rate R_V must be the same in the two "pipes." (3) We can also relate v to v_0 (and to h) through Bernoulli's equation (Eq. 14-28).

Calculations: From Eq. 14-24,

$$R_V = av = Av_0$$

and thus

$$v_0 = \frac{a}{A}v.$$

Because $a \ll A$, we see that $v_0 \ll v$. To apply Bernoulli's equation, we take the level of the hole as our reference level for measuring elevations (and thus gravitational potential energy). Noting that the pressure at the top of the tank and at the bullet hole is the atmospheric pressure p_0 (because both places are exposed to the atmosphere), we write Eq. 14-28 as

$$p_0 + \tfrac{1}{2}\rho v_0^2 + \rho gh = p_0 + \tfrac{1}{2}\rho v^2 + \rho g(0). \qquad (14\text{-}39)$$

(Here the top of the tank is represented by the left side of the equation and the hole by the right side. The zero on the right indicates that the hole is at our reference level.) Before we solve Eq. 14-39 for v, we can use our result that $v_0 \ll v$ to simplify it: We assume that v_0^2, and thus the term $\frac{1}{2}\rho v_0^2$ in Eq. 14-39, is negligible relative to the other terms, and we drop it. Solving the remaining equation for v then yields

$$v = \sqrt{2gh}. \qquad \text{(Answer)}$$

This is the same speed that an object would have when falling a height h from rest.

Sample Problem 14-9 Build your skill

Many types of race cars depend on *negative lift* (or *downforce*) to push them down against the track surface so they can take turns quickly without sliding out into the track wall. Part of the negative lift is the *ground force*, which is a force due to the airflow beneath the car. As the race car in Fig. 14-22*a* moves forward at 27.25 m/s, air is forced to flow over and under the car (Fig. 14-22*a*). The air forced to flow under the car enters through a vertical cross-sectional area $A_0 = 0.0330\ \text{m}^2$ at the front of the car (Fig. 14-22*b*) and then flows beneath the car where the vertical cross-sectional area is $A_1 = 0.0310\ \text{m}^2$. Treat this flow as steady flow through a stationary horizontal pipe that decreases in cross-sectional area from A_0 to A_1 (Fig. 14-22*c*).

(a) At the moment it passes through A_0, the air is at atmospheric pressure p_0. At what pressure p_1 is the air as it moves through A_1?

KEY IDEAS (1) Because the flow is steady, we can apply Bernoulli's equation (Eq. 14-28) to the flow. To be consistent with the given subscripts, we write the equation as

$$p_0 + \tfrac{1}{2}\rho v_0^2 + \rho gy = p_1 + \tfrac{1}{2}\rho v_1^2 + \rho gy, \qquad (14\text{-}40)$$

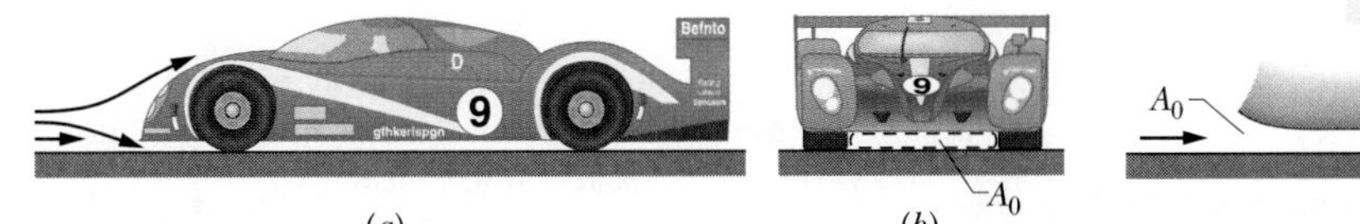

FIG. 14-22 (*a*) Air flows above and below a race car. (*b*) The flow beneath the car enters through vertical cross-sectional area A_0. (*c*) The flow is then constrained as in a pipe that narrows to vertical cross-sectional area A_1.

where ρ is the air density and y is the distance above the ground of the flowing air. (2) Because all the air entering through cross-sectional area A_0 flows through cross-sectional area A_1, the volume flow rate R_V through the two areas must be the same.

Calculations: From Eq. 14-24, we can write

$$A_0 v_0 = A_1 v_1,$$

or

$$v_1 = v_0 \frac{A_0}{A_1}. \tag{14-41}$$

Substituting Eq. 14-41 into Eq. 14-40 and rearranging give us

$$p_1 = p_0 - \tfrac{1}{2}\rho v_0^2 \left(\frac{A_0^2}{A_1^2} - 1 \right). \tag{14-42}$$

The speed of the air as it enters A_0 at the front of the car is equal to 27.25 m/s, the speed of the car as it moves forward through the air. Substituting this speed, the air density $\rho = 1.21\ \text{kg/m}^3$, and the values for A_0 and A_1 into Eq. 14-42, we find

$$p_1 = p_0 - \tfrac{1}{2}(1.21\ \text{kg/m}^3)(27.25\ \text{m/s})^2 \left(\frac{(0.0330\ \text{m}^2)^2}{(0.0310\ \text{m}^2)^2} - 1 \right)$$

$$= p_0 - 59.838\ \text{Pa} \approx p_0 - 59.8\ \text{Pa}. \quad \text{(Answer)}$$

Thus, the air pressure beneath the car is 59.8 Pa less than atmospheric pressure.

(b) If the horizontal cross-sectional area of the car is $A_h = 4.86\ \text{m}^2$, what is the magnitude of the net vertical force $F_{\text{net},y}$ on the car due to the air pressures above and below the car?

KEY IDEA The pressure on a surface is the force per unit area, as given by Eq. 14-4 ($p = F/A$).

Calculations: Here we are concerned with the top and bottom surfaces of the car, where we take both surfaces to have area A_h. Above the car the air is at atmospheric pressure p_0 and presses down on the car with a vertical component

$$F_{y,\text{above}} = -p_0 A_h.$$

Below the car, the air is at pressure $p_1 = p_0 - 59.838$ Pa and presses up on the car with a vertical component

$$F_{y,\text{below}} = (p_0 - 59.838\ \text{Pa})A_h.$$

The net vertical force is then

$$\begin{aligned} F_{\text{net},y} &= F_{y,\text{below}} + F_{y,\text{above}} \\ &= (p_0 - 59.838\ \text{Pa})A_h - p_0 A_h \\ &= -(59.838\ \text{Pa})(4.86\ \text{m}^2) = -291\ \text{N}. \quad \text{(Answer)} \end{aligned}$$

Danger of drafting: This net downward force, which is due to the reduced air pressure beneath the car, is the ground effect acting on the car. It is about 30% of the total negative lift that helps hold the car on the track. Without negative lift, a car must greatly slow for turns or else it will slide outward into the track wall. In a race, a driver can reduce the air drag on his car by closely following another car, a procedure known as *drafting*. However, the leading car disrupts the steady flow of air under the trailing car, eliminating the ground effect on the trailing car. If the trailing driver does not anticipate that elimination and slow accordingly, sliding into the track wall may be unavoidable.

REVIEW & SUMMARY

Density The **density** ρ of any material is defined as the material's mass per unit volume:

$$\rho = \frac{\Delta m}{\Delta V}. \tag{14-1}$$

Usually, where a material sample is much larger than atomic dimensions, we can write Eq. 14-1 as

$$\rho = \frac{m}{V}. \tag{14-2}$$

Fluid Pressure A **fluid** is a substance that can flow; it conforms to the boundaries of its container because it cannot withstand shearing stress. It can, however, exert a force perpendicular to its surface. That force is described in terms of **pressure** p:

$$p = \frac{\Delta F}{\Delta A}, \tag{14-3}$$

in which ΔF is the force acting on a surface element of area ΔA. If the force is uniform over a flat area, Eq. 14-3 can be written as

$$p = \frac{F}{A}. \tag{14-4}$$

The force resulting from fluid pressure at a particular point in a fluid has the same magnitude in all directions. **Gauge pressure** is the difference between the actual pressure (or *absolute pressure*) at a point and the atmospheric pressure.

Pressure Variation with Height and Depth Pressure in a fluid at rest varies with vertical position y. For y measured positive upward,

$$p_2 = p_1 + \rho g(y_1 - y_2). \quad (14\text{-}7)$$

The pressure in a fluid is the same for all points at the same level. If h is the *depth* of a fluid sample below some reference level at which the pressure is p_0, Eq. 14-7 becomes

$$p = p_0 + \rho gh, \quad (14\text{-}8)$$

where p is the pressure in the sample.

Pascal's Principle A change in the pressure applied to an enclosed fluid is transmitted undiminished to every portion of the fluid and to the walls of the containing vessel.

Archimedes' Principle When a body is fully or partially submerged in a fluid, a buoyant force $\vec{F}_b$ from the surrounding fluid acts on the body. The force is directed upward and has a magnitude given by

$$F_b = m_f g, \quad (14\text{-}16)$$

where m_f is the mass of the fluid that has been displaced by the body (that is, the fluid that has been pushed out of the way by the body).

When a body floats in a fluid, the magnitude F_b of the (upward) buoyant force on the body is equal to the magnitude F_g of the (downward) gravitational force on the body. The **apparent weight** of a body on which a buoyant force acts is related to its actual weight by

$$\text{weight}_{\text{app}} = \text{weight} - F_b. \quad (14\text{-}19)$$

Flow of Ideal Fluids An **ideal fluid** is incompressible and lacks viscosity, and its flow is steady and irrotational. A *streamline* is the path followed by an individual fluid particle. A *tube of flow* is a bundle of streamlines. The flow within any tube of flow obeys the **equation of continuity:**

$$R_V = Av = \text{a constant}, \quad (14\text{-}24)$$

in which R_V is the **volume flow rate,** A is the cross-sectional area of the tube of flow at any point, and v is the speed of the fluid at that point. The **mass flow rate** R_m is

$$R_m = \rho R_V = \rho Av = \text{a constant}. \quad (14\text{-}25)$$

Bernoulli's Equation Applying the principle of conservation of mechanical energy to the flow of an ideal fluid leads to **Bernoulli's equation:**

$$p + \tfrac{1}{2}\rho v^2 + \rho gy = \text{a constant} \quad (14\text{-}29)$$

along any tube of flow.

QUESTIONS

1 *The teapot effect.* Water poured slowly from a teapot spout can double back under the spout for a considerable distance before detaching and falling. (The water layer is held against the underside of the spout by atmospheric pressure.) In Fig. 14-23, in the water layer inside the spout, point a is at the top of the layer and point b is at the bottom of the layer; in the water layer outside the spout, point c is at the top of the layer and point d is at the bottom of the layer. Rank those four points according to the gauge pressure in the water there, most positive first.

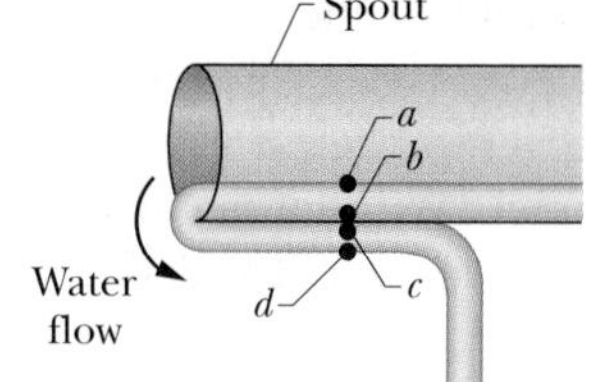

FIG. 14-23 Question 1.

2 Figure 14-24 shows a tank filled with water. Five horizontal floors and ceilings are indicated; all have the same area and are located at distances L, $2L$, or $3L$ below the top of the tank. Rank them according to the force on them due to the water, greatest first.

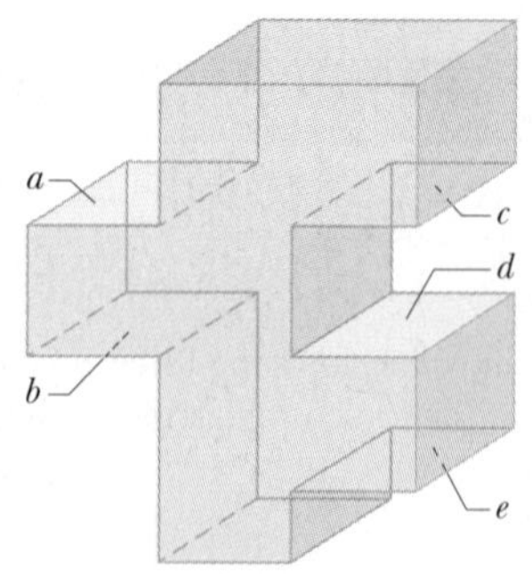

FIG. 14-24 Question 2.

3 We fully submerge an irregular 3 kg lump of material in a certain fluid. The fluid that would have been in the space now occupied by the lump has a mass of 2 kg. (a) When we release the lump, does it move upward, move downward, or remain in place? (b) If we next fully submerge the lump in a less dense fluid and again release it, what does it do?

4 Figure 14-25 shows four situations in which a red liquid and a gray liquid are in a U-tube. In one situation the liquids cannot be in static equilibrium. (a) Which situation is that? (b) For the other three situations, assume static equilibrium. For each, is the density of the red liquid greater than, less than, or equal to the density of the gray liquid?

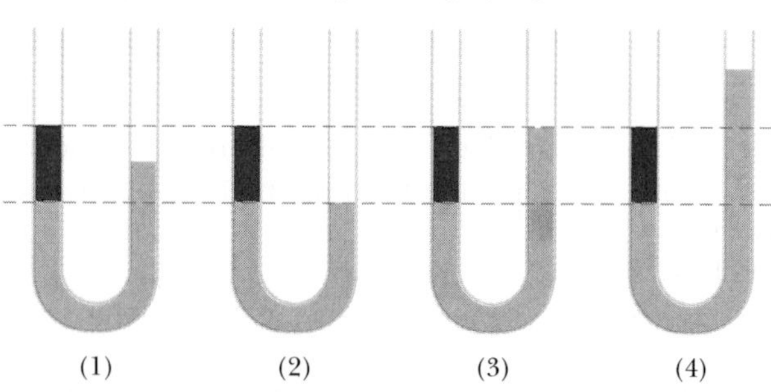

FIG. 14-25 Question 4.

5 A boat with an anchor on board floats in a swimming pool that is somewhat wider than the boat. Does the pool water level move up, move down, or remain the same if the anchor is (a) dropped into the water or (b) thrown onto the surrounding ground? (c) Does the water level in the pool move upward, move downward, or remain the same if, instead, a cork is dropped from the boat into the water, where it floats?

6 Figure 14-26 shows three identical open-top containers filled to the brim with water; toy ducks float in two of them. Rank the

FIG. 14-26 Question 6.

containers and contents according to their weight, greatest first.

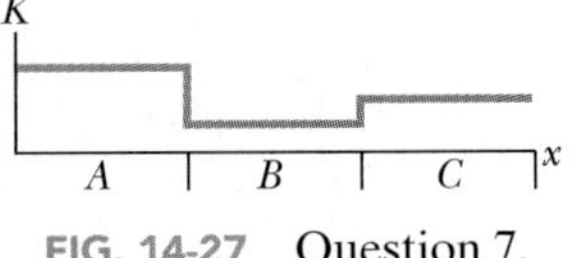

FIG. 14-27 Question 7.

7 Water flows smoothly in a horizontal pipe. Figure 14-27 shows the kinetic energy K of a water element as it moves along an x axis that runs along the pipe. Rank the three lettered sections of the pipe according to the pipe radius, greatest first.

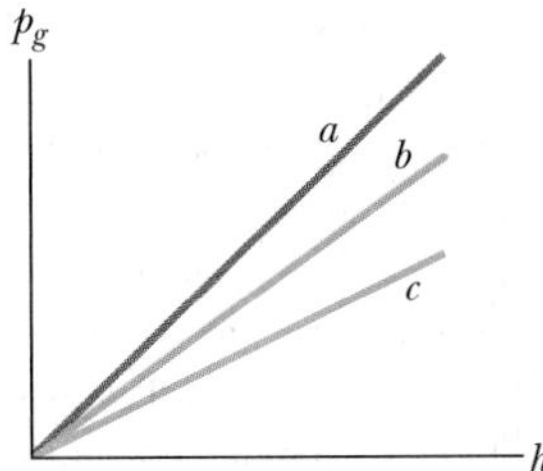

FIG. 14-28 Question 8.

8 The gauge pressure p_g versus depth h is plotted in Fig. 14-28 for three liquids. For a rigid plastic bead fully submerged in each liquid, rank the plots according to the magnitude of the buoyant force on the bead, greatest first.

9 Figure 14-29 shows four arrangements of pipes through which water flows smoothly toward the right. The radii of the pipe sections are indicated. In which arrangements is the net work done on a unit volume of water moving from the leftmost section to the rightmost section (a) zero, (b) positive, and (c) negative?

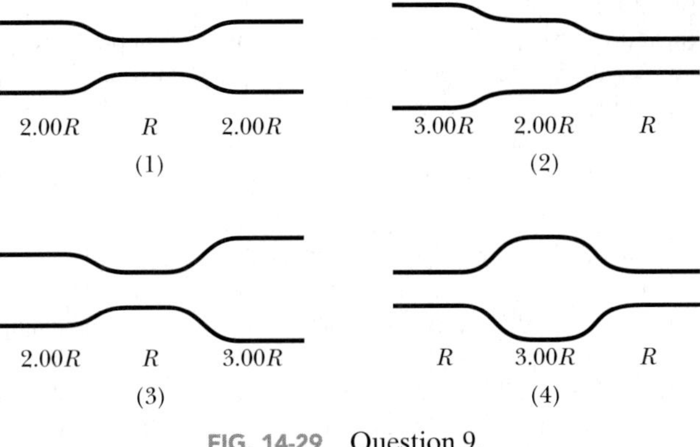

FIG. 14-29 Question 9.

10 A rectangular block is pushed face-down into three liquids, in turn. The apparent weight W_{app} of the block versus depth h in the three liquids is plotted in Fig. 14-30. Rank the liquids according to their weight per unit volume, greatest first.

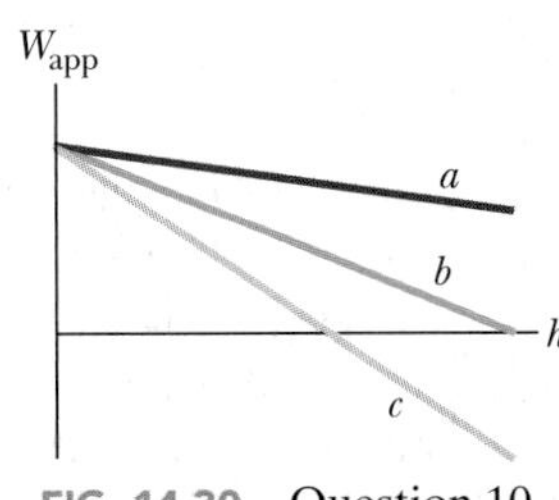

FIG. 14-30 Question 10.

PROBLEMS

GO Tutoring problem available (at instructor's discretion) in *WileyPLUS* and WebAssign

SSM Worked-out solution available in Student Solutions Manual

WWW Worked-out solution is at

ILW Interactive solution is at

http://www.wiley.com/college/halliday

• – ••• Number of dots indicates level of problem difficulty

 Additional information available in *The Flying Circus of Physics* and at flyingcircusofphysics.com

sec. 14-3 Density and Pressure

•**1** Find the pressure increase in the fluid in a syringe when a nurse applies a force of 42 N to the syringe's circular piston, which has a radius of 1.1 cm. SSM

•**2** Three liquids that will not mix are poured into a cylindrical container. The volumes and densities of the liquids are 0.50 L, 2.6 g/cm^3; 0.25 L, 1.0 g/cm^3; and 0.40 L, 0.80 g/cm^3. What is the force on the bottom of the container due to these liquids? One liter = 1 L = 1000 cm^3. (Ignore the contribution due to the atmosphere.)

•**3** An office window has dimensions 3.4 m by 2.1 m. As a result of the passage of a storm, the outside air pressure drops to 0.96 atm, but inside the pressure is held at 1.0 atm. What net force pushes out on the window? SSM

•**4** You inflate the front tires on your car to 28 psi. Later, you measure your blood pressure, obtaining a reading of 120/80, the readings being in mm Hg. In metric countries (which is to say, most of the world), these pressures are customarily reported in kilopascals (kPa). In kilopascals, what are (a) your tire pressure and (b) your blood pressure?

•**5** A fish maintains its depth in fresh water by adjusting the air content of porous bone or air sacs to make its average density the same as that of the water. Suppose that with its air sacs collapsed, a fish has a density of 1.08 g/cm^3. To what fraction of its expanded body volume must the fish inflate the air sacs to reduce its density to that of water? ILW

•**6** A partially evacuated airtight container has a tight-fitting lid of surface area 77 m^2 and negligible mass. If the force required to remove the lid is 480 N and the atmospheric pressure is 1.0×10^5 Pa, what is the internal air pressure?

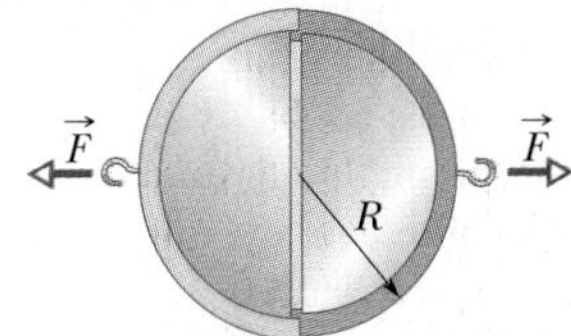

FIG. 14-31 Problem 7.

••**7** In 1654 Otto von Guericke, inventor of the air pump, gave a demonstration before the noblemen of the Holy Roman Empire in which two teams of eight horses could not pull apart two evacuated brass hemispheres. (a) Assuming the hemispheres have (strong) thin walls, so that R in Fig. 14-31 may be considered both the inside and outside radius, show that the force $\vec{F}$ required to pull apart the hemispheres has magnitude $F = \pi R^2 \Delta p$, where Δp is the difference between the pressures outside and inside the sphere. (b) Taking R as 30 cm, the inside pressure as 0.10 atm, and the outside pressure as 1.00 atm, find the force magnitude the teams of horses would have had to exert to pull apart the hemispheres. (c) Explain why one team of horses could have proved the point just as well if the hemispheres were attached to a sturdy wall.

sec. 14-4 Fluids at Rest

•**8** Calculate the hydrostatic difference in blood pressure between the brain and the foot in a person of height 1.83 m. The density of blood is 1.06×10^3 kg/m^3.

•9 At a depth of 10.9 km, the Challenger Deep in the Marianas Trench of the Pacific Ocean is the deepest site in any ocean. Yet, in 1960, Donald Walsh and Jacques Piccard reached the Challenger Deep in the bathyscaph *Trieste.* Assuming that seawater has a uniform density of 1024 kg/m^3, approximate the hydrostatic pressure (in atmospheres) that the *Trieste* had to withstand.

•10 The maximum depth d_{max} that a diver can snorkel is set by the density of the water and the fact that human lungs can function against a maximum pressure difference (between inside and outside the chest cavity) of 0.050 atm. What is the difference in d_{max} for fresh water and the water of the Dead Sea (the saltiest natural water in the world, with a density of 1.5×10^3 kg/m^3)?

•11 Crew members attempt to escape from a damaged submarine 100 m below the surface. What force must be applied to a pop-out hatch, which is 1.2 m by 0.60 m, to push it out at that depth? Assume that the density of the ocean water is 1024 kg/m^3 and the internal air pressure is at 1.00 atm. SSM

•12 The plastic tube in Fig. 14-32 has a cross-sectional area of 5.00 cm^2. The tube is filled with water until the short arm (of length d = 0.800 m) is full. Then the short arm is sealed and more water is gradually poured into the long arm. If the seal will pop off when the force on it exceeds 9.80 N, what total height of water in the long arm will put the seal on the verge of popping?

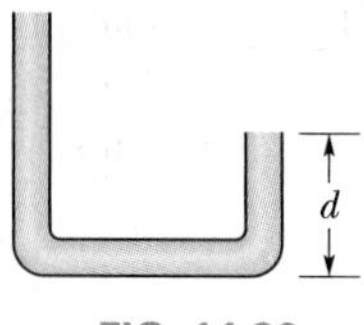

FIG. 14-32 Problems 12 and 75.

•13 What gauge pressure must a machine produce in order to suck mud of density 1800 kg/m^3 up a tube by a height of 1.5 m?

•14 *The bends during flight.* Anyone who scuba dives is advised not to fly within the next 24 h because the air mixture for diving can introduce nitrogen to the bloodstream. Without allowing the nitrogen to come out of solution slowly, any sudden air-pressure reduction (such as during airplane ascent) can result in the nitrogen forming bubbles in the blood, creating the *bends*, which can be painful and even fatal. Military special operation forces are especially at risk. What is the change in pressure on such a special-op soldier who must scuba dive at a depth of 20 m in seawater one day and parachute at an altitude of 7.6 km the next day? Assume that the average air density within the altitude range is 0.87 kg/m^3.

•15 *Giraffe bending to drink.* In a giraffe with its head 2.0 m above its heart, and its heart 2.0 m above its feet, the (hydrostatic) gauge pressure in the blood at its heart is 250 torr. Assume that the giraffe stands upright and the blood density is 1.06×10^3 kg/m^3. In torr (or mm Hg), find the (gauge) blood pressure (a) at the brain (the pressure is enough to perfuse the brain with blood, to keep the giraffe from fainting) and (b) at the feet (the pressure must be countered by tight-fitting skin acting like a pressure stocking). (c) If the giraffe were to lower its head to drink from a pond without splaying its legs and moving slowly, what would be the increase in the blood pressure in the brain? (Such action would probably be lethal.)

•16 In Fig. 14-33, an open tube of length L = 1.8 m and cross-sectional area A = 4.6 cm^2 is fixed to the top of a cylindrical barrel of diameter D = 1.2 m and height H = 1.8 m. The barrel and tube are filled with water (to the top of the tube). Calculate the ratio of the hydrostatic force on the bottom of the barrel to the gravitational force on the water contained in the barrel. Why is that ratio not equal to 1.0? (You need not consider the atmospheric pressure.)

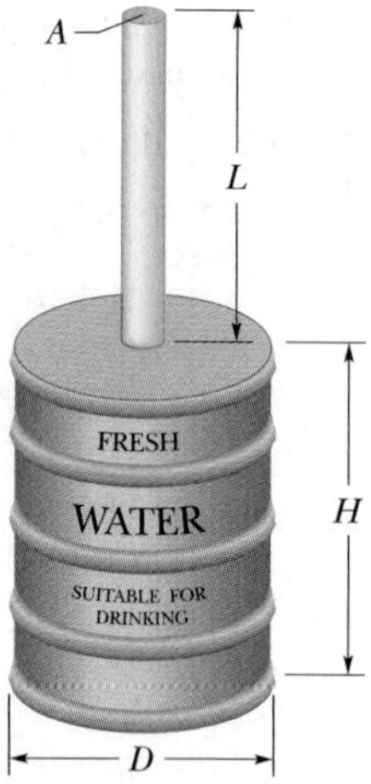

FIG. 14-33 Problem 16.

•17 *Blood pressure in Argentinosaurus.* (a) If this long-necked, gigantic sauropod had a head height of 21 m and a heart height of 9.0 m, what (hydrostatic) gauge pressure in its blood was required at the heart such that the blood pressure at the brain was 80 torr (just enough to perfuse the brain with blood)? Assume the blood had a density of 1.06×10^3 kg/m^3. (b) What was the blood pressure (in torr or mm Hg) at the feet?

•18 *Snorkeling by humans and elephants.* When a person snorkels, the lungs are connected directly to the atmosphere through the snorkel tube and thus are at atmospheric pressure. In atmospheres, what is the difference Δp between this internal air pressure and the water pressure against the body if the length of the snorkel tube is (a) 20 cm (standard situation) and (b) 4.0 m (probably lethal situation)? In the latter, the pressure difference causes blood vessels on the walls of the lungs to rupture, releasing blood into the lungs. As depicted in Fig. 14-34, an elephant can safely snorkel through its trunk while swimming with its lungs 4.0 m below the water surface because the membrane around its lungs contains connective tissue that holds and protects the blood vessels, preventing rupturing.

FIG. 14-34 Problem 18.

••19 Two identical cylindrical vessels with their bases at the same level each contain a liquid of density 1.30×10^3 kg/m^3. The area of each base is 4.00 cm^2, but in one vessel the liquid height is 0.854 m and in the other it is 1.560 m. Find the work done by the gravitational force in equalizing the levels when the two vessels are connected. SSM

••20 *g-LOC in dogfights.* When a pilot takes a tight turn at high speed in a modern fighter airplane, the blood pressure at the brain level decreases, blood no longer perfuses the brain, and the blood in the brain drains. If the heart maintains the (hydrostatic) gauge pressure in the aorta at 120 torr (or mm Hg) when the pilot undergoes a horizontal centripetal acceleration of $4g$, what is the blood pressure (in torr) at the brain, 30 cm radially inward from the heart? The perfusion in the brain is small enough that the vision switches to black and white and narrows to "tunnel vision" and the pilot can undergo *g*-LOC ("*g*-induced loss of consciousness"). Blood density is 1.06×10^3 kg/m^3.

••21 In analyzing certain geological features, it is often appropriate to assume that the pressure at some horizontal *level of compensation*, deep inside Earth, is the same over a large region and is equal to the pressure due to the gravitational force on the overlying material. Thus, the pressure on the level of compensation is given by the fluid pressure formula. This model requires, for one thing, that mountains have *roots* of continental rock extending into the denser mantle (Fig. 14-35). Consider a mountain of height $H = 6.0$ km on a continent of thickness $T = 32$ km. The continental rock has a density of 2.9 g/cm^3, and beneath this rock the mantle has a density of 3.3 g/cm^3. Calculate the depth D of the root. (*Hint:* Set the pressure at points a and b equal; the depth y of the level of compensation will cancel out.)

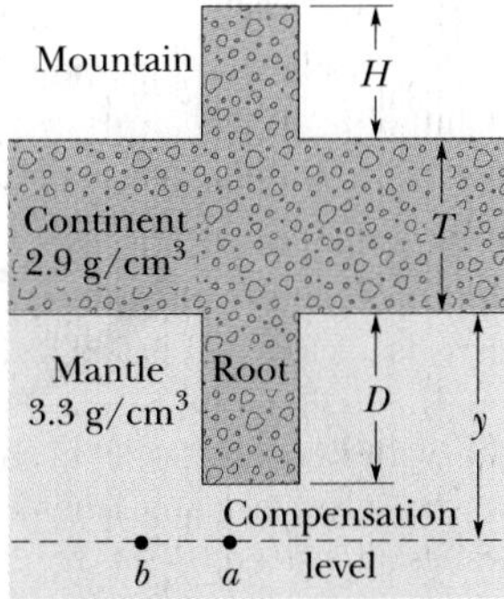

FIG. 14-35 Problem 21.

••22 The L-shaped tank shown in Fig. 14-36 is filled with water and is open at the top. If $d = 5.0$ m, what is the force due to the water (a) on face A and (b) on face B?

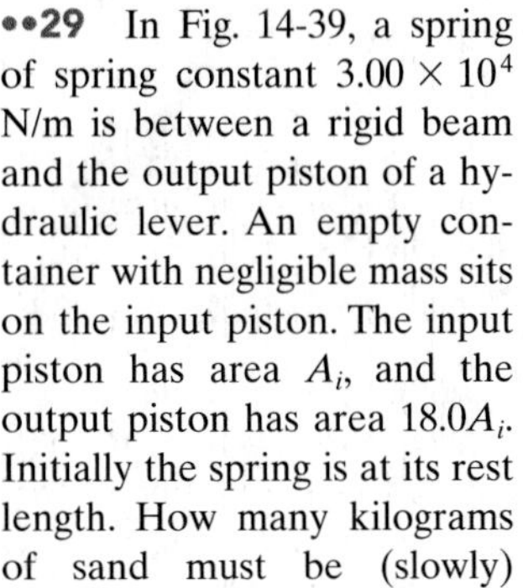

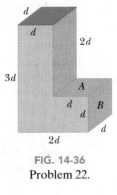

FIG. 14-36 Problem 22.

••23 A large aquarium of height 5.00 m is filled with fresh water to a depth of 2.00 m. One wall of the aquarium consists of thick plastic 8.00 m wide. By how much does the total force on that wall increase if the aquarium is next filled to a depth of 4.00 m? GO

•••24 In Fig. 14-37, water stands at depth $D = 35.0$ m behind the vertical upstream face of a dam of width $W = 314$ m. Find (a) the net horizontal force on the dam from the gauge pressure of the water and (b) the net torque due to that force about a line through O parallel to the width of the dam. (c) Find the moment arm of this torque. GO

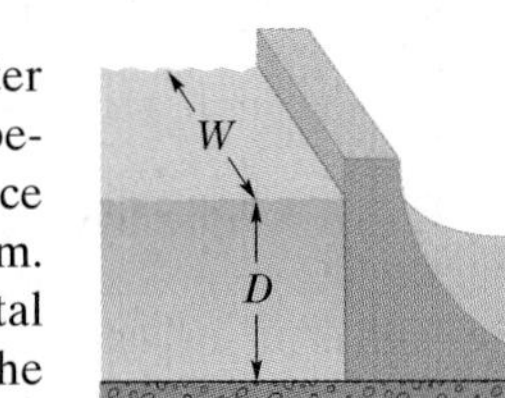

FIG. 14-37 Problem 24.

sec. 14-5 Measuring Pressure

•25 In one observation, the column in a mercury barometer (as is shown in Fig. 14-5*a*) has a measured height h of 740.35 mm. The temperature is -5.0°C, at which temperature the density of mercury ρ is 1.3608×10^4 kg/m^3. The free-fall acceleration g at the site of the barometer is 9.7835 m/s^2. What is the atmospheric pressure at that site in pascals and in torr (which is the common unit for barometer readings)?

•26 To suck lemonade of density 1000 kg/m^3 up a straw to a maximum height of 4.0 cm, what minimum gauge pressure (in atmospheres) must you produce in your lungs?

••27 What would be the height of the atmosphere if the air density (a) were uniform and (b) decreased linearly to zero with height? Assume that at sea level the air pressure is 1.0 atm and the air density is 1.3 kg/m^3. SSM

sec. 14-6 Pascal's Principle

•28 A piston of cross-sectional area a is used in a hydraulic press to exert a small force of magnitude f on the enclosed liquid. A connecting pipe leads to a larger piston of cross-sectional area A (Fig. 14-38). (a) What force magnitude F will the larger piston sustain without moving? (b) If the piston diameters are 3.80 cm and 53.0 cm, what force magnitude on the small piston will balance a 20.0 kN force on the large piston?

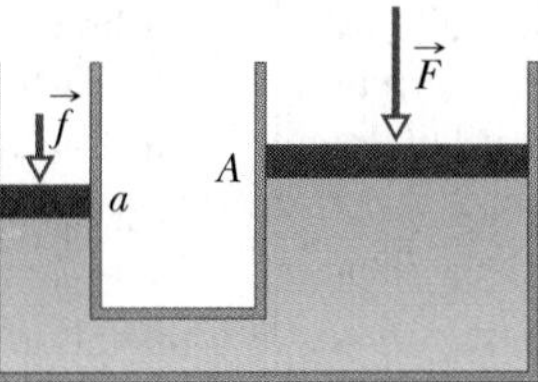

FIG. 14-38 Problem 28.

••29 In Fig. 14-39, a spring of spring constant 3.00×10^4 N/m is between a rigid beam and the output piston of a hydraulic lever. An empty container with negligible mass sits on the input piston. The input piston has area A_i, and the output piston has area $18.0A_i$. Initially the spring is at its rest length. How many kilograms of sand must be (slowly) poured into the container to compress the spring by 5.00 cm?

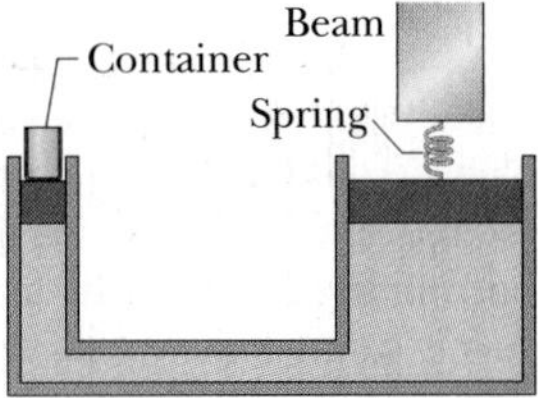

FIG. 14-39 Problem 29.

sec. 14-7 Archimedes' Principle

•30 In Fig. 14-40, a cube of edge length $L = 0.600$ m and mass 450 kg is suspended by a rope in an open tank of liquid of density 1030 kg/m^3. Find (a) the magnitude of the total downward force on the top of the cube from the liquid and the atmosphere, assuming atmospheric pressure is 1.00 atm, (b) the magnitude of the total upward force on the bottom of the cube, and (c) the tension in the rope. (d) Calculate the magnitude of the buoyant force on the cube using Archimedes' principle. What relation exists among all these quantities?

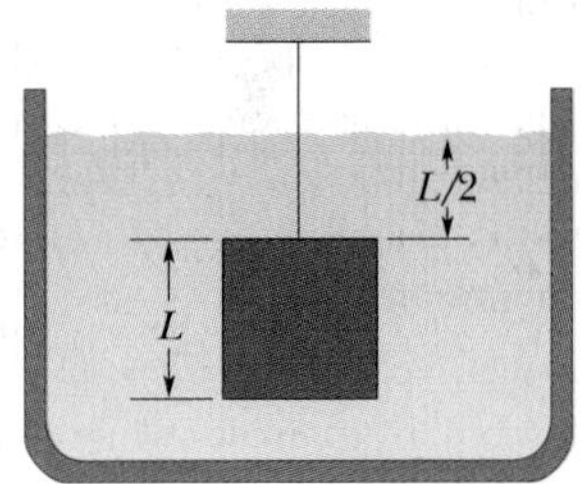

FIG. 14-40 Problem 30.

•31 An iron anchor of density 7870 kg/m^3 appears 200 N lighter in water than in air. (a) What is the volume of the anchor? (b) How much does it weigh in air? SSM

•32 A boat floating in fresh water displaces water weighing 35.6 kN. (a) What is the weight of the water this boat displaces when floating in salt water of density 1.10×10^3 kg/m^3? (b) What is the difference between the volume of fresh water displaced and the volume of salt water displaced?

•33 Three children, each of weight 356 N, make a log raft by lashing together logs of diameter 0.30 m and length 1.80 m. How many logs will be needed to keep them afloat in fresh water? Take the density of the logs to be 800 kg/m^3.

•34 A 5.00 kg object is released from rest while fully submerged in a liquid. The liquid displaced by the submerged object has a mass of 3.00 kg. How far and in what direction does the object move in 0.200 s, assuming that it moves freely and that the drag force on it from the liquid is negligible?

•35 A block of wood floats in fresh water with two-thirds of its volume V submerged and in oil with $0.90V$ submerged. Find the density of (a) the wood and (b) the oil. SSM

••36 A flotation device is in the shape of a right cylinder, with a height of 0.500 m and a face area of 4.00 m^2 on top and bottom, and its density is 0.400 times that of fresh water. It is initially held fully submerged in fresh water, with its top face at the water surface. Then it is allowed to ascend gradually until it begins to float. How much work does the buoyant force do on the device during the ascent?

••37 A hollow sphere of inner radius 8.0 cm and outer radius 9.0 cm floats half-submerged in a liquid of density 800 kg/m³. (a) What is the mass of the sphere? (b) Calculate the density of the material of which the sphere is made. SSM WWW

••38 *Lurking alligators.* An alligator waits for prey by floating with only the top of its head exposed, so that the prey cannot easily see it. One way it can adjust the extent of sinking is by controlling the size of its lungs. Another way may be by swallowing stones (*gastrolithes*) that then reside in the stomach. Figure 14-41 shows a highly simplified model (a "rhombohedron gater") of mass 130 kg that roams with its head partially exposed. The top head surface has area 0.20 m². If the alligator were to swallow stones with a total mass of 1.0% of its body mass (a typical amount), how far would it sink?

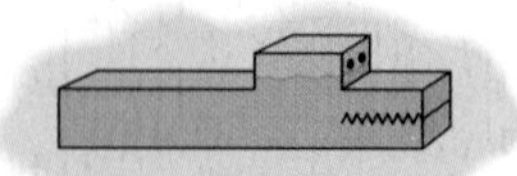

FIG. 14-41 Problem 38.

••39 What fraction of the volume of an iceberg (density 917 kg/m³) would be visible if the iceberg floats (a) in the ocean (salt water, density 1024 kg/m³) and (b) in a river (fresh water, density 1000 kg/m³)? (When salt water freezes to form ice, the salt is excluded. So, an iceberg could provide fresh water to a community.)

••40 A small solid ball is released from rest while fully submerged in a liquid and then its kinetic energy is measured when it has moved 4.0 cm in the liquid. Figure 14-42 gives the results after many liquids are used: The kinetic energy K is plotted versus the liquid density ρ_{liq}, and $K_s = 1.60$ J sets the scale on the vertical axis. What are (a) the density and (b) the volume of the ball? GO

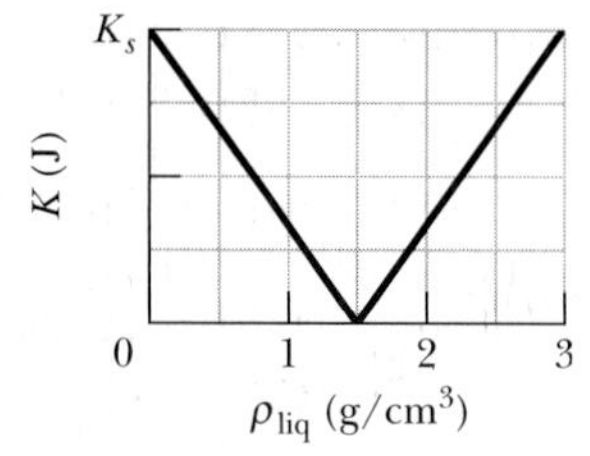

FIG. 14-42 Problem 40.

••41 A hollow spherical iron shell floats almost completely submerged in water. The outer diameter is 60.0 cm, and the density of iron is 7.87 g/cm³. Find the inner diameter. ILW

••42 In Fig. 14-43*a*, a rectangular block is gradually pushed face-down into a liquid. The block has height d; on the bottom and top the face area is $A = 5.67$ cm². Figure 14-43*b* gives the apparent weight W_{app} of the block as a function of the depth h of its lower face. The scale on the vertical axis is set by $W_s = 0.20$ N. What is the density of the liquid?

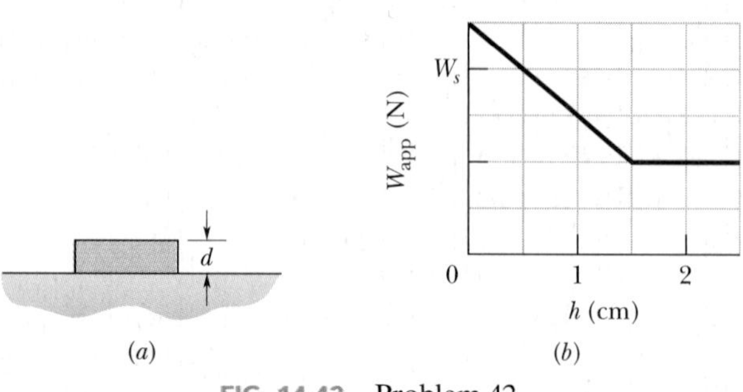

FIG. 14-43 Problem 42.

••43 An iron casting containing a number of cavities weighs 6000 N in air and 4000 N in water. What is the total volume of all the cavities in the casting? The density of iron (that is, a sample with no cavities) is 7.87 g/cm³.

••44 Suppose that you release a small ball from rest at a depth of 0.600 m below the surface in a pool of water. If the density of the ball is 0.300 that of water and if the drag force on the ball from the water is negligible, how high above the water surface will the ball shoot as it emerges from the water? (Neglect any transfer of energy to the splashing and waves produced by the emerging ball.)

••45 The volume of air space in the passenger compartment of an 1800 kg car is 5.00 m³. The volume of the motor and front wheels is 0.750 m³, and the volume of the rear wheels, gas tank, and trunk is 0.800 m³; water cannot enter these two regions. The car rolls into a lake. (a) At first, no water enters the passenger compartment. How much of the car, in cubic meters, is below the water surface with the car floating (Fig. 14-44)? (b) As water slowly enters, the car sinks. How many cubic meters of water are in the car as it disappears below the water surface? (The car, with a heavy load in the trunk, remains horizontal.)

FIG. 14-44 Problem 45.

••46 A block of wood has a mass of 3.67 kg and a density of 600 kg/m³. It is to be loaded with lead (1.14×10^4 kg/m³) so that it will float in water with 0.900 of its volume submerged. What mass of lead is needed if the lead is attached to (a) the top of the wood and (b) the bottom of the wood?

••47 When researchers find a reasonably complete fossil of a dinosaur, they can determine the mass and weight of the living dinosaur with a scale model sculpted from plastic and based on the dimensions of the fossil bones. The scale of the model is 1/20; that is, lengths are 1/20 actual length, areas are $(1/20)^2$ actual areas, and volumes are $(1/20)^3$ actual volumes. First, the model is suspended from one arm of a balance and weights are added to the other arm until equilibrium is reached. Then the model is fully submerged in water and enough weights are removed from the second arm to reestablish equilibrium (Fig. 14-45). For a model of a particular *T. rex* fossil, 637.76 g had to be removed to reestablish equilibrium. What was the volume of (a) the model and (b) the actual *T. rex*? (c) If the density of *T. rex* was approximately the density of water, what was its mass?

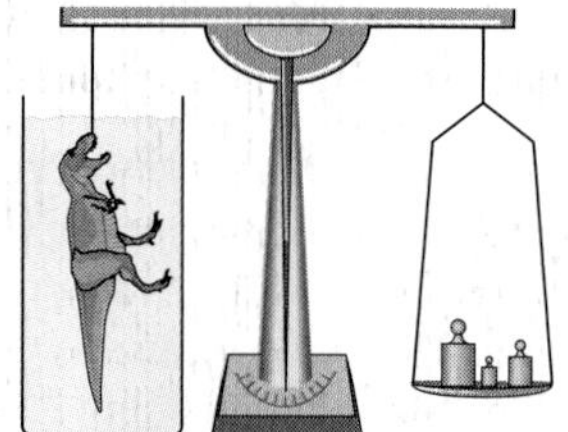

FIG. 14-45 Problem 47.

•••48 Figure 14-46 shows an iron ball suspended by thread of negligible mass from an upright cylinder that floats partially submerged in water. The cylinder

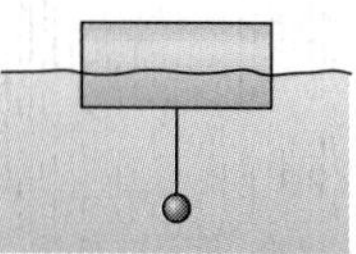

FIG. 14-46 Problem 48.

has a height of 6.00 cm, a face area of 12.0 cm^2 on the top and bottom, and a density of 0.30 g/cm^3, and 2.00 cm of its height is above the water surface. What is the radius of the iron ball? GO

sec. 14-9 The Equation of Continuity

•49 A garden hose with an internal diameter of 1.9 cm is connected to a (stationary) lawn sprinkler that consists merely of a container with 24 holes, each 0.13 cm in diameter. If the water in the hose has a speed of 0.91 m/s, at what speed does it leave the sprinkler holes? SSM

•50 Two streams merge to form a river. One stream has a width of 8.2 m, depth of 3.4 m, and current speed of 2.3 m/s. The other stream is 6.8 m wide and 3.2 m deep, and flows at 2.6 m/s. If the river has width 10.5 m and speed 2.9 m/s, what is its depth?

•51 *Canal effect.* Figure 14-47 shows an anchored barge that extends across a canal by distance d = 30 m and into the water by distance b = 12 m. The canal has a width D = 55 m, a water depth H = 14 m, and a uniform water-flow speed v_i = 1.5 m/s. Assume that the flow around the barge is uniform. As the water passes the bow, the water level undergoes a dramatic dip known as the canal effect. If the dip has depth h = 0.80 m, what is the water speed alongside the boat through the vertical cross sections at (a) point a and (b) point b? The erosion due to the speed increase is a common concern to hydraulic engineers.

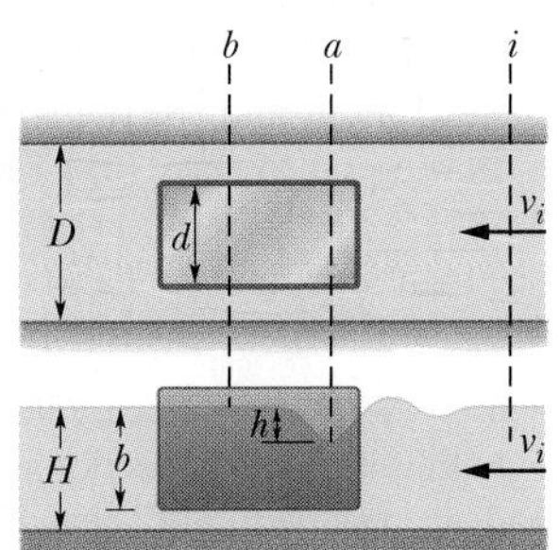

FIG. 14-47 Problem 51.

•52 Figure 14-48 shows two sections of an old pipe system that runs through a hill, with distances $d_A = d_B$ = 30 m and D = 110 m. On each side of the hill, the pipe radius is 2.00 cm. However, the radius of the pipe inside the hill is no longer known. To determine it, hydraulic engineers first establish that water flows through the left and right sections at 2.50 m/s. Then they release a dye in the water at point A and find that it takes 88.8 s to reach point B. What is the average radius of the pipe within the hill?

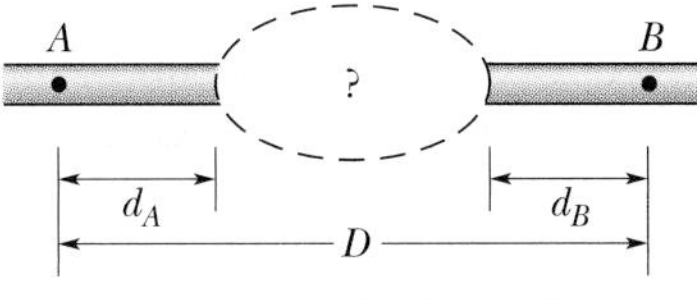

FIG. 14-48 Problem 52.

••53 Water is pumped steadily out of a flooded basement at a speed of 5.0 m/s through a uniform hose of radius 1.0 cm. The hose passes out through a window 3.0 m above the waterline. What is the power of the pump? SSM

••54 The water flowing through a 1.9 cm (inside diameter) pipe flows out through three 1.3 cm pipes. (a) If the flow rates in the three smaller pipes are 26, 19, and 11 L/min, what is the flow rate in the 1.9 cm pipe? (b) What is the ratio of the speed in the 1.9 cm pipe to that in the pipe carrying 26 L/min?

sec. 14-10 Bernoulli's Equation

•55 Water is moving with a speed of 5.0 m/s through a pipe with a cross-sectional area of 4.0 cm^2. The water gradually descends 10 m as the pipe cross-sectional area increases to 8.0 cm^2. (a) What is the speed at the lower level? (b) If the pressure at the upper level is 1.5×10^5 Pa, what is the pressure at the lower level? SSM

•56 The intake in Fig. 14-49 has cross-sectional area of 0.74 m^2 and water flow at 0.40 m/s. At the outlet, distance D = 180 m below the intake, the cross-sectional area is smaller than at the intake and the water flows out at 9.5 m/s. What is the pressure difference between inlet and outlet?

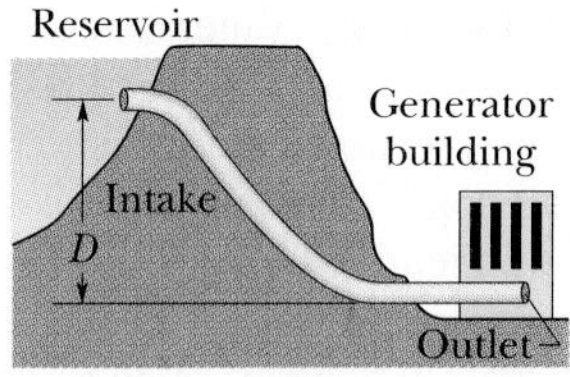

FIG. 14-49 Problem 56.

•57 A water pipe having a 2.5 cm inside diameter carries water into the basement of a house at a speed of 0.90 m/s and a pressure of 170 kPa. If the pipe tapers to 1.2 cm and rises to the second floor 7.6 m above the input point, what are the (a) speed and (b) water pressure at the second floor? ILW

•58 Models of torpedoes are sometimes tested in a horizontal pipe of flowing water, much as a wind tunnel is used to test model airplanes. Consider a circular pipe of internal diameter 25.0 cm and a torpedo model aligned along the long axis of the pipe. The model has a 5.00 cm diameter and is to be tested with water flowing past it at 2.50 m/s. (a) With what speed must the water flow in the part of the pipe that is unconstricted by the model? (b) What will the pressure difference be between the constricted and unconstricted parts of the pipe?

•59 A cylindrical tank with a large diameter is filled with water to a depth D = 0.30 m. A hole of cross-sectional area A = 6.5 cm^2 in the bottom of the tank allows water to drain out. (a) What is the rate at which water flows out, in cubic meters per second? (b) At what distance below the bottom of the tank is the cross-sectional area of the stream equal to one-half the area of the hole? SSM

•60 Suppose that two tanks, 1 and 2, each with a large opening at the top, contain different liquids. A small hole is made in the side of each tank at the same depth h below the liquid surface, but the hole in tank 1 has half the cross-sectional area of the hole in tank 2. (a) What is the ratio ρ_1/ρ_2 of the densities of the liquids if the mass flow rate is the same for the two holes? (b) What is the ratio R_{V1}/R_{V2} of the volume flow rates from the two tanks? (c) At one instant, the liquid in tank 1 is 12.0 cm above the hole. If the tanks are to have *equal* volume flow rates, what height above the hole must the liquid in tank 2 be just then?

•61 How much work is done by pressure in forcing 1.4 m^3 of water through a pipe having an internal diameter of 13 mm if the difference in pressure at the two ends of the pipe is 1.0 atm?

••62 In Fig. 14-50, water flows through a horizontal pipe and then out into the atmosphere at a speed v_1 = 15 m/s. The diameters of the left and right sections of the pipe are 5.0 cm and 3.0 cm. (a) What volume of water flows into the atmosphere during a 10 min period? In the left section of the pipe, what are (b) the speed v_2 and (c) the gauge pressure? GO

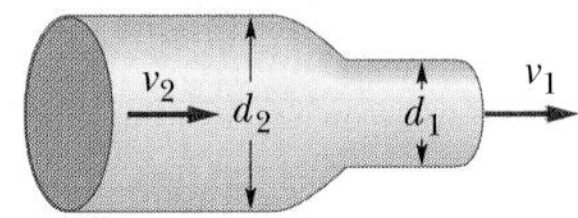

FIG. 14-50 Problem 62.

••63 In Fig. 14-51, the fresh water behind a reservoir dam has depth D = 15 m. A horizontal pipe 4.0 cm in diameter passes through the dam at depth d = 6.0 m. A plug secures the pipe opening. (a) Find the magnitude of the frictional force between plug and pipe wall. (b) The plug is removed. What water volume exits the pipe in 3.0 h? ILW

••64 Fresh water flows horizontally from pipe section 1 of cross-sectional area A_1 into pipe section 2 of cross-sectional area A_2. Figure 14-52 gives a plot of the pressure difference $p_2 - p_1$ versus the inverse area squared A_1^{-2} that would be expected for a volume flow rate of a certain value if the water flow were laminar under all circumstances. The scale on the vertical axis is set by $\Delta p_s = 300$ kN/m². For the conditions of the figure, what are the values of (a) A_2 and (b) the volume flow rate?

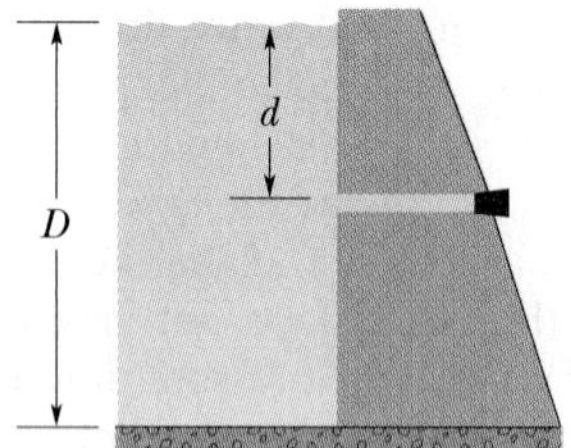

FIG. 14-51 Problem 63.

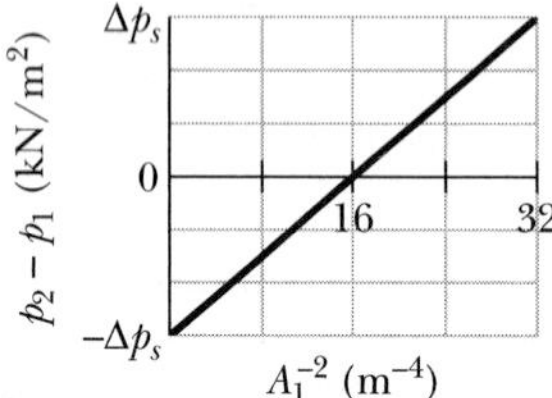

FIG. 14-52 Problem 64.

••65 Figure 14-53 shows a stream of water flowing through a hole at depth $h = 10$ cm in a tank holding water to height $H = 40$ cm. (a) At what distance x does the stream strike the floor? (b) At what depth should a second hole be made to give the same value of x? (c) At what depth should a hole be made to maximize x?

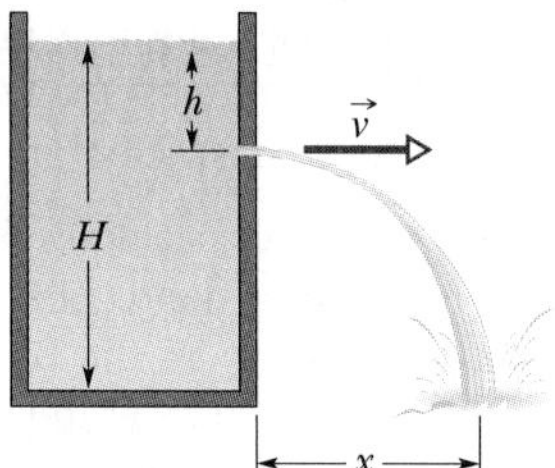

FIG. 14-53 Problem 65.

••66 In Fig. 14-54, water flows steadily from the left pipe section (radius $r_1 = 2.00R$), through the middle section (radius R), and into the right section (radius $r_3 = 3.00R$). The speed of the water in the middle section is 0.500 m/s. What is the net work done on 0.400 m³ of the water as it moves from the left section to the right section? 

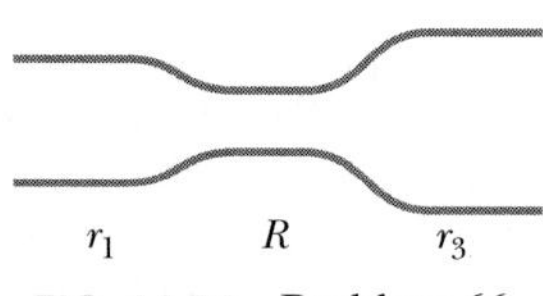

FIG. 14-54 Problem 66.

••67 A *venturi meter* is used to measure the flow speed of a fluid in a pipe. The meter is connected between two sections of the pipe (Fig. 14-55); the cross-sectional area A of the entrance and exit of the meter matches the pipe's cross-sectional area. Between the entrance and exit, the fluid flows from the pipe with speed V and then through a narrow "throat" of cross-sectional area a with speed v. A manometer connects the wider portion of the meter to the narrower portion. The change in the fluid's speed is accompanied by a change Δp in the fluid's pressure, which causes a height difference h of the liquid in the two arms of the manometer. (Here Δp means pressure in the throat minus pressure in the pipe.) (a) By applying Bernoulli's equation and the equation of continuity to points 1 and 2 in Fig. 14-55, show that

$$V = \sqrt{\frac{2a^2\,\Delta p}{\rho(a^2 - A^2)}},$$

where ρ is the density of the fluid. (b) Suppose that the fluid is fresh water, that the cross-sectional areas are 64 cm² in the pipe and 32 cm² in the throat, and that the pressure is 55 kPa in the pipe and 41 kPa in the throat. What is the rate of water flow in cubic meters per second? SSM WWW

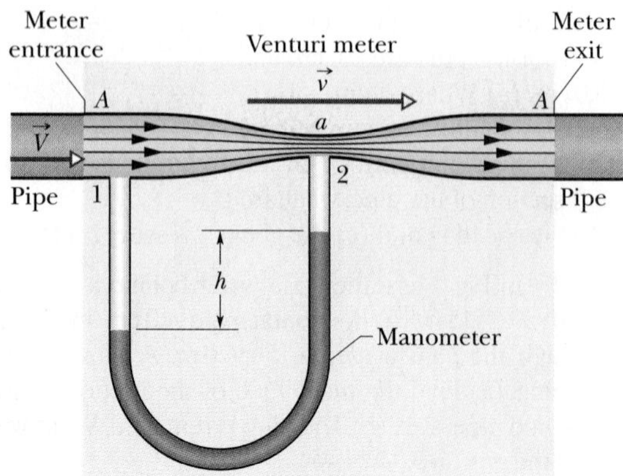

FIG. 14-55 Problems 67 and 68.

••68 Consider the venturi tube of Problem 67 and Fig. 14-55 without the manometer. Let A equal $5a$. Suppose the pressure p_1 at A is 2.0 atm. Compute the values of (a) the speed V at A and (b) the speed v at a that make the pressure p_2 at a equal to zero. (c) Compute the corresponding volume flow rate if the diameter at A is 5.0 cm. The phenomenon that occurs at a when p_2 falls to nearly zero is known as cavitation. The water vaporizes into small bubbles.

••69 A liquid of density 900 kg/m³ flows through a horizontal pipe that has a cross-sectional area of 1.90×10^{-2} m² in region A and a cross-sectional area of 9.50×10^{-2} m² in region B. The pressure difference between the two regions is 7.20×10^3 Pa. What are (a) the volume flow rate and (b) the mass flow rate?

••70 A pitot tube (Fig. 14-56) is used to determine the airspeed of an airplane. It consists of an outer tube with a number of small holes B (four are shown) that allow air into the tube; that tube is connected to one arm of a U-tube. The other arm of the U-tube is connected to hole A at the front end of the device, which points in the direction the plane is headed. At A the air becomes stagnant so that $v_A = 0$. At B, however, the speed of the air presumably equals the airspeed v of the plane. (a) Use Bernoulli's equation to show that

$$v = \sqrt{\frac{2\rho g h}{\rho_{air}}},$$

where ρ is the density of the liquid in the U-tube and h is the difference in the liquid levels in that tube. (b) Suppose that the tube contains alcohol and the level difference h is 26.0 cm. What is the plane's speed relative to the air? The density of the air is 1.03 kg/m³ and that of alcohol is 810 kg/m³.

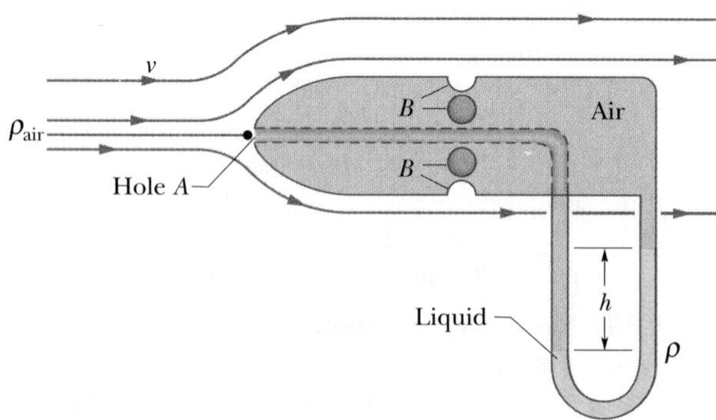

FIG. 14-56 Problems 70 and 71.

••71 A pitot tube (see Problem 70) on a high-altitude aircraft measures a differential pressure of 180 Pa. What is the aircraft's airspeed if the density of the air is 0.031 kg/m³?

•••72 A very simplified schematic of the rain drainage system for a home is shown in Fig. 14-57. Rain falling on the slanted roof runs off into gutters around the roof edge; it then drains through downspouts (only one is shown) into a main drainage pipe M below the basement, which carries the water to an even larger pipe below the street. In Fig. 14-57, a floor drain in the basement is also connected to drainage pipe M. Suppose the following apply:

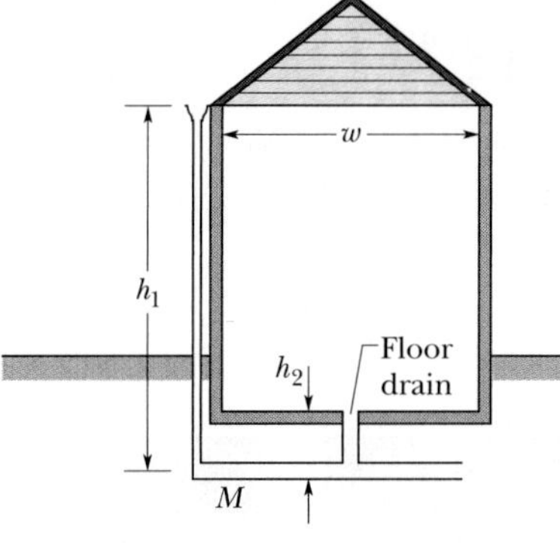

FIG. 14-57 Problem 72.

1. the downspouts have height $h_1 = 11$ m,
2. the floor drain has height $h_2 = 1.2$ m,
3. pipe M has radius 3.0 cm,
4. the house has side width $w = 30$ m and front length $L = 60$ m,
5. all the water striking the roof goes through pipe M,
6. the initial speed of the water in a downspout is negligible,
7. the wind speed is negligible (the rain falls vertically).

At what rainfall rate, in centimeters per hour, will water from pipe M reach the height of the floor drain and threaten to flood the basement?

Additional Problems

73 A glass ball of radius 2.00 cm sits at the bottom of a container of milk that has a density of 1.03 g/cm³. The normal force on the ball from the container's lower surface has magnitude 9.48×10^{-2} N. What is the mass of the ball?

74 When you cough, you expel air at high speed through the trachea and upper bronchi so that the air will remove excess mucus lining the pathway. You produce the high speed by this procedure: You breathe in a large amount of air, trap it by closing the glottis (the narrow opening in the larynx), increase the air pressure by contracting the lungs, partially collapse the trachea and upper bronchi to narrow the pathway, and then expel the air through the pathway by suddenly reopening the glottis. Assume that during the expulsion the volume flow rate is 7.0×10^{-3} m³/s. What multiple of the speed of sound v_s (= 343 m/s) is the airspeed through the trachea if the trachea diameter (a) remains its normal value of 14 mm and (b) contracts to 5.2 mm?

75 Figure 14-32 shows a modified U-tube: the right arm is shorter than the left arm. The open end of the right arm is height $d = 10.0$ cm above the laboratory bench. The radius throughout the tube is 1.50 cm. Water is gradually poured into the open end of the left arm until the water begins to flow out the open end of the right arm. Then a liquid of density 0.80 g/cm³ is gradually added to the left arm until its height in that arm is 8.0 cm (it does not mix with the water). How much water flows out of the right arm? SSM

76 Caught in an avalanche, a skier is fully submerged in flowing snow of density 96 kg/m³. Assume that the average density of the skier, clothing, and skiing equipment is 1020 kg/m³. What percentage of the gravitational force on the skier is offset by the buoyant force from the snow?

77 Figure 14-58 shows a *siphon,* which is a device for removing liquid from a container. Tube ABC must initially be filled, but once this has been done, liquid will flow through the tube until the liquid surface in the container is level with the tube opening at A. The liquid has density 1000 kg/m³ and negligible viscosity. The distances shown are $h_1 = 25$ cm, $d = 12$ cm, and $h_2 = 40$ cm. (a) With what speed does the liquid emerge from the tube at C? (b) If the atmospheric pressure is 1.0×10^5 Pa, what is the pressure in the liquid at the topmost point B? (c) Theoretically, what is the greatest possible height h_1 that a siphon can lift water?

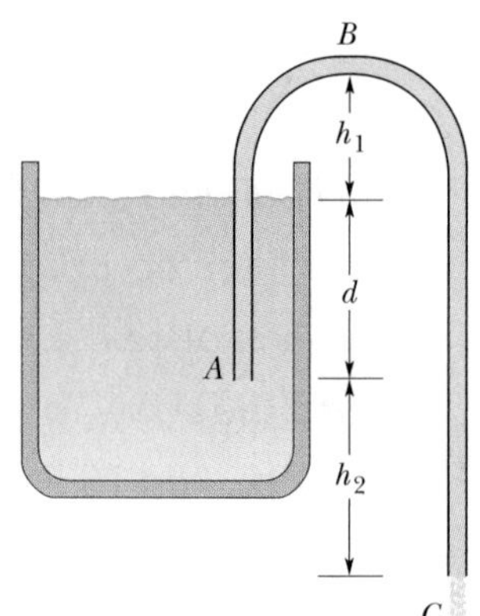

FIG. 14-58 Problem 77.

78 Suppose that your body has a uniform density of 0.95 times that of water. (a) If you float in a swimming pool, what fraction of your body's volume is above the water surface?

Quicksand is a fluid produced when water is forced up into sand, moving the sand grains away from one another so they are no longer locked together by friction. Pools of quicksand can form when water drains underground from hills into valleys where there are sand pockets. (b) If you float in a deep pool of quicksand that has a density 1.6 times that of water, what fraction of your body's volume is above the quicksand surface? (c) In particular, are you submerged enough to be unable to breathe?

79 If a bubble in sparkling water accelerates upward at the rate of 0.225 m/s² and has a radius of 0.500 mm, what is its mass? Assume that the drag force on the bubble is negligible.

80 What is the acceleration of a rising hot-air balloon if the ratio of the air density outside the balloon to that inside is 1.39? Neglect the mass of the balloon fabric and the basket.

81 A tin can has a total volume of 1200 cm³ and a mass of 130 g. How many grams of lead shot of density 11.4 g/cm³ could it carry without sinking in water?

82 A simple open U-tube contains mercury. When 11.2 cm of water is poured into the right arm of the tube, how high above its initial level does the mercury rise in the left arm?

83 An object hangs from a spring balance. The balance registers 30 N in air, 20 N when this object is immersed in water, and 24 N when the object is immersed in another liquid of unknown density. What is the density of that other liquid?

84 In an experiment, a rectangular block with height h is allowed to float in four separate liquids. In the first liquid, which is water, it floats fully submerged. In liquids A, B, and C, it floats with heights $h/2$, $2h/3$, and $h/4$ above the liquid surface, respectively. What are the *relative densities* (the densities relative to that of water) of (a) A, (b) B, and (c) C?

85 About one-third of the body of a person floating in the Dead Sea will be above the waterline. Assuming that the human body density is 0.98 g/cm³, find the density of the water in the Dead Sea. (Why is it so much greater than 1.0 g/cm³?)

15 Oscillations

If a tall building sways slowly in a wind, the occupants may not even notice the motion, but if the swaying repeats more than 10 times per second, it becomes annoying and may even cause motion sickness. One reason is that when a person is standing, the head tends to sway even more than the feet, setting off motion sensors in the balancing region of the inner ear. Various mechanisms are employed to decrease a building's sway. For example, the large ball (5.4×10^5 kg) seen in this photograph hangs on the 92nd floor of one of the world's tallest buildings.

How can the ball counter the building's sway?

The answer is in this chapter.

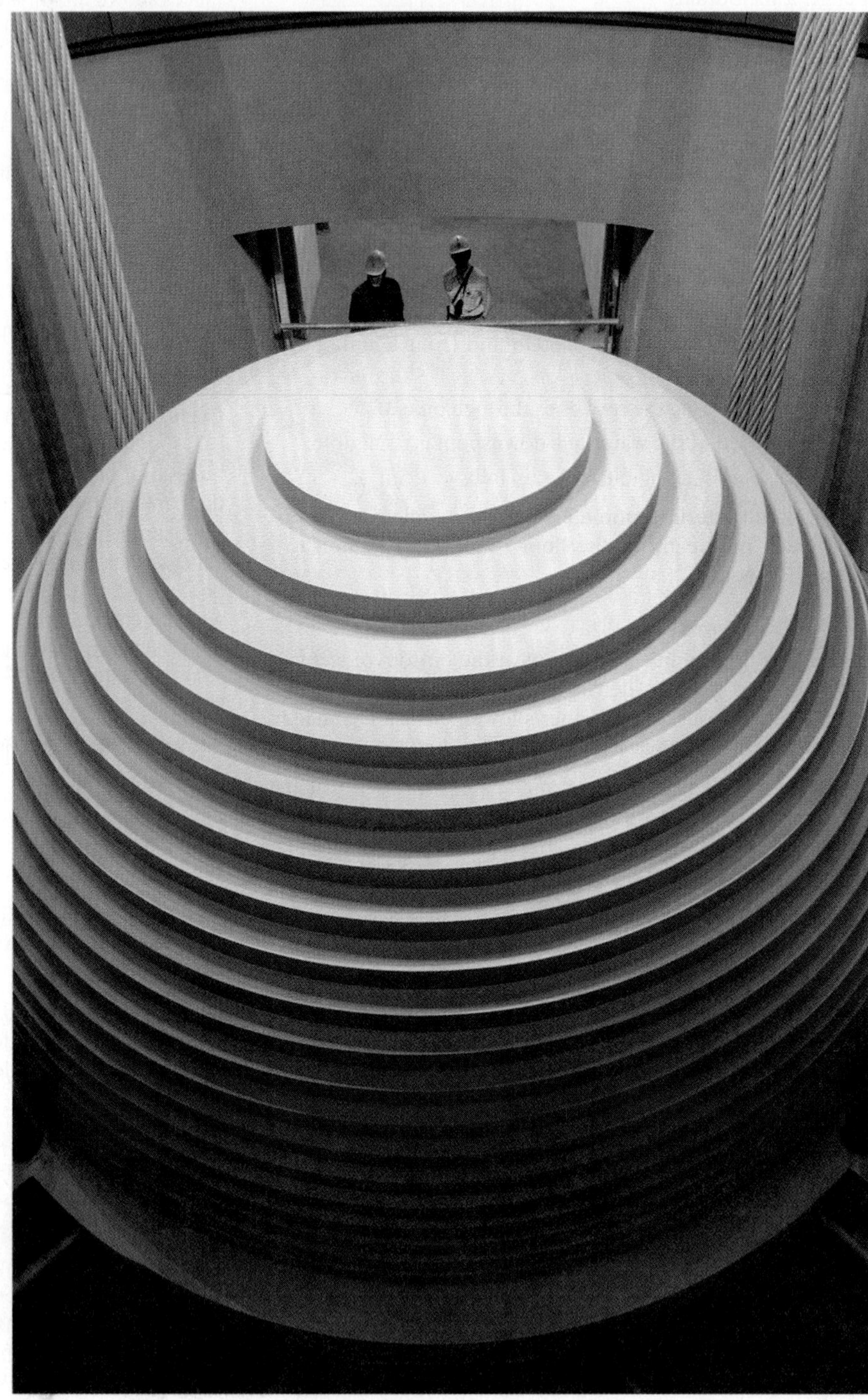

REUTERS/Richard Chung/Landov LLC

15-1 WHAT IS PHYSICS?

Our world is filled with oscillations in which objects move back and forth repeatedly. Many oscillations are merely amusing or annoying, but many others are financially important or dangerous. Here are a few examples: When a bat hits a baseball, the bat may oscillate enough to sting the batter's hands or even to break apart. When wind blows past a power line, the line may oscillate ("gallop" in electrical engineering terms) so severely that it rips apart, shutting off the power supply to a community. When an airplane is in flight, the turbulence of the air flowing past the wings makes them oscillate, eventually leading to metal fatigue and even failure. When a train travels around a curve, its wheels oscillate horizontally ("hunt" in mechanical engineering terms) as they are forced to turn in new directions (you can hear the oscillations).

When an earthquake occurs near a city, buildings may be set oscillating so severely that they are shaken apart. When an arrow is shot from a bow, the feathers at the end of the arrow manage to snake around the bow staff without hitting it because the arrow oscillates. When a coin drops into a metal collection plate, the coin oscillates with such a familiar ring that the coin's denomination can be determined from the sound. When a rodeo cowboy rides a bull, the cowboy oscillates wildly as the bull jumps and turns (at least the cowboy hopes to be oscillating).

The study and control of oscillations are two of the primary goals of both physics and engineering. In this chapter we discuss a basic type of oscillation called *simple harmonic motion.*

15-2 | Simple Harmonic Motion

Figure 15-1*a* shows a sequence of "snapshots" of a simple oscillating system, a particle moving repeatedly back and forth about the origin of an x axis. In this section we simply describe the motion. Later, we shall discuss how to attain such motion.

One important property of oscillatory motion is its **frequency,** or number of oscillations that are completed each second. The symbol for frequency is f, and its SI unit is the **hertz** (abbreviated Hz), where

$$1 \text{ hertz} = 1 \text{ Hz} = 1 \text{ oscillation per second} = 1 \text{ s}^{-1}. \tag{15-1}$$

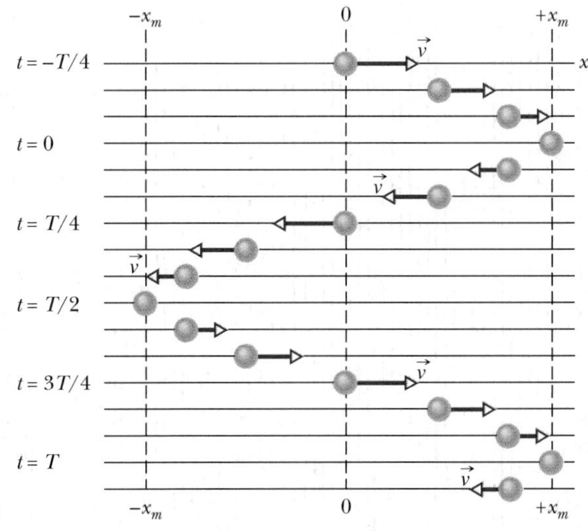

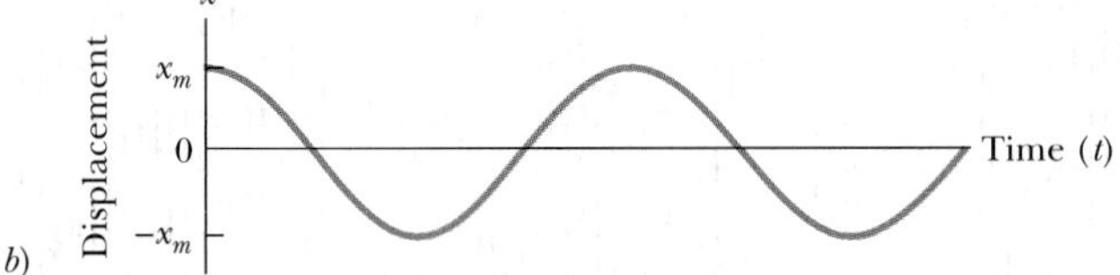

FIG. 15-1 (*a*) A sequence of "snapshots" (taken at equal time intervals) showing the position of a particle as it oscillates back and forth about the origin of an x axis, between the limits $+x_m$ and $-x_m$. The vector arrows are scaled to indicate the speed of the particle. The speed is maximum when the particle is at the origin and zero when it is at $\pm x_m$. If the time t is chosen to be zero when the particle is at $+x_m$, then the particle returns to $+x_m$ at $t = T$, where T is the period of the motion. The motion is then repeated. (*b*) A graph of x as a function of time for the motion of (*a*).

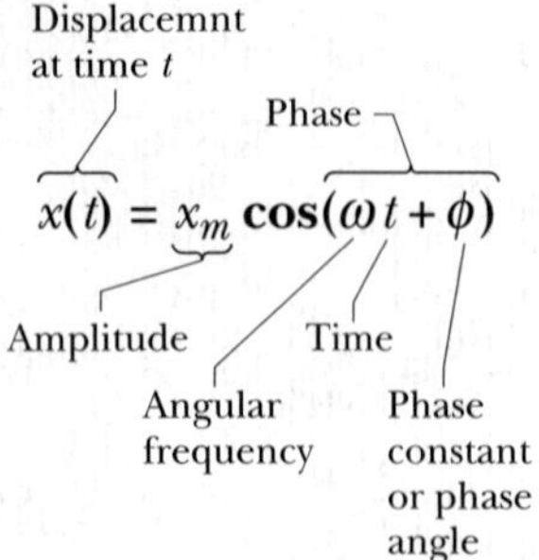

FIG. 15-2 A handy reference to the quantities in Eq. 15-3 for simple harmonic motion.

Related to the frequency is the **period** T of the motion, which is the time for one complete oscillation (or **cycle**); that is,

$$T = \frac{1}{f}. \tag{15-2}$$

Any motion that repeats itself at regular intervals is called **periodic motion** or **harmonic motion.** We are interested here in motion that repeats itself in a particular way—namely, like that in Fig. 15-1a. For such motion the displacement x of the particle from the origin is given as a function of time by

$$x(t) = x_m \cos(\omega t + \phi) \quad \text{(displacement)}, \tag{15-3}$$

in which x_m, ω, and ϕ are constants. This motion is called **simple harmonic motion** (SHM), a term that means the periodic motion is a sinusoidal function of time. Equation 15-3, in which the sinusoidal function is a cosine function, is graphed in Fig. 15-1b. (You can get that graph by rotating Fig. 15-1a counterclockwise by 90° and then connecting the successive locations of the particle with a curve.) The quantities that determine the shape of the graph are displayed in Fig. 15-2 with their names. We now shall define those quantities.

The quantity x_m, called the **amplitude** of the motion, is a positive constant whose value depends on how the motion was started. The subscript m stands for *maximum* because the amplitude is the magnitude of the maximum displacement of the particle in either direction. The cosine function in Eq. 15-3 varies between the limits ± 1; so the displacement $x(t)$ varies between the limits $\pm x_m$.

The time-varying quantity $(\omega t + \phi)$ in Eq. 15-3 is called the **phase** of the motion, and the constant ϕ is called the **phase constant** (or **phase angle**). The value of ϕ depends on the displacement and velocity of the particle at time $t = 0$. For the $x(t)$ plots of Fig. 15-3a, the phase constant ϕ is zero.

To interpret the constant ω, called the **angular frequency** of the motion, we first note that the displacement $x(t)$ must return to its initial value after one period T of the motion; that is, $x(t)$ must equal $x(t + T)$ for all t. To simplify this analysis, let us put $\phi = 0$ in Eq. 15-3. From that equation we then can write

$$x_m \cos \omega t = x_m \cos \omega(t + T). \tag{15-4}$$

The cosine function first repeats itself when its argument (the phase) has increased by 2π rad; so Eq. 15-4 gives us

$$\omega(t + T) = \omega t + 2\pi$$

or

$$\omega T = 2\pi.$$

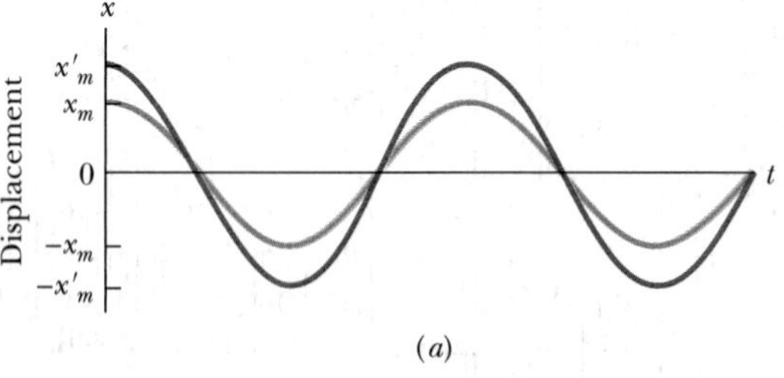

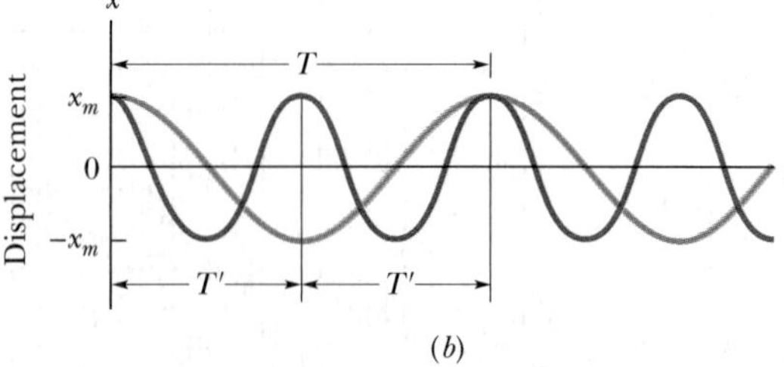

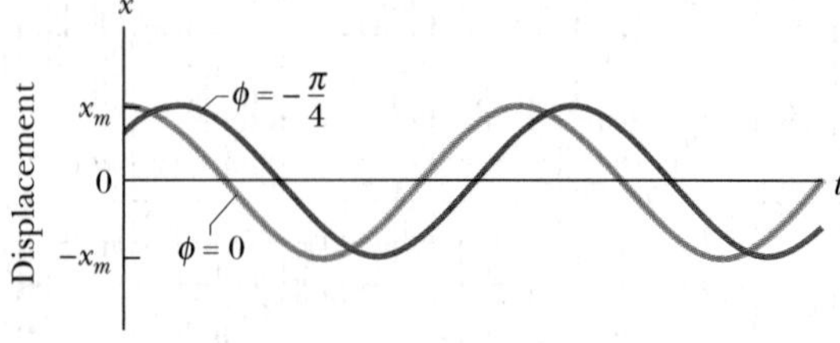

FIG. 15-3 In all three cases, the blue curve is obtained from Eq. 15-3 with $\phi = 0$. (a) The red curve differs from the blue curve *only* in that the red-curve amplitude x'_m is greater (the red-curve extremes of displacement are higher and lower). (b) The red curve differs from the blue curve *only* in that the red-curve period is $T' = T/2$ (the red curve is compressed horizontally). (c) The red curve differs from the blue curve *only* in that for the red curve $\phi = -\pi/4$ rad rather than zero (the negative value of ϕ shifts the red curve to the right).

Thus, from Eq. 15-2 the angular frequency is

$$\omega = \frac{2\pi}{T} = 2\pi f. \tag{15-5}$$

The SI unit of angular frequency is the radian per second. (To be consistent, then, ϕ must be in radians.) Figure 15-3 compares $x(t)$ for two simple harmonic motions that differ either in amplitude, in period (and thus in frequency and angular frequency), or in phase constant.

CHECKPOINT 1 A particle undergoing simple harmonic oscillation of period T (like that in Fig. 15-1) is at $-x_m$ at time $t = 0$. Is it at $-x_m$, at $+x_m$, at 0, between $-x_m$ and 0, or between 0 and $+x_m$ when (a) $t = 2.00T$, (b) $t = 3.50T$, and (c) $t = 5.25T$?

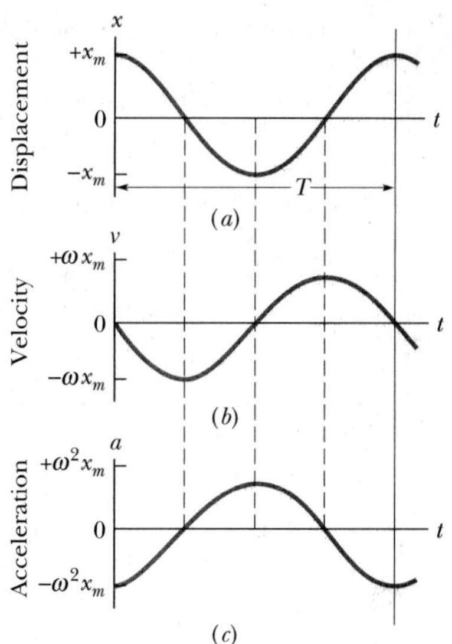

FIG. 15-4 (*a*) The displacement $x(t)$ of a particle oscillating in SHM with phase angle ϕ equal to zero. The period T marks one complete oscillation. (*b*) The velocity $v(t)$ of the particle. (*c*) The acceleration $a(t)$ of the particle.

The Velocity of SHM

By differentiating Eq. 15-3, we can find an expression for the velocity of a particle moving with simple harmonic motion; that is,

$$v(t) = \frac{dx(t)}{dt} = \frac{d}{dt}[x_m \cos(\omega t + \phi)]$$

or

$$v(t) = -\omega x_m \sin(\omega t + \phi) \quad \text{(velocity)}. \tag{15-6}$$

Figure 15-4*a* is a plot of Eq. 15-3 with $\phi = 0$. Figure 15-4*b* shows Eq. 15-6, also with $\phi = 0$. Analogous to the amplitude x_m in Eq. 15-3, the positive quantity ωx_m in Eq. 15-6 is called the **velocity amplitude** v_m. As you can see in Fig. 15-4*b*, the velocity of the oscillating particle varies between the limits $\pm v_m = \pm\omega x_m$. Note also in that figure that the curve of $v(t)$ is *shifted* (to the left) from the curve of $x(t)$ by one-quarter period; when the magnitude of the displacement is greatest (that is, $x(t) = x_m$), the magnitude of the velocity is least (that is, $v(t) = 0$). When the magnitude of the displacement is least (that is, zero), the magnitude of the velocity is greatest (that is, $v_m = \omega x_m$).

The Acceleration of SHM

Knowing the velocity $v(t)$ for simple harmonic motion, we can find an expression for the acceleration of the oscillating particle by differentiating once more. Thus, we have, from Eq. 15-6,

$$a(t) = \frac{dv(t)}{dt} = \frac{d}{dt}[-\omega x_m \sin(\omega t + \phi)]$$

or

$$a(t) = -\omega^2 x_m \cos(\omega t + \phi) \quad \text{(acceleration)}. \tag{15-7}$$

Figure 15-4*c* is a plot of Eq. 15-7 for the case $\phi = 0$. The positive quantity $\omega^2 x_m$ in Eq. 15-7 is called the **acceleration amplitude** a_m; that is, the acceleration of the particle varies between the limits $\pm a_m = \pm\omega^2 x_m$, as Fig. 15-4*c* shows. Note also that the acceleration curve $a(t)$ is shifted (to the left) by $\frac{1}{4}T$ relative to the velocity curve $v(t)$.

We can combine Eqs. 15-3 and 15-7 to yield

$$a(t) = -\omega^2 x(t), \tag{15-8}$$

which is the hallmark of simple harmonic motion:

In SHM, the acceleration is proportional to the displacement but opposite in sign, and the two quantities are related by the square of the angular frequency.

Thus, as Fig. 15-4 shows, when the displacement has its greatest positive value, the acceleration has its greatest negative value, and conversely. When the displacement is zero, the acceleration is also zero.

PROBLEM-SOLVING TACTICS

Tactic 1: ***Phase Angles*** Note the effect of the phase angle ϕ on a plot of $x(t)$. When $\phi = 0$, $x(t)$ has a graph like that in Fig. 15-4*a*, a typical cosine curve. Increasing ϕ shifts the curve leftward along the t axis. (You might remember this with the symbol $\uparrow\!\!\leftarrow\phi$, where the up arrow indicates an increase in ϕ and the left arrow indicates the resulting shift in the curve.) Decreasing ϕ shifts the curve rightward, as in Fig. 15-3*c* for $\phi = -\pi/4$.

Two plots of SHM with different phase angles are said to have a *phase difference;* each is said to be *phase-shifted* from the other, or *out of phase* with the other. The curves in Fig. 15-3*c*, for example, have a phase difference of $\pi/4$ rad; that is, one curve is phase-shifted from the other by $\pi/4$ rad.

Because SHM repeats after each period T and the cosine function repeats after each 2π rad, one period T represents a phase difference of 2π rad. In Fig. 15-4, $x(t)$ is phase-shifted to the right from $v(t)$ by one-quarter period, or $-\pi/2$ rad; it is shifted to the right from $a(t)$ by one-half period, or $-\pi$ rad. A phase shift of 2π rad causes a curve of SHM to coincide with itself; that is, it looks unchanged.

15-3 | The Force Law for Simple Harmonic Motion

Once we know how the acceleration of a particle varies with time, we can use Newton's second law to learn what force must act on the particle to give it that acceleration. If we combine Newton's second law and Eq. 15-8, we find, for simple harmonic motion,

$$F = ma = -(m\omega^2)x. \tag{15-9}$$

This result—a restoring force that is proportional to the displacement but opposite in sign—is familiar. It is Hooke's law,

$$F = -kx, \tag{15-10}$$

for a spring, the spring constant here being

$$k = m\omega^2. \tag{15-11}$$

We can in fact take Eq. 15-10 as an alternative definition of simple harmonic motion. It says:

> Simple harmonic motion is the motion executed by a particle subject to a force that is proportional to the displacement of the particle but opposite in sign.

The block–spring system of Fig. 15-5 forms a **linear simple harmonic oscillator** (linear oscillator, for short), where "linear" indicates that F is proportional to x rather than to some other power of x. The angular frequency ω of the simple harmonic motion of the block is related to the spring constant k and the mass m of the block by Eq. 15-11, which yields

$$\omega = \sqrt{\frac{k}{m}} \quad \text{(angular frequency).} \tag{15-12}$$

By combining Eqs. 15-5 and 15-12, we can write, for the **period** of the linear oscillator of Fig. 15-5,

$$T = 2\pi\sqrt{\frac{m}{k}} \quad \text{(period).} \tag{15-13}$$

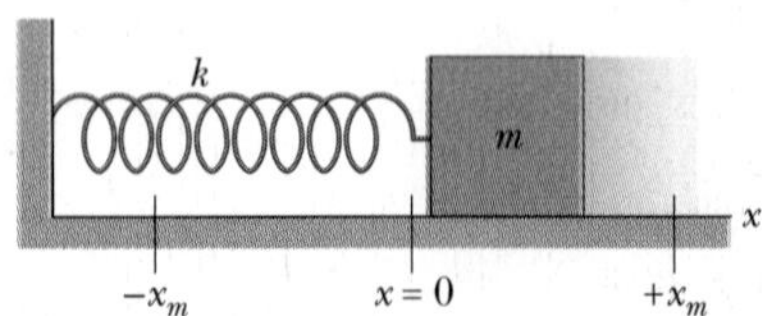

FIG. 15-5 A linear simple harmonic oscillator. The surface is frictionless. Like the particle of Fig. 15-1, the block moves in simple harmonic motion once it has been either pulled or pushed away from the $x = 0$ position and released. Its displacement is then given by Eq. 15-3.

Equations 15-12 and 15-13 tell us that a large angular frequency (and thus a small period) goes with a stiff spring (large k) and a light block (small m).

Every oscillating system, be it a diving board or a violin string, has some element of "springiness" and some element of "inertia" or mass, and thus resembles a linear oscillator. In the linear oscillator of Fig. 15-5, these elements are located in separate parts of the system: The springiness is entirely in the spring, which we assume to be massless, and the inertia is entirely in the block, which we assume to be rigid. In a violin string, however, the two elements are both within the string, as you will see in Chapter 16.

CHECKPOINT 2 Which of the following relationships between the force F on a particle and the particle's position x implies simple harmonic oscillation: (a) $F = -5x$, (b) $F = -400x^2$, (c) $F = 10x$, (d) $F = 3x^2$?

Sample Problem 15-1

A block whose mass m is 680 g is fastened to a spring whose spring constant k is 65 N/m. The block is pulled a distance $x = 11$ cm from its equilibrium position at $x = 0$ on a frictionless surface and released from rest at $t = 0$.

(a) What are the angular frequency, the frequency, and the period of the resulting motion?

KEY IDEA The block–spring system forms a linear simple harmonic oscillator, with the block undergoing SHM.

Calculations: The angular frequency is given by Eq. 15-12:

$$\omega = \sqrt{\frac{k}{m}} = \sqrt{\frac{65\text{ N/m}}{0.68\text{ kg}}} = 9.78\text{ rad/s}$$
$$\approx 9.8\text{ rad/s.} \qquad \text{(Answer)}$$

The frequency follows from Eq. 15-5, which yields

$$f = \frac{\omega}{2\pi} = \frac{9.78\text{ rad/s}}{2\pi\text{ rad}} = 1.56\text{ Hz} \approx 1.6\text{ Hz.} \qquad \text{(Answer)}$$

The period follows from Eq. 15-2, which yields

$$T = \frac{1}{f} = \frac{1}{1.56\text{ Hz}} = 0.64\text{ s} = 640\text{ ms.} \qquad \text{(Answer)}$$

(b) What is the amplitude of the oscillation?

KEY IDEA With no friction involved, the mechanical energy of the spring–block system is conserved.

Reasoning: The block is released from rest 11 cm from its equilibrium position, with zero kinetic energy and the elastic potential energy of the system at a maximum. Thus, the block will have zero kinetic energy whenever it is again 11 cm from its equilibrium position, which means it will never be farther than 11 cm from that position. Its maximum displacement is 11 cm:

$$x_m = 11\text{ cm.} \qquad \text{(Answer)}$$

(c) What is the maximum speed v_m of the oscillating block, and where is the block when it has this speed?

KEY IDEA The maximum speed v_m is the velocity amplitude ωx_m in Eq. 15-6.

Calculation: Thus, we have

$$v_m = \omega x_m = (9.78\text{ rad/s})(0.11\text{ m})$$
$$= 1.1\text{ m/s.} \qquad \text{(Answer)}$$

This maximum speed occurs when the oscillating block is rushing through the origin; compare Figs. 15-4*a* and 15-4*b*, where you can see that the speed is a maximum whenever $x = 0$.

(d) What is the magnitude a_m of the maximum acceleration of the block?

KEY IDEA The magnitude a_m of the maximum acceleration is the acceleration amplitude $\omega^2 x_m$ in Eq. 15-7.

Calculation: So, we have

$$a_m = \omega^2 x_m = (9.78\text{ rad/s})^2(0.11\text{ m})$$
$$= 11\text{ m/s}^2. \qquad \text{(Answer)}$$

This maximum acceleration occurs when the block is at the ends of its path. At those points, the force acting on the block has its maximum magnitude; compare Figs. 15-4*a* and 15-4*c*, where you can see that the magnitudes of the displacement and acceleration are maximum at the same times.

(e) What is the phase constant ϕ for the motion?

Calculations: Equation 15-3 gives the displacement of the block as a function of time. We know that at time $t = 0$, the block is located at $x = x_m$. Substituting these *initial conditions,* as they are called, into Eq. 15-3 and canceling x_m give us

$$1 = \cos\phi. \qquad (15\text{-}14)$$

Taking the inverse cosine then yields

$$\phi = 0\text{ rad.} \qquad \text{(Answer)}$$

(Any angle that is an integer multiple of 2π rad also satisfies Eq. 15-14; we chose the smallest angle.)

(f) What is the displacement function $x(t)$ for the spring–block system?

Calculation: The function $x(t)$ is given in general form by Eq. 15-3. Substituting known quantities into that equation gives us

$$x(t) = x_m\cos(\omega t + \phi)$$
$$= (0.11\text{ m})\cos[(9.8\text{ rad/s})t + 0]$$
$$= 0.11\cos(9.8t), \qquad \text{(Answer)}$$

where x is in meters and t is in seconds.

Sample Problem 15-2

At $t = 0$, the displacement $x(0)$ of the block in a linear oscillator like that of Fig. 15-5 is -8.50 cm. (Read $x(0)$ as "x at time zero.") The block's velocity $v(0)$ then is -0.920 m/s, and its acceleration $a(0)$ is $+47.0\ \text{m/s}^2$.

(a) What is the angular frequency ω of this system?

KEY IDEA With the block in SHM, Eqs. 15-3, 15-6, and 15-7 give its displacement, velocity, and acceleration, respectively, and each contains ω.

Calculations: Let's substitute $t = 0$ into each to see whether we can solve any one of them for ω. We find

$$x(0) = x_m \cos \phi, \tag{15-15}$$

$$v(0) = -\omega x_m \sin \phi, \tag{15-16}$$

and

$$a(0) = -\omega^2 x_m \cos \phi. \tag{15-17}$$

In Eq. 15-15, ω has disappeared. In Eqs. 15-16 and 15-17, we know values for the left sides, but we do not know x_m and ϕ. However, if we divide Eq. 15-17 by Eq. 15-15, we neatly eliminate both x_m and ϕ and can then solve for ω as

$$\omega = \sqrt{-\frac{a(0)}{x(0)}} = \sqrt{-\frac{47.0\ \text{m/s}^2}{-0.0850\ \text{m}}}$$

$$= 23.5\ \text{rad/s}. \quad \text{(Answer)}$$

(b) What are the phase constant ϕ and amplitude x_m?

Calculations: We know ω and want ϕ and x_m. If we divide Eq. 15-16 by Eq. 15-15, we find

$$\frac{v(0)}{x(0)} = \frac{-\omega x_m \sin \phi}{x_m \cos \phi} = -\omega \tan \phi.$$

Solving for $\tan \phi$, we find

$$\tan \phi = -\frac{v(0)}{\omega x(0)} = -\frac{-0.920\ \text{m/s}}{(23.5\ \text{rad/s})(-0.0850\ \text{m})}$$

$$= -0.461.$$

This equation has two solutions:

$$\phi = -25° \quad \text{and} \quad \phi = 180° + (-25°) = 155°.$$

(Normally only the first solution here is displayed by a calculator.) To choose the proper solution, we test them both by using them to compute values for the amplitude x_m. From Eq. 15-15, we find that if $\phi = -25°$, then

$$x_m = \frac{x(0)}{\cos \phi} = \frac{-0.0850\ \text{m}}{\cos(-25°)} = -0.094\ \text{m}.$$

We find similarly that if $\phi = 155°$, then $x_m = 0.094$ m. Because the amplitude of SHM must be a positive constant, the correct phase constant and amplitude here are

$$\phi = 155° \quad \text{and} \quad x_m = 0.094\ \text{m} = 9.4\ \text{cm}. \quad \text{(Answer)}$$

PROBLEM-SOLVING TACTICS

Tactic 2: ***Identifying SHM*** In linear SHM the acceleration a and displacement x of the system are related by an equation of the form

$$a = -(\text{a positive constant})x,$$

which says that the acceleration is proportional to the displacement from the equilibrium position but is in the opposite direction. Once you find such an expression for an oscillating system, you can immediately compare it with Eq. 15-8, identify the positive constant as being equal to ω^2, and so quickly get an expression for the angular frequency of the motion. With Eq. 15-5 you then can find the period T and the frequency f.

In some problems you might derive an expression for the force F as a function of displacement x. If the motion is linear SHM, the force and displacement are related by

$$F = -(\text{a positive constant})x,$$

which says that the force is proportional to the displacement but is in the opposite direction. Once you have found such an expression, you can immediately compare it with Eq. 15-10 and identify the positive constant as being k. If you know the mass that is involved, you can then use Eqs. 15-12, 15-13, and 15-5 to find the angular frequency ω, period T, and frequency f.

15-4 | Energy in Simple Harmonic Motion

In Chapter 8 we saw that the energy of a linear oscillator transfers back and forth between kinetic energy and potential energy, while the sum of the two—the mechanical energy E of the oscillator—remains constant. We now consider this situation quantitatively.

The potential energy of a linear oscillator like that of Fig. 15-5 is associated entirely with the spring. Its value depends on how much the spring is stretched or compressed—that is, on $x(t)$. We can use Eqs. 8-11 and 15-3 to find

$$U(t) = \tfrac{1}{2}kx^2 = \tfrac{1}{2}kx_m^2 \cos^2(\omega t + \phi). \tag{15-18}$$

Caution: A function written in the form $\cos^2 A$ (as here) means $(\cos A)^2$ and is *not* the same as one written $\cos A^2$, which means $\cos(A^2)$.

The kinetic energy of the system of Fig. 15-5 is associated entirely with the block. Its value depends on how fast the block is moving—that is, on $v(t)$. We can use Eq. 15-6 to find

$$K(t) = \tfrac{1}{2}mv^2 = \tfrac{1}{2}m\omega^2 x_m^2 \sin^2(\omega t + \phi). \quad (15\text{-}19)$$

If we use Eq. 15-12 to substitute k/m for ω^2, we can write Eq. 15-19 as

$$K(t) = \tfrac{1}{2}mv^2 = \tfrac{1}{2}kx_m^2 \sin^2(\omega t + \phi). \quad (15\text{-}20)$$

The mechanical energy follows from Eqs. 15-18 and 15-20 and is

$$\begin{aligned} E &= U + K \\ &= \tfrac{1}{2}kx_m^2 \cos^2(\omega t + \phi) + \tfrac{1}{2}kx_m^2 \sin^2(\omega t + \phi) \\ &= \tfrac{1}{2}kx_m^2 [\cos^2(\omega t + \phi) + \sin^2(\omega t + \phi)]. \end{aligned}$$

For any angle α,

$$\cos^2 \alpha + \sin^2 \alpha = 1.$$

Thus, the quantity in the square brackets above is unity and we have

$$E = U + K = \tfrac{1}{2}kx_m^2. \quad (15\text{-}21)$$

The mechanical energy of a linear oscillator is indeed constant and independent of time. The potential energy and kinetic energy of a linear oscillator are shown as functions of time t in Fig. 15-6*a*, and they are shown as functions of displacement x in Fig. 15-6*b*.

You might now understand why an oscillating system normally contains an element of springiness and an element of inertia: The former stores its potential energy and the latter stores its kinetic energy.

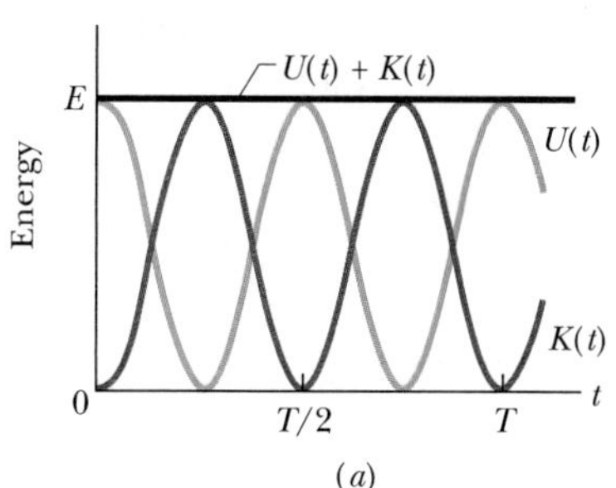

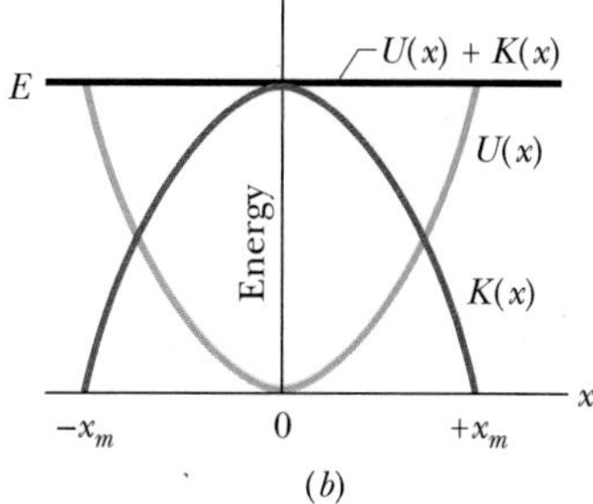

FIG. 15-6 (*a*) Potential energy $U(t)$, kinetic energy $K(t)$, and mechanical energy E as functions of time t for a linear harmonic oscillator. Note that all energies are positive and that the potential energy and the kinetic energy peak twice during every period. (*b*) Potential energy $U(x)$, kinetic energy $K(x)$, and mechanical energy E as functions of position x for a linear harmonic oscillator with amplitude x_m. For $x = 0$ the energy is all kinetic, and for $x = \pm x_m$ it is all potential.

CHECKPOINT 3 In Fig. 15-5, the block has a kinetic energy of 3 J and the spring has an elastic potential energy of 2 J when the block is at $x = +2.0$ cm. (a) What is the kinetic energy when the block is at $x = 0$? What is the elastic potential energy when the block is at (b) $x = -2.0$ cm and (c) $x = -x_m$?

Sample Problem 15-3

The huge ball that appears in this chapter's opening photograph hangs from four cables and swings like a pendulum when the building sways in the wind. When the building sways—say, eastward—the massive pendulum does also but delayed enough so that as it finally swings eastward, the building is swaying westward. Thus, the pendulum's motion is out of step with the building's motion, tending to counter it.

Many other buildings have other types of *mass dampers*, as these anti-sway devices are called. Some, like the John Hancock building in Boston, have a large block oscillating at the end of a spring and on a lubricated track. The principle is the same as with the pendulum: The motion of the oscillator is out of step with the motion of the building.

Suppose the block has mass $m = 2.72 \times 10^5$ kg and is designed to oscillate at frequency $f = 10.0$ Hz and with amplitude $x_m = 20.0$ cm.

(a) What is the total mechanical energy E of the spring–block system?

KEY IDEA The mechanical energy E (the sum of the kinetic energy $K = \tfrac{1}{2}mv^2$ of the block and the potential energy $U = \tfrac{1}{2}kx^2$ of the spring) is constant throughout the motion of the oscillator. Thus, we can evaluate E at any point during the motion.

Calculations: Because we are given amplitude x_m of the oscillations, let's evaluate E when the block is at position $x = x_m$, where it has velocity $v = 0$. However, to evaluate U at that point, we first need to find the spring constant k. From Eq. 15-12 ($\omega = \sqrt{k/m}$) and Eq. 15-5 ($\omega = 2\pi f$), we find

$$\begin{aligned} k &= m\omega^2 = m(2\pi f)^2 \\ &= (2.72 \times 10^5 \text{ kg})(2\pi)^2(10.0 \text{ Hz})^2 \\ &= 1.073 \times 10^9 \text{ N/m}. \end{aligned}$$

We can now evaluate E as

$$\begin{aligned} E &= K + U = \tfrac{1}{2}mv^2 + \tfrac{1}{2}kx^2 \\ &= 0 + \tfrac{1}{2}(1.073 \times 10^9 \text{ N/m})(0.20 \text{ m})^2 \\ &= 2.147 \times 10^7 \text{ J} \approx 2.1 \times 10^7 \text{ J}. \qquad \text{(Answer)} \end{aligned}$$

(b) What is the block's speed as it passes through the equilibrium point?

Calculations: We want the speed at $x = 0$, where the potential energy is $U = \frac{1}{2}kx^2 = 0$ and the mechanical energy is entirely kinetic energy. So, we can write

$$E = K + U = \tfrac{1}{2}mv^2 + \tfrac{1}{2}kx^2$$

$$2.147 \times 10^7 \text{ J} = \tfrac{1}{2}(2.72 \times 10^5 \text{ kg})v^2 + 0,$$

or

$$v = 12.6 \text{ m/s}. \qquad \text{(Answer)}$$

Because E is entirely kinetic energy, this is the maximum speed v_m.

15-5 | An Angular Simple Harmonic Oscillator

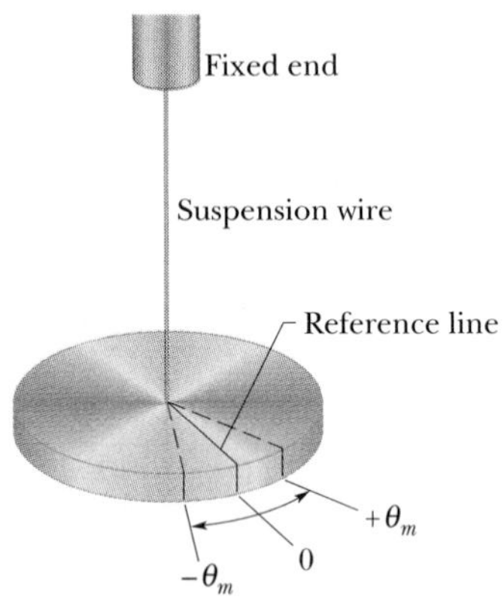

FIG. 15-7 A torsion pendulum is an angular version of a linear simple harmonic oscillator. The disk oscillates in a horizontal plane; the reference line oscillates with angular amplitude θ_m. The twist in the suspension wire stores potential energy as a spring does and provides the restoring torque.

Figure 15-7 shows an angular version of a simple harmonic oscillator; the element of springiness or elasticity is associated with the twisting of a suspension wire rather than the extension and compression of a spring as we previously had. The device is called a **torsion pendulum,** with *torsion* referring to the twisting.

If we rotate the disk in Fig. 15-7 by some angular displacement θ from its rest position (where the reference line is at $\theta = 0$) and release it, it will oscillate about that position in **angular simple harmonic motion.** Rotating the disk through an angle θ in either direction introduces a restoring torque given by

$$\tau = -\kappa\theta. \qquad (15\text{-}22)$$

Here κ (Greek *kappa*) is a constant, called the **torsion constant,** that depends on the length, diameter, and material of the suspension wire.

Comparison of Eq. 15-22 with Eq. 15-10 leads us to suspect that Eq. 15-22 is the angular form of Hooke's law, and that we can transform Eq. 15-13, which gives the period of linear SHM, into an equation for the period of angular SHM: We replace the spring constant k in Eq. 15-13 with its equivalent, the constant κ of Eq. 15-22, and we replace the mass m in Eq. 15-13 with *its* equivalent, the rotational inertia I of the oscillating disk. These replacements lead to

$$T = 2\pi\sqrt{\frac{I}{\kappa}} \quad \text{(torsion pendulum)}, \qquad (15\text{-}23)$$

which is the correct equation for the period of an angular simple harmonic oscillator, or torsion pendulum.

PROBLEM-SOLVING TACTICS

Tactic 3: Identifying Angular SHM When a system undergoes angular simple harmonic motion, its angular acceleration α and angular displacement θ are related by an equation of the form

$$\alpha = -(\text{a positive constant})\theta.$$

This equation is the angular equivalent of Eq. 15-8 ($a = -\omega^2 x$). It says that the angular acceleration α is proportional to the angular displacement θ from the equilibrium position but is in the direction opposite the displacement. If you have an expression of this form, you can identify the positive constant as being ω^2, and then you can determine ω, f, and T.

You can also identify angular SHM if you have an expression for the torque τ in terms of the angular displacement, because that expression must be in the form of Eq. 15-22 ($\tau = -\kappa\theta$) or

$$\tau = -(\text{a positive constant})\theta.$$

This equation is the angular equivalent of Eq. 15-10 ($F = -kx$). It says that the torque τ is proportional to the angular displacement θ from the equilibrium position but tends to rotate the system in the opposite direction. If you have an expression of this form, then you can identify the positive constant as being the system's torsion constant κ. If you know the rotational inertia I of the system, you can then determine T.

Sample Problem 15-4

Figure 15-8*a* shows a thin rod whose length L is 12.4 cm and whose mass m is 135 g, suspended at its midpoint from a long wire. Its period T_a of angular SHM is measured to be 2.53 s. An irregularly shaped object, which we call object X, is then hung from the same wire, as in Fig. 15-8*b*, and its period T_b is found to be 4.76 s. What is the rotational inertia of object X about its suspension axis?

KEY IDEA The rotational inertia of either the rod or object X is related to the measured period by Eq. 15-23.

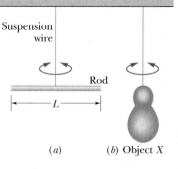

FIG. 15-8 Two torsion pendulums, consisting of (*a*) a wire and a rod and (*b*) the same wire and an irregularly shaped object.

Calculations: In Table 10-2*e*, the rotational inertia of a thin rod about a perpendicular axis through its midpoint is given as $\frac{1}{12}mL^2$. Thus, we have, for the rod in Fig. 15-8*a*,

$$I_a = \tfrac{1}{12}mL^2 = (\tfrac{1}{12})(0.135 \text{ kg})(0.124 \text{ m})^2$$
$$= 1.73 \times 10^{-4} \text{ kg}\cdot\text{m}^2.$$

Now let us write Eq. 15-23 twice, once for the rod and once for object X:

$$T_a = 2\pi\sqrt{\frac{I_a}{\kappa}} \quad \text{and} \quad T_b = 2\pi\sqrt{\frac{I_b}{\kappa}}.$$

The constant κ, which is a property of the wire, is the same for both figures; only the periods and the rotational inertias differ.

Let us square each of these equations, divide the second by the first, and solve the resulting equation for I_b. The result is

$$I_b = I_a\frac{T_b^2}{T_a^2} = (1.73 \times 10^{-4} \text{ kg}\cdot\text{m}^2)\frac{(4.76 \text{ s})^2}{(2.53 \text{ s})^2}$$
$$= 6.12 \times 10^{-4} \text{ kg}\cdot\text{m}^2. \qquad \text{(Answer)}$$

15-6 | Pendulums

We turn now to a class of simple harmonic oscillators in which the springiness is associated with the gravitational force rather than with the elastic properties of a twisted wire or a compressed or stretched spring.

The Simple Pendulum

If an apple swings on a long thread, does it have simple harmonic motion? If so, what is the period T? To answer, we consider a **simple pendulum,** which consists of a particle of mass m (called the *bob* of the pendulum) suspended from one end of an unstretchable, massless string of length L that is fixed at the other end, as in Fig. 15-9*a*. The bob is free to swing back and forth in the plane of the page, to the left and right of a vertical line through the pendulum's pivot point.

The forces acting on the bob are the force $\vec{T}$ from the string and the gravitational force $\vec{F}_g$, as shown in Fig. 15-9*b*, where the string makes an angle θ with the vertical. We resolve $\vec{F}_g$ into a radial component $F_g \cos\theta$ and a component $F_g \sin\theta$ that is tangent to the path taken by the bob. This tangential component produces a restoring torque about the pendulum's pivot point because the component always acts opposite the displacement of the bob so as to bring the bob back toward its central location. That location is called the *equilibrium position* ($\theta = 0$) because the pendulum would be at rest there were it not swinging.

From Eq. 10-41 ($\tau = r_\perp F$), we can write this restoring torque as

$$\tau = -L(F_g \sin\theta), \tag{15-24}$$

where the minus sign indicates that the torque acts to reduce θ and L is the moment arm of the force component $F_g \sin\theta$ about the pivot point. Substituting Eq. 15-24 into Eq. 10-44 ($\tau = I\alpha$) and then substituting mg as the magnitude of F_g, we obtain

$$-L(mg\sin\theta) = I\alpha, \tag{15-25}$$

where I is the pendulum's rotational inertia about the pivot point and α is its angular acceleration about that point.

We can simplify Eq. 15-25 if we assume the angle θ is small, for then we can approximate $\sin\theta$ with θ (expressed in radian measure). (As an example, if $\theta = 5.00° = 0.0873$ rad, then $\sin\theta = 0.0872$, a difference of only about 0.1%.) With that approximation and some rearranging, we then have

$$\alpha = -\frac{mgL}{I}\theta. \tag{15-26}$$

This equation is the angular equivalent of Eq. 15-8, the hallmark of SHM. It tells us

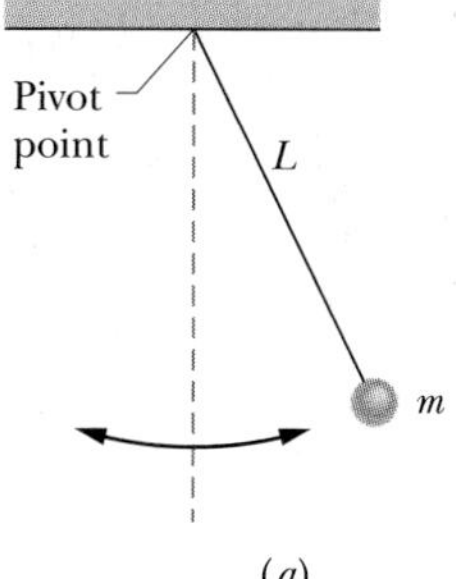

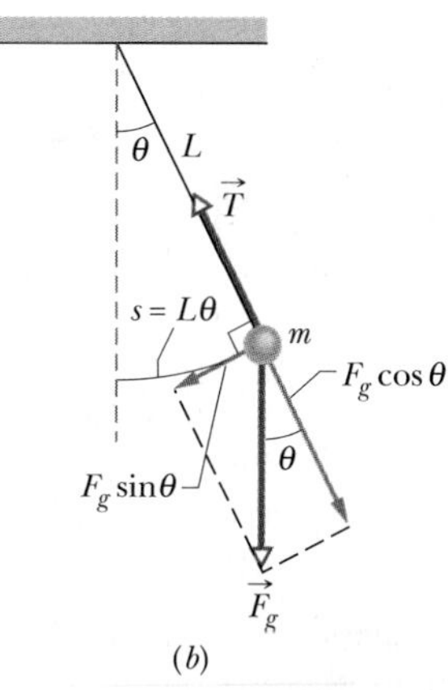

FIG. 15-9 (*a*) A simple pendulum. (*b*) The forces acting on the bob are the gravitational force $\vec{F}_g$ and the force $\vec{T}$ from the string. The tangential component $F_g \sin\theta$ of the gravitational force is a restoring force that tends to bring the pendulum back to its central position.

that the angular acceleration α of the pendulum is proportional to the angular displacement θ but opposite in sign. Thus, as the pendulum bob moves to the right, as in Fig. 15-9*a*, its acceleration *to the left* increases until the bob stops and begins moving to the left. Then, when it is to the left of the equilibrium position, its acceleration to the right tends to return it to the right, and so on, as it swings back and forth in SHM. More precisely, the motion of a *simple pendulum swinging through only small angles* is approximately SHM. We can state this restriction to small angles another way: The **angular amplitude** θ_m of the motion (the maximum angle of swing) must be small.

Comparing Eqs. 15-26 and 15-8, we see that the angular frequency of the pendulum is $\omega = \sqrt{mgL/I}$. Next, if we substitute this expression for ω into Eq. 15-5 ($\omega = 2\pi/T$), we see that the period of the pendulum may be written as

$$T = 2\pi\sqrt{\frac{I}{mgL}}. \tag{15-27}$$

All the mass of a simple pendulum is concentrated in the mass m of the particle-like bob, which is at radius L from the pivot point. Thus, we can use Eq. 10-33 ($I = mr^2$) to write $I = mL^2$ for the rotational inertia of the pendulum. Substituting this into Eq. 15-27 and simplifying then yield

$$T = 2\pi\sqrt{\frac{L}{g}} \quad \text{(simple pendulum, small amplitude).} \tag{15-28}$$

We assume small-angle swinging in this chapter.

The Physical Pendulum

A real pendulum, usually called a **physical pendulum,** can have a complicated distribution of mass, much different from that of a simple pendulum. Does a physical pendulum also undergo SHM? If so, what is its period?

Figure 15-10 shows an arbitrary physical pendulum displaced to one side by angle θ. The gravitational force $\vec{F}_g$ acts at its center of mass C, at a distance h from the pivot point O. Comparison of Figs. 15-10 and 15-9*b* reveals only one important difference between an arbitrary physical pendulum and a simple pendulum. For a physical pendulum the restoring component $F_g \sin\theta$ of the gravitational force has a moment arm of distance h about the pivot point, rather than of string length L. In all other respects, an analysis of the physical pendulum would duplicate our analysis of the simple pendulum up through Eq. 15-27. Again (for small θ_m), we would find that the motion is approximately SHM.

If we replace L with h in Eq. 15-27, we can write the period as

$$T = 2\pi\sqrt{\frac{I}{mgh}} \quad \text{(physical pendulum, small amplitude).} \tag{15-29}$$

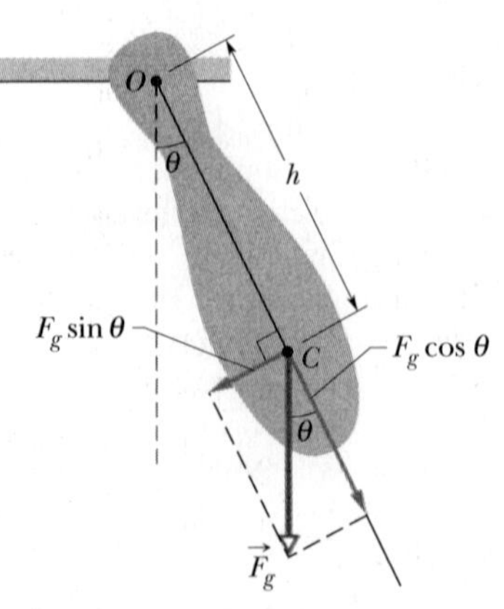

FIG. 15-10 A physical pendulum. The restoring torque is $hF_g \sin\theta$. When $\theta = 0$, center of mass C hangs directly below pivot point O.

As with the simple pendulum, I is the rotational inertia of the pendulum about O. However, now I is not simply mL^2 (it depends on the shape of the physical pendulum), but it is still proportional to m.

A physical pendulum will not swing if it pivots at its center of mass. Formally, this corresponds to putting $h = 0$ in Eq. 15-29. That equation then predicts $T \to \infty$, which implies that such a pendulum will never complete one swing.

Corresponding to any physical pendulum that oscillates about a given pivot point O with period T is a simple pendulum of length L_0 with the same period T. We can find L_0 with Eq. 15-28. The point along the physical pendulum at distance L_0 from point O is called the *center of oscillation* of the physical pendulum for the given suspension point.

Measuring g

We can use a physical pendulum to measure the free-fall acceleration g at a particular location on Earth's surface. (Countless thousands of such measurements have been made during geophysical prospecting.)

To analyze a simple case, take the pendulum to be a uniform rod of length L, suspended from one end. For such a pendulum, h in Eq. 15-29, the distance between the pivot point and the center of mass, is $\frac{1}{2}L$. Table 10-2e tells us that the rotational inertia of this pendulum about a perpendicular axis through its center of mass is $\frac{1}{12}mL^2$. From the parallel-axis theorem of Eq. 10-36 ($I = I_{com} + Mh^2$), we then find that the rotational inertia about a perpendicular axis through one end of the rod is

$$I = I_{com} + mh^2 = \tfrac{1}{12}mL^2 + m(\tfrac{1}{2}L)^2 = \tfrac{1}{3}mL^2. \qquad (15\text{-}30)$$

If we put $h = \frac{1}{2}L$ and $I = \frac{1}{3}mL^2$ in Eq. 15-29 and solve for g, we find

$$g = \frac{8\pi^2 L}{3T^2}. \qquad (15\text{-}31)$$

Thus, by measuring L and the period T, we can find the value of g at the pendulum's location. (If precise measurements are to be made, a number of refinements are needed, such as swinging the pendulum in an evacuated chamber.)

CHECKPOINT 4 Three physical pendulums, of masses m_0, $2m_0$, and $3m_0$, have the same shape and size and are suspended at the same point. Rank the masses according to the periods of the pendulums, greatest first.

Sample Problem 15-5

In Fig. 15-11a, a meter stick swings about a pivot point at one end, at distance h from the stick's center of mass.

(a) What is the period of oscillation T?

KEY IDEA The stick is not a simple pendulum because its mass is not concentrated in a bob at the end opposite the pivot point—so the stick is a physical pendulum.

Calculations: The period for a physical pendulum is given by Eq. 15-29, for which we need the rotational inertia I of the stick about the pivot point. We can treat the stick as a uniform rod of length L and mass m. Then Eq. 15-30 tells us that $I = \frac{1}{3}mL^2$, and the distance h in Eq. 15-29 is $\frac{1}{2}L$. Substituting these quantities into Eq. 15-29, we find

$$T = 2\pi\sqrt{\frac{I}{mgh}} = 2\pi\sqrt{\frac{\frac{1}{3}mL^2}{mg(\frac{1}{2}L)}} = 2\pi\sqrt{\frac{2L}{3g}} \qquad (15\text{-}32)$$

$$= 2\pi\sqrt{\frac{(2)(1.00\text{ m})}{(3)(9.8\text{ m/s}^2)}} = 1.64\text{ s}. \qquad \text{(Answer)}$$

Note the result is independent of the pendulum's mass m.

(b) What is the distance L_0 between the pivot point O of the stick and the center of oscillation of the stick?

Calculations: We want the length L_0 of the simple pendulum (drawn in Fig. 15-11b) that has the same period as the physical pendulum (the stick) of Fig. 15-11a.

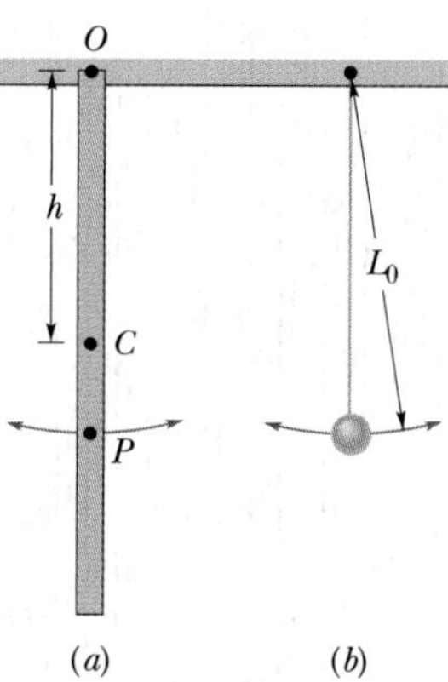

FIG. 15-11 (a) A meter stick suspended from one end as a physical pendulum. (b) A simple pendulum whose length L_0 is chosen so that the periods of the two pendulums are equal. Point P on the pendulum of (a) marks the center of oscillation.

Setting Eqs. 15-28 and 15-32 equal yields

$$T = 2\pi\sqrt{\frac{L_0}{g}} = 2\pi\sqrt{\frac{2L}{3g}}.$$

You can see by inspection that

$$L_0 = \tfrac{2}{3}L = (\tfrac{2}{3})(100\text{ cm}) = 66.7\text{ cm}. \qquad \text{(Answer)}$$

In Fig. 15-11a, point P marks this distance from suspension point O. Thus, point P is the stick's center of oscillation for the given suspension point.

Sample Problem 15-6 Build your skill

A competition diving board sits on a fulcrum about one-third of the way out from the fixed end of the board (Fig. 15-12*a*). In a running dive, a diver takes three quick steps along the board, out past the fulcrum so as to rotate the board's free end downward. As the board rebounds back through the horizontal, the diver leaps upward and toward the board's free end (Fig. 15-12*b*). A skilled diver trains to land on the free end just as the board has completed 2.5 oscillations during the leap. With such timing, the diver lands as the free end is moving downward with greatest speed (Fig. 15-12*c*). The landing then drives the free end down substantially, and the rebound catapults the diver high into the air.

Figure 15-12*d* shows a simple but realistic model of a competition board. The board section beyond the fulcrum is treated as a stiff rod of length L that can rotate about a hinge at the fulcrum, compressing an (imaginary) spring under the board's free end. If the rod's mass is $m = 20.0$ kg and the diver's leap lasts $t_{fl} = 0.620$ s, what spring constant k is required of the spring for a proper landing?

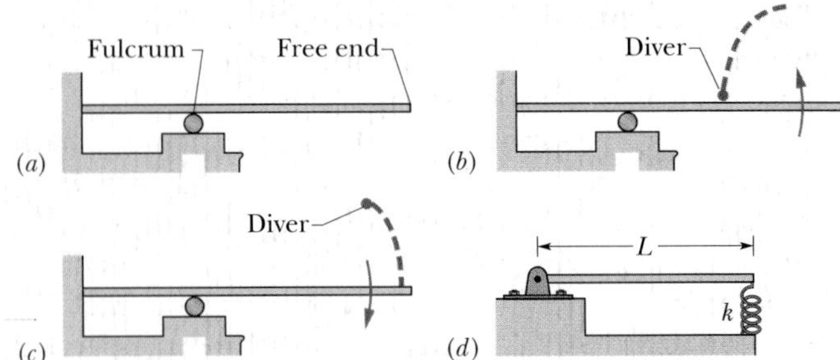

FIG. 15-12 (*a*) A diving board. (*b*) The diver leaps upward and forward as the board moves through the horizontal. (*c*) The diver lands 2.5 oscillations later. (*d*) A spring-oscillator model of the oscillating board.

KEY IDEA If the rod is in SHM, then the acceleration and displacement of the oscillating end of the rod must be related by an expression in the form of Eq. 15-8 ($a = -\omega^2 x$). If so, we shall be able to find ω and then the desired k from the expression.

Torque and force: Because the rod rotates about the hinge as the free end oscillates, we are concerned with a torque $\vec{\tau}$ on the rod about the hinge. That torque is due to the force $\vec{F}$ on the rod from the spring. Because $\vec{F}$ varies with time, $\vec{\tau}$ must also. However, at any given instant we can relate the magnitudes of $\vec{\tau}$ and $\vec{F}$ with Eq. 10-39 ($\tau = rF\sin\phi$). Here we have

$$\tau = LF\sin 90°, \tag{15-33}$$

where L is the moment arm of force $\vec{F}$ and 90° is the angle between the moment arm and the force's line of action. Combining Eq. 15-33 with Eq. 10-44 ($\tau = I\alpha$) gives us

$$I\alpha = LF, \tag{15-34}$$

where I is the rod's rotational inertia about the hinge and α is its angular acceleration about that point. From Eq. 15-30, the rod's rotational inertia I is $\frac{1}{3}mL^2$.

Now let us mentally erect a vertical x axis through the oscillating right end of the rod, with the positive direction upward. Then the force on the right end of the rod from the spring is $F = -kx$, where x is the vertical displacement of the right end.

Substituting these expressions for I and F into Eq. 15-34 gives us

$$\frac{mL^2\alpha}{3} = -Lkx. \tag{15-35}$$

Mixture: We now have a mixture of linear displacement x (vertically) and rotational acceleration α (about the hinge). We can replace α in Eq. 15-35 with the (linear) acceleration a along the x axis by substituting according to Eq. 10-22 ($a_t = \alpha r$) for tangential acceleration. Here the tangential acceleration is a and the radius of rotation r is L, so $\alpha = a/L$. With that substitution, Eq. 15-35 becomes

$$\frac{mL^2a}{3L} = -Lkx,$$

which yields

$$a = -\frac{3k}{m}x. \tag{15-36}$$

Equation 15-36 is, in fact, of the same form as Eq. 15-8 ($a = -\omega^2 x$). Therefore, the rod does indeed undergo SHM, and comparison of Eqs. 15-36 and 15-8 shows that

$$\omega^2 = \frac{3k}{m}.$$

Solving for k and substituting for ω from Eq. 15-5 ($\omega = 2\pi/T$) give us

$$k = \frac{m}{3}\left(\frac{2\pi}{T}\right)^2, \tag{15-37}$$

where T is the period of the board's oscillation. We want the time of flight t_{fl} to last for 2.5 oscillations of the board and thus also for 2.5 oscillations of our rod. Thus we want $t_{fl} = 2.5T$. Substituting this and given data into Eq. 15-37 leads to

$$k = \frac{m}{3}\left(\frac{2\pi}{t_{fl}}2.5\right)^2 \tag{15-38}$$

$$= \frac{(20.0\text{ kg})}{3}\left(\frac{2\pi}{0.620\text{ s}}2.5\right)^2$$

$$= 4.28 \times 10^3\text{ N/m}. \qquad \text{(Answer)}$$

This is the effective spring constant k of the diving board.

Diving skills: From Eq. 15-38, we see that a diver with a longer flight time t_{fl} requires a smaller spring constant k in order to land at the proper instant at the end of the board. The value of k can be increased by moving the fulcrum toward the free end or decreased by moving it in the opposite direction. A skilled diver trains to leap with a certain flight time t_{fl} and knows how to set the fulcrum position accordingly.

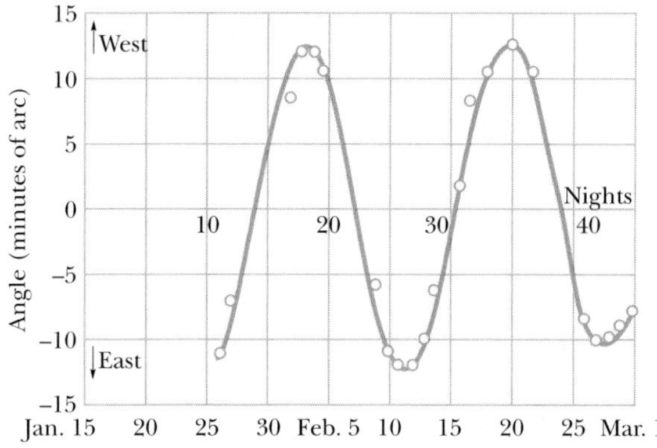

FIG. 15-13 The angle between Jupiter and its moon Callisto as seen from Earth. The circles are based on Galileo's 1610 measurements. The curve is a best fit, strongly suggesting simple harmonic motion. At Jupiter's mean distance from Earth, 10 minutes of arc corresponds to about 2×10^6 km. (Adapted from A. P. French, *Newtonian Mechanics*, W. W. Norton & Company, New York, 1971, p. 288.)

15-7 | Simple Harmonic Motion and Uniform Circular Motion

In 1610, Galileo, using his newly constructed telescope, discovered the four principal moons of Jupiter. Over weeks of observation, each moon seemed to him to be moving back and forth relative to the planet in what today we would call simple harmonic motion; the disk of the planet was the midpoint of the motion. The record of Galileo's observations, written in his own hand, is still available. A. P. French of MIT used Galileo's data to work out the position of the moon Callisto relative to Jupiter. In the results shown in Fig. 15-13, the circles are based on Galileo's observations and the curve is a best fit to the data. The curve strongly suggests Eq. 15-3, the displacement function for SHM. A period of about 16.8 days can be measured from the plot.

Actually, Callisto moves with essentially constant speed in an essentially circular orbit around Jupiter. Its true motion—far from being simple harmonic—is uniform circular motion. What Galileo saw—and what you can see with a good pair of binoculars and a little patience—is the projection of this uniform circular motion on a line in the plane of the motion. We are led by Galileo's remarkable observations to the conclusion that simple harmonic motion is uniform circular motion viewed edge-on. In more formal language:

> Simple harmonic motion is the projection of uniform circular motion on a diameter of the circle in which the circular motion occurs.

Figure 15-14*a* gives an example. It shows a *reference particle* P' moving in uniform circular motion with (constant) angular speed ω in a *reference circle.* The radius x_m of the circle is the magnitude of the particle's position vector. At any time t, the angular position of the particle is $\omega t + \phi$, where ϕ is its angular position at $t = 0$.

The projection of particle P' onto the x axis is a point P, which we take to be a second particle. The projection of the position vector of particle P' onto the x axis gives the location $x(t)$ of P. Thus, we find

$$x(t) = x_m \cos(\omega t + \phi),$$

which is precisely Eq. 15-3. Our conclusion is correct. If reference particle P' moves in uniform circular motion, its projection particle P moves in simple harmonic motion along a diameter of the circle.

Figure 15-14*b* shows the velocity $\vec{v}$ of the reference particle. From Eq. 10-18 ($v = \omega r$), the magnitude of the velocity vector is ωx_m; its projection on the x axis is

$$v(t) = -\omega x_m \sin(\omega t + \phi),$$

which is exactly Eq. 15-6. The minus sign appears because the velocity component of P in Fig. 15-14*b* is directed to the left, in the negative direction of x.

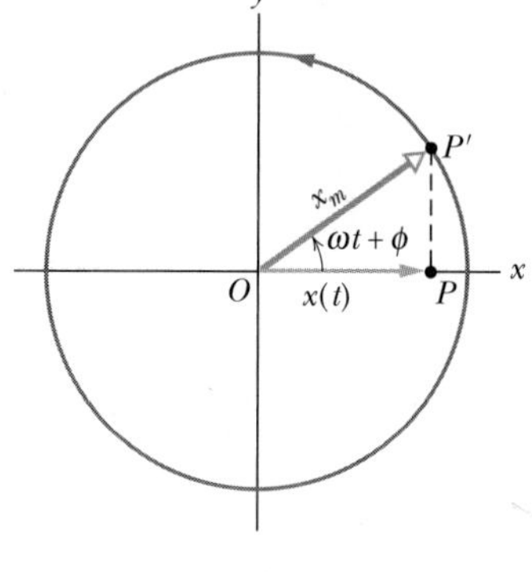

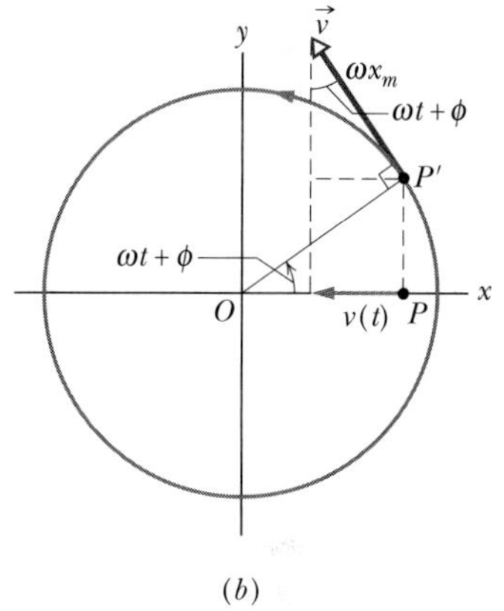

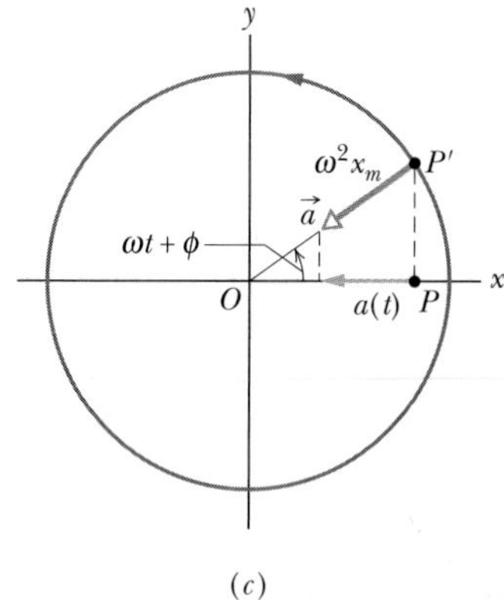

FIG. 15-14 (*a*) A reference particle P' moving with uniform circular motion in a reference circle of radius x_m. Its projection P on the x axis executes simple harmonic motion. (*b*) The projection of the velocity $\vec{v}$ of the reference particle is the velocity of SHM. (*c*) The projection of the radial acceleration $\vec{a}$ of the reference particle is the acceleration of SHM.

Figure 15-14c shows the radial acceleration $\vec{a}$ of the reference particle. From Eq. 10-23 ($a_r = \omega^2 r$), the magnitude of the radial acceleration vector is $\omega^2 x_m$; its projection on the x axis is

$$a(t) = -\omega^2 x_m \cos(\omega t + \phi),$$

which is exactly Eq. 15-7. Thus, whether we look at the displacement, the velocity, or the acceleration, the projection of uniform circular motion is indeed simple harmonic motion.

15-8 | Damped Simple Harmonic Motion

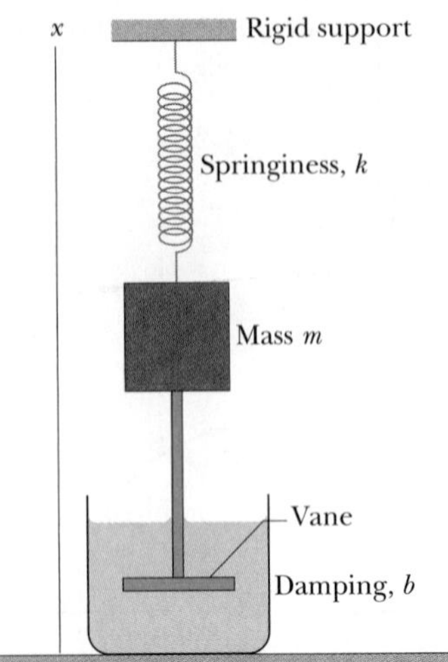

FIG. 15-15 An idealized damped simple harmonic oscillator. A vane immersed in a liquid exerts a damping force on the block as the block oscillates parallel to the x axis.

A pendulum will swing only briefly underwater, because the water exerts on the pendulum a drag force that quickly eliminates the motion. A pendulum swinging in air does better, but still the motion dies out eventually, because the air exerts a drag force on the pendulum (and friction acts at its support point), transferring energy from the pendulum's motion.

When the motion of an oscillator is reduced by an external force, the oscillator and its motion are said to be **damped.** An idealized example of a damped oscillator is shown in Fig. 15-15, where a block with mass m oscillates vertically on a spring with spring constant k. From the block, a rod extends to a vane (both assumed massless) that is submerged in a liquid. As the vane moves up and down, the liquid exerts an inhibiting drag force on it and thus on the entire oscillating system. With time, the mechanical energy of the block–spring system decreases, as energy is transferred to thermal energy of the liquid and vane.

Let us assume the liquid exerts a **damping force** $\vec{F}_d$ that is proportional to the velocity $\vec{v}$ of the vane and block (an assumption that is accurate if the vane moves slowly). Then, for components along the x axis in Fig. 15-15, we have

$$F_d = -bv, \tag{15-39}$$

where b is a **damping constant** that depends on the characteristics of both the vane and the liquid and has the SI unit of kilogram per second. The minus sign indicates that $\vec{F}_d$ opposes the motion.

The force on the block from the spring is $F_s = -kx$. Let us assume that the gravitational force on the block is negligible relative to F_d and F_s. Then we can write Newton's second law for components along the x axis ($F_{\text{net},x} = ma_x$) as

$$-bv - kx = ma. \tag{15-40}$$

Substituting dx/dt for v and d^2x/dt^2 for a and rearranging give us the differential equation

$$m\frac{d^2x}{dt^2} + b\frac{dx}{dt} + kx = 0. \tag{15-41}$$

The solution of this equation is

$$x(t) = x_m e^{-bt/2m} \cos(\omega' t + \phi), \tag{15-42}$$

where x_m is the amplitude and ω' is the angular frequency of the damped oscillator. This angular frequency is given by

$$\omega' = \sqrt{\frac{k}{m} - \frac{b^2}{4m^2}}. \tag{15-43}$$

If $b = 0$ (there is no damping), then Eq. 15-43 reduces to Eq. 15-12 ($\omega = \sqrt{k/m}$) for the angular frequency of an undamped oscillator, and Eq. 15-42 reduces to Eq. 15-3 for the displacement of an undamped oscillator. If the damping constant is small but not zero (so that $b \ll \sqrt{km}$), then $\omega' \approx \omega$.

We can regard Eq. 15-42 as a cosine function whose amplitude, which is $x_m e^{-bt/2m}$, gradually decreases with time, as Fig. 15-16 suggests. For an undamped

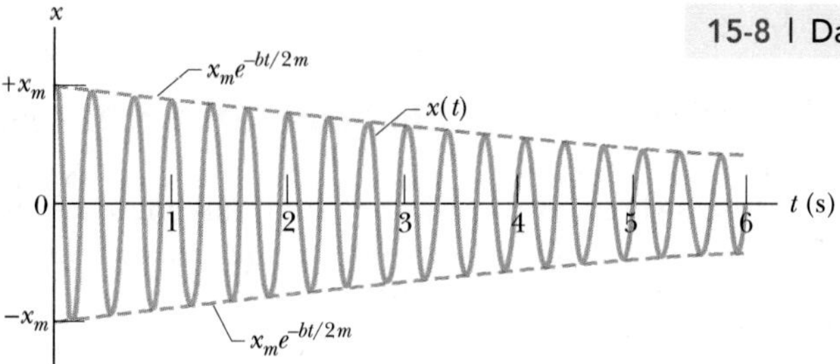

FIG. 15-16 The displacement function $x(t)$ for the damped oscillator of Fig. 15-15, with values given in Sample Problem 15-7. The amplitude, which is $x_m e^{-bt/2m}$, decreases exponentially with time.

oscillator, the mechanical energy is constant and is given by Eq. 15-21 ($E = \frac{1}{2}kx_m^2$). If the oscillator is damped, the mechanical energy is not constant but decreases with time. If the damping is small, we can find $E(t)$ by replacing x_m in Eq. 15-21 with $x_m e^{-bt/2m}$, the amplitude of the damped oscillations. By doing so, we find that

$$E(t) \approx \tfrac{1}{2}kx_m^2 e^{-bt/m}, \tag{15-44}$$

which tells us that, like the amplitude, the mechanical energy decreases exponentially with time.

CHECKPOINT 5 Here are three sets of values for the spring constant, damping constant, and mass for the damped oscillator of Fig. 15-15. Rank the sets according to the time required for the mechanical energy to decrease to one-fourth of its initial value, greatest first.

Set 1	$2k_0$	b_0	m_0
Set 2	k_0	$6b_0$	$4m_0$
Set 3	$3k_0$	$3b_0$	m_0

Sample Problem 15-7

For the damped oscillator of Fig. 15-15, $m = 250$ g, $k = 85$ N/m, and $b = 70$ g/s.

(a) What is the period of the motion?

KEY IDEA Because $b \ll \sqrt{km} = 4.6$ kg/s, the period is approximately that of the undamped oscillator.

Calculation: From Eq. 15-13, we then have

$$T = 2\pi\sqrt{\frac{m}{k}} = 2\pi\sqrt{\frac{0.25\text{ kg}}{85\text{ N/m}}} = 0.34\text{ s.} \quad \text{(Answer)}$$

(b) How long does it take for the amplitude of the damped oscillations to drop to half its initial value?

KEY IDEA The amplitude at time t is displayed in Eq. 15-42 as $x_m e^{-bt/2m}$.

Calculations: The amplitude has the value x_m at $t = 0$. Thus, we must find the value of t for which

$$x_m e^{-bt/2m} = \tfrac{1}{2}x_m.$$

Canceling x_m and taking the natural logarithm of the equation that remains, we have $\ln\frac{1}{2}$ on the right side and

$$\ln(e^{-bt/2m}) = -bt/2m$$

on the left side. Thus,

$$t = \frac{-2m\ln\frac{1}{2}}{b} = \frac{-(2)(0.25\text{ kg})(\ln\frac{1}{2})}{0.070\text{ kg/s}}$$
$$= 5.0\text{ s.} \quad \text{(Answer)}$$

Because $T = 0.34$ s, this is about 15 periods of oscillation.

(c) How long does it take for the mechanical energy to drop to one-half its initial value?

KEY IDEA From Eq. 15-44, the mechanical energy at time t is $\frac{1}{2}kx_m^2 e^{-bt/m}$.

Calculations: The mechanical energy has the value $\frac{1}{2}kx_m^2$ at $t = 0$. Thus, we must find the value of t for which

$$\tfrac{1}{2}kx_m^2 e^{-bt/m} = \tfrac{1}{2}(\tfrac{1}{2}kx_m^2).$$

If we divide both sides of this equation by $\frac{1}{2}kx_m^2$ and solve for t as we did above, we find

$$t = \frac{-m\ln\frac{1}{2}}{b} = \frac{-(0.25\text{ kg})(\ln\frac{1}{2})}{0.070\text{ kg/s}} = 2.5\text{ s.} \quad \text{(Answer)}$$

This is exactly half the time we calculated in (b), or about 7.5 periods of oscillation. Figure 15-16 was drawn to illustrate this sample problem.

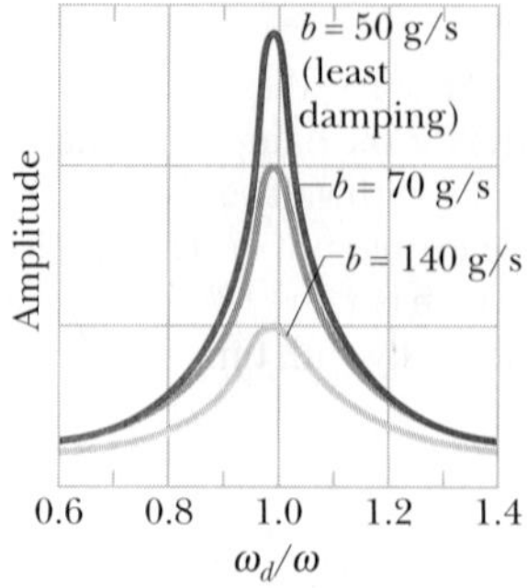

FIG. 15-17 The displacement amplitude x_m of a forced oscillator varies as the angular frequency ω_d of the driving force is varied. The curves here correspond to three values of the damping constant b.

15-9 | Forced Oscillations and Resonance

A person swinging in a swing without anyone pushing it is an example of *free oscillation.* However, if someone pushes the swing periodically, the swing has *forced,* or *driven, oscillations. Two* angular frequencies are associated with a system undergoing driven oscillations: (1) the *natural* angular frequency ω of the system, which is the angular frequency at which it would oscillate if it were suddenly disturbed and then left to oscillate freely, and (2) the angular frequency ω_d of the external driving force causing the driven oscillations.

We can use Fig. 15-15 to represent an idealized forced simple harmonic oscillator if we allow the structure marked "rigid support" to move up and down at a variable angular frequency ω_d. Such a forced oscillator oscillates at the angular frequency ω_d of the driving force, and its displacement $x(t)$ is given by

$$x(t) = x_m \cos(\omega_d t + \phi), \tag{15-45}$$

where x_m is the amplitude of the oscillations.

How large the displacement amplitude x_m is depends on a complicated function of ω_d and ω. The velocity amplitude v_m of the oscillations is easier to describe: it is greatest when

$$\omega_d = \omega \quad \text{(resonance)}, \tag{15-46}$$

a condition called **resonance.** Equation 15-46 is also *approximately* the condition at which the displacement amplitude x_m of the oscillations is greatest. Thus, if you push a swing at its natural angular frequency, the displacement and velocity amplitudes will increase to large values, a fact that children learn quickly by trial and error. If you push at other angular frequencies, either higher or lower, the displacement and velocity amplitudes will be smaller.

Figure 15-17 shows how the displacement amplitude of an oscillator depends on the angular frequency ω_d of the driving force, for three values of the damping coefficient b. Note that for all three the amplitude is approximately greatest when $\omega_d/\omega = 1$ (the resonance condition of Eq. 15-46). The curves of Fig. 15-17 show that less damping gives a taller and narrower *resonance peak.*

All mechanical structures have one or more natural angular frequencies, and if a structure is subjected to a strong external driving force that matches one of these angular frequencies, the resulting oscillations of the structure may rupture it. Thus, for example, aircraft designers must make sure that none of the natural angular frequencies at which a wing can oscillate matches the angular frequency of the engines in flight. A wing that flaps violently at certain engine speeds would obviously be dangerous.

Resonance appears to be one reason buildings in Mexico City collapsed in September 1985 when a major earthquake (8.1 on the Richter scale) occurred on the western coast of Mexico. The seismic waves from the earthquake should have been too weak to cause extensive damage when they reached Mexico City about 400 km away. However, Mexico City is largely built on an ancient lake bed, where the soil is still soft with water. Although the amplitude of the seismic waves was small in the firmer ground en route to Mexico City, their amplitude substantially increased in the loose soil of the city. Acceleration amplitudes of the waves were as much as 0.20g, and the angular frequency was (surprisingly) concentrated around 3 rad/s. Not only was the ground severely oscillated, but many intermediate-height buildings had resonant angular frequencies of about 3 rad/s. Most of those buildings collapsed during the shaking (Fig. 15-18), while shorter buildings (with higher resonant angular frequencies) and taller buildings (with lower resonant angular frequencies) remained standing.

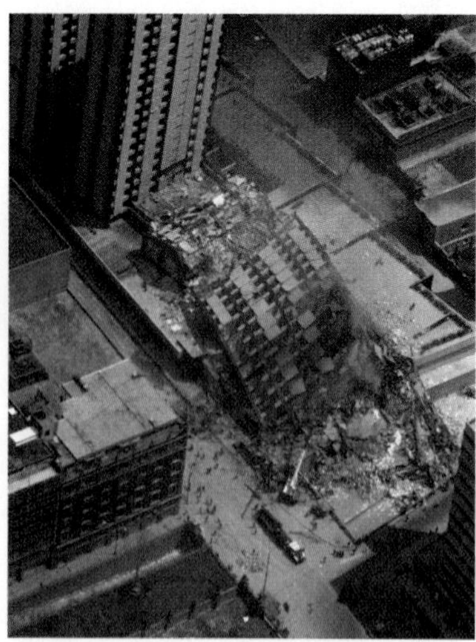

FIG. 15-18 In 1985, buildings of intermediate height collapsed in Mexico City as a result of an earthquake far from the city. Taller and shorter buildings remained standing. *(John T. Barr/Getty Images News and Sport Services)*

REVIEW & SUMMARY

Frequency The *frequency* f of periodic, or oscillatory, motion is the number of oscillations per second. In the SI system, it is measured in hertz:

$$1 \text{ hertz} = 1 \text{ Hz} = 1 \text{ oscillation per second} = 1 \text{ s}^{-1}. \quad (15\text{-}1)$$

Period The *period* T is the time required for one complete oscillation, or **cycle.** It is related to the frequency by

$$T = \frac{1}{f}. \quad (15\text{-}2)$$

Simple Harmonic Motion In *simple harmonic motion* (SHM), the displacement $x(t)$ of a particle from its equilibrium position is described by the equation

$$x = x_m \cos(\omega t + \phi) \quad \text{(displacement)}, \quad (15\text{-}3)$$

in which x_m is the **amplitude** of the displacement, the quantity $(\omega t + \phi)$ is the **phase** of the motion, and ϕ is the **phase constant.** The **angular frequency** ω is related to the period and frequency of the motion by

$$\omega = \frac{2\pi}{T} = 2\pi f \quad \text{(angular frequency)}. \quad (15\text{-}5)$$

Differentiating Eq. 15-3 leads to equations for the particle's SHM velocity and acceleration as functions of time:

$$v = -\omega x_m \sin(\omega t + \phi) \quad \text{(velocity)} \quad (15\text{-}6)$$

and

$$a = -\omega^2 x_m \cos(\omega t + \phi) \quad \text{(acceleration)}. \quad (15\text{-}7)$$

In Eq. 15-6, the positive quantity ωx_m is the **velocity amplitude** v_m of the motion. In Eq. 15-7, the positive quantity $\omega^2 x_m$ is the **acceleration amplitude** a_m of the motion.

The Linear Oscillator A particle with mass m that moves under the influence of a Hooke's law restoring force given by $F = -kx$ exhibits simple harmonic motion with

$$\omega = \sqrt{\frac{k}{m}} \quad \text{(angular frequency)} \quad (15\text{-}12)$$

and

$$T = 2\pi\sqrt{\frac{m}{k}} \quad \text{(period)}. \quad (15\text{-}13)$$

Such a system is called a **linear simple harmonic oscillator.**

Energy A particle in simple harmonic motion has, at any time, kinetic energy $K = \frac{1}{2}mv^2$ and potential energy $U = \frac{1}{2}kx^2$. If no friction is present, the mechanical energy $E = K + U$ remains constant even though K and U change.

Pendulums Examples of devices that undergo simple harmonic motion are the **torsion pendulum** of Fig. 15-7, the **simple pendulum** of Fig. 15-9, and the **physical pendulum** of Fig. 15-10. Their periods of oscillation for small oscillations are, respectively,

$$T = 2\pi\sqrt{I/\kappa} \quad \text{(torsion pendulum)}, \quad (15\text{-}23)$$

$$T = 2\pi\sqrt{L/g} \quad \text{(simple pendulum)}, \quad (15\text{-}28)$$

$$T = 2\pi\sqrt{I/mgh} \quad \text{(physical pendulum)}. \quad (15\text{-}29)$$

Simple Harmonic Motion and Uniform Circular Motion Simple harmonic motion is the projection of uniform circular motion onto the diameter of the circle in which the circular motion occurs. Figure 15-14 shows that all parameters of circular motion (position, velocity, and acceleration) project to the corresponding values for simple harmonic motion.

Damped Harmonic Motion The mechanical energy E in a real oscillating system decreases during the oscillations because external forces, such as a drag force, inhibit the oscillations and transfer mechanical energy to thermal energy. The real oscillator and its motion are then said to be **damped.** If the **damping force** is given by $\vec{F}_d = -b\vec{v}$, where $\vec{v}$ is the velocity of the oscillator and b is a **damping constant,** then the displacement of the oscillator is given by

$$x(t) = x_m e^{-bt/2m} \cos(\omega' t + \phi), \quad (15\text{-}42)$$

where ω', the angular frequency of the damped oscillator, is given by

$$\omega' = \sqrt{\frac{k}{m} - \frac{b^2}{4m^2}}. \quad (15\text{-}43)$$

If the damping constant is small ($b \ll \sqrt{km}$), then $\omega' \approx \omega$, where ω is the angular frequency of the undamped oscillator. For small b, the mechanical energy E of the oscillator is given by

$$E(t) \approx \tfrac{1}{2}kx_m^2 e^{-bt/m}. \quad (15\text{-}44)$$

Forced Oscillations and Resonance If an external driving force with angular frequency ω_d acts on an oscillating system with *natural* angular frequency ω, the system oscillates with angular frequency ω_d. The velocity amplitude v_m of the system is greatest when

$$\omega_d = \omega, \quad (15\text{-}46)$$

a condition called **resonance.** The amplitude x_m of the system is (approximately) greatest under the same condition.

QUESTIONS

1 The acceleration $a(t)$ of a particle undergoing SHM is graphed in Fig. 15-19. (a) Which of the labeled points corresponds to the particle at $-x_m$? (b) At point 4, is the velocity of the particle positive, negative, or zero? (c) At point 5, is the particle at $-x_m$, at $+x_m$, at 0, between $-x_m$ and 0, or between 0 and $+x_m$?

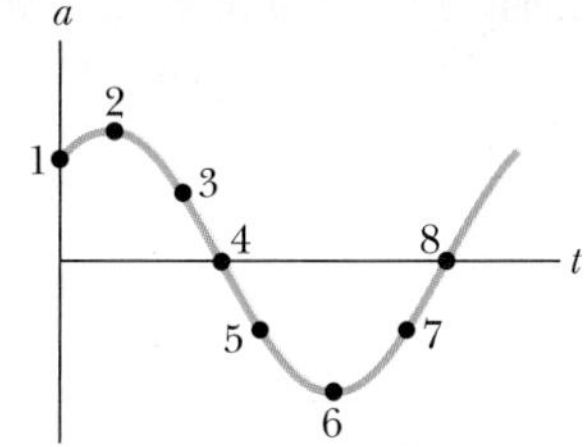

FIG. 15-19 Question 1.

2 Which of the following relationships between the acceleration a and the displacement x of a particle involve SHM: (a) $a = 0.5x$, (b) $a = 400x^2$, (c) $a = -20x$, (d) $a = -3x^2$?

3 Which of the following describe ϕ for the SHM of Fig. 15-20a:

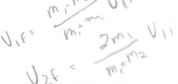

(a) $-\pi < \phi < -\pi/2$,
(b) $\pi < \phi < 3\pi/2$,
(c) $-3\pi/2 < \phi < -\pi$?

4 The velocity $v(t)$ of a particle undergoing SHM is graphed in Fig. 15-20*b*. Is the particle momentarily stationary, headed toward $-x_m$, or headed toward $+x_m$ at (a) point *A* on the graph and (b) point *B*? Is the particle at $-x_m$, at $+x_m$, at 0, between $-x_m$ and 0, or between 0 and $+x_m$ when its velocity is represented by (c) point *A* and (d) point *B*? Is the speed of the particle increasing or decreasing at (e) point *A* and (f) point *B*?

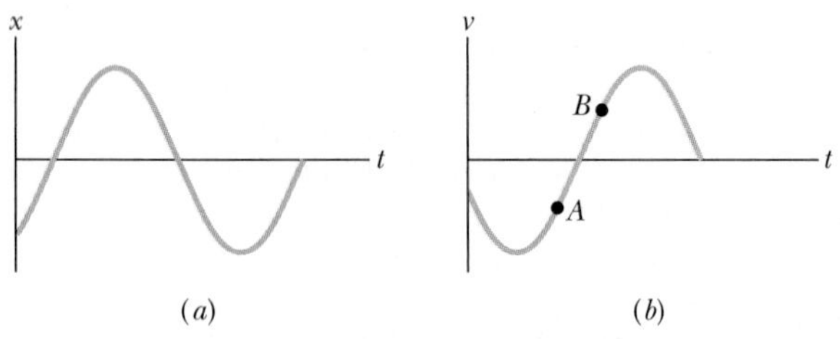

FIG. 15-20 Questions 3 and 4.

5 Figure 15-21 shows the $x(t)$ curves for three experiments involving a particular spring–box system oscillating in SHM. Rank the curves according to (a) the system's angular frequency, (b) the spring's potential energy at time $t = 0$, (c) the box's kinetic energy at $t = 0$, (d) the box's speed at $t = 0$, and (e) the box's maximum kinetic energy, greatest first.

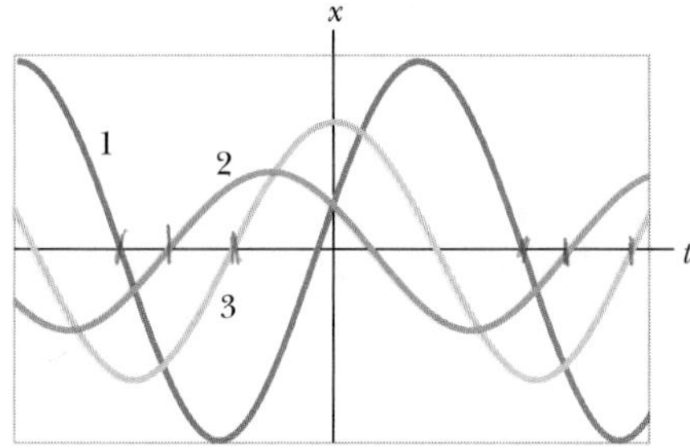

FIG. 15-21 Question 5.

6 Figure 15-22 gives, for three situations, the displacements $x(t)$ of a pair of simple harmonic oscillators (*A* and *B*) that are identical except for phase. For each pair, what phase shift (in radians and in degrees) is needed to shift the curve for *A* to coincide with the curve for *B*? Of the many possible answers, choose the shift with the smallest absolute magnitude.

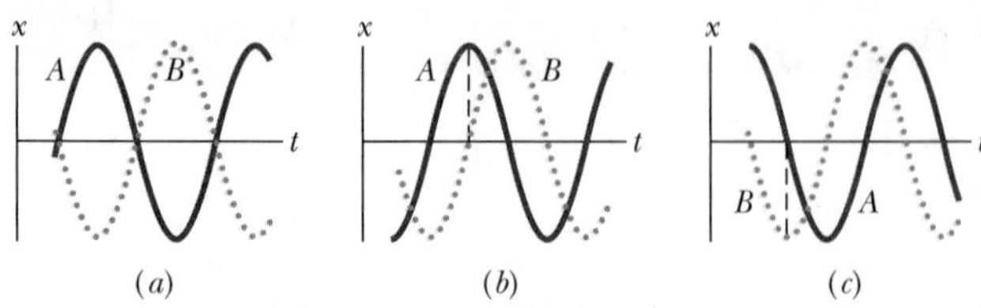

FIG. 15-22 Question 6.

7 You are to complete Fig. 15-23*a* so that it is a plot of velocity *v* versus time *t* for the spring–block oscillator that is shown in Fig. 15-23*b* for $t = 0$. (a) In Fig. 15-23*a*, at which lettered point or in what region between the points should the (vertical) *v* axis intersect the *t* axis? (For example, should it intersect at point *A*, or maybe in the region between points *A* and *B*?) (b) If the block's velocity is given by $v = -v_m \sin(\omega t + \phi)$, what is the value of ϕ? Make it positive, and if you cannot specify the value (such as $+\pi/2$ rad), then give a range of values (such as between 0 and $\pi/2$).

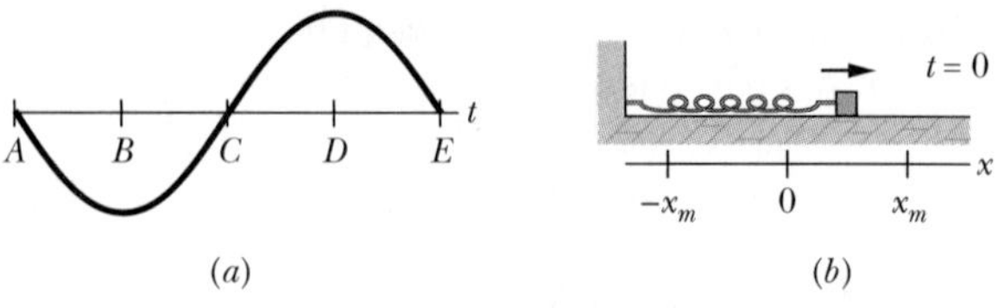

FIG. 15-23 Question 7.

8 You are to complete Fig. 15-24*a* so that it is a plot of acceleration *a* versus time *t* for the spring–block oscillator that is shown in Fig. 15-24*b* for $t = 0$. (a) In Fig. 15-24*a*, at which lettered point or in what region between the points should the (vertical) *a* axis intersect the *t* axis? (For example, should it intersect at point *A*, or maybe in the region between points *A* and *B*?) (b) If the block's acceleration is given by $a = -a_m \cos(\omega t + \phi)$, what is the value of ϕ? Make it positive, and if you cannot specify the value (such as $+\pi/2$ rad), then give a range of values (such as between 0 and $\pi/2$).

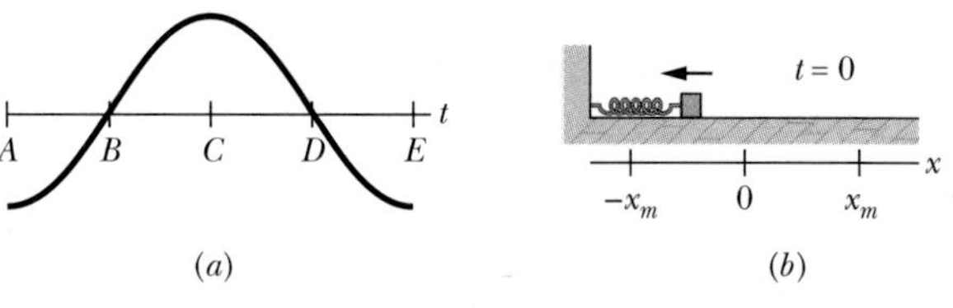

FIG. 15-24 Question 8.

9 In Fig. 15-25, a spring–block system is put into SHM in two experiments. In the first, the block is pulled from the equilibrium position through a displacement d_1 and then released. In the second, it is pulled from the equilibrium position through a greater displacement d_2 and then released. Are the (a) amplitude, (b) period, (c) frequency, (d) maximum kinetic energy, and (e) maximum potential energy in the second experiment greater than, less than, or the same as those in the first experiment?

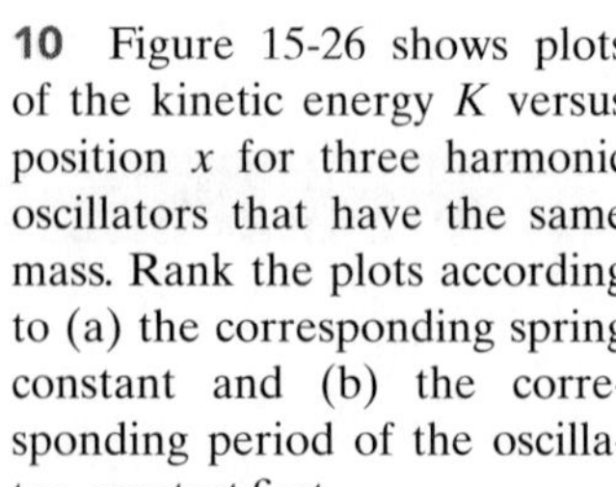

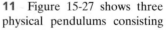

FIG. 15-25 Question 9.

10 Figure 15-26 shows plots of the kinetic energy *K* versus position *x* for three harmonic oscillators that have the same mass. Rank the plots according to (a) the corresponding spring constant and (b) the corresponding period of the oscillator, greatest first.

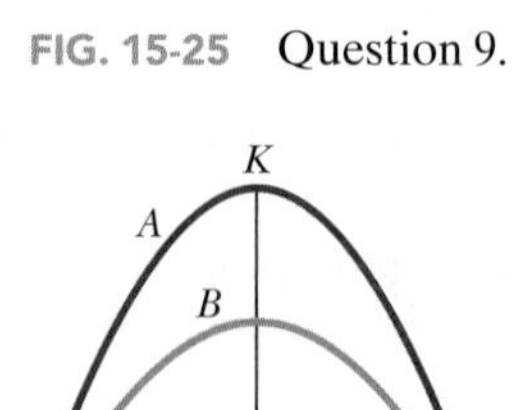

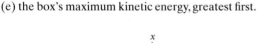

FIG. 15-26 Question 10.

11 Figure 15-27 shows three physical pendulums consisting of identical uniform spheres of the same mass that are rigidly connected by identical rods of negligible mass. Each pendulum is vertical and can pivot about suspension point *O*. Rank the pendulums according to period of oscillation, greatest first.

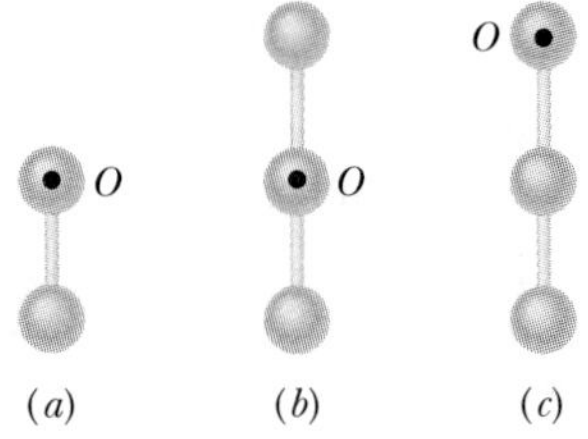

FIG. 15-27 Question 11.

12 You are to build the oscillation transfer device shown in Fig. 15-28. It consists of two spring–block systems hanging from a flexible rod. When the spring of system 1 is stretched and then released, the resulting SHM of system 1 at frequency f_1 oscillates the rod. The rod then exerts a driving force on system 2, at the same frequency f_1. You can choose from four springs with spring constants k of 1600, 1500, 1400, and 1200 N/m, and four blocks with masses m of 800, 500, 400, and 200 kg. Mentally determine which spring should go with which block in each of the two systems to maximize the amplitude of oscillations in system 2.

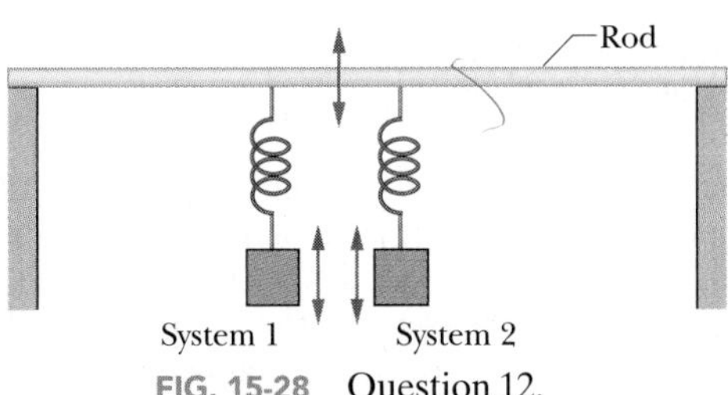

FIG. 15-28 Question 12.

PROBLEMS

GO Tutoring problem available (at instructor's discretion) in *WileyPLUS* and WebAssign
SSM Worked-out solution available in Student Solutions Manual
WWW Worked-out solution is at
• – ••• Number of dots indicates level of problem difficulty
ILW Interactive solution is at
http://www.wiley.com/college/halliday
 Additional information available in *The Flying Circus of Physics* and at flyingcircusofphysics.com

sec. 15-3 The Force Law for Simple Harmonic Motion

•1 What is the maximum acceleration of a platform that oscillates at amplitude 2.20 cm and frequency 6.60 Hz?

•2 A particle with a mass of 1.00×10^{-20} kg is oscillating with simple harmonic motion with a period of 1.00×10^{-5} s and a maximum speed of 1.00×10^3 m/s. Calculate (a) the angular frequency and (b) the maximum displacement of the particle.

•3 In an electric shaver, the blade moves back and forth over a distance of 2.0 mm in simple harmonic motion, with frequency 120 Hz. Find (a) the amplitude, (b) the maximum blade speed, and (c) the magnitude of the maximum blade acceleration. SSM

•4 A 0.12 kg body undergoes simple harmonic motion of amplitude 8.5 cm and period 0.20 s. (a) What is the magnitude of the maximum force acting on it? (b) If the oscillations are produced by a spring, what is the spring constant?

•5 An object undergoing simple harmonic motion takes 0.25 s to travel from one point of zero velocity to the next such point. The distance between those points is 36 cm. Calculate the (a) period, (b) frequency, and (c) amplitude of the motion.

•6 An automobile can be considered to be mounted on four identical springs as far as vertical oscillations are concerned. The springs of a certain car are adjusted so that the oscillations have a frequency of 3.00 Hz. (a) What is the spring constant of each spring if the mass of the car is 1450 kg and the mass is evenly distributed over the springs? (b) What will be the oscillation frequency if five passengers, averaging 73.0 kg each, ride in the car with an even distribution of mass?

•7 An oscillator consists of a block of mass 0.500 kg connected to a spring. When set into oscillation with amplitude 35.0 cm, the oscillator repeats its motion every 0.500 s. Find the (a) period, (b) frequency, (c) angular frequency, (d) spring constant, (e) maximum speed, and (f) magnitude of the maximum force on the block from the spring. SSM

•8 An oscillating block–spring system takes 0.75 s to begin repeating its motion. Find (a) the period, (b) the frequency in hertz, and (c) the angular frequency in radians per second.

•9 A loudspeaker produces a musical sound by means of the oscillation of a diaphragm whose amplitude is limited to 1.00 μm. (a) At what frequency is the magnitude a of the diaphragm's acceleration equal to g? (b) For greater frequencies, is a greater than or less than g? SSM

•10 What is the phase constant for the harmonic oscillator with the position function $x(t)$ given in Fig. 15-29 if the position function has the form $x = x_m \cos(\omega t + \phi)$? The vertical axis scale is set by $x_s = 6.0$ cm.

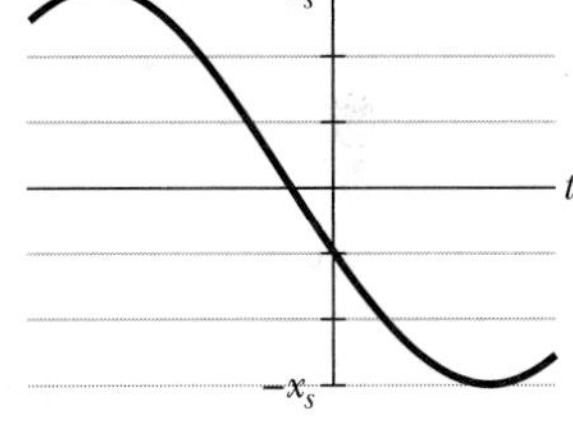

FIG. 15-29 Problem 10.

•11 The function $x = (6.0 \text{ m}) \cos[(3\pi \text{ rad/s})t + \pi/3 \text{ rad}]$ gives the simple harmonic motion of a body. At $t = 2.0$ s, what are the (a) displacement, (b) velocity, (c) acceleration, and (d) phase of the motion? Also, what are the (e) frequency and (f) period of the motion?

•12 What is the phase constant for the harmonic oscillator with the velocity function $v(t)$ given in Fig. 15-30 if the position function $x(t)$ has the form $x = x_m \cos(\omega t + \phi)$? The vertical axis scale is set by $v_s = 4.0$ cm/s.

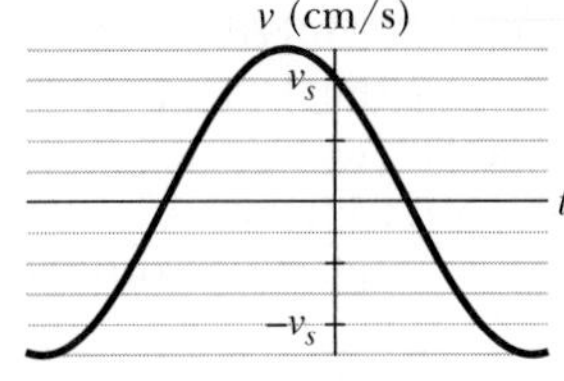

FIG. 15-30 Problem 12.

•13 In Fig. 15-31, two identical springs of spring constant 7580 N/m are attached to a block of mass 0.245 kg. What is the frequency of oscillation on the frictionless floor?

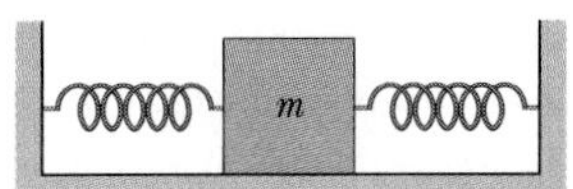

FIG. 15-31 Problems 13 and 23.

••14 Figure 15-32 shows block 1 of mass 0.200 kg sliding to the right over a frictionless elevated surface at a speed of 8.00 m/s. The block undergoes an elastic collision with stationary block 2, which is attached to a spring of spring constant 1208.5 N/m. (Assume that the spring does not affect the collision.) After the collision, block 2 oscillates in SHM with a period of 0.140 s, and block 1 slides off the opposite end of the elevated surface, landing a distance d from the base of that surface after falling height $h = 4.90$ m. What is the value of d?

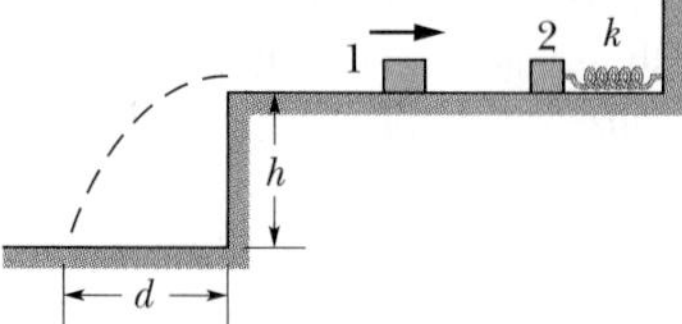

FIG. 15-32 Problem 14.

••15 An oscillator consists of a block attached to a spring ($k = 400$ N/m). At some time t, the position (measured from the system's equilibrium location), velocity, and acceleration of the block are $x = 0.100$ m, $v = -13.6$ m/s, and $a = -123$ m/s^2. Calculate (a) the frequency of oscillation, (b) the mass of the block, and (c) the amplitude of the motion. ILW

••16 At a certain harbor, the tides cause the ocean surface to rise and fall a distance d (from highest level to lowest level) in simple harmonic motion, with a period of 12.5 h. How long does it take for the water to fall a distance $0.250d$ from its highest level?

••17 A block is on a horizontal surface (a shake table) that is moving back and forth horizontally with simple harmonic motion of frequency 2.0 Hz. The coefficient of static friction between block and surface is 0.50. How great can the amplitude of the SHM be if the block is not to slip along the surface? SSM WWW

••18 Two particles execute simple harmonic motion of the same amplitude and frequency along close parallel lines. They pass each other moving in opposite directions each time their displacement is half their amplitude. What is their phase difference?

••19 Two particles oscillate in simple harmonic motion along a common straight-line segment of length A. Each particle has a period of 1.5 s, but they differ in phase by $\pi/6$ rad. (a) How far apart are they (in terms of A) 0.50 s after the lagging particle leaves one end of the path? (b) Are they then moving in the same direction, toward each other, or away from each other? SSM

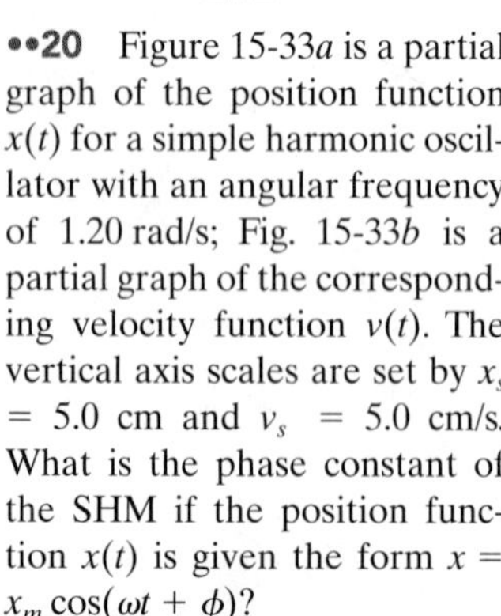

••20 Figure 15-33a is a partial graph of the position function $x(t)$ for a simple harmonic oscillator with an angular frequency of 1.20 rad/s; Fig. 15-33b is a partial graph of the corresponding velocity function $v(t)$. The vertical axis scales are set by $x_s = 5.0$ cm and $v_s = 5.0$ cm/s. What is the phase constant of the SHM if the position function $x(t)$ is given the form $x = x_m \cos(\omega t + \phi)$?

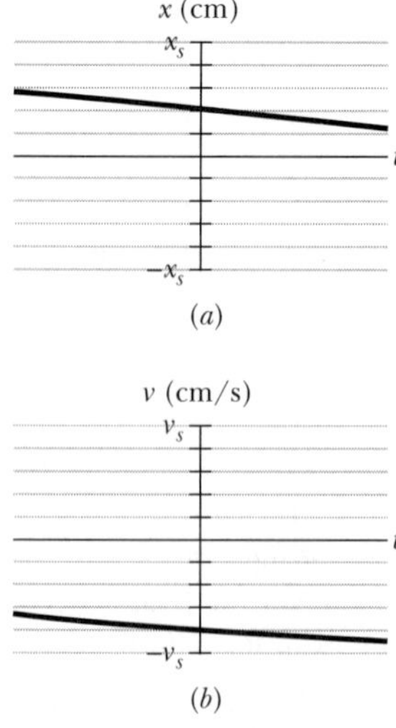

FIG. 15-33 Problem 20.

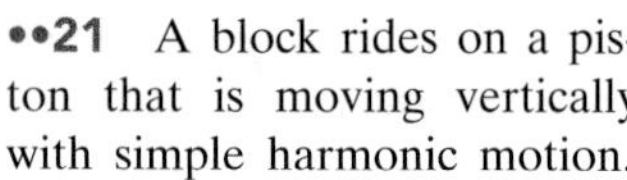

••21 A block rides on a piston that is moving vertically with simple harmonic motion. (a) If the SHM has period 1.0 s, at what amplitude of motion will the block and piston separate? (b) If the piston has an amplitude of 5.0 cm, what is the maximum frequency for which the block and piston will be in contact continuously?

••22 A simple harmonic oscillator consists of a block of mass 2.00 kg attached to a spring of spring constant 100 N/m. When $t = 1.00$ s, the position and velocity of the block are $x = 0.129$ m and $v = 3.415$ m/s. (a) What is the amplitude of the oscillations? What were the (b) position and (c) velocity of the block at $t = 0$ s?

••23 In Fig. 15-31, two springs are attached to a block that can oscillate over a frictionless floor. If the left spring is removed, the block oscillates at a frequency of 30 Hz. If, instead, the spring on the right is removed, the block oscillates at a frequency of 45 Hz. At what frequency does the block oscillate with both springs attached? ILW

•••24 In Fig. 15-34, two blocks ($m = 1.8$ kg and $M = 10$ kg) and a spring ($k = 200$ N/m) are arranged on a horizontal, frictionless surface. The coefficient of static friction between the two blocks is 0.40. What amplitude of simple harmonic motion of the spring–blocks system puts the smaller block on the verge of slipping over the larger block? GO

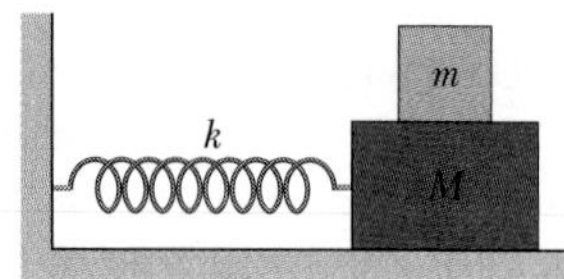

FIG. 15-34 Problem 24.

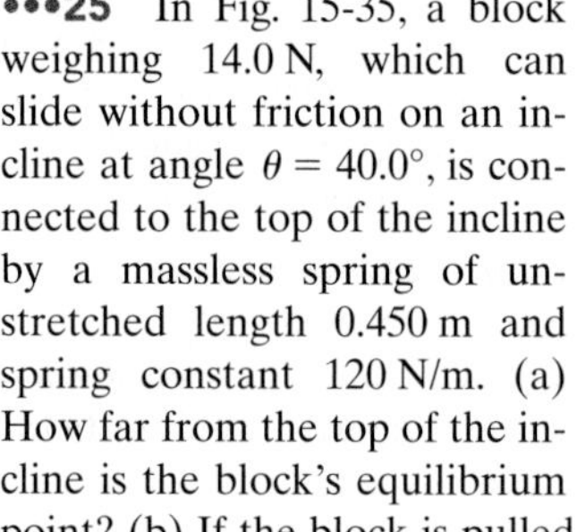

•••25 In Fig. 15-35, a block weighing 14.0 N, which can slide without friction on an incline at angle $\theta = 40.0°$, is connected to the top of the incline by a massless spring of unstretched length 0.450 m and spring constant 120 N/m. (a) How far from the top of the incline is the block's equilibrium point? (b) If the block is pulled slightly down the incline and released, what is the period of the resulting oscillations? GO

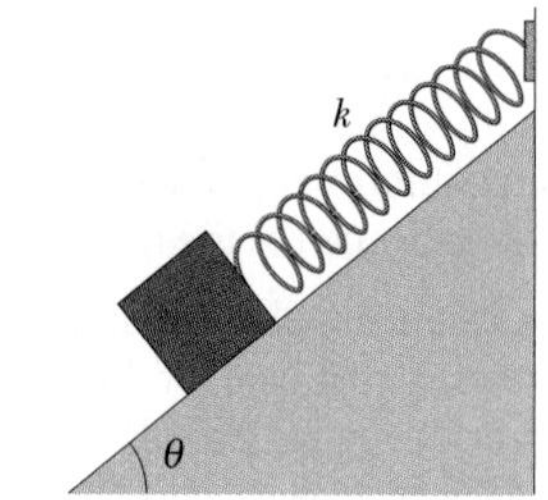

FIG. 15-35 Problem 25.

•••26 In Fig. 15-36, two springs are joined and connected to a block of mass 0.245 kg that is set oscillating over a frictionless floor. The springs each have spring constant $k = 6430$ N/m. What is the frequency of the oscillations?

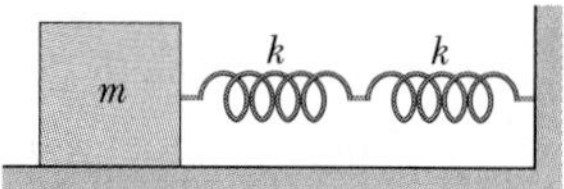

FIG. 15-36 Problem 26.

sec. 15-4 Energy in Simple Harmonic Motion

•27 Find the mechanical energy of a block–spring system having a spring constant of 1.3 N/cm and an oscillation amplitude of 2.4 cm. SSM

•28 An oscillating block–spring system has a mechanical energy of 1.00 J, an amplitude of 10.0 cm, and a maximum speed of 1.20 m/s. Find (a) the spring constant, (b) the mass of the block, and (c) the frequency of oscillation.

•29 When the displacement in SHM is one-half the amplitude x_m, what fraction of the total energy is (a) kinetic energy and (b) potential energy? (c) At what displacement, in terms of the amplitude, is the energy of the system half kinetic energy and half potential energy? SSM

•**30** Figure 15-37 gives the one-dimensional potential energy well for a 2.0 kg particle (the function $U(x)$ has the form bx^2 and the vertical axis scale is set by $U_s = 2.0$ J). (a) If the particle passes through the equilibrium position with a velocity of 85 cm/s, will it be turned back before it reaches $x = 15$ cm? (b) If yes, at what position, and if no, what is the speed of the particle at $x = 15$ cm?

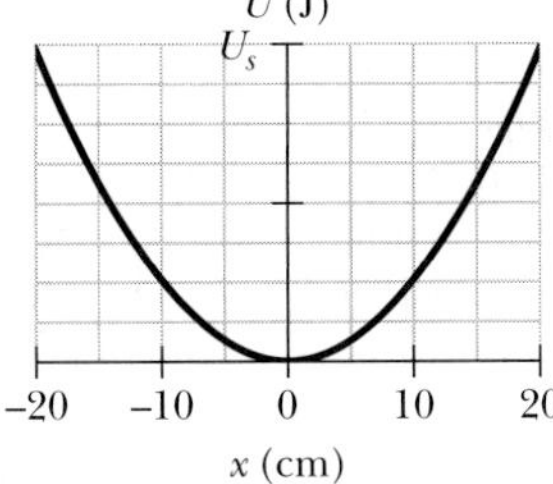

FIG. 15-37 Problem 30.

•**31** A 5.00 kg object on a horizontal frictionless surface is attached to a spring with $k = 1000$ N/m. The object is displaced from equilibrium 50.0 cm horizontally and given an initial velocity of 10.0 m/s back toward the equilibrium position. What are (a) the motion's frequency, (b) the initial potential energy of the block–spring system, (c) the initial kinetic energy, and (d) the motion's amplitude? ILW

•**32** Figure 15-38 shows the kinetic energy K of a simple harmonic oscillator versus its position x. The vertical axis scale is set by $K_s = 4.0$ J. What is the spring constant?

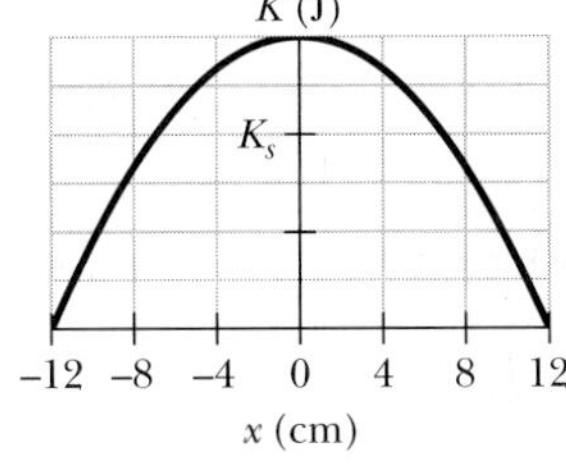

FIG. 15-38 Problem 32.

••**33** A 10 g particle undergoes SHM with an amplitude of 2.0 mm, a maximum acceleration of magnitude 8.0×10^3 m/s^2, and an unknown phase constant ϕ. What are (a) the period of the motion, (b) the maximum speed of the particle, and (c) the total mechanical energy of the oscillator? What is the magnitude of the force on the particle when the particle is at (d) its maximum displacement and (e) half its maximum displacement?

••**34** If the phase angle for a block–spring system in SHM is $\pi/6$ rad and the block's position is given by $x = x_m \cos(\omega t + \phi)$, what is the ratio of the kinetic energy to the potential energy at time $t = 0$?

••**35** A block of mass $M = 5.4$ kg, at rest on a horizontal frictionless table, is attached to a rigid support by a spring of constant $k = 6000$ N/m. A bullet of mass $m = 9.5$ g and velocity $\vec{v}$ of magnitude 630 m/s strikes and is embedded in the block (Fig. 15-39). Assuming the compression of the spring is negligible until the bullet is embedded, determine (a) the speed of the block immediately after the collision and (b) the amplitude of the resulting simple harmonic motion. GO

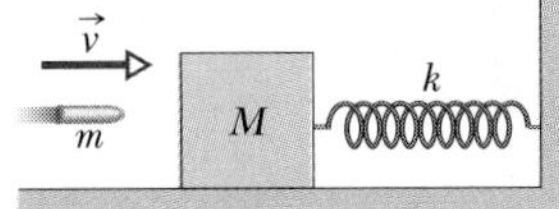

FIG. 15-39 Problem 35.

••**36** In Fig. 15-40, block 2 of mass 2.0 kg oscillates on the end of a spring in SHM with a period of 20 ms. The position of the block is given by $x = (1.0 \text{ cm}) \cos(\omega t + \pi/2)$. Block 1 of mass 4.0 kg slides toward block 2 with a velocity of magnitude 6.0 m/s, directed along the spring's length. The two blocks undergo a completely inelastic collision at time $t = 5.0$ ms. (The duration of the collision is much less than the period of motion.) What is the amplitude of the SHM after the collision? GO

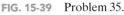

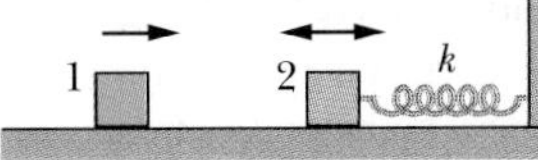

FIG. 15-40 Problem 36.

•••**37** A massless spring hangs from the ceiling with a small object attached to its lower end. The object is initially held at rest in a position y_i such that the spring is at its rest length. The object is then released from y_i and oscillates up and down, with its lowest position being 10 cm below y_i. (a) What is the frequency of the oscillation? (b) What is the speed of the object when it is 8.0 cm below the initial position? (c) An object of mass 300 g is attached to the first object, after which the system oscillates with half the original frequency. What is the mass of the first object? (d) How far below y_i is the new equilibrium (rest) position with both objects attached to the spring?

sec. 15-5 An Angular Simple Harmonic Oscillator

•**38** A 95 kg solid sphere with a 15 cm radius is suspended by a vertical wire. A torque of 0.20 N · m is required to rotate the sphere through an angle of 0.85 rad and then maintain that orientation. What is the period of the oscillations that result when the sphere is then released?

••**39** The balance wheel of an old-fashioned watch oscillates with angular amplitude π rad and period 0.500 s. Find (a) the maximum angular speed of the wheel, (b) the angular speed at displacement $\pi/2$ rad, and (c) the magnitude of the angular acceleration at displacement $\pi/4$ rad. SSM WWW

sec. 15-6 Pendulums

•**40** Suppose that a simple pendulum consists of a small 60.0 g bob at the end of a cord of negligible mass. If the angle θ between the cord and the vertical is given by

$$\theta = (0.0800 \text{ rad}) \cos[(4.43 \text{ rad/s})t + \phi],$$

what are (a) the pendulum's length and (b) its maximum kinetic energy?

•**41** (a) If the physical pendulum of Sample Problem 15-5 is inverted and suspended at point P, what is its period of oscillation? (b) Is the period now greater than, less than, or equal to its previous value?

•**42** In Sample Problem 15-5, we saw that a physical pendulum has a center of oscillation at distance $2L/3$ from its point of suspension. Show that the distance between the point of suspension and the center of oscillation for a physical pendulum of any form is I/mh, where I and h have the meanings assigned to them in Eq. 15-29 and m is the mass of the pendulum.

•**43** In Fig. 15-41, the pendulum consists of a uniform disk with radius $r = 10.0$ cm and mass 500 g attached to a uniform rod with length $L = 500$ mm and mass 270 g. (a) Calculate the rotational inertia of the pendulum about the pivot point. (b) What is the distance between the pivot point and the center of mass of the pendulum? (c) Calculate the period of oscillation. SSM

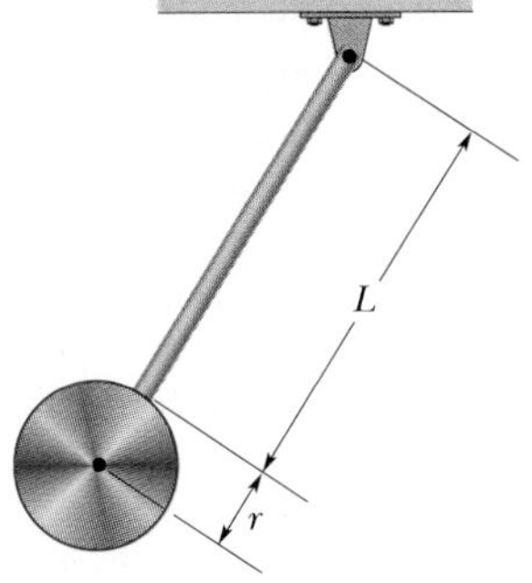

FIG. 15-41 Problem 43.

•**44** A physical pendulum consists of a meter stick that is pivoted at a small hole drilled through the stick a distance d from the 50 cm mark. The period of oscillation is 2.5 s. Find d. ILW

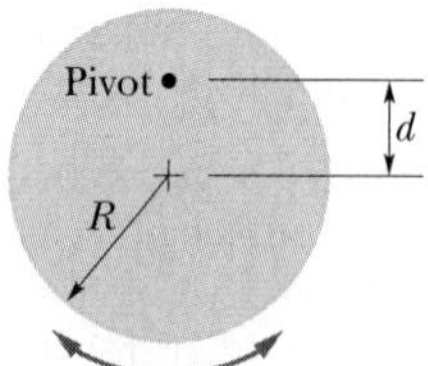

FIG. 15-42 Problem 45.

•45 In Fig. 15-42, a physical pendulum consists of a uniform solid disk (of radius $R = 2.35$ cm) supported in a vertical plane by a pivot located a distance $d = 1.75$ cm from the center of the disk. The disk is displaced by a small angle and released. What is the period of the resulting simple harmonic motion?

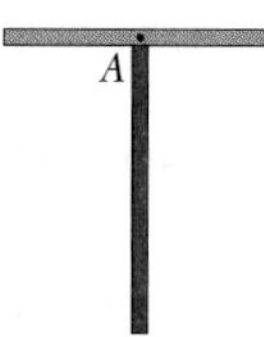

FIG. 15-43 Problem 46.

•46 A physical pendulum consists of two meter-long sticks joined together as shown in Fig. 15-43. What is the pendulum's period of oscillation about a pin inserted through point A at the center of the horizontal stick?

•47 A performer seated on a trapeze is swinging back and forth with a period of 8.85 s. If she stands up, thus raising the center of mass of the *trapeze + performer* system by 35.0 cm, what will be the new period of the system? Treat *trapeze + performer* as a simple pendulum.

••48 A thin uniform rod (mass = 0.50 kg) swings about an axis that passes through one end of the rod and is perpendicular to the plane of the swing. The rod swings with a period of 1.5 s and an angular amplitude of 10°. (a) What is the length of the rod? (b) What is the maximum kinetic energy of the rod as it swings?

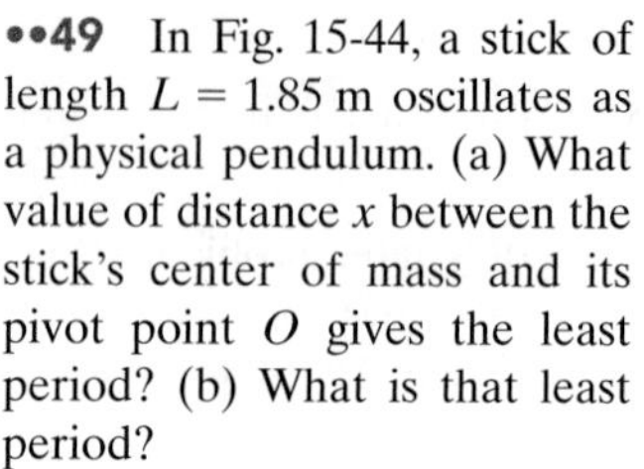

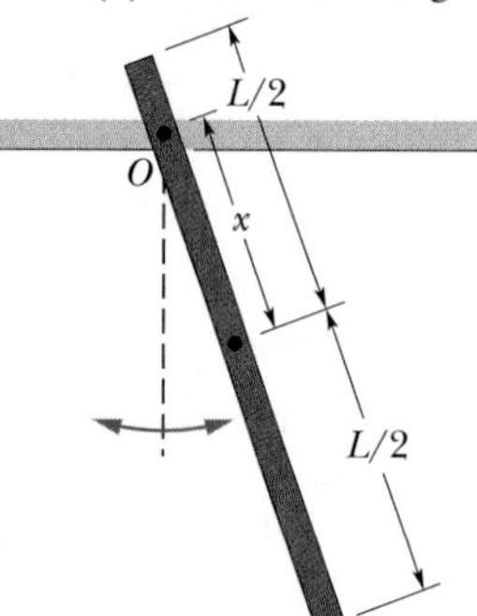

FIG. 15-44 Problem 49.

••49 In Fig. 15-44, a stick of length $L = 1.85$ m oscillates as a physical pendulum. (a) What value of distance x between the stick's center of mass and its pivot point O gives the least period? (b) What is that least period?

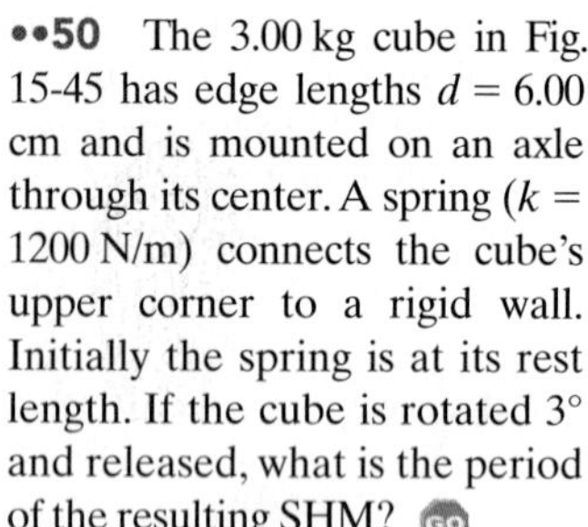

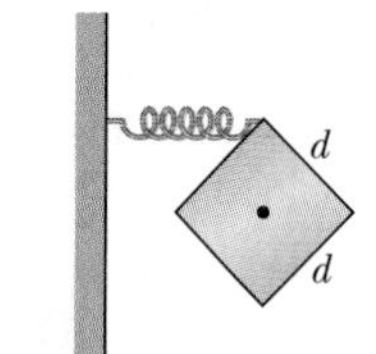

FIG. 15-45 Problem 50.

••50 The 3.00 kg cube in Fig. 15-45 has edge lengths $d = 6.00$ cm and is mounted on an axle through its center. A spring (k = 1200 N/m) connects the cube's upper corner to a rigid wall. Initially the spring is at its rest length. If the cube is rotated 3° and released, what is the period of the resulting SHM? GO

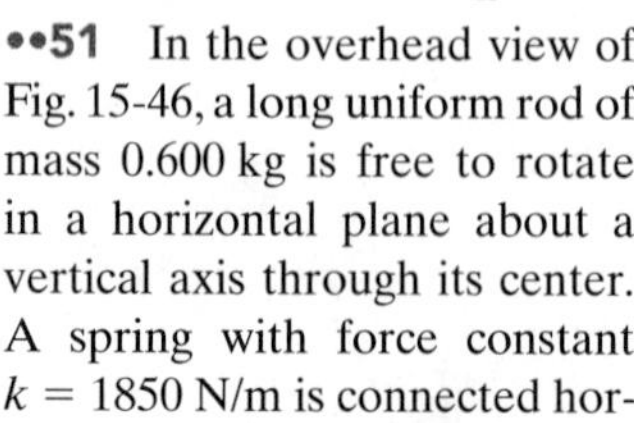

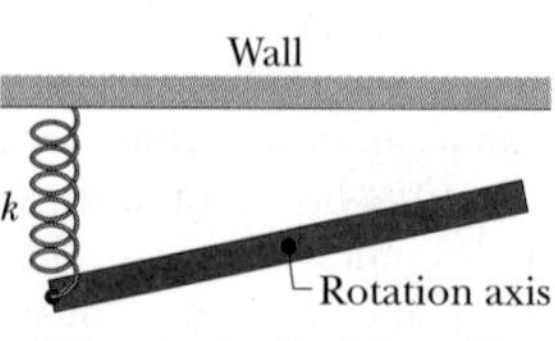

FIG. 15-46 Problem 51.

••51 In the overhead view of Fig. 15-46, a long uniform rod of mass 0.600 kg is free to rotate in a horizontal plane about a vertical axis through its center. A spring with force constant $k = 1850$ N/m is connected horizontally between one end of the rod and a fixed wall. When the rod is in equilibrium, it is parallel to the wall. What is the period of the small oscillations that result when the rod is rotated slightly and released? SSM ILW

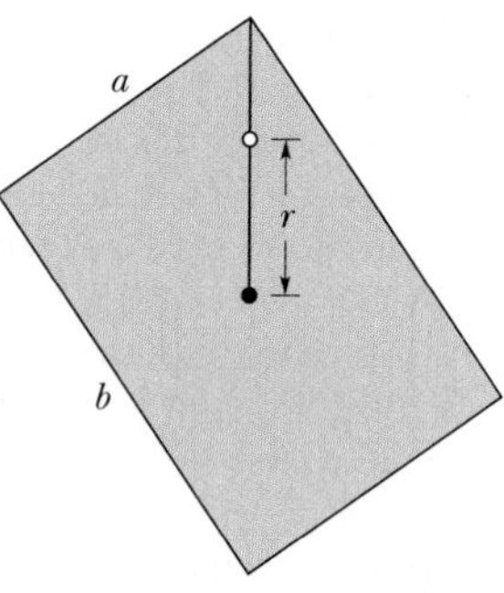

FIG. 15-47 Problem 52.

••52 A rectangular block, with face lengths $a = 35$ cm and $b = 45$ cm, is to be suspended on a thin horizontal rod running through a narrow hole in the block. The block is then to be set swinging about the rod like a pendulum, through small angles so that it is in SHM. Figure 15-47 shows one possible position of the hole, at distance r from the block's center, along a line connecting the center with a corner. (a) Plot the period of the pendulum versus distance r along that line such that the minimum in the curve is apparent. (b) For what value of r does that minimum occur? There is actually a line of points around the block's center for which the period of swinging has the same minimum value. (c) What shape does that line make?

••53 The angle of the pendulum of Fig. 15-9b is given by $\theta = \theta_m \cos[(4.44 \text{ rad/s})t + \phi]$. If at $t = 0$, $\theta = 0.040$ rad and $d\theta/dt = -0.200$ rad/s, what are (a) the phase constant ϕ and (b) the maximum angle θ_m? (*Hint:* Don't confuse the rate $d\theta/dt$ at which θ changes with the ω of the SHM.)

••54 In Fig. 15-48a, a metal plate is mounted on an axle through its center of mass. A spring with $k = 2000$ N/m connects a wall with a point on the rim a distance $r = 2.5$ cm from the center of mass. Initially the spring is at its rest length. If the plate is rotated by 7° and released, it rotates about the axle in SHM, with its angular position given by Fig. 15-48b. The horizontal axis scale is set by $t_s = 20$ ms. What is the rotational inertia of the plate about its center of mass? GO

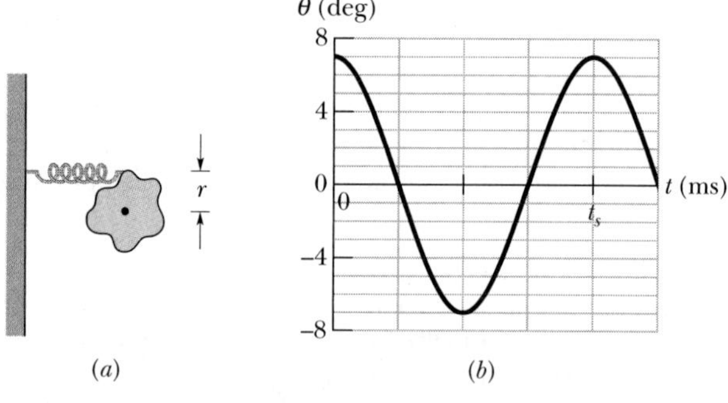

FIG. 15-48 Problem 54.

•••55 A pendulum is formed by pivoting a long thin rod about a point on the rod. In a series of experiments, the period is measured as a function of the distance x between the pivot point and the rod's center. (a) If the rod's length is $L = 2.20$ m and its mass is $m = 22.1$ g, what is the minimum period? (b) If x is chosen to minimize the period and then L is increased, does the period increase, decrease, or remain the same? (c) If, instead, m is increased without L increasing, does the period increase, decrease, or remain the same?

•••56 In Fig. 15-49, a 2.50 kg disk of diameter $D = 42.0$ cm is supported by a rod of length $L = 76.0$ cm and negligible mass that is pivoted at its end. (a) With the massless torsion spring unconnected, what is the period of oscillation? (b) With the torsion spring connected, the rod is vertical at equilibrium.

What is the torsion constant of the spring if the period of oscillation has been decreased by 0.500 s?

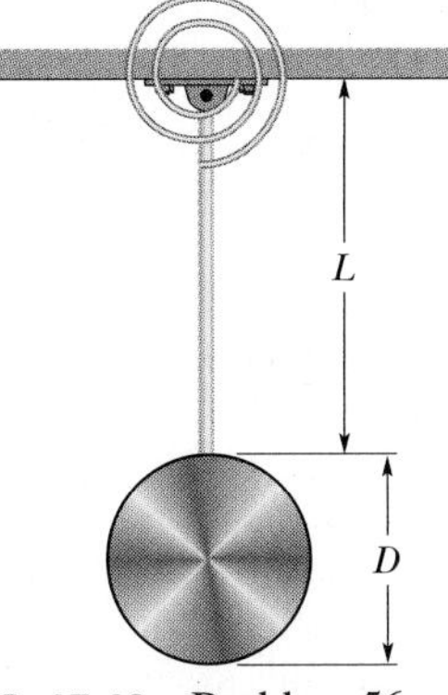

FIG. 15-49 Problem 56.

sec. 15-8 Damped Simple Harmonic Motion

•57 In Fig. 15-15, the block has a mass of 1.50 kg and the spring constant is 8.00 N/m. The damping force is given by $-b(dx/dt)$, where $b = 230$ g/s. The block is pulled down 12.0 cm and released. (a) Calculate the time required for the amplitude of the resulting oscillations to fall to one-third of its initial value. (b) How many oscillations are made by the block in this time? SSM WWW

•58 In Sample Problem 15-7, what is the ratio of the amplitude of the damped oscillations to the initial amplitude at the end of 20 cycles?

•59 The amplitude of a lightly damped oscillator decreases by 3.0% during each cycle. What percentage of the mechanical energy of the oscillator is lost in each cycle?

••60 The suspension system of a 2000 kg automobile "sags" 10 cm when the chassis is placed on it. Also, the oscillation amplitude decreases by 50% each cycle. Estimate the values of (a) the spring constant k and (b) the damping constant b for the spring and shock absorber system of one wheel, assuming each wheel supports 500 kg.

sec. 15-9 Forced Oscillations and Resonance

•61 For Eq. 15-45, suppose the amplitude x_m is given by

$$x_m = \frac{F_m}{[m^2(\omega_d^2 - \omega^2)^2 + b^2\omega_d^2]^{1/2}},$$

where F_m is the (constant) amplitude of the external oscillating force exerted on the spring by the rigid support in Fig. 15-15. At resonance, what are the (a) amplitude and (b) velocity amplitude of the oscillating object?

•62 Hanging from a horizontal beam are nine simple pendulums of the following lengths: (a) 0.10, (b) 0.30, (c) 0.40, (d) 0.80, (e) 1.2, (f) 2.8, (g) 3.5, (h) 5.0, and (i) 6.2 m. Suppose the beam undergoes horizontal oscillations with angular frequencies in the range from 2.00 rad/s to 4.00 rad/s. Which of the pendulums will be (strongly) set in motion?

••63 A 1000 kg car carrying four 82 kg people travels over a "washboard" dirt road with corrugations 4.0 m apart. The car bounces with maximum amplitude when its speed is 16 km/h. When the car stops, and the people get out, by how much does the car body rise on its suspension?

Additional Problems

64 A block is in SHM on the end of a spring, with position given by $x = x_m \cos(\omega t + \phi)$. If $\phi = \pi/5$ rad, then at $t = 0$ what percentage of the total mechanical energy is potential energy?

65 Figure 15-50 gives the position of a 20 g block oscillating in SHM on the end of a spring. The horizontal axis scale is set by $t_s = 40.0$ ms. What are (a) the maximum kinetic energy of the block and (b) the number of times per second that maximum is reached? (*Hint:* Measuring a slope will probably not be very accurate. Find another approach.)

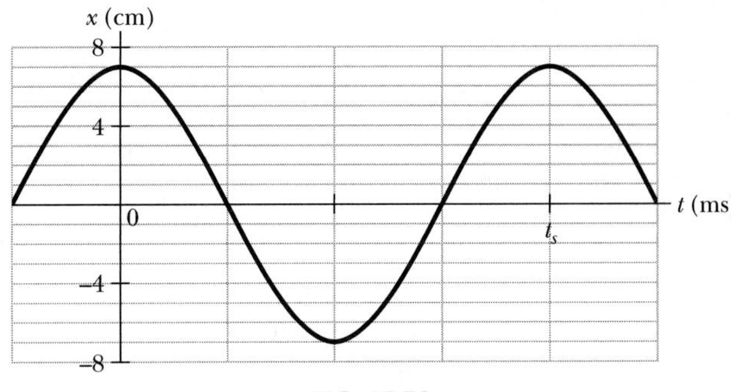

FIG. 15-50 Problems 65 and 66.

66 Figure 15-50 gives the position $x(t)$ of a block oscillating in SHM on the end of a spring ($t_s = 40.0$ ms). What are (a) the speed and (b) the magnitude of the radial acceleration of a particle in the corresponding uniform circular motion?

67 Figure 15-51 shows the kinetic energy K of a simple pendulum versus its angle θ from the vertical. The vertical axis scale is set by $K_s = 10.0$ mJ. The pendulum bob has mass 0.200 kg. What is the length of the pendulum?

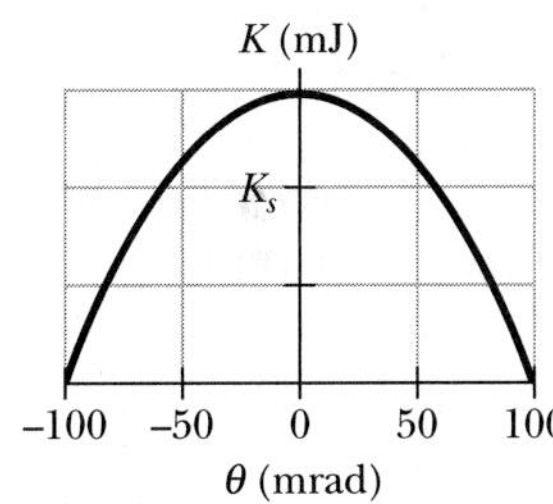

FIG. 15-51 Problem 67.

68 Although California is known for earthquakes, it has large regions dotted with precariously balanced rocks that would be easily toppled by even a mild earthquake. The rocks have stood this way for thousands of years, suggesting that major earthquakes have not occurred in those regions during that time. If an earthquake were to put such a rock into sinusoidal oscillation (parallel to the ground) with a frequency of 2.2 Hz, an oscillation amplitude of 1.0 cm would cause the rock to topple. What would be the magnitude of the maximum acceleration of the oscillation, in terms of g?

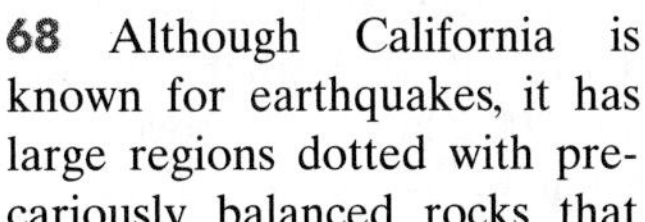

69 A 4.00 kg block is suspended from a spring with $k = 500$ N/m. A 50.0 g bullet is fired into the block from directly below with a speed of 150 m/s and becomes embedded in the block. (a) Find the amplitude of the resulting SHM. (b) What percentage of the original kinetic energy of the bullet is transferred to mechanical energy of the oscillator?

70 A 55.0 g block oscillates in SHM on the end of a spring with $k = 1500$ N/m according to $x = x_m \cos(\omega t + \phi)$. How long does the block take to move from position $+0.800x_m$ to (a) position $+0.600x_m$ and (b) position $-0.800x_m$?

71 A loudspeaker diaphragm is oscillating in simple harmonic motion with a frequency of 440 Hz and a maximum displacement of 0.75 mm. What are the (a) angular frequency, (b) maximum speed, and (c) magnitude of the maximum acceleration?

72 The tip of one prong of a tuning fork undergoes SHM of frequency 1000 Hz and amplitude 0.40 mm. For this tip, what is the magnitude of the (a) maximum acceleration, (b) maximum velocity, (c) acceleration at tip displacement 0.20 mm, and (d) velocity at tip displacement 0.20 mm?

73 A flat uniform circular disk has a mass of 3.00 kg and a radius of 70.0 cm. It is suspended in a horizontal plane by a vertical wire attached to its center. If the disk is rotated 2.50 rad about the wire, a torque of 0.0600 N · m is required to maintain that orientation. Calculate (a) the rotational inertia of the disk about the wire, (b) the torsion constant, and (c) the angular frequency of this torsion pendulum when it is set oscillating.

74 A uniform circular disk whose radius R is 12.6 cm is suspended as a physical pendulum from a point on its rim. (a) What is its period? (b) At what radial distance $r < R$ is there a pivot point that gives the same period?

75 What is the frequency of a simple pendulum 2.0 m long (a) in a room, (b) in an elevator accelerating upward at a rate of 2.0 m/s^2, and (c) in free fall? SSM

76 A particle executes linear SHM with frequency 0.25 Hz about the point $x = 0$. At $t = 0$, it has displacement $x = 0.37$ cm and zero velocity. For the motion, determine the (a) period, (b) angular frequency, (c) amplitude, (d) displacement $x(t)$, (e) velocity $v(t)$, (f) maximum speed, (g) magnitude of the maximum acceleration, (h) displacement at $t = 3.0$ s, and (i) speed at $t = 3.0$ s.

77 A 50.0 g stone is attached to the bottom of a vertical spring and set vibrating. If the maximum speed of the stone is 15.0 cm/s and the period is 0.500 s, find the (a) spring constant of the spring, (b) amplitude of the motion, and (c) frequency of oscillation.

78 A 2.00 kg block hangs from a spring. A 300 g body hung below the block stretches the spring 2.00 cm farther. (a) What is the spring constant? (b) If the 300 g body is removed and the block is set into oscillation, find the period of the motion.

79 The end point of a spring oscillates with a period of 2.0 s when a block with mass m is attached to it. When this mass is increased by 2.0 kg, the period is found to be 3.0 s. Find m.

80 A 0.10 kg block oscillates back and forth along a straight line on a frictionless horizontal surface. Its displacement from the origin is given by

$$x = (10 \text{ cm}) \cos[(10 \text{ rad/s})t + \pi/2 \text{ rad}].$$

(a) What is the oscillation frequency? (b) What is the maximum speed acquired by the block? (c) At what value of x does this occur? (d) What is the magnitude of the maximum acceleration of the block? (e) At what value of x does this occur? (f) What force, applied to the block by the spring, results in the given oscillation?

81 A 3.0 kg particle is in simple harmonic motion in one dimension and moves according to the equation

$$x = (5.0 \text{ m}) \cos[(\pi/3 \text{ rad/s})t - \pi/4 \text{ rad}],$$

with t in seconds. (a) At what value of x is the potential energy of the particle equal to half the total energy? (b) How long does the particle take to move to this position x from the equilibrium position?

82 A massless spring with spring constant 19 N/m hangs vertically. A body of mass 0.20 kg is attached to its free end and then released. Assume that the spring was unstretched before the body was released. Find (a) how far below the initial position the body descends, and the (b) frequency and (c) amplitude of the resulting SHM.

83 The piston in the cylinder head of a locomotive has a stroke (twice the amplitude) of 0.76 m. If the piston moves with simple harmonic motion with an angular frequency of 180 rev/min, what is its maximum speed? SSM

84 A wheel is free to rotate about its fixed axle. A spring is attached to one of its spokes a distance r from the axle, as shown in Fig. 15-52. (a) Assuming that the wheel is a hoop of mass m and radius R, what is the angular frequency ω of small oscillations of this system in terms of m, R, r, and the spring constant k? What is ω if (b) $r = R$ and (c) $r = 0$?

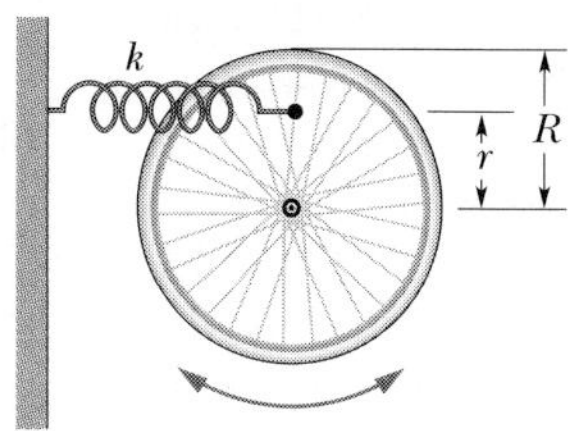

FIG. 15-52 Problem 84.

85 The scale of a spring balance that reads from 0 to 15.0 kg is 12.0 cm long. A package suspended from the balance is found to oscillate vertically with a frequency of 2.00 Hz. (a) What is the spring constant? (b) How much does the package weigh?

86 A uniform spring with $k = 8600$ N/m is cut into pieces 1 and 2 of unstretched lengths $L_1 = 7.0$ cm and $L_2 = 10$ cm. What are (a) k_1 and (b) k_2? A block attached to the original spring as in Fig. 15-5 oscillates at 200 Hz. What is the oscillation frequency of the block attached to (c) piece 1 and (d) piece 2?

87 In Fig. 15-53, three 10 000 kg ore cars are held at rest on a mine railway using a cable that is parallel to the rails, which are inclined at angle $\theta = 30°$. The cable stretches 15 cm just before the coupling between the two lower cars breaks, detaching the lowest car. Assuming that the cable obeys Hooke's law, find the (a) frequency and (b) amplitude of the resulting oscillations of the remaining two cars.

FIG. 15-53 Problem 87.

88 A simple pendulum of length 20 cm and mass 5.0 g is suspended in a race car traveling with constant speed 70 m/s around a circle of radius 50 m. If the pendulum undergoes small oscillations in a radial direction about its equilibrium position, what is the frequency of oscillation?

89 A vertical spring stretches 9.6 cm when a 1.3 kg block is hung from its end. (a) Calculate the spring constant. This block is then displaced an additional 5.0 cm downward and released from rest. Find the (b) period, (c) frequency, (d) amplitude, and (e) maximum speed of the resulting SHM. SSM

90 A block weighing 20 N oscillates at one end of a vertical spring for which $k = 100$ N/m; the other end of the spring is attached to a ceiling. At a certain instant the spring is stretched 0.30 m beyond its relaxed length (the length when no object is attached) and the block has zero velocity. (a) What is the net force on the block at this instant? What are the (b) amplitude and (c) period of the resulting simple harmonic motion? (d) What is the maximum kinetic energy of the block as it oscillates?

91 A 1.2 kg block sliding on a horizontal frictionless surface is attached to a horizontal spring with $k = 480$ N/m. Let x be the displacement of the block from the position at which the spring is unstretched. At $t = 0$ the block passes through $x = 0$ with a speed of 5.2 m/s in the positive x direction. What are the (a) frequency and (b) amplitude of the block's motion? (c) Write an expression for x as a function of time. SSM

92 A simple harmonic oscillator consists of an 0.80 kg block attached to a spring ($k = 200$ N/m). The block slides on a horizontal frictionless surface about the equilibrium point $x = 0$ with a total mechanical energy of 4.0 J. (a) What is the amplitude of the oscillation? (b) How many oscillations does the block complete in 10 s? (c) What is the maximum kinetic energy attained by the block? (d) What is the speed of the block at $x = 0.15$ m?

93 An engineer has an odd-shaped 10 kg object and needs to find its rotational inertia about an axis through its center of mass. The object is supported on a wire stretched along the desired axis. The wire has a torsion constant $\kappa = 0.50$ N · m. If this torsion pendulum oscillates through 20 cycles in 50 s, what is the rotational inertia of the object?

94 A grandfather clock has a pendulum that consists of a thin brass disk of radius $r = 15.00$ cm and mass 1.000 kg that is attached to a long thin rod of negligible mass. The pendulum swings freely about an axis perpendicular to the rod and through the end of the rod opposite the disk, as shown in Fig. 15-54. If the pendulum is to have a period of 2.000 s for small oscillations at a place where $g = 9.800$ m/s^2, what must be the rod length L to the nearest tenth of a millimeter?

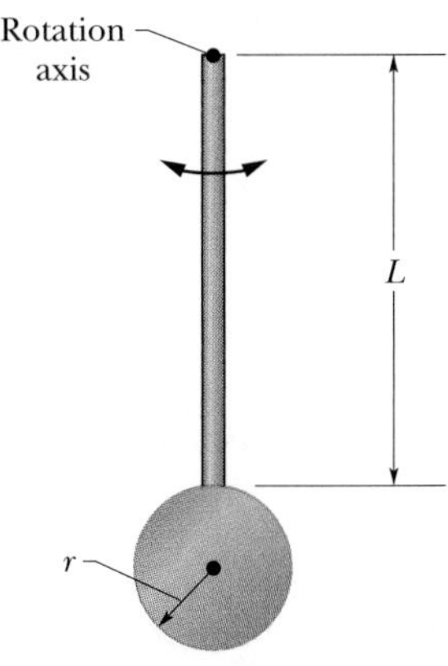

FIG. 15-54 Problem 94.

95 A block sliding on a horizontal frictionless surface is attached to a horizontal spring with a spring constant of 600 N/m. The block executes SHM about its equilibrium position with a period of 0.40 s and an amplitude of 0.20 m. As the block slides through its equilibrium position, a 0.50 kg putty wad is dropped vertically onto the block. If the putty wad sticks to the block, determine (a) the new period of the motion and (b) the new amplitude of the motion.

96 When a 20 N can is hung from the bottom of a vertical spring, it causes the spring to stretch 20 cm. (a) What is the spring constant? (b) This spring is now placed horizontally on a frictionless table. One end of it is held fixed, and the other end is attached to a 5.0 N can. The can is then moved (stretching the spring) and released from rest. What is the period of the resulting oscillation?

97 A 4.00 kg block hangs from a spring, extending it 16.0 cm from its unstretched position. (a) What is the spring constant? (b) The block is removed, and a 0.500 kg body is hung from the same spring. If the spring is then stretched and released, what is its period of oscillation?

98 A damped harmonic oscillator consists of a block ($m = 2.00$ kg), a spring ($k = 10.0$ N/m), and a damping force ($F = -bv$). Initially, it oscillates with an amplitude of 25.0 cm; because of the damping, the amplitude falls to three-fourths of this initial value at the completion of four oscillations. (a) What is the value of b? (b) How much energy has been "lost" during these four oscillations?

99 A common device for entertaining a toddler is a *jump seat* that hangs from the horizontal portion of a doorframe via elastic cords (Fig. 15-55). Assume that only one cord is on each side in spite of the more realistic arrangement shown. When a child is placed in the seat, they both descend by a distance d_s as the cords stretch (treat them as springs). Then the seat is pulled down an extra distance d_m and released, so that the child oscillates vertically, like a block on the end of a spring. Suppose you are the safety engineer for the manufacturer of the seat. You do not want the magnitude of the child's acceleration to exceed $0.20g$ for fear of hurting the child's neck. If $d_m = 10$ cm, what value of d_s corresponds to that acceleration magnitude?

FIG. 15-55 Problem 99.

100 What is the phase constant for SMH with $a(t)$ given in Fig. 15-56 if the position function $x(t)$ has the form $x = x_m \cos(\omega t + \phi)$ and $a_s = 4.0$ m/s^2?

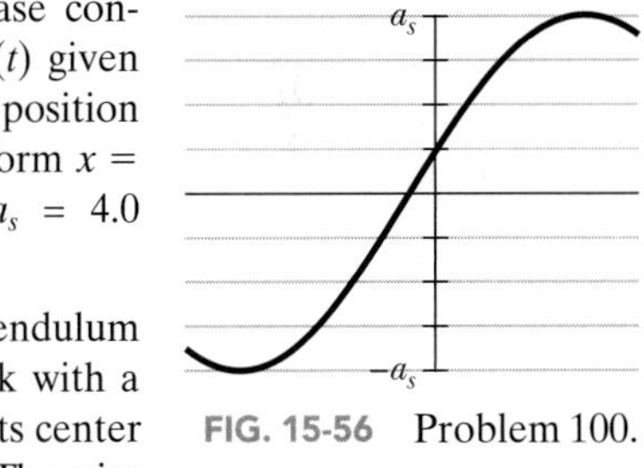

FIG. 15-56 Problem 100.

101 A torsion pendulum consists of a metal disk with a wire running through its center and soldered in place. The wire is mounted vertically on clamps and pulled taut. Figure 15-57a gives the magnitude τ of the torque needed to rotate the disk about its center (and thus twist the wire) versus the rotation angle θ. The vertical axis scale is set by $\tau_s = 4.0 \times 10^{-3}$ N · m. The disk is rotated to $\theta = 0.200$ rad and then released. Figure 15-57b shows the resulting oscillation in terms of angular position θ versus time t. The horizontal axis scale is set by $t_s = 0.40$ s. (a) What is the rotational inertia of the disk about its center? (b) What is the maximum angular speed $d\theta/dt$ of the disk? (*Caution:* Do not confuse the (constant) angular frequency of the SHM with the (varying) angular speed of the rotating disk, even though they usually have the same symbol ω. *Hint:* The potential energy U of a torsion pendulum is equal to $\frac{1}{2}\kappa\theta^2$, analogous to $U = \frac{1}{2}kx^2$ for a spring.)

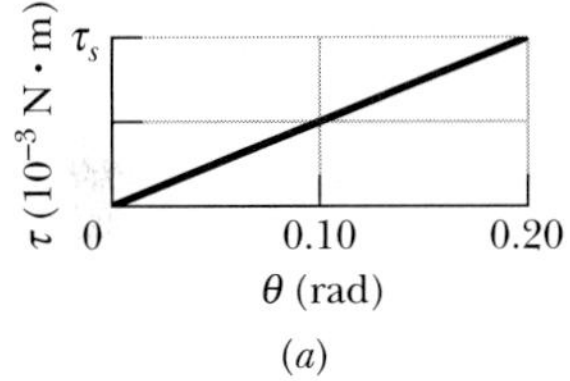

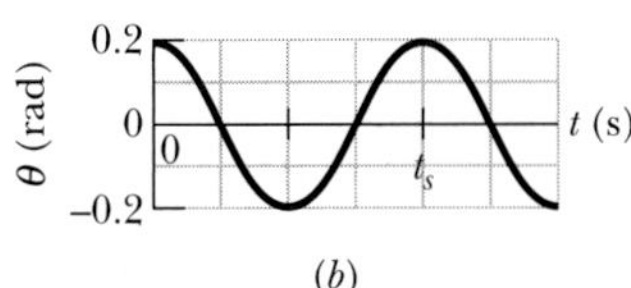

FIG. 15-57 Problem 101.

102 A spider can tell when its web has captured, say, a fly because the fly's thrashing causes the web threads to oscillate. A spider can even determine the size of the fly by the frequency of the oscillations. Assume that a fly oscillates on the *capture thread* on which it is caught like a block on a spring. What is the ratio of oscillation frequency for a fly with mass m to a fly with mass $2.5m$?

103 For a simple pendulum, find the angular amplitude θ_m at which the restoring torque required for simple harmonic motion deviates from the actual restoring torque by 1.0%. (See "Trigonometric Expansions" in Appendix E.)

104 A simple harmonic oscillator consists of a block attached to a spring with $k = 200$ N/m. The block slides on a frictionless surface, with equilibrium point $x = 0$ and amplitude 0.20 m. A graph of the block's velocity v as a function of time t is shown in Fig. 15-58. The horizontal scale is set by $t_s = 0.20$ s. What are (a) the period of the SHM, (b) the block's mass, (c) its displacement at $t = 0$, (d) its acceleration at $t = 0.10$ s, and (e) its maximum kinetic energy?

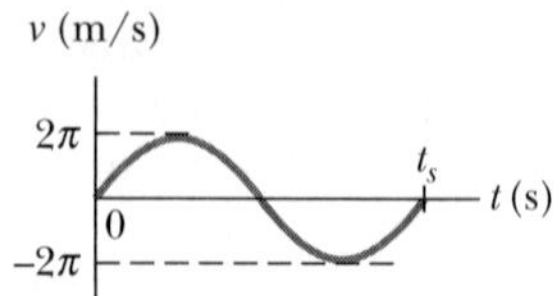

FIG. 15-58 Problem 104.

105 A simple harmonic oscillator consists of a 0.50 kg block attached to a spring. The block slides back and forth along a straight line on a frictionless surface with equilibrium point $x = 0$. At $t = 0$ the block is at $x = 0$ and moving in the positive x direction. A graph of the magnitude of the net force $\vec{F}$ on the block as a function of its position is shown in Fig. 15-59. The vertical scale is set by $F_s = 75.0$ N. What are (a) the amplitude and (b) the period of the motion, (c) the magnitude of the maximum acceleration, and (d) the maximum kinetic energy?

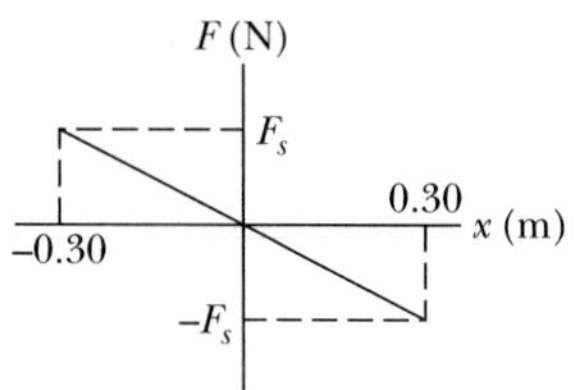

FIG. 15-59 Problem 105.

106 In Fig. 15-60, a solid cylinder attached to a horizontal spring ($k = 3.00$ N/m) rolls without slipping along a horizontal surface. If the system is released from rest when the spring is stretched by 0.250 m, find (a) the translational kinetic energy and (b) the rotational kinetic energy of the cylinder as it passes through the equilibrium position. (c) Show that under these conditions the cylinder's center of mass executes simple harmonic motion with period

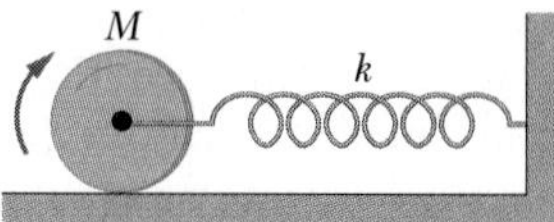

FIG. 15-60 Problem 106.

$$T = 2\pi\sqrt{\frac{3M}{2k}},$$

where M is the cylinder mass. (*Hint:* Find the time derivative of the total mechanical energy.)

107 A block weighing 10.0 N is attached to the lower end of a vertical spring ($k = 200.0$ N/m), the other end of which is attached to a ceiling. The block oscillates vertically and has a kinetic energy of 2.00 J as it passes through the point at which the spring is unstretched. (a) What is the period of the oscillation? (b) Use the law of conservation of energy to determine the maximum distance the block moves both above and below the point at which the spring is unstretched. (These are not necessarily the same.) (c) What is the amplitude of the oscillation? (d) What is the maximum kinetic energy of the block as it oscillates?

108 A 2.0 kg block executes SHM while attached to a horizontal spring of spring constant 200 N/m. The maximum speed of the block as it slides on a horizontal frictionless surface is 3.0 m/s. What are (a) the amplitude of the block's motion, (b) the magnitude of its maximum acceleration, and (c) the magnitude of its minimum acceleration? (d) How long does the block take to complete 7.0 cycles of its motion?

109 The vibration frequencies of atoms in solids at normal temperatures are of the order of 10^{13} Hz. Imagine the atoms to be connected to one another by springs. Suppose that a single silver atom in a solid vibrates with this frequency and that all the other atoms are at rest. Compute the effective spring constant. One mole of silver (6.02×10^{23} atoms) has a mass of 108 g.

110 In Fig. 15-61, a 2500 kg demolition ball swings from the end of a crane. The length of the swinging segment of cable is 17 m. (a) Find the period of the swinging, assuming that the system can be treated as a simple pendulum. (b) Does the period depend on the ball's mass?

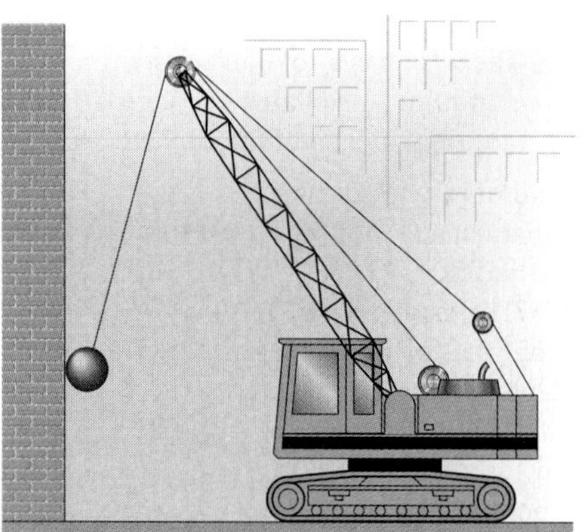
FIG. 15-61 Problem 110.

111 The center of oscillation of a physical pendulum has this interesting property: If an impulse (assumed horizontal and in the plane of oscillation) acts at the center of oscillation, no oscillations are felt at the point of support. Baseball players (and players of many other sports) know that unless the ball hits the bat at this point (called the "sweet spot" by athletes), the oscillations due to the impact will sting their hands. To prove this property, let the stick in Fig. 15-11*a* simulate a baseball bat. Suppose that a horizontal force $\vec{F}$ (due to impact with the ball) acts toward the right at *P*, the center of oscillation. The batter is assumed to hold the bat at *O*, the pivot point of the stick. (a) What acceleration does the point *O* undergo as a result of $\vec{F}$? (b) What angular acceleration is produced by $\vec{F}$ about the center of mass of the stick? (c) As a result of the angular acceleration in (b), what linear acceleration does point *O* undergo? (d) Considering the magnitudes and directions of the accelerations in (a) and (c), convince yourself that *P* is indeed the "sweet spot."

112 A 2.0 kg block is attached to the end of a spring with a spring constant of 350 N/m and forced to oscillate by an applied force $F = (15\text{ N})\sin(\omega_d t)$, where $\omega_d = 35$ rad/s. The damping constant is $b = 15$ kg/s. At $t = 0$, the block is at rest with the spring at its rest length. (a) Use numerical integration to plot the displacement of the block for the first 1.0 s. Use the motion near the end of the 1.0 s interval to estimate the amplitude, period, and angular frequency. Repeat the calculation for (b) $\omega_d = \sqrt{k/m}$ and (c) $\omega_d = 20$ rad/s.

Waves—I

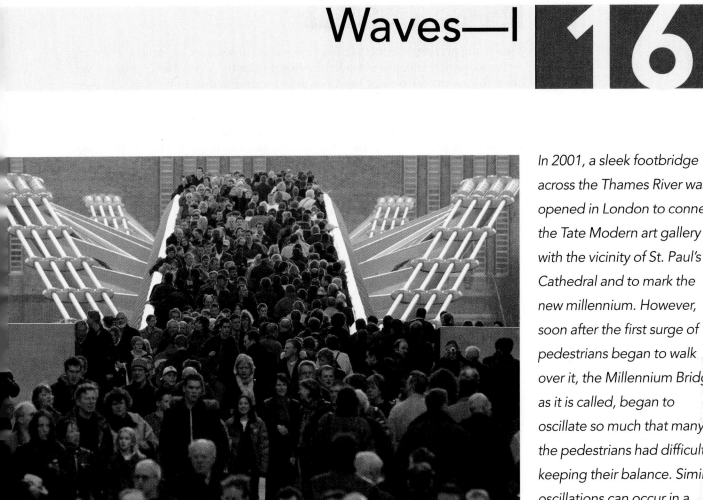

Sion Touhig/Getty Images News and Sport Services

Bruno Vincent/Reportage/Getty Images, Inc.

In 2001, a sleek footbridge across the Thames River was opened in London to connect the Tate Modern art gallery with the vicinity of St. Paul's Cathedral and to mark the new millennium. However, soon after the first surge of pedestrians began to walk over it, the Millennium Bridge, as it is called, began to oscillate so much that many of the pedestrians had difficulty keeping their balance. Similar oscillations can occur in a mosh pit on a standard wood floor, especially if the participants are pogoing or body slamming.

What causes such oscillations, which can be the nightmare of structural engineers?

The answer is in this chapter.

16-1 WHAT IS PHYSICS?

One of the primary subjects of physics is waves. To see how important waves are in the modern world, just consider the music industry. Every piece of music you hear, from some retro-punk band playing in a campus dive to the most eloquent concerto playing on the Web, depends on performers producing waves and your detecting those waves. In between production and detection, the information carried by the waves might need to be transmitted (as in a live performance on the Web) or recorded and then reproduced (as with CDs, DVDs, or the other devices currently being developed in engineering labs worldwide). The financial importance of controlling music waves is staggering, and the rewards to engineers who develop new control techniques can be rich.

This chapter focuses on waves traveling along a stretched string, such as on a guitar. The next chapter focuses on sound waves, such as those produced by a guitar string being played. Before we do all this, though, our first job is to classify the countless waves of the everyday world into basic types.

16-2 | Types of Waves

Waves are of three main types:

1. ***Mechanical waves.*** These waves are most familiar because we encounter them almost constantly; common examples include water waves, sound waves, and seismic waves. All these waves have two central features: They are governed by Newton's laws, and they can exist only within a material medium, such as water, air, and rock.
2. ***Electromagnetic waves.*** These waves are less familiar, but you use them constantly; common examples include visible and ultraviolet light, radio and television waves, microwaves, x rays, and radar waves. These waves require no material medium to exist. Light waves from stars, for example, travel through the vacuum of space to reach us. All electromagnetic waves travel through a vacuum at the same speed $c = 299\,792\,458$ m/s.
3. ***Matter waves.*** Although these waves are commonly used in modern technology, they are probably very unfamiliar to you. These waves are associated with electrons, protons, and other fundamental particles, and even atoms and molecules. Because we commonly think of these particles as constituting matter, such waves are called matter waves.

Much of what we discuss in this chapter applies to waves of all kinds. However, for specific examples we shall refer to mechanical waves.

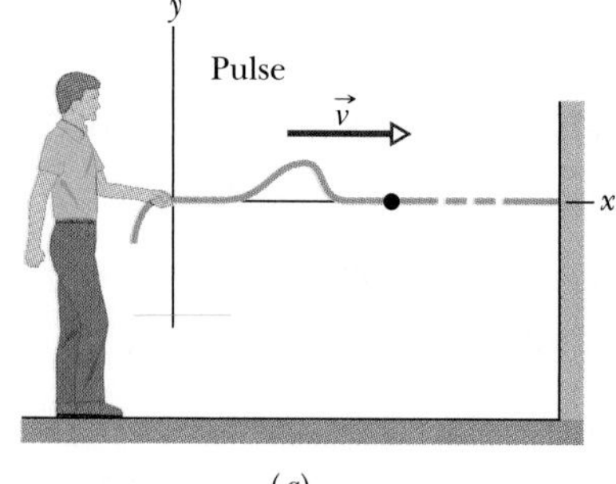

(*a*)

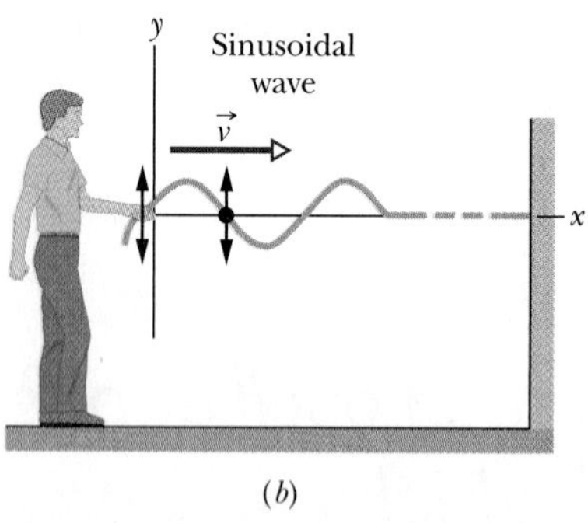

(*b*)

FIG. 16-1 (*a*) A single pulse is sent along a stretched string. A typical string element (marked with a dot) moves up once and then down as the pulse passes. The element's motion is perpendicular to the wave's direction of travel, so the pulse is a *transverse wave*. (*b*) A sinusoidal wave is sent along the string. A typical string element moves up and down continuously as the wave passes. This too is a transverse wave.

16-3 | Transverse and Longitudinal Waves

A wave sent along a stretched, taut string is the simplest mechanical wave. If you give one end of a stretched string a single up-and-down jerk, a wave in the form of a single *pulse* travels along the string, as in Fig. 16-1*a*. This pulse and its motion can occur because the string is under tension. When you pull your end of the string upward, it begins to pull upward on the adjacent section of the string via tension between the two sections. As the adjacent section moves upward, it begins to pull the next section upward, and so on. Meanwhile, you have pulled down on your end of the string. As each section moves upward in turn, it begins to be pulled back downward by neighboring sections that are already on the way down. The net result is that a distortion in the string's shape (the pulse) moves along the string at some velocity $\vec{v}$.

If you move your hand up and down in continuous simple harmonic motion, a continuous wave travels along the string at velocity $\vec{v}$. Because the motion of your hand is a sinusoidal function of time, the wave has a sinusoidal shape at any given instant, as in Fig. 16-1*b*; that is, the wave has the shape of a sine curve or a cosine curve.

We consider here only an "ideal" string, in which no friction-like forces within the string cause the wave to die out as it travels along the string. In addition, we assume that the string is so long that we need not consider a wave rebounding from the far end.

One way to study the waves of Fig. 16-1 is to monitor the **wave forms** (shapes of the waves) as they move to the right. Alternatively, we could monitor the motion of an element of the string as the element oscillates up and down while a wave passes through it. We would find that the displacement of every such oscillating string element is *perpendicular* to the direction of travel of the wave, as indicated in Fig. 16-1*b*. This motion is said to be **transverse,** and the wave is said to be a **transverse wave.**

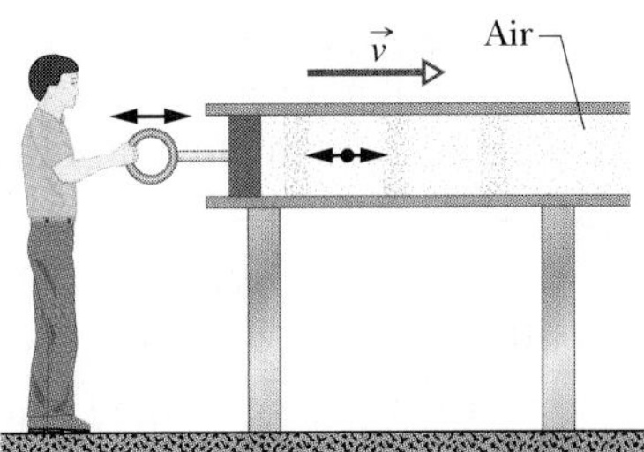

FIG. 16-2 A sound wave is set up in an air-filled pipe by moving a piston back and forth. Because the oscillations of an element of the air (represented by the dot) are parallel to the direction in which the wave travels, the wave is a *longitudinal wave.*

Figure 16-2 shows how a sound wave can be produced by a piston in a long, air-filled pipe. If you suddenly move the piston rightward and then leftward, you can send a pulse of sound along the pipe. The rightward motion of the piston moves the elements of air next to it rightward, changing the air pressure there. The increased air pressure then pushes rightward on the elements of air somewhat farther along the pipe. Moving the piston leftward reduces the air pressure next to it. As a result, first the elements nearest the piston and then farther elements move leftward. Thus, the motion of the air and the change in air pressure travel rightward along the pipe as a pulse.

If you push and pull on the piston in simple harmonic motion, as is being done in Fig. 16-2, a sinusoidal wave travels along the pipe. Because the motion of the elements of air is parallel to the direction of the wave's travel, the motion is said to be **longitudinal,** and the wave is said to be a **longitudinal wave.** In this chapter we focus on transverse waves, and string waves in particular; in Chapter 17 we focus on longitudinal waves, and sound waves in particular.

Both a transverse wave and a longitudinal wave are said to be **traveling waves** because they both travel from one point to another, as from one end of the string to the other end in Fig. 16-1 and from one end of the pipe to the other end in Fig. 16-2. Note that it is the wave that moves from end to end, not the material (string or air) through which the wave moves.

Sample Problem 16-1

Seismic waves are waves that travel either through Earth's interior or along the ground. Seismology stations are set up mainly to record seismic waves generated by earthquakes, but they also record seismic waves generated by any large release of energy near Earth's surface, such as an explosion. As the seismic waves travel past a station, they oscillate a recording pen and the pen traces out a graph. Figure 16-3*a* is one of the graphs created by seismic waves from the mysterious sinking of the Russian submarine *Kursk* in August 2000. The first oscillations of the pen are marked with an arrow and were of small amplitude. Much stronger oscillations began about 134 s later.

From this, analysts concluded that the first seismic waves were generated by an onboard explosion, possibly a torpedo that failed to launch when fired. The explosion presumably breached the hull, started a fire, and sank the submarine. The later, much stronger seismic waves were generated after the submarine was sunk and were possibly generated when the fire caused several of the powerful missiles on board to explode simultaneously.

(*a*)

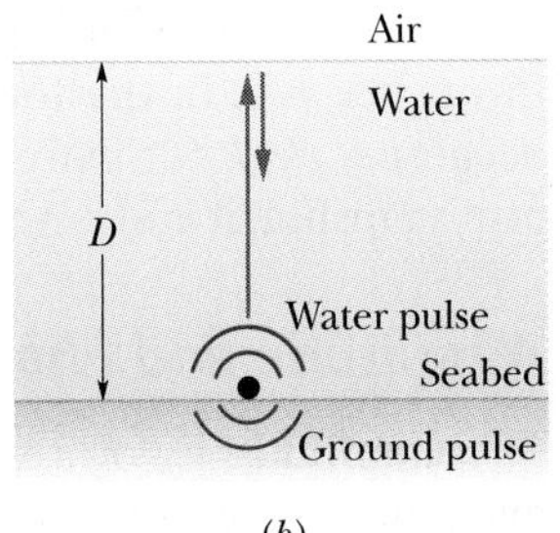

(*b*)

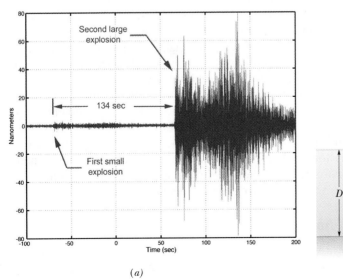

FIG. 16-3 (*a*) Graph made by a recording device as seismic waves from the *Kursk* passed the device. Amplitude is plotted vertically; time increases rightward. *(Courtesy of Jay Pulli/BBN Technologies)* (*b*) With the submarine sitting at depth *D*, the large explosion sent pulses both into the ground and up through the water.

These stronger waves arrived at seismology stations as pulses separated by a time interval Δt of about 0.11 s. To what depth D did the submarine sink?

KEY IDEA The speed of a sound is equal to the distance traveled divided by the travel time.

Calculations: Let us assume that the stronger explosion occurred after the *Kursk* was sitting on the ocean floor (the seabed). That explosion sent a pulse into the seabed and a pulse upward through the water (Fig. 16-3*b*). The pulse traveling through the water "bounced" several times between the water surface and the seabed. Each time it hit the seabed, it sent another pulse into the ground, and seismology stations detected those ground pulses as they arrived one after another. Thus, the time Δt between any two detected ground pulses was equal to the round-trip time for the water pulse to travel up to the water surface and back to the seabed. From Eq. 2-2 ($v_{avg} = \Delta x/\Delta t$), we can relate the speed v of the water pulse to the round-trip distance $2D$ and the round-trip time Δt as

$$v = \frac{2D}{\Delta t},$$

or

$$D = \frac{v\,\Delta t}{2}. \tag{16-1}$$

Waves travel through seawater at about 1500 m/s. Substituting this value and $\Delta t = 0.11$ s in Eq. 16-1, we find that the seismic data predicted the sunken *Kursk* was at a depth of approximately

$$D = \frac{(1500 \text{ m/s})(0.11 \text{ s})}{2}$$

$$= 82.5 \text{ m} \approx 83 \text{ m}. \qquad \text{(Answer)}$$

In fact, the wreck was discovered at a depth of about 115 m.

16-4 | Wavelength and Frequency

To completely describe a wave on a string (and the motion of any element along its length), we need a function that gives the shape of the wave. This means that we need a relation in the form $y = h(x, t)$, in which y is the transverse displacement of any string element as a function h of the time t and the position x of the element along the string. In general, a sinusoidal shape like the wave in Fig. 16-1*b* can be described with h being either a sine or cosine function; both give the same general shape for the wave. In this chapter we use the sine function.

Imagine a sinusoidal wave like that of Fig. 16-1*b* traveling in the positive direction of an x axis. As the wave sweeps through succeeding elements (that is, very short sections) of the string, the elements oscillate parallel to the y axis. At time t, the displacement y of the element located at position x is given by

$$y(x, t) = y_m \sin(kx - \omega t). \tag{16-2}$$

Because this equation is written in terms of position x, it can be used to find the displacements of all the elements of the string as a function of time. Thus, it can tell us the shape of the wave at any given time and how that shape changes as the wave moves along the string.

The names of the quantities in Eq. 16-2 are displayed in Fig. 16-4 and defined next. Before we discuss them, however, let us examine Fig. 16-5, which shows five "snapshots" of a sinusoidal wave traveling in the positive direction of an x axis. The movement of the wave is indicated by the rightward progress of the short arrow pointing to a high point of the wave. From snapshot to snapshot, the short arrow moves to the right with the wave shape, but the string moves *only* parallel to the y axis. To see that, let us follow the motion of the red-dyed string element at $x = 0$. In the first snapshot (Fig. 16-5*a*), this element is at displacement $y = 0$. In the next snapshot, it is at its extreme downward displacement because a *valley* (or extreme low point) of the wave is passing through it. It then moves back up through $y = 0$. In the fourth snapshot, it is at its extreme upward displacement because a *peak* (or extreme high point) of the wave is passing through it. In the fifth snapshot, it is again at $y = 0$, having completed one full oscillation.

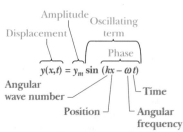

FIG. 16-4 The names of the quantities in Eq. 16-2, for a transverse sinusoidal wave.

Amplitude and Phase

The **amplitude** y_m of a wave, such as that in Fig. 16-5, is the magnitude of the maximum displacement of the elements from their equilibrium positions as the wave passes through them. (The subscript m stands for maximum.) Because y_m is a magnitude, it is always a positive quantity, even if it is measured downward instead of upward as drawn in Fig. 16-5a.

The **phase** of the wave is the *argument* $kx - \omega t$ of the sine in Eq. 16-2. As the wave sweeps through a string element at a particular position x, the phase changes linearly with time t. This means that the sine also changes, oscillating between $+1$ and -1. Its extreme positive value ($+1$) corresponds to a peak of the wave moving through the element; at that instant the value of y at position x is y_m. Its extreme negative value (-1) corresponds to a valley of the wave moving through the element; at that instant the value of y at position x is $-y_m$. Thus, the sine function and the time-dependent phase of a wave correspond to the oscillation of a string element, and the amplitude of the wave determines the extremes of the element's displacement.

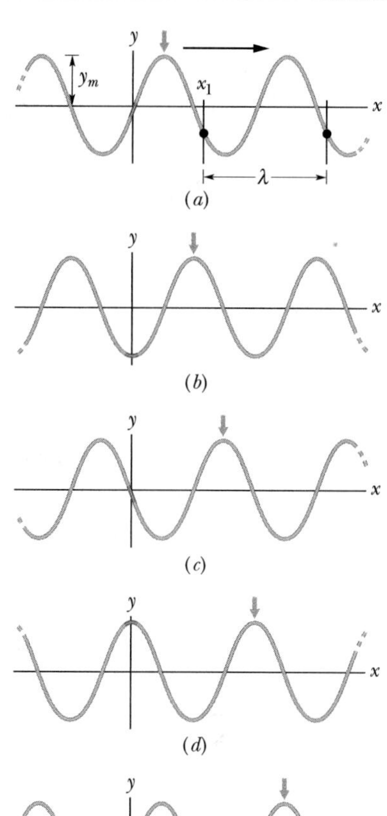

FIG. 16-5 Five "snapshots" of a string wave traveling in the positive direction of an x axis. The amplitude y_m is indicated. A typical wavelength λ, measured from an arbitrary position x_1, is also indicated.

Wavelength and Angular Wave Number

The **wavelength** λ of a wave is the distance (parallel to the direction of the wave's travel) between repetitions of the shape of the wave (or *wave shape*). A typical wavelength is marked in Fig. 16-5a, which is a snapshot of the wave at time $t = 0$. At that time, Eq. 16-2 gives, for the description of the wave shape,

$$y(x, 0) = y_m \sin kx. \tag{16-3}$$

By definition, the displacement y is the same at both ends of this wavelength—that is, at $x = x_1$ and $x = x_1 + \lambda$. Thus, by Eq. 16-3,

$$\begin{aligned} y_m \sin kx_1 &= y_m \sin k(x_1 + \lambda) \\ &= y_m \sin(kx_1 + k\lambda). \end{aligned} \tag{16-4}$$

A sine function begins to repeat itself when its angle (or argument) is increased by 2π rad, so in Eq. 16-4 we must have $k\lambda = 2\pi$, or

$$k = \frac{2\pi}{\lambda} \quad \text{(angular wave number)}. \tag{16-5}$$

We call k the **angular wave number** of the wave; its SI unit is the radian per meter, or the inverse meter. (Note that the symbol k here does *not* represent a spring constant as previously.)

Notice that the wave in Fig. 16-5 moves to the right by $\frac{1}{4}\lambda$ from one snapshot to the next. Thus, by the fifth snapshot, it has moved to the right by 1λ.

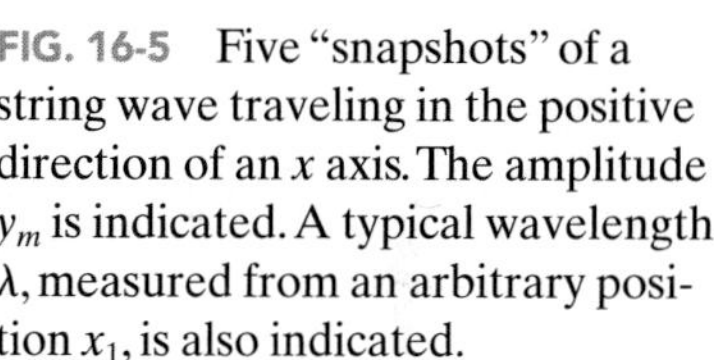

Period, Angular Frequency, and Frequency

Figure 16-6 shows a graph of the displacement y of Eq. 16-2 versus time t at a certain position along the string, taken to be $x = 0$. If you were to monitor the string, you would see that the single element of the string at that position moves up and down in simple harmonic motion given by Eq. 16-2 with $x = 0$:

$$\begin{aligned} y(0, t) &= y_m \sin(-\omega t) \\ &= -y_m \sin \omega t \qquad (x = 0). \end{aligned} \tag{16-6}$$

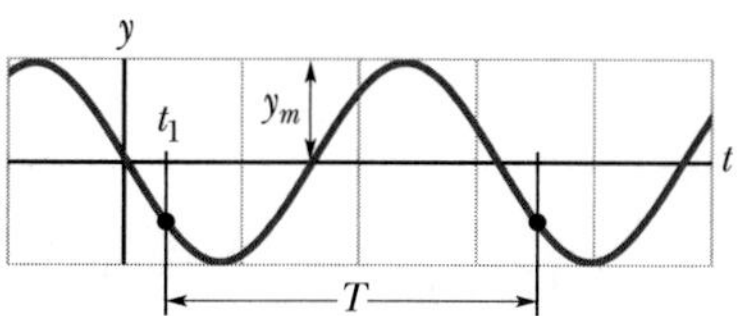

FIG. 16-6 A graph of the displacement of the string element at $x = 0$ as a function of time, as the sinusoidal wave of Fig. 16-5 passes through the element. The amplitude y_m is indicated. A typical period T, measured from an arbitrary time t_1, is also indicated.

Here we have made use of the fact that $\sin(-\alpha) = -\sin \alpha$, where α is any angle. Figure 16-6 is a graph of this equation; it *does not* show the shape of the wave.

We define the **period** of oscillation T of a wave to be the time any string element takes to move through one full oscillation. A typical period is marked on the graph of Fig. 16-6. Applying Eq. 16-6 to both ends of this time interval and equating the results yield

$$\begin{aligned} -y_m \sin \omega t_1 &= -y_m \sin \omega(t_1 + T) \\ &= -y_m \sin(\omega t_1 + \omega T). \end{aligned} \tag{16-7}$$

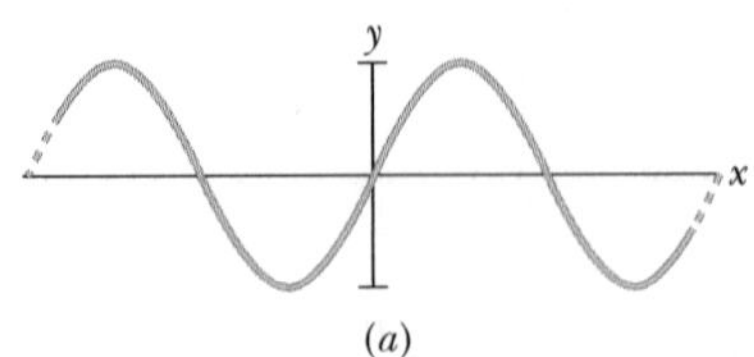

(a)

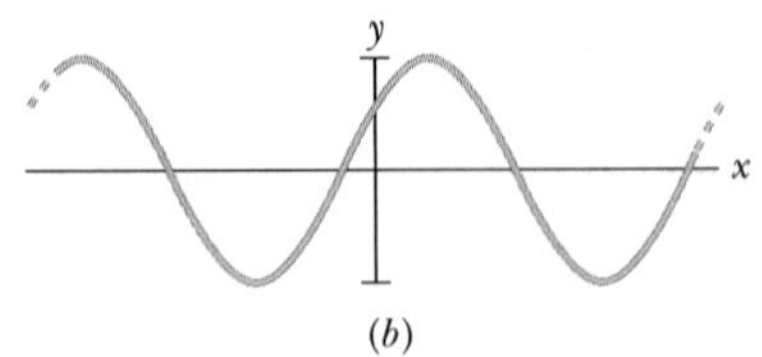

(b)

FIG. 16-7 A sinusoidal traveling wave at $t = 0$ with a phase constant ϕ of (a) 0 and (b) $\pi/5$ rad.

This can be true only if $\omega T = 2\pi$, or if

$$\omega = \frac{2\pi}{T} \quad \text{(angular frequency).} \tag{16-8}$$

We call ω the **angular frequency** of the wave; its SI unit is the radian per second.

Look back at the five snapshots of a traveling wave in Fig. 16-5. The time between snapshots is $\frac{1}{4}T$. Thus, by the fifth snapshot, every string element has made one full oscillation.

The **frequency** f of a wave is defined as $1/T$ and is related to the angular frequency ω by

$$f = \frac{1}{T} = \frac{\omega}{2\pi} \quad \text{(frequency).} \tag{16-9}$$

Like the frequency of simple harmonic motion in Chapter 15, this frequency f is a number of oscillations per unit time—here, the number made by a string element as the wave moves through it. As in Chapter 15, f is usually measured in hertz or its multiples, such as kilohertz.

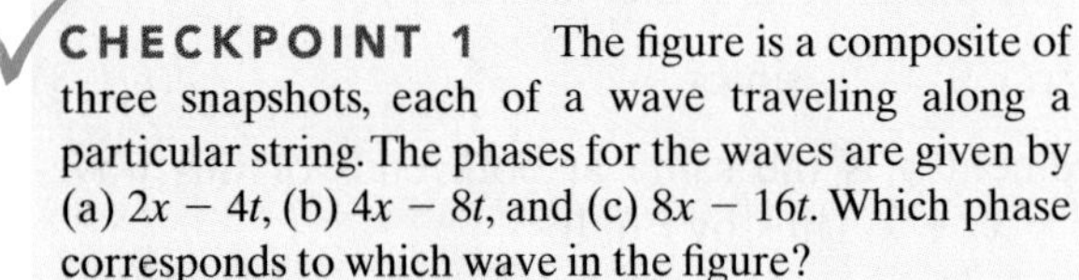
CHECKPOINT 1 The figure is a composite of three snapshots, each of a wave traveling along a particular string. The phases for the waves are given by (a) $2x - 4t$, (b) $4x - 8t$, and (c) $8x - 16t$. Which phase corresponds to which wave in the figure?

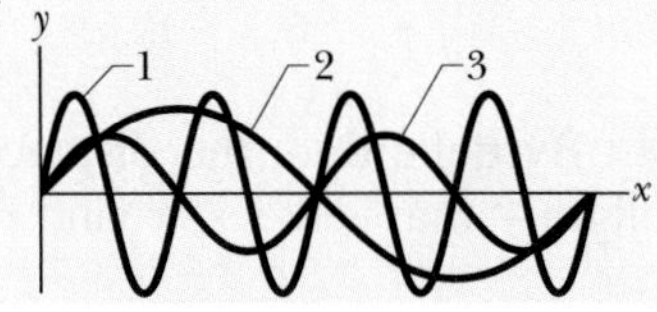

Phase Constant

When a sinusoidal traveling wave is given by the wave function of Eq. 16-2, the wave near $x = 0$ looks like Fig. 16-7a when $t = 0$. Note that at $x = 0$, the displacement is $y = 0$ and the slope is at its maximum positive value. We can generalize Eq. 16-2 by inserting a **phase constant** ϕ in the wave function:

$$y = y_m \sin(kx - \omega t + \phi). \tag{16-10}$$

The value of ϕ can be chosen so that the function gives some other displacement and slope at $x = 0$ when $t = 0$. For example, a choice of $\phi = +\pi/5$ rad gives the displacement and slope shown in Fig. 16-7b when $t = 0$. The wave is still sinusoidal with the same values of y_m, k, and ω, but it is now shifted from what you see in Fig. 16-7a (where $\phi = 0$).

16-5 | The Speed of a Traveling Wave

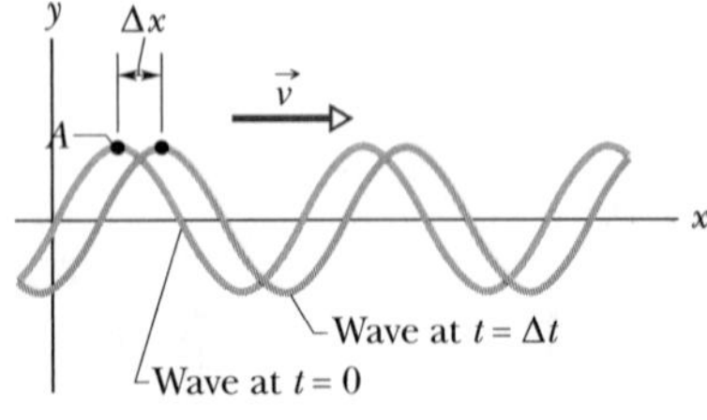

FIG. 16-8 Two snapshots of the wave of Fig. 16-5, at time $t = 0$ and then at time $t = \Delta t$. As the wave moves to the right at velocity $\vec{v}$, the entire curve shifts a distance Δx during Δt. Point A "rides" with the wave form, but the string elements move only up and down.

Figure 16-8 shows two snapshots of the wave of Eq. 16-2, taken a small time interval Δt apart. The wave is traveling in the positive direction of x (to the right in Fig. 16-8), the entire wave pattern moving a distance Δx in that direction during the interval Δt. The ratio $\Delta x/\Delta t$ (or, in the differential limit, dx/dt) is the **wave speed** v. How can we find its value?

As the wave in Fig. 16-8 moves, each point of the moving wave form, such as point A marked on a peak, retains its displacement y. (Points on the string do not retain their displacement, but points on the wave *form* do.) If point A retains its displacement as it moves, the phase in Eq. 16-2 giving it that displacement must remain a constant:

$$kx - \omega t = \text{a constant.} \tag{16-11}$$

Note that although this argument is constant, both x and t are changing. In fact, as t increases, x must also, to keep the argument constant. This confirms that the wave pattern is moving in the positive direction of x.

To find the wave speed v, we take the derivative of Eq. 16-11, getting

$$k\frac{dx}{dt} - \omega = 0$$

or

$$\frac{dx}{dt} = v = \frac{\omega}{k}. \tag{16-12}$$

Using Eq. 16-5 ($k = 2\pi/\lambda$) and Eq. 16-8 ($\omega = 2\pi/T$), we can rewrite the wave speed as

$$v = \frac{\omega}{k} = \frac{\lambda}{T} = \lambda f \quad \text{(wave speed).} \tag{16-13}$$

The equation $v = \lambda/T$ tells us that the wave speed is one wavelength per period; the wave moves a distance of one wavelength in one period of oscillation.

Equation 16-2 describes a wave moving in the positive direction of x. We can find the equation of a wave traveling in the opposite direction by replacing t in Eq. 16-2 with $-t$. This corresponds to the condition

$$kx + \omega t = \text{a constant}, \tag{16-14}$$

which (compare Eq. 16-11) requires that x *decrease* with time. Thus, a wave traveling in the negative direction of x is described by the equation

$$y(x, t) = y_m \sin(kx + \omega t). \tag{16-15}$$

If you analyze the wave of Eq. 16-15 as we have just done for the wave of Eq. 16-2, you will find for its velocity

$$\frac{dx}{dt} = -\frac{\omega}{k}. \tag{16-16}$$

The minus sign (compare Eq. 16-12) verifies that the wave is indeed moving in the negative direction of x and justifies our switching the sign of the time variable.

Consider now a wave of arbitrary shape, given by

$$y(x, t) = h(kx \pm \omega t), \tag{16-17}$$

where h represents *any* function, the sine function being one possibility. Our previous analysis shows that all waves in which the variables x and t enter into the combination $kx \pm \omega t$ are traveling waves. Furthermore, all traveling waves *must* be of the form of Eq. 16-17. Thus, $y(x, t) = \sqrt{ax + bt}$ represents a possible (though perhaps physically a little bizarre) traveling wave. The function $y(x, t) = \sin(ax^2 - bt)$, on the other hand, does *not* represent a traveling wave.

CHECKPOINT 2 Here are the equations of three waves: (1) $y(x, t) = 2\sin(4x - 2t)$, (2) $y(x, t) = \sin(3x - 4t)$, (3) $y(x, t) = 2\sin(3x - 3t)$. Rank the waves according to their (a) wave speed and (b) maximum speed perpendicular to the wave's direction of travel (the transverse speed), greatest first.

Sample Problem 16-2

A wave traveling along a string is described by

$$y(x, t) = 0.00327\sin(72.1x - 2.72t), \tag{16-18}$$

in which the numerical constants are in SI units (0.00327 m, 72.1 rad/m, and 2.72 rad/s).

(a) What is the amplitude of this wave?

KEY IDEA Equation 16-18 is of the same form as Eq. 16-2,

$$y = y_m \sin(kx - \omega t), \tag{16-19}$$

so we have a sinusoidal wave. By comparing the two equations, we can find the amplitude.

Calculation: We see that

$$y_m = 0.00327 \text{ m} = 3.27 \text{ mm}. \quad \text{(Answer)}$$

(b) What are the wavelength, period, and frequency of this wave?

Calculations: By comparing Eqs. 16-18 and 16-19, we see that the angular wave number and angular frequency are

$$k = 72.1 \text{ rad/m} \quad \text{and} \quad \omega = 2.72 \text{ rad/s}.$$

We then relate wavelength λ to k via Eq. 16-5:

$$\lambda = \frac{2\pi}{k} = \frac{2\pi \text{ rad}}{72.1 \text{ rad/m}} = 0.0871 \text{ m} = 8.71 \text{ cm}. \qquad \text{(Answer)}$$

Next, we relate T to ω with Eq. 16-8:

$$T = \frac{2\pi}{\omega} = \frac{2\pi \text{ rad}}{2.72 \text{ rad/s}} = 2.31 \text{ s}, \qquad \text{(Answer)}$$

and from Eq. 16-9 we have

$$f = \frac{1}{T} = \frac{1}{2.31 \text{ s}} = 0.433 \text{ Hz}. \qquad \text{(Answer)}$$

(c) What is the velocity of this wave?

Calculation: The speed of the wave is given by Eq. 16-13:

$$v = \frac{\omega}{k} = \frac{2.72 \text{ rad/s}}{72.1 \text{ rad/m}} = 0.0377 \text{ m/s} = 3.77 \text{ cm/s}. \qquad \text{(Answer)}$$

Because the phase in Eq. 16-18 contains the position variable x, the wave is moving along the x axis. Also, because the wave equation is written in the form of Eq. 16-2, the *minus* sign in front of the ωt term indicates that the wave is moving in the *positive* direction of the x axis. (Note that the quantities calculated in (b) and (c) are independent of the amplitude of the wave.)

(d) What is the displacement y at $x = 22.5$ cm and $t = 18.9$ s?

Calculation: Equation 16-18 gives the displacement as a function of position x and time t. Substituting the given values into the equation yields

$$\begin{aligned} y &= 0.00327 \sin(72.1 \times 0.225 - 2.72 \times 18.9) \\ &= (0.00327 \text{ m}) \sin(-35.1855 \text{ rad}) \\ &= (0.00327 \text{ m})(0.588) \\ &= 0.00192 \text{ m} = 1.92 \text{ mm}. \qquad \text{(Answer)} \end{aligned}$$

Thus, the displacement is positive. (Be sure to change your calculator mode to radians before evaluating the sine.)

Sample Problem 16-3

In Sample Problem 16-2d, we showed that at $t = 18.9$ s the transverse displacement y of the element of the string at $x = 0.255$ m due to the wave of Eq. 16-18 is 1.92 mm.

(a) What is u, the transverse velocity of the same element of the string, at that time? (This velocity, which is associated with the transverse oscillation of an element of the string, is in the y direction. Do not confuse it with v, the constant velocity at which the *wave form* travels along the x axis.)

KEY IDEAS The transverse velocity u is the rate at which the displacement y of the element is changing. In general, that displacement is given by

$$y(x, t) = y_m \sin(kx - \omega t). \qquad (16\text{-}20)$$

For an element at a certain location x, we find the rate of change of y by taking the derivative of Eq. 16-20 with respect to t while treating x as a constant. A derivative taken while one (or more) of the variables is treated as a constant is called a *partial derivative* and is represented by the symbol $\partial/\partial x$ rather than d/dx.

Calculations: Here we have

$$u = \frac{\partial y}{\partial t} = -\omega y_m \cos(kx - \omega t). \qquad (16\text{-}21)$$

Next, substituting numerical values from Sample Problem 16-2, we obtain

$$\begin{aligned} u &= (-2.72 \text{ rad/s})(3.27 \text{ mm}) \cos(-35.1855 \text{ rad}) \\ &= 7.20 \text{ mm/s}. \qquad \text{(Answer)} \end{aligned}$$

Thus, at $t = 18.9$ s, the element of the string at $x = 22.5$ cm is moving in the positive direction of y with a speed of 7.20 mm/s.

(b) What is the transverse acceleration a_y of the same element at that time?

KEY IDEA The transverse acceleration a_y is the rate at which the transverse velocity of the element is changing.

Calculations: From Eq. 16-21, again treating x as a constant but allowing t to vary, we find

$$a_y = \frac{\partial u}{\partial t} = -\omega^2 y_m \sin(kx - \omega t).$$

Comparison with Eq. 16-20 shows that we can write this as

$$a_y = -\omega^2 y.$$

We see that the transverse acceleration of an oscillating string element is proportional to its transverse displacement but opposite in sign. This is completely consistent with the action of the element itself—namely, that it is moving transversely in simple harmonic motion. Substituting numerical values yields

$$\begin{aligned} a_y &= -(2.72 \text{ rad/s})^2(1.92 \text{ mm}) \\ &= -14.2 \text{ mm/s}^2. \qquad \text{(Answer)} \end{aligned}$$

Thus, at $t = 18.9$ s, the element of string at $x = 22.5$ cm is displaced from its equilibrium position by 1.92 mm in the positive y direction and has an acceleration of magnitude 14.2 mm/s^2 in the negative y direction.

PROBLEM-SOLVING TACTICS

***Tactic 1:* Evaluating Large Phases** Sometimes, as in Sample Problems 16-2d and 16-3, an angle much greater than 2π rad (or 360°) crops up and you are asked to find its sine or cosine. Adding or subtracting an integral multiple of 2π rad to such an angle does not change the value of any of its trigonometric functions. In Sample Problem 16-2d, for example, we used the angle -35.1855 rad in a sine function. Adding $(6)(2\pi \text{ rad})$ to this angle yields

$$-35.1855 \text{ rad} + (6)(2\pi \text{ rad}) = 2.51361 \text{ rad},$$

an angle of less than 2π rad that has the same trigonometric functions as -35.1855 rad (Fig. 16-9). As an example, the sine of both 2.51361 rad and -35.1855 rad is 0.588.

Your calculator will reduce such large angles for you automatically. *Caution:* Do not round off large angles if you intend to take their sines or cosines. In taking the sine of a very large angle, you are throwing away most of the angle and taking the sine of what is left over. If, for example, you were to round -35.1855 rad to -35 rad (a change of 0.5% and normally a reasonable step), you would be changing the sine of the angle by 27%. Also, if you change a large angle from degrees to radians, be sure to use an exact conversion factor (such as $180° = \pi$ rad) rather than an approximate one (such as $57.3° \approx 1$ rad).

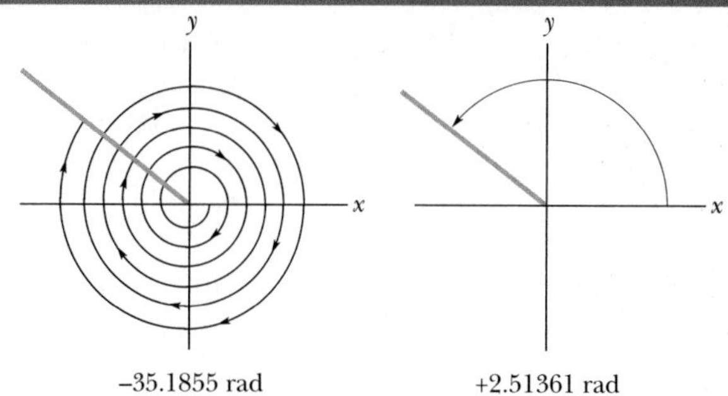

FIG. 16-9 These two angles are different, but all their trigonometric functions are identical.

16-6 | Wave Speed on a Stretched String

The speed of a wave is related to the wave's wavelength and frequency by Eq. 16-13, but *it is set by the properties of the medium.* If a wave is to travel through a medium such as water, air, steel, or a stretched string, it must cause the particles of that medium to oscillate as it passes. For that to happen, the medium must possess both mass (so that there can be kinetic energy) and elasticity (so that there can be potential energy). Thus, the medium's mass and elasticity properties determine how fast the wave can travel in the medium. Conversely, it should be possible to calculate the speed of the wave through the medium in terms of these properties. We do so now for a stretched string, in two ways.

Dimensional Analysis

In dimensional analysis we carefully examine the dimensions of all the physical quantities that enter into a given situation to determine the quantities they produce. In this case, we examine mass and elasticity to find a speed v, which has the dimension of length divided by time, or LT^{-1}.

For the mass, we use the mass of a string element, which is the mass m of the string divided by the length l of the string. We call this ratio the *linear density* μ of the string. Thus, $\mu = m/l$, its dimension being mass divided by length, ML^{-1}.

You cannot send a wave along a string unless the string is under tension, which means that it has been stretched and pulled taut by forces at its two ends. The tension τ in the string is equal to the common magnitude of those two forces. As a wave travels along the string, it displaces elements of the string by causing additional stretching, with adjacent sections of string pulling on each other because of the tension. Thus, we can associate the tension in the string with the stretching (elasticity) of the string. The tension and the stretching forces it produces have the dimension of a force—namely, MLT^{-2} (from $F = ma$).

We need to combine μ (dimension ML^{-1}) and τ (dimension MLT^{-2}) to get v (dimension LT^{-1}). A little juggling of various combinations suggests

$$v = C\sqrt{\frac{\tau}{\mu}}, \tag{16-22}$$

in which C is a dimensionless constant that cannot be determined with dimensional analysis. In our second approach to determining wave speed, you will see that Eq. 16-22 is indeed correct and that $C = 1$.

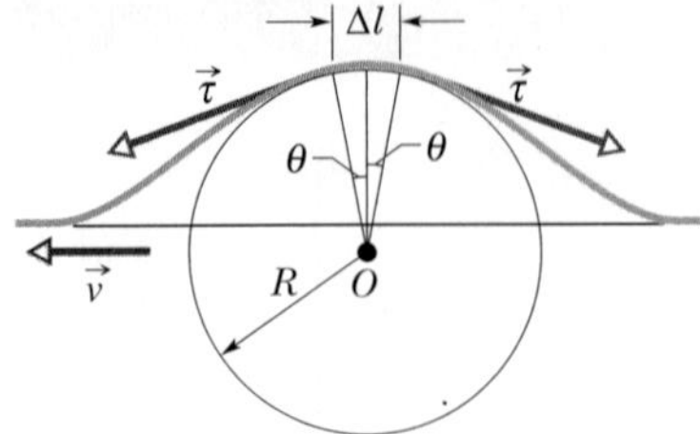

FIG. 16-10 A symmetrical pulse, viewed from a reference frame in which the pulse is stationary and the string appears to move right to left with speed v. We find speed v by applying Newton's second law to a string element of length Δl, located at the top of the pulse.

Derivation from Newton's Second Law

Instead of the sinusoidal wave of Fig. 16-1*b*, let us consider a single symmetrical pulse such as that of Fig. 16-10, moving from left to right along a string with speed v. For convenience, we choose a reference frame in which the pulse remains stationary; that is, we run along with the pulse, keeping it constantly in view. In this frame, the string appears to move past us, from right to left in Fig. 16-10, with speed v.

Consider a small string element of length Δl within the pulse, an element that forms an arc of a circle of radius R and subtending an angle 2θ at the center of that circle. A force $\vec{\tau}$ with a magnitude equal to the tension in the string pulls tangentially on this element at each end. The horizontal components of these forces cancel, but the vertical components add to form a radial restoring force $\vec{F}$. In magnitude,

$$F = 2(\tau \sin \theta) \approx \tau(2\theta) = \tau \frac{\Delta l}{R} \quad \text{(force)}, \tag{16-23}$$

where we have approximated $\sin \theta$ as θ for the small angles θ in Fig. 16-10. From that figure, we have also used $2\theta = \Delta l/R$. The mass of the element is given by

$$\Delta m = \mu \, \Delta l \quad \text{(mass)}, \tag{16-24}$$

where μ is the string's linear density.

At the moment shown in Fig. 16-10, the string element Δl is moving in an arc of a circle. Thus, it has a centripetal acceleration toward the center of that circle, given by

$$a = \frac{v^2}{R} \quad \text{(acceleration)}. \tag{16-25}$$

Equations 16-23, 16-24, and 16-25 contain the elements of Newton's second law. Combining them in the form

$$\text{force} = \text{mass} \times \text{acceleration}$$

gives

$$\frac{\tau \, \Delta l}{R} = (\mu \, \Delta l) \frac{v^2}{R}.$$

Solving this equation for the speed v yields

$$v = \sqrt{\frac{\tau}{\mu}} \quad \text{(speed)}, \tag{16-26}$$

in exact agreement with Eq. 16-22 if the constant C in that equation is given the value unity. Equation 16-26 gives the speed of the pulse in Fig. 16-10 and the speed of *any* other wave on the same string under the same tension.

Equation 16-26 tells us:

The speed of a wave along a stretched ideal string depends only on the tension and linear density of the string and not on the frequency of the wave.

The *frequency* of the wave is fixed entirely by whatever generates the wave (for example, the person in Fig. 16-1*b*). The *wavelength* of the wave is then fixed by Eq. 16-13 in the form $\lambda = v/f$.

CHECKPOINT 3 You send a traveling wave along a particular string by oscillating one end. If you increase the frequency of the oscillations, do (a) the speed of the wave and (b) the wavelength of the wave increase, decrease, or remain the same? If, instead, you increase the tension in the string, do (c) the speed of the wave and (d) the wavelength of the wave increase, decrease, or remain the same?

Sample Problem 16-4

In Fig. 16-11, two strings have been tied together with a knot and then stretched between two rigid supports. The strings have linear densities $\mu_1 = 1.4 \times 10^{-4}$ kg/m and $\mu_2 = 2.8 \times 10^{-4}$ kg/m. Their lengths are $L_1 = 3.0$ m and $L_2 = 2.0$ m, and string 1 is under a tension of 400 N. Simultaneously, on each string a pulse is sent from the rigid support end, toward the knot. Which pulse reaches the knot first?

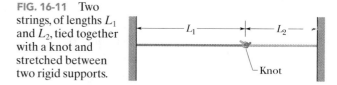

KEY IDEAS

1. The time t taken by a pulse to travel a length L is $t = L/v$, where v is the constant speed of the pulse.
2. The speed of a pulse on a stretched string depends on the string's tension τ and linear density μ, and is given by Eq. 16-26 ($v = \sqrt{\tau/\mu}$).
3. Because the two strings are stretched together, they must both be under the same tension τ (= 400 N).

Calculations: Putting these three ideas together gives us, as the time for the pulse on string 1 to reach the knot,

$$t_1 = \frac{L_1}{v_1} = L_1\sqrt{\frac{\mu_1}{\tau}} = (3.0\text{ m})\sqrt{\frac{1.4 \times 10^{-4}\text{ kg/m}}{400\text{ N}}}$$
$$= 1.77 \times 10^{-3}\text{ s.}$$

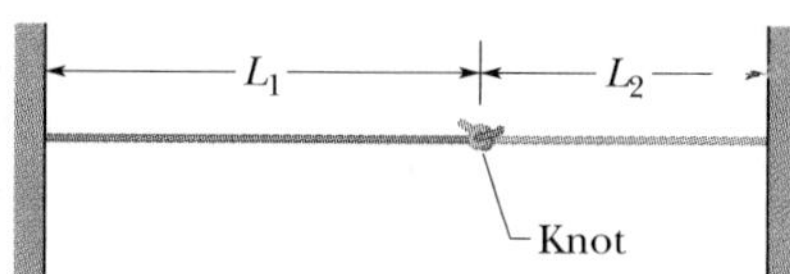

FIG. 16-11 Two strings, of lengths L_1 and L_2, tied together with a knot and stretched between two rigid supports.

Similarly, the data for the pulse on string 2 give us

$$t_2 = L_2\sqrt{\frac{\mu_2}{\tau}} = 1.67 \times 10^{-3}\text{ s.}$$

Thus, the pulse on string 2 reaches the knot first.

Now look back at the second key idea. The linear density of string 2 is greater than that of string 1, so the pulse on string 2 must be slower than that on string 1. Could we have guessed the answer from that fact alone? No, because from the first key idea we see that the distance traveled by a pulse also matters.

16-7 | Energy and Power of a Wave Traveling Along a String

When we set up a wave on a stretched string, we provide energy for the motion of the string. As the wave moves away from us, it transports that energy as both kinetic energy and elastic potential energy. Let us consider each form in turn.

Kinetic Energy

A string element of mass dm, oscillating transversely in simple harmonic motion as the wave passes through it, has kinetic energy associated with its transverse velocity $\vec{u}$. When the element is rushing through its $y = 0$ position (element b in Fig. 16-12), its transverse velocity—and thus its kinetic energy—is a maximum. When the element is at its extreme position $y = y_m$ (as is element a), its transverse velocity—and thus its kinetic energy—is zero.

Elastic Potential Energy

To send a sinusoidal wave along a previously straight string, the wave must necessarily stretch the string. As a string element of length dx oscillates transversely, its length must increase and decrease in a periodic way if the string element is to fit the sinusoidal wave form. Elastic potential energy is associated with these length changes, just as for a spring.

When the string element is at its $y = y_m$ position (element a in Fig. 16-12), its length has its normal undisturbed value dx, so its elastic potential energy is zero. However, when the element is rushing through its $y = 0$ position, it has maximum stretch and thus maximum elastic potential energy.

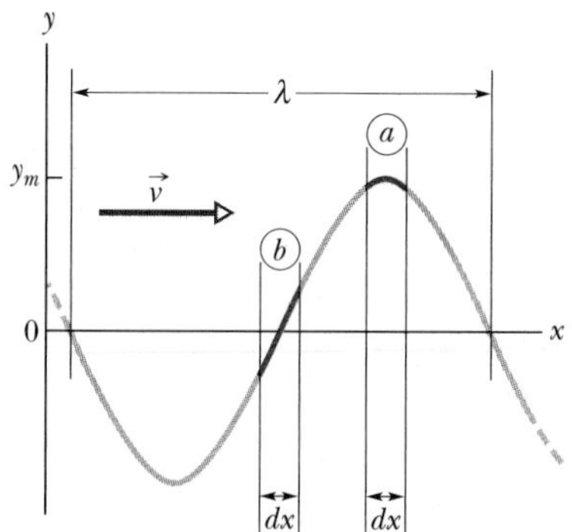

FIG. 16-12 A snapshot of a traveling wave on a string at time $t = 0$. String element a is at displacement $y = y_m$, and string element b is at displacement $y = 0$. The kinetic energy of the string element at each position depends on the transverse velocity of the element. The potential energy depends on the amount by which the string element is stretched as the wave passes through it.

Energy Transport

The oscillating string element thus has both its maximum kinetic energy and its maximum elastic potential energy at $y = 0$. In the snapshot of Fig. 16-12, the regions of the string at maximum displacement have no energy, and the regions at zero displacement have maximum energy. As the wave travels along the string, forces due to the tension in the string continuously do work to transfer energy from regions with energy to regions with no energy.

Suppose we set up a wave on a string stretched along a horizontal x axis so that Eq. 16-2 describes the string's displacement. We might send a wave along the string by continuously oscillating one end of the string, as in Fig. 16-1*b*. In doing so, we continuously provide energy for the motion and stretching of the string—as the string sections oscillate perpendicularly to the x axis, they have kinetic energy and elastic potential energy. As the wave moves into sections that were previously at rest, energy is transferred into those new sections. Thus, we say that the wave *transports* the energy along the string.

The Rate of Energy Transmission

The kinetic energy dK associated with a string element of mass dm is given by

$$dK = \tfrac{1}{2}\, dm\, u^2, \tag{16-27}$$

where u is the transverse speed of the oscillating string element. To find u, we differentiate Eq. 16-2 with respect to time while holding x constant:

$$u = \frac{\partial y}{\partial t} = -\omega y_m \cos(kx - \omega t). \tag{16-28}$$

Using this relation and putting $dm = \mu\, dx$, we rewrite Eq. 16-27 as

$$dK = \tfrac{1}{2}(\mu\, dx)(-\omega y_m)^2 \cos^2(kx - \omega t). \tag{16-29}$$

Dividing Eq. 16-29 by dt gives the rate at which kinetic energy passes through a string element, and thus the rate at which kinetic energy is carried along by the wave. The ratio dx/dt that then appears on the right of Eq. 16-29 is the wave speed v, so we obtain

$$\frac{dK}{dt} = \tfrac{1}{2}\mu v \omega^2 y_m^2 \cos^2(kx - \omega t). \tag{16-30}$$

The *average* rate at which kinetic energy is transported is

$$\left(\frac{dK}{dt}\right)_{\text{avg}} = \tfrac{1}{2}\mu v \omega^2 y_m^2 \,[\cos^2(kx - \omega t)]_{\text{avg}}$$

$$= \tfrac{1}{4}\mu v \omega^2 y_m^2. \tag{16-31}$$

Here we have taken the average over an integer number of wavelengths and have used the fact that the average value of the square of a cosine function over an integer number of periods is $\frac{1}{2}$.

Elastic potential energy is also carried along with the wave, and at the same average rate given by Eq. 16-31. Although we shall not examine the proof, you should recall that, in an oscillating system such as a pendulum or a spring–block system, the average kinetic energy and the average potential energy are equal.

The **average power,** which is the average rate at which energy of both kinds is transmitted by the wave, is then

$$P_{\text{avg}} = 2\left(\frac{dK}{dt}\right)_{\text{avg}} \tag{16-32}$$

or, from Eq. 16-31,

$$P_{\text{avg}} = \tfrac{1}{2}\mu v \omega^2 y_m^2 \quad \text{(average power)}. \tag{16-33}$$

The factors μ and v in this equation depend on the material and tension of the string. The factors ω and y_m depend on the process that generates the wave. The dependence of the average power of a wave on the square of its amplitude and also on the square of its angular frequency is a general result, true for waves of all types.

Sample Problem 16-5

A string has linear density $\mu = 525$ g/m and tension $\tau = 45$ N. We send a sinusoidal wave with frequency $f = 120$ Hz and amplitude $y_m = 8.5$ mm along the string. At what average rate does the wave transport energy?

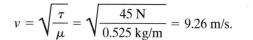

KEY IDEA The average rate of energy transport is the average power P_{avg} as given by Eq. 16-33.

Calculations: To use Eq. 16-33, we first must calculate angular frequency ω and wave speed v. From Eq. 16-9,

$$\omega = 2\pi f = (2\pi)(120 \text{ Hz}) = 754 \text{ rad/s}.$$

From Eq. 16-26 we have

$$v = \sqrt{\frac{\tau}{\mu}} = \sqrt{\frac{45 \text{ N}}{0.525 \text{ kg/m}}} = 9.26 \text{ m/s}.$$

Equation 16-33 then yields

$$P_{avg} = \tfrac{1}{2}\mu v \omega^2 y_m^2$$
$$= (\tfrac{1}{2})(0.525 \text{ kg/m})(9.26 \text{ m/s})(754 \text{ rad/s})^2(0.0085 \text{ m})^2$$
$$\approx 100 \text{ W}. \qquad \text{(Answer)}$$

16-8 | The Wave Equation

As a wave passes through any element on a stretched string, the element moves perpendicularly to the wave's direction of travel. By applying Newton's second law to the element's motion, we can derive a general differential equation, called the *wave equation*, that governs the travel of waves of any type.

Figure 16-13*a* shows a snapshot of a string element of mass dm and length ℓ as a wave travels along a string of linear density μ that is stretched along a horizontal x axis. Let us assume that the wave amplitude is small so that the element can be tilted only slightly from the x axis as the wave passes. The force $\vec{F}_2$ on the right end of the element has a magnitude equal to tension τ in the string and is directed slightly upward. The force $\vec{F}_1$ on the left end of the element also has a magnitude equal to the tension τ but is directed slightly downward. Because of the slight curvature of the element, these two forces produce a net force that causes the element to have an upward acceleration a_y. Newton's second law written for y components ($F_{net,y} = ma_y$) gives us

$$F_{2y} - F_{1y} = dm\, a_y. \qquad (16\text{-}34)$$

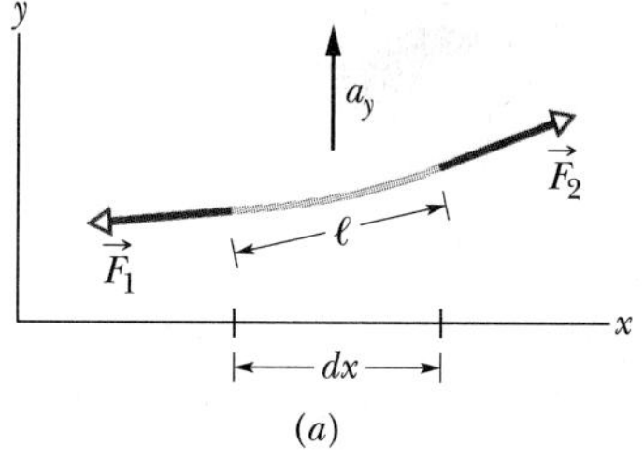

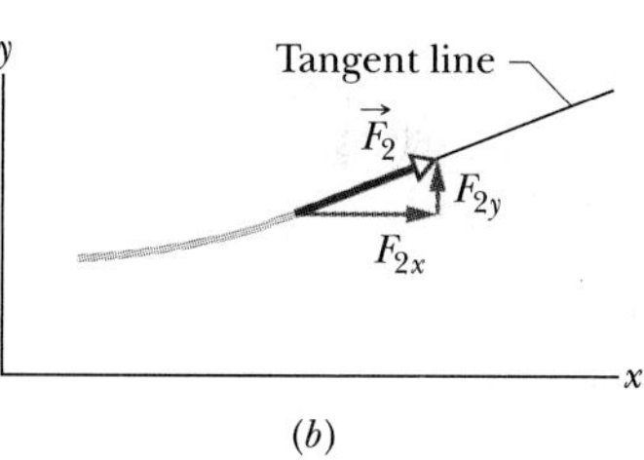

FIG. 16-13 (*a*) A string element as a sinusoidal transverse wave travels on a stretched string. Forces $\vec{F}_1$ and $\vec{F}_2$ act at the left and right ends, producing acceleration $\vec{a}$ having a vertical component a_y. (*b*) The force at the element's right end is directed along a tangent to the element's right side.

Let's analyze this equation in parts.

Mass. The element's mass dm can be written in terms of the string's linear density μ and the element's length ℓ as $dm = \mu\ell$. Because the element can have only a slight tilt, $\ell \approx dx$ (Fig. 16-13*a*) and we have the approximation

$$dm = \mu\, dx. \qquad (16\text{-}35)$$

Acceleration. The acceleration a_y in Eq. 16-34 is the second derivative of the displacement y with respect to time:

$$a_y = \frac{d^2y}{dt^2}. \qquad (16\text{-}36)$$

Forces. Figure 16-13*b* shows that $\vec{F}_2$ is tangent to the string at the right end of the string element. Thus we can relate the components of the force to the string slope S_2 at the right end as

$$\frac{F_{2y}}{F_{2x}} = S_2. \qquad (16\text{-}37)$$

We can also relate the components to the magnitude F_2 ($= \tau$) with

$$F_2 = \sqrt{F_{2x}^2 + F_{2y}^2}$$

or

$$\tau = \sqrt{F_{2x}^2 + F_{2y}^2}. \qquad (16\text{-}38)$$

However, because we assume that the element is only slightly tilted, $F_{2y} \ll F_{2x}$ and therefore we can rewrite Eq. 16-38 as

$$\tau = F_{2x}. \qquad (16\text{-}39)$$

Substituting this into Eq. 16-37 and solving for F_{2y} yield

$$F_{2y} = \tau S_2. \qquad (16\text{-}40)$$

Similar analysis at the left end of the string element gives us

$$F_{1y} = \tau S_1. \qquad (16\text{-}41)$$

We can now substitute Eqs. 16-35, 16-36, 16-40, and 16-41 into Eq. 16-34 to write

$$\tau S_2 - \tau S_1 = (\mu\, dx)\frac{d^2y}{dt^2},$$

or

$$\frac{S_2 - S_1}{dx} = \frac{\mu}{\tau}\frac{d^2y}{dt^2}. \tag{16-42}$$

Because the string element is short, slopes S_2 and S_1 differ by only a differential amount dS, where S is the slope at any point:

$$S = \frac{dy}{dx}. \tag{16-43}$$

First replacing $S_2 - S_1$ in Eq. 16-42 with dS and then using Eq. 16-43 to substitute dy/dx for S, we find

$$\frac{dS}{dx} = \frac{\mu}{\tau}\frac{d^2y}{dt^2},$$

$$\frac{d(dy/dx)}{dx} = \frac{\mu}{\tau}\frac{d^2y}{dt^2},$$

and

$$\frac{\partial^2 y}{\partial x^2} = \frac{\mu}{\tau}\frac{\partial^2 y}{\partial t^2}. \tag{16-44}$$

In the last step, we switched to the notation of partial derivatives because on the left we differentiate only with respect to x and on the right we differentiate only with respect to t. Finally, substituting from Eq. 16-26 ($v = \sqrt{\tau/\mu}$), we find

$$\frac{\partial^2 y}{\partial x^2} = \frac{1}{v^2}\frac{\partial^2 y}{\partial t^2} \quad \text{(wave equation).} \tag{16-45}$$

This is the general differential equation that governs the travel of waves of all types.

16-9 | The Principle of Superposition for Waves

It often happens that two or more waves pass simultaneously through the same region. When we listen to a concert, for example, sound waves from many instruments fall simultaneously on our eardrums. The electrons in the antennas of our radio and television receivers are set in motion by the net effect of many electromagnetic waves from many different broadcasting centers. The water of a lake or harbor may be churned up by waves in the wakes of many boats.

Suppose that two waves travel simultaneously along the same stretched string. Let $y_1(x, t)$ and $y_2(x, t)$ be the displacements that the string would experience if each wave traveled alone. The displacement of the string when the waves overlap is then the algebraic sum

$$y'(x, t) = y_1(x, t) + y_2(x, t). \tag{16-46}$$

This summation of displacements along the string means that

Overlapping waves algebraically add to produce a **resultant wave** (or **net wave**).

This is another example of the **principle of superposition,** which says that when several effects occur simultaneously, their net effect is the sum of the individual effects.

Figure 16-14 shows a sequence of snapshots of two pulses traveling in opposite directions on the same stretched string. When the pulses overlap, the resultant pulse is their sum. Moreover,

Overlapping waves do not in any way alter the travel of each other.

16-10 | Interference of Waves

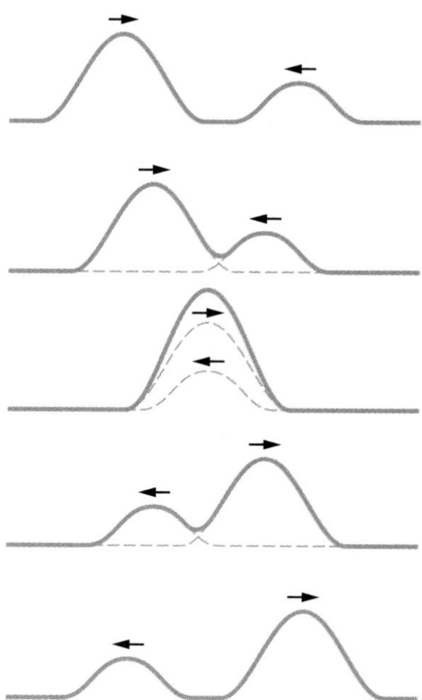

FIG. 16-14 A series of snapshots that show two pulses traveling in opposite directions along a stretched string. The superposition principle applies as the pulses move through each other.

Suppose we send two sinusoidal waves of the same wavelength and amplitude in the same direction along a stretched string. The superposition principle applies. What resultant wave does it predict for the string?

The resultant wave depends on the extent to which the waves are *in phase* (in step) with respect to each other—that is, how much one wave form is shifted from the other wave form. If the waves are exactly in phase (so that the peaks and valleys of one are exactly aligned with those of the other), they combine to double the displacement of either wave acting alone. If they are exactly out of phase (the peaks of one are exactly aligned with the valleys of the other), they combine to cancel everywhere, and the string remains straight. We call this phenomenon of combining waves **interference,** and the waves are said to **interfere.** (These terms refer only to the wave displacements; the travel of the waves is unaffected.)

Let one wave traveling along a stretched string be given by

$$y_1(x, t) = y_m \sin(kx - \omega t) \tag{16-47}$$

and another, shifted from the first, by

$$y_2(x, t) = y_m \sin(kx - \omega t + \phi). \tag{16-48}$$

These waves have the same angular frequency ω (and thus the same frequency f), the same angular wave number k (and thus the same wavelength λ), and the same amplitude y_m. They both travel in the positive direction of the x axis, with the same speed, given by Eq. 16-26. They differ only by a constant angle ϕ, the phase constant. These waves are said to be *out of phase* by ϕ or to have a *phase difference* of ϕ, or one wave is said to be *phase-shifted* from the other by ϕ.

From the principle of superposition (Eq. 16-46), the resultant wave is the algebraic sum of the two interfering waves and has displacement

$$y'(x, t) = y_1(x, t) + y_2(x, t)$$
$$= y_m \sin(kx - \omega t) + y_m \sin(kx - \omega t + \phi). \tag{16-49}$$

In Appendix E we see that we can write the sum of the sines of two angles α and β as

$$\sin \alpha + \sin \beta = 2 \sin \tfrac{1}{2}(\alpha + \beta) \cos \tfrac{1}{2}(\alpha - \beta). \tag{16-50}$$

Applying this relation to Eq. 16-49 leads to

$$y'(x, t) = [2y_m \cos \tfrac{1}{2}\phi] \sin(kx - \omega t + \tfrac{1}{2}\phi). \tag{16-51}$$

As Fig. 16-15 shows, the resultant wave is also a sinusoidal wave traveling in the direction of increasing x. It is the only wave you would actually see on the string (you would *not* see the two interfering waves of Eqs. 16-47 and 16-48).

> If two sinusoidal waves of the same amplitude and wavelength travel in the *same* direction along a stretched string, they interfere to produce a resultant sinusoidal wave traveling in that direction.

Displacement

$$y'(x,t) = [2y_m \cos \tfrac{1}{2}\phi] \sin(kx - \omega t + \tfrac{1}{2}\phi)$$

Magnitude gives amplitude — Oscillating term

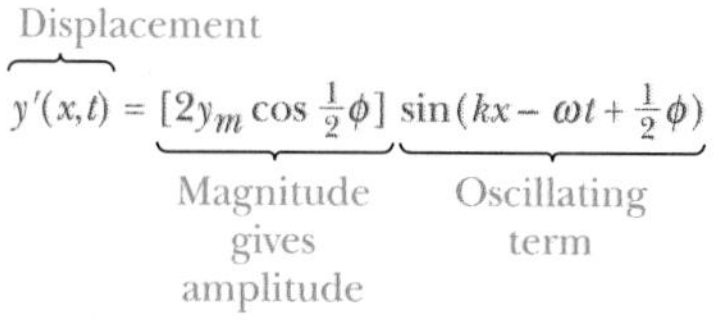

FIG. 16-15 The resultant wave of Eq. 16-51, due to the interference of two sinusoidal transverse waves, is also a sinusoidal transverse wave, with an amplitude and an oscillating term.

The resultant wave differs from the interfering waves in two respects: (1) its phase constant is $\frac{1}{2}\phi$, and (2) its amplitude y'_m is the magnitude of the quantity in the brackets in Eq. 16-51:

$$y'_m = |2y_m \cos \tfrac{1}{2}\phi| \quad \text{(amplitude)}. \tag{16-52}$$

If $\phi = 0$ rad (or 0°), the two interfering waves are exactly in phase, as in Fig. 16-16*a*. Then Eq. 16-51 reduces to

$$y'(x, t) = 2y_m \sin(kx - \omega t) \quad (\phi = 0). \tag{16-53}$$

This resultant wave is plotted in Fig. 16-16*d*. Note from both that figure and Eq. 16-53 that the amplitude of the resultant wave is twice the amplitude of either

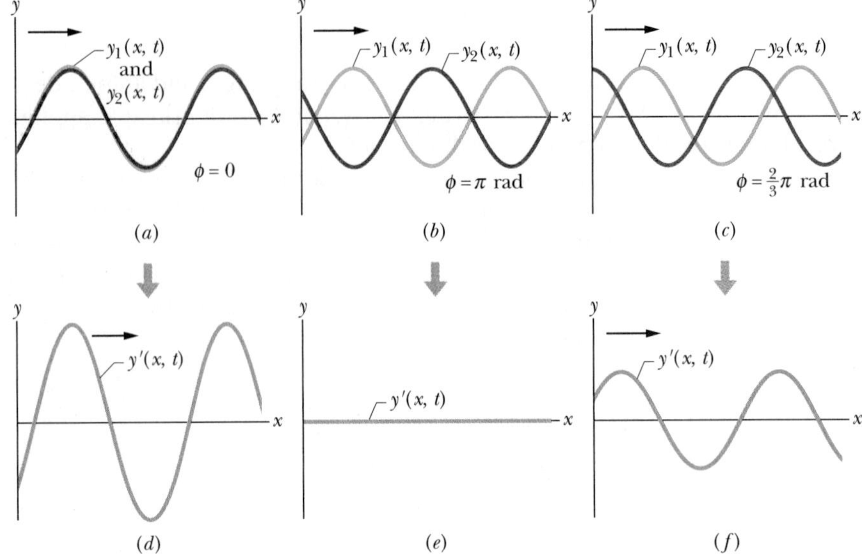

FIG. 16-16 Two identical sinusoidal waves, $y_1(x, t)$ and $y_2(x, t)$, travel along a string in the positive direction of an x axis. They interfere to give a resultant wave $y'(x, t)$. The resultant wave is what is actually seen on the string. The phase difference ϕ between the two interfering waves is (*a*) 0 rad or 0°, (*b*) π rad or 180°, and (*c*) $\frac{2}{3}\pi$ rad or 120°. The corresponding resultant waves are shown in (*d*), (*e*), and (*f*).

interfering wave. That is the greatest amplitude the resultant wave can have, because the cosine term in Eqs. 16-51 and 16-52 has its greatest value (unity) when $\phi = 0$. Interference that produces the greatest possible amplitude is called *fully constructive interference.*

If $\phi = \pi$ rad (or 180°), the interfering waves are exactly out of phase as in Fig. 16-16*b*. Then $\cos \frac{1}{2}\phi$ becomes $\cos \pi/2 = 0$, and the amplitude of the resultant wave as given by Eq. 16-52 is zero. We then have, for all values of x and t,

$$y'(x, t) = 0 \qquad (\phi = \pi \text{ rad}). \tag{16-54}$$

The resultant wave is plotted in Fig. 16-16*e*. Although we sent two waves along the string, we see no motion of the string. This type of interference is called *fully destructive interference*.

Because a sinusoidal wave repeats its shape every 2π rad, a phase difference $\phi = 2\pi$ rad (or 360°) corresponds to a shift of one wave relative to the other wave by a distance equivalent to one wavelength. Thus, phase differences can be described in terms of wavelengths as well as angles. For example, in Fig. 16-16*b* the waves may be said to be 0.50 wavelength out of phase. Table 16-1 shows some other examples of phase differences and the interference they produce. Note that when interference is neither fully constructive nor fully destructive, it is called *intermediate interference*. The amplitude of the resultant wave is then intermediate between 0 and $2y_m$. For example, from Table 16-1, if the interfering waves

TABLE 16-1

Phase Difference and Resulting Interference Types[a]

Phase Difference, in			Amplitude of Resultant Wave	Type of Interference
Degrees	Radians	Wavelengths		
0	0	0	$2y_m$	Fully constructive
120	$\frac{2}{3}\pi$	0.33	y_m	Intermediate
180	π	0.50	0	Fully destructive
240	$\frac{4}{3}\pi$	0.67	y_m	Intermediate
360	2π	1.00	$2y_m$	Fully constructive
865	15.1	2.40	$0.60y_m$	Intermediate

[a]The phase difference is between two otherwise identical waves, with amplitude y_m, moving in the same direction.

have a phase difference of 120° ($\phi = \frac{2}{3}\pi$ rad = 0.33 wavelength), then the resultant wave has an amplitude of y_m, the same as that of the interfering waves (see Figs. 16-16*c* and *f*).

Two waves with the same wavelength are in phase if their phase difference is zero or any integer number of wavelengths. Thus, the integer part of any phase difference *expressed in wavelengths* may be discarded. For example, a phase difference of 0.40 wavelength is equivalent in every way to one of 2.40 wavelengths, and so the simpler of the two numbers can be used in computations.

CHECKPOINT 4 Here are four possible phase differences between two identical waves, expressed in wavelengths: 0.20, 0.45, 0.60, and 0.80. Rank them according to the amplitude of the resultant wave, greatest first.

Sample Problem 16-6

Two identical sinusoidal waves, moving in the same direction along a stretched string, interfere with each other. The amplitude y_m of each wave is 9.8 mm, and the phase difference ϕ between them is 100°.

(a) What is the amplitude y'_m of the resultant wave due to the interference, and what is the type of this interference?

KEY IDEA These are identical sinusoidal waves traveling in the *same direction* along a string, so they interfere to produce a sinusoidal traveling wave.

Calculations: Because they are identical, the waves have the *same amplitude.* Thus, the amplitude y'_m of the resultant wave is given by Eq. 16-52:

$$y'_m = |2y_m \cos \tfrac{1}{2}\phi| = |(2)(9.8 \text{ mm}) \cos(100°/2)| = 13 \text{ mm}. \quad \text{(Answer)}$$

We can tell that the interference is *intermediate* in two ways. The phase difference is between 0 and 180°, and, correspondingly, the amplitude y'_m is between 0 and $2y_m$ (= 19.6 mm).

(b) What phase difference, in radians and wavelengths, will give the resultant wave an amplitude of 4.9 mm?

Calculations: Now we are given y'_m and seek ϕ. From Eq. 16-52,

$$y'_m = |2y_m \cos \tfrac{1}{2}\phi|,$$

we now have

$$4.9 \text{ mm} = (2)(9.8 \text{ mm}) \cos \tfrac{1}{2}\phi,$$

which gives us (with a calculator in the radian mode)

$$\phi = 2 \cos^{-1} \frac{4.9 \text{ mm}}{(2)(9.8 \text{ mm})} = \pm 2.636 \text{ rad} \approx \pm 2.6 \text{ rad}. \quad \text{(Answer)}$$

There are two solutions because we can obtain the same resultant wave by letting the first wave *lead* (travel ahead of) or *lag* (travel behind) the second wave by 2.6 rad. In wavelengths, the phase difference is

$$\frac{\phi}{2\pi \text{ rad/wavelength}} = \frac{\pm 2.636 \text{ rad}}{2\pi \text{ rad/wavelength}} = \pm 0.42 \text{ wavelength. (Answer)}$$

16-11 | Phasors

We can represent a string wave (or any other type of wave) vectorially with a **phasor.** In essence, a phasor is a vector that has a magnitude equal to the amplitude of the wave and that rotates around an origin; the angular speed of the phasor is equal to the angular frequency ω of the wave. For example, the wave

$$y_1(x, t) = y_{m1} \sin(kx - \omega t) \tag{16-55}$$

is represented by the phasor shown in Fig. 16-17*a*. The magnitude of the phasor is the amplitude y_{m1} of the wave. As the phasor rotates around the origin at angular speed ω, its projection y_1 on the vertical axis varies sinusoidally, from a maximum of y_{m1} through zero to a minimum of $-y_{m1}$ and then back to y_{m1}. This variation corresponds to the sinusoidal variation in the displacement y_1 of any point along the string as the wave passes through that point.

When two waves travel along the same string in the same direction, we can represent them and their resultant wave in a *phasor diagram*. The phasors in Fig.

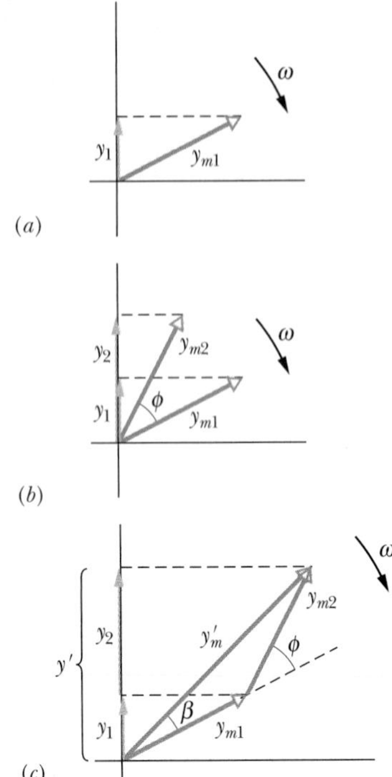

FIG. 16-17 (a) A phasor of magnitude y_{m1} rotating about an origin at angular speed ω represents a sinusoidal wave. The phasor's projection y_1 on the vertical axis represents the displacement of a point through which the wave passes. (b) A second phasor, also of angular speed ω but of magnitude y_{m2} and rotating at a constant angle ϕ from the first phasor, represents a second wave, with a phase constant ϕ. (c) The resultant wave is represented by the vector sum y'_m of the two phasors.

16-17b represent the wave of Eq. 16-55 and a second wave given by

$$y_2(x, t) = y_{m2} \sin(kx - \omega t + \phi). \tag{16-56}$$

This second wave is phase-shifted from the first wave by phase constant ϕ. Because the phasors rotate at the same angular speed ω, the angle between the two phasors is always ϕ. If ϕ is a *positive* quantity, then the phasor for wave 2 *lags* the phasor for wave 1 as they rotate, as drawn in Fig. 16-17b. If ϕ is a negative quantity, then the phasor for wave 2 *leads* the phasor for wave 1.

Because waves y_1 and y_2 have the same angular wave number k and angular frequency ω, we know from Eqs. 16-51 and 16-52 that their resultant is of the form

$$y'(x, t) = y'_m \sin(kx - \omega t + \beta), \tag{16-57}$$

where y'_m is the amplitude of the resultant wave and β is its phase constant. To find the values of y'_m and β, we would have to sum the two combining waves, as we did to obtain Eq. 16-51. To do this on a phasor diagram, we vectorially add the two phasors at any instant during their rotation, as in Fig. 16-17c where phasor y_{m2} has been shifted to the head of phasor y_{m1}. The magnitude of the vector sum equals the amplitude y'_m in Eq. 16-57. The angle between the vector sum and the phasor for y_1 equals the phase constant β in Eq. 16-57.

Note that, in contrast to the method of Section 16-10:

We can use phasors to combine waves *even if their amplitudes are different.*

Sample Problem 16-7

Two sinusoidal waves $y_1(x, t)$ and $y_2(x, t)$ have the same wavelength and travel together in the same direction along a string. Their amplitudes are $y_{m1} = 4.0$ mm and $y_{m2} = 3.0$ mm, and their phase constants are 0 and $\pi/3$ rad, respectively. What are the amplitude y'_m and phase constant β of the resultant wave? Write the resultant wave in the form of Eq. 16-57.

KEY IDEAS (1) The two waves have a number of properties in common: Because they travel along the same string, they must have the same speed v, as set by the tension and linear density of the string according to Eq. 16-26. With the same wavelength λ, they must have the same angular wave number k $(= 2\pi/\lambda)$. Also, with the same wave number k and speed v, they must have the same angular frequency ω $(= kv)$.

(2) The waves (call them waves 1 and 2) can be represented by phasors rotating at the same angular speed ω about an origin. Because the phase constant for wave 2 is *greater* than that for wave 1 by $\pi/3$, phasor 2 must *lag* phasor 1 by $\pi/3$ rad in their clockwise rotation, as shown in Fig. 16-18a. The resultant wave due to the interference of waves 1 and 2 can then be represented by a phasor that is the vector sum of phasors 1 and 2.

Calculations: To simplify the vector summation, we drew phasors 1 and 2 in Fig. 16-18a at the instant when phasor 1 lies along the horizontal axis. We then drew lagging phasor 2 at positive angle $\pi/3$ rad. In Fig. 16-18b we shifted phasor 2 so its tail is at the head of phasor 1. Then we can draw the phasor y'_m of the resultant wave from the tail of phasor 1 to the head of phasor 2. The phase constant β is the angle phasor y'_m makes with phasor 1.

To find values for y'_m and β, we can sum phasors 1 and 2 directly on a vector-capable calculator (by adding a vector of magnitude 4.0 and angle 0 rad to a vector of

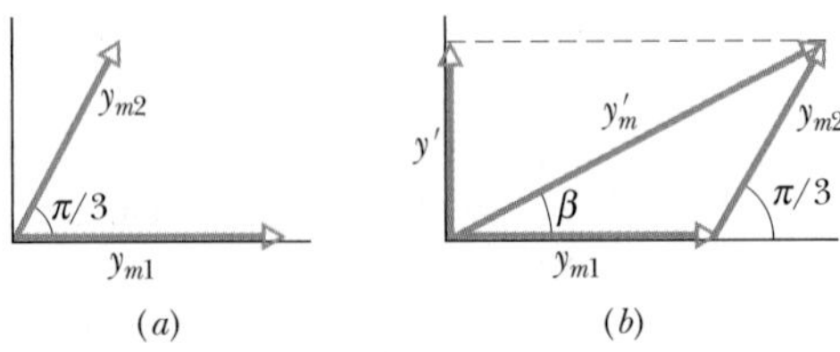

FIG. 16-18 (a) Two phasors of magnitudes y_{m1} and y_{m2} and with phase difference $\pi/3$. (b) Vector addition of these phasors at any instant during their rotation gives the magnitude y'_m of the phasor for the resultant wave.

magnitude 3.0 and angle $\pi/3$ rad) or we can add the vectors by components. For the horizontal components we have

$$y'_{mh} = y_{m1} \cos 0 + y_{m2} \cos \pi/3$$
$$= 4.0 \text{ mm} + (3.0 \text{ mm}) \cos \pi/3 = 5.50 \text{ mm}.$$

For the vertical components we have

$$y'_{mv} = y_{m1} \sin 0 + y_{m2} \sin \pi/3$$
$$= 0 + (3.0 \text{ mm}) \sin \pi/3 = 2.60 \text{ mm}.$$

Thus, the resultant wave has an amplitude of

$$y'_m = \sqrt{(5.50 \text{ mm})^2 + (2.60 \text{ mm})^2}$$
$$= 6.1 \text{ mm} \qquad \text{(Answer)}$$

and a phase constant of

$$\beta = \tan^{-1} \frac{2.60 \text{ mm}}{5.50 \text{ mm}} = 0.44 \text{ rad.} \qquad \text{(Answer)}$$

From Fig. 16-18*b*, phase constant β is a *positive* angle relative to phasor 1. Thus, the resultant wave *lags* wave 1 in their travel by phase constant $\beta = +0.44$ rad. From Eq. 16-57, we can write the resultant wave as

$$y'(x, t) = (6.1 \text{ mm}) \sin(kx - \omega t + 0.44 \text{ rad}). \qquad \text{(Answer)}$$

16-12 | Standing Waves

In Section 16-10, we discussed two sinusoidal waves of the same wavelength and amplitude traveling *in the same direction* along a stretched string. What if they travel in opposite directions? We can again find the resultant wave by applying the superposition principle.

Figure 16-19 suggests the situation graphically. It shows the two combining waves, one traveling to the left in Fig. 16-19*a*, the other to the right in Fig. 16-19*b*. Figure 16-19*c* shows their sum, obtained by applying the superposition principle graphically. The outstanding feature of the resultant wave is that there are places along the string, called **nodes,** where the string never moves. Four such nodes are marked by dots in Fig. 16-19*c*. Halfway between adjacent nodes are **antinodes,** where the amplitude of the resultant wave is a maximum. Wave patterns such as that of Fig. 16-19*c* are called **standing waves** because the wave patterns do not move left or right; the locations of the maxima and minima do not change.

> If two sinusoidal waves of the same amplitude and wavelength travel in *opposite* directions along a stretched string, their interference with each other produces a standing wave.

To analyze a standing wave, we represent the two combining waves with the equations

$$y_1(x, t) = y_m \sin(kx - \omega t) \tag{16-58}$$

and

$$y_2(x, t) = y_m \sin(kx + \omega t). \tag{16-59}$$

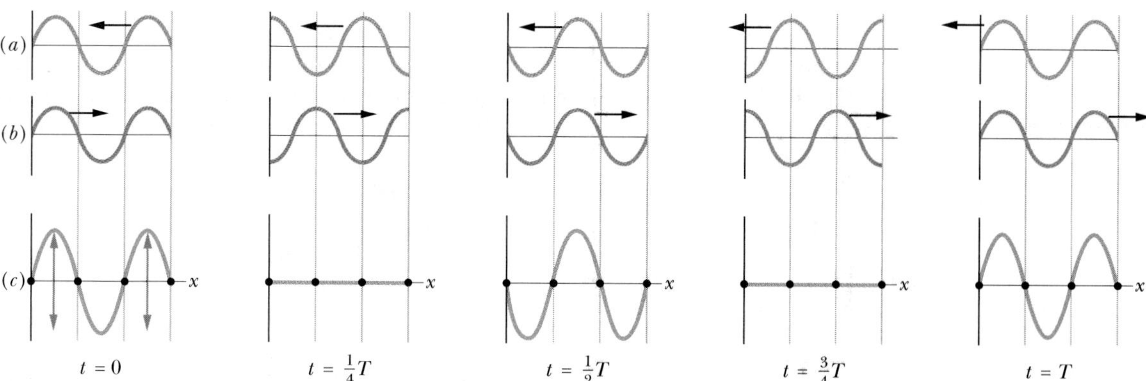

FIG. 16-19 (*a*) Five snapshots of a wave traveling to the left, at the times *t* indicated below part (*c*) (T is the period of oscillation). (*b*) Five snapshots of a wave identical to that in (*a*) but traveling to the right, at the same times *t*. (*c*) Corresponding snapshots for the superposition of the two waves on the same string. At $t = 0, \frac{1}{2}T$, and T, fully constructive interference occurs because of the alignment of peaks with peaks and valleys with valleys. At $t = \frac{1}{4}T$ and $\frac{3}{4}T$, fully destructive interference occurs because of the alignment of peaks with valleys. Some points (the nodes, marked with dots) never oscillate; some points (the antinodes) oscillate the most.

Displacement

$y'(x,t) = [2y_m \sin kx]\cos \omega t$

Magnitude gives amplitude at position x

Oscillating term

FIG. 16-20 The resultant wave of Eq. 16-60 is a standing wave and is due to the interference of two sinusoidal waves of the same amplitude and wavelength that travel in opposite directions.

The principle of superposition gives, for the combined wave,

$$y'(x,t) = y_1(x,t) + y_2(x,t) = y_m \sin(kx - \omega t) + y_m \sin(kx + \omega t).$$

Applying the trigonometric relation of Eq. 16-50 leads to

$$y'(x,t) = [2y_m \sin kx] \cos \omega t, \tag{16-60}$$

which is displayed in Fig. 16-20. This equation does not describe a traveling wave because it is not of the form of Eq. 16-17. Instead, it describes a standing wave.

The quantity $2y_m \sin kx$ in the brackets of Eq. 16-60 can be viewed as the amplitude of oscillation of the string element that is located at position x. However, since an amplitude is always positive and $\sin kx$ can be negative, we take the absolute value of the quantity $2y_m \sin kx$ to be the amplitude at x.

In a traveling sinusoidal wave, the amplitude of the wave is the same for all string elements. That is not true for a standing wave, in which the amplitude *varies with position.* In the standing wave of Eq. 16-60, for example, the amplitude is zero for values of kx that give $\sin kx = 0$. Those values are

$$kx = n\pi, \quad \text{for } n = 0, 1, 2, \ldots. \tag{16-61}$$

Substituting $k = 2\pi/\lambda$ in this equation and rearranging, we get

$$x = n\frac{\lambda}{2}, \quad \text{for } n = 0, 1, 2, \ldots \quad \text{(nodes)}, \tag{16-62}$$

as the positions of zero amplitude—the nodes—for the standing wave of Eq. 16-60. Note that adjacent nodes are separated by $\lambda/2$, half a wavelength.

The amplitude of the standing wave of Eq. 16-60 has a maximum value of $2y_m$, which occurs for values of kx that give $|\sin kx| = 1$. Those values are

$$\begin{aligned} kx &= \tfrac{1}{2}\pi, \tfrac{3}{2}\pi, \tfrac{5}{2}\pi, \ldots \\ &= (n + \tfrac{1}{2})\pi, \quad \text{for } n = 0, 1, 2, \ldots. \end{aligned} \tag{16-63}$$

Substituting $k = 2\pi/\lambda$ in Eq. 16-63 and rearranging, we get

$$x = \left(n + \frac{1}{2}\right)\frac{\lambda}{2}, \quad \text{for } n = 0, 1, 2, \ldots \quad \text{(antinodes)}, \tag{16-64}$$

as the positions of maximum amplitude—the antinodes—of the standing wave of Eq. 16-60. The antinodes are separated by $\lambda/2$ and are located halfway between pairs of nodes.

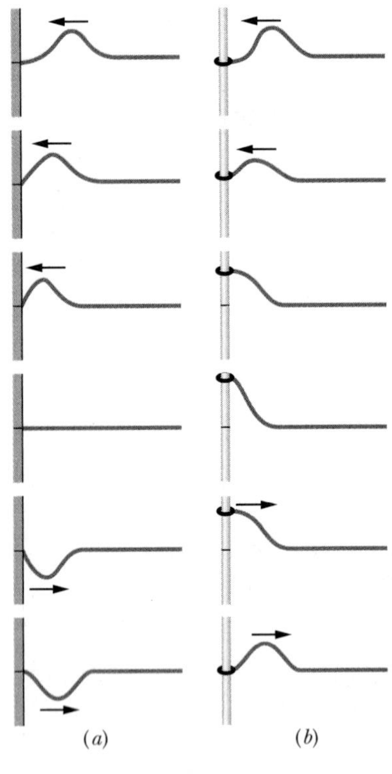

FIG. 16-21 (*a*) A pulse incident from the right is reflected at the left end of the string, which is tied to a wall. Note that the reflected pulse is inverted from the incident pulse. (*b*) Here the left end of the string is tied to a ring that can slide without friction up and down the rod. Now the pulse is not inverted by the reflection.

Reflections at a Boundary

We can set up a standing wave in a stretched string by allowing a traveling wave to be reflected from the far end of the string so that the wave travels back through itself. The incident (original) wave and the reflected wave can then be described by Eqs. 16-58 and 16-59, respectively, and they can combine to form a pattern of standing waves.

In Fig. 16-21, we use a single pulse to show how such reflections take place. In Fig. 16-21*a*, the string is fixed at its left end. When the pulse arrives at that end, it exerts an upward force on the support (the wall). By Newton's third law, the support exerts an opposite force of equal magnitude on the string. This second force generates a pulse at the support, which travels back along the string in the direction opposite that of the incident pulse. In a "hard" reflection of this kind, there must be a node at the support because the string is fixed there. The reflected and incident pulses must have opposite signs, so as to cancel each other at that point.

In Fig. 16-21*b*, the left end of the string is fastened to a light ring that is free to slide without friction along a rod. When the incident pulse arrives, the ring moves up the rod. As the ring moves, it pulls on the string, stretching the string and producing a reflected pulse with the same sign and amplitude as the incident pulse. Thus, in such a "soft" reflection, the incident and reflected pulses reinforce each other, creating an antinode at the end of the string; the maximum displacement of the ring is twice the amplitude of either of these pulses.

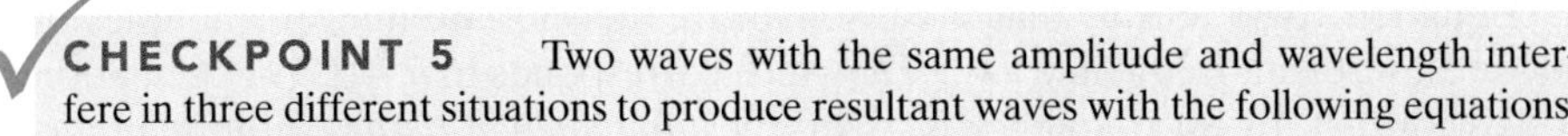

CHECKPOINT 5 Two waves with the same amplitude and wavelength interfere in three different situations to produce resultant waves with the following equations:

(1) $y'(x, t) = 4 \sin(5x - 4t)$
(2) $y'(x, t) = 4 \sin(5x) \cos(4t)$
(3) $y'(x, t) = 4 \sin(5x + 4t)$

In which situation are the two combining waves traveling (a) toward positive x, (b) toward negative x, and (c) in opposite directions?

16-13 | Standing Waves and Resonance

Consider a string, such as a guitar string, that is stretched between two clamps. Suppose we send a continuous sinusoidal wave of a certain frequency along the string, say, toward the right. When the wave reaches the right end, it reflects and begins to travel back to the left. That left-going wave then overlaps the wave that is still traveling to the right. When the left-going wave reaches the left end, it reflects again and the newly reflected wave begins to travel to the right, overlapping the left-going and right-going waves. In short, we very soon have many overlapping traveling waves, which interfere with one another.

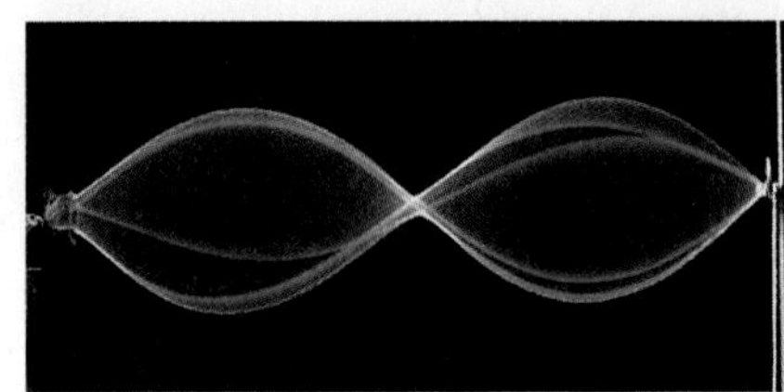

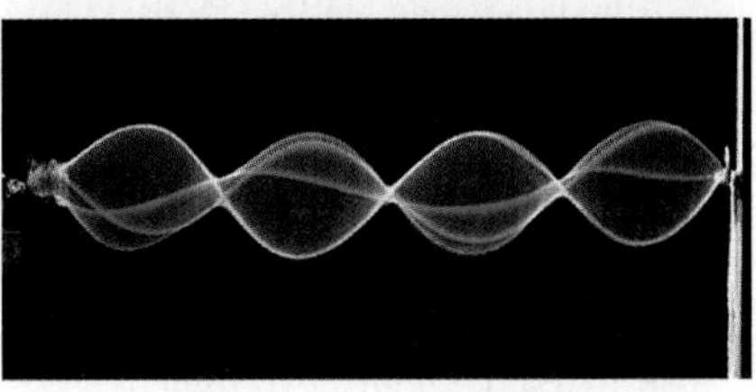

FIG. 16-22 Stroboscopic photographs reveal (imperfect) standing wave patterns on a string being made to oscillate by an oscillator at the left end. The patterns occur at certain frequencies of oscillation. *(Richard Megna/Fundamental Photographs)*

For certain frequencies, the interference produces a standing wave pattern (or **oscillation mode**) with nodes and large antinodes like those in Fig. 16-22. Such a standing wave is said to be produced at **resonance,** and the string is said to *resonate* at these certain frequencies, called **resonant frequencies.** If the string is oscillated at some frequency other than a resonant frequency, a standing wave is not set up. Then the interference of the right-going and left-going traveling waves results in only small (perhaps imperceptible) oscillations of the string.

Let a string be stretched between two clamps separated by a fixed distance L. To find expressions for the resonant frequencies of the string, we note that a node must exist at each of its ends, because each end is fixed and cannot oscillate. The simplest pattern that meets this key requirement is that in Fig. 16-23*a*, which shows the string at both its extreme displacements (one solid and one dashed, together forming a single "loop"). There is only one antinode, which is at the center of the string. Note that half a wavelength spans the length L, which we take to be the string's length. Thus, for this pattern, $\lambda/2 = L$. This condition tells us that if the left-going and right-going traveling waves are to set up this pattern by their interference, they must have the wavelength $\lambda = 2L$.

A second simple pattern meeting the requirement of nodes at the fixed ends is shown in Fig. 16-23*b*. This pattern has three nodes and two antinodes and is said to be a two-loop pattern. For the left-going and right-going waves to set it up, they must have a wavelength $\lambda = L$. A third pattern is shown in Fig. 16-23*c*. It has four nodes, three antinodes, and three loops, and the wavelength is $\lambda = \frac{2}{3}L$. We could continue this progression by drawing increasingly more complicated patterns. In each step of the progression, the pattern would have one more node and one more antinode than the preceding step, and an additional $\lambda/2$ would be fitted into the distance L.

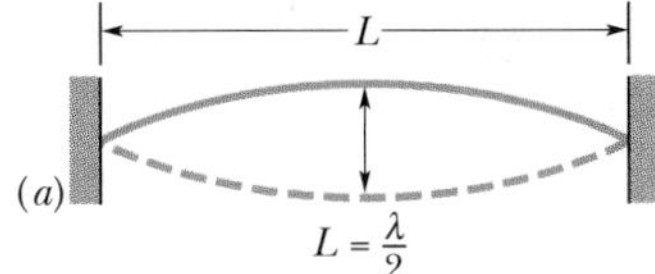

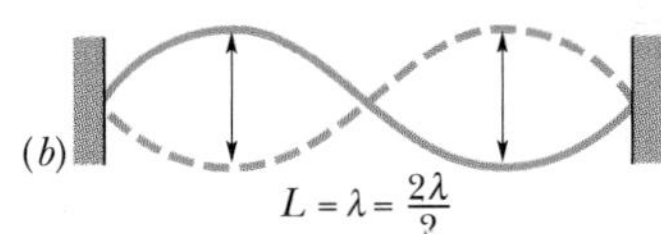

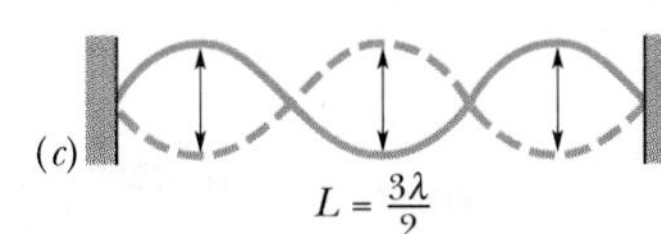

FIG. 16-23 A string, stretched between two clamps, is made to oscillate in standing wave patterns. (*a*) The simplest possible pattern consists of one *loop,* which refers to the composite shape formed by the string in its extreme displacements (the solid and dashed lines). (*b*) The next simplest pattern has two loops. (*c*) The next has three loops.

Thus, a standing wave can be set up on a string of length L by a wave with a wavelength equal to one of the values

$$\lambda = \frac{2L}{n}, \quad \text{for } n = 1, 2, 3, \ldots. \tag{16-65}$$

The resonant frequencies that correspond to these wavelengths follow from

FIG. 16-24 One of many possible standing wave patterns for a kettledrum head, made visible by dark powder sprinkled on the drumhead. As the head is set into oscillation at a single frequency by a mechanical oscillator at the upper left of the photograph, the powder collects at the nodes, which are circles and straight lines in this two-dimensional example. *(Courtesy Thomas D. Rossing, Northern Illinois University)*

Eq. 16-13:

$$f = \frac{v}{\lambda} = n\frac{v}{2L}, \quad \text{for } n = 1, 2, 3, \ldots . \tag{16-66}$$

Here v is the speed of traveling waves on the string.

Equation 16-66 tells us that the resonant frequencies are integer multiples of the lowest resonant frequency, $f = v/2L$, which corresponds to $n = 1$. The oscillation mode with that lowest frequency is called the *fundamental mode* or the *first harmonic*. The *second harmonic* is the oscillation mode with $n = 2$, the *third harmonic* is that with $n = 3$, and so on. The frequencies associated with these modes are often labeled f_1, f_2, f_3, and so on. The collection of all possible oscillation modes is called the **harmonic series**, and n is called the **harmonic number** of the nth harmonic.

For a given string under a given tension, each resonant frequency corresponds to a particular oscillation pattern. Thus, if the frequency is in the audible range, you can hear the shape of the string. Resonance can also occur in two dimensions (such as on the surface of the kettledrum in Fig. 16-24) and in three dimensions (such as in the wind-induced swaying and twisting of a tall building).

Footbridges and Mosh Pits

When the Millennium Bridge over the Thames River was first opened to pedestrians, the bridge did not initially exhibit oscillations. The footsteps of the pedestrians produced vertical and horizontal forces on the bridge that tended to set up the second harmonic on the bridge (which is much like the second harmonic on a string), but the pedestrians were few and their walking was uncoordinated. However, once their number exceeded a critical value, the second harmonic suddenly became noticeable and walking became difficult. To keep their balance, many pedestrians began to time their steps to the bridge's swaying, which then became even worse and led to the bridge being closed until damping devices (see Sample Problem 15-3) could be added.

Similar structural oscillations occur when an audience on a football stadium deck or in a concert hall begins to sway or stomp in a coordinated fashion. Perhaps the worse situation can occur in a mosh pit on a lightweight suspended floor. When the crowded dancers begin to pogo, in which their coordinated jumps are in time with the music's beat, they can set up resonance in the floor, which typically has a resonant frequency of 2 Hz. Resonance can then quickly build as more people are forced to coordinate their motion with the oscillations, which might become large enough to break the floor and cause it to collapse. To avoid that possibility, modern building codes commonly require that suspended dance floors be built with resonant frequencies no lower than 5 Hz.

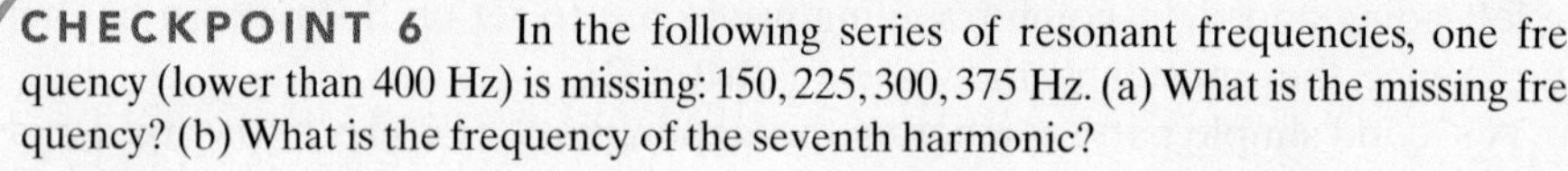

CHECKPOINT 6 In the following series of resonant frequencies, one frequency (lower than 400 Hz) is missing: 150, 225, 300, 375 Hz. (a) What is the missing frequency? (b) What is the frequency of the seventh harmonic?

Sample Problem 16-8 Build your skill

Figure 16-25 shows a pattern of resonant oscillation of a string of mass $m = 2.500$ g and length $L = 0.800$ m and that is under tension $\tau = 325.0$ N. What is the wavelength λ of the transverse waves producing the standing-wave pattern, and what is the harmonic number n? What is the frequency f of the transverse waves and of the oscillations of the moving string elements? What is the maximum magnitude of the transverse velocity u_m of the element oscillating at coordinate $x = 0.180$ m (note the x axis in the figure)? At what point during the element's oscillation is the transverse velocity maximum?

KEY IDEAS (1) The traverse waves that produce a standing-wave pattern must have a wavelength such that an integer number n of half-wavelengths fit into the length L of the string. (2) The frequency of those waves and of the oscillations of the string elements is given by Eq. 16-66 ($f = nv/2L$). (3) The displacement of a string el-

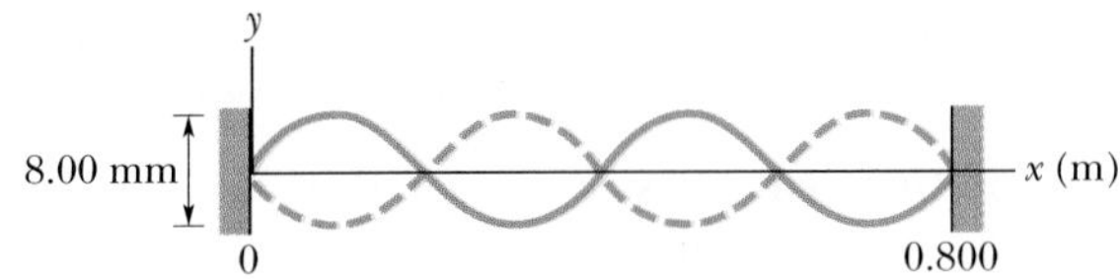

FIG. 16-25 Resonant oscillation of a string under tension.

ement as a function of position x and time t is given by Eq. 16-60:

$$y'(x,t) = [2y_m \sin kx] \cos \omega t. \qquad (16\text{-}67)$$

Wavelength and harmonic number: In Fig. 16-25, the solid line, which is effectively a snapshot (or freeze frame) of the oscillations, reveals that 2 full wavelengths fit into the length $L = 0.800$ m of the string. Thus, we have

$$2\lambda = L,$$

or

$$\lambda = \frac{L}{2}. \qquad (16\text{-}68)$$

$$= \frac{0.800\ \text{m}}{2} = 0.400\ \text{m}. \qquad \text{(Answer)}$$

By counting the number of loops (or half-wavelengths) in Fig. 16-25, we see that the harmonic number is

$$n = 4. \qquad \text{(Answer)}$$

We reach the same conclusion by comparing Eqs. 16-68 and 16-65 ($\lambda = 2L/n$). Thus, the string is oscillating in its fourth harmonic.

Frequency: We can get the frequency f of the transverse waves from Eq. 16-13 ($v = \lambda f$) if we first find the speed v of the waves. That speed is given by Eq. 16-26, but we must substitute m/L for the unknown linear density μ. We obtain

$$v = \sqrt{\frac{\tau}{\mu}} = \sqrt{\frac{\tau}{m/L}} = \sqrt{\frac{\tau L}{m}}$$

$$= \sqrt{\frac{(325\ \text{N})(0.800\ \text{m})}{2.50 \times 10^{-3}\ \text{kg}}} = 322.49\ \text{m/s}.$$

After rearranging Eq. 16-13, we write

$$f = \frac{v}{\lambda} = \frac{322.49\ \text{m/s}}{0.400\ \text{m}}$$

$$= 806.2\ \text{Hz} \approx 806\ \text{Hz}. \qquad \text{(Answer)}$$

Note that we get the same answer by substituting into Eq. 16-66:

$$f = n\frac{v}{2L} = 4\frac{322.49\ \text{m/s}}{2(0.800\ \text{m})}$$

$$= 806\ \text{Hz}. \qquad \text{(Answer)}$$

Now note that this 806 Hz is not only the frequency of the waves producing the fourth harmonic but also the frequency of the string elements that oscillate vertically in Fig. 16-25. It is also the frequency of the sound you would hear from the string.

Transverse velocity: The displacement y' of the string element located at coordinate x is given by Eq. 16-67 as a function of time t. The term $\cos \omega t$ contains the dependence on time and thus provides the "motion" of the standing wave. The term $2y_m \sin kx$ sets the extent of the motion—that is, the amplitude. The greatest amplitude occurs at an antinode, where $\sin kx$ is $+1$ or -1 and thus the greatest amplitude is $2y_m$. From Fig. 16-25, we see that $2y_m = 4.00$ mm, which tells us that $y_m = 2.00$ mm.

We want the transverse velocity—the velocity of a string element parallel to the y axis. To find it, we take the time derivative of Eq. 16-67:

$$u(x,t) = \frac{\partial y'}{\partial t} = \frac{\partial}{\partial t}[(2y_m \sin kx) \cos \omega t]$$

$$= [-2y_m\omega \sin kx] \sin \omega t. \qquad (16\text{-}69)$$

Here the term $\sin \omega t$ provides the variation with time and the term $-2y_m\omega \sin kx$ provides the extent of that variation. We want the absolute magnitude of that extent:

$$u_m = |-2y_m\omega \sin kx|.$$

To evaluate this for the element at $x = 0.180$ m, we first note that $y_m = 2.00$ mm, $k = 2\pi/\lambda = 2\pi/(0.400\ \text{m})$, and $\omega = 2\pi f = 2\pi(806.2\ \text{Hz})$. Then the maximum speed of the element at $x = 0.180$ m is

$$u_m = \left| -2(2.00 \times 10^{-3}\ \text{m})(2\pi)(806.2\ \text{Hz}) \times \sin\left(\frac{2\pi}{0.400\ \text{m}}(0.180\ \text{m})\right)\right|$$

$$= 6.26\ \text{m/s}. \qquad \text{(Answer)}$$

To determine when the string element has this maximum speed, we could investigate Eq. 16-69. However, a little thought can save a lot of work. The element is undergoing simple harmonic motion and must come to a momentary stop at its extreme upward position and extreme downward position. It has the greatest speed as it zips through the midpoint of its oscillation, just as a block does in a block–spring oscillator.

PROBLEM-SOLVING TACTICS

Tactic 2: Harmonics on a String When you need to obtain information about a certain harmonic on a stretched string of given length L, first draw that harmonic (as in Fig. 16-23). If you are asked about, say, the fifth harmonic, you need to draw five loops between the fixed support points. That would mean that five loops, each of length $\lambda/2$, occupy the length L of the string. Thus, $5(\lambda/2) = L$, and $\lambda = 2L/5$. You can then use Eq. 16-13 ($f = v/\lambda$) to find the frequency of the harmonic.

Keep in mind that the wavelength of a harmonic is set only by the length L of the string, but the frequency depends also on the wave speed v, which is set by the tension and the linear density of the string via Eq. 16-26.

REVIEW & SUMMARY

Transverse and Longitudinal Waves Mechanical waves can exist only in material media and are governed by Newton's laws. **Transverse** mechanical waves, like those on a stretched string, are waves in which the particles of the medium oscillate perpendicular to the wave's direction of travel. Waves in which the particles of the medium oscillate parallel to the wave's direction of travel are **longitudinal** waves.

Sinusoidal Waves A sinusoidal wave moving in the positive direction of an x axis has the mathematical form

$$y(x, t) = y_m \sin(kx - \omega t), \tag{16-2}$$

where y_m is the **amplitude** of the wave, k is the **angular wave number,** ω is the **angular frequency,** and $kx - \omega t$ is the **phase.** The **wavelength** λ is related to k by

$$k = \frac{2\pi}{\lambda}. \tag{16-5}$$

The **period** T and **frequency** f of the wave are related to ω by

$$\frac{\omega}{2\pi} = f = \frac{1}{T}. \tag{16-9}$$

Finally, the **wave speed** v is related to these other parameters by

$$v = \frac{\omega}{k} = \frac{\lambda}{T} = \lambda f. \tag{16-13}$$

Equation of a Traveling Wave Any function of the form

$$y(x, t) = h(kx \pm \omega t) \tag{16-17}$$

can represent a **traveling wave** with a wave speed given by Eq. 16-13 and a wave shape given by the mathematical form of h. The plus sign denotes a wave traveling in the negative direction of the x axis, and the minus sign a wave traveling in the positive direction.

Wave Speed on Stretched String The speed of a wave on a stretched string is set by properties of the string. The speed on a string with tension τ and linear density μ is

$$v = \sqrt{\frac{\tau}{\mu}}. \tag{16-26}$$

Power The **average power** of, or average rate at which energy is transmitted by, a sinusoidal wave on a stretched string is given by

$$P_{avg} = \tfrac{1}{2}\mu v \omega^2 y_m^2. \tag{16-33}$$

Superposition of Waves When two or more waves traverse the same medium, the displacement of any particle of the medium is the sum of the displacements that the individual waves would give it.

Interference of Waves Two sinusoidal waves on the same string exhibit **interference,** adding or canceling according to the principle of superposition. If the two are traveling in the same direction and have the same amplitude y_m and frequency (hence the same wavelength) but differ in phase by a **phase constant** ϕ, the result is a single wave with this same frequency:

$$y'(x, t) = [2y_m \cos \tfrac{1}{2}\phi] \sin(kx - \omega t + \tfrac{1}{2}\phi). \tag{16-51}$$

If $\phi = 0$, the waves are exactly in phase and their interference is fully constructive; if $\phi = \pi$ rad, they are exactly out of phase and their interference is fully destructive.

Phasors A wave $y(x, t)$ can be represented with a **phasor.** This is a vector that has a magnitude equal to the amplitude y_m of the wave and that rotates about an origin with an angular speed equal to the angular frequency ω of the wave. The projection of the rotating phasor on a vertical axis gives the displacement y of a point along the wave's travel.

Standing Waves The interference of two identical sinusoidal waves moving in opposite directions produces **standing waves.** For a string with fixed ends, the standing wave is given by

$$y'(x, t) = [2y_m \sin kx] \cos \omega t. \tag{16-60}$$

Standing waves are characterized by fixed locations of zero displacement called **nodes** and fixed locations of maximum displacement called **antinodes.**

Resonance Standing waves on a string can be set up by reflection of traveling waves from the ends of the string. If an end is fixed, it must be the position of a node. This limits the frequencies at which standing waves will occur on a given string. Each possible frequency is a **resonant frequency,** and the corresponding standing wave pattern is an **oscillation mode.** For a stretched string of length L with fixed ends, the resonant frequencies are

$$f = \frac{v}{\lambda} = n\frac{v}{2L}, \quad \text{for } n = 1, 2, 3, \ldots. \tag{16-66}$$

The oscillation mode corresponding to $n = 1$ is called the *fundamental mode* or the *first harmonic*; the mode corresponding to $n = 2$ is the *second harmonic*; and so on.

QUESTIONS

1 Figure 16-26*a* gives a snapshot of a wave traveling in the direction of positive x along a string under tension. Four string elements are indicated by the lettered points. For each of those elements, determine whether, at the instant of the snapshot, the element is moving upward or downward or is momentarily at rest. (*Hint:* Imagine the wave as it moves through the four string elements, as if you were watching a video of the wave as it traveled rightward.)

Figure 16-26*b* gives the displacement of a string element located at, say, $x = 0$ as a function of time. At the lettered times, is the element moving upward or downward or is it momentarily at rest?

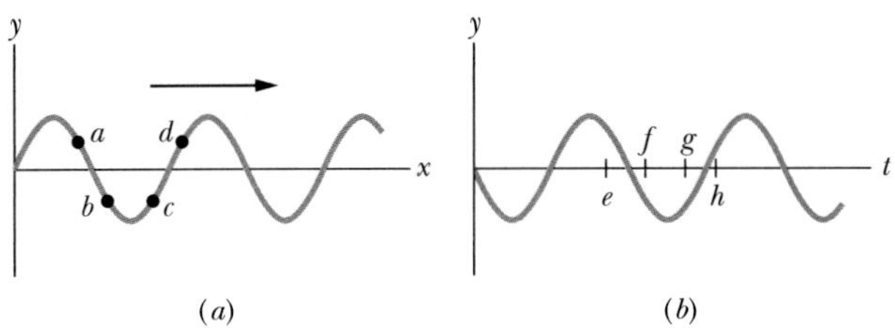

FIG. 16-26 Question 1.

2 Figure 16-27 shows three waves that are *separately* sent along a string that is stretched under a certain tension along an x axis. Rank the waves according to their (a) wavelengths, (b) speeds, and (c) angular frequencies, greatest first.

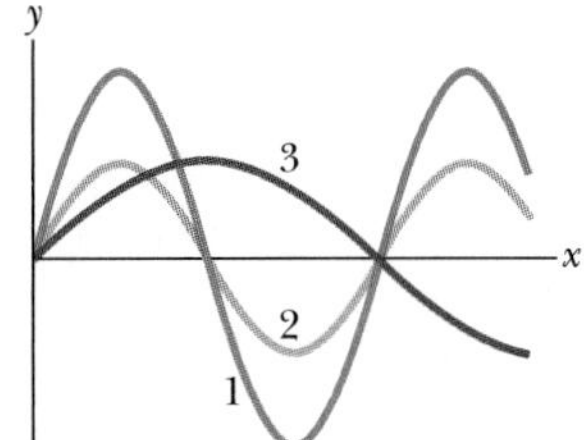

FIG. 16-27 Question 2.

3 The following four waves are sent along strings with the same linear densities (x is in meters and t is in seconds). Rank the waves according to (a) their wave speed and (b) the tension in the strings along which they travel, greatest first:

(1) $y_1 = (3\text{ mm})\sin(x - 3t)$, (3) $y_3 = (1\text{ mm})\sin(4x - t)$,

(2) $y_2 = (6\text{ mm})\sin(2x - t)$, (4) $y_4 = (2\text{ mm})\sin(x - 2t)$.

4 In Fig. 16-28, wave 1 consists of a rectangular peak of height 4 units and width d, and a rectangular valley of depth 2 units and width d. The wave travels rightward along an x axis. Choices 2, 3, and 4 are similar waves, with the same heights, depths, and widths, that will travel leftward along that axis and through wave 1. Right-going wave 1 and one of the left-going waves will interfere as they pass through each other. With which left-going wave will the interference give, for an instant, (a) the deepest valley, (b) a flat line, and (c) a flat peak $2d$ wide?

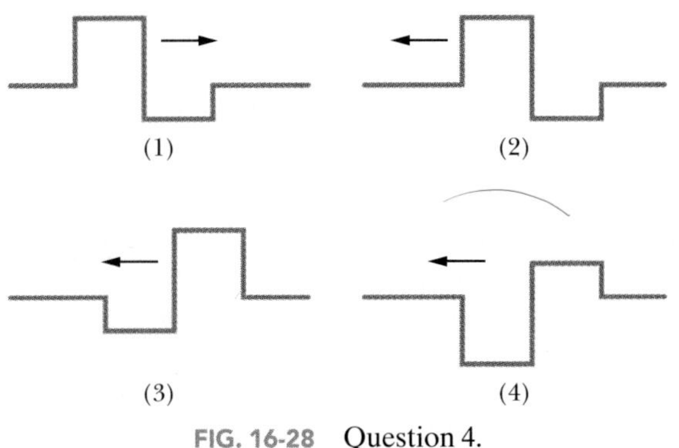

FIG. 16-28 Question 4.

5 A sinusoidal wave is sent along a cord under tension, transporting energy at the average rate of $P_{\text{avg},1}$. Two waves, identical to that first one, are then to be sent along the cord with a phase difference ϕ of either 0, 0.2 wavelength, or 0.5 wavelength. (a) With only mental calculation, rank those choices of ϕ according to the average rate at which the waves will transport energy, greatest first. (b) For the first choice of ϕ, what is the average rate in terms of $P_{\text{avg},1}$?

6 The amplitudes and phase differences for four pairs of waves of equal wavelengths are (a) 2 mm, 6 mm, and π rad; (b) 3 mm, 5 mm, and π rad; (c) 7 mm, 9 mm, and π rad; (d) 2 mm, 2 mm, and 0 rad. Each pair travels in the same direction along the same string. Without written calculation, rank the four pairs according to the amplitude of their resultant wave, greatest first. (*Hint:* Construct phasor diagrams.)

7 If you start with two sinusoidal waves of the same amplitude traveling in phase on a string and then somehow phase-shift one of them by 5.4 wavelengths, what type of interference will occur on the string?

8 If you set up the seventh harmonic on a string, (a) how many nodes are present, and (b) is there a node, antinode, or some intermediate state at the midpoint? If you next set up the sixth harmonic, (c) is its resonant wavelength longer or shorter than that for the seventh harmonic, and (d) is the resonant frequency higher or lower?

9 Figure 16-29 shows phasor diagrams for three situations in which two waves travel along the same string. All six waves have the same amplitude. Rank the situations according to the amplitude of the net wave on the string, greatest first.

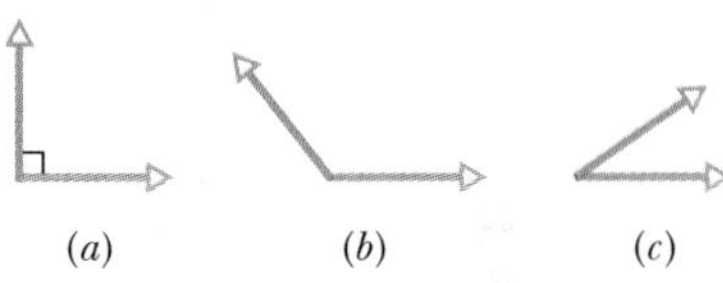

FIG. 16-29 Question 9.

10 (a) If a standing wave on a string is given by

$$y'(t) = (3\text{ mm})\sin(5x)\cos(4t),$$

is there a node or an antinode of the oscillations of the string at $x = 0$? (b) If the standing wave is given by

$$y'(t) = (3\text{ mm})\sin(5x + \pi/2)\cos(4t),$$

is there a node or an antinode at $x = 0$?

11 Strings A and B have identical lengths and linear densities, but string B is under greater tension than string A. Figure 16-30 shows four situations, (a) through (d), in which standing wave patterns exist on the two strings. In which situations is there the possibility that strings A and B are oscillating at the same resonant frequency?

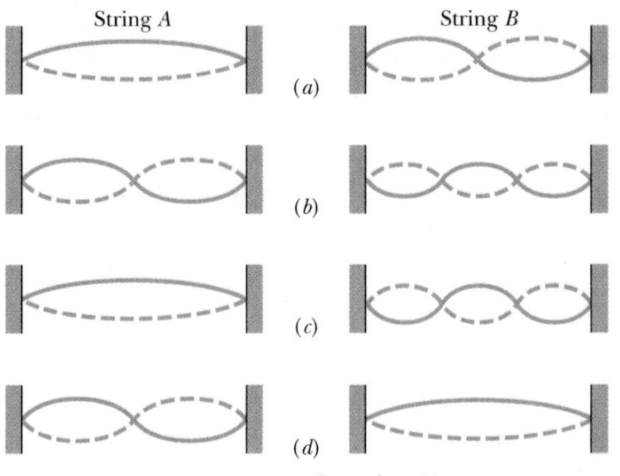

FIG. 16-30 Question 11.

PROBLEMS

GO Tutoring problem available (at instructor's discretion) in *WileyPLUS* and WebAssign
SSM Worked-out solution available in Student Solutions Manual
WWW Worked-out solution is at
ILW Interactive solution is at
http://www.wiley.com/college/halliday
• – ••• Number of dots indicates level of problem difficulty
Additional information available in *The Flying Circus of Physics* and at flyingcircusofphysics.com

sec. 16-5 The Speed of a Traveling Wave

•1 A wave has an angular frequency of 110 rad/s and a wavelength of 1.80 m. Calculate (a) the angular wave number and (b) the speed of the wave.

•2 A sand scorpion can detect the motion of a nearby beetle (its prey) by the waves the motion sends along the sand surface (Fig. 16-31). The waves are of two types: transverse waves traveling at $v_t = 50$ m/s and longitudinal waves traveling at $v_l = 150$ m/s. If a sudden motion sends out such waves, a scorpion can tell the distance of the beetle from the difference Δt in the arrival times of the waves at its leg nearest the beetle. If $\Delta t = 4.0$ ms, what is the beetle's distance?

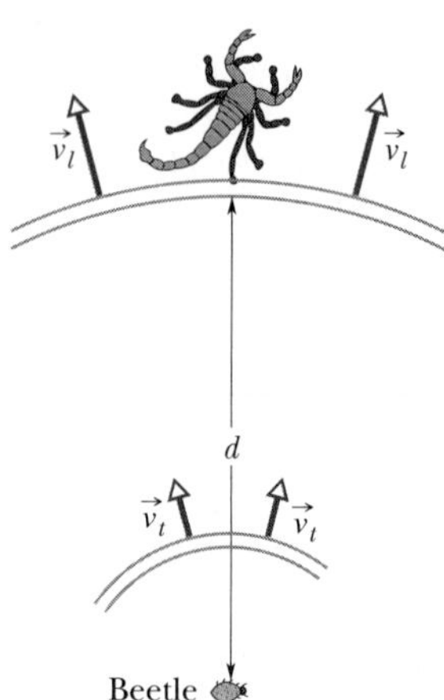

FIG. 16-31 Problem 2.

•3 A sinusoidal wave travels along a string. The time for a particular point to move from maximum displacement to zero is 0.170 s. What are the (a) period and (b) frequency? (c) The wavelength is 1.40 m; what is the wave speed?

•4 *A human wave.* During sporting events within large, densely packed stadiums, spectators will send a wave (or pulse) around the stadium (Fig. 16-32). As the wave reaches a group of spectators, they stand with a cheer and then sit. At any instant, the width w of the wave is the distance from the leading edge (people are just about to stand) to the trailing edge (people have just sat down). Suppose a human wave travels a distance of 853 seats around a stadium in 39 s, with spectators requiring about 1.8 s to respond to the wave's passage by standing and then sitting. What are (a) the wave speed v (in seats per second) and (b) width w (in number of seats)?

FIG. 16-32 Problem 4.

•5 If $y(x, t) = (6.0 \text{ mm}) \sin(kx + (600 \text{ rad/s})t + \phi)$ describes a wave traveling along a string, how much time does any given point on the string take to move between displacements $y = +2.0$ mm and $y = -2.0$ mm?

••6 Figure 16-33 shows the transverse velocity u versus time t of the point on a string at $x = 0$, as a wave passes through it. The scale on the vertical axis is set by $u_s = 4.0$ m/s. The wave has form $y(x, t) = y_m \sin(kx - \omega t + \phi)$. What is ϕ? (*Caution:* A calculator does not always give the proper inverse trig function, so check your answer by substituting it and an assumed value of ω into $y(x, t)$ and then plotting the function.)

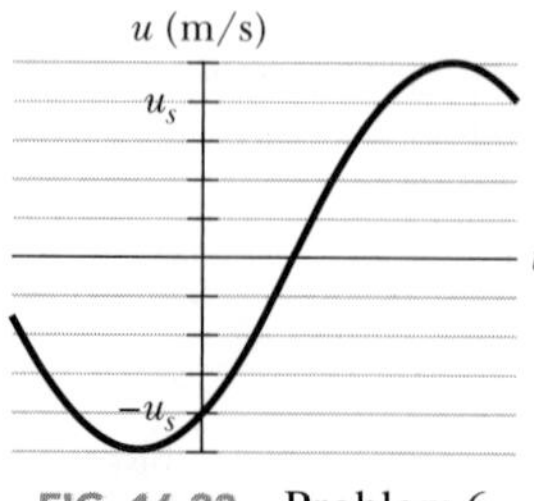

FIG. 16-33 Problem 6.

••7 A sinusoidal wave of frequency 500 Hz has a speed of 350 m/s. (a) How far apart are two points that differ in phase by $\pi/3$ rad? (b) What is the phase difference between two displacements at a certain point at times 1.00 ms apart? ILW

••8 The equation of a transverse wave traveling along a very long string is $y = 6.0 \sin(0.020\pi x + 4.0\pi t)$, where x and y are expressed in centimeters and t is in seconds. Determine (a) the amplitude, (b) the wavelength, (c) the frequency, (d) the speed, (e) the direction of propagation of the wave, and (f) the maximum transverse speed of a particle in the string. (g) What is the transverse displacement at $x = 3.5$ cm when $t = 0.26$ s?

••9 A transverse sinusoidal wave is moving along a string in the positive direction of an x axis with a speed of 80 m/s. At $t = 0$, the string particle at $x = 0$ has a transverse displacement of 4.0 cm from its equilibrium position and is not moving. The maximum transverse speed of the string particle at $x = 0$ is 16 m/s. (a) What is the frequency of the wave? (b) What is the wavelength of the wave? If the wave equation is of the form $y(x, t) = y_m \sin(kx \pm \omega t + \phi)$, what are (c) y_m, (d) k, (e) ω, (f) ϕ, and (g) the correct choice of sign in front of ω?

••10 The function $y(x, t) = (15.0 \text{ cm}) \cos(\pi x - 15\pi t)$, with x in meters and t in seconds, describes a wave on a taut string. What is the transverse speed for a point on the string at an instant when that point has the displacement $y = +12.0$ cm?

••11 A sinusoidal wave moving along a string is shown twice in Fig. 16-34, as crest A travels in the positive direction of an x axis by distance $d = 6.0$ cm in 4.0 ms. The tick marks along the axis are separated by 10 cm; height $H = 6.00$ mm. If the wave equation is of the form $y(x, t) = y_m \sin(kx \pm \omega t)$, what are (a) y_m, (b) k, (c) ω, and (d) the correct choice of sign in front of ω?

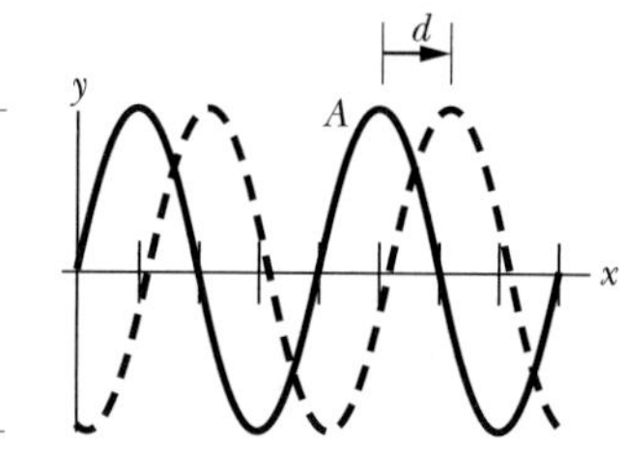

FIG. 16-34 Problem 11.

••12 A sinusoidal wave travels along a string under tension. Figure 16-35 gives the slopes along the string at time $t = 0$. The scale of the x axis is set by $x_s = 0.80$ m. What is the amplitude of the wave? GO

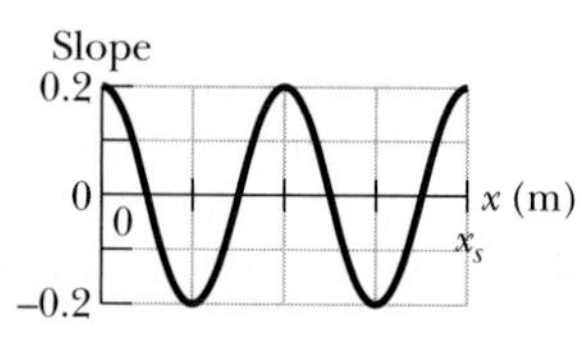

FIG. 16-35 Problem 12.

••13 A sinusoidal transverse wave of wavelength 20 cm travels along a string in the positive direction of an x axis. The displacement y of the string particle at $x = 0$ is given in Fig. 16-36 as a function of time t. The scale of the vertical axis is set by $y_s = 4.0$ cm. The wave equation is to be in the form $y(x, t) = y_m \sin(kx \pm \omega t + \phi)$. (a) At $t = 0$, is a plot of y versus x in the shape of a positive sine function or a negative sine function? What are (b) y_m, (c) k, (d) ω, (e) ϕ, (f) the sign in front of ω, and (g) the speed of the wave? (h) What is the transverse velocity of the particle at $x = 0$ when $t = 5.0$ s? GO

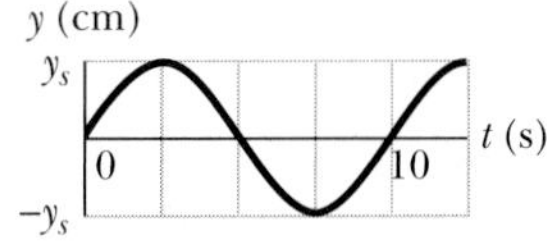

FIG. 16-36 Problem 13.

sec. 16-6 Wave Speed on a Stretched String

•14 The tension in a wire clamped at both ends is doubled without appreciably changing the wire's length between the clamps. What is the ratio of the new to the old wave speed for transverse waves traveling along this wire?

•15 What is the speed of a transverse wave in a rope of length 2.00 m and mass 60.0 g under a tension of 500 N? SSM

•16 The heaviest and lightest strings on a certain violin have linear densities of 3.0 and 0.29 g/m. What is the ratio of the diameter of the heaviest string to that of the lightest string, assuming that the strings are of the same material?

•17 A stretched string has a mass per unit length of 5.00 g/cm and a tension of 10.0 N. A sinusoidal wave on this string has an amplitude of 0.12 mm and a frequency of 100 Hz and is traveling in the negative direction of an x axis. If the wave equation is of the form $y(x, t) = y_m \sin(kx \pm \omega t)$, what are (a) y_m, (b) k, (c) ω, and (d) the correct choice of sign in front of ω? SSM WWW

•18 The speed of a transverse wave on a string is 170 m/s when the string tension is 120 N. To what value must the tension be changed to raise the wave speed to 180 m/s?

•19 The linear density of a string is 1.6×10^{-4} kg/m. A transverse wave on the string is described by the equation

$$y = (0.021 \text{ m}) \sin[(2.0 \text{ m}^{-1})x + (30 \text{ s}^{-1})t].$$

What are (a) the wave speed and (b) the tension in the string?

•20 The equation of a transverse wave on a string is

$$y = (2.0 \text{ mm}) \sin[(20 \text{ m}^{-1})x - (600 \text{ s}^{-1})t].$$

The tension in the string is 15 N. (a) What is the wave speed? (b) Find the linear density of this string in grams per meter.

••21 A sinusoidal transverse wave is traveling along a string in the negative direction of an x axis. Figure 16-37 shows a plot of the displacement as a function of position at time $t = 0$; the scale of the y axis is set by $y_s = 4.0$ cm. The string tension is 3.6 N, and its linear density is 25 g/m. Find the (a) amplitude, (b) wavelength, (c) wave speed, and (d) period of the wave. (e) Find the maximum transverse speed of a particle in the string. If the wave is of the form $y(x, t) = y_m \sin(kx \pm \omega t + \phi)$, what are (f) k, (g) ω, (h) ϕ, and (i) the correct choice of sign in front of ω? SSM ILW

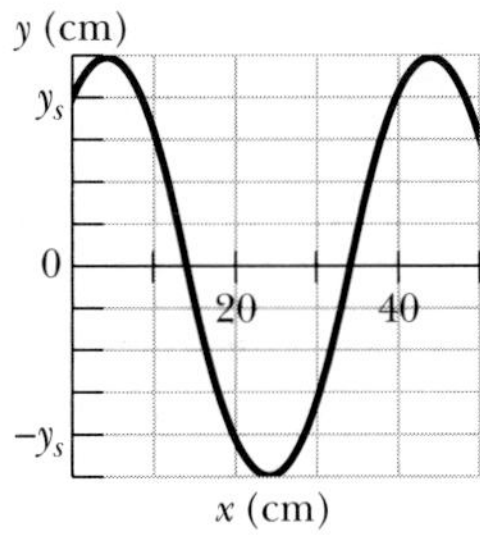

FIG. 16-37 Problem 21.

••22 A sinusoidal wave is traveling on a string with speed 40 cm/s. The displacement of the particles of the string at $x = 10$ cm is found to vary with time according to the equation $y = (5.0 \text{ cm}) \sin[1.0 - (4.0 \text{ s}^{-1})t]$. The linear density of the string is 4.0 g/cm. What are (a) the frequency and (b) the wavelength of the wave? If the wave equation is of the form $y(x, t) = y_m \sin(kx \pm \omega t)$, what are (c) y_m, (d) k, (e) ω, and (f) the correct choice of sign in front of ω? (g) What is the tension in the string?

••23 A 100 g wire is held under a tension of 250 N with one end at $x = 0$ and the other at $x = 10.0$ m. At time $t = 0$, pulse 1 is sent along the wire from the end at $x = 10.0$ m. At time $t = 30.0$ ms, pulse 2 is sent along the wire from the end at $x = 0$. At what position x do the pulses begin to meet? ILW

•••24 In Fig. 16-38a, string 1 has a linear density of 3.00 g/m, and string 2 has a linear density of 5.00 g/m. They are under tension due to the hanging block of mass $M = 500$ g. Calculate the wave speed on (a) string 1 and (b) string 2. (*Hint:* When a string loops halfway around a pulley, it pulls on the pulley with a net force that is twice the tension in the string.) Next the block is divided into two blocks (with $M_1 + M_2 = M$) and the apparatus is rearranged as shown in Fig. 16-38b. Find (c) M_1 and (d) M_2 such that the wave speeds in the two strings are equal.

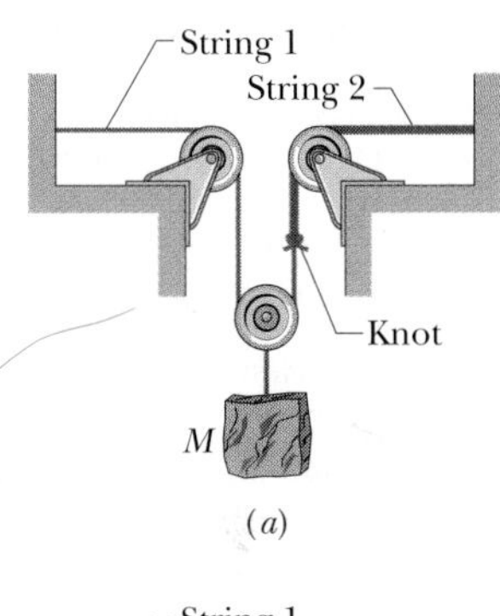

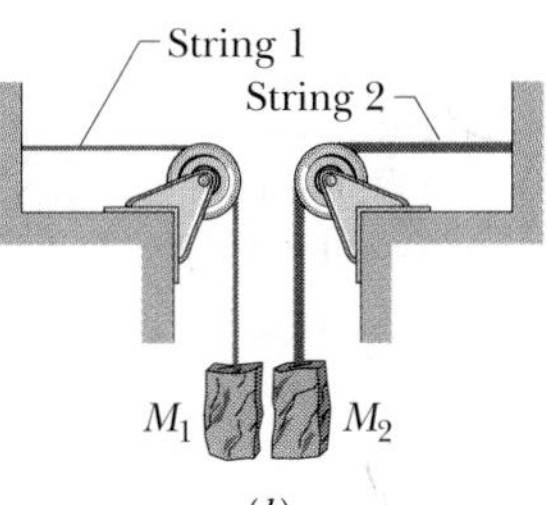

FIG. 16-38 Problem 24.

•••25 A uniform rope of mass m and length L hangs from a ceiling. (a) Show that the speed of a transverse wave on the rope is a function of y, the distance from the lower end, and is given by $v = \sqrt{gy}$. (b) Show that the time a transverse wave takes to travel the length of the rope is given by $t = 2\sqrt{L/g}$.

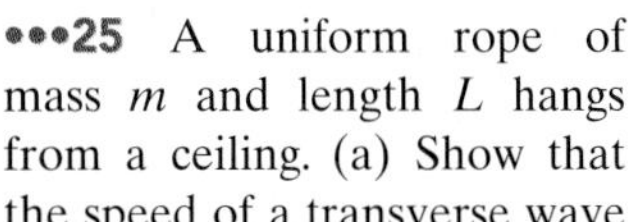

sec. 16-7 Energy and Power of a Wave Traveling Along a String

•26 A string along which waves can travel is 2.70 m long and has a mass of 260 g. The tension in the string is 36.0 N. What must be the frequency of traveling waves of amplitude 7.70 mm for the average power to be 85.0 W?

••27 A sinusoidal wave is sent along a string with a linear density of 2.0 g/m. As it travels, the kinetic energies of the mass elements along the string vary. Figure 16-39a gives

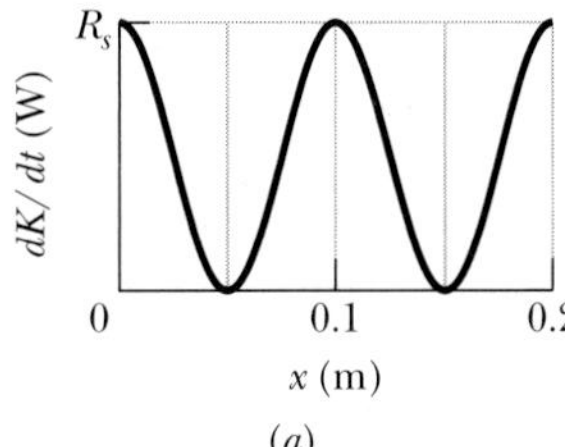

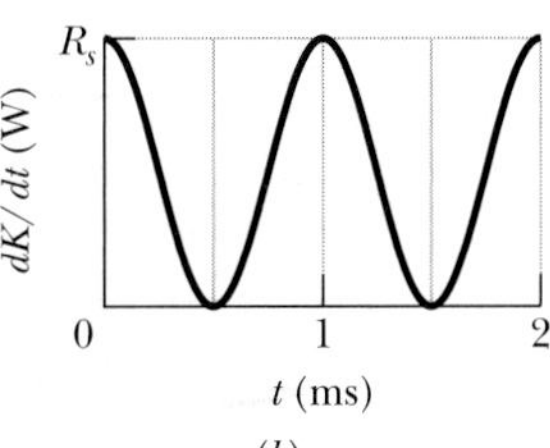

FIG. 16-39 Problem 27.

the rate dK/dt at which kinetic energy passes through the string elements at a particular instant, plotted as a function of distance x along the string. Figure 16-39b is similar except that it gives the rate at which kinetic energy passes through a particular mass element (at a particular location), plotted as a function of time t. For both figures, the scale on the vertical (rate) axis is set by $R_s = 10$ W. What is the amplitude of the wave? GO

sec. 16-8 The Wave Equation

•28 Use the wave equation to find the speed of a wave given by

$$y(x, t) = (3.00 \text{ mm}) \sin[(4.00 \text{ m}^{-1})x - (7.00 \text{ s}^{-1})t].$$

••29 Use the wave equation to find the speed of a wave given by

$$y(x, t) = (2.00 \text{ mm})[(20 \text{ m}^{-1})x - (4.0 \text{ s}^{-1})t]^{0.5}.$$

•••30 Use the wave equation to find the speed of a wave given in terms of the general function $h(x, t)$:

$$y(x, t) = (4.00 \text{ mm})\, h[(30 \text{ m}^{-1})x + (6.0 \text{ s}^{-1})t].$$

sec. 16-10 Interference of Waves

•31 Two identical traveling waves, moving in the same direction, are out of phase by $\pi/2$ rad. What is the amplitude of the resultant wave in terms of the common amplitude y_m of the two combining waves? SSM

•32 What phase difference between two identical traveling waves, moving in the same direction along a stretched string, results in the combined wave having an amplitude 1.50 times that of the common amplitude of the two combining waves? Express your answer in (a) degrees, (b) radians, and (c) wavelengths.

••33 Two sinusoidal waves with the same amplitude of 9.00 mm and the same wavelength travel together along a string that is stretched along an x axis. Their resultant wave is shown twice in Fig. 16-40, as valley A travels in the negative direction of the x axis by distance $d = 56.0$ cm in 8.0 ms. The tick marks along the axis are separated by 10 cm, and height H is 8.0 mm. Let the equation for one wave be of the form $y(x, t) = y_m \sin(kx \pm \omega t + \phi_1)$, where $\phi_1 = 0$ and you must choose the correct sign in front of ω. For the equation for the other wave, what are (a) y_m, (b) k, (c) ω, (d) ϕ_2, and (e) the sign in front of ω?

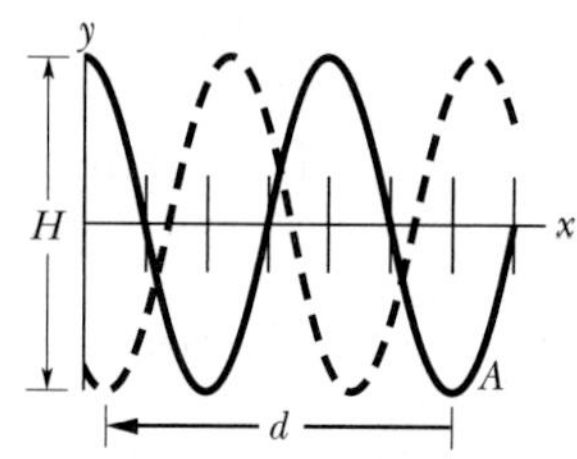

FIG. 16-40 Problem 33.

•••34 A sinusoidal wave of angular frequency 1200 rad/s and amplitude 3.00 mm is sent along a cord with linear density 2.00 g/m and tension 1200 N. (a) What is the average rate at which energy is transported by the wave to the opposite end of the cord? (b) If, simultaneously, an identical wave travels along an adjacent, identical cord, what is the total average rate at which energy is transported to the opposite ends of the two cords by the waves? If, instead, those two waves are sent along the *same* cord simultaneously, what is the total average rate at which they transport energy when their phase difference is (c) 0, (d) 0.4π rad, and (e) π rad?

sec. 16-11 Phasors

•35 Two sinusoidal waves of the same frequency travel in the same direction along a string. If $y_{m1} = 3.0$ cm, $y_{m2} = 4.0$ cm, $\phi_1 = 0$, and $\phi_2 = \pi/2$ rad, what is the amplitude of the resultant wave? SSM

••36 Two sinusoidal waves of the same frequency are to be sent in the same direction along a taut string. One wave has an amplitude of 5.0 mm, the other 8.0 mm. (a) What phase difference ϕ_1 between the two waves results in the smallest amplitude of the resultant wave? (b) What is that smallest amplitude? (c) What phase difference ϕ_2 results in the largest amplitude of the resultant wave? (d) What is that largest amplitude? (e) What is the resultant amplitude if the phase angle is $(\phi_1 - \phi_2)/2$?

••37 Two sinusoidal waves of the same period, with amplitudes of 5.0 and 7.0 mm, travel in the same direction along a stretched string; they produce a resultant wave with an amplitude of 9.0 mm. The phase constant of the 5.0 mm wave is 0. What is the phase constant of the 7.0 mm wave?

••38 Four waves are to be sent along the same string, in the same direction:

$$y_1(x, t) = (4.00 \text{ mm}) \sin(2\pi x - 400\pi t)$$
$$y_2(x, t) = (4.00 \text{ mm}) \sin(2\pi x - 400\pi t + 0.7\pi)$$
$$y_3(x, t) = (4.00 \text{ mm}) \sin(2\pi x - 400\pi t + \pi)$$
$$y_4(x, t) = (4.00 \text{ mm}) \sin(2\pi x - 400\pi t + 1.7\pi).$$

What is the amplitude of the resultant wave?

••39 These two waves travel along the same string:

$$y_1(x, t) = (4.60 \text{ mm}) \sin(2\pi x - 400\pi t)$$
$$y_2(x, t) = (5.60 \text{ mm}) \sin(2\pi x - 400\pi t + 0.80\pi \text{ rad}).$$

What are (a) the amplitude and (b) the phase angle (relative to wave 1) of the resultant wave? (c) If a third wave of amplitude 5.00 mm is also to be sent along the string in the same direction as the first two waves, what should be its phase angle in order to maximize the amplitude of the new resultant wave? GO

sec. 16-13 Standing Waves and Resonance

•40 A 125 cm length of string has mass 2.00 g and tension 7.00 N. (a) What is the wave speed for this string? (b) What is the lowest resonant frequency of this string?

•41 What are (a) the lowest frequency, (b) the second lowest frequency, and (c) the third lowest frequency for standing waves on a wire that is 10.0 m long, has a mass of 100 g, and is stretched under a tension of 250 N? SSM WWW

•42 String A is stretched between two clamps separated by distance L. String B, with the same linear density and under the same tension as string A, is stretched between two clamps separated by distance $4L$. Consider the first eight harmonics of string B. For which of these eight harmonics of B (if any) does the frequency match the frequency of (a) A's first harmonic, (b) A's second harmonic, and (c) A's third harmonic?

•43 A string fixed at both ends is 8.40 m long and has a mass of 0.120 kg. It is subjected to a tension of 96.0 N and set oscillating. (a) What is the speed of the waves on the string? (b) What is the longest possible wavelength for a standing wave? (c) Give the frequency of that wave. SSM

•44 Two sinusoidal waves with identical wavelengths and amplitudes travel in opposite directions along a string with a speed of 10 cm/s. If the time interval between instants when the string is flat is 0.50 s, what is the wavelength of the waves?

•45 A nylon guitar string has a linear density of 7.20 g/m and is under a tension of 150 N. The fixed supports are distance D = 90.0 cm apart. The string is oscillating in the standing wave pattern shown in Fig. 16-41. Calculate the (a) speed, (b) wavelength, and (c) frequency of the traveling waves whose superposition gives this standing wave. ILW

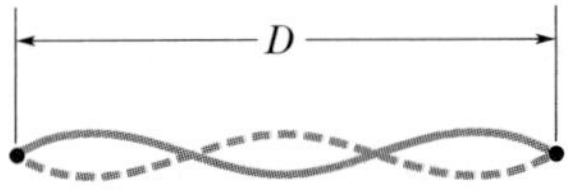

FIG. 16-41 Problem 45.

•46 A string under tension τ_i oscillates in the third harmonic at frequency f_3, and the waves on the string have wavelength λ_3. If the tension is increased to $\tau_f = 4\tau_i$ and the string is again made to oscillate in the third harmonic, what then are (a) the frequency of oscillation in terms of f_3 and (b) the wavelength of the waves in terms of λ_3?

•47 A string that is stretched between fixed supports separated by 75.0 cm has resonant frequencies of 420 and 315 Hz, with no intermediate resonant frequencies. What are (a) the lowest resonant frequency and (b) the wave speed? SSM ILW

•48 If a transmission line in a cold climate collects ice, the increased diameter tends to cause vortex formation in a passing wind. The air pressure variations in the vortexes tend to cause the line to oscillate (*gallop*), especially if the frequency of the variations matches a resonant frequency of the line. In long lines, the resonant frequencies are so close that almost any wind speed can set up a resonant mode vigorous enough to pull down support towers or cause the line to *short out* with an adjacent line. If a transmission line has a length of 347 m, a linear density of 3.35 kg/m, and a tension of 65.2 MN, what are (a) the frequency of the fundamental mode and (b) the frequency difference between successive modes?

•49 One of the harmonic frequencies for a particular string under tension is 325 Hz. The next higher harmonic frequency is 390 Hz. What harmonic frequency is next higher after the harmonic frequency 195 Hz?

••50 A rope, under a tension of 200 N and fixed at both ends, oscillates in a second-harmonic standing wave pattern. The displacement of the rope is given by

$$y = (0.10\text{ m})(\sin \pi x/2) \sin 12\pi t,$$

where $x = 0$ at one end of the rope, x is in meters, and t is in seconds. What are (a) the length of the rope, (b) the speed of the waves on the rope, and (c) the mass of the rope? (d) If the rope oscillates in a third-harmonic standing wave pattern, what will be the period of oscillation?

••51 A string oscillates according to the equation

$$y' = (0.50\text{ cm}) \sin\left[\left(\frac{\pi}{3}\text{ cm}^{-1}\right)x\right] \cos[(40\pi\text{ s}^{-1})t].$$

What are the (a) amplitude and (b) speed of the two waves (identical except for direction of travel) whose superposition gives this oscillation? (c) What is the distance between nodes? (d) What is the transverse speed of a particle of the string at the position x = 1.5 cm when $t = \frac{9}{8}$ s?

••52 A standing wave pattern on a string is described by

$$y(x, t) = 0.040\,(\sin 5\pi x)(\cos 40\pi t),$$

where x and y are in meters and t is in seconds. For $x \geq 0$, what is the location of the node with the (a) smallest, (b) second smallest, and (c) third smallest value of x? (d) What is the period of the oscillatory motion of any (nonnode) point? What are the (e) speed and (f) amplitude of the two traveling waves that interfere to produce this wave? For $t \geq 0$, what are the (g) first, (h) second, and (i) third time that all points on the string have zero transverse velocity?

••53 Two waves are generated on a string of length 3.0 m to produce a three-loop standing wave with an amplitude of 1.0 cm. The wave speed is 100 m/s. Let the equation for one of the waves be of the form $y(x, t) = y_m \sin(kx + \omega t)$. In the equation for the other wave, what are (a) y_m, (b) k, (c) ω, and (d) the sign in front of ω? SSM WWW

••54 For a certain transverse standing wave on a long string, an antinode is at $x = 0$ and an adjacent node is at x = 0.10 m. The displacement $y(t)$ of the string particle at $x = 0$ is shown in Fig. 16-42, where the scale of the y axis is set by y_s = 4.0 cm. When t = 0.50 s, what is the displacement of the string particle at (a) x = 0.20 m and (b) x = 0.30 m? What is the transverse velocity of the string particle at x = 0.20 m at (c) t = 0.50 s and (d) t = 1.0 s? (e) Sketch the standing wave at t = 0.50 s for the range x = 0 to x = 0.40 m.

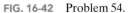

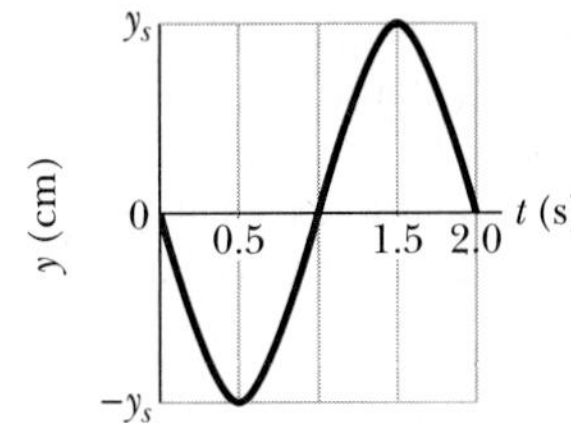

FIG. 16-42 Problem 54.

••55 A generator at one end of a very long string creates a wave given by

$$y = (6.0\text{ cm}) \cos \frac{\pi}{2}[(2.00\text{ m}^{-1})x + (8.00\text{ s}^{-1})t],$$

and a generator at the other end creates the wave

$$y = (6.0\text{ cm}) \cos \frac{\pi}{2}[(2.00\text{ m}^{-1})x - (8.00\text{ s}^{-1})t].$$

Calculate the (a) frequency, (b) wavelength, and (c) speed of each wave. For $x \geq 0$, what is the location of the node having the (d) smallest, (e) second smallest, and (f) third smallest value of x? For $x \geq 0$, what is the location of the antinode having the (g) smallest, (h) second smallest, and (i) third smallest value of x?

••56 Two sinusoidal waves with the same amplitude and wavelength travel through each other along a string that is stretched along an x axis. Their resultant wave is shown twice in Fig. 16-43, as the antinode A travels from an extreme upward displacement to an extreme downward displacement in 6.0 ms. The tick marks along the axis are separated by 10 cm; height H is 1.80 cm. Let the equation for one of the two waves be of the form $y(x, t) = y_m \sin(kx + \omega t)$. In the equation for the other wave, what are (a) y_m, (b) k, (c) ω, and (d) the sign in front of ω? GO

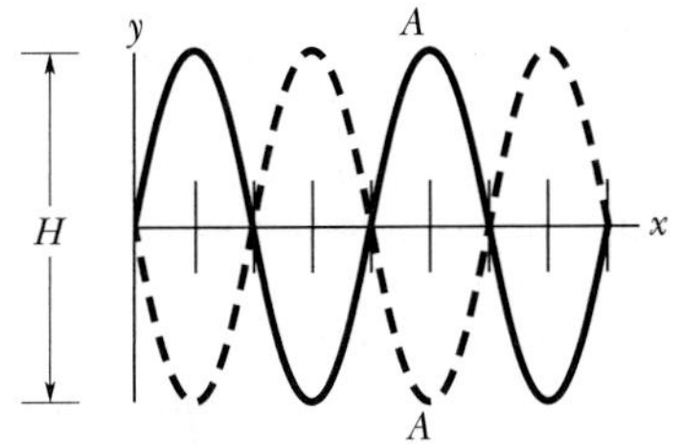

FIG. 16-43 Problem 56.

••57 The following two waves are sent in opposite directions on a horizontal string so as to create a standing wave in a vertical plane:

$$y_1(x, t) = (6.00 \text{ mm}) \sin(4.00\pi x - 400\pi t)$$

$$y_2(x, t) = (6.00 \text{ mm}) \sin(4.00\pi x + 400\pi t),$$

with x in meters and t in seconds. An antinode is located at point A. In the time interval that point takes to move from maximum upward displacement to maximum downward displacement, how far does each wave move along the string?

••58 In Fig. 16-44, a string, tied to a sinusoidal oscillator at P and running over a support at Q, is stretched by a block of mass m. Separation $L = 1.20$ m, linear density $\mu = 1.6$ g/m, and the oscillator frequency $f = 120$ Hz. The amplitude of the motion at P is small enough for that point to be considered a node. A node also exists at Q. (a) What mass m allows the oscillator to set up the fourth harmonic on the string? (b) What standing wave mode, if any, can be set up if $m = 1.00$ kg?

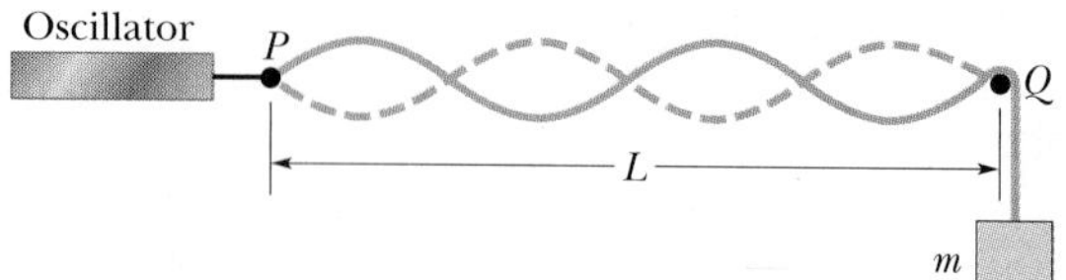

FIG. 16-44 Problems 58 and 60.

•••59 In Fig. 16-45, an aluminum wire, of length $L_1 = 60.0$ cm, cross-sectional area 1.00×10^{-2} cm^2, and density 2.60 g/cm^3, is joined to a steel wire, of density 7.80 g/cm^3 and the same cross-sectional area. The compound wire, loaded with a block of mass $m = 10.0$ kg, is arranged so that the distance L_2 from the joint to the supporting pulley is 86.6 cm. Transverse waves are set up on the wire by an external source of variable frequency; a node is located at the pulley. (a) Find the lowest frequency that generates a standing wave having the joint as one of the nodes. (b) How many nodes are observed at this frequency?

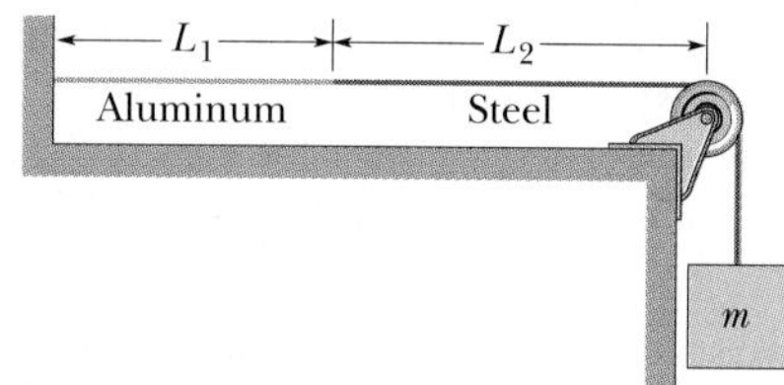

FIG. 16-45 Problem 59.

•••60 In Fig. 16-44, a string, tied to a sinusoidal oscillator at P and running over a support at Q, is stretched by a block of mass m. The separation L between P and Q is 1.20 m, and the frequency f of the oscillator is fixed at 120 Hz. The amplitude of the motion at P is small enough for that point to be considered a node. A node also exists at Q. A standing wave appears when the mass of the hanging block is 286.1 g or 447.0 g, but not for any intermediate mass. What is the linear density of the string? GO

Additional Problems

61 Three sinusoidal waves of the same frequency travel along a string in the positive direction of an x axis. Their amplitudes are y_1, $y_1/2$, and $y_1/3$, and their phase constants are 0, $\pi/2$, and π, respectively. What are the (a) amplitude and (b) phase constant of the resultant wave? (c) Plot the wave form of the resultant wave at $t = 0$, and discuss its behavior as t increases. SSM

62 Figure 16-46 shows the displacement y versus time t of the point on a string at $x = 0$, as a wave passes through that point. The scale of the y axis is set by $y_s = 6.0$ mm. The wave has form $y(x, t) = y_m \sin(kx - \omega t + \phi)$. What is ϕ? (*Caution:* A calculator does not always give the proper inverse trig function, so check your answer by substituting it and an assumed value of ω into $y(x, t)$ and then plotting the function.)

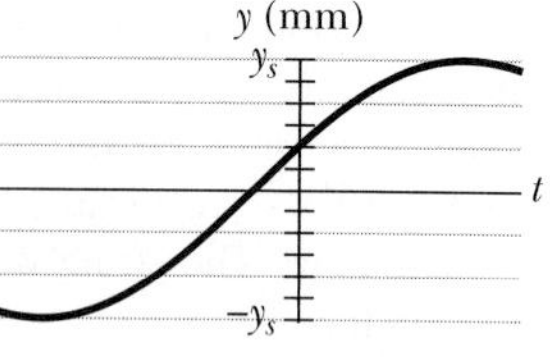

FIG. 16-46 Problem 62.

63 Two sinusoidal waves, identical except for phase, travel in the same direction along a string, producing a net wave $y'(x, t) = (3.0 \text{ mm}) \sin(20x - 4.0t + 0.820 \text{ rad})$, with x in meters and t in seconds. What are (a) the wavelength λ of the two waves, (b) the phase difference between them, and (c) their amplitude y_m?

64 Figure 16-47 shows transverse acceleration a_y versus time t of the point on a string at $x = 0$, as a wave written in the form $y(x, t) = y_m \sin(kx - \omega t + \phi)$ passes through that point. The scale of the vertical axis is set by $a_s = 400$ m/s^2. What is ϕ? (*Caution:* A calculator does not always give the proper inverse trig function, so check your answer by substituting it and an assumed value of ω into $y(x, t)$ and then plotting the function.)

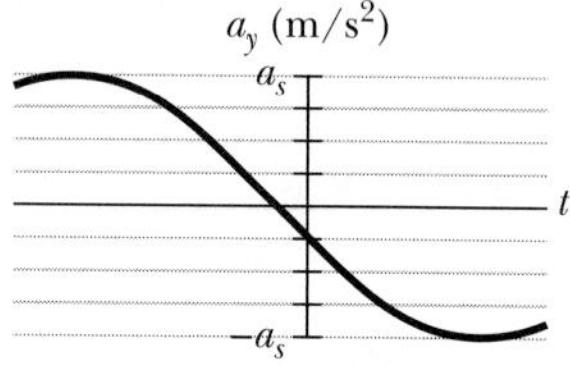

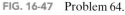

FIG. 16-47 Problem 64.

65 At time $t = 0$ and at position $x = 0$ m along a string, a traveling sinusoidal wave with an angular frequency of 440 rad/s has displacement $y = +4.5$ mm and transverse velocity $u = -0.75$ m/s. If the wave has general form $y(x, t) = y_m \sin(kx - \omega t + \phi)$, what is phase constant ϕ?

66 A single pulse, given by $h(x - 5.0t)$, is shown in Fig. 16-48 for $t = 0$. The scale of the vertical axis is set by $h_s = 2$. Here x is in centimeters and t is in seconds. What are the (a) speed and (b) direction of travel of the pulse? (c) Plot $h(x - 5t)$ as a function of x for $t = 2$ s. (d) Plot $h(x - 5t)$ as a function of t for $x = 10$ cm.

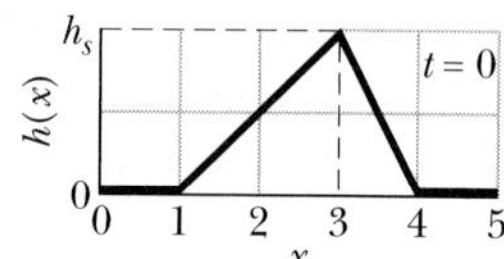

FIG. 16-48 Problem 66.

67 A transverse sinusoidal wave is generated at one end of a long, horizontal string by a bar that moves up and down through a distance of 1.00 cm. The motion is continuous and is repeated regularly 120 times per second. The string has linear density 120 g/m and is kept under a tension of 90.0 N. Find the maximum value of (a) the transverse speed u and (b) the transverse component of the tension τ.

(c) Show that the two maximum values calculated above occur at the same phase values for the wave. What is the transverse displacement y of the string at these phases? (d) What is the maximum rate of energy transfer along the string? (e) What is the transverse displacement y when this maximum transfer occurs? (f) What is the minimum rate of energy transfer along the string? (g) What is the transverse displacement y when this minimum transfer occurs?

68 Two sinusoidal 120 Hz waves, of the same frequency and amplitude, are to be sent in the positive direction of an x axis

that is directed along a cord under tension. The waves can be sent in phase, or they can be phase-shifted. Figure 16-49 shows the amplitude y' of the resulting wave versus the distance of the shift (how far one wave is shifted from the other wave). The scale of the vertical axis is set by y'_s = 6.0 mm. If the equations for the two waves are of the form $y(x, t) = y_m \sin(kx \pm \omega t)$, what are (a) y_m, (b) k, (c) ω, and (d) the correct choice of sign in front of ω?

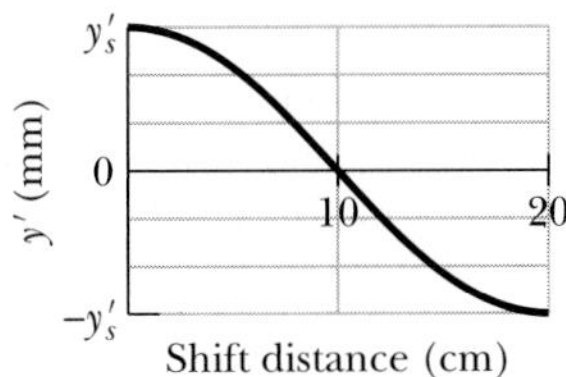

FIG. 16-49 Problem 68.

69 A sinusoidal transverse wave of amplitude y_m and wavelength λ travels on a stretched cord. (a) Find the ratio of the maximum particle speed (the speed with which a single particle in the cord moves transverse to the wave) to the wave speed. (b) Does this ratio depend on the material of which the cord is made? SSM

70 A sinusoidal transverse wave traveling in the positive direction of an x axis has an amplitude of 2.0 cm, a wavelength of 10 cm, and a frequency of 400 Hz. If the wave equation is of the form $y(x,t) = y_m \sin(kx \pm \omega t)$, what are (a) y_m, (b) k, (c) ω, and (d) the correct choice of sign in front of ω? What are (e) the maximum transverse speed of a point on the cord and (f) the speed of the wave?

71 A sinusoidal transverse wave traveling in the negative direction of an x axis has an amplitude of 1.00 cm, a frequency of 550 Hz, and a speed of 330 m/s. If the wave equation is of the form $y(x,t) = y_m \sin(kx \pm \omega t)$, what are (a) y_m, (b) ω, (c) k, and (d) the correct choice of sign in front of ω?

72 Two sinusoidal waves of the same wavelength travel in the same direction along a stretched string. For wave 1, y_m = 3.0 mm and $\phi = 0$; for wave 2, y_m = 5.0 mm and $\phi = 70°$. What are the (a) amplitude and (b) phase constant of the resultant wave?

73 A wave has a speed of 240 m/s and a wavelength of 3.2 m. What are the (a) frequency and (b) period of the wave?

74 When played in a certain manner, the lowest resonant frequency of a certain violin string is concert A (440 Hz). What is the frequency of the (a) second and (b) third harmonic of the string?

75 A 120 cm length of string is stretched between fixed supports. What are the (a) longest, (b) second longest, and (c) third longest wavelength for waves traveling on the string if standing waves are to be set up? (d) Sketch those standing waves.

76 The equation of a transverse wave traveling along a string is

$$y = 0.15 \sin(0.79x - 13t),$$

in which x and y are in meters and t is in seconds. (a) What is the displacement y at x = 2.3 m, t = 0.16 s? A second wave is to be added to the first wave to produce standing waves on the string. If the wave equation for the second wave is of the form $y(x, t) = y_m \sin(kx \pm \omega t)$, what are (b) y_m, (c) k, (d) ω, and (e) the correct choice of sign in front of ω for this second wave? (f) What is the displacement of the resultant standing wave at x = 2.3 m, t = 0.16 s?

77 A 1.50 m wire has a mass of 8.70 g and is under a tension of 120 N. The wire is held rigidly at both ends and set into oscillation. (a) What is the speed of waves on the wire? What is the wavelength of the waves that produce (b) one-loop and (c) two-loop standing waves? What is the frequency of the waves that produce (d) one-loop and (e) two-loop standing waves? SSM

78 Energy is transmitted at rate P_1 by a wave of frequency f_1 on a string under tension τ_1. What is the new energy transmission rate P_2 in terms of P_1 (a) if the tension is increased to $\tau_2 = 4\tau_1$ and (b) if, instead, the frequency is decreased to $f_2 = f_1/2$?

79 The equation of a transverse wave traveling along a string is

$$y = (2.0 \text{ mm}) \sin[(20 \text{ m}^{-1})x - (600 \text{ s}^{-1})t].$$

Find the (a) amplitude, (b) frequency, (c) velocity (including sign), and (d) wavelength of the wave. (e) Find the maximum transverse speed of a particle in the string.

80 Oscillation of a 600 Hz tuning fork sets up standing waves in a string clamped at both ends. The wave speed for the string is 400 m/s. The standing wave has four loops and an amplitude of 2.0 mm. (a) What is the length of the string? (b) Write an equation for the displacement of the string as a function of position and time.

81 In an experiment on standing waves, a string 90 cm long is attached to the prong of an electrically driven tuning fork that oscillates perpendicular to the length of the string at a frequency of 60 Hz. The mass of the string is 0.044 kg. What tension must the string be under (weights are attached to the other end) if it is to oscillate in four loops?

82 *Body armor.* When a high-speed projectile such as a bullet or bomb fragment strikes modern body armor, the fabric of the armor stops the projectile and prevents penetration by quickly spreading the projectile's energy over a large area. This spreading is done by longitudinal and transverse pulses that move *radially* from the impact point, where the projectile pushes a cone-shaped dent into the fabric. The longitudinal pulse, racing along the fibers of the fabric at speed v_l ahead of the denting, causes the fibers to thin and stretch, with

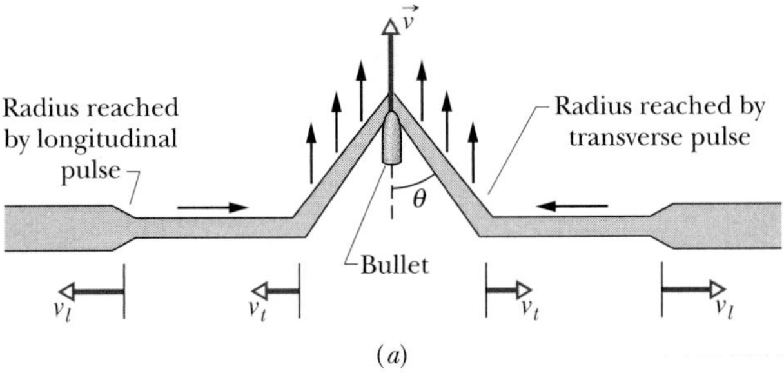

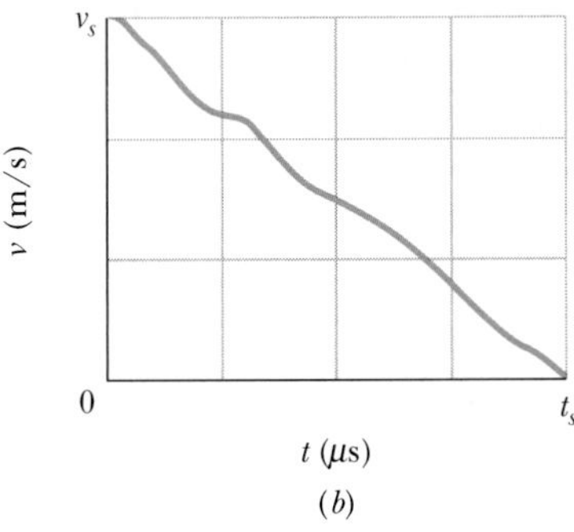

FIG. 16-50 Problem 82.

material flowing radially inward into the dent. One such radial fiber is shown in Fig. 16-50a. Part of the projectile's energy goes into this motion and stretching. The transverse pulse, moving at a slower speed v_t, is due to the denting. As the projectile increases the dent's depth, the dent increases in radius, causing the material in the fibers to move in the same direction as the projectile (perpendicular to the transverse pulse's direction of travel). The rest of the projectile's energy goes into this motion. All the energy that does not eventually go into permanently deforming the fibers ends up as thermal energy.

Figure 16-50b is a graph of speed v versus time t for a bullet of mass 10.2 g fired from a .38 Special revolver directly into body armor. The scales of the vertical and horizontal axes are set by $v_s = 300$ m/s and $t_s = 40.0$ μs. Take $v_l = 2000$ m/s, and assume that the half-angle θ of the conical dent is 60°. At the end of the collision, what are the radii of (a) the thinned region and (b) the dent (assuming that the person wearing the armor remains stationary)?

83 (a) What is the fastest transverse wave that can be sent along a steel wire? For safety reasons, the maximum tensile stress to which steel wires should be subjected is 7.00×10^8 N/m^2. The density of steel is 7800 kg/m^3. (b) Does your answer depend on the diameter of the wire?

84 (a) Write an equation describing a sinusoidal transverse wave traveling on a cord in the positive direction of a y axis with an angular wave number of 60 cm^{-1}, a period of 0.20 s, and an amplitude of 3.0 mm. Take the transverse direction to be the z direction. (b) What is the maximum transverse speed of a point on the cord?

85 A wave on a string is described by

$$y(x, t) = 15.0 \sin(\pi x/8 - 4\pi t),$$

where x and y are in centimeters and t is in seconds. (a) What is the transverse speed for a point on the string at $x = 6.00$ cm when $t = 0.250$ s? (b) What is the maximum transverse speed of any point on the string? (c) What is the magnitude of the transverse acceleration for a point on the string at $x = 6.00$ cm when $t = 0.250$ s? (d) What is the magnitude of the maximum transverse acceleration for any point on the string?

86 A standing wave results from the sum of two transverse traveling waves given by

$$y_1 = 0.050 \cos(\pi x - 4\pi t)$$

and

$$y_2 = 0.050 \cos(\pi x + 4\pi t),$$

where x, y_1, and y_2 are in meters and t is in seconds. (a) What is the smallest positive value of x that corresponds to a node? Beginning at $t = 0$, what is the value of the (b) first, (c) second, and (d) third time the particle at $x = 0$ has zero velocity?

87 In a demonstration, a 1.2 kg horizontal rope is fixed in place at its two ends ($x = 0$ and $x = 2.0$ m) and made to oscillate up and down in the fundamental mode, at frequency 5.0 Hz. At $t = 0$, the point at $x = 1.0$ m has zero displacement and is moving upward in the positive direction of a y axis with a transverse velocity of 5.0 m/s. What are (a) the amplitude of the motion of that point and (b) the tension in the rope? (c) Write the standing wave equation for the fundamental mode. SSM

88 A certain transverse sinusoidal wave of wavelength 20 cm is moving in the positive direction of an x axis. The transverse velocity of the particle at $x = 0$ as a function of time is shown in Fig. 16-51, where the scale of the vertical axis is set by $u_s = 5.0$ cm/s. What are the (a) wave speed, (b) amplitude, and (c) frequency? (d) Sketch the wave between $x = 0$ and $x = 20$ cm at $t = 2.0$ s.

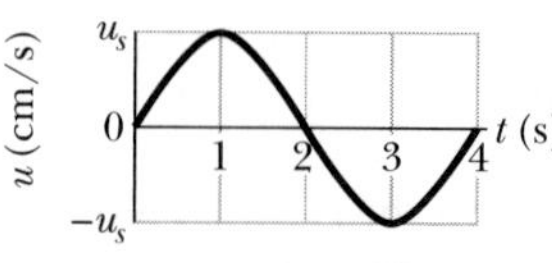

FIG. 16-51 Problem 88.

89 The type of rubber band used inside some baseballs and golf balls obeys Hooke's law over a wide range of elongation of the band. A segment of this material has an unstretched length ℓ and a mass m. When a force F is applied, the band stretches an additional length $\Delta\ell$. (a) What is the speed (in terms of m, $\Delta\ell$, and the spring constant k) of transverse waves on this stretched rubber band? (b) Using your answer to (a), show that the time required for a transverse pulse to travel the length of the rubber band is proportional to $1/\sqrt{\Delta\ell}$ if $\Delta\ell \ll \ell$ and is constant if $\Delta\ell \gg \ell$. SSM

90 Two waves,

$$y_1 = (2.50 \text{ mm}) \sin[(25.1 \text{ rad/m})x - (440 \text{ rad/s})t]$$

and $$y_2 = (1.50 \text{ mm}) \sin[(25.1 \text{ rad/m})x + (440 \text{ rad/s})t],$$

travel along a stretched string. (a) Plot the resultant wave as a function of t for $x = 0$, $\lambda/8$, $\lambda/4$, $3\lambda/8$, and $\lambda/2$, where λ is the wavelength. The graphs should extend from $t = 0$ to a little over one period. (b) The resultant wave is the superposition of a standing wave and a traveling wave. In which direction does the traveling wave move? (c) How can you change the original waves so the resultant wave is the superposition of standing and traveling waves with the same amplitudes as before but with the traveling wave moving in the opposite direction? Next, use your graphs to find the place at which the oscillation amplitude is (d) maximum and (e) minimum. (f) How is the maximum amplitude related to the amplitudes of the original two waves? (g) How is the minimum amplitude related to the amplitudes of the original two waves?

91 Two waves are described by

$$y_1 = 0.30 \sin[\pi(5x - 200)t]$$

and

$$y_2 = 0.30 \sin[\pi(5x - 200t) + \pi/3],$$

where y_1, y_2, and x are in meters and t is in seconds. When these two waves are combined, a traveling wave is produced. What are the (a) amplitude, (b) wave speed, and (c) wavelength of that traveling wave?

92 The speed of electromagnetic waves (which include visible light, radio, and x rays) in vacuum is 3.0×10^8 m/s. (a) Wavelengths of visible light waves range from about 400 nm in the violet to about 700 nm in the red. What is the range of frequencies of these waves? (b) The range of frequencies for shortwave radio (for example, FM radio and VHF television) is 1.5 to 300 MHz. What is the corresponding wavelength range? (c) X ray wavelengths range from about 5.0 nm to about 1.0×10^{-2} nm. What is the frequency range for x rays?

93 A traveling wave on a string is described by

$$y = 2.0 \sin\left[2\pi\left(\frac{t}{0.40} + \frac{x}{80}\right)\right],$$

where x and y are in centimeters and t is in seconds. (a) For $t = 0$, plot y as a function of x for $0 \leq x \leq 160$ cm. (b) Repeat (a) for $t = 0.05$ s and $t = 0.10$ s. From your graphs, determine (c) the wave speed and (d) the direction in which the wave is traveling.

Waves—II

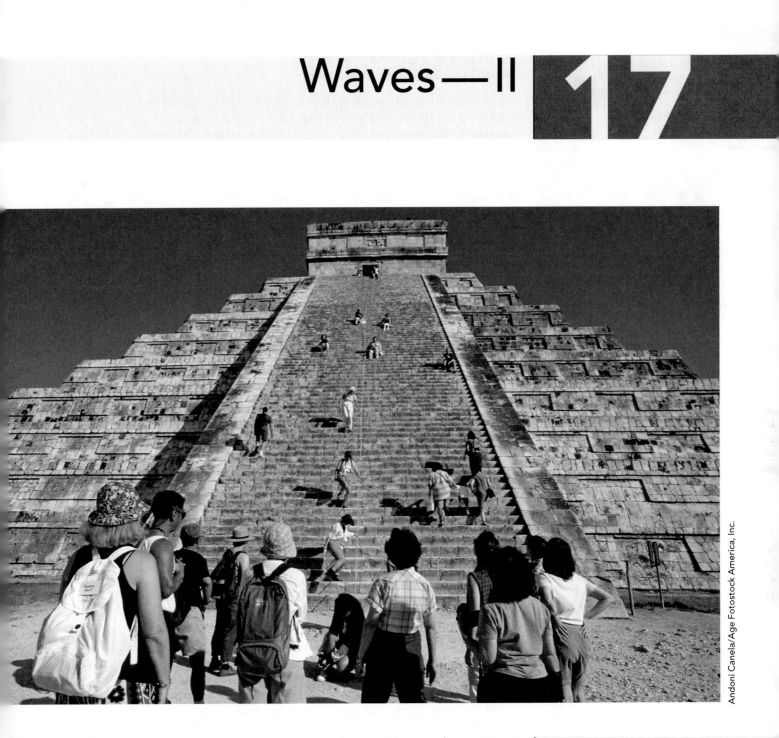

Andoni Canela/Age Fotostock America, Inc.

Echoes can be enchanting in certain outdoor settings and annoying in rooms where they make speech unintelligible, but they always mimic the source of the sound. For example, the sound of a handclap returns as the sound of a handclap. However, an echo in front of these steps up the side of a pyramid in the Mayan ruins at Chichen Itza, Mexico, is remarkably different because the handclap returns as a musical note descending in frequency.

What causes this musical echo, said to be a chirped echo?

The answer is in this chapter.

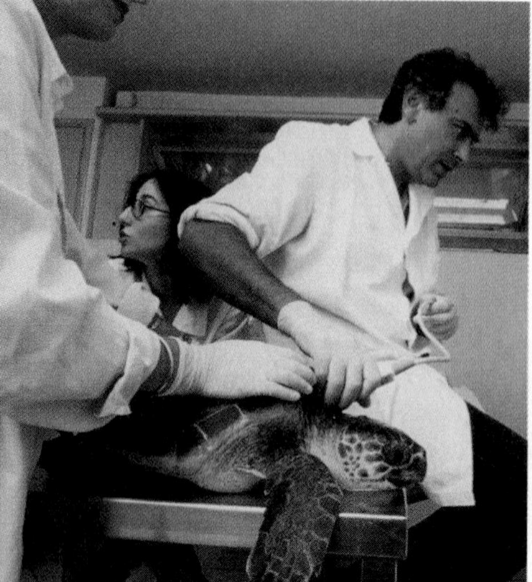

FIG. 17-1 A loggerhead turtle is being checked with ultrasound (which has a frequency above your hearing range); an image of its interior is being produced on a monitor off to the right. *(Mauro Fermariello/SPL/Photo Researchers)*

17-1 WHAT IS PHYSICS?

The physics of sound waves is the basis of countless studies in the research journals of many fields. Here are just a few examples. Some physiologists are concerned with how speech is produced, how speech impairment might be corrected, how hearing loss can be alleviated, and even how snoring is produced. Some acoustic engineers are concerned with improving the acoustics of cathedrals and concert halls, with reducing noise near freeways and road construction, and with reproducing music by speaker systems. Some aviation engineers are concerned with the shock waves produced by supersonic aircraft and the aircraft noise produced in communities near an airport. Some medical researchers are concerned with how noises produced by the heart and lungs can signal a medical problem in a patient. Some paleontologists are concerned with how a dinosaur's fossil might reveal the dinosaur's vocalizations. Some military engineers are concerned with how the sounds of sniper fire might allow a soldier to pinpoint the sniper's location, and, on the gentler side, some biologists are concerned with how a cat purrs.

To begin our discussion of the physics of sound, we must first answer the question "What *are* sound waves?"

17-2 | Sound Waves

As we saw in Chapter 16, mechanical waves are waves that require a material medium to exist. There are two types of mechanical waves: *Transverse waves* involve oscillations perpendicular to the direction in which the wave travels; *longitudinal waves* involve oscillations parallel to the direction of wave travel.

In this book, a **sound wave** is defined roughly as any longitudinal wave. Seismic prospecting teams use such waves to probe Earth's crust for oil. Ships carry sound-ranging gear (sonar) to detect underwater obstacles. Submarines use sound waves to stalk other submarines, largely by listening for the characteristic noises produced by the propulsion system. Figure 17-1 suggests how sound waves can be used to explore the soft tissues of an animal or human body. In this chapter we shall focus on sound waves that travel through the air and that are audible to people.

Figure 17-2 illustrates several ideas that we shall use in our discussions. Point S represents a tiny sound source, called a *point source,* that emits sound waves in all directions. The *wavefronts* and *rays* indicate the direction of travel and the spread of the sound waves. **Wavefronts** are surfaces over which the oscillations due to the sound wave have the same value; such surfaces are represented by whole or partial circles in a two-dimensional drawing for a point source. **Rays** are directed lines perpendicular to the wavefronts that indicate the direction of travel of the wavefronts. The short double arrows superimposed on the rays of Fig. 17-2 indicate that the longitudinal oscillations of the air are parallel to the rays.

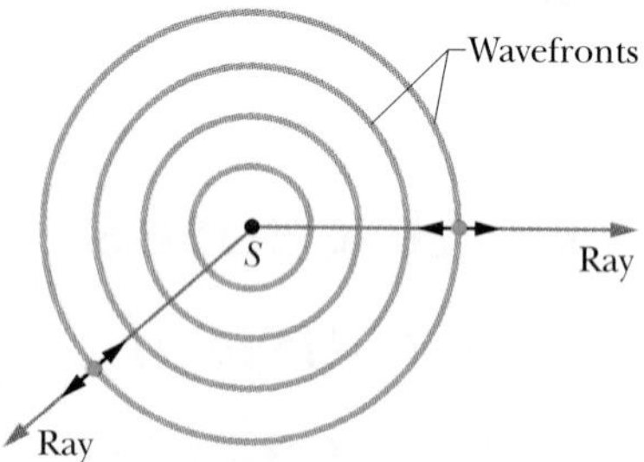

FIG. 17-2 A sound wave travels from a point source S through a three-dimensional medium. The wavefronts form spheres centered on S; the rays are radial to S. The short, double-headed arrows indicate that elements of the medium oscillate parallel to the rays.

Near a point source like that of Fig. 17-2, the wavefronts are spherical and spread out in three dimensions, and there the waves are said to be *spherical.* As the wavefronts move outward and their radii become larger, their curvature decreases. Far from the source, we approximate the wavefronts as planes (or lines on two-dimensional drawings), and the waves are said to be *planar.*

17-3 | The Speed of Sound

The speed of any mechanical wave, transverse or longitudinal, depends on both an inertial property of the medium (to store kinetic energy) and an elastic property of the medium (to store potential energy). Thus, we can generalize Eq. 16-26, which gives the speed of a transverse wave along a stretched string, by writing

$$v = \sqrt{\frac{\tau}{\mu}} = \sqrt{\frac{\text{elastic property}}{\text{inertial property}}}, \tag{17-1}$$

where (for transverse waves) τ is the tension in the string and μ is the string's linear density. If the medium is air and the wave is longitudinal, we can guess that the inertial property, corresponding to μ, is the volume density ρ of air. What shall we put for the elastic property?

In a stretched string, potential energy is associated with the periodic stretching of the string elements as the wave passes through them. As a sound wave passes through air, potential energy is associated with periodic compressions and expansions of small volume elements of the air. The property that determines the extent to which an element of a medium changes in volume when the pressure (force per unit area) on it changes is the **bulk modulus** B, defined (from Eq. 12-25) as

$$B = -\frac{\Delta p}{\Delta V/V} \quad \text{(definition of bulk modulus).} \tag{17-2}$$

Here $\Delta V/V$ is the fractional change in volume produced by a change in pressure Δp. As explained in Section 14-3, the SI unit for pressure is the newton per square meter, which is given a special name, the *pascal* (Pa). From Eq. 17-2 we see that the unit for B is also the pascal. The signs of Δp and ΔV are always opposite: When we increase the pressure on an element (Δp is positive), its volume decreases (ΔV is negative). We include a minus sign in Eq. 17-2 so that B is always a positive quantity. Now substituting B for τ and ρ for μ in Eq. 17-1 yields

$$v = \sqrt{\frac{B}{\rho}} \quad \text{(speed of sound)} \tag{17-3}$$

as the speed of sound in a medium with bulk modulus B and density ρ. Table 17-1 lists the speed of sound in various media.

The density of water is almost 1000 times greater than the density of air. If this were the only relevant factor, we would expect from Eq. 17-3 that the speed of sound in water would be considerably less than the speed of sound in air. However, Table 17-1 shows us that the reverse is true. We conclude (again from Eq. 17-3) that the bulk modulus of water must be more than 1000 times greater than that of air. This is indeed the case. Water is much more incompressible than air, which (see Eq. 17-2) is another way of saying that its bulk modulus is much greater.

TABLE 17-1

The Speed of Sound[a]

Medium	Speed (m/s)
Gases	
Air (0°C)	331
Air (20°C)	343
Helium	965
Hydrogen	1284
Liquids	
Water (0°C)	1402
Water (20°C)	1482
Seawater[b]	1522
Solids	
Aluminum	6420
Steel	5941
Granite	6000

[a]At 0°C and 1 atm pressure, except where noted.

[b]At 20°C and 3.5% salinity.

Formal Derivation of Eq. 17-3

We now derive Eq. 17-3 by direct application of Newton's laws. Let a single pulse in which air is compressed travel (from right to left) with speed v through the air in a long tube, like that in Fig. 16-2. Let us run along with the pulse at that speed, so that the pulse appears to stand still in our reference frame. Figure 17-3*a* shows the situation as it is viewed from that frame. The pulse is standing still, and air is moving at speed v through it from left to right.

Let the pressure of the undisturbed air be p and the pressure inside the pulse be $p + \Delta p$, where Δp is positive due to the compression. Consider an element of air of thickness Δx and face area A, moving toward the pulse at speed v. As this element enters the pulse, the leading face of the element encounters a region of higher pressure, which slows the element to speed $v + \Delta v$, in which Δv is negative. This slowing is complete when the rear face of the element reaches the pulse, which requires time interval

$$\Delta t = \frac{\Delta x}{v}. \tag{17-4}$$

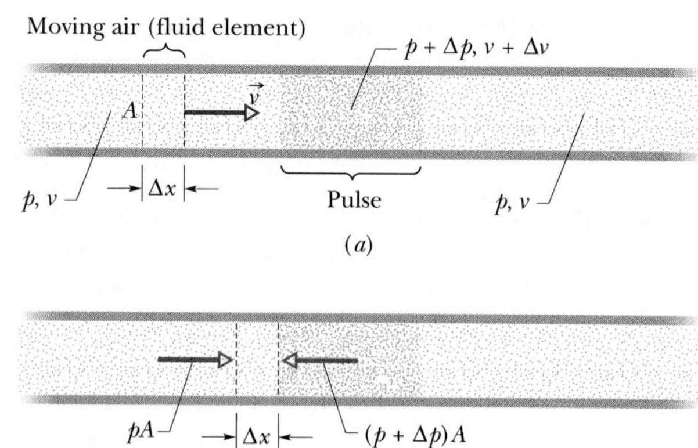

FIG. 17-3 A compression pulse is sent from right to left down a long air-filled tube. The reference frame of the figure is chosen so that the pulse is at rest and the air moves from left to right. (*a*) An element of air of width Δx moves toward the pulse with speed v. (*b*) The leading face of the element enters the pulse. The forces acting on the leading and trailing faces (due to air pressure) are shown.

Let us apply Newton's second law to the element. During Δt, the average force on the element's trailing face is pA toward the right, and the average force on the leading face is $(p + \Delta p)A$ toward the left (Fig. 17-3*b*). Therefore, the average net force on the element during Δt is

$$\begin{aligned} F &= pA - (p + \Delta p)A \\ &= -\Delta p\, A \qquad \text{(net force).} \end{aligned} \tag{17-5}$$

The minus sign indicates that the net force on the air element is directed to the left in Fig. 17-3*b*. The volume of the element is $A\,\Delta x$, so with the aid of Eq. 17-4, we can write its mass as

$$\Delta m = \rho\,\Delta V = \rho A\,\Delta x = \rho A v\,\Delta t \qquad \text{(mass).} \tag{17-6}$$

The average acceleration of the element during Δt is

$$a = \frac{\Delta v}{\Delta t} \qquad \text{(acceleration).} \tag{17-7}$$

Thus, from Newton's second law ($F = ma$), we have, from Eqs. 17-5, 17-6, and 17-7,

$$-\Delta p\, A = (\rho A v\,\Delta t)\frac{\Delta v}{\Delta t},$$

which we can write as

$$\rho v^2 = -\frac{\Delta p}{\Delta v/v}. \tag{17-8}$$

The air that occupies a volume $V\,(= Av\,\Delta t)$ outside the pulse is compressed by an amount $\Delta V\,(= A\,\Delta v\,\Delta t)$ as it enters the pulse. Thus,

$$\frac{\Delta V}{V} = \frac{A\,\Delta v\,\Delta t}{Av\,\Delta t} = \frac{\Delta v}{v}. \tag{17-9}$$

Substituting Eq. 17-9 and then Eq. 17-2 into Eq. 17-8 leads to

$$\rho v^2 = -\frac{\Delta p}{\Delta v/v} = -\frac{\Delta p}{\Delta V/V} = B.$$

Solving for v yields Eq. 17-3 for the speed of the air toward the right in Fig. 17-3, and thus for the actual speed of the pulse toward the left.

Sample Problem 17-1

When a sound pulse, as from a handclap, is produced at the foot of the stairs at the Mayan pyramid shown in the chapter's opening photograph, the sound waves reflect from the steps in succession, the closest (lowest) one first (Fig. 17-4*a*) and the farthest (highest) one last (Fig. 17-4*b*). The depth and height of the steps are $d =$ 0.263 m, and the speed of sound is 343 m/s. The paths taken by the sound waves to and from the steps near the bottom of the stairs are approximately horizontal. The slanted paths taken by the sound waves to and from the steps near the top are approximately 45° to the horizontal. At what frequency f_{bot} do the echo pulses arrive at the listener from the bottom steps? At what frequency f_{top} do they arrive from the top steps a short time later?

KEY IDEAS (1) The frequency f at which the pulses return to the listener is the inverse of the time Δt between successive pulses. (2) The time interval Δt required by sound to travel a given distance L is related to the speed of sound v by $v = L/\Delta t$.

Calculations: In Fig. 17-4*a* at the bottom of the stairs, the sound wave that reflects from the higher step travels a distance $L = 2d$ more than the sound wave that reflects from the lower step. (The higher wave must travel twice across the step's depth.) So, the arrivals of the echo pulses at the listener are separated by the time interval

$$\Delta t_{bot} = \frac{L}{v} = \frac{2d}{v} \tag{17-10}$$

$$= \frac{2(0.263 \text{ m})}{343 \text{ m/s}} = 1.533 \times 10^{-3} \text{ s}.$$

The frequency f_{bot} at which the pulses arrive at the listener is

$$f_{bot} = \frac{1}{\Delta t_{bot}} \tag{17-11}$$

$$= \frac{1}{1.533 \times 10^{-3} \text{ s}} = 652 \text{ Hz}. \qquad \text{(Answer)}$$

The time interval Δt_{bot} is too short for a listener to distinguish the individual pulses. Instead, the frequency f_{bot} is brought to consciousness—the listener hears a musical note of frequency 652 Hz.

In Fig. 17-4*b* at the top of the stairs, the slanted approach and return of the sound waves means that the wave reflected from the higher step travels a distance $L = 2\sqrt{2}d$ more than the wave that reflects from the lower step. (The travel is twice along the hypotenuse of a right triangle with equal legs of length d.) So, now the arrivals of the echo pulses at the listener are separated by the time interval

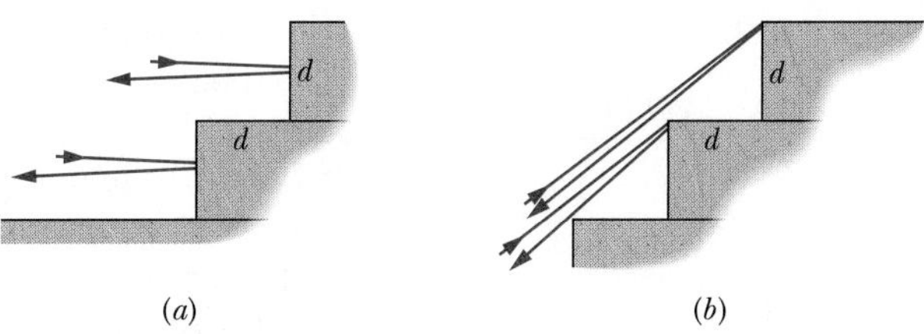

FIG. 17-4 Sound waves reflect from (*a*) the bottom steps and (*b*) the top steps of a tall flight of stairs.

$$\Delta t_{top} = \frac{L}{v} = \frac{2\sqrt{2}d}{v} \tag{17-12}$$

$$= \frac{2\sqrt{2}(0.263 \text{ m})}{343 \text{ m/s}} = 2.168 \times 10^{-3} \text{ s},$$

and the frequency that is brought to consciousness is

$$f_{top} = \frac{1}{\Delta t_{top}}$$

$$= \frac{1}{2.168 \times 10^{-3} \text{ s}} = 461 \text{ Hz}. \qquad \text{(Answer)}$$

Thus, a handclap in front of the stairs produces an echo that begins with a frequency of 652 Hz and ends with a frequency of 461 Hz. You might be able to hear such a musical echo from other stairs or even from a picket fence if you stand alongside it.

17-4 | Traveling Sound Waves

Here we examine the displacements and pressure variations associated with a sinusoidal sound wave traveling through air. Figure 17-5*a* displays such a wave traveling rightward through a long air-filled tube. Recall from Chapter 16 that

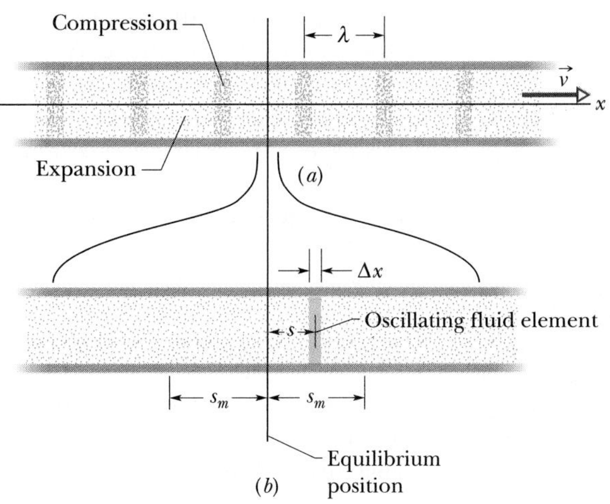

FIG. 17-5 (*a*) A sound wave, traveling through a long air-filled tube with speed v, consists of a moving, periodic pattern of expansions and compressions of the air. The wave is shown at an arbitrary instant. (*b*) A horizontally expanded view of a short piece of the tube. As the wave passes, an air element of thickness Δx oscillates left and right in simple harmonic motion about its equilibrium position. At the instant shown in (*b*), the element happens to be displaced a distance s to the right of its equilibrium position. Its maximum displacement, either right or left, is s_m.

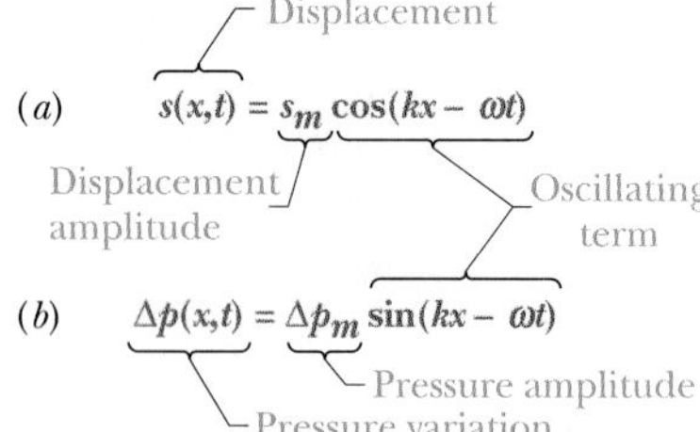

FIG. 17-6 (*a*) The displacement function and (*b*) the pressure-variation function of a traveling sound wave consist of an amplitude and an oscillating term.

we can produce such a wave by sinusoidally moving a piston at the left end of the tube (as in Fig. 16-2). The piston's rightward motion moves the element of air next to the piston face and compresses that air; the piston's leftward motion allows the element of air to move back to the left and the pressure to decrease. As each element of air pushes on the next element in turn, the right–left motion of the air and the change in its pressure travel along the tube as a sound wave.

Consider the thin element of air of thickness Δx shown in Fig. 17-5*b*. As the wave travels through this portion of the tube, the element of air oscillates left and right in simple harmonic motion about its equilibrium position. Thus, the oscillations of each air element due to the traveling sound wave are like those of a string element due to a transverse wave, except that the air element oscillates *longitudinally* rather than *transversely*. Because string elements oscillate parallel to the y axis, we write their displacements in the form $y(x, t)$. Similarly, because air elements oscillate parallel to the x axis, we could write their displacements in the confusing form $x(x, t)$, but we shall use $s(x, t)$ instead.

To show that the displacements $s(x, t)$ are sinusoidal functions of x and t, we can use either a sine function or a cosine function. In this chapter we use a cosine function, writing

$$s(x, t) = s_m \cos(kx - \omega t). \qquad (17\text{-}13)$$

Figure 17-6*a* labels the various parts of this equation. In it, s_m is the **displacement amplitude**—that is, the maximum displacement of the air element to either side of its equilibrium position (see Fig. 17-5*b*). The angular wave number k, angular frequency ω, frequency f, wavelength λ, speed v, and period T for a sound (longitudinal) wave are defined and interrelated exactly as for a transverse wave, except that λ is now the distance (again along the direction of travel) in which the pattern of compression and expansion due to the wave begins to repeat itself (see Fig. 17-5*a*). (We assume s_m is much less than λ.)

As the wave moves, the air pressure at any position x in Fig. 17-5*a* varies sinusoidally, as we prove next. To describe this variation we write

$$\Delta p(x, t) = \Delta p_m \sin(kx - \omega t). \qquad (17\text{-}14)$$

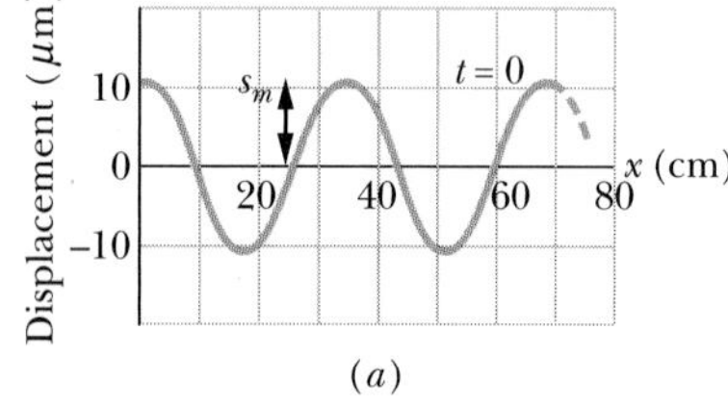

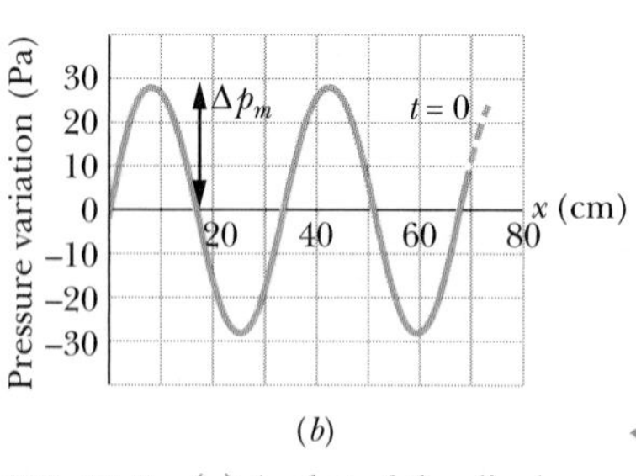

FIG. 17-7 (*a*) A plot of the displacement function (Eq. 17-13) for $t = 0$. (*b*) A similar plot of the pressure-variation function (Eq. 17-14). Both plots are for a 1000 Hz sound wave whose pressure amplitude is at the threshold of pain; see Sample Problem 17-2.

Figure 17-6*b* labels the various parts of this equation. A negative value of Δp in Eq. 17-14 corresponds to an expansion of the air, and a positive value to a compression. Here Δp_m is the **pressure amplitude,** which is the maximum increase or decrease in pressure due to the wave; Δp_m is normally very much less than the pressure p present when there is no wave. As we shall prove, the pressure amplitude Δp_m is related to the displacement amplitude s_m in Eq. 17-13 by

$$\Delta p_m = (v\rho\omega)s_m. \qquad (17\text{-}15)$$

Figure 17-7 shows plots of Eqs. 17-13 and 17-14 at $t = 0$; with time, the two curves would move rightward along the horizontal axes. Note that the displacement and pressure variation are $\pi/2$ rad (or 90°) out of phase. Thus, for example, the pressure variation Δp at any point along the wave is zero when the displacement there is a maximum.

CHECKPOINT 1 When the oscillating air element in Fig. 17-5*b* is moving rightward through the point of zero displacement, is the pressure in the element at its equilibrium value, just beginning to increase, or just beginning to decrease?

Derivation of Eqs. 17-14 and 17-15

Figure 17-5*b* shows an oscillating element of air of cross-sectional area A and thickness Δx, with its center displaced from its equilibrium position by distance s. From Eq. 17-2 we can write, for the pressure variation in the displaced element,

$$\Delta p = -B\frac{\Delta V}{V}. \tag{17-16}$$

The quantity V in Eq. 17-16 is the volume of the element, given by

$$V = A\,\Delta x. \tag{17-17}$$

The quantity ΔV in Eq. 17-16 is the change in volume that occurs when the element is displaced. This volume change comes about because the displacements of the two faces of the element are not quite the same, differing by some amount Δs. Thus, we can write the change in volume as

$$\Delta V = A\,\Delta s. \tag{17-18}$$

Substituting Eqs. 17-17 and 17-18 into Eq. 17-16 and passing to the differential limit yield

$$\Delta p = -B\frac{\Delta s}{\Delta x} = -B\frac{\partial s}{\partial x}. \tag{17-19}$$

The symbols ∂ indicate that the derivative in Eq. 17-19 is a *partial derivative,* which tells us how s changes with x when the time t is fixed. From Eq. 17-13 we then have, treating t as a constant,

$$\frac{\partial s}{\partial x} = \frac{\partial}{\partial x}[s_m\cos(kx - \omega t)] = -ks_m\sin(kx - \omega t).$$

Substituting this quantity for the partial derivative in Eq. 17-19 yields

$$\Delta p = Bks_m\sin(kx - \omega t).$$

Setting $\Delta p_m = Bks_m$, this yields Eq. 17-14, which we set out to prove.

Using Eq. 17-3, we can now write

$$\Delta p_m = (Bk)s_m = (v^2\rho k)s_m.$$

Equation 17-15, which we also wanted to prove, follows at once if we substitute ω/v for k from Eq. 16-13.

Sample Problem 17-2

The maximum pressure amplitude Δp_m that the human ear can tolerate in loud sounds is about 28 Pa (which is very much less than the normal air pressure of about 10^5 Pa). What is the displacement amplitude s_m for such a sound in air of density $\rho = 1.21$ kg/m^3, at a frequency of 1000 Hz and a speed of 343 m/s?

KEY IDEA The displacement amplitude s_m of a sound wave is related to the pressure amplitude Δp_m of the wave according to Eq. 17-15.

Calculations: Solving that equation for s_m yields

$$s_m = \frac{\Delta p_m}{v\rho\omega} = \frac{\Delta p_m}{v\rho(2\pi f)}.$$

Substituting known data then gives us

$$s_m = \frac{28 \text{ Pa}}{(343 \text{ m/s})(1.21 \text{ kg/m}^3)(2\pi)(1000 \text{ Hz})}$$
$$= 1.1 \times 10^{-5} \text{ m} = 11\ \mu\text{m}. \quad \text{(Answer)}$$

That is only about one-seventh the thickness of this page. Obviously, the displacement amplitude of even the loudest sound that the ear can tolerate is very small.

The pressure amplitude Δp_m for the *faintest* detectable sound at 1000 Hz is 2.8×10^{-5} Pa. Proceeding as above leads to $s_m = 1.1 \times 10^{-11}$ m or 11 pm, which is about one-tenth the radius of a typical atom. The ear is indeed a sensitive detector of sound waves.

17-5 | Interference

Like transverse waves, sound waves can undergo interference. Let us consider, in particular, the interference between two identical sound waves traveling in

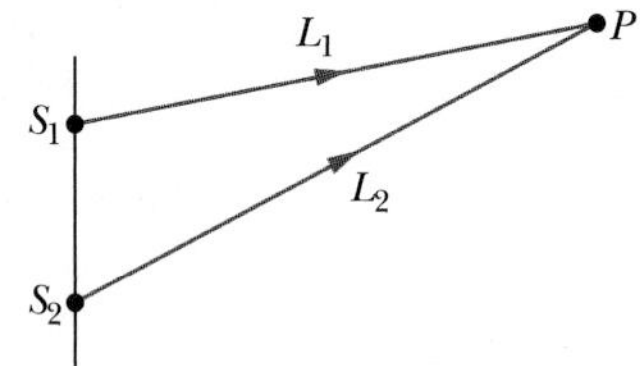

FIG. 17-8 Two point sources S_1 and S_2 emit spherical sound waves in phase. The rays indicate that the waves pass through a common point P.

the same direction. Figure 17-8 shows how we can set up such a situation: Two point sources S_1 and S_2 emit sound waves that are in phase and of identical wavelength λ. Thus, the sources themselves are said to be in phase; that is, as the waves emerge from the sources, their displacements are always identical. We are interested in the waves that then travel through point P in Fig. 17-8. We assume that the distance to P is much greater than the distance between the sources so that we can approximate the waves as traveling in the same direction at P.

If the waves traveled along paths with identical lengths to reach point P, they would be in phase there. As with transverse waves, this means that they would undergo fully constructive interference there. However, in Fig. 17-8, path L_2 traveled by the wave from S_2 is longer than path L_1 traveled by the wave from S_1. The difference in path lengths means that the waves may not be in phase at point P. In other words, their phase difference ϕ at P depends on their **path length difference** $\Delta L = |L_2 - L_1|$.

To relate phase difference ϕ to path length difference ΔL, we recall (from Section 16-4) that a phase difference of 2π rad corresponds to one wavelength. Thus, we can write the proportion

$$\frac{\phi}{2\pi} = \frac{\Delta L}{\lambda}, \tag{17-20}$$

from which

$$\phi = \frac{\Delta L}{\lambda} 2\pi. \tag{17-21}$$

Fully constructive interference occurs when ϕ is zero, 2π, or any integer multiple of 2π. We can write this condition as

$$\phi = m(2\pi), \quad \text{for } m = 0, 1, 2, \ldots \quad \text{(fully constructive interference).} \tag{17-22}$$

From Eq. 17-21, this occurs when the ratio $\Delta L/\lambda$ is

$$\frac{\Delta L}{\lambda} = 0, 1, 2, \ldots \quad \text{(fully constructive interference).} \tag{17-23}$$

For example, if the path length difference $\Delta L = |L_2 - L_1|$ in Fig. 17-8 is equal to 2λ, then $\Delta L/\lambda = 2$ and the waves undergo fully constructive interference at point P. The interference is fully constructive because the wave from S_2 is phase-shifted relative to the wave from S_1 by 2λ, putting the two waves *exactly in phase* at P.

Fully destructive interference occurs when ϕ is an odd multiple of π, a condition we can write as

$$\phi = (2m + 1)\pi, \quad \text{for } m = 0, 1, 2, \ldots \quad \text{(fully destructive interference).} \tag{17-24}$$

From Eq. 17-21, this occurs when the ratio $\Delta L/\lambda$ is

$$\frac{\Delta L}{\lambda} = 0.5, 1.5, 2.5, \ldots \quad \text{(fully destructive interference).} \tag{17-25}$$

For example, if the path length difference $\Delta L = |L_2 - L_1|$ in Fig. 17-8 is equal to 2.5λ, then $\Delta L/\lambda = 2.5$ and the waves undergo fully destructive interference at point P. The interference is fully destructive because the wave from S_2 is phase-shifted relative to the wave from S_1 by 2.5 wavelengths, which puts the two waves *exactly out of phase* at P.

Of course, two waves could produce intermediate interference as, say, when $\Delta L/\lambda = 1.2$. This would be closer to fully constructive interference ($\Delta L/\lambda = 1.0$) than to fully destructive interference ($\Delta L/\lambda = 1.5$).

Sample Problem 17-3 Build your skill

In Fig. 17-9a, two point sources S_1 and S_2, which are in phase and separated by distance $D = 1.5\lambda$, emit identical sound waves of wavelength λ.

(a) What is the path length difference of the waves from S_1 and S_2 at point P_1, which lies on the perpendicular bisector of distance D, at a distance greater than D from the sources? What type of interference occurs at P_1?

Reasoning: Because the waves travel identical distances to reach P_1, their path length difference is

$$\Delta L = 0. \quad \text{(Answer)}$$

From Eq. 17-23, this means that the waves undergo fully constructive interference at P_1.

(b) What are the path length difference and type of interference at point P_2 in Fig. 17-9a?

Reasoning: The wave from S_1 travels the extra distance D $(= 1.5\lambda)$ to reach P_2. Thus, the path length difference is

$$\Delta L = 1.5\lambda. \quad \text{(Answer)}$$

From Eq. 17-25, this means that the waves are exactly out of phase at P_2 and undergo fully destructive interference there.

(c) Figure 17-9b shows a circle with a radius much greater than D, centered on the midpoint between sources S_1 and S_2. What is the number of points N around this circle at which the interference is fully constructive?

Reasoning: Imagine that, starting at point a, we move clockwise along the circle to point d. As we move to point d, the path length difference ΔL increases and so the type of interference changes. From (a), we know that the path length difference is $\Delta L = 0\lambda$ at point a. From (b), we know that $\Delta L = 1.5\lambda$ at point d. Thus, there must be one point along the circle between a and d at which $\Delta L = \lambda$, as indicated in Fig. 17-9b. From Eq. 17-23, fully constructive interference occurs at that point. Also, there can be no other point along the way from point a to point d at which fully constructive interference occurs, because there is no other integer than 1 between 0 and 1.5.

We can now use symmetry to locate the other points of fully constructive interference along the rest of the circle. Symmetry about line cd gives us point b, at which $\Delta L = 0\lambda$. Also, there are three more points at which $\Delta L = \lambda$. In all we have

$$N = 6. \quad \text{(Answer)}$$

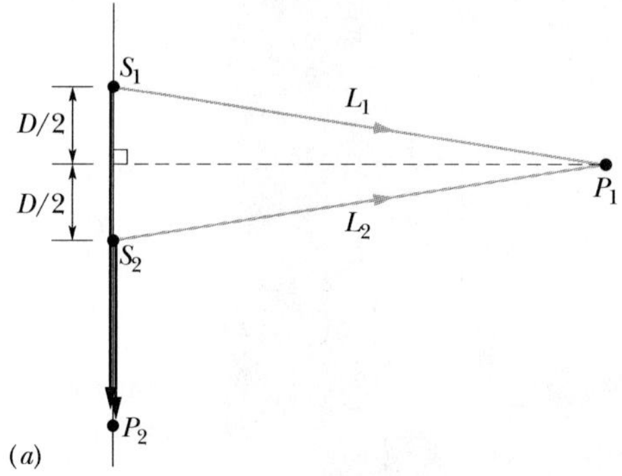

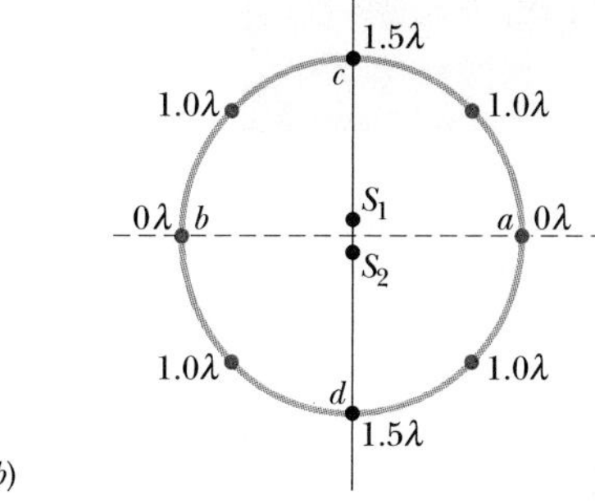

FIG. 17-9 (a) Two point sources S_1 and S_2, separated by distance D, emit spherical sound waves in phase. The waves travel equal distances to reach point P_1. Point P_2 is on the line extending through S_1 and S_2. (b) The path length difference (in terms of wavelength) between the waves from S_1 and S_2, at eight points on a large circle around the sources.

17-6 | Intensity and Sound Level

If you have ever tried to sleep while someone played loud music nearby, you are well aware that there is more to sound than frequency, wavelength, and speed. There is also intensity. The **intensity** I of a sound wave at a surface is the average rate per unit area at which energy is transferred by the wave through or onto the surface. We can write this as

$$I = \frac{P}{A}, \quad (17\text{-}26)$$

where P is the time rate of energy transfer (the power) of the sound wave and A is the area of the surface intercepting the sound. As we shall derive shortly, the intensity I is related to the displacement amplitude s_m of the sound wave by

$$I = \tfrac{1}{2}\rho v \omega^2 s_m^2. \quad (17\text{-}27)$$

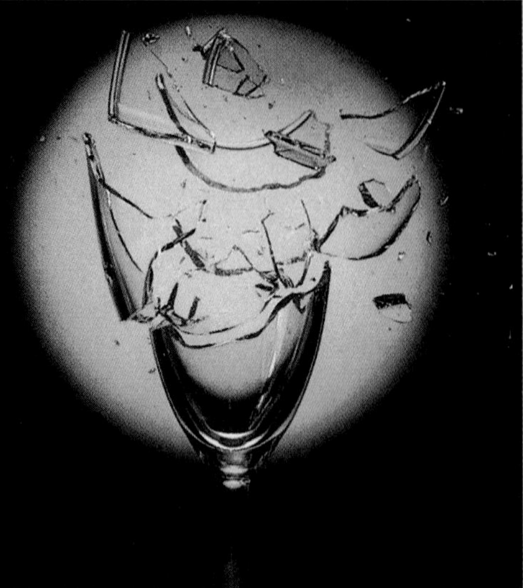

Sound can cause the wall of a drinking glass to oscillate. If the sound produces a standing wave of oscillations and if the intensity of the sound is large enough, the glass will shatter. *(Ben Rose/The Image Bank/Getty Images)*

Variation of Intensity with Distance

How intensity varies with distance from a real sound source is often complex. Some real sources (like loudspeakers) may transmit sound only in particular directions, and the environment usually produces echoes (reflected sound waves) that overlap the direct sound waves. In some situations, however, we can ignore echoes and assume that the sound source is a point source that emits the sound *isotropically*—that is, with equal intensity in all directions. The wavefronts spreading from such an isotropic point source S at a particular instant are shown in Fig. 17-10.

Let us assume that the mechanical energy of the sound waves is conserved as they spread from this source. Let us also center an imaginary sphere of radius r on the source, as shown in Fig. 17-10. All the energy emitted by the source must pass through the surface of the sphere. Thus, the time rate at which energy is transferred through the surface by the sound waves must equal the time rate at which energy is emitted by the source (that is, the power P_s of the source). From Eq. 17-26, the intensity I at the sphere must then be

$$I = \frac{P_s}{4\pi r^2}, \tag{17-28}$$

where $4\pi r^2$ is the area of the sphere. Equation 17-28 tells us that the intensity of sound from an isotropic point source decreases with the square of the distance r from the source.

CHECKPOINT 2 The figure indicates three small patches 1, 2, and 3 that lie on the surfaces of two imaginary spheres; the spheres are centered on an isotropic point source S of sound. The rates at which energy is transmitted through the three patches by the sound waves are equal. Rank the patches according to (a) the intensity of the sound on them and (b) their area, greatest first.

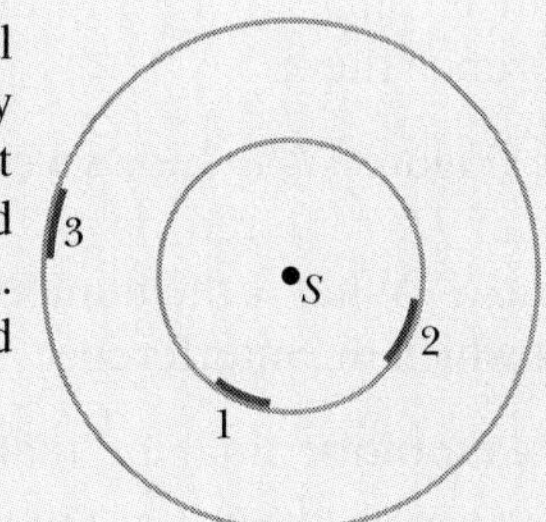

The Decibel Scale

You saw in Sample Problem 17-2 that the displacement amplitude at the human ear ranges from about 10^{-5} m for the loudest tolerable sound to about 10^{-11} m for the faintest detectable sound, a ratio of 10^6. From Eq. 17-27 we see that the intensity of a sound varies as the *square* of its amplitude, so the ratio of intensities at these two limits of the human auditory system is 10^{12}. Humans can hear over an enormous range of intensities.

We deal with such an enormous range of values by using logarithms. Consider the relation

$$y = \log x,$$

in which x and y are variables. It is a property of this equation that if we *multiply* x by 10, then y increases by 1. To see this, we write

$$y' = \log(10x) = \log 10 + \log x = 1 + y.$$

Similarly, if we multiply x by 10^{12}, y increases by only 12.

Thus, instead of speaking of the intensity I of a sound wave, it is much more convenient to speak of its **sound level** β, defined as

$$\beta = (10 \text{ dB}) \log \frac{I}{I_0}. \tag{17-29}$$

Here dB is the abbreviation for **decibel,** the unit of sound level, a name that was chosen to recognize the work of Alexander Graham Bell. I_0 in Eq. 17-29 is a

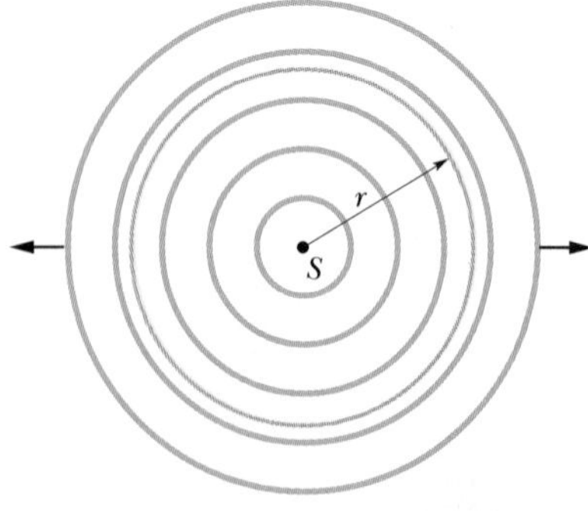

FIG. 17-10 A point source S emits sound waves uniformly in all directions. The waves pass through an imaginary sphere of radius r that is centered on S.

standard reference intensity ($= 10^{-12}$ W/m^2), chosen because it is near the lower limit of the human range of hearing. For $I = I_0$, Eq. 17-29 gives $\beta = 10 \log 1 = 0$, so our standard reference level corresponds to zero decibels. Then β increases by 10 dB every time the sound intensity increases by an order of magnitude (a factor of 10). Thus, $\beta = 40$ corresponds to an intensity that is 10^4 times the standard reference level. Table 17-2 lists the sound levels for a variety of environments.

TABLE 17-2

Some Sound Levels (dB)

Hearing threshold	0
Rustle of leaves	10
Conversation	60
Rock concert	110
Pain threshold	120
Jet engine	130

Derivation of Eq. 17-27

Consider, in Fig. 17-5*a*, a thin slice of air of thickness dx, area A, and mass dm, oscillating back and forth as the sound wave of Eq. 17-13 passes through it. The kinetic energy dK of the slice of air is

$$dK = \tfrac{1}{2}\, dm\, v_s^2. \qquad (17\text{-}30)$$

Here v_s is not the speed of the wave but the speed of the oscillating element of air, obtained from Eq. 17-13 as

$$v_s = \frac{\partial s}{\partial t} = -\omega s_m \sin(kx - \omega t).$$

Using this relation and putting $dm = \rho A\, dx$ allow us to rewrite Eq. 17-30 as

$$dK = \tfrac{1}{2}(\rho A\, dx)(-\omega s_m)^2 \sin^2(kx - \omega t). \qquad (17\text{-}31)$$

Dividing Eq. 17-31 by dt gives the rate at which kinetic energy moves along with the wave. As we saw in Chapter 16 for transverse waves, dx/dt is the wave speed v, so we have

$$\frac{dK}{dt} = \tfrac{1}{2}\rho A v \omega^2 s_m^2 \sin^2(kx - \omega t). \qquad (17\text{-}32)$$

The *average* rate at which kinetic energy is transported is

$$\left(\frac{dK}{dt}\right)_{\text{avg}} = \tfrac{1}{2}\rho A v \omega^2 s_m^2 [\sin^2(kx - \omega t)]_{\text{avg}}$$
$$= \tfrac{1}{4}\rho A v \omega^2 s_m^2. \qquad (17\text{-}33)$$

To obtain this equation, we have used the fact that the average value of the square of a sine (or a cosine) function over one full oscillation is $\frac{1}{2}$.

We assume that *potential* energy is carried along with the wave at this same average rate. The wave intensity I, which is the average rate per unit area at which energy of both kinds is transmitted by the wave, is then, from Eq. 17-33,

$$I = \frac{2(dK/dt)_{\text{avg}}}{A} = \tfrac{1}{2}\rho v \omega^2 s_m^2,$$

which is Eq. 17-27, the equation we set out to derive.

Sample Problem 17-4

An electric spark jumps along a straight line of length $L = 10$ m, emitting a pulse of sound that travels radially outward from the spark. (The spark is said to be a *line source* of sound.) The power of the emission is $P_s = 1.6 \times 10^4$ W.

(a) What is the intensity I of the sound when it reaches a distance $r = 12$ m from the spark?

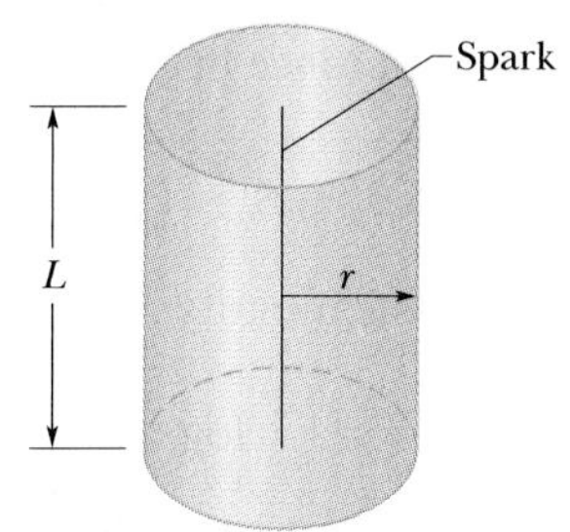

FIG. 17-11 A spark along a straight line of length L emits sound waves radially outward. The waves pass through an imaginary cylinder of radius r and length L that is centered on the spark.

KEY IDEAS (1) Let us center an imaginary cylinder of radius $r = 12$ m and length $L = 10$ m (open at both ends) on the spark, as shown in Fig. 17-11. Then the intensity I at the cylindrical surface is the ratio P/A, where P is the time rate at which sound energy passes through the surface and A is the surface area. (2) We assume that the principle of conservation of energy applies to the sound energy. This means that the rate P at which energy is transferred through the cylinder must equal the rate P_s at which energy is emitted by the source.

Calculations: Putting these ideas together and noting that the area of the cylindrical surface is $A = 2\pi rL$, we have

$$I = \frac{P}{A} = \frac{P_s}{2\pi rL}. \tag{17-34}$$

This tells us that the intensity of the sound from a line source decreases with distance r (and not with the square of distance r as for a point source). Substituting the given data, we find

$$I = \frac{1.6 \times 10^4\ \text{W}}{2\pi(12\ \text{m})(10\ \text{m})}$$
$$= 21.2\ \text{W/m}^2 \approx 21\ \text{W/m}^2. \quad \text{(Answer)}$$

(b) At what time rate P_d is sound energy intercepted by an acoustic detector of area $A_d = 2.0\ \text{cm}^2$, aimed at the spark and located a distance $r = 12$ m from the spark?

Calculations: We know that the intensity of sound at the detector is the ratio of the energy transfer rate P_d there to the detector's area A_d:

$$I = \frac{P_d}{A_d}. \tag{17-35}$$

We can imagine that the detector lies on the cylindrical surface of (a). Then the sound intensity at the detector is the intensity I (= 21.2 W/m²) at the cylindrical surface. Solving Eq. 17-35 for P_d gives us

$$P_d = (21.2\ \text{W/m}^2)(2.0 \times 10^{-4}\ \text{m}^2) = 4.2\ \text{mW}. \quad \text{(Answer)}$$

Sample Problem 17-5

Many veteran rockers suffer from acute hearing damage because of the high sound levels they endured for years while playing music near loudspeakers or listening to music on headphones. Some, like Ted Nugent, can no longer hear in a damaged ear. Others, like Peter Townshend of the Who, have a continuous ringing sensation (tinnitus). Recently, many rockers, such as Lars Ulrich of Metallica (Fig. 17-12), began wearing special earplugs to protect their hearing during performances. If an earplug decreases the sound level of the sound waves by 20 dB, what is the ratio of the final intensity I_f of the waves to their initial intensity I_i?

FIG. 17-12 Lars Ulrich of Metallica is an advocate for the organization HEAR (Hearing Education and Awareness for Rockers), which warns about the damage high sound levels can have on hearing. *(Tim Mosenfelder/ Getty Images News and Sport Services)*

KEY IDEA For both the final and initial waves, the sound level β is related to the intensity by the definition of sound level in Eq. 17-29.

Calculations: For the final waves we have

$$\beta_f = (10\ \text{dB}) \log \frac{I_f}{I_0},$$

and for the initial waves we have

$$\beta_i = (10\ \text{dB}) \log \frac{I_i}{I_0}.$$

The difference in the sound levels is

$$\beta_f - \beta_i = (10\ \text{dB}) \left(\log \frac{I_f}{I_0} - \log \frac{I_i}{I_0} \right). \tag{17-36}$$

Using the identity

$$\log \frac{a}{b} - \log \frac{c}{d} = \log \frac{ad}{bc},$$

we can rewrite Eq. 17-36 as

$$\beta_f - \beta_i = (10\ \text{dB}) \log \frac{I_f}{I_i}. \tag{17-37}$$

Rearranging and then substituting the given decrease in sound level as $\beta_f - \beta_i = -20$ dB, we find

$$\log \frac{I_f}{I_i} = \frac{\beta_f - \beta_i}{10\ \text{dB}} = \frac{-20\ \text{dB}}{10\ \text{dB}} = -2.0.$$

We next take the antilog of the far left and far right sides of this equation. (Although the antilog $10^{-2.0}$ can be evaluated mentally, you could use a calculator by keying in 10^-2.0 or using the 10^x key.) We find

$$\frac{I_f}{I_i} = \log^{-1}(-2.0) = 0.010. \quad \text{(Answer)}$$

Thus, the earplug reduces the intensity of the sound waves to 0.010 of their initial intensity, which is a decrease of two orders of magnitude.

17-7 | Sources of Musical Sound

Musical sounds can be set up by oscillating strings (guitar, piano, violin), membranes (kettledrum, snare drum), air columns (flute, oboe, pipe organ, and the digeridoo of Fig. 17-13), wooden blocks or steel bars (marimba, xylophone), and many other oscillating bodies. Most common instruments involve more than a single oscillating part.

Recall from Chapter 16 that standing waves can be set up on a stretched string that is fixed at both ends. They arise because waves traveling along the string are reflected back onto the string at each end. If the wavelength of the waves is suitably matched to the length of the string, the superposition of waves traveling in opposite directions produces a standing wave pattern (or oscillation mode). The wavelength required of the waves for such a match is one that corresponds to a *resonant frequency* of the string. The advantage of setting up standing waves is that the string then oscillates with a large, sustained amplitude, pushing back and forth against the surrounding air and thus generating a noticeable sound wave with the same frequency as the oscillations of the string. This production of sound is of obvious importance to, say, a guitarist.

We can set up standing waves of sound in an air-filled pipe in a similar way. As sound waves travel through the air in the pipe, they are reflected at each end and travel back through the pipe. (The reflection occurs even if an end is open, but the reflection is not as complete as when the end is closed.) If the wavelength of the sound waves is suitably matched to the length of the pipe, the superposition of waves traveling in opposite directions through the pipe sets up a standing wave pattern. The wavelength required of the sound waves for such a match is one that corresponds to a resonant frequency of the pipe. The advantage of such a standing wave is that the air in the pipe oscillates with a large, sustained amplitude, emitting at any open end a sound wave that has the same frequency as the oscillations in the pipe. This emission of sound is of obvious importance to, say, an organist.

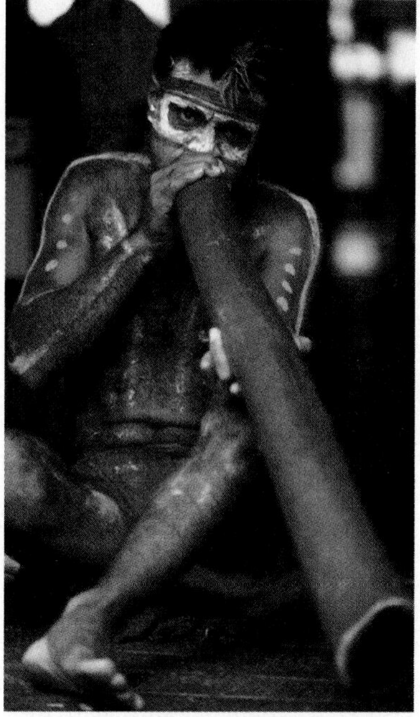

FIG. 17-13 The air column within a digeridoo ("a pipe") oscillates when the instrument is played. *(Alamy Images)*

Many other aspects of standing sound wave patterns are similar to those of string waves: The closed end of a pipe is like the fixed end of a string in that there must be a node (zero displacement) there, and the open end of a pipe is like the end of a string attached to a freely moving ring, as in Fig. 16-21*b*, in that there must be an antinode there. (Actually, the antinode for the open end of a pipe is located slightly beyond the end, but we shall not dwell on that detail.)

The simplest standing wave pattern that can be set up in a pipe with two open ends is shown in Fig. 17-14*a*. There is an antinode across each open end, as required. There is also a node across the middle of the pipe. An easier way of representing this standing longitudinal sound wave is shown in Fig. 17-14*b*—by drawing it as a standing transverse string wave.

The standing wave pattern of Fig. 17-14*a* is called the *fundamental mode* or *first harmonic*. For it to be set up, the sound waves in a pipe of length L must have a wavelength given by $L = \lambda/2$, so that $\lambda = 2L$. Several more standing sound wave patterns for a pipe with two open ends are shown in Fig. 17-15*a* using string wave representations. The *second harmonic* requires sound waves of wavelength $\lambda = L$, the *third harmonic* requires wavelength $\lambda = 2L/3$, and so on.

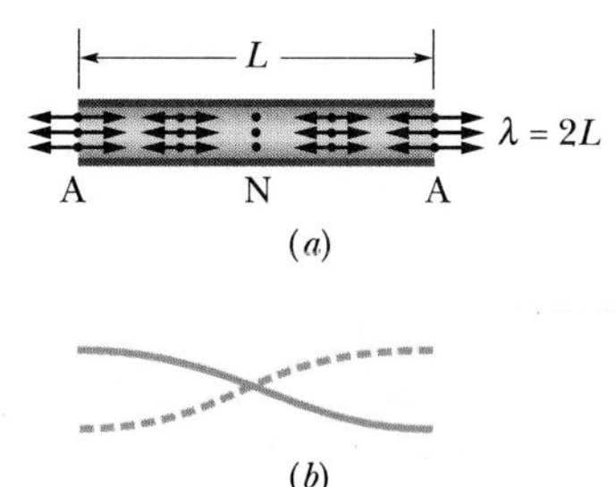

FIG. 17-14 (*a*) The simplest standing wave pattern of displacement for (longitudinal) sound waves in a pipe with both ends open has an antinode (A) across each end and a node (N) across the middle. (The longitudinal displacements represented by the double arrows are greatly exaggerated.) (*b*) The corresponding standing wave pattern for (transverse) string waves.

More generally, the resonant frequencies for a pipe of length L with two open ends correspond to the wavelengths

$$\lambda = \frac{2L}{n}, \quad \text{for } n = 1, 2, 3, \ldots, \tag{17-38}$$

where n is called the *harmonic number*. Letting v be the speed of sound, we write the resonant frequencies for a pipe with two open ends as

$$f = \frac{v}{\lambda} = \frac{nv}{2L}, \quad \text{for } n = 1, 2, 3, \ldots \quad \text{(pipe, two open ends).} \tag{17-39}$$

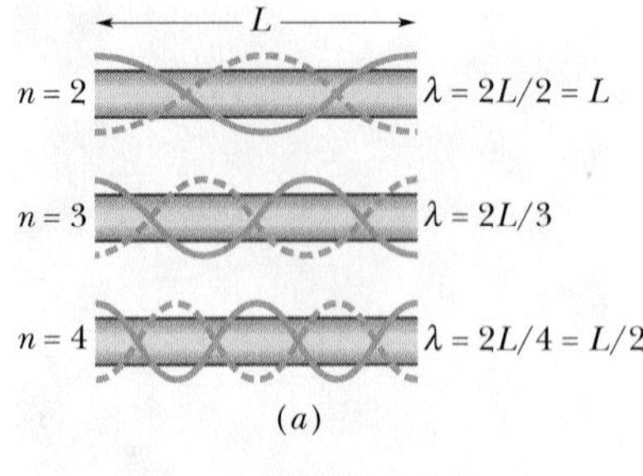

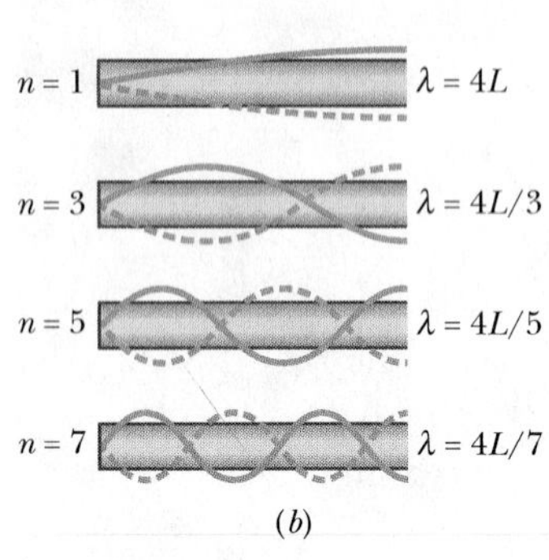

FIG. 17-15 Standing wave patterns for string waves superimposed on pipes to represent standing sound wave patterns in the pipes. (a) With *both* ends of the pipe open, any harmonic can be set up in the pipe. (b) With only *one* end open, only odd harmonics can be set up.

Figure 17-15b shows (using string wave representations) some of the standing sound wave patterns that can be set up in a pipe with only one open end. As required, across the open end there is an antinode and across the closed end there is a node. The simplest pattern requires sound waves having a wavelength given by $L = \lambda/4$, so that $\lambda = 4L$. The next simplest pattern requires a wavelength given by $L = 3\lambda/4$, so that $\lambda = 4L/3$, and so on.

More generally, the resonant frequencies for a pipe of length L with only one open end correspond to the wavelengths

$$\lambda = \frac{4L}{n}, \qquad \text{for } n = 1, 3, 5, \ldots, \tag{17-40}$$

in which the harmonic number n *must be an odd number*. The resonant frequencies are then given by

$$f = \frac{v}{\lambda} = \frac{nv}{4L}, \qquad \text{for } n = 1, 3, 5, \ldots \quad \text{(pipe, one open end).} \tag{17-41}$$

Note again that only odd harmonics can exist in a pipe with one open end. For example, the second harmonic, with $n = 2$, cannot be set up in such a pipe. Note also that for such a pipe the adjective in a phrase such as "the third harmonic" still refers to the harmonic number n (and not to, say, the third possible harmonic).

The length of a musical instrument reflects the range of frequencies over which the instrument is designed to function, and smaller length implies higher frequencies. Figure 17-16, for example, shows the saxophone and violin families, with their frequency ranges suggested by the piano keyboard. Note that, for every instrument, there is overlap with its higher- and lower-frequency neighbors.

In any oscillating system that gives rise to a musical sound, whether it is a violin string or the air in an organ pipe, the fundamental and one or more of the higher harmonics are usually generated simultaneously. Thus, you hear them together—that is, superimposed as a net wave. When different instruments are played at the same note, they produce the same fundamental frequency but different intensities for the higher harmonics. For example, the fourth harmonic of middle C might be relatively loud on one instrument and relatively quiet or even missing on another. Thus, because different instruments produce different net waves, they sound different to you even when they are played at the same note. That would be the case for the two net waves shown in Fig. 17-17, which were produced at the same note by different instruments.

CHECKPOINT 3 Pipe A, with length L, and pipe B, with length $2L$, both have two open ends. Which harmonic of pipe B has the same frequency as the fundamental of pipe A?

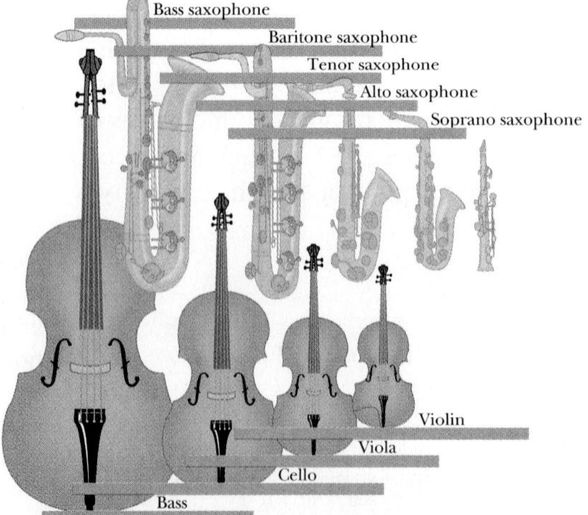

FIG. 17-16 The saxophone and violin families, showing the relations between instrument length and frequency range. The frequency range of each instrument is indicated by a horizontal bar along a frequency scale suggested by the keyboard at the bottom; the frequency increases toward the right.

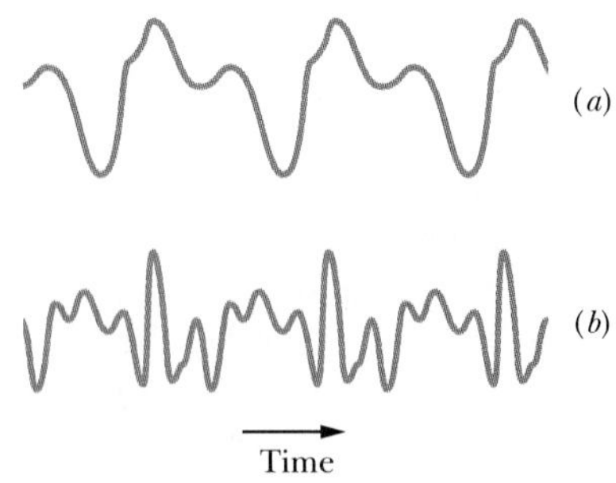

FIG. 17-17 The wave forms produced by (a) a flute and (b) an oboe when played at the same note, with the same first harmonic frequency.

Sample Problem 17-6

Weak background noises from a room set up the fundamental standing wave in a cardboard tube of length $L =$ 67.0 cm with two open ends. Assume that the speed of sound in the air within the tube is 343 m/s.

(a) What frequency do you hear from the tube?

KEY IDEA With both pipe ends open, we have a symmetric situation in which the standing wave has an antinode at each end of the tube. The standing wave pattern (in string wave style) is that of Fig. 17-14*b*.

Calculation: The frequency is given by Eq. 17-39 with $n = 1$ for the fundamental mode:

$$f = \frac{nv}{2L} = \frac{(1)(343 \text{ m/s})}{2(0.670 \text{ m})} = 256 \text{ Hz}. \quad \text{(Answer)}$$

If the background noises set up any higher harmonics, such as the second harmonic, you also hear frequencies that are *integer* multiples of 256 Hz.

(b) If you jam your ear against one end of the tube, what fundamental frequency do you hear from the tube?

KEY IDEA With your ear effectively closing one end of the tube, we have an asymmetric situation—an antinode still exists at the open end, but a node is now at the other (closed) end. The standing wave pattern is the top one in Fig. 17-15*b*.

Calculation: The frequency is given by Eq. 17-41 with $n = 1$ for the fundamental mode:

$$f = \frac{nv}{4L} = \frac{(1)(343 \text{ m/s})}{4(0.670 \text{ m})} = 128 \text{ Hz}. \quad \text{(Answer)}$$

If the background noises set up any higher harmonics, they will be *odd* multiples of 128 Hz. That means that the frequency of 256 Hz (which is an even multiple) cannot now occur.

17-8 | Beats

If we listen, a few minutes apart, to two sounds whose frequencies are, say, 552 and 564 Hz, most of us cannot tell one from the other. However, if the sounds reach our ears simultaneously, what we hear is a sound whose frequency is 558 Hz, the *average* of the two combining frequencies. We also hear a striking variation in the intensity of this sound—it increases and decreases in slow, wavering **beats** that repeat at a frequency of 12 Hz, the *difference* between the two combining frequencies. Figure 17-18 shows this beat phenomenon.

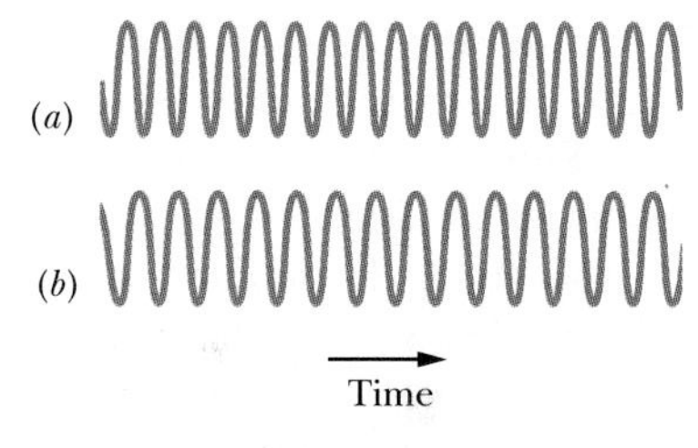

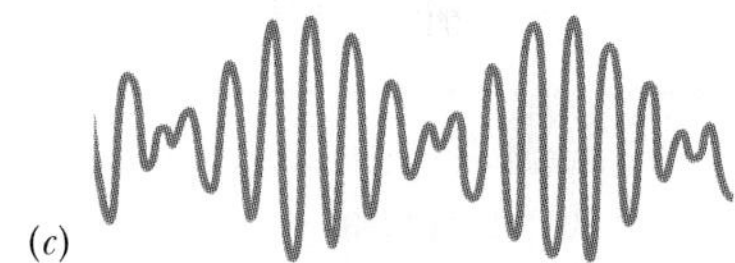

FIG. 17-18 (*a*, *b*) The pressure variations Δp of two sound waves as they would be detected separately. The frequencies of the waves are nearly equal. (*c*) The resultant pressure variation if the two waves are detected simultaneously.

Let the time-dependent variations of the displacements due to two sound waves of equal amplitude s_m be

$$s_1 = s_m \cos \omega_1 t \quad \text{and} \quad s_2 = s_m \cos \omega_2 t, \tag{17-42}$$

where $\omega_1 > \omega_2$. From the superposition principle, the resultant displacement is

$$s = s_1 + s_2 = s_m(\cos \omega_1 t + \cos \omega_2 t).$$

Using the trigonometric identity (see Appendix E)

$$\cos \alpha + \cos \beta = 2 \cos[\tfrac{1}{2}(\alpha - \beta)] \cos[\tfrac{1}{2}(\alpha + \beta)]$$

allows us to write the resultant displacement as

$$s = 2s_m \cos[\tfrac{1}{2}(\omega_1 - \omega_2)t] \cos[\tfrac{1}{2}(\omega_1 + \omega_2)t]. \tag{17-43}$$

If we write

$$\omega' = \tfrac{1}{2}(\omega_1 - \omega_2) \quad \text{and} \quad \omega = \tfrac{1}{2}(\omega_1 + \omega_2), \tag{17-44}$$

we can then write Eq. 17-43 as

$$s(t) = [2s_m \cos \omega' t] \cos \omega t. \tag{17-45}$$

We now assume that the angular frequencies ω_1 and ω_2 of the combining waves are almost equal, which means that $\omega \gg \omega'$ in Eq. 17-44. We can then regard Eq. 17-45 as a cosine function whose angular frequency is ω and whose amplitude (which is not constant but varies with angular frequency ω') is the absolute value of the quantity in the brackets.

A maximum amplitude will occur whenever $\cos \omega' t$ in Eq. 17-45 has the value $+1$ or -1, which happens twice in each repetition of the cosine function. Because $\cos \omega' t$ has angular frequency ω', the angular frequency ω_{beat} at which beats occur is $\omega_{beat} = 2\omega'$. Then, with the aid of Eq. 17-44, we can write

$$\omega_{beat} = 2\omega' = (2)(\tfrac{1}{2})(\omega_1 - \omega_2) = \omega_1 - \omega_2.$$

Because $\omega = 2\pi f$, we can recast this as

$$f_{beat} = f_1 - f_2 \quad \text{(beat frequency).} \tag{17-46}$$

Musicians use the beat phenomenon in tuning instruments. If an instrument is sounded against a standard frequency (for example, the note called "concert A" played on an orchestra's first oboe) and tuned until the beat disappears, the instrument is in tune with that standard. In musical Vienna, concert A (440 Hz) is available as a telephone service for the city's many musicians.

Sample Problem 17-7

When an emperor penguin returns from a search for food, how can it find its mate among the thousands of penguins huddled together for warmth in the harsh Antarctic weather? It is not by sight, because penguins all look alike, even to a penguin.

The answer lies in the way penguins vocalize. Most birds vocalize by using only one side of their two-sided vocal organ, called the *syrinx*. Emperor penguins, however, vocalize by using both sides simultaneously. Each side sets up acoustic standing waves in the bird's throat and mouth, much like in a pipe with two open ends. Suppose that the frequency of the first harmonic produced by side A is $f_{A1} = 432$ Hz and the frequency of the first harmonic produced by side B is $f_{B1} = 371$ Hz. What is the beat frequency between those two first-harmonic frequencies and between the two second-harmonic frequencies?

KEY IDEA The beat frequency between two frequencies is their difference, as given by Eq. 17-46 ($f_{beat} = f_1 - f_2$).

Calculations: For the two first-harmonic frequencies f_{A1} and f_{B1}, the beat frequency is

$$f_{beat,1} = f_{A1} - f_{B1} = 432 \text{ Hz} - 371 \text{ Hz}$$
$$= 61 \text{ Hz.} \quad \text{(Answer)}$$

Because the standing waves in the penguin are effectively in a pipe with two open ends, the resonant frequencies are given by Eq. 17-39 ($f = nv/2L$), in which L is the (unknown) length of the effective pipe. The first-harmonic frequency is $f_1 = v/2L$, and the second-harmonic frequency is $f_2 = 2v/2L$. Comparing these two frequencies, we see that, in general,

$$f_2 = 2f_1.$$

For the penguin, the second harmonic of side A has frequency $f_{A2} = 2f_{A1}$ and the second harmonic of side B has frequency $f_{B2} = 2f_{B1}$. Using Eq. 17-46 with frequencies f_{A2} and f_{B2}, we find that the corresponding beat frequency is

$$f_{beat,2} = f_{A2} - f_{B2} = 2f_{A1} - 2f_{B1}$$
$$= 2(432 \text{ Hz}) - 2(371 \text{ Hz})$$
$$= 122 \text{ Hz.} \quad \text{(Answer)}$$

Experiments indicate that penguins can perceive such large beat frequencies (humans cannot). Thus, a penguin's cry can be rich with different harmonics and different beat frequencies, allowing the voice to be recognized even among the voices of thousands of other, closely huddled penguins.

17-9 | The Doppler Effect

A police car is parked by the side of the highway, sounding its 1000 Hz siren. If you are also parked by the highway, you will hear that same frequency. However, if there is relative motion between you and the police car, either toward or away from each other, you will hear a different frequency. For example, if you are driving *toward* the police car at 120 km/h (about 75 mi/h), you will hear a *higher* frequency (1096 Hz, an *increase* of 96 Hz). If you are driving *away from* the police car at that same speed, you will hear a *lower* frequency (904 Hz, a *decrease* of 96 Hz).

These motion-related frequency changes are examples of the **Doppler effect.** The effect was proposed (although not fully worked out) in 1842 by Austrian physicist Johann Christian Doppler. It was tested experimentally in

1845 by Buys Ballot in Holland, "using a locomotive drawing an open car with several trumpeters."

The Doppler effect holds not only for sound waves but also for electromagnetic waves, including microwaves, radio waves, and visible light. Here, however, we shall consider only sound waves, and we shall take as a reference frame the body of air through which these waves travel. This means that we shall measure the speeds of a source S of sound waves and a detector D of those waves *relative to that body of air.* (Unless otherwise stated, the body of air is stationary relative to the ground, so the speeds can also be measured relative to the ground.) We shall assume that S and D move either directly toward or directly away from each other, at speeds less than the speed of sound.

If either the detector or the source is moving, or both are moving, the emitted frequency f and the detected frequency f' are related by

$$f' = f\frac{v \pm v_D}{v \pm v_S} \quad \text{(general Doppler effect)}, \tag{17-47}$$

where v is the speed of sound through the air, v_D is the detector's speed relative to the air, and v_S is the source's speed relative to the air. The choice of plus or minus signs is set by this rule:

When the motion of detector or source is toward the other, the sign on its speed must give an upward shift in frequency. When the motion of detector or source is away from the other, the sign on its speed must give a downward shift in frequency.

In short, *toward* means *shift up,* and *away* means *shift down.*

Here are some examples of the rule. If the detector moves toward the source, use the plus sign in the numerator of Eq. 17-47 to get a shift up in the frequency. If it moves away, use the minus sign in the numerator to get a shift down. If it is stationary, substitute 0 for v_D. If the source moves toward the detector, use the minus sign in the denominator of Eq. 17-47 to get a shift up in the frequency. If it moves away, use the plus sign in the denominator to get a shift down. If the source is stationary, substitute 0 for v_S.

Next, we derive equations for the Doppler effect for the following two specific situations and then derive Eq. 17-47 for the general situation.

1. When the detector moves relative to the air and the source is stationary relative to the air, the motion changes the frequency at which the detector intercepts wavefronts and thus changes the detected frequency of the sound wave.
2. When the source moves relative to the air and the detector is stationary relative to the air, the motion changes the wavelength of the sound wave and thus changes the detected frequency (recall that frequency is related to wavelength).

Detector Moving, Source Stationary

In Fig. 17-19, a detector D (represented by an ear) is moving at speed v_D toward a stationary source S that emits spherical wavefronts, of wavelength λ and fre-

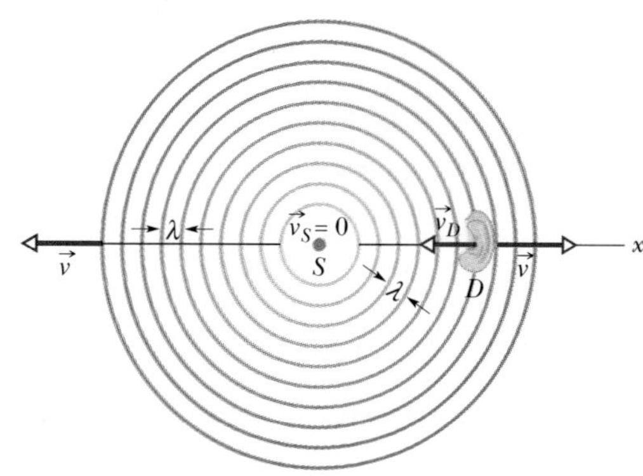

FIG. 17-19 A stationary source of sound S emits spherical wavefronts, shown one wavelength apart, that expand outward at speed v. A sound detector D, represented by an ear, moves with velocity $\vec{v}_D$ toward the source. The detector senses a higher frequency because of its motion.

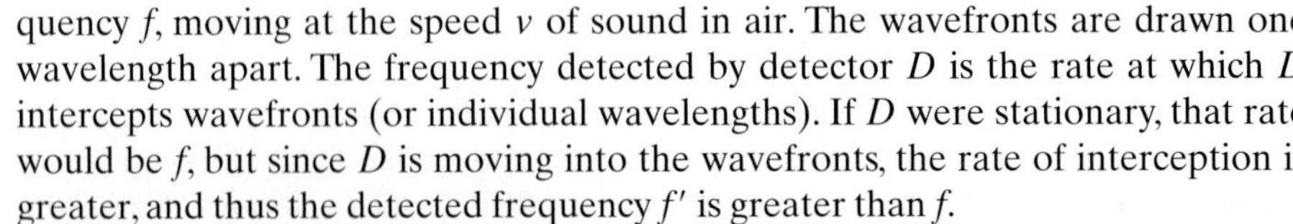

quency f, moving at the speed v of sound in air. The wavefronts are drawn one wavelength apart. The frequency detected by detector D is the rate at which D intercepts wavefronts (or individual wavelengths). If D were stationary, that rate would be f, but since D is moving into the wavefronts, the rate of interception is greater, and thus the detected frequency f' is greater than f.

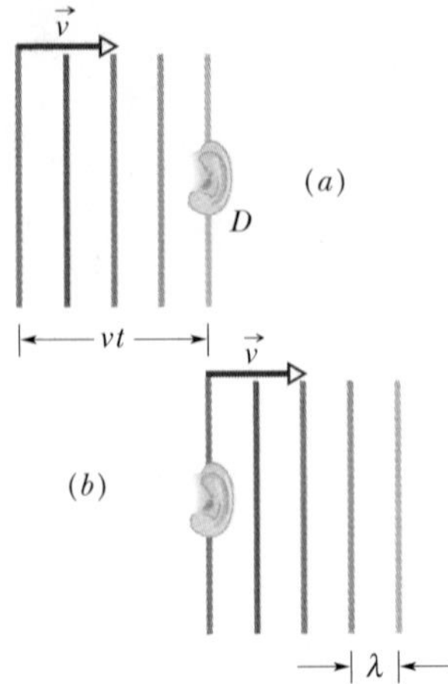

FIG. 17-20 The wavefronts of Fig. 17-19, assumed planar, (*a*) reach and (*b*) pass a stationary detector *D*; they move a distance *vt* to the right in time *t*.

Let us for the moment consider the situation in which D is stationary (Fig. 17-20). In time t, the wavefronts move to the right a distance vt. The number of wavelengths in that distance vt is the number of wavelengths intercepted by D in time t, and that number is vt/λ. The rate at which D intercepts wavelengths, which is the frequency f detected by D, is

$$f = \frac{vt/\lambda}{t} = \frac{v}{\lambda}. \tag{17-48}$$

In this situation, with D stationary, there is no Doppler effect—the frequency detected by D is the frequency emitted by S.

Now let us again consider the situation in which D moves in the direction opposite the wavefront velocity (Fig. 17-21). In time t, the wavefronts move to the right a distance vt as previously, but now D moves to the left a distance v_Dt. Thus, in this time t, the distance moved by the wavefronts relative to D is $vt + v_Dt$. The number of wavelengths in this relative distance $vt + v_Dt$ is the number of wavelengths intercepted by D in time t and is $(vt + v_Dt)/\lambda$. The *rate* at which D intercepts wavelengths in this situation is the frequency f', given by

$$f' = \frac{(vt + v_Dt)/\lambda}{t} = \frac{v + v_D}{\lambda}. \tag{17-49}$$

From Eq. 17-48, we have $\lambda = v/f$. Then Eq. 17-49 becomes

$$f' = \frac{v + v_D}{v/f} = f\frac{v + v_D}{v}. \tag{17-50}$$

Note that in Eq. 17-50, $f' > f$ unless $v_D = 0$ (the detector is stationary).

Similarly, we can find the frequency detected by D if D moves away from the source. In this situation, the wavefronts move a distance $vt - v_Dt$ relative to D in time t, and f' is given by

$$f' = f\frac{v - v_D}{v}. \tag{17-51}$$

In Eq. 17-51, $f' < f$ unless $v_D = 0$. We can summarize Eqs. 17-50 and 17-51 with

$$f' = f\frac{v \pm v_D}{v} \quad \text{(detector moving, source stationary).} \tag{17-52}$$

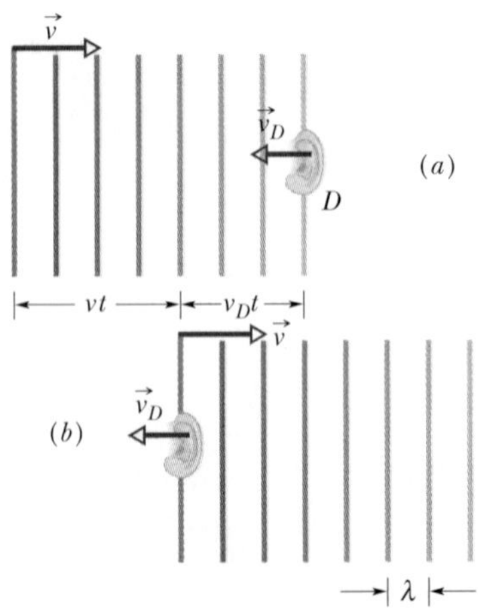

FIG. 17-21 Wavefronts traveling to the right (*a*) reach and (*b*) pass detector *D*, which moves in the opposite direction. In time *t*, the wavefronts move a distance *vt* to the right and *D* moves a distance v_Dt to the left.

Source Moving, Detector Stationary

Let detector D be stationary with respect to the body of air, and let source S move toward D at speed v_S (Fig. 17-22). The motion of S changes the wavelength of the sound waves it emits and thus the frequency detected by D.

To see this change, let T $(= 1/f)$ be the time between the emission of any pair of successive wavefronts W_1 and W_2. During T, wavefront W_1 moves a distance vT and the source moves a distance v_ST. At the end of T, wavefront W_2 is emitted. In the direction in which S moves, the distance between W_1 and W_2, which is the wavelength λ' of the waves moving in that direction, is $vT - v_ST$. If D detects those waves, it detects frequency f' given by

$$f' = \frac{v}{\lambda'} = \frac{v}{vT - v_ST} = \frac{v}{v/f - v_S/f}$$

$$= f\frac{v}{v - v_S}. \tag{17-53}$$

Note that f' must be greater than f unless $v_S = 0$.

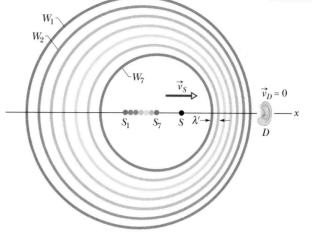

FIG. 17-22 A detector D is stationary, and a source S is moving toward it at speed v_S. Wavefront W_1 was emitted when the source was at S_1, wavefront W_7 when it was at S_7. At the moment depicted, the source is at S. The detector senses a higher frequency because the moving source, chasing its own wavefronts, emits a reduced wavelength λ' in the direction of its motion.

In the direction opposite that taken by S, the wavelength λ' of the waves is $vT + v_S T$. If D detects those waves, it detects frequency f' given by

$$f' = f\frac{v}{v + v_S}. \tag{17-54}$$

Now f' must be less than f unless $v_S = 0$.

We can summarize Eqs. 17-53 and 17-54 with

$$f' = f\frac{v}{v \pm v_S} \quad \text{(source moving, detector stationary)}. \tag{17-55}$$

General Doppler Effect Equation

We can now derive the general Doppler effect equation by replacing f in Eq. 17-55 (the source frequency) with f' of Eq. 17-52 (the frequency associated with motion of the detector). The result is Eq. 17-47 for the general Doppler effect.

That general equation holds not only when both detector and source are moving but also in the two specific situations we just discussed. For the situation in which the detector is moving and the source is stationary, substitution of $v_S = 0$ into Eq. 17-47 gives us Eq. 17-52, which we previously found. For the situation in which the source is moving and the detector is stationary, substitution of $v_D = 0$ into Eq. 17-47 gives us Eq. 17-55, which we previously found. Thus, Eq. 17-47 is the equation to remember.

CHECKPOINT 4 The figure indicates the directions of motion of a sound source and a detector for six situations in stationary air. For each situation, is the detected frequency greater than or less than the emitted frequency, or can't we tell without more information about the actual speeds?

	Source	Detector		Source	Detector
(a)	→	• 0 speed	(d)	←	←
(b)	←	• 0 speed	(e)	→	←
(c)	→	→	(f)	←	→

Sample Problem 17-8 Build your skill

Bats navigate and search out prey by emitting, and then detecting reflections of, ultrasonic waves, which are sound waves with frequencies greater than can be heard by a human. Suppose a bat emits ultrasound at frequency $f_{be} = 82.52$ kHz while flying with velocity $\vec{v}_b = (9.00 \text{ m/s})\hat{\text{i}}$ as it chases a moth that flies with velocity $\vec{v}_m = (8.00 \text{ m/s})\hat{\text{i}}$. What frequency f_{md} does the moth detect? What frequency f_{bd} does the bat detect in the returning echo from the moth?

KEY IDEAS The frequency is shifted by the relative motion of the bat and moth. Because they move along a single axis, the shifted frequency is given by Eq. 17-47 for the general Doppler effect. Motion *toward* tends to shift the frequency *up*, and motion *away* tends to shift the frequency *down*.

Detection by moth: The general Doppler equation is

$$f' = f\frac{v \pm v_D}{v \pm v_S}. \tag{17-56}$$

Here, the detected frequency f' that we want is the frequency f_{md} detected by the moth. On the right side of the equation, the emitted frequency f is the bat's emission frequency $f_{be} = 82.52$ kHz, the speed of sound is $v = 343$ m/s, the speed v_D of the detector is the moth's speed $v_m = 8.00$ m/s, and the speed v_S of the source is the bat's speed $v_b = 9.00$ m/s.

These substitutions into Eq. 17-56 are easy to make. However, the decisions about the plus and minus signs can be tricky. Think in terms of *toward* and *away*. We have the speed of the moth (the detector) in the numerator of Eq. 17-56. The moth moves *away* from the bat, which tends to lower the detected frequency. Because the speed is in the numerator, we choose the minus sign to meet that tendency (the numerator becomes smaller). These reasoning steps are shown in Table 17-3.

We have the speed of the bat in the denominator of Eq. 17-56. The bat moves *toward* the moth, which tends to increase the detected frequency. Because the speed is in the denominator, we choose the minus sign to meet that tendency (the denominator becomes smaller).

With these substitutions and decisions, we have

$$f_{md} = f_{be}\frac{v - v_m}{v - v_b}$$
$$= (82.52 \text{ kHz})\frac{343 \text{ m/s} - 8.00 \text{ m/s}}{343 \text{ m/s} - 9.00 \text{ m/s}}$$
$$= 82.767 \text{ kHz} \approx 82.8 \text{ kHz}. \qquad \text{(Answer)}$$

Detection of echo by bat: In the echo back to the bat, the moth acts as a source of sound, emitting at the frequency f_{md} we just calculated. So now the moth is the source (moving *away*) and the bat is the detector (moving *toward*). The reasoning steps are shown in Table 17-3. To find the frequency f_{bd} detected by the bat, we write Eq. 17-56 as

$$f_{bd} = f_{md}\frac{v + v_b}{v + v_m}$$
$$= (82.767 \text{ kHz})\frac{343 \text{ m/s} + 9.00 \text{ m/s}}{343 \text{ m/s} + 8.00 \text{ m/s}}$$
$$= 83.00 \text{ kHz} \approx 83.0 \text{ kHz}. \qquad \text{(Answer)}$$

Some moths evade bats by "jamming" the detection system with ultrasonic clicks.

TABLE 17-3

Bat to Moth		Echo Back to Bat	
Detector	Source	Detector	Source
moth	bat	bat	moth
speed $v_D = v_m$	speed $v_S = v_b$	speed $v_D = v_b$	speed $v_S = v_m$
away	toward	toward	away
shift down	shift up	shift up	shift down
numerator	denominator	numerator	denominator
minus	minus	plus	plus

17-10 | Supersonic Speeds, Shock Waves

If a source is moving toward a stationary detector at a speed equal to the speed of sound—that is, if $v_S = v$—Eqs. 17-47 and 17-55 predict that the detected frequency f' will be infinitely great. This means that the source is moving so fast that it keeps pace with its own spherical wavefronts, as Fig. 17-23*a* suggests. What happens when the speed of the source *exceeds* the speed of sound?

For such *supersonic* speeds, Eqs. 17-47 and 17-55 no longer apply. Figure 17-23*b* depicts the spherical wavefronts that originated at various positions of the source. The radius of any wavefront in this figure is vt, where v is the speed of sound and t is the time that has elapsed since the source emitted that wavefront. Note that all the wavefronts bunch along a V-shaped envelope in the two-dimensional drawing of Fig. 17-23*b*. The wavefronts actually extend in three dimensions, and the bunching actually forms a cone called the *Mach cone*. A *shock wave* is said to exist along the surface of this cone, because the bunching of wavefronts causes an abrupt rise and fall of air pressure as the surface

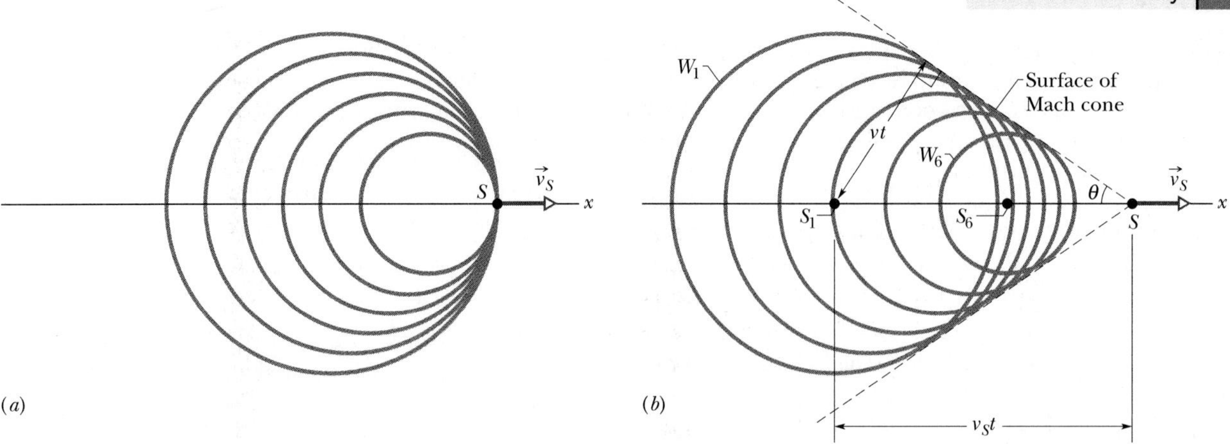

FIG. 17-23 (*a*) A source of sound *S* moves at speed v_S equal to the speed of sound and thus as fast as the wavefronts it generates. (*b*) A source *S* moves at speed v_S faster than the speed of sound and thus faster than the wavefronts. When the source was at position S_1 it generated wavefront W_1, and at position S_6 it generated W_6. All the spherical wavefronts expand at the speed of sound v and bunch along the surface of a cone called the Mach cone, forming a shock wave. The surface of the cone has half-angle θ and is tangent to all the wavefronts.

passes through any point. From Fig. 17-23*b*, we see that the half-angle θ of the cone, called the *Mach cone angle*, is given by

$$\sin\theta = \frac{vt}{v_S t} = \frac{v}{v_S} \quad \text{(Mach cone angle)}. \tag{17-57}$$

The ratio v_S/v is called the *Mach number.* When you hear that a particular plane has flown at Mach 2.3, it means that its speed was 2.3 times the speed of sound in the air through which the plane was flying. The shock wave generated by a supersonic aircraft (Fig. 17-24) or projectile produces a burst of sound, called a *sonic boom,* in which the air pressure first suddenly increases and then suddenly decreases below normal before returning to normal. Part of the sound that is heard when a rifle is fired is the sonic boom produced by the bullet. A sonic boom can also be heard from a long bullwhip when it is snapped quickly: Near the end of the whip's motion, its tip is moving faster than sound and produces a small sonic boom—the *crack* of the whip.

FIG. 17-24 Shock waves produced by the wings of a Navy FA 18 jet. The shock waves are visible because the sudden decrease in air pressure in them caused water molecules in the air to condense, forming a fog. *(U.S. Navy photo by Ensign John Gay)*

REVIEW & SUMMARY

Sound Waves Sound waves are longitudinal mechanical waves that can travel through solids, liquids, or gases. The speed v of a sound wave in a medium having **bulk modulus** B and density ρ is

$$v = \sqrt{\frac{B}{\rho}} \quad \text{(speed of sound)}. \tag{17-3}$$

In air at 20°C, the speed of sound is 343 m/s.

A sound wave causes a longitudinal displacement s of a mass element in a medium as given by

$$s = s_m \cos(kx - \omega t), \tag{17-13}$$

where s_m is the **displacement amplitude** (maximum displacement) from equilibrium, $k = 2\pi/\lambda$, and $\omega = 2\pi f$, λ and f being the wavelength and frequency, respectively, of the sound wave.

The sound wave also causes a pressure change Δp of the medium from the equilibrium pressure:

$$\Delta p = \Delta p_m \sin(kx - \omega t), \tag{17-14}$$

where the **pressure amplitude** is

$$\Delta p_m = (v\rho\omega)s_m. \tag{17-15}$$

Interference The interference of two sound waves with identical wavelengths passing through a common point depends on their phase difference ϕ there. If the sound waves were emitted in phase and are traveling in approximately the same direction, ϕ is given by

$$\phi = \frac{\Delta L}{\lambda} 2\pi, \tag{17-21}$$

where ΔL is their **path length difference** (the difference in the distances traveled by the waves to reach the common point). Fully constructive interference occurs when ϕ is an integer multiple of 2π,

$$\phi = m(2\pi), \quad \text{for } m = 0, 1, 2, \ldots, \tag{17-22}$$

and, equivalently, when ΔL is related to wavelength λ by

$$\frac{\Delta L}{\lambda} = 0, 1, 2, \ldots. \tag{17-23}$$

Fully destructive interference occurs when ϕ is an odd multiple of π,

$$\phi = (2m + 1)\pi, \quad \text{for } m = 0, 1, 2, \ldots, \tag{17-24}$$

and, equivalently, when ΔL is related to λ by

$$\frac{\Delta L}{\lambda} = 0.5, 1.5, 2.5, \ldots. \tag{17-25}$$

Sound Intensity The **intensity** I of a sound wave at a surface is the average rate per unit area at which energy is transferred by the wave through or onto the surface:

$$I = \frac{P}{A}, \tag{17-26}$$

where P is the time rate of energy transfer (power) of the sound wave and A is the area of the surface intercepting the sound. The intensity I is related to the displacement amplitude s_m of the sound wave by

$$I = \tfrac{1}{2}\rho v \omega^2 s_m^2. \tag{17-27}$$

The intensity at a distance r from a point source that emits sound waves of power P_s is

$$I = \frac{P_s}{4\pi r^2}. \tag{17-28}$$

Sound Level in Decibels The *sound level* β in *decibels* (dB) is defined as

$$\beta = (10\text{ dB}) \log \frac{I}{I_0}, \tag{17-29}$$

where I_0 ($= 10^{-12}$ W/m^2) is a reference intensity level to which all intensities are compared. For every factor-of-10 increase in intensity, 10 dB is added to the sound level.

Standing Wave Patterns in Pipes Standing sound wave patterns can be set up in pipes. A pipe open at both ends will resonate at frequencies

$$f = \frac{v}{\lambda} = \frac{nv}{2L}, \quad n = 1, 2, 3, \ldots, \tag{17-39}$$

where v is the speed of sound in the air in the pipe. For a pipe closed at one end and open at the other, the resonant frequencies are

$$f = \frac{v}{\lambda} = \frac{nv}{4L}, \quad n = 1, 3, 5, \ldots. \tag{17-41}$$

Beats *Beats* arise when two waves having slightly different frequencies, f_1 and f_2, are detected together. The beat frequency is

$$f_{\text{beat}} = f_1 - f_2. \tag{17-46}$$

The Doppler Effect The *Doppler effect* is a change in the observed frequency of a wave when the source or the detector moves relative to the transmitting medium (such as air). For sound the observed frequency f' is given in terms of the source frequency f by

$$f' = f\frac{v \pm v_D}{v \pm v_S} \quad \text{(general Doppler effect)}, \tag{17-47}$$

where v_D is the speed of the detector relative to the medium, v_S is that of the source, and v is the speed of sound in the medium. The signs are chosen such that f' tends to be *greater* for motion toward and *less* for motion away.

Shock Wave If the speed of a source relative to the medium exceeds the speed of sound in the medium, the Doppler equation no longer applies. In such a case, shock waves result. The half-angle θ of the Mach cone is given by

$$\sin\theta = \frac{v}{v_S} \quad \text{(Mach cone angle)}. \tag{17-57}$$

QUESTIONS

1 In Fig. 17-25, three long tubes (A, B, and C) are filled with different gases under different pressures. The ratio of the bulk modulus to the density is indicated for each gas in terms of a basic value B_0/ρ_0. Each tube has a piston at its left end that can send a sound pulse through the tube (as in Fig. 16-2). The three pulses are sent simultaneously. Rank the tubes according to the time of arrival of the pulses at the open right ends of the tubes, earliest first.

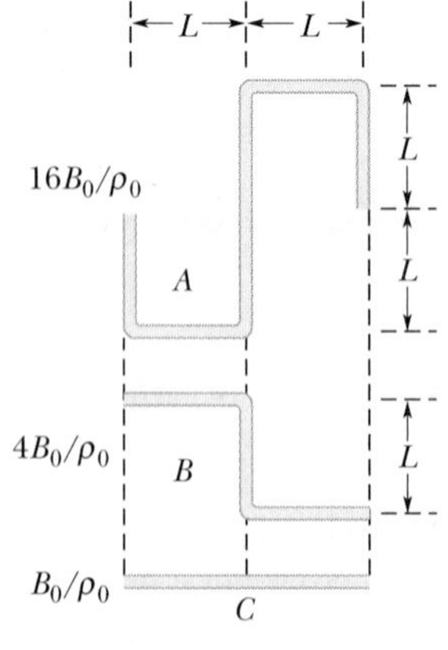

FIG. 17-25 Question 1.

2 In Fig. 17-26, two point sources S_1 and S_2, which are in phase, emit identical sound waves of wavelength 2.0 m. In terms of wavelengths, what is the phase difference between the waves arriving at point P if (a) $L_1 = 38$ m and $L_2 = 34$ m, and (b) $L_1 = 39$ m and $L_2 = 36$ m? (c) Assuming that the source separation is much smaller than L_1 and L_2, what type of interference occurs at P in situations (a) and (b)?

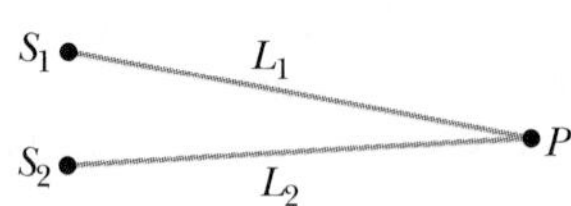

FIG. 17-26 Question 2.

3 In a first experiment, a sinusoidal sound wave is sent through a long tube of air, transporting energy at the average rate of $P_{\text{avg},1}$. In a second experiment, two other sound waves, identical to the first one, are to be sent simultaneously through the tube with a phase difference ϕ of either 0,

0.2 wavelength, or 0.5 wavelength between the waves. (a) With only mental calculation, rank those choices of ϕ according to the average rate at which the waves will transport energy, greatest first. (b) For the first choice of ϕ, what is the average rate in terms of $P_{avg,1}$?

4 Pipe A has length L and one open end. Pipe B has length $2L$ and two open ends. Which harmonics of pipe B have a frequency that matches a resonant frequency of pipe A?

5 For a particular tube, here are four of the six harmonic frequencies below 1000 Hz: 300, 600, 750, and 900 Hz. What two frequencies are missing from the list?

6 The sixth harmonic is set up in a pipe. (a) How many open ends does the pipe have (it has at least one)? (b) Is there a node, antinode, or some intermediate state at the midpoint?

7 In Fig. 17-27, pipe A is made to oscillate in its third harmonic by a small internal sound source. Sound emitted at the right end happens to resonate four nearby pipes, each with only one open end (they are *not* drawn to scale). Pipe B oscillates in its lowest harmonic, pipe C in its second lowest harmonic, pipe D in its third lowest harmonic, and pipe E in its fourth lowest harmonic. Without computation, rank all five pipes according to their length, greatest first. (*Hint:* Draw the standing waves to scale and then draw the pipes to scale.)

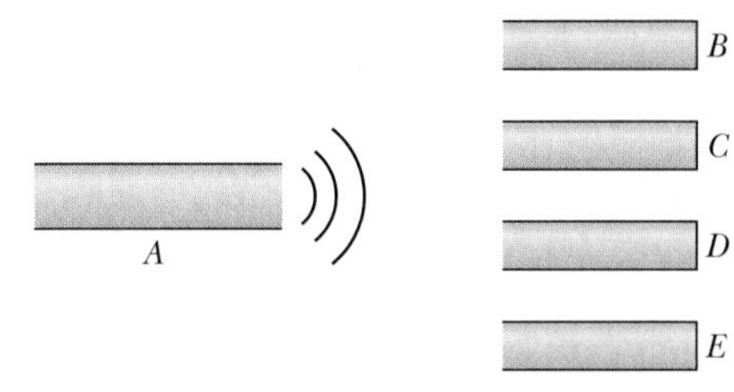

FIG. 17-27 Question 7.

8 Figure 17-28 shows a stretched string of length L and pipes a, b, c, and d of lengths L, $2L$, $L/2$, and $L/2$, respectively. The string's tension is adjusted until the speed of waves on the string equals the speed of sound waves in the air. The fundamental mode of oscillation is then set up on the string. In which pipe will the sound produced by the string cause resonance, and what oscillation mode will that sound set up?

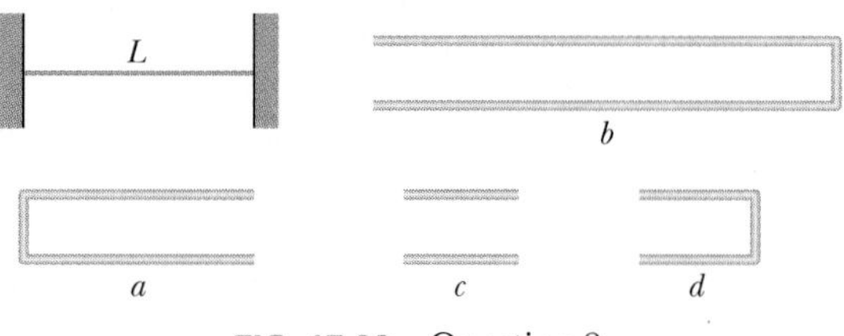

FIG. 17-28 Question 8.

9 Figure 17-29 shows a moving sound source S that emits at a certain frequency, and four stationary sound detectors. Rank the detectors according to the frequency of the sound they detect from the source, greatest first.

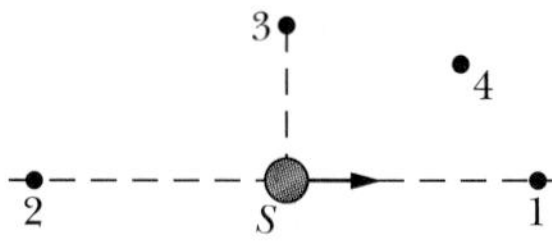

FIG. 17-29 Question 9.

10 A friend rides, in turn, the rims of three fast merry-go-rounds while holding a sound source that emits isotropically at a certain frequency. You stand far from each merry-go-round. The frequency you hear for each of your friend's three rides varies as the merry-go-round rotates. The variations in frequency for the three rides are given by the three curves in Fig. 17-30. Rank the curves according to (a) the linear speed v of the sound source, (b) the angular speeds ω of the merry-go-rounds, and (c) the radii r of the merry-go-rounds, greatest first.

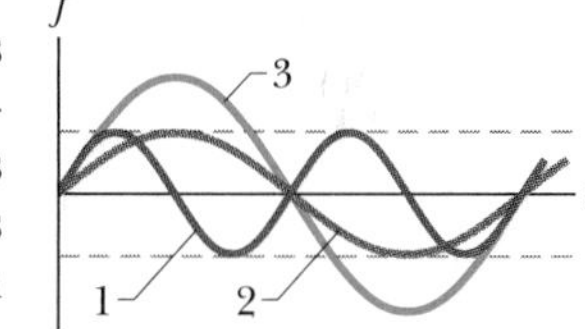

FIG. 17-30 Question 10.

PROBLEMS

GO Tutoring problem available (at instructor's discretion) in *WileyPLUS* and WebAssign
SSM Worked-out solution available in Student Solutions Manual
WWW Worked-out solution is at http://www.wiley.com/college/halliday
• – ••• Number of dots indicates level of problem difficulty
ILW Interactive solution is at http://www.wiley.com/college/halliday
Additional information available in *The Flying Circus of Physics* and at flyingcircusofphysics.com

Where needed in the problems, use

speed of sound in air = 343 m/s

and density of air = 1.21 kg/m^3

unless otherwise specified.

sec. 17-3 The Speed of Sound

•1 When the door of the Chapel of the Mausoleum in Hamilton, Scotland, is slammed shut, the last echo heard by someone standing just inside the door reportedly comes 15 s later. (a) If that echo were due to a single reflection off a wall opposite the door, how far from the door would that wall be? (b) If, instead, the wall is 25.7 m away, how many reflections (back and forth) correspond to the last echo?

•2 A column of soldiers, marching at 120 paces per minute, keep in step with the beat of a drummer at the head of the column. The soldiers in the rear end of the column are striding forward with the left foot when the drummer is advancing with the right foot. What is the approximate length of the column?

•3 Two spectators at a soccer game in Montjuic Stadium see, and a moment later hear, the ball being kicked on the playing field. The time delay for spectator A is 0.23 s, and for spectator B it is 0.12 s. Sight lines from the two spectators to the player kicking the ball meet at an angle of 90°. How far are (a) spectator A and (b) spectator B from the player? (c) How far are the spectators from each other?

•4 What is the bulk modulus of oxygen if 32.0 g of oxygen occupies 22.4 L and the speed of sound in the oxygen is 317 m/s?

••5 A stone is dropped into a well. The splash is heard 3.00 s later. What is the depth of the well? SSM WWW

••6 *Hot chocolate effect.* Tap a metal spoon inside a mug of water and note the frequency f_i you hear. Then add a spoonful of powder (say, chocolate mix or instant coffee) and tap again as you stir the powder. The frequency you hear has a lower value f_s because the tiny air bubbles released by the powder change the water's bulk modulus. As the bubbles reach the water surface and disappear, the frequency gradually shifts back to its initial value. During the effect, the bubbles don't appreciably change the water's density or volume or the sound's wavelength. Rather, they change the value of dV/dp—that is, the differential change in volume due to the differential change in the pressure caused by the sound wave in the water. If $f_s/f_i = 0.333$, what is the ratio $(dV/dp)_s/(dV/dp)_i$?

••7 Earthquakes generate sound waves inside Earth. Unlike a gas, Earth can experience both transverse (S) and longitudinal (P) sound waves. Typically, the speed of S waves is about 4.5 km/s, and that of P waves 8.0 km/s. A seismograph records P and S waves from an earthquake. The first P waves arrive 3.0 min before the first S waves. If the waves travel in a straight line, how far away does the earthquake occur? SSM ILW

••8 A man strikes one end of a thin rod with a hammer. The speed of sound in the rod is 15 times the speed of sound in air. A woman, at the other end with her ear close to the rod, hears the sound of the blow twice with a 0.12 s interval between; one sound comes through the rod and the other comes through the air alongside the rod. If the speed of sound in air is 343 m/s, what is the length of the rod?

sec. 17-4 Traveling Sound Waves

•9 Diagnostic ultrasound of frequency 4.50 MHz is used to examine tumors in soft tissue. (a) What is the wavelength in air of such a sound wave? (b) If the speed of sound in tissue is 1500 m/s, what is the wavelength of this wave in tissue? SSM

•10 The pressure in a traveling sound wave is given by the equation

$$\Delta p = (1.50 \text{ Pa}) \sin \pi[(0.900 \text{ m}^{-1})x - (315 \text{ s}^{-1})t].$$

Find the (a) pressure amplitude, (b) frequency, (c) wavelength, and (d) speed of the wave.

•11 If the form of a sound wave traveling through air is

$$s(x, t) = (6.0 \text{ nm}) \cos(kx + (3000 \text{ rad/s})t + \phi),$$

how much time does any given air molecule along the path take to move between displacements $s = +2.0$ nm and $s = -2.0$ nm?

•12 *Underwater illusion.* One clue used by your brain to determine the direction of a source of sound is the time delay Δt between the arrival of the sound at the ear closer to the source and the arrival at the farther ear. Assume that the source is distant so that a wavefront from it is approximately planar when it reaches you, and let D represent the separation between your ears. (a) If the source is located at angle θ in front of you (Fig. 17-31), what is Δt in terms of D and the speed of sound v in air? (b) If you are submerged in water and the sound source is directly to your right, what is Δt in terms of D and the speed of sound v_w in water? (c) Based on the time-delay clue, your brain interprets the submerged sound to arrive at an angle θ from the forward direction. Evaluate θ for fresh water at 20°C.

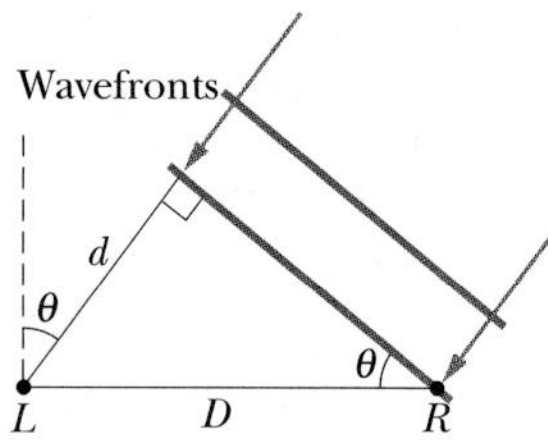

FIG. 17-31 Problem 12.

••13 A handclap on stage in an amphitheater sends out sound waves that scatter from terraces of width $w = 0.75$ m (Fig. 17-32). The sound returns to the stage as a periodic series of pulses, one from each terrace; the parade of pulses sounds like a played note. (a) Assuming that all the rays in Fig. 17-32 are horizontal, find the frequency at which the pulses return (that is, the frequency of the perceived note). (b) If the width w of the terraces were smaller, would the frequency be higher or lower?

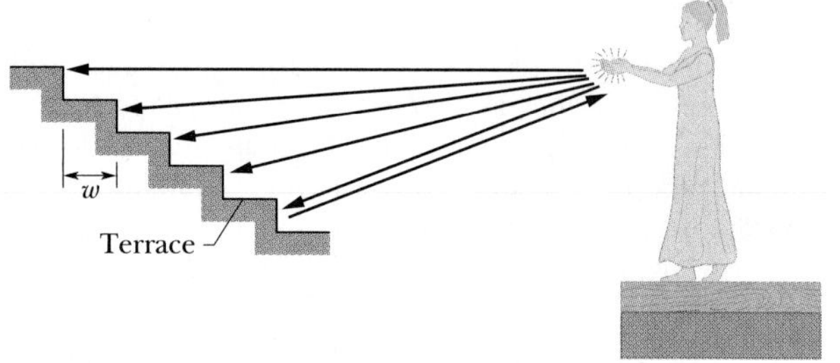

FIG. 17-32 Problem 13.

••14 Figure 17-33 shows the output from a pressure monitor mounted at a point along the path taken by a sound wave of a single frequency traveling at 343 m/s through air with a uniform density of 1.21 kg/m³. The vertical axis scale is set by $\Delta p_s = 4.0$ mPa. If the *displacement* function of the wave is written as $s(x, t) = s_m \cos(kx - \omega t)$, what are (a) s_m, (b) k, and (c) ω? The air is then cooled so that its density is 1.35 kg/m³ and the speed of a sound wave through it is 320 m/s. The sound source again emits the sound wave at the same frequency and same pressure amplitude. What now are (d) s_m, (e) k, and (f) ω?

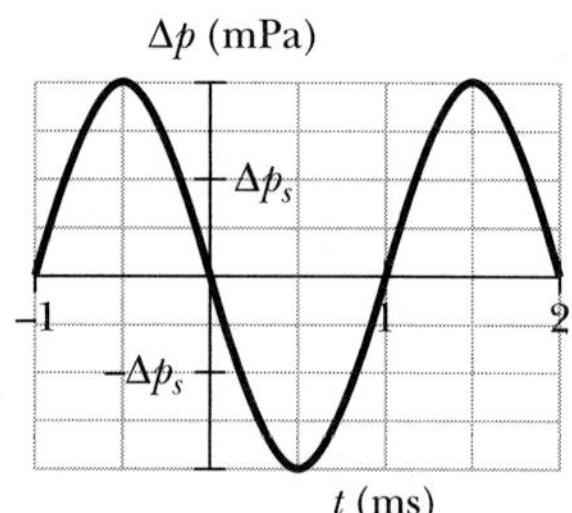

FIG. 17-33 Problem 14.

••15 A sound wave of the form $s = s_m \cos(kx - \omega t + \phi)$ travels at 343 m/s through air in a long horizontal tube. At one instant, air molecule A at $x = 2.000$ m is at its maximum positive displacement of 6.00 nm and air molecule B at $x = 2.070$ m is at a positive displacement of 2.00 nm. All the molecules between A and B are at intermediate displacements. What is the frequency of the wave?

sec. 17-5 Interference

•16 Two sound waves, from two different sources with the same frequency, 540 Hz, travel in the same direction at 330 m/s. The sources are in phase. What is the phase difference of the waves at a point that is 4.40 m from one source and 4.00 m from the other?

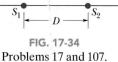

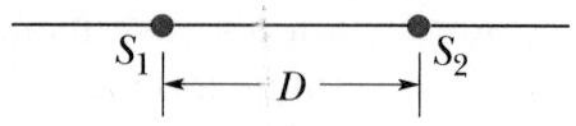

S_1 — D — S_2

FIG. 17-34 Problems 17 and 107.

••17 Figure 17-34 shows two isotropic point sources of sound, S_1 and S_2. The sources emit waves in phase at wavelength 0.50 m; they are separated by $D = 1.75$ m. If we move a sound detector along a large circle centered at the midpoint between the sources, at how many points do waves arrive at the detector (a) exactly in phase and (b) exactly out of phase?

••18 In Fig. 17-35, sound with a 40.0 cm wavelength travels rightward from a source and through a tube that consists of a straight portion and a half-circle. Part of the sound wave travels through the half-circle and then rejoins the rest of the wave, which goes directly through the straight portion. This rejoining results in interference. What is the smallest radius r that results in an intensity minimum at the detector?

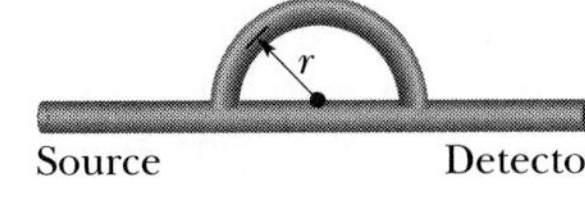

FIG. 17-35 Problem 18.

••19 In Fig. 17-36, two speakers separated by distance $d_1 = 2.00$ m are in phase. Assume the amplitudes of the sound waves from the speakers are approximately the same at the listener's ear at distance $d_2 = 3.75$ m directly in front of one speaker. Consider the full audible range for normal hearing, 20 Hz to 20 kHz. (a) What is the lowest frequency $f_{min,1}$ that gives minimum signal (destructive interference) at the listener's ear? By what number must $f_{min,1}$ be multiplied to get (b) the second lowest frequency $f_{min,2}$ that gives minimum signal and (c) the third lowest frequency $f_{min,3}$ that gives minimum signal? (d) What is the lowest frequency $f_{max,1}$ that gives maximum signal (constructive interference) at the listener's ear? By what number must $f_{max,1}$ be multiplied to get (e) the second lowest frequency $f_{max,2}$ that gives maximum signal and (f) the third lowest frequency $f_{max,3}$ that gives maximum signal? SSM

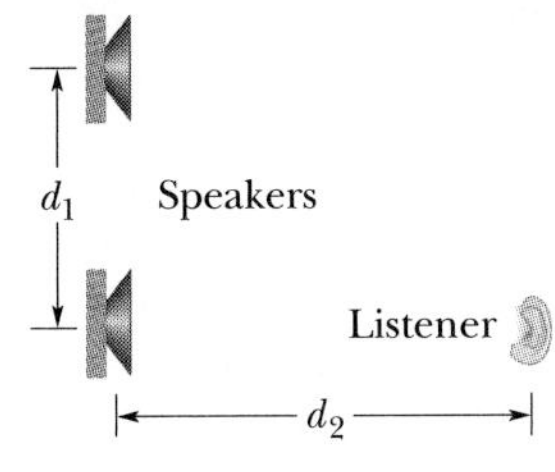

FIG. 17-36 Problem 19.

••20 In Fig. 17-37, sound waves A and B, both of wavelength λ, are initially in phase and traveling rightward, as indicated by the two rays. Wave A is reflected from four surfaces but ends up traveling in its original direction. Wave B ends in that direction after reflecting from two surfaces. Let distance L in the figure be expressed as a multiple q of λ: $L = q\lambda$. What are the (a) smallest and (b) second smallest values of q that put A and B exactly out of phase with each other after the reflections?

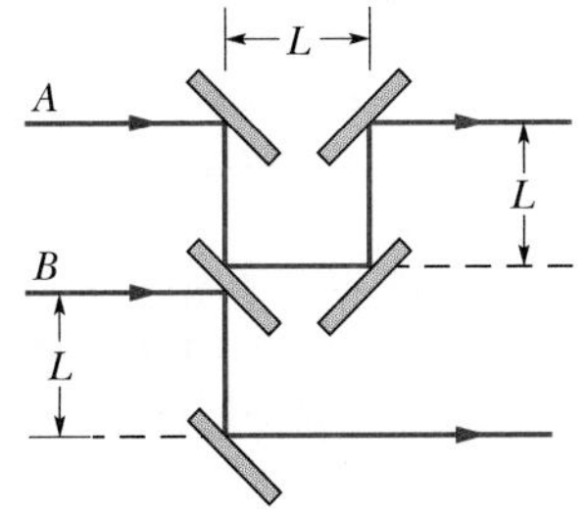

FIG. 17-37 Problem 20.

••21 Two loudspeakers are located 3.35 m apart on an outdoor stage. A listener is 18.3 m from one and 19.5 m from the other. During the sound check, a signal generator drives the two speakers in phase with the same amplitude and frequency. The transmitted frequency is swept through the audible range (20 Hz to 20 kHz). (a) What is the lowest frequency $f_{min,1}$ that gives minimum signal (destructive interference) at the listener's location? By what number must $f_{min,1}$ be multiplied to get (b) the second lowest frequency $f_{min,2}$ that gives minimum signal and (c) the third lowest frequency $f_{min,3}$ that gives minimum signal? (d) What is the lowest frequency $f_{max,1}$ that gives maximum signal (constructive interference) at the listener's location? By what number must $f_{max,1}$ be multiplied to get (e) the second lowest frequency $f_{max,2}$ that gives maximum signal and (f) the third lowest frequency $f_{max,3}$ that gives maximum signal? ILW

••22 Figure 17-38 shows four isotropic point sources of sound that are uniformly spaced on an x axis. The sources emit sound at the same wavelength λ and same amplitude s_m, and they emit in phase. A point P is shown on the x axis. Assume that as the sound waves travel to P, the decrease in their amplitude is negligible. What multiple of s_m is the amplitude of the net wave at P if distance d in the figure is (a) $\lambda/4$, (b) $\lambda/2$, and (c) λ?

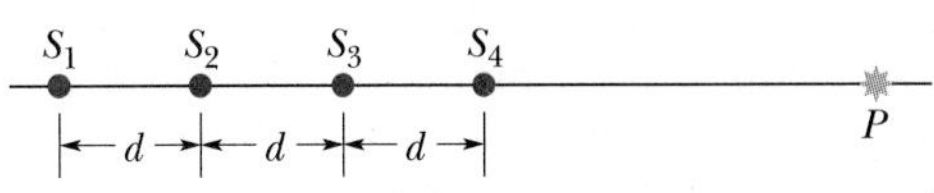

FIG. 17-38 Problem 22.

•••23 Figure 17-39 shows two point sources S_1 and S_2 that emit sound of wavelength $\lambda = 2.00$ m. The emissions are isotropic and in phase, and the separation between the sources is $d = 16.0$ m. At any point P on the x axis, the wave from S_1 and the wave from S_2 interfere. When P is very far away ($x \approx \infty$), what are (a) the phase difference between the arriving waves from S_1 and S_2 and (b) the type of interference they produce? Now move point P along the x axis toward S_1. (c) Does the phase difference between the waves increase or decrease? At what distance x do the waves have a phase difference of (d) 0.50λ, (e) 1.00λ, and (f) 1.50λ? GO

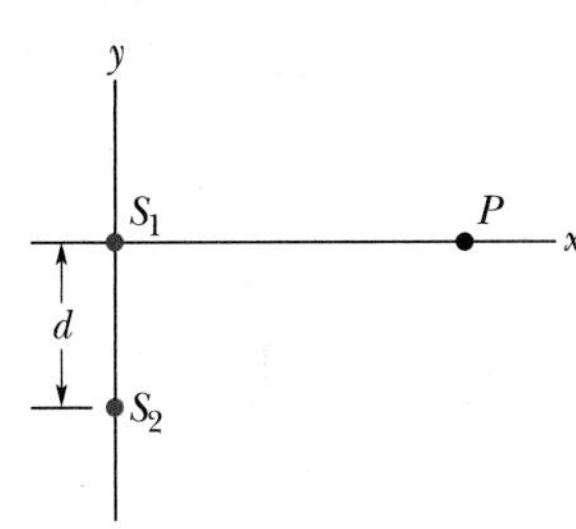

FIG. 17-39 Problem 23.

sec. 17-6 Intensity and Sound Level

•24 A 1.0 W point source emits sound waves isotropically. Assuming that the energy of the waves is conserved, find the intensity (a) 1.0 m from the source and (b) 2.5 m from the source.

•25 A source emits sound waves isotropically. The intensity of the waves 2.50 m from the source is 1.91×10^{-4} W/m^2. Assuming that the energy of the waves is conserved, find the power of the source. SSM

•26 Two sounds differ in sound level by 1.00 dB. What is the ratio of the greater intensity to the smaller intensity?

•27 A sound wave of frequency 300 Hz has an intensity of 1.00 μW/m^2. What is the amplitude of the air oscillations caused by this wave?

•28 The source of a sound wave has a power of 1.00 μW. If it is a point source, (a) what is the intensity 3.00 m away and (b) what is the sound level in decibels at that distance?

•29 A certain sound source is increased in sound level by 30.0 dB. By what multiple is (a) its intensity increased and (b) its pressure amplitude increased? SSM WWW

•30 Suppose that the sound level of a conversation is initially at an angry 70 dB and then drops to a soothing 50 dB. Assuming that the frequency of the sound is 500 Hz, determine the (a) initial and (b) final sound intensities and the (c) initial and (d) final sound wave amplitudes.

•**31** Male *Rana catesbeiana* bullfrogs are known for their loud mating call. The call is emitted not by the frog's mouth but by its eardrums, which lie on the surface of the head. And, surprisingly, the sound has nothing to do with the frog's inflated throat. If the emitted sound has a frequency of 260 Hz and a sound level of 85 dB (near the eardrum), what is the amplitude of the eardrum's oscillation? The air density is 1.21 kg/m^3.

•**32** Approximately a third of people with normal hearing have ears that continuously emit a low-intensity sound outward through the ear canal. A person with such *spontaneous otoacoustic emission* is rarely aware of the sound, except perhaps in a noise-free environment, but occasionally the emission is loud enough to be heard by someone else nearby. In one observation, the sound wave had a frequency of 1665 Hz and a pressure amplitude of 1.13×10^{-3} Pa. What were (a) the displacement amplitude and (b) the intensity of the wave emitted by the ear?

•**33** When you "crack" a knuckle, you suddenly widen the knuckle cavity, allowing more volume for the synovial fluid inside it and causing a gas bubble suddenly to appear in the fluid. The sudden production of the bubble, called "cavitation," produces a sound pulse—the cracking sound. Assume that the sound is transmitted uniformly in all directions and that it fully passes from the knuckle interior to the outside. If the pulse has a sound level of 62 dB at your ear, estimate the rate at which energy is produced by the cavitation. GO

••**34** *Party hearing.* As the number of people at a party increases, you must raise your voice for a listener to hear you against the *background noise* of the other partygoers. However, once you reach the level of yelling, the only way you can be heard is if you move closer to your listener, into the listener's "personal space." Model the situation by replacing you with an isotropic point source of fixed power P and replacing your listener with a point that absorbs part of your sound waves. These points are initially separated by $r_i = 1.20$ m. If the background noise increases by $\Delta\beta = 5$ dB, the sound level at your listener must also increase. What separation r_f is then required?

••**35** A point source emits 30.0 W of sound isotropically. A small microphone intercepts the sound in an area of 0.750 cm^2, 200 m from the source. Calculate (a) the sound intensity there and (b) the power intercepted by the microphone.

••**36** Two atmospheric sound sources A and B emit isotropically at constant power. The sound levels β of their emissions are plotted in Fig. 17-40 versus the radial distance r from the sources. The vertical axis scale is set by $\beta_1 = 85.0$ dB and $\beta_2 = 65.0$ dB. What are (a) the ratio of the larger power to the smaller power and (b) the sound level difference at $r = 10$ m? GO

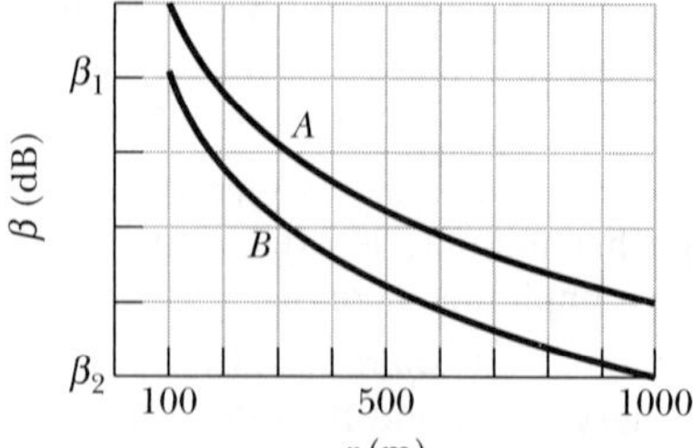

FIG. 17-40 Problem 36.

•••**37** A sound source sends a sinusoidal sound wave of angular frequency 3000 rad/s and amplitude 12.0 nm through a tube of air. The internal radius of the tube is 2.00 cm. (a) What is the average rate at which energy (the sum of the kinetic and potential energies) is transported to the opposite end of the tube? (b) If, simultaneously, an identical wave travels along an adjacent, identical tube, what is the total average rate at which energy is transported to the opposite ends of the two tubes by the waves? If, instead, those two waves are sent along the *same* tube simultaneously, what is the total average rate at which they transport energy when their phase difference is (c) 0, (d) 0.40π rad, and (e) π rad?

sec. 17-7 Sources of Musical Sound

•**38** The crest of a *Parasaurolophus* dinosaur skull contains a nasal passage in the shape of a long, bent tube open at both ends. The dinosaur may have used the passage to produce sound by setting up the fundamental mode in it. (a) If the nasal passage in a certain *Parasaurolophus* fossil is 2.0 m long, what frequency would have been produced? (b) If that dinosaur could be recreated (as in *Jurassic Park*), would a person with a hearing range of 60 Hz to 20 kHz be able to hear that fundamental mode and, if so, would the sound be high or low frequency? Fossil skulls that contain shorter nasal passages are thought to be those of the female *Parasaurolophus.* (c) Would that make the female's fundamental frequency higher or lower than the male's?

•**39** A violin string 15.0 cm long and fixed at both ends oscillates in its $n = 1$ mode. The speed of waves on the string is 250 m/s, and the speed of sound in air in 348 m/s. What are the (a) frequency and (b) wavelength of the emitted sound wave?

•**40** A sound wave in a fluid medium is reflected at a barrier so that a standing wave is formed. The distance between nodes is 3.8 cm, and the speed of propagation is 1500 m/s. Find the frequency of the sound wave.

•**41** In pipe A, the ratio of a particular harmonic frequency to the next lower harmonic frequency is 1.2. In pipe B, the ratio of a particular harmonic frequency to the next lower harmonic frequency is 1.4. How many open ends are in (a) pipe A and (b) pipe B?

•**42** Organ pipe A, with both ends open, has a fundamental frequency of 300 Hz. The third harmonic of organ pipe B, with one end open, has the same frequency as the second harmonic of pipe A. How long are (a) pipe A and (b) pipe B?

•**43** (a) Find the speed of waves on a violin string of mass 800 mg and length 22.0 cm if the fundamental frequency is 920 Hz. (b) What is the tension in the string? For the fundamental, what is the wavelength of (c) the waves on the string and (d) the sound waves emitted by the string? SSM ILW

•**44** The water level in a vertical glass tube 1.00 m long can be adjusted to any position in the tube. A tuning fork vibrating at 686 Hz is held just over the open top end of the tube, to set up a standing wave of sound in the air-filled top portion of the tube. (That air-filled top portion acts as a tube with one end closed and the other end open.) (a) For how many different positions of the water level will sound from the fork set up resonance in the tube's air-filled portion, which acts as a pipe with one end closed (by the water) and the other end open? What are the (b) least and (c) second least water heights in the tube for resonance to occur?

•**45** In Fig. 17-41, S is a small loudspeaker driven by an audio oscillator with a frequency that is varied from 1000 Hz to 2000 Hz, and D is a cylindrical pipe with two open ends and a length of 45.7 cm. The speed of sound in the air-filled pipe is 344 m/s. (a) At how many frequencies does the sound from the loudspeaker set up resonance in the pipe? What are the (b) lowest and (c) second lowest frequencies at which resonance occurs? SSM

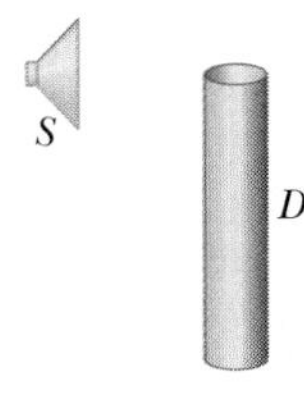

FIG. 17-41 Problem 45.

••46 One of the harmonic frequencies of tube A with two open ends is 325 Hz. The next-highest harmonic frequency is 390 Hz. (a) What harmonic frequency is next highest after the harmonic frequency 195 Hz? (b) What is the number of this next-highest harmonic?

One of the harmonic frequencies of tube B with only one open end is 1080 Hz. The next-highest harmonic frequency is 1320 Hz. (c) What harmonic frequency is next highest after the harmonic frequency 600 Hz? (d) What is the number of this next-highest harmonic?

••47 A violin string 30.0 cm long with linear density 0.650 g/m is placed near a loudspeaker that is fed by an audio oscillator of variable frequency. It is found that the string is set into oscillation only at the frequencies 880 and 1320 Hz as the frequency of the oscillator is varied over the range 500–1500 Hz. What is the tension in the string? SSM

••48 A tube 1.20 m long is closed at one end. A stretched wire is placed near the open end. The wire is 0.330 m long and has a mass of 9.60 g. It is fixed at both ends and oscillates in its fundamental mode. By resonance, it sets the air column in the tube into oscillation at that column's fundamental frequency. Find (a) that frequency and (b) the tension in the wire. GO

••49 A well with vertical sides and water at the bottom resonates at 7.00 Hz and at no lower frequency. (The air-filled portion of the well acts as a tube with one closed end and one open end.) The air in the well has a density of 1.10 kg/m^3 and a bulk modulus of 1.33×10^5 Pa. How far down in the well is the water surface?

••50 Pipe A, which is 1.20 m long and open at both ends, oscillates at its third lowest harmonic frequency. It is filled with air for which the speed of sound is 343 m/s. Pipe B, which is closed at one end, oscillates at its second lowest harmonic frequency. This frequency of B happens to match the frequency of A. An x axis extends along the interior of B, with $x = 0$ at the closed end. (a) How many nodes are along that axis? What are the (b) smallest and (c) second smallest value of x locating those nodes? (d) What is the fundamental frequency of B? GO

sec. 17-8 Beats

•51 The A string of a violin is a little too tightly stretched. Beats at 4.00 per second are heard when the string is sounded together with a tuning fork that is oscillating accurately at concert A (440 Hz). What is the period of the violin string oscillation?

•52 A tuning fork of unknown frequency makes 3.00 beats per second with a standard fork of frequency 384 Hz. The beat frequency decreases when a small piece of wax is put on a prong of the first fork. What is the frequency of this fork?

••53 Two identical piano wires have a fundamental frequency of 600 Hz when kept under the same tension. What fractional increase in the tension of one wire will lead to the occurrence of 6.0 beats/s when both wires oscillate simultaneously? SSM

••54 You have five tuning forks that oscillate at close but different frequencies. What are the (a) maximum and (b) minimum number of different beat frequencies you can produce by sounding the forks two at a time, depending on how the frequencies differ?

sec. 17-9 The Doppler Effect

•55 A state trooper chases a speeder along a straight road; both vehicles move at 160 km/h. The siren on the trooper's vehicle produces sound at a frequency of 500 Hz. What is the Doppler shift in the frequency heard by the speeder?

•56 An ambulance with a siren emitting a whine at 1600 Hz overtakes and passes a cyclist pedaling a bike at 2.44 m/s. After being passed, the cyclist hears a frequency of 1590 Hz. How fast is the ambulance moving?

•57 A whistle of frequency 540 Hz moves in a circle of radius 60.0 cm at an angular speed of 15.0 rad/s. What are the (a) lowest and (b) highest frequencies heard by a listener a long distance away, at rest with respect to the center of the circle? ILW

••58 A stationary motion detector sends sound waves of frequency 0.150 MHz toward a truck approaching at a speed of 45.0 m/s. What is the frequency of the waves reflected back to the detector?

••59 An acoustic burglar alarm consists of a source emitting waves of frequency 28.0 kHz. What is the beat frequency between the source waves and the waves reflected from an intruder walking at an average speed of 0.950 m/s directly away from the alarm? ILW

••60 A sound source A and a reflecting surface B move directly toward each other. Relative to the air, the speed of source A is 29.9 m/s, the speed of surface B is 65.8 m/s, and the speed of sound is 329 m/s. The source emits waves at frequency 1200 Hz as measured in the source frame. In the reflector frame, what are the (a) frequency and (b) wavelength of the arriving sound waves? In the source frame, what are the (c) frequency and (d) wavelength of the sound waves reflected back to the source?

••61 In Fig. 17-42, a French submarine and a U.S. submarine move toward each other during maneuvers in motionless water in the North Atlantic. The French sub moves at speed $v_F = 50.00$ km/h, and the U.S. sub at $v_{US} = 70.00$ km/h. The French sub sends out a sonar signal (sound wave in water) at 1.000×10^3 Hz. Sonar waves travel at 5470 km/h. (a) What is the signal's frequency as detected by the U.S. sub? (b) What frequency is detected by the French sub in the signal reflected back to it by the U.S. sub? GO

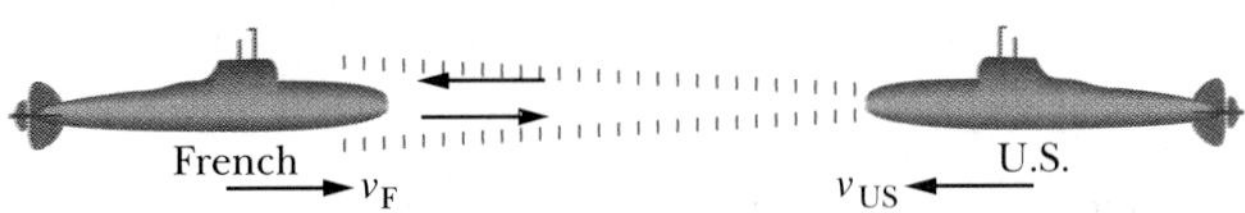

FIG. 17-42 Problem 61.

••62 A stationary detector measures the frequency of a sound source that first moves at constant velocity directly toward the detector and then (after passing the detector) directly away from it. The emitted frequency is f. During the approach the detected frequency is f'_{app} and during the recession it is f'_{rec}. If $(f'_{app} - f'_{rec})/f = 0.500$, what is the ratio v_s/v of the speed of the source to the speed of sound?

••63 A bat is flitting about in a cave, navigating via ultrasonic bleeps. Assume that the sound emission frequency of the bat is 39 000 Hz. During one fast swoop directly toward a flat wall surface, the bat is moving at 0.025 times the speed of sound in air. What frequency does the bat hear reflected off the wall?

••64 Figure 17-43 shows four tubes with lengths 1.0 m or 2.0 m, with one or two open ends as drawn. The third harmonic is set up in each tube, and some of the sound that escapes from them is detected by detector D, which moves directly away from the tubes. In terms of the speed of sound v, what speed must the detector have such that the detected frequency of the sound from (a) tube 1, (b) tube 2, (c) tube 3, and (d) tube 4 is equal to the tube's fundamental frequency?

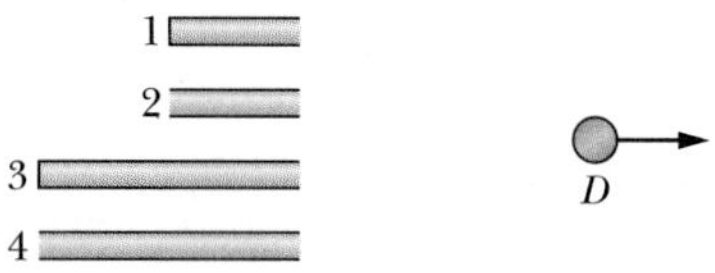

1 2 3 4 D

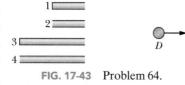

FIG. 17-43 Problem 64.

•••65 A girl is sitting near the open window of a train that is moving at a velocity of 10.00 m/s to the east. The girl's uncle stands near the tracks and watches the train move away. The locomotive whistle emits sound at frequency 500.0 Hz. The air is still. (a) What frequency does the uncle hear? (b) What frequency does the girl hear? A wind begins to blow from the east at 10.00 m/s. (c) What frequency does the uncle now hear? (d) What frequency does the girl now hear? SSM WWW

•••66 Two trains are traveling toward each other at 30.5 m/s relative to the ground. One train is blowing a whistle at 500 Hz. (a) What frequency is heard on the other train in still air? (b) What frequency is heard on the other train if the wind is blowing at 30.5 m/s toward the whistle and away from the listener? (c) What frequency is heard if the wind direction is reversed?

•••67 A 2000 Hz siren and a civil defense official are both at rest with respect to the ground. What frequency does the official hear if the wind is blowing at 12 m/s (a) from source to official and (b) from official to source? GO

sec. 17-10 Supersonic Speeds, Shock Waves

•68 The shock wave off the cockpit of the FA 18 in Fig. 17-24 has an angle of about 60°. The airplane was traveling at about 1350 km/h when the photograph was taken. Approximately what was the speed of sound at the airplane's altitude?

••69 A jet plane passes over you at a height of 5000 m and a speed of Mach 1.5. (a) Find the Mach cone angle (the sound speed is 331 m/s). (b) How long after the jet passes directly overhead does the shock wave reach you? SSM

••70 A plane flies at 1.25 times the speed of sound. Its sonic boom reaches a man on the ground 1.00 min after the plane passes directly overhead. What is the altitude of the plane? Assume the speed of sound to be 330 m/s.

Additional Problems

71 In Fig. 17-44, sound of wavelength 0.850 m is emitted isotropically by point source S. Sound ray 1 extends directly to detector D, at distance $L = 10.0$ m. Sound ray 2 extends to D via a reflection (effectively, a "bouncing") of the sound at a flat surface. That reflection occurs on a perpendicular bisector to the SD line, at distance d from the line. Assume that the reflection shifts the sound wave by 0.500λ. For what least value of d (other than zero) do the direct sound and the reflected sound arrive at D (a) exactly out of phase and (b) exactly in phase?

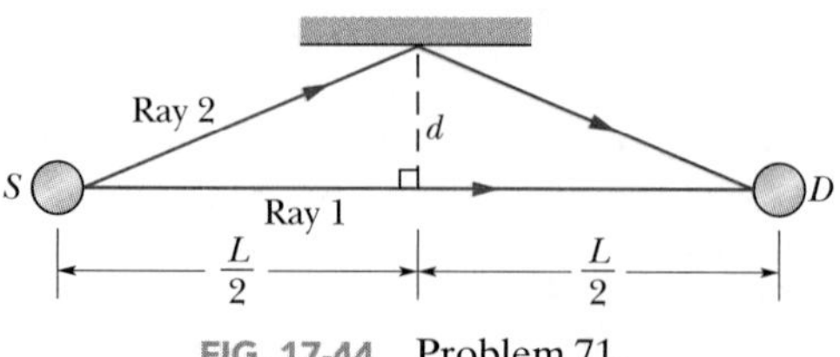

FIG. 17-44 Problem 71.

72 A detector initially moves at constant velocity directly toward a stationary sound source and then (after passing it) directly from it. The emitted frequency is f. During the approach the detected frequency is f'_{app} and during the recession it is f'_{rec}. If the frequencies are related by $(f'_{app} - f'_{rec})/f = 0.500$, what is the ratio v_D/v of the speed of the detector to the speed of sound?

73 Two sound waves with an amplitude of 12 nm and a wavelength of 35 cm travel in the same direction through a long tube, with a phase difference of $\pi/3$ rad. What are the (a) amplitude and (b) wavelength of the net sound wave produced by their interference? If, instead, the sound waves travel through the tube in opposite directions, what are the (c) amplitude and (d) wavelength of the net wave?

74 A sinusoidal sound wave moves at 343 m/s through air in the positive direction of an x axis. At one instant, air molecule A is at its maximum displacement in the negative direction of the axis while air molecule B is at its equilibrium position. The separation between those molecules is 15.0 cm, and the molecules between A and B have intermediate displacements in the negative direction of the axis. (a) What is the frequency of the sound wave?

In a similar arrangement, for a different sinusoidal sound wave, air molecule C is at its maximum displacement in the positive direction while molecule D is at its maximum displacement in the negative direction. The separation between the molecules is again 15.0 cm, and the molecules between C and D have intermediate displacements. (b) What is the frequency of the sound wave?

75 In Fig. 17-45, sound waves A and B, both of wavelength λ, are initially in phase and traveling rightward, as indicated by the two rays. Wave A is reflected from four surfaces but ends up traveling in its original direction. What multiple of wavelength λ is the smallest value of distance L in the figure that puts A and B exactly out of phase with each other after the reflections?

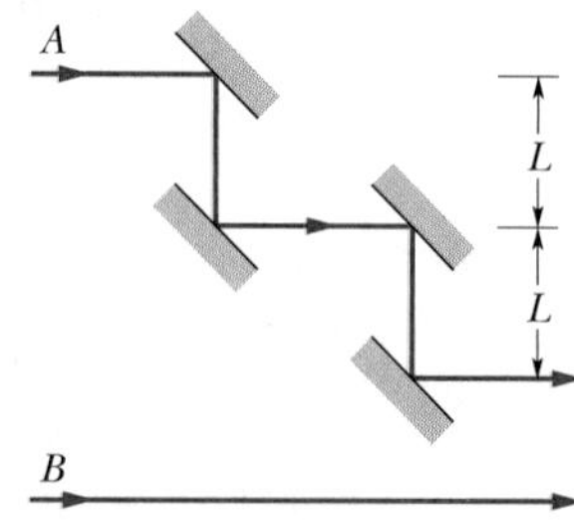

FIG. 17-45 Problem 75.

76 A trumpet player on a moving railroad flatcar moves toward a second trumpet player standing alongside the track while both play a 440 Hz note. The sound waves heard by a stationary observer between the two players have a beat frequency of 4.0 beats/s. What is the flatcar's speed?

77 A siren emitting a sound of frequency 1000 Hz moves away from you toward the face of a cliff at a speed of 10 m/s. Take the speed of sound in air as 330 m/s. (a) What is the frequency of the sound you hear coming directly from the siren? (b) What is the frequency of the sound you hear reflected off the cliff? (c) What is the beat frequency between the two sounds? Is it perceptible (less than 20 Hz)? SSM

78 A sound source moves along an x axis, between detectors A and B. The wavelength of the sound detected at A is 0.500 that of the sound detected at B. What is the ratio v_s/v of the speed of the source to the speed of sound?

79 A certain loudspeaker system emits sound isotropically with a frequency of 2000 Hz and an intensity of 0.960 mW/m^2 at a distance of 6.10 m. Assume that there are no reflections. (a) What is the intensity at 30.0 m? At 6.10 m, what are (b) the displacement amplitude and (c) the pressure amplitude?

80 At a certain point, two waves produce pressure variations given by $\Delta p_1 = \Delta p_m \sin \omega t$ and $\Delta p_2 = \Delta p_m \sin(\omega t - \phi)$. At this point, what is the ratio $\Delta p_r/\Delta p_m$, where Δp_r is the pressure amplitude of the resultant wave, if ϕ is (a) 0, (b) $\pi/2$, (c) $\pi/3$, and (d) $\pi/4$?

81 The sound intensity is 0.0080 W/m^2 at a distance of 10 m from an isotropic point source of sound. (a) What is the power of the source? (b) What is the sound intensity 5.0 m from the source? (c) What is the sound level 10 m from the source? SSM

82 The average density of Earth's crust 10 km beneath the continents is 2.7 g/cm^3. The speed of longitudinal seismic waves at that depth, found by timing their arrival from distant earthquakes, is 5.4 km/s. Use this information to find the bulk modulus of Earth's crust at that depth. For comparison, the bulk modulus of steel is about 16×10^{10} Pa.

83 Two identical tuning forks can oscillate at 440 Hz. A person is located somewhere on the line between them. Calculate the beat frequency as measured by this individual if (a) she is standing still and the tuning forks move in the same direction along the line at 3.00 m/s, and (b) the tuning forks are stationary and the listener moves along the line at 3.00 m/s.

84 You can estimate your distance from a lightning stroke by counting the seconds between the flash you see and the thunder you later hear. By what integer should you divide the number of seconds to get the distance in kilometers?

85 (a) If two sound waves, one in air and one in (fresh) water, are equal in intensity and angular frequency, what is the ratio of the pressure amplitude of the wave in water to that of the wave in air? Assume the water and the air are at 20°C. (See Table 14-1.) (b) If the pressure amplitudes are equal instead, what is the ratio of the intensities of the waves? SSM

86 Find the ratios (greater to smaller) of the (a) intensities, (b) pressure amplitudes, and (c) particle displacement amplitudes for two sounds whose sound levels differ by 37 dB.

87 Figure 17-46 shows an air-filled, acoustic interferometer, used to demonstrate the interference of sound waves. Sound source S is an oscillating diaphragm; D is a sound detector, such as the ear or a microphone. Path SBD can be varied in length, but path SAD is fixed. At D, the sound wave coming along path SBD interferes with that coming along path SAD. In one demonstration, the sound intensity at D has a minimum value of 100 units at one position of the movable arm and continuously climbs to a maximum value of 900 units when that arm is shifted by 1.65 cm. Find (a) the frequency of the sound emitted by the source and (b) the ratio of the amplitude at D of the SAD wave to that of the SBD wave. (c) How can it happen that these waves have different amplitudes, considering that they originate at the same source? SSM

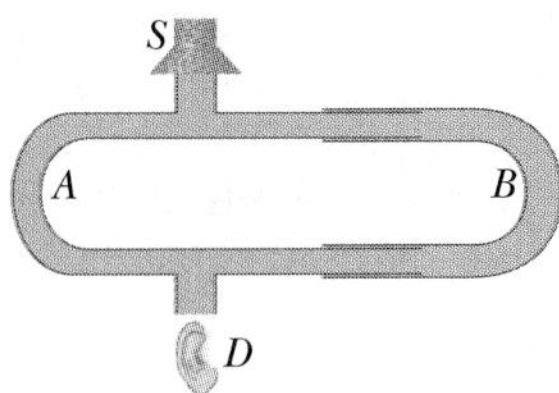

FIG. 17-46 Problem 87.

88 A bullet is fired with a speed of 685 m/s. Find the angle made by the shock cone with the line of motion of the bullet.

89 A sperm whale (Fig. 17-47a) vocalizes by producing a series of clicks. Actually, the whale makes only a single sound near the front of its head to start the series. Part of that sound then emerges from the head into the water to become the first click of the series. The rest of the sound travels backward through the spermaceti sac (a body of fat), reflects from the frontal sac (an air layer), and then travels forward through the spermaceti sac. When it reaches the distal sac (another air layer) at the front of the head, some of the sound escapes into the water to form the second click, and the rest is sent back through the spermaceti sac (and ends up forming later clicks).

Figure 17-47b shows a strip-chart recording of a series of clicks. A unit time interval of 1.0 ms is indicated on the chart. Assuming that the speed of sound in the spermaceti sac is 1372 m/s, find the length of the spermaceti sac. From such a calculation, marine scientists estimate the length of a whale from its click series.

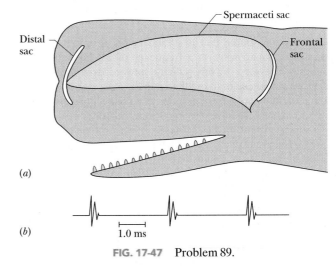

FIG. 17-47 Problem 89.

90 A continuous sinusoidal longitudinal wave is sent along a very long coiled spring from an attached oscillating source. The wave travels in the negative direction of an x axis; the source frequency is 25 Hz; at any instant the distance between successive points of maximum expansion in the spring is 24 cm; the maximum longitudinal displacement of a spring particle is 0.30 cm; and the particle at $x = 0$ has zero displacement at time $t = 0$. If the wave is written in the form $s(x, t) = s_m \cos(kx \pm \omega t)$, what are (a) s_m, (b) k, (c) ω, (d) the wave speed, and (e) the correct choice of sign in front of ω?

91 At a distance of 10 km, a 100 Hz horn, assumed to be an isotropic point source, is barely audible. At what distance would it begin to cause pain?

92 The speed of sound in a certain metal is v_m. One end of a long pipe of that metal of length L is struck a hard blow. A listener at the other end hears two sounds, one from the wave that travels along the pipe's metal wall and the other from the wave that travels through the air inside the pipe. (a) If v is the speed of sound in air, what is the time interval Δt between the arrivals of the two sounds at the listener's ear? (b) If $\Delta t = 1.00$ s and the metal is steel, what is the length L?

93 A pipe 0.60 m long and closed at one end is filled with an unknown gas. The third lowest harmonic frequency for the pipe is 750 Hz. (a) What is the speed of sound in the unknown gas? (b) What is the fundamental frequency for this pipe when it is filled with the unknown gas?

94 Four sound waves are to be sent through the same tube of air, in the same direction:

$$s_1(x, t) = (9.00 \text{ nm}) \cos(2\pi x - 700\pi t)$$
$$s_2(x, t) = (9.00 \text{ nm}) \cos(2\pi x - 700\pi t + 0.7\pi)$$
$$s_3(x, t) = (9.00 \text{ nm}) \cos(2\pi x - 700\pi t + \pi)$$
$$s_4(x, t) = (9.00 \text{ nm}) \cos(2\pi x - 700\pi t + 1.7\pi).$$

What is the amplitude of the resultant wave? (*Hint:* Use a phasor diagram to simplify the problem.)

95 Straight line AB connects two point sources that are 5.00 m apart, emit 300 Hz sound waves of the same amplitude, and emit exactly out of phase. (a) What is the shortest distance between the midpoint of AB and a point on AB where the interfering waves cause maximum oscillation of the air molecules? What are the (b) second and (c) third shortest distances?

96 A point source that is stationary on an x axis emits a sinusoidal sound wave at a frequency of 686 Hz and speed 343 m/s. The wave travels radially outward from the source, causing air molecules to oscillate radially inward and outward. Let us define a wavefront as a line that connects points where the air molecules have the maximum, radially outward displacement. At any given instant, the wavefronts are concentric circles that are centered on the source. (a) Along x, what is the adjacent wavefront separation? Next, the source moves along x at a speed of 110 m/s. Along x, what are the wavefront separations (b) in front of and (c) behind the source?

97 You are standing at a distance D from an isotropic point source of sound. You walk 50.0 m toward the source and observe that the intensity of the sound has doubled. Calculate the distance D.

98 On July 10, 1996, a granite block broke away from a wall in Yosemite Valley and, as it began to slide down the wall, was launched into projectile motion. Seismic waves produced by its impact with the ground triggered seismographs as far away as 200 km. Later measurements indicated that the block had a mass between 7.3×10^7 kg and 1.7×10^8 kg and that it landed 500 m vertically below the launch point and 30 m horizontally from it. (The launch angle is not known.) (a) Estimate the block's kinetic energy just before it landed.

Consider two types of seismic waves that spread from the impact point—a hemispherical *body wave* traveled through the ground in an expanding hemisphere and a cylindrical *surface wave* traveled along the ground in an expanding shallow vertical cylinder (Fig. 17-48). Assume that the impact lasted 0.50 s, the vertical cylinder had a depth d of 5.0 m, and each wave type received 20% of the energy the block had just before impact. Neglecting any mechanical energy loss the waves experienced as they traveled, determine the intensities of (b) the body wave and (c) the surface wave when they reached a seismograph 200 km away. (d) On the basis of these results, which wave is more easily detected on a distant seismograph?

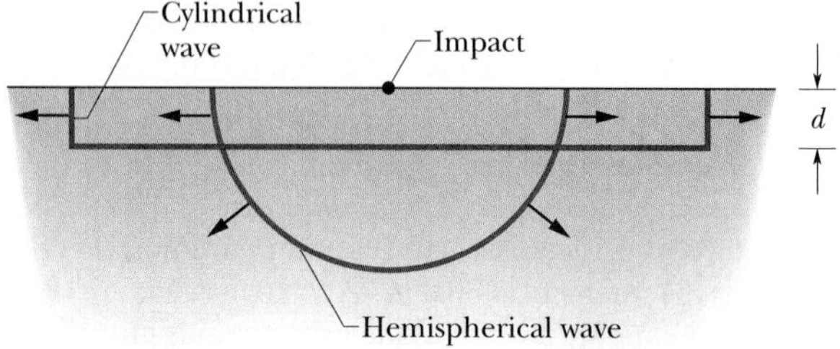

FIG. 17-48 Problem 98.

99 An avalanche of sand along some rare desert sand dunes can produce a booming that is loud enough to be heard 10 km away. The booming apparently results from a periodic oscillation of the sliding layer of sand—the layer's thickness expands and contracts. If the emitted frequency is 90 Hz, what are (a) the period of the thickness oscillation and (b) the wavelength of the sound?

100 Passengers in an auto traveling at 16.0 m/s toward the east hear a siren frequency of 950 Hz from an emergency vehicle approaching them from behind at a speed (relative to the air and ground) of 40.0 m/s. The speed of sound in air is 340 m/s. (a) What siren frequency does a passenger riding in the emergency vehicle hear? (b) What frequency do the passengers in the auto hear after the emergency vehicle passes them?

101 Ultrasound, which consists of sound waves with frequencies above the human audible range, can be used to produce an image of the interior of a human body. Moreover, ultrasound can be used to measure the speed of the blood in the body; it does so by comparing the frequency of the ultrasound sent into the body with the frequency of the ultrasound reflected back to the body's surface by the blood. As the blood pulses, this detected frequency varies.

Suppose that an ultrasound image of the arm of a patient shows an artery that is angled at $\theta = 20°$ to the ultrasound's line of travel (Fig. 17-49). Suppose also that the frequency of the ultrasound reflected by the blood in the artery is increased by a maximum of 5495 Hz from the original ultrasound frequency of 5.000 000 MHz. (a) In Fig. 17-49, is the direction of the blood flow rightward or leftward? (b) The speed of sound in the human arm is 1540 m/s. What is the maximum speed of the blood? (*Hint:* The Doppler effect is caused by the component of the blood's velocity along the ultrasound's direction of travel.) (c) If angle θ were greater, would the reflected frequency be greater or less? SSM

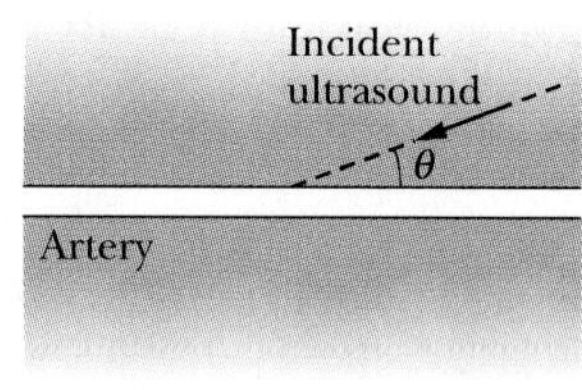

FIG. 17-49 Problem 101.

102 Pipe A has only one open end; pipe B is four times as long and has two open ends. Of the lowest 10 harmonic numbers n_B of pipe B, what are the (a) smallest, (b) second smallest, and (c) third smallest values at which a harmonic frequency of B matches one of the harmonic frequencies of A?

103 *Waterfall acoustics.* The turbulent impact of the water in a waterfall causes the surrounding ground to oscillate in a wide range of low frequencies. If the water falls freely (instead of hitting rock on the way down), the oscillations are greatest in amplitude at a particular frequency f_m. This fact suggests that acoustic resonance is involved and f_m is the fundamental frequency. The following table gives, for nine U.S. and Canadian waterfalls, measured values for f_m and for the length L of the water's free fall. Determine how to plot the data to get the speed of sound in the water of a waterfall. From the plot, find the speed of sound if waterfall resonance is effectively like that in a tube with (a) two open ends and (b) only one open end. The speed of sound in turbulent water filled with air bubbles can be about 25% less than the speed of 1400 m/s in still water. (c) From the answers to (a) and (b), determine how many open ends are effectively involved in waterfall resonance.

WATERFALL	1	2	3	4	5	6	7	8	9
f_m (Hz)	5.6	3.8	8.0	6.1	8.8	6.0	19	21	40
L (m)	97	71	53	49	35	24	13	11	8

104 A person on a railroad car blows a trumpet note at 440 Hz. The car is moving toward a wall at 20.0 m/s. Calculate the frequency of (a) the sound as received at the wall and (b) the reflected sound arriving back at the trumpeter.

105 A police car is chasing a speeding Porsche 911. Assume that the Porsche's maximum speed is 80.0 m/s and the police car's is 54.0 m/s. At the moment both cars reach their maximum speed, what frequency will the Porsche driver hear if the frequency of the police car's siren is 440 Hz? Take the speed of sound in air to be 340 m/s.

106 A sound wave travels out uniformly in all directions from a point source. (a) Justify the following expression for the displacement s of the transmitting medium at any distance r from the source:

$$s = \frac{b}{r}\sin k(r - vt),$$

where b is a constant. Consider the speed, direction of propagation, periodicity, and intensity of the wave. (b) What is the dimension of the constant b?

107 In Fig. 17-34, S_1 and S_2 are two isotropic point sources of sound. They emit waves in phase at wavelength 0.50 m; they are separated by $D = 1.60$ m. If we move a sound detector along a large circle centered at the midpoint between the sources, at how many points do waves arrive at the detector (a) exactly in phase and (b) exactly out of phase?

108 Suppose a spherical loudspeaker emits sound isotropically at 10 W into a room with completely absorbent walls, floor, and ceiling (an *anechoic chamber*). (a) What is the intensity of the sound at distance $d = 3.0$ m from the center of the source? (b) What is the ratio of the wave amplitude at $d = 4.0$ m to that at $d = 3.0$ m?

109 To search for a fossilized dinosaur embedded in rock, paleontologists can use sound waves to produce a computer image of the dinosaur. The image then guides the paleontologists as they dig the dinosaur out of the rock. (The technique is shown in the opening scenes of the movie *Jurassic Park.*) The basic idea of the detection technique is that a strong pulse of sound is emitted by a source (a seismic gun) at ground level and then detected by hydrophones that lie at evenly spaced depths in a bore hole drilled into the ground. The source and one hydrophone are shown in Fig. 17-50.

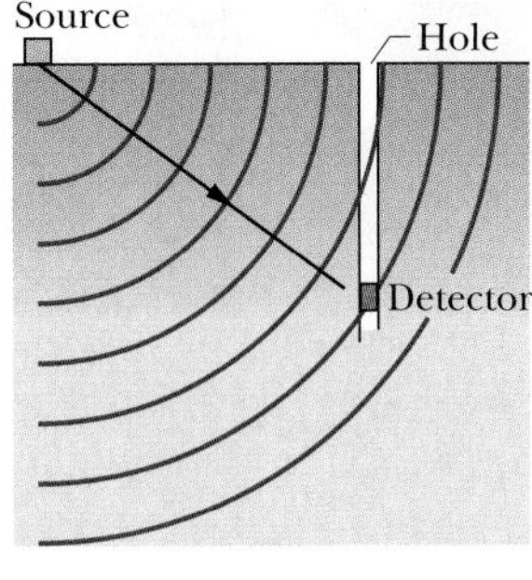

FIG. 17-50 Problem 109.

If the sound wave travels from the source to the hydrophone through only rock as in Fig. 17-50, it travels at a known speed V and takes a certain time T. If, instead, it travels through a fossilized bone along the way, it takes slightly more time because it travels more slowly in the bone than in the rock. By measuring the difference Δt between the expected and measured travel times, the distance d traveled in the bone can be determined. After this procedure is repeated for many locations of the source and hydrophones, a computer can transform the many computed distances d into an image of the fossil.

(a) Let the speed of sound through fossilized bone be $V - \Delta V$, where ΔV is small relative to V. Show that the distance d is given by

$$d \approx \frac{V^2\,\Delta t}{\Delta V}.$$

(b) For $V = 5000$ m/s and $\Delta V = 200$ m/s, what typical value of Δt can be expected if the sound passes along the diameter of a leg bone of an adult *T. rex*? (Estimate the bone's diameter.)

110 The period of a pulsating variable star may be estimated by considering the star to be executing *radial* longitudinal pulsations in the fundamental standing wave mode; that is, the star's radius varies periodically with time, with a displacement antinode at the star's surface. (a) Would you expect the center of the star to be a displacement node or antinode? (b) By analogy with a pipe with one open end, show that the period of pulsation T is given by $T = 4R/v$, where R is the equilibrium radius of the star and v is the average sound speed in the material of the star. (c) Typical white dwarf stars are composed of material with a bulk modulus of 1.33×10^{22} Pa and a density of 10^{10} kg/m^3. They have radii equal to 9.0×10^{-3} solar radius. What is the approximate pulsation period of a white dwarf?

111 A listener at rest (with respect to the air and the ground) hears a signal of frequency f_1 from a source moving toward him with a velocity of 15 m/s, due east. If the listener then moves toward the approaching source with a velocity of 25 m/s, due west, he hears a frequency f_2 that differs from f_1 by 37 Hz. What is the frequency of the source? (Take the speed of sound in air to be 340 m/s.)

112 A guitar player tunes the fundamental frequency of a guitar string to 440 Hz. (a) What will be the fundamental frequency if she then increases the tension in the string by 20%? (b) What will it be if, instead, she decreases the length along which the string oscillates by sliding her finger from the tuning key one-third of the way down the string toward the bridge at the lower end?

18 Temperature, Heat, and the First Law of Thermodynamics

Courtesy Nathan Schiff, Ph. D., USDA Forest Service, Center for Bottomland Hardwoods Research, Stoneville, MS

The fairly small Melanophila *beetles are known for a bizarre behavior: They fly toward forest fires and mate near them, and then the females fly into the still smoldering ruins to lay their eggs under burnt bark. This is the ideal environment for the larvae that hatch from the eggs, because the tree can no longer protect itself from the larvae by rosin or chemical means. If a beetle were at the periphery of a fire, detecting the fire would be easy, of course. However, these beetles can detect a fairly large fire from as far away as 12 km. They do this without seeing or smelling the fire.*

How can the beetles detect a distant fire?

The answer is in this chapter.

18-1 WHAT IS PHYSICS?

One of the principal branches of physics and engineering is **thermodynamics,** which is the study and application of the *thermal energy* (often called the *internal energy*) of systems. One of the central concepts of thermodynamics is *temperature,* which we begin to explore in the next section. Since childhood, you have been developing a working knowledge of thermal energy and temperature. For example, you know to be cautious with hot foods and hot stoves and to store perishable foods in cool or cold compartments. You also know how to control the temperature inside home and car, and how to protect yourself from wind chill and heat stroke.

Examples of how thermodynamics figures into everyday engineering and science are countless. Automobile engineers are concerned with the heating of a car engine, such as during a NASCAR race. Food engineers are concerned both with the proper heating of foods, such as pizzas being microwaved, and with the proper cooling of foods, such as TV dinners being quickly frozen at a processing plant. Geologists are concerned with the transfer of thermal energy in an El Niño event and in the gradual warming of ice expanses in the Arctic and Antarctic. Agricultural engineers are concerned with the weather conditions that determine whether the agriculture of a country thrives or vanishes. Medical engineers are concerned with how a patient's temperature might distinguish between a benign viral infection and a cancerous growth.

The starting point in our discussion of thermodynamics is the concept of temperature and how it is measured.

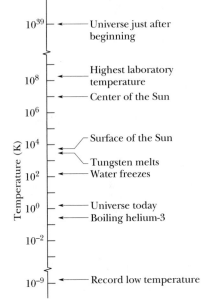

FIG. 18-1 Some temperatures on the Kelvin scale. Temperature $T = 0$ corresponds to $10^{-\infty}$ and cannot be plotted on this logarithmic scale.

18-2 | Temperature

Temperature is one of the seven SI base quantities. Physicists measure temperature on the **Kelvin scale,** which is marked in units called *kelvins.* Although the temperature of a body apparently has no upper limit, it does have a lower limit; this limiting low temperature is taken as the zero of the Kelvin temperature scale. Room temperature is about 290 kelvins, or 290 K as we write it, above this *absolute zero.* Figure 18-1 shows a wide range of temperatures.

When the universe began 13.7 billion years ago, its temperature was about 10^{39} K. As the universe expanded it cooled, and it has now reached an average temperature of about 3 K. We on Earth are a little warmer than that because we happen to live near a star. Without our Sun, we too would be at 3 K (or, rather, we could not exist).

18-3 | The Zeroth Law of Thermodynamics

The properties of many bodies change as we alter their temperature, perhaps by moving them from a refrigerator to a warm oven. To give a few examples: As their temperature increases, the volume of a liquid increases, a metal rod grows a little longer, and the electrical resistance of a wire increases, as does the pressure exerted by a confined gas. We can use any one of these properties as the basis of an instrument that will help us pin down the concept of temperature.

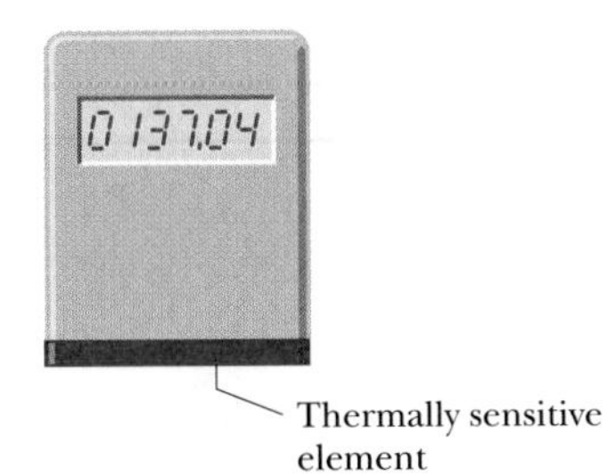

FIG. 18-2 A thermoscope. The numbers increase when the device is heated and decrease when it is cooled. The thermally sensitive element could be—among many possibilities—a coil of wire whose electrical resistance is measured and displayed.

Figure 18-2 shows such an instrument. Any resourceful engineer could design and construct it, using any one of the properties listed above. The instrument is fitted with a digital readout display and has the following properties: If you heat it (say, with a Bunsen burner), the displayed number starts to increase; if you then put it into a refrigerator, the displayed number starts to decrease. The instrument is not calibrated in any way, and the numbers have (as yet) no physical meaning. The device is a *thermoscope* but not (as yet) a *thermometer.*

Suppose that, as in Fig. 18-3*a*, we put the thermoscope (which we shall call body T) into intimate contact with another body (body A). The entire system is confined within a thick-walled insulating box. The numbers displayed by the

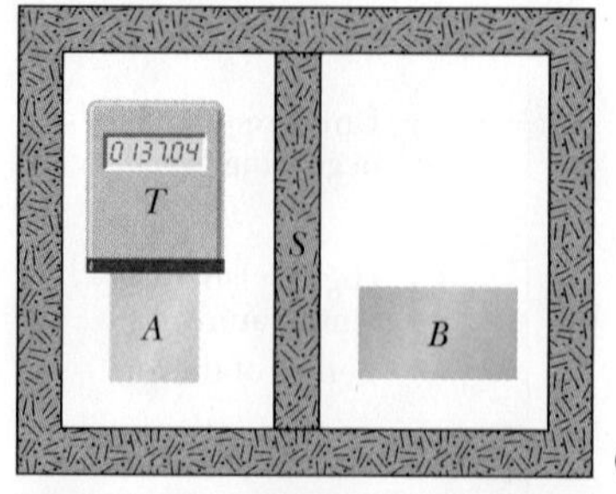

(a)

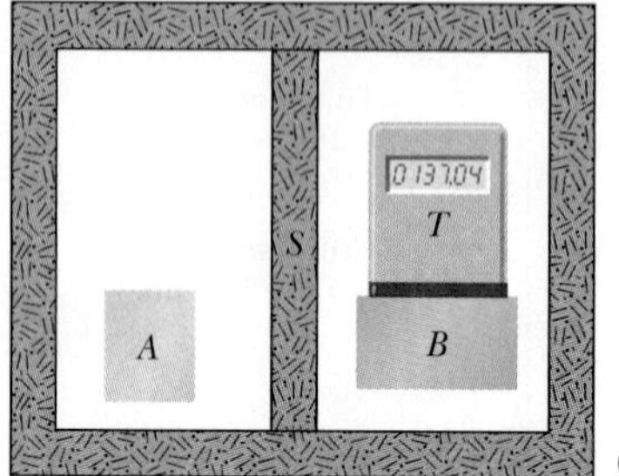

(b)

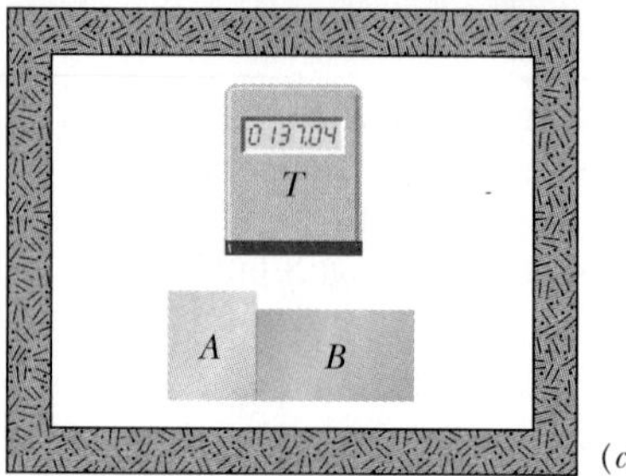

(c)

FIG. 18-3 (*a*) Body *T* (a thermoscope) and body *A* are in thermal equilibrium. (Body *S* is a thermally insulating screen.) (*b*) Body *T* and body *B* are also in thermal equilibrium, at the same reading of the thermoscope. (*c*) If (*a*) and (*b*) are true, the zeroth law of thermodynamics states that body *A* and body *B* are also in thermal equilibrium.

thermoscope roll by until, eventually, they come to rest (let us say the reading is "137.04") and no further change takes place. In fact, we suppose that every measurable property of body *T* and of body *A* has assumed a stable, unchanging value. Then we say that the two bodies are in *thermal equilibrium* with each other. Even though the displayed readings for body *T* have not been calibrated, we conclude that bodies *T* and *A* must be at the same (unknown) temperature.

Suppose that we next put body *T* into intimate contact with body *B* (Fig. 18-3*b*) and find that the two bodies come to thermal equilibrium *at the same reading of the thermoscope.* Then bodies *T* and *B* must be at the same (still unknown) temperature. If we now put bodies *A* and *B* into intimate contact (Fig. 18-3*c*), are they immediately in thermal equilibrium with each other? Experimentally, we find that they are.

The experimental fact shown in Fig. 18-3 is summed up in the **zeroth law of thermodynamics:**

> If bodies *A* and *B* are each in thermal equilibrium with a third body *T*, then *A* and *B* are in thermal equilibrium with each other.

In less formal language, the message of the zeroth law is: "Every body has a property called **temperature.** When two bodies are in thermal equilibrium, their temperatures are equal. And vice versa." We can now make our thermoscope (the third body *T*) into a thermometer, confident that its readings will have physical meaning. All we have to do is calibrate it.

We use the zeroth law constantly in the laboratory. If we want to know whether the liquids in two beakers are at the same temperature, we measure the temperature of each with a thermometer. We do not need to bring the two liquids into intimate contact and observe whether they are or are not in thermal equilibrium.

The zeroth law, which has been called a logical afterthought, came to light only in the 1930s, long after the first and second laws of thermodynamics had been discovered and numbered. Because the concept of temperature is fundamental to those two laws, the law that establishes temperature as a valid concept should have the lowest number—hence the zero.

18-4 | Measuring Temperature

Here we first define and measure temperatures on the Kelvin scale. Then we calibrate a thermoscope so as to make it a thermometer.

The Triple Point of Water

To set up a temperature scale, we pick some reproducible thermal phenomenon and, quite arbitrarily, assign a certain Kelvin temperature to its environment; that is, we select a *standard fixed point* and give it a standard fixed-point *temperature.* We could, for example, select the freezing point or the boiling point of water but, for technical reasons, we select instead the **triple point of water.**

Liquid water, solid ice, and water vapor (gaseous water) can coexist, in thermal equilibrium, at only one set of values of pressure and temperature. Figure 18-4 shows a triple-point cell, in which this so-called triple point of water can be achieved in the laboratory. By international agreement, the triple point of water has been assigned a value of 273.16 K as the standard fixed-point temperature for the calibration of thermometers; that is,

$$T_3 = 273.16 \text{ K} \quad \text{(triple-point temperature)}, \tag{18-1}$$

in which the subscript 3 means "triple point." This agreement also sets the size of the kelvin as 1/273.16 of the difference between absolute zero and the triple-point temperature of water.

Note that we do not use a degree mark in reporting Kelvin temperatures. It is 300 K (not 300°K), and it is read "300 kelvins" (not "300 degrees Kelvin"). The usual SI prefixes apply. Thus, 0.0035 K is 3.5 mK. No distinction in nomenclature is made between Kelvin temperatures and temperature differences, so we can write, "the boiling point of sulfur is 717.8 K" and "the temperature of this water bath was raised by 8.5 K."

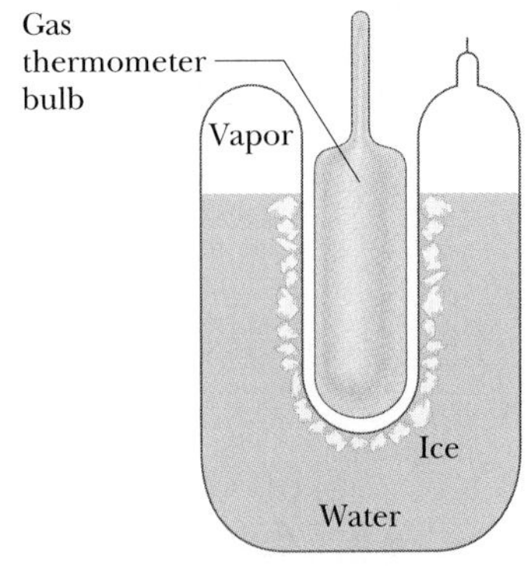

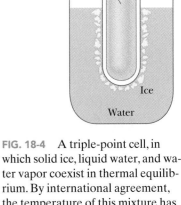

FIG. 18-4 A triple-point cell, in which solid ice, liquid water, and water vapor coexist in thermal equilibrium. By international agreement, the temperature of this mixture has been defined to be 273.16 K. The bulb of a constant-volume gas thermometer is shown inserted into the well of the cell.

The Constant-Volume Gas Thermometer

The standard thermometer, against which all other thermometers are calibrated, is based on the pressure of a gas in a fixed volume. Figure 18-5 shows such a **constant-volume gas thermometer;** it consists of a gas-filled bulb connected by a tube to a mercury manometer. By raising and lowering reservoir R, the mercury level in the left arm of the U-tube can always be brought to the zero of the scale to keep the gas volume constant (variations in the gas volume can affect temperature measurements).

The temperature of any body in thermal contact with the bulb (such as the liquid surrounding the bulb in Fig. 18-5) is then defined to be

$$T = Cp, \tag{18-2}$$

in which p is the pressure exerted by the gas and C is a constant. From Eq. 14-10, the pressure p is

$$p = p_0 - \rho gh, \tag{18-3}$$

in which p_0 is the atmospheric pressure, ρ is the density of the mercury in the manometer, and h is the measured difference between the mercury levels in the two arms of the tube.* (The minus sign is used in Eq. 18-3 because pressure p is measured *above* the level at which the pressure is p_0.)

If we next put the bulb in a triple-point cell (Fig. 18-4), the temperature now being measured is

$$T_3 = Cp_3, \tag{18-4}$$

in which p_3 is the gas pressure now. Eliminating C between Eqs. 18-2 and 18-4 gives us the temperature as

$$T = T_3\left(\frac{p}{p_3}\right) = (273.16\text{ K})\left(\frac{p}{p_3}\right) \quad \text{(provisional).} \tag{18-5}$$

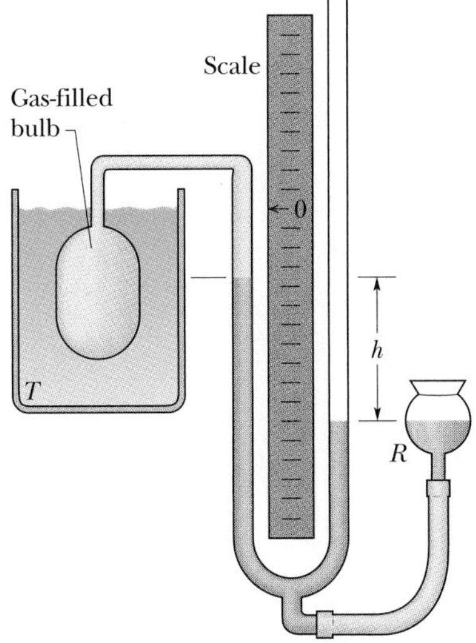

FIG. 18-5 A constant-volume gas thermometer, its bulb immersed in a liquid whose temperature T is to be measured.

We still have a problem with this thermometer. If we use it to measure, say, the boiling point of water, we find that different gases in the bulb give slightly different results. However, as we use smaller and smaller amounts of gas to fill the bulb, the readings converge nicely to a single temperature, no matter what gas we use. Figure 18-6 shows this convergence for three gases.

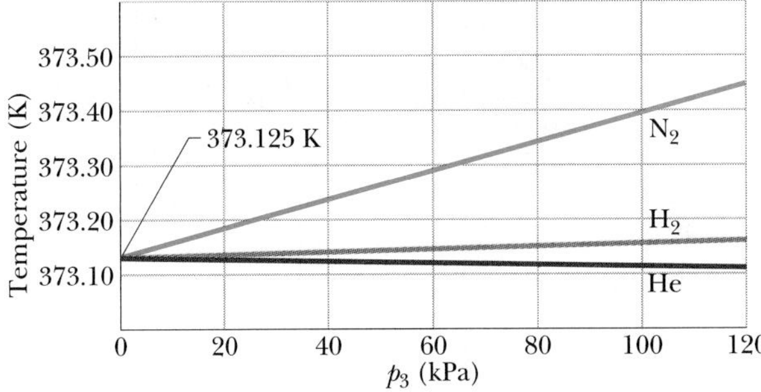

FIG. 18-6 Temperatures measured by a constant-volume gas thermometer, with its bulb immersed in boiling water. For temperature calculations using Eq. 18-5, pressure p_3 was measured at the triple point of water. Three different gases in the thermometer bulb gave generally different results at different gas pressures, but as the amount of gas was decreased (decreasing p_3), all three curves converged to 373.125 K.

*For pressure units, we shall use units introduced in Section 14-3. The SI unit for pressure is the newton per square meter, which is called the pascal (Pa). The pascal is related to other common pressure units by

$$1\text{ atm} = 1.01 \times 10^5\text{ Pa} = 760\text{ torr} = 14.7\text{ lb/in.}^2.$$

Thus the recipe for measuring a temperature with a gas thermometer is

$$T = (273.16\text{ K})\left(\lim_{\text{gas}\to 0} \frac{p}{p_3}\right). \tag{18-6}$$

The recipe instructs us to measure an unknown temperature T as follows: Fill the thermometer bulb with an arbitrary amount of *any* gas (for example, nitrogen) and measure p_3 (using a triple-point cell) and p, the gas pressure at the temperature being measured. (Keep the gas volume the same.) Calculate the ratio p/p_3. Then repeat both measurements with a smaller amount of gas in the bulb, and again calculate this ratio. Continue this way, using smaller and smaller amounts of gas, until you can extrapolate to the ratio p/p_3 that you would find if there were approximately no gas in the bulb. Calculate the temperature T by substituting that extrapolated ratio into Eq. 18-6. (The temperature is called the *ideal gas temperature.*)

18-5 | The Celsius and Fahrenheit Scales

So far, we have discussed only the Kelvin scale, used in basic scientific work. In nearly all countries of the world, the Celsius scale (formerly called the centigrade scale) is the scale of choice for popular and commercial use and much scientific use. Celsius temperatures are measured in degrees, and the Celsius degree has the same size as the kelvin. However, the zero of the Celsius scale is shifted to a more convenient value than absolute zero. If T_C represents a Celsius temperature and T a Kelvin temperature, then

$$T_C = T - 273.15°. \tag{18-7}$$

In expressing temperatures on the Celsius scale, the degree symbol is commonly used. Thus, we write 20.00°C for a Celsius reading but 293.15 K for a Kelvin reading.

The Fahrenheit scale, used in the United States, employs a smaller degree than the Celsius scale and a different zero of temperature. You can easily verify both these differences by examining an ordinary room thermometer on which both scales are marked. The relation between the Celsius and Fahrenheit scales is

$$T_F = \tfrac{9}{5}T_C + 32°, \tag{18-8}$$

where T_F is Fahrenheit temperature. Converting between these two scales can be done easily by remembering a few corresponding points, such as the freezing and boiling points of water (Table 18-1). Figure 18-7 compares the Kelvin, Celsius, and Fahrenheit scales.

We use the letters C and F to distinguish measurements and degrees on the two scales. Thus,

$$0°\text{C} = 32°\text{F}$$

TABLE 18-1

Some Corresponding Temperatures

Temperature	°C	°F
Boiling point of water[a]	100	212
Normal body temperature	37.0	98.6
Accepted comfort level	20	68
Freezing point of water[a]	0	32
Zero of Fahrenheit scale	≈ −18	0
Scales coincide	−40	−40

[a]Strictly, the boiling point of water on the Celsius scale is 99.975°C, and the freezing point is 0.00°C. Thus, there is slightly less than 100 C° between those two points.

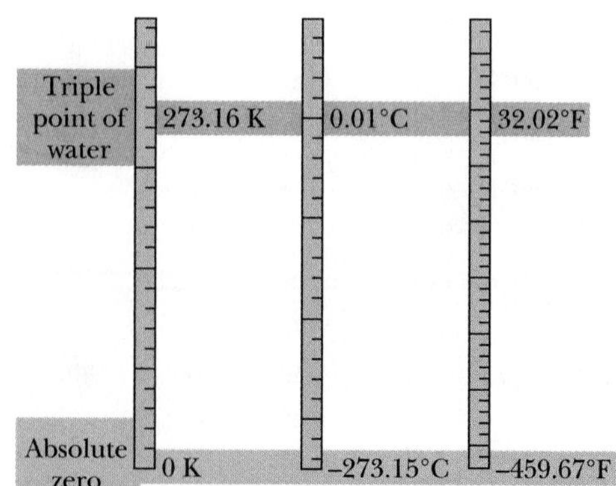

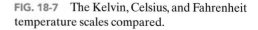

FIG. 18-7 The Kelvin, Celsius, and Fahrenheit temperature scales compared.

means that 0° on the Celsius scale measures the same temperature as 32° on the Fahrenheit scale, whereas

$$5\ \text{C}° = 9\ \text{F}°$$

means that a temperature difference of 5 Celsius degrees (note the degree symbol appears *after* C) is equivalent to a temperature difference of 9 Fahrenheit degrees.

CHECKPOINT 1 The figure here shows three linear temperature scales with the freezing and boiling points of water indicated. (a) Rank the degrees on these scales by size, greatest first. (b) Rank the following temperatures, highest first: 50°X, 50°W, and 50°Y.

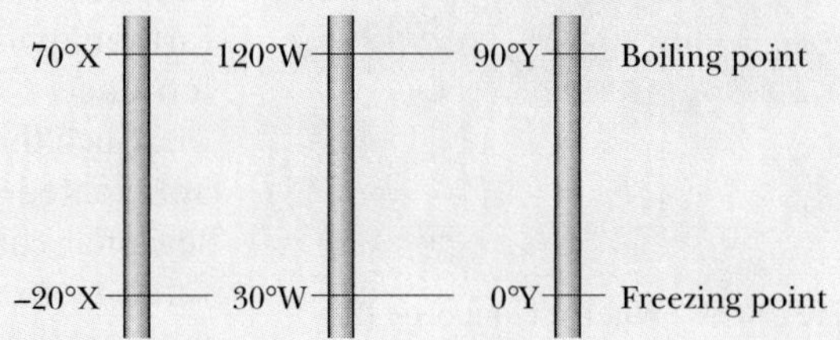

Sample Problem 18-1

Suppose you come across old scientific notes that describe a temperature scale called Z on which the boiling point of water is 65.0°Z and the freezing point is −14.0°Z. To what temperature on the Fahrenheit scale would a temperature of $T = -98.0$°Z correspond? Assume that the Z scale is linear; that is, the size of a Z degree is the same everywhere on the Z scale.

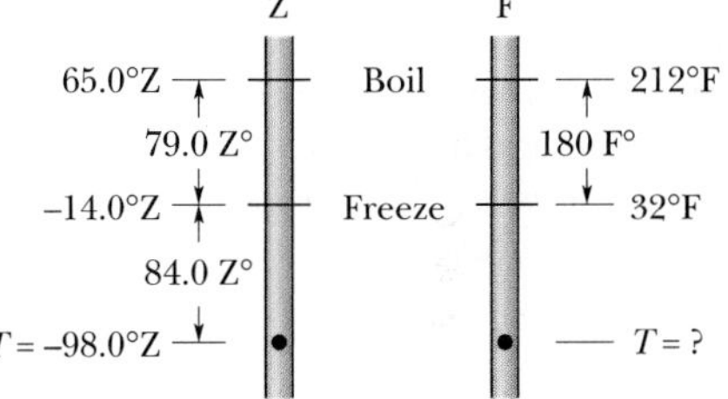

FIG. 18-8 An unknown temperature scale compared with the Fahrenheit temperature scale.

KEY IDEA A conversion factor between two (linear) temperature scales can be calculated by using two known (benchmark) temperatures, such as the boiling and freezing points of water. The number of degrees between the known temperatures on one scale is equivalent to the number of degrees between them on the other scale.

Calculations: We begin by relating the given temperature T to *either* known temperature on the Z scale. Since $T = -98.0$°Z is closer to the freezing point (−14.0°Z) than to the boiling point (65.0°Z), we use the freezing point. Then we note that T is *below this point* by −14.0°Z − (−98.0°Z) = 84.0 Z° (Fig. 18-8). (Read this difference as "84.0 Z degrees.")

Next, we set up a conversion factor between the Z and Fahrenheit scales to convert this difference. To do so, we use *both* known temperatures on the Z scale and the corresponding temperatures on the Fahrenheit scale. On the Z scale, the difference between the boiling and freezing points is 65.0°Z − (−14.0°Z) = 79.0 Z°. On the Fahrenheit scale, it is 212°F − 32.0°F = 180 F°. Thus, a temperature difference of 79.0 Z° is equivalent to a temperature difference of 180 F° (Fig. 18-8), and we can use the ratio (180 F°)/(79.0 Z°) as our conversion factor.

Now, since T is below the freezing point by 84.0 Z°, it must also be below the freezing point by

$$(84.0\ \text{Z}°)\frac{180\ \text{F}°}{79.0\ \text{Z}°} = 191\ \text{F}°.$$

Because the freezing point is at 32.0°F, this means that

$$T = 32.0°\text{F} - 191\ \text{F}° = -159°\text{F}. \quad \text{(Answer)}$$

PROBLEM-SOLVING TACTICS

Tactic 1: ***Temperature Changes*** Between the boiling and freezing points of water, there are (approximately) 100 kelvins and 100 Celsius degrees. Thus, a kelvin is the same size as a Celsius degree. From this or from Eq. 18-7, we then know that any temperature change is the same number whether expressed in kelvins or Celsius degrees. For example, a temperature change of 10 K is exactly equivalent to a temperature change of 10 C°.

Between the boiling and freezing points of water, there are 180 Fahrenheit degrees. Thus, 180 F° = 100 K, and a Fahrenheit degree must be 100/180, or 5/9, the size of a kelvin or Celsius degree. From this or from Eq. 18-8, we then know that any temperature change expressed in Fahrenheit degrees must be $\frac{9}{5}$ times that same temperature change expressed in either kelvins or Celsius degrees. For example, in Fahrenheit degrees, a temperature change of 10 K is (9/5)(10 K), or 18 F°.

Take care not to confuse a *temperature* with a temperature *change* or *difference*. A temperature of 10 K is certainly not the same as one of 10°C or 18°F but, as noted above, a temperature *change* of 10 K is the same as one of 10 C° or 18 F°. This distinction is very important in an equation containing a temperature T instead of a temperature change or difference such as $T_2 - T_1$: A temperature T by itself should generally be in kelvins and not degrees Celsius or Fahrenheit. In short, beware the "bare T."

FIG. 18-9 When a Concorde flew faster than the speed of sound, thermal expansion due to the rubbing by passing air increased the aircraft's length by about 12.5 cm. (The temperature increased to about 128°C at the aircraft nose and about 90°C at the tail, and cabin windows were noticeably warm to the touch.) *(Hugh Thomas/BWP Media/Getty Images News and Sport Services)*

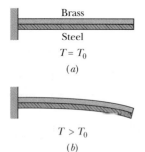

18-6 | Thermal Expansion

You can often loosen a tight metal jar lid by holding it under a stream of hot water. Both the metal of the lid and the glass of the jar expand as the hot water adds energy to their atoms. (With the added energy, the atoms can move a bit farther from one another than usual, against the spring-like interatomic forces that hold every solid together.) However, because the atoms in the metal move farther apart than those in the glass, the lid expands more than the jar and thus is loosened.

Such **thermal expansion** of materials with an increase in temperature must be anticipated in many common situations. When a bridge is subject to large seasonal changes in temperature, for example, sections of the bridge are separated by *expansion slots* so that the sections have room to expand on hot days without the bridge buckling. When a dental cavity is filled, the filling material must have the same thermal expansion properties as the surrounding tooth; otherwise, consuming cold ice cream and then hot coffee would be very painful. When the Concorde aircraft (Fig. 18-9) was built, the design had to allow for the thermal expansion of the fuselage during supersonic flight because of frictional heating by the passing air.

The thermal expansion properties of some materials can be put to common use. Thermometers and thermostats may be based on the differences in expansion between the components of a *bimetal strip* (Fig. 18-10). Also, the familiar liquid-in-glass thermometers are based on the fact that liquids such as mercury and alcohol expand to a different (greater) extent than their glass containers.

Linear Expansion

If the temperature of a metal rod of length L is raised by an amount ΔT, its length is found to increase by an amount

$$\Delta L = L\alpha\,\Delta T, \tag{18-9}$$

in which α is a constant called the **coefficient of linear expansion.** The coefficient α has the unit "per degree" or "per kelvin" and depends on the material. Although α varies somewhat with temperature, for most practical purposes it can be taken as constant for a particular material. Table 18-2 shows some coefficients of linear expansion. Note that the unit C° there could be replaced with the unit K.

The thermal expansion of a solid is like photographic enlargement except it is in three dimensions. Figure 18-11*b* shows the (exaggerated) thermal expansion of a steel ruler. Equation 18-9 applies to every linear dimension of the ruler, including its edge, thickness, diagonals, and the diameters of the circle etched on it and the circular hole cut in it. If the disk cut from that hole originally fits snugly in the hole, it will continue to fit snugly if it undergoes the same temperature increase as the ruler.

Brass

Steel

$T = T_0$

(*a*)

$T > T_0$

(*b*)

FIG. 18-10 (*a*) A bimetal strip, consisting of a strip of brass and a strip of steel welded together, at temperature T_0. (*b*) The strip bends as shown at temperatures above this reference temperature. Below the reference temperature the strip bends the other way. Many thermostats operate on this principle, making and breaking an electrical contact as the temperature rises and falls.

Volume Expansion

If all dimensions of a solid expand with temperature, the volume of that solid must also expand. For liquids, volume expansion is the only meaningful expansion parameter. If the temperature of a solid or liquid whose volume is V is increased by an amount ΔT, the increase in volume is found to be

$$\Delta V = V\beta\,\Delta T, \tag{18-10}$$

where β is the **coefficient of volume expansion** of the solid or liquid. The coefficients of volume expansion and linear expansion for a solid are related by

$$\beta = 3\alpha. \tag{18-11}$$

TABLE 18-2

Some Coefficients of Linear Expansion[a]

Substance	α (10^{-6}/C°)	Substance	α (10^{-6}/C°)
Ice (at 0°C)	51	Steel	11
Lead	29	Glass (ordinary)	9
Aluminum	23	Glass (Pyrex)	3.2
Brass	19	Diamond	1.2
Copper	17	Invar[b]	0.7
Concrete	12	Fused quartz	0.5

[a]Room temperature values except for the listing for ice.

[b]This alloy was designed to have a low coefficient of expansion. The word is a shortened form of "invariable."

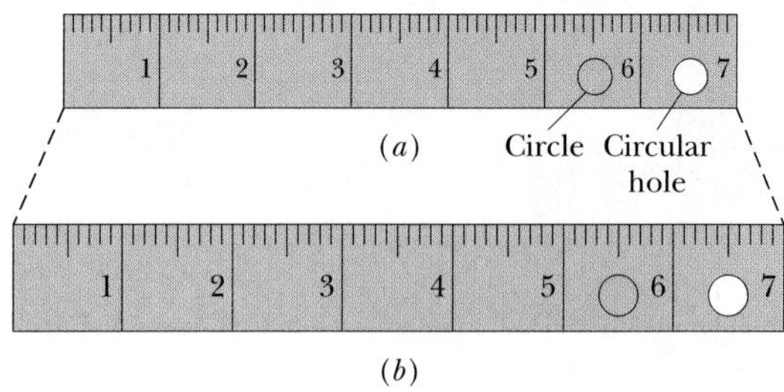

FIG. 18-11 The same steel ruler at two different temperatures. When it expands, the scale, the numbers, the thickness, and the diameters of the circle and circular hole are all increased by the same factor. (The expansion has been exaggerated for clarity.)

The most common liquid, water, does not behave like other liquids. Above about 4°C, water expands as the temperature rises, as we would expect. Between 0 and about 4°C, however, water *contracts* with increasing temperature. Thus, at about 4°C, the density of water passes through a maximum. At all other temperatures, the density of water is less than this maximum value.

This behavior of water is the reason lakes freeze from the top down rather than from the bottom up. As water on the surface is cooled from, say, 10°C toward the freezing point, it becomes denser ("heavier") than lower water and sinks to the bottom. Below 4°C, however, further cooling makes the water then on the surface *less* dense ("lighter") than the lower water, so it stays on the surface until it freezes. Thus the surface freezes while the lower water is still liquid. If lakes froze from the bottom up, the ice so formed would tend not to melt completely during the summer, because it would be insulated by the water above. After a few years, many bodies of open water in the temperate zones of Earth would be frozen solid all year round—and aquatic life could not exist.

CHECKPOINT 2 The figure here shows four rectangular metal plates, with sides of L, $2L$, or $3L$. They are all made of the same material, and their temperature is to be increased by the same amount. Rank the plates according to the expected increase in (a) their vertical heights and (b) their areas, greatest first.

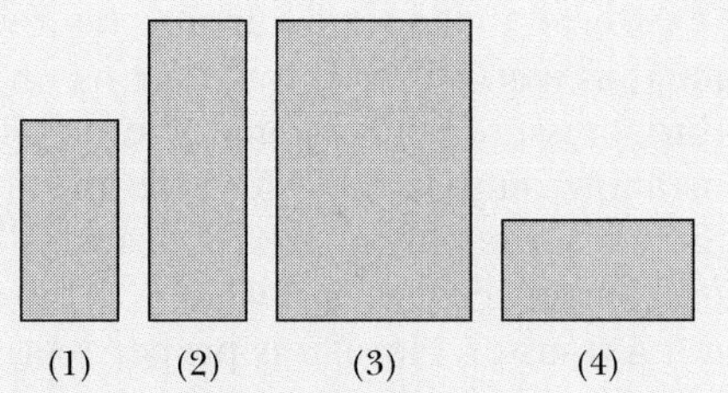

Sample Problem 18-2

On a hot day in Las Vegas, an oil trucker loaded 37 000 L of diesel fuel. He encountered cold weather on the way to Payson, Utah, where the temperature was 23.0 K lower than in Las Vegas, and where he delivered his entire load. How many liters did he deliver? The coefficient of volume expansion for diesel fuel is 9.50×10^{-4}/C°, and the coefficient of linear expansion for his steel truck tank is 11×10^{-6}/C°.

KEY IDEA The volume of the diesel fuel depends directly on the temperature. Thus, because the temperature decreased, the volume of the fuel did also, as given by Eq. 18-10 ($\Delta V = V\beta\,\Delta T$).

Calculations: We find

$$\Delta V = (37\,000 \text{ L})(9.50 \times 10^{-4}/\text{C°})(-23.0 \text{ K}) = -808 \text{ L}.$$

Thus, the amount delivered was

$$V_{del} = V + \Delta V = 37\,000 \text{ L} - 808 \text{ L} = 36\,190 \text{ L}. \quad \text{(Answer)}$$

Note that the thermal expansion of the steel tank has nothing to do with the problem. Question: Who paid for the "missing" diesel fuel?

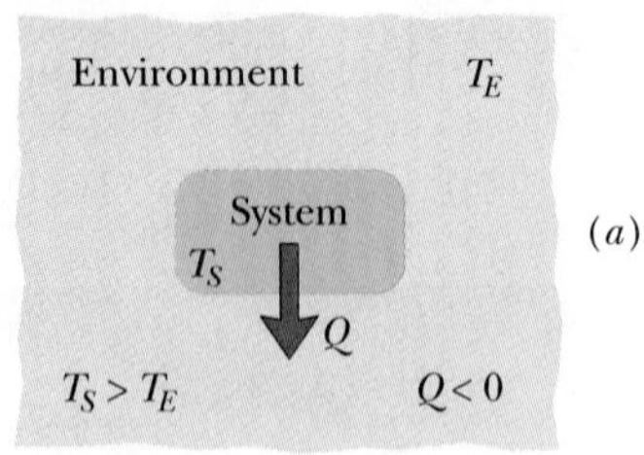

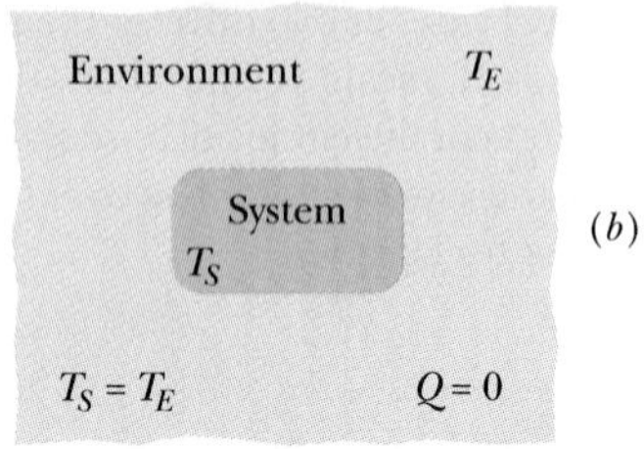

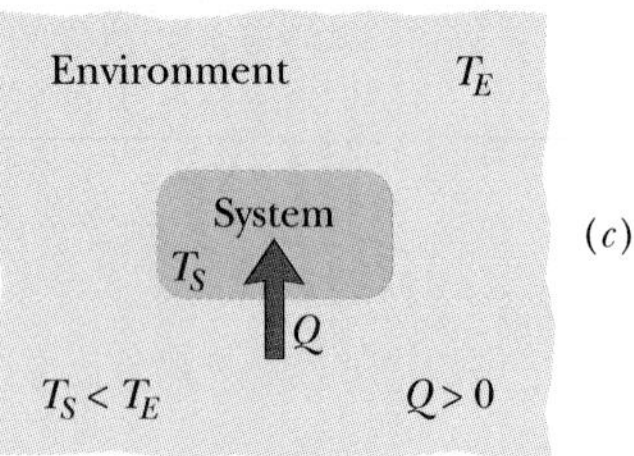

FIG. 18-12 If the temperature of a system exceeds that of its environment as in (*a*), heat Q is lost by the system to the environment until thermal equilibrium (*b*) is established. (*c*) If the temperature of the system is below that of the environment, heat is absorbed by the system until thermal equilibrium is established.

18-7 | Temperature and Heat

If you take a can of cola from the refrigerator and leave it on the kitchen table, its temperature will rise—rapidly at first but then more slowly—until the temperature of the cola equals that of the room (the two are then in thermal equilibrium). In the same way, the temperature of a cup of hot coffee, left sitting on the table, will fall until it also reaches room temperature.

In generalizing this situation, we describe the cola or the coffee as a *system* (with temperature T_S) and the relevant parts of the kitchen as the *environment* (with temperature T_E) of that system. Our observation is that if T_S is not equal to T_E, then T_S will change (T_E can also change some) until the two temperatures are equal and thus thermal equilibrium is reached.

Such a change in temperature is due to a change in the thermal energy of the system because of a transfer of energy between the system and the system's environment. (Recall that *thermal energy* is an internal energy that consists of the kinetic and potential energies associated with the random motions of the atoms, molecules, and other microscopic bodies within an object.) The transferred energy is called **heat** and is symbolized Q. Heat is *positive* when energy is transferred to a system's thermal energy from its environment (we say that heat is absorbed by the system). Heat is *negative* when energy is transferred from a system's thermal energy to its environment (we say that heat is released or lost by the system).

This transfer of energy is shown in Fig. 18-12. In the situation of Fig. 18-12*a*, in which $T_S > T_E$, energy is transferred from the system to the environment, so Q is negative. In Fig. 18-12*b*, in which $T_S = T_E$, there is no such transfer, Q is zero, and heat is neither released nor absorbed. In Fig. 18-12*c*, in which $T_S < T_E$, the transfer is to the system from the environment; so Q is positive.

We are led then to this definition of heat:

> Heat is the energy transferred between a system and its environment because of a temperature difference that exists between them.

Recall that energy can also be transferred between a system and its environment as *work W* via a force acting on a system. Heat and work, unlike temperature, pressure, and volume, are not intrinsic properties of a system. They have meaning only as they describe the transfer of energy into or out of a system. Similarly, the phrase "a $600 transfer" has meaning if it describes the transfer to or from an account, not what is in the account, because the account holds money, not a transfer. Here, it is proper to say: "During the last 3 min, 15 J of heat was transferred to the system from its environment" or "During the last minute, 12 J of work was done on the system by its environment." It is meaningless to say: "This system contains 450 J of heat" or "This system contains 385 J of work."

Before scientists realized that heat is transferred energy, heat was measured in terms of its ability to raise the temperature of water. Thus, the **calorie** (cal) was defined as the amount of heat that would raise the temperature of 1 g of water from 14.5°C to 15.5°C. In the British system, the corresponding unit of heat was the **British thermal unit** (Btu), defined as the amount of heat that would raise the temperature of 1 lb of water from 63°F to 64°F.

In 1948, the scientific community decided that since heat (like work) is transferred energy, the SI unit for heat should be the one we use for energy—namely, the **joule.** The calorie is now defined to be 4.1868 J (exactly), with no reference to the heating of water. (The "calorie" used in nutrition, sometimes called the Calorie (Cal), is really a kilocalorie.) The relations among the various heat units are

$$1 \text{ cal} = 3.968 \times 10^{-3} \text{ Btu} = 4.1868 \text{ J}. \tag{18-12}$$

18-8 | The Absorption of Heat by Solids and Liquids

Heat Capacity

The **heat capacity** C of an object is the proportionality constant between the heat Q that the object absorbs or loses and the resulting temperature change ΔT of the object; that is,

$$Q = C\,\Delta T = C(T_f - T_i), \tag{18-13}$$

in which T_i and T_f are the initial and final temperatures of the object. Heat capacity C has the unit of energy per degree or energy per kelvin. The heat capacity C of, say, a marble slab used in a bun warmer might be 179 cal/C°, which we can also write as 179 cal/K or as 749 J/K.

The word "capacity" in this context is really misleading in that it suggests analogy with the capacity of a bucket to hold water. *That analogy is false,* and you should not think of the object as "containing" heat or being limited in its ability to absorb heat. Heat transfer can proceed without limit as long as the necessary temperature difference is maintained. The object may, of course, melt or vaporize during the process.

Specific Heat

Two objects made of the same material—say, marble—will have heat capacities proportional to their masses. It is therefore convenient to define a "heat capacity per unit mass" or **specific heat** c that refers not to an object but to a unit mass of the material of which the object is made. Equation 18-13 then becomes

$$Q = cm\,\Delta T = cm(T_f - T_i). \tag{18-14}$$

Through experiment we would find that although the heat capacity of a particular marble slab might be 179 cal/C° (or 749 J/K), the specific heat of marble itself (in that slab or in any other marble object) is 0.21 cal/g · C° (or 880 J/kg · K).

From the way the calorie and the British thermal unit were initially defined, the specific heat of water is

$$c = 1\ \text{cal/g}\cdot\text{C}° = 1\ \text{Btu/lb}\cdot\text{F}° = 4190\ \text{J/kg}\cdot\text{K}. \tag{18-15}$$

Table 18-3 shows the specific heats of some substances at room temperature. Note that the value for water is relatively high. The specific heat of any substance actually depends somewhat on temperature, but the values in Table 18-3 apply reasonably well in a range of temperatures near room temperature.

CHECKPOINT 3 A certain amount of heat Q will warm 1 g of material A by 3 C° and 1 g of material B by 4 C°. Which material has the greater specific heat?

TABLE 18-3

Some Specific Heats and Molar Specific Heats at Room Temperature

Substance	Specific Heat (cal/g·K)	Specific Heat (J/kg·K)	Molar Specific Heat (J/mol·K)
Elemental Solids			
Lead	0.0305	128	26.5
Tungsten	0.0321	134	24.8
Silver	0.0564	236	25.5
Copper	0.0923	386	24.5
Aluminum	0.215	900	24.4
Other Solids			
Brass	0.092	380	
Granite	0.19	790	
Glass	0.20	840	
Ice (−10°C)	0.530	2220	
Liquids			
Mercury	0.033	140	
Ethyl alcohol	0.58	2430	
Seawater	0.93	3900	
Water	1.00	4180	

Molar Specific Heat

In many instances the most convenient unit for specifying the amount of a substance is the mole (mol), where

$$1\ \text{mol} = 6.02 \times 10^{23}\ \text{elementary units}$$

of *any* substance. Thus 1 mol of aluminum means 6.02×10^{23} atoms (the atom is the elementary unit), and 1 mol of aluminum oxide means 6.02×10^{23} molecules (the molecule is the elementary unit of the compound).

When quantities are expressed in moles, specific heats must also involve moles (rather than a mass unit); they are then called **molar specific heats.** Table 18-3 shows the values for some elemental solids (each consisting of a single element) at room temperature.

An Important Point

In determining and then using the specific heat of any substance, we need to know the conditions under which energy is transferred as heat. For solids and liquids, we usually assume that the sample is under constant pressure (usually atmospheric) during the transfer. It is also conceivable that the sample is held at constant volume while the heat is absorbed. This means that thermal expansion of the sample is prevented by applying external pressure. For solids and liquids, this is very hard to arrange experimentally, but the effect can be calculated, and it turns out that the specific heats under constant pressure and constant volume for any solid or liquid differ usually by no more than a few percent. Gases, as you will see, have quite different values for their specific heats under constant-pressure conditions and under constant-volume conditions.

Heats of Transformation

When energy is absorbed as heat by a solid or liquid, the temperature of the sample does not necessarily rise. Instead, the sample may change from one *phase,* or *state,* to another. Matter can exist in three common states: In the *solid state,* the molecules of a sample are locked into a fairly rigid structure by their mutual attraction. In the *liquid state,* the molecules have more energy and move about more. They may form brief clusters, but the sample does not have a rigid structure and can flow or settle into a container. In the *gas,* or *vapor, state,* the molecules have even more energy, are free of one another, and can fill up the full volume of a container.

To *melt* a solid means to change it from the solid state to the liquid state. The process requires energy because the molecules of the solid must be freed from their rigid structure. Melting an ice cube to form liquid water is a common example. To *freeze* a liquid to form a solid is the reverse of melting and requires that energy be removed from the liquid, so that the molecules can settle into a rigid structure.

To *vaporize* a liquid means to change it from the liquid state to the vapor (gas) state. This process, like melting, requires energy because the molecules must be freed from their clusters. Boiling liquid water to transfer it to water vapor (or steam—a gas of individual water molecules) is a common example. *Condensing* a gas to form a liquid is the reverse of vaporizing; it requires that energy be removed from the gas, so that the molecules can cluster instead of flying away from one another.

The amount of energy per unit mass that must be transferred as heat when a sample completely undergoes a phase change is called the **heat of transformation** L. Thus, when a sample of mass m completely undergoes a phase change, the total energy transferred is

$$Q = Lm. \tag{18-16}$$

When the phase change is from liquid to gas (then the sample must absorb heat) or from gas to liquid (then the sample must release heat), the heat of transformation is called the **heat of vaporization** L_V. For water at its normal boiling or condensation temperature,

$$L_V = 539 \text{ cal/g} = 40.7 \text{ kJ/mol} = 2256 \text{ kJ/kg}. \tag{18-17}$$

When the phase change is from solid to liquid (then the sample must absorb heat) or from liquid to solid (then the sample must release heat), the heat of transformation is called the **heat of fusion** L_F. For water at its normal freezing or melting temperature,

$$L_F = 79.5 \text{ cal/g} = 6.01 \text{ kJ/mol} = 333 \text{ kJ/kg}. \tag{18-18}$$

Table 18-4 shows the heats of transformation for some substances.

TABLE 18-4

Some Heats of Transformation

	Melting		Boiling	
Substance	Melting Point (K)	Heat of Fusion L_F (kJ/kg)	Boiling Point (K)	Heat of Vaporization L_V (kJ/kg)
Hydrogen	14.0	58.0	20.3	455
Oxygen	54.8	13.9	90.2	213
Mercury	234	11.4	630	296
Water	273	333	373	2256
Lead	601	23.2	2017	858
Silver	1235	105	2323	2336
Copper	1356	207	2868	4730

Sample Problem 18-3

(a) How much heat must be absorbed by ice of mass $m = 720$ g at $-10°$C to take it to liquid state at 15°C?

KEY IDEAS The heating process is accomplished in three steps: (1) The ice cannot melt at a temperature below the freezing point—so initially, any energy transferred to the ice as heat can only increase the temperature of the ice, until 0°C is reached. (2) The temperature then cannot increase until all the ice melts—so any energy transferred to the ice as heat now can only change ice to liquid water, until all the ice melts. (3) Now the energy transferred to the liquid water as heat can only increase the temperature of the liquid water.

Warming the ice: The heat Q_1 needed to increase the temperature of the ice from the initial value $T_i = -10°\text{C}$ to a final value $T_f = 0°\text{C}$ (so that the ice can then melt) is given by Eq. 18-14 ($Q = cm\,\Delta T$). Using the specific heat of ice c_{ice} in Table 18-3 gives us

$$\begin{aligned} Q_1 &= c_{\text{ice}}m(T_f - T_i) \\ &= (2220\ \text{J/kg}\cdot\text{K})(0.720\ \text{kg})[0°\text{C} - (-10°\text{C})] \\ &= 15\,984\ \text{J} \approx 15.98\ \text{kJ}. \end{aligned}$$

Melting the ice: The heat Q_2 needed to melt all the ice is given by Eq. 18-16 ($Q = Lm$). Here L is the heat of fusion L_F, with the value given in Eq. 18-18 and Table 18-4. We find

$$Q_2 = L_F m = (333\ \text{kJ/kg})(0.720\ \text{kg}) \approx 239.8\ \text{kJ}.$$

Warming the liquid: The heat Q_3 needed to increase the temperature of the water from the initial value $T_i = 0°\text{C}$ to the final value $T_f = 15°\text{C}$ is given by Eq. 18-14 (with the specific heat of liquid water c_{liq}):

$$\begin{aligned} Q_3 &= c_{\text{liq}}m(T_f - T_i) \\ &= (4190\ \text{J/kg}\cdot\text{K})(0.720\ \text{kg})(15°\text{C} - 0°\text{C}) \\ &= 45\,252\ \text{J} \approx 45.25\ \text{kJ}. \end{aligned}$$

Total: The total required heat Q_{tot} is the sum of the amounts required in the three steps:

$$\begin{aligned} Q_{\text{tot}} &= Q_1 + Q_2 + Q_3 \\ &= 15.98\ \text{kJ} + 239.8\ \text{kJ} + 45.25\ \text{kJ} \\ &\approx 300\ \text{kJ}. \qquad \text{(Answer)} \end{aligned}$$

Note that the heat required to melt the ice is much greater than the heat required to raise the temperature of either the ice or the liquid water.

(b) If we supply the ice with a total energy of only 210 kJ (as heat), what then are the final state and temperature of the water?

KEY IDEA From step 1, we know that 15.98 kJ is needed to raise the temperature of the ice to the melting point. The remaining heat Q_{rem} is then 210 kJ – 15.98 kJ, or about 194 kJ. From step 2, we can see that this amount of heat is insufficient to melt all the ice. Because the melting of the ice is incomplete, we must end up with a mixture of ice and liquid; the temperature of the mixture must be the freezing point, 0°C.

Calculations: We can find the mass m of ice that is melted by the available energy Q_{rem} by using Eq. 18-16 with L_F:

$$m = \frac{Q_{\text{rem}}}{L_F} = \frac{194\ \text{kJ}}{333\ \text{kJ/kg}} = 0.583\ \text{kg} \approx 580\ \text{g}.$$

Thus, the mass of the ice that remains is 720 g – 580 g, or 140 g, and we have

$$580\ \text{g water} \quad \text{and} \quad 140\ \text{g ice}, \quad \text{at } 0°\text{C}. \qquad \text{(Answer)}$$

Sample Problem 18-4

A copper slug whose mass m_c is 75 g is heated in a laboratory oven to a temperature T of 312°C. The slug is then dropped into a glass beaker containing a mass $m_w = 220$ g of water. The heat capacity C_b of the beaker is 45 cal/K. The initial temperature T_i of the water and the beaker is 12°C. Assuming that the slug, beaker, and water are an isolated system and the water does not vaporize, find the final temperature T_f of the system at thermal equilibrium.

KEY IDEAS (1) Because the system is isolated, the system's total energy cannot change and only internal transfers of thermal energy can occur. (2) Because nothing in the system undergoes a phase change, the thermal energy transfers can only change the temperatures.

Calculations: To relate the transfers to the temperature changes, we can use Eqs. 18-13 and 18-14 to write

$$\text{for the water:} \quad Q_w = c_w m_w (T_f - T_i); \quad (18\text{-}19)$$

$$\text{for the beaker:} \quad Q_b = C_b (T_f - T_i); \quad (18\text{-}20)$$

$$\text{for the copper:} \quad Q_c = c_c m_c (T_f - T). \quad (18\text{-}21)$$

Because the total energy of the system cannot change, the sum of these three energy transfers is zero:

$$Q_w + Q_b + Q_c = 0. \quad (18\text{-}22)$$

Substituting Eqs. 18-19 through 18-21 into Eq. 18-22 yields

$$c_w m_w (T_f - T_i) + C_b (T_f - T_i) + c_c m_c (T_f - T) = 0. \quad (18\text{-}23)$$

Temperatures are contained in Eq. 18-23 only as differences. Thus, because the differences on the Celsius and Kelvin scales are identical, we can use either of those scales in this equation. Solving it for T_f, we obtain

$$T_f = \frac{c_c m_c T + C_b T_i + c_w m_w T_i}{c_w m_w + C_b + c_c m_c}.$$

Using Celsius temperatures and taking values for c_c and c_w from Table 18-3, we find the numerator to be

$$(0.0923 \text{ cal/g} \cdot \text{K})(75 \text{ g})(312°\text{C}) + (45 \text{ cal/K})(12°\text{C}) + (1.00 \text{ cal/g} \cdot \text{K})(220 \text{ g})(12°\text{C}) = 5339.8 \text{ cal},$$

and the denominator to be

$$(1.00 \text{ cal/g} \cdot \text{K})(220 \text{ g}) + 45 \text{ cal/K} + (0.0923 \text{ cal/g} \cdot \text{K})(75 \text{ g}) = 271.9 \text{ cal/C°}.$$

We then have

$$T_f = \frac{5339.8 \text{ cal}}{271.9 \text{ cal/C°}} = 19.6°\text{C} \approx 20°\text{C}. \quad \text{(Answer)}$$

From the given data you can show that

$$Q_w \approx 1670 \text{ cal}, \quad Q_b \approx 342 \text{ cal}, \quad Q_c \approx -2020 \text{ cal}.$$

Apart from rounding errors, the algebraic sum of these three heat transfers is indeed zero, as Eq. 18-22 requires.

18-9 | A Closer Look at Heat and Work

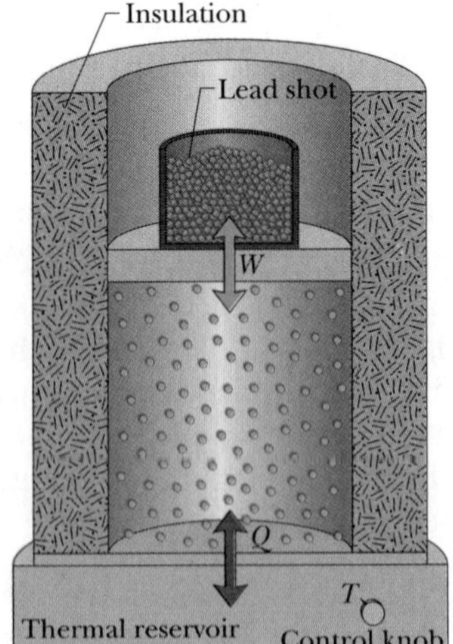

FIG. 18-13 A gas is confined to a cylinder with a movable piston. Heat Q can be added to or withdrawn from the gas by regulating the temperature T of the adjustable thermal reservoir. Work W can be done by the gas by raising or lowering the piston.

Here we look in some detail at how energy can be transferred as heat and work between a system and its environment. Let us take as our system a gas confined to a cylinder with a movable piston, as in Fig. 18-13. The upward force on the piston due to the pressure of the confined gas is equal to the weight of lead shot loaded onto the top of the piston. The walls of the cylinder are made of insulating material that does not allow any transfer of energy as heat. The bottom of the cylinder rests on a reservoir for thermal energy, a *thermal reservoir* (perhaps a hot plate) whose temperature T you can control by turning a knob.

The system (the gas) starts from an *initial state i*, described by a pressure p_i, a volume V_i, and a temperature T_i. You want to change the system to a *final state f*, described by a pressure p_f, a volume V_f, and a temperature T_f. The procedure by which you change the system from its initial state to its final state is called a *thermodynamic process.* During such a process, energy may be transferred into the system from the thermal reservoir (positive heat) or vice versa (negative heat). Also, work can be done by the system to raise the loaded piston (positive work) or lower it (negative work). We assume that all such changes occur slowly, with the result that the system is always in (approximate) thermal equilibrium (that is, every part of the system is always in thermal equilibrium with every other part).

Suppose that you remove a few lead shot from the piston of Fig. 18-13, allowing the gas to push the piston and remaining shot upward through a differential displacement $d\vec{s}$ with an upward force $\vec{F}$. Since the displacement is tiny, we can assume that $\vec{F}$ is constant during the displacement. Then $\vec{F}$ has a magnitude that is equal to pA, where p is the pressure of the gas and A is the face area of the piston. The differential work dW done by the gas during the displacement is

$$dW = \vec{F} \cdot d\vec{s} = (pA)(ds) = p(A\,ds) = p\,dV, \tag{18-24}$$

in which dV is the differential change in the volume of the gas due to the movement of the piston. When you have removed enough shot to allow the gas to change its volume from V_i to V_f, the total work done by the gas is

$$W = \int dW = \int_{V_i}^{V_f} p\,dV. \tag{18-25}$$

During the change in volume, the pressure and temperature of the gas may also change. To evaluate the integral in Eq. 18-25 directly, we would need to know how pressure varies with volume for the actual process by which the system changes from state i to state f.

There are actually many ways to take the gas from state i to state f. One way is shown in Fig. 18-14a, which is a plot of the pressure of the gas versus its volume and which is called a p-V diagram. In Fig. 18-14a, the curve indicates that the pressure decreases as the volume increases. The integral in Eq. 18-25 (and thus the work W done by the gas) is represented by the shaded area under the curve between points i and f. Regardless of what exactly we do to take the gas along the curve, that work is positive, due to the fact that the gas increases its volume by forcing the piston upward.

Another way to get from state i to state f is shown in Fig. 18-14b. There the change takes place in two steps—the first from state i to state a, and the second from state a to state f.

Step ia of this process is carried out at constant pressure, which means that you leave undisturbed the lead shot that ride on top of the piston in Fig. 18-13. You cause the volume to increase (from V_i to V_f) by slowly turning up the temperature control knob, raising the temperature of the gas to some higher value T_a. (Increasing the temperature increases the force from the gas on the piston, moving it upward.) During this step, positive work is done by the expanding gas (to lift the loaded piston) and heat is absorbed by the system from the thermal reservoir (in response to the arbitrarily small temperature differences that you create as you turn up the temperature). This heat is positive because it is added to the system.

Step af of the process of Fig. 18-14b is carried out at constant volume, so you must wedge the piston, preventing it from moving. Then as you use the control knob to decrease the temperature, you find that the pressure drops from p_a to its final value p_f. During this step, heat is lost by the system to the thermal reservoir.

For the overall process iaf, the work W, which is positive and is carried out only during step ia, is represented by the shaded area under the curve. Energy is transferred as heat during both steps ia and af, with a net energy transfer Q.

Figure 18-14c shows a process in which the previous two steps are carried out in reverse order. The work W in this case is smaller than for Fig. 18-14b, as is the

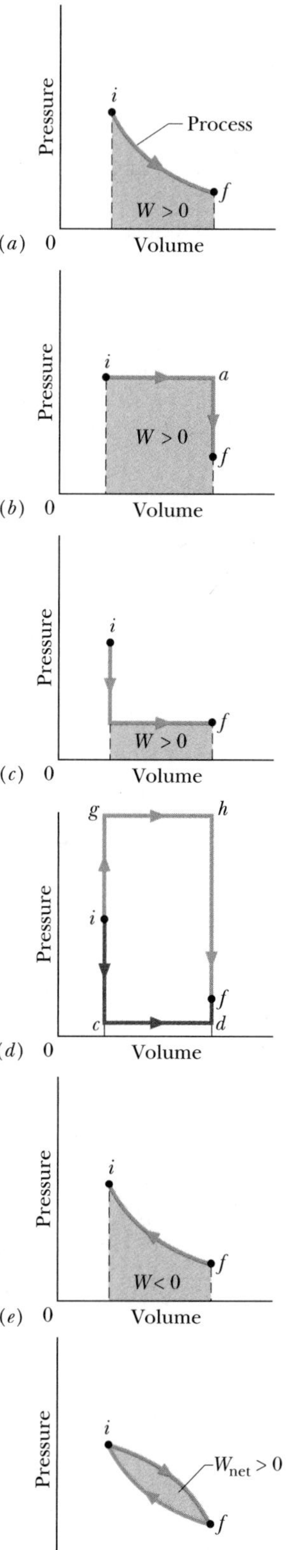

FIG. 18-14 (a) The shaded area represents the work W done by a system as it goes from an initial state i to a final state f. Work W is positive because the system's volume increases. (b) W is still positive, but now greater. (c) W is still positive, but now smaller. (d) W can be even smaller (path $icdf$) or larger (path $ighf$). (e) Here the system goes from state f to state i as the gas is compressed to less volume by an external force. The work W done *by* the system is now negative. (f) The net work W_{net} done by the system during a complete cycle is represented by the shaded area.

net heat absorbed. Figure 18-14*d* suggests that you can make the work done by the gas as small as you want (by following a path like *icdf*) or as large as you want (by following a path like *ighf*).

To sum up: A system can be taken from a given initial state to a given final state by an infinite number of processes. Heat may or may not be involved, and in general, the work W and the heat Q will have different values for different processes. We say that heat and work are *path-dependent* quantities.

Figure 18-14*e* shows an example in which negative work is done by a system as some external force compresses the system, reducing its volume. The absolute value of the work done is still equal to the area beneath the curve, but because the gas is *compressed,* the work done by the gas is negative.

Figure 18-14*f* shows a *thermodynamic cycle* in which the system is taken from some initial state i to some other state f and then back to i. The net work done by the system during the cycle is the sum of the *positive* work done during the expansion and the *negative* work done during the compression. In Fig. 18-14*f*, the net work is positive because the area under the expansion curve (i to f) is greater than the area under the compression curve (f to i).

CHECKPOINT 4 The p-V diagram here shows six curved paths (connected by vertical paths) that can be followed by a gas. Which two of the curved paths should be part of a closed cycle (those curved paths plus connecting vertical paths) if the net work done by the gas during the cycle is to be at its maximum positive value?

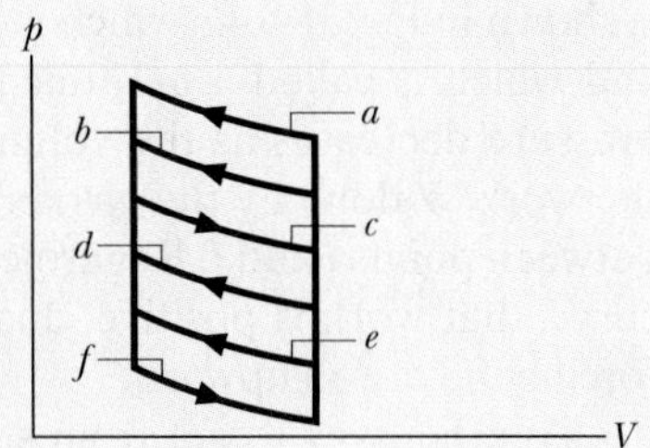

18-10 | The First Law of Thermodynamics

You have just seen that when a system changes from a given initial state to a given final state, both the work W and the heat Q depend on the nature of the process. Experimentally, however, we find a surprising thing. *The quantity $Q - W$ is the same for all processes.* It depends only on the initial and final states and does not depend at all on how the system gets from one to the other. All other combinations of Q and W, including Q alone, W alone, $Q + W$, and $Q - 2W$, are *path dependent;* only the quantity $Q - W$ is not.

The quantity $Q - W$ must represent a change in some intrinsic property of the system. We call this property the *internal energy* E_{int} and we write

$$\Delta E_{\text{int}} = E_{\text{int},f} - E_{\text{int},i} = Q - W \quad \text{(first law)}. \tag{18-26}$$

Equation 18-26 is the **first law of thermodynamics.** If the thermodynamic system undergoes only a differential change, we can write the first law as*

$$dE_{\text{int}} = dQ - dW \quad \text{(first law)}. \tag{18-27}$$

The internal energy E_{int} of a system tends to increase if energy is added as heat Q and tends to decrease if energy is lost as work W done by the system.

In Chapter 8, we discussed the principle of energy conservation as it applies to isolated systems—that is, to systems in which no energy enters or leaves the system. The first law of thermodynamics is an extension of that principle to

*Here dQ and dW, unlike dE_{int}, are not true differentials; that is, there are no such functions as $Q(p, V)$ and $W(p, V)$ that depend only on the state of the system. The quantities dQ and dW are called *inexact differentials* and are usually represented by the symbols $đQ$ and $đW$. For our purposes, we can treat them simply as infinitesimally small energy transfers.

systems that are *not* isolated. In such cases, energy may be transferred into or out of the system as either work W or heat Q. In our statement of the first law of thermodynamics above, we assume that there are no changes in the kinetic energy or the potential energy of the system as a whole; that is, $\Delta K = \Delta U = 0$.

Before this chapter, the term *work* and the symbol W always meant the work done *on* a system. However, starting with Eq. 18-24 and continuing through the next two chapters about thermodynamics, we focus on the work done *by* a system, such as the gas in Fig. 18-13.

The work done *on* a system is always the negative of the work done *by* the system, so if we rewrite Eq. 18-26 in terms of the work W_{on} done *on* the system, we have $\Delta E_{int} = Q + W_{on}$. This tells us the following: The internal energy of a system tends to increase if heat is absorbed by the system or if positive work is done *on* the system. Conversely, the internal energy tends to decrease if heat is lost by the system or if negative work is done *on* the system.

CHECKPOINT 5 The figure here shows four paths on a p-V diagram along which a gas can be taken from state i to state f. Rank the paths according to (a) the change ΔE_{int} in the internal energy of the gas, (b) the work W done by the gas, and (c) the magnitude of the energy transferred as heat Q between the gas and its environment, greatest first.

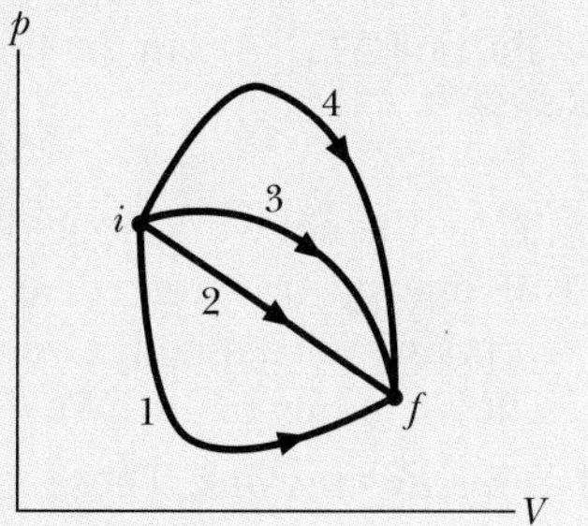

18-11 | Some Special Cases of the First Law of Thermodynamics

Here we look at four different thermodynamic processes to see what consequences follow when we apply the first law of thermodynamics. The results are summarized in Table 18-5.

1. ***Adiabatic processes.*** An adiabatic process is one that occurs so rapidly or occurs in a system that is so well insulated that *no transfer of energy as heat* occurs between the system and its environment. Putting $Q = 0$ in the first law (Eq. 18-26) yields

$$\Delta E_{int} = -W \quad \text{(adiabatic process).} \tag{18-28}$$

This tells us that if work is done *by* the system (that is, if W is positive), the internal energy of the system decreases by the amount of work. Conversely, if work is done *on* the system (that is, if W is negative), the internal energy of the system increases by that amount.

Figure 18-15 shows an idealized adiabatic process. Heat cannot enter or leave the system because of the insulation. Thus, the only way energy can be

TABLE 18-5

The First Law of Thermodynamics: Four Special Cases

The Law: $\Delta E_{int} = Q - W$ (Eq. 18-26)

Process	Restriction	Consequence
Adiabatic	$Q = 0$	$\Delta E_{int} = -W$
Constant volume	$W = 0$	$\Delta E_{int} = Q$
Closed cycle	$\Delta E_{int} = 0$	$Q = W$
Free expansion	$Q = W = 0$	$\Delta E_{int} = 0$

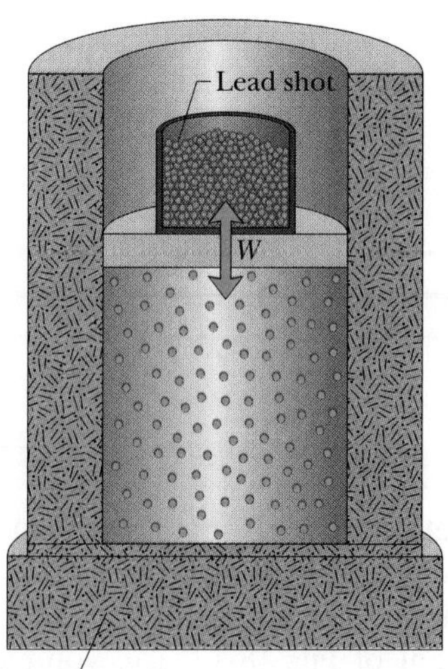

FIG. 18-15 An adiabatic expansion can be carried out by slowly removing lead shot from the top of the piston. Adding lead shot reverses the process at any stage.

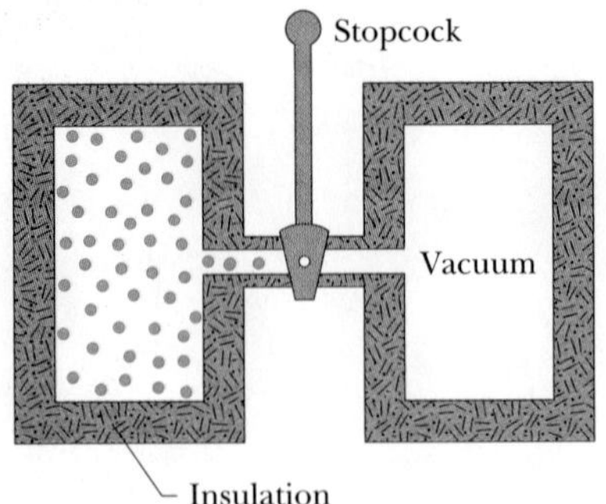

FIG. 18-16 The initial stage of a free-expansion process. After the stopcock is opened, the gas fills both chambers and eventually reaches an equilibrium state.

transferred between the system and its environment is by work. If we remove shot from the piston and allow the gas to expand, the work done by the system (the gas) is positive and the internal energy of the gas decreases. If, instead, we add shot and compress the gas, the work done by the system is negative and the internal energy of the gas increases.

2. ***Constant-volume processes.*** If the volume of a system (such as a gas) is held constant, that system can do no work. Putting $W = 0$ in the first law (Eq. 18-26) yields

$$\Delta E_{int} = Q \quad \text{(constant-volume process).} \tag{18-29}$$

Thus, if heat is absorbed by a system (that is, if Q is positive), the internal energy of the system increases. Conversely, if heat is lost during the process (that is, if Q is negative), the internal energy of the system must decrease.

3. ***Cyclical processes.*** There are processes in which, after certain interchanges of heat and work, the system is restored to its initial state. In that case, no intrinsic property of the system—including its internal energy—can possibly change. Putting $\Delta E_{int} = 0$ in the first law (Eq. 18-26) yields

$$Q = W \quad \text{(cyclical process).} \tag{18-30}$$

Thus, the net work done during the process must exactly equal the net amount of energy transferred as heat; the store of internal energy of the system remains unchanged. Cyclical processes form a closed loop on a p-V plot, as shown in Fig. 18-14*f*. We discuss such processes in detail in Chapter 20.

4. ***Free expansions.*** These are adiabatic processes in which no transfer of heat occurs between the system and its environment and no work is done on or by the system. Thus, $Q = W = 0$, and the first law requires that

$$\Delta E_{int} = 0 \quad \text{(free expansion).} \tag{18-31}$$

Figure 18-16 shows how such an expansion can be carried out. A gas, which is in thermal equilibrium within itself, is initially confined by a closed stopcock to one half of an insulated double chamber; the other half is evacuated. The stopcock is opened, and the gas expands freely to fill both halves of the chamber. No heat is transferred to or from the gas because of the insulation. No work is done by the gas because it rushes into a vacuum and thus does not meet any pressure.

A free expansion differs from all other processes we have considered because it cannot be done slowly and in a controlled way. As a result, at any given instant during the sudden expansion, the gas is not in thermal equilibrium and its pressure is not uniform. Thus, although we can plot the initial and final states on a p-V diagram, we cannot plot the expansion itself.

CHECKPOINT 6 For one complete cycle as shown in the p-V diagram here, are (a) ΔE_{int} for the gas and (b) the net energy transferred as heat Q positive, negative, or zero?

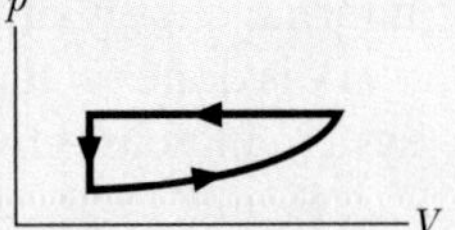

Sample Problem 18-5

Let 1.00 kg of liquid water at 100°C be converted to steam at 100°C by boiling at standard atmospheric pressure (which is 1.00 atm or 1.01×10^5 Pa) in the arrangement of Fig. 18-17. The volume of that water changes from an initial value of 1.00×10^{-3} m^3 as a liquid to 1.671 m^3 as steam.

(a) How much work is done by the system during this process?

KEY IDEAS (1) The system must do positive work because the volume increases. (2) We calculate the work W done by integrating the pressure with respect to the volume (Eq. 18-25).

Calculation: Because here the pressure is constant at 1.01×10^5 Pa, we can take p outside the integral. We then have

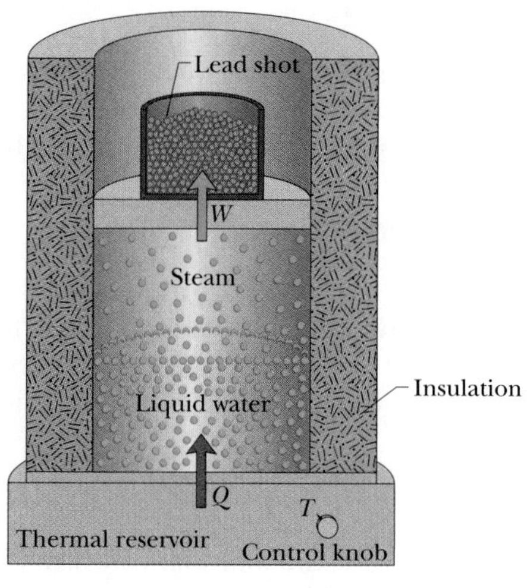

FIG. 18-17 Water boiling at constant pressure. Energy is transferred from the thermal reservoir as heat until the liquid water has changed completely into steam. Work is done by the expanding gas as it lifts the loaded piston.

$$\begin{aligned} W &= \int_{V_i}^{V_f} p\,dV = p\int_{V_i}^{V_f} dV = p(V_f - V_i) \\ &= (1.01 \times 10^5\ \text{Pa})(1.671\ \text{m}^3 - 1.00 \times 10^{-3}\ \text{m}^3) \\ &= 1.69 \times 10^5\ \text{J} = 169\ \text{kJ}. \qquad \text{(Answer)} \end{aligned}$$

(b) How much energy is transferred as heat during the process?

KEY IDEA Because the heat causes only a phase change and not a change in temperature, it is given fully by Eq. 18-16 ($Q = Lm$).

Calculation: Because the change is from liquid to gaseous phase, L is the heat of vaporization L_V, with the value given in Eq. 18-17 and Table 18-4. We find

$$\begin{aligned} Q &= L_V m = (2256\ \text{kJ/kg})(1.00\ \text{kg}) \\ &= 2256\ \text{kJ} \approx 2260\ \text{kJ}. \qquad \text{(Answer)} \end{aligned}$$

(c) What is the change in the system's internal energy during the process?

KEY IDEA The change in the system's internal energy is related to the heat (here, this is energy transferred into the system) and the work (here, this is energy transferred out of the system) by the first law of thermodynamics (Eq. 18-26).

Calculation: We write the first law as

$$\begin{aligned} \Delta E_{\text{int}} &= Q - W = 2256\ \text{kJ} - 169\ \text{kJ} \\ &\approx 2090\ \text{kJ} = 2.09\ \text{MJ}. \qquad \text{(Answer)} \end{aligned}$$

This quantity is positive, indicating that the internal energy of the system has increased during the boiling process. This energy goes into separating the H_2O molecules, which strongly attract one another in the liquid state. We see that, when water is boiled, about 7.5% (= 169 kJ/2260 kJ) of the heat goes into the work of pushing back the atmosphere. The rest of the heat goes into the system's internal energy.

18-12 | Heat Transfer Mechanisms

We have discussed the transfer of energy as heat between a system and its environment, but we have not yet described how that transfer takes place. There are three transfer mechanisms: conduction, convection, and radiation.

Conduction

If you leave the end of a metal poker in a fire for enough time, its handle will get hot. Energy is transferred from the fire to the handle by (thermal) **conduction** along the length of the poker. The vibration amplitudes of the atoms and electrons of the metal at the fire end of the poker become relatively large because of the high temperature of their environment. These increased vibrational amplitudes, and thus the associated energy, are passed along the poker, from atom to atom, during collisions between adjacent atoms. In this way, a region of rising temperature extends itself along the poker to the handle.

Consider a slab of face area A and thickness L, whose faces are maintained at temperatures T_H and T_C by a hot reservoir and a cold reservoir, as in Fig. 18-18. Let Q be the energy that is transferred as heat through the slab, from its hot face to its cold face, in time t. Experiment shows that the *conduction rate* P_{cond} (the amount of energy transferred per unit time) is

$$P_{\text{cond}} = \frac{Q}{t} = kA\,\frac{T_H - T_C}{L}, \qquad (18\text{-}32)$$

in which k, called the *thermal conductivity*, is a constant that depends on the

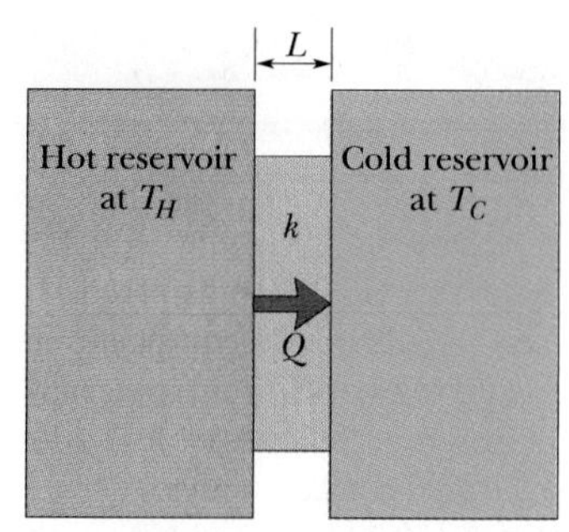

FIG. 18-18 Thermal conduction. Energy is transferred as heat from a reservoir at temperature T_H to a cooler reservoir at temperature T_C through a conducting slab of thickness L and thermal conductivity k.

TABLE 18-6

Some Thermal Conductivities

Substance	k (W/m·K)
Metals	
Stainless steel	14
Lead	35
Iron	67
Brass	109
Aluminum	235
Copper	401
Silver	428
Gases	
Air (dry)	0.026
Helium	0.15
Hydrogen	0.18
Building Materials	
Polyurethane foam	0.024
Rock wool	0.043
Fiberglass	0.048
White pine	0.11
Window glass	1.0

material of which the slab is made. A material that readily transfers energy by conduction is a *good thermal conductor* and has a high value of k. Table 18-6 gives the thermal conductivities of some common metals, gases, and building materials.

Thermal Resistance to Conduction (R-Value)

If you are interested in insulating your house or in keeping cola cans cold on a picnic, you are more concerned with poor heat conductors than with good ones. For this reason, the concept of *thermal resistance* R has been introduced into engineering practice. The R-value of a slab of thickness L is defined as

$$R = \frac{L}{k}. \tag{18-33}$$

The lower the thermal conductivity of the material of which a slab is made, the higher the R-value of the slab; so something that has a high R-value is a *poor thermal conductor* and thus a *good thermal insulator.*

Note that R is a property attributed to a slab of a specified thickness, not to a material. The commonly used unit for R (which, in the United States at least, is almost never stated) is the square foot–Fahrenheit degree–hour per British thermal unit ($\text{ft}^2 \cdot \text{F}^\circ \cdot \text{h/Btu}$). (Now you know why the unit is rarely stated.)

Conduction Through a Composite Slab

Figure 18-19 shows a composite slab, consisting of two materials having different thicknesses L_1 and L_2 and different thermal conductivities k_1 and k_2. The temperatures of the outer surfaces of the slab are T_H and T_C. Each face of the slab has area A. Let us derive an expression for the conduction rate through the slab under the assumption that the transfer is a *steady-state* process; that is, the temperatures everywhere in the slab and the rate of energy transfer do not change with time.

In the steady state, the conduction rates through the two materials must be equal. This is the same as saying that the energy transferred through one material in a certain time must be equal to that transferred through the other material in the same time. If this were not true, temperatures in the slab would be changing and we would not have a steady-state situation. Letting T_X be the temperature of the interface between the two materials, we can now use Eq. 18-32 to write

$$P_{\text{cond}} = \frac{k_2 A (T_H - T_X)}{L_2} = \frac{k_1 A (T_X - T_C)}{L_1}. \tag{18-34}$$

Solving Eq. 18-34 for T_X yields, after a little algebra,

$$T_X = \frac{k_1 L_2 T_C + k_2 L_1 T_H}{k_1 L_2 + k_2 L_1}. \tag{18-35}$$

Substituting this expression for T_X into either equality of Eq. 18-34 yields

$$P_{\text{cond}} = \frac{A(T_H - T_C)}{L_1/k_1 + L_2/k_2}. \tag{18-36}$$

We can extend Eq. 18-36 to apply to any number n of materials making up a slab:

$$P_{\text{cond}} = \frac{A(T_H - T_C)}{\Sigma\,(L/k)}. \tag{18-37}$$

The summation sign in the denominator tells us to add the values of L/k for all the materials.

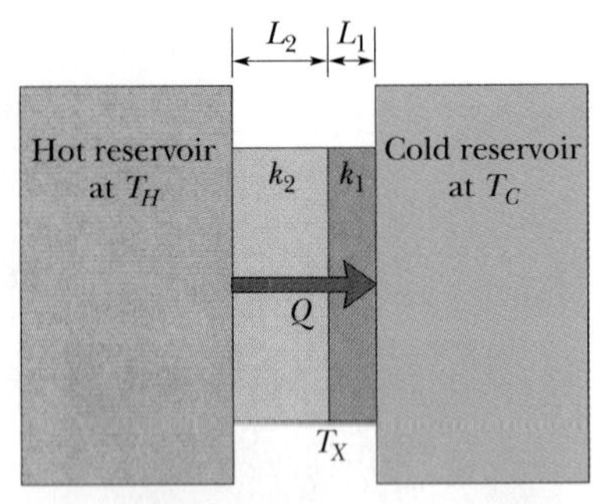

FIG. 18-19 Heat is transferred at a steady rate through a composite slab made up of two different materials with different thicknesses and different thermal conductivities. The steady-state temperature at the interface of the two materials is T_X.

CHECKPOINT 7 The figure shows the face and interface temperatures of a composite slab consisting of four materials, of identical thicknesses, through which the heat transfer is steady. Rank the materials according to their thermal conductivities, greatest first.

25°C, 15°C, 10°C, −5.0°C, −10°C

a, *b*, *c*, *d*

Convection

When you look at the flame of a candle or a match, you are watching thermal energy being transported upward by **convection.** Such energy transfer occurs when a fluid, such as air or water, comes in contact with an object whose temperature is higher than that of the fluid. The temperature of the part of the fluid that is in contact with the hot object increases, and (in most cases) that fluid expands and thus becomes less dense. Because this expanded fluid is now lighter than the surrounding cooler fluid, buoyant forces cause it to rise. Some of the surrounding cooler fluid then flows so as to take the place of the rising warmer fluid, and the process can then continue.

Convection is part of many natural processes. Atmospheric convection plays a fundamental role in determining global climate patterns and daily weather variations. Glider pilots and birds alike seek rising thermals (convection currents of warm air) that keep them aloft. Huge energy transfers take place within the oceans by the same process. Finally, energy is transported to the surface of the Sun from the nuclear furnace at its core by enormous cells of convection, in which hot gas rises to the surface along the cell core and cooler gas around the core descends below the surface.

Radiation

The third method by which an object and its environment can exchange energy as heat is via electromagnetic waves (visible light is one kind of electromagnetic wave). Energy transferred in this way is often called **thermal radiation** to distinguish it from electromagnetic *signals* (as in, say, television broadcasts) and from nuclear radiation (energy and particles emitted by nuclei). (To "radiate" generally means to emit.) When you stand in front of a big fire, you are warmed by absorbing thermal radiation from the fire; that is, your thermal energy increases as the fire's thermal energy decreases. No medium is required for heat transfer via radiation—the radiation can travel through vacuum from, say, the Sun to you.

The rate P_{rad} at which an object emits energy via electromagnetic radiation depends on the object's surface area A and the temperature T of that area in kelvins and is given by

$$P_{rad} = \sigma\varepsilon A T^4. \tag{18-38}$$

Here $\sigma = 5.6704 \times 10^{-8}\ \text{W/m}^2\cdot\text{K}^4$ is called the *Stefan–Boltzmann constant* after Josef Stefan (who discovered Eq. 18-38 experimentally in 1879) and Ludwig Boltzmann (who derived it theoretically soon after). The symbol ε represents the *emissivity* of the object's surface, which has a value between 0 and 1, depending on the composition of the surface. A surface with the maximum emissivity of 1.0 is said to be a *blackbody radiator,* but such a surface is an ideal limit and does not occur in nature. Note again that the temperature in Eq. 18-38 must be in kelvins so that a temperature of absolute zero corresponds to no radiation. Note also that every object whose temperature is above 0 K—including you—emits thermal radiation. (See Fig. 18-20.)

FIG. 18-20 A false-color thermogram reveals the rate at which energy is radiated by a cat. The rate is color-coded, with white and red indicating the greatest radiation rate. The nose is cool. *(Edward Kinsman/Photo Researchers)*

The rate P_{abs} at which an object absorbs energy via thermal radiation from its environment, which we take to be at uniform temperature T_{env} (in kelvins), is

$$P_{abs} = \sigma\varepsilon A T_{env}^4. \tag{18-39}$$

The emissivity ε in Eq. 18-39 is the same as that in Eq. 18-38. An idealized blackbody radiator, with $\varepsilon = 1$, will absorb all the radiated energy it intercepts (rather than sending a portion back away from itself through reflection or scattering).

Because an object will radiate energy to the environment while it absorbs energy from the environment, the object's net rate P_{net} of energy exchange due to thermal radiation is

$$P_{net} = P_{abs} - P_{rad} = \sigma\varepsilon A(T_{env}^4 - T^4). \tag{18-40}$$

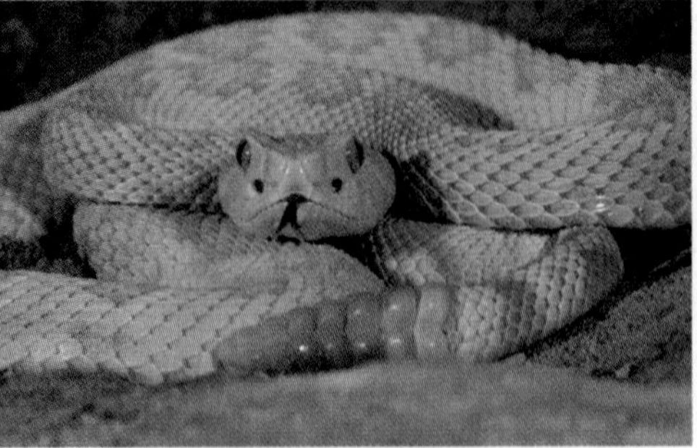

FIG. 18-21 A rattlesnake's face has thermal radiation detectors, allowing the snake to strike at an animal even in complete darkness. *(David A. Northcott/Corbis Images)*

P_{net} is positive if net energy is being absorbed via radiation and negative if it is being lost via radiation.

Let's now return to the story about the ability of a *Melanophila* beetle to detect a fairly large fire from a distance of 12 km without seeing or smelling it. A pair of organs along each side of the beetle's body can detect even low-level thermal radiation. Each organ contains about 70 small knob-like sensors that expand very slightly when they absorb thermal radiation from the fire; the expansion causes them to press down on sensory cells. Thus, the detector is a mechanism that transfers energy from the thermal radiation to the energy of a mechanical device. The beetle can locate the fire by orienting itself so that all four infrared-detecting organs are affected, and then it flies toward the fire so that the response of the organs increases.

Thermal radiation is also involved in the numerous medical cases of a *dead* rattlesnake striking a hand reaching toward it. Pits between each eye and nostril of a rattlesnake (Fig. 18-21) serve as sensors of thermal radiation. When, say, a mouse moves close to a rattlesnake's head, the thermal radiation from the mouse triggers these sensors, causing a reflex action in which the snake strikes the mouse with its fangs and injects its venom. The thermal radiation from a reaching hand can cause the same reflex action even if the snake has been dead for as long as 30 min because the snake's nervous system continues to function. As one snake expert advised, if you must remove a recently killed rattlesnake, use a long stick rather than your hand.

Sample Problem 18-6

Figure 18-22 shows the cross section of a wall made of white pine of thickness L_a and brick of thickness L_d $(= 2.0L_a)$, sandwiching two layers of unknown material with identical thicknesses and thermal conductivities. The thermal conductivity of the pine is k_a and that of the brick is k_d $(= 5.0k_a)$. The face area A of the wall is unknown. Thermal conduction through the wall has reached the steady state; the only known interface temperatures are $T_1 = 25°C$, $T_2 = 20°C$, and $T_5 = -10°C$. What is interface temperature T_4?

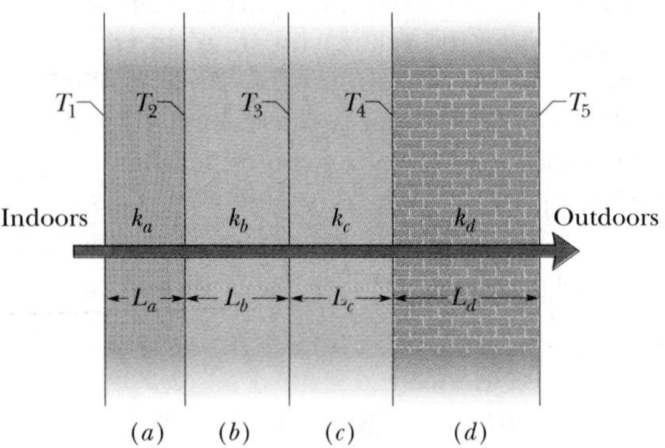

FIG. 18-22 A wall of four layers through which there is steady-state heat transfer.

KEY IDEAS (1) Temperature T_4 helps determine the rate P_d at which energy is conducted through the brick, as given by Eq. 18-32. However, we lack enough data to solve Eq. 18-32 for T_4. (2) Because the conduction is steady, the conduction rate P_d through the brick must equal the conduction rate P_a through the pine. That gets us going.

Calculations: From Eq. 18-32 and Fig. 18-22, we can write

$$P_a = k_a A \frac{T_1 - T_2}{L_a} \quad \text{and} \quad P_d = k_d A \frac{T_4 - T_5}{L_d}.$$

Setting $P_a = P_d$ and solving for T_4 yield

$$T_4 = \frac{k_a L_d}{k_d L_a}(T_1 - T_2) + T_5.$$

Letting $L_d = 2.0L_a$ and $k_d = 5.0k_a$, and inserting the known temperatures, we find

$$T_4 = \frac{k_a(2.0L_a)}{(5.0k_a)L_a}(25°C - 20°C) + (-10°C)$$
$$= -8.0°C. \qquad \text{(Answer)}$$

Sample Problem 18-7 Build your skill

During an extended wilderness hike, you have a terrific craving for ice. Unfortunately, the air temperature drops to only 6.0°C each night—too high to freeze water. However, because a clear, moonless night sky acts like a blackbody radiator at a temperature of $T_s = -23°C$, perhaps you can make ice by letting a shallow layer of water radiate energy to such a sky. To start, you thermally insulate a container from the ground by placing a poorly conducting layer of, say, foam rubber or straw beneath it. Then you pour water into the container, forming a thin, uniform layer with mass $m = 4.5$ g, top surface area $A = 9.0 \text{ cm}^2$, depth $d = 5.0$ mm, emissivity $\varepsilon = 0.90$, and initial temperature 6.0°C. Find the time required for the water to freeze via radiation. Can the freezing be accomplished during one night?

KEY IDEAS (1) The water cannot freeze at a temperature above the freezing point. Therefore, the radiation must first remove an amount of energy Q_1 to reduce the water temperature from 6.0°C to the freezing point of 0°C. (2) The radiation then must remove an additional amount of energy Q_2 to freeze all the water. (3) Throughout this process, the water is also absorbing energy radiated to it from the sky. We want a net loss of energy.

Cooling the water: Using Eq. 18-14 and Table 18-3, we find that cooling the water to 0°C requires an energy loss of

$$\begin{aligned} Q_1 &= cm(T_f - T_i) \\ &= (4190 \text{ J/kg}\cdot\text{K})(4.5 \times 10^{-3}\text{ kg})(0°\text{C} - 6.0°\text{C}) \\ &= -113 \text{ J}. \end{aligned}$$

Thus, 113 J must be radiated away by the water to drop its temperature to the freezing point.

Freezing the water: Using Eq. 18-16 ($Q = mL$) with the value of L being L_F from Eq. 18-18 or Table 18-4, and inserting a minus sign to indicate an energy loss, we find

$$\begin{aligned} Q_2 = -mL_F &= -(4.5 \times 10^{-3}\text{ kg})(3.33 \times 10^5 \text{ J/kg}) \\ &= -1499 \text{ J}. \end{aligned}$$

The total required energy loss is thus

$$Q_{tot} = Q_1 + Q_2 = -113 \text{ J} - 1499 \text{ J} = -1612 \text{ J}.$$

Radiation: While the water loses energy by radiating to the sky, it also absorbs energy radiated to it from the sky. In a total time t, we want the net energy of this exchange to be the energy loss Q_{tot}; so we want the power of this exchange to be

$$\text{power} = \frac{\text{net energy}}{\text{time}} = \frac{Q_{tot}}{t}. \qquad (18\text{-}41)$$

The power of such an energy exchange is also the net rate P_{net} of thermal radiation, as given by Eq. 18-40; so the time t required for the energy loss to be Q_{tot} is

$$t = \frac{Q}{P_{net}} = \frac{Q}{\sigma\varepsilon A(T_s^4 - T^4)}. \qquad (18\text{-}42)$$

Although the temperature T of the water decreases slightly while the water is cooling, we can approximate T as being the freezing point, 273 K. With $T_s = 250$ K, the denominator of Eq. 18-42 is

$$\begin{aligned} &(5.67 \times 10^{-8}\text{ W/m}^2\cdot\text{K}^4)(0.90)(9.0 \times 10^{-4}\text{ m}^2) \\ &\times [(250\text{ K})^4 - (273\text{ K})^4] = -7.57 \times 10^{-2}\text{ J/s}, \end{aligned}$$

and Eq. 18-42 gives us

$$\begin{aligned} t &= \frac{-1612 \text{ J}}{-7.57 \times 10^{-2}\text{ J/s}} \\ &= 2.13 \times 10^4 \text{ s} = 5.9 \text{ h}. \qquad \text{(Answer)} \end{aligned}$$

Because t is less than a night, freezing water by having it radiate to the dark sky is feasible. In fact, in some parts of the world people used this technique long before the introduction of electric freezers.

REVIEW & SUMMARY

Temperature; Thermometers Temperature is an SI base quantity related to our sense of hot and cold. It is measured with a thermometer, which contains a working substance with a measurable property, such as length or pressure, that changes in a regular way as the substance becomes hotter or colder.

Zeroth Law of Thermodynamics When a thermometer and some other object are placed in contact with each other, they eventually reach thermal equilibrium. The reading of the thermometer is then taken to be the temperature of the other object. The process provides consistent and useful temperature measurements because of the **zeroth law of thermodynamics:** If bodies A and B are each in thermal equilibrium with a third body C (the thermometer), then A and B are in thermal equilibrium with each other.

The Kelvin Temperature Scale In the SI system, temperature is measured on the **Kelvin scale,** which is based on the *triple point* of water (273.16 K). Other temperatures are then defined by use of a *constant-volume gas thermometer,* in which a sample of gas is maintained at constant volume so its pressure is proportional to its temperature. We define the *temperature T* as measured with a gas thermometer to be

$$T = (273.16\text{ K})\left(\lim_{\text{gas}\to 0}\frac{p}{p_3}\right). \qquad (18\text{-}6)$$

Here T is in kelvins, and p_3 and p are the pressures of the gas at 273.16 K and the measured temperature, respectively.

Celsius and Fahrenheit Scales The Celsius temperature scale is defined by

$$T_C = T - 273.15°, \qquad (18\text{-}7)$$

with T in kelvins. The Fahrenheit temperature scale is defined by

$$T_F = \tfrac{9}{5}T_C + 32°. \qquad (18\text{-}8)$$

Thermal Expansion All objects change size with changes in temperature. For a temperature change ΔT, a change ΔL in any linear dimension L is given by

$$\Delta L = L\alpha\,\Delta T, \qquad (18\text{-}9)$$

in which α is the **coefficient of linear expansion.** The change ΔV in the volume V of a solid or liquid is

$$\Delta V = V\beta\,\Delta T. \tag{18-10}$$

Here $\beta = 3\alpha$ is the material's **coefficient of volume expansion.**

Heat Heat Q is energy that is transferred between a system and its environment because of a temperature difference between them. It can be measured in **joules** (J), **calories** (cal), **kilocalories** (Cal or kcal), or **British thermal units** (Btu), with

$$1 \text{ cal} = 3.968 \times 10^{-3} \text{ Btu} = 4.1868 \text{ J}. \tag{18-12}$$

Heat Capacity and Specific Heat If heat Q is absorbed by an object, the object's temperature change $T_f - T_i$ is related to Q by

$$Q = C(T_f - T_i), \tag{18-13}$$

in which C is the **heat capacity** of the object. If the object has mass m, then

$$Q = cm(T_f - T_i), \tag{18-14}$$

where c is the **specific heat** of the material making up the object. The **molar specific heat** of a material is the heat capacity per mole, which means per 6.02×10^{23} elementary units of the material.

Heat of Transformation Heat absorbed by a material may change the material's physical state—for example, from solid to liquid or from liquid to gas. The amount of energy required per unit mass to change the state (but not the temperature) of a particular material is its **heat of transformation** L. Thus,

$$Q = Lm. \tag{18-16}$$

The **heat of vaporization** L_V is the amount of energy per unit mass that must be added to vaporize a liquid or that must be removed to condense a gas. The **heat of fusion** L_F is the amount of energy per unit mass that must be added to melt a solid or that must be removed to freeze a liquid.

Work Associated with Volume Change A gas may exchange energy with its surroundings through work. The amount of work W done *by* a gas as it expands or contracts from an initial volume V_i to a final volume V_f is given by

$$W = \int dW = \int_{V_i}^{V_f} p\,dV. \tag{18-25}$$

The integration is necessary because the pressure p may vary during the volume change.

First Law of Thermodynamics The principle of conservation of energy for a thermodynamic process is expressed in the **first law of thermodynamics,** which may assume either of the forms

$$\Delta E_{\text{int}} = E_{\text{int},f} - E_{\text{int},i} = Q - W \quad \text{(first law)} \tag{18-26}$$

or

$$dE_{\text{int}} = dQ - dW \quad \text{(first law)}. \tag{18-27}$$

E_{int} represents the internal energy of the material, which depends only on the material's state (temperature, pressure, and volume). Q represents the energy exchanged as heat between the system and its surroundings; Q is positive if the system absorbs heat and negative if the system loses heat. W is the work done *by* the system; W is positive if the system expands against an external force from the surroundings and negative if the system contracts because of an external force. *Q and W are path dependent; ΔE_{int} is path independent.*

Applications of the First Law The first law of thermodynamics finds application in several special cases:

adiabatic processes: $Q = 0, \quad \Delta E_{\text{int}} = -W$

constant-volume processes: $W = 0, \quad \Delta E_{\text{int}} = Q$

cyclical processes: $\Delta E_{\text{int}} = 0, \quad Q = W$

free expansions: $Q = W = \Delta E_{\text{int}} = 0$

Conduction, Convection, and Radiation The rate P_{cond} at which energy is *conducted* through a slab whose faces are maintained at temperatures T_H and T_C is

$$P_{\text{cond}} = \frac{Q}{t} = kA\,\frac{T_H - T_C}{L}, \tag{18-32}$$

in which A and L are the face area and length of the slab, and k is the thermal conductivity of the material.

Convection occurs when temperature differences cause an energy transfer by motion within a fluid. *Radiation* is an energy transfer via the emission of electromagnetic energy. The rate P_{rad} at which an object emits energy via thermal radiation is

$$P_{\text{rad}} = \sigma\varepsilon A T^4, \tag{18-38}$$

where σ ($= 5.6704 \times 10^{-8}$ W/m$^2 \cdot$K^4) is the Stefan–Boltzmann constant, ε is the emissivity of the object's surface, A is its surface area, and T is its surface temperature (in kelvins). The rate P_{abs} at which an object absorbs energy via thermal radiation from its environment, which is at the uniform temperature T_{env} (in kelvins), is

$$P_{\text{abs}} = \sigma\varepsilon A T_{\text{env}}^4. \tag{18-39}$$

QUESTIONS

1 Materials A, B, and C are solids that are at their melting temperatures. Material A requires 200 J to melt 4 kg, material B requires 300 J to melt 5 kg, and material C requires 300 J to melt 6 kg. Rank the materials according to their heats of fusion, greatest first.

2 Figure 18-23 shows three linear temperature scales, with the freezing and boiling points of water indicated. Rank the three scales according to the size of one degree on them, greatest first.

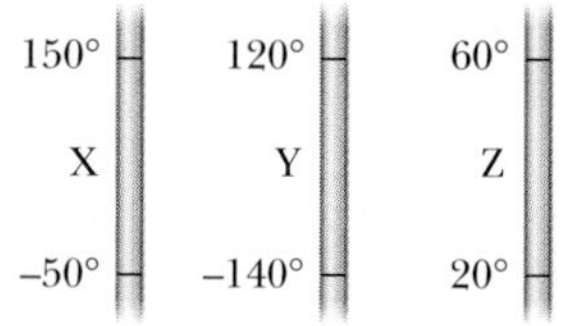

	X	Y	Z
boiling	150°	120°	60°
freezing	−50°	−140°	20°

FIG. 18-23 Question 2.

3 The initial length L, change in temperature ΔT, and change in length ΔL of four rods are given in the following table. Rank the rods according to their coefficients of thermal expansion, greatest first.

Rod	L (m)	ΔT (C°)	ΔL (m)
a	2	10	4×10^{-4}
b	1	20	4×10^{-4}
c	2	10	8×10^{-4}
d	4	5	4×10^{-4}

4 Figure 18-24 shows three different arrangements of materials 1, 2, and 3 to form a wall. The thermal conductivities are $k_1 > k_2 > k_3$. The left side of the wall is 20 C° higher than the right side. Rank the arrangements according to (a) the (steady state) rate of energy conduction through the wall and (b) the temperature difference across material 1, greatest first.

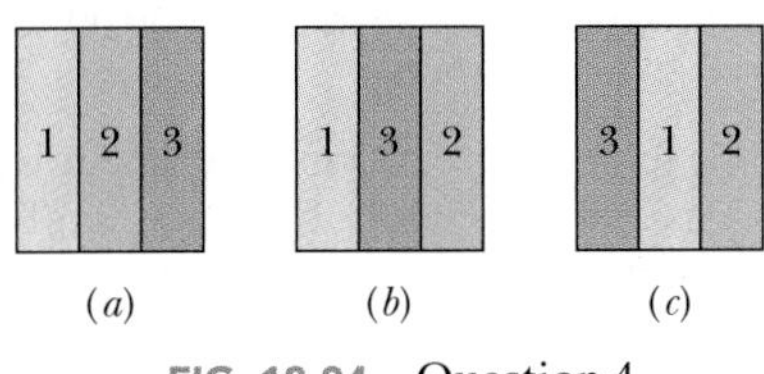

FIG. 18-24 Question 4.

5 Figure 18-25 shows two closed cycles on p-V diagrams for a gas. The three parts of cycle 1 are of the same length and shape as those of cycle 2. For each cycle, should the cycle be traversed clockwise or counterclockwise if (a) the net work W done by the gas is to be positive and (b) the net energy transferred by the gas as heat Q is to be positive?

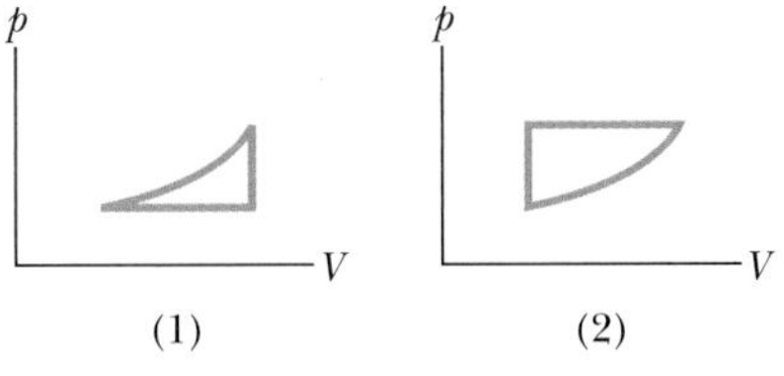

FIG. 18-25 Questions 5 and 6.

6 For which cycle in Fig. 18-25, traversed clockwise, is (a) W greater and (b) Q greater?

7 A hot object is dropped into a thermally insulated container of water, and the object and water are then allowed to come to thermal equilibrium. The experiment is repeated twice, with different hot objects. All three objects have the same mass and initial temperature, and the mass and initial temperature of the water are the same in the three experiments. For each of the experiments, Fig. 18-26 gives graphs of the temperatures T of the object and the water versus time t. Rank the graphs according to the specific heats of the objects, greatest first.

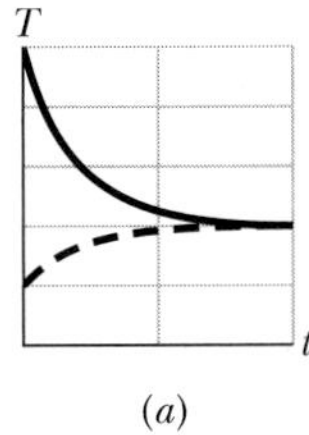

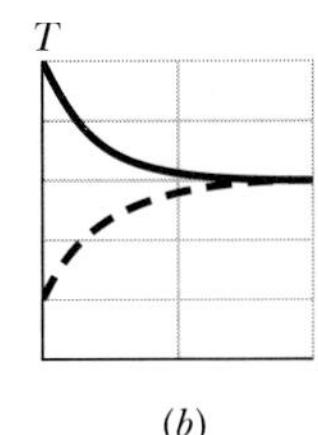

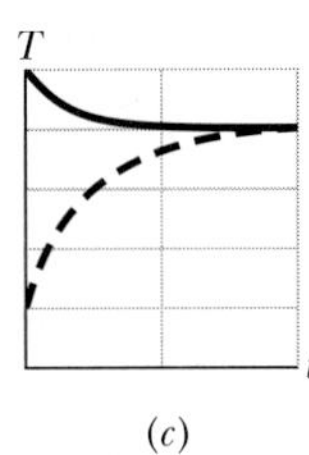

FIG. 18-26 Question 7.

8 A sample A of liquid water and a sample B of ice, of identical mass, are placed in a thermally insulated container and allowed to come to thermal equilibrium. Figure 18-27a is a sketch of the temperature T of the samples versus time t. (a) Is the equilibrium temperature above, below, or at the freezing point of water? (b) In reaching equilibrium, does the liquid partly freeze, fully freeze, or undergo no freezing? (c) Does the ice partly melt, fully melt, or undergo no melting?

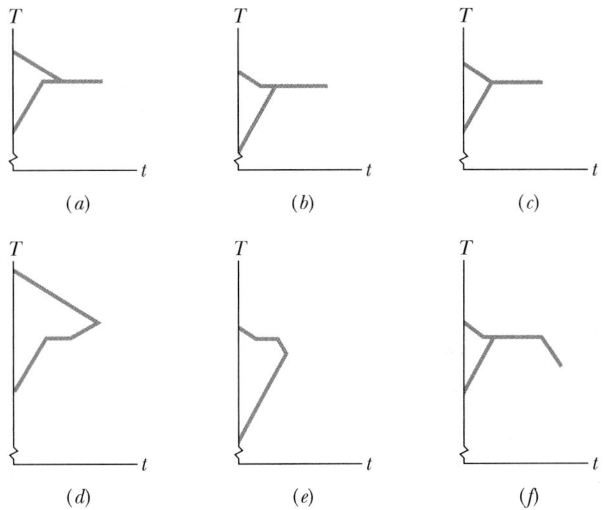

FIG. 18-27 Questions 8 and 9.

9 Question 8 continued: Graphs b through f of Fig. 18-27 are additional sketches of T versus t, of which one or more are impossible to produce. (a) Which is impossible and why? (b) In the possible ones, is the equilibrium temperature above, below, or at the freezing point of water? (c) As the possible situations reach equilibrium, does the liquid partly freeze, fully freeze, or undergo no freezing? Does the ice partly melt, fully melt, or undergo no melting?

10 A solid cube of edge length r, a solid sphere of radius r, and a solid hemisphere of radius r, all made of the same material, are maintained at temperature 300 K in an environment at temperature 350 K. Rank the objects according to the net rate at which thermal radiation is exchanged with the environment, greatest first.

11 Three different materials of identical mass are placed one at a time in a special freezer that can extract energy from a material at a certain constant rate. During the cooling process, each material begins in the liquid state and ends in the solid state; Fig. 18-28 shows the temperature T versus time t. (a) For material 1, is the specific heat for the liquid state greater than or less than that for the solid state? Rank the materials according to (b) freezing-point temperature, (c) specific heat in the liquid state, (d) specific heat in the solid state, and (e) heat of fusion, all greatest first.

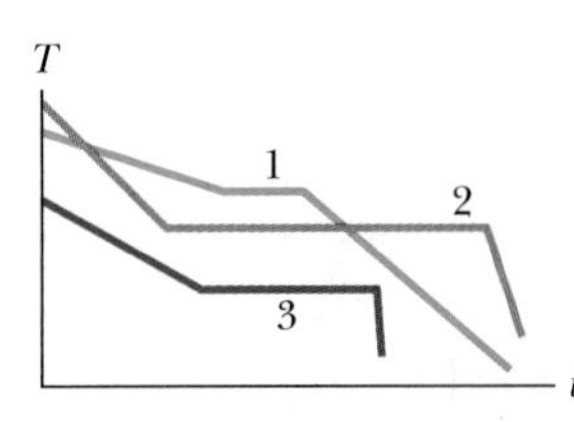

FIG. 18-28 Question 11.

PROBLEMS

Tutoring problem available (at instructor's discretion) in *WileyPLUS* and WebAssign
SSM Worked-out solution available in Student Solutions Manual
• – ••• Number of dots indicates level of problem difficulty
Additional information available in *The Flying Circus of Physics* and at flyingcircusofphysics.com
WWW Worked-out solution is at
ILW Interactive solution is at
http://www.wiley.com/college/halliday

sec. 18-4 Measuring Temperature

•1 A gas thermometer is constructed of two gas-containing bulbs, each in a water bath, as shown in Fig. 18-29. The pressure difference between the two bulbs is measured by a mercury manometer as shown. Appropriate reservoirs, not shown in the diagram, maintain constant gas volume in the two bulbs. There is no difference in pressure when both baths are at the triple point of water. The pressure difference is 120 torr when one bath is at the triple point and the other is at the boiling point of water. It is 90.0 torr when one bath is at the triple point and the other is at an unknown temperature to be measured. What is the unknown temperature?

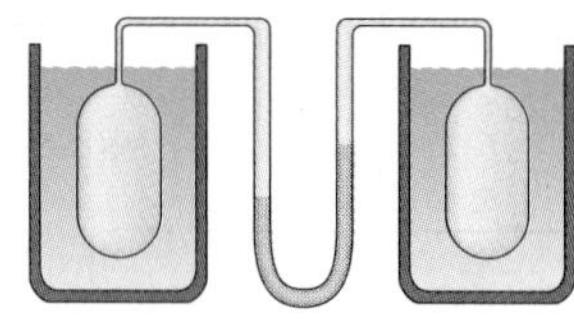
FIG. 18-29 Problem 1.

•2 Two constant-volume gas thermometers are assembled, one with nitrogen and the other with hydrogen. Both contain enough gas so that $p_3 = 80$ kPa. (a) What is the difference between the pressures in the two thermometers if both bulbs are in boiling water? (*Hint:* See Fig. 18-6.) (b) Which gas is at higher pressure?

•3 Suppose the temperature of a gas is 373.15 K when it is at the boiling point of water. What then is the limiting value of the ratio of the pressure of the gas at that boiling point to its pressure at the triple point of water? (Assume the volume of the gas is the same at both temperatures.)

sec. 18-5 The Celsius and Fahrenheit Scales

•4 (a) In 1964, the temperature in the Siberian village of Oymyakon reached −71°C. What temperature is this on the Fahrenheit scale? (b) The highest officially recorded temperature in the continental United States was 134°F in Death Valley, California. What is this temperature on the Celsius scale?

•5 At what temperature is the Fahrenheit scale reading equal to (a) twice that of the Celsius scale and (b) half that of the Celsius scale?

••6 On a linear X temperature scale, water freezes at −125.0°X and boils at 375.0°X. On a linear Y temperature scale, water freezes at −70.00°Y and boils at −30.00°Y. A temperature of 50.00°Y corresponds to what temperature on the X scale?

••7 Suppose that on a linear temperature scale X, water boils at −53.5°X and freezes at −170°X. What is a temperature of 340 K on the X scale? (Approximate water's boiling point as 373 K.) ILW

sec. 18-6 Thermal Expansion

•8 An aluminum flagpole is 33 m high. By how much does its length increase as the temperature increases by 15 C°?

•9 Find the change in volume of an aluminum sphere with an initial radius of 10 cm when the sphere is heated from 0.0°C to 100°C. SSM

•10 An aluminum-alloy rod has a length of 10.000 cm at 20.000°C and a length of 10.015 cm at the boiling point of water. (a) What is the length of the rod at the freezing point of water? (b) What is the temperature if the length of the rod is 10.009 cm?

•11 A circular hole in an aluminum plate is 2.725 cm in diameter at 0.000°C. What is its diameter when the temperature of the plate is raised to 100.0°C? ILW

•12 At 20°C, a brass cube has an edge length of 30 cm. What is the increase in the cube's surface area when it is heated from 20°C to 75°C?

•13 What is the volume of a lead ball at 30.00°C if the ball's volume at 60.00°C is 50.00 cm^3?

••14 When the temperature of a metal cylinder is raised from 0.0°C to 100°C, its length increases by 0.23%. (a) Find the percent change in density. (b) What is the metal? Use Table 18-2.

••15 An aluminum cup of 100 cm^3 capacity is completely filled with glycerin at 22°C. How much glycerin, if any, will spill out of the cup if the temperature of both the cup and the glycerin is increased to 28°C? (The coefficient of volume expansion of glycerin is 5.1×10^{-4}/C°.) SSM WWW

••16 At 20°C, a rod is exactly 20.05 cm long on a steel ruler. Both the rod and the ruler are placed in an oven at 270°C, where the rod now measures 20.11 cm on the same ruler. What is the coefficient of linear expansion for the material of which the rod is made?

••17 A steel rod is 3.000 cm in diameter at 25.00°C. A brass ring has an interior diameter of 2.992 cm at 25.00°C. At what common temperature will the ring just slide onto the rod? ILW

••18 When the temperature of a copper coin is raised by 100 C°, its diameter increases by 0.18%. To two significant figures, give the percent increase in (a) the area of a face, (b) the thickness, (c) the volume, and (d) the mass of the coin. (e) Calculate the coefficient of linear expansion of the coin.

••19 A vertical glass tube of length L = 1.280 000 m is half filled with a liquid at 20.000 000°C. How much will the height of the liquid column change when the tube is heated to 30.000 000°C? Take $\alpha_{glass} = 1.000\,000 \times 10^{-5}$/K and $\beta_{liquid} = 4.000\,000 \times 10^{-5}$/K. GO

••20 In a certain experiment, a small radioactive source must move at selected, extremely slow speeds. This motion is accomplished by fastening the source to one end of an aluminum rod and heating the central section of the rod in a controlled way. If the effective heated section of the rod in Fig. 18-30 has length d = 2.00 cm, at what constant rate must the temperature of the rod be changed if the source is to

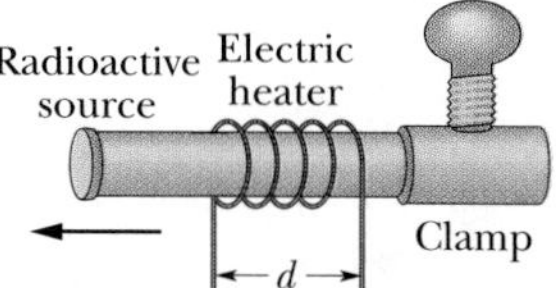

FIG. 18-30 Problem 20.

move at a constant speed of 100 nm/s?

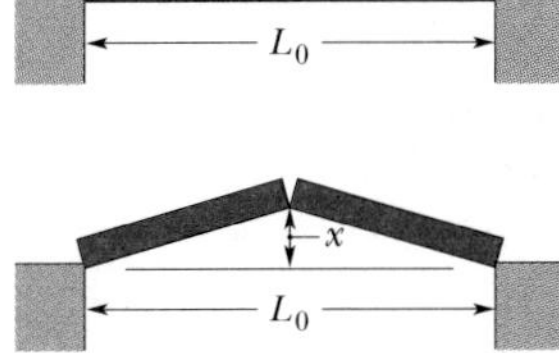

FIG. 18-31 Problem 21.

•••21 As a result of a temperature rise of 32 C°, a bar with a crack at its center buckles upward (Fig. 18-31). If the fixed distance L_0 is 3.77 m and the coefficient of linear expansion of the bar is 25×10^{-6}/C°, find the rise x of the center. SSM ILW

sec. 18-8 The Absorption of Heat by Solids and Liquids

•22 A certain substance has a mass per mole of 50.0 g/mol. When 314 J is added as heat to a 30.0 g sample, the sample's temperature rises from 25.0°C to 45.0°C. What are the (a) specific heat and (b) molar specific heat of this substance? (c) How many moles are present?

•23 A certain diet doctor encourages people to diet by drinking ice water. His theory is that the body must burn off enough fat to raise the temperature of the water from 0.00°C to the body temperature of 37.0°C. How many liters of ice water would have to be consumed to burn off 454 g (about 1 lb) of fat, assuming that burning this much fat requires 3500 Cal be transferred to the ice water? Why is it not advisable to follow this diet? (One liter = 10^3 cm^3. The density of water is 1.00 g/cm^3.)

•24 How much water remains unfrozen after 50.2 kJ is transferred as heat from 260 g of liquid water initially at its freezing point?

•25 Calculate the minimum amount of energy, in joules, required to completely melt 130 g of silver initially at 15.0°C. SSM

•26 One way to keep the contents of a garage from becoming too cold on a night when a severe subfreezing temperature is forecast is to put a tub of water in the garage. If the mass of the water is 125 kg and its initial temperature is 20°C, (a) how much energy must the water transfer to its surroundings in order to freeze completely and (b) what is the lowest possible temperature of the water and its surroundings until that happens?

•27 A small electric immersion heater is used to heat 100 g of water for a cup of instant coffee. The heater is labeled "200 watts" (it converts electrical energy to thermal energy at this rate). Calculate the time required to bring all this water from 23.0°C to 100°C, ignoring any heat losses. SSM

•28 What mass of butter, which has a usable energy content of 6.0 Cal/g (= 6000 cal/g), would be equivalent to the change in gravitational potential energy of a 73.0 kg man who ascends from sea level to the top of Mt. Everest, at elevation 8.84 km? Assume that the average g for the ascent is 9.80 m/s^2.

••29 What mass of steam at 100°C must be mixed with 150 g of ice at its melting point, in a thermally insulated container, to produce liquid water at 50°C? ILW

••30 A 150 g copper bowl contains 220 g of water, both at 20.0°C. A very hot 300 g copper cylinder is dropped into the water, causing the water to boil, with 5.00 g being converted to steam. The final temperature of the system is 100°C. Neglect energy transfers with the environment. (a) How much energy (in calories) is transferred to the water as heat? (b) How much to the bowl? (c) What is the original temperature of the cylinder?

••31 *Nonmetric version:* (a) How long does a 2.0×10^5 Btu/h water heater take to raise the temperature of 40 gal of water from 70°F to 100°F? *Metric version:* (b) How long does a 59 kW water heater take to raise the temperature of 150 L of water from 21°C to 38°C?

••32 Samples A and B are at different initial temperatures when they are placed in a thermally insulated container and allowed to come to thermal equilibrium. Figure 18-32a gives their temperatures T versus time t. Sample A has a mass of 5.0 kg; sample B has a mass of 1.5 kg. Figure 18-32b is a general plot for the material of sample B. It shows the temperature change ΔT that the material undergoes when energy is transferred to it as heat Q. The change ΔT is plotted versus the energy Q per unit mass of the material, and the scale of the vertical axis is set by $\Delta T_s = 4.0$ C°. What is the specific heat of sample A? GO

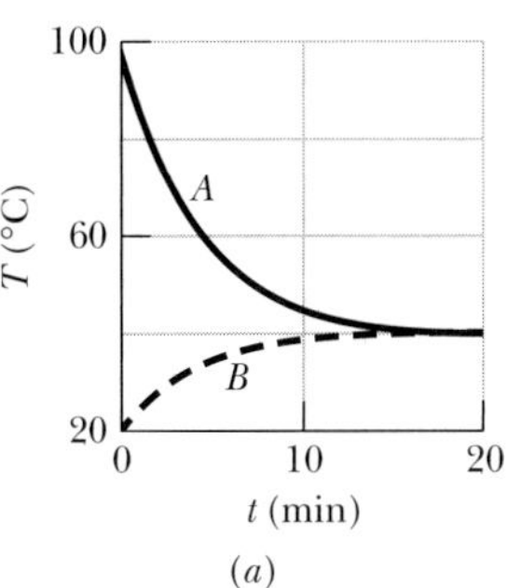

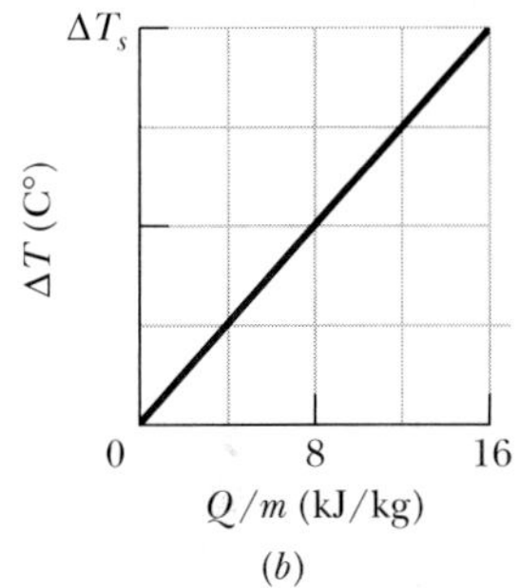

FIG. 18-32 Problem 32.

••33 In a solar water heater, energy from the Sun is gathered by water that circulates through tubes in a rooftop collector. The solar radiation enters the collector through a transparent cover and warms the water in the tubes; this water is pumped into a holding tank. Assume that the efficiency of the overall system is 20% (that is, 80% of the incident solar energy is lost from the system). What collector area is necessary to raise the temperature of 200 L of water in the tank from 20°C to 40°C in 1.0 h when the intensity of incident sunlight is 700 W/m^2?

••34 A 0.400 kg sample is placed in a cooling apparatus that removes energy as heat at a constant rate. Figure 18-33 gives the temperature T of the sample versus time t; the horizontal scale is set by $t_s = 80.0$ min. The sample freezes during the energy removal. The specific heat of the sample in its initial liquid phase is 3000 J/kg · K. What are (a) the sample's heat of fusion and (b) its specific heat in the frozen phase?

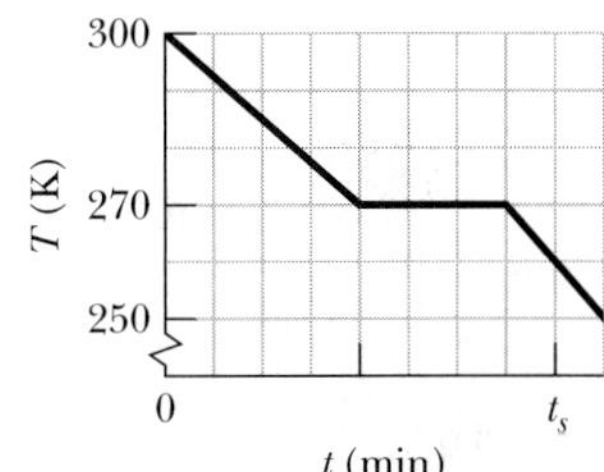

FIG. 18-33 Problem 34.

••35 An insulated Thermos contains 130 cm^3 of hot coffee at 80.0°C. You put in a 12.0 g ice cube at its melting point to cool the coffee. By how many degrees has your coffee cooled once the ice has melted and equilibrium is reached? Treat the coffee as though it were pure water and neglect energy exchanges with the environment.

••36 A 0.530 kg sample of liquid water and a sample of ice are placed in a thermally insulated container. The container also contains a device that transfers energy as heat from the liquid water to the ice at a constant rate P, until thermal equilibrium is reached. The temperatures T of the liquid water and the ice are given in Fig. 18-34 as functions of time t; the horizontal scale is set by $t_s = 80.0$ min. (a) What is rate P? (b) What is the initial mass of the ice in the container? (c) When thermal equilibrium is reached, what is the mass of the ice produced in this process?

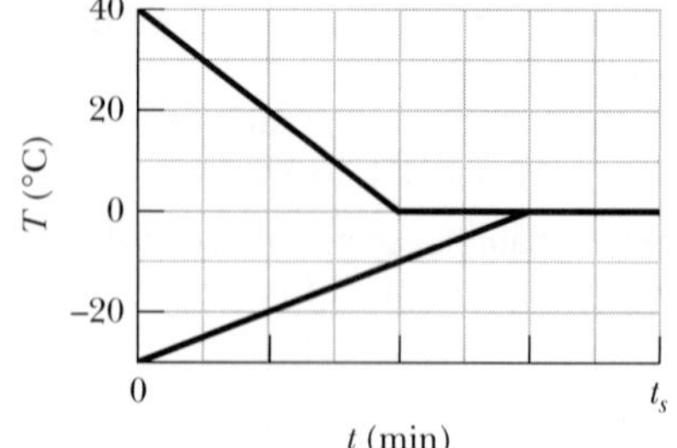

FIG. 18-34 Problem 36.

••37 Ethyl alcohol has a boiling point of 78.0°C, a freezing point of −114°C, a heat of vaporization of 879 kJ/kg, a heat of fusion of 109 kJ/kg, and a specific heat of 2.43 kJ/kg·K. How much energy must be removed from 0.510 kg of ethyl alcohol that is initially a gas at 78.0°C so that it becomes a solid at −114°C? GO

••38 The specific heat of a substance varies with temperature according to $c = 0.20 + 0.14T + 0.023T^2$, with T in °C and c in cal/g·K. Find the energy required to raise the temperature of 2.0 g of this substance from 5.0°C to 15°C.

••39 A person makes a quantity of iced tea by mixing 500 g of hot tea (essentially water) with an equal mass of ice at its melting point. Assume the mixture has negligible energy exchanges with its environment. If the tea's initial temperature is $T_i = 90°C$, when thermal equilibrium is reached what are (a) the mixture's temperature T_f and (b) the remaining mass m_f of ice? If $T_i = 70°C$, when thermal equilibrium is reached what are (c) T_f and (d) m_f?

•••40 *Icicles.* Liquid water coats an active (growing) icicle and extends up a short, narrow tube along the central axis (Fig. 18-35). Because the water–ice interface must have a temperature of 0°C, the water in the tube cannot lose energy through the sides of the icicle or down through the tip because there is no temperature change in those directions. It can lose energy and freeze only by sending energy up (through distance L) to the top of the icicle, where the temperature T_r can be below 0°C. Take $L = 0.12$ m and $T_r = -5°C$. Assume that the central tube and the upward conduction path both have cross-sectional area A. In terms of A, what rate is (a) energy conducted upward and (b) mass converted from liquid to ice at the top of the central tube? (c) At what rate does the top of the tube move downward because of water freezing there? The thermal conductivity of ice is 0.400 W/m·K, and the density of liquid water is 1000 kg/m³.

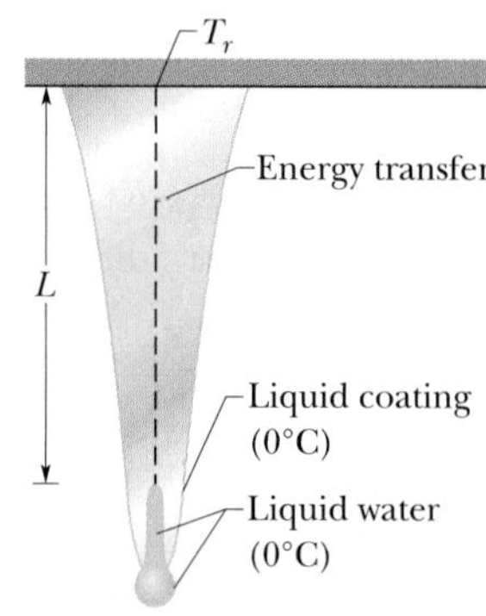

FIG. 18-35 Problem 40.

•••41 (a) Two 50 g ice cubes are dropped into 200 g of water in a thermally insulated container. If the water is initially at 25°C, and the ice comes directly from a freezer at −15°C, what is the final temperature at thermal equilibrium? (b) What is the final temperature if only one ice cube is used? SSM WWW

•••42 A 20.0 g copper ring at 0.000°C has an inner diameter of $D = 2.54000$ cm. An aluminum sphere at 100.0°C has a diameter of $d = 2.545\ 08$ cm. The sphere is placed on top of the ring (Fig. 18-36), and the two are allowed to come to thermal equilibrium, with no heat lost to the surroundings. The sphere just passes through the ring at the equilibrium temperature. What is the mass of the sphere?

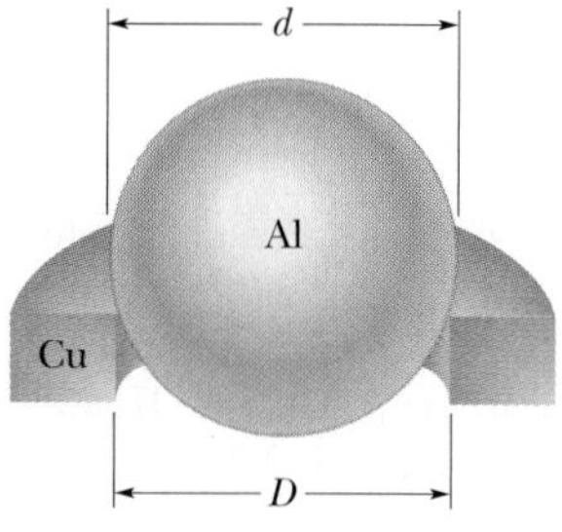

FIG. 18-36 Problem 42.

sec. 18-11 Some Special Cases of the First Law of Thermodynamics

•43 A gas within a closed chamber undergoes the cycle shown in the p-V diagram of Fig. 18-37. The horizontal scale is set by $V_s = 4.0$ m³. Calculate the net energy added to the system as heat during one complete cycle. SSM ILW

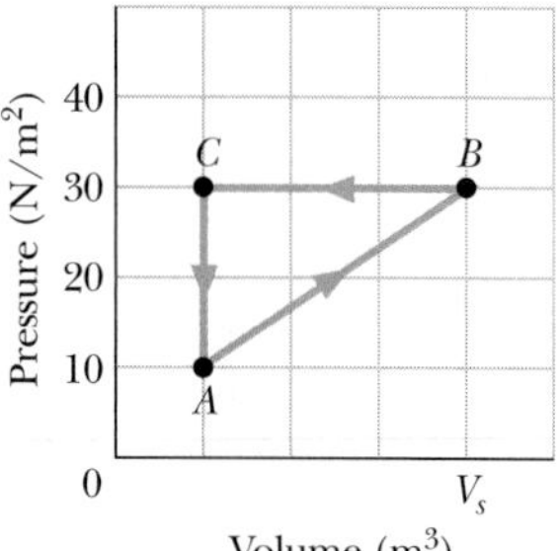

FIG. 18-37 Problem 43.

•44 Suppose 200 J of work is done on a system and 70.0 cal is extracted from the system as heat. In the sense of the first law of thermodynamics, what are the values (including algebraic signs) of (a) W, (b) Q, and (c) ΔE_{int}?

•45 In Fig. 18-38, a gas sample expands from V_0 to $4.0V_0$ while its pressure decreases from p_0 to $p_0/4.0$. If $V_0 = 1.0$ m³ and $p_0 = 40$ Pa, how much work is done by the gas if its pressure changes with volume via (a) path A, (b) path B, and (c) path C?

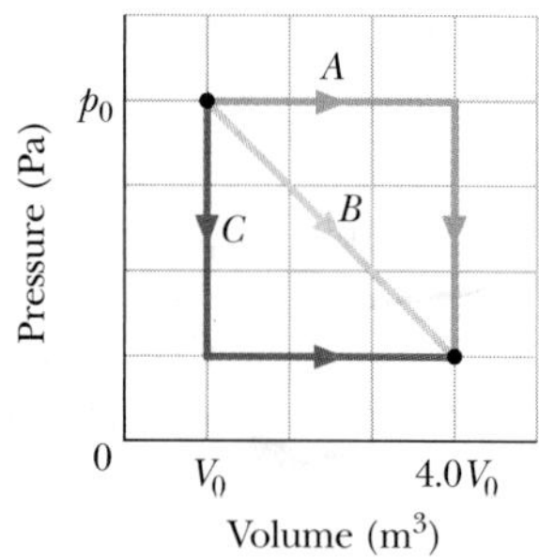

FIG. 18-38 Problem 45.

•46 A thermodynamic system is taken from state A to state B to state C, and then back to A, as shown in the p-V diagram of Fig. 18-39a. The vertical scale is set by $p_s = 40$ Pa, and the horizontal scale is set by $V_s = 4.0$ m³. (a)–(g) Complete the table in Fig. 18-39b by inserting a plus sign, a minus sign, or a zero in each indicated cell. (h) What is the net work done by the system as it moves once through the cycle $ABCA$?

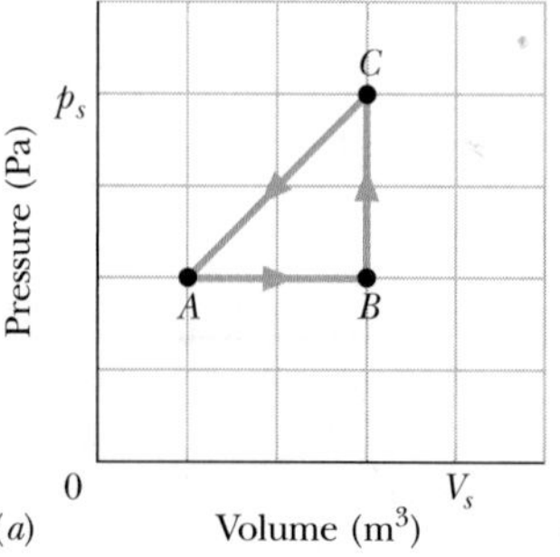

(b)

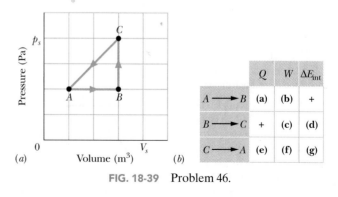

	Q	W	ΔE_{int}
A → B	(a)	(b)	+
B → C	+	(c)	(d)
C → A	(e)	(f)	(g)

FIG. 18-39 Problem 46.

••47 Figure 18-40 displays a closed cycle for a gas (the figure is not drawn to scale). The change in the internal energy of the gas as it moves from a to c along the path abc is -200 J. As it moves from c to d, 180 J must be transferred to it as heat. An additional transfer of 80 J to it as heat is needed as it moves from d to a. How much work is done on the gas as it moves from c to d? GO

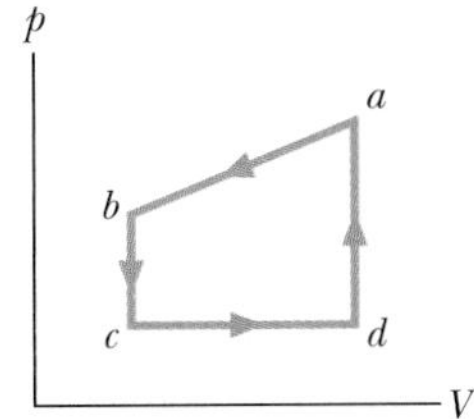

FIG. 18-40 Problem 47.

••48 A sample of gas is taken through cycle $abca$ shown in the p-V diagram of Fig. 18-41. The net work done is $+1.2$ J. Along path ab, the change in the internal energy is $+3.0$ J and the magnitude of the work done is 5.0 J. Along path ca, the energy transferred to the gas as heat is $+2.5$ J. How much energy is transferred as heat along (a) path ab and (b) path bc?

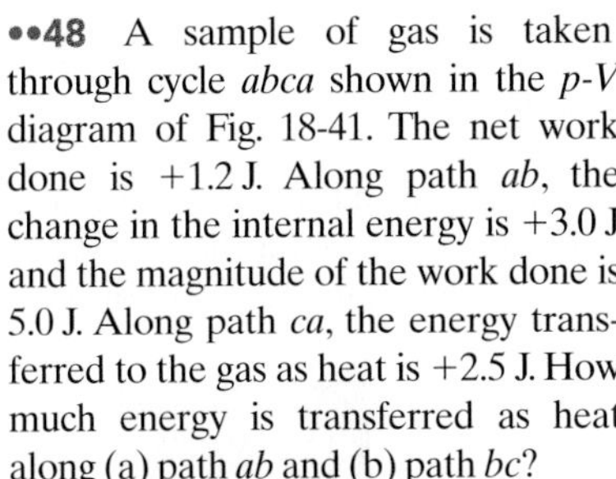

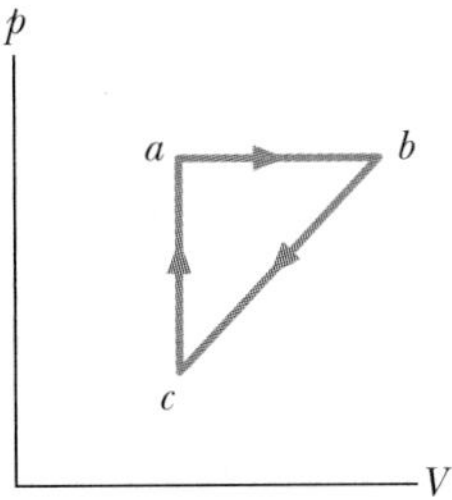

FIG. 18-41 Problem 48.

••49 When a system is taken from state i to state f along path iaf in Fig. 18-42, $Q = 50$ cal and $W = 20$ cal. Along path ibf, $Q = 36$ cal. (a) What is W along path ibf? (b) If $W = -13$ cal for the return path fi, what is Q for this path? (c) If $E_{\text{int},i} = 10$ cal, what is $E_{\text{int},f}$? If $E_{\text{int},b} = 22$ cal, what is Q for (d) path ib and (e) path bf? SSM WWW

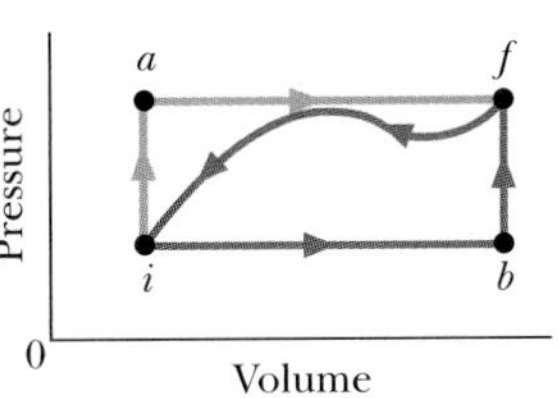

FIG. 18-42 Problem 49.

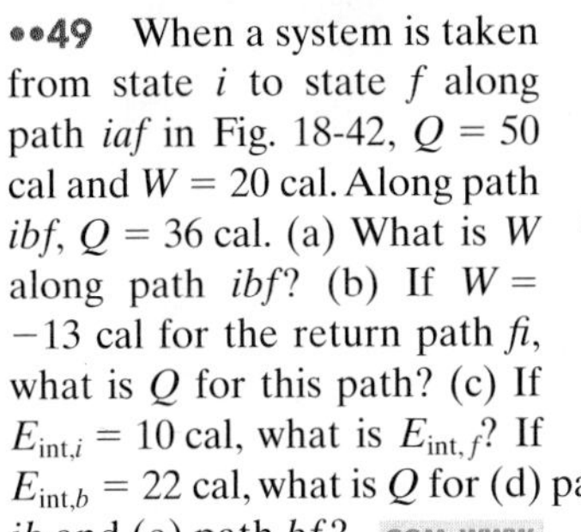

••50 Gas within a chamber passes through the cycle shown in Fig. 18-43. Determine the energy transferred by the system as heat during process CA if the energy added as heat Q_{AB} during process AB is 20.0 J, no energy is transferred as heat during process BC, and the net work done during the cycle is 15.0 J.

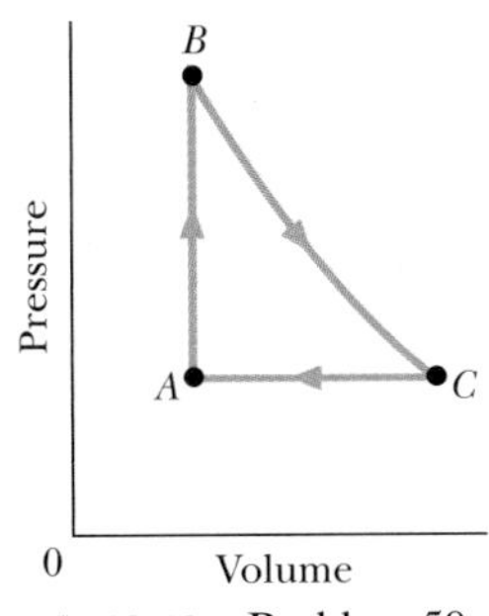

FIG. 18-43 Problem 50.

sec. 18-12 Heat Transfer Mechanisms

•51 Consider the slab shown in Fig. 18-18. Suppose that $L = 25.0$ cm, $A = 90.0\text{ cm}^2$, and the material is copper. If $T_H = 125°\text{C}$, $T_C = 10.0°\text{C}$, and a steady state is reached, find the conduction rate through the slab. SSM

•52 If you were to walk briefly in space without a spacesuit while far from the Sun (as an astronaut does in the movie *2001, A Space Odyssey*), you would feel the cold of space—while you radiated energy, you would absorb almost none from your environment. (a) At what rate would you lose energy? (b) How much energy would you lose in 30 s? Assume that your emissivity is 0.90, and estimate other data needed in the calculations.

•53 A cylindrical copper rod of length 1.2 m and cross-sectional area 4.8 cm^2 is insulated to prevent heat loss through its surface. The ends are maintained at a temperature difference of 100 C° by having one end in a water–ice mixture and the other in a mixture of boiling water and steam. (a) At what rate is energy conducted along the rod? (b) At what rate does ice melt at the cold end? ILW

•54 The ceiling of a single-family dwelling in a cold climate should have an R-value of 30. To give such insulation, how thick would a layer of (a) polyurethane foam and (b) silver have to be?

•55 A sphere of radius 0.500 m, temperature 27.0°C, and emissivity 0.850 is located in an environment of temperature 77.0°C. At what rate does the sphere (a) emit and (b) absorb thermal radiation? (c) What is the sphere's net rate of energy exchange?

••56 A solid cylinder of radius $r_1 = 2.5$ cm, length $h_1 = 5.0$ cm, emissivity 0.85, and temperature 30°C is suspended in an environment of temperature 50°C. (a) What is the cylinder's net thermal radiation transfer rate P_1? (b) If the cylinder is stretched until its radius is $r_2 = 0.50$ cm, its net thermal radiation transfer rate becomes P_2. What is the ratio P_2/P_1?

••57 In Fig. 18-44a, two identical rectangular rods of metal are welded end to end, with a temperature of $T_1 = 0°\text{C}$ on the left side and a temperature of $T_2 = 100°\text{C}$ on the right side. In 2.0 min, 10 J is conducted at a constant rate from the right side to the left side. How much time would be required to conduct 10 J if the rods were welded side to side as in Fig. 18-44b?

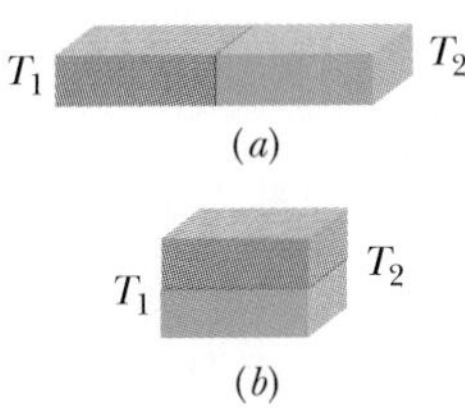

FIG. 18-44 Problem 57.

••58 Figure 18-45 shows the cross section of a wall made of three layers. The thicknesses of the layers are L_1, $L_2 = 0.700L_1$, and $L_3 = 0.350L_1$. The thermal conductivities are k_1, $k_2 = 0.900k_1$, and $k_3 = 0.800k_1$. The temperatures at the left and right sides of the wall are $T_H = 30.0°\text{C}$ and $T_C = -15.0°\text{C}$, respectively. Thermal conduction is steady. (a) What is the temperature difference ΔT_2 across layer 2 (between the left and right sides of the layer)? If k_2 were, instead, equal to $1.1k_1$, (b) would the rate at which energy is conducted through the wall be greater than, less than, or the same as previously, and (c) what would be the value of ΔT_2?

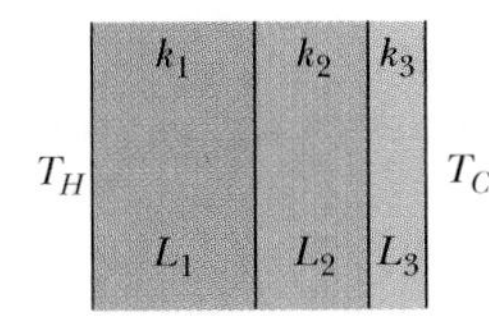

FIG. 18-45 Problem 58.

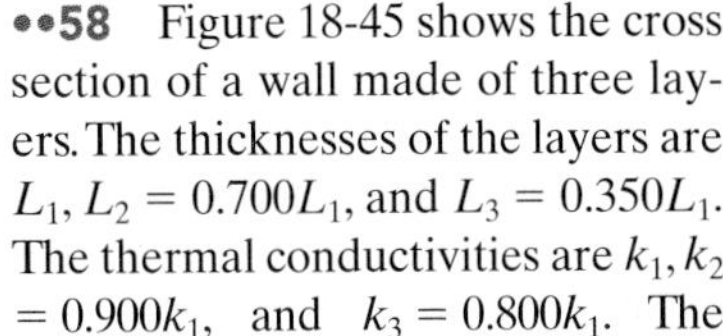

••59 (a) What is the rate of energy loss in watts per square meter through a glass window 3.0 mm thick if the outside temperature is $-20°\text{F}$ and the inside temperature is $+72°\text{F}$? (b) A storm window having the same thickness of glass is installed parallel to the first window, with an air gap of 7.5 cm between the two windows. What now is the rate of energy loss if conduction is the only important energy-loss mechanism?

••60 The giant hornet *Vespa mandarinia japonica* preys on Japanese bees. However, if one of the hornets attempts to invade a beehive, several hundred of the bees quickly form a compact ball around the hornet to stop it. They don't sting, bite, crush, or suffocate it. Rather they overheat it by quickly raising their body temperatures from the normal 35°C to 47°C or 48°C, which is lethal to the hornet but not to the bees (Fig. 18-46). Assume the following: 500 bees form a ball of radius $R = 2.0$

FIG. 18-46 Problem 60. (© Dr. Masato Ono, Tamagawa University)

cm for a time $t = 20$ min, the primary loss of energy by the ball is by thermal radiation, the ball's surface has emissivity $\varepsilon = 0.80$, and the ball has a uniform temperature. On average, how much additional energy must each bee produce during the 20 min to maintain 47°C?

••61 Figure 18-47 shows (in cross section) a wall consisting of four layers, with thermal conductivities $k_1 = 0.060$ W/m·K, $k_3 = 0.040$ W/m·K, and $k_4 = 0.12$ W/m·K (k_2 is not known). The layer thicknesses are $L_1 = 1.5$ cm, $L_3 = 2.8$ cm, and $L_4 = 3.5$ cm (L_2 is not known). The known temperatures are $T_1 = 30$°C, $T_{12} = 25$°C, and $T_4 = -10$°C. Energy transfer through the wall is steady. What is interface temperature T_{34}?

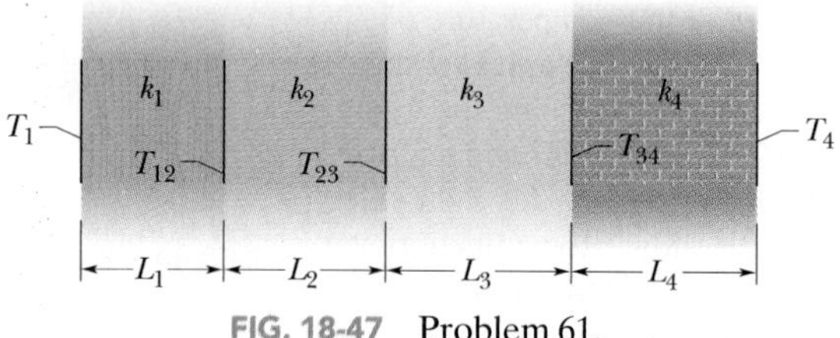

FIG. 18-47 Problem 61.

••62 *Penguin huddling.* To withstand the harsh weather of the Antarctic, emperor penguins huddle in groups (Fig. 18-48). Assume that a penguin is a circular cylinder with a top surface area $a = 0.34$ m² and height $h = 1.1$ m. Let P_r be the rate at which an individual penguin radiates energy to the environment (through the top and the sides); thus NP_r is the rate at which N identical, well-separated penguins radiate. If the penguins huddle closely to form a *huddled cylinder* with top surface area Na and height h, the cylinder radiates at the rate P_h. If $N = 1000$, (a) what is the value of the fraction P_h/NP_r and (b) by what percentage does huddling reduce the total radiation loss?

FIG. 18-48 Problem 62.
(*Alain Torterotot/Peter Arnold, Inc.*)

••63 Ice has formed on a shallow pond, and a steady state has been reached, with the air above the ice at −5.0°C and the bottom of the pond at 4.0°C. If the total depth of *ice + water* is 1.4 m, how thick is the ice? (Assume that the thermal conductivities of ice and water are 0.40 and 0.12 cal/m·C°·s, respectively.)

••64 *Leidenfrost effect.* A water drop that is slung onto a skillet with a temperature between 100°C and about 200°C will last about 1 s. However, if the skillet is much hotter, the drop can last several minutes, an effect named after an early investigator. The longer lifetime is due to the support of a thin layer of air and water vapor that separates the drop from the metal (by distance L in Fig. 18-49). Let $L = 0.100$ mm, and assume that the drop is flat with height $h = 1.50$ mm and bottom face area $A = 4.00 \times 10^{-6}$ m². Also assume that the skillet has a constant temperature $T_s = 300$°C and the drop has a temperature of 100°C. Water has density $\rho = 1000$ kg/m³, and the supporting layer has thermal conductivity $k = 0.026$ W/m·K. (a) At what rate is energy conducted from the skillet to the drop through the drop's bottom surface? (b) If conduction is the primary way energy moves from the skillet to the drop, how long will the drop last?

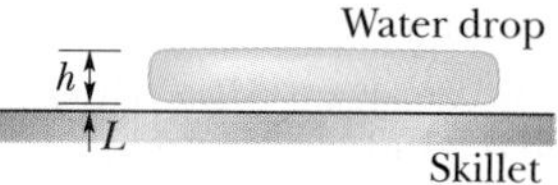

FIG. 18-49 Problem 64.

••65 A tank of water has been outdoors in cold weather, and a slab of ice 5.0 cm thick has formed on its surface (Fig. 18-50). The air above the ice is at −10°C. Calculate the rate of ice formation (in centimeters per hour) on the ice slab. Take the thermal conductivity of ice to be 0.0040 cal/s·cm·C° and its density to be 0.92 g/cm³. Assume no energy transfer through the tank walls or bottom. SSM

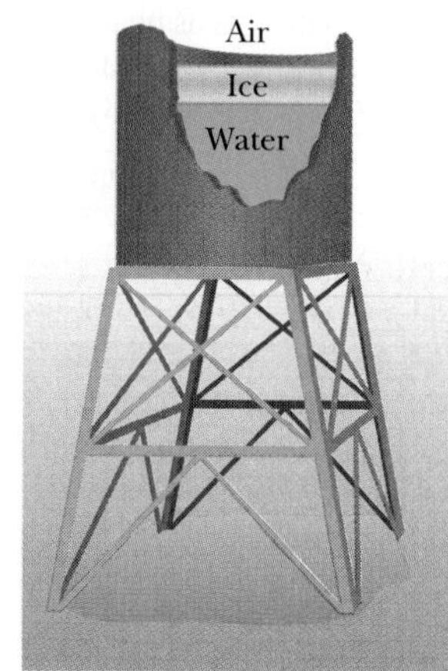

FIG. 18-50 Problem 65.

•••66 *Evaporative cooling of beverages.* A cold beverage can be kept cold even on a warm day if it is slipped into a porous ceramic container that has been soaked in water. Assume that energy lost to evaporation matches the net energy gained via the radiation exchange through the top and side surfaces. The container and beverage have temperature $T = 15$°C, the environment has temperature $T_{env} = 32$°C, and the container is a cylinder with radius $r = 2.2$ cm and height 10 cm. Approximate the emissivity as $\varepsilon = 1$, and neglect other energy exchanges. At what rate dm/dt is the container losing water mass?

Additional Problems

67 In the extrusion of cold chocolate from a tube, work is done on the chocolate by the pressure applied by a ram forcing the chocolate through the tube. The work per unit mass of extruded chocolate is equal to p/ρ, where p is the difference between the applied pressure and the pressure where the chocolate emerges from the tube, and ρ is the density of the chocolate. Rather than increasing the temperature of the chocolate, this work melts cocoa fats in the chocolate. These fats have a heat of fusion of 150 kJ/kg. Assume that all of the work goes into that melting and that these fats make up 30% of the chocolate's mass. What percentage of the fats melt during the extrusion if $p = 5.5$ MPa and $\rho = 1200$ kg/m³?

68 In a series of experiments, block B is to be placed in a thermally insulated container with block A, which has the same mass as block B. In each experiment, block B is initially at a certain temperature T_B, but temperature T_A of block A is changed from experiment to experiment. Let T_f represent the final temperature of the two blocks when they reach ther-

mal equilibrium in any of the experiments. Figure 18-51 gives temperature T_f versus the initial temperature T_A for a range of possible values of T_A, from $T_{A1} = 0$ K to $T_{A2} = 500$ K. The vertical axis scale is set by $T_{fs} = 400$ K. What are (a) temperature T_B and (b) the ratio c_B/c_A of the specific heats of the blocks?

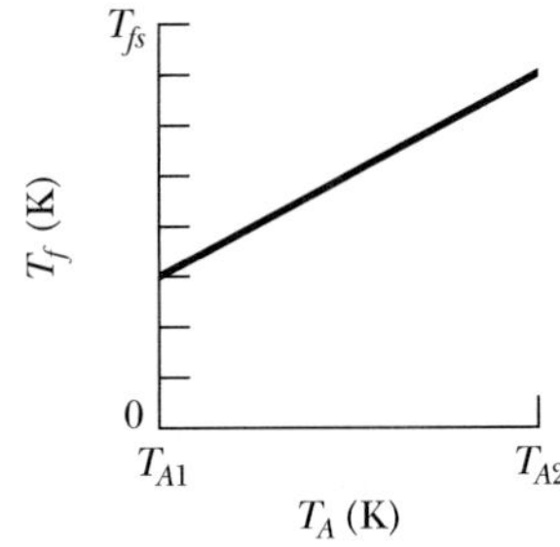

FIG. 18-51 Problem 68.

69 A 0.300 kg sample is placed in a cooling apparatus that removes energy as heat at a constant rate of 2.81 W. Figure 18-52 gives the temperature T of the sample versus time t. The temperature scale is set by $T_s = 30°C$ and the time scale is set by $t_s = 20$ min. What is the specific heat of the sample?

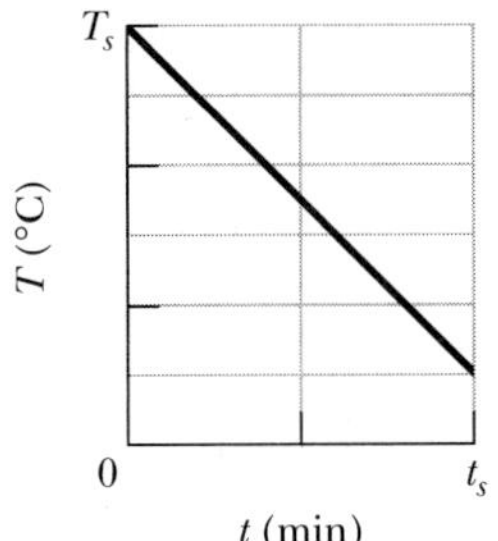

FIG. 18-52 Problem 69.

70 Calculate the specific heat of a metal from the following data. A container made of the metal has a mass of 3.6 kg and contains 14 kg of water. A 1.8 kg piece of the metal initially at a temperature of 180°C is dropped into the water. The container and water initially have a temperature of 16.0°C, and the final temperature of the entire system is 18.0°C.

71 What is the volume increase of an aluminum cube 5.00 cm on an edge when heated from 10.0°C to 60.0°C?

72 A copper rod, an aluminum rod, and a brass rod, each of 6.00 m length and 1.00 cm diameter, are placed end to end with the aluminum rod between the other two. The free end of the copper rod is maintained at water's boiling point, and the free end of the brass rod is maintained at water's freezing point. What is the steady-state temperature of (a) the copper–aluminum junction and (b) the aluminum–brass junction?

73 A sample of gas undergoes a transition from an initial state a to a final state b by three different paths (processes), as shown in the p-V diagram in Fig. 18-53, where $V_b = 5.00V_i$. The energy transferred to the gas as heat in process 1 is $10p_iV_i$. In terms of p_iV_i, what are (a) the energy transferred to the gas as heat in process 2 and (b) the change in internal energy that the gas undergoes in process 3? SSM

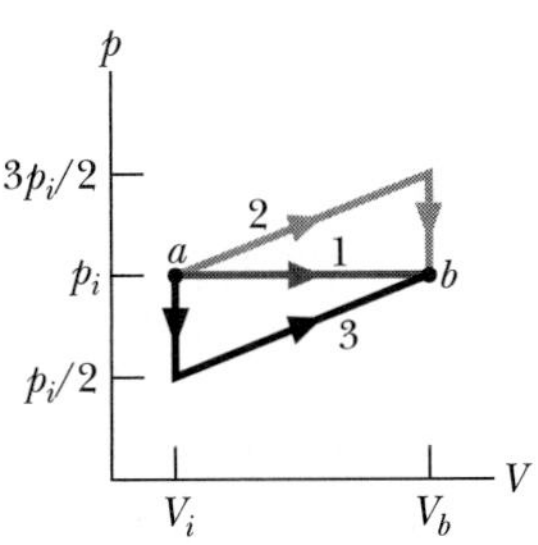

FIG. 18-53 Problem 73.

74 The average rate at which energy is conducted outward through the ground surface in North America is 54.0 mW/m², and the average thermal conductivity of the near-surface rocks is 2.50 W/m·K. Assuming a surface temperature of 10.0°C, find the temperature at a depth of 35.0 km (near the base of the crust). Ignore the heat generated by the presence of radioactive elements.

75 The temperature of a Pyrex disk is changed from 10.0°C to 60.0°C. Its initial radius is 8.00 cm; its initial thickness is 0.500 cm. Take these data as being exact. What is the change in the volume of the disk? (See Table 18-2.) SSM

76 In a certain solar house, energy from the Sun is stored in barrels filled with water. In a particular winter stretch of five cloudy days, 1.00×10^6 kcal is needed to maintain the inside of the house at 22.0°C. Assuming that the water in the barrels is at 50.0°C and that the water has a density of 1.00×10^3 kg/m³, what volume of water is required?

77 A sample of gas expands from an initial pressure and volume of 10 Pa and 1.0 m³ to a final volume of 2.0 m³. During the expansion, the pressure and volume are related by the equation $p = aV^2$, where $a = 10$ N/m⁸. Determine the work done by the gas during this expansion. SSM

78 (a) Calculate the rate at which body heat is conducted through the clothing of a skier in a steady-state process, given the following data: the body surface area is 1.8 m², and the clothing is 1.0 cm thick; the skin surface temperature is 33°C and the outer surface of the clothing is at 1.0°C; the thermal conductivity of the clothing is 0.040 W/m·K. (b) If, after a fall, the skier's clothes became soaked with water of thermal conductivity 0.60 W/m·K, by how much is the rate of conduction multiplied?

79 Figure 18-54 displays a closed cycle for a gas. From c to b, 40 J is transferred from the gas as heat. From b to a, 130 J is transferred from the gas as heat, and the magnitude of the work done by the gas is 80 J. From a to c, 400 J is transferred to the gas as heat. What is the work done by the gas from a to c? (*Hint:* You need to supply the plus and minus signs for the given data.)

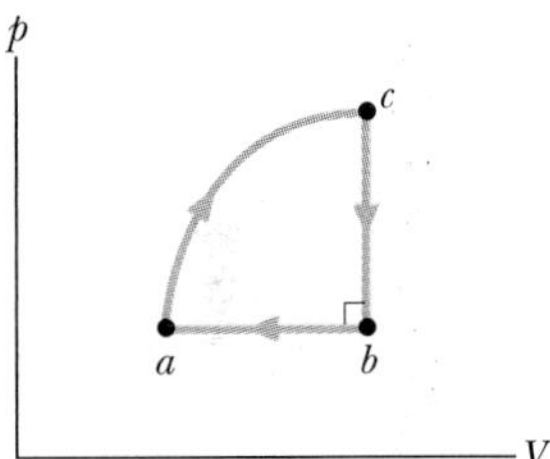

FIG. 18-54 Problem 79.

80 A glass window pane is exactly 20 cm by 30 cm at 10°C. By how much has its area increased when its temperature is 40°C, assuming that it can expand freely?

81 A 2.50 kg lump of aluminum is heated to 92.0°C and then dropped into 8.00 kg of water at 5.00°C. Assuming that the lump–water system is thermally isolated, what is the system's equilibrium temperature? SSM

82 Figure 18-55*a* shows a cylinder containing gas and closed by a movable piston. The cylinder is kept submerged in an ice–water mixture. The piston is *quickly* pushed down from position 1 to position 2 and then held at position 2 until the gas is again at the temperature of the ice–water mixture; it then is *slowly* raised back to position 1. Figure 18-55*b* is a p-V diagram for the process. If 100 g of ice is melted during the cycle, how much work has been done *on* the gas?

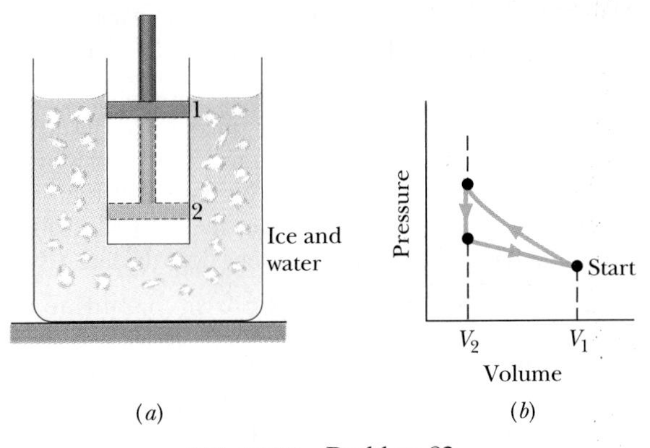

FIG. 18-55 Problem 82.

83 The temperature of a 0.700 kg cube of ice is decreased to $-150°C$. Then energy is gradually transferred to the cube as heat while it is otherwise thermally isolated from its environment. The total transfer is 0.6993 MJ. Assume the value of c_{ice} given in Table 18-3 is valid for temperatures from $-150°C$ to 0°C. What is the final temperature of the water? SSM

84 A steel rod at 25.0°C is bolted at both ends and then cooled. At what temperature will it rupture? Use Table 12-1.

85 Suppose that you intercept 5.0×10^{-3} of the energy radiated by a hot sphere that has a radius of 0.020 m, an emissivity of 0.80, and a surface temperature of 500 K. How much energy do you intercept in 2.0 min?

86 Three equal-length straight rods, of aluminum, Invar, and steel, all at 20.0°C, form an equilateral triangle with hinge pins at the vertices. At what temperature will the angle opposite the Invar rod be 59.95°? See Appendix E for needed trigonometric formulas and Table 18-2 for needed data.

87 It is possible to melt ice by rubbing one block of it against another. How much work, in joules, would you have to do to get 1.00 g of ice to melt?

88 A thermometer of mass 0.0550 kg and of specific heat 0.837 kJ/kg·K reads 15.0°C. It is then completely immersed in 0.300 kg of water, and it comes to the same final temperature as the water. If the thermometer then reads 44.4°C, what was the temperature of the water before insertion of the thermometer?

89 A recruit can join the semi-secret "300 F" club at the Amundsen–Scott South Pole Station only when the outside temperature is below $-70°C$. On such a day, the recruit first basks in a hot sauna and then runs outside wearing only shoes. (This is, of course, extremely dangerous, but the rite is effectively a protest against the constant danger of the cold.)

Assume that upon stepping out of the sauna, the recruit's skin temperature is 102°F and the walls, ceiling, and floor of the sauna room have a temperature of 30°C. Estimate the recruit's surface area, and take the skin emissivity to be 0.80. (a) What is the approximate net rate P_{net} at which the recruit loses energy via thermal radiation exchanges with the room? Next, assume that when outdoors, half the recruit's surface area exchanges thermal radiation with the sky at a temperature of $-25°C$ and the other half exchanges thermal radiation with the snow and ground at a temperature of $-80°C$. What is the approximate net rate at which the recruit loses energy via thermal radiation exchanges with (b) the sky and (c) the snow and ground?

90 A rectangular plate of glass initially has the dimensions 0.200 m by 0.300 m. The coefficient of linear expansion for the glass is 9.00×10^{-6}/K. What is the change in the plate's area if its temperature is increased by 20.0 K?

91 An athlete needs to lose weight and decides to do it by "pumping iron." (a) How many times must an 80.0 kg weight be lifted a distance of 1.00 m in order to burn off 1.00 lb of fat, assuming that that much fat is equivalent to 3500 Cal? (b) If the weight is lifted once every 2.00 s, how long does the task take?

92 Icebergs in the North Atlantic present hazards to shipping, causing the lengths of shipping routes to be increased by about 30% during the iceberg season. Attempts to destroy icebergs include planting explosives, bombing, torpedoing, shelling, ramming, and coating with black soot. Suppose that direct melting of the iceberg, by placing heat sources in the ice, is tried. How much energy as heat is required to melt 10% of an iceberg that has a mass of 200 000 metric tons? (Use 1 metric ton = 1000 kg.)

93 A sample of gas expands from $V_1 = 1.0$ m³ and $p_1 = 40$ Pa to $V_2 = 4.0$ m³ and $p_2 = 10$ Pa along path B in the p-V diagram in Fig. 18-56. It is then compressed back to V_1 along either path A or path C. Compute the net work done by the gas for the complete cycle along (a) path BA and (b) path BC.

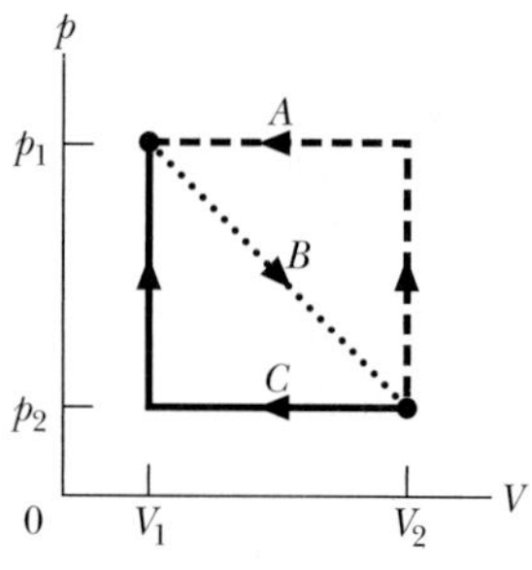

FIG. 18-56 Problem 93.

94 Soon after Earth was formed, heat released by the decay of radioactive elements raised the average internal temperature from 300 to 3000 K, at about which value it remains today. Assuming an average coefficient of volume expansion of 3.0×10^{-5} K^{-1}, by how much has the radius of Earth increased since the planet was formed?

95 Figure 18-57 displays a closed cycle for a gas. The change in internal energy along path ca is -160 J. The energy transferred to the gas as heat is 200 J along path ab, and 40 J along path bc. How much work is done by the gas along (a) path abc and (b) path ab?

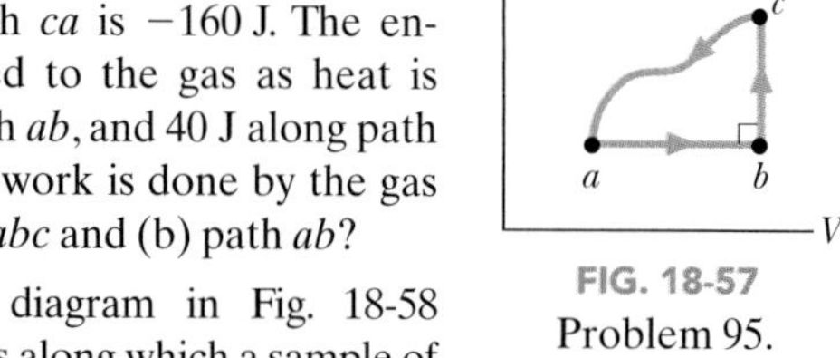

FIG. 18-57 Problem 95.

96 The p-V diagram in Fig. 18-58 shows two paths along which a sample of gas can be taken from state a to state b, where $V_b = 3.0V_1$. Path 1 requires that energy equal to $5.0p_1V_1$ be transferred to the gas as heat. Path 2 requires that energy equal to $5.5p_1V_1$ be transferred to the gas as heat. What is the ratio p_2/p_1?

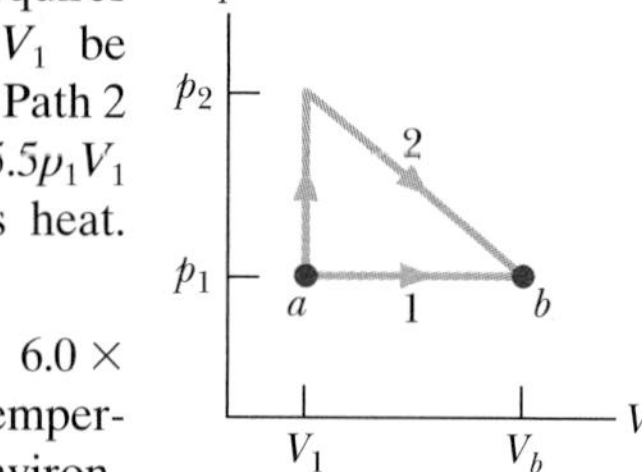

FIG. 18-58 Problem 96.

97 A cube of edge length 6.0×10^{-6} m, emissivity 0.75, and temperature $-100°C$ floats in an environment at $-150°C$. What is the cube's net thermal radiation transfer rate?

98 A *flow calorimeter* is a device used to measure the specific heat of a liquid. Energy is added as heat at a known rate to a stream of the liquid as it passes through the calorimeter at a known rate. Measurement of the resulting temperature difference between the inflow and the outflow points of the liquid stream enables us to compute the specific heat of the liquid. Suppose a liquid of density 0.85 g/cm³ flows through a calorimeter at the rate of 8.0 cm³/s. When energy is added at the rate of 250 W by means of an electric heating coil, a temperature difference of 15 C° is established in steady-state conditions between the inflow and the outflow points. What is the specific heat of the liquid?

99 An object of mass 6.00 kg falls through a height of 50.0 m and, by means of a mechanical linkage, rotates a paddle wheel that stirs 0.600 kg of water. Assume that the initial gravitational potential energy of the object is fully transferred to thermal energy of the water, which is initially at 15.0°C. What is the temperature rise of the water?

The Kinetic Theory of Gases

Tom Branch/Photo Researchers

When a container of cold champagne, soda pop, or any other carbonated drink is opened, a slight fog forms around the opening and some of the liquid sprays outward. In the photograph, the fog is the white cloud that surrounds the stopper, and the spray has formed streaks within the cloud.

What causes the fog?

The answer is in this chapter.

19-1 WHAT IS PHYSICS?

One of the main subjects in thermodynamics is the physics of gases. A gas consists of atoms (either individually or bound together as molecules) that fill their container's volume and exert pressure on the container's walls. We can usually assign a temperature to such a contained gas. These three variables associated with a gas—volume, pressure, and temperature—are all a consequence of the motion of the atoms. The volume is a result of the freedom the atoms have to spread throughout the container, the pressure is a result of the collisions of the atoms with the container's walls, and the temperature has to do with the kinetic energy of the atoms. The **kinetic theory of gases,** the focus of this chapter, relates the motion of the atoms to the volume, pressure, and temperature of the gas.

Applications of the kinetic theory of gases are countless. Automobile engineers are concerned with the combustion of vaporized fuel (a gas) in the automobile engines. Food engineers are concerned with the production rate of the fermentation gas that causes bread to rise as it bakes. Beverage engineers are concerned with how gas can produce the head in a glass of beer or shoot a cork from a champagne bottle. Medical engineers and physiologists are concerned with calculating how long a scuba diver must pause during ascent to eliminate nitrogen gas from the bloodstream (to avoid the *bends*). Environmental scientists are concerned with how heat exchanges between the oceans and the atmosphere can affect weather conditions.

The first step in our discussion of the kinetic theory of gases deals with measuring the amount of a gas present in a sample, for which we use Avogadro's number.

19-2 | Avogadro's Number

When our thinking is slanted toward atoms and molecules, it makes sense to measure the sizes of our samples in moles. If we do so, we can be certain that we are comparing samples that contain the same number of atoms or molecules. The *mole* is one of the seven SI base units and is defined as follows:

> One mole is the number of atoms in a 12 g sample of carbon-12.

The obvious question now is: "How many atoms or molecules are there in a mole?" The answer is determined experimentally and, as you saw in Chapter 18, is

$$N_A = 6.02 \times 10^{23}\ \text{mol}^{-1} \quad \text{(Avogadro's number)}, \tag{19-1}$$

where mol^{-1} represents the inverse mole or "per mole," and mol is the abbreviation for mole. The number N_A is called **Avogadro's number** after Italian scientist Amedeo Avogadro (1776–1856), who suggested that all gases occupying the same volume under the same conditions of temperature and pressure contain the same number of atoms or molecules.

The number of moles n contained in a sample of any substance is equal to the ratio of the number of molecules N in the sample to the number of molecules N_A in 1 mol:

$$n = \frac{N}{N_A}. \tag{19-2}$$

(*Caution:* The three symbols in this equation can easily be confused with one another, so you should sort them with their meanings now, before you end in "N-confusion.") We can find the number of moles n in a sample from the mass M_{sam} of the sample and either the *molar mass* M (the mass of 1 mol) or the molecular mass m (the mass of one molecule):

$$n = \frac{M_{\text{sam}}}{M} = \frac{M_{\text{sam}}}{mN_{\text{A}}}. \tag{19-3}$$

In Eq. 19-3, we used the fact that the mass M of 1 mol is the product of the mass m of one molecule and the number of molecules N_{A} in 1 mol:

$$M = mN_{\text{A}}. \tag{19-4}$$

PROBLEM-SOLVING TACTICS

Tactic 1: *Avogadro's Number of What?* In Eq. 19-1, Avogadro's number is expressed in terms of mol^{-1}, which is the inverse mole, or 1/mol. We could instead explicitly state the elementary unit involved in a given situation. For example, if the elementary unit is an atom, we might write $N_{\text{A}} = 6.02 \times 10^{23}$ atoms/mol. If, instead, the elementary unit is a molecule, then we might write $N_{\text{A}} = 6.02 \times 10^{23}$ molecules/mol. Being explicit about the unit is usually the better style of writing.

19-3 | Ideal Gases

Our goal in this chapter is to explain the macroscopic properties of a gas—such as its pressure and its temperature—in terms of the behavior of the molecules that make it up. However, there is an immediate problem: which gas? Should it be hydrogen, oxygen, or methane, or perhaps uranium hexafluoride? They are all different. Experimenters have found, though, that if we confine 1 mol samples of various gases in boxes of identical volume and hold the gases at the same temperature, then their measured pressures are nearly—though not exactly—the same. If we repeat the measurements at lower gas densities, then these small differences in the measured pressures tend to disappear. Further experiments show that, at low enough densities, all real gases tend to obey the relation

$$pV = nRT \quad \text{(ideal gas law)}, \tag{19-5}$$

in which p is the absolute (not gauge) pressure, n is the number of moles of gas present, and T is the temperature in kelvins. The symbol R is a constant called the **gas constant** that has the same value for all gases—namely,

$$R = 8.31 \text{ J/mol} \cdot \text{K}. \tag{19-6}$$

Equation 19-5 is called the **ideal gas law.** Provided the gas density is low, this law holds for any single gas or for any mixture of different gases. (For a mixture, n is the total number of moles in the mixture.)

We can rewrite Eq. 19-5 in an alternative form, in terms of a constant called the **Boltzmann constant** k, which is defined as

$$k = \frac{R}{N_{\text{A}}} = \frac{8.31 \text{ J/mol} \cdot \text{K}}{6.02 \times 10^{23} \text{ mol}^{-1}} = 1.38 \times 10^{-23} \text{ J/K}. \tag{19-7}$$

This allows us to write $R = kN_{\text{A}}$. Then, with Eq. 19-2 ($n = N/N_{\text{A}}$), we see that

$$nR = Nk. \tag{19-8}$$

Substituting this into Eq. 19-5 gives a second expression for the ideal gas law:

$$pV = NkT \quad \text{(ideal gas law)}. \tag{19-9}$$

(*Caution*: Note the difference between the two expressions for the ideal gas law—Eq. 19-5 involves the number of moles n, and Eq. 19-9 involves the number of molecules N.)

You may well ask, "What is an *ideal gas,* and what is so 'ideal' about it?" The answer lies in the simplicity of the law (Eqs. 19-5 and 19-9) that governs its macroscopic properties. Using this law—as you will see—we can deduce many properties of the ideal gas in a simple way. Although there is no such thing in nature as a truly ideal gas, *all real* gases approach the ideal state at low enough densities—that is, under conditions in which their molecules are far enough apart that they do not interact with one another. Thus, the ideal gas concept allows us to gain useful insights into the limiting behavior of real gases.

FIG. 19-1 A railroad tank car crushed overnight. *(Photo courtesy www.Houston.RailFan.net)*

The interior of the railroad tank car in Fig. 19-1 was being cleaned with steam by a crew late one afternoon. Because the job was unfinished at the end of their work shift, they sealed the car and left for the night. When they returned the next morning, they discovered that something had crushed the car in spite of its extremely strong steel walls, as if some giant creature from a grade B science fiction movie had stepped on it during a rampage that night.

With Eq. 19-9, we can explain what actually crushed the railroad tank car. When the car was being cleaned, its interior was filled with very hot steam, which is a gas of water molecules. The cleaning crew left the steam inside the car when they closed all the valves on the car at the end of their work shift. At that point the pressure of the gas in the car was equal to atmospheric pressure because the valves had been opened to the atmosphere during the cleaning. As the car cooled during the night, the steam cooled and much of it condensed, which means that the number N of gas molecules and the temperature T of the gas both decreased. Thus, the right side of Eq. 19-9 decreased, and because volume V was constant, the gas pressure p on the left side also decreased. At some point during the night, the gas pressure inside the car reached such a low value that the external atmospheric pressure was able to crush the car's steel walls. The cleaning crew could have prevented this accident by leaving the valves open, so that air could enter the car to keep the internal pressure equal to the external atmospheric pressure.

Work Done by an Ideal Gas at Constant Temperature

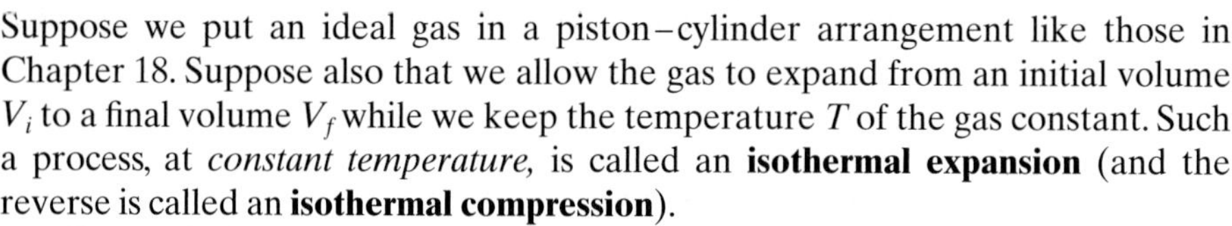

Suppose we put an ideal gas in a piston–cylinder arrangement like those in Chapter 18. Suppose also that we allow the gas to expand from an initial volume V_i to a final volume V_f while we keep the temperature T of the gas constant. Such a process, at *constant temperature,* is called an **isothermal expansion** (and the reverse is called an **isothermal compression**).

On a p-V diagram, an *isotherm* is a curve that connects points that have the same temperature. Thus, it is a graph of pressure versus volume for a gas whose temperature T is held constant. For n moles of an ideal gas, it is a graph of the equation

$$p = nRT\frac{1}{V} = (\text{a constant})\frac{1}{V}. \tag{19-10}$$

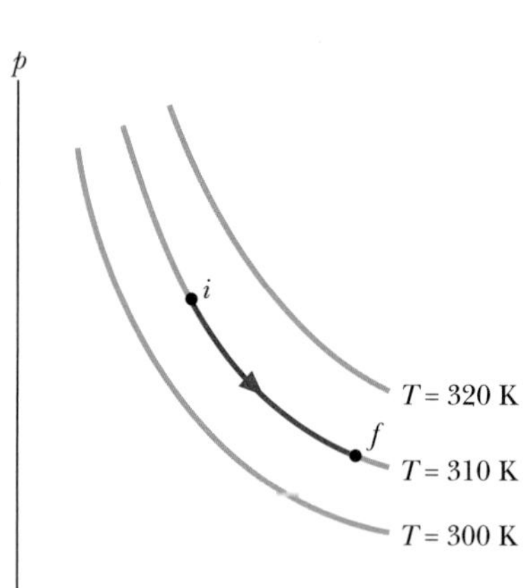

FIG. 19-2 Three isotherms on a p-V diagram. The path shown along the middle isotherm represents an isothermal expansion of a gas from an initial state i to a final state f. The path from f to i along the isotherm would represent the reverse process—that is, an isothermal compression.

Figure 19-2 shows three isotherms, each corresponding to a different (constant) value of T. (Note that the values of T for the isotherms increase upward to the right.) Superimposed on the middle isotherm is the path followed by a gas during an isothermal expansion from state i to state f at a constant temperature of 310 K.

To find the work done by an ideal gas during an isothermal expansion, we start with Eq. 18-25,

$$W = \int_{V_i}^{V_f} p\, dV. \tag{19-11}$$

This is a general expression for the work done during any change in volume of any gas. For an ideal gas, we can use Eq. 19-5 ($pV = nRT$) to substitute for p, obtaining

$$W = \int_{V_i}^{V_f} \frac{nRT}{V}\, dV. \tag{19-12}$$

Because we are considering an isothermal expansion, T is constant, so we can move it in front of the integral sign to write

$$W = nRT \int_{V_i}^{V_f} \frac{dV}{V} = nRT \Big[\ln V \Big]_{V_i}^{V_f}. \tag{19-13}$$

By evaluating the expression in brackets at the limits and then using the relationship $\ln a - \ln b = \ln(a/b)$, we find that

$$W = nRT \ln \frac{V_f}{V_i} \quad \text{(ideal gas, isothermal process).} \tag{19-14}$$

Recall that the symbol ln specifies a *natural* logarithm, which has base e.

For an expansion, V_f is greater than V_i, so the ratio V_f/V_i in Eq. 19-14 is greater than unity. The natural logarithm of a quantity greater than unity is positive, and so the work W done by an ideal gas during an isothermal expansion is positive, as we expect. For a compression, V_f is less than V_i, so the ratio of volumes in Eq. 19-14 is less than unity. The natural logarithm in that equation—hence the work W—is negative, again as we expect.

Work Done at Constant Volume and at Constant Pressure

Equation 19-14 does not give the work W done by an ideal gas during *every* thermodynamic process. Instead, it gives the work only for a process in which the temperature is held constant. If the temperature varies, then the symbol T in Eq. 19-12 cannot be moved in front of the integral symbol as in Eq. 19-13, and thus we do not end up with Eq. 19-14.

However, we can always go back to Eq. 19-11 to find the work W done by an ideal gas (or any other gas) during any process, such as a constant-volume process and a constant-pressure process. If the volume of the gas is constant, then Eq. 19-11 yields

$$W = 0 \quad \text{(constant-volume process).} \tag{19-15}$$

If, instead, the volume changes while the pressure p of the gas is held constant, then Eq. 19-11 becomes

$$W = p(V_f - V_i) = p\,\Delta V \quad \text{(constant-pressure process).} \tag{19-16}$$

CHECKPOINT 1 An ideal gas has an initial pressure of 3 pressure units and an initial volume of 4 volume units. The table gives the final pressure and volume of the gas (in those same units) in five processes. Which processes start and end on the same isotherm?

	a	b	c	d	e
p	12	6	5	4	1
V	1	2	7	3	12

Sample Problem 19-1

A cylinder contains 12 L of oxygen at 20°C and 15 atm. The temperature is raised to 35°C, and the volume is reduced to 8.5 L. What is the final pressure of the gas in atmospheres? Assume that the gas is ideal.

KEY IDEA Because the gas is ideal, its pressure, volume, temperature, and number of moles are related by the ideal gas law, both in the initial state i and in the final state f (after the changes).

Calculations: From Eq. 19-5 we can write

$$p_iV_i = nRT_i \quad \text{and} \quad p_fV_f = nRT_f.$$

Dividing the second equation by the first equation and solving for p_f yields

$$p_f = \frac{p_iT_fV_i}{T_iV_f}. \tag{19-17}$$

Note here that if we converted the given initial and final volumes from liters to the proper units of cubic meters, the multiplying conversion factors would cancel out of Eq. 19-17. The same would be true for conversion factors that convert the pressures from atmospheres to the proper pascals. However, to convert the given temperatures to kelvins requires the addition of an amount that would not cancel and thus must be included. Hence, we must write

$$T_i = (273 + 20) \text{ K} = 293 \text{ K}$$

and $$T_f = (273 + 35) \text{ K} = 308 \text{ K}.$$

Inserting the given data into Eq. 19-17 then yields

$$p_f = \frac{(15 \text{ atm})(308 \text{ K})(12 \text{ L})}{(293 \text{ K})(8.5 \text{ L})} = 22 \text{ atm}. \quad \text{(Answer)}$$

Sample Problem 19-2

One mole of oxygen (assume it to be an ideal gas) expands at a constant temperature T of 310 K from an initial volume V_i of 12 L to a final volume V_f of 19 L. How much work is done by the gas during the expansion?

KEY IDEA Generally we find the work by integrating the gas pressure with respect to the gas volume, using Eq. 19-11. However, because the gas here is ideal and the expansion is isothermal, that integration leads to Eq. 19-14.

Calculation: Therefore, we can write

$$W = nRT \ln \frac{V_f}{V_i}$$

$$= (1 \text{ mol})(8.31 \text{ J/mol} \cdot \text{K})(310 \text{ K}) \ln \frac{19 \text{ L}}{12 \text{ L}}$$

$$= 1180 \text{ J}. \quad \text{(Answer)}$$

The expansion is graphed in the p-V diagram of Fig. 19-3. The work done by the gas during the expansion is represented by the area beneath the curve *if*.

You can show that if the expansion is now reversed, with the gas undergoing an isothermal compression from 19 L to 12 L, the work done by the gas will be −1180 J. Thus, an external force would have to do 1180 J of work on the gas to compress it.

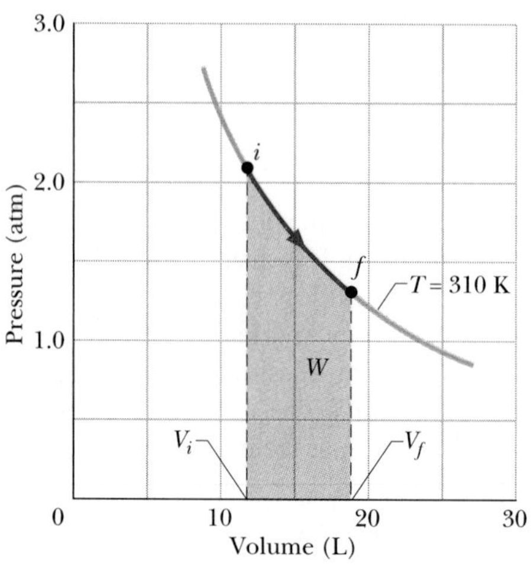

FIG. 19-3 The shaded area represents the work done by 1 mol of oxygen in expanding from V_i to V_f at a constant temperature T of 310 K.

19-4 | Pressure, Temperature, and RMS Speed

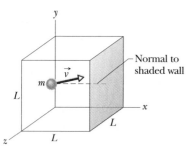

FIG. 19-4 A cubical box of edge length L, containing n moles of an ideal gas. A molecule of mass m and velocity $\vec{v}$ is about to collide with the shaded wall of area L^2. A normal to that wall is shown.

Here is our first kinetic theory problem. Let n moles of an ideal gas be confined in a cubical box of volume V, as in Fig. 19-4. The walls of the box are held at temperature T. What is the connection between the pressure p exerted by the gas on the walls and the speeds of the molecules?

The molecules of gas in the box are moving in all directions and with various speeds, bumping into one another and bouncing from the walls of the box like balls in a racquetball court. We ignore (for the time being) collisions of the molecules with one another and consider only elastic collisions with the walls.

Figure 19-4 shows a typical gas molecule, of mass m and velocity $\vec{v}$, that is about to collide with the shaded wall. Because we assume that any collision of a molecule with a wall is elastic, when this molecule collides with the shaded wall, the only component of its velocity that is changed is the x component, and that component is reversed. This means that the only change in the particle's momentum is along the x axis, and that change is

$$\Delta p_x = (-mv_x) - (mv_x) = -2mv_x.$$

Hence, the momentum Δp_x delivered to the wall by the molecule during the collision is $+2mv_x$. (Because in this book the symbol p represents both momentum and pressure, we must be careful to note that here p represents momentum and is a vector quantity.)

The molecule of Fig. 19-4 will hit the shaded wall repeatedly. The time Δt between collisions is the time the molecule takes to travel to the opposite wall and back again (a distance $2L$) at speed v_x. Thus, Δt is equal to $2L/v_x$. (Note that this result holds even if the molecule bounces off any of the other walls along the way, because those walls are parallel to x and so cannot change v_x.) Therefore, the average rate at which momentum is delivered to the shaded wall by this single molecule is

$$\frac{\Delta p_x}{\Delta t} = \frac{2mv_x}{2L/v_x} = \frac{mv_x^2}{L}.$$

From Newton's second law ($\vec{F} = d\vec{p}/dt$), the rate at which momentum is delivered to the wall is the force acting on that wall. To find the total force, we must add up the contributions of all the molecules that strike the wall, allowing for the possibility that they all have different speeds. Dividing the magnitude of the total force F_x by the area of the wall ($= L^2$) then gives the pressure p on that wall, where now and in the rest of this discussion, p represents pressure. Thus, using the expression for $\Delta p_x/\Delta t$, we can write this pressure as

$$p = \frac{F_x}{L^2} = \frac{mv_{x1}^2/L + mv_{x2}^2/L + \cdots + mv_{xN}^2/L}{L^2}$$

$$= \left(\frac{m}{L^3}\right)(v_{x1}^2 + v_{x2}^2 + \cdots + v_{xN}^2), \quad (19\text{-}18)$$

where N is the number of molecules in the box.

Since $N = nN_A$, there are nN_A terms in the second set of parentheses of Eq. 19-18. We can replace that quantity by $nN_A(v_x^2)_{avg}$, where $(v_x^2)_{avg}$ is the average value of the square of the x components of all the molecular speeds. Equation 19-18 then becomes

$$p = \frac{nmN_A}{L^3}(v_x^2)_{avg}.$$

However, mN_A is the molar mass M of the gas (that is, the mass of 1 mol of the gas). Also, L^3 is the volume of the box, so

$$p = \frac{nM(v_x^2)_{avg}}{V}. \quad (19\text{-}19)$$

For any molecule, $v^2 = v_x^2 + v_y^2 + v_z^2$. Because there are many molecules and because they are all moving in random directions, the average values of the squares of their velocity components are equal, so that $v_x^2 = \frac{1}{3}v^2$. Thus, Eq. 19-19 becomes

$$p = \frac{nM(v^2)_{avg}}{3V}. \quad (19\text{-}20)$$

The square root of $(v^2)_{avg}$ is a kind of average speed, called the **root-mean-square speed** of the molecules and symbolized by v_{rms}. Its name describes it rather well: You *square* each speed, you find the *mean* (that is, the average) of all these squared speeds, and then you take the square *root* of that mean. With $\sqrt{(v^2)_{avg}} = v_{rms}$, we can then write Eq. 19-20 as

$$p = \frac{nMv_{rms}^2}{3V}. \quad (19\text{-}21)$$

Equation 19-21 is very much in the spirit of kinetic theory. It tells us how the pressure of the gas (a purely macroscopic quantity) depends on the speed of the molecules (a purely microscopic quantity).

We can turn Eq. 19-21 around and use it to calculate v_{rms}. Combining Eq. 19-21 with the ideal gas law ($pV = nRT$) leads to

$$v_{rms} = \sqrt{\frac{3RT}{M}}. \quad (19\text{-}22)$$

Table 19-1 shows some rms speeds calculated from Eq. 19-22. The speeds are sur-

TABLE 19-1

Some RMS Speeds at Room Temperature (T = 300 K)[a]

Gas	Molar Mass (10^{-3} kg/mol)	v_{rms} (m/s)
Hydrogen (H_2)	2.02	1920
Helium (He)	4.0	1370
Water vapor (H_2O)	18.0	645
Nitrogen (N_2)	28.0	517
Oxygen (O_2)	32.0	483
Carbon dioxide (CO_2)	44.0	412
Sulfur dioxide (SO_2)	64.1	342

[a]For convenience, we often set room temperature equal to 300 K even though (at 27°C or 81°F) that represents a fairly warm room.

prisingly high. For hydrogen molecules at room temperature (300 K), the rms speed is 1920 m/s, or 4300 mi/h—faster than a speeding bullet! On the surface of the Sun, where the temperature is 2×10^6 K, the rms speed of hydrogen molecules would be 82 times greater than at room temperature were it not for the fact that at such high speeds, the molecules cannot survive collisions among themselves. Remember too that the rms speed is only a kind of average speed; many molecules move much faster than this, and some much slower.

The speed of sound in a gas is closely related to the rms speed of the molecules of that gas. In a sound wave, the disturbance is passed on from molecule to molecule by means of collisions. The wave cannot move any faster than the "average" speed of the molecules. In fact, the speed of sound must be somewhat less than this "average" molecular speed because not all molecules are moving in exactly the same direction as the wave. As examples, at room temperature, the rms speeds of hydrogen and nitrogen molecules are 1920 m/s and 517 m/s, respectively. The speeds of sound in these two gases at this temperature are 1350 m/s and 350 m/s, respectively.

A question often arises: If molecules move so fast, why does it take as long as a minute or so before you can smell perfume when someone opens a bottle across a room? The answer is that, as we shall discuss in Section 19-6, each perfume molecule may have a high speed but it moves away from the bottle only very slowly because its repeated collisions with other molecules prevent it from moving directly across the room to you.

Sample Problem 19-3

Here are five numbers: 5, 11, 32, 67, and 89.

(a) What is the average value n_{avg} of these numbers?

Calculation: We find this from

$$n_{avg} = \frac{5 + 11 + 32 + 67 + 89}{5} = 40.8. \quad \text{(Answer)}$$

(b) What is the rms value n_{rms} of these numbers?

Calculation: We find this from

$$n_{rms} = \sqrt{\frac{5^2 + 11^2 + 32^2 + 67^2 + 89^2}{5}} = 52.1. \quad \text{(Answer)}$$

The rms value is greater than the average value because the larger numbers—being squared—are relatively more important in forming the rms value. To test this, let us replace 89 in our set of five numbers by 300. The average value of the new set of five numbers (as you should show) is 2.0 times the previous average value. The rms value, however, is 2.7 times the previous rms value.

19-5 | Translational Kinetic Energy

We again consider a single molecule of an ideal gas as it moves around in the box of Fig. 19-4, but we now assume that its speed changes when it collides with other molecules. Its translational kinetic energy at any instant is $\frac{1}{2}mv^2$. Its *average* translational kinetic energy over the time that we watch it is

$$K_{avg} = \left(\tfrac{1}{2}mv^2\right)_{avg} = \tfrac{1}{2}m(v^2)_{avg} = \tfrac{1}{2}mv_{rms}^2, \tag{19-23}$$

in which we make the assumption that the average speed of the molecule during our observation is the same as the average speed of all the molecules at any given time. (Provided the total energy of the gas is not changing and provided we observe our molecule for long enough, this assumption is appropriate.) Substituting for v_{rms} from Eq. 19-22 leads to

$$K_{avg} = \left(\tfrac{1}{2}m\right)\frac{3RT}{M}.$$

However, M/m, the molar mass divided by the mass of a molecule, is simply Avogadro's number. Thus,

$$K_{avg} = \frac{3RT}{2N_A}.$$

Using Eq. 19-7 ($k = R/N_A$), we can then write

$$K_{avg} = \tfrac{3}{2}kT. \tag{19-24}$$

This equation tells us something unexpected:

At a given temperature T, all ideal gas molecules—no matter what their mass—have the same average translational kinetic energy—namely, $\tfrac{3}{2}kT$. When we measure the temperature of a gas, we are also measuring the average translational kinetic energy of its molecules.

CHECKPOINT 2 A gas mixture consists of molecules of types 1, 2, and 3, with molecular masses $m_1 > m_2 > m_3$. Rank the three types according to (a) average kinetic energy and (b) rms speed, greatest first.

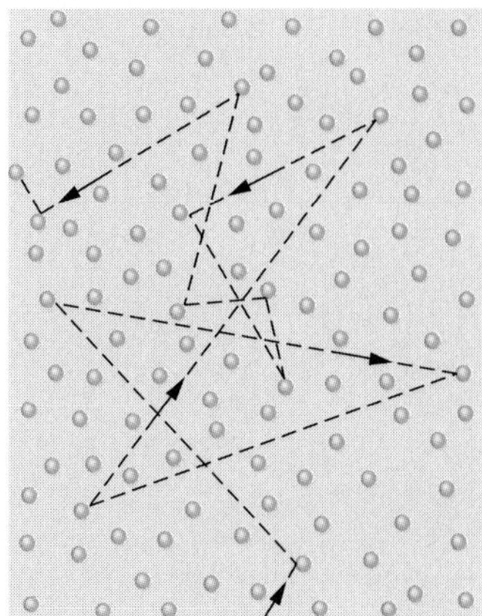

FIG. 19-5 A molecule traveling through a gas, colliding with other gas molecules in its path. Although the other molecules are shown as stationary, they are also moving in a similar fashion.

19-6 | Mean Free Path

We continue to examine the motion of molecules in an ideal gas. Figure 19-5 shows the path of a typical molecule as it moves through the gas, changing both speed and direction abruptly as it collides elastically with other molecules. Between collisions, the molecule moves in a straight line at constant speed. Although the figure shows the other molecules as stationary, they are also moving.

One useful parameter to describe this random motion is the **mean free path** λ of the molecules. As its name implies, λ is the average distance traversed by a molecule between collisions. We expect λ to vary inversely with N/V, the number of molecules per unit volume (or density of molecules). The larger N/V is, the more collisions there should be and the smaller the mean free path. We also expect λ to vary inversely with the size of the molecules—with their diameter d, say. (If the molecules were points, as we have assumed them to be, they would never collide and the mean free path would be infinite.) Thus, the larger the molecules are, the smaller the mean free path. We can even predict that λ should vary (inversely) as the *square* of the molecular diameter because the cross section of a molecule—not its diameter—determines its effective target area.

The expression for the mean free path does, in fact, turn out to be

$$\lambda = \frac{1}{\sqrt{2}\pi d^2 N/V} \quad \text{(mean free path)}. \tag{19-25}$$

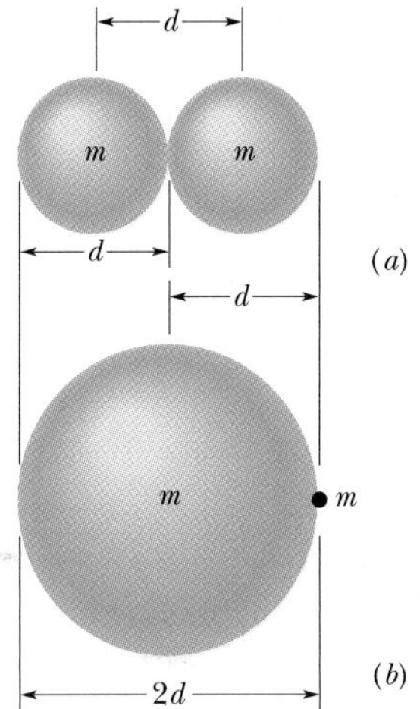

FIG. 19-6 (*a*) A collision occurs when the centers of two molecules come within a distance d of each other, d being the molecular diameter. (*b*) An equivalent but more convenient representation is to think of the moving molecule as having a *radius* d and all other molecules as being points. The condition for a collision is unchanged.

To justify Eq. 19-25, we focus attention on a single molecule and assume—as Fig. 19-5 suggests—that our molecule is traveling with a constant speed v and that all the other molecules are at rest. Later, we shall relax this assumption.

We assume further that the molecules are spheres of diameter d. A collision will then take place if the centers of two molecules come within a distance d of each other, as in Fig. 19-6*a*. Another, more helpful way to look at the situation is to consider our single molecule to have a *radius* of d and all the other molecules to be *points,* as in Fig. 19-6*b*. This does not change our criterion for a collision.

As our single molecule zigzags through the gas, it sweeps out a short cylinder of cross-sectional area πd^2 between successive collisions. If we watch this molecule for a time interval Δt, it moves a distance $v\,\Delta t$, where v is its assumed speed. Thus, if we align all the short cylinders swept out in interval Δt, we form a composite cylinder (Fig. 19-7) of length $v\,\Delta t$ and volume $(\pi d^2)(v\,\Delta t)$. The number of collisions that occur in time Δt is then equal to the number of (point) molecules that lie within this cylinder.

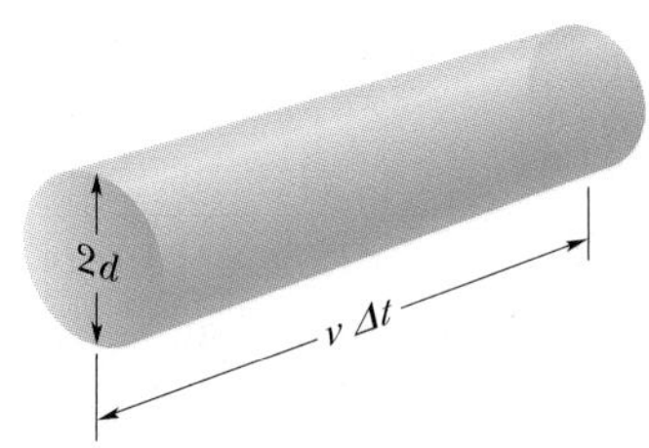

FIG. 19-7 In time Δt the moving molecule effectively sweeps out a cylinder of length $v\,\Delta t$ and radius d.

Since N/V is the number of molecules per unit volume, the number of molecules in the cylinder is N/V times the volume of the cylinder, or $(N/V)(\pi d^2 v\,\Delta t)$. This is also the number of collisions in time Δt. The mean free path is the length of the path (and of the cylinder) divided by this number:

$$\lambda = \frac{\text{length of path during } \Delta t}{\text{number of collisions in } \Delta t} \approx \frac{v\,\Delta t}{\pi d^2 v\,\Delta t\, N/V}$$

$$= \frac{1}{\pi d^2 N/V}. \tag{19-26}$$

This equation is only approximate because it is based on the assumption that all the molecules except one are at rest. In fact, *all* the molecules are moving; when this is taken properly into account, Eq. 19-25 results. Note that it differs from the (approximate) Eq. 19-26 only by a factor of $1/\sqrt{2}$.

The approximation in Eq. 19-26 involves the two v symbols we canceled. The v in the numerator is v_{avg}, the mean speed of the molecules *relative to the container*. The v in the denominator is v_{rel}, the mean speed of our single molecule *relative to the other molecules*, which are moving. It is this latter average speed that determines the number of collisions. A detailed calculation, taking into account the actual speed distribution of the molecules, gives $v_{rel} = \sqrt{2}\,v_{avg}$ and thus the factor $\sqrt{2}$.

The mean free path of air molecules at sea level is about 0.1 μm. At an altitude of 100 km, the density of air has dropped to such an extent that the mean free path rises to about 16 cm. At 300 km, the mean free path is about 20 km. A problem faced by those who would study the physics and chemistry of the upper atmosphere in the laboratory is the unavailability of containers large enough to hold gas samples that simulate upper atmospheric conditions. Yet studies of the concentrations of Freon, carbon dioxide, and ozone in the upper atmosphere are of vital public concern.

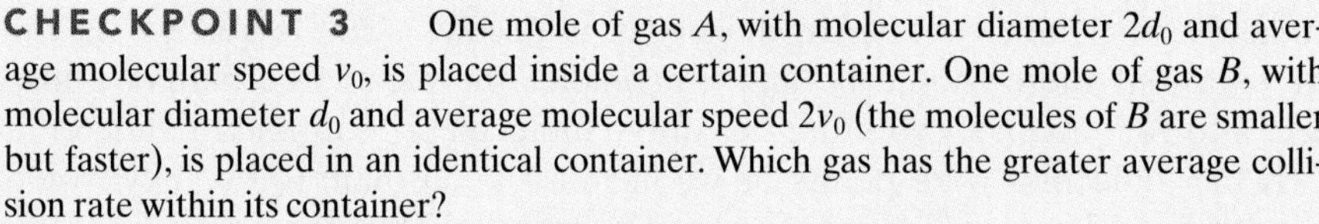

CHECKPOINT 3 One mole of gas A, with molecular diameter $2d_0$ and average molecular speed v_0, is placed inside a certain container. One mole of gas B, with molecular diameter d_0 and average molecular speed $2v_0$ (the molecules of B are smaller but faster), is placed in an identical container. Which gas has the greater average collision rate within its container?

Sample Problem 19-4

(a) What is the mean free path λ for oxygen molecules at temperature $T = 300$ K and pressure $p = 1.0$ atm? Assume that the molecular diameter is $d = 290$ pm and the gas is ideal.

KEY IDEA Each oxygen molecule moves among other *moving* oxygen molecules in a zigzag path due to the resulting collisions. Thus, we use Eq. 19-25 for the mean free path.

Calculation: We first need the number of molecules per unit volume, N/V. Because we assume the gas is ideal, we can use the ideal gas law of Eq. 19-9 ($pV = NkT$) to write $N/V = p/kT$. Substituting this into Eq. 19-25, we find

$$\lambda = \frac{1}{\sqrt{2}\pi d^2 N/V} = \frac{kT}{\sqrt{2}\pi d^2 p}$$

$$= \frac{(1.38 \times 10^{-23}\ \text{J/K})(300\ \text{K})}{\sqrt{2}\pi(2.9 \times 10^{-10}\ \text{m})^2(1.01 \times 10^5\ \text{Pa})}$$

$$= 1.1 \times 10^{-7}\ \text{m}. \qquad \text{(Answer)}$$

This is about 380 molecular diameters.

(b) Assume the average speed of the oxygen molecules is $v = 450$ m/s. What is the average time t between successive collisions for any given molecule? At what rate does the molecule collide; that is, what is the frequency f of its collisions?

KEY IDEAS (1) Between collisions, the molecule travels, on average, the mean free path λ at speed v. (2) The average rate or frequency at which the collisions occur is the inverse of the time t between collisions.

Calculations: From the first key idea, the average time between collisions is

$$t = \frac{\text{distance}}{\text{speed}} = \frac{\lambda}{v} = \frac{1.1 \times 10^{-7}\ \text{m}}{450\ \text{m/s}}$$

$$= 2.44 \times 10^{-10}\ \text{s} \approx 0.24\ \text{ns}. \qquad \text{(Answer)}$$

This tells us that, on average, any given oxygen molecule has less than a nanosecond between collisions.

From the second key idea, the collision frequency is

$$f = \frac{1}{t} = \frac{1}{2.44 \times 10^{-10}\ \text{s}} = 4.1 \times 10^9\ \text{s}^{-1}. \qquad \text{(Answer)}$$

This tells us that, on average, any given oxygen molecule makes about 4 billion collisions per second.

19-7 | The Distribution of Molecular Speeds

The root-mean-square speed v_{rms} gives us a general idea of molecular speeds in a gas at a given temperature. We often want to know more. For example, what fraction of the molecules have speeds greater than the rms value? What fraction have speeds greater than twice the rms value? To answer such questions, we need to know how the possible values of speed are distributed among the molecules. Figure 19-8*a* shows this distribution for oxygen molecules at room temperature ($T = 300$ K); Fig. 19-8*b* compares it with the distribution at $T = 80$ K.

In 1852, Scottish physicist James Clerk Maxwell first solved the problem of finding the speed distribution of gas molecules. His result, known as **Maxwell's speed distribution law,** is

$$P(v) = 4\pi\left(\frac{M}{2\pi RT}\right)^{3/2} v^2 e^{-Mv^2/2RT}. \tag{19-27}$$

Here M is the molar mass of the gas, R is the gas constant, T is the gas temperature, and v is the molecular speed. It is this equation that is plotted in Fig. 19-8*a*, *b*. The quantity $P(v)$ in Eq. 19-27 and Fig. 19-8 is a *probability distribution function:* For any speed v, the product $P(v)\,dv$ (a dimensionless quantity) is the fraction of molecules with speeds in the interval dv centered on speed v.

As Fig. 19-8*a* shows, this fraction is equal to the area of a strip with height $P(v)$ and width dv. The total area under the distribution curve corresponds to the fraction of the molecules whose speeds lie between zero and infinity. All molecules fall into this category, so the value of this total area is unity; that is,

$$\int_0^\infty P(v)\,dv = 1. \tag{19-28}$$

The fraction (frac) of molecules with speeds in an interval of, say, v_1 to v_2 is then

$$\text{frac} = \int_{v_1}^{v_2} P(v)\,dv. \tag{19-29}$$

Average, RMS, and Most Probable Speeds

In principle, we can find the **average speed** v_{avg} of the molecules in a gas with the following procedure: We *weight* each value of v in the distribution; that is, we mul-

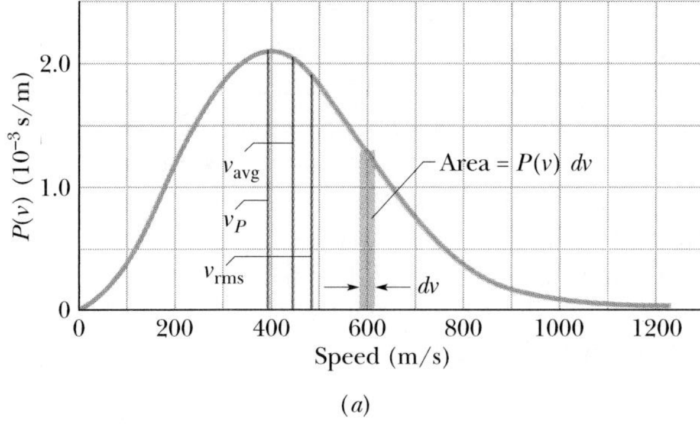

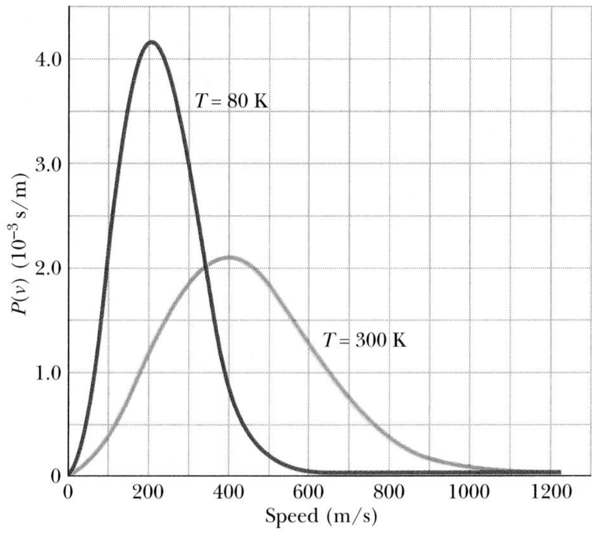

FIG. 19-8 (*a*) The Maxwell speed distribution for oxygen molecules at $T = 300$ K. The three characteristic speeds are marked. (*b*) The curves for 300 K and 80 K. Note that the molecules move more slowly at the lower temperature. Because these are probability distributions, the area under each curve has a numerical value of unity.

tiply it by the fraction $P(v)\,dv$ of molecules with speeds in a differential interval dv centered on v. Then we add up all these values of $v\,P(v)\,dv$. The result is v_{avg}. In practice, we do all this by evaluating

$$v_{avg} = \int_0^\infty v\,P(v)\,dv. \tag{19-30}$$

Substituting for $P(v)$ from Eq. 19-27 and using generic integral 20 from the list of integrals in Appendix E, we find

$$v_{avg} = \sqrt{\frac{8RT}{\pi M}} \quad \text{(average speed).} \tag{19-31}$$

Similarly, we can find the average of the square of the speeds $(v^2)_{avg}$ with

$$(v^2)_{avg} = \int_0^\infty v^2\,P(v)\,dv. \tag{19-32}$$

Substituting for $P(v)$ from Eq. 19-27 and using generic integral 16 from the list of integrals in Appendix E, we find

$$(v^2)_{avg} = \frac{3RT}{M}. \tag{19-33}$$

The square root of $(v^2)_{avg}$ is the root-mean-square speed v_{rms}. Thus,

$$v_{rms} = \sqrt{\frac{3RT}{M}} \quad \text{(rms speed),} \tag{19-34}$$

which agrees with Eq. 19-22.

The **most probable speed** v_P is the speed at which $P(v)$ is maximum (see Fig. 19-8*a*). To calculate v_P, we set $dP/dv = 0$ (the slope of the curve in Fig. 19-8*a* is zero at the maximum of the curve) and then solve for v. Doing so, we find

$$v_P = \sqrt{\frac{2RT}{M}} \quad \text{(most probable speed).} \tag{19-35}$$

A molecule is more likely to have speed v_P than any other speed, but some molecules will have speeds that are many times v_P. These molecules lie in the *high-speed tail* of a distribution curve like that in Fig. 19-8*a*. We should be thankful for these few, higher speed molecules because they make possible both rain and sunshine (without which we could not exist). We next see why.

Rain The speed distribution of water molecules in, say, a pond at summertime temperatures can be represented by a curve similar to that of Fig. 19-8*a*. Most of the molecules do not have nearly enough kinetic energy to escape from the water through its surface. However, small numbers of very fast molecules with speeds far out in the high-speed tail of the curve can do so. It is these water molecules that evaporate, making clouds and rain a possibility.

As the fast water molecules leave the surface, carrying energy with them, the temperature of the remaining water is maintained by heat transfer from the surroundings. Other fast molecules—produced in particularly favorable collisions—quickly take the place of those that have left, and the speed distribution is maintained.

Sunshine Let the distribution curve of Fig. 19-8*a* now refer to protons in the core of the Sun. The Sun's energy is supplied by a nuclear fusion process that starts with the merging of two protons. However, protons repel each other because of their electrical charges, and protons of average speed do not have enough kinetic energy to overcome the repulsion and get close enough to merge. Very fast protons with speeds in the high-speed tail of the distribution curve can do so, however, and for that reason the Sun can shine.

Sample Problem 19-5

A container is filled with oxygen gas maintained at room temperature (300 K). What fraction of the molecules have speeds in the interval 599 to 601 m/s? The molar mass M of oxygen is 0.0320 kg/mol.

KEY IDEAS

1. The speeds of the molecules are distributed over a wide range of values, with the distribution $P(v)$ of Eq. 19-27.
2. The fraction of molecules with speeds in a differential interval dv is $P(v)\,dv$.
3. For a larger interval, the fraction is found by integrating $P(v)$ over the interval.
4. However, the interval $\Delta v = 2$ m/s here is small compared to the speed $v = 600$ m/s on which it is centered.

Calculations: Because Δv is small, we can avoid the integration by approximating the fraction as

$$\text{frac} = P(v)\,\Delta v = 4\pi\left(\frac{M}{2\pi RT}\right)^{3/2} v^2 e^{-Mv^2/2RT}\,\Delta v.$$

The function $P(v)$ is plotted in Fig. 19-8*a*. The total area between the curve and the horizontal axis represents the total fraction of molecules (unity). The area of the thin gold strip represents the fraction we seek.

To evaluate frac in parts, we can write

$$\text{frac} = (4\pi)(A)(v^2)(e^B)(\Delta v), \qquad (19\text{-}36)$$

where

$$A = \left(\frac{M}{2\pi RT}\right)^{3/2} = \left(\frac{0.0320\text{ kg/mol}}{(2\pi)(8.31\text{ J/mol}\cdot\text{K})(300\text{ K})}\right)^{3/2} = 2.92\times10^{-9}\text{ s}^3/\text{m}^3$$

$$\text{and } B = -\frac{Mv^2}{2RT} = -\frac{(0.0320\text{ kg/mol})(600\text{ m/s})^2}{(2)(8.31\text{ J/mol}\cdot\text{K})(300\text{ K})} = -2.31.$$

Substituting A and B into Eq. 19-36 yields

$$\begin{aligned}\text{frac} &= (4\pi)(A)(v^2)(e^B)(\Delta v)\\ &= (4\pi)(2.92\times10^{-9}\text{ s}^3/\text{m}^3)(600\text{ m/s})^2(e^{-2.31})(2\text{ m/s})\\ &= 2.62\times10^{-3}. \qquad \text{(Answer)}\end{aligned}$$

Thus, at room temperature, 0.262% of the oxygen molecules will have speeds that lie in the narrow range between 599 and 601 m/s. If the gold strip of Fig. 19-8*a* were drawn to the scale of this problem, it would be a very thin strip indeed.

Sample Problem 19-6

The molar mass M of oxygen is 0.0320 kg/mol.

(a) What is the average speed v_{avg} of oxygen gas molecules at $T = 300$ K?

KEY IDEA To find the average speed, we must weight speed v with the distribution function $P(v)$ of Eq. 19-27 and then integrate the resulting expression over the range of possible speeds (0 to ∞).

Calculation: We end up with Eq. 19-31, which gives us

$$v_{avg} = \sqrt{\frac{8RT}{\pi M}} = \sqrt{\frac{8(8.31\text{ J/mol}\cdot\text{K})(300\text{ K})}{\pi(0.0320\text{ kg/mol})}} = 445\text{ m/s}. \qquad \text{(Answer)}$$

This result is plotted in Fig. 19-8*a*.

(b) What is the root-mean-square speed v_{rms} at 300 K?

KEY IDEA To find v_{rms}, we must first find $(v^2)_{avg}$ by weighting v^2 with the distribution function $P(v)$ of Eq. 19-27 and then integrating the expression over the range of possible speeds. Then we must take the square root of the result.

Calculation: We end up with Eq. 19-34, which gives us

$$v_{rms} = \sqrt{\frac{3RT}{M}} = \sqrt{\frac{3(8.31\text{ J/mol}\cdot\text{K})(300\text{ K})}{0.0320\text{ kg/mol}}} = 483\text{ m/s}. \qquad \text{(Answer)}$$

This result, plotted in Fig. 19-8*a*, is greater than v_{avg} because the greater speed values influence the calculation more when we integrate the v^2 values than when we integrate the v values.

(c) What is the most probable speed v_P at 300 K?

KEY IDEA Speed v_P corresponds to the maximum of the distribution function $P(v)$, which we obtain by setting the derivative $dP/dv = 0$ and solving the result for v.

Calculation: We end up with Eq. 19-35, which gives us

$$v_P = \sqrt{\frac{2RT}{M}} = \sqrt{\frac{2(8.31\text{ J/mol}\cdot\text{K})(300\text{ K})}{0.0320\text{ kg/mol}}} = 395\text{ m/s}. \qquad \text{(Answer)}$$

This result is also plotted in Fig. 19-8*a*.

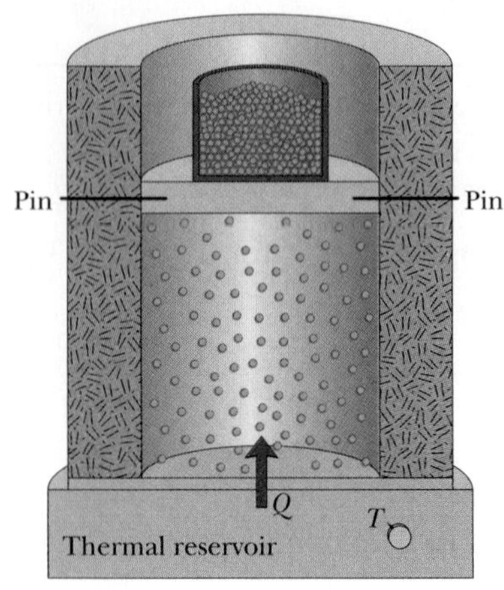

(a)

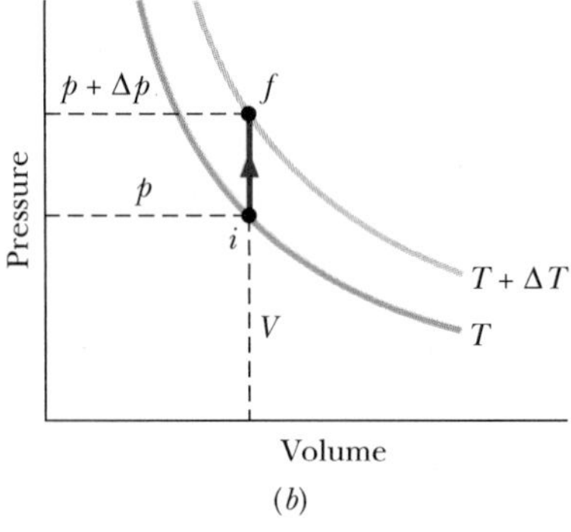

(b)

FIG. 19-9 (a) The temperature of an ideal gas is raised from T to $T + \Delta T$ in a constant-volume process. Heat is added, but no work is done. (b) The process on a p-V diagram.

19-8 | The Molar Specific Heats of an Ideal Gas

In this section, we want to derive from molecular considerations an expression for the internal energy E_{int} of an ideal gas. In other words, we want an expression for the energy associated with the random motions of the atoms or molecules in the gas. We shall then use that expression to derive the molar specific heats of an ideal gas.

Internal Energy E_{int}

Let us first assume that our ideal gas is a *monatomic gas* (which has individual atoms rather than molecules), such as helium, neon, or argon. Let us also assume that the internal energy E_{int} of our ideal gas is simply the sum of the translational kinetic energies of its atoms. (As explained by quantum theory, individual atoms do not have rotational kinetic energy.)

The average translational kinetic energy of a single atom depends only on the gas temperature and is given by Eq. 19-24 as $K_{avg} = \frac{3}{2}kT$. A sample of n moles of such a gas contains nN_A atoms. The internal energy E_{int} of the sample is then

$$E_{int} = (nN_A)K_{avg} = (nN_A)(\tfrac{3}{2}kT). \tag{19-37}$$

Using Eq. 19-7 ($k = R/N_A$), we can rewrite this as

$$E_{int} = \tfrac{3}{2}nRT \quad \text{(monatomic ideal gas)}. \tag{19-38}$$

The internal energy E_{int} of an ideal gas is a function of the gas temperature *only*; it does not depend on any other variable.

With Eq. 19-38 in hand, we are now able to derive an expression for the molar specific heat of an ideal gas. Actually, we shall derive two expressions. One is for the case in which the volume of the gas remains constant as energy is transferred to or from it as heat. The other is for the case in which the pressure of the gas remains constant as energy is transferred to or from it as heat. The symbols for these two molar specific heats are C_V and C_p, respectively. (By convention, the capital letter C is used in both cases, even though C_V and C_p represent types of specific heat and not heat capacities.)

Molar Specific Heat at Constant Volume

Figure 19-9*a* shows n moles of an ideal gas at pressure p and temperature T, confined to a cylinder of fixed volume V. This *initial state i* of the gas is marked on the p-V diagram of Fig. 19-9*b*. Suppose now that you add a small amount of energy to the gas as heat Q by slowly turning up the temperature of the thermal reservoir. The gas temperature rises a small amount to $T + \Delta T$, and its pressure rises to $p + \Delta p$, bringing the gas to *final state f*. In such experiments, we would find that the heat Q is related to the temperature change ΔT by

$$Q = nC_V\,\Delta T \quad \text{(constant volume)}, \tag{19-39}$$

where C_V is a constant called the **molar specific heat at constant volume.** Substituting this expression for Q into the first law of thermodynamics as given by Eq. 18-26 ($\Delta E_{int} = Q - W$) yields

$$\Delta E_{int} = nC_V\,\Delta T - W. \tag{19-40}$$

With the volume held constant, the gas cannot expand and thus cannot do any work. Therefore, $W = 0$, and Eq. 19-40 gives us

$$C_V = \frac{\Delta E_{int}}{n\,\Delta T}. \tag{19-41}$$

From Eq. 19-38, the change in internal energy must be

$$\Delta E_{int} = \tfrac{3}{2}nR\,\Delta T. \tag{19-42}$$

Substituting this result into Eq. 19-41 yields

$$C_V = \tfrac{3}{2}R = 12.5 \text{ J/mol} \cdot \text{K} \quad \text{(monatomic gas)}. \tag{19-43}$$

As Table 19-2 shows, this prediction of the kinetic theory (for ideal gases) agrees very well with experiment for real monatomic gases, the case that we have assumed. The (predicted and) experimental values of C_V for *diatomic gases* (which have molecules with two atoms) and *polyatomic gases* (which have molecules with more than two atoms) are greater than those for monatomic gases for reasons that will be suggested in Section 19-9.

We can now generalize Eq. 19-38 for the internal energy of any ideal gas by substituting C_V for $\frac{3}{2}R$; we get

$$E_{\text{int}} = nC_V T \quad \text{(any ideal gas)}. \tag{19-44}$$

This equation applies not only to an ideal monatomic gas but also to diatomic and polyatomic ideal gases, provided the appropriate value of C_V is used. Just as with Eq. 19-38, we see that the internal energy of a gas depends on the temperature of the gas but not on its pressure or density.

When an ideal gas that is confined to a container undergoes a temperature change ΔT, then from either Eq. 19-41 or Eq. 19-44 we can write the resulting change in its internal energy as

$$\Delta E_{\text{int}} = nC_V \, \Delta T \quad \text{(ideal gas, any process)}. \tag{19-45}$$

This equation tells us:

A change in the internal energy E_{int} of a confined ideal gas depends on the change in the gas temperature only; it does *not* depend on what type of process produces the change in the temperature.

TABLE 19-2

Molar Specific Heats at Constant Volume

Molecule	Example		C_V (J/mol · K)
Monatomic	Ideal		$\frac{3}{2}R = 12.5$
	Real	He	12.5
		Ar	12.6
Diatomic	Ideal		$\frac{5}{2}R = 20.8$
	Real	N_2	20.7
		O_2	20.8
Polyatomic	Ideal		$3R = 24.9$
	Real	NH_4	29.0
		CO_2	29.7

As examples, consider the three paths between the two isotherms in the p-V diagram of Fig. 19-10. Path 1 represents a constant-volume process. Path 2 represents a constant-pressure process (that we are about to examine). Path 3 represents a process in which no heat is exchanged with the system's environment (we discuss this in Section 19-11). Although the values of heat Q and work W associated with these three paths differ, as do p_f and V_f, the values of ΔE_{int} associated with the three paths are identical and are all given by Eq. 19-45, because they all involve the same temperature change ΔT. Therefore, no matter what path is actually taken between T and $T + \Delta T$, we can *always* use path 1 and Eq. 19-45 to compute ΔE_{int} easily.

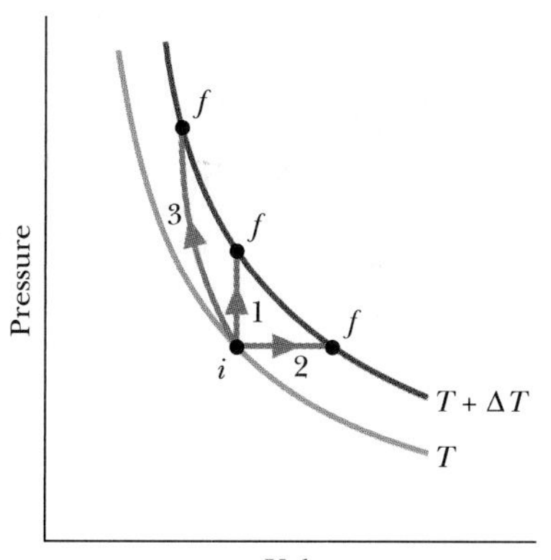

FIG. 19-10 Three paths representing three different processes that take an ideal gas from an initial state i at temperature T to some final state f at temperature $T + \Delta T$. The change ΔE_{int} in the internal energy of the gas is the same for these three processes and for any others that result in the same change of temperature.

Molar Specific Heat at Constant Pressure

We now assume that the temperature of our ideal gas is increased by the same small amount ΔT as previously but now the necessary energy (heat Q) is added with the gas under constant pressure. An experiment for doing this is shown in Fig. 19-11a; the p-V diagram for the process is plotted in Fig. 19-11b. From such experiments we find that the heat Q is related to the temperature change ΔT by

$$Q = nC_p \, \Delta T \quad \text{(constant pressure)}, \tag{19-46}$$

where C_p is a constant called the **molar specific heat at constant pressure.** This C_p is *greater* than the molar specific heat at constant volume C_V, because energy must now be supplied not only to raise the temperature of the gas but also for the gas to do work—that is, to lift the weighted piston of Fig. 19-11a.

To relate molar specific heats C_p and C_V, we start with the first law of thermodynamics (Eq. 18-26):

$$\Delta E_{\text{int}} = Q - W. \tag{19-47}$$

We next replace each term in Eq. 19-47. For ΔE_{int}, we substitute from Eq. 19-45. For Q, we substitute from Eq. 19-46. To replace W, we first note that since

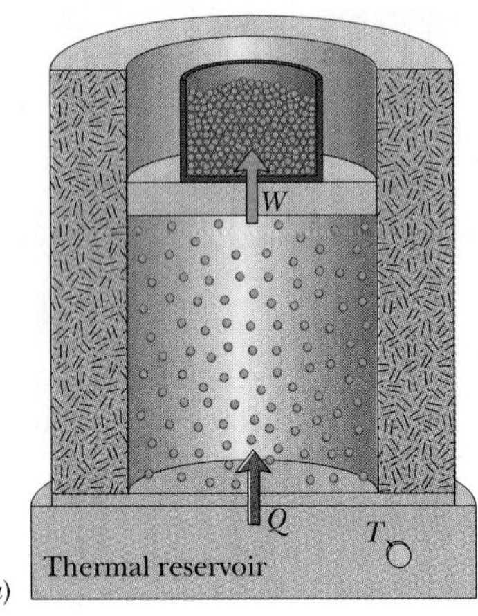

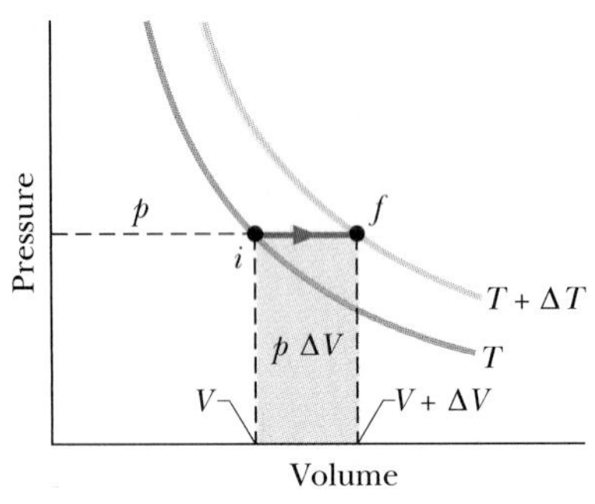

FIG. 19-11 (*a*) The temperature of an ideal gas is raised from T to $T + \Delta T$ in a constant-pressure process. Heat is added and work is done in lifting the loaded piston. (*b*) The process on a *p-V* diagram. The work $p\,\Delta V$ is given by the shaded area.

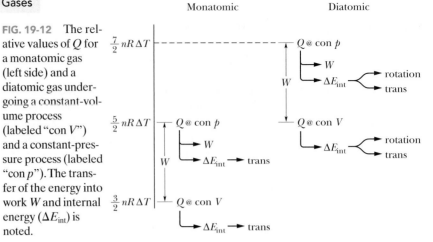

FIG. 19-12 The relative values of Q for a monatomic gas (left side) and a diatomic gas undergoing a constant-volume process (labeled "con V") and a constant-pressure process (labeled "con p"). The transfer of the energy into work W and internal energy (ΔE_{int}) is noted.

the pressure remains constant, Eq. 19-16 tells us that $W = p\,\Delta V$. Then we note that, using the ideal gas equation ($pV = nRT$), we can write

$$W = p\,\Delta V = nR\,\Delta T. \tag{19-48}$$

Making these substitutions in Eq. 19-47 and then dividing through by $n\,\Delta T$, we find

$$C_V = C_p - R$$

and then

$$C_p = C_V + R. \tag{19-49}$$

This prediction of kinetic theory agrees well with experiment, not only for monatomic gases but also for gases in general, as long as their density is low enough so that we may treat them as ideal.

The left side of Fig. 19-12 shows the relative values of Q for a monatomic gas undergoing either a constant-volume process ($Q = \frac{3}{2}nR\,\Delta T$) or a constant-pressure process ($Q = \frac{5}{2}nR\,\Delta T$). Note that for the latter, the value of Q is higher by the amount W, the work done by the gas in the expansion. Note also that for the constant-volume process, the energy added as Q goes entirely into the change in internal energy ΔE_{int} and for the constant-pressure process, the energy added as Q goes into both ΔE_{int} and the work W.

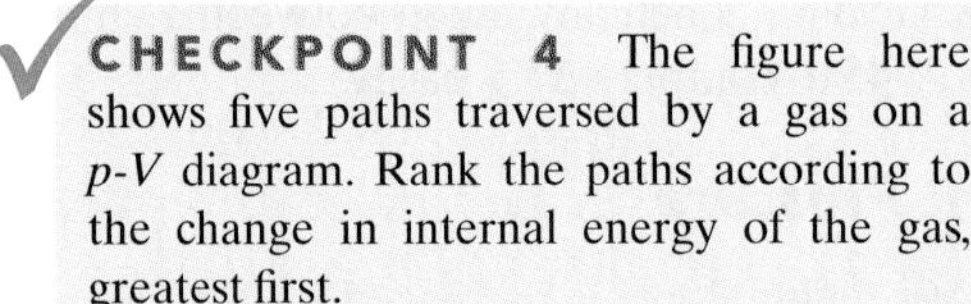

CHECKPOINT 4 The figure here shows five paths traversed by a gas on a *p-V* diagram. Rank the paths according to the change in internal energy of the gas, greatest first.

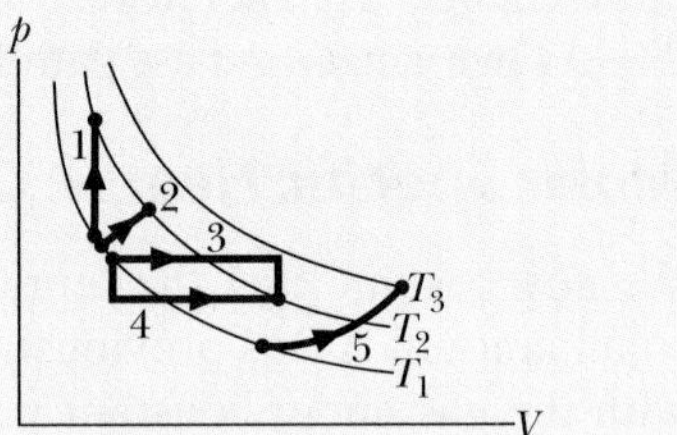

Sample Problem 19-7 Build your skill

A bubble of 5.00 mol of helium is submerged at a certain depth in liquid water when the water (and thus the helium) undergoes a temperature increase ΔT of 20.0 C° at constant pressure. As a result, the bubble expands. The helium is monatomic and ideal.

(a) How much energy is added to the helium as heat during the increase and expansion?

KEY IDEA Heat Q is related to the temperature change ΔT by a molar specific heat of the gas.

Calculations: Because the pressure p is held constant during the addition of energy, we use the molar specific heat at constant pressure C_p and Eq. 19-46,

$$Q = nC_p\,\Delta T, \tag{19-50}$$

to find Q. To evaluate C_p we go to Eq. 19-49, which tells us that for any ideal gas, $C_p = C_V + R$. Then from Eq. 19-43, we know that for any *monatomic* gas (like the helium here), $C_V = \frac{3}{2}R$. Thus, Eq. 19-50 gives us

$$\begin{aligned} Q &= n(C_V + R)\,\Delta T = n(\tfrac{3}{2}R + R)\,\Delta T = n(\tfrac{5}{2}R)\,\Delta T \\ &= (5.00 \text{ mol})(2.5)(8.31 \text{ J/mol}\cdot\text{K})(20.0 \text{ C}^\circ) \\ &= 2077.5 \text{ J} \approx 2080 \text{ J.} \qquad \text{(Answer)} \end{aligned}$$

(b) What is the change ΔE_{int} in the internal energy of the helium during the temperature increase?

KEY IDEA Because the bubble expands, this is not a constant-volume process. However, the helium is nonetheless confined (to the bubble). Thus, the change ΔE_{int} is the same as *would occur* in a constant-volume process with the same temperature change ΔT.

Calculation: We can now easily find the constant-volume change ΔE_{int} with Eq. 19-45:

$$\begin{aligned} \Delta E_{\text{int}} &= nC_V\,\Delta T = n(\tfrac{3}{2}R)\,\Delta T \\ &= (5.00 \text{ mol})(1.5)(8.31 \text{ J/mol}\cdot\text{K})(20.0 \text{ C}^\circ) \\ &= 1246.5 \text{ J} \approx 1250 \text{ J.} \qquad \text{(Answer)} \end{aligned}$$

(c) How much work W is done by the helium as it expands against the pressure of the surrounding water during the temperature increase?

KEY IDEAS The work done by *any* gas expanding against the pressure from its environment is given by Eq. 19-11, which tells us to integrate $p\,dV$. When the pressure is constant (as here), we can simplify that to $W = p\,\Delta V$. When the gas is *ideal* (as here), we can use the ideal gas law (Eq. 19-5) to write $p\,\Delta V = nR\,\Delta T$.

Calculation: We end up with

$$\begin{aligned} W &= nR\,\Delta T \\ &= (5.00 \text{ mol})(8.31 \text{ J/mol}\cdot\text{K})(20.0 \text{ C}^\circ) \\ &= 831 \text{ J.} \qquad \text{(Answer)} \end{aligned}$$

Another way: Because we happen to know Q and ΔE_{int}, we can work this problem another way: We can account for the energy changes of the gas with the first law of thermodynamics, writing

$$\begin{aligned} W &= Q - \Delta E_{\text{int}} = 2077.5 \text{ J} - 1246.5 \text{ J} \\ &= 831 \text{ J.} \qquad \text{(Answer)} \end{aligned}$$

The transfers: Let's follow the energy. Of the 2077.5 J transferred to the helium as heat Q, 831 J goes into the work W required for the expansion and 1246.5 J goes into the internal energy E_{int}, which, for a monatomic gas, is entirely the kinetic energy of the atoms in their translational motion. These several results are suggested on the left side of Fig. 19-12.

19-9 | Degrees of Freedom and Molar Specific Heats

As Table 19-2 shows, the prediction that $C_V = \frac{3}{2}R$ agrees with experiment for monatomic gases but fails for diatomic and polyatomic gases. Let us try to explain the discrepancy by considering the possibility that molecules with more than one atom can store internal energy in forms other than translational kinetic energy.

Figure 19-13 shows common models of helium (a *monatomic* molecule, containing a single atom), oxygen (a *diatomic* molecule, containing two atoms), and methane (a *polyatomic* molecule). From such models, we would assume that all three types of molecules can have translational motions (say, moving left–right and up–down) and rotational motions (spinning about an axis like a top). In addition, we would assume that the diatomic and polyatomic molecules can have oscillatory motions, with the atoms oscillating slightly toward and away from one another, as if attached to opposite ends of a spring.

To keep account of the various ways in which energy can be stored in a gas, James Clerk Maxwell introduced the theorem of the **equipartition of energy:**

Every kind of molecule has a certain number f of *degrees of freedom,* which are independent ways in which the molecule can store energy. Each such degree of freedom has associated with it—on average—an energy of $\frac{1}{2}kT$ per molecule (or $\frac{1}{2}RT$ per mole).

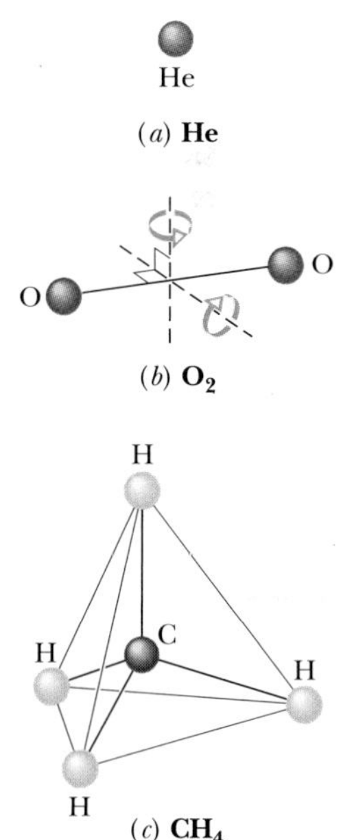

FIG. 19-13 Models of molecules as used in kinetic theory: (*a*) helium, a typical monatomic molecule; (*b*) oxygen, a typical diatomic molecule; and (*c*) methane, a typical polyatomic molecule. The spheres represent atoms, and the lines between them represent bonds. Two rotation axes are shown for the oxygen molecule.

Let us apply the theorem to the translational and rotational motions of the molecules in Fig. 19-13. (We discuss oscillatory motion in the next section.) For the translational motion, superimpose an xyz coordinate system on any gas. The molecules will, in general, have velocity components along all three axes. Thus, gas molecules of all types have three degrees of translational freedom (three ways to move in translation) and, on average, an associated energy of $3(\frac{1}{2}kT)$ per molecule.

For the rotational motion, imagine the origin of our xyz coordinate system at the center of each molecule in Fig. 19-13. In a gas, each molecule should be able

TABLE 19-3

Degrees of Freedom for Various Molecules

		Degrees of Freedom			Predicted Molar Specific Heats	
Molecule	Example	Translational	Rotational	Total (f)	C_V (Eq. 19-51)	$C_p = C_V + R$
Monatomic	He	3	0	3	$\frac{3}{2}R$	$\frac{5}{2}R$
Diatomic	O_2	3	2	5	$\frac{5}{2}R$	$\frac{7}{2}R$
Polyatomic	CH_4	3	3	6	$3R$	$4R$

to rotate with an angular velocity component along each of the three axes, so each gas should have three degrees of rotational freedom and, on average, an additional energy of $3(\frac{1}{2}kT)$ per molecule. *However,* experiment shows this is true only for the polyatomic molecules. According to *quantum theory,* the physics dealing with the allowed motions and energies of molecules and atoms, a monatomic gas molecule does not rotate and so has no rotational energy (a single atom cannot rotate like a top). A diatomic molecule can rotate like a top only about axes perpendicular to the line connecting the atoms (the axes are shown in Fig. 19-13*b*) and not about that line itself. Therefore, a diatomic molecule can have only two degrees of rotational freedom and a rotational energy of only $2(\frac{1}{2}kT)$ per molecule.

To extend our analysis of molar specific heats (C_p and C_V, in Section 19-8) to ideal diatomic and polyatomic gases, it is necessary to retrace the derivations of that analysis in detail. First, we replace Eq. 19-38 ($E_{int} = \frac{3}{2}nRT$) with $E_{int} = (f/2)nRT$, where f is the number of degrees of freedom listed in Table 19-3. Doing so leads to the prediction

$$C_V = \left(\frac{f}{2}\right)R = 4.16f \text{ J/mol}\cdot\text{K}, \tag{19-51}$$

which agrees—as it must—with Eq. 19-43 for monatomic gases ($f = 3$). As Table 19-2 shows, this prediction also agrees with experiment for diatomic gases ($f = 5$), but it is too low for polyatomic gases ($f = 6$ for molecules comparable to CH_4).

Sample Problem 19-8 Build your skill

We transfer 1000 J to a diatomic gas, allowing it to expand with the pressure held constant. The gas molecules rotate but do not oscillate. How much of the 1000 J goes into the increase of the gas's internal energy? Of that amount, how much goes into ΔK_{tran} (the kinetic energy of the translational motion of the molecules) and ΔK_{rot} (the kinetic energy of their rotational motion)?

KEY IDEAS

1. The transfer of energy as heat Q to a gas under constant pressure is related to the resulting temperature increase ΔT via Eq. 19-46 ($Q = nC_p\,\Delta T$).
2. Because the gas is diatomic with molecules undergoing rotation but not oscillation, the molar specific heat is, from Fig. 19-12 and Table 19-3, $C_p = \frac{7}{2}R$.
3. The increase ΔE_{int} in the internal energy is the same as would occur with a constant-volume process resulting in the same ΔT. Thus, from Eq. 19-45, $\Delta E_{int} = nC_V\,\Delta T$. From Fig. 19-12 and Table 19-3, we see that $C_V = \frac{5}{2}R$.
4. For the same n and ΔT, ΔE_{int} is greater for a diatomic gas than a monatomic gas because additional energy is required for rotation.

Increase in E_{int}: Let's first get the temperature change ΔT due to the transfer of energy as heat. From Eq. 19-46, substituting $\frac{7}{2}R$ for C_p, we have

$$\Delta T = \frac{Q}{\frac{7}{2}nR}. \tag{19-52}$$

We next find ΔE_{int} from Eq. 19-45, substituting the molar specific heat $C_V (= \frac{5}{2}R)$ for a constant-volume process and using the same ΔT. Because we are dealing with a diatomic gas, let's call this change $\Delta E_{int,dia}$. Equation 19-45 gives us

$$\Delta E_{int,dia} = nC_V\,\Delta T = n\tfrac{5}{2}R\left(\frac{Q}{\frac{7}{2}nR}\right) = \tfrac{5}{7}Q$$

$$= 0.71428Q = 714.3 \text{ J.} \qquad \text{(Answer)}$$

In words, about 71% of the energy transferred to the gas goes into the internal energy. The rest goes into the work required to increase the volume of the gas.

Increases in K: If we were to increase the temperature of a *monatomic* gas (with the same value of n) by the amount given in Eq. 19-52, the internal energy would change by a smaller amount, call it $\Delta E_{int,mon}$, because rotational motion is not involved. To calculate that smaller amount, we still use Eq. 19-45 but now we substitute the value of C_V for a monatomic gas—namely, $C_V = \frac{3}{2}R$. So,

$$\Delta E_{\text{int,mon}} = n\tfrac{3}{2}R\,\Delta T.$$

Substituting for ΔT from Eq. 19-52 leads us to

$$\Delta E_{\text{int,mon}} = n\tfrac{3}{2}R\left(\frac{Q}{n\frac{7}{2}R}\right) = \tfrac{3}{7}Q$$

$$= 0.42857Q = 428.6\text{ J}.$$

For the monatomic gas, all this energy would go into the kinetic energy of the translational motion of the atoms. The important point here is that for a diatomic gas with the same values of n and ΔT, the same amount of energy goes into the kinetic energy of the translational motion of the molecules. The rest of $\Delta E_{\text{int, dia}}$ (that is, the additional 285.7 J) goes into the rotational motion of the molecules. Thus, for the diatomic gas,

$$\Delta K_{\text{trans}} = 428.6\text{ J} \quad \text{and} \quad \Delta K_{\text{rot}} = 285.7\text{ J}. \qquad \text{(Answer)}$$

Sample Problem 19-9

A cabin of volume V is filled with air (which we consider to be an ideal diatomic gas) at an initial low temperature T_1. After you light a wood stove, the air temperature increases to T_2. What is the resulting change ΔE_{int} in the internal energy of the air in the cabin?

KEY IDEAS As the air temperature increases, the air pressure p cannot change but must always be equal to the air pressure outside the room. The reason is that, because the room is not airtight, the air is not confined. As the temperature increases, air molecules leave through various openings and thus the number of moles n of air in the room decreases. Thus, we *cannot* use Eq. 19-45 ($\Delta E_{\text{int}} = nC_V\,\Delta T$) to find ΔE_{int}, because it requires constant n. However, we *can* relate the internal energy E_{int} at any instant to n and the temperature T with Eq. 19-44 ($E_{\text{int}} = nC_VT$).

Calculations: From Eq. 19-44 we can then write

$$\Delta E_{\text{int}} = \Delta(nC_VT) = C_V\,\Delta(nT).$$

Next, using Eq. 19-5 ($pV = nRT$), we can replace nT with pV/R, obtaining

$$\Delta E_{\text{int}} = C_V\,\Delta\left(\frac{pV}{R}\right).$$

Because p, V, and R are all constants, this yields

$$\Delta E_{\text{int}} = 0, \qquad \text{(Answer)}$$

even though the temperature changes.

Why does the cabin feel more comfortable at the higher temperature? There are at least two factors involved: (1) You exchange electromagnetic radiation (thermal radiation) with surfaces inside the room, and (2) you exchange energy with air molecules that collide with you. When the room temperature is increased, (1) the amount of thermal radiation emitted by the surfaces and absorbed by you is increased, and (2) the amount of energy you gain through the collisions of air molecules with you is increased.

19-10 | A Hint of Quantum Theory

We can improve the agreement of kinetic theory with experiment by including the oscillations of the atoms in a gas of diatomic or polyatomic molecules. For example, the two atoms in the O_2 molecule of Fig. 19-13*b* can oscillate toward and away from each other, with the interconnecting bond acting like a spring. However, experiment shows that such oscillations occur only at relatively high temperatures of the gas—the motion is "turned on" only when the gas molecules have relatively large energies. Rotational motion is also subject to such "turning on," but at a lower temperature.

Figure 19-14 is of help in seeing this turning on of rotational motion and oscillatory motion. The ratio C_V/R for diatomic hydrogen gas (H_2) is plotted there against temperature, with the temperature scale logarithmic to cover several orders of magnitude. Below about 80 K, we find that $C_V/R = 1.5$. This result implies that only the three translational degrees of freedom of hydrogen are involved in the specific heat.

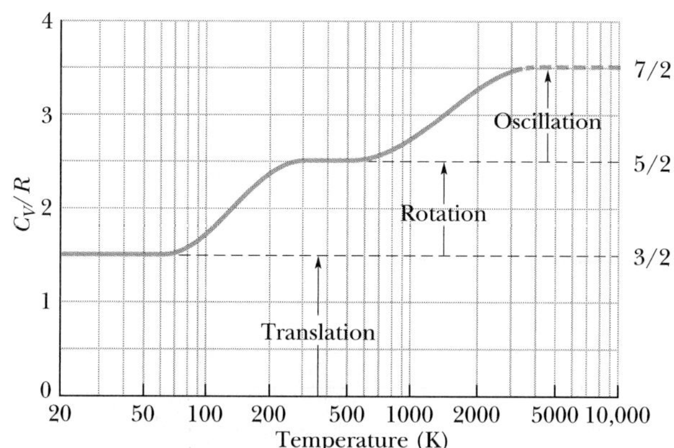

FIG. 19-14 C_V/R versus temperature for (diatomic) hydrogen gas. Because rotational and oscillatory motions begin at certain energies, only translation is possible at very low temperatures. As the temperature increases, rotational motion can begin. At still higher temperatures, oscillatory motion can begin.

As the temperature increases, the value of C_V/R gradually increases to 2.5, implying that two additional degrees of freedom have become involved. Quantum theory shows that these two degrees of freedom are associated with the rotational motion of the hydrogen molecules and that this motion requires a certain minimum amount of energy. At very low temperatures (below 80 K), the molecules do not have enough energy to rotate. As the temperature increases from 80 K, first a few molecules and then more and more obtain enough energy to rotate, and C_V/R increases, until all of them are rotating and $C_V/R = 2.5$.

Similarly, quantum theory shows that oscillatory motion of the molecules requires a certain (higher) minimum amount of energy. This minimum amount is not met until the molecules reach a temperature of about 1000 K, as shown in Fig. 19-14. As the temperature increases beyond 1000 K, more molecules have enough energy to oscillate and C_V/R increases, until all of them are oscillating and $C_V/R = 3.5$. (In Fig. 19-14, the plotted curve stops at 3200 K because there the atoms of a hydrogen molecule oscillate so much that they overwhelm their bond, and the molecule then *dissociates* into two separate atoms.)

19-11 | The Adiabatic Expansion of an Ideal Gas

We saw in Section 17-4 that sound waves are propagated through air and other gases as a series of compressions and expansions; these variations in the transmission medium take place so rapidly that there is no time for energy to be transferred from one part of the medium to another as heat. As we saw in Section 18-11, a process for which $Q = 0$ is an *adiabatic process.* We can ensure that $Q = 0$ either by carrying out the process very quickly (as in sound waves) or by doing it (at any rate) in a well-insulated container. Let us see what the kinetic theory has to say about adiabatic processes.

Figure 19-15*a* shows our usual insulated cylinder, now containing an ideal gas and resting on an insulating stand. By removing mass from the piston, we can allow the gas to expand adiabatically. As the volume increases, both the pressure and the temperature drop. We shall prove next that the relation between the pressure and the volume during such an adiabatic process is

$$pV^\gamma = \text{a constant} \qquad \text{(adiabatic process)}, \qquad (19\text{-}53)$$

in which $\gamma = C_p/C_V$, the ratio of the molar specific heats for the gas. On a p-V diagram such as that in Fig. 19-15*b*, the process occurs along a line (called an *adiabat*) that has the equation $p = (\text{a constant})/V^\gamma$. Since the gas goes from an initial state i to a final state f, we can rewrite Eq. 19-53 as

$$p_iV_i^\gamma = p_fV_f^\gamma \qquad \text{(adiabatic process)}. \qquad (19\text{-}54)$$

To write an equation for an adiabatic process in terms of T and V, we use the ideal gas equation ($pV = nRT$) to eliminate p from Eq. 19-53, finding

$$\left(\frac{nRT}{V}\right)V^\gamma = \text{a constant}.$$

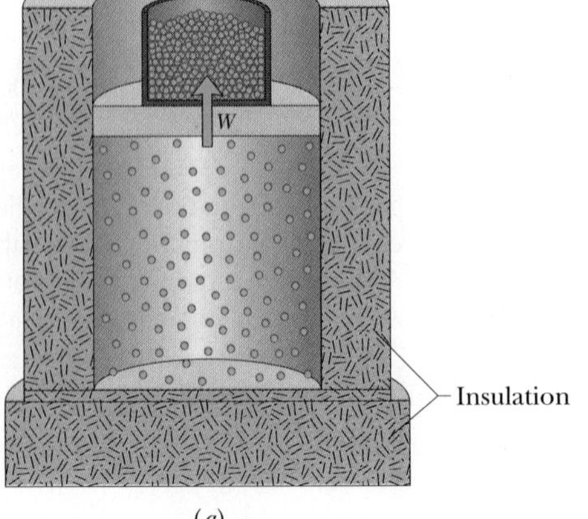

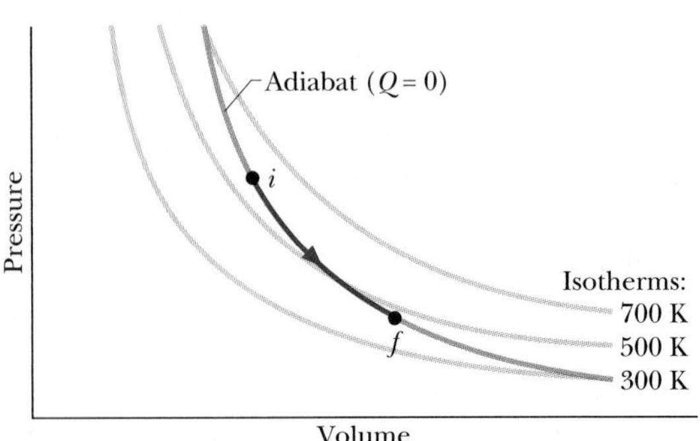

FIG. 19-15 (*a*) The volume of an ideal gas is increased by removing mass from the piston. The process is adiabatic ($Q = 0$). (*b*) The process proceeds from i to f along an adiabat on a p-V diagram.

Because n and R are constants, we can rewrite this in the alternative form

$$TV^{\gamma-1} = \text{a constant} \quad \text{(adiabatic process)}, \tag{19-55}$$

in which the constant is different from that in Eq. 19-53. When the gas goes from an initial state i to a final state f, we can rewrite Eq. 19-55 as

$$T_i V_i^{\gamma-1} = T_f V_f^{\gamma-1} \quad \text{(adiabatic process)}. \tag{19-56}$$

Understanding adiabatic processes allows you to understand why popping the cork on a cold bottle of champagne (as in this chapter's opening photograph) or the tab on a cold can of soda causes a slight fog to form at the opening of the container. At the top of any unopened carbonated drink sits a gas of carbon dioxide and water vapor. Because the gas pressure is greater than atmospheric pressure, the gas expands out into the atmosphere when the container is opened. Thus, the gas volume increases, but that means the gas must do work pushing against the atmosphere. Because the expansion is rapid, it is adiabatic, and the only source of energy for the work is the internal energy of the gas. Because the internal energy decreases, the temperature of the gas also decreases, which causes the water vapor in the gas to condense into tiny drops of fog. (Note that Eq. 19-56 also tells us that the temperature must decrease during an adiabatic expansion: V_f is greater than V_i, and so T_f must be less than T_i.)

Proof of Eq. 19-53

Suppose that you remove some shot from the piston of Fig. 19-15*a*, allowing the ideal gas to push the piston and the remaining shot upward and thus to increase the volume by a differential amount dV. Since the volume change is tiny, we may assume that the pressure p of the gas on the piston is constant during the change. This assumption allows us to say that the work dW done by the gas during the volume increase is equal to $p\,dV$. From Eq. 18-27, the first law of thermodynamics can then be written as

$$dE_{\text{int}} = Q - p\,dV. \tag{19-57}$$

Since the gas is thermally insulated (and thus the expansion is adiabatic), we substitute 0 for Q. Then we use Eq. 19-45 to substitute $nC_V\,dT$ for dE_{int}. With these substitutions, and after some rearranging, we have

$$n\,dT = -\left(\frac{p}{C_V}\right) dV. \tag{19-58}$$

Now from the ideal gas law ($pV = nRT$) we have

$$p\,dV + V\,dp = nR\,dT. \tag{19-59}$$

Replacing R with its equal, $C_p - C_V$, in Eq. 19-59 yields

$$n\,dT = \frac{p\,dV + V\,dp}{C_p - C_V}. \tag{19-60}$$

Equating Eqs. 19-58 and 19-60 and rearranging then give

$$\frac{dp}{p} + \left(\frac{C_p}{C_V}\right)\frac{dV}{V} = 0.$$

Replacing the ratio of the molar specific heats with γ and integrating (see integral 5 in Appendix E) yield

$$\ln p + \gamma \ln V = \text{a constant}.$$

Rewriting the left side as $\ln pV^{\gamma}$ and then taking the antilog of both sides, we find

$$pV^{\gamma} = \text{a constant}. \tag{19-61}$$

Free Expansions

Recall from Section 18-11 that a free expansion of a gas is an adiabatic process that involves no work done on or by the gas, and no change in the internal energy of the gas. A free expansion is thus quite different from the type of adiabatic process described by Eqs. 19-53 through 19-61, in which work is done and the internal energy changes. Those equations then do *not* apply to a free expansion, even though such an expansion is adiabatic.

Also recall that in a free expansion, a gas is in equilibrium only at its initial and final points; thus, we can plot only those points, but not the expansion itself, on a p-V diagram. In addition, because $\Delta E_{\text{int}} = 0$, the temperature of the final state must be that of the initial state. Thus, the initial and final points on a p-V diagram must be on the same isotherm, and instead of Eq. 19-56 we have

$$T_i = T_f \quad \text{(free expansion).} \tag{19-62}$$

If we next assume that the gas is ideal (so that $pV = nRT$), then because there is no change in temperature, there can be no change in the product pV. Thus, instead of Eq. 19-53 a free expansion involves the relation

$$p_iV_i = p_fV_f \quad \text{(free expansion).} \tag{19-63}$$

Sample Problem 19-10

In Sample Problem 19-2, 1 mol of oxygen (assumed to be an ideal gas) expands isothermally (at 310 K) from an initial volume of 12 L to a final volume of 19 L.

(a) What would be the final temperature if the gas had expanded adiabatically to this same final volume? Oxygen (O_2) is diatomic and here has rotation but not oscillation.

KEY IDEAS

1. When a gas expands against the pressure of its environment, it must do work.
2. When the process is adiabatic (no energy is transferred as heat), then the energy required for the work can come only from the internal energy of the gas.
3. Because the internal energy decreases, the temperature T must also decrease.

Calculations: We can relate the initial and final temperatures and volumes with Eq. 19-56:

$$T_iV_i^{\gamma-1} = T_fV_f^{\gamma-1}. \tag{19-64}$$

Because the molecules are diatomic and have rotation but not oscillation, we can take the molar specific heats from Table 19-3. Thus,

$$\gamma = \frac{C_p}{C_V} = \frac{\frac{7}{2}R}{\frac{5}{2}R} = 1.40.$$

Solving Eq. 19-64 for T_f and inserting known data then yield

$$T_f = \frac{T_iV_i^{\gamma-1}}{V_f^{\gamma-1}} = \frac{(310\text{ K})(12\text{ L})^{1.40-1}}{(19\text{ L})^{1.40-1}}$$

$$= (310\text{ K})(\tfrac{12}{19})^{0.40} = 258\text{ K}. \quad \text{(Answer)}$$

(b) What would be the final temperature and pressure if, instead, the gas had expanded freely to the new volume, from an initial pressure of 2.0 Pa?

KEY IDEA The temperature does not change in a free expansion.

Calculation: Thus, the temperature is

$$T_f = T_i = 310\text{ K}. \quad \text{(Answer)}$$

We find the new pressure using Eq. 19-63, which gives us

$$p_f = p_i\frac{V_i}{V_f} = (2.0\text{ Pa})\frac{12\text{ L}}{19\text{ L}} = 1.3\text{ Pa}. \quad \text{(Answer)}$$

PROBLEM-SOLVING TACTICS

Tactic 2: A Graphical Summary of Four Gas Processes
In this chapter we have discussed four special processes that an ideal gas can undergo. An example of each (for a monatomic ideal gas) is shown in Fig. 19-16, and some associated characteristics are given in Table 19-4, including two process names (isobaric and isochoric) that we have not used but that you might see in other courses.

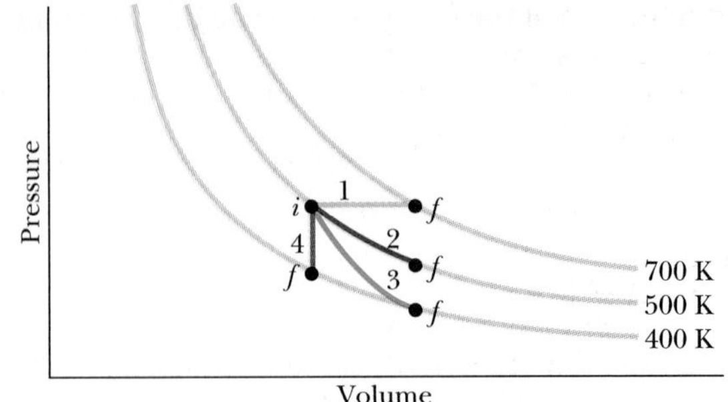

FIG. 19-16 A p-V diagram representing four special processes for an ideal monatomic gas.

TABLE 19-4

Four Special Processes

Path in Fig. 19-16	Constant Quantity	Process Type	Some Special Results ($\Delta E_{int} = Q - W$ and $\Delta E_{int} = nC_V \Delta T$ for all paths)
1	p	Isobaric	$Q = nC_p \Delta T; W = p \Delta V$
2	T	Isothermal	$Q = W = nRT \ln(V_f/V_i); \Delta E_{int} = 0$
3	$pV^\gamma, TV^{\gamma-1}$	Adiabatic	$Q = 0; \quad W = -\Delta E_{int}$
4	V	Isochoric	$Q = \Delta E_{int} = nC_V \Delta T; \quad W = 0$

CHECKPOINT 5 Rank paths 1, 2, and 3 in Fig. 19-16 according to the energy transfer to the gas as heat, greatest first.

REVIEW & SUMMARY

Kinetic Theory of Gases The *kinetic theory of gases* relates the *macroscopic* properties of gases (for example, pressure and temperature) to the *microscopic* properties of gas molecules (for example, speed and kinetic energy).

Avogadro's Number One mole of a substance contains N_A (*Avogadro's number*) elementary units (usually atoms or molecules), where N_A is found experimentally to be

$$N_A = 6.02 \times 10^{23} \text{ mol}^{-1} \quad \text{(Avogadro's number).} \tag{19-1}$$

One molar mass M of any substance is the mass of one mole of the substance. It is related to the mass m of the individual molecules of the substance by

$$M = mN_A. \tag{19-4}$$

The number of moles n contained in a sample of mass M_{sam}, consisting of N molecules, is given by

$$n = \frac{N}{N_A} = \frac{M_{sam}}{M} = \frac{M_{sam}}{mN_A}. \tag{19-2, 19-3}$$

Ideal Gas An *ideal gas* is one for which the pressure p, volume V, and temperature T are related by

$$pV = nRT \quad \text{(ideal gas law).} \tag{19-5}$$

Here n is the number of moles of the gas present and R is a constant (8.31 J/mol·K) called the **gas constant.** The ideal gas law can also be written as

$$pV = NkT, \tag{19-9}$$

where the **Boltzmann constant** k is

$$k = \frac{R}{N_A} = 1.38 \times 10^{-23} \text{ J/K}. \tag{19-7}$$

Work in an Isothermal Volume Change The work done *by* an ideal gas during an **isothermal** (constant-temperature) change from volume V_i to volume V_f is

$$W = nRT \ln \frac{V_f}{V_i} \quad \text{(ideal gas, isothermal process).} \tag{19-14}$$

Pressure, Temperature, and Molecular Speed The pressure exerted by n moles of an ideal gas, in terms of the speed of its molecules, is

$$p = \frac{nMv_{rms}^2}{3V}, \tag{19-21}$$

where $v_{rms} = \sqrt{(v^2)_{avg}}$ is the **root-mean-square speed** of the molecules of the gas. With Eq. 19-5 this gives

$$v_{rms} = \sqrt{\frac{3RT}{M}}. \tag{19-22}$$

Temperature and Kinetic Energy The average translational kinetic energy K_{avg} per molecule of an ideal gas is

$$K_{avg} = \tfrac{3}{2}kT. \tag{19-24}$$

Mean Free Path The *mean free path* λ of a gas molecule is its average path length between collisions and is given by

$$\lambda = \frac{1}{\sqrt{2}\pi d^2 N/V}, \tag{19-25}$$

where N/V is the number of molecules per unit volume and d is the molecular diameter.

Maxwell Speed Distribution The *Maxwell speed distribution* $P(v)$ is a function such that $P(v)\,dv$ gives the *fraction* of molecules with speeds in the interval dv at speed v:

$$P(v) = 4\pi\left(\frac{M}{2\pi RT}\right)^{3/2} v^2 e^{-Mv^2/2RT}. \tag{19-27}$$

Three measures of the distribution of speeds among the molecules of a gas are

$$v_{avg} = \sqrt{\frac{8RT}{\pi M}} \quad \text{(average speed),} \tag{19-31}$$

$$v_P = \sqrt{\frac{2RT}{M}} \quad \text{(most probable speed),} \tag{19-35}$$

and the rms speed defined above in Eq. 19-22.

Molar Specific Heats The molar specific heat C_V of a gas at constant volume is defined as

$$C_V = \frac{Q}{n\,\Delta T} = \frac{\Delta E_{int}}{n\,\Delta T}, \tag{19-39, 19-41}$$

in which Q is the energy transferred as heat to or from a sample of n moles of the gas, ΔT is the resulting temperature change of the gas, and ΔE_{int} is the resulting change in the internal energy of the gas. For an ideal monatomic gas,

$$C_V = \tfrac{3}{2}R = 12.5 \text{ J/mol·K}. \tag{19-43}$$

The molar specific heat C_p of a gas at constant pressure is defined to be

$$C_p = \frac{Q}{n\,\Delta T}, \qquad (19\text{-}46)$$

in which Q, n, and ΔT are defined as above. C_p is also given by

$$C_p = C_V + R. \qquad (19\text{-}49)$$

For n moles of an ideal gas,

$$E_{\text{int}} = nC_VT \quad \text{(ideal gas).} \qquad (19\text{-}44)$$

If n moles of a confined ideal gas undergo a temperature change ΔT due to *any* process, the change in the internal energy of the gas is

$$\Delta E_{\text{int}} = nC_V\,\Delta T \quad \text{(ideal gas, any process),} \qquad (19\text{-}45)$$

in which the appropriate value of C_V must be substituted, according to the type of ideal gas.

Degrees of Freedom and C_V We find C_V by using the *equipartition of energy* theorem, which states that every *degree of freedom* of a molecule (that is, every independent way it can store energy) has associated with it—on average—an energy $\frac{1}{2}kT$ per molecule ($=\frac{1}{2}RT$ per mole). If f is the number of degrees of freedom, then $E_{\text{int}} = (f/2)nRT$ and

$$C_V = \left(\frac{f}{2}\right)R = 4.16f \text{ J/mol}\cdot\text{K}. \qquad (19\text{-}51)$$

For monatomic gases $f = 3$ (three translational degrees); for diatomic gases $f = 5$ (three translational and two rotational degrees).

Adiabatic Process When an ideal gas undergoes a slow adiabatic volume change (a change for which $Q = 0$), its pressure and volume are related by

$$pV^\gamma = \text{a constant} \quad \text{(adiabatic process),} \qquad (19\text{-}53)$$

in which $\gamma (= C_p/C_V)$ is the ratio of molar specific heats for the gas. For a free expansion, however, $pV =$ a constant.

QUESTIONS

1 For a temperature increase of ΔT_1, a certain amount of an ideal gas requires 30 J when heated at constant volume and 50 J when heated at constant pressure. How much work is done by the gas in the second situation?

2 The dot in Fig. 19-17*a* represents the initial state of a gas, and the vertical line through the dot divides the p-V diagram into regions 1 and 2. For the following processes, determine whether the work W done by the gas is positive, negative, or zero: (a) the gas moves up along the vertical line, (b) it moves down along the vertical line, (c) it moves to anywhere in region 1, and (d) it moves to anywhere in region 2.

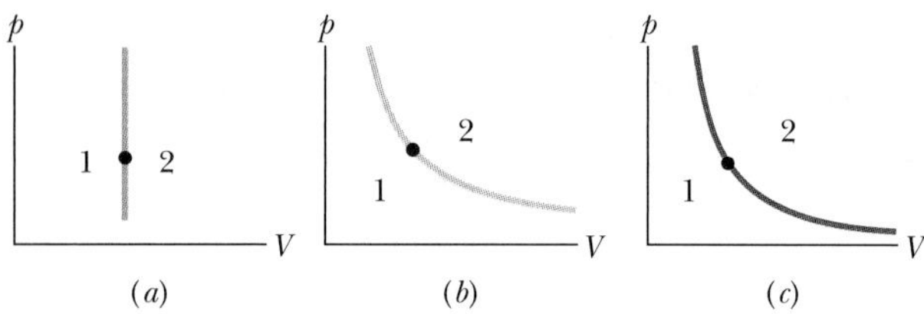

FIG. 19-17 Questions 2, 4, and 6.

3 For four situations for an ideal gas, the table gives the energy transferred to or from the gas as heat Q and either the work W done by the gas or the work W_{on} done on the gas, all in joules. Rank the four situations in terms of the temperature change of the gas, most positive first.

	a	b	c	d
Q	−50	+35	−15	+20
W	−50	+35		
W_{on}			−40	+40

4 The dot in Fig. 19-17*b* represents the initial state of a gas, and the isotherm through the dot divides the p-V diagram into regions 1 and 2. For the following processes, determine whether the change ΔE_{int} in the internal energy of the gas is positive, negative, or zero: (a) the gas moves up along the isotherm, (b) it moves down along the isotherm, (c) it moves to anywhere in region 1, and (d) it moves to anywhere in region 2.

5 A certain amount of energy is to be transferred as heat to 1 mol of a monatomic gas (a) at constant pressure and (b) at constant volume, and to 1 mol of a diatomic gas (c) at constant pressure and (d) at constant volume. Figure 19-18 shows four paths from an initial point to four final points on a p-V diagram. Which path goes with which process? (e) Are the molecules of the diatomic gas rotating?

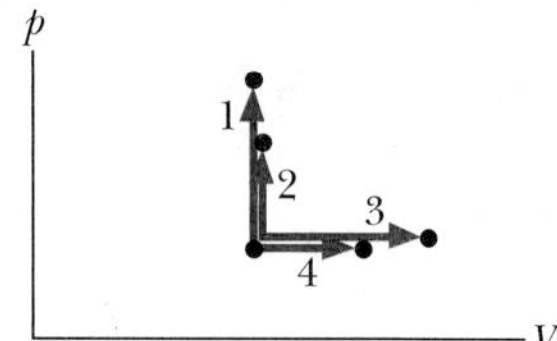

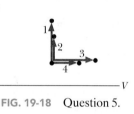

FIG. 19-18 Question 5.

6 The dot in Fig. 19-17*c* represents the initial state of a gas, and the adiabat through the dot divides the p-V diagram into regions 1 and 2. For the following processes, determine whether the corresponding heat Q is positive, negative, or zero: (a) the gas moves up along the adiabat, (b) it moves down along the adiabat, (c) it moves to anywhere in region 1, and (d) it moves to anywhere in region 2.

7 An ideal diatomic gas, with molecular rotation but not oscillation, loses energy as heat Q. Is the resulting decrease in the internal energy of the gas greater if the loss occurs in a constant-volume process or in a constant-pressure process?

8 In the p-V diagram of Fig. 19-19, the gas does 5 J of work when taken along isotherm ab and 4 J when taken along adiabat bc. What is the change in the internal energy of the gas when it is taken along the straight path from a to c?

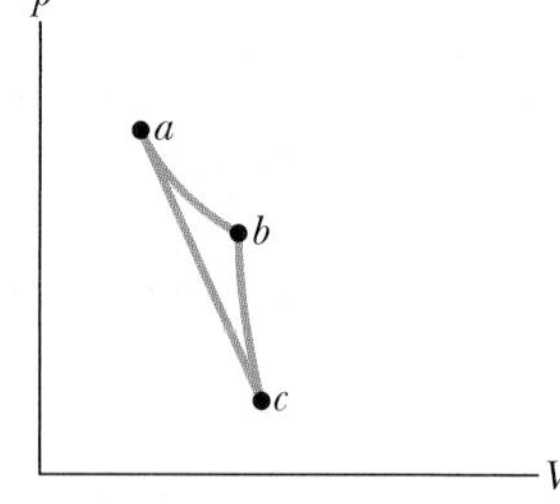

FIG. 19-19 Question 8.

9 (a) Rank the four paths of Fig. 19-16 according to the work done by the gas, greatest first. (b) Rank paths 1, 2, and 3 according to the change in the internal energy of the gas, most positive first and most negative last.

10 Does the temperature of an ideal gas increase, decrease, or stay the same during (a) an isothermal expansion, (b) an expansion at constant pressure, (c) an adiabatic expansion, and (d) an increase in pressure at constant volume?

PROBLEMS

GO Tutoring problem available (at instructor's discretion) in *WileyPLUS* and WebAssign

SSM Worked-out solution available in Student Solutions Manual

WWW Worked-out solution is at http://www.wiley.com/college/halliday

• – ••• Number of dots indicates level of problem difficulty

ILW Interactive solution is at http://www.wiley.com/college/halliday

Additional information available in *The Flying Circus of Physics* and at flyingcircusofphysics.com

sec. 19-2 Avogadro's Number

•1 Gold has a molar mass of 197 g/mol. (a) How many moles of gold are in a 2.50 g sample of pure gold? (b) How many atoms are in the sample?

•2 Find the mass in kilograms of 7.50×10^{24} atoms of arsenic, which has a molar mass of 74.9 g/mol.

sec. 19-3 Ideal Gases

•3 The best laboratory vacuum has a pressure of about 1.00×10^{-18} atm, or 1.01×10^{-13} Pa. How many gas molecules are there per cubic centimeter in such a vacuum at 293 K?

•4 Compute (a) the number of moles and (b) the number of molecules in 1.00 cm^3 of an ideal gas at a pressure of 100 Pa and a temperature of 220 K.

•5 An automobile tire has a volume of 1.64×10^{-2} m^3 and contains air at a gauge pressure (pressure above atmospheric pressure) of 165 kPa when the temperature is 0.00°C. What is the gauge pressure of the air in the tires when its temperature rises to 27.0°C and its volume increases to 1.67×10^{-2} m^3? Assume atmospheric pressure is 1.01×10^5 Pa.

•6 A quantity of ideal gas at 10.0°C and 100 kPa occupies a volume of 2.50 m^3. (a) How many moles of the gas are present? (b) If the pressure is now raised to 300 kPa and the temperature is raised to 30.0°C, how much volume does the gas occupy? Assume no leaks.

•7 Oxygen gas having a volume of 1000 cm^3 at 40.0°C and 1.01×10^5 Pa expands until its volume is 1500 cm^3 and its pressure is 1.06×10^5 Pa. Find (a) the number of moles of oxygen present and (b) the final temperature of the sample. SSM

•8 A container encloses 2 mol of an ideal gas that has molar mass M_1 and 0.5 mol of a second ideal gas that has molar mass $M_2 = 3M_1$. What fraction of the total pressure on the container wall is attributable to the second gas? (The kinetic theory explanation of pressure leads to the experimentally discovered law of partial pressures for a mixture of gases that do not react chemically: *The total pressure exerted by the mixture is equal to the sum of the pressures that the several gases would exert separately if each were to occupy the vessel alone.*)

•9 Suppose 1.80 mol of an ideal gas is taken from a volume of 3.00 m^3 to a volume of 1.50 m^3 via an isothermal compression at 30°C. (a) How much energy is transferred as heat during the compression, and (b) is the transfer *to* or *from* the gas?

•10 *Water bottle in a hot car.* In the American Southwest, the temperature in a closed car parked in sunlight during the summer can be high enough to burn flesh. Suppose a bottle of water at a refrigerator temperature of 5.00°C is opened, then closed, and then left in a closed car with an internal temperature of 75.0°C. Neglecting the thermal expansion of the water and the bottle, find the pressure in the air pocket trapped in the bottle. (The pressure can be enough to push the bottle cap past the threads that are intended to keep the bottle closed.)

••11 Suppose 0.825 mol of an ideal gas undergoes an isothermal expansion as energy is added to it as heat Q. If Fig. 19-20 shows the final volume V_f versus Q, what is the gas temperature? The scale of the vertical axis is set by $V_{fs} = 0.30$ m^3, and the scale of the horizontal axis is set by $Q_s = 1200$ J.

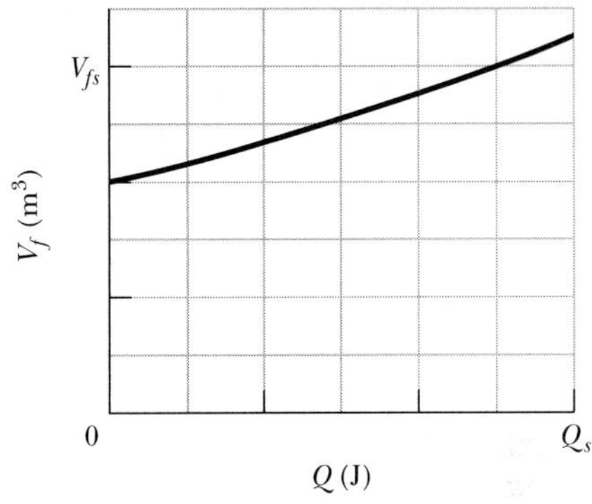

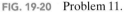

FIG. 19-20 Problem 11.

••12 In the temperature range 310 K to 330 K, the pressure p of a certain nonideal gas is related to volume V and temperature T by

$$p = (24.9 \text{ J/K})\frac{T}{V} - (0.00662 \text{ J/K}^2)\frac{T^2}{V}.$$

How much work is done by the gas if its temperature is raised from 315 K to 325 K while the pressure is held constant?

••13 Air that initially occupies 0.140 m^3 at a gauge pressure of 103.0 kPa is expanded isothermally to a pressure of 101.3 kPa and then cooled at constant pressure until it reaches its initial volume. Compute the work done by the air. (Gauge pressure is the difference between the actual pressure and atmospheric pressure.) SSM ILW WWW

••14 *Submarine rescue.* When the U. S. submarine *Squalus* became disabled at a depth of 80 m, a cylindrical chamber was lowered from a ship to rescue the crew. The chamber had a radius of 1.00 m and a height of 4.00 m, was open at the bottom, and held two rescuers. It slid along a guide cable that a diver had attached to a hatch on the submarine. Once the chamber reached the hatch and clamped to the hull, the crew could escape into the chamber. During the descent, air was released from tanks to prevent water from flooding the chamber. Assume that the interior air pressure matched the water pressure at depth h as given by $p_0 + \rho g h$, where $p_0 = 1.000$ atm is the surface pressure and $\rho = 1024$ kg/m^3 is the density of seawater. Assume a surface temperature of 20.0°C and a submerged water temperature of −30.0°C. (a) What is the air volume in the chamber at the surface? (b) If air had not been released from the tanks, what would have been the air volume in the chamber at depth $h = 80.0$ m? (c) How many moles of air were needed to be released to maintain the original air volume in the chamber? GO

••15 A sample of an ideal gas is taken through the cyclic process *abca* shown in Fig. 19-21. The scale of the vertical axis is set by $p_b = 7.5$ kPa and $p_{ac} = 2.5$ kPa. At point *a*, $T = 200$ K. (a) How many moles of gas are in the sample? What are (b) the temperature of the gas at point *b*, (c) the temperature of the gas at point *c*, and (d) the net energy added to the gas as heat during the cycle? GO

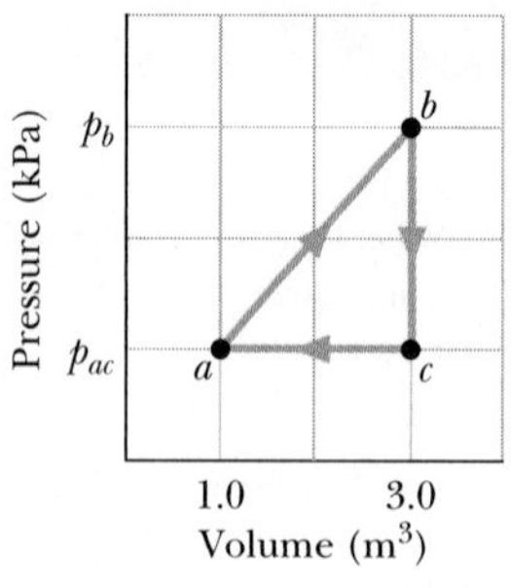

FIG. 19-21 Problem 15.

•••16 An air bubble of volume 20 cm³ is at the bottom of a lake 40 m deep, where the temperature is 4.0°C. The bubble rises to the surface, which is at a temperature of 20°C. Take the temperature of the bubble's air to be the same as that of the surrounding water. Just as the bubble reaches the surface, what is its volume?

•••17 Container A in Fig. 19-22 holds an ideal gas at a pressure of 5.0×10^5 Pa and a temperature of 300 K. It is connected by a thin tube (and a closed valve) to container B, with four times the volume of A. Container B holds the same ideal gas at a pressure of 1.0×10^5 Pa and a temperature of 400 K. The valve is opened to allow the pressures to equalize, but the temperature of each container is maintained. What then is the pressure in the two containers?

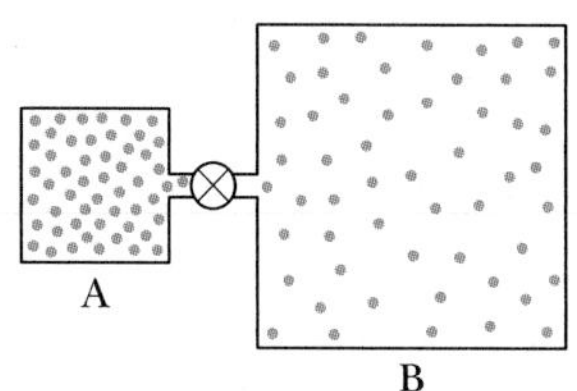

FIG. 19-22 Problem 17.

sec. 19-4 Pressure, Temperature, and RMS Speed

•18 Calculate the rms speed of helium atoms at 1000 K. See Appendix F for the molar mass of helium atoms.

•19 The lowest possible temperature in outer space is 2.7 K. What is the rms speed of hydrogen molecules at this temperature? (The molar mass is given in Table 19-1.) SSM

•20 Find the rms speed of argon atoms at 313 K. See Appendix F for the molar mass of argon atoms.

•21 (a) Compute the rms speed of a nitrogen molecule at 20.0°C. The molar mass of nitrogen molecules (N_2) is given in Table 19-1. At what temperatures will the rms speed be (b) half that value and (c) twice that value?

•22 The temperature and pressure in the Sun's atmosphere are 2.00×10^6 K and 0.0300 Pa. Calculate the rms speed of free electrons (mass 9.11×10^{-31} kg) there, assuming they are an ideal gas.

••23 A beam of hydrogen molecules (H_2) is directed toward a wall, at an angle of 55° with the normal to the wall. Each molecule in the beam has a speed of 1.0 km/s and a mass of 3.3×10^{-24} g. The beam strikes the wall over an area of 2.0 cm², at the rate of 10^{23} molecules per second. What is the beam's pressure on the wall?

••24 At 273 K and 1.00×10^{-2} atm, the density of a gas is 1.24×10^{-5} g/cm³. (a) Find v_{rms} for the gas molecules. (b) Find the molar mass of the gas and (c) identify the gas. (*Hint:* The gas is listed in Table 19-1.)

sec. 19-5 Translational Kinetic Energy

•25 Determine the average value of the translational kinetic energy of the molecules of an ideal gas at (a) 0.00°C and (b) 100°C. What is the translational kinetic energy per mole of an ideal gas at (c) 0.00°C and (d) 100°C?

•26 What is the average translational kinetic energy of nitrogen molecules at 1600 K?

••27 Water standing in the open at 32.0°C evaporates because of the escape of some of the surface molecules. The heat of vaporization (539 cal/g) is approximately equal to εn, where ε is the average energy of the escaping molecules and n is the number of molecules per gram. (a) Find ε. (b) What is the ratio of ε to the average kinetic energy of H_2O molecules, assuming the latter is related to temperature in the same way as it is for gases?

sec. 19-6 Mean Free Path

•28 The mean free path of nitrogen molecules at 0.0°C and 1.0 atm is 0.80×10^{-5} cm. At this temperature and pressure there are 2.7×10^{19} molecules/cm³. What is the molecular diameter?

•29 The atmospheric density at an altitude of 2500 km is about 1 molecule/cm³. (a) Assuming the molecular diameter of 2.0×10^{-8} cm, find the mean free path predicted by Eq. 19-25. (b) Explain whether the predicted value is meaningful. SSM

•30 At what frequency would the wavelength of sound in air be equal to the mean free path of oxygen molecules at 1.0 atm pressure and 0.00°C? Take the diameter of an oxygen molecule to be 3.0×10^{-8} cm.

••31 In a certain particle accelerator, protons travel around a circular path of diameter 23.0 m in an evacuated chamber, whose residual gas is at 295 K and 1.00×10^{-6} torr pressure. (a) Calculate the number of gas molecules per cubic centimeter at this pressure. (b) What is the mean free path of the gas molecules if the molecular diameter is 2.00×10^{-8} cm?

••32 At 20°C and 750 torr pressure, the mean free paths for argon gas (Ar) and nitrogen gas (N_2) are $\lambda_{Ar} = 9.9 \times 10^{-6}$ cm and $\lambda_{N_2} = 27.5 \times 10^{-6}$ cm. (a) Find the ratio of the diameter of an Ar atom to that of an N_2 molecule. What is the mean free path of argon at (b) 20°C and 150 torr, and (c) −40°C and 750 torr?

sec. 19-7 The Distribution of Molecular Speeds

•33 Ten particles are moving with the following speeds: four at 200 m/s, two at 500 m/s, and four at 600 m/s. Calculate their (a) average and (b) rms speeds. (c) Is $v_{rms} > v_{avg}$?

•34 The speeds of 22 particles are as follows (N_i represents the number of particles that have speed v_i):

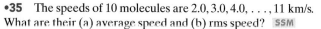

N_i	2	4	6	8	2
v_i (cm/s)	1.0	2.0	3.0	4.0	5.0

What are (a) v_{avg}, (b) v_{rms}, and (c) v_P?

•35 The speeds of 10 molecules are 2.0, 3.0, 4.0, . . . , 11 km/s. What are their (a) average speed and (b) rms speed? SSM

••36 Figure 19-23 gives the probability distribution for nitrogen gas. The scale of the horizontal axis is set by $v_s = 1200$ m/s. What are the (a) gas temperature and (b) rms speed of the molecules?

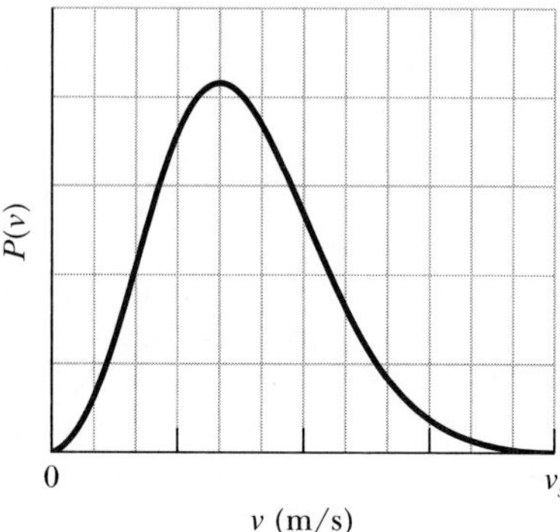

FIG. 19-23 Problem 36.

••37 At what temperature does the rms speed of (a) H_2 (molecular hydrogen) and (b)

O_2 (molecular oxygen) equal the escape speed from Earth (Table 13-2)? At what temperature does the rms speed of (c) H_2 and (d) O_2 equal the escape speed from the Moon (where the gravitational acceleration at the surface has magnitude 0.16g)? Considering the answers to parts (a) and (b), should there be much (e) hydrogen and (f) oxygen high in Earth's upper atmosphere, where the temperature is about 1000 K?

••38 Two containers are at the same temperature. The first contains gas with pressure p_1, molecular mass m_1, and rms speed v_{rms1}. The second contains gas with pressure $2.0p_1$, molecular mass m_2, and average speed $v_{avg2} = 2.0v_{rms1}$. Find the mass ratio m_1/m_2.

••39 A hydrogen molecule (diameter 1.0×10^{-8} cm), traveling at the rms speed, escapes from a 4000 K furnace into a chamber containing *cold* argon atoms (diameter 3.0×10^{-8} cm) at a density of 4.0×10^{19} atoms/cm^3. (a) What is the speed of the hydrogen molecule? (b) If it collides with an argon atom, what is the closest their centers can be, considering each as spherical? (c) What is the initial number of collisions per second experienced by the hydrogen molecule? (*Hint*: Assume that the argon atoms are stationary. Then the mean free path of the hydrogen molecule is given by Eq. 19-26 and not Eq. 19-25.)

••40 It is found that the most probable speed of molecules in a gas when it has (uniform) temperature T_2 is the same as the rms speed of the molecules in this gas when it has (uniform) temperature T_1. Calculate T_2/T_1.

••41 Figure 19-24 shows a hypothetical speed distribution for a sample of N gas particles (note that $P(v) = 0$ for speed $v > 2v_0$). What are the values of (a) av_0, (b) v_{avg}/v_0, and (c) v_{rms}/v_0? (d) What fraction of the particles has a speed between $1.5v_0$ and $2.0v_0$? SSM WWW

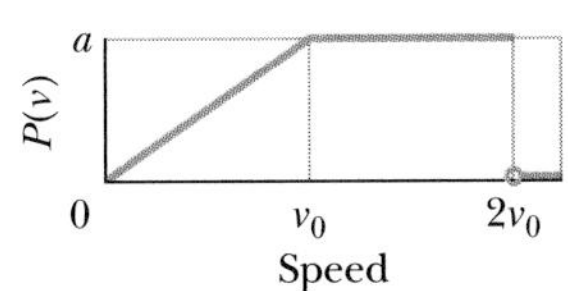

FIG. 19-24 Problem 41.

sec. 19-8 The Molar Specific Heats of an Ideal Gas

•42 What is the internal energy of 1.0 mol of an ideal monatomic gas at 273 K?

••43 The temperature of 2.00 mol of an ideal monatomic gas is raised 15.0 K at constant volume. What are (a) the work W done by the gas, (b) the energy transferred as heat Q, (c) the change ΔE_{int} in the internal energy of the gas, and (d) the change ΔK in the average kinetic energy per atom?

••44 Under constant pressure, the temperature of 2.00 mol of an ideal monatomic gas is raised 15.0 K. What are (a) the work W done by the gas, (b) the energy transferred as heat Q, (c) the change ΔE_{int} in the internal energy of the gas, and (d) the change ΔK in the average kinetic energy per atom?

••45 A container holds a mixture of three nonreacting gases: 2.40 mol of gas 1 with $C_{V1} = 12.0$ J/mol · K, 1.50 mol of gas 2 with $C_{V2} = 12.8$ J/mol · K, and 3.20 mol of gas 3 with $C_{V3} = 20.0$ J/mol · K. What is C_V of the mixture? SSM

••46 One mole of an ideal diatomic gas goes from a to c along the diagonal path in Fig. 19-25. The scale of the vertical axis is set by $p_{ab} = 5.0$ kPa and $p_c = 2.0$ kPa, and the scale of the horizontal axis is set by $V_{bc} = 4.0$ m^3 and $V_a = 2.0$ m^3. During the transition, (a) what is the change in internal energy of the gas, and (b) how much energy is added to the gas as heat? (c) How much heat is required if the gas goes from a to c along the indirect path abc? GO

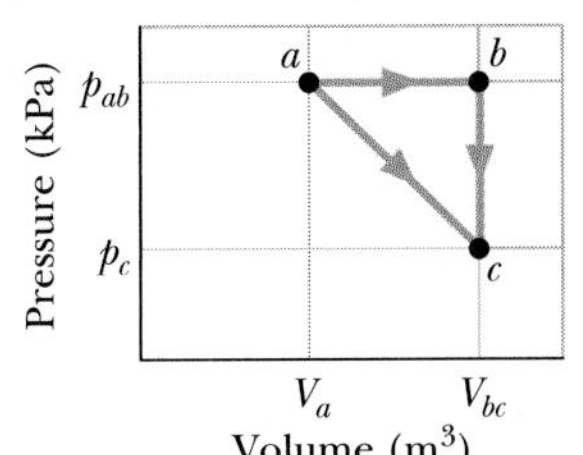

FIG. 19-25 Problem 46.

••47 The mass of a gas molecule can be computed from its specific heat at constant volume c_V. (Note that this is not C_V.) Take $c_V = 0.075$ cal/g · C° for argon and calculate (a) the mass of an argon atom and (b) the molar mass of argon. ILW

••48 When 20.9 J was added as heat to a particular ideal gas, the volume of the gas changed from 50.0 cm^3 to 100 cm^3 while the pressure remained at 1.00 atm. (a) By how much did the internal energy of the gas change? If the quantity of gas present was 2.00×10^{-3} mol, find (b) C_p and (c) C_V.

••49 The temperature of 3.00 mol of an ideal diatomic gas is increased by 40.0 C° without the pressure of the gas changing. The molecules in the gas rotate but do not oscillate. (a) How much energy is transferred to the gas as heat? (b) What is the change in the internal energy of the gas? (c) How much work is done by the gas? (d) By how much does the rotational kinetic energy of the gas increase? GO

sec. 19-9 Degrees of Freedom and Molar Specific Heats

•50 We give 70 J as heat to a diatomic gas, which then expands at constant pressure. The gas molecules rotate but do not oscillate. By how much does the internal energy of the gas increase?

•51 When 1.0 mol of oxygen (O_2) gas is heated at constant pressure starting at 0°C, how much energy must be added to the gas as heat to double its volume? (The molecules rotate but do not oscillate.) ILW

••52 Suppose 12.0 g of oxygen (O_2) gas is heated at constant atmospheric pressure from 25.0°C to 125°C. (a) How many moles of oxygen are present? (See Table 19-1 for the molar mass.) (b) How much energy is transferred to the oxygen as heat? (The molecules rotate but do not oscillate.) (c) What fraction of the heat is used to raise the internal energy of the oxygen?

••53 Suppose 4.00 mol of an ideal diatomic gas, with molecular rotation but not oscillation, experienced a temperature increase of 60.0 K under constant-pressure conditions. What are (a) the energy transferred as heat Q, (b) the change ΔE_{int} in internal energy of the gas, (c) the work W done by the gas, and (d) the change ΔK in the total translational kinetic energy of the gas? SSM WWW

sec. 19-11 The Adiabatic Expansion of an Ideal Gas

•54 Suppose 1.00 L of a gas with $\gamma = 1.30$, initially at 273 K and 1.00 atm, is suddenly compressed adiabatically to half its initial volume. Find its final (a) pressure and (b) temperature. (c) If the gas is then cooled to 273 K at constant pressure, what is its final volume?

•55 A certain gas occupies a volume of 4.3 L at a pressure of 1.2 atm and a temperature of 310 K. It is compressed adiabatically to a volume of 0.76 L. Determine (a) the final pressure and (b) the final temperature, assuming the gas to be an ideal gas for which $\gamma = 1.4$.

•56 We know that for an adiabatic process $pV^\gamma =$ a constant. Evaluate "a constant" for an adiabatic process involving exactly 2.0 mol of an ideal gas passing through the state having exactly $p = 1.0$ atm and $T = 300$ K. Assume a diatomic gas whose molecules rotate but do not oscillate.

••57 Figure 19-26 shows two paths that may be taken by a gas from an initial point i to a final point f. Path 1 consists of an isothermal expansion (work is 50 J in magnitude), an adiabatic expansion (work is 40 J in magnitude), an isothermal compression (work is 30 J in magnitude), and then an adiabatic compression (work is 25 J in magnitude). What is the change in the internal energy of the gas if the gas goes from point i to point f along path 2? GO

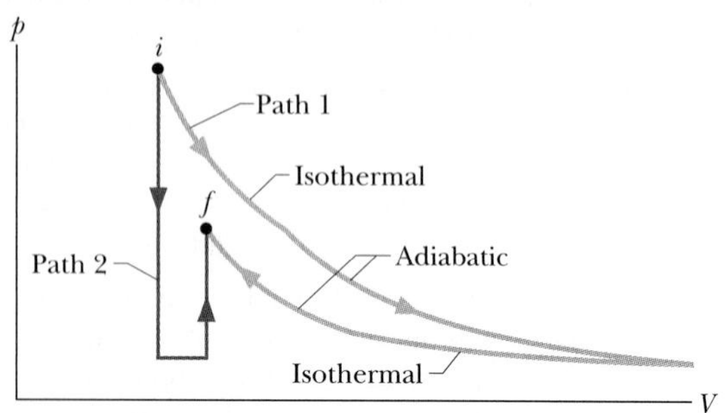

FIG. 19-26 Problem 57.

••58 *Adiabatic wind.* The normal airflow over the Rocky Mountains is west to east. The air loses much of its moisture content and is chilled as it climbs the western side of the mountains. When it descends on the eastern side, the increase in pressure toward lower altitudes causes the temperature to increase. The flow, then called a chinook wind, can rapidly raise the air temperature at the base of the mountains. Assume that the air pressure p depends on altitude y according to $p = p_0 \exp(-ay)$, where $p_0 = 1.00$ atm and $a = 1.16 \times 10^{-4}\ \text{m}^{-1}$. Also assume that the ratio of the molar specific heats is $\gamma = \frac{4}{3}$. A parcel of air with an initial temperature of $-5.00°C$ descends adiabatically from $y_1 = 4267$ m to $y = 1567$ m. What is its temperature at the end of the descent?

••59 A gas is to be expanded from initial state i to final state f along either path 1 or path 2 on a p-V diagram. Path 1 consists of three steps: an isothermal expansion (work is 40 J in magnitude), an adiabatic expansion (work is 20 J in magnitude), and another isothermal expansion (work is 30 J in magnitude). Path 2 consists of two steps: a pressure reduction at constant volume and an expansion at constant pressure. What is the change in the internal energy of the gas along path 2?

••60 *Opening champagne.* In a bottle of champagne, the pocket of gas (primarily carbon dioxide) between the liquid and the cork is at pressure of $p_i = 5.00$ atm. When the cork is pulled from the bottle, the gas undergoes an adiabatic expansion until its pressure matches the ambient air pressure of 1.00 atm. Assume that the ratio of the molar specific heats is $\gamma = \frac{4}{3}$. If the gas has initial temperature $T_i = 5.00°C$, what is its temperature at the end of the adiabatic expansion?

••61 The volume of an ideal gas is adiabatically reduced from 200 L to 74.3 L. The initial pressure and temperature are 1.00 atm and 300 K. The final pressure is 4.00 atm. (a) Is the gas monatomic, diatomic, or polyatomic? (b) What is the final temperature? (c) How many moles are in the gas?

•••62 An ideal diatomic gas, with rotation but no oscillation, undergoes an adiabatic compression. Its initial pressure and volume are 1.20 atm and 0.200 m³. Its final pressure is 2.40 atm. How much work is done by the gas? GO

•••63 Figure 19-27 shows a cycle undergone by 1.00 mol of an ideal monatomic gas. The temperatures are $T_1 = 300$ K, $T_2 = 600$ K, and $T_3 = 455$ K. For $1 \rightarrow 2$, what are (a) heat Q, (b) the change in internal energy ΔE_{int}, and (c) the work done W? For $2 \rightarrow 3$, what are (d) Q, (e) ΔE_{int}, and (f) W? For $3 \rightarrow 1$, what are (g) Q, (h) ΔE_{int}, and (i) W? For the full cycle, what are (j) Q, (k) ΔE_{int}, and (l) W? The initial pressure at point 1 is 1.00 atm ($= 1.013 \times 10^5$ Pa). What are the (m) volume and (n) pressure at point 2 and the (o) volume and (p) pressure at point 3?

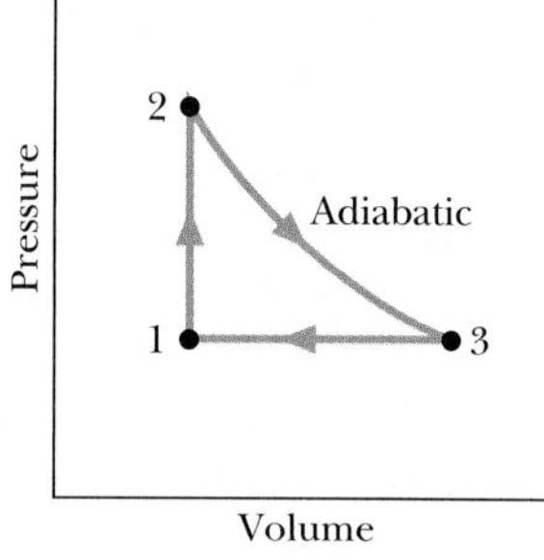

FIG. 19-27 Problem 63.

Additional Problems

64 In an interstellar gas cloud at 50.0 K, the pressure is 1.00×10^{-8} Pa. Assuming that the molecular diameters of the gases in the cloud are all 20.0 nm, what is their mean free path?

65 The temperature of 3.00 mol of a gas with $C_V = 6.00$ cal/mol · K is to be raised 50.0 K. If the process is at *constant volume,* what are (a) the energy transferred as heat Q, (b) the work W done by the gas, (c) the change ΔE_{int} in internal energy of the gas, and (d) the change ΔK in the total translational kinetic energy? If the process is at *constant pressure,* what are (e) Q, (f) W, (g) ΔE_{int}, and (h) ΔK? If the process is *adiabatic,* what are (i) Q, (j) W, (k) ΔE_{int}, and (l) ΔK?

66 Oxygen (O_2) gas at 273 K and 1.0 atm is confined to a cubical container 10 cm on a side. Calculate $\Delta U_g / K_{\text{avg}}$, where ΔU_g is the change in the gravitational potential energy of an oxygen molecule falling the height of the box and K_{avg} is the molecule's average translational kinetic energy.

67 The envelope and basket of a hot-air balloon have a combined weight of 2.45 kN, and the envelope has a capacity (volume) of $2.18 \times 10^3\ \text{m}^3$. When it is fully inflated, what should be the temperature of the enclosed air to give the balloon a *lifting capacity* (force) of 2.67 kN (in addition to the balloon's weight)? Assume that the surrounding air, at 20.0°C, has a weight per unit volume of 11.9 N/m³ and a molecular mass of 0.028 kg/mol, and is at a pressure of 1.0 atm. SSM

68 (a) An ideal gas initially at pressure p_0 undergoes a free expansion until its volume is 3.00 times its initial volume. What then is the ratio of its pressure to p_0? (b) The gas is next slowly and adiabatically compressed back to its original volume. The pressure after compression is $(3.00)^{1/3}p_0$. Is the gas monatomic, diatomic, or polyatomic? (c) What is the ratio of the average kinetic energy per molecule in this final state to that in the initial state?

69 The temperature of 2.00 mol of an ideal monatomic gas is raised 15.0 K in an adiabatic process. What are (a) the work W done by the gas, (b) the energy transferred as heat Q, (c) the change ΔE_{int} in internal energy of the gas, and (d) the change ΔK in the average kinetic energy per atom? SSM

70 During a compression at a constant pressure of 250 Pa, the volume of an ideal gas decreases from 0.80 m³ to 0.20 m³. The initial temperature is 360 K, and the gas loses 210 J as heat. What are (a) the change in the internal energy of the gas and (b) the final temperature of the gas?

71 At what frequency do molecules (diameter 290 pm) collide in (an ideal) oxygen gas (O_2) at temperature 400 K and pressure 2.00 atm? SSM

72 An ideal gas consists of 1.50 mol of diatomic molecules that rotate but do not oscillate. The molecular diameter is 250 pm. The gas is expanded at a constant pressure of 1.50×10^5 Pa, with a transfer of 200 J as heat. What is the change in the mean free path of the molecules?

73 An ideal monatomic gas initially has a temperature of 330 K and a pressure of 6.00 atm. It is to expand from volume 500 cm^3 to volume 1500 cm^3. If the expansion is isothermal, what are (a) the final pressure and (b) the work done by the gas? If, instead, the expansion is adiabatic, what are (c) the final pressure and (d) the work done by the gas?

74 An ideal gas with 3.00 mol is initially in state 1 with pressure $p_1 = 20.0$ atm and volume $V_1 = 1500$ cm^3. First it is taken to state 2 with pressure $p_2 = 1.50p_1$ and volume $V_2 = 2.00V_1$. Then it is taken to state 3 with pressure $p_3 = 2.00p_1$ and volume $V_3 = 0.500V_1$. What is the temperature of the gas in (a) state 1 and (b) state 2? (c) What is the net change in internal energy from state 1 to state 3?

75 An ideal gas undergoes an adiabatic compression from $p = 1.0$ atm, $V = 1.0 \times 10^6$ L, $T = 0.0°$C to $p = 1.0 \times 10^5$ atm, $V = 1.0 \times 10^3$ L. (a) Is the gas monatomic, diatomic, or polyatomic? (b) What is its final temperature? (c) How many moles of gas are present? What is the total translational kinetic energy per mole (d) before and (e) after the compression? (f) What is the ratio of the squares of the rms speeds before and after the compression?

76 An ideal gas, at initial temperature T_1 and initial volume 2.0 m^3, is expanded adiabatically to a volume of 4.0 m^3, then expanded isothermally to a volume of 10 m^3, and then compressed adiabatically back to T_1. What is its final volume?

77 A sample of ideal gas expands from an initial pressure and volume of 32 atm and 1.0 L to a final volume of 4.0 L. The initial temperature is 300 K. If the gas is monatomic and the expansion isothermal, what are the (a) final pressure p_f, (b) final temperature T_f, and (c) work W done by the gas? If the gas is monatomic and the expansion adiabatic, what are (d) p_f, (e) T_f, and (f) W? If the gas is diatomic and the expansion adiabatic, what are (g) p_f, (h) T_f, and (i) W? SSM

78 Calculate the work done by an external agent during an isothermal compression of 1.00 mol of oxygen from a volume of 22.4 L at 0°C and 1.00 atm to a volume of 16.8 L.

79 A steel tank contains 300 g of ammonia gas (NH_3) at a pressure of 1.35×10^6 Pa and a temperature of 77°C. (a) What is the volume of the tank in liters? (b) Later the temperature is 22°C and the pressure is 8.7×10^5 Pa. How many grams of gas have leaked out of the tank?

80 At what temperature do atoms of helium gas have the same rms speed as molecules of hydrogen gas at 20.0°C? (The molar masses are given in Table 19-1.)

81 Figure 19-28 shows a hypothetical speed distribution for particles of a certain gas: $P(v) = Cv^2$ for $0 < v \leq v_0$ and $P(v) = 0$ for $v > v_0$. Find (a) an expression for C in terms of v_0, (b) the average speed of the particles, and (c) their rms speed. SSM

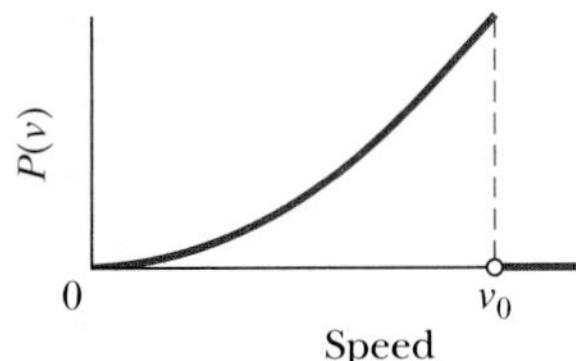

FIG. 19-28 Problem 81.

82 In an industrial process the volume of 25.0 mol of a monatomic ideal gas is reduced at a uniform rate from 0.616 m^3 to 0.308 m^3 in 2.00 h while its temperature is increased at a uniform rate from 27.0°C to 450°C. Throughout the process, the gas passes through thermodynamic equilibrium states. What are (a) the cumulative work done on the gas, (b) the cumulative energy absorbed by the gas as heat, and (c) the molar specific heat for the process? (*Hint:* To evaluate the integral for the work, you might use

$$\int \frac{a + bx}{A + Bx}\,dx = \frac{bx}{B} + \frac{aB - bA}{B^2}\ln(A + Bx),$$

an indefinite integral.) Suppose the process is replaced with a two-step process that reaches the same final state. In step 1, the gas volume is reduced at constant temperature, and in step 2 the temperature is increased at constant volume. For this process, what are (d) the cumulative work done on the gas, (e) the cumulative energy absorbed by the gas as heat, and (f) the molar specific heat for the process?

83 An ideal gas undergoes isothermal compression from an initial volume of 4.00 m^3 to a final volume of 3.00 m^3. There is 3.50 mol of the gas, and its temperature is 10.0°C. (a) How much work is done by the gas? (b) How much energy is transferred as heat between the gas and its environment? SSM

84 (a) What is the number of molecules per cubic meter in air at 20°C and at a pressure of 1.0 atm ($= 1.01 \times 10^5$ Pa)? (b) What is the mass of 1.0 m^3 of this air? Assume that 75% of the molecules are nitrogen (N_2) and 25% are oxygen (O_2).

85 Figure 19-29 shows a cycle consisting of five paths: AB is isothermal at 300 K, BC is adiabatic with work = 5.0 J, CD is at a constant pressure of 5 atm, DE is isothermal, and EA is adiabatic with a change in internal energy of 8.0 J. What is the change in internal energy of the gas along path CD?

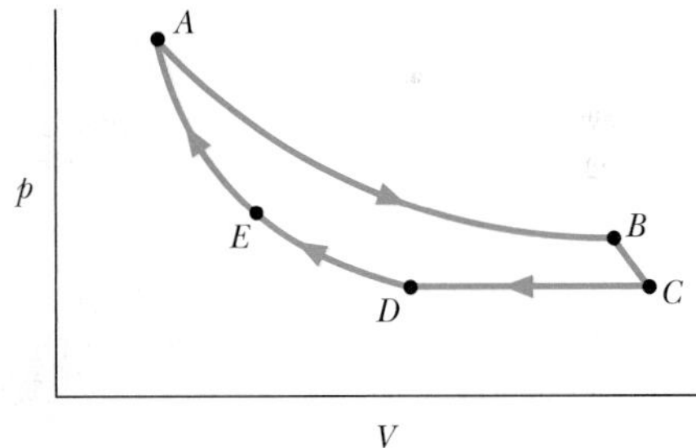

FIG. 19-29 Problem 85.

86 An ideal gas initially at 300 K is compressed at a constant pressure of 25 N/m^2 from a volume of 3.0 m^3 to a volume of 1.8 m^3. In the process, 75 J is lost by the gas as heat. What are (a) the change in internal energy of the gas and (b) the final temperature of the gas?

87 An ideal gas is taken through a complete cycle in three steps: adiabatic expansion with work equal to 125 J, isothermal contraction at 325 K, and increase in pressure at constant volume. (a) Draw a p-V diagram for the three steps. (b) How much energy is transferred as heat in step 3, and (c) is it transferred *to* or *from* the gas?

88 (a) What is the volume occupied by 1.00 mol of an ideal gas at standard conditions—that is, 1.00 atm ($= 1.01 \times 10^5$ Pa) and 273 K? (b) Show that the number of molecules per cubic centimeter (the *Loschmidt number*) at standard conditions is 2.69×10^9.

20 Entropy and the Second Law of Thermodynamics

Inflating a balloon with your breath and stretching a rubber band with your hands require effort because the rubber (or rubber-like material) resists being stretched. In most materials, the resistance to stretching is due to the forces that bind the atoms and molecules together. Because any stretching tends to separate the atoms and molecules, the binding forces resist the stretching. However, rubber is very different because it is elastic and stretching does not tend to increase the separation of the atoms and molecules. Thus, its resistance is not due to binding forces. Instead, it is due to a quantity that gives direction to the flow of time.

What causes a rubber band or balloon to resist stretching?

The answer is in this chapter.

Photo provided courtesy of Ronald P. Fowler, Jr., Flower Entertainment, www.flowerclown.com.

20-1 WHAT IS PHYSICS?

Time has direction, the direction in which we age. We are accustomed to one-way processes—that is, processes that can occur only in a certain sequence (the right way) and never in the reverse sequence (the wrong way). An egg is dropped onto a floor, a pizza is baked, a car is driven into a lamppost, large waves erode a sandy beach—these one-way processes are **irreversible,** meaning that they cannot be reversed by means of only small changes in their environment.

One goal of physics is to understand why time has direction and why one-way processes are irreversible. Although this physics might seem disconnected from the practical issues of everyday life, it is in fact at the heart of any engine, such as a car engine, because it determines how well an engine can run.

The key to understanding why one-way processes cannot be reversed involves a quantity known as *entropy*.

20-2 | Irreversible Processes and Entropy

The one-way character of irreversible processes is so pervasive that we take it for granted. If these processes were to occur *spontaneously* (on their own) in the wrong way, we would be astonished. Yet *none* of these wrong-way events would violate the law of conservation of energy.

For example, if you were to wrap your hands around a cup of hot coffee, you would be astonished if your hands got cooler and the cup got warmer. That is obviously the wrong way for the energy transfer, but the total energy of the closed system (*hands* + *cup of coffee*) would be the same as the total energy if the process had run in the right way. For another example, if you popped a helium balloon, you would be astonished if, later, all the helium molecules were to gather together in the original shape of the balloon. That is obviously the wrong way for molecules to spread, but the total energy of the closed system (*molecules* + *room*) would be the same as for the right way.

Thus, changes in energy within a closed system do not set the direction of irreversible processes. Rather, that direction is set by another property that we shall discuss in this chapter—the *change in entropy* ΔS of the system. The change in entropy of a system is defined in the next section, but we can here state its central property, often called the *entropy postulate:*

> If an irreversible process occurs in a *closed* system, the entropy S of the system always increases; it never decreases.

Entropy differs from energy in that entropy does *not* obey a conservation law. The *energy* of a closed system is conserved; it always remains constant. For irreversible processes, the *entropy* of a closed system always increases. Because of this property, the change in entropy is sometimes called "the arrow of time." For example, we associate the explosion of a popcorn kernel with the forward direction of time and with an increase in entropy. The backward direction of time (a videotape run backwards) would correspond to the exploded popcorn re-forming the original kernel. Because this backward process would result in an entropy decrease, it never happens.

There are two equivalent ways to define the change in entropy of a system: (1) in terms of the system's temperature and the energy the system gains or loses as heat, and (2) by counting the ways in which the atoms or molecules that make up the system can be arranged. We use the first approach in the next section and the second in Section 20-8.

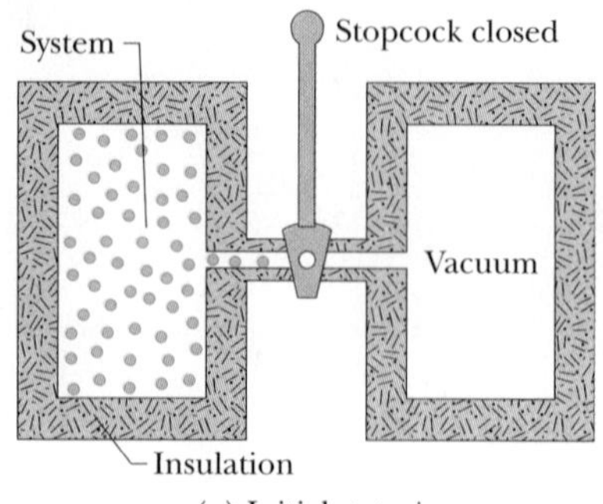

(*a*) Initial state *i*

Irreversible process

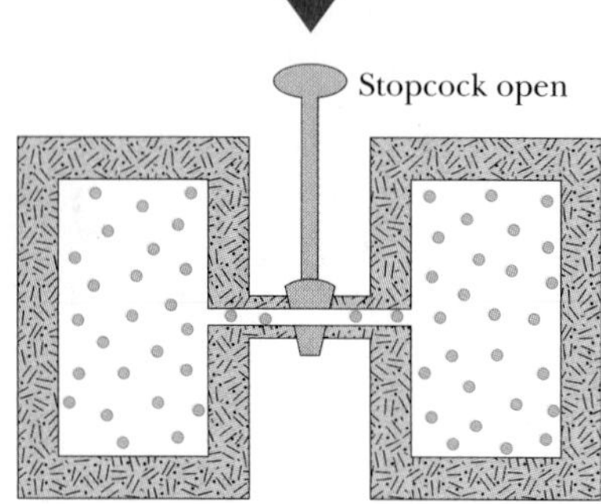

(*b*) Final state *f*

FIG. 20-1 The free expansion of an ideal gas. (*a*) The gas is confined to the left half of an insulated container by a closed stopcock. (*b*) When the stopcock is opened, the gas rushes to fill the entire container. This process is irreversible; that is, it does not occur in reverse, with the gas spontaneously collecting itself in the left half of the container.

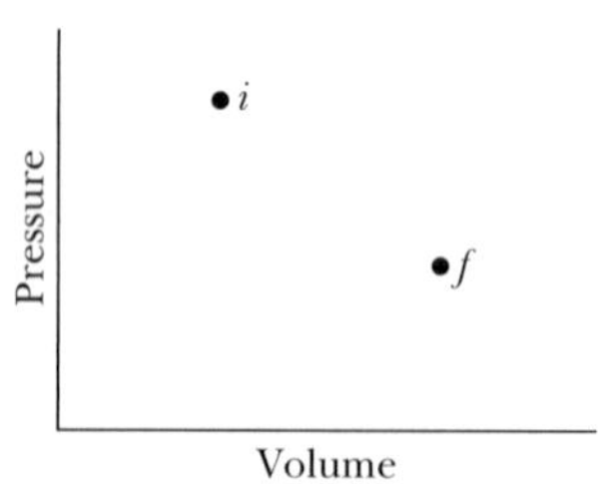

FIG. 20-2 A *p-V* diagram showing the initial state *i* and the final state *f* of the free expansion of Fig. 20-1. The intermediate states of the gas cannot be shown because they are not equilibrium states.

20-3 | Change in Entropy

Let's approach this definition of *change in entropy* by looking again at a process that we described in Sections 18-11 and 19-11: the free expansion of an ideal gas. Figure 20-1*a* shows the gas in its initial equilibrium state *i*, confined by a closed stopcock to the left half of a thermally insulated container. If we open the stopcock, the gas rushes to fill the entire container, eventually reaching the final equilibrium state *f* shown in Fig. 20-1*b*. This is an irreversible process; all the molecules of the gas will never return to the left half of the container.

The *p-V* plot of the process, in Fig. 20-2, shows the pressure and volume of the gas in its initial state *i* and final state *f*. Pressure and volume are *state properties,* properties that depend only on the state of the gas and not on how it reached that state. Other state properties are temperature and energy. We now assume that the gas has still another state property—its entropy. Furthermore, we define the **change in entropy** $S_f - S_i$ of a system during a process that takes the system from an initial state *i* to a final state *f* as

$$\Delta S = S_f - S_i = \int_i^f \frac{dQ}{T} \quad \text{(change in entropy defined).} \tag{20-1}$$

Here Q is the energy transferred as heat to or from the system during the process, and T is the temperature of the system in kelvins. Thus, an entropy change depends not only on the energy transferred as heat but also on the temperature at which the transfer takes place. Because T is always positive, the sign of ΔS is the same as that of Q. We see from Eq. 20-1 that the SI unit for entropy and entropy change is the joule per kelvin.

There is a problem, however, in applying Eq. 20-1 to the free expansion of Fig. 20-1. As the gas rushes to fill the entire container, the pressure, temperature, and volume of the gas fluctuate unpredictably. In other words, they do not have a sequence of well-defined equilibrium values during the intermediate stages of the change from initial equilibrium state *i* to final equilibrium state *f*. Thus, we cannot trace a pressure–volume path for the free expansion on the *p-V* plot of Fig. 20-2 and, more important, we cannot find a relation between Q and T that allows us to integrate as Eq. 20-1 requires.

However, if entropy is truly a state property, the difference in entropy between states *i* and *f* must depend *only on those states* and not at all on the way the system went from one state to the other. Suppose, then, that we replace the irreversible free expansion of Fig. 20-1 with a *reversible* process that connects states *i* and *f*. With a reversible process we can trace a pressure–volume path on a *p-V* plot, and we can find a relation between Q and T that allows us to use Eq. 20-1 to obtain the entropy change.

We saw in Section 19-11 that the temperature of an ideal gas does not change during a free expansion: $T_i = T_f = T$. Thus, points *i* and *f* in Fig. 20-2 must be on the same isotherm. A convenient replacement process is then a reversible isothermal expansion from state *i* to state *f*, which actually proceeds *along* that isotherm. Furthermore, because T is constant throughout a reversible isothermal expansion, the integral of Eq. 20-1 is greatly simplified.

Figure 20-3 shows how to produce such a reversible isothermal expansion. We confine the gas to an insulated cylinder that rests on a thermal reservoir maintained at the temperature T. We begin by placing just enough lead shot on the movable piston so that the pressure and volume of the gas are those of the initial state *i* of Fig. 20-1*a*. We then remove shot slowly (piece by piece) until the pressure and volume of the gas are those of the final state *f* of Fig. 20-1*b*. The temperature of the gas does not change because the gas remains in thermal contact with the reservoir throughout the process.

The reversible isothermal expansion of Fig. 20-3 is physically quite different from the irreversible free expansion of Fig. 20-1. However, *both processes have the same initial state and the same final state and thus must have the same change in entropy.* Because we removed the lead shot slowly, the intermediate states of the gas are equilibrium states, so we can plot them on a p-V diagram (Fig. 20-4).

To apply Eq. 20-1 to the isothermal expansion, we take the constant temperature T outside the integral, obtaining

$$\Delta S = S_f - S_i = \frac{1}{T}\int_i^f dQ.$$

Because $\int dQ = Q$, where Q is the total energy transferred as heat during the process, we have

$$\Delta S = S_f - S_i = \frac{Q}{T} \quad \text{(change in entropy, isothermal process).} \tag{20-2}$$

To keep the temperature T of the gas constant during the isothermal expansion of Fig. 20-3, heat Q must have been energy transferred *from* the reservoir *to* the gas. Thus, Q is positive and the entropy of the gas *increases* during the isothermal process and during the free expansion of Fig. 20-1.

To summarize:

> To find the entropy change for an irreversible process occurring in a *closed* system, replace that process with any reversible process that connects the same initial and final states. Calculate the entropy change for this reversible process with Eq. 20-1.

When the temperature change ΔT of a system is small relative to the temperature (in kelvins) before and after the process, the entropy change can be approximated as

$$\Delta S = S_f - S_i \approx \frac{Q}{T_{avg}}, \tag{20-3}$$

where T_{avg} is the average temperature of the system in kelvins during the process.

CHECKPOINT 1 Water is heated on a stove. Rank the entropy changes of the water as its temperature rises (a) from 20°C to 30°C, (b) from 30°C to 35°C, and (c) from 80°C to 85°C, greatest first.

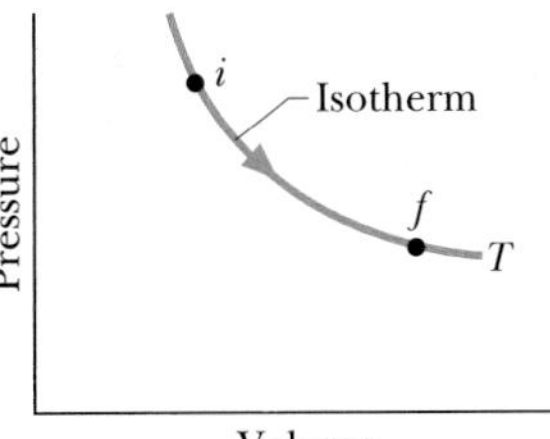

FIG. 20-4 A p-V diagram for the reversible isothermal expansion of Fig. 20-3. The intermediate states, which are now equilibrium states, are shown.

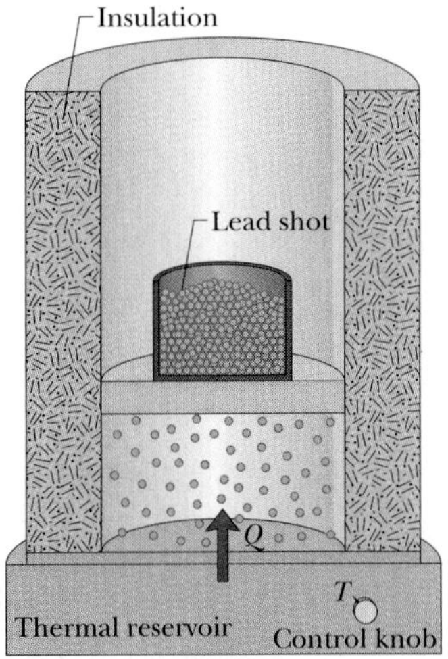

(*a*) Initial state i

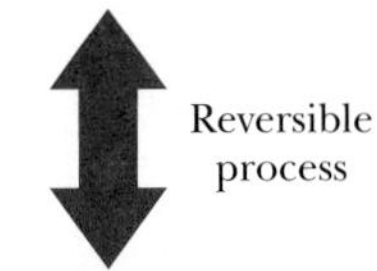

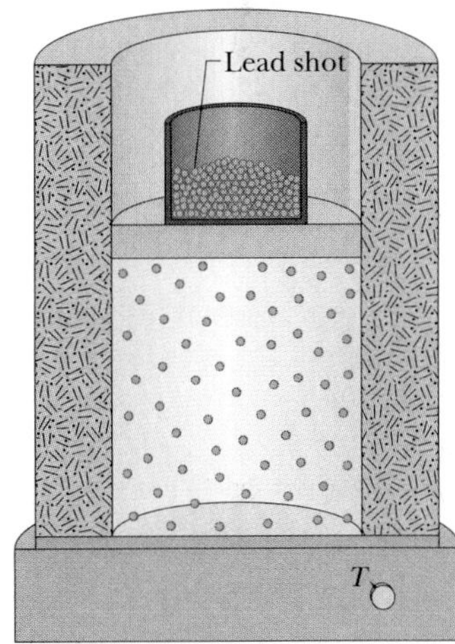

(*b*) Final state f

FIG. 20-3 The isothermal expansion of an ideal gas, done in a reversible way. The gas has the same initial state i and same final state f as in the irreversible process of Figs. 20-1 and 20-2.

Entropy as a State Function

We have assumed that entropy, like pressure, energy, and temperature, is a property of the state of a system and is independent of how that state is reached. That entropy is indeed a *state function* (as state properties are usually called) can be deduced only by experiment. However, we can prove it is a state function for the special and important case in which an ideal gas is taken through a reversible process.

To make the process reversible, it is done slowly in a series of small steps, with the gas in an equilibrium state at the end of each step. For each small step, the energy transferred as heat to or from the gas is dQ, the work done by the gas is dW, and the change in internal energy is dE_{int}. These are related by the first law of thermodynamics in differential form (Eq. 18-27):

$$dE_{\text{int}} = dQ - dW.$$

Because the steps are reversible, with the gas in equilibrium states, we can use Eq. 18-24 to replace dW with $p\,dV$ and Eq. 19-45 to replace dE_{int} with $nC_V\,dT$. Solving for dQ then leads to

$$dQ = p\,dV + nC_V\,dT.$$

Using the ideal gas law, we replace p in this equation with nRT/V. Then we divide each term in the resulting equation by T, obtaining

$$\frac{dQ}{T} = nR\frac{dV}{V} + nC_V\frac{dT}{T}.$$

Now let us integrate each term of this equation between an arbitrary initial state i and an arbitrary final state f to get

$$\int_i^f \frac{dQ}{T} = \int_i^f nR\frac{dV}{V} + \int_i^f nC_V\frac{dT}{T}.$$

The quantity on the left is the entropy change $\Delta S\ (= S_f - S_i)$ defined by Eq. 20-1. Substituting this and integrating the quantities on the right yield

$$\Delta S = S_f - S_i = nR\ln\frac{V_f}{V_i} + nC_V\ln\frac{T_f}{T_i}. \qquad (20\text{-}4)$$

Note that we did not have to specify a particular reversible process when we integrated. Therefore, the integration must hold for all reversible processes that take the gas from state i to state f. Thus, the change in entropy ΔS between the initial and final states of an ideal gas depends only on properties of the initial state (V_i and T_i) and properties of the final state (V_f and T_f); ΔS does not depend on how the gas changes between the two states.

CHECKPOINT 2 An ideal gas has temperature T_1 at the initial state i shown in the p-V diagram here. The gas has a higher temperature T_2 at final states a and b, which it can reach along the paths shown. Is the entropy change along the path to state a larger than, smaller than, or the same as that along the path to state b?

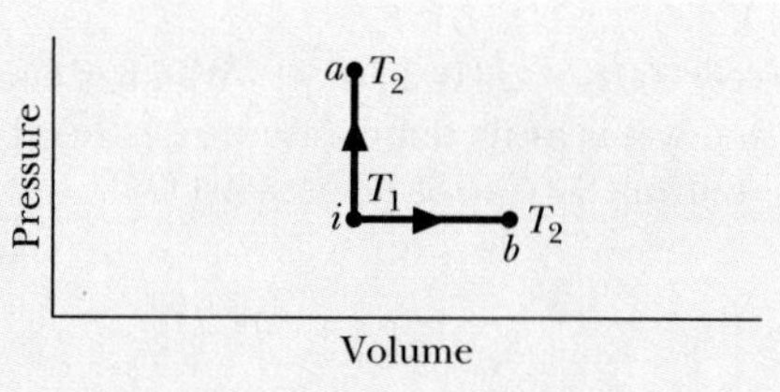

Sample Problem 20-1

Suppose 1.0 mol of nitrogen gas is confined to the left side of the container of Fig. 20-1*a*. You open the stopcock, and the volume of the gas doubles. What is the entropy change of the gas for this irreversible process? Treat the gas as ideal.

KEY IDEAS (1) We can determine the entropy change for the irreversible process by calculating it for a reversible process that provides the same change in volume. (2) The temperature of the gas does not change in the free expansion. Thus, the reversible process should be an isothermal expansion—namely, the one of Figs. 20-3 and 20-4.

Calculations: From Table 19-4, the energy Q added as heat to the gas as it expands isothermally at temperature T from an initial volume V_i to a final volume V_f is

$$Q = nRT\ln\frac{V_f}{V_i},$$

in which n is the number of moles of gas present. From Eq. 20-2 the entropy change for this reversible process is

$$\Delta S_{\text{rev}} = \frac{Q}{T} = \frac{nRT\ln(V_f/V_i)}{T} = nR\ln\frac{V_f}{V_i}.$$

Substituting $n = 1.00$ mol and $V_f/V_i = 2$, we find

$$\Delta S_{rev} = nR \ln \frac{V_f}{V_i} = (1.00 \text{ mol})(8.31 \text{ J/mol}\cdot\text{K})(\ln 2)$$
$$= +5.76 \text{ J/K}.$$

Thus, the entropy change for the free expansion (and for all other processes that connect the initial and final states shown in Fig. 20-2) is

$$\Delta S_{irrev} = \Delta S_{rev} = +5.76 \text{ J/K}. \quad \text{(Answer)}$$

Because ΔS is positive, the entropy increases, in accordance with the entropy postulate of Section 20-2.

Sample Problem 20-2 Build your skill

Figure 20-5*a* shows two identical copper blocks of mass $m = 1.5$ kg: block L at temperature $T_{iL} = 60°\text{C}$ and block R at temperature $T_{iR} = 20°\text{C}$. The blocks are in a thermally insulated box and are separated by an insulating shutter. When we lift the shutter, the blocks eventually come to the equilibrium temperature $T_f = 40°\text{C}$ (Fig. 20-5*b*). What is the net entropy change of the two-block system during this irreversible process? The specific heat of copper is 386 J/kg · K.

KEY IDEA To calculate the entropy change, we must find a reversible process that takes the system from the initial state of Fig. 20-5*a* to the final state of Fig. 20-5*b*. We can calculate the net entropy change ΔS_{rev} of the reversible process using Eq. 20-1, and then the entropy change for the irreversible process is equal to ΔS_{rev}.

Calculations: For the reversible process, we need a thermal reservoir whose temperature can be changed slowly (say, by turning a knob). We then take the blocks through the following two steps, illustrated in Fig. 20-6.

Step 1 With the reservoir's temperature set at 60°C, put block L on the reservoir. (Since block and reservoir are at the same temperature, they are already in thermal equilibrium.) Then slowly lower the temperature of the reservoir and the block to 40°C. As the block's temperature changes by each increment dT during this process, energy dQ is transferred as heat *from* the block to the reservoir. Using Eq. 18-14, we can write this transferred energy as $dQ = mc\,dT$, where c is the specific heat of copper. According to Eq. 20-1, the entropy change ΔS_L of block L during the full temperature change from initial temperature T_{iL} (= 60°C = 333 K) to final temperature T_f (= 40°C = 313 K) is

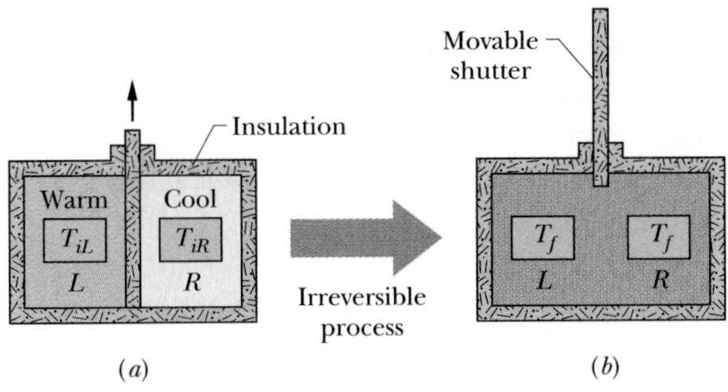

FIG. 20-5 (*a*) In the initial state, two copper blocks L and R, identical except for their temperatures, are in an insulating box and are separated by an insulating shutter. (*b*) When the shutter is removed, the blocks exchange energy as heat and come to a final state, both with the same temperature T_f.

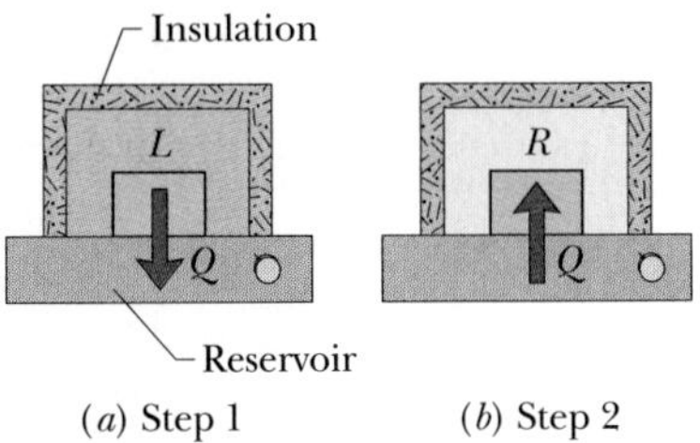

FIG. 20-6 The blocks of Fig. 20-5 can proceed from their initial state to their final state in a reversible way if we use a reservoir with a controllable temperature (*a*) to extract heat reversibly from block L and (*b*) to add heat reversibly to block R.

$$\Delta S_L = \int_i^f \frac{dQ}{T} = \int_{T_{iL}}^{T_f} \frac{mc\,dT}{T} = mc \int_{T_{iL}}^{T_f} \frac{dT}{T}$$
$$= mc \ln \frac{T_f}{T_{iL}}.$$

Inserting the given data yields

$$\Delta S_L = (1.5 \text{ kg})(386 \text{ J/kg}\cdot\text{K}) \ln \frac{313 \text{ K}}{333 \text{ K}}$$
$$= -35.86 \text{ J/K}.$$

Step 2 With the reservoir's temperature now set at 20°C, put block R on the reservoir. Then slowly raise the temperature of the reservoir and the block to 40°C. With the same reasoning used to find ΔS_L, you can show that the entropy change ΔS_R of block R during this process is

$$\Delta S_R = (1.5 \text{ kg})(386 \text{ J/kg}\cdot\text{K}) \ln \frac{313 \text{ K}}{293 \text{ K}}$$
$$= +38.23 \text{ J/K}.$$

The net entropy change ΔS_{rev} of the two-block system undergoing this two-step reversible process is then

$$\Delta S_{rev} = \Delta S_L + \Delta S_R$$
$$= -35.86 \text{ J/K} + 38.23 \text{ J/K} = 2.4 \text{ J/K}.$$

Thus, the net entropy change ΔS_{irrev} for the two-block system undergoing the actual irreversible process is

$$\Delta S_{irrev} = \Delta S_{rev} = 2.4 \text{ J/K}. \quad \text{(Answer)}$$

This result is positive, in accordance with the entropy postulate of Section 20-2.

20-4 | The Second Law of Thermodynamics

Here is a puzzle. We saw in Sample Problem 20-1 that if we cause the reversible process of Fig. 20-3 to proceed from (*a*) to (*b*) in that figure, the change in entropy of the gas—which we take as our system—is positive. However, because the process is reversible, we can just as easily make it proceed from (*b*) to (*a*), simply by slowly adding lead shot to the piston of Fig. 20-3*b* until the original volume of the gas is restored. In this reverse process, energy must be extracted as heat *from the gas* to keep its temperature from rising. Hence Q is negative and so, from Eq. 20-2, the entropy of the gas must decrease.

Doesn't this decrease in the entropy of the gas violate the entropy postulate of Section 20-2, which states that entropy always increases? No, because that postulate holds only for *irreversible* processes occurring in closed systems. The procedure suggested here does not meet these requirements. The process is *not* irreversible, and (because energy is transferred as heat from the gas to the reservoir) the system—which is the gas alone—is *not* closed.

However, if we include the reservoir, along with the gas, as part of the system, then we do have a closed system. Let's check the change in entropy of the enlarged system *gas + reservoir* for the process that takes it from (*b*) to (*a*) in Fig. 20-3. During this reversible process, energy is transferred as heat from the gas to the reservoir—that is, from one part of the enlarged system to another. Let $|Q|$ represent the absolute value (or magnitude) of this heat. With Eq. 20-2, we can then calculate separately the entropy changes for the gas (which loses $|Q|$) and the reservoir (which gains $|Q|$). We get

$$\Delta S_{\text{gas}} = -\frac{|Q|}{T}$$

and

$$\Delta S_{\text{gas}} = +\frac{|Q|}{T}.$$

The entropy change of the closed system is the sum of these two quantities: 0.

With this result, we can modify the entropy postulate of Section 20-2 to include both reversible and irreversible processes:

> If a process occurs in a *closed* system, the entropy of the system increases for irreversible processes and remains constant for reversible processes. It never decreases.

Although entropy may decrease in part of a closed system, there will always be an equal or larger entropy increase in another part of the system, so that the entropy of the system as a whole never decreases. This fact is one form of the **second law of thermodynamics** and can be written as

$$\Delta S \geq 0 \quad \text{(second law of thermodynamics)}, \tag{20-5}$$

where the greater-than sign applies to irreversible processes and the equals sign to reversible processes. Equation 20-5 applies only to closed systems.

In the real world almost all processes are irreversible to some extent because of friction, turbulence, and other factors, so the entropy of real closed systems undergoing real processes always increases. Processes in which the system's entropy remains constant are always idealizations.

Force Due to Entropy

To understand why rubber resists being stretched, let's write the first law of thermodynamics

$$dE = dQ - dW$$

for a rubber band undergoing a small increase in length dx as we stretch it between our hands. The force from the rubber band has magnitude F, is directed inward, and does work $dW = -F\,dx$ during length increase dx. From Eq. 20-2 ($\Delta S = Q/T$), small changes in Q and S at constant temperature are related by $dS = dQ/T$, or $dQ = T\,dS$. So, now we can rewrite the first law as

$$dE = T\,dS + F\,dx. \tag{20-6}$$

To good approximation, the change dE in the internal energy of rubber is 0 if the total stretch of the rubber band is not very much. Substituting 0 for dE in Eq. 20-6 leads us to an expression for the force from the rubber band:

$$F = -T\frac{dS}{dx}. \tag{20-7}$$

This tells us that F is proportional to the rate dS/dx at which the rubber band's entropy changes during a small change dx in the rubber band's length. Thus, you can *feel* the effect of entropy on your hands as you stretch a rubber band.

To make sense of the relation between force and entropy, let's consider a simple model of the rubber material. Rubber consists of cross-linked polymer chains (long molecules with cross links) that resemble three-dimensional zig-zags (Fig. 20-7). When the rubber band is at its rest length, the polymers are coiled up in a spaghetti-like arrangement. Because of the large disorder of the molecules, this rest state has a high value of entropy. When we stretch a rubber band, we uncoil many of those polymers, aligning them in the direction of stretch. Because the alignment decreases the disorder, the entropy of the stretched rubber band is less. That is, the change dS/dx in Eq. 20-7 is a negative quantity because the entropy decreases with stretching. Thus, the force on our hands from the rubber band is due to the tendency of the polymers to return to their former disordered state and higher value of entropy.

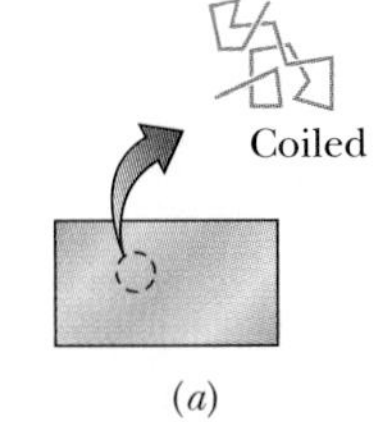

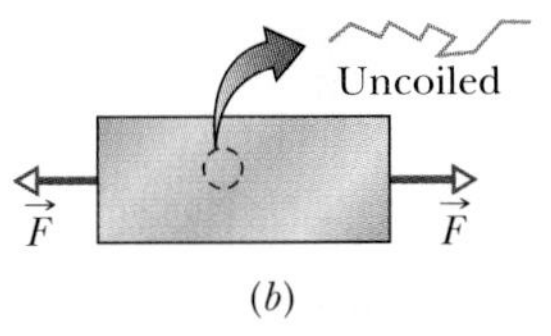

FIG. 20-7 A section of a rubber band (*a*) unstretched and (*b*) stretched, and a polymer within it (*a*) coiled and (*b*) uncoiled.

Sample Problem 20-3

The force of a stretched rubber band is given approximately by Hooke's Law of Eq. 7-21 ($F_x = -kx$), in which k is the *spring constant.* Suppose a rubber band with k = 50.0 N/m and at temperature T = 27°C is stretched by x = 1.2 cm. For a small additional stretching, at what rate dS/dx does the entropy of the rubber band decrease?

KEY IDEA The force of a stretched rubber is due to the change in the entropy of the polymers according to Eq. 20-7 ($F = -T\,dS/dx$).

Calculation: From Eq. 20-7, we know that the magnitude of the force is equal to $T|dS/dx|$. From Eq. 7-21, we know that the magnitude is also equal to $k|x|$. Thus,

$$T\left|\frac{dS}{dx}\right| = k|x|,$$

which yields

$$\left|\frac{dS}{dx}\right| = \frac{k|x|}{T} = \frac{(50.0\ \text{N/m})(0.012\ \text{m})}{(273\ \text{K} + 27\ \text{K})}$$
$$= 2.0 \times 10^{-3}\ \text{J/K}\cdot\text{m}. \qquad \text{(Answer)}$$

20-5 | Entropy in the Real World: Engines

A **heat engine,** or more simply, an **engine,** is a device that extracts energy from its environment in the form of heat and does useful work. At the heart of every engine is a *working substance.* In a steam engine, the working substance is water, in both its vapor and its liquid form. In an automobile engine the working substance is a gasoline–air mixture. If an engine is to do work on a sustained basis, the working substance must operate in a *cycle;* that is, the working substance must pass through a closed series of thermodynamic processes, called *strokes,* returning again and again to each state in its cycle. Let us see what the laws of thermodynamics can tell us about the operation of engines.

A Carnot Engine

We have seen that we can learn much about real gases by analyzing an ideal gas, which obeys the simple law $pV = nRT$. Although an ideal gas does not exist, any real gas approaches ideal behavior if its density is low enough. Similarly, we can study real engines by analyzing the behavior of an **ideal engine.**

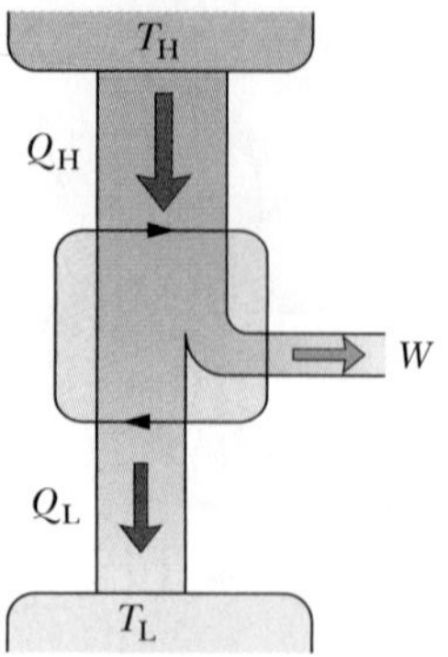

FIG. 20-8 The elements of a Carnot engine. The two black arrowheads on the central loop suggest the working substance operating in a cycle, as if on a p-V plot. Energy $|Q_H|$ is transferred as heat from the high-temperature reservoir at temperature T_H to the working substance. Energy $|Q_L|$ is transferred as heat from the working substance to the low-temperature reservoir at temperature T_L. Work W is done by the engine (actually by the working substance) on something in the environment.

In an ideal engine, all processes are reversible and no wasteful energy transfers occur due to, say, friction and turbulence.

We shall focus on a particular ideal engine called a **Carnot engine** after the French scientist and engineer N. L. Sadi Carnot (pronounced "car-no"), who first proposed the engine's concept in 1824. This ideal engine turns out to be the best (in principle) at using energy as heat to do useful work. Surprisingly, Carnot was able to analyze the performance of this engine before the first law of thermodynamics and the concept of entropy had been discovered.

Figure 20-8 shows schematically the operation of a Carnot engine. During each cycle of the engine, the working substance absorbs energy $|Q_H|$ as heat from a thermal reservoir at constant temperature T_H and discharges energy $|Q_L|$ as heat to a second thermal reservoir at a constant lower temperature T_L.

Figure 20-9 shows a p-V plot of the *Carnot cycle*—the cycle followed by the working substance. As indicated by the arrows, the cycle is traversed in the clockwise direction. Imagine the working substance to be a gas, confined to an insulating cylinder with a weighted, movable piston. The cylinder may be placed at will on either of the two thermal reservoirs, as in Fig. 20-6, or on an insulating slab. Figure 20-9 shows that, if we place the cylinder in contact with the high-temperature reservoir at temperature T_H, heat $|Q_H|$ is transferred *to* the working substance *from* this reservoir as the gas undergoes an isothermal *expansion* from volume V_a to volume V_b. Similarly, with the working substance in contact with the low-temperature reservoir at temperature T_L, heat $|Q_L|$ is transferred *from* the working substance *to* the low-temperature reservoir as the gas undergoes an isothermal *compression* from volume V_c to volume V_d.

In the engine of Fig. 20-8, we assume that heat transfers to or from the working substance can take place *only* during the isothermal processes *ab* and *cd* of Fig. 20-9. Therefore, processes *bc* and *da* in that figure, which connect the two isotherms at temperatures T_H and T_L, must be (reversible) adiabatic processes; that is, they must be processes in which no energy is transferred as heat. To ensure this, during processes *bc* and *da* the cylinder is placed on an insulating slab as the volume of the working substance is changed.

During the consecutive processes *ab* and *bc* of Fig. 20-9, the working substance is expanding and thus doing positive work as it raises the weighted piston. This work is represented in Fig. 20-9 by the area under curve *abc*. During the consecutive processes *cd* and *da*, the working substance is being compressed, which means that it is doing negative work on its environment or, equivalently, that its environment is doing work on it as the loaded piston descends. This work is represented by the area under curve *cda*. The *net work per cycle,* which is represented by W in both Figs. 20-8 and 20-9, is the difference between these two areas and is a positive quantity equal to the area enclosed by cycle *abcda* in Fig. 20-9. This work W is performed on some outside object, such as a load to be lifted.

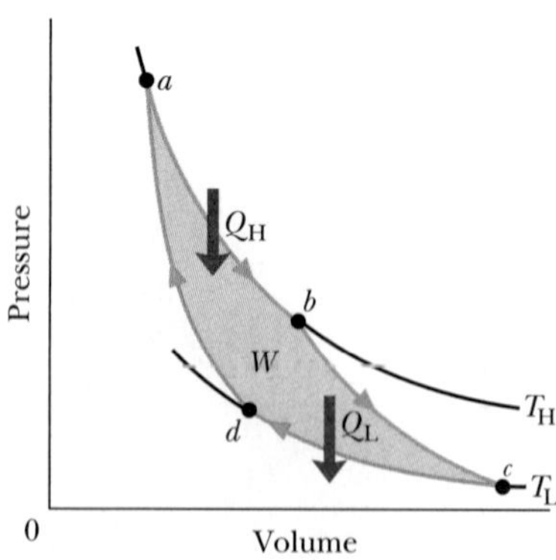

FIG. 20-9 A pressure–volume plot of the cycle followed by the working substance of the Carnot engine in Fig. 20-8. The cycle consists of two isothermal (*ab* and *cd*) and two adiabatic processes (*bc* and *da*). The shaded area enclosed by the cycle is equal to the work W per cycle done by the Carnot engine.

Equation 20-1 ($\Delta S = \int dQ/T$) tells us that any energy transfer as heat must involve a change in entropy. To illustrate the entropy changes for a Carnot engine, we can plot the Carnot cycle on a temperature–entropy (T-S) diagram as shown in Fig. 20-10. The lettered points *a*, *b*, *c*, and *d* in Fig. 20-10 correspond to the lettered points in the p-V diagram in Fig. 20-9. The two horizontal lines in Fig. 20-10 correspond to the two isothermal processes of the Carnot cycle (because the temperature is constant). Process *ab* is the isothermal expansion of the cycle. As the working substance (reversibly) absorbs energy $|Q_H|$ as heat at constant temperature T_H during the expansion, its entropy increases. Similarly, during the isothermal compression *cd*, the working substance (reversibly) loses energy $|Q_L|$ as heat at constant temperature T_L, and its entropy decreases.

The two vertical lines in Fig. 20-10 correspond to the two adiabatic processes of the Carnot cycle. Because no energy is transferred as heat during the two processes, the entropy of the working substance is constant during them.

The Work To calculate the net work done by a Carnot engine during a cycle, let us apply Eq. 18-26, the first law of thermodynamics ($\Delta E_{int} = Q - W$), to the working substance. That substance must return again and again to any arbitrarily selected state in the cycle. Thus, if X represents any state property of the working substance, such as pressure, temperature, volume, internal energy, or entropy, we must have $\Delta X = 0$ for every cycle. It follows that $\Delta E_{int} = 0$ for a complete cycle of the working substance. Recalling that Q in Eq. 18-26 is the *net* heat transfer per cycle and W is the *net* work, we can write the first law of thermodynamics for the Carnot cycle as

$$W = |Q_H| - |Q_L|. \quad (20\text{-}8)$$

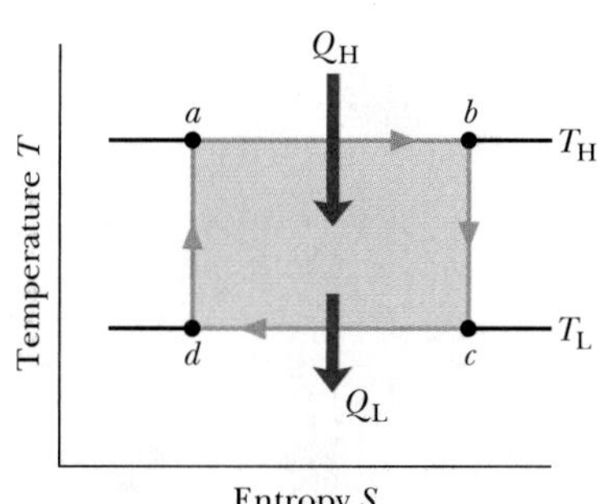

FIG. 20-10 The Carnot cycle of Fig. 20-9 plotted on a temperature–entropy diagram. During processes *ab* and *cd* the temperature remains constant. During processes *bc* and *da* the entropy remains constant.

Entropy Changes In a Carnot engine, there are *two* (and only two) reversible energy transfers as heat, and thus two changes in the entropy of the working substance—one at temperature T_H and one at T_L. The net entropy change per cycle is then

$$\Delta S = \Delta S_H + \Delta S_L = \frac{|Q_H|}{T_H} - \frac{|Q_L|}{T_L}. \quad (20\text{-}9)$$

Here ΔS_H is positive because energy $|Q_H|$ is *added to* the working substance as heat (an increase in entropy) and ΔS_L is negative because energy $|Q_L|$ is *removed from* the working substance as heat (a decrease in entropy). Because entropy is a state function, we must have $\Delta S = 0$ for a complete cycle. Putting $\Delta S = 0$ in Eq. 20-9 requires that

$$\frac{|Q_H|}{T_H} = \frac{|Q_L|}{T_L}. \quad (20\text{-}10)$$

Note that, because $T_H > T_L$, we must have $|Q_H| > |Q_L|$; that is, more energy is extracted as heat from the high-temperature reservoir than is delivered to the low-temperature reservoir.

We shall now use Eqs. 20-8 and 20-10 to derive an expression for the efficiency of a Carnot engine.

Efficiency of a Carnot Engine

The purpose of any engine is to transform as much of the extracted energy Q_H into work as possible. We measure its success in doing so by its **thermal efficiency** ε, defined as the work the engine does per cycle ("energy we get") divided by the energy it absorbs as heat per cycle ("energy we pay for"):

$$\varepsilon = \frac{\text{energy we get}}{\text{energy we pay for}} = \frac{|W|}{|Q_H|} \quad \text{(efficiency, any engine)}. \quad (20\text{-}11)$$

For a Carnot engine we can substitute for W from Eq. 20-8 to write Eq. 20-11 as

$$\varepsilon_C = \frac{|Q_H| - |Q_L|}{Q_H} = 1 - \frac{|Q_L|}{|Q_H|}. \quad (20\text{-}12)$$

Using Eq. 20-10 we can write this as

$$\varepsilon_C = 1 - \frac{T_L}{T_H} \quad \text{(efficiency, Carnot engine)}, \quad (20\text{-}13)$$

where the temperatures T_L and T_H are in kelvins. Because $T_L < T_H$, the Carnot engine necessarily has a thermal efficiency less than unity—that is, less than 100%. This is indicated in Fig. 20-8, which shows that only part of the energy extracted as heat from the high-temperature reservoir is available to do work, and the rest is delivered to the low-temperature reservoir. We shall show in Section 20-7 that no real engine can have a thermal efficiency greater than that calculated from Eq. 20-13.

Inventors continually try to improve engine efficiency by reducing the energy $|Q_L|$ that is "thrown away" during each cycle. The inventor's dream is to produce

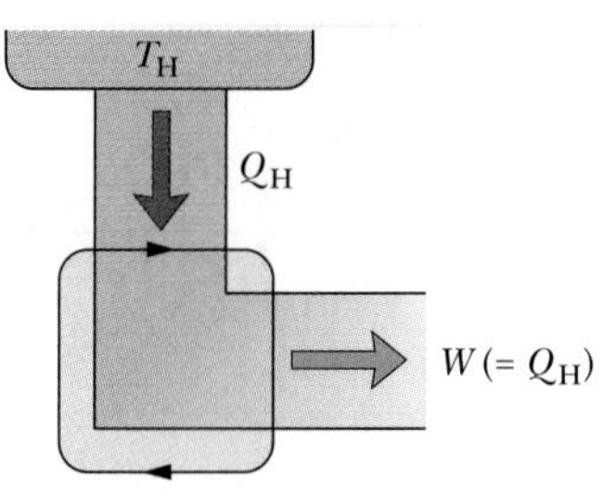

FIG. 20-11 The elements of a perfect engine—that is, one that converts heat Q_H from a high-temperature reservoir directly to work W with 100% efficiency.

FIG. 20-12 The North Anna nuclear power plant near Charlottesville, Virginia, which generates electric energy at the rate of 900 MW. At the same time, by design, it discards energy into the nearby river at the rate of 2100 MW. This plant and all others like it throw away more energy than they deliver in useful form. They are real counterparts of the ideal engine of Fig. 20-8. *(© Robert Ustinich)*

the *perfect engine,* diagrammed in Fig. 20-11, in which $|Q_L|$ is reduced to zero and $|Q_H|$ is converted completely into work. Such an engine on an ocean liner, for example, could extract energy as heat from the water and use it to drive the propellers, with no fuel cost. An automobile fitted with such an engine could extract energy as heat from the surrounding air and use it to drive the car, again with no fuel cost. Alas, a perfect engine is only a dream: Inspection of Eq. 20-13 shows that we can achieve 100% engine efficiency (that is, $\varepsilon = 1$) only if $T_L = 0$ or $T_H \to \infty$, impossible requirements. Instead, experience gives the following alternative version of the second law of thermodynamics, which says in short, *there are no perfect engines:*

> No series of processes is possible whose sole result is the transfer of energy as heat from a thermal reservoir and the complete conversion of this energy to work.

To summarize: The thermal efficiency given by Eq. 20-13 applies only to Carnot engines. Real engines, in which the processes that form the engine cycle are not reversible, have lower efficiencies. If your car were powered by a Carnot engine, it would have an efficiency of about 55% according to Eq. 20-13; its actual efficiency is probably about 25%. A nuclear power plant (Fig. 20-12), taken in its entirety, is an engine. It extracts energy as heat from a reactor core, does work by means of a turbine, and discharges energy as heat to a nearby river. If the power plant operated as a Carnot engine, its efficiency would be about 40%; its actual efficiency is about 30%. In designing engines of any type, there is simply no way to beat the efficiency limitation imposed by Eq. 20-13.

Stirling Engine

Equation 20-13 applies not to all ideal engines but only to those that can be represented as in Fig. 20-9—that is, to Carnot engines. For example, Fig. 20-13 shows the operating cycle of an ideal **Stirling engine.** Comparison with the Carnot cycle of Fig. 20-9 shows that each engine has isothermal heat transfers at temperatures T_H and T_L. However, the two isotherms of the Stirling engine cycle are connected, not by adiabatic processes as for the Carnot engine but by constant-volume processes. To increase the temperature of a gas at constant volume reversibly from T_L to T_H (process *da* of Fig. 20-13) requires a transfer of energy as heat to the working substance from a thermal reservoir whose temperature can be varied smoothly between those limits. Also, a reverse transfer is required in process *bc*. Thus, reversible heat transfers (and corresponding entropy changes) occur in all four of the processes that form the cycle of a Stirling engine, not just two processes as in a Carnot engine. Thus, the derivation that led to Eq. 20-13 does not apply to an ideal Stirling engine. More important, the efficiency of an ideal Stirling engine is lower than that of a Carnot engine operating between the same two temperatures. Real Stirling engines have even lower efficiencies.

The Stirling engine was developed in 1816 by Robert Stirling. This engine, long neglected, is now being developed for use in automobiles and spacecraft. A Stirling engine delivering 5000 hp (3.7 MW) has been built. Because they are quiet, Stirling engines are used on some military submarines.

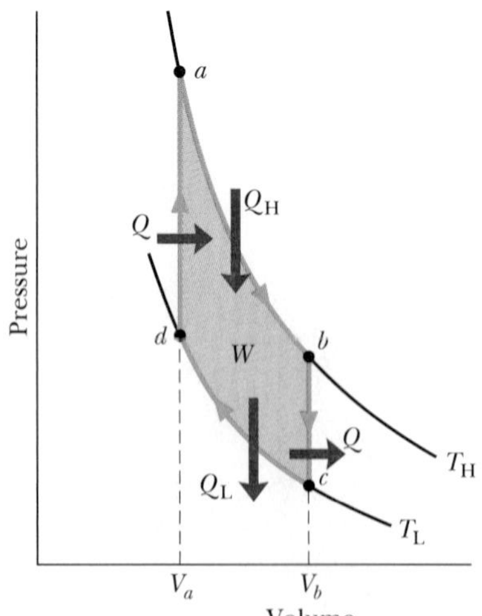

FIG. 20-13 A *p-V* plot for the working substance of an ideal Stirling engine, with the working substance assumed for convenience to be an ideal gas.

CHECKPOINT 3 Three Carnot engines operate between reservoir temperatures of (a) 400 and 500 K, (b) 600 and 800 K, and (c) 400 and 600 K. Rank the engines according to their thermal efficiencies, greatest first.

Sample Problem 20-4

Imagine a Carnot engine that operates between the temperatures $T_H = 850$ K and $T_L = 300$ K. The engine performs 1200 J of work each cycle, which takes 0.25 s.

(a) What is the efficiency of this engine?

KEY IDEA The efficiency ε of a Carnot engine depends only on the ratio T_L/T_H of the temperatures (in kelvins) of the thermal reservoirs to which it is connected.

Calculation: Thus, from Eq. 20-13, we have

$$\varepsilon = 1 - \frac{T_L}{T_H} = 1 - \frac{300\text{ K}}{850\text{ K}} = 0.647 \approx 65\%. \quad \text{(Answer)}$$

(b) What is the average power of this engine?

KEY IDEA The average power P of an engine is the ratio of the work W it does per cycle to the time t that each cycle takes.

Calculation: For this Carnot engine, we find

$$P = \frac{W}{t} = \frac{1200\text{ J}}{0.25\text{ s}} = 4800\text{ W} = 4.8\text{ kW}. \quad \text{(Answer)}$$

(c) How much energy $|Q_H|$ is extracted as heat from the high-temperature reservoir every cycle?

KEY IDEA For any engine, including a Carnot engine, the efficiency ε is the ratio of the work W that is done per cycle to the energy $|Q_H|$ that is extracted as heat from the high-temperature reservoir per cycle ($\varepsilon = W/|Q_H|$).

Calculation: Here we have

$$|Q_H| = \frac{W}{\varepsilon} = \frac{1200\text{ J}}{0.647} = 1855\text{ J}. \quad \text{(Answer)}$$

(d) How much energy $|Q_L|$ is delivered as heat to the low-temperature reservoir every cycle?

KEY IDEA For a Carnot engine, the work W done per cycle is equal to the difference in the energy transfers as heat: $|Q_H| - |Q_L|$, as in Eq. 20-8.

Calculation: Thus, we have

$$|Q_L| = |Q_H| - W = 1855\text{ J} - 1200\text{ J} = 655\text{ J}. \quad \text{(Answer)}$$

(e) By how much does the entropy of the working substance change as a result of the energy transferred to it from the high-temperature reservoir? From it to the low-temperature reservoir?

KEY IDEA The entropy change ΔS during a transfer of energy as heat Q at constant temperature T is given by Eq. 20-2 ($\Delta S = Q/T$).

Calculations: Thus, for the *positive* transfer of energy Q_H from the high-temperature reservoir at T_H, the change in the entropy of the working substance is

$$\Delta S_H = \frac{Q_H}{T_H} = \frac{1855\text{ J}}{850\text{ K}} = +2.18\text{ J/K}. \quad \text{(Answer)}$$

Similarly, for the *negative* transfer of energy Q_L to the low-temperature reservoir at T_L, we have

$$\Delta S_L = \frac{Q_L}{T_L} = \frac{-655\text{ J}}{300\text{ K}} = -2.18\text{ J/K}. \quad \text{(Answer)}$$

Note that the net entropy change of the working substance for one cycle is zero, as we discussed in deriving Eq. 20-10.

Sample Problem 20-5

An inventor claims to have constructed an engine that has an efficiency of 75% when operated between the boiling and freezing points of water. Is this possible?

KEY IDEA The efficiency of a real engine (with its irreversible processes and wasteful energy transfers) must be less than the efficiency of a Carnot engine operating between the same two temperatures.

Calculation: From Eq. 20-13, we find that the efficiency of a Carnot engine operating between the boiling and freezing points of water is

$$\varepsilon = 1 - \frac{T_L}{T_H} = 1 - \frac{(0 + 273)\text{ K}}{(100 + 273)\text{ K}} = 0.268 \approx 27\%.$$

Thus, the claimed efficiency of 75% for a real engine operating between the given temperatures is impossible.

PROBLEM-SOLVING TACTICS

Tactic 1: The Language of Thermodynamics A rich but sometimes misleading language is used in scientific and engineering studies of thermodynamics. You may see statements that say heat is added, absorbed, subtracted, extracted, rejected, discharged, discarded, withdrawn, delivered, gained, lost, transferred, or expelled, or that it flows from one body to another (as if it were a liquid). You may also see statements that describe a body as *having* heat (as if heat can be held or possessed) or that its heat is increased or decreased. You should always keep in mind what is meant by the term *heat* in science and engineering:

> Heat is energy that is transferred from one body to another body due to a difference in the temperatures of the bodies.

When we identify one of the bodies as being our system, any such transfer of energy into it is positive heat Q and out of it is negative heat Q.

The term *work* also requires close attention. You may see statements that say work is produced or generated, or combined with heat or changed from heat. Here is what is meant by the term *work*:

Work is energy that is transferred from one body to another body due to a force that acts between them.

When we identify one of the bodies as being our system of interest, any such transfer of energy out of the system is either positive work W done *by* the system or negative work W done *on* the system. Any such transfer of energy into the system is negative work done *by* the system or positive work done *on* the system. (The preposition that is used is important.) Obviously, this can be confusing—whenever you see the term *work,* read carefully.

20-6 | Entropy in the Real World: Refrigerators

A **refrigerator** is a device that uses work to transfer energy from a low-temperature reservoir to a high-temperature reservoir as the device continuously repeats a set series of thermodynamic processes. In a household refrigerator, for example, work is done by an electrical compressor to transfer energy from the food storage compartment (a low-temperature reservoir) to the room (a high-temperature reservoir).

Air conditioners and heat pumps are also refrigerators. The differences are only in the nature of the high- and low-temperature reservoirs. For an air conditioner, the low-temperature reservoir is the room that is to be cooled and the high-temperature reservoir is the (presumably warmer) outdoors. A heat pump is an air conditioner that can be operated in reverse to heat a room; the room is the high-temperature reservoir, and heat is transferred to it from the (presumably cooler) outdoors.

Let us consider an *ideal refrigerator:*

In an ideal refrigerator, all processes are reversible and no wasteful energy transfers occur as a result of, say, friction and turbulence.

Figure 20-14 shows the basic elements of an ideal refrigerator. Note that its operation is the reverse of how the Carnot engine of Fig. 20-8 operates. In other words, all the energy transfers, as either heat or work, are reversed from those of a Carnot engine. We can call such an ideal refrigerator a **Carnot refrigerator.**

The designer of a refrigerator would like to extract as much energy $|Q_L|$ as possible from the low-temperature reservoir (what we want) for the least amount of work $|W|$ (what we pay for). A measure of the efficiency of a refrigerator, then, is

$$K = \frac{\text{what we want}}{\text{what we pay for}} = \frac{|Q_L|}{|W|} \quad \text{(coefficient of performance, any refrigerator),} \tag{20-14}$$

where K is called the *coefficient of performance.* For a Carnot refrigerator, the first law of thermodynamics gives $|W| = |Q_H| - |Q_L|$, where $|Q_H|$ is the magnitude of the energy transferred as heat to the high-temperature reservoir. Equation 20-14 then becomes

$$K_C = \frac{|Q_L|}{|Q_H| - |Q_L|}. \tag{20-15}$$

Because a Carnot refrigerator is a Carnot engine operating in reverse, we can combine Eq. 20-10 with Eq. 20-15; after some algebra we find

$$K_C = \frac{T_L}{T_H - T_L} \quad \text{(coefficient of performance, Carnot refrigerator).} \tag{20-16}$$

For typical room air conditioners, $K \approx 2.5$. For household refrigerators, $K \approx 5$. Perversely, the value of K is higher the closer the temperatures of the two reservoirs are to each other. That is why heat pumps are more effective in temperate climates than in climates where the outside temperature is much lower than the desired inside temperature.

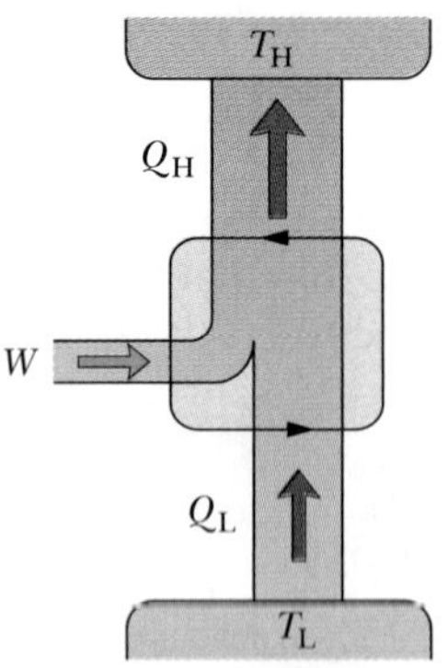

FIG. 20-14 The elements of a refrigerator. The two black arrowheads on the central loop suggest the working substance operating in a cycle, as if on a p-V plot. Energy is transferred as heat Q_L to the working substance from the low-temperature reservoir. Energy is transferred as heat Q_H to the high-temperature reservoir from the working substance. Work W is done on the refrigerator (on the working substance) by something in the environment.

It would be nice to own a refrigerator that did not require some input of work—that is, one that would run without being plugged in. Figure 20-15 represents another "inventor's dream," a *perfect refrigerator* that transfers energy as heat Q from a cold reservoir to a warm reservoir without the need for work. Because the unit operates in cycles, the entropy of the working substance does not change during a complete cycle. The entropies of the two reservoirs, however, do change: The entropy change for the cold reservoir is $-|Q|/T_L$, and that for the warm reservoir is $+|Q|/T_H$. Thus, the net entropy change for the entire system is

$$\Delta S = -\frac{|Q|}{T_L} + \frac{|Q|}{T_H}.$$

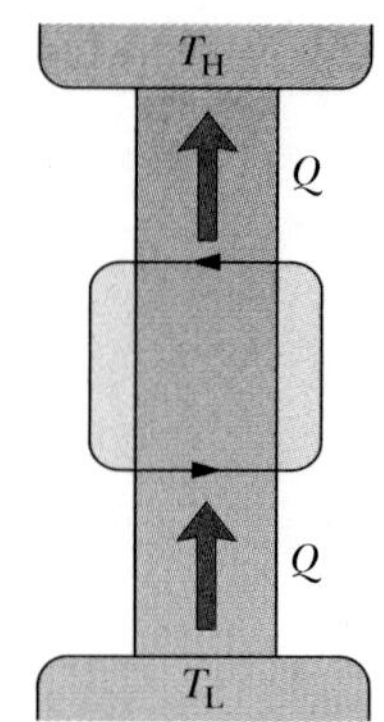

FIG. 20-15 The elements of a perfect refrigerator—that is, one that transfers energy from a low-temperature reservoir to a high-temperature reservoir without any input of work.

Because $T_H > T_L$, the right side of this equation is negative and thus the net change in entropy per cycle for the closed system *refrigerator + reservoirs* is also negative. Because such a decrease in entropy violates the second law of thermodynamics (Eq. 20-5), a perfect refrigerator does not exist. (If you want your refrigerator to operate, you must plug it in.)

This result leads us to another (equivalent) formulation of the second law of thermodynamics:

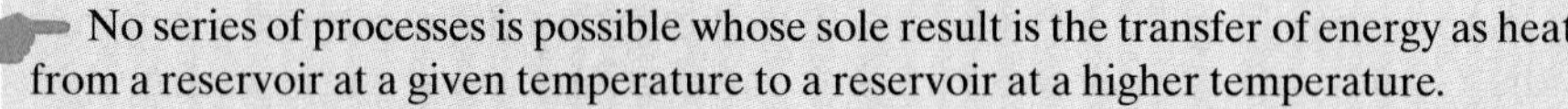

No series of processes is possible whose sole result is the transfer of energy as heat from a reservoir at a given temperature to a reservoir at a higher temperature.

In short, *there are no perfect refrigerators.*

CHECKPOINT 4 You wish to increase the coefficient of performance of an ideal refrigerator. You can do so by (a) running the cold chamber at a slightly higher temperature, (b) running the cold chamber at a slightly lower temperature, (c) moving the unit to a slightly warmer room, or (d) moving it to a slightly cooler room. The magnitudes of the temperature changes are to be the same in all four cases. List the changes according to the resulting coefficients of performance, greatest first.

20-7 | The Efficiencies of Real Engines

Let ε_C be the efficiency of a Carnot engine operating between two given temperatures. In this section we prove that no real engine operating between those temperatures can have an efficiency greater than ε_C. If it could, the engine would violate the second law of thermodynamics.

Let us assume that an inventor, working in her garage, has constructed an engine X, which she claims has an efficiency ε_X that is greater than ε_C:

$$\varepsilon_X > \varepsilon_C \quad \text{(a claim).} \tag{20-17}$$

Let us couple engine X to a Carnot refrigerator, as in Fig. 20-16*a*. We adjust the

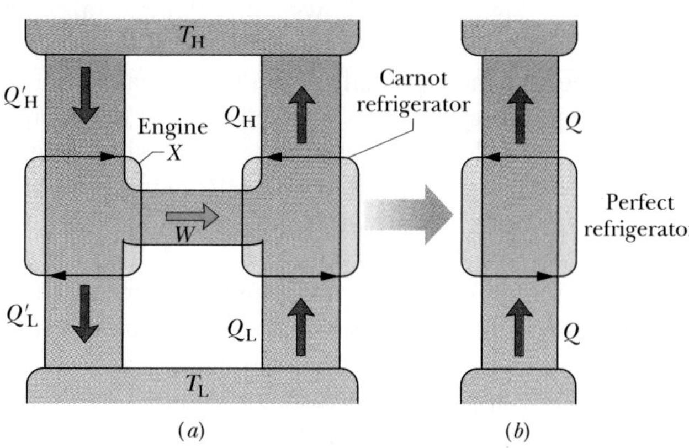

FIG. 20-16 (*a*) Engine X drives a Carnot refrigerator. (*b*) If, as claimed, engine X is more efficient than a Carnot engine, then the combination shown in (*a*) is equivalent to the perfect refrigerator shown here. This violates the second law of thermodynamics, so we conclude that engine X *cannot* be more efficient than a Carnot engine.

strokes of the Carnot refrigerator so that the work it requires per cycle is just equal to that provided by engine X. Thus, no (external) work is performed on or by the combination *engine* + *refrigerator* of Fig. 20-16*a*, which we take as our system.

If Eq. 20-17 is true, from the definition of efficiency (Eq. 20-11), we must have

$$\frac{|W|}{|Q'_H|} > \frac{|W|}{|Q_H|},$$

where the prime refers to engine X and the right side of the inequality is the efficiency of the Carnot refrigerator when it operates as an engine. This inequality requires that

$$|Q_H| > |Q'_H|. \quad (20\text{-}18)$$

Because the work done by engine X is equal to the work done on the Carnot refrigerator, we have, from the first law of thermodynamics as given by Eq. 20-8,

$$|Q_H| - |Q_L| = |Q'_H| - |Q'_L|,$$

which we can write as

$$|Q_H| - |Q'_H| = |Q_L| - |Q'_L| = Q. \quad (20\text{-}19)$$

Because of Eq. 20-18, the quantity Q in Eq. 20-19 must be positive.

Comparison of Eq. 20-19 with Fig. 20-16 shows that the net effect of engine X and the Carnot refrigerator working in combination is to transfer energy Q as heat from a low-temperature reservoir to a high-temperature reservoir without the requirement of work. Thus, the combination acts like the perfect refrigerator of Fig. 20-15, whose existence is a violation of the second law of thermodynamics.

Something must be wrong with one or more of our assumptions, and it can only be Eq. 20-17. We conclude that *no real engine can have an efficiency greater than that of a Carnot engine when both engines work between the same two temperatures.* At most, the real engine can have an efficiency equal to that of a Carnot engine. In that case, the real engine *is* a Carnot engine.

20-8 | A Statistical View of Entropy

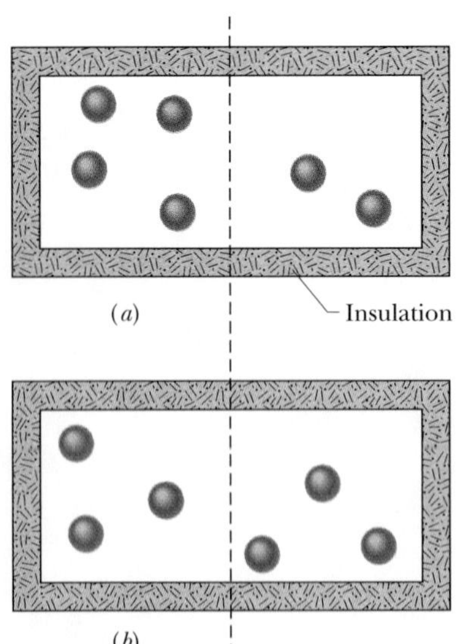

FIG. 20-17 An insulated box contains six gas molecules. Each molecule has the same probability of being in the left half of the box as in the right half. The arrangement in (*a*) corresponds to configuration III in Table 20-1, and that in (*b*) corresponds to configuration IV.

In Chapter 19 we saw that the macroscopic properties of gases can be explained in terms of their microscopic, or molecular, behavior. For one example, recall that we were able to account for the pressure exerted by a gas on the walls of its container in terms of the momentum transferred to those walls by rebounding gas molecules. Such explanations are part of a study called **statistical mechanics.**

Here we shall focus our attention on a single problem, one involving the distribution of gas molecules between the two halves of an insulated box. This problem is reasonably simple to analyze, and it allows us to use statistical mechanics to calculate the entropy change for the free expansion of an ideal gas. You will see in Sample Problem 20-7 that statistical mechanics leads to the same entropy change we obtained in Sample Problem 20 1 using thermodynamics.

Figure 20-17 shows a box that contains six identical (and thus indistinguishable) molecules of a gas. At any instant, a given molecule will be in either the left or the right half of the box; because the two halves have equal volumes, the molecule has the same likelihood, or probability, of being in either half.

Table 20-1 shows the seven possible *configurations* of the six molecules, each configuration labeled with a Roman numeral. For example, in configuration I, all six molecules are in the left half of the box ($n_1 = 6$) and none are in the right half ($n_2 = 0$). We see that, in general, a given configuration can be achieved in a number of different ways. We call these different arrangements of the molecules *microstates.* Let us see how to calculate the number of microstates that correspond to a given configuration.

Suppose we have N molecules, distributed with n_1 molecules in one half of the box and n_2 in the other. (Thus $n_1 + n_2 = N$.) Let us imagine that we distribute

TABLE 20-1

Six Molecules in a Box

Configuration Label	n_1	n_2	Multiplicity W (number of microstates)	Calculation of W (Eq. 20-20)	Entropy 10^{-23} J/K (Eq. 20-21)
I	6	0	1	$6!/(6!\ 0!) = 1$	0
II	5	1	6	$6!/(5!\ 1!) = 6$	2.47
III	4	2	15	$6!/(4!\ 2!) = 15$	3.74
IV	3	3	20	$6!/(3!\ 3!) = 20$	4.13
V	2	4	15	$6!/(2!\ 4!) = 15$	3.74
VI	1	5	6	$6!/(1!\ 5!) = 6$	2.47
VII	0	6	1	$6!/(0!\ 6!) = 1$	0
			Total = 64		

the molecules "by hand," one at a time. If $N = 6$, we can select the first molecule in six independent ways; that is, we can pick any one of the six molecules. We can pick the second molecule in five ways, by picking any one of the remaining five molecules; and so on. The total number of ways in which we can select all six molecules is the product of these independent ways, or $6 \times 5 \times 4 \times 3 \times 2 \times 1 = 720$. In mathematical shorthand we write this product as $6! = 720$, where 6! is pronounced "six factorial." Your hand calculator can probably calculate factorials. For later use you will need to know that $0! = 1$. (Check this on your calculator.)

However, because the molecules are indistinguishable, these 720 arrangements are not all different. In the case that $n_1 = 4$ and $n_2 = 2$ (which is configuration III in Table 20-1), for example, the order in which you put four molecules in one half of the box does not matter, because after you have put all four in, there is no way that you can tell the order in which you did so. The number of ways in which you can order the four molecules is $4! = 24$. Similarly, the number of ways in which you can order two molecules for the other half of the box is simply $2! = 2$. To get the number of *different* arrangements that lead to the (4, 2) split of configuration III, we must divide 720 by 24 and also by 2. We call the resulting quantity, which is the number of microstates that correspond to a given configuration, the *multiplicity* W of that configuration. Thus, for configuration III,

$$W_{\mathrm{III}} = \frac{6!}{4!\,2!} = \frac{720}{24 \times 2} = 15.$$

Thus, Table 20-1 tells us there are 15 independent microstates that correspond to configuration III. Note that, as the table also tells us, the total number of microstates for six molecules distributed over the seven configurations is 64.

Extrapolating from six molecules to the general case of N molecules, we have

$$W = \frac{N!}{n_1!\,n_2!} \quad \text{(multiplicity of configuration).} \tag{20-20}$$

You should verify that Eq. 20-20 gives the multiplicities for all the configurations listed in Table 20-1.

The basic assumption of statistical mechanics is

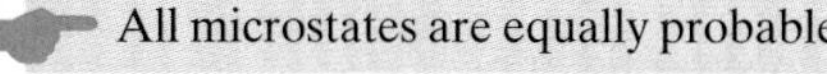
All microstates are equally probable.

In other words, if we were to take a great many snapshots of the six molecules as they jostle around in the box of Fig. 20-17 and then count the number of times each microstate occurred, we would find that all 64 microstates would occur equally often. Thus the system will spend, on average, the same amount of time in each of the 64 microstates.

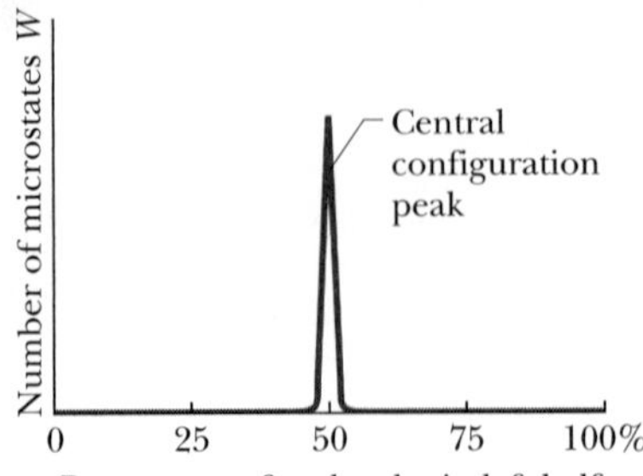

FIG. 20-18 For a *large* number of molecules in a box, a plot of the number of microstates that require various percentages of the molecules to be in the left half of the box. Nearly all the microstates correspond to an approximately equal sharing of the molecules between the two halves of the box; those microstates form the *central configuration peak* on the plot. For $N \approx 10^{22}$, the central configuration peak is much too narrow to be drawn on this plot.

Because all microstates are equally probable but different configurations have different numbers of microstates, the configurations are *not* all equally probable. In Table 20-1 configuration IV, with 20 microstates, is the *most probable configuration,* with a probability of $20/64 = 0.313$. This result means that the system is in configuration IV 31.3% of the time. Configurations I and VII, in which all the molecules are in one half of the box, are the least probable, each with a probability of $1/64 = 0.016$ or 1.6%. It is not surprising that the most probable configuration is the one in which the molecules are evenly divided between the two halves of the box, because that is what we expect at thermal equilibrium. However, it *is* surprising that there is *any* probability, however small, of finding all six molecules clustered in half of the box, with the other half empty.

For large values of N there are extremely large numbers of microstates, but nearly all the microstates belong to the configuration in which the molecules are divided equally between the two halves of the box, as Fig. 20-18 indicates. Even though the measured temperature and pressure of the gas remain constant, the gas is churning away endlessly as its molecules "visit" all probable microstates with equal probability. However, because so few microstates lie outside the very narrow central configuration peak of Fig. 20-18, we might as well assume that the gas molecules are always divided equally between the two halves of the box. As we shall see, this is the configuration with the greatest entropy.

Sample Problem 20-6

Suppose that there are 100 indistinguishable molecules in the box of Fig. 20-17. How many microstates are associated with the configuration $n_1 = 50$ and $n_2 = 50$, and with the configuration $n_1 = 100$ and $n_2 = 0$? Interpret the results in terms of the relative probabilities of the two configurations.

KEY IDEA The multiplicity W of a configuration of indistinguishable molecules in a closed box is the number of independent microstates with that configuration, as given by Eq. 20-20.

Calculations: For the (n_1, n_2) configuration (50, 50), that equation yields

$$W = \frac{N!}{n_1!\,n_2!} = \frac{100!}{50!\,50!} = \frac{9.33 \times 10^{157}}{(3.04 \times 10^{64})(3.04 \times 10^{64})} = 1.01 \times 10^{29}. \quad \text{(Answer)}$$

Similarly, for the configuration (100, 0), we have

$$W = \frac{N!}{n_1!\,n_2!} = \frac{100!}{100!\,0!} = \frac{1}{0!} = \frac{1}{1} = 1. \quad \text{(Answer)}$$

The meaning: Thus, a 50–50 distribution is more likely than a 100–0 distribution by the enormous factor of about 1×10^{29}. If you could count, at one per nanosecond, the number of microstates that correspond to the 50–50 distribution, it would take you about 3×10^{12} years, which is about 200 times longer than the age of the universe. Keep in mind that the 100 molecules used in this sample problem is a very small number. Imagine what these calculated probabilities would be like for a mole of molecules, say about $N = 10^{24}$. Thus, you need never worry about suddenly finding all the air molecules clustering in one corner of your room, with you gasping for air in another corner.

Probability and Entropy

In 1877, Austrian physicist Ludwig Boltzmann (the Boltzmann of Boltzmann's constant k) derived a relationship between the entropy S of a configuration of a gas and the multiplicity W of that configuration. That relationship is

$$S = k \ln W \quad \text{(Boltzmann's entropy equation).} \qquad (20\text{-}21)$$

This famous formula is engraved on Boltzmann's tombstone.

It is natural that S and W should be related by a logarithmic function. The total entropy of two systems is the *sum* of their separate entropies. The probability of occurrence of two independent systems is the *product* of their separate probabilities. Because $\ln ab = \ln a + \ln b$, the logarithm seems the logical way to connect these quantities.

Table 20-1 displays the entropies of the configurations of the six-molecule system of Fig. 20-17, computed using Eq. 20-21. Configuration IV, which has the greatest multiplicity, also has the greatest entropy.

When you use Eq. 20-20 to calculate W, your calculator may signal "OVERFLOW" if you try to find the factorial of a number greater than a few hundred. Fortunately, there is a very good approximation, known as **Stirling's approximation,** not for $N!$ but for $\ln N!$, which is exactly what is needed in Eq. 20-21. Stirling's approximation is

$$\ln N! \approx N(\ln N) - N \quad \text{(Stirling's approximation)}. \tag{20-22}$$

The Stirling of this approximation was an English mathematician and not the Robert Stirling of engine fame.

CHECKPOINT 5 A box contains 1 mol of a gas. Consider two configurations: (a) each half of the box contains half the molecules and (b) each third of the box contains one-third of the molecules. Which configuration has more microstates?

Sample Problem 20-7

In Sample Problem 20-1 we showed that when n moles of an ideal gas doubles its volume in a free expansion, the entropy increase from the initial state i to the final state f is $S_f - S_i = nR \ln 2$. Derive this result with statistical mechanics.

KEY IDEA We can relate the entropy S of any given configuration of the molecules in the gas to the multiplicity W of microstates for that configuration, using Eq. 20-21 ($S = k \ln W$).

Calculations: We are interested in two configurations: the final configuration f (with the molecules occupying the full volume of their container in Fig. 20-1b) and the initial configuration i (with the molecules occupying the left half of the container). Because the molecules are in a closed container, we can calculate the multiplicity W of their microstates with Eq. 20-20. Here we have N molecules in the n moles of the gas. Initially, with the molecules all in the left half of the container, their (n_1, n_2) configuration is $(N, 0)$. Then, Eq. 20-20 gives their multiplicity as

$$W_i = \frac{N!}{N!\,0!} = 1.$$

Finally, with the molecules spread through the full volume, their (n_1, n_2) configuration is $(N/2, N/2)$. Then, Eq. 20-20 gives their multiplicity as

$$W_f = \frac{N!}{(N/2)!\,(N/2)!}.$$

From Eq. 20-21, the initial and final entropies are

$$S_i = k \ln W_i = k \ln 1 = 0$$

and

$$S_f = k \ln W_f = k \ln(N!) - 2k \ln[(N/2)!]. \tag{20-23}$$

In writing Eq. 20-23, we have used the relation

$$\ln \frac{a}{b^2} = \ln a - 2 \ln b.$$

Now, applying Eq. 20-22 to evaluate Eq. 20-23, we find that

$$\begin{aligned} S_f &= k \ln(N!) - 2k \ln[(N/2)!] \\ &= k[N(\ln N) - N] - 2k[(N/2)\ln(N/2) - (N/2)] \\ &= k[N(\ln N) - N - N\ln(N/2) + N] \\ &= k[N(\ln N) - N(\ln N - \ln 2)] = Nk \ln 2. \end{aligned} \tag{20-24}$$

From Eq. 19-8 we can substitute nR for Nk, where R is the universal gas constant. Equation 20-24 then becomes

$$S_f = nR \ln 2.$$

The change in entropy from the initial state to the final is thus

$$\begin{aligned} S_f - S_i &= nR \ln 2 - 0 \\ &= nR \ln 2, \end{aligned} \quad \text{(Answer)}$$

which is what we set out to show. In Sample Problem 20-1 we calculated this entropy increase for a free expansion with thermodynamics by finding an equivalent reversible process and calculating the entropy change for *that* process in terms of temperature and heat transfer. In this sample problem, we calculate the same increase in entropy with statistical mechanics using the fact that the system consists of molecules. In short, the two, very different approaches give the same answer.

REVIEW & SUMMARY

One-Way Processes An **irreversible process** is one that cannot be reversed by means of small changes in the environment. The direction in which an irreversible process proceeds is set by the *change in entropy* ΔS of the system undergoing the process. Entropy S is a *state property* (or *state function*) of the system; that is, it depends only on the state of the system and not on the way in which the system reached that state. The *entropy postulate* states (in part): *If an irreversible process occurs in a closed system, the entropy of the system always increases.*

Calculating Entropy Change The **entropy change** ΔS for an irreversible process that takes a system from an initial state i to a final state f is exactly equal to the entropy change ΔS for *any reversible process* that takes the system between those same two states. We can compute the latter (but not the former) with

$$\Delta S = S_f - S_i = \int_i^f \frac{dQ}{T}. \quad (20\text{-}1)$$

Here Q is the energy transferred as heat to or from the system during the process, and T is the temperature of the system in kelvins during the process.

For a reversible isothermal process, Eq. 20-1 reduces to

$$\Delta S = S_f - S_i = \frac{Q}{T}. \quad (20\text{-}2)$$

When the temperature change ΔT of a system is small relative to the temperature (in kelvins) before and after the process, the entropy change can be approximated as

$$\Delta S = S_f - S_i \approx \frac{Q}{T_{\text{avg}}}, \quad (20\text{-}3)$$

where T_{avg} is the system's average temperature during the process.

When an ideal gas changes reversibly from an initial state with temperature T_i and volume V_i to a final state with temperature T_f and volume V_f, the change ΔS in the entropy of the gas is

$$\Delta S = S_f - S_i = nR \ln \frac{V_f}{V_i} + nC_V \ln \frac{T_f}{T_i}. \quad (20\text{-}4)$$

The Second Law of Thermodynamics This law, which is an extension of the entropy postulate, states: *If a process occurs in a closed system, the entropy of the system increases for irreversible processes and remains constant for reversible processes. It never decreases.* In equation form,

$$\Delta S \geq 0. \quad (20\text{-}5)$$

Engines An **engine** is a device that, operating in a cycle, extracts energy as heat $|Q_H|$ from a high-temperature reservoir and does a certain amount of work $|W|$. The *efficiency* ε of any engine is defined as

$$\varepsilon = \frac{\text{energy we get}}{\text{energy we pay for}} = \frac{|W|}{|Q_H|}. \quad (20\text{-}11)$$

In an **ideal engine,** all processes are reversible and no wasteful energy transfers occur due to, say, friction and turbulence. A **Carnot engine** is an ideal engine that follows the cycle of Fig. 20-9. Its efficiency is

$$\varepsilon_C = 1 - \frac{|Q_L|}{|Q_H|} = 1 - \frac{T_L}{T_H}, \quad (20\text{-}12, 20\text{-}13)$$

in which T_H and T_L are the temperatures of the high- and low-temperature reservoirs, respectively. Real engines always have an efficiency lower than that given by Eq. 20-13. Ideal engines that are not Carnot engines also have lower efficiencies.

A *perfect engine* is an imaginary engine in which energy extracted as heat from the high-temperature reservoir is converted completely to work. Such an engine would violate the second law of thermodynamics, which can be restated as follows: No series of processes is possible whose sole result is the absorption of energy as heat from a thermal reservoir and the complete conversion of this energy to work.

Refrigerators A refrigerator is a device that, operating in a cycle, has work W done on it as it extracts energy $|Q_L|$ as heat from a low-temperature reservoir. The coefficient of performance K of a refrigerator is defined as

$$K = \frac{\text{what we want}}{\text{what we pay for}} = \frac{|Q_L|}{|W|}. \quad (20\text{-}14)$$

A **Carnot refrigerator** is a Carnot engine operating in reverse. For a Carnot refrigerator, Eq. 20-14 becomes

$$K_C = \frac{|Q_L|}{|Q_H| - |Q_L|} = \frac{T_L}{T_H - T_L}. \quad (20\text{-}15, 20\text{-}16)$$

A *perfect refrigerator* is an imaginary refrigerator in which energy extracted as heat from the low-temperature reservoir is converted completely to heat discharged to the high-temperature reservoir, without any need for work. Such a refrigerator would violate the second law of thermodynamics, which can be restated as follows: No series of processes is possible whose sole result is the transfer of energy as heat from a reservoir at a given temperature to a reservoir at a higher temperature.

Entropy from a Statistical View The entropy of a system can be defined in terms of the possible distributions of its molecules. For identical molecules, each possible distribution of molecules is called a **microstate** of the system. All equivalent microstates are grouped into a **configuration** of the system. The number of microstates in a configuration is the **multiplicity** W of the configuration.

For a system of N molecules that may be distributed between the two halves of a box, the multiplicity is given by

$$W = \frac{N!}{n_1!\, n_2!}, \quad (20\text{-}20)$$

in which n_1 is the number of molecules in one half of the box and n_2 is the number in the other half. A basic assumption of **statistical mechanics** is that all the microstates are equally probable. Thus, configurations with a large multiplicity occur most often. When N is very large (say, $N = 10^{22}$ molecules or more), the molecules are nearly always in the configuration in which $n_1 = n_2$.

The multiplicity W of a configuration of a system and the entropy S of the system in that configuration are related by Boltzmann's entropy equation:

$$S = k \ln W, \quad (20\text{-}21)$$

where $k = 1.38 \times 10^{-23}$ J/K is the Boltzmann constant.

When N is very large (the usual case), we can approximate $\ln N!$ with *Stirling's approximation:*

$$\ln N! \approx N(\ln N) - N. \quad (20\text{-}22)$$

QUESTIONS

1 In four experiments, 2.5 mol of hydrogen gas undergoes reversible isothermal expansions, starting from the same volume but at different temperatures. The corresponding p-V plots are shown in Fig. 20-19. Rank the situations according to the change in the entropy of the gas, greatest first. (*Hint:* See Sample Problem 20-1.)

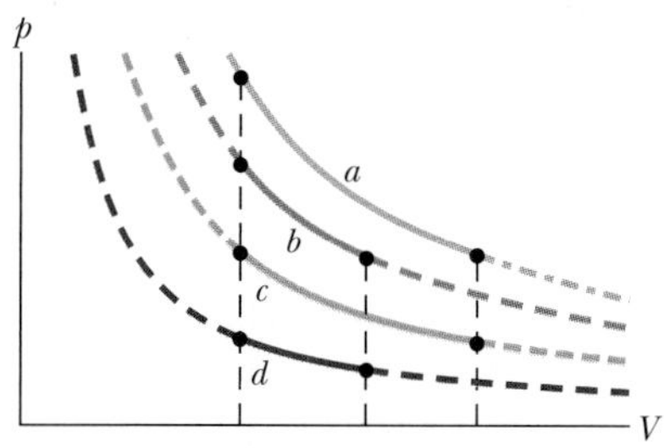

FIG. 20-19 Question 1.

2 In four experiments, blocks A and B, starting at different initial temperatures, were brought together in an insulating box (as in Sample Problem 20-2) and allowed to reach a common final temperature. The entropy changes for the blocks in the four experiments had the following values (in joules per kelvin), but not necessarily in the order given. Determine which values for A go with which values for B.

Block	Values			
A	8	5	3	9
B	−3	−8	−5	−2

3 Point i in Fig. 20-20 represents the initial state of an ideal gas at temperature T. Taking algebraic signs into account, rank the entropy changes that the gas undergoes as it moves, successively and reversibly, from point i to points $a, b, c,$ and d, greatest first.

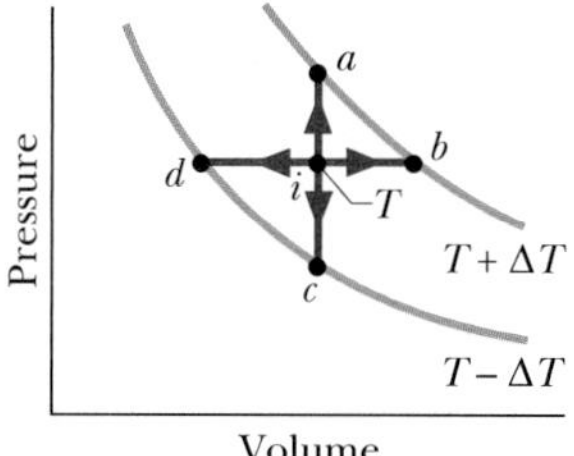

FIG. 20-20 Question 3.

4 An ideal monatomic gas at initial temperature T_0 (in kelvins) expands from initial volume V_0 to volume $2V_0$ by each of the five processes indicated in the T-V diagram of Fig. 20-21. In which process is the expansion (a) isothermal, (b) isobaric (constant pressure), and (c) adiabatic? Explain your answers. (d) In which processes does the entropy of the gas decrease?

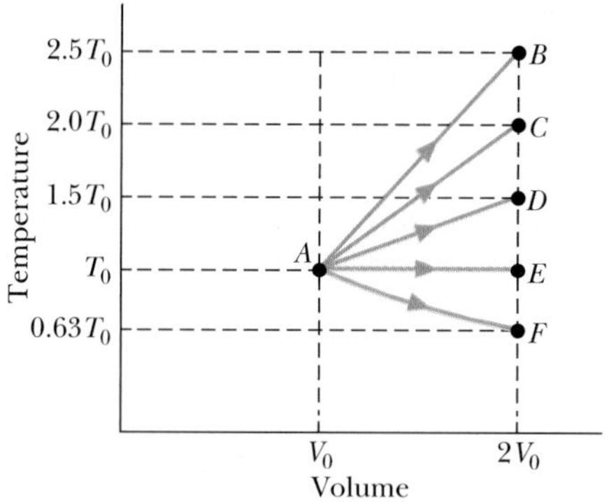

FIG. 20-21 Question 4.

5 A gas, confined to an insulated cylinder, is compressed adiabatically to half its volume. Does the entropy of the gas increase, decrease, or remain unchanged during this process?

6 Three Carnot engines operate between temperature limits of (a) 400 and 500 K, (b) 500 and 600 K, and (c) 400 and 600 K. Each engine extracts the same amount of energy per cycle from the high-temperature reservoir. Rank the magnitudes of the work done by the engines per cycle, greatest first.

7 An inventor claims to have invented four engines, each of which operates between constant-temperature reservoirs at 400 and 300 K. Data on each engine, per cycle of operation, are: engine A, $Q_H = 200$ J, $Q_L = -175$ J, and $W = 40$ J; engine B, $Q_H = 500$ J, $Q_L = -200$ J, and $W = 400$ J; engine C, $Q_H = 600$ J, $Q_L = -200$ J, and $W = 400$ J; engine D, $Q_H = 100$ J, $Q_L = -90$ J, and $W = 10$ J. Of the first and second laws of thermodynamics, which (if either) does each engine violate?

8 Does the entropy per cycle increase, decrease, or remain the same for (a) a Carnot refrigerator, (b) a real refrigerator, and (c) a perfect refrigerator (which is, of course, impossible to build)?

9 Does the entropy per cycle increase, decrease, or remain the same for (a) a Carnot engine, (b) a real engine, and (c) a perfect engine (which is, of course, impossible to build)?

10 A box contains 100 atoms in a configuration that has 50 atoms in each half of the box. Suppose that you could count the different microstates associated with this configuration at the rate of 100 billion states per second, using a supercomputer. Without written calculation, guess how much computing time you would need: a day, a year, or much more than a year.

PROBLEMS

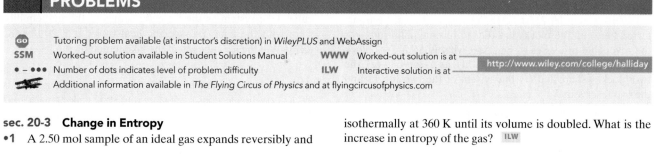

GO Tutoring problem available (at instructor's discretion) in *WileyPLUS* and WebAssign
SSM Worked-out solution available in Student Solutions Manual
WWW Worked-out solution is at http://www.wiley.com/college/halliday
• – ••• Number of dots indicates level of problem difficulty
ILW Interactive solution is at http://www.wiley.com/college/halliday
 Additional information available in *The Flying Circus of Physics* and at flyingcircusofphysics.com

sec. 20-3 Change in Entropy

•**1** A 2.50 mol sample of an ideal gas expands reversibly and isothermally at 360 K until its volume is doubled. What is the increase in entropy of the gas? ILW

•**2** How much energy must be transferred as heat for a reversible isothermal expansion of an ideal gas at 132°C if the entropy of the gas increases by 46.0 J/K?

•**3** Find (a) the energy absorbed as heat and (b) the change in entropy of a 2.00 kg block of copper whose temperature is increased reversibly from 25.0°C to 100°C. The specific heat of copper is 386 J/kg · K. ILW

•**4** (a) What is the entropy change of a 12.0 g ice cube that melts completely in a bucket of water whose temperature is just above the freezing point of water? (b) What is the entropy change of a 5.00 g spoonful of water that evaporates completely on a hot plate whose temperature is slightly above the boiling point of water?

•**5** Suppose 4.00 mol of an ideal gas undergoes a reversible isothermal expansion from volume V_1 to volume $V_2 = 2.00V_1$ at temperature $T = 400$ K. Find (a) the work done by the gas and (b) the entropy change of the gas. (c) If the expansion is reversible and adiabatic instead of isothermal, what is the entropy change of the gas? SSM

•**6** An ideal gas undergoes a reversible isothermal expansion at 77.0°C, increasing its volume from 1.30 L to 3.40 L. The entropy change of the gas is 22.0 J/K. How many moles of gas are present?

••**7** In an experiment, 200 g of aluminum (with a specific heat of 900 J/kg · K) at 100°C is mixed with 50.0 g of water at 20.0°C, with the mixture thermally isolated. (a) What is the equilibrium temperature? What are the entropy changes of (b) the aluminum, (c) the water, and (d) the aluminum–water system? SSM WWW

••**8** A 364 g block is put in contact with a thermal reservoir. The block is initially at a lower temperature than the reservoir. Assume that the consequent transfer of energy as heat from the reservoir to the block is reversible. Figure 20-22 gives the change in entropy ΔS of the block until thermal equilibrium is reached. The scale of the horizontal axis is set by $T_a = 280$ K and $T_b = 380$ K. What is the specific heat of the block?

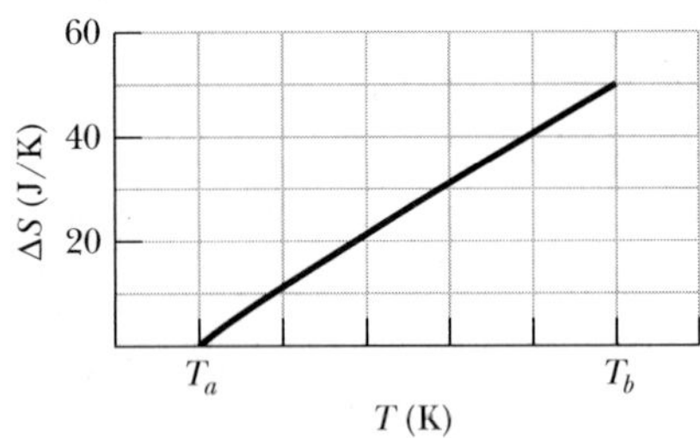

FIG. 20-22 Problem 8.

••**9** In the irreversible process of Fig. 20-5, let the initial temperatures of identical blocks L and R be 305.5 and 294.5 K, respectively, and let 215 J be the energy that must be transferred between the blocks in order to reach equilibrium. For the reversible processes of Fig. 20-6, what is ΔS for (a) block L, (b) its reservoir, (c) block R, (d) its reservoir, (e) the two-block system, and (f) the system of the two blocks and the two reservoirs?

••**10** A gas sample undergoes a reversible isothermal expansion. Figure 20-23 gives the change ΔS in entropy of the gas versus the final volume V_f of the gas. The scale of the vertical axis is set by $\Delta S_s = 64$ J/K. How many moles are in the sample?

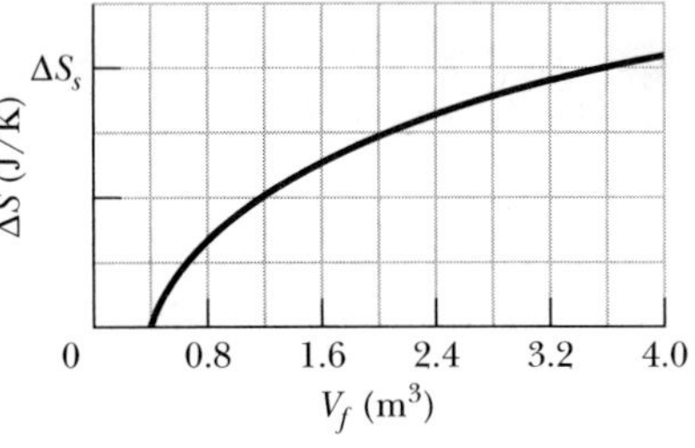

FIG. 20-23 Problem 10.

••**11** A 50.0 g block of copper whose temperature is 400 K is placed in an insulating box with a 100 g block of lead whose temperature is 200 K. (a) What is the equilibrium temperature of the two-block system? (b) What is the change in the internal energy of the system between the initial state and the equilibrium state? (c) What is the change in the entropy of the system? (See Table 18-3.) ILW

••**12** At very low temperatures, the molar specific heat C_V of many solids is approximately $C_V = AT^3$, where A depends on the particular substance. For aluminum, $A = 3.15 \times 10^{-5}$ J/mol · K^4. Find the entropy change for 4.00 mol of aluminum when its temperature is raised from 5.00 K to 10.0 K.

••**13** In Fig. 20-24, where $V_{23} = 3.00V_1$, n moles of a diatomic ideal gas are taken through the cycle with the molecules rotating but not oscillating. What are (a) p_2/p_1, (b) p_3/p_1, and (c) T_3/T_1? For path $1 \rightarrow 2$, what are (d) W/nRT_1, (e) Q/nRT_1, (f) $\Delta E_{int}/nRT_1$, and (g) $\Delta S/nR$? For path $2 \rightarrow 3$, what are (h) W/nRT_1, (i) Q/nRT_1, (j) $\Delta E_{int}/nRT_1$, (k) $\Delta S/nR$? For path $3 \rightarrow 1$, what are (l) W/nRT_1, (m) Q/nRT_1, (n) $\Delta E_{int}/nRT_1$, (o) $\Delta S/nR$?

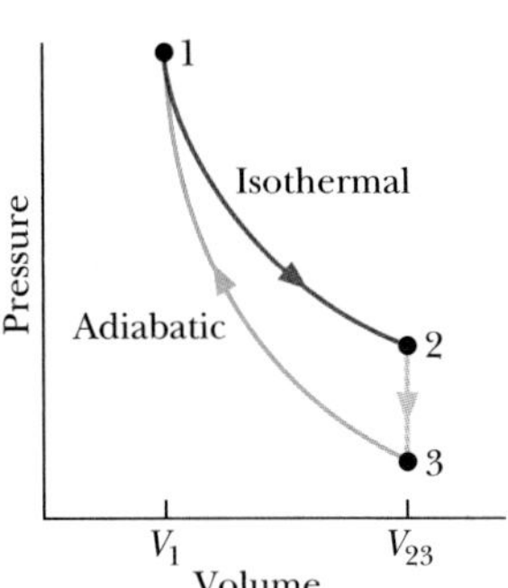

FIG. 20-24 Problem 13.

••**14** A 2.0 mol sample of an ideal monatomic gas undergoes the reversible process shown in Fig. 20-25. The scale of the vertical axis is set by $T_s = 400.0$ K and the scale of the horizontal axis is set by $S_s = 20.0$ J/K. (a) How much energy is absorbed as heat by the gas? (b) What is the change in the internal energy of the gas? (c) How much work is done by the gas? GO

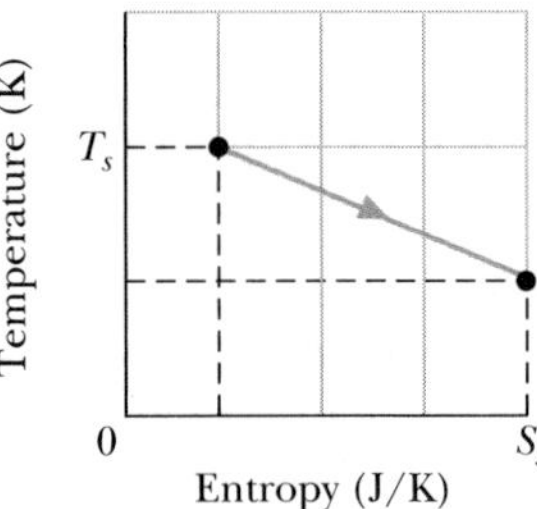

FIG. 20-25 Problem 14.

••**15** A 10 g ice cube at −10°C is placed in a lake whose temperature is 15°C. Calculate the change in entropy of the cube–lake system as the ice cube comes to thermal equilibrium with the lake. The specific heat of ice is 2220 J/kg · K. (*Hint:* Will the ice cube affect the lake temperature?)

••**16** (a) For 1.0 mol of a monatomic ideal gas taken through the cycle in Fig. 20-26, where $V_1 = 4.00V_0$, what is W/p_0V_0 as the gas goes from state a to state c along path abc? What is $\Delta E_{int}/p_0V_0$ in going (b) from b to c and (c) through one full cycle? What is ΔS in going (d) from b to c and (e) through one full cycle?

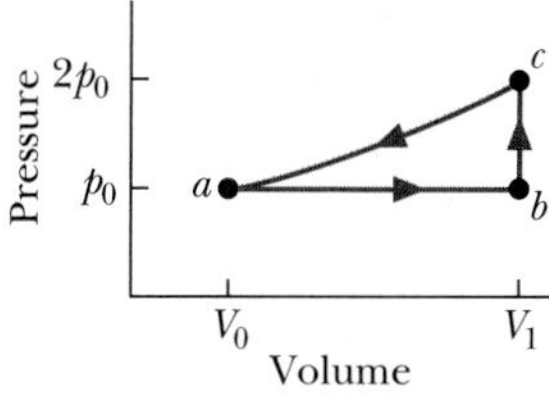

FIG. 20-26 Problem 16.

••17 A mixture of 1773 g of water and 227 g of ice is in an initial equilibrium state at 0.000°C. The mixture is then, in a reversible process, brought to a second equilibrium state where the water–ice ratio, by mass, is 1.00:1.00 at 0.000°C. (a) Calculate the entropy change of the system during this process. (The heat of fusion for water is 333 kJ/kg.) (b) The system is then returned to the initial equilibrium state in an irreversible process (say, by using a Bunsen burner). Calculate the entropy change of the system during this process. (c) Are your answers consistent with the second law of thermodynamics?

••18 An 8.0 g ice cube at −10°C is put into a Thermos flask containing 100 cm^3 of water at 20°C. By how much has the entropy of the cube–water system changed when equilibrium is reached? The specific heat of ice is 2220 J/kg · K.

•••19 Energy can be removed from water as heat at and even below the normal freezing point (0.0°C at atmospheric pressure) without causing the water to freeze; the water is then said to be *supercooled*. Suppose a 1.00 g water drop is supercooled until its temperature is that of the surrounding air, which is at −5.00°C. The drop then suddenly and irreversibly freezes, transferring energy to the air as heat. What is the entropy change for the drop? (*Hint:* Use a three-step reversible process as if the water were taken through the normal freezing point.) The specific heat of ice is 2220 J/kg · K. GO

•••20 An insulated Thermos contains 130 g of water at 80.0°C. You put in a 12.0 g ice cube at 0°C to form a system of *ice + original water*. (a) What is the equilibrium temperature of the system? What are the entropy changes of the water that was originally the ice cube (b) as it melts and (c) as it warms to the equilibrium temperature? (d) What is the entropy change of the original water as it cools to the equilibrium temperature? (e) What is the net entropy change of the *ice + original water* system as it reaches the equilibrium temperature? GO

•••21 Suppose 1.00 mol of a monatomic ideal gas is taken from initial pressure p_1 and volume V_1 through two steps: (1) an isothermal expansion to volume $2.00V_1$ and (2) a pressure increase to $2.00p_1$ at constant volume. What is Q/p_1V_1 for (a) step 1 and (b) step 2? What is W/p_1V_1 for (c) step 1 and (d) step 2? For the full process, what are (e) $\Delta E_{int}/p_1V_1$ and (f) ΔS? The gas is returned to its initial state and again taken to the same final state but now through these two steps: (1) an isothermal compression to pressure $2.00p_1$ and (2) a volume increase to $2.00V_1$ at constant pressure. What is Q/p_1V_1 for (g) step 1 and (h) step 2? What is W/p_1V_1 for (i) step 1 and (j) step 2? For the full process, what are (k) $\Delta E_{int}/p_1V_1$ and (l) ΔS?

•••22 Expand 1.00 mol of an monatomic gas initially at 5.00 kPa and 600 K from initial volume $V_i = 1.00$ m^3 to final volume $V_f = 2.00$ m^3. At any instant during the expansion, the pressure p and volume V of the gas are related by $p = 5.00 \exp[(V_i - V)/a]$, with p in kilopascals, V_i and V in cubic meters, and $a = 1.00$ m^3. What are the final (a) pressure and (b) temperature of the gas? (c) How much work is done by the gas during the expansion? (d) What is ΔS for the expansion? (*Hint:* Use two simple reversible processes to find ΔS.)

sec. 20-5 Entropy in the Real World: Engines

•23 A Carnot engine has an efficiency of 22.0%. It operates between constant-temperature reservoirs differing in temperature by 75.0 C°. What is the temperature of the (a) lower-temperature and (b) higher-temperature reservoir?

•24 In a hypothetical nuclear fusion reactor, the fuel is deuterium gas at a temperature of 7×10^8 K. If this gas could be used to operate a Carnot engine with $T_L = 100$°C, what would be the engine's efficiency? Take both temperatures to be exact and report your answer to seven significant figures.

•25 A Carnot engine operates between 235°C and 115°C, absorbing 6.30×10^4 J per cycle at the higher temperature. (a) What is the efficiency of the engine? (b) How much work per cycle is this engine capable of performing? SSM WWW

•26 A Carnot engine absorbs 52 kJ as heat and exhausts 36 kJ as heat in each cycle. Calculate (a) the engine's efficiency and (b) the work done per cycle in kilojoules.

•27 A Carnot engine whose low-temperature reservoir is at 17°C has an efficiency of 40%. By how much should the temperature of the high-temperature reservoir be increased to increase the efficiency to 50%?

••28 A 500 W Carnot engine operates between constant-temperature reservoirs at 100°C and 60.0°C. What is the rate at which energy is (a) taken in by the engine as heat and (b) exhausted by the engine as heat?

••29 Figure 20-27 shows a reversible cycle through which 1.00 mol of a monatomic ideal gas is taken. Volume $V_c = 8.00V_b$. Process bc is an adiabatic expansion, with $p_b = 10.0$ atm and $V_b = 1.00 \times 10^{-3}$ m^3. For the cycle, find (a) the energy added to the gas as heat, (b) the energy leaving the gas as heat, (c) the net work done by the gas, and (d) the efficiency of the cycle. SSM ILW

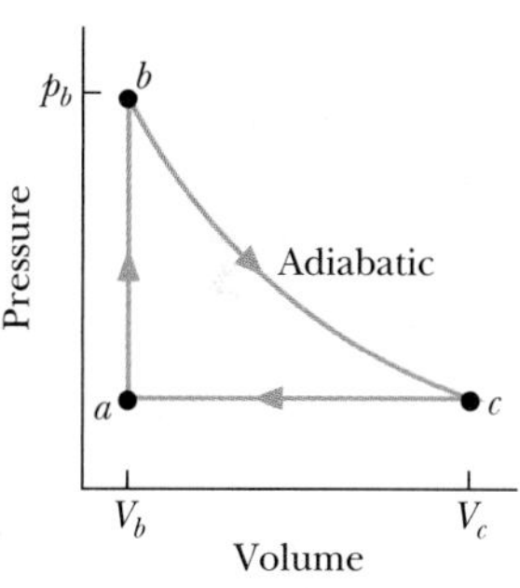

FIG. 20-27 Problem 29.

••30 A Carnot engine is set up to produce a certain work W per cycle. In each cycle, energy in the form of heat Q_H is transferred to the working substance of the engine from the higher-temperature thermal reservoir, which is at an adjustable temperature T_H. The lower-temperature thermal reservoir is maintained at temperature $T_L = 250$ K. Figure 20-28 gives Q_H for a range of T_H. The scale of the vertical axis is set by $Q_{Hs} = 6.0$ kJ. If T_H is set at 550 K, what is Q_H?

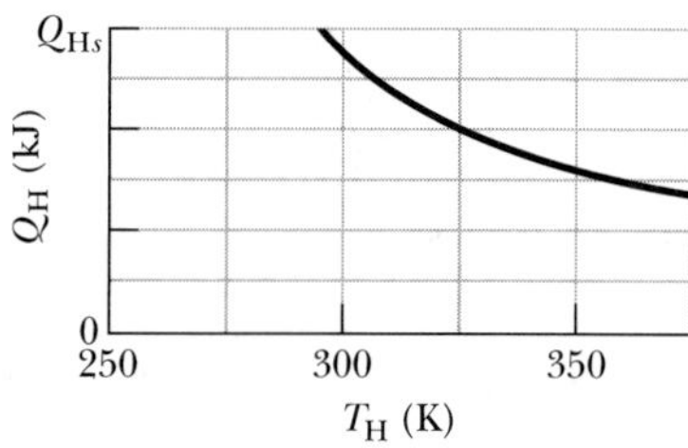

FIG. 20-28 Problem 30.

••31 Figure 20-29 shows a reversible cycle

through which 1.00 mol of a monatomic ideal gas is taken. Assume that $p = 2p_0$, $V = 2V_0$, $p_0 = 1.01 \times 10^5$ Pa, and $V_0 = 0.0225$ m^3. Calculate (a) the work done during the cycle, (b) the energy added as heat during stroke *abc*, and (c) the efficiency of the cycle. (d) What is the efficiency of a Carnot engine operating between the highest and lowest temperatures that occur in the cycle? (e) Is this greater than or less than the efficiency calculated in (c)? GO

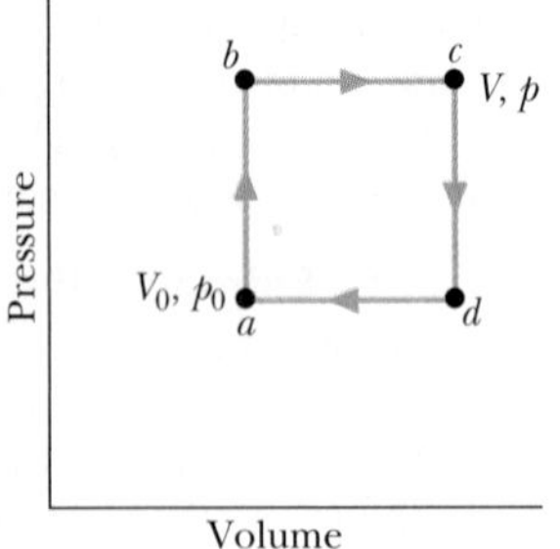

FIG. 20-29 Problem 31.

••32 An ideal gas (1.0 mol) is the working substance in an engine that operates on the cycle shown in Fig. 20-30. Processes *BC* and *DA* are reversible and adiabatic. (a) Is the gas monatomic, diatomic, or polyatomic? (b) What is the engine efficiency?

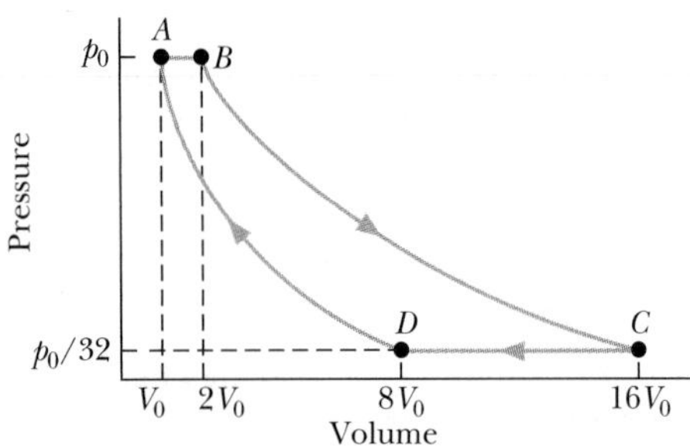

FIG. 20-30 Problem 32.

••33 The efficiency of a particular car engine is 25% when the engine does 8.2 kJ of work per cycle. Assume the process is reversible. What are (a) the energy the engine gains per cycle as heat Q_{gain} from the fuel combustion and (b) the energy the engine loses per cycle as heat Q_{lost}. If a tune-up increases the efficiency to 31%, what are (c) Q_{gain} and (d) Q_{lost} at the same work value?

••34 In the first stage of a two-stage Carnot engine, energy is absorbed as heat Q_1 at temperature T_1, work W_1 is done, and energy is expelled as heat Q_2 at a lower temperature T_2. The second stage absorbs that energy as heat Q_2, does work W_2, and expels energy as heat Q_3 at a still lower temperature T_3. Prove that the efficiency of the engine is $(T_1 - T_3)/T_1$.

•••35 The cycle in Fig. 20-31 represents the operation of a gasoline internal combustion engine. Volume $V_3 = 4.00V_1$. Assume the gasoline–air intake mixture is an ideal gas with $\gamma = 1.30$. What are the ratios (a) T_2/T_1, (b) T_3/T_1, (c) T_4/T_1, (d) p_3/p_1, and (e) p_4/p_1? (f) What is the engine efficiency?

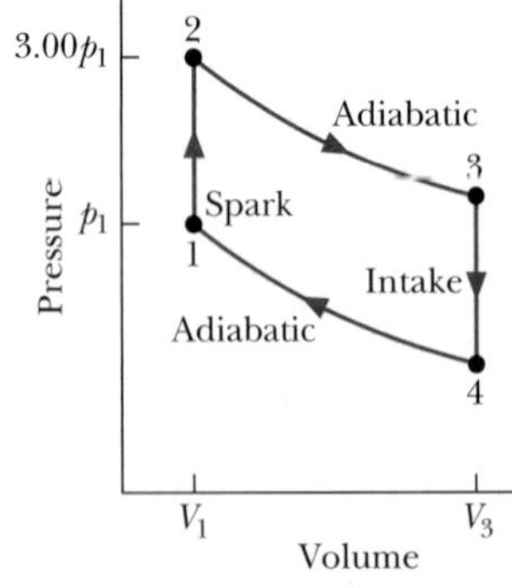

FIG. 20-31 Problem 35.

sec. 20-6 Entropy in the Real World: Refrigerators

•36 The electric motor of a heat pump transfers energy as heat from the outdoors, which is at −5.0°C, to a room that is at 17°C. If the heat pump were a Carnot heat pump (a Carnot engine working in reverse), how much energy would be transferred as heat to the room for each joule of electric energy consumed?

•37 A Carnot air conditioner takes energy from the thermal energy of a room at 70°F and transfers it as heat to the outdoors, which is at 96°F. For each joule of electric energy required to operate the air conditioner, how many joules are removed from the room? SSM

•38 To make ice, a freezer that is a reverse Carnot engine extracts 42 kJ as heat at −15°C during each cycle, with coefficient of performance 5.7. The room temperature is 30.3°C. How much (a) energy per cycle is delivered as heat to the room and (b) work per cycle is required to run the freezer?

•39 A heat pump is used to heat a building. The outside temperature is −5.0°C, and the temperature inside the building is to be maintained at 22°C. The pump's coefficient of performance is 3.8, and the heat pump delivers 7.54 MJ as heat to the building each hour. If the heat pump is a Carnot engine working in reverse, at what rate must work be done to run it? SSM

•40 How much work must be done by a Carnot refrigerator to transfer 1.0 J as heat (a) from a reservoir at 7.0°C to one at 27°C, (b) from a reservoir at −73°C to one at 27°C, (c) from a reservoir at −173°C to one at 27°C, and (d) from a reservoir at −223°C to one at 27°C?

••41 Figure 20-32 represents a Carnot engine that works between temperatures $T_1 = 400$ K and $T_2 = 150$ K and drives a Carnot refrigerator that works between temperatures $T_3 = 325$ K and $T_4 = 225$ K. What is the ratio Q_3/Q_1? GO

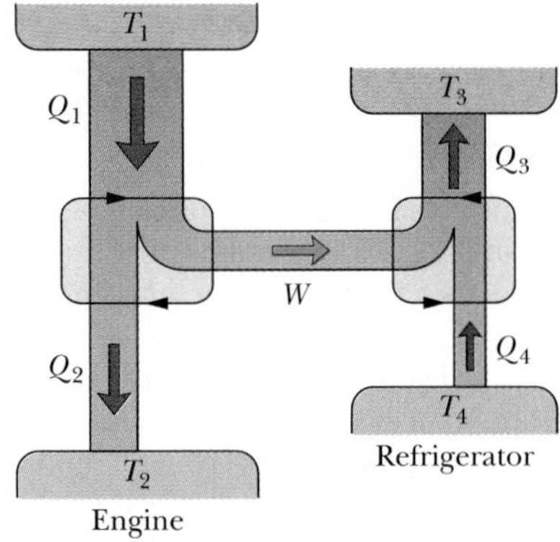

FIG. 20-32 Problem 41.

••42 (a) During each cycle, a Carnot engine absorbs 750 J as heat from a high-temperature reservoir at 360 K, with the low-temperature reservoir at 280 K. How much work is done per cycle? (b) The engine is then made to work in reverse to function as a Carnot refrigerator between those same two reservoirs. During each cycle, how much work is required to remove 1200 J as heat from the low-temperature reservoir?

••43 An air conditioner operating between 93°F and 70°F is rated at 4000 Btu/h cooling capacity. Its coefficient of performance is 27% of that of a Carnot refrigerator operating between the same two temperatures. What horsepower is required of the air conditioner motor? ILW

••44 The motor in a refrigerator has a power of 200 W. If the freezing compartment is at 270 K and the outside air is at 300 K, and assuming the efficiency of a Carnot refrigerator, what is the maximum amount of energy that can be extracted as heat from the freezing compartment in 10.0 min?

sec. 20-8 A Statistical View of Entropy

•45 Construct a table like Table 20-1 for eight molecules.

••46 A box contains N identical gas molecules equally divided between its two halves. For $N = 50$, what are (a) the multiplicity W of the central configuration, (b) the total number of microstates, and (c) the percentage of the time the system spends in the central configuration? For $N = 100$, what are (d) W of the central configuration, (e) the total number of microstates, and (f) the percentage of the time the system spends in the central configuration? For $N = 200$, what are (g) W of the central configuration, (h) the total number of microstates, and (i) the percentage of the time the system spends in the central configuration? (j) Does the time spent in the central configuration increase or decrease with an increase in N?

•••47 A box contains N gas molecules. Consider the box to be divided into three equal parts. (a) By extension of Eq. 20-20, write a formula for the multiplicity of any given configuration. (b) Consider two configurations: configuration A with equal numbers of molecules in all three thirds of the box, and configuration B with equal numbers of molecules in each half of the box divided into two equal parts rather than three. What is the ratio W_A/W_B of the multiplicity of configuration A to that of configuration B? (c) Evaluate W_A/W_B for $N = 100$. (Because 100 is not evenly divisible by 3, put 34 molecules into one of the three box parts of configuration A and 33 in each of the other two parts.) SSM WWW

Additional Problems

48 Figure 20-33 gives the force magnitude F versus stretch distance x for a rubber band, with the scale of the F axis set by $F_s = 1.50$ N and the scale of the x axis set by $x_s = 3.50$ cm. The temperature is 2.00°C. When the rubber band is stretched by $x = 1.70$ cm, at what rate does the entropy of the rubber band change during a small additional stretch?

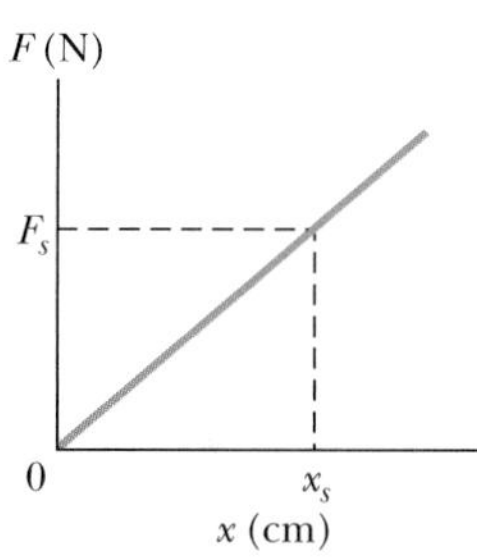

FIG. 20-33 Problem 48.

49 As a sample of nitrogen gas (N_2) undergoes a temperature increase at constant volume, the distribution of molecular speeds increases. That is, the probability distribution function $P(v)$ for the molecules spreads to higher speed values, as suggested in Fig. 19-8*b*. One way to report the spread in $P(v)$ is to measure the difference Δv between the most probable speed v_P and the rms speed v_{rms}. When $P(v)$ spreads to higher speeds, Δv increases. Assume that the gas is ideal and the N_2 molecules rotate but do not oscillate. For 1.5 mol, an initial temperature of 250 K, and a final temperature of 500 K, what are (a) the initial difference Δv_i, (b) the final difference Δv_f, and (c) the entropy change ΔS for the gas? SSM

50 A three-step cycle is undergone by 3.4 mol of an ideal diatomic gas: (1) the temperature of the gas is increased from 200 K to 500 K at constant volume; (2) the gas is then isothermally expanded to its original pressure; (3) the gas is then contracted at constant pressure back to its original volume. Throughout the cycle, the molecules rotate but do not oscillate. What is the efficiency of the cycle?

51 Suppose that a deep shaft were drilled in Earth's crust near one of the poles, where the surface temperature is −40°C, to a depth where the temperature is 800°C. (a) What is the theoretical limit to the efficiency of an engine operating between these temperatures? (b) If all the energy released as heat into the low-temperature reservoir were used to melt ice that was initially at −40°C, at what rate could liquid water at 0°C be produced by a 100 MW power plant (treat it as an engine)? The specific heat of ice is 2220 J/kg · K; water's heat of fusion is 333 kJ/kg. (Note that the engine can operate only between 0°C and 800°C in this case. Energy exhausted at −40°C cannot warm anything above −40°C.)

52 (a) A Carnot engine operates between a hot reservoir at 320 K and a cold one at 260 K. If the engine absorbs 500 J as heat per cycle at the hot reservoir, how much work per cycle does it deliver? (b) If the engine working in reverse functions as a refrigerator between the same two reservoirs, how much work per cycle must be supplied to remove 1000 J as heat from the cold reservoir?

53 A 600 g lump of copper at 80.0°C is placed in 70.0 g of water at 10.0°C in an insulated container. (See Table 18-3 for specific heats.) (a) What is the equilibrium temperature of the copper–water system? What entropy changes do (b) the copper, (c) the water, and (d) the copper–water system undergo in reaching the equilibrium temperature?

54 Suppose 0.550 mol of an ideal gas is isothermally and reversibly expanded in the four situations given below. What is the change in the entropy of the gas for each situation?

Situation	(a)	(b)	(c)	(d)
Temperature (K)	250	350	400	450
Initial volume (cm^3)	0.200	0.200	0.300	0.300
Final volume (cm^3)	0.800	0.800	1.20	1.20

55 A 0.600 kg sample of water is initially ice at temperature −20°C. What is the sample's entropy change if its temperature is increased to 40°C? SSM

56 What is the entropy change for 3.20 mol of an ideal monatomic gas undergoing a reversible increase in temperature from 380 K to 425 K at constant volume?

57 A three-step cycle is undergone reversibly by 4.00 mol of an ideal gas: (1) an adiabatic expansion that gives the gas 2.00 times its initial volume, (2) a constant-volume process, (3) an isothermal compression back to the initial state of the gas. We do not know whether the gas is monatomic or diatomic; if it is diatomic, we do not know whether the molecules are rotating or oscillating. What are the entropy changes for (a) the cycle, (b) process 1, (c) process 3, and (d) process 2?

58 Suppose 1.0 mol of a monatomic ideal gas initially at 10 L and 300 K is heated at constant volume to 600 K, allowed to expand isothermally to its initial pressure, and finally compressed at constant pressure to its original volume, pressure, and temperature. During the cycle, what are (a) the net energy entering the system (the gas) as heat and (b) the net work done by the gas? (c) What is the efficiency of the cycle?

59 A 2.00 mol diatomic gas initially at 300 K undergoes this cycle: It is (1) heated at constant volume to 800 K, (2) then allowed to expand isothermally to its initial pressure, (3) then compressed at constant pressure to its initial state. Assuming the gas molecules neither rotate nor oscillate, find (a) the net

energy transferred as heat to the gas, (b) the net work done by the gas, and (c) the efficiency of the cycle.

60 A 45.0 g block of tungsten at 30.0°C and a 25.0 g block of silver at −120°C are placed together in an insulated container. (See Table 18-3 for specific heats.) (a) What is the equilibrium temperature? What entropy changes do (b) the tungsten, (c) the silver, and (d) the tungsten–silver system undergo in reaching the equilibrium temperature?

61 A cylindrical copper rod of length 1.50 m and radius 2.00 cm is insulated to prevent heat loss through its curved surface. One end is attached to a thermal reservoir fixed at 300°C; the other is attached to a thermal reservoir fixed at 30.0°C. What is the rate at which entropy increases for the rod–reservoirs system?

62 An ideal refrigerator does 150 J of work to remove 560 J as heat from its cold compartment. (a) What is the refrigerator's coefficient of performance? (b) How much heat per cycle is exhausted to the kitchen?

63 A Carnot refrigerator extracts 35.0 kJ as heat during each cycle, operating with a coefficient of performance of 4.60. What are (a) the energy per cycle transferred as heat to the room and (b) the work done per cycle? SSM

64 Four particles are in the insulated box of Fig. 20-17. What are (a) the least multiplicity, (b) the greatest multiplicity, (c) the least entropy, and (d) the greatest entropy of the four-particle system?

65 A brass rod is in thermal contact with a constant-temperature reservoir at 130°C at one end and a constant-temperature reservoir at 24.0°C at the other end. (a) Compute the total change in entropy of the rod–reservoirs system when 5030 J of energy is conducted through the rod, from one reservoir to the other. (b) Does the entropy of the rod change? GO

66 An apparatus that liquefies helium is in a room maintained at 300 K. If the helium in the apparatus is at 4.0 K, what is the minimum ratio $Q_{\text{to}}/Q_{\text{from}}$, where Q_{to} is the energy delivered as heat to the room and Q_{from} is the energy removed as heat from the helium?

67 System A of three particles and system B of five particles are in insulated boxes like that in Fig. 20-17. What is the least multiplicity W of (a) system A and (b) system B? What is the greatest multiplicity W of (c) A and (d) B? What is the greatest entropy of (e) A and (f) B? SSM

68 Calculate the efficiency of a fossil-fuel power plant that consumes 380 metric tons of coal each hour to produce useful work at the rate of 750 MW. The heat of combustion of coal (the heat due to burning it) is 28 MJ/kg.

69 The temperature of 1.00 mol of a monatomic ideal gas is raised reversibly from 300 K to 400 K, with its volume kept constant. What is the entropy change of the gas?

70 Repeat Problem 69, with the pressure now kept constant.

71 Suppose that 260 J is conducted from a constant-temperature reservoir at 400 K to one at (a) 100 K, (b) 200 K, (c) 300 K, and (d) 360 K. What is the net change in entropy ΔS_{net} of the reservoirs in each case? (e) As the temperature difference of the two reservoirs decreases, does ΔS_{net} increase, decrease, or remain the same?

72 A Carnot engine whose high-temperature reservoir is at 400 K has an efficiency of 30.0%. By how much should the temperature of the low-temperature reservoir be changed to increase the efficiency to 40.0%?

73 A box contains N molecules. Consider two configurations: configuration A with an equal division of the molecules between the two halves of the box, and configuration B with 60.0% of the molecules in the left half of the box and 40.0% in the right half. For $N = 50$, what are (a) the multiplicity W_A of configuration A, (b) the multiplicity W_B of configuration B, and (c) the ratio $f_{B/A}$ of the time the system spends in configuration B to the time it spends in configuration A? For $N = 100$, what are (d) W_A, (e) W_B, and (f) $f_{B/A}$? For $N = 200$, what are (g) W_A, (h) W_B, and (i) $f_{B/A}$? (j) With increasing N, does f increase, decrease, or remain the same?

74 Suppose 2.00 mol of a diatomic gas is taken reversibly around the cycle shown in the T-S diagram of Fig. 20-34, where $S_1 = 6.00$ J/K and $S_2 = 8.00$ J/K. The molecules do not rotate or oscillate. What is the energy transferred as heat Q for (a) path 1 → 2, (b) path 2 → 3, and (c) the full cycle? (d) What is the work W for the isothermal process? The volume V_1 in state 1 is 0.200 m³. What is the volume in (e) state 2 and (f) state 3?

What is the change ΔE_{int} for (g) path 1 → 2, (h) path 2 → 3, and (i) the full cycle? (*Hint:* (h) can be done with one or two lines of calculation using Section 19-8 or with a page of calculation using Section 19-11.) (j) What is the work W for the adiabatic process?

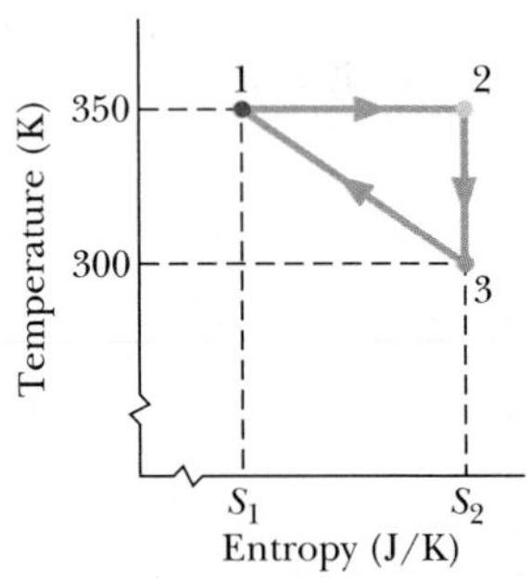

FIG. 20-34 Problem 74.

75 An inventor has built an engine X and claims that its efficiency ε_X is greater than the efficiency ε of an ideal engine operating between the same two temperatures. Suppose you couple engine X to an ideal refrigerator (Fig. 20-35a) and adjust the cycle of engine X so that the work per cycle it provides equals the work per cycle required by the ideal refrigerator. Treat this combination as a single unit and show that if the inventor's claim were true (if $\varepsilon_X > \varepsilon$), the combined unit would act as a perfect refrigerator (Fig. 20-35b), transferring energy as heat from the low-temperature reservoir to the high-temperature reservoir without the need for work.

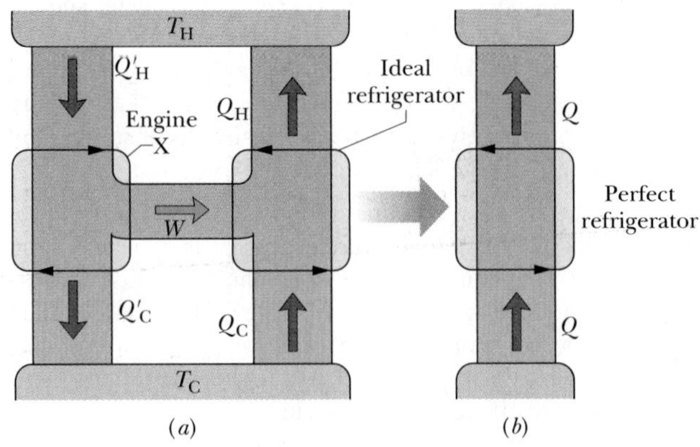

FIG. 20-35 Problem 75.

The International System of Units (SI)*

A

TABLE 1

The SI Base Units

Quantity	Name	Symbol	Definition
length	meter	m	". . . the length of the path traveled by light in vacuum in 1/299,792,458 of a second." (1983)
mass	kilogram	kg	". . . this prototype [a certain platinum–iridium cylinder] shall henceforth be considered to be the unit of mass." (1889)
time	second	s	". . . the duration of 9,192,631,770 periods of the radiation corresponding to the transition between the two hyperfine levels of the ground state of the cesium-133 atom." (1967)
electric current	ampere	A	". . . that constant current which, if maintained in two straight parallel conductors of infinite length, of negligible circular cross section, and placed 1 meter apart in vacuum, would produce between these conductors a force equal to 2×10^{-7} newton per meter of length." (1946)
thermodynamic temperature	kelvin	K	". . . the fraction 1/273.16 of the thermodynamic temperature of the triple point of water." (1967)
amount of substance	mole	mol	". . . the amount of substance of a system which contains as many elementary entities as there are atoms in 0.012 kilogram of carbon-12." (1971)
luminous intensity	candela	cd	". . . the luminous intensity, in a given direction, of a source that emits monochromatic radiation of frequency 540×10^{12} hertz and that has a radiant intensity in that direction of 1/683 watt per steradian." (1979)

*Adapted from "The International System of Units (SI)," National Bureau of Standards Special Publication 330, 1972 edition. The definitions above were adopted by the General Conference of Weights and Measures, an international body, on the dates shown. In this book we do not use the candela.

TABLE 2

Some SI Derived Units

Quantity	Name of Unit	Symbol	
area	square meter	m^2	
volume	cubic meter	m^3	
frequency	hertz	Hz	s^{-1}
mass density (density)	kilogram per cubic meter	kg/m^3	
speed, velocity	meter per second	m/s	
angular velocity	radian per second	rad/s	
acceleration	meter per second per second	m/s^2	
angular acceleration	radian per second per second	rad/s^2	
force	newton	N	$kg \cdot m/s^2$
pressure	pascal	Pa	N/m^2
work, energy, quantity of heat	joule	J	$N \cdot m$
power	watt	W	J/s
quantity of electric charge	coulomb	C	$A \cdot s$
potential difference, electromotive force	volt	V	W/A
electric field strength	volt per meter (or newton per coulomb)	V/m	N/C
electric resistance	ohm	Ω	V/A
capacitance	farad	F	$A \cdot s/V$
magnetic flux	weber	Wb	$V \cdot s$
inductance	henry	H	$V \cdot s/A$
magnetic flux density	tesla	T	Wb/m^2
magnetic field strength	ampere per meter	A/m	
entropy	joule per kelvin	J/K	
specific heat	joule per kilogram kelvin	$J/(kg \cdot K)$	
thermal conductivity	watt per meter kelvin	$W/(m \cdot K)$	
radiant intensity	watt per steradian	W/sr	

TABLE 3

The SI Supplementary Units

Quantity	Name of Unit	Symbol
plane angle	radian	rad
solid angle	steradian	sr

Some Fundamental Constants of Physics*

B

Constant	Symbol	Computational Value	Best (1998) Value	
			Value[a]	Uncertainty[b]
Speed of light in a vacuum	c	3.00×10^8 m/s	2.997 924 58	exact
Elementary charge	e	1.60×10^{-19} C	1.602 176 462	0.039
Gravitational constant	G	6.67×10^{-11} m^3/s$^2 \cdot$ kg	6.673	1500
Universal gas constant	R	8.31 J/mol $\cdot$ K	8.314 472	1.7
Avogadro constant	N_A	6.02×10^{23} mol^{-1}	6.022 141 99	0.079
Boltzmann constant	k	1.38×10^{-23} J/K	1.380 650 3	1.7
Stefan–Boltzmann constant	σ	5.67×10^{-8} W/m$^2 \cdot$ K^4	5.670 400	7.0
Molar volume of ideal gas at STP[d]	V_m	2.27×10^{-2} m^3/mol	2.271 098 1	1.7
Permittivity constant	ϵ_0	8.85×10^{-12} F/m	8.854 187 817 62	exact
Permeability constant	μ_0	1.26×10^{-6} H/m	1.256 637 061 43	exact
Planck constant	h	6.63×10^{-34} J $\cdot$ s	6.626 068 76	0.078
Electron mass[c]	m_e	9.11×10^{-31} kg	9.109 381 88	0.079
		5.49×10^{-4} u	5.485 799 110	0.0021
Proton mass[c]	m_p	1.67×10^{-27} kg	1.672 621 58	0.079
		1.0073 u	1.007 276 466 88	1.3×10^{-4}
Ratio of proton mass to electron mass	m_p/m_e	1840	1836.152 667 5	0.0021
Electron charge-to-mass ratio	e/m_e	1.76×10^{11} C/kg	1.758 820 174	0.040
Neutron mass[c]	m_n	1.68×10^{-27} kg	1.674 927 16	0.079
		1.0087 u	1.008 664 915 78	5.4×10^{-4}
Hydrogen atom mass[c]	$m_{^1H}$	1.0078 u	1.007 825 031 6	0.0005
Deuterium atom mass[c]	$m_{^2H}$	2.0141 u	2.014 101 777 9	0.0005
Helium atom mass[c]	$m_{^4He}$	4.0026 u	4.002 603 2	0.067
Muon mass	m_μ	1.88×10^{-28} kg	1.883 531 09	0.084
Electron magnetic moment	μ_e	9.28×10^{-24} J/T	9.284 763 62	0.040
Proton magnetic moment	μ_p	1.41×10^{-26} J/T	1.410 606 663	0.041
Bohr magneton	μ_B	9.27×10^{-24} J/T	9.274 008 99	0.040
Nuclear magneton	μ_N	5.05×10^{-27} J/T	5.050 783 17	0.040
Bohr radius	a	5.29×10^{-11} m	5.291 772 083	0.0037
Rydberg constant	R	1.10×10^7 m^{-1}	1.097 373 156 854 8	7.6×10^{-6}
Electron Compton wavelength	λ_C	2.43×10^{-12} m	2.426 310 215	0.0073

[a]Values given in this column should be given the same unit and power of 10 as the computational value.

[b]Parts per million.

[c]Masses given in u are in unified atomic mass units, where 1 u = $1.660\,538\,86 \times 10^{-27}$ kg.

[d]STP means standard temperature and pressure: 0°C and 1.0 atm (0.1 MPa).

*The values in this table were selected from the 1998 CODATA recommended values (www.physics.nist.gov).

C Some Astronomical Data

Some Distances from Earth

To the Moon*	3.82×10^{8} m	To the center of our galaxy	2.2×10^{20} m
To the Sun*	1.50×10^{11} m	To the Andromeda Galaxy	2.1×10^{22} m
To the nearest star (Proxima Centauri)	4.04×10^{16} m	To the edge of the observable universe	$\sim 10^{26}$ m

*Mean distance.

The Sun, Earth, and the Moon

Property	Unit	Sun		Earth	Moon
Mass	kg	1.99×10^{30}		5.98×10^{24}	7.36×10^{22}
Mean radius	m	6.96×10^{8}		6.37×10^{6}	1.74×10^{6}
Mean density	kg/m^3	1410		5520	3340
Free-fall acceleration at the surface	m/s^2	274		9.81	1.67
Escape velocity	km/s	618		11.2	2.38
Period of rotation[a]	—	37 d at poles[b]	26 d at equator[b]	23 h 56 min	27.3 d
Radiation power[c]	W	3.90×10^{26}			

[a]Measured with respect to the distant stars.

[b]The Sun, a ball of gas, does not rotate as a rigid body.

[c]Just outside Earth's atmosphere solar energy is received, assuming normal incidence, at the rate of 1340 W/m^2.

Some Properties of the Planets

	Mercury	Venus	Earth	Mars	Jupiter	Saturn	Uranus	Neptune	Pluto
Mean distance from Sun, 10^6 km	57.9	108	150	228	778	1430	2870	4500	5900
Period of revolution, y	0.241	0.615	1.00	1.88	11.9	29.5	84.0	165	248
Period of rotation,[a] d	58.7	−243[b]	0.997	1.03	0.409	0.426	−0.451[b]	0.658	6.39
Orbital speed, km/s	47.9	35.0	29.8	24.1	13.1	9.64	6.81	5.43	4.74
Inclination of axis to orbit	<28°	≈3°	23.4°	25.0°	3.08°	26.7°	97.9°	29.6°	57.5°
Inclination of orbit to Earth's orbit	7.00°	3.39°		1.85°	1.30°	2.49°	0.77°	1.77°	17.2°
Eccentricity of orbit	0.206	0.0068	0.0167	0.0934	0.0485	0.0556	0.0472	0.0086	0.250
Equatorial diameter, km	4880	12 100	12 800	6790	143 000	120 000	51 800	49 500	2300
Mass (Earth = 1)	0.0558	0.815	1.000	0.107	318	95.1	14.5	17.2	0.002
Density (water = 1)	5.60	5.20	5.52	3.95	1.31	0.704	1.21	1.67	2.03
Surface value of g,[c] m/s^2	3.78	8.60	9.78	3.72	22.9	9.05	7.77	11.0	0.5
Escape velocity,[c] km/s	4.3	10.3	11.2	5.0	59.5	35.6	21.2	23.6	1.1
Known satellites	0	0	1	2	60 + ring	31 + rings	21 + rings	11 + rings	1

[a]Measured with respect to the distant stars.

[b]Venus and Uranus rotate opposite their orbital motion.

[c]Gravitational acceleration measured at the planet's equator.

Conversion Factors D

Conversion factors may be read directly from these tables. For example, 1 degree = 2.778×10^{-3} revolutions, so $16.7° = 16.7 \times 2.778 \times 10^{-3}$ rev. The SI units are fully capitalized. Adapted in part from G. Shortley and D. Williams, *Elements of Physics,* 1971, Prentice-Hall, Englewood Cliffs, NJ.

Plane Angle

°	′	″	RADIAN	rev
1 degree = 1	60	3600	1.745×10^{-2}	2.778×10^{-3}
1 minute = 1.667×10^{-2}	1	60	2.909×10^{-4}	4.630×10^{-5}
1 second = 2.778×10^{-4}	1.667×10^{-2}	1	4.848×10^{-6}	7.716×10^{-7}
1 RADIAN = 57.30	3438	2.063×10^{5}	1	0.1592
1 revolution = 360	2.16×10^{4}	1.296×10^{6}	6.283	1

Solid Angle

1 sphere = 4π steradians = 12.57 steradians

Length

cm	METER	km	in.	ft	mi
1 centimeter = 1	10^{-2}	10^{-5}	0.3937	3.281×10^{-2}	6.214×10^{-6}
1 METER = 100	1	10^{-3}	39.37	3.281	6.214×10^{-4}
1 kilometer = 10^{5}	1000	1	3.937×10^{4}	3281	0.6214
1 inch = 2.540	2.540×10^{-2}	2.540×10^{-5}	1	8.333×10^{-2}	1.578×10^{-5}
1 foot = 30.48	0.3048	3.048×10^{-4}	12	1	1.894×10^{-4}
1 mile = 1.609×10^{5}	1609	1.609	6.336×10^{4}	5280	1

1 angström = 10^{-10} m
1 nautical mile = 1852 m
= 1.151 miles = 6076 ft

1 fermi = 10^{-15} m
1 light-year = 9.461×10^{12} km
1 parsec = 3.084×10^{13} km

1 fathom = 6 ft
1 Bohr radius = 5.292×10^{-11} m
1 yard = 3 ft

1 rod = 16.5 ft
1 mil = 10^{-3} in.
1 nm = 10^{-9} m

Area

METER^2	cm^2	ft^2	in.^2
1 SQUARE METER = 1	10^{4}	10.76	1550
1 square centimeter = 10^{-4}	1	1.076×10^{-3}	0.1550
1 square foot = 9.290×10^{-2}	929.0	1	144
1 square inch = 6.452×10^{-4}	6.452	6.944×10^{-3}	1

1 square mile = 2.788×10^{7} ft^2 = 640 acres
1 barn = 10^{-28} m^2

1 acre = 43 560 ft^2
1 hectare = 10^{4} m^2 = 2.471 acres

Volume

	METER^3	cm^3	L	ft^3	in.^{3}
1 CUBIC METER =	1	10^6	1000	35.31	6.102×10^4
1 cubic centimeter =	10^{-6}	1	1.000×10^{-3}	3.531×10^{-5}	6.102×10^{-2}
1 liter =	1.000×10^{-3}	1000	1	3.531×10^{-2}	61.02
1 cubic foot =	2.832×10^{-2}	2.832×10^4	28.32	1	1728
1 cubic inch =	1.639×10^{-5}	16.39	1.639×10^{-2}	5.787×10^{-4}	1

1 U.S. fluid gallon = 4 U.S. fluid quarts = 8 U.S. pints = 128 U.S. fluid ounces = 231 in.^{3}
1 British imperial gallon = 277.4 in.^{3} = 1.201 U.S. fluid gallons

Mass

Quantities in the colored areas are not mass units but are often used as such. For example, when we write 1 kg "=" 2.205 lb, this means that a kilogram is a *mass* that *weighs* 2.205 pounds at a location where g has the standard value of 9.80665 m/s^2.

	g	KILOGRAM	slug	u	oz	lb	ton
1 gram =	1	0.001	6.852×10^{-5}	6.022×10^{23}	3.527×10^{-2}	2.205×10^{-3}	1.102×10^{-6}
1 KILOGRAM =	1000	1	6.852×10^{-2}	6.022×10^{26}	35.27	2.205	1.102×10^{-3}
1 slug =	1.459×10^4	14.59	1	8.786×10^{27}	514.8	32.17	1.609×10^{-2}
1 atomic mass unit =	1.661×10^{-24}	1.661×10^{-27}	1.138×10^{-28}	1	5.857×10^{-26}	3.662×10^{-27}	1.830×10^{-30}
1 ounce =	28.35	2.835×10^{-2}	1.943×10^{-3}	1.718×10^{25}	1	6.250×10^{-2}	3.125×10^{-5}
1 pound =	453.6	0.4536	3.108×10^{-2}	2.732×10^{26}	16	1	0.0005
1 ton =	9.072×10^5	907.2	62.16	5.463×10^{29}	3.2×10^4	2000	1

1 metric ton = 1000 kg

Density

Quantities in the colored areas are weight densities and, as such, are dimensionally different from mass densities. See the note for the mass table.

	slug/ft^3	KILOGRAM/ METER^3	g/cm^3	lb/ft^3	lb/in.^{3}
1 slug per foot^3 =	1	515.4	0.5154	32.17	1.862×10^{-2}
1 KILOGRAM per METER^3 =	1.940×10^{-3}	1	0.001	6.243×10^{-2}	3.613×10^{-5}
1 gram per centimeter^3 =	1.940	1000	1	62.43	3.613×10^{-2}
1 pound per foot^3 =	3.108×10^{-2}	16.02	16.02×10^{-2}	1	5.787×10^{-4}
1 pound per inch^3 =	53.71	2.768×10^4	27.68	1728	1

Time

	y	d	h	min	SECOND
1 year =	1	365.25	8.766×10^3	5.259×10^5	3.156×10^7
1 day =	2.738×10^{-3}	1	24	1440	8.640×10^4
1 hour =	1.141×10^{-4}	4.167×10^{-2}	1	60	3600
1 minute =	1.901×10^{-6}	6.944×10^{-4}	1.667×10^{-2}	1	60
1 SECOND =	3.169×10^{-8}	1.157×10^{-5}	2.778×10^{-4}	1.667×10^{-2}	1

Speed

	ft/s	km/h	METER/SECOND	mi/h	cm/s
1 foot per second =	1	1.097	0.3048	0.6818	30.48
1 kilometer per hour =	0.9113	1	0.2778	0.6214	27.78
1 METER per SECOND =	3.281	3.6	1	2.237	100
1 mile per hour =	1.467	1.609	0.4470	1	44.70
1 centimeter per second =	3.281×10^{-2}	3.6×10^{-2}	0.01	2.237×10^{-2}	1

1 knot = 1 nautical mi/h = 1.688 ft/s   1 mi/min = 88.00 ft/s = 60.00 mi/h

Force

Force units in the colored areas are now little used. To clarify: 1 gram-force (= 1 gf) is the force of gravity that would act on an object whose mass is 1 gram at a location where g has the standard value of 9.80665 m/s^2.

	dyne	NEWTON	lb	pdl	gf	kgf
1 dyne =	1	10^{-5}	2.248×10^{-6}	7.233×10^{-5}	1.020×10^{-3}	1.020×10^{-6}
1 NEWTON =	10^5	1	0.2248	7.233	102.0	0.1020
1 pound =	4.448×10^5	4.448	1	32.17	453.6	0.4536
1 poundal =	1.383×10^4	0.1383	3.108×10^{-2}	1	14.10	1.410×10^2
1 gram-force =	980.7	9.807×10^{-3}	2.205×10^{-3}	7.093×10^{-2}	1	0.001
1 kilogram-force =	9.807×10^5	9.807	2.205	70.93	1000	1

1 ton = 2000 lb

Pressure

	atm	dyne/cm^2	inch of water	cm Hg	PASCAL	lb/in.^{2}	lb/ft^2
1 atmosphere =	1	1.013×10^6	406.8	76	1.013×10^5	14.70	2116
1 dyne per centimeter^2 =	9.869×10^{-7}	1	4.015×10^{-4}	7.501×10^{-5}	0.1	1.405×10^{-5}	2.089×10^{-3}
1 inch of water^a at 4°C =	2.458×10^{-3}	2491	1	0.1868	249.1	3.613×10^{-2}	5.202
1 centimeter of mercury^a at 0°C =	1.316×10^{-2}	1.333×10^4	5.353	1	1333	0.1934	27.85
1 PASCAL =	9.869×10^{-6}	10	4.015×10^{-3}	7.501×10^{-4}	1	1.450×10^{-4}	2.089×10^{-2}
1 pound per inch^2 =	6.805×10^{-2}	6.895×10^4	27.68	5.171	6.895×10^3	1	144
1 pound per foot^2 =	4.725×10^{-4}	478.8	0.1922	3.591×10^{-2}	47.88	6.944×10^{-3}	1

^{a}Where the acceleration of gravity has the standard value of 9.80665 m/s^2.

1 bar = 10^6 dyne/cm^2 = 0.1 MPa   1 millibar = 10^3 dyne/cm^2 = 10^2 Pa   1 torr = 1 mm Hg

Energy, Work, Heat

Quantities in the colored areas are not energy units but are included for convenience. They arise from the relativistic mass–energy equivalence formula $E = mc^2$ and represent the energy released if a kilogram or unified atomic mass unit (u) is completely converted to energy (bottom two rows) or the mass that would be completely converted to one unit of energy (rightmost two columns).

	Btu	erg	ft · lb	hp · h	JOULE	cal	kW · h	eV	MeV	kg	u
1 British thermal unit =	1	1.055×10^{10}	777.9	3.929×10^{-4}	1055	252.0	2.930×10^{-4}	6.585×10^{21}	6.585×10^{15}	1.174×10^{-14}	7.070×10^{12}
1 erg =	9.481×10^{-11}	1	7.376×10^{-8}	3.725×10^{-14}	10^{-7}	2.389×10^{-8}	2.778×10^{-14}	6.242×10^{11}	6.242×10^{5}	1.113×10^{-24}	670.2
1 foot-pound =	1.285×10^{-3}	1.356×10^{7}	1	5.051×10^{-7}	1.356	0.3238	3.766×10^{-7}	8.464×10^{18}	8.464×10^{12}	1.509×10^{-17}	9.037×10^{9}
1 horsepower-hour =	2545	2.685×10^{13}	1.980×10^{6}	1	2.685×10^{6}	6.413×10^{5}	0.7457	1.676×10^{25}	1.676×10^{19}	2.988×10^{-11}	1.799×10^{16}
1 JOULE =	9.481×10^{-4}	10^{7}	0.7376	3.725×10^{-7}	1	0.2389	2.778×10^{-7}	6.242×10^{18}	6.242×10^{12}	1.113×10^{-17}	6.702×10^{9}
1 calorie =	3.968×10^{-3}	4.1868×10^{7}	3.088	1.560×10^{-6}	4.1868	1	1.163×10^{-6}	2.613×10^{19}	2.613×10^{13}	4.660×10^{-17}	2.806×10^{10}
1 kilowatt-hour =	3413	3.600×10^{13}	2.655×10^{6}	1.341	3.600×10^{6}	8.600×10^{5}	1	2.247×10^{25}	2.247×10^{19}	4.007×10^{-11}	2.413×10^{16}
1 electron-volt =	1.519×10^{-22}	1.602×10^{-12}	1.182×10^{-19}	5.967×10^{-26}	1.602×10^{-19}	3.827×10^{-20}	4.450×10^{-26}	1	10^{-6}	1.783×10^{-36}	1.074×10^{-9}
1 million electron-volts =	1.519×10^{-16}	1.602×10^{-6}	1.182×10^{-13}	5.967×10^{-20}	1.602×10^{-13}	3.827×10^{-14}	4.450×10^{-20}	10^{-6}	1	1.783×10^{-30}	1.074×10^{-3}
1 kilogram =	8.521×10^{13}	8.987×10^{23}	6.629×10^{16}	3.348×10^{10}	8.987×10^{16}	2.146×10^{16}	2.497×10^{10}	5.610×10^{35}	5.610×10^{29}	1	6.022×10^{26}
1 unified atomic mass unit =	1.415×10^{-13}	1.492×10^{-3}	1.101×10^{-10}	5.559×10^{-17}	1.492×10^{-10}	3.564×10^{-11}	4.146×10^{-17}	9.320×10^{8}	932.0	1.661×10^{-27}	1

Power

	Btu/h	ft · lb/s	hp	cal/s	kW	WATT
1 British thermal unit per hour =	1	0.2161	3.929×10^{-4}	6.998×10^{-2}	2.930×10^{-4}	0.2930
1 foot-pound per second =	4.628	1	1.818×10^{-3}	0.3239	1.356×10^{-3}	1.356
1 horsepower =	2545	550	1	178.1	0.7457	745.7
1 calorie per second =	14.29	3.088	5.615×10^{-3}	1	4.186×10^{-3}	4.186
1 kilowatt =	3413	737.6	1.341	238.9	1	1000
1 WATT =	3.413	0.7376	1.341×10^{-3}	0.2389	0.001	1

Magnetic Field

	gauss	TESLA	milligauss
1 gauss =	1	10^{-4}	1000
1 TESLA =	10^{4}	1	10^{7}
1 milligauss =	0.001	10^{-7}	1

1 tesla = 1 weber/meter^2

Magnetic Flux

	maxwell	WEBER
1 maxwell =	1	10^{-8}
1 WEBER =	10^{8}	1

Mathematical Formulas E

Geometry

Circle of radius r: circumference $= 2\pi r$; area $= \pi r^2$.

Sphere of radius r: area $= 4\pi r^2$; volume $= \frac{4}{3}\pi r^3$.

Right circular cylinder of radius r and height h:
area $= 2\pi r^2 + 2\pi rh$; volume $= \pi r^2 h$.

Triangle of base a and altitude h: area $= \frac{1}{2}ah$.

Quadratic Formula

If $ax^2 + bx + c = 0$, then $x = \dfrac{-b \pm \sqrt{b^2 - 4ac}}{2a}$.

Trigonometric Functions of Angle θ

$$\sin\theta = \frac{y}{r} \qquad \cos\theta = \frac{x}{r}$$

$$\tan\theta = \frac{y}{x} \qquad \cot\theta = \frac{x}{y}$$

$$\sec\theta = \frac{r}{x} \qquad \csc\theta = \frac{r}{y}$$

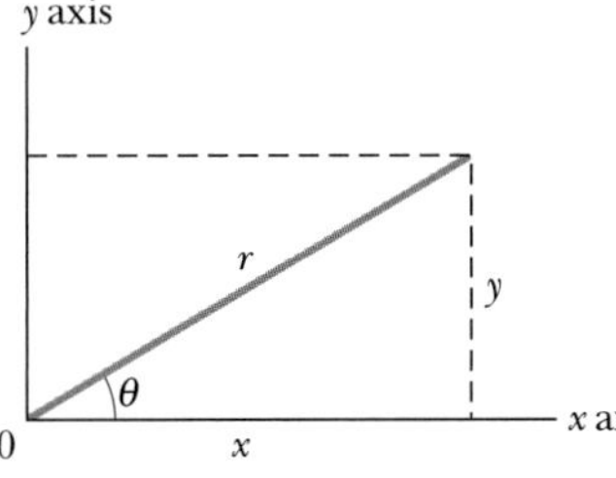

Pythagorean Theorem

In this right triangle,
$a^2 + b^2 = c^2$

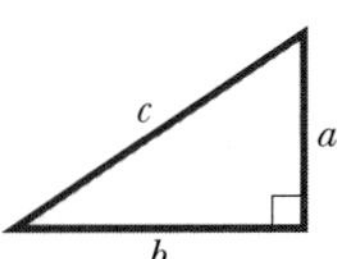

Triangles

Angles are A, B, C

Opposite sides are a, b, c

Angles $A + B + C = 180°$

$$\frac{\sin A}{a} = \frac{\sin B}{b} = \frac{\sin C}{c}$$

$c^2 = a^2 + b^2 - 2ab\cos C$

Exterior angle $D = A + C$

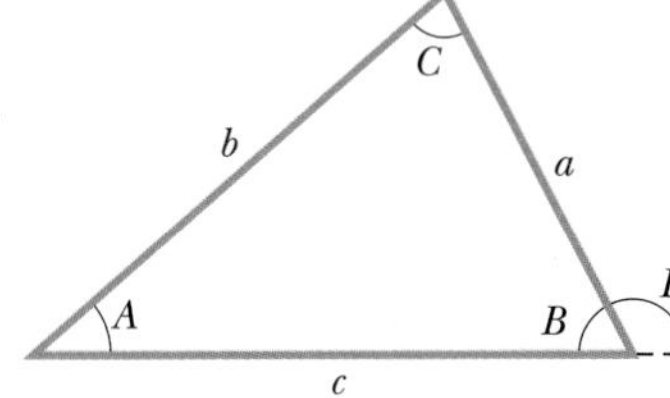

Mathematical Signs and Symbols

$=$ equals

$\approx$ equals approximately

$\sim$ is the order of magnitude of

$\neq$ is not equal to

$\equiv$ is identical to, is defined as

$>$ is greater than ($\gg$ is much greater than)

$<$ is less than ($\ll$ is much less than)

$\geq$ is greater than or equal to (or, is no less than)

$\leq$ is less than or equal to (or, is no more than)

$\pm$ plus or minus

$\propto$ is proportional to

Σ the sum of

x_{avg} the average value of x

Trigonometric Identities

$\sin(90° - \theta) = \cos\theta$

$\cos(90° - \theta) = \sin\theta$

$\sin\theta/\cos\theta = \tan\theta$

$\sin^2\theta + \cos^2\theta = 1$

$\sec^2\theta - \tan^2\theta = 1$

$\csc^2\theta - \cot^2\theta = 1$

$\sin 2\theta = 2\sin\theta\cos\theta$

$\cos 2\theta = \cos^2\theta - \sin^2\theta = 2\cos^2\theta - 1 = 1 - 2\sin^2\theta$

$\sin(\alpha \pm \beta) = \sin\alpha\cos\beta \pm \cos\alpha\sin\beta$

$\cos(\alpha \pm \beta) = \cos\alpha\cos\beta \mp \sin\alpha\sin\beta$

$$\tan(\alpha \pm \beta) = \frac{\tan\alpha \pm \tan\beta}{1 \mp \tan\alpha\tan\beta}$$

$\sin\alpha \pm \sin\beta = 2\sin\frac{1}{2}(\alpha \pm \beta)\cos\frac{1}{2}(\alpha \mp \beta)$

$\cos\alpha + \cos\beta = 2\cos\frac{1}{2}(\alpha + \beta)\cos\frac{1}{2}(\alpha - \beta)$

$\cos\alpha - \cos\beta = -2\sin\frac{1}{2}(\alpha + \beta)\sin\frac{1}{2}(\alpha - \beta)$

Binomial Theorem

$$(1+x)^n = 1 + \frac{nx}{1!} + \frac{n(n-1)x^2}{2!} + \cdots \qquad (x^2 < 1)$$

Exponential Expansion

$$e^x = 1 + x + \frac{x^2}{2!} + \frac{x^3}{3!} + \cdots$$

Logarithmic Expansion

$$\ln(1+x) = x - \tfrac{1}{2}x^2 + \tfrac{1}{3}x^3 - \cdots \qquad (|x| < 1)$$

Trigonometric Expansions (θ in radians)

$$\sin\theta = \theta - \frac{\theta^3}{3!} + \frac{\theta^5}{5!} - \cdots$$

$$\cos\theta = 1 - \frac{\theta^2}{2!} + \frac{\theta^4}{4!} - \cdots$$

$$\tan\theta = \theta + \frac{\theta^3}{3} + \frac{2\theta^5}{15} + \cdots$$

Cramer's Rule

Two simultaneous equations in unknowns x and y,

$$a_1x + b_1y = c_1 \quad \text{and} \quad a_2x + b_2y = c_2,$$

have the solutions

$$x = \frac{\begin{vmatrix} c_1 & b_1 \\ c_2 & b_2 \end{vmatrix}}{\begin{vmatrix} a_1 & b_1 \\ a_2 & b_2 \end{vmatrix}} = \frac{c_1b_2 - c_2b_1}{a_1b_2 - a_2b_1}$$

and

$$y = \frac{\begin{vmatrix} a_1 & c_1 \\ a_2 & c_2 \end{vmatrix}}{\begin{vmatrix} a_1 & b_1 \\ a_2 & b_2 \end{vmatrix}} = \frac{a_1c_2 - a_2c_1}{a_1b_2 - a_2b_1}.$$

Products of Vectors

Let $\hat{i}$, $\hat{j}$, and $\hat{k}$ be unit vectors in the x, y, and z directions. Then

$$\hat{i}\cdot\hat{i} = \hat{j}\cdot\hat{j} = \hat{k}\cdot\hat{k} = 1, \qquad \hat{i}\cdot\hat{j} = \hat{j}\cdot\hat{k} = \hat{k}\cdot\hat{i} = 0,$$

$$\hat{i}\times\hat{i} = \hat{j}\times\hat{j} = \hat{k}\times\hat{k} = 0,$$

$$\hat{i}\times\hat{j} = \hat{k}, \qquad \hat{j}\times\hat{k} = \hat{i}, \qquad \hat{k}\times\hat{i} = \hat{j}$$

Any vector $\vec{a}$ with components a_x, a_y, and a_z along the x, y, and z axes can be written as

$$\vec{a} = a_x\hat{i} + a_y\hat{j} + a_z\hat{k}.$$

Let $\vec{a}$, $\vec{b}$, and $\vec{c}$ be arbitrary vectors with magnitudes a, b, and c. Then

$$\vec{a}\times(\vec{b}+\vec{c}) = (\vec{a}\times\vec{b}) + (\vec{a}\times\vec{c})$$

$$(s\vec{a})\times\vec{b} = \vec{a}\times(s\vec{b}) = s(\vec{a}\times\vec{b}) \qquad (s = \text{a scalar}).$$

Let θ be the smaller of the two angles between $\vec{a}$ and $\vec{b}$. Then

$$\vec{a}\cdot\vec{b} = \vec{b}\cdot\vec{a} = a_xb_x + a_yb_y + a_zb_z = ab\cos\theta$$

$$\vec{a}\times\vec{b} = -\vec{b}\times\vec{a} = \begin{vmatrix} \hat{i} & \hat{j} & \hat{k} \\ a_x & a_y & a_z \\ b_x & b_y & b_z \end{vmatrix}$$

$$= \hat{i}\begin{vmatrix} a_y & a_z \\ b_y & b_z \end{vmatrix} - \hat{j}\begin{vmatrix} a_x & a_z \\ b_x & b_z \end{vmatrix} + \hat{k}\begin{vmatrix} a_x & a_y \\ b_x & b_y \end{vmatrix}$$

$$= (a_yb_z - b_ya_z)\hat{i} + (a_zb_x - b_za_x)\hat{j} + (a_xb_y - b_xa_y)\hat{k}$$

$$|\vec{a}\times\vec{b}| = ab\sin\theta$$

$$\vec{a}\cdot(\vec{b}\times\vec{c}) = \vec{b}\cdot(\vec{c}\times\vec{a}) = \vec{c}\cdot(\vec{a}\times\vec{b})$$

$$\vec{a}\times(\vec{b}\times\vec{c}) = (\vec{a}\cdot\vec{c})\vec{b} - (\vec{a}\cdot\vec{b})\vec{c}$$

Derivatives and Integrals

In what follows, the letters u and v stand for any functions of x, and a and m are constants. To each of the indefinite integrals should be added an arbitrary constant of integration. The *Handbook of Chemistry and Physics* (CRC Press Inc.) gives a more extensive tabulation.

1. $\dfrac{dx}{dx} = 1$
2. $\dfrac{d}{dx}(au) = a\dfrac{du}{dx}$
3. $\dfrac{d}{dx}(u + v) = \dfrac{du}{dx} + \dfrac{dv}{dx}$
4. $\dfrac{d}{dx}x^m = mx^{m-1}$
5. $\dfrac{d}{dx}\ln x = \dfrac{1}{x}$
6. $\dfrac{d}{dx}(uv) = u\dfrac{dv}{dx} + v\dfrac{du}{dx}$
7. $\dfrac{d}{dx}e^x = e^x$
8. $\dfrac{d}{dx}\sin x = \cos x$
9. $\dfrac{d}{dx}\cos x = -\sin x$
10. $\dfrac{d}{dx}\tan x = \sec^2 x$
11. $\dfrac{d}{dx}\cot x = -\csc^2 x$
12. $\dfrac{d}{dx}\sec x = \tan x \sec x$
13. $\dfrac{d}{dx}\csc x = -\cot x \csc x$
14. $\dfrac{d}{dx}e^u = e^u\dfrac{du}{dx}$
15. $\dfrac{d}{dx}\sin u = \cos u\dfrac{du}{dx}$
16. $\dfrac{d}{dx}\cos u = -\sin u\dfrac{du}{dx}$

1. $\int dx = x$
2. $\int au\,dx = a\int u\,dx$
3. $\int (u + v)\,dx = \int u\,dx + \int v\,dx$
4. $\int x^m\,dx = \dfrac{x^{m+1}}{m+1}\quad (m \neq -1)$
5. $\int \dfrac{dx}{x} = \ln |x|$
6. $\int u\dfrac{dv}{dx}\,dx = uv - \int v\dfrac{du}{dx}\,dx$
7. $\int e^x\,dx = e^x$
8. $\int \sin x\,dx = -\cos x$
9. $\int \cos x\,dx = \sin x$
10. $\int \tan x\,dx = \ln |\sec x|$
11. $\int \sin^2 x\,dx = \frac{1}{2}x - \frac{1}{4}\sin 2x$
12. $\int e^{-ax}\,dx = -\dfrac{1}{a}e^{-ax}$
13. $\int xe^{-ax}\,dx = -\dfrac{1}{a^2}(ax + 1)\,e^{-ax}$
14. $\int x^2e^{-ax}\,dx = -\dfrac{1}{a^3}(a^2x^2 + 2ax + 2)e^{-ax}$
15. $\int_0^\infty x^ne^{-ax}\,dx = \dfrac{n!}{a^{n+1}}$
16. $\int_0^\infty x^{2n}e^{-ax^2}\,dx = \dfrac{1\cdot3\cdot5\cdots(2n-1)}{2^{n+1}a^n}\sqrt{\dfrac{\pi}{a}}$
17. $\int \dfrac{dx}{\sqrt{x^2 + a^2}} = \ln(x + \sqrt{x^2 + a^2})$
18. $\int \dfrac{x\,dx}{(x^2 + a^2)^{3/2}} = -\dfrac{1}{(x^2 + a^2)^{1/2}}$
19. $\int \dfrac{dx}{(x^2 + a^2)^{3/2}} = \dfrac{x}{a^2(x^2 + a^2)^{1/2}}$
20. $\int_0^\infty x^{2n+1}\,e^{-ax^2}\,dx = \dfrac{n!}{2a^{n+1}}\quad (a > 0)$
21. $\int \dfrac{x\,dx}{x + d} = x - d\ln(x + d)$

E Properties of the Elements

All physical properties are for a pressure of 1 atm unless otherwise specified.

Element	Symbol	Atomic Number Z	Molar Mass, g/mol	Density, g/cm^3 at 20°C	Melting Point, °C	Boiling Point, °C	Specific Heat, $J/(g \cdot °C)$ at 25°C
Actinium	Ac	89	(227)	10.06	1323	(3473)	0.092
Aluminum	Al	13	26.9815	2.699	660	2450	0.900
Americium	Am	95	(243)	13.67	1541	—	—
Antimony	Sb	51	121.75	6.691	630.5	1380	0.205
Argon	Ar	18	39.948	1.6626×10^{-3}	−189.4	−185.8	0.523
Arsenic	As	33	74.9216	5.78	817 (28 atm)	613	0.331
Astatine	At	85	(210)	—	(302)	—	—
Barium	Ba	56	137.34	3.594	729	1640	0.205
Berkelium	Bk	97	(247)	14.79	—	—	—
Beryllium	Be	4	9.0122	1.848	1287	2770	1.83
Bismuth	Bi	83	208.980	9.747	271.37	1560	0.122
Bohrium	Bh	107	262.12	—	—	—	—
Boron	B	5	10.811	2.34	2030	—	1.11
Bromine	Br	35	79.909	3.12 (liquid)	−7.2	58	0.293
Cadmium	Cd	48	112.40	8.65	321.03	765	0.226
Calcium	Ca	20	40.08	1.55	838	1440	0.624
Californium	Cf	98	(251)	—	—	—	—
Carbon	C	6	12.01115	2.26	3727	4830	0.691
Cerium	Ce	58	140.12	6.768	804	3470	0.188
Cesium	Cs	55	132.905	1.873	28.40	690	0.243
Chlorine	Cl	17	35.453	3.214×10^{-3} (0°C)	−101	−34.7	0.486
Chromium	Cr	24	51.996	7.19	1857	2665	0.448
Cobalt	Co	27	58.9332	8.85	1495	2900	0.423
Copper	Cu	29	63.54	8.96	1083.40	2595	0.385
Curium	Cm	96	(247)	13.3	—	—	—
Darmstadtium	Ds	110	(271)	—	—	—	—
Dubnium	Db	105	262.114	—	—	—	—
Dysprosium	Dy	66	162.50	8.55	1409	2330	0.172
Einsteinium	Es	99	(254)	—	—	—	—
Erbium	Er	68	167.26	9.15	1522	2630	0.167
Europium	Eu	63	151.96	5.243	817	1490	0.163
Fermium	Fm	100	(237)	—	—	—	—
Fluorine	F	9	18.9984	1.696×10^{-3} (0°C)	−219.6	−188.2	0.753
Francium	Fr	87	(223)	—	(27)	—	—
Gadolinium	Gd	64	157.25	7.90	1312	2730	0.234
Gallium	Ga	31	69.72	5.907	29.75	2237	0.377
Germanium	Ge	32	72.59	5.323	937.25	2830	0.322
Gold	Au	79	196.967	19.32	1064.43	2970	0.131

Element	Symbol	Atomic Number Z	Molar Mass, g/mol	Density, g/cm^3 at 20°C	Melting Point, °C	Boiling Point, °C	Specific Heat, $J/(g \cdot °C)$ at 25°C
Hafnium	Hf	72	178.49	13.31	2227	5400	0.144
Hassium	Hs	108	(265)	—	—	—	—
Helium	He	2	4.0026	0.1664×10^{-3}	−269.7	−268.9	5.23
Holmium	Ho	67	164.930	8.79	1470	2330	0.165
Hydrogen	H	1	1.00797	0.08375×10^{-3}	−259.19	−252.7	14.4
Indium	In	49	114.82	7.31	156.634	2000	0.233
Iodine	I	53	126.9044	4.93	113.7	183	0.218
Iridium	Ir	77	192.2	22.5	2447	(5300)	0.130
Iron	Fe	26	55.847	7.874	1536.5	3000	0.447
Krypton	Kr	36	83.80	3.488×10^{-3}	−157.37	−152	0.247
Lanthanum	La	57	138.91	6.189	920	3470	0.195
Lawrencium	Lr	103	(257)	—	—	—	—
Lead	Pb	82	207.19	11.35	327.45	1725	0.129
Lithium	Li	3	6.939	0.534	180.55	1300	3.58
Lutetium	Lu	71	174.97	9.849	1663	1930	0.155
Magnesium	Mg	12	24.312	1.738	650	1107	1.03
Manganese	Mn	25	54.9380	7.44	1244	2150	0.481
Meitnerium	Mt	109	(266)	—	—	—	—
Mendelevium	Md	101	(256)	—	—	—	—
Mercury	Hg	80	200.59	13.55	−38.87	357	0.138
Molybdenum	Mo	42	95.94	10.22	2617	5560	0.251
Neodymium	Nd	60	144.24	7.007	1016	3180	0.188
Neon	Ne	10	20.183	0.8387×10^{-3}	−248.597	−246.0	1.03
Neptunium	Np	93	(237)	20.25	637	—	1.26
Nickel	Ni	28	58.71	8.902	1453	2730	0.444
Niobium	Nb	41	92.906	8.57	2468	4927	0.264
Nitrogen	N	7	14.0067	1.1649×10^{-3}	−210	−195.8	1.03
Nobelium	No	102	(255)	—	—	—	—
Osmium	Os	76	190.2	22.59	3027	5500	0.130
Oxygen	O	8	15.9994	1.3318×10^{-3}	−218.80	−183.0	0.913
Palladium	Pd	46	106.4	12.02	1552	3980	0.243
Phosphorus	P	15	30.9738	1.83	44.25	280	0.741
Platinum	Pt	78	195.09	21.45	1769	4530	0.134
Plutonium	Pu	94	(244)	19.8	640	3235	0.130
Polonium	Po	84	(210)	9.32	254	—	—
Potassium	K	19	39.102	0.862	63.20	760	0.758
Praseodymium	Pr	59	140.907	6.773	931	3020	0.197
Promethium	Pm	61	(145)	7.22	(1027)	—	—
Protactinium	Pa	91	(231)	15.37 (estimated)	(1230)	—	—
Radium	Ra	88	(226)	5.0	700	—	—
Radon	Rn	86	(222)	9.96×10^{-3} (0°C)	(−71)	−61.8	0.092
Rhenium	Re	75	186.2	21.02	3180	5900	0.134
Rhodium	Rh	45	102.905	12.41	1963	4500	0.243
Rubidium	Rb	37	85.47	1.532	39.49	688	0.364
Ruthenium	Ru	44	101.107	12.37	2250	4900	0.239
Rutherfordium	Rf	104	261.11	—	—	—	—

Element	Symbol	Atomic Number Z	Molar Mass, g/mol	Density, g/cm^3 at 20°C	Melting Point, °C	Boiling Point, °C	Specific Heat, J/(g·°C) at 25°C
Samarium	Sm	62	150.35	7.52	1072	1630	0.197
Scandium	Sc	21	44.956	2.99	1539	2730	0.569
Seaborgium	Sg	106	263.118	—	—	—	—
Selenium	Se	34	78.96	4.79	221	685	0.318
Silicon	Si	14	28.086	2.33	1412	2680	0.712
Silver	Ag	47	107.870	10.49	960.8	2210	0.234
Sodium	Na	11	22.9898	0.9712	97.85	892	1.23
Strontium	Sr	38	87.62	2.54	768	1380	0.737
Sulfur	S	16	32.064	2.07	119.0	444.6	0.707
Tantalum	Ta	73	180.948	16.6	3014	5425	0.138
Technetium	Tc	43	(99)	11.46	2200	—	0.209
Tellurium	Te	52	127.60	6.24	449.5	990	0.201
Terbium	Tb	65	158.924	8.229	1357	2530	0.180
Thallium	Tl	81	204.37	11.85	304	1457	0.130
Thorium	Th	90	(232)	11.72	1755	(3850)	0.117
Thulium	Tm	69	168.934	9.32	1545	1720	0.159
Tin	Sn	50	118.69	7.2984	231.868	2270	0.226
Titanium	Ti	22	47.90	4.54	1670	3260	0.523
Tungsten	W	74	183.85	19.3	3380	5930	0.134
Unnamed	Uuu	111	(272)	—	—	—	—
Unnamed	Uub	112	(285)	—	—	—	—
Unnamed	Uut	113	—	—	—	—	—
Unnamed	Unq	114	(289)	—	—	—	—
Unnamed	Uup	115	—	—	—	—	—
Unnamed	Uuh	116	—	—	—	—	—
Unnamed	Uus	117	—	—	—	—	—
Unnamed	Uuo	118	(293)	—	—	—	—
Uranium	U	92	(238)	18.95	1132	3818	0.117
Vanadium	V	23	50.942	6.11	1902	3400	0.490
Xenon	Xe	54	131.30	5.495×10^{-3}	−111.79	−108	0.159
Ytterbium	Yb	70	173.04	6.965	824	1530	0.155
Yttrium	Y	39	88.905	4.469	1526	3030	0.297
Zinc	Zn	30	65.37	7.133	419.58	906	0.389
Zirconium	Zr	40	91.22	6.506	1852	3580	0.276

The values in parentheses in the column of molar masses are the mass numbers of the longest-lived isotopes of those elements that are radioactive. Melting points and boiling points in parentheses are uncertain.

The data for gases are valid only when these are in their usual molecular state, such as H_2, He, O_2, Ne, etc. The specific heats of the gases are the values at constant pressure.

Source: Adapted from J. Emsley, *The Elements,* 3rd ed., 1998, Clarendon Press, Oxford. See also www.webelements.com for the latest values and newest elements.

G Periodic Table of the Elements

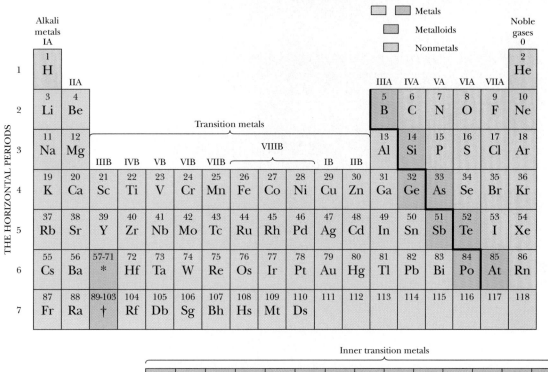

Metals
Metalloids
Nonmetals

THE HORIZONTAL PERIODS

Period	IA (Alkali metals)	IIA	IIIB	IVB	VB	VIB	VIIB	VIIIB	VIIIB	VIIIB	IB	IIB	IIIA	IVA	VA	VIA	VIIA	0 (Noble gases)
			Transition metals															
1	1 H																	2 He
2	3 Li	4 Be											5 B	6 C	7 N	8 O	9 F	10 Ne
3	11 Na	12 Mg											13 Al	14 Si	15 P	16 S	17 Cl	18 Ar
4	19 K	20 Ca	21 Sc	22 Ti	23 V	24 Cr	25 Mn	26 Fe	27 Co	28 Ni	29 Cu	30 Zn	31 Ga	32 Ge	33 As	34 Se	35 Br	36 Kr
5	37 Rb	38 Sr	39 Y	40 Zr	41 Nb	42 Mo	43 Tc	44 Ru	45 Rh	46 Pd	47 Ag	48 Cd	49 In	50 Sn	51 Sb	52 Te	53 I	54 Xe
6	55 Cs	56 Ba	57-71 *	72 Hf	73 Ta	74 W	75 Re	76 Os	77 Ir	78 Pt	79 Au	80 Hg	81 Tl	82 Pb	83 Bi	84 Po	85 At	86 Rn
7	87 Fr	88 Ra	89-103 †	104 Rf	105 Db	106 Sg	107 Bh	108 Hs	109 Mt	110 Ds	111	112	113	114	115	116	117	118

Inner transition metals

Lanthanide series *	57 La	58 Ce	59 Pr	60 Nd	61 Pm	62 Sm	63 Eu	64 Gd	65 Tb	66 Dy	67 Ho	68 Er	69 Tm	70 Yb	71 Lu
Actinide series †	89 Ac	90 Th	91 Pa	92 U	93 Np	94 Pu	95 Am	96 Cm	97 Bk	98 Cf	99 Es	100 Fm	101 Md	102 No	103 Lr

Elements 111, 112, 114, and 116 have been discovered but, as of 2003, have not yet been named. Evidence for the discovery of elements 113 and 115 has been reported. See www.webelements.com for the latest information and newest elements.

Answers

to Checkpoints and Odd-Numbered Questions and Problems

Chapter 1

P **1.** (a) 10^9 μm; (b) 10^{-4}; (c) 9.1×10^5 μm **3.** (a) 160 rods; (b) 40 chains **5.** (a) 4.00×10^4 km; (b) 5.10×10^8 km^2; (c) 1.08×10^{12} km^3 **7.** 1.9×10^{22} cm^3 **9.** 1.1×10^3 acre-feet **11.** 1.21×10^{12} μs **13.** (a) 1.43; (b) 0.864 **15.** (a) 495 s; (b) 141 s; (c) 198 s; (d) -245 s **17.** C, D, A, B, E; the important criterion is the consistency of the daily variation, not its magnitude **19.** 5.2×10^6 m **21.** (a) 1×10^3 kg; (b) 158 kg/s **23.** 9.0×10^{49} atoms **25.** (a) 1.18×10^{-29} m^3; (b) 0.282 nm **27.** 1750 kg **29.** 1.9×10^5 kg **31.** 1.43 kg/min **33.** (a) 22 pecks; (b) 5.5 Imperial bushels; (c) 200 L **35.** (a) 18.8 gallons; (b) 22.5 gallons **37.** (a) 11.3 m^2/L; (b) 1.13×10^4 m^{-1}; (c) 2.17×10^{-3} gal/ft^2; (d) number of gallons to cover a square foot **39.** 0.3 cord **41.** (a) 293 U.S. bushels; (b) 3.81×10^3 U.S. bushels **43.** 8×10^2 km **45.** 0.12 AU/min **47.** 3.8 mg/s **49.** 10.7 habaneros **51.** (a) yes; (b) 8.6 universe seconds **53.** (a) 3.88; (b) 7.65; (c) 156 ken^3; (d) 1.19×10^3 m^3 **55.** 1.2 m **57.** (a) 4.9×10^{-6} pc; (b) 1.6×10^{-5} ly **59.** (a) 3.9 m, 4.8 m; (b) 3.9×10^3 mm, 4.8×10^3 mm; (c) 2.2 m^3, 4.2 m^3

Chapter 2

CP **1.** b and c **2.** (check the derivative dx/dt) (a) 1 and 4; (b) 2 and 3 **3.** (a) plus; (b) minus; (c) minus; (d) plus **4.** 1 and 4 ($a = d^2x/dt^2$ must be constant) **5.** (a) plus (upward displacement on y axis); (b) minus (downward displacement on y axis); (c) $a = -g = -9.8$ m/s^2 **Q** **1.** (a) all tie; (b) 4, tie of 1 and 2, then 3 **3.** (a) negative; (b) positive; (c) yes; (d) positive; (e) constant **5.** (a) positive direction; (b) negative direction; (c) 3 and 5; (d) 2 and 6 tie, then 3 and 5 tie, then 1 and 4 tie (zero) **7.** (a) 3, 2, 1; (b) 1, 2, 3; (c) all tie; (d) 1, 2, 3 **9.** (a) D; (b) E **P** **1.** (a) +40 km/h; (b) 40 km/h **3.** 13 m **5.** (a) 0; (b) -2 m; (c) 0; (d) 12 m; (e) +12 m; (f) +7 m/s **7.** 1.4 m **9.** 128 km/h **11.** 60 km **13.** (a) 73 km/h; (b) 68 km/h; (c) 70 km/h; (d) 0 **15.** (a) -6 m/s; (b) $-x$ direction; (c) 6 m/s; (d) decreasing; (e) 2 s; (f) no **17.** (a) 28.5 cm/s; (b) 18.0 cm/s; (c) 40.5 cm/s; (d) 28.1 cm/s; (e) 30.3 cm/s **19.** -20 m/s^2 **21.** (a) m/s^2; (b) m/s^3; (c) 1.0 s; (d) 82 m; (e) -80 m; (f) 0; (g) -12 m/s; (h) -36 m/s; (i) -72 m/s; (j) -6 m/s^2; (k) -18 m/s^2; (l) -30 m/s^2; (m) -42 m/s^2 **23.** (a) +1.6 m/s; (b) +18 m/s **25.** (a) 3.1×10^6 s; (b) 4.6×10^{13} m **27.** 1.62×10^{15} m/s^2 **29.** (a) 30 s; (b) 300 m **31.** (a) 10.6 m; (b) 41.5 s **33.** (a) 3.56 m/s^2; (b) 8.43 m/s **35.** (a) 4.0 m/s^2; (b) $+x$ **37.** (a) -2.5 m/s^2; (b) 1; (d) 0; (e) 2 **39.** 40 m **41.** 0.90 m/s^2 **43.** (a) 15.0 m; (b) 94 km/h **45.** (a) 29.4 m; (b) 2.45 s **47.** (a) 31 m/s; (b) 6.4 s **49.** (a) 5.4 s; (b) 41 m/s **51.** 4.0 m/s **53.** (a) 20 m; (b) 59 m **55.** (a) 857 m/s^2; (b) up **57.** (a) 1.26×10^3 m/s^2; (b) up **59.** (a) 89 cm; (b) 22 cm **61.** 2.34 m **63.** 20.4 m **65.** (a) 2.25 m/s; (b) 3.90 m/s **67.** 100 m **69.** 0.56 m/s **71.** (a) 82 m; (b) 19 m/s **73.** (a) 2.00 s; (b) 12 cm; (c) -9.00 cm/s^2; (d) right; (e) left; (f) 3.46 s **75.** (a) 48.5 m/s; (b) 4.95 s; (c) 34.3 m/s; (d) 3.50 s **77.** 414 ms **79.** 90 m **81.** (a) 3.0 s; (b) 9.0 m **83.** 2.78 m/s^2 **85.** (a) 0.74 s; (b) 6.2 m/s^2 **87.** 17 m/s **89.** +47 m/s **91.** (a) 3.1 m/s^2; (b) 45 m; (c) 13 s **93.** (a) 1.23 cm; (b) 4 times; (c) 9 times; (d) 16 times; (e) 25 times **95.** 25 km/h **97.** 1.2 h **99.** $4H$ **101.** (a) 3.2 s; (b) 1.3 s **103.** (a) 10.2 s; (b) 10.0 m **105.** (a) 8.85 m/s; (b) 1.00 m **107.** (a) 2.0 m/s^2; (b) 12 m/s; (c) 45 m **109.** 3.75 ms **111.** (a) 5.44 s; (b) 53.3 m/s; (c) 5.80 m **113.** (a) 9.08 m/s^2; (b) $0.926g$; (c) 6.12 s; (d) $15.3T_r$; (e) braking; (f) 5.56 m

Chapter 3

CP **1.** (a) 7 m ($\vec{a}$ and $\vec{b}$ are in same direction); (b) 1 m ($\vec{a}$ and $\vec{b}$ are in opposite directions) **2.** c, d, f (components must be head-to-tail; $\vec{a}$ must extend from tail of one component to head of the other) **3.** (a) +, +; (b) +, −; (c) +, + (draw vector from tail of $\vec{d}_1$ to head of $\vec{d}_2$) **4.** (a) 90°; (b) 0° (vectors are parallel—same direction); (c) 180° (vectors are antiparallel—opposite directions) **5.** (a) 0° or 180°; (b) 90° **Q** **1.** Either the sequence $\vec{d}_2, \vec{d}_1$ or the sequence $\vec{d}_2, \vec{d}_2, \vec{d}_3$ **3.** yes, when the vectors are in same direction **5.** (a) yes; (b) yes; (c) no **7.** all but (e) **9.** (a) $+x$ for (1), $+z$ for (2), $+z$ for (3); (b) $-x$ for (1), $-z$ for (2), $-z$ for (3) **P** **1.** (a) 47.2 m; (b) 122° **3.** (a) -2.5 m; (b) -6.9 m **5.** (a) 156 km; (b) 39.8° west of due north **7.** (a) 6.42 m; (b) no; (c) yes; (d) yes; (e) a possible answer: $(4.30\text{ m})\hat{i} + (3.70\text{ m})\hat{j} + (3.00\text{ m})\hat{k}$; (f) 7.96 m **9.** (a) $(-9.0\text{ m})\hat{i} + (10\text{ m})\hat{j}$; (b) 13 m; (c) 132° **11.** 4.74 km **13.** (a) $(3.0\text{ m})\hat{i} - (2.0\text{ m})\hat{j} + (5.0\text{ m})\hat{k}$; (b) $(5.0\text{ m})\hat{i} - (4.0\text{ m})\hat{j} - (3.0\text{ m})\hat{k}$; (c) $(-5.0\text{ m})\hat{i} + (4.0\text{ m})\hat{j} + (3.0\text{ m})\hat{k}$ **15.** (a) -70.0 cm; (b) 80.0 cm; (c) 141 cm; (d) $-172°$ **17.** (a) 1.59 m; (b) 12.1 m; (c) 12.2 m; (d) 82.5° **19.** (a) 38 m; (b) $-37.5°$; (c) 130 m; (d) 1.2°; (e) 62 m; (f) 130° **21.** 5.39 m at 21.8° left of forward **23.** 2.6 km **25.** 3.2 **27.** (a) 7.5 cm; (b) 90°; (c) 8.6 cm; (d) 48° **29.** (a) $8\hat{i} + 16\hat{j}$; (b) $2\hat{i} + 4\hat{j}$ **31.** (a) $a\hat{i} + a\hat{j} + a\hat{k}$; (b) $-a\hat{i} + a\hat{j} + a\hat{k}$; (c) $a\hat{i} - a\hat{j} + a\hat{k}$; (d) $-a\hat{i} - a\hat{j} + a\hat{k}$; (e) 54.7°; (f) $3^{0.5}a$ **33.** (a) -18.8 units; (b) 26.9 units, $+z$ direction **35.** (a) -21; (b) -9; (c) $5\hat{i} - 11\hat{j} - 9\hat{k}$ **37.** (a) 12; (b) $+z$; (c) 12; (d) $-z$; (e) 12; (f) $+z$ **39.** 22° **41.** 70.5° **43.** (a) 3.00 m; (b) 0; (c) 3.46 m; (d) 2.00 m; (e) -5.00 m; (f) 8.66 m; (g) -6.67; (h) 4.33 **45.** (a) 27.8 m; (b) 13.4 m **47.** (a) 30; (b) 52 **49.** (a) -2.83 m; (b) -2.83 m; (c) 5.00 m; (d) 0; (e) 3.00 m; (f) 5.20 m; (g) 5.17 m; (h) 2.37 m; (i) 5.69 m; (j) 25° north of due east; (k) 5.69 m; (l) 25° south of due west **51.** (a) 103 km; (b) 60.9° north of due west **53.** (a) 140°; (b) 90.0°; (c) 99.1° **55.** (a) -83.4; (b) $(1.14 \times 10^3)\hat{k}$; (c) 1.14×10^3, θ not defined, $\phi = 0°$; (d) 90.0°; (e) $-5.14\hat{i} + 6.13\hat{j} + 3.00\hat{k}$; (f) 8.54, $\theta = 130°$, $\phi = 69.4°$ **57.** (a) 3.0 m^2; (b) 52 m^3; (c) $(11\text{ m}^2)\hat{i} + (9.0\text{ m}^2)\hat{j} + (3.0\text{ m}^2)\hat{k}$ **59.** (a) $+y$; (b) $-y$; (c) 0; (d) 0; (e) $+z$; (f) $-z$; (g) ab; (h) ab; (i) ab/d; (j) $+z$ **61.** (a) 0; (b) 0; (c) -1; (d) west; (e) up; (f) west **63.** Walpole (where the state prison is located) **65.** (a) $(9.19\text{ m})\hat{i}' + (7.71\text{ m})\hat{j}'$; (b) $(14.0\text{ m})\hat{i}' + (3.41\text{ m})\hat{j}'$ **67.** (a) $11\hat{i} + 5.0\hat{j} - 7.0\hat{k}$; (b) 120°; (c) -4.9; (d) 7.3 **69.** (a) $(-40\hat{i} - 20\hat{j} + 25\hat{k})$ m; (b) 45 m **71.** 4.1

Chapter 4

CP **1.** (draw $\vec{v}$ tangent to path, tail on path) (a) first; (b) third **2.** (take second derivative with respect to time) (1) and (3) a_x

and a_y are both constant and thus $\vec{a}$ is constant; (2) and (4) a_y is constant but a_x is not, thus $\vec{a}$ is not **3.** no **4.** (a) v_x constant; (b) v_y initially positive, decreases to zero, and then becomes progressively more negative; (c) $a_x = 0$ throughout; (d) $a_y = -g$ throughout **5.** (a) $-(4\text{ m/s})\hat{i}$; (b) $-(8\text{ m/s}^2)\hat{j}$ **Q 1.** (a) $(7\text{ m})\hat{i} + (1\text{ m})\hat{j} + (-2\text{ m})\hat{k}$; (b) $(5\text{ m})\hat{i} + (-3\text{ m})\hat{j} + (1\text{ m})\hat{k}$; (c) $(-2\text{ m})\hat{i}$ **3.** (a) all tie; (b) 1 and 2 tie (the rocket is shot upward), then 3 and 4 tie (it is shot into the ground!) **5.** decreases **7.** (a) all tie; (b) all tie; (c) 3, 2, 1; (d) 3, 2, 1 **9.** (a) 0; (b) 350 km/h; (c) 350 km/h; (d) same (nothing changed about the vertical motion) **11.** (a) 90° and 270°; (b) 0° and 180°; (c) 90° and 270° **13.** 2, then 1 and 4 tie, then 3 **P 1.** $(-2.0\text{ m})\hat{i} + (6.0\text{ m})\hat{j} - (10\text{ m})\hat{k}$ **3.** (a) 6.2 m **5.** $(-0.70\text{ m/s})\hat{i} + (1.4\text{ m/s})\hat{j} - (0.40\text{ m/s})\hat{k}$ **7.** (a) 7.59 km/h; (b) 22.5° east of due north **9.** (a) 0.83 cm/s; (b) 0°; (c) 0.11 m/s; (d) −63° **11.** (a) $(8\text{ m/s}^2)t\hat{j} + (1\text{ m/s})\hat{k}$; (b) $(8\text{ m/s}^2)\hat{j}$ **13.** (a) $(6.00\text{ m})\hat{i} - (106\text{ m})\hat{j}$; (b) $(19.0\text{ m/s})\hat{i} - (224\text{ m/s})\hat{j}$; (c) $(24.0\text{ m/s}^2)\hat{i} - (336\text{ m/s}^2)\hat{j}$; (d) −85.2° **15.** $(32\text{ m/s})\hat{i}$ **17.** (a) $(-1.50\text{ m/s})\hat{j}$; (b) $(4.50\text{ m})\hat{i} - (2.25\text{ m})\hat{j}$ **19.** (a) $(72.0\text{ m})\hat{i} + (90.7\text{ m})\hat{j}$; (b) 49.5° **21.** (a) 3.03 s; (b) 758 m; (c) 29.7 m/s **23.** 43.1 m/s (155 km/h) **25.** (a) 18 cm; (b) 1.9 m **27.** (a) 10.0 s; (b) 897 m **29.** (a) 1.60 m; (b) 6.86 m; (c) 2.86 m **31.** (a) 202 m/s; (b) 806 m; (c) 161 m/s; (d) −171 m/s **33.** 3.35 m **35.** 78.5° **37.** (a) 11 m; (b) 23 m; (c) 17 m/s; (d) 63° **39.** 4.84 cm **41.** (a) 32.3 m; (b) 21.9 m/s; (c) 40.4°; (d) below **43.** (a) ramp; (b) 5.82 m; (c) 31.0° **45.** 64.8° **47.** (a) yes; (b) 2.56 m **49.** (a) 2.3°; (b) 1.4 m; (c) 18° **51.** (a) 31°; (b) 63° **53.** the third **55.** (a) 75.0 m; (b) 31.9 m/s; (c) 66.9°; (d) 25.5 m **57.** (a) 12 s; (b) 4.1 m/s²; (c) down; (d) 4.1 m/s²; (e) up **59.** (a) 1.3×10^5 m/s; (b) 7.9×10^5 m/s²; (c) increase **61.** (a) 7.32 m; (b) west; (c) north **63.** $(3.00\text{ m/s}^2)\hat{i} + (6.00\text{ m/s}^2)\hat{j}$ **65.** 2.92 m **67.** 160 m/s² **69.** (a) 13 m/s²; (b) eastward; (c) 13 m/s²; (d) eastward **71.** 1.67 **73.** (a) 38 knots; (b) 1.5° east of due north; (c) 4.2 h; (d) 1.5° west of due south **75.** 60° **77.** 32 m/s **79.** (a) $(80\text{ km/h})\hat{i} - (60\text{ km/h})\hat{j}$; (b) 0°; (c) answers do not change **81.** (a) $(-32\text{ km/h})\hat{i} - (46\text{ km/h})\hat{j}$; (b) $[(2.5\text{ km}) - (32\text{ km/h})t]\hat{i} + [(4.0\text{ km}) - (46\text{ km/h})t]\hat{j}$; (c) 0.084 h; (d) 2×10^2 m **83.** (a) 2.7 km; (b) 76° clockwise **85.** 2.64 m **87.** (a) 2.5 m; (b) 0.82 m; (c) 9.8 m/s²; (d) 9.8 m/s² **89.** (a) −30°; (b) 69 min; (c) 80 min; (d) 80 min; (e) 0°; (f) 60 min **91.** (a) 62 ms; (b) 4.8×10^2 m/s **93.** (a) 6.7×10^6 m/s; (b) 1.4×10^{-7} s **95.** (a) 4.2 m, 45°; (b) 5.5 m, 68°; (c) 6.0 m, 90°; (d) 4.2 m, 135°; (e) 0.85 m/s, 135°; (f) 0.94 m/s, 90°; (g) 0.94 m/s, 180°; (h) 0.30 m/s², 180°; (i) 0.30 m/s², 270° **97.** (a) 6.79 km/h; (b) 6.96° **99.** (a) 16 m/s; (b) 23°; (c) above; (d) 27 m/s; (e) 57°; (f) below **101.** (a) 24 m/s; (b) 65° **103.** (a) 1.5; (b) (36 m, 54 m) **105.** (a) 0.034 m/s²; (b) 84 min **107.** (a) 44 m; (b) 13 m; (c) 8.9 m **109.** (a) 2.6×10^2 m/s; (b) 45 s; (c) increase **111.** (a) 45 m; (b) 22 m/s **113.** (a) 2.00 ns; (b) 2.00 mm; (c) 1.00×10^7 m/s; (d) 2.00×10^6 m/s **115.** (a) 4.6×10^{12} m; (b) 2.4×10^5 s **117.** 93° from the car's direction of motion **119.** (a) 8.43 m; (b) −129° **121.** (a) 63 km; (b) 18° south of due east; (c) 0.70 km/h; (d) 18° south of due east; (e) 1.6 km/h; (f) 1.2 km/h; (g) 33° north of due east **123.** 3×10^1 m **125.** (a) 14 m/s; (b) 14 m/s; (c) −10 m; (d) −4.9 m; (e) +10 m; (f) −4.9 m **127.** 67 km/h **129.** (a) from 75° east of due south; (b) 30° east of due north. For a second set of solutions, substitute west for east in both answers. **131.** (a) 11 m; (b) 45 m/s

Chapter 5

CP 1. c, d, and e ($\vec{F}_1$ and $\vec{F}_2$ must be head-to-tail, $\vec{F}_{net}$ must be from tail of one of them to head of the other) **2.** (a) and (b) 2 N, leftward (acceleration is zero in each situation) **3.** (a) equal; (b) greater (acceleration is upward, thus net force on body must be upward) **4.** (a) equal; (b) greater; (c) less **5.** (a) increase; (b) yes; (c) same; (d) yes **Q 1.** increase **3.** (a) 2 and 4; (b) 2 and 4 **5.** (a) 2, 3, 4; (b) 1, 3, 4; (c) 1, $+y$; 2, $+x$; 3, fourth quadrant; 4, third quadrant **7.** (a) 20 kg; (b) 18 kg; (c) 10 kg; (d) all tie; (e) 3, 2, 1 **9.** (a) increases from initial value mg; (b) decreases from mg to zero (after which the block moves up away from the floor) **11.** (a) M; (b) M; (c) M; (d) $2M$; (e) $3M$ **P 1.** (a) 1.88 N; (b) 0.684 N; (c) $(1.88\text{ N})\hat{i} + (0.684\text{ N})\hat{j}$ **3.** 2.9 m/s² **5.** (a) $(-32.0\text{ N})\hat{i} - (20.8\text{ N})\hat{j}$; (b) 38.2 N; (c) −147° **7.** (a) $(0.86\text{ m/s}^2)\hat{i} - (0.16\text{ m/s}^2)\hat{j}$; (b) 0.88 m/s²; (c) −11° **9.** 9.0 m/s² **11.** (a) 8.37 N; (b) −133°; (c) −125° **13.** (a) 108 N; (b) 108 N; (c) 108 N **15.** (a) 4.0 kg; (b) 1.0 kg; (c) 4.0 kg; (d) 1.0 kg **17.** (a) $-9.80\hat{j}$ m/s²; (b) $2.35\hat{j}$ m/s²; (c) 1.37 s; (d) $(-5.56 \times 10^{-3}\text{ N})\hat{j}$; (e) $(1.333 \times 10^{-3}\text{ N})\hat{j}$ **19.** (a) 42 N; (b) 72 N; (c) 4.9 m/s² **21.** (a) 11.7 N; (b) −59.0° **23.** (a) 0.022 m/s²; (b) 8.3×10^4 km; (c) 1.9×10^3 m/s **25.** 1.2×10^5 N **27.** (a) 494 N; (b) up; (c) 494 N; (d) down **29.** 1.5 mm **31.** (a) 46.7°; (b) 28.0° **33.** (a) 0.62 m/s²; (b) 0.13 m/s²; (c) 2.6 m **35.** (a) 1.18 m; (b) 0.674 s; (c) 3.50 m/s **37.** (a) 2.2×10^{-3} N; (b) 3.7×10^{-3} N **39.** 1.8×10^4 N **41.** (a) 31.3 kN; (b) 24.3 kN **43.** (a) 1.4 m/s²; (b) 4.1 m/s **45.** (a) 1.23 N; (b) 2.46 N; (c) 3.69 N; (d) 4.92 N; (e) 6.15 N; (f) 0.250 N **47.** (a) 2.18 m/s²; (b) 116 N; (c) 21.0 m/s² **49.** 6.4×10^3 N **51.** (a) 0.970 m/s²; (b) 11.6 N; (c) 34.9 N **53.** (a) 1.1 N **55.** (a) 3.6 m/s²; (b) 17 N **57.** (a) 4.9 m/s²; (b) 2.0 m/s²; (c) up; (d) 120 N **59.** (a) 0.735 m/s²; (b) down; (c) 20.8 N **61.** $2Ma/(a + g)$ **63.** (a) 0.653 m/s³; (b) 0.896 m/s³; (c) 6.50 s **65.** 81.7 N **67.** (a) 8.0 m/s; (b) $+x$ **69.** (a) 13 597 kg; (b) 4917 L; (c) 6172 kg; (d) 20 075 L; (e) 45% **71.** (a) 0; (b) 0.83 m/s²; (c) 0 **73.** (a) 0.74 m/s²; (b) 7.3 m/s² **75.** (a) rope breaks; (b) 1.6 m/s² **77.** 2.4 N **79.** (a) 4.6 m/s²; (b) 2.6 m/s² **81.** (a) 65 N; (b) 49 N **83.** (a) 11 N; (b) 2.2 kg; (c) 0; (d) 2.2 kg **85.** (a) 4.6×10^3 N; (b) 5.8×10^3 N **87.** (a) 4 kg; (b) 6.5 m/s²; (c) 13 N **89.** 195 N **91.** (a) 44 N; (b) 78 N; (c) 54 N; (d) 152 N **93.** 16 N **95.** (a) 1.8×10^2 N; (b) 6.4×10^2 N **97.** (a) $(5.0\text{ m/s})\hat{i} + (4.3\text{ m/s})\hat{j}$; (b) $(15\text{ m})\hat{i} + (6.4\text{ m})\hat{j}$ **99.** 16 N **101.** (a) 2.6 N; (b) 17° **103.** (a) 4.1 m/s²; (b) 836 N

Chapter 6

CP 1. (a) zero (because there is no attempt at sliding); (b) 5 N; (c) no; (d) yes; (e) 8 N **2.** ($\vec{a}$ is directed toward center of circular path) (a) $\vec{a}$ downward, $\vec{F}_N$ upward; (b) $\vec{a}$ and $\vec{F}_N$ upward **Q 1.** (a) same; (b) increases; (c) increases; (d) no **3.** (a) decrease; (b) decrease; (c) increase; (d) increase; (e) increase **5.** (a) upward; (b) horizontal, toward you; (c) no change; (d) increases; (e) increases **7.** At first, $\vec{f}_s$ is directed up the ramp and its magnitude increases from $mg \sin \theta$ until it reaches $f_{s,max}$. Thereafter the force is kinetic friction directed up the ramp, with magnitude f_k (a constant value smaller than $f_{s,max}$). **9.** (a) all tie; (b) all tie; (c) 2, 3, 1 **11.** 4, 3, then 1, 2, and 5 tie **P 1.** (a) 2.0×10^2 N; (b) 1.2×10^2 N **3.** (a) 1.9×10^2 N; (b) 0.56 m/s² **5.** 36 m **7.** (a) 11 N; (b) 0.14 m/s² **9.** (a) 6.0 N; (b) 3.6 N; (c) 3.1 N **11.** (a) 1.3×10^2 N; (b) no; (c) 1.1×10^2 N; (d) 46 N; (e) 17 N **13.** (a) 3.0×10^2 N; (b) 1.3 m/s² **15.** 2° **17.** (a) no; (b) $(-12\text{ N})\hat{i} + (5.0\text{ N})\hat{j}$ **19.** (a) 19°; (b) 3.3 kN **21.** (a) $(17\text{ N})\hat{i}$; (b) $(20\text{ N})\hat{i}$; (c) $(15\text{ N})\hat{i}$ **23.** 1.0×10^2 N **25.** 0.37 **27.** (a) 3.5 m/s²; (b) 0.21 N **29.** (a) 0; (b) $(-3.9\text{ m/s}^2)\hat{i}$; (c) $(-1.0\text{ m/s}^2)\hat{i}$ **31.** (a) 66 N; (b) 2.3 m/s² **33.** 4.9×10^2 N **35.** 9.9 s **37.** 2.3 **39.** (a) 3.2×10^2 km/h; (b) 6.5×10^2 km/h; (c) no **41.** 21 m **43.** 0.60 **45.** (a) 10 s; (b) 4.9×10^2 N; (c) 1.1×10^3 N **47.** 1.37×10^3 N **49.** (a) light; (b) 778 N;

(c) 223 N; (d) 1.11 kN **51.** 12° **53.** 2.2 km **55.** 1.81 m/s **57.** 2.6×10^3 N **59.** (a) 8.74 N; (b) 37.9 N; (c) 6.45 m/s; (d) radially inward **61.** (a) 69 km/h; (b) 139 km/h; (c) yes **63.** (a) 7.5 m/s²; (b) down; (c) 9.5 m/s²; (d) down **65.** (a) 27 N; (b) 3.0 m/s² **67.** (a) 35.3 N; (b) 39.7 N; (c) 320 N **69.** $g(\sin\theta - 2^{0.5}\mu_k \cos\theta)$ **71.** (a) 3.0×10^5 N; (b) 1.2° **73.** 147 m/s **75.** (a) 56 N; (b) 59 N; (c) 1.1×10^3 N **77.** (a) 275 N; (b) 877 N **79.** (b) 240 N; (c) 0.60 **81.** (a) 13 N; (b) 1.6 m/s² **83.** 0.76 **85.** (a) 3.21×10^3 N; (b) yes **87.** 3.4 m/s² **89.** (a) 84.2 N; (b) 52.8 N; (c) 1.87 m/s² **91.** (a) 222 N; (b) 334 N; (c) 311 N; (d) 311 N; (e) c, d **93.** (a) 6.80 s; (b) 6.76 s **95.** 3.4% **97.** (a) $\mu_k mg/(\sin\theta - \mu_k \cos\theta)$; (b) $\theta_0 = \tan^{-1}\mu_s$ **99.** (a) $v_0^2/(4g\sin\theta)$; (b) no **101.** (a) 30 cm/s; (b) 180 cm/s²; (c) inward; (d) 3.6×10^{-3} N; (e) inward; (f) 0.37 **103.** (a) 0.34; (b) 0.24 **105.** 0.18 **107.** 0.56 **109.** (a) 2.1 m/s²; (b) down the plane; (c) 3.9 m; (d) at rest

Chapter 7

CP **1.** (a) decrease; (b) same; (c) negative, zero **2.** (a) positive; (b) negative; (c) zero **3.** zero **Q** **1.** (a) positive; (b) negative; (c) negative **3.** all tie **5.** all tie **7.** *b* (positive work), *a* (zero work), *c* (negative work), *d* (more negative work) **9.** (a) *A*; (b) *B* **P** **1.** (a) 5×10^{14} J; (b) 0.1 megaton TNT; (c) 8 bombs **3.** (a) 2.9×10^7 m/s; (b) 2.1×10^{-13} J **5.** (a) 2.4 m/s; (b) 4.8 m/s **7.** 20 J **9.** 0.96 J **11.** (a) 1.7×10^2 N; (b) 3.4×10^2 m; (c) -5.8×10^4 J; (d) 3.4×10^2 N; (e) 1.7×10^2 m; (f) -5.8×10^4 J **13.** (a) 1.50 J; (b) increases **15.** (a) 62.3°; (b) 118° **17.** (a) 12 kJ; (b) −11 kJ; (c) 1.1 kJ; (d) 5.4 m/s **19.** (a) $-3Mgd/4$; (b) Mgd; (c) $Mgd/4$; (d) $(gd/2)^{0.5}$ **21.** 4.41 J **23.** 25 J **25.** (a) 25.9 kJ; (b) 2.45 N **27.** (a) 7.2 J; (b) 7.2 J; (c) 0; (d) −25 J **29.** (a) 6.6 m/s; (b) 4.7 m **31.** (a) 0.90 J; (b) 2.1 J; (c) 0 **33.** (a) 0.12 m; (b) 0.36 J; (c) −0.36 J; (d) 0.060 m; (e) 0.090 J **35.** (a) 0; (b) 0 **37.** 5.3×10^2 J **39.** (a) 42 J; (b) 30 J; (c) 12 J; (d) 6.5 m/s, $+x$ axis; (e) 5.5 m/s, $+x$ axis; (f) 3.5 m/s, $+x$ axis **41.** 4.00 N/m **43.** 4.9×10^2 W **45.** (a) 0.83 J; (b) 2.5 J; (c) 4.2 J; (d) 5.0 W **47.** 7.4×10^2 W **49.** (a) 1.0×10^2 J; (b) 8.4 W **51.** (a) 32.0 J; (b) 8.00 W; (c) 78.2° **53.** (a) 1×10^5 megatons TNT; (b) 1×10^7 bombs **55.** −6 J **57.** (a) 98 N; (b) 4.0 cm; (c) 3.9 J; (d) −3.9 J **59.** −37 J **61.** 165 kW **63.** (a) 1.8×10^5 ft · lb; (b) 0.55 hp **65.** (a) 797 N; (b) 0; (c) −1.55 kJ; (d) 0; (e) 1.55 kJ; (f) *F* varies during displacement **67.** (a) 1.20 J; (b) 1.10 m/s **69.** (a) 314 J; (b) −155 J; (c) 0; (d) 158 J **71.** (a) 23 mm; (b) 45 N **73.** 235 kW **75.** (a) 13 J; (b) 13 J **77.** (a) 0.6 J; (b) 0; (c) −0.6 J **79.** (a) 6 J; (b) 6.0 J

Chapter 8

CP **1.** no (consider round trip on the small loop) **2.** 3, 1, 2 (see Eq. 8-6) **3.** (a) all tie; (b) all tie **4.** (a) *CD*, *AB*, *BC* (0) (check slope magnitudes); (b) positive direction of *x* **5.** all tie **Q** **1.** (a) 12 J; (b) −2 J **3.** (a) 3, 2, 1; (b) 1, 2, 3 **5.** 2, 1, 3 **7.** +30 J **9.** (a) increasing; (b) decreasing; (c) decreasing; (d) constant in *AB* and *BC*, decreasing in *CD* **P** **1.** (a) 167 J; (b) −167 J; (c) 196 J; (d) 29 J; (e) 167 J; (f) −167 J; (g) 296 J; (h) 129 J **3.** (a) 4.31 mJ; (b) −4.31 mJ; (c) 4.31 mJ; (d) −4.31 mJ; (e) all increase **5.** 89 N/cm **7.** (a) 13.1 J; (b) −13.1 J; (c) 13.1 J; (d) all increase **9.** (a) 2.6×10^2 m; (b) same; (c) decrease **11.** (a) 2.08 m/s; (b) 2.08 m/s; (c) increase **13.** (a) 17.0 m/s; (b) 26.5 m/s; (c) 33.4 m/s; (d) 56.7 m; (e) all the same **15.** (a) 0.98 J; (b) −0.98 J; (c) 3.1 N/cm **17.** (a) 8.35 m/s; (b) 4.33 m/s; (c) 7.45 m/s; (d) both decrease **19.** (a) 2.5 N; (b) 0.31 N; (c) 30 cm **21.** (a) 4.85 m/s; (b) 2.42 m/s **23.** -3.2×10^2 J **25.** (a) no; (b) 9.3×10^2 N **27.** (a) 784 N/m; (b) 62.7 J; (c) 62.7 J; (d) 80.0 cm **29.** (a) 39.2 J; (b) 39.2 J; (c) 4.00 m **31.** (a) 35 cm; (b) 1.7 m/s **33.** (a) 2.40 m/s; (b) 4.19 m/s **35.** −18 mJ **37.** (a) 39.6 cm; (b) 3.64 cm **39.** (a) 2.1 m/s; (b) 10 N; (c) $+x$ direction; (d) 5.7 m; (e) 30 N; (f) $-x$ direction **41.** (a) −3.7 J; (c) 1.3 m; (d) 9.1 m; (e) 2.2 J; (f) 4.0 m; (g) $(4 - x)e^{-x/4}$; (h) 4.0 m **43.** (a) 5.6 J; (b) 3.5 J **45.** (a) 30.1 J; (b) 30.1 J; (c) 0.225 **47.** (a) −2.9 kJ; (b) 3.9×10^2 J; (c) 2.1×10^2 N **49.** 0.53 J **51.** (a) 1.5 MJ; (b) 0.51 MJ; (c) 1.0 MJ; (d) 63 m/s **53.** 1.2 m **55.** (a) 67 J; (b) 67 J; (c) 46 cm **57.** (a) 1.5×10^{-2} N; (b) $(3.8 \times 10^2)g$ **59.** (a) −0.90 J; (b) 0.46 J; (c) 1.0 m/s **61.** (a) 19.4 m; (b) 19.0 m/s **63.** 20 cm **65.** (a) 7.4 m/s; (b) 90 cm; (c) 2.8 m; (d) 15 m **67.** (a) 10 m; (b) 49 N; (c) 4.1 m; (d) 1.2×10^2 N **69.** 4.33 m/s **71.** (a) 5.5 m/s; (b) 5.4 m; (c) same **73.** (a) 109 J; (b) 60.3 J; (c) 68.2 J; (d) 41.0 J **75.** 3.7 J **77.** 15 J **79.** (a) 2.7 J; (b) 1.8 J; (c) 0.39 m **81.** 80 mJ **83.** (a) 7.0 J; (b) 22 J **85.** (a) 7.4×10^2 J; (b) 2.4×10^2 J **87.** 25 J **89.** 24 W **91.** −12 J **93.** (a) 8.8 m/s; (b) 2.6 kJ; (c) 1.6 kW **95.** (a) 300 J; (b) 93.8 J; (c) 6.38 m **97.** 738 m **99.** (a) −0.80 J; (b) −0.80 J; (c) +1.1 J **101.** (a) 2.35×10^3 J; (b) 352 J **103.** (a) −3.8 kJ; (b) 31 kN **105.** (a) 2.1×10^6 kg; (b) $(100 + 1.5t)^{0.5}$ m/s; (c) $(1.5 \times 10^6)/(100 + 1.5t)^{0.5}$ N; (d) 6.7 km **107.** (a) 5.6 J; (b) 12 J; (c) 13 J **109.** (a) 4.9 m/s; (b) 4.5 N; (c) 71°; (d) same **111.** (a) 1.2 J; (b) 11 m/s; (c) no; (d) no **113.** 54% **115.** (a) 2.7×10^9 J; (b) 2.7×10^9 W; (c) $\$2.4 \times 10^8$ **117.** (a) 5.00 J; (b) 9.00 J; (c) 11.0 J; (d) 3.00 J; (e) 12.0 J; (f) 2.00 J; (g) 13.0 J; (h) 1.00 J; (i) 13.0 J; (j) 1.00 J; (l) 11.0 J; (m) 10.8 m; (n) It returns to $x = 0$ and stops. **119.** (a) 3.7 J; (b) 4.3 J; (c) 4.3 J **121.** (a) 4.8 N; (b) $+x$ direction; (c) 1.5 m; (d) 13.5 m; (e) 3.5 m/s **123.** (a) 24 kJ; (b) 4.7×10^2 N **125.** (a) 3.0 mm; (b) 1.1 J; (d) yes; (e) ≈ 40 J; (f) no **127.** (a) 6.0 kJ; (b) 6.0×10^2 W; (c) 3.0×10^2 W; (d) 9.0×10^2 W **129.** 3.1×10^{11} W **131.** 880 MW **133.** (a) $v_0 = (2gL)^{0.5}$; (b) $5mg$; (c) $-mgL$; (d) $-2mgL$ **135.** because your force on the cabbage (as you lower it) does work

Chapter 9

CP **1.** (a) origin; (b) fourth quadrant; (c) on *y* axis below origin; (d) origin; (e) third quadrant; (f) origin **2.** (a)–(c) at the center of mass, still at the origin (their forces are internal to the system and cannot move the center of mass) **3.** (Consider slopes and Eq. 9-23.) (a) 1, 3, and then 2 and 4 tie (zero force); (b) 3 **4.** (a) unchanged; (b) unchanged (see Eq. 9-32); (c) decrease (Eq. 9-35) **5.** (a) zero; (b) positive (initial p_y down *y*; final p_y up *y*); (c) positive direction of *y* **6.** (No net external force; $\vec{P}$ conserved.) (a) 0; (b) no; (c) $-x$ **7.** (a) 10 kg · m/s; (b) 14 kg · m/s; (c) 6 kg · m/s **8.** (a) 4 kg · m/s; (b) 8 kg · m/s; (c) 3 J **9.** (a) 2 kg · m/s (conserve momentum along *x*); (b) 3 kg · m/s (conserve momentum along *y*) **Q** **1.** (a) 2 N, rightward; (b) 2 N, rightward; (c) greater than 2 N, rightward **3.** (a) *x* yes, *y* no; (b) *x* yes, *y* no; (c) *x* no, *y* yes **5.** b, c, a **7.** (a) one was stationary; (b) 2; (c) 5; (d) equal (pool player's result) **9.** (a) *C*; (b) *B*; (c) 3 **11.** (a) *c*, kinetic energy cannot be negative; *d*, total kinetic energy cannot increase; (b) *a*; (c) *b* **P** **1.** (a) −1.50 m; (b) −1.43 m **3.** (a) −0.45 cm; (b) −2.0 cm **5.** (a) 0; (b) 3.13×10^{-11} m **7.** (a) −6.5 cm; (b) 8.3 cm; (c) 1.4 cm **9.** $(-4.0\text{ m})\hat{i} + (4.0\text{ m})\hat{j}$ **11.** (a) 28 cm; (b) 2.3 m/s **13.** (a) $(2.35\hat{i} - 1.57\hat{j})$ m/s²; (b) $(2.35\hat{i} - 1.57\hat{j})t$ m/s, with *t* in seconds; (d) straight, at downward angle 34° **15.** 53 m **17.** 4.2 m **19.** (a) 7.5×10^4 J; (b) 3.8×10^4 kg · m/s;

(c) 39° south of due east **21.** (a) 5.0 kg · m/s; (b) 10 kg · m/s **23.** (a) 67 m/s; (b) $-x$; (c) 1.2 kN; (d) $-x$ **25.** 1.0 × 10³ to 1.2 × 10³ kg · m/s **27.** (a) 42 N · s; (b) 2.1 kN **29.** 5 N **31.** (a) 5.86 kg · m/s; (b) 59.8°; (c) 2.93 kN; (d) 59.8° **33.** (a) 2.39 × 10³ N · s; (b) 4.78 × 10⁵ N; (c) 1.76 × 10³ N · s; (d) 3.52 × 10⁵ N **35.** (a) 9.0 kg · m/s; (b) 3.0 kN; (c) 4.5 kN; (d) 20 m/s **37.** 9.9 × 10² N **39.** 3.0 mm/s **41.** 55 cm **43.** (a) $-(0.15\text{ m/s})\hat{i}$; (b) 0.18 m **45.** (a) 14 m/s; (b) −45° **47.** (a) $(1.00\hat{i} - 0.167\hat{j})$ km/s; (b) 3.23 MJ **49.** 3.1 × 10² m/s **51.** (a) 33%; (b) 23%; (c) decreases **53.** (a) 721 m/s; (b) 937 m/s **55.** (a) 4.4 m/s; (b) 0.80 **57.** (a) +2.0 m/s; (b) −1.3 J; (c) +40 J; (d) system got energy from some source, such as a small explosion **59.** 25 cm **61.** (a) 99 g; (b) 1.9 m/s; (c) 0.93 m/s **63.** (a) 1.2 kg; (b) 2.5 m/s **65.** −28 cm **67.** (a) 3.00 m/s; (b) 6.00 m/s **69.** (a) 0.21 kg; (b) 7.2 m **71.** (a) 433 m/s; (b) 250 m/s **73.** (a) 4.15 × 10⁵ m/s; (b) 4.84 × 10⁵ m/s **75.** 120° **77.** (a) 1.57 × 10⁶ N; (b) 1.35 × 10⁵ kg; (c) 2.08 km/s **79.** (a) 46 N; (b) none **81.** (a) 1.78 m/s; (b) less; (c) less; (d) greater **83.** (a) 1.92 m; (b) 0.640 m **85.** 28.8 N **87.** 1.10 m/s **89.** (a) 7290 m/s; (b) 8200 m/s; (c) 1.271 × 10¹⁰ J; (d) 1.275 × 10¹⁰ J **91.** (a) 1.0 kg · m/s; (b) 2.5 × 10² J; (c) 10 N; (d) 1.7 kN; (e) answer for (c) includes time between pellet collisions **93.** (a) $(7.4 \times 10^3\text{ N}\cdot\text{s})\hat{i} - (7.4 \times 10^3\text{ N}\cdot\text{s})\hat{j}$; (b) $(-7.4 \times 10^3\text{ N}\cdot\text{s})\hat{i}$; (c) 2.3 × 10³ N; (d) 2.1 × 10⁴ N; (e) −45° **95.** (a) 3.7 m/s; (b) 1.3 N · s; (c) 1.8 × 10² N **97.** 1.18 × 10⁴ kg **99.** +4.4 m/s **101.** (a) 1.9 m/s; (b) −30°; (c) elastic **103.** (a) 6.9 m/s; (b) 30°; (c) 6.9 m/s; (d) −30°; (e) 2.0 m/s; (f) −180° **105.** (a) 25 mm; (b) 26 mm; (c) down; (d) 1.6 × 10⁻² m/s² **107.** (a) 0.745 mm; (b) 153°; (c) 1.67 mJ **109.** (a) $(2.67\text{ m/s})\hat{i} + (-3.00\text{ m/s})\hat{j}$; (b) 4.01 m/s; (c) 48.4° **111.** 0.22% **113.** 190 m/s **115.** (a) 4.6 × 10³ km; (b) 73% **117.** (a) 50 kg/s; (b) 1.6 × 10² kg/s **119.** (a) −0.50 m; (b) −1.8 cm; (c) 0.50 m **121.** (a) 0.800 kg · m/s; (b) 0.400 kg · m/s **123.** 29 J **125.** 5.0 × 10⁶ N **127.** (a) 1; (b) 1.83 × 10³; (c) 1.83 × 10³; (d) all the same **129.** 5.0 kg **131.** 2.2 kg **133.** (a) 11.4 m/s; (b) 95.1° **135.** (a) 0; (b) 0; (c) 0

Chapter 10

CP **1.** b and c **2.** (a) and (d) ($\alpha = d^2\theta/dt^2$ must be a constant) **3.** (a) yes; (b) no; (c) yes; (d) yes **4.** all tie **5.** 1, 2, 4, 3 (see Eq. 10-36) **6.** (see Eq. 10-40) 1 and 3 tie, 4, then 2 and 5 tie (zero) **7.** (a) downward in the figure ($\tau_{net} = 0$); (b) less (consider moment arms) **Q** **1.** (a) c, a, then b and d tie; (b) b, then a and c tie, then d **3.** c, a, b **5.** larger **7.** (a) decrease; (b) clockwise; (c) counterclockwise **9.** all tie **P** **1.** 14 rev **3.** 11 rad/s **5.** (a) 4.0 rad/s; (b) 11.9 rad/s **7.** (a) 4.0 m/s; (b) no **9.** (a) 30 s; (b) 1.8 × 10³ rad **11.** (a) 3.00 s; (b) 18.9 rad **13.** 8.0 s **15.** (a) 44 rad; (b) 5.5 s; (c) 32 s; (d) −2.1 s; (e) 40 s **17.** (a) 3.4 × 10² s; (b) −4.5 × 10⁻³ rad/s²; (c) 98 s **19.** 6.9 × 10⁻¹³ rad/s **21.** (a) 20.9 rad/s; (b) 12.5 m/s; (c) 800 rev/min²; (d) 600 rev **23.** (a) 2.50 × 10⁻³ rad/s; (b) 20.2 m/s²; (c) 0 **25.** (a) 40 s; (b) 2.0 rad/s² **27.** (a) 3.8 × 10³ rad/s; (b) 1.9 × 10² m/s **29.** (a) 7.3 × 10⁻⁵ rad/s; (b) 3.5 × 10² m/s; (c) 7.3 × 10⁻⁵ rad/s; (d) 4.6 × 10² m/s **31.** (a) 73 cm/s²; (b) 0.075; (c) 0.11 **33.** 12.3 kg · m² **35.** 0.097 kg · m² **37.** (a) 1.1 kJ; (b) 9.7 kJ **39.** (a) 0.023 kg · m²; (b) 11 mJ **41.** 4.7 × 10⁻⁴ kg · m² **43.** (a) 49 MJ; (b) 1.0 × 10² min **45.** 4.6 N · m **47.** −3.85 N · m **49.** (a) 28.2 rad/s²; (b) 338 N · m **51.** 0.140 N **53.** 2.51 × 10⁻⁴ kg · m² **55.** (a) 6.00 cm/s²; (b) 4.87 N; (c) 4.54 N; (d) 1.20 rad/s²; (e) 0.0138 kg · m² **57.** (a) 4.2 × 10² rad/s²; (b) 5.0 × 10² rad/s **59.** (a) 19.8 kJ; (b) 1.32 kW **61.** 396 N · m **63.** 5.42 m/s **65.** 9.82 rad/s **67.** (a) 5.32 m/s²; (b) 8.43 m/s²; (c) 41.8° **69.** (a) 314 rad/s²; (b) 7.54 m/s²; (c) 14.0 N; (d) 4.36 N **71.** 6.16 × 10⁻⁵ kg · m² **73.** (a) 1.57 m/s²; (b) 4.55 N; (c) 4.94 N **75.** (a) 4.81 × 10⁵ N; (b) 1.12 × 10⁴ N · m; (c) 1.25 × 10⁶ J **77.** 30 rev **79.** 3.1 rad/s **81.** (a) 0.791 kg · m²; (b) 1.79 × 10⁻² N · m **83.** (a) 2.3 rad/s²; (b) 1.4 rad/s² **85.** 1.4 × 10² N · m **87.** 4.6 rad/s² **89.** (a) −67 rev/min²; (b) 8.3 rev **93.** 0.054 kg · m² **95.** (a) 5.92 × 10⁴ m/s²; (b) 4.39 × 10⁴ s⁻² **97.** 2.6 J **99.** (a) 0.32 rad/s; (b) 1.0 × 10² km/h **101.** (a) 7.0 kg · m²; (b) 7.2 m/s; (c) 71° **103.** (a) 1.4 × 10² rad; (b) 14 s **105.** (a) 221 kg · m²; (b) 1.10 × 10⁴ J **107.** 0.13 rad/s **109.** 6.75 × 10¹² rad/s **111.** (a) 1.5 × 10² cm/s; (b) 15 rad/s; (c) 15 rad/s; (d) 75 cm/s; (e) 3.0 rad/s **113.** 18 rad **115.** (a) 10 J; (b) 0.27 m

Chapter 11

CP **1.** (a) same; (b) less **2.** less (consider the transfer of energy from rotational kinetic energy to gravitational potential energy) **3.** (draw the vectors, use right-hand rule) (a) $\pm z$; (b) $\pm y$; (c) $-x$ **4.** (see Eq. 11-21) (a) 1 and 3 tie; then 2 and 4 tie, then 5 (zero); (b) 2 and 3 **5.** (see Eqs. 11-23 and 11-16) (a) 3, 1; then 2 and 4 tie (zero); (b) 3 **6.** (a) all tie (same τ, same t, thus same ΔL); (b) sphere, disk, hoop (reverse order of I) **7.** (a) decreases; (b) same ($\tau_{net} = 0$, so L is conserved); (c) increases **Q** **1.** (a) 1, 2, 3 (zero); (b) 1 and 2 tie, then 3; (c) 1 and 3 tie, then 2 **3.** (a) spins in place; (b) rolls toward you; (c) rolls away from you **5.** a, then b and c tie, then e, d (zero) **7.** D, B, then A and C tie **9.** (a) same; (b) increase; (c) decrease; (d) same, decrease, increase **P** **1.** (a) 0; (b) $(22\text{ m/s})\hat{i}$; (c) $(-22\text{ m/s})\hat{i}$; (d) 0; (e) 1.5 × 10³ m/s²; (f) 1.5 × 10³ m/s²; (g) $(22\text{ m/s})\hat{i}$; (h) $(44\text{ m/s})\hat{i}$; (i) 0; (j) 0; (k) 1.5 × 10³ m/s²; (l) 1.5 × 10³ m/s² **3.** 0.020 **5.** −3.15 J **7.** (a) $(-4.0\text{ N})\hat{i}$; (b) 0.60 kg · m² **9.** (a) 63 rad/s; (b) 4.0 m **11.** 4.8 m **13.** (a) $-(0.11\text{ m})\omega$; (b) −2.1 m/s²; (c) −47 rad/s²; (d) 1.2 s; (e) 8.6 m; (f) 6.1 m/s **15.** 0.50 **17.** (a) 13 cm/s²; (b) 4.4 s; (c) 55 cm/s; (d) 18 mJ; (e) 1.4 J; (f) 27 rev/s **19.** (a) $(6.0\text{ N}\cdot\text{m})\hat{j} + (8.0\text{ N}\cdot\text{m})\hat{k}$; (b) $(-22\text{ N}\cdot\text{m})\hat{i}$ **21.** $(-2.0\text{ N}\cdot\text{m})\hat{i}$ **23.** (a) $(50\text{ N}\cdot\text{m})\hat{k}$; (b) 90° **25.** (a) $(-1.5\text{ N}\cdot\text{m})\hat{i} - (4.0\text{ N}\cdot\text{m})\hat{j} - (1.0\text{ N}\cdot\text{m})\hat{k}$; (b) $(-1.5\text{ N}\cdot\text{m})\hat{i} - (4.0\text{ N}\cdot\text{m})\hat{j} - (1.0\text{ N}\cdot\text{m})\hat{k}$ **27.** (a) 9.8 kg · m²/s; (b) $+z$ direction **29.** (a) 0; (b) $(8.0\text{ N}\cdot\text{m})\hat{i} + (8.0\text{ N}\cdot\text{m})\hat{k}$ **31.** (a) 0; (b) −22.6 kg · m²/s; (c) −7.84 N · m; (d) −7.84 N · m **33.** (a) $(-1.7 \times 10^2\text{ kg}\cdot\text{m}^2\text{/s})\hat{k}$; (b) $(+56\text{ N}\cdot\text{m})\hat{k}$; (c) $(+56\text{ kg}\cdot\text{m}^2\text{/s}^2)\hat{k}$ **35.** (a) $48t\hat{k}$ N · m; (b) increasing **37.** (a) 1.47 N · m; (b) 20.4 rad; (c) −29.9 J; (d) 19.9 W **39.** (a) 4.6 × 10⁻³ kg · m²; (b) 1.1 × 10⁻³ kg · m²/s; (c) 3.9 × 10⁻³ kg · m²/s **41.** (a) 1.6 kg · m²; (b) 4.0 kg · m²/s **43.** (a) 3.6 rev/s; (b) 3.0; (c) forces on the bricks from the man transferred energy from the man's internal energy to kinetic energy **45.** (a) 267 rev/min; (b) 0.667 **47.** (a) 750 rev/min; (b) 450 rev/min; (c) clockwise **49.** 0.17 rad/s **51.** (a) 1.5 m; (b) 0.93 rad/s; (c) 98 J; (d) 8.4 rad/s; (e) 8.8 × 10² J; (f) internal energy of the skaters **53.** 3.4 rad/s **55.** 1.3 × 10³ m/s **57.** 11.0 m/s **59.** (a) 18 rad/s; (b) 0.92 **61.** 1.5 rad/s **63.** (a) 0.180 m; (b) clockwise **65.** 0.070 rad/s **67.** (a) 0.148 rad/s; (b) 0.0123; (c) 181° **69.** 0.041 rad/s **71.** 39.1 J **73.** (a) 6.65 × 10⁻⁵ kg · m²/s; (b) no; (c) 0; (d) yes **75.** (a) 0.333; (b) 0.111 **77.** (a) 58.8 J; (b) 39.2 J **79.** (a) 0.81 mJ; (b) 0.29; (c) 1.3 × 10⁻² N **81.** (a) $mR^2/2$; (b) a solid circular cylinder **83.** rotational speed would decrease; day would be about 0.8 s longer **85.** (a) 149 kg · m²; (b) 158 kg · m²/s; (c) 0.744 rad/s **87.** (a) 0; (b) 0; (c) $-30t^3\hat{k}$ kg · m²/s; (d) $-90t^2\hat{k}$ N · m; (e) $30t^3\hat{k}$ kg · m²/s;

(f) $90t^2\hat{k}$ N·m **89.** (a) 61.7 J; (b) 3.43 m; (c) no **91.** (a) 12.7 rad/s; (b) clockwise **93.** (a) $mvR/(I + MR^2)$; (b) $mvR^2/(I + MR^2)$ **95.** (a) 1.6 m/s²; (b) 16 rad/s²; (c) (4.0 N)$\hat{i}$ **97.** 0.47 kg·m²/s

Chapter 12

CP 1. *c, e, f* **2.** (a) no; (b) at site of $\vec{F}_1$, perpendicular to plane of figure; (c) 45 N **3.** *d* **Q 1.** *a* and *c* (forces and torques balance) **3.** (a) 12 kg; (b) 3 kg; (c) 1 kg **5.** (a) 1 and 3 tie, then 2; (b) all tie; (c) 1 and 3 tie, then 2 (zero) **7.** increase **9.** (a) at *C* (to eliminate forces there from a torque equation); (b) plus; (c) minus; (d) equal **P 1.** (a) 1.00 m; (b) 2.00 m; (c) 0.987 m; (d) 1.97 m **3.** 7.92 kN **5.** (a) 9.4 N; (b) 4.4 N **7.** (a) 1.2 kN; (b) down; (c) 1.7 kN; (d) up; (e) left; (f) right **9.** (a) 2.8×10^2 N; (b) 8.8×10^2 N; (c) 71° **11.** 74.4 g **13.** (a) 5.0 N; (b) 30 N; (c) 1.3 m **15.** 8.7 N **17.** (a) 2.7 kN; (b) up; (c) 3.6 kN; (d) down **19.** (a) 0.64 m; (b) increased **21.** 13.6 N **23.** (a) 1.9 kN; (b) up; (c) 2.1 kN; (d) down **25.** (a) 192 N; (b) 96.1 N; (c) 55.5 N **27.** (a) 6.63 kN; (b) 5.74 kN; (c) 5.96 kN **29.** 2.20 m **31.** (a) $(-80\text{ N})\hat{i} + (1.3 \times 10^2\text{N})\hat{j}$; (b) $(80\text{ N})\hat{i} + (1.3 \times 10^2\text{ N})\hat{j}$ **33.** (a) 445 N; (b) 0.50; (c) 315 N **35.** (a) 60.0°; (b) 300 N **37.** 0.34 **39.** (a) slides; (b) 31°; (c) tips; (d) 34° **41.** (a) 211 N; (b) 534 N; (c) 320 N **43.** (a) 6.5×10^6 N/m²; (b) 1.1×10^{-5} m **45.** (a) 866 N; (b) 143 N; (c) 0.165 **47.** (a) 0.80; (b) 0.20; (c) 0.25 **49.** (a) 1.4×10^9 N; (b) 75 **51.** (a) 1.2×10^2 N; (b) 68 N **53.** 76 N **55.** (a) 8.01 kN; (b) 3.65 kN; (c) 5.66 kN **57.** 71.7 N **59.** (a) $L/2$; (b) $L/4$; (c) $L/6$; (d) $L/8$; (e) $25L/24$ **61.** (a) 1.8×10^7 N; (b) 1.4×10^7 N; (c) 16 **63.** 0.29 **65.** 60° **67.** (a) 270 N; (b) 72 N; (c) 19° **69.** (a) 106 N; (b) 64.0° **71.** 2.4×10^9 N/m² **73.** (a) 88 N; (b) $(30\hat{i} + 97\hat{j})$ N **75.** (a) $a_1 = L/2$, $a_2 = 5L/8$, $h = 9L/8$; (b) $b_1 = 2L/3$, $b_2 = L/2$, $h = 7L/6$ **77.** (a) *BC, CD, DA*; (b) 535 N; (c) 757 N **79.** (a) 1.38 kN; (b) 180 N **81.** (a) $\mu < 0.57$; (b) $\mu > 0.57$ **83.** $L/4$ **85.** (a) $(35\hat{i} + 200\hat{j})$ N; (b) $(-45\hat{i} + 200\hat{j})$ N; (c) 1.9×10^2 N

Chapter 13

CP 1. all tie **2.** (a) 1, tie of 2 and 4, then 3; (b) line *d* **3.** (a) increase; (b) negative **4.** (a) 2; (b) 1 **5.** (a) path 1 (decreased *E* (more negative) gives decreased *a*); (b) less (decreased *a* gives decreased *T*) **Q 1.** Gm^2/r^2, upward **3.** *b* and *c* tie, then *a* (zero) **5.** $3GM^2/d^2$, leftward **7.** (a) positive *y*; (b) yes, rotates counterclockwise until it points toward particle *B* **9.** 1, tie of 2 and 4, then 3 **11.** *b, d,* and *f* all tie, then *e, c, a* **P 1.** 19 m **3.** $\frac{1}{2}$ **5.** $-5.00d$ **7.** 2.60×10^5 km **9.** 0.8 m **11.** (a) $M = m$; (b) 0 **13.** 8.31×10^{-9} N **15.** (a) $-1.88d$; (b) $-3.90d$; (c) $0.489d$ **17.** 2.6×10^6 m **19.** (a) 17 N; (b) 2.4 **21.** (a) 7.6 m/s²; (b) 4.2 m/s² **23.** 5×10^{24} kg **25.** (a) 9.83 m/s²; (b) 9.84 m/s²; (c) 9.79 m/s² **27.** (a) $(3.0 \times 10^{-7}\text{ N/kg})m$; (b) $(3.3 \times 10^{-7}\text{ N/kg})m$; (c) $(6.7 \times 10^{-7}\text{ N/kg}\cdot\text{m})mr$ **29.** (a) 0.74; (b) 3.8 m/s²; (c) 5.0 km/s **31.** (a) 0.0451; (b) 28.5 **33.** 5.0×10^9 J **35.** (a) 0.50 pJ; (b) −0.50 pJ **37.** (a) 1.7 km/s; (b) 2.5×10^5 m; (c) 1.4 km/s **39.** (a) 82 km/s; (b) 1.8×10^4 km/s **41.** -4.82×10^{-13} J **43.** 6.5×10^{23} kg **45.** 5×10^{10} stars **47.** (a) 6.64×10^3 km; (b) 0.0136 **49.** (a) 7.82 km/s; (b) 87.5 min **51.** (a) 1.9×10^{13} m; (b) $3.6R_P$ **55.** 0.71 y **57.** 5.8×10^6 m **59.** $(GM/L)^{0.5}$ **61.** (a) 2.8 y; (b) 1.0×10^{-4} **63.** (a) 3.19×10^3 km; (b) lifting **65.** (a) $r^{1.5}$; (b) r^{-1}; (c) $r^{0.5}$; (d) $r^{-0.5}$ **67.** (a) 7.5 km/s; (b) 97 min; (c) 4.1×10^2 km; (d) 7.7 km/s; (e) 93 min; (f) 3.2×10^{-3} N; (g) no; (h) yes **69.** 1.1 s **71.** (a) 1.0×10^3 kg; (b) 1.5 km/s **73.** $-0.044\hat{j}$ μN **75.** (a) 2.15×10^4 s; (b) 12.3 km/s; (c) 12.0 km/s; (d) 2.17×10^{11} J; (e) -4.53×10^{11} J; (f) -2.35×10^{11} J; (g) 4.04×10^7 m; (h) 1.22×10^3 s; (i) elliptical **77.** $0.37\hat{j}$ μN **79.** 29 pN **81.** 2.5×10^4 km **83.** (a) 2.2×10^{-7} rad/s; (b) 89 km/s **85.** 3.2×10^{-7} N **87.** (a) 0; (b) 1.8×10^{32} J; (c) 1.8×10^{32} J; (d) 0.99 km/s **89.** $GM_Em/12R_E$ **91.** (a) 1.4×10^6 m/s; (b) 3×10^6 m/s² **93.** $2\pi r^{1.5}G^{-0.5}(M + m/4)^{-0.5}$ **95.** 2.4×10^4 m/s **97.** −1.87 GJ **99.** (a) $GMmx(x^2 + R^2)^{-3/2}$; (b) $[2GM(R^{-1} - (R^2 + x^2)^{-1/2})]^{1/2}$ **101.** (a) Gm^2/R_i; (b) $Gm^2/2R_i$; (c) $(Gm/R_i)^{0.5}$; (d) $2(Gm/R_i)^{0.5}$; (e) Gm^2/R_i; (f) $(2Gm/R_i)^{0.5}$; (g) The center-of-mass frame is an inertial frame, and in it the principle of conservation of energy may be written as in Chapter 8; the reference frame attached to body *A* is noninertial, and the principle cannot be written as in Chapter 8. Answer (d) is correct. **103.** (a) 1.9×10^{11} m; (b) 4.6×10^4 m/s

Chapter 14

CP 1. all tie **2.** (a) all tie (the gravitational force on the penguin is the same); (b) $0.95\rho_0, \rho_0, 1.1\rho_0$ **3.** 13 cm³/s, outward **4.** (a) all tie; (b) 1, then 2 and 3 tie, 4 (wider means slower); (c) 4, 3, 2, 1 (wider and lower mean more pressure) **Q 1.** *b*, then *a* and *d* tie (zero), then *c* **3.** (a) moves downward; (b) moves downward **5.** (a) downward; (b) downward; (c) same **7.** *B, C, A* **9.** (a) 1 and 4; (b) 2; (c) 3 **P 1.** 1.1×10^5 Pa **3.** 2.9×10^4 N **5.** 0.074 **7.** (b) 26 kN **9.** 1.08×10^3 atm **11.** 7.2×10^5 N **13.** -2.6×10^4 Pa **15.** (a) 94 torr; (b) 4.1×10^2 torr; (c) 3.1×10^2 torr **17.** (a) 1.0×10^3 torr; (b) 1.7×10^3 torr **19.** 0.635 J **21.** 44 km **23.** 4.69×10^5 N **25.** 739.26 torr **27.** (a) 7.9 km; (b) 16 km **29.** 8.50 kg **31.** (a) 2.04×10^{-2} m³; (b) 1.57 kN **33.** five **35.** (a) 6.7×10^2 kg/m³; (b) 7.4×10^2 kg/m³ **37.** (a) 1.2 kg; (b) 1.3×10^3 kg/m³ **39.** (a) 0.10; (b) 0.083 **41.** 57.3 cm **43.** 0.126 m³ **45.** (a) 1.80 m³; (b) 4.75 m³ **47.** (a) 637.8 cm³; (b) 5.102 m³; (c) 5.102×10^3 kg **49.** 8.1 m/s **51.** (a) 3.0 m/s; (b) 2.8 m/s **53.** 66 W **55.** (a) 2.5 m/s; (b) 2.6×10^5 Pa **57.** (a) 3.9 m/s; (b) 88 kPa **59.** (a) 1.6×10^{-3} m³/s; (b) 0.90 m **61.** 1.4×10^5 J **63.** (a) 74 N; (b) 1.5×10^2 m³ **65.** (a) 35 cm; (b) 30 cm; (c) 20 cm **67.** (b) 2.0×10^{-2} m³/s **69.** (a) 0.0776 m³/s; (b) 69.8 kg/s **71.** 1.1×10^2 m/s **73.** 44.2 g **75.** 45.3 cm³ **77.** (a) 3.2 m/s; (b) 9.2×10^4 Pa; (c) 10.3 m **79.** 5.11×10^{-7} kg **81.** 1.07×10^3 g **83.** 6.0×10^2 kg/m³ **85.** 1.5 g/cm³

Chapter 15

CP 1. (sketch *x* versus *t*) (a) $-x_m$; (b) $+x_m$; (c) 0 **2.** a (*F* must have the form of Eq. 15-10) **3.** (a) 5 J; (b) 2 J; (c) 5 J **4.** all tie (in Eq. 15-29, *m* is included in *I*) **5.** 1, 2, 3 (the ratio *m*/*b* matters; *k* does not) **Q 1.** (a) 2; (b) positive; (c) between 0 and $+x_m$ **3.** a and b **5.** (a) all tie; (b) 3, then 1 and 2 tie; (c) 1, 2, 3 (zero); (d) 1, 2, 3 (zero); (e) 1, 3, 2 **7.** (a) between *D* and *E*; (b) between $3\pi/2$ rad and 2π rad **9.** (a) greater; (b) same; (c) same; (d) greater; (e) greater **11.** *b* (infinite period, does not oscillate), *c, a* **P 1.** 37.8 m/s² **3.** (a) 1.0 mm; (b) 0.75 m/s; (c) 5.7×10^2 m/s² **5.** (a) 0.50 s; (b) 2.0 Hz; (c) 18 cm **7.** (a) 0.500 s; (b) 2.00 Hz; (c) 12.6 rad/s; (d) 79.0 N/m; (e) 4.40 m/s; (f) 27.6 N **9.** (a) 498 Hz; (b) greater **11.** (a) 3.0 m; (b) −49 m/s; (c) -2.7×10^2 m/s²; (d) 20 rad; (e) 1.5 Hz; (f) 0.67 s **13.** 39.6 Hz **15.** (a) 5.58 Hz; (b) 0.325 kg; (c) 0.400 m **17.** 3.1 cm **19.** (a) $0.18A$; (b) same direction **21.** (a) 25 cm; (b) 2.2 Hz **23.** 54 Hz **25.** (a) 0.525 m; (b) 0.686 s **27.** 37 mJ **29.** (a) 0.75; (b) 0.25; (c) $2^{-0.5}x_m$ **31.** (a) 2.25 Hz; (b) 125 J; (c) 250 J; (d) 86.6 cm **33.** (a) 3.1 ms; (b) 4.0 m/s; (c) 0.080 J; (d) 80 N; (e) 40 N **35.** (a) 1.1 m/s; (b) 3.3 cm **37.** (a) 2.2 Hz; (b) 56 cm/s; (c) 0.10 kg; (d) 20.0 cm **39.** (a) 39.5 rad/s;

(b) 34.2 rad/s; (c) 124 rad/s^2 **41.** (a) 1.64 s; (b) equal **43.** (a) 0.205 kg·m^2; (b) 47.7 cm; (c) 1.50 s **45.** 0.366 s **47.** 8.77 s **49.** (a) 0.53 m; (b) 2.1 s **51.** 0.0653 s **53.** (a) 0.845 rad; (b) 0.0602 rad **55.** (a) 2.26 s; (b) increases; (c) same **57.** (a) 14.3 s; (b) 5.27 **59.** 6.0% **61.** (a) $F_m/b\omega$; (b) F_m/b **63.** 5.0 cm **65.** (a) 1.2 J; (b) 50 **67.** 1.53 m **69.** (a) 16.6 cm; (b) 1.23% **71.** (a) 2.8×10^3 rad/s; (b) 2.1 m/s; (c) 5.7 km/s^2 **73.** (a) 0.735 kg · m^2; (b) 0.0240 N · m; (c) 0.181 rad/s **75.** (a) 0.35 Hz; (b) 0.39 Hz; (c) 0 (no oscillation) **77.** (a) 7.90 N/m; (b) 1.19 cm; (c) 2.00 Hz **79.** 1.6 kg **81.** (a) 3.5 m; (b) 0.75 s **83.** 7.2 m/s **85.** (a) 1.23 kN/m; (b) 76.0 N **87.** (a) 1.1 Hz; (b) 5.0 cm **89.** (a) 1.3×10^2 N/m; (b) 0.62 s; (c) 1.6 Hz; (d) 5.0 cm; (e) 0.51 m/s **91.** (a) 3.2 Hz; (b) 0.26 m; (c) $x = (0.26\text{ m})\cos(20t - \pi/2)$, with t in seconds **93.** 0.079 kg·m^2 **95.** (a) 0.44 s; (b) 0.18 m **97.** (a) 245 N/m; (b) 0.284 s **99.** 50 cm **101.** (a) 8.11×10^{-5} kg · m^2; (b) 3.14 rad/s **103.** 14.0° **105.** (a) 0.30 m; (b) 0.28 s; (c) 1.5×10^2 m/s^2; (d) 11 J **107.** (a) 0.45 s; (b) 0.10 m above and 0.20 m below; (c) 0.15 m; (d) 2.3 J **109.** 7×10^2 N/m **111.** (a) F/m; (b) $2F/mL$; (c) 0

Chapter 16

CP **1.** a, 2; b, 3; c, 1 (compare with phase in Eq. 16-2, then see Eq. 16-5) **2.** (a) 2, 3, 1 (see Eq. 16-12); (b) 3, then 1 and 2 tie (find amplitude of dy/dt) **3.** (a) same (independent of f); (b) decrease ($\lambda = v/f$); (c) increase; (d) increase **4.** 0.20 and 0.80 tie, then 0.60, 0.45 **5.** (a) 1; (b) 3; (c) 2 **6.** (a) 75 Hz; (b) 525 Hz **Q** **1.** *a*, upward; *b*, upward; *c*, downward; *d*, downward; *e*, downward; *f*, downward; *g*, upward; *h*, upward **3.** (a) 1, 4, 2, 3; (b) 1, 4, 2, 3 **5.** (a) 0, 0.2 wavelength, 0.5 wavelength (zero); (b) $4P_{avg,1}$ **7.** intermediate (closer to fully destructive) **9.** *c*, *a*, *b* **11.** *d* **P** **1.** (a) 3.49 m^{-1}; (b) 31.5 m/s **3.** (a) 0.680 s; (b) 1.47 Hz; (c) 2.06 m/s **5.** 1.1 ms **7.** (a) 11.7 cm; (b) π rad **9.** (a) 64 Hz; (b) 1.3 m; (c) 4.0 cm; (d) 5.0 m^{-1}; (e) 4.0×10^2 s^{-1}; (f) $\pi/2$ rad; (g) minus **11.** (a) 3.0 mm; (b) 16 m^{-1}; (c) 2.4×10^2 s^{-1}; (d) minus **13.** (a) negative; (b) 4.0 cm; (c) 0.31 cm^{-1}; (d) 0.63 s^{-1}; (e) π rad; (f) minus; (g) 2.0 cm/s; (h) −2.5 cm/s **15.** 129 m/s **17.** (a) 0.12 mm; (b) 141 m^{-1}; (c) 628 s^{-1}; (d) plus **19.** (a) 15 m/s; (b) 0.036 N **21.** (a) 5.0 cm; (b) 40 cm; (c) 12 m/s; (d) 0.033 s; (e) 9.4 m/s; (f) 16 m^{-1}; (g) 1.9×10^2 s^{-1}; (h) 0.93 rad; (i) plus **23.** 2.63 m **27.** 3.2 mm **29.** 0.20 m/s **31.** $1.41y_m$ **33.** (a) 9.0 mm; (b) 16 m^{-1}; (c) 1.1×10^3 s^{-1}; (d) 2.7 rad; (e) plus **35.** 5.0 cm **37.** 84° **39.** (a) 3.29 mm; (b) 1.55 rad; (c) 1.55 rad **41.** (a) 7.91 Hz; (b) 15.8 Hz; (c) 23.7 Hz **43.** (a) 82.0 m/s; (b) 16.8 m; (c) 4.88 Hz **45.** (a) 144 m/s; (b) 60.0 cm; (c) 241 Hz **47.** (a) 105 Hz; (b) 158 m/s **49.** 260 Hz **51.** (a) 0.25 cm; (b) 1.2×10^2 cm/s; (c) 3.0 cm; (d) 0 **53.** (a) 0.50 cm; (b) 3.1 m^{-1}; (c) 3.1×10^2 s^{-1}; (d) minus **55.** (a) 2.00 Hz; (b) 2.00 m; (c) 4.00 m/s; (d) 50.0 cm; (e) 150 cm; (f) 250 cm; (g) 0; (h) 100 cm; (i) 200 cm **57.** 0.25 m **59.** (a) 324 Hz; (b) eight **61.** (a) $0.83y_1$; (b) 37° **63.** (a) 0.31 m; (b) 1.64 rad; (c) 2.2 mm **65.** 1.2 rad **67.** (a) 3.77 m/s; (b) 12.3 N; (c) 0; (d) 46.4 W; (e) 0; (f) 0; (g) ± 0.50 cm **69.** (a) $2\pi y_m/\lambda$; (b) no **71.** (a) 1.00 cm; (b) 3.46×10^3 s^{-1}; (c) 10.5 m^{-1}; (d) plus **73.** (a) 75 Hz; (b) 13 ms **75.** (a) 240 cm; (b) 120 cm; (c) 80 cm **77.** (a) 144 m/s; (b) 3.00 m; (c) 1.50 m; (d) 48.0 Hz; (e) 96.0 Hz **79.** (a) 2.0 mm; (b) 95 Hz; (c) +30 m/s; (d) 31 cm; (e) 1.2 m/s **81.** 36 N **83.** (a) 300 m/s; (b) no **85.** (a) 1.33 m/s; (b) 1.88 m/s; (c) 16.7 m/s^2; (d) 23.7 m/s^2 **87.** (a) 0.16 m; (b) 2.4×10^2 N; (c) $y(x, t) = (0.16\text{ m})\sin[(1.57\text{ m}^{-1})x]\sin[(31.4\text{ s}^{-1})t]$ **89.** (a) $[k\,\Delta\ell\,(\ell + \Delta\ell)/m]^{0.5}$ **91.** (a) 0.52 m; (b) 40 m/s; (c) 0.40 m **93.** (c) 2.0 m/s; (d) $-x$

Chapter 17

CP **1.** beginning to decrease (example: mentally move the curves of Fig. 17-7 rightward past the point at $x = 42$ m) **2.** (a) 1 and 2 tie, then 3 (see Eq. 17-28); (b) 3, then 1 and 2 tie (see Eq. 17-26) **3.** second (see Eqs. 17-39 and 17-41) **4.** *a*, greater; *b*, less; *c*, can't tell; *d*, can't tell; *e*, greater; *f*, less **Q** **1.** *C*, then *A* and *B* tie **3.** (a) 0, 0.2 wavelength, 0.5 wavelength (zero); (b) $4P_{avg,1}$ **5.** 150 Hz and 450 Hz **7.** *E*, *A*, *D*, *C*, *B* **9.** 1, 4, 3, 2 **P** **1.** (a) 2.6 km; (b) 2.0×10^2 **3.** (a) 79 m; (b) 41 m; (c) 89 m **5.** 40.7 m **7.** 1.9×10^3 km **9.** (a) 76.2 μm; (b) 0.333 mm **11.** 0.23 ms **13.** (a) 2.3×10^2 Hz; (b) higher **15.** 960 Hz **17.** (a) 14; (b) 14 **19.** (a) 343 Hz; (b) 3; (c) 5; (d) 686 Hz; (e) 2; (f) 3 **21.** (a) 143 Hz; (b) 3; (c) 5; (d) 286 Hz; (e) 2; (f) 3 **23.** (a) 0; (b) fully constructive; (c) increase; (d) 128 m; (e) 63.0 m; (f) 41.2 m **25.** 15.0 mW **27.** 36.8 nm **29.** (a) 1.0×10^3; (b) 32 **31.** 0.76 μm **33.** 2 μW **35.** (a) 5.97×10^{-5} W/m^2; (b) 4.48 nW **37.** (a) 0.34 nW; (b) 0.68 nW; (c) 1.4 nW; (d) 0.88 nW; (e) 0 **39.** (a) 833 Hz; (b) 0.418 m **41.** (a) 2; (b) 1 **43.** (a) 405 m/s; (b) 596 N; (c) 44.0 cm; (d) 37.3 cm **45.** (a) 3; (b) 1129 Hz; (c) 1506 Hz **47.** 45.3 N **49.** 12.4 m **51.** 2.25 ms **53.** 0.020 **55.** 0 **57.** (a) 526 Hz; (b) 555 Hz **59.** 155 Hz **61.** (a) 1.022 kHz; (b) 1.045 kHz **63.** 41 kHz **65.** (a) 485.8 Hz; (b) 500.0 Hz; (c) 486.2 Hz; (d) 500.0 Hz **67.** (a) 2.0 kHz; (b) 2.0 kHz **69.** (a) 42°; (b) 11 s **71.** (a) 2.10 m; (b) 1.47 m **73.** (a) 21 nm; (b) 35 cm; (c) 24 nm; (d) 35 cm **75.** 0.25 **77.** (a) 9.7×10^2 Hz; (b) 1.0 kHz; (c) 60 Hz, no **79.** (a) 39.7 μW/m^2; (b) 171 nm; (c) 0.893 Pa **81.** (a) 10 W; (b) 0.032 W/m^2; (c) 99 dB **83.** (a) 7.70 Hz; (b) 7.70 Hz **85.** (a) 59.7; (b) 2.81×10^{-4} **87.** (a) 5.2 kHz; (b) 2 **89.** 2.1 m **91.** 1 cm **93.** (a) 3.6×10^2 m/s; (b) 150 Hz **95.** (a) 0; (b) 0.572 m; (c) 1.14 m **97.** 171 m **99.** (a) 11 ms; (b) 3.8 m **101.** (a) rightward; (b) 0.90 m/s; (c) less **103.** (a) 5.5×10^2 m/s; (b) 1.1×10^3 m/s; (c) 1 **105.** 400 Hz **107.** (a) 14; (b) 12 **109.** (b) 0.8 to 1.6 μs **111.** 4.8×10^2 Hz

Chapter 18

CP **1.** (a) all tie; (b) 50°X, 50°Y, 50°W **2.** (a) 2 and 3 tie, then 1, then 4; (b) 3, 2, then 1 and 4 tie (from Eqs. 18-9 and 18-10, assume that change in area is proportional to initial area) **3.** *A* (see Eq. 18-14) **4.** *c* and *e* (maximize area enclosed by a clockwise cycle) **5.** (a) all tie (ΔE_{int} depends on *i* and *f*, not on path); (b) 4, 3, 2, 1 (compare areas under curves); (c) 4, 3, 2, 1 (see Eq. 18-26) **6.** (a) zero (closed cycle); (b) negative (W_{net} is negative; see Eq. 18-26) **7.** *b* and *d* tie, then *a*, *c* (P_{cond} identical; see Eq. 18-32) **Q** **1.** *B*, then *A* and *C* tie **3.** *c*, then the rest tie **5.** (a) both clockwise; (b) both clockwise **7.** *c*, *b*, *a* **9.** (a) *f*, because ice temperature will not rise to freezing point and then drop; (b) *b* and *c* at freezing point, *d* above, *e* below; (c) in *b* liquid partly freezes and no ice melts; in *c* no liquid freezes and no ice melts; in *d* no liquid freezes and ice fully melts; in *e* liquid fully freezes and no ice melts **11.** (a) greater; (b) 1, 2, 3; (c) 1, 3, 2; (d) 1, 2, 3; (e) 2, 3, 1 **P** **1.** 348 K **3.** 1.366 **5.** (a) 320°F; (b) −12.3°F **7.** −92.1°X **9.** 29 cm^3 **11.** 2.731 cm **13.** 49.87 cm^3 **15.** 0.26 cm^3 **17.** 360°C **19.** 0.13 mm **21.** 7.5 cm **23.** 94.6 L **25.** 42.7 kJ **27.** 160 s **29.** 33 g **31.** 3.0 min **33.** 33 m^2 **35.** 13.5 C° **37.** 742 kJ **39.** (a) 5.3°C; (b) 0; (c) 0°C; (d) 60 g **41.** (a) 0°C; (b) 2.5°C **43.** −30 J **45.** (a) 1.2×10^2 J; (b) 75 J; (c) 30 J

47. 60 J **49.** (a) 6.0 cal; (b) −43 cal; (c) 40 cal; (d) 18 cal; (e) 18 cal **51.** 1.66 kJ/s **53.** (a) 16 J/s; (b) 0.048 g/s **55.** (a) 1.23 kW; (b) 2.28 kW; (c) 1.05 kW **57.** 0.50 min **59.** (a) 1.7×10^4 W/m^2; (b) 18 W/m^2 **61.** −4.2°C **63.** 1.1 m **65.** 0.40 cm/h **67.** 10% **69.** 4.5×10^2 J/kg·K **71.** 0.432 cm^3 **73.** (a) $11p_1V_1$; (b) $6p_1V_1$ **75.** 4.83×10^{-2} cm^3 **77.** 23 J **79.** 3.1×10^2 J **81.** 10.5°C **83.** 79.5°C **85.** 8.6 J **87.** 333 J **89.** (a) 90 W; (b) 2.3×10^2 W; (c) 3.3×10^2 W **91.** (a) 1.87×10^4; (b) 10.4 h **93.** (a) −45 J; (b) +45 J **95.** (a) 80 J; (b) 80 J **97.** −6.1 nW **99.** 1.17 C°

Chapter 19

CP 1. all but c **2.** (a) all tie; (b) 3, 2, 1 **3.** gas A **4.** 5 (greatest change in T), then tie of 1, 2, 3, and 4 **5.** 1, 2, 3 ($Q_3 = 0$, Q_2 goes into work W_2, but Q_1 goes into greater work W_1 and increases gas temperature) **Q 1.** 20 J **3.** d, then a and b tie, then c **5.** (a) 3; (b) 1; (c) 4; (d) 2; (e) yes **7.** constant-volume process **9.** (a) 1, 2, 3, 4; (b) 1, 2, 3 **P 1.** (a) 0.0127 mol; (b) 7.64×10^{21} atoms **3.** 25 molecules/cm^3 **5.** 186 kPa **7.** (a) 0.0388 mol; (b) 220°C **9.** (a) 3.14×10^3 J; (b) from **11.** 360 K **13.** 5.60 kJ **15.** (a) 1.5 mol; (b) 1.8×10^3 K; (c) 6.0×10^2 K; (d) 5.0 kJ **17.** 2.0×10^5 Pa **19.** 1.8×10^2 m/s **21.** (a) 511 m/s; (b) −200°C; (c) 899°C **23.** 1.9 kPa **25.** (a) 5.65×10^{-21} J; (b) 7.72×10^{-21} J; (c) 3.40 kJ; (d) 4.65 kJ **27.** (a) 6.76×10^{-20} J; (b) 10.7 **29.** (a) 6×10^9 km **31.** (a) 3.27×10^{10} molecules/cm^3; (b) 172 m **33.** (a) 420 m/s; (b) 458 m/s; (c) yes **35.** (a) 6.5 km/s; (b) 7.1 km/s **37.** (a) 1.0×10^4 K; (b) 1.6×10^5 K; (c) 4.4×10^2 K; (d) 7.0×10^3 K; (e) no; (f) yes **39.** (a) 7.0 km/s; (b) 2.0×10^{-8} cm; (c) 3.5×10^{10} collisions/s **41.** (a) 0.67; (b) 1.2; (c) 1.3; (d) 0.33 **43.** (a) 0; (b) +374 J; (c) +374 J; (d) $+3.11 \times 10^{-22}$ J **45.** 15.8 J/mol·K **47.** (a) 6.6×10^{-26} kg; (b) 40 g/mol **49.** (a) 3.49 kJ; (b) 2.49 kJ; (c) 997 J; (d) 1.00 kJ **51.** 8.0 kJ **53.** (a) 6.98 kJ; (b) 4.99 kJ; (c) 1.99 kJ; (d) 2.99 kJ **55.** (a) 14 atm; (b) 6.2×10^2 K **57.** −15 J **59.** −20 J **61.** (a) diatomic; (b) 446 K; (c) 8.10 mol **63.** (a) 3.74 kJ; (b) 3.74 kJ; (c) 0; (d) 0; (e) −1.81 kJ; (f) 1.81 kJ; (g) −3.22 kJ; (h) −1.93 kJ; (i) −1.29 kJ; (j) 520 J; (k) 0; (l) 520 J; (m) 0.0246 m^3; (n) 2.00 atm; (o) 0.0373 m^3; (p) 1.00 atm **65.** (a) 900 cal; (b) 0; (c) 900 cal; (d) 450 cal; (e) 1200 cal; (f) 300 cal; (g) 900 cal; (h) 450 cal; (i) 0; (j) −900 cal; (k) 900 cal; (l) 450 cal **67.** 349 K **69.** (a) −374 J; (b) 0; (c) +374 J; (d) $+3.11 \times 10^{-22}$ J **71.** 7.03×10^9 s^{-1} **73.** (a) 2.00 atm; (b) 333 J; (c) 0.961 atm; (d) 236 J **75.** (a) monatomic; (b) 2.7×10^4 K; (c) 4.5×10^4 mol; (d) 3.4 kJ; (e) 3.4×10^2 kJ; (f) 0.010 **77.** (a) 8.0 atm; (b) 300 K; (c) 4.4 kJ; (d) 3.2 atm; (e) 120 K; (f) 2.9 kJ; (g) 4.6 atm; (h) 170 K; (i) 3.4 kJ **79.** (a) 38 L; (b) 71 g **81.** (a) $3/v_0^3$; (b) $0.750v_0$; (c) $0.775v_0$ **83.** (a) −2.37 kJ; (b) 2.37 kJ **85.** −3.0 J **87.** (b) 125 J; (c) to

Chapter 20

CP 1. a, b, c **2.** smaller (Q is smaller) **3.** c, b, a **4.** a, d, c, b **5.** b **Q 1.** a and c tie, then b and d tie **3.** b, a, c, d **5.** unchanged **7.** A, first; B, first and second; C, second; D, neither **9.** (a) same; (b) increase; (c) decrease **P 1.** 14.4 J/K **3.** (a) 5.79×10^4 J; (b) 173 J/K **5.** (a) 9.22 kJ; (b) 23.1 J/K; (c) 0 **7.** (a) 57.0°C; (b) −22.1 J/K; (c) +24.9 J/K; (d) +2.8 J/K **9.** (a) −710 mJ/K; (b) +710 mJ/K; (c) +723 mJ/K; (d) −723 mJ/K; (e) +13 mJ/K; (f) 0 **11.** (a) 320 K; (b) 0; (c) +1.72 J/K **13.** (a) 0.333; (b) 0.215; (c) 0.644; (d) 1.10; (e) 1.10; (f) 0; (g) 1.10; (h) 0; (i) −0.889; (j) −0.889; (k) −1.10; (l) −0.889; (m) 0; (n) 0.889; (o) 0 **15.** +0.76 J/K **17.** (a) −943 J/K; (b) +943 J/K; (c) yes **19.** −1.18 J/K **21.** (a) 0.693; (b) 4.50; (c) 0.693; (d) 0; (e) 4.50; (f) 23.0 J/K; (g) −0.693; (h) 7.50; (i) −0.693; (j) 3.00; (k) 4.50; (l) 23.0 J/K **23.** (a) 266 K; (b) 341 K **25.** (a) 23.6%; (b) 1.49×10^4 J **27.** 97 K **29.** (a) 1.47 kJ; (b) 554 J; (c) 918 J; (d) 62.4% **31.** (a) 2.27 kJ; (b) 14.8 kJ; (c) 15.4%; (d) 75.0%; (e) greater **33.** (a) 33 kJ; (b) 25 kJ; (c) 26 kJ; (d) 18 kJ **35.** (a) 3.00; (b) 1.98; (c) 0.660; (d) 0.495; (e) 0.165; (f) 34.0% **37.** 20 J **39.** 440 W **41.** 2.03 **43.** 0.25 hp **47.** (a) $W = N!/(n_1!\, n_2!\, n_3!)$; (b) $[(N/2)!\,(N/2)!]/[(N/3)!\,(N/3)!\,(N/3)!]$; (c) 4.2×10^{16} **49.** (a) 87 m/s; (b) 1.2×10^2 m/s; (c) 22 J/K **51.** (a) 78%; (b) 82 kg/s **53.** (a) 40.9°C; (b) −27.1 J/K; (c) 30.5 J/K; (d) 3.4 J/K **55.** 1.18×10^3 J/K **57.** (a) 0; (b) 0; (c) −23.0 J/K; (d) 23.0 J/K **59.** (a) 25.5 kJ; (b) 4.73 kJ; (c) 18.5% **61.** 0.141 J/K·s **63.** (a) 42.6 kJ; (b) 7.61 kJ **65.** (a) 4.45 J/K; (b) no **67.** (a) 1; (b) 1; (c) 3; (d) 10; (e) 1.5×10^{-23} J/K; (f) 3.2×10^{-23} J/K **69.** +3.59 J/K **71.** (a) 1.95 J/K; (b) 0.650 J/K; (c) 0.217 J/K; (d) 0.072 J/K; (e) decrease **73.** (a) 1.26×10^{14}; (b) 4.71×10^{13}; (c) 0.37; (d) 1.01×10^{29}; (e) 1.37×10^{28}; (f) 0.14; (g) 9.05×10^{58}; (h) 1.64×10^{57}; (i) 0.018; (j) decrease

Index

Figures are noted by page numbers in *italics*, tables are indicated by t following the page number.

W9-BOU-591

The Chris Farley SHOW

A BIOGRAPHY IN THREE ACTS

Tom Farley, Jr.
and Tanner Colby

VIKING

VIKING

Published by the Penguin Group

Penguin Group (USA) Inc., 375 Hudson Street, New York, New York 10014, U.S.A. • Penguin Group (Canada), 90 Eglinton Avenue East, Suite 700, Toronto, Ontario, Canada M4P 2Y3 (a division of Pearson Penguin Canada Inc.) • Penguin Books Ltd, 80 Strand, London WC2R 0RL, England • Penguin Ireland, 25 St Stephen's Green, Dublin 2, Ireland (a division of Penguin Books Ltd) • Penguin Books Australia Ltd, 250 Camberwell Road, Camberwell, Victoria 3124, Australia (a division of Pearson Australia Group Pty Ltd) • Penguin Books India Pvt Ltd, 11 Community Centre, Panchsheel Park, New Delhi – 110 017, India • Penguin Group (NZ), 67 Apollo Drive, Rosedale, North Shore 0632, New Zealand (a division of Pearson New Zealand Ltd) • Penguin Books (South Africa) (Pty) Ltd, 24 Sturdee Avenue, Rosebank, Johannesburg 2196, South Africa

Penguin Books Ltd, Registered Offices:
80 Strand, London WC2R 0RL, England

First published in 2008 by Viking Penguin,
a member of Penguin Group (USA) Inc.

10 9 8 7 6 5 4 3 2

Grateful acknowledgment is made for permission to reprint excerpts from the following copyrighted works:

"All You Need Is Love" by John Lennon and Paul McCartney. © 1967 Sony/ATV Tunes LLC. All rights administered by Sony/ATV Music Publishing, 8 Music Square West, Nashville, TN 37203. All rights reserved. Used by permission.

"Working for the Weekend" words and music by Paul Dean, Matthew Frenette and Michael Reno. © 1981 EMI April Music (Canada) Ltd., Dean of Music, EMI Blackwood Music Inc. and Duke Reno Music. All rights in the U.S.A. controlled and administered by EMI April Music Inc. and EMI Blackwood Music Inc. All rights reserved. International copyright secured. Used by permission.

Library of Congress Cataloging-in-Publication Data
Farley, Tom, 1961-
The Chris Farley show : a biography in three acts / Tom Farley, Jr. & Tanner Colby.
p. cm.
Includes index.
ISBN 978-0-670-01923-6
1. Farley, Chris, 1964-1997. 2. Comedians—United States—Biography. 3. Actors—United States—Biography. I. Colby, Tanner. II. Title.
PN2287.F33F37 2008
792.02'8092—dc22 2007040481

Printed in the United States of America
Set in Minion with AG Schoolbook
Designed by Daniel Lagin

For my brother, whom I love and miss;
my dad, who gave us all he had to give;
my kids, Mary Kate, Emma, and Tommy,
the most inspiring, funny, caring, and wonderful kids anyone could have;
and most important,
my wife and deepest friend, Laura, whom I love with all my heart

For Gus, a great writer of stories and chapter books

Nothing you can make that can't be made,
No one you can save that can't be saved,
Nothing you can do, but you can learn how to be you in time.
It's easy.
All you need is love.

—JOHN LENNON/PAUL McCARTNEY

CONTENTS

INTRODUCTION

Irish brothers share one of the strangest relationships on earth. We fight like hell among ourselves on a daily basis, but one word or action against one brother brings the wrath of God down upon you from the others. That was Chris and me. We were always competing, whether it was driveway basketball, touch football, or Monopoly. Most of the time, those games would end in a brawl. Nothing bloody, mind you. Drawing blood would bring the fury of Mom or Dad down on all of us. No, most of the time we'd strike a few blows and then run like hell. And let me tell you, nothing was more terrifying than being chased through the neighborhood by a crazy, mad Irish sibling who outweighed you by twenty-five pounds and had a brick in his hand!

But rare was the time that I wouldn't come running if Chris was in trouble. I was the older brother; that was my job. And, Chris being Chris, it was a job that put me in harm's way more times than I would have liked. One such time, when I was in eighth grade and Chris was in sixth, he got into a fight with a classmate. He tackled the kid and threw him to the ground, landing on top of him and breaking his collarbone. Word got around school that the kid's seventh-grade brother was gunning for Chris. Naturally, I had to step in. I put the word out that the brother would have

to get through me first. I found out later that day that the kid's brother was named Rocky. No shit: Rocky! The guy was massive (a future all-city lineman in high school, no less). No fight ensued, but I did learn that I possessed a real gift of what the Irish call "the gab." I talked my way out of it. It was my only defense, without which Chris would have certainly gotten me killed several times over. Life with Chris was exciting; he brought drama and danger into our lives. But no matter what he put you through, he could always just give you a look and make you laugh. Boy, did he make us laugh.

We always loved to tell "Chris stories." I've heard them from friends, relatives, teachers, coaches—even priests and nuns. You could be the funniest guy in the room just by describing some of the stuff Chris did. For every hilarious thing he did on camera, there were twenty things he did offscreen that just blew it away. He lived to make others laugh, and he was fearless about it. In the years since Chris passed away, there have been countless times when Chris's buddies would find themselves huddled together, sharing these crazy stories. At one time, I even thought that a collection of those stories would make a fantastic book. I still do. But I now believe that those funny stories alone would not paint the right picture of who this kid was. Chris had far too much depth and way too much pain. We all enjoyed Chris so much, and it's hard to put those things into words.

I began this project by listing all the people who either knew Chris the best or were there at the important moments in his life. I spoke to most of them and gave them assurances that this was a project that our family was behind all the way. I wanted them to be open and honest about their memories, opinions, and feelings about being part of what, for most, was an unforgettable relationship. I'm not sure I was totally prepared for the story that Tanner and I ended up with. The funny stories and outrageous moments are definitely in there, but what emerged was this amazing picture of the multifaceted character traits that Chris possessed. He was hilarious, yes, but he was also a very religious, very caring—and very troubled and addicted person. It's a sad story, no question about it. But it's Chris.

Soon after Chris died, I told my wife that my greatest fear was being sixty years old and trying hard to remember this kid who was my brother. I guess anyone who's lost someone close can say that. Being able to watch the fun movies and video clips only gets you so far; it's not the full picture. I'm pleased that this book will be something I can pick up when I'm older, remember Chris and his wild life, and be once more amazed that I had such an unbelievable person in my life.

The Chris Farley SHOW

ACT I

So that night I got blind drunk and threw up in my bed. Then I called Patrick the next day and said, "Man, this was great! When are we gonna do it again?" And I got blind drunk every weekend until I graduated high school.

Then I went on to college at Marquette University. I was away from home and that meant I could party every night. I did. Each year I got worse and worse. Freshman year I'd party Thursday, Friday, and Saturday. Sophomore year it'd be Wednesday, Thursday, Friday, and Saturday. Then it started on Tuesdays, and by the time I was a senior it was every night.

Every drug that I tried I couldn't wait to try more. Sophomore year I tried marijuana and fell completely in love with it. I went home and watched *Love Connection* and was like, "Ooooohhhh man . . ." You know? I couldn't understand why everyone wasn't getting high. It was the best way to live. My God, how boring it must be for you poor sober people. So I got high every day from that day on. I'd try psychedelics and I'd have a really bad trip and still couldn't wait to do it again. "Maybe this one'll be different." And that was the way it was. I just wanted to escape.

And so I remember reading about John Belushi in the book *Wired*. A lot of people read *Wired* and thought, "Man, that poor guy! I never wanna do drugs again!" But I was like, "Yeah! If that's what it takes, I wanna do it!" 'Cause I wanted to be like him in every way, like all those guys from that show. I thought that's what you had to do.

When I got outta school, I didn't know what I was gonna do with my life. I knew I didn't have much in the grades department, and so I was very fearful. A whole lot of fear. I remember drinking was the only time I felt, you know, good. I went and worked for my dad after school. I'd show up late and stuff like that. He was the boss, and so I was his screw-up son. I didn't get in too much trouble. He'd let it slide.

The one thing I knew was that I wanted to go into acting. I went down to Chicago to try to go into a place called Second City. I auditioned for that and got in pretty quickly, but I couldn't stop partying. They gave me a warning: "If you do it again, we're gonna kick you off the main stage."

I wanted to continue performing, so I only got high for the perfor-

Only this time, abandoning his popular, manic persona, he held the audience captive by simply standing at center stage and speaking in calm, measured tones. It was a Chris Farley that only a handful of close friends and family members ever knew. Dressed in a blue button-down shirt and stone-colored khakis, he paced a small circle, nervously fidgeting with his hands and running them through his slicked-back hair. At no point did he fall down or remove his pants. And there, alone on the stage that day with no crazy characters to hide behind, no wild-man stunts to impress, Chris gave a "motivational speech" quite unlike any he'd ever delivered on television.

CHRIS FARLEY:

Good to be here. Um, pretty nervous. I was here a couple times, so I know what it's like to be sitting where you are, full of fear and anxiety. Kinda how I'm feeling tonight! Heh heh heh!

Anyhoo.

I'm supposed to share my experience, strength, and hope with you, and so I'll start. I remember my first drink. I was seventeen years old, almost eighteen. My friend Patrick was a year above me and I admired him quite a bit, looked up to him. He was a great football player, all-state and everything like that. I went to a party with him one night. We went down in the basement. The guys started drinkin', and they went, "C'mon and take a drink, Chris."

So I took a shot, and I remember going, "Man, this sucks. I can't believe you do this."

"Just take it down like medicine," they said.

So I wolfed down about ten of 'em, no problem. And I remember hearing stuff like, "Man, I thought he was wild before, and now he's *really* gonna be a *wild* man!"

So that kind of planted in my cranium what I'd always wanted, and that was to fit in, or to be liked. Everyone seemed to love it. When I went upstairs, I remember the girls were like, "Great! Chris, *finally*!"

And I was like, "Yeah! Maybe I'll even get a chick now!"

CHAPTER 1

A Motivated Speaker

MIKE CLEARY, *friend, Edgewood High School:*

Freshman year of college we're heading out on a road trip to Milwaukee to see a big game. We're in the car. We've got the fifth of vodka, the gallon of orange juice. We're ready to get loaded and party. Just as we start to drive, Chris says, "Stop!"

We stop the car, and he pulls out a rosary. We have to sit there in the car and say one decade of the rosary—ten Hail Marys and an Our Father—before we can leave. Then he balls the rosary up in his hand, tosses it in the glove compartment, slams it shut, looks at all of us, and says, "Well, it's in God's hands now." And we hit the road.

On June 24, 1994, life for Chris Farley was good. He had just finished his fourth season on NBC's *Saturday Night Live* and was coming into his own as one of the most promising stars in American comedy. As the earnest, sad-sack Chippendales dancer and the swaggering, van-dwelling Motivational Speaker, Chris was bringing a kind of energy and anarchy to the show not seen since its seventies heyday.

Dana Carvey, Mike Myers, and Phil Hartman were stepping down as the show's reliable go-to players, and Chris was leading the charge—alongside Adam Sandler, Tim Meadows, David Spade, and Rob Schneider—in the next cycle of revitalizing and redefining the late-night institution for a new generation. Chris was also about to take on his first starring role in a feature film, *Tommy Boy*. The following year, *Tommy Boy* would open at the top of the North American box office and solidify Chris's status as a bankable movie star.

From his very first days onstage—starting in plays at summer camp and eventually at Chicago's Second City—Chris had possessed a singular talent for capturing and relating to an audience. In the words of *SNL* creator Lorne Michaels, "People liked Chris Farley, they trusted Chris Farley, and they thought they knew Chris Farley." In his lifetime, that likability translated to a huge following on television and three straight number one box-office hits. And since his death at the age of thirty-three, Chris's appeal persists. In the past ten years, *Saturday Night Live*'s *Best of Chris Farley* DVD has sold over a million copies, making it the second-best-selling title in the show's entire history. *Tommy Boy*, for its part, has gone on to become one of Paramount Studios' top-selling DVDs of all time.

But Chris's success had not come easily. His rise at *SNL* had been marred by a constant struggle with alcohol and drug addiction. High school and college drinking had given way, eventually, to cocaine and heroin use. Through the intervention of friends and family, Chris had attempted several different recovery programs, all of them eventually ineffective.

But on June 24, 1994, Chris was clean and sober and standing onstage in a large auditorium at Hazelden, a nationally renowned drug rehabilitation facility in Center City, Minnesota. Chris had visited Hazelden twice before, both times as a rather unwilling and uncooperative resident. This time he walked through its doors of his own free will, as a guest, invited to share his recovery experience with other addicts struggling in their own battles with the disease.

Chris's presence filled the auditorium; he knew how to work a room.

mance, on marijuana. Then afterward I couldn't wait to get ripped. I remember one time my director was giving me notes, and I drank a pint of Bacardi in about ten minutes, before he was done talking. He asked me a question, and I was slurring my words. He said, "Oh, you're no good. I'll talk to you tomorrow." But it was kind of tolerated. My lifestyle cradled it, because I didn't have to wake up in the morning. I could get blind drunk every night, and that's what I did.

Then I went to New York and started working on *Saturday Night Live*. That was, I thought, a dream come true. I'd read all about my idols and how they partied back then. I thought, man, this is gonna be great! I am gonna get *ripped*!

Well, that just wasn't the case. It wasn't hip anymore. I stuck out like a sore thumb, taking my clothes off at parties and making a fool of myself, which I had learned to do pretty good because I thought people would like me. Nobody's afraid of the fool. "Hi! C'mon, idiot! C'mon aboard!" I was totally full of fear. I'd do anything for you to like me, including doing things that I didn't want to do. As long as I had my substance, I was okay.

I went back to Second City after my first year on *Saturday Night Live* and took a bunch of acid and cocaine and a ton of liquor and went onstage and made a complete ass of myself. They booed. I remember during a blackout between scenes someone yelled, "Get the drunk off the stage!" That kinda rang true.

I had to cover my ass so they'd hire me back at *SNL* again next year. So I came here to the Shoemaker Unit at Hazelden. I hated every minute of it. I complied and kissed ass until the counselors went home and then screwed around and tried to make everybody on the unit laugh, and didn't take it serious one bit. Got outta there and thought I was cured. "All right, I did twenty-eight days sober, no problem."

So I got outta here, didn't go to meetings, didn't get a sponsor, didn't do anything that they told me. And guess what? I got back to New York and started doing a lot of drugs. I thought, if I don't drink in front of people, they're not gonna know I'm high. I thought I was fooling everyone, and I was fooling no one.

That Christmas, after a real bad bender, my apartment was just totally ripped out. I'd ripped apart drawers, everything was on the floor, because I'd been looking for something. "Oh, what's this? Is it in here? No?" *Crash!* "Where's this?! What's that?! Oh fuck!" So Christmas, coming home to surprise my parents, what a lovely gift I was. They put me in a detox for a couple of days.

The whole rest of that season I did the outpatient thing, which was a complete joke. I would comply with them and say, "I'm really trying hard." Meanwhile I always had a thing of urine in my pocket just in case they tested me. God, what an ass, asking my friends for their urine. "Kevin, yeah, you got that, uh, urine?" Jesus. Everyone knew I was using. I just remember a horrible dismay. I was crying all the time, because I could not stop. I couldn't imagine a life with sobriety, because drugs and alcohol were the only thing that was my friend. I knew I was in trouble.

I came back to Shoemaker. I decided to make sure they *knew* that I was trying. "By God, I'm your boy, boss. I'm *trying*. Pluggin' away!" So I screwed around and complied in treatment again, and didn't take it serious. I wasn't listening, and that's what you gotta do about this disease, because it's hell to stop.

I got outta there thinkin' I was cured. La di da! Didn't last even as long as I did the first time out, and by that time I had almost thrown in the towel. I went out to California to do some work the next summer. I got into another rehab out there. It was like every time I turned around I was in friggin' rehab. God, it sucks! But I kinda started takin' this one serious, because I was like, "I don't wanna come back here, man." The door was open just a little bit. I was sick of using, and I knew I was gonna be fired very soon. I didn't want that because *SNL* was everything I'd worked for.

They told me to go to Fellowship New York, a halfway house that had just opened. So I went there and this time I was gonna finish it, you know, give it a real shot. I was frightened of going to recovery meetings. Because what if I couldn't do it? That's what would really suck.

I was glad to be sober, but after ninety days people weren't patting me on the back anymore, sayin' "Good job on that sobriety! Go get 'em!"

People just expected it. And why shouldn't they expect me to be sober? I'm working for them. But I wanted the pats on the back, and they weren't doing that.

That ninety-day mark was a real tough one for me. After a bad day at read-through, the writers didn't write me into the show, and I was going back and forth. I used. I did five bags of heroin. Then I came back and told my boss. I thought if I was honest with him, you know? That's another manipulating tool. "I'm being honest with you, so you won't fire me, right? Because I'm *trying*. Can't you see I'm trying?" All that bullshit.

So, I lost my job for about a week. I kept begging and crying, the same manipulative things. Finally he gave me my job back, but he sent me to this place in Alabama, which was kind of like a boot camp. It was exactly what I needed, a good kick in the rear end. They told me stuff like "You're arrogant. You're complying." They made me cry every single day. They'd say that if you pick up drugs and alcohol, you're a baby. I didn't like to be called a baby. I didn't like to be called arrogant. I didn't like to be called all those things that I was.

It was around Christmas time, too. Man, what a horrible place to be over Christmas, you know? Hearing "Have yourself a merry little Christmas . . ." when I'm in a stinky hospital ward. But I did things in this treatment that I didn't do before, like making sure I made my bed every day. I practiced what I would be doing on the outside. I prayed to God in the morning to please keep me sober that day, and then I'd thank Him for keeping me sober every night.

So I got outta that thing in Alabama. I got a sponsor. I got a home group. I was reading from the Big Book. I went to a morning meeting every day at seven-thirty. I got involved, because I know I can't stay sober without these things, without going one hundred percent. But I can stay sober when I do. And sobriety's good, man. Sobriety's not carrying around urine jars—that's a real treat. It's not waking up in a horrible apartment with everything broken in it. I have a nice apartment now that's all taken care of. I make my bed every day. I do the things that I did in treatment. I

have a very healthy fear of getting high, and I have to take it serious, man. Because if I don't, I'm gonna use, and I cannot use again. I hate that shit. God, I hate it. I hate being a slave to that shit.

The ninety-day mark was a real kicker for me, again. I remember it was on St. Patrick's Day. I like to have an icy cold Guinness on St. Patrick's Day. I'm Irish! I have to drink, right? And I remember pacing back and forth in the rain outside a bar, crying. I was so scared, and I was just crying and crying and praying to God to help me. Then I stopped. I remembered that I don't have to drink. I called the halfway house, went to a meeting, and I did what I had to do. And today I have one year, six months, and six days. That's the most time I've ever had. And I can do this. I know I can do it.

We all can do it.

CHAPTER 2

Madison, Wisconsin

GREG MEYER, *friend:*

We were all sitting in the library one afternoon—me, Chris, Dan Healy, Mike Cleary, a bunch of guys. We're sitting at this table, and Chris is just cracking us up. Finally, he gets up to go to class, and as he's leaving somebody says, "He's going to be on *Saturday Night Live.*"

Everyone at the table just nodded. "Definitely."

Chris Farley's grandfather Donald Stephen Farley worked as an executive with the A&P supermarket company in Philadelphia, Pennsylvania. He lost everything he had in the 1930s Depression and returned to his family's home in Madison, Wisconsin. There he joined his brothers in a hardware business that sold machine parts and services. One of those services was laying asphalt roads, a lucrative field in the booming infrastructure build-out following World War II. Hanging out their shingle as Farley Oil, the brothers bought and sold road-paving contracts. They were middlemen, salesmen. They bid on contracts with

state and county officials and in turn brokered the services of the crews that laid the actual roads.

Donald's son Tom Farley, the second-youngest of six, applied for a special driver's license and began driving for the family business at the age of fourteen. Later, at Georgetown University in Washington, D.C., he discovered his calling in the game of politics. He soon found himself president of the campus Young Republicans and a frequent dinner guest of Wisconsin senator Joseph McCarthy.

During his senior year, Tom met Mary Anne Crosby, the daughter of an established Boston family and a student at Marymount College. Upon graduation, he moved back to Madison to attend law school at the University of Wisconsin, the first step in his plans to seek a career in law and a future in elected office. Mary Anne followed him to the Midwest, and they married in 1959.

The following year, Donald Farley suffered two massive heart attacks. He could no longer run the family business or support a family. With two parents, several siblings, a wife, and a new daughter depending on him, Tom had little choice. With only one year of law school remaining, he quit, shelved his dreams, and for the next thirty years plowed his expensive East Coast education and considerable personal charm into selling asphalt.

He sold a lot of it.

As a partner in Farley Oil—and later owner of his own company, Scotch Oil—Tom Farley was very successful. He became well known across the state, thriving in a business run entirely on his boisterous laugh and hearty handshake. His success gave him the means to provide for his family, which in the Irish Catholic tradition would soon grow quite large. Tom and Mary Anne's daughter, Barbara, was born in 1960; Tom Jr. a year after that. Two years later, on February 15, 1964, at 3:34 P.M., Mary Anne gave birth to her second son, Christopher Crosby Farley. He weighed eight pounds, fifteen ounces. Next came Kevin Farley in 1965, and then finally John, the youngest, in 1968.

Although Tom Farley, Sr., had grown up in a middle-class pocket of

Madison proper, when it came time to make a home of his own, he moved to the Village of Maple Bluff. Maple Bluff was, and is, an idyllic slice of affluent twentieth-century suburbia. Clustered on the eastern shore of Madison's Lake Mendota, it is home to the governor's executive mansion as well as the stately residence of one Oscar Mayer, proprietor of a local luncheon meat and hot dog concern. There, among Maple Bluff's tree-canopied lanes and rolling green lawns, Tom and Mary Anne raised their children. Over the next fifteen years they lived in four different homes, each one bigger than the one before. The last had a commanding lake-front view. Growing up, Chris would lack for little in the way of material comfort. The Farleys lived well. On paper, at least, it looked like the American Dream.

TOM FARLEY, *brother:*

When Chris came along, my grandmother insisted that my mom wasn't going to be able to handle three kids at once, so this Spanish woman came to help the family. My first memory is this woman coming into our lives because of Chris. I always remember that Chris got special attention.

KEVIN FARLEY, *brother:*

Maple Bluff was a great neighborhood. We were always outside playing, jumping in the leaves, riding our bikes, like kids do.

Chris was always popular, right off the bat. He always wanted to start up a game, get everyone together. We'd play kick the can or ghost in the graveyard, which was what we called hide-and-seek. I was the shy kid, and I was amazed at how he could make friends so easily. We changed schools a good bit, but no matter what school Chris went to, he always instantly had a new group of friends. Making people laugh was just instinctive. And also he looked to Dad. Dad was very outgoing. My parents always had parties, were very involved in the community. A lot of that carried over for Chris.

What I remember most from the earliest years are the Christmases we used to have. That was always a big event. Whenever the relatives came over we were sort of made to dress up and look nice, basically put on a show for the rest of the family, talking to all the aunts and uncles. Dad insisted on that.

TOM FARLEY:

It's been explained to me by more than a few therapists that we exhibited a typical Irish Family Syndrome. The father is the bullhorn and the head of the family, but not really the head of the family. It's really the mother who keeps everything together, and Mom always did. Our life was straight out of *Angela's Ashes*, only, you know, with plenty of money. Dad always drove the big Cadillac. We were certainly well off by Wisconsin standards, or at least gave the impression that we were. There was a point when we were all taken out of the parochial schools and sent to public schools for a year. Dad had some excuse that, looking back, didn't really hold water. But this was 1974, and Dad was in the oil business. He'd had a bad year and couldn't keep up with the tuition. But he always kept up appearances that everything was fine.

KEVIN FARLEY:

Dad loved politics. He ran for school board at one point, but didn't win. That was probably because he had all his kids in private school. They sort of hammered him on that. But we went out and put up signs for the race. Dad joined the board for Maple Bluff. It was a subdivision, but it had its own councils and so on. He enjoyed that immensely. He was a conservative man, politically, and very civic-minded.

TOM FARLEY:

Dad's voice was a sonic boom. All he'd say was, "*It's time to go to mass! Everybody in the car!*" and you'd scramble like it was a DEFCON 4 siren

from the Strategic Air Command. You didn't want to get on his bad side. He was very lenient, but with four hyperactive boys, somebody's got to crack the whip sometimes. And when the whip would crack, it would crack hard.

KEVIN FARLEY:

He was very strict, but if you could get a laugh out of him, you were okay. And Chris knew that. One time Chris walked into Mrs. Jennings's class at Edgewood Grade School and said, "Excuse me, Mrs. Jennings, where do I 'shit' down?"

She hit the roof and called my dad in for a conference. She told him what happened, and said Chris needed to be suspended. Chris was like, "I didn't say it, honest."

And Dad said, "Well, Chris says he didn't say that. And if my son says he didn't do it, then I believe him. You must have heard him wrong."

So she backed down. Then, on the way home, Dad turned to Chris and went, "You said it, didn't you?"

"Yeah."

"I knew it."

They both had a laugh over it, and that was it. He knew Chris had done it, but it was okay to laugh as long as nobody got hurt. Those kinds of incidents cropped up all the time.

As strict as Dad could be, when he decided it was time to have fun, it was time to have fun. We would pile into the station wagon and go shopping or out to mass. Sometimes we'd go out to the apple orchards to pick apples. The church bazaars my dad loved. He'd come in and say, "There's a church bazaar out in Lodi!"

And we'd go, "Aw, jeez . . ."

And then we'd all get in the car and go all the way out to Lodi for homemade pies and such at this bazaar out in a farm field somewhere. The rituals of our house when we were young all centered around the family. There was never a time when we wanted to rebel and get away from it.

JOHN FARLEY, *brother:*

Family dinners were very important. We had a dinner bell. Anything we were doing anywhere in the neighborhood, we could hear this giant bell outside our kitchen. We'd stop what we were doing—setting fires, whatever—our heads would pop up like deer and we'd run home.

There were actually two bells. There was our dinner bell at six-thirty, and there was also a giant whistle that would blow through the entire neighborhood at five o'clock. It wasn't from a factory. It wasn't the emergency broadcast system. It was just a whistle that the town of Maple Bluff had. Why it went off every day at five we still don't know. We assumed it meant it was time for all the families to start their cocktail hour.

KEVIN FARLEY:

Other than family, the one thing that was important to my parents was education, in particular a Catholic education. Some parents are really hard on good grades, but our parents cared more that we learned how to be good people, that we had big hearts and were kind. I don't know of any better guy in the world than my dad, just in terms of being a strong, moral person. He always stressed that in us.

TOM FARLEY:

If Dad instilled anything in Chris it was this love of the underdog, for the kid that's getting picked on. If we were driving down the road and you made a joke about some strange-looking homeless person out on the sidewalk, man, he'd lock those brakes up and the hand would come back. You didn't dare do that.

My dad was very Catholic, and in Catholicism that whole idea of right and wrong, good and evil is very important. Chris was very aware of that from an early age. It all stemmed from *The Exorcist*. The mere fact that we'd seen that movie brought the devil into our house, and that

started this whole superstition in Chris, not just of good and evil, but the literal, physical devil. He and I shared a bedroom for a time, and he was next to the closet; that just freaked the hell out of him. "Tommy, we gotta change beds," he'd say. "Tommy, please. The devil's in the closet."

KEVIN FARLEY:

Every night for months after *The Exorcist* came out, he'd just show up in our room with a sleeping bag and crash on the floor between Johnny and me. It was sort of an unspoken thing. If you asked him why, he'd say, "Shut up, okay? I'm just sleeping here." Chris was afraid of the dark, and he hated sleeping alone.

He was a very spiritual person, instinctively spiritual, and he'd always talk about it, so much so that he'd scare the crap out of you. As you grow up, even though you still call the devil by name, you begin to understand him as a spiritual idea, and a lot of people stop believing in the devil altogether, which, of course, is exactly what the devil wants. But Chris, he believed in the devil. He believed in hell, and it scared him.

TOM FARLEY:

He prayed to St. Michael the Archangel every night, because Michael was the one who'd thrown Lucifer out of heaven. It was more superstition than spirituality, to be honest. He read something once about the different ways your shoes land after you take them off means different kinds of luck. If your shoe was to one side, it was bad luck. If it was upright it was good luck, and so on. So every night I'd kick off my shoes, not caring where they landed, and Chris would say, "Tommy, pick up your shoe and set it right."

"No."

"C'mon."

"No."

"Please."

"Do it yourself."

And he'd get out of bed and go and move my shoes; he felt that strongly about it.

KEVIN FARLEY:

Growing up, Chris was wild and crazy and liked to have fun, and Tommy was more reserved. It really reflected, more than anything, the two sides of our dad. Dad would carry himself as this very professional gentleman, but he could also be this boisterous, crazy, laugh-out-loud kind of guy. And Tom and Chris were the two sides of that personality. To the extreme, really. John and I are somewhere in the middle.

TOM FARLEY:

Kevin was very focused, got decent grades. We called him Silent Sam, Steady Eddie. He just did his thing and did it well. John was the gopher. He was so much younger than the rest of us. He was always pleasing people, doing what it took to tag along. Still is to this day. As for myself, I was the brains in the family, which is really kind of sad. But I was Tom Farley, Jr., and everything that that entailed. My dad went to Georgetown, and so from day one the pressure was on me as the oldest son to live up to Dad's expectations.

The expectations for Chris were that there were no expectations. He just kind of marched to his own drummer. One day Chris said, "I want to join the hockey team." The next day he had a brand-new set of hockey gear, never mind that he couldn't really skate that well. So there was full support for him in whatever he wanted to do, but no real expectation that he should fail or succeed.

Chris and I were always together, but I was trying my best to toe the line and he was effortlessly crossing over the line, trampling it with no consequences; it annoyed the crap out of me. And because he was always so funny, my friends would want him to hang around. I hated that.

KEVIN FARLEY:

What was most important to Chris, really, was that he made people laugh. Chris was always the fat kid. Kids can be pretty mean, and humor was his only weapon, from grade school on. He wanted to be a football player, and that meant being part of the popular crowd. He used his humor to do that.

TED DONDANVILLE, *friend, Red Arrow Camp:*

I met Chris at summer camp, with all the other brothers. Tom was actually my counselor, and Johnny wound up being my best friend. You didn't forget Chris. Even if I'd never seen him again after camp, I'd remember him. During mass, if the priest made the mistake of asking for audience involvement, Chris was right there. His hand would shoot up, and then he'd figure out whether or not he had something to say.

DICK WENZELL, *play director, Red Arrow Camp:*

Red Arrow Camp was established in 1922, and was named after the Red Arrow Army, Second Division, from Wisconsin. It had originally been built as a logging camp in the nineteenth century. Some of the cabins date back to that time. It was a resident seven-week camp.

TIM HENRY, *friend, Red Arrow Camp:*

Chris always had some kind of stunt going. On Sundays they'd load all us Catholic boys into this old school bus and drive us into town. The girls' camps would come to Sunday mass, too. Now, you're never allowed to have candy at camp, but somehow one Sunday Chris has gotten ahold of these white tic tacs. He fills his mouth with them, and he's walking up the aisle for communion so prayerfully, and when he gets in front of the girl campers he rolls his eyes back like he's going to pass out and then he falls

and hits his mouth on the side of the bench and spits out all the tic tacs. They go clattering across this wooden bench, and Chris is yelling, "Oh my God! My teeth!" The girls were just aghast. We were all laughing hysterically.

HAMILTON DAVIS, *friend, Red Arrow Camp:*

It was anything for a laugh, absolutely anything. They gave away all these awards for good behavior and accomplishments and such. Chris didn't care.

TOM FARLEY:

He was our windup toy. You said it. He did it.

KEVIN FARLEY:

He didn't win a lot of the awards, but because he was so funny they'd put him in the camp play, and he was the star. Chris would always credit Dick Wenzell with encouraging that in him.

DICK WENZELL:

Chris was strictly a jock, but he had a lot of charisma. Once I got him onstage, his connection with the audience was unbelievable. Not only could he project to the audience, he could also receive from them. Chris could take whatever the audience gave to him and build on it. He just did it naturally. Visiting parents would comment on how magnetic he was. And this was when he was ten years old.

TED DONDANVILLE:

The camp play was really a bunch of skits strung together. One year they did a takeoff on *Snow White and the Seven Dwarfs*, and Chris played the villain. When they caught him, he told them they could do what they wanted to him as long as they didn't step on his blue suede shoes. Then he launched into this Elvis impersonation that brought the house down.

HAMILTON DAVIS:

Whatever the story was, they'd just drop him in there. He was such a crowd-pleaser that they didn't even have to have a part for him. Just put him up onstage. One time he did "Hound Dog" dressed up like Miss Piggy.

FRED ALBRIGHT, *counselor, Red Arrow Camp:*

When people ask "Where did Chris Farley get his start?" I say he got it at Red Arrow Camp. As a kid, he was just a miniature version of what he would become. Dick Wenzell used the expression "He was always onstage." And that was the case. A lot of it was a diversion, because down deep Chris was one of the most sensitive guys you'll ever meet. Even though he came across as this kind of rough, gruff, jovial guy, you could hurt his feelings with just a word or two. Incredibly sensitive guy.

MIKE CLEARY:

Chris was the very first guy I met at Edgewood. I grew up in Scarsdale, New York, and moved to Madison in high school. Edgewood can be a little clubby. For days, the rest of the kids didn't even come near me. Then one day I was sitting in the commons, getting ready for football practice. Chris came up and said, "You're the guy from New York? Hi, I'm Chris Farley." He was the first person to make me feel comfortable being there.

ROBERT BARRY, *friend:*
At school you always wanted to be around Chris. He was a blast, but his focus was always on you, talking *you* up, making *you* feel better. "This is my buddy Robert," he'd say. "He's all-state basketball." He's this. He's that. He's the greatest. It was never about himself.

TODD GREEN, *friend:*

We had the closest thing to what I would call a dream high school situation, where six or seven guys were as close as brothers and laughed their asses off every single day, and Chris was the glue that kept us together. He was such a pivotal part of our high school experience. Chris was the type of person who didn't see social class, or ethnicity, or anything like that. He came from a lot more money than most of us, but you would never know.

MIKE CLEARY:

One time I was visited by an old friend of mine from back east. He'd been a big football player, but he'd had this horrible car accident, and now he was in a wheelchair. A lot of kids could be uncomfortable around that, but Chris just embraced him. He spent the entire day making this kid feel welcome and totally at ease. And the thing is, he did it with no effort. His generosity was so commonplace that it was utterly unremarkable.

DAN HEALY, *friend:*

Chris made people feel good about themselves. Everyone was on a pedestal for some reason or another. He drew people together, naturally, and it was cathartic to be around him. To me, Chris would bat his eye and I would lose it laughing so hard my sides would hurt.

GREG MEYER:

People ask me what it was like going to school with Chris Farley, and I say, "You've seen him on *SNL*, right?"

"Yeah."

"Well, crank that up times ten."

JOHN FARLEY:

With the birth of the VCR, we memorized *Animal House*, *Stripes*, *Caddyshack*, *Meatballs*, *History of the World*, *High Anxiety*, and *Blues Brothers*. And I'm not talking about memorizing the lines. We memorized everything, every inch of footage. The foreground, the background, we memorized it all. And Chris pulled from that constantly.

One of Chris's favorite bits to do was to put his arms out like Frankenstein and make this monster voice, "Urgggh duugggh!" That was from an obscure scene in *Meatballs* with Spaz introducing himself to the cabin where, in the background, some fat camper was doing that Frankenstein thing. The whole thing was maybe half a second of film, maybe. But even that we had down. Take the original cast of *Saturday Night Live*, add in Mel Brooks, and you have our childhood.

NICK BURROWS, *guidance counselor/assistant football coach:*

Every time you'd walk down to the cafeteria, packed full of three-hundred-plus kids, all you had to do was listen for the roar of laughter and you'd know where Chris Farley was sitting. As I remember, Chris didn't really tell jokes. It was just who he was. He just *was funny,* being himself. People just liked hanging around him. I was his guidance counselor, and I liked hanging around him.

JOEL MATURI, *dean of discipline/head football coach:*

His antics were never mean or destructive. Chris did a lot of crazy things, but most of the stories of Chris being in trouble at Edgewood are, unfortunately, fabricated. I say "unfortunately" because they sound very entertaining many years later. There is one hilarious story about Chris and a nun that I know for a fact just isn't true.

NICK BURROWS:

We had a geometry teacher named Colonel McGivern. He was a retired air force colonel. Back then we had a lecture hall where all the kids would go for these huge group lectures while the colonel did equations and theorems on this big overhead projector. Well, one day Chris gets down in the aisle and belly-crawls up to the front of the room. He gets to the stage stairs, waits for Colonel McGivern to turn back to the projector, and then sneaks up and around, behind the curtain.

Now, Colonel McGivern had this thing called the Groaner of the Day, a really bad, corny joke that he'd use to end each day's lecture. He'd tell it, and the kids would all groan because it was so lame. So Chris waits, and just as Colonel McGivern delivers his punch line, Chris drops his pants and moons the entire audience, sticking his rear end out between two folds in the curtain. Well, the whole place erupts with laughter, and the colonel—who can't see Chris—stands there scratching his head, going, "Jeez, I didn't think it was that funny . . ." And then of course everyone really loses it.

Some sophomore girl gets offended, and she rats Chris out. I get a call from Joel Maturi, telling me that Chris has done this thing and needs to be punished.

"Nick, uh, are you familiar with the term 'hung a moon'?" he asks me.

"Sure, Coach."

"Well, that's what he did, and we need to get his parents over here and sort this out."

So I call Mr. and Mrs. Farley, and they come down to the office. I tell the Farleys about Chris hanging a moon and the colonel and the Groaner of the Day, the whole story. And Mrs. Farley just busts out laughing. She can't stop. Then I start laughing. Then Mr. Farley starts losing it, too. So here we all are dying laughing, waiting for the dean of discipline to come down.

Now, Joel Maturi is a real straight-arrow, buttoned-down kind of guy. We all straighten up as he comes in with his yellow legal pad where he's got the incident written down. "Mr. and Mrs. Farley," he says, very businesslike, sitting down, reading from his notes. "Ahem. Yes, it would appear, Mr. and Mrs. Farley, that your son has 'hung a moon' in his geometry class."

Mrs. Farley loses it again. She gets me laughing. Mr. Farley busts out again. And finally Maturi, as stiff as he is, he starts laughing. We're all roaring in my office. Finally, Maturi takes the legal pad, chucks it on the table, and says, "We're just going to forget about this one."

It should have been a suspension, but he threw it out because, quite frankly, we all thought it was too funny.

TOM FARLEY:

Chris would show up in Room 217, the detention hall, on a fairly regular basis, and Coach Maturi always seemed to be laughing under his breath, saying, "*. . . dammit, Farley.*"

When Chris was sorry, he was genuinely sorry. He'd be so guilty and remorseful, and he would always take his punishment. He knew it was the price to pay for getting the laugh. But before that apology would come, he had to get a laugh and you had to admit that it was funny.

Chris's bedroom was at the other end of the hallway from the bathroom, and lots of times he was just too lazy to get up and go. So what he did was he kept glasses in his bedroom. He'd pee in those, and then take them down to the bathroom and empty them out once he was ready to get up. Well, one time my mom found one of the glasses. We were all sitting at the dinner table the next day, and she had told my dad about it.

Dad was so furious he didn't know what to do. I mean, here was Chris peeing in our drinking glasses. Chris knew that Dad knew, and Dad knew that Chris knew that he knew, and there was dead silence at the table. Everyone was waiting for the other shoe to drop. Finally, Dad reaches over to take a sip of his water, and Chris goes, "You're not gonna drink out of *that* glass, are you?"

"Goddammit!"

Chris knew what was coming, but he had to get the laugh first.

KEVIN FARLEY:

Chris was such a natural talent that he was always being asked by the teachers to try out for the school plays. But he wouldn't have any part of it. "That's for pussies," he'd say.

DAN HEALY:

Madison is sports crazy. They'll watch anything played with a ball. It wasn't cool to do drama. In a perfect world I think Chris would have been about six-foot-three and played in the NFL. I remember when we started freshman football. It was a big deal at Edgewood. You knew making the football team was a key part of fitting in. After one of those first practices I heard this voice behind me say, "Well, my brother told me that if I can start on 'O,' then I'll probably start on 'D.' "

I turned around, and there was Chris. He was already pretty overweight, and he was wearing these saggy gray wool socks with his football uniform. Everyone else was in bright white athletic socks, and here's this chubby kid dressed kind of funny. I just thought, this poor kid actually thinks he's gonna play? But he did. And he was great.

KEVIN FARLEY:

Chris would play noseguard, and because I was his brother he'd want to hit me pretty hard in practice. My God, he hit me good. A couple of times I think I practically blacked out. "God, you really nailed me," I'd say.

"I know," he'd say. "I wanted to."

We were really competitive, and he was a really good football player. You couldn't move him because his legs were so powerful and he was so low to the ground. He was a really good interior lineman. He just didn't have the height or the NFL build that he needed.

JOEL MATURI:

Chris was like Rudy in that way, the kid in that story from Notre Dame. I mean that very honestly. He was all hustle. Chris would be the first one to jump into a drill, first one to volunteer for anything. We were not a great football program. I think his junior year we went 5–5, and his senior year we went 6–4. But Chris always thought the glass was half full. When other kids might have said something was impossible, Chris thought it was possible. He always *believed.* And that's why you loved him.

MIKE CLEARY:

Around Madison you played against teams with these huge, enormous guys who went on to Division One teams. They'd just steamroll right over you. It was so demoralizing. But with Chris on the field, you'd never let that get to you. He'd never let you forget about having fun, even when these future NFLers were grinding your face into the mud.

PAT O'GARA, *friend:*

When we played football, all the guys would go in to take showers. And of course all these sophomores and freshmen were nervous about shower-

ing with the older guys. Chris would be in there, in the showers, buck naked, curling his finger with a come-hither look at these kids going "Want some candy?" It'd scare the crap out of them. It was always an interesting time when Chris would hit the showers. He had a reputation of, well, exposing himself. All the time.

GREG MEYER:

He was naked a lot.

PAT O'GARA:

Wasn't ashamed at all. So, junior year, I was sitting in typing lab, practicing, and Chris was sitting next to me. I said to him, "Chris, I dare you to whip it out in front of this girl here."

I'm typing away, and Chris just pulls his pants down and lays it out. I don't think twice about it. To me, I'm just like, "Jesus, what a sick bastard."

And that was the end of it. Nothing happened. Then, about a month later, Coach Maturi says to me, "I understand that you dared Chris Farley to expose himself to this girl."

"Yes, sir," I said. "I said that."

"Well, she's been in therapy for about a month now because she keeps having flashbacks and has had a lot of psychological problems."

A prank like that would normally just get you disciplined somehow, but this girl and her parents were making a real issue out of it. Chris wound up getting expelled.

TOM FARLEY:

Dad, typically, said the school was overreacting. It was the school's fault, not Chris's. So rather than demoting him down to public school, he was sent off to private boarding school, La Lumiere in LaPorte, Indiana, for the rest of the semester. Senior year they let him back in.

GREG MEYER:

Chris was really, really pissed at himself, very disappointed to be leaving Edgewood. He'd do that thing out of nowhere where he'd smack his head and go, "Fuck! *Idiot!* Can't believe I did that."

KEVIN FARLEY:

When he would get into trouble, even as a kid, it was like something would take control of him that he couldn't help but cut up and make people laugh, even though he knew he'd get in trouble for it. And he'd look back at whatever stupid thing he'd just done and be like, "Why did I do that? Why did I get myself in detention just to get a laugh out of Dan Healy?"

NICK BURROWS:

The thing is, he was accepted by his friends and the other kids without all the crazy behavior. Quite frankly, he didn't need it. But he felt that he needed it. Someone at Marquette told me a story about a time on a Monday morning on campus. It was a dreary winter morning, and Chris all of a sudden just broke out running, dove into a snowbank, and started kicking his legs out in the air. Now, why would he feel compelled to do that?

GREG MEYER:

He was always insecure about his weight. He'd project this attitude of not caring to everyone, but among the inner circle of guys, he talked about it quite a lot. He said it was the worst thing in his life.

TOM FARLEY:

At some point, Chris started getting into his share of fights. We were at a basketball game in Stoughton, Wisconsin. I was there with a girl, and

word came up through the bleachers that Chris was down, just beaten and bloodied. Someone had called him a fatty, and that was enough for Chris to go off. We took him to the hospital and patched him up. Ruined my date.

HAMILTON DAVIS:

He was always fighting his weight, and I mean bad. He wanted to get with the guys who were lifting weights, would play any sport. He just wanted that weight off him so bad. We were both big guys. One summer he was a cookie, working in the kitchen, and I was a counselor, and we went on this diet together. We lost a bunch of weight, thirty pounds or so. He looked great. Chris always wanted to be mainstream, getting a girl and all that.

KIT SEELIGER, *girlfriend:*

I met Chris freshman year in high school. We had a lot of classes together, and we were pals. For Valentine's Day, Edgewood would do carnation sales. You could buy them for your girlfriend or whatever, and sophomore year Chris bought me one and gave me a box of candy. I think, as far as high school goes, we were boyfriend and girlfriend then. I was a late bloomer. I didn't have boys beating down my door, by any means, and maybe I felt safe to him. I'm sure that probably had something to do with it.

We talked on the phone a lot. Then as we got a little bit older we'd go out on dates and so forth. Chris was pretty insecure when it came to dating and girls, but the fact that we were friends beforehand made it a little bit easier. When his parents took him out of Edgewood and sent him to La Lumiere, that's when our relationship was probably the most serious. We talked on the phone probably every single night. We wrote long letters, back and forth to each other once or twice a week. We'd end each letter by writing, "I miss you so much," and it was a game to see how many *o*'s you could put in the word "so." It was all pretty innocent, really. For the

most part, we were just good friends. I know he told me things that he never told his guy friends. He would say he loved me, and things like that, things I know he never admitted to his friends that he'd said.

Then, the summer going into senior year, Chris broke up with me. I was shocked. But it was after a time when he had started standing me up. My mom couldn't stand Chris. She'd get so angry at the thought of my waiting around for him. But around the time we broke up, things were really taking a turn for the worse.

PAT O'GARA:

Toward the end of Chris's junior year, we were at a party one night at Pete Kay's house. Chris and I were downstairs in the basement all alone. I was drinking something. Chris had a bunch of beers, and he was just slamming them. I was like, "What are you doing? Slow down. This is ridiculous."

But he went and got drunk. I remember the next football practice on Monday he kept telling everyone, "O'Gara got me drunk!" He actually told one of the assistant coaches, who was like, "What the hell'd you do to Farley?"

That's how it all started.

TODD GREEN:

For the first two and a half years of high school, Chris was adamantly against drinking or doing drugs of any kind. Totally and completely.

ROBERT BARRY:

Chris didn't drink until he was a junior, but the minute he did it was all over. I don't ever remember a time when I could sit down with Chris and have "a beer" and have a conversation. It was balls to the wall, all the time.

KEVIN FARLEY:

The first time he got drunk, he came in, woke me up, and got into a fight with me. Then he woke up in the morning and didn't remember it, and I was like, "Whoa. What's wrong here?"

It was the first sip he ever had, and there was a complete personality change. Everybody was drinking in high school, but nobody I knew had the kind of reaction he had. Chris was a fun guy, an outgoing guy, and it turned him into kind of a monster. It was like pouring gasoline on a fire. And I don't think he liked it. He didn't like being that guy, but at the same time he craved it.

And nobody talked about it. You don't narc on your brother. All I knew was that everyone went out to the football games and went drinking. That's what you did. We just thought, hey! Chris is *so* crazy! And there were other guys just as hard-core as he was, passed out at parties and all that.

He got in trouble a couple of times, but for the most part it just slid by. We had these bedrooms downstairs where it was real easy to sneak out of the house. You could come and go, and you had plenty of time to clean yourself up before you went upstairs and saw the folks. I don't know how much they knew what was really going on. They didn't know. They didn't want to know.

TOM FARLEY:

Drinking was okay in our house. There was alcoholism on my mom's side of the family. On my dad's side, well, it was Wisconsin. You drank first and asked questions later. Our father was "a couple of tumblers of scotch a night" guy. From early on, we all knew what five o'clock meant. It was all right out there in the open.

The drinking age was eighteen, which effectively made it sixteen, so for us to go out and drink in high school was no big deal. All you did in Madison on Friday nights was cruise around, eat, drink, and then every-

one would end up at Rocky's Pizza. Whenever we went out, Dad would call us into the living room and dole out twenty-dollar bills to all of us. "C'mon in here!" he'd say. "Here you go. Buy a round for the boys on me!"

"But we're only seventeen."

"Be careful!"

We'd all walk out the door, head right down to Vic Pierce's liquor store, and charge a six-pack on Dad's account. You were ready to go. Chris was very generous, and Dad always made sure we had the means to be generous. It was part of being a Farley.

TODD GREEN:

Mr. Farley always made you feel like a million bucks as soon as you walked in the door of that house. It was always, "Hey! How ya doin'?!" All the Edgewood guys, we loved Mr. Farley, loved him.

DAN HEALY:

You heard all these stories about Mr. Farley being friendly with Joseph McCarthy and all these very powerful, conservative people in Wisconsin. He talked it and walked it.

TOM FARLEY:

My dad created an image for himself. He was always dressed in a custom-tailored blue blazer and slacks, perfectly starched shirt and tie. It was always, "Hey, Dad looks great! Check out his new blazer!" And, of course, no comment that the blazer was a size sixty-five because Dad was hitting four hundred pounds.

ROBERT BARRY:

He was an overwhelming personality. Just a loud, gregarious guy. His whole business was going around schmoozing people all over the state, and he was very good at it. The Farleys had this image they projected, living in Maple Bluff, having the status symbols. They would do a lot of different things to cover their problems up. But when you look at the family now, you can see how much of a façade it all was.

JOHN FARLEY:

Maple Bluff was a fantasyland, a place where there weren't any consequences. What we'd do in the Farley household, back in the eighties, back when everyone drank, was we had *fun.* We'd party and laugh. We'd put on little comedy sketches with each other. We weren't trying to be comedians; we were just doing what came naturally and what we liked to do, just joking around. Didn't know drinking was bad for you. Didn't have a clue. We grew up in a time and a place where at five o'clock every day, everyone would break out the cocktails. We thought it was normal. Put a gun to our heads, we thought it was normal.

TOM FARLEY:

We lived in a make-believe world. We were living with the elephant in the room—the literal elephant in the room—that no one wanted to talk about. My dad weighed six hundred pounds by the time he died. But Dad wasn't overweight. Dad didn't drink too much. Dad was just Dad. We didn't talk about it among ourselves, and we certainly didn't talk about it with anyone else.

Mom was really the only one to acknowledge what was going on with Chris. She was the voice in the wilderness, and for years there was nobody else on board with her.

"What's wrong with Chris?"

"Your mother and I are handling it."

And that was as far as it went for a long, long time. Nobody was willing to face the truth. Nobody confronted Chris about his problem, because doing so would have meant acknowledging that Dad had a problem—that we all had a problem.

CHAPTER 3

An Epiphany

DICK WENZELL:

As an actor, it took me a long time to learn how to let the character take me over, to lose myself in it. I think Chris figured that out very early in life, and he could hide himself by being anybody else that he wanted to be.

After Chris's expulsion and subsequent stint at La Lumiere prep school, he was allowed to return to Edgewood for the whole of his senior year. He graduated in the spring of 1982 with little direction beyond a vague sense that he might one day work in the family business. But Chris's father firmly believed that all of his children should receive a Catholic education, and so Chris found himself enrolled at Marquette, a Jesuit university in Milwaukee, Wisconsin.

Though Chris's grades did not make the cut for admission, they did qualify him for the university's Freshman Frontiers Program, a preliminary summer school in which students could earn their way into the incoming class. Together with high school friend Dan Healy, Chris entered the program, passed, and began regular course work in the fall.

It has been said that college is less a place for academic achievement and more a rite of passage through which to discover oneself. By that standard, Chris's years at Marquette were a ringing success. As always, he won friends easily, especially among the Marquette rugby team—the gung-ho, gonzo athletes who would become his ad hoc fraternity. Given the preferences of his father, and lacking any clear ones of his own, Chris started out in the business studies program. Bored and disinterested, he did not do well. However, as his understanding of his talents grew, he changed course and dedicated himself to a career in acting and comedy. Though his early attempts at performance were frequently clumsy, Chris kept at it with a dedication he had previously shown only in his football uniform.

As a Jesuit institution, Marquette teaches its students to pursue knowledge and personal growth not only for their own self-advancement but also for the glory of God and service to others. Chris's four years in Milwaukee would give him the opportunity to do precisely that.

JIM MURPHY, *friend:*

When he first came to Marquette, he was a very preppy guy. Super preppy. He would always have his hair combed and always had polo shirts and khakis and Top-Siders. Then over time he would take those Ralph Lauren button-down oxfords and he'd rip the sleeves off. Then, in the next phase, the khakis became army fatigues and the oxfords turned into flannel shirts. It was the same thing everyone did in college. We all kind of came in as one thing and left as another.

FR. MATT FOLEY, *friend:*

The first time I saw Chris was on the rugby pitch. I was the president of the rugby team, and a sophomore. He was a freshman. He was wearing some kind of obnoxious chartreuse-colored polo shirt with argyle shorts and gym shoes. I was drawn to him right away because I thought, oh, this poor soul is going to get his ass kicked. The guys on the rugby team are a

little rough, and here comes the preppy, portly kid. But he hustled and made many friends in no time because he was such a classic character.

One of the great things about Chris was that he was so very generous. Marquette would play Madison every year in rugby, so we went out there that fall. We'd been drinking all day, and Chris was like, "Let's all go over to my place!"

We get there, and Mr. Farley and Mrs. Farley are just the most gracious hosts. All these drunk, dirty rugby guys are running around their house—this is around one in the morning—and Mrs. Farley is going, "Oh, you boys! Let me make some sandwiches." And she just brought out all these beautiful sandwiches.

We rode back to Milwaukee in the back of a pickup, freezing our asses off. It had a camper top on it, but it wasn't very warm. The bars in Madison closed at one, but the bars at Marquette didn't close until three. You could close the bars in one town and still make it to close the bars in the other. That's the rugby mentality at its finest.

KEVIN FARLEY:

When I got to Marquette, Chris had only been there two years, and he was already a legend, flat out the funniest guy on campus, and that really grew out of the rugby parties. The Avalanche was a typical Milwaukee bar. That's where the rugby players would go party after their games. It had a great jukebox and a couple of pool tables. They had fifty-cent Red, White and Blue beers. When you finished your bottle, you threw it against the back wall of the bar.

Chris started doing this thing after the games: naked beer slides. Everyone in the bar would pour out their beer and he'd take off all his clothes and take a running start and slide across the bar like Pete Rose coming into home base. The Avalanche is long gone, but people still do it. It's something of a Marquette tradition. People have built a legend around it.

JIM MURPHY:

It got to a point where every team coming in to play Marquette had heard about the beer slides and wanted to see them. Chris would start to take his clothes off, going, "Aw, man. Why do I always have to do this?" But he'd kind of set the tone early, and he always had to live up to himself.

PAT FINN, *friend:*

Any time during a game that there was a lull, or any opportunity for a laugh, you'd just look over at Chris. One game, we file in for this line out. Everyone is waiting for the inbound pass. Chris kind of looks over and then pulls up his pants into his ass like a thong. The whole other team just turns and stares, like, "What the hell's with this guy?" Then, with the whole team distracted, Chris takes the pass and gets about twenty yards on the play. He was always hilarious.

DAN HEALY:

Freshman year of college, a group of us had done a road trip for some winter festival. It was freezing, snow everywhere, and we were goofing around, diving and playing in the snow. And while we were all screwing around, Chris just stopped cold. He stopped, and he turned to me and said, "I think I can make people laugh."

He'd had an epiphany, literally. He was starting to realize that he had this ability, a calling in life.

FR. MATT FOLEY:

He saw his talent as a gift from God—there's no doubt about that. I went away to seminary after his freshman year, but one thing I found fascinating about him in those two semesters was that he had a tremendous faith life, devoted and disciplined. He was not evangelical; he didn't preach about it, but it was something in the fiber of his soul.

At Marquette, there's the Joan of Arc Chapel, this beautiful chapel brought over stone by stone from Europe, and they have a daily mass. Inevitably, you would find Chris there, disheveled and partied out and just sort of scruffy. There's not a doubt in my mind that he was in church more than any other student on that campus, at least three or four times a week.

PAT FINN:

After I met Chris, we signed up for a class called the Philosophy of Humor. We thought, could there be an easier A? I don't know why we thought that, since we'd never gotten an A before. Truth be told, Chris got a D. But Father Nauss, who taught the class, gave us each a copy of "A Clown's Prayer." The last lines go, "Never let me forget that my total effort is to cheer people, make them happy, and forget momentarily all the unpleasantness in their lives. / And in my final moment, may I hear You whisper: 'When you made My people smile, you made Me smile.'" It meant a lot to us, and we kept it in our wallets.

There were times, for instance, when Chris and I'd be on the highway, going through a tollbooth. He'd do a bit in front of the tollbooth taker, and it'd make the guy laugh. At first you were kind of like, oh, that was a little weird. But on the other hand it was like, you know, he just made that guy's day. That guy's gonna go home and tell his wife, "Yeah, this big guy came through in a car today and did this thing with the steering wheel . . ."

One of the cool things about Chris, and one of the noble things about Chris, is that if he made somebody's day better, if he could ease the pain and sadness in the world just a bit, that was why he felt he was here.

MARK HERMACINSKI, *friend:*

The thing about Chris was that he always made all of us feel like *we* were the funny ones. He always listened and anything you'd throw out there, he'd bounce it back as a joke, so he made everyone feel like they were a

part of what was going on. Even if you're not a comedian, it's fun to sit around and laugh together, and he could pull the humor out of you.

MICHAEL PRICE, *dean, College of Communications:*

I was going through the student registrations for the spring semester of his sophomore year, and I saw that Chris hadn't preregistered. I called him in. He said he really didn't want to be there anymore. He didn't want to be in school, period. I said, "Chris, what do you want to do?"

"I want to be at Second City," he said. "It's a comedy company in Chicago."

"Well, I can certainly see you doing that," I said, "but why don't we talk it over with your folks? See what they say."

He seemed a little reluctant to even bring this subject up with them. His father wanted him to go into business. But Chris and his parents came down to my office. We talked about what Chris wanted. I said, "You guys talk it over, and I'll be right outside."

When I came back in, the decision was made. Chris would remain at Marquette. He could drop the business studies major, and he'd major in communications studies and minor in theater. Then, if he graduated, they would support his wishes to do what he wanted. Onward he went.

PAT FINN:

Chris and I took this professional speaking class together. The teacher would give you a topic, like a how-to speech or a "talk about a relative" speech. Then you would have to write a three- or eight-minute talk.

The first one was fine, but we both thought it was a little easy. We were sitting at lunch right before the next class, and one of us said, "Hey, why don't we make up each other's topics and go improvise it?"

So that's what we did. Our speech was supposed to be on our summer job. Chris decided that I'd spent the summer repairing air conditioners. I told Chris that he'd been a carny in the circus.

Chris got up with nothing but a blank sheet of paper and said, "Hi,

I'm Chris. This summer I spent three months as a carny with the circus." And he just started telling stories. How he fell in love with the fat lady, made out with a midget. The class stared at him. It was amazing to watch, because you knew it was complete fiction.

After those speech classes, Chris and I were determined. We wanted to "do comedy," but what that meant we didn't actually know. If you want to be a lawyer you go talk to somebody's uncle. But it wasn't like we could call up Bob Newhart and ask him. We'd sit and listen to *National Lampoon* albums. We'd watch *Saturday Night Live*, David Letterman. We'd do everything we could to see anybody do anything funny.

JIM MURPHY:

Chris and I both had these things that we were pretty passionate about, art and comedy, but we were at a Catholic liberal arts school in Wisconsin, not the place most conducive to learning these things. Fortunately Marquette actually did have a healthy theater program. The big thing for Chris was when he got a part in the school play. It was Sam Shepard's *Curse of the Starving Class*. It kind of got him going.

TOM FARLEY:

Growing up, we didn't have a lot of choices: you played sports. But at Marquette a dean said to Chris, "Why don't you try out for one of the plays?"

Chris was like, "Guys don't do that."

But he did it anyway and eventually found out that he wasn't an athlete who happened to be kind of funny. He was an actor who happened to be very athletic.

MICHAEL PRICE:

Dance was a part of the theater requirement. If you're going to be a theater minor or major, you gotta do that. So Chris took ballet. You know, he

moved well. He was large, but he was not a slouch. He was great. Sheila Reilly, the ballet teacher, had a dress code and it was pretty stringent, and I think Chris got by without having to abide by that. He may be one of the first students she ever allowed to wear sweats instead of tights.

KEVIN FARLEY:

It was a very formative time for him. After he moved out of the dorms, he and a bunch of his rugby pals lived in this big piece-of-shit house up on Nineteenth Street. They called it the Red House. It was a hellhole.

JOHN FARLEY:

Whenever I'd walk into the Red House I'd say, "Well, well, well. Looks like somebody forgot there's a rule against alcoholic beverages in fraternities on probation." It was just disgusting. Dad refused to go in for basic sanitary reasons. Everything had a touch of something on it. Odd smell, too.

JIM MURPHY:

One of our roommates was the Budweiser rep on campus, so everything was Budweiser, signs and cups everywhere. Chris's room was at the top of the stairs. Any time your parents would come, it was the first room they'd see. He was such a slob, food and clothes everywhere. There'd be all these fruit flies and it was, well, let's just say it was college.

MARK HERMACINSKI:

We put Farley close to the bathroom, but that didn't help much.

KEVIN FARLEY:

The bathroom was unspeakable. One of the things that made Chris a legend on that campus was his room, simply the fact that a human being could survive in there. People would come over just to look at it.

JIM MURPHY:

We were pretty poor, so we didn't turn on the heat until after Thanksgiving to try and save money. It was so cold. We had a rugby awards banquet there. We kind of cleaned it all up and everyone brought dates. The girls all came in dresses, and they were freezing. We had a bucket for ice so we could make cocktails. We put it out on Friday night. By Sunday night, the ice hadn't melted.

The other thing I remember about the Red House was every couple of months Chris's mom and dad—this being Wisconsin—would send him a twenty-five-pound summer sausage and a twenty-five-pound wheel of cheese. Any time you'd look in the refrigerator there was never anything inside it except this gigantic sausage and this huge wheel of cheese.

MARK HERMACINSKI:

The kitchen was all infested with flies and maggots. After two months we just closed the door. But we had a lot of fun. Since we had the campus Budweiser rep, they'd pull up the truck with ten, fifteen kegs and we'd have a party with five hundred of our closest friends. Chris always drew a crowd. Wherever he was, in Madison, on a rugby road trip, he'd have a crowd around him in minutes.

TOM FARLEY:

The place where Chris really learned how to be this galvanizing figure was at Red Arrow camp. It was like a graduate school in male bonding. Every

summer in college we all went back as counselors, and every summer all the kids and the other counselors, they'd rally around Chris.

FRED ALBRIGHT:

The kids loved him. How could you not? Every year, he was one of the most popular counselors. Chris had this very sensitive, sweet side to him, an empathetic side that helped him communicate with kids in this amazing way.

DICK WENZELL:

He could take any old boring activity and make it a fun, exciting experience. If you were eight years old, digging for earthworms with Chris was the most fun you ever had. He'd just hypnotize these kids. "We're digging for monsters, boys! Oh! Hey! I found a big one!" And he'd lead them off on this grand adventure. And that's the same thing he did with an audience.

RANDY HOPPER, *counselor, Red Arrow Camp:*

Chris came to life around those kids. He was the biggest kid there. We had this flag-football game called the Salad Bowl. When it was time for the big game, Chris would get up there in front of the boys and give them a pep talk that convinced them that this was the game of their lives. He'd get them all riled up. He'd be waving fistfuls of bacon, going, "Sooweee! Love them Hogs!" doing his big, high-school-coach motivational speech. In ten minutes, he'd have thirty kids so riveted and excited to be playing in this game that the Super Bowl would pale by comparison. His ability to connect with people was uncanny.

FRED ALBRIGHT:

He couldn't talk to girls, though. Around girls, Chris'd hide behind his jokes. He'd start flexing and doing this jokey, deep-voice, macho-man thing to try and hit on them. One night we're out at a bar near the camp and he goes up to these women and says, "Well, which one of you little ladies is gonna go home with me tonight?"

And one of the women looks at him and says, "Well, it can't be me, because you're my son's counselor."

Chris felt so small after that you just about could've balled him up and put him in a thimble.

He had one girl that he had met up north in Minocqua during the summer. She really adored Chris. They'd kid around, and it was like they were buddies, but Chris had this greater attraction to her. He would talk to me every day about how he'd had some moment with her. I see her now fairly often, and she didn't even realize that he was so completely enthralled with her.

TIM HENRY:

Being a camp counselor was all about working the girls from the other camps. If you played your cards right you could go out several nights a week, but the girls were only allowed to go out one night a week. So you always had one little honey from each camp.

RANDY HOPPER:

We'd all go out in the woods, have a big bonfire, and it was all about trying to score. That was the game. Chris was not too adept at that, not usually. But he would entertain everyone.

TIM HENRY:

It would always piss Tom off, because Chris didn't know when to stop. We'd be out in the woods, Tom would be working his magic on some girl, about to close the deal, and Chris would come barreling through the campsite fucking around. Then Tom would have to tell me to fuck off because I was laughing so hard.

TOM FARLEY:

One night, these two girls I knew from Georgetown had come all the way up from Chicago to visit. Chris was so out of control, and he wouldn't turn it off. These were nice girls, and one of them I was really trying to get serious with. They were getting annoyed, really offended. He wouldn't stop, and I couldn't take it anymore. I lost it. He was acting up and running around, and I grabbed him and I beat the shit out of him, just whaled on him, punching him in the face. He was crying so hard he couldn't even fight back. He was so scared. The next day he showed up at mass black and blue and bloodied all over.

TIM HENRY:

Chris so wanted to be with one of those girls, but he would always go back to what was safe. He would revert to the guy who's had too many beers and got silly. Most nights the parties would go really late. As the hours wore on, everybody would start to drift away and pair off out in the woods, and there would be Chris, alone by the campfire.

DAN HEALY:

I only stayed at Marquette for a year. After that, I transferred to University of Wisconsin. That sophomore year was a crucial time for Chris. I started to hear stories about him drinking alone in his room, which was weird. We were always big social drinkers, but never the sit-alone-in-your-room

kind of drinkers. The Marquette rugby team came into Madison one time. We all went out and drank, and then afterward I made a point of saying to Chris, "What's going on? I hear you're really going off the deep end." He backed away immediately. He didn't want anything to do with that conversation.

JIM MURPHY:

Then, during our junior year in the Red House, Chris read that book about John Belushi, *Wired*.

MARK HERMACINSKI:

Wired was the only book that Chris Farley read in college. The only one.

JIM MURPHY:

For our spring break that year, me, Chris, and this other guy went to L.A. We had a couch to crash on, and Chris was really developing an appetite for this career he wanted. We went around doing all the Hollywood tourist stuff, and the whole time Chris was like, "Jimmy, I really think I can do this."

I'm a huge fan of Buster Keaton and all those early physical comedians. One time I was trying to turn Chris on to Fatty Arbuckle. So I made him sit down and watch one of Arbuckle's films. At the end of it, the only thing Chris said to me was, "Wow, Jimmy, he did all of his own stunts." He fixated on this one thing about Arbuckle, and that was all he really took away from it. And that's sort of what happened with Belushi. When Chris read *Wired*, he just took completely the wrong thing away from it. You could tell that what he saw in Belushi and what you and I saw in Belushi were two different things. Chris wasn't blindly imitating Belushi, but reading that book validated all the addictions and impulses that Chris already had inside him.

Chris didn't smoke pot freshman or sophomore year, didn't do any

drugs at all, other than drinking. Then all of a sudden that year it clicked. I have this vivid memory of him in his room one day. While everybody else was going to class, he sat in this chair with a big red bong. He sat there doing bong hits and chugging Robitussin cough syrup. Back and forth. One after another.

At that point you could see where it was going. I'd try to explain to him, you know, that you can only get so high. It's like pouring water into a glass. You can pour in all you want. After a while, it's all just spilling over the sides. But it was the same thing no matter what he did. It was the same when he tried pot, or when he tried mushrooms, or when he tried comedy.

PAT FINN:

The first place Chris and I ever got up and performed together was during our senior year in a skinny bar in a bad neighborhood in downtown Milwaukee. It was called Wimpy's Hunt Club, and it had an open-mike night at midnight. The stage was literally twelve milk crates turned over in the corner. There were about nine people in the audience, all factory workers on break from the brewery. Tough crowd.

Chris and I went up there, and it was like when you're in middle school and you go on a date and you don't really know what you're supposed to do. It was that awkward, and that bad. We *bombed.* We signed up to do it again the next week. Then we found out that a bunch of Marquette students were going to be there. We chickened out and didn't go. Everyone got real mad at us.

Finally we signed up for the Follies, the school talent show. We got an actress, and we decided we'd do a parody of *The Dating Game.* Jim Murphy was the host, and another friend of ours, Seamus, was the third bachelor. We kept on meaning to get together and work, and we'd talk about it every once in a while, but we never had any idea what we were doing. Then about a week before the show the girl quit school and moved to Chicago, and we figured we'd just blow the thing off.

But the night of the Follies, I was over at Chris's house and we were

hanging out. About fifteen, twenty minutes into it, and we get a phone call from the stage manager. "Where the hell are you guys? You better get down here. People are pretty pissed. They want to see you—and they're screaming for Farley."

We hopped on our bikes and rode over. It was way bigger than we thought it was going to be. There were at least a couple thousand people there. The show had three more acts to go before it ended. "You're goin' after this singer," the stage manager told us.

Chris just goes, "All right, Jim, you're the emcee." He looked over at Seamus, who had overalls on. "Seamus, you're farmer guy." Then he flicked my collar up on my shirt, unbuttoned a button, and said, "And Finner, you're cool guy."

"Okay. What about you?" I asked.

He pulled out these nerdy glasses and said, "I'll be nerdy guy. Let's go."

Jim went out, made up some intro, brought out Seamus, and they did a funny little Q&A. Then I went out, doing this "cool guy" walk, hopped on my stool, and answered some questions. There was no girl, mind you, just the host and three male contestants, but that became part of the gag.

Then Jim said, "All right, let's bring out the next guy." The spotlight hit Chris coming out of the curtain. He ran as fast as he could and then tripped and slid across the entire length of the stage. The place went berserk. Then he went over to his stool, clumsily knocking it over. He finally clambered on and then fell right off. It was insane. The audience loved it.

As soon as it was over, Chris and I ran backstage, and I remember he just grabbed me by the shirt and he looked right in my eyes and said, "We're gonna be doing this for the rest of our lives. That was the greatest high I've ever felt in my life."

MICHAEL PRICE:

In the spring of Chris's senior year, we got one of those rare days when you can open windows, be outside, and throw a ball around. There was this white house on Kilborn Avenue where all these girls lived. Chris had a cherry smoke bomb. He lit it and put it on the open windowsill, thinking the smoke would drive everyone out and it'd be a good prank. But he forgot that when you light those things they twirl around and spin out of control. Well, it spun off and landed on their couch. And it burned. I mean, it really burned. Pretty soon the house was on fire, and it was spreading to the second floor. Chris figured he'd better get the heck out of there. So he took off with a friend, and they went down to Illinois, just across the border.

The next morning I got to work about seven, my phone rang and it was the police department. They were wondering if Chris was around. I called the Red House, and they said Chris wasn't there. I hung up, and then, about fifteen minutes later, two of the Red House guys showed up at my office. They said they couldn't talk to me on the phone because their line had been tapped. I said, "Oh, come on. You guys are outta your mind."

They told me where he was. I called him in Illinois. I talked to him and told him to come back. Then I called his folks in Madison. They asked me to give them the name of a lawyer in Milwaukee, so I did. Between the lawyer and his parents it was decided that Chris would come to my office, we would all meet there, and then he would go and turn himself in.

I called the police department, and I told them what the lawyer told me to say, that I knew where Chris was and he would come down there on his own the next day. Chris came back. The attorney took care of things. And after many weeks of delayed hearings and so on, Chris came away from it with a "dangerous use of firearms" charge, or something like that. He ended up with about thirty hours of community service, but he couldn't get his diploma.

* * *

In *Tommy Boy*, Chris's partying, rugby-playing alter ego graduated from Marquette in seven years. In the real world, Chris squeaked out in four and a half. As a result of the smoke-bomb incident, he was put on probation, and university policy did not allow students on probation to graduate. However, he was allowed to walk in the graduation ceremony with his classmates, complete his course work at the University of Wisconsin in the fall, and receive a Marquette diploma the following December.

Forced to return to Madison for school, Chris moved into an apartment downtown, close to many of his high school friends who had never left. Between finishing his classes and performing wherever and whenever he could, Chris took the only job offer he had. He went to work for his father.

MIKE CLEARY:

To understand Chris you have to understand something about Madison. Madisonians tend to be very educated, very literate, and upwardly mobile, but I would say that seventy-five percent of them have never seen the ocean. And I'm not kidding. Madison's got everything you need—that's the default mentality here. And Chris came back in large part because the family discouraged him from doing anything else.

TOM FARLEY:

Dad always wanted all of us home. It was almost like, "You can't make it out east, and so you don't need to try it." Kevin bought into that at first, and Chris did for a while, too. Dad had tried it with me. After I left Georgetown, I said, "Hey, all my buddies are going up to New York. That's where I want to go."

Dad sat me down and said, "You'll never make it."

"Watch me," I said. And I left.

Dad and I butted heads throughout our lives. If he said something was blue, I said it was red. Chris was the opposite. Everything he did was to

please Dad. At that point, he had done the plays at Marquette, but he really had no idea of how to go about making that into a possible career. Dad just said flat out, "No one's going to offer you a job. You'd better come work for me." So Chris went to work for Dad, but it wasn't a two-man job, so there wasn't a whole lot for him to do. The job was really a joke.

KEVIN FARLEY:

When Chris finally left and I came in to take over the job, we opened up the drawers of his desk and there was nothing in it but *Cracked* and *Mad* magazines.

JOHN FARLEY:

What did Chris do for my dad? Hell, what did my dad do? No one really knew. He'd take people to dinner, entertain them, and they'd buy stuff from him. That was our notion of work. Dad didn't really have to sell his product. He was selling roads. Everyone needs roads, and all roads are basically the same. Oil plus gravel equals road. The funniest, nicest, coolest guy was going to get the bid from the county, and there was no better guy to hang out with than my dad. Add Chris to the mix and they could sell anybody anything.

My grandfather was a salesman, and some days he had to go on four breakfasts, going town to town to town. Then he'd have to go to all these lunches, and then come home and have a dinner. It was the same for my dad. My dad knew every restaurant, every bar in Wisconsin. You'd drive by some place way the hell out in the middle of nowhere, and Dad would be like, "They can sauté a mushroom like nobody's business. Good cheeseburgers."

My dad's clients were these farmers, these down-home guys from rural towns who just happened to sit on the county highway commission handing out multimillion-dollar contracts to pave roads. The big thrill of their month was when Tom Farley would drive out from Madison and take them out for a schmooze and a steak dinner. You know those square

pats of butter they keep on the table? Farmers in Wisconsin eat those like appetizers, like a predinner mint. Just open 'em up and eat 'em. Sweet Jesus that's insane, but it's a very Wisconsin thing.

Sit. Eat. Talk. Drink. That was the business. It was about putting on a show, buying the round of scotches, prime rib for everyone. That's the key to who the Farleys are. We'd rather see a smile on someone's face, even if it meant hurting ourselves. I don't know why we did, but we did.

TOM FARLEY:

Chris was a great entertainer of clients, but for Dad to keep paying him twenty grand a year just to go to lunch was a bit much. So Chris started doing these open mikes at the student union. He bombed, failed miserably. He was getting up at the liberal, progressive University of Wisconsin and telling crude lesbian jokes. That went over like a fart in church. He'd get heckled and booed. Then he found improv at the Ark.

TODD GREEN:

Chris, Greg Meyer, and me all lived near each other downtown. And one night Chris said, "Guys, we gotta go to this thing, the Ark."

All through our childhood we always knew that Chris was going to do something. We just didn't know what. That night they were doing some skit and they needed audience participation. Chris started to get into it, and he completely stole the show from the performers. That was the start of the whole thing.

DENNIS KERN, *director, Ark Improvisational Theater:*

Chris always spoke fondly of his days at the Ark, and I was always very appreciative of that. We sort of took him in off the street—quite literally—and gave him a home. He showed up at the theater one night after a show and stumbled in through the door. He was so drunk he could

barely even form coherent sentences. He was just going on, like, "Wanna do . . . comedy . . . improv, I wanna—gotta do this . . ."

I could barely understand him. To be honest, I thought he might be retarded. I didn't think he'd remember anything that I told him, so I said, "Look, we're having a rehearsal tomorrow. Why don't you come by and join us then?" Then I showed him out the door, thinking that was the end of the whole episode.

The next day I got a call from my wife, who was at the theater. "Did you tell some big guy that he could come to our rehearsal today?" she asked.

"Yeah," I said, "but I didn't think he'd actually show up."

"Well, he's here. And he brought a case of beer."

CHAPTER 4

Attacking the Stage

BRIAN STACK, *cast member, Ark Improvisational Theater:*

Keith Richards said that the first time he heard rock and roll it was like the whole world went from black-and-white to Technicolor. That's how Chris always seemed to describe finding comedy.

As a city, Madison, Wisconsin, has something of a split personality. On the one hand, it's a typical Midwestern town with no shortage of beer, football, and competitive bratwurst eating. On the other hand, thanks largely to the University of Wisconsin, Madison carries with it a long history of liberal, even radical, politics. Wisconsin governor and U.S. senator Robert LaFollette launched his left-wing Progressive Party in Madison. And in the late sixties and early seventies, the university itself saw some of the country's most violent antiwar protests, culminating in the bombing of the school's Army Mathematics Research Center at Sterling Hall.

In such a hothouse political environment, a small but vibrant arts community was bound to spring up as well. It would still be a few years

before enterprising UW students hatched the *Onion*, the satirical newspaper that eventually found its way to Internet fame and glory. In Chris's day, if you lived in Madison and had a notion to seek a career in comedy, you went to the Ark.

Dennis and Elaine Kern founded the Ark Improvisational Theater in Madison in 1982. Both professional actors and directors, they had left New York City determined to do theater on their own terms and, God forbid, actually make a living at it. For its first two years, the Ark staged weekly shows at a local bar, Club de Wash, and offered classes in improv and acting. It quickly became recognized as a stepping-stone for those on their way to greater opportunities in neighboring Chicago. Joan Cusack, who had joined the cast of *Saturday Night Live* in 1985, was among the Ark's alumni.

As the theater grew more established, the Kerns purchased a defunct downtown building that had once housed a Brinks truck garage. They gutted it and installed a small one-hundred-seat theater. It was that converted garage that Chris Farley stumbled into late one night in August of 1986. For over twenty years, he had been a performer in search of a stage. He had finally found it.

JODI COHEN, *director/cast member, Ark Improvisational Theater:*

I was there the day that Chris came to audition. I forget what the scene was, but as part of it he fell out of his chair and—*smack*—landed on the ground. I thought he'd really had a heart attack. My first reaction was: Is the theater insured for this? What's going to happen? That's how convincing this fall was.

Elaine and I didn't want him in the company. He seemed like a wild card, and he didn't seem very focused. But Dennis said, "This guy is really talented. He should be in the group." We formed a company called Animal Crackers, and Chris performed with them.

BRIAN STACK:

I had never met anyone like him before. He had such incredible enthusiasm in everything he did. One thing that gets lost a lot is that when it came to the work, he was always very serious about it. He was always on time for rehearsal. In fact, he was usually there before everybody else. I never, ever remember him being late for a show.

I don't know if Chris had ADD, but people who have a lack of focus, when it comes to something they're passionate about, they hyperfocus. Chris was certainly like that when it came to acting. Our group's shows were mostly short-form, game-oriented improv with a lot of audience suggestions. But we also did some sketch-type stuff, and Chris was great with both of them. He was just a blast to work with from day one.

One thing that always amazed me was his ability to do things that if I had done them would have put me in the hospital, and then he'd get right up from them. He could slam into walls and slam down on the stage. He was such a natural athlete. He was almost like a ballet dancer. Even though Chris is known for being a great physical comedian, some of my favorite things were the subtle little characters he would do.

DENNIS KERN:

The Motivational Speaker appeared onstage for the first time at the Ark. It wasn't the same as on *Saturday Night Live*, but it was there in its infant form.

PAT O'GARA:

The Motivational Speaker actually started back in high school and was based heavily on our coach, Joel Maturi, who would go off on these inspirational speeches. He'd be prepping us for the game, briefing us on the other team's defense and all that, and Chris would be right there behind him, imitating him, making all these faces and forcing us all to laugh.

JOEL MATURI:

The Motivational Speaker is based in part on me; there is some truth to that. Mostly some of my mannerisms, the hiking up the pants, the spreading the legs and crouching down to get serious. I was pretty vocal with the pep talks and the Knute Rockne speeches. Those kinds of things. I think the more philosophical side of the character was actually based on his dad.

DENNIS KERN:

We taught Chris the basics of improv and scene work at the Ark, but the natural talent he had was already present. As a performer, he was just there in the moment, like Johnny Carson used to be on the *Tonight Show*. What Carson was so brilliant at was just reacting and responding naturally to the environment around him in a way that made you laugh. Chris had those same instincts. He just knew what to do.

BRIAN STACK:

He could do the same thing fifty times and somehow always make it funny. If a pretty woman walked by he would drop and start doing push-ups, starting out ". . . 198 . . . 199 . . . 200." I'd seen him do that lots of times. It shouldn't have been funny to me anymore, yet it always was. It's hard to explain why it was; it just was. You could videotape it and analyze it with a computer, like you would a golf swing, but you still wouldn't understand it, and you could never hope to replicate it.

One night after a show we went to this bar, and Chris was making this middle-aged couple in the bar laugh. He was dancing with the guy's wife and doing these cat-eye things with his hands. The husband was laughing so hard that he was actually falling off his bar stool, and he eventually said to Chris, "What's your name? I want to be sure and remember it. I've never laughed like this." It was strange. Everyone sort of sensed that there was just something unique about him. Chris wasn't famous, but it was the

same reaction he would get years later after he left Madison and became a movie star.

DENNIS KERN:

Chris and Brian Stack had just started rehearsals on *Cowboys No. 2*, a Sam Shepard play that we were going to put on. And Chris, meanwhile, had been taking trips down to Chicago here and there with his father. Then it became clear what all those trips to Chicago were for.

BRIAN STACK:

When Chris decided to leave, it was pretty upsetting. He loved the Ark, but he was bursting at the seams to get out, and Chicago was the first step. I think Dennis was happy that Chris was leaving to pursue his dream, but he seemed kind of angry on his last night.

DENNIS KERN:

I was happy for him, but at the same time I thought it was too soon. I thought that he needed to be more in contact with the source of his creativity before he went to try at the professional level. I always knew he would make it, but I don't know that he was grounded enough in the technique of acting to have something to hold on to. He was immensely talented, but that talent was sort of at the whim of whoever needed the next laugh.

TOM FARLEY:

The experience he had at the Ark told him he had to get out of Madison. Plus, he couldn't take the job at Scotch Oil anymore. As much as he wanted to please Dad, after a year of selling asphalt even Chris was like, "I gotta get out of here."

JOHN FARLEY:

Dad had made the ultimate sacrifice. He would have been a great lawyer, smartest man you ever met. Dreamed of going into politics. But he had given all that up to raise a family. And so he wanted everyone to stay in Madison, because that was what he'd sacrificed everything for. Years later, Chris had to cry for this scene when he filmed *Black Sheep*. So he turned to me and he said, "Johnny, make me cry."

I said, "Well, Dad's all alone in Wisconsin with two ladies. All his boys have gone and moved on with their lives."

"Shut up."

He really got angry that I had said it. Somehow it had triggered the wrong emotion.

MIKE CLEARY:

When Chris was working for his dad, he called me up one day and said, "I gotta talk to you about something."

"What's that?"

"Well, I have an opportunity to go to Chicago to study at Second City. What do you think?"

I really wasn't sure about taking risks like that. I said, "Chris, you need to just work with your dad. Establish a solid career and maybe do this stuff on the side." That was totally my mentality. Finally, I said, "Well, what does your dad say about it?"

And his exact words were, "My dad says I should definitely take the opportunity and go for it. He's gonna back me one hundred percent."

I said, "Well then there's no conversation here. You have to go."

TOM FARLEY:

We thought Chris would come running home in six months, and he never came back.

* * *

In June of 1987, Chris left for Chicago. He moved into a small apartment off Armitage Avenue, just north of Chicago's Old Town neighborhood. There he rejoined his Marquette rugby and acting friend Pat Finn.

The yellow porch light of Second City had led Chris to Chicago, but he quickly found that the doors of the renowned comedy institution did not immediately open for untrained unknowns fresh off the bus from Wisconsin. Forced to look elsewhere for a place to learn and perform, he found it at ImprovOlympic.

Today, ImprovOlympic has become an industry mainstay in its own right, producing a steady stream of bankable film and television stars, among them Mike Myers, Vince Vaughn, John Favreau, Andy Richter, Tina Fey, Steven Colbert, the Upright Citizens Brigade, and director Adam McKay, not to mention a healthy chunk of the writing staff at *Late Night with Conan O'Brien.*

But when Chris arrived, ImprovOlympic was still a fledgling outfit of vagabond comedians looking to make the funny anywhere they could. Teacher and director Charna Halpern had founded the group in 1981 with several goals in mind. Second City used improv as a means to create sketch comedy. Halpern wanted a curriculum in which improvised performance was the end in itself. At Second City, only a handful of seasoned performers trickled up to the main stage. Halpern gave ImprovOlympic a communitarian ethos, allowing even new and less-experienced students the chance to practice and learn in front of a paying audience.

In 1984, comedy guru Del Close joined Halpern's cause. As a director in Second City's early heyday, Close had trained and mentored a who's who of comedy, from John Belushi to Harold Ramis to Bill Murray. He was instrumental in shaping the forms and conventions of the Chicago school of improvisation. Perhaps his most notable contribution was the Harold, a long-form, fully improvised performance in which a whole cast works together off of a single audience suggestion to create a cohesive, continuous series of scenes.

For Close, the goal of improv was not to get laughs but rather to find the real, emotional truth of the characters that created those laughs. He

found the perfect instrument for that in Chris Farley. With Halpern's instruction and Close's inspiration, Chris began his comedy education in earnest. Some expressed doubts about his raw, unschooled talents, but those doubts quickly vanished. Chris, performing full throttle at night and bumbling through a comical parade of semiemployment by day, proved to everyone that he was destined for a life onstage.

PAT FINN:

After I graduated from Marquette, I went down to Chicago. Chris followed about a year behind me. We had no jobs, and we had no idea what we were doing. He moved into his place off Armitage, and we went from there down to Second City one day around two in the afternoon. We just kind of paced around in front of the theater, back and forth. In our minds, the scenario literally went something like this: Somebody up on the second floor would say, "What? We need two more people for the Second City main stage? Where are we going to find— Oh, wait! What about these two people out front? They look *hilarious.*"

That was about how far we'd thought things out. Then, after about ten minutes of pacing around, Joel Murray—Brian and Bill Murray's brother—walks by. He was at Second City at the time, and he and I had gone to the same grammar school, so I knew him a little.

Chris said, "There's Joel Murray. He's Bill's brother. You should talk to him."

"I don't know, Chris."

"You got to! C'mon. That's why we're here."

So Joel walked up, and I said, "Joel. Hi. I'm Pat Finn, from St. Joe's."

"Yeah. Little Finn," he said. "What's goin' on?"

"Um, nothin'. This is my friend Chris. We wanna get into comedy."

He just kind of looked at us. Chris's eyes had this look like the next thing out of Joel's mouth was going to be the keys to the kingdom. And, actually, it turns out it was.

JOEL MURRAY, *cast member, Second City:*

So one day here's Pat Finn, who I haven't seen since high school, standing there with this big guy. I could tell that the big guy was restraining all of his energy to just listen and be attentive to what I was saying. But I basically told them, "Go find Charna Halpern and Del Close at the ImprovOlympic and study with them, and then see if somebody'll let you paint the bathroom at Second City."

It was funny, knowing Chris later, just to watch him holding it in, trying not to be an idiot.

CHARNA HALPERN, *director/teacher, ImprovOlympic:*

ImprovOlympic didn't even have an actual theater at the time. We performed at Orphans, this bar on Lincoln Avenue. We had to be out by ten o'clock so the band could come on. We got kicked out of Orphans, and we moved around to like fourteen different spaces. It was an insane time. But I kept attracting these really brilliant people—Farley and Pat Finn, Mike Myers, Vince Vaughn, Jon Favreau, Andy Richter—and the shows kept getting better and better. But even though we were getting thrown out of these places, the audience was following us. It just kind of kept snowballing, getting bigger and bigger, and that was what it was like when Farley showed up.

BRIAN STACK:

I went down to Chicago to visit him at ImprovOlympic. He was taking classes, but Charna hadn't let him go up onstage to perform yet. After the show, he was pacing outside, and I could just see he had all this pent-up, frustrated energy that had nowhere to go. You could see how he was bursting at the seams, how he *needed* to get up onstage.

CHARNA HALPERN:

One night after a couple of weeks, Chris came up to me with Pat Finn and said, "Let me go onstage! Let me play tonight!"

"You?" I said. "God, no. You're definitely not ready."

He started getting violent. He was banging his fists on the wall above me, like, "Let me go! Let me go onstage! I'll be great! You'll see!"

"I'll tell you when you're ready to be onstage," I said.

He was just not hearing it. After a good seven minutes of his badgering, I finally got fed up myself. I said, "All right, I'll tell you what. You can go onstage and play tonight, but if you're bad you will never ever get on my stage again. Do you wanna take that chance?"

Before I even finished my sentence, he was bounding out into the room to tell the guys he was going on. Everyone was looking at me like, "Are you crazy?" But he got up there and was absolutely hilarious.

The good thing about it was that when he got back to class, he started to calm down. Once I'd let him go onstage, he'd lost that need to prove something. From then on he was really willing to learn and get better.

PAT FINN:

From that point on, he just committed a hundred and ten percent. We took classes with Charna. Then we got classes with Del. There were two improv teams that got assembled around that period. One was very cerebral—that wasn't us. We were the physical group, called Fish Shtick. People wanted to watch them because they were so smart and heady, but then they'd want to watch us because we were just off the wall.

BRIAN STACK:

Chris once said that Del Close told him to attack the stage like a bull and try and kill the audience with laughter.

NOAH GREGOROPOULOS, *cast member, ImprovOlympic:*

He was so big and emotional, very physical. He gave one guy a permanent scar on his forehead when he dove from the bar onto the stage in this overblown ninja thing. He landed on him, smashing his glasses into his head. His commitment was just past the point of safety.

TIM MEADOWS, *cast member, Second City:*

I was already touring for Second City, and I used to go back and perform at ImprovOlympic every now and then. One night I went up there and did a Harold with Chris's team. It's difficult when you're the new guy in a group, because they already have their dynamic and you don't know how you'll fit into it. But the very first time we performed together Chris was right there for me. I started a scene where I was hanging something up, and it was obvious to everyone in the audience that I was hanging up laundry on a clothesline. But Chris came out and said, "Doctor, what does the X-ray say?"

I just looked at him and said, "Well, it's not good."

And it got a big laugh because it was such a change from where people thought it was going.

PAT FINN:

When you get a suggestion in an improv set, usually one performer goes out and sets the stage based on the idea. And sometimes that person is out there for a while, just fumbling around. He doesn't know where he's going, and because he doesn't know, it's really difficult to step in and help him.

One night this girl walks out and puts a pretend briefcase down and goes, "I had such a great day today, honey. They made me partner at the law firm and they love me and somebody's gonna be interviewing me for *Newsweek . . .*"

And on and on and on. It was this long exposition, just going nowhere. We're all standing there at the back of the stage, thinking, how do we even enter this scene? She's giving us nowhere to go.

This goes about a minute or so. She's droning on and on, and finally Chris storms out of the back line and goes, "Sweet Lord, would you just shut up and bowl?!"

His instincts were near perfect. With one line he put the whole scene in a place and a context and established what the joke was. "My God, every time you bowl it's something different. You're a doctor. You're a lawyer. 'Look at me, I discovered something!' Bowl the *goddamned* ball."

JAMES GRACE, *cast member, ImprovOlympic:*

He was an amazing processor of information. He wasn't great at getting things started, but if you gave him anything, he would take it, internalize it, have a perspective on it, be affected by it, and ride it out for the scene.

CHARNA HALPERN:

He was an amazing listener onstage, like a sponge. You could just see him reflecting your idea through his facial expression and taking on your mood. He totally got it.

And he was in incredible shape. That always surprised people. I remember one football sketch he was in where his teammates were making fun of him for being out of shape. They'd say, "All right, fatty, drop and give me twenty." And Chris could do it, with no problem, clapping his hands in between each push-up, even. He was all muscle under there.

TIM HENRY:

We're at a bar in Chicago one night. It's ten degrees outside. Chris has got his English driving cap on, Timberland boots, and some cutoff sweatpants, and he's sporting these huge muttonchop sideburns. The after-

work crowd is there. A bunch of little honeys are at the bar, and Chris starts chatting them up. "Hey, how are you? What do you guys do?" he says.

They work in advertising or insurance or whatever, and they ask him, "So what do you do?"

Chris is standing there, sweating in ten-degree weather, and he goes, "Me? What do I do? I'm an aerobics instructor."

We're all laughing, 'cause we know he's winding up to have fun with them.

"Aerobics instructor?" they say. "Are you kidding me?"

And Chris, with one hand on the bar and one hand on the stool, defying all laws of physics, goes from standing stock still, leaps into a perfect backflip, and lands back right on his feet. Hat doesn't even come off his head.

"Yeah," he says. "You know, aerobics instructor."

TED DONDANVILLE:

I got hooked on Chris's shows very early. When I went back to Red Arrow Camp to be a counselor, Kevin and Johnny told me Chris was down in Chicago. I'd been thrown out of the University of Denver; I wasn't having a traditional college experience. So I started hanging out with Chris a lot. When you'd go and see him in these bars, you'd have to sit through an hour and a half of bullshit watching these kids learn how to do comedy. But however good or bad the shows were, Chris always had that moment, that one moment where lightning would strike and he'd just kill the audience.

PAT FINN:

ImprovOlympic was very young, and I think that's what made it, for lack of a better word, romantic. There were no agents coming to see you. There was nobody pitching a screenplay. It was just about the pure love of the game, going out every night and making people laugh.

JAMES GRACE:

It was like a wave of energy. We were doing five shows a week, one on Thursday and two on Fridays and Saturdays. We were just all constantly together all the time, performing, hanging out. Everybody who was there had come because they wanted to challenge themselves and push the boundaries of comedy. People were either rehearsing or performing every night of the week. You were consuming it all the time. And when you weren't rehearsing or performing, you were hanging out with people whom you rehearsed and performed with. Once you were on a team, that was basically your fraternity.

It was very collegiate, especially when it came to the drinking. I would say that if you took a clinical definition of alcoholism, then everybody there had a huge problem. Farley always did everything bigger than everybody else, but we were all out of control. One time I saw Pat Finn fall down two flights of stairs solely to make me laugh. That's what we did, outrageous things all across the board.

PAT FINN:

Chris must have had something like forty jobs during that time. One day he and I were walking down Armitage Avenue, and he was like, "Yeah, I worked there. I worked there."

And I was like, "When?"

"Well, I worked at the butcher shop for like an hour, and they fired me. Then I got a job at the hardware store the day after the butcher shop, but I was really tired 'cause I'd had to get up so early for the butcher shop, you know? So I fell asleep on some boxes in the back."

I said, "How could you fall asleep within hours of your first day on the job?"

"I don't know, but they were really pissed."

So he lost that job. Basically he'd lost every job up and down the street. Eventually, we'd go to church and he'd pull the little tags on the bulletin board that said "Need neighborhood workers" and stuff like that.

He got a job as a bouncer, but then one day he said to me, "Hey, I think I kinda got fired from that bouncer job at the bar."

"Jeez, Chris"—this was on Sunday—"you started it on Friday. What happened?"

Apparently a fight had broken out, so he—Chris, the bouncer—had left, because he didn't know what to do. And he'd caused the fight.

What had happened was Chris was checking IDs, and, goofing around, he goosed some girl in the butt. Her boyfriend thought it was somebody else, and he started shoving people and it broke out into a real melee, so Chris just kind of slipped out the front door.

All the other bouncers came out from inside and finally settled it. Then the owners came out and said, "My God, who's on the front door?"

At that moment, Chris came back around from this alley down the side of the bar. He saw the owners, panicked, turned down this alley, and yelled *"And stay out!"*

"Where the hell were you?" they asked.

"Where was *I*? Where the hell were *you*? There were like nine guys in the alley on top of me."

"What?"

"It's okay. I took care of it. But, man . . ."

"Oh Chris, we're so sorry."

So on Saturday he went back to bounce again. I asked him how that went.

"Well," he said, "normally you get your shift drink around eleven. But the girl behind the bar really liked me, so I got my shift drink at seven. Then I had another one. She was making those greyhounds that I like. Man, I had a lot of 'em."

"Were you okay?"

"Well, that's my question. I'm not sure. I passed out on the people in line while I was checking IDs, and all the bouncers had to take me across the street and put me into bed."

"Uh-huh."

"So do I just go over there to get a paycheck, or do I ask 'em to mail it? How does that work?"

"I don't know, Chris. You probably drank more than your paycheck."

"Yeah, you're probably right. I guess I just won't go over there for a while."

He didn't seem particularly fazed by it. It just kind of reinforced to him that he needed to find a way to make a living in comedy.

CHARNA HALPERN:

We got a pilot, an improv game show, similar to *Whose Line Is It Anyway?* I picked out a bunch of our best performers, of which Chris was one, and they flew us all out to L.A. We were doing some really smart work, and the producer just wanted us to dumb it down. "It's too smart," he kept saying, "too smart." And he started firing some of the best people. He wanted to bring in all these dick-joke stand-up comics.

So, one by one, my cast was getting fired. It was just a nightmare, which Del had warned me was going to happen. Chris could see what was going on. At one point in the rehearsal he said, "Look, I'm sorry, but I don't wanna get fired."

"Do what you gotta do," I said.

And so he hiked up his shorts into the crack of his ass and started jumping around doing the monkey dog boy dance, which is when you hold your crotch with one hand and put your fingers up your nose and just start jumping around being silly. And, oh, they were on the floor laughing, because that's the kind of dumb stuff they wanted. And he saved his job.

But Chris wasn't always a caricature of the fat guy. He did beautiful scenes. When he did serious scenes, oh my God, he could make you cry.

NOAH GREGOROPOULOS:

Chris's vision of himself was that everyone just wanted to see fatty fall down, so that was what he was going to give them. But there were plenty of other guys who could fall off a chair and eat in Roman proportions.

What Del Close liked about Chris wasn't necessarily what everybody else liked about him. Del felt that Chris was in touch with genuine emotions in a way not all improvisers allow themselves. That's what Del was really attracted to in Chris. He wanted to show Chris that he could be more than just a one-note performer.

CHARNA HALPERN:

After Chris was done working with me, I couldn't wait for him to get to Del, because that was the next level. I said to Del, "I can't wait for you to see this guy. I want to see what you think."

Del watched him, and after the show he turned to me, and the first thing he said was, "Oh, that's the next John Belushi."

CHAPTER 5

Whale Boy

JUDITH SCOTT, *cast member:*

If you think of the rest of the Second City cast as flat land, Chris was something that fell out of the sky and gave us shape. He might blow out a huge crater, like a meteor, or just collide with the ground, becoming this huge mountain. And by creating this landscape, he gave the rest of us the terrain on which to play.

Chris Farley spent a little over eighteen months studying and performing at ImprovOlympic. The young theater was rapidly becoming one of Chicago comedy's best-known training grounds, but at the time it remained just that: a place to learn. For actors seeking out professional opportunities and a professional paycheck, Second City was still the place to be. Since its founding in 1959, Second City had established itself as the nation's graduate school of comedy. In the early days, Robert Klein, Joan Rivers, and Alan Arkin all came across its stage. Over subsequent years, dozens of Hollywood stars matriculated there as well.

Chris auditioned for Second City's touring company in January 1989

and was offered a position. Most performers would have spent months or even years on the road before joining the main-stage ensemble; Chris made the move in a matter of weeks. Del Close had been offered the opportunity to direct Second City's spring revue, and he was given great latitude to mold the show and its cast to his own liking. The performer he liked most was Chris. Second City producer Joyce Sloane expressed reservations about the young performer's readiness and his outsized partying habits, but Charna Halpern insisted that working with Close was exactly the kind of discipline Chris needed. Sloane ultimately agreed.

And everyone agreed that Chris's potential was virtually without limit. During his eighteen months at Second City, Chris performed in three revues: *The Gods Must Be Lazy*, *It Was Thirty Years Ago Today*, which marked the theater's thirtieth anniversary, and *Flag Smoking Permitted in Lobby Only*. Also making the leap from the touring company at that time was Chris's friend Tim Meadows. Second City veterans David Pasquesi, Holly Wortell, Joe Liss, Judith Scott, and Joel Murray, as well as understudy Tim O'Malley, rounded out the cast. With each show, Chris's reputation grew. He created a number of characters and scenes that would go down as some of the best in the theater's history.

TIM O'MALLEY, *cast member:*

I was sitting in the main lobby at Second City. Chris came through the front door, all big and boisterous like he always did. He went upstairs and auditioned. What he did for his audition was he pretended he was late for whatever the scene was, took a running leap from stage left, and landed flat in the middle of the stage. They hired him right away, just on his energy and his commitment. Everyone was like, "You should have seen this guy's audition. He was fucking nuts."

TIM MEADOWS:

Me and Farley were the two new guys, both coming straight out of the touring company. We bonded over the fact that we didn't know what the fuck we were doing.

HOLLY WORTELL, *cast member:*

When we started rehearsals, Chris's inexperience showed. You had to come up with your own scene ideas, and he was not very good at that. One day we were throwing out these social, political, and cultural ideas, and Del said, "Chris, do you have anything?"

Chris went, "Um . . . yeah. I was thinking, um, that there's rich people . . . and, uh, there's poor people . . . you know . . . something about that."

And I was like, "What was *that*?"

TOM GIANAS, *director:*

Chris was a great writer, actually. He just worked on his feet. You gave him a premise, and he'd spin it into gold.

JOEL MURRAY:

The first time I had to improvise with him onstage, the suggestion we got from the audience was "the drunk tank." It was like, okay, there's a natural. I said to Chris, "Let's do a two-person scene where I'm your dad and I'm picking you up at the drunk tank, but in actuality I'm a drunk, too."

"Yeah," Chris said. "Let's go with that."

I didn't know anything about Chris and his father at that point; I just figured he was Irish so he'd know what I was talking about.

It was a great scene, and the emotions in it were so close to home in some ways, this drunk Irish father in his pajamas picking up his son. He's coming down on the boy, but everything he says—like "That time your

mother left me after I mowed down the hedges . . ."—reveals that he's a drunk, too.

So father and son have this meaningful talk about their drinking, and Chris is defending himself, like, "But I'm really *good* at it." Like his dad should be proud of him. Of course it winds up with Chris's character offering to drive home. And the dad, who's been drinking, says, "Yeah, yeah. I think that's a good idea."

And that became Chris's big sketch in the first show. What do you do with a drunken sailor? You have him play a drunken sailor. Del Close was in the audience, and he came backstage and said, "Yeah, well, that one's ready. Script that."

NATE HERMAN, *director:*

At Second City, a lot of the performers tend to be very verbal. But every once in a while a physical comedian comes along, and when an actor has that rough physicality in such a small setting it really tends to explode.

PAT FINN:

When Chris's first revue opened, he was an instant hit. There was a scene where he played a waiter. The people eating dinner were the heart of the scene, but Chris came out and got a huge laugh with "Can I get you something to eat?" That was it. He went over to the other side of the stage to make the drinks and sandwiches in the background, and every single head in the audience slowly turned to watch Chris. It was the oddest thing. Even if he was doing nothing, you wanted to watch him do nothing.

JOE LISS, *cast member:*

His mere presence would induce laughter. Anything he'd do on top of that was gravy.

DAVID PASQUESI, *cast member:*

Crowds loved him. I don't think you can find anyone who'll refute that. That's not an opinion.

HOLLY WORTELL:

On Friday and Saturday nights, we had a break between shows. Chris and I would always dash out and get something to eat, and we'd always run into about half the audience from the first show. We'd both been onstage for the past two hours, but everyone would come right up to Chris and say, "You were so great!"

Chris would go to great pains to say, "Hey, she was in the show, too. Wasn't she good?" And that was very sweet of him, but it seemed logical to me that people would notice him more.

JOEL MURRAY:

During that first show the cast, minus Holly, went away to Joyce Sloane's place on Lake Michigan. It was the dead of winter, and we'd brought a whole bunch of "Murray Brothers' Tea," this big thing of psilocybin mushroom tea. We wrote half the show that night, just from stuff we came up with screwing around. I've never laughed so hard in my life.

JOE LISS:

We were fucking around inside the house. I started doing this crazy English character. Tim was going on about how he couldn't feel his legs. The sun was coming up, and we couldn't find Chris. "Where's Chris?" "Let's find him!" We struck out of the house on an "expedition" to track him down. We ran out, and there he was, lying out on the ice of Lake Michigan. "Look," we said, "it's a whale!" "No! It's a boy!" *"It's a Whale Boy!"*

JOEL MURRAY:

The whole lake was frozen ice, these huge glacier formations, crazy stuff. And at one point, Farley was out there, shirt open, T-shirt over his head, diving like a seal onto the ice, doing these crazy belly flops, his stomach bright red. We riffed on this Whale Boy thing for hours, high on mushroom tea, laughing our asses off out there on the frozen ice of Lake Michigan.

JUDITH SCOTT:

We came back to the cabin, and as the drugs wore off we realized that we'd written a scene.

TIM MEADOWS:

We had trouble putting it on paper. When we got back we tried to improvise with this Whale Boy character, and there was just too much information. Plus, we all remembered it differently because we'd been tripping. So Nate Herman, our new director, sat down and said, "Each of you tell me your version of the story."

We did that, and then a few days later he came in with the whole thing. The story was that we had raised this Whale Boy as a real boy and hidden his true origin from him. The scene was his coming-out party into society. But the tension underneath was that his mother couldn't stand the fact that this was not her own son. Ultimately, we have to tell Chris the truth: that he's not our son, but in fact the product of a whale impregnated by radioactive human sperm and medical waste. Once he discovers that, Chris launches into this grand song and soliloquy about reclaiming his true identity.

HOLLY WORTELL:

We wanted him to be able to spout from his head like a whale, and so our stage manager took a football helmet, drilled a hole in it, and connected this water hose so Chris could squirt water out of it.

JOEL MURRAY:

He looked like a Hummel figurine. When he would come out like a choirboy and sing the Whale Boy song, the crowd would go nuts.

The last preview before opening night, he does the song and goes to exit through a door, and he pulls the doorknob off. There's nothing he can do, so he jumps into the crowd, and he breaks his foot. We're all like, "Suck it up, Farley. C'mon." But it turns out he really did break his foot, and he had a walking cast for opening night.

TIM O'MALLEY:

Ian Gomez took over, and Chris was out for a good eight weeks. But nobody could stop him from coming in and doing the improv sets. He'd come in and do those on a cast.

JOE LISS:

Chris was always "boy." He was Whale Boy. We did another sketch where he was Caramel Boy. Then of course there was *Tommy Boy*.

NATE HERMAN:

Paul Sills, one of the founders of Chicago improv, used to say, "Out there doesn't matter. This is the only place that matters. The stage is the only place you exist." So if the stage is the only place you feel real, it makes sense to make the whole world your stage.

ROBERT BARRY, *friend, Edgewood High School:*

Whatever Chris Farley did in the movies or on TV, that was just Chris Farley. He was never acting. Or, perhaps, Chris was always acting. That might be a better way of putting it.

CHARNA HALPERN:

He was always more himself onstage than off. He was more intelligent onstage. He'd step out there, and it was like a light would go on behind his eyes.

HOLLY WORTELL:

I remember we'd be in his apartment and he would be really calm. He'd talk about his feelings, things that he would never say around the guys. Then we would walk down the stairs and open that door out onto Wells Street and, literally, the second that door opened he turned into "Farley." It was like a jacket that he put on when he was leaving the house.

Chris would tell me a story about his day, like maybe a story about ordering a sandwich from the guy at the deli, and he'd do his voice and the guy's voice, acting out this little scene. But instead of just saying, "I told the guy, 'I want a tomato,'" Chris was really acting it up, like, "I *told* the guy, 'I *want* a *tomato!*'"

And so I asked him, "Chris, how come when you speak, you're imitating yourself?"

"What?"

I said, "When you tell that story about yourself, you're doing your own dialogue in a different voice. You're doing a character voice *for you.*"

And that kind of blew his mind.

Tom Farley and Mary Anne Crosby at a Georgetown University spring formal, 1956

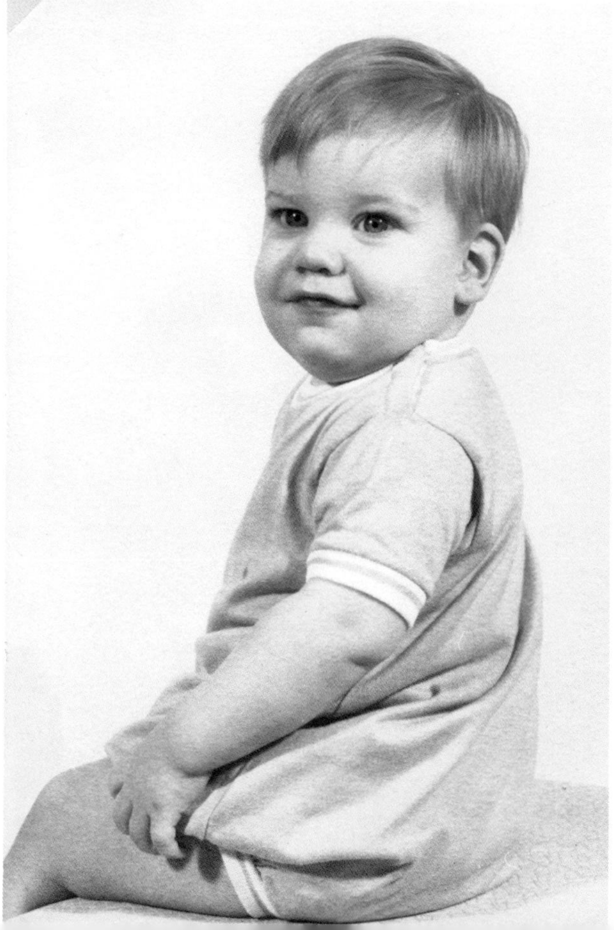

An early portrait of Chris, 1965

Hanging stockings on Christmas Eve, 1967, (from left) Chris, Tom Jr., Kevin, Mary Anne, and Barbara

A Farley Christmas card family portrait, 1968, (from left) Kevin, Chris, Tom Jr., Tom Sr., Mary Anne, Johnny, and Barbara

The Farley kids on summer vacation at Sagamore Beach, Cape Cod, 1969, (from left) Chris, Johnny, Barbara, Kevin, and Tom Jr.

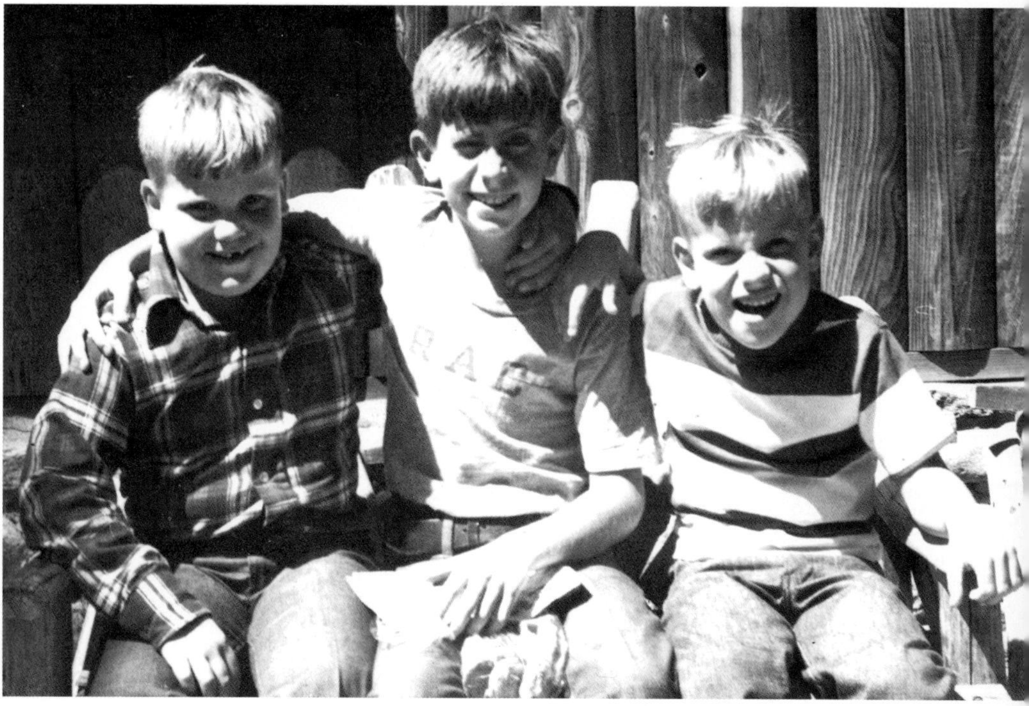

Chris and Kevin visiting brother Tom Jr. at Red Arrow Camp, 1971

Chris and Tom Jr. playing in a private plane owned by family friends visiting from Detroit, 1970

Chris at summer camp, 1973

Chris as a Cub Scout, photographed by den leader Mary Anne Farley, 1973

Chris (standing, top left) and his fellow cabin mates at Red Arrow Camp, 1974

A Farley family photo, taken at their home on Farwell Drive in the Village of Maple Bluff, 1972: Mary Anne, Tom Sr., Chris, Tom Jr., and John (standing); Barbara and Kevin (sitting)

Taking flight at age eleven

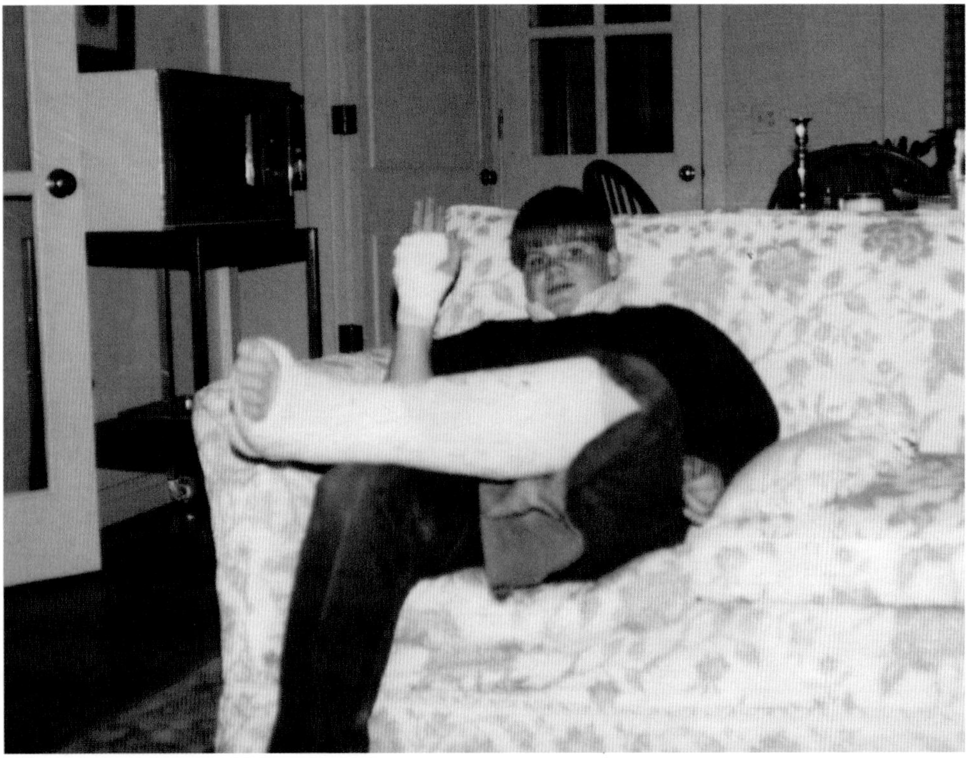

Sporting a broken leg and wrist from skating, 1977

Playing the class clown for friends in fifth grade at St. Patrick's School in Madison, 1974

Posing in a new suit for his eighth-grade graduation from Edgewood Campus School, 1978

Going out for the Madison Lakers youth hockey team. He played one year.

Noseguard, Edgewood varsity football team, 1980

Getting ready for his junior year homecoming dance with Greg Meyer (center), Robert Barry (second from right), and their dates

At a high school dance, senior year, with Todd Green (left) and Dan Healy

Fr. Matt Foley (bottom, far left) kneeling next to Chris and the rest of the gonzo athletes of the Marquette rugby team

Becoming an actor: backstage as "The Policeman" in a Marquette production of Sam Shepard's *Curse of the Starving Class*

The intrepid inhabitants of the legendary "Red House" at Marquette, with Jim Murphy (seated, bottom left) next to Chris, Dan Healy (perched, top left), and Mark Hermacinski (looking up, top center)
COURTESY OF JIM MURPHY

Enjoying college
COURTESY OF MARK HERMACINSKI

Leading his cabin on a woodland hike at Red Arrow Camp
COURTESY OF RED ARROW CAMP

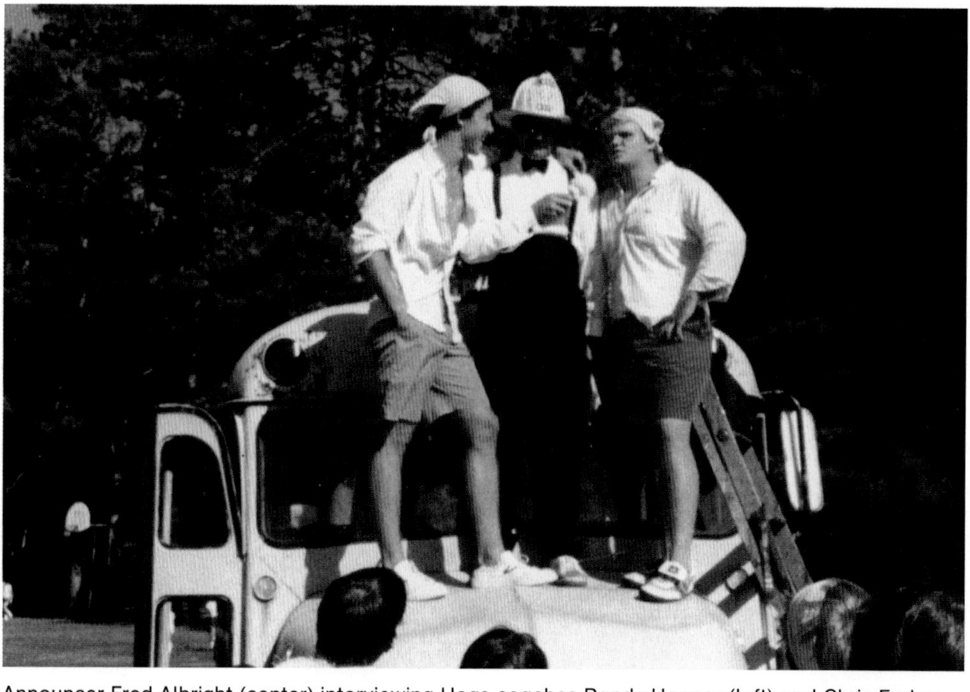

Announcer Fred Albright (center) interviewing Hogs coaches Randy Hopper (left) and Chris Farley (right) as they prepare to rally their team for Red Arrow Camp's annual "Salad Bowl," 1983

Performing in a Red Arrow counselor stunt night, with Fred Albright

The Farleys, minus Johnny, at Chris's "graduation" from Marquette University, 1986

Onstage at the Ark Improvisational Theater in Madison
COURTESY OF JIM MURPHY

The founding cast of FishShtick at Chicago's ImprovOlympic theater, including Pat Finn (standing, second from right), James Grace (seated, center), and Chris
© IO THEATER

The cast of Second City during Chris's first revue, *The Gods Must Be Lazy*: Tim Meadows and David Pasquesi (back row); Chris, Judith Scott, and Joe Liss (middle row); and Holly Wortell and Joel Murray (seated)

The original stage incarnation of Matt Foley, Motivational Speaker, (from left) Chris, Jill Talley, Bob Odenkirk, Holly Wortell, and Tim Meadows

TED DONDANVILLE:

Most of the people at Second City, after the show they'd all go out for drinks, and they'd just hang out with you like a normal person. But Chris and Joe Liss, they couldn't stop. The red nose and floppy shoes stayed on.

NATE HERMAN:

His greatest frustration was trying to beat that character that he was becoming. He really seemed like he was uncomfortable being the great, swaggering drunk; it didn't suit him. I always thought that inside there was this nice kid from Wisconsin going, "I'm really not comfortable doing this, but this is who I'm supposed to be."

TIM MEADOWS:

Chris used to say that he only had one character, and that was the fat, loud guy. But of course that wasn't true at all. In our last revue, we did a scene where the premise was basically that Chris and I were good friends, we're hanging out, and then his sister, played by Jill Talley, comes in. She and I really hit it off and I'm digging on her and Chris is *really* not okay with it.

JILL TALLEY, *cast member:*

The more Tim and I would milk our flirtation, the more Chris would amp up his reactions. Tim was giving me a foot rub, I was laughing at all of his jokes, and Chris was just fuming, bubbling under the surface, which he did so brilliantly.

Then Chris comes right out and tells Tim, "You can't date my sister."

"Why, because we're friends?"

"No, because you're black."

TIM MEADOWS:

The thing that I loved about Chris was that he was willing to be the racist in that scene. He starts off as my best friend, and then when I exit the stage for a moment he does this really subtle change where he confronts her and tells her he doesn't like it. Then I walk back onstage right as he says "nigger." Every night you could hear a pin drop as soon as he said it. People didn't expect that from Second City, and they certainly didn't expect it from Chris.

FR. MATT FOLEY:

When he was at Second City, he would call me late at night and I could tell he was using, that he was not doing well. He was really struggling; he was so damn lonely. He'd lost some of his anchors, and he was ashamed of his drug use.

TIM MEADOWS:

When we were in Chicago, we all drank, and we all did our share of other things. But one night after we did the first show we were getting notes from Del in the back. Farley went into the kitchen, got a bottle of wine, and just started guzzling it straight down. I remember watching him drink that bottle and thinking, holy shit.

Then, when we would go out drinking after the show, it would never end. His personality changed. He was a messy drunk. He would just get loud and in your face. I would go to one bar with him, and then he'd ask me if I was going to Burton Place, this bar that was open until five in the morning. I'd say, "No, I don't like that place." There was just a bad vibe in there, a lot of people who were involved in heavier drugs. I used to call it Satan's Den, and he would always tease me about that. "You wanna go over to Satan's Den with me?"

TIM O'MALLEY:

When you drank or got high with Chris, it was like corralling a tiger. I was doing coke on a regular basis, and I knew I was an addict. I don't know if I recognized Chris as being worse off than me, though. Guzzle and pour, guzzle and pour, slobber and puke. That was about it for us. We'd sleep all day. If we had rehearsal, we'd haul out of bed and make it there by eleven. We'd always try and get home by sunrise and get some sleep, but some days we were partying right up through to rehearsal, then try and get a nap and some food in before the show. Then, sometimes, we were just high for days.

TED DONDANVILLE:

The first hour of drinking with Chris was fun. The second hour was the best hour of your life. The rest of the night was pure hell.

DAVID PASQUESI:

He was taking it as far as someone could while still making it to work, and he was being rewarded all along the way. So there was no reason to stop.

TOM GIANAS:

Chris came in to rehearsal one day, and he was really out of it, kind of falling asleep. I took him aside and told him he had to go home. But Chris could always turn on the charm. He talked me into letting him stay. I always regretted that one moment. It was just a minor incident, but it was one instance where I had an opportunity to exert some discipline over him, and I didn't do it.

JOE LISS:

Second City sees itself as a family, and we were a pretty codependent family, too. There could have been a big intervention from the cast, but it didn't happen. We were all doing drugs.

The one thing we did crack down on was drinking during the show. That wasn't okay. One time I caught Chris in between sets. There's the greenroom backstage, and then there's another dressing room on the other side with a pass-through between them. I'm backstage, and I look down the pass-through and see Farley at his locker. He looks around, reaches into his locker, takes out a big tumbler, takes a big sip, and then hides it back in his locker.

I'm like, "You motherfucker."

So I wait for Chris to leave and go and get a big box of salt from the kitchen. I take the drink, fill it with salt, stir it up and let it dissolve, put it back, and head back to the other room and wait.

Chris comes back for the next show, reaches into his locker, gets the drink, takes another big sip of it and just does this beautiful spit take, spewing the drink all over. There was no more drinking during that show.

CHARNA HALPERN:

We had a couple of meetings with Chris where Del Close told him about the clinic where he went through aversion therapy and stopped drinking. It's a horrible process, like something out of *The Twilight Zone*. They shoot you up with this chemical and then they make you drink this watered-down alcohol until you vomit. Then they throw you back in your bed, and a couple of hours later they shoot you up and make you go through it again. They keep doing this until even the smell of alcohol makes you throw up. It sounds futuristic, but it works.

So Del told Chris about this therapy. "You need this," he said. "Try it. It works."

And Chris was like, "Nah, that's too permanent."

PAT FINN:

We were at a bar in Chicago once, Chris and me and his dad. Chris's dad goes, "Finner, you want a beer?"

"Oh yeah that'd be great, Mr. Farley."

"Christopher?"

"Uh, yeah. Thanks, Dad," Chris says.

"All right," he tells the bartender. "Two Old Styles for Finner, and a coupla Old Styles for Christopher. And I'll have a scotch. Tall glass. Rocks. Leave the bottle."

The guy just kind of stares at him. Mr. Farley stares back. Then he finally turns around to get the drinks.

I go, "Wow, Mr. Farley, really? What are we, in the Old West?"

"Look, Finner," he says. "I like scotch, and when I want a drink, and this bartender's talkin' to his gal pal down at the end of the bar, I'm not gonna wait for him. So I get a bottle, and I can have scotch whenever I want it. On top of that it's kind of fun to see what they charge me, because they never know how many shots are left in the bottle."

TOM GIANAS:

When his dad would come to the shows, they'd go to That Steak Joint, which was this steak restaurant right next door to the theater. They had this thing called the Trencherman's Cut, which was this ungodly cut of meat, just an unholy-size piece of Chicago beef. You would buy it for the table and eat it family style. When Chris and his dad would go, they'd each order one.

TIM MEADOWS:

His father used to tease him if he couldn't finish his. It was funny to see Chris when his parents came to town, because it was the only time you'd see him dressed up. The pressed shirt, the sport coat, the slacks. He'd have a haircut and a shave. He wanted everything to go right.

TIM O'MALLEY:

He was still a kid. His parents paid his rent for him. We told him to save some money and pay his own rent so he'd feel more responsible for himself. "But my dad wants to pay for everything," he'd say.

JOE LISS:

It was like they were still treating him like a kid at college, and he was embarrassed of it, or tired of it. He would get these care packages from his mother, and he would get visibly angered by them. She would send him food and slippers, new clothes from Brooks Brothers, and he'd be like, "Aw, dammit." But if you looked at his private life, you were like, "Chris, you need new clothes. You need a cleaning lady."

HOLLY WORTELL:

One time I tried to get Chris to open a bank account. He didn't have one. He would just cash his checks and have cash on hand. He was very generous, would buy everybody drinks, but he wouldn't save anything. I said to him, "Look, just walk with me up the street and open up an account."

"No."

"Chris, you can't just spend all your money. You're spending all your money on food, drugs, and alcohol. You have to put some away for rent and bills and savings."

And he wouldn't do it. He would not set foot in the bank with me.

JILL TALLEY:

We all took turns bringing him home. I remember being shocked that he didn't really have anything, nothing that you would associate with a home. It was really sparse, just a bare mattress and a big trash bag full of clothes. That was it.

JOEL MURRAY:

I had an apartment above Los Piñatas, this Mexican place by Second City. I always believed that Joyce Sloane had something to do with Farley all of a sudden having the apartment next to me. She knew that I would watch out for him.

He was like Pigpen from *Peanuts*. We had mice, too. The mice would come through my apartment and stop and give me this look, like "We're just on our way over to Farley's place."

JOYCE SLOANE, *producer:*

He always talked about how he wanted a girlfriend, and I said to him, "Do you think you could bring a girl up there? To that apartment? You gotta clean it up."

I tried to tell him that Bluto was just a character John Belushi played in *Animal House*; that wasn't what he was really like. When John's widow Judy was in town, I introduced him to her, and I said, "Do you think a lady like this would've been married to him if he was really that character that you saw?"

I thought that was a good thing to do, to let him know what John was really like. It didn't sink in, obviously. So I hired a cleaning woman for him to try and fix his place up. She came every week.

TIM O'MALLEY:

For all the trouble, everyone wanted to take Chris under his wing. The alumni that came in, like Bill Murray, would just adopt him immediately. It just oozed out of him: "Love me. Please love me." He got so much love. He got a ton of it. I don't know how he was even able to process it.

DAVID PASQUESI:

It was pretty obvious that there were problems for Chris from the word go. I don't think we realized that it was as detrimental as it was, but we could all see it. Then there was some odd behavior that was a bit obsessive-compulsive. We just thought it was weird, and we were right. That was some weird shit.

JOE LISS:

Before he put his pants on, he'd lick the tip of the belt, then the buckle, then he'd put on his pants. Before he went onstage, he'd touch the floor twice, touch the wing of the stage twice, and then go on. Every time.

BOB ODENKIRK, *cast member:*

I cannot express to you how much he licked everything. He'd open his wallet and lick everything inside it, the pictures, the money. He had to lick his shoelaces to tie them. He'd lick his finger and touch the stair, lick the finger, touch the stair, and do it all the way up the staircase. It was totally nuts.

HOLLY WORTELL:

One night I was onstage, and I said a cue line for Chris to enter. He should have been standing right at the door, but I could see him off in the greenroom, and that meant he had to go all the way down the corridor backstage and enter from the other side. I remember thinking, he's got to cover all that distance, touching every door twice, touching every stair twice, touching all the coats twice. It's going to be an hour before he gets there.

One day we were walking down the street, and he kept bending down and touching the sidewalk twice, touching the parking meters twice. I said to him, "Chris, why do you do that?"

And he said, "I'm just trying to even everything up."

JOE LISS:

He was superstitious, too. There was this Ouija board backstage. "Get that thing out of here!" he'd say. He was afraid of it. So, of course, someone took the Ouija board and put it on his pile of props in the back. "Who did this?!" he yelled. Then he took a towel—he wouldn't touch it with his hands—and he picked it up and threw it aside.

TIM MEADOWS:

I was surprised by Chris's faith. He'd go to mass all the time. I always admired that about him. I grew up in a Christian household, but once I got to Chicago I didn't go to church anymore. I just prayed whenever I wanted something.

HOLLY WORTELL:

We would do charity gigs sometimes where they would pay us a hundred dollars, which was *huge* for us; the most we ever made was $435 a week. But whenever we would do those charity shows, Chris was the only person in the cast who turned down the money. He would always say, "Please keep this and make sure it gets to the people you're helping."

JUDITH SCOTT:

He had a moral code, a sense of right and wrong. There was something in him that was correct and proper and Midwestern. If we were in a meeting and somebody's idea got shot down, Chris would immediately take up for that person and defend them.

HOLLY WORTELL:

In a show, you do the scenes, the director gives you notes on those scenes, and then you build on that the next night. I was so excited that I was go-

ing to get to work with Del Close, but what I came to learn was that Del *hated* women, didn't think they were funny, didn't want them in the show. At a certain point I was like, "What is this guy doing? He's not coming up with any ideas. He's not pushing us. He's not doing *anything.*" He would say to us, "No notes after the show tonight."

Then, after several weeks, Chris came to me and said, "Holly, I know I shouldn't be saying this, but I've just got to tell you something."

"What?"

"When Del says we aren't having notes after the show, we are. We're all going over to Joel's apartment, and he's giving us notes. Del says that you and Judith are evil witches and you're trying to ruin our show."

And what was happening was that Judith Scott, Joe Liss, and myself, who hadn't worked with Del over at the ImprovOlympic, would go home every night, not knowing what was going on, and Chris, Tim Meadows, Joel Murray, and David Pasquesi were all going over and working on the show with Del. For Chris to go behind Del's back and betray him was *huge.* Chris would never go against any authority figure, but he did it because he felt what Del was doing to Judith and me was wrong.

JILL TALLEY:

Once in a blue moon I'd make it to church, and I would always see Chris there. I always got the biggest kick out of watching him come back from taking communion, because it was like he knew everyone there, all the parishioners. He'd nod and smile at them, and they all really liked him. The priests really liked him, too. He was very much a part of that community. It was just a pious, quiet side of him that you'd never know unless you saw him there. Then, fourteen hours later, I'd be carrying him home from a bar and putting him to bed.

FR. MATT FOLEY:

Chris and I would sit and talk all night. He would ask me about God, about faith. His biggest questions always related to his struggles with evil,

with his addictions. Drugs were Satan, to Chris. Fighting that took a lot out of him.

HOLLY WORTELL:

When he was drunk, he might do or say something he shouldn't have done. But one day he told me, "I went to church and I confessed, and now it's over so I can do it again."

And I said, "Chris, I'm Jewish. But I'm pretty sure that's not how your religion works. That doesn't even make sense."

One night, Matt Foley came to see the show, and the three of us went out after. I said to Matt, "I have to ask you this. Chris says if you sin and go to church and confess and say your Hail Marys, then you can just do it again."

And Matt said, "No, of course you can't do it again. You have to try *not* to do it again."

I punched Chris in the arm and said, "See?"

FR. MATT FOLEY:

There are two ways to look at confession. One is to say, "I'm in a state of grace. God has blessed me, but in the midst of that greatness I have treated my brother or sister poorly and need to make atonement."

The flip side is to say, "I'm a sinner. I'm not worthy. I have to find Christ to find my way out of my natural state, which is sin."

I'm not sure which side Chris worked from, but I suspect it was the latter. "I am not worthy to receive You, but only say the word and I shall be healed." That's what Chris believed. And that's not wrong, but it needs to be balanced with an understanding of the good side.

Chris was grateful for each day of the life God had given him. He never saw himself as an equal of God, and many times we do. A lot of our humanism is very arrogant. It says, "I'm God. I'm equal to God." Chris was never like that. He stood in awe of God.

BOB ODENKIRK:

He saw things in a very simple way, good and evil, right and wrong. I was raised Catholic, too. I can remember being scared of the devil, of hell. It's real in your head.

One time at Second City, Chris was violently drunk at a party. He was picking up these chairs and throwing them across the room. Me and my girlfriend brought him back to his place, and he started throwing the furniture there, too, chucking it across the room. Then he just stopped, looked over at me, and said, "Odie, do you think Belushi's in heaven?"

It was sad, and chilling, and I didn't know how to answer him. It made me think that those myths that you live with as a kid, they don't always help you when you're an adult. They don't help you deal with grown-up things.

* * *

Even as Chris's personal troubles began to mount, his professional fortunes only grew. Every night the audience favored him more and more. His reputation began to draw interest from the talent scouts at *Saturday Night Live* as well as representatives at Brillstein-Grey, the powerhouse management firm in Beverly Hills. In the spring of 1990, Chris was signed by Brillstein's Marc Gurvitz, who would manage him for his entire career.

On March 24, 1990, *Saturday Night Live* flew Chris to New York to attend a taping of the show and to have an informal meeting with executive producer Lorne Michaels. It was the first solid indication of their interest in him and of things to come. At the same time, work had begun on Chris's third Second City revue, *Flag Smoking Permitted in Lobby Only*. For this show, producer Joyce Sloane brought in Tom Gianas, a young director from outside the Second City universe. Gianas in turn hired actress Jill Talley and *SNL* writer Bob Odenkirk to join the cast. It was a fortuitous meeting for all. At Gianas's request, Odenkirk helped Chris create the signature character that would take him to *Saturday Night Live* and beyond.

TOM GIANAS:

When you're lined up to direct a show at Second City, you just go in and watch the performers. You make notes, jot down inspirations. Night after night, I would go there to work and make notes, and every time I saw him onstage, he transformed me from a director into an audience member; I forgot to take notes. I would just sit back and laugh and laugh and laugh. That's never happened to me before or since.

I've worked with a lot of great people over the years, from Jack Black to Steve Carell, and this is something I can only say about Chris: From the moment he stepped onstage, the audience was completely invested in him. There was just a sense of "He's gonna cause trouble, and I want to be here when it happens."

When I arrived, he was doing the Motivational Speaker guy, but it wasn't as a motivational speaker yet, because that idea didn't exist. It was just *that guy,* in a million different contexts, usually a coach, or maybe an angry dad. It would destroy the audience every night. I said to Odenkirk, "We cannot open the show without that character."

And that's when Bob came back with a sketch about a family with pot-smoking kids who hire this Motivational Speaker, a guy who lives in a van down by the river, is thrice divorced, and uses the complete disaster of his life as an example of what not to become.

BOB ODENKIRK:

I sat down to write it, and the sketch came out pretty much whole the way it was done. I handed it to Chris, and watching where he took it was insane.

TIM O'MALLEY:

Chris could never remember his lines during rehearsal. He'd get so amped up with the energy of that character, doing the hips and the arm-pumping thing. He screwed it up every night for eight weeks. Odenkirk

and Pasquesi sat down with him and went over and over his lines. He was like, "This is hard. This is like learning the Our Father."

Then, on opening night, we were all worried he was going to screw it up—and he nailed it. When he came offstage, I said, "Why the hell did you finally get it tonight?"

"Big game," he said. "Coach is here." Del was in the audience.

FR. MATT FOLEY:

I was in the audience that night. When he said, "My name is Matt Foley, and I am a motivational speaker!" I was probably as red as a beet. I smiled and slid down a little further in my chair.

After the show we went and hung out in a bar for a time. Chris told me that he was never going to change the name, that it was always going to be Matt Foley. I'm honored by it still.

TOM GIANAS:

To this day, it's got to be the funniest thing I've ever seen. It never stopped making me laugh, and that's rare.

NATE HERMAN:

Chris was never captured in either movies or TV as good as he was onstage. He was too explosive. He just seems flat in all those movies. It's like watching a large wild animal in a cage.

BOB ODENKIRK:

The Matt Foley sketch was basically the same every night, but he was always on the edge of that character, forgetting his lines, making up new ones, changing the blocking. He was never content to mimic last night's performance, which a lot of actors do. Sometimes, you try to make the performance fun and surprising again for the people onstage, who might

be a little tired of it. Chris was doing a great show for the audience, but he was doing a completely different show for the rest of the cast. I can picture it in my head right now like it was yesterday. Night after night I'd stand there, four feet away from him, and just watch in complete awe.

HOLLY WORTELL:

I used to have to hold my cheeks with my hands because my face hurt from trying so hard not to laugh.

TIM MEADOWS:

Just watching him adjust his tie, or hitch up his pants, was enough to make you lose it. He'd get closer and closer to your face every night when he was saying his lines. I played the son. He started picking me up and tossing me up in the air, flipping me around. Then out of nowhere he'd kiss me. He just had a ball doing it.

JILL TALLEY:

He was so into the character that he'd be swinging his head around and his glasses would go flying off. Then he'd proceed to act like he couldn't see for five minutes, stumbling around looking for his glasses, and that would become the scene. Funny things would just happen organically. Even if he did the script word for word, it felt new every night.

BOB ODENKIRK:

I just remember thinking, no one else in the world will ever be able to do this character.

TIM MEADOWS:

After Farley left for *SNL*, Ian Gomez filled in for him. Every night he tried to do a different character to make that scene work, and he could never do it. Finally we said, "We gotta cut it."

TOM GIANAS:

The night that *SNL* came to scout him, he was nervous but confident. I remember saying to the cast, "The set's yours. You can put up whatever you want tonight." I wanted Chris's strongest pieces to go up so he'd have a good shot at it.

TIM MEADOWS:

We had a great show that night. And it was great for me. Chris and I worked so well together. That helped me get noticed, and I got hired at *SNL* about six months later.

JILL TALLEY:

He was nervous about *SNL*. He went back and forth with everyone. "What should I do?" He called his parents, his priest, the entire cast. We were all like, "What do you mean, 'What should I do?' You take it."

But he went round and round on it. Chris had his apartment, his bars, his church, and Second City right in this little four-block radius. That was his world, and I don't know that he ever really left it. He was scared to leave that behind, to leave his family, to leave everything he'd ever known.

KEVIN FARLEY:

I was at the airport when Chris left. Chris was crying, and Dad was crying. It was sad to watch. When you're from the Midwest, you don't really

ask for the spotlight. You just have your Sunday Packers game and that's about as exciting as it gets. But I think Dad knew, we all knew, that after this nothing would ever be the same.

PAT FINN:

I got married the Saturday of Chris's first night on the show, and he was all bummed out that he wasn't going to get to be in the wedding. He called me early in the week and was just apologizing profusely because he had to miss it. "Maybe I can make it," he said. "Get them to put me on next week."

"No! You're doin' the show!" I said.

"I know. I sorta got to. I mean, I shouldn't ask if I could skip it, right?"

"Chris. C'mon. It's *Saturday Night Live*!"

"No, but it's your wedding."

And that's the great thing about Chris: *SNL* was his dream, but if he could have skipped that first show to make my wedding, he would have.

The night of the ceremony, we were all at the Hilton. At ten-thirty, with everyone out on the dance floor and the wedding in full swing, about ten of us, including me in my tux and my wife in her wedding dress, snuck out and went over to the bar in the hotel and watched Chris make his debut on *Saturday Night Live*. It was so strange, so surreal. We'd all grown up with this show, and Chris was the first one we'd ever known to join those ranks. Just a few weeks before, he'd been hanging out in our apartments, and now he'd made it.

CHAPTER 6

Super Fan

CONAN O'BRIEN, *writer:*

When Chris first got to the show, I met him hanging out in the conference room outside Lorne's office. He was dressed kind of like a kid going to a job interview. We chatted for a bit. I liked him right away.

I came in and out of that conference room several times during the day, and Chris was still waiting. Lorne would do that to you, make you wait a long time. At the end of the day, I was feeling bad for him, so I said, "Hey, kid. I'll show you around the studio," and I led him on kind of a mock tour where I pretended to be in charge of everyone. Chris fell in and started playing along with me. After that I left and went home. I came back to work the next day, and Chris was still waiting outside Lorne's office.

He had this energy, even when he was sitting there waiting for his meeting, rocking back and forth in his ill-fitting sports jacket with his tie all pulled off to the side. He seemed really earnest about doing the show. You just had the feeling that he was going to be a lot of fun and he belonged here. It was like the show—and I

don't mean this to sound condescending—but it was like the show had been given this new golden retriever puppy.

From the day he arrived at *Saturday Night Live*, Chris Farley was already suffering comparisons to the other outrageous, larger-than-life figure in *SNL* history: John Belushi. When Chris died seven years later, eerily, at the same age as Belushi, those comparisons became gospel. In truth the two men shared far more differences than similarities. Still, in life and in death, Chris has borne the accusation of trying too hard to follow in Belushi's footsteps—an accusation with varying shades of truth. Yes, Chris looked up to and admired his predecessor, but whatever influence Belushi's ghost had on a young Chris Farley paled in comparison to the truly dominant forces in his life: his father, his family, and his faith. As far as drugs and alcohol went, Chris's bad habits were very much his own, seeded in his DNA and showing up at keg parties long before Belushi's demise. And if Chris followed Belushi in more positive ways, he was hardly alone.

In the comedy epidemic of the twentieth century, John Belushi was Patient Zero. The twin blockbuster successes of *Saturday Night Live* and *National Lampoon's Animal House* fundamentally changed the landscape of being funny. Movie studios began churning out huge blockbuster comedies like *Ghostbusters* and *Beverly Hills Cop*. Stars like Eddie Murphy, Mike Myers, and Jim Carrey beat a well-trod path from sketch-comedy cult status to Hollywood fame and fortune. Second City and ImprovOlympic grew from regional theaters into multiheaded corporate enterprises, churning out hundreds of aspiring comedians every year and spawning scores of other schools and venues across the country. Chris Farley and his friends were the first generation born into and weaned on that era. Their reverence for it and obsession with it was the common denominator that bound them together.

It all began in 1975 when producer Lorne Michaels assembled the original cast of *SNL* and took to the air live from New York every Satur-

day night. Following his departure in 1980, producer Dick Ebersol took over the show. Ebersol presided over some difficult years but also cultivated the stardom of Eddie Murphy and assembled the all-star cast of Billy Crystal, Christopher Guest, and Martin Short.

In 1985, Lorne Michaels returned. The show needed new direction, and he needed a job. After a rocky start, he went back to the drawing board in 1986 and assembled the cast—Dana Carvey, Phil Hartman, Jan Hooks, Nora Dunn, Jon Lovitz, Kevin Nealon, Victoria Jackson, and Weekend Update anchor Dennis Miller—that would breathe new life into the show. Mike Myers came aboard in '89, but otherwise no visible changes where made, or needed, for the rest of the eighties.

Then, in the fall of 1990, a slow transition began to take place. Nora Dunn and Jon Lovitz left; Chris Farley and Chris Rock entered. Far younger than the established cast, the two became fast friends and soon found themselves sharing an office. Farley and Rock were the only performers added that fall. Tim Meadows, Chris's Second City cast mate, would come on board at midseason.

Back in the writers' room, Jim Downey, a freshman writer in *SNL*'s early years, had assumed the reins of head writer and producer. At the core of the writing staff was a group that had led the resurgence from the show's mid-eighties nadir: Robert Smigel, Jack Handey, Bob Odenkirk, and Conan O'Brien. Meanwhile, Tom Schiller, Al Franken, Tom Davis, and Marilyn Suzanne Miller—also veterans of the show's original writing staff—had all come back for an additional go-round. Added to that was a very young team of stand-up comedians—Adam Sandler, David Spade, and Rob Schneider—whose age and sense of humor would ultimately bring about a generational shift at the show. Both on camera and off, *SNL* found itself with a varsity squad and a junior-varsity squad. It was an odd mix of talent, but it worked well. For a while.

Chris arrived in New York in October. His older brother, Tom, had lived in the city for many years, and together they found an apartment for Chris on Seventh Avenue, just north of Times Square and right around the corner from the show's Studio 8H in Rockefeller Center. The canyons of midtown Manhattan were a striking contrast to the cozy comforts of

Chicago's Old Town, but Chris soon discovered the Carnegie Deli, St. Malachy's Church on West Forty-ninth Street, and a fine Irish pub called The Fiddler's Green, all within a small walking radius. He had made his home again, scarcely able to believe what that new home was. As many latter-day *SNL* writers and performers have said, anyone who works at the show is a fan of the show, first and foremost. And Chris was surely that.

ROBERT SMIGEL, *writer/coproducer:*

I was a coproducer as well as a writer, and so I got to go with Lorne to Chicago to scout the Second City show. Hiring Chris was probably the easiest casting decision Lorne's ever had to make. In all the shows I scouted before or after, I'd never seen anybody leap out at you from the stage the way Chris did. Lorne hired him the next day.

JIM DOWNEY, *head writer/producer:*

There was so much buzz about Farley that our checking him out was almost pro forma. It was kind of automatic.

LORNE MICHAELS, *executive producer:*

I'd had something of a concern that maybe he was too big, personality-wise, to play on television. Theatrically, he was sort of playing to the back of the house. But after we saw him, there really wasn't much doubt.

ROBERT SMIGEL:

Lorne invited me to be in on his meeting with Chris. Chris showed up, and he was in full altar-boy mode, lots of "yes, sirs" and bright-eyed alertness. He was so transparently on his best behavior that you kind of had to laugh and wonder if it was inversely proportional to his worst behavior. Lorne talked about the show and what would be expected of him, and

Chris just kept sweetly nodding his head in agreement. Lorne had been told, at that point, about Chris's problems. I don't remember exactly what he said, but he told Chris, in so many words, that it wouldn't be tolerated. He even said something to the effect of "We don't want another Belushi."

LORNE MICHAELS:

It wasn't presented to us that Chris had any sort of problem, just that he was still a little young and liked to party too much.

TOM FARLEY:

All the cast and writers were sort of strolling in over the course of that first week. Chris immediately gravitated to this younger, newer crowd of writers and actors: Rob Schneider, Adam Sandler, and David Spade. They were coming on as writers. The only two new cast members were Chris and Chris Rock. They got all the press.

DAVID SPADE, *cast member:*

I had done four shows as a writer/performer. Then it was summer break, and when I got back Farley and Rock came on as featured players. Sandler came about six months later.

I met Chris the first day, walking over from the Omni Berkshire, where *SNL* had put us up. I saw him downstairs, and I'd heard about him. We talked and then we walked over to 30 Rock together. I thought he was funny. He was a nice Wisconsin dude, a genuine, sweet guy. I was out from Arizona. I'm not really a bad guy. We just gravitated to hanging out all the time and stayed buddies ever since.

MARCI KLEIN, *talent coordinator:*

I first met him the day he started. He was wearing this English driving cap and looking very Irish. He was very quiet and deferential, very nervous, like I was the person in charge or something, which I thought was funny, because I wasn't. He would get so nervous; that was one of the things that was really charming about him.

CHRIS ROCK, *cast member:*

We both got hired the same day, which was probably one of the greatest days of my life. We were the new guys, and they threw us together. The funny thing was that everyone was worried about *me*—I lived in Brooklyn and didn't want to move to Manhattan, because I couldn't park on the street and I couldn't get a cab. I said it in the *Live from New York* book: Two guys named Chris both get hired on the same day and share an office. One's a black guy from Bed-Stuy and one's a white guy from Madison, Wisconsin. Now, which one is going to OD?

KEVIN FARLEY:

When he got the show it was sort of strange, kind of scary in a way. Chris always liked the camaraderie of Second City, so a high-pressure situation like *Saturday Night Live* seemed mean and cutthroat. Dad was nervous. We all were. Chris could barely flush the toilet. How was he going to handle fame and television and New York City?

TOM FARLEY:

I was working at Bear Stearns in those days, at Forty-sixth and Park. Chris and I went out to lunch one day, and afterward I had to go back to work. We were standing on Park Avenue, right where it goes into the Met Life Building. Chris asked me about some good places to go shopping. I

pointed him in the direction of Fifth Avenue. Then he said, "Look, I don't have any money. I'm staying at this fancy hotel, and I've got this nice, big salary, but I haven't gotten paid yet."

So I went to a cash machine, got out $160, gave it to him, and went back to work. I got a call about a half hour later. "Uh, Tommy, I had a little problem," Chris said.

"What do you mean?" I asked.

"Well, I went up to Fifth Avenue like you told me and I was walking down the street and I saw these guys playing cards."

And I was like, oh no. "What did you do?" I said.

"Well, these guys were playing cards for money, and they were winning."

"Yeah, okay."

"And the guy looked at me and he said, 'Where's the card? Take a guess. How much you got, buddy?' And I said, 'I got $160.' And he said, 'Well, put it down.' "

"Let me guess, you lost the whole nut."

"Uh, yeah . . . can I get some more money?"

So I went and met him at the cash machine and said, "Don't do that again. Stay away from the guys playing three-card monte."

He said, "But I don't understand. How were all those other guys winning?"

"They were all part of the scam, Chris."

"Oh."

AL FRANKEN, *writer:*

Chris was very shy and self-effacing when he showed up. That never really changed too much, though he did get less shy. He was also one of these guys who came in with an incredible amount of respect for what had gone before him. He was just genuinely awed to be there, and wanted to know everything about the show. I don't want to name names, but some people would come in saying, "I was destined to be here! Get out of the way, old man!" That was not Chris.

CHRIS ROCK:

You could just tell he was funny. Normally you meet a guy, and you're automatically skeptical about him. You're basically not funny until proven otherwise. But there was something about Farley where you could tell he was funny when he said hi.

TOM FARLEY:

Chris called me up the first week he was there and said, "Hey, they're going to film me for the opening montage. Do you want to come down and watch?"

I said, "Sure. Where are you doing it?"

"Well, you can pick anywhere you want," he said, "so I was either going to do the steps of St. Patrick's Cathedral or McSorley's Ale House." Two very telling choices, and no question as to which one won out.

"Which is it going to be?" I asked.

"McSorley's."

So I met him down there at like two A.M. on a Tuesday night after it closed. We kept drinking beer after beer during the shoot. "Draining our props," as we called it. We were up till five in the morning.

My parents came into town for the first show. We had a blast showing them the whole New York thing. They stayed at the RIHGA Royal. We went to Gallagher's Steak House, which was Dad's favorite. Dad stayed in the hotel. He didn't go to the show. With his weight it was just too much trouble. So we all went to the dress rehearsal and then went back to the room to watch the live show together.

KEVIN FARLEY:

We were just excited as hell. It was sort of surreal, sitting there waiting for his first scene. Kyle MacLachlan was the host, and they did this *Twin Peaks* parody. Chris was the killer. He really didn't have anything to say, any real laugh lines in the sketch, but of course he found a way to milk a few

laughs by giving a look or drawing a word out. It's strange to see your brother on national television. I just sat there going, "This is weird."

TOM FARLEY:

I remember walking out of Rockefeller Plaza after the first show on our way to the after party. There were all these limos lined up. Chris said, "Wow! Look at all those limos! Ain't that something?" Pause. "I wonder if we can get a cab."

"Chris, who do you think these limos are for?" I said.

"What do you mean?"

I saw some guy with a clipboard and walked over to him. "Is one of these cars for Chris Farley?" I asked.

"Yes, sir," he said. "Right over there."

"Wow," Chris said, "How did you know that?"

"Because that's how it works here."

MIKE SHOEMAKER, *producer:*

The first show he didn't have much, because nobody knew him to write for him at that point. But that always takes a while. We knew he was going to hit big, and he did it pretty quickly. The "Chippendales" piece was only his fourth show.

JIM DOWNEY:

The thing that suggested "Chippendales" was less Farley and more Patrick Swayze being the host. You had a guy who was sort of built like—to the extent that I notice these things—like a male stripper. And he obviously could dance; that was how he'd come up in show business.

The second element was that nothing made me laugh more than the band Loverboy, whose big hit was "Working for the Weekend." So you had Patrick Swayze, male stripper, you had Loverboy—just add Farley.

KEVIN NEALON, *cast member:*

I played one of the judges, and my experience was the same as anyone who's seen it on television. I can't even think of the word to describe it. Incredulous, maybe? I did everything I could to keep a straight face.

JIM DOWNEY:

We didn't know it was going to be as popular as it was. You never do. In read-through Chris is just sitting fully clothed at the table while Lorne reads stage directions. We didn't know until he did it at dress.

MIKE MYERS, *cast member:*

I knew in rehearsal that a star was born.

DOUG ROBINSON, *agent:*

Adam Venit and I were agents at CAA. We'd known about Chris from Chicago, and we had been talking to Marc Gurvitz at Brillstein about signing him. At CAA, you have to get a consensus from the entire group of agents if you want to sign someone. All we did was show everyone a video of the "Chippendales" sketch, and it was done. We signed him right then.

BOB ODENKIRK:

I didn't like the fact that the first thing he became known for was that Chippendales thing, which I hated. Fucking lame, weak bullshit. I can't believe anyone liked it enough to put it on the show. Fuck that sketch. He never should have done it.

TOM DAVIS, *writer:*

When you get laughs like that, there's nothing wrong with what's going on onstage.

ROBERT SMIGEL:

It was a fantastic sketch—I'd say it's one of the funniest sketches in the history of the show—because of the way Downey wrote it. If the sketch had been written that a fat guy was trying out for the Chippendales and everyone was making fun of him or acting like it was crazy, then yes, it would have been just a cheap laugh at the expense of a fat guy. But the way it was constructed, with everyone sincerely believing that this guy has a shot, the judges studiously scribbling notes on his dance moves, that's what makes it original and completely hysterical.

JIM DOWNEY:

My overriding note to Chris was, "You're not at all embarrassed here. They're telling you, 'Our audience tends to prefer a more sculpted, lean physique as opposed to a fat, flabby one,' but your feelings are never hurt. You're processing that like it's good information. Like you're going to learn from this and take it to your next audition."

Of all the pieces I've done it's one of the most commented upon, and that's of course because of Farley. I can't take any credit for that except casting him. He was also very nimble and a good dancer, which made it impossible to feel like it was just a freak show.

Later in that show, Jack Handey had written a sketch about a mousetrap-building class. It was one of those group scenes where everybody has a very small part to play. Farley had only one little bit to do, but he had so won the audience over with "Chippendales" that he got the biggest reaction in the piece. They had already adopted him as their own.

CHRIS ROCK:

"Chippendales" was a weird sketch. I always hated it. The joke of it is basically, "We can't hire you because you're fat." I mean, he's a fat guy, and you're going to ask him to dance with no shirt on. Okay. That's enough. You're gonna get that laugh. But when he stops dancing you have to turn it in his favor. There's no turn there. There's no comic twist to it. It's just fucking mean. A more mentally together Chris Farley wouldn't have done it, but Chris wanted so much to be liked.

I wanted to be liked, but I had no problem saying something was racist and I wasn't doing it. Imagine if they'd had me in that sketch and then said at the end, "Oh, we can't hire you. You're a nigger." Would I have done a sketch like that? If I had, ten years later I'd want to shoot myself.

That was a weird moment in Chris's life. As funny as that sketch was, and as many accolades as he got for it, it's one of the things that killed him. It really is. Something happened *right then.*

TIM MEADOWS:

You had to prove yourself to get airtime and to get your sketches on. It was obvious after "Chippendales" that he was going to be one of the standouts. I can't remember too many read-throughs where he didn't have something to do.

TOM DAVIS:

During the early read-throughs, I was shocked when I heard him struggle to read scripts he was seeing for the first time. He struggled with reading in the way of someone who was not schooled well. But on the stage he was brilliant, absolutely brilliant. He was a star. I saw it the first day he walked in the office. There was some quality that shone brightly. Despite his shortcomings as someone who hadn't honed his skills yet, it was clear that he had the raw talent to work on that show successfully, which he did.

ALEC BALDWIN, *host:*

In the cast of *Saturday Night Live* you have people who've come from improv troupes, and you have people who've done a lot of stand-up comedy. You can distinguish the real actors from the stand-ups, and Chris was a good actor, a very good actor. He could have had a career for the rest of his life. Fat, thin, old, young, he was a really talented guy.

JIM DOWNEY:

Farley was like an old-school cast member. The first cast was a repertory company. They weren't comedians; they were funny actors, and they were called upon to do lots of different things. In the nineties, we got those with more of a stand-up background. And that's not to knock the stand-ups. Nobody has ever made me laugh harder than David Spade. But the danger is that a stand-up can do an absolutely devastating ten-minute audition, but that might not help you two years later when you need a Senator Harkin in your congressional hearings piece. As a general rule, we've always done better with the old Dan Aykroyd types.

LORNE MICHAELS:

I used to say that he was the son that John Belushi and Dan Aykroyd never had. When Chris was a kid, he used to tape his eyebrow up to try and figure out how Belushi did it.

TOM SCHILLER, *writer/short-film director:*

It was very shortly after he arrived that he asked me about Belushi. I could tell by the gleam in his eye that he was very excited to find out as much as he could. I don't recall exactly what I told him. I guess I told him a lot. But there was an immediate bond, and he wanted to be in one of my films to take part in that heritage.

MARILYN SUZANNE MILLER, *writer:*

Chris would come up to me and say, "You knew John. What was he like? How did he dress? Just tell me everything you knew about him." He just had this reverence, and it was always about real minutiae.

MIKE SHOEMAKER:

He always idolized Belushi, but when I looked at him I saw more of Aykroyd and Murray.

TIM MEADOWS:

Chris would always throw in bits or accents inspired by the older guys, more Bill Murray than anyone. He stole his "Da Bears" character from Aykroyd. Todd O'Connor, that's basically Aykroyd's Irwin Mainway, that cheesy salesman who sold all the dangerous toys.

ROBERT SMIGEL:

I had moved to Chicago in 1982 to take improv classes with the Second City Players Workshop. Today it seems you can take them anywhere, but back then that was the only place to go.

I was always a huge sports guy, and I wanted to see all the iconic Chicago sports landmarks, so not long after I moved there I went by myself to a Cubs game at Wrigley Field. I just walked up to the box office about a half hour before game time and said, "Give me the best ticket you've got." It was right behind the dugout, so I was pretty excited.

But during the game I noticed that the fans in the crappiest seats—the bleachers—were the ones having all the fun. I got a better look at them outside the ballpark after the game. They had their Cubs T-shirts on over their collared, button-down shirts; it was like a uniform. They all seemed to wear aviator sunglasses, which had been out of style for a good five years. And they all had these big walrus mustaches, which I think to

them were some sort of symbol of virility, probably passed on from Dick Butkus to Mike Ditka. They had a swagger to them, even though at the time all their teams pretty much sucked, which made the whole thing funnier.

Years later, at *Saturday Night Live*, there was a writers' strike that shut us down after February. Bob Odenkirk and I decided to go back to Chicago and do a stage show we'd talked about. The original sketch as we did it then was three of these guys sitting on a porch, talking about the Bears. It was more absurd, just guys talking about stuff and the conversation always going back to "and I'll tell ya who else will be riding high come January. A certain team known as . . . *da Bears.*" It was a hit in Chicago, but we never thought about doing it on *SNL*. Too local, we figured.

Then in 1991, Joe Mantegna hosted, and Bob suggested we do it. To make it more accessible, we set it in a sports bar and made it a parody of the Chicago sportswriters show, only with these ridiculous fans and their outrageous predictions given as sincere, cogent game analysis. Their idea of predicting an upset was to say that the Bears would only win by thirty points, instead of seven hundred.

Jim Downey is from Joliet, Illinois, and he was really insistent that the Chicago accents be dead-on. He felt that was really important to the joke. The only person in the cast who could really do the accent was Chris. Mike Myers could do a pretty decent accent, because he'd lived there awhile. Downey actually thought my accent was the best, and he wanted me to play the third guy.

So we did that original sketch with Mantegna. It did well, and that, I thought, was it. But it caught on in Chicago. A popular deejay named Jonathon Brandmeier played excerpts from it all the time. By spring the Bulls were headed for their first championship, and, coincidentally, George Wendt was hosting our season finale. It made sense to do another one. It was fine, but in Chicago it had a life of its own. "Da Bulls" was suddenly the rallying cry of the team. It's funny, because I had never thought of it as a catchphrase. In the scripts, it was always spelled "the Bulls"; the "da" just came out in the delivery.

That summer in Chicago, Michael Jordan was honored by Comic Re-

lief. George, Chris, and I were invited to do the Super Fans onstage with Jordan at this benefit at the Chicago Theater. Jordan had a lot of fun doing it and got comfortable with the idea that he could do comedy. That led to him hosting the season opener in the fall, where we did another one. It just took off from there. I think we did eight or nine in total.

It was fun to do on the show, but we had even more fun doing things at Chicago events. Chris and I started doing predictions every Thursday night for the local Chicago NBC affiliate. We'd tape them via satellite from 30 Rock in New York. Chris's managers never liked the idea of his doing these Chicago appearances if he wasn't getting paid, so he only did a few. I always thought that was ridiculous. Some of them were once-in-a-lifetime experiences.

When the Bears made the playoffs, in December of '91, Chris, George Wendt, and I actually stood at the fifty-yard line at Soldier Field and made a pregame speech. Half the fans went nuts and half were just weirded out, I think. Then they let us stay and watch the game on the field. I felt like a moron wearing my costume out there during the game, but I wasn't going to pass up the chance to watch from the sidelines.

At halftime they had a contest where they let kids try and make a field goal from a tee on the ten-yard line. Quite spontaneously, they asked us if we wanted to participate. We said sure, and suddenly on the PA it was, "Ladies and gentlemen, we direct your attention to our additional contestants."

We let Chris take the first kick. He lined up to make the field goal, took a running start and then slipped and nosedived right into the mud. The whole stadium went crazy. George and I didn't know what to do. We'd let the funniest guy go first, and now we had no way to top it. I figured the only thing I could do was to try and actually make the kick. So, with beer in hand, I lined up and kicked the ball and somehow I made it. The crowd loved it.

Then it was George's turn, and by now he'd figured out a topper. He had Chris snap the ball to me. Then instead of placing the ball down, we ran a "fake" and I lateraled to George. I "blocked" Chris, and George ran it in for a touchdown, as the announcer called it on the PA. That was just

so much fun. The stadium was packed, and everyone was cheering and going crazy for a fat guy running ten yards. It was one of those perfect moments in life, where all your hard work comes together in a way that's better than anything you could have imagined.

TIM MEADOWS:

It was a great time during those first couple of years. We were young. We all had money, we lived in New York, and we were on *Saturday Night Live*. We would always have dinner after the read-throughs. Saturday nights after the show we would hang out. We all played golf. We'd go to movies. We got invited to a lot of stuff, basketball games, baseball games. We all immediately bonded with each other.

CHRIS ROCK:

We just had fun. There was a lot of McDonald's, a lot of going out to eat, a lot of watching MTV. We'd watch the musical guests rehearse. "Let's run down and see Pearl Jam." We got to see a lot of the grunge stuff before it really took off; all those bands made their first appearance on *SNL*.

KEVIN NEALON:

More and more, the older cast members started to realize we had our own little west wing going on with Spade, Farley, and those other guys. You'd go over and there'd be *Playboy* magazines lying around and it'd be a bit of a mess. They'd all be talking about what models they'd gotten laid with, and then you went into Dana or Phil's office and they'd be talking about their 401(k).

DAVID SPADE:

We would just hang out and get in trouble. We shared an office, me and Chris. You'd have to go through our office to get to another little office for

Rock and Sandler. There was always Smigel around, or whoever else. We were all back there, screwing around and staying up late, trying to write and think of ideas and brainstorm and do whatever you gotta do to stay alive on that show.

MARCI KLEIN:

You just spend so many hours there, and so much time of your life. We were all sort of the same age. We were all always together. It's just inevitable that this group forms. We go to the party, then hang out at somebody's apartment, and the next thing I know, it's six A.M. and Chris and Adam are crashing in my living room.

TOM FARLEY:

The guys at *SNL*, they were like Chris's new rugby team. Not nearly as wild but a lot of fun, and they gave Chris that fraternity atmosphere that he always thrived in. And then, somewhere along that first year on the show, Chris started dating Lorne's assistant, this girl named Erin Maroney. Erin was great, just a lot of fun. And I thought it was the start of a whole new thing for Chris, a really good direction. It was the first time I'd seen him in a real relationship, you know, ever.

DAVID SPADE:

We'd always be outside Lorne's office waiting for whatever, so we had a lot of interaction with Erin. Chris would hide underneath her desk, and he'd pop out and surprise her when she got back. He flirted with her all the time, and it paid off. They started dating. Erin was very cute, very sweet.

She went back to Wisconsin with him once, and she'd tell hilarious stories about what it was like around that dinner table. She's kind of a preppy East Coast girl and there she is with seven Farleys fighting over chicken and steak, and the mother's passing around a "yuck bag" where they throw all their extra bones and corncobs. I don't think I ever would

have put those two together, but I think she saw through all the other stuff to see that he was a good guy and a fun guy, and as history shows sometimes that's enough.

TOM ARNOLD, *friend:*

Chris was in love with Erin. He talked about her all the time. She was very smart. Actually, she had the best line ever. They were all flying back from Los Angeles to New York, and Chris kept coming up and talking to Lorne, sucking up. So, after the third or fourth time he came by sucking up to Lorne, she turned to him and said, "Pace yourself, Tubby. It's a long flight." I think that's why he loved her. A funny woman who's attractive *and* smart, now that's what you want.

TODD GREEN, *friend, Edgewood High School:*

Chris really wanted to end up with an Irish Catholic girl that his family would approve of—and Erin fit the bill. Despite the fame, and despite the access he had to all these beautiful women, he really just longed for, as he would say, "a nice Irish gal." He was in love with the idea of them together.

TIM MEADOWS:

It was the first time I'd ever seen him in a serious relationship. It was surprising. I could see why he was attracted to her. They both had a similar sense of humor. She was also very cute, just had a really sweet personality, and she's really funny. I really liked that time of his life, because it was nice seeing him with someone who cared about him. He had someone who was actually going to make a home for him.

LORNE MICHAELS:

He saw in Erin someone who liked him for who he was. They would go to mass together on Sundays. He took her home to meet his parents, and I know she did the same. Her mother loved Chris. I think they just genuinely liked and cared for each other. But it was always more of a brother-sister thing than a boyfriend-girlfriend thing, from my perspective.

TIM MEADOWS:

Basically, that first year, we only had each other. And we all had one thing in common: We couldn't believe that we were doing this show. The people from *Saturday Night Live*, they were our heroes. It's like if you dreamed of being a Major League Baseball player and you got your chance to do it—only there's really only one baseball team in the country, and they've drafted you.

After the first season wrapped in May, we had this party at the restaurant downstairs at Rockefeller Center. A bunch of us went back up to the office for a little after party. Then, about four in the morning, Farley and I were both leaving at the same time. We were waiting for the elevator, and he said, "Timmy, can you believe this is our life?"

I said, "Chris, I think that every day."

Then he just grabbed me, and hugged me. He couldn't believe where his life was going.

CHAPTER 7

The Place in Alabama

LOVERBOY, *band:*

Everybody's workin' for the weekend.
Everybody wants a new romance.
Everybody's goin' off the deep end.
Everybody needs a second chance, oh.

Chris Farley's life was going in two wildly different directions at once. With "Chippendales," "Super Fans," and a number of solid turns in supporting roles, he had established himself as one of the standouts at *Saturday Night Live*. But his problem with drugs and alcohol had worsened severely. Around the office, Chris was always happy and hardworking, determined and focused. The sloppy, crazy party guy showed up at social functions here and there, but he was no longer a nightly fixture. Chris, knowing what was at stake, had learned how to hide his problem. And safely hidden, it festered.

The summer following his first year on the show, Chris returned to Chicago and dropped by Second City to see his old friends. He had clearly

been using and was in no condition to perform. The cast pleaded with him not to go onstage, but during the improv set he burst out from behind the curtain anyway. The audience greeted him with a huge cheer. But as he began to stumble his way through a scene, his inebriation was obvious to everyone. People began to boo. After one scene, an audience member yelled, "Get the drunk off the stage." Even in his failed stand-up comedy routines at Marquette, Chris had at least come off as affable and good-natured. He had never suffered wholesale rejection quite like this. The crowd's reaction cut deep.

On June 16, 1991, worried that the management at *SNL* would hear of the incident, Chris checked himself into the Hazelden recovery facility in Center City, Minnesota. It was the first time he sought treatment for his addiction.

Red Arrow camp counselor Fred Albright once observed, in his unique Wisconsin vernacular, that "Chris was extremely complex, and yet as easy to understand as a ripe watermelon." The doctors and therapists at Hazelden, in somewhat more sophisticated language, came to the same conclusion. Their diagnosis of Chris would come as no surprise to anyone who'd ever spent five minutes with the boy:

> *Chris's inclination is to be compliant. . . . Whenever Chris has been confronted for not meeting unit expectations he has apologized and assured that this will not happen again, [but he] does not appear to be willing to fully accept responsibility for his behavior. . . . Chris allows his fear to dictate the terms of his recovery. . . . Chris has identified that his use of humor serves the function of diverting attention from issues that may be painful. . . . that it's with humor that his family deals with conflict and pain. . . . Chris sees his life and his drinking as a benefit to his work as a comedian, and this may complicate his motivation to get help for these issues. . . . [Aftercare issues include] compulsive overeating, possible obsessive compulsive behavior. . . .*

Diagnosing the problem was easy. But unlocking it and treating it would prove to be a complex proposition indeed. After twenty-eight days, Chris checked out of Hazelden, but the program had accomplished little, and his destructive behaviors quickly resumed.

That fall, Chris returned to New York for the 1991–1992 season, his second, on *Saturday Night Live*. He moved from his apartment near Times Square to a place near Riverside Drive on Manhattan's Upper West Side. Along with Chris Rock, Chris was promoted from featured player to repertory player, and escalators in his contract pushed his salary from $4,500 to $6,500 per show. That year, Adam Sandler, David Spade, and Rob Schneider graduated from the writers' room to join Tim Meadows as full-fledged featured players.

Despite the unusual influx of so many new performers, *Saturday Night Live* still belonged very much to its senior members. Kevin Nealon took over the Weekend Update desk from the departing Dennis Miller; Mike Myers's long-running "Wayne's World" sketch had spawned a hit movie; and Phil Hartman and Dana Carvey would dominate the coming election year with their lauded takes on Arkansas governor Bill Clinton, President George H. W. Bush, and feisty contender Ross Perot.

Chris, always happy to be a team player, continued to do his best in small, supporting parts, such as Todd O'Connor in the "Super Fans" sketches and as Jack Germond in the *McLaughlin Group* parodies. He took on starring roles in now-classic sketches like Robert Smigel's "Schmitt's Gay." And veteran writers Jim Downey and Tom Davis saw a unique side of Chris underneath all of his wild-man antics and brought it brilliantly to life in Chris's signature piece, "The Chris Farley Show."

Through all of this, most of the cast and crew of *Saturday Night Live* failed to see the true nature of Chris's spiraling addiction. But over time, deep cracks began to appear in the wall he had built up between his personal and professional lives. It soon became readily apparent to all that Chris Farley was headed for a serious reckoning.

TOM DAVIS:

Saturday Night Live had really changed. The smell of marijuana no longer hit you when you stepped off the elevator; that sort of thing just wasn't tolerated anymore. However, Fiddler's Green was a bar down on Forty-eighth Street. It was the nearest watering hole to the seventeenth floor. Not literally. There was the Rainbow Room and one other really high-end restaurant that sort of frowned on long-haired, bearded comedy writers dashing in for a quick Rémy Martin. But Fiddler's Green was an Irish bar, like a real Irish bar. A lot of NBC people, mainly the union guys, they'd go there to drink. So it became a convenient place to duck out to. Around ten-thirty at night, when everyone else was eating candy bars and drinking coffee, you'd excuse yourself, grab your copy of *The Gulag Archipelago*, say you had to take a dump, and then sprint up Forty-eighth Street to get booze.

Chris discovered this fairly early on, and so I had the experience, on more than one occasion, of running into him there. I'd see him doing shots of tequila, literally throwing the shots back in a way that made me cringe. I like drinking, but you can't drink like that. He was going for oblivion. On one such night I told him, "Chris, don't go back to the office. Don't let them see you like this."

"Okay," he said.

Twenty minutes later we were both back in the office. He was obviously drunk in front of these younger writers, and it was funny to them. He would entertain them and they would all laugh. But if you were aware of what was going on, it wasn't so funny. He would slap himself so hard that you could see the mark on his face, and that would get a laugh from those writers, but I would see the mark on his face, and I just saw disaster.

DAVID SPADE:

We saw him drinking, but then everyone was drinking, so who cares? I started to notice it was a problem more when I'd leave him at night. I'm a

lightweight when it comes to that. I couldn't keep up, and he'd get angry when I'd leave him at the bar. He'd start getting mad that I wouldn't stay and drink with him. Like, really mad.

Then I started noticing that when we would walk down the street, strangers at sidewalk bars and restaurants would recognize us from the show and go, "Hey, Spade and Farley! What's up! Come and have a drink with us."

And Chris would go, "Okay."

I'd be like, "Are you kidding?"

Even the people at the table, who'd been half joking, they were like, "Wait, he's actually coming over?"

And so he would go drink with random people. Then the drugs kicked in and it escalated.

TIM MEADOWS:

In Chicago, everybody socialized and drank at a bar after the show. And that's just part of Chicago; there's a bar on every corner. In New York, he made more money and the drugs got harder. That was the big difference.

CHRIS ROCK:

He got high and we didn't, so he stayed away a lot.

STEVEN KOREN, *writer:*

I had a different perspective than most. I saw things that other people didn't see. I eventually became a writer, but when Chris got to the show I was still a receptionist. I would answer the phones until eight or nine at night and then stay late and write jokes to try and get them onto Weekend Update. Sometimes I was there until four, five in the morning. I'd be hunched over a desk in some dark office and I'd hear the elevator, and Chris would come in with some shady-looking characters, people I'd

never seen before. They'd go back into some office, turn on some music, and I wouldn't see or hear from them for a couple of hours. I didn't know what was going on, nor was it my place to say anything.

Then there was one incident where an office got messed up, and I got yelled at the next day. They thought I'd done it because I was the only person there that late. Turned out it was Chris and a bunch of these people. That was really the beginning. That was when I was like, oh, he's doing *other* things, beyond the norm.

KEVIN NEALON:

I remember seeing him being a little more sweaty in sketches. His moods would change more. When he first showed up he was that lovable Chris guy. Then you started to see sides of him that were a little more irritated, more impatient.

TIM MEADOWS:

Tom Davis was the first one at the show to say something.

TOM DAVIS:

And when Tom Davis is the one doing the intervention on you, that's when you know you're in trouble.

TIM MEADOWS:

He got me and Sandler and Spade and we all went over to Chris's dressing room. We sat him down and said, "You're hurting yourself, and you need to stop." We were young. We didn't know much about interventions. We just told him that we loved him and we didn't want to watch him doing this to himself, and that it was time to grow up.

TODD GREEN:

Chris and I had stayed close through college, and I moved to New York about nine months after he did. We tried to do an intervention in New York pretty early on. Me, Tom Farley, and Kevin Cleary all went to lunch at P.J. Clarke's. We tried to say to him that we loved him and we were worried about him, and he just didn't want to hear any of it. It was very confrontational. "Who are you guys to judge me?" That sort of thing. It wasn't easy. Kevin, Mike Cleary's brother, was the other big part of Chris's life in New York. Kevin died on 9/11, but in those days the three of us were always together.

MIKE CLEARY, *friend, Edgewood High School:*

Kevin was Chris's sounding board for the first three years of *SNL*. They were incredibly close. Kevin and I never went a day without talking to each other at least once, and so the few times we did go two, three days without talking, it was always because of Chris. And Kevin would never, ever talk about it. All he'd say was, "I had to deal with something."

TOM FARLEY:

Any time I was called in to bail out Chris, Kevin Cleary was there. I couldn't deal with Chris without Kevin, but I know Kevin bailed Chris out several times without me. He was the guy. He kept the pieces together.

Right before Christmas during Chris's second year on the show, Erin Maroney called me and said I needed to come down to Chris's apartment. He'd trashed the place. Kevin and I went over, and it was a mess. Chris stumbled out and was like, "What's up with you guys?" Trying to play it off like it was nothing.

Erin had also called Al Franken, so he came as well. Franken had a talk with him, but he was still pretty out of it. Franken also said he knew some people at Smithers, a rehab facility in New York, that he could get

Chris in there. But I called my dad, and he said, "Just get him home. Get two tickets and fly him home and we'll deal with it here."

I got him up to my apartment in Westchester. My wife drove us to the airport. This was on a Sunday. I got him on the plane, this little puddle jumper. He immediately said to the stewardess, "Gimme a screwdriver." So I gave him one. That's what they tell you to do. Give them anything they want to keep them pacified until you can get them to the rehab facility, to avoid a confrontation. So he drank the screwdriver and then just crashed out in his seat. There were some mechanical problems, and we sat out on the tarmac for an hour. He was out cold the whole time. Finally, they came on and said the flight wasn't going anywhere any time soon, and we all had to deplane back to the terminal. I woke him up and dragged him down the stairs out onto the tarmac. He looked around and said, "Damn, looks just like Westchester."

"Chris, it is Westchester. We haven't gone anywhere."

"Oh."

We went back in. There was no bar in this tiny airport, so there was nothing to do but sit. Chris started getting loud, like, "What's going on here? Why are we just sitting here?" And I kept trying to keep him quiet. We waited for another hour, and then finally they canceled the flight entirely. I called my wife. She came back to pick us up. After we got home, she took our daughter and said, "Look, I'm going to my sister's while you figure this out."

So she took off, and Chris went inside and just crashed in our bed to sleep it off. It was a long, long night. I called home and told my parents he'd be a day late. I took him back to the airport the next day, but I didn't go with him. I had to go back to work, so I just put him on the plane. I figured by that point he could fly home by himself.

Chris rolled off the plane in Madison, and of course the first thing out of my dad's mouth: "Hey! You look great! That a new blazer?" No acknowledgment of the problem. No discipline. No nothing. Refused to deal with it. It was only at my mother's insistence that he agreed to check him in to the local hospital for a few days, and then, after Christmas, after the incident at the hotel, they sent him back to Hazelden.

KEVIN FARLEY:

Around Christmas, me and my girlfriend were in Chicago, and we went out with Chris and Robert Smigel. They'd just done the halftime show at the Bears game at Soldier Field. We went out to the Chop House, and Chris was getting into it pretty hard. He was just ripping it up. Smigel eventually left, because Chris was too much to handle. I took him back to his room at the Westin Hotel. I said, "Chris, just go to bed. Let's just go to bed and sleep this off." And I left him in his room, assuming that he would stay there.

HOLLY WORTELL:

Chris called me up and said, "Come and meet me at Carly's." So I went and met him. That night he got really drunk really fast. People around us were starting to leave. It wasn't fun anymore. It was trouble. We got out onto Wells Street and I said, "Chris, where are you staying?"

He told me his hotel, which was downtown. Even if I'd told the cab-driver where to go, I doubt Chris could have made it back. So I went with him and got him up to his room. I didn't think I could leave him alone, so I said, "Can you call your brother? Where is he?"

"I don't know."

Then Chris got into the minibar, taking out all the little bottles and just shooting them. I yelled, *"Chris, stop it!"*

"No!"

He downed a few more of them, and then he looked up at me and sort of lunged at me. I stepped aside, but his momentum kept him careening forward. Now, we were about fifteen stories up in this high-rise hotel, and the room had these large picture windows that started about four feet up, went all the way to the ceiling and ran the whole width of the room. Under the window was this waist-high radiator. Chris ran smack into the plate-glass window and smashed right through it. His body was hanging out at a ninety-degree angle, and if not for this radiator that

caught him at his waist, he would have crashed right through and fallen to the street below.

He hung there for a few seconds. Then he lurched back in, and I screamed—his arm was sliced open all the way from his shoulder to his wrist.

And at that moment I sort of lost all my emotions. I said, "Chris, we have to go to the hospital right now." I walked over to the phone, I dialed the hotel operator, told them there'd been an accident in one of the rooms and they needed to call an ambulance. I grabbed Chris and said, "Let's go."

"I want to put my jacket on first," he said.

"No, let's go."

"I wanna put my jacket on!"

So I said okay, helped him put his jacket on, we walked out, got in the elevator, went downstairs, and there were three security guys waiting for us. They hadn't called an ambulance, because they said they didn't know exactly what the situation was. I said, "Well, someone's been hurt, and we need one right now! I could sue this whole hotel!"

I walked outside and hailed a cab. We got in, and I said, "Northwestern Hospital. Emergency room. Immediately."

We took off and drove as fast as we could to the hospital. I remember the last block was a one-way going the wrong way. The cabbie stopped and turned and looked at me, and I said, "*Just go.*"

He took off going the wrong way on this street, did a U-turn, and got us to the emergency room. I got Chris out, took him in, and did what I could to fill out the paperwork to get him admitted. Someone came up to me and said, "Isn't that the guy from *Saturday Night Live*?"

I said, "Yes. Please keep this quiet and tell the next shift to keep it quiet."

"Absolutely."

I sat there for a while. Then I asked if I could see him. They were calling in a specialist to look at him, but they said I could go in and visit first. I walked into his room, and the curtains were drawn on either side of his

bed. I pulled them aside and looked in and he was lying there. His arm was laid out and the muscle was just hanging open. With his other hand he was just lifting up the flap of skin and poking around and looking inside. I told him to stop. He was still inebriated, and obviously in some state of shock.

The doctors came in. I went back to the waiting room and called the family. Then, when they started to arrive, I went home.

KEVIN FARLEY:

Holly called, and I went to see him. He was sitting there, still hammered, with his arm all bandaged up. He had already tried to sneak out of the hospital that morning to go get a bottle of vodka. I looked him in the eye and said, "Chris, you can't do this anymore. You've got to sober up. It's time to sober up."

* * *

After slicing his arm open in Chicago, Chris agreed to go back to rehab, but it was a brief visit. He returned to New York—too soon, in the eyes of many—to get back to work on the rest of the *SNL* season. He joined an outpatient program, but used and manipulated the system beyond the point of any real effectiveness. Chris's near-death experience in the hotel may not have opened his eyes to the full extent of the problem, but it did open plenty of others'. When he returned to New York, his arm tied up in stitches and swathed in bandages, his friends and his family could ignore the problem no longer.

TOM FARLEY:

When I walked into his apartment that Christmas and saw that it was trashed, that was the first time I woke up to the reality of the problem. I was a big drinker, and what had brought my behavior under control was having a wife and settling down; she just wouldn't tolerate it. And here

Chris had this really nice girlfriend. They seemed to have a nice relationship. When I saw that she wasn't doing the same for him, that's when I realized his problem was of a whole different order.

Mom had known for a while. She saw it way before everyone, but after that night, seeing his apartment and putting him on the plane, that's when I started going over to Mom's camp.

JOHN FARLEY:

No one educated the rest of the family for a long time, and we didn't educate ourselves. I've got seven years sober now, but at the time? Recovery? What was that? I didn't know. I was chugging beers right alongside Chris, saying, "Gee, sucks for you." It was like Chris had a net thrown over him, and he was taken away from the party, but the party just kept on going.

KEVIN FARLEY:

The first step is to recognize it and to talk about it, and when Chris almost went out the window we at least started to talk about it. We all acknowledged that Chris had a problem. Except for Dad. He would never even mention the incident. And of course there was no discussion that Dad's drinking might be a problem. Never.

Then, it's one thing to talk about it. It's another thing to do something about it.

CHARNA HALPERN:

I got a phone call from Chris's manager, and he said, "I know that you have some kind of hold on Chris. You gotta talk to him. And you gotta talk to his family, because we don't know what to do anymore."

So I called Mr. Farley and said, "We've really got to help Chris. When he comes home, he can't drink. And you can't drink, either, because it's so much a part of your relationship."

And Mr. Farley said, "I don't have to quit drinking. I'm not the one with the problem. Del Close stopped drinking. If he managed to do it, Chris is gonna have to do it."

"Yes, but he can't," I said. "Maybe we can help him if we all stop drinking."

"I'm not gonna do that."

FR. MATT FOLEY:

You liked Mr. Farley. You really liked the guy, but his world was very black and white. He was very caring, but there was the right way and the wrong way with him. There was no middle ground.

Growing up, we were always told you can be critical inside the home, but don't ever bring it out in the street. That's an Irish Catholic thing, a clan thing. In Chris's case that aversion to dealing with matters openly would be even more multiplied, because if Chris had an eating and a drinking problem, that would mean somebody else in the room had an eating and a drinking problem.

TOM FARLEY:

Chris was sent back to rehab, but it was really more of a quick fix. Get him in, patch him up, and send him back on the road. Nothing was really accomplished by it. But when he went back to the show in January with his arm covered in bandages, it was a wake-up call for everyone else, too. Nobody could hide from the problem anymore.

TOM DAVIS:

The story was that he'd gone through a plate-glass door, not a high-rise hotel window. That story was accepted at the time, because, well, going through a plate-glass door is bad enough on its own.

Chris knew that he had fucked up and that he was going to have to

answer questions about what had happened. He was not unintelligent. He was very aware of all these forces swirling around him. Rather than have rumors floating around, he chose to reveal the problem to us, to go on the offensive about it and try and make a joke out of it.

He came into the writers' room and took his shirt off and went "Behold!" doing this shticky, horror-movie kind of thing. We all saw the Frankensteinian stitches up his arm—and nobody laughed. It was horrific. It was maybe the first time that those young guys didn't give Chris a laugh when he wanted it. We were all just like, "Holy shit . . . oh, fuck."

JIM DOWNEY:

Farley always reminded me of the lyrics to that old Irish wake song, "The Parting Glass": "All the harm e'er I've done, alas! it was to none but me." That was Farley. I don't think he knowingly ever hurt another person in his life, and, quite honestly, the drug problems are not remotely the first thing that I think of when it comes to Farley; I think of him being goofy and funny around the office. I didn't notice that it affected his work at all.

MIKE SHOEMAKER:

It was never a problem in the office, and it was never to the point where it was an embarrassment, ever. You knew that he was drinking, but he was always functional. And there were never problems Saturday night when the show went to air.

TIM MEADOWS:

The only time he would drink too much and it would be noticeable and annoying was at the party after the show every week. And he was never an angry drunk, which is kind of an easier drunk to diagnose. That person obviously needs help. But Chris would just get sloppy and tell you how

much he loved you, and he wanted to hug you and make you laugh. What I think, and this is just my opinion, is that he would drink when he got home after work every night, after read-through and after rehearsal.

TODD GREEN:

He may not have let on anything up at 30 Rock, but those times in New York, those were rough. He was over at Kevin Cleary's a lot, either trying to hold it together or coming off a binge. I know Erin dealt with a lot of the fallout, too.

KEVIN FARLEY:

Chris wanted to marry Erin. He even bought a ring. And she was just a super person, had a lot of patience. But she couldn't handle the ups and downs. I think she was probably scared to death. Who wouldn't be?

TOM FARLEY:

I couldn't tell you exactly when, but at some point the relationship was just over. She was staying with him a lot, and then she wasn't anymore. He'd gotten that ring, but I don't know if he ever even proposed. Ultimately that was never in the cards.

TODD GREEN:

Erin was so young then. I mean, she was a kid. But she really tried to help him. I remember going up to her parents' place in Westchester—me, Kevin Cleary, and her—and trying to figure out how we could help Chris. We just talked and looked at each other like, "What are we going to do?" Not a whole lot was accomplished.

It's funny. Mrs. Farley would call me and say, "If there's anything you can do to help him . . ." And, naturally, I would do anything, but you're so ill equipped at that age. The gravity of it is enormous and really beyond

your understanding. You just think, *I'm* his friend and *I'm* different and *I* can get through to him.

Labor Day after his second year on the show, Chris and I went up to Newport and stayed with his cousin. He was holding it together, but I knew he was using. He cleaned up a bit, and we went into town, where we ate at the Black Pearl, which is the restaurant where his dad and his mom went on their first date. Chris called his parents to tell them he was there, and he was really happy about it. He seemed leveled out, and I was trying to rationalize it in my head. Oh, he's okay. Things are better now. Things like that.

But when we got back to New York, I dropped him off at his apartment on the Upper West Side. As I pulled away, I looked back down the street. I saw him turn away from his front door and hop in a cab going downtown, and I knew where he was going.

TOM ARNOLD:

I'd seen a sketch Chris had done on *Saturday Night Live* where he impersonated me when I was married to Roseanne. It was very funny. I talked to some of the other guys from the show, and they said, "Oh, you have to meet Chris. You guys have a lot in common." So Roseanne and I went on and we hosted the show and I realized that Chris and I had an extra lot in common; I'd been sober since December 10, 1989.

Chris came out and we spent some time together. He worked on my HBO special, and he did stuff on *Roseanne*. As I got to know Chris, Lorne Michaels talked to me about him, saying he was afraid Chris had a problem, was kind of caught up in the Belushi thing.

In September of 1992, Chris was staying with me and Roseanne for a week. I had a show called *The Jackie Thomas Show*, and he came in and played my brother and, you know, I could tell. There were times when he didn't want to be around me at all. He disappeared a couple nights. We shot the show Friday night—he was great on the show, by the way—and after the taping I said, "Hey, why don't you come upstairs to my office real quick before you take off back to New York."

"Sure," he said.

He came up, and me and Rob Lowe and some other guys did an intervention on him. Everybody told him how much they loved him and how they were worried about him and how we wanted him to go to rehab right now. He said, "Well, I'd like to go back to New York first." Which, you know, is a classic addict thing to say. "I just want to go back and get my clothes and stuff." What they mean is: "I want to go see my dealer to get high before I go in."

We said, "Well, you know, when you go into rehab, they give you other drugs that kind of detox you from the drugs you're on now, so it'll get you fucked up."

"Oh, really?" he said. "So you're telling me if I go right now to this place, I can get fucked up?"

"Yeah, pretty much, for a couple days."

"Well, let's go."

So we all drove down there, and he went in.

TOM FARLEY:

After Tom Arnold's intervention, Chris finally seemed to get serious about it. He stayed at a facility out there, came back, and spent time at Hazelden's New York inpatient facility. Then he moved to this halfway house that had just been opened. It was on Sixteenth Street and Second Avenue. It was transitional housing for people trying to regain their footing in society. I wasn't even allowed in there. He might have had a room, but I believe it was just a bunch of cots in the basement, and they had their meetings upstairs. He was living there full-time and going to perform on national TV every week.

TODD GREEN:

Kevin and I met him over there a lot. At the time it was like, "Holy shit. This is where Chris is living?" But Chris liked it. He was glad to have it.

AL FRANKEN:

When he was living at Hazelden, he'd bring guys from the halfway house to see the show. Chris came to me a lot, because I often wrote about twelve-step stuff. I would talk to him about it and encourage him to go to meetings and stay on the program. I understood the ins and outs of the program, because I go to Al-Anon, which is for family members and friends of alcoholics. Chris knew I believed in the program, and he believed in it, too. And that's the aspect of this that people may not know. He *tried.* He really tried.

TODD GREEN:

It was a cycle. There were times when he was in really good shape, but other times there would be so much pressure to drink, and he'd ask me and Kevin to take him home; he was incapable of pulling himself out of a social situation that was bad.

ALEC BALDWIN:

I've done a lot of drug and alcohol rehabilitation work, privately, with friends over the years. Chris knew that, and so he really reached out to me for a brief period of time. I talked to him after a show and said, "Here's my phone number. I want you to call me."

And his persona offstage was exactly like the "Chris Farley Show" sketch. He was like, "Whoa, this is *your* phone number? And it's okay if I *call* it?"

"Yes, Chris. It's okay. Just call me, and we'll talk about whatever you're going through."

That lasted for a couple of months. He'd call and we'd talk. Then he'd go back to his old habits, and we'd stop.

DAVID SPADE:

Everyone wanted to take care of Chris. The hosts would hear about the problem, and they'd say, "Can I take you to lunch?" And I think Chris liked the attention of that more than the actual conversation. I used to do an imitation of Chris going, "Alec Baldwin, are you in recovery? We should go to lunch."

"Okay."

"Tom Hanks, you're in recovery, we should go to lunch."

"No, I'm not."

"Well . . . do you do drugs?"

"No."

"Are you afraid of the dark?"

"No."

"Have you ever seen a scary movie?"

"Yes."

"Me, too! Let's go talk about it."

FR. MATT FOLEY:

He wanted to find a nice woman, go and live in the suburbs, and start a family and be like everybody else. I think he felt if he did that he could escape it. But to stay in the environment he was in, he was never going to. They find you, those people. They always find you.

One time I came to visit New York, and Chris sent a driver to pick me up. The driver didn't know who I was. He's just going off about Chris's women, and how he always hooked Chris up with hot girls. All this wild and crazy stuff. Then halfway through the ride he says, "So what do you do?"

"I'm a priest."

Well, the conversation came to a dead halt after that. I'm not naïve, certainly, but it was a window into the kind of world Chris was in. How do you not find trouble in that world?

ALEC BALDWIN:

Chris's problem was that everywhere he went, people thought he was Falstaff. Chris was going to be the jolly fat man who would hoist a beer with you and snort a line with you. Everywhere he went someone was shoving a mug in his hand.

MICHAEL McKEAN, *cast member:*

While we were filming *Coneheads,* I was talking to Lorne, and one of us mentioned the fact that Chris was keeping clean at the time, and Lorne said, "Yeah, he's being good. That's the deal. We've already done the fat guy in the body bag."

TOM DAVIS:

I said to him once, "Chris, you don't want to die like Belushi, do you?"

And he said, "Oh, yeah, that'd be really cool."

And I actually started crying. I wept for him. He said, "Davis, you're crying."

"Yes, Chris. I'm crying for you."

He said, "Wow. Thanks, man."

I didn't say you're welcome.

TODD GREEN:

I never bought in to the fact that Chris was obsessed with Belushi. The press just seemed to make such a huge thing about it. I watched *Animal House* with Chris. I watched *Saturday Night Live* with Chris. Sure, we all thought Belushi was great, but I can't ever remember Chris being obsessed with the guy.

MIKE SHOEMAKER:

I think he romanticized Belushi's death, but that's not the same thing as having a death wish. Chris also wanted to be what Belushi couldn't be. He wanted to have the Chris Farley story be its own story. So I think he was of two minds on it.

LORNE MICHAELS:

Chris romanticized Belushi's life and his death, to a certain degree, but I told him there was nothing romantic about it. I said, "John missed most of the eighties, all of the nineties, and I don't think that was his intention." I was pretty brutal with Chris. I mean, we buried John.

KEVIN FARLEY:

When Chris finally cleaned up, the difference was that for the first time Lorne looked him in the face and said, "I will fire you. I'm not going through another Belushi, and I will fire you." And he meant it.

STEVEN KOREN:

Chris had really been doing well. He'd been clean for a while. Then we were doing the Glenn Close show just before Christmas. We were in the middle of this read-through, and Chris had written a sketch. I thought it was hilarious, but Chris didn't seem to think it went well. We had a half-hour break in the middle of the meeting, and Chris just didn't come back for the second half.

TOM FARLEY:

Chris left the show and went over to Hell's Kitchen and scored some heroin.

DAVID SPADE:

I found a bunch of bags of coke or heroin in a drawer in our office. I said, "What are these?" I didn't even know.

He said, "Get the fuck out of here!"

And he kicked me out of our office. Adam came in and asked what was going on. I said, "Farley's out of it, and there's some shit in there. I don't know what it is."

Chris was pretty good with Adam, so Adam said, "Let me talk to him."

Adam went in the office.

"Fuck you, Sandler! Get out of here!"

Adam came back out and said, "That didn't work too good."

I said, "Let's talk to Marci and see what she can do."

Marci came in and said, "Where's Chris?"

"Fuck you, Marci!"

She said, "I'm telling Lorne."

JIM DOWNEY:

Lorne said, "I think we have to fire him."

TOM ARNOLD:

Lorne called me. He said, "Chris relapsed. He's in his office, weeping. He's crying out for help so loudly that we can hear him out in the hallway."

LORNE MICHAELS:

It was a very adolescent cri de coeur, an attempt to play on everyone's sympathies. But as soon as I heard it was heroin, I was having none of it. I had been through it with John, and I wasn't doing it again.

MIKE SHOEMAKER:

I don't remember Chris actually being fired. He was suspended. But we never said "Empty your office." Because then what could Chris do but go on a binge? But it was severe, and there was an ultimatum attached. Either you come back clean or you don't come back. We never said, "You're outta here," because the problems always manifested when he was "out of here," because this was the only thing he cared about.

TODD GREEN:

The week he was kicked off, he watched the show with me and Kevin at Kevin's apartment. It killed him not to be on.

JIM DOWNEY:

I just sat there in the meeting with Chris, being somewhat cold about the whole thing. It was easier than being warm, and probably more effective. He would start to shake and cry, and I would just tune it out. It was a manipulation. He was trying to get your sympathy so you'd let it slide. He had plenty of people wanting to play that motherly role, and he needed people to say, "This is real simple: Fix the problem or you're out."

LORNE MICHAELS:

I basically just told him, "This is what it is, and I'm really, really disappointed. And you have to get some help, because this is not a problem that you can solve by yourself."

I don't know where I had heard about the place in Alabama, but I thought it was exactly what he needed. It was a real stripped-down, no-nonsense place. I had also seen enough of the Hollywood version of rehab where nothing actually happens. I wanted a place to get his undivided attention. That and the threat of losing the show were the only things that could do it.

TODD GREEN:

Kevin Cleary and I drove him to the airport. That was really hard. It was right before Christmas, and that was always an important time of year for Chris, to be with his friends and his family. No matter where any of us were, we always made it a point to be together in Madison for Christmas. So it was just heartbreaking knowing he was going away to spend it in Alabama at the kind of place he was going to.

We were driving to LaGuardia and it was really quiet in the car, and the song "Bad" by U2 came on. And while we were sitting there, listening to this song, Chris out of nowhere just asked, "What's this song about?" And it wasn't like Chris to ask something like that. Kevin and I were like, oh shit. We didn't really know what to say, so we just told him. "It's about trying to save a friend who dies from a heroin overdose."

"Oh."

Nobody said anything else. We just sat back and listened to the rest of the song. We got to the airport, and in those days you could still walk with someone all the way to the gate, so we went through security with him and walked through the terminal to meet his plane. After we said good-bye he walked over to the gate, and, right at the entrance to the jetway, he stopped and looked back at us for a moment. He had this deadly serious expression on his face. He gave us a thumbs-up, turned around, and walked onto the plane.

ACT II

CHAPTER 8

A Friendly Visitor

TOM SCHILLER:

He was a kind of secret, angelic being who tore too quickly through life, leaving a wake of laughter behind him. As corny as that sounds, it's the truth.

For the next three years, Chris Farley stayed clean and sober. At Lorne Michaels's behest, he had spent the entire Christmas break at a hard-core, locked-down rehab facility in Alabama. Unlike the celebrity resort and spa recovery units of Southern California, this joint was one step above prison, and it was staffed by, in the words of Tom Arnold, "a bunch of big black guys who didn't take any of Chris's shit." And it worked.

Chris's puppy-dog personality and endearing sense of humor had allowed him to weasel his way out of just about any difficult situation he'd faced in the past. But the people in Alabama weren't having any of it. And, finally, Chris wasn't having any of it, either. He realized he could no longer bullshit everyone, and he knew it was his last chance to stop bullshitting himself. He took the program seriously, took its message to

heart, and took a new direction in life when he returned to New York. He moved into an apartment on the Upper East Side in the same building as Dana Carvey. A year later he would move back downtown to a new apartment on Seventeenth Street, a place chosen specifically for its proximity to his old halfway house and its steady availability of meetings and support groups. As the hoary cliché goes, Chris was a changed man. He was calmer, more thoughtful, and more focused.

He was also funnier. Chris missed the first show of 1993, but he was soon back in full swing, and over the following year he would establish himself as the show's new breakout star. That February, the writers resurrected "The Chris Farley Show," this time with one of Chris's childhood idols, former Beatle Paul McCartney. The very next week Chris got to share the stage with returning *SNL* legend Bill Murray. The coming months brought some of Chris's most memorable characters, including the blustery Weekend Update commentator Bennett Brauer, the outlandish man-child Andrew Giuliani, a ravenous Gap Girl, and the titular heroine of Adam Sandler's "Lunchlady Land."

And on the second-to-last show of his third season, with Bob Odenkirk's blessing, Chris dusted off an old script lying around from his Second City days and brought it in to the weekly read-through. It was a hit, both at the table and on the air. That Saturday night, with one unforgettable performance, the phrase "van down by the river" assumed its permanent place in the national lexicon.

STEVEN KOREN:

Chris had been doing the Motivational Speaker character at Second City, but I didn't know what it was. Since Bob Odenkirk had already written it, they just needed a writer to babysit it through production, check the cue cards and all that. It was never anyone's favorite job to get. Little did I know.

So I was sitting there watching the rehearsal, making sure the camera angles were right, and I said to Chris, "You know, you're gonna hurt your voice talking like that. Are you sure you want to do the voice that way?"

He was like, "Don't worry, Steve. I got this one down."

That was a good lesson for a young writer: just trust the actors. When he did it live the place exploded.

DAVID SPADE:

In rehearsal, he'd done the thing with his glasses where he's like, "Is that Bill Shakespeare? I can't see too good." But he'd never done the twisting his belt and hitching up his pants thing. He saved that for the live performance, and so none of us had ever seen it. He knew that would break me. He started hitching up his pants, and I couldn't take it. And whenever the camera was behind him focusing on me, he'd cross his eyes. I was losing it.

Once we started laughing, Chris just turned it on more. And we're not supposed to do that. Lorne doesn't like it at all, but Chris loved to bust us up. Sometimes after the show he'd say, "All I'm trying to do is make you laugh. I don't care about anything else."

NORM MacDONALD, *writer/Weekend Update anchor:*

Lorne didn't like us cracking up on air. He didn't want it to be like *The Carol Burnett Show*. He hated that. When people crack up on *Saturday Night Live*, it's normally fake, because we've already done the sketches and rehearsed them so much. But it was always Chris's goal when it was live on air to make you laugh, to take you out of your character, and he always succeeded. You could never not laugh.

He would do little asides, especially to Sandler, even if Sandler wasn't in the sketch. One time Chris was in a Japanese game show sketch, and when he went to write down his answer for the game, he just took a big whiff of the Magic Marker and did a look to Sandler off camera. Sandler wasn't even in the sketch, but if you watch the tape you can hear him laughing offscreen.

MICHAEL McKEAN:

It was nice to share the stage with that kind of manic energy. For one thing, you knew the focus was elsewhere. No one was watching me. I could have sat down and eaten a sandwich during some of the sketches we did together.

CHRIS ROCK:

You never really shared the stage with him. It was always his stage, and deservedly so. The weird thing is that nobody got mad about that. There's a lot of competition on that show, but no one was competing with Farley. We'd all get upset if someone else had a sketch on and we didn't, but I can't think of one person who was ever upset about Chris getting a sketch on. No one ever complained.

ALEC BALDWIN:

Whenever I was watching Chris perform I would think, "How do I get where he's at? How do I get to be as funny and as honest and as warm?" There are comics that I've worked with who are the most self-involved bastards you've ever met in your life, and they can't fake the kind of decency Chris had. Chris was someone who was very vulnerable; it was a card he played. It was a tool in his actor's repertoire, and yet it was something totally genuine. Even when he plays Matt Foley, and he's hectoring people in this totally overbearing way, there's a tinge of the character's own neediness. Even underneath that, there's Chris.

KEVIN NEALON:

He was so fallible. People just felt for him. Women felt protective of him, because they could tell he wasn't watching out for himself. And men related to all his anxieties and imperfections.

LORNE MICHAELS:

One time we were in the studio, and Chevy Chase came by. Chris was practicing one of his pratfalls. He showed it to Chevy, and Chevy said, "What are you breaking your fall on?"

Chevy always had something to break his fall; you plan these things out. But Chris had watched Chevy and bought the illusion of it. How do you fall? You just fall on the ground and you don't mind the pain, because that's the price of doing it. So there was an honesty and a straightforwardness in him that people responded to.

NORM MacDONALD:

What I would do with Chris, when it came to writing a sketch, was just listen to him and observe him. There was this one thing he did. He'd tell a story—and I'm not doing this justice—but he'd tell a story like, "Anyways, Norm. Did I tell ya I seen my friend Bill the other day, and I says to him, I says, I look him right in the eye and I says to him, I says, I says to Bill, I says to him, get this, what I says to him is I says, get this, what I says, you won't believe what I says to him, I says . . ."

And of course the joke was that he'd never get to what he'd actually said to the guy. And Chris could keep this going for twenty, twenty-five minutes straight. He'd do it two hundred different ways. It would just get funnier and funnier and funnier. When you can reduce something to four words and be funny for twenty-five minutes without an actual joke or a punch line, that's genius. It's not even really comedy anymore. It's almost like music, like jazz variations.

I always liked comedians who just keep repeating things until nobody's laughing anymore, but then they take it so far that eventually it's funnier than it was in the beginning. There are only a couple of performers on the planet who can do that. Andy Kaufman could do it, and Chris Farley could do it.

So I had him do it on Weekend Update. Lorne had decided that the "I says to the guy" segment would last for thirty seconds, which I knew

would never work. At dress, I told Chris to do it for four minutes. So he did, and it was just like I thought. People weren't laughing for a while, but then right as he hit the four-minute mark it was really starting to kill. That's when I realized he should have done it for eight minutes.

But he never got to do it on air, because Lorne went ballistic on me that I'd let Chris go so far over time. I tried to explain to Lorne that it wasn't funny for thirty seconds, but Chris understood it completely.

ALEC BALDWIN:

There are people who are smart in a way that has no applicability to performance, but Chris's brains and his quickness inside of performance were amazing. He knew exactly how to scan a line, exactly what inflection to have, how to time it, what expression to make. A great performer is someone who puts together a half a dozen things in an instant, and Chris was one of the most skilled performers I've ever seen in that respect. And he knew that his opportunity would come. He wasn't sitting there, calculating how he was going to trump you or dominate the scene. He just patiently waited for his moment and then arrived fully in that moment.

STEVE LOOKNER, *writer:*

When it came to performing in your sketches, Chris was never some egotistical guy who was going to take your material and do it however he wanted to. He wanted to make sure he was getting the sketch the way you wanted it.

FRED WOLF, *writer:*

The highlight of my career, still, was the first sketch I got on at *Saturday Night Live*, this thing called "The *Mr. Belvedere* Fan Club." Chris had a big turn in that sketch where he played a crazy person obsessed with *Mr. Belvedere*. He brought down the house. Afterward he came up to me, saying,

"That was funny. Thanks for the good stuff." I couldn't believe that about him. To me, it was the other way around. I should have been thanking him.

DAVID MANDEL, *writer:*

He always went out of his way to make sure people knew what material was yours, that they were your jokes, and he was just the guy who said the lines.

IAN MAXTONE-GRAHAM, *writer:*

I worked on some of the Motivational Speaker sketches, because Bob Odenkirk was gone by then. Matt Foley was very much Chris's character, but Chris was also very loyal. We always had to call Bob up and read it to him over the phone and get his blessing.

SIOBHAN FALLON, *cast member:*

There was always an air of competition at *Saturday Night Live*. At read-through, people would purposefully not laugh at something even though it was funny, because they wanted something else to make it on the show. But Chris would laugh no matter what. If it was funny, he gave it a big, big laugh. He didn't discriminate. He was honest.

NORM MacDONALD:

I don't think Chris knew how to hate. I'd feel bad sometimes, because I'd be complaining and I'd go, "You know who sucks?" And I'd go off about so-and-so, some guy on the show. And Chris would immediately go, "I think he's funny, Norm. Why don't you like him?" So then I'd just feel like a jerk.

DAVID MANDEL:

The show was in a very weird spot at that time. During the election year, everything was Phil Hartman and Dana Carvey doing Clinton and Bush and Perot. Chris was a full cast member, and incredibly popular, but in those sketches he'd just do small, memorable turns as Joe Midwestern Guy. Al Franken and I wrote the sketch where Bill Clinton goes jogging and stops in at McDonald's. In that one Chris played Hank Holdgren from Holdgren Hardware in Fond du Lac, Wisconsin. In a lot of those small supporting roles I think you saw the road not taken for Chris. If he hadn't found comedy, you could totally see him being the friendly hardware-store guy.

TODD GREEN:

When Chris interacted with celebrities, the guest hosts, he would always introduce me by saying, "This is my friend Todd. We met in second grade and grew up together." He was proud that he was my friend, and he wanted to share that. I remember him regaling Glenn Close with stories of Madison, and you could see that she saw the genuineness in him. She just looked at him and said, "You really are an amazing guy."

I'm a huge, huge Beatles fan, and so when Paul McCartney was on the show, that was a really big deal. Ten years earlier, Chris and I had been listening to Beatles albums in our basements. He called me during the week at like two in the afternoon and said, "What're you doing?"

"What am I doing?" I said. "I'm working, like most people."

"You know what I'm doing?"

"What?"

And then he took the phone and held it up, and I could hear Paul McCartney singing "Yesterday."

"I'm just here, hanging with Paul McCartney," he said. Then he giggled and hung up the phone.

The night of the show, he said, "Listen, I want you guys to hang out in my dressing room tonight. I have a surprise for you."

So, Kevin and I wait and watch on the monitor in the dressing room. McCartney comes out and does the first song, and we watch him, wondering, "What's the surprise? Why didn't he come and get us?" Whatever. Didn't matter. It was one of his new songs. Then the second song comes and goes, another new one, and still no Chris. Just before the end of the show, when we're pretty sure Chris has forgotten about us, he barrels into the dressing room and says, "Okay, Greenie, you're on! Follow me!"

We go running down the hallway to the studio. Paul and Linda McCartney come out. Chris introduces me to them. I'm in a state of shock, and the four of us walk out to the stage together. Chris and I stop just short of the cameras, and Paul and Linda go out and he sings "Hey Jude."

And at that moment, Chris wasn't a member of the show anymore. It was just two buddies from Wisconsin who grew up on the Beatles, listening to Paul McCartney. Chris literally forgot that he had to go back onstage for the good-nights.

I think, deep down, all of the guys from Edgewood figured that one day we'd end up back in Madison and it would be just the way it was. I think even Chris believed that. Even ranked against all the fame and money and stardom, he felt the days back at Edgewood were the best days of our lives.

KEVIN FARLEY:

When you come to the conclusion that you're an alcoholic, and you go to these stupid meetings, they're filled with down-and-out people right off the street. I'd go to Madison, and I'd see Chris, who was on *Saturday Night Live*, had money, had fame. He'd go and drink coffee and talk with these regular folk, and he could talk to them more easily than he could talk to Mick Jagger or Paul McCartney. He felt more at ease with the average Joes.

What he loved was the honesty. Nobody is as honest as they are in one of those meetings, when they're admitting their faults, admitting that they're broken human beings. Contrast that with the *Saturday Night Live*

after party, where everyone wants you to think they're hot stuff. They're putting on airs, and it's all bullshit because we're all just broken people anyway. To witness people being honest about themselves and with themselves is a life-changing thing, because it's something you so rarely see. That's what's truly amazing about recovery, and amazing about how it changed Chris.

TOM FARLEY:

He was a lot more fun to be around. He was much, much funnier. You could have thoughtful, engaging discussions with him, and he wouldn't get mad or defensive. That was a huge difference from when he was drinking.

For Chris, being in recovery was a little like being at camp. That's how he treated it. Make your bed every morning for inspection, that sort of thing. And that carried over once he got out of rehab, too. As disgusting a slob as he was before, he was that clean and organized once he got sober. He turned into a neat freak.

BOB ODENKIRK:

After all the years of being in and out of rehab, I never thought that Chris could take it seriously. But one time I was at this party out in L.A., and I saw him turn down a beer. He was saying no, and he meant it. I thought, oh my God, *he figured it out.* He knows how dire this is, and he's really taking charge. It wasn't about pleasing everyone else. It was about him and his choice. I was really impressed, and I thought, wow, nothing is going to stop this guy.

TOM FARLEY:

Toward the end of Chris's *Saturday Night Live* run, my son was born, and he had to stay in intensive care for a week. One day I asked Chris to watch

the two older girls while my wife and I went to the hospital. We went down to his apartment, and we were ringing the buzzer, waiting for someone to let us in. There was no answer. I was starting to wonder where the new, reliable Chris was. Then, around the corner here he comes with these huge bags of Cheetos and ice cream and these enormous Barney dolls, walking down the street. It was just a great sight. Chris was so happy that we'd asked him to look after the girls, that we trusted him with that responsibility. He was so proud that he could be a better part of their lives.

KEVIN FARLEY:

Chris paved the way for the rest of us. When he went down to Alabama, I started to look at myself. I was doing the same stuff he was—coke, pot, drinking all the time. I saw where I was headed. I never went into rehab. I just walked into a meeting one day. That's when I realized we all are alcoholics, the whole family. My mother stopped drinking then, too, at the same time I did. We would go to meetings a lot together. We realized Dad was an alcoholic, and we saw the patterns very clearly once we'd changed our own. But Tommy, Johnny, and Dad were still drinking. Barb was the exception. She was never a drinker at all.

TOM FARLEY:

When I look back, or when people ask me what regrets I have, what I realize is that I always felt that Chris's problems were his own. I was still drinking, and I didn't take an active role in his recovery. It was his deal, and that was that. Then one day he asked me to come to his second-anniversary meeting, which he was going to lead. I said, "Great, where do you meet?" He gives me the address of this place down on Eleventh Avenue and Forty-something, the real fringe of Hell's Kitchen. You go down there and you think, Jesus, what am I walking into?

But that's where Chris liked to go. He had his choice between meetings on Park Avenue and in Hell's Kitchen, and he wanted to be with the

desperate, hard-luck cases to remind himself that his celebrity didn't put him above them in any way. He stood up there in front of them and said, "Look, I woke up the same way you guys did this morning, wondering if I was going to stay sober today. My disease is no different than yours." And it wasn't bullshit. He just seemed so wise and intelligent and in control. I just sat there thinking, this is my *brother*? I looked at him in a whole different light from then on.

NORM MacDONALD:

What's hard for a comedian is that they make a living on their anxieties and their self-doubts, but in real life they try and separate themselves from that. Chris didn't do that. He was absolutely honest in what he was.

CONAN O'BRIEN:

You got the sense with Chris that he wasn't punching a clock. And, obviously, that isn't always a fun way to live. But it was fun for everybody else.

DAVID MANDEL:

Emilio Estevez hosted the show in support of *Mighty Ducks 2*, which we were given a screening of. We saw it in a private screening room. No one really cared about *Mighty Ducks 2*, so only a couple of us were there.

Now, in *Mighty Ducks 2*—which, if you need your memory refreshed—they're training for the Junior Olympics, and they let some street kids from L.A. join the team. So now they have some black kids on the squad, and they do a giant musical montage where they take the rap song "Whoomp! There It Is" and change it to "Quack! There It Is." And of course it was the *actual musicians* who'd sung "Whoomp! There It Is." They'd sold out to Disney and done "Quack! There It Is."

It was so ridiculous that those of us in the room started clapping along and jumping up and down and dancing with the music in the mid-

dle of the screening room. The next thing you know, Farley's pants are down around his ankles, and he's standing up on a chair, smacking his ass in time to "Quack! There It Is." I have never seen anything funnier in my life. And yet when you look back sometimes you think, you know, maybe that was a cry for help.

SARAH SILVERMAN, *cast member:*

The cast was on a retreat, sitting around a campfire, and Chris sidled up to Jim Downey. I overheard him say in this little-boy voice, "Hey, Jim? Do you think it would help the show if I got *even fatter*?"

Jim said, in his parental voice, "No, Chris. I think you're fine."

Chris said, "Are you sure? 'Cause I will. For the show."

Chris was fucking around for sure, and seeing the back-and-forth of the conversation was hilarious, but there was an element of truth in it: He would do anything to be funny.

NORM MacDONALD:

I never thought about it as needing attention, because Chris laughed at everybody else, too. He loved Sandler and Spade and me, guys who were much less funny than he was. And he was always more generous in giving you a laugh than in taking one for himself.

His greatest love was just the act of laughter itself. As much as he made other people laugh, to watch Chris do it was the most beautiful thing you'd ever see. Nobody could laugh with as much unbridled glee. He'd just go into these paroxysms of mirth. If Chris laughed at one of your jokes, you felt like the king of the world.

STEVEN KOREN:

Chris was really smart. He knew exactly what he was doing. It's the same with Jim Carrey. He knows the exact degrees to which he's being big or small or clever. When they're that good, they know the difference between

being laughed at and laughed with. There's a definite awareness. I guess Chris was a victim of his own desire to make people laugh, but also I think his heart was so big that if he was the butt of the joke it was okay. He wanted to give people laughter so much that it was okay if it hurt him a little bit. It was a conscious decision, I think.

JAY MOHR, *cast member:*

No one was laughing at Chris. Everyone was laughing with him. Show me someone who was laughing at Chris Farley, and I'll show you a real cocksucker.

DAVID SPADE:

I would have to write sketches all week to try and stay alive on the show, and Chris would be written for, so he didn't write a lot, or read. So while I was busting my hump, he'd be bored behind me trying to amuse himself. One night he goes, "Davy, turn around."

"I'm busy," I say.

"Turn around."

"Dude, if this is Fat Guy in a Little Coat again, it's not funny anymore."

"It's not."

"Really?"

"I promise."

So I turn around and he's got my Levi's jacket on, and he goes, "*Fat Guy in a Little Coat* . . . give it a chance."

And the coat rips, and that's how we wound up putting it in *Tommy Boy.*

TIM HERLIHY, *writer:*

When comedians get together . . . I wouldn't call it one-upsmanship, but it is like a game. Who can be the funniest? When I knew Chris, he was

surrounded by the elite of comedy. Sandler's a huge, funny star, but you always knew Farley was going to top him. He was the funniest among a group of very funny, talented people. All of us who worked with him are richer for it. We're better writers, better performers.

FRED WOLF:

Comics are a pretty strange breed. Put all of us in a room and we can fight among ourselves and disagree with all our bitterness and neuroses. But when it came to Farley, it was unanimous: He was the best.

NORM MacDONALD:

What astonished me about Chris was that he could make *everyone* laugh. He could make a child laugh. He could make an old person laugh. A dumb person, a smart person. A guy who loved him, a guy who hated him.

IAN MAXTONE-GRAHAM:

He was a very funny, jovial presence in the office. He'd be very, very outgoing, and then he'd have this very cute, shy thing he'd do where he'd sort of retreat into himself. Hugely outgoing and hugely shy. That was the rhythm of his behavior. You can see that in some of his sketches.

BOB ODENKIRK:

Most of my memories are just of hanging out with Chris and him making me laugh *so hard.* But then, if Chris wasn't being silly, if he was just listening to you quietly, that was as funny as when he was worked up. "The Chris Farley Show" on *SNL*, that was Chris behaving himself.

JOHN GOODMAN, *host:*

"The Chris Farley Show," that was Chris.

MIKE SHOEMAKER:

That was Chris.

STEVEN KOREN:

That was him.

JACK HANDEY, *writer:*

He was basically playing himself.

TIM MEADOWS:

That's how he acted whenever he was around someone he admired. Until he got to know you, he really was that guy—shy and asking a lot of dumb questions but not wanting to be too intrusive. It was a very endearing quality.

TOM DAVIS:

I thought of that sketch originally. I thought, what the fuck are we going to do with this guy? He's just over the fucking top all the time. I buttonholed Downey and said, "Let's do 'The Chris Farley Show' and just have him talk as he really is so he doesn't go over the top."

JIM DOWNEY:

Farley was such a comedy nerd. He knew all the old shows, better than I did. He'd come up and say, "Do you remember that superheroes sketch

on the show where they were having that party?" Then he'd proceed to do the entire sketch for me, his version probably longer than the original. He'd finish that and be like, "You remember that?"

"Yes, Chris," I'd say, thinking this was all leading up to something significant. "What about it?"

"That was awesome."

So we decided to put that in a talk-show format, with some poor sap being trapped on a talk show with Farley asking him retarded questions. We submitted it, actually, as a joke at read-through. I thought it was too inside, and so would never make it on the air. But Lorne liked it immediately, and seemed to have big hopes for it.

At dress, Steve Koren was watching it, grinning ear to ear and laughing. And of course the audience loved it. I don't think it was just Farley being adorable. I guess there was just something universal about it, and I didn't appreciate the resonance it had.

The first one was with Jeff Daniels. Then we did Martin Scorsese. By the time Paul McCartney came around, I actually didn't want to bring it back at all. I just thought, what can you do that's different? If you just do the same thing over again just because people liked it, they might stop liking it. But Lorne insisted on doing it.

LORNE MICHAELS:

Actually, I think Chris was the one who was adamant about doing it with McCartney.

ALEC BALDWIN:

We were just dying. We couldn't believe how perfect it was. How hard is it to make Beatlemania funny again? How hard is it to make gooing over McCartney funny? We didn't know if that would work. But Chris came on, and we were sobbing with laughter it was so funny. It was going along, and then Chris says, "You remember that time you got arrested in Japan for pot?"

And McCartney just suddenly changed his tone. "Oh, those are the things I'd like to forget, Chris."

They played it perfectly.

JOHN GOODMAN:

The funniest bit that I ever saw him do was that McCartney interview. When I first met him he was like this kid who kept staring at me, just like that character did. I don't know why, but he seemed genuinely thrilled to meet me. It wasn't a celebrity type of thing, either; it was just that he couldn't get enough of other people, of their stories. He was endlessly curious.

BOB ODENKIRK:

I said once—and it was misquoted in that fucking *Live from New York* book—that Chris was like a child. How it reads in that book was that I meant Chris was like a little baby, which wasn't what I said at all. The whole quote—which they didn't include, because they're dicks—was, "Look, don't take this the wrong way, but Chris was *like* a child. He was like a child in his reverence and awe of the world around him." And he was. He was so respectful of everyone, like he always had something to learn from you.

JAY MOHR:

I learned from Chris how to have more fun. Nothing is that serious. Acting is really a ridiculous way to make a living. You're playing make-believe, and Chris never got away from that fact. Kids never come home and say, "I was over at Michael's house and we played make-believe. It was awful. We were in a spaceship and I had a helmet on and there were Martians and then we chased them through the woods—and it just wasn't my thing." No kid has ever said that. It's make-believe. You paint as you go. I

have a three-year-old son now. I open up his coloring book and say, "What color are these footprints going to be?"

"Green."

Great fucking idea. Green footprints, that was Chris.

MARILYN SUZANNE MILLER:

I would write songs and musical numbers for *SNL*, and when Kelsey Grammar was host I wrote a sketch called "Iron John: The Musical." Chris sang in that one, and he came up to me after the performance and said, "I love those musical things. I just love them. I really want to do more of them."

Musical numbers are very emotional things, and it's a very childlike desire to want to have that kind of honest, sincere outlet. He wanted to share as much of himself with the audience as he could. He was not a great singer, as I recall. During the rehearsals, beads of sweat would literally drip down his forehead as he was trying comically hard to hit all his cues and hit all his notes. He so much wanted to succeed. And when you see a young guy working like that, with sweat running down his forehead, that's kind of a wonderful thing. When the old cast would get laughs, we were practically counting them. We were very calculating. Chris wasn't a calculated performer. He was out there for the love.

IAN MAXTONE-GRAHAM:

They have a meeting with the host every Monday night, and there were always two jokes that we used to do every single week. Let's say Kevin Bacon was the host, and Tom Arnold was set to host the next week. Lorne would announce, "Everyone, this week's host is Kevin Bacon." And everyone would applaud.

Then Al Franken would say, " And next week: Tom Arnold." And everyone would applaud much louder. That was the little icebreaker they'd do to set the host at ease and poke fun at him a bit.

The other thing that would always happen was that at the very end of the meeting, Farley would jump in—kind of like that determined kid on the football team that's never won a game—he'd jump in and say, "C'mon, let's *do it* this time!" It was always very funny and sort of combined his childish eagerness with great comic timing and a great sense of the moment.

ALEC BALDWIN:

When you were on the set with Chris, he'd be giggling and pinching you and saying, "Where you want to go after the show, man?" It was like being in homeroom in high school. There's a quotient of people at *Saturday Night Live* for whom the show is like operating an elevator. We go up. We go down. What's the big deal? Then there were the people like Chris who made it their mission every week to make it the best show possible, and enjoy it.

ROBERT SMIGEL:

Just seeing Chris at the door of my office would put a smile on my face. He radiated this earnestness, and he really *believed* in the work that he did. Chris was also unique among comedians in being so open about his faith and spirituality. Most people in this industry are so caught up in being sarcastic or casually ironic that they're loath to admit that they actually care about anything. Admitting that you believe in God is the same as admitting that you like Bob Seger. Okay, even I'm not crazy about Seger. But I like Springsteen, and even Bruce is just too earnest for lots of comedy writers to give it up for.

CONAN O'BRIEN:

There are a lot of us in comedy who are a lot more Catholic than anybody knows. Our Catholicism is sort of under our skin. People were surprised at the depth of Chris's faith; to me it made perfect sense. A lot of people

think that they're mutually exclusive. How can you be dancing in a Chippendales thong and going to mass at the same time? But if you're Catholic you think, *of course* that's how it works.

TOM FARLEY:

Pretty early on, Chris told me he'd found this church, St. Malachy's. "They call it the Actors' Chapel," he said. He totally ate that up. In his mind it was this place where all these old Broadway stars and vaudevillians had come to mass and prayed and found guidance. The first time I went with him, these old ladies who were sort of scattered about the pews would break out with the most beautiful voices during the hymns. They'd really belt it out, and you could picture them singing in their old musicals and operas. It was really special for Chris in that way.

MSGR. MICHAEL CRIMMINS, *priest, St. Malachy's:*

He used to come on Tuesdays and Thursdays to the noon mass. He went to confession regularly. He'd bring his mother and his family to mass whenever they were in town.

Sister Theresa O'Connell was really his mentor and spiritual adviser. She knew him well. Unfortunately, she's passed on, and they kept their relationship very private. But she suggested to him that he volunteer through our Encore Friendly Visitors Program, and he did.

SIOBHAN FALLON:

Chris lived close to me, and we went to the same church, Holy Trinity on the Upper West Side. He'd go to St. Malachy's from work and Holy Trinity on the weekends. As he did everything big and great, he'd be in the back of church, praying intensely, bowed down in this dramatic position, practically kicking himself over whatever he'd done the night before. I'd say, "Hi, Chris."

And he'd say, "Well, God's gonna be mad at me this time!"

I have no doubt that Chris is sitting pretty up in heaven, entertaining everybody. He was a good guy. He's taken care of. You can say that addiction is a selfish disease, but Chris wasn't selfish. He always looked outside of himself. At that stage of someone's career, in your twenties and early thirties, you're so selfish and so self-consumed—especially actors. I think you'd be hard pressed to find anyone at that age who was thinking about anything other than getting themselves ahead. And so for Chris to be doing the work he was doing was amazing.

TIM MEADOWS:

A kid from the Make-A-Wish Foundation came to *SNL* once to meet Farley. I got to see that, but I had no idea that he was a part of this program at St. Malachy's. He never talked about it. I was one of his best friends, and I didn't know about it until after he'd passed away.

NORM MacDONALD:

It was amazing at the funeral to hear people talking. It was like, "My God, this is a person I never knew."

SR. PEGGY McGIRL, *executive assistant, Encore Community Services:*

Whenever he came here he was very regular and without any airs. He had a quiet way about him; he didn't like to have any attention focused on him at all. His main concern was just to be there to help the seniors. We have parties twice a year for people who are homebound, seniors and people with disabilities. One is in the spring and the other is around Halloween. If Chris was in town, he was always there.

KEITH HOCTER, *volunteer, Encore Community Services:*

I met Chris through the parties at St. Malachy's. We worked the door together. He was just extremely friendly, not at all hung up about who he was, and he was pretty famous by that point. He just showed up and did what the rest of us did, which was whatever the sisters told us to do.

For a lot of the seniors, this was their big social event of the year. The party would have about a hundred and fifty people. A lot of them were in wheelchairs and walkers, and back then the church didn't have an elevator. The party was in the basement, and the only way down was this old, narrow set of stairs. It had one of those side-rail lifts, but the thing never worked.

Chris and I and the other volunteers, we'd each grab a corner of the wheelchair, tip them back, and then just talk to them and keep them calm while we took them down. We never dropped anybody; I guess that's the first measure of success. And we never had anyone freak out on us, either. Half of our job was to get them down safely. The other was to make them feel good while we were carrying them down. Then, at the end of the evening, we would stay and, one by one, help carry them all back up to the street again.

JOHN FARLEY:

One time we were in Chicago, coming back from filming this HBO special where Chris had this quick little cameo. As we get out of the limo at Chris's apartment, there's an old woman standing on the corner, begging for a quarter to get on the bus. Two minutes later, she finds herself in the back of a limousine with a hundred dollars in her hand. And Chris tells the limo driver, "Take her anywhere she wants to go."

The next day the same driver comes back to pick Chris up, and he says, "Do you have any idea where that woman wanted me to take her? Please, let's not do that again."

AL FRANKEN:

Tony Hall is a former congressman from Dayton, Ohio. His son Matt had leukemia, and a mutual friend asked me to go and visit him at Sloan-Kettering. The second time I went there I said, "Who's your favorite cast member on *SNL*?" It was Chris. So I asked Chris if he'd come and visit him, and he did.

Matt just loved it. His parents are very Christian, especially his mom, and Chris and I were just swearing up a storm. Matt laughed. His mom didn't know whether to be happy or shocked or what.

After we'd spent a while with Matt and said good-bye, Chris went around and visited every single kid in the cancer ward. He stayed there and entertained every last one of them. Then at the end of the day, as we were walking out of the hospital, Chris just broke down and started sobbing. I think it was all sort of wrapped up with his own issues that he was dealing with at the time. I said to him, "Don't you see how much joy you bring to these people? Don't you see what you just did, how valuable that is?"

Chris went back and visited Matt again. When they had Matt's funeral, I went, and they had made a bulletin board of "Matt's Favorite Stuff." In the middle of it was a photo of him and Chris from that day.

FR. JOE KELLY, S.J., *priest, St. Malachy's:*

He believed that comedy was a ministry of its own. Anything that made people laugh was worthwhile. But at the same time, he wanted a little more than that. He was a bit of a disturbed guy. I'm talking personally, now. A bit of a disturbed guy.

He used to come up to my room here, just to sit and chat and talk about different things, especially about how important the Friendly Visitors Program was to him. He felt that without the program, without the work he did here, his life wouldn't have much meaning. Doing what he did here gave him a purpose outside of some of the trivial work he was doing in entertainment.

Anybody who's constantly making a fool of himself and getting

laughs out of those crazy facial gestures and so on, very often they're hiding something they don't want to face themselves. I think that was the case with Chris. He was a much deeper person than he let on. One gift he had was the ability to make people laugh. The other gift he had was himself. Just being the person he was was a gift for others. And I don't think he realized that for quite some time.

SR. PEGGY McGIRL:

We have a residence, a converted hotel, that is now a home for the homeless and the mentally ill, and Chris used to visit a man named Willie. He also spent time with another resident, a woman named Lola, but it was mostly Willie.

Willie was about seventy years old, and he had been homeless before coming to our residence. Chris took Willie out to dinner every week, and to famous restaurants. Chris treated him as an equal, always. He would take him to Broadway shows, take him out to ball games. If Chris was walking down the street on the way to his office, he'd stop in to see how Willie was doing. Whenever he had to go away for work, he'd send Willie postcards, and whenever he came back he always brought Willie a souvenir. They were friends for over five years.

TODD GREEN:

On the one-year anniversary of Chris's death, St. Malachy's was having a memorial mass for him, and I went with Tommy and Kevin Cleary. There was an elderly black guy there with a Chicago Bulls hat on. He was not quite homeless, but clearly one step away from it. He stood up to speak. He said his name was Willie, and he talked about Chris and about all the things he had done for him, all the time Chris had spent with him. Kevin, Tommy, and I, we just looked at each other—we had no idea.

The man spoke for a little while longer. Then he started to break down crying. He said, "This hat, this is the last thing Chris ever gave me, and I really miss him."

SR. PEGGY McGIRL:

After Chris passed, Willie became very quiet. Eventually, some time later, he moved back down south to be with his family, to let his family take care of him. Whatever problems had put him on the street and made him homeless, he overcame them and went back home.

When you receive love, it releases you from the things that trouble you. Just knowing that someone cares about you can give you strength and courage. And I always believed that it was Chris's love for Willie, and the things he did for Willie, that finally set him free.

CHAPTER 9

The Magic Sixty-six

FR. TOM GANNON, S.J., *friend:*

People always ask me what kind of guy Chris Farley was, and I say, "Go and see *Tommy Boy.*" That's Chris Farley.

In the summer of 1991, director Penelope Spheeris was working at Paramount Studios, preparing to shoot a feature-film version of "Wayne's World," Mike Myers's popular *Saturday Night Live* sketch about two kids with a cable-access television show. Lorne Michaels approached her about using Chris Farley in the movie. Chris was still a relative unknown. He had only one year of television under his belt, and Spheeris had no idea who he was. But, she says, "Lorne told me we should give him a part because he's going to be the biggest thing ever." On that recommendation she cast him sight unseen as a security guard at the movie's Alice Cooper concert. As film debuts go it was a small one, but Chris carried it off well.

Wayne's World, meanwhile, carried off buckets and buckets of money. The low-budget comedy shattered everyone's expectations, earning

$121 million at the box office. Paramount immediately ordered up *Wayne's World 2*, as well as another *SNL* franchise movie, *Coneheads*.

SNL alum Dan Aykroyd sat down with writer Tom Davis, and the two began drafting the story of Beldar and Prymatt Conehead as a movie. Aykroyd, impressed by what he'd seen in Chris on television, wrote a part for him as Ronnie, the love interest of young Connie Conehead. Over the following year Chris took on other supporting roles, as a roadie in *Wayne's World 2* and as a security guard in Adam Sandler's *Airheads*.

By the end of Chris's fourth year on *Saturday Night Live*, he was hands down a cast favorite, and Lorne Michaels felt that the young star was ready to carry his own feature—an original story, rather than another *SNL* franchise picture. Drawing on personal experience, Michaels laid out the central premise of the plot: a father who dies too soon and a son forced to take on responsibilities for which he is not prepared. *SNL* writers Bonnie and Terry Turner took on screenwriting duties, and Paramount bought the idea immediately, based largely on the casting of Chris Farley and Rob Lowe as brothers.

Very little of the original concept and script made it to the screen. The Turners' draft was largely discarded, and *SNL* writer Fred Wolf wrote the bulk of the final screenplay, receiving no credit for it. Wolf wrote most of the story on the fly while director Peter Segal and producers Bob Weiss and Michael Ewing scrambled at record speed to get the film off the ground.

The story that ultimately emerged was that of young Tommy Callahan, heir to a Midwestern auto-parts manufacturer. When Tommy's father dies, he has to make his dad's annual sales trip to save the company and his hometown. Joining him on the trip is his late father's assistant, Richard Hayden, played by the acerbic David Spade. Lorne Michaels had paired up the two close friends on a hunch that their complementary talents might make them the best comedy duo since John Belushi and Dan Aykroyd hit the screen in *Blues Brothers*, fifteen years earlier. Brian Dennehy took on the role of Big Tom Callahan, with Bo Derek as his bride. *Doc Hollywood*'s Julie Warner was cast as Chris's love interest, Michelle Brock. Aykroyd played the heavy, a competing auto-parts magnate

named Ray Zalinsky. And Kevin and John Farley also turned up in small roles as their older brother endeavored to give them a leg up in show business.

Few on set knew it at the time, but they were working with Chris at the single high point of his life. He was confident and self-assured, and it showed in his performance. Thanks in no small part to Chris's commitment, *Tommy Boy* lives on today as a minor classic, a staple of cable-TV comedy, and a brief glimpse of what might have been.

BOB WEISS, *producer:*

I got a call from John Goldwyn at Paramount saying, "We have this picture. Would you come in?"

"Do you have a script?" I asked.

"No."

"Whaddya got?"

"We've got thirty pages and a release date."

"Okay."

So I met with Lorne, and a lunch was set up at the Paramount commissary for me to meet with Chris.

He sat down at lunch, and he was really like an enthusiastic kid. He was just thrilled about being there and wasn't jaded in any way. How refreshing it was to have someone like that inside the Paramount commissary. You told him any piece of what was going on and it was "great." It was "cool." It was "exciting." And that's infectious. You can build on that.

Originally, there was another director involved. I acquainted him with the facts of our shooting schedule, and he left the picture; he was afraid his head would explode. Then Pete Segal agreed to come on board.

MICHAEL EWING, *associate producer:*

Tommy Boy, at that point, was called *Billy the III: A Midwestern*. It was changed because *Billy Madison* was being shot at the same time. Bonnie

and Terry Turner had written the first draft. It was a sweet script, but it was a bit of a mess—and it was a famous bit of mess, because now Chris Farley was attached to it and it was going to get made. I walked in to the office the first day, and Pete said, "Well, what do you think of the script?"

I said, "I think we have work to do."

FRED WOLF:

I shared an office at *SNL* with Spade, Sandler, and Rock. Farley was always coming over to our office to hang out. It almost seemed like he was a part of the office, too. I was the quiet guy who observed them, and so it seemed like I might be a good guy to bring a little of their sense of humor to the page.

Jim Downey and I were hired to do a polish of the Turners' script. We were working on it literally as the pages came in. Then Downey had to go back to the show, leaving me to do what amounted to a full rewrite. We all went up to Toronto, and I started commuting back and forth to the show, just like Chris and David did. I got married while the film was being shot, and at one point I was writing pages from my hotel room in Hawaii and faxing them in. It was crazy.

PETER SEGAL, *director:*

Even though there was no script, I really believed in Chris. One night before production started, I was driving him to the Palm in L.A. to meet Brian Dennehy, who was to play his father. This was at a time when *Saturday Night Live* was in its nadir. People were writing all the articles about "*Saturday Night Dead*" and so on. And here we were with a movie with a fixed start date and no finished script. Chris turned to me and said, "Pete, everyone expects us to fail."

"Yeah," I said. "I think you're right."

"Our only victory will be a success."

And at that point I knew we were bonded, because we were both taking a leap of faith in each other, that we were gonna go through this as

comrades. Fred Wolf had given us sixty-six pages. I called them the Magic Sixty-six, because that's all we had. We left for Toronto with very little to go on. It was like taking the pin out of a grenade and going jogging.

FRED WOLF:

It was my first movie. I didn't know enough to know that it was abnormal, though obviously it was. Pete and I would write until three or four in the morning, and then he'd have to be on the set by six. Then, while they were shooting what we'd written the night before, I'd be in the hotel room writing the scenes for the next day.

PETER SEGAL:

Fred and I would meet for dinner, or lie out on the floor of the hotel room with note cards, talking about things that'd happened to us. Fred would say, "I once left an oil can in the engine, and the hood flew up in my face."

And I'd say, "Great, put it down on a card. It's in."

Then I told him how I was once backing up at a gas station in Glendale and I hyperextended my car door on the cement post. So that went in. We just started building these stacks of cards.

After that, Fred and I would go to dinner and watch Chris and Dave interact, and we'd literally just start taking notes on things that they would say to each other. One day on the set, Chris came out with a new sport coat, and he said, "Does this coat make me look fat?"

And Spade said, "No, your face does."

I stopped what I was doing and said, "Wait, wait, wait. What was that? Say that again! That's gold! We gotta put that in the movie!"

Unfortunately, we ran into trouble with the start date. We were warned that if we started past a certain day in July we would run into the *Saturday Night Live* season and then we'd be splitting time with the show, which we ended up doing.

ERIC NEWMAN, *production assistant:*

Lorne had Paramount give us a plane that would shuttle Chris and David and me back and forth from Toronto to New York. It was my job to accompany them wherever they went. Mostly Chris. David didn't really need the accompaniment. We went back and forth twice a week. We'd shoot on Tuesday, fly down Wednesday morning, do the read-through, fly back up to Toronto, work for two days, then fly back down on Friday for blocking, do the show, then fly back to Toronto at four in the morning.

MICHAEL EWING:

To have Chris and David flying back and forth was actually lucky for us. We had a chance to write material so we'd have something to shoot when they got back. I would get calls at two in the morning from Pete and Fred. They would call and say, "You gotta call the casting director. We need a police officer for this new scene."

So I'd have to call casting, then wardrobe, then do everything to make sure that actor was ready first thing in the morning. The scene when Chris and David drive the car up to the airport, ditch it, and throw the guy the keys? That's that guy. He was hired at two-thirty in the morning and had to be on the set at sunrise. And that happened on a daily basis.

BOB WEISS:

It was rough, but the beautiful thing in *Tommy Boy* of course is the chemistry between Chris and David Spade. It wasn't like, "Hey, let's invent a comedy team." It's not that easy. There was just a hunch that their personal relationship would really pay off. And it did. A lot of what you see in that movie is who they really were. We started filming, and I was like, "Fuck, this is funny." They just hit home run after home run.

PETER SEGAL:

When *Tommy Boy* came out it was resoundingly dissed by every critic. But a couple of them, and one in the *L.A. Times*, said these guys were the new Laurel and Hardy. I was too young to really appreciate the Hope and Crosby *Road* movies, but there were certain comedy teams that were produced in the eighties, like Eddie Murphy and Nick Nolte, or Gene Wilder and Richard Pryor. It was kind of neat to think, God forbid, that these guys could become like them.

MICHAEL EWING:

We were having a reading of the script up in Canada. The studio had flown in for the occasion, and we knew we didn't have a third act. It was a fucking mess. It didn't go so well. But Chris was being his wonderful, boisterous self, kind of the life of the party, and at one point David Spade turned to him and said, "Chris!" Then he made this hand gesture like he was turning down his hearing aid a little bit. That was the first introduction that I had to them and their relationship.

DAVID SPADE:

We got close just by spending twenty-four/seven together. He trusted me enough to know I was a really great friend. And I was a huge fan—*fan*'s an odd word, but I really thought he was talented, and he knew that I genuinely believed that. During *Tommy Boy*, and even during *SNL*, I was always trying to come up with ways for him to score, to think of jokes for him, and that's a sacrifice you maybe don't see a lot of in show business. When we presented together at the Oscars, I came up with a punch line so I could set him up and he could get the laugh. Over time we just built up a mutual trust and respect.

ERIC NEWMAN:

Chris and David were somewhat competitive, which comes from being on *Saturday Night Live*. They all have that.

David was always looking to get out of work, wondering when he'd get days off. So we're on the plane one week, the three of us, and I get out a schedule. I have a copy, I hand David a copy, I'm taking David through it, and Chris goes "What is that?"

"Oh, David wanted a copy of the schedule," I say.

Chris rips my schedule out of my hand and says, "I get whatever David gets!"

Then he starts looking at it—sideways, because he can't figure it out—flips it upside down, can't figure it out, then drops it and abruptly pretends to fall asleep. It was hilarious. He'd acted out a little bit, and he knew it, so he tried to make it into a joke. He knew there were guys who became movie stars and became dicks, and he didn't want to become that guy.

LORNE MICHAELS:

I always said that while making the movies, Chris would put on thirty pounds and David would lose thirty pounds, but no matter what, the amount of weight in the frame stayed the same. Chris would get bigger, and you'd be saying "Get Spade a banana," because he was wasting away.

ROB LOWE, *costar:*

Chris and David were literally like an old married couple. They could be so petty with each other in ways that were so funny and unbelievable that you were never really sure when it was an act, which it often was, and when it crossed over to become real, which it often did.

The two instances I really remember were when they fought over me, like I was some girl. We were shooting in Toronto, and Spade and I had been at the gym at the same time and we ended up hanging out in the

Jacuzzi. Then, the next day, Chris said to Spade, "Where were you? I called you in your room."

"I was hanging out in the Jacuzzi with Rob."

"You were in the Jacuzzi with Rob?"

"Uh . . . yeah."

"Why didn't you guys call me?"

And it became this whole thing of who was in the Jacuzzi with me, and it just went on from there.

DAVID SPADE:

Then, one night after flying back from New York, Chris goes, "I got the flu, so when I land I'm going right to bed."

"Okay."

We got in to our hotel in Toronto. Chris was being cranky and grumpy, and he went to bed. Rob Lowe called, and I said, "Farley's crashing. You want to grab a drink?" He said sure. So we went down and we had one quick drink and went to bed, because we both had a six A.M. call.

The next day I'm sitting in makeup. Farley's staring at me in the mirror, biting his lip, which means there's a fight coming. He goes, "How's Rob Lowe?"

"He's all right."

"Huh? How's Rob Lowe?" And he kept saying it. "How's Rob Lowe?"

I said, "Uh, I don't get it."

"Where's your precious Rob Lowe?"

"Oh, you mean last night. Yeah, I had a drink with Rob Lowe."

"Oh, yeah. I heard all about it."

We'd just been together too much at that point. So we come to the set, it's twenty-five degrees and I'm huddling on the ground, waiting for the scene to start, trying to eat a tuna fish sandwich with my freezing fingers. Chris walks up and steps on the sandwich and my hand with his boot. I yell, "Ow, you motherfucker!"

And he goes, "*Huh?! How's Rob Lowe?*"

"*What?!*"

I throw my Diet Coke on him and he throws me into a wall and down the stairs and he comes to hit me and they yell, "*Action!*"

We both freeze in the middle of this fight, wait for our cue, and then open the door and walk in. I just stare at the other actors for a moment, and then I say, "Fuck this." And I walk out.

I just leave and go back to my room, and Farley goes, "What's *his* problem?"

Chris was actually jealous of Rob Lowe. He admitted it later. That's probably why I'm not married now; my first experience didn't work out.

MICHAEL EWING:

Best friends are always competitive, and comedians and actors are always competitive in a certain way. That's just part of it, that's part of the one-upmanship. And that carried over into their lives, with women and with friends. When you get people like Chris was, like Dave is, those are complicated relationships.

PETER SEGAL:

David had a boldness about calling out the elephant in the room where nobody else would. It was all playful, but it was the kind of humor that unless you knew Chris you would never go there. There was a lot of honesty in Dave's jokes toward Chris, and I think Chris appreciated the ballsiness of it. The guys from *SNL* all tell me that everyone felt Chris was the funniest guy. So for Dave to be the one to crack Chris up, well, that was like being the one to pluck the thorn from the lion's paw. He had a friend for life.

JULIE WARNER, *costar:*

David doesn't drink or do anything bad. He's very orderly, very much a grown-up at a young age. That stability is part of what attracted Chris to him. But I could see David's frustration. He knew that there was so much danger for Chris. He felt responsible, and it's too much responsibility for any one person. I've been around enough addicts to know that your greatest fear is to be abandoned. Chris knew David was never going to leave him, so there was safety there.

DAVID SPADE:

He trusted me. He also thought I was smart, whether I am or not. He would always ask my advice on a million things, and I would just try and help him in the best way I could. He had his problems, and it was a mounting pile of them as time went on. It was really hard for me. It scared me for myself. I wasn't even half as famous as he was, and I'm having my own weird stuff going on. Is this what it's like, being famous? Because it's getting harder and scarier to watch.

KEVIN FARLEY:

At the time of *Tommy Boy*, Chris was on his second year of sobriety. He already had a year under his belt. He was down to maybe 225 pounds, which was a pretty good weight for him. He could move really well in that movie, and it showed. That was probably the best he ever looked, and his sobriety was probably the best it had ever been. Everything was clicking on all levels.

MICHAEL EWING:

Every night after we'd finish shooting, I'd take him either to a meeting or to church to visit with his priest. Sometimes both. He would call the priest

also, sometimes late at night. He was completely straight during that whole shoot. He was dedicated to helping himself, and he was totally serious about it.

PETER SEGAL:

He was very superstitious, and I think it all tied in to habits that he felt would keep him on the straight and narrow. His habit of visiting his priest every day, it was as much for the routine as for the actual counseling.

DAVID SPADE:

He had to pull up his pant leg twice and tap the ground twice before every take, which over the course of an entire movie gets a little annoying, to say the least. So one time while he was tapping the ground like that I said, "You know, Chris, the good thing about the devil is that he won't come to you. You have to summon him like you're doing now."

"What are you talking about?" he said.

"You see, I like God. That's just my own thing. But you like to tap on the roof of hell and invite the devil to join you."

"*Shut the fuck up.* That's not what I'm doing."

"Well, there's no superstition in the Bible, Chris, which you'd know if you ever leafed through it. That's something the devil made up so people would invite him into their lives."

He was fucking stunned. I loved it. He didn't know what to do. Even though I was totally bullshitting him, it sort of sounded like it made sense, and he just stared at me, frozen.

PETER SEGAL:

And there was no joking about that stuff, either. Every time he smoked a cigarette he brought it out backwards to his lips and touched it, and then turned it around and put it in his mouth. If he turned around, did a 360-degree turn to the right, he'd have to unwind and do another 360 to the

left, like how you'd spin if you went around the corner and forgot something and came back for it. You'd ask him about it, and he'd say, "I gotta undo. I wound up to the right, and I gotta unwind to the left." There were all these habits. For some reason, I think they were comfort factors to Chris.

JULIE WARNER:

Chris looked everywhere for safety and support, and he felt very safe within the family he'd found on this film. Whatever feeling of acceptance he craved, I think he found it there.

Chris was very silly with me at first. He was like, "I can't believe they cast someone so pretty to be the love interest." He had this kind of goofy thing that he did around girls. It was like, "Oh, Julie, I can't even look at you. You're so pretty. I can't even talk to you."

It was put on, but it was obviously his defense for the fact that he couldn't have a real moment with you. That whole aw-shucks character he did on "The Chris Farley Show," that was all very deliberate. He'd say stuff to me like, "You were really good in *Doc Hollywood*. Especially in the naked part. You remember that time in *Doc Hollywood* when you were naked? You remember that? That was *awesome.*"

At the beginning it was funny. Then it went on for weeks, every time I saw him. I even said to David, "Is it always going to be like that?"

He said, "Chris can't talk to girls he thinks are pretty."

To be an actor, you have to make it real, play off the other person, listen and react. And Chris had that. We really connected when we were acting together. Interestingly, when Chris was on camera, it was the only time I could get him to look me directly in the eye.

PETER SEGAL:

We went to a club one night after work. Chris came in a three-piece suit, wearing his black horn-rimmed glasses. There were a lot of pretty women there, and Chris looked really tense. I said, "What's the matter?"

He was looking at these beautiful women at the bar, and he said, "Pete, beauty makes me angry. I can't tell whether I want to take 'em home or club 'em over the head with an empty wine bottle."

I've quoted that line to so many people over the years. It was hilarious, but you could tell that he was tortured by his own insecurities. He was the bravest guy in the room, yet fearful in ways that he would never let on.

MICHAEL EWING:

Chris was very self-deprecating, so self-deprecating. It would leave you always thinking, you know, *ouch*. You could see that there was some wound just below the surface, just a hair below the surface, and sometimes not even that deep. One time on the set, the crew was cracking up after the take, which happened a lot, and Chris walked by and said, "Yep, everybody likes it when fatty falls down." I was like, oh, there's the crux of it.

JULIE WARNER:

It's hard for me not to be able to connect with someone I'm working with. At a certain point during the shoot, Chris was doing his routine, and I stopped him and said, "Can I have lunch with you? Can I just sit and talk to you?"

So we had lunch in his trailer, and he made about forty jokes about "I can't believe you're in my trailer" and all that. But eventually it all just went away. All the shtick, it was just gone. We talked about football, about Madison. I think once he realized I was a safe person he stopped being so sheepish. He had an amazing ability to keep people at arm's length.

We had one kiss at the end of the movie. It was kind of a throwaway moment, but he talked about it all morning. I really wanted to put him at ease about it, make him feel like I was psyched about it. Because I wasn't unpsyched about it or anything, and I actually thought Chris was sexy.

Talent is sexy. Chris was a big guy, but he was cute. I hated that he didn't feel worthy of that. The first time we kissed the crew applauded and ribbed him a little bit. He was really embarrassed, but once he got past it he was fine.

There was something deeply lonely about him. Profoundly and deeply lonely. He was a man. He wanted that kind of companionship, and yet he did not know how to get it.

TOM FARLEY:

Even though our dad was incredibly proud of Chris's career, Chris always suspected that what Dad really wanted was for him to settle down with a wife and kids. It's like, no matter how successful you are, until you show that you can raise a good family you haven't really proved yourself. That was the struggle that Chris always went through, wanting to be a family guy like Dad was and yet wanting the success in his acting life, too. But very few people can make it work on both ends successfully. If Dad had had a choice, he would have been running for Congress or making deals on Wall Street with all his Georgetown buddies. He'd given that up. But Chris could never be content with his professional success, because he was living by Dad's barometer and not his own.

FRED WOLF:

We weren't trying to write a movie about Chris and his dad, but I think a lot of it just subconsciously worked its way in there.

PETER SEGAL:

In writing that movie, we pooled our emotional stories as well as our comedic stories. It was all done out of desperation, and then, ironically, there was serendipity to it. I think Chris brought a lot of his relationship with his dad to that movie; he tapped into those feelings when Big Tom

Callahan died. But it wasn't Chris's story. It wasn't Fred's story. It wasn't my story. It was everyone's story.

BOB WEISS:

When Terry and Bonnie Turner were writing it, Terry would say, "Well, this is like my father's story." And Lorne would say that it was like his father's story. Turns out it was a lot like Chris's father's story. It made me wonder, well, whose father is it? I mean, what's the deal? But that's why the film is so accessible. We all know the dynamic of trying to struggle under this giant paternal shadow. It's universal.

BRIAN DENNEHY, *costar:*

When I got on the set I figured my job was to be like Chris, not for Chris to be like me. I had to create this crazy, Rabelaisian character who would be an older, more settled version of Chris's character. There was never any conversation about it; it's just something Chris and I both understood. When we did the scene outside my office where we did that sumo thing, bumping into each other, that was my idea. I said to Peter, "It should be like two crazy-ass rednecks or sumo wrestlers who meet in a bar and start tussling like wild bears." Because that's what they are, these characters. All that came out of my watching Chris and thinking what it would mean to be his father.

I think that character loved his son and, to a certain extent, spoiled him. I think he also represented the ideal father for all of us. Psychiatrists might not call him the perfect father, but a lot of kids would.

KEVIN FARLEY:

Chris always used to say, "I'm only doing funny stuff to make one guy back in Madison laugh." And if you saw my dad around town, talking about Chris, he'd be gushing. He couldn't have been prouder. But like a lot of dads, he was a little reserved about actually showing that to us. You

want to have your dad say you did a good job. And Dad would do that. He'd go, "Good job, son." Really brusque and understated. But most families, especially Irish families, they just don't communicate that well.

BRIAN DENNEHY:

We all grow up with that necessity to be what our fathers want us to be, and probably, ultimately, failing. There's no question about that. My father's been dead now for twenty-five years, and there's not a day that goes by that I don't find myself thinking about it. Our relationship was unusual, my father's and mine. Our family was classically Irish Catholic in the sense that the family was unquestionably the most important aspect of all of our lives, and yet we were not close, if that makes any sense. There is an emotional distance between us that exists today.

Philip Larkin, a British poet whom I love very much, wrote a poem that really says it all: "They fuck you up, your mom and dad / They may not mean to, but they do. / They fill you with the faults they had / And add some extra, just for you." And that's not a criticism of parents, but I think it says that there's something inevitable about it. Your parents want you to know things that they've learned, but they can only do so much. You've pretty much got to learn it yourself.

PETER SEGAL:

Knowing more now about his relationship with his father, I recognize a lot of things in hindsight. Chris was a really good athlete. He idolized his coach, who was also a father figure. And I realized that the best way to work with Chris was not as a director but as a coach.

For example, take the day he shot the scene where he looks at his grades and says, "A D plus! I passed!!!" He wasn't getting it right, and he was so furious with himself. I told him to go out and run around the quad a couple of times and come back in. I said, literally, "Take a lap." There were times when he'd be so amped up with coffee and cigarettes that I'd have him drop and do twenty push-ups. I just needed him to work it off.

He'd go, take the lap, do the push-ups, then he'd come back and he'd look at me like I was about to put him into the game. And that was okay. Every actor is different, and that was how I had to deal with this particular person. In that instance, I was his coach. He would have walked through a brick wall for me if I'd asked him to.

ROB LOWE:

Pete Segal is a comedy mathematician. He really understands the timing and the beats in a way that a lot of other directors don't. And Chris's style was very wild and unstructured and, frankly, lacking in technique. So it was a good mix between the two of them.

JULIE WARNER:

I'm sure Paramount wasn't thrilled about the amount of money that was being spent, but Pete knew there was a gold mine there, and he was determined to get at it. He knew that the movie was only going to work if Chris was free to have fun, and that meant making sure he felt safe and not pressured. He gave Chris the trust and patience that I don't think he found anywhere else. Once Chris felt that safety, he was able to shine.

MICHAEL EWING:

What people responded to in the movie was the comedy, number one, but also this underlying heart that's woven through it. They're dancing at the wedding, and Brian Dennehy suddenly drops dead of a heart attack. I mean, what comedy has one of the main characters drop dead a third of the way through the movie? This is a comedy that people thought was just light and fun, but it also dealt with real things in a real way.

We were just a little movie, and by the end of the shoot Paramount didn't really want to spend any more money on music or anything. They said, "Here's the money to make the movie, and not a penny more."

Then Sherry Lansing saw the first screening. I was sitting across from

her, and as the lights came up I could see there were tears running down her face. We went outside; she gave Pete a big hug and gave me a hug and said, "My God, where did all that heart come from? That wasn't in the script."

It was just one of those rare things that happens in movies sometimes. It all came together. Then they approved the extra money to do a real score and everything.

FRED WOLF:

The critics totally missed the point of *Tommy Boy*, and, of course, history has proven them wrong. It's seen as this mini comic gem. A few years ago, *Time* magazine listed the "Top 10 Movies to Watch to Make Yourself Feel Better." It went all the way back to *Adam's Rib* and *Cocoanuts* with the Marx Brothers, and *Tommy Boy* was on that list. That was really great to see. I wanted to fax that to every movie critic in America.

ROB LOWE:

To this day, people stop me on the street and say they love *Tommy Boy*. It's the ultimate movie for fifteen-year-old boys. And if you compare *Tommy Boy* to what they're making today for fifteen-year-old boys, it's the fucking *Magnificent Ambersons*.

DAVID SPADE:

Looking back it feels like it was a big hit, even though it wasn't. It did all right. It just has nice memories about it. It's the most-talked-about movie that I've ever had any part of, certainly, and that's ninety percent because of Chris.

PETER SEGAL:

The premiere was very small. The movie was about to start and everyone had gone to their seats. I was nervous as hell, and I went into the men's room. Chris came in behind me. He said, "Well, this is it." He was nervous as hell, too. We knew we had been through a real war together. On the same side, but still a war. To this day it's the most difficult shoot that I've ever experienced.

We stayed there in the men's room and talked for a little while, knowing that the movie was starting. It was like that moment when you buckle yourself in to a roller coaster and you know that, as afraid as you are of going up that first hill, there's nothing you can do about it. I gave him a hug, and he said—and he was very adamant about it—he said, "Please don't leave me. Let's do this again. Promise me we'll do this again."

MICHAEL EWING:

I have a tradition that I get all of the actors to sign my movie poster for me. So one day I gave Chris the *Tommy Boy* poster. He took it, signed it, and handed it back to me. And what it said just cut straight to my heart, and really surprised me. What he wrote on the poster was: "Dear Michael, Don't give up on me. Chris."

CHAPTER 10

The Lost Boys

MARILYN SUZANNE MILLER, writer:

In the years I was back at *Saturday Night Live*, I so didn't belong there. But then of course, no one belonged there. The cast didn't belong there. The writers didn't belong there. And we didn't belong there with each other. The whole thing was a real marriage of hope.

Just two years earlier, during the run of the 1992 presidential election, *Saturday Night Live* had been at the top of its game, consistently funny and culturally relevant. But in the fall of 1994, as Chris Farley and David Spade flew back and forth from Toronto to film *Tommy Boy*, they returned each week to find the show slipping further and further into confusion and disrepair.

Cast stalwarts Dana Carvey, Jan Hooks, and Phil Hartman had all left. In their place, Lorne Michaels had hired a slew of actors and comedians, both young and old, known and unknown. In all, the cast swelled to seventeen members, more than double the original group of Not Ready for Primetime Players in 1975. But in spite of all the talent in the room

(or because of it), very little seemed to work. The cast was not a team. It was an odd collection of ill-fitting parts. There was little chemistry and no love lost among several of those sharing the stage. It was not a happy time.

Off-camera the changes were just as severe, and the process just as broken. The younger writers were coming to the fore, but the writing staff as a whole never gelled, especially with veterans like Al Franken and Marilyn Suzanne Miller feeling pushed out and stymied by the new generation. Caught in the midst of this chaos, and trying to manage it, was head writer Jim Downey. Downey's experience probably encapsulates best what everyone was going through: At the end of the year he was served with divorce papers on the same day he was fired.

With the show in a rut, Chris found himself in one, too. He put in a hilarious turn as a lost contestant on a Japanese game show, and he took on some of the show's political humor with his impression of House Speaker Newt Gingrich—a role that would even take him to the halls of Congress. But as far as memorable performances go, that fifth year added virtually nothing original to Chris's *SNL* legacy. The Motivational Speaker came back again (and again). So did the Gap Girls and the Super Fans. And as *Saturday Night Live* limped to the end of a particularly disappointing season, Chris's attentions drifted elsewhere.

MIKE SHOEMAKER:

It was a terrible year. Everyone was miserable. And once it starts getting bad, it almost has to get worse.

STEVE LOOKNER:

There was definitely a sense at the start of the '94–'95 season that we needed to make the show better. After every taping there was more discussion over what worked and what didn't, more of a conscious effort to pull things together. Nobody wanted to be the cast that brought *Saturday Night Live* to an end.

TIM HERLIHY:

The ratings plunged. It wasn't just a critical reaction; it was a popular one, too.

JIM DOWNEY:

My feeling was that the show had been running on vapors for a while, but the ratings had been crazily spiked by *Wayne's World*. It annoyed me that the network didn't care if the show sucked while the ratings were high. They only cared if the show sucked and the ratings were low.

DAVID MANDEL:

It was just very unclear what the show was supposed to be. When you look at the 1992 year, you had Carvey and Myers and Hartman, Jan Hooks and Kevin Nealon. Those guys were all-stars. Hartman used to put on a bald cap and play ten different characters with ten different voices in ten different sketches. So the beauty of adding a guy like Sandler to that group was that Sandler could go on Update and do his weird, funny thing and kill with it. Same with Chris. He could be a killer supporting part, like in the "Da Bears" sketches, then turn and have his own starring role, like in the "Chippendales" sketch. That was all you really needed of Farley in a given show. It was like a flavor of something. Jim Downey used to say something very interesting, and I will paraphrase it. He used to say that Farley and Sandler were like the special teams on a football team, the great kicker or the great punter, the guy you need to come on, do his thing, and then get off the field.

After the all-stars like Hartman left the show, it never seemed like a working cast so much as "Here's the Sandler sketch. Here's the Farley sketch. Here's the Spade sketch." All of a sudden, we were playing a football game with nothing but these special teams guys out on the field, and that's not a team that's going to play well for a whole four quarters.

JANEANE GAROFALO, *cast member:*

The system was flawed in a way that funneled the cream to the bottom and the mediocrity to the top. When we did the table read-throughs on Wednesdays, there were always funny sketches in there. Rarely did they hit the air. Downey was still there, but he wasn't spiritually there. I think there were some personal things going on in his life that he wasn't fully present, emotionally. He didn't have the reserves needed to manage the room. The system was just broken.

MARK McKINNEY, *cast member:*

People were clinging to the stuff that worked in a time without a lot of focus. It was really, really hard slogging. But I saw Chris as ensconced in a brotherhood of his own making with several of the writers. He was comfortable in a way that I never was.

JANEANE GAROFALO:

Chris had the luxury of not only being talented but also well liked. When he would come onstage, even just to take his mark during a commercial break, people would start cheering. It was clear that he was an audience favorite, and kind of the go-to guy for a laugh.

FRED WOLF:

All the writers wanted to get their stuff on the show, and you learned very quickly that there were guys that you could count on. You could ride their charisma onto the air. We would do that with Chris.

Ian Maxtone-Graham gave me this diagram he'd made of "Fred Wolf's Sketches for Chris Farley." There were three different dials on it. The first one was labeled "Chris is: Dry. Moist. Soaking Wet." And the dial was set on "Soaking Wet." The second one was labeled "Chris is: Quiet. Talking Loud. Screaming at the Top of His Lungs." And the dial was set on

"Screaming at the Top of His Lungs." The third one was labeled "Chris is saying: Gosh! Oh no! Oh, sweet mother of God!" And the dial was set to "Oh, sweet mother of God!"

It seemed like every sketch I wrote for a while had Chris getting soaking wet and screaming, at the top of his lungs, "Oh, sweet mother of God!" But I couldn't resist writing them, because they would always bring down the house.

ROBERT SMIGEL:

When we did the first "Motivational Speaker" sketch, I added something that I thought was really helpful at the time but that I somewhat regretted later. The sketch was pretty much word for word as Bob Odenkirk had written it at Second City, except for the ending. The stage version didn't really have a topper for Chris, other than "You'll have plenty of time to live in a van down by the river when you're . . . living in a van down by the river!" Chris was so powerful onstage that it carried you to the end. But TV flattens stuff out and I thought it needed something more, so I added the part where he's telling David Spade, "Ol' Matt's gonna be your shadow! Here's Matt, here's you! There's Matt, there's you!" And then he falls and smashes through the table.

It worked really well, but it inaugurated this trend of Chris being really clumsy and falling down a lot. There were several more "Motivational Speaker" sketches, and all of those ended with him crashing through something. Then the writers started having him fall through other stuff. He used to joke about it. "Everybody laughs when fatty falls down." Chris and I would laugh about how hacky it had become. I'd say, "Chris, give me a triple boxtop." And he'd do a certain kind of fall for me.

That sort of broad clumsiness was actually the opposite of what Chris's talents as a physical comedian were. What really struck me at Second City was how graceful and nimble and athletic he was, a brilliant physical performer who was also capable of really specific, subtle things. But a lot of that got buried in this succession of sketches with yelling and pratfalls. It was to Chris's detriment, and the show's.

JANEANE GAROFALO:

I think that the writers began to use him as a bit of a crutch, but that's not entirely the writers' fault. There's a natural instinct among a lot of comedians, particularly younger ones, to want to get a laugh. You want desperately to be liked, and sometimes the quickest route is to be loud and broad in your gestures. I think Chris did that in the beginning, and then, unfortunately, it stuck.

DAVID MANDEL:

As much as the writers used him in a certain way, he also liked working in that certain way. It was easy for him to default to the pratfalls and so on. He could power through a sketch just by hiking up his pants and playing with his hair. Those were stock Chris Farley moves. He also hadn't started wearing his glasses when he should, and he couldn't always read the cue cards. You'd write a quiet, subtler sketch, and he'd flub a line 'cause he'd miss the cue card. So maybe you didn't want to take a chance with him on that kind of sketch, and you'd default to something loud and physical.

There was never any sketch where we said, "This sketch isn't working. Let's have Farley walk in to be the joke." It was not a fallback move. But there were definitely a lot of sketches, especially in that last season, that could be reduced to: "Chris yells a lot."

MICHAEL McKEAN:

It paralleled Raymond Chandler's rule: Any time the action starts to slow up, just have a guy come through the door with a gun. That's how they used Chris. He would bring a lot of juice to what could have looked like lazy writing, and he saved a lot of bad sketches. There was this sketch with Deion Sanders—I mean, the comedy stylings of Deion Sanders, first of all—where this flying saucer lands and they keep sending men in to explore, and they all either get killed or anally probed. Then they send Chris in, and he comes out with his clothes in tatters, virtually naked, having

been anally violated. That's all there was to the sketch. In fact, I think I've probably embroidered it a little. But even with that, Chris gave it a shot, and he was funny.

JANEANE GAROFALO:

The Deion Sanders alien anal probing sketch, it was so embarrassing.

AL FRANKEN:

The show was always best when there was a balance between the writers and the performers, when both were operating at their peak level and working together. To some extent, Sandler and Spade and Schneider and those guys were not in sync with the writers, at least with my generation of writers. I was not thrilled with what was happening. But maybe it was just time for me to go.

MARILYN SUZANNE MILLER:

There was a quality among those guys, Rob Schneider, Adam, Chris, and Spade, that it was "our show." It was a very David Spade attitude, and it certainly excluded me. Also, I think they knew that, at the bottom line, we weren't loaded with respect for what they were coming up with.

For some reason the phrase "anal probe" found its way into virtually every sketch. Most of those didn't make it to air, but at the read-through table it seemed like "anal probe," "bitch," and "whore" had assumed the same status as "Good morning, how are you?" It was imbecilic and just as offensive as offensive could be.

TOM SCHILLER:

I think that the humor did change, and I didn't get into it that much. And that's because the times changed. But the stuff we were doing in the first five years of *SNL*, I wouldn't say it was necessarily so smart. When they

talk about this "dumbing down of comedy," I think comedy just keeps changing with the times, all the time. You can trace the evolution of vaudeville to *Ed Sullivan* to *Your Show of Shows* to *Laugh-In* to *Saturday Night Live*. And it just keeps evolving.

JOHN GOODMAN:

It's similar to what happened to the guys who took over *National Lampoon* after Doug Kenney and Henry Beard left, when it all fell to tits and racial slurs. Michael O'Donoghue used to say that comedy isn't a rapier; it's clubbing a baby seal. But you can only club that baby seal for so long.

TOM DAVIS:

They were taking their cues from *Animal House*, whereas we had taken our cues from Bob and Ray, Sid Caesar, and Johnny Carson. Comedy just takes these turns. But that show has to stay young. It doesn't matter if you like it or agree with it or think it's funny. It has to stay young.

MARILYN SUZANNE MILLER:

Chris was part of this gang, and he identified with this sort of gang spirit that they had. When he and Adam and Spade did those Gap Girls, it was kind of like the gang was getting together to play, only they were doing it on national television. They were like the Little Rascals, or the Lost Boys from Never-Never Land.

I remember being overwhelmed one night at some of the capers that were going on. All these overtly sexual—and, frankly, homoerotic—hijinks. Just constantly grabbing each other's asses—and much worse than that. I went into an office with Al Franken, and he explained to me that when a bunch of guys are marooned on an island together, as was the case with that show, you get this kind of behavior. It happens at boys' prep schools, on submarines. There was a sketch Jim Downey wrote on the old show, "The Adventures of Miles Cowperthwaite." It was about this

young boy trapped on a ship with all these pirates, and it was all about manly men being manly and doing manly things at sea to prove their manliness—and they all turn out to be gay. Everything these kids were doing was like that.

JIM DOWNEY:

It became more of the atmosphere of the show, because you had this critical mass of young guys. I always went to all-boys schools, so I have to admit it's something that makes me laugh, you know, when it's done right. Chris would burst through the double doors of the writers' room with his pants around his ankles and his privates tucked back between his thighs doing the thing from *Silence of the Lambs*. He'd start rubbing his breasts and saying, "Am I pretty?" It was just so balls out, so to speak. I mean, you had to give it up for that.

MIKE SHOEMAKER:

Comedy people, when we're alone and insulated, just get more and more shocking, and it doesn't play to the rest of the world. It's the same way to this day. I've seen worse before and since. A lot of it was disgusting, but in the context of this place it was always funny. We were just constantly thinking, oh, this is so damn funny, but if anybody saw it we'd all be arrested.

JIM DOWNEY:

It's hard not to laugh even if you think it's encouraging irresponsible behavior. Sometimes, to get Chris to stop doing something, we'd talk among ourselves while he was out of the room and agree not to laugh no matter how funny it got. Chris'd get perplexed, and eventually frustrated, because no one was laughing. Then he would just escalate more.

Farley liked to do this routine where he would jokingly hit on waitresses. He'd say, "Well, little lady, I've got a problem. I'm in from Moline,

Illinois—work with a grain elevator outfit out there—and I'm in town for a couple days on business. And darn it, if I don't use my whole expense account the home office'll be liable to cut me back. So, how's about you and me do this town up right." And so on, using all this weird, Jazz Age lingo. You'd be like, Chris, what the hell are you talking about?

One night we were at this Mexican restaurant in Midtown named Jose's. It was one of those places where you buzzed downstairs and they let you in and the entire restaurant was up on the second floor. One night, Farley was doing his goofy routine with the waitress all night, and she was kind of rolling her eyes, like, "Yeah, yeah, buddy."

The rest of us, I suppose, were not giving him enough attention, so he felt he had to take it up a notch. He jumped up, scooped her up in his arms, and ran down the stairs and out of the restaurant. I turned and looked out the window, and I saw him dashing up Fifty-fourth Street and getting into a cab with her. We all hung back, staying in the restaurant, like, "We're not going to bite. We can't give him the satisfaction." Then I said, "Jesus, we could all be sued." I was acting in loco parentis with these kids, so I ran downstairs after him. But Chris liked to do that, do big put-ons with strangers who didn't know who he was. In most cases people realized it was a joke and were happy to be a part of it.

NORM MacDONALD:

Chris would do things with girls, like a kid would do. He'd always be like, "You shure are purty. Can I touch your leg?" It was all for the comic effect of how you're *not* supposed to approach a girl. It was all harmless, but obviously because he had a lot of money, some extra came on the show and decided this amounted to sexual harassment.

JIM DOWNEY:

The second-to-last show of the '93–'94 season, I had written a piece about Bill Clinton called "Real Stories of the Arkansas Highway Patrol." We had to go upstate and do some outdoor filming. Some women were extras in

the piece, and one of them went up in the car with us. It was me and her and Schneider and Farley. It was a limo, with that wide space between the two rows and seats facing each other. Schneider and I were sitting together, and Farley was next to this girl. He was doing his usual "Hey there, little lady!" shtick. And he was poking her and hugging her, but if you knew Chris you knew it was all playful. I finally told him to knock it off—not because I thought it was assaultive behavior but because it was getting annoying.

Well, this girl went to the talent department and complained, hinting at some sort of legal action for what Chris had done. But Chris never did anything wrong. I know because I was sitting there, and as the producer of the show I never would have allowed it. My impression, honestly, was that she was mostly complaining about the size of her part. She thought she had several lines, and it actually wasn't a speaking role. I think we paid her for a speaking part instead of as an extra, and that was the end of it.

MIKE SHOEMAKER:

Nothing ever came of it. It was actually a very minor incident. It became a much bigger story in people's minds because of the prank that followed, more so than because of the incident itself.

JIM DOWNEY:

So the next week it's last show of the season. Farley came in, and we decided to have some fun with him. It was just completely random and totally unplanned. He came by, and I said to him very casually, "Chris, you know about the lawsuit, right?"

"What?" he said.

"You know, the sexual harassment suit. Anyway, you're not going to do any jail time. That's—don't worry about that. I mean, it's not one hundred percent you won't, but it's at least a sixty to seventy percent chance you won't do any jail time."

"Wh-what are you talking about?"

"You know. The girl from the limousine. Anyway, it's too early to tell, but NBC's lawyers are all over it."

He was really starting to shake and sweat. Then the other writers started gathering around. Mind you, I'd seen Farley do plenty of similar put-ons to other people, so in no way did I think this was unfair. And also, I thought that he needed to learn a lesson, that the kind of outlandish behavior he pulled in the limo can have consequences, even if it's harmless and well intentioned.

I said, "Now, Chris, I used to be a process server, so I know how this works. If you're walking down the street, for the next two . . . well, for the next several months, if you're walking down the street and someone approaches you, do not wait to find out who it is—you *run.* You flat out run."

And then Ian Maxtone-Graham chimed in, "Oh yeah, I was a process server for a whole summer. If they even *touch* you with the document, you've been served. If it touches anywhere on your person."

Eventually, everyone's getting in on this, giving Chris advice on how to hide out and things like that. I don't know what happened with Chris in the intervening days, but we went to the prop department and had them make up a subpoena, and I had one of the writers I knew from *Seinfeld* serve Farley with a lawsuit at the end-of-the-season party. He was devastated. A couple of people were coming up to me, saying, "C'mon, that's cruel. He's close to tears."

NORM MacDONALD:

Chris was just ashen, and the even crueler part was that they didn't let him in on the joke until an hour or two later. To make it that much worse, his mother was standing right there beside him when it happened. It was really terrible.

MICHAEL McKEAN:

It was a really shitty thing to do.

JIM DOWNEY:

And I was like, "Now, wait a minute. I've seen Chris put many a waitress through the paces before. He's a big boy." But finally I said, "Okay, let's end it." I went over and talked to him. It took me about a half hour to convince him that it was a put-on. As far as I heard, he was never mad about it, because he liked to put one over on other people, too. I talked to him a few days later and I reminded him, "You're a celebrity now, and people will be on the make. You should keep that fake subpoena as a reminder not to do anything that could be misconstrued."

And he said, "I don't have it. I burned it."

It was like he had to destroy the evidence of the whole thing.

FR. MATT FOLEY, friend:

Chris was very much a man's man. There were girls who were his friends, but anyone who was being honest would say he did some pretty inappropriate things with women. He was often mean to them. It was weird. It was the trust thing: Will you love me for who I am?

Chris used to say that every girl he went out with before he got famous looked like him with a wig on. Not to slam those women, but it's probably true. Then, all of a sudden, he's famous and these hot girls are all over him. So obviously, sexual issues, relationships, were very difficult things for him. I think he trusted God implicitly; I don't think he trusted people. "Why do these women want to go out with me?" He was very confused by that. He didn't trust them. He didn't know who to trust.

TIM MEADOWS:

That was something we talked about quite a bit. He'd always say, "How could any beautiful girl love my fat ass?"

MARILYN SUZANNE MILLER:

One of the real differences between John Belushi and Chris Farley was that John Belushi was married, whereas Chris was sort of the opposite of married. I wouldn't even put him in the category of "single." He wasn't single; he was the opposite of married.

FR. MATT FOLEY:

He went out with this girl named Lorri—her nickname was Kit Kat—this really hot girl. I was in New York one weekend and Chris told me, "I really like this Kit Kat girl." I saw her on the set. She was this five-foot-ten Victoria's Secret model, long legs, just hot. They clearly weren't going to talk about second-century world history together. Chris said, "What should I do? I don't know if she likes Spade or not. I want to ask her out, but I'm so confused."

I said, "Well, Chris, why don't you go to work today and ask David if it's okay if you ask her out?"

He did, David said it was okay and he asked her out. So here we are on a date, Chris, Kit Kat—and me, his priest. The next night we all went to a movie together. It was just bizarre as hell. It was like I was back in eighth grade.

DAVID SPADE:

Lorri lived directly across the street from me. I'd see her at the deli. She was this Victoria's Secret girl, who I eventually realized was one of the Victoria's Secret girls back from when I used to look at Victoria's Secret and found her quite striking. She was very friendly. I invited her to the show, and we started talking. I didn't have a whole lot of friends in New York outside of *SNL*, so it was nice to meet someone to hang out with.

We became friends and started dating. Chris would hang out with us. She thought he was funny, and I didn't mind that he'd come along. This happened a lot, and he'd always paw all over her, going, "You're so purdy,"

and all that. And I wouldn't get mad, but I was like, "Dude, be a little more respectful, to me and her. C'mon with that shit."

Then sometimes Lorri and I would go do stuff, and I'd say, "You know, Chris isn't doing anything this weekend. Can he come with us?"

LORRI BAGLEY, *girlfriend:*

Chris and I met because we were both best friends with David Spade. David and I had met, and we just clicked. We'd meet for breakfast, hang out after work. We even wound up living across the street from each other. Before Chris and I met, David would always say, "You're just like Chris. The two of you are the same person."

"I want to meet him," I'd say.

"No, if you do, you'll fall in love with him."

Men being men, I think David would have liked to date me, but for me it was just never like that. He asked me to go to the movies one night, and I said no, because I didn't want to be alone in a movie theater with him. So he said, "I'll bring Chris."

For a year, the three of us were just friends. We were like this fun-loving threesome that hung out together all the time. It was the most fun time of my life in New York. This was all about a year before *Tommy Boy*. I was a model, and I was doing Victoria's Secret shoots, and Fred Wolf said, "We have to have a pretty girl in the movie. It should be you." So I did the scene as the girl in the pool at the motel.

After *Tommy Boy*, we all kept hanging out just like before. Chris and I had never really been alone. We were always with David or a group. But Chris would always do these little things, like pulling my chair closer to his at the dinner table—little things that said, "She's mine." One night after my acting class, we were all at the Bowery Bar, just dancing and having fun. Sandler was there. It got late, people were going home, and Chris wanted to go out some more. Sandler looked at us and was like, "You guys are *baaaaad* . . ." He saw the connection. So Chris and I went out alone. That was the first night he kissed me. He was a very good kisser.

When it all first came up, Chris came to my apartment and said, "I

have to work with David. Until I finish *Saturday Night Live*, we can't see each other, because I can't go to work every day and have that kind of stress."

I said I understood, and we stayed apart for like three days. We just couldn't do it. David lived right across the street from me, on West Seventy-ninth, so that didn't make it easy. One night there was an after party for the show. Chris didn't go so he could come and see me, but it turned out David didn't go, either. He came home and saw Chris in the car out front waiting for me to come down.

I was getting ready to head out when David called me. "Is that Chris waiting for you downstairs?" he said.

"Um . . . yeah."

"You fucking bitch."

And he hung up the phone.

NORM MacDONALD:

It drove a wedge between them. Chris wasn't a ladies' man like Spade was. Chris wanted to fall in love and be married. Spade's the opposite. He's a real playboy. Chris decided that Spade had a million girlfriends, so he could have just this one.

TIM MEADOWS:

Spade dates nothing but hot girls, still to this day. But for Farley she was a coup.

DAVID SPADE:

And that was the part that ultimately kind of pissed me off. I had brought him into the mix. I should have just kept it the two of us, but I always made sure Chris was involved, because he didn't have anyone.

FRED WOLF:

I loved Lorri Bagley. She was great. It's really fun to walk around with celebrities and see everyone's reactions. But when we walked around and Lorri Bagley was part of it, she definitely did not detract from the excitement factor. She just had a stunning quality about her.

TODD GREEN:

She was so beautiful. Chris would just look at Kevin and me and shrug his shoulders like, "Can you *fucking* believe that I'm with this woman?"

We were playing golf down in Hilton Head. Chris was down there at some diet clinic and a bunch of us went down every year to play golf. Chris was actually not a bad golfer, but he wasn't having a good game. He was just getting frustrated. All day he kept muffing his drives and missing putts and getting more and more angry. Finally I said to him, "Farls, why are you so upset? You're dating Kitty Kat."

He just howled. He did that really deep, guttural laugh he had. And for the rest of the entire round, every time he missed a shot he'd just shrug and say, "Hey, I'm dating Kitty Kat."

She was flighty, but she really cared for Chris, and she genuinely loved him.

LORRI BAGLEY:

Chris would tell me stories about his life before he was sober, and I just couldn't picture it. He liked that. He liked that I couldn't even imagine that side of him. He was so organized, and so hardworking. He'd wake up every morning and make his bed, go to his meeting. He had the neatest, cleanest apartment.

And he was so romantic, always a gentleman. He would always walk on the street side of the sidewalk, always stand up when you left the table, and always stand up for you when you came back to sit down. He was very elegant that way, chivalrous, like someone from a different time.

Once we were meeting for dinner in New York. We were supposed to meet at a certain time, and I got there forty-five minutes late. He had been outside waiting for me the whole time, just so he could be there to open the door and make sure he could pay for the taxi. I mean, who does that? That's so much better than flowers.

Although when he did buy me flowers, that was always special, too. I was in this phase where I was always changing my hair color, and whatever my hair color was, he'd match the roses to it. I always loved that. He never got just red. One time the florist messed up and sent me plain red roses. He was so upset he called and bitched them out. He just hated to be typical. He wanted there to be thought behind everything he did.

Another time I was in Los Angeles, and we'd gotten into this huge fight. I said, "Okay, come out to L.A. and we'll work things out." I was staying at the Four Seasons. Every hour on the hour he sent something new. One hour it was flowers. The next hour it was a bottle of champagne. It went on for ten hours.

And the first night I spent with him, he got up to go downstairs to get water. I was lying there without any clothes on. He went to his closet and got out his robe and came and wrapped it around me, just so I would feel safe. He was a beautiful man.

* * *

While things were going well for Chris privately, *Saturday Night Live* had continued to suffer, and it was clear that major structural changes were needed. Early in the year, reporters had begun to take aim at the show's shortcomings, and by season's end the media had launched a full-fledged assault. Particularly derogatory was a *New York* magazine article by a reporter who had lived in and among the cast for several weeks. While the criticisms in the piece were not wholly without merit, its perspective was rather myopic, and its tone was unrelentingly foul. The magazine's cover featured Chris wearing a television on his head—the poster boy for the death of *Saturday Night Live*. And the headline of the piece, "Comedy Isn't Funny," wasn't exactly what Chris thought his legacy at the show would be.

TIM HERLIHY:

The stuff with the press that year was heartbreaking. Not only were they saying bad things, but Phil Hartman was saying things to *TV Guide*, and a lot of us were being misquoted here and there. The show was just being eviscerated.

FRED WOLF:

The worst hit piece was the *New York* magazine article. The guy who wrote that was living in our midst for at least half a season. He was around all the time. Then all this stuff came out and he just tore the show apart.

NORM MacDONALD:

The guy was really down on Chris in the article, but when Chris was telling stories in the writers' room, this reporter was on the floor. He was laughing like crazy. But the guy had the agenda to write this hit piece, and he was going to write it regardless. Even when he came there and found out that Chris was funny, it didn't matter to him. And then to have Chris go to a photo shoot where they put a TV on his head and to put him on the cover—to put a guy through all that, completely unknowing of what you're going to write about him, it was just low.

Later, when Chris filmed *Dirty Work* with me, he was saying he felt bad that he and Sandler had "ruined the show."

I said, "No, Chris. That's insane. They said that at the time, but you guys have all come out as the biggest comedy stars in the world."

MIKE SHOEMAKER:

The irony is that Farley and Sandler were the poster boys for the show's problems that year, and yet every week we'd do a show and they were the only ones getting laughs.

JIM DOWNEY:

What I didn't like was the opportunism of the press. It was a lot of late hits and piling on after the whistle. Basically, to be honest, I just wanted out.

MIKE SHOEMAKER:

At the end of the season, everything was in limbo. Nobody knew who or what was coming back, management included. Nobody thought the show would be canceled, but we thought we might be. There was never a time when Chris and those guys were officially fired. Everyone just kind of instinctively knew it was time to move on. All the writers just left, every single one of them.

DAVID MANDEL:

I'd love to say it was an ax coming down, a real housecleaning, which is what *SNL* needed. But it was more that we were all just exhausted, working like dogs, and people began drifting away. If there was an ax, it was a very passive-aggressive ax, which is *Saturday Night Live* in a nutshell.

KEVIN FARLEY:

I don't even know what word you would use to describe what happened at the end of that year. Weird? Crazy? The whole place runs on rumors and innuendo. But Chris had a lot of meetings with his managers, who told him he'd be fine stepping right into movies. I don't think Chris was fired, and he didn't exactly quit. He just never went back.

LORNE MICHAELS:

Chris's head had been turned by the exposure he'd gotten from the movies. Starting back when he was in the first *Wayne's World*, Chris was only

on the screen for a minute, but the audience clearly knew him and liked him and was invested in him. Gurvitz and Brillstein were pressuring him to get out there. There was a Chris Farley business now.

MIKE SHOEMAKER:

There's a sketch at the very end of Chris's last show, written by Fred Wolf. It's Chris and Adam and Jay Mohr and all those guys, playing themselves. They're at the zoo and they're screwing around, daring each other to jump into this polar bear pit. "I bet I can swim across the moat and back before the polar bear gets me." That sort of thing. It was the last sketch that those guys ever did on *Saturday Night Live*, and I always remember it as sort of being a metaphor for their leaving the show. Everybody leaps into the polar bear pit, and, one by one, they all get mauled and eaten alive.

CHAPTER 11

The Polar Bear Pit

NORM MacDONALD:

When Chris left *Saturday Night Live*, it seemed like he wasn't ready for Hollywood. There was the *Cable Guy* thing, the *Beverly Hills Ninja* thing. Hollywood was just ready to use a naïve guy in any way they could to make money. And Chris was naïve, but he certainly wasn't stupid. He saw what was happening, and it hurt him a lot. Perhaps because of his faith, Chris had great confidence in human beings and their capacity for being good. And they're not, really. Especially not in this town.

Paramount Studios released *Tommy Boy* on March 31, 1995. Despite a lukewarm critical reception, it opened number one at the box office and went on to gross a respectable $32 million. Suddenly there was a lot of growth potential in the Chris Farley business.

Paramount immediately ordered up, in essence, *Tommy Boy II*. No sequel ideas followed naturally out of Tommy Callahan's story, but that was no obstacle. The movie's basic formula was lifted, reupholstered, and

set down in vaguely different circumstances—and *Black Sheep* was born. This time, instead of playing the screw-up son of a successful father, Chris played the screw-up brother of a successful politician. David Spade no longer played an uptight assistant helping Chris not ruin a sales trip; he played an uptight assistant helping Chris not ruin a gubernatorial campaign.

Chris had signed a two-picture deal with Paramount, and the studio's interpretation of his contract prevented him from taking on any other films so long as they presented him with a "viable" project by a certain date. Fred Wolf was hired to write the screenplay, and on that certain date, under the threat of a lawsuit, he was compelled to turn in whatever script he had. Then, literally at the eleventh hour, *Wayne's World* director Penelope Spheeris was attached to direct.

Chris was a valuable commodity coming out of *Tommy Boy*, and had many options from which to choose. Paramount shut them all down, including a part in the Farrelly brothers' *Kingpin* and the lead in *The Cable Guy* (a project that would involve Chris in a wholly separate legal imbroglio).

And so Chris was shoehorned into the thankless role of Mike Donnelly, a warmhearted but hapless counselor at a community recreation center who's such a political nightmare he's got to be put under wraps during his brother's bid for the governor's mansion. Ever the optimist, Chris was determined to make the best of it. He hired old Red Arrow friend Ted Dondanville to be his personal assistant and constant companion, and then turned his attention to trying to improve the film, bringing in several writers to punch up the script. He also sought out Tim Matheson and Bruce McGill. Matheson and McGill had starred as Otter and D-Day, respectively, in *National Lampoon's Animal House*. Hoping their comedy talents would improve the film's prospects, Chris used his newfound clout to bring them in for supporting roles.

But no matter how hard Chris tried, *Black Sheep* was not going to be *Tommy Boy*. Despite the similarities, the film didn't have the same director or the same producers. Nor for that matter did it have the same stars.

The personal and professional chemistry of Chris Farley and David Spade had inspired *Tommy Boy* and come off beautifully on film. But the fabric of that relationship had begun to fray. Chris was receiving more attention, and more money, which would sow seeds of discontent in any partnership. And then there was the thing with the girl.

LORNE MICHAELS:

Black Sheep was an act of desperation by Paramount. Sherry Lansing felt that they missed it on *Tommy Boy*. They didn't know what they had; they hadn't marketed it well. Then after its release it got reevaluated. If nothing else, Sherry Lansing's son Jack said *Tommy Boy* was his favorite movie ever. Suddenly they wanted another. I kept saying, "We don't have one."

ERIC NEWMAN:

When Lorne was making *Wayne's World 2*, Mike Myers had written a script that the studio, for legal reasons, couldn't proceed with. So Mike, as is his style, dug his heels in and said he wasn't doing it. The reaction from Paramount was severe. They threatened litigation, and Mike found himself with no choice but to make the movie. That's probably why *Wayne's World 2* turned out the way that it did.

When there was a question about Chris doing *Black Sheep*, all the same people were involved, and it got really ugly again. Paramount was making threats. Chris's people were really angry, and they should have been.

DOUG ROBINSON, *agent:*

Our interpretation of the contract was that Chris owed Paramount one of his next two movies. Their interpretation was quite different, and they were really firm about Chris not doing *Cable Guy*, or *Kingpin*, which was another possible project we had lined up for him. Chris was being considered to play the Amish kid, the part eventually taken by Randy Quaid. We

really wanted him to do that. But Paramount was putting a lot of pressure on Chris, and he ultimately didn't want to fight it.

FRED WOLF:

I got a call from an executive at Paramount saying that I had to deliver a finished script by midnight on Sunday, the last day Chris was contractually allowed to get out of the movie. If I didn't have a finished script—any finished script—they were going to sue me. I sat down and wrote forty-five pages that weekend. Eric Newman met me at Paramount at around eleven forty-five. We made copies and distributed them to the people at Paramount. They had their script, and they forced Chris to do it.

DAVID SPADE:

Now, we're getting close to summer, and that's the only time *SNL* cast members can shoot movies. I ended up going back in the fall, and Chris didn't know at that point if he was going back or not.

But that summer Chris was also offered $3 million to do *Cable Guy*, and the Paramount deal was for way, way less, probably under a million. The thing was, I didn't owe Paramount anything. I didn't have a two-picture deal. I could say no.

So Chris comes to me at Au Bon Pain under 30 Rock on the way in to work. He sits me down and says, "Listen, I know they want you to do *Black Sheep*, and I owe them a movie and you don't. So when you read the script, if you don't like it then I'm free to go do *Cable Guy*."

"Right," I say.

"But if you say you want to do it, I have to do it, too."

"Okay."

"I read it. I wasn't crazy about it. You read it, and you decide."

And so I'm in a tough spot. If I say yes, Fred Wolf gets paid and gets a movie made, and so do I. If I say no, Fred and I don't have work, but Chris gets to go and do the other one.

I say, "Look, I'll read it, and I'll decide based on no reason other than whether or not I like it. And if I like it, I have to say yes."

"Fair enough."

I read it, and I thought it was actually pretty good. Coming off *Tommy Boy*, I thought Chris, Fred, and I could pull it off.

ERIC NEWMAN:

And so *Black Sheep* was concocted to preempt *Cable Guy*, but, unfortunately, at the same time, the *Cable Guy* deal was falling apart on its own due to the Jim Carrey thing.

BERNIE BRILLSTEIN, *manager/founder, Brillstein-Grey Entertainment:*

It was the worst. Endeavor sent me the script for *The Cable Guy*. You can't even imagine how different that script was from what got made. It was a simple, fun story. Gurvitz and I took it over to Columbia, to Mark Canton. They bought it with Chris attached. We were then going through all the preproduction, and Brad Grey and I got a call from Canton. "Please be over here at six o'clock. It's very important that we see you."

Now, when someone calls you for an important six o'clock meeting, it's never good news, and it's never to give you money, ever. We went over to Columbia and Canton said, "Somehow, the *Cable Guy* script made it to Jim Carrey."

"Somehow?" I said. "Did it fly over there or did you send it by cab?"

That started the meeting out on a bit of a hostile front.

"Jim Carrey wants to do it," he said, "and we want to make it our summer tent-pole movie."

Brad and I had brought Columbia this script, and, without our knowing, they had brought on Ben Stiller, Judd Apatow, and Jim Carrey, who wanted to turn it into a dark, black comedy. They were going to pay Carrey $20 million, and it was the first time anyone had broken the $20 million ceiling. I was very blunt. I said, "You just lost twenty million."

So now Canton says, "Here's what we'll do. We'll pay off Chris. You two can stay with the picture as producers, and we'll pay you."

It was a tough decision. It was a lot of money. It was our script, so I didn't feel bad about taking the money, honest to God. And they weren't going to budge off of Jim Carrey, so as long as Chris got paid, that was the best we could hope for, for him.

PENELOPE SPHEERIS, *director:*

One Sunday afternoon, I got a call from John Goldwyn at Paramount. He put Sherry Lansing on the phone, and they said, "Chris Farley wants to do this movie called *The Cable Guy*, and if we don't exercise our option on him by tomorrow morning, we lose our rights to hold him to another movie." They asked me if I would direct it.

I said, "Well, where's the script?"

"We don't have one," they said, "but we have a great idea."

They pitched me the idea. It didn't seem something I really wanted to do. Then they told me what they were going to pay me—and it was obscene. It was about two and a half or three million dollars. They needed to get the picture done that badly. So, I hate to sound crass, but I did it for the money. Plus, after doing *Wayne's World* I really did love Chris and really did want to work with him. Between those two things, I went for it.

DAVID SPADE:

And that's when the trouble started. I believe up until that point we would have had another *Tommy Boy* on our hands. But Penelope got paid more than all of us put together, because she'd done *Wayne's World*. So all the power went to her. The problem is, you have to give a lot of credit on *Wayne's World* to Mike and Dana. I'm sure she did something right, but as far as the funny is concerned, that's Mike and Dana, in my opinion.

So Penelope says, "I know how to make you guys funny." Which is the first red flag. Chris, Fred, and I knew what we needed to do. We just needed someone to shoot us, but she ripped forty pages out of the script

and said, "I'm going to work on this with my friend." All our complaints fell on deaf ears, and Fred got fired.

PENELOPE SPHEERIS:

There was one point in a meeting when we were discussing the script with the studio people. Fred came up with some stupid-ass idea, and I said, "I'm not going to do that."

Then everybody looked at Lorne, and Lorne said, "Well, Fred *is* the writer on the show." Parentheses: The writer is king on *Saturday Night Live*.

So I said, "Okay, you guys can take your two and a half million and shove it up your ass."

And I walked out. I couldn't believe I'd done it, but I was walking across the parking lot and I heard the click-click-click of Karen Rosenfeld's high heels, and she came up to me going, "Penelope, don't leave. Please, please."

They didn't care about me, mind you. They just didn't want to lose the director, any director, for fear of derailing the project and losing Chris.

ERIC NEWMAN:

A movie's like a train, a five-hundred-ton train, and once it leaves the station there's not a whole lot you can do about it. If you're Jerry Bruckheimer it's pretty easy to stand on principle and say, "This movie's not ready." (Not that he ever has.) But if you're not at that level, you work with what you have. Everyone thinks, okay, it'll all come out in the wash. The process will right the ship. And it never does. And so, despite the best efforts of Chris, David, Lorne, and Fred, *Black Sheep* is an entirely forgettable movie. It's a terrible movie. It's a really bad movie.

TED DONDANVILLE, *friend:*

Johnny Farley had moved to Chicago and started performing at Improv-Olympic, and he and I were drinking buddies. He told me Chris was looking for an assistant. So I just called Chris up and asked him for the job, and I got it. I wasn't hired for my secretarial efficiency but because that concrete wall seals off after you become famous. You can only trust the people on one side of it, and I was on that side. And of course I'd taken care of him when he was the poor, starving actor at Second City.

We flew out to Los Angeles on the Fourth of July of 1995, and I was his personal assistant from that day on. When we got there we moved into the Park Hyatt hotel. He had a nice suite, with a big living room area. He lived there for all of *Black Sheep*, and then for all the time we weren't on location for *Beverly Hills Ninja* and *Edwards & Hunt*. That was his home in L.A. Filming started about a week after we got to town.

Chris, because of his clout, got them to hire writers to work on punching up the script three or four nights a week. The guys who wrote *Edwards & Hunt*, which became *Almost Heroes*, they came in. Those guys would come over to Chris's hotel suite, we'd order food, and they'd sit around for hours going over the upcoming scenes, trying to make it better. Those sessions were a great time, a lot of fun to watch, these intellectual writers setting Chris up and giving him stuff to work on. He'd act it out a little and tweak it with them.

The process was more enjoyable than the actual product. Not a lot of that material got worked in, ultimately, because it was all last-second stuff. So Chris knew it was a piece of crap, but he was going to go down swinging.

TIM MATHESON, *costar:*

Chris was very positive. Always prepared. You could tell that all the principal performers were just doing it by rote, to fill an obligation, except for Chris. I didn't much believe in the movie. I just figured it had to be at least half as good as *Tommy Boy*, and that would be okay. But they just kept

adding and changing crap all the time, and never to make it better. It just got dumber.

FRED WOLF:

If you're going to do something in a slapdash manner, you need the captain of the ship to make sure it comes off right. I think we were all missing that.

On *Tommy Boy*, Pete Segal would call me in my room and say, "We're out here with Chris and Rob Lowe. He's washing him off after the cow-tipping scene. Do you have anything?"

And I knew that any time you had Chris dancing you had comedy gold. So I'd say, "Why not have him singing 'Maniac' from *Flashdance*?"

Then they were able to knock it out on the spot. On *Black Sheep*, the director wasn't speaking to me, and I was banned from the set. Penelope Spheeris fired me a total of three times. Chris rehired me twice, and Lorne Michaels a third time. I missed Pete Segal.

TED DONDANVILLE:

Chris liked the way that Penelope was very open about hearing his ideas. On the other hand, he didn't really trust her comedy chops. He had a fear of her not knowing what was funny and what wasn't, and he was worried about the lack of strong direction. Personally, though, they got along great. He liked the freedom she gave him.

PENELOPE SPHEERIS:

It was actually Chris's idea to get Tim Matheson and Bruce McGill for their parts, specifically because of *Animal House*. For me, I trusted him on it, and then I met the guys and they were just great. My first impulse was, "Who's going to believe Chris Farley and Tim Matheson as brothers?" But they really felt like brothers.

TIM MATHESON:

I went in for the audition, and to my surprise Chris was there. He was the star, and I was just coming in for this supporting part, but he was so gracious and so deferential and so flattering that you honestly would have thought it was *his* audition. But I got the feeling that he was responsible for my being there. I was very grateful for that, and I wanted to deliver.

He wanted to hear anything and everything I could remember about *Animal House*. He wanted to get my take on the whole experience and what it was like working with Belushi. I liked to make him laugh, so I told him as many stories as I could.

Chris had an innocence about him, a golly-shucks-gee kind of thing. John felt like an older brother. Chris felt like a younger brother. John had big designs, was always in charge. He grabbed that ball and ran with it. Chris was always "Golly-gee, what do you think?" Belushi was also a sweet guy, but Belushi was very aware of who he was, the impact he had on people, and the clout that it gave him, both in the industry and just over average people. John was very savvy. You got the feeling that Chris wasn't. Or, if he was, he chose to ignore it. He had a very salt-of-the-earth quality about him. When he introduced me to his brother, it was all about how great his brother was.

BRUCE McGILL, *costar:*

Toward the end of the movie, we all played in a golf tournament, so we spent about six hours out there together. You usually try and keep humiliating experiences down to ten, fifteen minutes, right? But a game of golf is just hours and hours of ego-bruising degradation, which breaks people down and opens them up. If you want to know something about a guy, go play golf with him.

And while we were out there we had a very protracted, involved three-way conversation about what had happened to John, and how he should avoid that for himself. It was fascinating to see how interested he was in Belushi and Belushi's demise—and how adamant he was that he would

not go down that road. Cut to two years later and he went exactly the same way. If there was ever a moth to a flame, whether it was conscious, unconscious, I don't know.

LORRI BAGLEY:

When Chris left *SNL*, he told me that the only goal he'd ever had in life was to be on that show, that his father had loved John Belushi on that show, and if he could make it there, he'd make his father happy. That's where the Belushi thing came from—his father.

BRUCE McGILL:

If Belushi made Chris's father laugh, well, there you go. It's positively Greek.

PENELOPE SPHEERIS:

My problem with *Black Sheep* was that then and to this day I find Chris Farley absolutely, brilliantly, hilariously funny. I don't think I've ever even smiled at anything David Spade's ever done. Chris was lovable and positive, and David was so bitter and negative. You take your pick.

I still have a recording of a message David left on my answering machine. He said, "You've spent this whole movie trying to cut my comedy balls off."

DAVID SPADE:

The main problem was that Penelope separated us. She had Chris go off and do one thing and me go off and do another. We kept saying, "Look, our characters just need to be together. We need to fight and bicker and do all that shit."

And Chris wasn't helping much, because he thought he should be

doing more dramatic stuff, that the movie should be more about his character and Tim Matheson's character and less about me. He even hinted that "would I mind" if I got paid not to be in it so they could make it more of a dramedy. And I don't think he meant it to be offensive to me. He just wanted to act and didn't want to keep doing fatty falls down. Personally, I thought it was too early; we needed more experience before we tried to do those things.

So they added a few scenes for Chris and Tim to be a little more serious, and they had another writer come in to work on the ending. And I was kind of on my own. I didn't have anyone to play off of. I didn't have Chris, and my humor is funny when I have someone to play off of.

ERIC NEWMAN:

Actors need rules, and those rules need to come from the director. Penelope clearly didn't get David, and she really allowed him to meander. Chris Farley alone is the comedy team of Costello and Costello. You needed the sharp-tongued straight man. You can pretend that you're just making a Chris Farley movie, but you're not. It's a Chris Farley/David Spade movie.

TIM MATHESON:

I sensed that there was something wrong with Chris and David, but I thought it was David not wanting to be the second banana.

PENELOPE SPHEERIS:

You could feel the tension between them, believe me.

LORRI BAGLEY:

I was staying with Chris in his hotel room, and at the end of the shoot he would come back and see me and vent out all of his tension. "You're the reason me and my best friend aren't talking." That sort of thing. It was hard.

TED DONDANVILLE:

It was about the girl, basically, and some underlying jealousy, too, which Spade actually handled very well. Spade was pissed off about how things went down with Kit Kat. I think it was one of those things where Spade thought she was his girl, but they weren't really dating to begin with.

DAVID SPADE:

Dating, not dating, whatever. I don't know what you want to call it. We were certainly hanging out a lot. I wasn't her boyfriend, but we were very close. And I didn't find out from them. It was an accident. For some reason, I wasn't supposed to know. So if she and I weren't dating, then why was I being kept in the dark? And whatever. The problems with Lorri were that I felt somewhat betrayed on both sides. I felt like, here's my friend. I always made sure he got to hang out with us because he said he had no one else to be with, and to have that bite me in the ass later didn't sit right.

TED DONDANVILLE:

I talked to Lorri about it a lot, and according to her, Chris was guilt-ridden over it. She says that she and David were never more than friends, but who knows how women revisit stuff.

LORRI BAGLEY:

Chris and I were just emotional, crazy people. Our relationship was always rocky. Up and down, like being on a ship at sea. David always used to say, "You two are going to kill each other." The thing with Chris, the thing we had problems with, was intimacy. Any time you got close to him, if he let down his guard and really let you in, then he'd push you back out. As close as you got the night before, you'd be pushed that much further away the next morning.

TED DONDANVILLE:

Whenever Lorri would visit L.A., there was a standard pattern to it. They'd start talking on the phone a lot, then she'd come out and he'd send me away. For a couple of days it'd be full-time, lovey-dovey, baby-talk heaven. Then for a couple of days they'd begin to resemble a calmer, everyday, normal couple. I'd be invited back in—the third wheel added to their little bicycle—and we'd all hang out together. Then it'd start to disintegrate into some crazy-ass fight. I'd get a call from Chris in his hotel suite: "Come up. Come through the bedroom." I'd go up and she'd be on the other side of his bedroom door, out in the suite, yelling and screaming, and Chris would be like, "Get her her own fucking room tonight and fly her back tomorrow."

She'd fly back to New York, they'd ignore each other for a month or so, then the phone calls would start and the whole thing would crank up again.

LORRI BAGLEY:

In every part of his life Chris had a role to play. His relationships with people were always predicated on "What do *you* want?" And then he would be that for them. I actually tried to break all those different roles down. We'd fight about it, but then in the end he'd feel better that he

didn't have to play some part. I always felt that he could only really be himself with me.

People would never believe that we were together. They didn't understand it, or questioned the reasons behind it. That was until they saw us together. They'd look back and forth between the two of us and say, "God, they're like the same person." One time he was going to do an interview. We were with his publicist in the car. She was like, "Chris, you're so calm." Then she looked over at me.

"Kitten makes me calm," he said. He always said to me, "You're the only girl I feel comfortable with. I've always been nervous and anxious around women, but not with you."

TED DONDANVILLE:

I wouldn't say Lorri *wasn't* attracted to the fame and success, but she and Chris were genuinely close. She will tell you that they had this famous romance for the ages. Fact of the matter is, Chris had other girlfriends here and there. But I will give Lorri credit for being the most important woman in his life. That is certainly true. Unfortunately, of all the girls I saw Chris go out with, I didn't think she was the healthiest or most stable.

DAVID SPADE:

Those two together, with Chris in a free fall, it was like nitroglycerine. I don't know if Chris was in love with her. I know he spent tons of time with her, and she was okay with the other women, or whatever his famous life brought him, because they were "soul mates." I was like, "Shit, is that how it works? I need me a soul mate. That's awesome."

LORRI BAGLEY:

Chris attracted control freaks. He made them feel wanted. And David is a control freak. If he's not in control of a situation, he'll just get out of it. It got ugly, and it was horrible for the two of them.

ERIC NEWMAN:

Every partnership has its problems. Costarring in a movie is hard. I think Chris and David really cared about each other, and they saw that they were good together. But I don't think the quality of that movie was in any way affected by the deterioration of their relationship. While their problems may have impaired the process a little bit, that movie, in its DNA, was a turd.

FRED WOLF:

I grew up in New York City but then later moved to Pennsylvania, to the town where they filmed *Deer Hunter*, so you know how dreary that was. I would take the bus into Pittsburgh to watch the Marx Brothers movies. My dream had always been to work with a comedy team, and here I was. I thought *Black Sheep* could have been a repeat of *Tommy Boy*, but the missing ingredient remained missing. The movie isn't atrocious. It opened bigger than *Tommy Boy*. Both of them were number one in the country, but the drop-off was a lot quicker because, ultimately, it wasn't the same kind of movie.

* * *

Black Sheep wrapped in late summer, and Chris moved back to Chicago, taking up residence at his new apartment in the John Hancock Center, which he had bought after leaving *Saturday Night Live*. Also located in the Hancock Center was the radio studio of Erich "Mancow" Muller, a popular Chicago morning deejay. Chris frequently popped in at the show on his way in or out of the building. Along with regular drop-ins at Second City and ImprovOlympic, the show gave Chris a stage whenever he needed one.

That fall, a small group of people coalesced around Chris, forming a sometime entourage and de facto inner circle. Ted Dondanville stayed on as his personal assistant. Kevin and Johnny Farley had both moved to Chicago to take classes and perform at Second City. The brothers had

never been very long apart, but now they found themselves all living in close proximity for the first time since high school. Chris also met a young woman, Jillian Seely. Seely herself had quit drinking several years before, and she was a great help to Chris. They attended recovery meetings together and became fast friends.

And naturally, any decent Chris Farley entourage needed to include a Roman Catholic priest. Sadly, Chris's longtime confidant Father Matt Foley had left Chicago to do four years of missionary work in the small town of Quechultenango, Mexico. The two friends spoke by telephone often, but Chris needed spiritual guidance closer to home. In the months between filming *Tommy Boy* and *Black Sheep*, he had gone to Bellarmine, a Jesuit retreat house in Barrington, Illinois, just outside of Chicago proper. There he met Father Tom Gannon, who would meet with him and talk to him on the phone regularly over the next two years.

Meanwhile, preproduction work was already under way on Chris's next film, *Beverly Hills Ninja.* He worked with the writers and producers on finalizing the script and took daily martial arts lessons from a teacher named Master Guo. For month after happy month, everything seemed fine.

JOHN FARLEY:

When I graduated from college, the family was driving back to Wisconsin from Colorado. I was the young sapling, had no clue what direction to take in life. I had Tommy on one side of me and Chris on the other. Tommy was saying, "Go into business." And Chris was going, "Go into comedy." They were kidding around, tugging back and forth on me.

I went back to Red Arrow Camp to be a counselor, basically piloted a ski boat all summer. Then I tried to get a job driving a Frito-Lay truck. They turned me down. Maybe it was my DUI. Maybe I was overqualified? But I doubt that. So then my mother was buying a new car, and I went with her. I thought, I like cars. Maybe I'll sell cars. I asked about it. One of the salesmen took me aside and said, very seriously, "Son, this isn't a job. This is a career. You're makin' a *career* move here."

"Wow. Thanks."

And I never went back. I figured, comedy, what the hell? I went to Chicago and did exactly what Chris did, started working at the Mercantile Exchange and started taking improv classes at night. Kevin got out of the asphalt business a year or two later when he saw how easily I'd gotten out of it.

KEVIN FARLEY:

I had worked with Dad for six years. In September of 1994, I packed my bags and moved down to Chicago. I lived on Johnny's couch, got a running job at the Chicago Board of Trade, took classes at Second City, and worked as a host there, seating people and doing dishes.

Second City has a business theater, which is upstairs and is pretty lucrative; you can hire players from Second City to perform at your corporate events and write material for you. Eventually, they thought I had a little talent, and they sent me out on these corporate gigs. I got to make a living doing that.

JOHN FARLEY:

I did the corporate thing a bit, too, because it paid well, but mostly I was in the touring company.

KEVIN FARLEY:

Everyone in our family is funny. Mom's hysterical. When you have a large family you want to have your own identity, so we all developed different senses of humor. We're very similar in our mannerisms, but all unique. Johnny is out there. His mind works in a really dark but funny way. I'm a little more goofy and silly. Chris was just outlandish, in your face and raucous. Tom is very cerebral, and dry. Only he never got up onstage with it.

JOHN FARLEY:

If you had to break down the Farley brothers, I'm Chevy Chase, Kevin is Dan Aykroyd, and Chris is John Belushi. And Tommy is Garrett Morris.

TED DONDANVILLE:

I'll be honest: Johnny and Kevin are often funnier than Chris in real life, in more normal ways anyway. Kevin Farley is the funniest guy in the world at a cocktail party. He tells stories, is very engaging. Tom and Johnny, too. The difference is when you put a spotlight on someone, there's a very different kind of funny you need to deliver, and that's where Chris was like Michael Jordan: He would always make the shot. But at the same time, a lot of people who'd meet Chris socially just didn't get him.

JILLIAN SEELY, *friend:*

I met Chris buying a cup of coffee. I really didn't know who he was. I remember he had an Elmer Fudd hat on, and he was wearing those electrician glasses he had. He looked like he was mentally retarded. He'd just finished *Black Sheep* and had moved into his apartment at the Hancock. I worked at a hair salon in the Bloomingdale's building at 900 North Michigan Avenue, and he was there in the building with Johnny. I was looking at Chris and he was looking at me, and we both started smiling and laughing. He asked me if I would marry him, and then he introduced himself. "Hi, I'm Chris Farley."

He asked me if I'd join him that night at a restaurant down the street. It was for John's birthday, I think. I showed up, brought some friends, and we all hung out and had a great time. We ended up going to a restaurant that was open really late and just laughing and talking all night. The next day he called me at work at around nine in the morning and said, "Hey, I noticed last night that you don't drink."

"Yeah, I quit a long time ago," I said.

"Me, too."

He told me a little bit about his problems, and then he asked if I would go to a meeting with him later that night. I said sure. We went to the meeting and then went out to dinner, and we just clicked. From that day on we just started hanging out all the time. We laughed our asses off together.

The thing that was great about being with Chris was that he started all of his conversations with "How was *your* day? What did *you* do?" Nobody does that anymore. That's why Chris was so different from most people. He was not selfish at all when it came to being a friend. We would stay up until three, four in the morning, opening ourselves up to each other, even when we were complete strangers to each other. To this day I don't know why. I've had friends who made me laugh and friends I could have really serious talks with, but I'd never had all of that in the same person like I did with Chris.

KEVIN FARLEY:

We always thought Jillian was super nice. They did hit it off right away.

JOHN FARLEY:

At the time we were busy setting up his apartment at the Hancock building. It was crazy, because the Hancock building is literally a retirement community. That and the studio for Jerry Springer. Chris was the only young person in the building. "Dad says it's the best place in town," Chris said. And maybe it was, back in the sixties, but the people who were hip when they moved in were the only ones still there.

KEVIN FARLEY:

Whenever Chris was in Chicago we would meet and go out to dinner at the Cheesecake Factory, or the Chop House, or Gibson's. Things were clicking with his career. It was a really good time. Nobody was worried about him.

TED DONDANVILLE:

At that point, sobriety was just part of his routine. It wasn't a chore or a burden. It was a balanced part of his life. We weren't hanging out at raging keggers or anything, but we'd go out to things where there was liquor served. People would buy him shots and he'd accept them graciously. Then he'd hand them to me and say, "Here, Ted. You do it."

Even when he was enraged or in a foul mood, he'd just go to a meeting, get himself together, and come back calmer. In the Second City days, drunk or sober, he was always a comedian without a stage, always fucking around. Now he was very much in control of himself. He didn't need to prove something to somebody all the time. He could turn on the comedian when he needed to be there.

JOHN FARLEY:

Chris and Ted, honest to God, were like Felix and Oscar. They were the Odd Couple. Just to watch them interact was hysterical. Chris lived to give him hell, and Ted was like, "Whatever."

Chris would hide things and then demand them from Ted, just for fun. They'd walk out the front door and Chris would go, "Where's the little, you know, my recording thing I need?"

"I didn't see it," Ted would say

"You didn't see it? Well, let's go back and look for it then!"

Then Chris would go back in and wait while Ted looked for the thing and say, "Look! Here it is under the couch!"

"Uh, okay."

"You idiot! Let's go!"

It was fun for Chris to beat up on him like you would a little brother, but Ted could ride out all of Chris's mood swings without even a blip in his pulse rate. He just didn't care. If you put Chris in real terms, he was a company making millions and millions of dollars, and Teddy was in charge. It was like, Holy Lord, this ship is headed for the rocks and nobody's at the wheel.

KEVIN FARLEY:

To be an assistant to a star like that, you've got a lot of people calling you all the time—agents, heads of studios. It's not an easy job. Chris would get frustrated with Ted, because oftentimes Ted wasn't as thorough as he needed to be. But they were friends, so that's why there was never really any employer-employee etiquette to be observed.

TED DONDANVILLE:

Chris was such a people pleaser that he'd give everyone what they wanted, always be so deferential. But he had just as much ego and just as much of a temper as anyone. All of that negative energy had to get channeled somewhere, and it got channeled to Kevin, Johnny, and me. He'd never let anyone else see that side of him, and so we'd take the brunt of it. But we also understood it for what it was, blowing off steam. Any outburst was immediately followed by a shower of apologies.

JOHN FARLEY:

Teddy was a good companion, and honest. He comes from more money than Chris or any of us had ever seen, so he didn't give two shits about Chris's money or his fame. He was just doing it for fun. He was probably the most trustworthy guy Chris could have had by his side. And we all had fun together. We'd go to Second City, work out at the gym. Chris was really into his martial arts training for *Beverly Hills Ninja*.

TED DONDANVILLE:

Master Guo had been a karate champion in Communist China and had defected. He didn't speak very good English, and he only weighed about a hundred and ten pounds, but he was an amazing teacher. He and Chris used to do this thing where they'd stand shoulder width apart, clasp one hand, and then push and pull, and the first one to have a foot pulled off

the ground would lose. Chris outweighed his teacher more than two to one, but the guy got Chris off his feet every time, without even trying. He was the real deal. But he was very impressed with Chris for what a fast learner he was. This was the football player in him coming back.

Thinking back on it, the martial arts training was something Chris lacked later on, namely a hobby, something to keep him occupied. It was a noteworthy time in that there was nothing too noteworthy about it. He was sober and happy and having a good time. It was never that way again.

JOHN FARLEY:

None of us saw it coming.

TED DONDANVILLE:

Chris was going to have a Christmas party at the Hancock, but first he had to go to New York to attend a screening of *Black Sheep*. For whatever reason, in the days before he left for New York, he started getting angry. He always had a temper, but this was a little more consistent, and more fierce. It was a gathering storm.

Then a rewrite of *Ninja* came back, and it really sucked. Following right on that, I was filming some of his training to send in to the screenwriters to come up with jokes, but the battery died on the camcorder halfway through the training session. Afterward, we went to look at it and it was all fucked up. Chris went into a rage, yelling and screaming and ranting about this goddamned script. He left Chicago really pissed off.

LORRI BAGLEY:

I picked him up in New York. The car came and got me, and I went to the airport to meet him. We were going to stop by the hotel and then go and have dinner. He got into the limousine, and as we drove off we started

talking about work. And while we were talking it was like a black cloud came over him. I saw the Chris I knew literally disappear, just vanish into this distant world. I said, "What's going on?" But he had checked out.

We went by the hotel and then got back in the car to go and have dinner. Chris was quiet for a moment. Then he turned to me and said, "Kitten, I'm drinking tonight."

ACT III

CHAPTER 12

Raising the White Flag

TED DONDANVILLE:

That first relapse, that was the big one. The rest were just dominoes.

Chris Farley had been sober for three years. At a time when his commitment had never seemed stronger, he gave it all up with one drink on the flight from Chicago to New York. As news of the relapse spread, Chris's friends and family all asked the same question: why? Some felt it was the gathering stress and anxiety over his career. Others felt that Chris had never successfully dealt with every aspect of his compulsive and addictive behaviors, most notably with regard to food. Still others felt that Chris's sobriety had always relied too much on external motivators, like the threat of losing his job at *Saturday Night Live*. Whatever precipitated the relapse, it happened. And it was devastating.

LORRI BAGLEY:

I started crying. "No, you're not," I said. "You're not going to do that, and you're not going to do it with me."

"I've already started. That wasn't water I was drinking in the hotel. It was vodka," he told me.

We went to dinner, and he started drinking martinis. After that, we went to the Rainbow Room. Steve Martin was there. Chris was acting like a madman. I thought I could get him through the night, call Ted the next morning and find his sponsor, and see where to take him.

I got him back to the hotel room. He was drinking and crying. Then he said he was going to this spot in Hell's Kitchen to get drugs. I said, "If you want to go there, I'm going with you, and then you need to take me home because I'm not going to play with that."

We got in the limo, went to this place, and he went inside while I waited outside. After a minute I got nervous and went in after him. He looked at me and said, "Get me out of here."

So he left without doing anything, and I took him back to the hotel. He just kept drinking and crying and talking about the voices in his head. He kept saying, "How do you turn off the voices in your head? They're in my head. How do I get them out?"

I finally fell asleep as the sun was rising. When I woke up I called his name, and he wasn't there. I waited and waited, not knowing what to do. About an hour and a half later, I started getting ready to go, and he came in with sunglasses on. I could tell he hadn't been to sleep, and he was way too calm and mellowed out from what he'd been the night before. He was on something.

I told him, "I thought that I could handle being around you if you started using again. I thought I'd never leave you no matter what, but I can't be around this. It's just too much." And I left and went home.

Enjoying the early days at *SNL*, (from left) Chris, Erin Maroney, the late Kevin Cleary, and Todd Green

COURTESY OF TODD GREEN

Chris's breakout role at *Saturday Night Live*: the much-adored, and much-maligned, Chippendales sketch, with host Patrick Swayze

© 1990 EDIE BASKIN

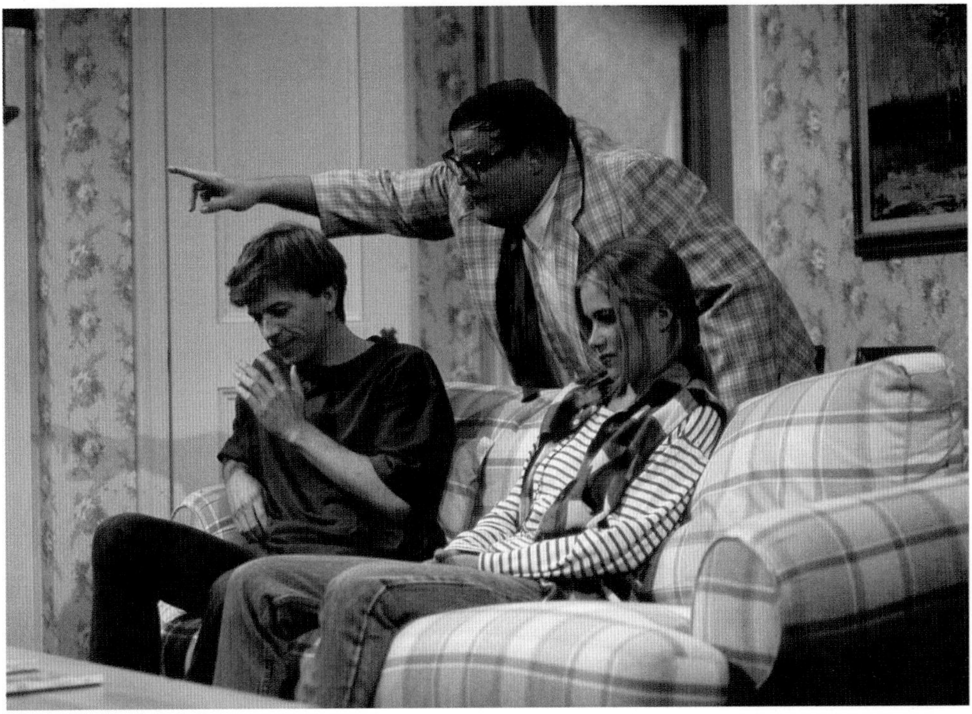

As Matt Foley, the thrice-divorced motivational speaker, berating cast mate David Spade (left) and host Christina Applegate

As the eponymous heroine of "Lunchlady Land," a rousing high school cafeteria anthem by Adam Sandler (right center)

"Awesome!" Basking in the glow of former Beatle Paul McCartney in an episode of "The Chris Farley Show"

Goofing around during rehearsal as the Blind Melon "Bee Girl," with Second City and *SNL* cast mate Tim Meadows

Riding high on the success of the "Super Fans" at Chicago's Soldier Field, with *SNL* writer Robert Smigel

Preparing for the big wedding scene in *Tommy Boy*, (from left) John Farley, Chris, Bo Derek, David Spade, and Kevin Farley

"Interestingly, when Chris was on camera, it was the only time I could get him to look me directly in the eye." —*Tommy Boy* costar Julie Warner

"Holy Schnike!" With Big Tom Callahan (Brian Dennehy) on the set of *Tommy Boy*

Tommy Callahan fumbling his way to Chicago as a phony airline attendant

Being coached by director Peter Segal on the set of *Tommy Boy*

Chris playing with his dinghy on the set of *Tommy Boy*

Escorting his mother (left) to the *Tommy Boy* premiere on the Paramount lot

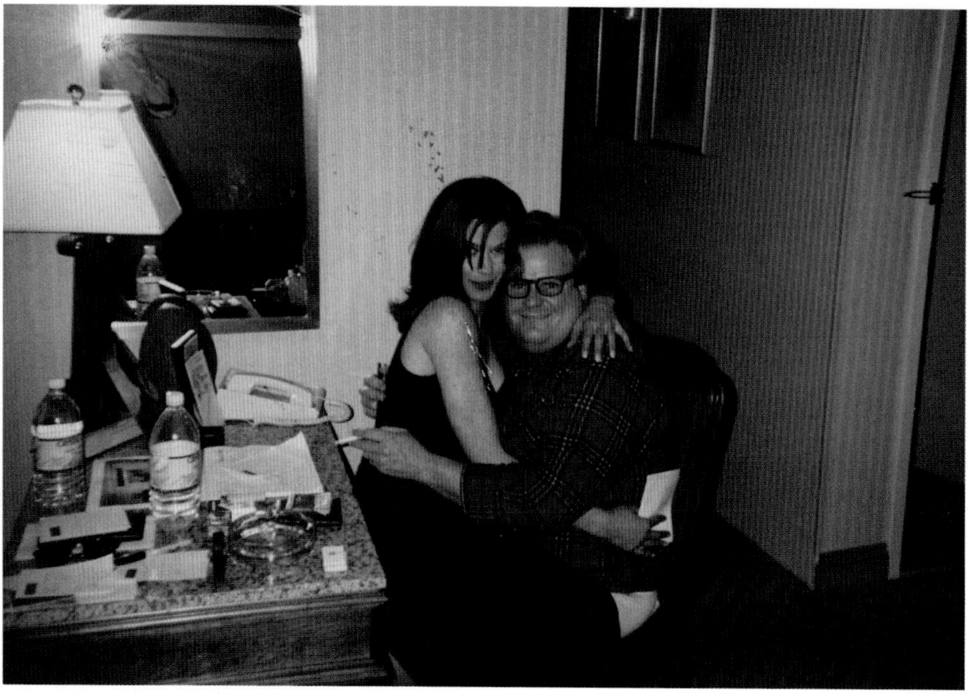

With Lorri "Kit Kat" Bagley
COURTESY OF LORRI BAGLEY

A prop photo used in the filming of *Black Sheep*, with Chris in character as kids' rec center counselor Mike Donnelly

Performing for the cameras—a contact sheet from a *Black Sheep* publicity photo shoot

Rough times on the *Black Sheep* set for Chris and David

© 1996 PARAMOUNT

On the set of *Beverly Hills Ninja*, (from left) Ted Dondanville, John Farley, Kevin Farley, Chris, director Dennis Dugan, and Chris Rock

© 2007 COURTESY OF THE FARLEY FAMILY ARCHIVES

Dear Mom and Dad 12/19/95

I am so sorry to have put you through so much worry. I cant explaine what happened I think the stress of my upcomming movie pluse the problums Im having with Ninja got to me. I Wanted to escape. I now know more than ever Booze is never the answer. I Cant tell you how awful I felt the guilt, the shame, the hard work to get those three yrs gone with a sip of that Booze. This is a wonderful place and Im getting stronger than ever. one good thing I know I will never let anything take away my sobriety again. not work, not relationships nothing. I will have a stronger sobriety because of this, because I know how wonderful it is. And so Im here getting stronger, Im only sorry Ill have to miss you at Christmas Ill miss you so much. please think of me because you can bet Ill be thinking of you. I love you Both with all my Hart.

Chris

A letter from Chris to his parents, written from rehab on the heels of his first relapse in three years

A moment of reflection while filming *Edwards & Hunt* (*Almost Heroes*)
© 2007 COURTESY OF THE FARLEY FAMILY ARCHIVES

With friend and Chicago radio personality Erich "Mancow" Muller
COURTESY OF PHILBIN FLASH

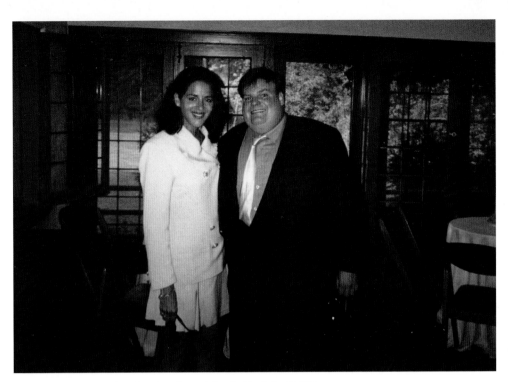

With Jillian Seely at Tim Meadows's wedding in the fall of 1997

COURTESY OF JILLIAN SEELY

With Adam Sandler, Chris Rock, and *SNL* writer Tim Herlihy, also at Meadows's wedding

COURTESY OF JILLIAN SEELY

LIKE HIS GOOFBALL CHARACTERS, HE IS SWEET, LIKABLE AND OUT OF CONTROL. HE SAYS HE WANTS TO CHANGE – HE DOESN'T WANT TO END UP LIKE HIS IDOL JOHN BELUSHI. IS HE KIDDING HIMSELF? Photograph by Dan Winters

Chris Farley: ON THE EDGE OF DISASTER

IT'S A SHIMMERY CALIFORNIA DAY, QUITE WARM, PERFECT FOR TRUE-LOVE romance or maybe a quick bonk, and Chris Farley, the comic and really big guy, is cruising up La Cienega Boulevard in a red Mustang convertible. The massive expanse of his forehead and the bristling parabola of his Fu Manchu are dewy with sweat. He's got that hair that shoots out, giving him a zonked, horned look. He's wearing a black linen suit, a shirt the color of flaming napalm and, over his eyes, a pair of highly electric, blue-tinted shades.

At a traffic light, a car pulls up next to his. Farley dips his head, peers inside and sees two "hot tamales," as he likes to call them. Dimly, an idea formulates. It boils down to this: If he plays his cards right, who knows what might develop?

"Hi, gals!" he booms, waving. "How ya doing, gals! Excuse me, gals! It's Rex Flexal here! Hey, do you know how to get to the Beverly Center?"

The girls look him over. It's an odd, suspended moment. They seem to recognize the big guy – this is Hollywood, after all, where just about anybody could be somebody – but they aren't sure. They ponder him, his hefty size, his bright happy cheeks, his otherworldly outfit, trying to make up their mind about what he is, what he stands for, where he could possibly have come from and what his prospects might be.

"What's your problem?" one of them says finally.

"You're a f---ing a--hole," the other one says.

Then they are off, at full speed, leaving Farley – who sometimes calls himself Rex Flexal – far behind.

I turn his way, expecting to see at least mild embarrassment. Instead, Farley wears a comical, stunned-carp look, like he can't believe what just happened. "Do you think they're lesbians?" he bellows in my direction. I find myself nodding vigorously. Then Farley scratches his head and lights a cigarette.

"All I wanted to do was strike up a conversation," he says. "What if I was in need of direction? I mean, what if I was really in need of direction?"

The question hangs in the air. It lingers. And then its specifics fade, leaving behind only the general idea of direction and whether Farley – this fellow in the blue-tinted shades and the flaming-napalm shirt, the man of so many unseemly rumors, mostly having to do with drugs, and drinking, and binge eating, and one or two sexual matters, not to mention the occasional spat with pal and frequent moviemaking partner David Spade – could perhaps use a tad more of it, that thing, direction.

AT LEAST PROFESSIONALLY, FARLEY NOW FINDS HIMSELF IN THE MOOD for change. The first time we meet, in Room 905 of the ultraswank Beverly Hills Four Seasons hotel, the first thing Farley says is, "David Mamet just wrote a script for me on the life and times of Roscoe Arbuckle. You know the man, the 1920s comic, Fatty Arbuckle? It's a tragic story. He was a huge star, bigger than Chaplin, but he was brought to his knees through this incident with this girl. She died. He was acquitted. It's a complicated story. It will be a departure from what I've previously done."

The details of the Fatty plot line aren't entirely clear to me, though I have heard the name before. But I do get Farley's point. In the trio of movies that made him a star – *Tommy Boy* and *Black Sheep*, both co-starring David Spade, and *Beverly Hills Ninja* – as well as during

Chris's problems go public four and a half months before his death

Trying to unwind during one of too many trips to rehab

Fumbling his way through the opening sketch of his catastrophic *Saturday Night Live* hosting appearance, with Tim Meadows (left) and Lorne Michaels

Tom Arnold (right) arriving for Chris's funeral at Our Lady Queen of Peace Church in Madison, Wisconsin

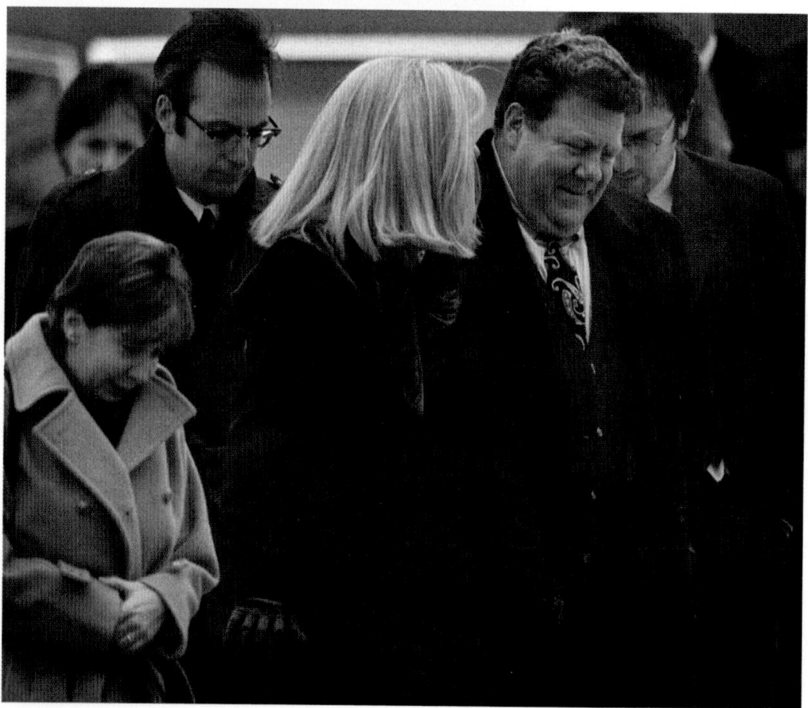

Holly Wortell, Bob Odenkirk, Bonnie Hunt (turned away), George Wendt, and Robert Smigel outside Our Lady Queen of Peace

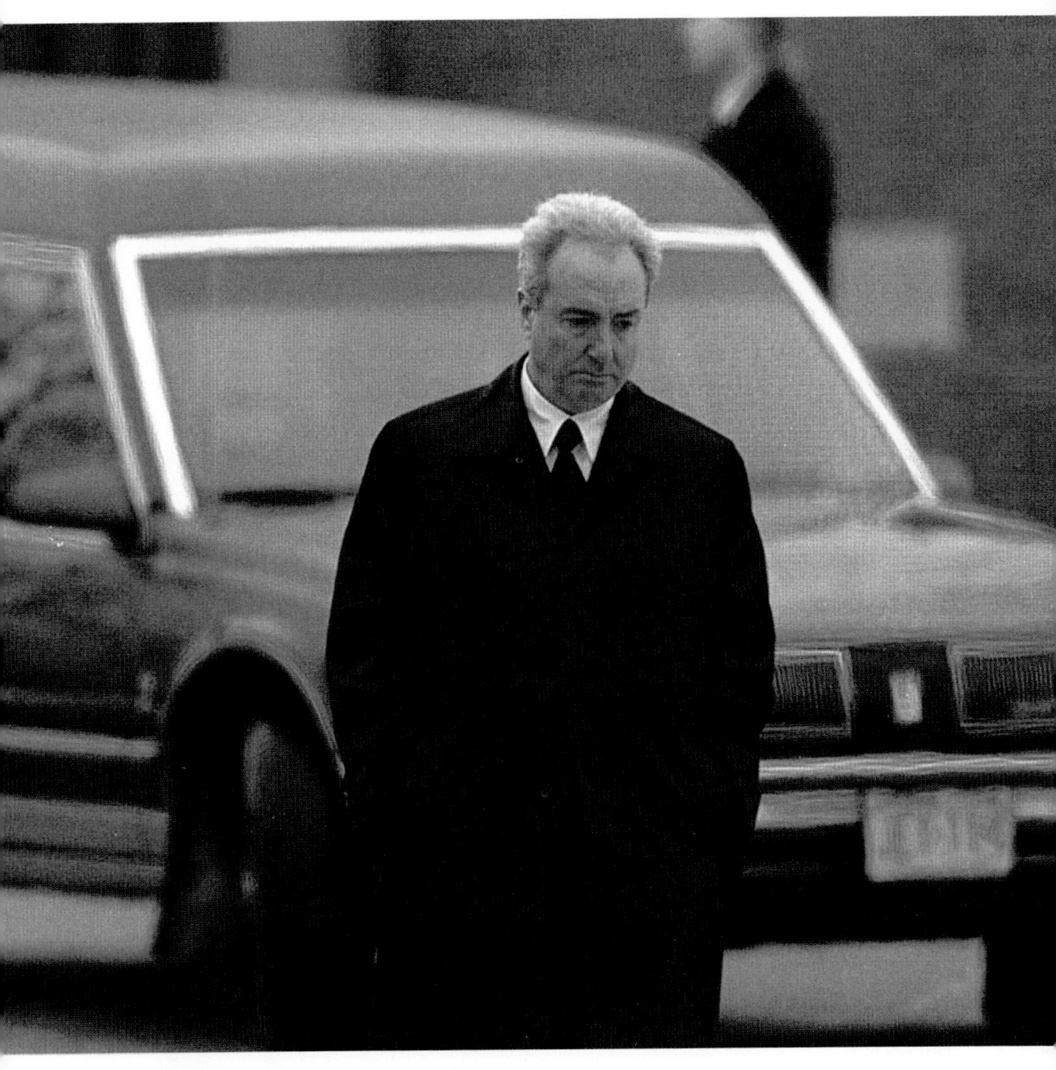

Lorne Michaels leaving the funeral service

TOM FARLEY:

Chris called me and said, "Tommy, you want to go and see a sneak preview of *Black Sheep* in New Jersey?" I said sure. He was staying at the Four Seasons, so I stopped up there. He looked fine to me. The limo was going to pick us up any second, and Chris was getting into the minibar, filling his pockets with the little bottles. I said, "Chris, what are you doing?"

"Oh, I'm just getting a couple of these for the limo driver," he said. "They like that."

And after three years of sobriety, I actually let myself think that was okay. I just didn't for the life of me think he could be lying. We got in the limo and drove out to Jersey.

We were up in the screening room, waiting for everyone to filter into the theater. At one point he said, "I gotta go to the bathroom." I did, too, so I went with him. We got to the men's room, and it was this small, janitor's closet kind of thing. I followed him in anyway. He was like, "What are you doing?"

"I gotta go to the bathroom," I said.

We're brothers, for God's sake. I'd been in the bathroom with him hundreds of times. But all of a sudden he's like, "Get out. I can't . . . I gotta go by myself."

I thought that was very strange, but I left him to it. Then we watched the movie, and we were driving somewhere else, and, same thing, "I gotta go to the bathroom." So we pull over and he goes in someplace to use the bathroom. And, of course, what he was doing was drinking. I certainly didn't see him give away any of his little bottles. I didn't put two and two together until the next day when I got a call from Dad, saying, "What the hell happened?"

"What do you mean?"

"Chris trashed his hotel room at the Four Seasons and did three thousand dollars' worth of damage."

And then it was back to rehab.

KEVIN FARLEY:

When I got the call in Chicago that he had relapsed, it was devastating; devastating to me, devastating to him, and devastating to everyone in the family. He took so much pride in his sobriety, more than any of the movies or work he'd ever done, more than any other success in his life. Those three years were his crowning achievement.

JILLIAN SEELY:

Chris and I had been hanging out every day, and then right around Christmas he disappeared off the face of the earth. He called me early in the morning on New Year's Eve. He said, "Hey, it's Chris."

"Hey," I said, "why haven't you called me?" I was pissed.

"I relapsed."

"You're lying," I said. "I don't believe you."

We had talked so much about his sobriety, and I was so confident in him that I really couldn't imagine it.

He said, "I really want to see you. Will you come and meet me at a meeting?"

There was a meeting that morning. I went with him. He told me about the relapse and the screening. He said that he'd fucking hated *Black Sheep*, that it was just *Tommy Boy II*, only worse.

TED DONDANVILLE:

That screening didn't help. He saw one shitty movie that he'd made, and then he really started worrying that, with *Ninja*, he was working on a second one.

* * *

Beverly Hills Ninja was a script that had been around, and around. Several stars had turned it down. Chris himself had passed on it a number of times. But negotiations for the film had taken a dramatic turn the previ-

ous summer, while filming for *Black Sheep* was still under way. Sensing Chris's impending stardom, *Ninja's* producers got very aggressive. They offered him an ungodly salary, and that changed the whole equation. Chris was still reluctant, and his managers were vehemently opposed, regardless of the payday. But Chris's dad counseled him otherwise, essentially saying, "You don't turn down that kind of money." Show business is not the asphalt business, but in this as in all things Chris listened to his father. He signed on to play Haru, an infant boy orphaned in Japan and raised to fulfill a prophecy as the Great White Ninja. Through a series of slapstick setups and wacky misadventures, Haru makes his way to California and solves a crime. Shakespeare it wasn't.

Chris had rationalized his taking the film by saying it would make a good kids' movie. Another big factor in the decision was simply his confidence. The story was one big, long pratfall, and Chris's abilities as a physical comedian had never failed to deliver huge laughs. But Chris's other major asset was his Midwestern Everyman appeal. Dressing him in martial arts garb and giving him hokey, Zen-sounding dialogue was not a good fit, and it flopped onscreen.

The project was not without its bright spots. It did turn out to be a successful children's movie. Chris Rock was struggling professionally at the time, and Farley used the movie to lend his *SNL* friend a helping hand. But all things considered, it was a serious detour.

Following his stint in rehab, Chris flew to Hollywood in January of 1996 and started production on the film. He stayed clean throughout the shoot, determined not to let the relapse derail his three years of hard-fought sobriety. But the change in him was obvious to everyone on the set. His anxiety was rapidly eclipsing his boisterous amiability, and the strength and serenity he'd possessed just a few weeks before had all but vanished.

ROB LOWE, *costar,* Tommy Boy*:*

At that point, Chris could have done almost anything, career-wise, and for him to do a movie where he offered himself up as "the fat guy" I felt

was a recipe for psychic disaster. I don't want to sound overly dramatic by saying that that movie killed him, but the decision to do it was Chris surrendering a creative part of himself. He was raising the white flag to easy Hollywood mediocrity. I know that he hated himself for saying yes.

BRAD JANKEL, *producer:*

I used to work at a company called Motion Picture Corporation of America. We'd had some success with the Farrelly brothers. We did *Dumb & Dumber* and *Kingpin*. We did *Bio-Dome* with Pauly Shore. Then we got ahold of *Beverly Hills Ninja*. If I recall properly, Dana Carvey had been attached to it at one point, and Chris held Dana in such high regard that he was open to looking at the script. Chris really was the first guy that we went to. We offered him a ton of money.

TED DONDANVILLE:

One day while we were filming *Black Sheep*, Chris's agents showed up. Agents always stand out because they're the only ones in L.A. who wear dark suits. They were really happy, really up. They came rushing in and met Chris in the trailer. *Beverly Hills Ninja* was a movie that he had rejected a number of times. "No, thanks. Pass." But this time the script came in with an offer for $6 million, which at the time was like a three hundred percent increase in what Chris was making. All of a sudden it was like, "Eh, maybe the script isn't so bad after all. . . ."

BERNIE BRILLSTEIN:

Marc Gurvitz and I, we told him not to do *Ninja*. It was about a fat guy in tights. Let's face it, who wants to see Chris like that? It was an embarrassing film. But the offer was $6 million, which even then was a lot of money. Chris called up and said, "I have to do it. My dad says I can't turn down that kind of money."

DOUG ROBINSON:

After his father weighed in, there wasn't much of a conversation. I remember sitting with Chris in his trailer at Paramount, telling him that there would be other big paydays down the road. It was one of those situations where you want to advise your client not to do it. But you could see that as soon as you told him the amount of money involved, the ship had sailed.

LORNE MICHAELS:

It was a bad decision in that no one was telling the truth and people had all kinds of different agendas. There are so many rationalizations: "It'll get your price up." "It's important to keep working." "Not every movie's a masterpiece." But Chris was an incredibly sensitive kid. No matter what he did, he always had some kind of hope for pride in his work, and so for Chris to do something that empty just didn't feel right.

LORRI BAGLEY:

It was fear. He took that movie out of fear. Chris's only goal in life was to be on *Saturday Night Live*, and after that was over he would always say, "I never imagined myself being a film actor." And I'd think, oh shit, he's out of his comfort zone. Movies and Hollywood and all that are out of his comfort zone, and things are going to get difficult.

KEVIN FARLEY:

Chris loved football, and he loved being part of a team. *Saturday Night Live*, Second City, those are team sports. Chris was competitive. He wanted to win, but he wanted the whole team to win. Winning doesn't mean anything if you can't share it with someone.

When you're a movie star, all the pressure is on you as an individual. But if you put too much pressure on one player, you're not going to win

ball games. On a film, clearly one guy is making the most money, and they're banking on him, win or lose. His career is on the line. In Hollywood, you're kind of alone.

TED DONDANVILLE:

The crew used to joke that when the *Beverly Hills Ninja* action figures came out, the Haru figure would come with its own Ted figure. Unlike most personal assistants, I was never out running errands and taking care of other things. Chris always wanted me to delegate that stuff to other people so I could stay, literally, right next to him all the time. We'd work all day and he'd be like, "No, c'mon. I'm taking you to dinner."

On that movie he was hung out to dry. That was a Chris Farley movie from start to finish. Except for those few Chris Rock scenes, every scene hinged on him.

BRAD JANKEL:

Farley wanted Chris Rock in the movie. He was very adamant about that. It was almost to the point where he wouldn't do the film if Rock wasn't involved. I think it was a bit of a life raft for him. I distinctly remember that Chris's best days on the set were the days that Rock was there.

Whatever reservations Chris may have had, to me he was just so appreciative, and grateful, and he really felt a personal responsibility that came with taking that much money. I had never experienced that with an actor before. He was so excited that we were hiring him, let along paying him so much, that during shooting he'd often say to me, "God, I hope I'm doing okay. Are you guys happy?"

And that kind of raised the bar for us. After he'd say those things to me, I'd walk away going, "Oh jeez, I hope *I'm* gonna deliver for *him.*"

Chris also helped develop the script. When we went into *Ninja*, we wanted it to be a really broad, adult movie, and, to his credit, Chris really took it more in the direction of being a movie for kids and families. That's what he wanted. And thank God he did. It wasn't a huge success, but

kids loved it. They really turned out, and that's where the movie made its money.

JASON DAVIS, *costar:*

The producer of *Beverly Hills Ninja* had done a movie that I was in, and I got cut out of that one. But he said I'd be perfect for this. I remember going to audition for it. I had to run into a wall and fall down. That was the audition. When they said I had the part, I got real excited. I got to play Chris Farley as a little kid. At the time, what could have been better? He was one of my heroes.

For my big scene, I was supposed to flip a stick around and accidentally hit this kid on the head. The director said, "You can hit him as hard as you want, because he has padding."

So we do the scene. We're twirling the sticks around, and I smack him on the head, only not on the part where he has padding. The kid goes down for the count. I just remember turning my head over to the right, looking at the director and seeing Chris going *"Oh, fuck"* and laughing his head off. He had this expression like, yup, he's just like me.

TED DONDANVILLE:

The role of the young Chris Farley is supposed to be this shitty little ninja, right? Well, Jason Davis literally was a shitty ninja. He was such a bad athlete and such a spastic little kid that it was funny to see him play this idiot. Chris saw this kid who was just a mess, and that amused him.

KEVIN FARLEY:

Chris liked that kid a lot. He always used to say, "That kid reminds me of me when I was little." Jason was kind of an out-of-control little guy. He'd go up and talk to anybody. Since Jason looked up to Chris so much, Chris kind of took him under his wing whenever he was around.

JASON DAVIS:

For the three days I got to work on the movie, I got to hang out with Chris a lot. Then, about a week later, he called me and said, "Hey, we should go out to lunch."

I thought, wow, that's really awesome. So we went out to lunch, and from then on we just really clicked. We'd hang out. We'd go and do stuff together.

What I loved about Chris was that he was the only person who ever understood me. For a period of time, if I needed to count on anyone, it was Chris. I was like eleven or twelve, and my mom wanted to send me away to fat camp. I was trying every which way to get out of it. I was praying for Chris to kidnap me and hold me for ransom so I wouldn't have to go. But Chris would talk to me about being a kid with a weight problem, and he really helped me a lot. I could call him and say, "My mom's being a bitch. I'm pissed off. I want to run away. I hate it here. What should I do?"

And he'd be like, "Relax. You're obviously going through denial. Your mother loves you. Don't run away. Besides, you'd have to stop at McDonald's every other block just to survive."

NANCY DAVIS, *Jason Davis's mother:*

Jason saw a side of Chris that maybe a lot of other people didn't see. Chris understood what he was going through, and touched Jason in this amazing way. He really did. When Jason was filming *Beverly Hills Ninja* with Chris, he was going through a hard time in his life. His father and I were getting divorced, and Jason had some similar issues with Chris, some weight issues, really deep father issues, and maybe Chris saw that in Jason. It was like he could just sense it, and he really made a point of helping him to deal with it. People are so afraid to talk about what's bothering them, and Chris really got him to open up about it. Jason's always been quirky and different, and Chris gave him the strength to think that was okay.

TED DONDANVILLE:

When I first started working for Chris, he'd been sober for about two and a half years. He was doing really well. He was very comfortable in his sobriety, and very strong. After that first relapse, he was very different. It changed him. He was still sober most of the time, but it was not the same strength and confidence you saw before. He'd be on edge about it. He used to never care if I drank around him. Now, if he found out I'd been out drinking, he'd get angry. So after that first relapse, there was a change. After the second, third, and fourth, they were all kind of the same.

BRAD JANKEL:

It was no secret that he was battling addiction. In preproduction he had it down. He didn't talk about it. He didn't need to talk about it. But then, during filming, he would struggle at times. And by struggle I mean that his actions were a little more overt. He would talk about his need to stay sober. There were a few more mood swings, but his commitment stayed the same. You could tell he was just trying to buckle down and get through it. I appreciated that he was open about it, and we all tried to support him.

TED DONDANVILLE:

The whole time we were shooting *Ninja*, there were no problems whatsoever. He had the first relapse before shooting, and then the second relapse came after—idle time.

BRAD JANKEL:

When he came back for reshoots, he was a different person. I wasn't there that day, but I got calls about all the problems. They couldn't make the film match. He'd put on weight, and he looked horrible. He wasn't the same old Chris. They kind of had to shoot around him different ways.

When we wrapped shooting, I could sense that he was starting to feel that the movie wouldn't turn out how he'd hoped. Chris went into it thinking he was making a good, earnest film for kids, but in the end he wasn't proud of it the way he was about other movies.

BERNIE BRILLSTEIN:

After the first screening of *Ninja*, I took Chris into the bathroom and he just cried on my shoulder. Cried and cried. It was one of the saddest things I've ever been through.

BRAD JANKEL:

I was disappointed, too. It didn't turn out as I had hoped. Thank God it played with the kids, because it missed with the older crowd. Not everything turns out the way you want; some things fail. I mean, you're talking to the guy who produced *Bio-Dome*.

There's a reason they give the Best Picture Oscar to the producer, because it's a collaborative effort and the producer is responsible for bringing all those collaborators together. But Chris felt like the whole thing was *his* failure, and it wasn't.

* * *

Principal photography on *Beverly Hills Ninja* wrapped in March of 1996. Without the incentive to stay clean for work, Chris relapsed a second time and returned to Hazelden in Minnesota. Other than the reshoots on *Ninja*, he spent the spring and early summer in Chicago, passing time with friends and family and working his twelve-step program. It was a frustrating time. Chris would maintain his sobriety for six weeks, two months at a stretch, then head angrily back to the starting line and begin again. It was not easy, but no one could say that Chris was not trying in earnest.

In May, he returned to Marquette to receive the speech department's Distinguished Young Alumnus Award, an honor of which he was exceed-

ingly proud. Then, in June, he joined David Spade at the MTV Movie Awards to accept the Best On-screen Duo award for *Tommy Boy*.

Chris went back to work that August, taking the lead role in *Edwards & Hunt*, a satirical spoof of Lewis and Clark's historic journey to find an overland route to the Pacific Ocean. Before the film's eventual release, the studio's marketing team found cause to change its name, and it has gone down in history as *Almost Heroes*.

Chris had been cast in the role of Bartholomew Hunt, master tracker and woodsman. *Spinal Tap* veteran Christopher Guest signed on to direct. The producers set about assembling a comedy ensemble to flesh out the exploring party, including Eugene Levy and Bokeem Woodbine. Several actors were considered for the lead role opposite Chris, and the part ultimately went to *Friends* star Matthew Perry.

Beverly Hills Ninja had been a bad choice, and *Black Sheep* was no choice at all, but at the time *Edwards & Hunt* was a good bet. On paper, the script recalled the absurdity of classic Mel Brooks, and it was widely thought to be one of the funniest unproduced scripts in Hollywood. As the cast and crew began shooting in the forests and small towns of northern California, everyone believed, somewhat prematurely, they had a bona fide hit in the works.

MARK NUTTER, *screenwriter:*

Tom Wolfe and I were working on a sitcom with our third partner, Boyd Hale. One Sunday, we decided to skip a trip to Anaheim to watch the L.A. Rams—it was raining—and had some beers at a bar in Venice instead. Boyd came up with the original concept of a Lewis and Clark comedy; the ideas flowed freely from there, and we wrote the script in a couple months. Our agent Rob Carlson also represented Steve Oedekerk, who had written for Jim Carrey, and with Steve's help the script made the rounds.

TOM WOLFE, *screenwriter:*

Steve was so hot at the time, because of his stuff with Jim Carrey, that *Edwards & Hunt* became a hot script. Turner Pictures ended up buying it. I don't remember exactly when Farley came on board, but I know he was first. For the other lead, we wanted Hugh Laurie, who's now the star of *House* but at the time was not well known. Turner wanted someone far more famous. I remember one of the executives saying, "Yeah, we could get a bunch of great actors and make a great little movie . . . but then what?"

And of course, from her way of looking at it, she was right. It would be much harder to market a "little" film with "unknown" actors. For a time Hugh Grant was interested, but he bowed out. I remember when Denise Di Novi brought up Matthew Perry for the first time. Being the self-loathing TV writers we were, we weren't that thrilled with a TV actor being the costar. But we were the writers, and this was our first film. The caterer had more power than we did.

DENISE DI NOVI, *producer:*

The script was brilliant. We even hired Christopher Guest to direct it. I've thought so many times about what went wrong. I always like to say I have the distinction of making the only unsuccessful Christopher Guest movie.

You never know with movies. It's kind of like alchemy; the chemistry just didn't work. It had all the right actors. It had a great director, a great script. But I think the tone of the comedy was very odd. It almost read better than it played. It wanted to be a quirky, British, *Black Adder* type of comedy, and here we had Chris Farley and Matthew Perry. It was just a weird combination.

TOM WOLFE:

We thought Chris Guest would bring the right sensibility to the script, but then the notes about changes came from the studio. They saw it from the beginning as a buddy comedy between Edwards and Hunt, and less like the ensemble comedy we saw it as. In one meeting, someone from Turner called *Lethal Weapon* "the greatest buddy film ever made," and I thought, oh shit, this is not a good sign.

TED DONDANVILLE:

They had to trim fat, and so they trimmed a lot of the funny stuff going on around Chris, which, ironically, would have helped Chris's performance. The movie would have been funnier with all the things the peripheral characters had to do, but of course those are the first to go; you have to build a movie around "the star." They cut the ensemble scenes first, Matthew Perry's second, and Chris's never.

BOKEEM WOODBINE, *costar:*

It was one of those things where they should have left well enough alone. Stuff was getting cut, and it was kind of bizarre. We shot the ending one way, and it was a pretty cool ending. Then they totally reshot the ending without my character there. That kind of shit happens sometimes. You don't get your feelings hurt, but the thing is, it affected the film, because it wasn't linear. At one point we're all together, fighting as this band against these conquistador bastards, and then in the next scene they all meet up and regroup—except for me, and it's like, "What happened to the brother?"

TED DONDANVILLE:

The movie came out after Chris died and when Matthew Perry had gone into rehab, and so a lot of the speculation was that drugs were to blame for the movie's failure, but that wasn't true.

The first weekend of filming, we were shooting in Redding, California, and he was drinking then. Denise Di Novi and the other producers, they found out about it, realized it was a major problem, and even threatened to shut down production. Chris thought he'd really blown it, that this time the personal shit had hit the professional fan.

So Chris had a sobriety bodyguard for the rest of the film, somebody who stayed with him the whole time. With that, Chris stayed pretty much okay for the rest of the shoot. As always, it was the work that kept him going.

DENISE DI NOVI:

That was the only incident. We traveled around a lot on this shoot, and whatever city we were in I would try and find a sponsor and a meeting for him to go to. I have to say it made me gain a lot of respect for people in recovery. No matter what teeny-tiny town we were in, within a couple of hours somebody would show up.

He also went to church every Sunday. No matter where we were on the road, he would find a church nearby and go. He got upset one week when we had to shoot on a Sunday. So instead of going to mass, he had a priest come from his church in Santa Monica and say mass for him on the set.

He was great with my kids. I remember I was sitting on the set one day with my son, who was about four and a half. We were playing a game for about a half an hour, and Chris was just watching us, staring. He said, "Boy, your son is so lucky to have you. I bet he's going to grow up to be a really great person."

And there was just such a sadness in that moment. He had a really big

heart, but there was a melancholy there. I never really understood where it came from.

TED DONDANVILLE:

During *Edwards & Hunt*, this girl from the Make-A-Wish Foundation who was dying of AIDS wanted to meet Matthew Perry. She showed up on the set, and everyone doted on her. But Chris noticed she had a brother off to the side, whom everyone was ignoring, and who had probably been ignored for a while, what with his sister taking up all the attention. Chris went and brought the brother into his trailer and goofed around with him and just gave him the greatest day of his life. And it was Chris who instigated it. It wasn't the kid coming over and asking to meet him. The boy ended up having a better Make-A-Wish day than his sister did, and he hadn't even made a wish.

LORRI BAGLEY:

If you even mentioned that, say, your friend's mother had died, tears would start to well up in Chris's eyes. He had such a big heart, and when you have a heart that big, you have to find ways to protect yourself. People who didn't know Chris that well just thought he was the most naïve little kid—and he was—but he knew that he was, and he knew how to use it. He knew exactly how to push your buttons, how to hurt your feelings, how to get you to feel sorry for him. Chris even said to me, "I know how to play people's games." And he did. He was always a hundred percent aware of what was going on and what he was doing.

ERIC NEWMAN:

He'd play dumb with people, but then later he'd recount something from their conversations that let you know he knew exactly what was going on. "The Chris Farley Show" character, the guy who asks the dumb questions,

that was the guy he could become when he wanted to. It was a defense mechanism. It protected him, and it made people feel better. I remember we were sitting once in Toronto with Jim Carrey. Carrey's father had just died, and Chris sat down with him and kind of took on this obsequious role. At the time I didn't really think much of it, but looking back, he did it to put Jim Carrey at ease, and Carrey is, by all accounts, a pretty uncomfortable guy out in the world.

TED DONDANVILLE:

He had a very high psychological intelligence. Anything that had to be learned in a book he probably shied away from, but he understood people very, very well.

Bob Timmons was a professional sobriety guy. He's the guy who would send out bodyguards to sit with Chris on the movie set. These "sober companions" got paid good money to essentially sit around the set and do nothing. They certainly got paid more than I did. Chris resented that, and he went through those guys like nothing. He would break them down, mentally. It was like the movie *Jeremiah Johnson*, where they only send out one guy at a time to try and kill Robert Redford, until he kills so many of them that they have to have respect for him.

On *Edwards & Hunt* there was a thief on the set who'd made off with some production equipment, and the producers were trying to figure out who'd done it. Gary was Chris's new sobriety watcher, and the producers came around after the theft, asking Chris, "How well do you know Gary?"

"Hmm," Chris shrugged, "not *very.*"

Things got a little uncomfortable for Gary after that. People stopped talking to him. Then a couple weeks later he left, never really knowing why. Then they sent out another guy. That's basically how it went. Chris would treat these guys like a new best friend, and then a few weeks later—either passively, aggressively, or passive-aggressively—they'd be gone, not really understanding why it didn't work out.

I don't know that those people helped so much. Ultimately, all those

professional celebrity recovery people have their own issues and their own agendas.

ERIC NEWMAN:

Maybe it's a recovery thing, but Chris attracted loonies.

JILLIAN SEELY:

There are some real wackos out in L.A., a lot of really disgusting people. There was one guy who wanted to be Chris's sponsor, but he charged $125 an hour, and this was back in 1996. I'd never heard of anything like that. You're someone's sponsor, but you charge them by the hour? What the fuck?

KEVIN FARLEY:

There's this whole community of recovery "professionals" who charge you money to watch you. They'll escort you to Hazelden and charge you three grand just for riding on a plane. And that really breaks with tradition. You're not supposed to take money for anything you do to help someone in recovery, and, if you do, it's only because those folks have that money to give.

If you go back to the way it all started, with Bill W. and Dr. Bob, they would go and visit hospitals and they would just find drunks, bring them to meetings and start talking to them. They never charged them money for it. You can make the argument that it costs $13,000 to spend a few weeks at such-and-such facility because they've got staff costs and real estate to keep up. But the whole idea of recovery is that it's free. I don't think it does any of these stars any good to try and deal with the problem inside that bubble. The whole idea is to break yourself down and destroy this illusion you have that you can handle the booze again. You go to a Chicago meeting, or a Madison meeting, and you get humbled pretty quickly.

JOHN FARLEY:

Chris needed a psychologist, not a guy punching a clock to sit around with him. I don't know how much Chris spent on them, but those guys charge exorbitant amounts of money. Jillian, Tim O'Malley, those guys didn't charge him anything, and they were the ones who were really helping him. But I think also the studios were stipulating that those guys be around; otherwise Chris couldn't get insured.

BOKEEM WOODBINE:

I saw those bodyguards hanging around Chris, but they were so cool about their shit that I didn't see them for what they were for the longest time. I didn't realize that Chris was going through the struggles he was having.

I remember one Friday we had wrapped for the week, and we were in somebody's trailer. I had a six-pack of Corona, and I was ready for the weekend. Chris had this trick where he knew how to open up a bottle with a cigarette lighter. I held up a beer and said, "Hey, Chris, could you open this for me?"

Everyone just stared at me. Chris said sure. He went to open it, and his hand was shaking. I thought he was making a joke, you know, being Chris. So I just started laughing, saying, "Look at this guy. He's hilarious." And everyone was looking at me like, what the fuck is wrong with you? And I wasn't getting it. It all went right over my head.

Well, about the third beer, Chris obliged me and opened it, but his hands were still shaking terribly. Then he left to go take a walk, smoke a cigarette or something, and Matthew Perry was like, "Bokeem, what the hell's wrong with you?"

"What do you mean?" I said.

"Chris is an alcoholic. He's having a hard time right now, and you're giving him fucking beer to open?"

"Shit, nobody sent me the memo. I didn't know."

Of course, if I had known, I never would have even brought the beer

around. But I was impressed by his dedication. I got the definite vibe that he was serious about not screwing up, and it seemed like his mental attitude was getting progressively healthier throughout the shoot. He was more and more upbeat every day we were working.

TED DONDANVILLE:

After *Edwards & Hunt*, when we weren't working, Chris would deliberately do things like going to health resorts or weight-loss clinics, places where he'd be safe and preoccupied with staying in shape. That year was definitely more up than down, and the downs were not so terrible. The relapses were smaller. He'd have a couple drinks, kick himself over it, talk it out, and be right back at a meeting the next day.

KEVIN FARLEY:

Chris was fighting it like crazy. He'd put together two, three months, and you'd think, okay, he's sober again. Then something would trigger it.

TODD GREEN:

Chris had been in L.A. full-time, so that Christmas I hadn't seen him in probably six or seven months. We all got together at some restaurant in Madison. At one point we kind of broke away and had our own little conversation, and Chris said, "Greenie, man, we're almost thirty-three years old and we're not married. God, that's weird." He just seemed really anxious.

I said, "Chris, what's wrong?"

"I just . . . I don't know. I'm playing these parts where I'm just the funny guy, and I'd like to do some more serious stuff."

I told him, "Why don't you come back to New York and do something onstage, Broadway or off Broadway?"

He really liked the idea. He seemed to take it seriously. Then we agreed that if the Packers went to the Super Bowl we would get together

there, since we'd both be going anyway. He was doing great, and we were all so proud of him. I walked away from that conversation thinking, Chris is back. We've got him back for good.

TIM O'MALLEY, *cast member, Second City:*

When Chris was at a weight-loss camp that November, he started to put together some time again. I met up with him again sometime that winter. I had about a year sober by that point. After Chris had left Second City, I'd stayed in Chicago, and my addiction just got worse. Eventually, when my brothers wouldn't come and bail me out of jail anymore, I cleaned up. Chris called me, and we started to go to meetings.

On January 20, 1997, we hooked up at a party at the Hancock for a recovery club that's here in Chicago. He had ninety-nine days. He was so happy, happiest I'd ever seen him. I watched him do backflips. He was doing actual backflips in the room. He was jumping and flipping and yelling, "I did it, Tim! Ninety-nine days! I did it! I did it!"

I was so thrilled for him that day. I was so fucking happy that I was crying.

He said, "I'm going to be here in Chicago for a while. If I keep checking in with you, will you help me?"

I took it as a request for sponsorship. He didn't specifically ask me to sponsor him or spell it out like that, but I could tell he wanted help. So for a few weeks he'd call me, I'd pick him up and take him to meetings. He started to get it again. He was happy and it was fun and we could crack jokes and make fun of each other just like we used to.

And then he stopped calling.

CHAPTER 13

The Devil in the Closet

JOHN FARLEY:

Chris said to me once, "You know what my dream is, Johnny?"

"What?" I said, thinking this was going to be some odd, Alice in Wonderland, through-the-rabbit-hole kind of thing. "What's your dream?"

"Here's what it is," he said. "It's me and Dad. We're both really skinny, and we're the coolest guys at the party, doing backflips all over the place and dancing up a storm to 'Twisting the Night Away.' That would be really cool."

On January 11, 1997, *Beverly Hills Ninja* opened in theaters nationwide. Despite a unanimous critical thumping, it earned over $12 million on its first outing, topping the weekly box office. Following *Tommy Boy* and *Black Sheep*, it was Chris's third-straight number-one film.

That January also marked Chris's third-straight month of sobriety. After staying clean during the principal photography of *Edwards & Hunt*, he'd relapsed again in September and, with varying degrees of failure,

cycled through three separate rehab facilities over the next two months. Then, in late October, Chris showed definite signs of improvement. When he celebrated his ninety-ninth day of sobriety in Chicago with Tim O'Malley, there was cause for hope.

But Chris's confidence was on the wane. In New Orleans, Todd Green was expecting to meet Chris at the Super Bowl to watch the Green Bay Packers take on the New England Patriots. When Chris didn't show, Todd called the Farley home in Madison, only to be told that the Super Bowl "wouldn't be good for Chris right now." Chris knew all too well what New Orleans's French Quarter would look like after a Packers win (or, for that matter, a Packers loss). He had chosen to watch the game at the home of a friend instead.

Despite making money, *Beverly Hills Ninja* was largely an embarrassment. It bombed with critics and disappointed even hard-core fans. Chris found himself at a professional crossroads. Hollywood had typecast him as the clown, and he had been fully complicit in that, playing the part whenever he was called upon to do so. But fatty could only fall down so many times. Fortunately, a project had arrived with the potential to take Chris in a new direction. Earlier that year, Bernie Brillstein had brought Chris together with screenwriter and playwright David Mamet, and together they'd agreed to collaborate on Chris's first dramatic film: a biopic of Fatty Arbuckle.

Roscoe "Fatty" Arbuckle was a silent-film star bigger in his day than Charlie Chaplin. He was on the receiving end of Hollywood's first-ever million-dollar contract. He was also on the losing end of Hollywood's first-ever sex scandal, being wrongly accused of sexually assaulting and fatally wounding a young woman. Arbuckle watched his career implode even as his innocence was proven in court. Brillstein was drawn to the story for its showbiz history and intrigue. Chris was drawn to it for the man himself. Arbuckle was a brilliant physical comedian who loathed his extra girth and outsized persona, despite having made it his professional stock-in-trade. After years of being made to play the crazy fat guy, Chris was being asked to play the guy behind the crazy fat guy. He was being asked to play himself, a role he rarely performed for anyone. Much like

Jackie Gleason's turn as Minnesota Fats in *The Hustler*, this was the role that would have fundamentally altered the course of Chris's career.

With the Arbuckle biopic ahead of him and ninety-nine days behind him, Chris was in good spirits. On the first weekend in March, the U.S. Comedy Arts Festival in Aspen, Colorado, was hosting a reunion of *Saturday Night Live* cast members, hosts, and writers. Several dozen stars from the show's history attended, from founding fathers Chevy Chase and Steve Martin to freshmen Molly Shannon and Cheri Oteri. For Chris to share that stage was an honor beyond anything he could have imagined growing up. It should have been one of the highlights of his career. It wasn't.

JOHN FARLEY:

I don't know what the hell happened. I remember everything had been fine in Chicago, but on the flight to Aspen he was acting strange. He may have relapsed that morning, or the night before. I just remember sitting on the plane, thinking, oh no.

CONAN O'BRIEN, *writer,* SNL*:*

When we were in Aspen, you could tell that the trolley was barely making it around the curves.

KEVIN FARLEY:

When I arrived he was already well into it, drinking and doing coke. From there it was just a total disaster. Spade really looked after him that weekend.

DAVID SPADE:

I went to meet him in his room to go to dinner with Lorne, and when I got to him he was already so messed up. We walked into the restaurant,

and it wasn't just Lorne. It was Lorne, Steve Martin, Dan Aykroyd, Chevy Chase, and Bernie Brillstein, all these people that Chris looked up to at this really nice, formal dinner. I said, real quick, "Hey, Chris, come over to the bathroom. I gotta tell you something." And I took him into the kitchen, out the back door into the alley, and I said, "We're getting the fuck out of here. You can't sit with these people in this condition."

These strangers showed up, and he started drinking with them. I tried to stay with him, but eventually I just had to go to bed. I was at lunch the next day, and he walked in. He was with the same people and obviously hadn't gone to bed. They were all wired, and Chris's eyes were rolling back. He said, "Davy. Davy, please stay with me. Don't leave me with these people."

JOHN FARLEY:

One day we had lunch at the restaurant on the top of the mountain. While we were eating, Chris started crying, saying, "I can't stop. I just can't stop." He was crying his eyes out right in the restaurant. Chris wore his heart on his sleeve; he didn't care one bit if he was crying in public, but people were starting to recognize him. We were like, okay, we've got to get Chris Farley off this mountain right now.

BERNIE BRILLSTEIN:

During the reunion, Chris was out onstage with about forty people from *SNL*. They were just telling stories, but Chris was crazed. I thought he was going to have a heart attack onstage. Finally, Dana Carvey quietly took him off.

CHEVY CHASE, *original cast member,* Saturday Night Live:

I read him the riot act that weekend. Everybody did. Chris was drunk and stoned and, on top of that, way overweight. I sat with him and I said, "Look, you're not John Belushi. And when you overdose or kill yourself,

you will not have the same acclaim that John did. You don't have the record of accomplishment that he had. You don't have the background that he had. And you don't have the same cultural status that he had. You haven't had the chance to get that far, and you're already screwing yourself up."

He kept saying, "I'm just trying to level out."

That's what he said he was doing with the drinking and the cocaine. It's so silly. It means if you took nothing you'd be level already. Why take all this shit that's killing you? And I told him that. I said, "I've experienced this. I've seen who dies. I've seen how far you think you can go, what you can take and what you can't. You're just going to end up being an overweight guy who could fall on his stomach and had one or two funny things in his career, but nothing that's ever really stood out. You'll be a blip in the *New York Times* obituaries page, and that'll be it. Is that what you want?"

BOB ODENKIRK, *cast member, Second City:*

I was at a party for *Mr. Show*. Somebody came in and said, "Chris is out back. He wants to talk to you."

There was this skanky deejay guy, the kind of guy who hangs out in these party vacation towns. He'd tried to pass David Cross some cocaine earlier, so I knew who he was. I go out back, and there's a limo. I go to the door and knock and the window rolls down. There's Chris, and he's packed in there with girls and hangers-on and this fucking scumbag who was pushing coke around. Chris is bloated and red-faced; he hasn't shaved. We talk for a few minutes, but there's really nothing to say at those times.

I'd seen Chris fucked up before, but this time he looked as bad as anyone has ever looked. It was a horrible thing to watch. It's one thing to shake your finger at a friend and say, "You're gonna kill yourself." It's another thing to look at him and know he's going to do it.

TOM FARLEY:

After Aspen, his managers said, "He's going to rehab, and we're serious this time. He's going away for thirteen weeks and he's not coming back—except to present at the Oscars."

KEVIN FARLEY:

Brillstein-Grey sent him back to the lock-up down south, but they thought it would be okay for him to go to the Oscars, under supervision, and present an award.

TOM FARLEY:

This woman who ran the facility said the only way they'd let Chris go was if he was there with someone from treatment. The next thing you know, *she's* the one who's going with him, and she made him pay her extra for her time, buy her first-class airfare, buy her a dress, and do the same for her daughter to accompany her. I don't think that helped. It just made him feel used.

KEVIN FARLEY:

I thought she was really unprofessional about the whole thing. It was her opportunity to go to the Oscars; she basked in the limelight for a little while. We were in the hotel, and she started rummaging through Chris's gift basket, looking at all the high-end cosmetics they put in there. And Chris was like, "What the fuck are you doing? Put that shit down. Don't you think I might want to give that to my mom?"

This woman just got way into it. "Ooh look, there's George Clooney!" Who gives a fuck? Why don't you do your job?

TOM FARLEY:

I didn't get it. Chris's managers were the ones busting him the hardest for fucking up at Aspen, and then two weeks later they were the same ones lobbying for him to come back and present at the Oscars. It was a money thing. The Oscars are exposure, and exposure means money. I guess they thought Chris needed it to help his career.

TOM ARNOLD:

Chris had a fear that his movies were starting to suck, and, you know, I know what that fear is like. But there's always options if you're talented.

BERNIE BRILLSTEIN:

A few months earlier, I'd taken him to New York to meet with David Mamet about the Fatty Arbuckle story. That story has always fascinated me, only because Arbuckle was innocent. Chris came to the meeting at a little restaurant down in the Village, and he was the good Chris, the well-behaved Chris, because he couldn't believe that David Mamet even wanted to meet him. Mamet loved him. It was a great meeting. He said yes before we got up from the table, and he wrote it for Chris. To this day I know that it would have changed his career.

TOM FARLEY:

As soon as he heard little bits and pieces about Arbuckle's life, he said, "This is me." It was the whole idea that nobody understands the real person underneath. "I'm going to tell them about the real Fatty Arbuckle, and maybe they'll understand the real Chris Farley."

ERICH "MANCOW" MULLER, *friend:*

Chris had all these pictures of clowns in his hallway. He said that they frightened and fascinated him, and that he found them sad. When he was drinking he would always talk like Burl Ives and sing old Burl Ives songs. He'd go, "A little, bitty tear let me down, spoiled my act as a clown." He'd sing that over and over and over.

FR. TOM GANNON, S.J., *friend:*

He felt his career was in trouble, and not just because of the drugs. The fatty-falls-down humor was beginning to take a toll. Sometime that year, he told me, "I can't keep this up. I can't keep falling down and walking into walls." But people wanted him to keep doing the same thing, because it assured them financial success.

BOB WEISS, *producer,* Tommy Boy:

Chris had an idea of reinventing himself in a certain way that didn't take into account very real forces in this industry, forces that can be tidal in nature.

FRED WOLF:

By that point, people were coming at Chris from every angle. They were trying to hire me in the hopes that they might make a deal with him. We went to dinner one night in New York, and he was telling me that he wanted to do movies like *Nothing in Common*, the Tom Hanks/Jackie Gleason movie. I was absolutely convinced that that was what he could do. We started throwing around some ideas, and we kept getting interrupted by fans coming up and saying, "Chris, I love you! You're so funny."

And then as they would walk away, Chris would sigh and say, "But that's all they want."

So we kept trying to have this very serious conversation about his career, but the fans just kept coming and coming and asking for lines from *SNL* or bits from *Tommy Boy*. They wouldn't leave him alone.

FR. TOM GANNON:

One night we were at Gibson's. People pretty much left him alone that night. But one couple came up and thanked him for his work and told him how much they loved him. Then they walked away and he turned to me and said, "They don't really love me. If they knew me, they wouldn't love me at all."

I said, "That's not true, Chris. People do love you. They don't love you the same way I do, or your family does, but they're sincere. You bring a lot of happiness into their lives."

He got a lot of that kind of attention, but he didn't get any nourishment from it, and so he felt he needed more of it all the time.

LORRI BAGLEY:

Chris would go to premieres and goof off on the red carpet, but then he'd complain that the business wouldn't take him seriously. I told him, "Chris, when you stop playing the clown, they'll stop treating you like the clown. They'll take you seriously when you take yourself seriously."

PETER SEGAL, *director,* Tommy Boy*:*

There were bidding wars for Chris on multiple projects, but most of them were not that good. He'd come to me with these scripts, and I'd turn them down. I kept saying, "No, Chris. That's not a good one for you or for me."

So there was a tension between us, because he thought I didn't want to work with him anymore. There was a long time when he wouldn't return my phone calls, and so I sat down and wrote him a long letter. I told him the reason I was turning these projects down was because I

believed his potential was so much greater. And I think he realized it, too. He eventually called me back to thank me for the letter.

But I really meant what I'd said. I thought he could win an Oscar one day. I know people might think I'm crazy saying that, looking at his brief career, but I really believed in his talent. It was way beyond what he was showing.

BRIAN DENNEHY, *costar,* Tommy Boy:

Myself, I never understood why you'd want to be the twentieth-best dramatic actor in the movie business when you were already the best comedian in the movie business. But there is this impulse that comedians have to do serious work.

Interestingly enough, I think with the right part and the right director Chris could have done it. There was a sadness and a vulnerability and a fear that existed in his face and in his eyes. Jackie Gleason had it, a sense that "the world can never take away the pain that I feel, pain that I know that I have, but that I don't fully understand." You can see a little bit of it in *Tommy Boy*, but he hadn't even really begun to explore it.

There are two ways to act, and some people are good enough to do both. One is to erect this very complicated, layered character around you in order to hide behind it, in order to disguise and protect yourself. It's a kind of architecture. You're creating a building. It may be a very impressive building, but it's still a fucking building.

The other way to act is to absolutely strip away everything that keeps you and your soul and your mind from the audience. You rip it away and say, "How much more of myself can I expose to help the audience understand this character?" It's more difficult, and it's more profound, because, ultimately, the real challenge of art is to understand more about yourself. And I think Chris could have done it. I think he would have done it, had he lived. But most comedians, in fact most actors, are not capable of that.

* * *

With *Tommy Boy, Black Sheep*, and *Beverly Hills Ninja*, Chris had joined the ranks of elite Hollywood stars who could "open" a film—a certain core audience could be counted on to turn out for any Chris Farley movie. Even if Chris wasn't thrilled with the reigning definition of "a Chris Farley movie," it was an enviable place to be, and a strong place from which to make a bold, smart career move.

But that spring, Chris's dance card was strangely empty. As a rule, studios take out short-term insurance policies on their lead actors to cover any possible interruptions in the production process. Many of those insurers were refusing to underwrite Chris's films until he could once again prove his dependability. And so, while the Arbuckle project plodded along at the glacial pace of most Hollywood development deals, Chris was having trouble getting even a typical Chris Farley movie off the ground.

In this troubled time one good project did come his way, a voice-over gig for a little animated movie called *Shrek*. In 1997, computer-animated movies were still in their infancy—Pixar's trendsetting *Toy Story* had opened only eighteen months before—and so there was little reason to believe that this fun sideline project would go on to spawn one of the most popular, highest-grossing film franchises of all time. Chris took it on almost as a lark.

Shrek was a popular children's book by William Steig about an ornery yet good-hearted ogre who lives alone in the woods, cast out from the world. Jeffrey Katzenberg, head of DreamWorks Animation, had procured the film rights. Chris was his first choice to play the title role. According to everyone involved, Chris Farley's Shrek was one of the funniest, most heartfelt performances he ever gave. Tragically, no one has ever heard it.

TERRY ROSSIO, *screenwriter:*

Chris was the number-one choice, and everyone was thrilled that he agreed to the project. For an animated feature his voice was perfect, very distinctive. Also, you know, Shrek kind of looked like Farley, or Farley looked like Shrek.

The recording sessions were essentially everybody in the booth rolling off our chairs onto the floor, laughing our asses off. I brought my daughter, who was twelve years old at the time, to one of the sessions at the Capitol Records building. It was her first time ever coming in with me to work, and she concluded I had the best job in the world, listening to funny people be funny.

ANDREW ADAMSON, *director:*

The character of Shrek is to some degree rebelling against his own vulnerabilities. And I think that's probably a reason Katzenberg went to Chris, because there was an aspect of that in him, covering vulnerability in humor and keeping people at arm's length. Within minutes of meeting Chris you saw his vulnerability. Sometimes he would switch on this very gruff persona, and you realized it was because he felt like he was exposing too much.

It didn't make the final film, but at one stage there was a moment in the script where Shrek was walking along, singing "Feeling Groovy," Simon and Garfunkel's "Fifty-ninth Street Bridge" song. Chris was just so into it. When we were recording, I kind of got the impression that he wasn't sure whether he was supposed to be doing a comedic take on the song or a sincere, heartfelt one. He was singing and putting himself out there in a way that was very touching. It made me see the longing in him to do something more genuine with his career. It made me feel bad, because we were in fact asking for a "funny" version. But that he was willing to give it to us, even though he felt so vulnerable about it, made it a very sad and touching moment.

TERRY ROSSIO:

We spoke about the essence or wellspring of Chris's humor; much of it was the humor of discomfort. He would occupy a space of discomfort until it became funny. Shrek, in the Chris Farley version of the story, was unhappy at his place in the world, unhappy to be cast as the villain. So for me,

Chris's comedic persona was key to the creation of the Shrek character—a guy who rejected the world because the world rejected him.

ANDREW ADAMSON:

After Chris died, we all had personal thoughts about whether we could use his voice track and find someone to impersonate him to finish the film. We definitely thought about whether that was the appropriate thing to do, but ultimately we felt that we weren't far enough along in developing the story and the character. The animation process depends a lot on the actor. His death was quite devastating, both personally and to the process of creating the film. We spent almost a year banging our heads against the wall until Mike Myers was able to come on board. Chris's Shrek and Mike's Shrek are really two completely different characters, as much as Chris and Mike are two completely different people.

TERRY ROSSIO:

They're both great in their own way. Mike created a very interesting character, a Shrek who has a sense of humor that's not that good, but it makes him happy. Chris's Shrek was born of frustration and self-doubt, an internal struggle between the certainty of a good heart and the insecurity of not understanding things.

ANDREW ADAMSON:

I always found Chris a very fun person to be around. Containing him in a recording booth was a great challenge, but he was a very down-to-earth guy on a certain level. We had an enjoyable relationship. The drug problem didn't impact his work at all, and to be honest, I had no idea it was happening. Everything I'd seen indicated that he had overcome those demons. He was going through rehab at the time and was very disciplined about it. Any other impressions I had were thirdhand and after the fact. I really felt like he was on an upward spiral.

* * *

And Chris was—that week. But the next week he was back on a downward one, and who could say where he was going the week after that? By the time he finished voicing *Shrek* in early May, Chris's ability to maintain his sobriety had all but vanished. His relapses started coming randomly, suddenly, and with alarming frequency.

One of Chris's counselors described him as having the most severe addictive personality he'd ever seen—this in several decades of helping patients. As Chris surrendered his hold on sobriety, his compulsive overeating ran rampant as well. Chris had fought a constant battle with his weight since childhood. Those who knew him well knew it was the bane of his existence. Given the severe health risks of obesity, Chris was doing almost as much damage to himself with food as he was with drugs and alcohol.

After presenting at the Oscars on March 24, Chris had returned to rehab in Alabama, emerging sober to work on *Shrek* in April and early May. Following yet another relapse, he returned to the outpatient program at Hazelden Chicago on May 19. It accomplished little. June and July were spent in and mostly out of rehab, and by August the situation was catastrophic.

Chris's relationship with Lorri Bagley, rocky and unstable in the best of times, was severely broken. It never ended, but the blowouts got bigger and more explosive, and the separations grew longer and longer. Friends who were active in Chris's recovery, like Jillian Seely and Tim O'Malley, did their best to keep him on the straight and narrow, but their efforts were increasingly frustrated. Chris would either insulate himself from his friends in order to use, or insulate himself in order not to use. He had so removed himself from his usual social networks that many assumed he was simply off somewhere else, stone sober and hard at work. Chris had never let the trappings of fame and success put any distance between him and his loved ones. But addiction finally succeeded where fame could not.

ROBERT BARRY, *friend, Edgewood High School:*

Toward the end Chris would go hang out with these Board of Trade guys in Chicago. They had tons of money and wanted to hang with celebrities. When he was in Chicago, Chris didn't call up Dan Healy or me or the Edgewood guys anymore. He'd call up those people. I never even visited his place in the Hancock.

FR. MATT FOLEY, friend:

For the last three years I had been living in Mexico, doing missionary work. I talked to Chris and his parents on a regular basis, but then Chris stopped returning my calls. One of the last times I saw him was on a trip to Chicago. We went to work out at a health club there by the John Hancock building. After that we were supposed to hang out all day, but he basically wanted to get rid of me. He didn't want me around because I would have told him he was full of shit.

JOEL MURRAY, *cast member, Second City:*

The people who loved him didn't want him to drink, so he couldn't be with us anymore. I'd invite him over to barbecues and stuff out in L.A., and I could tell that he had a whole other thing going on. It wasn't a celebrity, big-shot kind of thing; it was an "I gotta go do this stuff that I don't want to tell you about" kind of thing. He was the worst liar in the world, so he'd just kind of be evasive. Next thing you know he's hanging out with nefarious types who just want to wind up the comedy toy, and that's never good.

DAVID SPADE:

There's no shortage of those sorts of people. I've talked to Aykroyd about Belushi, and it's the same experience. Friends you've known for three days aren't friends I want to hang with.

I was working in TV, he was off doing his movies, and we'd just slowed down a little bit. It wasn't Lorri. That was done with, but we'd been a little bit on the outs, and because of that I got a lot of shit toward the end about "Why weren't you there for him?" But being that close, I dealt with it all the time. And in that situation, before the guy's dead, he's just kind of an asshole. Truth is, you get a junkie who's wasted all the time and moody and angry and trying to knock you around, you say, "Okay, you go do that, and I'll be over here." I think that's understandable.

TED DONDANVILLE:

Chris never had any animosity toward Spade at all; he had just respected Spade's decision to walk away for a while. But after being all alone on *Ninja* and *Edwards & Hunt*, Chris started to realize how much he needed his friend. It was like Mick Jagger after those first two solo albums—maybe it was good to have Keith Richards around.

TOM FARLEY:

I always told Chris, "You love humor, but look around at the people you're with when you're doing these drugs. These people have no humor in their lives. You keep this up and you will end up surrounded by people who are not your friends." And that's exactly what happened.

NORM MacDONALD:

Sometimes you'd see him with prostitutes. That was mostly at the very end, like when he hosted *SNL*. The amazing thing was how well he treated them. He really fell for them. He'd take them to dinner and treat them so sweetly. He'd treat them equal to any other person at the table. He'd introduce them to you as his girlfriend.

TIM O'MALLEY:

Escorts and strippers are just part of the deal when you're lonely and lost. It's like phone sex, trying to reach out and talk to somebody. Every phone book has a hundred phone numbers in it; you can always dig up someone to spend time with you.

I went into his apartment one night, and he said, "Yeah, I relapsed last night. I had a pizza, and I figured since I'd relapsed on my OA program I'd have a bottle of scotch, and then I went to the Crazy Horse and I spent eleven grand."

"Jesus, you were giving the girls five hundred a dance?"

"Yeah, how'd you know?"

"Because I know how it goes. You were trying to get some girl to come home with you by overtipping her, and those girls don't want anything but more money. First of all," I told him, "separate your food program from your alcohol problem. Food's not going to kill you tonight."

I hated the Overeaters Anonymous program for that, because if he relapsed on that he'd just go ahead and go the distance.

KEVIN FARLEY:

For Chris, by that point, every relapse meant going all the way. Some addicts will put a toe back in the water, but Chris would always dive back into the deep end. And that's what happened when he went to Hawaii.

DAVID SPADE:

I was at the Mondrian in L.A., and Chris was there. He was doing an interview, and he had one of his sobriety bodyguards with him. It was kind of sad, because I hadn't seen him in a while. He came over to my table—the bodyguard let him come over alone for a bit—he came over and he said, "Nobody cares about anything but *Tommy Boy*. Can we do another one? Can we do . . . something?"

"Of course. There's always scripts they want us to do. I didn't know if you wanted to do anything anymore."

"We gotta do it, because that's the only one that matters."

"Okay," I said. "Let's find something."

Then these two cute girls came over. They said, "Hey, come party with us. We're in town with Spanish *Playboy*." Or something ridiculous like that.

Chris said, "I can't."

"Oh, c'mon," they said. "Just come up to our room for a bit."

Chris looked at me. I said, "I'll cover for you. I can buy you about five minutes."

"Thanks, Davy."

He took off, and then the bodyguard came over and said, "Where's Chris?"

"He went to the bathroom."

"Which bathroom?"

"There's one in the hotel."

"You fucked this."

"Sorry."

It was the wrong thing to do, I know. But we'd had a really nice moment together, and I liked that. It proved that we were still close, could still be friends, and I wanted to help him out. But then they couldn't find Chris. He disappeared, and it just turned into chaos.

KEVIN FARLEY:

US magazine was doing a big feature article on him at the time, and Chris was spending his days with this reporter. Chris woke me up in the middle of the night and asked me if I wanted to come down and take a whirlpool with these girls he'd met from *Playboy*. He'd already relapsed and started drinking. I said no and went back to bed. I figured he'd play in the Jacuzzi and then go up to his room and sleep it off. But I got up the next morning and found out he'd relapsed hard, bought these girls plane tickets and

gone to Hawaii. When that *US* reporter showed up and there was no Chris, the shit hit the fan. Gurvitz had to put that fire out.

When I talked to Chris about it later, he didn't even remember going to Hawaii. He just woke up there. But when he called Dad from Hawaii, Dad was like, "Hey, you're on vacation!" The level of denial at that point was just crazy.

FR. TOM GANNON:

You cannot understand Chris Farley without grappling with the relationship between him and his father. That was the dominant force in his life. He talked to his father every day on the phone, and was constantly trying to please him. And I think he *did* please him. But the family, which looked so normal on the outside, was terribly dysfunctional.

ERIC NEWMAN:

If you were a shrink, you could retire on that family.

TIM O'MALLEY:

The first people we know as God are our parents. And if you don't get approval from your parents, eventually you can mature and find that from other places. But Chris was never able to do that. He was never able to find it from God or anyone else.

TOM ARNOLD:

Even when he was thirty years old, Chris would literally sit at his dad's feet and tell him stories. I don't think anything made him happier than to sit at the foot of his dad's recliner and tell him stories about show business, or food.

There were a couple other times where I went with Chris to the Taste

of Madison, which is this festival in the city square where every three feet there's a booth of a different kind of food. All the conversations Chris had with his dad that weekend were just "Hey, did you have that pork chop on a stick?" "Yeah, that was good. Did you get some of this?" You know, they were surface conversations, the kind I would have with my dad, the kind that don't get really deep. Because if you get deep it's pretty painful.

KEVIN FARLEY:

I think my dad was basically a happy guy, but he had an addiction to food and alcohol. And when you get to be six hundred pounds, you're in such a hole that what are you going to do to get out? And that's what depressed him. He was confused by it. He'd be like, "I don't know how I got this big. I don't know how this happened." I watched my dad's eating habits. Yes, he ate a lot, but was it proportional to the weight he gained? No way. Part of it had to be genetic.

My father was handicapped, and when you have someone in your family with an illness, you want to do what you can to make them feel better. It wasn't just Chris. We all wanted to make Dad happy, because we all knew he was on borrowed time.

JOHN FARLEY:

Then there's the other element to it, not wanting to get skinny or sober because he didn't want Dad to feel bad. Chris said that to me, that he should stay heavy for Dad.

LORRI BAGLEY:

Chris was very protective of his father. One night after I went with Chris to a meeting, he asked me if I wanted to meet his parents for dinner. When we were in the elevator going up to see them, Chris was like, "Look, my dad has this problem. Please don't stare at him."

A year later, the first time I spent the night with Chris, he showed me

a picture of his family from when he was a kid, and his father was so thin. I said, "What happened?" But Chris never really told me.

CHARNA HALPERN, *director/teacher, ImprovOlympic:*

I had a very intense night with him alone in my house once. We were listening to a Cat Stevens album, *Tea for the Tillerman*, and the song "Father and Son" came on. Chris started crying. Cried and cried and cried. He said, "I love my dad so much, and I don't want him to die."

I said, "He probably feels the same way about you. You're both in the same situation. You're both alcoholics. You're both overweight. Maybe you can help each other."

"Yeah, but we can't," he said. "It'll never happen."

HOLLY WORTELL, *cast member, Second City:*

His dad was of a different generation. They didn't go to see "headshrinkers." Chris told me that his father finally agreed to go with him to this weight-loss clinic once. They were sitting in a group therapy session, and everyone was going around the circle talking about their issues with food. His dad just stood up and said, "Let's go." They got up and went outside, and his dad said, "We're not like these people. They've got problems. That's not us. We're leaving."

FR. TOM GANNON:

They walked out, checked in to a resort on an island off the coast of Florida, took out a room, and proceeded to go on a binge together. With that kind of enabling, the kid didn't stand a chance. The father was in denial, but in all fairness, I don't think the brothers were straight with the father, either. Dad knew about the drinking but not so much about the drugs. The father never accepted that Chris was a drug addict until the very end, even though the two of them talked every day. So there was a lot of posturing going on.

TOM ARNOLD:

It's not his father's fault, what happened to Chris. It's not. Chris had access to every tool in the world. He went to the best treatment centers, had the best people being of service to him, reaching out to him.

You look at all the pieces of Chris's life, his father, his mother, his brothers, his life growing up, his work—everything. You look at all that and maybe some things are off or a little dysfunctional, but at the end of the day it's his responsibility. It's not like I didn't sit with him a dozen times where he looked me in the eye and knew what he had to do to stay sober. You can't blame your circumstances, and after a certain point you can't even blame your father. You can't blame him; you have to have compassion for him. It all comes down to you, and you've got to be a man about it.

LORRI BAGLEY:

Chris knew that to be himself, to be healthy, he'd have to pull away from the family, and he couldn't do it. He said he couldn't do it. But you have to cut the emotional umbilical cord at some point. Some American Indians have a ritual where you're not allowed to be a part of the tribe until you leave, go out in the wilderness, rename yourself, and come back. Then you're accepted as a man. But we don't have that in our culture. That's why families in the country are falling apart, and why women have to deal with all this Madonna/whore bullshit. It's because men don't grow up, and Chris never grew up.

ERICH "MANCOW" MULLER:

That May, Chris Rock was performing in Chicago. Farley called me and said, "I've broken out of prison. I'm out. I want to go see my boy Chris Rock!" Chris broke out of rehab to go to this show. I met him at his apartment, and I was begging him not to drink. I was sitting there, going, "No. No, Chris. Please."

He said, "Just a little splash." That's how it started off, a Coke with a splash of whiskey—and I mean just a drop. Then an hour later it turned into a glass of whiskey with a splash of Coke. We went to the concert to meet Tim Meadows and his wife, and I spent the whole night fighting him.

TIM MEADOWS:

We went backstage after the show to see Rock, and Farley was drunk, fooling around in front of these girls. We'd been talking about going out for dinner after the show, but Rock and I looked at each other, and I said, "I can't do it. I can't be around him anymore like this."

Rock said, "Yeah, I know what you mean. I'll take care of him tonight."

CHRIS ROCK:

He was so fucking drunk, drunk to the point where he was being rude and grabby with girls. He would go too far and you'd call him on it, and he'd give you his crying apology, the Farley Crying Apology. We probably had about four of those that night.

I remember dropping him off at his apartment. He wanted me to come up and see his place, and I just didn't have it in me. He was so fucked up. I just couldn't go up there. And as I drove away, I knew. It had gotten to that point. I knew that was the last time I'd ever see him alive.

JILLIAN SEELY:

I was waiting for Chris to pick me up for the Chris Rock show, and I got a phone call from him saying there weren't enough tickets and so I couldn't go. That was Sunday. Then Tuesday I got a call at nine o'clock at night from a nurse at the Northwestern psych ward. Hazelden had to send him to the hospital to get sober before they'd let him back into treatment.

Chris got on the phone. "I'm really scared," he said. "I totally relapsed on Sunday and went back to treatment, and they made me come here. Will you come and see me?"

So I went over to Northwestern. I went up to Chris's room, and I heard him go, "Hey, hey, in here."

He was in the bathroom blowing his cigarette smoke into the air vent. I looked down at this stainless-steel paper towel rack, and there were lines of cocaine on it. Chris had gotten one of the hospital staff to bring him coke in the detox ward.

I said, "I'm totally telling on you." I went out into the hallway and started yelling, "Chris is doing cocaine in his room!"

They came in and restrained him. He was screaming at me, "You're a fucking narc! I hate you!" It was like a scene out of a bad movie. It was horrible, really horrible.

KEVIN FARLEY:

The fact that Chris was able to score cocaine *inside* the detox ward was just insane. When you're famous there aren't any rules. That's when I knew things were getting bad. He was in a mental ward. You couldn't get any lower than that.

As a kid, when he watched *The Exorcist*, he was terrified of the idea that something evil could take over your body, possess you, and make you do things you can't control. Here he had this thing that was eating away at him from the inside, and he was powerless to stop it. And that scared the living shit out of him.

FR. TOM GANNON:

On the surface, the Farleys are a wonderful family. They're loving. They're supportive. They're there for one another. I didn't get to know the father. Met him once, maybe. Spoke to him on the phone a couple of times. And I suppose I have to be honest; I didn't care for him that much.

Whereas the mother is a lovely person, caught in the same vortex as the rest of them.

And therein lies the key to the problem: They didn't know how to manage Chris. When it's all said and done, I don't know that they were any more or less dysfunctional than any other family, but Chris's personality was so outsized that it sort of took over. It's that old story from his childhood, when the nuns said that Chris didn't know the difference between somebody laughing with him or laughing at him. That played out in the family as well. At what point do you draw a line that this bizarre behavior is too much to handle?

TOM FARLEY:

Nobody ever thought of the problem in terms of Chris's health or the idea that he could die. Mom maybe had some premonitions of disaster but didn't talk about it. No one talked about it—and that was the problem. My parents' reaction was always the same: "Chris is out of control." Or "Dammit, how could he do this? He's going to ruin his career."

And the motivations were always external, like getting fired from *Saturday Night Live*. It was always concern about the symptoms and never the disease, which none of us genuinely understood. But Chris wasn't "out of control"; he was sick. And his sickness was just so deep and so entrenched.

TOM ARNOLD:

It's harder for some people, and I don't know what it would've taken for Chris to really, truly hit bottom. The absolute worst I ever saw him was at a Planet Hollywood opening in Indianapolis that July. It was the bad Chris. I mean, he was just so fucked up. He had his shirt up over his head and people were taking pictures. Kevin was with him. I said to Kevin, "You better get him out of here. I'm gonna fuckin' tackle him, 'cause I have had it."

KEVIN FARLEY:

Jillian and I were trying to get him out of the bar, but he didn't want to leave. And at that point I couldn't control him. Either he's going to take a swing at me and we can get into a fight there in front of the cameras, or I can go home. We'd flown in on a private jet that night, so Jillian and I left and took it back together.

JILLIAN SEELY:

We were really quiet on the plane. We were both so sad that Chris had started drinking again. The next day I got a phone call around noon. I thought it would be Chris, calling from Indianapolis, confused and wondering why he'd been left behind and maybe having learned a bit of a lesson. But he was like, "Hey, what's going on? I'm back in Chicago. Want to get lunch?" The plane went right back for him and picked him up. No consequences for his actions at all.

But his behavior at the party made the *Enquirer* and the entertainment TV shows. And then that profile in *US* magazine came out a few weeks later. It was a pretty hard-core article.

TOM FARLEY:

That was the first time there had really been any public exposure of Chris's problems, which is pretty amazing when you look back on it. At that point, he was really staring at the abyss; it looked like he was going to lose it all. Brillstein-Grey went into damage control mode, trying to clean up the press.

They also sent Chris back to Promises in Malibu and made him start seeing this therapist in L.A. The sessions Chris had with this guy weren't really therapy sessions; it was more this guy telling Chris what he had to do, and why, if he wanted to save himself. He really got into Chris's noggin. He hit him in a weak spot, that superstitious thing that he always had. He was telling Chris there was this other side to him, this other being in-

side him that was bent on destruction. That really played to a lot of Chris's fears, and I don't think it was helpful at all. I think he just confused the boy.

FR. TOM GANNON:

Chris thought of his addiction in terms of good and evil, that drugs were the devil's way of controlling him, and I tried to steer him away from that way of thinking, because it isn't very helpful. Like many Irish Catholics, Chris's spirituality was sort of a mix between religion and superstition.

TOM FARLEY:

He told me that heroin was the devil. "I've seen the devil, Tommy." That's what he told me after he'd tried it.

LORRI BAGLEY:

Chris told me that every time you do heroin, you can feel it take a part of your soul.

KEVIN FARLEY:

Chris would talk about his addiction in those terms, because that was the vocabulary he had for it. A lot of people laugh at that concept, but I think it's as good a framework as any. What is a demon? A demon is something that wants you dead. And whatever was in possession of Chris certainly wanted him dead.

FR. MATT FOLEY:

Chris knew all too well that addiction was a disease. He and I had endless talks about it. He needed to separate himself from the shame that he felt. He needed to learn how to forgive himself and accept forgiveness from

others. But I can relate to his thinking. I struggle with temptation every day, as do we all. There is no blessing that comes out of drugs and alcohol, and in that sense they're evil.

TIM O'MALLEY:

They say that you should go back to your faith when you get sober, but it's up to the individual the role that their faith plays. How did I survive? How did I not run myself off the road when I was driving around in my underwear looking for crack? I'd have to say it was God. But a lot of people don't go back, because they feel so burned by the nuns and the priests.

I don't think Chris ever got a chance to really clarify or learn properly some of the ways to sort out your life. So I think he used religion and did the best he could with it, still trying to be a good Catholic boy using the garbage we were taught by the nuns, the angel on one shoulder and the devil on the other. It's a fifth-grader's view of spirituality.

FR. TOM GANNON:

Chris was caught in a transition in Catholicism between an old-church approach to faith and a newer way of thinking. The old view of spirituality was that life was like climbing a mountain. You have to fight onward and upward, climbing with your spiritual crampons until you reach the top—and that's perfection. You pass the trial and you pass the test and you get so many gold stars in your copybook. Then you come before the heavenly throne for judgment, and maybe you've got a couple of indulgences in your back pocket in case your accounting was wrong.

But that kind of faith only gets a person so far. Your spiritual life isn't like climbing a mountain, waiting to find God at the top. It's a journey, full of highs and lows, and God is there with you every step of the way, in the here and now and in the hereafter. The first approach is really a whole lot of smoke and mirrors. It's only the second one that allows a person to grow, but that second view is hard for people to get ahold of unless they get in touch with themselves.

Chris didn't feel that he was worthy of God's love. He felt he had to prove himself. Well, you're never going to get very far in any relationship with that kind of belief. Imagine if you had to prove yourself to your spouse every single day; that's not the way love works. In all of our talks, that was the one thing I really tried to work with him on, adjusting to this different idea of faith, but he never really moved from one to the other. It's hard. It takes a long time to come around to that way of thinking, and Chris just ran out of time.

CHAPTER 14

Fatty Falls Down

CHRIS FARLEY:

The notion of love is something that would be a wonderful thing. I don't think I've ever experienced it, other than the love of my family. At this point it's something beyond my grasp. But I can imagine it, and longing for it makes me sad.

In January of 1997, Chris Farley had the number-one movie in America and a ninety-day sobriety chip of which he could rightly be proud. A mere seven months later, all of that seemed an impossible memory.

On August 1, Chris's relapse at Planet Hollywood triggered a small whirlwind of negative press, capped by the September issue of *US* magazine with its profile, aptly titled "Chris Farley: On the Edge of Disaster." Not only had the reporter been witness to Chris's relapse and subsequent escape to Hawaii, Chris himself, in full heart-on-sleeve mode, had divulged a year's worth of personal therapy in just a few short interviews. Even Chris's friends and manager Marc Gurvitz had given frank assessments of Chris's condition. The piece was a public relations nightmare.

On August 8, Chris checked into Promises, an upscale recovery facil-

ity in Malibu, California. He stayed until the end of the month. On September 1, he checked out for a brief opportunity to go back to work, flying to Toronto to film a small cameo in Norm MacDonald's feature film *Dirty Work*, about a guy who goes into the revenge-for-hire business to help raise money for a friend's operation.

Chris stayed clean during his shooting days—he always did—but he would vanish at night, and in general he did not look well. Norm found Chris's behavior unusual; he had never seen his friend under the influence before. He openly questioned Chris as to whether or not it was a good idea for him to continue working, but Chris insisted that he was fine and that, after such a long professional drought, he was grateful for the opportunity. At that point, *Shrek* was an ongoing concern, and the Fatty Arbuckle biopic was still alive somewhere, but Chris's inability to get insured had effectively stalled his career. The only producer willing to give him a shot was Brian Grazer at Universal, who wanted Chris for a film called *The Gelfin*, which would begin filming in January with newly minted star Vince Vaughn.

On September 10, Chris's part in *Dirty Work* wrapped and it was back to Promises in Malibu. By this point he had cycled through a dozen rehab facilities in under twenty-two months, and the routine treatments had reached a point of diminishing returns. He knew the system better than most of the counselors assigned to his case. Institutions that once frightened Chris now merely bored him. The constant physical strain of using and drying out frayed his nerves. And the long, dark nights spent alone in strange beds ate away at what little reserve of humanity he had left.

Chris was convinced—or, more aptly, had convinced himself—that there had to be answers elsewhere. On his tenth day at Promises, fed up with the whole ordeal, he went down to the basement, flipped the master circuit breaker, and cut the power for the entire facility. Once the lights were restored, the security team found Chris lounging in the common room, naked, quietly leafing through old magazines. "You found me!" he proclaimed. The police were called, and he was asked to leave.

A now-familiar pattern played itself out, and by mid-October Chris swore he was ready to recommit himself to sobriety. He rented a house in

Los Angeles and asked Kevin to move in with him. Kevin now had four years sober on his own, and Chris thought his brother would be able to take him to meetings and keep him on track.

In truth, no one could keep Chris on track but Chris, but there was one place where he had managed to stay sober and happy: his old job at *Saturday Night Live*. Marc Gurvitz called *SNL* producer Lorne Michaels. They agreed that having Chris come back to host might help him in some way to deal with his problems. But the week Chris arrived in New York, he wasn't just having problems, he was having a full-blown meltdown. The results, broadcast live on national television for millions to see, were not pretty.

KEVIN FARLEY:

He was gasping for air by that point. I lost him for three days. It was the weekend before he was supposed to host *Saturday Night Live*, and he was just gone. I was staying at the house, and Marc Gurvitz was calling me, saying, "Where's Chris? We've got to get him on a plane. Lorne expects him in New York."

I told him I had no idea where Chris was. For all I knew, Chris could have died that weekend. He was with Leif Garrett, of all people. Leif fucking Garrett, and some other losers. When Chris finally showed up, high on heroin after three days missing, Leif came into the house and was like, "Your brother's so fucking funny, man." I almost took him out right there, but I was so sad and spent with the whole situation that I just didn't have the energy to punch him in the face.

That whole weekend was so sad and out of control. I think Chris planned it that way, to be gone right up until he had to leave, so that there would be no chance of anyone having an intervention and sending him back to rehab. He was planning on carrying the party right on to New York; it rolled right into *SNL*, and the result was a complete disaster.

LORNE MICHAELS:

The decision to have Chris come back and host wasn't made because he was red hot in show business and it would be great for the show. I think it might have been some desire of mine to help him get back in touch with a time in his life when he was happy. When he was at the show, he knew what the rules were, and I felt it might help him to come back.

TOM GIANAS, *writer:*

When he got back, you could tell that things were bad. He was very tense, distracted and, well, fucked up. I don't think Lorne expected that, otherwise he never would have let him host. We sure didn't expect it. I still thought he was sober. We went out to dinner Monday night after the pitch meeting, and we were running interference with waiters all night, trying to keep liquor and booze away from him. And that lasted the whole week.

MOLLY SHANNON, *cast member:*

He was just indulging in everything: girls, Chinese food, drugs, booze, cold syrup. Everything.

MARCI KLEIN, *talent coordinator:*

By Tuesday night, I knew he was out of control. I had heard that he'd been going up to some of the newer cast members and saying, "Hey, let's go out!" And a lot of the new cast, they really looked up to Chris, wanted to go with him and hang out with him. So I called a meeting with everyone. I met with them in the talent office, and I said, "Look, I know Chris, and I know what's going to happen. If he wants you to go out, you're not going. If any of you help him get drinks or liquor or anything, if you encourage him in any way, you are going to be a part of helping him die."

TIM MEADOWS:

That was a fucking rough week. I tried to hang out with him as much as I could whenever he was at the show, because Chris would never do anything in front of me. But he started all that shit, like, "I've got to . . . go to my car for something." He would make excuses to leave, and he'd come back and he'd be happy again. That whole week he had these total strangers hanging around.

TODD GREEN:

When Chris got to New York he shut everybody down. I wasn't able to see him. Kevin Cleary wasn't able to see him. I don't think Tommy could even get to him, and the rumor mill was cranking.

TOM FARLEY:

It was Tuesday night at about five o'clock. I was working up in Greenwich, and I knew he was coming out to New York. I called my wife, and I said, "This is weird. He's been here since Sunday, and he hasn't even called me."

She said, "Just get in your car and go down there."

So I drove down to the Waldorf-Astoria where he was staying, and he had this bodyguard sitting outside the door. I went in, and Chris was like, "What're you doing here?"

"I just wanted to see my brother," I said. "I haven't seen you in a while."

"Okay, great. I'm just about to head over to *Saturday Night Live*. They're writing tonight, and I'm going to go help come up with ideas."

So we hopped in this limo downstairs that was supposed to take us over to 30 Rock. Mike Shoemaker had one of the NBC pages in the car to babysit Chris. Chris got in and said, "We've got to make a stop. I've got to pick up some friends on 110th Street."

One Hundred and Tenth Street? Next thing I knew we pulled up at

this place in Harlem. Chris went up and came out of this building with two women, looking exactly like what they were. They got in the car, and Chris said, "Okay, let's go to *SNL*."

This NBC kid looked over at me like, "Are you shitting me?"

I just shrugged. So Chris took these two hookers up to *Saturday Night Live*. It was Marci Klein's birthday that night. They were all going out to the Havana Room later. Chris was hanging out with Tracy Morgan and Jim Breuer, talking to them about skits and stuff. I was stuck out in the lobby with these two girls, trying to make small talk. Then Chris went into Cheri Oteri's room to talk with her for a while. I knew where the evening was going, and I knew we weren't going to end up at Marci's birthday party. I got up and knocked on Cheri's door and said, "Chris, you know what, I'm outta here. I'm gonna get some sleep." And I left.

The next day, I called up Kevin Cleary and told him about Chris, that he was just a mess. Kevin said to come back down to the city, we'd take Chris to lunch and have yet another intervention. So I left work, drove in, Kevin and I went to the Waldorf at around eleven-thirty and called up to the room. Chris answered the phone. I said, "Hey, I'm down in the lobby. It's me and Kevin Cleary."

And as soon as he heard those two names he knew what was up. Kevin Cleary was the one person in the world Chris could never bullshit. Chris said, "Yeah, yeah, I'm just waking up. Give me twenty minutes."

We waited a half hour or so, called back up to his room, and there was a "do not disturb" on his phone line. I called the front desk and asked to be put through, but they said he'd left specific instructions not to be disturbed.

Kevin and I went out and had lunch and talked about what we could do, and that was it. I never saw him alive again.

NORM MacDONALD:

I had a sketch I wanted to do with him. The idea was that I'd play Fast Eddie Felson and he'd play Minnesota Fats and we'd play a really long game of pool, just like in *The Hustler*, only we'd never sink a single ball.

Chris told me to come to his hotel to work on the sketch, which was already kind of odd. Why wouldn't we just write in the office? But I went to the hotel and he had this guy guarding the door, and then I went inside and he had tumblers of vodka and orange juice and there were girls there. He kept going in the other room.

Working on a sketch with Chris was never anything but a great delight, but his mind wasn't working like it used to. There was all this self-doubt. He was very preoccupied.

I remember he said to me, "Sometimes, life seems so cruel." He kept saying that: Life seems so cruel.

I said, "Why do you say that, Chris? You're always laughing."

"I don't know," he said. "I just want to die."

I've heard people say that before, and it's sort of a meaningless phrase. Maybe he meant it; I don't know. I don't know how people talk under the influence.

ROBERT SMIGEL, *writer:*

I wasn't writing for the show full-time then. I was just doing the cartoons and occasionally faxing in a sketch idea. In this case, ESPN had asked me to do a "Super Fans" bit with George Wendt about Ditka becoming coach of the New Orleans Saints. When Chris hosted a few weeks later, I figured we could do something similar on the show, so I'd faxed in a version with Chris in it.

Chris's condition was obvious as soon as he showed up. You can't replace a host after the sketches are written and the sets are built, but that doesn't happen until Thursday. And by Tuesday everyone knew how bad it was. Granted, pulling the host even on a Tuesday would have been a huge crisis, but I thought that's what was needed: dire consequences and an inescapable, public humiliation. I thought Chris should be fired. That was the only thing that would teach him the same lesson that he'd learned when Lorne had suspended him and threatened to fire him four years earlier. I actually thought it was a gift opportunity, a chance to help a

drowning friend. But I wasn't a full-timer there, and I wasn't part of making that decision.

NORM MacDONALD:

It was shocking to everybody that Lorne let the show go forward.

LORNE MICHAELS:

I don't remember anybody threatening to resign over it. I don't think that the show should be used as therapy, but there was no way for us, for me anyway, to get through to him in a conversation. That didn't work anymore. If Chris couldn't get it any other way, he could at least watch it on television. The thing he most cared about was the show. I think there was a feeling that the process of doing this, succeeding at it or failing at it, could be brutally honest with him in a way that he was no longer getting in a lecture.

MIKE SHOEMAKER, *producer:*

The truth is you're so busy on that show that you just keep moving, and I think we all hoped that the old Chris would kick in and we'd be fine. I was also just mad at him. More than being worried, I was angry.

ROBERT SMIGEL:

That whole show is like watching a slow death march, but the cold open was just the worst.

TOM FARLEY:

I was in shock when I saw the opening sketch. It was Lorne and Tim Meadows talking about how Chris had kicked his drug problem and he'd

be in perfect shape to host. Then Chris bursts in, and he's obviously in terrible shape. They'd made jokes about Chris's drinking before, but that was when he was sober and doing well. This was not funny at all.

TIM MEADOWS:

When that sketch was pitched at read-through, nobody knew that by Saturday it would no longer be a sketch but the real, honest truth. At the time I didn't think it was in such bad taste, because that's what we'd always done. We'd always been able to make fun of ourselves.

ROBERT SMIGEL:

They cut that opening sketch and the monologue out of the syndicated version of the show. When you see the reruns, it starts with the opening credits and goes straight into a sketch. I believe it's the only time in the history of the show that's been done.

NORM MacDONALD:

He blew out his voice in dress, and so the live show was just awful. He was like a marathon runner stumbling to the finish line before it even began.

MARCI KLEIN:

I said to Lorne, "I think he's going to have a heart attack."

TOM GIANAS:

He did this incredibly physical scene with Molly Shannon, a Mary Katherine Gallagher scene, and it was on right before the Motivational Speaker scene, where he played the coach of a spinning class. He was just huffing and puffing his way through the scene with Molly, sweating like crazy.

And then, by the time he got to the Motivational Speaker scene, he was gone. It was really hard to watch.

NORM MacDONALD:

They did overuse him in the sketches, knowing what his condition was. In the heyday of *SNL*, back in the seventies, they'd get hosts who were completely drunk, and so they'd just write them out of the show. And this was a far more severe case than that.

I remember trying to stretch out Weekend Update. Chris had just done the Motivational Speaker, where they'd put him on a goddamned exercise bike just to make it that much worse. Lorne's usually pretty tight on time, but I just tried to draw Update out as much as I could to give Chris a chance to rest.

MOLLY SHANNON:

We didn't rehearse the Mary Katherine Gallagher sketch too much. And that's worrisome if you're doing a live performance that's very physical. I was throwing him into tables and breakaway walls. He was throwing me around. We did this big dance. Physically, I was a little scared, because he was so big and he'd been drinking. But I felt, for both the dress rehearsal and the live show, the second the camera was on he was completely there. I was amazed by his performance under the circumstances. With such little rehearsal, and without paying much attention in rehearsal, we put him in front of the audience, and he pulled it all together.

LORNE MICHAELS:

The last time I spoke with Chris was at the party at the end of the show. It was around four A.M. after that very bumpy week. We talked about the show. We talked about his health. I said all the things I'd said before about the way he was living and about taking care of himself. He listened. He

agreed. He beat himself up a little bit for my benefit. And I said in the most severe way I could, "You'll never get away with it."

"I know, I know."

We hugged and said how much we loved each other. Then after we hugged, he looked at me with that look we all know so well, that smiling-boy look, and he said, "I was funny tonight, wasn't I, boss?"

"Yeah, you were."

* * *

Following his week in New York, Chris returned to Los Angeles to find that Kevin had moved out, refusing to enable his brother's addiction any longer. Understanding Kevin's need to protect his own sobriety, Chris let it go without any confrontation. Other than a brief visit at Thanksgiving, the two brothers would never speak to each other again.

On November 1, Chris flew to Naples, Florida, where he checked into the Willows rehab facility. He checked out at the end of the month and left to go home to Madison for Thanksgiving. He drank in his car on the way to the airport.

After spending the holiday with his family, Chris returned to Chicago for a brief stopover. He was on his way to the Jesuit retreat house at Bellarmine, a place he often went to find peace and quiet and spiritual refuge. He never made it.

JOHN FARLEY:

Chris got back to Chicago, and he and Ted and I were in the car. "Let's go up to Bellarmine," Chris said.

I said, "I don't want to go to a Jesuit retreat house, Chris."

I didn't, and most people wouldn't. I mean, it's a nice place, but I wasn't in the mood for three days of silent meditation. But I really thought Chris should go, because then he'd be safe, and no one would have to worry about him. You always worried about him when he was in town. It was always a pit in your stomach. I said, "You go up to Bellarmine. You

can relax and everything'll be safe. You won't have these vultures around you."

"Please go," he said.

"I don't want to go."

So I walked away, thinking everything was fine. Then, later that night, I saw him coming around the corner in a bar. I thought, oh shit, he didn't go because I didn't go with him. That one stayed with me for a long time. Maybe if I'd taken him to the retreat everything would have been fine. Maybe maybe maybe.

TED DONDANVILLE:

I was supposed to go to Bellarmine with Chris, but then it was one drink, and then two, and then the trip was off. So now Chris needed to hide out, let people think he'd gone on this retreat so he could keep partying. He got a suite at the Four Seasons, right across the street from his apartment, and the party was on.

JILLIAN SEELY:

My hair salon was right downstairs from the Four Seasons. I went in that weekend, and the girls were like, "Chris just came down here in his pajamas with some random girl."

I went straight up to his room. It was a mess. I said to him, "Chris, what are you doing?"

He had the Big Book in his hands, and he was saying, "Jillian, let's get sober! I've got the book."

I said, "Chris, if you read that right now, you're going to go crazy. You can't be drunk and reading from this."

He got all freaked out and he went and grabbed a towel and threw it over the book to hide it, and then he said, "Come in the bathroom with me!" And he grabbed me and pulled me into the bathroom. He was just out of his mind.

Eventually, he calmed down a bit, and we talked a long time that night, about the rehabs, his father, his career. He told me about another movie he was doing. It was starting in February, with Vince Vaughn, called *The Gelfin*. He said, "Read this inside and out. I'm getting sober for the New Year, and you're coming to L.A. with me to do this movie."

He said to me, "Jillian, why don't you relapse with me and then we can go through treatment together and get better for the New Year?"

I said, "No. I'm not going to start drinking just so that I can go to treatment with you."

TED DONDANVILLE:

By the end he'd started thinking, oh, when I need to be sober I can go to rehab, and when I'm free to party, I can go party. He was picking his rehab spots based on which one was easiest, or which one was most comfortable. "Oh, I'm sick of this place, let's try this place." It was just not proper thinking. It was the thinking of an alcoholic.

LORRI BAGLEY:

I felt like I was supposed to save Chris, even though he told me that I wasn't. He said, "I know what I have to do, and the only way I can do it is with me and with God."

Chris knew what he had to do. He knew what he had to do to stay sober—and he chose not to do it. He told me why. He made me promise that I would never tell anyone why, and I never will. But maybe three months before he died, he was in a rehab in L.A. He called me late one night and told me why he wasn't going to stay sober anymore, and, at that point, we both knew what that meant. The thing is, Chris actually had great willpower, and great strength. Once he decided something, it was done.

JILLIAN SEELY:

Any idea that Chris wanted to die is bullshit. Chris was so full of life, and he had a boundless enthusiasm for everything and everyone. He enjoyed his life and savored it and was full of hope for the future.

When I saw him in those last weeks, he gave me the *Gelfin* script and told me he was getting sober and going back to work. When I picked him up to go to my Christmas party, I caught him practicing his karaoke in the mirror just to make sure he'd do a really good job. When he called his mom that night, he was telling her how happy and excited he was about taking her back to Ireland next year. And when I talked to him on the phone three days later and he said, "I'll call you back in an hour," I don't think he thought that was the last time anyone would ever hear from him.

TED DONDANVILLE:

I don't buy that it was a death wish, that it was a slow suicide. I just don't. You have to discount anything Chris might have said to people, especially to women. How was he trying to manipulate them? How was he trying to play on their sympathies?

The only thing was he said to me once, "Do you ever feel like you're doomed?" But I think that's something we all might say at some time or another. So I don't think you can look at what he was saying. You have to look at what he was doing. What he was doing was playing with fire and the consequences be damned, but he was also making a lot of plans for the future.

The binge that started at the Four Seasons lasted about four days, calling friends and picking up strangers and bringing them along. By the end of it I went up to the suite and there were all these food-service carts everywhere, ashtrays overflowing. After that Chris crashed for a few days, slept it off, and took it easy.

FR. TOM GANNON:

Chris called and asked if we could get together to talk. I said, "Sure, I'll come up to the apartment and we'll have mass together."

"I'd love that," he said.

So I went up, we had a long talk, I gave him confession, and we said a mass. Then we went out to dinner, came back to the apartment, and talked some more. He went on about his addiction and how bad he felt about where he was headed, both personally and professionally, and what he should do with his life. I had to be careful about bringing up his father, because he was always very sensitive when you did. I suggested he dedicate himself to going to daily mass, not because that would help with his addiction but because it might give him a safe, grounded place from which he could rededicate himself to treatment.

We both agreed that the rehab programs were getting him nowhere. I think he went to every rehab program known to man; he must have spent about half a million dollars on them. He had all the lingo down, but he didn't have the reality down. People have to internalize those twelve steps and make them their own, and Chris wasn't doing that.

I left around midnight. As I was driving home I just thought, this kid is going down the tubes. I had a deep foreboding. I came so close to turning the car around, going back, taking him to my place and keeping him there for a couple of days. But you can't do that. He's a grown man with his own free will, and what can you do?

TOM ARNOLD:

There was opportunity to cut Chris's money off at the end. You can commit somebody, legally commit them and cut off all their access to their funds. It came up with the people at Brillstein-Grey. They proposed it, but you have to get the family signed off on it. Ultimately it was his father's decision, and his father wouldn't go along.

TED DONDANVILLE:

Gurvitz wanted to send him away for a year, the most hard-core approach possible. But Chris's dad was like, "Chris is a grown-up. He can make his own decisions." And in a way his dad was right. If it wasn't Chris's decision to go, sending him there wouldn't accomplish anything.

TIM O'MALLEY:

By the time I got to Chris that December, everyone was telling me, "Forget it. We've tried. Just give up."

And I said, "You guys didn't give up on me, why should I give up on him?"

The last ten days of his life he called me every day. It was a slow, horrible thing. He'd call at five, six in the morning and plead with me to meet him at the Pump Room.

I'd say, "No, I am not going to meet you at a fucking bar. I will pick you up and take you to a meeting."

"I don't want to go to a meeting. Everybody recognizes me. I get bothered."

"Fine. I'll take you to a halfway house where people are so bottomed out that they don't care who's sitting next to them."

But he still wouldn't go. And it was the same thing every day. He'd call, we'd pray together. He kept saying, "Please, I need your help. I need your spiritual guidance."

I said, "Chris, all I got is what I got. I can't do anything for you unless you want to go to a meeting. You gotta start over, and you can start today."

"I can't start over."

"Yes, you can."

"No, I can't."

And it was the same conversation every day.

JILLIAN SEELY:

That Saturday, he asked me to come over, and we hung out. We made Christmas cookies together, went to a meeting. Then on Sunday he called me and I picked him up and we went to my Christmas party. We sang karaoke. I have a picture of the two of us that night as we walked into the club, and we were both sober. Then, by the end of the night, he had started drinking and someone snapped another picture of us. It's the last picture of the two of us together, and you can see the difference.

At the end of the party, I said, "Chris, it's time to go."

He was with a bunch of girls, and he was like, "No, no, I'm gonna stay here."

My friend and I told him he really needed to leave, and he got defensive, saying "You're not the boss of me," and all that. So we left, and he went out with all these people drinking. That was the last time I saw him.

TIM O'MALLEY:

Monday morning, I stopped by his apartment on my way to a meeting to see if he wanted to go. We got in an argument about this Fatty Arbuckle project. He was obsessed with doing it, but his managers had brought him into a meeting and told him he couldn't do it until he'd been sober for two years, otherwise no one would insure him. He didn't think that was fair. To me, that was the first time he'd been fired in his life, for real, where someone actually said no to him. I said, "Chris, this is good. It's good that you're going to let go of this."

"But it's going to get made without me." He had the script and he showed it to me, and he was like, "I have to do it."

"The Fatty Arbuckle movie is not a reality," I said. "It's just a script on your desk. You've got to learn how to not drink. Nothing else comes before that."

"But it's different for me. I'm famous."

"Bullshit. You're no different from me. You're just an Irish fucker who

can't stop drinking. This movie is not real. What's real is your torture. You've got to start from ground zero and fix it."

"I can't do it again."

"Yes, you can. I did it, and I had nothing. I had no career. I had no success. If I can do it, you can do it. You have even more to live for."

"But you're strong, and I'm weak."

"Fuck that. I'm as weak as you are."

"But my dad says . . ."

"Fuck your dad. If it were up to me you wouldn't do any work at all for a year. You stay here and you get sober and you work your steps and just get a grip on how to live."

And that's where I left it. That Fatty Arbuckle movie, that was the line in the sand. Either you get sober or you get dead.

TIM HENRY, *friend:*

On Tuesday, Johnny and Teddy and a bunch of guys from Chicago were meeting for lunch at Gibson's. Chris was late, and everyone was getting annoyed. We all had jobs to get back to. I ended up going to the Hancock to get him. Some mysterious girl was there, a joint burning in the ashtray. I was worried, but even though it's so obvious that the inevitable is next, you still don't believe that it's going to happen.

TED DONDANVILLE:

During lunch, Chris was adamant. "This is it," he said. "No more fucking around. We've got another couple weeks to party over Christmas, and then that's it. We're gonna get sober, rent the house in Beverly Hills, get to work on *The Gelfin*. No more fucking around."

He told me he wanted me to hire a trainer, a personal chef; he was going to get back in shape. And those plans were made. I'd rented the house. I was asking around to find a trainer and a chef. Chris had every intention of going back to work in January.

TIM HENRY:

I drove home that day and called Tommy and said, "Chris says he's cleaning up and getting serious after Christmas, but this is a new low."

"I know," Tommy said. "I get these calls all the time."

TOM FARLEY:

I asked Johnny after the fact, you know, "How could you sit there and drink with him?"

And Johnny was like, "What're you gonna do? Chris was already rolling when he got to the table." That's when Johnny just left. He couldn't take it anymore.

TED DONDANVILLE:

I know Johnny had a lot of guilt about what he could have done, should have done. But Chris knew the deal. And you have to remember, there was a physical fear when it came to standing up to Chris, not just an emotional one. He was bigger than you. Johnny said to him once, "You're sick. You've got to stop this." And Chris almost ripped his head off.

When Chris would relapse, all his friends in recovery would abandon him for the sake of their own sobriety. I understand it on one level, having now quit drinking myself, but in some ways it seems perverse. When he needed them the most, they were gone. Johnny and I were the only two people close to Chris who still drank, so we were the only ones around to look after him when he relapsed.

The problem was that even though Johnny and I were heavy drinkers—we could go eight, nine hours—there was always a point where Chris just wore us out. That night, we'd been drinking since Gibson's. It was around two in the morning, and we were at the Hunt Club. These guys wanted Chris to come and party with them at this place up in Lincoln Park. Of course Chris was up for it, but Johnny and I couldn't take it anymore. We had to get off. I said I was going home, and Chris told me to

get a room at the Ritz-Carlton, across the street from the Hancock, said he wanted me nearby. He told me to take care of the bill, and he took off with those people. And that was the last time I ever saw him alive.

JOHN FARLEY:

Chris was going to go all night, and I said, "I'm not doing this with you." I had to get away from it. It was making me ill. Chris and I had been living together at the Hancock, and the vibe had just gotten terrible. He wasn't sleeping at night, and it was a mess. There was stuff everywhere. I was like, eh, I shouldn't be here. That was the other big what if: if only I had stayed. But whatever he was going through, I thought he just needed to be left alone. Plus, I wasn't getting any sleep. So I went with Teddy and checked into the hotel.

TIM O'MALLEY:

Chris called me around five o'clock Wednesday morning. I said, "Chris, I'm sleeping. What is it?"

He said, "I really need your help, please."

I didn't know what to do. He had been calling me every day, and we'd been having the same endless conversation. He wanted me to meet him at the Pump Room, again. He said Joyce Sloane was going to be there. I told him that I wouldn't meet him at a bar. I said, "I'm coming downtown tonight for a meeting at six o'clock. Call me if you want to go." And he never called.

JOYCE SLOANE, *producer, Second City:*

We had a lunch date at the Pump Room, me and Chris and Holly Wortell. I had talked to him the night before to confirm the date, and the last thing he said to me was "I'll see you tomorrow."

HOLLY WORTELL:

Joyce and I met there at noon, and we waited and waited and waited.

JILLIAN SEELY:

I got a call from a friend around ten-thirty. She said, "I saw Chris out last night. He was in really bad shape."

So I called. He picked up on the speakerphone. I could tell he was out of it. I heard somebody laughing in the background. I said, "Who is that?"

"It's nobody. It's nobody," he said.

He asked if he could call me back. I knew he wasn't going to. I said, "Chris, do me a favor and just stay in tonight. Please do not go out."

"Okay, okay, I won't go out."

"Okay."

"I love you."

"I love you, too."

"I'll call you back in an hour."

CHAPTER 15

The Parting Glass

FR. MATT FOLEY:

That night I was in Xochitepec, this small mountain village in Mexico. It had no roads and no electricity, and it was about a seven-hour walk from my parish. I was on a journey with a missionary team. We would walk to a town and spend a day ministering to the people there. Then we'd sleep on the floor of the chapel, wake up the next day, and journey on to the next town.

Sleeping in the chapel that night, I dreamt that I was surrounded by my old Marquette rugby friends. We were all talking about someone we had lost. It was such a vivid dream that it woke me out of a deep sleep. I got up and walked outside. There was this big, full moon, and I just stood there and looked up at the sky for the longest time, trying to figure out what this dream meant.

The next day we celebrated mass and then walked several hours to our truck and drove to a village called Santa Cruz, where there was a phone line. There was a message telling us that we had to come home, immediately. And to be out on a journey like that and to get a message that you had to come home, it could only mean one thing. It meant that someone had died.

After Ted and Johnny went home, Chris stayed out, partying with his newfound companions. As Tuesday night surrendered to Wednesday morning, he left the Hunt Club, hit a few more bars, and eventually wound up at the home of a commodities broker in Chicago's Lincoln Park neighborhood, where the party was in full swing until sunrise and beyond. Chris's host offered to hire a call girl for him. According to tabloid accounts of her story, she arrived at the party around eleven A.M.

In midafternoon, having missed his lunch with Joyce Sloane and Holly Wortell, Chris went with the escort back to her apartment. There he joined her and a friend, smoking crack cocaine and snorting heroin for several hours. Chris called a car service to take them to dinner, but when the car arrived she suggested that he was too worn out to go to a restaurant and that they should just go back to his apartment in the Hancock Center. They did.

Once Chris got home, his brother John called and invited him to dinner with Ted Dondanville and friends at their hotel. Chris declined. He stayed in the apartment, where his binge continued into the night. At that point, Chris had been awake for four days, ever since Jillian Seely's party Sunday night. At ten-thirty Wednesday evening, he took Jillian's phone call, assuring her that everything was fine and that he would call her back.

Chris and the escort began arguing over money. Around three in the morning, she decided to leave, collected her things, and headed for the front door. Chris stood up to follow her and collapsed in the middle of his living room. She turned around, walked back over, and knelt down next to him. He was having trouble breathing. She stole his watch, took pictures of his body, stood up, and walked away. Before passing out, his last words to her were "Don't leave me."

On her way out, believing Chris to be safely unconscious, she stopped by a side table in the foyer. She took out a pen and a piece of paper and left him a short note, saying that he was just so much fun, and she'd had such a lovely time.

TED DONDANVILLE:

Johnny and I woke up on Thursday. I needed something out of Chris's apartment, but I didn't have my keys. Johnny had a set, so we checked out of the hotel and he came with me to the Hancock on his way to Second City.

JOHN FARLEY:

Teddy and I walked in and saw him on the floor. At first, of course, I thought he was joking. Then I realized he wasn't. I dropped to the ground beside him and started giving him CPR. That didn't work. I turned to Ted and told him to call 911. He ran to the phone and called for an ambulance. He told them it was for Chris Farley. That was a mistake.

JILLIAN SEELY:

I was at work and they pulled me aside and told me there were these reports going around. I didn't believe it. I went to the Hancock and tried to get the doorman to let me up, but he wouldn't. I kept calling different phones in the apartment. Eventually Ted picked up. I said, "Are they giving Chris CPR? Is he okay? Is he going to be okay?"

He said, "Jillian, it's not a good time. We'll call you later."

I sat in the lobby. Nobody was telling me what was happening. I started getting hysterical. All these news crews started showing up. The EMS teams were coming through the lobby, but they were coming through really slow, taking their time, like there was no more emergency.

JOHN FARLEY:

The paramedics arrived. They tried to revive him, but they couldn't do anything. When they pronounced Chris dead, Teddy knelt down and put a rosary in his hand.

TED DONDANVILLE:

The media latched onto it: "He died clutching a rosary. It's a sign!" But, no, I put it there. I told Johnny I'd handle everything with the paramedics and the cops, and he should go in the other room and call his parents before the story broke.

JOHN FARLEY:

I went into the back room and called my dad. That was horrible. That was the worst phone call anyone could ever make.

KEVIN FARLEY:

I was in my apartment in L.A. I had just rehearsed Tom Arnold's show that morning. I came home around noon, and there were all these messages on my machine. They all said there was something wrong with Chris, and I needed to call home. I called Dad immediately. He answered the phone, and I said, "What's wrong with Chris?"

There was a very long pause. "We lost him."

The room started spinning, and I hit my knees. I didn't believe it. My mind wouldn't even go there. I was lying there, half crumpled on the floor, when Tom came in the front door and said, "We're gonna help you."

PAT FINN, friend:

It was raining the day he died. It doesn't rain much in L.A., but this was one of those days where it just poured. My wife and I had gone out to lunch with our two little girls and we'd just come home. I was outside with the girls, playing in one of the puddles in the street. My wife came back out of the house and told me I needed to come inside. There were twenty-six messages on the answering machine.

LORRI BAGLEY:

I was at my girlfriend's house, visiting her new baby, and the phone rang. It's funny. People don't react to death like you see in the movies, with all the screaming and hysterics. It's not like that at all. It just doesn't compute, doesn't add up. You sit there, and you can't figure it out.

JOHN FARLEY:

I stayed in the back room, so I didn't see what was going on. Teddy was handling it.

TED DONDANVILLE:

The media reports all said that no illegal drugs were found in Chris's apartment, just a few prescription antidepressants. That's not exactly true. While the cops were sweeping the apartment, any time they came across something illegal, a baggie of cocaine maybe, they'd come over, quietly slip it to me, and say, "Here." Essentially, they got rid of the evidence. They were cops, but they were Chicago cops. Chris was dead. Anything illegal he'd been doing was beside the point. Let him rest in peace.

JOHN FARLEY:

They put him in a body bag and took him out. I went down the back way, where I could get to the garage without going by anybody; there were too many people in the front. I got in the car, pulled out of the garage, and then slipped right by them while they were waiting for me to come out the front door. It was pretty bad. It was a madhouse. They'd blocked off the whole street.

DAVID SPADE:

I was at a read-through for my show, *Just Shoot Me*, and Gurvitz called. He said, "I'm giving you about a twenty-minute head start on this, just so you know before the whole world does." I went back to the read-through and I fell apart. They took me in the other room, and I just couldn't stop bawling.

TOM FARLEY:

I was at a meeting at a friend's, talking about some business ventures. In his office, he had a TV with CNN on in the background, muted. I looked over and I saw Chris and David Spade in a clip from the movies. I said, "Oh, there's Chris. Turn it up." He turned the volume on, and just as he did they switched to this scene in front of the Hancock. My friend stood up, handed me his phone, and said, "Take as long as you need."

JIM DOWNEY, head writer/producer:

I was playing in the basement with my son, and my wife said, "There's a phone call for you." So we went upstairs, and as she handed me the phone I looked over and saw the TV, which was muted, and it was a montage of Farley. My son, who was about four years old, started laughing hysterically at what Chris was doing on the television; I put the phone to my ear, and Mike Shoemaker told me Chris was dead.

MIKE SHOEMAKER:

I realized we would need to choose a sketch to give out to the media as a clip. I remember sitting in Marci Klein's office, crying and thinking, what sketch would Chris want us to use? I picked "The Chris Farley Show" with Paul McCartney.

FRED WOLF:

My manager called me and said, "Chris Farley died." Five months later he called me and said, "Phil Hartman died." Thankfully, he hasn't had to call again.

FR. MATT FOLEY:

After we got the message, my team and I left Santa Cruz. It was a two-hour drive by truck on a dusty road back to Quechultenango. I was driving. It was a very quiet ride. We all knew that somebody in the truck had lost someone; we just didn't know which of us it was.

I finally pulled up to the parish house, we got out, and someone handed me a note saying that Chris had died; both Mrs. Farley and my sister had called. I was devastated. I had prayed so hard for him, and I had never given up on him. My sister had already booked my ticket home. I caught an all-night bus to Mexico City and flew out the next morning.

KEVIN FARLEY:

Brillstein-Grey got me a flight out that afternoon, and my manager drove me to the airport and got me on a plane. I flew into Chicago, met Johnny and Ted and Maria, my girlfriend at the time. I grabbed Johnny and hugged him. He looked like hell, like he'd come through a concentration camp. We got on a plane and flew into Madison together.

TOM FARLEY:

I drove home from my friend's office. I don't really remember the drive; I was just crying my eyes out. We had Laura's sister take the kids and we went out the next day.

JOHN FARLEY:

They had to do an autopsy, so that took a little while. Dad started taking care of all the arrangements for the funeral.

TODD GREEN:

Kevin Cleary and I flew back to Madison together. Mr. Farley called me in Kevin's hotel room, and he said, "Listen, I want all the Edgewood guys to be the pallbearers. You, Barry, Healy, Meyer, and the two Cleary boys. That's the way we want it. That's the way Chris would want it."

TIM HENRY:

Everybody had gone to the funeral home to meet with the priest and the funeral director, and Mrs. Farley asked if I would stay at the house while they were gone, just to watch the phones and be there if people came by. While I was waiting, a lot of neighbors came by, dropping off food and so forth. And of course, this being Wisconsin, several people brought over cases of beer. "Yeah, put that first twelve on the back porch so they'll get nice and cold!"

Then the phone rang. It was David Spade. He asked me to give Mrs. Farley his number and to tell her he'd called.

DAVID SPADE:

It was just very hard. Everybody takes it differently. I couldn't really talk to Johnny or Kevin; they reminded me too much of him. I talked to the family but didn't go to the funeral. I caught some shit for that, but it was my choice. I couldn't deal with it. I couldn't put myself through it, and that was selfish, but I didn't want to grieve in public. I've talked to Sandler and those guys, and they get it. They understand. I just don't like it that some people took that as meaning we weren't getting along.

TOM FARLEY:

Over the weekend, Kevin, Johnny, and I had to get Chris a shirt and a pair of socks to be buried in. We went to the big and tall store where Chris would shop when he was home; they had his measurements. We got him a white shirt, but instead of black socks, they had these red and green Ho-Ho-Ho socks with a little Santa Claus on them. I said, "I think Chris would want to be buried in these." So we bought those, and we all had a good laugh about it.

We went and delivered them to the funeral home, and they told us to take them around to the back entrance, which is where they actually prepare the bodies. That's when it really hit me again: I was delivering socks for my brother to be buried in. The whole week was just full of those moments, realizations like that.

JOEL MURRAY:

The morning of the service, David Pasquesi, Bonnie Hunt, Holly Wortell, and I ended up in a car together, driving up to Madison and telling stories. To a man, everyone in that car was saying, "Why isn't this his wedding? Why aren't we here for him to be marrying some nice local girl? What a party that would have been."

TOM FARLEY:

The funeral was two days before Christmas, and so everyone went through hell trying to get to Madison from all over. They already had holiday travel booked elsewhere and they had to change flights, and so many of the flights were already oversold. It was a nightmare.

JOHN GOODMAN:

I flew in through Chicago and the flight was late and I had to sprint for about a half mile through the terminal to make my connection—and me

sprinting is not cool. For a moment I thought maybe there would have to be two funerals. Then I landed in Madison and the taxi got lost and couldn't find the church. Finally I saw this huge mass of reporters on the street, and I told him to let me out and I just walked.

JOHN FARLEY:

I'll never forget the sight of John Goodman. The parking lot had been kept empty, and this massive bank of news crews had been cordoned off way back at the street. All of a sudden, you see this pack of reporters in a startled panic as Goodman just parts them like the Red Sea, elbowing them aside and yelling "Get outta the way!" He breaks through them and here he comes, trudging through the snow with these two massive, heavy suitcases under his arms and his big beige raincoat flapping in the freezing wind. John Goodman, that motherfucker, he loved Chris. Come hell or high water he was gonna make that funeral.

TIM MEADOWS:

Lorne was flying up by himself from Colorado. I met up with him at the Madison airport and we got a car. We were running really late. The service had already started, so I didn't see his body in the casket, and I didn't really want to.

TOM FARLEY:

We had an open casket at the church. We stood there in this receiving line that just stretched on and on forever. People were coming through, paying their respects, and then taking their seats in the church. After a while we just shut the casket and had people go to their seats, otherwise we never would have gotten to the ceremony.

TODD GREEN:

All the Edgewood guys were there. It was really hard for Mike and Kevin Cleary. They'd already lost their father and a brother, and Kevin had put so much into trying to save Chris. But the person I felt really sorry for was Kit Kat. She was there all by herself, and nobody talked to her.

LORRI BAGLEY:

I didn't really know where to be, or who to be with. I was having these heaving sobs, and this woman took me and let me sit next to her in the pew. Dan Aykroyd came over to me and said something. I don't remember what it was, but it immediately put me at ease. He knew what to say, because he'd been there before.

JILLIAN SEELY:

Everyone was saying, "You had to have seen it coming." But Chris was so full of life that you just wouldn't think that he would die. People were in shock. I was standing next to Adam Sandler, and I said, "This just doesn't seem real."

"Yeah," he said, "I keep expecting him to open up the coffin and be okay."

ALEC BALDWIN:

It's sad when something like that happens to anyone, but somehow it seemed sadder when it happened to Chris. Most of the people whom I've seen go down that path, they didn't have the humanity that he had.

KEVIN FARLEY:

A lot of people showed. Sandler, Chris Rock, and John Goodman. Al Franken and Norm Macdonald. I was just blown away by the life that

Chris had lived. There was a deep melancholy in the room, but you also felt this great love from everyone. He had touched so many people. As sad as I was, I was really proud of him.

PAT FINN:

I was one of the pallbearers, along with the Edgewood guys. Just walking the casket in was tough. When you're thirty-three you don't expect to be doing this. It's something you should be doing for your great-grandfather or something.

It was also strange because the room was filled with people whose names were synonymous with comedy and laughter, and yet the room was the exact opposite.

KEVIN FARLEY:

Nobody in that room could hold it together.

TIM MEADOWS:

I was sitting in front of Sandler and Rock and Rob Schneider. At one point I started crying, and Aykroyd came over and put his arm around me. After the funeral, Chris Rock just lost it. That was the first time I'd ever seen him cry. Sandler, too. That's how it was the whole day. We were a bunch of men who never cried, who never got emotional, who never showed that side of ourselves to each other. And we all just cried and cried uncontrollably all day.

ROBERT BARRY:

Mr. Farley was by far the worst of anyone. Chris was his life. Every Saturday night he'd line up a tumbler of Dewars Scotch, pull up in front of the TV, and laugh and laugh for hours at whatever Chris did. After the ceremony, the pallbearers and the family went back to put Chris in the mau-

soleum. Mr. Farley had his head in his hands, and he was just sobbing. "My boy's not supposed to be gone. Not before me."

PAT FINN:

We were there in the mausoleum, probably about fifteen or twenty of us, for the priests to say the last, final blessing. And I'll never forget the sight of Mr. Farley, getting up from his chair, which was tough for him to do, and putting his arms around the casket. He stood up, just this big bear of a man, and he reached around and he hugged the casket and he wouldn't let it go.

TED DONDANVILLE:

He stood up and raised his arm and with his big, open hand he slapped the coffin twice, loud and hard. *Boom! Boom!* It echoed across the room, sending a jolt through everyone. It was like a final send-off, a father's last good-bye.

FR. TOM GANNON:

I only really remember one thing from the funeral, and that was looking at the father and thinking he wouldn't last a year after Chris.

KEVIN FARLEY:

We all knew Dad wouldn't be too long to follow. I think even he knew it. He closed down the business, paid off the mortgage, made sure all his insurance was in order. When you wake up in the morning, what gets you through the day is your hopes and dreams for tomorrow. For Dad a lot of that was gone after Chris died. He couldn't find it again. They say that happens when you bury a child. I would have long talks with him, and he was just confused about the whole thing, wondering why it had happened, asking God why it had happened. It was such a shock that it left

him in a daze. For the rest of his life he just sat in his chair, staring out the window. But from that day until the day he died, a little over a year later, he never picked up the bottle again.

FRED WOLF:

I actually didn't go to Chris's funeral. My own father had had a heart attack and almost died. I'd gone up to Montana to visit him. I spent that day at the hospital, and I told him about Chris, that the funeral was going on right at that moment.

My father and I never knew how to talk to each other. He was an alcoholic, and our relationship was difficult. I didn't know him that well. One of the only really heartfelt conversations we ever had was that day, about Chris. My dad was saying how the things Chris did are so important for the world, that Chris may have been fighting these demons, but he helped a lot of people who were fighting those same demons feel better, if only for a little while. And I know that sounds sappy. It sounds like something you'd see sewn onto a quilt for sale in the window of some souvenir shop. But at the same time, there's a lot of truth in those quilts.

BOB ODENKIRK:

At the core of being funny is frustration, and even some anger, at the world. And Chris had so much constantly happening inside him that he was always being chased into that corner. He was always living inside that space, and that's why he was just funny *all* of the time. That was his choice. He made a lot of unhealthy choices, but that was the healthiest choice he could make to deal with the feelings that he had. You take some of the most intellectual comics in the world, and what's going on in their work, on a basic emotional level, is the same thing that was going on with Chris—his life was the purest expression of what it is to be a comedian.

DAN HEALY, friend:

It wasn't just that he made you laugh hysterically all the time; he did, but it was more who he was. I've struggled so many times to put into words exactly why Chris had such a huge impact on all of our lives. He had such a faith in other people. He believed in those basic things like goodness and right and wrong. When you were with him, he had this demeanor that simplified things for you. He let you take everything that was complicated in your life and just set it aside for a bit. And that was really the gift he gave us, honestly. Being with Chris reminded you that there was a time when you could still believe in all the things that he believed in. It reminded you of a time when you were lucky enough to look at the world through honest eyes.

IAN MAXTONE-GRAHAM, writer:

The week that Paul McCartney was the musical guest, someone said, "Let's do a 'Chris Farley Show.' "

Chris had done it twice before, with Jeff Daniels and with Martin Scorsese, and we all said, "Eh, it's been done. There's no new moves there."

But they decided to do it anyway. Franken and I were assigned to write it, and I was so glad that we hadn't persuaded people not to do it. It's so often the case that you write something you're not that excited about and then the performer brings something to it and you think, my God, I'm glad I was at work that day.

It played unbelievably sweet. It was so sweet that even now, ten years later, I get goose bumps just thinking about watching it. There's that one moment where Chris says to McCartney, "Remember when you were in the Beatles, and you did that album *Abbey Road*, and at the very end of the song it goes, 'And in the end, the love you take is equal to the love you make'? You remember that?"

"Yes."

"Um, is that true?"

And McCartney says, "Yes, Chris. In my experience it is. I've found that the more you give, the more you get."

And Chris is just like, "*Awesome!*"

And in that moment Chris isn't acting at all. It's really Chris, tapping into that quiet, needy part of himself. You see it up there on the stage. What you see in that sketch is the actual Chris Farley being happy that the actual Paul McCartney is telling him that there is an infinite amount of love in the world, and that someday that love will come back to him.

Epilogue

Christmas was always Chris Farley's favorite time of year, a time he made certain to be home in Madison, surrounded by family and childhood friends. But on December 22, 1997, he had come home to stay. Following the funeral service at Our Lady Queen of Peace Church, Chris was laid to rest in a mausoleum at Resurrection Cemetery, just down the road.

Although the last days of Chris's life in Chicago had been toxic and frenzied, at some point along the way he had arrested his downward spiral and paused to do his holiday shopping, picking out special presents for his parents and his siblings, handwriting personal notes to accompany each specially wrapped box. And so on Christmas Eve, only two days after burying their son, with the winter chill blowing in from the frozen lake outside, the Farley family sat around the tree in their living room and opened their final presents from Chris. He'd bought his mother two small ceramic clowns.

On January 2, 1998, the Office of the Cook County Medical Examiner issued an autopsy report in the case of Chris's death. It stated that he had died of opiate and cocaine intoxication, with coronary atherosclerosis as a significant contributing condition. Chris's body tested positive for

cocaine and morphine (metabolized heroin) as well as traces of marijuana and the prescription antidepressant Prozac. No alcohol was found in his system at the time, but his liver showed signs of significant damage from years of drinking. Blockages of fifty to ninety percent were found in his major coronary arteries from years of unhealthy eating. The report ruled his death an accident.

On May 29, 1998, *Edwards & Hunt* was released under the newly test-marketed name *Almost Heroes*. Chris's passing cast a shadow over the film, and costar Matthew Perry's own public struggle with addiction at the time didn't help much, either. The film's offbeat sense of humor failed to translate onscreen, and scenes of Chris's character acting drunk and out of control were particularly difficult to watch. Critics panned the film, lamenting the tragedy of its being the final installment of Chris's brief career. It earned a little over $6 million at the box office and quickly passed into history.

That summer, Chris was treated to one last curtain call. Norm MacDonald's *Dirty Work*, which featured Chris in a small cameo, was released on June 12. In it Chris played an ornery bar patron whose nose had been bitten off by a Saigon whore, providing some of the film's best laughs.

During the year after Chris's funeral, the health of his father, Tom Farley, Sr., deteriorated rapidly. Chris's death had forced him to stop drinking, but the damage had largely been done. Morbid obesity and liver failure left him severely debilitated, and soon his condition was exacerbated by a bad fall. In March 1999, he checked in to a hospital. After several days of constant vigil, Tom asked his family to go home and rest while he did the same. He died three hours later. He was sixty-three years old.

In the months immediately following Chris's death, several of his friends had made charitable donations to the family in Chris's name. Lorne Michaels of *Saturday Night Live*, meanwhile, issued a *Best of Chris Farley* home video, pledging a portion of the proceeds once again to the family. With this capital, Tom Farley, Sr., started the Chris Farley Foundation (www.thinklaughlive.com), a nonprofit organization to promote awareness and prevention of substance abuse problems. In its earliest in-

carnation, the foundation produced anti-drug public service announcements with past and current stars of *Saturday Night Live*.

From 1999 to 2003, the foundation hosted Comics Come Home in partnership with Comedy Central. An annual comedy event held in Madison, it featured the talents of David Spade, Dave Chappelle, Tom Arnold, Norm MacDonald, Bob Saget, and others. With the funds raised at Comics Come Home and other events, the Chris Farley Foundation works with high schools and colleges across the Midwest to develop programs and seminars aimed at educating kids on the dangers of drugs, primarily through the use of humor and strong communication skills. Today, Tom Farley, Jr., serves as the foundation president.

In late 2003, ImprovOlympic founder Charna Halpern petitioned the Hollywood Walk of Fame committee to give Chris Farley a star on Hollywood Boulevard. The organization hands out very few posthumous honors and almost none to those who pass away in less than rosy circumstances, as Chris had. But in April 2004, John Belushi was at long last honored by the organization, and Halpern seized on the precedent to lobby even more strongly on Chris's behalf. Michael Ewing, one of the producers of *Tommy Boy*, approached Paramount Studios to sponsor Chris's application. With the tenth anniversary DVD release of *Tommy Boy* just on the horizon, the right moment presented itself. Paramount helped push the nomination through and pegged the DVD's launch to the upcoming event.

On Friday, August 26, 2005, the Farley family joined several luminaries from Chris's life—Adam Sandler, David Spade, Chris Rock, Tom Arnold, Sarah Silverman, Peter Segal, Bernie Brillstein, and more—as they unveiled Chris's star, the 2,289th, on Hollywood's Walk of Fame. The tone was far more festive than somber. Mary Anne Farley accepted the award on Chris's behalf. Chris Rock declared that "every fat comedian working today owes Chris eighty bucks," and David Spade wistfully observed that if Chris were alive today, "he'd be working for Sandler, too." Chris's star is located at the corner of Hollywood Boulevard and Cosmo Street, directly in front of the theater for ImprovOlympic West.

In 2007, St. Malachy's Church in New York celebrated the second annual Father George Moore Awards. Moore was a pastor at St. Malachy's in the 1970s and 1980s and was a driving force in the efforts to save Times Square from the drugs and crime that had overtaken it. The George Moore Awards honor individuals who embody the clerygyman's commitment to community service and who have elevated mankind's spirit through their work in theater, television, film, music, or art. Chris, both as a famous movie star and as a humble parishioner, surely fit that description, and, on the tenth anniversary of his passing, St. Malachy's chose him to receive it.

On October 1, the Farleys joined Lorne Michaels, Alec Baldwin, Dan Aykroyd, and the entire current cast of *Saturday Night Live* for a special mass said in Chris's honor—with a sermon delivered by Father Matt Foley—followed by a dinner and awards ceremony at the landmark Broadway restaurant Sardi's. At the dinner, *SNL*'s Amy Poehler served as emcee, presenting a number of smaller awards to other members of the St. Malachy's community and introducing the evening's enertainment: a sketch from the *SNL* players, and young Broadway star Matthew Gumley performing "The Rose," a song Chris had often sung for the seniors he entertained at St. Malachy's social events.

In the midst of the proceedings, just before Mary Anne Farley accepted Chris's award from Lorne Michaels, *SNL* veteran Dan Aykroyd rose to say a few words. He did not go up to use the microphone and podium on the dais, but instead walked the aisles among the dinner guests, speaking off the cuff about his memories. He did a few spot-on impressions of Chris. He spoke of time spent together on the set of *Tommy Boy*, how the young star would come to his trailer and sit at his knee to hear stories about the old days. He spoke of Chris's faith, of his belief in using laughter to bring joy to those less fortunate. He spoke of Chris taking his God-given talent and turning it back out into the world to try and make it a better place. Concluding the speech, Aykroyd singled out all the actors, comedians, and other artists in attendance, and he challenged them to do the same.

ACKNOWLEDGMENTS

TOM FARLEY:

I want to thank everyone who poured their heart out in these interviews. I know well how talking about Chris can be both fun and painful, so I appreciate everyone who shared their memories and emotions in this book. I would especially like to acknowledge the following people who have provided endless help and support to me, the Chris Farley Foundation, and this project:

My beautiful wife, Laura; my fantastic kids, Mary Kate, Emma, and Tommy; my mom, Mary Anne Farley, and my sister, Barb; my brothers Kevin and John, the greatest, funniest guys I know; Fr. Matt Foley and Fr. Tom Gannon; the *SNL* family, who have been amazing to us every step of the way, especially Marci Klein, who understood Chris the second he walked into 30 Rock, and Lorne Michaels; Chris's buddies David Spade, Adam Sandler, Chris Rock, Robert Smigel, and Rob Schneider, who have all supported our foundation from the beginning; Chris's homeboys, Dan Healy, Mike Cleary, Greg Meyer, Todd Green, Robert Barry, and Pat O'Gara; Bob and Sue Krohn and the entire Red Arrow Camp family, who gave the Farley boys so much of our character and values; Marquette Uni-

versity and former dean of communications Michael Price, for believing in Chris and pointing him toward the stage; Madison, Wisconsin—our hometown and the greatest place on earth; the Second City, especially Joyce Sloan and Andrew Alexander—I feel like I'm home every time I walk in the door; the Second City gang: Holly Wortell, Tim Meadows, David Pasquesi, Joel Murray, Pat Finn, Tim O'Malley, and Tim Kazurinsky; Charna Halpern at ImprovOlympic, for all she's done for Chris and his lasting memory.

Humorology at the University of Wisconsin, year after year the most amazing group of young, talented, and philanthropic college students in the country; Jim Farley, my cousin, college roommate, and true friend; my good buddies Neil Lane, Nils Dahl, and John Plum; James Bonneville and Trevor Stebbins; Tim Henry and Don Beeby; Michael and Carol Lesser for helping to launch the Chris Farley Foundation; Shelly Dutch, who does more to help kids in recovery than any foundation I know; Cindy Grant, for her endless support for the foundation and all that we do; Tanner Colby, who now knows more about Chris than anyone alive (welcome to therapy, my friend).

And finally, Chris's closest friend and conscience, the late Kevin Francis Cleary.

TANNER COLBY:

I would like to thank, first and foremost, Tom Farley and the Farley family—Mary Anne, Barb, Kevin, and Johnny—for trusting me with their first, last, and only chance to do this project.

This book almost didn't happen, and credit for the fact that it did goes to the newest and greatest literary agency in the western world, Foundry Media. I owe an incalculable debt to Peter McGuigan, my agent, for picking up this ball and running with it—and sticking with it despite some rocky moments; Hannah Gordon, for bringing me to Foundry and fielding my near-daily queries and neurotic pesterings; Yfat Reiss Gendell, for her crack legal advice and perpetually sunny demeanor; as well as

Kristina Schulz, Stephanie Abou, and everyone else at Foundry who makes it feel like a second home.

On the day this book was purchased by Viking, it was remarked in the publishing blogosphere that Viking was "too good" an imprint for a book about Chris Farley, a comment I take some pride in. Chris's story is not what most people think it is or expect it to be, and I thank Wendy Wolf, a great editor and a wonderful collaborator, for seeing the story underneath and being its greatest advocate. Given her list of bestselling and prize-winning authors, it's an honor just to be stacked on the same shelf in the same office. I also want to thank Liz Parker, for proving that a good assistant can be your best friend in the whole world; Carolyn Coleburn and Ann Day, for plotting a PR campaign that every author should be lucky enough to have; Nancy Sheppard and Andrew Duncan, for the shrewdest of marketing strategies; Sharon Gonzalez, for ferreting out the last (?) mistakes and errors; Paul Buckley, for a book jacket that deserves to be framed; and Daniel Lagin, for a layout that begs to be read.

Months of collaboration were required to pull this off, and for those who helped I'm eternally grateful. I'd like to thank John and Kevin Farley, again, for lobbying where it was most needed; Ted Dondanville, for opening his Rolodex and, more reluctantly, his memories; Tom Davis, Todd Green, and Ian Maxtone-Graham, for their extracurricular help; Christie Tuite, for finding the elusive Jim Downey; Marc Liepis, for enduring far too many e-mails; Mike Bosze and Joey Handy, for their access to Broadway Video; Mike Shoemaker and Marci Klein, for access at *SNL*; Chris Osbrink and Tyson Miller at Callahan, for fielding my constant follow-ups; Julie Warner, for the same; Chris Saito, Susan Wright, and Brian Palagallo, for extra help at Paramount; Jillian Seely, Brian Stack, Lorri Bagley, Holly Wortell, Jim Murphy, and Mark Hermacinski, for their wonderful photographs; Edie Baskin, for hers; Jay Forman and Todd Levin, for poring over much longer drafts than this; Becky Poole, for doing the hard work I didn't want to do; Anna Thorngate, for a great edit; Shawn Coyne, for giving me a great start; Michelle Best, Father Baker, and everyone at St. Malachy's for honoring Chris; and all the agents, manag-

ers, publicists, their assistants, and their assistants' assistants, for helping us get the 130- plus interviews that make up this book.

And lastly, I'll always remember Mach Arom, for opening a door; Matt Atkatz, for his continuing friendship and patronage; Sheila Thibodaux and Marla Fredericks, for getting the money in on time; Richard Belzer, for being there at the beginning; Chris Meloni, for being, very simply, a great guy; Rex Reed, for his home, hospitality, and friendship; Mitch Glazer and Kelly Lynch, for their hospitality in Los Angeles; Laila Nabulsi, for being such a wonderful muse; Judy Belushi Pisano, for her inspiration and spirit; Jerry Daigle, for being a clutch player; Alan Donnes, for everything else; Mom and Dad, for always being there when I need them; Mason, Jenni, Gus, and Lena, for being family; and Ms. Emily Holland, for bringing me home.

NOTE ON SOURCES AND METHOD

The text of Chapter 1, "A Motivated Speaker," was transcribed and condensed from a speech given by Chris Farley at the Hazelden drug rehabilitation facility in Center City, Minnesota, in the summer of 1994. The Chris Farley quote at the opening of Chapter 13, "The Devil in the Closet," was taken from the article "Chris Farley: On the Edge of Disaster," which appeared in *US* magazine and was written by Erik Hedegaard. Some quotes from Tom Davis were drawn from his forthcoming memoir, *38 Years of Short-Term Memory Loss*. Other books that were helpful as general references include *Live from New York*, by Tom Shales and Jim Miller, and *Gasping for Airtime*, by Jay Mohr. Otherwise, all of the quotes and material in this book were drawn from original interviews conducted by the authors between the months of October 2005 and May 2007.

Given the confidential, anonymous nature of drug and alcohol rehabilitation, not to mention Chris Farley's nonexistent filing and organizational skills, very little hard documentation exists as to the exact dates and places of his attempts at treatment and his attendance at various recovery meetings. The facts and time lines presented here were drawn from personal notes kept by Mary Anne Farley over the course of Chris's life.

NOTE ON SOURCES AND METHOD

To every extent possible, the stories presented by the interviewees were checked against contemporary sources as well as the accounts of other eyewitnesses. In many instances, however, no such verification was possible. And, naturally, the opinions and recollections recounted by the participants vary wildly and often directly contradict one another. (Chris Farley was many different things to many different people; such was the nature of his personality.) We have endeavored to present all points of view—even some the authors and the Farley family do not agree with—in the belief that everyone's opinions have merit and deserve their day in court. Somewhere in this tangle of foggy recollection, iffy hindsight, and outright delusion lies the truth, and readers are invited to find it on their own.

THE -OGRAPHIES

FILMOGRAPHY

1992

Wayne's World, dir. Penelope Spheeris
as Security Guard

1993

Coneheads, dir. Steve Barron
as Ronnie the Mechanic

Wayne's World 2, dir. Stephen Surjik
as Milton

1994

Airheads, dir. Michael Lehmann
as Officer Wilson

1995

Billy Madison, dir. Tamra Davis
as Bus Driver (uncredited)

Tommy Boy, dir. Peter Segal
as Thomas "Tommy" Callahan III

1996

Black Sheep, dir. Penelope Spheeris
as Mike Donnelly

1997

Beverly Hills Ninja, dir. Dennis Dugan
as Haru

1998 (POSTHUMOUS RELEASE)

Almost Heroes, dir. Christopher Guest
as Bartholomew Hunt

Dirty Work, dir. Bob Saget
as Jimmy (uncredited)

SELECTED VIDEOGRAPHY: NOTABLE *SATURDAY NIGHT LIVE* APPEARANCES

1990–91 SEASON

09/29/90, Host: Kyle MacLachlan
"Tim Peaks," as Leo the killer

10/20/90, Host: George Steinbrenner
"Middle-Aged Man," as drinking buddy
"Weekend Update," as Tom Arnold

01/12/91, Host: Joe Mantegna
"Bill Swerski's Super Fans," as Todd O'Conner
"I'm Chillin," as B-Fats

01/19/91, Host: Sting
"Hedley & Wyche," as British toothpaste user

02/16/91, Host: Roseanne
"After the Laughter," as Tom Arnold

02/23/91, Host: Alec Baldwin
"The McLaughlin Group," as Jack Germond

05/18/91, Host: George Wendt
"Bill Swerski's Super Fans," as Todd O'Conner

1991–92 SEASON

09/28/91, Host: Michael Jordan
"Bill Swerski's Super Fans: Michael Jordan," as Todd O'Conner
"Schmitt's Gay," as house sitter

10/05/91, Host: Jeff Daniels
"The Chris Farley Show," as himself

11/16/91, Host: Linda Hamilton
"The Chris Farley Show," as himself (with Martin Scorsese)
"Schillervision: Secret Taste Test Gone Wrong," as angry dinner patron

11/23/91, Host: Macaulay Culkin
"Bill Swerski's Super Fans: Thanksgiving," as Todd O'Conner

12/14/91, Host: Steve Martin
"Not Gonna Phone It In Tonight," as himself

01/18/92, Host: Chevy Chase
"Bill Swerski's Super Fans: Quizmasters," as Todd O'Conner

05/09/92, Host: Tom Hanks
"Mr. Belvedere Fan Club," as Mr. Belvedere fan

1992–93 SEASON

12/05/92, Host: Tom Arnold
"Bill Clinton Visits McDonald's," as Hank Holdgren
"Bill Swerski's Super Fans: Hospital," as Todd O'Conner

02/13/93, Host: Alec Baldwin
"Gap Girls," as Cindy
"The Chris Farley Show," as himself (with Paul McCartney)

02/20/93, Host: Bill Murray
"Fond du Lac Men's Jazz Ensemble," as dancer
"The Whipmaster," as bartender

04/10/93, Host: Jason Alexander
"Weekend Update," as Bennett Brauer

05/08/93, Host: Christina Applegate
"Motivational Speaker," as Matt Foley
"Gap Girls," as Cindy

05/15/93, Host: Kevin Kline
"Weekend Update," as Bennett Brauer

1993–94 SEASON

10/02/93, Host: Shannen Doherty
"Relapsed Guy," as the relapsed guy

10/23/93, Host: John Malkovich
"Of Mice and Men," as Lenny/himself

10/30/93, Host: Christian Slater
"Motivational Speaker: Halloween," as Matt Foley

12/04/93, Host: Charlton Heston
"The Herlihy Boy House-Sitting Service," as Mr. O'Malley

11/11/93, Host: Sally Field
"Motivational Speaker: Santa Claus," as Matt Foley

01/08/94, Host: Jason Patric
"Cold Open: Giuliani Inauguration," as Andrew Giuliani
"The Herlihy Boy Dog-Sitting Service," as Mr. O'Malley

01/15/94, Host: Sara Gilbert
"Gap Girls," as Cindy
"Lunchlady Land," as the lunch lady

02/19/94, Host: Martin Lawrence
"Motivational Speaker: Scared Straight," as Matt Foley

03/12/94, Host: Nancy Kerrigan
"Pair Skating," as skating partner

03/19/94, Host: Helen Hunt
"Weekend Update," as Bennett Brauer

04/09/94, Host: Kelsey Grammer
"Yankees Game," as Andrew Giuliani
"Iron John: The Musical," as singer

04/16/94, Host: Emilio Estevez
"The Herlihy Boy Grandmother-Sitting Service," as Mr. O'Malley

05/14/94, Host: Heather Locklear
"Bill Swerski's Super Fans: Letter to Michael Jordan," as Todd O'Conner

1994–95 SEASON

09/24/94, Host: Steve Martin
"Clinton Auditions," as himself

11/19/94, Host: John Turturro
"It's a Wonderful Newt," as Newt Gingrich

12/03/94, Host: Roseanne
"A Woman Exploited," as Tom Arnold

12/10/94, Host: Alec Baldwin
"Japanese Game Show," as contestant

12/17/94, Host: George Foreman
"Motivational Speaker: George Foreman's Comeback," as Matt Foley

01/14/95, Host: Jeff Daniels
"Congressional Session," as Newt Gingrich

01/21/95, Host: David Hyde Pierce
"Little Women," as Toby Adams

02/18/95, Host: Deion Sanders
"Strange Visitors," as SWAT agent

03/25/95, Host: John Goodman
"Bill Swerski's Super Fans," as Todd O'Conner (with Brian Dennehy and Dan Aykroyd)

02/18/95, Host: Courtney Cox
"Motivational Speaker: Venezuela," as Matt Foley
"Gap Girls: Gapardy," as Cindy

05/11/95, Host: David Duchovny
"The Polar Bear Pit," as himself

1997–98 SEASON

10/25/97, Host: Chris Farley
"Cold Opening: I Can Do It," as himself
"Monologue," as himself
"Goth Talk," as rowdy friend
"Morning Latte," as Gil
"Mary Katherine Gallagher," as overweight classmate
"Motivational Speaker: Spinning Class," as Matt Foley
"El Niño," as El Niño
"Sally Jesse Raphael: Giant Baby," as the giant baby
"Monday Night Football Recording Session," as Hank Williams, Jr.
"Bill Swerski's Super Fans: Ditka Goes to New Orleans," as Todd O'Conner

Index

Farley, Tom, 1961-

The Chris Farley show.

DATE

$26.95

B
Farley
F

Farley, Tom, 1961-

The Chris Farley show.